Methods for General and Molecular Microbiology

3rd Edition

To my beloved parents

C. Narayana Reddy and C. Balamma

CONTENTS

REVIEWERS

Bratina Boni
Robert Brambl
John A. Breznak
Robert A. Britton
James W. Brown
D. B. Clewell
Thomas Corner
Frank Dazzo
Piet de Boer
Barry Ensch
Scott Ensign
Douglas Eveleigh
Heidi Goodrich-Blair
J. R. J. J. Graber
R. Gunsalus
Robert P. Hausinger
Robert J. Hawley
Ronald Hicks
John Hobbie
Julius H. Jackson
E. Kellenberger
S. Koval
J. R. Leadbetter
Paul Lepp

Tim Lilburn
Steven Lower
Margaret J. McFall-Ngai
R. Meganathan
N. C. Mishra
Tapan Mishra
Shanna Nesby
Jim Philp
A. Potter
Jorge Rodrigues
R. Sanford
T. M. Schmidt
John Scott
Uwe Sleytr
Loren R. Snyder
Kevin Sowers
Gerard J. Spahn
David Stahl
Bradley Stevenson
Murray Stewart
Joanne Whallon
J. Wood
Vincent Young

CONTRIBUTORS

LUIS A. ACTIS
Dept. of Microbiology, Miami University, Oxford, OH 45056

WILLIAM S. ADNEY
National Renewable Energy Laboratory, Chemical and
Biosciences Center, 1617 Cole Blvd., Golden, CO 80401

RUDOLF AMANN
Dept. of Molecular Ecology, Max Planck Institute for Marine
Microbiology, Celsiusstrasse 1, D-28359 Bremen, Germany

JOHN O. BAKER
National Renewable Energy Laboratory, Chemical and
Biosciences Center, 1617 Cole Blvd., Golden, CO 80401

W. EMMETT BARKLEY
Proven Practices, 6014 Conway Rd., Bethesda, MD 20817

SAM W. BAUSHKE
A126 Engineering Research Complex, Michigan State
University, East Lansing, MI 48824

TERRY J. BEVERIDGE
Dept. of Molecular & Cellular Biology, College of Biological
Science, University of Guelph, Guelph, Ontario, Canada
N1G 2W1

PAUL H. BLUM
School of Biological Sciences, University of Nebraska, E234
Beadle Center, Lincoln, NE 68588

JOHN A. BREZNAK
Michigan State University, East Lansing, MI 48824 [current
address: 15221 East Ridgeway Dr., Fountain Hills, AZ 85268]

DANIEL H. BUCKLEY
Dept. of Crop and Soil Sciences, Cornell University, Ithaca,
NY 14853

GARY CECCHINI
Molecular Biology Division, VA Medical Center, San
Francisco, and Dept. of Biochemistry and Biophysics,
University of California-San Francisco, San Francisco,
CA 94121

TODD A. CICHE
Dept. of Microbiology and Molecular Genetics, Michigan
State University, East Lansing, MI 48824-4320

A. JOHN CLUTTERBUCK
Institute of Biomedical and Life Sciences, Division of
Molecular Genetics, Anderson College, University of Glasgow,
56 Dumbarton Rd., Glasgow G11 6NU, Scotland

RALPH N. COSTILOW
Formerly Dept. of Microbiology, Michigan State University,
East Lansing, Michigan (Deceased)

JORGE H. CROSA
Dept. of Molecular Microbiology and Immunology, Oregon
Health Sciences University, Portland, OR 97201-3098

HERIBERT CYPIONKA
Institut für Chemie und Biologie des Meeres, Carl von
Ossietzky University Oldenburg, Carl-von-Ossietzky Str. 9-11,
Postfach 2503, 26111 Oldenburg, Germany

LACY DANIELS
Texas A&M Health Science Center, College of Pharmacy,
1010 West Avenue B, Kingsville, TX 78363

SHILADITYA DASSARMA
Center of Marine Biotechnology, University of Maryland
Biotechnology Institute, Columbus Ctr., Suite 236,
701 E. Pratt St., Baltimore, MD 21202

ROWLAND H. DAVIS
Dept. of Molecular Biology and Biochemistry, University of
California-Irvine, Irvine, CA 92697-3900

FRANS J. DE BRUIJN
Laboratoire Interactions Plantes Micro-organismes, B.P. 27,
31326 Castanet-Tolosan Cedex, France

STEPHEN R. DECKER
National Renewable Energy Laboratory, Chemical and
Biosciences Center, 1617 Cole Blvd., Golden, CO 80401

EDWARD F. DeLONG
Dept. of Civil and Environmental Engineering & Dept. of Biological Engineering, Massachusetts Institute of Technology, 48-427 MIT, Cambridge, MA 02139

CARLOS G. DOSORETZ
Faculty of Civil and Environmental Engineering, Technion-Israel Institute of Technology, Haifa 32000, Israel

CHRISTOPHE J. DOUADY
Genome Atlantic and Dept. of Biochemistry and Molecular Biology, Dalhousie University, 5850 College St., Halifax, Nova Scotia, Canada B3H 1X5 [Present address: Université Lyon I, Ecologie des Hydrosystèmes Fluviaux, 69622 Villeurbanne Cedex, France]

YVES F. DUFRÊNE
Unité de Chimie des Interfaces, Université Catholique de Louvain, Croix du Sud 2/18, B-1348 Louvain-la-Neuve, Belgium

GARY M. DUNNY
Dept. of Microbiology, University of Minnesota, 1340 Mayo, MMC196, 420 Delaware St. SE, Minneapolis, MN 55455

DAVID EMERSON
American Type Culture Collection, 10801 University Blvd., Manassas, VA 20110

PATRICIA L. FOSTER
Dept. of Biology, Jordan Hall 142, Indiana University, Bloomington, IN 47405

BERNHARD M. FUCHS
Dept. of Molecular Ecology, Max Planck Institute for Marine Microbiology, Celsiusstrasse 1, D-28359 Bremen, Germany

ROBERT L. GHERNA
American Type Culture Collection, Rockville, MD 20852

SHANA K. GOFFREDI
Environmental Science and Engineering, MC 138-78, California Institute of Technology, 1200 East California Blvd., Pasadena, CA 91125

ROBERT P. GUNSALUS
Dept. of Microbiology, Immunology, and Molecular Genetics and The Molecular Biology Institute, 1601 Molecular Sciences Bldg., University of California-Los Angeles, Los Angeles, CA 90095-1489

RICHARD S. HANSON
Dept. of Microbiology, University of Minnesota, 420 Delaware St. SE, Minneapolis, MN 55455

GEORGE HARAUZ
Dept. of Molecular & Cellular Biology, University of Guelph, 50 Stone Road East, Guelph, Ontario, Canada, N1G 2W1

BOB HARRIS
Dept. of Molecular & Cellular Biology, College of Biological Science, University of Guelph, Guelph, Ontario, Canada N1G 2W1

SYED A. HASHSHAM
A126 Engineering Research Complex, Michigan State University, East Lansing, MI 48824

GRAHAM F. HATFULL
Dept. of Biological Sciences, University of Pittsburgh, 376 Crawford Hall, 4249 5th Ave., Pittsburgh, PA 15260

ROBERT P. HAUSINGER
Dept. of Microbiology and Molecular Genetics, Michigan State University, East Lansing, MI 48824-4320

WILLIAM HENDRICKSON
Dept. of Microbiology and Immunology (M/C 790), University of Illinois, Chicago, IL 60612

MICHAEL E. HIMMEL
National Renewable Energy Laboratory, Chemical and Biosciences Center, 1617 Cole Blvd., Golden, CO 80401

ALAN G. HINNEBUSCH
Laboratory of Gene Regulation and Development, National Institute of Child Health and Human Development, National Institutes of Health, Bethesda, MD 20892

WILLIAM R. JACOBS, JR.
Howard Hughes Medical Institute, Dept. of Microbiology and Immunology, Albert Einstein College of Medicine, 1300 Morris Park Ave., Bronx, NY 10461

DEBORAH JACOBS-SERA
Dept. of Biological Sciences, University of Pittsburgh, Pittsburgh, PA 15260

JOHN L. JOHNSON
Formerly Virginia Polytechnic Institute and State University, Blacksburg, VA 24061 (Deceased)

DAVID M. KARL
Dept. of Oceanography, University of Hawaii, Honolulu, HI 96822

ARTHUR L. KOCH
Indiana University, Jordan Hall 142, 1001 E. Third St., Bloomington, IN 47405-6801

SUSAN F. KOVAL
Dept. of Microbiology and Immunology, University of Western Ontario, London, Ontario, Canada N6A 5C1

NOEL R. KRIEG
617 Broce Dr., Blacksburg, VA 24060

CHRISTOPHER J. KRISTICH
Dept. of Microbiology, University of Minnesota, 1340 Mayo, MMC196, 420 Delaware St. SE, Minneapolis, MN 55455

JEROME J. KUKOR
School of Environmental and Biological Sciences, Rutgers University, 216 Martin Hall, 88 Lipman Dr., New Brunswick, NJ 08901-8525

M. A. LACHANCE
Dept. of Biology, University of Western Ontario, London, Ontario, Canada N6A 5B7

JOSEPH S. LAM
Dept. of Microbiology, College of Biological Science, University of Guelph, 50 Stone Rd. East, Guelph, Ontario, Canada N1G 2W1

MICHELLE H. LARSEN
Howard Hughes Medical Institute, Dept. of Microbiology
and Immunology, Albert Einstein College of Medicine,
1300 Morris Park Ave., Bronx, NY 10461

JOHN R. LAWRENCE
National Water Research Institute, Saskatoon,
Saskatchewan, S7N 3H5 Canada

ROBERT E. MARQUIS
Dept. of Microbiology and Immunology, University of
Rochester Medical Center, School of Medicine and Dentistry,
601 Elmwood Ave., Box 672, Rochester, NY 14642-8672

TERENCE L. MARSH
Center for Microbial Ecology and Dept. of Microbiology
& Molecular Genetics, Michigan State University,
East Lansing, MI 48824

MATTHEW J. MARTON
Rosetta Inpharmatics LLC, a wholly owned subsidiary
of Merck & Co. Inc., Seattle, WA 98109

GEORGE A. MARZLUF
Dept. of Biochemistry, The Ohio State University,
Columbus, OH 43210-1292

R. MEGANATHAN
Dept. of Biological Sciences, 319 Montgomery Hall,
Northern Illinois University, DeKalb, IL 60115

CLAUDIA A. MICKELSON
EHS, Massachusetts Institute of Technology, N52-496,
77 Massachusetts Ave., Cambridge, MA 02139-4307

DIANNE MOYLES
Dept. of Molecular & Cellular Biology, College of Biological
Science, University of Guelph, Guelph, Ontario,
Canada N1G 2W1

SCOTT B. MULROONEY
Dept. of Microbiology and Molecular Genetics, Michigan
State University, East Lansing, MI 48824-4320

ROBERT G. E. MURRAY
Dept. of Microbiology and Immunology, University
of Western Ontario, London, Ontario, Canada N6A 5C1

LUCY M. MUTHARIA
Dept. of Molecular & Cellular Biology, University of Guelph,
488 Gordon St., New Science Complex, Guelph, Ontario,
Canada N1G 2W1

CINDY H. NAKATSU
Dept. of Agronomy, Purdue University, Young Hall, 302
Wood St., West Lafayette, IN 47907-2108

KRISHNAMURTHY NATARAJAN
School of Life Sciences, Jawaharlal Nehru University,
New Delhi 110067, India

CAMILLA L. NESBØ
Genome Atlantic and Dept. of Biochemistry and Molecular
Biology, Dalhousie University, 5850 College St., Halifax,
Nova Scotia, Canada B3H 1X5

T. R. NEU
Dept. of River Ecology, Helmholtz Centre for Environmental
Research—UFZ, Brueckstrasse 3A, 39114 Magdeburg,
Germany

J. R. PATEREK
Gas Technology Institute, 1700 South Mount Prospect Rd.,
Des Plaines, IL 60018

WILLIAM R. PEARSON
Dept. of Biochemistry and Molecular Genetics, Jordan Hall,
Box 800733, University of Virginia, Charlottesville,
VA 22908

JAKOB PERNTHALER
Limnologische Station, Institut für Pflanzenbiologie,
Universität Zürich, Seestrasse 187, CH-8802 Kilchberg,
Switzerland

JOSEPH E. PETERS
Dept. of Microbiology, Cornell University, 175A Wing Hall,
Ithaca, NY 14853

ALLEN T. PHILLIPS
Dept. of Molecular and Cell Biology, The Pennsylvania State
University, University Park, PA 16802

JANE A. PHILLIPS
College of Biological Sciences, University of Minnesota,
124 Gortner Ave., St. Paul, MN 55108

YAMINI RANGANATHAN
Dept. of Biological Sciences, 319 Montgomery Hall,
Northern Illinois University, DeKalb, IL 60115

C. A. REDDY
Dept. of Microbiology and Molecular Genetics, 6198
Biomedical & Physical Sciences Bldg., Michigan State
University, East Lansing, MI 48824-4320

CARL F. ROBINOW
Formerly University of Western Ontario, London,
Ontario, Canada (Deceased)

SILVIA ROSSBACH
Dept. of Biological Sciences, Western Michigan University,
Kalamazoo, MI 49008-5410

CHRISTINE E. SALOMON
Center for Drug Design, University of Minnesota, 7-241
PWB, MMC204, 516 Delaware St. SE, Minneapolis,
MN 55455

THOMAS M. SCHMIDT
Dept. of Microbiology and Molecular Genetics,
Michigan State University, East Lansing, MI 48824

IMKE SCHRÖDER
Dept. of Microbiology, Immunology, and Molecular Genetics,
1601 Molecular Sciences Bldg., University of California-Los
Angeles, Los Angeles, CA 90095-1489

J. SCOTT
Dept. of Public Health Sciences, Faculty of Medicine,
University of Toronto, 223 College St., Toronto, Ontario,
Canada M5T 1R4

JOHANNES SIKORSKI
DSMZ-Deutsche Sammlung von Mikroorganismen und
Zellkulturen GmbH, Inhoffenstrasse 7b, D-38124
Braunschweig, Germany

ROBERT M. SMIBERT
Formerly Virginia Polytechnic Institute and State University,
Blacksburg, Virginia (Deceased)

L. R. SNYDER
Dept. of Microbiology and Molecular Genetics, Michigan
State University, East Lansing, MI 48824

KEVIN R. SOWERS
Center of Marine Biotechnology, University of Maryland
Biotechnology Institute, Columbus Ctr., Suite 236,
701 E. Pratt St., Baltimore, MD 21202

G. DENNIS SPROTT
Institute for Biological Sciences, National Research Council of
Canada, 100 Sussex Dr., Ottawa, Ontario, Canada K1A 0R6

JANE TANG
American Type Culture Collection, 10801 University Blvd.,
Manassas, VA 20110

ANDREAS TESKE
Dept. of Marine Sciences, University of North Carolina
at Chapel Hill, 351 Chapman Hall, CB 3300, Chapel Hill,
NC 27599

R. G. THORN
Dept. of Biology, University of Western Ontario, London,
Ontario, Canada N6A 5B7

BRIAN J. TINDALL
DSMZ-Deutsche Sammlung von Mikroorganismen und
Zellkulturen GmbH, Inhoffenstrasse 7B, D-38124
Braunschweig, Germany

MARCELO E. TOLMASKY
Dept. of Biological Science, College of Natural Sciences
and Mathematics, California State University Fullerton,
Fullerton, CA 92834-6850

DON WALTHERS
Dept. of Microbiology and Immunology (M/C 790),
University of Illinois, 835 S. Wolcott Ave., Rm E610 MSB,
Chicago, IL 60612

BORIS WAWRIK
Biotechnology Center for Agriculture and the Environment,
School of Environmental and Biological Sciences, Rutgers
University, Foran Hall, 59 Dudley Rd., New Brunswick, NJ
08901-8520

TIMOTHY J. WELCH
Dept. of Molecular Microbiology and Immunology, Oregon
Health Sciences University, Portland, OR 97201-3098

WILLIAM B. WHITMAN
Dept. of Microbiology, University of Georgia, 527 Biological
Sciences Bldg., Athens, GA 30602-2605

WILLIS A. WOOD
The Agouron Institute, 1055 East Colorado Blvd., Pasadena,
CA 91106

GERBEN J. ZYLSTRA
Biotechnology Center for Agriculture and the Environment,
School of Environmental and Biological Sciences, Rutgers
University, Foran Hall, 59 Dudley Rd., New Brunswick,
NJ 08901-8520

PREFACE

Methods in General and Molecular Microbiology (MGMM) is a substantially revised, updated, and expanded version of its successful predecessor, *Methods in General and Molecular Bacteriology* (MGMB), published by ASM Press in 1994, with my colleague Philipp Gerhardt serving as the Editor-in-Chief. The objective of MGMM is to provide a comprehensive yet moderately priced book that will serve as a first source for traditional methods of microbiology as well as modern molecular biological methods commonly used with microbial cells. Both previous editions of this manual, MGMB and *Methods for General Bacteriology* (MGB, 1991), were popular not only in North America, but also in many other developed and developing countries worldwide. It is hoped that the audience for this edition will be even wider and will include not only "card-carrying" microbiologists, but also scientists in allied disciplines working with microbes as experimental models or as tools for various biotechnological applications. It is hoped that this manual will be on the bookshelf of every serious practitioner of microbiology in academic, industrial, governmental, and clinical laboratories and that it will serve as a rich resource of methods for the seasoned professional as well as for undergraduate and graduate students and postdoctorals.

The primary stimulus for launching this new edition of the manual (MGMM) came from the fact that over a decade had elapsed since publication of the previous edition (MGMB) and during this time not only new methods but also new areas of microbiology (such as community and genomic analysis) had emerged. Considering the fact that MGMB and its predecessors had been used widely around the globe, it was also felt that a new section on general methods in mycology would make this edition even more useful to its worldwide readership. Owing to this increased scope and the need to accommodate additional methodologies, this edition contains 47 total chapters (as compared to 31 chapters in MGMB). The section on systematics from the previous edition has been dropped: much of the still-relevant material in that section has been incorporated into other sections, whereas outdated material from that original section was omitted altogether. Inasmuch as this edition covers methods for microbes representing all the three domains of life, i.e., Bacteria, Archaea, and Eukarya, the title has been broadened from

MGMB (Bacteriology) to MGMM (Microbiology). MGMM is intended as a laboratory manual of methods to complement traditional microbiology textbooks and systematic treatises of general microbiology. However, aside from a chapter on bacteriophages, this manual does not cover viruses, algae, or protozoa.

MGMM is organized into sections corresponding to key subject areas of general bacterial and archaeal biology (microscopy, growth, metabolism, and molecular genetics), including a new section, Community and Genomic Analysis. This is followed by the new section on mycology and appendices on laboratory safety and culture preservation. Each section is divided into chapters, and each chapter has a table of contents to help the reader see the organization of the chapter and easily locate a specific topic of interest. A decimal numbering system is used throughout the manual to facilitate quick identification, cross referencing, and indexing. A comprehensive list of references is provided at the end of each chapter.

The editors refrained from imposing a single, rigid format on the authors other than requesting them to use MGMB as a guide in preparing their manuscripts. The editors also did not delineate in detail the topic assigned to an individual author(s). In spite of this, the chapters turned out to be remarkably consistent in their format and scope. For many of the chapter topics covered in this manual, the state of the art has so developed that entire books dedicated to the topic covered by a single chapter are currently available. Therefore, in many instances, the chapter authors have primarily described reliable methods that are widely applicable to basic studies on the topic covered. Each chapter presents sufficient background principles to understand the how and why of a given method, followed by a step-by-step description of the procedure. Common problems, precautions, and pitfalls of the methods are presented as appropriate. In many cases, commercial sources for equipment and materials are given. These suggestions, however, are not meant either to endorse or to exclude a particular product. Many of the commercial sources for products are also readily ascertained from catalogues published by various commercial firms, from annual buyer's guides of various journals, and from extensive information available on the Internet.

As mentioned above, two sections appear in this edition for the first time. Rapid advances in genomics and genome-based approaches have warranted the creation of the new section, Community and Genomic Analysis. The ability to determine complete genome sequences and metagenomes from microbial communities has revolutionized this field and has extended our ability to obtain valuable information on microbes that are yet to be cultured. Our rapidly expanding knowledge in this area has been put together in seven excellent chapters authored by leading researchers in this area.

The new section on Mycology includes a comprehensive chapter on general methods and three excellent chapters on filamentous fungi, focusing on their physiology, metabolism, and genetic methods and on the principles and practice of DNA microarray technology. Again, each chapter is written by authorities in their respective areas of specialization.

In each of the sections retained from MGMB, existing chapters have been revised (quite extensively in many cases) and a number of new chapters have been added. New authors, revising existing chapters from the previous edition, have provided a fresh perspective, but in most cases the original authors of the analogous former chapter are credited.

Notable new chapters in MGMM, added to sections from the previous edition, include "Laser Scanning Microscopy," "Computational Image Analysis and Reconstruction from Transmission Electron Micrographs," and "Atomic Force Microscopy" in the section Morphology and Ultrastructure; "Energetics, Stoichiometry, and Kinetics of Microbial Growth" and "General Methods To Investigate Microbial Symbioses" in the Growth section; "Bacterial Respiration," "Carbohydrate Fermentations," "Metabolism of Aromatic Compounds," and two chapters on plant polymer-degrading enzymes ("Cellulases, Hemicellulases, and Pectinases" and "Lignin and Lignin-Modifying Enzymes") in the section on Metabolism; and "Measuring Spontaneous Mutation Rates," "Genetics of Archaea," and "Genetic Manipulations Using Phages" in the section on Molecular Genetics.

It goes without saying that a project of this size could not have been successfully completed without the excellent cooperation and enormous effort on the part of the section editors and especially on the part of the authors, many of whom have revised their chapters more than once; the external referees, who provided rigorous and constructive critiques; and the dedicated publications staff of the ASM Press. Any corrections and constructive suggestions for improvement from the users of this new edition are welcome.

C. A. REDDY
Editor-in-Chief

ACKNOWLEDGMENTS

My sincere appreciation and gratitude to the following people for their valuable contributions to producing this book:

- The section editors for their wise counsel, understanding, cooperation, expertise, and able editing of the chapters in their respective sections
- The chapter authors for their dedication in spending countless hours of their time in writing, rewriting, and delivering the splendid scholarly chapters which are the heart of this book
- The host institutions of the chapter authors and editors for subsidizing the considerable amount of time and incidental costs involved in the preparation and editing of their manuscripts
- The reviewers for their expertise, time, and diligence in providing constructive critiques which contributed greatly to the quality of this book

- The staff of ASM Press, including Jeff J. Holtmeier, Director; Eleanor S. Tupper, Senior Production Editor; Steve Burdette, Amber Esplin, Yvonne Finnegan, Julie Winters, and Nancy Wachter for excellent copyediting; and Barbara Littlewood for indexing. I am particularly grateful to Gregory W. Payne, Senior Editor, ASM Press, for his invaluable time, counsel, friendship, encouragement, and help in completing this book.
- Sasikala C. Reddy for her understanding, encouragement, and support in every possible way during my efforts in getting this book completed

C. A. REDDY
Editor-in-Chief

MORPHOLOGY AND ULTRASTRUCTURE

I

Introduction to Morphology and Ultrastructure

T. J. BEVERIDGE

Microbiology and microscopy have enjoyed a strong synergistic union since the earliest times of scientific investigation. It is impossible for us to understand, today, how the crude lenses of Antonie van Leeuwenhoek or Robert Hooke could have achieved the resolution they did, or how the almost happenstance discovery of a fundamental differential stain by Christian Gram would eventually partition bacteria into two fundamental categories, gram-positive and gram-negative bacteria. Because the individual cells of prokaryotes are so minute, they cannot be seen without some sort of microscopy. Accordingly, there necessarily has been a fine history between microscopy and microbiology. Even the best light microscopes that are available today can discriminate only shape and form unless distinctive staining procedures are used for differentiation of fine structures and added contrast. Consequently, other kinds of microscopy, often using wavelengths and energies different from those of light, have been resorted to. Yet, light microscopy continues to be a hallmark of microbiology, and the first two chapters of this section cover the general methods employed. Chapter 1 outlines the basic components of a simple compound light microscope and their operation. Chapter 2 broaches the various manipulations to microorganisms before they can be viewed; these include the simple isolation of cells, chemical fixation, and both general and specific staining protocols for cells and constituent structures.

It cannot be emphasized too much how important are these tried and true optical techniques to microbiology. It saddens those of us who specialize in microscopic imaging to sometimes see how neglected and infrequently used is light microscopy in modern microbiological research laboratories. Some laboratories do not even possess a workable light microscope! Researchers in these laboratories seem confident in their ability to maneuver relatively small molecules of DNA as they manipulate the genetic message in cells and have little desire to actually visualize the cellular results. In these laboratories, possible altered phenotypes, microbial contaminants, and lethal or growth effects are rarely monitored by light microscopy during such refined molecular biological manipulations. Even more, by microscopy's absence, a clear tangible distance is established between the experimenter and that being experimented on, which lends a certain remoteness to the experimental process. We need to rediscover our traditional microscopic roots in microbiology, and chapters 1 and 2 are a required knowledge base for students and researchers alike.

Chapter 3 covers confocal laser scanning microscopy, which is fast becoming an essential part of microbiology's imaging armament. Good light microscopes combined with excellent coherent light sources (lasers) provide accurate resolution plus defined focal sections through relatively thick specimens. Specialized (but often user-friendly) computer software allows the easy manipulation of serial focal sections so that cells located in x, y, and z axial planes of the sample can be visualized and stacked to form three-dimensional images. Fluorescent probes, excited at specific wavelengths, provide exquisite regional discrimination of macromolecules within cellular space so that their location can be seen (e.g., within nucleoids, division septa, secretory apparatuses, endospores, etc.). Confocal laser scanning microscopes (CLSMs) remain relatively expensive instruments (compared to more simple light microscopes), but they are now readily available at the university, college, and (even) departmental levels. Because of their expense and technical sophistication, most CSLMs are operated by a dedicated technical staff member who will aid students and researchers alike. Much of our current knowledge of the growth and spatial distribution of cells within biofilms has come from the use of CSLMs.

It is safe to say that almost all of our knowledge of the fine structures in prokaryotes has been derived by using electron microscopes. Scanning electron microscopes (SEMs) have clarified the topography of biofilms, colonies, and (sometimes) individual cells. So-called environmental SEMs can give images of specimens while they are partially hydrated under high humidity. Transmission electron microscopes (TEMs), though, have been our most important tool in microbiology for determining the macromolecular structures of prokaryotes. Cell surfaces, juxtapositions of enveloping layers, distributions of cytoplasmic constituents, intramembranous inclusions, and cytoplasmic particles (e.g., polyphosphate, sulfur, and polyhydroxyalkoanate granules) have all been seen at high resolution, sometimes confirming less resolving pre-TEM light microscopic evidence.

Electron microscopy requires difficult and (often) extensive preprocessing before specimens can be viewed. Frequently these procedures can induce artifacts that are difficult to recognize for the uninitiated researcher. For this reason, chapter 4 outlines tried and true protocols, using more traditional avenues for TEM, which most microbiology laboratories should be capable of. Here, for even the difficult procedure of thin sectioning, cells can be grown, chemically fixed, stained, and embedded in plastic without specialized equipment. These samples can then be taken to in-house TEM facilities or shipped over larger distances. Because SEMs and TEMs are expensive to buy, operate, and maintain, these microscopes are usually clustered within a university facility containing ancillary equipment for the manipulation of your samples. Specimens can be thin sectioned, TEM grids can be prepared, and an operator can be supplied for viewing specimens. Remember, the technical staff members within these electron microscopy facilities are expert microscopists but they may not be experienced in microbiology. For this reason, it can be advantageous to network with an experienced microbiology structuralist. Over the last decade, a renaissance in microbiological ultrastructure has gradually occurred with the advent of cryogenic techniques. Here, cells are vitrified to preserve their native structures by ultrarapid freezing. One such cryogenic method, freeze substitution, is described in chapter 4 since it has brought remarkable clarity to a number of microbial structures, especially the enveloping layers of bacteria. With some effort and dedication, freeze substitution can be done in a typical microbiology laboratory.

However, more dedicated cryogenic techniques, such as those required for the production of frozen foils or frozen hydrated thin sections, need more dedicated technical skill and equipment (including cryoTEMs with electron energy filters) and are best left to the experts.

As previously mentioned, user-friendly computer software is an essential part of CSLMs and is included with them. However, computer manipulation of electron micrographs is more difficult and uses sophisticated Fourier transformation- or correlation averaging-based analysis to reduce the signal-to-noise ratio within the raw images. Chapter 5 is a short but comprehensive explanation of image enhancement in electron microscopy. It also includes the exciting new techniques of electron spectroscopic imaging (an electron-filtering method) and tomography (three-dimensional imaging by tilting the specimen) as applied to such small particles as ribosomal subunits so that rRNA can be seen as it folds its way around putative ribosomal proteins.

Scanning probe microscopy, (SPM) which is represented by atomic force microscopy (AFM) in chapter 6, is an entirely different type of microscopy in that no lenses are used. Here, a sharp tip on a bendable cantilever is dragged over a sample in a raster pattern with the tip forced up and down depending on the contours found on the specimen. The tip actually is in contact with the specimen and, depending on the physical attributes of the sample's surface, can detect the topography of macromolecules and even atoms. Unlike other forms of scanning probe microscopy, such as scanning tunneling microscopy, AFM can be done under water, which makes it a valuable microscopy for biological substances. Chapter 6 explains the basic setup of the atomic force microscope and how AFM is making inroads in microbiology. Another valuable trait of the microscope is that the tip, on a cantilever of a known spring constant, can probe the physical nature of the sample to determine such important traits as elasticity, viscosity, and adhesion. At one time atomic force microscopes were highly specialized and infrequently encountered in a research setting, but now they are an item that all research-intensive universities have.

It is one thing to discern the structures of microorganisms by looking at intact cells, but it is an entirely different and more demanding task to decipher the contours of their individual components. For this, the components usually have to be isolated and purified. Once these steps are accomplished, these cellular constituents can not only be visualized by microscopy (usually by using either TEM or AFM), they can also be used for biochemical analyses to determine their chemical makeups and possible activities. Chapter 7 describes the manner in which microbes can be broken or permeabilized so that their cellular constituents can be fractionated to purity. It also gives valuable advice on how to determine so-called biochemical signatures of structural components (e.g., the lipopolysaccharide of the gram-negative outer membrane), which is essential for tracking the various fractionations and (eventually) determining purity. Many of the described methods are used to obtain the components of cells that are eventually imaged by the TEM techniques described in chapter 4.

One of the more difficult aspects of any form of microscopy is to be able to unequivocally identify specific structures or macromolecules within a cell. Here, a specific probe to label such a constituent is an invaluable tool. One of the best probes is an antibody specific to the structure of interest. Chapter 8 describes the techniques used to obtain such probes and how to ensure their specificity. These probes can be polyvalent antibodies obtained from an animal after immunization or monoclonal antibodies developed through myeloma-spleen hybridomas. Highest specificity is via monoclonal antibodies with discrimination of a single epitope, but highest labeling capacity is through polyclonal antibodies capable of labeling many epitopes at a single time. Both factors have to be weighed against one another when determining the best probe for your sample. Once the immunological probe has been obtained, it must be labeled with a visual marker that can be detected by microscopy. For light microscopy, this label is usually a fluorescent marker (e.g., fluorescein; see chapters 2 and 3), but for electron microscopy an electron-dense marker (e.g., colloidal gold; see chapter 4) is needed. For AFM, the probe must be large enough to be visualized against the typical topography of the cell (e.g., a bacteriophage that is specific for the surface).

Together, the chapters in this section are an integration of the methods generally used in microbiology to visualize microorganisms and their constituent parts, along with the techniques required for component isolation and visual detection. Although these chapters make a correlated union, they will also be helpful when applied to some of the techniques described in other sections and chapters of this book. Indeed, the images that have been chosen to portray the various stains, cells, and structures in this section have also been chosen so as to help illustrate the microorganisms mentioned in other chapters of the book. One of the great pleasures of combining microbiology and microscopy is to enjoy the simplicity and beauty of microbial shape and form, but each image also possesses a wealth of scientific information. This marriage of science and visual enjoyment has its roots in the 20th century and continues today with more technical sophistication and interpretative skills.

1

Light Microscopy

R. G. E. MURRAY AND CARL F. ROBINOW

The light microscope is essential in any laboratory or classroom that is concerned with microbes, cell cultures, and the interactions of cellular life in nature or in experiments. Its primary functions are to assist in description and identification, the monitoring of growth or processes, the control of biological experiments, and the observation of microscopic phenomena. Thoughtful and well-directed microscopy provides basic information and an appreciation of the living world. Furthermore, judicious use can save the biologist from many a grievous error and will help in the anticipation of events or the recognition of novelty. All this can be accomplished at several levels of resolution and in a number of modes of preparation and presentation.

Light microscopy is made easy, interesting, and useful for bacteriological purposes if at least four different kinds of instruments are readily available, permanently set up for work, and maintained in good order. These are (i) a rugged, well-equipped, but uncomplicated microscope for "encounters of the first kind," (ii) a phase-contrast microscope, (iii) an optimally equipped and adjusted microscope for high resolution of detail and the best of photomicrography, and (iv) an instrument equipped for fluorescence microscopy. Many modern microscopes are equipped to combine two or more of these functions and are very satisfactory if the transitions are easy to accomplish. An additional instrument, a stereoscopic "dissecting" microscope, is valuable for undertaking the isolation of bacteria from nature on agar plates when colonies early in growth are small and close together.

Excellent microscopes are available. Those from first-rank manufacturers are all of comparable qualities and offer a range of accessories and support structures; the choice is likely to be based on convenience factors and personal preferences. Other choices may involve technical requirements, and it is always advisable to make direct comparisons of the possible instruments and the alternative accessories for a given material under given conditions of work. The really sophisticated instruments are expensive. With care, good work can be done with first-class optics fitted to a general-purpose or even a student microscope stand. However, it is hard to equal the modern instruments from the best manufacturers. What matters is an awareness that a good microscope (whatever its age), a discerning choice of components, and a knowledge of how to use them enhance the results and the pleasure derived from the work, save time, and improve accuracy.

The descriptions that follow are based on an old-fashioned basic microscope with separate light source, mirror, and adjustable and centerable components. The optical principles are no different for the enclosed modern instruments with substage illumination, which seem simpler to use but can suffer from similar problems. The use of poorly adjusted and poorly illuminated microscopes is so widespread that most scientists are no longer aware of what microscopes can attain and are astounded when given a demonstration of the best resolution obtainable with even a basic microscope when it is used in accord with the principles outlined below (see sections 1.1 and 1.2).

1.1. BASIC MICROSCOPY

The application of microscopy to the basic requirements of bacteriology is simple, and the questions answered with its help are as follows. Are there objects of bacterial size and staining properties? Are they consistent in size and shape? Are the bacteria gram positive or gram negative? Is there more than one kind of organism present? . . . And so on to more sophisticated questions. These primary questions can usually be answered without attaining the highest resolution, so a simplified procedure suffices as long as the user remembers that higher resolution demands improvement by adjustments at almost every step.

The beginner would be well advised to have at hand a fully illustrated manual for microscopy, of which there are many available (1, 3, 6, 7, 9, 10, 14, 16, 22).

1.1.1. Instrumentation

The basic microscope (Fig. 1) should be equipped with the following.

1. Oculars (eyepieces) of a magnification of at least ×10. Prolonged observation is more comfortable with the use of a binocular system. The oculars fit into the microscope tube, which may be fixed or adjustable in length (to 180 mm, or whatever tube length the manufacturer recommends). Observations on bacterial structure at high resolution are assisted by higher-magnification (×12 or ×15) compensating oculars (see section 1.2.2). Another circumstance in which high-power (15×) oculars assist vision is the scanning of growth on agar plates with a 10× objective.

2. Low-power, dry objectives (e.g., 10× and 40×). These are essential for looking at growing cultures and for identifying rewarding areas for study at high magnification, according to the nature of the preparation. Ideally, the objectives (including the oil-immersion objective) should all be parfocal, i.e., close to being in focus when exchanged by the nosepiece carrier.

3. An achromatic oil-immersion objective (90× to 100×). This is essential to bacteriologists for observing the finest details in specimens.

4. A stage with a mechanical slide-holding device or with simple spring clips. If a mechanical stage is provided, it should be easily removable so that petri plates may be examined at low power.

5. A substage condenser of the "improved Abbe type." This is essential for use with oil-immersion objectives, because it provides the quality and quantity of light they re-

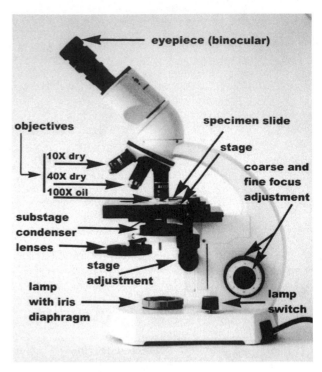

FIGURE 1 Basic light microscope and its parts. The assistance of R. Van Twest, Department of Microbiology, University of Guelph, is appeciated.

quire. It should have an iris diaphragm (aperture diaphragm) and a movable filter carrier incorporated below the lower lens. Some condensers have a movable top element (essential for work with high-power objectives) that tilts out of the way for purposes of low-power examination. A condenser that has three lenses, is aplanatic, and when under oil, has a numerical aperture (NA) of 1.3 to 1.4 is preferable when high-resolution as well as general use is expected (see section 1.2.1).

6. The light source in modern microscopes is built into the base to form a unit with a fixed mirror placed to illuminate the condenser through a plane glass in a mount incorporating an iris diaphragm (field diaphragm). Filters can be placed onto the opening when needed. The light itself is usually mounted in a centerable collar so that the bulb can be positioned to focus centrally in the axis of the condenser. Some specialized microscopes add the optical train for epi-illumination which directs the light to a beam splitter halfway up the tube. This beam splitter sends the beam down to the objective, which now plays the roles of both condenser and objective. Epi-illumination is most useful for fluorescence microscopy (23). There are also "inverted" microscopes with the objective below the stage and a long-focus condenser above it; this setup is useful for examining tissue culture bottles and the like.

Old-fashioned but eminently useful microscopes employ an external light source and, in a gimbal under the microscope, an adjustable two-sided mirror (flat on one side for normal work and concave on the other for low-magnification examination without the condenser). The simplest form of lamp, a frosted bulb behind a "bull's-eye" lens, is adequate for much work. However, the use of a lamp with a "collector" lens, capable of projecting a sharp image of the light source and provided with a filter carrier as well as an iris diaphragm (field diaphragm), results in greatly improved images of details in stained preparations, especially when Koehler illumination (see section 1.2.7) is used. A frosted-glass diffusing screen held in the filter carrier of the condenser can be of value for low-power microscopy, especially when a three-element condenser is used.

1.1.2. Operating Steps

Two basic kinds of bacteriological preparations are used for simple microscopy: a specimen under a coverslip, usually a wet mount (see chapter 2, section 2.2.1.1), or a stained smear on the slide surface (see chapter 2, sections 2.2.1 and 2.2.4). The former preparation is suitable for examination with all objectives. Start with a dry objective, either low power (10×) or "high dry" (40×), to see whether the specimen is rewarding and to select an area for study or to detect motility. Then use an oil immersion objective (90× to 100×) for more definitive observation. A stained smear is usually not suitable for study with dry objectives (although they may be used to locate the bacteria when the smear is very thin) because the smear must be covered with immersion oil to provide a clear and translucent image of the bacteria. The steps described below assume that a progression in levels of magnification will be used, although one can go directly to using an oil immersion objective.

Supposing that a slide preparation is to be examined by use of the simple illumination source described above, the steps in operating the basic microscope are as follows.

1. With the eyepiece and 10× objective in place, the condenser lens fully up and almost level with the stage, and the aperture diaphragm of the condenser fully open, tilt the mirror to provide illumination up the optical axis. Adjust the objective by using the coarse control wheel to focus the light field, and then center it with the mirror. Use the flat mirror for most microscopy; the concave mirror is used only for low-power microscopy when a condenser is not being used.

2. Place a suitable slide on the stage, and focus an appropriate objective to get a sharp (not necessarily well-illuminated) image of the specimen.

3. Improve the illumination by adjusting the positions of the condenser and the aperture diaphragm to get the best image and appropriate contrast. The simplest way to determine the optimum condenser position is to remove the eyepiece and look down the tube at the visible illumination of the back of the objective lens. While looking, adjust the position of the condenser to give the fullest and most homogeneous illumination of the objective. Any lack of concentration can be recognized and adjusted by tilting the mirror. Reduce glare (light scattering from lens mounts) by closing the aperture diaphragm of the condenser (if available) just enough to impinge on the bright disk. This will approximate "critical illumination," i.e., the light source is centered and focused in the plane of the specimen. This whole procedure applies to all objective lenses as long as they are first focused on the specimen. Alternatively, or to check that the critical illumination is properly adjusted, place a wire loop against the lamp or against the frosted glass of other illuminators (the effective light source) while the objective is in focus on the specimen. Then adjust the condenser to get the sharpest image of the wire loop on the specimen.

4. Replace the eyepiece to examine the specimen, and make any final adjustments for the best illumination and focus.

5. Turn the nosepiece to any other objective desired. Many microscope makers provide parfocal objectives (i.e., the focal position for one is nearly identical to that for all the others). Clearance is sufficient for the dry lenses. If the lenses are from different manufacturers or otherwise mismatched, it is wise to raise the tube a little before bringing into place and focusing the oil-immersion objective, which has a small focal distance (often 0.2 mm). The clearance will be even smaller when the specimen is mounted under a coverslip (which should be no. 1 thickness).

6. If going direct to oil immersion, apply a drop of oil over the specimen on the slide before placing it on the stage. If the specimen has been scanned for a rewarding area with a lower power, it is convenient to apply the oil drop to the slide in place with the nosepiece turned halfway to the correct objective before swinging the oil-immersion objective into position.

7. Using and focusing the oil immersion objective takes a little practice and care even if the objective has a spring-loaded front end to protect both the lens and slide from a ham-fisted operator. While looking from the side, lower the objective slowly onto the oil drop and down just a little more so that the lens almost touches the slide. Using the coarse focusing wheel and watching in the eyepiece, look for the specimen while raising the objective *gently*. When the specimen is found, use the fine focus wheel. Adjust the illumination as described above, if needed. Focus with care; contact between the objective and the slide can crack the slide and damage the lens mounting. Adjust for critical illumination as described in step 3 above.

8. On a binocular microscope, adjust the eyepiece tube that has a knurled ring for adjusting tube length so that both eyes are in exact focus.

9. If the image is hazy, moves, or is only dim and not sharp, there is something wrong. It may be necessary to clean the front lens of the objective (with lens paper *only*). A common cause of hazy images is bubbles entrapped in the immersion oil (take out the eyepiece and look down the tube; the bubbles are then easily identified). To remove the bubbles, raise the objective, turn the nosepiece, wipe off the oil with lens tissue (the bubbles usually come up onto the objective), and then turn back the objective and refocus. A poor image may also be caused by having accidentally closed the diaphragm, moved the lamp, or moved the mirror. See section 1.1.7 for other troubleshooting suggestions.

1.1.3. Operating Rules

Remember the following simple rules of microscopy and practice them.

Rule 1. The focused image requires the best possible quality and quantity of light.

Rule 2. If there is too much light, it is best to reduce it with a neutral-density filter or with a voltage controller. The light can also be moved farther away, but then the condenser must be refocused. With oil-immersion (large-aperture) objectives, avoid reducing the illuminating aperture (by lowering the condenser or closing the aperture diaphragm more than 10% of the width of the opening), although this stratagem is helpful for small-aperture, dry objectives and for use in looking at living, unstained cells.

Rule 3. If there is too little light for the objective used, either there is something in the way, the alignment has shifted, or a more appropriate illumination system (lamp and condenser) is needed.

Rule 4. The specimen must be part of a (nearly) homogeneous optical system and therefore must be mounted in oil, water, or another optically appropriate medium. No details can be discerned in dried, stained films of bacteria viewed with dry objectives.

Rule 5. Keep all lenses clean and free from dust, fingerprints, noseprints, and the grime from oily eyelashes (with lens paper *only*).

1.1.4. Comfortable Microscopy

Long hours at a microscope are needlessly tiring if the microscopist cannot sit comfortably and upright while looking into the instrument. Inclined eyepiece tubes are no help if the table or chair height is inappropriate and causes stooping and straining. Binocular viewing also contributes to restful study.

Chairs (including stools in teaching laboratories) should be adjustable. For the very tall, a block of wood can be cut and used on a table of fixed height to raise the microscope to just the right position. The whole object of these adjustments is to allow the microscopist to look down the tube with minimal deflection of the neck or back. A comfortable position minimizes strain and the misery that can result.

1.1.5. Immersion Oils

Immersion oils are designed for use with oil-immersion lenses and provide an environment of the correct refractive index for the front lens of the objective. They also generate a homogeneous system, approximating the refractive index of glass, and they clarify the dried cells of bacteriological slide preparations by permeating and embedding the stained cells.

The commercially formulated oils recommended by microscope manufacturers have been tested to be sure that they do not have acidic or solvent effects on lenses or their mountings; substitutes are not recommended and may be a false economy. Certain properties can be chosen for special purposes (5), such as very-high-viscosity oil to fill wide gaps or to stay in place on horizontal or inverted microscopes and low-fluorescence oil for fluorescence microscopy. For general purposes, use immersion oil of a fairly high viscosity (e.g., type B, high-viscosity oil of 1,250 cP from Cargille Laboratories Inc., Cedar Grove, N.J.), which stays more or less where it is placed and creeps less than low-viscosity oils. The only problem with this level of viscosity is that bubbles are easily entrained, but this problem is avoidable.

Oil immersion objectives should always be cleaned after use or, at least, at the end of a day's work. Use lens paper moistened with xylol or another appropriate solvent as recommended by the lens manufacturer. Ethanol and methanol are recommended by Wild-Leitz for its lenses, but xylol is also safe to use. The modern nonoxidizing oils are less of a problem than the traditional cedarwood oil (which hardens, oxidizes, and leaves annoying residues), but they still slowly alter and may also creep where not wanted. Take caution: modern immersion oils are safe, but before 1972 nondrying oils were often formulated with polychlorinated biphenyl compounds, which are now believed to be carcinogenic and toxic, so look out for old bottles!

Never use more oil than necessary; usually 1 drop will do. A good oil bottle will help to reduce mess; e.g., use a glass bottle with a broad base and an applicator (glass rod or wire loop) attached to the cap. Among the best is a double-chambered bottle in which the stopper is formed into a vessel for oil and the base is made to hold the solvent for cleaning. Squeeze bottles tend to generate bubbles unless used carefully.

Oiling the condenser to the slide, which is necessary for highest resolution with the oil-immersion objective (see section 1.2.1), can be a messy procedure. It demands a high-viscosity oil and some simple precautions both in mounting and in demounting of a preparation, as follows.

1. Place a drop of high-viscosity oil on the top lens of the condenser, and then lower the lens below the stage level. Avoid bubbles in the oil drop.

2. Place a small drop of oil on the underside of the slide below the area you wish to examine.

3. Set the slide down on the microscope stage with the drop central in the hole, and hold the slide there while adjusting the mechanical stage or clips to keep the slide in place.

4. Raise the condenser so that drop meets drop and until the top condenser lens is fully covered with oil.

5. Place a drop of oil over the specimen, and focus with the objective. Then adjust the condenser to give Koehler illumination (see section 1.2.7).

6. When demounting the preparation, lower the condenser before picking up the slide. Lateral movement or wide scanning may lead to oil's being deposited or creeping under the stage. Remember this possibility, and clean the stage underside when cleaning the top lens (with lens tissue!) after use.

7. Only condensers with integral or screw-on top elements are suitable for oil immersion. Movable "swing-out" top elements on some "universal" condensers (i.e., those that can be slid aside with a lever for low-power work) must be oiled with great care or not at all because subsequent cleaning is difficult.

8. Cleaning procedures for condensers after use are the same as these for objectives. Remove the condenser from its substage mount to avoid (or clean up) oil that has flowed around the condenser housing. Remove excess oil with lens tissue, and then clean the lenses with fresh lens tissue moistened with xylol or the appropriate solvent recommended by the microscope maker.

1.1.6. Care and Cleaning

Microscopes are remarkably tough and durable. With reasonable care and protection from the elements (particularly the hostile acid-laden air of some laboratories), a microscope will last a lifetime or more. Take care of the microscope as follows.

1. Protect the microscope from dust and grit. Put a dust cover over the microscope when it is not in use; the best cover is a rigid, transparent, and all-enclosing glass or plastic "bell." Do not allow dust to accumulate anywhere; it drifts into the lenses and mechanisms.

2. The moving parts, especially the rackwork and gears, should be cleaned and treated with new grease at long intervals (the grease for model-train gears from a hobby shop works well). Do not use thin oil on gears or bearing surfaces; the tube or condenser may then sink from its own weight.

3. Clean the stage regularly, and mop up any spills.

4. Keep the lenses clean, and especially, clean up after a session of oil immersion work. *Never use* a finger instead of an appropriate lens tissue (see below).

5. Do not attempt to repair an objective lens or the fine adjustments. These repairs are best left to professionals.

Good microscopes deserve a professional cleaning and adjustment every 20 years or so! Microscope dealers usually have or can recommend a repair facility.

6. Keep the tube closed at all times with an eyepiece, and keep all objective mounts filled or plugged, to minimize dust in the tube. This should not prevent you from removing an eyepiece to check the quality of illumination of the objective.

The "old soldiers" among microscopes can be cleaned, repaired if necessary, and put back to good use. The brass bodies on ancient ones (How elegant they are! How fine their lenses may be!) can be rubbed down with a lightly oiled cloth or one lightly moistened with tarnish remover and then a dry cloth. Modern black finishes should just be kept clean.

Clean lenses make for a dramatic improvement in image quality over that achieved with dirty lenses. Optical surfaces need special care in cleaning, and some general rules apply.

1. Keep a good supply of lens tissue (lens paper), a soft brush (artist's watercolor brush, 0.6 cm wide), and a nebulizer bulb with a short narrow rubber tube attached near the microscopy area. Keep them scrupulously clean and store them in a dust-free box.

2. Nonadherent dust can be easily removed by using the brush, puffing with the rubber bulb, or wiping with lens tissue after breathing on the optical surface.

3. Finger marks, other adherent grease, and dirt must be polished off with lens tissue doubled over a finger and barely moistened with xylol or the recommended solvent (see section 1.1.5). Do not flood lenses with solvents, and never use (without specific recommendation) alcohol, ether, or acetone for fear of penetrating the cement between lenses. If the lens has a raised mount, clean the edges with moistened lens tissue wrapped around an applicator stick.

4. If xylol is not effective, try using tissue moistened with distilled water.

5. The usual way of cleaning oil from objective and condenser lenses is to wipe most of it away with lens tissue and finish with tissue barely moistened with xylol. Oil and grease can also be efficiently removed with the surface of a freshly broken piece of polystyrene foam (common packing material) by pressing it against the objective front lens and rotating it (H. Pabst quoted by James [10]). The foam has lipophilic properties. Xylol dissolves the foam, so do not use both.

6. Never take an objective apart to clean the components, because the interlens distances are critical. Dust can be puffed from the back lens or lifted with a superclean brush.

7. Dust on oculars is a constant problem and produces dark spots which move around as the lens is rotated. The eyepieces are quite simple in construction (top and bottom lenses with a fixed diaphragm between), and both lenses unscrew. If cleaning of the external top and bottom surfaces does not remove a spot, there may be particles on the inner surfaces. Be careful of "bloomed" lens surfaces (those with antireflective coating which appears iridescent blue), and use a brush or tissue with care because some blooms are soft; try puffing first.

8. *Keep fingers off optical surfaces!*

1.1.7. Troubleshooting

The list of technical problems, causes, and remedies shown in Table 1 is taken (with permission) from the excellent book by James (10). For additional advice, consult other books (see References), an experienced microscopist, or an instrument technician to identify the problem. If the problem is optical, work along the light path in systematic order from the lamp and then through the condenser, specimen, objective, and ocular. Adjust at each step, and determine the effect of adjustment on the problem; this is where Table 1 is useful.

1.2. HIGH-RESOLUTION MICROSCOPY

The research microscope comes into its own for the resolution of fine detail at high magnification. Its use is not called for in determining the outcome of a Gram staining test, but it will help to settle questions about flagella and matters requiring the perception of fine detail; also, high resolution allows for the sharpest photomicrographs. For such purposes it is essential to pay close attention to the quality of the optical components. Achieving high resolution as well as freedom from chromatic and spherical aberration requires attention to basic principles (1, 13, 18).

1.2.1. Resolution Requirements

Resolution (R) is the shortest distance between points of detail which will still appear as a distinct gap in the images, visual or photographic. $R = \lambda/2NA$, where λ is the wavelength of the light used and NA is a measure of the light-gathering power of the objective. The purpose of a good research microscope is to help make R as small as possible. The smallest values of R require not only an objective of high NA but also a condenser of equivalent quality (see section 1.2.3) and the use of short-wavelength light. Unfortunately, light of the most effective wavelength (namely, UV light of about 365 nm) is not perceived by the eye and does not pass through glass. Indirect methods of microscopy, requiring a microscope with lenses (lamp, condenser, and objective) made of quartz (permeable in various degrees to UV light), take advantage of the short wavelength of UV light, as is the case for fluorescence microscopy. The best results with visible light are obtained in the green-yellow region of the spectrum because it is the wavelength to which the eyes are most sensitive (close to the mercury green spectral line) and for which microscope objectives are designed to transmit with a minimum of aberrations. So much for the λ part of the expression for resolution.

The NA of oil immersion objectives is usually 1.30 or 1.32, and the best ones may have an NA of 1.40. The NA represents the sine of one-half of the angle described by the cone of light admitted by the objective lens, multiplied by the refractive index of the medium through which light passes on its way to the lens. The maximum value of the sine function cannot exceed 1.0, but the refractive index of the medium through which light passes on the way from the condenser to the objective can be raised to that of optical glass (i.e., 1.5150) by the use of immersion oil. It is this fact that allows lenses of NAs of greater than 1.0 to be filled with light. Provided that certain other conditions are met, the resolving power of the costly objectives with NAs of 1.30 to 1.40 engraved on their glittering housings can be utilized to the fullest.

1.2.2. Objectives and Oculars

Large-aperture achromatic or apochromatic objectives are spherically correct for one or three colors, respectively, in the center of the field of view (where most of the work is done at high magnification). The two kinds of objectives

TABLE 1 Common problems in light microscopy and their causes and remedies[a]

Problem	Possible cause	Remedy(ies)
Coarse adjustment is too stiff	Mechanism adjustment was faulty	With many stands, adjust simply by moving the two coarse control knobs in opposite directions
	Dirt in rackwork	Clean and apply new grease
Tube or stage sinks spontaneously under its own weight (image drifts out of focus)	Incorrect adjustment of rackwork	As with first remedy
	Lubrication with too-thin oil	As with second remedy
	Faulty adjustment of focus control	As with first remedy
Focus adjustment stops before image is focused	Fine adjustment at the end of its travel	Bring a 10× objective into position with the revolving nosepiece, set the fine focus control at the middle of its range, and then focus with the coarse adjustment
Drift of focus with the slightest movement of the fine adjustment (especially with oil immersion objectives)	Objective insufficiently screwed into the revolving nosepiece	Seat the objective fully into the nosepiece
	Surface of the coverslip stuck to the objective by the layer of oil	Use less-viscous immersion oil; clip specimen firmly
Veiled, spotty image	Dirt or grease (i) on the eyepiece (spots move when the eyepiece is rotated in the tube), (ii) on the objective, (iii) on the coverslip (spots move when the specimen is shifted), or (iv) on any surface of the illumination apparatus	Clean where necessary
Sharply focused spots or specks in the image which change and disappear when the condenser is moved up and down	Dirt (i) on the light source or on the diffusing screen in front of it with critical illumination, (ii) on the cover plate of a built-in lamp, or (iii) on a filter near the cover plate with Koehler illumination	Clean where necessary; when the contaminated surface cannot be reached, change the focusing of the condenser slightly or tolerate the problem for the sake of resolution
Hazy image that cannot be brought sharply into focus	Wrong immersion medium (oil instead of air, air instead of oil, air bubble in oil)	Use correct immersion medium
	Transparent contamination on objective front lens	Clean where necessary
	Coverslips too thick or too thick a layer of mounting medium	Use a correct coverslip or an appropriate objective
	Irregularly distributed remnants of immersion oil on the coverslip when high-power dry objective is used	Clean the coverslip with dry cloth or paper tissue; beware of solvents, as some may weaken or dissolve mounting medium
	Slide upside down on the stage (only with high-power objectives)	Invert the slide; make sure a label is not stuck to the wrong side of the slide
Object field partially illuminated	Filter holder partially in the light path	Move the filter holder
	Objective not clicked into position	Reposition the objective
	Condenser (or swing-out lens) not in the optical axis	Realign the condenser
Object drifts diagonally during focusing	A lens is not centered in the optical axis	Check centering of the lenses at all points in the optical path
Object field unevenly illuminated	Mirror not correctly in position	Reposition the mirror
	Condenser not centered (with critical illumination)	Realign the condenser
	Irregularity in light source and/or diffusing screen (with critical illumination)	Move the condenser slightly up and down; use ground glass in front of the light source
Drift of cloud across the field; after this, image out of focus (oil immersion)	Air bubble in the immersion oil; oil in image space with a dry objective	Wipe off the oil from the specimen, and set up anew; clean the slide and objective carefully

(Continued on next page)

TABLE 1 *(Continued)*

Problem	Possible cause	Remedy(ies)
Sharply delineated bright spots in the image	Transverse reflections in the interior of the microscope (often sickle or ring shaped)	Try another eyepiece; use correct Koehler illumination
	Longitudinal reflections in the tube, causing round light spots	Use lenses with antireflective coatings; change the combination objective-eyepiece
Unsharp bright spots in the image	Contamination at a lens surface, on upper or underside of the object, or air in immersion oil of the condenser (differentiate as explained before)	When localization and removal of the contamination are not possible, reduce the effect by opening the condenser diaphragm somewhat more

[a]Reprinted from reference 10 with permission.

work almost equally well, price notwithstanding, especially when narrow-band-pass (interference) filters are used.

The objective best suited to the task in hand should be complemented by an ocular of commensurate quality. What is needed for apochromatic oil-immersion objectives are compensating eyepieces. "They are designed to correct the lateral chromatic error of magnification inherent in objective lenses of high corrections and high apertures"; experience amply proves the soundness of the additional advice of Shillaber (18) that "the compensating ocular should be of high power, preferably 15× or 20×. The ability of apochromatic objectives to take a high-power eyepiece should be fully utilized; otherwise, the fine detail that they are capable of bringing out in a photomicrograph is likely to be lost."

For those who wear glasses with major corrections, it is convenient and helpful to buy high-eyepoint oculars which are made so that glasses may be worn while observing.

1.2.3. Condensers

Another important station in the path of the light from lamp to object to eye is the substage condenser. There is no need for oiling the condenser in routine use that does not require the highest resolution; remember that objective NAs of 1.30 to 1.40 will be fully utilized only if the condenser NA is at least as high. Such condensers are usually also achromatic or aplanatic, and their properties should match the optical properties of the objectives described above.

Remember the role of the refractive index in the expression for NA: the full resolving power of an objective of high NA is utilized only when the medium through which light passes between the condenser and the lens system of the objective is homogeneous and glasslike all the way. Therefore, the maximum resolution of an oil-immersion lens requires oil between the condenser and the object slide as well as between the slide and the objective lens. The condenser iris diaphragm (aperture diaphragm) restricts the beam and thus affects the NA attained (see section 1.2.7).

1.2.4. Mirrors

A mirror is used to reflect the beam of light along the axis of the condenser. The adjustable mirror issued with old-style microscopes has a plane side for work with a condenser and a concave side for work at low powers without a condenser. The plane mirror is made by depositing metal on the underside of an optically flat disk, does well in everyday

work, and is durable. However, it is not the best for optimum illumination, because the front and back of the glass that covers the reflecting metal produce separate and overlapping reflections of the light source or of the opening of the field diaphragm when the Koehler procedure is followed, thus adding undesirable scattered light to the final image. Instead, use a "first-surface mirror," which lacks protective coverings and gives but a single reflection.

1.2.5. Filters

The light used for microscopy is modified in intensity or wavelength by the use of filters. Neutral-density filters (aluminized optical flats) allow the brightness to be adjusted for comfortable vision. Color filters restrict the wavelength used. Two kinds of color filters are available: expensive interference filters of narrow band-pass consist of two half-silvered glass plates facing each other across a precisely measured gap; Wratten filters consist of colored gelatin mounted between two 5-cm^2 flats of optically polished glass. Wratten filters deteriorate with time because air creeps in from the sides, but they come in a wide range of colors and densities and are valuable accessories. A red filter (Wratten no. 29) or an orange filter (Wratten no. 22) is helpful in making dense structures transparent or enhancing structures stained with blue dyes. A green filter (Wratten no. 11 or no. 58B) is the most generally useful because it gives the kind of light to which the human retina is maximally sensitive and because many commonly used histological stains are red. Their contrast is greatly enhanced when the light is green, the complementary color.

1.2.6. Light Sources

Optimum performance of a highly resolving objective depends on optimum illumination. The field of view must be evenly filled with light of a brightness comfortable for viewing or sufficient for photography. This requires a more complex lamp than that effective for ordinary observation. The light from the usual coiled filaments is unevenly distributed over their surface and therefore over their image unless diffused by ground glass. The best structureless, uniformly bright sources of light are small mercury or zirconium arcs and the less expensive conventional projection lamps with tungsten bead or ribbon filaments that can be imaged to fill the condenser. The Koehler system of illumination is designed to provide the appropriate quality of light and requires that the source be focused in the plane of the object being viewed. Therefore, the microscope lamp should be equipped with a lens system to project a sharp image of the light source into the entrance plane of the substage condenser. In front of this lens, there should be an iris diaphragm, the so-called field diaphragm (see below), which can restrict the area of illumination without interfering with the quality of the light. There should also be a carrier for filters.

1.2.7. Optimum (Koehler) Illumination

The Koehler system of illumination is useful for high-resolution microscopy because it provides appropriate illumination of the field and makes good use of high-quality optics. It is achieved by focusing an image of the filament or light source at the level of the condenser diaphragm (i.e., the lower focal plane of the condenser), when the condenser is in the correct position relative to the specimen. Under these conditions, an objective lens in focus on the specimen will be fully illuminated whatever the size of the source. Effectively, this applies to every part of a source, so that there is even illumination across the field of view even when a coil filament lamp is used. The principle can be applied usefully to both high-power dry and oil-immersion objectives.

With an eyepiece, objective, and condenser of matching optical qualities and with a first-surface mirror receiving filtered green light from a bright and homogeneous source, the research microscope is ready to be set up in the best way for work and needs only a rewarding specimen preparation to show its worth.

The steps allowing achievement of Koehler (nearly optimum) illumination for an oil-immersion objective are as follows.

1. With a specimen slide in place, switch on the lamp and adjust the condenser to give the smallest bright spot on the specimen. This approximates the focal position of the condenser.

2. Hold a small white card against the underside of the condenser, and observing in the mirror, focus the image of the lamp filament on the card and move the lamp so that the image covers the opening of the condenser. Alternatively, focus the image on the closed condenser (aperture) diaphragm. If the physical arrangement is awkward, use a small mirror for observation or approximate the focal position on a card placed over the mirror.

3. Insert appropriate filters in the light path, and fully open all diaphragms.

4. Oil the condenser with 1 drop of high-viscosity oil on the top lens, and raise it to meet a similar drop on the underside of the slide (see section 1.1.5).

5. If necessary, scan the preparation with a low-power objective to identify a rewarding area, using reasonable centration and illumination (section 1.1.1), and mark it by the coordinates provided on many mechanical stages.

6. Apply oil to the top of the slide (see section 1.1.5 for precautions), change to the oil-immersion objective, and focus on the specimen. A fuzzy or moving image may indicate bubbles in the oil; check by looking down the tube after removing the eyepiece.

7. Adjust the field (iris) diaphragm of the lamp to a minimum opening, and bring it into view with the mirror. Focus the image margin with the substage condenser in the plane of the specimen (i.e., so that the objective and condenser are both in focus). Readjust the field diaphragm so that its image just impinges on the field of view and centers exactly with the mirror. The optical alignment is then correct.

Note that if the margins of the field diaphragm give an uneven color fringe when white light is used, the condenser needs centration. To center it, turn the two adjustment knobs to give movement toward the red side, centering the aperture with the mirror, and stop when the color fringes are symmetrical.

8. Open the field diaphragm until it just impinges on the margin of the field of view. This prevents illumination of the specimen outside the field of view, which scatters light into the image plane. The field diaphragm does not affect resolution in the area illuminated and can be left in view.

9. Removing the ocular (which should be 12× or 15×), look down the tube and restrict the condenser aperture diaphragm to abolish glare or reflections. This restriction should not be more than 1/10 the diameter of the fully illuminated back lens; excess restriction lowers resolution.

10. Replace the eyepiece, and insert filters appropriate to observation or photography (section 1.2.5).

The same principles apply to adjustment of the illumination provided by modern integral substage systems or to incident-light systems (3). However, the centration of the light for these systems is attained by adjusting the position of the bulb in its mount with centering screws. The filament image is focused by positioning the lamp forward or backward in its sleeve, because the diaphragm and the lenses are fixed in place. The point of reference for alignment is usually an integral field diaphragm.

The same principles apply also to dry lenses, but oil immersion of the condenser is then unnecessary as long as the condenser is of adequate quality.

1.3. DARK-FIELD MICROSCOPY

Dark-field microscopy provides a useful means of looking at wet mounts of unstained specimens and detecting very small structures by reflected and diffracted light, revealed like motes of dust in a beam of sunlight or like planets and moons in a night sky. It is performed by illuminating the specimen with a hollow cone of light such that only the light diffracted by objects in the field of view is transmitted up the microscope tube to the eye or camera; the beams forming the cone of light focused on the specimen are at too low an angle to be captured by the objective. The result is a field of bright objects (spirochetes, bacteria, particles, organelles, etc.) against a dark background. It is an appropriate technique for all powers of objectives, but it has some requirements that impose the need for specialized condensers for use with oil-immersion objectives.

Low-power (10×) and high-dry (45×) objectives can be used in a dark-field mode by using an ordinary condenser equipped with a central patch stop in its filter holder. This patch stop allows light paths only in the periphery of the condenser and so attains a hollow cone of light effective at a particular condenser position. This means that the NA of the condenser must be considerably greater than that of the objective being used in order to attain an appropriate illumination angle and the condenser diaphragm must be completely open. The cone of light accepted by an objective limits the usefulness of patch stops, so they do not work for objectives with an NA greater than 0.60 to 0.65. A simple version of low-power dark-field microscopy with 10× and 45× objectives can also be attained by using a phase-contrast condenser (see section 1.4); it is useful but does not provide the highest-quality image. Few problems are encountered when dry objectives are used with a condenser particularly designed for dark-field microscopy, and an excellent image can be expected.

Oil-immersion objectives (which collect light at an angle proportional to the NA) require specialized condensers of high NA, which must be oiled to the slide so that an effective cone of light can be produced. These reflecting condensers consist of either a paraboloid mirror with a central stop or, more effectively, a central spherical reflecting surface which reflects much of the light entering the condenser to a cardioid (curved) peripheral mirror on the perimeter, reflecting a very low-angled beam in the form of a cone (refer to texts for optical diagrams).

However, when these condensers are used, most oil-immersion objectives do not provide an adequately dark ground without the presence of devices to restrict scatter from the margins and mountings of their lenses. Excellence requires an objective with an adjustable diaphragm in the back focal plane or restriction in the back focal plane by use of a funnel stop (a metal tube with a hole in the lower end) that fits inside the objective to block transmitted rays scattered from the periphery of the objective.

Intensive light is necessary to show very thin or very small objects. Sunlight from a heliostat has been used to show bacterial flagella in action.

It is sometimes difficult to get a good dark background when there is poor centration of the condenser relative to the objective. This requires some trial and error to gain symmetrical illumination. It is helpful to find the best setting for the condenser by using tongue or cheek scrapings mounted in water under a coverslip; these large epithelial cells are easy to find. All the precautions about oiling the condenser and slide (section 1.1.5) must be followed. But given a well-aligned microscope, all that matters then is the intensity, geometry, and quality of the cone of light with an appropriate objective.

The focus of the condenser is crucial in dark-field microscopy. The apex of the cone of light *must* be at the level of the specimen, which is also the plane of focus of the objective. Above this focal point is a dark conical space, in which the front lens of the objective is placed, symmetrical to the dark conical space formed below the specimen above the condenser. When the dark-field condenser is brought up to the slide, with immersion oil making a complete contact, a bright ring of light is projected onto the specimen. As the condenser is raised farther, the light comes to a bright intense spot, which must be close to the best focus. Because the large-aperture condensers for oil immersion microscopy have a very short focal length, it is important to use a thin slide (~0.8 mm thick) so that the condenser can focus through it on the specimen. Standard slides (~1.2 mm) may be too thick, thus making high-resolution dark-field microscopy impossible.

For many years, dark-field examination of exudates and media was the accepted method for demonstrating the presence of spirochetes. It is impressive to be able to see *Leptospira* spp. in low-power dark-ground images by using a 10× objective and 15× oculars, but it is even more so when they appear in all their glory under oil immersion at an effective NA of 1.3. It is a pity that these beautiful images are seldom seen nowadays because of the convenience of phase-contrast microscopy.

A procedure for oil-immersion dark-field microscopy is as follows.

1. See that the microscope and illumination system are fully aligned, as for bright-field microscopy.

2. Replace the condenser with the dark-field condenser, and apply 1 drop of high-viscosity oil to the top lens.

3. Make a wet mount of the specimen on a clean, thin slide. If the slide is very clean, it is helpful to make a small wax pencil mark under the coverslip for preliminary focus.

4. Raise the condenser to meet the specimen slide, and observe the ring of light. Focus this to give the smallest spot. An asymmetrical ring of light above or below focus indicates that the condenser is not properly illuminated.

5. Focus the oil-immersion objective on the specimen. The objective should be equipped in the body with a funnel stop or with an iris diaphragm controlled by a knurled ring. In the latter case, start with the diaphragm half closed and make final adjustments later for maximum darkness of the field.

6. Move the specimen slowly to a useful field, and make final adjustments in the condenser focus for maximum brightness of the reflections from the specimen.

1.4. PHASE-CONTRAST MICROSCOPY

Phase-contrast microscopy is a system for gaining contrast in a translucent specimen without the help of stains (4, 10, 17, 22) and has the advantage of using high-resolution optical components. It requires a specialized condenser and specialized objectives with appropriate annular modifications to the lens systems.

Stained specimens or biological material with opaque areas form images in a microscope because the various components transmit different amounts of light, so that amplitude and wavelength are modified by absorption and scattering in the specimen. Most cells and their components do not cause enough amplitude modification to give useful contrast in the image. However, their materials are translucent and have different refractive indices from neighboring structures and from the mounting medium, and these differences cause different degrees of phase retardation in the light beam passing through a structure compared with phases of beams that have passed through other material or only through the mounting medium. The purpose of the phase microscope is to take advantage of changes in phase, converting these into amplitude differences to form an image with enhanced contrast.

An instrument that is permanently set up for phase-contrast microscopy is preferable to one that provides alternating service with ordinary optics on the same stand. Phase-contrast objectives are not suitable for the best bright-field work, despite their high quality. It is better to have objectives dedicated to each purpose. Phase-contrast microscopy demands much more light than is adequate for the examination of stained specimens with ordinary optics. A green filter should be used to reduce unavoidable chromatic aberration.

One of the advantages of phase-contrast microscopy over the less-available dark-field microscopy is that the former uses the full aperture of the objective lens. Therefore, the best of phase-contrast objectives should be used. This in turn implies the use of a substage condenser with a suitably large aperture and of 12× to 15× compensating eyepieces, which fully utilize the fine resolution that the objectives are capable of giving. Medium-power (40× or 60×) oil-immersion phase-contrast objectives are available and provide crisp, rewarding images for work with large cells. For the very best results, the condenser and slide must be connected by oil when high-power (90× or 100×) objectives with NAs higher than 1.0 are used. However, an oiled condenser is not required for all applications. A dry condenser is adequate for observing the shapes and forms of bacteria and their spores and for checking motility.

Phase-contrast microscopy works because a phase annulus is inserted at the lower focal plane of the condenser (to generate a cone of light) and because a phase plate is incorporated in the back focal plane of the objective. An image of the annulus will thus be formed on the phase plate. Some light beams are diffracted by the specimen (e.g., a living cell) to go through all of the objective field, and some go directly through the specimen and the phase plate ring; both take part in forming an image. The phase ring in the objective is a flat, ring-form groove in an optical flat plate; its size fits the cone of the direct light beams generated by the condenser annulus. The groove is formed so that the phase ring is less-thick glass, causing at least a 0.25-nm-wavelength phase retardation compared with the phase of a direct (nondiffracted) beam. The phase difference of the diffracted and nondiffracted beams, some of which are also retarded by passing through a structure in the specimen, causes either destructive interference (dark contrast) or constructive interference (bright contrast) when the beams form the image. Therefore, it is essential to center the annulus plate in the condenser with respect to the analyzer plate in the back focal plane of the objective. This is achieved both by imaging with a focusing eyepiece (or "telescope") and by centering the visible rings by using the two adjusting knobs on the sides of the condenser. The bright image of the annulus should be completely enclosed by the gray annulus of the phase plate when centration and condenser positions are correct; complete filling of the annulus is not essential.

1.4.1. System Assembly

1. Set up the microscope as for Koehler illumination (section 1.2.7) without the annulus in place (use the bright-field position for the usual rotatable carrier in the condenser of a phase microscope), and adjust the phase objective to focus on a specimen.

2. Revolve the condenser carrier to bring the appropriate annulus into position below the condenser for the objective in use.

3. Insert a focusing telescope instead of the eyepiece and focus on the gray ring image of the phase plate, which has distinct inner and outer edges.

4. Look in the telescope for all or part of a very bright ring of light, the image of the condenser annulus, which has to be brought into register with the objective phase ring, seen as faint rings for each side of the groove in the phase plate. Achieve centering either by manipulating the condenser centering knobs or by nudging the annulus with a knurled ring on the carrier (systems vary according to manufacturer).

5. Replace the telescope with the eyepiece, adjust the fine focus, and regulate the light intensity (by using a transformer) for comfortable viewing. The image should not shift asymmetrically on focusing; if it does, recheck step 4.

6. If the image is unsatisfactory, check that all surfaces are clean, that the specimen is not too thick, that there are no bubbles in the oil if you are using oil immersion, and that the correct annulus is in place. The image quality deteriorates if any part of the direct beam falls outside of the annulus in the objective phase plate or is not fully concentric. Accurate focusing of the condenser is essential.

1.4.2. Specimen Preparation

The specimens for phase-contrast microscopy should be as thin as possible and must be mounted in a fluid or gel and under a coverslip to give a homogeneous background to the images. For general work, a clean standard slide and coverslip are satisfactory.

When intracellular detail of living organisms is wanted, attention must be paid to the refractive index of the medium in which the cells are mounted. This is most easily provided by dissolving gelatin or bovine serum albumin (15 to 30%) in the medium. According to the concentration used, this equalizes or brings closer together the refractive indices of the contents and medium, since phase retardation is proportional to the refractive index multiplied by the light path. Useful preparations can be made by mounting a water suspension on a slide that has a thin layer of agar dried on it and then applying a coverslip. The disturbing bright halos that usually surround dark-contrast images are then reduced or abolished, and the cells are more gray than black. As an additional benefit, the cells then show internal structures that are not easily discerned in life, e.g., nucleoids and granules other than the obvious lipid droplets, or developing endospores. When the mounting fluid and an adjacent structure are identical in refractive index, the structure will disappear; this technique of immersion refractometry can be used to measure the densities and masses of structures (15).

1.4.3. Phase Condenser for Low-Power Dark Field

The condenser phase plate is, essentially, a form of patch stop, and so, given suitable geometry, it should produce dark-field images with ordinary low-power objectives. The annulus for the oil immersion phase commonly gives a reasonable semblance of dark-field optics (section 1.3) with an ordinary 10× objective and sometimes with a 45× objective, after a little fiddling with centration and the position of the condenser.

1.5. INTERFERENCE MICROSCOPY

Interference microscopes have definite applications in bacteriology for discerning the structures of cells. Such microscopes are expensive, which makes them less available, and their operation and the interpretation of the images obtained are best learned by practice with experts. Like phase-contrast microscopes, they attain contrast in images of translucent specimens by detecting phase changes induced in light that traverses cell components with different masses and refractive indices. The two systems differ in that the interference microscope develops separate object and reference beams, rather than forming an image from the direct and diffracted elements of a single beam from a condenser annulus as in the phase-contrast microscope (10, 17, 20, 22).

Interference microscopes provide superior images of the internal morphological details of cells without the halos that surround cells when viewed by phase-contrast microscopy. A great advantage of the former is that phase changes may be measured to provide quantitative cytological data. Furthermore, the lack of an annulus avoids deterioration of the image by optical effects from cell structures. The lenses are of high quality, are used to their effective NAs, and will operate in either transmitted- or epi-illumination modes. The quality of light and the adjustment of the microscope to give critical or Koehler illumination at the outset are important.

The Nomarski-type interference microscope uses polarized light (white or monochromatic) from a filter at the source to fill the condenser through a Wollaston (birefringent) prism. The prism generates two polarized beams at right angles, each filling the lenses. The resulting cone of light traversing the specimen and illuminating the objective is a complex of these two beams; in effect, they are very close together, in parallel, and uniform. They act as object and reference beams according to what they traverse in the specimen. The beams are recombined above the objective by another Wollaston prism, allowing interference (constructive or destructive) to take place. The plane-polarized beam is recaptured by an "analyzer" polarizing filter, which is usually in a fixed orientation (this means that the polarizing filter at the source has to be rotated to give the appropriate orientation). If white light is used, with appropriate manipulation parts of the image will contain interference colors which are related to the amount and sign of the phase change. This information and photometric measurements will allow generation of mass data.

In some interference microscope systems (including the Nomarski type), the image has a pseudo-three-dimensional appearance, which is a consequence of an angle of shear between the beams.

1.6. FLUORESCENCE MICROSCOPY

Fluorescence microscopy employs all the principles of optics described for the research microscope (section 1.2). The differences in practice and design relate to generation and transmission of wavelengths of light suitable to the excitation of fluorochrome stains and of natural fluorescence in the specimen. The secondary emitted wavelengths are detected as an image of a fluorescing object. Because the excitation process usually requires short wavelengths in the near-UV or blue range, the lamp (a high-pressure mercury vapor arc lamp) and any lens between the lamp and object must be made of material (quartz) appropriate for the passage of that range of wavelengths. A first-surface mirror is essential (to avoid an interfering glass layer). The immersion oil for the objective and condenser must be nonfluorescent (e.g., a special synthetic formulation or sandalwood oil). Quartz slides and coverslips must be used. Good advice on details comes from specific (11, 23) and general texts.

Most important are the light source and the arrangement of filters in the light path, which vary in mechanical arrangement among manufacturers (who provide appropriate sets of filters for specific fluorochromes). The light must include wavelengths appropriate to the fluorochrome being used, and high-pressure mercury vapor arc lamps have the most inclusive spectral range. Before entering the lens train, the beam must be filtered to minimize heat transfers (i.e., remove the longer wavelengths) and then passed through another filter to isolate the frequency appropriate to activation of the fluorochrome. When the beam has passed through the objective, the light is still UV, so a barrier filter to stop the passage of eye-damaging frequencies is essential. This filter in the tube must allow the passage of the longer wavelengths arising from the excited fluorochrome. The choice of the latter may be governed by requirements of multiple colors when mixed probes are used.

1.6.1. Fluorescence Systems

The illumination systems for fluorescence microscopy take several forms: (i) transmitted light through a substage condenser; (ii) dark-field illumination through a specialized substage condenser; and (iii) incident illumination (epi-illumination) through a specialized objective, in which case the objective acts as both condenser and objective. The second and third systems have the advantage of an effective dark-field system, because the optical path is such that a direct beam does not go to the eye and there is minimal interference with the detection of the weaker fluorescent light from the specimen. In the third system, which is the best, the objective acts as the condenser and suffers no centration problem and the weak fluorescence suffers minimum attenuation from specimen thickness. The exciter beam is reflected down to the objective from a side or rear port in the microscope tube by a beam-splitting mirror or prism that reflects the exciting wavelength but transmits visible light back from the objective to the eye.

The manner of using the microscope and its filters may be unique to the instrument, so consult the manual provided by the microscope manufacturer.

1.6.2. Confocal Scanning Microscopy

The elaborate equipment (2, 19) required for confocal scanning microscopy, utilizing an intense beam of light from a laser, is designed to scan the sample by illuminating and imaging one very small area at a time in a single focal plane of the specimen. It is confocal because the scanning and the image are both attained through the objective. An epi-illumination system focuses a small spot of light at a plane in the specimen. This illuminated spot is imaged through a conjugate aperture, which accepts only the direct beams (but not the diffracted beams) for forming the image. The specimen is scanned through the objective by a moving beam producing a series of spots of light or aperture images. A raster scan allows the synthesis of a complete image in a detector system, and the image is displayed on a monitor. Only structure that is in focus will form an image. This imaging system will operate either with direct imaging by visible light or with fluorescence imaging by UV light.

There are two major forms of the scanning equipment. One form involves tandem multiple-aperture arrays, and the other involves a regulated movement of a very fine laser beam. The principle is straightforward, but the equipment is very specialized and requires its own instruction manuals and expert users.

The advantages of this method of microscopy are that it can be used at high magnification with the best of epi-illumination objectives (including oil-immersion) to study large objects that scatter a lot of light in other modes of microscopy. It can focus on the structure of a surface and, particularly, can allow examination of structure within large cells. As far as bacteria are concerned, a major use is the study of the interactions of bacteria with or within eukaryotic cells.

Information is increased by using confocal scanning microscopy as a form of fluorescence microscopy; photomultiplier imaging reduces the problems with dim signals, and the confocal system evades fluorescence interference from features above and below the plane of focus. Because images can be generated at known depths in the specimen, a computer correlation of a stacked series of images generates three-dimensional information about the specimen.

1.7. CELL MEASUREMENT

It is important to determine the magnification imposed on bacterial cells either in projections for drawings or on photomicrographs. Equally important, a range of dimensions may be needed as part of experiments or for the description of cells for purposes of classification. Although a number of measuring techniques may be applied, they all must be based on a measurement standard provided by a stage micrometer; the use of the stage micrometer is clearly described in texts (12, 18).

1.7.1. Stage Micrometer

A stage micrometer is a slide on which a number of lines have been ruled by a grating engine or deposited photographically to show a precise scale. Usually, there are 10 parallel lines at 0.1-mm (100 μm) spacing and 10 parallel lines at 0.01-mm (10 μm) spacing. Image the micrometer slide with the same care given to a specimen. Project the image onto the ground glass of a photomicrographic camera for direct measurement, or record the image on film under the same conditions used for photomicrographs. A number of measurements of the appropriate intervals are then used to generate either the magnification on the film, the length required for a bar scale to apply to micrographs, or a set of measurements of the cells under study.

1.7.2. Eyepiece Graticule Micrometer

An eyepiece graticule micrometer is an optically flat glass disk, about 12 mm in diameter, with an arbitrary but accurately proportioned scale engraved on it. The scale is usually in numbered units, each with 10 divisions. The graticule disk is placed in the plane of the intermediate image inside a Huygenian eyepiece, which has a diaphragm at the plane of the intermediate image where the graticule disk can be rested (access is obtained by unscrewing the top element of the eyepiece). In more complex eyepieces, the diaphragm is below the lower field lens, and so there are fewer problems with parallax.

Each graticule must be calibrated for each eyepiece-objective combination to be used. With a stage micrometer and an appropriate set of ruled lines in focus, rotate the eyepiece so that the graticule scale is aligned to the rulings and note a number of coincidence points. Then it is easy to derive a statistical measurement of the graticule unit in micrometers to two decimal places. Final direct measurements against the real specimen should not be considered more precise than one decimal place.

There are graticules ruled in squares for counting of cells (see chapter 2.1.4) or in circles for rough sizing of cells. When counting, relate the squares to the specimen area and count the cells lying within the squares and touching two of the sides. In this application, it is advantageous to have a special eyepiece allowing focus on the grid rulings by rotation of the top lens.

1.7.3. Eyepiece Screw Micrometer

The eyepiece screw micrometer is a specialized eyepiece (sometimes called a filar micrometer) for direct measurements after calibration against the rulings of a stage micrometer. The units are on a wheel (with or without a vernier reading scale) to be read against a fixed mark, and the wheel drives a hairline across the field, allowing a measurement in calibrated units by difference. As with an eyepiece graticule micrometer, the cursor line and the object must be in exact focus to minimize distortion by parallax.

1.8. PHOTOMICROGRAPHY

All types of light microscopy can make exact records on film, whatever the wavelength of light being used. The technical requirements are important, in all cases, so that the most faithful record may be obtained. There are excellent books available that amplify and explain the requirements (3, 4, 6, 8, 15, 18, 21). Chapters 3 through 5 provide information on digital imaging and image processing by computers.

Four types of photomicrography apparatuses are in general use, as follows.

1. A roll-film camera back and shutter (usually for 35-mm film), together with an integral device that includes the ocular and a beam-splitting prism with a side viewing arm to allow focusing, form the working unit. The whole device rests on the microscope tube.

2. Integral cameras have been developed by a number of microscope makers and come complete with an exposure meter and timing controls. These cameras are usually adequate and produce effective routine photomicrographs.

3. Digital and video cameras are available and used effectively for recording images. Both allow for storage and can provide images for projection, computer archiving, or image processing.

4. An old-fashioned light-tight bellows is sometimes used. It is located on a stand on which a microscope can also be placed. A light-excluding sleeve is fitted at the bottom of the bellows and meshes with its mate, which slips onto the top of the microscope tube. A plate carrier is fitted at the top of the bellows and can be made for either cut film or instant (Polaroid) negative film.

In all cases the success of photomicrography and the resulting photographic print depends on the following.

1. A first-class specimen preparation, appropriately mounted and exactly focused.

2. Good optics, properly aligned and optimally (Koehler) illuminated (section 1.2.7) to give the best possible image, and a suitable choice of color filters (section 1.2.5).

3. A camera-microscope assembly that is free from vibration.

4. Appropriate photographic materials and a processing technique appropriate to giving optimum grain size with adequate contrast and gray scale.

5. An appropriate final magnification, attained by printing with an enlarger.

The camera-microscope assembly can present some problems, because fixing photographic devices on the microscope tube tends to allow the transmission of vibrations. Partly, this is because pressing the shutter release removes the beam splitter before activating the shutter, and each action can generate persisting vibrations. When a shutter integral to the camera is used, it should be of the iris diaphragm type, and a focal-plane shutter must be avoided. Always use a cable release to actuate the shutter. Shutters for the timing of exposures are best put in the light path between lamp and microscope. Vibration is no problem with the bellows cameras as long as the light-tight collar around the eyepiece allows the camera and the microscope to be independent and not to touch each other. Vibrations must be kept to a minimum; a sturdy, heavy table that is not attached to a wall is a help.

Film for black-and-white photomicrography should be a panchromatic film of ASA 60 to 100 for general purposes (to keep the exposures at a manageable level). For phase-contrast photomicrography, a faster film (ASA 300 to 400) is needed because the light levels are much lower. An exposure meter allows repetition of values established by testing. If an old bellows-type photomicrographic camera is available, acquire it because it is the best apparatus for the highest-quality work, even though it is not the most convenient to use.

When exposing negatives and printing micrographs, think about the final magnification and the detail to be shown. Use film processing that enhances contrast with reasonably fine grain, as specified by the manufacturer. Grain in a developed negative dictates that you should not enlarge the film more than 2 to 2.5 times in printing. Good microscopy of most ordinary bacteria allows a total magnification of up to ×1,500, which is needed for fine detail and can be attained on the film (see section 1.6.1 for the determination of magnification). The larger images taken on a bigger area (4 by 5 in. [~10 by 13 cm]) using cut film in an old-fashioned camera, for which enlargements of 1.25 to 1.5 times are usually sufficient, are better than higher enlargements of much smaller images on 35-mm film. If the micrographs are not as sharp and as good as they appeared in the properly adjusted microscope, try again. Focusing is critical, whatever the camera used. The trick with bellows cameras is to use a plain glass insert and to focus on the image obtained with a focusing magnifier adjusted to focus on a mark on the inside of the glass. The traditional ground glass and black hood can still be used effectively.

Prints should be enlarged to a format that allows easy visibility of the important detail, with the object occupying most of the frame.

1.9. REFERENCES

Books on light microscopy and photomicrography, ranging from encyclopedias to paperbacks, are readily available. However, books can give only basic principles. Satisfying microscopy is best learned through direct instruction and practical experience, not a small part of which is the art of preparing worthwhile specimens. No single book is directed toward microscopy for bacteriologists, but the principles are the same for all users. The list presented below is representative of the books available.

1. **Barer, R.** 1968. *Lecture Notes on the Use of the Microscope.* Blackwell Scientific Publications, Oxford, United Kingdom.
   The advice of a master microscopist.
2. **Boyde, A.** 1990. Confocal optical microscopy, p. 185–204. In P. J. Duke and A. G. Michette (ed.), *Modern Microscopies.* Plenum Press, New York, NY.
   A description of equipment and applications.
3. **Bradbury, S., and B. Bracegirdle.** 1998. *Introduction to Light Microscopy.* Springer-Verlag, New York, NY.
4. **Bradbury, S., and P. Everett.** 1996. *Contrast Techniques in Light Microscopy.* Bios Scientific Publishers, Oxford, United Kingdom.
   Short handbook with simple explanations of lens systems.
5. **Cargille, J. I.** 1975. *Immersion Oil and the Microscope.* Technical reprint 10-1051. R. P. Cargille Laboratories, Cedar Grove, NJ.
   Another excellent booklet produced by a manufacturer.
6. **Culling, C. F. A.** 1974. *Modern Microscopy—Elementary Theory and Practice.* Butterworth & Co., London, United Kingdom.
   Another short paperback book.
7. **Delly, J. G.** 1998. *Photography through the Microscope.* Eastman Kodak Co., Rochester, NY.

An example of the booklets produced by manufacturers involved in aspects of microscopy. These booklets are obtainable from such firms and their agents in updated versions and are generally excellent.

8. **Engle, C. E. (ed.).** 1968. *Photography for the Scientist.* Academic Press, Inc., New York, NY.
   General aspects of scientific photography, including photomicrography.
9. **Hartley, W. G.** 1993. *The Light Microscope—Its Use and Development.* Senecio Publishing Co., Oxford, United Kingdom.
10. **James, J.** 1976. *Light Microscopic Techniques in Biology and Medicine.* Martinus Nijhoff Medical Division, Amsterdam, The Netherlands.
    A fine book on theory and practice, with emphasis on the latter. It provides good advice on special and advanced techniques, including phase-contrast, interference, dark-field, polarization, and fluorescence microscopy.
11. **McKinney, R. M., and W. B. Cherry.** 1985. Immunofluorescence microscopy, p. 891–897. In E. H. Lennette, A. Balows, W. J. Hausler, and H. J. Shadomy (ed.), *Manual of Clinical Microbiology,* 4th ed. American Society for Microbiology, Washington, DC.
12. **Mollring, F. K.** 1981. *Microscopy from the Very Beginning.* Carl Zeiss, Oberkochen, Germany.
    Another excellent booklet produced by a manufacturer.
13. **Pluta, M.** 1988 to 1993. *Advanced Light Microscopy,* Vol. 1–3. Polish Scientific Publisher/Elsevier, Amsterdam, The Netherlands.
14. **Quesnel, L. B.** 1971. Microscopy and micrometry, p. 1–103. In J. R. Norris and D. W. Ribbons (ed.), *Methods in Microbiology,* vol. 5A. Academic Press Ltd., London, United Kingdom.
15. **Quesnel, L. B.** 1972. Photomicrography and macrophotography, p. 276–358. In J. R. Norris and D. W. Ribbons (ed.), *Methods in Microbiology,* vol. 7B. Academic Press Ltd., London, United Kingdom.
    Quesnel's two articles are useful resources for optical details and practical advice on microscopy and photomicrography for microbiology in particular.
16. **Richardson, J. H.** 1991. *Handbook for the Light Microscope.* Noyes Publications, Park Ridge, NJ.
17. **Ross, K. F. A.** 1967. *Phase Contrast and Interference Microscopy for Cell Biologists.* Edward Arnold, London, United Kingdom.
    An excellent explanation of the theory and practice of phase-contrast and interference microscopy. Also practical discussion of photographic techniques applied to microscopy.
18. **Shillaber, C. P.** 1944. *Photomicrography in Theory and Practice.* John Wiley & Sons, Inc., New York, NY.
    Nothing is likely to replace this classic text, which deals exhaustively but readably with the properties of objective lenses, oculars, and condensers. It sets out the practice of good illumination, weighs the advantages of different mounting media, and deals with both theoretical and practical bench microscopy; however, it antedates phase microscopy.
19. **Shuman, H., J. M. Murray, and C. Di Lullo.** 1989. Confocal microscopy: an overview. *BioTechniques* 7:154–613.
    This review includes an example of three-dimensional reconstruction.
20. **Slayter, E. M.** 1970. *Optical Methods in Biology.* John Wiley & Sons, Inc., New York, NY.
    A source book for the theoretical bases of most forms of microscopy and for analytical processes including diffraction, spectroscopy, and related optical techniques. It is concerned with principles and not practice.
21. **Smith, R. F.** 1990. *Microscopy and Photomicrography—a Working Manual.* CRC Press, Inc., Boca Raton, FL.

A professionally illustrated procedural manual with minimal theory. Useful for a beginner with no experience.

22. **Spencer, M.** 1982. *Fundamentals of Light Microscopy.* Cambridge University Press, Cambridge, United Kingdom.
A useful general survey.

23. **Wang, Y.-L., and D. L. Taylor (ed.).** 1989. *Fluorescence Microscopy of Living Cells in Culture, part A. Fluorescent Analogs, Labelling Cells, and Basic Microscopy.* Academic Press, Inc., New York, NY.
A volume with helpful technical advice.

2

Sampling and Staining for Light Microscopy

TERRY J. BEVERIDGE, JOHN R. LAWRENCE, AND ROBERT G. E. MURRAY

Specific morphological details are required to characterize a microorganism; these are usually determined by means of light microscopy, but some require more sophisticated techniques such as laser scanning microscopy (chapter 3) or electron microscopy (chapter 4). Some of the light microscopy methods employed are time honored and trace their genesis to the early days of bacteriological science. This chapter is based on and is an updated version of a chapter in the earlier edition of this book, *Methods for General and Molecular Bacteriology* (12), to which R. G. E. Murray, R. A. Doetsch, and C. F. Robinow were major contributors; we recommend that you read the former chapter. Here, we have concentrated on the general means of characterizing bacteria by light microscopy because it includes the cytological approach to making the best of preparations for study and photomicrography. The methods are meticulously described in section 2.2.7, "Cytological Preparations," of the previous edition of this book.

The first approach to the study of natural populations is morphological classification, an assessment of relative numbers, and formation of an idea of the complexity of the bacterial community in advance of any attempt at cultivation. More specialized methods estimate the proportions of growing and nongrowing cells and the productivity of the populations (8, 13). Despite the problems in choice of method and accuracy of results (3), light microscopy provides most of the basic biological information about an ecosystem.

The establishment and maintenance of pure cultures (and even the struggle to maintain impure cultures of difficult bacteria) require control by microscopy. Frequently, contamination plagues the cultures maintained by those less practiced in bacteriology, so that regular microscopy and cultural control must be encouraged. The identification or description of bacteria inevitably requires the application of determinative staining methods, including those suitable for recognition of shape, special structures, behavior, or life cycles, as well as those for the measurement of cells. Digital image analysis techniques may also be applied to extract information regarding cell morphotypes, numbers, and biovolumes (9). The information gleaned from these techniques may be useful, but it is important to understand the methodological limitations because any laboratory manipulation may introduce some alteration of form and structure, albeit small in most cases. Despite shortcomings, microscopic observations of bacteria are necessary, but usually not sufficient, factors for identifying them. Nevertheless, errors in identification are often traceable to mistakes in judging the shape, Gram reaction, and motility of a new isolate.

2.1. SAMPLING

It is difficult to observe living bacteria directly in natural habitats, and information so obtained may be meager. There are several reasons for this.

- Not all environments contain large populations of bacteria per unit of mass. Organisms in marine and lake waters, for example, usually must be concentrated by filtration or centrifugation to obtain sufficient numbers for study. Waters are often low in nutrients, and minimal steady state and division often lead to extremely small cells (ultramicrocells). The presence of flocs or aggregates (ten to several hundred micometers in diameter) in most aquatic environments leads to localized high concentrations of bacteria. Better sites for productive growth may be provided by interfaces on suspended solids, surfaces of rocks, sands, sediments, and the surfaces of water plants; these sites adsorb and provide nutrients, making adhesion to surfaces advantageous and sometimes leading to the formation of biofilms.

- Certain environments may support too many bacteria per unit of mass and must be diluted for examination or cultivation. Sewage sludge contains an astonishing array of microorganisms intermixed with particulate materials. Certain soils and marine muds contain only a few bacteria mixed with much opaque colloidal matter, some of which may be hard to distinguish from bacterial forms.

- Most bacteria constituting the natural flora of an environment do not reveal particularly distinctive morphological features, such as the star-shaped *Prosthecomicrobium* species, sheaths of *Sphaerotilus* or *Leptothrix* spp., and trichomes of *Caryophanon* spp.

2.1.1. Air

The field of aerobiology encompasses both indoor and outdoor components and focuses on a wide range of microorganisms, including pathogenic and nonpathogenic varieties. Although the numbers in a unit of volume may be quite small, the simplest approach to air sampling is the use of open petri dishes containing a suitable nutrient agar near a suspected contamination source and is applicable when the concentration of organisms is relatively high. When large volumes of air have to be sampled and the organisms have to be characterized, bubble a defined volume (use a flowmeter in the system) through a flask of fluid medium, and then subject it to plate counting (7). Impaction and impingement samplers are also commercially available for sampling airborne organisms (26).

Filter sampling methods for bacteria (2) are derived from those used to sample particulates of industrial or public health importance; commercial filtering units are readily available. The pore size used for bacteria is usually 0.22 to 0.45 μm. Bacterial spores and vegetative cells are both counted (section 2.1.4) either by performing direct microscopic counts or by depositing a plate-sized (90-mm-diameter) filter on a nutrient surface for CFU counting after incubation. However, many vegetative bacterial cells die rapidly on an air filter because of excessive desiccation, and viable-cell counting can be inaccurate. It is also possible to do direct counting of bacteria on membrane filters by using epifluorescence microscopy (section 2.1.4).

2.1.2. Water

Since bacteria have optical properties similar to those of water and give minimal contrast with the use of ordinary transmission light microscopy, examine specimens by using phase-contrast or dark-field microscopy. Epifluorescence microscopy used with fluorescent nucleic acid stains (acridine orange, DAPI [4′, 6′-diamidino-2-phenylindole], etc.) is also an efficient method for detection of bacteria (see chapter 3). Direct microscopic examination of water samples generally reveals few bacteria (5, 11). Use filtration techniques to concentrate cells for estimation of the total number per unit of volume, as well as for direct visualization of different morphological types. Centrifuging water samples from most sources leads to a remarkably large gelatinous pellet, which includes polymers of diverse origins, entrapped particulates, and a diversity of microorganisms. Wet mounts of the pellets examined by phase-contrast microscopy or preparations stained with simple basic dyes or fluorescent stains reveal the biota for preliminary morphological classification (section 2.2.1).

As with air sampling (see above), membrane filters provide a rapid and effective means of sampling from fluids for direct microscopic counting of bacteria and viable-cell counting. Natural waters and other fluids (e.g., foods, oils, and solvent emulsions) can all be sampled, although each type of sample requires appropriate filters and methods. General overviews of membrane filtration can be found in most microbiology textbooks, and a comprehensive technical review is provided by Brock (2). A recommended filtration procedure for direct counting (33) is as follows.

1. Prestain polycarbonate filters (25-mm diameter; pore diameter, 0.2 μm) for 5 min in a 0.2% (wt/vol) solution of Irgalan black (color index, acid black no. 107; Union Carbide Corp., New York, NY) in 2% (vol/vol) acetic acid. (Prestained filters are usually commercially available and are preferred.)

2. Rinse the filter in cell-free distilled water and place it wet on a cell-free glass filter apparatus (Millipore Corp., Bedford, MA). Stained filters, dried after rinsing, can be stored for future use.

3. Fix the cells in 0.1% (vol/vol) aqueous glutaraldehyde.

4. Place the sample over the filter, and then add acridine orange (40% dye content) to a concentration of 0.1% (wt/vol) in 0.02 M Tris, pH 7.2 (at 20°C), to make a final concentration of 0.02% (wt/vol).

5. After staining for 3 min, draw the water sample-acridine orange through the filter membrane by suction.

6. Remove the membrane from the filter apparatus and place it over a drop of immersion oil (type 1, F, or A; Cargille Laboratories, Inc., Cedar Grove, NJ) on a glass microscope slide. Place another drop of oil on top of the membrane, followed by a glass coverslip. The steps must be performed rapidly to prevent the filter membrane from becoming dry. Examine the preparation by epifluorescence microscopy with a 100- or 200-W halogen lamp, a BG-12 excitation filter, an LP-510 barrier filter, and an FT-510 beam splitter. Bacteria on this filter will fluoresce green, and individual morphological types can be observed. Although acridine orange is easy to use (although a teratogen), it suffers from a tendency for nonspecific binding to organic matter, and depending on the sample source, other fluorescent stains should be considered.

Another, time-honored technique for scanning the biota that attach to surfaces, popularized by A. T. Henrici decades ago, involves suspending slides or coverslips for hours or days in the water being sampled and retrieving them for staining and microscopy. Many organisms from fresh and salt water are able to attach to surfaces.

2.1.3. Soil

One simple means of obtaining a selection (but by no means a complete one) of soil microorganisms for microscopic examination consists of placing the soil sample in the bottom of a watch glass or small dish and adding water *below* the sample by using a Pasteur pipette until a water surface is formed above the sample. Surface tension will have ensured the positioning of many soil cells at the air-water interface. A coverslip can be floated on the surface and can then be examined by microscopy. In situ incubation of glass slides will also provide an impression of the bacterial populations present.

2.1.4. Direct Total-Cell Counting

Direct counts can be made on bacterium-retaining filters either by the use of epifluorescence microscopy after appropriate staining of the cells on a portion of the filter membrane or by the use of phase microscopy on a clarified portion of the membrane (2). The counting is done for large numbers by using an appropriate eyepiece graticule ruled in squares (counting the cells within a square, counting a statistically sufficient number of squares, and relating the counts to the total area counted, the filtration area, and the volume filtered). For smaller numbers, use a statistically appropriate number of total fields of the objective in use (which has to be calibrated as to the area observed for subsequent relation of the counts to the filtration area and volume filtered). Use filters of low porosity (0.2 μm), because many bacteria in waters are small. Filters used for fluorescence microscopy should have minimum reflectance, should not fluoresce, and should be stained black (section 2.1.2). The size of the filter, e.g., 25 or 90 mm in diameter, must be appropriate for the volumes and cell concentrations involved and for the available filtering apparatus (e.g., the syringe attachment or vacuum system).

The filters must be made transparent for counting by phase or ordinary microscopy. Dry the filters after staining them with a basic dye (0.01 % [wt/vol] thionine). Clear a cut-out portion of the filter with a drop of immersion oil or xylene on the slide and on the membrane before covering with a coverslip. When doing filtrations in the field, dry the filter immediately after filtration; later, wet a cut-out portion for appropriate staining and then dry it for microscopy (as described above).

A simple method used for total counts is the acridine orange direct count by fluorescence microscopy (2, 4, 6), although newer stains such as the LIVE/DEAD Gram stain provided by Molecular Probes, Inc. (Eugene, OR), are also quite straightforward to use. For acridine orange counting, add the stain to the sample by bringing the acridine orange concentration to 0.01 to 0.1% (wt/vol) before filtration or by holding a small volume of acridine orange on the filter for 1 min after filtering the sample. Other fluorochromes, such as the fluorescein isothiocyanate, DAPI, and SYTO® series of nucleic acid stains, can also be used (more details are given in chapter 3). Diluent or rinsing waters should be of an appropriate quality and free of cells; e.g., samples from the sea require filter-sterilized artificial seawater medium. Formaldehyde (2%, wt/vol) may be used to prevent growth in samples. Published applications of the methods should be consulted, and consideration should be given to the sampling and counting strategies (2, 8, 15, 33). Also see chapter 1.

2.1.5. Direct Viable-Cell Counting

Viable-cell counting from a natural sample (without added formaldehyde) requires filtration of a set of appropriate dilutions and placing of the filters on a suitable nutrient agar for counting of the colonies formed after incubation. Viable-cell counts can also be estimated by using the frequency of dividing cells (2, 39) or by counting cells on a membrane filter that are incorporating a given substrate (2, 6, 49), and the proporton of live to dead cells can be estimated by correlating to total counts. The accuracy of these methods is not great.

The nalidixic acid method is representative of the approaches to viable-cell counts made by estimating synthetic activity in bacteria (8). Here, the drug inhibits DNA synthesis and division and the cells grow longer and more stainable. The procedure compares a fixed (2% [wt/vol] formaldehyde) aliquot of the sample with an unfixed aliquot to which is added 0.025% (wt/vol) yeast extract as a growth substrate plus 0.002% (wt/vol) nalidixic acid after an incubation period of 6 h. Filter the samples, and stain with 0.01% (wt/vol) acridine orange or fluorescein isothiocyanate for epifluorescence microscopy. Viable cells elongate when exposed to yeast extract and nalidixic acid, inducing nonseptate filaments which are enumerated as viable cells. Gram-positive cells are relatively insensitive to nalidixic acid; in some environmental samples, the nalidixic acid method may underestimate their numbers and one of the methods described below may be more appropriate. The technique can be modified or expanded as necessary, for example, with the use of in situ hybridization to allow assessment of the activity of specific species or groups of organisms (35).

Although acridine orange has been a traditional stain for nucleic acids, more stains are now available. DAPI is popular; it stains blue and penetrates into living cells. Ethidium bromide stains red but cannot enter cells to stain nucleic acids unless the semipermeability of the plasma membrane has been disrupted. A combination of DAPI and ethidium bromide can differentiate live (DAPI-stained, blue) from dead (ethidium bromide-stained, red) cells. A wide range of nucleic acid stains (the SYTO® series) is available from various suppliers (e.g., Molecular Probes, Inc.).

Certain companies have attempted to maximize the ease of differentiating between live and dead bacteria. The most popular is Molecular Probes' LIVE/DEAD BacLight® viability kit where two fluorescent reagents are mixed together to stain the cells (see section 2.3.5.3). There is no data available about how these two stains interact with archaea, but presumably, similar responses should be seen. There are not as many commercial kits available for determining yeast viability. Molecular Probes, Inc., also markets a LIVE/DEAD Yeast® kit which can be used for yeasts and a number of filamentous fungi. Here, a novel two-color fluorescent metabolic probe (FUN® 1) is used in combination with calcofluor white M2R, which is a general stain for many fungi. FUN 1 initially stains the cells a diffuse green, but this color is gradually metabolized into orange-red that is partitioned into the intravacuolar regions. Only living cells stain this way since the conversion from green to orange-red requires an intact cytoplasmic membrane and metabolic capacity. Another similar reagent (FUN® 2) is also available, and if this reagent is used, the initial diffuse green becomes yellow-orange. Cells that are not metabolically active remain green with both FUN 1 and FUN 2. These reagents have proven reliable for *Saccharomyces*, *Candida*, *Neurospora*, and *Aspergillus* spp. Additional information on these fluorescent stains for both bacteria and yeasts can be found on the company's website at www.probes.com (product information sheets MP 07007 and MP 07009).

Other fluorescent stains can also be used as vital dyes. For example, 0.02 to 0.05% (wt/vol) fluorescein diacetate is nonfluorescent when taken up by cells but is hydrolyzed by cellular esterases (38). The released fluorescein is easily detected by either transmitted- or epifluorescence microscopy, and the method may be applied to samples on filters as described above. Additional stains for metabolic activity (CTC) and specific enzyme activities (ELF-97) are discussed in chapter 3. A series of protocols for physiological characterization of bacterial cells is provided in Lisle et al. (37). In the application of fluorescent stains, consideration should be given to factors such as nonspecific binding, autofluorescence, and excitation-emission spectra and factors that may influence the efficiency of a fluor (see chapter 3 and reference 37).

2.1.6. Detection of Specific Bacteria

Recognition of specific bacteria in an environment by microscopy requires the application of either immunofluorescence techniques (see chapter 3), direct serological procedures (see chapter 8), or molecular methods by using species-specific RNA or DNA probes. The increasing power of molecular probes makes them the most popular and specific of all staining regimens today since probes for multiple receptors can be run on the same sample at the same time (e.g., Color Plate 1 and reference 48).

2.2. PREPARATION

Microscopy is effective and important in the primary characterization of organisms either present in the sample or growing in liquid media. Low-power microscopy can also be applied to colonies growing on agar plates, and valuable information about the arrangement and viability of cells within a colony can be obtained. Proper study of single colonies requires proper expertise and tools (often readily manufactured out of simple laboratory materials by an enterprising student) which are beyond the scope of this chapter. Excellent instructions were provided in the corresponding chapter of the previous edition of this book, and we refer you to it (12).

2.2.1. Living-Cell Suspensions

There are several approaches to the examination of living bacterial cells, either in a natural specimen or in culture. The initial preparation is usually a smear or film made by spreading a droplet with a wire loop over a few square centimeters of a microscope slide and then drying and fixing the slide (section 2.2.3) before simple staining (section 2.2.4). Cultures grown in liquid media can be used with the proviso that dilution with sterile medium (e.g., by adding a loopful to a drop of medium on a slide) is generally required for reducing the population for microscopic study to a useful concentration. To easily detect bacterial cells in liquid suspension, there should be more than 10^5 cells ml^{-1}. Cultures grown on solid media can be suspended on a microscope slide with a sterile loop in a drop of the liquid medium or a diluent (not tap water) to a faint turbidity. Collect cells that are in very low concentration either by centrifugation or accumulation on a membrane filter (e.g., a 0.2-µm-pore-size membrane filter in a syringe adapter), which is gently pressed onto a slide to transfer cells for simple staining (50). Remember that the age and conditions of a culture may affect the size and shape of cells, their surface components (flagella, pili, and capsules), their inclusions (sulfur, volutin and poly-β-hydroxyalkoanate granules, and endospores), and the selection of mutants. The culture conditions must be clearly defined in describing bacterial form and structure. Examination of living cells under conditions appropriate for growth is important to defining growth habit (e.g., mycelial or differentiated forms) and behavior (e.g., motility), as described for the following methods.

2.2.1.1. Wet Mounts

In preparing wet mounts of specimens, place a loopful or 1 drop (ca. 0.05 to 0.1 ml) of the sample on a clean and degreased (by prior heating for 20 min at 400°C) microscope slide and cover it with a glass coverslip. The latter should be 18 to 22 mm^2 and of no. 1 thickness (~0.15 mm thick) to allow an oil-immersion objective to focus through it. To prevent convection currents, drifting, and drying, seal the edges of the coverslip to the slide by applying Vaspar (a mixture of equal parts of petroleum and paraffin wax), birthday-candle wax (using the wick as a brush after melting the wax, not lighting it, in a pilot flame), or nail polish.

The motility of strictly aerobic bacteria can be observed only for a brief time in these preparations, because the bacteria cease moving once the oxygen is depleted. The inclusion of small air bubbles prolongs activity. Heat from the light source may interfere with motility if observations are to be made for a long period. A green filter is easy on the observer's eyes, and it utilizes the best corrections of the lenses.

Phase-contrast microscopy is recommended for examining wet mounts, and the thinnest possible film should be used for best results. Placing a piece of blotting paper over the coverslip before sealing assists in drawing off sufficient fluid to give a satisfactory thin film. Good views of cells can be obtained if a thin film of 0.75 to 1.5% (wt/vol) agar (for best results, use Noble agar or dialyzed agar) or 15 to 30% (wt/vol) gelatin is dried on the surface of the slide to absorb the drop of culture as it spreads under the applied coverslip. The cells are immobilized and immersed in a higher-refraction-index fluid and are close to the coverslip.

2.2.2. Negative Staining

Negative staining provides the simplest and often the quickest means of gaining information about cell shape, cell breakage, and refractile inclusions in cells such as sulfur and poly-β-hydroxyalkoanate granules and about endospores (16).

Procedure

1. Place a droplet of 7% (wt/vol) nigrosin on a coverslip (no. 1 thickness).
2. Mix a small sample of bacteria into the droplet.
3. Place another coverslip, rotated at 45°, over the droplet, and slide the two coverslips apart to form a thin film on each. Alternatively, spread the droplet over most of the coverslip with a loop until the fluid starts to dry and there are thicker and thinner parts of the film visible.
4. Let the film dry *completely*.
5. Place the coverslip on a slide, *film side down*, and fix it in place with several spots of birthday-candle wax along two edges.
6. Observe thin areas of the film under oil immersion. Only experience teaches how much of a culture to mix into how large a droplet and the appearance of an effectively spread film.

These "air-mounted" preparations reveal bacteria or fragments of them unstained and standing out brightly against a sepia background (Color Plate 2). They can be used to monitor the progress and extent of cell disruption and disintegration (see chapter 7). Negatively stained preparations should not be used for making cell length and width measurements (see below) because the capsule or slime layer outside the cell wall may exclude the nigrosin and because they display dried, unfixed, and partially collapsed cells. Furthermore, the negatively charged particles of colloidal nigrosin do not react with the bacterial surface because, at physiological pH, it also is negatively charged.

Consequently, the dark film dries so that the bacterium appears to be somewhat larger than in life, even if no capsule is present.

During storage, nigrosin solutions can acquire contaminating organisms. A little formaldehyde (0.5%) helps prevent growth and does not harm the solution. Flame sterilize loops, because the presence of dead bacteria leads to confusion. India ink also forms a sort of negative stain when added to a wet mount and can display large capsules effectively if the coverslip is pressed down tightly.

2.2.3. Fixation of Smears and Suspensions

A smear of bacteria (section 2.2.1) merely dried on a microscope slide can wash off during staining unless the cells are made to stick to the glass. Good preservation of structure usually requires a treatment to inactivate enzymes and to cross-link macromolecules. This fixation can be physical or chemical and usually results in the killing of the cells. (Remember that spores and some vegetative cells, e.g., those of *Mycobacterium* spp., are remarkably resistant to heat and chemical killing. Also, aerosol droplets can be formed and can spread to the surroundings, and fingers are easily contaminated.)

2.2.3.1. Heat Fixation

Heat fixation is the most common procedure used for stabilizing bacterial smears. Allow the smear to dry completely. Pass the underside of the slide several times over the flame of a Bunsen burner to induce adherence. The smear is now ready for simple staining (section 2.2.4), Gram staining (section 2.3.5), or other procedures. When considering morphological interpretations, remember that the drying and heating cause some shrinkage and distortion.

2.2.3.2. Chemical Fixation

Chemical fixation provides more accurate preservation of shape and structure, although it is more time-consuming than heat fixation. Useful procedures are as follows.

Fixation with an Aldehyde

Add formalin to a suspension to give a 5% (vol/vol) solution (i.e., ~1.7% [wt/vol] formaldehyde), and hold for a few minutes before making and drying a smear. This is a simple and adequate preparation for Gram staining. Alternatively, use 3% (vol/vol) glutaraldehyde in the same fashion.

Osmium Tetroxide Fixation of Wet Films

Prior to drying a smear, expose it to the fumes from aqueous 1% (wt/vol) OsO$_4$ for 2 min in a closed vessel. (*Caution:* osmium vapors are harmful, and this step must be done in a fume cabinet!) Cell arrangements will be better preserved if this step is included.

There is considerable advantage to microscopy on chemically fixed and lightly stained microorganisms on coverslips that are *mounted in water and not allowed to dry at any time in the processing*.

2.2.4. Simple Staining

Morphological studies of bacteria are generally done on heat-fixed smears that are stained with basic dyes. When a single dye is used, the process is referred to as simple staining. Simple staining of smears on slides is accomplished conveniently on a rack made of glass rods linked by short lengths of rubber tubing or of linked brass rods supporting slides over a sink or suitable vessel for disposal of stains and wash water. Coverslip preparations are best handled in Columbia staining dishes.

Procedure for Slides

1. Make a smear (section 2.2.1) of either living or chemically fixed cells (section 2.2.3.2).
2. Let the smear dry completely.
3. Heat fix (section 2.2.3) to gain adherence to the glass.
4. Flood the heat- or chemically fixed smear briefly with a solution of a basic dye such as crystal violet (for 10 s) or methylene blue (for 30 s), and then gently rinse with tap or distilled water and blot dry with absorbent paper.

Preparation of Staining Solutions

Crystal violet (Hucker formula)

```
Crystal violet . . . . . . . . . . . . . . . . . . . . . . . . . . . 2.0 g
Ethanol, 95% (vol/vol) . . . . . . . . . . . . . . . . . . 20 ml
Ammonium oxalate, 1% (wt/vol), aqueous . . . . 80 ml
```

Dissolve the crystal violet in the ethanol. Then add the ammonium oxalate solution, and allow it to stand for 48 h before use.

Methylene blue (Loeffler formula)

```
Methylene blue chloride . . . . . . . . . . . . . . . . . 1.6 g
Ethanol, 95% (vol/vol) . . . . . . . . . . . . . . . . . . 100 ml
Potassium hydroxide, 0.01% (wt/vol),
    aqueous . . . . . . . . . . . . . . . . . . . . . . . . . . . 100 ml
```

Prepare a saturated solution of methylene blue by adding the dye to the ethanol. Then add 30 ml of the supernatant solution to the potassium hydroxide.

If the cells stain lightly or not at all with one of the dyes, it may be necessary to stain them with carbol fuchsin (section 2.3.6.1); they may even be acid fast (section 2.3.6). A few seconds of exposure to this dye is usually enough for most bacteria.

2.2.5. Permanent Mounts

For extended preservation, the stained and dried film of bacteria on a coverslip should be cleared with xylene and then inverted on a drop of Canada balsam on a slide. Wipe excess balsam away with a tissue moistened with xylene. The balsam hardens in a day. Neutral mounting media consisting of a plastic (polystyrene) dissolved in a solvent (toluene or xylene) and containing a plasticizer (tricresyl phosphate) are marketed under such names as Permount (Fisher Scientific, Pittsburgh, PA). (*Caution: xylene, toluene, and many polystyrenes can be carcinogenic.*) Some of these media may be painted onto a stained film on a glass microscope slide without the need for a coverslip. Permount is neutral and does not become acid or discolor with age, nor does it tend to trap bubbles under the coverslip. Stained films of bacteria on coverslips, mounted in water and sealed with petroleum, clear nail polish, or candle wax, can be kept if evaporation is prevented. They remain useful for only a few days kept in a moist chamber, but they provide excellent material for photomicrography.

2.3. CHARACTERIZATION

The methods described below have been found satisfactory for revealing determinative characteristics for a large number of different bacteria. These techniques, however, are amenable to improvement to satisfy conditions for particular bacteria and the manner in which they have been grown. Cytologically rewarding preparations require great care in fixation, staining, handling, and microscopy. The following are sound routine methods, but some of the procedures, especially those for cytoplasmic inclusions, are most effective if applied to chemically fixed preparations with drying avoided at any stage.

2.3.1. Size

Cell lengths and widths cannot be measured with great precision because of certain unavoidable technical difficulties. The boundaries of living bacteria do not appear sharp when examined by phase-contrast or bright-field microscopy, and in phase microscopy they and internal details are obscured by halos. The halo problem can be reduced by mounting the organisms in a medium of a refractive index close to that of their cellular substance, which is done simply by using solutions of gelatin from 15 to 30% (wt/vol) (17), the best results being determined by experiment for each kind of organism and the detail to be observed. A less precise method to reduce the halos involves drying, ahead of time, a thin film of 15 to 30% (wt/vol) gelatin on slides and using these to make wet mounts. The fluid swells the gelatin, and the cells become embedded in it; this process immobilizes them under the coverslip and allows better phase relations between the organism and its environment. Similar preparations made on a thin layer of agar can also be useful (see section 2.2.1.1).

Cells in fixed and stained preparations all suffer some degree of distortion, depending on the technique used; hence, one obtains only an approximation of the true dimensions. Dried, stained films, even if prepared from formalin-fixed material, do not reveal the cell wall. Tannic acid-crystal violet staining (44) does reveal the cell wall and makes bacteria appear wider than they look when alive or dried, but this staining does not work well with gram-negative bacteria.

Microscopic measurements are made with either a calibrated ocular micrometer or a filar micrometer eyepiece. Calibration is attained by using a stage micrometer, which provides a set of engraved lines at defined intervals (chapter 1). After calibration, the stage micrometer is no longer required, but new calibrations must be made for each objective used. For repeated measurements and counting, consideration should be given to digital image analysis; an overview and examples are provided in reference 9.

2.3.2. Shape

For determinative and descriptive purposes, individual bacterial shapes are designated as straight, curved, spiral, coccobacillary, branching, pleomorphic, square ended, round ended, tapered, clubbed, or fusiform. Perhaps square, star shaped, stalked, and lobular should be added to these classical terms. Arrangements of individual bacteria are described as single, pairs, short chains (fewer than five bacteria), long chains (five or more bacteria), packets, tetrads, octets, clumps, filaments, and branching or mycelial forms.

In assigning the descriptive terms, it is assumed that natural groupings are not disturbed. For example, excessive shaking may break long chains or cause clumping of single cells. To minimize the breakup of long chains or other fragile associations, add formaldehyde to the culture to a final concentration of 1 to 2% (vol/vol). Centrifuge after 15 min, and resuspend the culture in water for making films or negative stains. This is highly recommended for streptococci.

The influence of culture age and medium composition also must be taken into account. When there are doubts, there is no substitute for examining early growth in the undisturbed state (section 2.2.1).

2.3.3. Mode of Division

The mode of division used by a given bacterium is ordinarily not seen at first glance, and continuous observation of the living organism in a suitable medium is required (41). Simple staining procedures may not adequately reveal details whether division is binary or ternary or takes place through budding or fragmentation. It may be necessary to use cell wall staining (44) and even electron microscopy (chapter 4) to determine this behavior.

2.3.4. Motility

Translational movement of bacteria by flagellar propulsion (swimming) may be observed in wet mounts (section 2.2.1.1) of specimens by use of, in most cases, the low-power or high-dry objectives. Bacteria vary in their translational velocity, and slow organisms must be differentiated from those showing only Brownian motion. To ascertain the presence of flagella in doubtful cases, as well as to determine flagellar distribution (polar, peritrichous, or lateral), staining procedures (section 2.3.10) and electron microscopy (chapter 4) may be required.

When a standard condenser and an oil-immersion objective are used, the light must be reduced by the aperture diaphragm and the condenser must be lowered to improve contrast. Much better resolution and contrast are attained with phase microscopy and (more dramatically) with dark-field microscopy. If the specialized condensers needed for the latter technique are not available, an adequate dark field is possible by using the stratagem of employing a phase microscope condenser, with the oil-immersion phase plate in place, for illuminating the specimen observed with standard (*not* phase) 10× or 45× objectives (chapter 1) and high-power eyepieces (12× to 16×). The condenser is not immersed, but its position for attaining the best dark field is discovered by trial. This trick makes it possible to see even spirochetes under low power!

Observation of living cells sometimes detects another motion effectively described as twitching (32), which is a sudden movement or change of position that is believed to be associated with the extension and retraction of fimbriae (pili) on the cells.

Some prokaryotic protists exhibit a peculiar type of translational movement known as gliding when in contact with a solid surface. Gliding is a comparatively slow, stately, intermittent progression parallel to the longitudinal axis of the organism. It is characterized by frequent directional changes and the absence of external locomotory organelles (32). Gliding is often not obvious on the primary medium of isolation but is usually facilitated on solid media containing very small amounts of the nutrients appropriate to the group of bacteria involved. It is best observed on an agar plate with a high-dry objective or with a low-power objective and high-power eyepiece. Single bacteria away from the margin of the colony and tracks or trails are often considered indicative of gliding motility. Young, active cultures of filamentous gliders form wavy, curly, or lacelike patterns. Translocation by gliding movement can occur at velocities of 10 to 15 μm/s.

Gliding should be differentiated from swarming, which is an expression of flagellar motility on agar under special conditions. Swarming is the active spreading of growth on rel-atively dry agar surfaces on which the bacteria move as groups or microcolonies (32) or, in some instances (e.g., *Proteus* species), as associations of filamentous swarmer cells. A continuously shifting pattern of organisms develops on the agar surface, ranging from short, tongue-like extensions and isolated comet-shaped groups cruising away from the mass to interlacing bands interspersed with empty areas. Swarming is being seen among an increasing number of bacteria as the conditions for induction are becoming known.

2.3.5. Gram Staining

Gram staining is the most important differential technique applied to bacteria (42). In theory, it should be possible to divide bacteria into two groups, gram positive and gram negative; in practice, there are instances when a given bacterium or archeon is gram variable (22–24). Numerous modifications of Gram staining have been published since the method was first developed by Christian Gram in 1884. The cells are stained with crystal violet and then treated with an iodine solution as a mordant (1). The purple-stained cells are then washed with ethanol. Gram-positive cells retain the stain (Color Plate 3), whereas gram-negative cells do not. To make the contrast between the two results obvious, the preparation is counterstained with a contrasting red dye (e.g., carbol fuchsin or safranin) so that gram-negative cells become red and are easily seen.

The Gram reaction is determined by the interaction of crystal violet and iodine and by the integrity and the structure of the cell wall, most particularly the molecular architecture of the peptidoglycan (murein) sacculus of bacteria (25, 27, 46). Cell walls of gram-positive cells do not allow the extraction of the crystal violet-iodine complex from the cytoplasm by a solvent. The crystal violet molecule is small enough to penetrate the interstices of the wall, but the dye-iodine complex is too large to exit (25). If the walls are broken or their structure is compromised by autolysis, exposure to lysozyme, or exposure to a wall-targeted antibiotic (e.g., penicillin), the complex is extracted during the decolorization step and the gram-positive bacteria stain as gram negative (Color Plate 3). Gram-positive bacteria that have died naturally stain as gram negative since autolysins have hydrolyzed their cell walls.

Archaea can also be stained by using the Gram reaction, but students *beware!* Their cell walls possess entirely different polymers and can be much simpler or more complex than those of bacteria; the staining response is not always unequivocal (1, 22–24).

The description of novel bacteria requires recording of the Gram reaction, but this description should also include the structural profile of the cell wall observed in sections by electron microscopy (chapter 4).

The smear of bacteria on the slide can be made directly for routine Gram staining purposes from a liquid or solid culture, as for simple staining (section 2.2.4), but is best made from a formalin-fixed, washed sample. The few extra minutes spent suspending the bacteria in 5% (vol/vol) formalin and concentrating and washing them by centrifugation pay off in good preservation of size and shape, good staining, and the absence of messy precipitates from liquid culture media. In either case, a thin smear should be used for air drying and heat fixation (thick clumps are hard to decolorize and can give a false positive).

Gram-negative bacteria may seem to be gram positive if the film is too thick and the decolorization is not completed. Gram-positive organisms, on the other hand, may seem to be gram negative if the film is overdecolorized; this happens particularly if the culture is in the late stationary phase of growth. Some *Bacillus* species are gram positive for only a few divisions after spore germination. It is advisable to prepare light films (faint turbidity) of young, actively growing cultures for best results, since older cultures tend to give variable reactions. A wise precautionary measure is to use known gram-positive and gram-negative organisms as controls.

It is important to standardize the Gram staining procedure, and two methods that give equivalent results are those of Hucker and of Burke (20, 21), as follows.

2.3.5.1. Hucker Staining Method
See references 6 and 20.

Solution A

Crystal violet (certified 90% dry content)	2.0 g
Ethanol, 95% (vol/vol)	20 ml

Solution B

Ammonium oxalate	0.8 g
Distilled water	80 ml

Mix A and B to obtain the crystal violet staining reagent. Store it for 24 h, and pass it through filter paper before use.

Mordant

Iodine	1.0 g
Potassium iodide	2.0 g
Distilled water	300 ml

Grind the iodine and potassium iodide in a mortar, and add water slowly with continuous grinding until the iodine is dissolved. Store the mordant in amber bottles.

Decolorizing Solvent
Ethanol, 95% (vol/vol)

Counterstain

Safranin O (2.5% [wt/vol] in 95 % [vol/vol] ethanol)	10 ml
Distilled water	100 ml

Procedure

1. Place the slide on a staining rack, and flood the smear with the crystal violet staining reagent for 1 min.
2. Wash the smear in a gentle and indirect stream of tap water for 2 s.
3. Flood the smear with the iodine mordant for 1 min.
4. Wash the smear in a gentle and indirect stream of tap water for 2 s; blot the film dry with absorbent paper.
5. Flood the smear with 95% ethanol for 30 s with agitation; blot the film dry with absorbent paper.
6. Flood the smear with the safranin counterstain for 10 s.
7. Wash the smear with a gentle and indirect stream of tap water until no color appears in the effluent, and then blot the film dry with absorbent paper.

2.3.5.2. Burke Staining Method
See reference 4.

Stain Solution A

Crystal violet (certified 90% dye content)	1.0 g
Distilled water	100 ml

Bicarbonate Solution B

Sodium bicarbonate	1.0 g
Distilled water	100 ml

Iodine Mordant

Iodine	1.0 g
Potassium iodide	2.0 g
Distilled water	100 ml

Prepare this mordant as for Hucker's method.

Decolorizing Solvent
Caution: this solvent is highly flammable.

Ethylether	1 volume
Acetone	3 volumes

Counterstain

Safranin O (85% dry content)	2.0 g
Distilled water	100 ml

Procedure

1. Flood the smear with solution A, add 2 or 3 drops of bicarbonate solution B, and let stand for 2 min.
2. Rinse off the stain with the iodine mordant, and then cover the smear with fresh iodine mordant for 2 min.
3. Wash off the mordant with a gentle and indirect stream of tap water for 2 s; blot *around* the stained area with absorbent paper, but do not allow the smear to dry.
4. Add the decolorizing solvent dropwise to the slanted slide until no color appears in the drippings (less than 10 s), and allow the smear to air dry.
5. Flood the smear for 5 to 10 s with the counterstain.
6. Wash the smear in a gentle and indirect stream of tap water until no color appears in the effluent; blot the smear dry with absorbent paper.

Several microbiological supply houses provide reliable kits which contain all of the reagents necessary for Gram staining; although these are more expensive than obtaining and making each reagent yourself, the convenience of a kit may be worthwhile (see section 2.5).

2.3.5.3. LIVE BacLight® and ViaGram® Methods
The LIVE BacLight® Gram stain kit is supplied by Molecular Probes, Inc. (Eugene, OR) and provides a one-step fluorescence analysis of the Gram state of living bacteria. The kit contains a mixture of green fluorescent SYTO9 and red fluorescent hexidium iodide (a nucleic acid stain). SYTO9 stains both live gram-positive and gram-negative bacteria. Hexidium iodide preferentially stains live gram-positive cells, and this stain displaces the SYTO9 dye. Accordingly, gram-negative bacteria fluoresce green, whereas gram positives fluoresce red. The Gram response can be distinguished between the two types in mixed cultures. Dead cells will give variable results, and it is best to use this method on actively growing exponential cultures. SYTO9 has an excitation-emission maximum of 490 nm/500 nm,

and hexidium iodide has an excitation-emission maximum of 480 nm/625 nm. Since this is a relatively new Gram staining method, it has not been tested on a large variety of bacteria and some inconsistencies with traditional Gram staining responses may still be found. So far, there are no data available about how the LIVE BacLight kits stain archaea.

ViaGram® Red+ Gram stain is another kit put out by Molecular Probes, Inc., and it has the advantage of also differentiating between live and dead cells based on their plasma membrane integrity. The kit contains three reagents, two nucleic acid stains for viability determination and a fluorescently labeled wheat germ agglutinin (WGA) for the determination of the Gram reaction. DAPI is a fluorescent blue stain used for cells with intact membranes (i.e., live cells), whereas SYTOX® Green can stain nucleic acids only if the plasma membrane has been disrupted (i.e., dead cells). The WGA is chemically attached to Texas Red®-X dye, and this entire complex selectively binds to most gram-positve bacteria. Once bacteria are stained, here are the possible results.

1. Gram negative, live: blue interior (DAPI)
2. Gram negative, dead: green interior (SYTOX Green)
3. Gram positive, live: blue interior (DAPI) and red surface (WGA-Texas Red)
4. Gram positive, dead: green interior (SYTOX Green) and red surface (WGA-Texas Red)

As with the LIVE BacLight® staining method, this is a relatively new method and a full range of bacteria have not yet been studied to ensure accurate correlation with more traditional Gram staining regimens. From our own experience, WGA (i.e., WGA-Texas Red) does not always bind to surfaces of gram-positive bacteria and clearly additional surface components such as capsules, S-layers, and sheaths could affect this new stain.

For more information on both of these new fluorescent stains, consult the product information pages (MP 07008 and MP 07022 at www.probes.com).

2.3.6. Acid-Fast Staining

A second important tinctorial property of certain bacteria is that of not being readily decolorized with acid-alcohol after staining with hot solutions of carbol fuchsin. Acid-fast bacteria include the actinomycetes, mycobacteria, and some relatives which, it is believed, react in this way because of limited permeability of the waxy components of the cell wall. Dormant endospores are also acid fast, but the cause is unknown.

Two methods are recommended for acid-fast staining.

2.3.6.1. Ziehl-Neelsen Staining Method
See reference 6.

Carbol Fuchsin Stain

Basic fuchsin	0.3 g
Ethanol, 95% (vol/vol)	10 ml
Phenol, heat-melted crystals	5 ml
Distilled water	95 ml

Dissolve the basic fuchsin in the ethanol, and then dissolve the phenol in the water. (*Caution:* phenol will harm skin and the fumes can be irritating.) Mix, and let the mixture stand for several days. Pass the mixture through filter paper before use.

Decolorizing Solvent

Ethanol, 95% (vol/vol)	97 ml
HCl (concentrated)	3 ml

Counterstain

Methylene blue chloride	0.3 g
Distilled water	100 ml

Procedure

1. Place a slide with an air-dried and heat-fixed smear on a slide carrier over a trough. Cut a piece of absorbent paper to fit the slide, and saturate the paper with the carbol fuchsin stain. Carefully heat the underside of the slide by passing a flame under the rack or by placing the slide on a hot plate until steam rises (*without boiling!*). Keep the preparation moist with stain and steaming for 5 min, repeating the heating as needed. (*Caution:* overheating causes spattering of the stain and may crack the slide.) Wash the film with a gentle and indirect stream of tap water until no color appears in the effluent.

2. Holding the slide with forceps, wash the slide with the decolorizing solvent. Immediately wash with tap water, as described above. Repeat the decolorizing and washing until the stained smear appears faintly pink.

3. Flood the smear with the methylene blue counterstain for 20 to 30 s, and wash it with tap water as described above.

4. Examine under oil immersion. Acid-fast bacteria appear red, and non-acid-fast bacteria (and other organisms) appear blue.

2.3.6.2. Truant Staining Method
See reference 51.

It is also possible to determine the acid fastness of an organism by using fluorescence techniques. A good example is the Truant technique. An advantage of this procedure is that slides of clinical specimens suspected of containing mycobacteria, for example, may be screened by using a 60× rather than a 100× objective; hence, an entire slide may be screened in a short time.

Fluorescent Staining Reagent

Auramine O, Cl 41000	1.50 g
Rhodamine B, Cl 749	0.75 g
Glycerol	75 ml
Phenol (heat-melted crystals)	10 ml
Distilled water	50 ml

Mix the two dyes with 25 ml of the water and the phenol. Add the remaining water and the glycerol, and mix again. Filter the resulting fluorescent staining reagent through glass wool.

Decolorizing Solvent

Ethanol, 70% (vol/vol)	99.5 ml
HCl (concentrated)	0.5 ml

Counterstain

Potassium permanganate	0.5 g
Distilled water	99.5 g

Procedure

1. Flood a lightly heat-fixed smear with the fluorescent staining reagent for 15 min.

2. Wash the slide with a gentle and indirect stream of distilled water until no color appears in the effluent.

3. Flood the smear with the decolorizing solvent for 2 to 3 min, and then wash the slide with distilled water as described above.

4. Flood the smear with the permanganate counterstain for 2 to 4 min.

5. Wash the slide with distilled water as described above, blot with absorbent paper, and dry.

Examine with a fluorescence microscope equipped with a BG-12 exciter filter and an OG-1 barrier filter. Acid-fast bacteria appear as brightly fluorescent yellow-orange cells on a dark background; non-acid-fast cells are dark.

2.3.7. Endospores

Mature, dormant endospores of bacteria, when viewed unstained, are sharp edged, even sized, and strongly refractile, shining brightly in a plane slightly above true focus. It is the core (protoplast) of the spore that is refractile; the surrounding cortex, coat, and exosporium appear dark and often difficult to discern. Electron microscopy is needed for determining the details of these peripheral structures. Do not assume that any highly refractile body within a bacterium is an endospore, particularly if the inclusion body is irregularly sized and if information concerning heat resistance is lacking; large poly-β-hydroxyalkoanate granules often appear in the cytoplasm of *Bacillus* spp. before sporulation and can be confusing.

2.3.7.1. Popping Test

A direct test is provided by a visible phenomenon (16) occurring in dormant endospores after immersion in an acid oxidizer (e.g., 0.1% wt/vol $KMnO_4$ in 0.3 N HNO_3). The spore cortex ruptures, and a portion of the spore protoplast including all of the nucleoplasm, now readily stainable with basic dyes, is herniated through the aperture. The dramatic suddenness of the event (after about 5 min in the solution) makes the term "popping test" appropriate. The test can be conveniently performed by mounting a dried film of spores on a coverslip in the reagent for 10 to 20 min; the consequent popping is visible with the oil immersion objective without staining, although staining makes the popping more easily visible.

2.3.7.2. Negative Staining

There are several methods for staining endospores in bacteria. The simplest method is the use of a negative stain (section 2.2.2), which yields lifelike preparations (16). In this method, mix a small loopful of 7% (wt/vol) aqueous nigrosin on a coverslip with an appropriate amount of culture, spread it into a thin film, and air dry. Invert the coverslip, place it film side down on a microscope slide (so the film remains in air when the coverslip is examined under oil), and maintain it in place with several spots of candle wax. Endospores appear as highly refractile spherical or ellipsoidal bodies both within the bacteria and free.

Endospores strongly resist staining by simple dyes until they germinate. Dormant endospores, once stained, are quite resistant to decolorization and hence are acid fast (section 2.3.6). A useful positive-staining method for bacterial endospores is that of Schaeffer-Fulton, as follows.

2.3.7.3. Schaeffer-Fulton Staining Method

In the Schaeffer-Fulton technique (47), 0.5% (wt/vol) aqueous malachite green is used instead of carbol fuchsin. Air dry the specimen on a glass slide, heat fix, cover with absorbent paper saturated with the dye, and place over a boiling-water bath for 5 min. Wash the slide in tap water, and counterstain the film with safranin for 30 s as in Gram staining (section 2.3.5). Then wash the slide and blot dry. Endospores appear bright green, and vegetative cells appear brownish red.

2.3.7.4. Dorner Staining Method

See reference 28.

1. Heat fix the sample on a glass slide, and cover it with a square of absorbent paper cut to fit the slide.

2. Saturate the paper with carbol fuchsin, and steam for 5 to 10 min, as described in section 2.3.6.

3. Remove the paper and decolorize the film with acid-alcohol (3 ml of concentrated HCl in 97 ml of 95% ethanol) for 1 min; rinse with tap water, and blot dry. Use forceps to handle the stain-covered slides.

4. Place a drop of 7% nigrosin (section 2.2.3) on the slide, cover with another slide, and draw the two apart to form a thin film of the nigrosin over the stained smear.

5. Examine under oil immersion. Vegetative cells appear colorless, the endospores appear red, and the background appears black to sepia.

2.3.8. Cysts

Bacterial cysts (e.g., as in *Azotobacter* spp.) stain weakly with simple stains and generally appear as spherical bodies surrounded by thick but poorly staining walls. The following method (52) is useful for demonstrating *Azotobacter* cysts, including forms developed prior to the appearance of a mature cyst.

Reagent

```
Glacial acetic acid . . . . . . . . . . . . . . . . . . . . . .     8.5 ml
Sodium sulfate (anhydrous) . . . . . . . . . . . . . .    3.25 g
Neutral red . . . . . . . . . . . . . . . . . . . . . . . . . .    200 mg
Light green S.F. yellowish . . . . . . . . . . . . . .    200 mg
Ethanol, 95% (wt/vol) . . . . . . . . . . . . . . . . . .     50 ml
Distilled water . . . . . . . . . . . . . . . . . . . . . . . .    100 ml
```

Add the chemicals and dyes to the water with continuous stirring for 15 min. Pass the mixture through a filter (pore diameter, 0.5 μm).

Procedure

Immerse the specimen in the reagent, and examine the wet preparation. Vegetative azotobacters are yellowish green; early encystment stages appear with a darker green cytoplasm somewhat receded from the outer cell wall, from which it is divided by a brownish red layer. In mature cysts, the central body appears dark green and is separated by the unstained intine from the outer, brownish red exine peripheral layers of the cyst.

2.3.9. Capsules and Slime Layers

Capsules and slime layers are produced by bacteria capable of forming them under specific culture conditions and are best demonstrated in wet preparations because the highly hydrated polymers constituting them are distorted and shrunk by drying and fixation.

The capsules of specific pathogens (e.g., pneumococci, *Klebsiella pneumoniae*, and meningococci) can be displayed effectively by using antisera specific for the capsule type, and this method can provide a presumptive identification. Suspend the clinical exudate or the organism in a drop of capsule-specific antiserum containing 0.01% (wt/vol) methylene blue under a coverslip. The cell is stained and the surrounding capsule becomes outlined by a dark blue line in a short time. For determinative purposes, this is known as the Neufeld Quellung (or capsule-swelling) test. Fluor-conjugated lectins may also be used to see capsules and more extensive exopolymer networks around cells and cell aggregates (chapter 3).

If antisera are not available, Duguid's method (see below) is the simplest and best staining method.

2.3.9.1. Duguid Negative Staining Method
See reference 29.

1. Place a large loopful of India ink on a clean slide, and mix in a loopful of culture; then place a glass coverslip over this in such a way that only part of the mixture is covered.

2. Press firmly down on the coverslip, using several thicknesses of absorbent paper, until the ink is sepia beneath the cover glass.

3. Examine with high-dry and oil immersion lens systems. Capsules appear as clear zones around the refractile organism and against the brownish black background full of dancing particles of India ink.

2.3.10. Flagella
Although Koch devised a technique for staining flagella over a century ago, no easy or constantly reliable method is yet available. Since the width of individual bacterial flagella lies below the limits of resolution for transmission light microscopy (flagella are 14 to 28 nm in diameter [53]), it is necessary to "tar and feather" the flagella to make them visible. NanoOrange® (Molecular Probes) is a new fluorescent protein stain that has also proven to be an effective method for screening bacteria for flagella (30).

Best results are obtained when using superclean slides; this reduces drying artifacts and precipitation of stains. Because flagella are sheared off easily by drying and mechanical forces, there should be a minimum of agitation while making the smear. More information about the form and disposition of flagella is obtained by electron microscopy (chapter 4).

Some methods for staining flagella are labeled as simple (31), but few are simple in practice. Leifson's method has enjoyed good success for many years.

2.3.10.1. Leifson Staining Method
See reference 36.

Solution A

Sodium chloride	1.5 g
Distilled water	100 ml

Solution B

Tannic acid	3.0 g
Distilled water	100 ml

Solution C

Pararosaniline acetate	0.9 g
Pararosaniline hydrochloride	0.3 g
Ethanol, 95% (vol/vol)	100 ml

Mix equal volumes of solutions A and B; then add 2 volumes of this mixture to 1 volume of solution C. The result-

ing dye solution may be kept under refrigeration for 1 to 2 months, but it should be used as soon as possible.

Procedure

1. Prepare an air-dried film on a slide. Using a wax glass-marking pencil, draw a rectangle around the film.

2. Flood dye solution onto the slide within the confines of the wax lines. Leave for 7 to 15 min, the best time to be determined experimentally.

3. As soon as a golden film develops on the dye surface and a precipitate appears throughout the film, remove the stain with gently flowing tap water. Air dry.

4. Examine under oil immersion. Bacterial bodies and flagella stain red.

2.3.11. Cytoplasmic Inclusions
Many bacteria grown under certain conditions produce deposits within the cytoplasm which are termed inclusions. Among these are deposits of fat, poly-β-hydroxyalkoanate (often referred to as poly-β-hydroxybutyrate), polyphosphate, starch-like polysaccharides, sulfur, and various crystals. Some of these are polymerized waste products, and others are food reserves. Several staining procedures can be used to reveal some of these inclusions, which are also evident by both phase-contrast and interference microscopy because of their refractility.

2.3.11.1. Staining of Poly-β-Hydroxyalkoanate

1. Prepare a heat-fixed film of the specimen on a slide and immerse in a filtered solution of 0.3% (wt/vol) Sudan black B made up in ethylene glycol. Stain for 5 to 15 min. Drain and air dry the slide.

2. Immerse and withdraw the slide several times in xylene, and blot dry with absorbent paper.

3. Counterstain for 5 to 10 s with 0.5% (wt/vol) aqueous safranin.

4. Rinse the slide with tap water, and blot dry.

5. Examine under oil immersion. Poly-β-hydroxyalkoanate inclusions appear as blue-black droplets, and cytoplasm appears pink.

2.3.11.2. Staining of Polyphosphate (Metachromatic or Volutin Granules)

1. Prepare a heat-fixed film of the specimen on a glass slide and stain for 10 to 30 s in a solution of Loeffler methylene blue (section 2.2.4) or 1% (wt/vol) toluidine blue.

2. Rinse the slide with tap water, blot dry, and examine under oil immersion. With methylene blue, polyphosphate granules appear as deep blue to violet spheres and the remaining cytoplasm appears light blue. Toluidine blue stains metachromatically, and the granules appear red in a blue cytoplasm.

2.3.11.3. Periodate-Schiff Staining of Glycogen-Like Polysaccharides
See reference 34.

1. Prepare a heat-fixed smear of the specimen on a glass slide.

2. Flood the slide for 5 min with periodate solution (20 ml of 4% [wt/vol] aqueous periodic acid, 10 ml of 0.2 M aqueous sodium acetate, 70 ml of 95% ethanol [protect the solution from light!]). Wash the slide with 70% ethanol.

3. Flood the slide for 5 min with a reducing solution containing 300 ml of ethanol, 5 ml of 2 M HCl, 10 g of potassium iodide, 5 ml of sodium thiosulfate pentahydrate, and

200 ml of distilled water. (Add the ethanol and then the HCl to the solution of potassium iodide and sodium thiosulfate in distilled water. Stir, allow the sulfur precipitate to settle, and then decant the supernatant for use.)

4. Wash the slide with 70% ethanol.

5. Stain for 15 to 45 min (the time to be determined experimentally) with the following Schiff reagent. (Dissolve 2 g of basic fuchsin in 400 ml of boiling distilled water, cool to 50°C, and filter through paper. Add 10 ml of 2 M HCl and 4 g of potassium metabisulfite to the filtrate. Stopper tightly, and allow to stand for 12 h in a cool, dark place. Add about 10 ml of 2 M HCl until the reagent, when dried on a glass slide, does not show a pink tint.) Store the reagent in the dark.

6. Wash the smear several times in a solution consisting of 2 g of potassium metabisulfite and 5 ml of concentrated HCl in 500 ml of distilled water.

7. Wash the slide with tap water, and counterstain the smear with a 0.002% (wt/vol) aqueous malachite green for 2 to 5 s.

8. Wash the slide in tap water, blot dry, and examine under oil immersion. Polysaccharides appear red, and cytoplasm appears green.

2.3.12. Nucleoids

The display of bacterial nucleoids requires special care in fixation and in mounting of the preparation for high-resolution microscopy. They can be seen in living cells (section 2.2.1) by phase-contrast microscopy as long as the refractive index of the surrounding medium is raised to a sufficient level (15 to 30% gelatin in a fluid mount; the exact concentration requires experimentation). Interference microscopy (chapter 1) may be used if it is available.

Fixing and staining of bacterial nucleoids to show reliable shape and form are difficult at the best of times (39), and the reader is referred to the excellent review article by Robinow and Kellenberger (43). The cytological methods given in the previous edition (12) are most reliable. Nucleoids can also be revealed by fluorescence methods (9, 18, 37), and more information can be found in chapter 3.

2.4. SPECIAL TECHNIQUES

A great number of techniques based on light microscopy are special to particular areas of bacteriology. Medical microbiology, in particular, has had to develop many methods to assist decisions about procedure, to attain an early presumptive diagnosis, to detect bacteria that are difficult to stain or that occur in small numbers, and to confirm the identity of a specific organism. These uses and attendant precautions are detailed in chapters of the *Manual of Clinical Microbiology*, in which the staining methods are separately summarized (6), as well as in other references in section 2.6.1. Here, we give examples of some special methods not routinely used.

2.4.1. Spirochetes

Spirochetes do not stain well with ordinary stains, but free-living spirochetes show up well negatively stained in nigrosin films (section 2.2.2). A number of spirochetes have a width near the limits of light microscope resolution; consequently, these bacteria are best observed in life by direct dark-field or phase-contrast microscopy. Spirochetes stained with Giemsa, a Romanowsky-type stain used for hematology (10), fluoresce bright golden yellow when ob-

served by dark-field microscopy. The following positive stain generally gives good results.

Fontana Silver Staining

Solution A (fixative)

Acetic acid	1 ml
Formalin	2 ml
Distilled water	100 ml

Solution B
Ethanol, absolute

Solution C (mordant)

Phenol	1 g
Tannic acid	5 g
Distilled water	100 ml

Solution D
Add a solution of 10% (wt/vol) aqueous ammonium hydroxide dropwise to a 0.5% (wt/vol) aqueous solution of silver nitrate until the precipitate that initially forms just redissolves.

Procedure

1. Using ultraclean slides, air dry a film of the specimen on a slide.

2. Fix the film in solution A for 1 to 2 min, and rinse it in solution B for 3 min.

3. Cover the film with solution C, and heat until steam rises for 30 s.

4. Wash the film with distilled water, and air dry.

5. Cover the film with solution D, and heat until steam rises and the film appears brown. Wash the film, and air dry.

6. Examine under oil immersion. Spirochetes appear brownish black.

2.4.2. Mycoplasmas

Mycoplasmas are the smallest free-living forms observable by light microscopy and are hard to recognize individually because of extreme polymorphism. They are parasitic in animal and plant tissues, and their growth and study require specialized techniques. The inexperienced would be well advised to seek expert help when faced with the need to recognize, cultivate, and identify mycoplasmas.

Mycoplasmas lack cell walls, and their cells are bounded only by a plasma membrane. The individual cells tend to be irregular, but a number of species and genera each have definite overall shapes (star shaped, round, goblet shaped, filamentous, and helical) which can be recognized by using phase-contrast or dark-field microscopy (chapter 1), usually assisted by electron microscopy of both negatively stained and sectioned preparations (chapter 4). Some species show a gliding motility on the surfaces of the tissues they infect. However, mycoplasmas growing in complex medium, tissue culture, or plant or animal tissues can be hard to distinguish among the large array of confusing artifacts and tissue components.

They grow slowly on appropriate nutrient agar plates, and colonies are microscopic, often ≤100 μm in diameter. Plates must be examined with a stereoscopic dissecting microscope (magnification, ×20 to ×60) by using oblique illumination to recognize the characteristic "fried-egg" colonies, which have a central plug penetrating the agar and a superficial spreading periphery. Colony margins may also be examined in situ at high magnification by placing a droplet of methylene blue onto the colony and covering it with a coverslip for examination by using transmitted light with high-power objectives.

The standard stained smears made by using heat-fixed films and Gram staining or staining with basic dyes are useless and uninformative. Stained preparations of mycoplasmas require careful chemical fixation, and staining with Giemsa (section 2.4.3.1) or Wright's (6) stain is preferred. The osmotic requirements for integrity of shape and form must be maintained until the chemical fixation is complete.

2.4.3. Rickettsiae

Most rickettsiae are obligate intracellular parasites that are rod shaped, occur in pairs, and show marked pleomorphism. Infected tissue samples may be examined directly by fluorescence microscopy with specific antisera labeled with fluorescein isothiocyanate. With the Gram stain, rickettsiae appear gram negative.

Rickettsiae may be seen in films of infected tissue by direct microscopic examination after staining by the Giemsa or Gimenez procedure (see below). Most rickettsiae appear as pleomorphic coccobacillary organisms varying in length from 0.25 to 2.0 μm. Pairs and short chains are most frequently observed. *Coxiella burnetii* is the smallest (0.25 by 1.0 μm), and it appears as a bipolarly staining rod in the cytoplasm of infected cells. The rickettsiae that cause typhus and spotted fever are generally larger (0.3 to 0.6 by 1.2 μm). Spotted-fever rickettsiae are found in the nuclei and cytoplasm of infected cells and often appear to be surrounded by a halo, sometimes mistaken for a capsule.

2.4.3.1. Giemsa Staining Method

Stock Solution

Giemsa powder	0.5 g
Glycerol	33 ml
Absolute methanol (acetone free)	33 ml

Dissolve the Giemsa powder in the glycerol at 55 to 60°C for 1.5 to 2 h. Add the methanol, mix thoroughly, allow to stand and decant. Store the sediment-free stock solution at room temperature. For use, dilute 1 part of the stock solution with 40 to 50 parts of distilled water or 0.5 M phosphate buffer at pH 6.8.

Procedure

1. Air dry the specimen on a glass slide, fix in absolute methanol for 5 min, and again air dry.
2. Cover the slide with freshly diluted Giemsa stain for 1 h.
3. Rinse the slide in 95% ethanol to remove any excess dye, and air dry.
4. Examine under oil immersion for the presence of basophilic intracytoplasmic bacteria which will be small, pleomorphic, and purple.

2.4.3.2. Gimenez Staining Method

Stock Solution

Basic fuchsin	10 g
Ethanol, 95% (vol/vol)	100 ml
Phenol, 4% (wt/vol) aqueous	250 ml
Distilled water	650 ml

Dissolve the dye in the ethanol, and then add the other ingredients. Allow the stock solution to stand at 37°C for 48 h. For use, dilute the stock solution 1:2.5 with phosphate buffer (pH 7.5) prepared by mixing 3.5 ml of 0.2 M NaH_2PO_4, 15.5 ml of 0.2 M Na_2HPO_4, and 19 ml of distilled water. Then filter the solution, which will keep for 3 to 4 days but should be filtered before each use.

Procedure

1. Heat fix the smear, and cover it with the staining solution for 1 to 2 min.
2. Wash the smear with tap water, and counterstain for 5 to 10 s with 0.8% (wt/vol) aqueous malachite green.
3. Wash the smear again with tap water, and make a second application of the malachite green. Rinse the slide thoroughly in tap water, and blot it dry.
4. Examine under oil immersion. Rickettsiae appear reddish against a green background.

2.4.4. *Legionella*

The members of the family *Legionellaceae* cause problems of recognition and detection because several species in the genera *Legionella*, *Fluoribacter*, and *Tatlockia* cause life-threatening disease, yet they and other species are common in freshwaters (including tap water) and inhabit some species of amoebae. Because they grow only on special media, a degree of suspicion of their presence is needed. The appearance of simple to pleomorphic bacilli in tissues and exudates or cultures from likely sources revealed by the nonspecific Gimenez (section 2.4.3.2) or Gram stain may suggest the use of the direct fluorescent-antibody test as a means of presumptive determination. This test involves the use of several group-specific polyvalent pools of fluorescent antibody. Once a group is identified, a species-specific fluorescent antibody can be used. False-positive reactions are rare and are usually laboratory induced (e.g., because of incorrect reagent or contaminations) when they occur. The commercial antibody is usually conjugated with fluorescein isothiocyanate. The procedure (45), which is an example of similar direct-fluorescent antibody tests for other organisms, is as follows.

Preparation

Make smears, impressions of the cut surfaces of tissues, or spreads of tissue scrapings on alcohol-cleaned slides. Distilled water and aqueous reagents must be filter sterilized, and the use of vessels and moist chambers (petri dish with moist filter paper) must be regulated to avoid any bacterial contamination of the slides from either other specimens or the environment, which may harbor members of the *Legionellaceae*.

Procedure

For a smear, air dry and gently heat fix on a slide. For a spread of tissue cells and impressions, fix in filter-sterilized 10% neutral formalin (add $CaCO_3$ to the bottle, and let it settle). Add enough of the appropriate conjugated antibody to spread over the entire surface of the smear. Use a moist chamber to prevent drying during the 20- to 30-min reaction at room temperature. Immerse the slide, first in buffered saline and then in distilled water. Drain, mount the smear in a drop of buffered glycerol (9 volumes of glycerol, 1 volume of 0.5 M $NaHCO_3$-Na_2CO_3 buffer [pH 9.0]), add a coverslip, and then seal with clear nail polish. Examine with a fluorescence microscope, preferably by epi-illumination (chapter 1), with appropriate exciting and barrier filters for the fluorochrome in use. Useful practical hints and a more extensive discussion of components are given by McKinney and Cherry (11).

2.5. SOURCES OF STAINING REAGENTS

Most of the laboratory requirements are available from the supply companies, but some stains are not widely stocked. E. Merck, Darmstadt, Germany, is a major source of biological stains, and this multinational company has acquired many of the suppliers of former times: BDH Chemicals, Harleco, Hopkin & Williams, J. T. and E. Gurr Co., and Raymond A. Lamb Co. A wide range of stains and other reagents are available through E. Merck subsidiaries. General staining reagents (e.g., Gram stain), glass slides, and coverslips, etc., are provided by a number of scientific suppliers. Merck subsidiaries and general suppliers are as follows.

BD, Franklin Lakes, NJ (supplier of BBL reagents)

BDH Chemicals, Toronto, Ontario, Canada

BDH Chemicals, Poole, Dorset, United Kingdom

Difco Laboratories, Detroit, MI

EM Science, Fort Washington, PA (in Canada, Cedarlane Labs Ltd., Hornby, Ontario)

Fisher Scientific Ltd., Pittsburgh, PA (in Canada, Nepean, Ontario)

Merck Japan Ltd., Tokyo, Japan

Sigma-Aldrich Canada, Oakville, Ontario

Many of the newer fluorescent stains can be obtained through Molecular Probes, Inc., Eugene, OR (www.probes.com).

2.6. REFERENCES

2.6.1. General References

1. **Beveridge, T. J.** 2001. Use of Gram stain in microbiology. *Biotech. Histochem.* **76:**111–118.
2. **Brock, T. D.** 1983. *Membrane Filtration: a User's Guide and Manual.* Science Tech Inc., Madison, WI.
   A complete guide and instruction in the use and interpretation of membrane filter techniques.
3. **Brock, T. D.** 1987. The study of microorganisms *in situ:* progress and problems, p. 1–17. *In* M. Fletcher, T. R. G. Gray, and J. G. Jones (ed.), *Ecology of Microbial Communities.* Cambridge University Press, Cambridge, United Kingdom.
   An assessment of the applications of counting techniques in microbial ecology.
4. **Clark, G.** 1973. *Staining Procedures Used by the Biological Stain Commission,* 3rd ed. The Williams & Wilkins Co., Baltimore, MD.
   A resource for methods used in association with H. A. Conn's Biological Stains (10).
5. **Hall, G. H., J. G. Jones, R. W. Pickup, and B. M. Simon.** 1990. Methods to study the bacterial ecology of freshwater environments. *Methods Microbiol.* **23:**181–210.
   A modern compilation of methods for bacteriological assessment of freshwaters.
6. **Hendrickson, D. A., and M. M. Krenz.** 1991. Reagents and stains, p. 1289–1314. *In* A. Balows, W. J. Hausler, Jr., K. L. Herrmann, H. D. Isenberg, and H. J. Shadomy (ed.), *Manual of Clinical Microbiology,* 5th ed. American Society for Microbiology, Washington, DC.
   A part of this chapter gives details of staining methods useful in clinical microbiology.
7. **Herbert, R. A.** 1990. Methods for enumerating microorganisms and determining biomass in natural environments. *Methods Microbiol.* **22:**1–39.
   A useful review and assessment of approaches to estimating biomass in environments.
8. **Karl, D. M.** 1986. Determination of *in situ* biomass, viability, metabolism, and growth, p. 85–176. *In* J. S. Poindexter and E. R. Leadbetter (ed.), *Bacteria in Nature,* vol. 2. Plenum Press, New York, NY.
   Descriptions and assessments of the methods applied to determination of biomass and viable counts in the field.
9. **Lawrence, J. R., D. R. Korber, and G. M. Wolfaardt.** 2001. Digital image analyses of microorganisms. *In* G. Bitton (ed.), *Encyclopedia of Environmental Microbiology,* John Wiley and Sons, New York, NY.
   A useful review with clear examples of the application of image-processing and analysis techniques.
10. **Lillie, R. D.** 1977. *H. A. Conn's Biological Stains,* 9th ed. The Williams & Wilkins Co., Baltimore, MD.
    An extensive compendium of stains and dyes with discussion of their chemical structures and applications. Also see reference 4.
11. **McKinney, R. M., and W. B. Cherry.** 1985. Immunofluorescence microscopy, p. 891-897. *In* E. H. Lennette, A. Balows, W. J. Hausler, Jr., and H. J. Shadomy (ed.), *Manual of Clinical Microbiology,* 4th ed. American Society for Microbiology, Washington, DC.
    A useful source of advice on the practice of immunofluorescence microscopy.
12. **Murray, R. G. E., R. N. Doetsch, and C. F. Robinow.** 1994. Determinative and cytological light microscopy, p. 21–41. *In* P. Gerhardt, R. G. E. Murray, W. A. Wood, and N. R. Kreig (ed.), *Methods for General and Molecular Bacteriology,* ASM Press, Washington, DC.
    A more expanded source for observation of live culures grown on agar, for chemical fixations, and for nucleoid visualization. We highly recommend that you read this chapter!
13. **Newell, S. Y., R. D. Fallon, and P. S. Tabor.** 1986. Direct microscopy of natural assemblages, p. 1–48. *In* J. S. Poindexter and E. R. Leadbetter (ed.), *Bacteria in Nature,* vol. 2. Plenum Press, New York, NY.
    A survey with useful references to studies in the field.
14. **Norris, J. R., and D. W. Ribbons (ed.).** 1971. *Methods in Microbiology,* vol. 5A. Academic Press, Inc., New York, NY.
    This is an old but useful volume; the material presented in the first part is particularly useful as a source of information supplementary to that presented here. The relevant chapters are as follows.

    I. Microscopy and micrometry, by L. B. Quesnel, p. 1–103.

    II. Staining bacteria, by J. R. Norris and H. Swain, p. 105–134.

    III. Techniques involving optical brightening agents, by A. M. Paton and S. M. Jones, p. 135–144.

    IV. Motility, by T. Iino and M. Enomoto, p. 145–163.
15. **Pickup, R. W.** 1991. Development of molecular methods for the detection of specific bacteria in the environment. *J. Gen. Microbiol.* **137:**1009–1019.
    A discussion of the strategies for sampling and detecting bacteria in the environment with special reference to genetically modified bacteria.
16. **Robinow, C. F.** 1960. Morphology of bacterial spores, their development and germination, p. 207–248. *In* I. C. Gunsalus and R. Y. Stanier (ed.), *The Bacteria,* vol. 1. Academic Press, Inc., New York, NY.
    A classic account of bacterial spores and their study using light microscopy. The same volume contains other illustrated articles on bacterial structure.

17. **Robinow, C. F.** 1975. The preparation of yeasts for light microscopy. *Methods Cell Biol.* **11**:1–22.
    Practical advice on using gelatin-agar slide cultures.
18. **Rosebrook, J. A.** 1991. Labeled-antibody techniques: fluorescent, radioisotopic, and immunochemical, p. 79–86. *In* A. Balows, W. J. Hausler, Jr., K. L. Herrmann, H. D. Isenberg, and H. J. Shadomy (ed.), *Manual of Clinical Microbiology*, 5th ed. American Society for Microbiology, Washington, DC.
    The applications of immunomicroscopy in clinical microbiology

2.6.2. Specific References

19. **Barer, R., R. F. A. Ross, and S. Thczk.** 1953. Refractometry of living cells. *Nature* (London) **171**:720–724.
20. **Bartholomew, J. W.** 1962. Variables influencing results, and the precise definition of steps in Gram staining as a means of standardizing the results obtained. *Stain Technol.* **37**:139–155.
21. **Bartholomew, J. W., and T. Mittwer.** 1952. The Gram stain. *Bacteriol. Rev.* **16**:1–29.
22. **Beveridge, T. J.** 1997. The response of S-layered bacteria to the Gram stain. *FEMS Microbiol. Rev.* **20**:2–10.
23. **Beveridge, T. J.** 1990. Mechanism of gram variability in select bacteria. *J. Bacteriol.* **172**:1609–1620.
24. **Beveridge, T. J., and S. Schultze-Lam.** 1997. The response of selected members of the *Archaea* to the Gram stain. *Microbiology* **142**:2887–2895.
25. **Beveridge, T. J., and J. A. Davies.** 1983. Cellular responses of *Bacillus subtilis* and *Escherichia coli* to the Gram stain. *J. Bacteriol.* **156**:846–858.
26. **Buttner, M. P., K. Willeke, and S. A. Grinshpun.** 1996. Sampling and analysis of airborne microorganisms, p. 629–640. *In* C. J. Hurst, G. R. Knudsen, M. McInerney, L. D. Stetzenbach, and M. V. Walter (ed.), *Manual of Environmental Microbiology.* ASM Press, Washington, DC.
27. **Davies, J. A., G. K. Anderson, T. J. Beveridge, and H. C. Clark.** 1983. Chemical mechanism of the Gram stain and synthesis of a new electron-opaque marker for electron microscopy which replaces the iodine mordant of the stain. *J. Bacteriol.* **156**:837–845.
    References 22 through 25 and 27 provide an up-to-date explanation of how the Gram stain works and why some prokaryotes stain in an unexpected way. Also see references 42 and 46.
28. **Dorner, W.** 1926. Un procédé simple pour la coloration des spores. *Lait* **6**:8–12.
29. **Duguid, J. P.** 1951. The demonstration of bacterial capsules and slime. *J. Pathol. Bacteriol.* **63**:673.
30. **Grossart, H.-P., G. F. Steward, J. Martinez, and F. Azam.** 2000. A simple, rapid method for demonstrating bacterial flagella. *Appl. Environ. Microbiol.* **66**:3632–3636.
31. **Heimbrook, M. E., W. L. L. Wang, and G. Campbell.** 1989. Staining bacterial flagella easily. *J. Clin. Microbiol.* **27**:2612–2615.
32. **Henrichsen, J.** 1972. Bacterial surface translocation: a survey and a classification. *Bacteriol. Rev.* **36**:478–503.
33. **Hobble, J. E., R. J. Daley, and S. Jasper.** 1977. Use of Nuclepore filters for counting bacteria by fluorescence microscopy. *Appl. Environ. Microbiol.* **33**:1225–1228.
34. **Hotchkiss, R. D.** 1948. A microchemical reaction resulting in the staining of polysaccharide structures in fixed tissue preparations. *Arch. Biochem.* **16**:131–141.
35. **Joux, F., and P. LeBaron.** 1997. Ecological implications of an improved direct viable count method for aquatic bacteria. *Appl. Environ. Microbiol.* **63**:3643–3647.
36. **Leifson, E.** 1951. Staining, shape, and arrangement of bacterial flagella. *J. Bacteriol.* **62**:377–389.
37. **Lisle, J. T., P. S. Stewart, and G. A. McFeters.** 1999. Fluorescent probes applied to physiological characterization of bacterial biofilms. p. 166–178. *In* R. J. Doyle (ed.), *Biofilms: Methods in Enzymology.* Academic Press, New York, NY.
38. **Lundgren, B.** 1981. Fluorescein diacetate as a stain of metabolically active bacteria in soil. *Oikos* **36**:17–22.
39. **Murray, R. G. E., and J. F. Whitfield.** 1956. The effects of the ionic environment on the chromatin structures of bacteria. *Can. J. Microbiol.* **2**:245–260.
40. **Newell, S. Y., and R. R. Christian.** 1981. Frequency of dividing cells as an estimator of bacterial productivity. *Appl. Environ. Microbiol.* **42**:23–31.
41. **Noller, E. C., and N. N. Durham.** 1968. Sealed aerobic slide culture for photomicrography. *Appl. Microbiol.* **16**:439–440.
42. **Popescu, A., and R. J. Doyle.** 1996. The Gram stain after more than a century. *Biotech. Histochem.* **71**:1415–1451.
43. **Robinow, C., and E. Kellenberger.** 1994. The bacterial nucleoid revisited. *Microbiol. Rev.* **58**:211–232.
44. **Robinow, C. E., and R. G. E. Murray.** 1953. The differentiation of cell wall, cytoplasmic membrane and cytoplasm of Gram-positive bacteria by selective staining. *Exp. Cell Res.* **4**:390–407.
45. **Rogers, F. G., and A. W. Pasculle.** 1991. *Legionella*, p. 442–453. *In* A. Balows, W. J. Hausler, Jr., K. L. Herrmann, H. D. Isenberg, and H. J. Shadomy (ed.), *Manual of Clinical Microbiology*, 5th ed. American Society for Microbiology, Washington, DC.
46. **Salton, M. R. J.** 1963. The relationship between the nature of the cell wall and the Gram stain. *J. Gen. Microbiol.* **30**:223–235.
47. **Schaeffer, A. B., and M. Fulton.** 1933. A simplified method of staining endospores. *Science* **77**:194.
48. **Seto, S., G. Layh-Schmitt, T. Kenri, and M. Miyata.** 2001. Visualization of the attachment organelle and cytoadherence proteins of *Mycoplasma pneumoniae* by immunofluorescence microscopy. *J. Bacteriol.* **183**:1621–1630.
49. **Tabor, P. S., and R. A. Neihof.** 1982. Improved method for determination of respiring individual microorganisms in natural waters. *Appl. Environ. Microbiol.* **43**:1249–1255.
50. **Traxler, R. W., and J. L. Arceneaux.** 1962. Method for staining cells from small inocula. *J. Bacteriol.* **84**:380.
51. **Truant, J. P., W. A. Brett, and W. Thomas.** 1962. Fluorescence microscopy of tubercle bacilli stained with auramine and rhodamine. *Henry Ford Hosp. Med. Bull.* **10**:287–296.
52. **Vela, G. R., and O. Wyss.** 1964. Improved stain for visualization of *Azotobacter* encystment. *J. Bacteriol.* **87**:476–477.
53. **Wilson, D. R., and T. J. Beveridge.** 1993. Bacterial flagellar filaments and their component flagellins. *Can. J. Microbiol.* **39**:451–472.

3

Laser Scanning Microscopy

J. R. LAWRENCE AND T. R. NEU

3.1. INTRODUCTION

Light microscopy provides a principal method for the observation and study of bacterial cells, aggregates, biofilms, and communities. Conventional light microscopy techniques, including phase contrast, dark field, differential interference contrast, and fluorescence, all provide effective means to observe bacterial cells in pure cultures and water samples, on surfaces, and in sediments, with or without staining. In addition to these methods, there have been substantial innovations in light microscopy through the integration of conventional microscope optics, a scanned laser beam, confocal optics, and digital image formation.

Confocal laser scanning microscopy (CLSM) with one-photon (1P) excitation became a practical technique with the introduction of the first commercially available models in the early 1980s. Laser scanning microscopes with two-photon (2P) excitation have been commercially available since 1997. Since that time, laser scanning microscopy (LSM) technologies have been applied in many fields. Valkenburg et al. (55) published the initial study using CLSM to examine the nucleoid of *Escherichia coli*, and Lawrence et al. (33) published an extensive study of bacterial biofilms.

In a conventional microscope, all the light passing through a specimen or emanating from it is imaged directly and simultaneously, forming an image. However, all light from the illuminated region returns to the eye or the detector, including light from above and below the focal plane. This stray light results in degradation of the image quality and interferes with three-dimensional (3-D) representation of the object. Both types of LSM, confocal or 1P and 2P or multiphoton systems, act to eliminate light not originating from the focal plane, resulting in a high-clarity image of the specimen that may be considered an optical thin section. The 1P and 2P optical microscopes may be used to view virtually any specimen or preparation suitable for conventional light and epifluorescence microscopy. However, 1P and 2P microscopy is most frequently used in conjunction with fluors or fluor-conjugated probes in the fluorescence mode to achieve optimal results. The option of using reflection imaging to reveal minerals or colloidal gold labels also exists but only for 1P excitation. The digital nature of the LSM images makes them amenable to image processing and analysis, allowing the user to obtain quantitative information or to make 3-D reconstructions of the material under study.

The ultimate goal in the application of LSM is to study and integrate physiological, biochemical, and molecular aspects of living microbial cells. This goal requires the development and application of noninvasive techniques to probe both cellular and extracellular processes of bacterial cultures and naturally occurring assemblages of bacteria, including aggregates and biofilms.

The purpose of this chapter is to provide a basis for understanding LSM and the approaches that may be used to apply high-resolution digital microscopy to the study of bacteria. Additional reviews of image analysis, digital imaging, and LSM for microbiology applications are also available (10, 27, 30, 32).

3.2. INSTRUMENTATION

Modern 1P and 2P systems are complicated, sensitive, and expensive with multiple components, including lasers, detectors, optical systems (glass or mechanical), step motors, and computer hardware and software. Many fundamental aspects of CLSM are covered in various books (15, 46), and readers are referred to these. The major manufacturers provide a variety of configurations, including "real-time" instruments, which provide video rate image collection but not high-resolution optical sectioning, and point scanning systems, which maximize lateral and axial resolution at the expense of image collection speed. The latter group may be subdivided into the 1P CLSM which uses classical physical barriers to remove out-of-focus light (pinholes or apertures) and those multiphoton or 2P systems where the optical sectioning is based on the physics of light absorption in the focal plane. All of these systems may use either a standard or an inverted light microscope as the fundamental platform; some LSM systems are designed to be easily exchanged between microscopes. Depending on the manufacturer and the intended applications, there are options including multiple laser attachments, multiple- or single-pinhole arrangements, and a selection of photomultiplier tubes (PMTs). PMTs may have various levels of light sensitivity and response to specific emission wavelengths. Additional equipment considerations include the host computer and software (both of which are usually predefined by the company) as well as an archiving system (e.g., compact disks [CDs]) and vibration isolation tables. The room itself should also be well isolated from vibration, be air conditioned, and have variable, controlled lighting levels.

In general, experience in light microscopy and fluorescence microscopy is a valuable starting point for LSM. Training is provided by the equipment manufacturers, and a variety of courses are offered worldwide for confocal 1P and 2P microscopy and the specific applications. The major focus of most courses is cell biology, although the information is of use and techniques are broadly transferable across disciplines.

Although some equipment alignments may be done by the user, in general these systems require technical adjustments by factory-trained engineers on at least an annual basis. In most of the current LSM systems, there are really no user-repairable parts. It is, however, important to do the following.

1. Establish user protocols for training and start-up and shutdown of the instrument.
2. Keep track of instrument use.
3. Keep a log of all the error messages displayed and their circumstances.
4. Check the alignment of the system periodically, especially during extended use, and always at the beginning of each session.
5. Log the results of alignment checks.
6. Periodically use reference samples and reference images relevant to the research area so as to assess instrument performance.
7. Ensure annual assessment and alignment to the manufacturer's specifications.

3.2.1. Lasers

Both 1P and 2P systems are equipped with a laser light source(s) that generates a range of excitation wavelengths. Commercially available lasers for confocal 1P microscopy include argon ion, helium-neon, argon-krypton, helium-cadmium, and UV varieties. The argon laser (25-, 50-, or 100-mW models with 5,000- to 10,000-h life spans) provides two main excitation lines at a 488-nm wavelength (the blue line) and a 514-nm wavelength (the green line) as well as minor lines with wavelengths ranging from 274 to 528 nm. To overcome limitations of the argon laser (i.e., limited excitation, little separation of excitation lines, and limited fluorochrome selection), most systems are equipped with additional helium-neon lasers (543- or 633-nm-wavelength line) or mixed-gas argon-krypton lasers (488-nm [blue]-, 568-nm [yellow]-, and 647-nm [red]-wavelength lines). The shift to alternate laser sources such as the Ar-Kr laser provides for excitation of up to three fluorochromes with little spectral emission overlap. The addition of a UV laser (e.g., 351- to 364-nm wavelength) further expands the excitation potential and the range of fluors. However, there are additional safety and operational considerations, for example, the need for quartz optics if excitation is performed using a 150-nm wavelength. A development to watch is the appearance of diode lasers that offer specific excitation lines. For example, a blue diode laser provides near-UV emission, allowing for excitation of DAPI (4',6'-diamidino-2-phenylindole) without a UV laser. 2P LSM systems employ a high-peak-power, infrared laser with an extremely short pulse in the femto- to picosecond range and a repetition rate of about 80 MHz. This laser produces the high photon density required to achieve the 2P effect of exciting the fluor only in the focal plane of the microscope lens. Current 1P and 2P systems use a fiber-optic system to connect the laser to the scan head of the microscope, greatly simplifying connection and alignment.

3.2.2. Scanning and Detection Systems

The core of the LSM system is the scan head, an integrated unit with galvanometric or acoustically controlled mirrors to scan the excitation wavelengths of the laser over the specimen. The laser light source is used to illuminate the specimen in a point-by-point fashion (point scanning) through a conventional objective lens, thereby exciting the fluorescent probes that have been applied to the specimen or creating a reflection image. The specimen may be scanned in a raster pattern at resolutions equal to 512 by 512, 512 by 768, or 1,024 by 1,024 pixels or at even higher resolution (new systems can do 4,096 pixels). Scan time may vary from a fraction of a second to several seconds. Note that the more rapid the scan, the lower the resolution of the image, whereas the longer the scan, the greater the photobleaching (destruction of a fluor by light, resulting in a loss of signal). When a specimen is scanned, one or more confocal pinholes allow only those fluorescence signals that arise from a focused xy plane to be detected by the PMT. The pinholes are positioned confocally in front of the incident laser light and in front of the photomultiplier detector, preventing fluorescent signals originating from above, below, or beside the point of focus from reaching the photodetector. In contrast to 1P systems, 2P microscopy achieves the selection of light from a focused xy plane because excitation occurs only in the true focal plane.

The selection of the fluorescence wavelengths to be detected may be accomplished by using a beam splitter and optical glass filter sets. However, it is also possible to select excitation-emission wavelengths without the need for optical lenses through the application of a filter-free, prism spectrophotometer-based system. This approach reduces the loss of signal and frees the user to optimize the detection of emission peaks, which can be very useful when more than one fluorescent stain or probe is used. Regardless of the excitation-emission system, a PMT detects the emitted light at each point and converts this into a digital grayscale image. Although custom PMTs are available for far-red imaging and potential exists for application of PMTs with various spectral ranges from 185 to 930 nm, the range of PMTs in commercial LSM systems has been limited. One critical limit is that the LSM PMTs generally have a response only to changes in image brightness of one order of magnitude. 1P and 2P systems may also incorporate either a fiber-optic light guide below the condenser and connected to one of the PMTs or a separate photodiode for imaging of nonconfocal transmitted laser images.

3.2.3. Microscopes and Objective Lenses

Additional details regarding objectives and microscope basics for light microscopy are found in chapter 1. Selection of the base microscope for LSM usually represents a compromise. Inverted microscopes offer advantages such as easy access to the sample for application of extraneous items such as microelectrodes, whereas the upright microscope offers the full range of objective lenses including water-immersible ones and gives the possibility of viewing the sample without a glass interface such as a coverslip. Some of the current 1P and 2P systems offer the convenience of being relatively easily switched between inverted- and upright-base microscopes (however, in practice loss of alignment may be a problem, especially with UV lasers).

The primary imaging tool is the objective lens, and its selection is based on the type of sample, sample preparation, required resolution, and base magnification. For 1P and 2P microscopes, the major limitation for all objective lenses is that the axial or z-dimension resolution is poor relative to the lateral resolution (Color Plate 4A and B). Older objective lenses are not corrected for imaging in the far red, and when wavelengths are extended into the UV range there may also be losses of transmission and serious image aberration. New lenses provided by most manufacturers and available for LSM are corrected from ~350 to 1,000 nm, overcoming these considerations. In general, one should use high-numerical-aperture (NA) oil or water immersion lenses, i.e., those with NAs of 1.2 to 1.4. Another critical factor in considering objective lenses is the working distance, i.e., that range of distances where the image formed by the lens is clearly focused. Working distances may be as small as <10 μm for some high-NA plan apochromat lenses, and some water immersion lenses (e.g., a 63×, 1.2-NA lens) have been designed for LSM applications allowing imaging through up to 220 μm of biological material. Application of these water immersion lenses, which are very expensive, assumes the use of fixed and stained specimens, optically appropriate mounting media, and high-quality coverslips (Corning Glass Works and NalgeNunc are suppliers of suitable coverslips and coverslip chambers) and proper adjustment of the objective lens as well as correction for the coverslip thickness and temperature. Generally, as the NA increases, the working distance of the objective lens decreases. The user should also consider that in LSM it is possible to zoom in while scanning, thereby increasing magnification without changing the objective lens.

This procedure can also increase resolution; e.g., a 20×, 0.75-NA lens will increase in resolution up to approximately four times during a zoom. In LSM the recorded digital image covers only the central part of the visual field of the objective lens. Errors in the objective lens increase with the distance from the center; thus, it is better to use a high-NA 60× lens and zoom than it is to use a similar 100× lens without zooming. Some of the issues associated with objective lens quality and corrections are illustrated in Color Plate 4. Color Plate 4A shows an image of 6-μm-diameter Focal Check beads (Molecular Probes, Eugene, OR) stained with green, red, and blue fluors and viewed with a 60×, 1.4-NA lens. The image shows a white peripheral ring around the beads, indicating good alignment and correction of the lens and LSM system. The loss of resolution when the image is formed by scanning in the xz direction is shown in Color Plate 4B. Here effects of light wavelength and point spread function error create distortion and misalignment of images from the three channels. Panel C shows the effect of using a 63×, 0.9-NA lens on the appearance of the same beads in the xy plane, and the further-reduced quality of the xz image is shown in panel D. When the same beads are viewed by using the outer edge of the 63× lens, the effect of the loss of correction with the distance from the central region of the scanned area is easily seen (Color Plate 4E). In panel F, the reader can see the loss of confocality (presence of multiple planes and bead colors) and chromatic correction in a 20×, 0.4-NA lens.

The thickness of an optical section obtained by confocal or 1P LSM is dependent on the NA of the objective lens and the diameter of the pinhole(s). In general, high-NA objectives such as those with NAs of 1.0 to 1.4 allow the use of lower pinhole or aperture settings, thus generating submicron optical sections. It is complicated to determine the exact thickness of an optical section. Neu et al. (44) used beads of known dimensions to show that a 40×, 0.55-NA water-immersible lens created an optical thin section that was 5 μm thick when the system was optimized for the pinhole and PMT, etc. In contrast, based on the calculations of Xiao and Kino (61), the theoretical thickness of an optical section taken with this lens was 4 μm.

Lenses with an NA lower than 0.5 will not provide confocal images (Color Plate 4F). However, there are lenses capable of extralong working distances (ELWDs) and water-immersible lenses with NAs of 0.5 to 0.9 that are particularly useful for imaging environmental samples. These lenses offer additional advantages, particularly for living specimens, e.g., no coverslip, long working distance, high NA, and superior brightness. There are a variety of high-NA water-immersible objectives supplied by the major manufacturers (i.e., Leica, Nikon, Olympus, and Zeiss).

For fixed specimens, factors such as the refraction properties of mounting and immersion media must be considered since they will diminish signal intensity and resolution. Therefore, both media should resemble each other in terms of their optical properties and be matched to the objective lens. See also chapter 1.

Although penetration of biological specimens is greatly enhanced by 1P and 2P microscopy, it is still a function of the transparency of the specimen, self-shading, diffraction by objects, and quenching of both the excitation and emission light by the specimen and the presence of the fluors. Solid surfaces may yield information from only the surface layer, whereas some biological materials such as bacterial biofilms may be imaged to a depth of about a millimeter. In 1P LSM applications, it is possible to effectively section

TABLE 1 Advantages and disadvantages of CLSM

Advantages	Disadvantages
Examination of fully hydrated samples up to several hundred micrometers thick (dependent upon the objective lens selected)	Fluorescence of background
Noninvasive optical sectioning with virtually no out-of-focus blur	Bleaching in out-of-focus area
xy, xz, and xt sectioning	Cell damage in out-of-focus area
Fluorescence mode and reflection mode available	Potential for chromatic aberration if UV laser is used
Simultaneous application of multiple probes and multichannel (4) imaging of digitally enhanced signals	Depth of laser penetration
Quantitative static and dynamic analyses of 3-D data sets, 3-D tomography, multicolor stereoscopic imaging, and computer animation	Low xz resolution

through several hundred micrometers with 63×, 0.9-NA water-immersible lenses and up to 1 mm (in exceptional cases) when using ELWD lenses, for example, 20× ELWD or 40× ELWD or 40×, 0.55-NA water-immersible lenses. Infrared illumination is less subject to scattering by biological specimens than visible light, which accounts for the greater penetration observed with 2P LSM systems. Sytsma et al. (52) were the first to apply 2P excitation in a study of oral biofilms where they reported that a dense, mixed-species biofilm of 100-μm thickness could be imaged after staining with acridine orange. However, the fluorescence decreased rapidly as a function of depth, and to counter this effect the excitation intensity had to be adjusted (52). A subsequent study indicated that in the same sample, there was a penetration of 70 μm for 2P excitation versus 15 μm for 1P excitation in oral biofilms stained with rhodamine B (57). A comparison of the advantages and disadvantages of 1P and 2P LSM systems is presented in Tables 1 and 2. Figure 1 shows a head-to-head comparison of 1P and 2P LSM imaging of a thick biofilm specimen stained with SYBR green (Molecular Probes), illustrating the increased resolution of the 2P system.

3.3. PERFORMANCE ASSESSMENT AND GENERAL OPERATION

There are a number of points to consider when setting up and operating an LSM system.

1. The system should be allowed to warm up for ~30 min prior to imaging. This warm-up is particularly critical with Ar-Kr systems to ensure a stable far-red (647-nm-wavelength) excitation line.

2. Internal mirrors and optical elements of the unit should be periodically (annually) aligned by the manufacturer.

3. Alignment must be confirmed prior to each use. Alignment is particularly critical for work with bacteria, where a misalignment of 0.5 μm cannot be ignored. Users should make up slides consisting of fluorescent beads of various intensities and Focal Check beads (Molecular Probes) for routine evaluation of image brightness and alignment. Focal Check beads allow assessment of images obtained from the same location but with different excitation-emission combinations (multiparameter imaging or colocalization studies), ensuring that they are in perfect register. Color Plate 4 is an image of Focal Check beads showing the appearance in the green, red, and far-red channels of a laser scanning microscope and the combined image from the three channels.

4. A variety of fluorescent beads ranging in diameter from 100 nm to 1 mm should be used to assess both image quality and resolution and to check the calibration of the instrument.

5. Adjustment of the computer monitor should be carried out periodically under normal working and lighting conditions. A monitor test image (both color and black and

TABLE 2 Advantages and disadvantages of 2P LSM

Advantages	Disadvantages
Excitation in the focal plane only with extremely small excitation volume (femtoliter)	Whole spectrum of infrared light is available only with 3 different mirror sets (700–1,200 nm)
Uncaging in extremely localized spots	Tuning of infared laser is necessary
No out-of-focus bleaching	Tuning determines the types of fluorochromes which can be excited
No out-of-focus cell damage	
No background fluorescence	
High depth penetration (0.5–2 mm)	
Inherent depth resolution	
Less scattering	Resolution is slightly lower than that with CLSM
No pinholes	
No expensive UV laser	
No UV photo damage	
No UV optics necessary	
Fewer filter problems	

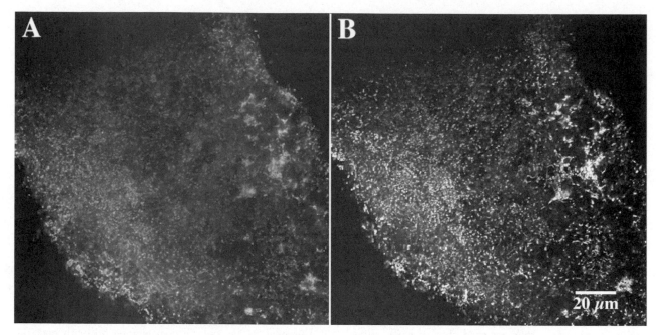

FIGURE 1 1P and 2P LSM images of a fluidized bed reactor biofilm which degrades EDTA and is growing on pumice. The specimen was stained with SYBR green, and serial sections were taken by using both LSM systems on the same sample sequentially. Images show enhanced resolution of the 2P system relative to that of the 1P system with a 39-μm-thick biofilm specimen.

white) should be displayed, and the brightness and contrast controls should be adjusted until all intensities are visible and well defined. The monitor image should also be centered and sized correctly.

6. Periodically check the step motor and calibration in the z dimension to ensure that the stage motion corresponds to that indicated by the software. This setting may also be assessed by using a range of sizes of fluorescent beads and coverslips of known thickness.

7. Ensure that the correct optical filters or settings are in place for the fluorochromes to be used. In the event that the microscope is equipped with a filter-free, prism spectrophotometer-based system, the software allows the selection of an infinite range of excitation-emission lines, optimization of the signal, and signal separation.

8. The PMTs should be adjusted when there is no incoming signal so that their black level is at the point where the boundary between scans is invisible against the dark background. Furthermore, the gain and black level of the PMTs should be adjusted to utilize the entire gray scale. Note that optimum brightness in an image usually occurs when only a few pixels of the image approach saturation (a gray level of 255 on a scale of 0 to 255 where 0 is black and 255 is white). Most of the commercial LSM systems include a digital indicator, so-called lookup tables (LUTs), where under- and oversaturated pixels appear in different colors, facilitating adjustments.

9. Ensure that the objective lens used for a particular analysis has sufficient magnification, resolving potential, and working distance. One also has to consider the nature of the sample, the presence or absence of a coverslip, and the type of mounting media, etc.

10. Place a sample on the microscope specimen stage and select a field for imaging using either standard epifluorescence or phase-contrast microscopy. The ideal field for initial setup should provide a fairly even signal response when viewed with epifluorescence or preliminarily scanned with LSM.

11. Switch to the 1P or 2P mode, and scan the sample by using a low zoom factor, with the PMT gain set at a low to midrange level and, for 1P, with the pinhole aperture about half open. Then, while scanning in real time, adjust the gain of the PMT so that no part of the image is saturated (white or other color depending on the LUT used).

12. Optimize the intensity of the incident laser light such that it provides a strong fluorescent signal while minimizing signal loss due to photobleaching. Laser intensity is usually controlled by using neutral-density filters or acoustic settings.

13. The size of the pinhole should be adjusted to achieve the smallest aperture possible. This may involve a sequence of adjustments with progressive reduction in aperture size (note that larger apertures produce thicker optical sections). With some instruments, setting the Airy disk for each objective lens to 1 sets the optimal pinhole.

14. By adjusting or balancing the PMT gain, laser intensity, and pinhole size, the user will also establish the optimal conditions for a specific combination of fluorochrome(s) and/or specimen. Various fluorochromes also have different quantum yields, and yield may change with the nature and intensity of the excitation light source. It is important to keep this in mind when fluorescent stains are added in combination and the specimen is imaged by using more than one excitation wavelength. Figure 2 illustrates the excitation-emission spectra of various fluors and their potential for interaction during imaging. Figure 2A illustrates the ideal case, and Fig. 2B shows the potential for signal overlap and bleed through with the use of more than one fluor in a sample.

The above list assumes proper preparation of both the samples and the system, each of which is required to achieve high-quality, information-packed images.

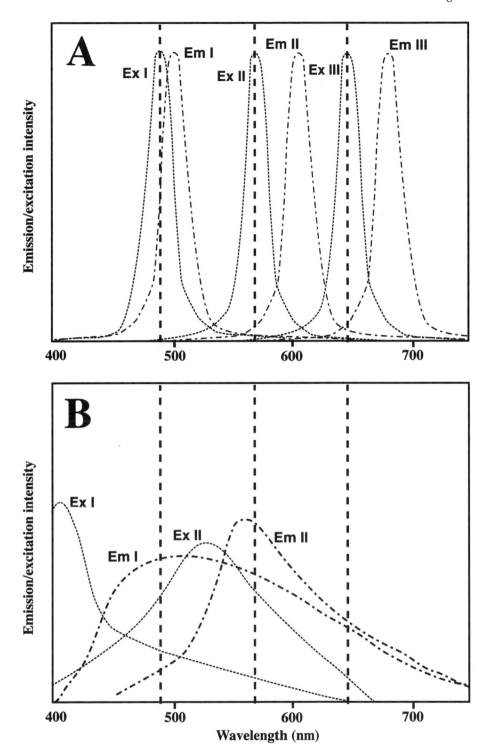

FIGURE 2 Graphs showing the nature of excitation-emission spectra for an ideal series of green, red, and far-red-emitting fluors (A) and problematic fluors where excitation (Ex) and emission (Em) peaks overlap, resulting in potential cross talk or bleed through of the signal into adjacent imaging windows (B).

3.4. SPECIAL CONSIDERATIONS FOR 2P IMAGING

The microscopic application of 2P excitation is still not fully explored and evaluations of basic protocols are necessary. These evaluations include, for example, the examina-

tion of suitable fluorochromes for microbiological samples. In principle, it is possible to use the same fluorochromes that have been developed for 1P excitation. However, in comparison to those for 1P excitation, the fluors for 2P excitation have a broader excitation cross-section. That means that the fluorochromes can be excited with a wider range of wavelengths. In addition, for 2P excitation usually only one excitation wavelength is available at a time, and consequently, for multiple staining the fluorochromes must have the same excitation cross-section while having different emission peaks. A further point to consider is the combination of 1P and 2P excitation. Observed differences in how they excite the same fluors indicate that they must be performed sequentially rather than simultaneously.

3.5. SAMPLE PREPARATION OPTIONS

In practice, one of the appealing aspects of fluorescence microscopy is the lack of a requirement for fixation, dehydration, and the air drying of samples required by some other techniques. As noted previously, one of the goals of this approach is to examine living bacterial cells, aggregates, and biofilms. In most instances, immobilization of the cells is required, and this may be carried out by using agar slide techniques or any one of the techniques outlined in the previous chapters on light microscopy (chapters 1 and 2) or in Caldwell et al. (12) or Lawrence et al. (32). These methods include light microscopy-friendly techniques such as continuous-flow slide culture and a variety of coupon-based approaches. However, it is also possible to examine cells and biofilms that have been fixed and embedded (e.g., Epon, Periplast, and TissueTek) or frozen and sectioned prior to being mounted on a slide and stained with a fluor or fluor-conjugated probe (see reference 34). Hydrophilic resins such as Nanoplast which stabilize structures and minimize artifacts may be employed with both transmission electron microscopy and LSM (19, 47). Simple techniques may be effectively used, e.g., the use of embedding and paraffin sections in combination with rRNA-targeted probes to study methanogens in anaerobic sludge granules (49) and cryosectioning to examine biofilms (62). A solution of 0.1% (wt/vol) agarose (39) or 20% (wt/vol) DNA-sequencing-grade acrylamide (14) for embedding of biofilms and aggregates for confocal microscopy can also be used. Advantages of gel embedding are that the sample may be either fixed and dehydrated or fully hydrated when examined and the gels are relatively nonfluorescent. A disadvantage is that the preparations may be relatively fragile.

3.5.1. Correlative Microscopy

Correlative microscopy is a name used when two or more microscopic techniques are employed to examine the same or parallel specimens. A number of methods should be considered when designing an analytical approach to particular specimens. Some authors have combined CLSM with conventional epifluorescence and transmission electron microscopy (35) or CLSM and scanning electron microscopy (29, 48), and others have combined CLSM with photon-counting microscopy (45).

3.6. FLUORESCENCE AND FLUORESCENT PROBES

Fluorescence is a near-perfect tool for the in vivo or in vitro, nondestructive, noninvasive study of biological samples including microorganisms. Fluors have unique and characteristic spectra for excitation and emission, which is a consequence of their electronic configurations (Fig. 2). For many fluors, there are published absorption-excitation and emission spectra which show the relative intensity of the fluorescence, in relation to the wavelength of light being applied (relative intensity is the vertical axis versus wavelength on the horizontal axis) (e.g., reference 21 and www.probes.com). Typical excitation-emission spectra are shown in Fig. 2. This spectral characteristic of fluors is fundamental to understanding their behavior and selecting specific fluors and fluor combinations for application in fluorescence and LSM studies. The signal emitted by a fluor often provides information about its environment, including its binding site. Fluors are used independently or as reporters conjugated to probes for specific macromolecules. These probes are considered in two classes, targeted (i.e., recognizing macromolecules) and responsive (i.e., changing the fluorescent signal as related to the environment of the fluor). Bright and Taylor (8) indicated that the use of fluorescence has the following major advantages.

1. Sensitivity, with detection of 10^{-11} to 10^{-12} M concentrations of a fluor possible.
2. Specificity: with proper controls, probes, and conditions, the identities and localization of a variety of macromolecules can be determined.
3. Spectroscopy and spectroscopic measurements may be used to obtain information about the environment of the fluor, e.g., by using pH-sensitive fluors.
4. Spatial resolution: the use of fluorescence allows analysis at the limits of light microscopy, e.g., the localization of specific carbohydrates (43, 44) or metals by using Newport Green (60).
5. Temporal resolution: the user can assess time-dependent processes related to metabolism, potentials (e.g., redox and electrical), and movement. For example, it is possible to determine diffusion rates (18, 31), visualize enzyme activities (56), and detect the rate of plasmid transfer (22).

For a fluor to be useful, it must meet certain criteria.

1. Have appropriate excitation and emission wavelengths.
2. Produce a detectable yield of visible light, i.e., a high quantum yield.
3. Resist photobleaching.
4. Be minimally toxic.
5. Be able to penetrate to the appropriate location.
6. Have a highly specific response to the target molecule(s).

In reality, there are no ideal fluors or fluorescently conjugated probe combinations and few if any ideal specimens. Therefore, a number of considerations must be taken into account when setting out to apply fluorescence imaging. Additional assessments must be considered when combinations of fluors and fluor-conjugated probes are used. These are critical assessments for 1P confocal microscopy, although there is extensive literature that provides basic information on expected excitation-emission spectra. On the other hand, for 2P or multiphoton laser microscopy the present literature is limited and practical evaluation is required to assess the excitation-emission spectra for any fluor prior to its use.

When applying a fluor or a fluor conjugate to an unknown sample, the user should assess the following aspects.

1. Is there autofluorescence or background fluorescence originating from the unstained sample? Selection of the appropriate filters may go a long way to limiting autofluorescence interference, as may the utilization of fluorescent reporter molecules which have maximal separation in terms of wavelength from autofluorescence wavelengths. Quenching solutions such as 0.1% (wt/vol) NaBH$_4$ (sodium

borohydride) used to reduce autofluorescence due to fixatives such as glutaraldehyde may have applications in other preparations. However, autofluorescence can also be a source of information; Lawrence et al. (28, 29) used the fluorescence of chlorophyll to identify and localize algae in microbial mats and river biofilms. Likewise, autofluorescence and LSM facilitated the examination of the distribution of photosynthetic bacteria in complex mat communities (58). Further, many bacteria may exhibit natural autofluorescence; for example, in methanogens autofluorescence is due to high levels of coenzyme F420.

2. What is the target molecule of interest (protein, carbohydrate, or nucleic acid), and what is the probe (i.e., antibody, lectin, rRNA, or DNA)?

3. Based on the presence or absence of autofluorescence and whether multiple probes are being applied, select the appropriate fluor based on excitation-emission spectra. See Fig. 2, which illustrates an ideal fluor combination where the excitation and emission spectra of the three fluors are well separated (Fig. 2A) and nonideal fluor combinations plus the nature of overlapping excitation-emission spectra, showing how tailing or bleed through may lead to conflicting signals during multiple staining (Fig. 2B).

4. Perform a practical assessment of fluor excitation-emission spectra on the 1P or 2P system to be used, covering all available excitation-emission combinations of the instrument. This step allows the assessment of such things as bleed through of the fluor signal into other emission wavelengths and the presence of other secondary excitation and emission peaks that may occur when 1P or 2P excitation is applied.

5. Assess the quantum yield or signal strength and the fade response curve of the fluor, i.e., is the fluor stable in this application or should a fade retardant be considered? Differential fading in dual- or multiple-fluor applications may create problems for quantitative imaging. For example, the results obtained using the LIVE/DEAD staining system (Molecular Probes) can be dramatically influenced by differential fading of the two components. Fading can be controlled through limiting observation and scan time and choosing the correct combination of zoom, pinhole, and PMT settings and laser intensity. Antifade reagents are recommended for many fluor applications such as fluorescence in situ hybridization (FISH). Florijn et al. (20) deal with these reagents, which include p-pheneylene-diamine, n-propylgallate, 2-mercapto-ethylamine, and 1,4-diazabocyclo-2,2,2-octane (DABCO). Note that fade retardants are highly toxic, must be used with caution, and may be ordered premixed from Molecular Probes or Citifluor (London, United Kingdom).

6. If a targeted conjugated probe is being used, then its specificity for target molecules in the sample must be assessed through the use of controls, including blanks; blinding of the probes and/or the reactive site; blocking of nonspecific binding sites; or inhibition of nonspecific binding. Elaborate protocols have been established for FISH (1) and for lectin use (44).

7. In addition, an important but understudied aspect is the influence of the fluor on the specificity of the conjugated probe, i.e., a fluorescein isothiocyanate-conjugated probe may exhibit specificities at variance with those for the same probe conjugated to tetramethyl rhodamine isothiocyanate (TRITC) or Cy5 (44). Assess the fluor for toxicity. Is it toxic in the range required to monitor changes in fluorescence and to observe a response in the test cells? Some fluorochromes such as CTC (5-cyano-2,3-ditolyl tetrazolium chloride; 1 to 20 mM) have demonstrated toxicity for microbial samples (13, 54). In contrast, other fluors (such as fluorescein and CellTracker) appear to have no negative effect when used on living bacteria (7, 11).

8. Although the above-described steps should result in an optimized protocol for the application of a single fluor or fluor conjugate, most applications with 1P or 2P excitation have colocalization or multiparameter imaging as the desired goal. Therefore, the implications of multiple fluors and fluor conjugates must also be assessed.

9. Fluor interactions may occur, and observations should assess the potential of "fluor A" to quench the fluorescence of "fluor B," resulting in attenuation of the signal. For example, the SYTO9 dye may quench dyes such as Cy3. Overlap of excitation-emission spectra creates problems with the separation of emission signals and unacceptable bleed through of the signal from fluor A into the observation channel for fluor B or fluor C (Fig. 2B). These problems can be avoided through sequential addition of stains or limitation of imaging to only one fluor at a time through sequential rather than simultaneous excitation-emission.

10. When the probe is a fluor conjugate, there exists the potential for one probe to interfere with the other or act as a target for other probes. The effects of incubation time, concentration, fluor labeling, the order of fluor addition, and probe interactions must all be studied.

A listing of fluors and fluor conjugates is provided in Table 3 illustrating the range of fluors available and their applications to the study of bacteria, aggregates, and biofilms.

3.7. LSM IMAGING

After the LSM system has been aligned and the correct fluor conjugate for the selected target in the specimen has been prepared for examination, there are a number of options available for imaging. The user may collect single images in the xy, xz plane or a series of xy or xz images creating an image stack for 3-D reconstruction and/or digital image analyses. It is also possible to collect images by using combinations of one, two, three, or more excitation and emission wavelengths by using 1P (including UV) or 2P systems. These images may be collected sequentially or simultaneously depending on the LSM system in use and the suitability of the fluors and fluor-conjugated probes selected. One may also collect images by using fluorescence, reflection (i.e., colloidal gold), and nonconfocal transmission modes.

3.7.1. Scanning in the xy Direction

1. After checking all alignments and optimizing the instrument as described in sections 3.3 and 3.6, the user may either switch between epifluorescence and LSM modes or scan continuously while examining the specimen and selecting locations for image collection. (Note that time is always of the essence in fluorescence microscopy; fluors decay at room temperature, in water, in light, and under observation.) Continuous scanning can be done with experience if the fluorescence signals are stable and resist bleaching. The actual taking or collection of images can be done in a random pattern or along a transect or consist of large-scale montages composed of a series of adjacent microscope fields. The latter approaches should be considered when collecting images for further analyses and statistical analysis of data (25, 28, 49). Once the appropriate location has been

TABLE 3 Fluorescent compounds and their application in microbial studies

Fluor	Application
Compounds for cell labeling and enumeration	Fluors which become fluorescent after binding to nucleic acids allow for visualization and enumeration of individual cells (24)
DAPI	
Acridine orange	
Hoechst 33258	
Hoechst 33342	
SYTO stains	
SYTO green	
SYTO blue	
SYTO orange	
SYTO red	
SYTO far-red	
SYBR green	
PicoGreen	
TOTO-1/TO-PRO-1	
YOYO-1/YO-PRO-1	
POPO-3	
SYBR green I/II	
ChemChrome V6	
Sytox stains	Used for fixed or membrane-damaged cells
Propidium iodide	Used for fixed or membrane-damaged cells
diOC$_6$	
Bis-oxonl	
Probe labels[a]	Suitable for conjugation to protein linkers on dextrans, lectins, nucleotides, oligonucleotides, and peptides and for creation of fluor-conjugated reporter molecules
Alexa series (488 and 546)	
Oregon Green (488) (difluorofluorescein)	
Fluorescein isothiocyanate (494)	
BODIPY FL (503)	
Carboxytetramethylrhodamine (550)	
Tetramethyl-rhodamine	
Texas Red (596)	
Cyanine dyes (Cy2, Cy3 [550], Cy5 [649])	
Compounds for identification	
FISH compounds	Detection of fluor-labeled oligonucleotide probe after hybridization
BacLight fluorescent Gram stain	Two-component fluorescent stain allowing in situ Gram reaction (pure cultures)
Green fluorescent protein	Tracking of green fluorescent protein-labeled cells
Immunostaining compounds	Fluorescent labels for antibody recognition of antigen
CellTracker	Labeled cells may be monitored for several generations based on partitioning of label to daughter cells (7)
Compounds for analysis of activity, viability, and metabolism	
RH-795	Analysis of cell membrane integrity and membrane potential
Rhodamine 123	Analysis of cell membrane integrity and membrane potential
Ethidium bromide	Analysis of membrane integrity and detection of spore germination (16)
diBAC$_4$(3)	Analysis of membrane integrity and detection of spore germination
LIVE/DEAD BacLight™ (propidium iodide-SYTO9)	Analysis of integrity of individual cell membranes and cell viability; see reference 14 for protocol
FISH compounds	Analysis of cellular activity and ribosome content
FISH-MAR[b] compounds	Analysis of ribosome content and response to substrate addition
In situ PCR compounds	Detection of specific mRNA activity
Sulfofluorescein diacetate	Detection of esterase activity

(Continued on next page)

TABLE 3 *Continued*

Fluor	Application
ELF-97	Detection of alkaline phosphatase activity; see reference 14 for protocol
Carboxyfluorescein diacetate	Detection of esterase activity
Triphenyl tetrazolium chloride	Detection of respiratory activity (dehydrogenase)
CTC	Detection of respiratory activity (dehydrogenase); see reference 14 for protocol
gfp gene green fluorescent protein gene	Analysis of gene expression of individual cells
Environmental reporters	
BCECF-AM	pH responsive
SNAFL-1	pH responsive
5- and 6-carboxyfluorescein	pH responsive
Newport Green	Sensitive to nickel concentrations; see reference 60 for details
gfp gene (green fluorescent protein gene)	Analysis of gene expression of individual cells may provide information on in situ conditions

[a]Numbers in parentheses are excitation maximums in nanometers.
[b]FISH-MAR, FISH microautoradiography.

found, stop scanning, select the appropriate zoom, and optimize pinhole, brightness, and contrast values. It may also be necessary to select whether to collect the image as a direct acquisition or through a mathematical filter. A number of mathematical filter options are available, including line averaging, frame averaging, and Kalman filtration. Kalman filtration (a running-average filter that mathematically averages sequentially collected images, thereby reducing noise level in the image but maintaining edge features) is extremely effective in reducing random noise and producing a clean image. Figure 3 illustrates the effect of mathematical filtration. Once these decisions have been made, then it is possible to start the collection of images.

2. For quantitative imaging purposes, it is necessary to maintain the microscope settings (i.e., gain, laser intensity, and aperture size) throughout the collection of images from an experimental series in order to create a valid comparative data set. Changes in any of the microscope settings will alter the section thickness, the brightness, and the apparent size of the objects being imaged.

3. All LSM images are grayscale; color is applied by using color LUTs which assign colors according to the corresponding grayscale value. Color is also assigned to indicate the excitation-emission combination that gave rise to the particular image. This assignment is arbitrary, but typically images are combined in the red-green-blue mode and stored in a tagged-image file format (TIFF) or a file format specific to the manufacturer (most often a modified TIFF). Most instruments will provide a display of each channel and, simultaneously, the combination of all channels. If all the probes had the same excitation wavelength but emitted at different wavelengths or there was minimal bleed through between channels, then simultaneous collection can be advised. In general, one is using a variety of fluors with different excitation-emission spectra, e.g., SYTO9, TRITC, Cy5, and autofluorescence (if present). In this case, the images for each channel should be collected sequentially. These images may then be overlaid in the operating software of the LSM system, or the overlaying may be done by using a variety of software packages (e.g., those of NIH IMAGE, Image J, or Imaris). To reduce bleaching and sample damage, start with the longest excitation wavelength and work towards the shortest.

3.7.2. Scanning in the *xz* Direction

During an *xy* scan, the laser beam is scanned over a defined area; however, the beam can also scan along a single line extending into the specimen (*xz* scan). When a line scan is repeated at many *z* levels, the resulting image (*z* scan) is a sagittal section through the sample. *z* scan images lack the resolution of *xy* images due to the nature of the corrections in standard objectives and the point spread function of the lens. Therefore, *xz* images are exaggerated in the *z* dimension and a single point will appear to have an elongated *z* axis (see beads in Color Plate 4B and D). This effect may be seen in Fig. 4B and C showing an *xy-xz* combination at the same location in a microbial biofilm. It is also possible to calculate *xz* images from *xy* serial sections by using various software packages.

Guidelines for *xz* scans are as follows.

1. Focus through the material to establish the top and bottom or beginning and end of the region to be scanned. Then take a single *xy* scan and use it as the basis to select the position and orientation of the *xz* line to be scanned.

2. For *xz* imaging, it may be necessary to increase brightness and contrast values and to adjust scanning time. Depending upon the instrument being used, this adjustment may not be possible or it can be done manually or automatically through the software. Typically, *xz* scans also benefit from the application of line averaging, or Kalman filtration.

3. It may be helpful to do a quick scan (to minimize photobleaching) through the site first in order to assess whether the selected section thickness and orientation are of value. Always remember that, depending on the objective lens, you may exceed the working distance if you select a large depth range for your sagittal section.

4. It is possible to create an *xz* series through a specimen; however, as noted previously, this can be done through calculation by using the *xy* series data set. Fig. 4 shows a typical *xy* and *xz* image pair taken through a biofilm stained with SYTO9 (Molecular Probes).

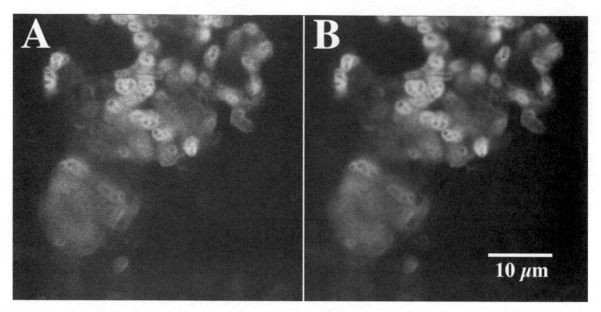

FIGURE 3   1P LSM images showing a direct single-scan image (A) and the influence of image averaging on the quality and signal-to-noise ratio of an LSM image (B).

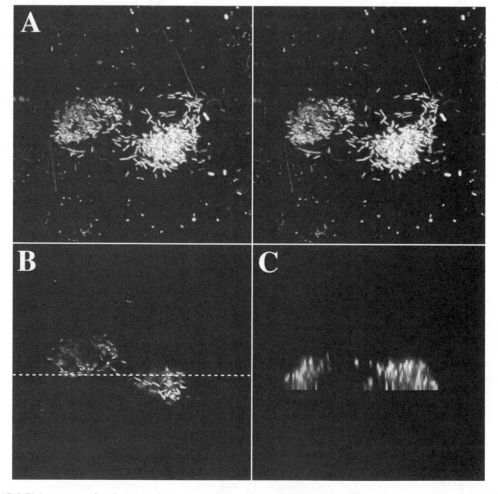

FIGURE 4   1P LSM micrographs showing the nature of the 3-D stereo pair (A) and a single *xy* (B) and a single *xz* (C) scan through a microbial biofilm stained with the nucleic acid stain SYTO9 (Molecular Probes).

3.8. 3-D IMAGING

The major advantage of LSM is the capacity to collect a series of images that allow the user to obtain 3-D spatial information. This capability follows from the optical sectioning capacity and the creation of a z series in perfect register. This series of xy images is referred to as a z series since it is taken along the z axis or commonly as an image stack. Serial optical sections can be used for a variety of 3-D reconstruction techniques (see below and Table 4). When an ideal z series is collected for display or analytical purposes, the sections should meet the following specific criteria.

1. They must be collected nondestructively where photobleaching is not a significant factor; however, in practice during collection of a series photobleaching can be compensated for by increasing brightness and contrast. Photobleaching can also be controlled through the application of fade-retardant solutions. The user should also consider the "total slice capacity" of the sample (this is a fixed number of sections that are taken from a location in a specimen before the effects of photobleaching become significant). This value may be best considered empirically through trial, error, and experience.

2. Sections must be adjacent, without oversampling or undersampling error. This requirement involves selection of the correct sectioning interval. Theoretically, with a high-NA objective, sections can be collected at intervals as small as 40 nm. In practice, optical sections are 0.5 to 1.5 μm in thickness, and this factor should be reflected in the optical sectioning interval, particularly in attempting to image bacteria. If the section interval is incorrect, oversampling or undersampling errors occur. These errors are most important in analysis of a series to extract depth information or in 3-D reconstruction. One quick test to assess sampling frequency is to project the series as a 3-D image; if there are visible discontinuities or steps in the image, then the sampling interval is incorrect.

3. Sections are collected within the working distance of the objective. As discussed above, a specific objective lens has a working distance that is a function of its magnification and NA, and this must be considered when obtaining a z series. (Again, empirical observation also serves to establish the working limits for specific samples.)

4. Sections must be homogeneous; minimize nonuniform bleaching by having the laser aligned. Also consider changing the fluor or the staining approach by using a more stable fluor or switch from a positive to a negative stain (11). Heterogeneity may also occur as a result of vignetting, which can be corrected by proper alignment or use of the proper objective and zoom combination.

5. Sections must be aligned and in register by ensuring the critical alignments as discussed above. Color Plate 4 shows the nature of misalignments that may be detected using Focal Check beads in multichannel mode. In this case, Color Plate 4A shows correct alignment whereas Color Plate 4C and D illustrate inherent problems with the objective lenses.

6. Sections must be calibrated so that image gray levels accurately represent concentrations or the dimensions of an object (see further discussion below).

7. Sections must be isotopic such that the xy pixels correspond to the z dimension. To collect an xz series, the procedure is essentially as described for the xy image collection.

3.9. MULTIPLE-PARAMETER IMAGING

The simultaneous or sequential acquisition of images by using a range of excitation wavelengths and selective detection of emission signals has a number of advantages, in particular by showing the identities and locations of bacterial cells and of bacterial macromolecules. A variety of probes such as fluor-conjugated antibodies, lectins, and nucleic acid stains, including rRNA probes, can be used to label bacteria and associated macromolecules. Although it is common to localize two to three signals, specific strategies allow collection of information on as many as seven different labels in a specimen (2). Several factors become critical in collecting multichannel images.

1. The nature of the emission and excitation optical filters or other mechanisms for selecting specific wavelengths of light. For example, a long-pass filter will allow all wavelengths greater than a set value to pass through and this may create problems for excitation and detection of other fluors. In contrast, a band-pass filter allows only specific wavelengths to pass, which can be desirable; however, it can result in insufficient light for detection of the fluor. For many applications, significant advantages may be found in LSM systems that do not use optical glass filters.

2. Misalignments in the optical pathway can cause the images from different channels to be offset. The presence of these misalignments in the xy or xz plane may be checked through imaging of Focal Check (6- or 15-μm-diameter) beads in the dual- or three-channel mode and overlaying of the resulting images. Any misalignment is evident in the overlay of the channels and should be corrected by following the alignment procedures outlined previously, arranging for a service call, or investing in high-quality objectives. Color Plate 4 illustrates the alignment and misalignments in three-channel images by using 6-μm-diameter Focal Check beads.

3. Axial chromatic aberration in the objective lens resulting in offset images. If the cause of misalignment is in the objective lens, realignment of the LSM system cannot correct the problem. For most lenses, maximum distortion occurs at the edge of the lens; 1.4-NA plan apochromat lenses provide the largest fully corrected area (cf. Color Plate 4A and E). Software solutions are also available that allow toggling of the digital images into alignment and cropping during the preparation of multichannel overlays.

4. Wavelength of light. The longer the wavelength of light, the thicker the optical section; thus, sections collected in the far red are thicker than those collected in the red and green (17). Fully corrected 350- to 1,000-nm-wavelength lenses minimize this problem.

5. Field curvature and lateral aberration, which result in objects at the edges of images appearing in slightly different locations. See points 2 and 3 regarding high-NA objective lenses and the use of Focal Check fluorescent beads (Color Plate 4).

6. Loss of signal and resolution with increasing wavelength of emission; thus, images originating in green and far red are of considerably different quality.

7. Superior results may be obtained by sequential addition and viewing of probes with the lasar scanning microscope or by simultaneous application of equimolar mixtures of probes labeled with different dyes.

8. It is advisable to include a color wheel in multiparameter images to facilitate interpretation of the combinations of probes in the specimen.

Amann et al. (2) present a method for multiple staining using rRNA-targeted fluorescent probes. Lawrence et al. (28) discuss multiple-parameter imaging and demonstrate the utility of nucleic acid staining to detect bacterial cells, lectin staining to assess bacterial exopolymers or glycoconjugates, and autofluorescence signals to detect phototrophic organisms. Lectins have tremendous potential as single probes, in combination with other fluors and probes, or in combination with one another. Some of the pitfalls of lectin staining have been presented by Neu et al. (44). Next, we present a brief protocol for the application of lectins to attached or immobilized cells or biofilms.

3.9.1. Lectin Staining

Labeled lectins have been applied in many pure culture studies to probe for cell surface structures. More recently, they have been used to image and characterize the exopolymeric substances of microbial aggregates and biofilms (28, 43, 59). Neu et al. (44) provide a fairly detailed discussion of their applications. A general protocol and considerations are outlined below.

1. Fluorescent lectins conjugated with different fluors such as fluorescein, fluorescein isothiocyanate, Oregon Green, TRITC, Texas Red, and Alexa are available (e.g., Sigma, St. Louis, MO, and Molecular Probes). Custom labeling (e.g., with Cy5 or Alexa) may be done by using commercial kits according to the instructions (Research Organics, Cleveland, OH, or Molecular Probes).

2. Lectins are employed alone or in combination for double and triple staining. In brief, they are dissolved at 100 µg ml^{-1} in filter-sterilized (0.2-µm-pore-size filter) water. Slides of 1 cm^2 carrying a biofilm are directly covered with 100 µl of lectin solution.

3. Next, samples are incubated in a humid chamber at $22 \pm 2°C$ for various times to assess the time required for complete staining. After determination of optimal staining time, the effect of lectin concentration (5 µg cm^{-2} to 100 µg cm^{-2}) needs to be determined.

4. After staining, all the slides are carefully rinsed with filter-sterilized water four times to remove unbound lectins. For each rinse, the wash water must be carefully added to the biofilm or bound cells and drawn off with filter paper.

5. The sample is then transferred to the laser scanning microscope stage. To avoid disturbance of the biofilm, no coverslip is used. If an upright microscope is employed, a 63×, 0.9-NA water-immersible lens is ideal for direct observation through the water droplet covering the biofilm. The lens must be carefully immersed into the droplet and slowly brought near the biofilm surface. For inverted microscopes, the substratum with the attached biofilm must be mounted upside down into a coverslip chamber (NalgeNunc International) by using spacers. This provides a water space to protect the biofilm or attached cells, and the specimen must then be examined with relatively-long-working-distance lenses. The same procedure of staining as described above can be applied to this inverted setup. Michael and Smith (40) also outline procedures for lectin application and the use of positive and negative controls. See also reference 57a. Color Plate 5 uses a three-color stereo pair to present the result of staining with a combination of three fluor-conjugated lectins. This can be best viewed by using stereo glasses. Color Plate 5 (panels B and C) presents high-magnification images of bacterial microcolonies triple-labeled with three lectins and showing detail of the hydrated exopolymer structure.

3.9.2. Imaging of the Metals Nickel and Cadmium

Wuertz et al. (60) presented a protocol for localization of the metals nickel and cadmium in biofilms. Conceptually the method is based on the observation that Newport Green at a concentration of 1 µM (22°C, pH 7) increases its fluorescence 16 times in the presence of 25 µM Ni. They concluded that the approach can be used to qualitatively demonstrate the presence of heavy metals such as Ni in defined systems.

1. Biofilms on 1-by-3-cm coupons were exposed to 100 µl of a 1 mM Ni solution for 1 h and subsequently washed two times with 100 µl of sterile distilled water.

2. Biofilms were then incubated with 100 µl of a 20 µM SYTO17 solution in the dark for 15 min. They were then washed three times with sterile distilled water.

3. Samples were then incubated in the dark for 15 min with 100 µl of a 2.4 µM solution of Newport Green.

4. Samples were observed by using the 488- and 543-nm-wavelength excitation lines of the laser scanning microscope for Newport Green and SYTO17, respectively. Either a 40×, 1.3-NA or a 100×, 1.3-NA plan neofluor oil immersion lens was used.

The method is suitable for qualitative detection and localization of nickel and perhaps cadmium in association with bacterial cells, colonies, and biofilms.

3.10. DYNAMIC IMAGING

As noted previously, one of the major advantages of fluorescence microscopy, including conventional epifluorescence and 1P and 2P LSM, is the capacity for temporal resolution, where time-dependent processes may be monitored by fluorescence. These processes would include diffusion and transport of molecules, changes in intra- and extracellular pH, cellular metabolism, kinetics of binding, and measurement of redox potential.

A number of approaches have been applied to study diffusion in bacterial biofilms and serve to illustrate the application of kinetic analyses using LSM systems. There are a number of fluor-conjugated probes that may be used to assess permeability and diffusion coefficients of bacterial cells and polymers; these include ficols, size-fractionated dextrans, and a range of fluorescent beads (10 nm to 15 um in diameter). Lawrence et al. (31) used 1P LSM to monitor the migration of fluor-conjugated dextrans to determine effective diffusion coefficients for biofilm systems. Microinjection and 1P LSM were used by De Beer et al. (18) to determine diffusion coefficients for biofilm materials. The more standard fluorescence recovery after photobleaching (FRAP) approach may also be applied to bacteria, aggregates, and biofilms. Birmingham et al. (4) used FRAP to monitor diffusion and binding of dextrans in oral biofilms. Laca et al. (26) used CLSM to monitor diffusion of protein in alginate beads. The details of the procedures and necessary equations are outlined in these and other papers (3, 6, 9).

Fluorescent probes such as fluorescein, 5- and 6-carboxyfluorescein, and dually labeled fluorescein-rhodamine dextrans, etc., exhibit pH-sensitive fluorescence and consequently have potential for in situ measurements. Caldwell et al. (10) presented a series of pseudocolor images of these gradients by using 1P LSM and the probe 5- and 6-carboxyfluorescein. In a more recent study of in situ pH, Vroom et al. (57) applied 2P LSM to detect pH gradients in mixed-species biofilms. In their application, they were able to demonstrate real-time imaging of pH gradients. Hunter and Beveridge (23a) also demonstrated pH gradients in biofilm regions and microcolonies; their study addressed concerns that pH-sensitive probes may be influenced by the presence of proteins and other biofilm constituents, indicating that these effects were not occurring in their system.

A common approach to the study of pH is the use of ratiometric probes such as 5- and 6-carboxyfluorescein and SNAFL-1 which exhibit no pH sensitivity when excited

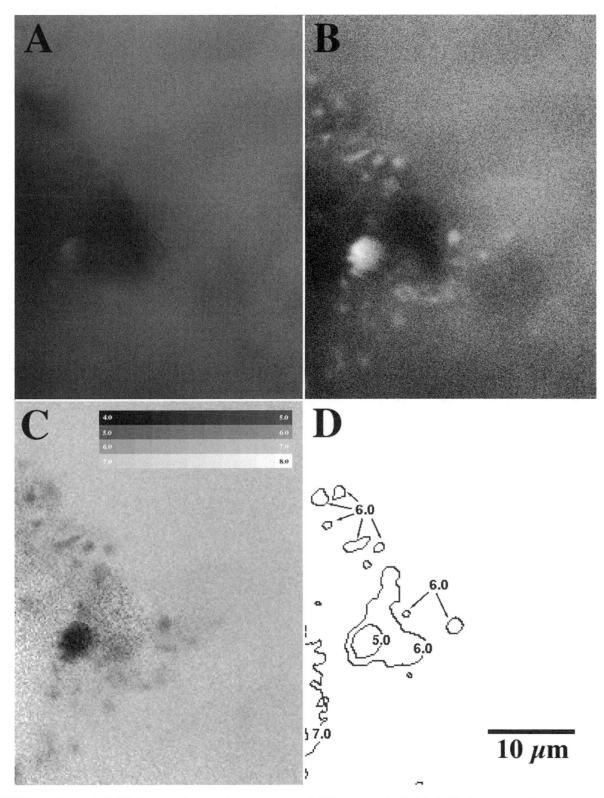

FIGURE 5 Series of 1P LSM images showing the monitoring of pH by using a dually labeled (pH-sensitive fluorescein and pH-insensitive rhodamine) 10,000-molecular-weight dextran in a microbial biofilm. (A) pH-sensitive imaging of fluorescein. (B) pH-insensitive fluorescence of the rhodamine. (C) Grayscale representation of the standard curve for pH versus the ratio of images A and B. (D) Contour map showing the distribution of pH levels within the microbial biofilm.

by 435-nm-wavelength light but are sensitive when excited by 490-nm-wavelength light. A ratio is taken of the pH-sensitive and pH-insensitive fluorescence signals, thereby eliminating effects such as variable concentration. When imaging biofilms, one needs to consider the lack of discrete boundaries within the biofilm and the presence of multiple environmental variables (e.g., Eh and ion concentration), as well as the potential for fluor sorption to macromolecules and fluor uptake into microbial cells (see, e.g., reference 23a). With bacterial cells, the major concern is effective loading of the fluor into the cells and its behavior in the presence of intracellular protein and other macromolecules. The calculation of pH from the ratio images is based on ratiometric imaging of pH-equilibrated cells, biofilm, or other specimens which have been stained in the same fashion as those under investigation. The cells used for equilibration may first be fixed in 70% (vol/vol) ethanol or rendered permeable by using valinomycin and nigericin, which act to equilibrate potassium and proton gradients across the cell membrane. These cells or biofilm materials are then placed in appropriate buffers (pH 5.0 to 8.0) with the fluor to obtain pH-equilibrated specimens that are analyzed by using LSM and digital image analysis to establish a standard calibration. Figure 5 illustrates the procedure, showing images taken in the green (pH responsive) and red (pH independent) wavelengths, the ratiometric image (with a standard grayscale calibration curve), and the resulting contour map of pH in the biofilm material.

The ratiometric method suffers from the drawback that quantitative pH imaging requires time-consuming, cumbersome calibration procedures as described above. However, many ratio probes such as 5- and 6-carboxyfluorescein and SNAFL-1 may be used in conjunction with a technique referred to as fluorescence lifetime imaging. This method was introduced ~10 years ago as an alternative method for achieving contrast in fluorescence images. It is based on detecting differences in the rates of decay of fluorescence of a molecule. This type of imaging also produces images that are independent of the concentration of the fluorescent probe and has the added advantage that straightforward calibrations of the fluor response in buffer are sufficient to create a valid standard curve. A number of publications have shown that this approach can be applied for determination of ion concentrations in bacterial cells and pH gradients in biofilms (51, 53, 57). The technique requires a modified 1P laser scanning microscopic equipped with an optical chopper to create nanosecond laser pulses and time-gated detection to collect the fluorescence emission following the short excitation pulse. The reference by Sanders et al. (51) provides a starting point for understanding the nature and application of fluorescence lifetime imaging of pH by using SNAFL-1 and 1P LSM.

3.11. HANDLING 3-D DATA SETS

There are many options for the presentation of 3-D data sets; these are summarized in Table 4. The selection of the option is dictated by personal choice, detection of specific information, and cost. The easiest but least informative presentation is the gallery in which all the images of a series are present sequentially (Fig. 6). Figure 7 shows the same image stack (or series) presented as a simple black-and-white stereo pair. In general, the use of color to code for depth information, create red-green anaglyph stereo images, and form three-color or multicolor stereo pairs (Color Plates 5 and 6) is an effective method, although publication costs may be inhibitory. Another option for the display of 3-D data sets is "simulated fluorescence," whereby the material is viewed as though it were illuminated from an oblique angle and the surface layer were fluorescent. The application of 3-D rendering through ray tracing or surface contour-based programs may also provide a useful presentation of 3-D data sets, allowing the viewer to examine the data set from various perspectives. For oral presentation, the evolution of computer animation through programs such as QuickTime allows the LSM user to present z series as live showings of images collected in movie format. This format

TABLE 4 Methods for the presentation of 3-D data sets

Method	Advantage	Disadvantage
Gallery	Simple presentation may allow specific marking of objects and depth	No true 3-D representation; limited data content
xyz projection	View of single section in xy and sections in xz and yz along user-defined line	Maximum-intensity projection in xy is not possible
Red-green anaglyph stereo projection	Comparatively simple to prepare; usually part of LSM software	Expense of color; requires special red-green glasses; up to 20% of people cannot see the stereo effect
Simulated fluorescence	Provides a type of 3-D view of a Z series	Presents only surface data
Stereo pair	Can be either black and white or color, can contain information from multiple probe signals, and can be very high quality	Requires special viewer; not oral presentation friendly; high expense; up to 20% of people cannot create the stereo effect
Polarized double projection	Effective display of data on computer screen	Requires special glasses and projector
3-D rendering	Detailed reconstruction of the data set; allows for rotation and fly through	Tends to be limited to external features; 3-D space-filling construction; extreme demand for computing time
Video	Animations can be transferred to video; easy to use	Requires additional equipment; not useful for publication
Movie	Audience can see Z series as collected; movie contains more visible information than the static presentation; easily appreciated	Requires laptop computer and special projection facility; not a publishable format at this time
Digital presentation	Allows use of single images or complex animation; easy for presentation	Not suitable for publication in most journals

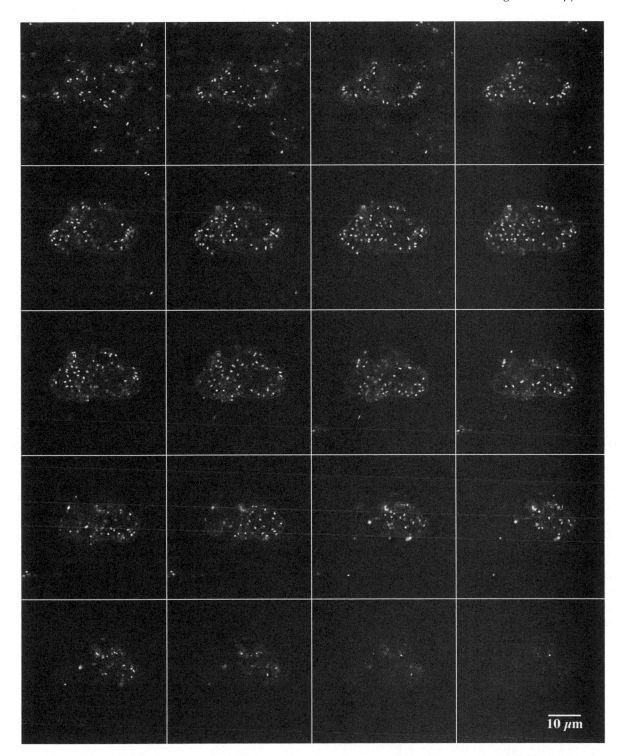

FIGURE 6 Gallery presentation of a Z series taken through a microbial biofilm stained with SYTO9.

also facilitates the projection of full-color rendering and animation of data sets, which provide a more information-rich presentation than static formats.

3.11.1. Quantification

Processing and analysis of digital images are major topics in their own right, and thus we can only touch upon various options available to LSM users with respect to quantification and deconvolution. See also references 30 and 32 and image-processing texts such as that by Russ (50).

Although there are many commercial and freeware systems for 3-D image analysis and processing, no single software package offers all the required features. Some users may prefer commercial systems which are offered by the

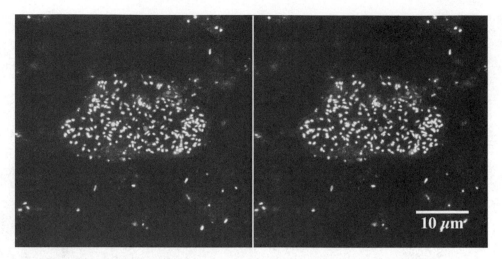

FIGURE 7   3-D stereo pair presentation of the Z series shown in Fig. 6.

LSM companies. In addition there are separate commercial 3-D imaging and image analysis systems, e.g., Imaris and VoxelShop (Bitplane, Switzerland), MicroVoxel (Indec Systems, Sunnyvale, CA), VoxelView (Vital Images, Fairfield, IA), VoxBlast (VayTek, Inc., Fairfield, IA), and ImageSpace (Molecular Dynamics, Sunnyvale, CA). On the other hand, freeware programs are available that provide image processing and analysis for single 2-D sections. The most widely distributed free software was NIH IMAGE, now replaced by Image J, available for MS Windows, Mac OS, Mac OS X, Linux, and the Sharp Zaurus PDA (http://rsb.info.nih.gov/ij/). Nevertheless, other tools have been developed and are available for download from the Internet, e.g., CELLSTAT (41), which is a program for UNIX workstations (http://www.cbm.biocentrum.dtu.dk), and ImageTool (36), which is a PC-based freeware image analysis package (http://ddsdx.uthscsa.edu/dig/itdesc.html) particularly useful for object-based analyses. COMSTAT (23) is an additional option written as a script in MATLAB (http://www.im.dtu.dk/comstat). Some consideration should be given to neural network systems for analysis of fluorescence images (5).

3.11.2. Deconvolution

As noted above, the major limitation of all light microscopic systems and particularly LSM is poor axial resolution. The principle solution to this limitation is image processing of the z series by computing of intensive restoration procedures or deconvolution. Deconvolution refers to the sharpening, or deblurring, of an image obtained using an optical microscope by removal of out-of-focus information, thereby mathematically producing an optical thin section. The most common algorithm used for deconvolution is the nearest-neighbor deconvolution algorithm; however, others such as blind deconvolution, iterative maximum likelihood estimation, and iterative constrained Tikhonov-Miller are included in certain software packages. The HUYGENS system (http://www.svi.nl) was designed to run on Silicon Graphics computers under the IRIX operating system, but is now available for a range of computer platforms. It is one of the most advanced packages and includes many image-processing features. Other specialized software packages are available for the removal of out-of-focus haze such as AUTO-DEBLUR (http://www.aqi.com), MICROTOME (http://www.vaytek.com), and EPR (http://www.scanalytics

.com). See, for example, Manz et al. (37). Another UNIX-based package for this purpose available over the Internet is XCOSM, which can be applied to LSM images or other types of fluorescent images (http://3dmicroscopy.wustl .edu/~xcosm). Regardless of the software package used, it is essential to have internal standards consisting of fluorescent beads of known dimensions in each z series to control and assess the deconvolution process.

3.12. LSM SAMPLING FREQUENCY

In LSM there are both 2-D- and 3-D-related considerations regarding sampling. In general, the Nyquist criterion indicates that the imaging system must sample with a spatial resolution of half that of the smallest object to be resolved. For 3-D imaging, it is necessary to sample at twice the spatial resolution of the optical system. In practical terms, this means x and y sampling at about 70-nm intervals and in the z axis, which has far poorer resolution, an interval of 150 nm is recommended. This sampling frequency creates a data set sufficiently continuous for correct image restoration (deconvolution) and representation of the sample area in subsequent processing and stereo presentation of the data. For some analytical purposes (e.g., cell counting and area measurements), the approach of sampling at discrete depths and locations for representative samples from a number of replicate locations provides a viable alternative (38).

Most microscopy studies rely upon relatively small areas of analysis; thus, typical data sets obtained may be prone to errors based on rare events. In practice, determinations of representative samplings for microscopic study are not easily made. However, Korber et al. (25) suggested analysis areas exceeding $>1 \times 10^5$ μm^2 and Møller et al. (42) examined contiguous areas of $>2.8 \times 10^5$ μm^2 for statistically representative results. Alternatively, the user may randomly select five or more replicate microscope fields (63× water-immersible lens) per treatment replicate, allowing application of analysis of variance to determine significant effects at P of <0.05 (28).

3.13. LSM IMAGE ARCHIVING AND PRINTING

Application of LSM techniques results in the creation of vast image and data sets; our facilities can produce several gigabytes per day. Thus, an important consideration is how to archive all this information. First, it is critical to have the most random-access memory possible and the largest hard drive(s) available for the operating computer. Second, options for longer-term information storage include CDs and DVDs; for temporary storage and transport, a variety of mini-hard drives and USB memory sticks are available. For relatively secure (significant concerns about the stability of CD media have been raised), portable, and relatively universal storage media, CDs and DVDs remain the best, most cost-effective recommendation. Other options may emerge, and the user must continually address this question.

Archiving represents another major hurdle that should be considered early in the process of developing an LSM-based research program. Images are stored in a variety of formats such as TIFF, GIFF, RAW, PICT, EPS, and JPEG and manufacturers' formats. Each of these has advantages and disadvantages. Some are dead ends which cannot be translated. Others offer various degrees of image fidelity and abilities to compress images, such as JPEG. In general, TIFF is the most universally used and can be opened by most software. Nevertheless, some of the advanced 3-D software packages use different image formats to analyze 3-D data sets, e.g., image cytometry standard, which in contrast to TIFF is designed for multidimensional image data.

Images may be printed for publication by using a variety of means including video printers, dye sublimation printers, slide printers, color laser printers, and digital printers. Increasingly, publishers can deal with the original digital file. The further development of the Internet and the way publications will be done in the future may allow more publication of animated 3-D multichannel LSM data.

3.14. REFERENCES

1. **Amann, R., R. Snaidr, M. Wagner, W. Ludwig, and K. H. Schleifer.** 1996. *In situ* visualization of high genetic diversity in a natural microbial community. *J. Bacteriol.* **178:**3496–3500.
   A useful example of application of multiple fluors.
2. **Amann, R. I., W. Ludwig, and K. H. Schleifer.** 1995. Phylogenetic identification and *in situ* detection of individual microbial cells without cultivation. *Microbiol. Rev.* **59:**143–169.
   Excellent review of both FISH and CLSM approaches with good color stereo images of microbial specimens.
3. **Axelrod, A., D. E. Koppel, J. Schlessinger, E. Elsen, and W. W. Webb.** 1976. Mobility measurement by analysis of fluorescence photobleaching recovery kinetics. *Biophys. J.* **16:**1055–1069.
   A more technical paper which provides the basis for using FRAP to calculate the mobility of molecules.
4. **Birmingham, J. J., N. P. Hughes, and R. Treloar.** 1995. Diffusion and binding measurements within oral biofilms using fluorescence photobleaching recovery methods. *Phil. Trans. R. Soc. Lond. B* **350:**325–343.
   A practical application of FRAP in microbial samples.
5. **Blackburn, N., A. Hagstrom, J. Wikner, R. Cuadros-Hansson, and P. K. Bjornsen.** 1998. Rapid determination of bacterial abundance, biovolume, morphology, and growth by neural network-based image analysis. *Appl. Environ. Microbiol.* **64:**3246–3255.
   An article for those with a greater interest in more sophisticated handling of images and data extraction.
6. **Blonk, J. C. G., A. Don, H. van Aalst, and J. J. Birmingham.** 1993. Fluorescence photobleaching recovery in the confocal scanning laser microscope. *J. Microsc.* **169:**363–374.

Technical paper providing more complex mathematical treatment of the data obtained through imaging FRAP.
7. **Boleti, H., D. M. Ojcius, and A. Dautry-Varsat.** 2000. Fluorescent labelling of intracellular bacteria in living host cells. *J. Microbiol. Methods* **40:**265–274.
   Technical information on the use of CellTracker with bacteria.
8. **Bright, G. R., and D. L. Taylor.** 1986. Imaging at low light level in fluorescence microscopy, p. 257–288. *In* D. L. Taylor, A. S. Waggoner, F. Lanni, and R. Birge (ed.), *Applications of Fluorescence in the Biomedical Sciences.* Liss Inc., New York, NY.
   An older book that still provides an excellent overview of and entry point into the very large cell biology literature which contains a vast array of vital information on fluors and techniques with potential for application in prokaryotic studies.
9. **Bryers, J. D., and F. Drummond.** 1998. Local macromolecule diffusion coefficients in structurally non-uniform bacterial biofilms using fluorescence recovery after photobleaching (FRAP). *Biotechnol. Bioeng.* **60:**462–473.
   Useful technical note on the FRAP method applied to microorganisms.
10. **Caldwell, D. E., D. R. Korber, and J. R. Lawrence.** 1992. Confocal laser microscopy and digital image analysis in microbial ecology. *Adv. Microb. Ecol.* **12:**1–67.
    The first comprehensive overview of applications and potential of 1P LSM and digital imaging focused on the area of microbial ecology; still provides a good starting point.
11. **Caldwell, D. E., D. R. Korber, and J. R. Lawrence.** 1992. Imaging of bacterial cells by fluorescence exclusion using scanning confocal laser microscopy. *J. Microbiol. Methods* **15:**249–261.
    Technical note describing details of the fluorescent negative staining technique for bacteria and biofilms.
12. **Caldwell, D. E., G. M. Wolfaardt, D. R. Korber, S. Karthikeyan, J. R. Lawrence, and D. K. Brannan.** 2001. Cultivation of microbial consortia and communities, p. 92–100. *In* C. J. Hurst, R. L. Crawford, G. R. Knudsen, M. J. McInerney, and L. D. Statzenbach (ed.), *Manual of Environmental Microbiology,* 2nd ed. ASM Press, Washington, DC.
    Overview; features many devices for cultivation of bacteria and biofilms that are compatible with microscopic studies.
13. **Choi, J.-W., B. F. Sherr, and E. B. Sherr.** 1999. Dead or alive? A large fraction of ETS-inactive marine bacterioplankton cells, as assessed by reduction of CTC, can become ETS-active with incubation and substrate addition. *Aquat. Microb. Ecol.* **18:**105–115.
    A good example of validation of a fluorescence technique for natural bacterial populations and important considerations in the process.
14. **Christensen, B. B., C. Sternberg, J. B. Andersen, L. Eberl, S. Møller, M. Givskov, and S. Molin.** 1999. Molecular tools for study of biofilm physiology. *Methods Enzymol.* **310:**20–42.
    Excellent overview of fluorescence-based techniques including those compatible with LSM.
15. **Conn, P. M.** 1999. *Methods in Enzymology,* vol. 307. *Confocal Microscopy.* Academic Press, New York, NY.
    A comprehensive compendium of papers on various areas of confocal microscopy with technical details and applications.
16. **Coote, P. J., C. M.-P. Billon, S. Pennel, P. J. McClure, D. P. Ferdinando, and M. B. Cole.** 1995. The use of confocal scanning laser microscopy (CSLM) to study the germination of individual spores of *Bacillus cereus. J. Microbiol. Methods* **21:**193–208.
    A good example of technical application of 1P LSM and fluorescent labeling.
17. **Cullander, C.** 1994. Imaging in the far-red with electronic light microscopy: requirements and limitations. *J. Microsc.* **176:**281–286.

18. **De Beer, D., P. Stoodley, and Z. Lewandowski.** 1997. Measurements of local diffusion coefficients in biofilms by microinjections and confocal microscopy. *Biotechnol. Bioeng.* **53:**151–158.
Excellent paper for those interested in calculation of diffusion from time course images of microinjected fluor-conjugated probes.

19. **Decho, A. W., and T. Kawaguchi.** 1999. Confocal imaging of in situ natural microbial communities and their extracellular polymeric secretions using Nanoplast resin. *BioTechniques* **27:**1246–1252.
Useful starting point for combined use of hydrophilic resin embedding and LSM.

20. **Florijn, R. J., J. Slats, H. J. Tanke, and A. K. Raap.** 1995 Analysis of antifading reagents for fluorescence microscopy. *Cytometry* **19:**177–182.

21. **Haugland, R. P.** 1998. *Handbook of Fluorescent Probes and Research Chemicals.* Molecular Probes Inc., Eugene, OR.
Comprehensive guide to the fluorescent products provided by Molecular Probes; includes much useful background information, excitation-emission spectra, and literature.

22. **Hausner, M., and S. Wuertz.** 1999. High rates of conjugation in bacterial biofilms as determined by quantitative in situ analysis. *Appl. Environ. Microbiol.* **65:**3710–3713.
A novel application of 1P LSM for in situ analyses of single-cell molecular events.

23. **Heydorn, A., A. T. Nielsen, M. Hentzer, C. Sternberg, M. Givskov, B. K. Ersböll, and S. Molin.** 2000. Quantification of biofilm structures by the novel computer program COMSTAT. *Microbiology* **146:**2395–2407.
Paper that provides a good entry point into the use of statistical considerations in imaging of microbial biofilms.

23a.**Hunter, R. C., and T. J. Beveridge.** 2005. Application of a pH-sensitive fluoroprobe (C-SNARF-4) for pH microenvironment analysis in *Pseudomonas aeruginosa* biofilms. *Appl. Environ. Microbiol.* **71:**2501–2510.
A good example of ratiometric applications of fluorescent pH reporters and useful controls for these studies.

24. **Kepner, R. L., and J. R. Pratt.** 1994. Use of fluorochromes for direct enumeration of total bacteria in environmental samples: past and present. *Microb. Rev.* **58:**603–615.
Excellent overview of the literature to 1994 on fluorochromes, including applications and problems.

25. **Korber, D. R., J. R. Lawrence, M. J. Hendry, and D. E. Caldwell.** 1993. Analysis of spatial variability within mot⁺ and mot⁻ *Pseudomonas fluorescens* biofilms using representative elements. *Biofouling* **7:**339–358.
Paper describes application of geostatistical methods to biofilm research and establishment of sampling requirements; also provides images of montages of large biofilm areas.

26. **Laca, A., L. A. Garcia, F. Argüeso, and M. Diaz.** 1999. Protein diffusion in alginate beads monitored by confocal microscopy. The application of wavelets for data reconstruction and analysis. *J. Ind. Microbiol. Biotechnol.* **23:**155–165.

27. **Lawrence, J. R., and T. R. Neu.** 1999. Confocal laser scanning microscopy for analysis of microbial biofilms. *Methods Enzymol.* **310:**131–144.
A brief overview of how to do 1P LSM with microbial biofilms and flocs; also examines specimen handling and upright versus inverted microscopes.

28. **Lawrence, J. R., T. R. Neu, and G. D. W. Swerhone.** 1998. Application of multiple parameter imaging for the quantification of algal, bacterial and exopolymer components of microbial biofilms. *J. Microbiol. Methods* **32:**253–261.
Useful techniques paper describing how to use autofluorescence, nucleic acid staining, and lectin staining with digital imaging to quantify major biofilm components.

29. **Lawrence, J. R., G. D. W. Swerhone, and Y. T. J. Kwong.** 1998. Natural attenuation of aqueous metal contamination by an algal mat. *Can. J. Microbiol.* **44:**825–832.
The paper provides a useful example of combining reflectance, fluorescence, and autofluorescence techniques.

30. **Lawrence, J. R., G. M. Wolfaardt, and T. R. Neu.** 1998. The study of biofilms using confocal laser scanning microscopy, p. 431–465. In M. H. F. Wilkinson and F. Schut (ed.), *Modern Microbiological Methods Series*, vol. •. *Digital Analysis of Microbes. Imaging, Morphometry, Fluorometry and Motility Techniques and Applications.* John Wiley and Sons, Sussex, United Kingdom.
A comprehensive overview of applications of 1P LSM to microbial biofilms.

31. **Lawrence, J. R., G. M. Wolfaardt, and D. R. Korber.** 1994. Monitoring diffusion in biofilm matrices using scanning confocal laser microscopy. *Appl. Environ. Microbiol.* **60:**1166–1173.
The first application of 1P LSM to determine diffusion and penetration of size-fractionated probes in complex biofilms.

32. **Lawrence, J. R., D. R. Korber, G. M. Wolfaardt, D. E. Caldwell and T. R. Neu.** 2001. Analytical imaging and microscopy techniques, p. 39–61. In C. J. Hurst, R. L. Crawford, G. R. Knudsen, M. J. McInerney, and L. D. Stetzenbach (ed.), *Manual of Environmental Microbiology*, 2nd ed. ASM Press, Washington, DC.
Comprehensive overview of digital microscopy as applied to a variety of environmental samples.

33. **Lawrence, J. R., D. R. Korber, B. D. Hoyle, J. W. Costerton, and D. E. Caldwell.** 1991. Optical sectioning of microbial biofilms. *J. Bacteriol.* **173:**6558–6567.
The first comprehensive study of biofilm architecture applying 1P LSM and digital image analyses.

34. **Lisle, J. T., P. S. Stewart, and G. A. McFeters.** 1999. Fluorescent probes applied to physiological characterization of bacterial biofilms. *Methods Enzymol.* **310:**166–178.
A comprehensive overview with practical protocols for application of a number of fluorescent reporters to microorganisms.

35. **Liss, S. N., I. G. Droppo, D. T. Flannigan, and G. G. Leppard.** 1996. Floc architecture in wastewater and natural riverine systems. *Environ. Sci. Technol.* **30:**680–686.
A useful example of the correlative microscopy approach.

36. **Liu, J., F. B. Dazzo, O. Glagoleva, B. Yu, and A. K. Jain.** 2001. CMEIAS: a computer-aided system for the image analysis of bacterial morphotypes in microbial communities. *Microb. Ecol.* **41:**173–194.
Excellent publication providing information on ImageTool and examples of the application of digital imaging and software routines in microbial ecology.

37. **Manz, W., G. Arp, G. Schumann-Kindel, U. Szewzky, and J. Reitner.** 2000. Widefield deconvolution epifluorescence microscopy combined with fluorescence in situ hybridization reveals the spatial arrangement of bacteria in sponge tissue. *J. Microbiol. Methods* **40:**125–134.
Excellent coverage of deconvolution approaches, with examples and color illustrations using microbial specimens.

38. **Manz, W., K. Wendt-Potthoff, T. R. Neu, U. Szewzyk, and J. R. Lawrence.** 1999. Phylogenetic composition, spatial structure, and dynamics of lotic bacterial biofilms investigated by fluorescent in situ hybridization and confocal laser scanning microscopy. *Microb. Ecol.* **37:**225–237.

39. **Massol-Deya, A., J. Whallon, R. F. Hickey, and J. Tiedje.** 1995. Channel structures in aerobic biofilms of fixed film reactors treating contaminated groundwater. *Appl. Environ. Microbiol.* **61:**767–777.

40. **Michael, T., and C. M. Smith.** 1995. Lectins probe molecular films in biofouling: characterization of early films on non-living and living surfaces. *Mar. Ecol. Progr. Ser.* **119:**229–236.
One of the first applications of lectins in complex habitats.

41. **Møller, S., C. S. Kristensen, L. K. Poulsen, J. M. Carstensen, and S. Molin.** 1995. Bacterial growth on surfaces: automated image analysis for quantification of growth rate-related parameters. *Appl. Environ. Microbiol.* **61:**741–748.

42. **Møller, S., D. R. Korber, G. M. Wolfaardt, S. Molin, and D. E. Caldwell.** 1997. The impact of nutrient composition on a degradative biofilm community. *Appl. Environ. Microbiol.* **63:**2432–2438.

43. **Neu, T. R.** 2000. In situ cell and glycoconjugate distribution of river snow as studied by confocal laser scanning microscopy. *Aquat. Microb. Ecol.* **21:**85–95.
Useful application publication illustrating information derived from 1P LSM based study.

44. **Neu, T. R., G. D. W. Swerhone, and J. R. Lawrence.** 2001. Assessment of lectin-binding analysis for in situ detection of glycoconjugates in biofilm systems. *Microbiology* **147:**299–313.
Excellent example of an assessment of fluors and probes for use with 1P LSM, with additional information on lectin applications.

45. **Palmer, R., Jr., B. Applegate, R. Burlage, G. Sayler, and D. White.** 1998. Heterogeneity of gene expression and activity in bacterial biofilms, p. 609–612. *In* A. Rhoda, M. Pazzagli, L. J. Kricka, and P. E. Stanley (ed.), *Bioluminescence and Chemical Luminescence: Perspectives for the 21st Century. Proceeding of the 10th International Symposium on Bioluminescence and Chemiluminescence.* John Wiley and Sons, Chichester, United Kingdom.
Useful example of correlative microscopy.

46. **Pawley, J. B. (ed.).** 2006. *Handbook of Biological Confocal Microscopy.* Plenum Press, New York, NY.
An excellent technical summary of all aspects of confocal microscopy.

47. **Perret, D., G. G. Leppard, M. Muller, N. Belzile, R. DeVitre, and R. Buffle.** 1991. Electron microscopy of aquatic colloids: non-perturbing preparation of specimens in the field. *Water Res.* **25:**1333–1343.
Describes the use of Nanoplast embedding techniques for delicate microbial structures.

48. **Podda, F., P. Zuddas, A. Minacci, M. Pepi, and F. Baldi.** 2000. Heavy metal coprecipitation with hydrozincite [$Zn_5(CO_3)_2(OH)_6$] from mine waters caused by photosynthetic microorganisms. *Appl. Environ. Microbiol.* **66:**5092–5098.

49. **Rocheleau, S., C. W. Greer, J. R. Lawrence, C. Cantin, L. Laramee, and S. Guiot.** 1999. Differentiation of *Methanosaeta concilii* and *Methanosarcina barkeri* in anaerobic mesophilic granular sludge by fluorescent in situ hybridization and confocal scanning laser microscopy. *Appl. Environ. Microbiol.* **65:**2222–2229.
Excellent example of combined application of embedding, rRNA probes, 1P LSM, and digital image analyses.

50. **Russ, J. C. (ed.).** 2002. *The Image Processing Handbook.* CRC Press, Boca Raton, FL.
Excellent comprehensive and understandable publication on all aspects of image processing.

51. **Sanders, R., A. Draaijer, H. C. Gerritsen, P. M. Houpt, and Y. K. Levine.** 1995. Quantitative pH imaging in cells using confocal fluorescence lifetime imaging microscopy. *Anal. Biochem.* **227:**302–308.
Publication provides a basis for starting fluorescence lifetime imaging using 1P LSM systems.

52. **Sytsma, J., J. M. Vroom, C. J. de Grauw, and H. C. Gerritsen.** 1998. Time-gated fluorescence lifetime imaging and microvolume spectroscopy using two-photon excitation. *J. Microsc.* **191:**39–51.
Provides current application of 2P LSM in conjunction with fluorescence lifetime imaging.

53. **Szmacinski, H., and J. R. Lakowicz.** 1993. Optical measurements of pH using fluorescent lifetimes and phase-modulation fluorometry. *Anal. Chem.* **65:**1668–1674.
Excellent starting point for understanding application of fluorescence life-time imaging.

54. **Ulrich, S., B. Karrasch, H. G. Hoppe, K. Jeskulke, and M. Mehrens.** 1996. Toxic effects on bacterial metabolism of the redox dye 5-cyano-2,3-ditolyl tetrazolium chloride. *Appl. Environ. Microbiol.* **62:**4587–4593.
Good critical evaluation of the application of fluorescent probes to assess bacterial metabolism.

55. **Valkenburg, J. A., C. L. Woldringh, G. J. Brakenhoff, H. T. van der Voort, and N. Nanninga.** 1985. Confocal scanning light microscopy of the *Escherichia coli* nucleoid: comparison with phase-contrast and electron microscope images. *J. Bacteriol.* **161:**478–483.
First application of confocal 1P LSM to study bacterial structure.

56. **Van Ommen Kloeke, F., and G. G. Geesey.** 1999. Localization and identification of populations of phosphatase-active bacterial cells associated with activated sludge flocs. *Microb. Ecol.* **38:**201–214.
Example of the application of fluorescence imaging of alkaline phosphatase activity in flocs.

57. **Vroom, J. M., K. J. de Grauw, H. C. Gerritsen, D. J. Bradshaw, P. D. Marsh, G. K. Watson, J. J. Birmingham, and C. Allison.** 1999. Depth penetration and detection of pH gradients in biofilms by two-photon excitation microscopy. *Appl. Environ. Microbiol.* **65:**3502–3511.
First substantial publication describing the use of 2P LSM to image microbial biofilms.

57a. **Wigglesworth-Cooksey, B., and K. E. Cooksey.** 2005. Use of fluorophore-conjugated lectins to study cell-cell interactions in model marine biofilms. *Appl. Environ. Microbiol.* **71:**428–435.
A good example of more extensive applications of lectins in marine habitats.

58. **Wiggli, M., A. Smallcombe, and R. Bachofen.** 1999. Reflectance spectroscopy and laser confocal microscopy as tools in an ecophysiological study of microbial mats in an alpine bog pond. *J. Microbiol. Meth.* **34:**173–182.
Useful example of LSM imaging using reflectance, autofluorescence, and staining to examine environmental samples.

59. **Wolfaardt, G. M., J. R. Lawrence, R. D. Robarts, and D. E. Caldwell.** 1998. In situ characterization of biofilm exopolymers involved in the accumulation of chlorinated organics. *Microb. Ecol.* **35:**213–223.
Application of fluor-conjugated lectins in complex biofilm system with quantification and analyses.

60. **Wuertz, S., E. Muller, R. Spaeth, P. Pfleiderer, and H.-C. Flemming.** 2000. Detection of heavy metals in bacterial biofilms and microbial flocs with the fluorescent complexing agent Newport Green. *J. Ind. Microbiol. Biotechnol.* **24:**116–123.
A good practical example of the application of LSM and calibration of fluorescence imaging.

61. **Xiao, G. Q., and G. S. Kino.** 1987. A real time confocal scanning optical microscope. *Proc. Soc. Photo Opt. Instrum. Eng.* **809:**107–113.
Presents a useful, simple equation for estimation of optical section thickness.

62. **Yu, F. P., G. M. Callis, P. S. Stewart, T. Griebe, and G. A. McFeters.** 1994. Cryosectioning of biofilms for microscopic examination. *Biofouling* **8:**85–91.
An early example of this now-standard method.

4

Electron Microscopy

TERRY J. BEVERIDGE, DIANNE MOYLES, AND BOB HARRIS

Microbiologists are dedicated to the study of very small cells. Today, as molecular biology is becoming an increasingly important research tool for microbiology, microscopy (especially high-resolution microscopy) is becoming an even more important implement for the identification of subtle cellullar changes and cloned products. Unfortunately, this emphasis on molecular biology has, somehow, deterred many younger workers from recognizing the importance of microscopy in their molecular studies. Imagine! It is not unusual to enter a modern microbiological laboratory today and discover that no light microscope can be found, or to enter a department of microbiology and find that no electron microscopy (EM) facility exists. Yet, viruses, bacteria, archaea, and most eucaryotic microorganisms cannot be seen as individual cells (or particles) without the aid of, at least, a light microscope. Few dedicated microbiological EM facilities remain, and microbiologists are becoming more and more dependent on generic college, university, or outside facilities that possess little experience in microbes at the levels of both specimen processing and the interpretation of results. It is analogous to taking a sore tooth to an auto mechanic to be repaired instead of to a dentist. Our own experience suggests that this neglect of both light microscopy and EM in present-day studies is having grave consequences for such disparate but important aspects as culture contamination and proper identification of molecular alterations in cells. For these reasons, microscopy in one form or another is inescapable. Researchers who do not routinely examine their cultures under a microscope run the very real risk of drawing conclusions based on results from contaminated cultures. Chapters 1 and 2 have explained the tried and true uses and advantages of light microscopy, and this chapter deals with EM. It is our hope that this chapter will explain some of the mysteries of EM as well as make the various techniques more user friendly to researchers who have lost the skills and recognized the importance of its use.

Of all the research techniques available to microbiologists, EM is the only one that can accurately give shape and form to cells and their component parts and identify positions of ultrastructural components (such as ribosomes, bilayers, and plasmids) and the viruses that afflict cells. There has been a renaissance of interest in EM since new preparatory procedures and equipment have made possible extreme resolution with well-preserved material. The chromosomes and ribosomes are found to be dispersed throughout most of the cytoplasm, and enveloping layers are more complicated than was first suspected. The accurate identification and determination of macromolecular distributions have provided a better understanding of the integrative steps required in cellular metabolic processes and of their alteration during changes in growth or during division. EM assists the forma-

tion of a conceptual bridge between cellular functions, the chemistry of the parts, and the behavior of the macromolecules that make up structure (Fig. 1 and 2). All students of biochemistry, molecular biology, biotechnology, industrial processes, and environmental studies really do require a microscopic view at one time or another. Accordingly, we also aim our chapter at these casual users as well as at neophytes at EM. Simple, routine types of EM are described together with hints based on experience of how best to go about preparing samples and maintaining equipment. The chapter provides a guide to the kind of information that can be obtained from basic techniques and suggests combinations of techniques for more complete information.

The first level of ultrastructural information is provided by transmission electron microscopy (TEM) and scanning

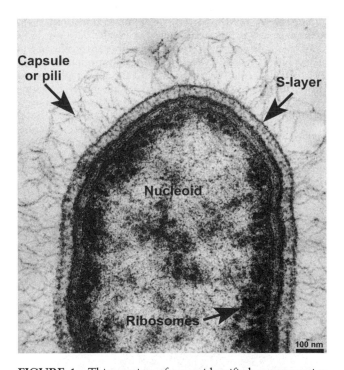

FIGURE 1 Thin section of an unidentified gram-negative bacterium fixed via the glutaraldehyde-osmium tetroxide protocol (section 4.2.2.1) showing the gram-negative envelope, condensed nucleoid, and clustered ribosomes produced by this method. An S-layer and capsule or pili (fimbriae) are also present. With the thin sectioning technique, it is difficult to differentiate between a capsule and pili, and a negative stain would help in identification. This is a TEM image.

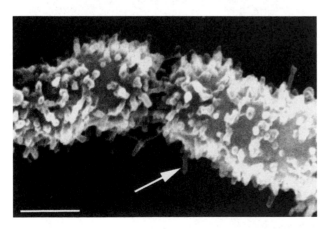

FIGURE 2 TEM image of a thin section of a freeze-substituted *E. coli* K-12 cell showing a well-preserved cytoplasm and cell envelope. Compare the cytoplasm and cell envelope of this cell with those shown in Fig. 1. (Reprinted from reference 8 with permission of the publisher.)

electron microscopy (SEM). These are tried and true techniques that have, over the last 40 years, deciphered the shapes, sizes, constituent arrangements, and cellular interactions of a great many bacteria (78, 81–83). Of the two microscopies, SEM is better for determining the surface appearance or topography of cells but usually does not have enough resolving power to elucidate more than general features (Fig. 3). SEM is a good technique for deciphering cell-cell associations, growth habits, and the topography of cell aggregates such as biofilms. TEM offers greater resolving power for a variety of preparations and is capable, with suitable material, of detecting the positions of cellular macromolecules (Fig. 1 and 2). Many procedures are suitable for TEM of bacteria, including negative staining, thin sectioning, and the use of shadowed replicas; most of this chapter concentrates on these procedures. It is important to

know what sort of information can be expected from these major techniques and to be aware of their limitations.

There are a number of EM books available and recommended (1–11), but nothing can replace the experience of the regular operators of a microbiological EM unit. The instruction from and the friendship of these highly qualified personnel should be nurtured. No chapter or textbook on EM can make you an expert microscopist; hands-on experience with guidance from someone who knows is essential.

Many new and complex derivatives of EM are available, all with impressive and complicated names, e.g., scanning transmission electron microscopy, energy-dispersive X-ray spectroscopy (EDS), electron energy loss spectroscopy (EELS), electron spectroscopic imaging (ESI), cryoTEM, and cryoSEM; often these techniques can be intermixed with one another (e.g., cryoTEM-EELS-ESI). Some microscopies cannot even be considered to involve typical lens-style microscopes, and they depend on the physical probing of specimens with sharp, pointed tips, thereby developing a tunneling current (scanning tunneling microscopy [STM]) or detecting atomic weak bonding forces (atomic force microscopy [AFM] and its derivatives). STM and AFM are described in more detail in chapter 6. To add to the confusion, new specimen-processing methods abound, e.g., cryofixation by freeze-plunging or freeze-slamming; freeze-substitution; the use of thin, frozen films; and cryosectioning. Preparative techniques allowing the localization of component macromolecules have also become crucial. All are highly specialized, requiring dedicated equipment and experienced personnel, but can be made available through collaborative ventures.

4.1. INSTRUMENTATION

4.1.1. Microscopy Procedures

Any electron microscope (Fig. 4) is an extremely complicated and expensive instrument. It is looked after by specialists and (usually) maintained, because of the expense of parts and labor, by service contracts with the manufacturer. Most instruments have multiple users to justify the capital expense and complex operating problems. EM facilities often provide thin sectioning, shadow casting, freeze-etching, and perhaps other kinds of assistance. Workers in charge of

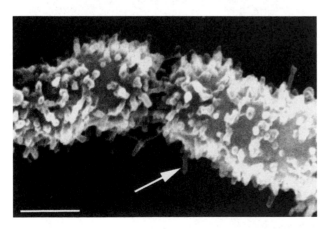

FIGURE 3 SEM image of a gold-sputtered *Chlorobium* sp. covered with special surface appendages called spinae (arrow). Bar = 500 nm. (Reprinted from reference 8 with permission of the publisher.)

FIGURE 4 Transmission electron microscope being operated by one of the authors (R. Harris) in the EDS mode to determine the elemental composition of a natural bacterial community.

these units establish rules and standards of operation, which protect the integrity of the equipment and the work that comes out of it, and are strict in enforcing them. Experience says that casual users of an electron microscope fit the description given by a museum guard of the museum's young patrons: "Lady, they will do *anything*." To this one must add, "And not tell you what it was!" To assist in troubleshooting and maintenance, each user must be considerate of the instrument (some parts are delicate, and it *does* matter which knobs you twist) and of the next user (a "golden rule") and must be scrupulously honest about anything done or not done.

Some procedural points to remember are as follows.

1. Get checked out on every instrument and procedure to know what to do and your limitations.

2. If equipment does not work, behaves strangely, or requires unusual force, *stop* the equipment in a safe position, think, and seek help.

3. Do not be tempted to undertake repairs. Own up and confess the problem.

4. Keep records, even if that is not the habit of the unit, and see that operating times are recorded and that every malfunction is reported.

5. Make notes of the steps in procedures, have them at hand when at work, and use a checklist, which should include the following: positions of switches and settings at start-up; start-up procedures; readiness requirements; alignment procedures for everyday operation for condenser adjustment, centering of the objective aperture, and astigmatism compensation; the setting for wide-field scanning; operating procedures and camera operation; warning signs and responses to them; standby and safety actions; the turn-off procedure; and positions of switches and settings after turn-off.

Good preparations, procedures, patience, and operational skills will be rewarded by good micrographs and will be appreciated by both the supervisor and the microscope.

4.1.2. Transmission Electron Microscope and Resolving Power

The value of a transmission electron microscope (Fig. 4) lies in its capacity to resolve objects that cannot be resolved by any form of light microscopy. The wavelength of electrons, which is shortened as the accelerating voltage is increased, permits the resolution of objects as small as 2 Å (0.2 nm or 0.0002 μm), whereas the limit of light microscopy is close to 0.2 μm. With biological specimens, resolutions of ~1.0 nm are achieved under only ideal conditions of operation, since cells are not infinitely thin and their substance contributes to "noise." The practical limit of real resolution, e.g., with thin biological membranes such as bacterial S-layers, is about 1.0 to 1.5 nm. Even then it takes real skill and a well-tuned instrument to do better than 2.0 to 2.5 nm. Thick specimens and support films increase electron scattering, and the resolution deteriorates accordingly.

It is useful to recall some of the limitations of TEM. Because specimens must be examined in a high vacuum to permit electron flow, cells need to be dried (unless they are held frozen [i.e., vitrified] in a special cryospecimen holder at extremely low temperatures). In addition, typical accelerating voltages (60 to 120 kV) subject the specimen to so much energy that chemical bonds (including covalent linkages) can be broken, molecular conformations can be altered, and substantial organic mass can be lost. Specimen exposure to the electron beam should be kept to a mini-

mum. Cell contrast is poor because carbon, hydrogen, oxygen, and nitrogen (which make up most cellular mass) scatter electrons poorly, and in comparison to resolution, cell contrast decreases with an increase in accelerating voltage (high kilovoltages are used to penetrate thick specimens). Consequently, the specimen must be as thin as possible and selectively stained or contrasted by using heavy-metal salts that increase the scattering power. Instruments in current use achieve enormous direct magnifications, but unless there is a need to resolve macromolecules, most applications are well served by magnifications of ×5,000 to ×15,000 (allowing subsequent photographic or computer enlargements of ×5 to ×10). Higher magnifications require more skill and a very well tuned instrument.

The value of TEM in microbiology is its outstanding ability to elucidate details of cell structure, provided that the preparative methods and the choice of EM technique are suitable to the structure involved, e.g., the use of heavy-metal salts to surround the bacterium and penetrate into surface irregularities supplies contrast by differentially impeding electrons, producing the effect of negative staining (section 4.2.2), analogous to the use of nigrosin in light microscopy (chapter 1). However, unless there are breaks in the cell wall or membrane to allow penetration of the metal salts, no intracytoplasmic features can be seen. Macromolecular arrangements and cellular components can be studied if cells are broken and fractionated (chapter 7) before negative staining (Fig. 5).

Microbes are too thick for adequate resolution of internal structures, but thin sectioning (into slices 50 to 100 nm thick) of chemically fixed and stained cells embedded in plastic allows their cytoplasmic arrangements to be discerned (Fig. 1 and 2). In this technique, heavy-metal stains of the sections are needed to *positively* stain the cells (as opposed to negative staining) and provide differential contrast for imaging. With bacteria, a thin section will show the arrangements of DNA, ribosomes, and cytoplasmic granules as well as the juxtapositioning of cell envelope layers (Fig.

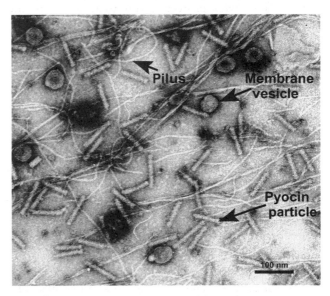

FIGURE 5 TEM image of a negatively stained preparation of isolated membrane vesicles, pili, and pyocin particles from *Pseudomonas aeruginosa*. Negative stains are particularly good for visualizing small particles for identification.

1 and 2). As mentioned previously, cellular elements (C, H, O, and N, etc.) do not scatter electrons proficiently. Most of the electrons of the electron beam pass cleanly through a specimen, and these unscattered electrons do not carry structural information. However, some electrons do have their trajectories altered as they pass through the specimen in the transmission electron microscope by elastic or inelastic scattering; this scattering by structures in the specimen provides information about the specimen. The elastic scattering is responsible for the high fidelity of TEM images.

Electron beams are usually generated by heating a pointed tungsten filament to a high temperature in the electron gun at the top of the of the transmission electron microscope column and focused by powerful symmetrical electromagnetic lenses that are stacked on top of one another in the column. More sophisticated electron guns can be outfitted with a lanthanum hexaboride (LaB_6) crystal to give a more brilliant coherent electron beam. Essentially, the column of a transmission electron microscope consists of condenser lenses (which focus the electron beam on a small area on the specimen) and magnifying lenses (functioning as objective, diffraction, intermediate, and projector lenses) immediately below. Electrons cannot be seen, and reliance is placed on a phosphorescent screen, which is excited by the incident electrons, to emit visible light. The emitted light intensity is low, and a darkened room is required to clearly see the images. Electron-sensitive photographic film or a charge-coupled device (CCD) camera is necessary to record the images.

4.1.3. Scanning Electron Microscope

In SEM, the electron beam rapidly scans the surface of the specimen to induce radiation of low-energy secondary electrons or higher-energy backscattered primary electrons. These electrons form an image on a cathode ray tube by synchronizing the movement of the electron beam in the microscope (and the information being collected) with the information displayed (during a raster) on the tube; the process is similar to the formation of a television picture. The specimen must first be chemically fixed, dried (best by critical-point drying or freeze-drying to avoid distortion), and given a thin metal coating to make it conductive (sputtered in vacuo with gold, platinum, or a 60:40 platinum-palladium mixture; the last of these gives a less granular metal layer). The resolution achieved by SEM has been markedly improved since its introduction (at best, ~30 nm in commercially available models) but is still lower than that routinely available by TEM. Newer low-voltage and field emission scanning electron microscopes are beginning to approach TEM resolution, and environmental scanning electron microscopes can view specimens under high relative humidities. A scanning electron microscope can display only the images of surfaces, but it does so in a three-dimensional mode that avoids the tedious preparation of the replicas of intact, fragmented, or frozen and fractured bacteria used in TEM for similar purposes. In bacteriology, SEM is most useful for discerning surface appendages and other surface structures (Fig. 3) and for defining shape and topographic relationships such as those in colonies or on surfaces of infected tissues (70–73).

Preparation of specimens for SEM involves fixation on a conductive surface such as graphite that can be affixed to the specimen holder. Heavy-metal coating done under vacuum prior to examination is usually required to avoid specimen charging and to give a strong surface signal for imaging. Because scanning electron microscopes are (usually)

not as expensive as transmission electron microscopes and are relatively easy to maintain and operate, and because their sample preparatory procedures are relatively straightforward, they are more frequently encountered than transmission electron microscopes by microbiologists. They are more likely to be used for a first look by students of microbial structure. Environmental scanning electron microscopes are becoming more readily available so that even hydrated specimens can be viewed, which gives a tremendous advantage over dried, coated specimens and provides a wonderful glimpse of microorganisms existing under quasinatural conditions (Fig. 6). Cryostages also provide a means of looking at frozen, hydrated specimens. Yet, resolving power and restriction to topography make scanning electron microscopes only partially useful for an encompassing EM analysis of microbial cells, and TEM must be, at some point, always considered. However, field emission and low-voltage systems are rapidly increasing the resolution gap between SEM and TEM. SEM remains a useful tool for looking at microorganisms in situ (e.g., to examine surfaces in natural environments for adhering microorganisms) (Fig. 6).

4.1.4. Specialized Electron Microscopes

This section is intended to identify and give a general explanation of highly sophisticated EM equipment and techniques but not procedures or methods for their use. The equipment and techniques have been chosen because they are important to research in microbiology. Special equipment and expert operators are required, so suitable problems must be taken to them. However, it is necessary to have a broad familiarity with them so as to know what they may be able to contribute to a research project.

4.1.4.1. STEMs

Scanning transmission electron microscopes (STEMs) combine some of the attributes of SEM and TEM; as in SEM, a narrow electron beam is scanned back and forth over the specimen; as in TEM, a transmitted signal can be recorded. The great advantage of a STEM is that a highly concen-

FIGURE 6 ESEM image of a wet and unaltered sample of a biofilm growing in a sulfur spring located near Ancaster, Ontario, Canada. Double arrows show the chains of cells (cyanobacteria) growing in the biofilm which are partially obscured by exopolymeric substances. The single arrow points to a break in the exopolymeric substances. This image was kindly supplied by S. Douglas of the NASA-Jet Propulsion Laboratory, Pasadena, CA.

trated and narrow electron beam can be directed onto and through a *thicker than normal* specimen. Because STEMs use a scanning mode, secondary and backscatter electron detectors can be placed above the specimen so that, at a flick of a switch, a STEM becomes a scanning electron microscope. Secondary electrons produce topographical images of the specimen, whereas the backscattered electrons provide density information. If an energy-dispersive X-ray spectrometer (see below for more details) is attached to a STEM, a compositional analysis of the elements in the specimen can be obtained. By selecting the energy line of a single element (e.g., Ca or Mg) and using a computer to combine the data with raster coordinates obtained in the STEM mode, a point map of the distribution of that element can be obtained (63, 64, 68, 69). Magnetosomes (Fe), polyphosphate (P) and sulfur (S) granules, and bacterium-associated minerals such as clays (Si) can be readily identified. The detection of toxic heavy metals complexed to bacteria from the environment (61) has provided some exciting examples of structural analysis by using STEMs.

4.1.4.2. Specialized Configurations for Spectroscopy and Other Types of Microscopy

There are three basic ways to obtain elemental analyses by EM: EDS (Fig. 4), wavelength-dispersive X-ray spectroscopy (WDS), and EELS. The application of EDS has already been mentioned (section 4.1.4.1). This can be done by SEM but is better done by scanning transmission electron microscopy because higher-resolution (elemental) point mapping can be achieved for single elements. A transmission electron microscope with an EDS detector will serve if no elemental imaging is required and if a full-spectrum analysis is satisfactory. As a specimen is irradiated by the electron beam, its constituent atoms emit "signature" X rays, and the X-ray emission spectrum is read by a detector placed above and to the side of the specimen. The X-ray

detector is separated from the high vacuum of the electron microscope column by a small borehole covered with a thin foil (usually of beryllium, germanium, or aluminum) which limits access to X rays with energies above those of the foil. For this reason, low-atomic-number elements (e.g., C, H, N, and O) are not easily resolved, although recent advances with plastic foils and "windowless" detectors are making the detection of such elements feasible. The borehole limits the number of X rays from regions other than the sample and the foil prevents the lowest-energy X rays from flooding the detector. EDS has been used to great advantage for the identification of fine-grained minerals formed on bacterial surfaces by biogeochemical means in the new field of geomicrobiology (Fig. 7) (61).

WDS is another technique for obtaining elemental analysis; it is restricted to scanning electron microscopes and relies on physical crystals (e.g., those of lithium fluoride or lead stearate) to separate and expand the X-ray emission spectrum focused on the detector. Analysis of the spectrum provides compositional analysis. Compared with EDS, WDS is rarely used on microbiological samples since it requires long counting times and can be used only if the material is very stable under the electron beam.

EELS requires a detector underneath the specimen (usually below the viewing screen) which is aligned to the optical axis of the electron beam and is, therefore, restricted to a transmission electron microscope or a STEM. Only the inelastically scattered electrons are used for EELS. These electrons have lost energy after atomic interaction with the specimen which results in lower energies and longer wavelengths; both of these traits provide elemental signatures of the atom(s) with which interaction occurred. Although these inelastic electrons lessen the quality of a TEM image by producing chromatic aberration, they are very useful for compositional analysis. Since the EELS detector operates in the vacuum of the electron microscope column without

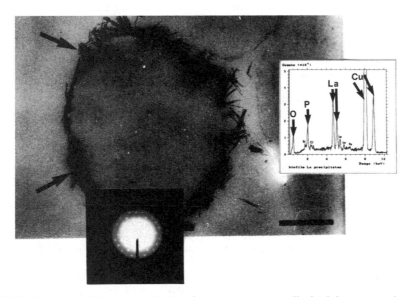

FIGURE 7 Unstained thin section of a *Pseudomonas aeruginosa* cell which has accumulated a lanthanum mineral phase on its surface. The lower inset uses SAED to show that it is a crystalline mineral phase due to the periodicity of the reflections in the diffractogram. The inset to the right is an EDS spectrum showing a high La concentration. The P and O could be due to the mineral, the cell, or the plastic. The high Cu is from the copper TEM grid. (From S. Langley and T. J. Beveridge, *Can. J. Microbiol.* **45**:616–622, 1999.)

metal foil filters (unlike the EDS detector), the technique can resolve a full spectrum of elements, including those of low atomic number. Compared with EDS, short counting times (10 s versus 100 s) are required but high doses of energy are required; highly beam-sensitive bacteriological samples should not be analyzed unless low-temperature (cryo-) stages are used to stabilize the specimen. However, so far, EELS is the best technique for analyses of elements of low atomic number.

One of the most exciting advances in TEM uses a modified version of EELS and is called ESI (62, 66). Two configurations are possible; one is a below-column spectrometer resembling that described above but with additional lenses and a filter attached. This configuration suffers from the disadvantage of a small sampling size, and the large number of lenses makes image acquisition difficult. Its advantage is that this ESI spectrometer can be added to most modern transmission electron microscopes. The other configuration requires a more dedicated instrument and relies on an in-column prism which allows for a larger sampling size and fewer lenses. Each ESI configuration separates the energy lines in the inelastic signal and allows the operator to choose a single elemental signal as the sole information signal to use in the formation of a TEM image. As a consequence, an elemental image at TEM resolution ($\sim$1.0 nm) is formed which has a much higher resolution than a point map formed by scanning transmission electron microscopy-EDS. This technique is complicated, requiring great expertise, and has only rarely been used in microbiology. Remarkably, the configuration of rRNA has been visualized in both small and large subunits of ribosomes by following the phosphorus line of the rRNA (Color Plate 7). By using the same phosphorus line, the bilayer profiles of membranes, such as plasma membranes and outer membranes, can also be imaged because of the phosphorus in phospholipids and lipopolysaccharides. Because of the need for extremely fine compositional analyses and because microbes are so small, ESI holds tremendous promise for TEM analysis in microbiology.

A comprehensive list of articles describing these methods of elemental analysis and their use in microbiology can be found in references 60 through 69.

Diffraction is another method for structural analysis but is applicable only to crystals and highly ordered macromolecular arrays. X-ray diffraction, which uses a beam of X rays, has been influential in determining molecular structure at the atomic level and in understanding the molecular conformations of proteins and nucleic acids. Like X rays, electrons diffract, and a transmission electron microscope can be used as an electron diffractometer by focusing incident electrons on extremely small areas. If the magnifying lenses are turned off in TEM for the viewing of a periodic object, an electron diffraction pattern will be obtained. Since the electron beam is concentrated on a small region of the specimen, this technique is called selected-area electron diffraction (SAED). Detection of high-level periodicity in most microbiological specimens is quite sensitive to electron bombardment, and SAED is restricted to biostructures with a high proportion of covalent bonding, to inorganic mineralized material such as magnetosomes (78), and to external mineral deposits (61). The combination of EDS or EELS with SAED is a most powerful technique for the analysis of fine-grained minerals associated with microorganisms (61).

A major challenge with imaging by EM is to obtain a good signal-to-noise ratio so that structure can be clearly seen. No matter what the resolving power of the microscope, there must be enough signal for our eyes to readily decipher an object and, hopefully, to see its macromolecular structure. In the early 1970s, optical diffractometers using lasers (as a coherent light source) and images of periodic structures (as diffraction grids) were used to painstakingly reconstruct images through mechanical means. The easy use of computers and the digitization of images has caused a major improvement in image processing. Selected negatives are digitized by using either a high-resolution television camera, a CCD camera, or a narrow-beam flat-bed scanner, and the information is then fed into a computer. Over the last few years, great advances have been made with CCD cameras with improved digitization. Two basic computer programs are used: Fourier analysis, which deals with highly periodic objects, and correlation averaging, which is used for more irregularly shaped objects (such as ribosomes) (Color Plate 7) or periodic objects with lattice defects. The structures of such diverse paracrystalline objects as viruses, bacteriophages, flagella, pili, S-layers, recombinant protein crystals, and parasporal bodies (Fig. 11 in chapter 5 shows a reconstruction of an S-layer) and such irregular particles as flagellar basal bodies and eucaryotic, archaeal, and bacterial ribosomes have been reconstructed by using computer-based techniques. Chapter 3 provides a more in-depth treatment of computer-based reconstruction techniques.

Cryo-transmission electron microscopes are relatively rare in microbiology, but they are rapidly revealing many secrets about microbial structure. These exotic microscopes are difficult to maintain, and the preparation of specimens is more of an art than a recipe. The specimen holder and specimen chamber are maintained at liquid-nitrogen or liquid-helium temperature so that snap-frozen specimens remain in a vitrified state (see section 4.2.4). Thin, frozen foils of small particles (e.g., membranes or ribosomes) (37d) or cryosections of microorganisms (37a–d) can be examined. Here, no heavy-metal stains can be used and we must rely on the actual mass and its differential organization within the section or foil for visualization against the overall mass of the frozen water in which the cells or particles are embedded. It is remarkable how well the cryoimages of bacteria compare with those from freeze-substitution or, even, conventional embedding. Subtle differences in shapes and dimensions of cell envelopes, ribosomes, and nucleoids are seen (37a, b, and c) but general shape and form is maintained. One of the most drastic differences is that a definite periplasmic space is seen in gram-positive bacteria (37b and c). New, very sophisticated cryo-transmission electron microscopes are now becoming available that are capable of doing tomography (producing a tilt series that can be used to produce a computer-generated three-dimensional image).

Scanning probe microscopy (SPM), which includes STM and AFM, is unlike any previously described microscopy. Here, no optical lenses are used, but instead, an atomically sharp tip (usually fabricated from silicon nitride [Si_3N_3]) attached to a cantilever of known rigidity is dragged over the surface of the specimen. The up and down movement of the tip provides topographical (z-axis) detail, and (as in SEM) as it is rastered over x,y coordinates, a three-dimensional image of the surface is generated (Fig. 5 in chapter 6). STM and AFM can provide atomic detail of hardy, robust inorganic samples (e.g., minerals and metal foils), but biological samples are orders of magnitude more difficult. Yet, since AFM can be done under water, it holds great promise for microbiology; chapter 6 provides an in-depth treatment of these new, exciting SPMs.

4.2. TECHNIQUES FOR TEM

This section describes a limited number of TEM methods that, from our experience, are considered useful to the worker who has need of basic structural information about microorganisms. Although we refer directly to procaryotes in our descriptions, most of the general principles are also suitable for eucaryotic microbes. Emphasis is placed on the use of negatively stained preparations and thin-sectioned materials for examination by TEM. No attempt is made to describe the operation and maintenance of the microscope (4, 6, 10) or other associated instruments. Materials that are required are mentioned (section 4.7), and their sources are given (section 4.8). It is assumed that instruments are functional and that experienced operators are available for assistance and instruction. The aim here is to supply the neophyte with helpful information on techniques and interpretation. General reading is provided in references 1 through 11, and more specialized sources of information are given in references 12 through 100.

4.2.1. Preparation

4.2.1.1. TEM Grids

See references 4 through 7 and 12.

Specimens in the form of thin sections, intact organisms, and cell components must be supported in the electron beam in accurate relation to the objective pole pieces of the microscope. This support has to be relatively transparent in the electron beam and is provided by an ultrathin film (section 4.2.1.2). To accomplish this, thin metal disks with a regular pattern of holes (grids) are usually covered with thin plastic or carbon films. Standard grids, of a diameter (2.3 or 3.0 mm) appropriate to the microscope used, are commercially available (see section 4.8, items *k*, *o*, *p*, *r*, *u*, *x*, *z*, and *cc*). Grids are made of several materials and are available in several mesh sizes, but the most commonly used grids are of copper and have a mesh size of 200 (mesh size indicates the lines per inch). The grids have a dull (matte) side and a shiny (bright) side. Before support films are applied, the grids must be clean and therefore must be treated with acetone and/or other solvents; however, supply houses can furnish precleaned grids.

4.2.1.2. Support Films for TEM Grids

Most thin sections as well as suspensions of bacterial cells for metal shadowing or for negative staining must be supported on thin films covering the holes in the grid. The film materials commonly used are nitrocellulose or collodion (Parlodion®), polyvinyl formol (Formvar®), and carbon. Films of the first two substances, which are plastics, are frequently stabilized by evaporative deposition of an additional thin layer of carbon. Formvar® is the most satisfactory plastic-film material and is used as follows.

1. Have available the materials listed in section 4.7.1.

2. Prepare a working solution of 0.2% (wt/vol) Formvar® in ethylene dichloride (chloroform can also be used). Mix thoroughly.

3. Degrease microscope slides (standard slides for light microscopy) by washing with soap (or detergent) and water, and rinse thoroughly in distilled or deionized water. Blot dry with lint-free tissue (lens tissue), which minimizes static charge buildup; never wipe completely dry. Cleaned slides may be stored indefinitely by being wrapped in folds of lint-free tissue.

4. Fill a Coplin jar two-thirds full with the working solution of 0.2% Formvar®.

5. Place two or three clean slides in the Coplin jar containing the Formvar® solution, immersed to three-quarters of their length. Allow them to stand for 3 to 5 min, withdraw one slide at a time with a slow, even motion, and hold the slide just above the solution for 10 s to allow excess to drain off in the solvent atmosphere. Now, completely withdraw the slide and place it standing on lint-free tissue for about 30 s allowing it to dry completely. With an alcohol-cleaned razor blade, score the Formvar® film around the entire perimeter of the slide. Be careful that small glass chips do not land on the Formvar® film!

6. Place a 6-in. (ca. 15-cm)-high, flat-bottomed, glass crystallizing dish on black paper. Fill the dish almost to the top with deionized water. Take a lint-free tissue, and drag it across the surface of the water to remove dust particles.

7. While holding the coated slide at a 45° angle over the dish, slowly immerse the slide into the water (breathing on dry Formvar® coatings helps to free them from the glass). The Formvar® film will slowly release from the glass surface where you have made your cuts and float off onto the water surface. Inspect each film visually to judge uniformity and thickness; they should appear gray to silver-gray, representing a thickness of ~60 nm according to interference color charts. The theoretically ideal thickness of 10 to 20 nm is rarely achieved. If uniformity or thickness of the films is unsatisfactory, repeat the above-decribed steps with appropriate modifications, as follows. If the thickness of the film is not uniform, then the slide was taken out of the solution too quickly or did not have enough time to drain in the solvent atmosphere. If the film is too thick, dilute the Formvar® solution with solvent. If film is too thin, increase the percentage of Formvar® to 0.3% and repeat the process.

8. Once satisfactory films are obtained on the water surface, use narrow-pointed forceps (Dumont no. 5) to carefully align the grids (dull side down) onto the films. To maintain a degree of consistency, avoid using film areas with obvious irregularities. Usually a film will accommodate 25 to 50 individual grids.

9. To retrieve the film-covered grids, cut a circular 9.0-cm-diameter filter paper (Whatman no. 1) into strips larger than the size of the film containing the grids and bend up one end of the filter paper strip (0.5 cm). While gripping the bent portion of the filter paper strip, hold the paper over the Formvar®-covered grids and lower it gently onto them until the paper has uniformly absorbed water. Just before the filter paper is completely wet, lift it quickly from the surface of the water (some filter papers are so "hairy" that the fibers damage the film; avoid these). Do not let the paper strip get completely wet or the film will separate from the paper and sink to the bottom of the dish.

10. Place the filter paper supporting the grids onto another piece of filter paper (grid side up) in a petri dish, and partially cover the dish with its lid while the contents dry.

11. After drying, lift the grids by their edges with narrow-pointed forceps. The plastic film will break cleanly around the perimeter of the grid and will be retained over the grid. Films that lift away from the filter paper as the grid is lifted or that do not break cleanly are too thick; some areas of the film may show this property while others are satisfactory. Films that are too thin will tear in the electron beam. Films that have large holes in them may also fail. Holes also result from too high an ambient humidity during drying.

Once in the microscope, because of ionization and consequent changes in electrostatic charge, plastic films expand and contract to cause specimen drift as the electron beam is intensified and decreased. Any movement will degrade the photographic image in proportion to the magnification used. Often, adjustments of beam intensity and a brief wait will establish a level of stability. Formvar® films prepared as described above are quite satisfactory for routine purposes using magnifications of up to ×20,000. For fine detail and higher magnifications, however, the stability of plastic films is improved by the deposition of a thin (2.0- to 3.0-nm) layer of carbon. Thin films of pure carbon avoid drift and improve resolution even more because of a more even and finer substructure, but they are fragile.

Formvar®-covered grids are usually placed within a carbon evaporator and are carbon coated by the heating of carbon electrodes in a 1.33- to 0.13-nPa vacuum. Evaporators are essential equipment of all EM laboratories; make sure that you have an operation manual and have been checked out on their use. Grids without Formvar® can also be carbon coated. Evaporate carbon onto a freshly cleaved mica surface under the same conditions; float the carbon film from the mica, and place grids on this film in the same way that Formvar® grids are prepared. Grids with only a carbon coating are *extremely fragile* and must be handled with care. They are especially useful for high resolution of subcellular components and viruses.

No matter how good the specimen preparation, accurate visualization of the specimen will clearly depend on the quality of the support film on the grid. For most routine investigations, Formvar®- and carbon-coated 200-mesh-size copper grids are recommended. The appropriate thickness of the coat can be gauged only by experience and by the silver-to-gray interference pattern of the raft of grids floating on water (as described above).

4.2.2. Negative Staining

4.2.2.1. Stain Choice

Negative staining is amongst the highest-resolving techniques that are easily available. It can be thought of as thin embedding since the sample is immersed in a thin film of a dilute heavy-metal salt, which is then dried down to a glassy consistency surrounding the object. Negative staining is especially good for visualizing subcellular particles such as ribosomes, cell wall or membrane fragments, viruses, and (in its most refined form) protein macromolecules. Most material does not have to be chemically fixed before a negative stain is applied, but some fragile specimens may require prefixation (see section 4.2.2.2, step 5). When objects are negatively stained, they appear white against the dark background of the stain. Since the heavy-metal solution penetrates into and around the contours of subunit arrangements, fine infrastructure can often be seen (see pyocins and flagella in Fig. 5).

Several different heavy-metal salts, usually in the range of 0.5 to 4.0% (wt/vol), have been used as negative stains. In our experience, ammonium molybdate is most useful because it spreads well, does not react with proteins or most other components in a suspension, and can be readily adjusted in tonicity as required. However, solutions of sodium or potassium phosphotungstate are perhaps the most widely used and are generally satisfactory at a neutral pH. Contrary to popular belief, phosphotungstic acid is almost never used, because of its very low pH; it is usually brought to neutral pH with sodium or potassium hydroxide to become a phosphotungstate salt. Uranium salts are also widely used because at pHs below 4.5 they dissociate into small uranyl (UO_2^{2+}) ions, which penetrate into surface irregularities and improve the definition of fine detail. At higher pHs, the uranyl ions polymerize into larger ions and can even precipitate from the solution. Uranyl acetate, which is generally available in the EM laboratory because it is also used for staining thin sections, is useful at concentrations of 1 to 2% (wt/vol), but its pH is low (~2.5 to 4.0). When uranyl acetate is used as a negative stain, both positive and negative staining may be seen on different areas of the same grid. In addition, uranyl acetate acts as a fixative for proteins and nucleic acids, can contract or round up mycoplasmas, disrupts some protein structures, and should be used only with well-washed preparations. Uranyl oxalate can be more useful than uranyl acetate because the oxalate anion is more tightly complexed to the uranyl cation and resists precipitation at more neutral pH. Uranyl formate, sodium silicotungstate, and various other salts are also sometimes used. Of all negative stains, uranium (uranyl) salts have the highest atomic number, 92, and provide the greatest specimen contrast. (*Caution*: salts of uranium are slightly radioactive and toxic. Care in both use and disposal is advised.)

Some workers recommend the addition of very small amounts of a wetting agent (such as peptone, glycerol, serum albumin, starch, bacitracin, or sucrose) to some stains to improve spreading (12). Not all investigators find this useful or necessary, especially if high magnifications are to be used and uncluttered backgrounds are desired.

4.2.2.2. Staining Procedures

There are many variations for negative staining (5, 10, 12). A standard procedure is a two-step method of first applying the specimen to the grid and then staining with 2% (wt/vol) aqueous ammonium molybdate, as follows.

1. Press a small piece of Parafilm® M (American National Can Co. [see section 4.8 for addresses of suppliers]) onto the laboratory bench surface, and then remove the tissue covering to provide a clean surface. Since the Parafilm® surface is hydrophobic, a small drop of the sample suspension can easily be applied with a Pasteur pipette. Also, next to the drop of suspension, place a small drop of 2% (wt/vol) ammonium molybdate onto the Parafilm® (for step 3).

2. Place a coated grid, with the film side down, on the drop of sample suspension. The surface tension of the meniscus spreads the suspended solids and allows them to adsorb to the grid. Large debris falls to the bottom of the drop and will not obscure the grid. Usually 10 to 60 s is enough time for particle adsorption. Small concentrations of detergents, organic solvents, or other surface-active agents can reduce adsorption and affect stainability. Trial and error will determine the optimum time for adsorption, which is also affected by the nature and charge of the film and the sizes and nature of the particles. To improve both adherence and distribution of the particles, the hydrophilicity of the support film can be improved by the use of coated grids that have been stored in a refrigerator or exposed to UV irradiation, to "glow discharge" in a high-vacuum evaporator, or to discharge from a Tesla coil. The exposures minimize electrostatic charges.

3. With narrow-pointed forceps, remove the grid. By touching the edge of the grid to a piece of filter paper (Whatman no. 1), remove the small drop still adhering to the grid. Quickly touch the film side of the grid to the small drop of 2% ammonium molybdate on the Parafilm®.

4. After 10 to 60 s, withdraw the grid from the drop of stain in the fashion described above. Staining times may require some experimentation, especially with different sorts of specimens.

5. Allow the sample to dry and examine it in the transmission electron microscope. For very delicate samples (such as bacterial S-layers), macromolecular arrangements can be perturbed by high surface tension during drying. This effect can be generally overcome by prefixation with 0.1 to 1.0% (vol/vol) glutaraldehyde in water or a suitable buffer. This prefixation is easily done by incorporating the fixative into step 2 (see above) and washing. (*Caution:* glutaraldehyde is volatile and toxic; prefixation should be done in a suitably vented fume cabinet or hood.)

There are several alternative procedures for negative staining of bacterial suspensions. One involves mixing equal volumes (or other ratios) of stain and suspension, placing the grid onto a drop of the mixture, and then withdrawing excess fluid with filter paper, either immediately or after an appropriate interval of time (e.g., 10 to 60 s). It is usually best to do this immediately after mixing, since interactions of stain and particles can, with time, produce confusing results. Sometimes, to supply a more uniform particle distribution, it is advantageous to pick up a thin film of the mixture in a platinum-iridium loop just bigger than the grid (e.g., ~3.2 mm) and to break this flat film over a grid resting on blotting paper. The grid is then allowed to dry and is examined. Occasionally, stain-suspension mixtures are sprayed onto grids, but this technique requires a special apparatus. This procedure is most useful for quantification of viruses or other small particles (12) and should be done in a special chamber for safety.

4.2.2.3. Washing

A useful accessory procedure after application of a suspension but before negative staining (in the two-step method) is to wash the grid to eliminate excessive debris, salt, and protein that can be precipitated by some stains and that can obscure detail. Washing can be done by touching the coated grid surface rapidly and repeatedly in succession to the surfaces of several fresh drops of distilled water or other fluid on Parafilm®.

In addition to water, various solutions may be tried as washes, but these can be risky. The presence of mineral salts and organic compounds, pH, and other conditions are sometimes essential for the maintenance of structure, but they can also produce an obscuring film as the specimen dries. In fact, during drying, tremendous concentration gradients can be produced, and these can denature proteins. Salts have a tendency to form large precipitates on the grid. Of all washing fluids other than water, 1% (wt/vol) aqueous ammonium acetate seems best since it is nonreactive with bacteria, mycoplasmas, membranes, bacteriophages, and most intracellular components and because any excess is volatile in the vacuum of the transmission electron microscope.

4.2.2.4. Pitfalls

For examination of negatively stained specimens, it is particularly important to balance contrast and resolution by using an appropriate accelerating voltage (usually 60 to 80 kV). Contamination of the microscope column, apertures, and specimen by evaporated stain or other substances is minimized by the use of a liquid nitrogen-cooled anticontamination device (or cold trap) and self-cleaning gold objective apertures in the microscope. Specimen drift is avoided by patience, experimentation with beam intensity, and the use of carbon-coated plastic or pure carbon support films. Beam damage to fine structure, as well as puddling of stain about or on the specimen, can be minimized by learning to achieve focus as quickly as possible with the lowest possible beam intensity and making the shortest possible photographic or CCD exposure that produces a good, accurate image. A short exposure at a relatively low magnification usually produces the best results; by trial and error, each microscopist will determine the magnifications most easily focused and producing the most consistent results.

Solutions of stains and wash solutions (buffers, salts, and even deionized, distilled water) may become contaminated with bacteria over time, and a quick examination of nigrosin films by light microscopy will soon confirm the condition of the solutions (chapter 2.2.3). Reduce the problem by keeping working stains in the refrigerator in a rack of small vials, which are filled as needed (with sterility precautions) from stock bottles. Contaminated solutions may be filtered but are best made up anew. A convenient method for storage and filtered delivery is the use of a plastic syringe with an attached 0.45- or 0.20-μm-pore-size Swinnex filter. Solutions of uranyl acetate require protection from light. This is done by covering the container with aluminum foil.

The distribution of negative stain and specimens on the support film may be uneven or unsatisfactory; it is therefore efficient to prepare two or more grids of the same material. A grid should be scanned thoroughly at low magnification. A satisfactory grid prepared by the two-step method will show a gradient of specimen and stain distributed from one side to the other (remember, fluid was withdrawn by touching *one edge* of the grid to the filter paper). If the particles are in low concentration in the original suspension or if drop withdrawal techniques are unsatisfactory, specimens and stain may be found on only one small area of the grid.

Ammonium molybdate produces less contrast than does phosphotungstate or uranyl acetate, and its use may require some practice. However, photographic negatives (with appropriate development) will have more contrast than anticipated and present a great range of printable tones. Digitized images from CCD cameras can be easily manipulated by computer for added contrast. Even molybdate tends to heavily surround most whole bacteria with an almost impenetrable mass of stain; shorter staining times and less concentrated (0.1 to 0.5%) solutions are often required for good results with these relatively large objects.

4.2.2.5. Expectations

Negative staining is the simplest and fastest EM method that can be used, but to obtain very good results it can take many different attempts with several combinations of staining and specimen concentrations. This can be frustrating, but do not give up; the results are well worth the effort!

What sort of structural information on bacteria can be obtained by negative-staining techniques? The size and shape of the cell can be ascertained. The presence of flagella (perhaps already suggested by motility studies) can be affirmed, and the distribution and fine structure of the flagella can be determined (Fig. 5 and 8). Other surface appendages such as pili (fimbriae) (Fig. 5) or the presence of S-layers or capsular materials can be revealed. Even the differences between walled and wall-less procaryotes (mycoplasmas or bacterial L forms) become apparent.

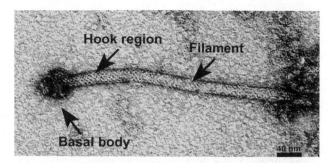

FIGURE 8 TEM image of a negatively stained flagellum isolated from *Vibrio cholerae*. Notice how the stain has penetrated between the flagellin subunits of the filament and the protein subunits of the hook (here artificially straightened). The flagellar rings of the basal body are also well differentiated. (From F. G. Ferris, T. J. Beveridge, M. L. Marseau-Day, and A. D. Larson, *Can. J. Microbiol.* **30**:322–333, 1984.)

Sometimes negative stains penetrate through the surface layers of bacteria so that membrane invaginations (e.g., mesosomes, which are considered to be artifacts) and the internal membranes of photosynthetic, methylotrophic, and nitrifying bacteria can be seen. Indeed, bacteria can be lysed and their component parts can be isolated (see chapter 7) so that they can be negatively stained (Fig. 5), and this is one of the most powerful methods of ultrastructural determination. The subunit arrangements of periodic structures, such as S-layers (chapter 5, Fig. 11), flagella, pili, spinae, and porins (after delipidation of outer membranes), are also revealed. Even minute structural differentiations within some appendages can be seen, e.g., the filament, hook, and basal-body rings of a flagellum (Fig. 8). With experience, gram-positive and gram-negative bacteria can be readily differentiated by their negatively stained envelope profiles.

Examples of the applications of negative staining may be found in references 77 through 85.

4.2.3. Thin Sectioning

See references 4 through 7, 10, and 13.

For conventional thin sectioning, bacteria must be chemically fixed, washed, dehydrated, and embedded in a liquid plastic that then solidifies with appropriate properties for the cutting of very thin sections. The sections are stained with solutions of heavy-metal salts to give sufficient contrast for examination by TEM (Fig. 1). The chemicals used in this technique (fixatives, stains, dehydration solvents, and plastics) can be very damaging to the specimen, and structural artifacts can be introduced. An experienced microscopist should help with the assessment of the initial preparations. Although many different combinations can be used, it is generally adequate to use sequential fixation in glutaraldehyde and osmium tetroxide, followed by dehydration and embedding in Epon 812®, LR White®, or Spurr's medium® (section 4.7.4). There are other choices of embedding plastics, but these three work reasonably well for routine purposes and are a good starting point. Once experience is gained with a sample and the way it reacts to fixation and embedding, modifications can be tried. *Caution:* chemical fixatives, heavy-metal stains, and plastic resins are chosen because of their high reactivity with and penetration into biological specimens. Remember, *microbiologists are made out of biological stuff too. These are highly toxic chem-*icals and should be treated with care! Surgical gloves and a well-vented fume cabinet (hood) with the glass door pulled down to protect the face are recommended.

4.2.3.1. Fixation with Glutaraldehyde and Osmium Tetroxide

1. To the cell suspension or liquid culture, add an equal volume of 5% (vol/vol) glutaraldehyde in aqueous 100 mM HEPES buffer (pH 6.8) containing 2 mM $MgCl_2$ (final glutaraldehyde concentration, 2.5%). Establish the pH of the growing culture immediately before fixation, and alter the pH of the fixative solution accordingly. (Sodium cacodylate at a 0.2 M concentration is another effective buffer often used for the fixative, but it contains arsenate, which vaporizes and is toxic. *Be careful!*) See sections 4.7 and 4.8 for sources and chemical formulations. Alternatively, centrifuge the cells and suspend the pellet in 2.5% glutaraldehyde in HEPES buffer. A wash in fresh medium or in an appropriate buffer may precede this step. For some procaryotes, especially some mycoplasmas, it may be preferable to make up the glutaraldehyde (perhaps at a different concentration) in fresh culture medium to balance osmolality or pH. (For easier manipulation and to reduce the number of centrifugations, it might be convenient to embed a concentrated slurry of cells in molten 2% [wt/vol] Noble agar or another highly refined gel and allow it to solidify. Cut into cubes of 1 mm^3, and process the cubes in small bottles through the reagents suggested below.)

2. Allow the cells to fix for 1 to 4 h at room temperature. The time is flexible; any time beyond 1 h is usually adequate, and bacteria held in glutaraldehyde for several days can even be shipped across the country if appropriately packaged to meet stringent transport regulations for dangerous chemicals. (The glutaraldehyde-fixed pellet may be enrobed in agar at this stage for easier handling; see steps 6 through 8.)

3. Wash the pellet twice in HEPES buffer, centrifuging between washes.

4. Suspend the washed pellet in 1% (wt/vol) buffered osmium tetroxide (OsO_4) for 1 to 4 h at room temperature. *Caution:* if the odor of osmium tetroxide is detectable, eyes may be subject to dangerous vapors. Persons wearing contact lenses are particularly vulnerable. For safety, a commercially available eyewash station (Nalge Co.) should be available, properly labeled, and strategically located.

5. Centrifuge and wash the pellet (now usually black due to a compound called osmium black) one to three times by suspension in the buffer, recentrifuging each time.

6. Make up and melt 2% (wt/vol) Noble® agar in deionized water, and cool to 50°C. Add a volume of agar equal to that of the pellet, and mix quickly with a warm Pasteur pipette.

7. With the pipette, quickly dispense the suspension onto an alcohol-cleaned glass slide, and allow it to solidify. The bacteria should be in high concentration and evenly distributed. With a clean razor blade, dice into 1-mm cubes.

8. Suspend the cubes for 15 min at room temperature first in deionized water to equilibrate them and then in 2% aqueous uranyl acetate for 1 to 2 h at room temperature. Wash with two to five changes of deionized water. The reason for replacing the buffer with water is that UO_2^{2+} will polymerize from the solution at pHs above 4.5; it is best to have an unbuffered system. Make sure that a phosphate-buffered system has not been used (unless the cells are well washed of the buffer), because uranyl and phosphate will rapidly complex together and precipitate from the solution.

4.2.3.2. Dehydration and Embedding with Plastic Medium

(This is a simple method for Spurr®'s resin or LR White® resin which is commonly used. For Epon 812®, follow the manufacturer's instructions.)

1. Wash the uranyl acetate-treated agar cubes once in 30% (vol/vol) ethanol for 10 min. (Use 2- to 3-ml volumes in 7-ml screw-cap glass vials.)

2. Replace with 50, 70, 80, and 95% (vol/vol) ethanol, successively, for 10 minutes each.

3. Replace with 100% (vol/vol) ethanol for 45 min, changing once at the 20-min point.

4. Replace with 1.0 ml of fresh 100% ethanol. Add 1.0 ml of complete plastic medium, and mix thoroughly but gently. Allow to stand, with occasional mixing, for 30 min or until cubes sink to the bottom of the vial. (The medium is most easily dispensed by using disposable plastic syringes.)

5. Replace the mixture with 2.0 ml of plastic medium, and allow to stand for 1 to 4 h until the cubes settle to the bottom. One change of the 100% plastic medium is recommended during this final infiltration step.

6. Insert typed labels into appropriately-sized Beem® capsules or flat embedding molds, and add to each a drop of complete plastic medium.

7. With a sharpened wooden applicator stick, place one cube (from step 5) into each Beem® capsule or mold. Fill the capsule with plastic medium. (Use 0.55 ml, so that the final block fits precisely into the microtome chuck.) Embeddings can also be done in gelatin capsules (available at a local drug store).

8. With caps *off* (to allow volatilization of any residual solvent), place the capsules in a 60 to 80°C oven for polymerization for the specified time (Epon 812®, 48 h; LR White®, 22 h; and Spurr®, 16 h; some plastics [e.g., LR White®] cure in the absence of oxygen and must be capped). Immediately upon removal, the blocks in the capsules may feel slightly spongy, but they will harden to optimal cutting quality after 2 to 4 h at room temperature.

4.2.3.3. Other Fixatives and Embedding Media

Many different fixatives have been devised for special purposes. These include permanganates, which have applications for thick-walled microorganisms such as yeasts and fungi, and other aldehydes (e.g., acrolein, formaldehyde, and crotonaldehyde). Principles of their use (including effects of pH, buffers, tonicity, concentration, temperature, and duration of fixation) are discussed in numerous books (4–7, 10, 13). Similarly, numerous embedding media are used (4–7, 10, 13), and their selection depends on their infiltration into cells, polymerization and cutting character, miscibility with water, and availability. Other embedding media in common use include Durcupan®, Araldite®, Maraglas®, Vestopal®, various methacrylates for special purposes, and numerous epoxy resins. The reader is referred to the literature and to local experts for advice on the use and properties of these media. The Lowicryls® are particularly useful for low-temperature embeddings and the preservation of antigenic sites for colloidal gold-antibody labeling (section 4.6.2) (14–17). *Caution*: Lowicryls can be highly allergenic, and surgical gloves must be worn.

4.2.3.4. Block Trimming

Before sectioning, remove the hardened blocks from Beem® or other capsules and trim them to the proper size and shape. Various methods of trimming have been described previously (4, 6, 10). Recent models of ultramicrotomes often incorporate trimming devices, and separate specimen trimmers are also available. The preparatory procedure is as follows.

1. With pliers or a Beem capsule press (E. F. Fullam), press firmly over the entire capsule to loosen the enclosed block. Now press from the pyramidal end, and proceed upward to force out the cured block. Alternatively, release blocks by cutting the capsule lengthwise on two sides with a razor blade. Mount the block in a holder, pyramidal end up, and place the holder in a microtome chuck.

2. Under a stereo-binocular microscope, use an alcohol-cleaned razor blade to carefully trim the top of the pyramid (which will be the front or cutting face when mounted in the ultramicrotome) until the cell mass has been reached. Cut down and away from this surface on all four sides at an angle of about 30° (the angle of the pyramid formed by the block in the capsule is about 60°), and leave a small, truncated pyramid (having a face about 1.0 mm^2) that encompasses the specimen to be sectioned. With care and practice, this may be adequate trimming—especially if the face is cut in the shape of a trapezoid. However, grooves inevitably left by the razor blade are rather large and run at right angles to the face; as a result, water may run onto the face during sectioning and interfere with good results. Better to further smooth the face in the microtome with a glass knife.

4.2.3.5. Thin Sectioning

Because procedures and equipment differ in each laboratory, no attempt is made here to describe the sectioning process. Descriptions of pitfalls are thoroughly described in TEM texts as well as in operating manuals available from the instrument makers (4, 6, 10).

4.2.3.6. Knives

Both glass and diamond knives can be used for cutting thin sections. Glass knives, which are less expensive, are usually prepared by the use of a knife-making apparatus, which is available in all TEM laboratories. The making of glass knives is described in texts (4, 6, 10) and in manuals accompanying the instruments. For sectioning, the choice of knife depends on financial limitations, operator preference and experience, and the embedding polymer. In some instances, glass knives are preferable and produce superior results. A diamond knife produces more consistent results as the operator becomes familiar with its edge and other qualities, especially when a standard embedding medium is consistently used. The angle of the edge of the diamond as the diamond sits in its holder must be specified when purchasing a diamond knife; an angle of 42 to 45° is satisfactory for most conventional embedding plastics.

4.2.3.7. Staining of Thin Sections

Even though most embedding procedures have a uranyl acetate-staining step, additional staining of the thin section is usually required for added contrast. The use of uranyl acetate followed by lead citrate is the most common staining procedure. For uranyl acetate staining, use the following procedure.

1. Place a drop of 1 to 2% (wt/vol) uranyl acetate on a sheet of clean dental wax or Parafilm®, and float a grid bearing a thin section, section surface down, on the surface of the stain.

2. Cover with a petri dish lid to maintain humidity and protect from dust, and stain for 5 to 7 min. Longer staining

times (up to 30 min) may be required for hard-to-stain structures such as mycobacterial cell walls.

3. Remove the grid with narrow-pointed tweezers, immerse three times in distilled water to wash, and dry on filter paper.

Some workers use uranyl acetate staining only, but the degree of contrast may be relatively feeble. Lead citrate (section 4.7.4) is commonly used to supply added contrast. For unknown reasons, uranyl acetate staining followed by lead citrate staining gives a much more intense stain than either alone. Lead citrate is conveniently stored in and dispensed from a plastic syringe, thus excluding air, since lead precipitates upon exposure to CO_2.

For staining with lead citrate, place a strip of dental wax or Parafilm in a large (6-in. [ca. 15-cm]-diameter) petri dish in which is placed a small (2-in.-diameter) dish containing a few pellets of NaOH. Add the stain as a single drop on the wax, and cover the larger dish with its top. Soon the internal atmosphere becomes alkaline because of the NaOH pellets and removes residual CO_2 which might affect staining. To stain the sections, float grids section side down on the drop of stain, cover the dish, and stain for 1 to 7 min. Remove the grid, and while holding it with tweezers, wash in 0.02 M NaOH and rinse several times in freshly boiled cooled water (boiling drives CO_2 from water). Touch the grid edge to filter paper to drain, and place the grid on filter paper to dry.

4.2.4. Cryofixation and Freeze-Substitution

See references 23 to 40.

Cryofixation has improved the preservation of the ultrastructures of microorganisms. Cryofixation relies on the ultrarapid freezing of cells so that they are fixed physically without the aid of toxic chemicals. The freezing requires specialized devices which either plunge the specimen into an ultracold liquid cryogen (usually propane or ethane held at the liquid-nitrogen temperature of $-196°C$) or slam the specimen onto an ultracold polished metal block (either copper or gold at $-196°C$ or colder). These devices do not have to be complicated and can be manufactured for a reasonable price by university machine shops (31, 32). Very expensive high-pressure freezing devices are also available for those who can afford them. Ice crystals are not formed by the ultrarapid freezing technique; rather, vitreous ice is formed which produces a "glass" around and within the cells. The resulting preservation of cells is so good that they can be thawed back to life. Detailed accounts of freezing theory are available (23, 28, 29, 37e–39).

Highly specialized cryoTEM laboratories may possess a cryostage on one of their microscopes, but they are rare, have been generally described in section 4.1.4.2, and will not be considered further here. Freeze-substitution, however, is a technique available to any TEM laboratory which can cryofix samples (Fig. 2). It has recently been shown to be especially good for the preservation of biofilm structure (36a). Once the samples are frozen, chemical fixatives are substituted into them at low temperatures so that the samples are chemically fixed without thawing. At the same time, the water (ice) is replaced with an organic solvent. Once substitution is complete, the specimen is ready for plastic resin infiltration and can, eventually, be thin sectioned like a conventional plastic block.

A simple plunging freeze-substitution method is as follows.

1. Wash cells free of culture medium in 100 mM HEPES buffer at pH 6.8 (other buffers and pHs can be used) by centrifugation (an Eppendorf centrifuge is satisfactory). Resuspend the pellet in a volume of molten 2% (wt/vol) Noble® agar equal to the volume of the pellet. Rapidly spread this suspension into a 20- to 30-μm-thick film over a cellulose-ester filter (Gelman Sciences, Ann Arbor, MI) by using the edge of a sterile glass microscope slide (the film is so thin that it can hardly can be seen, and the spreading requires a few tries to gain experience). Be sure that cellulose-ester filters are used, since other composites may dissolve during the substitution. While spreading the film or while waiting for the plunger to be available, ensure that a humid atmosphere is maintained over the agar film; drying artifacts are not wanted! High humidity can easily be produced by rigging a simple humidity chamber, consisting of the bottom of a large glass petri dish and a smaller-diameter glass beaker containing a large, damp (but not dripping) sponge. Force the sponge into the bottom of the beaker, and turn the beaker upside down over the agar film in the petri dish. Make sure that this humidity chamber has had time to equilibrate before putting the agar film inside.

2. Fill the cryogen reservoir with propane (or ethane) after it is cooled to liquid-nitrogen temperature. A 1-liter Dewar flask with a brass cylinder in the middle to hold the propane (the cryogen reservoir) can be easily manufactured. The top of the brass cylinder should hold 25 to 50 ml of propane for good heat exchange. If it can be stirred by a magnetic stirrer to reduce convection, so much the better. Propane is liquid at these low temperatures and is not volatile. *Caution:* once the freeze-plunging is finished, the propane will warm to room temperature and hence become volatile and *explosive!* Make sure that the freeze-plunging is done in a well-vented fume cabinet and that the fan in the cabinet is of a safe, sparkless variety. Turn the magnetic stirrer off and unplug it once the propane begins to warm up. All workers in the laboratory must be forewarned to turn off Bunsen burners, stoves, and heaters, etc., and not to operate electrical switches.

3. Immediately before plunging, cut small wedges from the agar film-cellulose-ester filter composite. These wedges must be pointed and resemble arrowheads ca. 5 mm long and 2 to 3 mm at their base. They are inserted into the plunger so that the pointed end hits the cryogen first. This will allow the specimen to be oriented for thin sectioning and ensure that the best frozen (pointed) region is sectioned.

4. For freeze-substitution, use 7-ml glass scintillation vials filled with 1 to 2 ml of substitution medium. Consistent results are obtained with 2% (wt/vol) osmium tetroxide and 2% (wt/vol) uranyl acetate in anhydrous acetone, although other media can be used (32). The anhydrous quality is an important criterion since all water (ice) must be completely replaced. Store the acetone with a hygroscopic molecular sieve in it (sodium aluminosilicate with a pore size of 0.4 nm; Sigma Chemicals), and the sieve should also be added to the substitution medium in the vials. Freeze the medium in each vial at $-196°C$ before use.

5. The time elapsed between spreading of the bacteria as an agar film onto the cellulose-ester filter, plunging, and transferring of the frozen cells onto the surface of the frozen substitution medium must be short, not exceeding 2 min. Seal vials by their screw caps, and hold at $-80°C$ in an ultracold freezer for 48 h. During this period, the specimen remains frozen, the substitution medium melts, the ice is replaced by acetone, and the cells become chemically fixed, stained, and dehydrated. The molecular sieve takes up the released water.

6. Allow the vials to come to room temperature, wash the specimen free of fixative with anhydrous acetone, and

infiltrate plastic resin into the specimen as described above (section 4.2.3.3). Section and stain the plastic embeddings as usual (sections 4.2.3.6 and 4.2.3.7).

4.2.5. Metal Shadowing

See references 1, 4, 5, 6, 10, 11, and 18 through 20.

Because of the growing popularity and availability of SEM, the use of direct metal shadowing of TEM preparations has declined. However, like SEM, metal shadowing is a method for eliciting information on the surface topography and dimensions of bacteria. Furthermore, because of the usually greater resolving power of TEM, these preparations provide greater resolution than those for SEM.

Shadowing was among the first techniques used to visualize bacteria by TEM. It is especially useful for determining the shapes of bacteria and viruses, the surfaces that coat them (e.g., S-layers and capsids), and the existence and placements of appendages (e.g., flagella, pili, and phage tail fibers). The vaporized metal hits the structures at an angle (10 to 45°), congeals on them, and mimics fine detail. Where surface elevations on the specimen limit exposure to the shadowing substance, little metal accumulates. Consequently, a gradation of metal shadow coats the specimen according to the elevations in its topography, resulting in a high-contrast, three-dimensional representation which can be seen by TEM. There are several methods of using shadowing (e.g., shadow casting, freeze-etching, and rotary shadowing); each has its particular advantage, and outlines of these methods follow.

4.2.5.1. Shadow Casting of Whole Mounts

See references 1, 5, 10, and 11.

Shadow casting involves the deposition of an electron-dense metal (e.g., platinum) at a specific angle onto the specimen while it is in a vacuum evaporator (4, 6). Shadow casting is especially useful for determining the topography of bacterial surfaces. Unlike negative stains, which can permeate through the various surface layers to the plasma membrane (and produce moiré patterns of superimposed structure), metal shadowing reveals only the upper surface of the specimen. In addition to delineating surface irregularities or periodic structures such as S-layers, the technique can be used to measure the vertical heights of structures in the specimen by measuring shadow lengths and calculating geometrically from the known shadow angle. This process can be aided by inclusion of known standards such as latex spheres (obtained from electron microscope suppliers) of a given size for purposes of calibration. Shadowing is conducted as follows.

1. Wash the specimen to be examined free from extraneous debris. The specimen can be chemically fixed first.
2. Place the material on a Formvar®-coated grid, and allow it to dry. Attach the grid to double-sided sticky tape on a glass microscope slide, together with a small piece of white paper to serve as a visual indicator of the amount of metal deposited during the evaporation.
3. Place the slide on the specimen support head of the vacuum evaporator.
4. Wrap Pt wire around the tungsten V-shaped filament held between the electrodes. The "V" is formed from standard 30-mil (ca. 0.2-cm)-diameter tungsten wire, and 7 to 10 mm of 10-mil (ca. 0.05-cm)-diameter Pt wire is wrapped around it. Before doing this, it is essential to heat the tungsten filament to white heat in vacuo and then reopen the evaporator and affix the Pt wire. (This procedure degasses the filament, improving the vacuum during subsequent

metal evaporation, and relieves strain produced in the wire during its bending, thus prolonging filament life.) Evaporators, such as Balzers freeze-etchers, are equipped with carbon and platinum evaporation sources. Simple but reliable sources consist of two graphite rods (one sharpened like a pencil) which touch one another, so that carbon is evaporated when a voltage of ~30 V is passed through them. For platinum evaporation, a small length of Pt wire is wrapped around one of the rods and a voltage is applied. More expensive and elaborate systems use electron beam guns as evaporation sources.

5. With the Pt wire wound around the apex of the tungsten filament, place this metal source approximately 12 cm above the grid at the desired angle. Establish a good vacuum (10 to 6 mm Hg) in the evaporator, turn up the controls to heat the metal quickly to its melting point, and hold it there until all is evaporated or until the shadow on the visual filter paper indicator appears correct. This judgment requires some practice and experience. Some evaporators have quartz monitors that measure the metal thickness by increased electrical conductance.

6. Sometimes it is useful to omit the Pt and to shadow with tungsten oxide so as to produce a minimal background and finer granularity than that obtained with Pt. To do this, heat the tungsten filament *in air* until it is coated with yellow tungsten oxide. Scrape off the oxide with a scalpel from all the wire except the apical few millimeters of the V, and proceed with the shadowing as described above.

4.2.5.2. Freeze Fracturing and Freeze-Etching

See references 21, 22, and 37.

Freeze fracturing is a difficult technique because it requires rapid freezing of a paste of cells to liquid-nitrogen temperature, and the frozen pellet must be maintained at a low temperature (−100 to −196°C) to preserve structure accurately. At these temperatures, the paste will fracture if hit with a knife, and the cleavage line will run through or around cells, thus exposing their constituents and surface structures. Membranes are frequently cleaved through their hydrophobic domain, so that intramembrane structure is revealed. The fracturing is done in vacuo, and the water in the exposed ice surface immediately begins to sublime, revealing structural components in the cells, their membranes, or the external milieu; this is the etching process. Freeze-etching is one of the few techniques besides thin sectioning that exposes the inside structures of bacteria (Fig. 9). However, the exposed structure in the frozen-cleaved surface of the pellet then has to be replicated as a carbon-supported metal replica of that surface. This means that a complex apparatus for freeze-cleaving, freeze-etching, shadow casting, and carbon deposition is required, all at controlled low temperature. Equipment allowing this series of critical steps in vacuo is provided by Balzers (BAL-TEC AG) and by Denton Vacuum Inc. The replica is examined by TEM and looks like a shadowed preparation.

Freeze fracturing (with or without etching) is a form of cryofixation, which was discussed in section 4.2.4. Both are carried out in sequence, starting with fracturing and followed by etching in a specialized apparatus, a freeze-etcher. Instead of plunging a thin bacterial film into a cryogen held at liquid-nitrogen temperature (as in freeze-substitution [section 4.2.4]), the paste of cells is formed into a 1-mm-diameter drop and placed onto a brass, copper, or (better because it is acid resistant) gold planchet before plunging. Propane and ethane held at liquid-nitrogen temperature are used as cryogens.

The frozen paste is placed onto an ultracold stage in the freeze-etcher, the stage temperature is set to −100 to

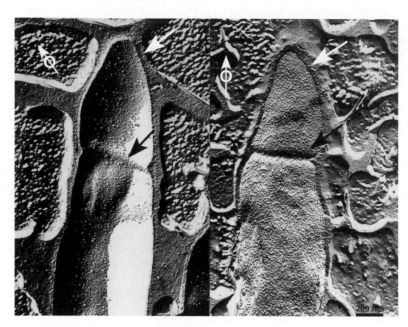

FIGURE 9 TEM image of reciprocal faces of two different dividing *Bacillus subtilis* cells showing the convex (right side) and concave (left side) fracture planes of the plasma membrane after freeze fracturing and -etching. The black arrows point to the growing septa, and the white arrows point to the edges of the cell walls. The white arrows with the circles denote the shadow direction of the platinum.

$-150°C$, and the chamber is evacuated to ca. 10^{-6} mm Hg. Once the vacuum and stage temperature are stable, the frozen drop of cell paste is cleaved with a liquid nitrogen-cooled microtome knife (usually a razor blade mounted on a rotary arm designed to strike the specimen). The fracture follows the lines of least bond energy, so that the fracture travels over cells and through cells and frequently splits membranes (e.g., the plasma membrane) (Fig. 9).

Initially, the fracture surface through the cell paste is smooth and flat. As the ice sublimes, the etching exposes ribosomes and DNA where the fracture has cut through cytoplasm, and of course, it also exposes the edges and (sometimes) the surfaces of walls and membranes (Fig. 9). Etching increases the topography of the entire fracture plane and provides more informative interfaces; if deep enough, it will also show the external wall surface of a whole cell that was embedded in ice. Generally, etching times of 20 to 60 s are used, after which the surface is shadowed with Pt as explained in section 4.2.5.1. Etching times can be decreased and deep etching can be attained if the microtome arm is cooled to $-196°C$ and placed over (but not touching) the fracture surface as a cold trap to speed sublimation.

The surfaces and interfaces now revealed in the ice can be visualized only by making a replica. This is done by shadow casting (section 4.2.5.1) a thin metal layer (Pt) at an angle to the fractured surface, which gives relief to show structure, and by supporting this thin layer with a continuous layer of carbon (i.e., a Pt-carbon replica now sits on top of the etched fracture face and precisely mimics its structure). Carbon coating is done in the same way that TEM grids are coated (section 4.2.1.2), except that it is performed in the freeze-etcher. Finally, the ice and organic constituents of the frozen cell mass (which are too thick for TEM) must be removed to provide the final replica for TEM, as follows.

Let air into the chamber, remove the specimen from the cold stage, and slowly lower it into a small volume of deionized water. Here, the paste will melt and the Pt-carbon replica will float off onto the water surface. Cleanse the replica of adhering cellular debris by floating it onto concentrated sulfuric acid in a small petri dish. (*Caution*: this is a strong acid!) This dish should be left for at least 60 min but can be left overnight without harm. Using a loop 4 to 5 mm in diameter, made out of 10-mil-diameter Pt wire and mounted in a suitable holder (a bacteriological inoculation loop holder works fine), carefully bring the loop under the replica to pick it up and transfer it into a dish of deionized water. The acid in the loop and the water in the dish mix violently and boil, but surprisingly, the replica remains intact and floats on the surface if the Pt-to-C ratio of the replica is correct. Using the loop, transfer the replica into a 5% (wt/vol) sodium hypochlorite solution (diluted from commercial bleach solution) for 1 to 2 h to oxidize the remaining organics and remove the charred debris from the replica.

Wash the replica free of salts and hypochlorite by transferring it two to five times into fresh water by using the loop. Surprisingly, this is the most tricky step because the cleaned replica can roll up on itself. If this happens, a quick transfer into concentrated sulfuric acid and then back into water will often generate a violent straightening. Some suppliers sell fluids which can be used to straighten replicas; these are organic solvents with high vaporization points which, when dropped onto the floating rolled-up replica in water, violently alter the surface tension and flatten the replica.

Mount the replica on a 200- or 400-mesh-size *uncoated* copper grid; the 200-mesh-size grid is best because there is more viewing space, but it may not give enough support to the replica. Mounting can be difficult because the replica will always prefer to stay at the surface of a water drop and not adhere to the grid surface. Using the loop, pick up the replica in a thin film of water, ensuring that the replica is in the center of the film. Now, taking the grid (held by narrow-pointed tweezers at the edge and bent at a right angle to the edge held by the tweezers), slowly move it through the center of the loop so that the film of water breaks, leaving the replica attached to the grid. Some beads of water will remain on the grid, and the replica may be floating on them. With pointed wedges of filter paper, carefully absorb the water and center the replica on the grid.

The technique of freeze-etching and the production of good replicas can be difficult, but they are well worth the trouble (Fig. 9). A good replica can be beautiful!

4.2.5.3. Rotary Shadowing of DNA and RNA

The spreading of nucleic acids onto fluid interfaces to make preparations for EM after rotary metal shadowing has come into widespread use. It is an essential method in many aspects of molecular biology and is well described in appropriate texts (18–20). Single- and double-stranded nucleic acids are too thin to be seen clearly by usual TEM methods, but once spread and dried, the polymers have enough height to be shadowed with Pt. In fact, the metal adds to their bulk and makes them easier to see. They are even more apparent when a rotary stage is used in the evaporator, which spins the specimen during shadowing, adding even more Pt to their mass and ensuring that all sides of the polymer are covered. Other than this, the process is similar to that outlined in section 4.2.5.1.

4.3. IMAGE ACQUISITION

4.3.1. Film

The working and reference material of the electron microscopist consists of electron micrographs, so the proper taking and processing of photographs is essential. Most microscopes are designed for sheet or roll film or both, although CCD cameras are becoming popular in modern transmission electron microscopes. Roll film (35 mm) is cheap, effective, and easily stored. A single roll usually provides ~40 exposures. Sheet film is larger (usually 10 by 8 cm) and consists of a photographic emulsion on a plastic (Estar) base. Since sheet film is larger, not as much can be stored in the microscope (e.g., only 20 to 40 exposures).

The major problem with all of these films is that the emulsion and the backing absorb and retain water vapor from the air. Therefore, it is wise to keep film in a series of aluminum desiccators with solid, lightproof lids, which are tapped to take a vacuum needle valve. Fill a tray in the bottom of the desiccator with phosphorus pentoxide. (*Caution*: this substance reacts violently with moisture, especially the moisture on fingers!) A modest vacuum and the pentoxide are effective in drying the film after a few days of storage. This drying prevents long pump-down times in the electron microscope as a result of outgassing.

Thirty-five-millimeter fine-grain positive film (Kodak no. 5302) is satisfactory for routine purposes when the electron microscope is equipped for roll film. For plate cameras, use Kodak electron image film no. 4489 or SO-163. It is important to first review the recommendations and data supplied by the manufacturer of any film considered for use. Experience will then indicate an appropriate routine exposure, development, and degree of contrast desired in the negative. Negatives can be easily scratched, and it is best to keep them in transparent sleeves for protection, handling, and easy scanning.

4.3.2. Printing

Skill in focusing the transmission electron microscope to achieve sharp negatives is obtained by experience, but it is helpful to recall that, at high magnifications, the exact focus is not necessarily the best for printing. In-focus negatives often do not have enough contrast for easy viewing; slightly underfocus images may make better prints. For publication, micrographs should have more contrast than normal since the published print loses contrast. It is therefore a good habit to take approximately three "through-focus" pictures of each field desired in the transmission electron microscope, attempting to have exact focus near the middle step.

The printing of electron micrographs requires more than ordinary attention to detail. A high-quality enlarger is essential. Automated photosensor exposure devices are helpful for reproducibility, but nothing can replace an experienced worker. Multigrade resin-coated papers used with appropriate enlarger filters give some control over the contrast of the print. It is not always necessary to print all negatives; this takes time and can be expensive. Negatives can be used in a slide projector and displayed on a screen at appropriate magnification. For this purpose, less sophisticated early-model projectors are often better since they are more easily adapted to strips or rolls of film. In fact, projection of negatives is a very good way to perform accurate measurements and statistical counts. A slide or negative of a measuring ruler will give a magnification factor for the projector and can be used for calibration each time the projector is used.

4.3.3. CCD Cameras

The addition of a digital CCD camera to an electron microscope has proven to be of great advantage. It provides the microscopist with the ability to display a "live" image on a computer monitor, perform measurements, count particles, add text, and print the resulting image and data all in one session. Digital images can be sent to off-campus investigators via the Internet, allowing much faster interpretations and diagnoses to take place. Image quality, although not equal to that obtained with photographic film, is certainly adequate for most purposes.

An ink-jet printer and photo-grade paper are sufficient for images to be displayed in reports, on posters, and in publications. CCD cameras with 1,000-by-1,000-pixel resolution are available at a moderate price, can be situated in the 35-mm-camera port or under the fluorescent viewing screen of a transmission electron microscope, and are easily installed and integrated into an existing system. Images can be stored on removable media such as a compact disk. As the life of a compact disk is not known, it is wise to take a few photographs on film for archival purposes.

The main advantage of digital imaging is the amount of time saved. As the price of electron microscope plate film doubles and triples, the CCD option becomes more attractive.

4.4. INTERPRETATION

Interpretation is not just seeing what is there; the EM image must also be related to what is already known about bacterial anatomy. Therefore, it is useful to have some knowledge of the field, and it is assumed that the student will have been exposed, through courses or some reading, to the general features of microbial ultrastructure. Unfortunately, pictorial atlases in which one can readily compare different genera are rare or hard to find. Specific references on interpretation, including some special reviews (74–100), are listed in section 4.9.3.

In a more precise sense, interpretation may be construed as getting the most information out of an electron micrograph, e.g., the details of fine periodic structure or macromolecular arrangements such as those in flagella or accessory protein layers of the cell envelope. For such advanced work, acquisition of a satisfactory micrograph is only the first step, and then methods of image enhancement and reconstruction can be used (section 4.9.2.10 and chapter 5). However, remember that it is rarely necessary to go this far; general information (about size, cell shape, mode of division, enveloping layers, flagellation, and internal membranes) is also useful and important to recognize and record (78, 81, 82).

4.5. SELECTION OF THE BEST MICROSCOPY TECHNIQUE

The proper selection of TEM technique is usually taken for granted, but it requires experience and is a source of confusion to students. Inexperience, today common with most microbiologists, usually dictates a shotgun approach that often includes the most sophisticated techniques. Don't "aim for the stars" right from the start. It is best to start with light microscopy (chapter 2) to gain preliminary information; e.g., is the culture pure, what is the shape of the bacterium, is it gram negative or gram positive, and is it motile? Light microscopy with stains or phase contrast will answer these simple questions. All EM takes time and effort and uses complicated, expensive equipment. Simple procedures are usually best, at least initially. Figure 10 describes a good stepwise approach for visualizing bacteria and their structures by TEM. It is a flow diagram that starts at a simple level and works toward more exotic techniques. Obviously, not all structures and techniques can be included, but there is enough information to assess the utility of methods and the times required to accomplish each individual step.

4.6. MACROMOLECULE IDENTIFICATION TECHNIQUES

The identities and positions of bacterial macromolecules, either within the cytoplasm or on the surfaces of cells, are often matters of importance to research and require appropriate TEM methods; e.g., it can be essential to know the cellular sites of enzymes, selected molecular species, and virulence factors or to monitor the synthesis and incorporation of cellular components. This objective requires the use of specialized electron-dense probes that can both search out and identify (with great precision) the molecules of interest, which can be present in low copy numbers. Several different techniques are used, and many of them require the production of specific antibodies (chapter 8.3 and 8.4); most are lengthy, and some are both technically complicated and difficult.

4.6.1. PCF
See refernces 41 through 44.

Polycationized ferritin (PCF) is a large (~11-nm) protein multimer with iron atoms in its core, which is readily seen by EM. It is a useful probe to determine electronegative sites on bacteria. Ferritin is purified from mammalian spleens and is chemically modified so that its overall charge is electropositive. The size of PCF does not allow it to penetrate into microbes, but it attaches to surfaces to reveal electronegative sites. PCF can be used on unstained whole mounts (43), in negative stains (42), in freeze-etchings (42), or in thin sections of prelabeled cells (Fig. 11) (41, 44). Since most bacterial surfaces are anionic (77, 83, 85), extremely low, nonsaturating concentrations of PCF can be used to probe for surface sites with the most electronegativity (43). Furthermore, because most bacterial capsules are also anionic, PCF can be used to stabilize and stain this hard-to-preserve material (Fig. 11) (41). PCF can also be used to probe for electronegative sites on isolated material such as gram-positive walls (43, 44). Other, smaller electropositive probes (e.g., cytochrome c) and electronegative probes (e.g., polyglutamic acid) are available, but they are not electron dense and require sophisticated methods of detection (79, 85). PCF is available through most supply houses (section 4.8).

4.6.2. Colloidal Gold-Antibody Probes
See references 45 through 59.

Colloidal gold can be made as gold sols to specific particle sizes (2 to 1,000 nm) in the laboratory (49, 53, 59) and is also available from commercial sources (section 4.8). Since colloidal gold is inexpensively and readily available from a number of commercial suppliers, it is not worth the complications to synthesize your own. The gold is reduced from chlorauric acid ($HAuCl_4$) to its colloidal metallic form, which is extremely electron dense. There is good attachment of colloidal gold to highly specific recognition molecules such as antibodies, protein A, protein G, avidin, and lectins. Commercial suppliers also provide colloidal gold with these probes. Once coated with these organic molecules, the gold particle becomes a highly specific marker for TEM. In fact, gold particles of different sizes and coated with different recognition molecules can be used to probe the same sample; the size of the gold particle is used to distinguish the different macromolecules. Because microbes are so small, gold particles of 5 to 50 nm are usually used.

4.6.2.1. Probe Production
Colloidal gold by itself has no specificity and is therefore not a complete probe; it must be coated with highly specific recognition molecules, most often antibody (chapter 8.3 and 8.4). In general, strong adsorption of proteins to gold surfaces occurs at pHs slightly above the isoelectric points of the proteins (i.e., their zwitterion is preferred, and interfacial tension is maximal). However, immunoglobulin G (IgG) molecules (even when affinity purified) can be a problem since they display a spectrum of isoelectric points. An alkaline pH of ca. 9.0 is best. Because IgG can aggregate

---

FIGURE 10 Flow diagram for the structural analysis of a microorganism. From the top down, the techniques go from simple to complex, and each has a section reference. The times given for each procedure suppose that all equipment is in place and ready for use. These techniques are for TEM, but SEM can also be appropriate for assessing general shape, growth habitat, and extracellular associations.

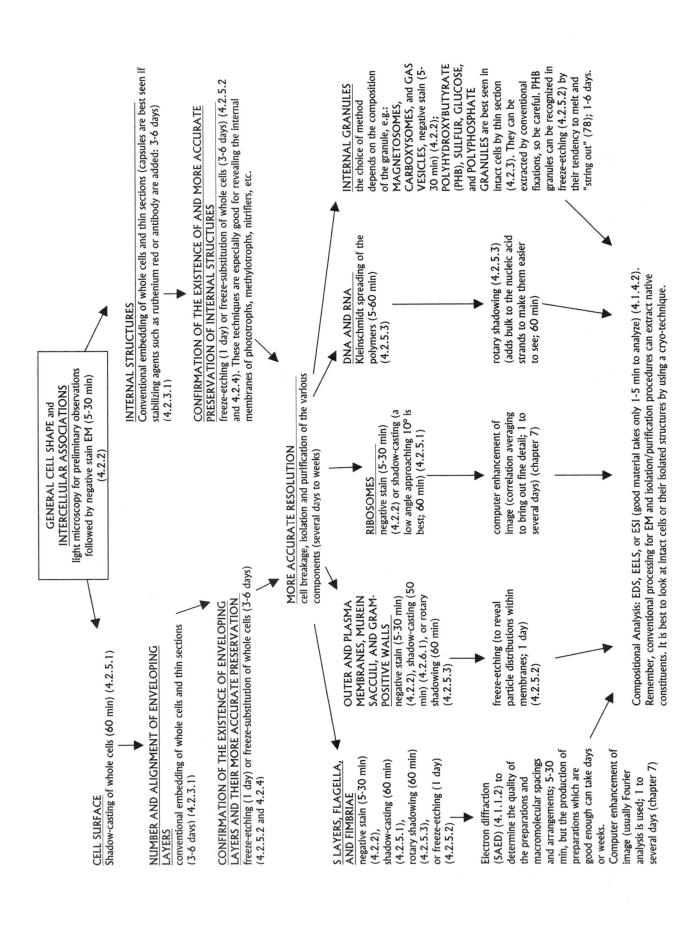

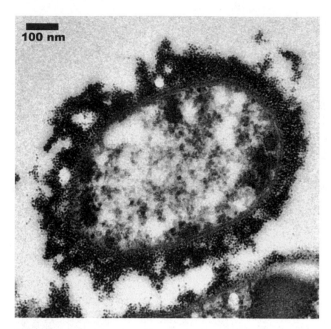

FIGURE 11 TEM image of a *Mannheimia haemolytica* cell expressing an acidic capsule that has been thin sectioned after labeling with PCF (electron-dense particles surrounding the bacterium) to both stabilize and label the capsule.

when subjected to an excess of salt or when concentrated, dilute buffers should be used (e.g., 2 mM Borax buffer [pH 9.01]) and the antibody concentration should never exceed 1 mg ml^{-1}. Centrifuge the IgG solution at 100,000 × *g* for 1 h before use to remove aggregates. Before adsorbing the antibody to the gold, some workers determine the minimal protecting concentration of the antibody as a guide (49). This is the amount of antibody that will protect a given volume of colloidal gold against flocculation when optimal pH and electrolyte concentration are used. For adsorption of the antibodies to colloidal gold particles, follow DeMey's method (49), which is useful for both IgG and IgM antibody classes. Of the two immunoglobulin classes, IgG binds more efficiently to the gold. Monoclonal antibodies have a narrower isoelectric point profile than do polyclonal antibodies. Carry out all reactions at room temperature, as follows.

1. Suspend the stabilized gold particles in 20 mM Tris-HCl buffer containing 1% (wt/vol) bovine serum albumin (BSA), 150 mM NaCl, and 20 mM NaN$_3$ (to discourage bacterial contamination) and adjust to pH 8.2 to 9.0. (For the higher pHs, use a Borax buffer, or for more physiological conditions, use phosphate-buffered saline at pH 6.8 to 7.6.) The colloidal gold concentration is usually adjusted to 20 to 80 μg ml^{-1}. Since the reaction area of the gold particle depends on particle size and since there may be a range of sizes in the gold sol, trial and error will optimize the reaction conditions. Without suitable analytical equipment, it is difficult to accurately measure the gold concentration by light absorption since the gold particles both absorb and scatter light at 520 nm. To help estimate the gold concentration, it is useful to know that 70 mg ml^{-1} of 15-nm gold particles gives an optical density at 520 nm of 0.25. Usually, small gold particles require more protein than larger particles do.

2. Centrifuge at low speed to remove gold aggregates (for 5-nm gold, use 4,800 × *g* for 20 min; for 15-nm gold, use 250 × *g* for 20 min).

3. Stir the gold sol rapidly, and slowly add the antibody. The antibody concentration is usually between 20 and 80 μg ml^{-1} of sol and must be above the minimal protecting concentration (49). After 2 min, add enough of a 10% BSA–20 mM Tris-HCl (pH 8.2 to 9.0) solution to bring the reaction mixture back to 1% BSA. This depends on the volume of antibody added. The BSA blocks the remaining adsorption sites on the gold particles that are not occupied by antibody and will also stop aggregation of the gold particles.

4. Centrifuge to pellet the antibody-coated particles (uncoated gold and unadsorbed antibody will stay in the supernatant fluid). Use 60,000 × *g* for 1 h for 5-nm gold and 15,000 × *g* for 15-nm gold.

5. Carefully remove the supernatant, resuspend the pellet in a few drops of 1% BSA buffer, and store overnight at 4°C. This period of storage ensures that the antibody-gold linkage becomes stable. Resuspend, and centrifuge the following morning.

6. Resuspend 5-nm gold to an A$_{520}$ of 0.250 and 15-nm gold to an A$_{520}$ of 0. 50. As long as NaN$_3$ is in the suspension, the suspension can be stored at 4°C for several weeks. The addition of glycerol to 50% (vol/vol) allows it to be frozen. Not all preparations freeze well. To be safe, freeze a small portion for 1 to 2 days and check its activity after thawing.

Because there is never absolute control over gold particle size, there will always be a small range of sizes. If there is a strict size requirement, the gold can be centrifuged through a 10 to 30% (wt/vol) isopycnic sucrose gradient (use a Beckman SW41 centrifuge rotor for 45 min at 41,000 rpm) to separate the particles into specific sizes. Make sure to wash or dialyze the sucrose from the gold before use.

There are also probe molecules other than antibody that can be used to target the gold (48–51, 59). These are attached to the gold particles as outlined above for antibody and include the following.

1. *Protein A*, specific for IgG and to a lesser extent IgM antibodies, and *protein G*, specific for IgG but also with good affinity for IgM. Because of their high specificity to antibodies, protein A and protein G are frequently used in the indirect-labeling technique (section 4.6.2.3) since they are also used in affinity chromatography (chapter 8.3.5). Protein A is a constituent of *Staphylococcus aureus* cell walls which has high affinity for IgGs and has been used extensively. Protein G, which is extracted from group G streptococcal cell walls (48), also has high affinity for IgG but seems to react more strongly to IgM than protein A does. Since monoclonal antibodies are frequently of the IgM class, protein G is a powerful addition to the immunolabeling arsenal.

2. *Lectins*. Lectins are specific for a range of carbohydrates (49, 52, 59); e.g., concanavalin A is specific for α-D-mannoside and α-D-glucoside, and wheat germ agglutinin is specific for *N*-acetyl-β-D-glucosamine.

3. *Avidin*. Avidin is found in hen egg white and is highly specific for *d*-biotin (vitamin H) (49, 52, 59).

4.6.2.2. Direct Antigen Labeling

Colloidal gold coated with antibody is a very specific electron-dense probe used to find a specific antigen. Bacterial surface components contain important antigens (e.g., lipopolysaccharides, outer membrane proteins, and capsule polymers), and these can be located on intact cells. Because antibodies have different affinities for their antigens and because they can exhibit nonspecific adsorption, each time a new colloidal gold-antibody probe is produced from different serum, the labeling conditions may have to be modified. A general recipe follows.

1. Float Formvar®- and carbon-coated *nickel* grids on a bacterial suspension (section 4.2.2.2). Process them one at a time.

2. Blot dry, and float on a 1 to 3% (wt/vol) BSA solution in phosphate-buffered saline (pH 7.4) for 15 min. Nonspecific adsorption of many proteins can occur, and BSA acts as a non-electron-dense blocking agent. Alternatively, use a 0.3% (wt/vol) solution of skim milk powder, which contains many proteins. Use the same droplet technique described in section 4.2.2.2 for negative staining.

3. Wash the grid by serial transfer onto 3 to 5 drops of 0.1% BSA (or skim milk) in phosphate-buffered saline.

4. Blot dry, and float the grid on the colloidal gold-antibody suspension for 10 to 30 min.

5. Wash as described in step 3. The final wash should be on droplets of deionized water to reduce salting-out of the phosphate-buffed saline in the transmission electron microscope. Too much washing in water can reduce the labeling, and too little washing produces severe salting and forms precipitates during drying. It is best to follow the washing by looking at a grid after each wash in the transmission electron microscope. This means that several grids should be ready to process.

6. Whole mounts of the cells without further staining can be suitable to observe the surface-labeling pattern, but a negative stain can often help. One percent (wt/vol) ammonium molybdate, pH 7.0 to 7.4 (section 4.2.2.1), is best because it has low contrast and the gold can easily be seen. Do the staining carefully and quickly so as not to disturb the labeling pattern. Do not use uranyl acetate, since its low pH can dissociate the gold probe from its binding site.

The most difficult aspect of using a gold probe is finding the best concentration for accurate labeling of the antigen. Even with blocking agents, there can be a high level of background labeling, and this makes the accuracy of the probe questionable. Finding the best probe concentration is difficult since every antibody preparation has a different avidity to an antigen. Trial and error is the only method, and steps 1 through 5 must be repeated until the best concentration is found. For this reason, controls are very important and must be processed through all steps 1 to 5. Uncoated (without antibody) gold and gold coated with nonspecific antibody should not label the specimen. Specific antibody (without the gold) should compete for the antigen sites with the gold-antibody probe. Steps 1 through 5 can also be used for lectin-coated gold.

It is possible to label more than one surface antigen and to distinguish one from another by using two different-sized colloidal gold particles (e.g., 5- and 15-nm gold). Each particle size is coated with a separate antibody. The two probes can be mixed for step 4, or they can be used one after the other. The more probes used, the greater is the difficulty in obtaining optimal labeling.

4.6.2.3. Indirect Antigen Labeling

For convenience, it is often best to use indirect gold labeling since a single gold probe can be used against a wide battery of antigens. For example, protein A-colloidal gold will bind to IgG and will recognize specific IgG that has adsorbed to the cell. Care must be taken to wash unbound IgG away before treatment with the gold probe. This is an indirect labeling method because the gold probe is not directly linked to the antigen but is only indirectly linked by the underlying IgG. Therefore, the actual gold particle is distanced from the antigen site by the span of both IgG and protein A molecules, but this distance is usually insignificant for most labeling studies. If the specific antibody is a

monoclonal antibody, there is a good chance it will be an IgM antibody. Protein A has a low binding efficiency for IgM, and it is preferable to use protein G, which has a higher affinity for this class of antibody. If it is possible to link biotinyl residues to the desired molecule, then avidin (usually streptavidin)-gold can be used as the probe (49).

For indirect labeling, the same general approach outlined in section 4.6.2.2 is used but for step 4 the specific antibody (not linked to gold) is used alone. After being washed with 3 to 5 drops of 0.1% BSA in phosphate-buffered saline, the grid is floated on protein A-gold (or protein G-gold) for 10 to 30 min and then washed as before. If appropriate, the grid can be negatively stained.

4.6.2.4. Thin-Section Antigen Labeling

Sometimes mounts of whole cells will not provide enough information about how labeled surface antigen components are arranged with respect to the underlying enveloping layers (e.g., labeled outer membrane proteins compared with the peptidoglycan layer). Label these cells as described in section 4.6.2.2 or 4.6.2.3, and fix or embed them as described in section 4.2.3.1. Since the cells are labeled before fixation, the gold probe is chemically fixed to the cell surface (e.g., the outer membrane), and this positioning survives embedding and curing. Thin sections show the position of the label with respect to the underlying envelope layers.

The difficulties will be greater if the intent is to label antigen constituents inside cells or to monitor the location of a component over time as it migrates through cellular space. The cells cannot be lysed to show their internal constituents, since this destroys the constituents' locations and cellular associations. One answer is to gold label thin sections of fixed cells after they have been cut and mounted on nickel grids by using either the direct or indirect method (Fig. 12). This is amongst the most difficult of all TEM procedures, and students should beware! Controls, to ensure labeling specificity, are essential.

Chemical fixation denatures bacterial constituents by altering their molecular folding and by increasing their covalent bonding. Dehydration and embedding in plastics remove all free cellular water and, presumably, most hydration shells surrounding bacterial macromolecules; thus, folding is again altered. All of these changes will affect the retention of antigenicity. This is not so important when surfaces are prelabeled with immunogold before fixation (as described above), but care must be taken for postlabeling of thin sections. Once again, trial and error is the best approach.

Osmium tetroxide is a strong oxidizing agent and should not be used in this fixation. Formaldehyde is a preferred fixative, but it is not a bifunctional agent and will not cross-link. Glutaraldehyde is bifunctional and is not as damaging as osmium tetroxide, but at high concentrations it reduces antigenicity. Uranyl acetate, because of its low pH and binding character, also denatures macromolecules and reduces antigenicity. For dehydration, ethanol is not as damaging as acetone or propylene oxide. Most of the usual plastics (e.g., Epon 812®, Vestopal®, and Araldite®) do not preserve antigenicity very well; LR White® and Lowicryl K4M® are better. It seems like a no-win situation, but there are reasonable chances of success. Use the following procedure.

Fixation. Use a mixture of 2% (wt/vol) formaldehyde (made fresh from paraformaldehyde; see section 4.7.4) and 0.1% (vol/vol) glutaraldehyde in 50 mM HEPES buffer (pH 7.0) for 60 min. Wash with buffer a few times to remove unbound fixative. (Proceed as described in section

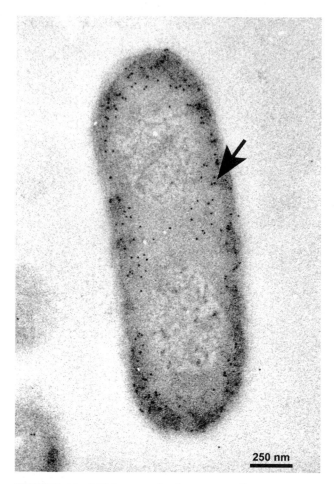

250 nm

FIGURE 12 TEM image of a thin-sectioned *E. coli* cell that has been section stained. The cell has been stained by the indirect protein A-colloidal gold method to detect the presence and location of penicillin-binding protein 3 from *Pseudomonas aeruginosa*, which has been cloned into it. The arrow points to a colloidal gold particle, and many more particles can be seen around the cell periphery.

4.2.3.1, but do not fix with osmium tetroxide or perform en bloc staining with uranyl acetate.)

Dehydration. Use a graded ethanol series, and repeat twice with 100% (vol/vol) ethanol, as described in sections 4.2.3.2 and 4.2.3.3.

Embedding. Infiltrate by using an ethanol-LR White® graded series (50 to 100% plastic medium as described in sections 4.2.3.2 and 4.2.3.3). Cure the block at 60°C for 22 h. Cut thin sections as usual, but do not stain with uranyl acetate or lead citrate. Lowicryl K4M® (16, 17) can also be used for embedding and has the advantage that embedding and curing with UV light can be carried out at −80°C to help retain antigenicity. (*Caution:* Lowicryl can induce severe allergic responses, so wear surgical gloves!)

Labeling sections with gold probe. Cut enough sections to make a number of *nickel* grids for trials and controls. Float the grids section side down on 1 to 3% (wt/vol)-BSA (or skim milk) in phosphate-buffed saline for 15 min to block nonspecific reactions. Wash with 2 to 5 drops of 0.1% BSA in phosphate-buffered saline. Float

the grid on the gold probe suspension for 30 to 60 min, and wash as before. Check by TEM to see how well the section is labeled (there has been no heavy-metal staining so far, and only the gold will be seen). Repeat the labeling on a fresh grid until a satisfactory labeling pattern is seen and the background is low. Controls with uncoated gold and nonspecific antibody-gold should be negative. Remember that only the antigens exposed on the surface of the section will be labeled and then only if they are not denatured; unless the antigen is present in a high copy number, expect only low labeling counts which may be only slightly above background and control counts. Statistical analysis of the counts can be helpful (49).

Staining of labeled sections can be tricky, and it is best to work out optimal gold labeling on unstained sections first. Preserving gold-labeling patterns after staining with lead citrate is unsatisfactory; instead, use 1 to 2% (wt/vol) uranyl acetate to stain the sections. The labeling efficiency of the gold may be reduced (and this is another reason for knowing how unstained sections label). Keep the period of uranyl acetate staining as short as possible for best results.

Sections can also be indirectly labeled with protein A- or protein G-gold (Fig. 12). For this labeling, the specific antibody is first adsorbed to the thin section, the grid is washed, and the indirect gold probe is adsorbed onto the first antibody. Sometimes, because gold counts are so low on thin sections, it is useful to try to amplify the gold labeling. This is also an indirect technique and uses a second intermediate antibody between the first (specific) adsorbed antibody and the gold probe; e.g., if the first specific antibody is a rabbit IgG, once it is adsorbed to the section, antibodies directed against rabbit IgG (maybe goat antibodies) would be used to decorate the first antibody at several places along its length. Next, anti-goat–colloidal gold (or protein A-gold) would be used to decorate all of the adsorbed goat antibodies. This three-step procedure amplifies the initial single antibody to become labeled with several gold particles and makes the antigenic site easier to see.

4.7. MATERIALS

This section provides a catalog of useful items for EM. Commercial sources for most of the materials are indicated by the letters in parentheses, which refer to the list in section 4.8.

4.7.1. Plastic Support Films

Formvar® powder (*a, k, o, p, r, u, x, z, cc*)

Formvar® solutions (*o, r, cc*)

Ethylene dichloride (*b, e, m, n, y, aa*)

Clean glass microscope slides

Lint-free photowipe tissues (*a, k, o, p, r, u, x, t, cc*)

Large glass crystallizing dish with straight sides and flat bottom. The best crystallizing dish is one measuring 200 by 80 mm (culture dish; Vitro) (*w, dd*). The top edge must be even and smooth. The bottom and sides can be permanently covered with black electrical tape to permit visualization of the floating films against a black background.

Coplin glass staining jars for microscope slides (*f, m, w, dd*)

Pasteur pipettes and rubber bulbs

Razor blades (alcohol washed)

Filter paper, Whatman no. 1, 9.0-cm diameter (*f, m*)

Petri dishes, standard, 9.0-cm diameter

Scissors

Grid tweezers, narrow-pointed no. 5 Dumont or equivalent (a, k, o, p, r, u, x, z, cc).

Copper or nickel grids, precleaned (a, k, o, p, r, u, x, z, cc)

Light source, fluorescent, adjustable

4.7.2. Carbon Films
High-power vacuum evaporator and accessories (d, g, q)

4.7.3. Negative Staining
Coated grids, prepared as described in section 4.2.1.2 or obtained from commercial sources (a, l, p, r, u, x, cc)

Sticky tape (Time Tape) (f, m, w)

Glass microscope slides

Zerostat gun (i, w) to discharge static charges

Pasteur pipettes

Ammonium acetate (b, e, m, n, q, y, aa)

Ammonium molybdate (b, e, m, n, q, y, aa)

Parafilm® (c, m)

Potassium or sodium phosphotungstate (b, e, m, n, q, y, aa)

Uranyl acetate (b, e, k, m, n, p, q, u, y, aa, cc)

4.7.4. Fixing, Embedding, Sectioning, and Staining
Glutaraldehyde (a, k, o, p, u, x, z, cc)

Paraformaldehyde. To generate formaldehyde, add 0.8 g of paraformaldehyde (b, e, m, n, q, y, aa) to 10 ml of water and heat to 60 to 70°C for 10 min. A white suspension is formed. With rapid stirring, add 1 to 3 drops of 1 M NaOH until the suspension clears. This gives 8% (wt/vol) formaldehyde. Dilute to 2% with buffer for fixation. *The entire procedure should be done in a fume hood to avoid fumes.* Cool to room temperature, and use immediately. Holding life is about 4 to 8 h.

Sodium cacodylate buffer. This may be purchased as buffer or as a kit with premeasured ingredients (cc). Otherwise, make up sodium cacodylate (a, k, o, p, u, x, z, cc) to 0.2 M in water and adjust to pH 7.2 with HCl. *Cacodylate is toxic; avoid contact and inhalation of arsenical vapor; use a fume hood.*

HEPES buffer. This may be purchased from a number of chemical supply houses (b, e, m, n, q, y, aa). It should be made up in deionized water to a stock solution of 1 M (pH 7.2) and diluted as needed.

Safety goggles

Osmium tetroxide (a, k, o, p, u, x, z, cc)

Noble® agar (h)

Uranyl acetate (a, k, o, p, u, x, z, cc)

Tryptone (h)

Trypticase soy broth (h)

Glass vials, screw cap, 1 dram (w, dd)

Respirator mask (v)

Disposable plastic containers, 4.5 oz (ca. 130 ml) (f, l, m)

Epon 812® embedding resin. Epon 812 (a, k, o, p, u, x, z, cc) is an epoxy-based resin and is one of the most commonly used embedding plastics. It relies on the polymer's strained epoxide ring, which can be broken at 60°C to reanneal with those of adjacent polymers to form a solid plastic block. The reaction is controlled by the addition of nadic methyl anhydride (a chemical "hardener" which increases bonding) and dodecylsuccinic anhydride (a chemical "softener" which decreases bonding). The reaction is catalyzed by tridimethylaminoethyl phenol. Epon 812 is quite viscous and can be difficult to infiltrate into microbes, especially fungi and yeasts, but it has good thin sectioning qualities and shows a small grain size when examined by TEM. Various lot numbers from chemical suppliers may have subtly different curing characteristics and may require some fine adjustments to the basic recipe. The major manufacturer of Epon 812 no longer makes this plastic resin, but stocks are still available. Some resins are very similar, e.g., Polybed 812® (z). Chemicals are usually stored at 4°C for best shelf life; remember to warm them to room temperature before opening so that water vapor does not condense in them. A procedure for preparing Epon 812® is given below. In separate, disposable 250-ml beakers, make the following.

Mixture A

Epon 812 . . . . 40 g
Dodecylsuccinic anydride . . . . 46.6 g

Mixture B

Epon 812 . . . . 42 g
Nadic methyl anhydride . . . . 33.5 g

Mix these ingredients thoroughly for ~15 min. Weigh out 6 g of mixture A into screw-cap bottles (20-ml scintillation vials work well), and then add 4 g of mixture B to each bottle (some of mixture B will probably be left over). Mix thoroughly for ~15 min, and then freeze at −20°C until ready for use. When required, take a bottle from the freezer, warm to room temperature, and add 0.3 ml of tridimethylaminoethyl phenol. Stir for ~30 min before use of this complete mixture.

Spurr® low-viscosity resin. This is supplied as a kit or as ingredients (a, k, o, p, u, x, z, cc) to be used as follows to make the stock mixture.

Stock mixture

ERL-4206 (vinylcyclohexene dioxide), 30 g or 27.0 ml

DER-736 (diglycidyl ether of propylene glycol epoxy resin), 18 g or 22.8 ml

Nonenylsuccinic anhydride, 78 g or 76.5 ml

Working in a chemical hood or wearing an appropriate mask, add the three components in the above-listed order to a disposable container, mixing for 5 min on a magnetic stirrer. This stock mixture, which lacks accelerator, can be stored tightly covered at −4°C for 3 to 5 months. Allow ample time to equilibrate to room temperature upon removal and before opening to avoid water condensation that may destroy the mixture.

Complete mixture

To make the complete mixture, put 22 ml of the stock mixture into a disposable container, add 0.24 ml of the accelerator S-I (dimethylaminoethanol), and stir for several minutes at room temperature. For each sample, allow at least 6 ml of complete mixture. This volume will take care of mixtures in steps 4 to 6 of the dehydration and embedding procedure and will provide for the filling of three embedding capsules. The useful life of the complete mixture is 2 days.

LR White® resin. This resin (*t, o, p, r, u, cc*) is an acrylic-based, hydrophilic plastic which comes as a fluid. It is cured at 60°C for 22 h and is simple to use. Follow the manufacturer's instructions for curing; the surface of the plastic block may be sticky because of a thin layer of partially cured resin which was exposed to air (oxygen). If this is a problem, a vacuum oven can be used. Gelatin capsules are recommended for polymerization. The advantages of LR White resin are its low viscosity (much lower than those of epoxy-based resins) for good infiltration, its low toxicity, and its simplicity of use. This resin works well for immunogold labeling (section 4.6.2.4). It comes in hard, medium, and soft grades.

Beem® capsules (*a, k, o, p, u, x, z, cc*)

Drying oven

Centrifuge, Eppendorf, and appropriate conical tubes with caps

Pliers

Razor blades, single edge

Ultramicrotome

Plate glass, special for glass knives, cut in appropriate strips for knife maker

Glass-knife maker (*s*)

Diamond knife (*j*), an expensive alternative to the glass knife

Dental wax

Glass weighing bottles

Plastic tissue culture dish, 35 by 100 mm

NaOH pellets

Petri dishes, glass or plastic

Lead citrate. This is available ready-made from commercial sources (*a, k, o, p, u, x, z, cc*). It can also be prepared as follows. To a 50-ml volumetric flask, add the following.

Lead nitrate . . . . 1.33 g
Sodium citrate . . . . 1.76 g
Distilled water . . . . 30.00 g

Shake intermittently over 30 min, and allow to stand at room temperature for 30 min. Add 8 ml of freshly prepared 1 M NaOH and dilute to 50 ml with water.

4.8. COMMERCIAL SOURCES

a. Agar Aids Ltd., 66a Cambridge Rd., Stansted CM24 8DA, England

b. Aldrich Chemical Co. Inc., 1001 West Saint Paul Ave., Milwaukee, WI 53233

c. American National Can Co., Greenwich, CT 06836. (Items from this company can be obtained through major scientific suppliers such as Fisher Scientific Co. [*m*].)

d. Balzers/BAL-TEC AG, Postfach 75, FL-9496 Balzers, Furstentum, Liechtenstein

e. Boehringer Mannheim Canada Ltd., 200 Micro Blvd., Laval, Quebec H7V 3Z9, Canada

f. Curtin-Matheson Scientific, Inc., 10727 Tucker St., Beltsville, MD 20705

g. Denton Vacuum Inc., Cherry Hill Industrial Center, Cherry Hill, NJ 08003

h. Difco Laboratories, P.O. Box 1958A, Detroit, MI 48232

i. Dishwasher, Inc., 1407 North Providence Rd., Columbia, MO 65201 (or local photo stores)

j. Dupont Instruments, Biomedical Division, Newtown, CT 06470

k. Electron Microscopy Sciences, P.O. Box 251, Fort Washington, PA 19034

l. Falcon Labware Division, 1950 Williams Dr., Oxnard, CA 93030

m. Fisher Scientific Co., 1 Reagent Ln., P.O. Box 375, Fairlawn, NJ 07410

n. Fluka Chemika-Biochemika, Fluka Chemie Ag, Industriestrasse 25, CH-9470 Buchs, Switzerland

o. Ernest E. Fullam, Inc., P.O. Box 444, Schenectady, NY 12301

p. J. B. EM Services Inc., P.O. Box 693, Pointe-Claire, Dorval, Quebec H9R 4S8, Canada

q. J. T. Baker Inc., 22 Red School Ln., Phillipsburg, NJ 08865

r. Ladd Research Industries, Inc., P.O. Box 901, Burlington, VT 05401

s. LKB Instruments, Inc., 12221 Parklawn Dr., Rockville, MD 20852

t. London Resin Co., Ltd., P.O. Box 34, Basingstoke, Hampshire RG25 2EX, England

u. Marivac Ltd., 5821 Russell St., Halifax, Nova Scotia B3K I X5, Canada

v. Mine Safety Appliances Co., Pittsburgh, PA 15208

w. Nalge Co., Division of Sybion Corp., Rochester, NY 14602

x. Ted Pella Co., P.O. Box 510, Tustin, CA 92680

y. Pierce, P.O. Box 117, Rockford, IL 61105

z. Polysciences, Inc., Paul Valley Industrial Park, Warrington, PA 18976

aa. Sigma Chemical Co., P.O. Box 14508, St. Louis, MO 63178

bb. Sorg Paper Co., Middletown, OH 45042

cc. SPI Supplies, P.O. Box 342, West Chester, PA 19380

dd. Tousimis Research Corp., P.O. Box 2189, Rockville, MD 20852

ee. Wheaton Scientific (Vitro), 1000 North 10th St., Millville, NJ 08332

(Most suppliers have Web pages that can be easily accessed for more information.)

4.9. REFERENCES

Many general and, even, specific references are relatively old. This does not necessarily reflect the absence of advancement in the field of EM, but it says, more, that the tried and true methods still work quite well for the beginning student and that the more advanced techniques require very specialized equipment and many years of experience.

4.9.1. General References

1. **Aldrich, H. C., and W. J. Todd (ed.).** 1986. *Ultrastructure Techniques for Microorganisms.* Plenum Press, New York, NY.
   One of the few books dedicated to EM techniques used on microorganisms, including chapters on conventional techniques (such as chemical fixation and embedding and freeze-etching) and also on more recent techniques (such as cryopreparation, image analysis, and immunolabeling). This book deserves a place in every microbiology EM laboratory.
2. **Duke, P. J., and A. G. Michette (ed.).** 1990. *Modern Microscopies: Techniques and Applications.* Plenum Press, New York, NY.
   Most chapters in this book are dedicated to X-ray and nuclear magnetic resonance microscopies, but there are a few worthwhile chapters on cryoTEM and image reconstruction.

3. **Dykstra, M. J.** 1992. *Biological Electron Microscopy: Theory, Techniques, and Troubleshooting.* Plenum Press, New York, NY.
   An up-to-date general overview of TEM and how it is used in biology. There is little information dedicated to microbiology.

4. **Glauert, A. M. (ed.).** 1972–1991. *Practical Methods in Electron Microscopy,* vol. 1–15. American Elsevier Publishing Co., Inc., New York, NY.
   Glauert, a distinguished electron microscopist with an interest in bacterial structure, has edited this remarkable series on state-of-the-art EM techniques. Most university libraries have the entire set, and it is well worthwhile to look through it. Some of the volumes are available as paperbacks. In particular, Glauert's part 1 of vol. 3, titled Fixation, Dehydration, and Embedding of Biological Specimens, is a handy 200-page paperback book which describes a full range of fixatives, buffers, stains, and plastics. Very good to have for laboratory recipes.

5. **Hajibagheri, M. A. N. (ed.).** 1999. *Electron Microscopy Methods and Protocols.* Humana Press, Totowa, NJ.
   A soft-cover book edited by Hajibagheri with chapters by many of the experts in EM. Techniques from negative staining to cryoTEM to colloidal gold labeling are covered.

6. **Hayat, M. A.** 1986. *Basic Techniques for Transmission Electron Microscopy.* Academic Press, Inc., New York, NY.
   This 400-page paperback book is filled with helpful hints and general recipes for processing of biological samples for TEM. It deserves a place in every EM laboratory.

7. **Hayat, M. A. (ed.).** 2000. *Principles and Techniques of Electron Microscopy. Biological Applications.* Cambridge University Press, Cambridge, United Kingdom.
   You cannot do EM in biology without running into Hayat's name. During the 1970s through the 1990s, he published a very successful multivolume set which deals with a full range of important techniques. Like Glauert's series, this set is available in most university libraries and should be examined. Here is a single book dedicated to biology that is full of useful information; it is 543 pages long.

8. **Koval, S. F., and T. J. Beveridge.** 2000. Electron microscopy, p. 276–287. *In* J. Lederberg (ed.), *Encyclopedia of Microbiology.* Academic Press, San Diego, CA.
   A good general overview of the various types of microscopy and how microscopy pertains to microbiology. Nicely illustrated with electron micrographs.

9. **Moses, N., P. S. Handley, H. J. Busscher, and P. G. Rouxhet (ed.).** 1991. *Structural and Physico-Chemical Methods for Microbial Cell Surface Analysis.* VCH Publishing, New York, NY.
   Although techniques other than EM are described, this book is dedicated to the elucidation of bacterial surfaces and surveys the most up-to-date methodology.

10. **Royal Microscopical Society.** *Royal Microscopical Society Microscopy Handbooks.* Oxford University Press, New York, NY.
    This is a series of short, soft-covered books dedicated to microscopy. No. 3 (P. J. Goodhew, Specimen Preparation for Transmission Electron Microscopy, 1984), no. 8. (S. K. Chapman, Maintaining and Monitoring the Transmission Electron Microscope, 1988), and no. 20 (D. Chescoe and P. J. Goodhew, The Operation of the Transmission Electron Microscope and the Scanning Electron Microscope, 1990) are particularly good volumes to look at for basic operation of a transmission or scanning electron microscope and for troubleshooting. No. 43 (M. Hoppert and A. Holzenburg, Electron Microscopy in Microbiology, 1998) is particularly recommended because it details how to use SEM and TEM to visualize microorganisms.

11. **Sommerville, J., and U. Scheer (ed.).** 1987. *Electron Microscopy in Molecular Biology: a Practical Approach.* IRL Press, Oxford, United Kingdom.
    This book does not deal with microorganisms but offers several methods for visualizing nucleic acids and proteins once they have been extracted from cells.

4.9.2. Specific References by Subject

4.9.2.1. Grids, Support Films, and Negative Staining
(See also references 4 through 7 and 10.)

12. **Harris, R. J.** 1997. *Royal Microscopical Society Handbook, no. 35. Negative Staining and Cryoelectron Microscopy: the Thin Film Techniques.* BIOS Scientific Publishing Ltd., Oxford, United Kingdom.
    Another of the Royal Microscopical Society handbooks that is right up to date and written by one of the experts who has good microbiological experience.

4.9.2.2. Sectioned Preparations
(See references 4 through 7.)

13. **Hobot, J. A.** 1990. New aspects of bacterial ultrastructure as revealed by modern acrylics for electron microscopy. *J. Struct. Biol.* **104:**169–177.
    A discourse, with an emphasis on bacteria, that explains the newer acrylic plastics that can be cured from −80 to +60°C.

4.9.2.3. Low-Temperature Embedding Resins

14. **Acetarin, J.-D., E. Carlemalm, and W. Villiger.** 1986. Developments of new Lowicryl resins for embedding biological specimens at even lower temperatures. *J. Microsc.* **143:**81–88.

15. **Armbruster, B. L., E. Carlemalm, R. Chiovetti, R. M. Caravito, J. A. Hobot, E. Kellenberger, and W. Villiger.** 1982. Specimen preparation for electron microscopy using low temperature embedding resins. *J. Microsc.* **126:**77–85.

16. **Carlemalm, E., M. Garavito, and W. Villiger.** 1982. Resin development for electron microscopy and an analysis of embedding at low temperature. *J. Microsc.* **126:**123–143.

17. **Carlemalm, E., W. Villiger, J. A. Hobot, J.-D. Acetarin, and E. Kellenberger.** 1985. Low temperature embedding with Lowicryl resins: two new formulations and some applications. *J. Microsc.* **140:**55–63.

4.9.2.4. Visualizing Macromolecules by Using Metal Shadowing

18. **Coggins, L. W.** 1987. Preparation of nucleic acids for electron microscopy, p. 1–29. *In* J. Sommerville and U. Scheer (ed.), *Electron Microscopy in Molecular Biology: a Practical Approach.* IRL Press, Oxford, United Kingdom.

19. **Glenney, J. R., Jr.** 1987. Rotary metal shadowing for visualizing rod-shaped proteins, p. 167–178. *In* J. Sommerville and U. Scheer (ed.), *Electron Microscopy in Molecular Biology: a Practical Approach.* IRL Press, Oxford, United Kingdom.

20. **Zentgraf, H., C. T. Bock, and M. Schrenk.** 1987. Chromatin spreading, p. 81–100. *In* J. Sommerville and U. Scheer (ed.), *Electron Microscopy in Molecular Biology: a Practical Approach.* IRL Press, Oxford, United Kingdom.

4.9.2.5. Freeze Fracturing and Freeze-Etching

21. **Chapman, R. L., and L. A. Staehelin.** 1986. Freeze-fracture (-etch) electron microscopy, p. 213–240. *In* H. C. Aldrich and W. J. Todd (ed.), *Ultrastructure Techniques for Microorganisms.* Plenum Press, New York, NY.
    Although old, this up-to-date chapter is devoted to microorganisms.

22. **Hui, S. W. (ed.).** 1989. *Freeze-Fracture Studies of Membranes.* CRC Press, Inc., Boca Raton, FL.
    Dedicated to the preparation of membranes for freeze fracture and the interpretation of their fracture planes.

4.9.2.6. Cryotechniques

Since one of the most recent advances in the preparation of biological specimens for EM is low-temperature ultrarapid freezing, a more extensive bibliography is given. The references below range from general books on the principles involved in rapid freezing and the formation of vitreous ice to specific journal papers which deal with bacteria. Freeze-substitution and thin frozen films are covered.

23. **Adrian, M., J. Dubochet, J. Lepault, and A. W. McDowall.** 1984. Cryo-electron microscopy of viruses. *Nature* (London) **308:**32–36.
    Thin, frozen films of virus particles exemplify exquisite detail against a background of vitreous ice.
24. **Amako, K., Y. Meno, and A. Takade.** 1988. Fine structures of the capsules of *Klebsiella pneumoniae* and *Escherichia coli* K-1. *J. Bacteriol.* **170:**4960–4962.
25. **Amako, K., K. Murata, and A. Umeda.** 1983. Structure of the envelope of *Escherichia coli* observed by the rapid-freezing and substitution fixation method. *Microbiol. Immunol.* **27:**95–99.
26. **Amako, K., K. Okada, and S. Miake.** 1984. Evidence for the presence of a capsule in *Vibrio vulnificus. J. Gen. Microbiol.* **130:**2741–2743.
27. **Amako, K., and A. Takade.** 1985. The fine structure of *Bacillus subtilis* revealed by the rapid-freezing and substitution-fixation method. *J. Electron Microsc.* **34:**13–17.
    References 24 to 27 are some of the earliest and best freeze-substitution studies on bacteria. Amako was one of the first to apply the technique to bacteria.
28. **Beckett, A., and N. D. Read.** 1986. Low-temperature scanning electron microscopy, p. 45–86. *In* H. C. Aldrich and W. J. Todd (ed.), *Ultrastructure Techniques for Microorganisms.* Plenum Press, New York, NY.
    A helpful chapter on the advantages of low temperatures for SEM.
29. **Carlemalm, E., W. Villiger, J.-D. Acetarin, and E. Kellenberger.** 1985. Low temperature embedding, p. 147–164. *In* M. Willer, R. P. Becker, A. Boyde, and J. J. Wolosewick (ed.), *The Science of Biological Specimen Preparation for Microscopy and Microanalysis, 1985.* Scanning Electron Microscopy, Inc., AMF O'Hare, IL.
    This Biozentrum group in Basel, Switzerland, developed the concept of low-temperature embedding and the use of Lowicryl resins.
30. **Dubochet, J., A. W. McDowall, B. Menge, E. N. Schmid, and K. G. Lickfeld.** 1983. Electron microscopy of frozen-hydrated bacteria. *J. Bacteriol.* **155:**381–390.
    Shows some of the first and best frozen thin sections of bacteria.
31. **Graham, L. L., and T. J. Beveridge.** 1990. Evaluation of freeze-substitution and conventional embedding protocols for routine electron-microscopic processing of eubacteria. *J. Bacteriol.* **172:**2141–2149.
32. **Graham, L. L., and T. J. Beveridge.** 1990. Effect of chemical fixatives on accurate preservation of *Escherichia coli* and *Bacillus subtilis* structure in cells prepared by freeze-substitution. *J. Bacteriol.* **172:**2150–2159.
33. **Graham, L. L., R. Harris, W. Villiger, and T. J. Beveridge.** 1991. Freeze-substitution of gram-negative eubacteria: general cell morphology and envelope profiles. *J. Bacteriol.* **173:**1623–1633.
    References 31 and 32 detail the biochemical preservation of bacteria and relate it to ultrastructural detail by comparing conventional methods with freeze-substitution. Reference 33 applies the freeze-substitution technique to a wide range of gram-negative bacteria. These articles will assist any microbiologist contemplating the use of freeze-substitution.
34. **Hobot, J. A.** 1991. Low temperature embedding techniques for studying microbial cell surfaces, p. 127–150. *In* N. Mozes, P. S. Handley, H. J. Busscher, and P. G. Rouxhet (ed.), *Structural and Physico-Chemical Methods for Microbial Cell Surface Analysis.* VCH Publishing, New York, NY.
    A good review of the current techniques that use low-temperature methacrylates to better preserve antigenicity and structure.
35. **Hobot, J. A., E. Carlemahn, W. Villiger, and E. Kellenberger.** 1984. Periplasmic gel: new concept resulting from the reinvestigation of bacterial cell envelope ultrastructure by new methods. *J. Bacteriol.* **160:**143–152.
36. **Hobot, J. A., W. Villiger, J. Escaig, M. Maeder, A. Ryter, and E. Kellenberger.** 1985. Shape and fine structure of nucleoids observed on sections of ultrarapidly frozen and cryosubstituted bacteria. *J. Bacteriol.* **162:**960–971.
    References 35 and 36 profoundly altered our perception of gram-negative periplasmic spaces (i.e., the concept of a "periplasmic gel") and the cytoplasmic distribution of the bacterial chromosome.
36a. **Hunter, R. C. and T. J. Beveridge.** 2005. High-resolution visualization of *Pseudomonas aeruginosa* PAO1 biofilms by freeze-substitution transmission electron microscopy. *J. Bacteriol.* **187:**7619–7630.
    Here is an article that shows the power of preservation that the technique of freeze-substitution has on bacterial biofilms. Exopolymeric substances, O side chains of lipopolysaccharide, and cells are all exquisitely maintained.
37a. **Matias, V. R. F., A. Al-Amoudi, J. Dubochet, and T. J. Beveridge.** 2003. Cryo-transmission electron microscopy of frozen-hydrated sections of *Escherichia coli* and *Pseudomonas aeruginosa. J. Bacteriol.* **185:**6112–6118.
    An example of cryoTEM and cryosections of Escherichia coli and Pseudomonas aeruginosa emphasizing the structure of the gram-negative envelope and the mass distribution.
37b. **Matias, V. R. F., and T. J. Beveridge.** 2005. Cryo-electron microscopy reveals native polymeric cell wall structure in *Bacillus subtilis* 168 and the existence of a periplasmic space. *Mol. Microbiol.* **56:**240–251.
37c. **Matias, V. R. F., and T. J. Beveridge.** 2006. Native cell wall organization shown by cryo-electron microscopy confirms the existence of a periplasmic space in *Staphylococcus aureus. J. Bacteriol.* **188:**1011–1021.
    Both references 37b and c give examples of how gram-positive bacteria look after they have been vitrified by freezing and cryosectioned. Both Bacillus subtilis and Staphylococcus aureus are shown to have a periplasmic space.
37d. **Renelli, M., V. Matias, R. Y. Lo, and T. J. Beveridge.** 2004. Characterization of DNA-containing membrane vesicles of *Pseudomonas aeruginosa* PAO1 and their genetic transformation potential. *Microbiology* **150:**2161–2169.
    An example of cryoTEM and frozen foils and one of the few articles that shows the mass distribution of intact outer and inner membrane vesicles while in their frozen vitrified state.
37e. **Robards, A. W., and U. B. Sleytr.** 1985. *Practical Methods in Electron Microscopy, vol. 10. Low Temperature Methods in Biological Electron Microscopy.* Elsevier Science Publishing, Amsterdam, The Netherlands.
38. **Roos, N., and A. J. Morgan.** 1990. *Royal Microscopical Society Handbook, no. 21. Cryopreparation of Thin Biological Specimens for Electron Microscopy. Methods and Applications.* Oxford University Press, New York, NY.
39. **Steinbrecht, R. A., and K. Zierold (ed.).** 1987. *Cryotechniques in Biological Electron Microscopy.* Springer-Verlag KG, Berlin, Germany.
    References 37b, c, and e, 38, and 39 are books and chapters which describe the principles involved in rapid freezing, the properties of different ice forms, freeze-substitution, frozen thin sections, and thin frozen films.
40. **Umeda, A., Y. Ueki, and K. Amako.** 1987. Structure of the *Staphylococcus aureus* cell wall determined by the freeze-substitution method. *J. Bacteriol.* **169:**2482–2487.

Other than Bacillus subtilis, *few gram-positive vegetative cells have been subjected to freeze-substitution. Here we have* Staphylococcus aureus.

4.9.2.7. PCF

41. **Jacques, M., M. Gottschalk, B. Foiry, and R. Higgins.** 1990. Ultrastructural study of surface components of *Streptococcus suis. J. Bacteriol.* **172:**2833–2838.
One of the more recent PCF studies on capsules.

42. **Sára, M., and U. B. Sleytr.** 1987. Charge distribution of the S layer of *Bacillus stearothermophilus* NRS 1536/3c and importance of charged groups for morphogenesis and function. *J. Bacteriol.* **169:**2804–2809.
Shows both negative stains and freeze-etching on PCF-labeled cell surfaces.

43. **Sonnenfeld, E. M., T. J. Beveridge, and R. J. Doyle.** 1985. Discontinuity of charge on cell wall poles of *Bacillus subtilis. Can. J. Microbiol.* **31:**875–877.
Limiting concentrations of PCF are used to locate the most anionic sites on the cell wall.

44. **Sonnenfeld, E. M., T. J. Beveridge, A. Koch, and R. J. Doyle.** 1985. Asymmetric distribution of charge on the cell wall of *Bacillus subtilis. J. Bacteriol.* **163:**1167–1171.
Shows the use of PCF for labeling electronegative sites on Bacillus subtilis *walls in thin section, before and after neutralization of charges. Eventually carboxyl and phosphoryl groups are shown to be the major wall anions.*

4.9.2.8. Immunogold Labeling

The ability to precisely label distinct antigens in bacteria has had a profound influence on the elucidation of microbial ultrastructure; for the first time in microbiology, it is possible to determine the exact locations of macromolecules and even distinct domains (epitopes) of macromolecules. Because of the importance of the immunogold labeling technique, this extensive bibliography includes recipes, reviews, and descriptions of the use of immunogold labeling in bacteriology. See also the references in chapter 8.

45. **Acker, G., D. Bitter-Suermann, U. Meier-Dieter, H. Peters, and H. Mayer.** 1986. Immunocytochemical localization of enterobacterial common antigen in *Escherichia coli* and *Yersinia enterocolitica. J. Bacteriol.* **168:**348–356.

46. **Acker, G., and C. Kammerer.** 1990. Localization of enterobacterial common antigen immunoreactivity in the ribosomal cytoplasm of *Escherichia coli* cells cryosubstituted and embedded at low temperature. *J. Bacteriol.* **172:**1106–1113.
Immunogold studies using conventionally and cryoprepared samples with a distinct E. coli *antigen.*

47. **Behnke, O., T. Ammitzboll, J. Jessen, M. Klokker, K. Nflausen, J. Tranum-Jensen, and L. Olsson.** 1986. Non specific binding of protein-stabilized gold sols as a source of error in immunocytochemistry. *Eur J. Cell Biol.* **41:**326–338.

48. **Bendayan, M., and S. Garzon.** 1988. Protein-G-gold complex: comparative evaluation with protein-A-gold for high-resolution immunocytochemistry. *J. Histochem. Cytochem.* **36:**597–607.
Shows that protein G is better for labeling of IgM antibodies, which make up the widest spectrum of monoclonal antibodies.

49. **DeMey, J.** 1984. Colloidal gold as a marker and tracer in light and electron microscopy. *EMSA Bull.* **14:**54–66.
A useful review including some of the most frequently used recipes.

50. **Dürrenberger, M., M.-A. Bjornsti, T. Uetz, J. A. Hobot, and E. Kellenberger.** 1988. Intracellular localization of histone-like protein HU in *Escherichia coli. J. Bacteriol.* **170:**4757–4768.
This is one of the most carefully controlled studies of thin-section labeling dealing with bacteria. See also reference 54.

51. **Dürrenberger, M. B.** 1989. Removal of background label in immunocytochemistry with the apolar Lowicryls by using washed protein A-gold-precoupled antibodies in a one-step procedure. *J. Electron Microsc. Tech.* **11:**109–116.
References 47, 49, and 51 describe some of the problems to be aware of with colloidal gold labeling.

52. **Hayat, M. A. (ed.).** 1991. *Colloidal Gold: Principles, Methods and Applications,* vol. 3. Academic Press, Inc., New York, NY.

53. **Hicks, D., and R. S. Molday.** 1984. Analysis of cell labeling for scanning and transmission electron microscopy, p. 203–219. *In* J. R. Revel, T. Barnard, and G. H. Haggis (ed.), *The Science of Biological Specimen Preparation for Microscopy and Microanalysis.* Scanning Electron Microscopy, Inc., AMF O'Hare, IL.
References 52 and 53 review the methodology used in immunolabeling. As an encompassing book on the topic, reference 50 is up to date.

54. **Hobot, J. A., M.-A. Bjornsti, and E. Kellenberger.** 1987. Use of on-section immunolabeling and cryosubstitution for studies of bacterial DNA distribution. *J. Bacteriol.* **169:**2055–2064.
This goes together with reference 50.

55. **Kellenberger, E., M. Dürrenberger, W. Villiger, E. Carlemalm, and M. Wurtz.** 1987. The efficiency of immunolabel on Lowicryl sections compared to theoretical predictions. *J. Histochem. Cytochem.* **35:**959–965.
A good explanation of why Lowicryl resins are useful for immunolabeling.

56. **Roth, J.** 1986. Post embedding cytochemistry with gold-labeled reagents: a review. *J. Microsc.* **143:**125–137.
Roth and Bendayan are two of the pioneers in immunogold labeling. References 48 and 56 are, then, by the experts.

57. **Slot, J. W., and H. J. Geuze.** 1984. A new method of preparing gold probes for multiple-labeling cytochemistry. *Eur. J. Cell Biol.* **38:**87–93.

58. **Slot, J. W., and H. J. Geuze.** 1991. Sizing of protein A-colloidal gold probes for immunoelectron microscopy. *J. Cell Biol.* **90:**533–536.
Recipes for making and sizing colloidal gold.

59. **Smit, J., and W. J. Todd.** 1986. Colloidal gold labels for immunocytochemical analysis of microbes, p. 469–517. *In* H. C. Aldrich and W. J. Todd (ed.), *Ultrastructure Techniques for Microorganisms.* Plenum Press, New York, NY.
One of the few review chapters which deals with the immunolabeling of microorganisms.

4.9.2.9. Compositional Analysis

Accurate compositional analysis requires exquisitely preserved structure, and cryopreservation is becoming the preferred preparatory method.

60. **Aldrich, H. C.** 1986. X-ray microanalysis, p. 517–525. *In* H. C. Aldrich and W. J. Todd (ed.), *Ultrastructure Techniques for Microorganisms.* Plenum Press, New York, NY.
A chapter dedicated to EDS applied to microorganisms.

61. **Beveridge, T. J., M. N. Hughes, H. Lee, K. T. Leung, R. K. Poole, I. Savvaidis, S. Silver, and J. T. Trevors.** 1996. Metal-microbe interactions. *Adv. Microb. Physiol.* **38:**177–243.
EDS and many other techniques are discussed with an emphasis on metal ions and geomicrobiology.

62. **Budd, P. M., and P. J. Goodhew.** 1988. *Royal Microscopical Society Handbook, no. 16. Light Element Analysis in the Transmission Electron Microscope: WEDS and EELS.* Oxford University Press, New York, NY.

63. **Chang, C.-F., H. Shuman, and A. P. Somlyo.** 1986. Electron probe analysis, X-ray mapping and electron energy-loss spectroscopy of calcium, magnesium, and monovalent

ions in log-phase and in dividing *Escherichia coli* B cells. *J. Bacteriol.* **167**:935–939.
One of the few compositional studies of the natural electrolyte concentrations in E. coli. It applies EDS and EELS with a field emission electron source.

64. **Johnstone, K., D. J. Ellar, and T. C. Appleton.** 1980. Location of metal ions in *Bacillus megaterium* spores by high-resolution electron probe X-ray microanalysis. *FEMS Microbiol. Lett.* **7**:97–101.
One of the first EDS studies applied to bacteria. In this case, endospores are analyzed.

65. **Morgan, J. A.** 1985. *Royal Microscopical Society Handbook*, no. 15. *X-Ray Microanalysis in Electron Microscopy for Biologists.* Oxford University Press, New York, NY.
References 8, 60, 61, 65, and 67 have good overviews of the topic. The physics and theory are emphasized in references 65 and 67, but references 8, 60, and 61 may be better to start with.

66. **Ottensmeyer, F. P.** 1984. Electron energy loss analysis and imaging in biology, p. 340–343. *In* G. W. Bailey (ed.), *Proceedings of the 42nd Annual Meeting of the Electron Microscopy Society of America.* San Francisco Press, Inc., San Francisco, CA.
A short synopsis of ESI by its pioneering scientist.

67. **Sigee, D. C., J. Morgan, A. T. Sumner, and A. Warley.** 1993. *X-Ray Microanalysis in Biology.* Cambridge University Press, New York, NY.

68. **Stewart, M., A. P. Somlyo, A. V. Somlyo, H. Shuman, J. A. Lindsay, and W. G. Murrell.** 1980. Distribution of calcium and other elements in cryosectioned *Bacillus cereus* T spores, determined by high-resolution scanning electron probe X-ray microanalysis. *J. Bacteriol.* **143**:481–491.

69. **Stewart, M., A. P. Somlyo, A. V. Somlyo, H. Shuman, J. A. Lindsay, and W. G. Murrell.** 1981. Scanning electron probe X-ray microanalysis of elemental distribution in freeze-dried cryosections of *Bacillus coagulans* spores. *J. Bacteriol.* **147**:670–674.
References 68 and 69 are two meticulous compositional studies on bacterial endospores. The researchers used frozen sections to point map the elements within sectioned spores by EDS.

4.9.2.10. Image Processing

Chapter 5 is dedicated to image processing and shows the power of the technique. For this reason, no references will be given here. Image processing has been used to clarify several bacterial structures such as ribosomes, pili, flagella, spinae, type III secretion systems, and S-layers. The latter have been the most studied because of their paracrystallinity and planar symmetry, and three-dimensional structure is available for several arrays. Also see reference 2.

4.9.2.11. SEM

70. **Hayat, M. A. (ed.).** 1974. *Principles and Techniques of Scanning Electron Microscopy. Biological Applications.* Van Nostrand Reinhold Co., New York, NY.
A multivolume set, full of good information, that was started in 1974. See volumes 1, 3, and 6 for SEM of spores, microorganisms, and bacteriophages.

71. **Joy, D. C.** 1984. Resolution in the low-voltage SEM, p. 444–449. *In* G. W. Bailey (ed.), *Proceedings of the 42nd Annual Meeting of the Electron Microscopy Society of America.* San Francisco Press, Inc., San Francisco, CA.
Describes the advantages of using low-voltage SEM. Also see the article "Low Voltage SEM," by J. Pawley, which immediately precedes this one.

72. **Passmore, S. M., and B. Bole.** 1976. Scanning electron microscopy of microbial colonies, p. 19–29. *In* R. Fuller and D. W. Lovelock (ed.), *Microbial Ultrastructure. The*

Use of the Electron Microscope. Academic Press, Inc., New York, NY.
An old reference but still useful.

73. **Yoshii, A., J. Tokumaga, and J. Tawara.** 1975. *Atlas of Scanning Electron Microscopy in Microbiology.* Igaku Shoin, Ltd., Tokyo, Japan.

4.9.2.12. STM and AFM

STM and AFM are the newest types of microscopies and have revealed startling atomic detail of surfaces and molecular detail of certain organic films. They are becoming more and more used in microbiology for both imaging and exploration of the mechanical properties of microorganisms. For this reason, chapter 6 is dedicated to so-called AFM, which is a type of SPM, and no references will be given here.

4.9.3. Specific References on Interpretation

It is all very well to read about and learn the many techniques used in EM and to process bacteria for ultrastructural analysis, but it is an additional problem to interpret the images produced by the techniques. It is best to consult with local EM experts, but remember that most biological electron microscopists deal with eucaryotic cells! Procaryotic structuralists around the world are glad to help, so send good micrographs to them and seek advice. In the meantime, here is a list of articles to help in interpretation. Many are older articles, but they have withstood the test of time quite well.

74. **Bayer, M. E.** 1991. Zones of membrane adhesion in the cryofixed envelope of *Escherichia coli. J. Struct. Biol.* **107**:268–280.
Adhesion zones, or "Bayer's patches," are thought to be zones where secretory proteins and outer membrane constituents are trafficked. Here is an article by the researcher who first discovered them.

75. **Bayer, M. E., and M. H. Bayer.** 1994. Periplasm, p. 447–464. *In* J.-M. Ghuysen and R. Hakenbeck (ed.), *Bacterial Cell Wall.* Elsevier Science B. V., Amsterdam, The Netherlands.
Many researchers would suggest that this often neglected cell wall constituent, the periplasm, is amongst the most important regions of the cell. Also see references 79, 83, 85, 88, 89, and 91.

76. **Bayer, M. E., and M. H. Bayer.** 1994. Biophysical and structural aspects of the bacterial capsule. *ASM News* **60**:192–198.
Capsules are amongst the most difficult structures to preserve because most of their mass is water. This reference gives a nice general overview of them.

77. **Beveridge, T. J.** 1981. Ultrastructure, chemistry, and function of the bacterial wall. *Int. Rev. Cytol.* **72**:229–317.
A review that correlates bacterial surface chemistry with a structural perspective and discusses what it means in a functional context. Somewhat dated but still worthwhile.

78. **Beveridge, T. J.** 1989. The structure of bacteria, p. 1–65. *In* E. R. Leadbetter and J. S. Poindexter (ed.), *Bacteria in Nature: a Treatise on the Interaction of Bacteria and Their Habitats*, vol. 3. Plenum Press, New York, NY.
One of the most comprehensive compilations of information on bacterial structure, it is useful reading before interpreting initial EM images.

79. **Beveridge, T. J.** 1999. Structures of gram-negative cell walls and their derived membrane vesicles. *J. Bacteriol.* **181**:4725–4733.

80. **Beveridge, T. J.** 1999. The ultrastructure of Gram-positive cell walls, p. 3–10. *In* V. Fischetti, R. Novick, J. Ferretti, D. Potnoy, and J. Rood (ed.), *Gram-Positve Pathogens.* ASM Press, Washington, DC.

81. **Beveridge, T. J.** 1999. Bacterial cells. *In* R. Atlas (ed.), *Encyclopedia of Life Sciences.* [Online.] Macmillan

References Ltd., London, United Kingdom. http://www.els.net.

82. **Beveridge, T. J.** 1999. Archaeal cells. *In* R. Atlas (ed.), *Encyclopedia of Life Sciences.* [Online.] Macmillan References Ltd., London, United Kingdom. http://www.els.net.

83. **Beveridge, T. J.** 1999. Bacterial cell wall. *In* R. Atlas (ed.), *Encyclopedia of Life Sciences.* [Online.] Macmillan References Ltd., London, United Kingdom. http://www.els.net.
If your university subscribes to the Encyclopedia of Life Sciences, *references 81 to 83 should be readily accessible to you and are a good starting point for differentiating bacteria from archaea and for getting an understanding of cell wall structure. Many other good, up-to-date microbiology topics are also covered; in fact, ~3,000 of them will be covered once the online site is completed!*

84. **Beveridge, T. J., and J. W. Costerton (ed.).** 1988. Shape and form. *Can. J. Microbiol.* **34:**363–420.
A compilation of short reviews by experts in the field of bacterial structure. The topics range from structural design strategies to growth and taxis. This compilation occurs in a special edition of the journal dedicated to R. G. E. Murray, who is one of the masters of bacterial structure.

85. **Beveridge, T. J., and L. L. Graham.** 1991. Surface layers of bacteria. *Microbiol. Rev.* **55:**684–705.
This is a review of the modern perception of the structure of bacterial surfaces. Cryotechniques, diffraction, image processing, immunogold labeling, and probes for charge are all covered.

86. **Bohrmann, B., W. Villiger, R. Johansen, and E. Kellenberger.** 1991. Coralline shape of the bacterial nucleoid after cryofixation. *J. Bacteriol.* **173:**3149–3158.
A wonderful blend of light microscopy and cryoTEM is used to define the shape of the bacterial nucleoid.

87. **Costerton, J. W.** 1979. The role of electron microscopy in the elucidation of bacterial structure and function. *Annu. Rev. Microbiol.* **33:**459–479.
A good overview which has stood the test of time.

88. **Graham, L. L., T. J. Beveridge, and N. Nanninga.** 1991. Periplasmic space and the concept of the periplasm. *Trends Biochem. Sci.* **16:**328–329.

89. **Graham, L. L., R. Harris, W. Villiger, and T. J. Beveridge.** 1991. Freeze-substitution of gram-negative eubacteria: general cell morphology and envelope profiles. *J. Bacteriol.* **173:**1623–1633.

90. **Hale, C. A., A. C. Rhee and P. A. De Boer.** 2000. Zip A-induced bundling of FtsZ polymers mediated by an interaction between C-terminal domains. *J. Bacteriol.* **182:**5153–5166.
There is increasing interest in the division machinery in gram-negative cells and the way in which various proteins form a primative cytoskeletal complex at the constriction site. Here is a recent reference that combines TEM and fluorescent light microscopy.

91. **Hobot, J. A., E. Carlemalm, W. Villiger, and E. Kellenberger.** 1984. Periplasmic gel: a new concept resulting from the reinvestigation of bacterial cell envelope ultrastructure by new methods. *J. Bacteriol.* **160:**143–152.
References 79, 85, 88, and 91 redefine the concept of the periplasm and the periplasmic space and show the impact freeze-substitution has had on bacterial structure.

92. **Jones, C. J., and S.-I. Aizawa.** 1991. The bacterial flagellum and flagellar motor: structure, assembly and function. *Adv. Microb. Physiol.* **32:**109–172.
There are more-recent reviews on flagella but this one combines TEM, biophysics, and molecular biology and is by a Japanese group that has performed many good structural studies on this fascinating motility organelle.

93. **Krell, P. J., and T. J. Beveridge.** 1987. The structure of bacteria and molecular biology of viruses. *Int. Rev. Cytol. Suppl.* **17:**15–88.
An overview of procaryotic and viral structures which was aimed at advanced university undergraduate studies.

94. **Nanninga, N.** 1998. Morphogenesis of *Escherichia coli. Microbiol. Mol. Biol. Rev.* **62:**110–129.
This is a review that encompasses most gram-negative structures and describes their functionality while at the same time contemplating what we know and what we still need to know. All students should read this one.

95. **Remsen, C. C.** 1982. Structural attributes of membranous organelles in bacteria. *Int. Rev. Cytol.* **76:**195–223.
Like reference 77, this one is rather old but has textbook structural information in it.

96. **Robinow, C., and E. Kellenberger.** 1994. The bacterial nucleoid revisited. *Microbiol. Rev.* **58:**211–232.
Our list of references would not be complete without one from these two experts Robinow and Kellenberger, who, together, have done more than any other researcher to describe and define that structure that clearly separates procaryotes from eucaryotes, the nucleoid. Light microscopy and TEM at their best are described in this review.

97. **Sleytr, U. B., and T. J. Beveridge.** 1999. Bacterial S-layers. *Trends Microbiol.* **7:**253–260.
A recent review by two researchers who have marveled at S-layers for over 30 years; it is dedicated to the expansive structural literature on procaryotic S-layers.

98. **Stoltz, J. F. (ed.).** 1991. *Structure of Phototrophic Prokaryotes.* CRC Press, Inc., Boca Raton, FL.
A review of the structure of cyanobacteria, the purple and green bacteria, and their internal, photosynthetic membranes.

99. **van Iterson, W. (ed.).** 1984. *Benchmark Papers in Microbiology,* vol. 17. *Inner Structures of Bacteria.* Van Nostrand Reinhold Co., Inc., New York, NY.

100. **van Iterson, W. (ed.).** 1984. *Benchmark Papers in Microbiology,* vol. 18. *Outer Structures of Bacteria.* Van Nostrand Reinhold Co., Inc., New York, NY.
References 99 and 100 are two wonderful volumes in which van Iterson has compiled the significant papers showing the progressive advance of the understanding of bacterial ultrastructure. Only the significant features of each paper are printed, and these are followed by editorial comments by an experienced microscopist.

5

Computational Image Analysis and Reconstruction from Transmission Electron Micrographs

GEORGE HARAUZ

Personal computers have become an essential fixture of every office and laboratory. Almost everyone has had some familiarity with using a computer to send messages or analyze and display laboratory data, including image data, at one point or another. It has become commonplace to render and submit figures for publication electronically. In this chapter, some specialized applications are described in which computers are essential for manipulating and enhancing images obtained by using a transmission electron microscope, particularly to understand the tertiary and quaternary structures of biological macromolecules and their complexes. In order to be able to employ these approaches fruitfully, a certain facility and intuition for mathematics are required. More commonly, one would enlist the assistance of a specialist with the appropriate skills and computational equipment. Thus, it is not possible, in a single chapter, to present a "how to do it" scheme of even selected aspects of specimen preparation, transmission electron microscopy (TEM) imaging, or computational analysis of the images. Instead, the conceptual principles of some techniques that are commonly encountered in the literature are outlined in a simplified way in order to help readers with a biological background understand them. A glossary at the end (section 5.10) redefines terms in boldface.

This chapter must be read with some knowledge of the basic principles of TEM of biological material. Chapter 4 is a great place to start, and there are numerous excellent monographs that are recommended for reference on this subject. The most recent edition of the book by Hayat (6) gives a good encyclopedic overview of conventional techniques for preparation of biological specimens for TEM, viz., thin sectioning. Of relevance here are its two chapters summarizing the variety and properties of negative and pos-

itive stains, and an earlier book by Hayat and Miller (7) is also recommended in this regard. Of course, a TEM facility should also have references with alternative protocols, e.g., those by Nasser Hajibagheri (10) and Hoppert and Holzenburg (8), the latter being aimed specifically at microbiologists. Researchers who are interested in TEM of biological macromolecular specimens per se should consult Harris (5) and Sommerville and Scheer (12). Finally, the computational techniques for analyzing and reconstructing macromolecular structures from TEM images, the subject of this chapter, are reviewed in monographs by Frank (1, 2), Häder (4), and Misell (9). Perhaps the best source for an introduction to computational image analysis is the opus magnum by Russ (11). It is richly illustrated with examples of electron micrographs of specimens from the biological and materials sciences and is accompanied by a general purpose software package called the image-processing tool kit.

5.1. MACROMOLECULAR STRUCTURE DETERMINATION BY TEM

One of the fundamental tenets of molecular biology and biochemistry is that the structure of a macromolecule, such as a protein, determines its function. The two most commonly known methods of determining the structures of soluble proteins are X-ray crystallography (diffractometry) and nuclear magnetic resonance (NMR) (3). However, neither method is universally applicable. The former approach requires three-dimensional (3-D) crystals of suitable size (~0.5 mm or larger) and high regularity, and this condition often represents a bottleneck in structural analyses. NMR is generally applicable to soluble proteins of <20 kDa, which must be highly concentrated and enriched with isotopes such as ^{13}C and ^{15}N. This goal is achieved by expressing the proteins in recombinant form in bacteria grown in minimal media with, say, [^{13}C]glucose as the sole carbon source. In contrast, TEM represents a *direct* approach to visualizing macromolecules and their modes of assembly and thus serves as an alternative or complement to X-ray crystallography or NMR (3). The resolutions of structures thus obtained are usually in an "intermediate" range (1 to 5 nm), but "high" resolution (0.3 to 1 nm) has been achieved with certain proteins arranged in highly ordered 2-D arrays (i.e., 2-D crystals) (21, 22). In general, though, TEM is an excellent way to investigate the quaternary or domain arrangement of fairly large macromolecules greater than about 10 nm in size. (Sometimes this is called blobology.)

One significant advantage of TEM of biological macromolecules is that only small amounts of material are required, about 1/1,000 of that needed for other biophysical approaches. The sample must be as biochemically homogeneous as possible and can be prepared for imaging in one of many different ways. Generally, the specimen support is a standard TEM grid that is coated with a layer of fenestrated (or holey) plastic, which itself supports an ultrathin carbon film (2 to 5 nm thick) to which the purified macromolecules are allowed to adsorb. Contrast is achieved by negative staining, usually with 2% (wt/vol) uranyl acetate, although one should always experiment with other stains as well (see chapter 4). If the macromolecule or complex is large enough (>10 nm in diameter), it can be imaged in frozen-hydrated form. Then the specimen support is merely the fenestrated plastic film, whose holes contain a thin layer of sample. The grid is rapidly frozen by plunging into liquid propane or ethane maintained at liquid-nitrogen temperatures so that vitreous (noncrystalline) ice is formed to trap the macromolecules in their native state. Stains are not used, and contrast is achieved instrumentally by varying the degree of focus (5, 9, 23).

In any structural biology project, there are limitations to TEM which must be dealt with for high-resolution work. There are potential specimen preparation artifacts, especially due to dehydration- and/or radiation-induced structural alterations. There are contrast limitations necessitating the use of heavy atom salts or metal shadowing. There are resolution limitations due to stain and image "noise." Moreover, TEM images are only 2-D and the structure being investigated is 3-D. These points will all be illustrated below. The driving force behind many recent advances in TEM technology has been to solve problems such as these. The technique of cryoTEM allows one to image freeze-dried or frozen-hydrated specimens trapped in their native state. One routinely performs low-electron-dose TEM, meaning that the specimen thus irradiated only during actual image formation, not during focusing which is done on an adjacent region. There are various "flavors" of TEM involving the use of combinations of elastically and inelastically scattered electrons to enhance contrast without the need for stains and/or high-resolution elemental mapping by either dark-field TEM, scanning TEM, or electron spectroscopic imaging (chapter 4). One can note here that scanning probe microscopies (including atomic force microscopy) are also viable and fruitful imaging techniques (chapter 6). All of these considerations are important to TEM of biological macromolecules in general and dictate requirements for specialized instruments that might not necessarily be found in every center or university. For present purposes, suffice it to say that one should strive to obtain the best possible TEM image of the biological macromolecule that one can. In these days when computer technology is changing rapidly, and when any computer purchased is soon out of date, the only truism is the principle of garbage in, garbage out. Get good data.

5.2. THE NECESSITY FOR COMPUTATIONAL IMAGE ANALYSIS

Once good micrographs are available, then computerized image analysis can be used to ameliorate some inherent limitations of TEM of biological macromolecules. The first limitation is that the images are fairly noisy in a visual sense, thus limiting the resolution of structural detail that can be interpreted. The sources of the noise include the uneven background of the carbon support or ice film, the irregular distribution of negative stain, and radiation-induced structural alterations of the structure under investigation. Although experimental approaches such as low-dose imaging are designed to minimize noise, it can never be entirely eliminated. Thus, one of the major themes of computational processing and analysis of macromolecular images is that of increasing the signal-to-noise ratio by averaging similar images of different individual complexes. This process of forming "characteristic projections" with an enhanced signal-to-noise ratio and better resolution of structural detail will be illustrated below. Secondly, the TEM image represents a 2-D *projection* of a 3-D object. This concept will also be exemplified later. In order to get the 3-D structure, one requires a large number of different views (i.e., from different orientations) of the object. Computers are then essential for determination of the relative angular orientations of different projections and for reconstructing an image of the 3-D structure from the 2-D data that are available.

For purposes of illustration in this chapter, two complementary approaches to analyzing macromolecular images have been chosen. The first is called **single-particle analysis**, where the purified protein or complex is imaged as isolated single particles in random orientations on a carbon film or in vitreous ice by low-electron-dose TEM. Two good recent reviews of single-particle analysis for asymmetric complexes are those by Czarnota et al. (14) and Radermacher (25). The single-particle approach is the method of choice for probing the structures of highly symmetric structures such as icosahedral viruses (31). Usually, resolution achievements can approach 1 nm in 3-D for asymmetric molecules and even better for symmetric ones such as viral capsids. It is thought that there is no inherent limitation in the technique to achieving even atomic resolution (33).

The second approach that will be briefly described is **2-D electron crystallography**, which can be applied when the specimen is arranged in the form of a 2-D crystal (e.g., a monolayer of protein). Again, the best-quality TEM images are obtained, and crystallographic computational approaches can be used to form a projection map of the unit cell at generally higher resolution than is achievable by single-particle analysis. Combining projection maps at different orientations yields the 3-D structure; reference 32 provides a nice schematic illustration of this process. The computational procedures for electron crystallography were originally developed with specimens such as bacterial surface arrays (e.g., S-layers). Instructive reviews with microbiological samples include references 9, 16, 29, and 32. Electron crystallography is increasingly being applied to determine the native 3-D structures of numerous membrane proteins in situ (20, 35); secondary structure elements can generally be defined. Indeed, atomic resolution can also be obtained, as first achieved for an archaeal bacteriorhodopsin (21) and shortly thereafter for other proteins such as bacterial porins (22).

5.3. COMPUTER AND SOFTWARE REQUIREMENTS

As indicated above, personal computers are an essential component of any laboratory, and few researchers work without them. In the computer's memory, an "image" is simply a large set of numbers indicating degrees of darkness and lightness. Any point in a digital image is called a pixel, meaning picture element, and has a certain size (with respect to the object) associated with it. For high-resolution work, one requires fine sampling, i.e., pixels spanning a small distance. Modern transmission electron microscopes can capture images digitally directly, or special photographic film can be scanned at high resolution. In popular computing, one hears of images being scanned at a certain "dpi", meaning "dots per inch." Scanners with resolutions of 600 dpi or better are commonplace and inexpensive. It is more appropriate here to think of scanning TEM film with pixel sizes of about 5 μm, at which point the size of the silver grains of the emulsion becomes the limiting factor. With respect to an image of a protein complex, pixel sizes should be no larger than about 0.5 nm at the object level. (To convert pixel sizes from the object level to the micrograph level, one simply needs to use the magnification factor.) Thus, the new 4,000 dpi scanners would have a pixel size of 6.35 μm, more appropriate for our purposes. One important point to note is that TEM image data are *not in color*, but instead are represented by *shades of gray*, also called gray levels, i.e., numbers ranging from 0 (dark) to 4,095 (light), say. Image data require substantial memory for programs to be able to manipulate them quickly (of the order of many hundreds of megabytes), significant hard drive space for storage (of the order of gigabytes), and a good-quality display that can render many colors. The more powerful the computer, the better.

Although there is a wide range of general software for manipulating image data, the specialized software that is directly applicable to TEM images is more limited (see section 5.9; italic letters in parentheses refer to items listed there). Some examples are Northern Eclipse 6.0 (*a*) and Scion Image (*b*), which run on personal computers with a recent version of the Windows operating system. The Scion program is adapted from the NIH *Image* software package that runs on Macintoshes (*c*), and both packages can be downloaded for free. The suites of computer programs for high-resolution analysis of macromolecular structure are quite complex, however, and have been developed on UNIX workstations. Two common packages are IMAGIC-5 (*d*) (34) and SPIDER (*e*) (18) and comprise both single-particle analysis and 2-D electron crystallography. All of the work presented here was performed by using IMAGIC-5. A new package called EMAN (*f*) (24) is based partly on IMAGIC-5 and focuses on single-particle analysis, but it is noteworthy because it is downloadable for free and provides extensive guidance for novice users. The SEMPER system is partly of historical importance because many techniques for analysis of TEM images of bacterial S-layers were developed by using it, but it remains current and easy to use and is of very general applicability, and a modern version is now available that can run on almost any computer platform (*g*) (27). Finally, structural molecular biologists are ultimately interested in atomic models of their proteins. A powerful suite of programs for manipulating molecular models and

incorporating them into 3-D reconstructions from TEM data is Insight 2000 (*h*). One should always be careful not to use any computer program as a "black box" without understanding some of the underlying principles behind its development.

5.4. WHAT IS SINGLE-PARTICLE ANALYSIS?

It is best first to describe what is meant by single-particle analysis. In Fig. 1, a hypothetical macromolecular complex is shown. This structure consists of 12 spheres (representing the protein subunits) arranged in what is called D6 symmetry (19, 26). One way of representing this 3-D structure is by shaded surface representations as shown in Fig. 1a. Here, we get a visual impression of the macromolecular complex as a "real" 3-D object from the depth cues given by the simulated light source. This sort of representation is common in computer graphics, especially in the entertainment industry. However, the representation in Fig. 1b is that of 2-D *projections*, which is what we get in a TEM image. When the macromolecular complex is imaged from a particular direction, its mass at each point along that direction is summed (figuratively speaking) and recorded on the micrograph. The degree of darkness on the image at any point is proportional to the amount of macromolecular mass above that point. A good analogy is that of a chest X ray—the details of both front and back ribs, and all the organs in between, are *projected* onto a 2-D film and appear superimposed.

As shown by the various images in Fig. 1 (remember that these are all of the *same* 3-D object), if the macromolecular complex lies in a certain way on its support in the transmission electron microscope, then the electron beam will impinge on it from a certain direction. In negatively stained preparations, this complex will tend to lie on the carbon support film on its flat side. Thus, we would usually see the top-down views of the structure, and other projection views such as side ones would occur less frequently. In frozen-hydrated preparations where the sample is effectively being imaged in a solution, a far greater variety of orientations is

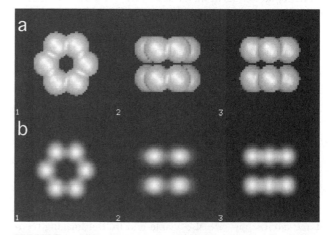

FIGURE 1 Simulated macromolecular complex, consisting of 12 spheres arranged in D6 symmetry, i.e., with one sixfold rotational axis of symmetry (the top-down view) and two twofold rotational axes of symmetry (the side views). In panel a, the 3-D structure is depicted via shaded surface representations. Here, the brighter regions act as a depth cue. In panel b, 2-D projections of the 3-D structure are shown, in which brighter regions represent a greater amount of molecular mass.

seen. It is far easier to prepare macromolecules for TEM by negative staining, and this method should be used first before proceeding to cryoTEM of vitreous ice-embedded material.

The effects of noise on the image are depicted in Fig. 2. Here, we have added randomly generated numbers with a Gaussian distribution to the gray level of each pixel of a set of images of the macromolecular complex lying in the top-down orientation. The simulated molecules are obscured by the noise, as is the case in TEM, and it is difficult to assess by visual inspection which projection view is represented. For this reason, computational single-particle analysis was developed. Consider that each of the images in Fig. 2 represents a 2-D top-down projection of a different individual macromolecular complex. However, since the preparation was biochemically pure to begin with, and since each individual macromolecular complex has the same 3-D structure, each image in Fig. 2 would be identical if they were noise free.

The computer programs that have been written for single-particle analysis do a number of complicated tasks. First of all, they select automatically (or allow one to select interactively) the images of different macromolecular complexes from a transmission electron micrograph that contains views of many of them. Then the programs align each macromolecular image with respect to a reference to facilitate comparison and sort out into smaller subgroups those that represent the same projection (the characteristic projections mentioned above). The sorting (sometimes called clustering) is done via a multivariate statistical analysis. Finally, we average together the images of macromolecular complexes viewed in the same orientation. (This process is simple arithmetic.) Since the macromolecular structural information is constant, it reinforces itself, whereas the noise, which is random, tends to cancel itself out. The effect of averaging is shown in Fig. 3. With increasing numbers of images being averaged, the final image of the macromolecular complex becomes clearer. In other words, the signal-to-noise ratio is increased.

In single-particle analysis, minimally hundreds and preferably thousands of images of a macromolecular complex are required in order to be able to get large enough subgroups representing the same 2-D projection for averaging. When a number of averages of different 2-D projections are obtained,

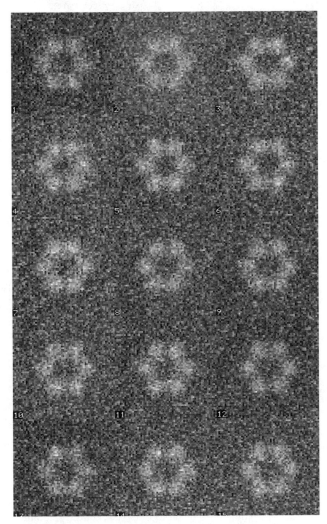

FIGURE 2 In transmission electron micrographs of biological macromolecules, noise obscures structural detail severely. Here, the top-down 2-D projection from Fig. 1b has been duplicated a number of times to represent different individual macromolecular complexes lying in the same orientation on a support film. Random noise has been added to each point (pixel) in each image. Thus, the numbered images portrayed here are all different because the noise is different in each, even though the complex's structural information is the same.

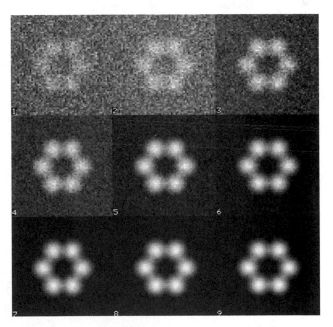

FIGURE 3 When noisy images like those in Fig. 2 are averaged together, the signal-to-noise ratio increases roughly according to the square root of the number of images being averaged. The constant structural information in each image is reinforced, whereas the random noise cancels itself out. The number of images being averaged in each panel is as follows: panel 1, 1; panel 2, 2; panel 3, 9; panel 4, 25; panel 5, 100; panel 6, 225; panel 7, 400; panel 8, 625; and panel 9, 900. Each image must represent the same 2-D orientation and the macromolecule must be in the same place in each image in order for the average to make sense. Thus, in single-particle analysis, a large population of images of different macromolecular complexes is aligned in order to facilitate comparison and sorted into homogeneous subgroups for averaging. Details of these processes are found in references 2, 14, 25, and 33.

they can be used to reconstruct a 3-D image of the structure. To return to the chest X-ray analogy, in order to see how the organs and bones in the chest are really arranged with respect to one another, one needs an X ray from front to back, side to side, and all the oblique views in between. In fact, this angular sampling is what medical X-ray computer-assisted tomography (CAT scanning) achieves.

In single-particle analysis of TEM images, we do not know a priori how the different views are related to one another—remember, we do not know the macromolecule's 3-

D structure yet. The methods for determining the relative angles amongst different projections are mathematically quite complex; some approaches for asymmetric objects are described in references 14, 25, and 33, and the strategy used for icosahedral viruses is reviewed in reference 31. Once the angles relating different projections are known, then a **3-D reconstruction from projections** can be performed. The computational principles of reconstruction from projections are depicted in Fig. 4 for a simple example and are used in many fields such as medicine as well as in structural biology.

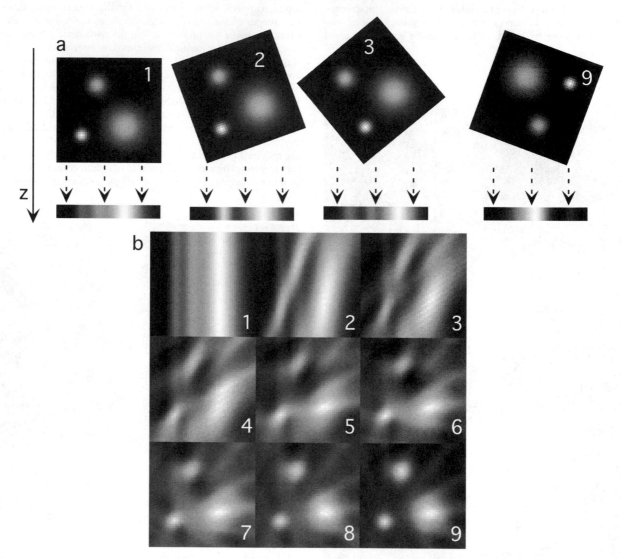

FIGURE 4 Principles of reconstruction from projections. (a) In this example, a 2-D image is projected into a 1-D image by summing all the density values along a line at a certain angle. (b) The process of reconstruction by back projection involves smearing back the densities of each 1-D projection along the direction in which the projection was collected into a blank 2-D image and adding together all the back projections possible. Here, the reconstructions from 1, 2, 3, ..., 9 different 1-D projections (i.e., collected at different angles of 0, 20, 40, ..., 160°) are shown. In panel b, image 1, the density from each point in a 1-D projection was back projected at an angle of 0° into the 2-D image that will form the reconstruction. The angle 0° is the one at which projection 1 was collected. In panel b, image 2, the 1-D density data are back projected into the 2-D image at the angle 20° associated with this projection and added to the data back projected from projection 1. On it goes, with images 3 and 4, etc., showing the cumulative results of back projecting density data from different angles and adding them to the previously back-projected information. A reasonable representation of the original object, a collection of three disks, is already obtained from nine projections. Reconstruction quality improves with increasing numbers of input projections that span all possible views. The reconstruction approach shown here is called tomography, meaning that 2-D slices of a 3-D object are reconstructed. Tomography requires a specialized geometry in which the 3-D object is tilted about an axis by a certain angle, e.g., using a tilt stage in a transmission electron microscope. In single-particle analysis, such a geometry generally does not apply and 3-D volume images are reconstructed directly (i.e., not slice by slice) from 2-D projection images.

5.5. SINGLE-PARTICLE ANALYSIS OF A MULTISUBUNIT ENZYME COMPLEX: A PHOSPHOENOLPYRUVATE SYNTHASE FROM THE ARCHAEON *STAPHYLOTHERMUS MARINUS*

The deep-sea archaeon *Staphylothermus marinus* possesses a phosphoenolpyruvate synthase that is unusual in that it is a large protein complex of molecular mass 2.25 MDa, comprising 24 identical subunits (13, 23). (The corresponding enzyme from mesophilic eubacteria is a simpler dimer.) Thus, it was of interest to find out how this enzyme was arranged, and its size (about 20 nm in diameter) made it suitable for TEM investigation. In Fig. 5, a transmission electron micrograph of a negatively stained preparation of the enzyme lying on a carbon support is shown, along with a set of individual smaller images of different complexes extracted from the larger micrographs. The latter images were the input to the programs used for single-particle analysis.

Two transmission electron micrographs of a frozen-hydrated preparation (i.e., embedded in vitreous ice) are

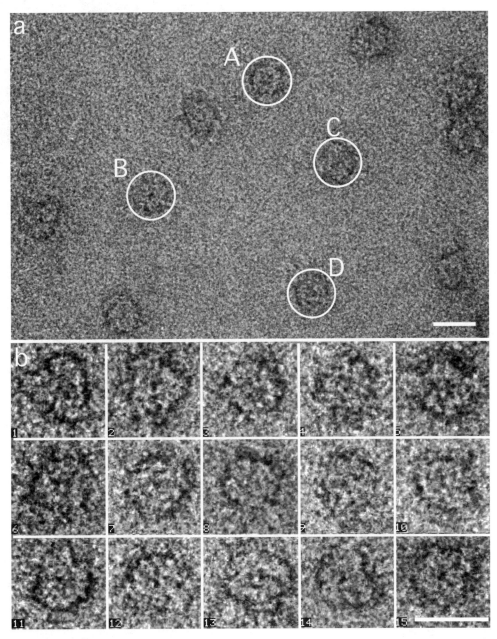

FIGURE 5 Homogeneous preparation of the 2.25-MDa phosphoenolpyruvate synthase from *S. marinus*, adsorbed to a thin carbon support, negatively stained with 2% (wt/vol) uranyl acetate to provide contrast, and imaged by TEM. (a) Field of view showing a number of different individual complexes. The scale bar represents 20 nm. (b) A set of smaller images of different individual complexes, extracted from larger micrographs like those in panel a. The scale bar represents 20 nm. These smaller images will be processed by computer programs that perform single-particle analysis to align, sort, average, and reconstruct them. In these images, the lighter regions represent proteinaceous material whereas the darker areas represent negative stain that embeds it. Some individual complexes are circled and denoted by A, B, C, and D.

shown in Fig. 6. Here, there is neither stain nor an underlying carbon support, and the individual complexes can be seen partly because of a phase-contrast effect. The images in Fig. 6 are of the same region of the specimen grid but differ in that they were recorded at different levels of defocus, i.e., at levels away from the true focus of the instrument. The concept of focus is somewhat less intuitive in TEM than in light microscopy; biological electron microscopists generally defocus (i.e., underfocus) to get a better-looking image. In cryoTEM, one achieves higher contrast in terms of ability to distinguish the macromolecules within the ice by defocusing, but at the cost of resolution of detail. In Fig. 6, one image is slightly blurrier than the other, but the macromolecular complexes stand out a bit better. As in Fig. 5, single macromolecular images were selected from the larger micrographs and subjected to single-particle analysis. Moreover, the micrographs of the frozen-hydrated preparations at different defocus levels were combined via an approach called **contrast transfer function correction** to maximize the amount of structural information obtained (23). This process is very difficult to explain without using mathematics. Suffice it to say that structural detail of different sizes is not recorded with the same efficiency at different defocus levels. We can predict theoretically how the blurring occurs at each level. Given a pair of (or even more) images recorded at different defocus levels, we can extract the best information from each micrograph and put complementary data together to yield the least blurry image possible.

Partial results of single-particle analyses of both frozen-hydrated and negatively stained preparations are shown in Fig. 7. The image averages in Fig. 7a to c show projection views (amongst many others obtained) of the frozen-hydrated macromolecular complex along axes that portray twofold, threefold, and fourfold rotational symmetry. The image average in Fig. 7d shows the projection view of the negatively stained macromolecular complex along its threefold rotational symmetry axis, its most stable orientation in this kind of preparation. For all preparations available in this study (23), projection views at different orientations such as those in Fig. 7 were then combined to give 3-D reconstructions. In Fig. 8, surface representations of various 3-D reconstructions of the complex are depicted, showing the underlying octahedral arrangement of this 24-meric complex and the overall consistency (yet subtle differences) obtained from different TEM preparations. The resolution of structural detail was better with the frozen-hydrated-preparation data, as is usually the case. Finally, an atomic model of the enzyme was generated and 24 subunits were fit into the 3-D reconstruction obtained by single-particle analysis, as shown in Color Plate 8. This exercise provided some useful insights into how this enzyme complex achieved structural stability under extreme conditions of temperature (23). Presently, single-particle analysis in conjunction with cryoTEM is an established tool for probing the structures of numerous macromolecular complexes, and its greatest notable recent success has been the elucidation of the architecture of prokaryotic ribosomal subunits (reviewed in references 17 and 33).

5.6. WHAT IS ELECTRON CRYSTALLOGRAPHY?

In electron crystallography, the macromolecule of interest is already arranged in the form of a 2-D array or crystal, one protein thick. In the simple example shown in Fig. 9a, each repeating motif in the 2-D crystal is a top-down view of the

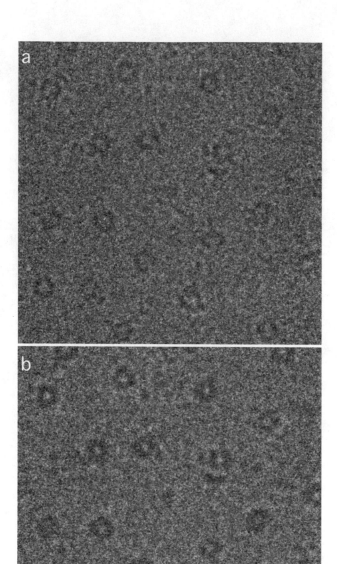

FIGURE 6 Homogeneous preparation of the 2.25-MDa phosphoenolpyruvate synthase from *S. marinus*, encased in a thin layer of vitreous ice and thereby trapped in a native, hydrated state. The sample was kept at liquid-nitrogen temperature in the transmission electron microscope during imaging. Panels a and b represent the same region of the specimen, but the images were recorded at different defocus values. The image in panel a is closer to the true focus position of the microscope than the one in panel b, and is somewhat crisper but shows less contrast. This effect is due to the contrast transfer function, which changes with defocus (5, 9, 23). Single-particle analyses of such frozen-hydrated material also incorporate contrast transfer function correction (e.g., reference 23). In these cryoTEM images, the darker regions represent the proteins. These images, like those in Fig. 5, were captured in digital form directly from the cryo-transmission electron microscope at a pixel size of 0.344 nm at the object level (23). Each panel in this figure is 600 by 600 pixels in size and thus spans a region on the specimen that is 206.4 nm in extent.

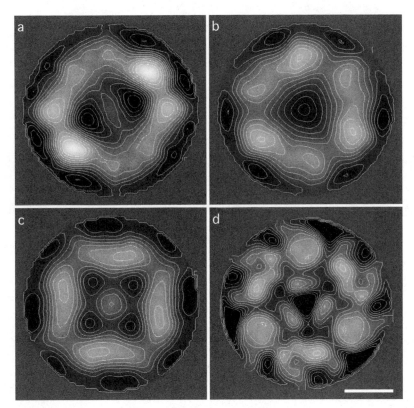

FIGURE 7   In cryoTEM images of frozen-hydrated preparations, biological macromolecules can exhibit a greater variety of orientations than those adsorbed to a support film. Single-particle analyses of contrast transfer function-corrected cryo-transmission electron micrographs (as those in Fig. 6) of the S. *marinus* phosphoenolpyruvate synthase yield projection averages viewed along twofold (a), threefold (b), and fourfold (c) rotational symmetry axes, consistent with an underlying octahedral shape. Single-particle analyses of TEM images of negatively stained preparations (as those in Fig. 5) yield primarily the threefold rotationally symmetric view (d) because this is the stablest way in which this complex can lie on the support film. In all of these images, contrast is such that lighter regions represent proteinaceous material. The scale bar represents 5 nm for all panels. Panels a through c are reproduced from reference 23 by permission of Academic Press.

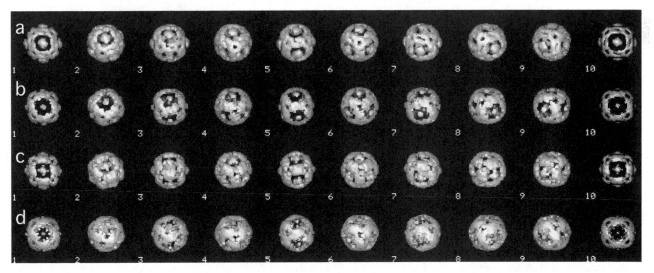

FIGURE 8   Projection averages at many different orientations, including those in Fig. 7, were used to reconstruct the 3-D structure of the archaeal phosphoenolpyruvate synthase from negatively stained preparations (1,028 input images) (a) and frozen-hydrated preparations (b through d) at high defocus (2,467 input images) (b) and low defocus (5,419 input images) (c) and after contrast transfer function correction (3,776 input images) (d). Here, the reconstructed 3-D objects are represented by shaded surfaces and are viewed from different orientations. This particular macromolecular complex has an octahedral shape and thus has fourfold, threefold, and twofold axes of rotational symmetry. There are always subtle differences among reconstructions from different preparations, and one must be cautious not to overinterpret details that might be spurious.

simulated structure first seen in Fig. 1. Again, with added noise in Fig. 9b (as in a TEM image), individual motifs are obscured. In electron crystallography, a mathematical tool called the Fourier transform allows us to convert the images in Fig. 9a and 9b into an alternative representation in which bright central spots indicate that we have *repeating* structural information (Fig. 9c and 9d, respectively). By selectively filtering these data as in Fig. 9d, i.e., zeroing out all

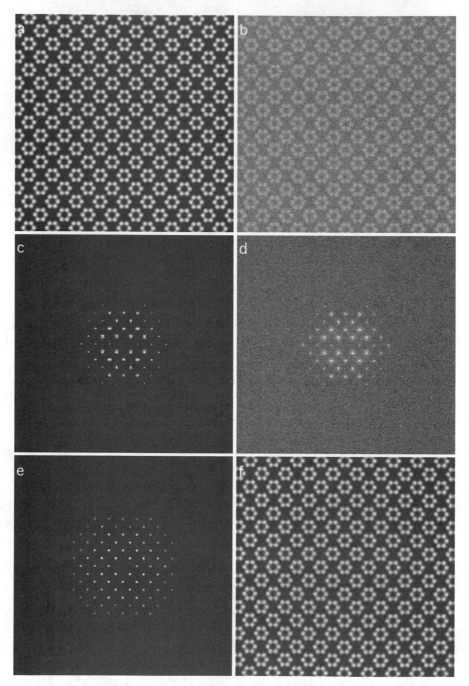

FIGURE 9 Basic principles of 2-D electron crystallography in general and of Fourier filtering in particular. (a) Simulated macromolecular complex, as shown in Fig. 1, arranged in a 2-D planar array (one complex thick). (b) Random noise has been added to each point in the image. (c) A Fourier transform of the image in panel a reveals a set of bright spots representative of the underlying regularity of the structure. Crystallographers are able to extract a great deal of information from the intensities and patterns of these spots. (d) A Fourier transform of the noisy image in panel b also reveals a set of bright spots, but now there is a diffuse background due to the noise, and the spots farther away from the origin are obscured. (e) The image in panel d can be filtered by computationally selecting only those values around the bright peaks and discarding all other values which represent primarily image noise. The bright peaks are arranged in a pattern that can be predicted by the way the repeating motifs are arranged with respect to one another, i.e., distances and angles between them. (f) By inverse Fourier transformation of the image in panel e, an improved image of the crystalline array is obtained for comparison with the original noisy image in panel b and the noise-free image in panel a.

pixels in the Fourier transform not associated with repeated information (Fig. 9e), and then doing an inverse Fourier transform, we obtain a less noisy image of the planar array (Fig. 9f). The jargon used to describe this process is **Fourier filtering**, or sometimes **image reconstruction**, which should not be confused with **3-D reconstruction from projections** that has already been mentioned. In practice, some fine tuning is performed by using computational alignment algorithms similar to those used in single-particle analysis, a scheme known as **correlation averaging**. The reader is referred to reference 16 for an overview of these approaches. In order to get a 3-D structure of the repeating motif of the crystal, one would tilt the specimen physically in the transmission electron microscope and obtain images at different angles of tilt, i.e., the different orientations (Fig. 10). Each image of the tilted crystal would be processed as those in Fig. 9, to provide a good 2-D projection, and the 3-D structure would be reconstructed from the set of 2-D projections as shown in Fig. 4. The angles of tilt are already known for each view. A good conceptual illustration of this tomographic reconstruction process is found in reference 32.

5.7. ELECTRON CRYSTALLOGRAPHIC ANALYSIS OF A NATURALLY OCCURRING 2-D PROTEIN ARRAY: AN S-LAYER FROM THE EUBACTERIUM *AEROMONAS SALMONICIDA*

S-layers, the proteinaceous surface arrays found on some gram-positive and gram-negative eubacteria and many archaea (28), are naturally occurring 2-D crystals that are amenable to image analysis by electron crystallography. We use here as an example the S-layer from *Aeromonas salmonicida* (Fig. 11a) (30). The S-layer is a virulence factor for this gram-negative fish pathogen, and as in all S-layers, the proteinaceous subunits self-assemble into oblique (p2), square (p4), or hexagonal (p3 or p6) motifs, depending on the particular protein being produced. The prefix "p" refers to the patterns of symmetry evident in the array (19, 26).

In this example, it is difficult to see in the untreated electron micrograph (Fig. 11a) what the individual repeating motifs look like, but after filtering the Fourier transform

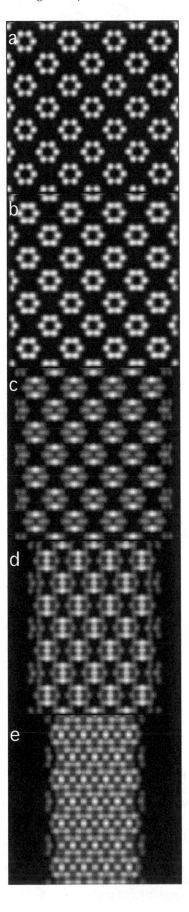

FIGURE 10 Specimens in the transmission electron microscope can be tilted about a single axis (or sometimes about two orthogonal axes) at defined angles and then imaged. In modern transmission electron microscopes, this entire process is computer controlled and automated. Here, a partial tilt series around a vertical axis of a segment of a simulated 2-D crystal is shown at tilt angles of 0° (a), 15° (b), 30° (c), 45° (d), and 60° (e). In practice, most transmission electron microscope specimen stages are limited to tilt angles of ±60°, where 0° represents the untilted specimen, but one would collect data at finer tilt increments such as every 3 or 5°. The limitation is the cumulative electron dose that the specimen can withstand before being distorted or destroyed. The object being tilted can be reconstructed tomographically as shown in Fig. 4. The Venetian blind effect of specimen grid bars makes it difficult to go any higher. This effect is already visible here via the confusing superimposition of repeating motifs at 60°. Since the object is not imaged fully from all possible orientations, the resolution of the final 3-D reconstruction is limited (the so-called missing-cone effect).

FIGURE 11 Example of 2-D electron crystallography. (a) S-layer from the eubacterium *Aeromonas salmonicida* with p4 symmetry (with unit cell dimensions $a = b = 12$ nm), negatively stained with 2% (wt/vol) uranyl acetate and imaged by TEM. The various possible planar symmetry groups are defined in references such as 16, 19, and 26. The distance between the center of each repeating motif and its neighbor along the lattice line (i.e., not diagonally) is thus 12 nm. (b) Fourier transform of the image in panel a. The arrangement of peaks is not as good as that in the simulated image in Fig. 9, reflecting the underlying imperfections of the 2-D array. (c) After filtering of the image in panel b and inverse Fourier transformation, an improved image of the bacterial surface array is obtained. The regional imperfections of the crystal, in terms of degree of staining and distortion, are evident. (d) Magnified view of a region of the filtered crystal. In panel a, the proteinaceous material is dark and the negative stain is light, as it would appear on the original electron micrograph recorded on film. In panels b through d, the contrast has been inverted.

shown in Fig. 11b, an image with enhanced signal-to-noise ratio as shown in Fig. 11c and d is obtained. Such analyses have become almost routine for investigating bacterial S-layers (e.g., references 16 and 28). Because S-layers are highly sensitive to damage by the electron beam, long exposures necessary for techniques such as selected-area electron diffraction (chapter 4) are usually avoided, making image reconstruction the preferred method of obtaining high-resolution, noise-reduced images. Most work has been done on negatively stained preparations like that in Fig. 11a, but some cryoTEM of frozen-hydrated specimens has been reported (see reference 30). More recently, some image reconstruction has been achieved on S-layers probed by scanning tunneling or atomic force microscopies (chapter 6).

Finally, electron crystallography is, at present, the most productive method for determining the structures of membrane proteins, and a tremendous amount of research activity is presently under way in this field (20, 35). Many of these examples, including various membrane channels (22), are of direct interest to structural microbiologists. In addition to membrane proteins, soluble proteins (such as enzymes) can be induced to form planar arrays, and electron crystallography can be used to solve their structures (15).

5.8. CONCLUDING REMARKS

TEM, particularly cryoTEM, has become a powerful tool for analyzing the tertiary and quaternary conformations of biological macromolecules and their complexes. Computational image analysis and reconstruction are integral tools in this process. This chapter has illustrated two approaches to manipulating and processing macromolecular TEM images that one would commonly find in the structural and molecular microbiological literature.

5.9. SOFTWARE SOURCES

a. Northern Eclipse 6.0. EMPIX Imaging, 3075 Ridgeway Dr., Unit #13, Mississauga, Ontario, L5L 5M6, Canada. http://www.empix.com.

b. Scion Image. Scion Corporation, 82 Worman's Mill Court, Suite H, Frederick, MD 21701. http://www .scioncorp.com.

c. NIH Image. Research Services Branch, National Institute of Mental Health, National Institute of Neurological Disorders and Stroke, c/o NIMH Public Inquiries, 6001 Executive Blvd., Room 8184, MSC 9663, Bethesda, MD 20892-9663. http://rsb.info.nih.gov/ nih-image.

d. IMAGIC-5. Michael Schatz, Image Science GmbH, Heilbronner Straße 10, D-10711 Berlin, Germany. http://www.ImageScience.de.

e. SPIDER. J. Frank, Wadsworth Center for Laboratories and Research, New York State Department of Health, Empire State Plaza, Albany, NY 12201-0509. http://www .wadsworth.org/spider_doc/spider/docs/spider.html.

f. EMAN. S. Ludtke, Verna and Marrs McLean Department of Biochemistry, Baylor College of Medicine, Houston, TX 77030. http://blake.bcm.tmc.edu/eman.

g. Image Objects (formerly SEMPER). Synoptics Ltd., Beacon House, Nuffield Rd., Cambridge, CB4 1TF, United Kingdom. http://www.synoptics.co.uk.

h. Insight 2000. Accelrys Inc. (formerly Molecular Simulations Incorporated), 9685 Scarnton Rd., San Diego, CA 92121-3752. http://www.accelrys.com.

5.10. GLOSSARY

2-D electron crystallography Electron beams in standard transmission electron microscopes cannot penetrate thick samples like whole bacteria or 3-D crystals of proteins. Occasionally, 2-D crystals of proteins that are one molecule thick can be produced, which are suitable for viewing by TEM because now the electrons can get through to form an image. These images can be manipulated computationally by using the same conceptual tools that X-ray crystallographers use for 3-D crystals.

3-D reconstruction from projections A TEM image is like an X-ray image in that 3-D information is *projected* into a 2-D image. By combining computationally the data from many 2-D images in which the object is viewed from different orientations, the 3-D internal density distribution can be calculated, i.e., reconstructed.

Contrast transfer function correction A transmission electron micrograph is blurry in a way defined by the contrast transfer function, which changes with the level of defocus at which the image was recorded. The information from two or more micrographs, recorded at different defocus levels, can be combined to give a minimally blurry result. A great deal of physics and mathematics are involved.

Correlation averaging In 2-D electron crystallography, the first step is Fourier filtering, which essentially averages the information from all repeating units. However, protein crystals are never perfect. Each repeating unit (unit cell) is

somewhat misplaced from where it should be, ideally. Correlation averaging is a tool used to reposition each unit cell computationally to give a better averaged image.

Fourier filtering The Fourier transform is a useful mathematical tool that can be used to show the degree of repeating information in any signals, including images. By selecting only this repeated information and inverse transforming, the level of noise can be reduced significantly because it is random. One occasionally reads about the fast Fourier transform, which is simply an algorithm for rapidly computing a Fourier transform for images of certain sizes (i.e., numbers of pixels).

Image reconstruction An imprecise term often used to denote Fourier filtering.

Single-particle analysis An electron micrograph of a protein sample will portray many protein macromolecules (or complexes). Each macromolecule possesses the same 3-D structure, but each looks different in the electron micrograph because it lies in a certain orientation and its image is noisy. Single-particle analysis combines all of the images of different individual macromolecules together to reduce noise and obtain the 3-D structure.

5.11. REFERENCES

The reference lists here are not exhaustive and are provided as reasonable places to start gathering more information. The choice of general references reflects as much what I had in hand (or rather, on shelf) as any other criterion. Specific references could number in the thousands.

5.11.1. General References

1. **Frank, J. (ed.).** 1992. *Electron Tomography: Three-Dimensional Imaging with the Transmission Electron Microscope.* Plenum Press, New York, NY.
2. **Frank, J.** 1996. *Three-Dimensional Electron Microscopy of Macromolecular Assemblies.* Academic Press, New York, NY. *Both references 1 and 2 are good, albeit specialized, overviews of computerized 3-D electron microscopy, the first one edited and the second one written by a pioneer in the field. Reference 1 describes 3-D reconstruction techniques that are applicable to any structures that can be imaged by TEM, even those as large as organelles and whole cells. Reference 2 is devoted to single-particle analysis like that described here.*
3. **Glasel, J. A., and M. P. Deutscher (ed.).** 1995. *Introduction to Biophysical Methods for Protein and Nucleic Acid Research.* Academic Press, San Diego, CA. *An overview in one volume of the myriad techniques for probing the structures of biological macromolecules.*
4. **Häder, D.-P. (ed.).** 2001. *Image Analysis: Methods and Applications,* 2nd ed. CRC Press, Boca Raton, FL. *A compendium of papers describing various applications of image analysis in the biological sciences, including electron microscopy.*
5. **Harris, J. R.** 1997. *Negative Staining and Cryoelectron Microscopy.* Bios Scientific Publishers, Oxford, United Kingdom. *A lot of excellent examples of high-quality TEM images of biological macromolecules are presented here. A good place to see what to aim for in one's own work, especially for cryoTEM.*
6. **Hayat, M. A.** 2000. *Principles and Techniques of Electron Microscopy. Biological Application,* 4th ed. Cambridge University Press, Port Chester, United Kingdom. *General, standard reference that should be in every TEM lab.*
7. **Hayat, M. A., and S. E. Miller.** 1990. *Negative Staining.* McGraw-Hill, Toronto, Canada. *More detailed description of a common preparation technique, of interest to the experienced practitioner.*

8. **Hoppert, M., and A. Holzenburg.** 1998. *Electron Microscopy in Microbiology.* Bios Scientific Publishers, Oxford, United Kingdom.
A good resource written specifically for microbiologists.

9. **Misell, D. L.** 1978. *Practical Methods in Electron Microscopy,* vol. 7. *Image Analysis, Enhancement, and Interpretation.* North-Holland, Amsterdam, The Netherlands.
An excellent classic in the field and a good review of techniques such as helical reconstruction.

10. **Nasser Hajibagheri, M. A. (ed.).** 1999. *Electron Microscopy Methods and Protocols.* Humana Press, Totowa, NJ.
Another general reference that every TEM lab should have available; includes techniques for macromolecules as well.

11. **Russ, J. C.** 1998. *The Image Processing Handbook,* 3rd ed. CRC Press, Boca Raton, FL.
Although image processing is encountered by everyone daily, most books on the subject are written by and for physicists, engineers, and computer scientists and combine a great deal of mathematics with trivial examples such as models' faces. This book helps nonspecialists get an appreciation of the tricks of the trade, presents a variety of real microscopic examples, and is accompanied by software to help put principles into practice.

12. **Sommerville, J., and U. Scheer (ed.).** 1987. *Electron Microscopy in Molecular Biology: a Practical Approach.* IRL Press, Oxford, United Kingdom.
A standard reference book on preparing biological macromolecules, including DNA.

5.11.2. Specific References

13. **Cicicopol, C., J. Peters, A. Lupas, Z. Cejka, S. Müller, R. Golbik, G. Pfeifer, H. Lilie, A. Engel, and W. Baumeister.** 1999. Novel molecular architecture of the multimeric archaeal PEP-synthase homologue (MAPS) from *Staphylothermus marinus. J. Mol. Biol.* **290:**347–361.
References 13 and 23 are intended to provide background for the specific example used here to illustrate single-particle analysis of a large macromolecular complex.

14. **Czarnota, G. J., D. R. Beniac, N. A. Farrow, G. Harauz, and F. P. Ottensmeyer.** 2001. Three-dimensional electron microscopy of biological macromolecules: quaternion-assisted angular reconstitution of single particles, p. 275–294. *In* D. Häder (ed.), *Image Analysis: Methods and Applications,* 2nd ed. CRC Press, Boca Raton, FL.
References 14 and 25 describe two slightly different approaches to angular reconstitution, i.e., the computational determination of relative angular orientations amongst different projections of asymmetric macromolecules.

15. **Ellis, M. J., and H. Hebert.** 2001. Structure analysis of soluble proteins using electron crystallography. *Micron* **32:**541–550.

16. **Engelhardt, H.** 1988. Correlation averaging and 3-D reconstruction of 2-D crystalline membranes and macromolecules. *Methods Microbiol.* **20:**357–413.
This review does for Fourier filtering and correlation averaging what references 14, 25, 31, and 33 do for single-particle analysis—it brings together the important concepts for a biologically oriented reader without obscuring them by excessive use of equations.

17. **Frank, J., P. Penczek, R. K. Agrawal, R. A. Grassucci, and A. B. Heagle.** 2000. Three-dimensional cryoelectron microscopy of ribosomes. *Methods Enzymol.* **317:**276–291.
The research groups of Frank and van Heel were pivotal both in developing techniques of single-particle analysis and in elucidating the structures of ribosomal subunits to high resolution in work spanning well over the past two decades. Reference 33 ranks also with references 2, 14, 25, and 31 in terms of reviewing single-particle analysis and cryoTEM and summarizes this group's work on ribosomes. Reference 17 is dedicated to reviewing contemporary progress in structural ribosomology.

18. **Frank, J., M. Radermacher, P. Penczek, J. Zhu, Y. Li, M. Ladjadj, and A. Leith.** 1996. SPIDER and WEB: processing and visualization of images in 3D electron microscopy and related fields. *J. Struct. Biol.* **116:**190–199.
References 18, 24, 27, and 34 describe software packages available for analysis of TEM images of macromolecules.

19. **Hammond, C.** 1992. *Introduction to Crystallography.* Oxford University Press and Royal Microscopical Society, Oxford, United Kingdom.
This is handbook 19 of the series produced by the Royal Microscopical Society, other volumes of which are noted in chapter 4. References 19 and 26 provide the reader with explanations of symmetry groups found in macromolecular crystals, including planar ones.

20. **Hasler, L., J. B. Heymann, A. Engel, J. Kistler, and T. Walz.** 1998. 2-D crystallization of membrane proteins: rationales and examples. *J. Struct. Biol.* **121:**162–171.
References 20 and 35 review some of the exciting advances in the field of structure determination of membrane proteins by cryoTEM.

21. **Henderson, R., J. M. Baldwin, T. A. Ceska, F. Zemlin, E. Beckmann, and K. H. Downing.** 1990. Model for the structure of bacteriorhodopsin based on high-resolution electron cryo-microscopy. *J. Mol. Biol.* **213:**899–929.
One of the classic milestones of biological macromolecular TEM.

22. **Jap, B. K., P. J. Walian, and K. Gehring.** 1991. Structural architecture of an outer membrane channel as determined by electron crystallography. *Nature* **350:**167–170.
A good example of the power of electron crystallography in elucidating membrane protein structure, of direct interest to structural and molecular microbiologists.

23. **Li, W., F. P. Ottensmeyer, and G. Harauz.** 2000. Quaternary organization of the *Staphylothermus marinus* phosphoenolpyruvate synthase: angular reconstitution from cryoelectron micrographs with molecular modelling. *J. Struct. Biol.* **132:**226–240.

24. **Ludtke, S. J., P. R. Baldwin, and W. Chiu.** 1999. EMAN: semiautomated software for high-resolution single-particle reconstructions. *J. Struct. Biol.* **128:**82–97.

25. **Radermacher, M.** 2001. Three-dimensional reconstruction of single particles in electron microscopy, p. 295–327. *In* D. Häder (ed.), *Image Analysis: Methods and Applications,* 2nd ed. CRC Press, Boca Raton, FL.

26. **Rhodes, G.** 1993. *Crystallography Made Crystal Clear: a Guide for Users of Macromolecular Models.* Academic Press, San Diego, CA.

27. **Saxton, W. O.** 1996. SEMPER: distortion compensation, selective averaging, 3-D reconstruction, and transfer function correction in a highly programmable system. *J. Struct. Biol.* **116:**230–236.

28. **Sleytr, U. B., and T. J. Beveridge.** 1999. Bacterial S-layers. *Trends Microbiol.* **7:**253–260.
An overall review of bacterial surface arrays, including TEM.

29. **Stewart, M.** 1990. Electron microscopy of biological macromolecules: frozen hydrated methods and computer image processing, p. 9–39. *In* P. J. Duke and A. G. Michette (ed.), *Modern Microscopies—Techniques and Applications,* Plenum Press, New York, NY.
Quick "big picture" review.

30. **Stewart, M., T. J. Beveridge, and T. J. Trust.** 1986. Two patterns in the *Aeromonas salmonicida* A-layer may reflect a structural transformation that alters permeability. *J. Bacteriol.* **166:**120–127.
This reference is intended to provide background to the specific example used here to illustrate 2-D electron crystallography.

31. **Thuman-Commike, P., and W. Chiu.** 2000. Reconstruction principles of icosahedral virus structure determination using electron cryomicroscopy. *Micron* **31**:687–711.
Of interest to virologists; a good recent review of single-particle analysis of highly symmetric objects.

32. **Unwin, N., and R. Henderson.** 1984. The structure of proteins in biological membranes. *Sci. Am.* **250:** 78–94.
A popular review intended for a general audience and of value for a beginner because of the clarity of the illustrations.

33. **van Heel, M., B. Gowen, R. Matadeen, E. V. Orlova, R. Finn, T. Pape, D. Cohen, H. Stark, R. Schmidt, M. Schatz, and A. Patwardhan.** 2000. Single-particle electron cryomicroscopy: towards atomic resolution. *Q. Rev. Biophys.* **33**:307–369.

34. **van Heel, M., G. Harauz, E. Orlova, R. Schmidt, and M. Schatz.** 1996. The next generation of the IMAGIC image processing system. *J. Struct. Biol.* **116**:17–24.

35. **Walz, T., and N. Grigorieff.** 1998. Electron crystallography of two-dimensional crystals of membrane proteins. *J. Struct. Biol.* **121**:142–161.

6

Atomic Force Microscopy

YVES F. DUFRÊNE

Light microscopy and electron microscopy have long been recognized as key tools in microbiological science (see chapters 1 and 4, respectively). Light microscopy is useful for counting and identifying microbial cells as well as for determining their general morphological details. Because the resolution is limited to 200 to 500 nm, i.e., the wavelength of the light source, information at the (supra)molecular level is not accessible. High-resolution images of microbial samples can be obtained by electron microscopy, which uses high-energy electrons instead of light as the incident beam. Elegant techniques have been developed for transmission electron microscopy (TEM) such as the use of freeze fracturing and surface replication to visualize, for example, cell surface layers and the use of negative staining for studying purified structures such as flagella and fimbriae (chapter 7). These approaches are limited by the requirement of vacuum conditions during the analysis, i.e., native, hydrated samples cannot be directly investigated unless sophisticated cryoTEM methods are employed. Furthermore, electron microscopy does not easily allow quantitative height measurements.

The advent of atomic force microscopy (AFM) (7), one technique in a family of new microscopies called scanning probe microscopies, has recently opened a wide range of novel applications for microbiologists. The technique provides three-dimensional (3-D) images of the surface ultrastructure with unprecedented resolution, in real time, under physiological conditions, and with minimal sample preparation. In addition, force measurements make it possible to probe physical properties such as molecular interactions and mechanical properties. Hence, the combination of AFM imaging and force measurements provides new avenues to study biological phenomena such as molecular recognition, protein folding, and cell adhesion (20a).

The general principle of AFM is surprisingly simple (Fig. 1). AFM imaging is performed, not by using an incident beam as in other classical microscopies, but by sensing the force between a very sharp probe and the sample surface. Thus, an AFM image is generated by recording the force changes as the probe (or sample) is scanned in the x and y directions. The sample is mounted on a piezoelectric scanner which ensures 3-D positioning with high resolution (Fig. 1). The force is monitored by attaching the probe to a soft cantilever, which acts as a spring, and measuring the bending, or deflection, of the cantilever. The larger the cantilever deflection, the higher the force that will be experienced by the probe. Most instruments today use an optical method to measure the cantilever deflection with high resolution; a laser beam is focused on the free end of the cantilever, and the position of the reflected beam is detected by a position-sensitive detector (photodiode).

This chapter focuses on the use of AFM in microbiology. Rather than providing an exhaustive review of the literature in this area, the chapter emphasizes methods and gives recommendations for reproducible, reliable experiments. Selected data are also presented to highlight the various applications offered by the technique for microbiology (for reviews of papers published in this area, see references 17 through 20 and 22). Section 6.1 concentrates on sample preparation procedures: selection of appropriate substrates and immobilization protocols available for isolated macromolecules, cell surface layers, and whole cells. Section 6.2 deals with the various aspects of AFM imaging: different imaging modes together with common problems and artifacts, imaging parameters, and imaging environments. Section 6.3 focuses on AFM force measurements: the principle of force-distance curves and their application to probe molecular interactions and mechanical properties. It is hoped that this contribution will be useful to microbiologists to evaluate the advantages and limitations of AFM in their specific fields and to define appropriate procedures that will lead them to successful experiments.

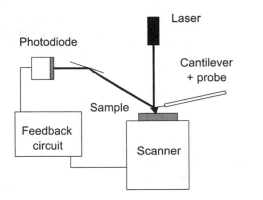

FIGURE 1 General design of an atomic force microscope.

6.1. SAMPLE PREPARATION

6.1.1. Substrates

Sample preparation is a key factor that determines the success of an AFM experiment. One point of particular importance is that the sample must be well attached to an appropriate solid substrate. For many applications, the substrate should typically be flat and easy to handle, prepare, and clean. Mica is certainly the most widely used substrate for biological applications. This layered mineral has strong 2-D bonds but is only weakly bound in the third dimension. As a result, it is easily cleaved, for instance by using an adhesive tape, to produce clean surfaces that are atomically flat over large areas. Glass, in the form of glass coverslips, is another suitable substrate for AFM studies. The fact that it is transparent may be an important advantage for experiments in which AFM and light microscopy are combined. Furthermore, its surface is fairly smooth and can be chemically modified for specific applications (see below). Glass coverslips, however, are always coated with organic contaminants and particles that should be removed before use. This removal can be achieved by washing in concentrated acidic solution (HCl-HNO$_3$, 3:1) followed by ultrasonication in several Milli-Q (Millipore) water solutions.

Both mica and glass have a hydrophilic surface. For some applications, best results may be obtained by using hydrophobic substrates, such as highly oriented pyrolytic graphite, a layered material from which flat, clean surfaces can easily be prepared by cleavage. Substrates may also be prepared by coating solids with a thin layer of metal such as gold. Gold-coated surfaces prepared by thermal evaporation are relatively flat and can be easily functionalized with self-assembled monolayers of organic alkanethiols (for details, see section 6.3.2). The functionalized surfaces can then be used as such or as substrates to covalently attach biomolecules (section 6.1.2). Finally, isoporous polymer membranes can be used to immobilize large objects such as whole cells (section 6.1.4).

In the following sections, various immobilization procedures are described. The protocols differ according to the type of sample investigated, i.e., macromolecules, cell surface layers, or whole cells.

6.1.2. Macromolecules

Since its invention, AFM has proved useful in imaging the structure and the conformation of isolated biomolecules, including polysaccharides, proteins, and nucleic acids. The literature describing sample preparation protocols and imaging conditions for biomolecules is abundant and not specific to microbiology (14, 27, 33, 48, 53, 66). Therefore, only the general features of these procedures will be mentioned.

Different methods are available to attach macromolecules to a substrate. For AFM in air, a simple immobilization procedure consists of depositing a drop of an aqueous solution containing the macromolecule of interest onto the substrate and letting the drop evaporate (38). Alternatively, the molecules can be sprayed from aqueous solution or in the presence of glycerol onto the substrate (46). Physical adsorption in the presence of appropriate electrolytes is another approach, which offers the advantage of allowing the sample to be directly imaged under liquid without the need for air drying. Addition of certain divalent cations may sometimes promote the attachment.

When firm, irreversible attachment is desired, macromolecules can be covalently bound to the substrate by using chemically derivatized alkanethiol or silane monolayers. In the first case, gold substrates are functionalized with alkanethiols (for details, see section 6.3.2) terminated by carboxyl functions which are then reacted with amino groups of proteins by using 1-ethyl-3-(3-dimethylaminopropyl) carbodiimide (EDC) and N-hydroxysuccinimide (NHS) (59). Alternatively, NHS-terminated monolayers can be directly formed by immersing the gold substrate in a solution of dithio-bis-succinimidylundecanoate (73). The other approach is based on the attachment of silane molecules such as 3-aminopropyltriethoxysilane (APTES) onto glass or mica to create a positive surface charge which can then bind anionic molecules (43).

6.1.3. Cell Surface Layers

Because of their well-organized structure and high rigidity, microbial cell surface layers made of 2-D protein crystals (i.e., S-layers) (Fig. 2) are ideal model systems for high-resolution imaging and for monitoring of structural changes in membrane proteins or other biological processes. AFM yields structural information directly under physiological conditions, which makes it a complementary tool to X-ray and electron crystallography (57).

During the past decade, various immobilization strategies have been established for AFM of cell surface layers. The procedure is often based on physical adsorption from aqueous solution in the presence of appropriate electrolytes. The optimal solution composition differs according to the type of specimen. Purple membranes from the archeon *Halobacterium* contain bacteriorhodopsin, a light-driven proton pump, in the form of highly ordered 2-D lattices. Appropriate buffer conditions for adsorption are as follows (52, 54): the pH may vary from 4 to 10; 150 mM KCl is added with 10 mM morpholineethanesulfonic acid (MES), HEPES, or Tris for adjustment of the pH. The purple membranes are diluted in the buffer to a concentration of about 50 μg/ml, and a drop of this solution is deposited onto freshly cleaved mica. After adsorption for 10 to 30 min, samples are gently washed with buffer to remove membranes that are not firmly attached. By using this procedure, individual bacteriorhodopsin molecules in the membrane were imaged at subnanometer resolution (Fig. 2). Furthermore, force-induced conformational changes were directly visualized (50, 57).

To adsorb in a defined orientation specimens that carry surface charges in an uneven distribution, such as purple

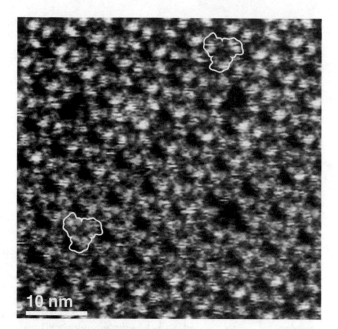

FIGURE 2 High-resolution AFM image, obtained with the specimen in aqueous solution, of purple membrane adsorbed to freshly cleaved mica (57). A fully reversible, force-induced conformational change was observed: at the top of the image, the force applied to the atomic force microscope stylus was 100 pN. During scanning of the surface line by line, the force was increased until it reached 150 pN at the bottom of the image. White outlines indicate bacteriorhodopsin trimers. (Reprinted from reference 57 with permission of the publisher.)

membranes, the surface properties of the substrate can be modified by coating with a polycation such as poly-L-lysine (53). The procedure involves dipping a mica or glass substrate into a 1- to 5-mg/ml solution of poly-L-lysine for a few minutes, rinsing with water, and drying. Before adsorbtion of the specimen, the quality of the coating should be checked, i.e., one should ensure that the substrate surface is not covered with polycation aggregates.

OmpF porin is a channel-forming protein of the outer membrane of *Escherichia coli* that forms 2-D crystals. The sample preparation (63) consists of preincubating freshly cleaved mica with 20 μl of buffer made of 100 mM NaCl, 20 mM HEPES, and 2 mM MgCl$_2$ (pH 7) for 5 min. After removal of the buffer, 1 to 3 μl of sample solution is added. This solution contains porin trimers dissolved at 1 mg/ml in octylpolyoxyethylene, mixed with dimyristoyl phosphatidylcholine at a lipid-to-protein ratio (wt/wt) of 0.2. After 10 min of contact time, the sample is washed with buffer to remove membranes that are not firmly adsorbed. The presence of Mg^{2+} is essential to compensate for the negative surface charges of both the sample and the substrate at pH ~7. This protocol allowed high-resolution imaging of porin OmpF 2-D crystals, with a lateral resolution of 1 nm and a vertical resolution of 0.1 nm (64). The contours determined by AFM agreed well with 3-D structural information from X-ray crystallography. In addition, voltage- and pH-dependent conformational changes of the extracellular loops of the protein were demonstrated (49).

For *E. coli* aquaporin Z, a membrane channel protein, the adsorption protocol involves first putting a drop of ad-sorption buffer made of 10 mM Tris-HCl (pH 7.5), 150 mM KCl, and 25 mM MgCl$_2$ onto freshly cleaved mica (65). A volume of 3 μl of protein crystal solution (0.1 mg/ml) is then injected into the drop. After 2 h of contact time, the sample is rinsed with a recording buffer made of 10 mM Tris-HCl, pH 7.5, and 150 mM KCl. AFM images revealed 2-D crystals with p42(1)2 and p4 symmetry (65). Imaging both crystal types before and after cleavage of the N termini allowed the cytoplasmic surface to be identified. Flexibility mapping and volume calculations identified the longest loop at the extracellular surface; this loop exhibited a reversible force-induced conformational change.

An elegant method has been developed to investigate S-layers from *Bacillus* species in conditions relevant to their native state (60, 75). The proteins are recrystallized on a lipid monolayer in a Langmuir-Blodgett trough, and the composite lipid-S-layer structure is then deposited onto a flat substrate. Under these conditions, S-layers attach to the lipid film with their inner face, which corresponds to the orientation found in the living organism. The protocol is as follows. A 1-mg/ml chloroform-methanol (9:1, vol/vol) solution of the lipid 1,2-dipalmitoyl-*sn*-3-phosphatidylethanolamine (DPPE) is spread onto a subphase of 1 mM borate buffer (pH 9.0; 10 mM CaCl$_2$) in a Langmuir-Blodgett trough. After evaporation of the solvent, the lipid molecules are compressed to a surface pressure of 30 mN/m. The clear supernatant of the S-layer protein solution is carefully injected beneath the lipid monolayer into the subphase. After 18 to 20 h of recrystallization at 20°C, 1-by-1-cm silicon wafers are carefully placed onto the liquid surface and removed after 15 min by being lifted horizontally from the surface with a pair of tweezers. Samples are rinsed twice with Milli-Q water and mounted into the AFM liquid cell while avoiding dewetting.

AFM imaging in buffer solution may be facilitated by covalently immobilizing the sample on the substrate. This type of procedure has been developed for the hexagonally packed intermediate (HPI) layer of *Deinococcus radiodurans* (35). First, silanization of glass substrates is performed with the NH$_2$-terminated reagent, silane APTES. Glass coverslips are covered with 30 ml of a 2% (wt/vol) solution of APTES in 95% (vol/vol) aqueous acetone and shaken for 3 min. After removal of the solution, the substrates are covered with 30 ml of acetone and shaken for 5 min and the acetone is removed. The last step is repeated 10 times, after which coverslips are put into an oven at 100°C for 1 h. The sample is then immobilized onto the derivatized glass surface by means of the photoactivatable heterobifunctional cross-linker N-5-azido-2-nitrobenzoyloxysuccinimide (ANB-NOS). Glass substrates are covered with 30 ml of 0.1 M Na$_2$CO$_3$, pH 9.0, containing 2 mg of ANB-NOS dissolved in 1 ml of dioxane and are shaken for 4 h. The excess reagent is removed by washing two to three times with four different solutions: Na$_2$CO$_3$ and 200 μl of butylamine, Na$_2$CO$_3$, water, and acetone. Then, a volume of 1 to 2 μl of the protein solution (1 to 10 mg/ml) is placed between two derivatized coverslips which are squeezed between two glass disks so as to bring the hydrophilic sample into close contact with the hydrophobic cross-linker. The azide is activated with UV irradiation at 366 nm at a distance of 10 cm for 3 min, and the sample is thoroughly rinsed with water or buffer to remove excess protein. The above-described immobilization strategy made it possible to image the HPI layer surface in buffer solution with a lateral resolution of 1 nm, the images being consistent within a few angstroms with data obtained by electron microscopy (36). It was also shown that the inner surface of the HPI layer had a protruding core with a central pore exhibiting two conformations, one with and the other without a central plug. In addition, individual pores were observed to switch from one state to the other, indicating that conformational changes can be detected at subnanometer resolution (51).

6.1.4. Whole Cells

There are several difficulties associated with the imaging of whole microbial cells. First, as opposed to animal cells, microbial cells have a well-defined shape and have no tendency to spread over substrates. As a result, the contact area between cells and the substrate is very small, often leading to cell detachment by the scanning probe. A second problem is the vertical motion of the sample (or probe), which is typically limited to a few micrometers. This makes it difficult to image the surfaces of large objects, such as microbial cells. For those reasons, immobilization by means of simple adsorption procedures is often inappropriate for investigating whole microbial cells.

For AFM studies related to bioadhesion issues, one immobilization method consists in exploiting the natural ability of microorganisms to firmly attach to solid substrates by means of extracellular polymeric substances. The sample preparation is dependent on the type of organisms and substrates investigated. It typically involves incubating the substrate with a cell suspension for a given period of time, resulting in the formation of a biofilm of adhering cells that can then be imaged by AFM (6).

For imaging of single cells, pretreatments such as air drying or chemical fixation can be used to promote cell attachment. The air-drying method typically involves placing a 5- to 50-μl droplet of a concentrated cell suspension on a 1-by-1-cm^2 substrate, mica or glass, and allowing drying in air (4, 40). Best results are obtained when using submonolayer cell coverages, which can be achieved by rinsing the substrate (three times in Milli-Q water) prior to air drying. Sometimes, covalent bonding of bacteria to silanized glass can be used since it does not appreciably affect the cell viability (10). Glass slides are soaked for 10 min in a 10% (wt/vol) silane solution, made of either 3-aminopropyltrimethoxysilane or 3-aminopropyldimethoxysilane in methanol, and then rinsed with copious amounts of methanol followed by Milli-Q water to remove excess silane. Then, a bacterial suspension (10^8 cells/ml) containing 100 mM EDC and 40 mM NHS is put in contact with silanized slides and allowed to shake (125 rpm) overnight. After rinsing with water, samples are either air dried by gently blowing air across the slide (for imaging in air) or kept hydrated (for imaging under water).

The above-described treatments may cause significant rearrangement or denaturation of the surface molecules, yielding a surface which is no longer representative of the native, hydrated state. To circumvent this problem, immobilization procedures that do not involve drying or chemical fixation have been developed, namely, immobilization in agar gels (Fig. 3A) and in porous membranes (Fig. 3B). In the first method (25), 5 μl of a highly concentrated cell suspension is deposited onto a clean coverglass. About 200 μl of 3% (wt/vol) molten agar (50°C) is dropped onto the cell layer. Another piece of coverglass is quickly put on top of the agar before hardening to create a flat surface. After 30 min, the solidified sample is turned upside down and the lower coverglass is pulled from the agar surface. Most cells are localized on that side of the agar disk, and weakly captured cells can be easily removed by washing with water. The cell-free side of the disk is carefully dried on a sheet of

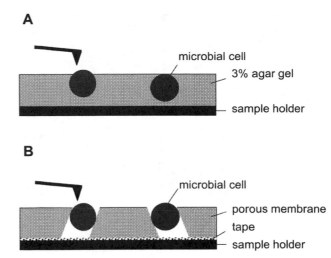

FIGURE 3 Schematic illustration of two immobilization methods for AFM investigation of native, single microbial cells. (A) Agar gel; (B) porous membrane.

tissue and then placed onto the sample holder. The use of such a soft, deformable immobilization matrix enabled in situ investigation of the growth process of *Saccharomyces cerevisiae* cells and direct visualization of micrometer-sized bud scars (25).

The second method consists of trapping spherical cells in a filter membrane with a pore size comparable to the dimensions of the cell (21, 31, 37, 70). Typically, a concentrated cell suspension (10 ml; 10^6 cells per ml) is gently sucked through an isopore polycarbonate membrane (Millipore) with a pore size slightly smaller than the cell size. After cutting of the filter (1 by 1 cm), the lower part is carefully dried on a sheet of tissue and the specimen is then attached to the sample holder by using a small piece of adhesive tape. This procedure offers two important advantages, i.e., it is fairly simple and straightforward, and it does not involve a macromolecular system such as agar, thus preventing the risk of cell surface contamination. An illustration of this immobilization procedure is given in Fig. 4, which shows an AFM height image, recorded under water, of two dividing daughter cells of *Lactococcus lactis*. Cells are trapped into a pore of the membrane, allowing repeated imaging without cell detachment or cell damage.

6.2. IMAGING

6.2.1. Imaging Modes

A number of AFM imaging modes are available. The most widely used is the contact mode, in which sample topography can be measured in different ways. In the constant-height mode, one simply records the cantilever deflection while the sample is scanned horizontally, i.e., at constant height. Minimizing large deflections, thus maintaining the applied force at small values, is often necessary to prevent sample damage. This is achieved in the constant-deflection mode, in which the sample height is adjusted to keep the deflection of the cantilever constant by using a feedback loop (Fig. 1). The feedback output is used to display a true height image. In many cases, small cantilever deflections do occur because the feedback loop is not perfect and the

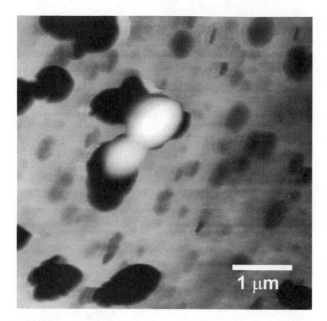

FIGURE 4   AFM height image (z range = 750 nm) recorded under water showing *Lactococcus lactis* cells trapped by a pore of a polycarbonate membrane. By trapping the cells mechanically into a porous membrane, single cells can be imaged under physiological conditions without any pretreatment.

resulting error signal can be used to generate a so-called deflection image.

For microbial specimens, both height and deflection images are often useful because they yield complementary information. The height image provides quantitative height measurements, thus allowing an accurate measure of surface roughness, the height of surface features, or the thickness of biological layers. The deflection signal is more sensitive to fine surface details than the height signal because the frequency response is much higher. This characteristic makes it valuable to reveal the surface ultrastructures of curved or rough samples such as microbial cells, for which conventional height images are often of poor resolution. Although lateral dimensions are accurately measured in the deflection imaging mode, this is not the case for vertical variations. Hence, one should keep in mind that deflection images do not provide quantitative height measurements. The complementarity of both height and deflection imaging modes is illustrated in Fig. 5 for the surfaces of native, dormant spores of the filamentous fungus *Aspergillus oryzae* (71). In the height image (Fig. 5A), one can see that the spore surface is rough and heterogeneous. The vertical cross section shows a ~35-nm step height, reflecting the thickness of the outer cell wall layer, which is in good agreement with electron microscopy data. The high-resolution deflection image recorded for the outer layer surface (Fig. 5B) shows well-ordered structures called rodlets, that are 10 nm in diameter and assembled in parallel to form fascicles interlaced with different orientations.

Because AFM can be operated in aqueous solution, it can be used to visualize in real time how cell surfaces interact with external agents or change during cell growth. For instance, real-time imaging was used to monitor the digestion of *S. cerevisiae* cell walls (3). Images of the cell surface

recorded as a function of time following addition of protease revealed the progressive formation of large depressions, about 500 nm in diameter and 50 nm in depth, reflecting the erosion of the mannoprotein outer layer. In other work, AFM was combined with thin-section TEM to investigate the changes in the cell walls of *Staphylococcus aureus* cells as they grow and divide (70). Nanoscale perforations were seen around the septal annulus at the onset of division and found to merge with time to form a single, larger perforation. These holes were suggested to reflect so-called murosomes, i.e., cell wall structures possessing high autolytic activity and which digest peptidoglycan. After cell separation, concentric rings were observed on the surface of the new cell wall and suggested to reflect newly formed peptidoglycan. Besides that of fungi and bacteria, it is worth noting that AFM may also be used to reveal the surface architecture of diatoms and viruses (15).

Accordingly, these data show distinct advantages of AFM over classical electron microscopy: (i) the ability to measure critical dimensions on the nanoscale such as the thickness of hydrated cell wall layers or the size of supramolecular structures and (ii) the possibility of probing surface ultrastructure in a hydrated state and in real time. Great advances are also being made with cryoelectron microscopy, in which hydrated specimens can be examined in a vitrified state. This is briefly described in chapter 4 in section 4.1.4.2.

Several other AFM imaging modes have been developed. Although their use in microbiology has been limited so far, they are briefly considered here because of their great potential for imaging surface topography at minimal applied forces and for mapping physical properties. Dynamic imaging modes consist of oscillating the probe or sample at a given frequency. In tapping mode AFM (TMAFM), or intermittent contact mode, the probe is excited externally and the amplitude and phase of the cantilever are monitored near the resonance frequency of the cantilever. TMAFM has been used to record high-resolution images of purple membranes without detectable deformation (47), to image *Shewanella putrefaciens* bacteria adhering to iron-coated and uncoated silica surfaces (26), to examine biofilms formed by *Pseudomonas* sp. on hematite and goethite minerals (24), and to observe changes in bacterial cell morphology resulting from adhesion-modifying chemicals (10). Force modulation microscopy is another dynamic mode which measures the amplitude and phase shift of the cantilever while the sample or the probe is vibrated, thereby allowing estimation of spatial viscoelasticity variations. Further information on the different AFM imaging modes can be found in the literature (14, 44, 48).

Time resolution is a crucial limitation of current AFM imaging modes. For instance, recording a high-resolution image of a 2-D protein crystal typically takes about 30 s, but for cells this procedure may take minutes, meaning that many dynamic processes occurring at short time scales are not accessible. Interestingly, novel scanning probe instruments are currently being developed with increased imaging rates (millisecond resolution), suggesting that within a few years microscopists should have new possibilities for probing molecular and cellular dynamics (32). Another issue is the alteration of the specimen by the tip during scanning. Here also, significant progress is being made to improve existing imaging modes. In particular, the recent active resonance control in TMAFM makes it possible to reduce tip-specimen interactions, thus yielding images of biological specimens with improved resolution (32).

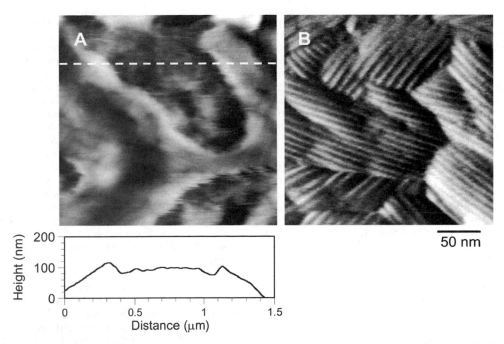

FIGURE 5 AFM height (A) and deflection (B) images recorded under water for the surfaces of dormant spores of the fungus *Aspergillus oryzae*. The height image (A) together with the vertical cross section (taken along the dashed line) reveal the presence of a heterogeneous outer layer ~35 nm thick. The deflection image (B) shows that the outer layer is made of rodlet structures 10 ± 1 nm in diameter.

6.2.2. Scanners, Cantilevers, and Probes

By applying appropriate voltages, piezoelectric materials expand in some directions and contract in others in a defined way. This is the basic idea behind the piezoelectric scanner (Fig. 1), which is used to position the sample in the x, y, and z directions with high accuracy and precision. Scan ranges are typically up to 100 μm laterally and a few micrometers vertically. One problem associated with the piezoscanner is that the expansion (i.e., the response) is not exactly linear with the applied voltage and shows hysteresis when the effects of increasing and decreasing applied voltages are compared. For large displacements, this nonlinearity can be important and be a source of inaccuracies.

Today, most AFM cantilevers and probes are made of silicon (Si) or silicon nitride (Si_3N_4) by using microfabrication techniques. Because the cantilever deflection is usually monitored by means of a laser beam, the backside of the cantilever is often coated with a thin layer of gold to enhance its reflectivity. The probe shape is either conical (Si) or pyramidal (Si_3N_4), with apex radii of curvature of about 10 nm (Si) and 20 to 50 nm (Si_3N_4). For topographic imaging, cantilevers are often used as received. However, organic contaminants on the probe may substantially influence the image quality. The use of a mixture of concentrated sulfuric acid and hydrogen peroxide (piranha solution) is an easy and effective way to remove this organic contamination (41). For biological specimens, best results are generally obtained with cantilevers exhibiting small spring constants, i.e., in the range of 0.01 to 0.10 N/m. Note that actual spring constants may substantially differ from values quoted by the manufacturer. Therefore, when accurate knowledge of the force is required, such as in quantitative force measurements, spring constants must be determined experimentally (13, 67).

The size and shape of the AFM probe play an important role in determining the resolution and the contrast of the images. At first glance, it is surprising to see that (sub)nanometer resolution is routinely acquired on reconstituted surface layers by using commercial probes with 10- to 50-nm radii of curvature. It is thought that nanoscale protrusions or asperities extending irregularly from the probe are responsible for the high resolution.

A common source of artifacts in imaging is the broadening of the surface features due to the finite size of the AFM probe. This effect is important when imaging height variations in the range of the probe radius of curvature, i.e., 10- to 50-nm. In this case, the sample interacts with the sides of the probe and the resulting image will be a combination of the real sample topography and of the probe geometry. The main consequence is that surface depressions appear smaller while protrusions appear broader. By using image-processing methods, the probe-sample interaction can be modeled and an image resembling more closely the actual surface may be restored (76). Another common problem is the shadowing or multiplication of small structures produced by multiple probe effects. These are due to the presence of multiple asperities on the probe apex which usually originate from contamination. If this problem is encountered during the course of an experiment, it is recommended that the probe be changed before proceeding further.

6.2.3. Imaging Parameters

In the constant-deflection mode, the sample height is adjusted to keep the deflection of the cantilever constant by using a feedback loop. To achieve accurate height adjustments, the feedback loop can be varied in different ways by using various gain parameters (for more details on feedback gain parameters, see reference 48). Most AFM systems use at least two gain controls, i.e., proportional and integral gains. It is often recommended to use a proportional gain higher than the integral gain. The gains should be high enough so that the probe can track accurately every surface feature; however, a too-high gain will result in oscillations and instabilities.

The operator should optimize the rate at which the image is acquired. An AFM image is obtained by raster scanning the probe over the sample, i.e., the sample or probe is moved along a line in the so-called fast scanning direction and then moves up to another scan line in the slow scanning direction. The scan rate is defined as the rate in the fast scanning direction and thus depends on the scanning frequency and on the image size. The scan rate should not be too high when nanometer resolution is desired because at high scan rates the probe may not respond faithfully to nanoscale features. On the other hand, low scan rates may cause sample damage when imaging soft, fragile samples. Therefore, scan rates are often in the range of 0.5 to 2 μm/s.

The force acting between the probe and the sample is another key parameter to control in order to obtain reliable, high-resolution images. Strong imaging forces may dramatically reduce the image resolution and cause molecular damage or displacement. When imaging in air, a layer of water condensation and other contamination often covers both the probe and the sample, forming a meniscus pulling the two together. The resulting strong attractive force, usually 10 to 100 nN, makes high-resolution imaging difficult and sometime causes sample damage. The capillary force can be eliminated by performing the imaging in aqueous solution. By selecting appropriate buffer conditions, it is generally possible to maintain an applied force in the range of 0.1 to 0.5 nN.

In addition to the sample surface topography, lateral forces (friction) between the probe and the sample may significantly contribute to the contrast of an AFM image. To assess the contribution of lateral forces to the apparent topographic contrast, it is advisable to compare images obtained by forward and backward scanning. Identical forward and backward images indicate that frictional effects are likely to be minimal. It may also be useful to change the scanning angle, say by 90°, to evaluate whether the observed contrast is affected by the scanning direction.

6.2.4. Imaging Environment

For most microbiological applications, water is the most relevant factor in the imaging environment. Imaging under water requires the use of a liquid cell, which is provided with most commercial microscopes. Depending on the cell design, the sample can be attached to the sample holder either mechanically or by using a small piece of double-sided adhesive tape or glue. When using the second option, keep in mind the possible risk of sample contamination. While mounting the wet sample into the cell, it is important to avoid passing the sample through the air-water interface and so also avoid contact with air bubbles, which may reorganize or denature the structures of interest. Some liquid cells use an O-ring to seal the sample and probe holder. However, best results can be obtained by leaving the system open. In this case, it is strongly recommended to protect the scanner head with a piece of parafilm to avoid contact with

the aqueous solution when filling the space between the sample and the probe holder.

High-resolution imaging requires vibrational and thermal stability. External vibration isolation, especially in the 10- to 100-Hz frequency range, can be accomplished by mounting the microscope stage on a mechanically isolated platform, such as an optical table with air bearings or heavy stone plinths hanging from elastic cords. To minimize thermal drift, it is important to wait at least 10 min so that the imaging force can be maintained constant and at low values. If a longer time is required to achieve thermal stability, one should keep in mind the possible risk of surface alteration due to contamination or cell lysis.

Another important factor for achieving (sub)molecular resolution is to use an adequate liquid composition because it will directly affect the forces between the probe and the sample. For macromolecules such as polysaccharides (38) and DNA (28), excellent results have been obtained by imaging under alcohols. For ordered microbial layers, aqueous buffers are generally used, the buffer composition having a profound influence on the image resolution. For instance, for purple membrane and OmpF porin surfaces (55), it was possible to improve the spatial resolution by changing the pH and electrolyte concentration, best results being obtained with 150 mM KCl or NaCl (in 10 mM Tris-HCl, pH 7.6). Accordingly, a general recommendation for microscopists investigating new microbial samples is to investigate the effects of pH and ionic strength on the image quality in order to define an optimal imaging environment.

6.3. FORCE MEASUREMENTS

6.3.1. Principle of Force-Distance Curves

Measuring the force acting between the AFM probe and the sample is important in defining the imaging force and thus optimizing the image resolution. As we shall see, AFM force measurements also provide microbiologists and biophysicists with the opportunity to probe the sample physical properties. Remarkably, the force sensitivity of the instrument, in the piconewton range, allows the forces associated with single biomolecules to be measured.

Force measurements are performed by monitoring, at a given (x,y) location, the cantilever vertical deformation, or deflection, as a function of the vertical displacement of the piezoelectric scanner (z). This process yields a raw voltage-displacement curve which can be converted into a force-distance curve as follows. By using the slopes of the curves in the region where the probe and the sample are in contact, the voltage can be converted into a cantilever deflection value. In order to minimize the possible effects of repulsive surface forces and/or sample deformation, it is recommended to consider the slope of the retraction curve. The cantilever deflection is then converted into a force (F) by using Hooke's law: $F = -k \times d$, where k is the cantilever spring constant and d is the cantilever deflection. The curve can be corrected by plotting F as a function of $(z - d)$. The zero separation distance is then determined as the position of the vertical linear parts of the curve in the contact region. Interestingly, spatially resolved mapping can be performed by recording multiple force-distance curve arrays in the (x,y) plane (for more details on the principle of force-distance curves, see references 9 and 11).

Different parts may be distinguished in a force-distance curve (Fig. 6). At long probe-sample separation distances, the force experienced by the probe is null. As the probe ap-

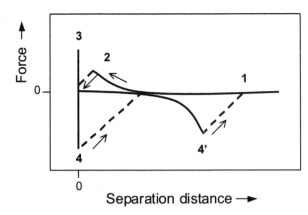

FIGURE 6 AFM force measurements: different parts of a force-distance curve. Labels indicate the approach (1, 2, 3) and retraction (3, 4, 4') parts.

proaches the surface, the cantilever may bend upwards due to repulsive forces until the probe jumps into contact when the gradient of attractive forces exceeds the spring constant plus the gradient of repulsive forces. Upon retraction of the probe from the surface, the curve often shows a hysteresis referred to as the adhesion pull-off force, which can be used to estimate the surface energies of solids or the binding forces between complementary biomolecules. In the presence of long, flexible molecules, an attractive force, referred to as an elongation force, may develop nonlinearly due to macromolecular stretching. The next sections show how the different portions of force-distance curves can be used in microbiology for probing molecular interactions, measuring sample elasticity, and stretching single molecules.

6.3.2. Probing Molecular Interactions

Cellular events such as microbial adhesion, microbial aggregation, and molecular recognition play a pivotal role in the natural environment, in medicine, and in biotechnological processes. Understanding the molecular basis of these phenomena requires knowledge of the physical properties of cell surfaces and of their molecular interactions. Until recently, these properties were not directly accessible to study.

The approach portion of force-distance curves can be used to characterize van der Waals and electrostatic forces, solvation, and hydration and steric and bridging forces associated with polymer-covered surfaces (9). By measuring electrostatic forces, at different pH values, between standard silicon nitride probes and the surfaces of purple membranes, the surface charge density of the latter was estimated to be -0.05 C/m^2 (8). In the same way, the surface potential of 2-D bacteriorhodopsin crystals could be determined and the isoelectric point was found to be about 4 (29). In the presence of 20 mM KCl, long-range repulsion forces were measured at the surfaces of purple membranes and OmpF porins by using silicon nitride probes, reflecting the electrostatic double-layer repulsion (55). These forces could be adjusted by changing the pH or the electrolyte concentration to optimize the image resolution. No repulsion force was found for the HPI layers, indicating that the surface charge of the latter was either too small to be detected or positive. Force-distance curves can detect subtle differences in cell surface composition. For instance, interaction forces between silicon nitride probes and confluent monolayers of isogenic strains of *E. coli* mutants were affected by the lengths of core

lipopolysaccharide molecules on the cell surface and by the production of the capsular polysaccharide, i.e., colanic acid (61).

Reproducible, quantitative force measurements require detailed control of the probe surface chemistry. The surface chemistries of most commercial probes are fairly complex, and the probe surface is often contaminated with gold and other materials. Probes of well-defined chemical compositions can be designed by functionalizing the surface with organic monolayers terminated by specific functional groups (e.g., OH and CH$_3$). The most common method is based on the formation of self-assembled monolayers of alkanethiols on gold surfaces. The procedure involves coating, by thermal evaporation, microfabricated cantilevers with a thin adhesive layer (Cr or Ti), followed by a 15- to 100-nm-thick Au layer, immersing the coated cantilevers in dilute (0.1 to 1 mM) ethanol solutions of the selected alkanethiol, rinsing with ethanol, and drying by using a gentle nitrogen flow. Although the protocol is fairly simple, it is important to validate the quality of the surface modification. This can be done by treating model substrates (glass, mica) in the same way as the probes and characterizing them by means of surface analysis techniques (e.g., contact angle measurements or X-ray photoelectron spectroscopy) (16). Another important point is to use the functionalized probes immediately after they are prepared in order to minimize surface contamination.

Chemically functionalized probes can be used to map physicochemical properties, such as surface hydrophobicity and surface charge. By using OH- and CH$_3$-terminated probes, patterns of nanoscale rodlets were visualized under physiological conditions on the surfaces of spores of *Phanerochaete chrysosporium* (16). Multiple (1,024) force-distance curves recorded over 500-by-500-nm areas at the spore surface, either in deionized water or in 0.1 M NaCl solutions, always showed no adhesion for both OH- and CH$_3$-terminated probes. Control experiments indicated that the lack of adhesion is due to the hydrophilic nature of the spore surface. Following a similar approach, COOH probes allow the mapping of surface charges and local isoelectric points at cell surfaces (2).

An alternative approach consists of immobilizing microbial cells directly on the AFM probe. In this case, the imaging capability of the atomic force microscope cannot be used, meaning that cell surface morphology cannot be controlled while measuring forces. A typical procedure is to adsorb poly(ethyleneimine) from a 1% (wt/vol) aqueous solution onto the probe (58). After rinsing of the probe with deionized water, a cell pellet is transferred onto the poly(ethyleneimine)-coated probe by using a micromanipulator and the coated probe is treated with a drop of 2.5% (vol/vol) glutaraldehyde at 4°C for 1 to 2 h and finally rinsed in water. This method has been used to measure the forces between *E. coli*-coated probes and solids of different surface hydrophobicities (58). Both attractive forces and cell adhesion behavior were promoted by substrate surface hydrophobicity, pointing to the role of hydrophobic interactions. In the above-described method, denaturation of cell surface molecules is likely to occur. This problem can be avoided by using probes coated with native living cells (42). To this end, glass beads are incubated in a solution of APTES (5 to 10% in acetone) for 6 h and the cells are then linked to the amino-functionalized beads by spinning a cell-bead mixture at 7,000 × g for 2 min. A single bacterium-coated bead is then attached to a cantilever with epoxy. By using this approach, the forces between living *Shewanella oneidensis* bacteria and goethite were measured quantitatively (42).

Finally, AFM probes can also be functionalized with biomolecules in order to measure intermolecular forces between individual ligands and receptors (30, 39). Such "nanobioprobes" were used to measure the specific interactions between cell surface lectins and mannose residues involved in the flocculation (aggregation) of the yeast *Saccharomyces carlsbergensis* (69). To measure these interactions at the single-molecule level, gold-coated AFM probes were functionalized with thiol-terminated hexasaccharide molecules, a strategy which provides enhanced accessibility of the carbohydrate moieties to the lectin binding sites. The force curves recorded for *S. carlsbergensis* in flocculating conditions and carbohydrate-coated probes showed adhesion forces of 121 ± 53 pN, reflecting the specific interaction between individual cell surface lectins and glucose residues. In the presence of mannose, the adhesion was strongly reduced, indicating that mannose had blocked the lectin receptor sites, thus confirming the specificity of the measured adhesion force. It is believed that such nanobioprobes have important promise in clinical microbiology and pathogenesis to investigate the interactions between microbial pathogens and host cells and localize cell surface receptors, which may eventually help in developing new therapeutic approaches.

6.3.3. Measuring Cell Surface Elasticity

The mechanical properties of microbial cell walls and surface appendages play an important role in controlling events such as cell growth, cell division, and cell adhesion. By pressing, in a controlled way, the AFM probe onto cell wall components or whole cells, it is possible to measure quantitatively sample elasticity.

So far, both the AFM probe and the sample were assumed to be nondeformable. In practice, most microbiological samples are fairly soft and can be significantly deformed by the probe, resulting in a contact region which is no longer a vertical line such as that represented in Fig. 6 but a curve. By subtracting the approach curves obtained for a soft sample and for a rigid reference material (substrate), it is possible to generate a so-called force-indentation curve, the shape of which provides direct information on the sample's elastic properties (for methodological details, see reference 74).

Multiple force-indentation curves recorded in buffer solution for a gram-negative magnetic bacterium, *Magnetospirillum gryphiswaldense*, were used to determine the effective compressibility of the cell wall, i.e., ~42 mN/m (5). By using approach curves, significant differences in the surface softnesses of the fibrillated *Streptococcus salivarius* HB strain and the nonfibrillated *S. salivarius* HBC12 strain were demonstrated (72). In agreement with AFM and electron microscopy images, a long-range repulsion was measured on the fibrillated strain and attributed to the compression of the soft cell surface fibrils.

The AFM "depression technique" is an original approach, complementary to the indentation method, for probing the elastic properties of isolated bacterial cell wall components (77, 78). The main features of the procedure are as follows. A drop of a solution containing the cell wall components (archeal sheaths and peptidoglycan sacculi, etc.) is placed onto a hard substrate and allowed to dry. The substrate is made of GaAs or Si_3N_4 and contains grooves that are narrow compared to the width or the length of the material to be investigated so that single wall components

bridging one or more grooves can be obtained. The force measurement consists of placing the probe at the midpoint of the unsuspended wall component region and increasing or decreasing the force as a linear function of time. Comparison of the cantilever deflections for the specimen and the hard substrate allows an accurate determination of the specimen elasticity. Applying this approach to the proteinaceous sheath of the archeon *Methanospirillum hungatei* GP1 yielded an elastic modulus of 2×10^{10} to 4×10^{10} N/m², indicating that the sheath could withstand an internal pressure of 400 atm, well beyond that needed for a eubacterial envelope to withstand turgor pressure (77). For murein sacculi of gram-negative bacteria in the hydrated state, elastic moduli of 2.5×10^7 N/m² were measured (78).

Interestingly, lateral variations of elasticity on single cells can be detected by using spatially resolved force maps. To this end, arrays of force-distance curves are recorded in parallel with topographic images. Each curve is then converted into a force-indentation curve to extract local elastic modulus (or Young's modulus) values. Applied to budding yeast cells, this approach yielded Young's modulus values of 6.1 and 0.6 MPa for the bud scar and surrounding cell surface, respectively (68). This result showed that the bud scar is 10 times stiffer than the surrounding cell wall, a finding which is consistent with the accumulation of chitin in this area.

6.3.4. Stretching Single Molecules

How stiff is a single polysaccharide molecule? What are the forces required to unfold a protein? What are the interactions responsible for the anchorage of proteins in cell membranes? How strong are the forces driving the formation of supramolecular assemblages? These questions, of great relevance in cellular and molecular microbiology, can now be addressed by means of AFM force spectroscopy.

When the AFM probe is pulled away from a sample that is made of long, flexible macromolecules, the retraction curves often show attractive elongation forces, developing nonlinearly, which reflect the stretching of the macromolecules (Fig. 6). In such experiments, referred to as single-molecule force spectroscopy experiments, it is common to present the force curve as the positive pulling force versus the extension (e.g., see Fig. 8). Two models from statistical mechanics are often used to describe the elasticity of long, flexible molecules, i.e., the worm-like chain (WLC) and the freely jointed chain (FJC) models (34). The WLC model describes the polymer as an irregular, curved filament, which is linear on the scale of the persistence length, a parameter which represents the stiffness of the molecule. In the FJC model, the polymer is considered as a series of rigid, orientationally independent statistical (Kuhn) segments, connected through flexible joints. An extended FJC (FJC+) model has been developed in which Kuhn segments can stretch and align under force.

In recent years, force spectroscopy experiments have provided new insight into the nanomechanical properties of single DNA, protein, and polysaccharide molecules (for reviews, see references 12, 23, and 34). For dextran, a polysaccharide found in bacterial and yeast species, the deformation at low forces was dominated by entropic forces and was described by an extended FJC model (45, 62). At strong forces, the dextran filaments underwent a distinct conformational change corroborated by molecular dynamic calculations. The polymer stiffened, and the segment elasticity was dominated by the bending of bond angles. Although the biological relevance of this behavior remains unclear, it may significantly improve the polymer ductibility.

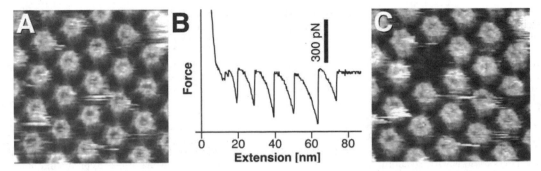

FIGURE 7 Combining AFM imaging and force spectroscopy to unzip proteins from the HPI layer of *Deinococcus radiodurans* (56). (A) Control AFM topograph of the inner surface of the HPI layer. (B) Force-extension curve recorded for this inner surface region showing a saw-tooth pattern with six force peaks of about 300 pN. (C) The same inner surface area imaged after recording of the force curve; a molecular defect the size of a hexameric HPI protein complex has clearly been created. (Reprinted from reference 56 with permission of the publisher.)

Another elegant study involving a microbial specimen is the unzipping of bacterial proteins from the HPI layer of *Deinococcus radiodurans* by using combined AFM imaging and force spectroscopy (56). Force-extension curves recorded for the inner surface of the HPI layer showed saw-tooth patterns with six force peaks of about 300 pN (Fig. 7). This behavior, well-fitted with a WLC model, was attributed to the sequential pulling out of the protomers of the hexameric HPI protein complex. After recording of the force curve, a molecular defect the size of a hexameric complex was clearly visualized by means of high-resolution imaging. Hence, AFM allows for correlation of force measurements with resulting structural changes.

Stretching macromolecules directly on living cells is very challenging due to the complex and dynamic nature of the surfaces. The forces between living *Shewanella oneidensis* bacteria and goethite, both commonly found in Earth near-surface environments, were measured quantitatively (42). Energy values derived from these measurements showed that the affinity between *S. oneidensis* and goethite rapidly increases by two to five times under anaerobic con-ditions in which electron transfer from the bacterium to the mineral is expected. Specific signatures in the force-extension curves were well-fitted with a WLC model and suggested to reflect the unfolding of a 150-kDa putative iron reductase.

Long, flexible macromolecules were stretched at the surface of living fungal spores (71). As can be seen in Fig. 8, force-extension curves recorded for germinating spores of *Aspergillus oryzae* showed single or multiple attractive forces of 400- ± 100-pN magnitude, along with characteristic elongation forces and rupture lengths ranging from 20 to 500 nm. Elongation forces were well fitted with an FJC+ model by using fitting parameters that were consistent with the stretching of individual dextran molecules (45, 62). The measured forces, attributed to the stretching of long cell surface polysaccharides, may be of great biological relevance. Incubating the spores of *A. oryzae* in liquid medium for a few hours leads to their germination and, in turn, to their aggregation. Hence, one may expect that the elastic properties of long surface macromolecules may play a significant role in modulating spore aggregation and other biointerfacial processes. Macromolecular stretching was also performed on the surface of *Pseudomonas putida* in solvents spanning a range of polarity and ionic strengths to learn about the macromolecule elasticity and adhesion. Shorter lengths were observed at higher ionic strength, indicating that the polymer chains were less extended in the presence of salt (1).

6.4. REFERENCES

1. **Abu-Lail, N. I., and T. A. Camesano.** 2002. Elasticity of *Pseudomonas putida* KT2442 surface polymers probed with single-molecule force microscopy. *Langmuir* **18:**4071–4081.
2. **Ahimou, F., F. A. Denis, A. Touhami, and Y. F. Dufrêne.** 2002. Probing microbial cell surface charges by atomic force microscopy. *Langmuir* **18:**9937–9941.
3. **Ahimou, F., A. Touhami, and Y. F. Dufrêne.** 2003. Real-time imaging of the surface topography of living yeast cells by atomic force microscopy. *Yeast* **20:**25–30.
4. **Amro, N. A., L. P. Kotra, K. Wadu-Mesthrige, A. Bulychev, S. Mobashery, and G.-Y. Liu.** 2000. High-resolution atomic force microscopy studies of the *Escherichia coli* outer membrane: structural basis for permeability. *Langmuir* **16:**2789–2796.

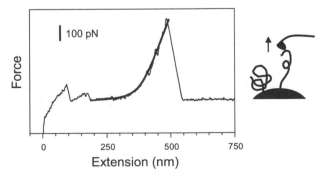

FIGURE 8 Stretching cell surface macromolecules. Typical force-extension curve recorded under water between a silicon nitride probe and the surface of a germinating *Aspergillus oryzae* spore (thin line, starting at 0 mm extension). Elongation forces were well described by an extended freely jointed chain model (thick line, starting at 200 mm extension) with parameters consistent with values reported for the elastic deformation of single dextran and amylose polysaccharides (71).

5. **Arnoldi, M., C. M. Kacher, E. Bäuerlein, M. Radmacher, and M. Fritz.** 1998. Elastic properties of the cell wall of *Magnetospirillum gryphiswaldense* investigated by atomic force microscopy. *Appl. Phys. A* **66:**S613–S617.

6. **Beech, I. B., C. W. S. Cheung, D. B. Johnson, and J. R. Smith.** 1996. Comparative studies of bacterial biofilms on steel surfaces using atomic force microscopy and environmental scanning electron microscopy. *Biofouling* **10:**65–77.

7. **Binnig, G., C. F. Quate, and C. Gerber.** 1986. Atomic force microscope. *Phys. Rev. Lett.* **56:**930–933.

8. **Butt, H.-J.** 1992. Measuring local surface charge densities in electrolyte solutions with a scanning force microscope. *Biophys. J.* **63:**578–582.

9. **Butt, H.-J., M. Jaschke, and W. Ducker.** 1995. Measuring surface forces in aqueous electrolyte solution with the atomic force microscope. *Bioelectrochem. Bioenerg.* **38:**191–201.

10. **Camesano, T. A., M. J. Natan, and B. E. Logan.** 2000. Observation of changes in bacterial cell morphology using tapping mode atomic force microscopy. *Langmuir* **16:**4563–4572.

11. **Cappella, B., and G. Dietler.** 1999. Force-distance curves by atomic force microscopy. *Surf. Sci. Rep.* **34:**1–104.

12. **Clausen-Schaumann, H., M. Seitz, R. Krautbauer, and H. E. Gaub.** 2000. Force spectroscopy with single biomolecules. *Curr. Opin. Chem. Biol.* **4:**524–530.

13. **Cleveland, J. P., S. Manne, D. Bocek, and P. K. Hansma.** 1993. A nondestructive method for determining the spring constant of cantilevers for scanning force microscopy. *Rev. Sci. Instrum.* **64:**403–405.

14. **Colton, R. J., A. Engel, J. E. Frommer, H. E. Gaub, A. A. Gewirth, R. Guckenberger, J. Rabe, W. M. Heckel, and B. Parkinson (ed.).** 1998. *Procedures in Scanning Probe Microscopies.* John Wiley & Sons Ltd., Chichester, England.

15. **Crawford, S. A., M. J. Higgins, P. Mulvaney, and R. Wetherbee.** 2001. Nanostructure of the diatom frustule as revealed by atomic force and scanning electron microscopy. *J. Phycol.* **37:**543–554.

16. **Dufrêne, Y. F.** 2000. Direct characterization of the physicochemical properties of fungal spores using functionalized AFM probes. *Biophys. J.* **78:**3286–3291.

17. **Dufrêne, Y. F.** 2001. Application of atomic force microscopy to microbial surfaces: from reconstituted cell surface layers to living cells. *Micron* **32:**153–165.

18. **Dufrêne, Y. F.** 2002. Atomic force microscopy, a powerful tool in microbiology. *J. Bacteriol.* **184:**5205–5213.

19. **Dufrêne, Y. F.** 2003. Recent progress in the application of atomic force microscopy imaging and force spectroscopy to microbiology. *Curr. Opin. Microbiol.* **6:**317–323.

20. **Dufrêne, Y. F.** 2004. Using nanotechniques to explore microbial surfaces. *Nat. Rev. Microbiol.* **2:**451–460.

20a. **Dufrêne, Y. F.** 2003. Atomic force microscopy provides a new means for looking at microbial cells. *ASM News* **69:**438–442.

21. **Dufrêne, Y. F., C. J. P. Boonaert, P. A. Gerin, M. Asther, and P. G. Rouxhet.** 1999. Direct probing of the surface ultrastructure and molecular interactions of dormant and germinating spores of *Phanerochaete chrysosporium. J. Bacteriol.* **181:**5350–5354.

22. **Firtel, M., and T. J. Beveridge.** 1995. Scanning probe microscopy in microbiology. *Micron* **26:**347–362.

23. **Fisher, T. E., P. E. Marszalek, and J. M. Fernandez.** 2000. Stretching single molecules into novel conformations using the atomic force microscope. *Nat. Struct. Biol.* **7:**719–724.

24. **Forsythe, J. H., P. A. Maurice, and L. E. Hersman.** 1998. Attachment of a *Pseudomonas* sp. to Fe(III)-(hydr)oxide surfaces. *Geomicrobiology* **15:**293–308.

25. **Gad, M., and A. Ikai.** 1995. Method for immobilizing microbial cells on gel surface for dynamic AFM studies. *Biophys. J.* **69:**2226–2233.

26. **Grantham, M. C., and P. M. Dove.** 1996. Investigation of bacterial-mineral interactions using fluid tapping mode atomic force microscopy. *Geochim. Cosmochim. Acta* **60:**2473–2480.

27. **Hansma, H. G., and J. H. Hoh.** 1994. Biomolecular imaging with the atomic force microscope. *Ann. Rev. Biophys. Biomol. Struct.* **23:**115–139.

28. **Hansma, H. G., J. Vesenka, C. Siegerist, G. Kelderman, H. Morrett, R. L. Sinsheimer, V. Elings, C. Bustamante, and P. K. Hansma.** 1992. Reproducible imaging and dissection of plasmid DNA under liquid with the atomic force microscope. *Science* **256:**1180–1184.

29. **Hartley, P., M. Matsumoto, and P. Mulvaney.** 1998. Determination of the surface potential of two-dimensional crystals of bacteriorhodopsin by AFM. *Langmuir* **14:**5203–5209.

30. **Hinterdorfer, P., W. Baumgartner, H. J. Gruber, K. Schilcher, and H. Schindler.** 1996. Detection and localization of individual antibody-antigen recognition events by atomic force microscopy. *Proc. Natl. Acad. Sci. USA* **93:**3477–3481.

31. **Holstein, T. W., M. Benoit, G. V. Herder, G. Wanner, C. N. David, and H. E. Gaub.** 1994. Fibrous minicollagens in *Hydra* nematocysts. *Science* **265:**402–404.

32. **Hörber, J. K. H., and M. J. Miles.** 2003. Scanning probe evolution in biology. *Science* **302:**1002-1005.

33. **Ikai, A.** 1996. STM and AFM of bio/organic molecules and structures. *Surf. Sci. Rep.* **26:**261–332.

34. **Janshoff, A., M. Neitzert, Y. Oberdörfer, and H. Fuchs.** 2000. Force spectroscopy of molecular systems—single molecule spectroscopy of polymers and biomolecules. *Angew. Chem. Int. Ed.* **39:**3213–3237.

35. **Karrasch, S., and A. Engel.** 1998. AFM imaging of HPI layers in buffer solution, p. 433–439. *In* R. J. Colton, A. Engel, J. E. Frommer, H. E. Gaub, A. A. Gewirth, R. Guckenberger, J. Rabe, W. M. Heckel, and B. Parkinson (ed.), *Procedures in Scanning Probe Microscopies.* John Wiley & Sons Ltd., Chichester, England.

36. **Karrasch, S., R. Hegerl, J. Hoh, W. Baumeister, and A. Engel.** 1994. Atomic force microscopy produces faithful high-resolution images of protein surfaces in an aqueous environment. *Proc. Natl. Acad. Sci. USA* **91:**836–838.

37. **Kasas, S., and A. Ikai.** 1995. A method for anchoring round shaped cells for atomic force microscope imaging. *Biophys. J.* **68:**1678–1680.

38. **Kirby, A. R., A. P. Gunning, and V. J. Morris.** 1996. Imaging polysaccharides by atomic force microscopy. *Biopolymers* **38:**355–366.

39. **Lee, G. U., L. A. Chrisey, and R. J. Colton.** 1994. Direct measurement of the forces between complementary strands of DNA. *Science* **266:**771–773.

40. **Lister, T. E., and P. J. Pinhero.** 2001. In vivo atomic force microscopy of surface proteins on *Deinococcus radiodurans. Langmuir* **17:**2624–2628.

41. **Lo, Y. S., N. D. Huefner, W. S. Chan, P. Dryden, B. Hagenhoff, and T. P. Beebe.** 1999. Organic and inorganic contamination on commercial AFM cantilevers. *Langmuir* **15:**6522–6526.

42. **Lower, S. K., M. F. Hochella, and T. J. Beveridge.** 2001. Bacterial recognition of mineral surfaces: nanoscale interactions between *Shewanella* and α-FeOOH. *Science* **292:**1360–1363.

43. **Lyubchenko, Y. L., A. A. Gall, L. S. Shlyakhtenko, R. E. Harrington, P. I. Oden, B. L. Jacobs, and S. M. Lindsay.** 1992. Atomic force microscopy imaging of double stranded DNA and RNA. *J. Biomol. Struct. Dyn.* **9:**589–606.

44. Magonov, S. N., and M.-H. Whangbo (ed.). 1996. *Surface Analysis with STM and AFM.* VCH, New York, NY.

45. Marszalek, P. E., A. F. Oberhauser, Y.-P. Pang, and J. M. Fernandez. 1998. Polysaccharide elasticity governed by chair-boat transitions of the glucopyranose ring. *Nature* 396:661–664.

46. McIntire, T. M., and D. A. Brant. 1997. Imaging of individual biopolymers and supramolecular assemblies using noncontact atomic force microscopy. *Biopolymers* 42:133–146.

47. Möller, C., M. Allen, V. Elings, A. Engel, and D. J. Müller. 1999. Tapping mode atomic force microscopy produces faithful high-resolution images of protein surfaces. *Biophys. J.* 77:1150–1158.

48. Morris, V. J., A. R. Kirby, and A. P. Gunning (ed.). 1999. *Atomic Force Microscopy for Biologists.* Imperial College Press, London, England.

49. Müller, D. J., and A. Engel. 1999. Voltage and pH-induced channel closure of porin OmpF visualized by atomic force microscopy. *J. Mol. Biol.* 285:1347–1351.

50. Müller, D. J., G. Büldt, and A. Engel. 1995. Force-induced conformational change of bacteriorhodopsin. *J. Mol. Biol.* 249:239–243.

51. Müller, D. J., W. Baumeister, and A. Engel. 1996. Conformational change of the hexagonally packed intermediate layer of *Deinococcus radiodurans* monitored by atomic force microscopy. *J. Bacteriol.* 178:3025–3030.

52. Müller, D. J., M. Amrein, and A. Engel. 1997. Adsorption of biological molecules to a solid support for scanning probe microscopy. *J. Struct. Biol.* 119:172–188.

53. Müller, D. J., A. Engel, and M. Amrein. 1997. Preparation techniques for the observation of native biological systems with the atomic force microscope. *Biosens. Bioelectron.* 12:867–877.

54. Müller, D. J., G. Büldt, and A. Engel. 1998. Preparation and observation of purple membranes by AFM, p. 425–428. *In* R. J. Colton, A. Engel, J. E. Frommer, H. E. Gaub, A. A. Gewirth, R. Guckenberger, J. Rabe, W. M. Heckel, and B. Parkinson (ed.), *Procedures in Scanning Probe Microscopies.* John Wiley & Sons Ltd., Chichester, England.

55. Müller, D. J., D. Fotiadis, S. Scheuring, S. A. Müller, and A. Engel. 1999. Electrostatically balanced subnanometer imaging of biological specimens by atomic force microscope. *Biophys. J.* 76:1101–1111.

56. Müller, D. J., W. Baumeister, and A. Engel. 1999. Controlled unzipping of a bacterial surface layer with atomic force microscopy. *Proc. Natl. Acad. Sci. USA* 96:13170–13174.

57. Müller, D. J., J. B. Heymann, F. Oesterhelt, C. Möller, H. Gaub, G. Büldt, and A. Engel. 2000. Atomic force microscopy of native purple membrane. *Biochim. Biophys. Acta* 1460:27–38.

58. Ong, Y.-L., A. Razatos, G. Georgiou, and M. M. Sharma. 1999. Adhesion forces between *E. coli* bacteria and biomaterial surfaces. *Langmuir* 15:2719–2725.

59. Patel, N., M. C. Davies, M. Hartshorne, R. J. Heaton, C. J. Roberts, S. J. B. Tendler, and P. M. Williams. 1997. Immobilization of protein molecules onto homogeneous and mixed carboxylate-terminated self-assembled monolayers. *Langmuir* 13:6485–6490.

60. Pum, D., M. Weinhandl, C. Hödl, and U. B. Sleytr. 1993. Large-scale recrystallization of the S-layer of *Bacillus coagulans* E38-66 at the air/water interface and on lipid films. *J. Bacteriol.* 175:2762–2766.

61. Razatos, A., Y.-L. Ong, M. M. Sharma, and G. Georgiou. 1998. Molecular determinants of bacterial adhesion monitored by atomic force microscopy. *Proc. Natl. Acad. Sci. USA* 95:11059–11064.

62. Rief, M., F. Oesterhelt, B. Heymann, and H. E. Gaub. 1997. Single molecule force spectroscopy on polysaccharides by atomic force microscopy. *Science* 275:1295–1297.

63. Schabert, F. A. 1998. Imaging of reconstituted OmpF porin in solution using AFM, p. 429–432. *In* R. J. Colton, A. Engel, J. E. Frommer, H. E. Gaub, A. A. Gewirth, R. Guckenberger, J. Rabe, W. M. Heckel, and B. Parkinson (ed.), *Procedures in Scanning Probe Microscopies.* John Wiley & Sons Ltd., Chichester, England.

64. Schabert, F. A., C. Henn, and A. Engel. 1995. Native *Escherichia coli* OmpF porin surfaces probed by atomic force microscopy. *Science* 268:92–94.

65. Scheuring, S., P. Ringler, M. Borgnia, H. Stahlberg, D. J. Müller, P. Agre, and A. Engel. 1999. High resolution AFM topographs of the *Escherichia coli* water channel aquaporin Z. *EMBO J.* 18:4981–4987.

66. Shao, Z., J. Mou, D. M. Czajkowsky, J. Yang, and J.-Y. Yuan. 1996. Biological atomic force microscopy: what is achieved and what is needed. *Adv. Phys.* 45:1–86.

67. Tortonese, M., and M. Kirk. 1997. Characterization of application specific probes for SPMs. *SPIE* 3009:53–60.

68. Touhami, A., B. Nysten, and Y. F. Dufrêne. 2003. Nanoscale mapping of the elasticity of microbial cells by atomic force microscopy. *Langmuir* 19:4539–4543.

69. Touhami, A., B. Hoffmann, A. Vasella, F. A. Denis, and Y. F. Dufrêne. 2003. Aggregation of yeast cells: direct measurement of discrete lectin-carbohydrate interactions. *Microbiol. SGM* 149:2873–2878.

70. Touhami, A., M. Jericho, and T. J. Beveridge. 2004. Atomic force microscopy of cell growth and division in *Staphylococcus aureus*. *J. Bacteriol.* 186:3286–3295.

71. van der Aa, B. C., R. M. Michel, M. Asther, M. T. Zamora, P. G. Rouxhet, and Y. F. Dufrêne. 2001. Stretching cell surface macromolecules by atomic force microscopy. *Langmuir* 17:3116–3119.

72. van der Mei, H. C., H. J. Busscher, R. Bos, J. de Vries, C. J. P. Boonaert, and Y. F. Dufrêne. 2000. Direct probing by atomic force microscopy of the cell surface softness of a fibrillated and non-fibrillated oral streptococcal strain. *Biophys. J.* 78:2668–2674.

73. Wagner, P., M. Hegner, P. Kernen, F. Zaugg, and G. Semenza. 1996. Covalent immobilization of native biomolecules onto Au(111) via N-hydroxysuccinimide ester functionalized self-assembled monolayers for scanning probe microscopy. *Biophys. J.* 70:2052–2066.

74. Weisenhorn, A. L., M. Khorsandi, S. Kasas, V. Gotzos, and H.-J. Butt. 1993. Deformation and height anomaly of soft surfaces studied with an AFM. *Nanotechnology* 4:106–113.

75. Wetzer, B., D. Pum, and U. B. Sleytr. 1997. S-layer stabilized solid supported lipid bilayers. *J. Struct. Biol.* 119:123–128.

76. Wilson, D. L., K. S. Kump, S. J. Eppell, and R. E. Marchant. 1995. Morphological restoration of atomic force microscopy images. *Langmuir* 11:265–272.

77. Xu, W., P. J. Mulhern, B. L. Blackford, M. H. Jericho, M. Firtel, and T. J. Beveridge. 1996. Modeling and measuring the elastic properties of an archaeal surface, the sheath of *Methanospirillum hungatei*, and the implication for methane production. *J. Bacteriol.* 178:3106–3112.

78. Yao, X., M. Jericho, D. Pink, and T. Beveridge. 1999. Thickness and elasticity of gram-negative murein sacculi measured by atomic force microscopy. *J. Bacteriol.* 181:6865–6875.

7

Cell Fractionation

SUSAN F. KOVAL AND G. DENNIS SPROTT

This chapter presents techniques for the fractionation of cellular components, including organelles and appendages, beginning with external surfaces and ending with internal components. The structural and physiological diversity of prokaryotes makes it impossible to present a single universal method for cell fractionation or to summarize all of the approaches developed to fit each bacterium. Extra effort was made to provide methods for the isolation of components from both bacterial and archaeal phylogenetic lineages. In an era of molecular biology, genomics, and proteomics, many researchers may not have experience in breakage and fractionation procedures (excluding those for nucleic acids), so it is hoped that such individuals will find a revisit to some of the germane, and also sometimes older, literature rewarding.

Large variations in genes exist even within the subgroupings of bacteria and often lead to changes in structure; these structural variations can affect cell fractionation. Indeed, a rational fractionation scheme can be developed only on the basis of solid knowledge of a particular organism. Since the information needed is often lacking for newly isolated or poorly studied bacteria, some preliminary exploration may be essential. It is important, for example, to determine at what stage in the growth cycle and in which growth medium the component of interest is produced optimally. Such factors can influence not only the yield, but also the chemical structures of cell components.

It is recommended that the isolation of any cellular component be monitored by electron microscopy (chapter 4) at all stages of the procedure. Electron microscopy often provides the only adequate means of monitoring the fractionation and the state of intactness of the cell component. This is especially important during the preliminary exploration of a bacterium.

Glassware must be carefully cleaned and sometimes sterilized to avoid contamination from stray microbial cells and enzymes, especially if the fraction is to be stored. Also, it is preferable to store samples at −20°C or below, provided that the component resists freeze-thaw damage. Samples can be stored for short periods of time at 4°C, with the growth of microbial contaminants inhibited temporarily with poisons such as 0.05% (wt/vol) sodium azide. Containment facilities may be needed to comply with safety regulations if pathogenic bacteria are used, as well as to protect against oxidative inactivation of specific components of anaerobic prokaryotes.

The fractionation of cytoplasmic or organelle-associated proteins is not discussed, since gel electrophoresis and the chromatography of soluble proteins are described in chapter 21. It should be appreciated, however, that many of the techniques described in this chapter, from cell breakage to centrifugation theory, will have similar applications.

7.1. CELL BREAKAGE

Because of differences in the architectures of prokaryotic cell surfaces, there is perhaps no greater area of variability, or importance, than the initial choice of breakage method.

This choice will depend on how easily the microorganism is broken and on how the method of breakage affects the structural intactness and functional activity of the isolated component(s). The factors pertinent to cell fractionation, often dictating the approach needed, are the differences in wall structures and the bonding forces within walls (and associated surface materials) responsible for maintaining cell shape.

Fundamental differences exist in the cell envelope structures of bacteria and archaea, which influence the choice of cell breakage method. The Gram staining reaction subdivides bacteria into gram-negative, gram-positive, and gram-variable species on the basis of wall structure (chapter 2.3.5.), and this staining may be useful as a first step in assigning a method for breakage. In general, the gram-positive cocci have thick peptidoglycan layers; this characteristic and their shear-resistant shape make them the most resilient to mechanical breakage. The Gram reaction has limited value for archaea because their cell walls can be so fundamentally different in structure and chemistry from those of bacteria as to defy conventional interpretation. Archaea, which include the methanogenic, extremely halophilic, and sulfur-dependent groupings, have at least five different envelope structures (71). Archaea that have only a proteinaceous S-layer as the sole wall component are relatively fragile (although the S-layer of some hyperthermophiles may be covalently bonded). Archaea with lysozyme-insensitive pseudomurein may be more difficult to break. *Thermoplasma* and *Methanoplasma* species resemble the mycoplasmas with respect to the complete absence of a wall structure.

The extent of breakage can be assessed by comparing the cell counts for the suspension before and after treatment. For routine assessments, it is adequate to view wet mounts by phase-contrast microscopy or by negative staining with nigrosin (chapter 2.2.2.). Microscopy, providing direct observation of the numbers and intactness of cells (chapter 2.1.4. and 2.1.5.), has an advantage over techniques which monitor the loss of cell viability, since viability may be lost without a concomitant release of internal contents. Alternatively, it is possible to make measurements to reflect the release of cytoplasmic contents into the supernatant following centrifugation (e.g., 8,000 × g for 15 min). Markers for cytoplasm include specific cytoplasmic enzyme activities, ions such as K^+, and UV-absorbing cofactors or nucleic acids.

7.1.1. Mechanical Breakage
Mechanical methods for cell breakage are the most universally successful. Each method, however, must be carefully assessed in light of the species studied and the cell fraction to be isolated.

7.1.1.1. Pressure Shearing
Pressure shearing using the French pressure cell is probably the most widely used and useful method of cell breakage, particularly for gram-negative bacteria (including the cyanobacteria) and gram-positive bacteria (particularly the

gram-positive bacilli). Gram-positive cocci are not broken readily by this device, and ballistic disintegration or the enzymatic digestion of the cell wall is more effective for breaking them. The French pressure cell is effective for many archaea, although *Methanobrevibacter* and *Thermoplasma* spp. are particularly resistant. An apparatus capable of achieving higher pressures, such as the EmulsiFlex™ high-pressure homogenizers sold by Avestin (www.avestin.com), can overcome these difficulties. The EmulsiFlex™ series, similar in principle to the French pressure cell, have capacities ranging from small scale (0.5 to 3.5 ml) to continuous flow (5 to <1,200 liters/h) depending on the model chosen. The pump operating these units is driven with compressed air or, in the case of the larger units, with electrical power.

The French pressure cell consists of a steel cylinder, piston, and pressure relief valve fitted with a replaceable nylon ball. Additional equipment includes a loading stand and a 10-ton (89,000 N) motor-driven hydraulic press which can deliver a constant force over a range of piston speeds. The Aminco pressure cell, accessories, and a suitable hydraulic pump can be obtained from SLM Instruments, Inc. (Urbana, IL), or Canberra Canada Ltd. (Kanata, Ontario). Alternatively, the pump, which is the most expensive part of the apparatus, can be purchased through a local supplier of hydraulic equipment, although commercial units may require some modification by a machine shop to fit the cell unit safely and satisfactorily. Aminco cells are available with maximum capacities of 3.7, 35, or 40 ml of sample. Both the 35- and 40-ml-capacity cells can be fitted with a one-way return valve for continuous operation with larger volumes of sample or for repeated processing. The larger cells are recommended for most applications, since they are more easily controlled than the smaller ones.

To use a French pressure cell, load the cell suspension (up to 30% wet cell paste by volume) into the pressure cell, taking care to avoid a gas pocket. Close the relief valve, and place the entire assembly into the hydraulic press. Lower the hydraulic press piston onto the cell piston, and allow it to reach the maximum force desired. Then slowly open the relief valve to the point where the piston moves downward while maintaining pressure. The cells are broken by the high shear force generated as the suspension passes through the small orifice of the relief valve.

This method has several advantages. Heating can be minimized by precooling the press cylinder and piston; heat is generated only when the cells pass through the relief valve. The effluent temperature will rise by 5 to 10°C, but the effluent can be rapidly cooled if it is collected in a metal tube or beaker in an ice bath. Anaerobiosis may be maintained by flushing the chamber with argon or N_2 while loading the cell suspension with a syringe preflushed with N_2 and collecting the effluent in a serum vial continuously flushed with N_2. Breakage is virtually instantaneous, and the broken suspension is not subjected to additional shear forces that could damage subcellular particles. If a satisfactory hydraulic press is used, very reproducible conditions can be obtained. The extent of breakage is little influenced by the density of the cell suspension, the growth phase at which the cells are harvested, or the breakage medium in which the cells are suspended. Multiple passes are done as needed.

Breakage Procedure for *Escherichia coli* Cells

Harvest cultures, wash them once, and resuspend them at 0.01 to 0.1 times the original culture volume in 0.01 M HEPES, pH 7.4, at 0°C. Other buffers such as phosphate may be used, but Tris buffer or chelating agents should be avoided because they cause outer membrane damage. Add a small amount of pancreatic RNase and DNase I (~0.1 mg/ml; omit Mg^{2+} to encourage ribosome dissociation), and break the suspension with a French pressure cell operated at a cell pressure of 16,000 lb/in^2 (1 lb/in^2 = 6.895 kPa). After breakage, add $MgCl_2$ to give a final concentration of 1 mM, and remove unbroken cells by centrifugation at 5,000 × g for 5 min. The purpose of the Mg^{2+} is to activate DNase. RNase acts in the absence of Mg^{2+}. Omission of the DNase allows for the isolation of short strands of DNA by this breakage method; viscosity is generally not a problem because of DNA shearing.

7.1.1.2. Ultrasonic Disintegration

A variety of ultrasonic probe devices are available to break both prokaryotic and eukaryotic cells. Programmable ultrasonic liquid processors are now available (e.g., Sonicator™ from Misonix Inc., Farmingdale, NY, and Sonic Dismembranator 500™ from Fisher Scientific). Operators will find the timer-programming feature convenient, although older, nonprogrammable sonicators (e.g., MSE and Braun types) function just as efficiently in most laboratories. The rapid vibration of the probe tip produces high-intensity sound waves, which generate microscopic gas bubbles (especially on small nucleation particles such as bacteria). Creation of these transient cavities (cavitation) is thought to create high shear gradients by microstreaming (59). This breakage method is tempting to use, since the probe devices are relatively inexpensive and effective. However, there are inherent disadvantages.

The major disadvantage is that breakage is not instantaneous; a cell suspension must be treated for 30 s to several minutes before a reasonable proportion of the cells are broken. During this time, subcellular particles released from broken cells are subjected to the same high shear forces as the unbroken cells. Membrane vesicles can be degraded into small lipoprotein fragments, which can no longer be sedimented by ultracentrifugation, and extensive redistribution of membrane proteins among various membranes can occur (e.g., between the inner and outer membranes of gram-negative bacteria).

In addition, it is difficult to control the temperature of the sample during breakage. Foaming can cause protein denaturation, and cavitation promotes oxidation of oxygensensitive enzymes and unsaturated lipids. These problems can be minimized by subjecting the sample to several short bursts (15 to 30 s) rather than one continuous treatment, with cooling periods between the bursts. Covering the sample with argon during treatment may minimize oxidation. It is difficult to obtain reproducible breakage, since the effectiveness of the probe depends upon sample viscosity and the size, shape, and composition of the sample vessel (e.g., a glass beaker is far more effective than a plastic one). *Caution*: wear ear protection, unless the probe is enclosed in a sound-absorbing cabinet. Avoid holding the vessel in the fingers, because ultrasonic transmission is effective through glass.

Although ultrasonic treatment is not recommended for primary cell breakage, it may be useful for small-scale preparation of walls or envelopes, for distinguishing soluble from particulate proteins, or for the initial identification of S-layers (by electron microscopy). In these procedures, very short bursts of ultrasonic energy are required so that subcellular organelles are not damaged. For the processing of difficult cells, glass beads (10 to 50 μm in size) (see section

7.1.1.3) can be added to the liquid to facilitate breakage. *Staphylococcus* species should be pretreated with lysostaphin. Ultrasonic treatment is useful for lysing spheroplasts (as in the procedure of Osborn and Munson [109] for the separation of plasma and outer membranes of gram-negative bacteria), for dispersing clumps of subcellular organelles, and for suspending centrifuge pellets.

7.1.1.3. Ballistic Disintegration

The term ballistic disintegration describes a variety of methods in which prokaryotes are broken by the shear forces developed when a suspension of cells together with small glass or plastic beads is shaken or agitated violently. Many commercial devices have been used, and they are detailed in reviews by Salton (120) and by Hughes et al. (59).

The Mickle shaker has been replaced by the MSK Tissue Disintegrator™ (B. Braun, Melsungen, Germany) that is available from many laboratory supply houses in the United States. The Braun disintegrator has a 65-ml sample container that is shaken horizontally at 2,000 to 4,000 oscillations/min. Liquid CO_2 delivered to the sample container provides cooling. Most samples can be disintegrated in 3 to 5 min at a temperature of <4°C. The Braun disintegrator is often the method of choice for the isolation of cell walls and particulate enzymes from gram-positive bacteria. The following procedure (156) can be used with most microorganisms.

Mix a cell suspension (30 ml, containing 20 to 50 mg [dry weight]/ml) in a suitable buffer with 20 ml of glass beads (Ballotini #12; 100 to 200 μm) in the sample container (an air space of at least 20% of the container volume is essential for breakage), and pass liquid CO_2 through the apparatus for 0.5 min to cool the container. Shake the container at 3,000 strokes/min for 3.5 to 5.0 min depending on the species. The beads may be removed by low-speed centrifugation or by passage through coarse-grade sintered glass. Because glass beads may liberate alkali when shaken violently, with potential damage to alkali-sensitive components, they must be acid washed, usually with nitric acid, before use. Suitable glass beads are supplied by Sigma Chemical (St. Louis, MO). (Some workers prefer to use plastic beads [118], to minimize enzyme denaturation.) When this procedure is used for the isolation of cell walls, the sample should be treated promptly as described below to inactivate autolytic enzymes (156).

Bacterial endospores can be broken with the Braun disintegrator by shaking dry spores with dry, acid-washed glass beads (5 g of dry spores plus 15 g of beads) as described above and then suspending the sample in buffer (137). Beads with a 470-μm diameter are appropriate; acid-washed beads with diameters of 425 to 600 μm are available from Sigma.

The Bead-Beater™ is sold by Biospec Products (Bartlesville, OK) as an apparatus for homogenizing and disrupting microorganisms. To use this apparatus, mix glass beads with up to 20 g (dry weight) of cells (500-μm diameter for yeast, fungi, and algae; 100 to 150 μm in diameter for bacteria) in a 30- or 60-ml chamber. A Teflon rotor within the chamber is driven by a conventional blender motor and provides breakage by bombardment with the beads. An external jacket may be filled with ice water to minimize heating during the 2 to 3 min normally needed to achieve breakage. For small samples (~1 ml), a mini Bead-Beater is available.

7.1.1.4. Solid Shearing

Cells may be broken by grinding a cell paste or lyophilized cells with abrasives such as alumina by using an agate mortar and pestle; they also may be broken by forcing a frozen suspension or paste through a small hole or slit. Hughes et al. (59) have provided a complete description of these methods.

7.1.1.5. Freezing and Thawing

Freezing and thawing have long been used to release high-M_r DNA from a variety of cells (121). The efficiency of the process depends on the organism and the operator. A washed cell pellet is frozen in liquid N_2 and ruptured by grinding the frozen cells with a pestle in a precooled mortar. This procedure of lysis has been used for the isolation of DNA from *Methanothermus fervidus* (152) and several other methanogens (62) and involves further treatment with sodium dodecyl sulfate (SDS).

7.1.2. Chemical Lysis

7.1.2.1. Detergent Uses and Actions

Detergents have multiple uses in cell fractionation. (i) When nonmembranous organelles such as ribosomes and nucleoids are sought, detergents provide a gentle means of lysing the cells once the integrity of the peptidoglycan (gram-positive bacteria) or the outer membrane plus peptidoglycan (gram-negative bacteria) has been damaged. (ii) Detergents are used to selectively solubilize the plasma membranes of gram-negative bacteria while leaving the outer membranes intact (section 7.4.2). (iii) Detergents also can be used to remove membrane contamination from ribosomes, polysomes, cell walls of gram-positive bacteria, or other cell fractions after cell breakage. (iv) Extraction and recovery of proteins expressed in *E. coli* can be simplified by the use of kits containing chemical lysis agents that gently disrupt the cell wall. These kits provide a simple, rapid alternative to mechanical methods for releasing expressed target proteins in preparation for purification or other applications. Harsh conditions associated with pressure disruption and sonication are avoided. One example is BugBuster™. This protein extraction reagent from Novagen (www.novagen.com) is a proprietary formulation of a mixture of nonionic detergents advertised not to denature soluble proteins. A companion nuclease can be purchased that reduces the viscosity of the extract. The low-viscosity clarified extract is fully compatible with affinity chromatography resins for fusion proteins. Sigma also supplies a comparable reagent called CelLytic B™.

Detergents are amphipathic molecules with both hydrophilic and hydrophobic regions and are sparingly soluble in water. At very low concentrations, detergents form true solutions in water. As the concentration is increased, additional molecules of detergent aggregate to form micelles. Here, the hydrophilic regions are exposed to water and the hydrophobic regions are shielded from water on the inside of the micelle. The concentration at which micelles begin to form, as the amount of added detergent is increased, is the critical micelle concentration (CMC). The CMC and the size and shape of the detergent micelle are characteristic of each detergent. Membranes undergo various degrees of disintegration at different detergent concentrations. At low concentrations, lysis or rupture is seen first; at intermediate detergent-to-membrane ratios (0.1 to 1 mg/mg of membrane lipid), membrane proteins may be selectively extracted; and at higher ratios (2 mg/mg of lipid), the membrane may be solubilized by the formation of soluble micelles of lipid-detergent, protein-detergent, and lipid-protein-detergent (54). Early reviews by Helenius et al. (54)

and Zulauf et al. (160) summarize the properties of many of the detergents used in cell fractionation.

Detergents may be grouped into three classes, which differ in micelle properties, protein binding, and response to other solutes. These classes are ionic detergents (cationic, anionic, and zwitterionic), nonionic detergents, and bile salts. A variety pack of detergents may be purchased from Calbiochem, La Jolla, CA, to explore suitability for specific applications.

7.1.2.2. Ionic Detergents

The most commonly used ionic detergents are SDS (also known as sodium lauryl sulfate), sodium N-lauroylsarcosine (Sarkosyl), alkyl benzene sulfonates (common household detergents), and quaternary amine salts such as cetyltrimethylammonium bromide. Ionic detergents tend to form small micelles (molecular mass [M_r], ~10 kDa) and exhibit rather high CMCs (the CMC of SDS is about 0.2% [wt/vol] at 20°C in dilute buffers). The CMC and the micelle solubility of ionic detergents are strongly influenced by the ionic strength of the solution and the nature of the counterions present. For example, a 10% (wt/vol) solution of SDS is soluble down to about 17°C, whereas a similar solution of Tris dodecyl sulfate is soluble at 0°C. Potassium dodecyl sulfate is soluble only at elevated temperatures, and K^+ must be excluded from all buffers when this detergent is used. The guanidinium salt of SDS is also insoluble.

Detergents such as SDS, which have a hydrophilic group that is strongly ionized, are not affected by pH and are not precipitated by 5% (wt/vol) trichloroacetic acid. Ionic detergents bind strongly to proteins, and for SDS, this strong binding usually results in unfolding and irreversible denaturation of proteins. This is the basis for their use in polyacrylamide gel electrophoresis (chapter 15.5.1). Although ionic detergents can pass through dialysis membranes, they bind strongly to proteins and cannot be removed completely by dialysis.

7.1.2.3. Nonionic Detergents

Commercial nonionic detergents include the polyoxyethylene detergents such as Triton X-100, Nonidet P-40, Brij 58, and Tween 80 (54). These detergents are not pure but a mixture of related compounds. In general, these detergents have high-molecular-mass (50-kDa or greater) micelles and low CMCs (0.1% or less), which limits their usefulness in procedures such as gel filtration and electrophoresis and makes them difficult to remove by dialysis. Although their properties are not strongly affected by pH or ionic strength, they may be precipitated by 5% (wt/vol) trichloroacetic acid. The advantage of these detergents is that they bind primarily to hydrophobic sites on the surfaces of proteins and thus do not cause extensive denaturation or loss of biological activity. Since nonionic detergents are uncharged, they do not interfere with separations that are based on the charge character of the cellular component. As noted in section 7.4.2, the outer membrane resistance of many gram-negative bacteria to nonionic (124) or weakly ionic (36) detergents allows the application of these detergents to the isolation of cell surface components.

There are a number of nonionic detergents consisting of acyl derivatives of sugars that have been synthesized for scientific rather than commercial applications. Examples include octyl-β-D-glucopyranoside, octyl-β-D-thioglycopyranoside, and dodecyl-β-D-maltoside. These are available from supply houses such as Calbiochem and Sigma. These detergents are similar to Triton X-100 in their membrane protein-solubilizing

properties, but they have several advantages. They are available as pure compounds of defined structures, and they generally have much higher CMCs, which makes them more useful in chromatography and facilitates removal by dialysis or gel filtration. They are available in different acyl chain lengths, which in turn results in ranges in properties such as CMC, the ease of dialysis, and micelle M_r. High cost is a major disadvantage.

7.1.2.4. Bile Salts

Bile salts are sterol derivatives, such as sodium cholate, deoxycholate, and taurocholate. Because of the poor packing ability of the bulky sterol nucleus, these detergents form small micelles (often of just a few molecules), and unlike those for other detergents, the micelle M_r is a function of concentration. As these detergents are salts of very insoluble weak acids with pK_as in the range of 6.5 to 7.5, they must be used at an alkaline pH. To avoid solubility problems, stock solutions are often prepared by dissolving the free acid in excess NaOH. Ionic composition, pH, and total detergent concentration affect their usefulness and must all be maintained constant when these detergents are used.

For the release of genomic DNA from rod-shaped halobacteria, sodium taurocholate may be preferred to sodium deoxycholate, since it dissolves more readily in the presence of high salt concentrations. The lysis of cells by bile salts is effective for only select members of the archaea and bacteria (70).

7.1.2.5. Problems with Detergents

In excess, detergents act by forming mixed micelles with lipids or by binding to proteins, forming soluble protein-detergent complexes. Since most detergents are used at concentrations well above the CMC, the ratio of detergent to protein or lipid is much more important than the actual detergent concentration. As a general rule, one must use at least 2 to 4 mg of detergent per mg of sample protein to ensure a satisfactory excess of detergent. For example, if 2% (vol/vol) Triton X-100 is used for membrane solubilization, the maximum protein content of the sample should be less than 5 to 10 mg/ml.

Triton X-100, one of the most useful nonionic detergents, presents problems in protein assays because it has an aromatic moiety that prevents the measurement of A_{280} and because it (and other nonionic or cationic detergents) forms a cloudy precipitate in chemical protein assays. Proteins can be assayed in the presence of these detergents by the modification of the procedure of Lowry et al., described by Peterson (113), in which an excess of SDS is added to the sample. This addition leads to the formation of stable mixed micelles of the detergents that do not interfere with the assay.

Triton X-100 and other similar nonionic detergents are soluble in ethanol-water mixtures, and proteins dissolved in these detergents may be freed from the detergents by ethanol precipitation as follows. Place the sample on ice, and add 2 volumes of ice-cold absolute ethanol with stirring. Allow the sample to stand overnight in a freezer, and collect the protein precipitate by centrifugation. Efficient precipitation requires a protein concentration of at least 0.2 mg/ml. Dilute samples may be concentrated by using an ultrafiltration apparatus (Amicon Corp., Beverly, MA) with a filter with an appropriate pore size, although this procedure is limited by the possibility that detergent micelles will also be concentrated. Triton detergents can be removed by an SM-2 resin (Bio-Rad Laboratories, Richmond, CA) (58).

SDS can be removed from samples by acetone precipitation as follows. Stir 6 volumes of anhydrous acetone into the sample at room temperature, and recover the precipitate by centrifugation. Wash the precipitate several times with acetone-water (6:1, vol/vol). Since the precipitate is often waxy and difficult to work with, it may be dispersed in water with a Potter-Elvehjem homogenizer and then lyophilized.

An alternative method for SDS removal, used often in DNA purification, is precipitation of the potassium salt of SDS made by the dropwise addition of potassium acetate to a final concentration of 0.3 to 0.5 M.

7.1.2.6. Dithiothreitol

Dithiothreitol will induce the lysis of the sheathed methanogens in the absence of an osmotic protectant, providing that the pH is moderately alkaline (133). The procedure can be scaled up to allow the isolation of cellular organelles or enzymes from *Methanospirillum hungatei* or *Methanosaeta concilii* but is generally ineffective with most other methanogens (100, 133).

7.1.3. Osmotic Lysis

Osmotic lysis in hypotonic solution may be achieved for some bacteria without the need of first weakening the cell wall. Extreme halophiles such as *Halobacterium*, as well as many moderate halophiles, require salts to maintain structural integrity and lyse in deionized water. The lytic susceptibility of a new culture should therefore be tested by suspension in low-ionic-strength solutions. Because of its simplicity, osmotic lysis is recommended as a first option for lysis, provided that the component to be isolated is stable under these conditions. Gas vesicles have been isolated from *Halobacterium salinarum* by using salt dilution (130).

Environmental cations can affect susceptibility to lysis in water (86, 95). For example, 18 marine and 2 terrestrial gram-negative isolates were found to give a range of lytic responses in distilled water; some lysed after exposure to Mg^{2+}-containing solutions and others were sensitized by washing in 0.5 M NaCl (e.g., *Pseudomonas aeruginosa*). Others resisted lysis in distilled water even after washing with NaCl (e.g., *E. coli*). Many methanogenic bacteria that have S-layers as the sole wall component (e.g., *Methanococcus voltae* and *Methanocaldococcus jannaschii*) and moderate halophiles (e.g., *Methanosarcina mazei*) lyse in water.

A general procedure for osmotic lysis in low-ionic-strength solutions, from Laddaga and MacLeod (86), is as follows.

1. Harvest exponential-growth-phase cells at 8,000 × *g* for 10 to 20 min, and wash them in an equal volume of 0.5 M NaCl up to three times (the outer membranes of gram-negative bacteria can sometimes be removed with further washes in 0.5 M sucrose).

2. Resuspend the cell pellet in deionized water.

Extraordinary resistance to lysis may require individualized procedures. The sulfur-dependent, wall-less thermophile *Thermoplasma acidophilum* grows optimally at pH 2 and can be lysed by raising the pH to 6 or 7. The organism otherwise resists mechanical breakage, nonionic detergents, pronase, and trypsin and is osmotically stable (87).

A procedure combining chemical and osmotic techniques is effective with a wide variety of methanogens, as follows (100).

1. Centrifuge a 10-ml sample of cell culture (0.3 to 0.5 mg [dry weight]/ml), freeze overnight at −20°C as cell paste, and thaw.

2. Resuspend thawed cells in 10-ml ABE buffer (50 mM ammonium bicarbonate–50 mM disodium EDTA, pH 8.0). Chelators of divalent cations (e.g., EDTA) facilitate lysis in cases where the cell wall has cationic cross bridging.

3. Add pronase to give 40 µg/ml, and incubate for 1 h at 37°C. If DNA is to be isolated, nucleases in the pronase (Roche, Indianapolis, IN) must be inactivated by autodigestion in 50 mM Tris-HCl (pH 8.0) for 1 h at 37°C. Note that proteinase K cannot be substituted for pronase with some strains.

4. Add dithiothreitol (200 µg/ml) for 15 min at 23°C.

5. Finally, add SDS (1 mg/ml), and heat the mixture at 55°C for 15 min to promote lysis.

Note that not all steps are necessary for all strains; the protease digestion step is likely to be unnecessary for most applications, excluding nucleic acid isolations, and can often be omitted.

7.1.4. Enzymatic Lysis

Osmotic lysis may be achieved in the many cases where osmotically sensitive cell forms can be generated by enzymatic activity, as described next in section 7.2.

7.2. GENERATION OF PROTOPLASTS AND SPHEROPLASTS

The strength of most bacterial walls resides in the murein (peptidoglycan) component, which may be damaged or destroyed by specific enzymes (e.g., lysozyme), leading to osmotically sensitive cells. The surrounding medium is generally hypotonic (or can be made so by dilution) so that lysis results, the extent of which should be monitored by phase-contrast microscopy (chapter 1). Lysis may be prevented by adding 0.3 to 0.8 M sucrose and 0.5 to 10 mM Mg^{2+} to the reaction mixture to generate round and osmotically sensitive, intact cells. These sensitive forms are protoplasts (from gram-positive cells), free of wall material but with the plasma membrane remaining, or spheroplasts (from gram-negative types), with both outer and plasma membranes remaining. Protoplasts and spheroplasts may not be perfectly round, and it is important to check their morphology with that of untreated cells. In addition to sucrose and Mg^{2+}, other osmotic stabilizers may be used (e.g., polyethylene glycol, sodium succinate, and certain salts from 0.2 to 0.8 M).

The enzymes that degrade murein can be divided into three classes: carbohydrases, acetylmuramyl-L-alanine amidases, and peptidases (66). Those most commonly used in forming protoplasts and spheroplasts are the muramidase subgroup of the carbohydrases, notably lysozyme. Structural variations in the murein can lead to differences in sensitivity to murein-lytic enzymes. For example, the extent of O acetylation of murein directly affects its susceptibility to egg white lysozyme (31). Since the extent of O acetylation may increase during the stationary phase (65), it is generally recommended that exponential cultures be used.

Osmotically sensitive forms may also be generated by interactions with antibiotics. Penicillins (β-lactam antibiotics) cause specific inhibition of murein biosynthesis and lead to unbalanced growth and lysis of β-lactamase-negative bacteria in a hypotonic environment. In the presence of hypertonic sucrose, osmotically sensitive spheres are formed from many rod-shaped gram-negative (spheroplasts) and gram-positive (protoplasts) bacteria, such as *Bacillus subtilis* and *Bacillus megaterium*. For the technique to work effectively, rapidly growing mid-exponential-phase cells are used. The method, with pros and cons for its use, has been described in detail by Kaback (68). *Staphylococcus aureus* forms protoplasts during growth in media containing sublethal doses of D-cycloserine in 1 M sucrose (147).

7.2.1 Protoplasts from Gram-Positive Bacteria

A procedure for generating protoplasts from gram-positive bacteria that are sensitive to lysozyme is as follows.

1. Wash the bacteria in a buffer of pH 6 to 8, and resuspend (1 to 10 mg [dry weight]/ml) in buffer containing hypertonic sucrose (usually 0.2 to 1.0 M) and lysozyme (Sigma) at 0.1 to 1.0 mg/ml. Treatment for 30 min at 20°C should result in complete protoplast formation. The enzyme may be used over a broad pH range, with optimum pHs of 6 to 7, and over a wide temperature range from 4 to 60°C (30, 119). Lysozyme can be inhibited by the presence of salts in excess of 0.2 M (132).

Many gram-positive bacteria are not sensitive to hen egg white lysozyme but may be sensitive to muramidases of broader substrate ranges (Table 1). These include lysostaphin from "*Staphylococcus staphylolyticus*," mutanolysin from *Streptomyces globisporus* (Sigma), and a fungal muramidase from *Chalaropsis* sp. (Miles Laboratories, Inc., Elkhart, IN). The latter enzyme, unlike egg white lysozyme, is active against murein acetylated at the C-6 position of muramic acid (52). Caution should be used when working with these enzymes since, unlike egg white lysozyme, they are often contaminated with proteases, lipases, or nucleases.

Methods can sometimes be found to cause the murein of insensitive strains to become sensitive to lysozyme. For example, *Mycobacterium smegmatis* becomes sensitive to lysozyme after prolonged exposure to DL-methionine (157). Protoplasts of *Bacillus megaterium* can be formed following incubation for 2 h at 30°C in nutrient broth containing 100 U of penicillin per ml. The treated cells are washed once in 0.85% (wt/vol) saline and resuspended for ~1 h in 0.3 M sucrose containing 500 μg of lysozyme per ml (96). Partially lysed and weakened cells of cyanobacteria (*Oscillatoria* spp.) can be made by treatment with ampicillin before lysozyme (148).

Clinical isolates of *Staphylococcus* spp. can be made more sensitive to lysostaphin by growth in P medium lacking glucose (44).

7.2.2. Spheroplasts from Gram-Negative Bacteria

Virtually all gram-negative bacteria have a murein structure that is sensitive to lysozyme. However, since their outer membranes are impermeable to lysozyme, their integrity must be compromised before lysozyme will penetrate to the murein. This can be done in various ways (Table 1). The basic lysozyme-EDTA procedure for forming spheroplasts and causing the lysis of a wide variety of gram-negative bacteria (68) is as follows.

1. Harvest bacteria in mid to late exponential growth phase, and wash twice with 10 mM Tris-HCl (pH 8.0) at 4°C. Resuspend the cells (1 g [wet weight]/80 ml), and stir at room temperature in 30 mM Tris-HC1 (pH 8.0) containing 20% (wt/wt) sucrose.

2. Add 10 mM potassium EDTA (pH 7.0) and 0.5 mg of lysozyme per ml (final concentrations), and incubate for 30 min at room temperature. Add lysozyme slowly under the surface of the liquid to decrease aggregation of the spheroplasts.

3. If lysis is required, dilute the sucrose either directly by dialysis or by centrifugation and resuspension of the pellet.

Some modifications to this method are described elsewhere (109), as is an efficient procedure for the production of viable spheroplasts from *E. coli* (97). This procedure is based on Ca^{2+} pretreatment, glycerol stabilization, EDTA incubation, and heat shock and utilizes much less lysozyme than standard procedures. Factors influencing the efficiency of spheroplasting include growth medium (cells grown in a rich medium are more suitable than cells from defined media), harvesting temperature (0°C versus room temperature), factors which might destroy the permeability barrier to sucrose, and the concentration of lysozyme.

7.2.3. Osmotically Sensitive Forms from Archaea

The formation of osmotically sensitive forms in archaea is less well understood, but has been achieved in selected cases (Table 1). Spheroplasts of *Halobacterium* spp. may be formed by resuspending the cells in 0.1 M 2-(N-morpholino) ethanesulfonic acid (MES) buffer (pH 7.0) containing 0.5 M sucrose, 0.25 M NaCl, and 0.01 M $MgCl_2$ (63). Also, *Methanospirillum* spp. and *Methanosaeta concilii* form spheroplasts when exposed to dithiothreitol at alkaline pH (see section 7.1.2.6) (13).

Strains of *Methanosarcina barkeri*, which have cell walls composed largely of protein, form metabolically active protoplasts when incubated at 37°C with pronase (2 mg/ml) and 0.5 M sucrose (67). During growth, some bacteria enter a lytic phase, often as a result of nutrient depletion. In these cases, it may be possible to prevent lysis by incorporating sucrose directly into the growth medium. This concept may be used to form protoplasts during growth limitation of *Methanobacterium bryantii* in a medium deficient in nickel and ammonium but containing 20 mM $MgCl_2$ to prevent lysis (61), and to form protoplasts from *Methanosarcina barkeri* cultures stabilized by 0.3 M sucrose or glucose (27). Finally, a pseudomurein endopeptidase has been discovered in *Methanobacterium wolfei* and found to be effective in preparing protoplasts of *Methanothermobacter thermoautotrophicus* stabilized with 0.8 M sucrose (74). The usefulness of this enzyme is limited by its O_2 sensitivity, but it was the first enzyme known to hydrolyze pseudomurein. It is possible to purify this enzyme aerobically and recover activity by incubating under reducing conditions with $Na_2S \cdot 9H_2O$ or dithiothreitol (104).

7.3. CENTRIFUGATION

The techniques of differential centrifugation and density gradient centrifugation are widely used in the purification of components from cell lysates. Detailed discussions of both the theoretical and practical aspects of preparative centrifugation are provided in comprehensive reviews (37, 115, 117, 140).

The sedimentation rate of particles subjected to a centrifugal force is proportional to the force applied, and also to the properties of the suspension fluid, as shown below for a spherical particle:

$$v = \frac{d^2(p_p - p_l)g}{18n}$$

where v is the sedimentation rate, d is the particle diameter, p_p is the particle density, p_l is the liquid density, g is the centrifugal force, and n is the viscosity of the liquid. At a constant centrifugal force and viscosity, the sedimentation rate is proportional to the size of the particle and to the difference between the densities of the particle and the liquid.

TABLE 1 Some commonly used methods to lyse bacteria and archaea to form osmotically sensitive cells in hypotonic solutions

Method	Species	Reference
Lysozyme	*Bacillus megaterium*	132
	Bacillus subtilis	132
	Micrococcus lysodeikticus	132
Lysozyme-EDTA-Tris (pH 8)	*Salmonella enterica* serovar Typhimurium	68
	Escherichia coli	68
	Enterobacteria	68
	Pseudomonas aeruginosa	68
	Micrococcus denitrificans	68
	Azotobacter vinelandii	68
	Bacillus subtilis	68
	Bacillus megaterium	68
	Clostridium thermoaceticum	68
	Aquaspirillum serpens	23
	411 gram-positive and gram-negative bacteria	146
	Thiobacillus strain A2	92
	Wolinella succinogenes	78
	Yersinia pestis	105
NaCl-sucrose-lysozyme	5 *Alteromonas* spp.	86
	7 *Vibrio* spp.	86
	2 *Cytophaga* spp.	86
	4 *Pseudomonas* spp.	86
	3 *Alcaligenes* spp.	86
EDTA-Tris (pH 8.6)	*Pseudomonas aeruginosa*	40
	Alcaligenes faecalis	43
	Escherichia coli	43
Mutanolysin ± 1M NaCl	*Streptococcus mutans*	129
	Streptococcus sanguis	129
	Streptococcus salivarius	129
Lysoamidase	*Staphylococcus aureus*	114
Penicillin G	Various gram-positive and gram-negative bacteria	68
Lysozyme-penicillin[a]	*Fusobacterium varium*	21
	Enterococcus faecium	21
Serine protease	*Methanobacterium formicicum*	20
Pronase	*Methanosarcina barkeri*	67
Dithiothreitol (pH >8)	*Methanospirillum* spp.	13
	Methanosaeta concilii	13
Pseudomurein endopeptidase	*Methanothermobacter thermoautotrophicus*	74
Nalidixic acid-glycine incubation, followed by freeze-thaw	*Escherichia coli* K-12	159
	Aeromonas hydrophila	159
	Staphylococcus epidermidis	159

[a] Lysozyme susceptibility follows growth in medium containing penicillin G.

Simple differential centrifugation (section 7.3.1) is a technique employed to differentially pellet cell fractions. Density gradient centrifugation can be based on either the size difference among particles (rate-zonal centrifugation [section 7.3.2]), or the density difference among particles (isopycnic separation [section 7.3.3]).

7.3.1. Differential Centrifugation

In differential centrifugation (pelleting), samples are centrifuged for a given time at a given speed, resulting in a supernatant and a pellet fraction. This technique is useful for the separation of particles with very different sedimentation velocities and thus for the isolation of partially purified fractions. For example, centrifugation for 5 to 10 min at 3,000 to 5,000 × g will pellet intact bacteria, leaving most cell fragments (walls, membranes, and appendages) in the supernatant. Cell wall fragments and large membrane structures can be pelleted by centrifugation at 20,000 to 50,000 × g for 20 min, whereas centrifugation at 200,000 × g for 1 h (in an ultracentrifuge) is required to pellet small membrane vesicles or ribosomes. For pelleting in an ultracentrifuge, thick-walled tubes can be used without tube caps in fixed-angle rotors. For ultracentrifugation of radioactive or biohazardous material, polycarbonate bottles with a three-piece cap assembly are available for general-purpose fixed-angle rotors and are used frequently for differential centrifugation where band recovery is not a problem.

Low-speed centrifugation may be performed anaerobically by using standard anaerobic methodologies and glass centrifuge tubes modified to accept serum bottle closures (134).

7.3.2. Rate-Zonal Centrifugation

Zonal centrifugation is effective for separating subcellular structures having similar buoyant densities but differing in shape or particle size, e.g., ribosomal subunits, classes of polysomes, and various forms of DNA molecules. Centrifugation is carried out either in swinging-bucket rotors or specially designed zonal rotors, and a shallow linear gradient (usually sucrose) is employed to prevent convection in the tubes or the chamber of the zonal rotor during centrifugation. The preparation of linear gradients is described in section 7.3.4. A density range is chosen such that during the separation the particle densities are always greater than the density of the liquid. The sample is applied as a zone or narrow band at the top of the gradient, and the run is terminated before the separating particles reach the bottom of the tube. Times and speeds must be determined empirically. For subcellular particles, a gradient of 15 to 40% (wt/wt) sucrose is commonly used, and centrifugation at 100,000 × g for 1 to 4 h is sufficient for the separation of most subcellular particles.

The use of conventional sucrose or glycerol gradients results in an increase in the concentration of the gradient material, and therefore of the liquid viscosity, as the particle moves down the gradient. Consequently, the rate of movement can decrease as the particle moves downward even though the centrifugal force increases, and this decrease ultimately results in less resolution between separating zones. When a cesium chloride gradient (section 7.3.6.2) is used, the viscosity decreases with increasing concentration, resulting in an accelerating gradient and better separations.

7.3.3. Isopycnic Separations

Isopycnic separation is often used to separate particles and membrane fractions on the basis of buoyant density instead of sedimentation velocity. Membrane fragments derived from the same subcellular membrane may differ greatly in size and, hence, in sedimentation velocity but should have the same buoyant densities. The sample is centrifuged in a solute density gradient (sucrose gradients are commonly used for membranes and organelles with a density of less than 1.3 g/ml; cesium chloride is used for denser structures such as viruses) until an equilibrium state is reached, at which time each particle has migrated to a point in the gradient where the particle has the same density as the surrounding solution (Table 2). This point is the isopycnic position of the particle. Since the equilibrium density is an equilibrium position, it may be approached by sedimentation from less dense regions of the gradient, from more dense regions (by flotation), or from particles dispersed throughout the gradient (Fig. 1). Since the sedimentation velocity of a particle becomes progressively lower as the particle approaches the region in the gradient where it has the same density as the solution, very long centrifugation times are required in order to approach equilibrium. This is particularly true for small membrane vesicles such as chromatophores and plasma membrane fragments from cells broken with a French press, since the sedimentation velocity of these vesicles is low even in the absence of the gradient. If centrifugal forces in the order of 100,000 to 200,000 × g are used, at least 18 h will be required for a reasonable separation, and periods as long as 72 h are needed for critical separations.

7.3.4. Preparation and Fractionation of Density Gradients

Density gradients may be subdivided into two main categories according to the method of preparation: step (discontinuous) gradients and linear (continuous) gradients.

Step gradients can be prepared by hand layering solutions of different densities in the centrifuge tube. Sucrose gradients are prepared often as a series of five to seven layers. When wettable tubes such as polycarbonate ones are used, each layer can be added by allowing it to run down the side of the tube from a pipette held at an angle against the side of the tube. When nonwettable tubes such as polyallomer or polypropylene ones are used, the pipette tip must be kept in contact with the meniscus during the addition of the solution to avoid mixing. Alternatively, an easy way to layer solutions is to float on the bottom sucrose layer a thin disk of cork with a diameter just less than the diameter of the tube. The solutions are added dropwise just above the center of the cork disk. The solutions will run over and under the cork disk, without disturbing the heavier sucrose solutions underneath. The sample is layered on top, and centrifugation is started immediately. Before and during centrifugation, the steps in these preformed gradients are smoothed by diffusion. Bottom- and middle-loaded samples, first having been mixed

TABLE 2 Buoyant densities of fractions isolated from prokaryotic cells

Fraction	Buoyant density (g/ml)	Medium	Source	Reference
Plasma membrane	1.14–1.16	Sucrose	Gram-negative enterics	109
	1.17	Sucrose	*Methanospirillum hungatei*	135
Outer membrane	1.22	Sucrose	Gram-negative enterics	109
Protein sheath	1.35	Sucrose	*Methanospirillum hungatei*	135
Fimbriae (pili)	1.117–1.137	Sucrose	*Escherichia coli*	72
	1.3	CsCl	*Pseudomonas aeruginosa*	110
Spinae	1.37	CsCl	Marine pseudomonad	33
Flagella (intact)	1.30	CsCl	*Escherichia coli*	28
Endospores	1.27	Percoll	*Clostridium perfringens*	144
	1.21	Percoll	*Bacillus subtilis*	144

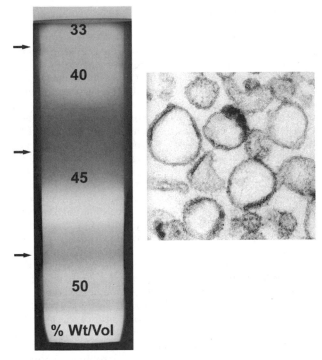

33

40

45

50

% Wt/Vol

FIGURE 1 Isolation of cytoplasmic membranes by discontinuous sucrose gradient centrifugation. A membrane fraction from *Methanospirillum hungatei* spheroplast lysate in 33% sucrose was loaded on 40 to 50% sucrose steps and centrifuged at 4°C for 48 h (110,000 × g, R_{max}) in a Beckman SW27 swinging-bucket rotor. A thin section of the upper band revealed closed vesicles bounded by the double-track membrane, typical of cytoplasmic membranes. See reference 135 for further details.

with gradient medium to make the sample solution denser than the remainder of the gradient, can float upward to find their own buoyant density.

Linear gradients of sucrose can be formed by allowing a step gradient to "age" in the centrifuge tube overnight at 4 to 8°C. A continuous gradient will form by diffusion. A variety of commercial gradient-forming devices are available, but these are unnecessary if the molecular mass of the gradient medium is less than 1,000 Da. Diffusion of higher-molecular-mass molecules such as Ficoll 400 is too slow, requiring a gradient maker for the preparation of linear gradients. Linear gradients are called preformed gradients for the obvious reason that they are formed prior to centrifugation. Continuous-density gradients may also be formed by centrifuging a solution, which initially is of the same density throughout the centrifuge tube, for sufficient time and at sufficient speed. During centrifugation, a gradient is formed as the density decreases at the top of the tube and increases toward the bottom. This type of gradient is known as a self-forming gradient but cannot be formed with all types of gradient media. It is very useful for cesium chloride gradients.

Unless one is following a specific protocol, a choice will have to be made on the gradient type for a particular separation. In the simplest terms, preformed gradients are used when the particles to be purified are large enough to reach their isopycnic positions in less time than it takes a gradient to self-form. This is very important in the case of large par-

ticles such as cells, cell organelles, and viruses. Shorter centrifugation times and weaker *g* forces can be used, which assists in the preservation of delicate structures. Smaller particles such as macromolecules and associated complexes require much longer centrifugation times to band, during which a gradient can be self-generated.

Various methods can be used for fractionating sucrose gradients. The gradients may be pumped out through a stainless steel cannula (e.g., a blunted syringe tip) inserted into the bottom of the tube by using a peristaltic pump. Alternatively, the bottom of the tube, or the side just below the band of interest, may be pierced with a needle and samples may be collected by gravity flow. Although simple to perform, these methods in which samples are collected starting from the bottom of the gradient can lead to cross contamination of upper bands with those appearing lower in the tube. In cases where the component of interest is the upper band, the gradient should be fractionated by upward displacement. The centrifuge tube is fitted with a tight-fitting rubber bung, pierced by a stainless steel cannula sufficiently long to extend to the bottom of the tube. A sucrose solution about 5% greater in density than the bottom of the gradient is pumped through the cannula, thus displacing the gradient through a short exit tube leading to a fraction collector. Alternatively, if bands are clearly visible, samples may be taken carefully from the top of the gradient by pipette.

7.3.5. Rotor, Tube, and Bottle Selection

There are basically four types of rotors used for centrifugal separations: (i) swinging bucket (swing out), (ii) fixed angle, (iii) vertical tube, and (iv) zonal. It is important to choose the correct rotor in order to achieve optimum separation results. This topic is discussed in references 37 and 117. The Beckman Coulter Internet site (http://www.beckmancoulter.com) is also an excellent source of information about the practical aspects of ultracentrifugation.

Swinging-bucket rotors are recommended for most preparative rate-zonal separation methods because of their relatively long path length. Vertical-tube rotors are also useful for rate-zonal separations, especially for large particles such as cells and subcellular organelles. The main advantage of vertical-tube rotors is the short time needed to achieve separation compared with that for other rotor types. Fixed-angle rotors are not normally used for rate-zonal separation. Beckman Coulter has published an excellent centrifuge product selection guide (BR-8101D) that includes a table for discontinued rotors and their recommended replacements. This guide may be useful when one is following a centrifugation procedure from the older literature. It is *extremely* dangerous to use older rotors that are not compatible with the newer models of ultracentrifuges or to exceed the maximum speed recommended for a particular rotor. Be aware that certain rotors may be derated (based on the total number of run hours) but still useable.

For isopycnic separations, sample particles will cease sedimenting when they reach their buoyant density and wall effects are much less important than they are for rate-zonal separation. Swinging-bucket, fixed-angle, and vertical-tube rotors may all be used for isopycnic separations; however, the density range and slope of an equilibrium gradient will depend upon the type of rotor used for isopycnic separation. Fixed-angle and vertical-tube rotors have some advantages over swinging-bucket rotors for these separations. Because the effective path length is shorter for these rotors than for swinging-bucket rotors, the time period of centrifugation is considerably shorter for the equilibrium banding of particles. The increased number of tubes in fixed-angle and vertical-tube rotors means that the sample capacity per rotor is much greater.

A wide variety of materials have been used for the manufacture of centrifuge tubes and bottles, each with its own distinct combination of properties to meet a variety of applications. It is important to choose a material that is compatible with the sample and the gradient medium. Available options include transparent, translucent, and opaque tubes; tubes that can be sliced, punctured, or sterilized and reused; and tubes resistant to a variety of chemical compounds. This selection can be overwhelming to the novice user. Thankfully, selection considerations are discussed in detail in the Beckman Coulter centrifuge product selection guide (BR-8101D). The tube size may vary somewhat from the actual filling capacity. Centrifuge tubes should be filled (to within ~3 mm of the tube rim) to prevent collapse of the tubes during centrifugation. An exception is thick-walled tubes that can be used uncapped and filled to half the maximum volume. The maximum filling volume of thick-walled tubes will depend on the tube angle of the rotor to be used. It is important to dry the outside of the tube before inserting it into the rotor or bucket. Any residual moisture on the outside of the tube can act as a lubricant under centrifugal force and contribute to the collapse of a tube, particularly with open-top tubes.

To use a tube closure or not: that is the question. Without a doubt, the most convenient tube closure is none at all. This applies to all tubes run in swinging-bucket rotors. Thick-walled polyallomer tubes can be run in all fixed-angle rotors without caps, provided that they are partially filled. Polycarbonate tubes are used like thick-walled polyallomer tubes, with or without caps, in fixed-angle rotors. Ultraclear, polypropylene, and polyethylene tubes can also be used without caps in fixed-angle rotors.

Tubes need to be capped for one of four reasons: (i) to prevent the collapse of thin-walled tubes during centrifugation, (ii) to contain radioactive or biohazardous samples, (iii) to maintain anaerobic conditions for oxygen-labile components, and (iv) to maintain sterility. Thin-walled polyallomer tubes must be capped when used in fixed-angle rotors. For swinging-bucket rotors, these thin-walled tubes should be filled to within 2 to 3 mm of the top, as tube caps are not used in swinging-bucket rotors. The standard tube cap assembly (with an O ring sealing system) is definitely one of the more tedious aspects of ultracentrifugation. Fortunately, Beckman Coulter has introduced alternatives. Polycarbonate bottles with three-piece cap assemblies are available for fixed-angle rotors. These are frequently used for differential centrifugation where the bottle does not need to be punctured for band recovery. Bottle assemblies can be filled to half the maximum volume. OptiSeal™ polyallomer tubes have the simplest closure; simply insert a tube plug and press, and an O ring seals securely against the tube's inner surface. Quick-Seal® tubes (polyallomer or ultraclear) have a 3-mm-long inlet through which the tube is filled. The top of the tube is then heat sealed by using a handheld or tabletop tube topper. Quick-Seal® tubes can be used in all swinging-bucket and most fixed-angle rotors and must be used in vertical rotors. Tube size must be compatible with the rotor and the sample volume. Many Beckman rotors can accommodate smaller tubes by using adapters or spacers. The Beckman g-Max system uses a combination of short polyallomer Quick-Seal® tubes and floating spacers to permit faster run times at maximum g forces.

7.3.6. Gradient Media

7.3.6.1. Sucrose and Ficoll

Sucrose concentrations in gradients are reported in the literature in several ways: density, percent sucrose by weight (wt/wt), percent sucrose per unit of volume (wt/vol, or grams/100 ml), or molar concentration of sucrose. In duplicating experiments reported in the literature, note how the gradient solutions were mixed: wt/wt or wt/vol. The most useful unit is percent sucrose by weight. Standard chemistry handbooks (e.g., *Handbook of Biochemistry and Molecular Biology*, [8]) contain tables defining both densities and viscosities of various concentrations of sucrose (percent) in water at different temperatures. For each density of sucrose, the percent wt/wt and wt/vol is given so conversions can be made. Sucrose gradients can be run at different temperatures (e.g., 4, 10, or 20°C), but remember that the density of the sucrose will vary according to temperature. This variation may affect band position.

When preparing sucrose solutions, one may assume that water or dilute buffers have a density of 1 g/ml. Therefore, to prepare a 54% (wt/wt) sucrose solution, for example, dissolve 54 g of sucrose in 46 ml of water. This will give 100 g of solution, and since the density of a 54% (wt/wt) sucrose solution is 1.2451 g/ml, the final volume will be 80.3 ml. Preparing solutions in this manner avoids the necessity of having to make viscous solutions up to fixed volumes. Sucrose allows for a density range of up to 1.35 g/ml. It is important to check the maximum density of the gradient material for each rotor. Since sucrose solutions are relatively viscous, linear gradients are generally employed.

Ficoll 400 (Amersham Pharmacia Biotech, Inc., Piscataway, NJ) is a hydrophilic polymer of sucrose with a molecular weight of 400,000. Solutions of up to 50% (wt/vol) Ficoll can be made, which allows a density range of up to 1.2 g/ml. Its advantages over sucrose for density gradient centrifugation lie in its lower osmotic pressure, which can be important in preserving the morphologies and activities of subcellular fractions.

7.3.6.2. Cesium Chloride and Potassium Bromide

Cesium chloride solutions have low viscosity so that preformed gradients of this solute are difficult to prepare and self-forming gradients are used. The sample is mixed with enough CsCl to provide a density equal to the average density of the subcellular particles. This homogeneous suspension is placed in the centrifuge, and the gradient is formed during centrifugation as a result of the sedimentation of the CsCl in the centrifugal field. CsCl allows a density range of up to 1.81 g/ml. KBr can also be used and is less expensive than CsCl (69).

7.3.6.3. Percoll

Percoll is a density gradient medium containing colloidal silica particles that are rendered nontoxic by a coating of polyvinyl-pyrrolidine (112). Percoll is recommended for separation of cells because it has low osmolarity, low viscosity, and large particles (30 nm in diameter) so that it does not permeate bacterial walls or membranes. It is used also to separate viruses and subcellular organelles. It can be purchased from Amersham Pharmacia Biotech as a sterile suspension which may be diluted as desired and used in gradients that are either preformed or self-formed during the centrifugation run to produce a density range of up to 1.3 g/ml. Since Percoll cannot penetrate walls or membranes, intact cells band at the buoyant densities of the intact cell. In contrast, gradient media with smaller particles (e.g., Metrizamide) permeate coatings above the membrane and thus indicate the buoyant densities of the protoplasts, as exemplified by lysozyme-sensitive spores (91). Density marker beads of differing densities and colors can be purchased (Amersham Pharmacia Biotech) for calibration. The use of marker beads is important for checking the shape of the gradient, as the rotor angle and the tube size affect the geometry (shape) of the gradient formed.

7.3.6.4. Metrizamide

Metrizamide, or 2(3-acetamido-5-*N*-methylacetamido-2,4,6-triiodobenzamido)-2-deoxy-D-glucose (Fluka, Buchs, Switzerland), is a density gradient medium which has a molecular weight of 789, a density of 2.17 g/ml, and very low viscosity. The last property is particularly useful in applications in which it is desirable to isolate membrane fractions by flotation to assess the association of the membrane with mutant proteins or products of genes cloned into multicopy vectors. Spheroplast lysates of *E. coli* are brought to a density of 1.29 g/ml with Metrizamide and placed into a centrifuge tube. The sample is layered over with a 1.27-g/ml Metrizamide solution. Rapid flotation of membranes occurs while the gradient is formed during centrifugation, leaving cytoplasmic proteins near the bottom of the tube. The short sedimentation distance and high force obtainable in a Beckman TL100 benchtop ultracentrifuge can be used to advantage (141).

7.4. ISOLATION OF CELL COMPONENTS

7.4.1. Appendages

7.4.1.1. Flagella

Intact flagella of most prokaryotes consist of three major components: filament, hook, and basal body (chapter 4, Fig. 8). The first two components are external to the cell wall and are relatively easy to isolate from other cell components. Flagellar filaments can be removed mechanically from cells by shear forces in a blender or Potter-Elvehjem homogenizer which removes filaments without damaging the cell walls of most organisms. Sheared flagella from some bacteria do retain the hook; however, experience has shown that this retention is rare. When flagella are released spontaneously from autolyzed cells, the hook often remains attached to the filament. Often, filaments can be sheared from cells, the cells can be placed in fresh medium, and regrowth of the filament can be monitored.

Note: check the culture supernatant by electron microscopy (negative stains, chapter 4.2.2 and chapter 4, Fig. 8) after harvesting for any filaments that have been removed during either growth or centrifugation. Often concentration and ultracentrifugation of the culture supernatant will produce a good yield.

Procedure for Isolation of Flagellar Filaments

1. Grow 1 to 3 liters of bacteria. Harvest cells by centrifugation, and resuspend cells in approximately 10% of the culture volume in 0.5 M HEPES or Tris buffer, pH 7.5. The density of the cell suspension is important for the removal of filaments.

2. Homogenize cells in a rotating-blade blender (Sorvall Omni-Mixer; Waring) at maximum speed for about 60 s. The volume of the cell suspension must at least be enough to cover the blades. For some bacteria, flagella can also be sheared by forcing of the cells through a 22-gauge needle and subsequent homogenization of cells in a Potter-Elvehjem Teflon glass tissue grinder.

3. Sediment deflagellated cells by differential centrifugation (6 to 10,000 × *g* for 15 min).

4. Centrifuge the cell-free supernatant at 100,000 × *g* in an ultracentrifuge for 90 min to pellet the filaments.

5. The sheared filaments can be further purified by isopycnic density gradient centrifugation using KBr (69). Resuspend the pellet of filaments from step 4 into 10 ml of buffer overnight at 4°C. Add 5 g of KBr (Sigma; ca. 99%), and dissolve the KBr with gentle stirring. Centrifuge at 210,000 × *g* for 24 h at 5°C (Beckman SW41Ti or SW50.1 rotor).

A single diffuse band of filaments usually occurs in the bottom third of the centrifuge tube. Remove the band from the top with a Pasteur pipette. KBr can be removed by dialysis against distilled water or buffer, and the filaments can be collected by ultracentrifugation at 100,000 × *g* for 90 min.

Procedure for Isolation of Intact Flagella

The DePamphilis and Adler (28) method can be used successfully with *E. coli* and *Bacillus subtilis*. Refer to the original paper for details of the procedure, which may be of use for work with other bacteria. In all steps, care must be taken not to break off the basal bodies from the filaments.

1. Grow 1 to 3 liters of bacteria, and harvest cells by centrifugation.

2. Gently resuspend cells in 50 ml of 20% (wt/wt) sucrose (final volume, ~70 ml).

3. To form protoplasts or spheroplasts, add in the following order 7 ml of 1 M Tris-HCl, (pH 7.8), 2 ml of 0.25% (wt/vol) lysozyme in 0.1 M Tris-HCl (pH 7.8) containing 0.2 M NaCl, and 6 ml of 0.1 M EDTA in 0.1 M Tris-HCl (pH 7.8). Mix the suspension after the addition of each reagent, and add the EDTA within 30 s after the lysozyme. Incubate at 30°C for 1.5 h with gentle shaking. A good yield of intact flagella depends on the formation of at least 80% protoplasts or spheroplasts.

4. Add 7 ml of 20% (vol/vol) Triton X-100. This addition should produce a clear, viscous lysate.

5. To free the flagellum preparation of DNA, add 0.8 ml of 1 M $MgCl_2$ followed by 1.5 mg of DNase I. Incubate the mixture at 30°C for 20 min. Do not add Mg^{2+} before the Triton X-100, since this prevents solubilization of the outer membrane of gram-negative bacteria.

6. Dilute the lysate immediately to 260 ml with cold 0.1 M Tris-HCl (pH 7.8) containing 0.5 mM EDTA (Tris-EDTA buffer). Rapidly pour 87 ml of cold-saturated $(NH_4)_2SO_4$ (in Tris-EDTA buffer) into the diluted lysate to give 25% of saturation. Stir the suspension slowly at 5°C for 2 h.

7. Centrifuge the suspension at 12,000 × *g* for 25 min in a fixed-angle rotor. Collect the viscous white material floating at the meniscus and adhering to the side of the tube. Discard the fluid.

8. Rinse the tube with Tris-EDTA buffer, and combine with recovered material to give a final volume of 40 ml.

9. Add 1 ml of 20% (vol/vol) Triton X-100 and dialyze the suspension at 5°C against 1 liter of Tris-EDTA buffer. The initially turbid suspension should clear completely upon dialysis.

10. Dilute the dialyzed preparation to 50 ml with Tris-EDTA buffer. To separate the intact flagella from other cellular debris, layer in duplicate tubes 25 ml of the diluted

preparation onto a gradient consisting of 2 ml of 20% (wt/wt) sucrose and 3 ml of 60% (wt/wt) sucrose (in Tris-EDTA). After centrifugation for 1 h a 40,000 × g in a swinging-bucket rotor (Beckman SW28), remove and discard the liquid above the sucrose layers.

11. Add 5 ml of Tris-EDTA buffer to the sucrose layers. Mix gently. Remove sucrose by dialysis overnight against Tris-EDTA buffer. Dilute the suspension to 25 ml with buffer, and add 0.15 ml of 20% (vol/vol) Triton X-100. Centrifuge at 4,000 × g for 10 min to remove unlysed cells and large aggregates. Retain the supernatant.

12. To further purify the intact flagella, use isopycnic density gradient centrifugation in CsCl. Dilute the supernatant material from step 11 to 27 ml with Tris-EDTA buffer. Add 12.1 g of CsCl (all at once), and dissolve rapidly. Centrifuge at 50,000 × g for 50 h (Beckman SW28 rotor). A band of flagella should be present near the center of the gradient. Collect this band, and remove the CsCl by dialysis against Tris-EDTA buffer. The appearance and number of bands in CsCl gradients and the purity and quality of the intact flagella recovered are markedly influenced by the rate of addition of CsCl and the age of the cells at harvesting.

Isolation of Intact Flagella from Archaea

Kalmokoff et al. (69) devised a procedure appropriate to *Methanococcus voltae*, whose cell envelope consists of only the plasma membrane and a protein surface array. This procedure is based on the temperature-induced phase separation of the nonionic detergent Triton X-114. At temperatures above 20°C, Triton X-114 separates into two phases: hydrophilic proteins separate into the upper aqueous phase, and hydrophobic integral membrane proteins separate into the lower detergent-rich phase. Since the envelopes of archaea contain only a few hydrophilic proteins, this technique allows a selective enrichment of the wall protein and flagella. For osmotically fragile methanogens, cells can be lysed by dilution in distilled water. Crude flagellar preparations are purified as described in reference 69. The procedure is not suitable for some archaea whose flagellar filaments are sensitive to Triton (142).

Procedure

1. Prepare spheroplasts (see section 7.2) from 1 liter of medium, if possible, and lyse them in distilled water with gentle mixing. Add DNase (1.5 mg) with 1 mM MgCl$_2$ to reduce viscosity.

2. Centrifuge the lysate at 22,000 × g for 30 min at 4°C, resuspend envelopes in 20 ml of cold 10 mM Tris-HCl (pH 7.5), and treat with 1% (vol/vol, final concentration) Triton X-114 for 30 min at 4°C with occasional mixing.

3. Incubate samples at 37°C to induce phase separation. Centrifuge at 300 × g for 3 min. Do not centrifuge at higher speeds, since the flagella will be pelleted in the lower detergent phase.

4. Remove the upper aqueous phase (~17 ml), and concentrate to 1 ml with an ultrafiltration apparatus (e.g., PM30 filter, model 8010; Amicon Corporation). Check the sample by electron microscopy for the presence of flagella; bacterial flagella are ~20 nm in diameter, and methanogen flagella are ~11 nm in diameter.

7.4.1.2. Fimbriae (Pili)

Fimbriae are proteinaceous, filamentous surface structures composed mostly of identical subunits (chapter 4, Fig. 1). They are narrower in diameter (< 10 nm; see chapter 4, Fig. 5) than flagella and have no basal structure. They are often found on flagellated cells, in which case isolation and purification will involve separation of these two kinds of appendages. They are very stable protein assemblies and are firmly attached, making them more resistant to shear forces than flagella.

Procedure for Removal of Fimbriae from Cells

1. Grow 1 liter of cells in appropriate medium and harvest by centrifugation. Resuspend the cells in 50 ml of 10 mM Tris-HCl (pH 7.2).

2. Remove fimbriae by blending for 2 min at maximum speed in a rotating-blade blender (e.g., Sorvall Omni-Mixer; Waring). See comments in section 7.4.1.1.

3. Remove cells by differential centrifugation (6,000 to 10,000 × g for 15 min), and retain the supernatant fluid ("shear supernatant") which contains the fimbriae.

Procedures for Purification of Fimbriae

Procedure 1

Fimbriae can be separated from other contaminating materials in the shear supernatant by differential centrifugation. A detailed description of the purification of gonococcal fimbriae is given in reference 53, whereas a method for the common fimbriae of nonflagellated *E. coli* is by precipitation in 0.1 M MgCl$_2$ (final concentration) (19). The fimbriae are collected by centrifugation, resuspension in buffer, and reprecipitation twice with 0.1 M MgCl$_2$.

Procedure 2

The shear supernatant (containing 0.5 M NaCl) of sheared *Pseudomonas aeruginosa* contains flagella and fimbriae, which are both precipitated by the addition of polyethylene glycol 6000 (110). After 18 h at 4°C, collect the precipitate by centrifugation and resuspend the pellet in 10% (wt/wt) (NH$_4$)$_2$SO$_4$ (pH 4). After 2 h at 4°C, the fimbriae form a precipitate while the flagella remain in suspension. Remove the flagella by repeating the (NH$_4$)$_2$SO$_4$ precipitation step. Dissolve the final pellet in water, dialyze to remove the (NH$_4$)$_2$SO$_4$, and subject the suspension to CsCl density gradient centrifugation.

Procedure 3

The flagella and fimbriae of *E. coli* precipitate upon the addition of (NH$_4$)$_2$SO$_4$ to give 20% saturation (72). Resuspend the pellet in buffer, and layer on top of a linear sucrose gradient (15 to 50% [wt/wt] sucrose). Remove contaminating flagella by incubating the fimbriae for 1 h at 37°C in 0.4% (wt/vol) SDS. Separate the dissociated flagella from the intact fimbriae by gel filtration on Sepharose CL-4B by using a buffer containing 0.4% SDS and 0.1% (wt/vol) EDTA. Only fimbriae elute in the void volume.

When appropriate, isolation is assisted by taking advantage of the buoyant density of fimbriae versus the densities of other structures (Table 2).

7.4.1.3. Spinae

Spinae are nonprosthecate, rigid, helical protein structures that extend outward from the cell surface of some bacteria (chapter 4, Fig. 3, and reference 32); they may coexist with flagella and require appropriate culture conditions for expression (34). They are readily detached from the cell surface by shear force, but the base may remain on the cell. Spinae are not efficiently detached from the cell by homogenization in a Potter-Elvehjem homogenizer. Their large size and characteristic density (Table 2) allow purification by differential and density gradient centrifugation. Here is a procedure for the isolation of spinae from a marine pseudomonad grown under conditions that repress flagellum synthesis (33).

1. Harvest cells from 1 liter of medium, and wash by centrifugation with resuspension in 100 ml of 0.15 M NaCl or an appropriate buffer according to the origin of the culture.

2. Homogenize the cells in a Waring type blender for 1 to 2 min, and sediment sheared cells by centrifugation.

3. Centrifuge the supernatant at 65,000 × g for 20 min to pellet the spinae. Spinae detached during growth may be recovered from the culture supernatant by retention on membrane filters (pore size, 0.45 μm; Millipore Corp., Bedford, MA).

4. Resuspend the spinae in a small volume of 0.15 M NaCl or buffer, and layer on a CsCl gradient (density, 1.3 to 1.4). Centrifuge at 60,000 × g for 90 min. Collect the band, and remove the CsCl by dialysis against 0.15 M NaCl or buffer.

7.4.2. Walls and Envelopes

The term cell wall is used to define most structural components external to the plasma membrane and should not be used in reference to the murein (peptidoglycan) component alone. Bacterial walls have proved to be sufficiently hardy to survive cell disruption procedures (section 7.1.1). Historically, one of the first successful bacterial fractionation techniques was the isolation of the cell walls from gram-positive bacteria (120). Gram-negative bacteria have more complex cell walls consisting of murein, periplasm, and outer membranes with or without surface embellishments (chapter 4, Fig. 1 and 2). It is obvious that the cell wall of a gram-negative bacterium cannot be isolated as a completely physiologically active wall, because the periplasmic components, which form part of the structure, are released during mechanical breakage. In addition, it has proved difficult to remove all of the plasma membrane from the murein and outer membranes of many gram-negative bacteria, probably due to adhesion zones between plasma and outer membranes. Therefore, this fraction as isolated from gram-negative bacteria is more accurately referred to as an envelope fraction. The cell walls of many archaea consist solely of a paracrystalline S-layer, and simple mechanical or osmotic lysis techniques usually result in the preparation of an envelope fraction.

Although many wall components can be isolated from whole cells, it is often advantageous to start with an envelope or wall fraction to reduce contamination with cytosol constituents.

Procedure for Obtaining Crude Cell Wall Fraction

Best separations of envelope or wall fractions from the cell contents are achieved following mechanical or osmotic breakage (section 7.1) followed by differential centrifugation (section 7.3.1.). Perform procedures at 0 to 4°C and as rapidly as possible to minimize enzymatic degradation of murein by autolysins, which is a definite problem with many *Bacillus* species. It may be necessary to heat the wall preparations at ~80°C after the first centrifugation or even to inactivate the enzymes by heating at 100°C for 5 to 10 min. RNase and DNase may be included during disruption of cells. Use an initial low-speed spin to remove unbroken cells, followed by centrifugation at 30,000 to 45,000 × g for 20 min to pellet the envelope fraction. Inevitably, some larger envelope fragments will sediment with the intact cells, but the low-speed spin is necessary to ensure the purity of the final fraction. The plasma membrane is disrupted by shear forces and usually reseals to form small vesicles

which require much higher centrifugal forces (~100,000 × g) to sediment. Thus, the crude wall or envelope fraction obtained at 30,000 to 45,000 × g does not usually contain much plasma membrane other than that present via adhesion zones. The choice of subsequent purification methods depends upon the purpose of the preparation and is considered separately for gram-positive and gram-negative bacteria.

Procedure for Isolation of Walls of Gram-Positive Bacteria

1. Wash the crude cell walls with saline (0.9% [wt/vol] NaCl) or buffers (50 mM Tris-HCl [pH 8] or 50 mM HEPES [pH 7.8]) to remove soluble proteins. If necessary, use 1 M NaCl to displace cytoplasmic contaminants (120), and subsequently wash the walls with distilled water to remove the salt. Avoid proteolytic enzymes to digest soluble proteins, because important wall proteins can be covalently bound to the peptidoglycan.

2. Suspend the washed envelopes in a minimum volume of cold water, and pour this suspension into 4% (wt/vol) SDS held in a boiling water bath to give a concentration of 30 mg (dry weight)/ml. Stir occasionally for 15 min or until the color is gone.

3. Recover the walls by centrifugation (45,000 × g for 15 min at 20°C), and wash the wall pellet repeatedly by suspension and centrifugation, first with 0.9% NaCl and then with water at 20°C to remove traces of SDS. The purified walls of gram-positive bacteria are white and devoid of any cellular pigments. The walls can be stored in suspension at −20 to −70°C or lyophilized. Verify the purity of the product by electron microscopy.

Procedure for Isolation of Envelopes of Gram-Negative Bacteria (Fig. 2)

1. Wash the crude envelope fraction several times with saline (0.9% NaCl) or buffers (50 mM Tris-HCl [pH 8] or 50 mM HEPES [pH 7.8]). It is advisable to include 1 mM MgCl₂ and/or CaCl₂. Mg^{2+} serves to maintain the integrity of the plasma membrane, and Ca^{2+} maintains the integrity of walls and S-layers; use will vary with intention.

2. Treat the preparation with RNase (100 μg/ml) for 2 to 3 h at 37°C, and wash again. For many gram-negative bacteria, this final preparation is suitably devoid of plasma membrane and cytosol contaminants, which should be verified by electron microscopy.

Selective solubilization of the plasma membrane by detergents provides a method for examining outer membrane proteins without degradation of the murein layer. The basis for selective solubilization is that the outer membrane of *E. coli* is insoluble in 2% (vol/vol) Triton X-100 in the presence of Mg^{2+} (e.g., 10 mM) (28, 125) or in Sarkosyl (36) whereas the plasma membrane is soluble. The fraction obtained should be referred to as the detergent-insoluble fraction rather than the wall since (especially for gram-negative bacteria) minor polypeptides, lipopolysaccharide (LPS), or lipids of the outer membrane may also be solubilized and thus a native outer membrane is not obtained. However, quite the opposite effect is obtained with *Treponema pallidum* (111), whose outer membrane has unusual properties and is solubilized by Triton X-114 while the plasma membrane remains intact (25).

Purification of the Murein-Outer Membrane Complex

Purification of the murein-outer membrane complex from the crude envelope fraction by density gradient separation

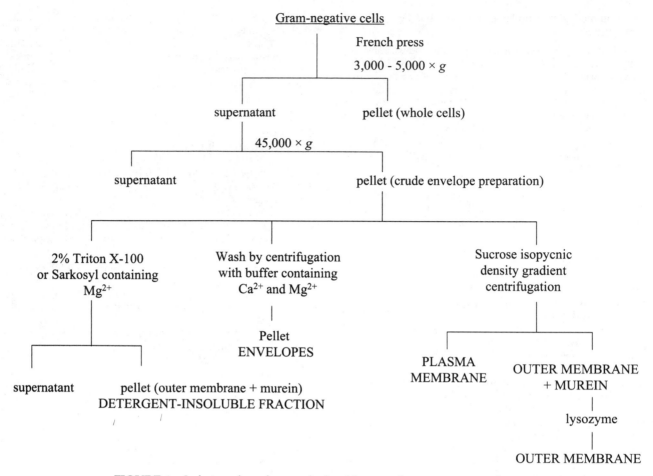

FIGURE 2   Isolation of envelopes and related fractions from gram-negative bacteria.

in sucrose is described by Schnaitman (124). This method attains separation of plasma and outer membranes, but remember that the murein is still associated with the outer membrane and has to be removed by lysozyme if purified outer membrane is required.

7.4.3. Wall Macromolecules

7.4.3.1. Murein and Pseudomurein

Most bacterial walls contain murein (peptidoglycan), whereas archaea have either pseudomurein (modified murein) or a completely different wall construction (71). For example, the archaea *Methanosarcina* and *Halococcus* possess complex anionic heteropolysaccharides and others have cell walls that consist solely of S-layers (section 7.4.3.4).

Murein can be prepared from bacteria either with or without prior cell disruption. On a small scale, the intact murein sacculus can be isolated from both gram-positive and gram-negative bacteria based on its insolubility in boiling SDS solutions. A general method (46) follows.

Procedure for Isolation of Murein Sacculi without Prior Cell Disruption

1. Suspend a wet bacterial paste or lyophilized bacteria in ice-cold water (2 to 4 mg [dry weight]/ml). Mix mechanically or with a hand homogenizer to obtain a homogeneous suspension, and add it dropwise to an equal volume of boiling 8% (wt/vol) SDS. Bring the mixture to a boil again as quickly as possible (to help inactivate autolysins), and boil for 30 min with continuous stirring.

2. Cool the sample, and leave it overnight at room temperature. Collect the insoluble wall material by centrifugation at $30,000 \times g$ for 15 min (for gram-positive bacteria) or $130,000 \times g$ for 60 min (for gram-negative bacteria). To avoid precipitation of SDS, do not refrigerate the samples while centrifuging.

3. Reextract the insoluble residue twice with boiling 4% SDS, until the murein sacculus appears free of other cellular constituents as judged by electron microscopy and chemical analysis.

4. Wash this preparation by resuspension and centrifugation, four times with water, twice with 2 M NaCl, and four times with water. The final residue may be frozen or lyophilized for storage.

A better separation of murein from the cytosol and membranes is often achieved after mechanical disruption of cells and extraction of the envelope fragments. This method was described in the previous section (7.4.2) for gram-positive bacteria. The same principles apply to gram-negative bacteria except that a larger starting culture volume is necessary to compensate for the smaller proportion of murein in these cells.

As a cautionary note, it should be remembered that in cultures of enteric bacteria and possibly other gram-negative bacteria grown to stationary phase, the outer membrane proteins may become cross-linked in a way which prevents their solubilization by boiling SDS. Care must be taken to use actively growing exponential-phase cultures.

7.4.3.2. Teichoic Acids

Complex polysaccharides, including murein, make up the major portion of the cell walls of the gram-positive bacteria. Often, secondary polymers of glycerol or ribitol teichoic acids (wall teichoic acids), teichuronic acids, or other polysaccharides are covalently linked to the murein strands. Lipoteichoic acids (membrane teichoic acids), also found in most gram-positive bacteria, are anchored through the lipid moiety to the outer face of the plasma membrane with the hydrophilic chains extending toward, and in some strains through, the cell wall (150). Isolation methods for these various accessory polymers have been reviewed critically elsewhere (47) and will not be addressed in detail here.

Teichoic acids may be released from the murein by acid or base hydrolysis of sensitive phosphodiester linkages between teichoic acid and the C-6 of muramic acid in the murein strand. Degradation of the polymer can be a problem, depending on the exact structure of the teichoic acid. If extensive damage occurs, use less harsh methods such as lysozyme or neuraminidase digestion of murein from a purified wall fraction (section 7.4.2). However, the enzymatic digestion may leave murein moieties close to the point of linkage to the teichoic acid polymer. Protocols for extraction by using trichloroacetic acid, acidic buffers, NaOH, or N,N'-dimethylhydrazine and by using enzymatic hydrolysis are given by Hancock and Poxton (47).

7.4.3.3. LPS

LPS is ubiquitous, with few exceptions, as a major surface component of gram-negative bacteria. Although there are common structural features found in almost all kinds of LPS, the structure varies so widely among different strains and species of bacteria that the choice of a suitable extraction method may be quite difficult. In addition, many individual species can produce more than one kind of LPS. A single strain of bacteria may produce multiple forms of LPS continuously during balanced growth, discontinuously as a result of genetic changes such as phase variation, or in response to physiological signals such as temperature, culture density, and nutrition. Some of the extraction procedures (38, 153) were developed before this heterogeneity was fully appreciated, and thus they may not yield a representative preparation. Because LPS is so frequently the subject of research and presents so many problems, an extensive discussion follows.

Most enteric bacteria and many *Pseudomonas* strains produce a large form of LPS termed the S (or smooth) form, which has three parts: (i) a lipid termed lipid A, which is a disaccharide of glucosamine containing both O-linked and N-linked fatty acids; (ii) a complex core oligosaccharide of about 10 sugars, which is attached at its reducing end to the lipid via the eight-carbon sugar 3-deoxy-D-manno-octulosonic acid (Kdo); and (iii) a polysaccharide called the O antigen, which is attached to the nonreducing end of the core oligosaccharide. The O antigen consists of a three-, four-, or five-sugar chain that is repeated about 15 to 20 times to form a long unbranched polymer. Since the O antigen polymers are not always the same length, the S form of some LPSs will appear after SDS-polyacrylamide gel electrophoresis (SDS-PAGE) as a ladder of discrete bands, each differing from the next by the size of a sugar chain. Strains which produce an S form LPS (S-LPS) also produce a smaller LPS species termed the R (or rough) form (R-LPS), which is very heterogeneous and contains only two parts, a lipid A and a varied population of core molecules which do not carry O antigens. In enteric bacteria, the S and R forms each represent about 50% of the total number of LPS molecules. The terms S and R correspond to smooth- and rough-colony designations given to strains which do and do not produce an O antigen, respectively. This LPS distinction is somewhat less applicable since it was discovered that the strains that exhibit smooth colonies produce a mixture of S- and R-LPS.

Many gram-negative bacteria (including members of genera such as *Haemophilus* and *Neisseria*) never produce S-LPS but instead produce molecules consisting only of lipid A and an antigenically complex, often heterogeneous oligosaccharide somewhat similar to the core present in R-LPS from enteric bacteria. These molecules are sometimes termed lipooligosaccharides to distinguish their properties from those of S-LPSs that bear O antigens. These molecules are similar in extraction properties to the R-LPSs from enteric bacteria.

The choice of methods for the isolation of LPSs from gram-negative bacteria depends mostly on the presence or absence of O side chains. The classical method for the isolation of S-LPS from gram-negative bacteria was described by Westphal and Jann (153) and utilizes hot aqueous phenol. This method is not successful in extracting R-LPS from whole cells. Galanos et al. (39) developed another method involving aqueous phenol, chloroform, and petroleum ether (PCP) for extraction of R-LPS. Another procedure described by Darveau and Hancock (26) for the simultaneous extraction of both S- and R-LPS has broader application. Although originally described for *P. aeruginosa*, it has been used successfully for other organisms. LPS isolated by any of the above-mentioned procedures is sufficiently pure for subsequent chemical analyses and SDS-PAGE (described below). LPS isolated by the phenol-water method preserves the reactivity of pseudomonad LPS with antibodies in Western immunoblots and is better in this regard than LPS isolated by the Darveau-Hancock method (26). Isolated LPS can be converted into a soluble uniform salt form (Ca^{2+}, Na^+, triethylammonium) by electrodialysis to the free-acid form (deionized) and neutralization with various bases (38).

Proteinase K Digestion of Bacterial Lysates and Analysis by SDS-PAGE

A breakthrough in the field of LPS analysis occurred when Hitchcock and Brown (56) devised the following simple procedure for analysis of LPS by SDS-PAGE of crude lysates of whole cells or outer membrane preparations.

1. Remove cells from an agar plate with a sterile swab, and resuspend them in 10 ml of phosphate-buffered saline (pH 7.2) to an optical density at 600 nm of 0.5 to 0.6. Because some bacteria produce a different LPS when grown on agar rather than in broth, and some also produce different LPSs at different stages of growth, it is recommended to standardize by using late-exponential-phase cells from a broth culture. Centrifuge cells from 10 ml of an overnight culture, and resuspend in phosphate-buffered saline to an optical density at 600 nm of 0.5 to 0.6. The volume of culture used may vary according to the organism under investigation. Use a portion (1.5 ml) of the cell suspension for LPS extraction, and sediment bacteria by centrifugation in

a microcentrifuge for 3 min. An equal amount (1 to 10 mg) of lyophilized cells may also be used.

2. Resuspend the pellet in 50 μl of SDS-PAGE lysis buffer (2% [wt/vol] SDS, 4% [vol/vol] 2-mercaptoethanol, 10% [vol/vol] glycerol, and 0.002% [vol/vol] bromophenol blue in 1 M Tris-HCl buffer [pH 6.8]), and heat at 100°C for 10 min. Add 25 μg of proteinase K, and incubate at 60°C for 1 h.

If multiple samples are to be prepared, enough proteinase K solution can be prepared for all; however, it cannot be stored after it is dissolved in lysis buffer. The digested samples may be frozen at this point. Prior to running the gel, mix 5 μl of digested sample with 10 μl of lysis buffer and load 3 to 6 μl of this dilution per 5-by-0.75-mm well (16-by-18-cm gels). Proteinase K is not stable during prolonged storage, and it should be replaced if a background of protein bands is visible on the gels.

A cleaner picture is obtained if proteinase K-digested samples (50 μl) are extracted with an equal volume of 90% (vol/vol) phenol at 65°C for 15 min (vortex every 5 min) (96). Centrifuge samples in a microcentrifuge for 10 min. Transfer the aqueous phase to a new microcentrifuge tube, and extract once with 10 volumes of ethyl ether to remove traces of phenol. Aspirate the upper ether phase, and mix the lower phase 1:1 with sample buffer as described above. Cleaner lanes are also obtained if the Hitchcock and Brown procedure is applied to envelope or outer membrane preparations (56).

Procedure for Tricine–SDS-PAGE

The Tricine-SDS gel system is modified from that of Lesse et al. (89) and gives vastly increased resolution (especially for R-LPS) over that obtained by the older glycine-SDS system (56).

1. The bottom buffer is 0.2 M Tris base (24.2 g/liter) adjusted to pH 8.9 with concentrated HCl. The top buffer is 0.1 M Tris base (12.1 g/liter), 0.1 M Tricine (17.9 g/liter), and 0.1% (wt/vol) SDS. A 4× gel buffer stock solution consists of 4 M Tris base (48.5 g/100 ml; heat to dissolve), adjusted to pH 8.45 with concentrated HCl, and 0.4% SDS. For the best results, add dry SDS to the top buffer and gel buffer stock just before use.

2. The running gel (per 20 ml) consists of 9 ml of 40% (wt/vol) acrylamide (final concentration, 18%), 3.6 ml of 2% (wt/vol) bis-acrylamide (final concentration, 0.36%), 5 ml of 4× gel buffer, 2.1 ml of glycerol, and 0.3 ml of water. Predissolved acrylamide and bis-acrylamide stock solutions can be obtained commercially. Degas the gel solution under a vacuum for 20 min, and add 4 μl of N,N,N′,N′-tetramethylethylenediamine and 40 μl of 10% (wt/vol) ammonium persulfate. After casting the gel, overlay it with water-saturated butanol. When polymerization is complete, wash out the butanol with 1× gel buffer. The gel can be stored overnight with an overlay of 1× gel buffer. For convenience, a stock solution of 10% ammonium persulfate can be prepared in advance, dispensed into small single-use vials, and stored frozen for up to 6 months.

3. The stacking gel (per 10 ml, 0.74 M Tris [final concentration]) contains 1.125 ml of 40% acrylamide (final concentration, 4.5%), 0.4 ml of 2% bis-acrylamide, 1.86 ml of gel buffer, and 6.62 ml of water. Degas the gel solution under a vacuum for 10 min, and add 12 μl of N,N,N′,N′-tetramethylethylenediamine and 120 μl of 10% ammonium persulfate. Wells should be rinsed well with top buffer before samples are added.

4. For a 16-cm-by-0.75-mm gel with 20.5-mm wells, perform electrophoresis at a constant current of 35 to 38 mA for about 4.5 h. The actual running time will depend upon the molecular size of the LPS and the separation desired. Cool the electrophoresis apparatus to room temperature with recirculating water.

Procedure for Silver Staining of LPS

Wear gloves for handling the gel and staining solution, both for safety and to prevent contamination. To prevent formation of a silver carbonate precipitate, it is important that the NaOH used be carbonate free and that the NH₄OH be fresh. It is convenient and inexpensive to purchase NaOH as a predissolved carbonate-free 1 N solution, so that bottles can be discarded after they have been opened a few times. Use deionized water, and agitate throughout all procedures. Because of the frequent decanting of washes, it is convenient to fix, wash, and stain gels in small baking dishes on a rocker platform adjacent to a vacuum aspiration apparatus.

1. Fix the gel overnight in 200 ml of freshly prepared 25% (vol/vol) isopropanol in 7% (vol/vol) acetic acid. Decant the solution, and oxidize the gel for 5 min in a freshly prepared solution consisting of 150 ml water, 1.05 g of periodic acid, and 4 ml of the 25% isopropanol–7% acetic acid fixing solution.

2. Wash the gel eight times, for 15 to 20 min each time, with 200 ml water. This extensive washing removes small aldehydes resulting from periodate oxidation.

3. Stain for 10 min in a freshly prepared solution consisting of 28 ml of 0.1 N NaOH, 1.25 ml of concentrated (29.4%) NH₄OH, 5 ml of freshly prepared 20% (wt/vol) silver nitrate, and 115 ml of water. If a dark brown precipitate forms in the staining solution, add concentrated NH₄OH dropwise until it just disappears. After staining, wash the gel four times, for 10 min each time, with 200 ml of water.

4. Develop in 250 ml of a solution that contains (per liter) 50 mg of citric acid and 0.5 ml of 37% (vol/vol) formaldehyde. Developing usually requires 5 to 10 min and can be stopped when the desired band intensity is reached by soaking the gel for 60 min in 200 ml of water containing 10 ml of 7% (vol/vol) acetic acid. If the gel is to be stored for a long period, repeat the fixation step to minimize band fading.

5. Record the image as a file by scanning or photographing the wet gel. Alternatively, the gel may be dried before photographing. Soak the gel overnight in water containing 5% (vol/vol) glycerol prior to drying on filter paper. Because of the high acrylamide content, dry the gel on a heated dryer with a high-power vacuum and a refrigerated or dry-ice trap to minimize cracking.

The Hitchcock-Brown method has several advantages that make it particularly suitable both for characterizing the LPS and for monitoring LPS extraction and purification procedures. Because it does not involve extraction and is independent of the size and complexity of the LPS, it gives an accurate picture of the distribution of various forms of LPS in a given culture or preparation. The procedure is useful for screening large numbers of clinical or environmental isolates. Depending on the branching pattern of the core carbohydrate, most LPS molecules yield either one or two aldehyde groups that are attached to the lipid and remain fixed to the gel after prolonged washing. Therefore the intensity of the silver stain reflects the number of molecules and not the number of sugar residues per molecule prior to cleavage; the silver stain pattern provides both qualitative and semiquantitative information on the distribution of different size classes of LPS in a sample. With minor modifications in gel strength and running time, the Hitchcock-Brown procedure can be used to characterize the polymerization numbers and size distribution of ladders of S-LPS and relative amounts of S- versus R-LPS, the chemotypes of LPS produced by various mutants, and the sizes and the properties of lipooligosaccharides. In addition to LPS visualized by silver staining, LPS prepared by the Hitchcock-Brown procedure and separated by SDS-PAGE is ideal for detection by Western immunoblotting, lectin binding, or autoradiography.

Procedure for Phenol-Water Extraction for S-LPS (153)
Dry, redistilled phenol can be purchased (molecular biology grade) and stored at 4°C for several months. This purchase avoids the hazards associated with redistilling of crystalline phenol and the instability of liquified phenol. To prepare 90% (wt/vol) phenol, melt by heating a 500-g bottle (cap loosened) of dry phenol to approximately 70°C in a microwave oven (or water bath) and add 61 ml of water while hot. *Caution:* do not attempt to weigh by removing crystals of phenol from the bottle. If lesser amounts are required, weigh 90 g of the liquified phenol and add 11 ml of water.

1. Thoroughly resuspend lyophilized bacteria to a concentration of 5% (wt/vol) in distilled water. For large amounts of LPS for carbohydrate analysis, resuspend 20 g of cells in 350 ml of water. For smaller amounts of LPS for column chromatography, SDS-PAGE, and Western immunoblotting, 1 to 2 g of lyophilized cells in 30 to 60 ml of distilled water is sufficient. However, higher yields of LPS extracted with phenol-water may sometimes be achieved from spheroplast lysates versus cells. As an alternative to using lyophilized cells, sensitive strains may be treated with EDTA-lysozyme and the extract may be digested with RNase and DNase (plus 20 mM MgCl$_2$) and then with trypsin. Purified LPS can also be extracted from RNase-treated envelopes (section 7.4.2) preextracted with chloroform-methanol to remove lipid (154) (use 50 ml of water/g of lipid-depleted lyophilized walls). This LPS preparation is usually free of contaminating nucleic acids.

2. Heat the aqueous suspension at 70°C in a water bath for 10 min.

3. Heat an equal volume of 90% (vol/vol or wt/vol) phenol to 70°C, and add it to the bacterial suspension. A glass Erlenmeyer flask is a suitable vessel for the phenol mixture.

4. Stir vigorously with a mechanical stirrer in a water bath at 70°C for 15 min.

5. Transfer the mixture to 50-ml centrifuge tubes (polycarbonate tubes must *not* be used), and cool on ice to about 10°C.

6. Centrifuge the emulsion at 10,000 × g for 30 min. Three layers are formed: an upper aqueous phase (containing the LPS), a lower phenol layer, and an insoluble residue (containing murein) at the interface of the two layers. Note that exceptions exist in which significant amounts of LPS are soluble in the phenol phase (2). This possibility must be checked before discarding the phenol phase.

7. Remove the upper aqueous layer with a Pasteur pipette, and retain it.

8. Extract the phenol phase and any insoluble residue again at 70°C with a volume of water equivalent to that removed in the aqueous phase. Combine the aqueous layers.

9. Dialyze the aqueous extract against distilled water for 2 to 3 days to remove the phenol.

10. Lyophilize and dissolve in 0.9% (wt/vol) NaCl. Avoid rotary evaporation when unbuffered (acidic) phenol is present, since structural damage such as the loss of O-acetyl groups may occur. Use saline rather than distilled water to dissolve the LPS as an aid for protein removal. Centrifuge at 27,000 × g for 30 min to remove any insoluble residue.

If the phenol extraction was performed on whole bacterial cells, about one half of the organic material in the crude extract will be RNA. This contamination can be avoided by extraction of cleaned walls (envelopes). LPS can be separated from the RNA by ultracentrifugation of the extract at 100,000 × g for 10 h (shorter times may result in partial recovery of LPS). The LPS should sediment to form a clear, gelatinous pellet. Experience has shown that phenol-extracted LPS of some bacteria does not sediment efficiently by ultracentrifugation. If the pellet has an opaque base, it is not free of contaminating material and must be washed with water by ultracentrifugation. Alternatively, before ultracentrifugation, treat the dialyzed suspension with RNase and DNase (1 μg/ml) and MgCl$_2$ (final concentration, 10 mM) at 37°C for 1 h. Resuspend the final pellets in a small volume of distilled water, and lyophilize.

If a low yield is obtained, LPS should be extracted from envelopes as mentioned above. The advantage of this approach is that the RNA has been digested prior to phenol extraction and the dialyzed aqueous extract can be lyophilized directly.

Procedure for Phenol-Chloroform-Petroleum Ether Extraction for R-LPS
The procedure described below (5) is modified from that of Galanos et al. (39) and gives a good yield of R-LPS from strains such as *E. coli* K-12 and from a variety of mutants that produce different chemotypes of LPS including deep-rough and heptose-deficient strains. The physical properties of purified R-LPS are determined by the size of the carbohydrate core. For example, lyophilized LPS from wild-type *E. coli* K-12 which contains a complete core is a fluffy powder which is readily dispersed in water, whereas a similar preparation of LPS from a deep-rough mutant which lacks the hexose outer core sugars and has only the inner heptose-Kdo region is a dense, waxy solid that can be dispersed in water only by heating and vigorous homogenization.

The efficiency of extraction of R-LPS by PCP and the purity of the final product are determined by the way the bacteria are treated prior to extraction. Unlike other extraction procedures, PCP extraction gives a poor yield with isolated outer membrane preparations and is most effective if the bacteria are intact. It is important that the bacteria are harvested in exponential phase and are free of broken or damaged cells, completely dry, and free of extraneous salt from the medium and that they are in the form of a loose, granular powder which is easily dispersed in the extraction mixture. The following procedure is designed for isolation of LPS from two 6-liter batches of *E. coli* K-12 in late exponential growth phase and yields about 120 mg of purified LPS or about 3 to 5% of the weight of the dry cells. It can be scaled down easily.

1. After harvesting, process cultures promptly through the drying procedure without overnight storage or freezing of cell pellets. Chill a 6-liter culture, and harvest the cells by centrifugation. Resuspend the pellets with a Potter-Elvehjem homogenizer to a total volume of 150 ml in cold, deionized water. Distribute the suspension into six 40-ml centrifuge tubes, and pellet the cells by centrifugation for 10 min at 12,000 × g. Suspend the cells as a thick paste in a minimum of cold water (no more than 1 ml per tube plus 1 to 2 ml for rinsing the tubes) by using a test tube as a homogenizer.

2. Transfer the cell paste into the bowl of a blender or Sorvall Omni-Mixer, add 90 ml of cold methanol, and homogenize for 30 s at maximum speed. The suspension should be smooth and free of clumps. Transfer to a Corex bottle, and centrifuge for 10 min at 8,000 × g. Discard the supernatant, and resuspend the pellet in 90 ml of cold acetone; homogenize with a blender as described above, and centrifuge as described above. Repeat the acetone wash step once more. Drain off all of the acetone, cover the mouth of the Corex bottle with a sheet of tissue or lens paper secured with a rubber band, place in a large lyophilizer jar, and freeze-dry overnight. The pellet will "pop" during drying, and the resulting chunks of dried bacteria can easily be crushed into a fine powder with a spatula and stored in a vacuum desiccator at room temperature until PCP extraction.

3. Caution: prepare PCP in a hood, and use gloves and extreme care for all steps. PCP (300 ml) is prepared by combining with vigorous stirring 40 ml liquified phenol (see above for preparation), 100 ml of chloroform, and 160 ml of petroleum ether (American Chemical Society grade; boiling range, 37 to 52°C). This generally forms a hazy emulsion. Add about 1 ml of melted dry phenol to break the emulsion, which will result in a few small drops of a separate phenol-rich phase that tends to adhere to the walls of the container. This phase can be removed by pouring the PCP through a piece of fluted Whatman filter paper in a glass funnel. The filtrate is a clear and stable single phase and can be stored for up to 1 month in a stoppered container at room temperature.

4. To extract the bacteria, combine the dried cells from two 6-liter batches (about 4 g [dry weight] total). Place in the bowl of a 100-ml Sorvall Omni-Mixer or other small blender (a beaker and a rotary probe homogenizer of the Braun type can also be used), and add 50 ml of PCP. Homogenize at maximum speed for about 30 s. The purpose is to suspend the bacteria completely without breaking the cells, and the suspension should have a smooth consistency resembling that of a milk shake. Transfer to a Corex glass bottle, and centrifuge for 5 min at 9,200 × g. Decant, and save the supernatant, which contains the bulk of the LPS. To increase the yield, remove the very rubbery pellet, break into pieces with a spatula, and homogenize it as described above with a second 50-ml portion of PCP. Centrifuge as described above, combine the supernatants, and filter through Whatman no. 1 paper in a small Büchner funnel.

5. It is necessary to concentrate the sample. A rotary evaporator is not recommended, since the solution may foam. Transfer portions of the sample into a 100-ml beaker, place on a hot plate or heating block which is just hot to the touch (about 50°C), and evaporate under a stream of N_2 gas until the entire sample has been concentrated to ~15 ml. Small flakes or a skin may form on the sides of the beaker. Cool, and add a small amount of water (no more than 2 ml) dropwise while stirring with a Teflon-coated stirring rod until no more waxy precipitate forms. Transfer to a Corex tube (some precipitate will remain on the sides of the beaker), and centrifuge at 12,000 × g for 10 min. Depending upon the LPS type and the extent to which the chloroform has been removed, it will either form a firm white pellet or float on top as a waxy cake. Remove the clear phenol liquid (aspirate with a hypodermic syringe with a long needle if necessary), and test to be sure that precipitation of the LPS is complete. Precipitation is complete if no more precipitate forms when a few drops of water or several volumes of methanol are added. Add about 10 ml of

methanol to the pellet or waxy cake in the Corex tube, and use an additional 10 ml to wash any precipitate from the sides of the precipitation beaker into the tube. Mix the methanol suspension in the Corex tube vigorously, and centrifuge at 12,000 × g for 15 min. Carefully aspirate the methanol from the fragile pellet, and wash the precipitate twice by suspending in 20 ml of methanol with a Teflon-coated stirring rod and centrifuging.

6. Drain off the methanol, and cap the tube with a piece of tissue or lens paper secured with a rubber band. Evaporate the solvent for about 2 h on a lyophilizer, and suspend the dry pellet as completely as possible in 24 ml of water containing 0.1 mM $MgCl_2$ by using a Teflon Potter-Elvehjem homogenizer. Depending upon the nature of the LPS, it may appear as an opalescent waxy suspension or a clear, very viscous solution. Centrifuge in an ultracentrifuge at 200,000 × g for 4 h. The pellets are sometimes transparent and invisible until the supernatant is poured off. These are suspended by thorough homogenization in ~25 ml of water, frozen, and lyophilized. It is important that the samples do not thaw during lyophilization and that they are completely dry before removing from the lyophilizer.

With some samples, the LPS cannot be removed from an aqueous solution by ultracentrifugation. If this occurs, the LPS can be recovered from aqueous solution by precipitation with 5 volumes of ethanol or acetone. This is somewhat less satisfactory, since it also precipitates any nucleic acids or proteins in the preparation.

To test for protein and nucleic acid content, dissolve the LPS at a concentration of 1 mg/ml in 0.1% (wt/vol) SDS solution. It may be necessary to boil the sample to dissolve it in the SDS. The protein content can be determined by a standard assay for protein (e.g., Lowry assay or A_{280}), and the nucleic acid content can be determined by measuring A_{260} versus a blank of SDS solution.

Darveau-Hancock Procedure for S- and R-LPS (26)

Day 1

1. Resuspend 500 mg of lyophilized bacterial cells in 15 ml of 10 mM Tris-HCl buffer (pH 8.0) containing 2 mM $MgCl_2$, 100 μg of pancreatic DNase 1 per ml (add 1.5 mg directly), and 25 μg of pancreatic RNase per ml (add 750 μl of 500 μg/ml stock). In this and all subsequent steps, the term Tris-HCl buffer refers to 10 mM Tris-HCl buffer (pH 8.0).

2. Pass this suspension through a French press (section 7.1.1.1) at 16,000 lb/in². Sonicate for two 30-s intervals at maximum output. Add an additional 100 μg of DNase per ml and 25 μg RNase per ml. Incubate the cell lysate at 37°C for 2 h with gentle shaking.

3. Add the following to the cell lysate: 5 ml of 0.5 M tetrasodium EDTA dissolved in Tris-HCl buffer, 2.5 ml of 20% (wt/vol) SDS in Tris-HCl buffer, and 2.5 ml of Tris-HCl buffer. Adjust the pH to 9.5. Vortex carefully, and incubate the lysate for 1 h at 37°C with gentle shaking. This incubation period is necessary for bacteria whose outer membrane may not dissociate as readily as that of *P. aeruginosa*.

4. Ultracentrifuge the suspension at 50,000 × g for 30 min at 15°C to sediment intact murein (23,500 rpm; Beckman type 40 rotor).

5. Remove the supernatant, and add pronase (from *Streptomyces griseus* [Sigma]) at 5 mg/25 ml to a final concentration of 200 μg/ml. Incubate overnight at 37°C with gentle shaking.

Day 2

6. If a precipitate forms after overnight incubation, centrifuge the mixture at 120 × g for 10 min. Discard the precipitate.

7. Precipitate LPS by adding 2 volumes of 0.375 M MgCl₂ in 95% ethanol and cooling to 0°C in a −20°C freezer or an ethanol–dry-ice bath. Do not allow the temperature of the mixture to go below −5°C. Centrifuge mixture at 12,000 × g for 15 min at 0°C (precool the rotor and centrifuge tubes).

8. Resuspend the pellet in 17.5 ml of Tris-HCl buffer. Add 2.5 ml of 20% (wt/vol) SDS and 5 ml of 0.5 M tetrasodium EDTA (both in 10 mM Tris-HCl). Sonicate for two 30-s intervals as in step 2.

9. Adjust to pH 8 by the dropwise addition of 2 M HCl.

10. Incubate the sample at 85°C for 30 min in a water bath, and then cool the sample to room temperature.

11. Adjust the pH to 9.5 with 2 M NaOH. Add pronase to 25 μg/ml (625 μg/25 ml), and incubate overnight at 37°C with gentle shaking.

Day 3

12. Precipitate the LPS as in step 7 with 2 volumes of 0.375 M MgCl₂ in ethanol. Centrifuge at 12,000 × g for 15 min at 0°C. If the supernatant remains turbid, add a further volume of 0.375 M MgCl₂ in 95% (vol/vol) ethanol and repeat the precipitation and centrifugation steps.

13. Resuspend the pellet in 15 ml of Tris-HCl buffer, sonicate as described above and centrifuge at approximately 400 × g for 5 min. For some bacteria, this pellet may contain a large amount of LPS. Resuspend the pellet in Tris-HCl buffer, dialyze against distilled water, and lyophilize.

14. Make up the supernatant from the low-speed centrifugation in step 13 to 25 ml with Tris-HCl buffer, and add MgCl₂ to 25 mM (125 mg/25 ml). Ultracentrifuge at 200,000 × g for 4 h at 15°C (Beckman type 70.1 Ti rotor). Resuspend the pellet in distilled water, and lyophilize.

7.4.3.4. S-Layers

Many prokaryotes in nature possess paracrystalline arrays of protein, forming surface layers (S-layers; chapter 4, Fig. 1) on their cell walls (79, 131). Electron microscopy has shown that S-layers have hexagonal, tetragonal, or oblique symmetry. S-layers are widely distributed in nearly every major taxonomic group of both *Bacteria* and *Archaea*. A notable exception is the absence of S-layers among the members of the *Enterobacteriaceae*. S-layers, when present, are usually the outermost component of the cell wall.

Most S-layers are composed of single homogeneous protein or glycoprotein species. The energy and information for assembly is contained within the individual protein monomers, and thus some S-layer proteins will self-assemble into the paracrystalline array. S-layers are noncovalently associated with the underlying cell wall (bacteria) or plasma membrane (most archaea) via a combination of ionic and hydrogen bonds or, in some cases, hydrophobic interactions. The same forces also maintain the integrity of the S-layer itself. The stability and conformation of the protein are often dependent upon divalent cations (most often Ca²⁺ and occasionally Mg²⁺), whereas the ability to integrate with the wall depends on ionic strength. A valuable feature of S-layers with regard to their isolation is that they usually remain associated with the underlying envelope components after mechanical or osmotic disruption of cells. Most techniques for the isolation and purification of S-layers thus involve the initial disruption of cells and subsequent differential centrifugation to separate cell envelope fragments. S-layers cover the entire cell surface, and their presence is not dependent on growth stage. Cells should be harvested at maximum cell density.

Reagents that are effective in removing S-layers from the cell surface often disrupt the subunit interactions of the array. Therefore, it should be specified whether the intact, paracrystalline S-layer or the component polypeptide is to be isolated. These will be considered separately below.

Isolation of Intact S-Layers

The isolation of intact S-layers is most successful with gram-positive bacteria and archaea. In gram-negative bacteria, the close association with the outer membrane makes this isolation nearly impossible (81, 101). Most S-layers are resistant to dissociation by the nonionic detergent Triton X-100, but the plasma membrane dissolves during Triton X-100 treatment of the cell walls of gram-positive bacteria (section 7.4.2). The contaminating plasma membrane fragments are then removed by washing. The remaining wall can be treated with lysozyme to remove peptidoglycan underlying the S-layer. For archaea having an S-layer as the sole wall component, with notable exceptions (e.g., *Methanoculleus marisnigri*), treatment of envelopes with 0.5% (vol/vol) Triton X-100 at 65°C for 30 min solubilizes the plasma membrane (80). The crude, intact S-layer is collected by centrifugation.

Procedure (12)

1. Grow 3 liters of cells in complex medium to mid exponential phase.

2. Prepare cell walls as described for gram-positive bacteria in section 7.4.2.

3. Treat the walls with 100 ml of 1% (vol/vol) Triton X-100 for 30 min at room temperature.

4. Wash walls by centrifugation three times with 250 ml of 0.05 M sodium phosphate buffer containing 1 mM MgCl₂.

5. Resuspend the final pellet in 250 ml of buffer containing 950 U (approximately 50,000 U/mg of protein; Sigma) of lysozyme per ml, and incubate at room temperature for 6 h.

6. Wash the residual intact S-layer with buffer by centrifugation at 45,000 × g for 30 min, and dialyze against distilled water.

7. The final S-layer preparation can be stored frozen at −20°C as a suspension, or lyophilized.

Some S-layers are remarkably stable in the presence of chemical perturbants. The intact hexagonal S-layer of *Deinococcus radiodurans* can be isolated following treatment of cells or outer membrane vesicles with 2% (wt/vol) SDS (both 20 and 60°C have been used) for 2 to 12 h (143), which dissolves the outer membrane away from the stable protein array. The inner perforate S-layer of *Lampropedia hyalina* is resistant to SDS at 24°C (6). The S-layered sheath of *Methanospirillum hungatei* is very resistant to heat or chemical solubilization (14). The S-layer of the extremely thermophilic archaeon *Thermoproteus tenax* is so resistant to chemical denaturants (76) that most of its constituent subunits appear to be covalently linked.

Isolation of S-Layer Proteins

Koval and Murray (81) and Messner and Sleytr (101) describe methods for the isolation of S-layer proteins. No single isolation procedure will apply to all because S-layers vary considerably in mechanical stability and ease of dispersion into subunits. The various procedures that have been used for the specific extraction of S-layer proteins are described below. Detergents can be used to solubilize S-layer proteins, but they usually result in nonspecific extraction of other wall-associated proteins.

General procedure for extraction of S-layer proteins with guanidine hydrochloride and urea

The chaotropic agents guanidine hydrochloride (guanidinium chloride) and urea have been used successfully with bacteria and archaea. Most S-layer proteins are readily solubilized from the underlying wall components by moderate concentrations of guanidine hydrochloride (1 to 2 M) or urea (4 to 6 M). The effective concentration must be determined for each individual S-layer by using SDS-PAGE and electron microscopy by negative staining to monitor the removal of the wall and the presence of subunits in extracts (chapter 4). The choice of chaotropic agent will require a trial of procedures, e.g., the S-layer proteins of *Aquaspirillum serpens* strains are more susceptible to proteolysis during extraction with guanidine hydrochloride than during extraction with urea (81). However, the use of urea is not compatible with some subsequent biochemical analyses of the polypeptide, such as N-terminal amino acid sequencing.

1. Grow 1 to 3 liters of cells in an appropriate medium to mid exponential phase, and prepare cell envelopes (section 7.4.2).

2. Treat the cell envelopes with 1% (vol/vol) Triton X-100 to remove contaminating plasma membrane. This is an optional step but reduces the total number of proteins seen by SDS-PAGE.

3. Wash the walls by centrifugation with 10 mM HEPES buffer (pH 7.4), resuspend the walls in the appropriate concentration of guanidine hydrochloride or urea, and incubate at 37°C for 2 h. If necessary, protease inhibitors (e.g., 1 mM phenylmethylsulfonyl fluoride) (9) may be included at this and subsequent stages.

4. Recover the envelopes by centrifugation, and reextract with the chaotropic agent.

5. Combine the extracts (i.e., supernatants), and dialyze at 4°C against several changes of distilled water with the appropriate cation. If the S-layer protein forms assembly products upon removal of the chaotropic agent, a precipitate will appear in the dialysis bag. Collect the precipitate by centrifugation at 6,000 × g for 15 min, and clean from soluble contaminants by washing and centrifugation. If a specific cation (e.g., Mg^{2+} or Ca^{2+}) is required for assembly and not included, amorphous aggregates will form upon removal of the chaotropic agent.

6. Concentrate the solubilized protein in a lyophilizer or an ultrafiltration apparatus with a PM 10 filter (Amicon Corp.). Store at −20°C as a concentrated solution or in a lyophilized state.

Procedure for extraction of S-layer proteins with metal-chelating agents

Because many S-layers, particularly those from gram-negative bacteria, are dependent on cations (Ca^{2+} or Mg^{2+}) for attachment and assembly, it is important to provide the necessary cations during growth and for the preparation and storage of cell envelopes. For some species, extraction of cell envelopes with 10 mM EDTA or EGTA will solubilize the S-layer protein. Follow the procedure described above, substituting metal-chelating agents for the chaotropic agents in steps 5 and 6. For S-layer proteins that strongly bind cations, dialysis against the chelating agent may be

necessary (81). Cation-dependent S-layer proteins from gram-negative bacteria are usually not capable of self-assembly in the absence of a suitable template (e.g., outer membrane vesicles or S-layer minus cell surfaces).

Other approaches

The synthesis of S-layer proteins is usually closely regulated, and they are not overproduced, but there are two examples to date where S-layer proteins have been detected in the supernatants of early-stationary-phase cultures. *Bacillus brevis* 47 secretes large amounts of its two S-layer proteins (158). *Campylobacter fetus* also secretes its S-layer protein (antigen a) into the culture medium in a soluble form, along with two to three additional antigens. This S-layer protein has also been isolated by treatment of cells with 0.2 M glycine hydrochloride at pH 2.2 (29). The S-layer protein of *A. vinelandii* is unusual in that it is removed from the cell surface by distilled-water washes (16). The S-layer of *A. vinelandii* is stabilized by Ca^{2+} and Mg^{2+}, but the stability is not as great as that of other S-layers such as those of *Aquaspirillum* spp., which require use of a metal-chelating agent for dissociation.

7.4.3.5. Exopolysaccharides

Exopolysaccharides cover the surfaces of many gram-negative and gram-positive bacteria (chapter 4, Fig. 6). They may form a capsule composed of high-molecular-weight polysaccharides attached to the cell surface (chapter 4, Fig. 1) and seen by light and electron microscopy, or they may produce slime either loosely attached to the cell surface or released into the culture fluid; these are called exopolymeric substances. Capsules are sometimes referred to as a glycocalyx, which is more appropriately considered a eukaryotic (often mammalian) structure. Growth conditions including carbon-nitrogen ratio, phase of growth, temperature, and growth in laboratory media can markedly affect the production of exopolysaccharide. Properties and structures of exopolysaccharides have been reviewed previously (60, 64).

Various methods are described to isolate the high- and low-molecular-weight exopolysaccharides (48). Bacteria may be grown and harvested from solid media in cases in which nondialyzable medium ingredients contaminate the final preparation. The following method modeled after that of Altman et al. (1) produces pure exopolysaccharides from *Actinobacillus pleuropneumonia* strains and is generally applicable for the isolation of capsules or slime. During isolation of exopolysaccharides, it is important to ensure that extracts are clarified by centrifugation. Because these polymers are very water soluble, haziness indicates impurities.

Procedure

1. Harvest bacteria, usually in late exponential to stationary phase, as a centrifuged pellet. Attempt to recover exopolysaccharides from the culture medium supernatant, unless their absence is established. Dialyze this supernatant against water, and concentrate by lyophilizing and dissolving in 0.9% (wt/vol) saline. Recover exopolysaccharide by precipitating with 5 volumes of 95% (vol/vol) ethanol. Dissolve the precipitate in distilled water. If hazy, remove impurities by centrifuging.

2. Wash the bacterial pellet from step 1 with 0.9% saline (the supernatant may be combined with the growth medium supernatant and processed as in step 1), and extract cells with phenol-water as described for LPS (section 7.4.3.3, steps 1 to 12). *Caution:* in rare cases, exopolysaccharides may be in the phenol phase (15). Remove impurities by centrifugation, if hazy.

3. Digest the exopolysaccharide preparation from steps 1 and 2 (plus 20 mM $MgCl_2$) with RNase and DNase followed by trypsin.

4. Dialyze against water, and recover by precipitating with 5 volumes of 95% ethanol.

5. Precipitate with 1% (wt/vol) cetyltrimethylammonium bromide. Store overnight at 4°C, and recover by centrifugation. Dissolve in 10% (wt/vol) NaCl, precipitate with ethanol, dialyze extensively against water, and lyophilize.

6. Obtain pure exopolysaccharide as the void-volume fraction following gel filtration on a column of Sephadex G-50 or as a homogeneous peak emerging in the salt gradient of a DEAE-Sephacel ion exchange column (Amersham Pharmacia Biotech).

7.4.4. Plasma Membranes

The plasma membrane was first isolated to obtain compositional information on its lipid (~25 to 35%) and protein components (see chapter 4, Fig. 9, to see protein particles in plasma membrane) and to study electron transport. Membrane vesicles were later shown to have advantages over bacterial cells for studies of active transport (68). In general, the murein portion of the cell wall is first removed to form protoplasts or spheroplasts (section 7.2.), which are then lysed to provide either the plasma membrane or the plasma plus outer membranes. When this is done osmotically, vesicles are produced with membranes oriented primarily as in the cell. When lysis is done by French pressure cell treatment (section 7.1.1), a high proportion of inside-out vesicles are generated (68, 75).

Orientation can be ascertained by electron microscopic observations of freeze-fractured preparations (chapter 4). DNase and RNase are added to remove DNA and RNA that tend to adhere to the membranes.

Osmotic lysis may not be the method of choice when special precautions are needed to preserve membrane structure and activity. For example, in an extreme halophile, *Halobacterium salinarum*, functional membrane vesicles were isolated following brief sonication of the cells suspended in 4 M NaCl–50 mM Tris-HCl (pH 7) (94). To prevent membrane solubilization in *Thermoplasma acidophilum* that occurs at pH >6, plasma membranes may be prepared by sonicating the cells in 0.05 M acetate buffer of pH 5 (87). As pointed out in section 7.1.1.2, the use of sonication can have serious disadvantages. Repeated freezing and thawing of osmotically sensitive cells can be used as an alternative breakage method for membrane isolation (99).

It may not be convenient, or possible, to remove the cell wall prior to breakage. Membranes may be isolated from French pressure cell lysates by sucrose density gradient centrifugation (section 7.3.6.1), although lower yields and lower purity may result for gram-negative bacteria. An isolation of plasma membranes from the archaeon *Methanospirillum hungatei* is illustrated in Fig. 1.

7.4.5. Membranes of Gram-Negative Bacteria

A total outer and inner membrane fraction can be obtained by lysis of EDTA-lysozyme spheroplasts (102, 109). Subsequent separation by sucrose density gradient centrifugation takes advantage of the fact that the outer membrane has a higher buoyant density than the plasma membrane (Table 2). Good separation of the plasma membrane from the outer membrane requires the absence of Mg^{2+} and maintenance of low ionic strength in all points of the isolation, assisted by the incorporation of 5 mM EDTA into the sucrose gradient (109). Tris or HEPES buffer can be used.

Procedure for Isolation of Membranes of Enteric Bacteria (109)

1. Grow 1 to 2 liters of bacteria to mid-exponential phase and prepare spheroplasts (section 7.2).

2. Lyse spheroplasts by dilution with water or brief sonication. (If the spheroplast preparation contains a significant number of intact rods, lysis by osmotic shock is advantageous since the unlysed cells can be removed by centrifugation at 1,200 × g for 15 min.)

3. Recover envelopes from the supernatant by centrifugation at approximately 100,000 × g for 2 h by using a fixed-angle ultracentrifugation rotor.

4. Resuspend pellets in 3 to 5 ml of 10 mM HEPES buffer (pH 7.4) with the aid of a glass tissue homogenizer. Avoid the presence of Mg^{2+} salts in the buffer, because this prevents subsequent separation of inner and outer membranes by isopycnic density gradient centrifugation.

5. Wash the envelope fraction twice and resuspend it in 0.5 to 2 ml of 10 mM HEPES buffer (pH 7.4).

6. Layer envelopes onto a discontinuous sucrose gradient (section 7.3.4) to separate outer from plasma membranes. For a large-scale preparation (from 1 to 2 liters of culture), use a Beckman SW28 rotor. Layer 6.3 ml each of 50, 45, 40, 35, and 30% (wt/wt) sucrose over a cushion (1.5 ml) of 55% sucrose. Make all sucrose solutions in 10 mM HEPES (pH 7.4) containing 5 mM EDTA. As much as 5 mg of membrane protein (in 2 to 3 ml) can be added to one tube. Centrifuge at 110,000 × g (R_{max}) for 18 to 40 h. Equilibrium is attained at 30 to 36 h, but adequate separation occurs by 18 h.

7. Collect fractions by piercing the bottom of the tube with an 18-gauge syringe needle and collecting counted numbers of drops in a series of test tubes to be read at A_{280}. Pool peak fractions, dilute ~4-fold with HEPES buffer (to a final sucrose concentration of 10% or less), and centrifuge at 100,000 × g for 2 h. Resuspend the washed pellets in a small volume of distilled water, again with the aid of a glass tissue homogenizer.

8. The identification of peaks as outer or plasma membrane is based on well-characterized markers: e.g., buoyant density (Table 2), enzyme activities, outer membrane KdO, and plasma membrane cytochromes.

EDTA is known to solubilize some outer membrane components (especially in *P. aeruginosa* [49]) or to result in reorganization of the outer membrane structure (93), and thus a native outer membrane may not be obtained. The French press method avoids the use of EDTA and is useable with large amounts of cells of a variety of species, for which spheroplasting methods may not be known.

Kaback (68) developed a procedure for preparation of membrane vesicles of *E. coli* for transport studies. In this procedure, lysis of spheroplasts is followed by differential centrifugations, with extensive homogenizations designed to aid in the fractionation of residual cells or partially lysed cells. Note that this procedure was developed for *E. coli* ML, a strain in which the spheroplast outer membrane is particularly sensitive to disruption and removal by EDTA. Thus, the vesicle preparation is essentially plasma membranes. With *E. coli* K-12 or smooth strains of *E. coli* or *Salmonella* spp., a large amount of outer membrane remains attached to the plasma membrane vesicles.

7.4.6. Specialized Membranes

7.4.6.1. Photosynthetic Membranes

The isolation of bacterial photosynthetic membranes has been one of the more demanding exercises in cell fractionation, especially with regard to defining the structural relationship of these intracytoplasmic membranes to the plasma membrane.

The culture conditions under which photosynthetic bacteria are grown have a profound effect on the composition and extent of intracytoplasmic membrane structure. Slight differences in light intensity and variations in temperature can influence both growth rate and pigmentation. For any organism, the conditions of culture must be defined precisely before isolation is attempted.

Many of the procedures described for isolation of photosynthetic membranes have as their goal the preparation of a photochemically active unit containing the bulk of the photopigments. Mechanical procedures for cell disintegration can produce a photochemically active preparation consisting of regular small vesicles. In view of the structural variety of intracytoplasmic membranes found in photosynthetic bacteria, some change in structure must have occurred. A structurally intact, intracytoplasmic membrane array may best be isolated by lysis of osmotically sensitive spheroplasts (section 7.2).

Sykes (140) has compiled an extensive description of procedures for isolation of photosynthetic membranes from the main groups of photosynthetic bacteria. The diversity and complexity of these membranous structures rule out the possibility of providing a single procedure for their preparation.

As an example, the isolation of chlorosomes from *Chlorobium limicola* strain 6230 is described here. The method of Schmidt (123) with slight modifications is used and results in structurally intact chlorosomes that are morphologically distinct from plasma membrane vesicles.

Procedure

1. Grow 2 to 5 liters of bacteria. Harvest cells by centrifugation, wash, and resuspend to 1/10 the original culture volume in 50 mM Tris-HCl buffer (pH 8).

2. Pass this suspension through a precooled French press (section 7.1.1) at 16,000 lb/in^2 in the presence of pancreatic DNase (100 µg/ml). Remove whole cells and large particles by centrifugation at 12,000 × g for 15 min.

3. Treat the supernatant with lysozyme (50 µg/ml) and EDTA (1 mM) for 30 to 45 min at 37°C.

4. Collect the chlorosome and plasma membrane vesicles by centrifugation at 100,000 × g for 2 h. Resuspend the pellet in Tris-HCl buffer to a final protein concentration of 5 to 7 mg/ml.

5. Briefly homogenize the suspension (using a Potter-Elvehjem Teflon-glass tissue grinder).

6. Layer 1 ml of the membrane homogenate onto a discontinuous sucrose gradient (25 to 55%, [wt/wt]) and centrifuge at 100,000 × g (R_{max}) for 18 h (Beckman SW28 rotor).

7. Remove bands from the top of the gradient with a Pasteur pipette. The chlorosomes are concentrated in the 25% sucrose layer. A pure plasma membrane fraction is found in the 35% layer. To obtain these purified fractions, it may be necessary to repeat the density gradient centrifugation two to three times.

7.4.6.2. Purple and Red Archaeal Membranes

Pigmented extreme halophiles (e.g., *Halobacterium salinarum*) have plasma membranes with differentiated areas that can be isolated as purple membrane sheets (85, 138). During growth under O_2-limited conditions in light, this membrane is largely purple as a result of the retinal protein chromophore known as bacteriorhodopsin. Biochemically the isolated purple membrane consists of one protein band with a molecular mass of 20 kDa on SDS-PAGE gels and ether lipids including sulfated lipids (84).

Cells grown aerobically produce mostly a red plasma membrane containing red C_{50} carotenoids called bacterioruberins. On SDS-PAGE gels, the red membrane yields at least six bands with molecular masses ranging from 10 to 60 kDa (the 20-kDa band is absent) and ether lipids which are nonsulfated (84).

Methods to isolate purple and red membranes have been summarized by Kates et al. (73) and are based on separation on sucrose density gradients following removal of stabilizing salts.

Procedure

1. Select growth conditions carefully to enrich for purple or red membranes as desired (73, 84).

2. Harvest cells from six 1.5-liter cell batches (10,000 × g for 20 min), and wash twice in salt solution (250 g of NaCl/liter, 9.8 g of MgSO$_4$/liter, 2 g of KCl/liter [pH 6.5]).

3. Resuspend in 20 ml of salt solution containing 10 mg of DNase, and stir for 20 min at room temperature. Dialyze at 4°C against 6 liters of distilled water for 6 h with two changes of water (one every 2 h).

4. Centrifuge the dialysate at 10,000 × g for 20 min to remove cell debris. Centrifuge the supernatant at 50,000 × g for 1.5 h. Suspend the pellet in 200 ml of distilled water, and centrifuge at 10,000 × g for 20 min. Discard the pellet, and recentrifuge the supernatant at 50,000 × g for 1.5 h.

5. Resuspend the membrane pellet in 18 ml of water, and layer on discontinuous sucrose density gradients consisting of 43 and 38% (wt/wt) steps. Centrifuge in a swinging-bucket rotor at 260,000 × g for 18 h. The red membrane appears at the top of the 38% step, and the purple membrane appears at the top of the 43% step.

6. Remove the membrane bands, and dialyze against water to remove sucrose prior to a second gradient of 43% and 38% steps for purple membrane purification or 38 and 30% steps for red membrane purification.

7.4.7. Periplasm

The locations of the periplasm in gram-positive and gram-negative bacteria are considerably different. In gram positives, the periplasm extends from the plasma membrane into the interstices of the cell wall. Gram-negative bacteria have a more localized periplasm with the consistency of a gel that fits between the outer and plasma membranes in a region called the periplasmic space (chapter 4, Fig. 2, and reference 57). The peptidoglycan (murein) layer resides in this space. Specific proteins in the periplasm account for 10 to 15% of the cell protein (55, 57), and a class of membrane-derived oligosaccharides (127) is also present. Periplasmic contents are released into the medium when the outer membrane of gram-negative bacteria is damaged or removed.

7.4.7.1. Chloroform Release of Periplasm

Chloroform release of periplasmic proteins (3) maintains activity of the periplasmic amino acid-binding proteins and appears to be little influenced by strain variation, thus providing a rapid procedure with which to compare these proteins among various strains. The mechanism of the chloroform effect is unknown, but it seems not to be the expected result of preferential loss of the outer membrane.

7.4.7.2. Osmotic Release of Periplasm

Periplasmic proteins also may be released by an osmotic shock for exponential-phase *E. coli* (106). This method can result in the necessity of centrifuging large volumes during scale up and recovering periplasmic contents from similar volumes of shock fluid. The procedure is as follows.

1. Harvest cells and wash twice with ~40 volumes of cold 10 mM Tris-HCl (pH 7.1) containing 30 mM NaCl. Resuspend 1 g (wet weight) of cells in 40 ml of 33 mM Tris-HCl (pH 7.1) at 24°C.
2. Dilute the cells with rapid stirring with 40 ml of 40% (wt/wt) sucrose in 33 mM Tris-HCl (pH 7.1). Then add sufficient 0.1 M disodium EDTA (pH 7.1) to give 0.1 mM EDTA.
3. Place the suspension on a rotary shaker at 180 rpm for 10 min at 24°C. Centrifuge the cells for 10 min at 13,000 × g in a refrigerated centrifuge.
4. Rapidly disperse the well-drained pellet in 80 ml of ice-cold 0.5 mM MgCl₂ solution (water has also been used) by using a rubber spatula. Gently stir the suspension in an ice bath for 10 min, and centrifuge.
5. Carefully remove the supernatant containing the osmotic shock proteins.

7.4.7.3. Calcium Release of Periplasm

Use of a high concentration of Ca^{2+} at a low temperature (7) is a third method rendering the outer membrane of gram-negative bacteria permeable to periplasmic proteins. Increased permeability results in enhanced DNA uptake in transformation assays and accounts for the ability to reconstitute active transport by adding osmotic shock proteins to a binding protein-deficient mutant of *E. coli* (7). The procedure is as follows.

1. Harvest ~2 × 10⁹ exponential-growth-phase cells. Wash the cells at room temperature with 5-ml portions of 50 mM Tris-HCl (pH 7.2) followed by 1 ml of 100 mM potassium phosphate buffer (pH 7.2).
2. Wash the cells at 0°C with 1 ml of 50 mM Tris-HCl (pH 7.2) containing 300 mM $CaCl_2$.
3. Resuspend the cells in 50 μl of the above-described Tris-Ca^{2+} buffer for reconstitution and transformation assays.

7.4.8. Nucleoid

The DNA-containing region of prokaryotic cells is generally referred to as the nucleoid (meaning nucleus like). Other terms are chromosome, nuclear body, nuclear region, DNA-plasm, and chromatin body. Nucleoids are visible in living cells by phase-contrast microscopy (chapter 1) or by bright-field microscopy after fixation and staining (chapter 2). The nucleoid has a lobular or coralline shape with clefts and is highly dispersed throughout the ribosome-containing cytoplasm (chapter 4, Fig. 2, and reference 18). The structural organization of the in vivo nucleoid is dependent on the constraints that fold and compact the DNA. As the DNA unfolds, the sedimentation and viscometric properties of the nucleoid change. Methods for the isolation of genomic DNA (chapter 23.2) do not retain the tertiary structure of the nucleoid. Investigations into the chemical basis of the structural organization of the nucleoid, as well as membrane attachment sites, require procedures for the isolation of intact nucleoids. Most research on isolated nucleoids has been carried out with *E. coli*.

Procedure for Isolation of Radiolabeled Nucleoids

The following procedure is from reference 139, with modifications (155).

1. Grow *E. coli* into the exponential phase in M9 medium supplemented with 0.2% (wt/vol) Casamino Acids.
2. Label cells (~2 × 10⁸ cells/ml) for 30 min at 37°C with 50 to 100 μCi of [³H]thymidine (10 to 50 Ci/mmol). Chill the culture quickly to 0°C, and harvest by centrifugation.
3. Resuspend cells in 0.2 ml of 4°C 10 mM Tris-HCl (pH 8.1), 10 mM sodium azide, 20% (wt/vol) sucrose, and 0.1 M NaCl.
4. Add 0.05 ml of freshly made 0.12 M Tris (pH 8.1) containing 50 mM EDTA and 4 mg of lysozyme/ml. Mix carefully and quickly.
5. After 30 s on ice, remove and add 0.25 ml of a solution containing 1% (wt/vol) Brij 58 plus 0.4% (wt/vol) deoxycholate, 2 M NaCl, and 10 mM EDTA to complete cell lysis. Slowly rotate the suspension until thoroughly mixed. Incubate at room temperature until the mixture has cleared (10 to 30 min). Brij 58 is a polyoxyethylene ether detergent (Sigma). The Na⁺ is important as a counterion to maintain DNA conformation.
6. Centrifuge the lysate for 5 min at 4,000 × g and 4°C. Some of the cellular DNA will precipitate.
7. Layer the supernatant fraction onto a 5-ml sucrose gradient (10 to 30%, wt/vol) containing 0.01 M Tris-HCl (pH 8.1), 1.0 M NaCl, 1 mM EDTA, and 1 mM mercaptoethanol. Centrifuge at 30,000 × g for 30 min in a Beckman SW50.1 rotor. The nucleoid sediments as a distinct band and, when removed from the gradient, remains compact. It is approximately 80% DNA by weight and includes a small amount of protein and nascent RNA.

The isolated nucleoid can be either relatively free of cell envelopes or associated with them depending on the conditions for cell lysis. Lysis in molar NaCl at 20 to 25°C results in nucleoids with few envelope components. Lysis at 4°C results in envelope-associated nucleoids. Isolation of completely envelope-free nucleoids requires a longer lysozyme treatment and use of an ionic detergent such as Sarkosyl (88, 155). Other macromolecules may bind to the nucleoid during cell lysis and must be recognized as artifacts. The effect of lysis and physiological conditions on the nucleoid is discussed by Korch et al. (77). Recently, a general lysis method (as in the above-described protocol) was compared to a procedure where *E. coli* spheroplasts were osmotically shocked to yield nucleoids free of envelope fragments (24). Similar structures were obtained by both methods, but the osmotic shock nucleoids remained relatively compact without the need for high ionic conditions compared to detergent-salt nucleoids.

7.4.9. Ribosomes

The cytoplasm of bacterial cells as observed by electron microscopy in thin sections is filled with an abundance of ribosomes (chapter 4, Fig. 2, and Color Plate 7). During ribosome isolation, all steps are conducted at 0 to 4°C to retard nuclease action, and it is prudent to sterilize all glassware and tubes. The RNase I-deficient *E. coli* strain D₁₀ is especially suitable for ribosome isolation (136). In some cases, for example, in ribosome isolation from *Bacillus subtilis*, the protease inhibitor phenylmethylsulfonylfluoride may be added as an ethanol solution to give a 1.0 mM final concentration prior to cell harvesting (128). The growth phase may be important as well, since *E. coli* ribosomes prepared for in vitro protein synthesis are most active for tRNA binding and elongation if prepared from cells early in the mid-exponential growth phase (116). Cells are generally broken by freeze-thawing and grinding with alumina (136), although breakage by sonication or French pressure cell treatment may be used also. Crude ribosomes can be purified by the ammonium sulfate procedure of Kurland (83) or on gradients of sucrose (116) or CsCl (45) as described previously. Readers are referred to these publications to choose the best method for their application.

Procedure for Ribosome Isolation from Frozen-Thawed Cells (116, 136)

1. Chill mid-exponential cells rapidly, and harvest them to obtain wet cell pellets. Freeze the wet pellets rapidly, and store them at $-80°C$.

2. Thaw the frozen cells (usually 25 to several hundred grams, wet weight) in a few milliliters of 10 mM Tris-HCl (pH 7.5) containing 6 mM $MgCl_2$, 30 mM NH_4Cl, 4 mM 2-mercaptoethanol, or 0.5 mM dithiothreitol.

3. Break the cells by grinding with a mortar and pestle for about 10 min with 2 to 2.5 g of alumina/g of wet cell paste. While continuing grinding, slowly add the above-described buffer solution (now containing 5 μg of DNase/ml) to a final volume of ~2 ml/g of cell paste.

4. Centrifuge the lysate to remove the alumina (16,000 × g for 10 min) and then to remove cell debris (27,000 × g for 45 min). A wash of the alumina pellet with the buffer solution may increase the ribosome yield by up to 60% (103). Combine these supernatants prior to the second low-speed centrifugation.

5. Harvest the crude ribosomes from the supernatant by centrifuging at 47,000 × g for 18 h. The use of higher centrifugation speeds (e.g., 150,000 × g for 3 h or 200,000 × g for 1 h) is reserved for cases where pressure-induced dissociation of the 70S ribosome is not a concern (116).

6. Rinse the ribosome pellets in buffer solution, and resuspend in the same buffer with gentle stirring over ~2 h. Clarify the suspension by centrifuging at 3,100 × g for 5 min. Yields are measured by monitoring A_{260}, with ~300 A_{260} units expected per gram of wet cells.

7. Gradient centrifugation or $(NH_4)_2SO_4$ fractionation may be used to purify these crude ribosomes. Archaea have 70S ribosomes, which can be isolated from methanogens (35, 98) and *Halobacterium salinarum* (98) by methods similar to those used for bacteria.

7.4.10. Endospores

Procedures for the production of high yields of free dormant spores by spore-forming bacteria generally involve growth in a moderately rich medium until the nutrients are depleted and autolysis occurs. Supplementation of the medium with minerals, especially Ca^{2+} (11), selection of a smooth-colony variant (11), and growth at above-optimal temperatures (10) may be necessary to produce maximally resistant spores. High aeration rates (on an agar surface, in a shaker flask, or in a sparged fermentor) are necessary to cultivate the aerobic spore formers and to prevent the accumulation of contaminating poly-hydroxyalkanoates. The release of endospores by mechanical means and high centrifugation forces should be avoided because spores can be fragile with these procedures. Dormant spores can be stored after lyophilization or by suspension in absolute alcohol or distilled water. Specific conditions are prescribed for producing spores from a variety of sporeformers (10, 11, 42, 51, 151).

Methods for isolating highly purified spores (91, 144) are based on their high density (Table 2). A given spore population may be heterogeneous with respect to density, resulting in more than one density band upon density gradient centrifugation. For example, spores of *Geobacillus stearothermophilus* separate in a gradient of a low medium (such as Metrizamide) into bands of germinated spores, lysozyme-resistant dormant spores, and lysozyme-susceptible dormant spores. The banding patterns are attributed to differences in the levels of penetration of the gradient medium into the spore integuments. In a gradient of Percoll, whose colloidal particles do not penetrate the spore integument, only one density band is obtained. Thus, Percoll provides a suitable medium to determine the wet density of the entire spore, with values comparing favorably with direct mass measurements. Metrizamide is suitable to determine the wet density of the spore protoplast in lysozyme-susceptible spores and to completely separate dormant spores from germinated spores, vegetative cells, parasporal crystals, and debris.

7.4.11. Gas Vesicles

Gas vesicles are found in aquatic prokaryotes ranging from cyanobacteria and purple sulfur bacteria to extremely halophilic archaea (148) and including certain others such as *Methanosarcina barkeri* strain FR1 (4) and *Bacillus megaterium* (90). Gas vesicles are found individually in cells or clustered together to form intracellular particles which appear as phase-light areas under phase-contrast microscopy or as bright refractile areas under bright-field microscopy (chapter 1). It is diagnostic of gas vesicles that they disappear during the application of pressure, generally of ~500 kPa, and are permeable to gases but not to water. The cylindrical structure of gas vesicles varies in dimensions among strains from 30 to 250 nm in width and 50 to 1,000 nm in length. At least eight *gvp* genes are required to form gas vesicles (90, 107), which consist of a main, extremely hydrophobic structural protein, GvpA, arranged in a linear crystalline array of ribs.

Gas vesicles are isolated from osmotic lysates of osmotically sensitive cells by a flotation method (149). A general simplified method follows, taken from details given for *Haloferax volcanii* (108).

1. Prepare spheroplasts or protoplasts for cells that do not lyse readily in hypotonic solutions.

2. Break them gently by osmotic means to prevent collapse of the gas vesicles. Perform lysis in 1 mM $MgSO_4$ containing 10 to 20 μg of DNase/ml to reduce viscosity for the subsequent flotation step. *Haloferax volcanii* is lysed by simply resuspending cells in 10 mM Tris-HCl (pH 7.2) and DNase.

3. Gas vesicles can be collected by flotation by using narrow, 4-mm-diameter tubes in a microcentrifuge operating for 20 min at 60 to 300 × g. Add 1% (vol/vol) Triton X-100 for 30 min at 37°C to aid removal of plasma membrane, and repeat flotation. Other detergents have been used, e.g., 0.1% SDS or 1% Tween 20; however, SDS or Sarkosyl can cause weakening of the vesicle.

4. Wash the surface gas vesicle layer several times with 10 mM Tris-HCl (pH 7.2) by flotation.

7.4.12. Inclusion Bodies

The isolation and analysis of glycogen (chapter 16.3.3.4) and poly-hydroxyalkanoate granules (chapter 16.4.4.10) are described elsewhere. These components can also be stained for light microscopy (chapter 2.3.11.3).

Inorganic polyphosphates (linear polymers of phosphate in anhydrous linkage) accumulate in many prokaryotes including *Methanosarcina* spp. (122) and have been observed upon addition of inorganic phosphate to cells previously phosphate starved (50, 82). The terms metachromatic, volutin, and Babes-Ernst granules refer to deposits of polyphosphates that can be stained for light microscopy (chapter 2.3.11.2).

Procedure for Isolation of Polyphosphate Inclusion Bodies (82)

1. Extract the cells (5 ml, 10^9 cells/ml) twice with 5-ml aliquots of cold 0.5 N perchloric acid for 5 min at 20°C, centrifuging to obtain acid-soluble polyphosphates of 2 to 10 residues of phosphoric acid.

2. Extract the residue with 5 ml of ethanol for 30 min at 20°C, and centrifuge.

3. Add 5 ml of ethanol-ether (3:1, vol/vol) to the cell residue from step 2, and heat in a water bath (fume hood) to approach boiling for 1 min. Centrifuge the cooled mixture to obtain lipid-depleted residue.

4. Extract the residue from step 3 twice with 5-ml aliquots of hot 0.5 N perchloric acid for 15 min at 70°C, and centrifuge to remove particulates. Separate nucleic acids from the longer polyphosphate chains found in the hot perchloric acid extract by adsorption of the nucleic acids onto charcoal (Norit A; BDH Chemicals, Poole, United Kingdom). The polymers may be sized by PAGE (22).

Molecular cloning of foreign proteins can result in the formation of insoluble intracellular aggregates of the expressed protein. These aggregates are also referred to as inclusion bodies that appear as amorphous granules in electron micrographs and as bright refractile particles in cells examined by phase-contrast light microsocpy. Lysis of bacteria containing inclusion bodies can generally be accomplished by the standard methods described in sections 7.1.1 and 7.1.4 (lysozyme). The inclusion bodies are readily isolated by low-speed centrifugation.

7.4.13. Magnetosomes

Magnetotactic bacteria (17) contain intracellular membrane-bound magnetic structures (magnetosomes) consisting of the magnetic iron mineral magnetite (Fe_3O_4) or greigite (Fe_3S_4). A single mineral particle typically 35 to 120 nm in size is deposited within a membrane vesicle located at a specific site in the cell, with multiple magnetosomes often arranged in linear arrays within the cytoplasm (41). Because known magnetotactic bacteria are microaerophilic or anaerobic, it is believed that magnetosomes facilitate migration to a suitable position in the oxygen gradient of aquatic habitats. Most of the limited numbers of magnetotactic bacteria isolated so far are poorly characterized with respect to growth conditions and physiology (126). Below is a procedure used to isolate magnetosomes from *Magnetospirillum magnetotacticum* (41).

Procedure for Isolation of Magnetosomes

1. Grow cells under microaerophilic conditions in liquid, defined medium containing 4 mM $NaNO_3$ and no organic nitrogen. Add 20 μg of iron with 20 μg of quinic acid to autoclaved and cooled medium.

2. Harvest the cells, and wash three times in buffer A (10 mM N-2-hydroxyethylpiperizine-N'-2-ethanesulfonic acid [pH 7.4], plus 10 μg of the protease inhibitor phenylmethylsulfonyl fluoride/ml).

3. Disrupt ~10^{12} cells in 30 ml of buffer A by French pressure cell treatment (approximately three passages at 18,000 lb/in²; 1lb/in² = 6.895 kPa).

4. Add 50 μg of DNase/ml, 100 μg of RNase/ml, and 10 μM $MgCl_2$, and incubate for 60 min at 23°C.

5. Place the disrupted cells in a centrifuge tube in the gap of a 2-kG radar magnet. Allow the black magnetic fraction 10 min to accumulate at the sides of the tube nearest the magnet. Remove the nonmagnetic fraction by aspiration.

6. Resuspend the magnetic fraction in 100-fold its volume of buffer A, and repeat the preceding step at least 10 times.

7. Resuspend the magnetosome fraction in 100-fold its volume of buffer A containing 1 M NaCl to remove electrostatically associated proteins.

8. Wash the purified magnetosomes again as in step 6 at least 10 times with buffer A.

A new type of magnet-sensitive structures (MSS) with noncrystalline organization of an iron-enriched matrix encompassed by an envelope was found in many archaea and bacteria. Both cells containing MSS and isolated MSS particles are attracted to a magnet (145).

Dr. Carl A. Schnaitman is gratefully acknowledged for those conceptual and written contributions that were carried over in this third edition from his first published version of "Cell Fractionation."

7.5. REFERENCES

1. **Altman, E., J.-R. Brisson, and M. B. Perry.** 1986. Structural studies of the capsular polysaccharide from *Haemophilus pleuropneumoniae* serotype 1. *Biochem. Cell Biol.* **64:**707–716.

2. **Altman, E., M. B. Perry, and J.-R. Brisson.** 1989. Structure of the lipopolysaccharide antigenic O-chain produced by *Actinobacillus pleuropneumoniae* serotype 4 (ATCC 33378). *Carbohydr. Res.* **191:**295–303.

3. **Ames, G. F.-L., C. Prody, and S. Kustu.** 1984. Simple, rapid, and quantitative release of periplasmic proteins by chloroform. *J. Bacteriol.* **160:**1181–1183.

4. **Archer, D. B., and N. R. King.** 1984. Isolation of gas vesicles from *Methanosarcina barkeri. J. Gen. Microbiol.* **130:**167–172.

5. **Austin, E. A., J.F. Graves, L. A. Hite, C. T. Parker, and C. A. Schnaitman.** 1990. Genetic analysis of lipopolysaccharide core biosynthesis by *Escherichia coli* K-12. *J. Bacteriol.* **172:**5312–5325.

6. **Austin, J. W., and R. G. E. Murray.** 1987. The perforate component of the regularly-structured (RS) layer of *Lampropedia hyalina. Can. J. Microbiol.* **33:**1039–1045.

7. **Bakau, B., J. M. Brass, and W. Boos.** 1985. Ca^{2+}-induced permeabilization of the *Escherichia coli* outer membrane: comparison of transformation and reconstitution of binding-protein-dependent transport. *J. Bacteriol.* **163:**61–68.

8. **Barber, E. J.** 1976. Viscosity and density tables, p. 415–418. *In* G. D. Fasman (ed.), *Handbook of Biochemistry and Molecular Biology*, 3rd ed., vol. 1. *Physical and Chemical Data.* CRC Press, Inc., Cleveland, OH.

9. **Barrett, A. J.** 1980. Introduction: the classification of proteinases. *Ciba Found. Symp.* **75:**1–13.

10. **Beaman, T. C., and P. Gerhardt.** 1986. Heat resistance of bacterial spores correlated with protoplast dehydration, mineralization, and thermal adaptation. *Appl. Environ. Microbiol.* **52:**1242–1246.

11. **Beaman, T. C., J. T. Greenamyre, T. R. Corner, H. S. Pankratz, and P. Gerhardt.** 1982. Bacterial spore heat resistance correlated with water content, wet density, and

protoplast/sporoplast volume ratio. *J. Bacteriol.* **150:**870–877.

12. **Beveridge, T. J.** 1979. Surface arrays on the wall of *Sporosarcina urea. J. Bacteriol.* **139:**1039–1048.

13. **Beveridge, T. J., G. B. Patel, B. J. Harris, and G. D. Sprott.** 1986. The ultrastructure of *Methanothrix concilii*, a mesophilic aceticlastic methanogen. *Can. J. Microbiol.* **32:**703–710.

14. **Beveridge, T. J., M. Stewart, R. J. Doyle, and G. D. Sprott.** 1985. Unusual stability of the *Methanospirillum hungatei* sheath. *J. Bacteriol.* **162:**728–737.

15. **Beynon, L. M., M. Moreau, J. C. Richards, and M. B. Perry.** 1991. Structure of the O-antigen of *Actinobacillus pleuropneumoniae* serotype 7 lipopolysaccharide. *Carbohydr. Res.* **209:**225–238.

16. **Bingle, W. H., J. L. Doran, and W. J. Page.** 1984. Regular surface layer of *Azotobacter vinelandii. J. Bacteriol.* **159:**251–259.

17. **Blakemore, R. P.** 1982. Magnetotactic bacteria. *Annu. Rev. Microbiol.* **36:**217–238.

18. **Bohrmann, B., W. Villiger, R. Johansen, and E. Kellenberger.** 1991. Coralline shape of the bacterial nucleoid after cryofixation. *J. Bacteriol.* **173:**3149–3158.

19. **Brinton, C. C., A. Buzzell, and M. A. Lauffer.** 1954. Electrophoresis and phage susceptibility on a filament-producing variant of the *E. coli* B bacterium. *Biochim. Biophys. Acta* **15:**533–542.

20. **Bush, J. W.** 1985. Enzymatic lysis of the pseudomurein-containing methanogen *Methanobacterium formicicum. J. Bacteriol.* **163:**27–36.

21. **Chen, W., K. Ohmiya, and S. Shimizu.** 1986. Protoplast formation and regeneration of dehydrodivanillin-degrading strains of *Fusobacterium varium* and *Enterococcus faecium. Appl. Environ. Microbiol.* **52:**612–616.

22. **Clark, J. E., H. Beegen, and H. G. Wood.** 1985. Isolation of polyphosphate from *Propionibacterium shermanii. Fed. Proc.* **44:**1079.

23. **Coulton, J. W., and R. G. E. Murray.** 1977. Membrane-associated components of the bacterial flagellar apparatus. *Biochim. Biophys. Acta* **465:**290–310.

24. **Cunha, S., T. Odijk, E. Suleymanoglu, and C. L. Woldringh.** 2001. Isolation of the *Escherichia coli* nucleoid. *Biochimie* **83:**149–154.

25. **Cunningham, T. M., E. M. Walker, J. N. Miller, and M. A. Lovett.** 1988. Selective release of the *Treponema pallidum* outer membrane and associated polypeptides with Triton X-114. *J. Bacteriol.* **170:**5789–5796.

26. **Darveau, R. P., and R. E. W. Hancock.** 1983. Procedure for isolation of bacterial lipopolysaccharides from both smooth and rough *Pseudomonas aeruginosa* and *Salmonella typhimurium* strains. *J. Bacteriol.* **155:**831–838.

27. **Davis, R. P., and J. E. Harris.** 1985. Spontaneous protoplast formation by *Methanosarcina barkeri. J. Gen. Microbiol.* **131:**1481–1486.

28. **De Pamphilis, M. L., and J. Adler.** 1971. Purification of intact flagella from *Escherichia coli* and *Bacillus subtilis. J. Bacteriol.* **105:**376–383.

29. **Dubreuil, J. D., S. M. Logan, S. Cubbage, D. NiEidhin, W. D. McCubbin, C. M. Kay, T. J. Beveridge, F. G. Ferris, and T. J. Trust.** 1988. Structural and biochemical analysis of a surface array protein of *Campylobacter fetus. J. Bacteriol.* **170:**4165–4173.

30. **Dunn, R. M., M. J. Munster, R. J. Sharp, and B. N. Dancer.** 1987. A novel method for regenerating the protoplasts of thermophilic bacilli. *Arch. Microbiol.* **146:**323–326.

31. **Dupont, C., and A. J. Clarke.** 1991. Dependence of lysozyme-catalysed solubilization of *Proteus mirabilis* peptidoglycan on the extent of O-acetylation. *Eur. J. Biochem.* **195:**763–769.

32. **Easterbrook, K. B.** 1989. Spinate bacteria, p. 1991–1993. *In* J.G. Holt (ed.), *Bergey's Manual of Systematic Bacteriology,* vol. 3. The Williams and Wilkins Co., Baltimore, MD.

33. **Easterbrook, K. B., and R. W. Coombs.** 1976. Spinin: the subunit protein of bacterial spinae. *Can. J. Microbiol.* **22:**438–440.

34. **Easterbrook, K. B., and S. Sperker.** 1982. Physiological controls of bacterial spinae production in complex medium and their value as indicators of spina function. *Can. J. Microbiol.* **28:**130–136.

35. **Elhardt, D., and A. Böck.** 1982. An *in vitro* polypeptide synthesizing system from methanogenic bacteria: sensitivity to antibiotics. *Mol. Gen. Genet.* **188:**128–134.

36. **Filip, C., G. Fletcher, J. L. Wulff, and C. F. Earhart.** 1977. Solubilization of the cytoplasmic membrane of *Escherichia coli* by the ionic detergent sodium lauryl sarcosinate. *J. Bacteriol.* **115:**717–722.

37. **Ford, T. C., and J. M. Graham.** 1991. *An Introduction to Centrifugation.* BIOS Scientific Publishers Ltd., Oxford, United Kingdom.

38. **Galanos, C., and O. Luderitz.** 1975. Electrodialysis of lipopolysaccharides and their conversion to uniform salt form. *Eur. J. Biochem.* **54:**603–610.

39. **Galanos, C., O. Lüderitz, and O. Westphal.** 1969. A new method for the extraction of R lipopolysaccharides. *Eur. J. Biochem.* **9:**245–249.

40. **Gilleland, H. E., Jr., J. D. Stinnett, I. L. Roth, and R. G. Eagon.** 1973. Freeze-etch study of *Pseudomonas aeruginosa*: localization within the cell wall of an ethylenediaminetetraacetate-extractable component. *J. Bacteriol.* **113:**417–432.

41. **Gorby, Y. A., T. J. Beveridge, and R. P. Blakemore.** 1988. Characterization of the bacterial magnetosome membrane. *J. Bacteriol.* **170:**834–841.

42. **Gould, G. W.** 1971. Methods for studying bacterial spores. *Methods Microbiol.* **6A:**361–381.

43. **Gray, G. W., and S. G. Wilkinson.** 1965. The effect of ethylenediaminetetraacetic acid on the cell walls of some gram-negative bacteria. *J. Gen. Microbiol.* **39:**385–399.

44. **Gruter, L., and R. Laufs.** 1991. Protoplast transformation of *Staphylococcus epidermidis. J. Microbiol. Methods* **13:**299–304.

45. **Hamilton, M. G.** 1971. Isodensity equilibrium centrifugation of ribosomal particles; the calculation of the protein content of ribosomes and other ribonucleoproteins from buoyant density measurements. *Methods Enzymol.* **20:**512–521.

46. **Hancock, I. C., and I. R. Poxton.** 1988. Isolation and purification of cell walls, p. 55–65. *In* I. C. Hancock and I. R. Poxton (ed.), *Bacterial Cell Surface Techniques.* John Wiley & Sons Ltd., Toronto, Canada.

47. **Hancock, I. C., and I. R. Poxton.** 1988. Teichoic acids and other accessory carbohydrates from Gram-positive bacteria, p. 79–88. *In* I. C. Hancock and I. R. Poxton (ed.), *Bacterial Cell Surface Techniques.* John Wiley & Sons Ltd., Toronto, Canada.

48. **Hancock, I. C., and I. R. Poxton.** 1988. Isolation of exopolysaccharides, p. 121–125. *In* I. C. Hancock and I. R. Poxton (ed.), *Bacterial Cell Surface Techniques.* John Wiley & Sons Ltd., Toronto, Canada.

49. **Hancock, R. E. W., and H. Nikaido.** 1978. Outer membranes of Gram-negative bacteria. XIX. Isolation from *Pseudomonas aeruginosa* PAO1 and use in reconstitution and definition of the permeability barrier. *J. Bacteriol.* **136:**381–390.

50. **Harold, F. M.** 1963. Accumulation of inorganic polyphosphate in *Aerobacter aerogenes.* I. Relationship to growth and nucleic acid synthesis. *J. Bacteriol.* **86:**216–221.

51. **Harwood, C. R., and S. M. Cutting.** 1990. *Molecular Biological Methods for Bacillus.* John Wiley & Sons, Inc., New York, NY.

52. **Hash, J. H., and M. V. Rothlauf.** 1967. The N,O-diacetylmuramidase of *Chalaropsis* species. *J. Biol. Chem.* **242:**5586–5590.

53. **Heckels, J. E., and M. Virji.** 1988. Detection and preparation of surface appendages, p. 67–72. *In* I. C. Hancock and I. R. Poxton (ed.), *Bacterial Cell Surface Techniques.* John Wiley & Sons Ltd., Toronto, Canada.

54. **Helenius, A., D. R. McCaslin, E. Fries, and C. Tanford.** 1979. Properties of detergents. *Methods Enzymol.* **56:**734–749.

55. **Heppel, L. A.** 1971. The concept of periplasmic enzymes, p. 223–247. *In* L. I. Rothfield (ed.), *Structure and Function of Biological Membranes.* Academic Press, Inc., New York, NY.

56. **Hitchcock, P. J., and T. M. Brown.** 1983. Morphological heterogeneity among *Salmonella* lipopolysaccharide chemotypes in silver-stained polyacrylamide gels. *J. Bacteriol.* **154:**269–277.

57. **Hobot, J. A., E. Carlemalm, W. Villiger, and E. Kellenberger.** 1984. Periplasmic gel: new concept resulting from the reinvestigation of bacterial cell envelope ultrastructure by new methods. *J. Bacteriol.* **160:**143–152.

58. **Holloway, P. W.** 1973. A simple procedure for the removal of Triton X-100 from protein samples. *Anal. Biochem.* **53:**304–308.

59. **Hughes, D. E., J. W. T. Wimpenny, and D. Lloyd.** 1971. The disintegration of microorganisms, p. 1–54. *In* J. R. Norris and D. W. Ribbons (ed.), *Methods in Microbiology,* vol. 5B. Academic Press, Inc., New York, NY.

60. **Jann, B., and K. Jann.** 1990. Structure and biosynthesis of the capsular antigens of *Escherichia coli,* p. 19–42. *Curr. Top. Microbiol. Immunol.* **150:**19–42.

61. **Jarrell, K. F., J. R. Colvin, and G. D. Sprott.** 1982. Spontaneous protoplast formation in *Methanobacterium bryantii. J. Bacteriol.* **149:**346–353.

62. **Jarrell, K. F., D. Faguy, A. M. Hebert, and M. L. Kalmokoff.** 1992. A general method of isolating high molecular weight DNA from methanogenic archaea (archaebacteria). *Can. J. Microbiol.* **38:**65–68.

63. **Jarrell, K. F., and G. D. Sprott.** 1984. Formation and regeneration of *Halobacterium* protoplasts. *Curr. Microbiol.* **10:**147–152.

64. **Jennings, H. J.** 1990. Capsular polysaccharides as vaccine candidates. *Curr. Top. Microbiol. Immunol.* **150:**97–127.

65. **Johannsen, L., H. Labischinski, B. Reinicke, and P. Giesbrecht.** 1983. Changes in the chemical structure of walls of *Staphylococcus aureus* grown in the presence of chloramphenicol. *FEMS Microbiol. Lett.* **16:**313–316.

66. **Jollès, P.** 1969. Lysozymes: a chapter of molecular biology. *Angew. Chem. Int. Ed.* **8:**227–294.

67. **Jussofie, A., F. Mayer, and G. Gottschalk.** 1986. Methane formation from methanol and molecular hydrogen by protoplasts of new methanogen isolates and inhibition by dicyclohexylcarbodiimide. *Arch. Microbiol.* **146:**245–249.

68. **Kaback, H. R.** 1971. Bacterial membranes. *Methods Enzymol.* **22:**99–120.

69. **Kalmokoff, M. L., K. F. Jarrell, and S. F. Koval.** 1988. Isolation of flagella from the archaebacterium *Methanococcus voltae* by phase separation with Triton X-114. *J. Bacteriol.* **170:**1752–1758.

70. **Kamekura, M., D. Oesterhelt, R. Wallace, P. Anderson, and D. J. Kushner.** 1988. Lysis of halobacteria in Bacto-Peptone by bile acids. *Appl. Environ. Microbiol.* **54:**990–995.

71. **Kandler, O., and H. König.** 1985. Cell envelopes of archaebacteria. *Bacteria* **8:**413–457.

72. **Karch, H., H. Leying, K.-H. Buscher, H.-P. Kroll, and W. Opferkuch.** 1985. Isolation and separation of physiochemically distinct fimbrial types expressed on a single culture of *Escherichia coli* O7:K1:H6. *Infect. Immun.* **47:**549–554.

73. **Kates, M., S. C. Kushwaha, and G. D. Sprott.** 1982. Lipids of purple membrane from extreme halophiles and of methanogenic bacteria. *Methods Enzymol.* **88:**98–111.

74. **Kiener, A., H. König, J. Winter, and T. Leisinger.** 1987. Purification and use of *Methanobacterium wolfei* pseudomurein endopeptidase for lysis of *Methanobacterium thermoautotrophicum. J. Bacteriol.* **169:**1010–1016.

75. **Kobayashi, H., J. van Brunt, and F. M. Harold.** 1978. ATP-linked calcium transport in cells and membrane vesicles of *Streptococcus faecalis. J. Biol. Chem.* **253:**2085–2092.

76. **König, H., and K. O. Stetter.** 1986. Studies on archaebacterial S-layers. *Syst. Appl. Microbiol.* **7:**300–309.

77. **Korch, C., S. Overb, and K. Kleppe.** 1976. Envelope-associated folded chromosomes from *Escherichia coli:* variations under different physiological conditions. *J. Bacteriol.* **127:**904–916.

78. **Kotzian, S., V. Kreis-Kleinschmidt, T. Krafft, O. Klimmek, J. M. Macy, and A. Kröger.** 1996. Properties of a *Wolinella succinogenes* mutant lacking periplasmic sulfide dehydrogenase (Sud). *Arch. Microbiol.* **165:**65–68.

79. **Koval, S. F.** 1988. Paracrystalline protein surface arrays on bacteria. *Can. J. Microbiol.* **34:**407–414.

80. **Koval, S. F., and K. F. Jarrell.** 1987. Ultrastructure and biochemistry of the cell wall of *Methanococcus voltae. J. Bacteriol.* **169:**1298–1306.

81. **Koval, S. F., and R. G. E. Murray.** 1984. The isolation of surface array proteins from bacteria. *Can. J. Biochem. Cell Biol.* **62:**1181–1189.

82. **Kulaev, I. S.** 1975. Biochemistry of inorganic polyphosphates. *Rev. Physiol. Biochem. Pharmacol.* **73:**131–157.

83. **Kurland, C. G.** 1971. Purification of ribosomes from *Escherichia coli. Methods Enzymol.* **20:**379–381.

84. **Kushwaha, S. C., M. Kates, and W. G. Martin.** 1975. Characterization and composition of the purple and red membrane from *Halobacterium cutirubrum. Can. J. Microbiol.* **53:**284–292.

85. **Kushwaha, S. C., M. Kates, and W. Stoeckenius.** 1976. Comparison of the purple membrane from *Halobacterium cutirubrum* and *Halobacterium halobium. Biochim. Biophys. Acta* **426:**703–710.

86. **Laddaga, R. A., and R. A. MacLeod.** 1982. Effects of wash treatments on the ultrastructure and lysozyme penetrability of the outer membrane of various marine and two terrestrial gram-negative bacteria. *Can. J. Microbiol.* **28:**318–324.

87. **Langworthy, T. A.** 1978. Membranes and lipids of extremely thermoacidophilic microorganisms, p. 11–30. *In* S. M. Friedman (ed.), *Biochemistry of Thermophily.* Academic Press, Inc., New York, NY.

88. **Leive, L.** 1965. Release of lipopolysaccharide by EDTA treatment of *Escherichia coli. Biochem. Biophys. Res. Commun.* **21:**290–296.

89. **Lesse, A. J., A. A. Campagnari, W. E. Bittner, and M. A. Apicella.** 1990. Increased resolution of lipopolysaccharides and lipooligosaccharides utilizing tricine-sodium dodecyl sulfate-polyacrylamide gel electrophoresis. *J. Immunol. Methods* **126:**109–117.

90. **Li, N., and M. C. Cannon.** 1998. Gas vesicle genes identified in *Bacillus megaterium* and functional expression in *Escherichia coli. J. Bacteriol.* **180:**2450–2458.

91. **Lindsay, J. A., T. C. Beaman, and P. Gerhardt.** 1985. Protoplast water content of bacterial spores determined by buoyant density sedimentation. *J. Bacteriol.* **163:**735–737.

92. **Lu, W., A. P. Wood, and D. P. Kelly.** 1983. An enzymatic lysis procedure for the assay of enzymes in *Thiobacillus* A2. *Microbios* **38**:171–176.

93. **Lugtenberg, B., and L. van Alphen.** 1983. Molecular architecture and functioning of the outer membrane of *Escherichia coli* and other Gram-negative bacteria. *Biochim. Biophys. Acta* **737**:51–115.

94. **MacDonald, R. E., and J. K. Lanyi.** 1975. Light-induced leucine transport in *Halobacterium halobium* envelope vesicles: a chemiosmotic study. *Biochemistry* **14**:2882–2889.

95. **MacLeod, R. A.** 1985. Marine microbiology far from the sea. *Annu. Rev. Microbiol.* **39**:1–20.

96. **Manoharan, R., E. Ghiamati, R. A. Dalterio, K. A. Britton, W. H. Nelson, and J. F. Sperry.** 1990. UV resonance Raman spectra of bacteria, bacterial spores, protoplasts and calcium dipicolinate. *J. Microbiol. Methods* **11**:1–15.

97. **Marvin, H. J. P., and B. Witholt.** 1987. A highly efficient procedure for the quantitative formation of intact and viable-lysozyme spheroplasts from *Escherichia coli*. *Anal. Biochem.* **164**:320–330.

98. **Matheson, A. T., M. Yaguchi, W. E. Balch, and R. S. Wolfe.** 1980. Sequence homologies in the N-terminal region of the ribosomal "A" proteins from *Methanobacterium thermoautotrophicum* and *Halobacterium cutirubrum*. *Biochim. Biophys. Acta* **626**:162–169.

99. **Mayer, F., A. Jussofie, M. Salzmann, M. Lübben, M. Rohde, and G. Gottschalk.** 1987. Immunoelectron microscopic demonstration of ATPase on the cytoplasmic membrane of the methanogenic bacterium strain GÖ1. *J. Bacteriol.* **169**:2307–2309.

100. **Meakin, S. A., J. H. E. Nash, W. D. Murray, K. J. Kennedy, and G. D. Sprott.** 1991. A generally applicable technique for the extraction of restrictable DNA from methanogenic bacteria. *J. Microbiol. Methods* **14**:119–126.

101. **Messner, P., and U. B. Sleytr.** 1988. Separation and purification of S-layers from Gram-positive and Gram-negative bacteria, p. 97–104. *In* I. Hancock and I. Poxton (ed.), *Bacterial Cell Surface Techniques*. John Wiley & Sons Ltd., Toronto, Canada.

102. **Miura, T., and S. Mizushima.** 1968. Separation by density gradient centrifugation of two types of membranes from spheroplast membranes of *Escherichia coli* K12. *Biochim. Biophys. Acta* **150**:159–161.

103. **Moore, P. B.** 1979. The preparation of deuterated ribosomal materials for neutron scattering. *Methods Enzymol.* **59**:639–655.

104. **Morii, H., and Y. Koga.** 1992. An improved assay for a pseudomurein-degrading enzyme from *Methanobacterium wolfei* and the protoplast formation of *Methanobacterium thermoautotrophicum* by the enzyme. *J. Ferment. Bioeng.* **73**:6–10.

105. **Nilles, M. L., A. W. Williams, E. Skrzypek, and S. C. Straley.** 1997. *Yersinia pestis* LcrV forms a stable complex with LcrG and may have a secretion-related regulatory role in the low-Ca^{2+} response. *J. Bacteriol.* **179**:1307–1316.

106. **Nossal, N. G., and L. A. Heppel.** 1966. The release of enzymes by osmotic shock from *Escherichia coli* in exponential phase. *J. Biol. Chem.* **241**:3055–3062.

107. **Offner, S., A. Hofacker, G. Wanner, and F. Pfeifer.** 2000. Eight of fourteen *gvp* genes are sufficient for formation of gas vesicles in halophilic archaea. *J. Bacteriol.* **182**:4328–4336.

108. **Offner, S., G. Wanner, and F. Pfeifer.** 1996. Functional studies of the *gvpACNO* operon of *Halobacterium salinarium* reveal that the GvpC protein shapes gas vesicles. *J. Bacteriol.* **178**:2071–2078.

109. **Osborn, M. J., and R. Munson.** 1974. Separation of the inner (cytoplasmic) and outer membranes of gram negative bacteria. *Methods Enzymol.* **31A**:642–653.

110. **Paranchych, W., P. A. Sastry, L. S. Frost, M. Carpenter, G. D. Armstrong, and T. H. Watts.** 1979. Biochemical studies on pili isolated from *Pseudomonas aeruginosa* strain PAO. *Can. J. Microbiol.* **25**:1175–1181.

111. **Penn, C. W., A. Cockayne, and M. J. Bailey.** 1985. The outer membrane of *Treponema pallidum*: biological significance and biochemical properties. *J. Gen. Microbiol.* **131**:2349–2357.

112. **Pertoft, H., T. C. Laurent, R. Seljelid, G. Akerstrom, L. Kagedal, and M. Hirtenstein.** 1979. The use of density gradients of Percoll^R for the separation of biological particles, p. 67–72. *In* H. Peeters (ed.), *Separation of Cells and Subcellular Elements*. Pergamon Press, Toronto, Canada.

113. **Peterson, G. L.** 1977. A simplification of the protein assay method of Lowry *et al.* which is more generally applicable. *Anal. Biochem.* **83**:346–356.

114. **Petrov, V. V., V. Y. Artzatbanov, E. N. Ratner, A. I. Severin, and I. S. Kulaev.** 1991. Isolation, structural and functional characterization of *Staphylococcus aureus* protoplasts obtained using lysoamidase. *Arch. Microbiol.* **155**:549–553.

115. **Price, C.A.** 1982. *Centrifugation in Density Gradients.* Academic Press, New York, NY.

116. **Rheinberger, H.-J., U. Geigenmüller, M. Wedde, and K. H. Nierhaus.** 1988. Parameters for the preparation of *Escherichia coli* ribosomes and ribosomal subunits active in tRNA binding. *Methods Enzymol.* **164**:658–670.

117. **Rickwood, D. (ed.).** 1984. *Centrifugation: a Practical Approach*, 2nd ed. IRL Press, Oxford, United Kingdom.

118. **Ross, J. W.** 1963. Continuous-flow mechanical cell disintegrator. *Appl. Microbiol.* **11**:33–35.

119. **Salton, M. R. J.** 1957. The properties of lysozyme and its action on microorganisms. *Bacteriol. Rev.* **21**:82–99.

120. **Salton, M. R. J.** 1974. Isolation of cell walls from Gram-positive bacteria. *Methods Enzymol.* **31**:653–667.

121. **Sambrook, J., E. F. Fritsch, and T. Maniatis.** 1989. *Molecular Cloning: a Laboratory Manual*, 2nd ed. Cold Spring Harbor Laboratory, Cold Spring Harbor, NY.

122. **Scherer, P. A., and H.-P. Bochem.** 1983. Ultrastructural investigation of 12 *Methanosarcinae* and related species grown on methanol for occurrence of polyphosphate-like inclusion. *Can. J. Microbiol.* **29**:1190–1199.

123. **Schmidt, K.** 1980. A comparative study on the composition of chlorosomes (chlorobium vesicles) and cytoplasmic membranes from *Chloroflex aurantiacus* strain OK-70-fl and *Chlorobium thiosulfatophilum* strain 6230. *Arch. Microbiol.* **124**:21–31.

124. **Schnaitman, C. A.** 1970. Protein composition of the cell wall and cytoplasmic membrane of *Escherichia coli*. *J. Bacteriol.* **104**:890–901.

125. **Schnaitman, C. A.** 1971. Solubilization of the cytoplasmic membrane of *Escherichia coli* by Triton X-100. *J. Bacteriol.* **108**:545–552.

126. **Schüler, D., and R. B. Frankel.** 1999. Bacterial magnetosomes: microbiology, biomineralization and biotechnological applications. *Appl. Microbiol. Biotechnol.* **52**:464–473.

127. **Schulman, H., and E. P. Kennedy.** 1979. Localization of membrane-derived oligosaccharides in the outer envelope of *Escherichia coli* and their occurrence in other gram-negative bacteria. *J. Bacteriol.* **137**:686–688.

128. **Sharrock, W. J., and J. C. Rabinowitz.** 1979. Fractionation of ribosomal particles from *Bacillus subtilis*. *Methods Enzymol.* **59**:371–382.

129. **Siegel, J. L., S. F. Hurst, E. S. Liberman, S. E. Coleman, and A. S. Bleiweis.** 1981. Mutanolysin-induced

spheroplasts of *Streptococcus mutans* are true protoplasts. *Infect. Immun.* **31**:808–815.

130. **Simon, R. D.** 1981. Morphology and protein composition of gas vesicles from wild type and gas vacuole defective strains of *Halobacterium salinarium* strain 5. *J. Gen. Microbiol.* **125**:103–111.

131. **Sleytr, W. B., and P. Messner.** 1988. Crystalline surface layers in procaryotes. *J. Bacteriol.* **170**:2891–2897.

132. **Spizizen, J.** 1962. Preparation and use of protoplasts. *Methods Enzymol.* **5**:122–134.

133. **Sprott, G. D., J. R. Colvin, and R. C. McKellar.** 1979. Spheroplasts of *Methanospirillum hungatei* formed upon treatment with dithiothreitol. *Can. J. Microbiol.* **25**:730–738.

134. **Sprott, G. D., and K. F. Jarrell.** 1981. K$^+$, Na$^+$, and Mg$^+$ content and permeability of *Methanospirillum hungatei* and *Methanobacterium thermoautotrophicum*. *Can. J. Microbiol.* **27**:444–451.

135. **Sprott, G. D., K. M. Shaw, and K. F. Jarrell.** 1983. Isolation and chemical composition of the cytoplasmic membrane of the archaebacterium *Methanospirillum hungatei. J. Biol. Chem.* **258**:4026–4031.

136. **Staechelin, T., and D. R. Maglott.** 1971. Preparation of *Escherichia coli* ribosomal subunits active in polypeptide synthesis. *Methods Enzymol.* **20**:449–456.

137. **Steinberg, W.** 1974. Properties and developmental roles of the lysyl- and tryptophanyl-transfer ribonucleic acid synthetase of *Bacillus subtilis*: common genetic origin of the corresponding spore and vegetative enzymes. *J. Bacteriol.* **118**:70–82.

138. **Stoeckenius, W., and R. Rowen.** 1967. A morphological study of *Halobacterium halobium* and its lysis in media of low salt concentration. *J. Cell Biol.* **34**:365–393.

139. **Stonington, O. G., and D. E. Pettijohn.** 1971. The folded genome of *Escherichia coli* isolated in a protein-DNA-RNA complex. *Proc. Natl. Acad. Sci. USA* **68**:6–9.

140. **Sykes, J.** 1971. Centrifugation techniques for the isolation and characterization of sub-cellular components from bacteria, p. 55–207. *In* J. R. Norris and D. W. Ribbons (ed.), *Methods in Microbiology*, vol. 5B. Academic Press, Inc., New York, NY.

141. **Thom, J. R., and L. L. Randall.** 1988. Role of the leader peptide of maltose-binding protein in two steps of the export process. *J. Bacteriol.* **170**:5654–5661.

142. **Thomas, N. A., S. L. Bardy, and K. F. Jarrell.** 2001. The archaeal flagellum: a different kind of prokaryotic motility structure. *FEMS Microbiol. Lett.* **25**:147–174.

143. **Thompson, B. G., R. G. E. Murray, and J. F. Boyce.** 1982. The association of the surface array and the outer membrane of *Deinococcus radiodurans. Can. J. Microbiol.* **28**:1081–1088.

144. **Tisa, L. S., T. Koshikawa, and P. Gerhardt.** 1982. Wet and dry bacterial spore densities determined by buoyant sedimentation. *Appl. Environ. Microbiol.* **43**:1307–1310.

145. **Vainshtein, M. B., N. E. Suzina, E. B. Kudryashova, E. V. Ariskina, and V. V. Sorokin.** 1998. On the diversity of magnetotactic bacteria. *Microbiology* **67**:670–676.

146. **Vandenbergh, P. A., R. E. Bawdon, and R. S. Berk.** 1979. Rapid test for determining the intracellular rhodanese activity of various bacteria. *Int. J. Syst. Bacteriol.* **29**:339–344.

147. **Virgilio, R., C. Gonzalez, N. Muñoz, T. Cabezon, and S. Mendoza.** 1970. *Staphylococcus aureus* protoplasting induced by D-cycloserine. *J. Bacteriol.* **104**:1386–1387.

148. **Walker, J. E., P. K. Hayes, and A. E. Walsby.** 1984. Homology of gas vesicle proteins in *Cyanobacteria* and *Halobacteria. J. Gen. Microbiol.* **130**:2709–2715.

149. **Walsby, A. E.** 1974. The isolation of gas vesicles from blue-green algae. *Methods Enzymol.* **31**:678–686.

150. **Ward, J. B.** 1981. Teichoic and teichuronic acids: biosynthesis, assembly, and location. *Microbiol. Rev.* **45**:211–243.

151. **Warth, A. D.** 1978. Relationship between the heat resistance of spores and the optimum and maximum growth temperature of *Bacillus* species. *J. Bacteriol.* **134**:699–705.

152. **Weil, C. F., D. S. Cram, B. A. Sherf, and J. N. Reeve.** 1988. Structure and comparative analysis of the genes encoding component C of methyl coenzyme M reductase in the extremely thermophilic archaebacterium *Methanothermus fervidus. J. Bacteriol.* **170**:4718–4726.

153. **Westphal, O., and K. Jann.** 1965. Bacterial lipopolysaccharides. *Methods Carbohydr. Chem.* **5**:83–91.

154. **Wilkinson, S. G., L. Galbraith, and G. A. Lightfoot.** 1973. Cell walls, lipids, and lipopolysaccharides of *Pseudomonas* species. *Eur. J. Biochem.* **33**:158–174.

155. **Worcel, A., and E. Burgi.** 1972. On the structure of the folded chromosome of *Escherichia coli. J. Mol. Biol.* **71**:127–147.

156. **Work, E.** 1971. Cell walls, p. 361–418. *In* J. R. Norris and D. W. Ribbons (ed.), *Methods in Microbiology*, vol. 5A. Academic Press, Inc., New York, NY.

157. **Yabu, K., and S. Takahashi.** 1977. Protoplast formation of selected *Mycobacterium smegmatis* mutants by lysozyme in combination with methionine. *J. Bacteriol.* **129**:1628–1631.

158. **Yamada, H., N. Tsukagoshi, and S. Udaka.** 1981. Morphological alterations of cell wall concomitant with protein release in a protein-producing bacterium, *Bacillus brevis* 47. *J. Bacteriol.* **148**:322–332.

159. **Yokomaku, D., N. Yamaguchi, and M. Nasu.** 2000. Improved direct viable count procedure for quantitative estimation of bacterial viability in freshwater environments. *Appl. Environ. Microbiol.* **66**:5544–5548.

160. **Zulauf, M., U. Fürstenberger, M. Grabo, P. Jäggi, M. Regenass, and J. P. Rosenbusch.** 1989. Critical micellar concentrations of detergents. *Methods Enzymol.* **172**:528–538.

8

Antigen-Antibody Reactions

LUCY M. MUTHARIA AND JOSEPH S. LAM

Reactions between antigens (Ags) and antibodies (Abs) are usefully exploited in many areas of life science research (1–4). Antibodies have many virtues as biological reagents, including specificity for an Ag, availability, and the usually visible secondary reactions due to the divalent nature of the Ab. The monoclonal Ab (MAb) technology developed by Köhler and Milstein (66) allows for the production of unlimited quantities of Abs against virtually any molecule. A MAb is known for its unique epitope specificity and reproducibility in Ag-Ab reactions. Although MAbs have obvious advantages over conventional polyclonal Abs, one should also consider the disadvantages of MAbs, i.e., high production costs and the requirement for highly trained personnel to produce and maintain the MAbs. In most cases, the generation of polyclonal Abs requires nothing more than an immunogen, a rabbit, and a syringe. The resulting polyclonal antisera will be adequate for most needs in microbiology and at a fraction of the cost of MAb production. As there is an immense volume of information concerning all aspects of Ag-Ab reactions in the literature, the objective of this chapter is to provide readers with simple and useful protocols and an introduction to some of the more novel techniques.

8.1. HANDLING ANIMALS

The success of Ab production depends on many factors, including Ag quality, dose, and route of immunization and animal host and health. With few exceptions, animals are used to generate Abs; therefore, the knowledge of animal handling, use, and welfare is of primary concern for anyone who needs to produce Abs.

8.1.1. Animal Care Guidelines

Institutional animal facilities and programs should be operated in accordance with the governmental requirements and recommendations as outlined in *Guide for the Care and Use of Laboratory Animals*, the Animal Welfare Acts, and other applicable federal regulations and policies. In most cases, the use of animals for immunization and the production of Abs can be classified in the acute-care and low-pain categories. However, it is necessary for the researchers to be aware of the established guidelines. In brief, all animals should be cared for, transported, and handled in accordance with the Animal Welfare Act or its equivalent. The procedures used should be designed and performed with due consideration of their relevance to human and animal health; the humane use of animals is imperative, and where necessary the minimum number of animals should be considered. Anesthesia is to be used to minimize distress and pain, and animals that are suffering chronic pain or distress that cannot be relieved should be painlessly euthanized. Appropriate living conditions should be provided, and investigators or personnel handling the animals must be trained and qualified. A copy of the *Guide for the Care and Use of Laboratory Animals* can be obtained from the Animal Care Committee of your institution or directly from the Animal Welfare Division of the National Institutes of Health in the United States.

8.1.2. Animals

Rabbits, mice, and rats are the laboratory animals most commonly used for Ab production. The preference for any particular species depends on the volume of serum required and can be influenced by the specificity of the generated serum. In general, rabbits are the animals of choice for polyclonal Ab production due to their size and, more importantly, their gentle natures and the ease of handling. Also, anti-rabbit immunoglobulin Abs (secondary Abs) conjugated to either enzymes, biotin, fluorescence reagents (chapters 2 and 3), or electron-dense colloidal gold particles (chapter 4) are commercially available. For MAb production, mice or rats are usually chosen because most of the myeloma cell lines available are of murine origin. Taking into consideration the points discussed in section 8.1.1, no individual should handle animals without some form of training. The details on how to handle each type of animal should be learned from experienced and qualified personnel. For detailed descriptions of the proper handling of animals, one can consult several review chapters on the subject (6, 8–10).

8.1.3. Age and Size

8.1.3.1. Rabbits

Many breeds of rabbits, weighing from 2 to 7 kg, are readily available from approved commercial sources. When large Ab quantities are required, larger animals should be considered. Lop-eared rabbits, which have a body weight of up to 7 kg and large ears and veins, allow for easy bleeding. However, the most frequently used type of rabbit is the New Zealand White, adults of which weigh 5 to 6 kg and have the advantage of white skin and easily visible veins. Generally, 8- to 10-week-old animals, ca. 2 kg, are purchased, and by the time they have been immunized and are ready to be bled, they have been kept for another 4 to 6 weeks. It is worth noting that female rabbits are usually more tame and easier to handle than male animals. It is not uncommon for rabbits to be kept for more than 6 months for continuous immunization and bleeding. However, unless rabbits are housed in open pens, after 6 to 8 months they may have outgrown the standard-sized cages used by most institutions. Long-term care can be expensive. As an alternative, several smaller breeds of rabbits such as the Dutch Belted or the San Juan strains can be used and sera can be pooled to obtain a larger volume. The latter choice will guarantee that the standard-sized cages will be adequate to meet long-term animal-holding requirements.

8.1.3.2. Mice

Many laboratory strains of mice are available from commercial suppliers, and some institutions may even breed the more common strains such as the Swiss White, which is often used for microbiological work. Because of their small size, one cannot expect to obtain more than 0.2 to 0.3 ml of blood unless by cardiac puncture. BALB/c mice are the source of myeloma cell lines, and for obvious reasons, these mice are used for immunization to obtain primed splenocytes for plasmacytoma fusion experiments for MAb production. Animals that are 5 to 8 weeks postweaning, weighing ~20 g, are usually purchased. After being housed for the duration of the immunization procedures, they are large enough to yield sufficient amounts of splenocytes.

8.1.3.3. Rats

Commonly available strains of rats are the Hooded Lister and the two albinos, Wistar and the Sprague-Dawley. Rats are stronger and more agile than mice; they need to be handled with caution. *Do not handle them without proper training.* Animal size depends on how long one is intending to keep them. Animals that are 5 to 8 weeks postweaning can weigh between 75 and 150 g. They are slightly small, but after they have been kept for the duration of the immunization procedures and are ready to be bled, they should have grown to sufficient size to give a good yield of blood.

A frequently asked question is how many animals are needed for the generation of sufficient Abs. If the amount of Ag is not limiting, then more than one animal should be used because even genetically identical animals will respond to the same Ag differently (3). Therefore, it is advis-

able to immunize more than one animal with the same Ag, and if they provide adequate titers of Ab you can pool the sera collected. For rabbits, use a minimum of two animals; for mice, use at least five; and for rats, use a minimum of three. Note that antisera should not be pooled when the objective of the study is to evaluate the responses of different animals to an Ag or to monitor the Ab response during the course of immunization.

8.1.3.4. Chickens

Hen egg yolk has been successfully used to generate Abs (5). Each egg may contain in excess of 100 mg of egg yolk immunoglobulin Y (IgY), and the Ab is easily purified by precipitation from the water-soluble fraction of the egg yolk. Because such large amounts of Abs can be obtained, egg yolk Abs have been used for passive immunization studies (5, 7).

8.1.4. Injection Route and Dosage

Many factors determine the immunogenicity of an Ag. The dose and injection route significantly influence the outcome. The routes of injection commonly used on the three types of animals described above include intradermal (i.d.), subcutaneous (s.c.), intramuscular (i.m.), intraperitoneal (i.p.), and intravenous (i.v.). Other routes of immunization such as ingestion, inhalation, and skin application are also possible but are not as common. The i.d., s.c., and i.m. routes are normally chosen due to slow Ag release, which can enhance a humoral response. i.p. injection allows an Ag to come into contact with the lymphatics sooner than the previously described methods and is easy in rodents—mice are often immunized by this method to prime their spleen cells for MAb production. However, this route is not recommended for rabbits due to their large size. i.v. injection is often used for administering a booster dose for rapid response to an Ag, for instance, when wishing to produce agglutinating Abs to microbial surface Ags. When this route is chosen, the researcher should be cautious to exclude air bubbles or adjuvants from the injecting material. The practice of injecting into the footpads of rabbits to get a slow Ag release should not be used since the footpad is a sensitive site and injection causes severe pain and discomfort. For all injections, the materials should be suspended in isotonic saline to avoid causing a burning sensation in the animal. Very low doses of Ag may cause either nonresponsiveness or tolerance, while very high doses can also cause tolerance. In general, an Ag dose of 50 to 500 µg can be administered to elicit a good response. Table 1 summarizes the maximum volumes that can be injected into animals via the various routes.

8.1.5. Immunization Protocols

Circulating Abs to a specific Ag do not appear in significant amounts until at least 7 days after immunization. Most of

TABLE 1 Volumes of Ag-Ab mixtures for injection

Animal	Maximum vol (ml) for route of injection:				
	i.d.	s.c.	i.m.	i.p.	i.v.[a]
Rabbit	0.1	0.4	2.0	5.0	1.0
Mouse	0.1	0.1	0.2	1.0	0.2
Rat	0.1	0.2	0.5	2.5	0.5

[a]Freund's adjuvants should not be used with i.v. injections.

the Abs in an early (or primary) response belong to the IgM class, while Abs from a secondary response are mostly IgG. However, if the Ag is pure carbohydrate, the secondary-response Abs will be mainly IgM. The amount of Ab formed after a second or booster injection of an Ag is usually much greater than that formed after the first injection; thus, for the production of high-titer Abs, the following schedules can be used (e.g., for rabbits).

1. *Test bleed.* To collect preimmune serum (this should be done before the first injection) for use as a control to detect cross-reactive Abs

2. *Day 1.* Inject 0.1 ml each (1:1 emulsified mixture of protein Ag with Freund's complete adjuvant or another suitable adjuvant; see section 8.2.4) at four sites s.c., usually at the dorsal part of the body near the shoulders or the hips. Freund's complete adjuvant should not be used more than once because repeat injections can cause granuloma formation and ulceration of the tissue at the injection site (44, 53). The reaction to only one injection with this adjuvant is usually mild.

3. *Day 4.* Repeat the day 1 injection steps, except now Freund's incomplete adjuvant should be used (see section 8.2.4). After this step, the animals are allowed to rest for 14 days to allow the primary response to subside to a baseline level; otherwise, a secondary response may not be achieved.

4. *Day 18.* Inject 1.0 ml (1:1 mixture with Freund's incomplete adjuvant) i.m. into the thigh muscles of one leg of the animal.

5. *Day 22.* Bleed, and collect serum.

6. *Day 25.* Repeat the above-described step of i.m. injection on another leg of the animal.

7. *Day 29.* Bleed, collect serum, and pool it with the previous sample.

I.m. injections can be kept up once a month or once every 2 weeks, and serum can be collected 3 to 4 days later. To avoid batch-to-batch variability, the immune sera collected at different times should be pooled, unless one is attempting to monitor the immune response throughout a time course. The general immunization schedule given above has been used successfully with a wide range of Ags, including soluble proteins, enzymes, whole-cell bacterial suspensions, lipopolysaccharide (LPS), and polysaccharide-protein conjugates. Variations can be made at some steps; for instance, an i.v. injection *without adjuvant* may be given as a booster dose (on days 18 and 25) to elicit a more rapid response before bleeding to collect antiserum. Adjuvants are not used for i.v. immunizations.

8.1.6. Blood Collection
Blood should be collected in a dry, sterile container without adding coagulating or anticoagulating agents.

8.1.6.1. Rabbits
Restrain rabbits by wrapping each animal with a large towel so that only the head and the ears are exposed. This is more efficient than the use of commercially built rabbit-holding boxes. If the latter are used, the animals may kick or bounce violently and may sustain severe injuries such as a dislocated spine. Alternately, one could anesthetize the animals lightly to calm them down for all bleeding steps. Routinely, blood is withdrawn from the marginal vein of the ear. The vein closer to the head can be occluded by using a paper clip or by holding firmly with the fingers. This procedure produces a better blood flow at the more proximal part of the vein for puncturing. A small cut to the dilated vein with the

tip of a fresh scalpel blade will open the vein up. The droplets of blood are collected in a suitable container, usually a screw-capped glass bottle. With sufficient practice, 10 to 20 ml of blood can easily be obtained. If the inflicted incision is small enough, the bleeding will stop rapidly when the procedure is completed and gentle pressure is applied to the cut by using cotton wool. The animal will heal within a short time. Alternatively, the rabbit can be lightly anesthetized; a 22-gauge needle can be inserted into the vein, bevel up; and blood can be allowed to drip from the needle into the collection container. If sterile blood is required, it can be withdrawn directly from the vein into a syringe and needle or into a Vacutainer (Becton Dickinson, Rutherford, NJ). The latter method will require proper instruction and practice to avoid causing hematoma formation in the veins.

Larger volumes of blood can be obtained by bleeding from the central ear artery or by using cardiac puncture on anesthetized animals. The animal dies after cardiac puncture and is heavily sedated prior to starting the procedure. Cardiac puncture is not recommended for routine use but is useful if the animal is to be exsanguinated. The following steps can be followed for bleeding from the central ear artery.

1. Fifteen minutes prior to bleeding, administer s.c. or i.m. a dose of anesthetic—either 0.125 ml of Innovar-Vet/kg of body weight (11) or 0.4 ml of a mixture containing 1 mg of oxymorphone and 0.5 mg of acepromazine/kg (or as advised by a veterinary or animal technician). The drugs will tranquilize the animals for up to 45 to 60 min.

2. Dilate the central ear artery by warming the ear with a lamp or gently rubbing it with one's hands.

3. Insert a 20- or 21-gauge needle into the central ear artery, bevel up, and allow blood to drip into the container. Again, for collection of sterile blood, aspirate the blood into a sterile syringe. To prevent hematoma formation, apply pressure over the puncture site until bleeding has ceased; it will take a few minutes since the puncture site is an artery.

8.1.6.2. Mice
Only a small volume of blood can be obtained from a mouse due to its small size. The easiest and most humane way is to put the animal under light anesthesia, dip the tail into a beaker of warm water to dilate the tail veins, and nick a tail vein (usually at about the 1 o'clock position relative to the dorsal part of the tail) with a scalpel. One can massage the tail lightly while withdrawing blood into a Pasteur pipette. Approximately 0.2 ml can be obtained this way. Alternatively, bleeding can be accomplished by withdrawing from the venous plexus, located in the orbit behind the eyeball. The latter procedure is a more delicate operation and should not be attempted without instruction. One to three milliliters of blood can be obtained from the heart or from the heart cavity immediately after the animal is killed.

8.1.6.3. Rats
Blood can usually be collected from the tail vein of a rat by the procedure described above for mice. Again, once the animal is killed, larger amounts can be obtained from the heart.

8.1.7. Serum Preparation
Freshly collected blood should be allowed to clot at room temperature for 30 min, followed by incubation at 37°C for another 30 min. The cell-free fluid or serum, which contains the Abs, is collected by centrifugation. Alternatively, incubate the clotted blood at 4°C overnight to allow maximum shrinkage of the clot before centrifugation and the

collection of serum. The low-temperature protocol yields more serum per milliliter of blood and protects the Abs from hydrolysis by the action of naturally occurring proteases. Aliquot the serum into 0.5- to 2-ml volumes, and store the aliquots frozen at −20°C. For short-term storage at 4°C, add an antimicrobial agent to the serum.

8.2. ANTIGEN PREPARATION

8.2.1. Particulate Ags

8.2.1.1. Bacterial Cells
Due to their immunological "foreignness" and complex chemical compositions, bacteria are naturally immunogenic and when injected into animals will induce an Ab response. Therefore, the use of whole-cell suspensions to immunize animals is common, provided that a protective or polyvalent Ab response against multiple components or antigenic determinants of the microorganism is desired (51). Effective immunizing doses of whole-cell suspensions (see section 8.1.5) range from 5×10^8 to 1×10^{10} cells/ml. The estimation of cell counts per milliliter can be achieved by standard methods including serial dilutions, plate counts, or optical density measurements at 600 nm (OD_{600}) of standard cell suspensions (e.g., an OD_{600} of ≥0.6 for *Escherichia coli* indicates a cell concentration of approximately 10^9/ml of the suspension). Alternatively, the cell suspension can be compared to McFarland standards (14), which use various concentrations of $BaSO_4$ suspended in water to give a rough estimate of bacterial concentration from 10^8 to 10^{10} cells/ml. Comparison of the test and reference suspensions is made easier if they are held in front of a white sheet of paper.

In some cases, vaccines with live, attenuated bacteria are necessary; however, infecting animal hosts with live microorganisms may have some unexpected effects. Apart from the fact that live bacteria may be more toxic and distressing to animals, they can undergo antigenic and phase variations due to the pressure of host defense mechanisms. Bacterial cells can be killed either by heating at 100°C for 30 min or by suspending the cells in formalin-saline overnight at a final formalin (formaldehyde) concentration of 0.3% (vol/vol). Cells should be centrifuged and washed twice in isotonic saline to remove free formalin. Killing should be ascertained before injection.

As a common practice, Ags should be divided into at least 10 doses and stored frozen with or without adjuvants to ensure the consistency of each dose. Repeated thawing and freezing of Ag should be avoided.

8.2.1.2. Sheep Erythrocytes as Carriers
Both carbohydrate and protein Ags can easily be adsorbed onto sheep red blood cells (SRBC) to render them agglutinable with antiserum specific for the coating Ags. Thus, hemagglutination is easy and convenient for assessing Ag-Ab reactions. For the same reasons, SRBC can easily be used as a carrier for purified carbohydrate Ags that by themselves are normally not very immunogenic in rabbits (17). By treating red blood cells (RBC) with 0.005% (wt/vol) tannic acid, low-M_r proteins can be efficiently adsorbed and the complexes can be used as immunogens or components in passive agglutination reactions (section 8.5.1.3).

Attachment of LPS Ags onto SRBC
1. Boil an overnight culture of gram-negative bacteria for 1 h; centrifuge at $2,000 \times g$ for 10 min to sediment cell debris. The supernatant contains a crude LPS preparation.

2. Mix 4.5 ml of bacterial supernatant with 6 ml of 2.5% (vol/vol) SRBC (bacteria/SRBC ratio of 3:4 [vol/vol]; *note:* commercially available SRBC are usually prepared as a 2.5% solution in citrate buffer and are usually stable at 4°C for at least 1 month). Isotonic solutions such as phosphate-buffered saline (PBS) and Alsever's solution (see recipe below) should be used for all manipulations of the SRBC suspension. Incubate at 37°C for 30 min with occasional shaking.

3. Sediment the SRBC by centrifugation at $200 \times g$ for 10 min, and wash the cells twice with 10 ml of saline or isotonic buffer.

4. Resuspend the pellet with 6 ml of PBS, and the coated SRBC are ready for use in injections (Table 1) or in hemagglutination assays.

The attachment of LPS onto SRBC has the advantage of presenting the Ags on highly immunogenic particulate carriers. In addition, the amount of LPS presented to the animals will be much smaller than a dose of pure LPS (LPS is endotoxic); thus, this type of immunogen will be relatively nontoxic to the animals while effective in eliciting a response to LPS. By treating RBC with 0.005% (wt/vol) tannic acid, low-M_r proteins can be adsorbed to the RBC surface.

Alsever's Solution (Citrate Saline Solution)
Alsever's solution is an isotonic, anticoagulation blood preservative that permits the storage of whole blood at refrigeration temperatures for 10 weeks or more.

Dextrose	20.50 g
Sodium citrate (dihydrate)	8.00 g
Citric acid (monohydrate)	0.55 g
Sodium chloride	4.20 g
Distilled water	to 1 liter

Tanning of SRBC for Coating with Protein Ags
1. Add 3 ml of 0.005% (wt/vol) tannic acid to a centrifuge tube containing 3 ml of 2.5% SRBC. Incubate at 37°C for 10 min.

2. Centrifuge the cells at $2,000 \times g$ for 5 min, and wash once in 5 ml of PBS. Centrifuge as before, and resuspend the pellet in 3 ml of PBS.

3. To each centrifuge tube containing 3 ml of tanned SRBC, add 3 ml of 0.3-mg/ml soluble protein, mix gently, and incubate at 37°C for 15 min.

4. Centrifuge, and wash twice as described above. Resuspend each pellet in 3 ml of PBS. These cells are now ready for injection or for passive hemagglutination.

8.2.1.3. Bacterial Cells as Carriers
Smooth LPS containing O Ag sugars is generally strongly immunogenic, whereas semirough and rough LPS (containing only core oligosaccharides as terminal substituents) or capsular polysaccharides are often weakly immunogenic. To elicit a response to epitopes of the core region, Ags such as pure oligosaccharides, lipid A, and rough LPS can be attached onto bacterial cells of a rough strain, for instance, *E. coli* J5, according to the following method by Galanos et al. (18) with modifications by Bogard et al. (12).

1. Prepare heat-killed cells of the rough strain at a concentration of 5×10^9 cells/ml in 1% (vol/vol) acetic acid.

2. Heat the cell suspension to 100°C for 1 h, wash three times in distilled water, and lyophilize.

3. Dissolve lipid A or rough LPS in 0.5% (vol/vol) triethylamide at a concentration of 1 mg/ml, and add the

lyophilized acid-treated bacteria to a final concentration of 1 mg/ml.

4. Stir slowly for 30 min at room temperature.

5. Dehydrate the mixture in vacuo with a SpeedVac centrifuge (Savant Instruments Inc., Hicksville, NY).

Capsular polysaccharides made up of homopolymers or containing sialic acids induce T-cell-independent immune responses and produce low levels of humoral Abs. To improve their immunogenicity, the polysaccharides can be covalently linked to carrier proteins to form glycoconjugates. These glycoconjugates induce T-cell-dependent immune responses with the production of serum with high Ab titers (26).

8.2.2. Soluble Ags

Complete Ags usually have high molecular masses, although some naturally occurring immunogens can be small (e.g., insulin [6 kDa] and ribonuclease [14 kDa]). Molecules with molecular masses of <3 to 5 kDa are not good immunogens. As a rule, polypeptides of >20 kDa should be reasonably immunogenic and should induce a T-cell-dependent response with an increased level of specific IgG. Smaller peptides can be conjugated to larger protein carriers (such as SRBC as described in section 8.2.1), or polymerized by treatment with a cross-linking bifunctional agent, such as 0.5% (vol/vol) glutaraldehyde (final concentration), for 1 h at room temperature. Here, the material is dialyzed against buffer to remove the excess cross-linking agent. The disadvantage of glutaraldehyde treatment is the irreversible modification of certain native epitopes.

The purity of soluble Ags from microorganisms depends on the extraction method used. Soluble cell products such as toxins and enzymes are usually purified by a combination of column chromatography methods including ion exchange, gel filtration, chromatofocusing, and affinity techniques. These methods are further enhanced by separation using high-pressure liquid chromatography (HPLC) and fast protein liquid chromatograpy systems. Chapter 7 specifically deals with the fractionation of Ags from cells. The higher the purity, the easier it will be to raise specific Abs to the Ag of interest. The following protocol outlines frequently used rapid methods for purifying Ags from polyacrylamide gels, provided the protein band of interest has been identified.

8.2.2.1. Purification

Protein Ags are usually separated by using the standard sodium dodecyl sulfate (SDS)-polyacrylamide gel electrophoresis (PAGE) method of Laemmli (reference 23 and chapter 7) or, if native protein is important for eliciting an Ab response to conformational epitopes, by omitting SDS. The more commonly used methods for identifying the location of the band of interest in the gel are either staining a vertical strip of the gel to locate the Ag band or staining the entire gel lightly. The former has the advantage of avoiding chemical fixation of the entire gel and denaturing the Ag of interest. Normally, 0.05% (wt/vol) Coomassie brilliant blue R-250 is used to stain the gel. The sensitivity is roughly 1 to 2 μg per band in a lane of standard width. If the more sensitive silver staining method has to be used to see the band, the quantity is insufficient for immunization.

Direct Use of Gel Slices

1. Excise the band of interest with a razor or scalpel blade. Before discarding the rest of the gel, stain it to check the accuracy of the excision.

2. Wash the gel slices in PBS or distilled water for a few minutes. Finely chop the slices with a razor blade, or repeatedly squeeze the gel pieces through the barrel of a standard 5-ml syringe with the help of a small amount of buffer. For the latter option, do a final squeezing step with a 21-gauge needle attached.

3. For immunization, the homogenate can be used directly or lyophilized and ground into a fine powder. However, it is advisable to electroelute (see below) the proteins from these chopped-up gel slices before injection, since acrylamide is not easily degradable in an animal host.

8.2.2.2. Electroelution

Gel slices can be placed into a small dialysis tube containing 1 ml of 0.2 M Tris-acetate (pH 7.4), 1.0% (wt/vol) SDS, and 100 mM dithiothreitol per 0.1 g (wet weight) of polyacrylamide gel. The dialysis tubing is then placed into a horizontal electrophoresis chamber. Alternatively, gel slices can be placed into an ISCO sample cup (Canberra Packard, Mississauga, Ontario, Canada) with a dialysis membrane attached to the bottoms of both chambers. The running buffer contains 50 mM Tris-acetate (pH 7.4), 0.1% SDS, and 0.5 mM sodium thioglycolate, and electroelution is run for 3 h at 100 V. The gel slices can be removed and stained with Coomassie blue to verify the removal of Ags. The proteins are then dialyzed against distilled water (to remove all detergents and other chemicals) and lyophilized, and the product can now be incorporated into an emulsion with an adjuvant for immunization (25).

8.2.3. Haptens

Ags with molecular masses of <10 kDa are usually weak immunogens, with the exceptions described previously. Thus, one needs to conjugate low-molecular-mass molecules with little or no immunogenicity onto carriers before immunizing animals in order to raise specific Abs. The carriers can be globular protein molecules, such as bovine serum albumin (BSA), keyhole limpet hemacyanin, or particulate Ags as described above (section 8.2.1). For more detail on the choice of peptide sequences and the strategy for the coupling of peptides to carriers, consult the manual by Harlow and Lane (4).

A common procedure for conjugating haptens to protein carriers uses glutaraldehyde, which cross-links molecules through their amino groups. The following procedure has been used successfully to conjugate cadmium-binding peptide (29) and can be used to conjugate other small peptides with 9 to 25 amino acids to BSA.

1. Add 2 ml of 0.2% (vol/vol) glutaraldehyde to a 2-ml solution containing 3 mg of your peptide plus 1 mg of BSA. To slow the rate of coupling, the pK of the buffer used in the reaction mixture should be above the pK of the peptide's amino groups (if known) so that NH_2 and not NH_3 will be targeted.

2. Incubate the mixture at room temperature for 2 h with gentle mixing. Dialyze for 24 h against several changes of 5 mM Tris-HCl (pH 8.0).

3. Concentrate the contents of the dialysis bag either by using a SpeedVac (Savant) or by sprinkling dry flakes of polyethylene glycol (PEG; molecular mass, 6 kDa or higher) onto the surface of the dialysis bag placed in a Pyrex dish. This can be accomplished in a matter of minutes if sufficient PEG flakes are used to totally cover the dialysis bag.

Alternatively, 1 to 2% glutaraldehyde can be used to cross-link the peptide's amino groups such that it polymerizes, increasing Ag size, but this may interfere with the epitopes of interest.

8.2.4. Adjuvants

With few exceptions, adjuvants are always used to enhance the Ab response to an Ag. Several reviews can be consulted for more specific detail on adjuvants (37, 43, 45, 55, 56). Table 2 summarizes commonly used adjuvants and their characteristics.

8.2.4.1. Freund's Adjuvant

Both complete and incomplete forms of Freud's adjuvant are readily available commercially, and here is how they are used.

1. To aqueous Ags (usually suspended in isotonic saline), add an equal volume of Freund's adjuvant (either the complete or the incomplete form). Mix ingredients vigorously until a thick emulsion is formed. To help generate the emulsion, take the mixture into a glass syringe fitted with an 18-gauge needle, forcing the material in and out of the syringe until the syringe plunger is almost impossible to move. Alternatively, mix the Ag and adjuvant in a vortex vigorously until a thick emulsion develops. The emulsion is deemed ready when a droplet of it does not disperse immediately when added to a saline solution.

2. Take up the volume needed for injection into a fresh syringe, add a needle of an appropriate size, and the immunogen is ready for use.

8.2.4.2. Alum

The following procedure can be used for alum (34).

1. Add 5 ml of 10% (wt/vol) aluminum potassium sulfate (dissolved in distilled water) to a 50-ml conical centrifuge tube. Then add 22.8 ml of 0.25 N NaOH dropwise to the tube while vortexing.
2. Allow the mixture to settle for 10 min at room temperature.
3. Sediment the Al(OH)$_3$ by centrifugation at 1,000 × g for 10 min. Wash with 50 ml of distilled water, and repeat the centrifugation. Discard the supernatant.
4. The pellet contains the Al(OH)$_3$, or alum adjuvant, which has a binding capacity of 50 to 200 μg of protein Ag.
5. Add an aliquot of Ag to alum, allow the two components to mix at room temperature for 10 min, centrifuge at 10,000 × g for 10 min, and check the supernatant for unbound Ags. The sediment is ready for injection.

8.2.4.3. ISCOMs

Immunostimulatory complexes (ISCOMs) are adjuvant systems composed of cholesterol, lipid, phospholipid, and saponins (46, 47, 55). The early formulations of ISCOMs used Quil A, a mixture of saponins (46), which showed various levels of toxicity in animals. A recent-generation ISCOM containing a 7:3 ratio of Quil A and Quil C (ISCOPREP™703) is nontoxic as an animal and human vaccine (50). Preparation of ISCOM with Quil A can be done according to the method of Morein et al. (46). Since Quil A behaves like a detergent in water and forms micelles at the critical micelle concentration of 0.03% (wt/vol), it easily forms a complex with Ags in which the Ag is present in an accessible and multimeric fashion. ISCOMs prepared from Quil A have successfully elicited humoral responses against viral products in animal protection studies. Thus far,

it is not known whether ISCOMs can be used as effectively with other microbial products. Here is how to prepare them.

1. Using 5-ml layers, prepare a discontinuous sucrose gradient ranging from 10 to 40% (wt/wt) (chapter 7) in TN buffer (0.05 M Tris, 0.1 M NaCl, pH 7.2) containing 0.2% Quil A (wt/vol).
2. Solubilize viral glycoproteins or other Ags with 2% (vol/vol) Triton X-100 in TN buffer, and then place 200 μg of the Ag in 200 μl of 8% (wt/wt) sucrose in TN buffer containing 1% Triton X-100 on top of the gradient.
3. Centrifuge at 150,000 × g for 4 h at 20°C in an ultracentrifuge using an SW50 Beckman rotor.
4. Collect 250-μl fractions, and assess the quantity of Ags either by protein assay or by the measurement of radioactivity if the Ag is labeled. The fractions in which Ags are detected can be pooled. To remove excess Quil A, the pooled fractions are centrifuged again in a 10 to 40% sucrose gradient as described above but *without* Quil A. Fractions are collected as before, pooled, dialyzed against 0.5 M ammonium acetate, pH 7.0, and lyophilized.
5. The amount of Quil A will be approximately 5% (wt/vol) in the ISCOM, i.e., <2 μg of QuilA/10 μg of protein. The effective dose of Ag in the ISCOM will be in the range of 5 to 10 μg. Note that in a mouse weighing 20 g, 10 to 50 μg of Quil A is toxic (48); however, nontoxic derivatives exist (50).

8.2.4.4. Liposomes

Liposomes are micelles that take the shape of concentric spheres consisting of phospholipid bilayers. Proteins and other Ags can be trapped inside the liposomes. Furthermore, depending on the nature of the interaction between Ag molecules and liposomes, hydrophilic epitopes can be well exposed by the liposomes. A commonly used mixture of lipids, composed of egg lecithin, cholesterol, and stearylamine in molar ratios of 7:2:1, forms positively charged liposomes (30). If negatively charged liposomes are desired, phosphatidic acid or dicetylphosphate replaces stearylamine. Negatively charged liposomes are reported to be superior to positively charged ones as adjuvants to elicit an Ab response to diptheria toxin (42). While there are many combinations of phospholipids that can be used, the following general protocol can be used (35).

1. Dissolve 75 mg of phosphatidylcholine and 11 mg of cholesterol in chloroform by using a round-bottomed flask. A thin film will develop after low-power vacuum rotary evaporation at 37°C for a short time or under a nitrogen atmosphere.
2. BSA liposomes can be prepared by adding 10 mg of BSA in 10 ml of PBS (0.15 M NaCl, 10 mM phosphate buffer, pH 7.4) to the flask with the thin film of lipid and incubating at 37°C for 1 h with gentle rotation. Alternatively, the mixture can be subjected to bath sonication for 30 min.
3. Wash the BSA liposomes three times with PBS (100,000 × g, 30 min) to remove unattached BSA. Resuspend the pellet in a small volume of saline or PBS. The amount of bound BSA can be assessed by a standard protein assay. The Ag-incorporated liposomes are ready for use.

8.2.4.5. Ribi Adjuvant Systems

Adjuvant systems are available from Ribi ImmunoChemical Research, Inc. (Hamilton, MT), and they contain a combination of mycobacterial cell walls, trehalose dimycolate, and

TABLE 2 Adjuvants and their characteristics

Adjuvant(s) (Source)	Characteristics	Reference(s)
Freund's complete adjuvant (FCA)	Most commonly used adjuvant Mix 1:1 (vol/vol) with Ag until thick emulsion is formed Serves as a depot for slow release of Ags Emulsion of oil (Bayol F), detergent, and an extract from mycobacteria Must never be used for i.v. injections Disadvantages: (i) elicits granuloma at site of injection if used repeatedly and (ii) is difficult to prepare as an emulsion	40
Freund's incomplete adjuvant	Contains the oil and detergent only without bacterial extract Mix 1:1 (vol/vol) with Ag as described above Do not use in i.v. injections Normally used to replace Freund's complete adjuvant in subsequent injections	40
Alum	Aluminum hydroxide salt (gel-like) which also facilitates slow release of Ags Attracts immunocompetent lymphocytes to the area of injection to induce an improved Ab response Elicits IgG1 and IgE responses in mice Is less toxic than Freund's complete adjuvant	16, 48, 53
Liposomes	Can be made as lecithin-cholesterol-dicetyl phosphate in a molar ratio of 7:2:1 Are useful to entrap Ags for slow release Advantages: (i) function as a depot of Ags, (ii) help Ags to retain T-cell dependency of native protein Disadvantage: entrapped Ags, not exposed on the outer lipid layer, may not be detected by immune system	19, 35, 42, 52
LPS	B-cell mitogen Stimulates natural (nonspecific) Abs if administered without Ag but stimulates specific Ab response when injected with an Ag Usually, the lipid A region is the immunopotentiating component; however, mannan O side chain sugars from *Klebsiella* strain 03 and *E. coli* 09 were shown to have adjuvant effects as well Toxicity is the major disadvantage	41, 49, 51
Muramyl dipeptide	Active component of mycobacterial extract used in Freund's complete adjuvant Is prepared in many forms and derivatives Stimulates specific Ab response to a wide variety of natural Ags such as bacteria, viruses, and fungi Is nontoxic	38
Bacterial toxins and proteins	Genetically detoxified derivatives of cholera toxin or heat-labile *E. coli* toxin Potent adjuvants for induction of mucosal, humoral, and cell-mediated immunity	13, 16
Quil A	Composed of saponin, a mixture of water-soluble triterpene glycosides from the South American tree *Quillaia saponaria* Advantage: is highly surface active and forms stable complexes known as ISCOMs with viral envelopes Has been reported to be superior to Freund's complete adjuvant in eliciting Ab response to human serum albumin. Use of 5–10 μg of Ags in the ISCOM for each dose will elicit better response than 100–200 μg of aggregated Ag without Quil A	43, 46, 47, 50, 55
Synthetic adjuvants, formulation 1 (e.g., SFA-1, Syntex)	A synthetic MDP muramyl dipeptide was shown to stimulate IgG2 to human serum albumin in mice Toxicity in animals and low viscosity; the latter means that Ag-Ab mixture will be easy to mix and to administer to animals	38, 43

(Continued on next page)

TABLE 2 (*Continued*)

Adjuvant(s) (Source)	Characteristics	Reference(s)
Adjuvant system (Ribi Immunochem Research, Inc., Hamilton, MT)	Composed of monophosphoryl lipid A From 5–50 μg of monophosphoryl lipid A was shown to augment humoral responses to both polysaccharide and protein Ags Nontoxic and has low viscosity; thus, Ag-adjuvant mixtures are easy to prepare	31, 32, 39
SBAS4 (SmithKline Beecham Adjuvant System 4)	System containing alum and 3-deacylated monophosphoryl lipid A Known to elicite humoral and cellular immune responses	36, 39, 54
Bacterial DNA	Synthesized as oligodeoxynucleotides containing methylated CpG motifs derived from bacterial DNA Also called immunostimulatory CpG ODNs Is a potent stimulator of cytokines leading to induction of mucosal, humoral, and cell-mediated immune responses Can be administered to potentiate immune responses to protein Ags	20, 28, 33

a nontoxic lipid A derivative (monophosphoryl lipid A). The adjuvant effect appears to be based on the ability of monophosphoryl lipid A to inactivate suppressor T cells while stimulating a polyclonal B-cell response (32, 36, 38, 39).

8.2.4.6. DNA Immunization

Nucleic acid (DNA or RNA) immunization and genetic vaccines introduced in the early 1990s are some of the most important discoveries and novel strategies in vaccine development (20, 21, 24, 27). The essential features of a DNA vaccine are a bacterial plasmid vector engineered to carry a DNA insert encoding the protein immunogen(s) of interest, a eukaryote gene promoter, and a poly(A) site to enable expression of the protein in mammalian cells. The vectors are usually maintained in and purified from *E. coli*. Bacterial DNA is a potent adjuvant that stimulates immune cells and Ag-presenting cells and induces B-cell proliferation and the secretion of Abs. The immunostimulatory properties are attributed to unmethylated CpG dinucleotide motifs found with high frequency in bacterial and viral DNA (28).

For immunization, the recombinant plasmid DNA is delivered directly into muscles of the host animal and the DNA is taken up by tissue and translated in vivo into protein Ags. DNA vaccines induce both humoral (Ab) and mucosal immune responses and elicit strong, sustained T-cell-dependent responses. DNA vaccines are typically administered to the host by injecting tissues or by using a gene gun that provides high-pressure contact delivery. Recent studies describe systemic and mucosal routes with or without adjuvants with the induction of mucosal and humoral responses (15, 28). There are advantages of using nucleic acid vaccines; e.g., a variety of well-characterized plasmid vectors are available, recombinant plasmids are easily prepared and stable, insert DNA (encoding the immunogen) is stable, and there is never a need to prepare the Ag for immunization (15, 20, 24, 28).

8.3. CONVENTIONAL ANTIBODY PREPARATION

The polyclonal serum is expressed from the blood clot by centrifugation, and approximately 52% of its volume can be collected as serum (section 8.1.7). The serum can be used directly in Ag-Ab reactions, absorbed to deplete it of nonspecific or cross-reactive Abs (section 8.3.1), or fractionated to purify Abs free of other serum proteins (sections 8.3.2 to 8.3.4). The two principal types of proteins in serum are globulins and albumin, which can be fractionated by various means. The most common purification methods involve ammonium sulfate precipitation, dialysis to get rid of the salts, and anion exchange chromatography for purification. This combination is tedious but inexpensive and can effectively isolate all IgG molecules from serum. Gel filtration may be used instead of anion exchange chromatography when IgM purification is desired. More recently, the use of affinity chromatography can dramatically speed up the process, but more importantly, it can help to target the specific desired Ab if purified Ag is incorporated onto an affinity matrix. Affi-Gel blue (Bio-Rad) replaces the salt precipitation step because the Cibacron Blue® dye of its matrix (2 mg/ml) has a strong affinity for binding albumin (at 11 mg/ml); therefore, it can be used as a one-step method to yield a relatively clean IgG fraction. HPLC and fast protein liquid chromatography methods involving a high-pressure pump and special columns provide better speed and resolution for the separation of Abs from other serum proteins; however, the equipment is expensive and may not be readily available.

8.3.1. Absorption of Cross-Reactive and Nonspecific Abs

In addition to the Ag-specific Abs, polyclonal immune sera contain a variety of Abs representative of immune responses to Ags that an animal is exposed to from its environment, gut, or foods. These so-called background, or nonspecific, Abs include those to Ags of bacteria, other microorganisms, and macro-Ags such as pollen and dust mites, etc. When tested against bacterial whole-cell lysates, these serum Abs interact with common and conserved antigenic determinants of bacterial surface molecules including LPS, capsular polysaccharide, flagella, and outer membrane proteins. To increase the sensitivity of sera and enhance the detection of specific Ags, cross-reacting or background Abs are removed from the polyclonal antisera by adsorption with whole bacteria to remove cross-reacting Abs to cell surface-localized Ags or bacterial cell lysates are absorbed onto nitrocellulose disks to deplete the sera of all cross-reacting Abs (61, 62).

Procedure

1. Lyse *E. coli* cells from a 100-ml overnight culture by incubating the cells in a lysis buffer (50 mM Tris [pH 8], 2 mM EDTA, 2% [wt/vol] SDS, 4% [vol/vol] β-mercaptoethanol, 1% [wt/vol] lysozyme) for 1 h at 37°C.

2. Centrifuge the cell lysate at 12,000 × g at 4°C for 30 min. Discard the pellet, and collect the supernatant.

3. Immerse five pieces of nitrocellulose (NC) membranes, 35 cm^2, in the cell lysate supernatant, and incubate for 1 h on a shaker at room temperature.

4. Wash the treated NC membranes three times (10 min each) in Tris-buffered saline (TBS; 10 mM Tris-HCl, [pH 8], 150 mM NaCl).

5. Block the membranes for 1 h with 3 to 5% (wt/vol) skim milk in TBS. The purpose of this step is to fully saturate any remaining binding sites on the NC membranes to avoid nonspecific binding of Abs to the membranes in subsequent adsorption steps.

6. Wash the membranes again three times for 5 min each with TBS. Now the membranes are ready to be used for adsorption of background Abs from the antiserum. If the membranes are not being used right away, they can be air dried; wrapped in plastic film, sealed in a plastic bag by using a bag sealer device, or sealed in a zip-lock bag; and stored at −20°C until use. Before use, remove the membranes from the freezer, allow them to thaw at room temperature for 20 min, and soak for 1 to 3 min in PBS.

7. To begin the adsorption process, immerse one membrane in 10 to 20 ml of diluted antiserum (1/100 in saline or PBS) and incubate for 1.5 h at room temperature. Adsorption is done in petri dishes. Use a rotary shaker to provide good mixing during the incubation.

8. After the incubation, remove the membrane and replace it with a second membrane. Repeat this process until all five membranes have been used.

9. Now the antiserum is ready for use for Western immunoblotting and should yield results with little to no background protein band reactivity. If not immediately used, the freshly adsorbed antiserum should be stored frozen in the presence of a small amount of a preservative such as sodium azide. We often collect the used serum, add a preservative, store it frozen, and reuse it for blots up to five times or until the intensity of the reaction against the immunizing Ag is appreciably decreased.

8.3.2. Ammonium Sulfate Precipitation

Ammonium sulfate is the most widely used salt for the precipitation of proteins. $(NH_4)_2SO_4$ has the advantage of a high level of solubility which is only minimally dependent on temperature; its solubility varies only ~3% between 0 and 25°C. In contrast, Na_2SO_4, which is indicated in some methods, is five times as soluble at 25 as at 0°C. Proteins are polyvalent ions, with surface charges that interact with water molecules through hydrogen bonding. As SO_4^{2-} ions also attract water molecules, they compete with proteins and, at high concentrations, will strip away the solvation layer. The proteins will then have an increased tendency to interact with one another and precipitate from the solution. The gamma globulin fraction, containing Abs, is obtained after repeated precipitations with $(NH_4)_2SO_4$ added to make a 33.3% saturation solution (see the recipe below). Unwanted proteins, including albumin, will remain in the solution. A higher yield can be achieved by a single 50%

saturation, although a small amount of albumin may be coprecipitated.

Procedure

1. To prepare a saturated stock solution of $(NH_4)_2SO_4$, add 100 g of the crystals for every 100 ml of distilled water and leave the mixture stirring for 1 to 2 days on a magnetic stir plate at room temperature. There should always be undissolved crystals on the bottom; otherwise, the solution is not saturated. Check and adjust to pH 7.0. Store at 4°C since Ab precipitation is usually performed at this temperature.

2. To 20 ml of serum (already cooled to 4°C), slowly add 10 ml of saturated $(NH_4)_2SO_4$ dropwise from the stock solution while the solution is stirred. Continue the stirring for at least 2 h at 4°C.

3. Sediment the pellet by centrifugation at 10,000 × g for 10 min.

4. Resuspend the pellet with a small volume of buffer (5 ml), e.g., PBS or TBS (50 mM Tris, 1.5 M NaCl) at pH 7.0, and then add 2.5 ml of saturated $(NH_4)_2SO_4$ as in step 2. Sediment the pellet by centrifugation as before.

5. Resuspend the pellet with 10 ml of buffer, and transfer it into a dialysis bag. Dialyze overnight at 4°C against two to three changes of 2-liter volumes of buffer. The dialyzed material can now be stored frozen or can be lyophilized prior to further steps such as anion exchange chromatography.

8.3.3. Ion Exchange Chromatography

For the purification of IgG, ion exchange chromatography is a natural second step after ammonium sulfate precipitation to reduce albumin. The most commonly used matrix for Ab purification is one to which an ionizable DEAE group is attached. DEAE-Sephacel (cellulose) and DEAE-Sepharose from Pharmacia are convenient, preswollen, and ready to use. They have high capacities, and a 10-ml column can be used to bind 100 to 200 mg of protein. At pH 8.0, both matrices are strongly cationic for binding Abs. Immunoglobulins have pIs in the range of 6 to 8; thus, they will be anionic at pH 8.0 and will bind to DEAE at low salt concentrations. Elution of Abs can easily be achieved either by increasing the concentrations of competing ions (e.g., Cl^-) in the column buffer or by lowering the pH. The recovery of Abs should be almost 100%.

Procedure

1. Pack and equilibrate a DEAE-Sephacel or DEAE-Sepharose column (1.5 by 10 cm) with 10 mM Tris buffer, pH 8.0, at room temperature. To avoid the formation of air bubbles which will diminish the efficiency of the column, the buffer and gel slurry should be allowed to equilibrate to room temperature before starting. The size of the gel bed will be determined by the amount of Abs to be bound. As described above, one can pack a 10-ml column for every 100 mg of protein to be bound. If chromatographic columns are not available, a plastic disposable syringe will suffice as long as some fiberglass is used as a bottom screen to prevent leakage of the slurry.

2. Wash the column with at least 5 column volumes of the 10 mM Tris buffer. A gravity feed with a head pressure of 100 to 200 mm will be sufficient (65) if a peristaltic pump is not available.

3. Apply the ammonium sulfate-precipitated and dialyzed Ab solution onto the column. If the dialyzed Abs have been lyophilized, reconstitute them in 10 mM Tris.

4. Wash the column with 2 to 5 bed volumes of Tris buffer. Collect these fractions to ensure that Abs are not lost.

5. Develop the column with a linear gradient of increasing concentrations of NaCl made from 100 ml of 10 mM Tris buffer in the first chamber and 100 ml of 0.3 M NaCl in 10 mM Tris buffer in the second chamber of a gradient maker. The usual total volume of the gradient is 10 to 20 times the bed volume. The gradient is allowed to run over 4 h, and 5-ml fractions are collected. Gradient makers can be purchased from commercial sources, and as long as both chambers are of the same size, linear gradients can easily be achieved. To mix the solutions, a magnetic bar is placed in each chamber and the gradient maker is placed on a magnetic plate. Mouse IgG2a is usually eluted earlier, and IgG1 is usually eluted later. Mouse IgG3 may not be stable at low ionic strength; thus, the concentration of the Tris buffer used for the purification of IgG3 should be increased to 50 or 100 mM. Alternatively, the column can be eluted with a stepwise salt gradient. A series of 50-ml volumes in 10 mM Tris (pH 8.0) are used. The first contains no salt, and then increasing salt concentrations of 50 mM NaCl, 100 mM NaCl, 150 mM NaCl, 200 mM NaCl, 250 mM NaCl, and 300 mM NaCl are added. Most of the Abs should be eluted by the time the 200 mM NaCl step is added.

6. The fractions are monitored by using a spectrophotometer fitted with a UV detector or by assaying for proteins. The purity of the eluted Ab can be assessed by standard agarose gel electrophoresis or SDS-PAGE.

8.3.4. Gel Filtration

Gel filtration chromatography separates proteins according to their molecular sizes. Since IgM has a molecular mass of ~900 kDa, it can be separated easily from other Abs with average molecular masses of 150 kDa. Many commercial gels are suitable for this purpose, including Sepharose 6B, Sephacryl S-500, and Sephadex G-200 (all from Pharmacia), Ultrogel AcA 22 (LKB), and Bio-Gel P-300 (Bio-Rad). Sephacryl (a mixture of dextran and acrylamide) and Ultrogel (a mixture of agarose and acrylamide) are more resistant to compression and easier to pack and can be run at higher pressures and flow rates than gels like Sephadex (65). The columns used are typically long, thin ones (15 to 30 mm in diameter). When packed and properly degassed, the gel generally occupies approximately 70% of the column and the space between the gel beads occupies the remaining 30%. Either TBS or PBS can be used at pH 7 to 8 at an NaCl concentration of ≥100 mM to prevent adsorption effects. Sodium azide at 10 mM may be added in running buffers to prevent microbial growth. For best resolution, the flow rate of buffers should be low, which means that it will take 1 to 3 days to perform a proper run. An obvious disadvantage besides the time this takes to perform is the dilution of the Abs being eluted. However, concentration of the fractions is easily achieved by lyophilization or concentration by using PEG (see section 8.2.3).

8.3.5. Affinity Chromatography

Ag covalently bound onto a chromatographic gel matrix provides specific binding to isolate the Ab of choice from serum. Some of the commonly known gels are CNBr-activated Sepharose 4B and Sepharose 6B (Pharmacia), which are ready for the direct coupling of Ags through the primary amino groups of the Ags. For obvious reasons, direct coupling of high-molecular-mass polysaccharides onto CNBr-activated gels is not possible until the polysaccharides are conjugated to protein carriers. More appropriately, the coupling of microbial polysaccharides can be achieved through the binding of their carboxyl groups to one of the following gels: AH-Sepharose 4B (Pharmacia), Affi-Gel 102 (Bio-Rad), and Aminoethyl Bio-Gel (Bio-Rad). Detailed descriptions of the principles and procedures of the coupling of ligands to affinity gels can be found in a document entitled *Affinity Chromatography: Principles and Methods*, which can be obtained from Pharmacia free of charge.

8.3.5.1. Affi-Gel Blue

In addition to the aforementioned activated gel matrices for preparing affinity columns, one product from Bio-Rad known as Affi-Gel blue is worth mentioning (59). Affi-Gel blue (50 to 100 mesh size) is a beaded, cross-linked agarose with covalently attached Cibacron Blue F3GA dye. There is ca. 1.9 mg of dye/ml of gel, and the dye has a very high affinity for albumin. It can bind ≥11 mg of albumin/ml of Affi-Gel blue. Thus, a single passage of serum through a minicolumn of this gel can yield a relatively pure immunoglobulin fraction with albumin virtually removed. This method is very rapid compared to the steps involved in ammonium sulfate precipitation, dialysis, and concentration to isolate immunoglobulins. Minicolumns with 5- to 10-ml volumes can be prepared in Econo-columns (Bio-Rad) or in syringes. To reconstitute the column, albumin can be removed by using 3.0 M potassium thiocyanate or 8 M urea (65), and then the column should be reequilibrated with running buffer of near-neutral pH.

8.3.5.2. Protein A-Agarose

Protein A is a cell wall component of *Staphylococcus aureus* with a strong affinity for the Fc region of IgG. Protein A has a molecular mass of 42 kDa and consists of four globular immunoglobulin-binding sites at the N-terminal end. Protein A interacts with the immunoglobulins of 65 mammalian species (69). Purified protein A can be covalently attached to gel beads such as agarose and used as an affinity gel for the purification of IgG. The binding of protein A to IgG is generally at least five times stronger than binding to IgM. However, protein A-agarose columns can be used to purify IgM if buffers of high ionic strength and high pH are used (60, 70). Protein A-agarose can be obtained from Pharmacia or Bio-Rad; the latter product is known as the MAPS (MAb purification system), which can be purchased as a kit with the column and all the necessary buffers. In general, IgG2a, IgG2b, and IgG3 can be easily purified by this method, and IgG1 and IgM can be purified when high pH and high ionic strength are employed (63). Elution of the Abs is usually achieved by using acidic buffers such as citrate or glycine-HCl buffer at pH 5.0 or lower. The buffer conditions described below for HPLC purification of MAbs can be used. Naturally, columns should be run at a much lower flow rate, for example, 2 to 5 ml/h, than those in HPLC systems (5 to 10 ml/min). In order to avoid clogging of the rather expensive protein A-agarose columns, the methods described for the pretreatment of ascitic fluid could be used. Samples should at least be passed through a 0.45 to 0.22 μm-pore-size filter before being applied to the column. Acid-eluted Abs should be neutralized with 1 M Tris, pH 8.0 (65), immediately after elution from the column to avoid any loss of Ab activity.

8.3.5.3. Protein G

Protein G is a cell wall component of alpha-hemolytic group C and G streptococci (57, 58). Native protein G has a strong affinity for the Fc regions of all IgG Abs, including IgG classes that have low affinity for protein A; however, this protein also binds albumin (57). Recombinant protein G in which the albumin site has been eliminated is now available for immunoglobulin purification and immunodetection procedures (58).

8.4. MAb PREPARATION

Since the first report by Köhler and Milstein (66) of cell fusion between a hypoxanthine-aminopterin-thymidine (HAT)-sensitive variant of MOPC-21 myeloma cells and spleen cells immunized with SRBC, the hybridoma technique has been well exploited by scientists in all life science disciplines. The main advantages of MAbs over conventional Abs include increased reproducibility and specificity, improved sensitivity, and the ability to produce unlimited amounts of immunoglobulins. These advantages usually outweigh the expense and the fact that the people involved in MAb production require a fair amount of training. However, one should consider the *real need* for MAbs before embarking on the process. For most intended purposes, there is nothing wrong with rabbit polyclonal Abs, which cost a fraction of what it would cost for MAbs. For consideration of the legal ramifications of the use of animals for MAb production, consult the review by Kuhlmann et al. (67). The following protocol is for the production of murine MAbs by cell fusion by using a special mixture of PEG and dimethyl sulfoxide.

8.4.1. Production Protocol

Most protocols used for this purpose are based on the protocol of Galfré and Milstein (64) with PEG as a fusogen instead of Sendai virus as used in earlier studies. The following is a brief outline of the procedure described by Lam et al. (68) for the production of MAbs against LPS; all reagents are from Sigma.

1. In a 50-ml sterile centrifuge tube with 10 ml of serum-free medium (either RPMI 1640 or Dulbecco's modified Eagle medium [DMEM]), mix 10^8 spleen cells from an immunized mouse (BALB/c) with 10^7 myeloma cells in 10 ml of the same medium. This ratio of 10:1 of the two cell types may be varied depending on personal experience.

2. Centrifuge at $400 \times g$ for 5 min. Resuspend the pellet with 20 ml of serum-free medium, split the cell suspension into four equal aliquots of 5 ml each, and put the aliquots into 15-ml sterile centrifuge tubes. The purpose of working with four smaller cell pellets is to allow better exposure of cells to the fusogen in the step below. Repeat step 2.

3. To one of the four cell pellets, dispense 0.5 ml of the fusogen (40% [wt/vol] PEG [molecular weight of 1,050], 4% [vol/vol] dimethylsulfoxide, 56% [vol/vol] DMEM or RPMI 1640) in small dropwise quantities *in exactly 2 min.* (The other three tubes with the cell pellets should be kept on ice until ready for use.)

4. Add 20 ml of serum-free medium over 3 to 5 min. Centrifuge at $400 \times g$ for 10 min.

5. Tap the tube gently on the bench to dislodge the cell pellet. Gently resuspend the cell pellet in 12.5 ml of medium containing 20% (vol/vol) fetal calf serum (FCS). It is all right to leave some clumps of cells. Repeat steps 3 through 5 for each of the remaining cell pellets.

6. Combine the four fused cell suspensions, each 12.5 ml, and transfer to a large (75-mm²) tissue culture flask.

Incubate overnight at 37°C and 5% CO_2. This allows the cells to recover from the PEG shock before being subjected to the selective medium containing HAT and 20% FCS. At this time, prepare feeder cells from the spleen and thymus of a nonimmunized mouse (e.g., a CD1 mouse); incubate the cell suspension (at 10^6 cells in medium containing 10% FCS) overnight at 37°C and 5% CO_2.

7. On the next day, add to the fused cells the feeder cells and 50 ml of medium containing a 2× concentration of HAT (HAT can be obtained as a 50× stock from most suppliers of tissue culture materials) to yield a density of 5 $\times 10^4$ to 5×10^5 myeloma cells/ml. This is the concentration at which myeloma cells grow best.

8. Feeder cells are added at a concentration of 10^5 cells/ml. Transfer aliquots of the cell suspensions onto 24- or 96-well tissue culture plates.

9. Leave the fused cells undisturbed for 10 to 14 days in an incubator set at 37°C with 5% CO_2 and 95% humidity. Hybrid clones that can survive the HAT selection will appear as plaques (or colonies of clones) on the bottoms of the wells. These plaques can be confirmed by viewing with an inverted microscope.

10. Remove culture supernatants, assay for specific activity against the immunizing Ag (e.g., by enzyme-linked immunosorbent assay [ELISA]), and identify the wells positive for activity.

11. Clone cells in positive wells by using the limiting dilutions (71). The medium used should contain 20% (vol/vol) FCS and hypoxanthine-thymidine obtained as a 50× stock from suppliers. Reclone cells in the same manner.

12. Culture supernatants from the twice-cloned cell lines usually contain microgram quantities of MAbs and are ready for use for most Ag-Ab reactions. In order to produce larger quantities (such as milligram-per-milliliter amounts) of MAbs, the method of ascites fluid production described below can be followed. Several systems have also been described for scaling up the production of MAbs from hybridoma cells grown in serum-free media by using shaker flasks, roller bottles, and hollow fiber and capillary systems (refer to the Invitrogen website at http://www.invitrogen.com for the GIBCO BRL *Guide to Hybridoma Technology*).

8.4.2. Isotyping Abs

Before a MAb can be further characterized or exploited, it is necessary to identify the Ab isotype, including its class and subclass and the presence of a κ- or λ-light chain. Commercially produced anti-mouse (and anti-rat) Ab reagents are readily available such that there is no longer a need to produce Abs to each of the immunoglobulin isotypes for identification. Depending on the commercial mouse Ab typing kit purchased, a simple double immunodiffusion test, a more sensitive passive agglutination test, a dot-blotting method with precoated nitrocellulose strips, or a highly sensitive ELISA method can be used. Table 3 summarizes the various commercial kits.

8.4.3. Ascites Fluid Production

The term ascites refers to an accumulation of fluid in the abdominal cavity. In medicine, ascites is usually associated with circulatory congestion due to disorders of the heart, lungs, kidney, or liver. For research purposes, ascites is induced in mice as a source of the production of highly concentrated MAbs. Ascites fluids like serum contain 10 to 20 mg of Ab/ml. This level of Abs is at least 100 to 1,000 times more than that obtained in tissue culture supernatants of hybridoma lines. Both Freund's incomplete adjuvant and

TABLE 3 Commercial sources of immunoglobulin isotype determination reagents

Commercial reagents or kits	Technique used	Remarks
Bio-Rad mouse typer isotyping kit	ELISA	Complete kit includes rabbit anti-mouse sera, goat anti-rabbit Ab-peroxidase or horseradish conjugates, substrate solutions, and anti-κ and anti-λ reagents
Monoclonal typing kit, mouse or rat, immunodiffusion (ICN Immunobiologicals)	Gel diffusion	Simple to use, but one has to wait 24–48 h before reading results
Monoclonal typing kit, mouse, hemagglutination (ICN Immunobiologicals)	Hemagglutination	More sensitive than gel immunodiffusion and can detect 1–20 μg of MAbs
Mouse MAb isotyping reagents (Sigma)	Choice of uses	6 reagents are provided to identify all classes and subclasses, but no anti-light chain reagents. The reagents can be used in either ELISA or immunodiffusion methods
Sigma ImmunoType kit	Dot blotting	Precoated strips and reagents for color development are provided

pristane, a defined mineral oil (see below; 3, 72, 73), can induce ascites production in mice. The latter has been particularly useful for priming BALB/c mice for the injection of myeloma cells to induce Ab production; ca. 5 to 10 ml of ascites fluid can be generated by this procedure. However, before choosing to use this route of Ab production, one should realize that both Freund's adjuvant and pristane cause immunosuppression yet induce tumor formation in the peritoneal cavities of the injected animals. Without doubt, this will cause stress to the animals; repeated injection of the animals over prolonged periods and more than one harvest of ascites fluids should be avoided to minimize stress. The procedure is outlined below.

1. For pristane priming, inject 0.1 ml of pristane (2,6,10,14-tetramethylpentadecane; Aldrich Chemicals; catalog no. T2-280-2) i.p. into a BALB/c mouse and, 1 week later, inject an additional 0.2 ml of pristane i.p. (Most protocols recommend one to two 0.5-ml doses for priming; however, we find that increasing the second dose to 0.5 ml or higher did not increase the volume of ascites fluid recovered. By using two smaller doses of pristane, animals are subjected to less stress and the survival time is longer than when 0.5 ml is used.)

2. Optimally, pristane priming should not be done more than 60 days prior to the injection of the hybridoma cells.

3. One week after the administration of the second dose, 10^6 hybridoma cells (0.1 ml) can be injected i.p. into the animal. Note that a solid tumor with little or no ascites fluids production can sometimes occur, as a result of either the pristane priming conditions or the low level of secretion that can be characteristic of certain cell lines (65). (When tumors are obtained, harvest the hybridoma cells from the peritoneal cavity and suspend them in culture medium containing 20% [vol/vol] FCS; store cells frozen in liquid nitrogen or inject 10^6 cells i.p. into a mouse primed with Freund's incomplete adjuvant.)

4. Once the abdominal area of the animal begins to enlarge, it should be watched closely to avoid overextension of the peritoneal cavity and skin. When the bulging of the abdomen is deemed to be large enough or accounts for 20% of the total body weight, the animal is euthanized and ascites fluid is drained aseptically with a 20-gauge needle and syringe.

5. Remove the needle and dispense the fluid into sterile, 15-ml, conical centrifuge tubes with caps. Sediment the cells by centrifugation at 1,000 × g for 10 min.

6. Separate the fluid containing the Abs from the pellet. Note that the pellet contains the healthiest hybridoma cells and they should be resuspended in hypoxanthine-thymidine- or HAT-containing DMEM to be propagated for freezing and storage. Cell lines suspected of containing contaminating microbes can be cleaned by passage through an animal by using this protocol. The normal white blood cells from the ascites fluid will eventually die and should not pose any problem to the purity of the hybridomas. The ascites fluid is now ready to be used for Ag-Ab reactions, with or without further purification.

Many undesirable components such as lipids, adjuvants, and fibrins are included in ascites fluid collected from the animals. These impurities can clog up affinity columns and the more expensive HPLC columns; thus, they must be removed before proceeding with purification. The following methods can be used separately or in combination to achieve this goal.

1. Add $CaCl_2$ to a final concentration of ca. 1 mM (per Bio-Rad instructions for Affi-Prep protein A columns). Allow the mixture to sit for 2 h at 4°C. Use a wooden applicator stick or other suitable probe to remove the fibrin-lipid clot. Centrifuge at 10,000 × g for 10 min at 4°C, collect the clear supernatant, and filter it through a 0.22- or 0.45-μm-pore-size sterile filter. The supernatant is now ready for injection into HPLC columns.

2. Ascites fluid can also be poured through a small Sephadex G-25 column to adsorb clogging materials. The amount of Sephadex to use is 1 ml/ml of ascites fluid. The Sephadex column should be preequilibrated with 50 mM Tris, adjusted to pH 8.6 with HCl. Again, filter the eluant with a sterile filter (as described above) before subjecting it to further purification (see protein A Sepharose protocols in the Pharmacia manual).

3. Glass wool can also be used to adsorb clogging agents. Dilute ascites fluid samples 1:1 with 50 mM Tris, pH 8.6, and apply them to a small column made of either syringes or Pasteur pipettes. A small volume may be retained by the glass wool.

8.4.4. Ab Purification

Since any of the methods described in section 8.4.3 can be used to purify MAbs, this section will be devoted to describing two rapid methods, HPLC and a less expensive affinity protocol, that have been used successfully to purify MAbs of both IgG and IgM isotypes (Fig.1).

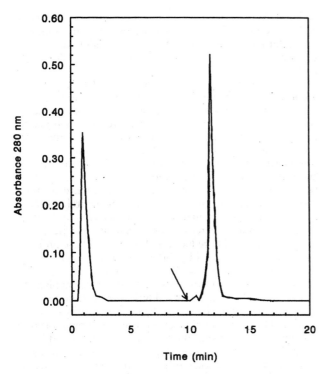

FIGURE 1 HPLC separation of IgM MAb from ascites fluid. The first peak represents unbound material, and the second peak represents pure IgM Abs eluted after the addition of elution buffer 1. The arrow indicates the time at which the elution buffer was added.

8.4.4.1. Protein A

While protein A has been used effectively to purify IgG Abs, it generally binds poorly to IgM and murine IgG1. When protein A-Sepharose affinity chromatography is used to purify IgM, antibody recovery is ca. 14 to 29% (63, 70). In recent years, many companies (e.g., Pharmacia, Bio-Rad, and Pierce) have reported that efficient binding between IgM and protein can be accomplished by using buffers of high pH and high ionic strength. Pharmacia uses 1.5 M glycine buffer, pH 8.9, containing 3 M NaCl, while other company protocols are proprietary (e.g., the Bio-Rad MAb purification system buffers). The following procedure can be used for purifying either IgM or murine IgG1 and a variation for purifying other isotypes is summarized in Table 4.

Procedure

1. Equilibrate a column (Bio-Rad Affi-Prep protein A preparative cartridge; binding capacity, ~35 mg of immunoglobulin) with binding buffer (see Table 4 for the recipe) for 12 min at 5 to 10 ml/min (15 column volumes).

Alternatively, a protein A Superose column (Pharmacia) can be used. The MAb purification system buffer from Bio-Rad could also be used; however, premade buffers are more expensive.

2. Dilute samples (1:5) with binding buffer, and filter through a 0.45- or 0.22-μm-pore-size filter if the ascites fluid has not been pretreated as described previously.

3. Load 0.2 to 1.0 ml of the diluted sample, and run it through the column at 5 to 10 ml/min. The flow rate may be reduced if the column cannot withstand the pressure of a high flow rate. All proteins that do not bind should elute within the first 5 min. To remove unbound proteins, run the column for 5 to 10 min so that the A_{280} remains at the baseline for at least 5 min.

4. Elute the immunoglobulin fraction with the first elution buffer (0.1 M citric acid buffer, pH 4, containing 0.15 M NaCl) for at least 5 min at the same flow rate as before. Repeat with the second elution buffer (0.1 M citric acid buffer, pH 3, containing 0.15 M NaCl) at 5 to 10 ml/min for 5 min.

5. Wash the column with PBS containing azide, and store and seal the column in the same buffer. Note that the eluted immunoglobulin should not be stored in the elution buffer, as the acidity may be deleterious to the IgM if the Ab is exposed to it for too long.

6. Dialyze the eluted IgM against PBS (2 4-liter volumes) overnight at 4°C.

7. Concentrate to the desired volume by placing the IgM solution into dialysis bags and sprinkling PEG flakes (molecular mass, 20 kDa) over the tubing. This process should not take any more than 30 to 60 min.

8. Assay for total protein content and analyze the sample by SDS-PAGE to ensure that only the heavy (~55 to 60 kDa)- and light (~28 kDa)-chain bands are seen (Fig. 2).

Figure 1 shows the purification of IgM MAb from ascites fluid by HPLC using this method.

8.4.4.2. DEAE Affi-Gel Blue

The DEAE Affi-Gel blue method is a one-step procedure for the isolation of IgG free from protease, nuclease, or albumin (59). Recovery can be as high as 77 to 80%. (Note that the DEAE Affi-Gel blue is not the same as the Bio-Rad Affi-Gel blue described in section 8.3.5.1.) IgG will be bound to the DEAE functional groups and therefore must be eluted.

Procedure

1. Preequilibrate a small column containing 5 to 10 ml of DEAE Affi-Gel blue gel with 20 mM Tris-HCl, pH 7.2.

2. Apply 1 to 5 ml of ascites fluid (pretreated as described above). Wash the column with 2 bed volumes of the same buffer.

3. Wash the column with 1 bed volume of buffer but now containing 25 mM NaCl to elute transferrin. Elute the

TABLE 4 Conditions for HPLC purification of various immunoglobulin isotypes

Isotype(s) of MAb	Binding buffer	Elution buffer 1	Elution buffer 2
IgM and murine IgG1	1.5 M glycine–3 M NaCl, pH 8.9	0.1 M citric acid–0.15 M NaCl, pH 4.0	0.1 M citric acid–0.15 M NaCl, pH 3.0
Murine IgG2a and IgG3	50 mM Tris-HCl, pH 8.6	0.1 M citric acid, pH 5.0	0.1 M citric acid, pH 4.0
Murine IgG2b	50 mM Tris-HCl, pH 8.6	0.1 M citric acid, pH 4.0	0.1 M citric acid, pH 3.0

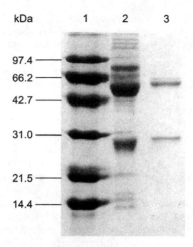

FIGURE 2 SDS-PAGE profiles of IgM MAb purification by HPLC. Lanes: 1, molecular mass standards; 2, ascites fluid before purification (note the large number of bands that are normally found in animal serum proteins or body fluids); and 3, purified IgM MAb eluted from the column (the second major peak in Fig. 1). Note the presence of only two major bands representing the heavy and light chains at apparent molecular masses of 60 and 28 kDa, respectively.

MAbs with 3 bed volumes of buffer containing 50 mM NaCl.

8.5. DETECTION OF REACTIONS

Ag-Ab reactions have wide application in all areas of biological, medical, and biochemical research where Abs are used as probes for the presence, structure, and even function of a given Ag. In many instances, there is a limited amount of Abs, Ags, or both and the economical use of these reagents becomes paramount. Furthermore, in diagnostic labs, screening systems must be easy to use routinely with a large number of samples yet be rapid, sensitive, and specific.

The classical serological methods, including immunoprecipitation, agglutination, immunodiffusion, and complement fixation, have traditionally been used in diagnostic laboratories. In recent years, these techniques have been replaced by highly sensitive and rapid immunoassay techniques such as ELISA and Western immunoblotting. The reaction mechanisms and application of both classical and modern techniques are discussed below.

8.5.1. Agglutination

Agglutination reactions are among the easiest of immunological tests to perform and evaluate and have long been used in bacterial identification and serological classification (74, 75, 76, 79, 80). Here, the Ag (generally a suspension of cells or Ag-coated particles) is mixed with the test Ab, and a positive reaction is observed as aggregation or clumping and is termed agglutination. The titer of an antiserum is the highest dilution of the serum that gives a visible aggregation reaction. If the Ag is heterologous to the Ab or if there is no Ag present, the suspensions remain unchanged.

For agglutination to occur, the Ag must have more than one epitope to allow the formation of large aggregates of cross-linked Ag-Ab complexes. Bacterial Ags such as LPS and its O Ag, carbohydrate-capsular Ags (K Ags), and flagella (H Ags), which contain repeated epitopes, and whole cells, which have large numbers of Ags, are ideal candidates for agglutination reactions. Ags used in agglutination reactions can be particulate (e.g., whole cells and Ag-coated particles; see section 8.2.1).

Diagnostic polyclonal antiserum should be extensively adsorbed with a bacterial strain deficient in the target Ag, e.g., a nonflagellated, unencapsulated, or O Ag-deficient strain (see section 8.3.1). Adsorption reduces nonspecific, cross-reactive activity and increases the specificity of the serum. The adsorbed polyclonal antiserum contains a mixture of Abs of different affinities to different determinants, and the combined activities allow detection of an organism bearing these Ags. The major problems encountered in the use of polyclonal antisera include batch-to-batch variation in Ab quality and residual nonspecific activity. In addition, while adsorption increases the specificity of the serum, there is considerable reduction of Ab titer. These problems are virtually eliminated by the use of highly specific MAbs in serotyping with exposed specific epitopes.

Despite the high affinities of MAbs, their application can be limited due to the fixed affinities of these reagents. This problem can be overcome by using mixtures of specific MAbs. They are invaluable in the serotyping of nontypeable and polyagglutinable bacterial isolates and in the identification of organisms expressing a specific antigenic determinant. The choice of polyclonal versus monoclonal antisera depends on the problem under investigation.

Bacterial agglutination techniques are routinely used for the detection of Abs to bacterial surface Ags, notably LPS, cell walls, and K and H Ags. Serological classification of bacteria is based on selective agglutination of the bacterial cells by Abs to these Ags (74, 76, 79, 80).

8.5.1.1. Qualitative Slide Agglutination

Slide agglutination assays can be performed on a slide, in tubes, or on microtiter plates.

1. Suspend a loopful of bacterial cells (or other particulate Ags) in 1.5 ml of normal saline (0.85% [wt/vol] NaCl).

2. Mix 25 μl or a loopful of the bacterial suspension with an equal volume of the diluted test antiserum on a glass slide.

3. Use normal cell suspensions mixed with normal saline or nonimmune serum as negative controls to detect the nonspecific agglutination of the cells.

4. Aggregation should be visible in about 1 min by eye or by low-power phase-contrast microscopy (chapter 1).

Note: (i) Do not use bacterial cells that autoagglutinate in normal saline or water in the absence of an Ab in these reactions. Minor autoagglutination can be overcome by using 0.1% (wt/vol) skim milk to block adherence or by suspending the cells in a 1:100 dilution of normal serum. These reagents, when used, should also be used in the controls. (ii) LPS O Ag-deficient strains may agglutinate with more than one serotype-specific serum. (iii) Heat serum at 56°C for 30 min to inactivate complement before testing for agglutination. (iii) Dilute serum to avoid a prozone effect, i.e., failure to agglutinate at high Ab concentrations.

8.5.1.2. Microtiter Plate Agglutination

1. Plate agglutination allows simultaneous testing of one or more antisera against one or more particulate Ags. Make serial doubling dilutions of the test serum or sera in normal saline or PBS, pH 7.4 (see section 8.5.4), containing 150 mM NaCl. To each serum dilution, add 0.005% (wt/vol)

Safranin O dye, making it easier to read the results. (Agglutinated cells appear as a pink mat.) Similarly, prepare a series of dilutions of nonimmune control serum, that is, serum obtained from a nonimmunized animal.

2. Dispense 25-μl volumes of each dilution of the test and control sera into two wells (duplicate series) of a 96-well microtiter plate.

3. Adjust the bacterial suspension to a concentration of 10^9 by adjusting it to an OD_{600} of 0.6 or by comparing it to the McFarlands standards (14).

4. Add equal volumes of the particulate Ag suspensions to the wells containing the sera. Gently rock the plate to mix the contents. Cover the plate with plastic food wrap to prevent evaporation.

5. Incubate the plates at room temperature for 50 to 60 min. Incubations at low temperatures may give indeterminate results or nonspecific clumping.

Record and compare the times of visible agglutination for different sera and particulate Ags. Record agglutination as ++++ to + for complete to slight agglutination and − for no agglutination. Express agglutination titers as the highest dilution of serum that gives a visible agglutination of cells.

8.5.1.3. Passive Agglutination

In the passive agglutination technique, specific Ags (or Abs) are adsorbed or chemically immobilized on erythrocytes (section 8.2.1), *Staphylococcus aureus* cells (77), or latex beads. The sensitized particles are then mixed with the test antisera (slide or microtiter assays), and agglutination is observed. The large size of the Ag- or Ab-coated particles enhances the visibility of the Ag-Ab reaction. This technique allows the use of extracted and soluble Ags rather than whole cells (which contain complex Ags) in the detection of homologous Abs by agglutination. Inhibition of passive agglutination by the addition of exogenous purified Ag to the reaction mixture can be used to confirm the specificity of the reaction. The *S. aureus* Cowan I strain (ATCC 12598) produces large amounts of cell wall-bound protein A that has high affinity for the Fc portion of IgG Abs from a range of animals (69). The immobilized Abs retain their Ag recognition properties. The IgG-coated cells are mixed with the serum or Ag preparation to detect the presence of the specific Ags (77).

Latex beads have several advantages over erythrocytes and *S. aureus* cells in passive agglutination assays. They are immunologically and chemically inert, can be stored for long periods without damage, have a uniform size, and are available in different colors. A mixture of different colored beads, each color sensitized by a specific Ag or Ab, can be used to detect the presence of different Abs (or Ags) in the same sample in an economical, time-saving manner (76, 78). Ags can be passively adsorbed or chemically coupled to surface-modified latex beads. Chemical coupling is described in detail by McIllmurray and Moody (79). The procedure described below is for passive agglutination of protein Ags to latex beads (Difco, Detroit, MI).

Procedure

1. Wash 1 ml of 10% (wt/vol) latex beads in 40 ml of the glycine-saline buffer (glycine, 14.0 g; NaCl, 17.0 g; NaOH, 0.7 g; sodium azide, 1.0 g; dissolve in 800 ml of distilled water and adjust to pH 8.2 with 1 M NaOH; make up to 1 liter). Centrifuge at 12,000 × *g* for 15 min, and repeat the wash.

2. Resuspend beads in 20 ml of the buffer. Resuspend the proteins in the buffer, and add 300 to 500 μg of protein to the beads. Place the flask on a rotary shaker at low speed, and mix the suspension at room temperature for 30 min.

3. Wash three times in 20 ml of the buffer.

4. Resuspend in 20 ml of a buffer containing 0.1% (wt/vol) skim milk or 1% (wt/vol) BSA (to block all remaining protein-binding sites) and incubate for 30 min. Centrifuge the beads, and resuspend in the above-described buffer containing 0.1% (wt/vol) sodium azide. At this point, the suspension should show no spontaneous agglutination of the sensitized particles. If clumping occurs, it may indicate that excess Ag was used in the coupling reaction and the process should be repeated using a lower Ag concentration. Store the suspension at 4°C in acid-clean glass or plastic containers having minimal protein-binding capacities.

5. Mix 25 μl of the activated latex beads and the test sample. Rock the mixture gently, and observe agglutination against a dark background.

8.5.2. Quellung Reaction

Encapsulated bacteria can rapidly be identified by the capsule-swelling Quellung reaction (also see chapter 2). In this assay, a drop of bacteria is mixed with a drop of antiserum or capsule-specific Ab and then mixed with an equal amount of methylene blue dye and the cells are observed with a light microscope. The binding of Ab to the capsule appears to increase the capsular volume possibly by stabilizing the structure. The fixed capsules are more visible and refractile and are easily identified because they show a sharply stained margin. This reaction is used in the identification of encapsulated *Streptococcus pneumoniae*, *Haemophilus influenzae* type b, and *Meningococcus* spp. strains (81).

8.5.3. Precipitation

The interaction of a multivalent *soluble* Ag with the homologous divalent Ab results in the formation of a cross-linked lattice of the two components. If the Ag has more than one Ab-binding site and there are sufficient quantities of both reagents to attain optimal proportions, the lattice is visible as a precipitate. Precipitation, however, occurs only at a narrow range of Ab and Ag concentrations, called the equivalence zone; in the excess of either component, the precipitate dissolves. The rate at which precipitation occurs is dependent on the ratio of Ab to Ag and the avidity of the interaction. Although precipitation reactions can be performed by using a solution (ring test), the most commonly used procedure is precipitation in an agar gel and is called immunodiffusion, which eliminates the necessity of finding the optimal proportions of Ag and Ab.

8.5.3.1. Radial Single and Ouchterlony Double Immunodiffusion

The detection of Ag-Ab complexes was made easier by the introduction of agar and agarose as diffusion matrices. These substances are chemically inert and do not influence the free diffusion of molecules. At low concentrations, agar and agarose form networks of large pores, allowing free movement of Ags and Abs. The rate of diffusion of each is a function of its M_r and structure. Molecules with similar structural properties will diffuse at similar rates and in this way form a concentrated front.

In radial single immunodiffusion, Ags diffuse from a well into an Ab-containing gel; precipitin rings are formed in the gel around the Ag well.

In the classical Ouchterlony double immunodiffusion method (84), Ag-containing wells are placed around a central

Ab-containing well and the components diffuse radially into the gel. Ab-Ag complexes form in the agar zone between the wells, and precipitation occurs at the point at which concentrations of the reagents attain equivalence. The diffusion of complex Ags against homologous polyclonal antiserum will result in the formation of a complex pattern of precipitin lines in the gel. If two adjacent wells containing a common Ag are tested against the homologous Ab, the precipitin line between the Ab and the Ag wells is continuous, representing a reaction of identity, (Fig. 3B, wells 1 and 3). Reactions of nonidentity are seen as precipitin lines that cross each other and represent Ab reactions to distinct Ags (Fig. 3C, wells 1 and 2). A line of partial identity is formed between two Ag wells that contain a related (continuous precipitin) but not identical (spur) Ag (Fig. 3A and C, wells 2 and 3). The spur points towards the Ag well possessing the distinct Ag determinants. The position of the precipitin lines is determined by the local concentration of the reactants in the gel and is a function of the original concentration in the well and of the diffusion rates of the reactants.

Double Immunodiffusion Procedure

Barbitone buffer (pH 8.2; ionic strength, 0.08)

Place 12 g of sodium barbital (5′5-diethylbarbituric acid–Na salt [Fisher Scientific]) in 800 ml of distilled water, and mix with 4.40 g of barbital (5′5-diethylbarbituric acid [Fisher Scientific]) in 150 ml of distilled water at 95°C. Adjust to pH 8.2 with 5 M NaOH, and bring to a final volume of 1 liter. Since sodium barbital and barbital are classified (barbiturate) substances, a license is required for the purchase of these reagents.

Gel preparation

The precoating gel is made by adding 0.5% (wt/vol) of agarose to distilled water and boiling to dissolve. To make the resolving gel, dissolve 2 g of electrophoresis-grade agarose (SeaKem; M_r, −0.16 to −0.19) in 50 ml of barbitone buffer and boil to dissolve the agarose. Add 50 ml of the buffer to make a 1% (wt/vol) buffered agarose solution. Store at 4°C for up to 4 weeks.

Making templates

Immunodiffusion assays can be performed on glass plates (such as a microscope slide) or on gel bond film (Mandel Scientific, Guelph, Canada).

1. Place the glass slide or gel bond film on a leveling table.

2. Pour a thin film of the precoating gel onto the plate or onto the hydrophilic side of the gel bond film (a drop of water will spread on this side). Let dry.

3. Pour 0.18 ml of the barbitone-buffered agarose/cm² onto the precoated plate or film to form the immunodiffusion template. Allow the agar to set.

4. Punch the required pattern of wells (each 20 mm in diameter and 1 cm apart) into the agar with a sharp gel punch.

5. Pipette 5 to 10 μl of the sample into the wells. Do not overfill the wells.

6. Develop the plates in a humid box for 12 to 48 h, and observe precipitin lines against a black surface. To enhance their visibility, soak the slides in 4% (wt/vol) tannic acid in saline.

Pressing, washing, and staining slides

1. Place the slide on a filter paper. Fill the wells with distilled water.

2. Press for 15 min by placing the slide under several pieces of Whatman no.1 filter paper, a glass plate, and several textbooks for weight.

3. Soak the plate in saline for 20 min. Repeat the washing and pressing process twice.

4. Soak the plate in distilled water for 20 min. Air dry the template.

5. Stain by placing the plate in Coomassie blue solution (see the recipe below) for 15 min.

6. Destain with destaining solution (as described below without the Coomassie dye). The staining is reversible, and it is possible to completely destain the gel. If this occurs, restain the gel.

7. Air dry the stained gel.

Staining solution

Coomassie brilliant blue	0.5 g
95% Ethanol	45 ml
Glacial acetic acid	45 ml
Distilled water	10 ml

Place the solution on a magnetic stirrer, and dissolve the Coomassie overnight before use. The stain is reusable.

8.5.3.2. Rocket Immunoelectrophoresis and CIE

In the rocket immunoelectrophoresis and counterimmunoelectrophoresis (CIE) techniques, immunoprecipitation in gels is accelerated by the active movement of Ags and Abs towards each other under the influence of an electrical field and sensitivity is increased by eliminating the dilution of reactants that occurs in more-traditional immunodiffusion assays. Rocket immunoelectrophoresis and CIE are analogous to the single radial and double Ouchterlony immunodiffusion assays, respectively.

In rocket immunoelectrophoresis the Ags are placed in wells at the cathode side of the gel and then electrophoresed into the Ab-containing agarose gel (83). Immunopre-

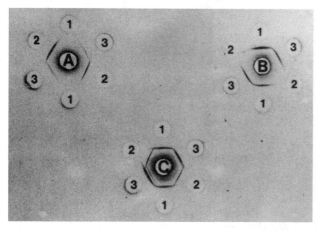

FIGURE 3 Double immunodiffusion profiles of antisera to proteases of *Aeromonas hydrophila*. Each well was filled with 10 μl of either an Ag or an antiserum preparation. Wells: A, rabbit antiserum to protease A; B, rabbit antiserum to protease B; C, a 1:1 (vol/vol) mixture of antiserum to protease A and antiserum to protease B, respectively. The Ags were placed into the outer wells. Well 1, protease B from strain 1; well 2, protease A from strain 2; well 3, extracellular products of strain 3. (Courtesy of K. Y. Leung and R. M. W. Stevenson.)

cipitates are observed as peaks radiating into the gel. The area below each precipitin peak is a function of the Ag concentration and is inversely proportional to the Ab concentration in the gel.

In CIE (82, 85), Ags and Abs are placed into paired wells (3-mm diameter) cut 1 cm apart in agar-coated slides. The Ags are placed in wells at the cathode side while the Abs are on the anode side. The gels, Ags, and Abs are all in barbital-Tris, pH 8.6, buffer. At that pH, most bacterial Ags have an overall negative charge because of ionized carboxylate groups and will migrate to the anode during electrophoresis. Abs are usually neutral or weakly charged and will drift (together with the water molecules) towards the cathode by electroendosmosis. The Abs and Ags, therefore, migrate towards each other in a concentrated fashion. Precipitation is detected in 20 to 60 min. CIE of complex Ag mixtures against homologous polyclonal antisera results in a complex pattern of precipitin lines.

Procedure for CIE

1. *Equipment.* Flat-bed electrophoresis chamber (water cooled if available) and power supply
2. *Buffer.* Barbituric acid-Tris (pH 8.6). This buffer is used to make the gels and to dissolve the Ags and Abs and is used for the cathode and anode wells.

Barbituric acid (0.02 M) . . . . 4.48 g/liter
Tris base (0.07 M) . . . . 8.86 g/liter
Calcium lactate (0.03 M) . . . . 0.108 g/liter

3. *Gel.* One percent (wt/vol) SeaKem ME agarose (M_r, −0.16 to −0.19). Dissolve the agar in the above-described buffer by boiling. Store at 4°C. Heat to redissolve, and keep at 56°C for use. (Warm antiserum to 56°C before addition to the agar for rocket immunoelectrophoresis.)
4. *Casting gels.* The gels are cast on the hydrophilic side of gel bond film. Use 0.18 ml of agarose/cm² of film. Punch paired Ag and Ab wells into the agar. Place the gel in the electrophoresis chamber, and load 2 to 4 μl of sample/well.
5. *Electrophoresis.* Perform the electrophoresis at 10 V/cm. Use five layers of Whatman no. 1 paper or J-cloth strips as wicks to connect the gel to the buffer during electrophoresis. Boil the strips in water before use to remove impurities.

8.5.3.3. XIE

Crossed (two-dimensional) immunoelectrophoresis (XIE) is a powerful and highly sensitive analytical technique that combines the separation of complex Ags by electrophoresis with immunoelectrophoresis (86). XIE has been used to investigate the immunogenicity of bacterial cell lysates and membrane Ags. Fig. 4 shows XIE of *Pseudomonas aeruginosa* cell lysates against homologous polyclonal rabbit serum.

In the first dimension, Ags are electrophoretically separated into zones on agarose or polyacrylamide gels on the basis of M_r. Gel strips containing the separated Ags are then placed along the cathode side of a rectangular slab of agarose gel containing the Abs; this is the second-dimension gel. An intermediate gel with or without additives is cast between the Ag- and Ab-containing gels. This gel can be used to stack the Ags or identify specific Ags. The Ags are then electrophoresed perpendicularly into the second-dimension gel, and upon interaction of the Ags with the homologous Abs, immunoprecipitates are formed in the second-dimension gel. A complex pattern of overlapping peaks is obtained when several Ags are electrophoresed into a polyclonal Ab-containing gel. The area below each pre-

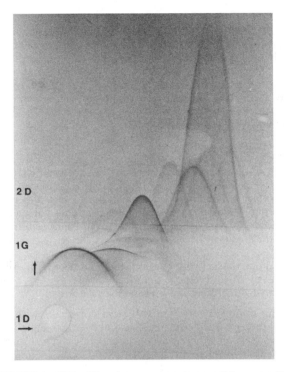

FIGURE 4 XIE of *Pseudomonas aeruginosa* cell lysate and homologous polyclonal antiserum. A 10-μl portion of the lysate was loaded in the first-dimension gel (1D), the intermediate gel (IG) contained NaCl, and the second-dimension gel (2D) contained antiserum.

cipitin peak is a function of Ag concentration and is inversely proportional to the Ab concentration in the second-dimension gel. The positions of specific Ags can be defined by incorporation of a monoclonal or specific Ab into the intermediate gel. The Ag will react with the Ab and form a precipitin in the intermediate gel.

The direct applications of XIE to investigate Ag structure as well as immunogenicity are exemplified by crossed affinoimmunoelectrophoresis (XAIE). In this form of XIE, a lectin such as concanavalin A is incorporated into the intermediate gel. The lectin will bind to and precipitate the respective glycosylated Ag as the Ags are electrophoresed from the first-dimension gel through the separating gel. The precipitin peak corresponding to that Ag will be identified by comparison of lectin-containing and lectin-deficient gels. The Agic and structural properties of immunogens can, therefore, be investigated by XAIE (86).

XAIE Procedure

Dissolve 1% (wt/vol) SeaKem agarose in barbituric-acid-Tris buffer (pH 8.6), and keep the agar at 56°C. Warm all sera or Ab solutions to 56°C before use.

First-dimension

1. Place a glass plate (10 by 10 cm) on a leveling table. Pour 18 ml of the 1% (wt/vol) barbituric-acid-Tris-buffered agarose onto the plate. Allow it to set.
2. Punch 10 sample wells (2-mm diameter) in the gel, and remove the gel plugs. The wells should be 1 cm apart and 2 cm from the edge of the plate.
3. Place the gel in the flat-bed electrophoresis chamber. Connect the gel to the buffer in the electrode wells with filter paper wicks. *Do not turn on the power source.*

4. Pipette 4-μl aliquots of the Ags into nine wells. Place 2 μl of a tracking dye (0.05% bromophenol blue)-indicator protein (0.1% BSA) mixture in the last well. Do not overfill the wells. Close the lid, and apply voltage.

5. Electrophorese samples until the tracking dye has moved 2.8 cm away from the well. Turn off the power supply.

6. Using a long, sharp blade, cut the sample lanes and trim to 1-by-6-cm strips (well to tracking dye front). This is the first-dimension gel.

Intermediate gel

1. Lift the first-dimension gel strips with the blade, and place on one end (on the *hydrophilic* face) of a gel bond film (5 by 6 cm). The films should be on a leveling table.

2. Place a metal bar 1 cm away from the first-dimension gel strip. This will be the space for the intermediate gel.

3. Add 0.1 ml of the prewarmed nonimmune serum into 1.5-ml 1% agarose at 56°C. Gently mix the gel, avoiding air bubbles.

4. Pour the gel into the intermediate gel space. Allow the gel to set, and remove the metal bar. This gives a continous first-dimension and intermediate gel strip.

The nonimmune serum will form complexes with and absorb nonspecific activities, thus improving the quality of the gel. Specific Abs, Ags, lectins, or saline can be incorporated into this gel, depending on the aim of the experiment.

Second-dimension gel

1. Add 0.5 ml of the test serum into 3.5 ml of the 1% (wt/vol) agarose at 56°C. Mix well by swirling the gel in the water bath, avoiding air bubbles.

2. Pour the gel onto the rest of the gel bond film, and allow it to set.

3. Place the gel in the electrophoresis unit with the first-dimension gel on the cathode side. Connect the wicks, close the lid, and electrophorese at 3 V/cm for 16 to 18 h.

4. Remove the plate and proceed to press, wash, and stain as described in section 8.5.3.1.

8.5.4. ELISA

The design and development of ELISA (89) provided a major revolution in diagnostic immunology. The application and versatility of ELISA were greatly facilitated by the design of the microtiter (96-well) plate. ELISA has become the most widely used immunoassay technique, with applications in routine diagnostic serology and clinical laboratories and in virtually all areas of biological sciences where Abs are used as research tools. In all immunoassays (including ELISA and Western and dot immunoblotting), the Ag-Ab reactions are monitored by enzyme activities rather than direct observations of immune complexes. The popularity of ELISA is due to its high sensitivity and specificity at low cost, its amplification systems allowing the detection of nanogram quantities, its adaptability to automation, its elimination of hazardous reagents, its use of color (chromogenic, bioluminescent, or fluorescent) reactions that can be evaluated visually, and its ability to simultaneously screen many samples in the same microtiter plate with ~50 μl in each well.

The two common requirements for all immunoassay procedures are (i) the effective immobilization of the Ag or Ab onto a suitable solid matrix without the loss of function, the adsorbed reagent then being used to capture the Ab or Ag from the test sample, and (ii) the availability of suitable reaction indicator or reporter systems for the detection of the bound Ag-Ab complexes. Immunoassays depend on the fact that enzymes or chemically active groups can be covalently linked to immunoglobulins and the resultant complex retains the former immunological or chemical activity. The high stability, specificity, and activity of the enzyme complexes are crucial to the high sensitivity seen in the immunoassay techniques.

8.5.4.1. Requirements

The sensitivity, reproducibility, and specificity of the ELISA depends on three major factors: (i) optimal concentration and stability of the immobilized or coating reagent, usually the Ag; (ii) reduction of nonspecific activities by blocking reagents; and (iii) choice of indicator system. These factors are discussed below because of their importance to the design and success of the assay. The reviews by Nakamura et al. (98), Tijseen (103), and Voller and Bidwell (104) will provide more detail.

8.5.4.1.1. Optimum Concentration of Ag Required To Coat the Matrix

The actual amount of Ag adsorbed to the plate is crucial to assay sensitivity. Generally, the optimal concentration range for Ags in the coating solution is 0.2 to 10 μg of purified Ags/ml, whereas complex mixtures and cell lysates require about 5 to 25 μg/ml. Too much Ag in the coating buffer results in protein complexes in solution which then bind to the matrix in "stacks" (rather than as layers) that are removed in subsequent washes, resulting in insufficient Ag coating. Too little Ag leads to the selective binding of high-affinity Abs so that low-affinity Abs are excluded from the assay, resulting in highly inaccurate results when hyperimmune serum and early-infection sera are compared. Consequently, several methods have been described previously for the estimation of the optimal Ag concentrations required to obtain the highest sensitivity reproducibly (97). The aim is to achieve an Ag concentration that is slightly below that needed to saturate all Ag-binding sites on the matrix. It is assumed that this concentration will give a one-layer distribution of the Ag on the matrix and allow optimal interaction with the Abs in subsequent steps.

8.5.4.1.2. Stability of the Ag

Ag adsorption is influenced by the binding capacity of the matrix and the stability of the bound Ag. The sensitivity of the ELISA is influenced by Ag concentrations, and therefore the matrix should have high adsorptive capacity without denaturing the Ag. There should also be minimal loss of Ag after adsorption.

Binding of Ags to microtiter plates is due mainly to hydrophobic interactions rather than size or ionic or charge interactions (87). Thus, the buffer systems used in the adsorption step reflect the chemical and structural properties of the Ag that determine its solubility rather than the effect of the buffer in Ag binding. For example, bicarbonate (pH 9.6) and phosphate (pH 7.2) buffers or water has been used in the adsorption of soluble Ags. Inorganic solvents (93, 94, 100), detergents (96, 100), and glutaraldehyde (105) have been used for the adsorption of lipidic, glycosylated, and integral membrane proteins. Detergents are reported to reduce the binding of some Ags to ELISA plates and should be avoided where possible (91). Integral membrane components, however, require detergents in the coating buffer at concentrations above the critical micellar concentrations for solubility. Several protocols have been used to enhance the immobilization of insoluble, complex, or small Ags on solid matrices. Some of these procedures include the use of polyvinyl chloride plates derivatized to contain aldehyde groups for glycoprotein adsorption (96) and avidin-coated plates to bind biotinylated peptides (90) and pretreatment of polystyrene tubes with 0.2% (vol/vol) glutaraldehyde for adsorption of small peptides (105). Improved Ag adsorption can be obtained following Ag denaturation, which may expose hydrophobic regions (88). $MgCl_2$ in the coating buffer enhances the adsorption of LPS (92), whereas preincubation of rough LPS with polymyxin increases its binding to microtiter plates (2, 102). In a novel procedure, HgCl is adsorbed onto plates precoated with BSA-glutathione for determination of the titers of HgCl-specific Abs (106).

The 96-well microtiter plates used in ELISA are made of polyvinyl chloride and polystyrene and are obtained from various commercial suppliers. Unlike other microtiter plates used for tissue culture or protein assays, the well surfaces of the ELISA plates are modified for high levels of protein binding. Improved binding of a specific Ag can be obtained by specific treatment of the wells before or after Ag adsorption.

8.5.4.1.3. Blocking Reagents

All immunoassays require blocking reagents to saturate the excess binding sites on the matrix, thereby preventing nonspecific adsorption of reactants and reducing background values. The best blocking reagent should have no affinity for the test Ags, the immunoglobulins, or the indicator system. Examples of blocking reagents include 0.1% (vol/vol) Tween 80, 3 to 10% (wt/vol) skim milk, 2 to 10% (vol/vol) fetal bovine serum, 1 to 3% (wt/vol) gelatin, 1 to 5% (wt/vol) BSA, and various combinations of these reagents. We have found that fetal bovine serum increases background levels when used as the blocking reagent due to the nonspecific binding of Abs or Ags to this reagent. Gelatin at a concentration above 0.5% may solidify at room temperature and will block the ELISA plate-washing apparatus; 5% skim milk is recommended for blocking, and 2% is recommended for all other steps. In laboratories that routinely use immunoassays, the cost of the blocking reagent could be an important consideration, and skim milk is inexpensive!

8.5.4.1.4. Indicator System

The sensitivity of the immunoassays is vastly increased by using labeled indicator systems that allow the detection of extremely low concentrations of immunoreactants. There are four main groups of indicator conjugates used in immunoassays, and a list of these follows.

1. *Immunoglobulin-enzyme conjugates.* The success of immunoglobulin conjugates depends on the bifunctional nature of the immunoglobulin molecule and the high stability and activity of the enzymes following conjugation. The enzymes are covalently conjugated to the immunoglobulin molecule without affecting Ag recognition and the binding properties of the Ab. The most commonly used conjugates are alkaline phosphatase (AP) and horseradish peroxidase (HRPO) because of their stability and low cost. Others include β-galactosidase, glucose oxidase, glucoamylase, carbonic anhydrase, and acetylcholinesterase (99, 103). The use of fluorescent probes and fluorometers for ELISA has also been described previously (99). The standard ELISA protocols use chromogenic substrates and calorimetric detection for quantification of Ag-Ab complexes. Recently, chemiluminescence for detection has become more widespread due to a demand for increased sensitivity and the quantification of compounds whose values fall below the

detection limit of the standard ELISA (95, 101, 107). The most commonly used immunoglobulin conjugates are the anti-Abs (second or reporter Abs). For example, goat anti-mouse IgG or IgM [or respective $F(ab')_2$] enzyme-conjugated Abs are used in all assays where the test Ab is from a mouse. The use of anti-species Abs eliminates the need to purify or label the test Ab for each assay. A large variety of highly or affinity-purified MAbs and high-affinity, enzyme-labeled anti-species (and anti-isotype) Abs are available commercially, e.g., from Cedarlane Laboratory (Hornby, Canada), Bio/CAN Scientific (Mississauga, Canada), Sigma, and Boehringer Mannheim.

2. *Ligand-receptor complexes.* Ligand-receptor complex systems make use of the specific and extremely high affinity of avidin (an egg white glycoprotein) or streptavidin complexes (from *Streptomyces avidii*) for biotin (a small, water-soluble vitamin). Each single molecule of avidin can bind four molecules of biotin with an affinity of 10^{-15}/M. The biotin-avidin complex is stable in the presence of high pH, proteolytic enzymes, and organic solvents. The test (or second) Ab [or respective $F(ab')_2$ fragments] is covalently labeled with biotin in a simple procedure (see section 8.5.5). The streptavidin complexes or streptavidin-biotin complexes conjugated to enzymes (HRPO or AP) are then used as the indicator (90). Streptavidin complexes are preferred over avidin because they have low background labeling due to a pI close to pH 7 and hence little nonspecific interaction with charged groups or glycoproteins. Also, streptavidin is not glycosylated and will not bind to tissue or cell lectins. In addition, the binding of one streptavidin–biotinylated-HRPO complex to a single biotin-labeled second Ab (or biotinylated Ag) greatly amplifies the ensuing positive reaction. Amplification obtained by this detection system allows the detection of very low levels of Ag-Ab activity, thereby increasing the sensitivity of the assay severalfold.

3. *Protein A conjugates.* Protein A is obtained by purification from *S. aureus* cell walls and has a well-known affinity for the Fc domain of IgG classes and subclasses from a diverse group of animals (69). The main disadvantage of protein A in immunoassays is that the binding to Abs is sensitive to changes in ionic strength and pH. Protein A may also show nonspecific binding to some bacterial Ags.

4. *Lectins.* Lectins are proteins that recognize and bind to specific carbohydrate groups. Biotinylated or enzyme-conjugated lectins, such as biotinylated concanavalin A, can be used to detect the binding of glycosylated moieties by immobilized Abs in Ag capture ELISA. These lectin conjugates can also be used to directly detect glycosylated molecules on Western immunoblots (110).

8.5.4.2. System Classification

The ELISA systems can be placed into two main classes based on the immobilized or matrix-bound component and the Ab on which the indicator molecule is localized.

1. *The immobilized or matrix-bound component.* If the Ag is adsorbed to the matrix, the ELISA is used to screen test sera for the presence of Abs that react with the immobilized Ag. This procedure is called Ag capture ELISA, and it has wide application in, e.g., the detection of Abs in sera from patients or animals that may have been immunized (vaccinated) with a specific Ag or that are suspected of having been exposed to or infected with a pathogen. If an Ab of known specificity is adsorbed the matrix, ELISA is used to screen test samples for the presence of Ags that react with

the immobilized Ab. This is also called Ag capture ELISA and has application in the diagnosis of ongoing infections in which the immobilized Ab is used to capture circulating Ags in serum. Another application is in the detection of specific contaminants in test samples, e.g., microorganisms or metals in water.

2. *The Ab on which the indicator molecule is localized.* In direct ELISA, the indicator molecule (enzyme, fluorochrome, or biotin) is directly conjugated to the first or primary Ab; i.e., the Ag-specific Ab is labeled with the indicator molecule. In these assays, the labeled primary Ab is added to plates coated with the test Ag. If the Ab recognizes determinants on the test Ag, the Ab will remain bound on the plates. The bound Ab is detected by the addition of the specific substrate (enzyme-labeled Ab) or can be visualized by fluorescence microscopy (fluorochrome-labeled Ab). Direct ELISA is a simple, one-step procedure. There are a few disadvantages to using direct ELISA: (i) every test Ab has to be labeled, and (ii) the sensitivity of the assay depends on how many Ab molecules are bound (or on the number of specific determinants) on the Ag.

In the indirect ELISA, the primary Ab is not labeled. The detector molecules are conjugated onto a second Ab, which is often an anti-species Ab. If the primary Ab is from a rabbit, the second Ab can be a goat or sheep anti-rabbit Ab. Indirect ELISA has several advantages. (i) The second Ab can be directed towards an immunoglobulin's Fc domain (anti-immunoglobulin class or isotype) or even a number of pooled immunoglobulins (anti-all isotypes). (ii) A single labeled anti-species second Ab can be used for all assays in which the primary Abs are from the same animal (e.g., an enzyme-labeled goat anti-rabbit immunoglobulin Ab can be used in all assays in which the primary Ab is generated in rabbits, regardless of the immunoglobulin class and Ag specificity of the primary Ab). (iii) Labeled anti-species Abs are available form many commercial suppliers (e.g., Sigma, BIO/CAN, Bio-Rad, Cedarlane, Amersham, and Pierce, etc.). (iv) The use of a second Ab increases the sensitivity of the assay; i.e., more than one second Ab molecule can potentially bind to the primary Ab (which is bound to the Ag), thereby amplifying the detection response.

8.5.4.3. Procedure

The following procedure describes an indirect, "checkerboard" ELISA that can be used to determine the optimal working concentrations of Ag and Ab. The optimal Ag concentration can then be used in assays to screen hybridoma supernatants or serum samples for specific Ab activity.

Reagents for ELISA

1. *Coating buffer.* 0.05 M carbonate-bicarbonate buffer, pH 9.6

> Solution A . . . . 21.2 g of Na_2CO_3/liter
> Solution B . . . . 16.8 g of $NaHCO_3$/liter

To prepare the working buffer, mix together 20 ml of solution A and 42.5 ml of solution B and adjust to 250 ml with distilled water. The pH should be 9.6. Add 0.05 g of NaN_3. Store at 4°C in a dark bottle for up to 4 weeks. Note that detergents or salts may be added to this buffer depending on the properties of the Ag. In addition, other buffers and solutions may be used as coating buffers.

2. *PBS*, pH 7.4

> Solution A . . . . 31.2 g of $NaH_2PO_4 \cdot 2H_2O$/liter
> Solution B . . . . 28.39 g of Na_2HPO_4 or 71.7 g of $Na_2HPO_4.H_2O$/liter

The working buffer is prepared as follows. Mix together 47.50 ml of solution A and 202.50 ml of solution B. Adjust to 800 ml with distilled water, add 8.75 g of NaCl, and make up to 1 liter. The pH should be 7.4.

3. *Wash buffer.* PBS containing 0.05% (vol/vol) Tween 20

4. *Blocking reagent.* PBS containing 5% (wt/vol) skim milk (Difco). This should be made fresh every day.

5. *Reagents for AP conjugates*

Substrate buffer. Sodium carbonate (0.05 M; pH 9.8) containing 10 mM $MgCl_2$. To prepare the substrate buffer, use the buffer stocks made for the coating buffer (step 1). Mix 27.5 ml of solution A with 35 ml of solution B, and make up to 250 ml with distilled water. Add 0.05 g of $MgCl_2 \cdot 6H_2O$. Check the pH before use. Store in a dark bottle at 4°C. Alternatively, use 1 M diethanolamine buffer (pH 9.8) containing 0.5 mM $MgCl_2$.

AP substrate. Dissolve 1 mg of *p*-nitrophenyl phosphate/ml in the buffer. The substrate can be obtained from Sigma in the form of 5-mg tablets. Store the substrate at −20°C.

Read the yellow color at A_{400}. Color development can be stopped by the addition of 10 μl of 3 M NaOH to the wells.

6. *Reagents for HRPO*

Substrate buffer. Citric acid buffer, pH 4.0, is made by dissolving 0.2 g of citric acid in 90 ml of distilled water. Adjust to pH 4.0 with 1 M NaOH. Store at room temperature.

HRPO substrate. Immediately before use, add to 10 ml of the citric acid buffer, 5 μl of 30% (vol/vol) H_2O_2 and 75 μl of 10-mg/ml ABTS [2,2 azino-bis(ethylbenzthiazoline)-6-sulfonic acid; Sigma].

Read the green color at A_{414}. Color development can be stopped by the addition of 0.08 M NaF to the wells.

Protocol for ELISA

1. *Ag coating or adsorption.* Make several dilutions of the Ag in the coating buffer at concentrations between 10 and 100 μg/ml for complex Ag mixtures or between 0.2 and 10 μg/ml for pure Ags. Mix very well. Distribute in triplicates in the vertical rows, 50 to 100 μl of each Ag dilution per well. Cover the plates with plastic food wrap, and incubate at 4°C for 16 h. Adsorption of Ags to the matrix increases with exposure time. Add appropriate protease inhibitors to the coating buffer when proteases are suspected. Alternatively, if the Ab reacts with denatured Ags, the sample can be boiled prior to adsorption.

2. *Washing.* Shake out the solution from the wells by using a strong flick of the wrist. Wash the plates by directing a jet of wash buffer from a wash bottle into each well. Wash three times with PBS-Tween, incubating 5 min each time. Rinse with PBS or distilled water. Shake out the buffer between each wash.

3. *Blocking.* Add 200 μl of 5% skim milk to each well. Incubate for 2 h at 37°C or 4 h at room temperature. Rinse three times with PBS or distilled water without incubation. (If not used immediately, allow plates to air dry at room temperature, wrap in food wrap, and store at −20°C.)

4. *Test Ab*. Prepare dilutions of the test Ab in 2% skim milk, and add them to the Ag-coated wells (volume equal to that of the Ag). Generally, use primary immune sera at a dilution of 1:10 and high-titer Ab sera at dilutions of 1:100 to 1:1,500 or higher. Aliquot the serum in horizontal rows such that each serum dilution is tested against all Ag dilutions, i.e., checkerboard titration. Incubate as in step 2 and repeat the washes.

5. *Second Ab*. Make a dilution of the labeled (conjugated) second Ab in 2% skim milk. Repeat the incubation and washing.

6. *Substrate*. Prepare the appropriate substrates for the conjugate used, and add to the wells in volumes equal to that of the coating Ag. Incubate the plates at 37°C, and take readings (at appropriate absorbance for the substrate) at 30, 45, and 60 min.

7. *Data analysis*. Plot the percent absorbance (calculated as the percentage of the maximum reading) against the Ag concentrations. From the graph, calculate the lowest Ag concentration that will give 50% of the maximum reading for the time point; this is the Ag concentration to use for subsequent ELISAs. The Ab titer at that Ag concentration can then be used in the positive controls. It is important that each plate contain the following control wells: (i) an Ag blank, (ii) a first-Ab blank, and (iii) homologous and a nonrelated Ag controls. The Ag-deficient control is used to zero the plate reader. Checkerboard ELISA can also be used to establish the optimum concentrations of conjugates and Ab. In each assay, only one parameter is varied at a time. Incubations can also be carried out at 37°C for 2 to 4 h at room temperature or 16 h at 4°C.

8.5.4.4. Chemiluminescence ELISA

Chemiluminescence is the production of visible light by a chemical reaction. Enzymes such as HRPO and AP, in the presence of hydrogen peroxide, catalyze the oxidation and excitation of the luminol substrate; the excited molecule decays to its ground state with concomitant emission of visible light (chemiluminescence). The chemiluminescence assay protocol based on the protocols of Lewkowich et al. (95) and Zhao and Lam (107) is similar to the standard calorimetric ELISA (section 8.5.4.3) with the following modifications. White, opaque microtiter plates (Corning Inc.) are used. Dilutions of 100 ng to 1 μg of pure Ag/ml or 500 ng to 10 μg of complex Ag mixtures/ml are used to coat the plates. Plates are washed rigorously (12 or more times) after each incubation step.

Detection of AP-Conjugated Abs

1. The assay is developed with 0.4 mM disodium 3-(4-methoxyspiro{1,2-dioxetane-3,2′-(5′-chloro)-tricyclo[3.3.1.1]decan}-4-yl)phenyl phosphate (CSPD) or CDP-Star® (Applied Biosystems, Bedford, MA) diluted 1:5 in substrate buffer (1 mM MgCl₂, 0.1 mM DEA, pH 9.5).

2. Incubate at room temperature for 20 to 40 min. The intensity of chemiluminescence emission can be increased by using chemiluminescence enhancers (Tropix, Bedford, MA).

3. Chemiluminescence is detected by using luminometers, e.g., the MicroLumat Plus microplate luminometer or the 1420-VICTOR² multilabel counter (Wallac, Montreal, Canada) using the chemiluminescence program.

Compared to calorimetric ELISA, chemiluminescence is clearly superior (95, 107).

8.5.5. Immunoblotting of Ags

Immunoblotting (Western blotting, electroblotting, and immunoelectroblotting) (113) has largely replaced other immunoelectrophoretic procedures. Here, the antigenic and electrophoretic properties of the molecule are characterized simultaneously. The technique can be applied to study any molecule that retains its functional antigenic structure following SDS-PAGE (22).

The initial step is separation of the antigenic mixtures by PAGE, separating the molecules on the basis of size (urea or SDS-PAGE), charge (isoelectric focusing), or both (two-dimensional gels [isoelectric focusing–SDS-PAGE]) (see chapter 17 for details). The separated components are then electrophoretically transferred onto a support matrix, often nitrocellulose. The immobilized Ags are allowed to react with the test Abs and then with the enzyme- or ligand-conjugated Abs and are visualized by the addition of specific chromogenic or chemiluminescent substrates in a process similar to the ELISA (108-111, 114). The substrates used in immunoblotting differ from those used in ELISA in that the product of the enzymatic reaction is insoluble and precipitates at the site of formation. The reaction is then observed as a labeled band on the blot (chromogenic assay) or film (chemiluminescence assay).

Although immunoblotting was originally used to investigate protein Ags, the development of nylon-backed nitrocellulose matrices with improved macromolecule binding has made it possible to study glycosylated proteins and lipids, lipoproteins, proteoglycans, viral Ags, and LPSs. Some Ags show reduced Ab binding following SDS-PAGE and immunoblotting. This problem can be reduced considerably by renaturation of antigenic sites in the blotted proteins (108). Conformational epitopes that are subject to alteration or dissociation following SDS-PAGE cannot be investigated by this method. Conformational epitopes may be indicated in ELISA, dot immunoblotting, or immunofluorescence. Because of the improved resolution of the Ab-binding molecules in test samples, immunoblotting is used in clinical laboratories for diagnosis and also for identification of false-positive ELISA results.

Assay Parameters

The parameters that should be optimized for immunoblotting are (i) the concentration of acrylamide that gives the best separation of the Ags and (ii) the efficiency of the binding of the Ags to the nitrocellulose.

Application

Immunoblotting has the same sensitivity as ELISA; however, it has several advantages. Adsorption of proteins onto nitrocellulose is not affected by detergents, less sample is required, and information on Ag structural properties is obtained. Nitrocellulose-bound molecules can also be chemically modified, and the nature of epitopes can be examined; e.g., periodate oxidation of the blots is used to differentiate periodate-sensitive and -insensitive antigenic determinants (114). Furthermore, immobilized Ags of interest can be excised and used in the immunization of animals to generate specific Abs (22; section 8.2.2).

Nitrocellulose

A wide variety of nitrocellulose matrices with improved Ag-binding capacities are available, including Hybond (Pierce, Amersham), Immobilon-P (Millipore), BioTrace HP (Gelman), and polyvinyl chloride (Millipore). Nitrocellulose is brittle and tears easily. When possible, the nylon-backed membranes, which are less fragile and have high molecule-binding capacities, should be used.

8.5.5.1. General Procedure

Similar methods can be found described in more detail in the Bio-Rad immunoblotting guide.

Reagents

1. *Transfer buffer*, pH 8.1 to 8.4:

Tris base . . . . 12.12 g
Glycine . . . . 57.48 g
Methanol . . . . 800 ml

Dissolve the Tris in 3.2 liters of distilled water, and add glycine and then methanol. Store at 4°C.

2. *Incubation and wash buffers.* To 10 mM PBS, pH 7.2, add 8.76 g of NaCl, 0.2 g of KCl, 1.44 g of $Na_2HPO_4 \cdot 2H_2O$, and 0.2 g of KH_2PO_4 in 1 liter of water. Alternatively, 10 mM TBS can be used.

3. *Wash buffers.* PBS or TBS containing 0.2% (vol/vol) Tween 20

4. *Blocking solutions.* Either 1% casein, 3% gelatin, 5% skim milk, or 3% BSA (all wt/vol) dissolved in PBS or TBS can be used. (BSA is expensive; skim milk and casein are cheaper.) If used, heat the buffer to dissolve the gelatin. Make stock solutions by autoclaving and storing in 100-ml volumes at 4°C. Heat to dissolve before use.

5. *Dilution buffers.* Dilute all Abs and conjugates in PBS or TBS containing 1% casein or 0.8 to 1.0% gelatin, 2% BSA, or 2% skim milk.

6. *Developing reagents.* Several reagents for blot development and premade kits are available commercially (e.g., from Bio-Rad). We find it is best to make the reagents as described below.

(i) HRPO conjugates

Solution A, 50 mg of 3,3'-diaminobenzidine tetrahydrochloride (Sigma) in 50 ml of PBS

Solution B, 30 mg of $CoCl_2$ in 50 ml of PBS

Pour first solution A and then solution B onto the blot. Add 1 ml of 30% H_2O_2 (stock). Agitate during the staining. 4-Chloro 1-napthol and aminoethylcarbazole are alternate reagents that are less sensitive than 3,3'-diaminobenzidine tetrahydrochloride but may show less background.

(ii) AP conjugates

Solution A, 300 mg of nitroblue tetrazolium (Sigma) in 10 ml of 70% dimethyl formamide (DMF; Sigma), i.e., 3 ml of water and 7 ml of DMF

Solution B, 150 mg of 5-bromo-4-chloro-3-indolyl phosphate-toluidine (Sigma) in 10 ml of DMF. Store 1-ml aliquots of solutions A and B at −20°C.

Bicarbonate buffer (pH 9.8), 8.4 mg of $NaHCO_3$ and 0.203 mg of $MgCl_2 \cdot 6H_2O$ in 1 liter of distilled water

Before use, add 1 ml each of solutions A and B to 100 ml of the bicarbonate buffer.

Protocol

It is important to remember the following.

1. Use prestained or biotinylated M_r standards to facilitate accurate identification of the immunolabeled bands.

2. Do not touch the nitrocellulose paper with bare hands.

3. Be economical; nitrocellulose is expensive.

4. The nitrocellulose membrane is always placed on the anode (positive) side of the chamber.

5. Remove all air bubbles from the sandwich, or the transfer will not be uniform.

6. Cool the buffer during high-voltage Ag transfer.

7. Silver stain the gel after blotting to confirm the efficiency of the transfer.

Sandwich Assembly

1. Cut a piece of nitrocellulose equal to the size of the gel to be blotted, float it carefully on distilled water or transfer buffer, and let it wet evenly. If parts of the nitrocellulose membrane do not wet after 15 min of soaking, discard it. Meanwhile, cut two pieces of Whatman filter paper. Soak the papers and the fiber pads in the transfer buffer.

2. Carefully remove one glass plate from the gel. Remove the stacking gel, and nick one corner of the gel for orientation.

3. Partially submerge one plate of the blot sandwich holder in transfer buffer in a dish, and lay a soaked fiber pad followed by a filter paper on the submerged plate. Next, place the gel on the filter paper and pour some buffer over the gel to reduce the chances of trapping air bubbles. Place the nitrocellulose membrane over the gel. Remove all air bubbles. Last, place the other filter paper and fiber pad on the sandwich.

4. Clamp the sandwich together. Place the sandwich in a transfer chamber with the gel side towards the cathode.

5. Connect the leads, and transfer at 100 V for 1 to 2 h with cooling or at 30 V for 16 h at room temperature or 4°C.

6. After transfer, remove the nitrocellulose membrane (Western blot or blot) and mark the corner corresponding to the nicked corner of the gel with a pencil. Place the blot in the blocking buffer, and block for 4 to 16 h at room temperature or 1 h at 37°C. At this point, the blot can be rinsed in PBS, air dried, and sealed in a plastic bag or wrapped in plastic food wrap at −20°C until use. *Do not fold the blot.*

8.5.5.2. Detection of Ags

8.5.5.2.1. Biotin-Streptavidin Systems (111)

All blocking and incubation steps for the biotin-streptavidin amplification method can be carried out for 1 h at 37°C, 2 to 4 h at room temperature, or 16 h at 4°C. The volumes of reagents or buffers used should be sufficient to cover the blot completely. Small dishes should be used at the Ab and reagent steps to save these reagents. Agitate the nitrocellulose blot throughout the following procedure.

1. Incubate the blot in the blocking solution.

2. Wash for 5 min in the wash buffer. Repeat twice more.

3. Place the blot in the test Ab and incubate. As a general estimation, use a dilution of Ab that gives an absorbance of 50% of the highest absorbance as determined by ELISA. Hybridoma culture supernatants are often used at a 1:5 dilution or undiluted.

4. Wash for 15 min in wash buffer. Repeat twice more.

5. Add the biotin-conjugated second Ab, diluted as indicated, for commercial preparations. Incubate. Repeat step 4.

6. Add a dilution of streptavidin-HRPO or streptavidin-biotin-HRPO complex, and incubate for 30 min at room temperature. Repeat step 4.

7. Place the immunoblots in developing solution. Wash the stained immunoblots extensively to stop color develop-

ment. Dry between paper towels, and store out of direct light.

8.5.5.2.2. Rapid-Detection Chromogenic and Chemiluminescent Enzyme Substrates

The rapid-detection protocol is a modification of the rapid immunodetection method using chemiluminesence (Millipore Corp., Bedford, MA). In this procedure, proteins are transferred onto either polyvinylidene difluoride (PVDF; Roche Diagnostics) or Immobilon™-P (Millipore) membranes, following the manufacturer's instructions for prewetting of the membranes. The membranes do not wet in PBS. PVDF and Immobilon-P membranes are hydrophobic, and they do not readily bind proteins except during electrophoretic transfer. They do require an incubation step in blocking buffer prior to reaction with the primary Ab. For all incubation and wash steps, good mixing of reagents with the membrane should be maintained.

Chromogenic Protocol

1. Following transfer, the PVDF blots are soaked briefly in methanol and completely air dried before immunodetection. The air-dried blots can also be wrapped in plastic film and stored at −20°C.
2. The blot is transferred directly (without a blocking step) into the primary Ab diluted in 1% skim milk (or no more than 1% casein). Incubate blots for 1.5 h at room temperature. Use of BSA or gelatin as blocking reagents results in high backgrounds and is not recommended.
3. Wash the blots for 1 to 2 min with PBS or with PBS-0.05% Tween 20. Wash two more times.
4. Incubate the washed blot in either AP- or HRPO-conjugated second Ab diluted in 1% skim milk. Incubate for 30 min.
5. Repeat the washing step as described above.
6. Develop either AP or HRPO chromogenic substrate (see section 8.5.5.1 above).

The entire developing process is completed in 3 h.

Rapid Detection by Chemiluminescence

1. By following steps 1 and 2 as described above for the chromogenic protocol, the Immobilon-P or PVDF blots are reacted with the primary Ab and the enzyme-conjugated secondary Abs. The Abs are used at one-half to one-fifth of the concentration used for detection by the chromogenic protocol.
2. Wash the blots (three 5-min washes) in PBS without Tween detergents.
3. Tranfer blots to a clean container (e.g., for small blots use a polystyrene weigh boat) containing a solution of the enhanced chemiluminescence substrate (ECL™; Amersham). Use the minimum volume required to cover the blot. Incubate for 1 to 3 min.
4. Drain off excess substrate (do not allow the blot to dry).
5. Using clean tweezers, lift and place the blot onto a clean glass plate and wrap (glass and blot) with plastic wrap. Gently remove all air bubbles between the blot and the plastic wrap.
6. Take the wrapped blot assembly into a dark room and place it, blot side up, in a film cassette.
7. Place a sheet of autoradiography film (Cronex medical film; Med-Tec Inc., Rexdale, Canada) onto the wrapped blot. Close the cassette. Leave at room temperature for 15 to 30 s, and expose the film to the generated chemiluminescence.

8. After exposure, remove the film and develop by immersing in Kodak GBX film developer solution (Amersham) until bands are visible on the film. To stop development, transfer the film to 0.2% (vol/vol) acetic acid. Rinse in water. If the bands are too dark or faint, place a new film in the cassette and repeat exposure for a shorter (seconds) or longer (up to 40 min) duration, respectively. The intensity of the chemiluminescence emitted is dependent on several factors, including the amount of the target Ag(s) on the blot, the affinity of the primary Ab for the target Ag, and the number or density of Ab-binding epitopes per Ag molecule.

Detection by chemiluminescence is two to five times more sensitive than detection by the chromogenic protocol.

8.5.5.2.3. Dye Staining of Ags

Proteins immobilized on nitrocellulose can be stained by amido black (112), India ink (19), fluorescein isothiocyanate (119), or the reversible Ponceau S stain (3). The transfer efficiency can be assessed by the following method.

1. Wash the blot four times for 10 min each in PBS containing 0.1% (vol/vol) Tween 20. Rinse well with water between these washes. The washing process is necessary to remove SDS, which interferes with staining.
2. Prepare 1 µl of Pelican India ink solution/ml in PBS-0.1% Tween 20. Mix for at least 1 h.
3. Stain the blot with the India ink for 1 to 18 h at room temperature with agitation.
4. Destain in several changes of distilled water. India ink will stain protein and nonprotein components.

8.5.5.3. Dot Blotting

Dot immunoblotting is an indirect ELISA conducted on a nitrocellulose matrix. The assay exploits the known ability of nitrocellulose to bind and immobilize a large variety of molecules in comparison with the ELISA plate matrix. In this procedure, 2-µl volumes of the test sample are dotted onto the nitrocellulose and allowed to dry. Dilute samples (e.g., from chromatographic columns) can be concentrated on the paper by repeated application, allowing the samples to dry after each application. The dot blot is then placed in blocking reagent for 4 to 16 h at room temperature or 1 h at 37°C. Process the dot blot as described in section 8.5.5.2. Whole-cell lysates, bacterial colonies, culture supernatants, and serological samples can be tested by this method. The major advantages of dot blots over ELISA are high sensitivity, easy execution, and rapid results. The Ag-immobilizing step of the ELISA is also eliminated. The disadvantage is that the results are not quantitative.

8.5.5.4. Detection of Glycoproteins

Several biotinylating agents can be used to label cell surfaces and soluble and immobilized proteins. Biotinylation is rapid and easy to perform, and the reaction is easily terminated by washing of the cells or by dialysis. The best results are obtained at high protein concentrations. The biotinylated molecules can be detected directly by the addition of streptavidin-enzyme complexes. Biotinylation may have an added advantage in Ag capture assays in which a second Ag-binding Ab may not be available. Biotinylated molecules can also be immobilized on avidin-coated plates.

The succinamide esters of biotin, such as sulfo-NHS-biotin (with or without a spacer arm), react with primary amine groups on the protein. The target amino acids are lysine residues. If the lysine residues are in the Ag recognition pocket of an Ab, there is potential for interference with the Ag's binding properties. The method described below uses the hydrazide derivative of biotin, which binds to aldehyde groups on carbohydrates (110, 111). This label does not seem to affect Ab activity and shows increased sensitivity compared to the biotin ester-labeled molecules.

Reagents

1. *Biotinylating reagent.* Biotin aminocaproyl hydrazide dissolved in dimethylformamide at 50 mg/ml

2. *Protein.* Prepare Ab solution at 3 mg/ml in 0.1 M sodium acetate buffer, pH 5.5

3. *Oxidizing reagent.* Dissolve 100 mM sodium metaperiodate ($NaIO_4$) in the sodium acetate buffer, pH 5.5. The periodate oxidizes carbohydrates to generate aldehyde groups, the target of the hydrazide group. Use of excess periodate will result in the production of carboxylic acid groups and reduce the efficiency of biotinylation

4. Add an equal amount of 80 mM sodium sulfite to quench the periodate

Biotinylation

1. Add 0.1 ml of the sodium metaperiodate reagent to 1 ml of the Ab solution. Incubate on ice for 30 min in the dark.

2. Add 0.25 ml of the sodium sulfite solution, and incubate at room temperature for 5 min to destroy all periodate.

3. Add 20 μl of biotin-LC-hydrazide to give a final concentration of 1 mM. Incubate for 15 min at room temperature.

4. Dialyze the labeled Ab extensively against PBS with 0.02% sodium azide. Omit the azide if peroxidase-coupled reagents will be used in the immunoassays.

5. Concentrate the Ab to 10 mg/ml and store at 4°C in azide, or add 1% BSA and store at −20°C.

Glycoproteins and polysaccharide-containing molecules can be detected by using lectin blots (110) or by using oxidation of periodate-sensitive epitopes on Ags immobilized on nitrocellulose membranes (114). Detection of periodate-sensitive epitopes is as follows.

1. For oxidation, incubate blots after transfer in 50 mM sodium acetate, pH 4.5, containing 10 mM periodic acid for 1 h in the dark. Incubate control blots in buffer without periodate.

2. Rinse in the sodium acetate buffer.

3. For reduction, incubate blots in 50 mM sodium borohydride in PBS for 15 min at room temperature.

4. Wash extensively in PBS. Proceed with step 1 of the blot immunostaining procedure described above.

8.5.6. Immunofluorescence Techniques

Fluorescent- and immunofluorescent-Ab techniques work on the same principle as the ELISA, except that in fluorescent-Ab assays the Abs are labeled with chemical compounds which emit a fluorescent signal upon activation by UV radiation of suitable wavelengths. The emitted fluorescence is observed by using a fluorescence microscope. The visual impact of the fluorescent dye is convincing evidence of an Ab-Ag interaction.

In the direct fluorescence assay, cells, tissues, or smears of the Ags are fixed on a glass slide and treated with the fluorochrome-labeled Ab and the resulting fluorescence is observed. The assay can also be performed with unfixed cells in a solution to study cell surface Ags. Great care must be taken to ensure the integrity of the cells during this procedure. See the excellent reviews by Johnson and Holborow (116) and McKinney (117).

Application

In pathology laboratories, immunofluorescence is still the most useful technique in terms of simplicity and speed for the screening of intracellular and extracellular pathogens in tissues, blood smears, and respiratory exudates. Although direct fluorescence assays using fluorochrome-labeled polyclonal Abs are used, most laboratories employ the indirect fluorescence technique (IFAT). The increased brightness of the labeled sample and the availability of commercial fluorochrome-labeled secondary Abs make the IFAT a very useful technique. The specificity of IFATs has been improved by the development of Ag-specific MAbs. One advantage of the IFAT compared with ELISAs is that the person reading the slides can often distinguish cross-reacting microorganisms in tissues by the morphology of the labeled cells, e.g., paired cocci as opposed to single coccal cells.

Another important area of application is the fluorescence-activated cell sorter technique (115). Cells are incubated with fluorescence-labeled immunoglobulins, and the labeled cells are fractionated on the basis of the intensity of the fluorescent signal or the type of signal when Abs conjugated to different fluorochromes are used. Fluorescence-labeled lectins have also been important in studies of cell surface glycosylation.

Limitations in Using the Technique

The best results are obtained by using an epifluorescence microscope to read the slides. This is expensive equipment for most laboratories. The method can be tedious for the screening of large numbers of samples and may not be sensitive. The diagnostic potential of the assay for early infections is also complicated by background noise due to low-level cross-reactivities and endogenous tissue fluorescence. Lastly, the interpretation of the fluorescent-antibody test results requires skill and is largely subjective, particularly for semiquantitative judgments and borderline-positive cases.

The sensitivity of fluorescence assays has been greatly improved by the development of fluorescent labels such as lanthanide chelate that upon activation emit a powerful signal and of highly sensitive fluorometers to quantitate these short-lived (microsecond) signals. The cost of the equipment is prohibitive for most labs. For these reasons fluorescent-antibody assays have been replaced in many areas of research and diagnostics by the highly sensitive and labor-saving ELISA as a screening procedure and by the more analytical immunogold labeling procedures (chapter 4.6.2) in immunocytochemistry. The most common application of the IFAT is in diagnostic laboratories where frozen tissue is examined for the presence of pathogens.

Fluorochromes

(Also see chapters 2 and 3.) The most commonly used fluorochromes are fluorescein isothiocyanate and tetramethyl rhodamine isothiocyanate (rhodamine). Abs and anti-Abs conjugated to these fluorochromes can be obtained commercially, and a general protocol for the labeling of Abs with fluorescein isothiocyanate and rhodamine can be found in reference 115. Because the fluorescence of fluorescein isothiocyanate is green (Color Plate 5) and that of rhodamine is red, it is possible to examine the distribution or presence of two different Ags on the same sample by using Abs of two or more specificities labeled with different fluorochromes (suppliers include Cedarlane Laboratories Ltd.,

Hornby, Canada; Molecular Probes Inc., Eugene, OR; and BD PharMingen, San Diego, CA).

Protocol

In this IFAT, the test Ab is unlabeled and a fluorochrome-labeled anti-species second Ab is used. Alternatively, the anti-Ab can be labeled with fluorescein, followed by an anti-fluorescein Ab conjugated to fluorescein isothiocyanate. In this method the fluorescein is a ligand. A more popular and sensitive procedure uses biotin-labeled anti-immunoglobulins followed by fluorochrome-labeled streptavidin or a streptavidin-biotinylated fluorochrome.

1. Prepare a suspension of the test cells, and apply 10 μl to the slide. Immediately aspirate all the suspension to leave a thin smear. Air dry and fix the smear by passing the slide over a flame, by heating in an oven at 60°C for 5 min, or by dipping for a few seconds in dry acetone or methanol at −20° C. Tissue extracts are smeared on a slide and fixed in the same manner. Acetone or methanol fixation will permeabilize membranes and allow cells in tissues or organelles to be detected more readily than by heat fixation. This treatment also allows binding of Abs to intracellular Ags that would otherwise not be accessible in the intact cell. For best results, use commercially available slides that have been treated for improved cell binding.

2. Add a drop of the test Ab onto the fixed sample, and incubate at room temperature for 5 min. Place the slides in a covered moist chamber (petri dish or sandwich box containing a wet paper towel) to avoid drying during the staining. Include positive and negative controls using specific and nonspecific antisera in the assay.

3. Rinse the slides with a gentle stream of PBS, pH 7.2, and then place them in a bath of PBS for 5 min. Repeat the process to replace the PBS twice.

4. Add a drop of the fluorochrome-labeled anti-species Ab. Incubate and wash as described above. Most fluorochrome-conjugated Abs are diluted between 1:16 and 1:32.

5. Shake the slides to remove most of the buffer. Add mounting fluid, consisting of 90 ml of glycerin in 0.5 M bicarbonate buffer, pH 9.6, since fluorescence is enhanced under alkaline conditions. The slides can then be stored in the dark at 4°C.

6. To score results, view the labeled cells under immersion oil (see chapter 1) by using a fluorescence microscope in a darkened room. Use the right filters for the fluorochrome (see chapters 2 and 3). Fluorescein isothiocyanate fluoresces apple green at 515 nm. Samples are scored as follows: no fluorescence is 0, faint fluorescence is +, and intense fluorescence is + + + +.

Problems and Troubleshooting

When everything fluoresces (including the negative controls), there is nonspecific binding of the fluorochrome-labeled Ab, which is used at very high concentrations in these assays. Centrifugation of the Ab before use, or preabsorption with unlabeled cells, will remove aggregates and "sticky" proteins. Also, some serum proteins will bind Abs in a non-specific manner and cells grown in serum should be washed well before staining. Here, too, the test Ab should be preabsorbed with unrelated cells to reduce non-specific background (section 8.3.1). This is especially important when polyclonal antiserum is used. Repeated freezing and thawing of the conjugated Ab and the use of high-salt-concentration buffers can result in complex formation which will be seen as bright fluorescent spots on the slide.

The labeled preparations are stable at 4°C in the presence of 0.02% (wt/vol) sodium azide to prevent contamination. If a low-level background persists, try preincubating the smear with a blocking reagent and diluting the antisera in PBS blocker (see section 8.5.4).

8.6. COMMERCIAL SOURCES AND THEIR URLS

Aldrich Chemical Co., Inc., Milwaukee, WI, http://www.sigmaaldrich.com

Amersham-Pharmacia Biotech, Uppsala, Sweden, http://www.amersham.com

Becton-Dickinson, Sparks, MD, http://www.bd.com or http://www.bdbiosciences.com

BIO/CAN Scientific, Mississauga, Canada, http://www.biocan.com

Bio-Rad Laboratories, Richmond, CA, http://www.bio-rad.com

Boehringer Mannheim Biochemica, http://biochem.boehringer-mannheim.com (select country)

Cedarlane Laboratories, Hornby, Canada, http://www.cedarlanelabs.com

Difco Laboratories, Detroit, MI, http://www.bd.com

Fisher Biotech, Fair Lawn, NJ, http://www.fishersci.ca or http://www.fishersci.com

Gelman Sciences Inc., Ann Arbor, MI, http://www.argus-inc.com//Gelman/Gelman.html or http://www.gelman.com

ICN Immunobiologicals, Costa Mesa, CA, http://www.icnbiomed.com

Invitrogen Life Technologies, http://www.invitrogen.com/

Mandel Scientific Company, Guelph, Canada, http://www.mandel.ca

Millipore, Bedford, MA, http://www.milliporecorp.com

Packard BioScience, http://www.packardbioscience.com

Pharmacia-LKB Biotechnology Inc., Uppsala, Sweden, http://www.pharmacia.com/

Pierce Chromatographic Specialities, Brockville, Canada, http://www.piercenet.com/

Pitman-Moore, Davis, CA, http://www.abcmedinc.com/catalog/cat5.htm

Ribi ImmunoChem Research, Inc. (a subsidiary of Corixa Corp.), Hamilton, MT, http://www.corixa.com

Savant Instruments Inc., Hicksville, NY, http://www.thermo.com/

Sigma Chemicals Co., St. Louis, MO, http://www.sigmaaldrich.com

Wallac, Montreal, Canada, http://www.wallac.de/1/default.asp

8.7. REFERENCES

8.7.1. General References

1. **Goers, J.** 1992. *Immunochemical Techniques Laboratory Manual*. Academic Press, New York, NY.
2. **Hancock, I. C., and I. R. Poxton.** 1988. *Bacterial Cell Surface Techniques. Modern Microbiological Methods*. Wiley Interscience, New York, NY.
   Provides specific protocols for studying immunochemistry of cell surface antigens.

3. **Harlow, E.** 1999. *Using Antibodies: a Laboratory Manual.* Cold Spring Harbor Laboratory Press, Cold Spring Harbor, NY.

4. **Harlow, E., and D. Lane.** 1988. *Antibodies, a Laboratory Manual.* Cold Spring Harbor Laboratory Press, Cold Spring Harbor, NY.
   References 3 and 4 are valuable reference books for labs using antibodies.

8.7.2. Specific References

8.7.2.1. Handling Animals

5. **Akita, E. M., and S. Nakai.** 1992. Immunoglobulin from egg yolk; isolation and purification. *J. Food Sci.* **57:**629–634.

6. **Herbert, W. J., and F. Kristensen.** 1986. Laboratory animal techniques for immunology, p. 133.1–133.36. *In* D. M. Weir, C. Blackwell, and L. A. Herzenberg (ed.), *Handbook of Experimental Immunology,* 4th ed. Blackwell Scientific Publications, Boston, MA.

7. **Lee, S. B., Y. Mine, and R. M. W. Stevenson.** 2000. Effects of hen egg yolk immunoglobulins in passive protection of rainbow trout against *Yersinia ruckeri. J. Agric. Food Chem.* **48:**110–115.

8. **Malik, V. S.** 1994. *Antibody Techniques.* Academic Press, New York, NY.

9. **Smelser, J. F.** 1985. Rabbits: a practical guide for the veterinary technician. *Vet. Tech.* **6:**121–128.

10. **Stewart, K. L., E. L. Johnstone, and J. A. Vecera.** 1988. The laboratory mouse and rat. *Vet. Tech.* **9:**264–271.

11. **Tillman, P., and C. Norman.** 1983. Droperidol-fentanyl as an aid to blood collection in rabbits. *Lab. Anim. Sci.* **33:**181–182.

8.7.2.2. Ag Preparation

12. **Bogard, C. W., Jr., D. L. Dunn, K. Abernthy, C. Kilgarriff, and P. C. Kung.** 1987. Isolation and characterization of murine monoclonal antibodies specific for gram-negative bacterial lipopolysaccharide: association of cross-genus reactivity with lipid A specificity. *Infect. Immun.* **55:**899–908.

13. **Bowman, C. C., and J. D. Clements.** 2001. Differential biological and adjuvant activities of cholera toxin and *Escherichia coli* heat-labile enterotoxin hybrids. *Infect. Immun.* **69:**1528–1535.

14. **Campbell, D. H., J. S. Garvey, N. E. Cremer, and D. H. Sussdorf.** 1970. *Methods in Immunology,* 2nd ed. W. A. Benjamin, Inc., New York, NY.

15. **Chowdhury, P. S., M. Gallo, and I. Pastan.** 2001. Generation of high titre antisera in rabbits by DNA immunization. *J. Immunol. Methods* **249:**147–154.

16. **Christodoulides, M., E. Rattue, and J. E. Heckles.** 2000. Effect of adjuvant on the immune response to a multiple Ag peptide (MAP) containing protective epitopes from *Neisseria meningitidis* class 1 porin. *Vaccine* **18:**131–139.

17. **Diano, M., A. Le Bivic, and M. Hirn.** 1987. A method for the production of highly specific polyclonal antibodies. *Anal. Biochem.* **166:**224–229.

18. **Galanos, C., O. Luderitz, and O. Westphal.** 1971. Preparation and properties of antisera against the lipid-A component of bacterial lipopolysaccharide. *Eur. J. Biochem.* **24:**116–122.

19. **Hancock, K., and V. C. W. Tsang.** 1983. India ink staining of proteins on nitrocellulose paper. *Anal. Biochem.* **133:**157–162.

20. **Hassan, U. A., A. M. Abai, D. R. Harper, B. W. Wren, and W. J. W. Morrow.** 1999. Nucleic acid immunization: concepts and techniques associated with third generation vaccines. *J. Immunol. Methods* **229:**1–22.

21. **Hassett, D. E., and J. L. Whitton.** 1996. DNA immunization. *Trends Microbiol.* **4:**307–312.

22. **Knudsen, K. A.** 1985. Proteins transferred to nitrocellulose for use as immunogen. *Anal. Biochem.* **147:**285–288.

23. **Laemmli, U. K.** 1970. Cleavage of structural proteins during the assembly of the head of bacteriophage T4. *Nature* **227:**680–685.

24. **Leitner, W. W., H. Ying, and N. P. Restifo.** 2000. DNA and RNA-based vaccines: principles, progress and prospects. *Vaccine* **18:**765–777.

25. **Leppard, K., N. Totty, M. Waterfield, E. Harlow, J. Jenkins, and L. Crawford.** 1981. Purification and partial amino acid sequence analysis of the cellular tumor antigen, p53, from mouse SV40-transformed cells. *EMBO J.* **2:** 1993–1999.

26. **Lindberg, A.** 1999. Glycoprotein conjugate vaccines. *Vaccine* **17:**S28–S36.

27. **Lipford, G. B., K. Heeg, and H. Wagner.** 1998. Bacterial DNA as immune cell activator. *Trends Microbiol.* **6:**496–500.

28. **McCluskie, M. J., and H. L Davis.** 1999. CpG DNA as mucosal adjuvants. *Vaccine* **18:**231–237.

29. **Rauser, W. E., A. A. Quesnel, J. S. Lam, and G. G. Southam.** 1988. An enzyme-linked immunosorbent assay for plant cadmium binding peptide. *Plant Sci.* **57:**37–43.

8.7.2.3. Adjuvants

30. **Allison, A. C., and G. Gregoriadis.** 1974. Liposomes as immunological adjuvants. *Nature* **252:**252.

31. **Baker, P. J., J. R. Hiernaux, M. B. Fauntleroy, B. Prescott, J. Cantrell, and J. A. Rubach.** 1988. Inactivation of suppressor T-cell activity by nontoxic monophosphoryl lipid A. *Infect. Immun.* **56:**1076–1083.

32. **Baker, P. J., J. R. Hiernaux, M. B. Fauntleroy, P. W. Stashak, B. Prescott, J. L. Cantrell, and J. A. Tudbach.** 1988. Ability of monophosphoryl lipid A to augment the antibody response of young mice. *Infect. Immun.* **56:**3064–3066.

33. **Branche, R., and G. Renoux.** 1972. Stimulation of rabies vaccine in mice by low doses of polyadenylic: polyuridylic complex. *Infect. Immun.* **6:**324–325.

34. **Chase, M. W.** 1967. Production of antiserum. *Methods Immunol. Immunochem.* **1:**197–209.

35. **Claassen, E., N. Kors, and N. van Rooijen.** 1987. Immunomodulation with liposome: the immune response elicited by liposomes with entrapped dichloromethyl-diphosphonate and surface associated antigen or hapten. *Immunology* **60:**509–515.

36. **De Becker, G., V. Moulin, B. Pajak, C. Bruck, M. Francotte, C. Thiriart, J. Urbain, and M. Moser.** 2000. The adjuvant monophosphoryl lipid A increases the function of antigen-presenting cells. *Int. Immunol.* **6:**807–815.

37. **Edelman, R.** 1980. Vaccine adjuvants. *Rev. Infect Dis.* **2:**370–383.

38. **Ellouz, F., A. Adam, R. Ciorbaru, and E. Ledcrer.** 1974. Minimal structural requirements for adjuvant activity of bacterial peptidoglycan derivatives. *Biochem. Biophys. Res. Commun.* **59:**1317–1325.

39. **Fitzgerald, T. J.** 1991. Syphilis vaccine: up-regulation of immunogenicity by cyclophosphamide, Ribi adjuvant, and indomethacin confers significant protection against challenge infection in rabbits. *Vaccine* **9:**266–272.

40. **Freund, J.** 1956. The mode of action of immunological adjuvants. *Adv. Tuberc. Res.* **7:**130–148.

41. **Gery, I., J. Kruger, and S. Spiesel.** 1972. Stimulation of B lymphocytes by endotoxin reactions of thymus-deprived mice and karyotypic analysis of dividing cells in mice bearing T_6T_6 thymus graft. *J. Immunol.* **108:**1088–1091.

42. Gregoriadis, G., D. Davis, and A. Davis. 1987. Liposomes as immunological adjuvants: antigen incorporation studies. *Vaccine* **5:**145–151.
43. Kenney, J. S., B. W. Hughes, M. P. Masada, and A. C. Allison. 1989. Influence of adjuvants on the quantity, affinity, isotype and epitope specificity of murine antibodies. *J. Immunol. Methods* **121:**157–166.
44. Kripke, M. L., and D. W. Weiss. 1970. Studies on the immune responses of Balb/c mice during tumor induction by mineral oil. *Intern. J. Cancer* **6:**422–430.
45. Moingeon, P., J. Haensler, and A. Lindberg. 2001. Towards the rational design of Th1 adjuvants. *Vaccine* **19:** 4363–4372.
46. Morein, B., B. Sundquist, S. Hoglund, K. Dalsgaard, and A. Osterhaus. 1984. ISCOM, a novel structure for antigenic presentation of membrane proteins from enveloped viruses. *Nature* **308:**457–459.
47. Morein, B., K. Lövgren, S. Höglund, and B. Sundquist. 1987. The ISCOM: an immunostimulating complex. *Immunol. Today* **8:**333–338.
48. Nicholson, K. G., D. A. J. Tyrrell, P. Harrison, C. W. Potter, R. Jennings, A. Clark, G. C. Scheld, J. M. Wood, R. Yelts, V. Seagroatt, A. Higgins, and S. G. Anderson. 1979. Clinical studies of mononvalent inactivated whole virus and subunit A/USSR/77 (H₁ H₁) vaccine serological and clinical reactions. *J. Biol. Stand.* **7:**123–136.
49. Ohta, M., N. Kido, T. Hasegawa, H. Ito, Y. Fujii, T. Arakawa, T. Komatsu, and N. Kato. 1987. Contribution of the mannan O side-chains to the adjuvant action of lipopolysaccharide. *Immunology* **60:**503–507.
50. Röonberg, B., M. Fekadu, and B. Morein. 1995. Adjuvant activity of non-toxic *Quillaja saponaria* Molin components for use in ISCOM matrix. *Vaccine* **13:**1375–1382.
51. Seppala, I. J. T., and O. Makela. 1984. Adjuvant effect of bacterial LPS and/or alum precipitation in response to polysaccharide and protein antigens. *Immunology* **53:**827–836.
52. Shek, P. N., and B. H. Sabiston. 1981. Immune response mediated by liposome associated protein antigens. I. Potentiation of the plaque forming cell response. *Immunology* **45:**349–356.
53. Taub, R. N., A. R. Krantz, and D. W. Dresser. 1970. The effect of localized injection of adjuvant material on the draining lymph node. I. Histology. *Immunology* **18:** 171–186.
54. Thoelen, S., N. De Clercq, and N. Tornieporth. 2001. A prophylatic hepatitis B vaccine with a novel adjuvant system. *Vaccine* **19:**2400–2403.
55. Sjölander, A., J. C. Cox, and I. A. Barr. 1998. ISCOMs: an adjuvant with multiple functions. *J. Leukoc. Biol.* **64:** 713–723.
56. Warren, H. S., F. R. Vogel, and L. A. Chedid. 1986. Current status of immunological adjuvants. *Annu. Rev. Immunol.* **4:**369–388.

8.7.2.4. Antibody Preparation

57. Bjorck, L., W. Karstern, G. Lindahl, and K. Wildeback. 1987. Streptococcal protein G, expressed by streptococci or *Escherichia coli*, has separate sites for human albumin and IgG. *Mol. Immunol.* **24:**1113–1122.
58. Bjorck, L., and G. Kronvall. 1984. Purification and some properties of streptococcal protein G, a novel IgG-binding reagent. *J. Immunol.* **133:**969–974.
59. Bruck, C., D. Portetelle, C. Gilneur, and A. Bollen. 1982. One step purification of mouse monoclonal antibodies from ascitic fluid by DEAE Affi-gel blue chromatography. *J. Immunol. Methods* **53:**313–319.
60. Cassone, A., A. Torosantucci, M. Boccanera, G. Pellegrini, C. Palma, and F. Malavasi. 1988. Production and characterisation of a monoclonal antibody to a cell surface, glucomannoprotein constituent of Candida albicans and other pathogenic Candida species. *J. Med. Microbiol.* **27:**233–238.
61. Creuzenet, C., and J. S. Lam. 2001. Topological and functional characterisation of WbpM, an inner-membrane UDP-GlcNAc C₆ dehydratase essential for lipopolysaccharide biosynthesis in *Pseudomonas aeruginosa*. *Mol. Microbiol.* **41:**1295–1310.
62. Davey, M. L., R. E. W. Hancock, and L. M. Mutharia. 1998. Influence of culture conditions on expression of the 40-kilodalton protein of *Vibrio anguillarum* serotype O2. *Appl. Environ. Microbiol.* **64:**138–146.
63. Ey, P. L., S. J. Prouse, and C. R. Jenkin. 1978. Isolation of pure IgG1, IgG2a and IgG2b immunoglobulins from mouse serum using protein A-Sepharose. *Immunochemistry* **15:**429–436.
64. Galfré, G., and C. Milstein. 1981. Preparation of monoclonal antibodies: strategies and procedure. *Methods Enzymol.* **73:**1–46.
65. Goding, J. W. 1986. *Monoclonal Antibodies: Principle and Practice*, 2nd ed. Academic Press, London, England.
66. Köhler, G., and C. Milstein. 1975. Continuous cultures of fused cells secreting antibody of predefined specificity. *Nature* **256:**495–499.
67. Kuhlmann, I., W. Kurth, and I. Ruhdel. 1989. Monoclonal antibodies: *in vivo* and *in vitro* production on a laboratory scale, with consideration of the legal aspects of animal protection. *Altern. Lab. Anim.* **17:**73–82.
68. Lam, J. S., L. A. MacDonald, M. Y. C. Lam, L. G. M. Duchesne, and G. G. Southam. 1987. Production and characterization of monoclonal antibodies against serotype strains of *Pseudomonas aeruginosa*. *Infect. Immun.* **55:**1051–1057.
69. Lindmark, R., K. Thoren-Tolling, and J. Sjoquist. 1983. Binding of immunoglobulins to protein A and immunoglobulin levels in mammalian sera. *J. Immunol. Methods* **62:**1–13.
70. Mariani, M., M. Cianfriglia, and A. Cassone. 1989. Is mouse IgM purification on protein A possible? *Immunol. Today* **10:**115–116. (Letter to the editor.)
71. Oi, V. T., and L. A. Herzenberg. 1980. Immunoglobulin-producing hybrid cell lines, p. 351–372. *In* B. B. Mishell and S. M. Shiigi (ed.), *Selected Methods in Cellular Immunology*. W. H. Freeman, San Francisco, CA.
72. Tung, A. S., S.-T. Ju, S. Sato, and A. Nisonoff. 1976. Production of large amounts of antibodies in individual mice. *J. Immunol.* **116:**676–681.
73. Wilner, M. A. E., H. D. Troutman, F. W. Trader, and I. W. McLean. 1963. Vaccine potentiation by emulsification with pure hydrocarbon compounds. *J. Immunol.* **91:** 210–229.

8.7.2.5. Agglutination

74. Edwards, P. R., and E. W. H. Ewing. 1962. *Identification of Enterobacteriaceae*, 2nd ed. Burgess Publishing Co., Minneapolis, MN.
75. Fung, J. C., and R. C. Tilton. 1985. Detection of bacterial antigens by counter immunoelectrophoresis, coagglutination and latex agglutination, p. 883–890. *In* E. H. Lennette, A. Balows, W. J. Hauser, Jr., and W. J. Shadomy (ed.), *Manual of Clinical Microbiology*, 4th ed. American Society for Microbiology, Washington, DC.
76. Handfield, S. G., A. Lane, and M. B. McIllmurray. 1987. A novel coloured latex test for detection and identification of more than one antigen. *J. Immunol. Methods* **97:** 153–158.

77. **Kronvall, G.** 1973. A rapid slide agglutination method for typing pneumococci by means of specific antibody adsorbed to protein A-containing staphylococci. *J. Med. Microbiol.* **6:**187–190.

78. **Lim, P. L., and K. H. Ko.** 1990. A tube latex test based on colour separation for detection of IgM antibodies to either of two different microorganisms. *J. Immunol. Methods* **135:**9–14.

79. **McIllmurray, M. B., and M. D. Moody.** 1986. Latex agglutination, p. 9–28. *In* R. B. Kohler (ed.), *Antigen Detection To Diagnose Bacterial Infections*, vol 1. CRC Press, Boca Raton, FL.

80. **Oakley, C. L.** 1971. Antigen-antibody reactions in microbiology, p. 174–217. *In* J. R. Norris and D. W. Ribbons (ed.), *Methods in Microbiology.* Academic Press, London, England.

8.7.2.6. Quellung Reaction

81. **Merill, C. W., J. M. Gwaltney, J. O. Hendley, and M. A. Sande.** 1973. Rapid identification of pneumococci. *N. Engl. J. Med.* **288:**510–512.

8.7.2.7. Immunodiffusion

82. **Bjerrum, O. J., and T. C. Bøg-Hansen.** 1975. Immunochemical gel precipitation techniques in membrane studies p. 378–426. *In* A. H. Maddy (ed.), *Biochemical Analysis of Membranes.* John Wiley and Sons, Inc., New York, NY.

83. **Laurell, C. B.** 1966. Quantitative estimation of proteins by electrophoresis in agarose gel containing antibodies. *Anal. Biochem.* **15:**45–52.

84. **Ouchterlony, Ö.** 1968. *Handbook of Immunodiffusion and Immunoelectrophoresis.* Ann Arbor Science Publishers, Ann Arbor, MI.

85. **Tilton, R.** 1983. Procedures for the detection of microorganisms by counter immunoelectrophoresis, p. 87–96. *In* J. D. Coonrod, L. J. Kunz, and M. J. Ferraro (ed.), *The Direct Detection of Microorganisms in Clinical Samples.* Academic Press, Orlando, FL.

86. **Weeke, B.** 1973. Crossed immunoelectrophoresis, p. 45–56. *In* N. H. Axelsen, J. Krøll, and B. Weeke (ed.), *A Manual of Quantitative Immunoelectrophoresis.* Blackwell Scientific Publications, Oxford, England.

8.7.2.8. ELISA

87. **Cantarero, L. A., J. E. Butler, and J. W. Osborne.** 1980. The adsorptive characteristic of proteins for polystyrene and their significance in solid phase immunoassays. *Anal. Biochem.* **105:**375–382.

88. **Conradie, J. D., M. Govender, and L. Visser.** 1983. ELISA solid phase: partial denaturation of coating antibody yields a more efficient solid phase. *J. Immunol. Methods* **59:**289–299.

89. **Engvall, E., and P. Perlmann.** 1971. Enzyme-linked immunosorbent assay (ELISA): quantitative assay of immunoglobulin. *Immunochemistry* **8:**871–879.

90. **Fischer, P. M., and M. E. H. Howden.** 1990. Direct, enzyme-linked immunosorbent assay of anti-peptide antibodies using capture of biotinylated peptides by immobilised avidin. *J. Immunoassay* **11:**311–327.

91. **Gardas, A., and A. Lewartowska.** 1988. Coating of proteins to polystyrene ELISA plates in the presence of detergents. *J. Immunol. Methods* **106:**251–255.

92. **Ito, J. I., Jr., A. C. Wunderlich, J. Lyons, C. E. Davis, D. G. Gurney, and A. I. Braude.** 1980. Role of magnesium in the enzyme-linked immunosorbent assay for lipopolysaccharides of rough *Escherichia coli* strain J5 and *Neisseria gonorrhoeae. J. Infect. Dis.* **142:**532–537.

93. **Julian, E., M. Cama, P. Martínez, and M. Luquin.** 2001. An ELISA for five glycolipids form the cell wall of *Mycobacterium tuberculosis*: Tween 20 interference in the assay. *J. Immunol. Methods* **251:**21–30.

94. **Lee, B. Y., D. Chatterjee, C. M. Bozic, P. J. Brennan, D. L. Cohn, J. D. Bales, S. M. Harrison, L. A. Androu, and I. M. Orme.** 1991. Prevalence of serum antibody to the type-specific glycopeptidolipid antigens of *Mycobacterium avium* in human immunodeficiency virus-positive and -negative individuals. *J. Clin. Microbiol.* **29:**1026–1029.

95. **Lewkowich, I. P., J. D. Campbell, and K. T. HayGlass.** 2001. Comparison of chemiluminescent assays and colorimetric ELISAs for quantification of murine IL-12, human IL-4 and murine IL-4: chemiluminescent substrates provide markedly enhanced sensitivity. *J. Immunol. Methods* **247:**111–118.

96. **Lutz, H. U., P. Stammler, and E. A. Fischer.** 1990. Covalent binding of detergent-solubilised membrane glycoproteins to "Chemobond" plates for ELISA. *J. Immunol Methods* **129:**211–220.

97. **Munoz, C., A. Nieto, A. Gaya, J. Martinez, and J. Vives.** 1986. New experimental criteria for optimisation of solid phase antigen concentration and stability in ELISA. *J. Immunol. Methods* **94:**137–144.

98. **Nakamura, R. M., A. Valler, and D. E. Bidwell.** 1986. Enzyme immunoassays: heterogeneous and homogenous systems, p. 27.1–27.2. *In* D. M. Weir, C. Blackwell, and L. A. Herzenberg (ed.), *Handbook of Experimental Immunology*, 4th ed., vol. 1. *Immunochemistry.* Blackwell Scientific Publications, Boston, MA.

99. **Pitzurra, L., E. Blast, A. Bartoli, P. Marconi, and F. Bistoni.** 1990. A rapid objective immunofluorescence microassay. Application for detection of surface and intracellular antigens. *J. Immunol. Methods* **135:**71–75.

100. **Reggiardo, Z., E. Vasquez, and L. Schnaper.** 1980. ELISA tests for antibodies against mycobacterial glycolipids. *J. Immunol. Methods* **34:**55–60.

101. **Rongen, H. A., H. M. van der Horst, A. J. van Oosterhout, A. Bult, and W. P. van Bennekom.** 1996. Application of xanthine oxidase-catalysed luminol chemiluminescence in a mouse interleukin-5 immunoassay. *J. Immunol. Methods* **197:**161–169.

102. **Scott, B. B., and G. R. Barclay.** 1987. Endotoxin-polymyxin complexes in an improved enzyme-linked immunosorbent assay for IgG antibodies in blood donor sera to gram-negative endotoxin core glycolipids. *Vox Sang.* **52:**272–280.

103. **Tijseen, P.** 1985. Chapter 1, p. 9–327. *In* R. H. Burdon and P. H. Van Knippenberg (ed.), *Laboratory Techniques in Biochemistry and Molecular Biology. Practice and Theory of Enzyme Immunoassays.* Elsevier Science Publishing Company, Inc., New York, NY.

104. **Voller, A., and D. E. Bidwell.** 1985. Enzyme immunoassays, p. 77–86. *In* W. P. Collins (ed.), *Alternate Immunoassays.* John Wiley & Sons, Inc., New York, NY.

105. **Weigand, K., C. Birr, and M. Suter.** 1981. The hexa- and pentapeptide extension of proalbumin. II. Processing of specific antibodies against the synthetic hexapeptide. *Biochim. Biophys. Acta* **670:**424–427.

106. **Wylie, D. E., L. D. Carlson, R. Carlson, F. W. Wagner, and S. M. Schuster.** 1991. Detection of mercuric ions in water by ELISA with a mercury-specific antibody. *Anal. Biochem.* **194:**381–387.

107. **Zhao, X., and J. S. Lam.** 2001. WaaP of *Pseudomonas aeruginosa* is a novel eukaryotic type protein-tyrosine kinase as well as a sugar kinase essential for the biosynthesis of core lipopolysaccharide. *J. Biol. Chem.* **277:**4722–4730.

8.7.2.9. Immunoblotting

108. **Birk, H. W., and H. Koepsell.** 1987. Reaction of monoclonal antibodies with plasma membrane proteins after binding to nitrocellulose: renaturation of antigenic sites and reduction of non-specific antibody binding. *Anal. Biochem.* **164:**12–22.

109. **Houston, B., and D. Peddie.** 1989. A method for detecting proteins immobilised on nitrocellulose membranes by *in situ* derivatization with fluorescein isothiocyanate. *Anal. Biochem.* **177:**263–267.

110. **Mutharia, L. M., and M. Steele.** 1995. Characterization of concanavalin A-binding glycoproteins from procyclic culture forms of *Trypanosoma congolense, T. simiae* and *T. brucei brucei. Parasitol Res.* **81:**245–252.

111. **O'Shannessy, D. J., P. J. Voorstad, and R. H. Quarles.** 1987. Quantitation of glycoproteins on electroblots using the biotin-streptavidin complex. *Anal. Biochem.* **163:**204–209.

112. **Schaffner, W., and C. Weissman.** 1973. A rapid, sensitive and specific method for the determination of protein in dilute solution. *Anal. Biochem.* **56:**502–514.

113. **Towbin, M., T. Staehlin, and J. Gordon.** 1979. Electrophoretic transfer of proteins from polyacrylamide gels to nitrocellulose sheets: procedure and some application. *Proc. Natl. Acad. Sci. USA* **76:**4350–4354.

114. **Woodward, M. P., W. W. Young, Jr., and R. A. Bloodgood.** 1985. Detection of monoclonal antibodies specific for carbohydrate epitopes using periodate oxidation. *J. Immunol. Methods* **78:**143–153.

8.7.2.10. Immunofluorescence

115. **Herzenberg, L. A., S. C. De Rosa, and L. A. Herzenberg.** 2000. Monoclonal antibodies and the FACS: complementary tools for immunobiology and medicine. *Immunol. Today* **21:**383–390.

116. **Johnson, G. D., and E. J. Holborow.** 1986. Preparation and use of fluorochrome conjugates, p. 18.1–18.20. *In* D. M. Weir, C. Blackwell, and L. A. Herzenberg (ed.), *Handbook of Experimental Immunology*, 4th ed., vol. 1. *Immunochemistry.* Blackwell Scientific Publications, Boston, MA.

117. **McKinney, R. M.** 1986. Immunofluorescence microscopy: reagents and technique, p. 35–49. *In* R .B. Kohler (ed.), *Antigen Detection To Diagnose Bacterial Infections*, vol. 1. *Methods.* CRC Press, Boca Raton, FL.

GROWTH

Introduction to Growth

JOHN A. BREZNAK

"The study of the growth of bacterial cultures . . . is the basic method of microbiology" (1). This quote, excerpted by Gerhardt to introduce this section in *Methods for General and Molecular Bacteriology*, is as true today as when written by Jacques Monod nearly 60 years ago. Despite the impressive impact that molecular biological methods have made on understanding the phylogenic diversity and physiological properties of microbes—especially those that have not yet yielded to cultivation in vitro—it is the study of the *growth* of microbes, either in the laboratory or in their natural habitat, that continues to provide some of the deepest insights into the role of particular microbes in the real world. Hence, most of the chapters in this section are concerned with microbial growth: its achievement, measurement, and interpretation in the laboratory, as well as thermodynamic relationships and physicochemical factors affecting growth.

Chapter 9 by Koch has been subject to minor updating since the last edition but remains largely intact, reflecting the fact that the mathematics of microbial growth, as well as the methods for its measurement, are fundamentally unchanged. It remains a cornerstone chapter of this section, important for beginning microbiologists and seasoned investigators as well, and it includes statistical and computer methods for growth measurement.

In chapter 10, dealing with microbial nutrition and media, Emerson and Tang of the American Type Culture Collection (ATCC) update and expand the former chapter by Cote and Gherna (also of the ATCC). Retained is an overview of macro- and micronutrient requirements of microbes and a description of defined and undefined media for cultivation of metabolically diverse microorganisms, from autotrophs to heterotrophs, and including both chemotrophs and phototrophs. Also retained are methods for sterilization of media and medium components, including heat-labile components. In chapter 11, Teske and Cypionka update the former chapter on enrichment and isolation written by Holt and Krieg. Notable additions include methods for enrichment and isolation of iron- and sulfide-oxidizing bacteria by using gradient culture systems, as well as methods for extinction and high-throughput culturing, single-cell isolation, and rRNA-based molecular methods to follow sought-after microbes from environmental samples through enrichment to pure culture.

In chapter 12 on culture techniques, Hashsham has revised, updated, and merged previous chapters by Krieg and Gerhardt (solid, liquid/solid, and semisolid culture) and Gerhardt and Drew (liquid culture) into one comprehensive and unified treatment encompassing batch, continuous, and specialized (e.g., fed-batch, immobilized-cell) cultivation at the laboratory scale. Hashsham and Bauschke follow this in chapter 13 with one of the most important new additions to this section: a discussion of the energetics, stoichiometry, and kinetics of microbial growth. This chapter reviews the chemical and thermodynamic underpinnings of microbial growth wherein materials (elements of the substrate[s] and their associated electrons) are converted into energy and cell material (biomass), emphasizing the importance of mass and redox balance for obtaining an accurate interpretation of the energetics of microbial growth on various substrates, or on a given substrate under various culture conditions. It ends with a discussion of the kinetics and modeling of batch and continuous (e.g., chemostat) culture systems.

Chapter 14 by this writer and Costilow required relatively minor revision from the previous edition, as the main physicochemical factors that affect microbial growth (temperature, pH, water activity, etc.) remain the same. However, a new addition to this chapter is a discussion of microaerophiles and methods for growth of oxygen-requiring, but at the same time oxygen-sensitive, microbes under hypoxic conditions.

The last two chapters are also important new inclusions in this section. Chapter 15 by Tindall et al. provides an extensive description of methods for phenotypic characterization of microbes. It is an update of a previous chapter by Smibert and Krieg that was included in a section entitled "Systematics" in the last edition. However, with reorganization of the current edition, and considering that many of the characterization methods depend on *growth* of the microbes in question, this section seemed a logical place for it. An important addition to this chapter is a discussion of principles of comparative systematics, thereby providing an intellectual context within which the results of phenotypic characterization tests can be interpreted effectively. Chapter 16 by Ciche and Goffredi is new to this edition and describes general methods to study the growth and activities of microbes on, or within, other organisms, i.e., microbial symbioses. In light of the fact that microbial symbioses are extremely diverse and include microbial associations with animals, plants, and even other microbes, this chapter focuses on general methods to establish that a "symbiosis" is actually occurring (i.e., a Koch's postulates approach to verification); methods to characterize "environmental" symbioses, wherein one or more of the microbial partners, or the host, have not yet been cultured individually in vitro; and model symbiotic systems, wherein the partners not only can be individually cultured in vitro but also are amenable to genetic manipulation. The emphasis here is on some well-studied animal-microbe symbioses as examples.

REFERENCES

1. **Monod, J.** 1949. The growth of bacterial cultures. *Annu. Rev. Microbiol.* **3:**371–394.

9

Growth Measurement

ARTHUR L. KOCH

Growing bacterial cultures and knowing their growth rate and density of growth are essential for microbial physiology. But obtaining this knowledge is not simple and is painstaking. Moreover, it is old-fashioned. Much modern biology avoids this care, and about half of the molecular biology successfully does without. However, it is truly needed for critical work in microbial physiology and hence for much of molecular biology. This circumstance means that much of the work and techniques presented here are old hat and the present chapter is anachronistic and is archaeology, but it is essential.

9.1. PRINCIPLES

Principles of bacterial growth are also discussed in chapter 13 of this volume and in references 1 through 8.

9.1.1. Definitions of Growth

To measure bacterial growth, precise definitions are needed. Probably the most basic definition of growth is based on the ability of individual cells to multiply, i.e., to initiate and complete cell division. This definition implies monitoring the increase in the total number of discrete bacterial particles. There are two basic ways to do this: by microscopic enumeration of the particles or by electronic enumeration of the particles passing through an orifice. This definition would also falsely include the fragmentation of nongrowing filamentous organisms as growth, although this kind of apparent growth cannot continue indefinitely.

A second definition of growth involves determining the increase in CFU. Since some cells may be dead or dying, this definition of growth may be different from the one based on the detection of discrete particles. In the long run, the increase in the number of organisms capable of indefinite growth is the only important consideration. This is the reason why colony counting and most-probable-number (MPN) methods of measurement are so important. Viable-counting methods, which seem so natural to a microbiologist, are really quite special in that cultures are diluted so that individual organisms cannot interact. For example, these methods cannot in principle be applied to obligately sexually reproducing organisms requiring male-female interaction or to colonial organisms such as myxobacteria that under certain conditions need to be part of a large mass of organisms that produces a sufficient amount of exoenzyme. Even when applied to procaryotes, there are special restrictions and limitations; e.g., exogenous CO_2 must be available in sufficient concentration, although it need not be supplied if many organisms are present (see chapter 6.5.4).

A third definition of growth is based on an increase in biomass. Macromolecular synthesis and increased capability for synthesis of cell components are the obvious basis for the measurement of growth by the bacterial physiologist, the biochemist, and the molecular biologist. From their point of view, cell division is an essential, but minor process that seldom limits growth; what usually limits growth is the rate at which enzymatic systems utilize resources to form biomass (cytoplasm).

A fourth definition of growth is based upon the action of the organisms in chemically changing their environment as a consequence of the increase in biomass.

9.1.2. Balanced Growth

The four definitions of growth mentioned above become synonymous under a single circumstance: balanced growth (17). An asynchronous culture can be said to be in balanced growth when all extensive properties increase proportionally with time. "Extensive" is a term from physical chemistry and refers to the properties of the system that change when there are altered amounts of substances of various kinds in the system. Thus, biomass, DNA, RNA, and cell number are extensive properties of a culture in a fixed volume, but temperature or constituent ratios (such as DNA or RNA per cell) are intensive properties that do not change during balanced growth. The application of this physicochemical principle to bacteriology lies in the thought that if a culture is grown for a long enough duration when growth is sparse enough not to alter the environment significantly, sooner or later the bacteria will come to achieve a stable growth state characteristic of any particular constant environment no matter what the condition of the cell was initially. Once this balanced growth state has been achieved, if the conditions remain the same the culture will remain in balanced growth indefinitely (until a nutrient is exhausted or the culture is altered by mutation and selection). The criteria given above are too stringent to be fully met, but it is readily possible to study cultures that are substantially held in balanced growth by maintaining growing cultures at low density by dilution and by using only a single measure (such as biomass measured turbidimetrically) to monitor growth. If the doubling time remains constant over an extended period, it is a good and useful assumption that growth is balanced.

Consider any extensive property (such as biomass, protein, DNA, or RNA) of a culture in balanced growth and call this property X. The rate of formation of X will be proportional to the amount of biomass, m. Call the proportionality constant C_1. Then,

$$dX/dt = C_1 m$$

Because growth is balanced, X/m is constant; call this constant C_2. Then,

$$dX/dt = (C_1/C_2)X$$

where (C_1/C_2) is a proportionality constant (equal to the specific growth rate, most usually designated by μ). Through the laws of calculus, this equation can be integrated and a boundary condition ($X = X_0$ when $t = 0$) can be imposed, yielding

$$X = X_0 e^{\mu t}$$

That is, for every cellular substance or even an extracellular product of the cells, after growth becomes balanced the increase of X will be exponential in time. If the ratio of one substance to every other substance is to remain constant, the same proportionality constant must apply for every choice of X. This common value of μ is called the **specific growth rate** or **growth rate constant**. Any substance that you wish to choose and call X will increase exponentially, but also any combination of substances, or, indeed, the rate of change of any substance will increase exponentially with the same specific growth rate, μ. Practically, this allows the rate of oxygen uptake or the production of a metabolite to be used as an index of growth.

Intensive properties, such as μ and the ratio of the concentrations of different cell constituents, must necessarily remain constant under conditions of balanced growth. In addition, the distribution of cell sizes stays constant. An average cell will have a constant rate of carrying out every cellular process, and newly arisen daughter cells will have a constant probability of being able to form a colony (i.e., the percentage of nonviable cells will remain constant). Therefore, under these special conditions, no matter what measure of growth is used (whether it is particle counting, colony formation, chemical determination of a cell substance, consumption, or excretion of a substance), the same specific growth rate will be obtained and that rate will be constant through time in a constant environment.

Such favorable conditions in principle occur when bacteria are cultivated under long-term continuous culture. Balanced growth should arise during chemostat growth that is limited by the rate of addition of a single nutrient and during turbidostat growth when the culture is mechanically diluted to maintain a constant amount of biomass in the growth vessel and growth is limited by the nature of nutrients and not their amount. Balanced cultures can be formed by repeated dilution under batch growth (6, 62), if carried out in such a way that the culture never goes into the lag phase or into the stationary phase.

9.1.3. Changes in Cell Composition at Different Growth Rates

Organisms respond to environmental conditions, both physical and chemical, by altering their own composition. These changes have been well documented for certain enteric bacteria (37, 46) but may occur with all procaryotes. In general, favorable growth conditions mean faster growth, which requires a higher concentration of ribosomes and associated proteins. In terms of gross composition, the most obvious change is an increase in RNA content. Also, under favorable growth conditions the cells can lay down reserve materials such as glycogen and poly-β-hydroxybutyric acid. These changes in composition lead to the possible pitfall of the microbiologist falsely relating one measure of growth to others when comparing growth in different environments.

9.1.4. Unbalanced Growth

Although balanced growth conditions lead to reproducible cultures, much of physiology deals with the responses of organisms to changes in their environment that lead to progressive changes in the organism during the ensuing "unbalanced growth." When a stationary-phase culture is inoculated into fresh medium, the properties of organisms change drastically through the course of the batch culture cycle. Though only well documented for certain enteric bacteria (1, 14), similar changes probably apply with all procaryotes. The exact sequence of changes in composition and morphology depends on the medium and on the age and condition of the inoculum. Culture cycle phenomena have relevance to growth measurements. These phenomena are particularly important in ecological studies, where the conditions under which organisms grow are critical and, to a large degree, uncontrollable. The changes in characteristics of cells are also involved in response to the fluctuations in natural conditions (37, 40).

A typical bacterial culture cycle progresses as follows. When a stationary-phase culture is diluted into rich medium, macromolecular synthesis accelerates. The components of the protein-synthesizing system (i.e., ribosomal proteins and RNA) are made first. Only after considerable macromolecular growth does cell division take place. During this lag phase, the average size of cells increases greatly. Later when the capacity of the medium to support balanced exponential rapid growth is exceeded, cellular processes slow down. Different processes slow differentially in a way to produce small, RNA-deficient cells. Finally, the cells largely die (20) and may eventually lyse.

Workers using the techniques from this book may employ intentional perturbations of growth. These perturbations can result from nutritional shifts or deficiencies, the use of growth inhibitors (particularly antibiotics), or the effects of radiation or other extreme physical conditions (e.g., high or low temperature and osmotic pressure). Their effects on growth can be quite complex, and the bacteriologist must be both cautious and critical.

9.1.5. Pitfalls in Growth Measurement

There are four classes of pitfalls in bacterial growth measurement. The first major pitfall is the tendency of most organisms toward either clumping or a having a filamentous habit of growth. This can occur even under mildly toxic conditions with bacteria that ordinarily divide regularly. The second pitfall is the differential viability of injured bacteria under different culture conditions. Repair processes may permit the recovery of viable cells under some but not other conditions (11). A third major pitfall is the possible development of resistant stages. Bacteria known to form resistant stages, such as spores, pose no problem since the controls and measurements to correct for such forms are well known; but when resistant stages are not suspected, error can arise. The fourth pitfall pertains to the way in which the inoculum is exposed to the new environment. Different results may be obtained if the concentration of an agent is raised gradually, if it is raised discontinuously, or if a high concentration is temporarily presented and then removed or lowered. The cell concentration at the time of challenge can be critical. Bacteria may have special ways, sometimes inducible and sometimes unknown, to protect themselves against toxic agents. The protection may be dependent on the number of organisms cooperating in detoxification.

The phenomena related to the fourth pitfall can be illuminated with a single example from work on the interaction of rifampin and *Escherichia coli* (42). Rifampin is a large molecule that can only very slowly enter the cell. Doses of rifampin can be detoxified by a sufficiently large number of cells, and the process appears to be inducible. Therefore, higher doses of drug can be tolerated when the cells have previously been gradually exposed to the antibiotic. Independent of this phenomenon, fast-growing bacteria are more resistant than slow-growing ones. This observation can be explained largely by the shorter time available for the drug to penetrate relative to the doubling time of the growing bacteria. For example, under conditions where the low dose by itself would not inhibit growth, a higher dose of antibiotic that is subsequently diluted can block growth. Thus, when growth is slowed by the brief high dose, the inhibition remains permanent because growth is slowed and now sufficient time is available to allow penetration of enough drug to maintain the blockage of growth.

9.1.6. Mycelial Growth

The evaluation of mycelial growth (the first pitfall) can be easier and also can be much more difficult than is the evaluation of "well-behaved" organisms that engage in binary fission and then promptly separate. The problems and methods for mycelial growth have been discussed by Calam (16) and Koch (39) and are briefly dealt with here. Filtering filamentous cells, with or without drying, is easier than filtering smaller nonfilamentous cells. In addition, the increase in the physical size of a mycelial colony of filamentous cells can be monitored. In the extreme case, a tube 1 m long containing a layer of nutrient agar is inoculated at one end, and the mycelial mass grows along the tube and may even be continued successively into other tubes (59). Such colonial growth is essentially one-dimensional, whereas growth of mycelial colonies on agar surfaces is two-dimensional. In liquid shake cultures (chapter 12.2.2) mycelial growth is three-dimensional. Under all three conditions the linear dimensions increase linearly (not exponentially) with time because of the nutrient limitation created by the cells.

The rate of increase in size of the colony in one, two, or three dimensions depends on the rate of elongation of the terminal hyphae that happen to be growing on the surface out from the edge of the colony, but the mobilization of resources into the mycelial mass depends on the surface area of the colony. Therefore, with shake cultures particularly, the results depend on the nature and size of inoculated fragments; with large fragments, growth becomes limited at an earlier stage by diffusion of nutrients into the mycelial mass. Thus, the major pitfall with the mycelial habit of growth is that the growth quickly deviates from exponential and depends on the geometry of growth and the nature of the inoculum. In shake cultures, the apparent growth may depend on the shape of the vessel and on the shaking speed, its character (circular or reciprocal), and the distance moved, because all of the above can affect the tendency of the mycelia to break into smaller pieces.

9.1.7. Cell Differentiation
The change of enteric bacteria from large RNA-rich forms in the exponential phase to small RNA-poor forms in the stationary phase has many of the aspects of differentiation. Microbiology, however, has much clearer examples of cell differentiation in the cases of transition of a rod to a coccus, of a vegetative cell to an endospore, and in the formation of exospores, cysts, buds, and prosthecae. The tendency to form filaments in certain circumstances can also be considered a differentiation. These changes and their reversal pose potential pitfalls for all approaches to growth measurement.

9.1.8. Cell Adsorption to Surfaces
Many bacteria naturally adhere to certain surfaces or can adapt and mutate to achieve a high avidity for solid surfaces including glass. Many experiments with chemostat culture have failed to achieve their primary goal because the organisms adhered to the vessel walls. Therefore, plastic or Teflon should be used whenever there is long-term contact of the organism with a culture vessel (37). Other approaches to minimize the effects of growth on vessel walls include the use of large culture volumes in large containers, the use of violent agitation, and the frequent subculture of the cells into fresh glassware. In addition, the use of detergents, vegetable oils, silicone coatings, and a high-ionic-strength medium may to some degree alleviate this problem.

This problem is particularly pertinent during dilution of the culture for measurement by highly sensitive means, such as microscopic and plate counts. It is therefore given further consideration in the discussions of those methods in sections 9.2.1, 9.3.1, and 9.3.2.

Currently, a great deal of research is being devoted to the study of biofilms. This is the flip side of what was said above since many organisms do bind to surfaces and many surfaces in nature have communities with multiple species. On many of these surfaces, polysaccharides are present, secreted by the cells. These help bacteria adhere.

9.1.9. Growth in Natural Environments
If the problem at hand is to measure growth under natural conditions, tremendous difficulties must be overcome (12). In nature, growth almost always involves mixed cultures with bacteria attached to each other or to solid particles. Six kinds of approaches have been used to deal with the measurement problem in these cases. First, $^{14}CO_2$ fixation has been used (72) to monitor biomass increase. Second, tritiated thymidine autoradiography has been used to identify cells engaged in DNA replication (13). Third, antibodies (60) or oligonucleotide probes (see chapter 39) tagged with fluorescent labels have been used to scan soil samples and identify particular organisms (60). Fourth, nucleic acids within cells have been stained with DAPI (4,6-diamidino-2-phenylindole) and acridine orange and then observed under the fluorescence microscope. Fifth, sophisticated mass spectrometry of the gases emanating from natural soil samples has been used to provide information about the growth of organisms within the sample (51). Sixth, agents that block cell division but not enlargement have been used. Cells that become bigger under such conditions can be considered to have been alive.

9.2. DIRECT COUNTS

9.2.1. Microscopic Enumeration
Microscopic enumeration in a counting chamber is a common technique that is quick and cheap and uses equipment readily available in the microbiological laboratory. Microscopic direct counts can also be made by membrane filter sampling and staining (chapter 2).

The counting chamber technique is subject to errors, but these can be overcome to a large degree by using improvements suggested by the work of Norris and Powell (53). The major difficulty in direct microscopic enumeration is the reproducibility of filling the counting chamber with fluid. In the technique recommended below, the thickness of the fluid filling is measured by focusing on bacteria attached to the top and bottom interfaces and measuring the distance between them with the micrometer scale on the microscope. The horizontal dimensions between the scribe marks in commercial chambers are quite accurate and cause no problem.

A second major difficulty is the adsorption of cells on the surfaces of glassware, including pipettes. The procedure described below avoids adsorption during the dilution process but desirably encourages it in the counting chamber. In the dilution process, it can be decreased by carrying out dilutions in high-ionic-strength medium (e.g., physiological saline or many minimal media without their carbon source) instead of water or in a solution of formaldehyde (any concentration between 0.5 and 5% is satisfactory) that has been neutralized with K_2HPO_4 together with a trace of anionic detergent such as sodium dodecyl sulfate (53). The formaldehyde stops growth and motility, and the K_2HPO_4-detergent combination prevents aggregation, but the detergent may lyse certain organisms even though it is used only in minute amounts. Alternatively, plastic containers and plastic pipette tips can be used for the dilution to decrease the loss by adsorption on surfaces. In the procedure described below, 0.1 N HCl is used as the final diluent to favor adsorption on glass surfaces of the chamber.

The Hawksley counting chamber (A70 Helber; Hawksley, Ltd., Lansing, England) is recommended in preference to the Petroff-Hausser counting chamber, and there are other varieties from other suppliers that will serve; for example, Preiser Scientific, Inc, or Arthur H. Thomas Co., Philadelphia, PA. The prime advantage of the former is that its optical path with an ordinary coverslip is short enough that the chamber can be used under an oil-immersion objective at high power. Although most counts are done under a high-dry objective, it is sometimes necessary to use the oil-immersion objective, either because there are too many cells or because they tend to clump. Both chambers are 20 μm deep with scribe marks defining counting areas of 50 by 50 μm.

One major advantage of microscopic examination is that one can gain additional information about the size and morphology of the objects counted. Oil-immersion and high-power objectives make counting more tedious, but critical distinctions can be made. The manufacturers of chambers supply alternative procedures. Additional discussion has been presented by Meynell and Meynell (6) and Postgate (58). For a discussion of statistical considerations, see section 9.7.

Procedure

1. Clean the chamber and the coverslip with water containing a small amount of anionic detergent; rinse with water and then alcohol, blot, and let air dry.

2. Make a preliminary estimation of the concentration of cells. Proceed to the next step if the concentration is less than 3×10^{11} cells per ml. Otherwise, make a primary dilution of the cell suspension in Norris-Powell diluent (5:1) prepared as follows. Add 5 ml of formalin (37% formaldehyde) to 1 liter of water. Adjust to pH 7.2 to 7.4 (indicator paper is sufficiently accurate) by adding solid K_2HPO_4; the amount of the phosphate needed will vary with the amount of formic acid in the formalin. Add a few milligrams of sodium dodecyl sulfate, and repeat until bubbles do not break immediately when air is passed through the solution with a Pasteur pipette.

3. Carry out a final single dilution of the sample in a ratio of at least 1:1 with 0.1 N HCl. This will kill the cells, had the formaldehyde not sufficed, and gives their surface a net positive charge so that they will not aggregate but will instead adsorb onto glass.

4. Immediately fill the hemocytometer counting chamber with approximately 5 μl of the diluted sample, using a Pipetteman, Eppendorf, Centaur, or other plastic-tipped pipette. Let the chamber stand for 1 to 2 min.

5. Examine with a phase-contrast microscope under a high-dry or oil immersion objective. Most of the cells will have attached to the bottom interface; a few cells will have attached to the top interface; and only a few cells will remain suspended and exhibit Brownian motion.

6. First, focus on the cells that have attached to the bottom interface, and read the markings on the dial of the focusing knob. On many microscopes of high quality, this dial reads directly in micrometers. Next, focus on the cells on the top interface. Note the distance between the top and bottom, and augment the difference by one bacterial diameter. This total distance will quite accurately measure the filling, which is nominally 20 μm. When gaining familiarity with the technique, make depth measurements in several well-separated regions of the chamber to find the uniformity of the depth of filling.

7. Count the cells lying within small squares. Optimally, the number in each small square should be in the range of 5 to 15. Score the cells that lie on the boundaries of a square if they are on the upper or right side but not if they are on the lower or left side. A hand tally counter is convenient to count the cells within a square. A second hand tally is convenient to keep track of the number of squares. At least 600 total organisms should be counted for accurate work (see below), but this need not be done with a single filling. Some authors recommend multiple fillings of the chamber, thus averaging the variability of the fillings; however, this is not necessary with the method described here, because the thickness of the filling is measured each time. It is best to count squares chosen in a systematic fashion, such as the four corner squares and the major diagonal squares. This prevents counting the same square twice and averages a possible geometric gradient of cells in the chamber.

8. Calculate the number of cells per ml of undiluted culture by use of the following formula:

$$\frac{\text{total bacteria counted} \times \text{dilution factor} \times 4 \times 10^8}{\text{number of small squares} \times \text{filling depth (in micrometers)}}$$

If the coverslip is precisely positioned, the volume of the filling on top of a small square is $50 \times 50 \times 20 \ \mu m^3 = 5 \times 10^4 \ \mu m^3 = 5 \times 10^{-8}$ ml. The reciprocal of this, 2×10^7/ml, is the usual factor in the formula quoted in the instructions supplied by the manufacturer; thus, the two formulas are the same if the filling depth is exactly 20 μm. For the procedure to be successful, only a few cells need be attached to each surface for the thickness measurement. It is most convenient, however, if most of the cells are on one surface for counting. Then, the final act of counting a small square is the focusing through the suspension to count the cells not attached and the ones on the upper interface that has fewer cells. Then, tally the square, refocus on the original surface, and move on to the next square.

9.2.2. Electronic Enumeration

The Coulter Counter (Coulter Electronics, Hialeah, FL), its commercial competitors, and particularly the laboratory-built versions have been important in the development of bacteriology over the last 40 years. Such instruments are used routinely in clinical hematology. They are also very useful in the enumeration of nonfilamentous yeasts and protozoa but not of mycelial or filamentous organisms. Although use of these counters has led to important concepts in bacteriology, the technique is difficult to apply in a valid way to estimate the volume of bacteria because of their small size and usually elongated shape. Attempts to improve and validate the technique for use in bacteriology have been made at the research level by people with backgrounds in physics or engineering. Evidence of the difficulties involved is the fact that many of the people who helped develop the technique for measuring distribution of cell volumes no longer use it. This article, therefore, only presents the principles, mentions the difficulties and the attempted solutions to these difficulties, and directs the reader to published literature. Then, the reader will be able to consider the applicability of the technique to a specific bacteriological problem.

The principle of electronic enumeration is as follows. A fixed volume of a diluted cell suspension is forced to flow through a very small orifice connecting two fluid compartments. Electrodes in each compartment are used to measure the electrical resistance of the system. Even though the medium conducts electricity readily, the orifice is so small that its electrical resistance is very high. Consequently, the electrical resistance of the rest of the electrical path is negligible by comparison. When a cell is carried through the orifice, the resistance further increases since the conductivity of the cell is lower than that of the medium. This change in resistance is sensed by a measuring circuit and converted into a voltage or current pulse. An electronics circuit similar to that used in counting radioactivity counts the pulses. A discriminator circuit eliminates very small pulses. An upper discriminator eliminates very high pulses, which might be due to dirt or other irrelevant particles. In advanced models, the pulses may be analyzed by size and the pulse sizes may be stored in a multichannel analyzer. Later, the data are recovered and plotted in a histogram, and the numbers, mean size, and standard deviation are calculated. All the data may be collected, and the discrimination against pulses that are too high or too low is carried out as the data are analyzed. The instrument needs some method of forcing an accurately known volume through the orifice during the counting period. This is usually done by displacing the fluid in contact with a mercury column past triggering electrodes that conduct effectively through the mercury but not through the diluent medium.

There are three major problem areas in electronic enumeration. First, some bacterial cells are very small (less than 0.4 μm^3), and the resistance pulses produced as they pass through the orifice are comparable to the noise generated by the turbulence that develops in the fluid flowing through the orifice. The discriminator dial on the instrument can be set to reject the turbulence noise, but then sample information, particularly about newly divided cells, may be lost. Blanks can be run, and their values can be subtracted, but blanks are particularly variable for small cell sizes. In addition, a pattern of turbulence can become established, remain for a while, and then be replaced with another pattern. Finally, the overall error increases when the blank has large statistical variation (section 9.7.3). There is little problem in the study of bacteria growing in rich medium in which even the newly formed cells are larger than the pulses produced by the turbulence or if interest is restricted to the relatively few large bacteria such as *Azotobacter agilis* and *Lineola longa*.

The second major problem in electronic counting results from the failure of cells to separate promptly from each other after cell division. This, and the tendency to form filaments and aggregates, can be minimized by careful choice of the organism and the conditions. The choice of *E. coli* for physiological and genetic studies was fortuitous because of the relatively small extent to which it remains as pairs or chains or forms aggregates. Various physical techniques such as mild ultrasound treatment or vigorous blending in a Vortex mixer can be used to try to separate pairs, disperse aggregates, or break up filaments. There is a related problem of coincidence (section 9.7.5), i.e., the passage of more than one cell through the orifice in a short enough time that a single larger cell is registered by the electronics. This problem can be dealt with (see below) by increasing the dilution so that the probability of coincidence is altered. Coincidence is an especially vexing problem when cell size distributions are to be accurately measured.

The third major problem in electronic counting is clogging of the orifice. The resistance change is a smaller proportion of the total orifice resistance when the orifice diameter is larger. Consequently, for small rod-shaped bacteria such as *E. coli*, orifices with diameters in the range of 12 to 30 μm must be used. The exact choice is a trade-off between an increased signal and the increased noise and chance of becoming clogged. Clogging is best prevented by ultrafiltration of all reagents. Alternatively, the diluent can be prepared and allowed to settle for a long time (months) in a siphon bottle so that the particulate-free solution can be withdrawn away from the bottom. Choosing solvents that do not tend to generate particulate matter can also be of help. Kubitschek (44) recommended 0.1 N HCl for this reason; moreover, it is entirely volatile and does not leach materials out of the glass, which later may form precipitates. However, the problems in practice are severe and become worse in some of the modifications needed to size bacterial cells more accurately.

Electronic counting has influenced the study of bacterial growth more because of its presumed ability to measure the size distribution of bacterial cells than because of its ability to enumerate them. In fact, it is very difficult to measure cell size accurately because of the nature of the resistance pulse generated by a cell passing through an orifice. Attempts have been made to overcome this difficulty by using a relatively long pore (100 μm) (43), but the resulting slower flow and longer path increase the chance of clogging. A second approach involves special hydrodynamic focusing of solutions in such a way that the bacterial cells pass very nearly down the center of the pore surrounded by fluid containing no particles (63, 65, 75).

Additional information about this technique can be found in references 2, 22, and 44 and in instruction manuals for the instruments.

9.2.3. Flow Cytometry

Flow cytometry has become an extremely powerful method for the studies of many aspects of the biology of eucaryotes, but the methods are only now coming into their own in the study of the biology of procaryotes. The delay was simply because the latter are smaller and require more development. The literature is extensive (50, 55, 71). The flow cytometry instruments that are now common in hospitals and research laboratories operate by forming a small-diameter stream of the sample suspension. They were originally designed for applications in immunology but now are commercially available from several manufacturers. The sample suspension is encased in a stream of fluid that is added in such a way that the stream of the sample is made still narrower in diameter. The flowing stream is examined with laser light of various frequencies and angles, and the output of the measurement circuits is used to detect when a particle passes through. The design permits the analysis of biomass by light-scattering methods and by staining of chemical components such as DNA with fluorescent dyes. The electronic circuits allow cells to be counted very rapidly. Growth can be monitored by measuring the increase in counts in sample counted for a fixed duration corresponding to a fixed volume. The problems for the application to procaryotes are in sensitivity and background noise. General applications in microbiology are described in references 23 and 61. Commercially available equipment has been used to study a variety of problems in ecology (15). Other applications are described in references 8 and 52.

Another instrumental approach (66) depends on flow directly on a microscope slide on an inverted microscope. It was developed in Norway by Steen and has the property that it is much cheaper and has been adapted to microbiological applications to access the level of many cellular functions.

9.3. COLONY COUNTS

Bacteriology really became an experimental science when Robert Koch listened to Fannie Hesse and developed the agar plate. This allowed not only the cloning of pure strains but also the enumeration of colonies arising from individual viable cells or CFU. Various colony count methods have been used: (i) **pour plates**, in which an aliquot sample of diluted cells is pipetted into an empty sterile petri dish, molten but cool (45°C) agar medium is poured onto the sample, and the contents are mixed by swirling and then allowed to harden; (ii) **spread plates**, in which the sample is pipetted onto the surface of solidified agar medium in a petri dish and the cells are distributed with a wire, glass, or Teflon spreader; (iii) **thin-layer plates**, in which the sample is pipetted into a tube containing a small volume (2.5 to 3.5 ml) of molten but cool soft (0.6 to 0.75%) agar medium, the mixture is poured onto hardened sterile agar medium in a petri dish, and this overlay is allowed to harden; (iv) **layered plates**, which are like the thin-layer plates except that an additional layer of agar medium is poured onto the newly congealed soft-agar medium containing the cells so that all colonies are subsurface; and (v) **membrane filter methods**, in which the diluted cells are filtered onto an appropriate presterilized membrane filter, which is then placed on an agar medium plate or onto blotter pads containing concentrated liquid medium. Sometimes it is necessary to carefully prewash the membranes and the pads with water or medium and then resterilize them.

The pour plate methods have variations in which the cells are grown in roller tubes or microtubes (56) and are examined with low-power microscopes when the colonies are small (57). There are many individual variations or techniques, sometimes resulting from historical accidents and sometimes resulting from special bacteriological circumstances. Automation of colony counting has put additional special restrictions on techniques but allows colonies on petri dishes to be counted rapidly with reduced operator error.

Two colony count methods are detailed below, and another is described briefly. The first is the spread plate, in which all colonies are surface colonies. It is chosen for presentation because it is reliable and because surface colonies are required to produce the proper color responses with many indicator agars. In many cases, different colors are given from subsurface colonies because the oxygenation is different; therefore, the acid production and reducing potential are different from those on the surface.

The second method is the layered plate. It is very useful because all colonies are subsurface and therefore much smaller and compact. They can be intensely colored. Many more colonies may be present, and yet the coincidence by fusion of colonies is small; this means that several thousand colonies per plate can be used to give statistically more meaningful results. This approach is especially recommended because the main difficulty with usual colony-counting methods is the lack of dynamic range. Rules have been issued that between 30 and 300 colonies are required. The lower limit is set by statistical accuracy, and the upper limit is set by coincidence limitations. This 10-fold range is inconvenient for many purposes, because in many cases one cannot predict the number within a factor of 10 when choosing the dilution factor. The extra care needed to prepare the overlay plates is justified because it allows one to count in the larger range of 30 to 2,000. A second major advantage is flexibility for nutrient supplementation. Minimal agar can be used to pour the basal layer in a large number of petri plates for indefinite storage. Stock supplies of the minimal soft agar can also be kept on hand. Then 10- to 50-fold excesses of needed special nutrients can be added to the aliquots used for the molten top agar. In some cases dyes and chromogenic substrates are added as needed to the soft agar to allow screening of the colonies. Toxic substances can be incorporated to measure frequencies of resistant mutants.

The third method is the pour plate. This method, although commonly employed, must be used with caution because the elevated temperature or the sudden change in temperature may kill some bacteria. The method also lacks the advantages of the foregoing two methods. Before routine use, test and compare pour plates with one of the other methods.

Several articles and books have been devoted to attempts to speed and automate growth measurement (2, 29). This is a field that is in so much flux that further discussion is not practical here.

Certain organisms are very sensitive to substances present in agar. Meynell and Meynell (6) presented an excellent discussion of these problems. Injured organisms may have additional special requirements, and the entire 26th symposium of the Society for General Microbiology (25) was devoted to these problems. Genetically defective organisms pose their own individualistic problems that can be research problems on their own, e.g., the ability of various repair mutants to form countable colonies (18).

The problem of quantifying the number of organisms in cultures of strict anaerobes is dealt with in references 19, 30, and 31 and in chapter 14.

9.3.1. Diluents

Bacterial cells often must be diluted from their original dense concentration to a sparse concentration suitable for observation in a microscope, measurement of cell numbers, analysis for genetic or metabolic properties, or washing preparatory to study. Whatever the use, the diluted cells must retain their original characteristics. The preservation of viability and metabolic activity is particularly important when measuring the CFU. Diluted cells are more likely to be harmed by an unfavorable environment than are cells in dense suspensions, in which the environment contains higher levels of materials leached from the cells. Consequently, care must be taken to use a suitable diluting solution.

Distilled or tap water alone should usually not be used, because they are osmotically hypotonic to all bacterial cells and unbuffered against pH change. Traces of heavy-metal contaminants and detergents in water supplies may also cause inactivation of cells in very dilute suspension. "Physiological saline" (0.85% NaCl) also is usually inadequate because it is isotonic only to mammalian cells and is unbuffered. Viability may be reduced by 50% or more when such diluents are used.

A common general-purpose diluent is phosphate-buffered saline with a small amount of protein, usually gelatin or peptone. The presence of the protein stabilizes fragile bacteria by binding metals and detergents, and the diluent is buffered at neutrality by the phosphate salts. Buffered saline with protein contains 8.5 g of NaCl, 0.3 g of anhydrous KH_2PO_4, 0.6 g of anhydrous Na_2HPO_4, and 0.1 g of gelatin or peptone per liter of distilled or deionized water. It is adjusted to pH 7.0 when necessary.

For critical or unusual situations, pretest the diluent for adverse effects and use a diluent adapted to the particular conditions. For plate counts, do not transfer bacteria growing rapidly at an incubator temperature into cold diluent. To minimize death or multiplication, do not suspend them in diluent for more than 30 min at room temperature. The growth medium devoid of the carbon source may be the best diluent for cultures as it may slow or arrest further growth. $MgCl_2$ (0.2 g/liter, sterilized and cooled separately as a 4% aqueous stock solution) may be added to preserve membrane integrity, especially with gram-negative bacteria. With anaerobic bacteria, the diluent should contain a reducing agent and be freed from dissolved oxygen (see chapter 14.6.2 and 14.6.4).

A detergent (e.g., 0.1% Tween 80) is sometimes suggested as a dispersing agent. However, this is effective mainly for mycobacteria and other cells with hydrophobic surfaces and may be harmful to other cells. Polymerized organic salts of sulfonic acids of the alkyl-aryl type are effective dispersing agents and apparently are not harmful to representative bacteria (48).

Selection and standardization of diluents are especially important in microbiological tests for safeguarding public consumption of food and water, which are detailed in standard methods manuals published by the American Public Health Association. A particularly good review of the literature and methods for using diluents is contained in *Standard Methods for the Examination of Dairy Products*, 16th ed. (49).

Meynell and Meynell (6) have also reviewed the use of diluents, and an entire symposium has dwelt on the death and survival of vegetative microorganisms (25). Further details of diluents are provided in section 9.2.1 and chapter 20.1.1.

9.3.2. Spread Plates

Prepare a suitable agar medium in ordinary glass (old-fashioned but ecologically sound) or plastic (modern and convenient) 9-cm petri plates. Various procedures have been suggested to pour and dry the plates. The following procedure is recommended.

1. Place the covered container of molten agar, shortly after removal from the autoclave and after addition of mildly thermolabile substances (sugars, dyes, antibiotics, and chromogenic substances), into a dishpan or other large vessel containing hot tap water (45 to 50°C). The agar cools quickly (temporarily warming the water further) but uniformly throughout the vessel and reaches a temperature plateau above its solidification point. For 1 liter of agar, 30 min is sufficient. With this procedure there is no gelling around the edges of the container.

2. Pour the agar into the petri plates. If the agar is sufficiently cooled, little condensation will form on the undersurface of the petri plate lid. Further, covering the plates with a sheet of newspaper can minimize condensation. If bubbles form on the agar surface, direct the flame of a Bunsen burner momentarily downward on them until they burst (be careful not to melt plastic petri plates). With 9-cm petri plates, 15 to 20 ml of agar is satisfactory, corresponding to a thickness of 0.24 to 0.32 cm. If thicker, the contrast will be less between the colonies and their background. If thinner plates are used, colonies may be small and present in reduced numbers, possibly because some nutrient is limiting or the plate becomes locally dry.

3. Dry the plates (to remove the water of condensation on the plate lids and water of syneresis on the agar) at room temperature overnight or for 24 h, depending on the relative humidity. Frequently the humidity is lower in incubation rooms or laminar flow hoods, so use them, although some modern incubation sites have controlled humidity.

4. Store the plates indefinitely at 4°C in closed plastic containers. Be sure to allow them to warm to room temperature before use.

5. Prepare a suitable dilution of the cell suspension based on all the information and hunches at your disposal. Calculate to get 100 to 200 colonies, but use the results of plates containing between 30 and 300. If the organism forms only small colonies, up to 500 may be counted. Make the dilutions with a diluent solution that does not favor adsorption to glass, if ordinary glassware and pipettes are to be used. High ionic strength, pH between 4 and 5 (if not harmful), and the presence of small amounts of anionic detergents (if not toxic) are helpful in this regard (section 9.3.1). Alternatively, use presterilized plastic vessels and pipette tips with a Pipetteman-type device.

6. The actual dilutions can be carried out in many ways. Historically, large volumes of diluent (99 ml) and 1-ml samples were used. As pipette and volumetric apparatuses have been improved, smaller volumes have been used, economizing on reagents and mixing time. Modern plastic-tipped semiautomatic pipettes allow very small samples of bacterial culture to be used, but then the problem is proper sterilization. This can be done in an autoclave or in a microwave oven. Commonly, presterilized tips are used. Good laboratory technique, however, depends on the continued sterility of the racks of pipette tips. When a microwave oven is used, the energy is absorbed by bacteria or other biological material much more readily than by the plastic (because the bacteria have dipoles and ions), so that the plastic remains cool but the cells and even spores are heated to lethal temperatures. For many purposes, 0.1-ml serological pipettes may be used to deliver a full 0.1-ml volume of culture into 0.9 or 4.9 ml of diluent (usually contained in tubes measuring 13 by 100 mm) or into 9.9 ml of diluent (usually contained in tubes measuring 16 by 150 mm). These dilution factors are about optimal for ease of making an accurate dilution and ease of mixing adequately (which is the most critical factor). Moreover, with this rule it is easy to check that the dilution was properly carried out and that no trivial error was introduced.

7. Pipette 0.1 ml of the final dilution onto the agar surface of the petri plate. Form two or three free-falling drops off center on the surface, and then blow the remaining fluid onto the surface. If using a pipetting device, push to the second stop of the pipettor. The cells may have a tendency to become immediately attached in situ, so do not delay in spreading the drops.

8. Sterilize a spreader (Fig. 1) by dipping it in 70% alcohol (do not use the autoclave), shaking off the excess alcohol, and igniting. A spreader can be made quickly from a glass Pasteur pipette by using a Bunsen burner with a wing tip.

9. Spread the plate. Try to achieve a uniform coverage as close to the edges as possible. The major difficulty with this method is learning the technique for uniform distribution of the cells. Therefore, after the plates have been incubated, examine the distribution of colonies on all of the plates to learn how uniform your spreading technique has become. Plates with a larger number of colonies are especially useful in this regard, even if they contain too many colonies to count. Also note whether the number of colonies is larger in the vicinity of the original droplets. If so, then the technique must be improved. The agar surface should be dry enough that the 0.1 ml delivered from the pipette is absorbed in 15 to 20 s by the agar. Some workers prefer to work with turntables, which speed the work of spreading the sample uniformly.

10. Incubate the inverted plates in a constant-temperature room or chamber with good temperature regulation (chapter 14.3). Incubating in closed containers is advantageous, but do not overfill the containers, as this will increase the time required for the temperature of the plates to become equilibrated to the temperature of the incubator. This is especially important when using temperature-sensitive mutants. Storing in closed containers also avoids the effects of any noxious gases that may be present in general-purpose constant-temperature rooms (e.g., acetic acid fumes from gel destaining). Opaque closed containers protect against inactivation of colored drugs and dyes by keeping out light. Finally, in closed containers the plates do not dry out and so can be retained for slow- or late-developing colonies (the container can also contain filter paper or paper towels that are kept moist to maintain high humidity).

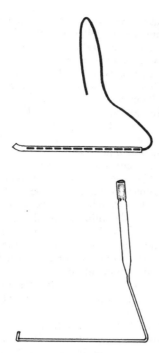

FIGURE 1 Spreaders. The one at the top is bent from a paper clip and fitted into a length of Teflon tubing. The one at the bottom is made by fusing the end of a Pasteur pipette and then bending it sharply in two places by using a Bunsen burner with a wing tip to define the spreading region. The handles of both are bent to the convenience of the user. Both spreaders have low heat capacity. The spreader at the top absorbs fewer bacteria because of its Teflon construction.

Another point of concern has to do with CO_2. Even organisms not usually isolated in a high-CO_2 atmosphere may have a CO_2 requirement. This may be particularly evident when single cells are spread at high dilution and incubated in normal laboratory air. Some of these considerations may be of small importance in any particular case, but all should be kept in mind.

11. Observe the plates before the colonies have fully matured. It is often possible to see that too many colonies are developing. To obtain reliable information with less coincidence correction, it may be possible to make fresh dilutions of a sample that had been retained in the cold or to count colonies under a dissecting microscope while they are still small.

12. Count the colonies. Depending on the circumstance, various types of illumination are advantageous and may not be obtainable with commercial colony counters. Experiment with various types of magnifying glasses that can be worn or clipped onto one's own glasses. Also try various lamps that have a magnifying lens as an integral part. The colonies may be enumerated by marking the bottom of the petri plate with a marking pen, or they may be counted by hand with an electronic counter, hand tally, or television-based scanning equipment. One technique that clearly marks the individual colonies is to stick the point of a colored pencil into the colony. Not all brands of colored pencils transfer color well; Eberhard Faber Mongol colored pencils do this well, but some colors work better than others (e.g., French Green no. 898). Several colors can be used for differential counting. If using an electronic scanning counter, pay careful attention to be sure that false-positive

counts are not registered as a result of dirt and imperfections in the petri dishes (which would not cause difficulty to a human eye).

9.3.3. Layered Plates

1. Prepare the petri plates with a base of 1.5% agar medium approximately 0.4 cm thick. Pour the plates on a level surface. Check the work area with a level carefully; if the surface is not true, use any shim material (metal strips, wood strips, or pieces of paper) to level a piece of plate glass, and pour the three layers of agar while the plate rests on this level surface.

Prepare small test tubes (13 by 100 mm) with 2.5 to 3 ml of soft 0.7% agar medium. It is convenient to prepare stock bottles with this strength of agar in the basal medium. Melt the agar in these bottles. (A microwave oven is the quick way, but experience is necessary to find the setting that melts but does not boil the agar explosively.) In any case, the agar must be thoroughly melted, or mixing cannot be uniform. Add nutrients, dyes, and inhibitors at this point, and then pipette aliquots into small test tubes (13 by 100 mm) previously placed in a water bath or heating block at 45°C. This pipetting can be done while the agar medium is still very hot, decreasing the chance of contamination. The agar will remain liquid at 45°C for several hours. It remains liquid for a longer time at 50°C, but this temperature is likely to cause some killing.

2. Dilute the sample as in the previous procedure (section 9.2.1).

3. Bring the base-medium petri plates from storage to at least 25°C (30°C is better).

4. Pipette 0.1 ml of the diluted sample near the inside lip of the tube. Do this on an identifiable side of the tube (e.g., the side with a trademark, or a side that has a frosted spot or an identifying mark made with a marking pen).

5. Immediately pour the soft agar out of the tube onto the base agar. Pour it over the side of the tube on which the sample has been pipetted. This will wash all of the organisms onto the base agar in such a way that they will be quantitatively transferred and uniformly mixed with the rest of the soft agar. They will not have a chance to become adsorbed to the glass of the test tube or to cluster in spots on the base agar.

6. Tilt and swirl the petri plate so that the melted agar covers the surface.

7. Place the petri plate on the level work area while the agar congeals.

8. Carry out the platings needed for the remainder of the experiment.

9. Pour or pipette 2.5 to 3 ml of additional soft agar onto the congealed agar surface, and distribute by rocking. Then, let this third layer congeal on the level surface.

10. Incubate the plates either upright or upside down, since contamination and degree of dryness are much less critical with this technique than with spread plates.

11. Examine and count the colonies. The subsurface colonies are compact and lens shaped with well-defined edges and are smaller than the usual surface colonies. This makes subsurface colonies a little more difficult to count, but the magnifying glass of a colony counter helps. Magnifying glasses on a headband also help, and these glasses have built-in prisms that reduce eyestrain. A dissecting microscope can also be used.

The extra trouble involved in the overlay technique is worthwhile when many colonies are present. The colonies should be uniformly distributed, and this can be checked by visual inspection. A fraction of a plate can be counted, and then the count can be prorated. A low-power dissecting binocular microscope allows virtually all errors due to coincidence to be eliminated because the colonies can be visualized in three dimensions. It is laborious to count under such conditions, but it can save a very carefully performed experiment from being incomplete and inconclusive.

9.3.4. Pour Plates

1. Prepare a number of tubes with 20 to 30 ml of thoroughly molten agar.

2. Place in a 45°C water bath or heated aluminum block. Allow adequate time for temperature equilibration.

3. Mix the sample aliquot. Do this well but without causing bubbles to form.

4. Pour the contents into a sterile petri dish, and allow them to set.

5. Incubate, examine, count, and calculate.

9.4. MOST PROBABLE NUMBERS

The concentration of viable cells can be roughly estimated by the MPN method (also called the **fraction-negative** or **endpoint-quantal method**), which involves the mathematical inference of the viable count from the fraction of multiple cultures that fail to show growth in a series of dilution tubes containing a suitable growth medium.

This method consists of making a number of replicate dilutions in a growth medium and recording the fraction of tubes showing bacterial growth. The tubes exhibiting no growth presumably failed to receive even a single cell that was capable of growth. Since the distribution of such cells must follow a Poisson distribution (see below), the mean number plated at this dilution can be calculated from the formula $P_0 - e^{-m}$, where m is the mean number and P_0 is the ratio of the number of tubes with no growth to the total number of tubes. The mean number, estimated by $-\ln P_0$ (ln designates the natural logarithm), is then simply multiplied by the dilution factor and by the volume inoculated into the growth tube to yield the viable count of the original sample. The MPN method is a very inefficient method from the point of view of statistics, because each tube corresponds to a small fraction of the surface of a petri dish in a plate count. Consequently, very many tubes or wells in a titer plate must be used, or the worker must be prepared to settle for a very approximate answer.

When is the MPN method of advantage? First, it can be used if there is no way to culture the bacterium on solidified medium. Second, it is preferred if the kinetics of growth are highly variable. Suppose some cells grow immediately and rapidly and end up making a large colony on solid agar that spreads over and obscures colonies of the organism of interest that form later. The small-colony formers may be more numerous but unmeasurable on plates because of the fewer but highly motile or rapidly growing bacteria. Third, it is preferred if other organisms not of interest are present in the sample and no selective method to eliminate them is available. The method has utility when the bacterium of interest produces some detectable product (e.g., a colored material, specific virus, antibiotic, or other metabolite). Then, even though contaminating organisms may overgrow the culture, the numbers of the bacterium in question can be estimated by the fraction of the tubes that fail to produce the characteristic product. Fourth, if agar and other solidifying materials have some factors (such as heavy metals) that may

alter the reliability of the count or interfere with the object of the experimental plan, the MPN method can be used to avoid these difficulties.

Modern developments in laboratory techniques can be used to speed the execution of the MPN method. Machines are available to fill the wells of plastic trays that have as many as 144 or more depressions. Scanning devices designed for other purposes can be used to aid in counting the number of wells with no growth. Similarly, automatic and semiautomatic pipettes can be used to fill small test tubes. Because these procedures make it possible to examine many more cultures, the classical tables of fixed numbers of tubes and fixed dilution series are obsolete and should be abandoned. A different approach and method of calculation is needed and is provided below.

The MPN method used at a single level of dilution when the mean number of bacterial cells capable of indefinite growth is 1.59 per tube is statistically the most accurate (24). This number would result in 20.8% of the tubes remaining sterile (Fig. 2). Consequently, if the expected number is known precisely and the assay is conducted for confirmation, all the available tubes should be seeded with a dilution that is expected to have the value 1.59. The accuracy falls

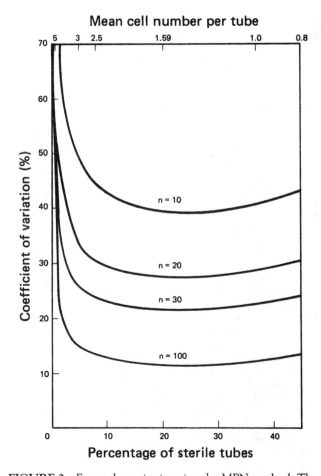

FIGURE 2 Errors that arise in using the MPN method. The CV for a single level of dilution is shown as a function of the percentage of sterile tubes (bottom abscissa) and the mean number of cells per tube (top abscissa). The respective optima occur at 20.8% and 1.59 mean cells per tube for a varying number of tubes (n) prepared at a single dilution level.

off rapidly with deviations from this optimum, particularly outside the range of an average of 1 to 2.5 cells per tube. Thus, the dilutions 10-fold away from the optimum contribute almost no information about the mean number of cells. Consequently, if the viable count is known within a narrow factor, the optimum strategy is to use a narrow range of dilution factors. If the uncertainty is greater, a wider range should be used.

Usually, prior knowledge is not available and growth tubes at several dilutions are needed. The simplest approach is to separate the dilution levels 10-fold. For this approach, little is lost by discarding the data from dilutions at which the percentage of sterile tubes lies outside the range of 8 to 36% or outside the range of 5 to 50% if a larger range of error is acceptable. The error from a dilution within the range can then be read or interpolated from Fig. 2. This method is simple but more wasteful than is necessary. Consequently, statistical methods have been developed to combine data from different dilution levels when specified numbers of tubes are run at each level.

Classically 10 cultures at each of three successive 10-fold dilutions have been used. Of the 1,000 possible outcomes of such tests, Halvorson and Ziegler (28) listed the 210 cases that could happen with P values of >0.01%. Finney (24) gave a good exposition of the statistical aspects and the approaches to calculate the MPN of replicates at each dilution. He pointed out that there could be a definite advantage in computation when carrying out error calculations if the dilution series extends over a range such that most dilution levels show some tubes with growth, but there is a higher dilution in which every tube shows no growth. Even then, the proper statistical calculations are cumbersome and very wasteful of time and material.

With modern computer programs (see Appendix, "The MPN Method"), it is easier, better, and more flexible to do away with tables and to calculate the MPN for an individually chosen and optimized arrangement of tubes and dilutions. Even for a standardized series of dilutions (for example, the data shown in Table 1), a computer program is of value since sometimes the possible outcomes obtained are not listed in the published tables.

Table 1 shows an example for the case of 10 tubes at each of three 10-fold dilutions. Thus, the numbers of tubes (t_1, t_2, and t_3) are each 10. The volumes of solution pipetted are $a_1 = 10$, $a_2 = 1$, and $a_3 = 0.1$. The data for the example are $g_1 = 8$, $g_2 = 5$, and $g_3 = 1$. Although 30 tubes are used, the accuracy is less than if all 30 tubes had been used at a single dilution (as given by the curve method $n = 30$ in Fig. 2) and would have been much poorer if a wider range had been chosen.

When designing an MPN measurement, one must choose first how many dilution tubes at each level and how many dilution levels are to be used for the analysis and how far apart they will be spaced. To do this, one must consider the expected titer and the confidence that one has in that estimate. The less sure one is of the titer of the sample, the wider is the range of dilution levels that one should test; conversely, if the titer can be quite accurately estimated or inferred, one should choose fewer levels. This flexibility, compared with existing tables, is really the point of using the computer program, because it allows the investigator to make optimum use of the MPN technique.

The program, as written, prints the output only on the screen. It is expected that the user will wish to modify the program to obtain a printed output instead. Printer directions have been omitted because format statements vary widely among computer systems. If the same volumes or the same number of levels are always used, the program can be customized to fix these parameters and to delete some of the requests that are made by the supplied program.

A statistician might properly point out that the program does not compute the error statistics. Although that feature could have been included in the program, it was omitted to make the program briefer, simpler, and more readily understandable and adaptable. The program is short enough that the user, it is hoped, will undertake the one-time-only job of keying it in. However, there are two even better reasons for not calculating the error directly from the statistical theory for the MPN. First, the most reliable data come from the level in which 20% of the cultures show no growth. The error can be quickly estimated with adequate accuracy by neglecting the rest of the data and using the level with the closest proportion to 20% and values read from Fig. 2. Second, as mentioned in section 9.2, the errors in pipetting and sampling are not included in such an error computation. Consequently, it is strongly recommended that several independent samples be taken and processed independently by MPN determinations. Consider an example in an ecological context in which the problem is to estimate the number of cells in a stream. It is far better to take samples from different locations within the stream or sample at a series of different times than to take one sample and analyze it with many tubes. With more original samples and fewer tubes in each individual MPN determination, the estimation of the mean and standard deviation (and standard error) of the independent MPN values will give an estimate of the combined error due to variation of the biological source, errors in the dilution procedure, and errors in the MPN determination itself.

9.5. BIOMASS MEASUREMENT

A measure of the mass of bacterial cell constituents frequently is used as the unit basis for measurement of a metabolic activity or the concentration of a morphological or chemical constituent; i.e., "biomass" and cell numbers provide the two basic parameters of bacterial growth.

The methods for measuring biomass seem obvious and straightforward, but in fact they are complicated if accuracy is sought. Furthermore, the results may be expressed in different ways, and, in some of these ways, the values may be more relative than absolute.

9.5.1. Wet Weight

A nominal wet weight of bacterial cells originally in liquid suspension is obtained by weighing a sample in a tared pan after separating and washing the cells by filtration or centrifugation. In either case, however, diluent is trapped in the interstitial (intercellular) space and adds to the total weight of the sample. The amount of interstitial diluent may be substantial. A mass of close-packed, rigid spheres contains in its interstice 27% of the volume of the cells. This is independent of size for uniform spheres, and a pellet of bacterial cells close packed by centrifugation may contain an interstice of 5 to 30%, depending on cell shape and amount of deformation.

A method for obtaining the actual wet weight of the cells themselves is to correct the nominal wet weight of a packed-cell pellet by subtracting the experimentally determined weight of diluent in the interstitial space. This is determined by use of a nonionic polymeric solute that permeates into the interstitial space but is too large to penetrate into bacterial cell walls, such as dextrans of very high average molecular weight. The method is described and discussed in chapter 20.1.1.7.

9.5.2. Dry Weight

A nominal dry weight (solids content) of bacterial cells originally in a liquid suspension is obtained by drying a measured wet weight or volume in an oven at 105°C to constant weight. The cells must be washed with water (possibly extracting cell components), or (better) a correction must be made for medium or diluent constituents that are dried along with the cells. Separating the cells by filtration poses particular problems. Oven drying may lose volatile components of the cells, and some degradation may occur, as evidenced by discoloration (particularly if a higher temperature is used). Some regain of moisture occurs during the transferring and weighing process in room atmosphere, so this should be done quickly within a fixed time for all replicate samples, especially if the relative humidity is high. It is best, of course, to use tared weighing vessels that can be sealed after drying.

Possibly more accurate determination can be made by drying the sample to constant weight in a desiccator vessel with P_2O_5, and under oil-pump vacuum at 80°C or by lyophilization (chapter 47.2.1). Results with the three methods were indistinguishable within 1% when compared for bacterial spores (10). An excellent discussion of dry-weight procedures and errors is given by Mallette (47).

The dry weight of cells may be expressed on a wet-weight basis (grams of solids per gram of wet cells) or on a wet-volume basis (grams of solids per cubic centimeter of wet cells or of cell suspension).

9.5.3. Water Content

The amount of cell water in fully hydrated cells is equal to the difference between the wet weight and the dry weight of the cells themselves, determined as described above.

Water content also can be determined relative to the humidity of the atmosphere or to the water activity of the solution in which the cells occur (chapter 14.2). Completely dried cells equilibrate with an atmosphere of controlled, known humidity in successively increasing increments up to saturation (70), resulting in a "sorption isotherm" curve (Fig. 3). The initial phase of water sorption, at very low humidities, represents tightly bound, monolayer-adsorbed water; the intermediate plateau phase represents loosely bound, multilayer-adsorbed water; and the terminal phase, at high humidities, represents bulk solution, "free" water. The total amount of cell water in fully hydrated cells is obtained at 100% humidity, the intercept for which occurs at a steep rise in the curve and consequently is difficult to determine precisely.

Another way to estimate the amount and distribution of cell water is to carry out isotope dilution measurements with a probe for water (e.g., ${}^3H_2O$), a probe that will not pass the cytoplasmic membrane (e.g., [^{14}C]sucrose), and a probe excluded from the cell wall (e.g., [^{14}C]dextran). See chapter 20.1.5 and reference 67.

The water content of cells may be expressed on a wet-weight basis (grams of H_2O per gram of wet cells), a dry-weight basis (grams of H_2O per gram of dry cells), or a wet-volume basis (grams of water per cubic centimeter of wet cells). Microbiologists perhaps most commonly use the wet-weight basis, but the dry-weight basis is less subject to experimental error. The two bases are correlated by the equation, $WC_{dry} = WC_{wet}/(1 - WC_{wet})$, where WC_{dry} and WC_{wet} are the dry and wet water content determinations.

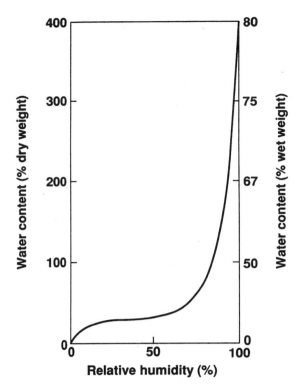

FIGURE 3 Typical water sorption isotherm curve for bacterial cells.

9.5.4. Volume

The volume of a mass of wet cells (or the average size of a single cell) is best obtained by the procedure described in chapter 20.1.4.

9.5.5. Wet and Dry Densities

The density (i.e., the specific gravity) of a bacterial cell may be obtained as either the wet density, based on the total of the solids and water contents, or the dry ("chemical") density, based on the solids content. Both densities are expressed in units of weight per unit of volume (grams per cubic centimeter). The wet density may be obtained simply by dividing the cell wet weight by the cell dry volume. Unfortunately, the dry volume is difficult to determine accurately. The cells first are completely dried, e.g., by lyophilization (chapter 47.2.1). Occluded gas and residual water vapor are removed by holding a fairly large mass (>2 g) of the dried cells under a high vacuum until the pressure becomes constant (e.g., at 0.01 mm Hg [ca. 1.33 mPa]). The volume of an inert and nonadsorbing gas (e.g., helium or nitrogen) displaced by the degassed cells then is measured in a volumetric adsorption apparatus. This procedure was exemplified with bacterial spores by Berlin et al. (1). The equipment may be obtained commercially (Fekrumeter; Gallard-Schlesinger, Inc., Carle Place, NY).

Wet and dry densities can also be determined by equilibrium buoyant density gradient centrifugation (26, 45, 69). For determination of the wet density of hydrated cells by this technique, gradients of Percoll (chapter 7) almost completely avoid variations in osmotic tonicity and water activity and have replaced earlier gradient media. Percoll gradients can be formed in the usual gradient maker or can be self-generated by prior centrifugation. Percoll is manufactured by the addition of polyvinylpyrrolidine to colloidal silica particles to create a stable suspension of dense particles that do not penetrate cell walls. Additional precision can be achieved by the use of colored beads of known densities to provide density markers (chapter 7.3) and laser-based machines to locate the bands (W. W. Baldwin and A. L. Koch, unpublished data). A major finding with the Percoll gradients is that some organisms maintain a constant density through the cell cycle (45) and others do not (21, 45).

Other substances such as Nycodenz and Metrizamide can be very useful under certain circumstances (chapter 7.3). They are derivatives of triiodobenzene and have a high density, and their contribution to the osmotic pressure can be significant; however, they afford a dense and transparent material for gradient construction.

9.6. LIGHT SCATTERING

Light-scattering methods are the techniques most generally used to monitor the growth of pure cultures. They can be powerful and useful, but they can lead to erroneous results. Their major advantages are that they can be performed quickly and nondestructively. However, they may give information about a quantity not of primary interest to the investigator. They give information mainly about macromolecular content (dry weight) and not about the number of cells. The physics and mathematics of light scattering are complex; still, without difficult physics, elementary considerations can give most of the needed answers.

The basic principle is that of Huygens and is the same one needed to understand the properties of lenses, such as those in microscopes. Electromagnetic radiation interacts with the electronic charges in all matter. When the light energy cannot be absorbed, a light quantum of the same energy (color) must be reradiated. This light can emerge in any direction. This means that all atoms in a physical body serve as secondary sources of light. Photons that happen to go in the direction of the original wave will stay in phase with the wave arriving directly from the light source, but light reirradiated from different points in the body that go in other directions will differ in phase at an observation point. The phase will differ depending on the distance that the emitting photon must travel throughout the object, because light going through matter is slowed; the degree of slowing is measured by the index of refraction.

It is the above circumstance that controls how light is bent and focused in large bodies such as prisms, lenses, raindrops, or a pane of glass. In the last case, all the light scattered in every direction except straight ahead cancels, leaving a beam of light going in the original direction. However, it is slightly retarded relative to a light ray not going through the pane; this is the basis leading to indices of refraction greater than unity.

9.6.1. Turbidimetry

Bacteria in suspension are between the size limits of atoms and an object such as a window pane. Consequently, most of the scattered light is deviated only slightly; i.e., it is directed almost but not quite in the same direction as the incident beam (Fig. 4). The light scattered from an atom or a very small particle depends inversely on the fourth power of the wavelength of light. Such particles therefore scatter blue light more strongly and transmit red light more efficiently. This is the reason that the sky appears blue, smoke

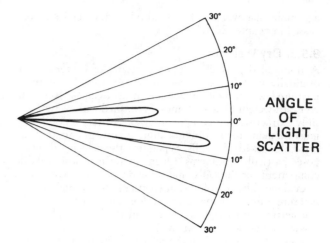

FIGURE 4 Light-scattering patterns from randomly oriented bacteria showing the angular distribution of light scattered by a beam traversing from left to right through a suspension of cells with the size and physical properties of *Escherichia coli* grown in minimal medium. The upper pattern shows the distribution if the cells are ellipsoidal with an axial ratio of 4:1. The lower pattern shows the distribution for spherical cells.

and many virus suspensions are bluish, and sunsets are orange and red. For a pane of glass, there is no light left uncanceled in any but the original direction, so the light retains its original hue and the pane appears transparent. For bacteria, the light scattered away from the original direction most nearly approximates an inverse second power (34). Therefore, suspensions of bacteria (like a cloud in the sky) are intermediate and appear white and turbid. Approximations of the turbidity of bacterial suspensions may be made by visual comparisons with **McFarland turbidity standards** consisting of $BaSO_4$ suspensions (see chapter 15). This standard was developed in 1907. It can be useful, and almost no equipment is required. A piece of newspaper is sufficient. A 0.5 McFarland standard should be prepared and quality controlled prior to beginning susceptibility testing. If tightly sealed to prevent evaporation and stored in the dark, the standard can be stored for up to 6 months. The McFarland standards are used to adjust the turbidity of the inoculum for antibiotic susceptibility testing.

The common practice in bacteriology is to use a colorimeter or spectrophotometer to measure turbidity directly. Ideally, such instruments measure only the primary beam of light that passes through the sample and reaches the photocell without deviation (Fig. 5A). Usually, the measurement is made of the light intensity relative to that which reaches the photocell when a sample of the suspending medium has replaced the cell suspension. From the above discussion, an ideal photometer must be designed with a narrow beam and a small detector so that only the light scattered in the forward direction reaches the photocell. That is, the instrument must have well-collimated optics. Such an ideal instrument gives larger apparent absorbance values than do simpler instruments with poorly collimated optics, since in the latter instruments a large percentage of the light scattered by the suspension is still intercepted by the phototube (Fig. 5B). Therefore, the measuring system responds as if there were less light scattered than is actually deviated from the forward direction.

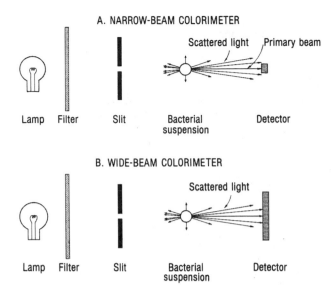

A. NARROW-BEAM COLORIMETER

Lamp   Filter   Slit   Bacterial   Detector
                          suspension

Scattered light   Primary beam

B. WIDE-BEAM COLORIMETER

Lamp   Filter   Slit   Bacterial   Detector
                          suspension

Scattered light

FIGURE 5 Schematic designs for filter-type "colorimeter" instruments. (A) Narrow-beam instrument. Almost all the scattered light deviates enough to escape falling on the photodetector. (B) Wide-beam instrument. A variable fraction of the scattered light falls on the detector, and thus the sensitivity is lowered.

The larger the number of bacteria in the light path, the lower the intensity of the light that goes through the sample without deviation. At low turbidities this is a simple geometric relationship, since the intensity of the unscattered light decreases exponentially as the number of bacteria increases. The geometric relationship between turbidity and bacterial numbers can be deduced by considering a suspension of bacteria that reduces the light intensity to 1/10 of the original intensity. Now consider two such identical suspensions in two cuvettes in a row such that the light beam goes through both suspensions. By how much would the light intensity be reduced after it goes through both tubes? The answer is 0.1 × 0.1, or 0.01, of the intensity of the incident beam. In the ideal case the same relationship holds when the concentration of bacteria is doubled and only one tube is used. When this is expressed mathematically, the intensity of the unscattered light (I) is equal to the intensity of the incident light (I_0) multiplied by $10^{-W/W_{10}}$, where W_{10} is that concentration of bacteria which gives a 10-fold decrease in the light intensity. That is:

$$I = I_0 \cdot 10^{-W/W_{10}}$$

Therefore, if $W = W_{10}$, I would be 0.1 of I_0. If this concentration of bacteria were doubled, I would be 0.01 of I_0. By taking base 10 logarithms (log) of both sides of the equation, this equation can be rewritten as follows:

$$-\log (I/I_0) = \log (I_0/I) = W/W_{10}$$

A similar law, the Beer-Lambert law, works for the absorption of light by colored samples. It could be derived in the same way. Therefore, most instruments have a scale that directly reads the quantity log I_0/I. This is called **absorbance** (A) or, on older instruments, **optical density** (OD). Optical density is the more general term and can be used for turbid as well as light-absorbing solutions, but most workers prefer the symbol A. From the above equation, ideally, a plot of A

of bacterial cultures versus cell numbers yields a straight line going through the origin. This could be written

$$A = K \cdot W$$

where K is the slope (and is I/W_{10}) and A has replaced log I_0/I.

As shown in Fig. 6, the actual absorbance becomes increasingly lower than predicted by the formula. More light gets through the turbid suspension than expected, for two reasons: first, because light scattered from one bacterium is rescattered by another bacterium so that the light is redirected back into the phototube; and second, because the bacteria interfere with the Brownian motion of one another so that they become more evenly distributed and scatter less light away from the direction of this beam (like the pane of glass considered above).

These considerations allow one to grasp some practical implications about bacterial turbidimetry. The results of both theoretical (34, 35) and experimental (27, 34, 36, 37) studies show that dilute suspensions of most bacteria, independently of cell size, have nearly the same absorbance per unit dry-weight concentration. However, very different absorbances are found per particle or per CFU with different sizes of bacterial cells. An approximate rule has been proposed that the dry-weight concentration is directly proportional to the absorbance (34). This rule applies to both

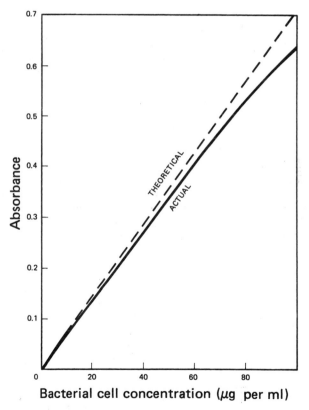

FIGURE 6 Absorbance (at 420 nm) as a function of bacterial cell concentration. The dashed line shows the theoretical relationship. The solid line shows an actual result obtained with 15 dilutions of a bacterial culture, whose concentrations were measured by dry-weight determinations and whose plot was obtained by the best-fitting quadratic curve that was forced to pass through the origin.

cocci and rods and is a first approximation to a more precise rule, i.e., that the light scattered out of the primary beam is proportional to the four-thirds power of the average volume of cells in the culture. Objects smaller than bacteria, such as suspensions of viruses, do not obey these rules. The rules also do not apply when the suspensions of small bacteria have a bluish cast. Larger objects (e.g., filaments and aggregates of bacteria) may still appear cloudy and not necessarily colored but may not obey the rule. When the rule applies, the proportionality constant relating the absorbance measurement to the dry-weight concentration has the same value for any good, well-collimated photometer. In such instruments, the phototube is at a considerable distance from the sample and/or the instruments are so well collimated that a narrow undeviated beam passes through and only the detector intercepts this narrow beam. The Zeiss instruments possess this design character. Other instruments can be assessed by examination of their optical design, or by comparison of the same suspension of cells within a short period when measured in an instrument from Carl Zeiss, Inc., Thornwood, NY, or other narrow-beam instrument and in the instrument under consideration. As time has passed, spectrophotometers with gratings instead of prisms and solid-state detectors instead of phototubes have made them simple, stable, and high resolution in terms of wavelength, but not necessarily of narrow light beam. The latter is a quality essential for turbidity measurements, so some care is needed in selecting a spectrophotometer.

If the instrument or the size of organism does not meet these criteria, establishing a standard curve on either a relative or an absolute basis can allow valid measurements to be made in many cases. Difficulty arises only as physical or biological conditions vary with the samples to be compared, since the experimenter cannot be sure that the same standard curve applies.

The above considerations apply only to dilute suspensions. Deviation from Beer's law must be applied after the absorbance exceeds about 0.3 when light of 500 nm or shorter is used (see the example in Fig. 6). A good discussion of this problem was given by Kavenagh (32). Failure to take this deviation into consideration is the most common difficulty encountered in the bacteriological literature that involves turbidimetry. One way to avoid this problem is to restrict culture densities and use only dilute cultures for measurement (i.e., to keep the absorbance below 0.3). However, the photometric measurements at this level are less accurate; an additional dilution may introduce significant pipetting error; for such small absorbance measurements, it is also critical that the cuvettes must be well matched and scrupulously clean; and each cuvette must be precisely placed in the spectrophotometer. It is usually better to make measurements at higher densities and correct for deviations from Beer's law.

9.6.1.1. Procedure

The important point about turbidimetric methods is that there is no set procedure. One uses the equipment at hand and seeks useful results. The most costly spectrophotometers increase the spectrophotometric precision and the stability and accuracy of wavelength settings. However, the main advantage of such instruments for the present purpose is that they have a larger dynamic range. Lower absorbances have validity when measured on good equipment because of electronic stability and increased reproducibility in positioning the cuvettes and because the cuvettes have plane parallel faces and are of accurate dimension. Higher absorbances, for which very little light reaches the phototube, have validity when measured on such equipment because there is less stray light. Moreover, the electronics can be compensated by a dark-current (zero) adjustment. Finally, the premier instruments may be read more accurately.

An important distinction among various types of equipment for turbidity measurements is that between single- and double-beam photometers. The critical test of the instrument is to obscure the phototube with the shutter or to place an opaque object in the light path and monitor the kinetics of decreases in transmission. In cheaper instruments, there will be a precipitous fall, which is followed by a much slower but significant fall that may continue for a very long time. Similarly, when the shutter is subsequently opened or the opaque object is removed from cheaper instruments, after the transmission has risen quickly it then rises further very slowly but never achieves a truly stable value. These drifts result from changes in phototube response because of memory to prior illumination conditions. Single-phototube, double-beam instruments avoid this because the same photodetector alternately measures the blank and the sample. However, simple single-beam instruments can be used with very little error if the phototube is alternately exposed to blanks, sample, and dark-current measurements in a routine standardized way with fixed timing.

There is a second reason for adopting a routine timing procedure for measuring cell turbidity, even with the best spectrophotometer. This is because turbidity fluctuates depending on the alignment of rod-shaped particles. The longer the rods, the more severe the alignment fluctuations will be in recently stirred cultures. For organisms with moderate ratios of length to width, such as *E. coli*, one approach is to mix the culture in the cuvette, place it in the spectrophotometer, wait for a definite period (e.g., 45 s), and then record the reading. Waiting for a long time in hopes of constancy is not recommended. This is because the cuvette will heat up from the light source, creating convection currents that cause partial vertical alignment of rod-shaped organisms. Another approach for rod-shaped organisms is to make the measurements in a slowly flowing system, where rod-shaped organisms will be oriented by the laminar flow (32). The orientation is greater nearer the walls than the center of the cuvette because the gradient of velocity is greater there. Of course, the procedure must be calibrated under the same flowing conditions with organisms of the same axial ratio.

9.6.1.2. Calibration

The procedure adopted for a given culture and instrument must be calibrated in some way to allow for comparison with other work. It is not valid to report turbidity measurements in absorbance units at a specified wavelength, for reasons to be itemized below, although this is commonly done. On the other hand, if a narrow-beam spectrophotometer is used, if the cells have a size range from 0.4 to 2 μm^3, and if reporting the measurement in terms of dry-weight concentration is acceptable, the conversion factors and deviations from Beer's law given below and elsewhere (36, 38) can be used.

To calibrate any adopted procedure, prepare a concentrated suspension of the organism grown under the appropriate conditions. Carry out a series of dilutions of the well-mixed suspension, and make measurements as quickly as possible so that growth need not be stopped or interfered with. Alternatively, place the concentrated suspension in an ice bath and make dilutions and measurements and/or introduce an agent such as chloramphenicol to stop protein synthesis. These are the effective ways to minimize biomass changes while measurements are made. In any case, make a repeat of the first dilution from the stock after all other measurements have been made. The first and last measurements with the same dilution factor must check. Plot the data as they are obtained on a working graph similar to that shown in Fig. 6, but use an abscissa expressed as the percent concentration relative to the undiluted stock. A feeling for the shape of the curve is achieved with this working graph. Inspect the graph after each point is added; this allows the detection of observational errors. Repeat particular dilutions when error is suspected. Make additional dilutions to define the parts of the curve where there is more curvature. Repeating a dilution can give an immediate estimation of error, but an additional intermediate dilution defines the curvature better. There need be no significant pipetting errors since the stock can be adjusted such that dilutions need not be high and relatively large volumes of solution can be pipetted into volumetric glassware. This working graph is sufficient to permit accurate estimates of growth of a culture, since only the relative change in biomass with time is needed for calculating specific growth rates and doubling times. However, for many other uses an absolute instead of a relative measure is needed. In addition, if biomass is measured on an absolute basis, the work is reportable in a way that allows accurate replication in other laboratories. Therefore, it is suggested that calibration be done on an absolute basis, as follows.

For absolute calibration it is necessary to measure accurately the stock suspension content by some other measure of bacterial biomass or numbers. This may be done by dry-weight determinations (47) (section 9.5.2) or by chemical determinations of a nucleic acid (chapter 18.6), protein (chapter 18.7.5 to 10), or total nitrogen (chapter 18.7.11). This must be done with adequate replication and quality control. Use many independent dilutions for viable counts, and use many independent fillings and adequate counts in the counting chamber for microscopic counts. Accuracy at this stage is to be emphasized, since the value obtained will pervade future work.

Prepare a graph that includes all the data that relate meter readings on the photometric instruments to cell number or mass, such as shown in Fig. 6. If computer facilities are available, calculate the best linear, quadratic, or cubic fit to the data. Programs available at most computing centers and available statistical software programs for PCs and Macintosh computers carry out an F-test to define which of the types of curves fit the data most efficiently. Simple programs with whatever computer facilities are available can then calculate the best-fitting curve. Programmable pocket calculators are adequate.

In work from this laboratory, we calibrated bacterial suspensions by using the narrow-beam Cary and Zeiss spectrophotometers. We also used and tested a range of commercial instruments with bacterial cells of various sizes (36). Over the size range from 0.4 to 2 μm^3, the results were very nearly the same when expressed on a dry-weight basis. Since the same curve applied to all the bacteria tested, the conversion formulas no doubt will apply to other unpigmented procaryotes whose mean size is within these limits. When the absorbance range is limited to the region below 1.1, a quadratic when passed through the origin fitted the data adequately: $A_\lambda = a_\lambda W - b_\lambda W^2$, where W is the dry-weight concentration of bacteria and a_λ and b_λ are constants

that apply at a given wavelength λ. The empirical equations for two convenient wavelengths are as follows:

$$A_{420} = 7.114 \times 10^{-3} W - 7.702 \times 10^{-6} W^2$$

$$A_{660} = 2.742 \times 10^{-3} W - 0.138 \times 10^{-6} W^2$$

The coefficient of the first power W (a_λ) decreases with increasing wavelength. The coefficient can be calculated at other wavelengths since it depends inversely on the wavelength raised to approximately the second power, actually the 2.11 power. Consequently, a_λ can be calculated at other intermediate wavelengths from:

$$a_\lambda = 2.439/\lambda^{2.11}$$

There is no rule to estimate b_λ at other wavelengths, although Kavenagh (32) measured this quantity at a number of wavelengths. This constant depends more critically on cell shape, cell internal heterogeneity, and photometer design. In fact, the quadratic fit is only marginally satisfactory at higher turbidities. For this reason and for photometric reasons, it is best to work at absorbances less than 1. Increasing the wavelength decreases b_λ, decreases multiple scattering of light into the detector, and improves linearity.

For routine calculation, the equations given above can be solved to yield W in the units of micrograms of dry weight per milliliter:

$$W = 461.81[1 - (1 - 0.60881A_{420})^{1/2}]$$

and

$$W = 9,929[1 - (1 - 0.07347A_{660})^{1/2}]$$

The latter expression can be expanded with no loss of accuracy to:

$$W = 364.74A_{660} + 6.7A_{660}^2$$

All three expressions can be used for optical densities of up to 1.

In the absence of pigments, the choice of wavelengths should be based on the following considerations. The lower the wavelength, the larger the turbidity. Therefore, lower cell concentrations can be directly measured. Conversely, low sensitivity is desirable for some purposes because more dense suspensions are of interest and an extra step of dilution can then be avoided. An additional reason for using a higher wavelength is that many growth media contain substances that absorb near the blue end of the visible spectrum and therefore appear yellow or brown. Then, 660 nm is a desirable wavelength because the medium blank is not very different from a water blank. Any higher wavelengths would be inconvenient because a different phototube would be required in many instruments. Chemical changes due to bacterial action, air oxidation, or variation in autoclaving alter the light transmission in complex media at shorter wavelengths. In any case, it is important when using a new medium that the supernatant of a culture of high density be compared with the original growth medium to see whether pigments are introduced or removed by bacterial growth.

Photometric methods can be used with colored organisms. It may be possible to choose a suitable wavelength to avoid absorption bands. Alternatively, choose a wavelength at which the absorption by cellular pigments is maximal, and report the biomass using the pigments as an indicator. The latter procedure has the difficulty that in many cases the pigment content varies with culture conditions.

If a narrow-beam spectrophotometer is not available for routine use, the above formula for dry weight can be used if there is temporary access to a narrow-beam instrument. Simply carry out the calibration procedure, but instead of the other methods listed above, also measure the turbidity of a known dilution on a narrow-beam instrument and use the formulas given above.

With wide-beam instruments the measured turbidity is less than in the narrow-beam instruments when a cuvette of the same size and shape is used; therefore, the sensitivity is lower (see above). This is a minor factor in most work. It is more significant that the calibration may change every time the phototube or the light bulb is changed; in some instruments, simply removing and replacing the housing compartment for the measuring cuvette cause measurable changes. Frequent restandardization of a wide-beam instrument against a narrow-beam instrument is recommended. As an example, the reading on an instrument of the old DU type (Beckman Instruments, Inc., Fullerton, CA) exhibited dramatic changes in sensitivity as thermospacers are inserted around the cuvette compartment in some models (36). This is not unexpected, since the addition of thermospacers lengthens the working distance and hence the instrument more closely approaches the quality of a narrow-beam instrument with its greater sensitivity.

Special quality control must be used with spectrophotometers in which gratings are used to disperse the spectrum. As the grating replicas age, they become warped and an increased amount of stray light hits the phototube.

9.6.1.3. Klett-Summerson Colorimeter

Of the many colorimeters and relatively inexpensive spectrophotometers, the Klett-Summerson colorimeter must be singled out for special comment. It was one of the first photoelectric instruments, it is rugged and stable, and many workers have used it for more than 80 years. It is still being used in many laboratories. It is stable because it is a double-beam instrument so that fluctuations in the light source are canceled out. It is not sensitive, since it has a simple CdS sensor, which produces only enough current so that a sensitive galvanometer is needed. The detector is far enough from the cuvettes that less of the scattered light is trapped than in many other (and more modern) instruments.

The Klett colorimeter is calibrated on a different scale from that used for more modern instruments: an absorbance of 1 on other instruments corresponds to 500 on the Klett scale. It reads so sluggishly that the turbidity of rod-shaped bacteria becomes stable by the time the dial can be properly adjusted; by then, the cells have achieved random orientation. Take care in reading the instrument dial and adjusting the dial to achieve a null reading on the galvanometer. Stand in a reproducible position so that parallax error is avoided. The major fault of this instrument is that there is no dark-current (zero) adjustment. Light can pass around the contents of the cuvette by being scattered on instrument surfaces and reflected through the glass walls of the cuvette. Test this (and other instruments without dark-current adjustments) by filling a cuvette with a highly absorbing material having an essentially infinite optical density (0% transmittance). In the Klett colorimeter, the reading should be beyond the printed scale of the instrument and in fact should be right on the vertical line that forms the left-hand side of the letter R in COPYRIGHT. Frequently the Klett colorimeter is out of calibration, and then there is an adventitious instrument error, which creates additional deviations to the Beer's law relationship. Special adjustment, repair, or replacement of the light source is needed. The problem can be partially alleviated, in some cases, by reblackening the internal surfaces.

9.6.2. Nephelometry

Although most of the light scattered by bacteria is nearly in the forward direction, instruments that measure the light scattered at 90° from the primary beam have also been used to measure bacterial concentrations. Instruments designed to make right-angle measurements of light scattering are called nephelometers (from the Greek *nephelos*, which means cloud). Bacterial concentration may be measured with this principle in a spectrofluorometer when the dials are set so that the excitation monochromator and the emission monochromator are at the same wavelength.

Nephelometric measurements can be ultrasensitive, so that very low concentrations of bacterial cells can be measured. There are two difficulties in making routine measurements. The first is caused by false signals from particulate impurities in the medium; this can be partially overcome by ultrafiltration of all reagents. The second is that it is necessary to standardize the instrument repeatedly (because the light intensity from the lamp can vary with time or during a series of measurements). Compared with a spectrophotometer, a different kind of standardization is needed with a nephelometer or spectrofluorometer. The operator sets the dark-current (zero) adjustment so that the scale reads zero with a reagent blank. Then, the gain is set to give a fixed response from a standard scattering material. That is, it is necessary to run a one-point standard curve routinely. Since a suspension of bacteria would not be stable, a secondary standard such as a piece of opal glass must be used. One can also use a suspension of glass beads, such as those used in rupturing bacteria and other biological material. Commercial preparations are also available. A tube can be prepared and sealed and used to calibrate the instrument. Then, a standard curve can be prepared relating meter readings after standardization to bacterial concentration. The standard curves are not linear at higher bacterial densities.

Although various specialized instruments have been made and used over the years to measure the light scattered at different angles, these instruments are of limited value because most of the light that is scattered from particles in the size range of bacteria is nearly in the forward direction (Fig. 4). If this light is to be sensed, the instrument must be capable of orientation to within a very few degrees of the direction of the incident light without intercepting undeviated light from the primary beam. This means that very well collimated light is necessary. Today, this is readily achieved with a laser. Instruments that use a low-angle range between 2 and 12° from the beam are available, but none is commercially available at a reasonable price. Measurements at higher angles contain information about the amount of cell material, but they also include information about internal structure and distribution of material within the cell. If it is known which factors are important and whether elongated particles are oriented or randomly distributed in space, light-scattering measurement over a range of angles can give information about the state of aggregation of the protoplasm, such as whether the ribosomes are in polysomes or as monosomes within the living cell (7); the thickness of the cell envelope (33, 35, 41, 73); and the distribution of cell mass from the center of the cell (33, 73).

Wyatt (73, 74) has developed an apparatus to measure the angular dependency of the light-scattering signal from 30 to 150°. In this angular range, different organisms, different treatments of a culture, and cultures in different phases of growth give characteristic patterns. These may be of use as a fingerprint technique for the diagnosis and study of drug action. For the reasons stated above, the light-scattering signals in the high-angle range cannot be used reliably for growth measurement because they are sensitive to details of subcellular structure.

Light-scattering measurements made over a range of angles from 0 to 20° are relatively independent of these three factors but are more critically dependent on overall size. Measurement in this range could nondestructively yield the biomass concentrations, the number concentration, the average axial ratio, and a measure of average biomass distribution around the center of the cell sizes for bacteria in balanced growth.

9.7. STATISTICS, CALCULATIONS, AND CURVE FITTING

Several aspects relevant to the measurement of bacterial growth are almost never included in courses in basic microbiology or statistics taken by undergraduate microbiology majors. For the former, this is because the needed material is a little beyond the scope of the courses, and for the latter, it is because the counts of colonies or particles is an all-or-none event and of lesser utility for most users of statistics. The following description is intended to make these mathematical tools readily available.

9.7.1. Population Distributions

The **binomial distribution** describes the chance of occurrence of two alternative events. For example, it provides the answer to the question, "What is the chance of having five boys in a family of nine, assuming that births of boys represent 0.56 of the total live births?" The numerical answer is $P_5 = 0.2600$. The relevant formula is as follows:

$$P_r = n! \, p^r \, (1 - p)^n / (n - r)! \, r!$$

where p is the chance of a specified response on a single try, n is the total number of trials, and r is the number of a specified kind of responses and would vary from 0 to n. The symbol ! means factorial; i.e., $n! = n \, (n - 1)(n - 2) \dots 1$. The formula shows that in different families of nine children more families would have five boys than any other number. For example, $P_4 = 0.2044$, $P_5 = 0.2600$, and $P_6 = 0.2207$. The plot of P against r is an example of a distribution histogram. The binomial distribution provides a way to estimate the mean and standard deviation of p from experimental data. Assuming that there are data on only a single family with n children and that there are by chance r boys in that family, statistical theory shows that the best estimate of p is r/n. Numerically, for the example where $r = 5$ and $n = 9$, p is $5/9 = 0.5556$, and its standard deviation is:

$$\sqrt{p(1 - p) / n} = \sqrt{(5/9)(4/9)/9} = 0.1656$$

This result is not very reliable because the coefficient of variation (CV) is nearly 30% (CV = 0.1656 × 100%/0.5556 = 29.81%). Looking for a family that had many more children and applying the same formula could improve the precision in estimating p. It would be not only easier but also better to pool the census data of a number of families. Suppose 50,000 boys are counted in 90,000 children in a large pool of families. Then, the same formula yields $p = 0.5556$, but now CV is 0.2981%. Not only is this more precise because large numbers are

counted, but also many families, for genetic and sociological reasons, may have different values of p. *It may be desirable to estimate the value of p that applies generally to the entire population.*

The formula for the distribution is cumbersome to calculate when the numbers are large. An important contribution of Gauss was to rewrite the binomial distribution for this large-number case. Thus, the **Gaussian distribution** applies as a generalization of the binomial distribution for the case in which the numbers involved are so large that they can be treated as a continuous distribution instead of one with discrete variables. The variables of the Gaussian distribution, replacing n and p, are the population mean (m) and standard deviation (s). The formula then becomes:

$$P_x = \frac{1}{\sqrt{2\pi}\sigma} e^{-(x-m)^2 / 2\sigma^2}$$

where e is 2.71828.

In this formula the continuous quantity (x) replaces the positive integer variable (r) as the measurement of response. This distribution, like the binomial, can be mathematically manipulated so that one can go from data to estimations of these two parameters. The Gaussian distribution is also called the **normal distribution**, partly because it is symmetrical about the mean and partly because it is so frequently observed.

For data that follow a Gaussian distribution, the estimate of the mean is called m and is given by $m = \Sigma x/n$. The estimate of the standard deviation is called s and is given by:

$$s = \sqrt{\frac{\Sigma x^2 - (\Sigma x)^2}{n - 1}}$$

The coefficient of variation, CV, is given by s/m.

The other limiting distribution of the binomial is the **Poisson distribution**. It applies for the case where n is very large and p is very small but the product np is finite. The best estimate of np is N, the observed number of total responses of a specified kind. This distribution would be useful if, for example, boys occurred very rarely (say, 1 in 10,000) but families were large (say, 100,000 children). Then, an average family would contain $N = 10$ boys and the standard deviation would be:

$$n\sqrt{p(1-p)/n} = \sqrt{n(N/n)(1 - N/n)} =$$
$$\sqrt{10,000 - (10/10,000)(1 - 10/10,000)} = 3.1621$$

A keen simplification, due to Poisson, was to assume that N/n is much smaller than 1. Then, the formula for the binomial distribution simplifies to a one-parameter distribution:

$$p = \frac{e^{-m} m^r}{m!}$$

The best estimate of the mean is $m = N$, and the best estimate of the standard deviation is s. Note that this is not very different from the binomial distribution, since $10^{.5} = 3.1623$ is not very different from 3.1621. Note that again n and p are replaced by different symbols, in this case by a single one, m. The important point is that *the count of the number of discrete objects provides not only the best estimate of the mean value (i.e., N = x = m) but also an estimate of the precision of the estimate ($\sqrt{N} = s = \sigma$).*

Consequently, in the enumeration of objects it does not matter how they are subdivided. Two replicate plates with 200 colonies on each are no better or worse from the point of view of Poisson statistics than is one plate with 400 colonies. In both cases the standard deviation, s, of the measurement is $\sqrt{400} = 20$ and the CV is $\sqrt{400}/400 = 5\%$. Therefore, to get the best estimate from a group of plates from the same or different dilutions of the same sample, simply add the total counts on all the plates and divide by the total volume used of the original solution. The standard deviation is the square root of the total count divided by the plated volume of solution.

As an example, imagine that duplicate plates were made at dilutions of both 10^{-5} and 10^{-6} with counts of 534 and 580 and of 32 and 60, respectively. The total count is 1,206. If 0.1 ml of these dilutions were plated, 2.2×10^{-6} ml would be the total volume of original culture used to make the four plates. Therefore, the best estimate of the concentration is $1,206/2.2 \times 10^{-6} = 5.46 \times 10^{8}$/ml. The standard deviation is $\sqrt{1,206}/(2.2 \times 10^{-6}) = 0.16 \times 10^{8}$/ml.

Justification for treating the results at the two different dilutions separately and not pooling them, as done above, depends on other kinds of errors being larger. Then, by comparing the results at different levels of dilution, some estimate is obtained of the variability due to the additional pipetting operation and to other sources of error that are not included in the calculation of the Poisson sampling error. Although this variability is of interest, sources of error could be more directly measured by carrying out independent dilutions from the same cell suspension. As an example, imagine that 0.1-ml aliquots were plated on single plates from each of 12 independent 10^{-5} fold dilutions and that the following set of colony counts were obtained: 534, 580, 760, 643, 565, 498, 573, 476, 555, 634, 514, and 693. The sum is 7,026, and the Poisson standard deviation is $\sqrt{7,026} = 83.8$. The mean of the numbers is 585.5, and the standard deviation by the Gaussian formula is 81.2. Calculation of the bacterial count of the original suspension together with the two different estimates of error yields the following count: $7,026/1.2 \times 10^{-5}$ ml $= 5.85 \times 10^{8}$/ml; Gaussian error $= \pm81.2/1.2 \times 10^{-5} = \pm0.81 \times 10^{8}$/ml; and Poisson error $= \pm83.8/1.2 \times 10^{-5} = \pm0.07 \times 10^{8}$/ml. The comparison of these two estimates of error suggests that considerable error is due to sources other than random sampling. An attempt should be made to find and reduce these sources of error. Until that is done, it is necessary to make many independent dilutions and use Gaussian statistics. *Until the source and magnitude of the experimental error are found, the Poisson error is irrelevant.*

The same point can be made in another way from this example. Imagine that only one plate had been made, say the first one, in which case only the Poisson error would be available for consideration. The count then would be $5.34 \times 10^{8} \pm 0.23 \times 10^{8}$, and the real error would be underestimated fourfold. It is therefore cautioned not to rely on Poisson statistics until their use has been justified for the conditions actually in use. Instead, make a comparison on at least several occasions with a Gaussian statistic measurement of error as indicated above.

9.7.2. Statistical Tests

Much of the statistics taught in elementary courses is concerned with whether a body of data is consistent with a hypothesis. Usually the **probability** (P) that the observed de-

viations from the hypothesis could occur by chance is computed. If P is small, but not very small, the hypothesis could still be false although improbable. If P is very small, the hypothesis can, with some confidence, be said to be true. These statistical tests are generally made on the assumption that the data follow a Gaussian distribution. In many cases in bacteriology, this assumption should be questioned and other appropriate statistical tools should be used.

The standard deviation has been defined above. This is frequently confused with another term, the **standard error**, also called **standard deviation of the mean**. The standard deviation measures the deviation of an individual measurement from the mean of many measurements. The standard error measures the mean of all the data observed from the mean of a hypothetical data base containing an infinite number of observations and is a measure of how close the average is to the "true" mean value.

The only statistical test mentioned in the text below is Student's **t test**. This applies to the difference in the means of two groups. The difference is divided by the standard error of the combined data. Thus, the **t** value measures the difference in the means by using the standard error as its unit of measurement. This ratio is compared with values given in tables. Use of the tables generates P values but requires a knowledge of the number of measurements and whether potential deviations can occur on both sides or only one side of the mean. The bigger that t is, the smaller P becomes; if P is not very small, the hypothesis that the two populations were identical may not be rejected.

In recent years, the **analysis of variance**, which is a subbranch of statistics, has been elaborated so that it now can be applied to many problems and replaces many of the more specialized techniques used previously. The availability of packaged programs for various computers means that it requires work, but much less work than previously, to learn, to use, and to apply statistical methods when they are appropriate in bacteriology.

9.7.3. Error Propagation

The accuracy of an estimate depends on the accuracy of its component measurements. The Poisson error of a colony count and the error of the dilution procedure both contribute to the error in the estimated concentration of organisms of the original undiluted suspension. Additional errors can only further blur the results or make them less precise. Even though errors in one part of the estimate may compensate for errors in another part, on the average random errors will make them larger. When errors in one measurement are independent of (uncorrelated with) errors in another measurement, the overall error can be calculated by two rules for **propagation of errors**, as follows.

1. If two quantities (x and y) are to be added or subtracted, the standard deviation(s) of the combined quantities is as follows:

$$S_{x+y} = S_{x-y} = \sqrt{S_x^2 + S_y^2}$$

2. If two quantities are to be multiplied or divided, the CV of the combined quantities is:

$$CV_{xy} = CV_{x/y} = \sqrt{CV_x^2 + CV_y^2}$$

As an example, apply the second rule to estimate the overall error in a single plate count containing colonies from the series of 10^5-fold dilutions. Assume that the dilutions were performed in five steps of 10-fold each and that the pipetting error of a single 10-fold dilution has a CV of 0.02. Then, the overall error of the five dilution steps is $\sqrt{5} \times 0.02$. This result is obtained by the repeated use of the second rule. It then must be combined with the Poisson error. Since the best estimate of the Poisson error CV is $1/\sqrt{585.5}$, then the overall CV is as follows:

$$CV = \sqrt{1/585.5 + 5(0.02^2)} = \sqrt{0.001709 + 0.00200} = 6.1\%$$

This 6.1% error is composed of a Poisson counting error of $4.1\% = 100/\sqrt{585.5}$ and an error due to the cumulative pipetting errors of $2\% \times \sqrt{5} = 4.47\%$. The rule to combine them gives a value smaller than their sum ($4.1 + 4.47 = 8.57\%$) but larger than the largest contributor to the error.

Two important experimental considerations derive from this example. First, *there is no reason for increasing the accuracy of one part of an experiment unless other sources of error comparable to it are also decreased or eliminated.* Second, *if an operation is to be done many times, it is worthwhile to devise a way to do it accurately and then obviate the need to carry out elaborate statistical calculations.* In the previous section, the pipetting error was neglected because it was assumed that pipetting can be, and was, done accurately. This is a reasonable thing to do if the CV of this error is smaller than the Poisson counting error. As an example, imagine that each pipetting operation had been carried out with an accuracy of 1% instead of 2%. Then, the overall pipetting error would have been $1\% \times \sqrt{5} = 2.23\%$ and consequently the overall total CV would have been $\sqrt{4.1\%^2 + 2.23\%^2} = 4.7\%$, only a little bit larger than the Poisson counting error by itself.

Similar logic follows for cases in which blank values and background values are to be subtracted (in which case the first rule for propagation of errors applies) or in which the instrument calibration factors are used to multiply observed values (and the second rule applies). In the measurement of controls used repeatedly in the calculation of data, errors should be reduced by repetitions or by more accurate measurement than for individual experimental values so that the control factors do not contribute significantly to the overall error of measurement.

9.7.4. Ratio Accuracy

There is a very powerful and general statistical method applicable to diverse experimental situations varying between large natural ecosystems at one extreme and a drop of culture on an electron microscope grid at the other. This method is to add a known number of reference particles, which may be bacteria, ferritin particles, polystyrene spheres, abortively transduced bacteria, plasmids, viruses, gold beads, etc. After mixing takes place, samples are taken and the ratio of the number of objects of interest to the number of reference particles is determined by appropriate means. The method can be illustrated for microscopic smears containing a class of recognizable organisms of unknown number and polystyrene beads of known concentration. The smear must be prepared from a known volume of cellular suspension and a known volume of suspension of beads of known concentration. The concentration of the beads in the final suspension is multiplied by the ratio of the counts of the unknown cells relative to those of the reference beads to calculate the concentration

of unknown cells. The second rule for the propagation of error applies in this case. If the concentration of the reference particles is known without error in the original stock solution, the coefficient of variation of the unknown particles is given by:

$$CV = \sqrt{\frac{1}{N_u} + \frac{1}{N_r}}$$

where N_u and N_r are the counts of the unknown and reference, respectively. To minimize the number of total counts following the first argument in the previous action, N_u should be about equal to N_r. Then, the CV will be about $\sqrt{2} = 1.4$-fold larger than if a very large number of reference cells (or unknown cells) had been counted.

9.7.5. Coincidence Correction

Coincidence corrections must be applied when too many colonies are on a plate or too many cells are on a square of a counting chamber or if too many radioactive decays are recorded by a radioactivity counter in a unit of time. For the case of a colony count, assume that if two cells initially are closer together than a distance r, they will be counted as a single colony. Let N_t be the true count and N_a be the actual count, and assume that the radius of the petri dish is R. Consider a single cell; the chance that another cell on the plate is within a distance r is $N_t r^2/R^2$; thus, the count is decreased by $N_t r^2/R^2$. The number of colonies not counted will be $N_a N_t r^2/R^2$. Therefore,

$$N_t = N_a + N_a N_t r^2/R^2 = N_a(1 + cN_t)$$

where $c = r^2/R^2$. It is usually convenient to substitute N_a for N_t on the far-right-hand side when the correction is small. From this formula it is clear why a fourfold reduction in colony size reduces the coincidence correction at a given count by 16-fold. This is the basis of the use of layered plates (section 9.3.3), which makes the colonies smaller and thus reduces coincidence.

9.7.6. Exponential-Growth Calculations

Under constant conditions after a long enough time when cell-cell interaction is small, growth measured in any manner is expected to proceed according to $X = X_o e^{\mu t}$ This can be written in any of the following equivalent ways:

$$\ln X = \ln X_0 + \mu t$$
$$\log X = \log X_0 + \mu t/2.303$$
$$X = X_0 2^{t/T_2}$$

In the last equation, T_2 is the **doubling time** and can be calculated from the following:

$$T_2 = (\ln 2)/\mu = 0.6931/\mu$$

Many symbols other than μ have been used for the specific growth rate, including a, k, and λ. Knowing these other symbols is important because they are used without definition in many papers. Confusion arises with μ, which designates the specific growth rate in the literature on continuous culture (chapter 13.2.2) and in microbial ecology. This is the usage employed here. Unfortunately, μ is also used to symbolize the number of doublings per hour in the literature on cell physiology. The latter usage differs from the former usage by a factor of $\ln 2 = 0.6931$. Although any time unit could be used, reciprocal hours appear to be nearly standard. The doubling time (T_2) is reported in the literature in either minutes or hours.

9.7.7. Plotting and Fitting Exponential Data

There are several alternative ways to fit data to the exponential-growth model that are equally valid, but differ in their precision and in the additional information yielded to the experimenter. The most simple conceptually is to look up the natural logarithms of the concentration of cells (or the dry weight of biomass, or other measurement of an extensive property of bacteria) and plot them on ordinary arithmetic graph paper against the time that the measurements were made. Then, draw a straight line through the data points as close to the points as possible, and determine its slope by the rise-over-fall method. If the time scale is in hours, then the slope is in units of reciprocal hours. At this point, ask two questions: How appropriate is a straight line to the data? Is the line that has been drawn a good summary of the data?

Common (base 10) logarithms may be used, in which case the slope must be multiplied by $2.303 = \ln 10$ to obtain μ. Base 2 logarithms also may be used, in which case multiply by $0.6931 = \ln 2$. There is an advantage in using base 2 logarithms: the reciprocal of the slope gives the doubling time directly. Whatever the base of logarithm used, plot both the characteristic and mantissa numbers on the arithmetic scale of the graph paper. Do not make the all-too-frequent blunder of mixing them (i.e., $\log_{10}$ of 4×10^8 cells $= 8.6 \neq 8.4$)!

It usually is more convenient to use semilogarithmic graph paper. Paper that has six divisions between darker lines on the arithmetic scale and two cycles on the logarithmic scale is recommended. Define the abscissa (x axis) according to hours or days on the major lines so that the minor lines represent an even number of minutes or hours. Mark each of the three unit labels with the appropriate powers of 10 on the ordinate (y axis). The printed scales may be multiplied by a constant, but it is invalid to add or subtract a constant from the logarithmic scale. Find the point corresponding to the amount of biomass, cell numbers, or other measure of extent of growth on the y axis, and mark the point exactly. It is useful to make a small point mark for exactness and surround it with a larger circle for better visualization. It also is useful, when light-scattering methods are used, to plot each point as soon as it is obtained. This frequently shows when errors have been made, whether they are biological, instrumental, arithmetical, or human. If the error is detected immediately, one has the opportunity to restart a culture, remeasure the culture, or replot a point. Once all the data have been plotted, draw the best straight line to fit the points.

The mathematical procedure that does this best is called the least-squares fit, which minimizes the square of the vertical distance of all points to the proposed line. This can be approximated visually by mentally noting the distance from the few points that are farthest from the position of a proposed straight line by moving a transparent ruler. By readjusting the ruler and remembering that the actual distances, even though the graph paper is semilogarithmic, are to be squared, one can do a quite accurate job of drawing very nearly the line that would be generated by the mathematical least-squares procedure.

Figure 7 shows an actual growth curve with an "eyeballed line," which is essentially the same as the computer-fitted line. This example is drawn from a carefully executed experiment with accurate data over a period permitting a 30-fold increase in cell mass. Note that the datum points fit close to the line. This implies that if overt or gross errors were made during the experiment, they were corrected.

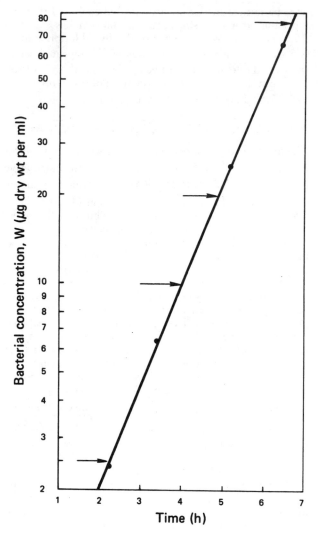

FIGURE 7 Growth curve plotted on semilogarithmic paper. See the text for conditions. The horizontal arrows are drawn to the curve at values of bacterial concentrations that are convenient powers of 2 apart, which facilitates estimations of the doubling time.

This example was chosen to show how sensitive growth rate measurements are to the accuracy of the data and to show the way the line is drawn.

Note that there are many kinds of semilogarithmic graph paper classified by how many cycles (powers of 10) they span. The smaller the number of cycles, the more spread out the points are and the easier it is to define the line accurately. Note also that there is only one type of semilogarithmic paper. The same paper is used whether the user chooses to work and think with natural, common, or base 2 logarithms.

Use the graph to find the doubling time, since this is the measure of growth rate that has the most intuitive appeal to bacteriologists. The doubling time is easily found by measuring the time for the bacterial concentration to change twofold. Put a tick mark on the graph where the line crosses some major division on the logarithmic scale of the paper; make another on the division twofold higher or lower. Read both times off the abscissa and subtract them to obtain the doubling time. Alternatively, better precision can be obtained by using fourfold, eightfold, or 2^N multiples of mass and then dividing the time difference by 2, 3, or N to obtain the resulting time difference for one doubling. In this way, the doubling time with better precision can be obtained. The specific growth rate or other measures of growth rate can then be calculated.

9.7.8. Least-Squares Fitting

It is very important to measure and use the best current estimation of the growth rate and extent of growth in carrying out a physiological experiment. For this reason the last section was written and is pertinent, but no one in a well-equipped laboratory will use these methods. Those to be presented are easier and more accurate and provide additional information. Using balanced growing cultures and measuring the doubling time is of the utmost importance. It is so important for achieving reliable results and conclusions that in this section several methods are provided for computing online these estimates both before an experiment proper is started as the inoculum is growing and while it is going on. It is for this reason that I have developed several convenient ways so that while carrying out experiments, the doubling time information is estimated at the time it is needed. The easiest and most flexible procedure is readily used by anyone with a computer that has an Excel® spreadsheet. It would take only a few moments to prepare a model spreadsheet copied from the appendix entitled "Cell Growth." It would then be recopied and used to insert data as it is obtained during an actual experiment. The section following is recommended to be read even though it gives an alternative method because it provides an understanding of what is being accomplished by any of these methods including the simple graphical one utilizing only a piece of semilogarithmic graph paper.

9.7.8.1. By Pocket Calculator

The least-squares procedure is more precise than visual fitting of exponential growth data, and there is no reason not to make the calculation this way since pocket calculators with linear-regression capability are readily available at low cost and are ubiquitous. Table 1 shows the data for Fig. 7 treated on a pocket calculator with least-squares capability. By simply following the instruction manual, one can use any calculator with linear-regression functions for this purpose.

The first step is to accumulate the sums, sums of squares, and sums of cross-products. Take care to calculate the natural logarithm (ln) of the cell count (or other measure of cell substance) and to use it as the Y variable. Key the time in for the X variable. Take care to convert the time to decimal units before executing the cumulation procedure ($\Sigma+$). Some calculators convert hours and minutes into decimal hours; thus, if 10.30 is keyed in and this is thought of as 10 h and 30 min, a function available on some calculators converts this to decimal time, 10.50, which is then suitable for regression. Be careful if time goes past noon (or midnight) to add 12 or 24 h to the time. Alternatively, calculate the elapsed time from the start of the growth run or use an elapsed-time counter that reads in decimal time during the execution of the experiment. Many clocks available for laboratory use actually read in minutes and decimals of minutes. A clock can also be watched and measurements can be made at the hour, the half-hour, or at any regular time interval.

It is important to use time as the X variable. This is because the formulas used in the linear regression have been

---

TABLE 1 Sample data for exponential growth by graphical estimation or pocket calculator computation

Sample data. These data were obtained with a subculture of *E. coli* ML308 growing in 0.2% glucose-M9 medium at 28°C

As recorded		As converted	
Time (h:min) (1)	A_{420} (2)	Time (min) (3)	$W(\mu g/ml)$ (4)
2:13	0.017	133	2.39
3:23	0.045	203	6.36
3:57	0.071	237	10.09
5:10	0.174	310	25.14
6:23	0.439	383	66.50

Graphical estimation. From Fig. 6 (which is the plot of data in column 1 versus data in column 4), determine the time at any value of W and also at twice that value (e.g., at $W = 10$ and 20 μg/ml):

Time (h:min)	$W(\mu g/ml)$
4:00	10
4:52	20

The time difference is 52 min; $T_2 = 52$ min; and the ratio of the W values is 2.

Alternatively, determine the time at any chosen value of W_0 and the time that the culture reaches $2^N W_0$. For example, if $W_0 = 2.5$ μg/ml and $N = 5$, then $W_5 = 2.5 \times 2^5 = 80$ μg/ml. The time when the culture had $W_0 = 2.5$ μg/ml as read from the graph is 2:15 p.m., and the time when $W_5 = 80$ μg/ml as read from the graph is 6:38 p.m. The difference in time, 263 min, is the time for five generations of growth, and therefore $T_2 = 263$ min/5 = 52.6 min.

Pocket calculator computation. Obtain the regression of the logarithms of data in column 4 against data in column 3 by the following steps:

1. Clear memory
2. Key 2.39, ln, key 133, $\Sigma+$
3. Key 6.36, ln, key 203, $\Sigma+$
4. Key 10.09, ln, key 237, $\Sigma+$
5. Key 25.14, ln, key 3 10, $\Sigma+$
6. Key 66.50, ln, key 383, $\Sigma+$
7. Key linear regression. Read ln $W_0 = -0.85456$.
8. Key X↔Y. Read = 0.0132 1/min.
9. These results can be expressed as ln $W = -0.85456 + 0.013213t$, or $W = 0.4255 e^{+0.013213t}$, when t is expressed in minutes.
10. To reexpress the specific growth rate on a per-hour basis, multiply by 60 to give 0.793/h.
11. To reexpress in terms of doubling time, divide into ln 2 to give $T_2 = 52.47$ min.
12. For an error calculation (see text), $T_2 = 52.47 \pm 0.65$ min and CV = 1.24%.

derived on the assumption that all error in the measurement is in the Y variable and that none of it is due to error in the X variable. For growth measurements, the time can be precisely defined (if necessary, to the second), and the major error is in the measurement of the number of cells or other indices of biomass.

The calculator derives the best estimate of the intercept and slope by the formulas applicable to the linear equation $Y = a_0 + a_1X$, i.e.:

$$a_0 = \frac{(\sum X^2)(\sum Y) - (\sum X)(\sum XY)}{n(\sum X^2) - (\sum X)^2}$$

$$a_1 = \frac{n(\sum XY)\sum - (\sum X)(\sum Y)}{n(\sum X^2) - (\sum X)^2}$$

where a_0 is the estimate of the value of Y when X is zero, a_1 is the slope of the regression, and n is the number of observations. For the case of exponential growth, the value of a_0 is the natural logarithm of the initial cell density. This is of lesser interest than the slope (a_1), which is the best estimate of the specific growth rate (μ).

A very real problem arises with the usual use of a linear regression in its unweighted form. To see this more clearly, consider a hypothetical situation in which the problem does not arise. Imagine that growth is measured by viable counts and that dilutions are made at every time point in such a way that precisely 400 colonies developed on each plate. In this case the coefficient of variation is $\sqrt{400}/400 = 5\%$ and is constant throughout the growth period, even though growth may have increased the number of cells per unit volume of culture many orders of magnitude during the experiment. This means that each value of ln N is known with a standard deviation of ln (1.05) = 0.05 above and ln (0.95) = −0.05 below the standard mean value and is thus constant throughout the experiment. Constancy of the error in the Y variable was another of the basic assumptions made in deriving the least-squares unweighted linear regression formula.

Now imagine that growth is monitored turbidimetrically without diluting samples for measurement. At low turbidities the coefficient of variation is high because of the difficulty of matching cuvettes and balancing the blank. At an apparent absorbance of 0.4, the relative error is at a minimum; at higher absorbances, the photometric precision falls and the scales cannot be read accurately. Consequently, the coefficient of variation rises. In addition, there are errors involved in correction at high absorbance for deviations from Beer's law and for problems resulting from stray light.

For these reasons, after logarithms have been taken, the standard deviation is not constant but falls to a minimum at intermediate absorbances and then becomes larger. Therefore, the data at intermediate absorbances have higher accuracy but do not determine the slope of the line set by the experimental data as well as is done by points at the low and high ends that span a longer stretch of time. One can overcome this problem and use the data in a balanced way by carrying out a weighted regression. To do so, it is necessary to estimate the error of each datum in one way or another. Ordinarily this involves a great deal of experimental and computational work and is not routinely done. Alternatively, make turbidimetric measurements by using techniques that are very precise so that the errors are very small. Additionally, take more datum points and exercise more care in regions at the low and high ends of the curves.

If the data are treated by pocket calculator (or by computer), no arithmetic mistakes are likely to be made. However, human copying error can easily invalidate the results. For this reason, it is strongly recommended that a graph also be made so that the results of the computation can be compared and qualitatively checked with the graphical results. Alternatively, the error of the slope should be calculated. The error of the slope of a linear regression is given by:

$$s_{a_1} = s_\mu = \frac{(\sum Y^2 - a_0 \sum Y - a_1 \sum XY)n}{(n-2)(\sum X^2)/n}$$

Except for $\sum Y^2$ most of the summations in this formula were also needed by the formula for calculating a_0 and a_1. Fortunately, $\sum Y^2$ is needed for calculating correlation coefficients and for calculating the standard deviation of the Y values. For this reason, most modern pocket calculators and statistical computer packages calculate everything needed for the error calculation when the data pair is entered (with the $\Sigma+$ key or its equivalent). With $\sum Y^2$ accumulated, it is no great difficulty to program the programmable versions of the hand calculators to calculate the error of the slope. The error given in the appendix entitled "Cell Growth" was calculated this way. One could also calculate errors of the intercept, of the means of X and of Y, and of the confidence interval. However, the error of the specific growth rate is the crucial value and worth the effort to calculate. The standard error of the doubling time can be calculated to sufficient accuracy by assuming that μ or T_2 (= ln $2/\mu$) have the same coefficient of variation. Algebraic manipulation shows that the standard deviation of the doubling time is the standard deviation of the specific growth rate multiplied by ln 2 and divided by μ^2; that is,

$$s_T 2 = s_\mu \ln 2/\mu^2$$

The error in the slope of the regression has some of the same properties as does the Poissonian $\sqrt{N}$ as an estimate of the error of a count. It tells about the internal accuracy of the measurement but does not assess external sources of variation.

The specific growth rate can be measured with precision within a run better than the reproducibility of the specific growth rate from run to run or of the reproducibility of growth rate of subcultures of a bacterial strain from day to day. The sources of this variation for a particular organism and culture conditions are unknown at present, but this is an important field of research for the future. However, these run-to-run variations must be taken into consideration for the practical purposes of making and using growth rate measurement to study many aspects of microbiology.

Most often, the reason for measuring growth rates is to measure the effect of some treatment. This requires comparison of growth with and without treatment. Ultimately, comparison requires replicate control cultures and replicate experimental cultures. Then, the standard t test of the specific growth rates of the two groups suffices to test the probability that the groups are not different. Other statistical procedures should be used when it is evident that there is a difference and the effect of the treatment is to be quantitated (see any statistics text, for example, reference 64). The real value of the calculation of s_μ is in allowing the experimenter to replicate the treatments and the controls less often once the relationship of the internally measured error to the external measured error has been measured enough times to instill confidence. If the external error is always small compared with that of the error in a single growth run, then a single run will have more significance—when backed up with an estimate of the error of the slope from previous experiments.

Another way to increase the precision of the measurement of the effect of a treatment when the variation from subculture to subculture is important is the "subject as its own control" method. Start by monitoring the growth of an untreated culture. When the growth rate has been determined, treat the culture and monitor growth to remeasure the specific growth rate. This procedure isolates the treatment as the sole variable between the two halves of the growth run. It is a smart move to run control experiments in which the treatment is only a sham treatment.

9.7.9. Nonexponential Growth

Except under exponential batch culture or continuous culture, growth does not stay balanced (if this is ever achieved) for very long. Even under continuous culture conditions there can be long-term changes that cause deviations from a constant state of balanced growth. So in all cases, the assumption of exponential growth is a limited one from the point of view of the biological growth process, fully independent of the difficulties and imprecision of making measurements. It becomes necessary, therefore, to limit the range of measurement. For batch cultures, this means excluding low- and high-density data or conducting the experimental measurements only over a limited range of growth density. Both of these approaches are attempts to obtain balanced growth by restricting the part of the total growth process considered. All such choices and judgments are difficult, and they are impossible to justify from a statistical point of view.

APPENDIX I: THE MPN METHOD

GENERAL FORMULA FOR MPN COMPUTATION

$$\Sigma a_i g_i / (1 - e^{-a_i n}) = a_i t_i$$

where the values of a_i are the volume tested, the values of t_i are the total number of tubes, and the values of g_i are the number of tubes showing growth in level i. The MPN value is symbolized in this equation by n. Various values for MPN are tried systematically by the fitting program. Through the computational procedure, the one that satisfies the above formula is selected.

SAMPLE DATA FOR MPN COMPUTATION

Level (i)	Vol (a_i)	Total no. (t_i)	No. positive (g_i)
1	10	10	8
2	1	10	5
3	0.1	10	1

BASIC COMPUTER PROGRAM FOR MPN DETERMINATION

The program with slight modification should work within any BASIC program and can be easily modified for other computational programs.

```
'MPN, 2004
DIM D(15,5):TV=0:DEFSNG A-Z:CLS
INPUT "DILUTION FACTOR TO TEST SOLUTION ";DIL
INPUT "NUMBER OF LEVELS TESTED ";I
```

```
PRINT "VOLUME, # OF TUBES, # WITH
GROWTH"
FOR J=1 TO I:INPUT V,N,G
  IF J=1 THEN NN=N
  D(J,1)=J:D(J,2)=V:D(J,3)=N:D(J,4)=G
  IF D(J,3)=0 THEN D(J,3)=NN
  TV=TV+D(J,2)*D(J,3)
NEXT J
X1=.001:X2=.8
X=X1
GOSUB "SUM"
Y1=Z
X=X2
GOSUB "SUM"
Y2=Z
"AAA"
X=X2-(X2-X1)*Y2/(Y2-Y1)
GOSUB "SUM"
Y=Z:PRINT X;Y
IF ABS(Z)<.0001 THEN "DDD"
IF Y*Y2>0 THEN "CCC"
X1=X2:YI=Y2
"BBB":
X2=X:Y2=Y:GOTO "AAA"
"CCC":
Y1=Y1/2:GOTO "BBB"
"DDD":
PRINT:PRINT "VOLUME, # OF TUBES, WITH GROWTH"
FOR J=1 TO I:PRINT D(J,2),D(J,3),D(J,4)
NEXT J
PRINT
PRINT "MOST PROBABLE NUMBER =";X
PRINT "ORIGINAL TITER =";X*DIL
END
"SUM"
ZZ=0
FOR J=1 TO I
  ZZ=ZZ+D(J,2)*D(J,4)/(1-EXP(-X*D(J,2)))
  TRON X
NEXT J
Z=ZZ-TV:PRINT TV, Z,
RETURN
```

PROCEDURE FOR BASIC COMPUTER PROGRAM TO CALCULATE MPN VALUES

1. Install the above program in a PC, Macintosh, or another microcomputer. The code can be copied from the above in a few minutes and then stored. Because it is written in a very primitive form of BASIC (Microsoft VBA®), it should run with substantially any dialect of BASIC, such as Quick Basic, Zbasic, FutureBasic, or VisualBasic. In any case, test the program first with the sample data given above under "Sample Data for MPN Computation"; the relevant formula is also given above.

2. Run the program. The program first asks for the dilution factor. This is the fold dilution between the experimental culture or field sample and the solution actually pipetted into the MPN tubes. Key this number in, and then key **<enter>** (not **<return>**). The computer then asks for the number of levels of dilution used in the sample. Key this number in, and again key **<enter>**. For the example shown in Table 1, this number is 3.

3. Next, the program requests sets of three numbers about a given level used in the procedure: key the volume of the diluted test suspension placed in the tubes, the number of

tubes inoculated at that level, and the observed number of positive cultures. Key these numbers separated by commas, and then key **<enter>**. So for the first level, enter 10, 10, 8 **<enter>**; for the second, 1, 10, 5 **<enter>**; and for the third, 0.1, 10, 1 **<enter>**. The first two must be followed by commas and the last with the **<enter>** command. If the **<enter>** key is used instead of the comma, the computer can still use the data and will request the remaining data. If the same number of tubes is used at all levels, it is necessary to **<enter>** the number of tubes only at the first level and then the middle entry can be omitted, but in this case be sure to put two commas between the volume of sample and the number of positive cultures.

4. After the data entry for the last level is completed, the computer uses an interpolation routine (the Illinois version of the regula falsi algorithm) to find the value that best fits the data according to the formula given above and prints the results to the screen. The reiterative process is shown on the screen as the computer finds successively better values of the MPN. For the example of Table 1, the final estimate of MPN is 0.27.

The program, as written, prints the output only on the screen. It is expected that the user will wish to modify the program to obtain a printed output instead. Printer directions have been omitted because format statements vary widely among computer systems. As another variation, if the same volumes or the same number of levels are always used, the program can be customized to fix these parameters and to delete some of the requests that are made by the supplied program.

APPENDIX II: CELL GROWTH: ASSESSED BY COMPUTER AND SPREADSHEET

In the first edition of this manual, only the pocket calculator routine was provided to calculate growth parameters. In the second edition, a decade later, a BASIC program written so as to be especially user friendly was given. For this third edition, given the progress in desktop computers and some predictions for more changes in the next decade, yet another approach is presented. This is the use of spreadsheets in Excel® that work on PC as well as Macintosh computers and also work in a variety of the vintage Excel® applications. It also works with the VBA program in Microsoft Office®. It is arranged in such a way that it can be conveniently used during the execution of a physiological experiment without curtailing other uses of the computer. This facility is possible because the growth data of each culture are accumulated and followed in their own spreadsheet. This means that the program can be hidden and the computer can be used for other purposes most of the time. When appropriate, each spreadsheet is brought to the front

and a new time point of data can be recorded and intermediate results or terminal output can be noted, stored, and printed. The user-friendly feature is that the investigator can study and monitor many independent cultures simultaneously. A second feature allows estimation of the culture density at a chosen time, as an interpolation and/or an extrapolation of the data.

For use, copy the spreadsheet as shown below to be a master file. Include the formulae in the several cells as specified in the legend. This effort should take a quarter of an hour. Make a copy, and then make another copy from which the sample data have been deleted and store it. Now insert the time and absorbance measurement as given in the example. If you have made no mistakes, the spreadsheet should generate the entire table except for the last line. In that last line (row 20) insert any time you choose and the program will calculate the dry weight and the optical density of the culture at that time. In the example, I have chosen the time of an earlier observation. It should be noted that the estimated optical density is not exactly what had been experimentally observed. This is because all the recorded data (both before and after) were used to interpolate what the optical density would have been.

For microbiological studies, recall the stored version and make copies for each culture in your day's experiment and label each. As the growth of these cultures is followed, insert data for each observed time point. After the third one in a given spreadsheet, the program will automatically calculate a growth rate per hour, a CV in percentage, and the doubling time in minutes.

If a clock that gives the time in hours and decimals is used, key this time in the "Hours" column and 0 or nothing in the "Minutes" column. If the growth runs continue past noon, use time expressed on a 24-h basis. If time goes past midnight or for several days it will be necessary to key in the hours as time passed from midnight of the day before the run started.

For the extrapolation or interpolation, it is possible to insert a time, still as hours and minutes, into columns A and B of row 20. The computer then computes the amount of growth at this chosen time employing the values of a_0 (intercept) and a_1 (exponential growth rate constant) from the last data listed. This part of the program of Excel® can either extrapolate or interpolate the biomass. This is a feature that allows the experimenter to estimate when another phase of the experimental program should be initiated such as diluting the culture or adding a drug. There are many modifications that the user may wish to make. It is suggested, however, that one tests the example and then a sample of one's own experimental data before modifying the program.

TABLE A1   Sample data and output, *E. coli* ML308, 28°C, M9 Med, 10/25/2003, Run B

Hours	Minutes	OD	Time (h)	Dry wt	Growth (1/h)	CV	Doubling time (min)	Slope	Intercept
2	13	0.0172	2.2167	2.5293					
3	23	0.0453	3.3833	6.6904					
3	57	0.071	3.95	10.528					
5	10	0.1745	5.1667	26.312	0.7939	0.2016	52.387	2.212	0.4459
6	23	0.4396	6.3833	69.484	0.7899	0.205	52.648	2.2033	0.4517

SETUP OF THE EXCEL® SPREADSHEET

The spreadsheet, if copied into an empty spreadsheet and then stored as a master, can be copied for use for each new experiment. By copying and storing it, the growth rate can be found very easily.

In Row 4 insert:

```
D4  =IF(ISBLANK($C4)," ",$A4+$B4/60)
E4  =IF(ISBLANK(C4)," ",481.81*
    (1-SQRT(1-0.60881*C4)))
F4  =IF(ROW($F4)<7," ",IF(ISBLANK($C4)," ",
    LN(INDEX(LOGEST($E$4:$E4,$D$4:
    $D4,TRUE,FALSE),1))))
G4  =IF(ROW($G4)<7," ",IF(ISBLANK($C4)," ",
    INDEX(LOGEST($E$4:$E4,$D$4:$D4,
    TRUE,FALSE),2)/INDEX(LOGEST($E$4:$E4
    ,$D$4:$D4,TRUE,FALSE),1)))
H4  =IF(ROW($H4)<7," ",
    IF(ISBLANK(C4)," ",(LN(2)*60)/F4))
```

Use the Fill command of Excel® to copy each of these lines from Row 4 to Row 19.

The numerical values in the above computer lines assume that the optical density of the cultures was measured at 420 μm. If this is not so, change the constants in E4 and C20 based on the information in section 9.6.1.2.

To allow extrapolation and interpolation of the amount of growth at a chosen time using the last estimate of fitted growth parameters, install the following formulae:

```
In C20 =7.114 × 10⁻³*E20 — 7.702 × 10⁻⁶*
       E20*E20
In D20 =A20+B20/60
In E20 =$K20 *($J20)^($D20)
In J20 =INDIRECT(CONCATENATE("R",
       19COUNTBLANK($C$4:$C$19),
       "C10"),FALSE)
In K20 =INDIRECT(CONCATENATE("R",
       19COUNTBLANK($C$4:$C$19),
       "C11"),FALSE)
```

For sample data and output, see Table A1.

9.8. REFERENCES

9.8.1. General References

1. **Dawson, P. S. S. (ed.).** 1974. *Microbial Growth.* Halsted Press, New York, NY.
   A collection of papers that earlier dominated the study of physiology of bacterial growth.
2. **Gall, L. S., and W. A. Curby.** 1979. *Instrumental Systems for Microbiological Analysis of Body Fluids.* CRC Press, Inc., Boca Raton, FL.
3. **Kavenagh, F. (ed.).** 1963. *Analytical Microbiology,* vol. 1. Academic Press, Inc., New York, NY.
4. **Kavenagh, F. (ed.).** 1972. *Analytical Microbiology,* vol. 2. Academic Press, Inc., New York, NY.
5. **Meadows, P., and S. J. Pirt (ed.).** 1969. Microbial growth. *Symp. Soc. Gen. Microbiol.* **19:**1–450.
6. **Meynell, G. G., and E. Meynell.** 1970. *Theory and Practice in Experimental Bacteriology,* 2nd ed. Cambridge Press, Cambridge, United Kingdom.
7. **Miller, J. H.** 1972. *Experiments in Molecular Genetics.* Cold Spring Harbor Laboratory, Cold Spring Harbor, NY.
8. **Pirt, S. J.** 1975. *Principles of Microbe and Cell Cultivation.* Blackwell Scientific Publications, Oxford, United Kingdom.

9.8.2. Specific References

9. **Amann, R. I., B. J. Binder, R. J. Olson, S. W. Chisholm, R. Devereux, and D. A. Stahl.** 1990. Combination of 16S rRNA targeted oligonucleotide probes with flow cytometry for analyzing mixed microbial populations. *Appl. Environ. Microbiol.* **56:**1919–1925.
10. **Black, S. H., and P. Gerhardt.** 1962. Permeability of bacterial spores. IV. Water content, uptake, and distribution. *J. Bacteriol.* **83:**960–987.
11. **Bridges, B. A.** 1976. Survival of bacteria following exposure to ultraviolet and ionizing radiation. *Symp. Soc. Gen. Microbiol.* **26:**183–208.
12. **Brock, T. D.** 1967. Bacterial growth in the sea; direct analysis by thymidine autoradiography. *Science* **155:**81–83.
13. **Brock, T. D.** 1971. Microbial growth rates in nature. *Bacteriol. Rev.* **35:**39–58.
14. **Buchanan, R. E.** 1918. Life phases in a bacterial culture. *J. Infect. Dis.* **23:**109–125.
    This classic has been reprinted in reference 1.
15. **Button, D. K., and B. R. Robertson.** 1989. Kinetics of bacterial processes in natural aquatic systems based on biomass determined by high-resolution flow cytometry. *Cytometry* **10:**558–563.
16. **Calam, C. T.** 1969. The evaluation of mycelial growth, p. 567–591. *In* J. R. Norris and D. W. Ribbons (ed.), *Methods in Microbiology,* vol. 1. Academic Press, Inc., New York, NY.
17. **Campbell, A.** 1957. Synchronization of cell division. *Bacteriol. Rev.* **21:**263–272.
18. **Capaldo, F. N., and S. D. Barbour.** 1975. The role of the *rec* genes in the viability of *Escherichia coli* K12, p. 405–418. *In* P. C. Hanawalt and R. B. Setlow (ed.), *Molecular Mechanisms for Repair of DNA,* part A. Plenum Publishing Corp., New York, NY.
19. **Cato, E. P., C. S. Cufflins, L. V. Holdeman, J. L. Johnson, W. E. C. Moore, R. M. Smibert, and K. D. S. Smith.** 1970. *Outline of Culture Methods in Anaerobic Bacteriology.* Virginia Polytechnic Institute and State University, Blacksburg.
20. **Clifton, C. E.** 1966. Aging of *Escherichia coli. J. Bacteriol.* **92:**905–912.
21. **Dicker, D. T., and M. L. Higgins.** 1987. Cell cycle changes in the buoyant density of exponential-phase cells of *Streptococcus faecium. J. Bacteriol.* **169:**1200–1204.
22. **Drake, J. F., and H. M. Tsuchlya.** 1973. Differential counting in mixed cultures with Coulter counters. *Appl. Microbiol.* **26:**9–13.
23. **Edwards, C., J. Porter, J. R. Saunders, J. Diaper, J. A. W. Morgan, and R. W. Pickup.** 1992. Flow cytometry and microbiology. *SGM Q.* **19:**105–108.
24. **Finney, D. J.** 1964. *Statistical Method in Biological Assay,* 2nd ed., p. 570–586. Hafner Publishing Co., New York, NY.
25. **Gray, T. R. G.** 1976. Survival of vegetative microbes in soil. *Symp. Soc. Gen. Microbiol.* **26:**327–364.
26. **Guerrero, R., C. Pedros-Alió, T. M. Schmidt, and J. Mas.** 1985. A survey of buoyant density of microorganisms in pure cultures and natural samples. *Microbiologia* **1:**53–65.
27. **Gunther, H. H., and F. Berhgter.** 1971. Bestimmung der Trockenmasse von Zellsuspensionen durch Extinktionsmessungen. *Z. Allg. Mikrobiol.* **11:**191–197.
28. **Halvorson, H. O., and N. R. Ziegler.** 1933. Application of statistics to problems in bacteriology. 1. A means of determining bacterial population by the dilution method. *J. Bacteriol.* **25:**101–121.

29. **Heden, C.-G., and T. Illeni.** 1975. *Automation in Microbiology and Immunology.* John Wiley & Sons, Inc., New York, NY.

30. **Holdernan, L. V., E. P. Cato, and W. E. C. Moore (ed.).** 1977. *Anaerobe Laboratory Manual,* 4th ed. Virginia Polytechnic Institute and State University, Blacksburg.

31. **Hungate, R. E.** 1971. A roll tube method for cultivation of strict anaerobes, p. 117–132. *In* J. R. Norris and D. W. Ribbons (ed.), *Methods in Microbiology,* vol. 3a. Academic Press, Inc., New York, NY.

32. **Kavenagh, F.** 1972. Photometric assaying, p. 43–121. *In* F. Kavenagh (ed.), *Analytical Microbiology.* Academic Press, Inc., New York, NY.

33. **Kerker, M., D. D. Coke, H. Chew, and P. J. McNulty.** 1978. Light scattering by structural spheres. *J. Opt. Soc. Am.* **68:**592–601.

34. **Koch, A. L.** 1961. Some calculations on the turbidity of mitochondria and bacteria. *Biochim. Biophys. Acta* **51:**429–441.

35. **Koch, A. L.** 1968. Theory of angular dependence of light scattered by bacteria and similar-sized biological objects. *J. Theor. Biol.* **18:**133–156.

36. **Koch, A. L.** 1970. Turbidity measurements of bacterial cultures in some available commercial instruments. *Anal. Biochem.* **38:**252–259.

37. **Koch, A. L.** 1971. The adaptive responses of *Escherichia coli* to a feast and famine existence. *Adv. Microb. Physiol.* **6:**147–217.

38. **Koch, A. L.** 1975. Lag in adaptation to lactose as a probe to the timing of permease incorporation into the cell membrane. *J. Bacteriol.* **124:**435–444.

39. **Koch, A. L.** 1975. The kinetics of mycelial growth. *J. Gen. Microbiol.* **89:**209–216.

40. **Koch, A. L.** 1976. How bacteria face depression, recession, and derepression. *Perspect. Biol. Med.* **20:**44–63.

41. **Koch, A. L., and E. Ehrenfeld.** 1968. The size and shape of bacteria by light scattering measurements. *Biochim. Biophys. Acta* **165:**262–273.

42. **Koch, A. L., and G. H. Gross.** 1979. Growth conditions and rifampin susceptibility. *Antimicrob. Agents Chemother.* **15:**220–228.

43. **Kubitschek, H. E.** 1964. Apertures for Coulter counters. *Rev. Sci. Instrum.* **35:**1598–1599.

44. **Kubitschek, H. E.** 1969. Counting and sizing microorganisms with the Coulter counter, p. 593–610. *In* J. R. Norris and D. W. Ribbons (ed.), *Methods in Microbiology,* vol. 1. Academic Press, Inc., New York, NY.

45. **Kubitschek, H. E.** 1987. Buoyant density variation during the cell cycle in microorganisms. *Crit. Rev. Microbiol.* **14:**73–97.

46. **Maaløe, O., and N. O. Kjeldgaard.** 1966. *Control of Macromolecular Synthesis.* W. A. Benjamin, Inc., New York, NY.

47. **Mallette, M. F.** 1969. Evaluation of growth by physical and chemical means, p. 521–566. *In* J. R. Norris and D. W. Ribbons (ed.), *Methods in Microbiology,* vol. 1. Academic Press, Inc., New York, NY.

48. **Marquis, R. E., and P. Gerhardt.** 1959. Polymerized organic salts of sulfonic acid used as dispersing agents in microbiology. *Appl. Microbiol.* **7:**105–108.

49. **Marshall, R. T.** 1993. Media, p. 89–101. *In* R. T. Marshall (ed.), *Standard Methods for the Examination of Dairy Products,* 16th ed. American Public Health Association, Washington, DC.

50. **Melamed, M. R., T. Lindmo, and M. L. Mendelsohn (ed.).** 1990. *Flow Cytometry and Sorting,* 2nd ed. Wiley-Liss, New York, NY.

51. **Mitruka, B. M.** 1976. *Methods of Detection and Identification of Bacteria.* CRC Press, Inc., Cleveland, OH.

52. **Nir, R., Y. Yisraeli, R. Lamed, and E. Sajar.** 1990. Flow cytometry sorting of viable bacteria and yeasts according to O-galactosidase activity. *Appl. Environ. Microbiol.* **56:**3861–3866.

53. **Norris, K. P., and E. O. Powell.** 1961. Improvements in determining total counts of bacteria. *J. R. Microsc. Soc.* **80:**107–119.

54. **Novick, A., and L. Szilard.** 1951. Experiments with the chemostat on the spontaneous mutations of bacteria. *Proc. Natl. Acad. Sci. USA* **36:**708–719.

55. **Ormerod, M. G. (ed.).** 2000. *Flow Cytometry: a Practical Approach,* 3rd ed. Oxford University Press, New York, NY.

56. **Perfiliv, B. V., and D. R. Gabe.** 1969. *Capillary Methods of Investigation of Microorganisms.* University of Toronto Press, Toronto, Canada.

57. **Postgate, J. R.** 1967. Viability measurements and the survival of microbes under minimal stress. *Adv. Microb. Physiol.* **1:**2–23.

58. **Postgate, J. R.** 1969. Viable counts and viability, p. 611–628. *In* J. R. Norris and D. W. Ribbons (ed.), *Methods in Microbiology,* vol. 1. Academic Press, Inc., New York, NY.

59. **Ryan, F. J., G. W. Beadle, and E. L. Tatum.** 1943. The tube method of measuring the growth rate of neurospora. *Am. J. Bot.* **30:**784–799.

60. **Schmidt, E. L., R. O. Bantkole, and B. B. Bohlool.** 1968. Fluorescent-antibody approach to study of rhizobia in soil. *J. Bacteriol.* **95:**1987–1992.

61. **Shapiro, H. M.** 2003. *Practical Flow Cytometry,* 4th ed. Wiley-Liss, New York, NY.

62. **Shehata, T. E., and A. G. Marr.** 1971. Effect of nutrient concentration on the growth of *Escherichia coli. J. Bacteriol.* **107:**210–216.

63. **Shuler, M. L., R. Arls, and H. M. Tsuchlya.** 1972. Hydrodynamic focusing and electronic cell-sizing techniques. *Appl. Microbiol.* **24:**384–388.

64. **Sokal, R. R., and F. J. Rohlf.** 1969. *Biometry.* W. H. Freeman & Co., San Francisco, CA.

65. **Spielman, L., and S. L. Goren.** 1968. Improving resolution in Coulter counting by hydrodynamic focusing. *J. Colloid Interface Sci.* **26:**175–182.

66. **Steen, H. B.** 1990. Flow cytometric studies of microorganisms, p. 605–622. *In* M. R. Melamed, T. Linmo, and M. L. Mendelsohn (ed.), *Flow Cytometry and Sorting,* 2nd ed. Wiley-Liss, New York, NY.

67. **Stock, J. B., B. Rauch, and S. Roseman.** 1977. Periplasmic spaces in *Salmonella typhimurium* and *Escherichia coli. J. Biol. Chem.* **252:**7850–7861.

68. **Taney, T. A., J. R. Gerke, and J. F. Pagano.** 1963. Automation of microbiological assays, p. 219–247. *In* F. Kavenagh (ed.), *Analytical Microbiology.* Academic Press, Inc., New York, NY.

69. **Tisa, L. S., T. Koshikawa, and P. Gerhardt.** 1982. Wet and dry bacterial spore densities determined by buoyant sedimentation. *Appl. Environ. Microbiol.* **43:**1307–1310.

70. **Troller, J. A., and J. H. B. Christian.** 1978. *Water Activity and Food.* Academic Press, Inc., New York, NY.

71. **Van Dilla, M. A., P. N. Dean, O. D. Laerum, and M. R. Melamed.** 1985. *Flow Cytometry: Instrumentation and Data Analysis.* Academic Press, Orlando, FL.

72. **Vollenwelder, R. A.** 1974. *A Manual on Methods for Measuring Primary Production in Aquatic Environments.* IBP handbook no. 12. Blackwell Scientific Publications, Oxford, United Kingdom.

73. **Wyatt, P. J.** 1973. Differential light scattering techniques for microbiology, p. 183–263. *In* J. R. Norris and D. W. Ribbons (ed.), *Methods in Microbiology*, vol. 8. Academic Press, Inc., New York, NY.

74. **Wyatt Technology.** 1988. *Literature on Model F and Model B Dawn Instruments*. Wyatt Technology, Santa Barbara, CA.

75. **Zimmerman, U., J. Schultz, and G. Pilwat.** 1973. Transcellular ion flow in *Escherichia coli* B and electrical sizing of bacterias. *Biophys. J.* **13:**1005–1013.

10

Nutrition and Media[†]

DAVID EMERSON AND JANE TANG

[†]This chapter is an updated and significantly revised version of an earlier chapter in *Methods for General and Molecular Bacteriology* by Rosalie J. Cote and Robert L. Gherna (see reference 9).

10.1. INTRODUCTION

Nutritional diversity is the hallmark of prokaryotes. As a result, the media requirements for prokaryotes reflect the diverse array of organic and inorganic energy sources that these organisms can utilize. To illustrate this point, at the ATCC there are approximately 18,000 strains of prokaryotes that require over 2,000 different kinds of media to maintain; in comparison, there are approximately 36,000 strains of fungi but these metabolically diverse eukaryotes require only about 100 different kinds of media for growth. The preparation of the appropriate medium is where every microbial investigation begins, whether it is the routine use of *Escherichia coli* to develop a clone library or an attempt to isolate a novel anaerobic hyperthermophile from a deep-sea hydrothermal vent. Every medium shares in common the necessity of providing an organism's energy requirements, sources of carbon, minerals, and any supplementary nutrients, as well as an appropriate atmosphere and a stable pH.

The roots of general microbiology lie in the seemingly simple concept pioneered by Beijerinck and Winogradsky in the late 1800s of using a selective medium to enrich from the environment a particular group of organisms. For example, medium that lacks a source of combined nitrogen can be used to select for organisms capable of fixing nitrogen, or medium that contains methanol as the sole energy source can be used to select for methylotrophic bacteria. While selective enrichment may be "seemingly simple," more than 100 years later we are still conceiving of new combinations of organic and/or inorganic compounds to discover whole new metabolic classes of organisms. Recent intersections between geology and microbiology, as well as environmental engineering and microbiology, have provided even more creative approaches for developing selective media for novel organisms; see, for example, Amend and Shock (2). This chapter places some emphasis on the nutritional needs of some of the broad groups of prokaryotes that have been discovered through using selective media as an enrichment tool.

Other important aspects of nutrition that are considered in this chapter are requirements for vitamins, trace minerals, other growth factors, buffers to control pH, sterilization, and preparation of solid media. A section on quality control of media is also included. In general, specific media formulations are not included in this chapter. The best general sources for media formulations are the websites of the large culture collections; for example, the ATCC, the DSMZ, or the JCM, where complete listings of media formulations are available free of charge. These can be searched either by organism or by media name. The URLs for these sites are listed under the sources section (section 10.9.2). In addition, the *Handbook of Microbiological Media* (4) has a large number of media formulations.

10.2. TERMINOLOGY

A sometimes bewildering array of terms are encountered to describe the general properties of different media. The following is a glossary of media descriptions.

Defined or synthetic medium: a medium in which the chemical composition is known.

Complex or undefined medium: a medium in which the chemical composition is not known or defined. Examples include yeast extract or peptone.

Minimal medium: a medium that is designed to provide a given organism with only the essential nutritional requirements that it cannot synthesize by itself. Generally all the components are defined.

Mineral medium: a defined medium that contains only essential minerals, trace elements, vitamins (if necessary), and a simple energy source, for example, glucose.

Auxotroph medium: a medium, usually defined, that contains an essential growth factor (for example, an amino acid) which an organism requires. Auxotroph media are often used in the selection of genetically altered microbes.

Selective medium: a medium designed either to select for growth of cells possessing a specific trait (for example, antibiotic resistance) or to inhibit a group of organisms. Examples include adding Brilliant Green to inhibit gram-positive bacteria and select for gram-negative organisms.

Enrichment medium: a medium designed to promote the growth of organisms with a specific physiological trait from a mixed population (for example, sulfate reduction). Enrichment can also be achieved by adding a specific supplement to an otherwise nonselective medium. For example, adding cysteine to enrich for *Francisella*, which specifically promotes the growth of this organism.

Differential medium: a medium designed to distinguish a particular microbe or group of microbes based on nutritional properties. For example, MacConkey medium differentiates between lactose fermenters and nonfermenters.

Bioassay medium: a medium designed to determine the concentration of an essential growth compound by quantifying the growth of a microbe that requires that compound in the medium. A standard curve is developed by growing the organism on a range of concentrations, from very limiting to nonlimited, of the compound.

10.3. BASIC NUTRITIONAL REQUIREMENTS

The essential nutritional requirements for prokaryotes can be divided into two classes: macronutrients and micronutrients. The macronutrients serve as the primary energy sources for growth as well as the chemical building blocks for cellular structures. The micronutrients, vitamins, minerals, and other cofactors, can provide essential functionality to enzymes and other parts of the cellular machinery.

10.3.1. Macronutrients

More than 30 elements are considered essential for cell growth; however, there are six nonmetals (carbon, oxygen, hydrogen, nitrogen, sulfur, and phosphorus) and two metals (potassium and magnesium) that constitute 98% of the dry weight of prokaryotic organisms (15). These elements must be present in relatively high concentrations in all growth media.

10.3.1.1. Carbon

The majority of known prokaryotes are heterotrophic; that is, they use carbon-containing organic compounds to satisfy their carbon and energy requirements. The different classes of organic compounds are shown below (see Table 3, in section 10.8.1). Some bacteria are very versatile in their ability to utilize a wide range of compounds as sole sources of carbon and energy, e.g., pseudomonads, whereas others are limited in their ability to decompose certain organic substances, e.g., methylotrophs.

Autotrophy, or the ability to fix CO_2 as the principal source of carbon for cellular materials, is also widespread throughout the prokaryotic world. Autotrophs require a source of CO_2 in the gas phase; however, this is often supplemented by inclusion of bicarbonate salts in the liquid phase. CO_2-bicarbonate or bicarbonate-carbonate mixtures can also act as a buffering system for media.

10.3.1.2. Nitrogen

Nitrogen is the next most abundant constituent in the cell, making up 8 to 15% of the dry weight (15, 25). It is the major noncarbon constituent of proteins, nucleic acids, peptidoglycan, and coenzymes. Microorganisms can obtain nitrogen either from inorganic sources such as salts of ammonium (NH_4^+) or nitrate (NO_3^-) or by taking up N-rich organic compounds such as amino acids. Unique to all of life are the nitrogen-fixing or diazotrophic prokaryotes that contain nitrogenase and are capable of fixing atmospheric nitrogen gas (N_2) as a source of nitrogen.

10.3.1.3. Sulfur

Sulfur is an absolute requirement for microorganisms because of its structural role in building amino acids (cysteine and methionine), coenzymes, and a number of vitamins (biotin, thiamine, and lipoic acid) (5, 25). Microorganisms generally acquire sulfur from inorganic sulfate salts or from organic sulfur-containing amino acids. In addition, sulfur compounds are involved in the energy metabolism of quite a large number of microbes, either serving as electron acceptors for anaerobic respiration, or as oxidizable substrates (see section 10.8.6.2.)

10.3.1.4. Phosphorus

Phosphorus, usually in the form of phosphate, is needed for ATP synthesis and serves as a cellular constituent for nucleotides, phospholipids, and coenzymes, as well as cell wall components of gram-positive bacteria (teichoic acids). Organisms obtain phosphorus from inorganic phosphate ions or organic sources, such as glycerophosphate or phosphate esters.

10.3.1.5. Potassium and Magnesium

Potassium is required by all organisms, and it is the principal inorganic cation in the cell. It is a cofactor for several enzymes, as well as being needed for carbohydrate metabolism and transport processes. Among the halophilic archaea it plays a key role in osmoregulation (15). Potassium requirements are normally met by the addition of potassium salts (sulfate, phosphate, or chloride) to the medium.

Magnesium stabilizes ribosomes, cell membranes, and nucleic acids. A number of enzymes also use magnesium as an essential cofactor, for example, ATPase. Magnesium sulfate or chloride salts are the usual form added to the growth medium to satisfy this requirement.

10.3.1.6. Others

Sodium is necessary for certain marine bacteria, phototrophs, and some anaerobes. Although calcium is not a requirement for growth, it stabilizes cell walls and plays a role in the heat stability of bacterial endospores (25).

10.3.2. Micronutrients

The micronutrients (minerals, vitamins, and other cofactors) are essential for optimal growth; however, the absolute quantities that are required are often very small (micro- or nanomolar concentrations, or less). For this reason, they are normally added to growth media from concentrated stock solutions. Ideally, in characterizing a microbe its specific vitamin requirements should be determined; however, this is tedious and often not done; it is simply easier to add a stock solution of vitamins. One common problem in growing microbes is that growth may seem to diminish upon serial transfers of the culture; often this is a result of some essential micronutrient being diluted out of the original culture medium.

10.3.2.1. Minerals

There are a number of elements that, although required in tiny amounts (milligrams per liter), are critical to the overall nutrition of microbes because of their role as cofactors in a wide variety of enzymatic reactions. These elements include iron, manganese, cobalt, copper, molybdenum, zinc, nickel, chloride, and boron. Some organisms require selenium and tungsten. In the case of complex media, these elements are usually abundant enough in the organic supplements so it is not necessary to add them to the media separately. An important exception to this can be iron. Due to its prevalence in respiratory enzymes and cytochromes it can potentially be growth limiting, especially in complex media in which the overall organic carbon content is low.

In most mineral salts media it is recommended to add minerals from a stock solution. In Table 1 a recipe for ATCC's trace mineral supplement (based on Wolfe's mineral solution) is provided along with the primary function of each element (9, 38). Many stock mineral solutions call for the use of a chelating agent, such as nitrilotriacetic acid or EDTA, to prevent precipitation of the relatively high concentrations of metals in the solution. Stock solutions of minerals should always be filter sterilized. While the heat of autoclaving does not harm the metals themselves, it can cause precipitation reactions. After the mineral solution is diluted into the final growth medium (usually 1:100 or 1:1,000), the solutions can be autoclaved without harm.

10.3.2.2. Vitamins

Growth factors are specific organic compounds that a cell may not be able to synthesize and are usually required in very small (parts per million) amounts. There are three main classes of growth factors: (i) amino acids for certain

TABLE 1 Trace mineral supplements

Ingredients	Concn (g per liter)	Function
EDTA	0.5	Chelating agent
$MgSO_4 \cdot 7H_2O$	3.0	Cocatalyst, contributes to pigmentation of some organisms
$MnSO_4 \cdot H_2O$	0.5	For enzymes reacting with active oxygen species, and water-splitting enzymes of photosystem II
NaCl	1.0	Electrolyte
$FeSO_4 \cdot 7H_2O$	0.1	For enzymes, electron carriers, and oxidoreductases
$Co(NO_3)_2 \cdot 6H_2O$	0.1	For making cobalamin (B_{12})
$CaCl_2$	0.1	Cocatalyst; membrane stability
$ZnSO_4 \cdot 7H_2O$	0.1	For different enzyme classes; dehydrogenases, polymerases, anhydrase, proteinases
$CuSO_4 \cdot 5H_2O$	0.01	For enzymes reacting with oxygen; oxygenases, oxidases, superoxide dismutase, and certain nitrite reductases
$AlK(SO_4)_2 \cdot 12H_2O$	0.01	Function unknown, probably for enzymatic activities
H_3BO_3	0.01	Uptake of Ca^{2+} and Mg^{2+}; increase retention of Ca^{2+} in cells; synthesis of certain quorum signal compounds
$Na_2MoO_4 \cdot 2H_2O$	0.01	For oxidoreductases and nitrogenases
Na_2SeO_3	0.001	Found in tRNA and selenocysteine of some oxidoreductases
$NiCl_2 \cdot 6H_2O$	0.02	For Ni-tetrapyrrole of methanogens, in urease, hydrogenase, CO dehydrogenase
$Na_2WO_4 \cdot 2H_2O$	0.01	Substitute for Mo in some oxidoreductases, but cannot be replaced by Mo

proteins, (ii) purines and pyrimidines for nucleic acids and coenzymes, and (iii) vitamins for coenzymes or prosthetic groups of certain enzymes (9).

Vitamins play catalytic roles within the cells and usually serve as components of coenzymes or as prosthetic groups of enzymes. There are certain bacteria that have complex vitamin requirements. The lactic acid bacteria, which include *Streptococcus*, *Leuconostoc*, and *Lactobacillus*, are well known for this, and they are used in vitamin assay procedures. For growth in complex media, yeast extract (see section 10.4.1.3) is often added as a supplement that is rich in vitamins and can fulfill the requirements for most prokaryotes. In defined and mineral media, vitamins are normally added from a stock solution. Stock solutions of vitamins should be made in ultrapure water and filter sterilized. They are normally added to growth media after the medium has been autoclaved. Some vitamins are light sensitive (e.g., riboflavin and vitamin B_6), so stock solutions should be kept at 4°C in the dark. It is recommended that vitamin stock solutions be made fresh every 6 months.

Below is a group of commonly used vitamins, with some detail as to their functions and properties. This information is largely drawn from an earlier edition of this book (9). Table 2 summarizes this list of vitamins and their functions.

10.3.2.2.1. PABA
p-Aminobenzoic acid (PABA) serves as a precursor for tetrahydrofolic acid biosynthesis; thus, it plays a part in the metabolism of thymine, methionine, serine, the purine bases, and vitamin B_{12}. The presence of these compounds may decrease or eliminate the requirement of a bacterium for PABA in the medium.

10.3.2.2.2. Folic Acid Group
Folic acid is primarily found in cells as tetrahydrofolate. It functions as a cofactor in the synthesis of thymine, purine bases, serine, methionine, and pantothenic acid. It plays a role in one-carbon (C_1) metabolism and transfer. Folic acid is only slightly soluble in water, but the ammonium salt is very soluble. It is stable to autoclaving.

10.3.2.2.3. Biotin
Biotin is responsible for biosynthetic reactions that require carbon dioxide fixation and release, as well as the synthesis of fatty acids. Free biotin is only slightly soluble in water, but its salts are freely water soluble. Biotin is stable to autoclaving and acids but can be oxidized to the sulfoxide and sulfone. For bacteria that can convert the oxidation products to biotin, these compounds can promote their growth. Oxidation is a problem only in extremely diluted solutions (<1 µg/ml) and if no reducing agents are present.

10.3.2.2.4. Nicotinic Acid and Its Derivatives
Nicotinic acid serves as the precursor of NAD and NADP, which are involved in redox reactions. Nicotinic acid and its amide are stable to autoclaving at neutral pH. Some organisms (*Haemophilus influenzae*) are unable to synthesize NAD from nicotinic acid, and thus, they require the addition of NAD (factor V) as the coenzyme. NAD is destroyed by autoclaving, acids, and alkalis, so it must be sterilized by filtration and added aseptically to the separately autoclaved medium.

10.3.2.2.5. Pantothenic Acid and Related Compounds
Pantothenic acid is the precursor of coenzyme A and a component of the acyl carrier protein. It also plays a part in the metabolism of fatty acids. A few bacteria utilize the intact coenzyme as a growth factor, but most others require pantothenic acid or pantotheine for growth. The requirement of the individual forms of these related compounds reflects the differences in the synthetic and transport abilities of the bacteria. Free pantothenic acid and many of its salt forms are very hygroscopic; thus, it is marketed as the calcium

TABLE 2 Vitamins and their functions

Vitamin	Functions
p-Aminobenzoic acid	Precursor of tetrahydrofolic acid; involved in one-carbon transfer
Biotin	Carbon dioxide fixation and release; fatty acid biosynthesis
Coenzyme M	Involved in methane formation
Cyanocobalamin (B_{12})	Molecular-rearrangement reactions; synthesis of deoxyribose
Folic acid	One-carbon metabolism
Lipoic acid	Transfer of acyl groups
Pantothenic acid	Precursor of coenzyme A; metabolism of fatty acids
Pyridoxine (B_6)	Deamination/transamination
Nicotinic acid	Precursor of NAD and NADP; involved in redox reactions
Riboflavin (B_2)	Precursor of FMN and FAD; involved in redox reactions
Vitamin K	Precursor of menaquinone
Thiamine (B_1)	Transketolase; decarboxylation

salt, which is only slightly hygroscopic, freely soluble in water, and autoclavable. Like its more complex derivatives, pantothenic acid is readily hydrolyzed, so its growth-promoting activity is destroyed by acid or alkaline hydrolysis.

10.3.2.2.6. Riboflavin and Its Derivatives
Riboflavin 5'-phosphate (flavin mononucleotide) and flavin adenine dinucleotide are the major coenzyme forms of riboflavin. They serve as cofactors in redox reactions. Riboflavin is slightly soluble in cold water but is easily soluble by warming. It is destroyed by visible light, especially at neutral pH or above; thus, it is important to store the solutions in the dark. Riboflavin is stable to autoclaving.

10.3.2.2.7. Thiamine
A large number of bacteria require thiamine, its precursors, or its coenzyme form, thiamine pyrophosphate (cocarboxylase) to grow. Thiamine pyrophosphate is involved in the decarboxylation of 2-oxo acids and in the transketolase reaction. At pH 5.0 or above, thiamine is cleaved into its component moieties in aqueous solutions when autoclaved; thus, the solutions need to be filter sterilized or autoclaved at pH below 5.0 if the intact form of this vitamin is required.

10.3.2.2.8. Vitamin B_6 Group
The vitamin B_6 group is comprised of pyridoxine, pyridoxal, pyridoxamine, and their phosphorylated derivatives. They are precursors of pyridoxalphosphate and serve as coenzymes in amino acid transformation (deamination and transamination) reactions. Pyridoxal is preferred by some bacteria, and adequate amounts of it are formed during autoclaving if sufficient pyridoxine is included in the medium. Some bacteria require pyridoxamine 5'-phosphate for growth; but for most auxotrophs the phosphorylated forms are inactive probably due to the lack of the necessary transport systems for these bacteria.

All three forms of the vitamin are stable to autoclaving; however, under the temperatures and pressures of autoclaving pyridoxal and pyridoxamine may react with many naturally occurring compounds and become inactive. Thus, it is crucial that these vitamins be filter sterilized. All forms of this vitamin are sensitive to light, especially at alkaline pH.

10.3.2.2.9. Vitamin B_{12}
Cyanocobalamin (vitamin B_{12}) functions in methyl (CH_3) transfer reactions. The coenzyme form (cyanocobalamin coupled to adenine nucleoside) plays a part in a number of carbon rearrangement reactions and in the biosynthesis of ribonucleotide reductase. Vitamin B_{12} itself is relatively stable to light and autoclaving. While vitamin B_{12} is synthesized by many bacteria, it is one of the vitamins most commonly added as a supplement to minimal media.

10.3.2.2.10. Lipoic Acid
Lipoic acid is essential for propagating members of the lactic acid bacteria. Its function is in the transfer of acyl groups and in the irreversible oxidative decarboxylation of 2-oxo acids. Acetate can alleviate the requirement of lipoic acid for many lipoate auxotrophs. Lipoic acid is stable to acid and autoclaving but is labile to oxidizing agents.

10.3.2.2.11. Vitamin K
Vitamin K and related compounds are the precursors of menaquinones, which function as carriers for redox reactions. A few organisms (e.g., *Porphyromonas levii*, *Prevotella melaninogenica*, and *Prevotella nigrescens*) need vitamin K or related compounds to grow (13, 23). Vitamin K_3 (menadione; 2-methyl-1,4-naphthoquinone) is destroyed by light, alkali, and reducing agents. Vitamin K_1 (2-methyl-3-phytyl-1,4-naphthoquinone) is more heat stable and biologically reactive for some bacteria. Both forms are fat soluble, and they can be added to media from an ethanol stock solution. Vitamin K stock solutions should be prepared fresh at monthly intervals and stored at refrigeration temperature in the dark.

10.3.2.2.12. Purines and Pyrimidines
The relationships of nucleosides between folic acid and vitamin B_{12} were mentioned above (sections 10.3.2.2.2 and 10.3.2.2.9). Some bacteria do not have the ability to synthesize purines and pyrimidines, for example, *Bacteroides fragilis* ATCC 29766 and *Vibrio cholerae* ATCC 14033. In this case, the nucleosides need to be added to the mineral-based media in milligram concentrations. The purine salts are more soluble than the free bases. Guanine and xanthine can be dissolved in a minimal amount of hydrochloric acid; the other compounds are soluble in hot water.

10.4. COMPLEX NUTRITIONAL COMPONENTS

Complex media contain many undefined ingredients, i.e., crude extracts of animals and protein digests. These highly nutritious substances promote bacterial growth by providing an excess of both micro- and macronutrients (9). Complex media are used routinely to propagate a variety of microorganisms and are especially useful when maximal growth rates and yields are the objective. The most common undefined components are listed below (4, 9).

10.4.1. Peptones, Extracts, and Hydrolysates

Peptones are hydrolyzed proteins that result from enzymatic hydrolysis, or acidic/alkali digestion resulting in individual amino acids and small peptides. Casein is the most common protein used for forming peptones, but more complex animal and plant hydrolysates can serve as raw materials for these digests as well. The protein source and the method and degree of digestion are important in determining the suitability of a particular digest in a specific medium. Although medium manufacturers rigidly evaluate the digestion products, lot-to-lot variations are expected, especially with respect to those undefined factors that promote the growth of certain very fastidious organisms. Peptones and hydrolysates are usually added to media at a concentration of 0.5 to 1.0%.

10.4.1.1. Acid Hydrolysates

Acid hydrolysates use heat and low pH to digest the proteins. As a result tryptophan and, to a lesser degree, serine and threonine are destroyed in the process. Further treatment during the manufacture of acid hydrolysates also limits the vitamin content of the end products. In addition, neutralization of the final products results in a high salt content. Among the common forms of acid-hydrolyzed casein are Acidicase and Casamino Acids. These and other acid-hydrolyzed caseins consist predominately of free amino acids but are deficient in tryptophan and cystine. They are used in assay media that are low in these growth factors, as well as in media for toxin production.

10.4.1.2. Enzymatic Digests

Enzymatic digests best conserve the intact nutritional content of the original protein source. Common enzymatic digests of casein include Casitone, Trypticase peptone, and tryptone. They consist mostly of free amino acids and small peptides and are used frequently in growth media for heterotrophs. Since the digests are low in carbohydrates, they can be used in fermentation test media for certain sugars.

Examples of enzymatic digests of meat protein are peptone, proteose peptone, Myosate, and Thiotone. They consist of a general mixture of amino acids and peptides suitable for growing most heterotrophs. Peptones with the designation "proteose" are hydrolyzed in such a way to have a high peptide content and are recommended for cultivating fastidious organisms. Thiotone is a digest high in sulfur amino acids and can be used in test media for detecting hydrogen sulfide.

Enzymatic digests of plant protein include Phytone, Soytone, soya peptone, and soy peptone. They are high in carbohydrate content and vitamins and can support rapid growth of most bacteria; however, acid production from the carbohydrates may also cause a rapid decline in culture density in the stationary phase, especially under fermentative conditions.

10.4.1.3. Extracts of Meat and Yeast

Extracts of meat and yeast contain large amounts of amino acids, vitamins, and trace elements. Plant extracts contain high concentrations of carbohydrates. They are often included in complex media as a source of nutrition. The commonly encountered meat extracts are beef extracts and brain heart infusion, which are made by boiling the substrate for a prescribed time (20 min at 100°C). They are free from fermentable carbohydrates but contain glycolic acid, lactic acid, and creatinine, which can serve as carbon sources. Yeast extract is produced by the autolytic reaction of the proteolytic enzymes from the yeast cells and consists of a concentrated solution of hydrolyzed yeast protein, which is high in water-soluble vitamins and carbohydrates as well. Commercially available powdered forms may not be equivalent in the growth-promoting potential to freshly prepared autolysates of yeast. The latter have been shown to be more effective in media for cultivating mycoplasmas, and preparation of fresh yeast autolysate was shown to be essential for the culture of acetogenic spirochetes from the termite gut (22).

10.4.1.4. Mixed Hydrolysates

Mixed hydrolysates are processed yeast or meat extracts to supplement the nutritional aspects of peptones by providing some amino acids, carbohydrates, nucleic acid fractions, organic acids, vitamins, and trace minerals lost during their manufacture. Mixed hydrolysates include polypeptone (casein and meat), tryptose (mixed enzymatic sources), and Biosate (yeast extract and casein digest). These are generally added to media at final concentrations of 0.3 to 0.5%.

10.4.2. Other Organic Nutrients

10.4.2.1. Fatty Acids

Certain groups of microorganisms can utilize fatty acids as carbon and energy sources, and some groups are unable to synthesize certain fatty acids required for their lipids; these must be provided in the medium. For example, members of *Mycobacterium* need oleic acid included in their propagation media (Middlebrook 7H10 agar, ATCC medium 173).

Several of the volatile fatty acids can be substituted for rumen fluid to cultivate the microorganisms present in the rumen. Volatile fatty acids can be prepared as a stock mixture and added to the propagation media. It should be remembered that fatty acids can be toxic even at low concentrations; this is especially true for unsaturated fatty acids. Thus, it may be necessary to include a detoxifying compound in the medium. Bovine serum albumin is commonly used for this purpose.

10.4.2.2. Others

Some *Streptococcus pneumoniae* strains and oral treponemes require choline to grow. This organic compound serves as a precursor for certain cellular lipids. In *S. pneumoniae*, the choline residues make up the lipoteichoic cell membrane. Purines and pyrimidines are the building blocks of nucleic acids and may be required in auxotrophic strains that are unable to synthesize these compounds.

10.4.3. Supplements

Many fastidious organisms grow in media that contain additional undefined supplements. These substances include growth factors, blood components, and other compounds or extracts serving as sources for vitamins, amino acids, or undefined growth factors.

10.4.3.1. Blood and Serum

Many clinical isolates grow on media containing blood or blood components. Routine cultivation media for human or animal pathogens contain 5 to 15% defibrinated blood. The most common source of blood is from sheep, although rabbit blood and horse blood are also used occasionally for certain fastidious microbes. A few organisms, such as members of the genus *Haemophilus*, may be inhibited by fresh sheep blood. Instead, chocolate medium (GC medium, ATCC medium 814) in which the blood is heated to 70 to 80°C for 15 min is used. In addition, species of *Haemophilus* and *Neisseria* require X factor (heme) and V factor (NAD).

Serum can serve as an alternate to blood. Horse and bovine sera are widely used in cultivating fastidious organisms, i.e., *Leptotrichia*, *Mycobacterium*, *Streptobacillus*, and *Treponema*. Fresh rabbit serum is crucial for cultivating *Borrelia* species. Lyophilized pooled rabbit serum is needed for cultivating species of *Leptospira*. Certain members of the *Mollicutes* (*Mycoplasma*, *Spiroplasma*, *Mesoplasma*, and pleuropneumonia-like organisms [mycoplasmas]) require serum as a source for sterols to grow.

It is important to ensure that the blood and serum used in formulating bacteriological media are free of pathogens and potential health risks, especially in light of the recent bovine spongiform encephalopathy outbreaks. Serum may have to undergo heat treatment for inactivating certain proteins to prevent a possible inhibiting effect on microbes. Media containing blood and serum are stable and usable if stored properly under refrigeration.

10.4.3.2. Rumen Fluid

To cultivate some ruminal bacteria, complex media may have to be supplemented with rumen fluid. Rumen fluid can be obtained from universities and colleges with animal science or veterinary programs that have access to fistulated cows (this is a not a stressful experience for the cow). Although the exact content of the rumen fluid is not known and varies with the diet of the animal, it serves as a supplement of amino acids, peptides, vitamins, fatty acids, carbohydrates, and other intermediate metabolites. Results may vary from batch to batch as well as with age. A mixture of volatile fatty acids can replace rumen fluid to fulfill the nutritional requirements of some ruminal organisms (*Ruminobacter*, *Ruminococcus*, and rumen treponemes) (9).

10.4.3.3. Soil Extract

Many soil microbes grow better and sporulate more readily in media containing soil extracts, although the composition is unknown and the reasons for the effects are not clear (7, 9). One "standard" soil extract uses African violet soil supplemented with sodium bicarbonate. This has been used as a supplement to cultivate *Azotobacter*, *Rhizobium*, and some *Bacillus* species.

10.4.3.4. Other Supplements

A number of undefined supplements are frequently added to mimic the natural habitat from where the organisms were isolated. For example, tomato juice is included in media to cultivate lactic acid bacteria (i.e., *Lactobacillus*, *Leuconostoc*, *Pediococcus*, *Bifidobacterium*, and *Propionibacterium*). V-8 canned vegetable juice is supplemented in media for propagating species of *Actinomadura* and *Microspora*, as well as organisms isolated from plants (*Pseudomonas tolaasii*) (9). Rabbit dung is another example. The pellets are often used as a bait in the enrichment of myxobacteria (33), and the aqueous extracts have proven useful in isolating planktonic *Herpetosiphon* in freshwater (32). Dry pellets of rabbit dung are boiled in water for 20 min, and the filtrate is adjusted to pH 7.2 before adding to enrichment media for isolation of appropriate organisms.

10.5. NONNUTRITIONAL MEDIA COMPONENTS

10.5.1. Agar

One of the major innovations in the growth of microbes was the use of agar as a solidifying agent for culture media, an idea suggested by Fannie Eilshemius Hesse, the wife of an associate of Robert Koch (5). The ability to cultivate bacteria on solid media revolutionized almost every aspect of microbiology. With the exception of some marine bacteria, agar is not digested or used as a nutrient source by most microbes.

Agar is the most commonly used gelling agent extracted from species of red algae. It is a sulfuric acid ester of the linear gelactan (26). A valuable property of agar is that it melts at 80 to 95°C but does not solidify until cooled down to 35 to 40°C. Typically 15 g of agar per liter is used for solid culture media, and lower concentrations can be used to produce soft or semisolid media. Most manufacturers sell different grades of agar that vary in the amount of impurities, such as trace elements and residual organic matter, that they contain. Noble agar is generally the highest grade and most expensive. Agarose, typically used for gel electrophoresis, is an even more purified form of agar that can be used if it is suspected that impurities are a problem; however, while agarose has a higher gelling strength than agar, it is also much more expensive.

Other different types of solidifying agents include gelatin, gellan gum, Gelrite, and silica gels. Gelatin can be digested by many organisms, so it is most useful when identifying bacteria with proteolytic activity by observing liquefaction of the gelatin medium. Silica gel is an inorganic solidifying agent, useful in culturing autotrophs in the complete absence of organic substances. Silica gel can also be used in testing the ability of heterotrophs to utilize certain organic compounds as single carbon sources, as well as in vitamin assays. Gelrite (Kelco Division of Merck & Co., Inc., San Diego, CA) or Phytagel (Sigma Chemical Co.) has a higher melting temperature than agar, so it is used for culturing thermophiles that grow at temperatures close to 100°C, which will liquefy the agar. Unlike agar, Gelrite requires a divalent cation (Mg^{2+} or Ca^{2+}) to solidify and it solidifies very rapidly at about 90°C, making it somewhat difficult to work with. One solution to this is to mix agar and Gelrite in a 1:3 ratio; this results in a solution that is easier to work with but that remains solid at temperatures of up to 80 to 90°C. Gellan gum has been shown to be effective for culturing novel soil microorganisms on solid medium (19). It requires a divalent cation, e.g., $CaCl_2$, for solidification.

10.5.2. Reducing Agents

While reducing agents do not contribute directly to an organism's nutrition, they are a necessity for the optimal growth of a number of strict anaerobes by lowering the redox potential of the medium to a point compatible with the stability and activity of many cellular enzymes (see chapter 14). Sodium sulfide is probably the most commonly used reducing agent. Liquid stock solutions are stable for months when stored under N_2 and can also be repeatedly frozen and thawed. Sulfide reduces growth media rapidly. It is important to remember that hydrogen sulfide gas is extremely toxic and is generated rapidly if sodium sulfide solutions come in contact with acid. Likewise, stocks of sodium sulfide should be filter sterilized and not autoclaved. Another quite commonly used reducing agent is cysteine. Both cysteine and sulfide can be used as sulfur sources by a number of prokaryotes; sulfide has an advantage in enrichment media of not being used as an energy source for fermentative microbes. Another very effective reducing agent for methanogens is 2-mercaptoethanesulfonic acid or coenzyme M, which is, itself, a key intermediate in methane formation. A stock solution of 5% (wt/vol) coenzyme M can be made in deionized water and filter sterilized. The stock solution is diluted 1:100 into the culture medium. Coenzyme M takes some time to fully reduce the culture medium; between 30 and 60 min is common; however, it has proven a very effective reducing agent for a wide variety of methanogens (D. Boone, personal communication). Other reducing agents that are used in more specialized applications include iron sulfide (FeS), titanium citrate, and hydrogen in the presence of palladium chloride ($PdCl_2$) catalyst; these are discussed in more detail in chapter 14.

10.5.3. Redox Indicators

Redox indicators are dyes that either change or lose color depending on the redox potential of the medium. Important factors to consider with any redox indicator are its midpoint potential and potential toxicity. The former consideration will guide the selection of the most appropriate redox indicator when a particular redox potential for a medium is desired. These dyes are not measures of specific redox potentials. Instead, they indicate whether the redox potential of the medium is above or below their mid-point potential. The most commonly utilized redox indicator is resazurin. At neutral pH, this dye is blue under aerobic conditions. However, once the O_2 is gassed out of the medium and it is heated either by boiling or autoclaving, an irreversible reaction leads to the formation of resorufin, which is pink. Upon addition of a reducing agent, such as sodium sulfide, capable of poising the redox around -110 mV or less, the resorufin ($E_0 = -51$ mV) becomes clear. Reduced forms of resazurin are sensitive to photo-oxidation and may produce toxic intermediates; therefore, resazurin should not be used in anaerobic phototrophic media that will be illuminated. Redox indicators are discussed in more detail in chapter 14.

10.5.4. Buffers

Many metabolic reactions coupled to cell growth result in the consumption or release of acidic or basic products, which can result in a substantial shift in the pH of the medium. For example, most fermentations result in the production of organic acids which can easily reduce the pH of a culture medium to below pH 5. The most common way to alleviate this problem is to add a buffering agent to the medium. The important things to consider in choosing a buffer are (i) its buffering range, (ii) its potential toxicity, (iii) its capacity to bind metals or otherwise interact with other media components, and (iv) the potential to be metabolized by the organisms that are growing in it. In chapter 14, there is an excellent description of the chemistry and buffering capacity of buffers, and so this chapter highlights only some of the buffers commonly used in media preparation.

10.5.4.1. Inorganic Buffers

10.5.4.1.1. Phosphates

Historically, phosphate is probably the most commonly used buffering system, made by mixing the monobasic salts of either KH_2PO_4 or NaH_2PO_4 with their dibasic forms, K_2HPO_4 or Na_2HPO_4, to provide buffering capacity in the range from pH 6 to 8. The exact pH depends upon the ratio of monobasic to dibasic salts. These buffers can cause precipitation reactions with metals, principally Ca^{2+}, Fe^{3+}, and Mg^{2+}, especially above pH 7.0, resulting in these metals no longer being available for microbial nutrition.

10.5.4.1.2. Bicarbonate-CO_2

The bicarbonate-CO_2 buffering system is the most widespread in nature, and as a result it is commonly used for methanogens, sulfate-reducing bacteria, and anoxygenic photosynthetic bacteria, as well as a variety of other lithotrophic prokaryotes. It is especially advantageous for autotrophic organisms since it provides both buffering capacity and a source of C. Bicarbonate buffers are more cumbersome to prepare than most other buffers; however, once procedures are established, they quickly become routine and require little more time than using more conventional laboratory buffers. Typically a calculated concentration of sodium bicarbonate (usually between 10 and 50 mM) is added to the medium, either before or after autoclaving, and then the pH is adjusted by gassing the medium with CO_2 after autoclaving. Using the equations in chapter 14, it is possible to calculate the ratios of bicarbonate:CO_2 that should be used, or this can be done empirically by gassing a known volume of medium with a known flow rate of CO_2 for different times until the final pH is achieved. Bicarbonate buffers must be used in a contained culture vessel, otherwise the CO_2 will escape to the atmosphere, causing the pH to rise as the bicarbonate dissociates to equilibrate the [CO_2]. This is especially important if the bicarbonate is added prior to autoclaving. In general, autoclaving is not recommended for bicarbonate buffers, because at high concentrations enough CO_2 may be driven out of the solution to pressurize the headspace and create an explosion hazard in glass bottles. An advantage of this buffering system is that it can be used over the range of pH 6 to 8 with few problems, such as metal precipitation reactions.

10.5.4.2. Organic Buffers

10.5.4.2.1. Tris

Tris-hydrochloride is a commonly used biological buffer. Its best buffering capacity is between pH 7.5 and 8.5, so if it is to be used at pH 7.0 a concentration in the range of 50 mM should be used. Tris is prepared by either adjusting the pH of the Tris-hydrochloride with sodium hydroxide to the desired pH or mixing a ratio of Tris-HCl with a basic form, e.g., tris[hydroxymethyl]aminomethane, to achieve the desired pH. The Sigma catalog lists the ratios of these buffers for different pHs. Tris buffers are temperature sensitive; for every degree Celsius above 25°C, Tris buffers decrease by approximately 0.025 pH unit; for every degree Celsius below 25°C, there is an approximate increase in pH of 0.03 unit. Tris buffers are also known to react with some metal cations, e.g., Ni^{2+} and Ca^{2+}.

10.5.4.2.2. Good's Buffers

The zwitterionic, organic buffers originally synthesized by Good et al. (14) have continuously grown in popularity for use in microbiological media. There are a number of these available from commercial sources, and together they can cover the pH range from about pH 5.5 to 11. Three of the more common examples are MES [2-(N-morpholino) ethanesulfonic acid], with a pH range of 5.5 to 6.7; HEPES [N-(2-hydroxyethyl)piperazine-N'-(4-butanesulfonic acid)], with a pH range of 6.8 to 8.2; and TAPS {N-[Tris(hydroxymethyl)methyl]-3-aminopropanesulfonic acid}, with a useful pH range of 7.7 to 9.1. These acidic buffers are simply added to the medium prior to autoclaving at an appropriate concentration, and the pH is adjusted with sodium or potassium hydroxide. They are chemically inert and nontoxic. Their only drawback is that they are relatively expensive, especially if large volumes of a medium are required.

10.5.4.2.3. Others

Organic acids, for example acetate and citrate, can be used as buffering agents at low pH, 4.0 to 5.5, when there are no good inorganic or synthetic buffers available. These have the obvious drawback that they are common microbial food sources. Protein and amino acids also have buffering capacity; in fact most complex media do not have specific buffers added, in part due to the buffering capacity of the proteinaceous components of these media.

10.6. PREPARATION OF MEDIA

10.6.1. Sterilization

After the ingredients of a given medium are weighed out and dissolved, either together or separately, they must be sterilized. While this step seems easy, most microbiologists have experienced the frustration of carefully mixing a complex set of ingredients, adjusting the pH, dispensing it into culture flasks, and autoclaving the medium, only to find upon opening the autoclave an hour later a precipitated mess. To avoid this outcome, thought must be given to the potential reactions of media ingredients at the heat and pressure in an autoclave, especially as a function of pH. If precipitation problems are encountered, or suspected, it is best to either filter sterilize the medium or heat sterilize the components separately and add them together after autoclaving. In most cases the pH of the media is adjusted prior to sterilization; however, it is sometimes necessary to check and readjust the pH after heat sterilization. It is also important to sterilize the medium as soon as it is prepared, to minimize the opportunity for growth of microbial contaminants. Several methods used for sterilization are as follows.

10.6.1.1. Autoclaving

Autoclaving utilizes high-pressure steam to kill microorganisms. This is probably the most commonly used method of sterilization, exposing the media to steam at 115 lbs/in^2 and 121°C for a minimum of 15 min. It is important that the internal temperature of the liquid in the autoclave reaches 121°C for the set time period of sterilization. Thus, longer running times are required for large loads and/or containers of larger volume.

Although autoclaving is a convenient and effective method of sterilization, many formulations cannot tolerate the high heat and pressure that contribute to undesirable chemical reactions (e.g., precipitation and browning) and breaking down of components. In most instances these reactions can be avoided by physically separating the reactants. Here are a few general rules to follow when autoclaving.

- Sterilize glucose separately from amino acids, peptones, or compounds containing phosphate.
- Autoclave phosphates separately from amino acids, peptones, or other mineral salt components.
- Alkaline media (pH, >7.6) should not be autoclaved. Prepare the medium at neutral pH and adjust it to the desired alkalinity after it has cooled down from autoclaving.
- Acidic media (pH, <6.0) also should not be autoclaved. Prepare the agar and the acidic ingredients separately in double strength (i.e., one liter's worth of ingredients dissolved in 500 ml of water, and 15 or 20 g of agar in 500 ml of water), and then autoclave them. After cooling down to about 50°C, combine the two solutions aseptically before pouring to plates or distributing into tubes.
- If precipitation problems are encountered with an agar-containing mineral salts solution, try autoclaving the agar separately from mineral salt solutions. There may be calcium and magnesium salts present in certain brands of agar which precipitate with other mineral salts.

The sterilizing effectiveness of the autoclave machines should be routinely monitored. This is usually performed by using a commercially available product, Kilit (BBL), which contains spores of *Bacillus stearothermophilus*.

10.6.1.2. Filtration

For heat-labile and volatile liquids, filtration is the method of choice for sterilization. A number of filtration systems are available commercially for different volumes of solution. The membrane filters are convenient to use, and their compositions need to be compatible with the chemical nature of the solutions to be sterilized. Most media and components are compatible with cellulose acetate or nitrocellulose filters. Nylon or Teflon membranes should be used for solvents.

Most bacteriological media are passed through filters with a pore size of 0.2 μm. Most microorganisms are trapped by this pore size; however, smaller organisms (i.e., *Pseudomonas diminuta*, mycoplasmas, rickettsias, and viruses) pass through these filters. If this is a concern, 0.1-μm-pore-size filters can be used instead.

If carbohydrates are to be used in sole-carbon source testing procedures, it is recommended to filter sterilize them, rather than autoclave. As mentioned above, vitamin and mineral stock solutions should be filter sterilized.

10.6.1.3. Tyndallization

Tyndallization method alternates heating and cooling by exposing the solutions to steam, or heating to 100°C in a water bath, for 30 min each day for three successive days. In this case the vegetative cells of heat-resistant endospore-forming bacteria (e.g., *Bacillus*) are killed by heating at 100°C. The endospores germinate when the medium cools down, and they in turn are killed by the subsequent heat treatment. This method is useful for solutions which contain powdery substances and is most often used with media containing elemental sulfur, which turns into an unworkable mass when autoclaved.

10.7. QUALITY CONTROL TESTING

For routine maintenance and characterization media, especially in clinical and quality control laboratories, it is important to ensure that media perform properly. Media purchased from commercial companies usually have gone through quality control procedures; however, it is always important to record the lot number of any individual medium in case problems arise. For media that are produced in-house, a testing protocol should be established. The protocol should include standard operating procedures for testing, the expected results, and criteria for acceptance.

An important aspect of quality control includes quality control organisms that will produce a positive or negative reaction on the media to be tested. These organisms should be obtained from a reputable source. For example, the ATCC has many quality control organisms, and every effort is made to minimize transfers and authenticate their identities and properties.

Frozen stocks (in 10% [vol/vol] glycerol or 5% [vol/vol] dimethyl sulfoxide) of the quality control organisms should be made and stored. After three passages of the working stock culture, it is necessary to go to another frozen stock culture. Minimizing transfers is crucial in the performance of these quality control organisms.

A good source of information on quality control of media is the *Difco Manual* (3), which provides the cultural reactions to each of its media and lists the organisms used for testing as well as positive and negative results. The incubation conditions are also included. This manual can serve as a guide to establish a media quality control program.

10.8. MEDIA FOR SPECIFIC GROUPS

In this section we consider media designed for some major metabolic groups of prokaryotes. This discussion is representative but by no means exhaustive; references and sources with more detail are provided for different groups.

10.8.1. Media for Heterotrophic Microbes

In general when we think of a bacterial growth medium, we think of a medium containing an organic energy and carbon source. If an organic compound can be made available to microbes, there is a good chance that there is a prokaryote that can extract energy from it. Major classes of organic compounds that are used for growth are listed in Table 3.

When a complex medium is used, several of these organic sources may be available simultaneously, whereas defined or minimal media are generally designed to provide only a single primary energy source. In general, for routine growth of a microbe in the laboratory a complex medium is best, yielding the most rapid growth rates and highest cell yields. Many standard microbiological media can be purchased premixed from commercial sources. This makes media preparation easy and fast and ensures a quality product. For more specific physiological or genetic studies more selective or defined media are often used. It is also important to remember that genetic drift or alteration is a real and often rapid phenomenon, so organisms selected for a specific trait should always be maintained on a medium that selects for that trait.

10.8.2. Phototrophs

10.8.2.1. Oxygenic Phototrophs

Cyanobacteria are nutritionally among the real minimalists of the microbial world, requiring little other than a few mineral salts and light to grow. Most are capable of synthesizing all of their vitamins, although a few require vitamin B_{12}. Despite these minimal media requirements, a number of cyanobacteria have proven quite challenging to grow, in part because they can be quite susceptible to toxic compounds that contaminate media components. For example, a number of marine strains, especially *Trichodesmium* spp., are exquisitely sensitive to Cu, and so in some cases extraordinary care needs to be taken in preparing media and trace elements solutions (see Waterbury [37] for a much more detailed discussion of media preparation for cyanobacteria). Light requirements are generally met by cool or warm white fluorescent lamps; lighting intensities should try to mimic those of the isolate's natural environment.

10.8.2.2. Anoxygenic Phototrophs

A large number of microbes are capable of carrying out photosynthesis under anaerobic conditions using reduced S compounds or H_2 in place of H_2O as a source of electrons for the incorporation of CO_2 into cell biomass. Two major groups of obligate anaerobic photosynthetic bacteria are the purple sulfur photosynthetic bacteria represented by the family *Chromatiaceae* and the green sulfur bacteria of the family *Chlorobiaceae*. Interestingly, both of these groups of organisms are found primarily in aquatic habitats including anoxic water columns, littoral sediments, or microbial mats and have similar media requirements; however, phylogenetically they are very distinct. The *Chlorobiaceae* form their own phylum within the domain *Bacteria*, while the *Chromatiaceae* are members of the gamma-proteobacteria. Both groups of organisms can be grown with a mineral salts medium that contains trace elements, vitamin B_{12}, and

TABLE 3 Major classes of organic growth compounds

Compound	Examples	Comments
Carbohydrates	Glucose, maltose, ribose	Excellent energy sources, readily soluble
Polysaccharides	Cellulose, starch	May be difficult to solubilize
Proteins/amino acids	Peptones, albumin	Excellent energy sources
Lipids	Palmitic acid, stearic acid	Rich energy sources, although limited range of organisms; solubility may be a problem
Organic acids	Acetate, butyrate	pH of medium must be adjusted, buffers are recommended
Alcohols	Ethanol, sorbitol	Some are toxic at higher concentrations
Aromatic hydrocarbons	Benzoate, toluene	Low solubility, volatility, toxicity
Aliphatic hydrocarbons	Hexadecane	Low solubility, volatility

sodium sulfide as a source of sulfide. An important physiological trait of most genera of the green sulfur bacteria is a preference for growth under low light intensity. See the chapter by Overmann (28) in *The Prokaryotes*, for an up-to-date and detailed discussion of the cultivation and lifestyle of the *Chlorobiaceae*, and the chapter by Pfennig and Trüper (29) for a discussion of cultivation of *Chromatiaceae*, especially vis-à-vis light requirements.

10.8.3. Biodegradation

The capacity for microbes to degrade and in many cases mineralize xenobiotics and organic pollutants is one of their most impressive traits. This is a large and ever-growing field and so is dealt with only in a cursory way here. Classes of toxicant compounds include aromatic hydrocarbons (e.g., benzene, toluene, and xylene), aliphatic hydrocarbons (e.g., gasoline), chlorinated aromatic hydrocarbons (e.g., polychlorinated biphenyls), and chlorinated aliphatics (e.g., tetrachloroethylene). Aside from their inherent toxicity, the property that makes many of these compounds recalcitrant is their insolubility and/or volatility. This has to be taken into account in preparing the media. In some cases the compound of interest may need to be dissolved in a potent organic solvent, for example, hexadecane, and added to the medium. In the case of toluene, the compound is sparingly soluble and quite volatile; therefore, it should be maintained in vapor phase in a properly sealed culture vessel to prevent its depletion by volatilization. Again, some of these compounds are themselves organic solvents, and so they can degrade or partition into rubber stoppers, requiring the use of Teflon-coated caps. For enrichment studies, often a mineral medium is used so that the only carbon and energy source available is the organic compound in question. In the case of anaerobic organisms capable of dechlorination, the chlorinated compound can serve as the electron acceptor and the organism may use an organic compound, for example, pyruvate, as the electron donor; thus, the dechlorination is the result of anaerobic respiration (10).

10.8.4. Methanogenesis

Methanogens, which constitute a large and well-studied group of *Archaea*, are unique in the microbial world for their ability to form methane. Despite being metabolically limited only to methane production, methanogens are a physiologically and morphologically varied lot, and there is a large literature devoted to them (34, 36, 40). Methanogens are among the strictest of anaerobes, meaning that they require the addition of a reducing agent (see section 10.5.2) to the growth medium. It is best to assume that they cannot tolerate any exposure to oxygen, so a redox indicator should be included as well. Specially adapted culture vessels have been designed for growing methanogens, the most notable being the Balch tube, an 18- by 150-mm glass tube with a serum vial-type lip that can accommodate a thick butyl rubber stopper and be sealed with an aluminum crimp cap. This allows the tubes to be pressurized with the CO_2–H_2 gas mixtures that many of these organisms require for optimal growth. Although methanogens are comprised of over a hundred different species, metabolically they grow on a limited range of substrates. The classic substrates are CO_2 plus H_2. Another well-studied group are the aceticlastic methanogens that can utilize acetate. Other potential substrates for methanogens are methanol, ethanol, methylated amines, and methylated sulfides. In every case, CH_4 is the primary product. It is important to remember when using the latter set of soluble substrates that CH_4 will build up in the headspace of the culture vessel. Glass media vessels should be vented regularly with a needle to prevent the buildup of potentially explosive concentrations of gas. An anaerobic gassing station is essential for preparing methanogen medium with prepurified 100% N_2 and 80% H_2/20% CO_2 or 70% N_2/30% CO_2. The gas must be completely oxygen free; if good-quality gas is not available, traces of O_2 can be removed by passing the gas stream over hot copper turnings or purchasing a commercially available oxygen trap. See the paper by Sowers and Noll (34) for a more detailed discussion of preparing a laboratory for the growth of methanogens.

10.8.5. Methane-Oxidizing Bacteria

The methane-oxidizing or methanotrophic bacteria are a ubiquitous group of organisms capable of oxidizing methane in the presence of air for growth (18). They are commonly found in aquatic and terrestrial environments often associated with aerobic-anaerobic interfaces, where they can take advantage of the methane that is supplied by methanogenesis in the anaerobic environment. These organisms are unique in being able to incorporate the formaldehyde that is formed as an intermediate of methane oxidation directly into cell carbon; thus, true methanotrophs do not require an exogenous supply of organic matter.

A much larger group of bacteria, the methylotrophs are capable of growing on methanol by a similar mechanism. Most methanotrophs are obligate, requiring either methane or methanol for growth; on the other hand methylotrophs that are unable to grow on methane are generally metabolically diverse and often are able to grow on a variety of organic compounds as well. Methanotrophs can be grown on a mineral salts medium with trace metals; some require vitamins, others do not. Methanol is the preferred electron donor, since the alcohol can easily be dissolved in the medium. Methane is necessary, especially in the enrichment stage, to select for true methanotrophs and must be added as a gas by bubbling a stoppered tube for a short time with a gas mixture. Consult a more specialized text (18) for advice on specific methane-containing gas mixtures to use for enrichment and cultivation of different methanotrophs. It is important to remember that methane (natural gas) is very flammable.

10.8.6. Sulfur Metabolism

10.8.6.1. Sulfate Reduction

Sulfate reduction is the most common mechanism for carbon mineralization in anaerobic marine sediments, making it, on a global scale, an important pathway for both sulfur and carbon cycling. The sulfate-reducing bacteria (SRB) are very diverse; however, known isolates are concentrated in the delta-proteobacteria. There are several excellent reviews on the feeding and care of SRB (20, 30). In general, these organisms require a source of sulfate, an organic electron donor, and strictly anaerobic conditions, although addition of a reducing agent is usually not necessary. Sulfate is generally provided by simply dissolving sodium sulfate into the growth medium. Organic acids like acetate or lactate are commonly used as electron donors for SRB, but a wide variety of alcohols and fatty acids can also be used. Selectivity for certain species can be achieved by using a specific electron donor. Some SRB, for example, *Desulfovibrio desulfuricans* and *Desulfovibrio vulgaris*, are capable of lithotrophic growth on SO_4 by coupling H_2 oxidation or formate oxidation to sulfate reduction.

10.8.6.2. Sulfur Disproportionation

This is a newly discovered metabolism that results in the disproportionation of either elemental sulfur or thiosulfate to sulfide and sulfate. In the case of sulfur the reaction is not energetically favorable by itself; however, if there is Fe(III) present in the medium that can react with the sulfide that is produced and remove it from the medium as FeS, the low concentration of the sulfide product drives the reaction to be thermodynamically favorable. Thus far, only one organism, *Desulfocapsa sulfexigens*, has been described as a true chemolithoautotroph that grows by sulfur disproportionation; however, it is likely that others await discovery (12). Several sulfate reducers are known to also be able to grow by disproportionation of thiosulfate, although these organisms require a source of organic carbon as well (30).

10.8.6.3. Sulfur Oxidation

The oxidation of different reduced forms of S is, in most cases, a thermodynamically favorable reaction; thus, there is a broad range of lithotrophic bacteria and archaea that can derive energy by using reduced S compounds as electron donors (20). The most commonly used form of sulfur for growth of these is thiosulfate, since the sodium salt is readily soluble at neutral pH and can be autoclaved. Other sulfur compounds are tetrathionate, which can be made soluble in acidic stock solutions, and elemental sulfur, which is completely insoluble and must be sterilized by tyndallization (10.6.1.3). Hydrogen sulfide (H_2S) gas may also be used. Please remember that H_2S is very poisonous and it is also critical to keep acids away from sodium sulfide solutions, since the acidification causes the sulfide to rapidly form volatile H_2S gas.

10.8.7. Dissimilatory Metal Reduction

10.8.7.1. Iron Reduction

Over the past decade a growing number of anaerobic bacteria have been recognized that can utilize ferric (Fe) oxides as an electron acceptor for carrying out anaerobic respiration (24). Two organisms that are well-known for carrying out this reaction are *Geobacter metallireducens* and *Shewanella oneidensis*; these organisms represent the extremes of metabolic diversity that are associated with these processes as well. *G. metallireducens* has a limited metabolic repertoire, consisting of electron acceptors or donors that it can use for growth. *S. oneidensis* is a facultative anaerobe and is metabolically very diverse, capable of utilizing a wide variety of electron donors and acceptors. However, when respiring on Fe(III) the two organisms have quite similar requirements: both require a source of amorphous Fe oxide, an electron donor, and buffering capacity. There are a variety of sources of Fe(III) that have been used (21). Chelated forms of Fe(III), including ferric citrate, Fe-pyrophosphate, and Fe-nitrilotriacetic acid are the easiest to work with and can all be purchased as powders that can be dissolved in the culture medium. Especially during enrichment studies it is probably best to work with more naturally occurring forms of Fe oxides, including synthetically produced Fe oxide, or naturally produced ferrihydrite (35); these are insoluble forms that require that the cells attach to the oxides during growth.

While acetate is the most oft-used electron donor for the growth of Fe-reducing bacteria, a wide range of organic compounds have been shown to be coupled to Fe reduction, including short- and long-chain fatty acids, alcohols, and some aromatic compounds. In addition some strains can oxidize H_2 and formate as well and grow as chemolithoautotrophs. Addition of extracellular quinones, principally as the humic acid analog, anthraquinone-2,6-disulfonate (AQDS) has also been shown to stimulate Fe reduction by acting as an intermediary electron shuttle (8).

10.8.7.2. Manganese Reduction

Manganic oxides are also excellent electron acceptors for anaerobic respiration, although relatively less common than iron. The most common Mn(IV) oxidation state does not exist in a chelated form, so Mn minerals such as colloidal Mn(IV) or vernadite (MnO_2) must be prepared synthetically and added as the insoluble oxides to the medium (21). The carbon sources used may be similar to those used for Fe reduction.

10.8.7.3. Arsenic and Selenium

Arsenic and selenium are metals known for their toxicity and, while naturally occurring, are most often associated in more concentrated forms with industrial processes such as mining and the production of semiconductors. The oxidized form of arsenic is arsenate, As(V), which can be reduced to arsenite, As(III). The oxidized forms of selenium are selenate, Se(VI), and selenite, Se(IV), which can be reduced to elemental selenium, Se(0). At present there are at least five species of bacteria that are known to respire anaerobically by coupling oxidation of organic acids to Se reduction (27). Other bacteria that can reduce Se(VI) or Se(IV) are known, but they are not able to grow via metal reduction. *Sulfurospirillum barnesii* is also able to respire on As(V), and several other species have also recently been shown to be capable of dissimilatory As reduction. The oxidized forms of both As and Se are available as salts and require no special handling. It is important to remember that the soluble forms are toxic.

10.8.8. Fe Oxidizers

Prokaryotes that can utilize Fe(II) as an energy source are another unusual group. Physiologically these organisms fall into two categories, acidophiles and neutrophiles. Fe(II) is quite stable in the presence of air at a pH of 4 or less, and thus, acidophilic Fe oxidizers largely avoid competition with abiotic oxidation. At neutral pH, chemical oxidation of Fe(II) predominates unless the oxygen concentration is kept very low, which is the case in the oxic-anoxic boundary areas where these organisms grow. Interestingly, acidophilic Fe oxidizers, such as *Acidothiobacillus ferrooxidans* and *Leptospirillum ferrooxidans*, cannot tolerate the presence of organic compounds, including agar, in their medium. As a result these organisms are normally maintained in a liquid mineral salts medium. Clever plating methods have been employed that use acidophilic heterotrophs to scavenge organic matter in the agar; these are overlayered with a thin hard layer of agar that can then support the growth of the Fe oxidizers (17).

Neutrophilic Fe oxidizers can be grown in media designed to create opposing gradients of Fe(II) and O_2; the microbes grow as a band at the oxic-anoxic interface (11). The most commonly employed method is to make a synthetic FeS solution which is mixed with 1% agarose to make a thin, solid layer in the bottom of a tube. This is overlayed with a semisolid mineral salts medium (containing 0.15% agarose) with an air space on top, which naturally causes Fe(II) to diffuse up from the bottom and O_2 to diffuse down from the air. The lithotrophic Fe oxidizers grow at the interface, and the gel prevents the cells from sinking out of the interface zone. $FeCO_3$ or $FeCl_2$ can be substituted for FeS in these media.

10.8.9. Nitrifying Bacteria

The nitrifiers are another group of phylogenetically diverse bacteria that share a quite limited lithotrophic metabolism, the ability to gain energy from the oxidation of ammonia or nitrite (1, 6). The ammonia oxidizers, consisting of the beta-proteobacteria *Nitrosomonas* and related genera, grow in mineral media with NH_4Cl or $(NH_4)_2SO_4$ as an energy source. The nitrite oxidizers, consisting of the alpha-proteobacteria *Nitrobacter* spp. and of the *Nitrospira* spp., which form a separate phylum, grow on $NaNO_2$ or KNO_2 as an energy source. Many of these organisms have slow doubling times (12 to 24 h). Ammonia oxidation results in acidification of the medium, so buffering is required, usually with HEPES or $CaCO_3$, and pH indicators can also be added as an indicator that growth is occurring. All known nitrifiers can grow autotrophically; thus, the addition of carbonate to medium can be a source of carbon. Some strains have been shown to be capable of incorporating organic compounds, thus carrying out mixotrophic growth; however, in some cases excess organic compounds can be toxic to these organisms.

10.8.10. Halophiles

Halophilic bacteria and archaea require the addition of salt to the medium, from 10 to 30% by weight, depending on the organism. In most cases, NaCl is the primary form of salt; however, many halophilic archaea have requirements for Mg salts as well. These organisms are nutritionally diverse and in general do not have any other extraordinary nutritional requirements beyond the addition of salt. The high salt concentrations can present problems when working with agar-solidified medium, as the salt causes the agar to be softer than normal. NaCl also affects the pH of the medium due to the high ionic strength.

10.8.11. Oligotrophs

While the physiological concept of microbes capable of growing under, or requiring, very low nutrient concentrations is not new, in the past few years good progress has been made in using very low nutrient media to enrich and grow bacteria that had previously been thought to be unculturable. A nice example of this is the enrichment and isolation of members of the Sar11 clade of marine bacteria. This clade had been identified through analysis of 16S rRNA clone libraries as ubiquitous in the open ocean; however, these organisms resisted attempts at cultivation. By using very dilute seawater amended with micromolar or smaller amounts of nitrogen and phosphorous and organic substrates, Rappe et al. were successfully able to cultivate members of Sar11 (31). A similar approach that uses a dilute gel-encapsulated medium has been used to cultivate a number of novel bacteria from the Sargasso Sea and from soil (39). Another recent study used progressively diluted nutrient media to aid in the isolation of previously uncultured, but numerically prevalent, ultramicrobacteria from freshwater lakes (16). The secret to the success of these types of studies is in designing a growth medium that carefully simulates both qualitatively and quantitatively the natural habitat of the organisms under study.

10.9. SOURCES

10.9.1. Sources of Manufacture

Becton Dickinson Microbiology Systems, 7 Loveton Circle, Sparks, MD 21152-0999. (Tel) 800-638-8663.

www.BD.com

Bellco Glass, Inc. 340 Edrudo Rd., Vineland, NJ 08360. (Tel) 800-257-7043.

www.bellcoglass.com (Good source for anaerobic culturing glassware)

BioMerieux, Inc., 595 Anglum Rd., Hazelwood, MO 63042-2320. (Tel) 800-638-4835.

www.biomerieux-vitek.com

Difco Laboratories: see Becton Dickinson Microbiology Systems.

Gibson Laboratories, Inc., 1040 Manchester St., Lexington, KY 40508. (Tel) 800-477-4763.

www.gibsonlabs.com

Oxoid Inc., Suite 100, 1926 Merivale Rd., Nepean, Ontario K2G 1E8, Canada. (Tel) 613-226-1318.

www.oxoid.co.uk

Raven Biologics Laboratories, 8607 Park Dr., P.O. Box 27261, Omaha, NE 68127. (Tel) 800-728-5702.

www.ravenlabs.com

Remel Inc., 12076 Santa Fe Dr., P.O. Box 14428, Lenexa, KS 66215. (Tel) 800-255-6730.

www.remelinc.com

Sigma, P.O. Box 14508, St. Louis, MO 63178. (Tel) 800-521-8956.

www.sigma-aldrich.com

10.9.2. Culture Collections

These sites have search capacities for finding media recipes.

American Type Culture Collection (ATCC)

http://www.atcc.org

Deutsche Sammlung von Mikroorganismen und Zellkulturen GmbH (DSMZ)

http://www.dsmz.de

Japan Collection of Microorganisms (JCM)

http://www.jcm.riken.go.jp

10.9.3. Others

Cyanobacteria

http://www-cyanosite.bio.purdue.edu/

Provides media recipes and information on culturing a wide range of cyanobacteria including prochlorophytes.

Methanogens

http://methanogens.pdx.edu/

A culture collection of methanogens; provides media recipes and detailed information on cultivation.

Biodegradation

http://umbbd.ahc.umn.edu/

Does not provide media recipes, but does have a wealth of information on the metabolic pathways of organisms that break down most toxic organic compounds, and includes links to the culture collections where specific media recipes can be found.

10.10. REFERENCES

1. **Abeliovich, A.** 2002. The nitrite-oxidizing bacteria. *In* M. Dworkin (ed.), *The Prokaryotes: An Evolving Electronic Resource for the Microbiological Community.* Springer-Verlag, New York, NY.

2. **Amend, J. P., and E. L. Shock.** 2001. Energetic of overall metabolic reactions of thermophilic and hyperthermophilic Archaea and Bacteria. *FEMS Microbiol. Rev.* **25:** 175–243.

3. **Anonymous.** 1998. *Difco Manual,* 11th ed. Becton Dickinson Microbiology System, Sparks, MD.

4. **Atlas, R. M.** 1996. *Handbook of Microbiological Media,* 2nd ed. CRC Press, Boca Raton, FL.

5. **Black, J. G.** 1999. *Growth and Culturing of Bacteria.* John Wiley & Sons, Inc., New York, NY.

6. **Bock, E., and M. Wagner.** 2002. Oxidation of inorganic nitrogen compounds as an energy source. *In* M. Dworkin (ed.), *The Prokaryotes: An Evolving Electronic Resource for the Microbiological Community,* 3rd ed. Springer-Verlag, New York, NY.

7. **Bridson, E. Y.** 1978. Natural and synthetic culture media for bacteria, p. 91–281. *In* M. Rechcigl (ed.), *CRC Handbook Series in Food and Nutrition,* vol. III. CRC Press, Cleveland, OH.

8. **Coates, J. D., D. J. Ellis, E. Roden, K. Gaw, E. L. Blunt-Harris, and D. R. Lovley.** 1998. Recovery of humics-reducing microorganisms from a diversity of sedimentary environments. *Appl. Environ. Microbiol.* **64:**1504–1509.

9. **Cote, R. J., and R. L. Gherna.** 1994. Nutrition and media, p. 155–178. *In* P. Gerhardt, R. G. E. Murray, W. A. Wood, and N. R. Krieg (ed.), *Methods for General and Molecular Bacteriology.* ASM Press, Washington, DC.

10. **DeWeerd, K. A., L. Mandelco, R. S. Tanner, C. R. Woese, and J. M. Suflita.** 1990. *Desulfomonile tiedjei* gen. nov. and sp. nov., a novel anaerobic, dehalogenating, sulfate-reducing bacterium. *Arch. Microbiol.* **154:**23–30.

11. **Emerson, D., and C. Moyer.** 1997. Isolation and characterization of novel iron-oxidizing bacteria that grow at circumneutral pH. *Appl. Environ. Microbiol.* **63:**4784–4792.

12. **Finster, K., W. Liesack, and B. Thamdrup.** 1998. Elemental sulfur and thiosulfate disproportionation by *Desulfocapsa sulfoexigens* sp. nov., a new anaerobic bacterium isolated from marine surface sediment. *Appl. Environ. Microbiol.* **64:**119–125.

13. **Gibbons, R. J., and J. B. MacDonald.** 1960. Hemin and vitamin K compounds as required factors for the cultivation of certain strains of *Bacteroides melaninogenicus.* *J. Bacteriol.* **80:**164–170.

14. **Good, N. E., D. Winget, W. Winter, T. N. Connolly, S. Izawa, and R. M. M. Singh.** 1966. Hydrogen ion buffers for biological research. *Biochemistry* **5:**467–477.

15. **Gottschal, J. C., W. Harder, and R. A. Prins.** 2002. Principles of enrichment, isolation, cultivation, and preservation of bacteria. *In* M. Dworkin (ed.), *The Prokaryotes: An Evolving Electronic Resource for the Microbiological Community,* 3rd ed. Springer-Verlag, New York, NY.

16. **Hahn, M. W., H. Lunsdorf, Q. Wu, M. Schauer, M. G. Hofle, J. Boenigk, and P. Stadler.** 2003. Isolation of novel ultramicrobacteria classified as actinobacteria from five freshwater habitats in Europe and Asia. *Appl. Environ. Microbiol.* **69:**1442–1451.

17. **Hallberg, K. B., and D. B. Johnson.** 2001. Biodiversity of acidophilic prokaryotes. *Adv. Appl. Microbiol.* **49:**37–84.

18. **Hanson, R. S.** 1998. Ecology of methylotrophic bacteria, p. 137–162. *In* R. S. Burlage, R. Atlas, D. Stahl, G. Geesey, and G. Saylor (ed.), *Techniques in Microbial Ecology.* Oxford Press, New York, NY.

19. **Janssen, P. H., P. S. Yates, B. E. Grinton, P. M. Taylor, and M. Sait.** 2002. Improved culturability of soil bacteria and isolation in pure culture of novel members of the divisions *Acidobacteria, Actinobacteria, Proteobacteria,* and *Verrucomicrobia. Appl. Environ. Microbiol.* **68:**2391–2396.

20. **Kelly, D. P., and A. P. Wood.** 1998. Microbes of the sulfur cycle, p. 31–57. *In* R. S. Burlage, R. A. Atlas, D. Stahl, G. Geesey, and G. Saylor (ed.), *Techniques in Microbial Ecology.* Oxford University Press, New York, NY.

21. **Kostka, J. E., and K. H. Nealson.** 1998. Isolation, cultivation and characterization of iron- and manganese-reducing bacteria, p. 58–78. *In* R. S. Burlage, R. A. Atlas, D. Stahl, G. Geesey, and G. Saylor (ed.), *Techniques in Microbial Ecology.* Oxford Press, New York, NY.

22. **Leadbetter, J. R., T. M. Schmidt, J. R. Graber, and J. A. Breznak.** 1999. Acetogenesis from H_2 plus CO_2 by spirochetes from termite guts. *Science* **283:**686–689.

23. **Lev, M.** 1959. The growth-promoting activity of compounds of the vitamin K group and analogues for a rumen strain of *Fusiformes nigrescens. J. Gen. Microbiol.* **20:**697–703.

24. **Lovley, D. R.** 2000. Fe(III) and Mn(IV) reduction, p. 3–30. *In* D. R. Lovley (ed.), *Environmental Microbe-Metal Interactions.* ASM Press, Washington, DC.

25. **Madigan, M. T., J. M. Martinko, and J. Parker.** 2003. *Brock Biology of Microorganisms.* Prentice-Hall, Upper Saddle River, NJ.

26. **Matsuhashi, T.** 1998. Agar, p. 335–375. *In* S. Dumitriu (ed.), *Polysaccharides: Structural Diversity and Functional Versatility.* Marcel Dekker, New York, NY.

27. **Oremland, R. S., P. R. Dowdle, S. Hoeft, J. P. Sharp, J. K. Schaefer, L. G. Miller, J. S. Blum, R. L. Smith, N. S. Bloom, and D. Wallschlaeger.** 2000. Bacterial dissimilatory reduction of arsenate and sulfate in meromictic Mono Lake, California. *Geochim. Cosmochim. Acta* **64:**3073–3084.

28. **Overmann, J.** 2002. The family *Chlorobiaceae. In* M. Dworkin (ed.), *The Prokaryotes: An Evolving Electronic Resource for the Microbiological Community,* 3rd ed. Springer-Verlag, New York, NY.

29. **Pfennig, N., and H. G. Truper.** 2002. The family *Chromatiaceae. In* M. Dworkin (ed.), *The Prokaryotes: An Evolving Electronic Resource for the Microbiological Community,* 3rd ed. Springer-Verlag, New York, NY.

30. **Rabus, R., T. Hansen, and F. Widdel.** 2001. Dissimilatory sulfate- and sulfur-reducing prokaryotes. *In* M. Dworkin (ed.), *The Prokaryotes: An Evolving Electronic Resource for the Microbiological Community,* 3rd ed. Springer-Verlag, New York, NY.

31. **Rappe, M. S., S. A. Connon, K. L. Vergin, and S. J. Giovannoni.** 2002. Cultivation of the ubiquitous marine bacterioplankton clade. *Nature* **418:**630–633.

32. **Reichenbach, H.** 2002. The genus *Herpetosiphon. In* M. Dworkin (ed.), *The Prokaryotes: An Evolving Electronic Resource for the Microbiological Community,* 3rd ed. Springer-Verlag, New York, NY.

33. **Reichenbach, H., and M. Dworkin.** 2002. The Myxobacteria. *In* M. Dworkin (ed.), *The Prokaryotes: An Evolving Electronic Resource for the Microbiological Community,* 3rd ed. Springer-Verlag, New York, NY.

34. **Sowers, K. R., and K. M. Noll.** 1995. Techniques for anaerobic growth, p. 15–48. *In* H. J. Schreier and K. R. Sowers (ed.), *Archaea: A Laboratory Manual, Methanogens.* Cold Spring Harbor Laboratory Press, Plainview, NY.

35. **Straub, K. L., M. Hanzlik, and B. E. Buchholz-Cleven.** 1998. The use of biologically produced ferrihydrite for the isolation of novel iron-reducing bacteria. *Syst. Appl. Microbiol.* **21:**442–449.

36. **Tumbula, D. L., T. L. Bowen, and W. B. Whitman.** 1995. Growth of methanogens on solidified medium, p. 49–56. *In* H. J. Schreier and K. R. Sowers (ed.), *Archaea: A Laboratory Manual, Methanogens.* Cold Spring Harbor Laboratory Press, Plainview, NY.

37. **Waterbury, J. B.** 2002. The cyanobacteria—isolation, purification, and identification. *In* M. Dworkin (ed.), *The Prokaryotes: An Evolving Electronic Resource for the Microbiological Community*, 3rd ed. Springer-Verlag, New York, NY.

38. **Wolin, E. A., M. J. Wolin, and R. S. Wolfe.** 1963. Formation of methane by bacterial extracts. *J. Biol. Chem.* **238:**2882–2886.

39. **Zengler, K., G. Toledo, M. Rappe, J. Elkins, E. J. Mathur, and M. Keller.** 2002. Cultivating the uncultured. *Proc. Natl. Acad. Sci. USA* **99:**15681–15686.

40. **Zinder, S. H.** 1998. Methanogens, p. 113–136. *In* R.S. Burlage, R. Atlas, D. Stahl, G. Geesey, and G. Saylor (ed.), *Techniques in Microbial Ecology.* Oxford Press, New York, NY.

11

Enrichment and Isolation

ANDREAS TESKE, HERIBERT CYPIONKA, JOHN G. HOLT, AND NOEL R. KRIEG

Since the last edition of this book, the development of molecular biological approaches has revolutionized microbiology. The possibility to analyze microbial communities without the necessity of cultivating their members has demonstrated an overwhelming microbial diversity. In 1987, 12 divisions were known in the bacterial domain. Within 10 years this has increased to 36, and more are coming (88). While about 6,000 species are described today, estimates based on reassociation kinetics of DNA isolated from natural environments come to 13,000 species already in 30 g of forest soil (197) or even 1 billion prokaryotic species worldwide based on a DNA-DNA homology of 70% within a species (209). At the same time it became increasingly clear that our success in cultivation of microbes from natural environments is minute. Usually less than 1% of the prokaryotes present in a natural environment grow on laboratory media (13). Therefore, many people came to the conclusion that microbial ecology should abandon cultivation attempts and rely on molecular biology only. Most of the microbes in natural environments were regarded as "unculturable." However, the huge numbers of prokaryotes in natural environments and their diversity are a fact. Mother Nature has cultivated them where they are. Obviously, they are not unculturable per se. It is only our techniques that are not efficient. With improved cultivation techniques and more sensitive analysis of cultures the success can be increased, although cultivation of 100% of a natural community is not feasible. In the meantime it has turned out that cultures of bacteria are still required to understand their properties and interactions. And just the molecular techniques in themselves turned out to be very helpful for cultivation experiments. They can help to study the composition of enrichment cultures in order to track a member of interest. In the future, studies on mixed cultures will be possible, where some members can be analyzed without separating them from their natural-environment mates. A new section (11.5) focuses on the use of enrichment and cultivation procedures with molecular methods, such as whole-cell fluorescent in situ hybridization (FISH) or denaturing gradient gel electrophoresis (DGGE), to monitor the progress of an enrichment, to evaluate the presence of contaminants, and to identify new isolates.

Enrichments make it possible to assess the differential effects of environmental factors imposed on mixed microbial populations and permit the selection of organisms capable of attacking or degrading particular substrates or of thriving under unusual conditions.

Successful isolation of a given organism into pure culture requires a sufficiently high proportion of that organism in the mixed population. Isolation is easiest when the organism is the numerically dominant member of the population. Enrichment methods are designed to increase the relative numbers of a particular organism by favoring its growth, its survival, or its spatial separation from other members of the population. Biophysical enrichments make use of such conditions as growth temperature, heat treatment, sonic oscillation, or UV irradiation to kill or inhibit the rest of the population. Such methods may also take advantage of some physical property of the desired organism, such as its size, motility, surface charge, or the presence of certain antigens, which can allow the organism to be preferentially separated from the rest of the community. Biochemical enrichments employ toxic agents to kill or inhibit the rest of the population without affecting the desired organism; alternatively, they may provide nutrient sources that can be used preferentially by a particular component of the mixed population. Biological enrichments may make use of specific hosts for selective growth of a particular organism, or they may take advantage of some pathogenic property, such as invasiveness, which the rest of the population does not possess. In many enrichment procedures several physical, chemical, and/or biological methods may be used in combination to achieve a maximum effect.

Most enrichments are carried out in closed systems, such as batch cultures in a flask or tube, where the concentrations of nutrients and metabolic products in the culture vessel continually change during bacterial growth. Open systems have also been used for enrichment. For example, the use of a chemostat (chapters 12.2.5 and 13.2) enables one to provide a constant environment for cultivation of bacterial cells by continuously supplying a growth-limiting nutrient and continually removing metabolic products. Alteration of the dilution rate controls the concentration of the growth-limiting nutrient, which differentially affects the growth rate of the various organisms in mixed culture, making it possible for one or another member of the mixed population to become predominant; for an example, see section 11.2.11.2. The use of gradients (11.2.14) is another approach that allows the organisms to find optimum conditions instead of forcing them into a fixed situation.

Bacteria are usually isolated from enrichment cultures by spatially separating the organisms in or on a solid medium and subsequently allowing them to grow into colonies. For organisms that cannot grow on solid media, dilution to extinction by allowing separation of cells into individual tubes of a liquid medium can be used. Because such ordinary isolation methods do not absolutely ensure purity, more difficult methods may be needed whereby an individual bacterial cell from a mixed population is spatially isolated under a microscope before being cultured into a clone.

This chapter is designed to demonstrate the multiplicity and in many instances the considerable ingenuity of enrichment and isolation methods for bacteria by presenting specific, selected examples. The wide-ranging examples for enrichment and isolation methods in this chapter in the previous edition were retained and expanded by new approaches, such as gradient culture, new electron acceptors (metal oxides and humics, halogenated organics, and disproportionating sulfur species), oligotrophic cultivation in seawater, and high-throughput cultivation.

Although many different methods are described, this chapter is not intended to be a comprehensive compendium for all bacterial taxa. The general references (1–11) given at the end of this chapter and those given for enrichment and isolation of mutant cells in chapter 28 should be consulted for additional methods and organisms.

Following the procedures described here will usually lead to the decribed microorganisms. One should, however, keep in mind that we—in contrast to Mother Nature—so far have cultivated only a small percentage of the prokaryotes on earth. Any variation of an enrichment procedure might give rise to a different outcome. Thermodynamics can give a guideline for the detection of new organisms. In principle, any situation with free energy could provide the energetical basis for microbial growth. In many cases, like disproportionation of elemental sulfur, the concentrations of substrates and products have to be controlled by further reactions in order to allow growth. Whenever an exergonic reaction is possible, there is a good chance to find a microbe that can use it for growth.

11.1. BIOPHYSICAL ENRICHMENT

11.1.1. Low-Temperature Incubation (139)

Low temperature retards the growth of many mesophilic bacteria but not psychrophilic and psychrotrophic bacteria. Therefore, incubation of enrichment cultures at 0 to 5°C can favor the growth of the latter organisms.

11.1.1.1. Psychrophiles

Inoue (92) isolated psychrophilic bacteria from Antarctic soil by spreading dilutions of the soil over plates of a glucose-yeast extract-peptone agar and incubating them for 14 to 24 days, with the temperature maintained throughout at below 5°C. In this manner nine strains of obligately psychrophilic bacteria with maximum growth temperatures of ca. 20°C were obtained (93).

11.1.1.2. *Listeria monocytogenes*

It may be difficult to isolate *Listeria monocytogenes* from material heavily contaminated with other bacteria (e.g., feces, silage, sewage, and clinical specimens from the cervix, vagina, meconium, nasopharynx, or tissues). Such materials may be enriched by the following procedure (174). Add a 1- to 2-g sample of the suspected material to each of two flasks containing 100 ml of sterile enrichment broth. Examples of suitable enrichment broths are TN medium (section 11.6.74) and PTN medium (section 11.6.57). Incubate one flask at 37°C and the other at 4°C. Plate 0.1-ml samples from the flask incubated at 37°C onto Tryptose agar (Difco) daily for 7 days. Incubate the plates at 37°C for 24 h, and examine the plates under oblique lighting for the presence of distinctive blue-green colonies. Plate 0.1-ml samples from the flask incubated at 4°C onto Tryptose agar at 7-day intervals for a period of up to 2 months. Incubate and examine the plates as described for the broth held at 37°C. Although the incubation at 37°C may give results more quickly, the cold enrichment usually gives a higher proportion of positive cultures. The reason why cold incubation enriches for *L. monocytogenes* has been variously attributed to an ability to survive longer than many other bacteria at 4°C (215), an ability to multiply at 4°C (179), and release of the organisms from an intracellular location (59).

11.1.1.3. Marine Luminescent Bacteria

Luminescent bacteria can be isolated from fresh herring (39). Although the isolates are psychrophilic, pure cultures can be obtained within a 2-week course. Place a fresh herring in a bowl, covered with salt water (3%) by half. Put the bowl into a refrigerator at 8 to 10°C for 2 days (not longer, not warmer, otherwise a bad smell will develop!). Light-emitting pinpoints can be detected, often on the sideline organ. The room has to be perfectly darkened, and the eye has to adapt to the darkness for at least 5 min. To isolate pure cultures, light-emitting pinpoints are transferred to agar medium (section 11.6.41) by means of sterile toothpicks. A freshly grown plate can be bright enough for reading a watch nearby.

11.1.2. High-Temperature Incubation

High-temperature incubation favors the selection of thermophilic bacteria because the growth of other organisms is inhibited at high temperature. The definition of "thermophilic" varies depending on the particular organisms being studied. The term usually refers to organisms that are unable to grow at temperatures below ca. 45 to 50°C; however, it has also been applied to other organisms, for example, *Campylobacter* species (e.g., *Campylobacter jejuni*, *C. coli*, and *C. lari*) that can be cultured at 42°C but not 25°C. Another term, "extreme thermophiles," is usually applied to organisms that can grow at temperatures above 70°C.

11.1.2.1. Thermophiles in Milk

For milk thermophiles, plate serial dilutions of milk in standard methods agar (section 11.6.66) and incubate the cultures at 55°C (14).

11.1.2.2. *Thermus* Species

Thermus aquaticus strains have an optimum growth temperature of 70°C, a maximum of 79°C, and a minimum of 40°C. *T. ruber* strains have an optimum temperature of 60°C, a maximum of 70°C, and a minimum of 37°C (214). For enrichment (33, 214), inoculate 0.5- to 1.0-ml samples of microbial mats, water from hot springs, thermally polluted water, or hot tap water into 10-ml portions of *Thermus* medium (basal salts medium [section 11.6.10] plus 0.1% pancreatic digest of casein and 0.14% yeast extract). Incubate in covered water bath shakers for 1 to 3 days at 70 to 75°C for *T. aquaticus* or 55 to 65°C for *T. ruber*. For isolation, streak turbid cultures onto the *Thermus* medium solidified with 2 to 3% agar. Wrap the petri dishes in household plastic wrap to avoid drying out during the incubation period (32), and incubate them aerobically at the appropriate optimum temperature. Colonies of *T. aquaticus* are yellow to pale or colorless. Colonies of *T. ruber* are red (214).

11.1.2.3. *Thermoplasma* Species

Thermoplasma species are best enriched and isolated from coal refuse piles (28, 58). Filter a liquid sample through a membrane filter with a pore size of 0.45 μm and then through a second filter with a pore size of 0.22 μm. Incubate aerobically in *Thermoplasma* isolation medium (section 11.6.72) at 55°C for 4 to 6 weeks or until turbidity develops. The organism can be isolated either by performing dilution to extinction in liquid medium (section 11.4.3) or by obtaining colonies on medium solidified with 10% (wt/vol) hydrolyzed starch of the type used for gel electrophoresis (175). Adjust the pH to ca. 3 immediately before pouring the plates. Incubate the plates within a sealed chamber in a humid atmosphere consisting of ca. 60% air and 40% CO_2 (vol/vol). *Thermoplasma* colonies appear within 7 days (or more), are small and colorless to brownish, and have a characteristic "fried-egg" appearance when viewed under a dissecting microscope (175).

11.1.3. High-Temperature Treatment

Treatment of a sample at an appropriate high temperature may select heat-resistant bacteria from a mixture of microorganisms. The heat treatment may be relatively mild (as in pasteurization), which even certain vegetative cells can survive, or much more severe, which only sporeformers can survive.

11.1.3.1. Thermoduric Organisms in Milk (14)

Pasteurize a sample of raw milk by heating it to 62.8°C and holding it at this temperature for 30 min. Cool, add dilutions to standard methods agar (section 11.6.66), and incubate the plates at 32°C. For instance, *Microbacterium lacticum* can be isolated with the same heat treatment and subsequent plating on yeast extract milk agar (section 11.6.82) with incubation at 32°C for 7 days.

11.1.3.2. Mesophilic Sporeformers

Heat water samples or soil suspensions at 80°C for 10 min, and then streak onto plating media. Isolation media for *Bacillus* species differ according to the special nutritional needs of the species (e.g., use of uric acid as a substrate for *B. fastidiosus*), and the reader is referred to references 48 and 212 for details. For *Clostridium* species, streak a roll tube of prereduced chopped-meat agar (see chapter 15.4.12) or streak a plate of freshly prepared agar medium and incubate it in an anaerobe jar.

Another approach is to add some of the sample in which spores are suspected to a suitable growth medium, heat at 80°C for 10 min, and then incubate the broth for about 1 day to permit growth to occur before streaking onto solid media. For *Bacillus* species, nutrient broth containing 0.5% yeast extract is usually a suitable growth medium; for

Clostridium species, prereduced starch broth (PY broth [see chapter 15.4.31] + 1.0% soluble starch) is often satisfactory.

If small numbers of spores are present in the samples, the proportion may be increased by adding the sample to a suitable sporulation medium and incubating the culture for various periods (2 to 21 days) prior to the heat treatment. For most *Bacillus* species, a suitable sporulation medium is nutrient broth (nutrient agar [see chapter 15.4.28] without the agar component) plus 0.5% yeast extract, 7×10^{-4} M $CaCl_2$, 1×10^{-3} M $MgCl_2$, and 5×10^{-5} M $MnCl_2$. Also see chapter 4.4.9. For most *Clostridium* species, no one medium is optimum for production of spores, but chopped-meat medium (see chapter 15.4.12) with and without glucose often supports sporulation (86).

The heat treatment employed may have to be modified for some types of sporeformers, since the endospores of some bacterial strains are not as heat resistant as others. Try a lower temperature (e.g., 75 or 70°C) if satisfactory results are not obtained at the usual temperature of 80°C.

11.1.3.3. Thermophilic Sporeformers (14)

Thermophilic sporeformers may be present, for example, in various sweetening agents used in ice cream. Initially prepare a 20% (wt/vol or vol/vol) solution of the sweetening agent (beet or cane sugar, lactose, cerelose, invert syrup, corn syrup, maple syrup, liquid sugar, or honey), and heat to 100°C for 5 min to kill nonsporeformers. Dilute with sterile water to give a final concentration of 13.3%.

For thermophilic, aerobic flat-sour organisms, plate 2-ml portions of the heated, diluted sugar solution in glucose-tryptone agar (section 11.6.29). Incubate in a humid chamber at 55°C for 48 h. Characteristic surface colonies are round, are 2 to 5 mm in diameter, have a typical opaque central spot, and are surrounded by a yellow halo in the medium. Subsurface colonies are compact and may be pinpoint in size; these should be subcultured and streaked onto glucose-tryptone agar.

For thermophilic, anaerobic hydrogen sulfide producers, add samples of the heated, diluted sugar solution to deep tubes of sulfite agar (1% tryptone or Trypticase, 0.1% sodium sulfite, and 2% agar). The tubes of medium should be melted and cooled to 55°C prior to inoculation. Allow the tube contents to solidify, and incubate at 55°C for 72 h. Sulfide producers will form blackened spherical areas in the medium.

11.1.4. Drying

A number of bacteria that normally reside in the soil and produce spores, cysts, or other dormant forms are resistant to dehydration. A soil sample is air dried to kill many of the vegetative cells and is then plated onto a suitable agar medium. This method is useful for members of the genera *Bacillus*, *Azotobacter*, and *Arthrobacter* (48).

11.1.5. Motility

Motility has been used as the basis for several ingenious methods for enrichment and isolation. The principle is that bacteria that swim in liquid media or swarm or glide across solid media may be able to outdistance other microorganisms.

11.1.5.1. Swarming Motility of *Clostridium tetani* (181)

Inoculate the specimen (soil, animal feces, or clinical material) onto a small area of a freshly prepared plate of blood agar (see chapter 15.4.9). Incubate the plate at 37°C in an anaerobic jar for 1 day, and examine the agar surface carefully for evidence of swarming (a thin layer of growth that has spread outward from the inoculated area). It may be helpful to scrape the surface of the medium with a needle to verify the occurrence of swarming. Suspend some growth from the edge of the swarming area in broth, and streak onto solid media containing 5% agar to obtain isolated colonies.

11.1.5.2. Swimming Motility of Treponemes (80, 83, 167–169)

The following method can be used for enrichment of cultivable *Treponema* species, many of which occur as part of the normal flora of the oral cavity and intestinal tract of humans and animals. Cut a well into the center of a thick layer of suitable agar medium (section 11.6.58) contained in either petri dishes or beakers. The well should be at least 7 mm deep and 2 to 10 mm in diameter and should not be cut to the bottom of the dish. Inoculate the well with the specimen (sample from the oral cavity, intestinal contents, or feces). Inoculate large wells (diameter, 10 mm) with up to 0.2 ml of sample; inoculate small wells (diameter, 2 mm) by stabbing ca. 2 mm obliquely into one side of the well. Do not allow any of the sample to be deposited on the surface of the agar. Immediately place the plate in an anaerobic jar, and incubate at 37°C for 4 to 7 days. In contrast to most other bacteria, treponemes can migrate through agar media. Look for treponemal growth occurring as a "haze" in the medium at some distance from the well. Subculture a sample from the outermost portions of this hazy region into a suitable prereduced semisolid medium (e.g., broth containing 0.15% agar). Since more than one kind of treponeme may be present, streak the subculture onto a roll tube of prereduced medium to obtain isolated colonies. These appear as hazy, whitish, dense areas in the medium.

A similar approach has been successfully used for the isolation of anaerobic marine spirochetes (43).

11.1.5.3. Swimming Motility of *Spirillum volutans* (91, 118, 166)

Obtain mixed cultures of the giant microaerophilic bacterium *Spirillum volutans* by preparing a hay infusion with stagnant pond water. After a surface scum develops, examine samples taken just beneath the scum. Look for very large spirilla (1.4 to 1.7 μm in diameter and up to 60 μm long) with bipolar flagellar fascicles that are clearly visible by dark-field microscopy. Enrich the culture by inoculating some of the hay infusion into Pringsheim soil medium (section 11.6.55) and incubating at room temperature. Even with this enrichment, *S. volutans* will be vastly outnumbered by other bacteria. However, it can be isolated by using a capillary tube (166), as follows. Soften the center of a short section of a sterile, cotton-plugged piece of 5-mm-diameter glass tubing in a flame. Pinch the tubing with square-ended forceps until it is almost closed. Reheat the flattened portion, and draw it out rapidly to form a long capillary tube 15 to 30 cm long and 0.1 to 0.3 mm wide, oval in cross section. Seal the ends of the capillary in a flame. Break the capillary near one end with sterile forceps, and draw up 10 to 20 cm of sterile Pringsheim soil medium (supernatant). Then, dip the capillary into the enrichment culture, and draw up another 2 to 4 cm, making sure that no air space occurs between the sterile medium and the culture. Seal the tip of the capillary, leaving a small air space. Mount the capillary on the stage of a ×100 microscope. *S. volutans* is often able to swim faster than the other bacteria

in the enrichment culture and thus can reach the distal end of the capillary first. As soon as some spirilla reach the distal end, break the capillary behind them, expel the spirilla into a tube of semisolid CHSS medium (section 11.6.21), and incubate at 30°C. Confirm the purity of the cultures by phase-contrast microscopy.

11.1.5.4. Gliding Motility of Cytophagas, Flexibacters, and Myxobacters

Spread dilutions of the specimen (soil, water, or animal dung which has been in contact with the soil) onto agar media that have a low nutrient concentration (e.g., 0.1% tryptone or Trypticase, or 1/10-strength nutrient agar). Another procedure is to smear terrestrial plant leaf material, algal fronds, or marine plants on nutrient-poor media. Cytophagas and flexibacters are able to migrate on the surface of solid media; their colonies can be recognized as thin, often nearly translucent colonies with fingerlike projections, which develop far beyond the streak of deposition line. Subculture from these colonies. The incorporation of penicillin G (15 U/ml) and chloramphenicol (5 μg/ml) into the agar media often helps to suppress the growth of other bacteria (206). Many of the fruiting myxobacters are bacteriolytic and can be enriched on bacterium-rich substrates to the point at which they produce characteristic fruiting bodies, which can then be transferred to agar media. Place natural materials such as dung from herbivorous animals, decaying plant material, or bark onto moistened paper contained in petri dishes or household plastic crisper boxes, and incubate for up to 3 weeks.

Myxococci can be enriched on sterilized, urine-free rabbit dung pellets (preferably from rabbits fed on non-antibiotic-containing feeds) placed on moistened soil in a petri dish (135). It is best to moisten the substrate with a solution of cycloheximide (30 to 50 mg per liter of distilled water) to inhibit fungi. Keep the preparations moist while incubating. Check periodically for the presence of characteristic fruiting bodies with the aid of a dissecting microscope. Transfer material from the fruiting body to an agar medium by using a sterile fine metal or glass needle (135). As with other gliding bacteria, pure cultures can be obtained by repeatedly removing cells from the leading edge of the colony and inoculating them onto the center of a fresh plate of agar medium. Bacteriolytic species can be inoculated onto an agar medium containing a suspension of heat-killed bacteria such as *Escherichia coli*.

11.1.5.5. Gliding Motility of *Beggiatoa* Species (45, 66, 147, 186)

Beggiatoa species form colorless filamentous chains of cells, which exhibit gliding motility on surfaces. The organisms occur in aerobic freshwater or marine environments rich in H_2S; the cells can oxidize the sulfide to elemental sulfur, which is deposited intracellularly.

For enrichment, first prepare extracted hay by the following procedure. Cut dried hay into small pieces, and extract by boiling the hay in a large volume of water. Change the water three times during the extraction. The final wash should have an amber color. Drain the hay, and place it on trays at 37°C to dry. The enrichment medium consists of 0.8% (wt/vol) of the dried, extracted hay in tap water, distributed in 70-ml volumes into 125-ml cotton-stoppered Erlenmeyer flasks. Sterilize by autoclaving. Inoculate the flasks with 5-ml portions of mud containing decaying plant materials, as from small ponds, lakes, or streams. Incubate the flasks at ca. 25°C for 10 days.

Effective enrichments are indicated by a strong odor of H_2S and the development of a white film on the surface of the medium and on the submerged upper walls of the flasks. Examine the film for the characteristic filaments of *Beggiatoa* organisms. To free the filaments of contaminants, remove a tuft of filaments with a loop needle or microforceps and place it into a vial of sterile basal salts solution (section 11.6.10) plus 5 mM neutralized sodium sulfide. Tease the tuft apart, swirl, and transfer to another wash bath. Repeat this process five times. Place the washed filaments on the dried surface of a 1.6% agar plate for about 1 min to absorb excess fluid, and then place them on the surface of plates containing a medium consisting of 0.1% yeast extract, 0.1% sodium acetate, and 2 mM sodium sulfide. View the plates after 24 h with a dissecting microscope, and use a fine sterile needle to transfer to a fresh plate those filaments growing from the mass that are free of contaminants. The filaments that grow away from the central inoculum are cut out on agar blocks and are used as inoculum for new agar plates. These procedures have led to the isolation of heterotrophic freshwater *Beggiatoa* strains (162, 187).

Marine *Beggiatoa* strains require different enrichment and isolation procedures (147). Plates for the isolation of marine *Beggiatoa* strains use a modified J3 basal medium, based on aged seawater (section 11.6.40). J3 basal medium is amended to produce an isolation medium (J-TS) by adding the following sterile stocks, with final concentrations in parentheses: (i) 7.5 ml of 200 mM $Na_2S_2O_3$ (2 mM); (ii) 3.75 ml of freshly neutralized 200 mM Na_2S (1 mM); this is autoclaved as a basic solution, which is quite stable against auto-oxidation, and then neutralized with an equimolar quantity of sterile HCl just prior to use; (iii) 15 ml of 1 M $NaHCO_3$ (20 mM); to make this stock, autoclave 8.4 g of $NaHCO_3$ (dry) and add 100 ml of sterile water when cool.

Immediately after solidification, agar plates are incubated in a bell jar for 24 h or more under anoxic conditions (99.5% N_2, 0.5% CO_2), with desiccant present to absorb water evaporating from the surface of the medium. The medium is buffered by the bicarbonate in conjunction with the level of atmospheric CO_2. After inoculation with a tuft of *Beggiatoa* spp., plates are placed in a micro-oxic atmosphere (0.5% CO_2; 0.2% O_2; balance N_2). Exposing the medium and bacteria to full air for approximately 20 min every day or two, as needed for inoculation or single-filament isolations from agar plates, poses no problem to the success of the technique. Pure cultures resulting from repeated single-filament isolations are maintained in sulfide-oxygen gradient media (section 11.2.14.2).

11.1.6. Filterability

The ability of some bacteria to pass through a membrane filter has been used to advantage for enrichments. In oligotrophic environments many bacteria are very small and may be able to pass filters with 0.45-μm pore size. The initial filtration of samples containing *Thermoplasma* species has already been mentioned (section 11.1.2.3). The small size and plastic properties of these wall-less cells allow them to pass through the pores of a 0.45-μm-pore-size membrane filter. During enrichment of bdellovibrios, the samples are filtered before being placed with potential prey cells (section 11.3.1). A somewhat different approach consists of placing a sterile filter on top of a suitable agar medium in a petri dish and inoculating the surface of the filter with the sample containing the desired organism. If the organism is small, motile and, preferably, flexible, it may be able to migrate through the filter pores into the agar below, where it can be isolated after growth by removing the filter and subculturing from the agar. This method is particularly useful for organisms that grow slowly and are easily overgrown by contaminants during isolation attempts. The principle is exemplified by enrichment procedures for spirochetes such as cultivable, anaerobic *Treponema* spp. and facultatively anaerobic *Spirochaeta* spp. and for thin, flagellated organisms such as *Aquaspirillum gracile* and *Serpens flexibilis*.

11.1.6.1. Spirochetes

The following method can be used for enrichment of cultivable *Treponema* species from the oral cavity and intestinal tract of humans and animals (86, 180). Prepare a petri dish of a suitable agar medium such as RGCA-SC medium (section 11.6.58). Place a membrane filter (pore size, 0.15 μm) on the surface of the agar. Place an O-ring (25 to 30 mm in diameter) that has been lightly coated with vacuum grease on top of the filter. Place several drops of the diluted specimen (sample from the oral cavity, intestinal contents, or feces) onto the center of the O-ring. Incubate the petri dish in an anaerobic jar at 37°C for 1 to 2 weeks. Treponemes are small enough to migrate through the filter pores and penetrate the underlying agar, where they grow as a haze in the medium. Remove the O-ring and membrane filter from the agar surface, and remove a plug of agar from the hazy region with a Pasteur pipette. Examine the plug by dark-field microscopy for treponemes and contaminants. Subculture and purify as described in section 11.1.5.2. The inclusion of polymyxin B (800 U/ml) and nalidixic acid (800 U/ml) in the agar medium often helps to suppress the growth of contaminants.

For enrichment of facultatively anaerobic *Spirochaeta* strains (43), place a membrane filter with a pore size of 0.3 or 0.45 μm on the surface of a suitable agar medium, such as GYPT medium (section 11.6.31). Place a drop of pond water or water-mud slurry on the disc near the center. Incubate the plate aerobically at 22 to 30°C for 12 to 18 h. Remove the filter from the medium, and continue incubating the plate. Within the agar medium, the spirochetes form a subsurface veil-like growth that expands toward the periphery of the plate and away from colonies of contaminating bacteria that may also have passed through the filter pores. Obtain pure cultures by serial dilution in deep tubes of the GYPT medium: *Spirochaeta* colonies resemble a cotton ball or veil-like growth with a denser center.

11.1.6.2. Thin, Flagellated Bacteria

Aquaspirillum gracile is the thinnest of the aerobic, freshwater spirilla, having a cell diameter of 0.2 to 0.3 μm. It can be isolated by the following procedure (44). Place a membrane filter with a pore size of 0.45 μm on the surface of the isolation agar (section 11.6.4), and deposit 0.5 ml of pond or stream water in the center of the filter. Incubate the plates for 1.5 to 2.0 h at room temperature. Remove the filter, and continue the incubation for 3 days or longer. *A. gracile* is thin enough to penetrate the pores of the filter into the underlying agar. Look for spreading, semitransparent areas of growth within the agar medium, and subculture from these areas. Other small bacteria (small vibrios, cocci, or short rods) may also pass through the filter to form small colonies on the surface of the isolation medium; these colonies can be distinguished easily from the subsurface, spreading, semitransparent growth typical of *A. gracile*.

Serpens flexibilis can be found in the sediment of eutrophic freshwater ponds. The cells are rod shaped with a diameter of 0.3 to 0.4 μm and move through agar gels in a characteristic serpentine manner. For enrichment, place a sterile membrane filter with a pore size of 0.3 to 0.45 μm on the surface of the isolation agar (85). Deposit a small amount of pond water-mud slurry on the center of the filter. Incubate for 6 to 12 h at 30°C, remove the filter, and incubate the plate for 2 to 4 days. The organism grows as a subsurface veil; remove a sample from the edge of this veil, and streak onto a second plate of sterile medium to obtain a pure culture.

11.1.7. Visible Illumination for Phototrophs

Unlike chemotrophic bacteria, phototrophic bacteria can use light as a source of energy. This ability has been used for enrichment of these bacteria: with light as the sole energy source, the growth of phototrophs will be favored over that of chemotrophs. However, chemotrophs may still be able to grow to some extent by using organic compounds produced by the phototrophs.

11.1.7.1. Unicellular Cyanobacteria

For freshwater cyanobacteria (formerly called blue-green algae), inoculate samples from ponds, streams, or reservoirs into tubes of BG-11 medium (section 11.6.12). For cyanobacteria from marine environments, use MN medium (section 11.6.44); some grow poorly in this medium, so ASN-III medium (section 11.6.6) is sometimes preferable. Wash all glassware well, and then rinse successively in tap water, concentrated nitric acid, and deionized water. Incubate enrichment cultures in an illuminated water bath at 35°C; this temperature inhibits the growth of most eucaryotic algae. Use a light intensity of 2,000 to 3,000 lx (186 to 279 ft-candles, or approximately the illumination provided by a 55-W incandescent desk lamp at a distance of 10 to 12 in. [25 to 30 cm]); some cyanobacteria may require lower intensities (500 lx or less). Examine the enrichment cultures periodically until there is evidence of development of cyanobacteria. Then, streak onto media solidified with 2% bacteriological-grade agar. Incubate plates under illumination (<500 lx) at 25°C in air or in an atmosphere slightly enriched with CO_2; clear plastic boxes such as household vegetable crispers make good chambers to help prevent evaporation. Use a dissecting microscope to detect the compact, deeply pigmented colonies of nonmotile cyanobacteria. Restreaking several times may be necessary to eliminate nonphototrophic bacterial contaminants. For motile cyanobacteria, enrich by placing a small patch of culture material at one side of a petri dish of agar medium. Illuminate the dish from the opposite side. The cyanobacteria will respond phototactically by gliding across the agar toward the region of higher light ginsity. When some of the organisms have reached the opposite side, subculture them to a new plate and repeat the procedure until nonphototrophic bacterial contaminants have been eliminated. This technique has also been used effectively with some filamentous cyanobacteria (183).

For slow-growing cyanobacteria such as members of the order *Pleurocapsales*, primary cultures are best established by direct isolation of colonies on solid media rather than by preliminary cultivation in a liquid medium (207). During transport of rock chips, mollusk shells, or macroalgae from intertidal zones to the laboratory, keep the samples in closed bottles or tubes containing a damp piece of filter paper; do not submerge the samples in seawater, since this promotes development of contaminants. In the laboratory, suspend material scraped from the natural substrates in sterile liquid medium, and streak several plates directly from the suspension in addition to preparing a liquid enrichment culture. If the suspension contains many contaminants, wash it repeatedly in sterile medium by low-speed centrifugation to reduce the level of contamination before streaking plates.

Confirm the purity of cultures of cyanobacteria by microscopic observation and also by inoculating into complex media and incubating in the dark at 30°C, where the cyanobacteria do not grow.

See section 11.1.12.1 for a further enrichment technique for cyanobacteria.

11.1.7.2. Purple Nonsulfur Bacteria (29)

Although termed purple nonsulfur bacteria, a number of these species can in fact use sulfide as an electron donor for growth, but they do so only when the sulfide is maintained at low, nontoxic concentrations. Consequently, the organisms are ordinarily cultured with organic electron donors. Rather specific enrichment for particular species is possible, depending on the carbon source provided. For enrichment, only substrates that cannot be fermented by nonphototrophic organisms are provided in the enrichment medium.

The purple nonsulfur bacteria can be readily isolated from freshwater and marine sediments and are less frequently isolated from field, lawn, or garden soils. Use a basal enrichment medium (section 11.6.8), and supplement it with a suitable carbon source. (The particular carbon source used depends on the genus and species desired; consult references 3 and 8 for specific compounds.) Place 0.1 g of the specimen into a screw-cap tube, and fill the tube completely with the enrichment medium, freshly made to minimize oxygenation. Tighten the cap so that there is no air space at the top, and incubate the culture at ca. 25°C. Illuminate the culture continuously with incandescent light (not fluorescent light); use a 50- or 75-W lamp at a distance of 40 to 60 cm. Make sure that the lamp does not heat the cultures. Look for development of turbidity with a brown, yellow, or pink tinge in 3 to 7 days. Transfer a drop of culture to a second tube of medium for a secondary enrichment. For purification, use an agar medium containing 1% yeast extract or peptone, 0.2 g sodium malate, and 1.5% agar. Use the shake tube method (section 11.4.2), or use pour plates and incubate in an anaerobic jar. Illuminate the cultures during incubation. Many of the purple nonsulfur bacteria also grow in the dark under an air atmosphere or under microaerobic conditions, although the colonies are less highly pigmented than when grown anaerobically in the light.

11.1.7.3. Green Sulfur Bacteria

The green sulfur phototrophs are not as easy to cultivate as the green nonsulfur forms, although massive developments ("blooms") of them are often readily visible to the eye, particularly in marine and brackish environments. Unless inocula from such blooms are available, a useful enrichment approach is to set up a **Winogradsky column.** This method provides anaerobic conditions and a long-lasting supply of H_2S, and successive blooms of the sulfur phototrophs (and many other microorganisms as well) usually result.

To prepare a Winogradsky column, obtain mud from freshwater, brackish, or marine environments (for example, mud from the edge of a freshwater pond or stream lake or from a salt marsh). Mix 3 parts of mud with 1 part of $CaSO_4$ · H_2O. Add some insoluble organic material such as finely shredded filter paper or small pieces of roots from aquatic plants. If paper is used, also add a small amount of NH_4MgPO_4. Pour the mixture into a tall glass cylinder (at least 5 cm in diameter) to a height of at least 15 cm, stirring to avoid air pockets. Hold the cylinder in the dark for 2 to 3 days to minimize the development of oxygenic photosynthetic organisms. Expose the cylinder to incandescent light or to diffuse daylight at 18 to 25°C during subsequent incubation. Anaerobic decomposition of the organic material in the column (with concomitant production of CO_2, alcohols, fatty acids, hydroxy acids, organic acids, and amines) and the formation of H_2S from the $CaSO_4$ will provide an appropriate array of microhabitats in which the sulfur phototrophs can thrive and form distinctive purple, red, or green patches on the sides of the glass column or layers or bands in the water column above the sediment-water interface.

Use Pasteur pipettes to obtain organisms from the glass surfaces or from the distinctive layers for microscopic examination, isolation, or further enrichment in liquid cultures. Alternatively, sequentially remove portions of the sediment with a spoon or spatula to expose the various zones of growth.

Enrichment can also be accomplished in defined liquid media. Select for specific sulfur phototrophs by varying the wavelength and intensity of the illumination used, the temperature of incubation, and/or the type and concentration of the electron donor. For a clear, detailed exposition of the many intricacies of such enrichment and the subsequent isolation of sulfur phototrophs into pure culture, see the very useful essay by Van Niel (199).

11.1.8. Visible Illumination To Enhance Pigment Production

Some bacteria respond to visible light by producing colored pigments, which can greatly aid the differentiation and isolation of these organisms. For example, *Brevibacterium* spp. produce a distinctive orange pigment only when exposed to visible light. Incubate plates in the light during the period of active growth, not after colonies have developed fully.

11.1.9. Sonic Oscillation Selection of *Sporocytophaga myxococcoides*

Enrichment of the microcyst-forming cellulolytic cytophaga *Sporocytophaga myxococcoides* takes advantage of a sonication-resistant form in the organism's life cycle (112). Place 100 ml of sporocytophaga medium (section 11.6.64) into a 500-ml Erlenmeyer flask, and inoculate with ca. 0.1 g of soil, mud, or plant material. After incubation at 30°C for 7 to 10 days with moderate agitation, remove 5 ml of culture, subject it to sonic oscillation for 15 to 30 s, and then use it to inoculate a secondary enrichment flask. After ca. 5 to 7 days, look for a distinctive yellow hue in the flask, and use phase-contrast microscopy to look for both microcysts and vegetative cells on and around the cellulose fibers. Subject a portion of this culture to sonic oscillation, and then use the pour plate method to obtain isolated colonies, as described in section 11.2.12.5.

11.1.10. Use of Antiserum Agglutination or Magnetic Beads Coated with Antibodies

The use of antiserum may enrich for the minority organism in mixed cultures. In a study of a culture in which two species of marine spirilla were present—a large and a small organism—the smaller organism was found to be greatly predominant. This made it impossible to obtain isolated

colonies of the larger organism by plating; moreover, none of a variety of selection methods was applicable. To enrich for the larger spirillum, the smaller organism was isolated and used to immunize a rabbit. When the resulting antiserum was added to the mixed culture, the smaller spirilla agglutinated and settled to the bottom of the tube. The supernatant, now containing a high proportion of the large spirilla, was used to obtain isolated colonies.

Magnetic beads coated with antibodies may be used to trap target organisms for which antibodies are available (47). This technique was used to isolate *Azospirillum* species from natural soils (81). Polyclonal antibodies raised in rabbits against whole cells of a mixture of five strains of *Azospirillum* were used to coat magnetic beads. These were then mixed with natural samples and used to capture the beads and bacteria on a magnet. The samples were washed before being used for cultivation experiments. The lectin concavalin A was used to isolate coliform bacteria from freshwater and sewage (159). The immunomagnetic separation method was successfully combined with matrix-assisted laser desorption ionization–time of flight mass spectroscopy (127).

11.1.11. Cell Density

The ability of some bacteria to float or to sink in aqueous environments or to form a band during density gradient centrifugation has been used to separate these organisms from contaminants.

11.1.11.1. Flotation on Water for *Lampropedia hyalina* (140)

The strictly aerobic, tablet-forming coccus *Lampropedia hyalina* forms a characteristic hydrophobic pellicle on the surface of liquid media. If complex media are inoculated with material from habitats rich in organic matter, the lampropedias often form contaminated pellicles containing their very distinctive cells. If such pellicles develop, transfer them with a Pasteur pipette to a plate containing soil seeded with starch or wheat grains covered with water. After incubation, pellicles containing the tablets of *L. hyalina* (as shown by microscopy) can be transferred to agar media for purification.

11.1.11.2. Swirling for *Achromatium oxaliferum* (111)

The relatively high density of cells of *Achromatium oxaliferum* is caused by the accumulation of intracellular $CaCO_3$, and this property can be exploited to concentrate the organism from natural samples. Place a small amount of sediment from a suspected source in the bottom of a large beaker, and cover it with ca. 1 cm of slightly alkaline water. Tilt the beaker, and gently swirl the contents to separate the achromatia as a thin, white deposit directly above the sediment. Transfer the cells to another beaker with a Pasteur pipette, and repeat the process until the achromatia are free of contaminants.

11.1.11.3. Density Gradient Centrifugation

The buoyant density of bacteria in pure culture and in samples from natural aquatic environments has been studied by density gradient centrifugation in Percoll gradients, and the average density of a representative bacterium is 1.080 pg μm^{-3} (78). There do not seem to be significant differences in density among bacteria; however, the density of a given bacterium can change by as much as 7% from the average as a result of the formation of inclusions such as sulfur and phosphorus storage materials or the formation of capsules and/or gas vesicles. Such inclusions may be responsible for the occurrence of two or more bands of organisms when samples from natural environments are subjected to density gradient centrifugation (78). Therefore, the density gradient technique may be useful in enriching for organisms with different types of inclusions.

Another enrichment application of density gradient centrifugation is the use of Urografin gradients to completely separate endospores from vegetative cells in sporulating cultures of *Bacillus* spp. (191).

For further information on density gradient centrifugation, see chapter 7.3.

11.1.11.4. Sampling the Neuston Layer for *Nevskia ramosa*

Nevskia ramosa is abundant in the epineustic layer on the surface films on calm water bodies. To isolate *Nevskia*, samples are taken with a sterile loop needle from the surface film and transferred to unshaken enrichment cultures with a medium free of combined nitrogen compounds (section 11.6.50) (188). At room temperature (20 to 24°C) *Nevskia ramosa* develops typical rosettes that consist of flat, binary branching polysaccharide stalks with cells in the tips. For isolation the cells can be streaked out on agar. Addition of ammonium salts to the medium results in submersed growth.

11.1.12. Radiation Resistance

Some bacteria are resistant to high doses of radiation such as UV light, X-rays, and gamma-radiation, and this can be used for selection purposes by killing the less resistant contaminants that may be present in a mixed culture. The following examples illustrate this principle.

11.1.12.1. UV Irradiation for Cyanobacteria

It is often difficult to obtain cultures of cyanobacteria free from bacterial contaminants, which frequently penetrate and live in the gelatinous sheaths that surround the cells and filaments of cyanobacteria (section 11.1.7.1). However, by treating a suspension with UV light for an appropriate period, it is possible to kill the contaminating bacteria and yet recover viable cyanobacteria. Success has been reported with the following method (75). Place a dilute suspension of cyanobacteria in a quartz chamber and irradiate with 275-nm UV light from a quartz-jacketed mercury vapor lamp. Agitate the suspension by continuous stirring during the incubation. At periodic intervals during the irradiation, remove samples and prepare a large number of dilution cultures from each sample. In the dilution cultures that show growth of the cyanobacteria, test for bacterial contamination by performing microscopic examination and by inoculating a variety of bacteriological media. One disadvantage of this method is that the final pure culture may contain cyanobacteria in which mutations have occurred because of the UV light treatment.

11.1.12.2. X-Ray and Gamma-Radiation Resistance for Deinococci (141, 142)

Members of the family *Deinococcaceae*, although not sporeformers, are highly resistant to X-rays and gamma-irradiation, and this property can be used to select for them. Samples suspected of containing deinococci (such as ground meat, fish, fecal samples, and sawdust) can be irradiated with 1.0 to 1.5 Mrad of gamma-radiation, killing most or all other vegetative cells. A few endospores of contaminating bacteria may survive, but if, before irradiation, the sample suspension is incubated for a time in a medium that allows endospores to germinate, this decreases the number of surviving sporeformers. After irradiation, plate the samples onto a suitable medium such as TGYM medium (section 11.6.69) and incubate at 25 to 30°C. Look for colonies that are pink, orange-red, or red.

Deinococci can also be selected from soil by preparing a slurry of the soil and exposing the supernatant in a petri dish to UV light at 600, 900, and 1,200 J m^{-2} (142).

11.1.13. Magnetic Field for Magnetotactic Bacteria (138)

A number of aquatic bacteria contain small enveloped crystals of the iron oxide magnetite, which enable the cells to become oriented in a magnetic field. The bacteria are able to move along the magnetic lines of force and, in natural bodies of water, also move downward, since the Earth's geomagnetic field has a downward component. Magnetotactic bacteria can be enriched from natural samples, such as pond water and sewage, by the following procedure. Place the water and sediment in a 2-liter beaker wrapped in opaque material and covered with plastic wrap. Attach a stirring-bar magnet to the beaker, with the south pole (use the north pole in the Southern hemisphere) touching the side of the beaker and positioned about 5 to 8 cm above the sediment. The magnetotactic cells will congregate near the magnet and can be removed with a pipette.

A simple "racetrack" for selection of magnetotactic bacteria can be constructed (216). Briefly, prepare a capillary tube from a Pasteur pipette and seal the small end in a flame. Fill the sealed capillary with filter-sterilized water (preferably water collected from the surface layer of sediment from which the inoculum is to be taken) by means of a syringe and needle. The large, nonsealed end of the capillary forms a well whose bottom can be plugged with a tiny wad of cotton. Place a drop of sediment containing magnetotactic bacteria into the well, and position a stirring-bar magnet near the opposite, sealed end of the capillary. Magnetotactic bacteria migrate from the well through the cotton plug, travel along the capillary, and arrive at the sealed end in a few minutes. The migration can be followed by dark-field microscopy. Harvest the accumulated cells by aseptically breaking the capillary near the sealed end and removing the contents with a narrow sterile pipette.

Some magnetotactic bacteria have been isolated in pure culture (26, 30, 130, 171).

11.1.14. Isoelectric Focusing of Cells

Isoelectric focusing (IEF) can be used to separate cells based on their surface charges (97) (Fig. 1). In a pH gradient of 2 to 10 and an electric field of 11.5 V cm^{-1}, cells are transported to the place where their net charge is compensated. By this technique a mixture of cells from *Chlorobium limicola* 6230, *Pseudomonas stutzeri* DSM 50227, and *Micrococcus luteus* DSM 20030 were successfully separated. A density gradient of Ficoll prevented convective currents in the system. The method was also tested with a concentrated mixture of bacteria from a shallow eutrophic lake and yielded up to 10 different bands. Species composition in each IEF band was analyzed by PCR and DGGE. Each IEF band exhibited a different species composition. After the separation of cells by IEF, three times more 16S rRNA signals could be detected by DGGE than in the unfractionated natural bacterial community. At the same time, the IEF fractions were

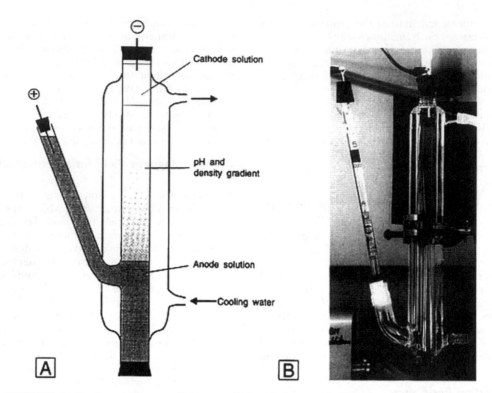

FIGURE 1 Isoelectric focusing. Schematic (A) and photograph (B) of apparatus for isoelectric focusing of whole bacterial cells. The cathode (−) and anode (+) consist of platinum wires which are connected to a power supply (97).

enriched for certain species and could be used in subsequent cultivation experiments.

11.2. BIOCHEMICAL ENRICHMENT

11.2.1. Alkali Treatment for Mycobacteria (24)

Add a maximum of 10 ml of the suspected sputum sample to a sterile, disposable, plastic 50-ml conical centrifuge tube with a leakproof and aerosol-free plastic screw cap. Add an equal volume of N-acetyl-L-cysteine-sodium hydroxide solution (section 11.6.49). Tighten the screw cap, and mix well in a Vortex mixer for a maximum of 30 s. Allow to stand at room temperature for 15 min. The N-acetyl-L-cysteine functions as a mucolytic agent; it converts the thick sputum to a thin, watery consistency. An extra pinch of the powder may be needed to liquefy highly mucoid sputum samples. The function of the sodium hydroxide is to destroy many of the contaminants present in the sputum; the mycobacteria are relatively resistant to the alkaline treatment. Fill the centrifuge tube to within 1 cm of the top with sterile 0.067 M phosphate buffer to neutralize the action of the sodium hydroxide. Centrifuge at 3,600 × g for 15 min to concentrate the mycobacteria. Carefully decant the supernatant into a splash-proof container. Add 1 to 2 ml of sterile water or buffer to suspend the sediment, and use this suspension to inoculate suitable culture media (for an example, see section 11.6.37).

CAUTION: Biosafety level 2 practices, containment equipment, and facilities are recommended for culturing

sputum samples, provided that aerosol-generating manipulation of such specimens are conducted in a class I or II biological safety cabinet. Biosafety level 3 practices, containment equipment, and facilities are recommended for activities involving the propagation and manipulation of cultures that may contain *Mycobacterium tuberculosis* (164). See chapter 46 for biosafety methods.

11.2.2. Incubation at Alkaline pH for Vibrios and *Sporosarcina ureae* (24, 49)

To select for *Vibrio cholerae* from a fluid stool sample or rectal swab, directly streak the specimen heavily onto a selective medium such as TCBS agar (section 11.6.68). If there are only small numbers of *V. cholerae* in the specimen, inoculate 20-ml portions of alkaline peptone water (section 11.6.3), incubate for 5 h at 35°C, and then streak this culture heavily onto the TCBS agar. The alkalinity of the peptone water (pH 8.4) and the TCBS (pH 8.6) inhibits the growth of most contaminants. On the TCBS agar, colonies of *V. cholerae* are yellow (sucrose fermenting) and oxidase positive.

Sporosarcina ureae in soil samples can also be isolated on alkaline agar media (pH 8.5). Streak the medium (tryptic soy-yeast extract agar with urea [section 11.6.75]) with a dilution of the soil, incubate at 30°C, and check colonies microscopically for the presence of packets or tetrads of cocci, which may be motile. To test for endospore formation, transfer suspected colonies to nutrient agar to which has been added 20 ml of a 10% (wt/vol) solution of urea (previously sterilized by filtration) and 50 mg of $MnSO_4 \cdot H_2O$

per liter. Incubate at a temperature below 25°C. Endospores are round and refractile by phase-contrast microscopy and are located centrally or laterally within the cocci.

11.2.3. Acid Treatment for *Legionella* Species

Isolation of *Legionella* species, particularly those from the environment, can sometimes be facilitated by acidification of samples to pH 2.2, which kills contaminants more quickly than it does the *Legionella* species. The following procedure is based on the isolation of *Legionella shakespearei* from water taken from an evaporative cooling tower (201). Pass 1 liter of water suspected of containing *Legionella* bacteria through a 0.45-μm-pore-size membrane filter. Cut the filter into small pieces with sterile scissors, and suspend the pieces in a tube containing 20 ml of sterile distilled water. Cap the tube, and shake vigorously by hand for 1 min to suspend the bacteria. Centrifuge 10 ml of the suspension at 4,000 rpm for 30 min, and then remove and discard 9 ml of the supernatant. Acidify the remaining 1 ml with 9 ml of pH 2.2 KCl-HCl buffer (add 5.3 ml of 0.2 N HCl and 25 ml of 0.2 N KCl to 100 ml of distilled water, adjust the pH to 2.2, and sterilize by filtration). Allow to stand for 5 min. Spread 0.1-ml portions onto the surface of BCYE agar plates (section 11.6.11) supplemented with glycine (3 g/liter), polymyxin B sulfate (79,200 IU/liter), vancomycin (5 mg/liter), and cycloheximide (80 mg/liter). Incubate the plates at 35°C for 7 days in a vessel containing a highly humid air atmosphere. Look for colonies that exhibit a "ground-glass" appearance when viewed under obliquely transmitted light. Presumptive identification of an isolate as being a member of the genus *Legionella* is based on demonstration of small (diameter, 0.5 μm), aerobic, gram-negative rods that require iron and cysteine for growth.

11.2.4. Incubation at Low pH

Most bacteria are inhibited by highly acidic conditions, but some are able to thrive, e.g., certain bacteria that live on fruit, are used in cheese ripening, or oxidize reduced-sulfur compounds to sulfuric acid. Incubation at low pH values can be very useful for the selection and isolation of these organisms.

11.2.4.1. Lactobacilli (126)

For lactobacilli in cheddar cheese, streak dilutions of the cheese on modified Rogosa's medium (section 11.6.46). This medium has a pH of 5.35 because of an acetic acid/acetate buffer system. At this pH, lactobacilli such as *Lactobacillus casei* and *L. planiarum* form colonies, but the common dairy organism *Lactococcus lactis* does not.

11.2.4.2. *Thiobacillus thiooxidans* (203)

Inoculate shallow layers of *Thiobacillus thiooxidans* medium (section 11.6.73) with samples from soil, mud, or water (marine mud is the most reliable source). Look for a drop in pH to 2.0 after 3 to 4 days or more, which virtually ensures the predominance of this acid-tolerant organism. Purify by streaking on the solidified medium.

11.2.4.3. *Frateuria* Species (190)

Members of the genus *Frateuria* are commonly found on fruit and can be enriched on *Frateuria* isolation medium (section 11.6.26), which has an initial pH of 4.5. Inoculate the medium with fruit samples, and incubate at 30°C. Streak plates of GYC agar (section 11.6.30) with material from the enrichment. GYC agar contains insoluble $CaCO_3$, and clear zones will form around the *Frateuria* colonies.

11.2.5. Inhibition by Toxic Metals

The salts of heavy metals such as tellurium, thallium, and selenium can, at low concentrations, exert an inhibitory effect on many bacteria. The use of a particular metal salt at an appropriate concentration can allow some bacteria to grow while inhibiting the growth of others. Some examples follow.

11.2.5.1. Tellurite Inhibition for Corynebacteria and Certain Streptococci

Potassium tellurite inhibits gram-negative bacteria and most gram-positive bacteria when used at a suitable concentration. To select for corynebacteria such as *Corynebacterium diphtheriae*, use potassium tellurite at a concentration of 0.0375%, as in cystine tellurite blood agar (section 11.6.22). Colonies of corynebacteria are gray or black as a result of reduction of the tellurite. For selection of certain streptococci (*Streptococcus mitis* and *S. salivarius*) and enterococci, use potassium tellurite at a concentration of 0.001%, as in mitis-salivarius agar (section 11.6.43). *S. mitis* forms tiny blue colonies, *S. salivarius* forms larger blue "gumdrop" colonies, and enterococci form small blue-black colonies.

11.2.5.2. Thallium Inhibition for Mycoplasmas and Enterococci (23, 105)

Use thallous acetate at a concentration of 0.023%, as in E agar (section 11.6.24) and E broth (section 11.6.25), to select for mycoplasmas such as *Mycoplasma pneumoniae* from the respiratory tract. Extract specimens collected on swabs into 2 ml of soybean-casein digest broth (section 11.6.62) containing 0.5% bovine serum albumin. Inoculate 0.1-ml amounts of the suspension into biphasic E medium (section 11.6.24) and also onto plates of E agar. Incubate the cultures at 37°C aerobically in sealed containers. Examine the plates at intervals for up to 30 days with a dissecting microscope (magnification, ×20 to ×60) for the appearance of minute colonies (diameter 10 to 100 μm) with a typical "fried-egg" appearance. Examine the biphasic cultures microscopically by looking through the side of the tube for spherules (fluid medium colonies); also observe the cultures for a decrease in pH (yellowing of the phenol red indicator). Inoculate E agar plates from the biphasic medium to obtain isolated colonies. For principles of biphasic media, see chapter 12.1.

Thallous acetate has also been used at a concentration of 0.1%, as in thallous acetate agar (section 11.6.70) to select for enterococci.

11.2.5.3. Selenite Inhibition for Salmonellae

Use sodium hydrogen selenite at a concentration of 0.4%, as in selenite F broth (section 11.6.60), to temporarily suppress the growth of coliforms while allowing salmonellae to grow. In addition to directly streaking stool samples onto selective and nonselective agar media, inoculate selenite F broth heavily (ca. 1 g or 1 ml of stool specimen in 8 to 10 ml of broth). Incubate this enrichment at 35 to 37°C for 12 to 16 h, and then streak onto the plating media.

11.2.6. Phenylethanol Inhibition for Gram-Positive Cocci

Use phenylethanol in agar media at a concentration of 0.25%, as in phenylethyl alcohol agar (section 11.6.54), to inhibit the growth of gram-negative bacteria, particularly *Proteus* species, when these occur in mixed culture with gram-positive cocci. For example, one application is the isolation of coagulase-positive staphylococci from a stool specimen.

11.2.7. Dye Inhibition for Gram-Negative Bacteria, Mycobacteria, and Arthrobacters

Gram-positive bacteria are generally inhibited by lower concentrations of triphenylmethane dyes (such as crystal violet, basic fuchsin, brilliant green, and malachite green) than are gram-negative bacteria. For example, the presence of malachite green at a concentration of 1:4,000,000 in media inhibits the growth of *Bacillus subtilis* and at 1:1,000,000 inhibits the growth of staphylococci; however, concentrations of 1:30,000 to 1:40,000 are required to inhibit *E. coli* or *Salmonella enterica* serovar Typhi (72). Brilliant green is used in media for the confirmed test for coliforms, as in brilliant green lactose bile broth (section 11.6.17), and in several selective media for members of the Enterobacteriaceae, such as salmonella-shigella agar (section 11.6.59) and brilliant green agar (section 11.6.16). Crystal violet is also used for selection of members of the Enterobacteriaceae, as in violet red bile agar (section 11.6.78) and MacConkey agar (section 11.6.38). Mycobacteria are very resistant to dyes; malachite green is often incorporated into media used for isolation of *M. tuberculosis* to inhibit the growth of contaminants, as, for example, in Lowenstein-Jensen medium (section 11.6.37). Methyl red has been used to select for soil arthrobacters (section 11.6.5); the dye inhibits other gram-positive bacteria but not arthrobacters (79).

11.2.8. Salt Inhibition

The growth of some bacteria is not inhibited by high concentrations of NaCl, and some bacteria actually require a high NaCl concentration for growth. For instance, the red extreme halophiles such as *Halobacterium* species need at least 15% NaCl. Regardless of whether it is required or merely tolerated, NaCl has proven to be useful as a selective agent for certain groups of bacteria.

11.2.8.1. Staphylococci

Most *Staphylococcus* species grow in media containing 10% NaCl. For selection of staphylococci from foods and miscellaneous environments, mannitol salt agar (section 11.6.39), which contains 7.5% NaCl, is a useful medium. The growth of many other genera of bacteria is suppressed at this concentration of salt.

11.2.8.2. Halobacteria (76)

The archaeobacterial genera *Halobacterium* and *Halococcus* occur in heavily salted proteinaceous materials (such as salted fish), in salterns, and in the Dead Sea and other highly saline lakes. They can often be isolated from samples of solar salt. They require a high concentration of NaCl for good growth (ca. 25% NaCl) and cannot grow with less than ca. 15% NaCl. In fact, they are killed by even brief exposures to salt concentrations of less than ca. 15%. The use of NaCl concentrations of 20% or greater makes their selection from natural sources a simple matter. For an example of a suitable enrichment medium, see section 11.6.32. Incubate inoculated flasks at 37°C with agitation, and subsequently obtain isolated colonies by streaking from the enrichment cultures onto plates of medium solidified with 2% agar. Incubate the plates at 37°C for 3 to 14 days in plastic bags to prevent excessive drying. Look for the development of pink or red colonies.

11.2.8.3. *Halomonas* Species (205)

Halomonads are eubacteria that tolerate both high and low salt concentrations; all known species grow in NaCl concentrations from 0.2 to 25%, depending on the type of medium used (205). They have been isolated from a wide variety of saline environments, including solar salterns, the Dead Sea, manganese nodules from the ocean, underground salt formations, and the Antarctic. They can be isolated on high-salt-containing media such as CAS medium (section 11.6.19). They form white to yellow colonies, not pink or red as do the archaeal halophiles.

11.2.9. Bile Inhibition for Enteric Bacteria

Bile or bile salts are often incorporated into culture media as selective agents for intestinal bacteria. There are some exceptions to this selectivity; for example, the plague bacillus *Yersinia pestis* can grow in pure bile even though it is not an intestinal organism (146). However, *Y. pestis* is closely related to *Y. pseudotuberculosis*, an intestinal organism, so perhaps the bile tolerance is not surprising. Nevertheless, the rule holds sufficiently well to make bile an important selective agent for gram-negative enteric rods and for enterococci. For selection of members of the family Enterobacteriaceae, bile or bile salts are incorporated into such media as MacConkey agar (section 11.6.38), salmonella-shigella agar (section 11.6.59), violet red bile agar (section 11.6.78), and many others. For enterococci, bile esculin agar (section 11.6.13) has proved to be an excellent selective and differential medium.

11.2.10. Antibiotic Inhibition

Antibiotics provide one of the easiest ways to devise a selective medium for strains of a particular species or genus. It makes no difference whether the organisms have medical importance or not; for instance, they may merely be harmless soil or water organisms. Obtain several reference strains of the bacterial taxon, and test them against a wide spectrum of individual antibiotics. Identify antibiotics to which all the strains are resistant, and incorporate several of these antibiotics into an appropriate sterile agar medium. When samples from nature are streaked onto the medium, the desired organisms, if present, will be able to grow, whereas many of the contaminant organisms will be suppressed by the antibiotics. A few specific examples follow.

(i) Because mycoplasmas lack cell walls, they are resistant to even very high concentrations of penicillin concentrations that inhibit most other bacteria. For instance, the antibiotic is used at a concentration of 194 U/ml in mycoplasma isolation media such as E agar (section 11.6.24).

(ii) Penicillin is useful in the isolation of *Bordetella pertussis* from the nasopharynx because it helps to suppress the growth of members of the normal flora while permitting *B. pertussis* to form characteristic pearl-gray ("mercury-drop") colonies in ca. 4 days. The clinical specimen (obtained by means of a nasopharyngeal swab) is streaked onto plates of Bordet-Gengou agar (section 11.6.15) containing penicillin (0.5 U/ml).

(iii) Antibiotics are necessary when attempting to isolate *Campylobacter jejuni* or other *Campylobacter* species from stool specimens, in order to suppress the members of the normal flora that otherwise would rapidly outgrow the campylobacters. One example of a medium suitable for isolation of *C. jejuni* is Modified Campy BAP (section 11.6.45), which contains cephalothin, polymyxin B, trimethoprim, vancomycin, and amphotericin B.

(iv) Three antibiotics—vancomycin, colistin, and nystatin—aid in the selection of *Neisseria meningitidis* from nasopharyngeal samples (to detect healthy carriers of the organism) and of *Neisseria gonorrhoeae* from clinical specimens (urethral exudates, cervical swabs, etc.). These antibiotics are incorporated into Thayer-Martin agar (section 11.6.71) and help to suppress the growth of members of the normal flora. The vancomycin suppresses gram-positive contaminants, the colistin acts mainly against gram-negative contaminants, and the nystatin is an antifungal agent.

Table 1 lists some other examples of bacteria for which antibiotics serve as selective agents. Do not overlook the use of cycloheximide, nystatin, and other antifungal agents as general inhibitors of fungi in selective media, especially when the source of the inoculum is the soil.

11.2.11. Dilute Media

Many bacteria are residents of soil and aquatic habitats low in nutrients and have difficulty growing in rich media. Also, many potential contaminants cannot compete in dilute media, so that the shortage of nutrients becomes a selective factor. Below are some examples of the use of dilute media for isolation of various soil or water bacteria.

11.2.11.1. Caulobacters (158)

Caulobacters are prosthecate bacteria that can grow at levels of nutrients that do not support good growth of many contaminants. To samples of water from ponds, streams, or lakes or to samples of tap water add 0.01% peptone and incubate aerobically at 20 to 25°C in bottles or flasks loosely covered with paper or aluminum foil. Examine the surface film daily by phase-contrast microscopy. When prosthecate bacteria occur in a relative proportion of ca. 1 in 10 or 20 cells (usually in about 4 days), streak the surface film onto plates of tap water agar containing 0.05% peptone and incubate at 10°C. The low peptone level allows the caulobacters to form tiny colonies but does not allow heavy overgrowth by other bacteria. After 4 days or more of incubation, select microcolonies of caulobacters under a dissecting microscope and transfer as patches to a richer medium (0.2% peptone, 0.1% yeast extract, 0.02% $MgSO_4 \cdot 7 H_2O$, 1.0% agar). After 2 days of incubation, prepare wet mounts from the patches to detect caulobacters. Obtain isolated colonies by streaking growth from the patches onto fresh plates. For marine caulobacters, add 0.01% peptone to samples of stored seawater and incubate at 13°C for ca. 7 days. When microscopic observation indicates the development of a suitable proportion of prosthecate bacteria, streak the surface film onto plates of seawater agar containing 0.05% peptone and incubate at 25°C. Subculture microcolonies to fresh plates as patches, and later purify by streaking.

11.2.11.2. Aquatic Spirilla (213)

In dilute media, aerobic chemoheterotrophic spirilla (the genera *Aquaspirillum* and *Oceanospirillum*) can often compete successfully with other bacteria for the nutrients present. For freshwater spirilla, add 1% peptone or yeast autolysate to samples of source water (e.g., water from stagnant ponds) and incubate at room temperature for ca. 7 days or until the spirilla become numerous. Then, add part

TABLE 1 Examples of the use of antibiotics in selective media

Bacterium	Antibiotic	Concn	Medium (chapter section)	Reference(s)
Agrobacterium biovar 1	Penicillin G	97,500 U/liter	*Agrobacterium* biovar 1 isolation medium (11.6.1)	106
	Streptomycin	30 mg/liter		
	Cycloheximide	250 mg/liter		
	Tyrothricin	1 mg/liter		
	Bacitracin	6,500 U/liter		
Agrobacterium biovar 2	Cycloheximide	250 mg/liter	*Agrobacterium* biovar 2 isolation medium (11.6.2)	106
	Bacitracin	100 mg/liter		
	Tyrothricin	1 mg/liter		
Brucella spp.	Bacitracin	25 U/ml	Brilliant green lactose bile broth (11.6.17) plus antibiotics	53, 65
	Polymyxin B	5 U/ml		
	Cycloheximide	100 μg/ml		
	Vancomycin	20 μg/ml		
	Nalidixic acid	5 pg/ml		
	Nystatin	1 U/ml		
Micrococcus spp. (from skin)	Furazolidone	20 mg/liter	*Frateuria* enrichment medium (11.6.26)	56, 108, 212
Nocardia spp.	Demeclocycline[a]	5 μg/ml	Diagnostic Sensitivity Test Agar (Oxoid) plus antibiotics	150
	or methacycline	10 μg/ml		
Pseudomonas spp.	Cycloheximide	900 μg/ml	*Pseudomonas* isolation medium (11.6.56)	77
	Nitrofurantoin	10 g/ml		
	Nalidixic acid	23 μg/ml		
Spirochaeta spp.	Rifampin	2 μg/ml	*Sphaerotilus* medium (11.6.63)	44

[a]Formerly known as demethylchlortetracycline.

of this initial culture to an equal part of the source water, and sterilize by autoclaving. Inoculate this mixture from the unsterilized portion of the original culture. After incubation and further development of the spirilla, dilute a portion of the second culture with more source water, sterilize the mixture, and inoculate it from the unsterilized portion. Continue to deplete the nutrients in this manner until the spirilla predominate. Obtain colonies by streaking onto plates of MPSS agar (section 11.6.48).

Another approach depends on low levels of nitrogen sources (213). Supplement samples of source water with 1% calcium malate or lactate, and incubate at room temperature for ca. 1 week. Make a serial transfer into sterile source water containing 1% of the carbon source, and incubate. Continue in this manner (three or four serial transfers) until the spirilla predominate. With this method it is important not to add a nitrogen source such as NH_4Cl in order to prevent overgrowth of the spirilla by contaminants. For marine spirilla, mix the seawater sample with an equal volume of Giesberger base medium (section 11.6.28) supplemented with 1% calcium lactate (213). After incubation, remove a portion of the original culture and incubate. Continue to successively deplete the nitrogen content by repeating this procedure until the spirilla become the predominating organisms. Obtain isolated colonies by streaking onto MPSS agar (section 11.6.48).

Another way to enrich for spirilla is to use a continuous-culture system that provides low levels of nutrients. For instance, in chemostat experiments with a mixture of a marine spirillum and a pseudomonad (95), the growth rate of the spirillum exceeded that of the pseudomonad when the dilution rate of the chemostat was decreased to the point at which the limiting carbon and energy source (lactate) fell below 10 mg/liter. In other experiments the growth rate of the spirillum exceeded that of *E. coli* when the lactate concentration fell below 5 mg/liter (96). In similar experiments with a freshwater spirillum and a pseudomonad, when the lactate concentration fell below ca. 0.09 mg/liter the pseudomonad was eliminated from the chemostat as a nongrowing population (129). The more efficient scavenging ability of the spirillum for lactate may be attributable to a lower K_m and a higher V_{max} of the transport system for lactate and also to a higher surface-to-volume ratio for the spirillum (129).

Under starvation conditions, spirilla appear to have a survival advantage (128). This may be related to their ability to form intracellular reserves of poly-β-hydroxybutyrate under conditions of prior growth with limiting levels of carbon and energy sources. The role of poly-β-hydroxybutyrate in bacterial survival has also been reported for other types of bacteria (60). Consequently, starvation conditions should be considered as a possible way to select for bacteria that form this polymer.

11.2.11.3. *Sphacrotilus* Species

The sheathed bacteria of the genus *Sphaerotilus* occur in streams contaminated with sewage or organic matter and form slimy tassels attached to submerged surfaces. They can also be isolated from rivers, open drains, or ditches where there is no initial evidence of their presence. The enrichment procedure takes advantage of the ability of *Sphaerotilus* species to grow at very low nutrient concentrations. To 50-ml volumes of *Sphaerotilus* medium (section 11.6.63), add 25-ml volumes of the water sample or 1-, 5-, and 10-ml portions of settled sewage or the settled liquor from various stages of sewage treatment. Incubate at 22 to 25°C for 5

days. Examine microscopically for evidence of filamentous growth after day 2. Obtain pure cultures by selecting a filament from the enrichment broth and streaking it onto plates of a solid medium (0.05% meat extract, 1.5% agar). Incubate plates for 24 h at 25°C, and examine under a dissecting microscope for the typical curling filaments of *Sphaerotilus* species. Transfer isolates to a Trypticase-glycerol broth (Trypticase [BBL Microbiology Systems, Cockeysville, MD], 5 g; glycerol, 5 g; distilled water, 1,000 ml [pH 7.0 to 7.2]), and incubate for up to 2 weeks. *Sphaerotilus* species form a heavy surface pellicle in 2 to 3 days, and the underlying broth remains clear. If turbidity develops, contamination has occurred. Even without development of turbidity, *Sphaerotilus* species should be reisolated from the surface pellicle by an additional streaking onto meat extract agar to ensure purity. To confirm that the isolated organisms form a sheath, place a small piece of slime growth on a slide in a drop of water, apply a coverslip, and press down on the coverslip with blotting paper. Place a very small drop of 1% crystal violet solution at the edge of the coverslip so that it flows into the preparation by capillary action. After 30 s, press again with blotting paper to remove excess dye and observe with a bright-phase oil-immersion lens. Both the cells and the sheath should be clearly visible.

11.2.11.4. Marine Oligotrophic Bacteria Growing on Natural Seawater Medium

The natural dilute medium that sustains a great diversity of microbial life is natural freshwater or seawater. Novel oligotrophic marine bacteria have been isolated on unamended seawater as the medium (52). Freshly collected seawater is filtered through a 0.2-μm-pore-diameter Supor membrane and is immediately autoclaved. In order to restore the bicarbonate buffer lost during autoclaving, the seawater is sparged with sterile CO_2 for at least 6 h, followed by sterile air for at least 12 h. Acid-washed polycarbonate containers are used for media whenever possible. Before use, the media are checked for sterility by directly counting cells stained with 4',6-diamidino-2-phenylindole (DAPI) as described earlier (198), except that 1% formaldehyde is used. Isolations on seawater medium often use dilution approaches (extinction culturing; see section 11.4.3.2.), to obtain numerically dominant members of the bacterioplankton community. In this way, several cosmopolitan lineages of oligotrophic marine bacteria, including the SAR11 cluster within the alpha-proteobacteria, were brought into pure culture years after their molecular detection by 16S rRNA sequencing of bacterioplankton samples.

11.2.12. Special Substrates

If a culture medium contains a sole carbon or nitrogen source or a selective electron acceptor that can be used only by a particular species or group of bacteria, the growth of that species or group will be favored over that of other organisms that may be present. However, be aware that other organisms may be able to grow to some extent by using products synthesized by the favored organisms; thus, this method is not absolutely selective but may merely increase the proportion of the desired organism in the population.

11.2.12.1. Tryptophan for Pseudomonads (117)

To enrich for pseudomonads capable of using tryptophan as the sole carbon and nitrogen source, inoculate a 250-ml Erlenmeyer flask containing 40 ml of tryptophan medium (section 11.6.76) with ca. 0.1 g of soil. Incubate with shaking at 25°C for 5 to 7 days. Transfer 0.1 ml to a second flask of medium, and incubate for 2 to 3 days. After a further serial transfer, obtain pure cultures by streaking onto tryptophan medium solidified with 15 g of agar per liter. After 1 to 3 days of incubation, subculture isolated colonies to agar slants.

11.2.12.2. N₂ for *Azospirillum* and *Azotobacter* Species

Azospirillum brasilense and *A. lipoferum* are microaerophilic nitrogen fixers associated with the roots of a variety of plants; they also occur in soil (61, 62). Place washed root pieces 5 to 8 mm long, macerated with a forceps, into nitrogen-free semisolid medium (Nfb medium [section 11.6.51]). Alternatively, inoculate the medium with a loopful of soil. Incubate with agitation for 40 h at 32°C; then, test the enrichment culture for acetylene-reducing activity (see chapter 15.3.11). Be careful not to disturb the dense subsurface pellicle that forms in the medium, since this may stop nitrogenase activity. If the culture reduces acetylene, enrich further by a serial transfer to fresh Nfb medium. Examine by phase-contrast microscopy, and look for plump, curved, motile rods ca. 1 μm wide and filled with intracellular granules. Then, streak onto plates of Nfb medium (solidified with 1.5% agar) containing 20 mg of yeast extract per liter. After 1 week, look for small, white, dense colonies and transfer to Nfb semisolid medium. For final purification, streak the Nfb culture onto BMS agar (section 11.6.14). Look for the development of typical pink, often wrinkled colonies.

Unlike *Azospirillum* spp., which fix nitrogen only at low oxygen tension, *Azotobacter* spp. have mechanisms to protect oxygen-labile nitrogenase from oxygen and can fix nitrogen aerobically. Inoculate 0.1 g of soil into 100 ml of nitrogen-free *Azotobacter* medium (section 11.6.7) contained in a 1-liter flask. Incubate at 30°C on a shaking machine. Observe the culture microscopically at periodic intervals for the development of large, ovoid cells that are 2 μm or more in diameter. Prepare a secondary enrichment culture, and purify by obtaining isolated colonies on nitrogen-free agar medium (*Azotobacter* medium solidified with 1.5% agar).

See also section 11.2.12.14 for a discussion of the use of unusual carbon sources for isolation of nitrogen fixers.

11.2.12.3. H₂ for *Aquaspirillum autotrophicum* (16)

Collect bacteria on the surface of membrane filters (pore size, 0.45 μm) by filtering water samples taken at different depths (e.g., 3 and 7 m) from eutrophic lakes. Place the filters, bacterium side up, onto the surface of mineral agar plates (section 11.6.42). Incubate under an atmosphere of 60% H₂, 30% air, and 10% CO₂ at 30°C. Subculture growth that develops on the filter surface to mineral agar plates. Obtain pure cultures by repeated streaking on mineral agar. Microscopically, look for spirilla 0.6 to 0.8 μm wide with bipolar tufts of flagella. Confirm autotrophy by demonstrating that growth is relying on CO₂ as the only carbon source.

11.2.12.4. Methanol for Hyphomicrobia (20)

Members of the genus *Hyphomicrobium* are able to use one-carbon compounds such as methanol or methylamine as sole carbon sources. Add the inoculum (5.0 ml of pond or ditch water, or 0.3 g of mud or soil—all preferably with a low organic content) to stoppered bottles (75- to 125-ml capacity), and add nonsterile *Hyphomicrobium* medium (section 11.6.33) through which nitrogen has been bubbled to provide anoxic conditions. Fill the bottles completely with the medium. Incubate in the dark at 30°C. Be sure that the bottles remain completely filled by adding fresh medium if necessary. Hyphomicrobia generally develop in ca. 8 days. Monitor their development by phase-contrast microscopy: look for rod-shaped cells with pointed ends or oval, egg-shaped, or bean-shaped forms, which produce filamentous outgrowths (prosthecae) that vary in length and may show branching. Prepare a secondary enrichment, this time with sterile medium. Finally, streak onto solidified *Hyphomicrobium* medium and incubate aerobically to obtain isolated colonies.

11.2.12.5. Cellulose for Cellulolytic Cytophagas

Various methods and media have been devised for enrichment of cellulolytic cytophagas (163). A plate culture method for enrichment is as follows. Place autoclaved, round, filter paper discs, such as are used in chemistry laboratories, on the surface of plates of ST6CX agar (section 11.6.65). Inoculate the surface of the filter papers with soil, rotting plant material, or drops from water samples. For soil, inoculate the filter papers in a regular pattern at different places with a few grains of soil by tamping the soil down on the filter with a glass rod or distributing it with a swab. Incubate the plates at 25 to 30°C. After 4 to 5 days and periodically up to 1 to 2 weeks, look for glassy, translucent, yellow to orange spots on the paper. Make transfers as early as possible from the margins of the areas of cellulose decomposition to fresh plates of medium with several small pieces of filter paper. Look for slender, flexible cytophaga cells by phase-contrast microscopy in the areas of cellulolysis. Obtaining pure cultures from enrichments can be difficult; for the various methods and their limitations, see reference 163.

11.2.12.6. Agar for Agarolytic Cytophagas

Some facultatively anaerobic marine cytophagas are able to hydrolyze and ferment agar; this trait is a valuable selective feature that can be used for their enrichment. Fill glass-stoppered bottles or screw-cap tubes to the top with non-sterile Veldkamp medium (section 11.6.77). Inoculate with marine mud from areas with decaying algae, stopper the bottles, and incubate them in the dark at 30°C. Look for development of turbidity, gas formation, and a drop in pH in 3 to 7 days. Isolate colonies by the shake tube method (section 11.4.2) with sterile Veldkamp medium containing 2% agar. Look for the development of colonies which, on microscopic examination, are composed of cells that exhibit the flexing movements characteristic of cytophagas. Subculture the colonies to media containing 1% agar and 1% yeast extract to demonstrate softening or liquefaction of the agar.

For other methods of obtaining agarolytic cytophagas, see reference 163.

11.2.12.7. Lactate for Propionibacteria

For most fermentative bacteria, lactate is an end product of fermentation, not a beginning substrate. The relatively unusual ability of propionibacteria to ferment lactate to propionate and CO₂ provides a basis for selection of these anaerobic organisms. Swiss-type cheeses are a good source of propionibacteria because these bacteria are the ripening agents. Fill a screw-cap culture tube (25-ml capacity) with a freshly boiled and cooled medium consisting of 4% sodium lactate and 1% yeast extract. Add ca. 0.2 g of CaCO₃ to the tube, and inoculate with a small piece of Swiss-type cheese.

Tighten the screw cap, and incubate the tube at 30°C. Look for the development of reddish brown turbidity in ca. 5 to 7 days. Obtain isolated colonies by streaking the culture onto sterile lactate-yeast extract agar and incubating anaerobically in a CO_2-enriched atmosphere, such as that provided by a GasPak (BBL).

11.2.12.8. Ammonium or Nitrite for Nitrifying Bacteria (136, 160)

Use the basal nitrification medium described in section 11.6.52. For ammonium oxidizers, which oxidize ammonium to nitrite, supplement the medium with 0.5 g of $(NH_4)_2SO_4$ per liter. For nitrite oxidizers, which oxidize nitrite to nitrate, use 0.5 g of $NaNO_2$ per liter. Prepare a 1:10 dilution of soil, and inoculate each flask of the enrichment medium with 1.0 ml of the dilution. Incubate at 28°C. At weekly intervals, test for disappearance of the ammonium or nitrite by removing samples of the enrichment culture to a spot plate. For ammonium, test 3-drop samples with Nessler's solution (see chapter 15.5.11); an orange or yellow color indicates ammonium. For nitrite, add 3 drops of solution A and 3 drops of solution B (see chapter 15.5.13) to 0.5-ml samples; a red color indicates nitrite. Compare the intensities of the colors with those obtained with a set of dilutions from standard solutions of ammonium or nitrite, and replace the amount of substrate lost from the enrichment culture by adding fresh substrate from a sterile stock solution. If no nitrite is present in the enrichment culture, test for the formation of nitrate by adding a small amount of zinc dust after the nitrite test reagents; the zinc will reduce any nitrate present to nitrite, which will then yield a red color. Continue to incubate the enrichment culture, and replace ammonium or nitrite at weekly intervals until a population of nitrifiers is built up as indicated by microscopic observation.

The isolation of nitrifiers into pure culture is extremely difficult. Completely inorganic media must be used, and the colonies formed are tiny, ca. 100 μm in diameter. Moreover, nitrifiers grow very slowly and are therefore often overgrown by contaminants. Using sterile enrichment medium as a diluent, prepare a series of 10-fold dilutions from the enrichment. Place 1.0-ml samples of each dilution into petri dishes; then, add silica gel medium prepared with double-strength enrichment medium as described in chapter 12.1.1.3. Mix the culture dilution with the medium immediately, and allow the medium to solidify. Incubate the plates in a humid atmosphere at 28°C, and examine periodically under a microscope for the development of tiny colonies. With a Pasteur pipette freshly drawn out in a flame to a fine tip, subculture well-isolated colonies into sterile medium. Determine the ability of the subcultures to use ammonium or nitrite, and also test for contaminants by streaking samples onto organic media such as nutrient agar (no growth should occur). Purify any apparently pure cultures in silica gel medium several more times, each time selecting well-isolated colonies under the microscope, testing for the ability of subcultures to use ammonium or nitrite and for the presence of contaminants.

11.2.12.9. Nitrate plus Organic Acids for Pseudomonads

Some members of the aerobic genus *Pseudomonas* can be enriched and isolated by taking advantage of their ability to use nitrate as a source of cellular nitrogen, together with their ability to use the salts of various organic acids as carbon and energy sources (184). Add 0.1 g of soil or mud or 0.1 ml of pond or river water to 3 ml of succinate-salts

medium (section 11.6.67). Incubate at 30°C to enrich for members of the fluorescent group of pseudomonads, such as *Pseudomonas putida*, *P. fluorescens*, or *P. aeruginosa*. Incubate at 41°C to select for *P. aeruginosa* (*P. putida* and *P. fluorescens* cannot grow at this temperature). Enrich for *P. acidovorans* by substituting glycolate, muconate, or norleucine for succinate in the enrichment medium and incubating at 30°C. Some pseudomonads can use not only oxygen as a terminal electron acceptor but also nitrate, and these pseudomonads can often be selected by increasing the nitrate concentration of the enrichment medium to 1.0% and by filling screw-cap tubes completely with the medium to establish oxygen-limiting conditions, which favor nitrate respiration. After preparing secondary enrichment cultures of pseudomonads, obtain isolated colonies by streaking agar plates (use the appropriate enrichment media solidified with 15 g of agar per liter). Incubate the plates aerobically at 30°C.

11.2.12.10. Sulfate plus Organic Acids for Sulfate Reducers

Many sulfate-reducing bacteria are able to oxidize lactate in the presence of sulfate, with the latter being reduced to H_2S. Inoculate screw-cap tubes or stoppered bottles with soil, mud, water, or fecal material, and fill the vessels completely with lactate medium (section 11.6.36). Incubate at 30°C. A blackening of the medium indicates sulfate reduction. Prepare a secondary enrichment before streaking onto solid media. To select for spore-forming species (*Desulfotomaculum* or *Desulfosporosinus*), heat enrichment cultures at 70°C for 10 min before streaking solid media. Obtain isolated colonies of the sulfate reducers by streaking plates of Iverson medium (section 11.6.34) and incubating them under a hydrogen atmosphere for 7 to 10 days at 30°C. To enrich for *Desulfotomaculum acetoxidans* (210), which cannot use lactate but can use acetate, use the enrichment medium of Widdel and Pfennig (section 11.6.79), which contains acetate as the oxidizable substrate. Use phase-contrast microscopy to look for development of motile straight or slightly curved rods that are 1.0 to 1.5 μm wide and 3.5 to 9.0 μm long and contain spores and also bright refractile areas. Use the shake tube method (section 11.4.2) to obtain isolated colonies. After 3 weeks of incubation, pick colonies and heat them at 70°C for 10 min; *D. acetoxidans* is a spore-former and can be selected by this heat treatment.

11.2.12.11. Elemental Sulfur for *Desulfuromonas acetoxidans*

Desulfuromonas acetoxidans obtains energy for growth by anaerobic sulfur respiration, with acetate, ethanol, or propanol serving as the carbon and energy source. Add samples of anaerobic, sulfide-containing water or mud from freshwater or marine sources to sterile screw-cap bottles. Fill the bottles with Pfennig and Biebl medium (section 11.6.53), in which flowers of sulfur are suspended; leave a small air bubble. Tighten the caps, and incubate the bottles at 28°C for 2 weeks with agitation (e.g., on a rotary shaker at 150 rpm). The action of the glass beads in the medium will gradually grind the sulfur to a very fine suspension. Positive enrichments can be recognized by a strong odor of H_2S. After two serial transfers in Pfennig and Biebl medium supplemented with a vitamin solution (section 11.6.79), allow the sulfur to settle and examine the supernatant for a faint turbidity of cells. Observe microscopically for small rods, 0.4 to 0.7 μm by 1 to 4 μm, some of which may be motile. Obtain isolated colonies by the shake tube procedure (section 11.4.2). To get a fine, homogeneous distribution of sulfur in the agar tubes, add 3 drops of an autoclaved polysulfide solution (10 g of $Na_2S \cdot 9 H_2O$ and 3 g of flowers of sulfur dissolved in 15 ml of distilled water) per 50 ml of the agar medium. Look for development of pink to ochre colonies. Because the sulfide formed by *D. acetoxidans* eventually causes inhibition of growth (no more than 0.1 g of H_2S can be tolerated), an alternative enrichment method has been devised in which the sulfur medium is inoculated not only with the mud or water sample but also with a pure culture of a green sulfur bacterium (of the family *Chlorobiaceae*). The latter organism continuously consumes the H_2S formed by *D. acetoxidans* by reoxidizing it to elemental sulfur, allowing fast-growing and highly enriched cultures which can be directly used to isolate pure cultures (155).

11.2.12.12. Thiosulfate, Sulfite, or Elemental Sulfur for Bacteria that Carry Out Disproportionation of Intermediate Sulfur Compounds

Thiosulfate, sulfite, or elemental sulfur can be dismutated by several sulfur bacteria (21, 195). The sulfur compounds are transformed to sulfide plus sulfate. This process represents the only known case of fermentation of inorganic compounds and appears to play an important role in the sulfur cycle of marine or freshwater sediments. The free energy is very low, and only few bacteria like *Desulfovibrio dismutans* can use it as the sole energy source. In the case of elemental sulfur, the end product sulfide has to be removed (e.g., by precipitation with amorphous ferric hydroxide) in order to obtain an exergonic reaction. Since many of the bacteria carrying out disproportionation of sulfur compounds do not grow autotrophically, 1 mM acetate is added to Widdel and Pfennig medium (section 11.6.79) as the carbon source. The formation of sulfate from intermediate sulfur compounds (10 mM sodium thiosulfate, 5 mM sodium sulfite, or a few grams per liter elemental sulfur) under anoxic conditions in the dark gives proof for the process of disproportionation.

11.2.12.13. Fe and Mn Oxides for Iron- and Manganese-Reducing Bacteria

Microbial oxidation of organic matter coupled to the reduction of Fe(III) and Mn(IV) is an important mechanism for organic matter oxidation in a variety of aquatic sediments, submerged soils, and in aquifers. Iron- and manganese-reducing bacteria are generally isolated using slight modifications of standard anaerobic techniques (22, 137). Culture media can be prepared with the classical approach (90) of boiling the media under a stream of anoxic gas to remove dissolved oxygen and then dispensing into tubes or bottles under anaerobic conditions. Alternatively, aerobic media may be dispensed into individual tubes or bottles and then the media can be vigorously bubbled with anoxic gas to strip dissolved oxygen from the media (121). Both media preparation approaches appear to yield similar organisms. Reducing agents such as Fe(II) (typically supplied at 1 to 3 mM as ferrous chloride), cysteine (0.25 to 1 mM), or sulfide (0.25 to 1 mM) can be added to dispensed media from anoxic stocks just prior to inoculation. Once Fe(III) reduction begins, the Fe(II) formed serves as protection against oxygen contamination.

Poorly crystalline Fe(III) oxide is typically the insoluble Fe(III) oxide of choice for culturing. An advantage of using poorly crystalline Fe(III) oxide as the electron acceptor is that most Fe(III)-reducing microorganisms convert the poorly crystalline Fe(III) oxide to the magnetic mineral magnetite during reduction. This is visually apparent as the reddish, nonmagnetic Fe(III) oxide is transformed into a black, highly magnetic precipitate (122). Reduction of the Mn(IV) oxide is also visually apparent in bicarbonate-buffered media because reduction of the dark Mn(IV) oxide results in its dissolution and concomitant accumulation of rhodochrosite, a white Mn(II) carbonate mineral.

An alternative electron acceptor that can be used for the recovery of Fe(III)- and Mn(IV)-reducing microorganisms is the humics analog, AQDS, which is typically provided at 5 mM. All of the Fe(III)-reducing microorganisms that have been evaluated can reduce AQDS, whereas microorganisms that do not reduce Fe(III) cannot reduce AQDS (123–125). Recovery of AQDS-reducing microorganisms either through enrichment and isolation procedures or dilution-to-extinction approaches yield organisms that also can reduce Fe(III) (51). The reduction of AQDS to AHQDS is visually apparent as the conversion of the relatively colorless AQDS to the orange AHQDS.

A variety of media have been successfully employed for the enrichment and isolation of iron- and manganese-reducing bacteria. Examples of a freshwater and a marine medium are described in section 11.6.35. They have a bicarbonate-carbon dioxide buffer system, and the headspace gas typically contains 20% carbon dioxide to establish an initial pH of ca. 6.8.

Iron- and manganese-reducing bacteria in environments can be enumerated with standard most-probable-number (MPN) culturing techniques using variations of media described above. Enumerations typically use Fe(III) or AQDS as the electron acceptor with the understanding that the Fe(III)-reducing microorganisms recovered are likely to have the ability to reduce Mn(IV) as well. Poorly crystalline Fe(III) oxide or Fe(III)-NTA is preferred over Fe(III)-citrate, which promotes the growth of fermentative microorganisms. One successful approach has been to add a combination of poorly crystalline Fe(III) oxide (100 mmol/liter) and 4 mM NTA to provide a supply of chelated Fe(III). Iron- and manganese-reducing bacteria also can be counted in plate counts in which Fe(III)-NTA or AQDS has been added as the electron acceptor. Clearing zones develop around bacteria reducing Fe(III)-NTA, and growth with AQDS as the electron acceptor results in the formation of orange colonies or zones.

11.2.12.14. Chlorinated Compounds for Dehalogenating Bacteria

Halogenated organics, for example, tetrachloroethene (perchloroethene [PCE]), trichlorethene, or chlorinated benzenes, are commonly used organic solvents, fungicides, insect repellents, and major intermediates in the manufacture of various chemicals which have been released into the environment and have become major groundwater pollutants. Although these compounds appear to be strictly anthropogenic and have no natural source, bacteria have been isolated that can degrade them by reductive dechlorination. PCE appears to be resistant to aerobic degradation but can undergo reductive dechlorination to ethene or ethane. The anaerobe "*Dehalococcoides ethenogenes*" (proposed name; 195), a deeply branching bacterium within the *Chloroflexi* phylum, dechlorinates PCE and other chloroethenes to ethene; hydrogen is required as the electron donor, and mixed organic substrates from bacterial extracts are required as the carbon and nutrient sources (131–133). The thiosulfate- and sulfate-reducing bacterium *Desulfomonile tiedjei* strain DCB-1 is able to reductively dechlorinate 3-chlorobenzoate and 3,5-chlorobenzoate to benzoate (177) and uses hydrogen and formate as electron donor for this process (138a).

These two well-described dehalogenating bacteria are phylogenetically and physiologically very different; there are no uniform isolation procedures. However, a common denominator is the use of low concentrations of halogenated compounds, to avoid toxicity and growth inhibition that affect different members of the dehalogenating enrichment cultures and consortia, and the dehalogenating bacteria themselves. Less than 1 mM 3-chlorobenzoate is used for isolation and maintenance of *Desulfomonile tiedjei* DCB-1 (177); chlorinated ethenes between 0.3 to 0.7 mM were used for "*Dehalococcoides ethenogenes*" (131–133). In both cases, hydrogen serves as electron donor for reductive dehalogenation; hydrogen is most likely provided by fermentative bacteria, or by methanol-disproportionating methanogens in dehalogenating enrichments (133). Another common denominator in the isolation procedures is the requirement for complex organic carbon sources; 5 to 20% (vol/vol) clarified rumen fluid for strain DCB-1 (177), and filter-sterilized anaerobic digester sludge supernatant (25%, vol/vol) plus extracts from a PCE-butyrate grown mixed culture (5%, vol/vol) for "*Dehalococcoides ethenogenes*" (131–133). These complex media allow the enrichment and isolation of dehalogenating bacteria from organic-rich inocula, such as anaerobic digestor sludge, without placing arbitrary bottlenecks on their substrate and vitamin needs.

Interestingly, the extract of a *Dehalococcoides* mixed culture is necessary to transfer and regrow a pure culture of "*Dehalococcoides ethenogenes*." The mixed culture likely contains unidentified growth factors produced by other microbial community members that are required for pure culture growth of "*Dehalococcoides ethenogenes*." For a description of the media and complex substrates involved in the isolation of "*Dehalococcoides ethenogenes*," see section 11.6.23.

11.2.12.15. Toluene for Toluene Oxidizers

To enrich for toluene-oxidizing bacteria, use cotton-stoppered flasks containing basal inorganic medium A or B (section 11.6.9). Inoculate with a small quantity of moist soil previously exposed to toluene vapor (i.e., incubated in a closed chamber containing a beaker of water saturated with toluene) for several days, or use fresh soil. Incubate the flasks at 25 to 30°C for 1 to 3 weeks in a closed chamber containing a beaker of water saturated with toluene. When growth occurs, obtain isolated colonies by streaking onto solidified medium and incubating the plates at 25 to 30°C in a toluene-containing atmosphere.

11.2.12.16. Bacterial Cells as Substrate for Myxococci

Myxococcus species lyse the cells of other bacteria by means of bacteriolytic enzymes and use the compounds liberated from the bacteria for growth. Take advantage of this in isolating myxococci from samples of soil, water, or plant material (134, 154, 178). Obtain bacterial cells for use as a substrate (*Enterobacter aerogenes* is suitable and convenient) by removing an entire 4-mm colony from the surface of an agar medium or by centrifuging the bacteria from a broth culture and washing them several times. Make a streak or smear of the bacteria ca. 1 cm wide and 4 cm long on the surface of a plate of water-agar medium (1.5% agar in distilled water). The smear of cells should be sufficiently thick to be barely visible to the eye. At one end of the smear, place two or three particles of soil or bits of plant material (e.g., bark or leaf). After 2 to 3 days and at daily intervals thereafter, examine the plates with a dissecting microscope for evidence of dissolution of the bacterial smear near the added parti-

cles. Also look for trails of gliding bacteria and development of fruiting bodies (usually yellow, orange, or pink) on and at the sides of the smear. Transfer the fruiting bodies found the greatest distance from the smear to an agar medium such as 0.2% tryptone or Casitone (Difco Laboratories, Detroit, MI) and 1.5% agar; crush each fruiting body in a drop of sterile water between two slides before streaking, in order to liberate the myxospores. The incorporation of cycloheximide (25 μg/ml) in both the nonnutritive agar and the subsequent plating medium will retard the growth of fungi, thereby aiding the isolation of the myxococci.

11.2.12.17. Unusual Carbon Sources for Nitrogen Fixers

The use of unusual carbon sources in nitrogen-free enrichment media can be exploited for the isolation of many of the N₂-fixing species (192, 196). The use of mannitol in *Azotobacter* medium (section 11.6.7) is a case in point. Very few contaminants are able to metabolize mannitol, which gives azotobacters a selective advantage. Ethanol is also a good selective carbon source to use for azotobacters.

Table 2 lists other examples of N₂ fixers that can be enriched and isolated by using specific compounds as sole carbon sources in nitrogen-free enrichment media.

11.2.13. Chemotactic Attraction

Actinoplanetes are members of the actinomycete family *Actinoplanaceae* and produce sporangia, which release motile spores. They were first isolated by using baiting techniques (section 11.3.5), but better recovery was achieved by using their attraction to chloride ions (152). A special chamber can be constructed (152), or a plastic tissue culture tray can be used. If using the latter, cut a 3-mm-deep channel in the connecting bridge between two wells. Place a soil

TABLE 2 Some unusual carbon sources for free-living nitrogen fixers[a]

Bacterium	Carbon source
Azotobacter vinelandii	L-Rhamnose
	Ethylene glycol
	Erythritol
	D-Arabitol
Azotobacter beijerinckii	L-Tartrate
	o-Hydroxybenzoate
	D-Glucuronate
	D-Galacturonate
Azotobacter armeniacus	Caprylate
Azomonas spp.	Benzoate
Beijerinckia indica	Citrate
Beijerinckia mobilis	Formate
	Benzoate
Derxia gummosa	Methane
	Methanol

[a]See references 27, 170, 192, and 196.

sample (0.5 g or less) in the bottom of both wells. Cover the soil with sterile water to a level of 2 mm above the bottom of the connecting channel, and let stand at 30°C for 1 h. Place a 1-μl capillary (Micro-caps; Drummond Scientific Co., Broomall, PA) filled with sterile phosphate buffer (5 to 10 mM [pH 6.8] containing 2 mM KCl), keeping the tip of the capillary 1 mm below the surface of the water. Let stand for 1 h, remove the capillary, wash with a few drops of sterile water, and blow the contents into 1 ml of sterile water or buffer. Spread a portion of the suspension on the surface of a plating medium such as casein starch agar (section 11.5.20). After incubation, most of the colonies, especially the pigmented ones, should be actinoplanetes.

Chemotactic responses can be used for selective enrichment of phototrophic consortia from freshwater lakes and sediments (74). Holes of 3 mm in diameter are drilled into one side of a 100-ml flat Meplats glass bottle. The bottle is filled with 20 ml of sediment slurry or an enrichment culture and gassed with N_2/CO_2 to maintain anoxic conditions. Stock solutions of various carbon and sulfur sources are diluted to a final concentration of 1 mM using centrifuged and filtered (0.2-μm-pore-width membrane filters) supernatant of the same culture. Capillaries (length, 50 mm; inside diameter, 0.1 by 1.0 mm; capacity, 5 μl; Vitro Dynamics, Rockaway, NJ) were filled with these solutions by capillary action and sealed at one end with plasticine (Idena, Berlin, Germany). Multiple capillaries can be used per substrate. Each capillary is inserted into a hole so that its end extends into the bacterial suspension, and then it is fixed with plasticine (Fig. 2). After incubation for 1 h under temperature and light conditions suitable for the target organism, the capillaries are removed, their open ends are sealed with plasticine, and the trapped consortia are counted by light microscopy. For the phototrophic consortia studied, sulfide, thiosulfate, 2-oxoglutarate, and citrate were the most powerful chemotactic attractants (74).

11.2.14. Gradient Culture

Enrichments are often incubated in homogeneously mixed systems. This provides for the bacteria only one fixed set of concentrations, which are often increased compared to the natural environment. Gradients are easily prepared by adding diffusible substrates in an agar or agarose plug. The developing diffusion patterns represent dynamic systems, in which the bacteria can find their optimum layer. Often steady states which are sustained by the bacterial activity are reached.

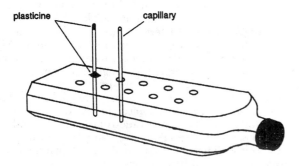

FIGURE 2 Capillary enrichment method, based on chemotactic attraction of motile bacteria to different carbon and sulfur sources in glass microcapillaries (74).

Autotrophic, oxygen-respiring bacteria that use Fe(II) and sulfide as electron donors have to compete with aerobic oxidation of their energy source. Such bacteria can be enriched and isolated by the gradient tube method. In principle, an agarose plug containing an oxygen-sensitive electron source [Fe(II) or sulfide] is placed at the bottom of a gradient tube; the plug is overlaid with a semisolid mineral salts–bicarbonate-buffered medium and a headspace of air. At this point the gradient (neglecting convective mixing) is theoretically a "step-gradient," i.e., all of the electron donor is below the interface of the agar plug and the overlay agar. The air headspace reservoir in the top of the tube constitutes an oxygen reservoir. Molecular diffusion and nonbiological reaction between electron donor and oxygen gradually alter the gradient shapes over a few days.

11.2.14.1. Gradient Culture for Neutrophilic Fe-Oxidizing Bacteria

Neutrophilic Fe-oxidizing bacteria can be cultured with a modified version of Kucera's and Wolfe's gradient enrichment for *Gallionella* (110). Briefly, an agarose plug containing an oxygen-sensitive electron source is placed anaerobically at the bottom of a gradient tube; the plug is overlaid with a semisolid mineral salts–bicarbonate-buffered medium and a headspace of air. For Fe-oxidizing bacteria, FeS, $FeCO_3$, pyrite, and different Fe(II) minerals and salts can be used (63, 64). FeS is prepared as described previously (72). The electron donors are mixed 1:1 with modified Wolfe's mineral medium in a flask (82) (section 11.6.47), and 1% (wt/vol) agarose (Pharmacia) is added. In a separate flask, the semisolid overlayer is prepared by adding 0.15% (wt/vol) agarose to the modified Wolfe's mineral medium. Both mixtures are autoclaved at 121°C for 20 min. A 0.75-ml aliquot of the molten FeS agarose mixture is added to the presterilized tube, and the agarose forms a solid plug. Sodium bicarbonate (5 mM final concentration; 1 M sterile stock solution) and 1 ml of vitamins (filter sterilized) are added to the slush agar overlayer while it is still molten. Then 3.75 ml of this mixture is placed on the surface of the FeS agar plug. While still molten, the overlayer is bubbled with filter-sterilized CO_2 dispensed through a cannula at a flow rate of 1.8 ml/s for 4 s, and the final pH is between 6.2 and 6.4. The tubes are capped with butyl rubber septa or stoppers and are allowed to sit for 6 to 24 h prior to inoculation. The tubes are inoculated by removing the top, inserting a pipette tip containing 10 to 15 μl of inoculant into the gel almost to the FeS plug, and expelling the contents of the pipette as the tip is withdrawn. The tubes are incubated in the dark at a constant temperature (maximum, 21°C) which should be cool enough to prevent the agar from liquefying (64). Other potential growth substrates can also be tested with these gradient tubes. For enrichment and isolation of Mn-oxidizing bacteria, the FeS is replaced with 0.5, 5, or 10 mM $MnCl_2$; thiosulfate (2.5 mM) and carbon substrates can be used in the same manner.

Bacteria ususally grow in a distinct band at the interface between the reduced electron donor that diffuses upward from the bottom plug and the oxygen that diffuses downward from the headspace. This growth pattern is shown in Fig. 3, with FeS-based growth of Fe-oxidizing bacteria. The four tubes on the left contain cultures of Fe-oxidizing bacteria, and the single tube on the far right represents an abiotic control. In culture tubes, a distinct band of cells develops at ca. 1 cm from the air-medium interface at the tops of the tubes. The milky region that can be seen clearly at the tops of the tubes is comprised principally of Fe oxide particles,

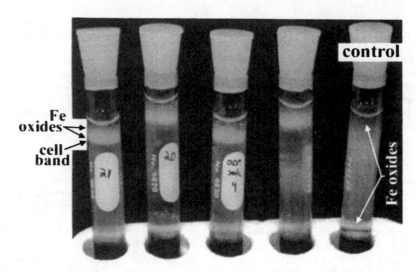

FIGURE 3 FeS-based growth of Fe-oxidizing bacteria in gradient tubes. The four tubes on the left contain cultures of Fe-oxidizing bacteria, and the single tube on the far right represents an abiotic control. In culture tubes, a distinct band of cells develops at ca. 1 cm from the air-medium interface at the tops of the tubes. The milky region that can be seen clearly at the tops of the tubes is comprised principally of Fe oxide particles, which directly overlie the band of cell growth. In the control tube, Fe oxides develop from ca. 1 cm from the top of the tube to ca. 3 cm from the FeS-medium interface at the bottom (63).

which directly overlie the band of cell growth. In the control tube, Fe oxides develop from ca. 1 cm from the top of the tube to ca. 3 cm from the FeS-medium interface at the bottom (63).

11.2.14.2. Gradient Culture for Sulfide-Oxidizing Bacteria (147)

A classical example is gradient culture of marine sulfide-oxidizing *Beggiatoa* strains (147). This culture method is used to maintain pure cultures and clones from individual *Beggiatoa* filaments that are picked from enrichments of *Beggiatoa*, as described in section 11.1.5.5. The gradient is prepared in a marine seawater medium (J3) (section 11.6.40). First a 4-ml quantity of J3 medium (pH 8.4; 1.5% agar; $NaHCO_3$ concentration lowered to 2.0 mM) supplemented with freshly neutralized Na_2S is solidified in the bottom of a 16- by 150-mm screw-cap tube. An initial sulfide concentration of 8 mM in this agar plug has proven satisfactory for all marine *Beggiatoa* isolates tested. The sulfide plug is then overlaid with 8.0 ml of semisolid J3 medium (0.25% agar; $NaHCO_3$ concentration lowered to 2.0 mM; no sulfide or thiosulfate).

Aging new gradient media for 2 to 3 days prior to inoculation establishes a sulfide-oxygen interface that is quite stable in both position and rates of nutrient flux. The interface is located near the top of the agar column, but the extent of sulfide and oxygen overlap is roughly 6 to 7 mm in uninoculated medium, compared with 0.2 mm or less in *Beggiatoa* cultures (148). Whether inoculated at the surface of this medium or stabbed throughout the upper few centimeters, the filaments rapidly proliferate at the sulfide-oxygen interface, forming a marked layer or "plate," which attains a maximum thickness of approximately 1 mm. Gliding motility and negative chemotactic responses allow these bacteria to track this interface as it slowly descends due to the gradual depletion of the sulfide reservoir.

11.2.15. Cyclic AMP and *N*-Acyl Homoserine Lactones as Growth Inducers

Extracellular signal compounds, such as cyclic AMP (cAMP) and *N*-acyl homoserine lactone, trigger a wide range of physiological responses in microbial cells. In gram-negative bacteria, *N*-acyl homoserine lactones are major autocrine molecular signal compounds, which trigger processes as diverse as resuscitation from the lag phase (25), bioluminescence (103), and virulescence (189). In enterobacteria, cAMP is involved in the regulation of the majority of genes expressed under starvation (173), including those coding for high-affinity sugar transport systems (67). The addition of extracellular cAMP has been shown to prevent substrate-accelerated death in starved laboratory cultures (42). Furthermore, the transcription factor σ^S (the gene product of *rpoS*) is involved in the transition to stationary phase. cAMP, in a complex with the cAMP receptor protein, acts as a negative regulator of the transcription of *rpoS* (119). It is therefore possible that the addition of extracellular cAMP hinders bacterial cells from entering the protective stationary phase but rather maintains them in a nutrient-scavenging state more favorable for cultivation. The ability of natural bacteria to acquire cAMP from the environment may obviate the need to accumulate cAMP intracellularly in preparation for recovery from starvation; marine planktonic bacteria express high-affinity uptake systems for cAMP and despite the low ambient concentrations (1 to 35 pM) are capable of increasing intracellular cAMP to levels (up to 2.8 μM) typical for *E. coli* and other prokaryotes (15). These observations make cAMP and *N*-acyl homoserine lactones promising candidates for improving the cultivation success of natural bacterioplankton.

Additions of cAMP and homoserine lactones in low concentration (10 μM) to liquid mineral media, amended with diverse substrate monomers and polymers, have improved the cultivation efficiency of heterotrophic bacteria from the Baltic Sea (34) and from a eutrophic lake (36). An alpha-proteobacterial bacterium that was isolated from the highest positive dilution of a cAMP-amended MPN dilution series showed lower doubling times and higher growth rates in the presence of cAMP (34). A wide phylogenetic range of bacteria is stimulated by cAMP additions, as shown by the high diversity of bacterial isolates obtained from cAMP-amended MPN cultures during a seasonal study of a freshwater lake. Hence, the composition of the cAMP-stimulated, cultivated fraction appears to be controlled by the physiological state of individual bacteria and the specific incubation conditions chosen rather than the overall composition of the bacterioplankton community that remains highly diversified and quite stable throughout the seasonal cycle (36).

11.3. BIOLOGICAL ENRICHMENT

11.3.1. Bacterial Parasitism by Bdellovibrios (185)

The tiny vibrios called bdellovibrios are widely distributed in soil and water and are capable of attaching to a wide variety of gram-negative bacteria, penetrating the cell wall, and multiplying within the periplasmic space, with consequent lysis of the host bacteria. The method for isolating bdellovibrios resembles that used for bacteriophages in many respects, with the difference that most bdellovibrios have a broad host range. Suspend 50 g of a soil sample in 500 ml of tap water, and shake vigorously for 1 h. Centrifuge the suspension for 5 min at 2,000 $\times$ g to remove the larger particles. Pass the supernatant through membrane filters of decreasing pore size: 3.0, 1.2, 0.8, 0.65, and 0.45 μm. Mix 0.5 ml of the final filtrate with 0.5 ml of a suspension (ca. 5 $\times 10^{10}$ cells per ml) of the host bacterium (for example, *Enterobacter aerogenes* or *Pseudomonas fluorescens*). Add the mixture to 4 ml of molten semisolid YP medium (section 11.6.83), mix, and pour over the surface of a plate of solid YP medium. After overnight incubation, examine the plates for plaques (areas of lysis). If plaques form within 24 h, they are attributable to bacteriophages rather than to bdellovibrios. Mark such plaques so that they will not be confused with plaques formed by bdellovibrios, which take at least 2 days to appear and grow larger with time. Cut out plaques suspected to be caused by bdellovibrios, suspend them in YP solution, and prepare a dilution series to be applied to lawns of host bacteria to obtain plaques that are well isolated. Examine one of the plaques by phase-contrast microscopy; look for tiny, highly motile vibrios ca. 0.3 μm wide. Suspend material from a plaque in YP broth, pass it through a membrane filter with a pore diameter of 0.45 μm, dilute the filtrate, and plate it onto lawns of the host bacteria. After three successive plaque isolations, the bdellovibrio strains will represent the descendants of a single bdellovibrio cell.

11.3.2. Plant Symbiosis by Rhizobia (12)

The nodules found on the roots of legumes represent a natural enrichment system for symbiotic nitrogen-fixing bacteria of the genera *Rhizobium* and *Bradyrhizobium*. Obtain nodulated roots of alfalfa, soybeans, or red clover, and wash the soil from them. Remove a nodule from the root, leaving a small portion of the root attached to the nodule. Use a camel hair brush to remove any soil still adhering to the nodule while holding the nodule under running water. Submerge the nodule in a 1:1,000 solution of $HgCl_2$ for 3 to 6 min; move the nodule around occasionally with sterile forceps. Transfer the nodule to 75% ethanol, and agitate it in the solution for several minutes. Remove it to sterile water, and agitate it for several minutes. Add 1 ml of sterile water to each of six sterile petri dishes. Transfer the nodule to the first dish, and crush it with sterile forceps. Mix the exudate with the water. Transfer one or two loopfuls of the suspension to the sterile water in the second dish, and mix. Continue to serially dilute in this manner for the remaining dishes. To each of the dilutions, add molten yeast extract mannitol agar (section 11.6.81) at 45°C. Incubate the solidified plates at room temperature, and subculture from well-isolated colonies.

A baiting technique for enrichment of rhizobia, similar to the enrichment of plant-pathogenic bacteria (section 11.3.5), can also be used. Plant susceptible legume species in soil naturally infested with the rhizobia being sought, and isolate the bacteria from the nodules that form, in the manner described above.

11.3.3. Animal Parasitism (70)

Inoculation of a host animal with a mixed culture containing a pathogen can select for the latter. The pathogen will predominate in the infected animal, often occurring in pure culture in the blood and tissues. Nonpathogenic contaminants are inhibited or destroyed by the defense mechanisms of the animal. For instance, if 1 ml of emulsified sputum containing *Streptococcus pneumoniae* and other bacteria is injected intraperitoneally into a mouse, a pure culture of the pneumococci can be obtained 4 to 6 h later by inserting a sterile, sharp-tipped capillary pipette through the abdomen and collecting some of the peritoneal fluid for cultivation on blood agar plates. A number of other animal pathogens can be similarly enriched by inoculating a host animal. For instance, *Borrelia* spp. can be selected by injecting the blood or tissue suspensions from infected arthropods into young or suckling mice (104), and pathogenic *Leptospira* spp. can be selected by injecting contaminated soil, mud, or water into weanling hamsters or guinea pigs (98).

11.3.4. Plant Parasitism

Plant-pathogenic bacteria can be enriched from their locus of infection by planting a susceptible plant in the area thought to be infected by the pathogen. If the "bait" becomes infected, the pathogen can be isolated from the diseased portions of the plant by appropriate isolation techniques for that pathogen.

11.3.5. Baiting for Actinoplanetes

The chemotactic attraction to chloride has already been mentioned as an enrichment method for actinoplanetes (section 11.2.13). However, actinoplanetes were first isolated in enrichments for aquatic molds, in which baits of various types were floated on water covering soil samples (55). Most of the water molds and actinoplanetes produce motile sporangiospores which are released from the soil-borne mycelium and will colonize the bait. For this type of enrichment of actinoplanetes, place the soil in a petri dish and cover it with charcoal-treated water. Place baits such as pollen grains (e.g., from *Pinus* species), seeds, human hair, or pieces of leaves on the surface of, or partially submerged in, the water. Incubate the plates for a few days, and periodically check the bait with a dissecting microscope for the characteristic mycelium and stalked sporangia of the actinoplanetes. The colonized baits can then be transferred with forceps to the surface of a suitable medium such as casein-starch agar (section 11.6.20).

11.3.6. Uncultured Symbionts and Parasites

Many bacteria and archaea that cannot (yet) be grown in pure laboratory culture can be maintained as symbionts or parasites of a specific host organism. In this way, maintaining the host organism is an indirect way of maintaining the "uncultured" symbiont as a natural enrichment. For example, a sponge of the genus *Axinella* harbors natural enrichments of the coldwater marine crenarchaeota (group I), which are abundant throughout the marine water column. The association between sponge and archaeon is specific, is maintained over time when the sponge is kept in an aquarium, and is quantitatively significant; up to 5% of the total rRNA in the sponge tissue is archaeal (161). Numerous specific associations between eukaryotic hosts and prokaryotic symbionts (38) give evidence of natural microbial enrichment and cultivation strategies whose biochemical and physiological underpinnings are essentially unknown. A few well-studied cases, for example the *Euprymna scolopes-Vibrio fischeri* symbiosis, show the complexity of host-symbiont interactions that hold clues to future cultivation strategies.

11.4. ISOLATION

Methods for the isolation of bacteria into pure cultures have benefited from substantial advances in novel micromanipulation and dilution techniques. Traditionally, isolation is done by obtaining individual colonies in or on a solidified nutrient medium by using either a streak plate or pour plate method, or by deep-agar dilution series. A single plating or dilution series does not guarantee purity. Also, obtaining a single colony does not always ensure purity, since colonies can arise from aggregates of cells as well as from individual cells. In slime producers, contaminants may be enmeshed in the chains of filaments formed by these organisms. Enrichment may be performed with selective media, but it is best to use nonselective media for purification because contaminants are more likely to grow and be detected on such media. Even with nonselective media, it is best not to pick (subculture) colonies too soon, because slow-growing contaminants may not yet have made their presence known. A pure culture should yield colonies that appear similar, and microscopic observation of the culture should reveal cells that are reasonably similar in appearance, particularly in regard to cell diameter and Gram reaction. There are, of course, some exceptions to these criteria; for example, colonies growing from a pure culture may exhibit smooth-rough variation; coccoid bodies, cysts, and spores may occur together in pure cultures of various organisms; and some organisms may show Gram stain variability. Nevertheless, the criteria are generally useful and apply in most cases.

Automatization and high-throughput approaches are changing this methodological landscape, in particular its reliance on enrichment steps and colony formation. For example, recent efforts to culture the uncultured (or not-yet-cultured) majority of marine bacteria rely on encapsulation of individual cells in gel microdroplets, and subsequent sorting of these encapsulated cells and microcolonies by flow cytometry, before parallel culturing in diverse oligotrophic media (218). Another approach uses a piezoelectric micropipettor for dispensing microdroplets of prediluted inoculum into microtiter plates for cultivation, with one or a few cells per well (35). In this way, individual microbial cells do not have to go through the highly selective pressures of the enrichment stage, before they have a chance to be isolated in pure culture. Basic principles, such as dilution, continue to be used, but the outcome, in terms of numbers of isolates, is going to grow beyond the capability of a human experimentator. The large numbers of isolates that are obtained by automated dilution methods are characterized by high-throughput sequencing and FISH analysis, rather than phenotypic analysis (35, 218). These methods can in principle be adapted to diverse, demanding media and complex cultivation regimes, with the result that the typical microbial isolate in 20 years may be isolated and identified exclusively by integrated, automatic procedures.

11.4.1. Spatially Streaking or Spreading on Solid Medium

There are many methods for streaking plates of solid media (streak plates), but the one illustrated in Fig. 4 almost invariably yields well-isolated colonies, even when done by a novice. Alternatively, spread dilutions of a mixed culture onto the surface of plates of solid media (for details, see chapter 9.3.2). For anaerobes, plates streaked or spread under an air atmosphere can subsequently be incubated in an anaerobic jar. Solid media for anaerobes should be freshly prepared and streaked within 4 h to avoid accumulating too much dissolved oxygen. Even so, it takes some time for an anaerobe jar to remove oxygen and establish anaerobic conditions; the use of roll tubes containing prereduced media eliminates this difficulty entirely (86, 89, 90). Such tubes are prepared by spinning sealed tubes of melted prereduced media so that the agar solidifies on the walls of the tubes as a thin layer. The method of streaking a roll tube is illustrated in Fig. 5. A roll tube can also be inoculated by adding a dilute suspension of cells and then rotating the tube to spread the cells over the surface. (Also see chapter 14.6.4.4.)

11.4.1.1. Isolation of Microcolony-Forming Bacteria on Filters

Growth of bacterial cells on thin agar slides or membrane filters and subsequent visualization of the microcolonies by light or epifluorescence microscopy can be used as an alternative to conventional spread plating. Advantages of the microcolony growth technique are rapid handling and the possibility of observing both dividing and nondividing cells on the same microscopic slide. Some bacteria from environmental samples are able to perform only a few cell divisions before their growth ceases; such organisms would never develop visible macrocolonies by conventional spread plating. The technique has been adapted for use with liquid media under aerobic and anaerobic conditions (85a). Basically, cells from a preculture are harvested and serially diluted in phosphate-buffered saline (PBS) (per liter: $Na_2HPO_4 \cdot 2H_2O$, 3.6 g; $NaH_2PO_4 \cdot H_2O$, 1.1 g; NaCl, 7.3 g; pH adjusted to 7.2 to 7.4 with NaOH). Samples are filtered onto black polycarbonate membrane filters (pore size, 0.2 μm; 25-mm diameter), which are immediately removed from the filter towers and placed on the surface of a PBS solution to avoid desiccation of the cells. The membrane filters are then mounted with the bacterial side towards a circular glass coverslip with 25-mm diameter (Menzel, Braunschweig, Germany) coated with a thin layer (ca. 0.1 mm) of silicone oil with a viscosity index of 10^5 m^2s^{-1} (Wacker-Chemie, Munich, Germany). In order to obtain a smooth layer of silicone oil, the silicone-coated coverslips are heated prior to use (ca. 24 h at 105°C). After the filters are mounted on the coverslips, the filter sandwiches are immediately placed, filter side downwards, on the PBS solution again to prevent the hydrophobic silicone oil from embedding the cells. The silicone-mounted membrane filters

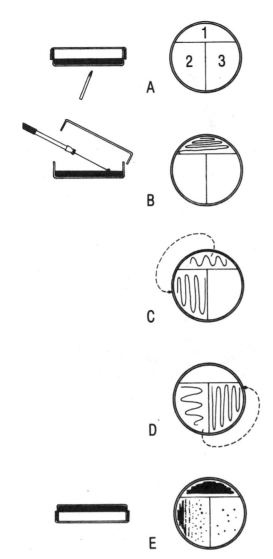

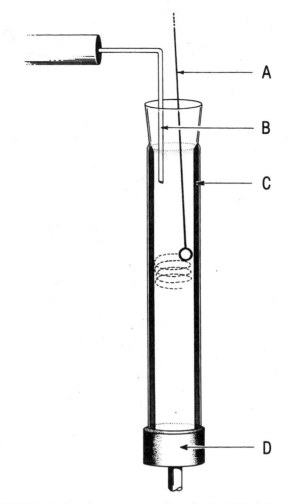

FIGURE 5 Streaking an anaerobic roll tube (86). (A) Loop needle (platinum or stainless steel; nichrome will cause oxidation of the medium). (B) Gassing cannula for continuous purging of the tube with oxygen-free gas. (C) Prereduced agar medium coating the inner wall of the tube. (D) Motor-driven tube holder for rotating the roll tube during streaking. Insert the needle with a loopful of inoculum to the bottom of the tube, press the loop flat against the agar, and draw it upward. After streaking in this manner for one-fourth of the way up the tube, turn the loop so that it is perpendicular to the agar (as shown), and continue to streak upward to the top. Remove the gas cannula, replace the rubber stopper in the tube, and incubate the culture in a vertical position.

FIGURE 4 Useful streak plate method for obtaining well-isolated colonies. (A) With a glass marker pencil, draw a "T" on the bottom of the petri dish to divide the plate into three sections. (B) Streak a loopful of culture lightly back and forth on the surface of the agar over section 1 as shown. Raise the lid of the dish just enough to allow the streaking to be done, and then replace it. Flame sterilize the loop, and allow it to cool (15 s). (C) Draw the loop over section 1 as shown, and immediately streak back and forth over section 2. Flame the needle, and allow it to cool. (D) Draw the loop over section 2 as shown, and then streak back and forth over section 3. (E) Incubate the dish in an inverted position as shown, to prevent drops of condensed water on the lid from falling onto the agar surface. Section 1 will develop the heaviest amount of growth, and section 2 or 3 will usually have well-isolated colonies.

can now be incubated floating on the surface or submerged within a suitable liquid medium (depending on the target of the enrichment), without disturbing the position of the individual cells on the filter (Fig. 6).

An interesting filter cultivation technique succeeded in cultivation of previously "unculturable" bacteria in a simulated natural habitat (101). In brief, bacteria of a marine sediment community were separated from sediment particles, serially diluted, mixed with warm agar made with seawater, and placed into a diffusion chamber. The chamber was formed by two polycarbonate membranes (pore size, 0.03 μm) which restricted movement of cells, but allowed diffusional exchange of chemicals and nutrients between the chamber and the environment. The first membrane was affixed to the base of the chamber; the agar with microorganisms was poured in, and the top was sealed with another membrane. The sealed chambers were placed on the surface of marine sediment and were kept in an aquarium. A gap was left between the agar and the top membrane, which was filled with seawater in the aquarium and allowed to observe

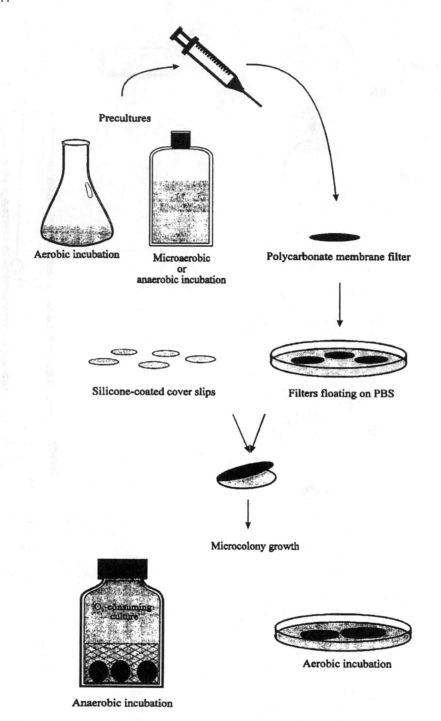

FIGURE 6 Membrane filter design and setups for aerobic and anaerobic incubation. Bacterial cells from precultures are filtered onto 0.2-μm-pore-size polycarbonate membrane filters and placed in PBS solution to avoid desiccation. Coverslips are covered with a thin layer of silicone oil. The membrane filters are mounted on the coverslip with the bacteria towards the silicone oil. The filter sandwich can be incubated either floating on a liquid surface or submerged within a liquid growth medium. For aerobic incubation, the filters can also be placed directly on a growth medium without being mounted on the coverslips (85a).

the undisturbed agar surface with freshly formed microcolonies after peeling off the membrane. Individual microcolonies were subcultured using the diffusion chamber technique. In this way, an average of ca. 22% of total counted cells were successfully cultivated, including several novel strains that were not closely related to known cultured bacteria. These cultivations were not possible in parallel experiments using conventional agar plates (101).

11.4.2. Serially Diluting in Solidified Medium

The simplest method for preparing a pour plate is to inoculate a tube of sterile, melted agar medium (cooled to 46 to 50°C) with a loopful of the sample, mix, pour the inoculated medium into a petri dish, and allow it to solidify. However, the sample often has to be diluted to produce well-isolated colonies, and the best approach is to use a series of 10-fold dilutions of the sample. Add 1.0-ml portions of each dilution to petri dishes, add 15 to 20 ml of the melted agar medium, mix by rotating the dishes several times, and allow the plates to solidify. For anaerobes, dilutions of the sample can be mixed with the melted, cooled (to 45°C) prereduced medium in roll tubes just before coating the walls of the tubes.

A disadvantage of diluting in agar media is that many of the isolated colonies are submerged in the agar and can be removed only by being dug out with a sterile instrument or punched out with a sterile Pasteur pipette. Another disadvantage is that the bacteria to be isolated must be able to withstand temporarily the 45 to 50°C temperature of the molten agar.

The agar shake technique (Fig. 7) has often been used for the isolation of anaerobic bacteria, especially phototrophic and sulfate-reducing bacteria (199). For this method, prepare a series of sterile tubes held in a water bath at 45°C and half fill them with molten agar medium at the same temperature. Inoculate the first tube of the series with a few drops of the mixed culture, and mix gently with the medium. Then transfer 1/10 (vol/vol) of the medium to the second tube of the series. Place the first tube in a vertical position, and immerse the bottom part in cold water to solidify the agar. Mix the contents of the second tube, transfer 1/10 of the medium to the third tube, and cool the second tube. Proceed in this manner for the remaining tubes of the series. Then, overlay the solidified agar in each tube with a melted, sterile mixture (1:1) of paraffin and paraffin oil (mineral oil) to a depth of ca. 2 cm. This seals the medium from the air. During solidification of the wax, the plug may contract, resulting in incomplete sealing; in this case apply mild local heating and tap the tube to remove any air bubbles under the seal. After the bacteria have grown, select a tube that contains well-isolated colonies, and remove the agar from the tube by first melting and discarding the paraffin seal and then inserting a sterile capillary pipette between the glass wall of the tube and the agar. Push the tip of the pipette down to the bottom of the tube, and apply air pressure to push the column of agar out of the tube into a sterile dish. Dissect the agar to remove the desired colonies.

11.4.3. Serially Diluting to Extinction in Liquid Medium

Serial dilution to extinction in liquid medium is useful when the desired organism cannot grow on solid media. A prerequisite is that the organism desired must be the predominant member of the mixed population.

11.4.3.1. MPN Dilutions

Serial dilution is often performed in multiple (at least three) parallel series of decimal dilutions, to combine isolations of numerically dominant organisms with MPN counts of prokaryotes. Tubes are filled with 9 ml of sterile medium and are inoculated with a 1-ml homogenized sample (Fig. 8 and 9). Several 10-fold dilution steps are performed, sufficient that the last 1 or 2 dilutions do not contain growing prokaryotes. Tubes showing growth after incubation (for slow-growing organisms this may take weeks to months) are recorded and used to calculate the MPN of viable cells in the original sample, according to published statistical tables and programs (100, 107). If necessary, the highest dilutions showing growth can subsequently be used as inoculum for new dilution series, with the ultimate purpose of bringing the desired, numerically dominant organism into pure culture. In this way, MPN dilutions are a powerful tool to isolate microorganisms that are numerically dominant and ecologically significant; frequently, these key community members cannot be isolated in enrichment culture due to quick overgrowth by "laboratory weeds," strains that are a minor component of the natural community but which respond enthusiastically to the laboratory media and conditions used (68). This approach relies on the assumption that one or a few organisms in a natural community or in an enrichment are numerically dominant, by at least 1 order of magnitude, given the large confidence intervals of MPN counts. However, a highly diversified microbial community may consist of numerous physiological and phylogenetic

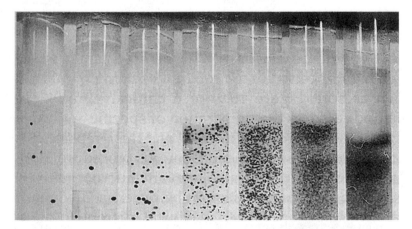

FIGURE 7 Colonies of oxygen-sensitive anaerobes using the agar shake technique. A dilution series was established from right to left, eventually yielding well-isolated colonies. The tubes are sealed with a sterile mixture of paraffin and mineral oil to maintain anaerobic conditions (photo courtesy of Norbert Pfennig [126a]).

groups, none of which is dominant; in this case, bacterial isolates from the highest MPN dilutions will necessarily represent a randomized selection of these nondominant groups, as far as they are compatible with the media used (34).

MPN dilutions tend to underestimate population densities of complex natural microbial communities and have to be regarded as conservative minimum estimates. For example, comparisons of MPN counts and rate measurements for sulfate-reducing bacteria indicated discrepancies of 2 to 3 orders of magnitude (193). Several approaches can improve MPN counts. MPN counts in sediments can be amended with sterilized natural sediment, which serves as an attachment matrix for microbial growth, and also as a source of complex substrates. In this way, cells that naturally grow on particles or in complex microbial associations have a better chance of regrowing in the highest dilutions that they reach. Further, MPN counts can be scored by activity measurements or detection of microbial reaction products, instead of relying on visible growth only (202). Also, MPNs can be prepared with a "background culture." For example, MPNs for fermenting and syntrophic bacteria were performed with a background culture of the sulfate reducer *Desulfomicrobium norvegicum* (ca. 10^6 cells/ml of medium) that lowers the redox potential and promotes syntrophic growth by scavenging hydrogen.

11.4.3.2. Extinction Culturing

Dilution series should allow for a sufficient margin of safety, to ensure that the dilutions are actually carried out to extinction. In other words, high dilutions that contain potentially single bacterial cells must not hit the upper end of the dilution series. Prepare a dilution of the mixed culture such that, when aliquots are added to a large number of vials of growth medium, the mean number of bacteria inoculated per vial will be <0.05. In other words, if 100 ml of the dilution contained a total of 5 bacteria, and if 1.0-ml aliquots were inoculated into 100 tubes, the mean number of bacteria inoculated per tube would be 0.05. On incubation, most of the vials would show no growth, but the few that did would be likely to have received only a single bacterium (P = 0.975). The smaller the mean number of bacteria inoculated per vial, the greater is the probability that the growth in the vial arose from a single bacterium. It is therefore imperative that most of the culture vials inoculated exhibit no growth, so that the few tubes that do exhibit growth will have a high likelihood of having been inoculated with a single cell.

Dilution to extinction was used to isolate novel oligotrophic marine bacteria on natural seawater medium (41). Natural communities of microorganisms were diluted to a known number, from 1 to 10 per tube, and then examined for microbial growth by flow cytometry, which is effective for counting very dilute populations of cells. By this method, bacterioplankton culturability from 2 to 60% was found for marine waters around Alaska and The Netherlands. Percent culturability is determined by the equation for estimation of culturability, $V = -\ln(1 - p)/X$. The theoretical number of pure cultures is estimated by the equation $u = -n(1 - p)\ln(1 - p)$, where u is an estimation of the expected number of pure cultures, n is the number of inoculated wells, V is the estimated culturability, p is the proportion of wells positive for growth (wells positive for growth divided by total inoculated wells), and X is the initial inoculum of cells added per well. To calculate the error, the exact lower and upper 95% confidence limits for the binomial proportion (p) are determined by using the SAS

package version 6.12 (SAS Institute Inc.). Next, these exact limits are put into the culturability equation and pure culture equation in place of the term p to give the exact lower and upper 95% confidence limits (CI_{95}) for percent culturability and the theoretical number of pure cultures (41, 52).

11.4.3.3. High-Throughput Culturing

High-throughput culturing, in conjunction with dilution to extinction, or MPN counts, is based on diluting liquid samples into multiwell plates. This protocol (52) can be adapted for different physiological groups of bacteria.

Cell counts of the liquid sample are determined by DAPI staining and epifluorescence microcopy. To determine viable cell counts, inocula of 50 or 100 μl are applied to spread plates of a suitable heterotrophic growth medium. Inoculum samples are diluted into liquid medium and distributed as 1-ml aliquots into 48-well non-tissue-culture treated polystyrene plates (Becton Dickinson, Franklin Lakes, NJ) to a final average inoculum ranging from one to five cells per well. At least one control plate is made for each sample collection by distribution of 1-ml aliquots of uninoculated medium. The extinction cultures are incubated under the desired conditions and time, and the agar cultures are incubated until colonies are large enough to count.

A cell array is made from each 48-well plate to examine wells for growth. Two hundred microliters from each well in the plate are filtered into the corresponding chamber of a 48-array filter manifold of custom design (HyTec Plastics, Corvallis, OR). Cells are then DAPI stained and vacuum filtered onto a 48- by 60-mm 0.2-μm-pore-diameter white polycarbonate membrane (cut from 8- by 10-in sheets; Whatman Nucleopore, Newton, MA). The membrane is laid on an oiled 75- by 50-mm slide (Corning Glass Works, Corning, NY) and covered with a 48- by 60-mm cover glass (Erie Scientific, Portsmouth, NH). The diameter of each sector of the array is 2 mm, which enables the detection of a culture with a cell titer as low as 1.3×10^3 cells/ml when 200 μl of sample is filtered. The array is then scored for growth by epifluorescence microscopy. Cell titers are estimated by counting five random fields within each positive sector.

Positive cultures from individual wells are split threefold. An aliquot is taken for molecular identification by PCR, RFLP, FISH, and sequencing; a second aliquot is transferred to fresh medium; a third aliquot is transferred to storage in 7% dimethyl sulfoxide (DMSO) and/or 10% glycerol in liquid nitrogen.

A new automated high-throughput approach (35) is based on the generation of very small droplets using a pressure pulse-driven microdispensing system (MicroDrop Autodrop microdispenser system, version 5.0; manufactured by MicroDrop GmbH, Norderstedt, Germany). The device works like an ink jet printer. Its core consists of a glass capillary (volume, 25 μl) which is surrounded by a piezoactuator. A voltage pulse causes the piezoactuator to contract, thereby creating a pressure pulse in the liquid inside the capillary. At the nozzle of the glass capillary, the pressure wave is transformed into a highly accelerated motion which leads to the expulsion of a small droplet. The droplet diameter was 55 to 75 μm, depending on the voltage (100 to 150 V), pulse duration (26 to 32 μs), drop frequency (150 Hz), and the geometry of the micropipette. The diameters of the droplets were determined under stroboscopic illumination by a video camera integrated into the Microdrop system. Droplet generation takes place with volume variation of less than 1%. The

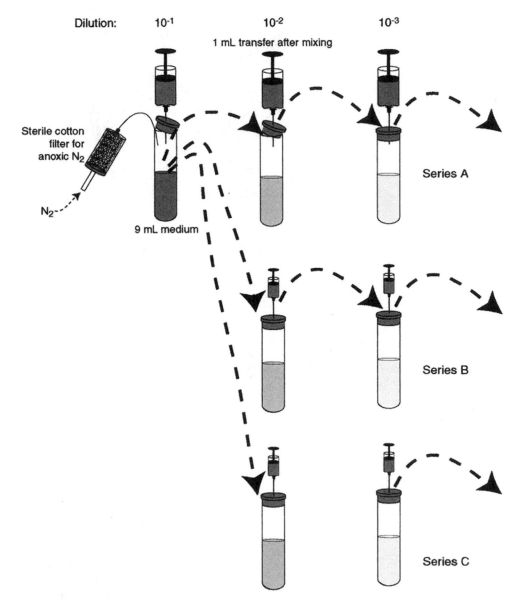

Dilution: 10^{-1} 10^{-2} 10^{-3}

1 mL transfer after mixing

Sterile cotton
filter for
anoxic N_2

N_2

9 mL medium

Series A

Series B

Series C

FIGURE 8 (A) MPN dilution series, here shown with Hungate tubes in three parallel dilution series. The inoculum is transferred anaerobically using N_2-flushed syringes (H. Cypionka).

capillary is connected to a gas pump; by decreasing or increasing the pressure on the liquid within the capillary, it can be filled with fluid or drained. The droplets generated are dispensed automatically at predetermined positions with the aid of an XYZ-positioning system (Fig. 10). Automated dispensing permits an inoculation of 96 samples in less than 1 min. Prior to the inoculation of the microtiter plates, the glass capillary is sterilized by repeated flushing with 70% ethanol, then sterile distilled water, and finally twice with the desired liquid sample. The liquid sample is prediluted such that each droplet harbors only one or a few bacterial cells.

This method can be used for high-throughput culturing, and also for the estimation of cultivation efficiency, the fraction of cells in a natural microbial community that can be cultured. After incubation, the number of microplate wells showing microbial growth is scored, and the fraction, p, of positive wells calculated (wells positive for growth/

total inoculated wells). The number of culturable cells per well, x, and the corresponding CI_{95} are calculated from p and the total number of inoculated wells, n, using formulas based on a binomial distribution:

$$x = -\ln\ (1 - p)$$
$$CI_{95} = 1.96 \times \sqrt{(p/\ [n(1 - p)])}$$

The method allows the determination of cultivation efficiencies with significantly lower statistical uncertainty and smaller confidence intervals than MPN dilutions. In a direct comparison, both approaches yielded similar isolates from lake water (35).

11.4.4. Washing Filamentous Bacteria

Filamentous bacteria can be separated from smaller, unicellular contaminants by washing of filaments in a grid. This

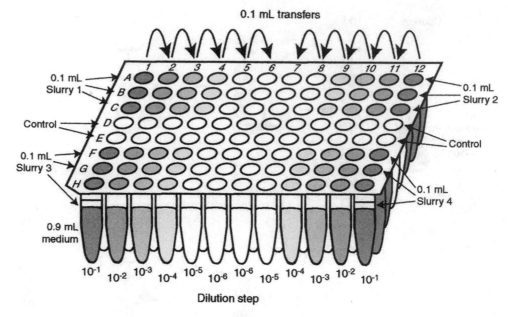

FIGURE 9 MPN counts for anaerobes with a microtiter plate. Inside an anaerobic glove box, four different samples are diluted down to $1:10^6$ on a deep-well microtiter plate. For each sample, a row remains uninoculated as a control. The wells are sealed with a capmat, and the plate is stored in an anoxic bag with an oxygen-consuming catalyst such as Merck Anaerocult A and an oxygen indicator such as phenol red (H. Cypionka).

method has been used successfully for cultivation of filamentous sulfate-reducing bacteria, genus *Desulfonema* (211), but can be adapted for the isolation of other filamentous bacteria, too. Briefly, a copper grid (an electron microscopy grid works well) is placed into a narrowed glass tube and attached to the glass wall with resin. The narrowed outlet of the glass tube (below the filter) is connected to a bent glass tube as outlet, via a flexible tube. After solidification of the resin, the apparatus is sterilized with 70% ethanol, dried with filtered air, and filled with sterile medium. The surface of the medium is kept ca. 1 to 2 cm above the grid and matches the level of the outlet (Fig. 11). Suspensions of filamentous bacteria taken from an enrichment or a natural population are placed on the grid. The filaments remaining on the grid are washed by rinsing with ca. 20 ml of medium. After the washing step, the flexible connection tube is closed and squeezed, to resuspend the filamentous bacteria from the grid, for subsequent transfer into fresh growth media.

11.4.5. Isolating Single Cells

A review of the various methods used for single-cell isolation may be found in reference 99. For occasional needs, the following procedure described by Lederberg (113) is useful. On the back of a clean microscope slide, draw a grid of 5-mm squares with India ink. Sterilize the face of the slide in a flame. After the slide cools, coat it with paraffin (mineral) oil to a depth of ca. 0.5 mm. (It is not necessary to sterilize the oil.) Heat 4-mm glass tubing in a flame, and draw it out to form a capillary with a terminal diameter of ca. 0.1 mm. Attach rubber tubing to the opposite end. Dilute the culture to a density of 10^6 to 10^7 bacteria per ml in medium, and draw up the suspension into the capillary by applying suction to the rubber tubing. Deposit a drop of suspension from the capillary at the center of each square, under the

oil. The drops will adhere to the glass and will flatten out to a diameter of 0.1 to 0.2 mm. Scan the flattened drops by phase-contrast or dark-field microscopy, and determine

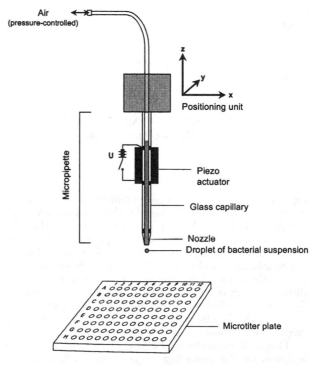

FIGURE 10 Schematic illustration of the Microdrop® device for automated inoculation of 170-μl aliquots of a bacterial suspension into a 96-well microtiter plate (35).

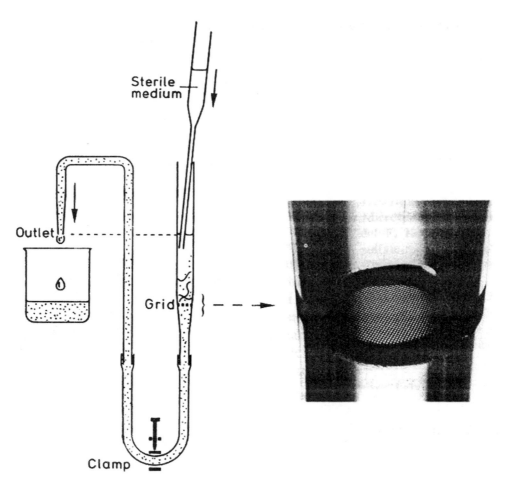

FIGURE 11 Apparatus for washing of filamentous bacteria (211).

which drops contain only a single cell of the desired type of organism. Frequently, such a single cell can be recovered by repeatedly flushing the drop in and out of a capillary pipette containing sterile medium. Another procedure is to add a small amount of sterile medium to the drop, incubate the slide in a container of oil until a clone develops, and remove some of the cells of the clone with a capillary pipette. Since the growth conditions for the clone are semi-anaerobic, this procedure may not work for strictly aerobic organisms.

When many isolations are required, a mechanical micromanipulator should be used to pick individual cells (73). A commercial micromanipulator (Eppendorf Model 5171) equipped with a pressure device (Eppendorf Model 5246 plus or Eppendorf CellTram Oil) is mounted onto an inverse phase-contrast microscope and used according to the manufacturer's instructions (Micromanipulator 5171: Operating Manual; CellTram Oil: Operating Manual; Transjector 5246: Operating Manual). Bacteria are isolated with a sterile capillary tube (Bactotip; Eppendorf), which possesses preferably a beveled tip (angle, 45°) with an opening of ca. 5 to 10 μm at the anterior end. Bacteria are diluted in PBS buffer, and the suspension is spread on a sterile microscopic coverslip as a thin film. A small volume (ca. 0.1 to 0.2 μl) solution is sucked into the bactotip. Bringing the opening of the Bactotip close to the surface of

a distinct bacterial cell (attached to the glass plane) causes a droplet of the solution to flow out of the tip and moisten the bacterial cell. The cells detach from the glass surface and are suspended in the droplet, which is then aspirated into the Bactotip and transferred to fresh medium (Fig. 12). The method has been tested with mixed laboratory cultures of bacterial species and with unicellular protists (73).

An ingenious new technique of optical trapping (aptly also called "optical tweezers") has been developed to isolate and manipulate microscopic particles, including bacteria, while they are being viewed through a high-resolution light microscope (18, 19). The technique is based on the use of a single strongly focused infrared laser beam at very low intensity so that a particle or a microbial cell becomes trapped from a suspension into the focus spot in a horizontally movable chamber; when the chamber is moved, the trapped particle is dragged along to another position in the chamber. After a single cell is optically trapped, it is dragged from the observation compartment containing the cell suspension into a culture compartment containing growth medium, which is then shut off or removed so that the single-cell culture can be separately observed and eventually subcultured (Fig. 13). The process is monitored by inverse microscopy. The technique thus enables a single cell type to be isolated from a predominant or mixed

Spread bacteria culture on a microscopic slide

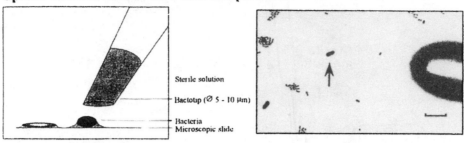

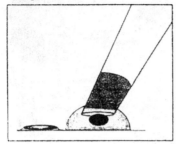

Suspend bacterium with sterile solution

Take up suspension

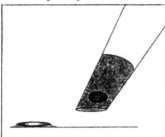
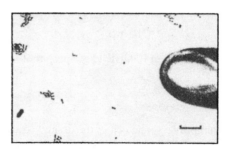

FIGURE 12 Isolation of a single bacterial cell of *Bacillus cereus* (arrow, larger cells) with the Bactotip method. The smaller cells are *Burkholderia* sp. (Left) Schematic drawing; (right) corresponding phase-contrast micrographs. Bar, 10 μm (73).

population without prior enrichment. The method has been used successfully to culture novel hyperthermophilic archaea identified in a mixture by rRNA whole-cell hybridization, as described below (87). This method works best with metabolically active, growing, and dividing cells; for example, cells from preincubated laboratory enrichments. Here, an individual cell has a reasonable chance to grow on laboratory media after it has been separated from its original milieu and microbial community. Survival and growth of an individually picked microorganism are less likely if it has been picked from a natural habitat with complex ecophysiological niches and microbial community interactions which cannot be mimicked easily in laboratory culture. Similar considerations apply to the mechanical separation of individual prokaryotic and eukaryotic microorganisms by micromanipulator (73).

11.5. rRNA-BASED METHODS IN CONJUNCTION WITH PHENOTYPIC CHARACTERIZATION

The rapidly expanding databases for rRNA sequences allow increasingly detailed cross-referencing to physiological properties and culture requirements of bacteria and archaea. New identification, cultivation, and isolation approaches combine rRNA-based methods with phenotypic characterizations.

11.5.1. rRNA Sequencing as a Guide for Enrichment and Isolation

The 16S rRNA identification of cultured bacteria has direct implications for phenotypic characterizations and can guide tests for unexpected or previously overlooked

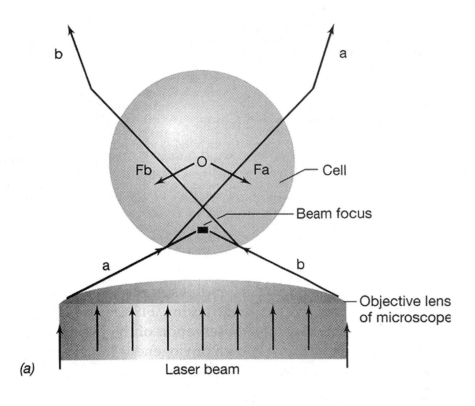

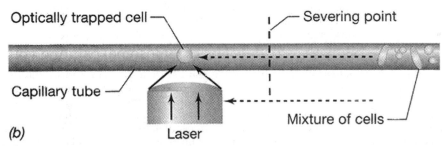

FIGURE 13 Optical tweezers for the isolation of bacteria. (a) Principle of the optical tweezers. A strongly focused laser beam on an object as small as a prokaryotic cell creates downward forces (Fa and Fb) on the cell, which allow the cell to be dragged in any direction as long as the beam force remains on it. (b) Isolation. The laser beam can lock onto a single cell present in a mixture in a capillary tube and drag the optically trapped cell away from the other cells. In the example used here, the desired cell is dragged from right to left. Once the desired cell is far enough from the other cells, the capillary is severed, the laser is turned off, and the cell is flushed into a tube of sterile medium (126a).

physiological and biochemical properties of misidentified strains on the basis of phylogenetic affiliation. One example of many is the reidentification of an alleged *Flavobacterium* isolate as a facultatively anaerobic, nitrate-reducing, thiosulfate-oxidizing strain of *Pseudomonas stutzeri* (182). By the same token, the phylogenetic affiliation of a microorganism can give hints for the isolation strategy. This strategy does not work if a new phylotype is not closely related to cultured prokaryotes; without a cultured model organism, a suitable cultivation strategy cannot be inferred. As a caveat, inferences for bacteria that are members of physiologically and phylogenetically well-characterized groups are not always valid. A case in point is the sulfate-reducing bacterium *Desulfospira joergensii*, a close relative of the acetate-oxidizing sulfate-reducing

genus *Desulfobacter* within the delta-proteobacterial sulfate-reducing bacteria. Despite its 16S rRNA position, *Desulfospira joergensii* does not oxidize acetate (71).

11.5.2. Uses of Whole-Cell FISH

A widely applicable method at the interface of phenotypic and phylogenetic characterization is whole-cell FISH with rRNA-targeted, fluorescently labeled oligonucleotides that hybridize with phylogenetically informative sequence motifs on their target rRNA molecules (13). Essentially, cells are fixed in buffered formaldehyde solution, collected on filters, or immobilized on glass slides, dehydrated by a series of ethanol washes, and incubated in a buffered solution containing the labeled probe and specific concentrations of formamide as a denaturant.

After washing away nonhybridized probe, the cells can be coated in mounting medium and are ready for viewing. Modifications of this approach aim at improved cell wall permeability for probes (in particular high-molecular-weight probes) and at better signal-to-noise ratio with multiple fluorescent labels or enzymatically amplified fluorescence signals. For detailed protocols, see reference 153. To the extent that particular phylotypes are congruent with phenotypic traits, phylogenetically related microorganisms with specific phenotypic properties can be identified in environmental samples, enrichments, and cocultures. The specificity of this "phylogenetic staining" method can be adjusted from domain-specific stains (bacteria versus archaea) to species differentiation, as long as species-specific rRNA sequence motifs exist. The continuously growing sequence databases necessitate continuous probe development and improvement, as older probes often turn out to be less specific or comprehensive than expected at the time when they were designed. Even domain-specific probes that target highly conserved sites have to be revised based on new sequence information, as demonstrated for general bacterial probes (57).

rRNA-targeted FISH can be used to stain specific microbial populations in natural samples and enrichments and to monitor them throughout successive enrichment and isolation steps. The phylogenetic target range of the probe can indicate the likely physiological properties and culture requirements of the target bacterium, as shown in the isolation of a new *Desulfovibrio* strain (102). Enrichment conditions for enigmatic microorganisms without cultured relatives, such as the hyperthermophilic *Korarchaeota*, can be improved empirically as soon as new organisms are visualized by FISH (40). FISH identification, in conjunction with physical cell separation by optical tweezers, was used for the isolation of a novel, morphologically conspicuous hyperthermophilic archaeon from a mixed enrichment culture (87). As shown by FISH, this archaeon represented a small percentage of the total archaeal population in the enrichment culture and could therefore not be isolated using serial dilution. However, the characteristic clusters formed by this archaeon could be picked by optical tweezers from the archaeal background population (87).

For structure-function analyses of natural microbial communities, FISH can be combined with microautoradiography to reveal substrate uptake patterns of different phylogenetic groups in mixed bacterial cultures, enrichments, and environmental samples (54, 114, 151). In brief, suspended cells are incubated with tritiated carbohydrates, short-chain fatty acids or amino acids, fixed with formaldehyde, and then filtered down and immobilized on glass slides for FISH hybridization with group-specific probes. After hybridization, the samples are washed, counterstained with DAPI, and incubated in photographic emulsion before development, fixation, and viewing. This method has allowed substantial progress in determining substrate uptake patterns and preferences for major phylogenetic groups of marine bacterioplankton (54). Such in situ analyses of nutrient specificity and uptake may open the way for the future laboratory cultivation of not-yet-culturable microorganisms in nature.

11.5.3. Molecular Monitoring of the Path from Environmental Samples to Enrichments and Pure Cultures

Molecular approaches can monitor the continuously changing microbial communities during an enrichment and isolation process. An enrichment can be regarded as a competition experiment in which selective physical, chemical, and biological factors trigger a succession of microbial community changes that lead to the stepwise elimination of most community members except a small group or even a single strain that is best adapted to the selective factors. The often complex community changes during the enrichment process can be monitored by several molecular approaches; here we focus on DGGE.

Since its introduction into microbial ecology (144), DGGE has emerged as a versatile screening method that allows the rapid identification of microorganisms with identical 16S rRNA sequences and the differentiation of isolates with divergent rRNA sequences. In brief, DGGE separates PCR-amplified 16S rRNA gene fragments of identical length and molecular weight, which would be impossible by agarose gel electrophoresis. Homologous 16S rRNA gene fragments separate on the basis of their melting domain stability, as they migrate through a polyacrylamide gel containing a gradient of DNA-denaturing substances, urea and formamide, in increasing concentrations. The 16S rRNA gene fragments melt into single-stranded conformation at different positions in the denaturing gradient, depending on their secondary structure and primary sequence. Identical fragments from identical or very closely related microorganisms denature simultaneously and result in DNA bands at the same position in the gradient. In single-stranded conformation, the DNA strands would separate from each other completely if they were not clamped together at one end by a GC-rich segment attached to one of the PCR primers that were used to generate them. Single-stranded fragments held together by GC clamps stop their migration. The electrophoresis is continued until all fragments have migrated into the denaturant gradient, have denatured at their respective gradient positions, and remain at their different positions as discrete bands. DGGE is sufficiently sensitive to recognize single-base-pair differences. Depending on the size of the DGGE gel and the number of gel lanes, several dozen samples can be analyzed in parallel, for example PCR products from environmental samples, enrichments, and pure cultures. Individual DGGE bands of mixed bacterial populations or cocultures can be excised from the acrylamide gel, reamplified by PCR, and sequenced for a detailed phylogenetic identification; this information can be used to design media for pure culture isolation of individual strains (Fig. 14) (194). DGGE can be used efficiently to monitor growth and changing species composition of microbial enrichment cultures (37). DGGE is extremely useful in tracking specific microorganisms in environmental sample series over spatial and temporal gradients, in particular when the microorganisms represented by each band and their physiological properties and preferences are well-known and can be understood in their environmental context (69). In this way, DGGE works as a molecular proxy for microbial population shifts and indicates the response of the community to specific factors that govern the growth and composition of a microbial community or an enrichment. Detailed descriptions of the method and a review of field applications have been published (143, 145).

Molecular tools can be used as auxiliary methods to check for the presence of contaminants in allegedly pure cultures. Whole-cell FISH with species- or strain-specific oligonucleotide probes (13) or DGGE (145) can identify contaminants within the limitations of microscopy or PCR sensitivity. They may not detect a contaminant that is very low in abundance, but they combine identification with detection of a contaminant.

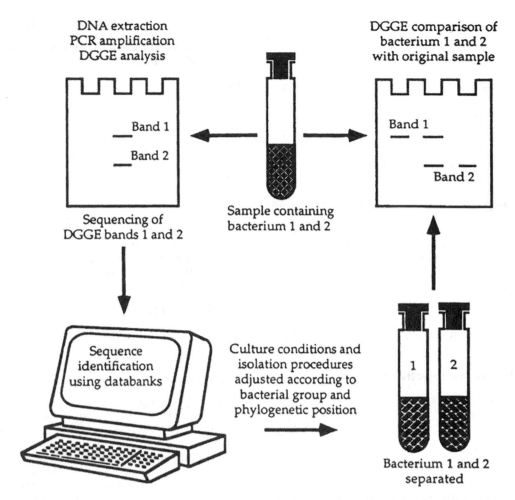

FIGURE 14 Use of DGGE in the analysis of mixed cultures and enrichments. After DNA isolation from the sample or mixed culture, PCR, and DGGE, the sequences of the DGGE bands are determined and compared to the sequence information available from databases, to identify the components of the microbial mixture, and subsequently to select culture conditions to separate and isolate the bacterial species in pure culture. For verification, the DGGE bands of the pure culture are compared to the DGGE pattern of the original sample (A. Teske).

11.6. MEDIA AND REAGENTS

11.6.1. *Agrobacterium* Biovar I Isolation Medium (172)

Mannitol	10.0 g
$NaNO_3$	4.0 g
$MgCl_2$	2.0 g
Calcium propionate	1.2 g
$MgHPO_4 \cdot 3H_2O$	0.2 g
$MgSO_4 \cdot 7H_2O$	0.1 g
$NaHCO_3$	0.075 g
Magnesium carbonate	0.075 g
Agar	20.0 g
Distilled water	1,000 ml

Adjust pH to 7.1 with 1 N HCl, autoclave, and cool. Add the following compounds aseptically to give the indicated final concentrations (in milligrams per liter).

Berberine	275
Sodium selenite	100
Penicillin G (1,625 U/mg)	60
Streptomycin sulfate	30
Cycloheximide	250
Tyrothricin	1
Bacitracin (65 U/mg)	100

11.6.2. *Agrobacterium* Biovar 2 Isolation Medium (149)

m-Erythritol	5.0 g
$NaNO_3$	2.5 g
K_2HPO_4	0.1 g
$CaCl_2$	0.2 g
NaCl	0.2 g
$MgSO_4 \cdot 7H_2O$	0.2 g
Fe-EDTA solution (0.65%, wt/vol)	2.0 ml
Biotin	2.0 µg
Agar	18.0 g
Distilled water	1,000 ml

Adjust pH to 7.0 with 1 N NaOH, autoclave, and cool. Add the following compounds aseptically to give the indicated final concentrations (in milligrams per liter).

Cycloheximide	250
Bacitracin	100
Tyrothricin	1
Sodium selenite	100

11.6.3. Alkaline Peptone Water for *Vibrio cholerae*

Peptone	10.0 g
NaCl	5.0 g
Distilled water	1,000 ml

Adjust pH to 8.4 with 1 N NaOH and sterilize by autoclaving at 121°C for 15 min.

11.6.4. *Aquaspirillum gracile* Isolation Medium (44)

Peptone	5.0 g
Yeast extract	0.5 g
Tween 80 (sorbitan monooleate polyoxyethylene)	0.02 g
K_2HPO_4	0.1. g
Agar	10.0 g
Distilled water	1,000 ml

Adjust to pH 7.2. Boil to dissolve the agar. Sterilize at 121°C for 15 min.

11.6.5. *Arthrobacter* Selective Agar (79)

Trypticase soy agar (BBL)	4.0 g
Yeast extract (Difco)	2.0 g
NaCl	20.0 g
Cycloheximide	0.1 g
Methyl red	160.0 mg
Distilled water	1,000 ml

Sterilize a stock solution of the methyl red, and add it aseptically to the rest of the autoclaved medium to give the indicated final concentration. Adjust the pH to that of the soil being sampled.

11.6.6. ASN-III Medium (165)

NaCl	25.0 g
$MgCl_2 \cdot 6H_2O$	2.0 g
KCl	0.5 g
$NaNO_3$	0.75 g
$K_2HPO_4 \cdot 3H_2O$	0.02 g
$MgSO_4 \cdot 7H_2O$	3.5 g
$CaCl_2 \cdot 2H_2O$	0.5 g
Citric acid	0.003 g
Ferric ammonium citrate	0.003 g
EDTA, disodium magnesium salt	0.0005 g
Na_2CO_3	0.02 g
Trace metal mix A5 (see BG-11 medium [section 11.6.12])	1.0 ml
Deionized water	1,000 ml

After autoclaving and cooling, the pH of the medium should be 7.5.

11.6.7. *Azotobacter* Medium

Mannitol	2.0 g
K_2HPO_4	0.5 g
$MgSO_4 \cdot 7H_2O$	0.2 g
$FeSO_4 \cdot 7H_2O$	0.1 g
Distilled water	1,000 ml

Adjust pH to 7.3 to 7.6. Sterilize at 121°C for 15 min.

11.6.8. Basal Enrichment Medium for Members of the *Rhodospirillaceae* (199)

Solution A

$NaHCO_3$	2.0 g
Distilled water	25.0 ml

Sterilize by filtration (positive pressure).

Solution B

NH,Cl	1.0 g
KH_2PO_4	0.5 g
$MgCl_2$	0.5 g
(NaCl for organisms from brackish or marine environments; 20–30 g)	
Trace-metal solution (see below)	1.0 ml
Distilled water	975 ml

Sterilize at 121°C for 15 min.

Solution C

$Na_2S \cdot 9H_2O$	3.0 g
Distilled water	200 ml

Sterilize in a flask with a Teflon-covered magnetized stirring bar at 121°C for 15 min. When cool, add 1.5 ml of sterile 2 M H_2SO_4 with stirring.

To prepare medium, combine solutions A and B. Adjust the pH to 7 with sterile Na_2CO_3 and H_3PO_4 as required. At the time of inoculation of primary enrichment cultures, add 1 ml of solution C per 100 ml of medium. For subsequent transfers, when members of the *Rhodospirillaceae* have established themselves, omit the Na_2S and substitute $MgSO_4$ for the $MgCl_2$ in the medium.

Trace element solution

EDTA, disodium salt	500 mg
$FeSO_4 \cdot 7H_2O$	200 mg
$ZnSO_4 \cdot 7H_2O$	10 mg
$MnCl_2 \cdot 4H_2O$	3 mg
H_3BO_3	30 mg
$CoCl_2 \cdot 6H_2O$	20 mg
$CuCl_2 \cdot 2H_2O$	1 mg
$NiCl_2 \cdot 6H_2O$	2 mg
$Na_2MoO_4 \cdot 2H_2O$	3 mg
Deionized water	1,000 ml

Dissolve the EDTA in a portion of the water. Separately dissolve the other ingredients in water and add them to the EDTA solution. Adjust the solution to ca. pH 3 and bring to a final volume of 1,000 ml.

11.6.9. Basal Inorganic Media A and B (50)

Medium A

Solution 1

$(NH_4)_2SO_4$	1.2 g
$CaCl_2 \cdot 2H_2O$	0.1 g
$MgSO_4 \cdot 7H_2O$	0.1 g

Ferric citrate . . . . 0.002 g
Distilled water . . . . 1,000 ml

Sterilize at 121°C for 15 min.

Solution 2

K_2HPO_4 . . . . 0.2. g
KH_2PO_4 . . . . 0.1 g
Distilled water . . . . 200 ml

Sterilize at 121°C for 15 min.

To prepare medium, combine solutions 1 and 2 aseptically. For a solid medium, use 20 g of agar per liter.

Medium B

K_2HPO_4 . . . . 0.8 g
KH_2PO_4 . . . . 0.2 g
$CaSO_4 \cdot 2H_2O$ . . . . 0.05 g
$MgSO_4 \cdot 7H_2O$ . . . . 0.5 g
$FeSO_4 \cdot 7H_2O$ . . . . 0.01 g
$(NH_4)_2SO_4$ . . . . 1.0 g
Distilled water . . . . 1,000 ml

Sterilize at 121°C for 15 min. For a solid medium, use 20 g of agar per liter.

11.6.10. Basal Salts Medium for *Thermus* Species
(32, 33)

Solution 1

Nitrilotriacetic acid . . . . 1.0 g
$CaSO_4 \cdot 2H_2O$ . . . . 0.6 g
$MgSO_4 \cdot 7H_2O$ . . . . 1.0 g
NaCl . . . . 0.08 g
KNO_3 . . . . 1.03 g
$NaNO_3$ . . . . 6.89 g
Na_2HPO_2 . . . . 1.11 g
Distilled water . . . . 1,000 ml

Solution 2

$FeCl_3$ solution . . . . 0.028 g
Distilled water . . . . 1,000 ml

Solution 3

$MnSO_4 \cdot H_2O$ . . . . 0.22 g
$ZnSO_4 \cdot 7H_2O$ . . . . 0.05 g
H_3BO_3 . . . . 0.05 g
$CaSO_4$ . . . . 0.0016 g
$Na_2MoO_4 \cdot 2H_2O$ . . . . 0.0025 g
$CoCl_2 \cdot 6H_2O$ . . . . 0.0046 g
Distilled water containing 0.5 ml of H_2SO_4
  per liter . . . . 1,000 ml

To prepare 1 liter of basal salts medium, combine 100 ml of solution A, 10 ml of solution B, and 10 ml of solution C, adjust the pH to 8.2, and bring to a final volume of 1,000 ml with distilled water. Sterilize by autoclaving. To make the complete medium for *Thermus* species, add 1.0 g of tryptone and 1.0 g of yeast extract per liter of basal salts medium before autoclaving. For a solid medium, add 15 g of agar per liter, boil to dissolve the agar, and then autoclave.

11.6.11. BCYE Agar for *Legionella* Species

Add 10.0 g of ACES buffer (*N*-2-acetamido-2-aminoethanesulfonic acid) to 500 ml of distilled water, and dissolve by heating in a water bath at 45 to 60°C. Mix this solution with 440 ml of distilled water to which 40 ml of 1.0 N KOH has been added (see note below). Add the following ingredients: activated charcoal (Norit SG: no. C5510; Sigma Chemical Co., St. Louis, MO), 2.0 g; yeast extract, 10.0 g; and agar, 17.0 g. Dissolve by boiling, autoclave at 121°C for 15 min, and cool to 50°C. Aseptically add L-cysteine · HCl · H_2O solution (0.4 g in 10 ml of distilled water, sterilized by filtration), followed by ferric PP_i solution (0.25 g of soluble ferric PP_i in 10 ml of distilled water, sterilized by filtration). The ferric PP_i must be kept dry and stored in the dark until used; it is not usable if the color changes from green to yellow or brown. Do not use heat over 60°C to dissolve the ferric PP_i; a 50°C water bath is satisfactory. Check the pH and adjust if necessary (see below). Dispense 20ml portions of the complete medium into petri dishes; swirl the medium between pouring plates to keep the charcoal particles suspended. The pH of the final solid medium should be 6.9 ± 0.05 at room temperature. The pH is critical since legionellas do not grow below pH 6.8 or above pH 7.0.

Note: When checking the pH, hold the bulk medium at 50°C while pouring one plate and checking its pH. When necessary, adjust the bulk medium with either 1.0 N KOH or 1.0 N HCl. Note that the pK_a of ACES buffer is influenced by temperature (0.02/°C); consequently, this must be considered with all pH determinations.

11.6.12. BG-11 Medium (165)

$NaNO_3$ . . . . 1.5 g
$K_2HPO_4 \cdot 3H_2O$ . . . . 0.04 g
$MgSO_4 \cdot 7H_2O$ . . . . 0.075 g
$CaCl_2 \cdot 2H_2O$ . . . . 0.036 g
Citric acid . . . . 0.006 g
Ferric ammonium citrate . . . . 0.006 g
EDTA, disodium magnesium salt . . . . 0.001 g
Na_2CO_3 . . . . 0.02 ml
Trace metal mix A5 (see below) . . . . 1.0 ml
Deionized water . . . . 1,000 ml

After autoclaving and cooling, the pH of the medium should be 7.4.

Trace metal mix A5

H_3BO_3 . . . . 2.86 mg/ml
$MnCl_2 \cdot 4H_2O$ . . . . 1.81 mg/ml
$ZnSO_4 \cdot 7H_2O$ . . . . 0.222 mg/ml
$Na_2MoO_4 \cdot 2H_2O$ . . . . 0.439 mg/ml
$CuSO_4 \cdot 5H_2O$ . . . . 0.079 mg/ml
$Co(NO_3)_2 \cdot 6H_2O$ . . . . 0.0494 mg/ml

11.6.13. Bile Esculin Agar (BBL or Difco)

Solution A

Beef extract . . . . 3.0 g
Peptone . . . . 5.0 g
Agar . . . . 15.0 g
Distilled water . . . . 400 ml

Solution B

```
Oxgall . . . . . . . . . . . . . . . . . . . . . . . . . 40.0 g
Distilled water . . . . . . . . . . . . . . . . . . . . 400 ml
```

Solution C

```
Ferric citrate . . . . . . . . . . . . . . . . . . . . . . . . 0.5 g
Distilled water . . . . . . . . . . . . . . . . . . . . . . 100 ml
```

Combine the three solutions, and heat to 100°C for 10 min. Sterilize at 121°C for 15 min. Cool to 50°C. Add aseptically 100 ml of a 1% solution of esculin (sterilized by filtration). Dispense into sterile tubes (for slants).

11.6.14. BMS Agar (Potato Agar) (61)

```
Washed, peeled, sliced potatoes . . . . . . . .    200 g
L-Malic acid . . . . . . . . . . . . . . . . . . . . . . .    2.5 g
KOH . . . . . . . . . . . . . . . . . . . . . . . . . . .    2.0 g
Raw cane sugar . . . . . . . . . . . . . . . . . . .    2.5 g
Vitamin solution (see section 11.6.45) . . . .    1.0 ml
Bromothymol blue (0.5% alcoholic
    solution) . . . . . . . . . . . . . . . . . . . . . .    2 drops
Agar . . . . . . . . . . . . . . . . . . . . . . . . . .    15.0 g
Distilled water . . . . . . . . . . . . . . . . . . . 1,000 ml
```

Place the potatoes in a gauze bag. Boil in 1,000 ml of water for 30 min; then, filter through cotton and save the filtrate. Dissolve the malic acid in 50 ml of water, and add 2 drops of bromothymol blue. Add KOH until the malic acid solution is green (pH 7.0). Add this solution together with the cane sugar, vitamins, and agar to the potato filtrate. Make up the volume to 1,000 ml with distilled water. Boil to dissolve the agar. Sterilize at 121°C for 15 min.

11.6.15. Bordet-Gengou Agar, Modified (116)

Place 125 g of washed, peeled, and sliced potatoes in a gauze bag. Submerge in a mixture of 10.0 ml of glycerol and 500 ml of water. Boil until the potatoes are soft; then, strain through the gauze into the water-glycerol mixture. Allow the fluid to stand in a tall cylinder until the supernatant is relatively clear. Decant the supernatant, and make up to 1,000 ml with distilled water. Add 5.6 g of NaCl. Heat. Add 22.5 g of agar, and dissolve by boiling with constant stirring. Sterilize at 121°C for 15 min, and cool to 45°C. Add 200 ml of defibrinated sheep blood (see chapter 15.4.9) aseptically, and mix. Dispense into petri dishes. (Bordet-Gengou Agar Base is available from BBL or Difco.)

11.6.16. Brilliant Green Agar (BBL or Difco)

```
Pancreatic digest of casein USP . . . . . . . . . .    5.0 g
Peptic digest of animal tissue USP . . . . . . . .    5.0 g
Yeast extract . . . . . . . . . . . . . . . . . . . . . .    3.0 g
NaCl . . . . . . . . . . . . . . . . . . . . . . . . . . .    5.0 g
Lactose . . . . . . . . . . . . . . . . . . . . . . . . .    10.0 g
Sucrose . . . . . . . . . . . . . . . . . . . . . . . . .    10.0 g
Phenol red . . . . . . . . . . . . . . . . . . . . . . .    0.08 g
Brilliant green . . . . . . . . . . . . . . . . . . . 0.0125 g
Agar . . . . . . . . . . . . . . . . . . . . . . . . . .    20.0 g
Distilled water . . . . . . . . . . . . . . . . . . . 1,000 ml
```

Adjust pH to 6.9. Boil to dissolve agar. Sterilize at 121°C for 15 min.

11.6.17. Brilliant Green Lactose Bile Broth (available from BBL or Difco as Brilliant Green Bile 2%)

Solution A

```
Peptone . . . . . . . . . . . . . . . . . . . . . . . . . . . 10.0 g
Lactose . . . . . . . . . . . . . . . . . . . . . . . . . . . 10.0 g
Distilled water . . . . . . . . . . . . . . . . . . . . .    500 ml
```

Solution B

```
Oxgall . . . . . . . . . . . . . . . . . . . . . . . . . . . 20.0 g
Distilled water . . . . . . . . . . . . . . . . . . . . .    200 ml
```

Mix solutions A and B. Make up to 975 ml with distilled water. Adjust pH to 7.4. Add 13.3 ml of a 0.1 aqueous solution of brilliant green. Make up the final volume to 1,000 ml with distilled water. Dispense into tubes containing inverted gas vials, and sterilize at 121°C for 15 min.

11.6.18. Brucella Agar (BBL or Difco)

```
Pancreatic digest of casein . . . . . . . . . . . . . . 10.0 g
Peptic digest of animal tissue . . . . . . . . . . . . 10.0 g
Glucose . . . . . . . . . . . . . . . . . . . . . . . . . . .  1.0 g
Yeast autolysate . . . . . . . . . . . . . . . . . . . . .  2.0 g
NaCl . . . . . . . . . . . . . . . . . . . . . . . . . . . .  5.0 g
Sodium bisulfite . . . . . . . . . . . . . . . . . . . . .  0.1 g
Agar . . . . . . . . . . . . . . . . . . . . . . . . . . . . 15.0 g
Distilled water . . . . . . . . . . . . . . . . . . . . . 1,000 ml
```

Adjust pH to 7.0. Boil to dissolve agar. Sterilize at 121°C for 15 min.

11.6.19. CAS Medium (205)

```
Yeast extract . . . . . . . . . . . . . . . . . . . . . .    1.0 g
Casamino Acids (Difco, not "vitamin
    free") . . . . . . . . . . . . . . . . . . . . . . . . .    7.5 g
Proteose Peptone no. 3 (Difco) . . . . . . . . . .    5.0 g
Sodium citrate . . . . . . . . . . . . . . . . . . . . .    3.0 g
MgSO₄ · 7H₂O . . . . . . . . . . . . . . . . . . . . .     20 g
Fe(NH₄)₂(SO₄)₂ . . . . . . . . . . . . . . . . . . .    5.0 mg
K₂HPO₄ . . . . . . . . . . . . . . . . . . . . . . . . .    7.5 g
NaCl . . . . . . . . . . . . . . . . . . . . . . . . . . .   80.0 g
Distilled water . . . . . . . . . . . . . . . . . . . . 1,000 ml
```

Adjust pH to 8.0 with NaOH. Store only in the dark; discard if crystals form.

11.6.20. Casein-Starch Agar (152)

```
Soluble starch . . . . . . . . . . . . . . . . . . . . . .     10 g
Casein . . . . . . . . . . . . . . . . . . . . . . . . . . .    1.0 g
K₂HPO₄ . . . . . . . . . . . . . . . . . . . . . . . . .    0.5 g
MgSO₄ . . . . . . . . . . . . . . . . . . . . . . . . . .    5.0 g
Agar . . . . . . . . . . . . . . . . . . . . . . . . . . . . 15.0 g
Distilled water . . . . . . . . . . . . . . . . . . . . . 1,000 ml
```

Adjust pH to 7.0 to 7.5. Boil to dissolve agar. Sterilize at 121°C for 15 min.

11.6.21. CHSS Medium for *Spirillum volutans*

```
Acid-hydrolyzed casein, vitamin free, salt free
    (ICN Nutritional Biochemicals,
    Cleveland, OH) . . . . . . . . . . . . . . . . . . .    2.5 g
```

Succinic acid (free acid) . . . . 1.0 g
(NH$_4$)$_2$SO$_4$ . . . . 1.0 g
MgSO$_4$ · 7H$_2$O . . . . 1.0 g
NaCl . . . . 0.1 g
FeCl$_3$ · 6H$_2$O . . . . 0.002 g
MnSO$_4$ · H$_2$O . . . . 0.002 g
Distilled water . . . . 1,000 ml

Adjust pH to 7.0 with 2 N KOH. For semisolid medium, add 1.5 g of agar and boil to dissolve the agar. Sterilize at 121°C for 20 min.

11.6.22. Cystine Tellurite Blood Agar (116)

Heart infusion agar (blood agar base; see
    chapter 15.4.35) . . . . 500 ml
Agar . . . . 2.5 g

Adjust pH to 7.4. Boil to dissolve agar. Sterilize at 121°C for 15 min. Cool to 56°C. Add the following ingredients aseptically.

Defibrinated rabbit or sheep blood (see
    chapter 15.4.9) . . . . 25 ml
0.3% potassium tellurite solution (sterilized
    by autoclaving) . . . . 75 ml
L-Cystine, powdered . . . . 22 mg

Stir while dispensing medium into petri dishes to keep the cystine suspended.

11.6.23. "Dehalococcoides ethenogenes" Medium

To enrich for PCE degraders in a methanol-utilizing anaerobic enrichment culture, PCE was fed every 2 days at increasing doses, from 3.5 μmol/liter of culture medium to 0.55 mmol (133). The pure culture isolation and propagation of the PCE-degrader "Dehalococcoides ethenogenes" requires, in addition to a basal mineral medium with different specific additions, anaerobic digester sludge supernatant and an extract from a PCE-butyrate-grown mixed culture. The basal mineral medium is purged with N$_2$, dispensed into 27-ml crimp-top tubes (Bellco, Vineland, NJ), sealed with aluminum-crimp Teflon-coated butyl rubber stoppers (Wheaton; Millville, NJ) and autoclaved at 121°C for 55 min.

Basal Medium

NH$_4$Cl . . . . 0.5 g
K$_2$HPO$_4$ . . . . 0.4 g
MgCl$_2$ · 6H$_2$O . . . . 0.1 g
CaCl$_2$ · 2H$_2$O . . . . 0.05 g
Resazurin . . . . 0.001 g
Trace element solution (217) . . . . 10 ml
Deionized H$_2$O . . . . 1,000 ml

The headspaces are flushed with sterile 70% N$_2$–30% CO$_2$, and receive the following additions before inoculation (final medium volume, ca. 10 ml per tube): 2 mM Na$_2$S · 9H$_2$O; 12 mM NaHCO$_3$; acetate (2 mM final concentration); 0.5% (vol/vol) vitamin B$_{12}$ solution (0.05 mg/liter); filter-sterilized anaerobic digester sludge supernatant (25%, vol/vol), filter-sterilized mixed culture extract (5%, vol/vol).

The size of the Dehalococcoides inoculum is 2% (vol/vol). Culture tubes are incubated upside down in an incubator-shaker at 150 rpm, with H$_2$ added to the headspace under overpressure (67 kPa, ca. 47.5 mmol/liter) immediately after inoculation. After several doses of PCE are consumed, NaHCO$_3$ is added to neutralize the HCl produced by the dechlorination process, and H$_2$ is added to replenish the headspace gas. The quantity of PCE added (0.3, 0.5, or 0.7 mmol/liter) is estimated directly from the syringe volume delivered because a few hours are needed for PCE to dissolve and to equilibrate with the headspace. See Maymó-Gatell et al. (133) for detailed GC quantification protocols of PCE, and dehalogenation products.

Anaerobic digestor sludge supernatant

Several liters of anaerobic sewage digester sludge are collected from a wastewater treatment facility, to ensure uniformity. The sludge is initially clarified by centrifugation for 30 min at 8,000 rpm (10,400 × g, in a Sorvall RC-2B centrifuge), and samples are frozen at −20°C until use. Thawed samples are then centrifuged for 30 min at 18,000 rpm (39,000 × g). The resulting liquid is neutralized to pH 7.0 with 1 M HCl and then prefiltered through a 0.8-μm-pore-size Supor-800 membrane filter (Gelman Sciences, Ann Arbor, MI). The sludge supernatant is then purged with N$_2$ and filter sterilized in an anaerobic chamber through a double 0.8- and 0.2-μm-pore-size Acrodisc PF filter (Gelman) into sterile vials. Alternatively, the sludge supernatant can be lyophilized (133).

Mixed culture extract

A mixed Dehalococcoides culture grown on butyrate-PCE is centrifuged for 20 min at 14,460 × g. The pellet is resuspended in distilled water so that the cells are concentrated 50-fold, and the suspension is frozen at −20°C until use. After the suspension is thawed at room temperature, it is passed through a French pressure cell at 20,000 lb/in^2 and centrifuged at 34,800 × g for 20 min. The pellet is discarded, and the supernatant is then purged with N$_2$ for 10 min, filter sterilized inside an anaerobic chamber by using a 25-mm-diameter Acrodisc combined filter (pore sizes, 0.8 and 0.2 μm; Gelman), and transferred into an autoclaved vial with a 70% N$_2$–30% CO$_2$ atmosphere. NaHCO$_3$ (final concentration, 1 liter) and Na$_2$S · 9H$_2$O (final concentration, 0.5 g/liter) are then added to buffer and reduce the extract before freezing and storage at −20°C until use. After thawing, the extract is kept at 4°C, and it is discarded after 4 weeks (133).

11.6.24. E Agar (116)

Papaic digest of soy meal USP . . . . 20.0 g
NaCl . . . . 5.0 g
Agar . . . . 10.0 g
Deionized water . . . . 1,000 ml

Heat to dissolve ingredients. Cool, and adjust pH to 7.4 with NaOH. Dispense, and sterilize at 121°C for 15 min. Cool to 50°C. To 65 ml of solution, add aseptically 10 ml of yeast dialysate (see below), 25 ml of horse serum, 2 ml of penicillin (10,000 U/ml), and 1 ml of 3.3% thallous acetate (sterilized by filtration). (CAUTION: thallium salts are poisonous!) Dispense 5-ml amounts into petri dishes (10 by 35 mm), and incubate overnight at room temperature before use.

Yeast Dialysate

Suspend 450 g of active dried yeast in 1,250 ml of distilled water at 40°C. Heat at 121°C for 5 min. Dialyze against

1 liter of distilled water at 4°C for 2 days. Discard the dialysis sac and its contents. Sterilize the dialysate at 121°C for 15 min. Store in a freezer.

Biphasic medium. For a biphasic medium, aseptically dispense 3-ml amounts of E agar into sterile screw-cap tubes (16 by 125 mm). After the medium solidifies, overlay it with 3 ml of E broth (section 11.6.24). Store at room temperature.

11.6.25. E Broth (116)

Papaic digest of soy meal USP	20.0 g
NaCl	5.0 g
Glucose	10.0 g
Phenol red, 2% aqueous solution	2.0 ml
Deionized water	1,000 ml

Adjust pH to 7.6. Dispense and sterilize at 121°C for 15 min. After the broth cools, add the same supplements as for E agar (section 11.6.23).

11.6.26. *Frateuria* Enrichment Medium (190)

Glucose	10.0 g
Ethanol	5.0 g
Yeast extract	5.0 g
Peptone	3.0 g
Acetic acid	0.3 ml
10% potato extract	1,000 ml

11.6.27. Furazolidone (FTO) Agar (56, 108, 204)

Peptone	10 g
Yeast extract	5.0 g
NaCl	5.0 g
Glucose	1.0 g
Agar	12 g
Distilled water	1,000 ml

Adjust pH to 7.0, boil to dissolve agar and autoclave at 121°C for 15 min. Cool to 48°C, and add 100 ml of a 0.02% acetone solution of furazolidone with slow stirring. Leave flask open or loosely covered in a water bath for 3 to 5 min to allow the acetone to evaporate.

11.6.28. Giesberger Base Medium (213)

NH_4Cl	1.0 g
K_2HPO_4	0.5 g
$MgSO_4$	0.5 g
Distilled water	1,000 ml

Adjust to pH 7. Sterilize at 121°C for 15 min. (*Note:* for marine organisms, substitute seawater for the distilled water.)

11.6.29. Glucose-Tryptone Agar (14)

Tryptone (Difco) or Trypticase (BBL)	10.0 g
Glucose	5.0 g
Agar	15.0 g
Bromocresol purple	0.04 g
Distilled water	1,000 ml

Adjust pH to 6.7. Boil to dissolve agar. Sterilize at 121°C for 15 min.

11.6.30. GYC Agar (190)

Glucose	50.0 g
Yeast extract	10.0 g
$CaCO_3$	25.0 g
Distilled water	1,000 ml

Boil to dissolve agar. Sterilize by autoclaving at 121°C for 15 min.

11.6.31. GYPT Medium (43)

Cellobiose	0.2 g
Trypticase (BBL)	0.1 g
Yeast extract	0.1 g
L-Cysteine	0.05 g
Rifampin	0.2 mg
Agar	1.0 g
Tris-HC1 buffer (1 M)	5.0 ml
Distilled water	20.0 ml
Seawater	75.0 ml

Replace the seawater with distilled water if attempting the isolation of freshwater strains. Sterilize the rifampin by filtration, and add aseptically to the autoclaved medium. Add 0.1 mg of resazurin to 100 ml of the medium if the medium is to be prereduced.

11.6.32. Halophile Medium (176)

Casamino Acids (Difco)	7.5 g
Yeast extract	10.0 g
Trisodium citrate	3.0 g
KCl	2.0 g
$MgSO_4 \cdot 7H_2O$	20.0 g
$FeCl_3$	0.023 g
NaCl	250 g
Distilled water	1,000 ml

Dissolve the solutes in 800 ml of the distilled water, and adjust the pH to 7.5 to 7.8 with 1 N KOH. Autoclave the medium at 120°C for 5 min, and filter to remove the precipitate. Adjust the pH to 7.4 with 1 N HCl, and make the medium up to 1,000 ml. Sterilize by autoclaving. For a solid medium, add 20 g of agar per liter, boil to dissolve the agar, and then autoclave. Dispense agar medium into petri dishes at 60 to 70°C to prevent premature solidification.

11.6.33. *Hyphomicrobium* Medium (20)

K_2HPO_4	1.74 g
$NaH_2PO_4 \cdot H_2O$	1.38 g
$(NH_4)_2SO_4$	0.5 g
$MgSO_4 \cdot 7H_2O$	0.2 g
$CaCl_2 \cdot 2H_2O$	0.025 mg
$FeCl_2 \cdot 4H_2O$	3.5 mg
Methanol	5.0 ml
KNO_3	5.0 g
Trace element solution (see below)	0.5 ml
Deionized water	1,000 ml

Adjust pH to 7.0 with NaOH. Do not sterilize when using the medium for enrichment. For sterile medium for purification, sterilize at 121°C for 15 min prior to adding the methanol. For solid medium to be used for isolation, add 15 g of agar per liter and omit the KNO_3; sterilize at 121°C for 15 min prior to adding the methanol. Remove the oxygen from all liquid media by bubbling nitrogen through them just before use; the KNO_3 in these media serves as the electron acceptor under anaerobic conditions.

Trace element solution

$ZnSO_4 \cdot 7H_2O$	50 mg
$MnCl_2 \cdot 4H_2O$	400 mg
$CoCl_2 \cdot 6H_2O$	1 mg
$CaSO_4 \cdot 5H_2O$	0.4 mg
H_3BO_3	2.0 mg
$Na_2MoO_4 \cdot 2H_2O$	500 mg
Distilled water	1,000 ml

11.6.34. Iverson Medium (94)

Trypticase soy agar (dehydrated; BBL)	40.0 g
Agar	5.0 g
Sodium lactate, 0.4% solution	600 ml
$MgSO_4 \cdot 7H_2O$	2.0 g
Ferrous ammonium sulfate	0.5 g
Distilled water	400 ml

Adjust pH to 7.2 to 7.4. Boil to dissolve agar. Sterilize at 121°C for 15 min.

11.6.35. Iron- and Manganese-Reducer Medium

Freshwater Medium

$NaHCO_3$	2.50 g
NH_4Cl	0.25 g
$NaH_2PO_4 \cdot H_2O$	0.60 g
KCl	0.10 g
Vitamin solution	10.0 ml
Mineral solution	10.0 ml
Deionized water	900 ml

Bring solution to a final volume of 1 liter. The medium is dispensed, sparged with an 80:20 mixture of $N_2:CO_2$ gas, and then autoclaved.

Marine Medium

NaCl	20.0 g
KCl	0.67 g
$NaHCO_3$	2.50 g
Vitamin solution	10.0 ml
Mineral solution	10.0 ml
RST minerals stock (50×)	20.0 ml
Salt stock*	50.0 ml
Deionized water	900 ml

*Add salt solution aseptically and anaerobically after autoclaving.

RST Minerals Stock 50×

NH_4Cl	5.0 g
KCl	0.5 g
KH_2PO_4	0.5 g
$MgSO_4 \cdot 7H_2O$	1.0 g
$CaCl_2 \cdot 2H_2O$	0.1 g
Deionized water	100 ml

Salt Stock

$MgCl_2 \cdot 6H_2O$	21.2 g
$CaCl_2 \cdot 2H_2O$	3.04 g
Deionized water	100 ml

Vitamin Solution

Biotin	2.0 mg
Folic acid	2.0 mg
Pyridoxine HCl	10.0 mg
Riboflavin	5.0 mg
Thiamine	5.0 mg
Nicotinic acid	5.0 mg
Pantothenic acid	5.0 mg
B_{12}	0.1 mg
p-Aminobenzoic acid	5.0 mg
Thioctic acid	5.0 mg
Distilled water	1,000 ml

Mineral Solution

Trisodium nitrilotriacetic acid	1.5 g
$MgSO_4$	3.0 g
$MnSO_4 \cdot H_2O$	0.5 g
NaCl	1.0 g
$FeSO_4 \cdot 7H_2O$	0.1 g
$CaCl_2 \cdot 2H_2O$	0.1 g
$CoCl_2 \cdot 6H_2O$	0.1 g
$ZnCl_2$	0.13 g
$CuSO_4 \cdot 5H_2O$	0.01 g
$AlK(SO_4)2 \cdot 12H_2O$	0.01 g
H_3BO_3	0.01 g
Na_2MoO_4	0.025 g
$NiCl_2 \cdot 6H_2O$	0.024 g
$Na_2WO_4 \cdot 2H_2O$	0.025 g
Distilled water	1,000 ml

Preparation of Fe(III) and Mn(IV) Forms

Poorly crystalline Fe(III) oxide

Dissolve $FeCl_3 \cdot 6H_2O$ in water to provide a final concentration of 0.4 M. Stir continually while slowly adjusting the pH to 7.0 dropwise with 10 M NaOH. It is extremely important not to let the pH rise above pH 7 even momentarily during the neutralization step because this will result in an Fe(III) oxide that is much less available for microbial reduction. Continue to stir for 30 min once pH 7 is reached and recheck the pH to be sure it has stabilized at pH 7. To remove dissolved chloride, centrifuge the suspension at 5,000 rpm for 15 minutes. Discard the supernatant, resuspend the Fe(III) oxide in water, and centrifuge. Repeat six times. On the last wash, resuspend the Fe(III) oxide to a final volume of approximately 400 ml, and after determining iron content, adjust Fe(III) concentration to approximately 1 mol per liter. Typically, Fe(III) oxide is added to individual tubes of media to provide 100 mmol per liter.

Fe(III)-citrate

Prior to the addition of any of the media constituents, heat 800 ml of water on a stirring hot plate to near boiling. Add Fe(III)-citrate [typically 13.7 g to provide a final concentration of ca. 50 mM Fe(III)]. Once the ferric citrate is dissolved, quickly cool the medium to room temperature in an ice bath. Adjust pH to 6.0 using 10 N NaOH. When the pH approaches 5.0, add the NaOH dropwise. Add medium constituents as outlined above. Bring to a final volume of 1 liter. Do not expose this medium to direct sunlight to prevent photoreduction of the Fe(III).

Fe(III) nitrilotriacetic acid

To make a stock of 100 mM Fe(III)-NTA, dissolve 1.64 g of $NaHCO_3$ in 80 ml of water. Add 2.56 g of $C_6H_6NO_6Na_3$ (sodium nitrilotriacetic acid) and then 2.7 g of $FeCl_3 \cdot 6H_2O$. Bring the solution up to 100 ml. Sparge the solution with N_2 gas and filter sterilize into a sterile, anaerobic serum bottle. Do not autoclave. Typically, 100 mM Fe(III)-NTA stock is added to individual tubes of media to provide a final concentration of 5 or 10 mmol of Fe(III).

Goethite

Prepare a 0.4 M $FeCl_3 \cdot 6H_2O$ solution. With continual stirring, adjust the pH to between 11 and 12 with 10 M NaOH solution. The suspension will become very thick. Ensure continual stirring and rinse the pH electrode frequently. The color of this suspension will turn to ochre as goethite is formed. One week at room temperature followed by 16 h at 90°C is sufficient to convert the Fe(III) to goethite. The suspension should be washed to remove chloride, as described above for poorly crystalline Fe(III) oxide. The formation of goethite should be confirmed by X-ray diffraction analysis. The Fe(III) oxide also should be tested with extractants (120, 157) to ensure that it does not contain poorly crystalline Fe(III) oxide.

Hematite

Hematite is readily available from chemical supply companies as "Ferric Oxide."

Manganese oxide

To 1 liter of a solution containing 80 mM NaOH and 20 mM $KMnO_4$ slowly add 1 liter of 30 mM $MnCl_2$ with mixing. Wash the manganese oxide precipitate, as described above for poorly crystalline Fe(III) oxide, to lower the dissolved chloride concentration.

11.6.36. Lactate Medium

Yeast extract	1.0 g
Sodium lactate	4.0 g
NH_4Cl	0.5 g
K_2HPO_4	1.0 g
$MgSO_4 \cdot 7H_2O$	0.2 g
$CaCl_2 \cdot 2H_2O$	0.1 g
$FeSO_4 \cdot 7H_2O$	0.1 g
$NaSO_4$	0.5 g
(NaCl, for marine organisms)	(20–30 g)
Distilled water	1,000 ml

Sterilize at 121°C for 15 min.

11.6.37. Lowenstein-Jensen Medium (BBL or Difco)

K_2HPO_4	2.4 g
$MgSO_4 \cdot 7H_2O$	0.24 g
Magnesium citrate	0.60 g
Asparagine	3.6 g
Potato flour	30.0 g
Glycerol	12.0 ml
Distilled water	600 ml
Homogenized whole eggs	1,000 ml
Malachite green, 2% aqueous solution	200 ml

Dissolve the salts and asparagine in the water. Add the glycerol and potato flour, and autoclave at 121°C for 30 min. Cleanse whole eggs, not more than 1 week old, by scrubbing with 5% soap solution. Allow to stand for 30 min in soap solution, and rinse thoroughly in cold running water. Immerse the eggs in 70% ethanol for 15 min, remove, and break into a sterile flask. Homogenize by shaking with sterile glass beads. Filter through four layers of sterile gauze. Add 1 liter of the homogenized eggs to the flask of cooled potato-salt mixture. Add the malachite green. Mix well, and dispense into sterile, screw-cap tubes (20 by 150 mm) 6 to 8 ml per tube. Slant the tubes, and inspissate them at 85°C for 50 min. Incubate for 48 h at 37°C to check sterility. Store the tubes in a refrigerator with their caps tightly sealed.

11.6.38. MacConkey Agar (BBL or Difco)

Peptone (Difco) or Gelysate (BBL)	17.0 g
Proteose Peptone (Difco) or Polypeptone (BBL)	3.0 g
Lactose	10.0 g
NaCl	5.0 g
Crystal violet	0.001 g
Agar	13.5 g
Distilled water	1,000 ml

Adjust pH to 7.1. Boil to dissolve agar. Sterilize at 121°C for 15 min.

11.6.39. Mannitol Salt Agar (BBL or Difco)

Beef extract	1.0 g
Proteose Peptone no. 3 (Difco) or Polypeptone (BBL)	10.0 g
NaCl	75.0 g
Mannitol	10.0 g
Phenol red	0.025 g
Agar	15.0 g
Distilled water	1,000 ml

Adjust pH to 7.4. Boil to dissolve agar. Sterilize at 121°C for 15 min. Cool to 55°C (45°C may not prevent premature solidification), and pour into petri dishes.

11.6.40. Marine Basal J3 Medium for Lithoautotrophic Marine *Beggiatoa* spp.

Solution 1

Aged natural seawater (salinity 3.2–3.5%), 500 ml prefiltered (Whatman #1 or Gelman GF/F) and filtered (0.45 μm)

Solution 2

Distilled water	200 ml
Agar	9.0 g

Solution 3

NH_4NO_3	0.06 g
Trace elements (SL8) (156)	0.75 ml
Mineral stock	50 ml

Mineral stock

K_2HPO_4	0.52 g
Na_2MoO_4	0.05 g

FeCl$_3$ · 6H$_2$O .... 0.29 g
Na$_2$S$_2$O$_5$ (sodium pyrosulfite) .... 0.75 g
Phenol red .... 10 ml
                    of a sterile solution (0.5%, Gibco)
Distilled water .... 1,000 ml

Autoclave solutions 1, 2, and 3 separately in Erlenmeyer flasks. After cooling to 50°C, aseptically combine in the solution 2 vessel (volume, >750 ml). This procedure ensures a marine basal medium (J3) free of precipitates. Then, supplement with 0.2 ml of Va vitamin solution.

Va vitamin solution

B$_{12}$ .... 1 mg
Thiamine .... 200 mg
Biotin .... 1 mg
Folic acid .... 1 mg
para-Aminobenzoic acid .... 10 mg
Nicotinic acid .... 100 mg
Inositol .... 1 mg
Calcium pantothenate .... 100 mg
Distilled water .... 1,000 ml

11.6.41. Marine Medium for Luminescent Bacteria (39)

Yeast extract .... 5.0 g
Tryptone .... 10 g
Glycerol .... 3 ml
NaCl .... 20 g
MgSO$_4$ .... 1 g
Tris .... 6 g
Distilled water .... 1,000 ml

Adjust pH to 7.5. Sterilize at 121°C for 15 min. Add 20 g of agar per liter for a solid medium.

11.6.42. Mineral Agar (16)

Na$_2$HPO$_4$ · 12H$_2$O .... 9.0 g
KH$_2$PO$_4$ .... 1.5 g
MgSO$_4$ · 7H$_2$O .... 0.2 g
NH$_4$Cl .... 1.0 g
Ferric ammonium citrate .... 0.005 g
CaCl$_2$ · 2H$_2$O .... 0.010 g
Trace elements solution (see below) .... 3.0 ml
Double-distilled water .... 1,000 ml

Adjust pH to 7.1. Sterilize at 121°C for 15 min. (For a solid medium, add 17 g of agar per liter, boil to dissolve the agar, and sterilize by autoclaving.) After the medium has cooled to 45 to 56°C, add sufficient NaHCO$_3$ solution (sterilized by filtration with positive pressure) to give a final concentration of 0.5 g of NaHCO$_3$ per liter.

Trace element solution

ZnSO$_4$ · 7H$_2$O .... 10 mg
MnCl$_2$ · 4H$_2$O .... 3 mg
H$_3$BO$_3$ .... 30 mg
CoCl$_2$ · 6H$_2$O .... 20 mg
CuCl$_2$ · 6H$_2$O .... 0.79 mg
NiCl$_2$ · 6H$_2$O .... 2 mg
Na$_2$MoO$_4$ · 2H$_2$O .... 3 mg
Double-distilled water .... 1,000 ml

11.6.43. Mitis-Salivarius Agar (46) (Difco)

Tryptone (Difco) .... 10.0 g
Proteose Peptone no. 3 (Difco) .... 5.0 g
Proteose Peptone (Difco) .... 5.0 g
Glucose .... 1.0 g
Sucrose .... 50.0 g
K$_2$HPO$_4$ .... 4.0 g
Trypan blue .... 0.075 g
Crystal violet .... 0.0008 g
Agar .... 15.0 g
Distilled water .... 1,000 ml

Adjust pH to 7.0. Boil to dissolve agar. Sterilize at 121°C for 15 min. Cool to 50 to 55°C. Add 1.0 ml of 0.1 potassium tellurite (sterilized by filtration), mix, and dispense into petri dishes.

11.6.44. MN Medium (165)

NaNO$_3$ .... 0.75 g
K$_2$HPO$_4$ · 3H$_2$O .... 0.02 g
MgSO$_4$ · 7H$_2$O .... 0.038 g
CaCl$_2$ · 2H$_2$O .... 0.018 g
Citric acid .... 0.003 g
Ferric ammonium citrate .... 0.003 g
EDTA, disodium magnesium salt .... 0.0005 g
Na$_2$CO$_3$ .... 0.02 g
Trace metal mix A5 (see BG-11 medium;
    section 11.5.12) .... 1.0 ml
Seawater .... 750 ml
Deionized water .... 250 ml

After autoclaving and cooling, the pH of the medium should be 8.3.

11.6.45. Modified Campy BAP (31)

Prepare brucella agar (section 11.6.18), sterilize by autoclaving, and cool to 50°C. Aseptically add defibrinated whole blood (10%, vol/vol; see chapter 15.4.9). Add the following antimicrobial agents from filter-sterilized stock solutions to give the indicated final concentrations.

Cephalothin .... 15 μg/ml
Polymyxin B .... 2.5 IU/ml
Trimethoprim .... 5 μg/ml
Vancomycin .... 10 μg/ml
Amphotericin B .... 2 μg/ml

11.6.46. Modified Rogosa Medium (126)

Solution A

MgSO$_4$ · 7H$_2$O .... 11.5 g
MnSO$_4$ · 4H$_2$O .... 2.8 g
FeSO$_4$ · 4H$_2$O .... 0.08 g
Distilled water .... 100 ml

Solution B

Yeast extract .... 6.0 g
Diammonium hydrogen citrate .... 2.4 g
KH$_2$PO$_4$ .... 7.2 g
Glucose .... 24.0 g
Tween 80 .... 1.2 g
Distilled water .... 100 ml

Solution C. Add 6.0 ml of solution A to 100 ml of solution B. Heat gently until the ingredients are dissolved. Then, add 60 ml of 4 M sodium acetate-acetic acid buffer (pH 5.37), and make up to a final volume of 200 ml with distilled water. The final pH should be 5.0.

Solution D

Separated raw milk (adjusted to pH 8.5)	1,000 ml
Trypsin	5 g
Chloroform	10 ml

Incubate at 37°C for 24 h, steam for 20 min, filter while hot, and adjust the pH to 6.65 with glacial acetic acid (ca. 0.5 ml per liter).

To prepare complete medium, add 19 g of agar to 700 ml of solution D, and dissolve by autoclaving at 121°C for 20 min. While the mixture is hot, add 185 ml of solution C (previously warmed to 50°C). Make up to 1,000 ml with hot solution D. A sample diluted with 3 parts of warm water should have a pH of 5.35 ± 0.05 at 30°C. Dispense the medium in 10-ml quantities into tubes, and store in a refrigerator without sterilizing. When preparing plates, melt the medium with as little heating as possible to avoid darkening and the formation of a precipitate.

11.6.47. Modified Wolfe Mineral Medium (82)

NH_4Cl	1.0 g
$MgSO_4 \cdot H_2O$	0.2 g
$CaCl_2 \cdot 2H_2O$	0.1 g
$KH_2PO_4 \cdot 3H_2O$	0.05 g
	(1/10 of original concentration)
Distilled water (see note)	1,000 ml

The low phosphate concentration allows autoclaving of the medium without precipitation of earth alkali phosphates, and prevents iron phosphate precipitation in culture medium for Fe oxidizers.

11.6.48. MPSS Agar

Peptone (Difco)	5.0 g
Succinic acid (free acid)	1.0 g
$(NH_4)_2SO_4$	1.0 g
$FeCl_2 \cdot 6H_2O$	0.002 g
$MnSO_4 \cdot H_2O$	0.002 g
Agar	15.0 g
Distilled water (see note)	1,000 ml

Adjust to pH 7.0 with KOH. Boil to dissolve agar. Sterilize at 121°C for 15 min. (*Note*: for marine organisms, substitute seawater for the distilled water.)

11.6.49. *N*-Acetyl-L-Cysteine-Sodium Hydroxide Reagent (70)

Combine equal volumes of 4% NaOH and 2.94% sodium citrate · 3H$_2$O. Dissolve 0.5% *N*-acetyl-L-cysteine powder in this solution. Use within 24 h. *Note*: it is best to use distilled water for preparation of the solutions to minimize chances of adding acid-fast tap water contaminants to the specimens.

11.6.50. *Nevskia* Medium (188)

Sodium lactate solution (50%)	5.5 g
K_2HPO_4	5.0 g

$MgSO_4 \cdot 7H_2O$	0.05 g
$CaCl_2$	0.015 g
Trace element solution (Widdel and Pfennig Medium [11.6.79])	1 ml
Vitamin solution (Widdel and Pfennig Medium [11.6.79])	1 ml

Adjust pH to 7.0. Sterilize at 121°C for 15 min. Add 20 g of agar per liter for a solid medium. To obtain submersed instead of surface-film growth, add 0.5 g of NH$_4$Cl per liter.

11.6.51. Nfb Medium (61)

L-Malic acid	5.0 g
K_2HPO_4	0.5 g
$MgSO_4 \cdot 7H_2O$	0.2 g
NaCl	0.1 g
$CaCl_2$	0.02 g
Trace metal solution (see below)	2.0 ml
Bromothymol blue (5% alcoholic solution)	2.0 ml
Fe EDTA (1.64% solution)	4.0 ml
Vitamin solution (see below)	1.0 ml
KOH	4.0 g
Agar	1.75 g
Distilled water	1,000 ml

Adjust pH to 6.8. Boil to dissolve agar. Sterilize at 121°C for 15 min. Cool to 45 to 50°C, and dispense 4-ml amounts into 6-ml serum vials with rubber diaphragms.

Trace metal solution

$Na_2MoO_4 \cdot 2H_2O$	0.2 g
$MnSO_4 \cdot H_2O$	0.235 g
H_3BO_3	0.28 g
$CuSO_4 \cdot 5H_2O$	0.008 g
$ZnSO_4 \cdot 7H_2O$	0.024 g
Distilled water	200 ml

Vitamin solution

Biotin	10 mg
Pyridoxine · HCl	20 mg
Distilled water	100 ml

Heating to nearly boiling is required to dissolve the biotin.

11.6.52. Nitrification Medium (160)

Na_2HPO_4	13.5 g
$MgSO_4 \cdot 7H_2O$	0.1 g
$FeCl_3 \cdot 6H_2O$	0.014 g
$CaCl_2 \cdot 2H_2O$	0.18 g
Distilled water	1,000 ml

Place 75-ml amounts of medium into 250-ml Erlenmeyer flasks, and sterilize at 121°C for 15 min. For nitrite oxidizers, add 0.5 g of NaNO$_2$ per liter prior to sterilization. For ammonium oxidizers, sterilize a stock solution of $(NH_4)_2SO_4$ separately from the basal medium and add aseptically to give a final concentration of 0.5 g/liter.

11.6.53. Pfennig and Biebl Sulfur Medium (155)

KH_2PO_4	1.0 g
NH_4Cl	0.3 g
$MgSO_4 \cdot 7H_2O$	1.0 g

MgCl$_2$ · 6H$_2$O .... 2.0 g
(NaCl, for marine organisms) .... (20 g)
CaCl$_2$ · 2H$_2$O .... 0.1 g
Trace element solution (see below) .... 10 ml
2 M H$_2$SO$_4$ solution .... 2.0 ml
Distilled water .... 1,000 ml

Sterilize by autoclaving. When cool, add the following components from sterile stock solutions.

Sodium acetate .... 0.5 g
NaHCO$_3$ (sterilized by filtration under
  positive pressure) .... 4.0 g
Na$_2$S · 9H$_2$O .... 0.3 g
Biotin .... 20 µg

Adjust the pH to 7.8. Grind highly purified flowers of sulfur in a mortar together with distilled water, and sterilize at 112 to 115°C for 30 min; decant the excess water. For every 50 ml of medium, add a pea-sized amount of the sulfur. Also add several glass beads to each bottle of medium.

Trace element solution

EDTA disodium salt .... 500 mg
FeSO$_4$ · 7H$_2$O .... 200 mg
ZnSO$_4$ · 7H$_2$O .... 10 mg
MnCl$_2$ · 4H$_2$O .... 3 mg
H$_3$BO$_3$ .... 30 mg
CoCl$_2$ · 6H$_2$O .... 20 mg
CuCl$_2$ · 2H$_2$O .... 1 mg
NiCl$_2$ · 6H$_2$O .... 2 mg
Na$_2$MoO$_4$ · 2H$_2$O .... 3 mg
Deionized water .... 1,000 ml

11.6.54. Phenylethyl Alcohol Agar (BBL or Difco)

Pancreatic digest of casein USP .... 15.0 g
Papaic digest of soya meal USP .... 5.0 g
NaCl .... 5.0 g
Phenylethyl alcohol .... 2.5 g
Agar .... 15.0 g
Distilled water .... 1,000 ml

Adjust pH to 7.3. Boil to dissolve the agar, and sterilize at 118°C for 15 min.

11.6.55. Pringsheim Soil Medium (166)
Place one wheat or barley grain in a large test tube, and cover with 3 to 4 cm of garden soil. Fill the tube almost to the top with tap water. Sterilize the medium at 121°C for 30 min.

11.6.56. *Pseudomonas* Isolation Medium (77)
Trypticase soy agar (BBL) is supplemented with the following (per liter).

Basic fuchsin .... 9.0 mg
Cycloheximide .... 0.9 g
2,3,5-Triphenyl tetrazolium chloride .... 0.14 g
Nitrofurantoin (see below) .... 10.0 mg
Nalidixic acid (see below) .... 23.0 mg

Add appropriate amounts of stock solutions of nitrofurantoin and nalidixic acid, sterilized by filtration, to the autoclaved medium to give the indicated final concentrations.

11.6.57. PTN Medium for *Listeria* Species (208)

Potassium thiocyanate .... 3.75 g
Nutrient broth, hydrated
  (Oxoid no. 2) .... q.s. to 1,000.00 ml
Nalidixic acid solution .... 19.0 ml

Autoclave the nutrient broth containing the potassium thiocyanate at 121°C for 15 min. Cool, and add the nalidixic acid solution aseptically to the sterile medium.

Nalidixic acid solution. Dissolve 0.1 g of crystalline nalidixic acid in 10 ml of 1 N NaOH; when dissolved, add 9 ml of sterile distilled water.

11.6.58. RGCA-SC Medium (86)

Glucose .... 0.0248 g
Cellobiose .... 0.0248 g
Soluble starch .... 0.05 g
(NH$_4$)$_2$SO$_4$ .... 0.1 g
Resazurin solution (0.025%) .... 0.4 ml
Distilled water .... 20 ml
Salts solution (chapter 15.4.31) .... 50 ml
Rumen fluid (Randolph Biologicals,
  Houston, TX) .... 30 ml
Cysteine hydrochloride .... 0.05 g
Hemin solution (chapter 15.4.31) .... 1.0 ml
Vitamin K$_1$ solution (chapter 15.4.31) ..... 0.02 ml

Prepare in a manner similar to that described for PY broth (chapter 15.4.31). Dispense 10-ml amounts into tubes containing 0.2 g of agar, and autoclave. To prepare the final medium, melt the contents of two tubes and cool to 50°C. To each tube, aseptically add the following.

Sterile inactivated rabbit serum (heated at
  60°C for 4 h) .... 1.5 ml
Cocarboxylase solution, 0.025%
  (sterilized by filtration) .... 0.2 ml
Sterile prereduced PY broth
  (chapter 15.4.31) .... 1.0 ml

Mix, and pour the contents of both tubes into a petri dish. Allow to solidify. Use immediately, or store in an anaerobic jar until needed.

11.6.59. *Salmonella-Shigella* (SS) Agar
(BBL or Difco)

Beef extract .... 5.0 g
Peptone .... 5.0 g
Lactose .... 10.0 g
Bile salts mixture .... 8.5 g
Sodium citrate .... 8.5 g
Sodium thiosulfate .... 8.5 g
Ferric citrate .... 1.0 g
Brilliant green .... 0.33 g
Neutral red .... 0.025 g
Agar .... 13.5 g
Distilled water .... 1,000 ml

Adjust pH to 7.0. Heat to boiling to dissolve agar. Do not sterilize by autoclaving. Cool to 42 to 45°C and dispense into petri dishes.

11.6.60. Selenite F Broth (115) (BBL or Difco)

Polypeptone (BBL) or tryptone	5.0 g
Lactose	4.0 g
Na$_2$HPO$_4$	10.0 g
Sodium hydrogen selenite	4.0 g
Distilled water	1,000 ml

Adjust pH to 7.0. Use immediately without sterilization, or place tubes in flowing steam for 30 min and store until needed.

11.6.61. Serpens Isolation Medium (85)

Yeast extract	0.2 g
Peptone	0.1 g
Hay extract (see below)	10.0 ml
Agar	1.0 g
Distilled water	90 ml

Prepare hay extract by boiling hay in 100 ml of water for 15 min and clarifying the mixture by centrifugation. Adjust pH to 7.0 with KOH before sterilization.

11.6.62. Soybean-Casein Digest Broth, also known as Trypticase Soy Broth (BBL) or Tryptic Soy Broth (Difco)

Pancreatic digest of casein USP	17.0 g
Papaic digest of soy meal USP	3.0 g
NaCl	5.0 g
K$_2$HPO$_4$	2.5 g
Glucose	2.5 g
Distilled water	1,000 ml

Adjust pH to 7.3. Sterilize at 118 to 121°C for 15 min.

11.6.63. Sphaerotilus Medium (17)

Sodium lactate	100 mg
NH$_4$Cl	1.7 mg
KH$_2$PO$_4$	8.5 mg
K$_2$HPO$_4$	21.5 mg
Na$_2$HPO$_4$ · 7H$_2$O	34.4 mg
MgSO$_4$ · 7H$_2$O	22.5 mg
CaCl$_2$	27.5 mg
FeCl$_3$ · 6H$_2$O	0.25 mg
Distilled water	1,000 ml

Adjust pH, if necessary, to 7.1 to 7.2. Dispense 50-ml volumes into French square bottles. Sterilize at 116°C for 15 min.

11.6.64. Sporocytophaga Medium

Whatman Chromedia 11	10 g
KNO$_3$	0.5 g
MgSO$_4$ · 7H$_2$O	0.2 g
CaCl$_2$ · 2H$_2$O	0.1 g
FeCl$_3$	0.02 g
Distilled water	1,000 ml

11.6.65. ST6CX Medium for Cellulolytic Cytophagas (163)

(NH$_4$)$_2$SO$_4$	1.0 g
MgSO$_4$ · 7H$_2$O	1.0 g
CaCl$_2$ · 2H$_2$O	1.0 g

MnSO$_4$ · H$_2$O	0.1 g
FeCl$_3$ · 6H$_2$O	0.2 g
Yeast extract (Difco)	0.02 g
Agar	10.0 g
Distilled water	1,000 ml

Boil to dissolve agar. Sterilize at 121°C for 15 min. From a separately autoclaved stock solution, add sufficient K$_2$HPO$_4$ to give a final concentration of 0.1%. From a filter-sterilized stock solution, add sufficient cycloheximide to give a final concentration of 25 μg/ml. Also add 1.0 ml of the following filter-sterilized trace element solution per liter of medium.

Trace element solution

MnCl$_2$ · 4H$_2$O	100 mg
CoCl$_2$	20 mg
CuSO$_4$	10 mg
Na$_2$MoO$_4$ · 2H$_2$O	10 mg
ZnCl$_2$	20 mg
LiCl$_2$	5 mg
SnCl$_2$ · 2H$_2$O	5 mg
H$_3$BO$_3$	10 mg
KBr	20 mg
KI	20 mg
20 mg EDTA, Na-Fe^{3+} salt (trihydrate)	8 g

11.6.66. Standard Methods Agar (14) (BBL or Difco)

Pancreatic digest of casein USP	5.0 g
Yeast extract	2.5 g
Glucose	1.0 g
Agar	15.0 g
Distilled water	1,000 ml

Adjust pH to 7.0. Boil to dissolve agar. Sterilize at 121°C for 15 min.

11.6.67. Succinate-Salts Medium

Sodium succinate	4.0 g
KNO$_3$	0.5 g
K$_2$HPO$_4$	0.5 g
MgSO$_4$ · 7H$_2$O	0.2 g
CaCl$_2$ · 2H$_2$O	0.1 g
FeSO$_4$ · 7H$_2$O	0.2 g

Adjust to pH 7.0. Sterilize at 121°C for 15 min.

11.6.68. TCBS Agar (7) (BBL)

Sodium thiosulfate	10.0 g
Sodium citrate	10.0 g
Oxgall	5.0 g
Sodium cholate	3.0 g
Sucrose	20.0 g
Pancreatic digest of casein USP	5.0 g
Peptic digest of animal tissue USP	5.0 g
Yeast extract	5.0 g
NaCl	10.0 g
Iron citrate	1.0 g
Thymol blue	0.04 g
Bromothymol blue	0.04 g
Agar	14.0 g
Distilled water	1,000 ml

The final pH should be 8.6. Heat with agitation, and boil for 1 min. Cool to 45 to 50°C, and dispense into petri dishes. Do not autoclave.

11.6.69. TGYM Medium (142)

Tryptone	5.0 g
Yeast extract	3.0 g
Glucose	1.0 g
DL-Methionine	1.0 g
Agar	15.0 g
Distilled water	1,000 ml

Adjust pH to 7.0. Boil to dissolve agar. Sterilize by autoclaving at 121°C for 15 min.

11.6.70. Thallous Acetate Agar

Thallous acetate (CAUTION: thallium salts are poisonous!)	1.0 g
Peptone	10.0 g
Yeast extract	10.0 g
Glucose	10.0 g
Agar	13.0 g
Distilled water	1,000 ml

Adjust pH to 6.0. Boil to dissolve agar, and sterilize at 118°C for 15 min. Cool to 45 to 50° C, and add aseptically 10 ml of triphenyl tetrazolium chloride (1% aqueous solution, sterilized by filtration). Dispense into petri dishes.

11.6.71. Thayer-Martin Agar

This medium is made most conveniently by using commercial concentrates and solutions or by using prepared media (BBL or Difco).

Solution A

Pancreatic digest of casein USP	15.0 g
Peptic digest of animal tissue USP	15.0 g
Cornstarch	2.0 g
K₂HPO₄	8.0 g
KH₂PO₄	2.0 g
NaCl	10.0 g
Agar (see note)	20.0 g
Distilled water	1,000 ml

Boil to dissolve agar. Sterilize at 121°C for 15 min. Cool to 50°C. Note: Finegold et al. (44) recommend increasing the agar to 30 g.

Solution B

Hemoglobin, dry	20.0 g
Distilled water	1,000 ml

Add dry power gradually to a little water to make a smooth paste; then, gradually add the rest of the water. Sterilize at 121°C for 15 min.

Solution C

Vitamin B₁₂	0.010 g
L-Glutamine	10.0 g
Adenine	1.0 g
Guanine hydrochloride	0.03 g
p-Aminobenzoic acid	0.013 g
L-Cystine	1.10 g
Glucose	100.0 g
NAD	0.250 g
Thiamine PPᵢ (cocarboxylase)	0.100 g
Ferric nitrate	0.020 g
Thiamine hydrochloride	0.003 g
Cysteine hydrochloride	25.9 g
Distilled water	1,000 ml

Sterilize by filtration.

Solution D

Vancomycin	30 mg
Colistin	75 mg
Nystatin	125,000 U
Distilled water	100 ml

Aseptically combine 1,000 ml of solution A, 1,000 ml of solution B, 20 ml of solution C, and 20 ml of solution D. Dispense into petri dishes.

11.6.72. Thermoplasma Isolation Medium (58)

KH₂PO₄	3.0 g
MgSO₄ · 7H₂O	1.02 g
CaCl₂ · H₂O	0.25 g
(NH₄)₂SO₄	0.2 g
Yeast extract	1.0 g
Deionized water	1,000 ml

Adjust to pH 2.0 with 10 N H₂SO₄. After autoclaving, add 25 ml of separately sterilized 40% glucose solution to give a final glucose concentration of 1.0%.

11.6.73. Thiobacillus thiooxidans Medium (203)

Na₂S₂O₃ · 5H₂O	10.0 g
KH₂PO₄	4.0 g
K₂HPO₄	4.0 g
MgSO₄ · 7H₂O	0.8 g
NH₄Cl	0.4 g
Trace metal solution (see below)	10.0 ml
Deionized water	1,000 ml

Sterilize at 121°C for 15 min. Adjust to pH 3.5 to 4.0 with sterile H₂SO₄. For a solid medium, add 15 g of agar per liter before adjusting the pH. Boil to dissolve the agar, sterilize by autoclaving, and cool to 45 to 50°C. Adjust pH to 3.5 to 4.0 with sterile H₂SO₄.

Trace metal solution

EDTA	50.0 g
ZnSO₄ · 7H₂O	22.0 g
CaCl₂	5.54 g
MnCl₂ · 4H₂O	5.06 g
FeSO₄ · 7H₂O	4.99 g
(NH₄)₆Mo₇O₂₄ · 4H₂O	1.10 g
CuSO₄ · 5H₂O	1.57 g
CoCl₂ · 6H₂O	1.61 g
Distilled water	1,000 ml

Adjust to pH 6.0 with KOH.

11.6.74. TN Medium for *Listeria* Species (109)

Glucose	2.0 g
Thallous acetate	2.0 g
Nalidixic acid solution	7.6 ml
Nutrient broth, hydrated	
(Oxoid no. 2)	q.s. to 1,000 ml

Combine ingredients and sterilize by autoclaving at 121°C for 15 min.

Nalidixic acid solution

Dissolve 0.1 g of crystalline nalidixic acid in 10 ml of 1 N NaOH; when dissolved, add 9 ml of distilled water.

11.6.75. Tryptic Soy-Yeast Extract Agar with Urea

Tryptic Soy Broth, dehydrated (Difco)	27.5 g
Yeast extract (Difco)	5.0 g
Glucose	5.0 g
Agar	15.0 g
Distilled water	1,000 ml

Adjust pH to 8.5 with NaOH before autoclaving. After sterilization, add sufficient urea solution (sterilized by filtration) to give a final urea concentration of 1%.

11.6.76. Tryptophan Medium (117)

$MgSO_4 \cdot 7H_2O$	0.2 g
K_2HPO_4	1.0 g
$MnCl_2 \cdot 4H_2O$	0.002 g
$FeSO_4 \cdot 7H_2O$	0.05 g
$CaCl_2$	0.02 g
$Na_2MoO_4 \cdot 2H_2O$	0.001 g
L-Tryptophan	1.0 g
Distilled water	1,000 ml

11.6.77. Veldkamp Medium (200)

NaCl	30.0 g
KH_2PO_4	1.0 g
NH_4Cl	1.0 g
$MgCl_2 \cdot 6H_2O$	0.5 g
$CaCl_2$	0.04 g
$NaHCO_3$	5.0 g
$Na_2S \cdot 9H_2O$	0.1 g
Ferric citrate, 0.004 M solution	5.0 ml
Trace element solution (see below)	2.0 ml
Powdered agar	5.0 g
Yeast extract	0.3 g
Distilled water	1,000 ml

Adjust to pH 7.0. Use without sterilization for primary enrichment. For a solid medium to be used for pour plates, prepare the medium at double strength with the omission of the powdered agar, and adjust the pH. Sterilize by filtration (positive pressure). Prepare a 4% solution of agar in freshly distilled boiling water, and sterilize by briefly autoclaving. Bring both the double-strength broth and the molten agar to 45 to 50°C, and combine equal volumes aseptically.

Trace element solution

H_3BO_3	2.8 g
$MnSO_4 \cdot H_2O$	2.1 g

$Cu(NO_3)_2 \cdot 3H_2O$	0.2 g
$Na_2MoO_4 \cdot 2H_2O$	0.75 g
$CoCl_2 \cdot 6H_2O$	0.2 g
$Zn(NO_3)_2 \cdot 6H_2O$	0.25 g
Deionized water	1,000 ml

11.6.78. Violet Red Bile Agar (14) (BBL or Difco)

Yeast extract	3.0 g
Peptone (Difco) or Gelysate (BBL)	7.5 g
Bile salts mixture	1.5 g
Lactose	10.0 g
NaCl	5.0 g
Neutral red	0.03 g
Crystal violet	0.002 g
Agar	15.0 g
Distilled water	1,000 ml

Adjust to pH 7.4. Boil to dissolve agar. Cool to ca. 45°C, and use for pour plates. After the inoculated medium has solidified, overlay it with more medium to prevent surface growth and spreading of colonies.

11.6.79. Widdel and Pfennig Medium (210)

Sodium acetate	1.23 g
Na_2SO_4	2.84 g
KH_2PO_4	0.68 g
$MgCl_2$	0.19 g
NH_4Cl	0.32 g
$CaCl_2$	0.07 g
Trace element solution (see below)	10 ml
Distilled water	1,000 ml

Sterilize at 121°C for 15 min. Then, add the following ingredients from sterile stock solutions.

$FeCl_2$ (from acidified stock solution)	0.00025 g
$NaHCO_3$ (sterilized by positive pressure	
filtration)	1.68 g
Na_2S (sterilized by filtration)	0.117 g
Vitamin solution (sterilized by filtration;	
see below)	5.0 ml
Vitamin B_{12} (sterilized by filtration)	20 μg

Adjust the pH of the medium to 7.1 with sterile H_3PO_4. After inoculating media, add 0.0315 g of $Na_2S_2O_4 \cdot 2H_2O$ per liter from a freshly prepared stock solution sterilized by filtration. For marine organisms, add 20 g of NaCl and 1.14 g of $MgCl_2$ per liter of medium.

Trace element solution

$ZnSO_4 \cdot 7H_2O$	10 mg
$MnCl_2 \cdot 4H_2O$	3 mg
H_3BO_3	30 mg
$CoCl_2 \cdot 6H_2O$	20 mg
$CuCl_2 \cdot 2H_2O$	1 mg
$NiCl_2 \cdot 6H_2O$	2 mg
$Na_2MoO_4 \cdot 2H_2O$	3 mg
Deionized water	1,000 ml

Vitamin solution

Biotin	0.2 mg
Niacin	2.0 mg

Thiamine	1.0 mg
p-Aminobenzoic acid	1.0 mg
Pantothenic acid	0.5 mg
Pyridoxine · HCl	5.0 mg
Distilled water	100 ml

Dissolve ingredients. Sterilize by filtration. Store at 4°C.

11.6.80. Wiley and Stokes Alkaline Medium

Yeast extract	20.0 g
(NH$_4$)$_2$SO$_4$	10.0 g
Tris buffer	15.7 g
Distilled water	1,000 ml

Prepare stock solutions of each ingredient separately and sterilize by autoclaving. (The pH of the Tris buffer solution should be 9.0.) Combine the ingredients aseptically from the sterile stock solutions. The final pH should be 8.7. For a solid medium, incorporate 15 g of agar per liter.

11.6.81. Yeast Extract Mannitol Agar

Mannitol	10.0 g
K$_2$HPO$_4$	0.5 g
MgSO$_4$ · 7H$_2$O	0.2 g
NaCl	0.1 g
CaCO$_3$	3.0 g
Yeast extract	0.2 g
Agar	15.0 g
Distilled water	1,000 ml

Boil to dissolve agar. Sterilize at 121°C for 15 min.

11.6.82. Yeast Extract Milk Agar

Yeast extract	3.0 g
Peptone	5.0 g
Fresh whole or skim milk	10.0 ml
Agar	15.0 g
Distilled water	1,000 ml

Dissolve the yeast extract and peptone by steaming, and adjust the cooled medium to pH 7.4. Add agar and milk, and autoclave at 121°C for 20 min. While hot, filter through paper pulp and adjust pH to 7.0 at 50°C. Dispense, and autoclave at 121°C for 15 min. Final pH should be 7.2.

11.6.83. YP Medium (185)

Yeast extract	3.0 g
Peptone	0.6 g
Distilled water	1,000 ml

Adjust pH to 7.2. Sterilize at 121°C for 15 min. For semisolid YP medium to be used as an overlay, add 6.0 g of agar per liter; add 19.0 g of agar per liter for a solid medium. Boil to dissolve agar, and sterilize by autoclaving.

11.7 REFERENCES

11.7.1. General References

1. **Aaronson, S.** 1970. *Experimental Microbial Ecology.* Academic Press, Inc., New York, NY.
   Contains a wealth of detailed methods for enrichment and isolation of a great variety of bacteria.

2. **Balows, A., H. G. Trüper, M. Dworkin, W. Harder, and K.-H. Schleifer (ed.).** 1992. *The Prokaryotes. A Handbook on the Biology of Bacteria: Ecophysiology, Isolation, Identification, Applications,* vol. 1–4, 2nd ed. Springer-Verlag, New York, NY.
   A comprehensive treatment of the prokaryotic world including isolation and enrichments of all types. Some chapters are classics on methodology and techniques, such as "Gram-negative mesophilic sulfate-reducing bacteria" by F. Widdel and F. Bak, vol. IV, p. 3352–3378.

3. **Collins, V. G.** 1969. Isolation, cultivation and maintenance of autotrophs, p. 1–52. *In* J. R. Norris and D. W. Ribbons (ed.), *Methods in Microbiology,* vol. 3B. Academic Press, Inc., New York, NY.
   A comprehensive treatment of the principles and techniques for enrichment and isolation of photo- and chemoautotrophs.

4. **Knudsen, G. R., M. J. McInerney, L. D. Stetzenbach, and R. L. Crawford (ed.).** 2001. *Manual of Environmental Microbiology,* 2nd ed. ASM Press, Washington, DC.
   With a general methodology section on microbial cultivation, identification, and community analysis, and chapters on specialized methodology for microbiological sampling, activity measurements, and specific microorganisms of environmental interest.

5. **Krieg, N. R., and J. G. Holt (ed.).** 1984. *Bergey's Manual of Systematic Bacteriology,* vol. 1. The Williams & Wilkins Co., Baltimore, MD.

6. **Lebeda, D. P.** 1990. *Isolation of Biotechnological Organisms from Nature.* McGraw-Hill Publishing Co., New York, NY.
   Covers both prokaryotic and eukaryotic microorganisms and presents selected techniques for isolation of organisms of potential biotechnological importance.

7. **Murray, P. (ed.).** 1999. *Manual of Clinical Microbiology,* 7th ed. ASM Press, Washington, DC.
   Principles and methods for enrichment, isolation, and identification of pathogenic bacteria, rickettsias, viruses, and fungi.

8. **Sneath, P. H. A., N. S. Mair, M. E. Sharpe, and J. G. Holt (ed.).** 1986. *Bergey's Manual of Systematic Bacteriology,* vol. 2. The Williams & Wilkins Co., Baltimore, MD.

9. **Staley, J. T., M. P. Bryant, N. Pfenning, and J. G. Holt (ed.).** 1989. *Bergey's Manual of Systematic Bacteriology,* vol. 3. The Williams & Wilkins Co., Baltimore, MD.

10. **Veldkamp, H.** 1970. Enrichment cultures of prokaryotic organisms, p. 305–361. *In* J. R. Norris and D. W. Ribbons (ed.), *Methods in Microbiology,* vol. 3A. Academic Press, Inc., New York, NY.
    Emphasizes the theoretical aspects of enrichment cultures and presents methods for the enrichment of specific organisms.

11. **Williams, S. T., M. E. Sharpe, and J. G. Holt (ed.).** 1989. *Bergey's Manual of Systematic Bacteriology,* vol. 4. The Williams & Wilkins Co., Baltimore, MD.
    References 5, 8, 9, and 11 contain descriptions of all the genera of prokaryotes with discussions of their enrichment and isolation.

11.7.2. Specific References

12. **Allen, O. N.** 1957. *Experiments in Soil Bacteriology,* 3rd ed. Burgess Publishing Co., Minneapolis, MN.

13. **Amann, R. I., W. Ludwig, and K.-H. Schleifer.** 1995. Phylogenetic identification and in situ detection of individual microbial cells without cultivation. *Microbiol. Rev.* **59:**143–169.

14. **American Public Health Association.** 1960. *Standard Methods for the Examination of Dairy Products,* 11th ed. American Public Health Association, New York, NY.

15. **Ammerman, J. W., and F. Azam.** 1981. Dissolved cyclic adenosine monophosphate (cAMP) in the sea and uptake of cAMP by marine bacteria. *Mar. Ecol. Prog. Ser.* **5:**85–89.

16. **Aragno, M., and H. G. Schlegel.** 1978. *Aquaspirillum autotrophicum*, a new species of hydrogen-oxidizing, facultatively autotrophic bacteria. *Int. J. Syst. Bacteriol.* **28:** 112–116.

17. **Armbruster, E. H.** 1969. Improved technique for isolation and identification of *Sphaerotilus*. *Appl. Microbiol.* **17:**320–321.

18. **Ashkin, A., and J. M. Dziedzic.** 1987. Optical trapping and manipulation of viruses and bacteria. *Science* **235:** 1517–1520.

19. **Ashkin, A., J. M. Dziedzic, and Y. Yamane.** 1987. Optical trapping and manipulation of single cells using infrared laser beams. *Nature* (London) **330:**769–771.

20. **Attwood, M. M., and W. Harder.** 1962. A rapid and specific enrichment procedure for *Hyphomicrobium* spp. *Antonie Leeuwenhoek J. Microbiol. Serol.* **38:**369–378.

21. **Bak, F., and H. Cypionka.** 1987. A novel type of energy metabolism involving fermentation of inorganic sulphur compounds. *Nature* **326:**891–892.

22. **Balch, W. E., G. E. Fox, L. J. Magrum, C. R. Woese, and R. S. Wolfe.** 1979. Methanogens: reevaluation of a unique biological group. *Microbiol. Rev.* **43:**260–296.

23. **Barnes, E. M.** 1956. Methods for the isolation of faecal streptococci (Lancefield Group D) from bacon factories. *J. Appl. Bacteriol.* **18:**193–203.

24. **Baron, E. J., and S. M. Finegold.** 1990. *Bailey & Scott's Diagnostic Microbiology*, 8th ed. C. V. Mosby Co., St. Louis, MO.

25. **Batchelor, S. E., M. Cooper, S. R. Chhabra, L. A. Glover, G. S. A. B. Stewart, P. Williams, and J. I. Prosser.** 1997. Cell density-regulated recovery of starved biofilm populations of ammonia-oxidizing bacteria. *Appl. Environ. Microbiol.* **63:**2281–2286.

26. **Bazylinski, D. A., R. B. Frankel, and H. W. Jannasch.** 1988. Anaerobic magnetite production by a marine, magnetotactic bacterium. *Nature* (London) **334:**518–519.

27. **Becking, J.-H.** 1984. Genus *Beijerinckia* Derx 1950, p. 315. *In* N. R. Krieg and J. G. Holt (ed.), *Bergey's Manual of Systematic Bacteriology*, vol. 1. The Williams & Wilkins Co., Baltimore, MD.

28. **Belly, R. T., B. B. Bohlool, and T. D. Brock.** 1973. The genus *Thermoplasma*. *Ann. N. Y. Acad. Sci.* **225:**94–107.

29. **Biebl, H., and N. Pfennig.** 1981. Isolation of members of the family *Rhodospirillaceae*, p. 267–273. *In* M. P. Starr, H. Stolp, H. G. Trüper, A. Balows, and H. G. Schlegel (ed.), *The Prokaryotes. A Handbook on Habitats, Isolation, and Identification of Bacteria*. Springer-Verlag, New York, NY.

30. **Blakemore, R. P., D. Maratea, and R. S. Wolfe.** 1979. Isolation and pure culture of a freshwater magnetic spirillum in chemically defined medium. *J. Bacteriol.* **140:**720–729.

31. **Blaser, M. J., I. D. Berkowitz, R. M. LaForce, J. Cravens, L. B. Reller, and W.-L. L. Wang.** 1979. Campylobacter enteritis: clinical and epidemiologic features. *Ann. Intern. Med.* **91:**179–184.

32. **Brock, T. D.** 1984. Genus *Thermus* Brock and Freeze 1969, p. 333–339. *In* N. R. Krieg and J. G. Holt (ed.), *Bergey's Manual of Systematic Bacteriology*, vol. 1. The Williams & Wilkins Co., Baltimore, MD.

33. **Brock, T. D., and H. Freeze.** 1969. *Thermus aquaticus* gen. n. and sp. n., a nonsporulating extreme thermophile. *J. Bacteriol.* **98:**289–297.

34. **Bruns, A., H. Cypionka, and J. Overmann.** 2002. Cyclic AMP and acyl homoserine lactones increase the cultivation efficiency of heterotrophic bacteria from the central Baltic Sea. *Appl. Environ. Microbiol.* **68:**3978–3987.

35. **Bruns, A., H. Hoffelner, and J. Overmann.** 2003. A novel approach for high throughput cultivation assays and

the isolation of planktonic bacteria. *FEMS Microbiol. Ecol.* **45:**161–171.

36. **Bruns, A., U. Nübel, H. Cypionka, and J. Overmann.** 2003. Effect of signal compounds and incubation conditions on the culturability of freshwater bacterioplankton. *Appl. Environ. Microbiol.* **69:**1980–1989.

37. **Buchholz-Cleven, B. E. E., B. Rattunde, and K. L. Straub.** 1997. Screening for genetic diversity of isolates of anaerobic Fe(II)-oxidizing bacteria using DGGE and whole-cell hybridization. *Syst. Appl. Microbiol.* **20:**301–309.

38. **Buchner, P.** 1965. *Endosymbiosis of Animals with Plant Microorganisms*. Interscience Publishers, New York, NY.

39. **Bukatsch, F.** 1968. Anreicherung und Isolierung mariner Leuchtbakterien. *Zentbl. Bakteriol. 1. Abt. Suppl. Heft* **1:** 399–406.

40. **Burggraf, S., P. Heyder, and N. Eis.** 1997. A pivotal Archaea group. *Nature* **385:**780.

41. **Button, D. K., F. Schut, P. Quang, R. Martin, and B. R. Robertson.** 1993. Viability and isolation of marine bacteria by dilution culture: theory, procedures, and initial results. *Appl. Environ. Microbiol.* **59:**881–891.

42. **Calcott, P. H., and J. R. Postgate.** 1972. On substrate-accelerated death in *Klebsiella aerogenes*. *J. Gen. Microbiol.* **70:**115–122.

43. **Canale-Parola, E.** 1984. Genus I. *Spirochaeta* Ehrenberg 1835, p. 41. *In* N. R. Krieg and J. G. Holt (ed.), *Bergey's Manual of Systematic Bacteriology*, vol. 1. The Williams & Wilkins Co., Baltimore, MD.

44. **Canale-Parola, E., S. L. Rosenthal, and D. G. Kupfer.** 1966. Morphological and physiological characteristics of *Spirillum gracile* sp. n. *Antonie Leeuwenhoek J. Microbiol. Serol.* **32:**113–124.

45. **Cataldi, M. S.** 1940. Aislamiento de *Beggiatoa alba* en cultivo puro. *Rev. Inst. Bacteriol. Dept. Nac. Hig.* (Buenos Aires) **9:**393–423.

46. **Chapman, G. H.** 1944. The isolation of streptococci from mixed cultures. *J. Bacteriol.* **48:**113–114.

47. **Christensen, B., T. Torsvik, and T. Lien.** 1992. Immunomagnetically captured thermophilic sulfate-reducing bacteria from North Sea oil waters. *Appl. Environ. Microbiol.* **58:**1244–1248.

48. **Claus, D., and R. C. W. Berkeley.** 1986. Genus *Bacillus* Cohn 1872, p. 1114–1120. *In* P. H. A. Sneath, N. S. Mair, M. E. Sharpe, and J. G. Holt (ed.), *Bergey's Manual of Systematic Bacteriology*, vol. 2. The Williams & Wilkins Co., Baltimore, MD.

49. **Claus, D., and F. Fahmy.** 1984. Genus *Sporosarcina* Kluyver and Van Niel 1936, p. 1204. *In* P. H. A. Sneath, N. S. Mair, M. E. Sharpe, and J. G. Holt (ed.), *Bergey's Manual of Systematic Bacteriology*, vol. 2. The Williams & Wilkins Co., Baltimore, MD.

50. **Claus, D., and N. Walker.** 1964. The decomposition of toluene by soil bacteria. *J. Gen. Microbiol.* **36:**107–122.

51. **Coates, J. D., D. J. Ellis, E. Roden, K. Gaw, E. L. Blunt-Harris, and D. R. Lovley.** 1998. Recovery of humics-reducing bacteria from a diversity of sedimentary environments. *Appl. Environ. Microbiol.* **64:**1504–1509.

52. **Connon, S. A., and S. J. Giovannoni.** 2002. High-throughput methods for culturing micro-organisms in very-low-nutrient media yield diverse new marine isolates. *Appl. Environ. Microbiol.* **68:**3878–3885.

53. **Corbel, M. J., and W. J. Brinley-Morgan.** 1984. Genus *Brucella* Meyer and Shaw 1920, p. 377–382. *In* N. R. Krieg and J. G. Holt (ed.), *Bergey's Manual of Systematic Bacteriology*, vol. 1. The Williams & Wilkins Co., Baltimore, MD.

54. **Cottrell, M. T., and D. L. Kirchman.** 2000. Natural assemblages of marine Proteobacteria and members of the *Cytophaga-Flavobacter* cluster consuming low- and high-

molecular-weight dissolved organic matter. *Appl. Environ. Microbiol.* **66:**1692–1697.

55. **Couch, J. N.** 1949. A new group of organisms related to *Actinomyces. J. Elisha Mitchell Sci. Soc.* **65:**315–318.

56. **Curry, J. C., and G. E. Borovian.** 1976. Selective medium for distinguishing micrococci from staphylococci in the clinical laboratory. *J. Clin. Microbiol.* **4:**455–457.

57. **Daims, H., A. Brühl, R. Amann, K.-H. Schleifer, and M. Wagner.** 1999. The domain-specific probe EUB338 is insufficient for the detection of all bacteria: development and evaluation of a more comprehensive probe set. *Syst. Appl. Microbiol.* **22:**434–444.

58. **Darland, G., T. D. Brock, W. Samsonoff, and S. F Conti.** 1970. A thermophilic, acidophilic mycoplasma isolated from a coal refuse pile. *Science* **170:**1416–1418.

59. **Davis, B. D., R. Dulbecco, H. N. Eisen, H. S. Ginsberg, W. B. Wood, and M. McCarty.** 1973. *Microbiology,* 2nd ed. Harper & Row, Hagerstown, MD.

60. **Dawes, E. A., and P. J. Senior.** 1973. The role and regulation of energy reserve polymers in microorganisms. *Adv. Microb. Physiol.* **10:**135–266.

61. **Döbereiner, J., and V. L. D. Baldani.** 1979. Selective infection of maize roots by streptomycin-resistant *Azospirillum lipoferum* and other bacteria. *Can. J. Microbiol.* **25:**1264–1268.

62. **Döbereiner, J., I. E. Marriel, and M. Nery.** 1976. Ecological distribution of *Spirillum lipoferum* Beijerinck. *Can. J. Microbiol.* **22:**1461–1473.

63. **Edwards, K. J., D. R. Rogers, C. O. Wirsen, and T. M. McCollom.** 2003. Isolation and characterization of novel psychrophilic, neutrophilic, Fe-oxidizing, chemolithoautotrophic alpha- and gamma-proteobacteria from the deep sea. *Appl. Environ. Microbiol.* **69:**2906–2913.

64. **Emerson, D., and C. Moyer.** 1997. Isolation and characterization of novel iron-oxidizing bacteria that grow at circumneutral pH. *Appl. Environ. Microbiol.* **63:**4784–4792.

65. **Farrell, I. D.** 1974. The development of a new selective medium for the isolation of *Brucella abortus* from contaminated sources. *Res. Vet. Sci.* **16:**280–286.

66. **Faust, L., and R. S. Wolfe.** 1961. Enrichment and cultivation of *Beggiatoa alba. J. Bacteriol.* **81:**88–106.

67. **Ferenci, T.** 1996. Adaptation to life at micromolar nutrient levels: the regulation of *Escherichia coli* glucose transport by endoinduction and cAMP. *FEMS Microbiol. Rev.* **18:**301–317.

68. **Ferris, M. J., A. L. Ruff-Roberts, E. D. Kopcynski, M. M. Bateson, and D. M. Ward.** 1996. Enrichment culture and microscopy conceal diverse thermophilic *Synechococcus* populations in a single hot spring microbial mat habitat. *Appl. Environ. Microbiol.* **62:**1045–1050.

69. **Ferris, M. J., and D. M. Ward.** 1997. Seasonal distributions of dominant 16S rRNA-defined populations in a hot spring microbial mat examined by denaturing gradient gel electrophoresis. *Appl. Environ. Microbiol.* **63:**1375–1381.

70. **Finegold, S. M., W. J. Martin, and E. G. Scott.** 1978. *Bailey and Scott's Diagnostic Microbiology,* 5th ed. The C. V. Mosby Co., St. Louis, MO.

71. **Finster, K., W. Liesack, and B. J. Tindall.** 1997. *Desulfospira joergensii,* gen. nov., sp. nov., a new sulfate-reducing bacterium isolated from marine surface sediment. *Syst. Appl. Microbiol.* **20:**201–208.

72. **Freeman, B. A.** 1977. *Burrows' Textbook of Microbiology,* 21st ed. The W. B. Saunders Co., Philadelphia, PA.

73. **Fröhlich, J., and H. König.** 1999. Rapid isolation of single microbial cells from mixed natural and laboratory populations with the aid of a micromanipulator. *Syst. Appl. Microbiol.* **22:**249–257.

74. **Fröstl, J. M., and J. Overmann.** 1998. Physiology and tactic response of the phototrophic consortium "*Chlorochromatium aggregatum.*" *Arch. Microbiol.* **169:**129–135.

75. **Gerloff, G. C., G. P. Fitzgerald, and F. Skoog.** 1950. The isolation, purification and culture of blue-green algae. *Am. J. Bot.* **37:**216–218.

76. **Gibbons, N. E.** 1969. Isolation, growth and requirements of halophilic bacteria, p. 169–183. *In* J. R. Norris and D. W. Ribbons (ed.), *Methods in Microbiology,* vol. 3B. Academic Press, Inc., New York, NY.

77. **Grant, M. A., and J. G. Holt.** 1977. Medium for the selective isolation of members of the genus *Pseudomonas* from natural habitats. *Appl. Environ. Microbiol.* **33:**1222–1224.

78. **Guerrero, R., S. Pedrós-Alió, T. N. Schmidt, and J. Mas.** 1985. A survey of buoyant density of microorganisms in pure cultures and natural samples. *Microbiologia* **1:**53–65.

79. **Hagedorn, C., and J. G. Holt.** 1975. A nutritional and taxonomic survey of *Arthrobacter* soil isolates. *Can. J. Microbiol.* **21:**353–361.

80. **Hampp, E. G.** 1957. Isolation and identification of spirochaetes obtained from unexposed canals of pulp-involved teeth. *Oral Surg. Oral Med. Oral Pathol.* **101:**1100–1104.

81. **Han, S. O., and P. B. New.** 1998. Isolation of *Azospirillum* spp. from natural soils by immunomagnetic separation. *Soil Biol. Biochem.* **30:**975–980.

82. **Hanert, H. H.** 1992. The genus *Gallionella.* p. 4082–4088. *In* A. Balows, H. G. Trüper, M. Dworkin, W. Harder, and K.-H. Schleifer (ed.), *The Prokaryotes. A Handbook on the Biology of Bacteria: Ecophysiology, Isolation, Identification, Applications.* 2nd ed. Springer-Verlag, New York, NY.

83. **Hanson, A. W.** 1970. Isolation of spirochaetes from primates and other mammalian species. *Br. J. Vener. Dis.* **46:**303–306.

84. **Hanson, A. W., and G. R. Cannefax.** 1964. Isolation of *Borrelia refringens* in pure culture from patients with condylomata acuminata. *J. Bacteriol.* **88:**111–113.

85. **Hespell, R. B.** 1984. Genus *Serpens* IIespell 1977, p. 373. *In* N. R. Krieg and J. G. Holt (ed.), *Bergey's Manual of Systematic Bacteriology,* vol. 1. The Williams & Wilkins Co., Baltimore, MD.

85a.**Højberg, O., S. J. Binnerup, and J. Sørensen.** 1997. Growth of silicone-immobilized bacteria on polycarbonate membrane filters, a technique to study microcolony formation under anaerobic conditions. *Appl. Environ. Microbiol.* **63:**2920–2924.

86. **Holdeman, L. V., E. P. Cato, and W. E. C. Moore (ed.).** 1977. *Anaerobe Laboratory Manual,* 4th ed. Virginia Polytechnic Institute and State University, Blacksburg.

87. **Huber, R., S. Burggraf, T. Mayer, S. M. Barns, P. Rossnagel, and K. O. Stetter.** 1995. Isolation of a hyperthermophilic archaeum predicted by in-situ RNA analysis. *Nature* **376:**57–58.

88. **Hugenholtz, P., B. M. Goebel, and N. R. Pace.** 1998. Impact of culture-independent studies on the emerging phylogenetic view of bacterial diversity. *J. Bacteriol.* **180:**4765–4774.

89. **Hungate, R. E.** 1950. The anaerobic cellulolytic bacteria. *Bacteriol. Rev.* **14:**1–49.

90. **Hungate, R. E.** 1969. A roll tube method for cultivation of strict anaerobes, p. 117–132. *In* J. R. Norris and D. W. Ribbons (ed.), *Methods in Microbiology,* vol. 3B. Academic Press, Inc., New York, NY.

91. **Hylemon, P. B., J. S. Wells, Jr., J. H. Bowdre, T. O. MacAdoo, and N. R. Krieg.** 1973. Designation of *Spirillum volutans* Ehrenberg 1832 as type species of the genus *Spirillum* Ehrenberg 1832 and designation of the neotype strain of *S. volutans. Int. J. Syst. Bacteriol.* **23:**20–27.

92. **Inoue, K.** 1976. Quantitative ecology of microorganisms of Syowa Station in Antarctica and isolation of psychrophiles. *J. Gen. Appl. Microbiol.* **22:**143–150.

93. **Inoue, K., and K. Komagata.** 1976. Taxonomy study on obligately psychrophilic bacteria isolated from Antarctica. *J. Gen. Appl. Microbiol.* **22:**165–176.

94. **Iverson, W. P.** 1966. Growth of *Desulfovibrio* on the surface of agar media. *Appl. Microbiol.* **14:**529–534.

95. **Jannasch, H.** 1967. Enrichments of aquatic spirilla in continuous culture. *Arch. Mikrobiol.* **69:**165–173.

96. **Jannasch, H.** 1968. Competitive elimination of *Enterobacteriaceae* from seawater. *Appl. Microbiol.* **16:**1616–1618.

97. **Jaspers, E., and J. Overmann.** 1997. Separation of bacterial cells by isoelectric focusing, a new method for analysis of complex microbial communities. *Appl. Environ. Microbiol.* **63:**3176–3181.

98. **Johnson, R. C., and S. Faine.** 1984. Genus I. *Leptospira* Noguchi 1917, p. 64. *In* N. R. Krieg and J. G. Holt (ed.), *Bergey's Manual of Systematic Bacteriology*, vol. 1. The Williams & Wilkins Co., Baltimore, MD.

99. **Johnstone, K. I.** 1969. The isolation and cultivation of single organisms, p. 455–471. *In* J. R. Norris and D. W. Ribbons (ed.), *Methods in Microbiology*, vol. 1. Academic Press, Inc., New York, NY.

100. **Jones, J. G.** 1979. *A Guide to Methods for Estimating Microbial Numbers and Biomass in Fresh Water.* Scientific publication no. 39. Freshwater Biological Association, Ambleside, United Kingdom.

101. **Kaeberlein, T., K. Lewis, and S. S. Epstein.** 2002. Isolating "uncultivable" microorganisms in pure culture in a simulated natural environment. *Science* **296:**1127–1129.

102. **Kane, M. D., L. K. Poulsen, and D. A. Stahl.** 1993. Monitoring the enrichment and isolation of sulfate-reducing bacteria by using oligonucleotide hybridization probes designed from environmentally derived 16S rRNA sequences. *Appl. Environ. Microbiol.* **59:**682–686.

103. **Kaplan, H., and E. Greenberg.** 1985. Diffusion of autoinducer is involved in regulation of the *Vibrio fischeri* luminescence system. *J. Bacteriol.* **163:**1210–1214.

104. **Kelly, R. T.** 1984. Genus IV Borrelia Swellengrebel 1907, p. 58. *In* N. R. Krieg and J. G. Holt (ed.), *Bergey's Manual of Systematic Bacteriology*, vol. 1. The Williams & Wilkins Co., Baltimore, MD.

105. **Kenny, G. E.** 1974. Mycoplasma, p. 333–337. *In* E. H. Lennette, E. H. Spaulding, and J. P. Truant (ed.), *Manual of Clinical Microbiology*, 2nd ed. American Society for Microbiology, Washington, DC.

106. **Kersters, K., and J. De Ley.** 1984. Genus III. *Agrobacterium* Conn, p. 247. *In* N. R. Krieg and J. G. Holt (ed.), *Bergey's Manual of Systematic Bacteriology*, vol. 1. The Williams & Wilkins Co., Baltimore, MD.

107. **Klee, A. J.** 1993. A computer program for the determination of most probable number and its confidence limits. *J. Microbiol. Methods* **18:**91–98.

108. **Kocur, M.** 1986. Genus I. *Micrococcus* Cohn 1872, p. 1005. *In* P. H. A. Sneath, N. S. Mair, M. E. Sharpe, and J. G. Holt (ed.), *Bergey's Manual of Systematic Bacteriology*, vol. 2. The Williams & Wilkins Co., Baltimore, MD.

109. **Kramer, P. A., and D. Jones.** 1969. Media selective for *Listeria monocytogenes*. *J. Appl. Bacteriol.* **32:**381–394.

110. **Kucera, S., and R. S. Wolfe.** 1957. A selective enrichment method for *Gallionella ferruginea*. *J. Bacteriol.* **74:**344–349.

111. **la Rivière, J. W M., and K. Schmidt.** 1989. Genus *Achromatium* 1893, p. 2132. *In* J. T. Staley, M. P. Bryant, N. Pfennig, and J. G. Holt (ed.), *Bergey's Manual of Systematic Bacteriology*, vol. 3. The Williams & Wilkins Co., Baltimore, MD.

112. **Leadbetter, E. R.** 1963. Growth and morphogenesis of *Sporocytophaga myxococcoides*. *Bacteriol. Proc.* **1963:**42.

113. **Lederberg, J.** 1954. A simple method for isolating individual microbes. *J. Bacteriol.* **88:**258–259.

114. **Lee, N., P. H. Nielsen, K. H. Andreasen, S. Juretschko, J. L. Nielsen, K.-H. Schleifer, and M. Wagner.** 1999. Combination of fluorescent in-situ hybridization and microautoradiography—a new tool for structure function analyses in microbial ecology. *Appl. Environ. Microbiol.* **65:**1289–1297.

115. **Leifson, E.** 1936. New selenite enrichment media for the isolation of typhoid and paratyphoid (*Salmonella*) bacilli. *Am. J. Hyg.* **24:**423–432.

116. **Lennette, E. H., A. Balows, W. J. Hausler, Jr., and J. P. Truant (ed.).** 1980. *Manual of Clinical Microbiology*, 3rd ed. American Society for Microbiology, Washington, DC.

117. **Lichstein, H. C., and E. L. Oginsky.** 1965. *Experimental Microbial Physiology*. W. H. Freeman and Co., San Francisco, CA.

118. **Linn, D. M., and N. R. Krieg.** 1971. Occurrence of two organisms in the type strain of *Spirillum lunatum*: rejection of the name *Spirillum lunatum* and characterization of *Oceanospirillum maris* subsp. *williamsiae* and an unclassified vibrioid bacterium. *Int. J. Syst. Bacteriol.* **28:**131–138.

119. **Loewen, P. C., B. Hu, J. Strutinsky, and R. Sparling.** 1998. Regulation in the rpoS region of *Escherichia coli*. *Can. J. Microbiol.* **44:**707–717.

120. **Lovley, D. R., and E. J. P. Phillips.** 1987. Rapid assay for microbially reducible ferric iron in aquatic sediments. *Appl. Environ. Microbiol.* **53:**1536–1540.

121. **Lovley, D. R., and E. J. P. Phillips.** 1988. Novel mode of microbial energy metabolism: organic carbon oxidation coupled to dissimilatory reduction of iron or manganese. *Appl. Environ. Microbiol.* **54:**1472–1480.

122. **Lovley, D. R., J. F. Stolz, G. L. Nord, and E. J. P. Phillips.** 1987. Anaerobic production of magnetite by a dissimilatory iron-reducing microorganism. *Nature* **330:**252–254.

123. **Lovley, D. R., J. D. Coates, E. L. Blunt-Harris, E. J. P. Phillips, and J. C. Woodward.** 1996. Humic substances as electron acceptors for microbial respiration. *Nature* **382:**445–448.

124. **Lovley, D. R., J. L. Fraga, E. L. Blunt-Harris, L. A. Hayes, E. J. P. Phillips, and J. D. Coates.** 1998. Humic substances as a mediator for microbially catalyzed metal reduction. *Acta Hydrochim. Hydrobiol.* **26:**152–157.

125. **Lovley, D. R., K. Kashefi, M. Vargas, J. M. Tor, and E. L. Blunt-Harris.** 2000. Reduction of humic substances and Fe(III) by hyperthermophilic microorganisms. *Chem. Geol.* **169:**289–298.

126. **Mabbitt, L. A., and M. Zielinika.** 1956. The use of a selective medium for the enumeration of lactobacilli in cheddar cheese. *J. Appl. Bacteriol.* **18:**95–101.

126a. **Madigan, M. E., J. M. Martinko, and J. Parker.** 2000. *Brock Biology of Microorganisms*, 9th ed. Prentice-Hall, Upper Saddle River, NJ.

127. **Madonna, A. J., F. Basile, E. Furlong, and K. J. Vorhees.** 2001. Detection of bacteria from biological mixtures using immunomagnetic separation combined with matrix-assisted laser desorption/ionization time-of-flight mass spectrocopy. *Rapid Commun. Mass Spectrom.* **15:**1068–1074.

128. **Matin, A., C. Veldhuis, V. Stegeman, and M. Veenhuis.** 1979. Selective advantage of a *Spirillum* sp. in a carbon-limited environment. Accumulation of poly-8-hydroxybutyric acid and its role in starvation. *J. Gen. Microbiol.* **112:**349–355.

129. **Matin, A., and H. Veldkamp.** 1978. Physiological basis of the selective advantage of a Spirillum sp. in a carbon-limited environment. *J. Gen. Microbiol.* **106:**187–197.

130. **Matsunaga, T., T. Sakaguchi, and R. Tadokoro.** 1991. Magnetite formation by a magnetic bacterium capable of growing aerobically. *Appl. Microbiol. Biotechnol.* **35:**651–655.

131. **Maymó-Gatell, X., T. Anguish, and S. H. Zinder.** 1999. Reductive dechlorination of chlorinated ethenes and 1,2-dichlorethane by "*Dehalococcoides ethenogenes*" 195. *Appl. Environ. Microbiol.* **65:**3108–3113.

132. **Maymó-Gatell, X., Y.-T. Chien, J. M. Gossett, and S. H. Zinder.** 1997. Isolation of a bacterium that reductively dechlorinates tetrachloroethene to ethene. *Science* **276:**1568–1571.

133. **Maymó-Gatell, X., V. Tandoi, J. M. Gossett, and S. H. Zinder.** 1995. Characterization of an H2-utilizing enrichment culture that reductively dechlorinates tetrachloroethene to vinyl chloride and ethene in the absence of methanogenesis and acetogenesis. *Appl. Environ. Microbiol.* **61:**3928–3933.

134. **McCurdy, H. D.** 1963. A method for the isolation of myxobacteria in pure culture. *Can. J. Microbiol.* **8:**282–285.

135. **McCurdy, H. D.** 1989. Order *Myxococcales* Tchan, Pochon, and Prevot 1948, p. 2142. *In* J. T. Staley, M. P. Bryant, N. Pfenning, and J. G. Holt (ed.), *Bergey's Manual of Systematic Bacteriology*, vol. 3. The Williams & Wilkins Co., Baltimore, MD.

136. **Meiklejohn, J.** 1950. The isolation of *Nitrosomonas europaea* in pure culture. *J. Gen. Microbiol.* **4:**185–191.

137. **Miller, T. L., and M. J. Wolin.** 1974. A serum bottle modification of the Hungate technique for cultivating obligate anaerobes. *Appl. Microbiol.* **27:**985–987.

138. **Moench, T. T.** Genus "*Bilophococcus*," p. 1889. *In* J. T. Staley, M. P. Bryant, N. Pfenning, and J. G. Holt (ed.), *Bergey's Manual of Systematic Bacteriology*, vol. 3. The Williams & Wilkins Co., Baltimore, MD.

138a. **Mohn, W. W., and J. M. Tiedje.** 1990. Strain DCB-1 conserves energy for growth from reductive dechlorination coupled to formate oxidation. *Arch. Microbiol.* **153:**267–271.

139. **Morita, R. Y.** 1975. Psychrophilic bacteria. *Bacteriol. Rev.* **39:**144–167.

140. **Murray, R. G. E.** 1984. Genus *Lampropedia* Schroeter 1886, p. 405. *In* N. R. Krieg and J. G. Holt (ed.), *Bergey's Manual of Systematic Bacteriology*, vol. 1. The Williams & Wilkins Co., Baltimore, MD.

141. **Murray, R. G. E.** 1986. Genus I. *Deinococcus* Brooks and Murray 1981, p. 1039. *In* P. H. A. Sneath, N. S. Mair, M. E. Sharpe, and J. G. Holt (ed.), *Bergey's Manual of Systematic Bacteriology*, vol. 2. The Williams & Wilkins Co., Baltimore, MD.

142. **Murray, R. G. E.** 1992. The family *Deinococcaceae*, p. 3733–3744. *In* A. Balows, H. G. Trüper, M. Dworkin, W. Harder, and K.-H. Schleifer (ed.), *The Prokaryotes. A Handbook on the Biology of Bacteria: Ecophysiology, Isolation, Identification, Applications*, 2nd ed. Springer-Verlag, New York, NY.

143. **Muyzer, G., T. Brinkhoff, U. Nübel, C. Santegoeds, H. Schäfer, and C. Wawer.** 1998. Denaturing gradient gel electrophoresis (DGGE) in microbial ecology, vol. 3.4.4, p. 1–27. *In* A. D. L. Akkermans, J. D. van Elsas, and F. J. de Bruijn (ed.), *Molecular Ecology Manual*. Kluwer Academic Publishers, Dordrecht, The Netherlands.

144. **Muyzer, G., E. C. De Waal, and A. G. Uitterlinden.** 1993. Profiling of complex microbial populations by denaturing gradient gel electrophoresis analysis of polymerase chain reaction-amplified genes coding for 16S rRNA. *Appl. Environ. Microbiol.* **59:**695–700.

145. **Muyzer, G., and K. Smalla.** 1998. Application of denaturing gradient gel electrophoresis (DGGE) and temperature gradient gel electrophoresis (TGGE) in microbial ecology. *Antonie Leeuwenhoek* **173:**127–141.

146. **Myrivk, Q. N., N. N. Pearsall, and R. S. Weiser.** 1974. *Fundamentals of Medical Bacteriology and Mycology.* Lea & Febiger, Philadelphia, PA.

147. **Nelson, D. C.** 1992. The genus *Beggiatoa*, p. 3171–3180. *In* A. Balows, H. G. Trüper, M. Dworkin, W. Harder, and K.-H. Schleifer (ed.), *The Prokaryotes. A Handbook on the Biology of Bacteria: Ecophysiology, Isolation, Identification, Applications*, 2nd ed. Springer-Verlag, New York, NY.

148. **Nelson, D. C., B. B. Jørgensen, and N. P. Revsbech.** 1986. Growth pattern and yield of a chemoautotrophic *Beggiatoa* sp. in oxygen-sulfide microgradients. *Appl. Environ. Microbiol.* **52:**225–233.

149. **New, P. B., and A. Keer.** 1972. Biological control of crown gall: field measurements and glass-house experiments. *J. Appl. Bacteriol.* **35:**279–287.

150. **Orchard, V. A., M. Goodfellow, and S. T. Williams.** 1977. Selective isolation and occurrence of nocardiae in soil. *Soil Biol. Biochem.* **9:**233–238.

151. **Ouverney, C., and J. A. Fuhrman.** 1999. Combined microautoradiography-16S rRNA probe technique for determination of radioisotope uptake by specific microbial cell types in situ. *Appl. Environ. Microbiol.* **65:**1746–1752.

152. **Palleroni, N. J.** 1980. A chemotactic method for the isolation of *Actinoplanaceae*. *Arch. Microbiol.* **128:**53–55.

153. **Pernthaler, J., F.-O. Glöckner, W. Schönhuber, and R. Amann.** 2001. Fluorescent in situ hybridization with rRNA-targeted oligonucleotide probes, p. 207–226. *In* J. Paul (ed.), *Methods in Microbiology*, Academic Press, New York, NY.

154. **Peterson, J. E.** 1969. Isolation, cultivation and maintenance of the myxobacteria, p. 185–210. *In* J. R. Norris and D. W. Ribbons (ed.), *Methods in Microbiology*, vol. 3B. Academic Press, Inc., New York, NY.

155. **Pfennig, N., and H. Biebl.** 1976. *Desulfuromonas acetoxidans* gen. nov. and sp. nov., a new anaerobic, sulfur-reducing, acetate-oxidizing bacterium. *Arch. Microbiol.* **110:**3–12.

156. **Pfennig, N., and H. Biebl.** 1981. The dissimilatory sulfur-reducing bacteria, p. 941–947. *In* M. P. Starr, H. Stolp, H. G. Trüper, A. Balows, and H. G. Schlegel (ed.), *The Prokaryotes*, vol 1. Springer, Berlin, Germany.

157. **Phillips, E. J. P., and D. R. Lovley.** 1987. Determination of Fe(III) and Fe(II) in oxalate extracts of sediment. *Soil Sci. Soc. Am. J.* **51:**938–941.

158. **Poindexter, J. S.** 1964. Biological properties and classification of the *Caulobacter* group. *Bacteriol. Rev.* **28:**231–295.

159. **Porter, J., and R. W. Pickup.** 1998. Separation of natural populations of coliform bacteria from freshwater and sewage by magnetic-bead cell sorting. *J. Microbiol. Methods* **33:**221–226.

160. **Pramer, D. A., and E. L. Schmidt.** 1964. *Experimental Soil Microbiology.* Burgess Publishing Co., Minneapolis, MN.

161. **Preston, C., Y. W. Ke, T. F. Molinski, and D. F. DeLong.** 1996. A psychrophilic crenarcheon inhabits a marine sponge: *Cenarchaeum symbiosum* gen. nov., sp. nov. *Proc. Natl. Acad. Sci. USA* **93:**6241–6246.

162. **Pringsheim, E. G.** 1964. Heterotrophism and species concepts in *Beggiatoa. Am. J. Bot.* **51:**898–913.

163. **Reichenbach, H.** 1989. Genus 1. *Cytophaga* Winogradsky 1929, p. 2015–2050. *In* J. T. Staley, M. P. Bryant, N. Pfenning, and J. G. Holt (ed.), *Bergey's Manual of Systematic Bacteriology*, vol. 3. The Williams & Wilkins Co., Baltimore, MD.

164. **Richardson, J. H., and W. E. Barkley.** 1988. *Biosafety in Microbiological and Biomedical Laboratories*, 2nd ed. U.S. Department of Health and Human Services, HHS publication no. (NIH) 88-8395, U.S. Government Printing Office, Washington, DC.

165. **Rippka, R., J. Derueiles, J. B. Waterbury, M. Herdman, and R. Y. Stanier.** 1979. Generic assignments, strain histories and properties of pure cultures of cyanobacteria. *J. Gen. Microbiol.* **111**:1–61.

166. **Rittenberg, B. T., and S. C. Rittenberg.** 1962. The growth of *Spirillum volutans* in mixed and pure cultures. *Arch. Mikrobiol.* **42**:138–153.

167. **Rosebury, T.** 1962. *Microorganisms Indigenous to Man.* McGraw-Hill Book Co., New York, NY.

168. **Rosebury, T., and G. Foley.** 1942. Isolation and pure cultivation of the smaller mouth spirochaetes by an improved method. *Proc. Soc. Exp. Biol. Med.* **47**:368–374.

169. **Rosebury, T., J. B. McDonald, S. A. Ellison, and S. G. Engel.** 1951. Media and methods for separation and cultivation of oral spirochaetes. *Oral Surg. Oral Med. Oral Pathol.* **4**:68–85.

170. **Sampalo, M.-J. A. M., E. M. R. da Silva, J. Döbereiner, M. G. Yates, and F. O. Pedrosa.** 1981. Autography and methylotrophy in *Derxia gummosa*, p. 447. *In* A. H. Gibson and W. E. Newton (ed.), *Current Perspectives in Nitrogen Fixation*, Canberra, Australia, Dec. 1–5, 1980. Elsevier/North Holland Biomedical Press, Amsterdam, The Netherlands.

171. **Schleifer, K.-H., D. Schüler, S. Spring, N. Weizenegger, R. Amann, W. Ludwig, and M. Köhler.** 1991. The genus *Magnetospirillum* gen. nov.: description of *Magnetospirillum gryphiswaldense* sp. nov. and transfer of *Aquaspirillum magnetotacticum* to *Magnetospirillum magnetotacticum* comb. nov. *Syst. Appl. Microbiol.* **14**:379–385.

172. **Schroth, M. N., J. P. Thompson, and D. C. Hildebrand.** 1965. Isolation of *Agrobacterium tumefaciens*-*A. radiobacter* group from soil. *Phytopathology* **55**:645–647.

173. **Schultz, J. E., G. I. Latter, and A. Matin.** Differential regulation by cyclic AMP of starvation protein synthesis in *Escherichia coli. J. Bacteriol.* **170**:3903–3909.

174. **Seeliger, H. P. R., and D. Jones.** 1986. Genus *Listeria* Pirie 1940, p. 1235–1245. *In* P. H. A. Sneath, N. S. Mair, M. E. Sharpe, and J. G. Holt (ed.), *Bergey's Manual of Systematic Bacteriology*, vol. 2. The Williams & Wilkins Co., Baltimore, MD.

175. **Segerer, A. H., and K. O. Stetter.** 1992. The genus *Thermoplasma*, p. 712–718. *In* A. Balows, H. G. Trüper, M. Dworkin, W. Harder, and K.-H. Schleifer (ed.), *The Prokaryotes. A Handbook on the Biology of Bacteria: Ecophysiology, Isolation, Identification, Applications*, 2nd ed. Springer-Verlag, New York, NY.

176. **Sehgal, S. N., and N. E. Gibbons.** 1960. Effect of some metal ions on the growth of *Halobacterium cutirubrum. Can. J. Microbiol.* **6**:165–169.

177. **Shelton, D. R., and J. M. Tiedje.** 1984. Isolation and partial characterization of bacteria in an anaerobic consortium that mineralizes 3-chlorobenzoic acid. *Appl. Environ. Microbiol.* **48**:840–848.

178. **Singh, B.** 1947. Myxobacteria in soils and composts: their distribution, number and lytic action on bacteria. *J. Gen. Microbiol.* **1**:1–10.

179. **Slack, J. M., and I. S. Snyder.** 1978. *Bacteria and Human Disease.* Year Book Medical Publishers, Chicago, IL.

180. **Smibert, R. M., and R. L. Claterbaugh.** 1972. A chemically-defined medium for *Treponema* strain PH-7 isolated from the intestine of a pig with swine dysentery. *Can. J. Microbiol.* **18**:1073–1078.

181. **Smith, L. D., and V. R. Dowell.** 1974. Clostridium, p. 376–380. *In* E. H. Lennette, E. H. Spaulding, and J. P. Truant (ed.), *Manual of Clinical Microbiology*, 2nd ed. American Society for Microbiology, Washington, DC.

182. **Sorokin, D. Y., A. Teske, L. A. Robertson, and J. G. Kuehnen.** 1999. Anaerobic oxidation of thiosulfate to tetrathionate by obligately heterotrophic bacteria, belonging to the *Pseudomonas stutzeri* group. *FEMS Microbiol. Ecol.* **30**:113–123.

183. **Stanier, R. Y., R. Kunisawa, M. Mandel, and G. Cohen-Bazire.** 1971. Purification and properties of unicellular blue-green algae (order *Chroococcales*). *J. Bacteriol. Rev.* **35**:171–205.

184. **Stanier, R. Y., N. J. Palleroni, and M. Doudoroff.** 1966. The aerobic pseudomonads: a taxonomic study. *J. Gen. Microbiol.* **43**:159–271.

185. **Stolp, H., and M. P. Starr.** 1963. *Bdellovibrio bacteriovorus* gen. et sp. n., a predatory, ectoparasitic, and bacteriolytic microorganism. *Antonie Leeuwenhoek J. Microbiol. Serol.* **29**:217–248.

186. **Strohl, W. R.** 1989. Genus I. *Beggiatoa* Trevisan 1842, p. 2095–2096. *In* J. T Staley, M. P. Bryant, N. Pfennig, and J. G. Holt (ed.), *Bergey's Manual of Systematic Bacteriology*, vol. 3. The Williams & Wilkins Co., Baltimore, MD.

187. **Strohl, W. R., and J. M. Larkin.** 1978. Enumeration, isolation, and characterization of *Beggiatoa* from freshwater sediments. *Appl. Environ. Microbiol.* **36**:755–770.

188. **Stürmeyer, H., J. Overmann, H.-D. Babenzien, and H. Cypionka.** 1998. Ecophysiological and phylogenetic studies of *Nevskia ramosa* in pure culture. *Appl. Environ. Microbiol.* **64**:1890–1894.

189. **Swift, S., M. J. Lynch, L. Fish, D. F. Kirke, J. M. Tomas, G. S. A. B. Steward, and P. Williams.** 1999. Quorum sensing-dependent regulation and blockade of exoprotease production in *Aeromonas hydrophila. Infect. Immun.* **67**:5192–5199.

190. **Swings, J., J. De Ley, and M. Gillis.** 1984. Genus 111. *Frateuria* Swings et al. 1980, p. 211. *In* N. R. Krieg and J. G. Holt (ed.), *Bergey's Manual of Systematic Bacteriology*, vol. 1. The Williams & Wilkins Co., Baltimore, MD.

191. **Tamir, H., and C. Gilvarg.** 1966. Density gradient centrifugation for the separation of sporulating forms of bacteria. *J. Biol. Chem.* **241**:1085–1090.

192. **Tchan, Y.-T., and P. B. New.** 1984. Genus 1. *Azotobacter* Beijerinck 1901, p. 220. *In* N. R. Krieg and J. G. Holt (ed.), *Bergey's Manual of Systematic Bacteriology*, vol. 1. The Williams & Wilkins Co., Baltimore, MD.

193. **Teske, A., N. B. Ramsing, K. Habicht, M. Fukui, J. Küver, B. B. Jørgensen, and Y. Cohen.** 1998. Sulfate-reducing bacteria and their activities in cyanobacterial mats of Solar Lake, Sinai (Egypt). *Appl. Environ. Microbiol.* **64**:2943–2951.

194. **Teske, A., P. Sigalevich, Y. Cohen, and G. Muyzer.** 1996. Molecular identification of bacteria from a coculture by denaturing gradient gel electrophoresis of 16S ribosomal DNA fragments as a tool for isolation in pure cultures. *Appl. Environ. Microbiol.* **62**:4210–4215.

195. **Thamdrup, B., K. Finster, J. W. Hansen, and F. Bak.** 1993. Bacterial disproportionation of elemental sulfur coupled to chemical reduction of iron or manganese. *Appl. Environ. Microbiol.* **59**:101–108.

196. **Thompson, J. P., and V. B. D. Skerman.** 1979. *Azotobacteraceae: The Taxonomy and Ecology of the Aerobic Nitrogen-Fixing Bacteria.* Academic Press Ltd., London, United Kingdom.

197. **Torsvik, V., J. Goksoyr, and F. L. Daae.** 1990. High diversity in DNA of soil bacteria. *Appl. Environ. Microbiol.* **56**:782–787.

198. **Turley, C. M.** 1993. Direct estimates of bacterial numbers in seawater samples without incurring cell loss due to sample storage, p. 143–147. *In* P. F. Kemp, B. F. Sherr, E. B. Sherr, and J. J. Cole (ed.), *Handbook of Methods in Aquatic Microbial Ecology.* Lewis Publishers, Boca Raton, FL.

199. **Van Niel, C. B.** 1971. Techniques for the enrichment, isolation, and maintenance of the photosynthetic bacteria. *Methods Enzymol.* **23**:3–28.

200. **Veldkamp, H.** 1961. A study of two marine agar-decomposing, facultatively anaerobic myxobacteria. *J. Gen. Microbiol.* **26:**331–342.
201. **Verma, U. K., D. J. Brenner, W. L. Thacker, R. F. Benson, G. Vesey, J. B. Kurtz, P. J. L. Dennis, A. G. Steigerwalt, J. S. Robinson, and C. W. Moss.** 1992. *Legionella shakespearei* sp. nov., isolated from cooling tower water. *Int. J. Syst. Bacteriol.* **42:**404–407.
202. **Vester, F., and K. Ingvorsen.** 1998. Improved most-probable-number method to detect sulfate-reducing bacteria with natural media and a radiotracer. *Appl. Environ. Microbiol.* **64:**1700–1707.
203. **Vishniac, W., and M. Santer.** 1957. The thiobacilli. *Bacteriol. Rev.* **21:**195–213.
204. **Von Rheinbaben, K. E., and R. M. Hodlak.** 1981. Rapid distinction between micrococci and staphylococci with furazolidone agars. *Antonie Leeuwenhoek J. Microbiol. Serol.* **47:**41–51.
205. **Vreeland, R. H.** 1992. The family *Halomonadaceae*, p. 3181–3188. *In* A. Balows, H. G. Trüper, M. Dworkin, W. Harder, and K.-H. Schleifer (ed.), *The Prokaryotes. A Handbook on the Biology of Bacteria: Ecophysiology, Isolation, Identification, Applications*, 2nd ed. Springer-Verlag, New York, NY.
206. **Warke, G. M., and S. A. Dhala.** 1968. Use of inhibitors for selective isolation and enumeration of cytophagas from natural substrates. *J. Gen. Microbiol.* **51:**43–48.
207. **Waterbury, J. B., and R. Y. Stanier.** 1978. Patterns of growth and development in pleurocapsalean cyanobacteria. *Microbiol. Rev.* **42:**2–44.
208. **Watkins, J., and K. P. Sleath.** 1981. Isolation and enumeration of *Listeria monocytogenes* from sewage, sewage sludge and river water. *J. Appl. Bacteriol.* **50:**1–9.
209. **Whitman, W. B., D. C. Coleman, and W. J. Wiebe.** 1998. Prokaryotes: the unseen majority. *Proc. Natl. Acad. Sci. USA* **95:**6578–6583.
210. **Widdel, E., and N. Pfennig.** 1977. A new anaerobic, sporing, acetate-oxidizing sulfate-reducing bacterium, *Desulfotomaculum* (emend.) *acetoxidans. Arch. Microbiol.* **112:**119–122.
211. **Widdel, F.** 1983. Methods for enrichment and pure culture isolation of filamentous gliding sulfate-reducing bacteria. *Arch. Microbiol.* **134:**282–285.
212. **Wiley, W. R., and J. L. Stokes.** 1962. Requirement of an alkaline pH and ammonia for substrate oxidation by *Bacillus pasteurii. J. Bacteriol.* **84:**730–734.
213. **Williams, M. A., and S. C. Rittenberg.** 1957. A taxonomic study of the genus *Spirillum* Ehrenberg. *Int. Bull. Bacteriol. Nomencl. Taxon.* **7:**49–111.
214. **Williams, R. A. D., and M. S. Da Costa.** 1992. The genus *Thermus* and related microorganisms, p. 3745–3753. *In* A. Balows, H. G. Trüper, M. Dworkin, W. Harder, and K.-H. Schleifer (ed.), *The Prokaryotes. A Handbook on the Biology of Bacteria: Ecophysiology, Isolation, Identification, Applications*, 2nd ed. Springer-Verlag, New York, NY.
215. **Wilson, G. S., and A. A. Miles.** 1964. *Topley and Wilson's Principles of Bacteriology and Immunity*, 5th ed. The Williams & Wilkins Co., Baltimore, MD.
216. **Wolfe, R. A., R. K. Thauer, and N. Pfennig.** 1987. A 'capillary racetrack' for isolation of magnetotactic bacteria. *FEMS Microbiol. Ecol.* **45:**31–35.
217. **Zeikus, J. G.** 1977. The biology of methanogenic bacteria. *Bacteriol. Rev.* **41:**514–541.
218. **Zengler, K., G. Toledo, M. Rappe, J. Elkins, E. J. Mathur, J. M. Short, and M. Keller.** 2002. Cultivating the uncultured. *Proc. Natl. Acad. Sci. USA* **99:**15681–15686.

12

Culture Techniques[†]

SYED A. HASHSHAM

The purpose of this chapter is to describe the main culture techniques for microbial growth. Many techniques are available including solid, semisolid, biphasic, immobilized, and liquid cultures. Factors governing the choice of the technique used for microbial cell cultures include the mass of microbial cells or their by-products desired, the mode of growth sought (aerobic respiration, anaerobic respiration, or fermentative), safety issues (e.g., production of aerosols), and experimental objectives (protein production, modeling, observation, etc.). For most biotechnology objectives (e.g., large-scale production of antibiotics), several culture techniques are combined in series to accomplish the task. This chapter first describes the solid, semisolid, biphasic, membrane surface, and immobilized culture techniques (section 12.1). This is followed by laboratory scale liquid-culture techniques including specialized liquid cultures such as synchronous and dialysis cultures (section 12.2). References for further reading are also included.

[†]This chapter is a revised and combined version of previous chapter 9 (written by Noel R. Krieg and Philipp Gerhardt) and previous chapter 10 (written by Philipp Gerhardt and Stephen W. Drew) in *Methods for General and Molecular Bacteriology* (MGMB).

12.1. SOLID, SEMISOLID, MEMBRANE SURFACE, IMMOBILIZED-CELL, AND BIPHASIC CULTURES

Culture media prepared in the solid state, in the form of firm gels, have been used in bacteriology since adopted by Robert Koch. The most important features of solidified media stem from their ability to enable separated colonies to arise from individual cells in a population diluted into or onto a solidified medium. Thus, the streak plate is a simple but effective technique for isolating pure cultures of bacteria and the pour plate, spread plate, and layered plate are similarly valuable for enumerating viable bacteria. Other single-cell or single-colony transfer techniques that rely on solid culture are multiple-point inoculation with velveteen and the auxanographic method. Solid media are also used in mass culture, bioautography, and physiological studies of bacterial cells. Semisolid, biphasic, and immobilized cultures are variations of the solid-culture method with respect to the consistency and type of the gelling agent used. The following sections describe each of the above, their applications, and the main type of solidifying agents currently in use.

12.1.1. Solid Cultures and Solidifying Agents

Solid culture is one of the most useful techniques in the isolation and cultivation from single cells. The maximum density of bacterial cells is obtained by growth as a colony on a solid surface. If the inoculum is highly diluted, each cell has the potential to develop into a distinct colony; such a procedure may be desirable even for mass culture if uniformity of colony characteristics is required before harvesting of the solid culture (for example, if one wants to harvest cells only when the colony texture becomes smooth). More frequently, the inoculum is not diluted and is spread evenly over the surface so that confluent solid growth results. The solid surface usually is that of an agar or otherwise solidified medium. However, a number of techniques have been described in which colonial growth is obtained on a membrane over a reservoir of liquid or solidified medium (see section 12.1.3). In solid culture the cells are already concentrated, so there is no need to use a centrifuge or other means of harvesting the cells. Solid cultures are also relatively free from macro- and micromolecular compounds of the nutrient medium. Certain cell forms, e.g., fruiting bodies of myxobacteria and endospores of certain *Bacillus* species, are better produced on a solid medium (sometimes this is the only medium from which they can be obtained). Solid culture, however, is limited in the extent to which it can be scaled up (e.g., one can effectively produce gram amounts of cells, but dekagram amounts become difficult and hectogram or kilogram amounts are impossible to achieve in the laboratory). Solid cultures are also not homogeneous in physiological properties of the cells. For example, the cells at the top of an aerobic colony (which is likely to be 1,000 cells deep) are nutrient starved but oxygen rich, whereas the opposite situation prevails at the bottom of the colony. In addition, solid cultures often yield a small number of cells from a given amount of medium. The following paragraphs describe some of the commonly used solidifying agents.

12.1.1.1. Agar

Agar is the most commonly used solidifying agent for microbiological media. It mainly consists of two polysaccharides, agarose [the gelling component, a repeating sequence of (1-4)-linked 3,6 anhydro-α-L-galactose and (1-3)-linked β-D-galactose] and agaropectin (the ionic, nongelatinous portion of agar [72]). Agaropectin may contain methylated or unmethylated galactoside units, sulfate, and pyruvate in varying amounts as well as D-glucuronic acid (72). Agarose makes up about 60% of the mixture (9). Agar is extracted from certain red marine macroalgae. When first extracted, agar is contaminated by algal cell debris and various impurities, most of which must be removed before the product is suitable for microbiological purposes. The key properties of agar that make it a versatile gelling agent for microbiological work are the following: it is not enzymatically degraded by most bacterial species; agar gels are stable up to 85°C or higher, yet molten agar does not gel until cooled to ca. 40°C; and agar gels have a high degree of transparency. Agar is available in various commercial grades, but for most culturing purposes "bacteriological" grade is satisfactory. Lower grades of commercial agar may contain troublesome impurities, e.g., starches, fatty acids, Cu^{2+}, or bleaching agents that are toxic; elevated levels of Ca^{2+} and Mg^{2+} that can cause precipitates; hydrogen peroxide (85); and thermoduric spores that resist the usual autoclaving procedures. "Special," "Select," "Noble," and "Purified" grades contain decreased levels of impurities and thus are suitable for electrophoretic, nutritional, enrichment and isolation, genetic, recombinant DNA, serological, and other special applications. Whenever an agar is used for a special purpose, pretest it to ensure effectiveness, check with the manufacturer's technical service for suitability and analysis, and if possible, use a single control lot number. Trace amounts of agar can sometimes inhibit PCR and nucleic acid sequencing (31, 103, 104).

To prepare an agar-solidified medium, first adjust the pH of the liquid medium to the desired value and then add the granular agar. Bacteriological or higher grades of agar do not alter the pH of the medium appreciably. For solid medium, use 15 to 20 g of agar per liter; for isolating highly motile organisms, even higher concentrations of agar may be required. For semisolid medium, use 1 to 4 g/liter, depending on the consistency required. Different brands and grades of agar may require different concentrations to achieve a particular degree of firmness, and the instructions of the manufacturer should always be consulted in this regard.

After adding the agar to a liquid medium, heat the mixture to boiling and dissolve the agar completely. If the medium is being heated over a flame or on a hot plate, stir it constantly during heating to prevent the agar from settling to the bottom of the container, where it can caramelize and char; then, bring the medium to a rolling boil for 1 min or so, being careful to avoid having it foam up and over the edge of the container. Alternatively, the agar can be dissolved by placing the mixture in a steam cabinet or microwave oven for an appropriate period. After the agar is melted, mix to ensure uniformity in concentration throughout the medium. Dispense the molten medium into tubes, flasks, or bottles, and sterilize in an autoclave.

When an agar medium with a pH of 6.0 or less is required, initially prepare and sterilize the medium at a pH greater than 6.0; otherwise, the agar will be hydrolyzed during heating and will fail to solidify adequately when the medium is later cooled. Once the medium has been sterilized and cooled to 45 to 50°C, add sufficient sterile acid aseptically to achieve the final pH value desired. It is useful to prepare an extra portion of medium to experiment with in order to determine the correct amount of acid to add to the main batch. Alternatively, autoclave agar separately at $2\times$ strength in water (or a neutral mixture of some of the medium components) and mix with the autoclaved, acidic main medium components when cooled.

For preparing tubes of slanted medium, place the hot tubes from the autoclave in a tilted position and allow them to cool and solidify. For preparing petri dishes of medium, first cool the molten medium to 45 to 50°C in a water bath and then dispense the medium aseptically into the bottom halves of the sterile dishes (usually 15 to 20 ml per dish), taking care to replace the tops after each dish is poured. Cover the dishes with newspaper or other insulation to prevent condensation of moisture underneath their tops as the agar solidifies. Stacking the poured plates also discourages condensation under the lids.

Notes

1. After the agar solidifies, drops of moisture (water of syneresis) may form on the surface. This moisture should be evaporated before the plates are inoculated. Usually, storage of the dishes in an inverted position overnight at room temperature results in sufficient drying of the agar. For more rapid drying, invert the dishes on the shelf of a 45°C gravity-convection incubator or in a sterile hood and adjust the agar-containing halves so that they are slightly ajar. Incubate in this manner until the water of syneresis disappears. For some procedures, such as in the use of velveteen replicators, dishes that have been dried more extensively may be required.

2. For storage of agar dishes under conditions that prevent severe drying, place them inverted in closed polyethylene bags (the bags in which plastic petri dishes are packaged make excellent storage bags for media). Agar dishes can be stored in this way at room temperature for 4 to 5 weeks and under refrigeration for even longer periods. Media containing blood or other heat-labile components should always be refrigerated.

3. Store all culture media, liquid and solid, in the dark. Storage under illumination, especially sunlight (as near a window), may result in photochemical generation of hydrogen peroxide or other toxic forms of oxygen, which can render the media inhibitory for the growth of bacteria (85, 100). Microaerophiles appear to be particularly sensitive to media that have been subjected to illumination (40).

12.1.1.2. Carrageenan

Carrageenan, long used in the food and dairy industry, has also been used as a solidifying agent for microbiological media. Also known as Irish moss or vegetable gelatin, carrageenan is extracted from marine macroalgae (14, 20, 44). There are several types of carrageenan: kappa, lambda, mu, and iota (94). All are complex polysaccharides composed of several types. Kappa carageenan, for example, consists of alternating units of mostly sulfated 3-linked D-galactosyl and 4-linked D-galactosyl that may also be sulfated. The potassium salt of kappa carrageenan is capable of forming rigid transparent gels, which can be an effective substitute for agar in many microbiological media (47, 101). Carrageenan is considerably less expensive than agar. Like agar gels, carrageenan gels can be used for streaking or spreading inocula. Carrageenan is not degraded by most species of bacteria, and gels stable to temperatures of 60°C can be prepared. However, carrageenan does have certain limitations. One is the high temperature at which liquid carrageenan media must be dispensed into petri dishes (55 to 60°C), which may preclude its use in pour plates for enumerating certain species and precludes its use in making media containing blood. Because of its high viscosity, it is difficult to dispense

molten carrageenan media by pipetting. For gels stable during incubation to 45°C, 2.0% carrageenan is used; for gels stable to 60°C, 2.4% carrageenan is used (47). Variability may exist between lots of carrageenan (101). Some investigators have found carrageenan to be unsuitable for semisolid media, whereas others have found it to be satisfactory (10, 47). In contrast to agar, carrageenan may cause alterations in the pH of some media during preparation and sterilization; therefore, determine whether and to what magnitude such pH changes occur for any given kind of medium. It may be necessary to compensate for such changes by preparing the medium at a different initial pH value than desired finally or by increasing the buffering capacity of the medium. Media containing a phosphate buffering system seem to be particularly likely to exhibit a large decrease in pH (47).

12.1.1.3. Silica Gel

Silica gel is an inorganic solidifying agent used in media for solid culture of autotrophic bacteria in the complete absence of organic substances (25, 41, 62, 98). Such inorganic media can be supplemented with various organic compounds to study the ability of heterotrophic bacteria to use these compounds as sole carbon sources. Vitamin requirements can also be determined by the use of silica gel media. Silica gel media can be prepared according to the methods of Funk and Krulwich (25) or Sommers and Harris (86) or more recent approaches (62, 98).

12.1.1.4. Gellan Gum

Gellan gum is an exopolysaccharide produced by the gram-negative bacterium *Sphingomonas paucimobilis* (5, 30, 38, 73, 92). After processing of the native product, gellan gum has relatively better clarity than bacteriological-grade agar at much lower concentrations (91). Gelrite and Phytogel are some of its trade names (91). Initial applications demonstrated a clear advantage of gellan gum for culturing and quantifying thermophilic bacteria at temperatures to 120°C and pressures to 265 atm (26,851 kPa) (15, 46, 83). It was also shown to be noninhibitory in PCR applications (70, 76). Prepare solid media solidified with gellan gum in accordance with the manufacturer's or other published directions or after appropriate experimentation.

12.1.1.5. Pluronic Polyol

Pluronic polyol, also known as reverse agar, liquefies on cooling to ca. 4°C and gels into a semirigid clear jelly at temperatures of 11 to 32°C depending on the polyol concentration (27). It is a block copolymer of polypropylene oxide and ethylene oxide and is a nontoxic solidifying agent. Its unusual solidifying characteristics have been used to isolate heat-sensitive bacteria and to enrich for denitrifying, sulfate-reducing, and methanogenic microbes (27). It is especially useful for isolation of fungal colonies (71).

12.1.1.6. Gelatin

Gelatin was the first solidifying agent used (by Robert Koch) for microbiological media, but it was soon replaced by agar because gelatin is liquefied by some species and melts at a relatively low temperature. At the 12% concentration effective for solidification, it melts at 28 to 30°C. Gelatin presently is used mainly as a specific hydrolysis test substrate in systematics. However, mixed with Noble agar it was found to be more suitable for the isolation of oral spirochete colonies than agar alone (12).

12.1.1.7. Starch

Starch of the type used for gel electrophoresis has been used as a solidifying agent in culture media for thermoacidophilic Archaea such as Thermoplasma, Acidianus, and Sulfolobus species (78, 79). For Thermoplasma species, starch plates have been reported to be superior to agar and gellan gum plates (78). Media containing 10 to 12% (wt/vol) starch are boiled to dissolve the polymer, poured like agar into petri dishes, and allowed to solidify and to dry overnight at 4°C (79). Starch also plays a critical role in medium formulations for the isolation of various pathogens and mammalian microflora (37, 61, 68).

12.1.1.8. Isubgol

Isubgol is a mucilaginous husk derived from the seeds of Plantago ovata (36). Its use as an economical gelling agent has been reported for both plant tissue culture (3, 7) and microbiological growth (36, 74). It requires a higher concentration than does agar.

12.1.2. Semisolid Culture and Applications

Semisolid media contain a low concentration of gelling agent (e.g., 0.1 to 0.4% agar) and have a soft, jelly-like consistency. Such media are useful for cultivating microaerophilic bacteria and for studying various aspects of motility and chemotaxis. These applications are briefly described below.

12.1.2.1. Microaerophiles

Microaerophilic bacteria must be grown under low levels of oxygen. They cannot grow under an atmosphere containing high levels of oxygen, e.g., 21% oxygen present in air. Campylobacter jejuni, a microaerophile, grows best in a 6% oxygen atmosphere. Helicobacter pylori living in mammalian or avian gut is also a microaerophile (39). Certain nitrogen-fixing bacteria that have a respiratory type of metabolism may grow as aerobes when supplied with a source of fixed nitrogen (such as ammonium sulfate) but can grow only as microaerophiles in nitrogen-deficient media. For example, Azospirillum species grow best in a 1% oxygen atmosphere under nitrogen-fixing conditions; the nitrogenase complex would be inactivated by excess oxygen. The general subject of microaerophily has been reviewed by Krieg and Hoffman (40).

The key factor for growing microaerophiles is the availability of "low" concentrations of O_2, as opposed to completely aerobic conditions or total lack of oxygen. To obtain a low concentration of O_2, microaerophiles may sometimes be grown deeper in a solid medium because oxygen diffusion is limited under such conditions. Diffusivity of oxygen through agar is reported to be in the range of 7.60×10^{-8} to 2.45×10^{-9} m^2s^{-1} at 30°C (53). Thus, if tubes of semisolid media are stored for long periods, oxygen may eventually diffuse deeply into the media. Placing such tubes in a boiling-water bath for several minutes may help drive off the dissolved oxygen.

The microaerophile C. jejuni can be easily cultivated in tubes of semisolid brucella medium (brucella broth containing 0.15 to 0.3% agar). The nitrogen-fixing Azospirillum species grow well in a semisolid nitrogen-deficient malate medium, and large crops of these cells for physiological studies can be obtained by using 1-liter Roux bottles containing 150 ml of culture medium stratified with 0.05% agar (59). This medium forms an 8-mm-deep layer on the wide, flat bottom of the Roux bottle. The agar concentration is high enough to allow the microaerophile to grow under aerobic conditions but low enough to permit the culture to be poured or pipetted from the Roux bottle after incubation into a sterile vessel. Sterile phosphate buffer is then added to the vessel, and the cells are collected by centrifugation and washed.

12.1.2.2. Motility

Semisolid media can also be used to determine whether a bacterial strain is motile. Inoculate a tube of motility medium (broth containing 0.4% agar) with a straight needle to one-half the depth of the tube. During growth, motile bacteria migrate from the line of inoculation to form diffuse turbidity in the surrounding medium; nonmotile bacteria grow only along the line of inoculation.

A modification of this method is to place a piece of open-ended glass tubing into the semisolid medium prior to sterilization. Inoculate only the upper part of the medium within the glass tubing. If the organism is motile, it will migrate downward, emerge from the bottom of the tubing, and then migrate upward to the surface of the medium, where it will grow extensively. This method is also helpful for selecting highly motile organisms for use in preparing H antigens for immunization. However, the method is suitable only for facultative organisms capable of metabolism under anaerobic conditions, such as members of the family Enterobacteriaceae. Strictly aerobic organisms such as Pseudomonas species do not migrate downward to the bottom of the glass tubing. Strategies to measure the rate of motility, e.g., by attaching latex particles to the cell, are also available (52). The subject of motility has been reviewed by Harshey (32).

12.1.2.3. Chemotaxis

Semisolid media are also useful in chemotaxis studies. For example, a semisolid medium containing an oxidizable carbon and energy source can be used to investigate positive chemotaxis in Escherichia coli (2), Vibrio cholerae (8), Salmonella enterica (87), and many other species. It is one of the most extensively studied subjects with reviews (2, 49, 93), methods of measurements (60), and models (88). When a petri dish containing the medium is inoculated heavily at its center, the bacteria migrate outward as an expanding ring of cells that consumes all of the oxidizable substrate as the cells move. When two oxidizable carbon sources are provided in the medium, two rings which migrate at different rates are formed. In contrast to capillary-tube methods for studying chemotaxis, the oxygen supply on the agar plate is never exhausted; thus, the migrating rings form only in response to self-created substance gradients and not in response to self-created oxygen gradients.

Negative chemotaxis also has been studied by using semisolid media (95). E. coli cells are suspended in the medium at a concentration sufficient to give visible turbidity, and the inoculated medium is dispensed into a petri dish. After the medium gels, a plug of hard agar (2% agar) containing the suspected repellent compound is inserted into the medium. If the compound is a repellent, the surrounding bacteria migrate away from the plug and leave a clear zone that becomes visible in ca. 30 min.

The use of plates of semisolid media also makes it possible to select nonchemotactic mutants (95). In the case of positive chemotaxis, nonchemotactic mutants (and also nonmotile mutants) fail to respond to a self-created substrate gradient and remain near the center of the petri plate, where they can be removed for subculturing and purification. In the case of negative chemotaxis, the nonchemotactic mutants can be isolated from the clear zone surrounding the hard agar plug.

12.1.3. Membrane Surface Cultures

Solid culture of bacteria can also be accomplished on the surface of a dialysis or microfiltration membrane in contact with an underlying reservoir of liquid or solidified medium. The principles and development of such colonial growth with membranes have been reviewed from time to time (26, 43, 66, 77). This technique is particularly useful for viable counts of dilute concentrations of cells in air or water after membrane filtration, for studying conjugational transfer of plasmids (97), and for studying microbial interactions (57).

12.1.4. Bioautography

An example of membrane surface culture is bioautography. It is a version of paper chromatography in which the growth of bacteria is used as a highly sensitive indicator for locating the positions of certain compounds on a paper chromatogram. The method has the advantage of specifically detecting compounds with biological activity, which chemical or radioisotopic detection systems lack. The method is particularly applicable to locating the position of growth factors on chromatograms of spent culture media or cell extracts when the concentration of the growth factors is so low as to preclude the use of ordinary detection systems. For example, the location of as little as 5 to 10 ng of folic acid on a chromatogram can be determined by use of bioautography. For a good example of bioautography applied to the detection of growth factors in cell extracts, see reference 84. Bioautography is also widely used by the pharmaceutical industry for the detection of antibiotic agents on paper chromatograms (69, 102). For bioautography of antibiotic agents, the agar medium should be nutritionally complete to allow the indicator organism to grow throughout the dish except at regions of the chromatogram where the antibiotic substance is located. More recent approaches have disks with multiple antibiotics that are easier to handle and use.

12.1.5. Immobilized-Cell Cultures

Immobilization of cells is important in many areas including maximization of product formation, minimization of cell mass, water distribution systems, groundwater remediation, and medical devices. Product maximization by immobilized cells has been practiced for many decades; e.g., in the "quick vinegar" process, acetic acid bacteria are immobilized as a colony film on a column of wood shavings, through which an ethanol solution is trickled downward while air is passed upward, yielding an acetic acid solution. Once the bacterial film is established, little further growth occurs and bacterial maintenance energy is derived from the oxidation of ethanol to acetate. The process can be operated continuously over periods of weeks or months. Similar examples exist for the production of kojic acid, using *Aspergillus oryzae* NRRL484 (99), or laccase from *Coriolus versicolor* IFO4937, using more sophisticated membrane surface liquid reactors (34). Immobilized cell culture technology has been reviewed for various applications (see reference 58 and other articles in the same issue). Laboratory scale studies often employ model surfaces and reactors that mimic the relevant environmental surface characteristics and conditions and allow intrusive or nondestructive observation of the immobilized cells.

12.1.6. Biphasic Cultures

Biphasic culture consists of a thick layer of solidified nutrient medium overlaid with a thin layer of nutrient broth (96). Populations in excess of 10^{11} cells per ml can be obtained in this way, and the technique is apparently applica-

ble to any type of bacterium. Biphasic cultures are found to be more efficient in growing many pathogens (6, 80) including *Mycobacterium* (1) and *Arcobacter* spp. (17). To prepare such a system, partially fill the container with hot medium containing 2 to 3% agar. After the agar is solidified, overlay it aseptically with a small volume of broth, inoculate, and incubate. If the bacteria are aerobic, clamp the container on a mechanical shaker to provide aeration and agitation of the broth during incubation. The culture is confined to the liquid overlay but has diffusional access to the reservoir of nutrients in the solidified base, and consequently the culture becomes densely concentrated. Since the movement of nutrients is dependent on diffusion, the agar base should be limited to about 5 cm in depth. The ratio of solid to liquid should be at least 4:1 and less than 10:1; the lower ratio provides a better yield, and the higher ratio provides a greater concentration of cells. An Erlenmeyer flask is a convenient container, but indentations at its base are helpful to hold the agar in place if a mechanical shaker is used; rectangular containers also are useful in this way. Also, use a greater percentage of agar if breakup occurs during shaking.

If all of the medium components are incorporated into the agar base, the overlay can be distilled water. After an overnight equilibration period, the resulting clear diffusate in the overlay can be inoculated. For bacteria such as gonococci, which normally must be grown in a turbid medium enriched with blood and starch and which yield sparse populations, a relatively clean and dense population of cells can be harvested from such a diffusate overlay (28). A variety of bacteria have been preserved in the short term by subculturing in a biphasic medium.

12.2. LABORATORY SCALE LIQUID CULTURES

Laboratory scale liquid cultures provide one of the most common techniques to grow and study the behavior of microorganisms. Containers for such cultures vary from simple culture tubes to large carboys and reactors. The setup depends on the need to maintain sterility, sampling, electron acceptor conditions, substrate addition and product removal, monitoring, and mixing regimes. The following sections describe the basic types of systems used for liquid cultures both in batch and continuous systems. It also describes some of the special types of liquid cultures that require dialysis bags or large animal hosts.

12.2.1. Range of Electron Acceptors and Donors and Carbon Sources

In all cases of microbial growth, it is important to ensure that the electron acceptor, electron donor, carbon source, and trace element requirements are met as per the needs of the particular type of microorganism. The section on energetics and stoichiometry of chapter 13 describes the theoretical aspects of microbial growth focusing on the electron acceptor and donor and the carbon and nitrogen sources. In addition, phosphorus and trace elements must also be added in sufficient quantities. Phosphorus requirements are generally met by adding one-fifth of the amount of nitrogen required using the stoichiometric equation (see chapter 13.1.10). Requirements for media preparation and trace elements vary considerably among the range of microorganisms known today. For example, the range of electron acceptors for *Shewanella oneidensis* MR-1 and *Shewanella putrefaciens* strain 200 includes O_2, NO_3^-, NO_2^-, NO,

Fe^{3+}, Mn^{4+}, Mn^{3+}, SO_3^{2-}, $S_2O_3^{2-}$, S, dimethyl sulfoxide, arsenate, chromate, vanadate, neptunium, fumarate, selenium, and technitium with the capability to use several organic acids, sugars, and hydrogen as electron donors. Table 1 lists a number of common electron acceptors and donors for some of the common types of microorganisms. Establishing the media constituents for growing a given type of microorganism, especially if it is a relatively less studied organism, should begin by a good literature search and evaluation with respect to the objective.

12.2.2. Batch Culture Vessels

12.2.2.1. Culture Tubes

12.2.2.1.1. Aerobic
The lipless Pyrex glass culture tube (usually 16 by 150 mm), plugged with nonabsorbent cotton or plastic foam, is the most convenient and widely used container for batch liquid culture of bacteria. However, for aerobic cultivation, test tubes usually provide only minimally effective conditions of oxygen supply. Fortunately, many bacteria are facultatively anaerobic and most microbiological uses of culture tubes do not require optimum growth conditions. To improve oxygen supply, increase the surface-to-volume ratio of the liquid medium by reducing the volume and slanting the tubes, or preferably, mount them on a rotary shaking machine to induce a vortex. Also increase the availability of air by using a small and loosely packed cotton plug, or a plastic or stainless-steel cap.

12.2.2.1.2. Anaerobic
The use of thick-walled culture tubes (Fig. 1 [Bellco Biotechnology, Inc., Vineland, NJ]) is now common for many types of anaerobic applications ranging from isolation of dehalorespiring microorganisms to methanogenic activity measurements to demonstration of photoautotrophic activity (4). Larger-volume serum bottles (up to 1,000 ml) may also be employed for ease of use and larger biomass and sample needs. The use of anaerobe culture tubes and serum bottles requires (i) proper sealing by special butyl rubber stoppers that can be punctured without causing leaks, (ii) purging of the headspace with appropriate oxygen-free gases, and (iii) proper sealing of the tubes or bottles with aluminum crimp caps. Laboratories engaged in anaerobic culture systems routinely assemble a system of gas cylinders, regulators, gassing and sparging stations, and cannulas for the preparation of media and sampling needs.

12.2.2.2. Shake Flasks

12.2.2.2.1. Aerobic
An Erlenmeyer flask (capacity, 100 to 2,000 ml) is commonly used as a container for producing masses of cells in

TABLE 1 Electron acceptors and donors and carbon sources for selected microorganisms

Electron acceptor	Electron donor	Carbon source	Examples	Applications or context
O_2	Organic compounds	Organic compounds	Heterotrophs	Wastewater treament, remediation, many others
O_2	NH_4^+	CO_2	Nitrosomonas	NH_4^+ oxidation to NO_2^-
O_2	NO_2^-	CO_2	Nitrobacter	NO_2^- oxidation to NO_3^-
O_2	HS^-/H_2S, S	CO_2	Thiobacilli	Concrete corrosion
O_2	H_2	CO_2	H_2-oxidizing bacteria	NO_3^- removal (because they can grow on NO_3^-)
O_2	Fe^{2+}	CO_2	Thiobacillus ferrooxidans	Iron oxidation
O_2	CH_4	CH_4	Methanotrophs	TCE^a remediation
NO_3^-	Organic compounds, some can use H_2	Organic compounds, CO_2	Denitrifiers	NO_3^- removal
NO_3^-	HS^-/H_2S, S	CO_2	Thiobacillus denitrificans	NO_3^- removal
AsO_4^{3-}	HS^-/H_2S	CO_2	Strain MLMS-1	Arsenic removal
SeO_3^{2-}	Organic compounds, H_2	Organic compounds, CO_2	Shewanella sp., Bacillus selenitireducens	Selenium nanosphere production
SO_4^{2-}	Lactate, acetate, H_2	Organic compounds	Sulfate-reducing bacteria	Sewer, anaerobic processes
Fe^{3+}	Acetate	Acetate	Desulfuromonas	Acid mine drainage
Mn^{4+}	Acetate	Acetate	Desulfuromonas	
CO_2	H_2	CO_2	Methanogens, acetogens	Anaerobic digestion
Organic compounds	Organic compounds	Organic compounds	Clostridia, lactic acid bacteria	Fermentation, probiotics
NO_2^-	NH_4^+	CO_2	Anaerobic ammonia oxidizers	Anammox process (NH_4^+ and NO_2^- removal)
PCE^b/TCE	Acetate, formate, H_2	Acetate, CO_2	Dehalococcoides	Remediation
Cr^{6+}	Lactate	Lactate	Shewanella	Remediation
U^{6+}	Organic compounds, H_2	Organic compounds, CO_2	Geobacter sp., Shewanella sp.	Uranium mobility

[a]TCE, trichloroethene.
[b]PCE, tetrachloroethene.

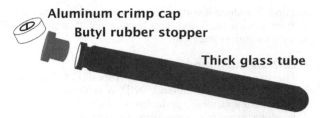

FIGURE 1 Thick glass culture tube with butyl rubber stoppers and aluminum crimp caps useful in isolation and culturing of anaerobic microorganisms.

the laboratory or as the first stage in a scale-up study of an industrial process involving batch liquid culture of bacteria. Consequently, maximally effective conditions of air (oxygen) supply are sought for obligately or facultatively anaerobic types of bacteria. Aerobic flask culture must be carried out with shaking of the flask to facilitate mass transfer of oxygen (as well as other gases and nutrients) at two levels: from the gas phase across the gas-liquid interface into the liquid phase, and from the liquid phase across the liquid-cell interface into the cell. Two major factors influence the ability of flask culture techniques to meet the oxygen requirements of the cells: gas exchange through the flask closure, and the liquid surface area available for oxygen transport from the gas phase.

The closure design for culture flasks must allow adequate gas exchange between the external environment and the flask interior, yet must maintain asepsis. Morton-type caps provide an adequate aseptic seal and good gas exchange, but they occasionally spin off a flask at high rotational speed. Foam plastic plugs also are effective aseptic closures for shake flask cultivation and provide moderate gas exchange, but the plugs must be sized to fit into the neck of the flask snugly yet without severe compression. Furthermore, foam plugs may contain plasticizers or other toxic chemicals, which may require thorough washing of the plugs prior to their use as closures for biological culture.

The rate at which dissolved oxygen is consumed in a liquid culture of bacteria is determined by the cell density and growth rate. The demand for dissolved oxygen is created by cell consumption and is met by continuous diffusion of oxygen from the gas phase to the liquid phase and thence into the cell. The interfacial surface area between gas and liquid often controls the flux or volumetric rate of oxygen transfer. A large interfacial surface area results in high volumetric rates of oxygen transfer and therefore allows rapid growth without oxygen limitation. The interfacial surface area can be maximized by maintaining low ratios of liquid volume to flask volume and by vigorously shaking the flasks. Oxygen limitation due to limited diffusion can usually be avoided by preparing flasks with liquid volumes of no more than 20% of flask volumes and by shaking the flasks at 200 to 350 rpm on a rotary shaker or at 150 to 250 strokes per min on a reciprocating shaker. During incubation of the inoculated flasks at an appropriate temperature, rotary shaking should cause the liquid meniscus to rise to approximately two-thirds of the flask height. Reciprocal shaking, such as is found in most water bath shakers, should be sufficiently vigorous to cause significant breaking (turbulence) of the liquid wave as it moves from side to side in the flask, but not so great as to wet the closure (24). Oxygen transfer is further enhanced by baffling. Baffled flasks have indentations in the side and/or bottom surface, which act to break vortex formation in flasks on a rotary shaking machine and to induce turbulence. Baffled flasks with good design of the baffles are available commercially (e.g., Wheaton Science Products).

12.2.2.2.2. Anaerobic

Flasks or bottles of any other convenient shape are used to cultivate anaerobic bacteria on an intermediate size scale. Two main principles prevail: the nutrient medium must be prereduced to an E_h of -150 mV or lower, and air must be removed from the immediate environment of the culture. Strict anaerobes require special precautions to remove all traces of oxygen. These are described more fully elsewhere in this manual.

A few practical tips apply to mass cultivation of anaerobes. Fill the container nearly full to minimize the effect of the overlying gas phase. Use as large an inoculum as possible, distributed from the bottom upward in a column of prereduced medium. Displace the gas phase with oxygen-free nitrogen or argon, plus 5 to 10% CO_2, and allow growth to begin with the medium quiescent. Once growth has progressed actively throughout the medium, provide agitation by means of a magnetic stirring bar or by purging nitrogen or an inert gas (e.g., helium or argon) through a tube with its outlet at the bottom of the container. Use a water trap to permit the escape of introduced and metabolic gases without the entrance of air. *Caution:* some anaerobic bacteria produce hydrogen or methane gas, which may cause flammable or even explosive conditions, and most anaerobes produce CO_2 and other gases, which will increase the internal pressure dangerously if adequate venting is not provided.

12.2.3. Batch Cultures

Growing microorganisms in batch culture vessels is one of the most common techniques employed in microbiology and biotechnology. In batch cultures, substrate concentration decreases from an initial value that is supplied to a final concentration remaining after the growth has slowed down significantly. Concentration of biomass or cell increases from an initial value that is determined by the inoculum size and liquid volume to a final concentration after growth. The volumes of batch cultures may vary from a few milliliters in small test tubes to thousands of liters in very large fermentors. Batch cultures invariably serve as an intermediate step between growth as a colony on a plate to small flasks to carboys to large fermentors. It should be noted that handling and manipulating large glass carboys during medium preparation, sterilization, and inoculation are hazardous, and shaking of filled glass carboys for aerobic cultivation is risky, although the use of polypropylene or polycarbonate plastic carboys lessens these concerns. Gas dispersion in any carboy is poor, even with stirrers, so that adequate aeration is unlikely. Although preparation of media in plastic carboys is sometimes necessary for large-scale work, cultivation in carboys should be avoided.

The following subsections describe some of the special types of batch cultures that require better control of one or more parameters related to substrate characteristics, microorganism growth stages, or feeding patterns.

12.2.3.1. pH-Controlled Batch Culture

Controlled batch culture requires instrumentation that monitors an environmental parameter and triggers an addition to the fermentor so that the indicator parameter is maintained at a steady value. An example of environmental parameter control in a fermentation system is the maintenance of pH at a constant value by the automatic addition of acid or base. The basic components of a pH control system are the pH electrode(s) with its shielded connector cable, the pH meter, an endpoint titrator, an acid or base reservoir with flexible tubing for connection to the fermentor, and a single peristaltic pump or solenoid valve. All components of the fermentor system, including pH and dissolved-oxygen electrodes, must be designed for steam sterilization or be compatible with other sterilization methods, such as ethylene oxide or formaldehyde-steam sparging. The pH meter and pH titrator must be interconnected and therefore should be purchased from the same manufacturer to ensure equipment match. Although a peristaltic pump can allow foolproof addition of acid or base against relatively high fermentor back-pressure, a simple tubing-pinch solenoid valve for control of gravity feed of acid or base to the fermentor is usually sufficient for laboratory scale operation. Fermentor manufacturers usually provide their own pH control systems.

12.2.3.2. Synchronous Batch Culture

Culture synchrony occurs when all of the cells divide at nearly the same instant, thus mimicking individual cell growth. For synchronous cultures, a plot of the logarithm of cell numbers versus time resembles a stair step rather than a straight line. Biochemical events associated with specific times in the individual cell cycle can be systematically studied in a synchronous culture. Although development of culture synchrony is relatively straightforward, maintenance of synchrony over long periods is a difficult task requiring precise control.

Culture synchrony can be achieved by periodically varying a critical environmental condition. The technique forces the synchronization of cell multiplication by interrupting, promoting, or retarding metabolic function in a cyclic manner. The population will gradually synchronize its response to these periodic disturbances of metabolic activity and will ultimately synchronize its growth pattern. The technique, however, requires severe disturbance of normal metabolic activity and is therefore of limited use in the study of growth cycle-linked bacterial physiology. One exception to this limitation is bacterial spore germination as a means of initiating culture synchrony. Step or pulse changes in the culture environment or addition of a specific chemical germinant can trigger spore germination and may closely model natural occurrences.

A generally useful method of synchronization is based on physical selection of a homogeneous fraction from a heterogeneous population of vegetative cells (21). This approach is often termed selection synchrony and avoids most of the problems of metabolic disturbance during synchronization by physically selecting cells that are in similar states of the cell growth cycle.

Kubitschek (42) and Poole (65) showed that the cell volume of *E. coli* increases linearly during the cell cycle while its cell mass increases exponentially. The observation that cell volume is lowest just after cell division suggests that centrifugation in a density gradient might allow recovery of a population of new daughter cells from an asynchronous culture. However, the observation that the volume increase is linear while the mass increase is exponential means that cells which are just ready to divide or cells which have just divided will have the greatest cell density despite cell size differences. Density gradient centrifugation

will result in cell fractions in which the most dense fraction contains both young (daughter) cells and mature (ready-to-divide) cells, while the least dense fraction will contain a homogeneous population of cells that have progressed through a common fraction of their cell cycles.

Selection synchrony through density gradient centrifugation cannot supply new daughter cells for direct study of cell cycle-linked physiology, because these fractions will always be contaminated with mature cells ready to divide. However, density gradient centrifugation can supply an adequate inoculum for synchronous culture growth. Cell cycle-linked physiology may then be studied by direct sampling of the synchronous culture. The procedure described below is presented as a general guideline for development of synchronous cultures and is based on the technique described by Mitchison and Vincent (54). This technique may be used as an initial guide for development of a synchronous cultivation technique specifically designed for the organism in use (48).

Synchronization Procedure

Delay synchronization experiments until reproducible batch cultivation conditions can be established, including determination of asynchronous culture kinetics. When these prerequisites are met, establish batch cultivation conditions so that two to five cell mass doublings occur during logarithmic growth.

1. Prepare 500 ml of sterilized medium in several culture vessels (250-ml Erlenmeyer flasks are convenient). Inoculate one half of the flasks, and incubate these under appropriate conditions. Store the remaining sterile flasks under identical conditions.

2. Harvest the cells from the batch culture at the mid-exponential growth phase. Rapidly cool the culture to 0 to 4°C by swirling the flasks in an ice bath. Harvest the cells by refrigerated centrifugation at 10,000 × g for 10 min.

3. Resuspend the sedimented cells in 2 ml of appropriate ice-cold buffer (0.1 M potassium phosphate; pH 7.0) by vigorous agitation with a Vortex mixer. If the cells tend to aggregate, use mild sonication or mild homogenization with a blender or tissue grinder to prepare a suspension of discrete cells. For cultures that are particularly difficult to suspend, use 0.01% (wt/vol) Tween 80 in the suspending buffer before mechanical or sonic treatment.

4. Rapidly, but carefully, layer the suspended cells on a sterile, precooled density gradient prepared from Ficoll, sucrose, Percoll, or another appropriate medium as dictated by the cell system. For example, to prepare an exponential sucrose gradient in a discontinuous manner, layer sterile ice-cold solutions of increasing sucrose concentration into presterilized, precooled (0°C) centrifuge tubes. Place 10 ml of 35% (wt/vol) sucrose in phosphate buffer into the bottom of a sterile centrifuge tube. Sequentially layer 10 ml each of sucrose-buffer solutions containing 26.5, 25.5, 24.5, 22.0, 19.0, and 15.0% (wt/vol), respectively, onto the 35% sucrose cushion (the buffers may all be prepared from the 35% sucrose stock solution). Filter sterilize the sucrose solutions prior to use in forming the gradient. All solutions must be ice-cold and preaerated (for aerobic cells).

5. Carefully centrifuge the cell-charged sealed centrifuge tubes in a precooled centrifuge at 2,500 × g for 15 to 20 min at 0°C. For more precise separations, adjust the time and speed of centrifugation so that the optically dense band of bacteria moves no more than two-thirds of the way down the centrifuge tube; speeds and times vary somewhat depending on the culture being handled.

6. Inoculate prewarmed flasks of growth medium with 0.5 ml directly from the lightest-density fraction of cells.

7. Carefully monitor the optical density of the newly inoculated culture flasks for a definite stepwise increase in optical density, an indication of synchronous growth.

Culture synchrony should be maintained for two to three cycles. A study over longer periods requires the reestablishment of synchrony through the procedure described above. Decay of synchrony occurs beyond a few cycles and should be carefully monitored and documented.

12.2.3.3. Fed-Batch or Sequencing Batch Culture

In a batch culture, the transition from exponential growth to stationary phase may occur for a variety of reasons, including the depletion of an essential nutrient substrate or the buildup of a toxic metabolite product. When the transition results from nutrient depletion, growth will continue if fresh medium is added. The medium addition rate and the culture volume must be increased exponentially to maintain a constant rate of exponential growth. This technique for exponential growth maintenance (18, 45) is usually called fed-batch culture. Periodic removal of culture volume to allow additional feeding is called extended-batch culture or repeated fed-batch culture. True fed-batch operation requires that the volume of the liquid medium in the fermentor increase during the fermentation. This requirement places an upper limit on the culture time based on feed rate and leads to changing conditions of aeration and agitation effectiveness. The fed-batch mode allows substantial improvements in cell mass or product productivity over an ordinary batch operation. The technique of fed-batch culture may also be used to supply large quantities of a potentially toxic substrate while maintaining a low concentration of the substrate in the medium.

12.2.4. Turbidostats

In a turbidostat, cell concentration (biomass) is monitored and maintained at a constant level by adjusting the feed rate of fresh nutrients (50, 55). The substrate need not be present in a limiting amount but, rather, is usually present in excess. The system may be operated over a wide range of biomass concentrations near the critical dilution rate as long as all medium components are in excess. However, it is most stable when the specific growth rate of the culture is near the maximum value, μ_{max}, for the particular medium in use. The term "turbidostat" includes any technique that holds the cell concentration constant and includes monitoring techniques based on cellular metabolism. Since certain metabolic functions (such as oxygen uptake, carbon dioxide evolution, and, in some cases, pH change) are intimately linked to cell growth rate and ultimately to the specific growth rate, these parameters may be used as control variables for medium replacement.

The precision with which the biomass can be controlled has historically been the weakest aspect of turbidostat cultivation. Most of the older methods of monitoring cell density rely on optical monitoring with some type of photoelectric sensor measuring scattered or transmitted light. These methods suffer from interference by bubbles and foam produced during aeration, wall growth on the fermentor and, even more critically, the optical device. Bubble interference can be eliminated by using an external flow-through optical cell. However, foaming remains a problem. Wall growth on optical surfaces can be partially remedied by wiping the surface, but the many designs for this have proved cumbersome and prone to failure.

12.2.5. Chemostats

Chemostats are completely mixed reactors with inflow of liquid containing the substrate and nutrients at a certain rate (volume per unit of time) and outflow of liquid at the same rate containing a much lower concentration of substrate and nutrient left over after use, and by-products and biomass produced from the biochemical reaction. Mixing in a chemostat is critical to minimize physical and chemical concentration gradients and population heterogeneity. Hence, they are sometimes referred to as continuous stirred tank reactors. Mixing or agitation is achieved by mechanical means or by gas sparging. Inadequate agitation allows the development of regions of poor mixing ("dead spots") and results in uncontrolled local concentration gradients. The physiological result of poor mixing is a highly heterogeneous cell population. For example, inadequate agitation of aerobic cultures can result in insufficient contact between the gas phase and the liquid medium. As a result, some regions of the reactor or vessel may be anaerobic. If the organism is facultative, the cell population will have a wide range of respiratory characteristics and the cell yield will be reduced. The operational volume of chemostats typically ranges from a few milliliters to thousands of liters, with typical laboratory units having operating volumes of 1 to 20 liters. Aerobic chemostats require lines for aeration, feeding, sampling, and a mechanism to mix the liquid. Often both aeration and mechanical mixing are practiced to avoid foam and dead space. Anaerobic chemostats need a water break at the effluent line to avoid exposure to air (Fig. 2). To maintain sterility and growth in the feed, the nutrient (nitrogen, phosphorus, and trace elements) line is often separated from the main line supplying the fermentable substrate and buffer. If intermittent feeding and/or withdrawal is needed, the system could be automated with the help of timers and automatic controllers.

Although it is possible to build a chemostat in a machine shop, the necessary attention to detail of design for mixing, aeration, sterilization, and maintenance of asepsis warrants the purchase of commercially available units. Chemostats with better control may also be the best option for more sophisticated studies related to, say, replication, gene expression patterns, and stress response. Commercial fermentors are capable of adequate aeration and agitation of relatively dense cultures and are usually equipped with a multiport head plate through which pH probes, dissolved-oxygen probes, and sampling or feed lines can be introduced into the fermentor. They can be quite expensive depending on the number of vessels, volume, probes, and desired controls.

Alternatives to commercial fermentors are also abundant and are in more common use. Chemostats of sizes varying from a few milliliters to tens of liters can be built in-house by the appropriate choice of vessel, mixing and aeration devices, peristaltic or syringe pumps for feeding and withdrawal, monitoring probes, temperature control systems, sampling ports, and controls and timers for automated operation. Laboratory-built systems are generally less sophisticated (unless the focus of building such a system is to push the limits of the existing commercially available systems) and may require somewhat more time for operation and maintenance. The main drawback of commercial units, however, is their high cost, requiring their purchase as specialized items instead of as routine laboratory equipment. Irrespective of the type of equipment or system used,

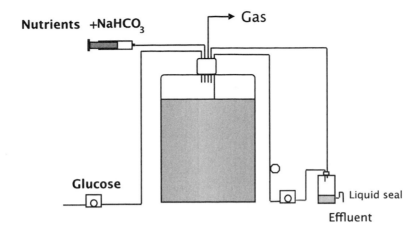

FIGURE 2 Schematic of an anaerobic continuous stirred tank reactor with separate pumps for supplying substrate and nutrients, effluent pump with water seal, and gas collection line. For smaller systems with multiple reactors, syringe pumps with multiple ports can be used.

the operation of a chemostat requires consideration of the following parameters in some form.

Preparation and operation of the chemostat require that all points of entry to or withdrawal from the system be designed for aseptic operation to prevent contamination. Most chemostats have a sample and inoculation line that penetrates the chemostat head plate and extends to within a few centimeters of the bottom of the vessel. Note that positive pressure on the fermentor must always be maintained. Chemostat continuous cultivation is always preceded by transient batch cultivation, during which time the cell mass accumulates at the expense of the substrate. Start the continuous culture while the batch cultivation is in the exponential phase of growth; this minimizes oscillations due to nutritional step-up and avoids inadvertent washout due to physiological lag. Start the dilution at a rate less than the desired operational dilution rate, and then increase to the operational dilution rate within one residence time, the average time that a cell remains in the vessel. This minimizes oscillations from toxic substrates. Chemostat response to toxic substrates is discussed in more detail by Pirt (63).

12.2.5.1. Sterilization of Liquid Media

Continuous cultivation requires the use of rather large quantities of medium and produces equally large volumes of product culture. For daily maintenance, it is advisable to prepare enough medium for 20 h of operation to allow flexibility in the scheduling of reservoir transfer. Since most laboratory scale operations require autoclaving of the medium as the means of sterilization, the required volumes of medium will probably be most conveniently prepared in carboys of appropriate size. (*Caution:* exercise extreme care in handling large volumes of hot liquids. Use autoclavable plastic carboys rather than glass ones.) For example, a 1-liter liquid working volume chemostat maintained so that the bacterial culture doubles every hour (dilution rate, 0.69 h^{-1}) would require 13.8 liters of fresh medium per 20-h interval. Small volumes of medium may be easily autoclaved and stored in standard 20-liter plastic carboys. The preparation of medium in small volumes (10 to 14 liters) allows adequate heat sterilization without excessive deterioration. The size of the chemostat to be operated is dictated by the need for sampling, cell mass, number of replicates or experimental reactors needed, and operational flexibility.

Standard procedures for the sterilization of liquid media are described elsewhere in this manual. Filter sterilization may be necessary to prevent destruction of heat-labile medium components. However, most simple media for cultivation of bacteria can be autoclaved prior to use, but larger volumes of medium require longer autoclaving times. Vessels containing approximately 10 liters of liquid should be autoclaved at 121°C for 30 to 90 min, depending on medium constituents. Media containing solids such as cornmeal flour or soy flour may require up to 90 min for complete sterilization of a 10-liter volume. However, media containing only dissolved components are usually sterile after 30 min at 121°C. All vessels should be vented during autoclaving to allow equilibration of pressure. The medium volume in culture reservoirs should not exceed 75% of the total reservoir volume, to minimize the chance of boil-over during autoclaving. Furthermore, it is essential that the gas space in the reservoir be vented during autoclaving.

12.2.5.2. Sterilization of Air

The air or gas supply to sparged chemostats must be sterilized prior to injection into the vessel. This is most easily done by physical removal of airborne microorganisms with a fibrous-medium filter. The filtration medium must be changed regularly to ensure adequate performance. A supplementary or backup air sterilization filter may be needed for long-duration cultivation as a precaution against fouling of the duty filter.

12.2.5.3. Sampling, Back Contamination, and Temperature Control

Sampling is best accomplished with an independent sample line system. Alternatively, samples may be taken from the effluent line, although this procedure is not recommended when a small sample is needed, since the lumen of the effluent tube may become coated with adherent bacteria which may break away and result in nonrepresentative samples. If an external sampling system is used in place of sample collection from the effluent line, the sample size must be kept below 5% of the reactor working volume so that steady-state conditions will not be disturbed greatly. Finally, the sample line must be purged of its entire contents before samples are collected for analysis. Some chemostats are equipped with a sterile air purge which displaces the sample line contents back into the fermentor after a sample is taken. If the latter system is used, purging prior to sample collection is not necessary.

The aeration and agitation conditions of continuous culture result in the production of an aerosol containing large numbers of the cultured bacterium. As a result, the medium feed line is subject to back contamination and must be fitted with a medium "break" tube, which acts as an aseptic seal between the culture vessel and the medium reservoir.

Most commercial chemostats for laboratory-scale operation are equipped with integral heat exchangers. The temperature is controlled by a temperature control unit to maintain a constant temperature inside the vessel. In some cases, the vessel is jacketed or submerged in a constant-temperature bath as a means of temperature control when accuracy is required.

12.2.5.4. Aerobic Operation

The agitation speed for adequate oxygen transfer in a chemostat is subject to change as the culture density (cell concentration) or operating conditions (volumetric airflow rate, temperature, and liquid volume) change. The operator should adjust the agitation speed and airflow rate so that the dissolved-oxygen concentration never falls below 30% of the initial saturated dissolved oxygen concentration (82). Maintenance of dissolved oxygen concentrations above 30% of the saturation value at ambient temperatures allows growth of most bacterial cultures under conditions in which oxygen is the growth-limiting substrate and keeps aerobic cells growing exponentially.

Some basic properties of oxygen supply and demand must be kept in mind when providing for adequate aeration in the cultivation of aerobic bacteria whether in a fermentor, a shake flask, or a test tube. Oxygen is quite insoluble; for example, there are only about 9 ppm (0.0009%) in water at 20°C in equilibrium with air. As the temperature is raised, oxygen becomes even less soluble. Oxygen solubility is directly proportional to the partial pressure of oxygen in the gas phase but is substantially independent of the total pressure and the presence of other gases. Bacteria utilize only dissolved (not gaseous) oxygen. Oxygen is so insoluble that only a small reservoir of it exists in solution at any given time. Consequently the rate of dissolved-oxygen supply must at least equal the rate of oxygen demand of the culture. Fortunately, cell respiration proceeds at a rate that is independent of the dissolved-oxygen concentration as long as it remains above a critical concentration, which is considerably below the saturation value. These basic properties of oxygen supply and demand are further developed and exemplified in a classic review on aeration and agitation in microbial culture by Finn (23) and in another excellent discussion in the monograph by Pirt (63).

Agitation requirements for adequate oxygen transfer are almost always well in excess of the agitation requirements for mixing of highly soluble nutrients. This rule also applies to agitation conditions established for optimum dissolution of other sparingly soluble substrates, such as immiscible hydrocarbons and steroids. For oxygen transfer, both agitation speed and volumetric airflow rate can be varied to establish the desired oxygen concentration in the fermentation medium. Instrumentation for dissolved-oxygen monitoring and for control of airflow rate is useful for the cultivation of aerobic organisms and allows the operator to control dissolved oxygen concentration as an independent parameter (19). Dissolved-oxygen control systems are available from all of the major fermentor manufacturers and usually allow control of both agitation and airflow rate in a sequential manner.

Most well-aerated and agitated cultures produce relatively stable foam at some point in the culture cycle. If foams are allowed to develop unchecked, they may wet the air filters and lead to back contamination of the culture as well as to foam spillage. Foams may be controlled by mechanical foam breakers or by the addition of chemical antifoam agents. Many commercial units are equipped with mechanical foam breakers as standard or optional equipment. Current designs for mechanical foam breakers are adequate for all but a very few culture conditions.

Chemical antifoam agents provide a less expensive means of foam control when intermittent or light use is expected (11). However, antifoam agents must be added to the medium and therefore contribute to the overall medium composition. Some antifoam agents such as vegetable (corn oil or cottonseed oil) and animal (lard oil) ones may be metabolized by the culture and therefore contribute to the carbon substrate pool. On the other hand, nonmetabolizable antifoam agents, such as the silicone antifoams, may be toxic at high concentrations. However, their action as surfactant materials requires their use in only very low concentrations. Effectiveness and toxicity should be determined on an individual basis. Antifoam agents can be added directly to the medium before sterilization or to the fermentation system through a feed line in the head plate (the antifoam agent, its reservoir, and the feed line must be sterilized prior to use). Automatic control units for antifoam addition are usually available commercially.

12.2.5.5. Anaerobic Operation

The primary concern in system design for anaerobic cultivation in chemostats is the exclusion of oxygen. Use a large inoculum (10 to 20% of total operating volume) to allow rapid establishment of the culture and to reduce the sensitivity of the system to leaks of oxygen from the external environment. If pH is to be controlled or nutrients are to be added in a continuous or semicontinuous fashion, keep the flexible tubing connecting the reservoirs to the vessel as short as possible, since oxygen permeates natural or silicone rubber tubing; use butyl rubber instead. Tygon tubing is also relatively oxygen impermeable but softens during autoclaving, so it must be clamped or wired in place. Finally, fit the gas exit of the fermentor with a water-gas trap to prevent back diffusion of oxygen into the fermentor head space.

In all cases, prereduce the medium by adding a reducing agent and sparge the vessel with oxygen-free gas prior to introducing the prereduced medium. Prepare and maintain acid, base, or nutrient solutions under oxygen-deficient conditions for addition to the vessel during operation. Absolute exclusion of oxygen from acid, base, or other solutions to be added to the fermentor in small quantities is not necessary since the reducing characteristics of an active anaerobic culture adequately cope with very slight additions of oxygen through the feed systems.

For the cultivation of anaerobic microbes, take special care to ensure establishment and maintenance of the chemically reduced state when autoclaving large volumes of prereduced medium. Connect vessels containing prereduced media to an oxygen-free gas supply immediately upon removal from the autoclave. As the steam in the gas headspace of the vessel begins to condense, a partial vacuum will form. Sparging with oxygen-free gas allows release of the vacuum formed during cooling without contamination of the medium by oxygen in the air.

12.2.6. Special Systems

12.2.6.1. Dialysis Culture

Dialysis is a process for separation of solute molecules by means of their unequal diffusion through a semipermeable membrane as a result of a concentration gradient (77, 89). The process is applied to the growth and maintenance of living cells by a technique called dialysis (or diffusion or perfusion) culture. To use the technique of dialysis culture, a membrane is positioned between a culture chamber and a dialysate reservoir. For dialysis culture to be effective, the volume of the reservoir must be larger than that of the culture or else the reservoir must be replenishable. Further, the permeability and area of the membrane must be sufficient to permit useful diffusion in rate and amount. Such a system of dialysis culture can be operated in vitro or in vivo and batchwise, continuously, or a combination of these modes.

Specific advantages and reasons for using dialysis for bacterial culture include (i) prolongation of the exponential growth phase in the batch cycle, which allows the attainment of very high densities of viable cells; (ii) extension of the maximum stationary phase in batch culture, thus permitting increased production of secondary metabolites associated with this phase; (iii) relief of product inhibition control by removal of metabolite products, thus enabling their greater production in batch or continuous culture; (iv) establishment of a steady-state population with mainly maintenance metabolism, thus immobilizing the cells for prolonged production of a metabolite product; (v) production of metabolites free from cells and, conversely, production of cells free from medium macromolecules; (vi) means to study and recover a cell population placed in an in situ or in vivo environment, such as in an ecological or animal system; and (vii) the capability for study of molecular interactions between separated populations of cells.

There are two basic principles underlying dialysis culture. First, it provides a means for achieving substrate-limited growth, i.e., fed-batch culture. Substrate in the dialysate reservoir diffuses through the membrane into the culture chamber, driven by a concentration gradient that results as the substrate is used for metabolism and growth. Second, dialysis culture provides a means for lowering the concentration of a diffusible metabolite product inhibitory to growth; the product in the culture chamber diffuses through the membrane and is diluted in the larger dialysate reservoir, thus relieving the feedback inhibition by the product that normally regulates its production. In the usual dialysis culture system, nutrient supply and product withdrawal occur concurrently; i.e., exchange dialysis occurs.

Three main types of membranes, differing in porosity, are applicable to dialysis culture. Dialysis membranes have a nominal pore size on the order of 10 nm in diameter so that they exclude cells and macromolecules but allow small molecules, such as the nutrients for bacterial growth, to pass. Filter membranes (or membrane filters) have a nominal porosity on the order of 100 nm, so they also exclude cells, but they allow macromolecules as well as small molecules to pass. Sheets of membrane are commercially available in a wide range of porosities and materials. Factors to consider for microbiological use include autoclavability, pore size, inertness, and permeability (which is governed by porosity, void space, and thickness). Solution transport membranes have no pores and pass gases because of their solubility in the membrane material itself (e.g., oxygen solubility in silicone rubber or polycarbonate). An interface between two physically different phases may also be used to separate the culture and dialysate (interface dialysis culture), such as in a liquid/solid biophasic system.

12.2.6.1.1. In Vitro Systems

Dialysis culture is often first attempted in the laboratory by suspending a membrane sac containing the culture in a tube, flask, or carboy of medium. The simplest such arrangement is to use a length of dialysis membrane tubing that is intussuscepted so as to form a double-walled tube containing culture in the annular space thus formed. The advantages of shake flasks are combined with those of membrane dialysis culture in a unit assembled from flanged Pyrex glass pipe. This and other flanged flask designs (89) enable the use of sheet membrane of any type. Scale-up of dialysis culture to fermentors is possible by carrying out growth in a separate culture circuit that is connected with a separate dialysate circuit by means of an intermediate dialyzer (26). Various types of hollow-fiber dialyzers have also been used (67). The design of the system can be optimized for a given culture situation. Stieber et al. (90) have developed and reviewed various modes of operation of an in vitro system.

12.2.6.1.2. Animal Systems

The term "diffusion chamber" is often used to describe a small dialysis culture unit that can be implanted within a living experimental animal (e.g., in the peritoneum or rumen or beneath the skin). Among a number of systems, the best are made with a nondegradable filter membrane sealed on both sides of an inert plastic cylinder (like a drum) through which sampling access can be provided. Bacteria introduced into diluent within the chamber grow entirely on diffusible nutrients from the host. Such systems have been used in bacteriology to study immune reactions and the growth of fastidious pathogens, e.g., *Mycobacterium leprae*, *Treponema pallidum*, and *Neisseria gonorrhoeae*.

Although not common, examples do exist whereby a microbial growth system is designed to be in contact with the bloodstream of a live animal (67). Such an ex vivo hemodialysis culture system enables the bacteria to grow entirely on the nutrients of the blood, yet separate from the macromolecular and cellular defense mechanisms, if a dialysis type of membrane of appropriate molecular exclusion is used. The host-parasite reactions of many facultatively aerobic (but no obligately anaerobic) bacteria were studied in this manner (29).

12.2.6.2. Product Removal Culture Systems

The product removal and consequent cell-concentrating feature of dialysis culture can be accomplished much more efficiently by other systems involving membranes in which the driving force is greater than a concentration gradient, such as a pressure gradient for microfiltration and an electrical gradient for electrodialysis. However, these systems are more complicated in design, tend to foul the membrane more than dialysis does, and have specialized applicability beyond the scope of this book. An example of a continuous microfiltration culture system is given in reference 56, and an example of an electrodialysis system is given in reference 33. Product removal and relief of product feedback inhibition of bacterial growth can also be attained by a variety of other means. These include extraction by use of nonaqueous solvents, biphasic aqueous systems, and catalytic membranes; sorption by solvents; vacuum evaporation and membrane pervaporation; and precipitation. These extractive culture systems are reviewed in a publication edited by Mattiasson and Holst (51).

12.2.6.3. High-Density Batch Culture Strategies

Ordinary liquid-culture methods produce relatively low densities of bacterial cell mass. Usual maximum growth densities of aerobic or facultatively aerobic bacteria in batch systems are on the order of 0.1 g (dry weight) of cell per liter in quiescent test tubes. Compared to this, shake flasks and aerated and agitated fermentors can achieve 10 and 100 times more biomass, respectively. Cell counts and other direct indices of growth increase similarly by decades. Maximum densities of anaerobic bacteria are generally about 1 decade less than those of aerobes.

Further large increases to still-higher cell densities can be attained by strategically using one or more of the critical limiting factors in batch growth discussed earlier in this chapter. High-density culture (concentrated culture) strategies for aerobic bacteria have resulted in concentrations as high as 150 g (dry weight) of cells/liter (26), but the maximum practical concentration is limited by oxygen transfer capability to about 70 g (dry weight) of cells/liter. Increasing the cell density makes more sense than making replicates or increasing the vessel volume. Optimization of a process for high cell density is desirable whether the process is for bacterial biomass production, metabolic-product formation, or substrate consumption.

The key principle underlying high-density culture is to prevent the depletion of an essential nutrient and relieve the feedback inhibition of bacterial growth by limiting the accumulation of toxic metabolic products. This is done by (i) supplying essential nutrients in increasing amounts to meet but not exceed growth needs, (ii) removing toxic products as they are formed, and (iii) controlling physicochemical growth factors at optimum levels. These are best accomplished in a fermentor system and are usually applied to aerobic bacteria, but they are also applicable to anaerobes. A summary review of the theoretical and historical background of high-density culture was presented by Shiloach et al. (82).

12.2.6.3.1. Aerobes

High-density culture of aerobes is best exemplified with *E. coli* and other enteric bacteria, which oxidize glucose and other carbon and energy sources largely to CO_2 via aerobic respiration in the presence of an adequate oxygen supply, but incompletely to organic acids, particularly acetic acid, via fermentation in the presence of an inadequate oxygen supply. It is this accumulation of acetic acid by fermentation as the result of oxygen depletion that mainly inhibits growth.

To keep aerobic cells growing exponentially in a fermentor, the dissolved-oxygen supply should be increased as growth increases and kept above 30% saturation; for a culture of 10 g (dry weight) of cells/liter, this amounts to 300 mmol of oxygen per liter per 30 min (82). This enormous oxygen demand can only partially be met by maximizing the airflow and agitation rates in a fermentor. To further increase the dissolved-oxygen concentration, oxygen-enriched air must be provided by adding pure oxygen from a liquid-oxygen cylinder. Dissolved-oxygen concentration is monitored with a probe (19). The increases in the oxygen-enriched airflow and agitation rates corresponding to the increase in growth are best controlled by a computer system (22).

Increasing the addition of essential nutrients as exponential growth progresses (fed-batch culture) is also usually essential and a primary consideration for high-density culture of aerobes. Glucose (or another carbon and energy source) must usually be maintained at a low level to eliminate toxic-product formation, e.g., acetic acid by *E. coli*. The system must be monitored with a glucose or acetate probe and controlled by a computer (to <2 g of glucose per liter or <1.5 g of acetate per liter for *E. coli*).

Removing a toxic metabolic product as it is formed is another common strategy for attaining a high density of bacterial cells. Although essential for anaerobes, product removal alone should usually be a secondary strategy for aerobes if bacterial biomass production is the objective: it is better to limit product formation than to correct it. In dialysis culture, product removal is coupled with nutrient supply; this combination strategy is especially applicable for small-scale laboratory use by use of the biphasic shake flask system.

Maintenance of pH at an optimum level is necessary for high-density culture and has become commonplace in fermentor systems with provision for a pH probe and either manual or computer control by the addition of ammonium hydroxide, which also maintains the nitrogen supply necessary for high-density culture.

Control of temperature is also usual in fermentor systems. For high-density culture, it is crucial not to exceed the optimum temperature, beyond which growth rates fall off precipitously. Indeed, a lower growth temperature (for *E. coli*, 22°C) has been used to lower the growth rate and so to extend the exponential phase (81), resulting in about a doubled yield. A practical compromise between these considerations is to use a moderately lower growth temperature, e.g., 30°C for *E. coli*.

12.2.6.3.2. Anaerobes

High-density cultures of anaerobic organisms are most common in two broad areas: production of probiotics (e.g., lactic acid bacteria [35, 75]) and environmental biotechnology (e.g., anaerobic granules containing more than 10^{11} cells/gram of granular sludge [16]). In addition, most high-density anaerobic cultures require specialized operation of the reactor itself in addition to the removal of any inhibitory products that may be produced. Membrane-based cell recycle systems (13), upflow anaerobic sludge blanket reactors (64), and numerous attached-growth or immobilized-cell technologies are examples of increasing the cell density by reactor operation and modifications.

12.3. REFERENCES

1. **Abe, C.** 1997. Rapid diagnosis of tuberculosis. *Kekkaku* **72:**659–672. (In Japanese.)
2. **Adler, J.** 1966. Chemotaxis in bacteria. *Science* **153:**708–716.
3. **Babbar, S. B., and N. Jain.** 1998. 'Isubgol' as an alternative gelling agent in plant tissue culture media. *Plant Cell Rep.* **17:**318–322.
4. **Balch, W. E., and R. S. Wolfe.** 1976. New approach to the cultivation of methanogenic bacteria: 2-mercaptoethanesulfonic acid (HS-CoM)-dependent growth of *Methanobacterium ruminantium* in a pressurized atmosphere. *Appl. Environ. Microbiol.* **32:**781–791.
5. **Banik, R. M., B. Kanari, and S. N. Upadhyay.** 2000. Exopolysaccharide of the gellan family: prospects and potential. *World J. Microbiol. Biotechnol.* **16:**407–414.
6. **Bannur, M., R. P. Fule, A. M. Saoji, and V. L. Jahagirdar.** 1995. Study of bacteraemia using conventional and biphasic culture methods. *Indian J. Pathol. Microbiol.* **38:**147–151.
7. **Bhattacharya, P., S. Dey, and B. C. Bhattacharyya.** 1994. Use of low-cost gelling agents and support matrices for

industrial-scale plant-tissue culture. *Plant Cell Tissue Organ Cult.* **37:**15–23.

8. **Boin, M. A., M. J. Austin, and C. C. Hase.** 2004. Chemotaxis in *Vibrio cholerae. FEMS Microbiol. Lett.* **239:** 1–8.

9. **Bridson, E. Y., and A. Brecker.** 1970. Design and formulation of culture media, p. 229-295. *In* J. R. Norris and D. W. Ribbons (ed.), *Methods in Microbiology*, vol. 3A. Academic Press, Inc., New York, NY.

10. **Bromke, B. J., and M. Furiga.** 1991. Carrageenan is a desirable substitute for agar in media growing *Trichomonas vaginalis. J. Microbiol. Methods* **13:**61–65.

11. **Bryant, J.** 1970. Anti foam agents, p. 187–203. *In* J. R. Norris and D. W. Ribbons (ed.), *Methods in Microbiology*, vol. 2. Academic Press, Inc., New York, NY.

12. **Chan, E. C. S., A. DeCiccio, R. McLaughlin, A. Klitorinos, and R. Siboo.** 1997. An inexpensive solid medium for obtaining colony-forming units of oral spirochetes. *Oral Microbiol. Immunol.* **12:**372–376.

13. **Chang, H. N., I. K. Yoo, and B. S. Kim.** 1994. High-density cell-culture by membrane-based cell recycle. *Biotechnol. Adv.* **12:**467–487.

14. **Chiovitti, A., G. T. Kraft, A. Bacic, D. J. Craik, S. L. A. Munro, and M. L. Liao.** 1998. Carrageenans from Australian representatives of the family Cystocloniaceae (Gigartinales, Rhodophyta), with description of Calliblepharis celatospora sp. nov., and transfer of Austroclonium to the family Areschougiaceae. *J. Phycol.* **34:**515–535.

15. **Deming, J. W., and J. A. Baross.** 1986. Solid medium for culturing black smoker bacteria at temperatures to 120°C. *Appl. Environ. Microbiol.* **51:**238–243.

16. **Diaz, E., R. Amils, and J. Sanz.** 2003. Molecular ecology of anaerobic granular sludge grown at different conditions. *Water Sci. Technol.* **48:**57–64.

17. **Dickson, J. S., T. R. Manke, I. V. Wesley, and A. L. Baetz.** 1996. Biphasic culture of *Arcobacter* spp. *Lett. Appl. Microbiol.* **22:**195–198.

18. **Dunn, I. J., and J. R. Mor.** 1975. Variable volume continuous culture. *Biotechnol. Bioeng.* **17:**1805–1822.

19. **Elsworth, R.** 1972. The value and use of dissolved oxygen measurement in deep culture. *Chem. Eng.* **258:**63–71.

20. **Epifanio, E. C., R. L. Veroy, F. Uyenco, G. J. B. Cajipe, and E. C. Laserna.** 1981. Carrageenan from *Eucheuma striatum* (Schmitz) in bacteriological media. *Appl. Environ. Microbiol.* **41:**155–158.

21. **Evans, J. B.** 1975. Preparation of synchronous cultures of *Escherichia coli* by continuous flow size selection. *J. Gen. Microbiol.* **91:**188–190.

22. **Fass, R., T. R. Clem, and J. Shiloach.** 1989. Use of a novel air separation system in a fed batch fermentative culture of *Escherichia coli. Appl. Environ. Microbiol.* **55:**1305–1307.

23. **Finn, R. K.** 1954. Agitation aeration in the laboratory and in industry. *Bacteriol. Rev.* **18:**254–274.

24. **Freedman, D.** 1970. The shaker in bioengineering, p. 175–185. *In* J. R. Norris and D. W. Ribbons (ed.), *Methods in Microbiology*, vol. 2. Academic Press, Inc., New York, NY.

25. **Funk, H. B., and T. A. Krulwich.** 1964. Preparation of clear silica gels that can be streaked. *J. Bacteriol.* **88:**1200–1201.

26. **Gallup, D. M., and P. Gerhardt.** 1963. Dialysis fermentor systems for concentrated culture of microorganisms. *Appl. Microbiol.* **11:**506–512.

27. **Gardener, S., and J. G. Jones.** 1984. A new solidifying agent for culture media which liquefies on cooling. *J. Gen. Microbiol.* **130:**731–733.

28. **Gerhardt, P., and C. G. Hedn.** 1960. Concentrated culture of gonococci in clear liquid medium. *Proc. Soc. Exp. Biol. Med.* **105:**49–51.

29. **Gerhardt, P., J. M. Quarles, T. C. Beaman, and R. C. Belding.** 1977. Ex vivo hemodialysis culture of microbial and mammalian cells. *J. Infect. Dis.* **135:**42–50.

30. **Giavasis, I., L. M. Harvey, and B. McNeil.** 2000. Gellan gum. *Crit. Rev. Biotechnol.* **20:**177–211.

31. **Gibb, A. P., and S. Wong.** 1998. Inhibition of PCR by agar from bacteriological transport media. *J. Clin. Microbiol.* **36:**275–276.

32. **Harshey, R. M.** 2003. Bacterial motility on a surface: many ways to a common goal. *Annu. Rev. Microbiol.* **57:** 249–273.

33. **Hongo, M., Y. Nomura, and M. Iwahara.** 1986. Novel method of lactic acid production by electrodialysis fermentation. *Appl. Environ. Microbiol.* **52:**314–319.

34. **Hoshino, K., M. Yuzuriha, S. Morohashi, S. Kagaya, and M. Taniguchi.** 2002. Production of laccase by membrane-surface liquid culture with nonwoven fabric of Coriolus versicolor. *Biol. Syst. Eng.* **830:**108–120.

35. **Ishizaki, A.** 2003. Advanced continuous fermentation for anaerobic microorganism. *Ferment. Biotechnol.* **862:**21–35.

36. **Jain, N., S. Gupta, and S. B. Babbar.** 1997. Isubgol as an alternative gelling agent for microbial culture media. *J. Plant Biochem. Biotechnol.* **6:**129–131.

37. **Jenkins, J. A., and P. W. Taylor.** 1995. An alternative bacteriological medium for the isolation of *Aeromonas* spp. *J. Wildl. Dis.* **31:**272–275.

38. **Jin, H., N. K. Lee, M. K. Shin, S. K. Kim, D. L. Kaplan, and J. W. Lee.** 2003. Production of gellan gum by *Sphingomonas paucimobilis* NK2000 with soybean pomace. *Biochem. Eng. J.* **16:**357–360.

39. **Kelly, D. J.** 2001. The physiology and metabolism of *Campylobacter jejuni* and *Helicobacter pylori. J. Appl. Microbiol.* **90:**16S–24S.

40. **Krieg, N. R., and P. S. Hoffman.** 1986. Microaerophily and oxygen toxicity. *Annu. Rev. Microbiol.* **40:**107–130.

41. **Kriukov, V. R.** 1981. Development of hydrogen bacteria on hard surfaces. *Mikrobiologiia* **50:**299–304.

42. **Kubitschek, H. E.** 1987. Buoyant density variation during the cell cycle in microorganisms. *Crit. Rev. Microbiol.* **14:**73–97.

43. **Landwall, P., and T. Holme.** 1977. Removal of inhibitors of bacterial growth by dialysis culture. *J. Gen. Microbiol.* **103:**345–352.

44. **Laserna, E. C., F. Uyenco, E. Epifanio, R. L. Veroy, and G. J. B. Cajipe.** 1981. Carrageenan from *Eucheuma striatum* (Schmitz) in media for fungal and yeast cultures. *Appl. Environ. Microbiol.* **42:**174–175.

45. **Lim, H. C., B. J. Chen, and C. C. Creagan.** 1977. An analysis of extended and exponentially fed batch cultures. *Biotechnol. Bioeng.* **19:**425–433.

46. **Lin, C. C., and J. L. E. Casida.** 1984. Gelrite as a gelling agent in media for growth of thermophilic microorganisms. *Appl. Environ. Microbiol.* **47:**427–429.

47. **Lines, A. D.** 1977. Value of the K$^+$ salt of carrageenan as an agar substitute in routine bacteriological media. *Appl. Environ. Microbiol.* **34:**637–639.

48. **Lloyd, D. L., J. C. Edwards, and A. H. Chagla.** 1975. Synchronous cultures of micro organisms: large scale preparation by continuous flow size selection. *J. Gen. Microbiol.* **88:**153–158.

49. **Lux, R., and W. Shi.** 2004. Chemotaxis-guided movements in bacteria. *Crit. Rev. Oral. Biol. Med.* **15:**207–220.

50. **Markx, G. H., C. L. Davey, and D. B. Kell.** 1991. The permittistat: a novel type of turbidostat. *J. Gen. Microbiol.* **137:**735–743.

51. **Mattiasson, B., and O. Holst (ed.).** 1991. *Extractive Bioconversions.* Marcel Dekker, Inc., New York, NY.

52. **McBride, M. J.** 2004. *Cytophaga flavobacterium* gliding motility. *J. Mol. Microbiol. Biotechnol.* **7:**63–71.

53. **Miller, C. W., M. H. Nguyen, M. Rooney, and K. Kailasapathy.** 2003. Novel apparatus to measure diffusion in gel type foods. *Food Australia* 9:432–435.

54. **Mitchison, J. W., and W. S. Vincent.** 1965. Preparation of synchronous cell cultures by sedimentation. *Nature* 205:987–989.

55. **Munson, R. J.** 1970. Turbidostats, p. 349–376. *In* J. R. Norris and D. W. Ribbons (ed.), *Methods in Microbiology*, vol. 2. Academic Press, Inc., New York, NY.

56. **Nipkow, A., J. G. Zeikus, and P. Gerhardt.** 1989. Microfiltration cell recycle pilot system for continuous thermoanaerobic production of exoamylase. *Biotechnol. Bioeng.* 34:1075–1084.

57. **Nordbring Hertz, B., M. Veenhuis, and W. Harder.** 1984. Dialysis membrane technique for ultrastructural studies of microbial interactions. *Appl. Environ. Microbiol.* 47:195–197.

58. **Norton, S., and J. C. Vuillemard.** 1994. Food bioconversions and metabolite production using immobilized cell technology. *Crit. Rev. Biotechnol.* 14:193–224.

59. **Okon, Y., S. L. Albrecht, and R. H. Burris.** 1976. Carbon and ammonia metabolism of *Spirillum lipoferum*. *J. Bacteriol.* 128:592–597.

60. **Olson, M. S., R. M. Ford, J. A. Smith, and E. J. Fernandez.** 2004. Quantification of bacterial chemotaxis in porous media using magnetic resonance imaging. *Environ. Sci. Technol.* 38:3864–3870.

61. **Papapetropoulou, M., G. Rodopoulou, and E. Giannoulaki.** 1995. Improved glutaminate-starch-penicillin agar for the isolation and enumeration of *Aeromonas hydrophila* from seawater by membrane filtration. *Pathol. Biol.* (Paris) 43:622–627.

62. **Parkinson, S. M., M. Wainwright, and K. Killham.** 1989. Observations on oligotrophic growth of fungi on silica-gel. *Mycol. Res.* 93:529–534.

63. **Pirt, J. S.** 1975. *Principles of Microbe and Cell Cultivation.* John Wiley & Sons, Inc., New York, NY.

64. **Pol, L. W. H., S. I. D. Lopes, G. Lettinga, and P. N. L. Lens.** 2004. Anaerobic sludge granulation. *Water Res.* 38:1376–1389.

65. **Poole, R. K.** 1977. Fluctuations in buoyant density during the cell cycle of *Escherichia coli* K12: significance for the preparation of synchronous cultures by age selection. *J. Gen. Microbiol.* 98:177–186.

66. **Portner, R., and I. H. Mark.** 1998. Dialysis cultures. *Appl. Microbiol. Biotechnol.* 50:403–414.

67. **Quarles, J. M., R. C. Belding, T. C. Beaman, and P. Gerhardt.** 1974. Hemodialysis culture of *Serratia marcescens* in a goat artificial kidney fermentor system. *Infect. Immun.* 9:550–558.

68. **Rabe, L. K., and S. L. Hillier.** 2003. Optimization of media for detection of hydrogen peroxide production by *Lactobacillus* species. *J. Clin. Microbiol.* 41:3260–3264.

69. **Ramirez, A., R. Gutierrez, G. Diaz, C. Gonzalez, N. Perez, S. Vega, and M. Noa.** 2003. High-performance thin-layer chromatography-bioautography for multiple antibiotic residues in cow's milk. *J. Chromatogr. B Anal. Technol. Biomed. Life Sci.* 784:315–322.

70. **Rath, P. M., and D. Schmidt.** 2001. Gellan gum as a suitable gelling agent in microbiological media for PCR applications. *J. Med. Microbiol.* 50:108–109.

71. **Reeslev, M., and A. Kjoller.** 1995. Comparison of biomass dry weights and radial growth-rates of fungal colonies on media solidified with different gelling compounds. *Appl. Environ. Microbiol.* 61:4236–4239.

72. **Roehrig, K. L.** 1984. *Carbohydrate Biochemistry and Metabolism.* The AVI Publishing Company, Westport, CT.

73. **Rule, P. L., and A. D. Alexander.** 1986. Gellan gum as a substitute for agar in leptospiral media. *J. Clin. Microbiol.* 23:500–504.

74. **Sahay, S.** 1999. The use of psyllium (isubgol) as an alternative gelling agent for microbial culture media. *World J. Microbiol. Biotechnol.* 15:733–735.

75. **Schiraldi, C., V. Adduci, V. Valli, C. Maresca, M. Giuliano, M. Lamberti, M. Carteni, and M. De Rosa.** 2003. High cell density cultivation of probiotics and lactic acid production. *Biotechnol. Bioeng.* 82:213–222.

76. **Schmidt, D., and P. M. Rath.** 2003. Faster genetic identification of medically important aspergilli by using gellan gum as gelling agent in mycological media. *J. Med. Microbiol.* 52:653–655.

77. **Schultz, J. S., and P. Gerhardt.** 1969. Dialysis culture of microorganisms: design, theory, and results. *Bacteriol. Rev.* 33:1–47.

78. **Segerer, A. H., and K. O. Stetter.** 1992. The genus Thermoplasma, p. 712–718. *In* A. Balows, H. G. Thiper, M. Dworkin, W. Harder, and K. H. Schleifer (ed.), *The Prokaryotes. A Handbook on the Biology of Bacteria: Ecophysiology, Isolation, Identification, Applications*, 2nd ed. Springer Verlag KG, Berlin, Germany.

79. **Segerer, A. H., and K. O. Stetter.** 1992. The order Sulfolobales, p. 684–701. *In* A. Balows, H. G. Thiper, M. Dworkin, W. Harder, and K. H. Schleifer (ed.), *The Prokaryotes. A Handbook on the Biology of Bacteria: Ecophysiology, Isolation, Identification, Applications*, 2nd ed. Springer Verlag KG, Berlin, Germany.

80. **Shadowen, R. D., and C. V. Sciortino.** 1989. Improved growth of *Campylobacter pylori* in a biphasic system. *J. Clin. Microbiol.* 27:1744–1747.

81. **Shiloach, J., and S. Bauer.** 1975. High yield growth of *E. coli* at different temperatures in a bench scale fermentor. *Biotechnol. Bioeng.* 17:227–239.

82. **Shiloach, J., M. V. d. Walle, J. B. Kaufman, and R. Fass.** 1991. High density growth of microorganisms for protein production, p. 33–46. *In* M. D. White, S. Reuveny, and A. Shafferman (ed.), *Biologicals from Recombinant Microorganisms and Animal Cells*. VCH Publishers, New York, NY.

83. **Shungu, D., M. Valiant, V. Tutlane, E. Weinberg, B. Weissberger, L. Koupal, H. Gadebusch, and E. Stapley.** 1983. Gelrite as an agar substitute in bacteriological media. *Appl. Environ. Microbiol.* 46:840–845.

84. **Sirotnak, F. M., G. J. Donati, and D. J. Hutchison.** 1963. Folic acid derivatives synthesized during growth of *Diplococcus pneumoniae*. *J. Bacteriol.* 85:658–665.

85. **Sneath, P. H. A.** 1955. Failure of *Chromobacterium violaceum* to grow on nutrient agar, attributed to hydrogen peroxide. *J. Gen. Microbiol.* 13:i.

86. **Sommers, L. E., and R. F. Harris.** 1968. Routine preparation of silica gel media using silicate solutions of varying pH. *J. Bacteriol.* 95:1174.

87. **Stecher, B., S. Hapfelmeier, C. Muller, M. Kremer, T. Stallmach, and W. D. Hardt.** 2004. Flagella and chemotaxis are required for efficient induction of *Salmonella enterica* serovar Typhimurium colitis in streptomycin-pretreated mice. *Infect. Immun.* 72:4138–4150.

88. **Stelling, J.** 2004. Mathematical models in microbial systems biology. *Curr. Opin. Microbiol.* 7:513–518.

89. **Stieber, R. W.** 1979. Dialysis Continuous Processes for Microbial Fermentations: Mathematical Models, Computer Simulations, and Experimental Tests. Ph.D. thesis. Michigan State University, East Lansing.

90. **Stieber, R. W., and P. Gerhardt.** 1979. Dialysis continuous process for ammonium lactate fermentation: improved mathematical model and use of deproteinized whey. *Appl. Environ. Microbiol.* 37:487–495.

91. **Sutherland, I. W.** 1999. Microbial polysaccharide products. *Biotechnol. Genet. Eng. Rev.* 16:217–229.

92. **Sworn, G., G. R. Sanderson, and W. Gibson.** 1995. Gellan gum fluid gels. *Food Hydrocolloids* 9:265–271.

93. **Szurmant, H., and G. W. Ordal.** 2004. Diversity in chemotaxis mechanisms among the bacteria and archaea. *Microbiol. Mol. Biol. Rev.* **68:**301–319.

94. **Towle, G. A., and R. L. Whistler.** 1973. Hemicellulose and gums, p. 198–248. *In* L. R. Miller (ed.), *Phytochemistry*, vol. 1. Van Nostrand Reinhold Co., New York, NY.

95. **Tso, W., and J. Adler.** 1974. Negative chemotaxis in *Escherichia coli. J. Bacteriol.* **118:**560–576.

96. **Tyrrell, E. A., R. E. MacDonald, and P. Gerhardt.** 1958. Biphasic system for growing bacteria in concentrated culture. *J. Bacteriol.* **75:**1–4.

97. **VanElsas, J. D., J. M. Govaert, and J. A. v. Veen.** 1987. Transfer of plasmid pFT30 between bacilli in soil as influenced by bacteria population dynamics and soil conditions. *Soil Biol. Biochem.* **19:**639–647.

98. **Wainwright, M., and A. Al-Talhi.** 1999. Selective isolation and oligotrophic growth of Candida on nutrient-free silica gel medium. *J. Med. Microbiol.* **48:**1130.

99. **Wakisaka, Y., T. Segawa, K. Imamura, T. Sakiyama, and K. Nakanishi.** 1998. Development of a cylindrical apparatus for membrane-surface liquid culture and production of kojic acid using *Aspergillus oryzae* NRRL484. *J. Ferment. Bioeng.* **85:**488–494.

100. **Waterworth, P. M.** 1969. The action of light on culture media. *J. Clin. Pathol.* **22:**273–277.

101. **Watson, N., and D. Apirion.** 1976. Substitute for agar in solid media for common usages in microbiology. *Appl. Environ. Microbiol.* **31:**509–513.

102. **Weinstein, M. J., and G. H. Wagman (ed.).** 1978. *Antibiotics. Isolation, Separation and Purification.* Elsevier Scientific Publishing Co., New York, NY.

103. **Yamaguchi, Y., S. Nimbari, H. Obata, T. Ookawara, H. Eguchi, T. Kurotsu, and K. Suzuki.** 2002. Effects of agarose and LB medium on dye-terminator DNA sequencing. *Yakugaku Zasshi-J. Pharm. Soc. Jpn.* **122:**495–498.

104. **Yamaguchi, Y., S. Nimbari, T. Ookawara, K. Oishi, H. Eguchi, and K. Suzuki.** 2002. Inhibitory effects of agarose gel and LB medium on DNA sequencing. *BioTechniques* **33:**282.

13

Energetics, Stoichiometry, and Kinetics of Microbial Growth[†]

SYED A. HASHSHAM AND SAM W. BAUSHKE

Materials, electrons, and **energy** are three essential components needed to support the growth of all microorganisms (Fig. 1). Materials serve as building blocks for the additional cell mass and include carbon, nitrogen, phosphorus, and trace elements. Carbon is obtained from the vast number of organic compounds (by heterotrophic microorganisms) or from CO_2 (by autotrophic microorganisms). Nitrogen is obtained from ammonium, the most common nitrogen source, or from nitrate, nitrite, dinitrogen, or other nitrogen oxides. Electrons are needed to reduce most of these materials from a relatively oxidized state that is commonly found in nature to a reduced state that is characteristic of cell material. Electrons come from an electron-rich compound, which is termed electron donor. To quantitatively relate reactants and products, it is often helpful to represent compounds in terms of their **electron equivalents (e$^-$ eq)**. For example, 1 mol of glucose can be represented as 24 e$^-$ eq (see section 13.1.1 for more details). Energy is needed to perform the re-

duction of the materials and to polymerize the building blocks together in the form of useful enzymes, structural materials, and information containing molecules such as nucleic acids. Energy is obtained by a transfer of electrons from the electron donor to an electron acceptor (a compound that is electron deficient). Thus, there are two separate needs for the electron donor, one to provide electrons to reduce the materials and the other to transfer electrons to the acceptor to release energy.

The study of energy transactions in chemical or biochemical reactions, e.g., during the transfer of electrons from an electron donor to an acceptor is known as **energetics**. The study of the quantitative relationships among the electron donor, electron acceptor, cell mass, nitrogen, and other reactants and products is known as **stoichiometry**. In addition to energetics and stoichiometry, it is also important to know the rate at which the enzymes present in the cell mass may carry out the biological reactions. The study of rates is known as **kinetics**. Kinetics plays an important role in determining the size of the reaction vessel in which the process is being carried out or in determining the outcome of competition between two populations or guilds.

[†]The section on kinetics in this chapter was previously written by Philipp Gerhardt and Stephen W. Drew.

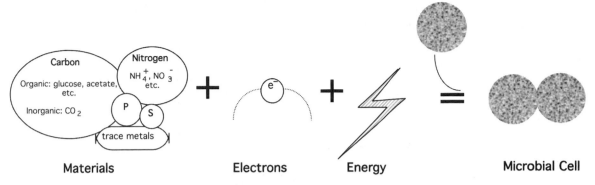

FIGURE 1 Simplified depiction of the three essential components needed to support the growth of all microorganisms.

The above description of energetics, stoichiometry, and kinetics highlights only the main aspects of microbial growth in a greatly simplified manner. In reality, the process of forming another microbial cell is extremely complex, especially at the molecular level of resolution. Reactions occurring within a single cell that ultimately enable the cell to grow and divide are only beginning to be understood and modeled through some very large-scale efforts (for an example, see http://microbialcellproject.org/). Yet at a coarser resolution, the growth and performance of many different types of cells, populations, and guilds can be modeled to obtain predictive equations for various applications. (A guild is defined as a group of many different types of populations capable of utilizing similar substrates.)

This chapter introduces concepts related to energetics, stoichiometry, and kinetics of microbial growth. It consists of two parts: the first part focuses on energetics and stoichiometry, and the second part focuses on kinetics of microbial growth. The section on energetics and stoichiometry includes some basic concepts such as computation of oxidation states and half-reactions that are critical to formulate a stoichiometric equation. Illustrative examples, when appropriate, are also presented. The section on kinetics presents equations governing cell growth, focusing on batch cultures and chemostats. All analyses and statements in this chapter assume the existence of a single cell (or spore that could transform into a vegetative state) that contains the enzymatic capacity and information necessary to put the building blocks, electrons, and energy into a copy of itself.

13.1. ENERGETICS AND STOICHIOMETRY

Quantitative studies related to growth of microorganisms often seek to determine the amount of cell mass that is expected from a given amount of material (electron donor, electron acceptor, nitrogen, etc.) or aim to compute how much donor (or acceptor) will be needed to consume a certain amount of acceptor (or donor). This requires writing a **stoichiometric equation, R**, that relates electron donor, electron acceptor, carbon source, nitrogen source, cell mass produced, and other by-products of donor oxidation and acceptor reduction in a quantitative manner. It combines principles from both redox chemistry and thermodynamics and serves as the basis for evaluating the quantitative relationships among the reactants and products. To write R, the process of microbial growth is viewed as a set of two biochemical reactions, namely the **energy-yielding reaction**,

R_e, and the **cell synthesis reaction, R_c** (Fig. 2). An energy-yielding reaction itself is constructed by combining an electron **donor half-reaction (R_d)** with an electron **acceptor half-reaction (R_a)**.

For simplicity, all analysis is carried out with 1 e⁻ eq of donor. Considering that R_d represents the source of electrons, a fraction of the donor, represented by f_e^0, must be transferred to the acceptor reaction to yield energy. The remaining fraction, represented by f_s^0 (which also equals $1 - f_e^0$), is transferred to the synthesis reaction to reduce the materials used for cell synthesis. Thus, 1 e⁻ eq of donor is distributed between the acceptor half-reaction and cell synthesis half-reaction. The overall stoichiometric equation R is written so that the donor half-reaction (R_d) seems to divide the 1 e⁻ eq between R_a and R_c in the ratio f_e^0 to f_s^0. In the form of an equation, the above statement can be written as $R = f_e^0 R_a + f_s^0 R_c - R_d$. This is depicted in Fig. 2 by arrows marked with f_e^0, representing the fraction of electrons transferred from the donor to the acceptor for energy, and f_s^0, representing the electrons transferred from the donor to the materials for cell synthesis. The total electron donor is equal to $f_s^0 + f_e^0$, which is equal to 1 e⁻ eq.

A cell synthesis reaction represents the assembly of raw materials from the environment into cell macromolecules. In a cell synthesis reaction, the Gibbs standard free energy at pH 7, $\Delta G_r^{0'}$, obtained from the energy-yielding reaction and electrons (f_s^0) from the electron donor are used to reduce carbon, nitrogen, and other cell constituents into an oxidation state of the materials found in the cell and to assemble these constituents into cellular macromolecules. If the donor is an organic compound, it may also serve as the source of carbon for cell synthesis. If the donor is an inorganic compound (e.g., NH_4^+, H_2, H_2S, etc.), the source of cell carbon is CO_2. In Fig. 2, the two sources of carbon are depicted by dotted lines emanating either from the donor or from CO_2 and ending at the source of carbon for cell synthesis. The conversion of carbon to cell material is a complex process. For simplicity of computing the energy required to convert the carbon source to cell carbon, this conversion is assumed to take place in two steps via pyruvate: (i) conversion of carbon source to pyruvate and (ii) conversion of pyruvate to cell carbon. The basis and need for this assumption are described in more detail in reference 20.

Considering that approximately 13% of cell weight (dry weight) is nitrogen, the synthesis of cells requires a significant amount of nitrogen. Depending on the source of nitrogen,

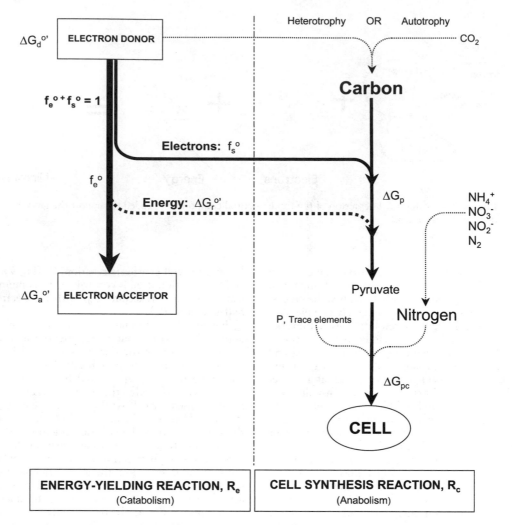

FIGURE 2 The overall process of microbial growth in terms of electron donor, electron acceptor, carbon and nitrogen sources, and energy for synthesis reaction.

electrons may also be needed to reduce it. NH_4^+ is the most common source of nitrogen. It is at the same oxidation state as cell nitrogen (i.e., -3). Hence, it does not require electrons to be reduced to the oxidation state of cell nitrogen. Other more oxidized nitrogen sources (e.g., NO_3^-, NO_2^-, and N_2) must be reduced to the level of NH_4^+ by addition of electrons from the donor. This is one of the reasons for a somewhat lower cell yield on relatively oxidized nitrogen sources. The impact of nitrogen source on the process of writing the stoichiometric equation R is incorporated by having separate cell synthesis half-reactions for each nitrogen source.

The overall process of writing a stoichiometric equation R described above involves a number of concepts. These are listed below and described in the sections that follow (13.1.1 to 13.1.11).

- oxidation state of reactants and products
- balanced reduction half-reaction normalized to 1 e⁻ eq
- Gibb's standard free energy at pH 7 for a reduction half-reaction ($\Delta G^{0'}$)
- common electron acceptor half-reactions (R_a)

- common electron donor half-reactions (R_d)
- electron donor-acceptor pair as energy-yielding reactions (R_e)
- model cell synthesis half-reactions as a function of nitrogen source (R_c)
- assumed efficiency of energy transfer (ε)
- fraction of electron donor used for cell synthesis (f_s^0)
- automated assignment of source of carbon for cell synthesis
- the overall stoichiometric equation (R)

The discussion in these sections is intended to be an introduction to the process of formulating an overall stoichiometric equation. For further details, the reader is referred to the original references (3, 20, 24).

13.1.1. Oxidation State of Reactants and Products

During microbial growth, elements such as carbon, nitrogen, oxygen, sulfur, etc., are often transformed from one oxidation state to another. An electron donor gets oxidized, and an electron acceptor gets reduced. Thus, the ability to evaluate the oxidation states of reactants and products is the first step in predicting the role of the compound as an electron donor or acceptor. Consider, for example, the following transformations: glucose ($C_6H_{12}O_6$) to CO_2; trichloroethene (TCE, C_2HCl_3) to ethene (C_2H_4); NH_4^+ to NO_3^-; NO_3^- to N_2; O_2 to H_2O; and Cr^{6+} to Cr^{3+}. All the above transformations are known to support microbial growth. Their role as electron donor or acceptor can be tentatively assigned by computing the oxidation state of carbon, nitrogen, oxygen, or chromium, respectively. The oxidation state of a given element in a compound is determined by the following rules.

- All elements when combined with themselves (e.g., O_2, H_2, and N_2) have an oxidation state of 0.
- Except when combined with itself, oxygen has an oxidation state of -2. However, in peroxides (H_2O_2 and Na_2O_2), the oxidation state of oxygen is -1.
- Except when combined with itself, hydrogen has an oxidation state of $+1$. However, in metal hydrides (NaH, LiH, etc.), the oxidation state of H is -1.
- The sum of oxidation states of all the elements in a species or group is the net charge on that group.
- Nitrogen and sulfur when present in an organic compound can be assumed to be in an oxidation state of -3 and -2, respectively, unless specified otherwise. This is only to simplify the calculation of the need for electrons and should not be used as a universal rule for the oxidation state of these compounds in microbial cells.

Following the above rules, the average oxidation state of each carbon in $C_6H_{12}O_6$ can be computed as follows. We begin by representing the unknown oxidation state, in this case that of carbon, as x. As per the above rules, the oxidation state of hydrogen is known to be $+1$ and that of oxygen is known to be -2. Thus, the sum of oxidation states of all elements in glucose can be written as $6x + 12(+1) + 6(-2)$. This must be equal to the charge on the compound, which in the case of glucose is zero. Solving for x yields the

average oxidation state of carbon in glucose as zero. Similarly, in CO_2 the oxidation state of carbon can be computed as $+4$ because each oxygen atom is at an oxidation state of -2. Thus, the transformation of $C_6H_{12}O_6$ to CO_2 is an oxidation involving a change in the oxidation state of each carbon atom from 0 to $+4$. Because there are 6 carbon atoms in glucose, a total of 24 e^- eq may be donated by 1 mol of glucose. This can also be readily seen by writing the balanced half-reaction for complete glucose oxidation, as follows: $C_6H_{12}O_6 + 6 H_2O \rightarrow 6 CO_2 + 24 H^+ + 24 e^-$.

A change in oxidation state combined with energy yield and its utilization for growth determines the role of the compound as a donor or acceptor. A substance being oxidized may have the potential to serve as an electron donor if it yields energy when combined with an acceptor to support microbial growth. Conversely, a substance being reduced may have the potential to serve as an electron acceptor if it yields energy when combined with an electron donor. Figure 3 depicts the oxidation states for a number of key elements present in various compounds of interest.

Considering the above examples of transformations, it can be easily shown that TCE transformation to ethene is a reductive process. Each carbon in TCE is at an oxidation state of $+1$, which is computed from the knowledge that the oxidation state of Cl in TCE is -1 and that of H is $+1$. Thus, the transformation of TCE to ethene has the potential to serve as an electron acceptor. The transformation of NH_4^+ to NO_3^-, however, is an oxidation process because nitrogen changes its oxidation state from -3 to $+5$. It involves a transfer of 8 electrons. Thus, the conversion of NH_4^+ to NO_3^- may potentially serve as a donor reaction. The remaining transformations (NO_3^- to N_2, O_2 to H_2O, and Cr^{6+} to Cr^{3+}) are all reductive processes involving a transfer of 5, 2, and 3 electrons, respectively. Each has the potential to serve as an electron acceptor. The following example highlights the use of oxidation state in tentatively predicting the possible role of a given compound in an energy-yielding reaction.

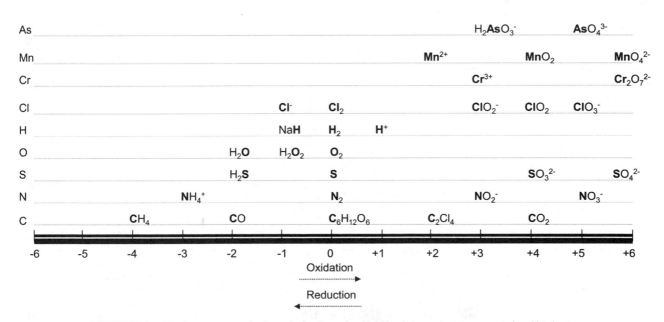

FIGURE 3 Oxidation state of selected elements (in boldface) in various compounds of biological interest.

13.1.1.1. Example 1

H_2S is known to be transformed to SO_4^{2-} and H^+ by microorganisms that grow on the topmost surface of concrete inside sewer pipes and cause corrosion of concrete due to the acid production. How many electrons are transferred in this process? What is the possible role of H_2S in this microbially mediated transformation process?

The oxidation state of S in H_2S must be -2 because H is at $+1$ and there is zero charge on the compound. The oxidation state of S in SO_4^{2-} is $+6$ because O must be at an oxidation state of -2 and the sum of oxidation states in SO_4^{2-}, i.e., $x + 4(-2)$, must be equal to the charge, -2. The change in the oxidation state of S from -2 to $+6$ indicates that the transformation of H_2S to SO_4^{2-} is an oxidation reaction involving 8 electrons. The transformation is expected to serve as an electron donor reaction because S is getting oxidized. Further evidence for this role may be obtained by computing the energy available when H_2S is combined with an appropriate electron acceptor (O_2 in this case).

13.1.2. Balanced Reduction Half-Reaction Normalized to 1 e⁻ eq

The information gathered from the above exercise related to oxidation states can be used to write half-reactions. Half-reactions are reactions involving electrons as a substrate or as a product. If a reaction does not contain electrons, it cannot be called a half-reaction. Each half-reaction can be written either as a reduction half-reaction or an oxidation half-reaction. Reduction half-reactions always have electrons as reactants on the left-hand side of the equation. Similarly, oxidation half-reactions always have electrons as product on the right-hand side of the reaction. The availability of balanced reduction half-reactions normalized to 1 e⁻ eq for the donor, acceptor, and cell synthesis greatly simplifies the process of writing the stoichiometric equation R and subsequent quantitative analysis. Normalization of the half-reactions to 1 e⁻ eq simplifies the procedure of obtaining R to a simple combination of the half-reactions for donor (R_d), acceptor (R_a), and cell synthesis (R_c) in a manner governed by f_s^0. All half-reactions, by convention, are written as reduction half-reactions. This is critical for correct use of the approach for obtaining R and f_s^0 described here (20). A number of half-reactions are listed in Table 1 (organic) and Table 2 (inorganic) with their associated Gibbs standard free energies (adopted from various sources such as references 20 and 24 and Thermodyn, an XL-based program on energetics [K. Hanselmann {http://www.microeco.unizh.ch/}], often with some modifications to present them as reduction half-reactions normalized to 1 e⁻ eq as needed in the approach described here). Any balanced reduction half-reaction normalized to 1 e⁻ eq can be obtained by following the steps described below.

1. Write the oxidized chemical species on the left-hand side and the reduced chemical species on the right-hand side.

2. Add electrons on the left-hand side because the half-reaction is a reduction half-reaction.

3. Add other species as needed to balance the carbon, nitrogen, phosphorus, etc.

4. Balance the oxygen by adding water (and not O_2).

5. Balance the hydrogen with H^+.

6. Balance the charge by changing the number of e⁻.

7. In the final step, the balanced half-reaction is divided by an integer (the multiplication coefficient of the e⁻) to convert the half-reaction to one electron equivalent.

13.1.2.1. Example 2

Arsenic compounds in drinking water sources are a health hazard around the world. Both chemical and biological means of transforming and removing arsenic compounds from drinking water are of interest. Strain MLMS-1 is a chemoautotrophic arsenate (AsO_4^{3-}) respirer, which grows by oxidizing sulfide to sulfate while reducing arsenate to arsenite ($H_2AsO_3^-$) (7). Write a balanced reduction half-reaction normalized to 1 e⁻ eq for the electron acceptor used by this organism.

Because O is at an oxidation state of -2 in both arsenate and arsenite, arsenic is at $+5$ oxidation state in arsenate and at $+3$ oxidation state in arsenite. Hence, arsenate is the oxidized species and arsenite is the reduced species. The following seven steps follow the procedure listed above to obtain the reduction half-reaction normalized to 1 e⁻ eq for the above transformation.

1. $AsO_4^{3-} = H_2AsO_3^-$
2. $AsO_4^{3-} + e^- = H_2AsO_3^-$
3. $AsO_4^{3-} + e^- = H_2AsO_3^-$
4. $AsO_4^{3-} + e^- = H_2AsO_3^- + H_2O$
5. $AsO_4^{3-} + 4H^+ + e^- = H_2AsO_3^- + H_2O$
6. $AsO_4^{3-} + 4H^+ + 2e^- = H_2AsO_3^- + H_2O$
7. $1/2\ AsO_4^{3-} + 2H^+ + e^- = 1/2\ H_2AsO_3^- + 1/2\ H_2O$

13.1.3. Gibbs Standard Free Energy at pH 7 for a Reduction Half-Reaction ($\Delta G^{0'}$)

The Gibbs standard free energy of a half-reaction modified for pH 7 ($\Delta G^{0'}$) for the donor and acceptor half-reactions is needed to compute the available free energy ($\Delta G_r^{0'}$) from the energy-yielding reaction (R_e) and to determine the fraction of electron donor, f_s^0, used for the cell synthesis reaction. These values in turn are used to write the stoichiometric equation, R. Tables 1 and 2 list the associated $\Delta G^{0'}$ for many half-reactions of interest. Values of $\Delta G^{0'}$ for any reaction can be computed from the free energies of formation (G_f^0) of reactants and products by using the following equation (10).

$$\Delta G^{0'} = \Sigma G_f^0 \text{ (products)} - \Sigma G_f^0 \text{ (reactants)} \quad (1)$$

Excellent resources to obtain G_f^0 values exist, e.g., references 20 and 24, *Brock Biology of Microorganisms* (10), the online version of the CRC's *Handbook of Chemistry and Physics* (http://www.hbcpnetbase.com/), and Thermodyn (see above). Some of these values are reproduced here for easy reference (Table 3).

13.1.3.1. Example 3

1,1,1-Trichloroethane (1,1,1-TCA) is a chlorinated solvent and a groundwater contaminant in the United States (14). Recently, it has been shown to support microbial growth, serving as an electron acceptor (22). You are interested in computing the Gibbs standard free energy ($\Delta G^{0'}$) for the transformation of 1,1,1-TCA to chloroethane so that the computed value can be used with the donor half-reaction to determine the amount of energy available. You have written the reduction half-reaction normalized to 1 e⁻ eq for the transformation of 1,1,1-TCA to chloroethane. Calculate the $\Delta G^{0'}$ for the half-reaction representing the conversion of 1,1,1-TCA to chloroethane.

First we write the balanced reduction half-reaction of 1,1,1-TCA transformation to chloroethane normalized to 1 e⁻ eq, following the procedure described in sections 13.1.1 and 13.1.2.

TABLE 1 Reduction half-reactions normalize to 1 e^- for selected organic compounds and the associated values of $\Delta G^{0'}$ and theoretical values of f_s^{0a}

Reaction no.	Reduced compounds	Substrates and products	$\Delta G^{0'}$ (kJ/e^- eq)	Computed f_s^0 values with acceptor:					
				O_2	Fe^{3+}	NO_3^-	MnO_2	SO_4^{2-}	CO_2
1	Glyoxylate	$1/4\ CO_2 + 1/4\ HCO_3^- + H^+ + e^- = 1/4\ HCOHCOO^- + 1/2\ H_2O$	49.47	0.77	0.77	0.76	0.71	0.44	0.41
2	Formaldehyde	$1/4\ CO_2 + H^+ + e^- = 1/4\ CH_2O + 1/4\ H_2O$	46.6	0.75	0.75	0.74	0.69	0.40	0.36
3	Glyceraldehyde	$1/4\ CO_2 + H^+ + e^- = 1/2\ CH_2OHCHOHCHO + 1/4\ H_2O$	42.66	0.73	0.72	0.72	0.66	0.34	0.30
4	Glucose	$1/4\ CO_2 + H^+ + e^- = 1/24\ C_6H_{12}O_6 + 1/4\ H_2O$	41.35	0.72	0.72	0.71	0.65	0.33	0.28
5	Glycine	$1/6\ CO_2 + 1/6\ HCO_3^- + 1/6\ NH_4^+ + H^+ + e^- = 1/6\ CH_2NH_2COOH + 1/2\ H_2O$	39.8	0.71	0.71	0.70	0.64	0.30	0.26
6	Formate	$1/2\ HCO_3^- + H^+ + e^- = 1/2\ HCOO^- + 1/2\ H_2O$	39.19	0.71	0.70	0.70	0.64	0.29	0.25
7	Glycerol	$3/14\ CO_2 + H^+ + e^- = 1/14\ CH_2OHCHOHCH_2OH + 3/14\ H_2O$	38.88	0.71	0.70	0.70	0.63	0.29	0.24
8	Ethylene glycol	$1/5\ CO_2 + H^+ + e^- = 1/10\ CH_2OHCH_2OH + 1/5\ H_2O$	38.25	0.70	0.70	0.69	0.63	0.28	0.23
9	Oxalacetate	$1/5\ CO_2 + 1/5\ HCO_3^- + H^+ + e^- = 1/10\ COO^-COCH_2COO^- + 1/2\ H_2O$	37.83	0.70	0.69	0.69	0.63	0.27	0.22
10	Methanol	$1/6\ CO_2 + H^+ + e^- = 1/6\ CH_3OH + 1/6\ H_2O$	36.84	0.70	0.69	0.68	0.62	0.26	0.21
11	Pyruvate	$1/5\ CO_2 + 1/10\ HCO_3^- + H^+ + e^- = 1/10\ CH_3COCOO^- + 2/5\ H_2O$	35.09	0.69	0.68	0.67	0.61	0.23	0.18
12	α-Ketoglutarate	$3/16\ CO_2 + 1/8\ HCO_3^- + H^+ + e^- = 1/16\ COO^-CH_2CH_2COCOO^- + 7/16\ H_2O$	34.25	0.67	0.67	0.66	0.59	0.22	0.16
13	Malate	$1/6\ CO_2 + 1/6\ HCO_3^- + H^+ + e^- = 1/2\ COO^-CHOHCH_2COO^- + 5/12\ H_2O$	34.17	0.67	0.66	0.66	0.59	0.22	0.16
14	Acetaldehyde	$1/5\ CO_2 + H^+ + e^- = 1/10\ CH_3CHO + 3/10\ H_2O$	33.6	0.67	0.66	0.65	0.58	0.20	0.15
15	Isocitrate	$3/18\ CO_2 + 3/18\ HCO_3^- + H^+ + e^- = 1/18\ COO^-CH_2CHCOO^-CHOHCOO^- + 4/9\ H_2O$	33.45	0.66	0.65	0.65	0.58	0.20	0.15
16	Citrate	$1/6\ CO_2 + 1/6\ HCO_3^- + H^+ + e^- = 1/18\ (COO^-)CH_2COH(COO^-)CH_2COO^- + 4/9\ H_2O$	33.08	0.66	0.65	0.65	0.58	0.19	0.14
17	α-Ketobutyrate	$1/8\ CO_2 + 1/16\ HCO_3^- + H^+ + e^- = 1/16\ CH_3CH_2COCOO^- + 1/2\ H_2O$	32.43	0.65	0.64	0.64	0.57	0.18	0.13
18	Lactate	$1/6\ CO_2 + 1/12\ HCO_3^- + H^+ + e^- = 1/12\ CH_3CHOHCOO^- + 1/3\ H_2O$	32.29	0.65	0.64	0.64	0.56	0.18	0.13
19	Alanine	$1/6\ CO_2 + 1/12\ HCO_3^- + 1/12\ NH_4^+ + 11/12\ H^+ + e^- = 1/12\ CH_3CHNH_2COO^- + 5/12\ H_2O$	31.37	0.64	0.63	0.62	0.55	0.16	0.11
20	Ethanol	$1/6\ CO_2 + H^+ + e^- = 1/12\ CH_3CH_2OH + 1/4\ H_2O$	31.18	0.64	0.63	0.62	0.55	0.16	0.11
21	β-Hydroxybutyrate	$3/18\ CO_2 + 1/18\ HCO_3^- + H^+ + e^- = 1/13\ CH_3CHOHCH_2COO^- + 1/3\ H_2O$	31.01	0.63	0.62	0.62	0.55	0.16	0.11
22	Glutamate	$1/6\ CO_2 + 1/9\ HCO_3^- + 1/18\ NH_4^+ + H^+ + e^- = 1/18\ COOHCH_2CH_2CHNH2COO^- + 4/9\ H_2O$	30.93	0.63	0.62	0.62	0.54	0.16	0.10
23	Acetoacetate	$3/16\ CO_2 + 1/16\ HCO_3^- + H^+ + e^- = 1/16\ CH_3COCH_2COO^- + 3/8\ H_2O$	30.72	0.63	0.62	0.62	0.54	0.15	0.10
24	Succinate	$1/7\ CO_2 + 1/7\ HCO_3^- + H^+ + e^- = 1/14(CH_2)_2(COO^-)_2 + 3/7\ H_2O$	29.09	0.61	0.60	0.60	0.52	0.13	0.07
25	Caproate	$5/32\ CO_2 + 1/32\ HCO_3^- + H^+ + e^- = 1/32\ CH_3(CH_2)_4COO^- + 11/32\ H_2O$	27.82	0.60	0.59	0.58	0.50	0.11	0.06
26	Butyrate	$3/20\ CO_2 + 1/20\ HCO_3^- + H^+ + e^- = 1/20\ CH_3CH_2CH_2COO^- + 7/20\ H_2O$	27.74	0.59	0.58	0.58	0.50	0.11	0.05
27	Propionate	$1/7\ CO_2 + 1/14\ HCO_3^- + H^+ + e^- = 1/14\ CH_3CH_2COO^- + 5/14\ H_2O$	27.63	0.59	0.58	0.58	0.50	0.10	0.05
28	Acetate	$1/8\ CO_2 + 1/8\ HCO_3^- + H^+ + e^- = 1/8\ CH_3COO^- + 3/8\ H_2O$	27.4	0.59	0.58	0.58	0.50	0.10	0.05
29	Benzoate	$1/5\ CO_2 + 1/30\ HCO_3^- + H^+ + e^- = 1/30\ C_6H_5COO^- + 13/30\ H_2O$	27.34	0.59	0.58	0.57	0.50	0.10	0.05
30	Palmitate	$15/19\ CO_2 + 1/92\ HCO_3^- + H^+ + e^- = 1/92\ CH_3(CH_2)_{14}COO^- + 31/92\ H_2O$	27.26	0.59	0.58	0.57	0.50	0.10	0.05

(Continued on next page)

TABLE 1 Reduction half-reactions normalized to 1 e⁻ for selected organic compounds and the associated values of $\Delta G^{0\prime}$ and theoretical values of f_s^{0a} (Continued)

Reaction no.	Reduced compounds	Substrates and products	$\Delta G^{0\prime}$ (kJ/e⁻ eq)	Computed f_s^0 values with acceptor:					
				O₂	Fe³⁺	NO₃⁻	MnO₂	SO₄²⁻	CO₂
31	Methane	1/8 CO₂ + H⁺ + e⁻ = 1/8 CH₄ + 1/4 H₂O	23.53	0.55	0.54	0.53	0.45	0.05	N/A
32	Dimethyl sulfide	1/2 (CH₃)2SO + H⁺ + e⁻ = 1/2 (CH₃)2S + 1/2 H₂O	−22.19	0.21	0.20	0.19	0.10		
33	Hexachloroethane	1/2 C₂Cl₆ + e⁻ = 1/2 C₂Cl₄ + Cl⁻	−28.03	0.18	0.17	0.16	0.07		
34	Dichlorophenol	1/2 C₆H₄Cl₂OH + 1/2 H⁺ + e⁻ = 1/2 C₆H₅ClOH + 1/2Cl⁻	−33	0.16	0.15	0.14	0.05		
35	Vinyl chloride	1/2 C₂H₃Cl + 1/2 H + e⁻ = 1/2 C₂H₄ + 1/2 Cl⁻	−35.81	0.15	0.13	0.13	0.04		
36	Tetrachloroethene	1/2 C₂Cl₄ + 1/2 H + e⁻ = 1/2 C₂HCl₃ + 1/2 Cl⁻	−47.79	0.10	0.09	0.08			
37	11-Dichloroethane	1/2 C₂H₄Cl₂ + 1/2 H + e⁻ = 1/2 C₂H₅Cl + 1/2 Cl⁻	−49.21	0.09	0.08	0.07			
38	Trichloroethene	1/2 C₂HCl₃ + 1/2 H + e⁻ = 1/2 C₂H₂Cl₂ + 1/2 Cl⁻	−51.14	0.09	0.07	0.07			
39	Carbon tetrachloride	1/2 CCl₄ + 1/2 H + e⁻ = 1/2 CHCl₃ + 1/2 Cl⁻	−57.4	0.06	0.05	0.05			
40	111-Trichloroethane	1/2 C₂H₃Cl₃ + 1/2 H+ + e⁻ = 1/2 C₂H₄Cl₂ + 1/2 Cl⁻	−113						

^{a}Computed assuming NH₄⁺ as nitrogen source, C₅H₇O₂N as the cell formula, and ε = 0.6. The values of f_s^0 are computed assuming the following end points for the acceptors: O₂ to H₂O, Fe³⁺ to Fe²⁺, NO₃⁻ to N₂, MnO₂ to Mn²⁺, SO₄²⁻ to HS⁻, and CO₂ to CH₄. These f_s^0 values are provided to save time in computation and do not necessarily imply the existence of an organism that can use the corresponding electron donor-acceptor pair. Data from references 3, 20, and 24.

$$1/4\ C_2H_3Cl_3 + 1/2\ H^+ + e^-$$
$$= 1/4\ C_2H_5Cl + 1/2\ Cl^- \tag{2}$$

Then, we obtain the free energies of formation (G_f^0) for $C_2H_3Cl_3$, H^+, e^-, C_2H_5Cl, and Cl^- from reference 20. The following values (in kilojoules/mole) were obtained from reference 20: −69.04 for 1,1,1-TCA aqueous state [aq], −39.87 for H^+ (aq) at pH 7, 0 for e^- (G_f^0 for e^- is always zero), −36.83 for chloroethane (aq), and −131.26 for Cl^- (aq). Note that unless a reactant or product is known to participate in the reaction in other forms, the values selected for G_f^0 should be for the aqueous state.

Using equation 1, the value of $\Delta G^{0\prime}$ for the above half-reaction is computed as:

$$\Delta G^{0\prime} = \Sigma G_f^0\ (\text{products}) - \Sigma\ G_f^0\ (\text{reactants})$$
$$= [1/4\ (-36.83) + 1/2\ (-131.26)] - [1/4(-69.04) + 1/2\ (-39.87) + (0)]$$
$$= -37.64\ \text{kJ/e}^-\ \text{eq}$$

It should be noted that the superscript "0" in $\Delta G_r^{0\prime}$ implies that the free energy is computed for standard conditions and the prime in ΔG_r^0 implies that the Gibbs standard free energy (originally ΔG_r^0) has been corrected for pH 7 (i.e., H^+ concentration is assumed to be 10^{-7} M instead of 1 M). This is to compare the energies at a pH that is relevant for most, but not all, biological reactions. There may be instances where this correction may be necessary for other values of pH (e.g., in biological reactions occurring at much lower pH, as in acid mine drainage). In such cases, the value of G_f^0 for H^+ corresponding to the pH of interest should be used. Because G_f^0 for H^+ at pH 0 is 0, the change in G_f^0 per pH unit is −5.69 kJ/mol.

13.1.4. Commonly Used Electron Acceptor Half-Reactions (R_a)

As mentioned previously, the compound that accepts the electrons from the donor in an energy-yielding reaction is known as the electron acceptor. It is the compound that the organism is said to "breathe." All aerobes, including most chemo-organotrophs and chemolithotrophs, can use oxygen as an electron acceptor. In the absence of oxygen, nitrate-reducing organisms or denitrifiers use NO_3^- and other nitrogen oxides as their electron acceptor; SO_4^{2-} is the major electron acceptor for sulfate-reducing organisms; and CO_2 is the electron acceptor for most methanogens and acetogens. These compounds can serve only as electron acceptors, because the corresponding elements are at their most oxidized states (0 for O_2, +5 for N, +6 for S, and +4 for CO_2). In addition, organic matter can be used by fermenters as a source of electron donor as well as a compound that can accept electrons, because carbon in organic compounds can exist at oxidation states ranging from −4 to less than +4.

Besides the above well-known acceptors, there are many other relatively less studied electron acceptors in nature, e.g., Fe^{3+}, Mn^{3+}, Cr^{6+}, etc. Over the past 2 decades, many more compounds have been added to the list of electron acceptors. These include several chlorinated compounds, such as chlorobenzoate (4), perchloroethene, TCE, dichloroethene, vinyl chloride (11), 1,1,1-TCA (22), chloropropanes, halophenols, etc., and inorganic ions, such as chlorate, perchlorate, arsenate, selenate, uranium(IV), etc. The half-reactions for many of these acceptors are listed in Tables 1 and 2.

Note that the term electron acceptor is reserved for the compounds whose reduction releases energy which must be

TABLE 2 Reduction half-reactions normalized to 1 e^- eq for selected inorganic compounds and the associated values of $\Delta G^{0'}$ and theoretical values of f_s^{0a}

Reaction no.	Reduced-oxidized compounds	Substrates and products	$\Delta G^{0'}$ (kJ/e⁻ eq)	Computed f_s^0 values with acceptor:					
				O_2	Fe^{3+}	NO_3^-	MnO_2	SO_4^{-2}	CO_2
1	Sulfite-sulfate	$1/2\ SO_4^{-2} + H^+ + e^- = 1/2\ SO_3^{-2} + 1/2\ H_2O$	50.30	0.26	0.25	0.25	0.21	0.08	0.07
2	Hydrogen-H⁺	$H^+ + e^- = 1/2\ H_2$	39.87	0.24	0.24	0.23	0.19	0.05	0.04
3	Ammonium-nitrogen	$1/6\ N_2 + 4/3\ H^+ + e^- = 1/6\ NH_4^+$	26.70	0.22	0.22	0.21	0.16	0.02	0.01
4	Sulfide-sulfur	$1/2\ S + 1/2\ H^+ + e^- = 1/2\ HS^-$	26.05	0.22	0.21	0.21	0.16	0.02	0.01
5	Thiosulfate-sulfate	$1/4\ SO_4^{-2} + 5/4\ H^+ + e^- = 1/8\ S_2O_3^{-2} + 5/8\ H_2O$	23.58	0.22	0.21	0.21	0.16	0.01	0.01
6	Methane-carbon dioxide	$1/8\ CO_2 + H^+ + e^- = 1/8\ CH_4 + 1/4\ H_2O$	23.12	0.22	0.21	0.21	0.16	0.01	0.00
7	Sulfide-sulfate	$1/8\ SO_4^{-2} + 9/8\ H^+ + e^- = 1/8\ HS^- - 1/2\ H_2O$	21.23	0.21	0.21	0.20	0.15	0.01	
8	Sulfide-sulfate	$1/8\ SO_4^{-2} + 19/16\ H^+ + e^- = 1/16\ H_2S + 1/16\ HS^- + 1/2\ H_2O$	20.85	0.21	0.21	0.20	0.15	0.00	
9	Sulfide-thiosulfate	$1/8\ S_2O_3^{-2} + H^+ + e^- = 1/4\ HS^- + 3/8\ H_2O$	20.26	0.21	0.20	0.20	0.15	0.00	
10	Sulfur-sulfate	$1/6\ SO_4^{-2} + 4/3\ H^+ + e^- = 1/6\ S + 2/3\ H_2O$	19.15	0.21	0.20	0.20	0.15	N/A	
11	Sulfide-sulfite	$1/6\ SO_3^{-2} + 5/4\ H^+ + e^- = 1/12\ H_2S + 1/12\ HS^- + 1/2\ H_2O$	11.03	0.20	0.19	0.18	0.13		
12	Arsenite-arsenate	$1/2\ AsO_4^{-3} + 2\ H^+ + e^- = 1/2\ H_2AsO_3^- + 1/2\ H_2O$	−14.47	0.15	0.14	0.14	0.08		
13	Cuprous-cupric	$Cu^{+2} + e^- = Cu^+$	−15.44	0.15	0.14	0.13	0.08		
14	Te-Te(OH)₃	$1/4\ Te(OH)_3 + 3/4\ H^+ + e^- = 1/4\ Te + 3/4\ H_2O$	−24.12	0.13	0.12	0.12	0.05		
15	Selenium-selenite	$1/4\ SeO_3^{-2} + 3/2\ H^+ + e^- = 1/4\ Se + 3/4\ H_2O$	−26.05	0.13	0.12	0.11	0.05		
16	Selenium-selenate	$1/6\ SeO_4^{-2} + 4/3\ H^+ + e^- = 1/6\ Se + 2/3\ H_2O$	−31.84	0.11	0.10	0.10	0.04		
17	Ammonium-nitrite	$1/6\ NO_2^- + 4/3\ H^+ + e^- = 1/6\ NH_4^+ + 1/3\ H_2O$	−32.93	0.11	0.10	0.10	0.03		
18	Ammonium-nitrate	$1/8\ NO_3^- + 5/4\ H^+ + e^- = 1/8\ NH_4^+ + 3/8\ H_2O$	−35.11	0.11	0.10	0.09	0.03		
19	Nitrite-nitrate	$1/2\ NO_3^- + H^+ + e^- = 1/2\ NO_2^- + 1/2\ H_2O$	−41.65	0.09	0.08	0.08	0.01		
20	Selenite-selenate	$1/2\ SeO_4^{-2} + H^+ + e^- = 1/2\ SeO_3^{-2} + 1/2\ H_2O$	−43.42	0.09	0.08	0.07	0.01		
21	Manganous-manganese dioxide	$1/2\ MnO_2 + 2H^+ + e^- = 1/2\ Mn^{+2} + H_2O$	−45.35	0.08	0.07	0.07	N/A		
22	Nitrogen-nitrate	$1/5\ NO_3^- + 6/5\ H^+ + e^- = 1/10\ N_2 + 3/5\ H_2O$	−72.20	0.02	0.01	N/A			
23	Ferrous-ferric	$Fe^{+3} + e^- = Fe^{+2}$	−74.27	0.01	N/A				
24	Mercury-mercuric	$1/2\ Hg^{+2} + e^- = 1/2\ Hg$	−77.19	0.00					
25	Water-oxygen	$1/4\ O_2 + H^+ + e^- = 1/2\ H_2O$	−78.72	N/A					
26	Nitrogen-nitrite	$1/3\ NO_2^- + 4/3\ H^+ + e^- = 1/6\ N_2 + 2/3\ H_2O$	−92.56						
27	Chloride-chlorate	$1/6\ ClO_3^- + H^+ + e^- = 1/6\ Cl^- + 1/2\ H_2O$	−100.34						
28	Nitrogen-nitrous oxide	$1/2\ N_2O + H^+ + e^- = 1/2\ N_2 - 1/2\ H_2O$	−126.40						

ᵃComputed assuming NH_4^+ as nitrogen source, $C_5H_7O_2N$ as cell formula, and $\varepsilon = 0.6$. Note that the values of f_s^0 are provided to save time in computation. They do not necessarily imply the existence of an organism that can use the corresponding electron acceptor-acceptor pair. Data from references 3, 20, and 24.

TABLE 3 Free energies of formation, G_f^0, for selected substances

Substance	G_f^0 (kJ mol^{-1})	Substance	G_f^0 (kJ mol^{-1})
CO_2	-394.4	PO_4^{3-}	-1026.55
CO	-137.34	$H_2PO_4^-$	-1130.2
O_2	0	Se^0	0
H_2	0	SeO_4^{2-}	-441.3
H_2O	-237.17	H_2Se	-77.09
HCO_3^-	-586.85	SeO_4^{2-}	-439.95
CO_3^{2-}	-527.90	$MgSO_4$	-1199.5
H_2CO_3	-623.16	PbS	-92.59
N_2	0	CuS	-49.02
NH_3	-26.57	Cu^{2+}	$+64.94$
NH_4^+	-79.37	Cu^+	$+50.28$
NO	$+86.57$	MoS_2	-225.42
NO_2	$+51.95$	MoO_4^{2-}	-836.3
NO_3^-	-111.34	ZnS	-198.60
NO_2^-	-37.2	Zn^{+2}	-147.1
N_2O	$+104.18$	Mn^{3+}	-82.12
S^0	0	Mn^{+2}	-227.93
SO_3^{2-}	-486.6	MnO_2	-456.71
SO_4^{2-}	-744.6	MnO_4^-	-506.57
HSO_3^-	-527.7	$MnSO_4$	-955.32
HSO_4^-	-755.9		
H_2O_2	-134.1	Aspartate	-700.4
$S_2O_3^{2-}$	-513.4	Crotonate	-277.4
H_2S	-27.87	Cysteine	-339.8
HS^-	$+12.05$	Fructose	-915.38
S^{2-}	$+85.8$	Fumarate	-604.21
Cl^-	-131.2	Gluconate	-1128.3
ClO_2^-	17.2	Glycerate	-658.1
ClO_4^-	-8.5	Guanine	$+46.99$
ClO_{3-}	-8.0	Glycolate	-530.95
ClO^-	-36.8	Lactose	-1515.24
$Cr_2O_7^{2-}$	-1301.1	Mannitol	-942.61
CrO_4^{2-}	-727.8	Methane	-50.75
Fe^{2+}	-78.87	Oxalate	-674.04
Fe^{3+}	-4.6	Phenol	-47.6
$FeCO_3$	-673.23	n-Propanol	-175.81
$FeSO_4$	-829.62	Ribose	-757.3
FeS_2	-150.84	Sucrose	-370.90
U^{+4}	-531.9	Urea	-203.76
U^{+3}	-476.2	Valerate	-344.34
Electron (e^-)	0, always		
H^+	0 at pH 0; -39.83 at pH 7 (-5.69 per pH unit)		
OH^-	57.3 at pH 14; -198.76 at pH 7; -237.57 at pH 0		

captured by the enzymatic machinery of the microorganisms to support growth. Many other compounds may use electrons (e.g., for the reduction of materials for cell synthesis or cometabolic reductive transformation of contaminants), but those compounds are not referred to as electron acceptors. Thus, availability of energy when combined with a donor and the subsequent use of energy in the cell synthesis reaction defines the term electron acceptor and not the acceptance of electrons alone.

13.1.5. Commonly Used Electron Donor Half-Reactions (R_d)
Similar to the electron acceptor, the compound that donates the electrons to the acceptor in an energy-yielding

reaction is known as the electron donor. It is the compound that is said to be the energy source for the microorganisms (although it is more accurate to state that energy is provided by the donor-acceptor pair). The array of compounds that can donate electrons is much larger (in thousands) than the number of electron acceptors, which is probably less than 100 at present. Many organic compounds (e.g., sugars, proteins, acids, and alcohols) and some inorganic compounds (e.g., NH_4^+, H_2, Fe^{2+}, Mn^{2+}, S, etc.) can serve as electron donors. During heterotrophic growth (i.e., when the donor is organic), the donor also serves as the carbon source. However, there are many denitrifying and sulfate-reducing bacteria that can oxidize an inorganic donor, especially H_2. In such cases, CO_2 serves as the carbon source. CO_2 is also the carbon source during autotrophic growth (i.e., when the donor is an inorganic compound) because the donor can only serve as a source of electrons. Nitrifiers are an example of autotrophic bacteria using NH_4^+ as donor and O_2 as acceptor. Similarly, iron- and sulfur-oxidizing bacteria, or the arsenate-respiring bacteria that use sulfide as the donor (see Example 2) are all autotrophic, using CO_2 as a carbon source.

13.1.6. Electron Acceptor-Donor Pair as Energy-Yielding Reaction (R_e)

The energy-yielding reaction can be obtained by subtracting the donor half-reaction (R_d) from the acceptor half-reaction (R_a). In the form of an equation, this is written as $R_e = R_a - R_d$. Because the identity of donor or acceptor may not always be known a priori, evaluation of their role proceeds by assuming that one of the half-reactions is a donor (R_d) and the other is an acceptor (R_a). The $\Delta G_r^{0'}$ is computed as $\Delta G_a^{0'} - \Delta G_d^{0'}$ under the assumed roles. If the computed $\Delta G_r^{0'}$ is negative, the assumed roles are said to be correct and it is concluded that energy is available and can potentially be used to support growth. A positive $\Delta G_r^{0'}$, on the other hand, implies that the assumed roles are incorrect and energy is not available for growth from the pair under the conditions tested. When the donor half-reaction is reversed to obtain $-R_d$, the sign of the Gibbs standard free energy for the donor half-reaction also reverses (i.e., it becomes the negative of whatever value was for $\Delta G_d^{0'}$). In addition, in an energy-yielding reaction (R_e), electrons must cancel out each other. Under no circumstances will an energy-yielding reaction have electrons as the substrate or product. Only half-reactions can have electrons. The following examples illustrate the process of writing an energy-yielding reaction.

13.1.6.1. Example 4

Evaluate if the oxygen half-reaction (O_2/H_2O) and the glucose half-reaction (glucose/CO_2) can be combined to serve as a source of energy for microbial growth.

First, the half-reactions for O_2 and glucose are written as they are listed in Tables 1 and 2, i.e., reduction half-reactions normalized to 1 e^- eq. The associated $\Delta G^{0'}$ for each half-reaction is also written as such. A tentative role to each half-reaction is assigned (say, R_a to O_2 half-reaction and R_d to glucose half-reaction).

R_a: $1/4\ O_2 + H^+ + e^- = 1/2\ H_2O$ -78.72 kJ/e^- eq **(3)**

R_d: $1/4\ CO_2 + H^+ + e^- = 1/24\ C_6H_{12}O_6 + 1/4\ H_2O$
$$+41.35 \text{ kJ/}e^- \text{ eq} \quad \textbf{(4)}$$

Then, the assumed donor half-reaction is reversed to obtain $-R_d$ (equation 4a). The sign of $\Delta G^{0'}$ for the donor half-

reaction is also reversed. The reversed donor half-reaction is added to equation 3 as follows to obtain equation 5:

$-R_d$: $1/24\ C_6H_{12}O_6 + 1/4\ H_2O$

$\underline{R_a\text{: } 1/4\ O_2 + H^+ + e^-}$

R_e: $1/24\ C_6H_{12}O_6 + 1/4\ O_2$

$$\begin{aligned} &= 1/4\ CO_2 + H^+ + e^- & -41.35 \text{ kJ/}e^- \text{ eq} \quad \textbf{(4a)} \\ &= 1/2\ H_2O & -78.72 \text{ kJ/}e^- \text{ eq} \quad \textbf{(3)} \\ \hline &= 1/4\ CO_2 + 1/4\ H_2O & -120.07 \text{ kJ/}e^- \text{ eq} \quad \textbf{(5)} \end{aligned}$$

Note that the energies for the two half-reactions are also added up. Because $\Delta G_r^{0'}$ (equation 5) is negative, indicating release of energy, the reaction can serve as an energy-yielding reaction and the assumed roles for glucose as electron donor and O_2 as electron acceptor were correct. In an energy-yielding reaction, both donor and acceptor must end up on the left-hand side and the products of their oxidation (CO_2 from glucose) or reduction (H_2O from O_2) must end up on the right-hand side of the reaction.

13.1.6.2. Example 5

Anaerobic ammonia oxidation (anammox) is a recently observed chemolithotrophic process whereby ammonia is oxidized to dinitrogen gas anaerobically with nitrite as an electron acceptor. The potential for existence of such a process was predicted a long time ago based on energetics. At present, many species including "*Candidatus* Scalindua wagneri," "*Candidatus* Scalindua brodae," "*Candidatus* Scalindua sorokinii," "*Candidatus* Brocadia anammoxidans," and "*Candidatus* Kuenenia stuttgartiensis" are known to obtain energy for growth by anaerobic ammonia oxidation. Compute the amount of energy available for the above donor-acceptor pair.

First, the half-reactions for NO_2^- and NH_4^+ (both transforming to N_2) are selected from Table 1. The associated $\Delta G^{0'}$ value for each half-reaction is also written as such. Then, the known roles are assigned to each half-reaction (R_a to NO_2^- half-reaction and R_d to NH_4^+ half-reaction).

R_a: $1/3\ NO_2^- + 4/3\ H^+ + e^- = 1/6\ N_2 + 2/3\ H_2O$
$$-92.56 \text{ kJ/}e^- \text{ eq} \quad \textbf{(6)}$$

R_d: $1/6\ N_2 + 4/3\ H^+ + e^- = 1/3\ NH_4^+$
$$26.70 \text{ kJ/}e^- \text{ eq} \quad \textbf{(7)}$$

The donor reaction is then reversed to obtain equation 7a, which is added to equation 6 as follows:

$-R_d$: $1/3\ NH_4^+ = 1/6\ N_2 + 4/3\ H^+ + e^-$
$$-26.70 \text{ kJ/}e^- \text{ eq} \quad \textbf{(7a)}$$

$\underline{R_a\text{: } 1/3\ NO_2^- + 4/3\ H^+ + e^- = 1/6\ N_2 + 2/3\ H_2O}$
$$-92.56 \text{ kJ/}e^- \text{ eq} \quad \textbf{(6)}$$

R_e: $1/3\ NH_4^+ + 1/3\ NO_2^- = 1/3\ N_2 + 2/3\ H_2O$
$$-118.26 \text{ kJ/}e^- \text{ eq} \quad \textbf{(8)}$$

Note that the availability of energy alone does not guarantee that microorganisms will be able to use the released energy for growth. Metabolic enzymes must also be available to capture the released energy. For less commonly known donor-acceptor pairs, the availability of enzymes to capture the available energy can be shown only experimentally. For example, the use of energy for growth from the C-Cl bond breakage was unknown until 1987 when Dolfing and Tiedje

(4) demonstrated that the energy released during the reduction of chlorobenzoate to benzoate can be captured for growth. Since then, many more microorganisms including *Dehalobacter restrictus* (8), *Dehalococcoides ethenogenes* (12), and *Dehalospirillum multivorans* have been shown to respire halogenated compounds. The subject of dehalorespiration was recently reviewed by Smidt and de Vos (21).

13.1.7. Automated Assignment of Source of Carbon for Cell Synthesis

In nature, carbon is available as reduced organic carbon with an average oxidation state varying from -4 to $+3$ or higher but not $+4$ (all organic compounds) or as CO_2 (including HCO_3^-, CO_3^{2-}), which is the most oxidized state of carbon ($+4$). Most compounds in the cell providing structure, function, or information and especially carbon- and nitrogen-containing compounds are found in the reduced form compared to the environment. Consider for example, the elemental formula for cell material of *Escherichia coli* (17), which has 50 to 54% C, 8% H, 20% O, and 14% N (all dry weight basis). This gives a formula for cell material as $CH_{1.9}O_{0.3}N_{0.2}$. From this formula, the oxidation state of cell carbon can be computed as -0.7, which is relatively reduced compared to that in a typical substrate such as glucose, where the average oxidation state of C is 0. Because carbon is also the most abundant element in microbial cells, constituting up to 50 to 55% of the dry cell mass, the need for electrons to reduce carbon is also substantial. This is especially true when the carbon source is more oxidized, e.g., CO_2/HCO_3^-. This is why for the same amount of donor, the cell yield is much lower when CO_2 is used as the carbon source (autotrophy) than when an organic compound is used as the carbon source (heterotrophy).

The incorporation of the carbon source into cell carbon and its effect on cell synthesis are taken into account by computing f_s^0 for each carbon source and by writing the half-reactions in a manner that automatically reflects the source of carbon in the overall stoichiometric equation. If the donor is an organic carbon source, the half-reaction for the donor combined with the energy and synthesis reactions ensures that the organic donor appears on the left-hand side of the stoichiometric equation R as a source of electrons as well as cell carbon. If the donor is an inorganic compound, the combination of half-reactions ensures that CO_2 and/or HCO_3^- appear on the left-hand side of R as the source of cell carbon. Thus, in the process of writing R, no active decision is required to correctly portray the carbon source. Exceptions may exist to this generalization, for which an inorganic donor (e.g., H_2) and an organic carbon source (e.g., acetate or formate) are necessary for growth. Dependence of f_s^0 on the source of carbon is described in section 13.1.10.

13.1.8. Model Cell Synthesis Half-Reactions as a Function of Nitrogen Source (R_c)

After carbon, nitrogen is the next most abundant element in microbial cells, constituting approximately 13% of the dry cell mass. In the cell, it is generally present in the average oxidation state of -3 (as NH_3/NH_4^+). Nitrogen in the environment may exist at an oxidation state varying from -3 (NH_3/NH_4^+) to $+5$ (NO_3^-). Thus, if the nitrogen source available to the cells is other than NH_3/NH_4^+, additional electrons will be needed to reduce it to the oxidation state of NH_3/NH_4^+. NO_3^- and NO_2^- may be used as nitrogen sources in the absence of NH_4^+, and N_2 may also serve as a source of nitrogen if a mechanism to fix N_2 is available to the microorganism. Other oxides of nitrogen can also be used as a source of nitrogen. From the perspective of the need for electrons to reduce the nitrogen source, NO_3^- is the most expensive and NH_3 is the least expensive among the four nitrogen sources. Similar to the lower yield observed when CO_2 is used as the carbon source, a somewhat lower growth on the more oxidized nitrogen sources is also expected and observed.

It is much easier to account for the variability in cell yield due to various nitrogen sources by writing separate cell synthesis half-reactions (R_c), one for each type of nitrogen source as shown in Table 4. These equations are written along the lines of R_a and R_d. When writing R, one of these four cell synthesis half-reactions is chosen corresponding to the nitrogen source available for growth. From the coefficients associated with the nitrogen source, it can be seen that the number of moles of electrons needed to make 1 mol of cell material will be 28 for NO_3^-, 26 for NO_2^-, and 23 for N_2 compared to only 20 for NH_4^+, illustrating the effect of nitrogen source on cell yield.

13.1.9. Assumed Efficiency of Energy Transfer (ϵ)

One of the key assumptions in the process of writing R is related to the efficiency of energy transfer, ε, by the cellular machinery, i.e., how much of the $\Delta G_r^{0'}$ can really be utilized and how much of it is lost as heat. The value of ε is assumed to be 0.6 (i.e., 40% of the energy is lost), unless demonstrated otherwise. This assumption has important implications for the cell yield. An efficiency much lower than 0.6 (which is possible in many cases) will result in a much lower cell yield. For many compounds, often the efficiency, ε, is unknown. The value of ε combined with a quantitative evaluation of the electrons and free energy also allows the calculation of the fraction of electrons used for cell synthesis (f_s^0) and those needed for energy generation (f_e^0) as described below. For a detailed description of the rationale for using 0.6, the reader is referred to reference 20.

13.1.10. Fraction of Electron Donor Used for Cell Synthesis (f_s^0)

The fraction of electron donor that is used by the cell synthesis reaction (f_s^0) must be known to write the stoichiometric equation. The value of f_s^0 depends upon the choice of electron donor and acceptor, carbon and nitrogen sources, and the efficiency of the enzymes that the microorganism has. The role of each one of these components in

TABLE 4 Half-reactions representing cell synthesis using NH_4^+, NO_3^-, NO_2^-, or N_2 as N source

N source	Cell synthesis half-reaction, R_c
NH_4^+	$1/20\ NH_4^+ + 1/5\ CO_2 + 1/20\ HCO_3^- + H^+ + e^- = 1/20\ C_5H_7O_2N + 9/20\ H_2O$
NO_3^-	$1/28\ NO_3^- + 5/28\ CO_2 + 29/28\ H^+ + e^- = 1/28\ C_5H_7O_2N + 11/28\ H_2O$
NO_2^-	$1/26\ NO_2^- + 5/26\ CO_2 + 27/26\ H^+ + e^- = 1/26\ C_5H_7O_2N + 10/26\ H_2O$
N_2	$1/46\ N_2 + 5/23\ CO_2 + H^+ + e^- = 1/23\ C_5H_7O_2N + 8/23\ H_2O$

determining f_s^0 is incorporated in the following equation, which is obtained by considering the amount of energy needed to synthesize an equivalent of cells, the amount of energy available from the donor-acceptor pair, and the losses. To compute f_s^0, Rittmann and McCarty (20) assumed pyruvate as the key intermediate to which all carbon sources must be converted (albeit theoretically in many cases) before being transformed to macromolecules of the cell. The following paragraphs present the procedure to compute f_s^0. For the rationale behind this assumption and the use of terms in this equation, the original reference should be consulted (20).

The fraction of electron donor (in terms of e^- eq) used for cell synthesis (f_s^0) is computed as follows:

$$f_s^0 = \frac{1}{1+A} \tag{9}$$

where A is computed as follows:

$$A = -\frac{\dfrac{\Delta G_p}{\varepsilon^n} + \dfrac{\Delta G_{pc}}{\varepsilon}}{\varepsilon \, \Delta G_r} \tag{10}$$

All the factors needed to compute A are related to the free energies associated with the donor, acceptor, or cell synthesis half-reactions and the related efficiency of transfer in the following manner.

ΔG_p is the difference between the free energy of the pyruvate half-reaction and that of the carbon source. It is equal to 113.8 kJ/e^- eq for autotrophic reactions, i.e., when CO_2 is the carbon source and is computed as $(35.09 - \Delta G_d^{0'})$ for heterotrophic reactions where donor and carbon source are the same.

ΔG_{pc} is the amount of energy needed to convert pyruvate to cell carbon, estimated by McCarty (13). It also includes the effects of various nitrogen sources by using a different value for each nitrogen source: 18.8 kJ/e^- eq for NH_4^+, 13.5 kJ/e^- eq for NO_3^-, 14.5 kJ/e^- eq for NO_2^-, and 16.4 kJ/e^- eq for N_2. In all cases, the cell formula is assumed to be $C_5H_7O_2N$. Combining the effects of different nitrogen sources into ΔG_{pc} amounts to an assumption that electrons to reduce the oxidized nitrogen sources come from pyruvate.

ΔG_r is the energy available from the electron donor-acceptor pair, i.e., $\Delta G_a^{0'} - \Delta G_d^{0'}$. Under standard conditions and pH 7, this is the same as $\Delta G_r^{0'}$.

ε is the efficiency of energy transfer, assumed to be 0.6 (but it may vary for different organisms and substrates).

n is a factor that addresses the fact that some carbon sources may actually release energy when converted to pyruvate and others may require energy. Thus, it is +1 when ΔG_p is positive and −1 when ΔG_p is negative.

The following example illustrates the use of the above equation for computing f_s^0.

13.1.10.1. Example 6

You are interested in removing NO_3^- from groundwater using denitrifying bacteria and decide to use methanol, which is also the carbon source, as the donor. Assume that no NH_4^+ is present, so NO_3^- is also the nitrogen source. Compute the value of f_s^0 expected for the denitrifying organisms during the above process.

First, we compute ΔG_p using $\Delta G_d^{0'}$ of the methanol half-reaction.

$$\Delta G_p = (35.09 - \Delta G_d^{0'})$$
$$= 35.09 - 36.84$$
$$= -1.75 \text{ kJ/}e^- \text{ eq}$$

Because ΔG_p is negative, $n = -1$. ΔG_{pc} is dictated by the nitrogen source, which in this case is NO_3^-. Therefore, the value of ΔG_{pc} is equal to 13.5 kJ/e^- eq.

Next, we obtain ΔG_r by computing $\Delta G_a^{0'} - \Delta G_d^{0'}$. The acceptor reaction is NO_3^- to N_2, implying that $\Delta G_a^{0'}$ is −72.20 kJ/e^- eq. The donor reaction is the same as above, and $\Delta G_d^{0'}$ is 36.84 kJ/e^- eq. Therefore,

$$\Delta G_r = (-72.20 - 36.84)$$
$$= -109.04 \text{ kJ/reaction}$$

Using an ε of 0.6 in equation 10, we can obtain the value of A as follows:

$$A = -\frac{\dfrac{-1.75}{0.6^{-1}} + \dfrac{13.5}{0.6}}{0.6(-109.04)} = 0.33$$

The value of f_s^0 is computed using equation 9 as:

$$f_s^0 = \frac{1}{1+0.33} = 0.75$$

This implies that f_e^0 which is equal to $(1 - f_s^0)$, is 0.25.

Figure 4 depicts the variability in f_s^0 for the organic and inorganic donors listed in Tables 1 and 2 when combined with a number of common electron acceptors. Several points are worth noting including the following: (i) the value of f_s^0 is generally much lower for autotrophic growth than for heterotrophic growth; (ii) oxygen, iron, and nitrate have higher f_s^0 than sulfate and CO_2, with manganese dioxide having f_s^0 closer to the former set of acceptors; and (iii) certain chlorinated compounds, even when acting as donor with high-energy-yielding acceptors, have relatively low f_s^0 values.

13.1.11. The Overall Stoichiometric Equation (R)

The stoichiomteric equation, R, serves many needs. Using this equation we can calculate (i) the amount of cells expected from a given substrate, (ii) the amount of oxygen or any other electron acceptor needed to accomplish the conversion, (iii) the amount of nitrogen required to make sure only the donor is limiting and not other key nutrients, and (iv) the amount of other products generated, etc. As mentioned earlier, the available electrons from the donor are apportioned into the energy-yielding reaction (f_e^0) and the cell synthesis reaction (f_s^0). For the purpose of writing the stoichiometric equation, it is easier to start with 1 e^- eq of the electron donor and divide it into two portions: one portion is used by the acceptor to yield energy, and the other portion is used in the cell synthesis reaction to reduce carbon and nitrogen (if it is other than NH_4^+/NH_3) and synthesize the macromolecules. In other words, the electron donor half-reaction (R_d) supplies 1 e^- eq, a fraction of which (f_e^0) goes to the acceptor half-reaction (R_a) and the remaining fraction (f_s^0) goes to the synthesis half-reaction (R_c). Using R_d, R_a, and R_c, this is written mathematically as

$$R = f_e^0 R_a + f_s^0 R_c - R_d \tag{11}$$

To obtain R, the acceptor half-reaction is multiplied by f_e^0, the cell synthesis half-reaction is multiplied by f_s^0, the donor half-reaction is multiplied by −1, and the three resulting equations are added algebraically. The following example illustrates the main steps.

13.1.11.1. Example 7

Write the overall stoichiometric equation for the oxidation of methanol to CO_2 (methanol acts as an electron donor and also as a carbon source), with NO_3^- as an electron acceptor and also serving as the nitrogen source. Use the value of f_s^0 as 0.75 (thus, $f_e^0 = 0.25$) computed above.

The half-reactions needed to write the stoichiometric equation are methanol to CO_2 half-reaction, NO_3^- to N_2 half-reaction, and cell synthesis half-reaction with NO_3^- as the nitrogen source. They are reproduced below:

R_a: $1/5\ NO_3^- + 6/5\ H^+ + e^- = 1/10\ N_2 + 3/5\ H_2O$

R_c: $1/28\ NO_3^- + 5/28\ CO_2 + 29/28\ H^+ + e^-$
$\qquad = 1/28\ C_5H_7O_2N + 11/28\ H_2O$

R_d: $1/6\ CO_2 + H^+ + e^- = 1/6\ CH_3OH + 1/6\ H_2O$

We multiply R_a by 0.25 (f_e^0), R_c by 0.75 (f_s^0), and R_d by -1; rewrite the resulting equations; and add them as shown below:

$f_e^0 R_a$: $0.05\ NO_3^- + 0.3\ H^+ + 0.25\ e^-$
$\qquad = 0.025\ N_2 + 0.15\ H_2O$

$f_s^0 R_c$: $0.027\ NO_3^- + 0.134\ CO_2 + 0.777\ H^+ + 0.75\ e^-$
$\qquad = 0.027\ C_5H_7O_2N + 0.29\ H_2O$

$-R_d$: $0.167\ CH_3OH + 0.167\ H_2O$
$\qquad = 0.167\ CO_2 + H^+ + e^-$

$0.167\ CH_3OH + 0.077\ NO_3^- + 0.077\ H^+ = 0.025\ N_2$
$\quad + 0.033\ CO_2 + 0.278\ H_2O + 0.027\ C_5H_7O_2N$ (12)

Note that no electrons remain in the final equation because 1 e^- eq from the donor is exactly divided between the synthesis and the energy-yielding reactions in the ratio 0.75:0.25.

Equation 12 is the stoichiometric equation, R. It may be used to compute the quantitative relationship among the reactants and products including cell mass. It is helpful to first calculate the molecular mass of each reactant or product. Thus, the molecular mass of methanol is $(1 \times 12 + 4 \times 1 + 1 \times 16) = 32$ and that of the cell of the given formula is $(12 \times 5) + (1 \times 7) + (16 \times 2) + 14 = 113$. Using the above equation, it can be computed that 3.05 g ($= 0.027 \times 113$) of cells are expected from 5.34 g ($= 0.167 \times 32$) of methanol oxidized to CO_2. Expressed as a ratio (of grams of cells produced per grams of substrate consumed), this is also known as the true yield coefficient Y. A value of 0.57 for the yield coefficient Y on the basis of methanol is typical of the values observed for aerobic and denitrifying processes. Equation 12 also indicates that for each 5.34 g of methanol oxidized, 4.77 g of NO_3^- (0.077×62) or 1.08 g of $NO_3^- - N$ (0.077×14) will be consumed.

Notes:

1. The true yield, Y in grams of cells per gram of organic matter expressed as chemical oxygen demand can also be computed from f_s^0 using the formula:

$$Y = f_s^0\ (M_c \text{ g of cells/mol of cells})/$$
$$[(n_e\ e-\text{ eq/mol cells})(8 \text{ g of COD}/e^-\text{ eq donor})]$$

where COD is chemical oxygen demand. It is defined as the amount of oxygen required to oxidize 1 g of any substrate, M_c is the empirical formula weight of cells (113 for $C_5H_7O_2N$), and n_e is the number of e^- eq per mol of cells (20 for NH_4^+, 28 for NO_3^-, 26 for NO_2^-, and 23 for N_2).

The last term (8 g of COD in 1 e^- eq of donor) is a constant which is true for all donors because 1 e^- eq of any organic donor requires 8 g of oxygen to be completely oxidized to CO_2. The above equation can be translated to the following four equations, one for each nitrogen source.

$$Y = 0.706\ f_s^0 : NH_4^+ \text{ as nitrogen source}$$
$$= 0.504\ f_s^0 : NO_3^- \text{ as nitrogen source}$$
$$= 0.543\ f_s^0 : NO_2^- \text{ as nitrogen source}$$
$$= 0.614\ f_s^0 : N_2 \text{ as nitrogen source}$$

Using the f_s^0 values for each nitrogen source and the above equations, the yield coefficient Y has the following relationship:

$$Y_{nitrate} = (0.79 \pm 0.06)\ Y_{ammonium}$$
$$Y_{nitrate} = (0.83 \pm 0.05)\ Y_{ammonium}$$
$$Y_{dinitrogen} = (0.91 \pm 0.05)\ Y_{ammonium}$$

Thus, the yield coefficient Y for denitrifiers growing on methanol with ammonium as nitrogen source will be approximately 0.72 (obtained by dividing the value of Y on nitrate, which is 0.57, by 0.79, the ratio between the yield coefficients for nitrate and ammonium). The above relationships were obtained by computing f_s^0 and Y for the four nitrogen sources over a range of donors (those listed in Tables 1 and 2) and acceptors (those shown in Fig. 4). It can be used to obtain approximate values. These relationships also highlight the effect of nitrogen source on cell yield.

2. Energy is also needed to maintain the cellular functions. This is known as maintenance energy and comes from the energy-yielding reaction. It is modeled by increasing the f_e^0 by a fraction, m, and representing it by f_e (thus, $f_e^0 + m = f_e$) and decreasing the f_s^0 by the same fraction and representing it by f_s (thus $f_s^0 - m = f_s$). The fraction m is computed from an equation that is a function of the rate of cell decay and cell age. The fraction m effectively reduces the amount of donor used for cell synthesis by an amount that is equal to the amount of donor required for maintenance (20).

3. In the reduction half-reactions used in the above analysis, all reactants and products except H^+ are at 1 M concentration. In practice, this is generally not the case; the reactants and products may be at concentrations other than 1 M, and their concentrations may be continuously changing with time due to microbial activity. In most cases, the change in $\Delta G^{0'}$ due to a change in reactant or product concentration will not be significant enough to change the conclusions about the potential for growth. However, in certain instances, it is possible that the reaction is thermodynamically favorable (meaning negative $\Delta G^{0'}$) under one set of reactant and product concentrations but unfavorable (meaning positive $\Delta G^{0'}$) under another. The effect of change in concentration is taken into account by the Nernst equation (20) whose general form is

$$\Delta G = \Delta G^{0'} + RT \ln Q \qquad (13)$$

where R, T, and Q are the gas constant (8.314 J mol^{-1} K^{-1}), temperature (in kelvin), and reaction quotient, respectively. The reaction quotient, Q, depends on the reaction of interest (20).

4. If the source of nitrogen is other than those mentioned above (NH_4^+, NO_3^-, NO_2^-, and N_2), a separate cell synthesis equation may be written corresponding to that source

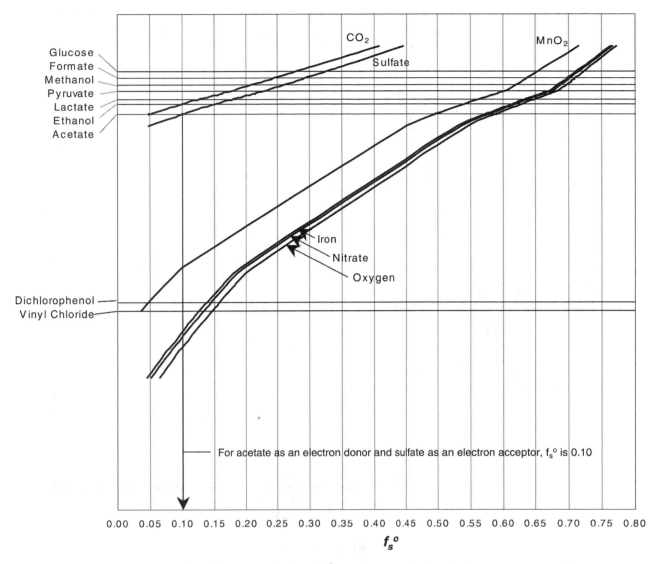

FIGURE 4 Trends in the computed value of f_s^0 as a function of selected electron acceptors and electron donors for NH_4^+ as the nitrogen sources and $C_5H_7O_2N$ as the cell formula. (Part 1) Organic donors (heterotrophy); (Part 2) Inorganic donors (autotrophy).

of nitrogen. Similarly, the cell synthesis equation R_c in the above analysis uses $C_5H_7O_2N$ as the cell formula. Significant variability in cell formula exists due to the variability in the type of microorganism and growth substrates. If a better estimate for the cell formula is known and such a theoretical analysis is still useful, writing a new R_c corresponding to the known cell formula may be more appropriate.

5. The effect of source of phosphorus, sulfur, and other trace elements on electron donor requirement is neglected in this analysis. Phosphorus constitutes only up to 2.5% of the cell mass under normal conditions, and most of it is present as phosphate, the oxidized state of phosphorus. Sulfur and other trace elements which may be present in a reduced state constitute even lower fractions of the cell mass. These are generally excluded from the quantitative analyses related to assess the need for electrons and the electron donor because the complexity of equations is increased without significant improvement in the energy, electrons, and material

balance. Experimentally, it is of course critical to ensure that these elements are present in excess.

13.2. KINETICS OF MICROBIAL GROWTH

A single cell grows into two cells with the input of raw materials (carbon, nitrogen, phosphorus, nutrients, etc.), electrons, and energy. For the purpose of modeling, it is important to assume that only one of the raw materials is limiting, called the limiting substrate (or just substrate) and that all other required substrates and trace elements are in excess. In practice, it is possible that more than one substrate is limiting. Such a scenario requires a more complex modeling approach that is outside the scope of this chapter. The limiting substrate is generally the electron donor, but it could be any one of the raw materials. The concentration of this limiting substrate can be represented as S (in milligrams per liter) at any time t, while the initial concentration can be represented

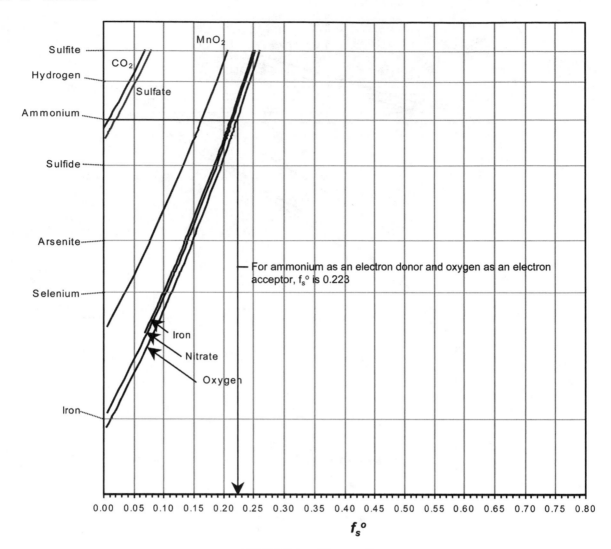

FIGURE 4   *(Continued)*

as S^0 at the start of the reaction, t_0. The initial concentration of cells can be represented as X^0 (in milligrams per liter) while the cell concentration at any time t can be represented as X. Some materials are utilized during the time the microorganisms grew, implying the existence of a rate at which substrate was utilized. This is called the rate of substrate utilization, r_{su}. The concentration of the substrate (S), biomass (X), change in the concentration of substrate (dS), change in the concentration of biomass (dX), change in time (dt), rate of change of biomass concentration (dX/dt), and rate of change of substrate concentration (dS/dt) need to be related to each other for a certain system (reactor vessel, plate colony, biofilm, etc.) in order to predict a target unknown parameter when all others are known.

Such predictive equations can be written for a single substrate, a group of microorganisms using a similar substrate (known as a guild, e.g., in fermentation with mixed population), or a collection of guilds each utilizing a different substrate but together performing an overall function (e.g., in anaerobic digestion). Often the same approach is applied for modeling even if a large number of different types of microorganisms utilize a collection of different types of substrates (e.g., in the case of aerobic bacteria utilizing complex carbon compounds as the donor substrate and oxygen as the acceptor substrate, only one of them will be assumed to be limiting either the donor or the acceptor). Thus, the growth of a single species on a single substrate becomes a special case of the more complex situations. Note also that there are some special cases that require additional steps to obtain appropriate modeling equations. These include diauxic growth, synchronous culture growth, and effect of toxic compounds. When a single species is growing on multiple substrates, there is a chance that the substrates will become limiting one after another; this is called diauxic growth, the modeling of which requires additional considerations beyond the scope of the current discussion. Similarly, there is also a scenario in which the growth of the cells is synchronized to divide at the same time. This is known as synchronous culture growth and has to be modeled differently from the growth discussed above, which is for asynchronous growth. These exercises are introductory because most models may need to be developed much beyond what is presented here. Many books and journal articles are available on this subject.

It so happens that a batch system is easier to operate but more complex to model. The reason is that in a batch system, substrate concentration keeps changing. Figure 5 shows that the specific growth rate (μ) is a function of the substrate concentration (S). Thus, in a batch system, unless the substrate is in great excess, microorganisms grow at different specific growth rates because S is changing with time. To calculate the total amount of growth (and thus the concentration of biomass after time t) or the new substrate concentration after some of it has been utilized, the total growth needs to be integrated mathematically. This is in contrast to a chemostat where a single substrate concentration, S, exists at all times, resulting in a constant specific growth rate (μ) at all times, no integration is required, and the analytical solution is much simpler.

The techniques used to measure growth monitor average values and obscure the fact that all bacterial cultures are grossly heterogeneous. They describe the growth of a population rather than of individual cells. Although all members of a particular population may be genetically identical, the individual members vary with respect to doubling time, age, composition, metabolic characteristics, and size. The magnitudes of these variations are influenced by the environment and can often be minimized by careful design and development of the system for cultivation. Most of the systems described here are designed to minimize heterogeneity within the population of a pure culture.

In such systems, the growth behavior of a bacterial population can be predicted by the simple relationship:

$$\frac{dX}{dt} = \mu X \qquad (14)$$

where dX is the increase in amount of cell mass (biomass), dt is the time interval, X is the amount of biomass, and μ is the specific growth rate, representing the rate of growth per unit amount of biomass and having dimensions of reciprocal time ($1/t$).

If μ is constant, integration of equation 14 shows that X will increase exponentially with time as follows:

$$\ln \frac{X}{X_0} = \mu t \qquad (15)$$

where ln is the natural logarithm (to the base $e = 2.303$) and X_0 is the amount of biomass when $t = 0$.

It follows by rearrangement of equation 15 that the final biomass concentration X is

$$X = X_0 e^{\mu t} \qquad (16)$$

Bacterial growth that follows this relationship is called **exponential or logarithmic growth**.

The relationship between the biomass doubling time (t_d) and the specific growth rate is found by letting $X = 2X_0$ at $t = t_d$ in equation 15 and solving the equation as follows:

$$t_d = \frac{\ln 2}{\mu} = \frac{0.693}{\mu} \qquad (17)$$

Equations 14 to 17 predict the growth of bacteria in simple systems in which the factors influencing growth are constant but do not allow prediction of deviation from constant-growth (steady-state) conditions. The techniques for continuous cultivation very nearly establish steady-state conditions of theoretically infinite duration, whereas the techniques for batch cultivation allow significant changes in the environment during the time course of cultivation. In batch culture, equations 14 to 17 will apply without adjustment to the value of μ only during the portion of the growth cycle in which the changes in the growth environment have no influence on population growth (i.e., during the exponential growth phase).

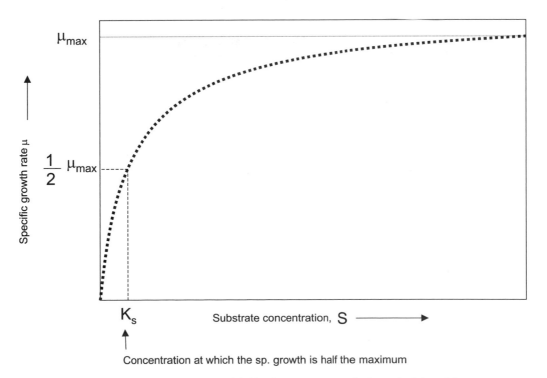

FIGURE 5 Specific growth rate and half velocity constant, the basis for Monod kinetics.

13.2.1. Modeling Batch Systems

13.2.1.1. Principles
A batch culture system is one in which nothing (with the frequent exception of the gas phase) is added to or removed from the culture vessel after a medium of appropriate composition is inoculated with living cells; this is called a **closed system**. It follows that a batch system can support cell multiplication for only a limited time and with progressive changes in the original medium and environment.

13.2.1.1.1. Normal Growth Cycle
Figure 6 shows an idealized normal growth cycle (curve) for a simple, homogeneous batch culture of bacteria. The normal growth cycle assumes asynchronous binary fission of the cells. Growth proceeds through a **lag phase**, during which cell mass increases but cell numbers do not, and into a **growth phase**, which usually is characterized by an exponential increase in numbers and follows the relationships of equations 14 to 17. Ultimately, changes in the chemical or physical environment result in a phase of no net increase in numbers, the **maximum stationary phase**. Cells in the stationary phase still require an energy source for the maintenance of viability. If the availability of an energy source in a batch culture is limited, there ensues a **death phase**, which is often characterized by an exponential decrease in the number of living cells.

The lag phase may be brought about by the shock of rapid change in the culture environment. In fresh medium, the length of the lag phase depends on the size of the inoculum, the age of the inoculum, and the changes in nutrient composition and concentration experienced by the cells. A small volume of inoculum transferred to a large volume of fresh medium may result in outward diffusion of vitamins, cofactors, and ions which are required for many intracellular enzyme activities. If cells are inoculated from a rich medium to a minimum medium, the lag time may be affected by inoculum size as a result of carryover of trace nutrients from the original medium.

The age of the inoculum influences the lag in a fresh medium as a result of toxic materials accumulated and essential nutrients depleted within the cells during their prior growth. Both a positive and a negative effect on the length of the lag phase in fresh medium can occur. In general, an increasing inoculum age lengthens the lag phase, especially when cells are inoculated from a nutritionally complex medium into a simple one.

Finally, changes in nutrient composition and concentration between the inoculum culture and the fresh medium may trigger the control and regulation of enzyme activities within the cells or of morphological differentiation, such as spore formation. If the cells are transferred from a simple medium to a richer medium, both time and nutrients will be expended to allow an increase in the concentration of enzymes essential for metabolism. When cells are inoculated from a rich to a less rich medium, the cells may resume exponential growth immediately but at a lower rate.

The fact that constant exponential growth can occur for even a limited time in batch culture shows that the growth rate can be virtually unaffected by changes in substrate concentrations over wide ranges. Under these conditions the culture is said to be in **balanced growth**, whose rate can be described by a single numerical value, μ. Eventually, the culture will deviate from constant exponential growth and can no longer be described only by the value of μ, even though it is possible to calculate this value for the case of limitation

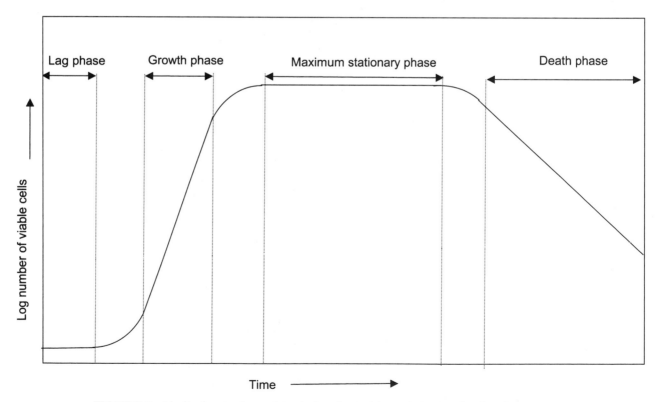

FIGURE 6 Idealized normal growth cycle for a bacterial population in a batch culture system.

by a single nutrient substrate. In a classic article (15), Monod described the relationship between substrate concentration and bacterial growth in simple systems at steady state as follows:

$$\mu = \frac{\mu_{max}S}{K_s + S} \quad (18)$$

where μ is the specific growth rate, μ_{max} is the maximum value of μ obtained when S is much greater than K_s, and K_s is the saturation constant equivalent to a Michaelis-Menten constant. K_s is the substrate concentration at which the specific growth rate is one-half the maximum growth rate, μ_{max}. S is the instantaneous or steady-state concentration of a single limiting nutrient. Both K_s and S have the units of milligrams per liter (or mass per unit volume) because both are concentrations.

Several alternative models for growth response to substrate concentration exist (15). Equation 18 may be used to describe growth response to substrate limitation only under steady-state conditions with uncomplicated bacterial systems. When exponential growth ceases as a result of substrate limitation, conditions are no longer steady state for cell mass accumulation or substrate concentration, and equation 14 must be used with equations 18 and 19 to adequately model the response of the culture to diminishing substrate. Simultaneous solution of equations 14, 18, and 19 yields a somewhat bulky model for batch growth, which is at best a rough estimation for very simple systems. This is why models of batch reactors are more complex than those of chemostats.

The maximum population in a batch culture can be estimated from experimental data relating the increase in cell number or mass to the corresponding decrease in substrate concentration (assuming that concentration of substrate is the only thing that limits growth):

$$\frac{(X - X_0)}{(S_0 - S)} = Y \quad (19)$$

where X and S are the cell and substrate concentrations at time t, X_0 and S_0 are the cell and substrate concentrations, respectively, at an earlier time, t_0, and Y is the overall yield coefficient. Equation 19 accurately describes the relationship between cell concentration and substrate concentration during exponential growth. If exponential growth continues unabated until the stationary phase is reached and the substrate is completely consumed during exponential growth, the maximum cell number or concentration will be given by equation 19. This estimation assumes that the yield coefficient is constant throughout the growth cycle and neglects substrate consumption during the lag and stationary phases. In fact, the yield coefficient cannot be assumed to be constant for other than constant exponential growth conditions at a single specific growth rate. It follows that the prediction of a maximum population from equation 19 will lead to an overestimation. The concept of yield coefficient is discussed in greater detail later in this chapter. Prediction of the time required to attain maximum population density in batch culture requires simplifying assumptions (1, 5).

13.2.1.1.2. Aberrant Growth Cycles

Aberrations from the normal batch growth cycle are common. Morphological change in the culture (such as an increase in opacity, cellular refractive index, individual cell size, or cell aggregation) can lead to an apparent change in the growth cycle if growth is determined by optical measurement. For example, stationary-phase bacterial cells are often more transparent than those growing in the exponential phase. As a result, a growth curve plotted as the logarithm of absorbance versus time may show an apparent decrease in stationary-phase cell concentration compared with that attained at the end of exponential growth. In this case, the aberrant growth cycle represents a change in the morphological characteristics of the cell rather than a change in the number of cells.

Growth curves plotting the log of cell number versus time occasionally show an unusually rapid increase in cell number just after the lag phase, which then settles into a slower increase. The unusual burst of growth may in fact be an indication of partial or complete culture synchrony. This **synchronized growth** may result, for example, from culture acclimation to a new nutritional environment. Such culture synchrony degenerates rapidly, and asynchronous growth usually dominates within two generations. However, special techniques have been devised to synchronize cell divisions in a growing population as a way to mimic individual cell growth.

Arithmetic (linear) growth occurs when the supply of a critical nutrient is regulated by an arithmetic process (such as by dropwise addition or by diffusion) and this process becomes limiting. For example, limited diffusion of air through the cotton plug in a test tube or of nutrients through a membrane in a dialysis culture (chapter 12) may cause a shift from exponential to arithmetic growth. Although most bacterial cultures reproduce by binary fission as individual cells, some (e.g., actinomycetes) reproduce as filamentous extensions. If growth occurs primarily through extension of the tips of the hyphae, the increase in cell mass with respect to time will be arithmetic.

Filamentous bacteria often grow as pellets in liquid culture; when this occurs, biomass increases more slowly than the classical exponential rate and is proportional to the cube of time. Microbial growth in pellets may be severely affected by diffusion of nutrients into the pellet and by diffusion of metabolic products out from the pellet, a possibility that is not taken into account by equation 18.

Bacterial cells in complex environments often metabolize usable substrates in a sequential manner. That is, the presence of certain substrates may lead to repression of the enzymes for metabolism of other substrates. In this instance, only when the concentration of the repressing substrate has been reduced through bacterial consumption can the enzymes for metabolism of other substrates be elaborated. This regulation of bacterial physiology leads to an aberrant growth cycle that shows one or more intermediate but transient stationary phases. This response to a changing environment is termed **diauxic growth**. A classical example of diauxic growth is that of *E. coli* growing in the presence of both glucose and lactose (Fig. 7). Rapid growth on glucose occurs first. At the point of glucose exhaustion, an inflection in the biomass curve occurs and there may even be a decline in biomass. A new enzyme system for metabolism of lactose is induced during this lag, and biomass accumulation at the expense of lactose then continues.

13.2.2. Continuous-Culture Systems Modeling

Continuous cultivation differs from batch cultivation in that a fresh supply of medium is added continuously at the same rate that culture is withdrawn (an **open system**). If the culture is well mixed, a sample that is representative of both the cell population and the substrate concentration within the fermentor can be withdrawn. The technique of

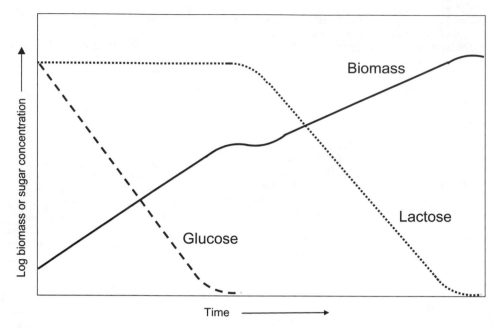

FIGURE 7 Idealized diauxic growth of a bacterial population (biomass) in a batch culture system with two usable substrates (glucose and lactose).

continuous cultivation theoretically allows continuous exponential growth of the culture in a system requiring constant addition of fresh medium and constant withdrawal of culture so that the culture volume and cell concentration remain constant with time. In its broadest sense, continuous cultivation does not require a constant cell concentration, but most literature accounts of quantitative study of continuous cultivation have dealt with such a situation and the principles developed below deal only with this situation.

The term "steady state" is often applied to continuous cultivation and means literally that no change in status occurs during the time span studied. In reality, this definition is too broad, since practical application of the continuous-cultivation theory often results in changes in some parameters while others remain constant. A continuous culture is therefore defined at steady state by an invariant biomass concentration with respect to the time span of observation. In contrast, a batch culture may have a steady-state dissolved oxygen concentration maintained by constant replacement, but it is not a continuous culture because the biomass concentration changes with time.

Continuous cultivation at steady state is possible only when all factors contributing to the accumulation of biomass are exactly balanced by all factors contributing to the loss of biomass from the system. This is shown by the following general material balance on the bacterial cells:

Cells accumulated within the system = cells added to the system − cells removed from the system + cells produced through growth − cells consumed through death

The balance is shown mathematically as follows:

$$\frac{dX}{dt} = \frac{QX_0}{V} - \frac{QX}{V} + \mu X - bX \qquad (20)$$

where Q is the medium flow rate to and from the fermentor (in liters per hour); V is the liquid volume within the fermentor (in liters); X_0 is the cell concentration in the feed in grams per liter (X_0 is generally zero for most laboratory scale systems, but for the sake of completeness of mass balance, it is included here); X is the cell concentration in the fermentor, again in grams per liter; μ is the specific growth rate, b is the specific death rate (in hours^{-1}), and dX/dt is the rate of change in cell mass (in grams per liter-hour).

At steady state, $dX/dt = 0$. The volume of a true continuous culture is fixed in theory and undergoes negligible variation in practice. Therefore, the flow rates to and from the fermentor must be identical. Finally, the specific death rate is almost always much lower than the specific growth rate, so the death term (bX) may be ignored. (Exceptions to this rule may occur at very low growth rates, in the presence of toxic substances, or in an extreme biophysical environment.) Since the feed to the fermentor is usually sterile such that $X_0 = 0$, at steady state, the following applies:

$$\mu = \frac{Q}{V} = D \qquad (21)$$

That is, the specific rate of growth of the population within the fermentor is determined by the dilution rate, D, where $D = Q/V$.

Many types of continuous cultures are possible (1, 5). The following discussion is limited to one of the most widely used types of continuous culture, the **chemostat**, which achieves steady state by precise control of the availability of a single growth-limiting substrate. The mathematical principles underlying the chemostat also apply to the **turbidostat**, which achieves steady state by actually removing the cell mass and replacing it with fresh medium at the same rate as cell growth. However, turbidostats are better suited to continuous-cultivation studies of growth at or near the maximum specific growth rate, μ_{max}.

Although many references on continuous cultures now exist, background reading on continuous culture should start with the classic papers published in 1950 by Monod

(16) and Novick and Szilard (18). Luedeking (9) has also presented a very useful discussion of continuous-culture theory and practice. The article describes graphical design and analysis of continuous-culture systems, avoiding the requirement for an accurate model of growth response to changing environmental properties (such as the Monod model). An empirical approach to continuous cultivation based on batch data has its pitfalls, which were discussed by Luedeking; however, the approach is still useful for obtaining approximate design criteria.

13.2.2.1. Chemostat

Deviations from simple chemostat theory abound, but most are the result of (i) product formation, which may require a sophisticated model of yield as a function of μ; (ii) imperfect mixing in the fermentor, which may allow a stable dilution rate that is higher than the critical dilution rate; (iii) wall growth, which has an effect similar to imperfect mixing but is more pronounced; and (iv) idiosyncrasies of physiological response. Pirt (19) has discussed the first three of these deviations.

In its simplest form, chemostat continuous cultivation can be described as a collection of reaction steps in which the rate of growth of a culture is determined by the lowest rate of nutrient metabolism. Although a growing culture requires the metabolism of many different nutrients, the growth rate of an ideal culture at any given instant is determined by the rate of metabolism of a single limiting nutrient. Monod (15, 16) found that growth rate dependence on substrate concentration could be predicted by an equation whose form is essentially a Michaelis-Menten-type function, as shown in equation 18. Monod's equation is most easily applied to experimental conditions of steady state. For usual purposes, a steady state exists when the substrate concentration does not fluctuate with time and the resulting cell population in the continuous-culture vessel is constant with time. When a chemostat is operated with a sterile feed ($X_0 = 0$) and without recycle, the specific growth rate is numerically equal to the dilution rate (equation 21). This identity is forced by controlling the availability of a limiting nutrient through the addition of fresh medium.

A limiting nutrient balance can be written for a chemostat:

input − output − consumed = accumulation

The mathematical expression for this, similar to equation 20 for cell mass, is as follows:

$$\frac{dS}{dt} = DS_0 - DS - \frac{\mu X}{Y_{x/s}} \qquad (22)$$

where D is the dilution rate ($D = Q/V$); S_0 and S are the limiting substrate concentrations in the feed and chemostat, respectively; μ is the specific growth rate of the culture in the chemostat; X is the cell mass (dry weight) in the fermentor; $Y_{x/s}$ is the overall yield coefficient (cells formed/substrate consumed); and dS/dt is the rate of change of substrate concentration in the fermentor. The overall yield term is a composite that includes contributions for both growth and maintenance. No products other than cells are assumed. If product formation does occur, an additional consumption term must be added. At steady state, equation 22 then becomes

$$D(S_0 - S) = \frac{\mu X}{Y_{x/s}} \qquad (23)$$

Substitution of equation 21 into equation 23 gives

$$X = Y_{x/s}(S_0 - S) \qquad (24)$$

Note that equation 24 was presented earlier without derivation for use with batch systems (equation 19). The overall growth yield is assumed to be dependent only on the limiting nutrient concentration and independent of specific growth rate. *Equations 21 and 24 are the steady-state equations for usual chemostat continuous cultivation.*

A model expressing the specific growth rate as a function of substrate concentration must be assumed before biomass concentration, substrate concentration, and specific growth rate can be related to define a stable set of operating conditions. Equation 18 provides an adequate model for many situations. Substitution of equation 21 into equation 18 yields

$$D = \frac{D_c S}{(K_s + S)} \qquad (25)$$

where D_c is the critical dilution rate corresponding to the maximum specific growth rate, μ_{max}. *Operation of the chemostat at dilution rates above D_c will result in complete washout of the culture.*

Equation 25 can be arranged to give

$$S = \frac{DK_s}{(D_c - D)} \qquad (26)$$

Substitution of equation 26 into equation 23 results in

$$X = Y_{x/s}\left(S_0 - \frac{DK_s}{D_c - D}\right) \qquad (27)$$

Equation 27 relates the steady-state biomass concentration to the dilution rate. Figure 8 depicts the generalized system response to changes in the dilution rate, as predicted by equations 26 and 27.

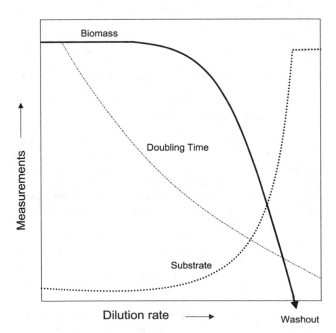

FIGURE 8 Generalized effects of changes in the dilution rate on culture variables in a chemostat system for continuous cultivation, where the dilution rate approaches μ_{max} at washout.

The Monod model (equation 18) is one of several models relating growth rate and substrate concentration. The assumptions implicit with this model are quite simple and do not adequately describe all systems. An excellent discussion of the fundamental theory of chemostat continuous culture is presented by Tempest (23)

13.2.2.2. Growth Yield Calculations

The concept of growth yield is developed from a material balance of a stable cultivation system. Monod originally defined the yield constant, Y, in terms of mass units:

$$Y = \frac{\text{grams (dry weight) of cells formed}}{\text{grams of substrate consumed}} \quad (28)$$

The yield from substrate is often indicated by $Y_{x/s}$, indicating a ratio of the respective concentrations. The convention of mass ratios for expressing the yield of cells on the basis of carbon substrate consumption is fairly well accepted. Yields can be easily calculated by determining the production of cell mass (dry weight) over a given period of the cultivation and dividing by the mass of carbon substrate consumed during the same period.

If the yield is to be expressed on the basis of ATP produced or oxygen consumed during growth, the cell yield can be defined as follows:

$$Y_{x/\text{ATP}} = \frac{\text{grams (dry weight) of cells formed}}{\text{moles of ATP formed}} \quad (29)$$

$$Y_{x/O_2} = \frac{\text{grams (dry weight) of cells formed}}{\text{moles of oxygen consumed}} \quad (30)$$

Under many conditions, the yield with respect to ATP synthesis or oxygen consumption is relatively constant for many organisms. For example, the yield of cells per mole of ATP synthesized under conditions of energy substrate limitation and fairly high growth rate is approximately 10 ± 2 g of cells (dry weight) formed per mol of ATP (2).

However, the yield of cells per mole of ATP is not constant for all bacteria and is generally variable if the energy substrate does not limit growth or if the growth rate of the culture is significantly lower than the maximum rate. Also, carbon and energy sources may be consumed not only for cellular growth but also for product formation. Furthermore, energy substrates are involved in cell maintenance as well as in cell growth. Since cells can use energy substrates for endogenous respiration without growth, maintenance energy requirements will contribute to the consumption of energy substrates without a concomitant increase in cell dry weight. Although the influence of endogenous metabolism on the calculation of cell growth yield is minor when the culture growth rate is near its maximum value with the energy substrate as the limiting nutrient, consumption of energy substrates for maintenance can be quite significant at growth rates lower than the maximum or in an extreme growth environment.

Since meaningful calculation of yield values requires careful and precise control of the cultivation conditions, yield data are best obtained from continuous-culture systems. In this way, the growth-limiting substrate and the specific rate of growth, μ, can be specified and controlled. Calculation of the overall growth yield from continuous-culture data assumes steady-state substrate and cell concentrations and therefore is most easily accomplished in a chemostat. The medium for chemostat operation must be designed so that a single nutrient limits the rate of cell growth. This can be easily accomplished by ensuring that all nutrients, other than the one on which yields will be based, are in excess. An insight into this medium design can be obtained from an elemental analysis of the bacteria to be grown in the chemostat. The medium must supply all of the components of the cell and can therefore be based on elemental analysis. The ultimate level of growth will be determined by the concentration of the limiting nutrient. Since cell concentration in a chemostat will automatically adjust to match the rate of limiting-nutrient addition, the ratios of medium components are more important than their absolute amounts. Media for the determination of overall growth yield should therefore be designed such that the ratios of elements supplied by the medium to the element chosen as the limiting nutrient are considerably higher than predicted by taking the ratio of elements from elemental analysis.

The calculations described below require the establishment of steady-state conditions in terms of cell concentration and substrate concentration. Make sure that this condition is met by monitoring them both over a long enough period to allow displacement of one reactor volume of medium. Whenever conditions are changed (such as increasing or decreasing the rate of medium addition to a chemostat), allow sufficient time for a return to steady state. This can be ensured by allowing at least four turnovers of fermentor liquid volume. Since a continuous reactor operates on a dilution principle, a step of pulse change in operating conditions would cause deviation from the preexistent steady state. This would approach a new steady-state value at a rate predicted by the dilution of the culture; that is, four mean residence times would be necessary for a close asymptotic approach to the new steady-state equilibrium value. For example, consider a 2-liter chemostat in which the rate of medium addition and withdrawal is 0.5 liter/h with a dilution rate of $FIV = D = 0.25 \text{ h}^{-1}$. The time necessary for four turnovers of medium volume is calculated by dividing 4 by the dilution rate. In this case, the reactor medium volume will have been completely replaced four times after 16 h $(4/0.25 \text{ h}^{-1})$. That is, at least 16 h must elapse after changing the growth conditions of a continuous culture prior to the achievement of steady-state conditions.

In reality, the time necessary to obtain a close asymptotic approach to the new steady-state value may be dependent on the way in which operating conditions have been changed. If the change in operating conditions involves a nutritional step-down, for instance, four mean residence times should be adequate. If the change involved a nutritional step-up under conditions of rapid growth, the system may require more than four mean residence times to achieve steady state. In the latter case the rapidly growing cells may require elaboration of new enzymes for metabolism of the added nutrients.

Once steady-state continuous cultivation is achieved, the overall growth yield is determined as shown below:

$$Y_{x/s} = \frac{(X - X_0)}{(S_0 - S)} = \frac{\text{grams (dry weight) of cell produced}}{\text{grams of substrate used}} \quad (31)$$

where X is grams of cells (dry weight) per liter of effluent medium, X_0 is grams of cells (dry weight) per liter of influent medium (usually zero), S is grams of substrate per liter of effluent medium, and S_0 is grams of substrate per liter of influent medium. If the medium added to the fermentor is sterile, the value of X_0 is zero and the yield becomes $Y_{x/s} = X/(S_0 - S)$.

To determine the overall growth yield, first determine the cell dry weight by quantitatively filtering a precisely measured volume of culture effluent through a preweighed 0.2-μm-pore-size membrane filter. Wash the collected cells with buffer or distilled water to remove excess medium. Dry the membrane filter and cells to constant weight in an oven whose temperature is not greater than 105°C. Lower oven temperatures may be required to prevent extensive browning and therefore weight loss of the sample. *Caution:* many filters lose weight to a variable extent after washing and drying. Therefore, several samples should be filtered and weighed, and several filters should be washed, dried, and weighed as controls. It is more accurate to centrifuge and wash cells as described above from a larger volume of medium and then to resuspend in a small volume of water, dry, and weigh.

Overall growth yield is related to maintenance and growth requirements for limiting substrates that act as energy sources according to the following equation:

$$\frac{1}{Y_{x/s}} = \frac{m}{\mu} + \frac{1}{Y_G} \qquad (32)$$

where μ is the specific growth rate (in reciprocal hours), m is the specific rate of substrate uptake for cellular maintenance (in reciprocal hours), $Y_{x/s}$ is the overall yield, and Y_G is the growth-specific yield. The values of m and Y_G can be estimated by plotting $1/Y_{x/s}$ versus $1/\mu$. (Remember that μ = D only for simple continuous culture without cell recycle.) If the data form a straight-line relationship, the intercept will be the value of $1/Y_G$ and the slope of the line will be the value of μ.

Although growth yields are conventionally expressed in terms of mass of cells formed per mass or moles of substrate used, this mixture of terms is often confusing. Herbert (6) has suggested that yields should be expressed in terms of gram-atoms of cellular carbon formed per gram-atom of element consumed from the limiting substrate. Adoption of this convention would standardize reporting procedures and eliminate much of the confusion in interpreting yield data. Expression of growth in terms of gram-atoms of cellular carbon formed does not require a determination of the complete elementary composition of the cells. Rather, only the carbon content of the cells need be determined to allow use of this convention.

13.2.2.3. Mean Cell Residence Time, θ, Versus Dilution Rate, D

Mean cell residence time signifies the amount of time the microorganisms spend in a given vessel, on an average. It is represented by θ and is defined as the total biomass in the system divided by the total biomass wasted daily, intentionally or otherwise. Mathematically, assuming Q as the flow rate in liters or cubic meters per day, X as the biomass concentration in milligrams per liter, and V as the liquid volume of the vessel in liters, the total biomass in the system can be represented as $V \cdot X$ and the total biomass wasted daily can be represented by $Q \cdot X$. Thus,

$$\theta = \frac{VX}{QX} \qquad (33)$$

which for chemostat, is the same as V/Q or the inverse of D (because D is Q/V). When cell mass is recycled or added to a vessel, then the equation takes a different form, although the basic definition of total biomass in the system divided by total biomass wasted daily still applies. It is also important to note that at steady state, the total biomass wasted must be equal to the total biomass being produced, or else biomass accumulation or washout will occur.

13.3. REFERENCES

1. **Bailey, J. E., and D. F. Ollis.** 1986. *Biochemical Engineering Fundamentals*, 2nd ed. McGraw Hill Book Co., New York, NY.
2. **Bauchop, T., and S. R. Elsden.** 1960. The growth of micro-organisms in relation to their energy supply. *J. Gen. Microbiol.* **23:**457–469.
3. **Criddle, C. S., L. M. Alvarez, and P. L. McCarty.** 1991. Microbial processes in porous media, p. 639–691. *In* J. Bear and M. Y. Corapcioglu (ed.), *Transport Processes in Porous Media.* Kluwer Academic Publishers, Dordrecht, The Netherlands.
4. **Dolfing, J., and J. M. Tiedje.** 1987. Growth yield increase linked to ATP production and growth in an anaerobic bacterium, strain DCB-1. *Arch. Microbiol.* **153:**264–266.
5. **Drew, S. W.** 1981. Liquid culture, p. 151–178. *In* P. Gerhardt, R. G. E. Murray, R. N. Costilow, E. W. Nester, W. A. Wood, N. R. Krieg, and G. B. Phillips (ed.), *Manual of Methods for General Bacteriology.* American Society for Microbiology, Washington, DC.
6. **Herbert, D.** 1976. Stoichiometric aspects of microbial growth, p. 1–30. *In* A. C. R. Dean, D. C. Ellwood, C. G. T. Evans, and J. Melling (ed.), *Continuous Culture 6: Applications and New Fields.* Ellis Horwood Ltd., Chichester, England.
7. **Hoeft, S. E., T. R. Kulp, J. F. Stolz, J. T. Hollibaugh, and R. S. Oremland.** 2004. Dissimilatory arsenate reduction with sulfide as electron donor: experiments with Mono lake water and isolation of strain MLMS-1, a chemoautotrophic arsenate respirer. *Appl. Environ. Microbiol.* **70:**2741–2747.
8. **Holliger, C., D. Hahn, H. Harmsen, W. Ludwig, W. Schumacher, B. Tindall, F. Vazquez, N. Weiss, and A. J. B. Zehnder.** 1998. *Dehalobacter restrictus* gen. nov. and sp. nov., a strictly anaerobic bacterium that reductively dechlorinates tetra- and trichloroethene in an anaerobic respiration. *Arch. Microbiol.* **169:**313–321.
9. **Luedeking, R. (ed.).** 1976. *Fermentation Process Kinetics*, vol. 1. Academic Press, Inc., New York, NY.
10. **Madigan, M. T., J. M. Martinko, and J. Parker.** 2002. *Brock Biology of Microorganisms*, 10th ed. Prentice Hall, Upper Saddle River, NJ.
11. **Maymó-Gatell, X., Y. Chien, J. M. Gossett, and S. H. Zinder.** 1997. Isolation of a bacterium that reductively dechlorinates tetrachloroethene to ethene. *Science* **276:**1568–1571.
12. **Maymó-Gatell, X., I. Nijenhuis, and S. H. Zinder.** 2001. Reductive dechlorination of cis-1,2-dichloroethene and vinyl chloride by "*Dehalococcoides ethenogenes*." *Environ. Sci. Technol.* **35:**516–521.
13. **McCarty, P. L.** 1971. Energetics and bacterial growth. *In* S. D. Faust and J. V. Hunter (ed.), *Organic Compounds in Aquatic Environments.* Marcel Dekker, New York, NY.
14. **McCarty, P. L.** 1997. Microbiology: breathing with chlorinated solvents. *Science* **276:**1521–1522.
15. **Monod, J.** 1949. The growth of bacterial cultures. *Annu. Rev. Microbiol.* **3:**371–394.
16. **Monod, J.** 1950. La technique de culture continue: theorie et applications. *Ann. Inst. Pasteur* (Paris) **79:**390–410.
17. **Neidhardt, F. C., R. Curtiss III, J. L. Ingraham, E. C. C. Lin, K. B. Low, B. Magasanik, W. S. Reznikoff,**

M. Riley, M. A. Schaechter, and E. H. Umbarger (ed.). 1996. Escherichia coli *and* Salmonella: *Cellular and Molecular Biology.* ASM Press, Washington, DC.

18. **Novick, A., and L. Szilard.** 1950. Experiments with the chemostat on spontaneous mutations of bacteria. *Proc. Natl. Acad. Sci. USA* **36:**708–719.

19. **Pirt, J. S.** 1975. *Principles of Microbe and Cell Cultivation.* John Wiley & Sons, Inc., New York, NY.

20. **Rittmann, B. E., and P. L. McCarty.** 2001. *Environmental Biotechnology: Principles and Applications,* 1st ed. McGraw-Hill Higher Education, New York, NY.

21. **Smidt, H., and W. M. de Vos.** 2004. Anaerobic microbial dehalogenation. *Annu. Rev. Microbiol.* **58:**43–73.

22. **Sun, B. L., B. M. Griffin, H. L. Ayala-del-Rio, S. A. Hashsham, and J. M. Tiedje.** 2002. Microbial dehalorespiration with 1,1,1-trichloroethane. *Science* **298:**1023–1025.

23. **Tempest, D. W. (ed.).** 1970. *The Continuous Cultivation of Microorganisms. 1. Theory of the Chemostat,* vol. 2. Academic Press, Inc., New York, NY.

24. **Thauer, R. K., K. Jungermann, and K. Decker.** 1977. Energy conservation in chemotrophic anaerobic bacteria. *Bacteriol. Rev.* **41:**100–180.

14

Physicochemical Factors in Growth

JOHN A. BREZNAK AND RALPH N. COSTILOW†

Some physicochemical factors affecting microbial growth are controlled by the constituents of the culture medium (hydrogen ion activity, water activity, osmotic pressure, and viscosity). Others are controlled by the external environment (temperature, oxygen, light, hydrostatic pressure, and magnetic-field strength). The oxidation-reduction potential is controlled by both the medium and the environment. All of these factors can influence the growth rate, cell yield, metabolic pattern, and chemical composition of bacteria. The control of hydrogen ion activity, temperature, and oxygen supply is critical with every bacterial culture, whereas the control of oxidation-reduction potential is of major importance in culturing obligate anaerobes. Light quality and quantity are critical for growth of phototrophs. This chapter addresses practical aspects of application and control of physicochemical factors.

14.1. HYDROGEN ION ACTIVITY (pH)

pH is a measure of the mean hydrogen ion activity of a solution (a_{H+}) and is defined as follows: $pH = -\log_{10} a_{H+}$. The activity of the H^+ ion is the product of its molar concentration and its activity coefficient, which, in turn, is a function of the total ionic strength of the solution (μ). In dilute solutions, where μ ranges from 0 to 0.25, the activity coefficient of the H^+ ion (at 25°C) ranges from 1 to 0.85, respectively. Most routinely used microbiological media can be considered ionically dilute solutions. For example, in a typical glucose salts medium, μ is about 0.02. Therefore, the activity coefficient of the H^+ ion is close to unity, and hence a_{H+} is essentially equal to the hydrogen ion concentration, $[H^+]$. Consequently, $pH \approx -\log [H^+]$ for most microbiological media.

The pH scale is typically thought of as ranging from 0 ($[H^+] = 1$ M) to 14 ($[H^+] = 10^{-14}$ M), although values outside this range are possible. Most known microbes grow over a relatively narrow range of pH, usually best near

---

†Deceased.

neutrality (pH 7.0). However, an ever-increasing number of extremophiles continue to be recognized and isolated from nature (78, 79). Many of these grow optimally at very low pH (acidophiles) or very high pH (alkaliphiles). Some of these organisms also grow at high temperatures (thermo- and hyperthermophiles) or high salinities (halophiles) or both. Others grow optimally at low temperatures (psychrophiles). Accordingly, for precise definition of the pH optimum for growth of such bacteria, it is important to keep in mind the factors that affect the activity coefficient of the hydrogen ion and hence the a_{H+}. Among these factors are temperature, ionic strength, ion charge, dielectric constant, and the physical size of various ions in the solution. In practice, the influence of such factors on pH is taken into account by standardizing the pH meter with a pH reference solution whose composition and temperature are as close as possible to those of the solution being measured (e.g., the microbial culture fluid).

14.1.1. pH Measurement

For accurate pH measurements of virtually all types of solutions, use a pH meter with a glass electrode. Two types of glass electrode systems may be used. One consists of a pH-measuring electrode paired with a separate pH reference electrode (Hg/Hg_2Cl_2 or $Ag/AgCl$), both of which are put into the solution being measured. Another system consists of a single electrode containing elements of both the pH-measuring and reference electrode, referred to as a pH combination electrode. This type of electrode is popular, because its generally compact size facilitates the measurement of small volumes contained in test tubes. Epoxy electrodes (pH-measuring, pH reference, and pH combination types) are also available and are generally more durable than glass electrodes. However, epoxy electrodes are not recommended for use with certain organic solvents; consult with the manufacturer for the solvent tolerance of such electrodes if this is a concern. So-called "microelectrodes," with tip diameters in the micrometer range, have become widely used for making measurements on tiny samples (e.g., insect guts) or for fine-scale resolution of physicochemical gradients present in natural samples (e.g., sediments). Microelectrodes capable of measuring the pH, E_h, certain ions other than protons (e.g., NO_3^-), various dissolved gases (e.g., O_2, H_2, N_2O, and others), and other physicochemical parameters (e.g., temperature, flow velocity, etc.) are available from Unisense, Aarhus, Denmark (www.unisense.com) and Diamond General Development Corp., Ann Arbor, MI (www.diamondgeneral.com). See chapter 17.2 for a further description of hydrogen ion and other ion-selective electrode systems, procedures used for calibration of pH meters, and proper care of electrodes. Also see references 31, 68, and 100 for details on measurement and control of pH.

When accurate pH measurements are desired, observe the following precautions.

pH Measurement

1. Standardize the pH meter with a buffer (a substance that resists change in pH; see below) whose temperature is similar to that of the sample. The pH of buffers changes with temperature; e.g., the pH of standard phosphate buffer is 6.98 at 0°C, 6.88 at 20°C, and 6.84 at 37°C. See the *Handbook of Chemistry and Physics* (99) for pH values of standard buffers at different temperatures. If pH measurements are to be made on numerous samples at different temperatures, a pH meter with a temperature compensation control is desirable, so calibration with standard buffer need be done at only one temperature. In any case, keep in mind that an average error of only 0.003 pH unit per degree Celsius exists at 1 pH unit from standardization; this is often less than the required accuracy for measurements.

2. Standardize the pH meter with a buffer that is near the expected pH of the sample and of similar ionic strength. If accurate readings are to be made across a fairly wide range of pH values (e.g., 3 pH units), a meter with a slope control is desirable and is used after two-point calibration. In this procedure, the pH meter is standardized with a buffer, the pH of which is at or just outside one end of the expected range; the electrode(s) is placed in a second standard buffer representing the other end of the expected range; and then the pH meter readout is adjusted to the appropriate pH by using the slope control. A frequently encountered sample of relatively high ionic strength is seawater. The pH of such samples should be made after standardization with a buffer consisting of 0.4186 M NaCl, 0.0596 M $MgCl_2$, 0.02856 M Na_2SO_4, 0.005 M $CaCl_2$, 0.02 M Tris, and 0.02 M Tris-HCl. The pH of this buffer is 8.835 at 5°C, 8.517 at 15°C, 8.224 at 25°C, and 7.953 at 35°C (100).

3. If the sample must be stirred (e.g., with a magnetic stirrer) during measurement, stir the calibrating buffer in the same manner, since stirring often alters the signal of the pH electrode.

4. Determine the pH of media after sterilization, not before. Autoclaving of media frequently results in a significant change in the pH owing to the escape of CO_2 or the precipitation of alkaline-earth phosphates. Even filtration may have a significant effect on the pH of media.

5. Just prior to use, check the pHs of media that have been stored.

6. Some pH electrodes (those with linen fiber junctions) do not give accurate pH readings with Tris buffers. Be sure that the electrodes used with Tris buffers are recommended for such use by the manufacturer, e.g., Orion no.71-03 Tris calomel pH electrodes with ceramic junctions. In addition, reference electrodes may be especially sensitive toward chelators (e.g., EDTA) or sulfides. For such instances, probes with salt-bridged reference electrodes should be used.

In instances when accuracy is not required, such as in the preparation of routine media, the pH may be measured by use of pH indicator dye solutions or pH paper. By proper selection of either, the pH can be estimated to within 0.2 pH unit. Indicator solutions and papers, and their corresponding color standards for various pH ranges, are available from most scientific supply companies. Common pH indicators and their useful pH ranges are listed in Table 1. Also see chapter 15.5.15.

A number of pH indicators are frequently incorporated into culture media to demonstrate pH changes during growth of microbes. Select the appropriate indicator for the pH range of interest, and make sure that it does not inhibit growth of the organism. All of these pH indicators are poorly soluble in water but are effective at very low concentrations (less than 0.01%) in media.

14.1.2. pH Buffers

Many organisms consume or produce significant amounts of acidic or basic ions during growth. This commonly occurs during anaerobic fermentation and during aerobic oxidation of the salts of organic acids. Unless acid production and consumption are balanced, large changes in pH may occur

TABLE 1 pH indicators and their useful pH range

Indicator[a]	pH range	pK$_a$	Color	
			Acid	Alkaline
Thymol blue	8.0–9.6	8.9	Yellow	Blue
Metacresol purple	7.4–9.0	8.3	Yellow	Purple
Cresol red	7.2–8.8	8.3	Yellow	Red
Phenol red	6.8–8.4	7.9	Yellow	Red
Bromthymol blue	6.0–7.6	7.0	Yellow	Blue
Bromcresol purple	5.2–6.8	6.3	Yellow	Purple
Methyl red	4.4–6.0	5.2	Yellow	Red
Bromcresol green	3.8–5.4	4.7	Yellow	Blue
Bromphenol blue	3.0–4.9	4.0	Yellow	Blue
Thymol blue	1.2–2.8	1.5	Red	Yellow

[a]Prepare 0.04% solutions by solubilizing 0.1 g in the smallest possible volume (10 to 30 ml) of 0.01 N NaOH, and then dilute to 250 ml.

during growth, and growth may be significantly inhibited. The pH of cultures may be controlled to some extent by incorporating pH buffers into the medium. Continuous pH control may be achieved by the automatic addition of acid or base (section 14.1.3).

pH buffers usually are mixtures of weak acids and their conjugate bases. In complex media, the acidic and basic groups of organic molecules (such as proteins, peptides, and amino acids) provide some buffering capacity. Owing to the variety of compounds present in such media, there may be a fair degree of buffering action over a wide range of pH. However, the buffering capacity at any given pH varies widely with the types and concentrations of organic molecules present. Consider the following in selecting a buffer for use: (i) the pH desired; (ii) possible inhibitory or toxic effects of the buffer; (iii) possible utilization of the buffer by the culture; and (iv) possible binding of di- and trivalent metal ions by the buffer.

Weak acids and bases buffer most effectively at the pH where they are 50% dissociated. This pH is equal to the pK$_a$ (the negative logarithm of the dissociation constant) of the acid or base. The effective range of a buffer is within 1 pH unit of its pK$_a$. A number of compounds that have been used as buffers in growth media are listed in Table 2, along with their pK$_a$ values. The Good buffers, originally developed by N. E. Good et al. (33), have greatly expanded the range of buffers useful in biological systems. The Good buffers have the advantages of a pK$_a$ between 6.0 and 8.5, high aqueous solubility, impermeability by biological membranes, minimum interaction with mineral cations, enzymatic and hydrolytic stability, little or no participation in biochemical reactions, minimum absorbance between 240 and 700 nm, and usually no toxicity. In addition, they show minimum effects due to ionic composition of solutions, concentration, and temperature. A disadvantage is that they are slightly more expensive than some of the other buffers.

TABLE 2 pK$_a$ values of chemical compounds used in buffers

Compound	pK$_a$ at:	
	25°C	20°C
Citric acid	3.13, 4.77, 6.40	
Phthalic acid	2.89, 5.51	
Barbituric acid	4.01	
Oxalic acid	4.19	
Succinic acid	4.16, 5.61	
Acetic acid	4.75	
Monobasic phosphate	7.21	
Tris(hydroxymethyl)aminomethane	8.08	
Boric acid	9.24	
Glycine	9.87	
The Good buffers:		
MES [2-(N-morpholino)-ethanesulfonic acid]		6.15
ADA (N-2-acetamidoiminodiacetic acid)		6.60
PIPES [piperazine-N,N'-bis(2-ethanesulfonic acid)]		6.80
ACES (N-2-acetamido-2-aminoethanesulfonic acid)		6.90
BES [N,N-bis-(2-hydroxyethyl)2-aminoethanesulfonic acid]		7.15
MOPS [3-(N-morpholino)propanesulfonic acid]		7.20
TES [3-tris(hydroxymethyl)methyl-2-aminoethanesulfonic acid]		7.50
HEPES (N-2-hydroxyethylpiperazine-N'-2-ethanesulfonic acid)		7.55
TAPSO {N-[tris(hydroxymethyl)methyl]-3-amino-2-hydroxy-propanesulfonic acid}	7.60	
HEPPS (N-2-hydroxyethylpiperazine-N'-3-propanesulfonic acid)		8.00
TRICINE [N-tris(hydroxymethyl)methylglycine]		8.15
BICINE [N,N-bis(2-hydroxyethyl)glycine]	8.30	
TAPS {N-[tris(hydroxymethyl)methyl]-3-aminopropanesulfonic acid}	8.40	

Note that the salts of citric, phthalic, and succinic acids buffer over a wide range of pH because these acids have more than one dissociable hydrogen ion and hence more than one pK_a.

Buffers containing the compounds in Table 2 are prepared by one of the following two general procedures.

1. Prepare a solution of the buffer salt at twice the desired concentration, and while the solution is being stirred, titrate it with either NaOH or HCl to the desired pH by using a pH meter. Dilute to final volume with distilled or deionized water. For example, to prepare 1 liter of 0.1 M Tris-hydrochloride buffer at pH 7.6, make 500 ml of a 0.2 M solution of Tris, titrate with HCl to pH 7.6, and then dilute to 1 liter. Dilution has only a small effect on the final pH. If this effect is important, make a final correction before adding the last few milliliters of water during dilution.

2. Prepare equimolar concentrations of the acidic and basic components of the buffer system. With the aid of a magnetic stirrer and a pH meter, titrate one solution with the other until the desired pH is obtained. If the desired pH of the buffer is near the pK_a of the acid component, the ratio of the volume of acidic solution to basic solution is about 1:1; if it is 1 pH unit above the pK_a, the ratio is about 1:10; and if it is 1 pH unit below the pK_a, the ratio is about 10:1. For example, to prepare about 100 ml of 0.1 M sodium phosphate buffer at pH 7.0, mix solutions of 0.1 M NaH_2PO_4 (monobasic) and 0.1 M Na_2HPO_4 (dibasic) in approximately equal proportions and then adjust to pH 7.0 by adding more of the appropriate solution as determined with a pH meter. To prepare 100 ml of the phosphate buffer at pH 6.0, titrate 95 ml of the monobasic solution with the dibasic solution.

Although the Good buffers are becoming increasingly popular, the most commonly used buffers in microbiological growth media are phosphate, Tris-hydrochloride, citrate, and acetate. These four systems buffer over almost the entire range of pH permitting bacterial growth. Solutions used for their preparation and the appropriate volumes required to buffer at various pH values are given in Table 3. As noted above, the actual pH of any buffer is temperature dependent. Tables for preparing buffers of other types and at different temperatures are available elsewhere (22, 32, 68, 99).

TABLE 3 Formulas for buffers frequently used in microbiological media[a]

pH	Acetate buffer[b]	Citrate-phosphate buffer[c]	Phosphate buffer[d]	Tris-hydrochloride buffer[e]
		Value of x to be used in buffer formulas		
4.0	41.0	30.7		
4.2	36.8	29.4		
4.4	30.5	27.8		
4.6	25.5	26.7		
4.8	20.0	25.2		
5.0	14.8	24.3		
5.2	10.5	23.3		
5.4	8.8	22.2		
5.6	4.8	21.0		
5.8		19.7	46.0	
6.0		17.9	43.85	
6.2		16.9	40.75	
6.4		15.4	36.75	
6.6		13.6	31.25	
6.8		9.1	25.5	
7.0		6.5	19.5	
7.2			14.0	44.2
7.4			9.5	41.4
7.6			6.5	38.4
7.8			4.25	32.5
8.0			2.65	26.8
8.2				21.9
8.4				16.5
8.6				12.2
8.8				8.1
9.0				5.0

[a]Data adapted from reference 32.
[b]Formula: x ml of 0.2 M acetic acid + (50 − x) ml of 0.2 M sodium acetate, diluted to 100 ml.
[c]Formula: x ml of 0.1 M citric acid + (50 − x) ml of 0.2 M Na_2HPO_4, diluted to 100 ml.
[d]Formula: x ml of 0.2 M NaH_2PO_4 + (50 − x) ml of 0.2 M Na_2HPO_4, diluted to 100 ml.
[e]Formula: 50 ml of 0.2 M Tris + x ml of 0.2 M HCl, diluted to 100 ml.

When cultures are incubated in atmospheres enriched with CO_2, the CO_2-bicarbonate equilibrium is an important consideration and, in fact, can be exploited as a buffer system itself at a pH between 6.0 and 8.0. CO_2 dissolves in and reacts with water to form carbonic acid, which readily ionizes to form a proton and a bicarbonate anion as follows.

$$CO_2 \text{ (gas)} \overset{K_s}{\leftrightarrow} CO_2 \text{ (dissolved)} \overset{+H_2O}{\underset{-H_2O}{\leftrightarrow}} H_2CO_3 \overset{K_{a1}}{\leftrightarrow} H^+ + HCO_3^-$$

The formation of carbonic acid and, hence, the acidity created by it are dependent on the concentration of dissolved CO_2, which, in turn, is dependent on the concentration of CO_2 in the gas phase in accord with Henry's law. K_s is the solubility constant of CO_2 (which is temperature dependent). Inasmuch as the concentration of H_2CO_3 in solution is usually very low, the dissociation constant K_{a1} usually refers to the reaction,

$$CO_2 \text{ (dissolved)} + H_2O \leftrightarrow H^+ + HCO_3^-$$

The Henderson-Hasselbach equation for this system is

$$pH = pK' + \log [HCO_3^-]/[CO_2 \text{ (dissolved)}]$$

where pK' (the negative logarithm of K_{a1} for the reaction shown above) normally takes into account the activity of H_2O (which is 1 by definition and is close to 1 for dilute solutions), which can itself be considered a constant at a given temperature. The pK' value is about 6.4 at 25°C. Therefore, the molar ratio of dissolved CO_2 to HCO_3^- is about 1:1 at this pH. CO_2-bicarbonate buffers are effective over a pH range of about 5.4 to 7.4. If the pH of the medium is below 5.0, essentially no bicarbonate is present, but an increase in CO_2 concentration in the atmosphere will still result in a decrease in pH as a result of the formation of carbonic acid. In an atmosphere of 50 to 100% CO_2, the medium must contain a substantial amount of bicarbonate (e.g., $NaHCO_3$) to maintain a pH level near neutrality. The high salt concentration that results may be toxic to some bacteria.

After appropriate substitution, the Henderson-Hasselbach equation can be solved to determine the molarity of bicarbonate needed in a medium for buffering at a given pH when the atmosphere contains a certain percentage of CO_2. For this calculation, use the following equation:

$$\log [HCO_3^-] = pH - pK' + \log [P \cdot \alpha \cdot (\%CO_2) \cdot (5.87 \times 10^{-7})]$$

where P is the atmospheric pressure in millimeters of Hg; α (often referred to as the Bunsen absorption coefficient) is the volume of gas (normalized to the volume it would occupy at a 0°C and 760 mm Hg) that is absorbed by a unit volume of water (at the temperature of the measurement) under a gas pressure of 760 mm Hg; $\%CO_2$ is the volume of CO_2 per unit volume of gas phase × 100; and 5.87×10^{-7} is a factor that converts α to moles per liter at 760 mm Hg (94). If atmospheric pressure varies between 720 and 760 mm Hg (96.0 and 101.3 kPa), the variation has little effect on the calculated $[HCO_3^-]$. Therefore, in most instances the following simplified equation can be used (calculated for $P = 740$ mm Hg):

$$\log [HCO_3^-] = pH - pK' + \log [\alpha \cdot (\%CO_2) \cdot (4.35 \times 10^{-4})]$$

When consulting published values for α, use tabulated values that take into account the contribution of solvent (in this case water) vapor pressure to the total gas pressure of 760 mm Hg; otherwise, a small error will be introduced in the calculation. This can be seen from Table 4, which lists values for α that do (column B) and do not (column A) consider the vapor pressure of water and which also lists values for pK' at various temperatures. As an example, the following is a calculation of the concentration of $NaHCO_3$ required to buffer a medium at an initial pH of 7.0 at 30°C under a gas phase containing 20% CO_2:

$$\log [HCO_3^-] = pH - pK' + \log [\alpha \cdot (\%CO_2) \cdot (4.35 \times 10^{-4})]$$

$$\log [HCO_3^-] = 7.0 - 6.348 + \log [(0.642) \cdot (20) \cdot (4.35 \times 10^{-4})] = -1.601$$

$$[HCO_3^-] = 0.025 \text{ M}$$

However, values for α and pK' are also influenced by the presence of other ions in solution. Therefore, in practice slight adjustments may have to be made in the amount of HCO_3^- included in the liquid phase, or in the concentration of CO_2 used in the gas phase, to achieve the desired pH of the medium before inoculation. Adjustment of the pH before inoculation can also be made by addition of sterile Na_2CO_3 or HCl. For growth of alkaliphiles, advantage can be taken of the second dissociation of bicarbonate to yield a proton and the carbonate anion:

$$HCO_3^- \overset{K_{a2}}{\leftrightarrow} H^+ + CO_3^{2-}$$

The pK of this reaction at 25°C is 10.33 and ranges from 10.56 at 5°C to 10.16 at 100°C. Hence, a buffer system of $NaHCO_3$–Na_2CO_3 can be prepared, as described above, that will facilitate control of pH between 9 and 11.5. In this case, inclusion of CO_2 in the gas phase is unnecessary, as virtually all of the relevant carbonate species will be present in solution. More detailed information on CO_2-bicarbonate-carbonate equilibria and buffering capacity can be found in the classic manual for manometric methods (94) and in Stumm and Morgan (87). The latter includes detailed discussion of the effect of other ions on the equilibrium and the solubility of the carbonate species.

The pH of media used for growing microbes that produce large amounts of acid (e.g., lactic acid bacteria) may be very

TABLE 4 Values of α (CO_2 solubility) and pK' at various temperatures

Temp (°C)	α^a		pK'^b
	A	B	
20	0.878	0.859	6.392
25	0.759	0.738	6.365
30	0.665	0.642	6.348
35	0.592	0.565	6.328
40	0.530	0.494	6.312

^{a} Milliliters of CO_2 (reduced to 0°C and 760 mm Hg) that will dissolve in 1 ml of pure water at a gas pressure of 1 atm (760 mm Hg) at the stated temperature. The values for α in column A do not consider the contribution of water vapor pressure to the total gas pressure (23, 55). The values in column B have been recalculated from q values in references 23 and 55 and do consider the contribution of water vapor pressure to the total gas pressure.

^{b} $pK' = -\log K_{a1}$ for the reaction: CO_2 (dissolved) $+ H_2O \overset{K_{a1}}{\leftrightarrow} H^+ + HCO_3^-$. Data from reference 94.

difficult to control with soluble buffers. However, the acid can be neutralized as fast as it is formed by inclusion of finely ground $CaCO_3$ (chalk) in the medium. In fact, a good way of visualizing acid production by colonies growing on agar plates is to include 0.3% finely ground chalk in the medium. Make sure that the chalk is well suspended in the molten agar when the plate is poured. Alternatively, pour a chalk-free base layer first and, when solidified, pour a relatively thin chalk-containing top layer; this keeps the suspended $CaCO_3$ close to colonies developing on the surface. Colonies that produce acid will be surrounded by a clear zone. The acid produced in liquid cultures can also be withdrawn by use of a dialysis or other product removal culture system (see chapter 12.2.6.1 and 12.2.6.2) or can be neutralized by the addition of alkali with a system under automatic pH control (see below).

14.1.3. Continuous Control of pH

In instances when the pH of a culture must be controlled within a narrow range, an automatic pH control system should be used. Such a system should include a pH meter, steam-sterilizable pH electrodes, a controller and controls for setting the desired pH limit, and pumps for the addition of acid and/or base. Most such systems are linked to a recorder to provide a continuous record. Automatic systems, including steam-sterilizable pH electrodes, are available from a number of sources (e.g., New Brunswick Scientific Co., Inc., Edison, NJ [www.nbsc.com]; The VirTis Co., Inc., Gardiner, NY [www.virtis.com]; and Cole-Parmer Instrument Co., Vernon Hills, IL [www.coleparmer.com].

Check pH control systems at frequent intervals for accuracy by using a separate pH meter. Problems that may be encountered are (i) improper or inadequate grounding of the system; (ii) establishment of pH control points (i.e., upper and lower limits) that are too narrow, resulting in almost constant addition of acid and base and hence significant dilution of the culture; (iii) aging and deterioration of the pH glass electrode as a result of repeated steam sterilization; and (iv) contamination of the reference electrode diaphragm by continuous exposure to medium constituents or metabolic products of the bacteria. The most common contaminants of the diaphragms are proteins and silver sulfide precipitate. The latter is formed by reaction of the silver chloride in the reference electrode with sulfur-containing substances in the medium. Remove the proteins by soaking the electrode in a protease (e.g., pronase) solution, and remove the silver sulfide deposits with an acidic solution of thiourea. For further discussion of automatic pH control and sterilization of electrodes, see chapter 12.2.3.1 and reference 68.

14.2. WATER ACTIVITY AND OSMOTIC PRESSURE

Water must be available to cells for metabolism and growth. However, the mere presence of water in a medium does not ensure its availability, which is determined by the water activity (a_w) of the medium. The a_w represents the mole fraction of the total water molecules that are available and is equal to the ratio of the vapor pressure of the solution to that of pure water (p/p_0). This ratio is equal to the fractional relative humidity (%RH/100) of the atmosphere above the medium at equilibrium. Temperature variation within the range in which most bacteria grow has little effect on the a_w of the medium.

The osmotic pressure (π) of a solution, expressed in atmospheres, is related to a_w as follows:

$$\pi = [RT/V_w] \cdot \ln a_w$$

where R is the gas constant (0.0821 liter atm mol^{-1} K^{-1}), T is the absolute temperature in kelvins (K) (= degrees Celsius + 273), and V_w is the volume of 1 mol of water. For more-detailed presentations of these relationships, see references 14, 36, 75, 76, 80, and 92.

The minimum a_w values at which microbes grow vary widely, but the optimum values for nonmarine species is typically greater than 0.99. By contrast, most marine bacteria are somewhat halophilic and require sodium ions; they grow best in the presence of 0.5 to 1.0 M NaCl, at which a_w is approximately equal to 0.98. Extreme halophiles are largely represented by some *Archaea* and grow best in 3 to 5 M NaCl ($a_w \approx 0.8$), which is near saturation. Species of so-called xerophilic fungi are capable of growth on relatively dry substrates such as dried fruits, candy, and cereal, whose a_w is about 0.7. For any given microbe, however, variations in a_w will almost certainly affect growth rates, cell composition, and metabolic activity. Many solutes have very specific effects at concentrations below those required to limit the availability of water. Therefore, determinations of growth-limiting a_w values should be conducted by using media in which a_w has been adjusted with more than one solute. The a_w and osmotic pressure of microbiological media are best controlled by adding nonnutrient solutes such as sodium chloride, potassium chloride, sodium sulfate, or mixtures of such salts. Scott (80) has detailed methods for calculating the a_w of nutrient media and for calculating the required concentrations of salt(s) to attain a given a_w value.

For considerations of osmotic pressure relative to bacterial permeability and transport, see chapter 20.4. See references 12, 14, 80, and 92 for reviews of the physiological effects of a_w on bacteria.

14.2.1. Measurement of Water Activity

A number of methods have been developed for the measurement of relative humidity or a_w directly. Special instruments and probes for this purpose are available from Omnimark Instrument Corp., Tempe, AZ [www.omniwww.com]; Rotronic Instrument Corp., Huntington, NY [www.rotronic-usa.com]; Omega Engineering Inc., Stamford, CT [www.omega.com]; GE General Eastern Instruments, Billerica, MA [www.globalspec.com/Supplier/Catalog/GEGeneralEastern]; and Newport Scientific Inc., Jessup, MD [www.newport-scientific.com]. The advantages and disadvantages of these instruments have been reviewed (76, 92).

The results of a study (54) of methods for determining a_w indicated that there is considerable variability among the values obtained by different procedures and that the significance of values reported beyond the second decimal place is doubtful. A statistical analysis (91) of measurements made with the electronic Sina-Scope instrument (now Novasina and marketed through Omnimark, above) over an a_w range of 0.755 to 0.967 indicated a variation of less than ±0.01. The accuracy for such instruments is currently stated to be ± 0.01.

14.2.2. Measurement of Osmotic Pressure

The most commonly used instruments for measuring osmotic pressure are osmometers and are based on measurement of vapor pressure (e.g., Discovery Diagnostics, Claremont, Ontario, Canada [www.discovery-diagnostics.com]) or freezing point depression (e.g., Advanced Instruments Inc., Norwood, MA [www.aicompanies.com]). Although some of these instruments are calibrated for direct readout, several standard solutions should be run simultaneously with the measured solutions.

14.3. TEMPERATURE

The temperature of incubation dramatically affects the growth rate of microbes, because it affects the rates of all cellular reactions. In addition, temperature may affect the metabolic pattern, nutritional requirements, and composition of bacterial cells.

The growth rates of most microbes respond to temperature in a manner similar to that shown in Fig. 1. The useful range of temperature at or near the optimum is usually quite narrow, and the maximum temperature for growth is only a few degrees (3 to 5°C) above the optimum. By contrast, the minimum temperature for growth may be 20 to 40°C below the optimum. However, determination of the minimum

temperature for growth is complicated by the fact that a shift of cells to lower temperatures is often accompanied by a period of no growth, whose duration is extended the lower the temperature. Hence, it is frequently necessary to incubate a bacterial culture for a prolonged period to measure the growth rate near the minimum temperature. For example, one *Escherichia coli* culture required over 41 h for one generation at 8°C compared to 21 min at the optimum temperature plateau of 36 to 42°C (45). The minimum temperature for growth is best determined by growing a culture somewhat above the minimum and then shifting it to progressively lower temperatures (46).

Over a limited range of temperature below the optimum, the changes in growth rates of a microbe are comparable to the responses of chemical reaction rates to temperature. Therefore, if the logarithm of the growth rate is plotted against the reciprocal of the absolute temperature (i.e., an **Arrhenius plot** [Fig. 1]), a linear slope is observed in a limited range (45, 74, 75). For most bacteria, the temperature coefficient (Q_{10}) for the growth rate in this limited range is about 2; i.e., the growth rate doubles with a 10°C increase in temperature. The slope of such an Arrhenius plot becomes essentially zero around the optimum growth temperatures and approaches infinity near the maximum temperature for growth.

Generally, microbes not only grow more slowly but also die more rapidly at temperatures above the maximum for growth. Consequently, incubation of a culture at temperatures above the microbe's optimum requires precise temperature control. To be safe, incubate a culture below its optimum temperature to an extent determined by the variability of the incubator.

The temperature within ordinary convection-type air incubators may vary by several degrees Celsius, even more if opened frequently. Incubators equipped with circulating fans or water jackets maintain a more constant temperature, but variations in temperature of 1°C are not uncommon. Nevertheless, for routine cultivation of microbes, an array of air incubators, including models designed for low-temperature incubation, are commercially available from scientific supply companies.

For accurate measurements of the effect of temperature on bacterial growth rate, however, use a stirred water bath whose temperature is controlled by a thermostat with a sensing element that is immersed in the circulating water. Such baths are available commercially and include models for high- and low-temperature incubation. For accurate measurements, also be sure to use an accurately calibrated thermometer. To assess the temperature response of a microbe, measure growth rates at a number of carefully controlled temperatures. Plot these data as illustrated in Fig. 1. If the general shapes of the resulting curves vary greatly from the normal pattern, repeat the determinations while closely monitoring the stability of the temperature of the water bath.

Practical procedures and precautions relative to temperature control of bacterial growth are as follows.

Temperature Control

Measure temperatures at various positions in air incubators by using thermometers inserted into stoppered water-filled flasks or tubes.

Air convection incubators may vary widely in temperature. Use circulation fans, and check to see that they are operating efficiently. Position cultures in the incubators so that "dead" air spaces are not created. Do not open incubators more frequently than absolutely necessary.

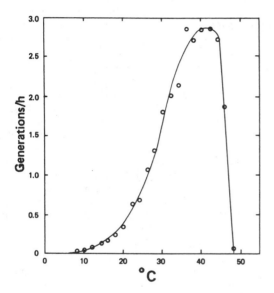

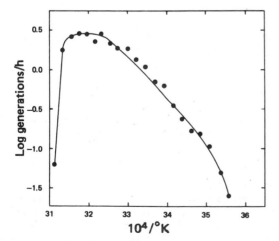

FIGURE 1 Effect of temperature on the generation time of *E. coli*, as depicted by an arithmetic plot (top) and by a semilogarithmic Arrhenius plot (bottom). Data from reference 45.

Surface cooling due to evaporation of media in air incubators must be controlled in critical studies of temperature effects. When possible, use tightly stoppered culture vessels. Where free air exchange is necessary, as for aerobic bacteria, use incubators with provision to maintain high relative humidities. This is exceedingly important at high incubation temperatures, at which loss in culture volume could present a serious problem, particularly with shaken cultures.

Use 50% polyethylene glycol antifreeze or light silicone oil in baths used for high-temperature or low-temperature incubation.

Temperature gradient incubators have been reviewed by Patching and Rose (74). These incubators provide temperatures that exceed the entire temperature range of growth for a bacterium. Different designs allow for incubation of organisms with various oxygen requirements. If large numbers of determinations are to be made, the construction and use of a gradient incubator may be desirable. A commercial source for a temperature gradient incubator is not known to us, but one can be constructed in a machine shop. A recent design has been described by Elsgaard and Jørgensen (27).

14.4. PRESSURE

High hydrostatic pressure forces volume changes and can have deleterious effects on cell macromolecules. High pressure can also inhibit or accelerate enzymatic reactions that are accompanied by an increase or decrease in volume, respectively (1). Little or no effect on most species has been observed at pressures ranging from 0.101 to 5 MPa (0.101 MPa = 1 atm = 15 lb/in^2 = 101,325 N/m^2 = 760 mm Hg = 1.01325 bar). However, at 5 MPa a detectable diminution in the growth rate of $E.$ $coli$ is observed (61). At 69 MPa, protein synthesis in $E.$ $coli$ ceases owing to inhibition of polysome formation and translocation (46). On the other hand, with increased exploration of the deep sea, barophilic (also known as piezophilic [1]) $Bacteria$ and $Archaea$ have been recognized that show optimum growth at hydrostatic pressures significantly greater than that of atmospheric (107). Some bacteria isolated from the Mariana Trench, the deepest ocean bottom in the world, have even proven to be obligate barophiles. For example, strain MT41, a γ-proteobacterium related to $Colwellia$, grows optimally at 69 MPa and also grows at 103 MPa, a pressure close to that at the depth of its origin (10 km). However, this strain does not grow at 35 MPa or less, which are pressures found at average depths of the sea (24, 108). Other obligately barophilic γ-proteobacteria have been isolated from this same habitat (42).

Studies of bacteria growing under high pressures are limited and currently represent a rather specialized domain of research, so details on the methods used are not described here. However, Marquis (60) has described experimental procedures for applying high pressures to cultures, and Jannasch and colleagues (49, 89) have described equipment that allows sampling, manipulation, and even isolation of cultures under high hydrostatic pressure without decompression. Recently, Nauhaus et al. (69) described a device for incubating marine sediments containing anaerobic methane oxidizing consortia under high hydrostatic pressures. The effects of hydrostatic pressure on the physiology and metabolism of bacteria are discussed in references 48, 60, and 61.

14.5. OXYGEN AND OTHER GASES

Gases frequently constitute substrates for bacteria, whether serving as an oxidizable energy source (e.g., H_2, CH_4, and CO), a terminal electron acceptor of aerobic respiration (e.g., O_2), or a source of nitrogen (e.g., N_2). Consequently, the metabolism and growth rates of bacteria are often dependent on the concentration of gas in solution. Among the most prevalent concerns in this regard are aerobic and facultative bacteria, whose growth rate and yield may depend critically on the concentration of oxygen in solution. Hence, the following discussion focuses on oxygen as an example, although the principles are also true for virtually any gas that dissolves in water.

Unlike most nutrients, oxygen is relatively insoluble in water (<10 mg/liter) and may quickly become limiting in liquid bacterial cultures unless special precautions are taken to ensure that it is supplied and dissolved continuously during growth. Principal factors to consider in supplying dissolved oxygen to liquid cultures (aeration) are as follows.

Aeration of Liquid Cultures

For availability of gaseous oxygen to the gas-liquid interface, the opening of the culture vessel must be sufficiently large and stoppered or covered with porous material so as to allow maximum exchange of the atmospheres inside and outside the vessel.

Large gaseous surface-to-liquid volume ratios in static cultures can be maintained only by use of very shallow liquid volumes in flat-bottom vessels such as Erlenmeyer or Fernbach flasks. Either rotary or reciprocal shaking of flasks greatly increases the surface area of a given volume of medium and hence increases the efficiency of mass transfer. This effect can be further enhanced by creation of additional turbulence in the flask, either by using flasks with "dimples" or "baffles" in the side of the flask or by placing a stainless-steel coil spring in the flask. The oxygen-solution rate obtained in a shaken flask decreases rapidly as the volume of culture increases and as the density of cells increases.

With large volumes of cultures (>0.5 liter), it is usually necessary to force air through (sparge) the liquid. The efficiency of oxygen dissolution in sparged cultures depends primarily on the number and smallness of the air bubbles and on the time of contact between the bubbles and the liquid. The use of gas spargers with small pores results in small bubbles but may induce foaming. The time of contact can be maximized by using culture vessels with high height-to-diameter ratios, by vigorously stirring the cultures, and by using baffles in the vessel.

Finally it should be remembered that, according to Henry's law, the mass of gas that dissolves in a definite mass of liquid at a given temperature is directly proportional to the partial pressure of that gas. Therefore, if agitated liquid cultures are incubated under (or sparged with) a gas phase containing O_2 at concentrations greater than atmospheric (i.e., >21% [vol/vol]), the amount of dissolved oxygen that can be provided to growing cells is increased. Gas solubilities are often expressed in terms of an Ostwald coefficient, L, which is the ratio of the volume of gas absorbed to the volume of the absorbing liquid, both measured at the same temperature. The partial pressure of the gas must be designated, and tabulated values are usually reported for the pure gas at 1 atm (101.325 kPa). Multiplication of L by the decimal fraction of a particular gas in the gas phase (e.g., 0.21 for O_2

in air) yields the equilibrium concentration of that gas in water, as milliliters of gas per milliliter of H_2O, at the specified temperature, T. Further multiplication by $(273.15)(1,000)/(22.4)$(absolute temperature in kelvins) expresses the equilibrium concentration as micromoles of gas per milliliter of water, or millimolar.

Table 5 gives Ostwald coefficients for various gases at five different temperatures. Using this table and the information above, the concentration of O_2 in water in equilibrium with air at 1 atm and 25°C is calculated to be 0.0065 ml of O_2/ml of H_2O, or 0.267 mM. A more extensive tabulation of Ostwald coefficients and other constants for various gases is given by Wilhelm et al. (103). See chapter 12 for more details on liquid cultures. The principles of aeration are also reviewed in references 75 and 96.

14.5.1. Dissolved-Oxygen Measurement

Dissolved oxygen is most conveniently measured polarographically with an oxygen electrode. The principles and operating procedures involved are described in chapter 17.2.3. Procedures for using such an instrument to determine the rates of oxygen solution in cultures are described in detail in references 40, 75, and 96.

14.5.2. Measurement of Oxygen Absorption Rate

Oxygen absorption rate (OAR) is expressed in millimoles of O_2 per liter per minute and provides information on the oxygenation efficiency of an aeration system. Such rates can be measured polarographically by using an oxygen electrode as mentioned above. In the absence of such equipment, the OAR can also be determined chemically by measuring the rate of oxidation of sulfite to sulfate by dissolved oxygen in the presence of a copper catalyst. However, the actual OAR in broth cultures may be quite different from that observed in sodium sulfite solutions. Procedures for measuring OAR in media in the presence or absence of biomass are described elsewhere (15, 75, 96). The following is a simple method (20) for measuring OAR by sulfite oxidation.

1. Instead of a growth medium, use the appropriate volume of a sodium sulfite solution containing 0.001 M copper sulfate, and operate the system at the same temperature and under the same conditions of aeration to be used for cultures. The appropriate concentration of sodium sulfite to use depends on the aeration conditions under study (see below).

2. At various intervals after initiating aeration, pipette duplicate 5-ml samples of the sulfite solution into 2.5- by 25-cm test tubes each containing a small pellet of dry ice. The rising CO_2 stirs the sample during titration and blankets the sample with gaseous CO_2 to prevent further oxida-

tion by air. If partial freezing occurs, warm the sample before completing the titration.

3. Add a few drops of a 10% starch solution, and titrate to a permanent blue end point with a freshly standardized solution of iodine. Standardize the iodine solution against a standard sodium thiosulfate solution. The normality of the iodine solution should be about one-fifth that of the sulfite solution.

4. Calculate the OAR as follows.

$$OAR = \frac{(\text{ml of titration difference} \times \text{normality of iodine})}{4} \times \frac{1,000 \text{ ml}}{5 \text{ ml}} \times \frac{1}{\text{min}}$$

The concentrations of sulfite and iodine most appropriate for any given efficiency of aeration may be calculated by substitution in this equation. For example, if the OAR is about 3.0, the use of a 0.15 N iodine solution would result in a titration difference of about 4 ml for a 10-min interval of oxygen absorption by the solution. The initial sodium sulfite concentration should be about five times the iodine concentration, or 0.75 N.

14.5.3. Hypoxia and Microaerophiles

A number of microorganisms capable of respiring with O_2 as terminal electron accceptor grow better (or will only grow) under hypoxic atmospheres, in which oxygen is present at concentrations substantially less than atmospheric, e.g., from 1 to 10% O_2 (vol/vol). Such microbes are termed microaerophiles. The physiological basis for this phenotype appears to reside in the sensitivity of cells to reactive oxygen species (ROS) (e.g., superoxide anion, hydrogen peroxide, and hydroxyl radical) generated by their own metabolism in the presence of O_2 or formed abiotically in culture media. Endogenous sources of ROS can be diminished by incubation of cultures under hypoxia. Formation of exogenous ROS can be diminished by preparation of media under hypoxic (or anoxic) conditions and by avoiding exposure of media to light; and residual ROS can be minimized by inclusion in media of agents that quench ROS, e.g., catalase, superoxide dismutase, pyruvate, and sodium metabisulfite in combination with an iron salt. See reference 53 for a more detailed discussion of this issue. Examples of microaerophiles include species of *Campylobacter* (98), *Helicobacter* (84), *Spirillum* (52, 73), and *Magnetospirillum* (85). Some bacteria may be considered conditional microaerophiles in that they exhibit a microaerophilic phenotype when grown under certain conditions, e.g., *Teredinibacter turnerae* and *Azospirillum* species, when growing under

TABLE 5 Ostwald coefficients for various gases

Temp (°C)	Ostwald coefficient[a]							
	O_2	H_2	N_2	CO	CH_4	C_2H_2	C_2H_4	C_2H_6
20	0.0334	0.0194	0.0169	0.0249	0.0367	1.108	0.1275	0.0514
25	0.0311	0.0191	0.0159	0.0233	0.0340	1.013	0.1162	0.0453
30	0.0293	0.0190	0.0151	0.0221	0.0316	0.935	0.1068	0.0405
35	0.0277	0.0189	0.0145	0.0211	0.0296	0.870	0.0990	0.0367
40	0.0265	0.0189	0.0140	0.0202	0.0280	0.816	0.0925	0.0337

[a]Values are in milliliters per milliliter of H_2O at a partial pressure of 1 atm of the pure gas and at the specified temperature.

N_2-fixing conditions, to prevent O_2-inactivation of nitrogenase (25, 37). A number of *Pseudomonas* and *Bacillus* species also exhibit a microaerophilic phenotype when growing in low-substrate media (62).

Considerations relating to the cultivation of microbes under hypoxia include not merely creating a hypoxic gas mixture or a liquid medium containing a low amount of dissolved O_2 (both of which are fairly easy to accomplish), but continually replenishing the O_2 as it is used up by cells. Taken together, these considerations have led to three general strategies for cultivation of microbes in hypoxia, as follows. (i) For motile microaerophiles that can tolerate O_2 to the extent that they can initiate growth in air-equilibrated medium (ca. 250 to 300 μM dissolved O_2), cells can be inoculated into *unshaken* flasks or serum bottles of liquid medium or "soft agar" medium under an atmosphere of air. This allows cells to migrate to, and grow within, their preferred position below the air-medium interface, within the O_2 gradient that is eventually created by the downwardly diffusing O_2 and their consumption of it. If cells cannot tolerate air-equilibrated media, then media can be deoxygenated by placing tubes in a boiling water bath for about 10 min (or using them soon after being removed from an autoclave) and quickly cooling them prior to inoculation. (ii) A gas mixture containing a low concentration of O_2 can be prepared and sparged through, or swept over, a liquid culture that is shaken or stirred. (iii) Cells can be grown under a hypoxic gas mixture of sufficient volume that the O_2 concentration is not significantly changed by the microbes' consumption of it during growth.

Strategy (i) has the advantage of simplicity and may even facilitate harvesting of cells (by aspiration of cells accumulated within a band). However, the growth rate of cells may be slow and cell yields may be limited, because although the cells may position themselves in a band at a zone of optimum dissolved O_2, a band of cells in a static culture is likely to become diffusion limited for oxidizable nutrients. Periodic swirling to disperse cells in the band may overcome this diffusion limitation, if tolerated by the culture.

Strategy (ii) requires only an appropriate gas mixture and sealed culture vessels equipped with ports for inflow and outflow of the (filter-sterilized) gas mixture. A gas mixture containing a low concentration of O_2 can be custom-prepared by most commercial suppliers of compressed gases. These are convenient but are considerably more expensive than preparing the desired mixture in the laboratory—especially so if mixtures containing different O_2 concentrations are being used. Gas mixtures containing variable amounts of O_2 are easily prepared by diluting compressed air with N_2 by means of a gas proportioner. For example, compressed air contains about 21% (vol/vol) O_2 (balance, almost entirely N_2): hence, a mixture of compressed air flowing at a rate of 50 ml/min with pure N_2 flowing at a rate of 450 ml/min will yield a blended gas containing 2.1% O_2 delivered at a rate of 500 ml/min.

Gas proportioners are units fitted with two or more graduated glass tubes through which individual gases flow, each containing a glass or stainless steel spherical float. The rate of flow of gas through the tubes is adjustable by needle valves and is indicated by the height of the float. The individual gases are then merged in a common mixing tube (usually an empty metal tube containing a wire brush or a coiled pipe cleaner to facilitate mixing) before issuing from the outflow port of the unit. The glass tubes and companion floats come in different sizes, depending on the range of gas flow rate desired. Therefore, one should have some idea of the proportions in which the gases are likely to be mixed and the total combined gas flow rate desired before purchasing such a device. This will depend on the rate of O_2 consumption per milliliter of culture, the volume of cultures, and the anticipated number of cultures to be grown at any one time. Keep in mind that outflow from a single gas proportioner can be connected to a manifold capable of delivery to several culture vessels, with optional additional flow controllers regulating the flow rate to the individual vessels. An alternative to the tube-float-type gas proportioners are mass flow blenders. These are electronically controlled and give digital readouts of individual and combined gas flow rates. Mass flow controllers and blenders are less susceptible to variations in temperature and back-pressure than are the tube-float units, but they are more expensive. Gas proportioners and mass flow blenders are available from a number of commercial sources, including Matheson Gases & Equipment, Chicago, IL (www.matheson-trigas.com); Cole-Parmer Instrument Co., Vernon Hills, IL (www.coleparmer.com); and Sable Systems International, Inc., Henderson, NV (www.sablesys.com).

Whatever device is used for mixing, it should be calibrated (or its precalibration should be checked) at a stated back-pressure of source gas (5 or 10 lb/in^2 back-pressure is typical). A fairly constant back-pressure can be achieved by using a good quality, two-stage regulator for the compressed gas tanks, or by using in-line pressure regulators between the compressed gas tank and the gas proportioner/blender. Individual or combined gas outflows are then easily measured by using mass flow detectors or simple "soap-bubble"-type glass flowmeters. The latter are available in sizes of up to 1,000 ml from Alltech Associates, Inc., Deerfield, IL (www.alltechWEB.com); and calibration involves measuring (with a stopwatch) the rate at which a bubble film is pushed up the graduated column by gas issuing from the gas proportioner/blender. A bubble solution can be purchased (e.g., "Snoop" gas leak detector, from Alltech) or prepared by making a dilute solution of dishwashing detergent in water. Sable Systems (above) also sells equipment for measuring the O_2 and CO_2 concentration of gas mixtures down to 1 ppm by volume (ppmv) (0.1 Pa). Low concentrations of O_2 (10 to 1,000 ppmv [1 to 100 Pa]) can also be measured colorimetrically, by using the pyrogallol method (6).

Strategy (iii) can be achieved a number of ways. An old-fashioned device to accomplish this is the "candle jar," which consists of a gas-tight jar containing the plates and/or tubes to be incubated and a small candle. The candle is lit just before the jar is sealed and allowed to burn until extinguished, a process that consumes some of the oxygen in the jar and also produces CO_2, the latter often required by, or stimulatory to, microaerophiles. Although somewhat useful for aerotolerant microaerophiles, the candle jar leaves an atmosphere that is still quite rich in O_2 (ca. 17% O_2 and 3% CO_2 [97]). Hence, it may not be suitable for microaerophiles like *Campylobacter*, which grow best at 5 to 6% O_2, except perhaps during primary isolation from feces when accompanied by other O_2-consuming colonies on plates (97). A better method involves the use of large desiccator jars, aluminum pressure cookers, or polycarbonate GasPak® jars (BD Diagnostic Systems, Sparks, MD [www.bd.com]) containing, or retrofitted with, at least one port that can be used to vacuum evacuate and fill the gas phase with (or dilute the ambient gas phase to) a defined hypoxic atmosphere. A simple procedure for doing this is described in chapter 15.2.65. An alternative to evacuation and filling of the headspace, which can be tedious if many jars are to be processed, is to include in each jar a CampyPak® or CampyPak Plus® H_2 plus CO_2-generating envelope (BD Diagnostics), which is activated by addition of water just before the jar is sealed. O_2 is then removed by reaction with liberated H_2 on the surface of palladium catalyst pellets (included as an integral component of the disposable CampyPak Plus system, or present in screen mesh baskets in GasPak jars) to form water. The resulting atmosphere will contain 5 to 15% O_2 and 5 to 12% CO_2, the exact concentrations depending on the volume of initial gas phase in the jar, which in turn will depend on the number of plates or tubes present. According to BD Diagnostics technical service, a single CampyPak envelope used in a 2.5-liter capacity GasPak jar containing 1 to 12 plates of medium will result in a final gas phase containing 6 to 9% O_2, 5.5 to 7% CO_2, and 0.2% residual H_2. Note that catalyst pellets become deactivated with repeated use but can be reactivated by being heated to 160°C for 1 h. However, they become permanently deactivated by sulfide, which is produced by some microaerophiles, e.g., *Campylobacter jejuni*.

An excellent system for strategy (iii) is a hypoxic glove box (Coy Lab Products Inc., Grass Lake, MI [www .coylab.com]). A glove box greatly facilitates both incubation and manipulations under hypoxia (e.g., sample preparation, dilution, plating, etc.). The system employs an O_2 sensor connected to solenoid valves that admit O_2 (e.g., compressed air) when the O_2 concentration falls below a set limit and dilute the gas phase (e.g., with N_2) if the O_2 rises above the set point. O_2 concentrations between 0 and 100% can be controlled to within ±1%; and controls for CO_2 concentration, humidity, and temperature are also available. The main advantages of this system are the relatively large reservoir of hypoxic gas phase (ca. 30 ft^3 in large models) and precise, automated control of a defined O_2 concentration. However, these systems are relatively expensive and hence are designed for long-term dedicated application. The Coy-made model used in my (J.A.B.'s) laboratory is a large, two-person vinyl glove box set at 2% O_2 and also maintained at 5% CO_2 via connections to compressed gas tanks containing 5% CO_2 (balance, air) and 5% CO_2 (balance, N_2). Included in the glove box is a compact rotary shaker, a Spectronic 20 colorimeter, an external footswitch-operated incandescent device for heat sterilization of inoculation loops, and an assortment of other supplies.

14.5.4. CO_2 and Capnophiles

Growth of a number of microorganisms requires, or is stimulated by, elevated levels of CO_2 in culture media, e.g., fermentative members of the order Cytophagales (77), members of the genus *Bacteroides* (81), and others. Such organisms are sometimes referred to as capnophiles (the prefix derived from the Greek noun *capnos*, meaning "smoke"). The term grew out of the original description of the gliding bacterium *Capnocytophaga*, which exhibited CO_2-dependent growth (56). It has come to be applied to heterotrophic microbes that require CO_2 mainly for carboxylation reactions leading to fermentative end products such as succinate, whose 4-carbon skeleton is derived by carboxylation of pyruvate or phosphoenolpyruvate. The term is not typically applied to autotrophs or to microbes such as acetogens or methanogens that use CO_2 as a major electron acceptor for anaerobic respiration. For capnophiles, CO_2 is supplied at concentrations of 5 to 20% (vol/vol) of the gas phase, and it should be accompanied by sufficient $NaHCO_3$ in the liquid phase to buffer the

medium at the optimum pH for growth, as described in section 14.1.2.

14.6. ANAEROBIOSIS

There is often no sharp line of demarcation between oxic and anoxic conditions in environments. Accordingly, there are bacteria that require oxygen (aerobes), that grow best at lower-than-atmospheric oxygen tensions (microaerophiles, above), that grow either with or without oxygen (facultative anaerobes), and that can grow only in the absence of oxygen (anaerobes).

In practice, bacteria that are unable to grow on or near the surface of solid or semisolid media in air at atmospheric pressure are considered to be anaerobic. Nonstringent (aeroduric, aerotolerant) anaerobes are able to grow on the surface of agar plates with low but significant levels of oxygen in the atmosphere (34), whereas stringent (obligate) anaerobes die, or their growth is inhibited, almost immediately on exposure to such an environment. Driving most of the oxygen out of a liquid medium by boiling or by repeated evacuation and flushing with an oxygen-free gas may not be sufficient to permit the growth of stringent anaerobes, especially if growth is to be initiated from inocula of low cell density. This is because stringent anaerobes require not only the absence of oxygen but also a low oxidation-reduction potential of the medium.

See references 17, 41, 44, 47, 57, 58, 63, and 65 for general principles and methods for growing anaerobic bacteria.

14.6.1. Oxidation-Reduction Potential (E_h)

The oxidation-reduction (redox) potential (E_h) provides a useful scale for measuring the degree of anaerobiosis (44). Simply stated, the E_h is a measure of the tendency of a solution to donate or receive electrons (i.e., to become oxidized or reduced, respectively). Measurements of E_h are expressed in units of electrical potential relative to the potential of the hydrogen electrode, which is assigned a value of zero at a pH of zero at 25°C and under an H_2 pressure of 1 atm. Under these conditions, the E_h of the hydrogen electrode is equal to E_H (the potential of the standard hydrogen electrode [zero voltage]). E_0 is the standard redox potential of a 50% reduced substance, based on the standard hydrogen electrode. At pH 7.0, the E_h of the hydrogen electrode at 25°C under 1 atm H_2 is -0.413 V, and it is designated E_0', the prime denoting pH 7.0 (90). Likewise, E_0' is the standard redox potential of any 50% reduced substance at pH 7.0, based on the standard hydrogen electrode. E_0' values are typically tabulated for compounds written as "oxidized form/reduced form"; so hydrogen would be written H^+/H_2, and oxygen would be written O_2/H_2O. The more positive the E_0' value, the stronger the *oxidizing* properties of the redox compound; the more negative the E_0' value, the stronger are its *reducing* properties. However, the E_h of a complex solution such as a bacterial culture, although a real and measurable quantity, actually represents the net E_h of a multitude of individual redox reactions, not all of which may be in equilibrium with the cell or freely reversible. As Morris (67) describes it, to speak of the E_h of a bacterial culture ". . . is conceptually the equivalent of discussing the strength of a complex mixture of different acids, bases and buffers." Nevertheless, E_h measurement and its control are of practical importance for the culture of anaerobes.

Oxygen is an oxidizing agent, and in normal laboratory media dissolved oxygen is almost always the agent primarily responsible for raising the E_h. Hence, media with positive

E_h values resulting from dissolved oxygen are unfavorable to the growth of most anaerobes, probably owing to the partial reduction of the O_2 therein to toxic oxygen species (i.e., O_2^-, H_2O_2, or OH^-) by the cells or by oxidizable medium components. By contrast, positive E_h values created by the presence of other chemicals in a medium may not affect the growth of even stringent anaerobes (71, 95). Therefore, although it may be difficult, if not impossible, to define specific tolerances to E_h for various anaerobic microbes, most stringent anaerobes will not initiate growth in media unless the E_h is ≤ -200 mV (67); and some methanogenic *Archaea* will not initiate growth unless the E_h is ≤ -330 mV (44). These considerations underscore one fundamental requirement during preparation of media for growth of stringent anaerobes: exclusion of O_2.

The concentration of oxygen in pure water at 30°C under an atmosphere of air is about 1.48×10^{20} molecules per liter (44). If all but the last molecule of O_2 were removed from 1 liter of water, the E_h should be about $+480$ mV based on theoretical calculations (44). Pirt (75) has found, however, that oxygen behaves anomalously in solution. Based on empirical observations, he estimated that an aqueous solution at 25°C containing 10^8 molecules O_2 per ml (i.e., less than 1 molecule per bacterium in a dense culture) has an E_h of -140 mV: lower than the theoretical estimate, but still more positive than desired for cultivation of stringent anaerobes. These examples illustrate the difficulty of creating strongly reducing conditions merely by removing most of the oxygen from liquid media. Therefore, a second fundamental feature of media preparation is to include a reducing agent to lower and poise the E_h in a range that permits growth of the more stringent anaerobes. In this respect, reducing agents act in a manner analogous to that of buffers, which poise the pH of a medium. This is discussed further in section 14.6.2 below.

14.6.1.1. Electrometric Measurement of E_h

The E_h of a solution is most accurately measured electrometrically. Detailed procedures and precautions for such determinations are described in a review by Jacob (47), and a system for doing so with a liquid culture of an anaerobe is diagrammed by Smith and Pierson (82). However, performance of such measurements is not practical during routine preparation of media and so is not discussed here.

14.6.1.2. Dyes Sensitive to E_h

E_h-sensitive dyes are used widely to estimate the E_h of media and of cultures, especially of anaerobes. The most useful dyes are those that are reversibly oxidized and reduced, are colored in the oxidized or reduced state, and are nontoxic. Each dye becomes reduced at a particular E_h, and the E_h at which it is 50% oxidized or reduced at pH 7.0 is its standard redox potential, E_0'. Various useful dyes and their standard redox potentials are listed in Table 6. The total range of E_h covered (from completely oxidized to completely reduced) by a redox dye having a two-electron transition is about 120 mV at constant pH. For example, methylene blue ($E_0' = 11$ mV) is almost completely oxidized (full color) at an E_h of 71 mV and almost completely reduced (colorless) at an E_h of -49 mV (41). However, the midpoint potential (E_0) of dyes varies with the pH. For example, the E_h values at which methylene blue is 50% reduced are 101, 11, and -50 mV at pH 5.0, 7.0, and 9.0, respectively (47). The exact change in E_0 with pH is variable among dyes. However, the E_0 for most dyes increases (become more positive) by 30 to 60 mV per unit of decrease in pH, and vice versa.

TABLE 6 Standard redox potentials of various dyes at pH 7.0 and 30°C[a]

Dye	E_0'(mV)
Methylene blue	11
Toluidine blue	−11
Indigo tetrasulfonate	−46
Resorufin[b]	−51
Indigo trisulfonate	−81
Indigo disulfonate	−125
Indigo monosulfonate	−160
1,5-Anthraquinone sulfate	−200
Phenosafranine	−252
Benzyl viologen	−359

[a]Data from reference 47, where a more complete list is to be found.
[b]Formed from resazurin by reduction.

A number of redox dyes are toxic to certain bacteria even at very low concentrations. Unless a dye is known to be nontoxic, use it only in control tubes or flasks of medium handled in exactly the same manner as the medium used for cultures. In such instances, use the dyes at the lowest possible concentration, because the dye may alter the E_h of media in which the redox potential is weakly poised (47). Resazurin is a widely used redox dye, because it is generally nontoxic to bacteria and is effective at concentrations of 1 to 2 μg per ml. When incorporated into media, resazurin (which is blue) first undergoes an irreversible reduction step to resorufin, which is pink at pH values near neutrality. This first reduction step can occur when media are heated under an O_2-free atmosphere. The second reduction step to dihydroresorufin (which is colorless) has an E_0' of -51 mV, so the resorufin/dihydroresorufin redox system becomes totally colorless at an E_h of about -110 mV (41). This usually requires the addition of a reducing agent to the medium (see below). When a very low E_h must be ascertained, phenosafranine ($E_0' = -252$ mV) may be incorporated into the medium, but it is often inhibitory. Titanium(III) citrate is both a reducing agent ($E_0' = -480$ mV) and an E_h indicator (109); it becomes colorless when completely oxidized (its use is described below along with that of other reducing agents). It should be kept in mind, however, that although redox dyes facilitate the preparation and use of anoxic media for anaerobes, they are poor indicators of the actual E_h of the medium. They only indicate that some minimum E_h has been achieved, i.e., resazurin incorporated into a medium at pH 7.0 will be converted to the colorless dihydroresorufin whether the E_h is -110, -200, or -300 mV.

14.6.2. Reducing Agents

Reducing agents are commonly added to anoxic media to depress and poise the redox potential at optimum levels for the growth of stringent anaerobes. Such agents must be nontoxic at the concentration used and create an E_h low enough in the medium for growth of the particular organism under study. The reducing agents most widely used in anaerobe cultures are listed in Table 7. Most of these agents have an E_0' low enough to completely reduce resazurin, but only those with an E_0' of ≤ -300 mV are likely to promote growth of stringent anaerobes. Most of these agents owe their reducing character to the presence of a reduced sulfur moiety (S^{2-}, HS^-, or R-SH). However, for organisms that

TABLE 7 Chemical reducing agents for anoxic media

Agent[a]	E_0' (mV)	Reference(s)	Conc in media
Ascorbic acid	+58	101	0.05%
Sodium thioglycolate	−140	101	0.05%
$Na_2S \cdot 9H_2O$[b]	−243	101	0.05%
FeS (amorphous hydrated)	<−270	13	4 μg/ml
Cysteine · HCl	−325	101	0.05%
Dithiothreitol	−330	19	1 mM
H_2 (+$PdCl_2$)	−413	2	Variable[c]
Titanium(III) citrate[d]	−480	50, 109	1–4 mM

[a]Stock solutions may be autoclaved and stored under O_2-free gas.

[b]At pH 7, about one-half the added sulfide exists as gaseous H_2S and one-half exists as HS^-.

[c]Insoluble $PdCl_2$ powder is included in the medium at ca. 330 μg/ml and acts as a catalyst for reduction of the medium by H_2, which can be included in the gas phase. For H_2-consuming bacteria such as methanogens, which are usually grown under a gas phase of H_2/CO_2 (80/20, vol/vol), the H_2 thereby acts both as a substrate and as a medium-reducing agent.

[d]Prepared from commercial 20% solutions of $TiCl_3$ as described in reference 50. Purchase $TiCl_3$ in small volumes under N_2, and prepare stock solutions of the citrate salt as needed.

are inhibited by such compounds, titanium(III) citrate or H_2 (+$PdCl_2$) may be tried. Amorphous FeS may also be suitable, as its solubility product is so low (3.7×10^{-19}) that little free sulfide will exist in the medium (13). O_2-consuming preparations of bacterial membranes may also be incorporated into liquid media as a type of reducing agent. Such preparations are available commercially (Oxyrase, Inc., Mansfield, OH [www.oxyrase.com]) and may be useful for growth of anaerobes that are sensitive to other reducing agents.

For maximum effectiveness, prepare stock solutions of reducing agents under O_2-free gas and with O_2-free water, sterilize them, and store them under anoxic conditions (see procedures for preparation of reduced media below). Add appropriate concentrations of the reducing agent to the medium just prior to use.

Facultative anaerobes may also be used to reduce the E_h of media for anaerobes. Growth of a bacterium such as *E. coli* in a medium prior to inoculation will scavenge residual oxygen and reduce the E_h to low levels if the culture is completely protected from oxygen. The *E. coli* cells may then be heat killed prior to inoculation with the anaerobe under study. It is necessary to be sure that the facultative bacterium used does not interfere with the growth of the desired anaerobe by depleting essential nutrients or by producing toxic products. This procedure has been used to isolate methanogens, which are among the most stringent of all anaerobes (83). The principle is also evidenced in infectious disease, in which infection of a contaminated wound by staphylococci predisposes it to subsequent infection by clostridia.

14.6.3. Techniques for Nonstringent Anaerobes

Procedures for preparing media and for cultivating and transferring nonstringent anaerobes (including many clostridia) are not difficult (88). All operations can be conducted in air by taking a few precautions to prevent excessive exposure of media and cells to oxygen. Some suggested

procedures and practices for aerotolerant anaerobes are as follows.

Cultivation of Aerotolerant Anaerobes

1. Prepare and dispense media in containers that provide for a small surface-to-volume ratio of the medium (e.g., 16-by 150-mm to 18- by 150-mm test tubes, Florence flasks, round-bottom flasks). Flasks are filled only half-way to avoid boiling over during autoclaving, but afterwards they are filled to the neck with homologous medium in companion flasks while still warm and allowed to cool before inoculation.

2. Screw-cap bottles or vials nearly completely filled with freshly autoclaved media are excellent for the prolonged incubation of anaerobes that produce little or no gas (e.g., photosynthetic bacteria).

3. Making liquid media slightly viscous, by incorporating 0.05 to 0.1% agar, reduces convection currents and is useful for test tube cultures.

4. A thick layer (ca. 1 cm) of sterile, molten vaspar (a mixture of 1 part petroleum jelly to 1 part paraffin) may be poured on the surface of inoculated media in test tubes. When solidified, the vaspar discourages diffusion of oxygen into the liquid. A disadvantage is that it is a nuisance to clean tubes that have contained vaspar, because it is insoluble in water.

5. Autoclave media without a reducing agent present, if possible, and add the agent from a sterile stock solution after the medium has cooled to about 45°C.

6. Avoid storage of media for prolonged periods at any temperature. Never store media in a refrigerator, because the solubility of oxygen in water increases as the temperature decreases. If storage is desired or necessary, store media under an anoxic atmosphere, e.g., N_2 or Ar.

7. Include resazurin (1 mg/liter) in the medium. If the top one-third of the medium is pink when ready for use, boil and cool the medium before inoculation.

8. When possible, use a fairly large inoculum (2 to 10%, vol/vol) of actively growing cells.

9. If cells are to be diluted before inoculation, use the growth medium or a freshly autoclaved diluent that contains a reducing agent.

10. Vegetative cells of nonstringent anaerobes (e.g., *Clostridium acetobutylicum*) or spores of stringent anaerobes (e.g., *Clostridium aminovalericum*) may be spread on the surface of a freshly prepared agar medium containing a reducing agent in a petri dish, provided that the dish is subsequently incubated in a jar or chamber free from oxygen. For this purpose, one of the following systems may be used.

Use a vacuum desiccator or other vacuum jar only for reasonably aerotolerant anaerobes. Evacuate three times to a partial vacuum (about 500 mm Hg), refilling each time with the anoxic gas desired. Use nitrogen, argon, or helium. If possible, include 5 to 10% carbon dioxide in the gas phase (with an appropriate amount of $NaHCO_3$ in the medium), as some anaerobes require CO_2 for growth. Methods for mixing gases and including $CO_2/NaHCO_3$ mixtures are described above.

Use sealed jars or pouches with gas generators. The system most commonly used is the GasPak jar or pouch (BD Diagnostics, above) with an H_2/CO_2-generating envelope. Evacuation of jars is not absolutely required, although anoxic conditions are attained more rapidly if jars are purchased with a vented lid to allow preliminary flushing of the assembled jar with an O_2-free gas. Some of the gas-generating envelopes require activation by the addition of water; others are entirely self-contained and include a catalyst and redox indicator. After activation, the H_2/CO_2 generator is inserted into the jar, which is then quickly sealed. The slow diffusion of oxygen through pouches or out of the walls of the polycarbonate jars is counteracted by the relatively large amount of reductant (H_2) generated by the gas generator envelope. If a redox indicator is not present (either in the medium or in the gas generation unit), methylene blue indicator strips may be inserted in the transparent containers before sealing and observed for decolorization. If such systems are used for the cultivation of anoxygenic phototrophs and the jars are placed in front of incandescent light bulbs, the temperature inside the jars may drift several degrees above ambient owing to the strong absorption of infrared light by CO_2. Therefore, temperature control may be necessary.

14.6.4. Hungate Technique for Stringent Anaerobes

Many anaerobes found in gastrointestinal tracts, sewage sludge, sediments, and other anoxic habitats require anoxic media of very low redox potential to initiate growth. Therefore, special precautions must be taken to protect media and cells from even brief exposure to oxygen. The fundamentals of two general procedures for doing this are described below and in section 14.6.5.

A roll-tube technique was described in 1950 by R. E. Hungate (43), who pioneered it for the isolation and maintenance of pure cultures of stringent anaerobic bacteria. Although many modifications of the so-called "Hungate technique" have evolved, the basic aspects of its execution have remained essentially unchanged. Its major advantages are that it requires little special apparatus and allows the use of defined, O_2-free atmospheres for cultivating specific groups such as the methanogens (106). Clear, well-illustrated descriptions of the various modifications for cultivation and preparation of anoxic media and solutions have been published (4, 5, 16, 17, 44, 57, 58, 65, 101, and 106). Many of the culture tubes, bottles, rubber stoppers, and aluminum seals required for the Hungate technique are commercially available from Bellco Glass Inc., Vineland, NJ (www.bellcoglass.com).

Even the most detailed descriptions of the Hungate technique or its modifications are frequently difficult to follow without demonstration. It is best to visit a laboratory where the technique is in use and observe the manipulations performed by someone skilled in their execution. Therefore, only the basic steps and simple procedures used in the technique are described here.

14.6.4.1. Removal of Oxygen from Gases

The gases that are used to replace air (generally N_2, Ar, CO_2, or H_2, or mixtures of these) must be treated to remove traces of oxygen. This treatment may be accomplished in one of the following ways.

Pass the gas through a heated (350 to 400°C) Pyrex or Vycor glass column (2 to 3 cm [inside diameter] by 20 to 30 cm) containing copper filings, which provide a large surface area for "scrubbing" out traces of O_2 (44). Upon trapping O_2, the copper will begin to turn black owing to formation of copper oxide ($2Cu^0 + O_2 \rightarrow 2CuO$). Cu^0 can be regenerated by purging the column with gas containing at least 3% H_2 until the characteristic red-orange color of reduced copper returns to the filings. (Caution: do not use pure H_2, unless all O_2 is swept from the column first, or an explosion might result.) Since water vapor is formed during the regeneration procedure ($H_2 + CuO \rightarrow Cu^0 + H_2O$), divert the humid outflow gas from the column to a vented fume hood to avoid wetting the downstream gas lines. If copper filings are unavailable, cupric oxide "wire" can be used after it is reduced to Cu^0 as described above. This system is very effective for removing trace amounts of O_2, even at fairly high flow rates (2 to 3 liters/min).

Bubble the gas through a solution of titanium(III) citrate, prepared as described in footnote d of Table 7. Use a gas washing bottle. The solution will become colorless when it is completely oxidized, and it must then be replaced.

14.6.4.2. Preparation of Prereduced Media

Combine the heat-stable ingredients of the medium (omitting the reducing agent) in a round-bottom boiling flask, and add a glass boiling chip. Boil the solution gently while passing a stream of oxygen-free gas over the surface by means of a gassing cannula. For a gassing cannula, use a sterile cotton-filled glass syringe barrel (such as that illustrated in Fig. 2) equipped with a bent 18- or 19-gauge needle 4 to 6 in. long (Popper & Sons, New Hyde Park, NY [www.popperandsons.com]). The vessel should have a relatively long thin neck to discourage reentry of air, and a rubber stopper may be placed loosely in the mouth of the vessel to hold the gassing needle in place yet allow venting of the exit gas and steam. If the medium is to be used soon after autoclaving, a reducing agent may be added at this time provided that it is heat stable; otherwise, it should be added just before inoculation. Continue the gassing while the medium cools to about 50 to 60°C, at which time the medium can be dispensed or the flask can be stoppered.

The choice of gas or gas mixture depends on the medium being used (e.g., CO_2 should be included in the gas mixture

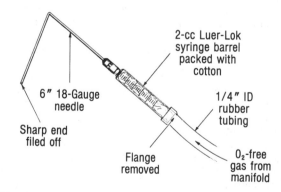

FIGURE 2 Gassing cannula used for the Hungate technique. At least two are needed: one for the vessel to be inoculated or filled with medium, and one for the vessel containing the inoculum or the medium to be dispensed. After assembly, autoclave the cotton-filled glass syringe and needle, dry in a drying oven at 100°C, allow to cool, and connect to butyl rubber tubing. Thereafter, flame the needle to sterilize it before inserting it into a vessel. This procedure also permits a constant check that gas is flowing through the needle, since the issuing gas should make a visible dent in the flame. Tubes or flasks must be constantly gassed when open.

if the medium is to be buffered with a $CO_2/NaHCO_3$ buffering system; see above) and the anaerobic microbe species to be cultivated. If H_2 is to be included in the gas phase for cultivation of H_2-consuming anaerobes, boil and cool the medium with N_2 in place of H_2, switching to H_2 when ready to dispense the medium into culture tubes (below) or after tubes are stoppered (by vacuum evacuation and flushing with the H_2-containing gas mix, below). Dispense the medium into tubes or bottles equipped with stoppers (section 14.6.4.4) while exercising the following precautions.

1. Maintain a constant flow of oxygen-free gas over the surface of the medium during transfer, and flush each tube or bottle with the same gas before and during transfer, up to the time at which the tube or bottle is stoppered while simultaneously withdrawing the gassing needle (below).

2. Fill the pipette used for transferring medium with the gas prior to drawing the medium into it. Use a rubber propipette bulb for transfer or a pistol-shaped, pump-driven autopipetter. Do not draw gas or liquid into the pipette faster than gas is entering the vessel.

3. After transfer, stopper the tube or bottle without allowing the entrance of air. This is a critical step and requires some practice. Place the stopper alongside the needle in the mouth of the tube and continue gassing for several seconds. Withdraw the needle smoothly and rapidly while pushing in the stopper. A right-handed person can hold the butt of the tube against his/her stomach by wrapping the third, fourth, and fifth fingers of the left hand around it, while manipulating the stopper with the left thumb and index finger. The right hand is then free to manipulate the gassing syringe and cannula. On removing the gassing cannula, the stopper need only be seated snug enough at first to prevent entry of air. Then, the gassing cannula can be hung up and both hands can be used to seat the stopper firmly with a twisting motion, but avoiding excessive force, especially if tubes other than thick-walled Bellco-type anaerobe culture tubes are used (section 14.6.4.4).

The time required to displace air from a culture tube or bottle with a gassing cannula before it is stoppered depends on the flow rate of gas and the headspace volume of the vessel. As a general rule, gas each tube for at least 10 s with a per-minute flow rate equal to 25 times the anticipated final headspace volume. For example, assume that 10 ml of medium is dispensed into a 20-ml capacity tube, leaving 10 ml of headspace, and that the tube is then gassed for 10 s with a probe delivering 250 ml of O_2-free gas per min. The fraction, x, of original atmospheric oxygen left in the tube will be e^{-kt}, where k is the dilution rate (25/min) and t is 0.17 min. Thus, $x = e^{-(25)(0.17)} = 0.0143$. Therefore, the amount of O_2 left in the tube after 10 s will be 21% O_2 (in air) $\times$ 0.0143 = 0.29% O_2, which is a total of about 1.2 μmol at room temperature (and is probably much lower, since this calculation assumes *complete* mixing of the gas phase and does not consider the gassing of the tube that took place before and during medium addition). Inasmuch as reducing agents are usually incorporated into media at a final concentration of about 1 mM, the tube will contain in total 10 μmol of reducing agent, or at least a 10-fold molar excess over O_2.

Alternatively, the flask of oxygen-free medium may be stoppered as described above and taken into an anoxic glove bag or glove box (below) for dispensing into tubes or bottles. This is much easier and much quicker. However, the stopper of the flask must be clamped or wired in place to prevent popping out and having the medium bubble over

during entry into the glove box, which is usually through an entry chamber that is vacuum evacuated and flushed with anoxic gas several times. After dispensing within a glove bag (which typically has a gas phase of 5 to 10% H_2 and 0 to 20% CO_2 (balance, N_2)), the gas phase in tubes and bottles may have to be changed to that desired for growth of the organism. This is easily done before autoclaving by using a manifold connected to a vacuum and a source of appropriate gas mixture. Diagrams of such manifolds are shown in Balch and Wolfe (5) and Widdel and Bak (101). It should also be mentioned that an alternative to deoxygenating a medium by boiling under an O_2-free gas is to alternately vacuum evacuate the batch of medium and bring it back to atmospheric pressure by flushing with an O_2-free gas. For this procedure, the medium is prepared in a thick-walled vacuum flask and stirred with a stirring bar during evacuation and flushing, which is controlled by manipulation of a 3-way stopcock (with one line connected to the vessel, one to the vacuum source, and one to the source of anoxic gas) and is monitored by a vacuum-pressure gauge mounted in the line between the 3-way stopcock and the vessel. A simple water faucet aspirator attached to a fast-flowing faucet works surprisingly well as a vacuum source, allowing evacuation down to -30 in. Hg (ca. -100 kPa) or more. This is sufficient to induce the visible liberation of dissolved gas bubbles from the medium during the evacuation phases.

For sterilization, the stoppers of tubes or bottles must be secured in position. If normal (i.e., tapered) stoppers are used, they must be clamped in place. A special press for holding an entire rack of stoppered culture tubes is available commercially (Bellco) or easily constructed in a machine shop. Clamps for single tubes may be constructed from wire coat hangers (106). Hungate-type anaerobe tubes equipped with a flanged butyl rubber stopper and a screw cap with a 9-mm opening to hold the stopper in place (no. 2047-16125; Bellco) do not require a clamp. Tubes or bottles with serum bottle necks and equipped with butyl rubber, mushroom-shaped serum stoppers may also be used (65); the stoppers may be held in place with aluminum seals. An aluminum-sealed tube assembly with mushroom-shaped stoppers was developed by Balch and Wolfe (5) and is available commercially (no. 2048-18150; Bellco). Rubber materials (stoppers, septa, gas tubing, etc.) should be butyl rubber if possible, as butyl rubber is fairly impermeable to oxygen. Hence, the headspace of sealed culture vessels will not be overly contaminated with O_2 diffusing out of air-equilibrated stoppers. If culture vessels are to be filled and sealed within an anoxic glove bag, preincubate stoppers in the glove bag for at least 24 h to allow traces of O_2 to diffuse out of the rubber before use.

14.6.4.3. Inoculation and Transfer

When either Hungate tubes (44, 58) or tubes or bottles with serum bottle necks (5, 65) are used, inoculation and transfer of cultures can be accomplished with sterile disposable glass pipettes, platinum loops, or sterile disposable plastic syringes fitted with 24- or 25-gauge hypodermic needles that are at least 7/8 (0.875) in. long. Use of the latter is easiest, as it avoids the need to remove and reinsert stoppers aseptically. Before use, flush syringes several times with sterile, O_2-free gas by inserting the syringe needle into the bore of an 18- or 19-gauge gassing cannula that has been briefly flamed with a Bunsen burner. Inoculate tubes possessing conventional tapered rubber stoppers by swabbing the stopper-tube junction with 100% ethanol, flaming the area, and then inserting the inoculating needle into the side of the stopper at a 20- to 40-degree angle from the long axis of the tube. The specimen or culture from which the inoculum is taken should also be manipulated the same way. For such inoculation and transfers, anaerobe tubes or serum bottles with mushroom-shaped stoppers are much easier to work with, as inoculation can be done through the flat top of the stoppers.

If stoppers must be removed for inoculation or transfer, loosen them first by pushing up around the perimeter the stopper while still maintaining an air-tight seal. The stopper-vessel junction is then ethanol-swabbed and flamed briefly before inserting a sterile gassing cannula into the mouth of the vessel as the stopper is removed completely (with fingertips or with sterile forceps). After inoculation, the stopper is briefly flamed and then reinserted as described above. During inoculation, the stopper may be placed on an ethanol-swabbed and flamed glass surface (e.g., the surface of an inverted glass beaker) until ready to be reinserted.

14.6.4.4. Roll Tubes, Shake Tubes, and Bottle Plates

Instead of petri dishes, roll tubes or shake tubes are often used for the isolation of single colonies and for the estimation of viable populations of stringent anaerobes. Prepare agar medium as above, dispense it into anoxic tubes, autoclave it, and cool it to 45°C while the stoppers are still clamped in place. If the tubes of medium have been prepared in advance, be sure that stoppers are clamped before boiling or steaming to melt the agar. Add reducing agent if necessary. Transfer samples of specimens, or dilutions made in prereduced medium or other diluent, to tubes as described above, and mix with the molten medium (but avoid frothing). To prepare a roll tube, roll the tubes horizontally under a cold-water tap until the molten agar solidifies as a shell on the inside wall of the tube. Try to coat the walls of the tube uniformly. A mechanical spinner that simplifies the procedure is available commercially (no. 7790-44125; Bellco).

Prepare shake tubes in a similar manner, except that after the inoculum is added to the molten agar, invert the tube several times to disperse the cells homogeneously throughout, and then allow the agar to solidify as an agar deep (with the tube held in an upright position) or as a slant. The latter yields a larger surface area, through which a needle or bent-tip Pasteur pipette can be used to make uncontaminated picks of isolated colonies. Each shake tube or roll tube is, in effect, a pour plate in a tube. Shake tubes to be used for picking colonies should be prepared with ca. 0.9 to 1% agar. Agar at this concentration is slightly more transparent than 1.5% agar medium, and it is easier to pick colonies from, as it allows controlled lateral movement of an already inserted picking needle or the finely drawn-out (through a flame) tip of a Pasteur pipette. Shake tubes should be incubated upside down (mouth side down) so that the water of syneresis that exudes from the agar during incubation does not accumulate on the agar surface through which colonies will be picked.

Roll tubes, prepared without inoculating the molten agar beforehand, may be subsequently streaked with an inoculating loop by starting at the bottom and, while simultaneously gassing, rotating the tube as the streaking loop is drawn straight up. This yields a spiral inoculation line with the more well-isolated colonies near the mouth facilitating subsequent picking. A device for rotating a tube for streaking is also available commercially (no. 7790-33333; Bellco), and the technique is clearly depicted by Wolfe (106).

A rubber-stoppered bottle plate is the anoxic analog of a streak plate (39), and special bottles for this purpose are available commercially (no. 2535-50020; Bellco). These bottles have a compartment for collection of the water of syneresis, which might otherwise smear the streaked agar surface. However, normal prescription bottles can also be used with care (93).

14.6.5. Anoxic Chambers for Stringent Anaerobes

Anoxic vinyl glove bags or acrylic glove boxes are efficient chambers in which to isolate and manipulate anaerobic microbes (2, 3). These types of chambers have been used in many laboratories. Even many extremely oxygen-sensitive methanogenic archaea can be safely handled in such chambers, with only minor modifications (5, 26, 51). Metcalf et al. (63) designed an "intrachamber" that can accommodate over 150 petri plates and enables plating of *Methanosarcina* species (and probably many other types of stringent anaerobes) at efficiencies comparable to those of traditional protocols. The intrachamber is housed within a large anoxic glove bag and has independent vacuum evacuation and gas-flushing lines, the latter including a line to an anoxic gas mixture containing 0.1% H_2S, which serves as a reducing agent and sulfur source. Anoxic glove bags and glove boxes are available from several commercial sources, including Coy Laboratory Products, Inc. (above) and Plas-Labs, Lansing, MI (www.plaslabs.com).

The primary advantages of anoxic chambers are that they facilitate the use of standard microbiological techniques, including spread plating, replica plating, and antibiotic susceptibility testing; they facilitate the preparation and distribution of anoxic media; and they require no special training to operate. However, a sizable initial investment is required, and a significant amount of laboratory space is occupied. In addition, anoxic chambers require constant supervision to ensure anaerobiosis. There is some inconvenience in working with gloves, and it is necessary to anticipate the need for media well in advance of use. Nevertheless, the combined use of the Hungate techniques and a properly functioning anoxic chamber makes it possible to conduct almost any kind of experiment with even the most stringent anaerobic microbes.

Considerations and Precautions for Use of Anoxic Chambers

Chambers that can be heated to serve as incubators are available. However, at 30 or 37°C chamber temperature, hands become warm and perspiration makes it difficult to remove the rubber gloves. Even with a chamber at room temperature, it is helpful to use cotton or nylon gloves under the rubber gloves when working for long periods. For some purposes, it may be preferable to have a small incubator inside the chamber. Alternatively, inoculated petri dishes or other cultures may be placed in jars, which can be sealed and removed to an external incubator (5).

Gas mixtures used in the chamber should contain at least 5% but no more than 10% H_2, which, with palladium pellets as a catalyst, continually removes traces of O_2 down to a concentration of 0 to 5 ppm. A commonly used gas mixture is 5% CO_2–10% H_2–85% N_2. If the microbes under study require CO_2 and are to be incubated in the glove bag, necessary precautions should be taken to ensure that the CO_2 does not become limiting. Gas mixtures may be purchased, but they are expensive. A gas-mixing apparatus is recommended, as described above (section 14.5.3).

Open flames cannot be used in the chamber, as there is no O_2 to support combustion. Install an electric incinerator (e.g., Steri-Loop) or a hot-wire incandescent flaming device operated by an external foot switch (Coy Laboratory Products). The latter device may be used to sterilize inoculating loops, the mouths of culture vessels, and other surfaces.

Installation of a forced-air filter such as those used in germfree hoods (Standard Safety Equipment Co., McHenry, IL [www.standardsafety.com]) will help prevent contamination.

Change the palladium catalyst at frequent intervals. Once a week is usually adequate, but twice weekly or more may be necessary when there is heavy usage. Unless poisoned by H_2S, the catalyst can be regenerated indefinitely by heating at 160°C for 2 h. Cool and return the catalyst to the chamber promptly after regeneration. Check the catalyst for activity occasionally by directing a stream of H_2 over the cold catalyst, which should heat up quickly. *Caution:* never expose hot catalyst to hydrogen, because a violent reaction may occur.

H_2S irreversibly poisons the catalyst. Whenever possible, grow H_2S-producing cultures in closed containers and either open them outside the glove box, or flush the container with anoxic gas before bringing them into the box to open them. To continually scrub H_2S from the chamber atmosphere, fashion a wire screen mesh envelope out of window screen stock and fill it with activated charcoal (8 to 12 mesh). Place this envelope in the palladium catalyst tray, between the fan and the catalyst, to trap any H_2S. This treatment reduces the H_2S level in a chamber from 100 to 1 ppm in 30 min (5, 64).

Keep the humidity at a low level. Place a relative-humidity indicator and a large tray of silica gel in the chamber. When the humidity exceeds 50%, change the silica gel. Regenerate the gel by heating at 160°C for 1 h.

Incubate materials (e.g., plasticware, rubber stoppers, syringes, needles, etc.) in the chamber at least overnight before use, to allow dissolved oxygen to diffuse out and be removed by the catalyst. When possible, keep agar plates in plastic bags and keep media in screw-cap tubes or bottles to minimize evaporation.

Keep the anoxic chamber under positive pressure. There is some gas exchange through common plastics and even more through polycarbonate and silicone plastics or rubber. If the chamber is used frequently, the concentration of H_2 should remain high enough in the chamber as a result of opening and flushing the entry lock. However, it may be necessary to add the gas mixture about twice a month to maintain good positive pressure. Loss of pressure is readily observed in flexible vinyl chambers, and a rapid loss is a sign of leaks (see below). The H_2 level should be monitored. Open, empty serum vials held in the chamber may be periodically sealed and removed, and the gas therein may be analyzed for H_2 by gas chromatography. Alternatively, specific gas monitors are available and can be placed in the chamber to provide continuous measurement of H_2 and O_2 levels. The O_2 level should be kept below 5 ppm.

Watch for leaks in the system. Whenever the positive pressure is lost unusually quickly or when the humidity stays high, it is likely that there is a leak. A leak is also indicated when an E_h indicator such as resazurin or phenosafranine in reduced media fails to remain colorless. The most common place for a leak to develop is in a rubber glove or the glove cuff, or around the seams that secure the gloves to the chamber. Leaks can be found by filling the chamber to a sufficient positive pressure to cause the gloves to turn inside out and protrude outward from the chamber. "Snoop" leak detector (above) or a dilute solution of soapy water can then be spread over parts of the gloves and chamber to locate leaks by bubble formation, but this is somewhat messy. Electronic leak detectors (e.g., Coy Laboratory Products) are also available.

Sterile, molten agar medium prepared outside the anoxic chamber may be brought into the chamber for pouring plates while still warm, but the vessel should be tightly sealed (under sterile anoxic gas) during entry; otherwise, the vacuum phase of the entry lock may cause the medium to boil and erupt. If not added previously, a reducing agent should be added to the medium before pouring plates. Alternatively, finely powdered $PdCl_2$ (ca. 300 μg/ml) may be incorporated into agar media that are poured into petri dishes and allowed to solidify outside the glove box. When subsequently brought into the glove box, reduction of the medium to a colorless dihydroresorufin end point will occur from the H_2 normally present as 5 to 10% of the anoxic chamber gas (2).

14.7. LIGHT

Light is of primary importance in the cultivation of photosynthetic bacteria. Their light-absorbing photopigments include bacteriochlorophylls and carotenoids in the anoxygenic phototrophs and chlorophylls, carotenoids, and phycobilins in the oxygenic phototrophs. As a group, the phototrophs absorb light in virtually all regions of the visible spectrum, as well as invisible (to the human eye) light in the near-infrared region. Absorption of near-infrared wavelengths is primarily by bacteriochlorophylls *a* and *b* of the purple bacteria (72). A group of *Archaea* (members of the family *Halobacteriaceae*) possess carotenoids and bacteriorhodopsin (a retinal-containing pigment) that absorb light strongly in the 500- to 650-nm range. The latter pigment functions as a light-driven proton pump, enabling cells to synthesize ATP by a nontraditional photophosphorylation reaction (86). Similar pigments, proteorhodopsins, have recently been discovered in as-yet-uncultured marine *Proteobacteria* (7, 8).

For photosynthetic growth of such organisms in the laboratory, selection of appropriate light sources and measurement of the quality and quantity of light used for illumination are critical. Excellent discussions of the production and measurement of photosynthetically usable light (or photosynthetically active radiation, PAR) have been given by Carr (18) and by Alex Ryer in a free, downloadable version of the Light Measurement Handbook © 1997 (sponsored by International Light; www.intl-light .com/handbook); only the general considerations are dealt with here.

14.7.1. Light Measurement

In most microbiological literature, the wavelength of light is expressed in nanometers (1 nm = 10^{-9} m = 10 Å). Luminous flux, also called irradiance, is the emission of a light source that actually illuminates an object. It is best expressed in terms of absolute energy such as ergs per square centimeter per second or watts per square meter, where $1 \text{ erg}\cdot\text{cm}^{-2}\cdot\text{s}^{-1} = 10^{-3} \text{ W}\cdot\text{m}^{-2}$. Irradiance is most accurately measured by means of a radiometer or quantum sensor. The latter (e.g., those manufactured by LI-COR; see below) report irradiance as micromoles quanta per square meter per second. However, this quantity is interconvertible with watts per square meter by the relationship $1 \text{ W}\cdot\text{m}^{-2} = 1 \text{ J}\cdot\text{m}^{-2}\cdot\text{s}^{-1}$. All that is required is to know how much energy change (e.g., $\Delta G^{0'}$, the Gibbs standard free energy change at pH 7, expressed in joules and designated $\Delta G^{0'}_{h\nu}$ referring to radiant energy) occurs when 1 μmol of light quanta is absorbed. This can be calculated (72) for any given wavelength of light by the relationship,

$$\Delta G^{0'}_{h\nu} = N_A \cdot h \cdot c \cdot \lambda^{-1}$$

where N_A is the Avogadro constant, 6.023×10^{23} quanta·mol^{-1}; h is Planck's constant, 6.63×10^{-34} J·s; c is the speed of light, 2.99×10^8 m·s^{-1}; and λ is the wavelength of light in meters. For example, the energy in 1 mol of quanta of light of wavelength 870 nm (= 870×10^{-9} m; i.e., the long wavelength absorption maximum of bacteriochlorophyll a) is 137.2 kJ per mole or 0.137 J per μmol.

Radiometers and quantum sensors, including portable hand-held and immersible models are available from LI-COR Biosciences, Lincoln, NE (www.licor.com/env/); Gigahertz-Optik Inc., Newburyport, MA (www.gigahertz-optik.com); and Hotek Technologies (www.hotektech.com). These can be purchased with photocells sensitive to specific ranges of wavelengths of light or broad-range sensitivity. For measurement of polychromatic light used in photosynthesis studies, use a photocell (with any necessary filters) with a relative spectral response that is similar to that of the in vivo spectrum of the organism. In reporting photosynthesis studies, be sure to state the light source and the type and spectral sensitivity of the photocell used for irradiance measurement.

14.7.2. Light Sources and Filters

Tungsten filament incandescent lightbulbs have broad emission spectra, with considerable emission in the near-infrared region (800 to 1,500 nm) and a maximum at about 900 nm. They should be used to illuminate bacteriochlorophyll-containing (anoxygenic) phototrophs. However, the maximum emission of tungsten bulbs depends on the temperature of the bulb, which in turn depends on its wattage. At higher wattages, the emission spectrum is shifted 10 to 30 nm toward the UV. The irradiance also decreases with time as the filament deteriorates, and hence it is important to check the irradiance intensity periodically. Emission spectra for various bulbs can be obtained from the manufacturer of the bulb.

A disadvantage of tungsten bulbs is their heat emission, and caution must be exercised to control the incubation temperature of illuminated cultures. A water filter (e.g., a relatively thin, water-containing, transparent bottle) placed between the light source and the culture can accomplish this, but keep in mind that water absorbs strongly in the near-infrared region. For monochromatic light $I = I_0 e^{-\alpha}$, where I_0 is the original intensity of radiation, I is the intensity after passing through 1 cm of water, and α is the absorption coefficient. Table 8 shows the values of α for water with respect to various wavelengths of light.

Fluorescent lamp emissions range from about 400 to 700 nm, with maximum intensities between 550 and 650 nm. However, fluorescent lamps do not emit much light in the

TABLE 8 Absorption coefficients (α) for water at 20°C[a]

Wavelength (nm)	α (cm^{-1})
760	0.026
970	0.460
1,190	1.05
1,450	26.0
1,940	114

[a]Data from reference 21.

near-infrared region and so are mainly useful for chlorophyll-containing (oxygenic) phototrophs.

Absorption and interference filters can be used to provide a selected range of wavelength of illumination. Popular among these are the various Kodak "Wratten"-type filters, which consist of water-soluble dyes suspended in gelatin. These filters are sandwiched between glass plates for routine use. Such filters can, for example, be used to selectively enrich anoxygenic phototrophs by blocking the lower wavelengths of light absorbed by chlorophylls of oxygenic phototrophs and to enrich for the near-infrared-absorbing anoxygenic phototrophs containing bacteriochlorophyll b (in vivo absorbing maxima out to 1,020 to 1,040 nm).

Regardless of the light source, light intensity is most easily controlled by altering the number of lamps or by inserting neutral filters between the light source and the culture. Neutral filters consist of partially darkened photographic plates or layers of wire mesh screen or cheesecloth.

14.8. DIFFUSION GRADIENTS

Gel-stabilized, one- and two-dimensional diffusion gradient systems have been used to study the responses of individual cultures and microbial communities to gradients of pH, salt concentration, and redox potential and have been reviewed by Wimpenny et al. (104, 105). They can be as simple as a tube with an agar plug in the bottom containing a substance that diffuses upward into an overlying layer of soft agar medium, or they can consist of a double-agar plate in which the first layer (containing a diffusible substance of interest) is poured and allowed to solidify with plates on an incline and the second layer is poured with the plate flat. This yields a gradient of the substance of interest across the surface of the plate. A more elaborate system was designed by Emerson et al. (30) and consists of a square arena containing a semisolid growth medium into which substances diffuse from reservoirs located on each side. Although their use is not yet widespread, gradient culture systems continue to hold great potential for studying the behavior of bacteria in an in vitro system that more closely mimics natural habitats, in which diffusion gradients are pervasive. For example, they have been used to study the oxidation of sulfide by *Beggiatoa* at the oxic-anoxic (sulfidic) interface (70) and to isolate novel strains of bacteria capable of oxidizing Fe(II) at neutral pH (29). The square diffusion chamber mentioned above was used to isolate the first extremely halophilic archaeon capable of oxidizing aromatic compounds (28).

14.9. MAGNETIC FIELDS

Some motile bacteria that possess magnetite- or greigite-containing magnetosomes, e.g., *Magnetospirillum* (formerly *Aquaspirillum*) *magnetotacticum* (59), align with the Earth's geomagnetic field, which has a strength of about 1 G. As a consequence, they show a biased swimming behavior, and both north-seeking and south-seeking forms are known. This behavior has been termed magnetotaxis and is thought, in part, to help these bacteria (many of which are microaerophilic) orient toward aquatic sediments where dissolved-oxygen concentrations are lower than in surface waters (11, 85). By contrast, the growth of various nonmagnetotactic bacteria is stimulated or inhibited, depending on the field strength (50 to 900 G) and frequency of a pulsed magnetic field (66). Equipment for exposing bacteria to magnetic fields is described in references 11 and 85. A microscope equipped with Helmholtz coils to observe magnetotactic behavior is described in reference 10.

14.10. VISCOSITY

Viscosity is a physical property of cultivation media not usually considered in the cultivation of microbes. However, the isolation and growth of some microbes is facilitated by using liquid media whose viscosity has been increased by incorporation of a thickening agent, either for the purpose of creating a gel-stabilized diffusion gradient (above) or to provide a substratum which favors the growth of certain large gliding bacteria (e.g., Desulfonema [102]). Spirochetes are a group of coiled bacteria that possess internal (periplasmic) flagella and are capable of swimming through solutions of viscosities that readily immobilize other flagellated bacteria (9, 35). The possible advantages of such ability in nature have been discussed by Harwood and Canale-Parola (38), who also comment on how this property has been used for selective isolation of spirochetes. Agents that can be used for imparting viscosity to liquid media include agar at low concentrations (0.1 to 0.3%, wt/vol), polyvinylpyrrolidone, methylcellulose, and Ficoll, which can be purchased from most chemical suppliers, as well as precipitated aluminum phosphate, which settles to form an artificial sediment (102).

14.11. REFERENCES

1. **Abe, F., and K. Horikoshi.** 2001. The biotechnological potential of piezophiles. *Trends Biotechnol.* **19:**102–108.
2. **Aranki, A., and R. Freter.** 1972. Use of anaerobic glove boxes for the cultivation of strictly anaerobic bacteria. *Am. J. Clin. Nutr.* **25:**1329–1334.
3. **Aranki, A., S. A. Syed, E. B. Kenney, and R. Freter.** 1969. Isolation of anaerobic bacteria from human gingiva and mouse cecum by means of a simplified glove box procedure. *Appl. Microbiol.* **17:**568–576.
4. **Balch, W. E., G. E. Fox, L. J. Magrum, C. R. Woese, and R. S. Wolfe.** 1979. Methanogens: reevaluation of a unique biological group. *Microbiol. Rev.* **43:**260–296.
5. **Balch, W. E., and R. S. Wolfe.** 1976. New approach to the cultivation of methanogenic bacteria: 2-mercaptoethanesulfonic acid (HS-CoM)-dependent growth of *Methanobacterium ruminantium* in a pressurized atmosphere. *Appl. Environ. Microbiol.* **32:**781–791.
6. **Baughn, A. D., and M. H. Malamy.** 2004. The strict anaerobe *Bacteroides fragilis* grows in and benefits from nanomolar concentrations of oxygen. *Nature* **427:**441–444.
7. **Béjà, O., L. Aravind, E. V. Koonin, M. T. Suzuki, A. Hadd, L. P. Nguyen, S. B. Jovanovich, C. M. Gates, R. A. Feldman, J. L. Spudich, E. N. Spudich, and E. F. DeLong.** 2000. Bacterial rhodopsin: evidence for a new type of phototrophy in the sea. *Science* **289:**1902–1906.
8. **Béjà, O., E. N. Spudich, J. L. Spudich, M. Leclerc, and E. F. DeLong.** 2001. Proteorhodopsin phototrophy in the ocean. *Nature* **411:**786–789.
9. **Berg, H. C., and L. Turner.** 1979. Movement of microorganisms in viscous environments. *Nature* **278:**349–351.
10. **Blakemore, R. P.** 1981. Magnetic navigation in bacteria. *Sci. Am.* **245:**58–65.
11. **Blakemore, R. P.** 1982. Magnetotactic bacteria. *Annu. Rev. Microbiol.* **36:**217–238.
12. **Bremer, E., and R. Krämer.** 2000. Coping with osmotic challenges: osmoregulation through accumulation and release of compatible solutes in bacteria, p. 79–97. *In* G. Storz and R. Hengge-Aronis (ed.), *Bacterial Stress Responses.* ASM Press, Washington, DC.
13. **Brock, T. D., and K. O'Dea.** 1977. Amorphous ferrous sulfide as a reducing agent for culture of anaerobes. *Appl. Environ. Microbiol.* **33:**254–256.
14. **Brown, A. D.** 1990. *Microbial Water Stress Physiology.* John Wiley & Sons, Inc., New York, NY.
15. **Brown, D. E.** 1970. Aeration in the submerged culture of microorganisms, p. 127–174. *In* J. R. Norris and D. W. Ribbons (ed.), *Methods in Microbiology*, vol. 2. Academic Press, Inc., New York, NY.
16. **Bryant, M. P.** 1972. Commentary on the Hungate technique for culture of anaerobic bacteria. *Am. J. Clin. Nutr.* **25:**1324–1328.
17. **Bryant, M. P., and L. A. Burkey.** 1953. Cultural methods and some characteristics of some of the more numerous groups of bacteria in the bovine rumen. *J. Dairy Sci.* **36:**205–217.
18. **Carr, N. G.** 1970. Production and measurement of photosynthetically useable light, p. 205–212. *In* J. R. Norris and D. W. Ribbons (ed.), *Methods in Microbiology*, vol. 2. Academic Press, Inc., New York, NY.
19. **Cleland, W. W.** 1964. Dithiothreitol, a new protective reagent for SH groups. *Biochemistry* **3:**480–482.
20. **Corman, J., H. M. Tsuchlya, H. J. Koepsell, R. G. Benedict, S. E. Kelley, V. H. Feger, R. G. Dworschak, and R. W. Jackson.** 1957. Oxygen absorption rates in laboratory and pilot plant equipment. *Appl. Microbiol.* **5:**313–318.
21. **Curcio, J. A., and C. C. Petty.** 1951. The near infrared absorption spectrum of liquid water. *J. Opt. Soc. Am.* **41:**302–304.
22. **Datta, S. P., and A. K. Grzybowski.** 1961. pH and acid-base equilibria, p. 19–58. *In* C. Long, E. J. King, and W. M. Speery (ed.), *Biochemists' Handbook.* D. Van Nostrand Co., Inc., New York, NY.
23. **Dean, J. A.** 1979. *Lange's Handbook of Chemistry*, 12th ed. McGraw-Hill Book Co., New York, NY.
24. **DeLong, E. F., D. G. Franks, and A. A. Yayanos.** 1997. Evolutionary relationships of cultivated psychrophilic and barophilic deep-sea bacteria. *Appl. Environ. Microbiol.* **63:**2105–2108.
25. **Distel, D. L., W. Morrill, N. MacLaren-Toussaint, D. Franks, and John Waterbury.** 2002. *Teredinibacter turnerae* gen. nov., sp. nov., a dinitrogen-fixing, cellulolytic, endosymbiotic γ-proteobacterium isolated from the gills of wood-boring molluscs (Bivalvia: Teredinidae). *Int. J. Syst. Evol. Microbiol.* **52:**2261–2269.
26. **Edwards, T., and B. C. McBride.** 1975. New method for the isolation and identification of methanogenic bacteria. *Appl. Microbiol.* **29:**540–545.
27. **Elsgaard, L., and L. W. Jørgensen.** 2002. A sandwich-designed temperature-gradient incubator for studies of microbial temperature responses. *J. Microbiol. Methods* **49:**19–29.
28. **Emerson, D., S. Chauhan, P. Oriel, and J. A. Breznak.** 1994. *Haloferax* sp. D1227, a halophilic Archaeon capable

of growth on aromatic compounds. *Arch. Microbiol.* **161:** 445–452.

29. **Emerson, D., and C. Moyer.** 1997. Isolation and characterization of novel iron-oxidizing bacteria that grow at circumneutral pH. *Appl. Environ. Microbiol.* **63:**4784–4792.

30. **Emerson, D., R. M. Worden, and J. A. Breznak.** 1994. A diffusion gradient chamber for studying microbial behavior and separating microorganisms. *Appl. Environ. Microbiol.* **60:**1269–1278.

31. **Galster, H.** 1991. *pH Measurement: Fundamentals, Methods, Applications, Instrumentation.* Weinheim, New York, NY.

32. **Gomori, G.** 1955. Preparation of buffers for use in enzyme studies. *Methods Enzymol.* **1:**138–146.

33. **Good, N. E., G. D. Winget, W. Winter, T. N. Connolly, S. Izawa, and R. M. M. Singh.** 1966. Hydrogen ion buffers for biological research. *Biochemistry* **5:**467–477.

34. **Gordon, J., R. A. Holman, and J. W. McLeod.** 1953. Further observations on production of hydrogen peroxide by anaerobic bacteria. *J. Pathol. Bacteriol.* **66:**527–537.

35. **Greenberg, E. P., and E. Canale-Parola.** 1977. Relationship between cell coiling and motility of spirochetes in viscous environments. *J. Bacteriol.* **131:**960–969.

36. **Griffin, D. M.** 1981. Water and microbial stress. *Adv. Microb. Ecol.* **5:**91–136.

37. **Hartmann, A., and J. I. Baldani.** 2003. The genus *Azospirillum.* In M. Dworkin (ed.), *The Prokaryotes,* electronic version, release 3.12. Springer-Verlag, New York, NY.

38. **Harwood, C. S., and E. Canale-Parola.** 1984. Ecology of spirochetes. *Annu. Rev. Microbiol.* **38:**161–192.

39. **Hermann, M., K. M. Noll, and R. S. Wolfe.** 1986. Improved agar bottle plate for isolation of methanogens or other anaerobes in a defined gas atmosphere. *Appl. Environ. Microbiol.* **51:**1124–1126.

40. **Hitchman, M. L.** 1978. *Measurement of Dissolved Oxygen. Chemical Analysis,* vol. 49. John Wiley & Sons, Inc., New York, NY.

41. **Holdeman, L. V., E. P. Cato, and W. E. C. Moore.** 1977. *Anaerobe Laboratory Manual,* 4th ed. Virginia Polytechnic Institute and State University, Blacksburg.

42. **Horikoshi, K.** 1998. Barophiles: deep-sea microorganisms adapted to an extreme environment. *Curr. Opin. Microbiol.* **1:**291–295.

43. **Hungate, R. E.** 1950. The anaerobic mesophilic cellulolytic bacteria. *Bacteriol. Rev.* **14:**1–49.

44. **Hungate, R. E.** 1969. A roll tube method for cultivation of strict anaerobes, p. 117–132. In J. R. Norris and D. W. Ribbons (ed.), *Methods in Microbiology,* vol. 3B. Academic Press, Inc., New York, NY.

45. **Ingraham, J. L.** 1958. Growth of psychrophilic bacteria. *J. Bacteriol.* **76:**75–80.

46. **Ingraham, J. L., and A. G. Marr.** 1996. Effect of temperature, pressure, pH and osmotic stress on growth, p. 1570–1578. In F. C. Neidhardt, R. Curtis III, J. L. Ingraham, E. C. C. Lin, K. B. Low, B. Magasanik, W. S. Reznikoff, M. Riley, M. Schaechter, and H. E. Umbarger (ed.), *Escherichia coli* and *Salmonella: Cellular and Molecular Biology,* vol. 2. ASM Press, Washington, DC.

47. **Jacob, H. E.** 1970. Redox potential, p. 92–123. In J. R. Norris and D. W. Ribbons (ed.), *Methods in Microbiology,* vol. 2. Academic Press, Inc., New York, NY.

48. **Jannasch, H. W., and C. D. Taylor.** 1984. Deep-sea microbiology. *Annu. Rev. Microbiol.* **38:**487–514.

49. **Jannasch, H. W., C. O. Wirsen, and C. D. Taylor.** 1982. Deep-sea bacteria: isolation in the absence of decompression. *Science* **216:**1315–1317.

50. **Jones, G. A., and M. D. Pickard.** 1980. Effect of titanium(III) citrate as a reducing agent on growth of rumen bacteria. *Appl. Environ. Microbiol.* **39:**1144–1147.

51. **Jones, W. J., W. B. Whitman, R. D. Fields, and R. S. Wolfe.** 1983. Growth and plating efficiency of methanococci on agar media. *Appl. Environ. Microbiol.* **46:**220–226.

52. **Krieg, N. R.** 1984. The genus *Spirillum* Ehrenberg 1832, 38[AL], p. 90–93. In N. R. Krieg and J. G. Holt (ed.), *Bergey's Manual of Systematic Bacteriology,* vol. 1. The Williams & Wilkins Co., Baltimore, MD.

53. **Krieg, N. R., and P. S. Hoffman.** 1986. Microaerophily and oxygen toxicity. *Annu. Rev. Microbiol.* **40:**107–130.

54. **Labuza, T. P., K. Acott, S. R. Tatini, R. Y. Lee, J. Fink, and W. McCall.** 1976. Water activity determination: a collaborative study of different methods. *J. Food Sci.* **41:**910–917.

55. **Lange, N. A.** 1961. *Handbook of Chemistry,* 10th ed. McGraw-Hill Book Company, Inc., New York, NY.

56. **Leadbetter, E. R., S. C. Holt, and S. S. Socransky.** 1979. *Capnocytophaga:* new genus of gram-negative gliding bacteria. I. General characteristics, taxonomic considerations and significance. *Arch. Microbiol.* **122:**9–16.

57. **Ljungdahl, L. G., and J. Wiegel.** 1986. Working with anaerobic bacteria, p. 84–96. In A. L. Demain and N. A. Solomon (ed.), *Manual of Industrial Microbiology and Biotechnology.* American Society for Microbiology, Washington, DC.

58. **Macy, J. M., J. E. Snellen, and R. E. Hungate.** 1972. Use of syringe methods for anaerobiosis. *Am. J. Clin. Nutr.* **25:**1318–1323.

59. **Maratea, D., and R. P. Blakemore.** 1981. *Aquaspirillum magnetotacticum* sp. nov., a magnetic spirillum. *Int. J. Syst. Bacteriol.* **31:**452–455.

60. **Marquis, R. E.** 1976. High-pressure microbial physiology. *Adv. Microb. Physiol.* **14:**159–241.

61. **Marquis, R. E., and P. Matsamura.** 1978. Microbial life under pressure, p. 105–158. In D. J. Kushner (ed.), *Microbial Life in Extreme Environments.* Academic Press, Inc., New York, NY.

62. **Mazumder, R., H. C. Pinkart, P. S. Alban, T. J. Phelps, and R. E. Benoit.** 2000. Low-substrate regulated microaerophilic behavior as a stress response of aquatic and soil bacteria. *Curr. Microbiol.* **41:**79–83.

63. **Metcalf, W. W., J. K. Zhang, and R. S. Wolfe.** 1998. An anaerobic, intrachamber incubator for growth of *Methanosarcina* spp. on methanol-containing solid media. *Appl. Environ. Microbiol.* **64:**768–770.

64. **Miguel, A. H., D. F. S. Natusch, and R. L. Tanner.** 1976. Adsorption and catalytic conversion of thiol vapors by activated carbon and manganese dioxide. *Atmos. Environ.* **10:**145–150.

65. **Miller, T. L., and M. J. Wolin.** 1974. A serum bottle modification of the Hungate technique for cultivating obligate anaerobes. *Appl. Microbiol.* **27:**985–987.

66. **Moore, R. L.** 1979. Biological effects of magnetic fields: studies with microorganisms. *Can. J. Microbiol.* **25:**1145–1151.

67. **Morris, J. G.** 1975. The physiology of obligate anaerobiosis. *Adv. Microb. Physiol.* **12:**169–246.

68. **Munro, A. L. S.** 1970. Measurement and control of pH values, p. 39–89. In J. R. Norris and D. W. Ribbons (ed.), *Methods in Microbiology,* vol. 2. Academic Press, Inc., New York, NY.

69. **Nauhaus, K., A. Boetius, M. Krüger, and F. Widdel.** 2002. *In vitro* demonstration of anaerobic oxidation of methane coupled to sulphate reduction in sediment from a marine gas hydrate area. *Environ. Microbiol.* **4:**296–305.

70. **Nelson, D. C.** 1989. Physiology and biochemistry of filamentous sulfur bacteria, p. 219–238. In H. G. Schlegel and B. Bowien (ed.), *Autotrophic Bacteria.* Science Tech Pub., Madison, WI.

71. **Onderdonk, A. B., J. Johnston, J. W. Mayhew, and S. L. Gorbach.** 1976. Effect of dissolved oxygen and E_h on *Bacteroides fragilis* during continuous culture. *Appl. Environ. Microbiol.* **31:**168–172.

72. **Overmann, J., and F. Garcia-Pichel.** 2000. The phototrophic way of life. *In* M. Dworkin (ed.), *The Prokaryotes*, electronic version, release 3.2. Springer-Verlag, New York, NY.

73. **Padgett, P. J., W. H. Cover, and N. R. Krieg.** 1982. The microaerophile *Spirillum volutans*: cultivation on complex liquid and solid media. *Appl. Environ. Microbiol.* **43:**469–477.

74. **Patching, J. W., and A. H. Rose.** 1970. The effects and control of temperature, p. 23–38. *In* J. R. Norris and D. W. Ribbons (ed.), *Methods in Microbiology*, vol. 2. Academic Press, Inc., New York, NY.

75. **Pirt, S. J.** 1975. *Principles of Microbe and Cell Cultivation*. John Wiley & Sons, Inc., New York, NY.

76. **Prior, B. A.** 1979. Measurement of water activity in foods: a review. *J. Food Prot.* **42:**668–674.

77. **Reichenbach, H.** 1999. The order Cytophagales. *In* M. Dworkin (ed.), *The Prokaryotes*, electronic version, release 3.0. Springer-Verlag, New York, NY.

78. **Rothschild, L. J., and R. L. Mancinelli.** 2001. Life in extreme environments. *Nature* **409:**1092–1101.

79. **Schlegel, H. G., and H. W. Jannasch.** 1992. Prokaryotes and their habits, p. 75–125. *In* A. Balows, H. G. Trüper, M. Dworkin, W. Harder, and K.-H. Schleifer (ed.), *The Prokaryotes*, vol. 1, 2nd ed. Springer-Verlag, New York, NY.

80. **Scott, W. J.** 1957. Water relations of food spoilage microorganisms. *Adv. Food Res.* **3:**84–123.

81. **Smith, C. J., E. R. Rocha, and B. J. Paster.** 2003. The medically important *Bacteroides* spp. in health and disease. *In* M. Dworkin (ed.), *The Prokaryotes*, electronic version, release 3.12. Springer-Verlag, New York, NY.

82. **Smith, M. V., and M. D. Pierson.** 1979. Effect of reducing agents on oxidation-reduction potential and the outgrowth of *Clostridium botulinum* type E spores. *Appl. Environ. Microbiol.* **37:**978–984.

83. **Smith, P. H., and R. E. Hungate.** 1958. Isolation and characterization of *Methanobacterium ruminantium* n. sp. *J. Bacteriol.* **75:**713–718.

84. **Solnick, J. V., J. L. O'Rourke, P. Vandamme, and A. Lee.** 2003. The genus *Helicobacter*. *In* M. Dworkin (ed.), *The Prokaryotes*, electronic version, release 3.13. Springer-Verlag, New York, NY.

85. **Spring, S., and D. A. Bazylinski.** 2000. Magnetotactic bacteria. *In* M. Dworkin (ed.), *The Prokaryotes*, electronic version, release 3.4. Springer-Verlag, New York, NY.

86. **Stoeckenius, W., and R. A. Bogomolni.** 1982. Bacteriorhodopsin and related pigments of halobacteria. *Annu. Rev. Biochem.* **52:**587–616.

87. **Stumm, W., and J. J. Morgan.** 1996. *Aquatic Chemistry*, 3rd ed. John Wiley & Sons, Inc., New York, NY.

88. **Sutter, V. L., D. M. Citron, and S. M. Finegold.** 1980. *Wadsworth Anaerobic Bacteriology Manual*, 3rd ed. C. V. Mosby Co., St. Louis, MO.

89. **Taylor, C. D., and H. W. Jannasch.** 1976. A subsampling technique for measuring growth of bacterial cultures under high hydrostatic pressure. *Appl. Environ. Microbiol.* **32:**355–359.

90. **Thauer, R. K., K. Jungermann, and K. Decker.** 1977. Energy conservation in chemotrophic anaerobic bacteria. *Bacteriol. Rev.* **41:**100–180.

91. **Troller, J. A.** 1977. Statistical analysis of a_w measurements obtained with the Sina scope. *J. Food Sci.* **42:**86–90.

92. **Troller, J. A., and J. H. B. Christian.** 1978. *Water Activity and Food*, Academic Press, Inc., New York, NY.

93. **Uffen, R. L., and R. S. Wolfe.** 1970. Anaerobic growth of purple nonsulfur bacteria under dark conditions. *J. Bacteriol.* **104:**462–472.

94. **Umbreit, W. W., R. H. Burris, and J. F. Stauffer.** 1964. *Manometric Techniques*, 4th ed. Burgess Publishing Co., Minneapolis, MN.

95. **Walden, W. C., and D. J. Hentges.** 1975. Differential effects of oxygen and oxidation-reduction potential on the multiplication of three species of anaerobic intestinal bacteria. *Appl. Microbiol.* **30:**781–785.

96. **Wang, D. L C., C. L. Cooney, A. L. Demain, P. Durmill, A. E. Humphrey, and M. D. Lilly.** 1979. *Fermentation and Enzyme Technology*, p. 157–193. John Wiley & Sons, Inc., New York, NY.

97. **Wang, W.-L. L., N. W. Luechtefeld, M. J. Blaser, and B. Reller.** 1982. Comparison of CampyPak II with standard 5% oxygen and candle jars for growth of *Campylobacter jejuni* from human feces. *J. Clin. Microbiol.* **16:**291–294.

98. **Wassenaar, T. M., and D. G. Newell.** 2001. The genus *Campylobacter*. *In* M. Dworkin (ed.), *The Prokaryotes*, electronic version, release 3.5. Springer-Verlag, New York, NY.

99. **Weast, R. C. (ed.).** 1989. *CRC Handbook of Chemistry and Physics*, 70th ed., p. D-144–D-150 and D-161–D-165. CRC Press, Inc., Boca Raton, FL.

100. **Westcott, C. C.** 1978. *pH Measurements*. Academic Press, Inc., New York, NY.

101. **Widdel, F., and F. Bak.** 1992. Gram-negative mesophilic sulfate-reducing bacteria, p. 3352–3378. *In* A. Balows, H. G. Trüper, M. Dworkin, W. Harder, and K.-H. Schleifer (ed.), *The Prokaryotes*, 2nd ed. Springer-Verlag, New York, NY.

102. **Widdel, F., G.-W. Kohring, and F. Mayer.** 1983. Studies on dissimilatory sulfate-reducing bacteria that decompose fatty acids. III. Characterization of the filamentous gliding *Desulfonema limicola* gen. nov. sp. nov., and *Desulfonema magnum* sp. nov. *Arch. Microbiol.* **134:**286–294.

103. **Wilhelm, E., R. Battino, and R. J. Wilcock.** 1977. Low-pressure solubility of gases in liquid water. *Chem. Rev.* **77:**219–262.

104. **Wimpenny, J. W. T., and D. E. Jones.** 1988. One-dimensional gel-stabilized model systems, p. 1–30. *In* J. W. T. Wimpenny (ed.), *CRC Handbook of Laboratory Model Systems for Microbial Ecosystems*, vol. II. CRC Press, Inc., Boca Raton, FL.

105. **Wimpenny, J. W. T., P. Waters, and A. Peters.** 1988. Gel-plate methods in microbiology, p. 229–251. *In* J. W. T. Wimpenny (ed.), *CRC Handbook of Laboratory Model Systems for Microbial Ecosystems*, vol. 1. CRC Press, Inc., Boca Raton, FL.

106. **Wolfe, R. S.** 1971. Microbial formation of methane. *Adv. Microb. Physiol.* **6:**107–146.

107. **Yayanos, A. A.** 1986. Evolutional and ecological implications of the properties of deep-sea barophilic bacteria. *Proc. Natl. Acad. Sci. USA* **83:**9542–9546.

108. **Yayanos, A. A., A. S. Dietz, and R. Van Boxtel.** 1981. Obligately barophilic bacterium from the Mariana Trench. *Proc. Natl. Acad. Sci. USA* **78:**5212–5215.

109. **Zehnder, A. J. B., and K. Wuhrmann.** 1976. Titanium (III) citrate as a nontoxic oxidation-reduction buffering system for the culture of obligate anaerobes. *Science* **194:**1165–1166.

15

Phenotypic Characterization and the Principles of Comparative Systematics

BRIAN J. TINDALL, JOHANNES SIKORSKI, ROBERT A SMIBERT, AND NOEL R. KRIEG

During a long-term study of a group of strains, stock cultures may be subject to genetic variation. Therefore, the cultures of an organism that are used near the end of a study may not contain quite the same organism as those used earlier. If cultures are properly preserved at the beginning of the study, one can always have a source of the organism with its original characteristics. It is also advantageous to have a working culture of a strain as well as a seed stock, which may be drawn upon should the working culture become contaminated. Chapter 47 of this volume outlines a number of preservation methods. In practice, however, only a few methods would be routinely used in a research laboratory, with freezing probably being the most popular. When freezing cultures, it is worthwhile remembering that standard flip-top (Eppendorf) microcentrifuge tubes are not suitable, due to the difficulty of opening them without contamination. Avoiding plastic altogether in mechanical freezers is to be recommended. Small glass vials (such as those used in autosamplers) with screw caps can easily be sterilized and opened with a minimal risk of contamination. Experience in the Deutsche Sammlung von Mikroorganismen und Zellkulturen with a glass capillary system for storing a wide variety of strains in liquid nitrogen (106) has shown that this method works well for both aerobes and anaerobes. In addition, a single (20-μl-culture-volume) capillary is opened per strain, and this practice further avoids the risk of contamination, which may be encountered when reopening vials.

In the past decades, our approach to characterizing or identifying prokaryotes has changed considerably. The approach which one takes is often dependent on the nature of the strain(s) being characterized. If it is suspected that a strain is novel, it may be prudent to initially determine the 16S rRNA gene sequence. Based on this information, it is usually possible to establish which known genera and species show a degree of similarity to the new isolate. It should be remembered that 16S rRNA gene sequences alone do not allow one to assign a new strain to one of several species in which the 16S rRNA gene sequence similarities among the strains approach 100%. On the other hand, when dealing with environments where familiar strains may be encountered (and where the taxon is well characterized), then it is often possible to rely on the results of the phenotypic tests. Irrespective of the approach, the scope of the methods used should be greater the more comprehensive the characterization or identification of the organism needs to be. Despite the advent of molecular methods (including whole-genome sequencing), it is also evident that prokaryotes do not always give up their secrets easily and a (phenotypically or epigenetically) well-characterized strain is almost a prerequisite for helping to make sense of the genomic data. In characterizing or identifying prokaryotes at the phenotypic level, certain general characteristics are of primary importance for determining the major group to which the new isolate is most likely to belong. The investigator should determine whether the organism is phototrophic, chemoautotrophic, or chemoheterotrophic and whether it is aerobic, anaerobic, microaerophilic, or facultative. Similarly, certain morphological features should be determined: the Gram reaction and the cell shape (e.g., rod, coccus, vibrioid, helix, or other). The presence of special morphological features (e.g., endospores, exospores, true branching, acid fastness, sheaths, cysts, stalks or other appendages, fruiting bodies, gliding motility, or budding division) is valuable. The arrangement of the cells, e.g., cocci occurring in chains or in irregular clusters, may be helpful. Three physiological tests are of primary importance: the oxidase test, the catalase test, and the determination of whether sugars are oxidized or fermented. The presence of special physiological characteristics (e.g., nitrogen-fixing ability, the ability to degrade cellulose, bioluminescence, the requirement for seawater for growth, the production of pigments, or a heme requirement) may narrow the field of possibilities considerably. The increasing interest in unusual environments means that determining the ranges of temperatures, pHs, and salinities for growth as well as the optima should also be a routine part of comprehensive characterization.

It is desirable to use an orderly approach to the characterization and identification of a strain, one that is based on common sense and uses only the tests that are pertinent. Avoid a "shotgun" approach, in which all sorts of tests are performed in the desperate hope that some of them may be helpful. Studies of complete genomes indicate that organisms with larger genomes may be more versatile than those with smaller genomes (134). The consequences of these observations may also have an impact on the scope of tests used in characterizing a novel strain. The table of contents and the major keys given in *Bergey's Manual of Systematic Bacteriology* (4, 10, 17, 18, 20) will be of great assistance, as will the general descriptions of the various genera and species of prokaryotes. Clearly, this type of approach is helped by access to the 16S rRNA gene data. However, it should be remembered that some groups of taxa with a low level of 16S rRNA gene similarity may show a relatively low degree of physiological diversity (e.g., the anaerobic clostridia) while others may show greater physiological diversity with a relatively small degree of difference in 16S rRNA genes (e.g., the genera and *Nitrobacter, Bradyrhizobium, Afipia,* and *Rhodopseudomonas*). After the initial possibilities have become more restricted, consult the detailed keys and tables from various sources (many of which are listed in references 1–4, 7, 8, 10–12, and 15–20) for the identification of genera and species. In the final steps of identification, compare the new strain with, preferably, either the type strain of the suspected species or some other named strain whose inclusion in the species has been firmly established.

In many cases, especially with pathogenic prokaryotes, one may be able to make a rapid presumptive identification of a genus or species by the use of antisera. However, such tests are often not absolutely specific, and additional physiological tests are usually required for confirmation of the identification. Agglutination of gram-negative enteric rods, for example, by polyvalent *Salmonella* serum is presumptive evidence for a *Salmonella* strain, but this identification must always be confirmed by a number of biochemical tests.

In performing phenotypic characterization tests, the organisms used for the inoculation of test media should be from fresh transfers and in good physiological condition. Old stock cultures, which may contain mostly dead or dying cells, are not satisfactory. Incubate the inoculated test media under conditions that are optimum for the organisms or for the characteristic being tested, e.g., optimum temperature, pH, gaseous conditions, and ionic conditions. Furthermore, apply characterization methods not only to the organism to be tested but, wherever possible, also in parallel to strains known to be positive and negative for the characteristic, as a check on the reliability of the reagents and media.

The methods described in this chapter are based primarily on methods developed for the characterization and identification of organisms which have usually been isolated on nutrient-rich media (and often at neutral pH). There is increasing evidence that the number of strains which may be isolated from a given environment may increase if less-nutrient-rich media are used (37, 220). Some of these strains may not grow in the media developed for the tests described in this chapter, and due consideration should be given to working with lower concentrations of substrates and/or other components in the media, as well as other effects (289). Other problems which may be encountered include the isolation of obligate acidophiles or alkaliphiles, in which shifts in pH required to detect a positive reaction may be minimal due to the buffering capacity of the growth medium. Naturally, at extremes of pH the standard indicators given in the individual recipes will also not work, and alternatives are to be found in chapter 11, Table 1. Despite tremendous advances in genomics, proteomics, and transcriptomics, there appears to be no easy molecular tool for determining the optimum temperature, salt concentration, or pH, although specific changes in the gene sequences of proteins may reflect adaptation to extremes of any of these three environmental parameters. In this sense, there is still a need to isolate (see chapter 11) and characterize prokaryotes from the diversity of natural environments in which they can be found (210). Combining the information gained using the methods described in this chapter together with other methods of analysis described in other chapters in this book and appreciating prokaryote diversity at different levels are some of the major goals of prokaryote systematics.

At the end of this chapter is listed a selection of general references (1–20) that provide identification schemes for various groups of bacteria, further information on methodology, and additional characterization methods. During the coming years the publication of the newer volumes of *Bergey's Manual of Systematic Bacteriology* will certainly provide additional information. A section dealing with characterization and systematics would not be complete without a brief overview of the objectives of systematics.

15.1. AN INTRODUCTION TO SYSTEMATICS

Systematics is not always fully appreciated. While there is often a tendency to equate systematics with taxonomy, it is more appropriate to regard the latter as a part of the former.

In simple terms, systematics may be considered to be one of the most elementary and comprehensive sciences, because it is an essential aspect in comparing any two elements in biology, whether it be genes, proteins, biochemical pathways, or organisms themselves. Thus, systematics continues to play an important role in an age when it is becoming increasingly easy to sequence whole genomes or to characterize the genes of populations of organisms.

Systematics has been defined in many ways but may be succinctly described as "the cradle of comparative biology." Taxonomy can be clearly defined as encompassing characterization, classification, and nomenclature. The characterization of an organism is no longer bounded by methodological barriers, and it is now possible to fully sequence the whole genome of a strain, to study individual genes, or to examine the genetic information by using amplified fragment length polymorphism, random amplification of polymorphic DNA (RAPD), and G+C content analysis. However, genes do not exist on their own, and it is becoming increasingly clear that the study of the biochemical pathways of an organism, the roles structural elements (proteins and lipopolysaccharides, etc.) may play in morphology, or the chemical composition of the cell should be correlated with the underlying genetic information. Classification is the arrangement of prokaryotes into groups. Different forms of classification may have different goals. Organisms may be grouped according to their pathogenic potentials (biological safety levels) or in an arrangement based on more complex theories (i.e., the course of evolution). Nomenclature is the naming of those groups. When the groups are species, genera, and families, etc., then the way they are named or the links between names which have been used in the past are governed by an International Code of Nomenclature (147). Identification is often considered to be a part of taxonomy but is concerned with comparing unknown organisms with organisms which have already been classified. As such, identification can be carried out only once a taxonomy has been established. Typically, identification protocols have the goal of quickly assigning an organism to a known group by using the minimum number of methods. In contrast, a novel organism should be characterized as fully as possible in order for subsequent identification systems to have a reliable basis on which to work. The more reliable the characterization and classification, the greater chance one will have of being able to pick identification methods which both are accurate and have a long-term future.

Nomenclature is regulated in prokaryote systematics by an official system of registering (or indexing) those names which may be used in prokaryote taxonomy via a centralized system. The formal term for registering a name is "valid publication [of a name]." This system was unique to the International Code of Nomenclature of Bacteria (but virology has followed this principle) and was introduced in 1980 to combat uncertainties in the application of some 40,000 names, which had accumulated over the previous ~200 years. With the introduction of the Approved Lists of Bacterial Names (240), only 2,000+ names made the grade into the modern system. While nomenclature is formally regulated, it is important to be aware that taxonomy is not regulated by the code. Thus, the code recognizes a "valid published species name," but the term "validly published species" has no meaning, despite being frequently met with in some of the literature. In order for a name to be validly published, the proposal for the name must be accompanied by a number of criteria (147, 247, 248). Among these criteria is the designation of a nomenclatural type. Just as

physics has reference points for the meter or the kilogram, so too does biology, with the nomenclatural types being the reference points for a taxon. In the cases of the species and subspecies, these reference points are represented by type strains (147). Given the central importance of type strains, it is important that they be made available as widely as possible. The best course of action would appear to be to deposit type strains in a suitable culture collection, which should be able to maintain the distribution of the strain in the future (83, 141). Clearly, depositing a type strain in such a way that it is not easy to access is counterproductive to the principle behind the deposit of type strains, that of making them widely and easily available for comparative purposes.

Organisms were initially described by a variety of observations, although the only way of observing organisms in natural samples was either by studying mass accumulations (as in Winogradsky columns), by using microscopy, by studying the results of spoilage, or by studying organisms in connection with a disease. Pure-culture techniques for aerobes became available at the end of the 19th century, with pure cultures of anaerobes being available at the beginning of the 20th century. Staining methods were also developed at this time, with Christian Gram (88) developing the Gram stain. The concept of the significance of the biochemical properties of a microorganism was formulated by Orla-Jensen (203, 204), and den Dooren de Jong (63) was also instrumental in developing this concept further (208). Thus, until the 1960s most microbiologists relied on morphological aspects together with biochemical and physiological properties. The discovery of the structure of DNA during the 1950s was also paralleled by the development of methods for the study of the chemical compositions of cells (gas chromatography, nuclear magnetic resonance [NMR], mass spectrometry, and gel electrophoresis, etc.). This development laid the basis for the further development, over the next 50 years, of the modern spectrum of methods which we know today, many of which are covered in this book. While the immediate question to most is what methods do we use, one must first examine what we try to achieve in characterizing and classifying prokaryotes.

The ultimate goal of taxonomy is to find the "correct order" which is present in nature. Although this is often called the "natural system," the concept of "natural" has, over the centuries, meant different things to different people (41). Prior to Darwin, the term natural would have referred to the order which was a result of the act of creation. During the 1940s, this term became associated with a particular philosophical approach outlined by J. S. L. Gilmour, later to become known by the term phenetic, and finally, the term natural was associated with evolution. Unfortunately, some confusion has arisen around the term phenetic because of the popular belief that it has nothing to do with evolution and that it deals only with phenotypic data. This is not so, and the proper definition by Cain and Harrison (41) makes clear and unambiguous reference to the fact that it deals with "overall similarity" and may include both phenotypic and genotypic data. Harrison (97) noted that the analogy of genotypic to genetic versus phenotypic to phenetic was unfortunate and misleading. This misconception persists most strongly in prokaryotic systematics. Based on the conceptual work of Gilmour, which has been elaborated by Sneath and Sokal (246, 249, 251), the phenetic approach made a clear distinction between the exact ancestor-descendant relationships (phyletic lineages) and those groups which have arisen as the result of evolution (phyletic groups). Under certain circumstances (largely those in which homoplasy, convergent evolution, parallelism, gene transfer, or gene conversion does not play a role), a phenetic system may also reflect the true course of evolution. In contrast to the phenetic approach (which has flavored much of modern microbial systematics), the cladistic approach has been much favored in botany and zoology, where bitter and often acrimonious debates have raged over the years on the merits of one system over another. Briefly, the cladistic approach (also known as phylogenetic systematics) evaluates the significance of individual characters and their contribution to delineating the evolutionary history of the organisms concerned. The term clade was coined by Huxley (112), but the major contribution to cladistic evaluation is usually attributed to Hennig (who used the term Phylogenetische Systematik [102]). This methodology centered largely on the use of morphological data. Debates over the advantages of one method over another included topics such as the "naturalness" of taxa and the "information content" of the classifications (73, 74).

One of the present problems is that the term phylogenetic has also come to be associated with a number of meanings. Cowan (54) noted different meanings being associated with the terms phylogeny and phylogenetic classification. However, he refers to Constance (51), who, writing in 1964, had not been caught up by the tendency to equate cladistics with Hennigian phylogenetic systematics. In a modern context, phylogenetics has become associated with the study of gene or protein sequences. The paradox arose in the early days of the use of S_{ab} values for evaluating 16S rRNA gene catalogues that this method was taken to be phylogenetic (301, 302), whereas Sneath (245) maintained that it was not cladistic and primarily phenetic. It is interesting that one author makes the distinction between phenetic and phylogenetic methods based on whether a program will allow one to mask a sequence (66). Sneath (245) has indicated that it is sometimes not easy to determine when "phylogenetic" means evolutionary, cladistic, or simply genomic. The confusion over how a particular term was, is, or will be used will no doubt continue, so it is probably best to avoid those terms which cause confusion altogether and simply divide the methods of classification into three: overall similarity, "character analysis" (273), and a combination of both (164). Hillis et al. (104) have also provided a good overview of how some of the methods used for inferring phylogenies from sequences may be classified.

Numerical taxonomy is, in its simplest form, the use of computers in the taxonomic process. In microbiology, the term is often (and very mistakenly) taken to mean "the use of computers in the evaluation of phenotypic data, based on phenetic philosophy." In areas such as gene or protein sequence analysis, the basic elements of numerical taxonomy are evident in many alignment programs and in the principle underlying simple BLAST searches. One of the basic principles of numerical taxonomy is that given enough data, the correct natural groups will be found. Certainly in microbiology the use of a wide range of biochemical and physiological test data has not proven to be the optimal solution, particularly across higher taxa. However, this should not be taken to mean that such data do not reflect evolution. If we mistakenly assume that all cocci are a monophyletic group, then it is not the data that are incorrect but rather our assumptions that led us to these conclusions. Siefert and Fox (236) have indicated how complex the interrelationships are between rods and cocci in relation to 16S rRNA gene sequence-based groupings. Anyone familiar with the principles of numerical taxonomy will have noticed an interesting trend as full genome sequences became available, with the sheer flood of data being taken as the solution. However, that this is not the case is now fully evident, and it is becoming increasingly difficult to find more than about 100 genes which can be used to cover all taxa (154). When different genes give different results, gene transfer is the simple answer (65), although this may well be an oversimplification (140). While zoologists may be content to classify all insects or all vertebrates, prokaryote systematics generally deals with all prokaryotes and in an evolutionary context attempts to extend this study back to the origin of life itself. It is the latter, the origin of life and the root of the tree of life, which may well be the most elusive element of all (27). In the absence of a root, it becomes difficult to determine whether major groups such as the *Bacteria* and the *Archaea* are monophyletic, monophyletic groups being the product of rooted trees.

A central element in any comparative work is the comparison of like with like. This concept has as much a biological as a philosophical basis. The idea of like with like benefited from a formal definition proposed by Owen, the concept of homology. Despite the widespread usage of the term homology in the post-Darwinian era, Owen was a firm opponent of Darwin's theory and his concept of homology was not formulated with an evolutionary basis. Nevertheless, this concept has become incorporated into an evolutionary context, although there is continued and regular debate on the exact use, meaning, and definition of the term. Certainly, in gene or protein sequence-based work there is a tendency to equate similarity (of the aligned sequences) with homology. While it is true that homologous proteins will show a degree of similarity, it is also true that some proteins may be similar but are not the same (i.e., homologues). Until recently, BLAST searches were called "homology searches." In reality, they determine the most similar sequence(s) in the database without being able to confirm to what extent the sequences being composed represent the same proteins or genes (i.e., homologues). This problem was highlighted by Margoliash et al. (161), but the debate about the misinterpretation of similarity as equating with homology regularly resurfaces. Patterson (212, 213) has outlined the problem, and Fitch (77) has also recently aired his views on this topic. It is also becoming increasingly clear how difficult it is to decide exactly what is orthologous, paralogous, or xenologous at the gene or protein sequence level. Lerat et al. (154) have indicated that it is not easy to unambiguously distinguish the three in closely related taxa and have proposed the term synologue. This term has also been used by Gogarten (85) in another context. The problem of identifying homology was clearly articulated by Sneath and Sokal (251), who proposed using the term isologue and developed the concept of operational homology. It is likely that the problem will continue to be with us for some time to come. Irrespective of how one sees the problem, homology may have slightly different interpretations at different levels. However, two elements should always be taken into consideration: those of "common evolutionary history" and "structure and function" (77). The latter is clearly not a one-dimensional concept, and the determination of homology should not be reduced to a simple linear, statistical relationship (301, 302). Clearly the correct use of terminology in the literature as well as correctly communicating these concepts to younger scientists, yet to be tainted by misuse, would help to reduce the confusion.

It is often claimed that sequence data are the only means of establishing any evolutionary dimension for prokaryotes, with the work of Zuckerkandl and Pauling (308) being the quoted source. However, these authors were well aware that certain aspects of the phenotype may be useful in determining at least parts of the evolutionary tree while also acknowledging that one must consider the results from individual sequences in the context of the evolutionary history of the whole organism. It is this history which may not be so easy to decipher. In other cases, one gene may not be in accord with others (25) or different treatments of the raw data may give different results (152), and it is evident that the original goal to test an evolutionary hypothesis (302) is often forgotten. Clearly, grouping all anoxygenic phototrophs into a single (monophyletic) group is not consistent with 16S rRNA gene groupings. However, the distribution of genes coding for (bacterio)chlorophyll is proving to be a tantalizing evolutionary puzzle, with some groups of organisms consistently having such genes (cyanobacteria, heliobacteria, and chlorobiaceae) while others, such as the alpha-, beta-, and gammaproteobacteria, seem to have a patchwork distribution. Clearly, in some groups phototrophic growth is a characteristic of all of the members and in others it is not. Morphology is as unpredictable; spirochete morphology is group specific, but cocci may be found across the breadth of modern taxa. The presence of various biochemical pathways such as methanogenesis may define fairly large, coherent groups, while the ability for nitrate reduction seems to surface at random. Given the fact that the phenotype of an organism is the result of the underlying genetic information, such problems will also be reflected at the genetic level. In some cases genes may be present but expressed either only to a low degree or not at all.

Wilson et al. (300) stated:

> We have already alluded to the observation that amino acid and nucleotide sequences evolve at fairly steady rates that seem virtually independent of rates of organismal evolution. Molecular evolutionists were slow to recognise this surprising and intriguing fact. They had assumed that organismal evolution depends on sequence evolution in proteins and expected a simple relationship between the two types of evolution. In particular it was expected that morphologically conservative organisms should have experienced slower macromolecular evolution than organisms that had evolved unusually rapidly at the morphological level. To date, however, there is no convincing evidence that proteins or nucleic acids of conservative creatures are conservative in regard to their amino acid or nucleotide sequences.

It is the regularity of biological systems which has arisen as the result of evolution that remains to be documented. Tracking those patterns in nature at one level can help to reinforce the information gathered at another. In the absence of a wealth of data on biochemical pathways and their end products or the structural components of prokaryotes, little progress could have been made in elucidating prokaryote genomes. Using data from annotated genomes can, in return, help to locate postulated new pathways or to confirm the molecular infrastructures underlying the results of simple biochemical tests. The presence of genes for bacteriochlorophyll synthesis does not confirm that they are expressed to any appreciable degree. Systems biology seeks to integrate the whole and steers a course similar to a broadly based approach to systematics, in which information from different levels (gene sequence, protein sequence, biochemical pathway, and end product) can be integrated to provide

a better appreciation of its significance. Simpson advocated taking a similar course in zoology (239). In this respect, the methods employed in prokaryote systematics will continue to expand, and some methods may fall by the wayside (186, 288). However, just as understanding the functioning of the ribosome requires an understanding of the three-dimensional interactions of the various RNAs and proteins (177), so too is there a need for the continued reference to the wider range of properties of an organism.

The methods described in this chapter deal primarily with the biochemical, physiological, and chemical properties of prokaryotes. Genetic aspects are covered in chapter 11.5 and sections IV and V. One chapter deals specifically with accessing data gained by genomic methods (chapter 35), and another deals with the thorny question of phylogenetic trees (chapter 36). The reader is also referred to more comprehensive reference works, such as *The Prokaryotes*, *Bergey's Manual of Systematic Bacteriology*, and specific chapters in these volumes (10, 17, 18, 20, 36, 118, 137, 156, 257). The *International Journal of Systematic Bacteriology*, now renamed the *International Journal of Systematic and Evolutionary Microbiology*, is one of the most important journals dealing with modern prokaryotic systematics. The reader is also referred to the List of Prokaryotic Names with Standing in Nomenclature, compiled by Jean Euzéby (www.bacterio.cict.fr/), which covers all validly published names and includes references to the original publications, as well as a wealth of additional information.

15.2. ROUTINE TESTS

15.2.1. Acetamide Hydrolysis

To test for acetamide hydrolysis, streak the surface of an acetamide agar slant (section 15.4.2) with a sample from a dilute suspension. Incubate for up to 7 days. Red or magenta signals a positive test; no color change signals a negative test.

Examples: positive, *Pseudomonas aeruginosa*; negative, *Pseudomonas fluorescens*.

15.2.2. Acid-Fast Reaction

See chapter 2.3.6 for two staining methods for the acid-fast reaction. A positive reaction indicates the presence of a member of the genus *Mycobacterium* (which is strongly acid fast), *Nocardia*, or *Rhodococcus* (members of the last two genera are partially acid fast) (17).

15.2.3. Acidification of Carbohydrates

15.2.3.1. Method 1

Incorporate a pH indicator into the medium during its preparation. For a list of the most commonly used indicators and their properties, see Table 8. In cases in which an indicator may be toxic to the bacteria or may be reduced during their growth, add drops of the indicator (0.02 to 0.04% alcoholic solution) to the culture after it has grown to maximum turbidity.

A relatively high concentration of sugars, e.g., 1 to 2% (wt/vol), is usually added to the basal medium. On the other hand, the peptone concentration in the basal medium should be kept as low as possible while still allowing good growth to minimize buffering and alkali formation. This is particularly important for characterizing aerobic organisms, which produce comparatively little acidity when oxidizing sugars. For instance, in determining the acidification of sugar media by various *Aquaspirillum* species, the peptone level in the medium should not exceed 0.2% and a pH indicator, such as phenol red, that will change color with only slight acidity should be used (114).

In general, sugar-containing media to be used for testing acidification can be sterilized by autoclaving at 115 to 118°C for 10 to 15 min. However, heating is known to alter some sugars, especially in an alkaline solution, or to convert one sugar to another (169). Consequently, when using xylose, lactose, sucrose, arabinose, trehalose, rhamnose, or salicin, it is prudent to sterilize the complete sugar-containing medium by filtration or to sterilize a 5 or 10% stock solution of the sugar separately by filtration and then add appropriate amounts aseptically to portions of the autoclaved, cooled basal medium (1).

Some organisms, such as *Neisseria* spp., show acidity best when grown in a semisolid medium.

Care should be taken to ensure that the basal medium to which the carbohydrates are added does not contain fermentable or oxidizable sugars, and control media without any added carbohydrates should always be included during testing of cultures. Moreover, basal media should not contain the oxidizable salts of organic acids, such as malate, pyruvate, or succinate, because oxidation of these substances by the test organism will cause an alkaline reaction that may obscure acidification due to carbohydrate oxidation or fermentation.

With cultures grown in defined media with an ammonium salt or nitrate as the major or sole nitrogen source, the utilization of the ammonium salt will cause a decrease in pH and the utilization of the nitrate will cause an increase in pH (168). This effect may lead to incorrect conclusions about the acidification of sugar media. For instance, almost all of the acidity produced in cultures of the aerobe *Azospirillum lipoferum* grown in a defined fructose-containing medium is due to the utilization of ammonium sulfate and not to the production of organic acids from the fructose (84).

15.2.3.2. Method 2

To provide a more quantitative measurement of acidification, use a pH meter to measure the pH of the culture after it has grown to maximum turbidity. Use a long, thin, combination pH electrode, which can be inserted directly into a culture tube. Affix a flat rubber washer with a diameter larger than that of the culture tube to the upper portion of the electrode to prevent the electrode tip from striking the bottom of the culture tube when the electrode is inserted. When determining the pHs of a number of cultures of pathogenic bacteria, dip the electrode into a beaker of distilled water between cultures. This water should later be autoclaved. When finished with the electrode, rinse it with a suitable disinfectant (e.g., 3% hydrogen peroxide or 70% ethanol).

For anaerobes cultured in prereduced PY broth (section 15.4.31) containing carbohydrates, the oxygen-free carbon dioxide used to purge the tube during inoculation will usually lower the pH of the medium to 6.2 to 6.4. Consequently, pHs of PY carbohydrate broth cultures are usually interpreted to indicate acidification when they are 5.5 to 6.0 (weak acid) or below 5.5 (strong acid). For PY broth cultures containing arabinose, ribose, or xylose, a pH of 5.7 or below usually indicates acidification, because the sterile medium purged with carbon dioxide may have a pH of 5.9 after 1 to 2 days of incubation (3). In any case, the pH of cultures in PY broth lacking any carbohydrate should be determined as a control, because acids may be formed from peptones in certain cases.

15.2.4. Agar Corrosion and Digestion

Colonies of agar-corroding or agar-pitting bacteria appear as if they are in a shallow pit in the surface of the agar medium. The term corrosion is used rather than digestion because the agar seems not to be softened or liquefied. Examples are *Eikenella corrodens*, *Bacteroides ureolyticus* (this organism is related to members of the genus *Campylobacter* [280]), *Kingella kingae*, and various *Moraxella* species.

Several species of gliding bacteria are able to digest agar by means of agarolytic enzymes. For example, the colonies of *Nannocystis exedens* may deeply corrode the surface of the medium and may even penetrate the medium, producing holes or tunnels in the agar. *Chondromyces* and *Polyangium* species can pit, erode, and penetrate agar media. *Vibrio alginolyticus* and some *Cytophaga* species can soften or totally liquefy agar gels, whereas others, such as *Cytophaga fermentans*, merely produce gelase fields and craters on agar plates.

15.2.5. Ammonia from Arginine

Inoculate arginine broth (see sections 15.4.3 and 15.4.4). Also inoculate a control lacking arginine. After incubation for 2 to 3 days, test samples of the culture on a spot plate with Nessler's solution (section 15.5.11). A positive test is indicated by yellow or orange compared with the control.

Examples: positive, *Enterococcus faecalis*; negative, *Streptococcus salivarius*.

15.2.6. Arginine Dihydrolase Activity

15.2.6.1. Method 1

The following method is widely used to distinguish among members of the family *Enterobacteriaceae*. Inoculate Møller's broth base (section 15.4.26) supplemented with 1% L-arginine monohydrochloride (or 2% of the DL form). Also inoculate a control lacking arginine. After inoculation, overlay the broth with 10 mm of sterile mineral (paraffin) oil. Examine daily for 4 days. A positive test is indicated by violet or reddish violet due to an increase in pH. Weak reactions are indicated by bluish gray.

Examples: positive, *Enterobacter cloacae*; negative, *Proteus vulgaris*.

15.2.6.2. Method 2

The following method is suitable for a wide variety of facultatively anaerobic bacteria. Inoculate Thornley's semisolid medium containing arginine (see section 15.4.36) and also a control lacking arginine. After stab inoculation, seal the medium with a layer of sterile melted petrolatum and incubate. A positive test is indicated by a change from yellow-orange to red within 7 days due to an increase in pH.

15.2.6.3. Method 3 (258)

The following method is suitable for aerobic and facultative bacteria and is based on the direct measurement of the disappearance of arginine. Make a dense suspension of the bacteria in 0.033 M phosphate buffer (pH 6.8) (see section 15.5.18). Purge 4 ml of the suspension by bubbling nitrogen through the suspension for several minutes, and add 1 ml of 0.001 M L-arginine monohydrochloride. After purging again, stopper the tubes, incubate them for 2 h, and heat them at 100°C for 15 min. After removing the cells by centrifugation, determine the concentration of arginine in the supernatant by the method of Rosenberg et al. (224) as follows. Mix 1 ml of the sample with 1 ml of 3 N NaOH, 2 ml of developing solution (see section 15.5.4), and 6 ml of water; read the tubes at 30 min against a blank prepared without arginine by using a colorimeter equipped with a green filter (540 nm); and compare the readings with those obtained with an uninoculated control containing arginine. A positive test is indicated by the disappearance of some or all of the arginine.

15.2.7. Aromatic Ring Cleavage (207)

Grow the culture in chemically defined medium containing an aromatic substrate such as 0.1% sodium p-hydroxybenzoate as the carbon source. (Also grow it on a yeast extract agar without the aromatic substrate to determine whether the enzymes are constitutive or inducible.) Scrape growth from the agar, and suspend it in 2 ml of 0.02 M Tris buffer (pH 8.0). Shake the tubes with 0.5 ml of toluene, and add 0.2 ml of either 0.1 M catechol or 0.1 M sodium protocatechuate. A bright yellow color appearing in a few minutes indicates *meta*-type cleavage. If no color appears, shake the tubes for 1 h at 30°C. Add solid $(NH_4)_2SO_4$ to saturation, adjust the pH to approximately 10 by adding 2 drops of 5 N ammonium hydroxide, and then add 1 drop of freshly prepared 25% (wt/vol) sodium nitroprusside (nitroferricyanide). Deep purple indicates *ortho*-type cleavage.

Examples: *meta*-type cleavage, *Delftia acidovorans* (*Pseudomonas acidovorans*); *ortho*-type cleavage, *Pseudomonas fluorescens*; negative, *Escherichia coli*.

15.2.8. Arylsulfatase Activity

Aseptically add a sufficient amount of a filter-sterilized 0.08 M solution of tripotassium phenolphthalein disulfate to a sterile liquid medium to give a final concentration of 0.001 M. Media containing methionine as the sole source of sulfur are best for the synthesis of arylsulfatase; sulfur sources such as sulfate, sulfite, thiosulfate, or cysteine may repress synthesis (20a, 170). Inoculate the medium, incubate it for 7 days, and add 1 N NaOH or 1 M Na_2CO_3, drop by drop. A positive test is indicated by faint pink to light red. An uninoculated control similarly treated should remain colorless.

Examples: positive, *Proteus rettgeri*; negative, *Chromobacterium violaceum*.

15.2.9. Autotrophic Growth on Hydrogen (159)

Prepare the medium (see section 15.4.5) in either large-volume bottles, which can be fitted with aluminium seals and a rubber stopper, or in suitable small conical flasks (100 ml), which can be placed in an anaerobic jar (see section 15.2.65). The gas phase may be mixed either externally (in the case of aluminum seal type bottles) or internally by evacuation and flushing with gas as described in section 15.2.65. The principal gases are H_2, O_2, CO_2, and N_2, and they may be mixed in their relative proportions as outlined in section 15.2.65 (see also Table 4). CAUTION: hydrogen and oxygen are flammable gases, and the hydrogen-oxygen mixtures should not be brought into contact with open flames or sparks. When using conical flasks in an anaerobic jar, the mouths of the flasks must be stoppered with a gas-permeable material, such as commercially available silicon stoppers. Other forms of stoppers may encourage the growth of bacteria and fungi under the damp conditions generated in the jar.

For the chemolithotrophic growth of aerobes, incubate the culture under an atmosphere of 10% (vol/vol) O_2, 10% CO_2, 60% H_2, and 20% N_2. For heterotrophic growth, supplement the mineral medium with an appropriate carbon source (0.2% carbohydrate or 0.1% organic acid). For growth on nitrogen-free medium, omit NH_4Cl and incubate the culture under an atmosphere of 2% (vol/vol) O_2, 10% CO_2, 10% H_2, and 78% N_2 or heterotrophically under 2% (vol/vol) O_2 and 98% N_2 (160).

Examples: positive, *Xanthobacter flavus*; negative, *Escherichia coli*.

15.2.10. Bacitracin Sensitivity

Use sterile commercially available differentiation (not sensitivity) disks (Difco Laboratories, Detroit, MI, and BD [BBL] Microbiology Systems, Cockeysville, MD) or sterile filter paper disks impregnated with 0.04 U of bacitracin (Sigma Chemical Co., St. Louis, MO). Place a disk on an inoculated blood agar plate (see section 15.4.9), and incubate for 24 h. A positive test is indicated by a zone of growth inhibition around the disk.

Examples: positive, *Streptococcus pyogenes*; negative, other beta-hemolytic streptococci.

15.2.11. Bile Solubility

Centrifuge a 24-h culture grown in 10 ml of Todd-Hewitt broth (see section 15.4.37), and discard the supernatant into a flask of disinfectant. Suspend the cells in 0.5 ml of 0.067 M phosphate buffer (pH 7.0) (see section 15.5.18). Add 0.5 ml of 10% sodium deoxycholate, and incubate the mixture at 37°C for 15 to 30 min. If the suspension becomes clear, the test is positive.

Examples: positive, *Streptococcus pneumoniae*; negative, other alpha-hemolytic streptococci.

15.2.12. Bile Tolerance

15.2.12.1. Method 1

Streak a plate of bile agar (section 15.4.8). Compare the growth on the plates with that occurring in the absence of the ox gall.

Examples: positive (10 and 40% bile), *Enterococcus faecalis*; positive (10 but not 40% bile), *Streptococcus salivarius*; negative (neither 10 nor 40% bile), *Streptococcus dysgalactiae*.

15.2.12.2. Method 2

Use method 2 for anaerobes (7). Inoculate PY broth (section 15.4.31) containing 2% ox gall and 1% glucose. Inoculate tubes with a Pasteur pipette while flushing them with oxygen-free carbon dioxide. Stopper the tubes, and incubate. Observe the growth response, and compare it with that in the absence of the ox gall.

Examples: growth, *Prevotella oralis*; no growth, *Prevotella melaninogenica*.

15.2.13. CAMP Test (Beta-Hemolysis Accentuation)

The CAMP test (the acronym derives from the originating authors' names) has been used mainly in the identification of *Streptococcus agalactiae* and of *Listeria monocytogenes* and *Listeria seeligeri*, but it might be useful for other bacteria as well.

Make a single streak of a strain of beta-hemolysin-producing *Staphylococcus aureus* (e.g., ATCC 25923) across the middle of a plate of blood agar (containing sheep erythrocytes that have been washed to remove natural antibodies to hemolysins). Make a single streak of each test organism perpendicular to but not quite touching the *Staphylococcus aureus* streak (about 3 to 4 mm away). For a positive control, make a similar streak of a reference strain of *Streptococcus agalactiae*, *Listeria monocytogenes*, or *Listeria seeligeri*, and for a negative control, make a streak of a reference strain of *Streptococcus pyogenes* or *Listeria innocua* (32). Incubate the plate aerobically overnight. A positive test is indicated by the appearance of a crescent- or arrowhead-shaped zone of enhanced hemolysis at the juncture of the test organism and the *Staphylococcus aureus* streak.

The test has also been used in the identification of *Listeria ivanovii*; in this case, however, a strain of *Rhodococcus equi* (e.g., NCTC 1621) is substituted for *Staphylococcus aureus*. *Listeria ivanovii* gives a positive reaction, whereas *Listeria monocytogenes* and *Listeria seeligeri* give a negative reaction (32).

15.2.14. Carbon Dioxide Requirement

Some aerobic, microaerophilic, or facultatively anaerobic chemoheterotrophs are characterized by a requirement for 3 to 15% CO_2 (vol/vol) in the gaseous atmosphere in the culture vessel. Such bacteria are called capnophilic bacteria. The CO_2 is required to initiate growth; after the organisms have begun to grow in a liquid or semisolid medium, they usually can provide themselves with CO_2 from their own metabolism. However, when growing in direct contact with the gaseous atmosphere, as on the surface of a solid medium, they may continue to require a supply of exogenous CO_2.

Some examples of capnophilic bacteria are as follows: *Capnocytophaga ochracea*, facultative anaerobe that requires 5% CO_2 (109); *Campylobacter jejuni* and other *Campylobacter* species, microaerophiles that require 3 to 5% CO_2 (241); *Neisseria gonorrhoeae*, aerobe that requires 3 to 10% CO_2, (283); *Brucella ovis* and some strains of *Brucella abortus*, aerobes that require 5 to 10% CO_2 (52) (note that *Brucella ovis* and *Brucella abortus* may be treated as biovars within the species *Brucella melitensis*, although the International Committee on Systematics of Prokaryotes subcommittee dealing with this genus is not in full agreement with this solution); some strains of *Streptococcus pneumoniae*, *Streptococcus mutans*, and *Streptococcus anginosus* (which includes "*Streptococcus milleri*"), facultative anaerobes that require 5% CO_2 (94).

There are some bacteria that do not have an absolute requirement for exogenous CO_2 but whose growth is nevertheless markedly stimulated by CO_2. For instance, *Neisseria meningitidis* grows much better when 5 to 8% CO_2 is provided (283). Similarly, many streptococci grow better in the presence of an atmosphere enriched with CO_2 (94).

15.2.15. Casein Hydrolysis

Combine sterile (autoclaved) skim milk at 50°C with an equal volume of double-strength nutrient agar (see section 15.4.28) or other carbohydrate-free agar medium at 50 to 55°C. Incubate streaked plates for up to 14 days, and look for clear zones surrounding the growth. Confirm by flooding the plates with 10% HCl. *Note*: acid production from the lactose in the milk may in rare instances inhibit the casein hydrolysis and necessitate prior dialysis of the skim milk.

Examples: positive, *Paenibacillus polymyxa* (*Bacillus polymyxa*); negative, *Paenibacillus macerans* (*Bacillus macerans*).

15.2.16. Catalase Activity

15.2.16.1. Method 1

Inoculate a nutrient agar slant (see section 15.4.28) or other medium lacking blood. After incubation, trickle 1 ml of 3% hydrogen peroxide down the slant. Examine immediately and after 5 min for the evolution of bubbles, which indicates a positive test. Alternatively, add a few drops of 3% peroxide to colonies on a plate or to the heavy growth that

may occur at or near the surfaces of semisolid media. For broth cultures, add 0.5 ml of 3% hydrogen peroxide to 0.5 ml of culture and observe for continuous bubbling. *Note*: media containing blood must not be used for catalase tests, because blood contains catalase activity unless the blood has been heated (see section 15.2.16.3). Also note that some bacteria (e.g., certain lactic acid organisms) make a nonheme "pseudocatalase" in media containing little or no glucose (293). Pseudocatalase production can be prevented by incorporating 1% glucose into the medium. When anaerobes are tested for catalase activity, it is important to expose the cultures to air for 30 min before adding peroxide (7).

Examples: positive, *Staphylococcus epidermidis*; negative, *Lactococcus lactis*.

15.2.16.2. Method 2

The following method eliminates problems of penetration that might occur when peroxide is added to colonies on a plate or to growth on a slant. Scrape the growth from a slant or plate with a nonmetallic instrument, and suspend it in a drop of 3% hydrogen peroxide on a slide. Examine immediately and at 5 min for bubbles, either macroscopically or with a low-power microscope (or hand lens). If the catalase activity is weak, a coverslip placed over the wet-mount preparation can help to capture the bubbles.

15.2.16.3. Method 3

Use the following method for certain bacteria that can make catalase only if grown on a heme-containing medium, e.g., certain lactic acid bacteria (293). To sterile blood agar base (section 15.4.35) containing 1% glucose to inhibit pseudocatalase formation, add 5% (vol/vol) of a 1:1 mixture of defibrinated blood (section 15.4.9) and sterile water. Heat the medium at 100°C for 15 min to inactivate blood catalase, cool to 45 to 50°C, and dispense into plates. Test growth on the plates directly with 3% hydrogen peroxide, or use method 2.

15.2.17. Cellular Pigmentation

Cells of many prokaryotes are strongly pigmented, with colors ranging from black, red, and yellow to green. Defining a simple absorption spectrum for the living cells may provide useful information on the nature of the pigments. Most dual-beam scanning spectrophotometers, which measure light in the UV to visible range, cover the wavelengths from about 200 to 900 nm. In some cases, namely, those of bacteriochlorophyll *b*-producing members of the anoxygenic phototrophic bacteria, a spectrophotometer which measures to about 1,200 nm is needed in order to record the peak that occurs at about 1,015 to 1,035 nm. A number of methods have been described for measuring the absorption spectra of cells. One of the problems which needs to be countered is the light-scattering effect of intact cells.

15.2.17.1. Method 1

Take a liquid culture, or make a suspension of the cells in a suitable medium. Filter the cell suspension through a glass microfiber filter (Whatman GF/C). Filter either water or the medium used for suspending the organisms through a second, "blank" filter. Either alter a cuvette (plastic disposable cuvettes are usually suitable) to take a strip of the filter, or cut the filters so that they fit inside of an intact cuvette. Place the blank filter in the reference beam of the spectrophotometer and the filter with the bacterial suspension in the measuring beam.

15.2.17.2. Method 2

Suspend a small volume of a liquid culture or colonies from a plate in a 50% solution of sucrose. Suspend the cells well in order to avoid aggregates. Use 50% sucrose as the reference blank.

15.2.17.3. Method 3

If an ultrasonic unit is available, this unit may be used to disrupt the cells. Be careful not to overheat the cells, which may have an adverse effect on some of the natural pigment protein complexes.

Various pigments absorb at characteristic wavelengths. Carotenoids typically absorb in the region of 400 to 600 nm (31, 230). Various bacteriochlorophylls (in intact cells) absorb over the range from 590 to 1,100 nm. Flexirubins are characterized by the fact that they undergo a reversible shift in the absorption spectrum at different pHs (230). At neutral pH, flexirubins are yellow, but at alkaline pHs they are blue. This method should not be confused with a test for carotenoid-like pigments first described by Molisch (174). In the presence of concentrated sulfuric acid, carotenoid pigments become blue. More information can be obtained by extracting the pigments and subjecting them to more-detailed analysis. Methods suitable for the extraction of pigments that are soluble in organic solvents are given below (see section 15.3.5).

15.2.18. Cellulase Activity

15.2.18.1. Method 1 (111)

The following method of preparing cellulose gives the best form of native cellulose for testing cellulolytic activity; for other methods, see references 15 and 16. Incorporate finely divided cellulose into appropriate carbohydrate-free agar media. To prepare the cellulose, wet grind 3% (wt/vol) Whatman no. 1 filter paper in a pebble mill as follows. Place 30 g of the paper (torn into small pieces) into a porcelain jar (ca. 4-liter capacity) with 1 liter of water; add enough flint pebbles (porcelain balls are not as satisfactory) that the liquid just covers them; roll the jar for 24 h at 74 rpm or until a very fine state of suspension has been achieved and the suspension has become viscous; stop before the viscosity decreases and copper-reducing substances appear. (To test for the latter, remove 1 ml of the suspension and test as described in section 15.2.42 with 1 ml of Benedict's reagent [section 15.5.1].) A positive test is indicated by the appearance of clear zones around the colonies on the cellulose agar. Long periods of incubation may be required.

Examples: positive, *Cellulomonas* species; negative, *Escherichia coli*.

15.2.18.2. Method 2

Method 2 is less sensitive than method 1. Place a strip of Whatman no. 1 filter paper in tubes of carbohydrate-free broth before autoclaving. A portion of the strip should extend above the level of the broth. A positive test is indicated by partial or complete disintegration of the paper strip during the growth of the culture.

15.2.19. Citrate Utilization

15.2.19.1. Method 1

Prepare a dilute suspension of the organisms in sterile water or saline. Make a single streak up a slant of Simmons' citrate agar (section 15.4.34). Incubate for up to 7 days. Blue indicates the utilization of citrate as the sole carbon source (i.e., a positive test).

Examples: positive, *Enterobacter aerogenes*; negative, *Escherichia coli*.

15.2.19.2. Method 2

From a dilute suspension, streak the entire surface of a slant of Christensen citrate agar (section 15.4.13). Incubate for up to 7 days. A positive test is indicated by red or magenta. *Note*: a positive reaction indicates that citrate is used but not necessarily as the sole carbon source; i.e., an organism could give a positive reaction on Christensen agar and a negative reaction on Simmons' citrate agar.

15.2.20. Coagulase Activity

Mix one loopful of growth from an agar slant, 0.1 ml of broth culture, or a single colony from an agar plate with 0.5 ml of undiluted rabbit plasma or plasma diluted 1:4 with saline. Incubate at 37°C, and examine at 4 and 24 h. A positive test is indicated by a solid clot or a loose clot suspended in the plasma. Granular or ropey formations are inconclusive.

Examples: positive, *Staphylococcus aureus*; negative, *Staphylococcus epidermidis*.

15.2.21. Coccoid Bodies

Use phase-contrast microscopy to examine cultures that are held static (not shaken during growth) for up to 4 weeks. Coccoid bodies may occur at as early as 2 or 3 days. A positive test is indicated by a predominance of round, refractile forms which have diameters greater than those of the original cells and which lack a thickened cell wall. Many of the forms have a discrete, dark peripheral region of cytoplasm in an otherwise empty-appearing cell. Coccoid bodies occur among certain spirilla, vibrios, and campylobacters.

Examples: positive, *Aquaspirillum itersonii*; negative, *Aquaspirillum serpens*.

15.2.22. Colonies

Measure the colony diameter in millimeters, describe the pigmentation, and describe the form, elevation, and margin as indicated in Fig. 1. Low-power microscopy (or a hand lens) may be necessary for observation of the margin. Also indicate whether the colonies are smooth (shiny, glistening surface), rough (dull, bumpy, granular, or matte surface), or mucoid (slimy or gummy appearance). Record the opacity of the colonies (transparent, translucent, or opaque) and their texture when tested with a needle: butyrous (butter-like texture), viscous (gummy), or dry (brittle or powdery). Describe the colonies from both young and old cultures. The medium, age of the culture, gaseous conditions, exposure to illumination, and other culture conditions may affect the colony characteristics.

15.2.23. Cysts and Microcysts

Examine cultures daily by phase-contrast microscopy. Initially rod-shaped cells become spherical or ovoid with thickened walls in older cultures. These forms may be optically dense, or they may be refractile. They do not have the heat resistance of endospores (section 15.2.76), except for the cyst-like forms made by certain *Nocardia* species, but they are extremely resistant to desiccation. These forms of *Azotobacter* strains are termed cysts. Those in the mycobacteria (order *Myxococcales*) are termed microcysts; here, the term cyst refers to the sporangium, if any, which contains the microcysts. These forms of *Nocardia* strains are termed microcysts or chlamydospores.

The following methods can be used to compare the levels of desiccation resistance of cyst-enriched cultures of an

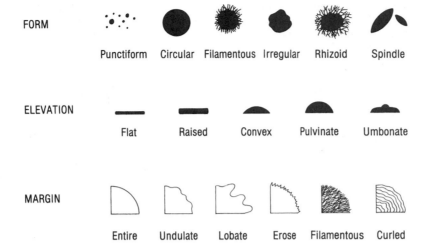

FIGURE 1 Diagram illustrating the various forms, elevations, and margins of bacterial colonies. (Adapted by permission of the McGraw-Hill Book Co. from reference 214.)

organism to those of cultures lacking cysts. See also chapter 2.3.8 for methods of staining cysts.

15.2.23.1. Method 1 (143)

Spread suspensions of the cells onto plates of media containing 2% agar, and incubate until the agar has dried completely to a thin film. Cut the agar film into pieces with sterile scissors, and transfer each piece aseptically to a 30-ml screw-cap vial half full of silica gel desiccant. Each vial should contain a sterile paper disk or funnel to separate the agar film from the desiccant. Store vials in the dark at room temperature. At various periods, remove a piece of agar film from a vial, place it upon fresh agar culture medium, and incubate for up to 7 days to see whether growth occurs.

15.2.23.2. Method 2 (226)

Prepare a suspension of the cells with sterile distilled water, and then prepare a series of 10-fold dilutions of this suspension. Transfer 100-μl portions of each dilution into sterile, 1.5-ml Eppendorf microcentrifuge tubes. Place the open tubes in a sterile petri dish, and incubate at 30°C until dry. Immediately add 100 μl of sterile distilled water to some of the tubes, then agitate vigorously to suspend the cells, and plate onto an appropriate agar medium to determine the initial colony count. After various storage periods, repeat the hydration and plating procedure with the remaining tubes to estimate the percentage of surviving cells.

15.2.24. Dye Tolerance

The ability to grow in the presence of low concentrations of various dyes added to culture media has been used to characterize and differentiate bacteria, especially those that have few other readily determinable phenotypic characteristics. For example, *Wolinella curva* can be differentiated from *Wolinella recta* by its ability to grow in the presence of Janus green, basic fuchsin, methyl orange, indulin scarlet, and safranin (266, 267).

Some dyes that have been used in dye tolerance tests for characterizing bacteria are listed in Table 1. In some instances, dye tolerance has been determined by soaking paper strips in a dye solution, placing the strips on inoculated agar plates, and noting whether growth occurs around the strips after incubation. For instance, taxonomic studies

of *Bordetella* species have used paper strips soaked in basic fuchsin (0.0005%), methyl violet (0.00025%), pyronin G (0.000125%), safranin O (0.000125%), and thionine (0.000125%) (117).

15.2.25. Esculin Hydrolysis

Grow cultures in broth or agar medium supplemented with 0.01% esculin and 0.05% ferric citrate. A positive test occurs when the medium becomes brownish black as a result of a reaction between 6,7-dihydroxycoumarin and Fe^{3+}.

Examples: positive, *Enterococcus faecalis*; negative, *Streptococcus mitis*.

15.2.26. Esterase Activity

See also "Lipase Activity" (section 15.2.47).

The hydrolysis of ester linkages between fatty acids and 4-methylumbelliferone results in the liberation of the latter compound, which is fluorescent when exposed to UV light

TABLE 1 Some dyes that have been used in dye tolerance tests for characterizing bacteria

Dye	% Concn (wt/vol)	Reference
Alizarin red	0.032	267
Azure II	0.0025	266
Basic fuchsin	0.005	266
	0.002	52
Brilliant green	0.001	62
Crystal violet	0.0005	266
	0.0001	62
Indulin scarlet	0.05	266
Janus green	0.01	266
Malachite green	0.001	62
Methyl orange	0.032	266
Methylene blue	0.1	184
	0.3	185
Safranin	0.05	266
Thionine	0.002	52

at 362 to 365 nm. Commercially available substrates include 4-methylumbelliferyl (4-MU) acetate, 4-MU propionate, 4-MU butyrate, 4-MU heptanoate, 4-MU nonanoate, 4-MU oleate, 4-MU elaidate, and 4-MU palmitate.

15.2.26.1. Method 1 (282)

A rapid spot test for the detection of butyrate esterase can be done as follows. Dissolve 100 mg of 4-MU butyrate (Sigma) in 10 ml of dimethyl sulfoxide (DMSO) and 100 μl of Triton X-100 detergent (Sigma). Dilute this stock solution 1:10 in citrate buffer (0.1 M citric acid, 0.05 M $Na_2HPO_4 \cdot 12H_2O$ [pH 5.01]). This diluted reagent can be stored in 250-μl portions at −70°C for at least 1 month. To test a culture, place a few drops of the reagent onto a piece of filter paper. Streak two or three colonies from a culture of the test organism onto the paper. After 30 s, expose the paper to a UV source (362 to 365 mn, or the long wavelength of a Wood's lamp). A positive test is indicated by light blue fluorescence. A negative test has no fluorescence.

Examples: positive, *Moraxella catarrhalis* (*Branhamella catarrhalis*), *Staphylococcus aureus*; negative, *Neisseria sicca*, *Neisseria lactamica*.

15.2.26.2. Method 2 (89)

The following quantitative method for esterases has been applied to mycobacteria but can be adapted for use with other bacteria. Prepare 0.04 M stock solutions of 4-MU acetate, 4-MU propionate, 4-MU butyrate, and 4-MU heptanoate in DMSO. Prepare 0.02 M stock solutions of 4-MU oleate, 4-MU elaidate, and 4-MU palmitate in DMSO. Store at −20°C until needed. Immediately before use, add 0.2 to 9.8 ml of 0.2 M phosphate buffer. The pH of the buffer depends on the esterase and the species. For mycobacterial esterases, the optimum pH is 8.0 to 8.5; however, spontaneous hydrolysis of the substrates is reported to occur at these pHs. Therefore, buffer with a pH of 7.3 has been used instead (89). Reaction mixtures contain 0.1 ml of a suspension of the bacteria to be tested, 0.9 ml of phosphate buffer, and 1.0 ml of the buffered 4-MU ester. (Also prepare a blank containing 0.1 ml of distilled water instead of bacteria.) Incubate the mixtures at 37°C for the desired period, and then stop the reaction by adding 1.0 ml of glycine buffer (0.2 M solution of glycine, adjusted to pH 10.5 with 1.0 N NaOH). Measure the fluorescence in quartz cuvettes at 37°C with a spectrofluorimeter using an excitation wavelength of 362 nm and an emission wavelength of 450 nm.

15.2.27. Flagella

Use flagellum staining with light microscopy (see chapter 2.3.10) or electron microscopy (see chapter 4.2.2 and 4.2.5.1). Agitate the bacteria as little as possible during their preparation, because flagella may be easily broken off the cells. Flagellar stains with light microscopy are occasionally misleading when applied to polar-flagellated bacteria, because a fascicle of several flagella sometimes seems to be a single flagellum (298). Culture conditions may be important; e.g., some bacteria, such as *Vibrio parahaemolyticus*, form a single polar flagellum when grown in liquid media but also form lateral flagella of shorter wavelengths when grown on solid media (235). Some bacteria, such as *Listeria monocytogenes*, may not form flagella at certain temperatures but do form them at lower temperatures (128). In some instances, glucose-containing media inhibit the formation of flagella (21).

15.2.28. Fluorescent Pigment

Make a single streak across a plate of King's medium B (see section 15.4.7). Examine the plates with the covers removed

at 24, 48, and 72 h under UV light (below 260 nm). A shortwave lamp used for the examination of mineral specimens is suitable. Plates should not be reincubated after examination, because the UV illumination may be bactericidal. A positive test is indicated by a fluorescent zone in the agar surrounding the growth. Fluorescent pseudomonads produce a yellow-green pigment; certain *Azotobacter*, *Azomonas*, and *Beijerinckia* species may form green or white fluorescent pigments.

Examples: positive, *Pseudomonas fluorescens*; negative, *Escherichia coli*.

Methanogenic members of the *Archaea* also fluoresce under UV light. The fluorescence is strong enough in actively growing cells to allow individual cells to be seen with an epifluorescence microscope. The cofactor F_{420} is one of the major components responsible for this autofluorescence. Freshly prepared wet mounts are best examined immediately under a suitable epifluorescence microscope using a 40× or 100× objective. Cells will fluoresce bright green when excited and viewed with an appropriate filter. Members of the methanogenic genus *Methanothrix* (*Methanosaeta*) do not produce strong fluorescence. In natural samples, algae may also fluoresce red while some natural minerals may also show up as bright fluorescent objects.

15.2.29. Formate-Fumarate Requirement

Use the formate-fumarate requirement test to distinguish certain bacteria that, when grown anaerobically, oxidize formate and concomitantly reduce fumarate to succinate. Prepare formate-fumarate stock solution (section 15.5.6), and add 1 drop of the solution per ml of prereduced PY broth (section 15.4.31) when inoculating the medium under oxygen-free carbon dioxide. Compare the growth response with that occurring in medium lacking the formate-fumarate. If the growth is greatly stimulated by the formate-fumarate, the test is positive. In some cases, growth may not occur at all unless the supplement is added.

Examples: positive, *Bacteroides ureolyticus* (this organism is related to members of the genus *Campylobacter* [280]); negative, *Prevotella oralis*.

15.2.30. Gas Production from Sugars

15.2.30.1. Method 1

Before autoclaving the sugar broth, place a small vial (ca. 10 by 75 mm; also known as a Durham tube) in an inverted position in the tube. After autoclaving, the vial will become completely filled with medium. For thermolabile sugars, add filter-sterilized stock solutions aseptically after the base medium has been autoclaved and cooled; allow the tubes to stand for a day or so to allow diffusion of the sugar into the inverted vials. Inoculate the medium and incubate. A positive test is indicated by the accumulation of gas within the vial. *Note:* if media are stored in a refrigerator before use and then are inoculated and incubated, dissolved gases in the media may be liberated and accumulate in the vials, giving false-positive reactions. Sterile controls should be used to detect this occurrence. Also, if the organism forms only small amounts of carbon dioxide, detection may be difficult because of the great solubility of carbon dioxide and because of rapid diffusion into the air. Methods 2 and 3 are designed to eliminate this difficulty.

15.2.30.2. Method 2

Method 2 is the same as method 1 but with the following change. After inoculation, add ca. 1 ml of sterile mineral (paraffin) oil to prevent the diffusion of carbon dioxide into the air. Alternatively, use sugar-containing agar medium cooled to 45°C after autoclaving (or, for thermolabile sugars, add a filter-sterilized stock solution of the sugar to the base agar medium which has been melted and cooled to 45°C). Inoculate the tubes, and allow them to solidify. Overlay with ca. 6 mm of sterile nonnutrient 2% agar (i.e., agar prepared only with water) and then a thin layer of sterile mineral oil. Incubate the cultures. A positive test is indicated by cracks in the medium and upward movement of the agar overlay.

15.2.30.3. Method 3 (7)

Use method 3 only for anaerobes. Cool melted prereduced PY agar (section 15.4.30) containing 1% glucose to 45°C, and inoculate it with 2 to 3 drops of culture. Cover the tubes with sterile aluminum foil (do not stopper them), allow the agar to solidify, and incubate the tubes. A positive test is indicated by the formation of gas bubbles in the medium or breaks in the agar.

15.2.31. Gas Vacuole Production

Some organisms produce gas vacuoles (vesicles), which appear as bright regions within the cells. First described by a number of authors in the same year (22, 130, 260), these vacuoles may be easily distinguished from other refractive bodies in the cell by a simple test. If cultures are left to stand in the cold (4°C), the gas-vacuolated cells will usually form a pellicle at the surface of a liquid medium.

15.2.31.1. Method 1 (131, 285)

Method 1 is for large volumes of culture or for demonstration purposes. Take a thick-walled glass bottle and fill it with the culture to be examined for the presence of gas vacuoles until it is nearly full. Insert a close-fitting rubber stopper into the mouth of the bottle. Compress the contents of the bottle by giving the rubber stopper a sharp tap with a light hammer. The object is to rapidly, but briefly, increase the gas pressure in the bottle and collapse the (gas-filled) vacuoles. If gas vacuoles are present, the turbidity of the culture will rapidly decrease due to a lessening of the refraction caused by the gas vacuoles. Compare a sample of the culture which has not been added to the bottle and a sample from the bottle as wet-mount preparations under the microscope. Untreated cells which contain gas vacuoles will have bright inclusions, whereas the inclusions will be absent from those cells which have been pressurized in the bottle.

Examples: gas vacuoles present, *Prosthecomicrobium pneumaticum*, *Ancylobacter aquaticus* (some strains).

15.2.31.2. Method 2

Method 2 is suitable for small volumes. Take a 1-ml luer-lock syringe and fill it with the strain to be tested. Seal the end of the syringe. This may be done by fitting a small luer-fitting stop value or using a syringe needle inserted into a small rubber stopper. Give the end of the syringe a sharp tap. Examine the untreated and treated cultures as wet-mount preparations under the microscope. Untreated cells which contain gas vacuoles will have bright inclusions, whereas the inclusions will be absent from those cells which have been pressurized in the syringe. .

15.2.32. Gelatin Hydrolysis

15.2.32.1. Method 1
Use broth supplemented with 12% gelatin.

Method 1A
Incubate at 20 to 22°C for up to 6 weeks. A positive test is indicated by liquefaction of at least a portion of the medium. Evaporation should be prevented during prolonged incubation.

Examples: positive, *Proteus vulgaris*; negative, *Escherichia coli*.

Method 1B
Incubate at the optimum growth temperature for the organism if greater than 22°C, and then chill in a refrigerator along with an uninoculated control. If the test culture fails to solidify, the organism has hydrolyzed the gelatin. If the test culture does solidify, remove the culture and the control and incubate in a tilted position at room temperature or in a 30°C incubator for ca. 30 min. If the culture liquefies before the control, a weakly positive reaction is indicated. However, if the test culture takes as long as the control to liquefy, this indicates a negative reaction.

Method 1C
Method 1C is the same as method 1B but with 4% gelatin. This is a more sensitive method than method 1B.

15.2.32.2. Method 2

Method 2 is more sensitive than method 1A or 1B. Streak plates of agar medium supplemented with 0.4% gelatin. Incubate them at the optimum temperature for the organism. Flood the plates with gelatin-precipitating reagent (15% $HgCl_2$ in 20% [vol/vol] concentrated HCl). A positive test is indicated by clear zones around colonies.

15.2.32.3. Method 3

Method 3 is sensitive and provides a more readily observable reaction than the methods described above. Culture the organism in a tube of broth containing a charcoal-gelatin disk (section 15.5.3). If the charcoal granules become dispersed in the medium, the test is positive.

15.2.33. Glycosidase Activity

15.2.33.1. Method 1

The following method depends on the use of chromogenic substrates—in particular, glycosides in which the sugar moiety is linked by a glycosidic bond to *o*-nitrophenol or *p*-nitrophenol. For example, the substrate for β-galactosidase is *o*-nitrophenyl-β-D-galactopyranoside (ONPG). Numerous other substrates are commercially available; some examples may be found in Table 2. For some enzymes, either *para*- or *ortho*-nitrophenyl substrates are available.

Dissolve 60 mg of the chromogenic substrate in 10 ml of 0.1 M sodium phosphate buffer (pH 7.0), and sterilize by filtration. Aseptically add 1 volume of this solution to 4 volumes of sterile peptone broth (0.1% peptone, 0.05% NaCl), and dispense half a portion of the mixture into small, sterile, capped tubes. The mixture can be stored for up to 1 month in a refrigerator or up to 6 months at −20°C. It must not show yellow coloration before use.

Culture the bacterium to be tested on an agar-solidified medium, preferably in the presence of the appropriate glycoside substrate (e.g., lactose for β-D-galactosidase and maltose for α-D-glucosidase), because some glycosidases are inducible. Remove some of the growth from the surface of the agar, and suspend the cells to a dense concentration in a tube of the peptone broth-chromogenic substrate. Incubate the tube in a water bath at 37°C, and examine it at 1 h and at intervals for up to 24 h. A control lacking the cell suspension should also be prepared and incubated. A positive test is indicated by yellow.

Note: merely because cells hydrolyze a chromogenic substrate, do not assume that the cells contain the glycosidase;

TABLE 2 Chromogenic substrates for glycosidase activity

Enzyme	Substrate
α-D-Acetylgalactosaminidase	o-Nitrophenyl N-acetyl-α-D-galactoasaminide
β-D-Acetylgalactosaminidase	p-Nitrophenyl N-acetyl-β-D-galactoasaminide
α-D-Acetylglucosaminidase	o-Nitrophenyl N-acetyl-α-D-glucosaminide
β-D-Acetylglucosaminidase	p-Nitrophenyl N-acetyl-β-D-glucosaminide
α-L-Arabinofuranosidase	p-Nitrophenyl α-L-arabinofuranoside
α-L-Arabinopyranosidase	p-Nitrophenyl α-L-arabinopyranoside
α-L-Fucosidase	p-Nitrophenyl α-L-fucopyranoside
β-D-Fucosidase	p-Nitrophenyl β-D-fucopyranoside
β-L-Fucosidase	p-Nitrophenyl β-L-fucopyranoside
α-D-Galactosidase	p-Nitrophenyl α-D-galactopyranoside
β-D-Galactosidase	o-Nitrophenyl β-D-galactopyranoside
β-D-Galacturonidase	p-Nitrophenyl β-D-galacturonide
α-D-Glucosidase	p-Nitrophenyl α-D-glucopyranoside
β-D-Glucosidase	p-Nitrophenyl β-D-glucopyranoside
α-D-Mannosidase	p-Nitrophenyl α-D-mannopyranoside
β-D-Mannosidase	p-Nitrophenyl β-D-mannopyranoside
α-D-Rhamnosidase	p-Nitrophenyl α-D-rhamnopyranoside
β-D-Xylosidase	p-Nitrophenyl β-D-xylopyranoside

the enzyme may instead be a phosphoglycosidase, which acts only on the phosphorylated form of the glycoside. For instance, *Staphylococcus aureus* cells produce o-nitrophenol when incubated with ONPG; however, they do this by transporting the ONPG into the cell by means of a membrane-bound phosphoenolpyruvate phosphotransferase system which phosphorylates the ONPG; the phospho-ONPG is then hydrolyzed by a cytoplasmic phospho-β-galactosidase, yielding galactose-6-phosphate and o-nitrophenol. Therefore, *Staphylococcus aureus* is ONPG positive but β-galactosidase negative. Also note that the fact that a cell suspension hydrolyzes a chromogenic substrate (e.g., ONPG) does not necessarily mean that the cells can ferment or grow on the natural substrate (e.g., lactose). This is because some bacterial strains are cryptic: they possess the hydrolytic enzyme but lack an effective transport system for the substrate. For example, many strains of *Salmonella enterica* serovar Arizonae can produce detectable amounts of yellow o-nitrophenol from ONPG but are unable to ferment or grow on lactose.

Examples (for β-D-galactosidase): positive, *Escherichia coli*; negative, *Proteus vulgaris*.

15.2.33.2. Method 2 (158)

Method 2 is a qualitative spot method that depends on the use of fluorogenic substrates—in particular, glycosides in which the sugar moiety is linked by a glycosidic bond to 4-methylumbelliferone. For example, the substrate for β-galactosidase is 4-MU–β-D-galactopyranoside. Numerous other substrates are commercially available; a few examples are found in Table 3.

Dissolve 15 μmol of each substrate (use 5 μmol for 4-MU–β-D-galactopyranoside) in 0.2 ml of DMSO. Bring the volume of each solution up to 10 ml with Dulbecco phosphate-buffered saline (PBS; 0.8% NaCl, 0.02% KCl, 0.115% Na_2HPO_4, 0.02% KH_2PO_4 [pH 7.3]). Remove a sample of the colony to be tested with a glass rod, and rub it vigorously onto a Whatman no. 1 qualitative filter paper. Then cover the bacterial smear with 2 drops (0.067 ml) of substrate. As controls, (i) prepare another bacterial smear nearby on the filter paper and cover it with 2 drops of the DMSO in which the substrate is dissolved and (ii) place 2 drops of the substrate nearby on the filter paper (no bacteria). Incubate for 30 min. Cover the spots on the filter paper with 2 drops of 0.1 N NaOH to increase the intensity of fluorescence. Examine in a dark room under UV light (362 to 365 nm, or the long wavelength of a Woods lamp). A positive test is indicated by light blue fluorescence. A negative test has no fluorescence.

Examples (for β-D-galactosidase): positive, *Escherichia coli*; negative, *Proteus vulgaris*.

TABLE 3 Fluorogenic substrates for glycosidase activity

Enzyme	Substrate
β-D-Glucosidase	4-MU-β-D-glucopyranoside
β-D-Galactosidase	4-MU-β-D-galactopyranoside
β-D-Glucuronidase	4-MU-β-D-glucuronide
N-Acetyl-β-D-glucosaminidase	4-MU-2-acetamide-2-deoxy-β-D-glucopyranoside
N-Acetyl-β-D-galactosaminidase	4-MU-2-acetamide-2-deoxy-β-D-galactopyranoside
α-D-Mannosidase	4-MU-α-D-mannopyranoside
α-L-Arabinosidase	4-MU-α-L-arabinopyranoside

15.2.33.3. Method 3

The following quantitative method for glycosidases has been applied to mycobacteria but can be adapted for use with other bacteria (89). Prepare 0.04 M stock solutions of 4-MU glycosides in DMSO. Store at −20°C until needed. Immediately before using, add 0.2 to 0.8 ml of 0.2 M phosphate buffer. The pH of the buffer depends on the glycosidase and the bacterial species; for mycobacterial glycosidases, optimal pHs occur within the range of 5.5 to 7.0. Reaction mixtures contain 0.1 ml of a suspension of the bacteria to be tested, 0.1 ml of the buffered 4-MU substrate, and 0.8 ml of 0.2 M phosphate buffer. (Also prepare a blank containing 0.1 ml of distilled water instead of bacteria.) Incubate the mixtures at 37°C for the desired period, and then stop the reaction by adding 1.0 ml of sodium glycinate buffer (0.2 M solution of glycine, adjusted to pH 10.5 with 1.0 N NaOH). Measure fluorescence in quartz cuvettes at 37°C with a spectrofluorimeter using an excitation wavelength of 362 nm and an emission wavelength of 450 nm.

15.2.34. Gram Reaction

See chapter 2.3.5 for two Gram staining methods. The basis of the Gram stain has also been investigated at the electron microscope level (30, 57). Wiegel (294) has drawn attention to the distinction between the Gram reaction and cell wall structure. Gram-positive organisms are traditionally associated with the presence of a thick peptidoglycan layer, and strains that are negative for the Gram stain traditionally have a thin peptidoglycan layer. However, this pattern is an oversimplification. Certainly, some strains that have an incomplete, thick peptidoglycan layer may stain as gram negative. Members of the *Planctomycetes* do not produce peptidoglycan and stain as gram negative, as do some members of the *Archaea*. In contrast, some gram-positive members of the *Archaea* also do not produce peptidoglycan but have a thick, extracellular polymer, which effectively prevents the staining complex from diffusing out of the cell (30). Members of the *Arthrobacter* group are notorious for staining as gram positive in young cultures but as gram negative once a culture ages. Electron microscopy of thin sections and chemical analysis of the cell wall are additional, supporting methods of characterization which have not lost their significance.

Alternatives to the classical Gram staining method include the lysis of gram-negative cells by the aminopeptidase reaction or by the action of KOH.

15.2.34.1. KOH Test

The KOH test described by Gregersen (91) is a simple method which may be used in addition to the Gram stain. The lysis of cells when the cells are mixed with 3% KOH is taken to indicate that the cell wall structure is of a type corresponding to gram-negative organisms.

Place 1 or 2 drops of a 3% KOH solution onto a clean glass slide. Pick a single colony or several small colonies with a loop, and mix them with the KOH solution on the slide for 5 to 10 s. Then slowly raise the loop from the suspension. If the cells have lysed, then the solution will be viscous, will stick to the loop, and will have the appearance of slime, which is slowly drawn out as threads as the loop is raised. Viewing this reaction against a dark background may help. Cells which lyse are recorded as "positive," but this result usually indicates that they will be negative for the Gram stain. The test should be carried out with young cultures. This test may be used in conjunction with the aminopeptidase test.

15.2.34.2. Aminopeptidase Test

The aminopeptidase test is a development of the observations of Cerny (45) that the production of an alanine peptidase which could hydrolyze L-alanine-4-nitroanilidehydrochloride was restricted to gram-negative organisms. This test is available as a commercial test kit (EMD Merck aminopeptidase). Make a suspension of the colonies in about 200 μl of sterile distilled water. Dip the test strip into the suspension, and incubate it for 10 to 30 min at 37°C. The formation of a yellow color by the release of 4-nitroanilidehydrochloride is taken to indicate a positive reaction. This method cannot be used for strongly pigmented strains.

15.2.35. Hemolysis

15.2.35.1. Method 1

Streak a plate of blood agar (section 15.4.9). Incubate it at 37°C for 24 to 48 h. If beta-hemolysis is occurring, a clear, colorless zone surrounds the colonies and the erythrocytes in the zone are completely lysed. If alpha-hemolysis is occurring, an indistinct zone of partial destruction of the erythrocytes surrounds the colonies, often accompanied by a greenish to brownish discoloration of the medium. If alpha-prime-hemolysis is occurring, a small halo of intact or partially lysed erythrocytes is present adjacent to the colony, with a zone of complete hemolysis extending farther out into the medium; the colony can be confused with a beta-hemolytic colony when examined only macroscopically. *Note*: some bacteria, such as certain staphylococci, produce beta-hemolysis only if the inoculated plates are first incubated for 48 h at 37°C and then chilled in a refrigerator for 1 h ("hot-cold hemolysis").

Sharper zones of hemolysis may be obtained by the following modification. Prepare a plate of blood agar base (without added blood), and allow it to solidify. Pour an overlay of 10 ml of blood agar onto the blood agar base, and allow it to solidify. Streak the surface of the blood agar overlay with the test culture.

15.2.35.2. Method 2

Streak a plate of blood agar base (without added blood). Pour an overlay of 10 ml of blood agar (at 45 to 50°C) onto the streaked plate. Hemolysis (alpha or beta) may be more pronounced than that with method 1 because the colonies are submerged, offering some protection for hemolytic enzymes (hemolysins) that may be oxygen labile (such as streptolysin O).

15.2.36. Hippurate Hydrolysis

15.2.36.1. Method 1

Prepare a broth medium supplemented with 1.0% sodium hippurate. Mark the level in each tube after autoclaving. Inoculate the medium, and incubate it along with a sterile control tube for 4 days. If evaporation occurs, restore the volumes in the tubes to their original levels with distilled water at the end of the incubation period. Place 1.0 ml of the sterile control broth in a small tube, and add 0.10-ml increments of ferric chloride solution (12 g of $FeCl_36H_2O$ in 100 ml of 2% HCl). Agitate the tube after each addition. At first a precipitate of iron hippurate will form, but upon further addition of ferric chloride this precipitate will dissolve, leaving a clear solution. Determine the smallest total amount of ferric chloride that will allow the initial precipitate to dissolve. Centrifuge the broth culture, add 1.0 ml of the supernatant to another small tube, and then add an amount of ferric chloride solution equivalent to that determined previously for the control. A positive test is indicated by the formation of a heavy permanent precipitate of iron benzoate.

Examples: positive, *Streptococcus agalactiae;* negative, *Streptococcus pyogenes.*

15.2.36.2. Method 2 (113)

Use the following method for a more rapid test. Prepare a 1% solution of sodium hippurate in distilled water, and dispense 0.4-ml portions into tubes. Store the corked tubes in a freezer at −20°C until needed. To test an organism, thaw a tube and mix a large loopful of the growth from the surface of a solid medium into the hippurate solution to form a very cloudy suspension. Incubate the tube in a heating block or water bath at 37°C for 2 h. After incubation, add ca. 0.2 ml of ninhydrin reagent (3.5 g of ninhydrin in 100 ml of a 1:1 [vol/vol] mixture of acetone and butanol). Do not shake the tube (76). Continue incubation at 37°C for 10 min. A positive test is indicated by the formation of a deep purple color as a result of the glycine, which is formed from hippurate hydrolysis. A negative test shows no color reaction or only a faint tinge of purple. Controls containing known positive and negative organisms should be included.

15.2.37. Hydrogen Sulfide Production from Thiosulfate

15.2.37.1. Method 1

Use a deep tube of a solid or semisolid agar medium supplemented with sodium thiosulfate (0.008 to 0.03%) and 0.025% ferric ammonium citrate or 0.015% ferrous sulfate. Sulfide production usually occurs best under anaerobic conditions; therefore, the medium should be inoculated by stabbing. A positive test is indicated by blackening of the medium during growth, as a result of the formation of iron sulfide (see also section 15.2.80).

Examples: positive, *Proteus vulgaris;* negative, *Escherichia coli.*

15.2.37.2. Method 2 (223)

The following method has been used with microaerophiles that do not produce H$_2$S anaerobically. Use triple sugar iron agar slants (section 15.4.38) that have been freshly prepared so that there is water of syneresis at the junction of the slant with the butt. (If the water of syneresis is not present, a small amount of sterile water can be added to the slant.) Inoculate the entire surface of the slant, and incubate it under microaerobic conditions (e.g., 6% O$_2$, 5% CO$_2$, 89% N$_2$) for up to 7 days. A positive test is indicated by blackening of the liquid at the slant-butt junction; a negative test shows no blackening.

Examples: positive, *Campylobacter coli;* negative, *Campylobacter jejuni.*

15.2.38. Hydrogen Sulfide Production from Cysteine

Cut filter paper into strips of 5 by 20 mm, and soak them in a solution of 5% lead acetate. Autoclave the strips in cotton-stoppered test tubes, and dry them in a drying oven. Inoculate a tube of liquid medium containing 0.05% cysteine. Before incubation, suspend a lead acetate strip in the tube so that the end of the strip is approximately 1.3 cm above the level of the medium; the top of the strip can be held by the cotton plug or other tube closure. The strip must not be allowed to come into contact with the medium.

Prepare uninoculated controls in a similar manner. A positive test is indicated by blackening (as a result of the formation of lead sulfide) of the lower portion of the strip after incubation for several days.

Examples: positive, *Aeromonas hydrophila;* negative, *Aeromonas caviae.*

15.2.39. Indole Production

15.2.39.1. Method 1

Inoculate a liquid medium that is free of carbohydrate, nitrate, and nitrite and contains a source of tryptophan (e.g., 1% tryptone broth or medium supplemented with 0.1% L-tryptophan). Test cultures of different ages by the following method. Add 1 ml of xylene (or water-saturated toluene), agitate the culture vigorously to extract the indole, and then allow the xylene to form a layer at the top of the broth. With the tube held in a slanted position, trickle 0.5 ml of Ehrlich's reagent (section 15.5.5) down the wall of the tube to form a layer between the xylene and the broth. Allow the tube to stand for 5 min. A positive test is indicated by the formation of a red ring just below the xylene layer. A tube of uninoculated medium should also be tested as a control. *Note:* if the medium in which the organisms are growing contains a carbohydrate, synthesis of the enzyme tryptophanase, responsible for indole formation, may be repressed (35). Also, nitrite at levels of 0.75 mg/ml or higher may block the detection of indole (244); therefore, the presence of nitrate in the medium may lead to a false-negative test result if the organism can reduce the nitrate to nitrite.

Examples: positive, *Escherichia coli;* negative, *Enterobacter aerogenes.*

15.2.39.2. Method 2

Method 2 is not as sensitive as method 1; however, method 2 eliminates the necessity for using flammable and toxic solvents. Use Kovács' reagent (section 15.5.7) rather than Ehrlich's reagent, and omit extraction with xylene. Add 0.5 ml of the reagent to the broth culture, and shake the tube gently. A positive test is indicated by red.

15.2.39.3. Method 3

Cut filter paper into strips, and soak them in a warm saturated solution of oxalic acid. Crystals of the acid will form on the strips upon cooling. Dry the strips thoroughly (sterilization by heat seems unnecessary), and insert them aseptically into the inoculated culture tube before incubating the culture. The strip should not come into contact with the medium but rather should be bent so that it presses against the wall of the tube and remains near the mouth when the cotton stopper or screw cap is replaced. If the test is positive, the paper strip becomes pink or red during the growth of the culture.

15.2.40. Indoxyl Acetate Hydrolysis (173)

Add 50 μl of a freshly prepared 10% (wt/vol) solution of indoxyl acetate in acetone to a blank paper disk (0.25-in. [ca. 0.64-cm]-diameter concentration disks [Difco]). Allow the disk to air dry. Store dried disks in a tightly capped amber or opaque bottle at 4°C in the presence of silica gel desiccant; the shelf life of the disks is at least 1 year.

Place a colony or a large loopful of growth from an agar medium onto a test disk, and immediately add 1 drop of sterile distilled water. Incubate the mixture for 5 to 10 min. A positive test is indicated by the development of a dark blue color. For a negative test, the blue is absent.

Examples: positive, *Campylobacter jejuni;* negative, *Campylobacter lari.*

15.2.41. Iron-Porphyrin Compounds (61)

The following test detects the presence of iron-porphyrin compounds such as cytochromes, catalase, and nitrate reductase. Dissolve 1.0 g of benzidine · 2HCl in 20 ml of glacial acetic acid, add 30 ml of distilled water, heat the solution gently, cool, and add 50 ml of 95% ethanol. Prepare a 5% solution of hydrogen peroxide by diluting a 30% stock solution. To perform the test, flood the bacterial growth on an agar plate with the benzidine solution and then immediately add an approximately equal volume of the 5% hydrogen peroxide solution. The benzidine solution must come into contact with all of the microbial growth before the hydrogen peroxide is added. A positive test is indicated by the prompt development of a blue-green to deep blue coloration.

Examples: positive, *Staphylococcus aureus;* negative, *Lactococcus lactis.*

CAUTION: benzidine is a carcinogen and is rapidly absorbed through the skin. Upon ingestion it may produce nausea, vomiting, and liver and kidney damage. Consequently, all precautions pertinent to the use, handling, and subsequent disposal of carcinogens should be observed for this compound.

15.2.42. 2-Ketogluconate from Glucose Oxidation

Grow the culture in a medium such as Haynes' broth (section 15.4.17) containing 4% potassium gluconate (sterilized by filtration and added aseptically). After various periods of incubation, remove 1-ml samples and add 1 ml of Benedict's reagent (section 15.5.1). Heat the mixture in a boiling-water bath for 10 min, and cool rapidly. A positive test is indicated by a yellowish brown precipitate of cuprous oxide.

Examples: positive, *Pseudomonas aeruginosa;* negative, *Escherichia coli.*

15.2.43. 3-Ketolactose from Lactose Oxidation (29)

Grow the organism initially on slants of medium containing 1% yeast extract, 2% glucose, 2% $CaCO_3$ (the $CaCO_3$ does not dissolve), and 1.5 to 2.0% agar. With a loop needle, scrape growth from the slant and deposit it as a small heap, approximately 5 mm in diameter, on a plate of lactose agar (1% lactose, 0.1% yeast extract, 2% agar). Four to six different strains may be tested on the same plate. After incubation for 1 to 2 days, flood the plate with a shallow layer of Benedict's reagent (section 15.5.1). A positive test is indicated by the formation of a yellow ring of cuprous oxide around the growth, reaching a maximum intensity in 2 h (approximately 2 to 3 cm).

Examples: positive, *Rhizobium radiobacter* (note that members of the species *Agrobacterium tumefaciens* and *Agrobacterium radiobacter* are considered to be a single species and have been transferred to the genus *Rhizobium*); all other bacteria so far studied are negative.

15.2.44. Lactic Acid Optical Rotation (7, 43)

To 10 ml of the culture, and also to 10 ml of uninoculated medium, add 0.2 ml of 50% (vol/vol) H_2SO_4. Use 1 ml of the acidified culture to determine the concentration of lactic acid (milliequivalents per 100 ml) by gas chromatography (see section 15.3.1). This concentration must be at least 1 meq/100 ml to determine optical rotation by the method described below.

See section 15.5.8 for the preparation of perchloric acid solution, buffer, lactic dehydrogenase solution, and NAD solution.

Prepare the L-(+)-lactic acid working standard as follows. Add 0.0296 g of L-(+)-lactic acid (calcium salt) to 5.0 ml of acidified uninoculated medium. Add 5.0 ml of distilled water. Add 1.0 ml of this solution to 1.0 ml of perchloric acid solution. Centrifuge, and save the supernatant, which is the working standard [equivalent to 1.0 pmol of L-(+)-lactic acid per 0.1 ml].

To 1.0 ml of acidified culture, add 1.0 ml of distilled water and 2.0 ml of perchloric acid solution. Centrifuge, and save the supernatant, which is the sample to be tested for optical rotation.

To 1.0 ml of acidified, uninoculated medium, add 1.0 ml of distilled water and 2.0 ml of perchloric acid solution. Centrifuge, and save the supernatant, which is the blank stock solution.

Use seven glass tubes, 12 by 75 mm. Add 0.1 ml of the blank stock solution to tube 1. Add 0.02, 0.04, 0.06, 0.08, and 0.10 ml of the working lactic acid standard to tubes 2 through 6, respectively. Make up the volume in each tube to 0.10 ml by adding 0.08, 0.06, 0.04, 0.02, and 0.00 ml of blank stock solution, respectively. To tube 7, add 0.10 ml of the sample to be tested. Add 3.0 ml of buffer to each of the seven tubes, followed by 0.03 ml of lactic dehydrogenase solution. Then add 0.1 ml of NAD solution, cover each tube with Parafilm, and invert gently several times to mix (do not shake the tubes). Incubate at 30°C for 60 min, being careful to start timing from the moment the NAD is added to the first tube.

Read the absorbance of the contents of each tube. Use a wavelength of 340 nm if possible. If the instrument is not capable of reading at this wavelength, use a wavelength as near to 340 nm as possible; for example, if a Spectronic 20 spectrophotometer (Thermo Electron Corporation) is used, set the wavelength to 366 nm. Use tube 1 as the blank to set the instrument to 0.00 absorbance. Read the absorbance values, being careful to use the same sequence as that used for the addition of the NAD. Use the same cuvette for all readings; pour out the contents after making a reading, rinse out the cuvette, and fill it with the next solution to be read.

Plot micromoles of L-(+)-lactic acid (0.0, 0.2, 0.4, 0.6, 0.8, and 1.0, μmol, respectively, for tubes 1 through 6) versus absorbance on linear graph paper. Draw the line of best fit. By comparison with this standard curve, determine the micromoles of L-(+)-lactic acid in the test sample (tube 7). Multiply this value by four to obtain the milliequivalent of L-(+)-lactic acid per 100 ml of culture. Calculate the percentage of L-(+)-lactic acid by using the following formula:

$$\frac{\text{meq of L-(+)-lactic acid (from enzyme reactions) per 100 ml}}{\text{meq of total lactic acid (from gas chromatography) per 100 ml}} \times 100$$

15.2.45. β-Lactamase Activity (200)

The following method uses the chromogenic cephalosporin substrate nitrocefin (obtainable from Oxoid or from BD [BBL]). Dissolve 10 mg of nitrocefin powder in 1 ml of DMSO. Dilute the solution 1:20 with 0.1 M phosphate buffer (pH 7.0) to a concentration of 500 μg/ml. The solution is yellow to light orange and is stable for many weeks at 4 to 10°C. For use, add some of the solution to the well of a microtiter plate or small test tube. Prepare a heavy suspension of the organism to be tested, and mix it with the nitrocefin solution. A positive test is indicated by the development of a red color, usually within 10 to 30 min, although incubation for as long as 6 h is needed in some instances.

15.2.46. Lecithinase Activity

Streak a plate of egg yolk agar (section 15.4.16) to obtain well-isolated colonies. A positive test is indicated by a cloudy (opaque) zone within the medium around the colony (see also section 15.2.47).

Examples: positive, *Bacillus cereus, Clostridium perfringens*; negative, *Bacillus subtilis, Clostridium butyricum.*

15.2.47. Lipase Activity

The lipase activity methods detect primarily the hydrolysis of ester-linked long-chain fatty acids. Although method 2 may indirectly detect the metabolism of the hydrolyzed fatty acids, their utilization is best checked by gas chromatography of the fatty acid methyl esters obtained after transmethylation of the medium which has been extracted by apolar organic solvents.

There are indications that some methods claiming to detect the degradation of fatty acids by esterases detect only the hydrolysis of the ester linkage, with the release of acidic fatty acid side chains. Many stains cited in the literature (see method 3) may be detecting only a lowering of the pH of the medium caused by the release of the free fatty acids.

15.2.47.1. Method 1 (7)

Streak a plate of egg yolk agar (section 15.4.16). After incubation, remove the petri dish cover and examine the growth closely under oblique illumination. A positive test is indicated by an oily, iridescent sheen or a pearly layer over the colony and on the surface of the surrounding agar.

Examples: positive, *Clostridium sporogenes*; negative, *Clostridium butyricum.*

15.2.47.2. Method 2

Use the following method for palmitic, stearic, and oleic acid esters (237). Use an agar medium supplemented with 0.01% $CaCl \cdot H_2O$. Sterilize Tween 40 (palmitic acid ester), Tween 60 (stearic acid ester), or Tween 80 (oleic acid ester) by autoclaving for 20 min at 121°C. Add sterile Tween to the molten agar medium at 45 to 50°C to give a final concentration of 1% (vol/vol). Shake the medium until the Tween is completely dissolved. Dispense into petri dishes, allow to solidify, and streak the plates. If the test is positive, an opaque halo forms around the colonies. Microscopically, the halo consists of crystals of calcium soaps (calcium salts of fatty acids). If the fatty acids are metabolized, then upon longer incubation the calcium salts of the fatty acids may disappear or the calcium salts will appear as a halo beneath the colonies.

Examples: positive (Tween 80), *Pseudomonas aeruginosa*; negative, *Pseudomonas putida.*

15.2.47.3. Method 3

Use the following method for fats of various types (15). Prepare fat of the desired type stained with Nile blue sulfate (section 15.5.12). Add 1 ml of a melted fat emulsion to 20 ml of an agar growth medium which has been melted and cooled to 45 to 50°C. Mix well, and dispense into petri dishes. Allow to solidify, and streak the plates. A positive test is indicated by the formation of a blue halo around the colonies.

15.2.48. Lysine Decarboxylase Activity

See the arginine dihydrolase method 1 (section 15.2.6). Substitute 1% L-lysine dihydrochloride (or 2% of the DL form) for the arginine.

Examples: positive, *Serratia marcescens*; negative, *Proteus vulgaris.*

15.2.49. Lysozyme Resistance (1, 167)

The following test has been used to differentiate various aerobic actinomycetes. Inoculate several bits of a colony of the isolate to be tested into a tube of basal medium without lysozyme (control) and a tube of basal medium containing lysozyme (section 15.4.23). Incubate for up to 7 days at room temperature. A positive test (i.e., resistance to lysozyme) is indicated by good growth in the presence and absence of lysozyme. A negative test (i.e., susceptibility to lysozyme) is indicated by good growth only in the medium lacking lysozyme.

Examples: positive, *Nocardia asteroides*; negative, *Streptomyces griseus.*

15.2.50. Malonate Utilization

Inoculate a tube of malonate broth (section 15.4.24) from a young agar slant or broth culture. Incubate it at 37°C for 48 h. A positive test is indicated by a change from green to deep blue.

Examples: positive, *Klebsiella pneumoniae*; negative, *Escherichia coli.*

15.2.51. Meat Digestion by Anaerobes (7)

Inoculate prereduced chopped-meat broth (section 15.4.12) while purging the tube with oxygen-free carbon dioxide. Incubate for up to 21 days. If the test is positive, meat particles disintegrate, leaving a fluffy powder at the bottom of the tube.

Examples: positive, *Clostridium sporogenes*; negative, *Clostridium butyricum.*

15.2.52. Methyl Red Reaction

Inoculate a tube of methyl red–Voges-Proskauer broth (section 15.4.27). Incubate at 30°C for 5 days (in many cases, incubation at 37°C for 48 h may also be satisfactory). Add 5 to 6 drops of methyl red solution (dissolve 0.1 g of methyl red in 300 ml of 95% ethanol, and make up to 500 ml with distilled water). A positive test is indicated by bright red, indicating a pH of 4.2 or less. A negative test gives a yellow or orange color, and a weakly positive test gives a red-orange color.

Examples: positive, *Escherichia coli*; negative, *Enterobacter aerogenes.*

15.2.53. Milk Reactions

15.2.53.1. Method 1

Inoculate a tube of litmus milk (section 15.4.22). Positive reaction results for cultures are indicated as follows: acid reaction, pink; alkaline reaction, blue; reduction of litmus, milk appears white due to discoloration of the litmus (usually occurs in the bottom portion of tube); acid curd, a firm pink clot which does not retract and is soluble in alkali; rennet curd, a soft curd which retracts and expresses a clear grayish fluid (whey) and is insoluble in alkali; peptonization (proteolysis), the milk becomes translucent as a result of hydrolysis of the casein, which often begins at the top of the medium, and the medium frequently becomes alkaline.

15.2.53.2. Method 2 (7)

Use the following method for anaerobes. Inoculate a tube of prereduced milk medium (section 15.34.25) while the tube is being purged with oxygen-free carbon dioxide. Incubate the tube for up to 3 weeks. Check for curd formation and/or peptonization.

15.2.54. Motility: Swimming

Prepare wet-mount samples from broth cultures that are young (e.g., still in the exponential phase) and have been grown at two temperatures (e.g., 37 and 20°C). In some instances, glucose-containing media may inhibit the formation of flagella (21). For methods of light microscopy observation of motility, see chapter 2.2.1.1. Examine wet-mount samples immediately after they are prepared. For anaerobes, the central portion of the mount sample is best for examination. For aerobes, areas near air bubbles or near the edge of the coverslip are best. Not all the cells in the mount sample need to be motile. For a bacterial strain to be considered motile, at least one cell should be seen to alter its location in relation to at least two other cells (15). *Note:* motility must be distinguished from Brownian motion (random "jiggling about" of cells) and from the passive motion of cells being carried about by currents under the coverslip. Anaerobes may also be tested by drawing up some of the broth culture into a capillary tube and immediately sealing both ends of the capillary in a narrow oxygen flame (53). Examine the capillary with a high-power dry objective. Some bacteria, such as the members of the orders *Myxococcales* and *Cytophagales*, lack flagella but exhibit "gliding motility" when in contact with a solid surface. For methods of detection and observation of gliding motility, see chapter 11.1.5.4 and 11.1.5.5.

15.2.55. Motility: Twitching (103)

Twitching motility is a type of motility in which cells on an agar surface exhibit small, intermittent jerks covering only short distances and often changing the direction of movement, which is not regularly related to the long axis of the cell. The cells move predominantly singly, although slower-moving aggregates occur. Sometimes a small number of cells lying together may be seen to make a jump, in any direction, in a body. The maximum speed of cells is about 1 to 5.4 µm/min. Twitching motility is independent of flagella but has been associated with the presence of pili. The phenomenon has been reported for relatively few bacteria, e.g., species of the genera *Neisseria*, *Moraxella*, *Acinetobacter*, *Kingella*, and *Eikenella*; these species do not exhibit either swimming or gliding motility. Most authors study twitching motility by microscopy of cultures growing on thin agar plates poor in nutrients. The use of a coverslip may prevent cell movement because of the adherence of the cells to the glass. The presence of a thin film of water at the agar surface, as with incubation in a humidified atmosphere, facilitates twitching motility.

When streaked in a line across a plate of agar, cultures of a twitching strain often exhibit spreading zones that vary in width, appearance, and distribution along the sides of the central streak of growth. The spreading may occur along the entire streak or as a varying number of fan-shaped zones that may be unevenly distributed.

Twitching motility also has been reported for at least one species of gliding bacteria, i.e., *Agitococcus lubricus*, which glides across agar surfaces and exhibits intense twitching motility when colonies are covered with a drop of water and a coverslip (78).

15.2.56. Nalidixic Acid Susceptibility (7)

Susceptibility or resistance to nalidixic acid is widely used for the differentiation of *Campylobacter* species. Streak a plate of an appropriate medium (e.g., *Brucella* agar), and on the portion of the plate with the heaviest inoculum place a sterile paper disk impregnated with 30 µg of nalidixic acid. Such disks can be obtained commercially (e.g., nalidixic acid Sensi-Disc [BD BBL]) in a convenient dispenser. After incubation, organisms showing zones of growth inhibition of any size around the disk are considered to be susceptible.

Examples: susceptible, *Campylobacter jejuni*; resistant, *Campylobacter fetus*.

15.2.57. Nitrate Reduction and Denitrification

See section 15.5.13 for the preparation of reagents and cautions about their carcinogenic properties.

15.2.57.1. Method 1 (190)

The following method is quantitative for nitrite production from nitrate. Inoculate a suitable broth medium supplemented with 0.1% KNO_3 and 0.17% agar (the latter to give a semisolid consistency and promote semianaerobic conditions). Gently mix the inoculum with the medium to distribute it throughout the tube. Examine the cultures periodically during incubation for gas bubble formation, which is presumptive evidence for denitrification. Control cultures grown without nitrate should not exhibit gas and should not give a positive test for nitrite with the procedure given below. To detect the ability of the organisms to reduce nitrate to nitrite, remove 0.1-ml samples from cultures of various ages and add 2 ml of test reagent (section 15.5.13). Add water to make the final volume up to 4 ml, and allow the color to develop for 15 min. Estimate the intensity of the purple-red color visually on a scale from 0 to 5. The intensity of the color can also be estimated by using a spectrophotometer at 540 nm. The accumulation of nitrite with the increasing age of the culture indicates the reduction of nitrate to nitrite, whereas the initial formation of nitrite followed by a decrease in its level indicates a further ability to reduce the nitrite. If no red color develops, add up to 5 mg of zinc powder per ml of the mixture. If the nitrate in the medium has not been reduced, the zinc will reduce it chemically to nitrite and a red color will appear. If no color appears, the organism has reduced all of the nitrite and denitrification has probably occurred. In this case an earlier examination of the culture may give a positive reaction for nitrite. *Note*: zinc powder may become inactive with long storage, and its ability to reduce nitrate to nitrite should be checked periodically with uninoculated medium.

Examples: denitrification, *Pseudomonas aeruginosa*; reduction of nitrate to nitrite, *Escherichia coli*; negative, *Bacillus sphaericus*.

15.2.57.2. Method 2

The following method is qualitative for nitrite. Grow the organisms as described for method 1. Examine the cultures periodically for gas bubbles, which are presumptive evidence for denitrification. To cultures of different ages, add 1 ml of solution A and 1 ml of solution B (section 15.5.13). Pink or red indicates the presence of nitrite. If no red color occurs, test the culture with zinc powder as described in method 1 to see whether nitrate is present. If the presence of neither nitrite nor nitrate can be demonstrated, denitrification has probably occurred. The controls described for method 1 also apply to method 2.

An alternative way to demonstrate gas production is to grow the organism in nitrate broth (no agar) containing a small, inverted vial (see section 15.2.30). This allows for the visible accumulation of gas during denitrification.

15.2.57.3. Method 3 (190)

The following method allows for the detection of nitrous oxide formed during denitrification and provides stronger evidence for denitrification than method 1 or 2. Inoculate tubes or flasks of semisolid nitrate medium. Seal the vessels with serum bottle stoppers. Inject acetylene gas, generated as described in section 15.3.11, through the stopper to give a final concentration of 1% (vol/vol). The acetylene will prevent N_2O formed during denitrification from being further reduced to N_2 (26). After the culture has grown, remove gas samples with a syringe and determine the presence of N_2O by gas chromatography with a Porapak Q (80- to 100-mesh) column (274 cm by 4.8 mm) at 35°C with helium as the carrier gas and a thermal conductivity detector.

15.2.58. Nitrite Reduction

See section 15.5.13 for the preparation of reagents and cautions about their carcinogenic properties.

Inoculate a suitable broth medium supplemented with 0.01 to 0.10% $NaNO_3$ or KNO_3 and 0.17% agar (the latter to give a semisolid consistency and promote semianaerobic conditions). Gently mix the inoculum with the medium to distribute it throughout the tube. To cultures of different ages, add 1 ml of solution A and 1 ml of solution B (section 15.5.13). A positive test is indicated by the lack of a pink or red color. A negative test is indicated by pink or red.

Comments: nitrite at a concentration of 0.1% is toxic to some bacteria, so it may be necessary to use 0.01% or an even lower concentration. For instance, *Neisseria gonorrhoeae* gives positive results with 0.01% nitrate but negative results with 0.1% nitrite (132).

The type of medium used may also affect the results. For instance, *Kingella kingae* and *Kingella indologenes* reduce nitrite when a rich basal medium such as Mueller-Hinton broth with yeast extract is used, but when less rich media are used the two species are negative for nitrite reduction (252).

15.2.59. Nucleic Acid Hydrolysis

15.2.59.1. Method 1

Use the following method to test for both DNase and RNase. Dissolve DNA or RNA (Sigma) in distilled water at a concentration sufficient to give 0.2% when added to an agar growth medium. DNA is readily soluble in water. To dissolve RNA in water, slowly add 1 N NaOH but do not allow the pH to become greater than 5.0. Add the nucleic acid solution to the agar medium just prior to autoclaving (121°C for 15 min), and pour into plates when the medium has cooled to 50°C. Make a single streak of the organism across the solidified medium. After incubation, flood the plate with 1 N HCl. In a positive test, a clear zone around the growth indicates DNase or RNase activity.

Examples: positive (DNase), *Serratia marcescens*; negative (DNase), *Enterobacter aerogenes*.

15.2.59.2. Method 2

Use the following method to test for DNase only. Add 100 mg of toluidine blue O (Sigma-Aldrich, St. Louis, MO) per liter of DNA-containing test medium (see method 1) prior to autoclaving (232). Alternatively, add methyl green solution (section 15.5.10) to give a final concentration of 0.005% (243). For the toluidine blue medium, make a single streak of the test organism across the plate; for the methyl green medium, streak the plate to obtain isolated colonies. For a positive test with the toluidine blue plates, a bright rose-pink zone surrounds the growth; for a positive test with the methyl green plates, the colonies are surrounded by a colorless zone on an otherwise green plate. These methods provide a more readily observable reaction for DNase than method 1. *Note*: the two dye methods

should not be used with bacteria grown under anaerobic conditions, because the indicator system may be nullified (217).

15.2.59.3. Method 3 (68)

Use the following method as a more rapid test for DNase, although it may not detect very weak DNase activity. Prepare a test medium by combining 31.5 mg of DNA (Difco), 5.0 ml of 0.1 M $CaCl_2$, and 5.0 ml of 0.1 M $MgCl_2$ in 100 ml of Tris buffer (pH 7.4). Make a heavy (milky) suspension of the organism in 0.5 ml of this medium, and incubate it for 2 h at 37°C. Add 1.6 ml of methyl green solution (section 15.5.10) to 100 ml of distilled water, and then add 2 drops of this solution to the bacterial suspension. Mix the contents of the tube, and incubate for an additional 2 h at 37°C. A positive reaction is indicated by a colorless tube. A negative test has a blue-green tube.

15.2.60. O/129 Antibacterial Susceptibility

Place one or two crystals of 2,4-diamino-6,7-diisopropyl-pteridine phosphate (Sigma-Aldrich) on an inoculated agar plate. Alternatively, soak sterile paper disks (6 to 8 mm in diameter) in a solution of the reagent (0.1 g in 100 ml of acetone), dry the disks, and apply them to an inoculated agar plate. A positive test is indicated by a zone of growth inhibition around the crystals or disks.

Examples: positive, *Vibrio parahaemolyticus*; negative, *Escherichia coli*.

15.2.61. Optochin Susceptibility

Place a sterile, commercially available (Difco, BD [BBL]) optochin differentiation disk or a sterile paper disk impregnated with 0.02 ml of a 0.025% solution of optochin (ethylhydrocupreine hydrochloride [Sigma-Aldrich]) on an inoculated agar plate. A positive test is indicated by a zone of growth inhibition around the disk.

Examples: positive, *Streptococcus pneumoniae*; negative, other alpha-hemolytic streptococci.

15.2.62. Ornithine Decarboxylase Activity

See method 1 for arginine dihydrolase (section 15.2.6). Substitute 1% L-ornithine dihydrochloride (or 2% of the DL form) for the arginine.

Examples: positive, *Enterobacter aerogenes*; negative, *Proteus vulgaris*.

15.2.63. Oxidase Activity (268)

Place ashless filter paper (Whatman no. 40, quantitative grade) into a petri dish, and wet it with 0.5 ml of a 1% solution of *N,N,N,N-tetramethyl-p-phenylenediamine* (not HCl salt; Sigma-Aldrich) in certified-grade DMSO. (CAUTION: tetramethyl-p-phenylenediamine is a respiratory system and skin irritant and should be handled in a fume hood.) Pick up one large isolated colony with a cotton-tipped swab, and allow the inoculum to dry on the swab for about 5 s. Then tap the swab lightly 10 times onto the wet filter paper. A positive test is indicated by the development of a blue-purple color on the tip of the swab within 15 s. The reagent solution is stable under refrigeration for at least 1 month.

Examples: positive, *Pseudomonas aeruginosa*; negative, *Escherichia coli*.

15.2.64. Oxidation-Fermentation Catabolism

A test carbohydrate (usually glucose) should be prepared as a stock solution, sterilized by filtration, and added aseptically to an autoclaved semisolid basal medium at 45 to 50°C to give a final concentration of 0.5 to 1.0%. The most widely used basal medium is that of Hugh and Leifson (see section 15.4.19). If the organism cannot grow in this medium because of complex nutritional requirements, the further addition of 0.1% yeast extract or 2% blood serum may permit growth. In cases in which acid production may be masked by the production of alkaline substances from the peptone, a medium such as that devised by Board and Holding (see section 15.4.10) may be suitable. In cases in which the bromothymol blue indicator present in the medium is inhibitory to the organism being tested, the substitution of another indicator such as phenol red or bromocresol purple may be necessary. For marine bacteria, the modified oxidation-fermentation medium of Leifson (see section 15.4.21) is useful.

To perform the test, place the tubes of the semisolid carbohydrate medium in a boiling-water bath for a few minutes to remove most of the dissolved oxygen; cool the tubes quickly, and inoculate them by stabbing with a straight needle. After inoculation, overlay the medium in only one of the tubes with 10 mm of sterile melted petrolatum; this tube is the anaerobic tube. No overlay is used for the second tube (the aerobic tube). Prepare the following controls to detect nonspecific reactions: (i) inoculate similar media lacking the carbohydrate, and (ii) incubate sterile media containing the carbohydrate.

Interpret the results as follows. (i) If only the aerobic tube is acidified, the organism catabolizes the carbohydrate by oxidation. Growth should be evident in the aerobic tube. (ii) If both the aerobic and the anaerobic tubes are acidified, the organism is capable of fermentation. Growth should be evident in both tubes. (iii) If neither tube becomes acidified, the organism is unable to catabolize the carbohydrate. If no growth is evident, the medium may be lacking some nutrient required for the organism being tested.

Examples: fermentation with glucose, *Staphylococcus epidermidis*; oxidation with glucose, *Micrococcus luteus*; negative reaction with glucose, *Paucimonas lemoignei* (*Pseudomonas lemoignei*).

15.2.65. Oxygen Relationships

One of the most fundamental phenotypic characteristics of a bacterium is its relationship to O_2. Although various categories have been devised to describe the oxygen relationships of bacteria, the following four major categories are useful. Chapter 10 and 14 also deal with this topic. A useful list of redox indicators (10.5.3; 14.6.1), as well as the redox potentials of a number of chemical reducing agents (10.5.2; 14.6.2), are given in these chapters.

15.2.65.1. Aerobes

Aerobes can use O_2 as an electron acceptor at the terminus of an electron transport chain, tolerate a level of O_2 equivalent to or higher than that present in an air atmosphere (21%), and have a strictly respiratory type of metabolism. Respiration is an energy-yielding process that generates a proton motive force for oxidative phosphorylation by means of an electron transport chain, for which an exogenous terminal electron acceptor is required. It differs from fermentation, which is an energy-yielding process that does not involve an electron transport chain and an exogenous terminal electron acceptor but instead relies on substrate-level phosphorylation and an endogenously generated electron acceptor (e.g., pyruvate from glycolysis, which can be reduced to lactate).

Note that some aerobes may be capable of growing anaerobically with electron acceptors other than O_2. For example, *Pseudomonas aeruginosa* is an aerobe even though it can grow anaerobically by respiring with nitrate. A survey of the literature suggests that all aerobes produce respiratory lipoquinones (as part of the respiratory chain). See section 15.3.5.1 for details on analyzing respiratory quinones. However, respiratory lipoquinones are also found in anaerobes.

15.2.65.2. Anaerobes

Anaerobes are incapable of O_2-dependent growth. Although many anaerobes take up O_2, they do not use it for respiration but instead make toxic derivatives from it, such as H_2O_2 or $O_2^•$ by auto-oxidation of oxygen-labile metabolites, by reactions catalyzed by various oxidases, or by other means, and they cannot grow in the presence of an oxygen concentration equivalent to that present in an air atmosphere (21% oxygen).

Some anaerobes, such as *Clostridium perfringens*, have a strictly fermentative type of metabolism. Other anaerobes may carry out anaerobic respiration in which an exogenous terminal electron acceptor other than O_2 is required. For example, *Desulfobacter postgatei* has a strictly respiratory type of metabolism in which sulfate or other oxidized sulfur compounds serve as terminal electron acceptors, being reduced to H_2S. *Desulfococcus multivorans* can respire in a similar manner with sulfate or other oxidized sulfur compounds, but in the absence of such electron acceptors it can grow by fermenting lactate or pyruvate.

15.2.65.3. Facultative Anaerobes

Facultative anaerobes grow both in the absence of oxygen and in the presence of a level of oxygen equivalent to that in an air atmosphere; they also grow under anaerobic conditions by fermentation, not by respiration.

Some facultative anaerobes, such as *Escherichia coli* and *Staphylococcus aureus*, grow aerobically by respiration with oxygen and grow anaerobically by fermentation. Others, such as *Streptococcus pyogenes* and *Lactobacillus plantarum*, have a strictly fermentative type of metabolism under both aerobic and anaerobic conditions.

15.2.65.4. Microaerophiles

Microaerophiles are capable of oxygen-dependent growth and cannot grow in the presence of a level of oxygen equivalent to that present in an air atmosphere (21% oxygen), although oxygen-dependent growth does occur at lower oxygen levels.

Although most microaerophiles are capable of respiring with O_2, *Borrelia* species seem to have a strictly fermentative type of metabolism. The reason they require a low level of O_2 for growth is not known.

Some microaerophiles are capable not only of respiring with O_2 but also of respiring anaerobically with electron acceptors other than O_2. For example, *Wolinella recta* is an H_2- or formate-requiring organism that was originally thought to be an anaerobe that respired with fumarate or nitrate as the terminal electron acceptor. It was later recognized as a microaerophile when the discovery was made that it can use O_2 as an electron acceptor if this molecule is present at very low concentrations (201). Similar results have been reported for other putative anaerobes, i.e., *Wolinella curva*, *Bacteroides ureolyticus* (this organism is related to members of the genus *Campylobacter* [280]), and *Campylobacter gracilis* (*Bacteroides gracilis*) (93).

Some microaerophiles, such as *Campylobacter jejuni* and *Spirillum volutans*, use only O_2 as a terminal electron acceptor and are not able to respire anaerobically.

A clue to the oxygen relationships of an isolate may be obtained by the following procedure. Melt the contents of a narrow (e.g., 16 by 150 mm) tube filled to three-fifths of its capacity with an appropriate agar medium. Cool the medium to 45°C, and inoculate it with the isolate. Gently mix the inoculum with the medium to distribute it uniformly, and then allow the agar to solidify. Upon incubation, the following growth responses may occur.

1. *Growth occurs only at the surface of the medium.* Growth only on the surface suggests that the organism is aerobic. (However, a fermentable substrate should be present in the medium in case the organism is a facultative anaerobe that not only respires with O_2 but also grows anaerobically through a fermentative type of metabolism.)

2. *Growth occurs only in the bottom regions of the tube and not near the surface.* Growth only on the bottom suggests that the organism is anaerobic. (However, anaerobes that are extremely intolerant of O_2 may not be able to grow even in the bottom-most portion of the tube because of the presence of small amounts of dissolved O_2.)

3. *Growth occurs throughout the tube.* Growth throughout the tube suggests that the organism is a facultative anaerobe. (The medium should contain a fermentable substrate, but no terminal electron acceptors other than O_2 should be present, since some aerobes can respire anaerobically with terminal electron acceptors other than O_2.)

4. *Growth occurs only in the form of a disk a short distance below the surface.* Growth in a small area a short distance below the surface suggests that the organism is a microaerophile.

For motile bacteria, a semisolid medium (0.2% agar) may be preferable to a solid medium because it can be inoculated by stabbing with an inoculating needle after the agar has gelled. This method makes it unnecessary to mix the medium with the inoculum and thus avoids the incorporation of some dissolved O_2 into the medium. Motile organisms usually exhibit negative or positive aerotaxis, which results in their translocation to a zone in the tube where the oxygen level is optimal for growth.

More quantitative information about the oxygen relationships of an isolate can be obtained by incubating cultures (preferably inoculated onto the surface of agar media, where they are most susceptible to O_2 in the gas phase) under a series of atmospheres containing various O_2 levels. The most convenient way to do this is to use sealable vessels, which can be connected by a manifold to a vacuum pump and various gas cylinders. Use polycarbonate jars such as those that are used routinely for the cultivation of anaerobes, provided they have a vent to allow evacuation and refilling with gases. Alternatively, adapt ordinary aluminum pressure cookers, obtainable at a department store, by removing the rubber gasket from the lid, cleaning all traces of grease from the gasket and the lid, and then gluing the gasket back into place with a silicone rubber adhesive. After the glue has cured thoroughly (which may take several days), the vessel should be connected to a vacuum pump to confirm that it can maintain a vacuum without leaking.

To obtain a particular oxygen level, merely evacuate the appropriate amount of air from the jar and replace it with N_2. This is accomplished most conveniently by measuring the decrease in pressure in the vessel by means of a vacuum gauge or mercury manometer. For instance, suppose that, at a particular geographical location, the uncompensated barometric pressure is 29 in. of mercury (737 mm Hg). To obtain an atmosphere of 7% O_2, i.e., 7/21 of the 21% O_2 in an air atmosphere, pump air out of the jar until the manometer or gauge indicates a vacuum of 19.3 in. (14/21 of 29 in.) or 491 mm Hg (14/21 of 737 mm Hg). Then add N_2 to the vessel until the manometer or gauge reads 0 in. of vacuum (i.e., atmospheric pressure), and seal the jar. Further examples are as presented in Table 4.

When testing an isolate that requires CO_2, an appropriate amount of CO_2 should be added from a tank of pure, compressed CO_2 before the final filling with N_2. When testing an isolate that requires H_2 as an electron donor, such as *Campylobacter mucosalis*, H_2 can be added instead of N_2 during the final filling of the culture vessel. (CAUTION: gas mixtures containing O_2 and H_2 are highly explosive. Do not use H_2 near open flames or any source of sparks, and when evacuating vessels and adding H_2, enclose the vessels in wire netting or a metal mesh cage. Wear safety goggles when using H_2. When incubating culture vessels, make sure that the incubators have thermoregulators that are isolated from the incubation chamber since some thermoregulators can generate sparks. When opening the culture vessels after incubation, first make sure that there are no flames or sparks in the area that could ignite the escaping gas mixture.)

15.2.66. Peptidase Activity

The test for peptidase activity depends on the use of amino acid-naphthylamides, in which an amino acid moiety is linked by a peptide bond to β-naphthylamine. For example, the substrate for glycine aminopeptidase is glycine-β-naphthylamide. Quantitative estimation of enzyme activity is made by using standardized bacterial suspensions or

TABLE 4 Examples of air evacuation requirements to obtain various oxygen levels

O_2 level desired (%)	Required vacuum gauge reading (in.) (mm)
0^a	
1	7.6 (702)
3	24.9 (632)
5	22.1 (562)
7	19.3 (491)
9	16.6 (421)
12	12.4 (316)
15	8.3 (211)
18	4.1 (105)
21	No evacuation of vessel

[a]To achieve strictly anaerobic conditions, merely evacuating the vessel and refilling it with N_2 (even if this process is repeated several times) may not suffice. This is because culture media usually contain dissolved O_2, especially after they have been stored for several hours. This dissolved O_2 bleeds off slowly into the gas phase within the culture vessel; meanwhile, some bacteria that are extremely efficient at scavenging O_2 may be able to respire with it and grow, or some bacteria that are highly susceptible to O_2 may be killed or inhibited. One way to minimize dissolved O_2 in agar media is to use only media that are freshly prepared and that are inoculated immediately after they have solidified. An alternative method is to add to the vessel a palladium catalyst (of the type used in commercial anaerobic jar systems), evacuate the jar, and refill it with N_2 several times and finally with H_2; the H_2, with the catalyst, will then remove any residual O_2. Even this method will not work with anaerobes that are extremely intolerant of O_2, and more stringent methods, such as using prereduced media, may be necessary (see chapter 14.6.4).

cell extracts and measuring the rate of formation of β-naphthylamine by using a fluorimeter (287, 291). For qualitative purposes, the following methods may be useful. However, as noted by Westley et al. (291), the culture medium and the age of the cells, unless carefully standardized for a particular group of organisms, can influence the reproducibility of the results.

15.2.66.1. Method 1

The following method was originally used to differentiate mutants of *Salmonella enterica* serovar Typhimurium that were unable to hydrolyze L-alanine β-naphthylamide (171). Stain colonies grown on agar media by pouring over them a mixture containing 5 ml of 0.2 M Tris-HCl buffer (pH 7.5), 0.2 ml of a solution of amino acid-naphthylamide substrate (10 mg/ml in *N,N*-dimethylformamide), and 10 mg of Fast Garnet GBC (Sigma-Aldrich). Colonies hydrolyzing the amino acid-naphthylamide to β-naphthylamine develop a dark red color, usually in 1 to 2 min.

15.2.66.2. Method 2 (56)

Dissolve amino acid-naphthylamide substrates in 0.1 M Tris-phosphate buffer (pH 8.0 for most substrates, pH 7.2 for hydroxyproline–β-naphthylamide, and pH 7.6 for γ-glutamyl–β-naphthylamide) to a concentration of 10 to 3 M. Solubilize the crystalline solids initially in a small volume of 50% ethanol or *N,N*-dimethylformamide, and then add the appropriate amount of buffer to achieve the final concentration. Dispense 40-μl volumes of the substrate solutions into small wells or microtubes. Dry the wells or microtubes under a vacuum at a temperature below 50°C, and store at 4°C until use.

Suspend cells from a culture on an agar medium in physiological saline (0.85% NaCl) to a turbidity equivalent to a McFarland standard of 3 (section 15.5.9), and add 20 to 40 μl of cell suspension to each well or microtube. Incubate for 1 to 4 h in a humid chamber, and then add 20 to 40 μl of detector reagent A and 20 μl of detector reagent B (section 15.5.14). Controls are prepared in the same manner as the test wells but without any substrate. Incubate for 15 min. A positive test is indicated by orange. Negative tests show no color change compared with the controls. (To eliminate any residual yellow in negative reaction mixtures after reagent B is added, expose the wells or microtubes to direct sunlight for 2 to 4 min.)

Examples (with γ-glutamic acid–β-naphthylamide as the substrate): positive, *Neisseria meningitidis*, *Moraxella urethralis*; negative, *Neisseria gonorrhoeae*, *Neisseria sieca*, *Moraxella lacunata*.

Some substrates that are available commercially from biochemical supply houses are listed in Table 5.

CAUTION: the β-naphthylamine that results from the hydrolysis of these substrates is a carcinogen that causes bladder tumors. Consequently, appropriate precautions should be taken during the use, handling, and subsequent disposal of all materials containing this substance.

15.2.67. pH Optima

Prokaryotes are found in a wide range of habitats of different pHs. While using suitable buffers to determine the pH range for growth and the optimum pH may sound like an easy task, determining the maximum or optimum pH at values above about 8.0 is not a simple task. Sorokin (253) has recently drawn attention to this problem. While NaOH-, NaCO₃-, and NaHCO₃-based systems are particularly susceptible to pH changes due to the effects of atmospheric CO_2 entering

TABLE 5 Some substrates available commercially from biochemical supply houses[a]

L-Alanine β-naphthylamide
L-Arginine β-naphthylamide
L-Asparagine β-naphthylamide
L-Aspartic acid α-(β-naphthylamide)
L-Cystine di-β-naphthylamide
L-Glutamic acid γ-(α-naphthylamide)
L-Glutamic acid γ-(β-naphthylamide)
Glycine β-naphthylamide
L-Histidine β-naphthylamide
trans-4-Hydroxy-L-proline β-naphthylamide
L-Isoleucine β-naphthylamide
L-Leucine β-naphthylamide
L-Leucine 4-methoxy-β-naphthylamide
L-Lysine 3-naphthylamide
D-Methionine β-naphthylamide
L-Phenylalanine β-naphthylamide
L-Proline β-naphthylamide
L-Pyroglutamic β-naphthylamide
L-Serine β-naphthylamide
L-Tryptophan β-naphthylamide
L-Tyrosine β-naphthylamide
L-Valine β-naphthylamide

[a]CAUTION: the β-naphthylamine that results from the hydrolysis of these substrates is a carcinogen that causes bladder tumors. Consequently, all precautions pertinent to the use, handling, and subsequent disposal of all materials containing it should be observed.

the system and lowering the pH, these effects may be enhanced due to the production of CO_2 by the organisms themselves. In addition, the pH may be altered by the production of acids from sugars (if included in the medium) or by the action of deaminases or decarboxylases on amino acids (present in yeast extract or other proteinaceous digests). Such effects are to be found not only in NaOH-, NaCO₃-, or NaHCO₃-buffered systems but also in systems in which biochemical buffers, such as CAPSO [3-(cyclohexylamino)-1-hydroxypropane sulfonic acid], CHES [2-(*N*-cyclohexylamino) ethane sulfonic acid], HEPES, MES (morpholineethane sulfonic acid), and PIPES [piperazine-*N,N*′-bis (2-ethane sulfonic acid)], etc., are used. These effects are not trivial, and pHs should be determined both at the beginning of the incubation period and at the end. Chapters 10.5.4 and 14.1 give an excellent overview of buffers. Additional information may also be found in Dawson et al. (58).

15.2.68. Phenylalanine Deaminase Activity

Inoculate heavily a phenylalanine agar slant (section 15.4.29). After incubation, allow 4 to 5 drops of 10% ferric chloride solution to run down the slant. A positive test is indicated by green, which is due to the formation of phenylpyruvic acid.

Examples: positive, *Proteus vulgaris*; negative, *Escherichia coli*.

15.2.69. Phosphatase Activity

15.2.69.1. Method 1

Use the following method to test for both acidic and alkaline phosphatases. To an agar growth medium such as nutrient agar (section 15.4.28), melted and cooled to 45 to 50°C, aseptically add a sufficient amount of a filter-sterilized 1% solution of the sodium salt of phenolphthalein diphosphate (Sigma-Aldrich) to give a final concentration of 0.01%. Incubate streaked plates for 2 to 5 days. Place 1 drop of ammonia solution (specific gravity, 0.880 [28 to 30%1]) in the lid of the inverted plate, and replace the culture over it to allow the ammonia fumes to reach the colonies. If the test is positive, colonies become red, as a result of the presence of free phenolphthalein.

Examples: positive, *Staphylococcus aureus*; negative, *Micrococcus luteus*.

15.2.69.2. Method 2 (38)

Use method 2 to test only for acid phosphatase. Prepare a very dense suspension of cells in saline (0.85% NaCl). Add 0.3 ml of the suspension to 0.3 ml of a substrate solution consisting of 0.01 M citrate buffer (pH 4.8) containing 0.01 M disodium *p*-nitrophenyl phosphate (Sigma-Aldrich). Incubate the mixture for up to 6 h at 37°C. Then add 0.3 ml of 0.04 M glycine buffer, which has been adjusted to pH 10.5 with NaOH. Also use a control lacking the bacteria. A positive test is indicated by a yellow color due to acid phosphatase activity.

15.2.69.3. Method 3

Use method 3 to test for only alkaline phosphatase. This method is similar to method 2, but the *p*-nitrophenyl phosphate is dissolved in the glycine buffer instead of the citrate buffer. No further addition of glycine buffer is necessary to develop the yellow color.

15.2.70. PHB Presence (149, 205, 259)

Grow a culture of the organism under conditions conducive to the formation of poly-β-hydroxybutyrate (PHB). For aerobic bacteria, these conditions usually are nitrogen and also oxygen limitations of exponential growth in the presence of excess carbon and an energy source. PHB occurs as refractile intracellular granules.

15.2.70.1. Staining Procedures

Staining may be used as presumptive evidence for PHB granules; however, chemical analysis is required for a definitive test. The most specific staining procedure is performed as follows. Prepare and filter a 1% aqueous solution of Nile Blue A (Sigma-Aldrich) just before use. Stain a heat-fixed smear of the bacteria with this solution for 10 min at 55°C in a Coplin staining jar. Wash the slide briefly with tap water to remove excess stain, and place it in 8% aqueous acetic acid solution for 1 min. Wash with tap water, and blot dry. Remoisten with tap water, and place a coverslip over the smear. Place immersion oil on the coverslip, and examine the preparation under an epifluorescence microscope by using an exciter filter that provides an excitation wavelength of approximately 460 nm. If the test is positive, PHB granules exhibit a bright orange fluorescence. Other cell inclusion bodies such as polyphosphate do not fluoresce in this manner.

15.2.70.2. Chemical Analysis

Place 1 ml of the culture into a conical glass centrifuge tube, add 9 ml of alkaline hypochlorite reagent (see section 15.5.20), and incubate at room temperature for 24 h (259). Alternatively, add 8 ml of commercial 5% hypochlorite (Clorox bleach) to 2 ml of culture and incubate (202). The hypochlorite will digest most cell components but will not digest the PHB granules. Centrifuge the mixture to collect the granules, and discard the clear supernatant fluid. Wash the sediment by suspending it in 10 ml of distilled water, centrifuging, and discarding the supernatant fluid. Wash the sediment once more with water, twice with 10-ml portions of acetone, and finally twice with 10-ml portions of diethyl ether. Dry the sediment, and add 2 ml of concentrated sulfuric acid. Place the tube in a boiling-water bath for 10 min. Cool it to room temperature. Using a UV spectrophotometer and far-UV quartz cuvettes, determine the absorption spectrum of the sample at 5-nm increments from 215 to 255 nm against a blank of plain concentrated sulfuric acid. (The sample may have to be diluted 1:10 or more with sulfuric acid if its A_{235} is greater than 1.0.) A positive test is indicated by the occurrence of an absorption peak at 235 nm as a result of crotonic acid formed from PHB by the action of the sulfuric acid.

Examples: positive, *Paucimonas lemoignei* (*Pseudomonas lemoignei*); negative, *Pseudomonas aeruginosa*.

15.2.71. PHB Hydrolysis (258)

The hydrolysis of PHB is very useful for distinguishing among the members of the genus *Pseudomonas*; however, the PHB granule suspension required for the test cannot be purchased commercially and is tedious to prepare. Details of its preparation are given in section 15.5.17.

Prepare a plate of mineral agar, such as that described in section 15.3.2, lacking carbon sources. To another portion of the medium that has been melted and cooled to 45 to 50°C, add a sufficient amount of the sterile concentrated PHB granule suspension to give a final concentration of 0.25% (wt/vol) in the medium. Pour the granule-agar suspension over the surface of the plate of solidified medium to form a thin overlay.

After the overlay solidifies, inoculate a washed suspension of the organism to be tested onto a small area of the plate. (Several strains can be tested on a single plate.) Incubate the plate for several days. A positive test is indicated by the growth of the organism at the site of inoculation and the clearing of the surrounding medium as a result of the depolymerization of the PHB granules.

Examples: positive, *Paucimonas lemoignei* (*Pseudomonas lemoignei*); negative, *Delftia acidovorans* (*Pseudomonas acidovorans*).

15.2.72. Proteinase Activity

Various chromogenic substrates for detection and quantitative assays of proteinases are available commercially from biochemical supply houses. These substrates consist of a dye that is chemically conjugated to a particular protein; hydrolysis of the protein results in liberation of the dye. Examples of substrates are azoalbumin, azocasein, and Azocoll (a dye-impregnated cowhide powder). Assay methods involve incubating a solution of the substrate (or a suspension, in the case of insoluble substrates) with the supernatant fluid from a bacterial culture (most bacterial proteinases of interest are exocellular), removing the residual unhydrolyzed substrate, and measuring the amount of soluble dye that has been liberated.

A typical protocol with azoalbumin is as follows. To a test tube, add 0.3 ml of the test culture supernatant, 1.7 ml of distilled water, and then 0.1 ml of azoalbumin solution (5 mg/ml in 0.1 M Tris buffer [pH 7.5]). Also prepare a blank containing 2.0 ml of distilled water and 1.0 ml of azoalbumin solution. Incubate the test tubes for 1 h in a water bath at 30°C. Then add 2.0 ml of 8% trichloroacetic acid to stop the reaction and to precipitate the unhydrolyzed azoalbumin. Pour the mixture into a polypropylene centrifuge tube, and centrifuge at 10,000 × g or at a speed sufficient to sediment the precipitated protein. Using a pipette and a suction bulb, carefully transfer 2.0 ml of the clear supernatant fluid to a tube. Add 2.0 ml of 0.5 M NaOH to intensify the color. Read the A_{400} of the solution. Although the units of enzyme activity are arbitrary, it is often convenient to let each 0.001 increase in absorbance beyond the reading of the blank equal 1 U of activity.

Azocoll, an insoluble substrate, is used in the form of a suspension (e.g., 50 mg suspended in 5 ml of 100 mM phosphate buffer [pH 7.0; EMD Biosciences catalogue no. 194932 or 194933]). Instead of centrifuging the reaction mixture after incubation, remove the residual unhydrolyzed protein by filtration and measure the A_{520}.

See also sections 15.2.15, 15.2.32, 15.2.51, and 15.2.53.

15.2.73. Pyocyanin and Other Phenazine Pigments

Inoculate a slant of King's medium A (section 15.4.1). Examine the slant at 24, 48, and 72 h. A positive test is indicated by pigments that diffuse into the agar and stain it. In some cases, the pigments are only slightly water soluble and precipitate as crystals in the medium.

Examples: pyocyanin, stains the medium blue (produced by *Pseudomonas aeruginosa*); phenazine-L-carboxylic acid, stains orange-yellow and may crystallize in the colonies and the surrounding medium (produced by *Pseudomonas aureofaciens*—note that members of this species are now considered to be members of the species *Pseudomonas chlororaphis* and one should use strains which were classified as *Pseudomonas aureofaciens* in the past); oxychloraphin, stains the medium pale yellow (produced by *Pseudomonas chlororaphis*); chloraphin, only very sparingly soluble and accumulates as isolated green crystals in the growth at the butt of the slant (produced by *Pseudomanas chlororaphis*).

15.2.74. Pyrazinamidase Activity

15.2.74.1. Method 1 (as Applied to *Corynebacterium* spp.) (263)

Inoculate a tube of the test medium (section 15.4.32) with bacteria from an overnight growth on chocolate agar slants (ca. 5×10^9 cells). Incubate at 37°C overnight. To each butt, add 1.0 ml of freshly prepared 1% (wt/vol) aqueous ferrous ammonium sulfate. Refrigerate the tube for 1 to 4 h. A positive test is indicated by the appearance of a pink band in the agar, indicating the formation of pyrazinoic acid. A negative test shows no pink band.

For a rapid test, omit the Dubos broth base from the culture medium and decrease the agar concentration to 0.2%. Prepare a physiological saline (0.85% NaCl) suspension of the bacteria to be tested from growth on chocolate agar slants, and with a pipette, mix the suspension into the test medium in the top half of the tube. Incubate for 2 h in a 37°C water bath. Add 1 ml of ferrous ammonium sulfate solution (see above), and incubate for 1 to 5 min. A positive test is indicated by pink. A negative test shows no pink.

Examples: positive, *Corynebacterium xerosis*; negative, *Corynebacterium pseudotuberculosis*.

15.2.74.2. Method 2 (as Applied to *Yersinia* spp.) (125)

Inoculate slants of the test medium (section 15.4.32) with bacteria grown overnight on soybean-casein digest agar (section 15.4.35), and incubate for 48 h at 25 to 30°C. Flood the slant with 1 ml of ferrous ammonium sulfate solution (see method 1), and incubate for 15 min. A positive test is indicated by pink. A negative test shows no pink.

Examples: positive, *Yersinia intermedia*; negative, *Yersinia pseudotuberculosis*.

15.2.75. Sodium Chloride Optima

The determination of the range of NaCl concentrations for growth and the optimum concentration is not a trivial matter. In the past, a number of "marine bacteria" were described which were later shown to be very halotolerant (28). Increasing interest in marine microbiology makes this aspect of greater significance, and determining salinity optima may be coupled with examining strains for their internal compatible solutes or other, salt-dependent properties. Many bacilli and members of the genera *Staphylococcus* may be isolated in the absence of salt but may grow at a wide range of salinities.

15.2.76. Spores (Endospores)

For staining of spores, see chapter 2.3.7. The media used for the growth of cultures should contain adequate levels of Ca^{2+} and Mn^{2+}. Media with and without carbohydrates should be inoculated, and cultures of different ages should be examined. *Note*: large inclusion granules (e.g., PHB) may be mistaken for endospores.

If spore-like structures are observed, a heat test should be performed for confirmation. Inoculate a tube of broth with growth from solid or liquid media in which sporulation is suspected to occur. Take care not to touch the wall of the tube with the inoculum. Place the tube in a water bath at 80°C along with an uninoculated tube of broth containing a thermometer. Make sure the level of the water in the bath is higher than the level of the broth. Heat for 10 min; begin timing when the thermometer reaches 80°C. Cool the tube, and incubate at the optimum growth temperature to see if growth occurs. *Note*: some endospores, such as those that occur in certain strains of *Clostridium botulinum*, may be killed at 80°C, necessitating the use of lower temperatures for testing heat resistance (e.g., 75 or 70°C). Also, apply the popping test (chapter 2.3.7.1). Certain *Nocardia* species form chlamydospores or microcysts, which can survive at 80°C for several hours and yet are not true endospores.

15.2.77. Starch Hydrolysis

Prepare an agar medium on which the organism can grow well, and add 0.2% soluble starch before boiling. The medium should preferably contain no glucose since glucose may diminish starch hydrolysis (292). Sterilize the medium at 115°C for 10 min. Make a single streak of the organism across the center of a plate of the starch agar. After incubation, flood the plate with iodine solution (the iodine normally used for Gram staining is suitable). A positive test is indicated by a colorless area around the growth.

Examples: positive, *Bacillus subtilis*; negative, *Escherichia coli*.

15.2.78. Sulfur Production by Microorganisms

Sulfur may be produced by certain organisms by the oxidation of sulfide, for example, and may be found either as globules inside the cell or as similarly sized droplets outside the cells. Examine the cells or cell suspensions in wet mount preparations, under the microscope. Sulfur globules appear as refractile bodies with a bright center and a characteristic birefringent edge. When large amounts of elemental sulfur are produced in a culture, the culture will turn milky white, without necessarily an increase in the number of cells. Very finely divided sulfur is white, not yellow. Although sulfur is soluble in certain solvents (such as carbon disulfide), care should be taken that the cells do not lyse as a result of the action of the solvent.

Examples: sulfur produced internally, *Chromatium vinosum* (when grown on sulfide); sulfur produced externally, *Ectothiorhodospira mobilis* (when grown on sulfide).

15.2.79. Temperature Optima

Given the wide range of habitats which prokaryotes can colonize, from the Antarctic to deep-sea hydrothermal vents, it should be obvious that determining the temperature range for growth and the optimum temperature is a routine part of characterizing an organism. When large numbers of organisms are being handled, a preliminary screening of aerobic strains in the range 0 to 40°C can probably be easily undertaken by using microtiter plates and a plate reader. Many such readers have a computer interface and allow downloading of the optical density readings in a format for use in common spreadsheet programs. This feature allows one to plot the optical density in graphical format. The method is not suitable for slow-growing organisms. It is also advisable to incubate the microtiter plates in closed plastic containers. At higher temperatures, evaporation of the small volumes becomes a problem.

Despite modern digital displays, normal incubators are not necessarily always accurate and better results can be obtained by using either water baths or a suitable temperature gradient incubator. Small water baths are usually suitable for incubating liquid cultures, but they generally cannot be shaken. A temperature gradient incubator used to be manufactured by Toyo Kagaku Sangyo Co. Ltd. and continues to give good service in the Deutsche Sammlung von Mikroorganismen und Zellkulturen. This system has the advantages of forming a temperature gradient and allowing both aerobic and anaerobic cultures to be incubated with shaking. The tubes can be removed, and the optical density can be measured in a suitably modified spectrophotometer. A similar system, using a heating element and a thermoelectric Peltier element for cooling, has also been described recently (69).

15.2.80. Triple Sugar Iron Agar Reactions

Inoculate a slant of triple sugar iron agar (section 15.4.38) by using a straight needle. First, stab the butt down to the bottom, withdraw the needle, and then streak the surface of the slant. Use a cotton stopper or a loosely fitting closure to permit access to air. Read results after incubation at 37°C for 18 to 24 h. Three kinds of data may be obtained from the reactions.

15.2.80.1. Sugar Fermentations

Acid butt, alkaline slant (yellow butt, red slant): glucose has been fermented, but not sucrose or lactose

Acid butt, acid slant (yellow butt, yellow slant): lactose and/or sucrose has been fermented

Alkaline butt, alkaline slant (red butt, red slant): neither glucose, lactose, nor sucrose has been fermented

15.2.80.2. Gas Production

Gas production is indicated by bubbles in the butt. With large amounts of gas, the agar may be broken or pushed upward.

15.2.80.3. Hydrogen Sulfide Production from Thiosulfate

Hydrogen sulfide production is indicated by a blackening of the butt as a result of the reaction of H_2S with the ferrous ammonium sulfate to form black ferrous sulfide.

15.2.81. Urease Activity

15.2.81.1. Method 1

Inoculate the surface of a Christensen urea agar slant (section 15.34.14), and incubate. A positive test is indicated by red-violet.

Examples: positive, *Proteus vulgaris*; negative, *Escherichia coli*.

15.2.81.2. Method 2 (114)

The following method is useful for organisms that cannot grow on Christensen urea agar. Prepare the following solution: 0.1065% BES buffer [*N,N*-bis(2-hydroxyethyl)-2-aminoethanesulfonic acid (Sigma or Valeant Pharmaceuticals)]–2.0% urea–0.001% phenol red (pH 7.0). Sterilize by filtration, and aseptically dispense 2.0-ml volumes into small tubes. Centrifuge a broth culture of the organism, and suspend the cells in sterile distilled water or saline to a dense concentration. Add 0.5 ml of the cell suspension to a tube of the urea medium. A control medium lacking urea should be inoculated similarly. Incubate the tubes for 24 h. A positive test is indicated by red-violet.

15.2.82. Voges-Proskauer Reaction

Inoculate two tubes of methyl red–Voges-Proskauer broth (section 15.4.27). Incubate one tube at 37°C and the other at 25C. After 48 h, remove 1 ml of culture to another tube, add 0.6 ml of 5% (wt/vol) α-naphthol dissolved in absolute ethanol, and mix thoroughly. Then add 0.2 ml of 40% aqueous KOH. Mix well, and incubate the tube in a slanted position to increase the surface area of the medium (the reaction is dependent on oxygen). Examine after 15 and 60 min. A positive test is indicated by a strong red color that begins at the surface of the medium.

Examples: positive, *Enterobacter aerogenes*; negative, *Escherichia coli*.

15.2.83. X and V Factor Requirements

Inoculate a plate of soybean-casein digest agar (section 15.4.35) as follows. Moisten a sterile cotton swab in the suspension, squeeze out excess fluid by pressing and rolling the swab against the wall of the tube, and roll the swab across the entire surface of the plate (do not scour the surface of the agar). Use sterile paper strips impregnated with the X factor (heme) or the V factor (NAD). The strips are obtainable commercially (Difco; BBL). With flamed forceps, place an X strip on the surface of the seeded agar, flame the forceps, and then place a V strip on the agar approximately 2 cm from the X strip. Incubate the plates for 1 to 2 days, and interpret the results as follows. A requirement for either X or V factor, but not both, is indicated by growth only around the appropriate strip. A requirement for both factors is indicated by the occurrence of growth only between the two strips. If growth occurs over the entire plate, neither factor is required.

Examples: positive for both X and V, *Haemophilus influenzae*; positive for V only, *Haemophilus parainfluenzae*; negative for both X and V, *Escherichia coli*.

15.3. SPECIAL TESTS AND METHODS

15.3.1. Fermentation Products

15.3.1.1. Organic Acids and Alcohols by Gas Chromatography

See reference 7 and chapters 17.3.5 and 18.8.

The following procedures were designed for the identification of anaerobes, but they can also be used for aerobic and facultative organisms. Acidify 1 ml of culture medium containing glucose or another fermentable carbohydrate in a 12-by-75-mm tube by adding 0.2 ml of 50% (vol/vol) H_2SO_4; samples thus acidified can be stored in a refrigerator for future analysis. Add 0.4 g of NaCl and 1 ml of diethyl ether (it should be possible to substitute tert-butylmethyl ether). Stopper the tube with a cork stopper. Mix the contents of the tube by gently inverting the tube about 20 times to extract the free fatty acids into the ether phase. Centrifuge the mixture for a few minutes to break the emulsion, and remove the ether (upper) layer from the aqueous layer. It is advisable not to remove too much of the ether layer in order to avoid contaminating the ether with the water. Place the ether layer into a new 12-by-75-mm tube, and add anhydrous $CaCl_2$ (4 to 20 mesh) to about one-fourth the volume of the ether to remove traces of dissolved water. Inject 14 μl into the gas chromatograph to detect short-chain volatile fatty acids and alcohols. Uninoculated medium should be acidified, extracted, and examined as a control.

Pyruvic, lactic, fumaric, and succinic acids are nonvolatile and must first be converted to their methyl esters for detection by gas chromatography. Place 1 ml of the culture in a 12-by-75-mm tube (sealed with a Teflon-lined screw cap), and add 2 ml of methanol and 0.4 ml of 50% H_2SO_4 (2 ml of boron trifluoride methanol [obtainable from any chromatographic supply house] may be added in place of plain methanol). Heat the mixture at 60°C for 30 min. Add 1 ml of water and 0.5 ml of chloroform to the tube, stopper the tube tightly, and gently invert it about 20 times. Remove the chloroform (bottom) layer, and inject 14 μl into the gas chromatograph.

A gas chromatograph equipped with a thermal conductivity detector is recommended. A flame ionization detector may be used; however, it will not detect formic acid. Use a stainless steel, aluminum, or glass column measuring 6 ft (1.8 m) by 0.25 in. (6 mm). Pack the column with Supelco SP-1000, 100 to 200 mesh (Supelco Inc., Bellefonte, PA), or with 5% FFAP on Chromosorb-G. Both volatile and methylated fatty acids are detected with the same column. Set the column oven temperature at 135 to 145°C. The carrier gas is helium at a flow rate of 120 cm^3/min. If the instrument is equipped with separate injector and detector oven heaters, set these at 20 to 30°C higher than the column temperature. Franzmann et al. (79) reported the use of an Nukol wide bore capillary column for volatile fatty acids and alcohol analysis, using hydrogen as the carrier gas (linear velocity, 20 cm/s). The column temperature was programmed as follows: hold at 75°C for 3 min, and then heat at 32°C/min to 170°C and hold for 6 min isothermally at this temperature. Detection was by flame ionization detection. Most gas chromatographs are now equipped with computer-based data storage, integration, and evaluation programs. Prepare standards for volatile acids, alcohols, and nonvolatile acids (section 15.5.19), and extract them in the same manner as the culture samples. Such standards may also be purchased from chromatography supply houses.

The presence or absence, as well as the approximate amounts, of the various metabolic end products constitute patterns that are very helpful in identifying bacteria. More quantitative information can be obtained by using careful technique and standard curves based on known amounts of each acid or alcohol. The order of elution of alcohols and volatile fatty acids from the gas chromatograph column is as follows: ethanol, n-propanol, isobutanol, n-butanol, isopentanol, n-pentanol, acetic acid, formic acid, propionic acid,

isobutyric acid, n-butyric acid, isovaleric acid, n-valeric acid, isocaproic acid, n-caproic acid, and heptanoic acid. For methyl esters of the nonvolatile acids, the order of elution is as follows: pyruvic acid, lactic acid, oxalacetic acid, oxalic acid, methylmalonic acid, malonic acid, fumaric acid, and succinic acid.

15.3.1.2. Organic Acids and Alcohols by HPLC

See reference 242 and chapters 17.3.6 and 18.8.

Add 3 to 5 drops of 50% sulfuric acid to a culture grown in a broth medium with glucose or another fermentable carbohydrate. Pipette 1 ml of the culture into a Clin-Elute CE-1001 column (Varian, Harbor City, CA). Elute the organic acids from the column by adding 9 ml of tert-butylmethyl ether (in 3-ml portions). Add 1 ml of 0.2 M NaOH to the 9 ml of ether extract, and shake. Remove the aqueous layer containing the sodium salts of the fatty acids, and wash it once with tert-butylmethyl ether. Inject 20 μl of the aqueous extract into a high-pressure liquid chromatography (HPLC) analyzer equipped with a UV detector set at 214 nm to detect the double bond of the carboxyl group of the fatty acids. Bio-Rad offers a suitable organic acid column (model HPX-87H; Richmond, CA); other manufacturers also make organic acid columns. Elute organic acids from the column with 5% acetonitrile in 0.013 N sulfuric acid at a rate of 0.8 ml/min; the column temperature should be 60°C. Record peaks by using an integrator-recorder or a computer-based program.

Alcohols can be separated on the organic acid column; they are eluted with distilled water and detected with a refractive index monitor. Organic acid and alcohol standards are available commercially.

15.3.1.3. Hydrogen and Methane by Gas Chromatography

Analyze cultures grown in a medium containing glucose or other fermentable carbohydrates for the production of hydrogen and methane. Stopper the cultures with a rubber stopper, and incubate them. Crimp-top or screw-capped tubes, fitted with rubber stoppers, used for the growth of strict anaerobes are suitable. Use a syringe fitted with an on/off valve to which is attached the syringe needle. After the cultures have grown, remove gas samples from the culture tubes by pushing the needle of a syringe through the rubber stopper and into the gas phase of the culture. Use a 1- or 5-ml disposable plastic syringe with a 1.5-in. (3.8-cm) needle, and start the needle into the culture tube at a point where the rubber stopper and glass wall meet; push the needle through the stopper at an angle that will permit the tip of the needle to enter the gas phase of the culture. Withdraw 0.7 to 1.0 ml of the gas phase, close the valve, and withdraw the needle. Inject the contents of the syringe into the gas chromatograph after the valve fitted to the syringe has been opened.

The gas chromatograph must be equipped with a thermal conductivity detector. Use either nitrogen or argon as the carrier gas, with a flow rate of 25 to 30 cm^3/min. Use an aluminum or stainless steel column measuring 16 ft by 0.25 in., packed with silica gel (grade 12; 80 to 100 mesh; Supelco Inc.). Set the column oven temperature to 25 to 30°C; if the instrument is equipped with separate injector and detector ovens, set these at the same temperature as the column oven or a few degrees higher. Standards for gas analysis can be purchased from any chromatography supply house; natural gas can be used to determine the elution point of methane. For mixtures of hydrogen and methane, hydrogen will be eluted from the column before methane.

15.3.2. Substrate Utilization Patterns

15.3.2.1. Auxanographic Nutritional Method

Seed a molten basal agar medium (45 to 50°C) lacking carbon sources with a washed suspension of the organism to be tested. After the plates have solidified, dip the edges of sterile filter paper disks into sterile solutions of various carbon sources, allowing the fluid to be absorbed by capillary action. Such disks may be applied immediately to the surface of the seeded agar plates or may be dried and stored in a sterile container until needed. Place each disk near the periphery of the seeded plate, using three to four disks per plate, or alternatively, place the disks at suitable distances on larger sheets of seeded agar (e.g., in a Pyrex baking dish). After incubation, determine whether a carbon source has been used by looking for a halo of turbid growth around the disk. In some cases, growth occurs only as a line between two disks, indicating that the organism requires both carbon sources for growth. If growth occurs over the entire plate, the inoculum was probably too dense; use a more dilute inoculum to seed the agar.

15.3.2.2. Multipoint Inoculators

The widespread commercial availability of sterile, plastic microtiter plates (plastic dishes with 96 miniature wells, each having a volume of ca. 0.2 ml) has led to the development of miniaturized procedures for the determination of acid production from carbohydrates, gas production, and other biochemical reactions. One can make a multiple inoculator by simply inserting the pointed ends of 27-mm stainless steel pins into wax-filled wells or a microtiter plate (38). One can also make an inoculator by inserting pins or brads into a wood, plastic, or metal template so that they are in the exact pattern of the wells of a microtiter plate (81). If wires are used, they may be bent to form a number of coils (as in a spring), the number and diameter of the coils determining how much liquid culture is transferred. An inoculator made entirely of metal has the advantage of being autoclavable, but one can effectively sterilize the other inoculators by dipping the pinheads into alcohol and then flaming them for 1 to 2 s with the pinheads pointing upward (80). Charge an inoculator by lowering the device onto a master microtiter plate in which each well contains a different strain of bacteria; in this way, each pinhead becomes charged with ca. 0.006 ml of culture. Then lower the inoculator onto a microtiter plate with its wells filled with carbohydrate broth, thereby causing each pinhead to inoculate its particular strain of bacteria into a separate well of broth. Return the inoculator to the master plate for recharging, and use it to inoculate a second microtiter plate that has its wells filled with a different carbohydrate broth, etc. To determine gas production from carbohydrates, cover each inoculated well with a sterile overlay of a sealant such as 1 part of Amojell (American Oil Co.) and 1 part of mineral oil (82). Cover the inoculated plates with their lids, and incubate them. Acid production is determined by a change in the color of the pH indicator in the medium. Gas production is indicated by the accumulation of bubbles under the overlay.

Filling the wells of a microtiter plate with sterile media can be greatly facilitated if one uses a multichannel automatic pipette or similar apparatus. One can construct a homemade device such as that described by Wilkins et al. (296) as follows. Remove the tips and hubs from eight 18-gauge hypodermic needles, and braze the needles into holes in a 13.5-cm-long piece of 7-mm stainless steel tubing so

that the needle centers are the same distance apart as the microtiter wells (Fig. 2). Close off one end of the tubing, and braze a luer-lock connector onto the other end. After sterilizing the assembly, aseptically connect the luer-lock to a sterile automatic pipetting syringe (Cornwall type; Becton, Dickinson & Co., Rutherford, NJ) equipped with a sterile filling hose (Fig. 2). With the end of the filling hose immersed in a flask of sterile medium, depress the plunger of the adjustable syringe several times to fill the syringe. Then place the eight-needle manifold into a row of eight wells on a microtiter plate, and depress the plunger of the syringe to deliver 0.2 ml of medium simultaneously to each of the eight wells. Continue to fill rows of wells in this manner until all of the wells on the plate have been filled. Rinse the assembly by flushing it with sterile water (or boiling water, if an agar medium has been dispensed), and then flush it several times with the next medium to be dispensed.

A number of manufacturers now produce multitip pipetting aids. These may be in 8- or 12-channel format and may be of the fixed-volume, single-delivery type or of a multiple-delivery, stepper principle type. The disposable tips may be autoclaved in suitable containers. In many cases this technology may be effectively used, in conjunction with multiwell microtiter plates, to replace the multiple-point inoculators described above. However, the equipment described above may be more suitable if larger volumes are to be inoculated.

Biochemical tests other than those for acid and gas production can be performed with microtiter plates. For example, perform the indole test by adding 2 drops of Kovács' reagent to the well cultures, or for the Voges-Proskauer test add 1 loopful of α-naphthol followed by 1 loopful of KOH (81). Some tests may not be suitable for microtiter plates; e.g., in testing for urease, the ammonia produced by a urease-positive organism may diffuse to adjacent wells to give false-positive reactions. Pinhead inoculators may be used to inoculate the surfaces of diagnostic agar media contained in conventional petri dishes (80, 155). Also, inoculators with the pointed ends of brads or pins projecting from the block may be used to pierce the agar surface (189). However, when one deposits various strains on a single petri dish, problems of overgrowth, spreading, or mutual interference may occur. These problems can be eliminated if one uses a petri dish divided into compartments (250), mini ice cube trays (80), or microtiter plates (81, 295). Square, sterile, plastic multiwell dishes used to be manufactured by Sterilin in the United Kingdom under the name "repli dishes" (Bibby Sterilin; 10 mm by 10 cm, divided into a five-by-five format) and by Flow Laboratories (now Titertek). In the absence of such systems on the current market, it may be pos-

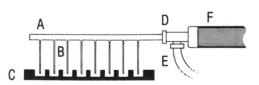

FIGURE 2 Diagram of apparatus for filling eight microtiter plate wells simultaneously. (A) Stainless steel tubing; (B) 18-gauge hypodermic needles brazed to tubing; (C) microtiter plate with wells; (D) Luer-Lok connector brazed to end of tubing; (E) filling hose attached to Cornwall syringe; (F) barrel and plunger of Cornwall syringe. By depressing the plunger, the eight wells can be filled simultaneously with sterile medium.

sible to substitute similarly formatted tissue culture multi-well dishes. Tiny slants, such as triple sugar iron agar slants, may be prepared in microtiter plates and inoculated by using a piercing inoculator (81).

For multipoint inoculation with anaerobes, dispense sterile semisolid medium into microtiter plate wells and incubate the plates with their lids in position overnight aerobically at 37°C to check for contamination and also to dry the plates slightly (295). Place the plates in an anaerobic glove box, and allow them to reduce for 2 days prior to inoculation. Inoculate the plates with a piercing inoculator within the glove box. Although each plate contains 96 wells, inoculate only 48 of them (in a checkerboard pattern) and leave 48 uninoculated to serve as controls to detect nonbiological pH changes or the possible migration of motile bacteria from one well to another. After inoculation, cover each plate with a sheet of sterile plastic tape (nontoxic plate sealer; Cooke Engineering Co.) by placing the tape with the adhesive side up on a rubber mat and pressing the inverted plate onto it. Roll the plates with the tape roller to ensure a firm seal on each well. To allow for the escape of gases during growth, punch a pinhole into the tape over each well by using the tape perforator. Incubate the plate within the anaerobic glove box for 3 days. Determine acid production by means of a narrow pH electrode (such as an S-30070-10 electrode [Sargent-Welch Co., Skokie, IL]) inserted into the medium of each well. Do not rinse the electrode between wells, because each insertion pushes the previous agar medium upward away from the electrode tip (295).

15.3.2.3. Velveteen for Multiple Inoculation

The multiple-inoculation method using velveteen is designed to determine the nutritional characteristics of a large number of strains. The procedure described below, in which the abilities of 267 strains of pseudomonads to use 146 different organic compounds as carbon and energy sources are determined (258), exemplifies the method.

Prepare a mineral agar base with the following composition: 40 ml of 1.0 M Na_2HPO_4-KH_2PO_4 buffer (pH 6.8), 20 ml of Hutner's vitamin-free mineral base (section 15.4.2.), 1.0 g of $(NH_4)_2SO_4$, 20 g of agar, and 940 ml of distilled water. Supplement the medium with 0.1 to 0.2% of the carbon source to be tested, and dispense the medium into petri dishes. Dry the sterile plates by incubating them at 37°C for 2 to 3 days, and store them at room temperature in plastic bags until used (up to 4 or 5 weeks of storage). Prepare a master plate by "patching" (depositing the inoculum in a small area) up to 23 different strains on a plate of mineral agar base containing 0.5% yeast extract. After growth occurs, prepare submaster plates by the velveteen replication method originally devised by Lederberg and Lederberg (151). In this method, a piece of sterilized velveteen cloth is affixed to a piston having a diameter slightly smaller than that of the petri dish. By gentle pressing of the velveteen onto the master plate and then onto a series of sterile plates, the plates become inoculated with the various strains in exactly the same pattern as the master plate. Up to nine submaster plates can be prepared from a single master plate. After growth occurs on the submaster plates, use each plate to print test plates for nine different carbon sources, followed by a terminal yeast extract agar plate to verify that transfer of all the strains has occurred. Read the plates at 48 and 96 h, and score the results as follows: 0, growth no greater than that on a control plate lacking any carbon source; +, good growth; ±, scanty growth but significantly

greater than that on the control plate; and + M, late growth of a few colonies in the area of the patch, presumed to arise from mutants in the inoculum. When strains with markedly different nutritional characteristics are being tested together on a plate, cross feeding (the diffusion of nutrients produced by one strain across to adjacent strains, resulting in a false-positive growth response by the latter strains) or interference between strains may occur. Consequently, follow up each initial random screening with a second screening with a different arrangement or pattern of strains on the master plate.

15.3.2.4. Commercial System Based on Nutritional Patterns

The nutritional system by Biolog (Hayward, CA) is based on the utilization of nutrients and other chemicals by bacteria. A 96-well microtiterplate with a different dehydrated nutrient in each well along with a tetrazolium redox indicator is used. A cell suspension is pipetted into the wells, and the plate is incubated for a few hours or overnight. Reduction of the colorless tetrazolium indicator to purple formazan shows which substrates have been oxidized. The method does not record growth but rather the ability of the cell suspension to metabolize the substrate. Sterile pipettes and saline for the cell suspension must be used. The Biolog GN2 plate is available for gram-negative aerobic bacteria, and the GP2 plate is used for gram-positive bacteria. Another plate, the MT plate, contains only the tetrazolium redox dye and a buffered medium; the user adds various substrates to the wells according to personal choice. The system allows computerized identification of many bacterial species. Manual data entry software that accesses all of the system's databases is available; alternatively, an automated plate reader can be used. There is also a computer program that allows users to develop their own databases.

15.3.3. PAGE Patterns

See also chapter 17.5.

Polyacrylamide gel electrophoresis (PAGE) patterns of water-soluble proteins can be useful for bacterial characterization and identification. Strains of a given species will have similar patterns, especially of the major proteins. Although the patterns are similar, they are not necessarily identical; for instance, some protein bands may be seen as dark bands for one strain and light ones for another strain. Some strains will yield distinctive patterns that help to identify the organism readily, whereas other organisms may show patterns that are not readily distinguishable from those of other species, and some additional tests may be needed to verify the identification of the bacteria. In an electrophoretic study of clostridia, Cato et al. (44) found that strains with more than 80% DNA-DNA binding usually produce identical patterns, strains with ca. 70% DNA-DNA binding produced patterns with some minor heterogeneity but overall similarity, and strains that were unrelated by DNA-DNA hybridization produced patterns with major differences. Other examples of the usefulness of PAGE can be found in references 192, 197, and 221. Highly reproducible patterns can be produced under rigorously standardized conditions and can be analyzed by rapid, computerized, numerical analysis (126). Detailed descriptions of this method can also be found in the articles of Jackman (115) and Pot et al. (219).

Both native (nondenaturing) gels and sodium dodecyl sulfate (SDS) gels can be used. Although an SDS gel yields more bands, this seems to offer no advantages and, in fact, makes visual comparison of the patterns more difficult. The following method has been reported to be useful for the rapid identification of various gram-positive and gram-negative isolates from human gingival crevice floras (178).

Resolving Gel (8.5%)

Add 11.3 g of acrylamide and 0.3 g of N,N'-methylenebisacrylamide to 50 ml of 0.375 M Tris-chloride buffer (pH 8.9). (The buffer can be prepared by dissolving 23.85 g of Tris preset-pH crystals [pH 8.8 at 25°C; Sigma] in 500 ml of distilled water.) Heat the mixture to dissolve the ingredients, and make up to a volume of 133 ml with pH 8.8 buffer. Filter the solution through rapid-flow filter paper, and add 0.06 ml of TEMED (N,N,N',N'-tetramethylethylenediamine). Degas the solution by passing 32 ml through a 50-ml syringe. Add a 0.3-ml portion of a freshly prepared aqueous 10% solution (wt/vol) of ammonium persulfate, pour the gels, and allow them to polymerize for 10 to 20 min. Rinse the tops of the gels with distilled water to remove any unpolymerized acrylamide.

Stacking Gel (4.7%)

Add 2.5 g of acrylamide and 0.07 g of N,N'-methylenebisacrylamide to 15 ml of 0.15 M Tris-chloride buffer (pH 7.0). (The buffer can be prepared by dissolving 5.82 g of Tris preset-pH crystals [pH 7.0 at 25°C; Sigma] in 250 ml of distilled water.) Store the solution in 10-ml amounts for no more than 1 week. Before use, add 7 drops of tracking dye (0.025% aqueous bromophenol blue) and 0.01 ml of TEMED to 10 ml of the solution. Degas, and add 0.1 ml of freshly prepared 10% aqueous ammonium persulfate. Add a comb to the resolving-gel plate, and pour the stacking gel. After the stacking gel has solidified (20 min), remove the comb and rinse the wells.

Electrode Buffer (0.025 M Tris–0.25 M Glycine Buffer)

Dissolve 5.4 g of Tris and 27.0 g of glycine in distilled water, and make up to a final volume of 1,000 ml.

Sample Preparation

Harvest the bacterial cells from a 5-ml broth culture by centrifuging at 8,000 × g for 10 min. To the unwashed cell pellet, add approximately 0.15 g of glass beads (74 to 110 μm in diameter; class IV, type C [Cataphore Division of Ferro Corp., Jackson, MS]) and 0.15 ml of Tris-chloride buffer (pH 7.0). Disrupt the cells by holding the tube against a Vortex mixer for 4 min. (Alternatively, the bacterial cells can be broken by using other methods.) Add a 1/3 volume of powdered sucrose to the sample. For gram-negative bacterial species, centrifuge the extracts at 8,000 × g and heat at 55°C for 5 min before adding the sucrose to reduce the background in the gels.

Electrophoresis

Add 30 μl of sample to a well, and operate the electrophoresis apparatus at room temperature at 150 V (constant) and 33 mA of initial current. Stop electrophoresis when the tracking dye reaches the bottom of the gel (about 2.5 h).

Gel Staining

Place the gels in 12% trichloroacetic acid for 20 min. Then incubate them for 10 min in 0.08% Coomassie blue stain (4.0 g of Brilliant Blue R [60% dye content], 900 ml of methanol, 400 ml of glacial acetic acid, 1,800 ml of distilled water). Destain the gels in 10% (vol/vol) glacial acetic acid. Gels can be saved by being placed into plastic Ziploc bags. Alternatively, commercially available silver staining kits

may be used. Gels may also be dried with a proper gel-drying apparatus. Gels may be either scanned or photographed with a digital imaging system and evaluated with appropriate software (e.g., GelCompar; BioNumerics, Sint-Martens-Latern, Belgium).

15.3.4. Fatty Acid (and Other Hydrophobic Side Chain) Profiles

See also chapter 17.3.5.6.

Long-chain fatty acids (9 to 20 carbons) found in the membranes of bacteria can be used to characterize and identify these organisms. By using a computer program and statistical analysis of the individual cellular fatty acids, the most probable identity of the isolate can be determined when the results are compared with information in databases in a library of known species. The accuracy of the system depends on using the recommended medium, incubation temperature, and time of incubation.

15.3.4.1. Gas Chromatography of Cellular Long-Chain Fatty Acids (181)

Suspend the growth from an agar medium in 0.5 ml of sterile distilled water, and place the suspension in a 13-by-100-mm tube with a Teflon-lined screw cap. Saponify the cells by adding 1 ml of NaOH-methanol reagent (15% NaOH in 50% aqueous methanol), and heat at 100°C for 30 min. Cool to room temperature, and add 1.5 ml of HCl-methanol reagent (25% HCl in methanol). Heat the mixture to 100°C for 15 min, and then cool it to room temperature. Extract the methyl esters of the fatty acids by adding 1.5 ml of tert-butylmethyl ether-hexane (1:1, vol/vol). Shake, and allow to stand for 1 to 2 min. Remove the aqueous (lower) layer with a Pasteur pipette, and discard it. Add 1 ml of phosphate buffer (pH 11.0) to wash the organic phase, shake, and allow to stand for 2 to 3 min. Remove and save ca. two-thirds of the organic (top) layer. Inject samples of this layer into a gas chromatograph equipped with a flame ionization detector. Use a fused silica capillary column (25 m by 0.2 mm [internal diameter]) containing cross-linked methylphenyl silicone (SE 54) as the stationary phase. Set the detector temperature to 300°C and the injector temperature to 250°C, and program the column oven temperature to increase from 170 to 300°C at a rate of 5°C/min and then maintain 300°C for 1 min. Peaks are recorded on an integrating recorder, and the retention times are compared with those of peaks obtained by using a standard mixture of methyl esters of fatty acids. This method is that commercially available as the MIDI system, based on the work of Miller (172) (see below). However, a variety of other methods of fatty acid hydrolysis can be found in the literature and the results obtained are not always directly comparable with those from the methods described above (142, 182).

15.3.4.2. MIDI Microbial Identification System for Identification of Bacteria by Gas Chromatography of Cellular Fatty Acids

The MIDI microbial identification system (Midi, Newark, DE) is an automated system for bacterial identification. In the past, most organisms in the system's computer libraries were clinical bacteria; however, the system can be used for bacteria from any source if users create their own libraries of data from known organisms. To identify an isolate, harvest cells from colonies, free the fatty acids from the cellular lipids by saponification, and then methylate the fatty acids to form the volatile methyl esters. These are extracted and automatically injected into a gas chromatograph for separation, identification, and quantification. A computer compares the fatty acid profile with other profiles in a database library of known organisms and selects the most likely name for the isolate. The system includes a gas chromatograph with automatic injection, an integrator, and a computer complete with data libraries for facultative, aerobic, and anaerobic bacteria. A custom library is also available for user-generated data on organisms not included in the available data libraries.

15.3.4.3. Simple Methods for Distinguishing Unresolved Peaks

In forsenic medicine, one would rarely rely on a single method for the identification of a compound. Peak identities would be confirmed by coupling the gas chromatograph to a mass spectrometer or a mass selective detector, or columns of different polarities would be used. In single-column systems, relatively apolar phases are the most popular, because of their thermal stability. Using a more-polar column usually reduces the upper temperature limit and increases the run times. More-polar columns generally also have the effects of altering the elution order of saturated and unsaturated fatty acids and delaying the elution of hydroxy fatty acids. In the popular MIDI system, some peaks are not unambiguously resolved and are recorded as "summed features." While to the layman this feature may appear to be trivial, the failure to distinguish the individual components in groups of peaks ("summed feature")—which may contain either a branched-chain hydroxy fatty acid or a straight-chain, unsaturated fatty acid (or even a mixture of both)—is a significant difference, because their presence reflects significant differences in the underlying biosynthetic pathways. In the absence of a mass spectrometer, simple methods are available which help to distinguish between some (but not all) of the summed features.

Some bacteria contain plasmalogens, which upon acid hydrolysis are released as dimethyl acetals (13). There is documented evidence that the retention times of these substances may overlap with those of hydroxylated fatty acids. In the case of the MIDI system, the choice of the correct identification database will help minimize incorrect identification of a peak (99, 176, 261, 262).

15.3.4.3.1. Method for Detecting Double Bonds

The original method for detecting double bonds used a methanolic solution of the fatty acid methyl ester, in which was suspended a small amount of platinum black catalyst, through which was bubbled hydrogen gas. This method is not suitable for small volumes, since the solvent would evaporate. Use a small, thick-walled glass bottle—a 50- or 100-ml Duran (Schott Glass) is suitable—and a cap with a hole in it. Either close the bottle with a specially made piece of glassware to which a stopcock is fitted, or take a suitably sized rubber stopper through which a standard glass stopcock is fitted. This will allow the bottle to be evacuated (with a vacuum pump) and flushed with hydrogen. Take about 200 µl of the fatty acid methyl esters released as described in 15.3.4.1. or 15.3.4.2 and transfer it to a small autosampler vial or a small glass tube which will fit into the Duran bottle. Remove the solvent in a gentle stream of nitrogen. Redissolve the fatty acid methyl esters in about 500 µl of methanol and add a small amount (tip of a fine spatula) of platinum black catalyst. Place the vial or glass tube in the Duran bottle, and close the bottle with the stopcock

adapter. It is convenient if a three-way valve is used to connect the bottle to a vacuum and a hydrogen supply. Evacuate and flush the bottle three to four times. Allow the reaction mixture to stand for 30 min, with occasional shaking. After 30 min, remove the reaction mixture and filter it using a 1-ml glass extraction column fitted with a glass-fiber frit. Wash the platinum black catalyst, remaining on the frit, twice with 2× 250 µl of methanol. Dry the hydrogenated fatty acids under a stream of nitrogen, and redissolve in 200 µl of tert-butylmethyl ether. Rerun the hydrogenated fatty acids under the same conditions as the original sample. Unsaturated fatty acids disappear from the sample, and the relative amounts of the corresponding saturated fatty acids increase as a result.

15.3.4.3.2. Method for Detecting Hydroxy Fatty Acids (163)

The presence of hydroxy groups may be conveniently detected by the formation of derivatives. While there are a variety of derivatizing agents on the market, a simple method involves the formation of the acetate derivatives. Take 200 µl of the fatty acid methyl esters released as described in 15.3.4.1 or 15.3.4.2, and dry them under a stream of nitrogen. Redissolve in 200 µl of acetic anhydride-pyridine (1:1, vol/vol), tightly close the vial, and heat for 30 min at 100°C. Other derivatizing agents which have different effects on the equivalent chain lengths of the fatty acids may also be used (163). Cool the samples, and rerun the acetylated fatty acids under the same conditions as the original sample. Acetylated hydroxy fatty acids increase their retention times.

15.3.4.4. Release of Ether-Linked Lipids

Ether-linked lipids have become synonymous with members of the Archaea. However, ether lipids are also found in a variety of members of the Bacteria. While the ether-linked side chains within members of the Archaea are always isoprenoid derivatives (and the stereochemistry of the glycerol backbone is opposite to that found in Bacteria and Eukarya), the ether-linked lipids in members of the Bacteria (145, 211, 225) reported to date are straight or branched chain, not isoprenoid derivatives. In the case of certain anaerobic members of the Bacteria, the ether lipids are present in the plasmalogen form. These lipids are acid labile and are usually detected as dimethyl acetals during the course of fatty acid methyl ester analysis (as in the MIDI system). Other forms of ether lipids (mono-, di-, or tetraethers) have usually been detected by other methods.

15.3.4.4.1. Method 1

Place 100 mg of freeze-dried cells in small glass screw-cap bottles. The screw caps should be fitted with Teflon-lined septa. Add 5 ml of methanol–tert-butylmethyl ether–concentrated sulfuric acid (50:50:1, vol/vol/vol) and heat overnight at 50°C. Cool, add 2 ml of 0.3% aqueous NaCl, mix, and allow the phases to separate. Remove and save the upper tert-butylmethyl ether phase, and reextract the aqueous methanol phase with 2.5 ml of tert-butylmethyl ether. Remove and combine the tert-butylmethyl ether phase with the first extract. If it is suspected that more-polar ether lipids are present (as in members of the genus Sulfolobus), extract the aqueous methanol phase twice with 2.5 ml of chloroform. This method may destroy some of the acid-labile hydroxydiethers. In addition, some di- and tetraether phospholipids are remarkably acid stable (notably, the phosphatidylethanolamine, phosphatidylserine, and phosphatidylcholine derivatives) and the di- or tetraethers are not released (136, 138, 180, 194).

15.3.4.4.2. Method 2

If acid-labile hydroxydiethers are suspected, substitute methanol containing 0.18% HCl for the *tert*-butylmethyl ether–concentrated sulfuric acid mixture (67, 254). Heat overnight at 50°C. The hydrolyzed ethers may be recovered by extraction with petroleum ether or hexane.

15.3.4.5. Thin-Layer Chromatography of Ether Lipids

Hydrolyzed ether lipids can be separated by thin-layer chromatography on silica gel (Macherey-Nagel; article no. 818 135). Diethers may be separated from any fatty acid methyl esters by using the solvent hexane–*tert*-butylmethyl ether (80:20, vol/vol). Hydroxylated fatty acids may run close to the diethers. Better separation of different classes of diethers (isoprenoid versus straight or branched chain) and monoethers can be achieved using double development in the same solvent. Di-, mono-, and tetraethers and hydroxydiethers may be separated using hexane–*tert*-butylmethyl ether–glacial acetic acid (25:25:1, vol/vol/vol). Chloroform-methanol-water (60:10:1, vol/vol/vol) is used for the separation of tetraether polyols (146). Ether lipids may be visualized by using either 5% ethanolic molybdophosphoric acid (heated at 150°C for 15 to 20 min) (section 15.6.6) or anisaldehyde-sulfuric acid (heated at 120°C for 15 min) (section 15.6.1.1). Monoethers with adjacent hydroxyl groups may be visualized with the periodate-Schiff reagent.

The long-chain diols reported for *Thermomicrobium roseum* (215, 216) have chromatographic and staining behaviors similar to those of monoether lipids.

15.3.4.6. Gas Chromatography of Ether Lipids

Di- and tetraether lipids may be separated with a high-temperature capillary column (BPX5; Scientific Glass Engineering), after conversion to their bis-trimethylsilyltrifluoracetamide (BTMSFA) derivatives, in 50 µl at 60°C for 1 h (191).

15.3.4.7. Analysis of Long-Chain Bases

Long-chain bases in members of the genera *Sphingobacterium* (304), *Sphingomonas* (303), *Chlorobium* (116), and *Deinococcus* (23, 24) have been described previously. In members of the genera *Sphingobacterium* and *Sphingomonas*, they are typically 2-amino-1,3-dihydroxy derivatives (i.e., sphingosines), while in *Deinococcus radiodurans* they are in the form of alkyl amines. Long-chain bases may be obtained by hydrolyzing whole cells (or lipid extracts) in aqueous methanolic 2 N HCl. Samples are heated at 100°C for 3 h and cooled. Adjust the pH to 12.0 with KOH, and extract twice with chloroform. The presence of long-chain bases may be determined by thin-layer chromatography or gas chromatography (148, 187, 198, 303, 304). Typically, for gas chromatography analysis, the extracts are dried and derivatization is carried out with 200 µl of bis(trimethylsilyl) trifluoracetamide either for 2 h at 70°C (198) or for 30 min at 60°C (148). After cooling, the solvent is evaporated and the sample is redissolved in hexane (200 to 300 µl) and analyzed with a capillary column, such as that used for the fatty acid methyl ester analysis (section 15.3.4.2); some modification to the run conditions may be required.

15.3.5. Lipids

Although all lipids may be extracted using the Bligh and Dyer method described in chapter 18.4, it is usually con-

venient to fractionate the lipids into the more-apolar lipids and the more-polar lipids by using a two-stage extraction method (270, 271). Although the more-apolar isoprenoids are routinely used for the extraction of respiratory lipoquinones (isoprenoid quinones) and polar lipids, it should be remembered that these isoprenoids (including a wide range of carotenoid pigments) are extracted with the respiratory lipoquinones and that chlorophylls and bacteriochlorophylls will be extracted along with more-polar pigments (e.g., carotenoids further modified by the addition of sugars) and the "classical" polar lipids. Although polar lipids are often reduced to the phospholipids, work over the past decades has shown that they may include phospholipids, glycolipids, and aminolipids, as well as a diversity of other components, which may be characteristic of certain higher taxonomic groupings. It is for this reason that it is better to analyze the polar lipid fraction as a whole and not to fractionate it (see chapter 18). It is generally convenient to carry out the extractions from freeze-dried cell material. Using a combination of respiratory lipoquinone analysis, polar lipids, and the nature of the polar-lipid side chains (fatty acids, long-chain bases, or isoprenoid side chains) provides a very useful method, which in conjunction with gene-sequencing techniques is a very powerful tool in the characterization and classification of prokaryotes.

15.3.5.1. Respiratory Lipoquinones

Respiratory lipoquinones are known to occur in a wide variety of prokaryotes, and they are now generally accepted as being of taxonomic significance. Benzoquinones, of which ubiquinones (coenzyme Q) are found in prokaryotes, are restricted to members of the *Alphaproteobacteria*, *Gammaproteobacteria*, and *Betaproteobacteria*. Napthoquinones, of which menaquinones (vitamin K_2) are found in a wide variety of prokaryotes, are by far the most widely distributed class of respiratory lipoquinones, found in members of both the *Bacteria* and the *Archaea*. The length of the isoprenoid side chain, the degree of saturation, and other forms of modification are also of taxonomic value. Methods which use heat or alkaline extraction are to be avoided (59, 181). The reason for avoiding these conditions is described in detail by Tindall (272).

Weigh out 100 mg of freeze-dried cell material. The extraction is most efficient when the dried cell material is relatively finely divided. Place the cell material in a small brown glass bottle (capacity, about 10 ml) which can be closed with a Teflon-lined screw cap. To stir the cell material, cut off a piece of a metal (magnetic) paper clip about 1 cm long and place it in the bottle. Add 3 ml of hexane-methanol (1:2, vol/vol), close the bottle with the screw cap, and stir at medium speed on a magnetic stirrer for 30 min. It is often recommended that the extraction be carried out under an inert gas and in the dark, but unless quantitative measurements are needed, these conditions do not appear to be important. After 30 min, remove the bottle from the stirrer, add 1 ml of ice-cold hexane, and mix briefly. Place the bottle on ice for about 15 min. Transfer the contents of the bottle to a glass centrifuge tube, and centrifuge briefly at about 2,000 to 3,000 rpm for 2 to 3 min to separate the methanol and cell debris from the upper hexane phase. Remove the upper hexane phase, and collect it in a narrow glass tube (Schott GL4) with a Teflon-lined screw cap. To the methanol and cell debris, add 2 ml of aqueous 0.3% NaCl and 2 ml of hexane. Mix the phases by using a Pasteur pipette, and then centrifuge as before. The upper hexane phase is collected and added to the first collected phase. If you wish to extract the polar lipids, do not discard the methanol–aqueous 0.3% NaCl and cell debris (see section 15.3.5.2). The hexane phase is dried under a gentle stream of nitrogen and then taken up in a small volume of a relatively volatile solvent, such as *tert*-butylmethyl ether.

Respiratory lipoquinones may be separated into their various classes (napthoquinones, benzoquinones, menathioquinones, and benzothiophene quinones, etc.). Although for separation on silica gel thin layers, hexane-diethyl ether (85:15, vol/vol) is the traditional solvent, hexane–*tert*-butylmethyl ether (80:20, vol/vol) is a suitable alternative and avoids some of the hazards associated with diethyl ether. Transfer the respiratory lipoquinone-containing material, in *tert*-butymethyl ether, to a 10-by-10-cm plastic-backed silica gel thin-layer plate containing a UV indicator (Macherey-Nagel no. 805 023, cut to 10 by 10 cm with a guillotine) as a thin line 1 cm from the bottom edge of the plate. Two samples may be conveniently run side by side on the same plate (each band being ~4 cm long). Use all of the extracted material to load the thin-layer chromatography plates. Label the thin-layer chromatography plates just below the upper edge with a soft black pencil. Run the plates until the solvent reaches the upper edge. Remove the plates from the chromatography tank, and allow them to dry briefly in the fume cupboard before detecting the respiratory lipoquinones by using brief irradiation under UV light (254 nm). Mark the UV-absorbing bands, and scrape them from the plates by using a fine spatula. Transfer the silica gel to a 1-ml glass column fitted with a glass-fiber filter. Compact the silica gel in the column, elute the respiratory lipoquinones twice with 0.5 ml of hexane-methanol (1:2, vol/vol), and collect the respiratory lipoquinones in small glass bottles suitable for an autosampler. Cool the samples on ice, and then add a few drops of aqueous 0.3% NaCl and allow them to stand for a few minutes so that the upper hexane phase can separate. Add about 0.5 ml of hexane, collect the upper hexane phase, dry under a stream of nitrogen, and redissolve in HPLC-grade methanol (~300 μl). Separating the respiratory lipoquinones into different classes has the advantage that compounds are not misidentified. Respiratory lipoquinones are generally eluted with an RP$_{18}$ column using a UV detector set at 269 nm. A Macherey-Nagel Nucleosil 125/2 120-3 C$_{18}$ column fitted with a guard column and using methanol–heptane (90:10, vol/vol) at a flow rate of 0.2 ml/min will separate Q7-Q11 or MK6-MK10 over a period of 30 min. Longer isoprenoid chain lengths can be separated by using longer elution times or by increasing the flow rate to 0.4 ml/min.

Additional information on mass spectrometry and NMR analyses can be found in articles by Collins (49, 50).

15.3.5.2. Polar Lipids
Either polar lipids may be extracted directly from freeze-dried cell material by using chloroform–methanol–0.3% aqueous NaCl, or the cell debris and methanol–aqueous 0.3% NaCl mixture obtained after respiratory lipoquinone extraction may be used (see section 15.3.5.1).

To the 100 mg of freeze-dried cell material in a small brown glass bottle fitted with a Teflon-lined screw cap, add a short length of a magnetic paper clip (see section 15.3.5.1), add 9.5 ml of chloroform–methanol–aqueous 0.3% NaCl, and stir overnight. If starting from the cell debris and methanol–aqueous 0.3% NaCl, add 2.5 ml of chloroform and 3 ml of methanol. Transfer all of the solvent and cell debris into a small brown glass bottle fitted with a

Teflon-lined screw cap. Check that the paper clip stirrer is still present or add a new one and stir overnight. To recover the polar lipids, first transfer the solvent and cell debris to glass centrifuge tubes. Centrifuge for 5 min at 3,000 rpm. Decant the solvent into clean glass containers (leaving the cell debris as a pellet behind), and add 2.5 ml of aqueous NaCl and 2.5 ml chloroform. Transfer the solvent mixture, which may start to separate into two phases, into glass centrifuge tubes. Centrifuge for 5 min at 3,000 rpm. The lower chloroform phase contains the polar lipids, which may be recovered by removing the lower layer. Dry under a stream of nitrogen, and redissolve in about 500 μl of chloroform-methanol (2:1, vol/vol).

This method does not efficiently extract all polar lipids from certain methanogenic archaea (193). In this, case substitute 5% trichloroacetic acid for the 0.3% NaCl in the initial extraction mixture. The polar lipids may be partitioned as described above, but the chloroform phase should be washed with 1.9 volumes of methanol–0.3% aqueous NaCl to remove excess trichloroacetic acid.

The polar-lipid mixture is spotted into one corner (1 cm from two adjacent edges) of an aluminum-backed 10-by-10-cm silica gel thin-layer chromatography plate (Macherey-Nagel article no. 818 135). The plates are developed in 20-by-10-by-5-cm (internal dimensions) glass chromatography tanks lined with Whatman no. 1 filter paper. Plates are run in the first direction by using chloroform-methanol-water (65:25:4, vol/vol/vol). The plates are then air dried for 30 min at room temperature and then run in the second solvent. Use a new chromatography chamber for the second dimension, lined with Whatman no. 1 filter paper and the solvent chloroform-methanol-glacial acetic acid-water (80:12:15:4, vol/vol/vol/vol).

In addition to comparison with standards and strains of known lipid compositions, the individual spots are identified by their reactions with specific stains. Consulting various reference works is essential (5, 6, 9, 13, 14). It is also possible to scrape areas from the plates which have not been sprayed, elute the lipids, and subject them to mass spectrometry. The various lipids are visualized by using the spray reagents listed in section 15.6. Plates may be either photographed, photocopied, or digitalized by using either a scanner or a digital camera.

15.3.6. Analysis of Pigments Soluble in Organic Solvents
Relatively apolar pigments (typically, carotenoids with few polar modifications) are extracted along with respiratory lipoquinones (section 15.3.5). More-polar pigments and bacteriochlorophyll are extracted along with the polar lipids. These extraction methods may be used to separate the pigments of different polarities, although the solvents traditionally used are a mixture of acetone and methanol, followed by partitioning into ether upon the addition of ether and 3% NaCl. Whatever the method used for extraction, bacteriochlorophyll and the majority of polar lipids may be removed by a simple saponification step. The lipid extract is dissolved in methanol, methanolic KOH is added to give a final KOH concentration of ~5%, and the pigments are placed in the dark under a nitrogen atmosphere. After standing overnight, the pigments are transferred into petroleum ether by adding 0.5 volumes of petroleum ether and 0.5 volumes of 3% NaCl. It is outside the scope of this chapter to describe pigment analysis in detail. However, the analysis of lipid pigments (including carotenoids and chlorophylls) by UV and visible-light spectroscopy with

suitable organic solvents (use glass or quartz cuvettes only), thin-layer chromatography, HPLC, mass spectrometry, and NMR have been reviewed recently (222, 230).

15.3.7. Polyamines

Polyamines are found in a variety of prokaryotes and are of taxonomic significance. In some cases, they are to be found in specific structures of the cell, such as covalently bound to the peptidoglycan of members of the *Sporomusa-Pectinatus-Selenomonas* group (119–124, 229, 264, 265). Various methods of analyzing the polyamine content of prokaryotes have been described, but two methods, based on HPLC separation, seem to be the most common. Gas chromatographic separation is of interest if gas chromatography-mass spectrometry analyses are to be carried out (195). Hamana and Matsuzaki reviewed the literature up to 1992 (92), but Hamana and colleagues have published extensively on this topic over the past 20-plus years. Busse and colleagues have also continued to use this method (40).

15.3.7.1. Method 1, Modified (40, 228)

Suspend 40 mg of freeze-dried cells in 1 ml of 0.2M $HClO_4$ in screw-cap tubes fitted with a Teflon-lined septum. Add 1 μmol of the internal standard, 1,8-diaminooctane (1 mM in 50 ml distilled water). Heat the samples at 100°C for 30 min. Cool, and sediment the cell material by centrifugation in a benchtop centrifuge (3,000 rpm for 5 min). Remove 200 μl of the supernatant, and add it to 300 μl of an aqueous 1% Na_2CO_3 solution in a screw-cap tube fitted with a Teflon-line septum. To this mixture is added 800 μl of a dansyl chloride solution (7.5 mg/ml in acetone). Heat for 20 min at 60°C, and cool. To bind excess dansyl chloride, add 100 μl of a proline solution (50 mg/ml in water) and incubate at 60°C for a further 10 min. Cool the sample on ice, and extract the dansylated polyamines with 100 μl of toluene. Use 10 μl of this mixture for injection into the HPLC column.

Dansylated polyamines are separated with an RP_{18} HPLC column (250-mm by 4-mm internal diameter) by using a linear gradient of acetone-water from 40:60 (vol/vol) to 85:15 (vol/vol) over 35 min followed by the final eluant for a further 15 min. The flow rate is 1 ml/min. Longer-chain polyamines may need elution with a higher percentage of acetonitrile (acetonitrile-water, 94:6 [vol/vol]) (306). The dansylated polyamines are detected by using a fluorescence detector at an excitation wavelength of 360 nm and a cutoff filter set to 450 nm. The column temperature is held at 40°C.

15.3.7.2. Method 2 (162)

Extract the cells using 1 M $HClO_4$, in the cold. Samples may be taken directly for ion exchange chromatography by using a Kyowa Seimitsu 62210F 4.6-by-80-mm column held at a temperature of 60°C. The eluant is 0.35 M potassium citrate containing 0.5 M KCl (first eluant), followed by 0.35 M potassium citrate containing 2.1 M KCl, at a flow rate of 1.3 ml/min. The polyamines are detected by post-column derivatization by using o-phthalaldehyde and fluorescence detection.

15.3.8. FTIR

The principle of Fourier transform infrared spectroscopy (FTIR) as a method of characterizing and identifying prokaryotes has been pioneered by Helm, Naumann, and colleagues (100, 101, 188). Essentially this method makes use of the characteristic signals of various chemical bonds in the infrared region for detecting similarities and differences between strains. The method has a number of variants, using either an FTIR spectrophotometer (transmission) or an FTIR microscope in transmission or reflected (Raman FTIR) mode (86, 87). Brucker has taken an interest in this method and supplies suitable FTIR equipment. Apart from the FTIR instrumentation, little else is needed. In the FTIR spectrophotometer method, an aqueous cell suspension is spread as a film on a suitable carrier. Older models used a potassium bromide disk, while newer models use a microtiter plate format. Some experience is needed to get the density and thickness of the film right. Once the film is dried, the measurements can be made automatically by the programmed FTIR spectrophotometer.

The use of FTIR microscopes allows one to observe colonies as they develop on an agar plate. In transmission mode, an infrared, transparent, sterile carrier is used to make an image of the colonies (cf. the velveteen method). The use of the FTIR microscope makes it possible to use smaller colonies than those needed for the original FTIR spectrophotometer method. If suitable software is available, the images of the colonies may be detected automatically. The alternative is to use the FTIR microscope to make Raman spectra of the colonies.

Although a complete infrared spectrum is recorded, the data have to be transformed, and not all regions of the infrared spectrum are useful. Various methods of data handling have been developed, and different wavelength windows may be used. These variations may lead to different groupings, according to how the data are handled. While this consequence may be considered to be negative, Tindall et al. (274) have taken advantage of this effect and used different spectral regions to detect those groups that are always recovered. Only those groups which are always recovered are taken to be significant.

15.3.9. MALDI-TOF Analysis of Whole Cells

Proteomics has seen a number of significant developments in the area of mass spectrometry. The principle of matrix-assisted laser desorption ionization–time of flight (MALDI-TOF) is fairly straightforward, with a laser being used to excite a protein sample held on a carrier and the protein once charged, accelerating into the mass spectrometer, its time taken to reach the detector being directly related to its mass and charge. If the sample does not absorb in the wavelength of the laser, then a suitable chromophore must be added. If whole cells are used, then proteins in the cells may be analyzed. Such patterns are reminiscent of protein SDS-PAGE methods in their ranges of proteins detected. However, it is also possible to select certain windows in the analysis. Like those for FTIR, the samples need little preparation, and the cost of consumables is very low. The method is still in the early stages of development as a method for prokaryote identification and characterization. The method is clearly one of potential, since it can be used for a variety of methods relevant to prokaryote identification, and the reader is referred to a number of reviews (75, 150, 233, 278). Only a brief overview can be given of some of the methods that are developing.

15.3.9.1. Whole-Cell MALDI-TOF Analysis: Lower-Molecular-Weight Region (500 Da to 3,500 Da) (135, 234)

A small amount of growth from a bacterial colony is placed on a suitable steel sample slide. To this is added 0.7 μl of the matrix solution, acetonitrile-water-methanol (1:1:1, vol/vol/vol), which contains 0.01 M 18-crown-6-ether and 0.1% formic acid, saturated with α-cyano-4-hydroxycinnamic acid (14 mg/ml). The samples are placed in the mass spectrometer and targeted with a 337-nm-wavelength laser. This method has been developed with Kratos and Waters-Micromass MALDI-TOF instruments. Data are evaluated by software supplied for the purpose. Although the mass region up to 30,000 Da may be scanned, the regions of interest lie between 500 and 3,500 Da.

15.3.9.2. Whole-Cell MALDI-TOF Analysis: Higher-Molecular-Weight Region (2,000 Da to 20,000 Da) (255, 276)

A small amount of growth from a bacterial colony is placed on a suitable steel sample slide. To this is added the matrix solution. One study has used 0.5 μl of a cell suspension mixed with 0.5 μl of ferulic acid (10 mg/ml) in a 0.1% trifluoracetic acid-acetonitrile mixture (7:3, vol/vol) (276). Another study has used 1 μl of 2,5-dihdyroxybenzoic acid (10 mg/ml) in water-acetonitrile containing 0.03% trifluoracetic acid (255). In both cases, samples were air dried. The samples are placed in the mass spectrometer and targeted with a 337-nm-wavelength laser. These methods have used Applied Biosystems Voyager DE model MALDI-TOF instruments. The mass region between 2,000 and 30,000 Da may be scanned, but the regions of interest lie between 2,000 and 15,000 Da. There is some indication that some of the peaks may result from ribosomal proteins. The study of Valentine et al. (276) concentrated on variations caused by growth conditions as well as the testing of an algorithm for identifying those peaks in the spectrum which allowed reliable identifcation irrespective of the growth conditions.

15.3.9.3. MALDI-TOF Analysis of DNA or RNA

MALDI-TOF analysis of DNA or RNA remains to be fully explored, and the reader is referred to two recent articles which illustrate the potential (233, 284). As in the case of whole-cell MALDI-TOF analysis, one of the advantages lies in the low cost of consumables. It would appear that the application of this method may also have further advantages over more conventional gel or capillary electrophoresis separation techniques currently used to analyze DNA or RNA.

15.3.10. Multitest Systems

Various systems that allow a large number of biochemical characterization tests to be performed conveniently and rapidly are commercially available. The number of such test systems has grown in recent years. Complete information about the methods can be obtained from the manufacturers. Charts, tables, coding systems, and characterization profiles designed to be used with the systems are also available from the manufacturers. Many of the systems now have computer programs that are used for bacterial identification. Table 6 provides a listing of some of these systems; although the list is not complete, it serves to indicate the diversity of multitest identification systems presently available. The reader should be aware that, in most cases, the purpose of these systems is the identification of clinical isolates and that they may not be reliable for environmental bacteria or isolates from nonclinical sources. In that case, investigators may have to develop their own charts and identification profiles by using species from culture collections as reference strains. The reader should see references 1 and 12 for additional information and information on the accuracy and reliability of these systems in clinical microbiology. It should be noted that no identification system is 100% accurate, although some systems are more reliable and accurate than others.

15.3.11. Nitrogenase Activity and Nitrogen Fixation

The development of the acetylene reduction method has made the testing of bacterial cultures for nitrogenase activity relatively easy in comparison with the use of $^{15}N_2$ incorporation methods, although the latter remain the most definitive. The principle of the acetylene reduction method is that the nitrogenase complex of enzymes is capable of reducing acetylene to ethylene. The organism to be tested is cultured under conditions conducive to nitrogenase formation. Acetylene gas is then added to the culture vessel, and the formation of ethylene is determined after a period of incubation. Controls involving media that contain ammonium ions, which repress nitrogenase synthesis (except in *Rhizobium* spp.), should exhibit little or no ethylene production in comparison. Comprehensive reviews of the acetylene reduction method and its application are available (95, 96, 218).

It is important to culture the organisms under conditions that are favorable for nitrogenase synthesis. This involves the use of media free from fixed nitrogen but containing suitable carbon and energy sources and also a source of molybdenum. Also, an appropriate gaseous atmosphere should be supplied. For anaerobes such as *Clostridium pasteurianum*, anaerobic conditions are essential. Anaerobic conditions are also the most favorable for nitrogenase synthesis by facultative anaerobes such as *Klebsiella* species. Many nitrogen fixers require oxygen for growth and energy production, but the levels of oxygen tolerance for growth under nitrogen-fixing conditions vary considerably among such organisms. For example, *Azospirillum* cells grow best under nitrogen-fixing conditions at ca. 1% oxygen and do not grow at all under an air atmosphere (21% oxygen), whereas *Azotobacter* cells can tolerate an air atmosphere. Even aerobic nitrogen fixers, however, fix nitrogen best when grown at oxygen levels lower than that of air.

15.3.11.1. Aerobic Nitrogen Fixers

Grow the organisms in a mineral salts medium (liquid or solid) in cotton-stoppered tubes or serum bottles. An example of such a medium is the Burk medium for *Azotobacter* spp. (section 15.4.5, or 15.4.11). It is sometimes desirable to add a low level of a source of fixed nitrogen (e.g., 0.0001 to 0.005% yeast extract) to initiate growth. At the end of the incubation period, replace the cotton stopper with a sterile serum bottle stopper. Generate the acetylene gas (CAUTION: flammable!) in a fume hood by using the following procedure (218). Add a lump (ca. 1 g) of calcium carbide to a tube half filled with 15 ml of water; stopper the tube immediately with a one-hole stopper fitted to a piece of rubber tubing, and immerse the distal end of the tubing in a beaker of water. After air has been flushed out of the tubing by the acetylene but while the acetylene is still being formed, use a syringe equipped with a 26-gauge hypodermic needle to withdraw acetylene from the tubing. If a large number of tests are to be performed, acetylene may be purchased in small "lecture" bottles. Inject the acetylene into the culture vessel through the serum bottle stopper to give 10% acetylene (vol/vol). After various periods of incubation, remove 1-ml gas samples from the culture vessel with a syringe and test for the presence of ethylene by gas chromatography (see below).

15.3.11.2. Facultative and Anaerobic Nitrogen Fixers

Bubble oxygen-free filtered nitrogen gas through sterile culture media in tubes, and inoculate the media while the

TABLE 6 Selected commercial multitest systems for identification of bacteria[a]

Manufacturer[b]	System	Microbe(s) identified	Description
Abbott Laboratories, Diagnostic Division	Quantum 11	*Enterobacteriaceae*, gram-negative nonfermenters, and other gram-negative bacteria	Cartridge of multicompartment wells containing dehydrated substrates for performing 20 characterization reactions. Incubation, 4–5 h
bioMerieux Vitek, Inc.	An-Ident	Anaerobes	Plastic strip of microtubes containing dehydrated substrates for 21 characterization reactions. Aerobic incubation, 4 h
	API 20E	*Enterobacteriaceae*, gram-negative nonfermenters, and other gram-negative bacteria	Reaction strip of microtubes containing dehydrated substrates allows for 20 characterization reactions. Incubation, 24–48 h
	API 20 Strep	Streptococci	Reaction strip allows for 20 characterization reactions. Incubation, 4 h
bioMérieux	Vitek 2 Card	Fermenting and nonfermenting rods	System includes 47 biochemical tests and 19 enzyme tests and allows rapid identification in about 10 h
bioMerieux Vitek, Inc.	API 20 A	Anaerobes	Reaction strip allows for 21 characterization reactions. Incubation (anaerobic conditions), 24 h
	Rapid ID 32 A	Anaerobes	System facilitates identification of anaerobes in 4 h
	API NH	*Neisseria* spp. and *Moraxella catarrhalis*	Reaction strip allows for characterization of 10 reactions. Incubation, 2 h
	API Rapid 20E	*Enterobacteriaceae*	Similar to API 20E but with unbuffered media and smaller microtubes; 21 characterization reactions. Incubation, 4 h
	API Coryne	*Corynebacterium* and coryneforms	Reaction strip allows for 12 characterization reactions. Incubation, 24–48 h
	API Listeria	*Listeria*	Reaction strip allows for 24-h identification of all *Listeria* species
	API Campy	*Campylobacter*	Reaction strip allows for 12 characterization reactions. Incubation, 24–48 h
	API 50 CH	General use	Reaction strip allows for 50 characterization reactions. Incubation, 24–48 h
	API 20 NE	Nonenteric gram-negative rods	Reaction strip allows for 20 characterization reactions. Incubation, 24–48 h
	API 10 S	Simple identification of gram-negative rods	Reaction strip allows for 10 characterization reactions. Incubation, 4 h
	API Staph	*Staphylococci*	Reaction strip allows for overnight characterization of staphylococci and micrococci
	ID 32 Staph	Staphylococci and micrococci from humans and animals	Helps in identification of staphylococci and micrococci from humans and animals; overnight incubation
	API Rapid ID 32 Strep	Members of the genus *Streptococcus* and related genera	Allows 4-h identification of streptococci and members of related genera
Austin Biological Laboratories	RIM-N	*Neisseria* spp., *Branhamella catarrhalis*	Rapid kit (containing solution tubes) to test for utilization of selected carbohydrates. Incubation, 30–60 min
Dade Behring Inc.	MicroScan	Gram-negative fermenters and nonfermenters, gram-positive cocci, *Listeria* spp., *Neisseria* spp., and anaerobes	Performs automated and direct MIC susceptibility testing and helps ensure detection of emerging and low-level antibiotic resistance in various pathogenic microbes
Difco/Pasco Laboratories	Tri Panel	Gram-negative and gram-positive bacteria	Microplate containing hydrated substrates for 30 characterization tests. Plate stored at −20°C. Incubation, 16–20 or 40–44 h, depending on organisms

(*Continued on next page*)

TABLE 6 (*Continued*)

Manufacturer	System	Microbe(s) identified	Description
Remel Inc.	RapID ANA II	Anaerobes (medically important)	Plastic tray with cavities containing dehydrated substrates for 19 characterization reactions. Incubation (aerobic conditions), 4 h
	RapID NH	*Neisseria*, *Haemophilus*, and *Moraxella* spp.	Plastic strip with cavities containing dehydrated substrates for 13 characterization tests. Incubation, 1–4 h
	RapID CB	*Corynebacterium* and other gram-positive coryneform rods	Plastic panel containing dehydrated substrates for characterization reactions. Incubation, 4 h
	RapID ONE	More than 70 medically important, gram-negative, oxidase-negative rods	Plastic panel containing dehydrated substrates for characterization reactions. Incubation, 4 h
	RapID NF	More than 70 medically important, gram-negative, oxidase-positive rods	Plastic panel containing dehydrated substrates for characterization reactions. Incubation, 4 h
	RapID SS/u	Urinary pathogens such as *E. coli*, *Klebsiella* spp., *Enterobacter* spp., and *Proteus* spp. and staphylococci	Plastic panel containing dehydrated substrates for 11 characterization reactions. Incubation, 2 h
	RapID STR	Group A, B, C, and G strepto-cocci, staphylococci, and some gram-positive rods	Plastic strips with cavities containing dehydrated substrates for 14 characterization tests. Incubation, 4 h
	Uni-N/F-Tek	Gram-negative nonfermenters	Used for identification of oxidase-positive, non-fermentative, gram-negative bacilli. Plastic plate with 11 pie-shaped wells and one center well containing hydrated agar media; allows 18 characterization reactions
BD-Diagnostic Systems	BBL Enterotube II	*Enterobacteriaceae*	A colony is picked onto the end of a long skewer that is drawn through a large tube containing hydrated agar cells, each with a different substrate. Allows identification in 18–24 h
	Oxi-Ferm Tube II	Gram-negative nonfermenters	Similar to Enterotube II; 9 characterization reactions. Incubation, 24–48 h
Sensititre	Sensititre	*Enterobacteriaceae* and nonfermenters	Plastic microtiter plate containing dehydrated, fluorogenic substrates for 32 characterization tests. Incubation, 5–18 h. Fluorometer required
bioMerieux Vitek	GNI+ card	*Enterobacteriaceae* and other gram-negative rods	Allows fairly accurate identification of clinical anaerobes in <5 h

*a*Although only selected commercially available kit systems are presented here, there are many other diagnostic kits on the market that are not mentioned here but for which the information is readily available on the Web. For example, a number of kit systems are available just for the identification of enterohemorrhagic *E. coli* alone. Similarly, quite a few systems are available for ubiquitous microbes such as streptococci, staphylococci, and *E. coli*.

*b*Locations of manufacturers are as follows: Abbott Laboratories, Abbott Park, IL; bioMerieux Vitek, Inc., Hazelwood, MO; bioMerieux, Marcy-l'Etoile, France; Austin Biological Laboratories, Austin, TX; Dade Behring Inc., Deerfield, IL; Difco/Pasco Laboratories, Wheat Ridge, CO; Remel Inc., Lenexa, KS; BD-Diagnostic Systems, Franklin Lakes, NJ; Sensititre (now Trek Diagnostics), Salem, NH.

tubes are being flushed with nitrogen. Seal the tubes with sterile serum bottle stoppers. Strictly anaerobic conditions may not be required for facultative bacteria such as *Klebsiella* spp. because the organisms themselves may use up small residual amounts of oxygen and create highly anaerobic conditions. For examples of media, see section 15.4.39 for *Klebsiella* medium, section 15.5.15 for *Clostridium pasteurianum* medium, and section 15.4.19 for *Bacillus* medium.

Alternatively, use 40-ml Pankhurst tubes (42). The Pankhurst tube (Fig. 3) consists of two test tubes, one larger (tube A) and one smaller (tube C), with a horizontal connecting tube (tube B) filled with nonabsorbent cotton as a gas filter. Dispense 5- to 10-ml volumes of culture medium into the larger tube, and temporarily stopper the tube with cotton. Sterilize the medium by autoclaving. Inoculate the cooled medium, and then aseptically seal both the larger and the smaller tubes with sterile serum bottle stoppers. Flush the assembly with nitrogen gas (this step may not be required [218]) by inserting hypodermic needles into both stoppers, one to serve as an inlet for the flushing gas and the other to serve as an exit. After flushing, remove the needles and inject 0.75 ml of a saturated solution of pyrogallol into the smaller tube, followed by 0.75 ml of a solution containing 10% (wt/vol) NaOH and 15% (wt/vol) K_2CO_3. These additions are soaked up by a little absorbent cotton previously placed in the bottom of the small tube as shown in Fig. 3. The reagents react with residual oxygen and use it up, creating anaerobic conditions. If the Pankhurst tube has not been flushed with nitrogen, 2 to 3 h is required to establish anaerobic conditions and ca. 12 ml of nitrogen

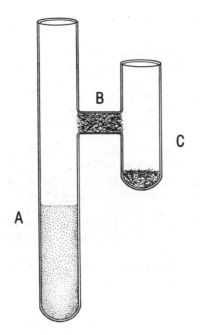

FIGURE 3 Pankhurst tube. (A) Larger test tube (1.9 by 15.7 cm) containing culture; (B) connecting tube (0.8 by 1.8 cm) with nonabsorbent cotton as gas filter; (C) smaller test tube (1.9 by 6.0 cm) with absorbent cotton to hold alkaline pyrogallol.

should be added at this time to replace the oxygen that has been removed (218).

After the culture has grown, inject acetylene through the serum bottle stopper to give a final concentration of 10% (vol/vol). Incubate the culture for various periods, and use a syringe to withdraw 1-ml gas samples for gas chromatography to determine the presence of ethylene.

15.3.11.3. Microaerophilic Nitrogen Fixers

Semisolid nitrogen-free media have been successfully used for demonstrating acetylene reduction by nitrogen-fixing spirilla; e.g., see section 15.4.5 or 15.4.6 for media for *Azospirillum* spp. Semisolid media can be incubated in an air atmosphere because they provide a stratified oxygen gradient from the surface of the medium downward. This allows microaerophilic bacteria to initiate growth at whatever oxygen level is most suitable. Growth begins as a thin band or pellicle several millimeters below the surface of the medium; as cell numbers increase, the pellicle becomes denser and migrates closer to the surface. The tube should not be agitated during growth or during subsequent addition of acetylene, since any disturbance of the pellicle may allow air to reach the organisms and inactivate the nitrogenase.

After growth has occurred, replace the cotton stopper of the tube or vial with a serum bottle stopper, and inject acetylene to give a final concentration of 10% (vol/vol). After incubation for various periods, remove 1-ml samples for gas chromatography (see below).

15.3.11.4. Gas Chromatography

See also chapter 17.3.5.

A gas chromatograph equipped with a hydrogen flame detector should be used. The column is usually ca. 5 ft (1.5 m) by 2 mm (inside diameter), filled with Porapak R or N

packing (Waters Associates). Keep the column oven temperature at 50°C, although any constant temperature between room temperature and 80°C will be satisfactory. Use nitrogen as the carrier gas, with a flow rate of ca. 50 cm³/min. Calibrate the chromatograph with dilutions of commercial ethylene (99% purity) in nitrogen.

Use plastic disposable syringes for the injection of gas samples and ethylene standards. Because of retention of some ethylene on the plastic surfaces, discard any syringe that has previously contained ethylene and replace it with a fresh syringe. Rubber serum bottle stoppers may also become contaminated with ethylene and should not be used again.

Hypodermic needles may punch out a hole in a serum bottle stopper or in the rubber septum of the gas chromatograph injector port. To prevent this, bend the tip of the needle with small pliers as shown in Fig. 4.

During chromatography with gas samples containing both acetylene and ethylene, the acetylene will take longer to pass through the Porapak column than the ethylene. Do not inject new samples into the gas chromatograph until the acetylene from the previous sample has passed from the column.

15.3.12. Antigen-Antibody Reactions

For a more detailed discussion of the principles on which tests of antigen-antibody reactions are based, see chapter 8 and reference 275.

15.3.12.1. Direct Bacterial Agglutination

Slides with ceramic rings ca. 14 mm in diameter are preferred for direct bacterial agglutination. However, glass plates marked off into 2.5-cm squares with a wax glass-marking pencil will suffice. Rehydrate the lyophilized antiserum as specified by the manufacturer. Place a drop of the antiserum into one of the rings, and place a drop of saline (0.85% NaCl) or normal rabbit serum in another ring. Mix a loopful of the growth from the surface of an agar medium into each drop, so that a homogeneous suspension of cells without lumps is obtained. The suspension should be visibly turbid but not extremely dense. Rock the slide in a circular fashion for 1 min to further mix the cells and liquid, being careful not to spill over the boundary of the ring. Examine

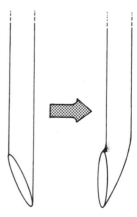

FIGURE 4 Diagram showing how the end of a hypodermic needle can be bent with a needle-nose pliers so that the needle will not punch a hole in a serum bottle stopper or in the rubber septum of a gas chromatograph injector port.

the slide by eye; for best visibility, hold the slide near a bright light and view it against a dark background. If the test is positive, the cells mixed with the antiserum show clumping (i.e., the suspension appears to be granular or, with strong reactions, even to curdle). Clumps are best seen by tilting the slide slightly so that fluid drains down toward the lower boundary of the ring. The saline or normal serum control should not show clumping. If it does, this clumping usually indicates a rough strain instead of a smooth strain. *Note*: the bacterium-antiserum mixture must not be allowed to dry, since no indication of agglutination can be obtained if drying occurs. If drying is a problem, repeat the test using a larger drop of antiserum.

CAUTION: after completion, discard the slide into a container of disinfectant (e.g., 1% Lysol). Wash hands and bench surface with disinfectant. This is important because during the mixing of cells with antiserum, small droplets may be scattered about on the slide and bench surfaces. The antiserum will not kill the bacteria.

15.3.12.2. Coagglutination

The coagglutination test is based on the use of *Staphylococcus aureus* cells that have an outer coating of protein A, which can combine directly with the Fc portion of immunoglobulin G (IgG) molecules from almost all mammals. When killed staphylococci are incubated with antibodies of the IgG class, the Fc portions of the antibodies bind to the protein A on the surface of the staphylococci. This leaves the Fab portions of the antibodies available for binding to a specific antigen. The antibody-coated staphylococci are suspended in saline and stored in a refrigerator until needed.

When the staphylococci are mixed with a saline suspension of bacterial cells for which the IgG antibodies are specific, the Fab portions of the IgG bind to the surfaces of the cells, causing visible clumping of the cells. A positive test is indicated by a granular appearance, whereas a negative test is indicated by a homogeneous, milky appearance.

An example of a commercially available coagglutination system is the Phadebact system for the identification of streptococci (Boule Diagnostics AB, Huddinge, Sweden; distributed in the United States by Remel, Lenexa, KS). In this system, stained staphylococcal cells are coated with antibodies against various Lancefield streptococcal grouping antigens (for example, the group A polysaccharide antigen characteristic of *Streptococcus pyogenes*). The coagglutination reaction produces readily visible clumps on a plastic-coated card, and no magnification is required. The procedure is simpler and much more rapid than that for the classical precipitin tests for Lancefield antigens (see below), but the manufacturer's directions should be strictly followed.

15.3.12.3. Latex Agglutination

In latex agglutination, tiny beads of polystyrene latex serve as an inert carrier of antibodies or antigens. The Fc portions of IgG molecules can bind spontaneously to latex; once the Fc portion of an IgG molecule attaches hydrophobically to the polystyrene surface, the attachment is virtually irreversible. Various soluble protein antigens, such as the proteins from a bacterial sonicate, may also bind spontaneously to latex. Latex beads can be obtained in various sizes but usually have an average diameter of 0.81 or 1.0 μm. They can be obtained with amino, carboxyl, and other functional chemical groups so that antigens can be covalently coupled to the beads; colored, fluorescent, and magnetic latex beads are also available. Latex beads and coupling reagents can be obtained from Difco and from Polyscience, Inc., Warrington, PA.

To prepare antibody-coated beads, add the inert latex particles (usually obtained as a 2.5% suspension) to antibodies of the IgG class and incubate the latex-antibody mixture for several hours (the time should be determined by the investigator). Then wash the particles, and suspend them in saline. Place a drop of the latex-antibody suspension on a slide (commercially available slides that have ceramic rings are convenient), and mix a drop of an antigen (either a soluble antigen or a whole-cell antigen) with the antibody-coated beads. A positive test is indicated by the agglutination or clumping of the particles; a negative test shows no change in the milky suspension.

An example of a latex slide agglutination test is the Campy slide test (BD [BBL]), which is used to confirm the identification of various *Campylobacter* species at the genus level. The test is based on the use of antibody-coated latex beads for the detection of certain cell wall antigens that are shared by the species (*Campylobacter jejuni*, *Campylobacter coli*, *Campylobacter fetus*, *Campylobacter lari*).

15.3.12.4. Hemagglutination

In hemagglutination, erythrocytes are used as carriers instead of bacterial cells or latex beads. Many polysaccharide and protein antigens can bind directly to the surfaces of erythrocytes. Treating erythrocytes with tannic acid stabilizes the cell membranes and gives the erythrocytes a longer shelf life.

To prepare tanned erythrocytes, pipette 3 ml of sheep blood into a 15-ml centrifuge tube, add 10 ml of physiological saline, mix, and centrifuge for 5 min at $500 \times g$. Wash the pellet three times in saline. After the final centrifugation, suspend the cells to a density of 0.5 ml of packed erythrocytes per 20 ml of saline. Add 2 ml of this cell suspension to 10 ml of distilled water. Read the optical density at 520 nm, and adjust it to between 0.4 and 0.6. Pipette 3 ml of the cell suspension into a centrifuge tube, and add 3 ml of 0.005% tannic acid (prepared in saline). Mix, and incubate at 37°C in a water bath for 10 min. Centrifuge at $500 \times g$ for 5 min. Suspend the cells in buffered saline (pH 7.2), wash them once, and suspend them in 3 ml of buffered saline.

To coat the tanned erythrocytes with a soluble antigen, add 4 ml of buffered saline to a centrifuge tube and add 1 ml of antigen solution and 1 ml of tanned erythrocytes. Incubate the mixture at room temperature for 10 min. Centrifuge it at $500 \times g$ for 5 min. Suspend the cells in 2 ml of a diluent consisting of a 1:100 dilution of normal rabbit serum in saline.

A hemagglutination test for antibodies, antigens, or other substances that react with coated erythrocytes can be done by using 13-by-100-mm tubes or a plastic microplate with round-bottom wells. A positive hemagglutination reaction is indicated by an even layer of erythrocytes coating the bottom of the tube; a negative test is indicated by a discrete button of cells at the bottom of the tube.

An interesting example of a hemagglutination test is the Staphyloslide test for the identification of *Staphylococcus aureus* (BD [BBL]). In this test, fibrinogen-coated sheep erythrocytes are used to detect a cell wall polypeptide called clumping factor, which is produced by most strains of *Staphylococcus aureus*. After the staphylococcal cells have been mixed with the coated erythrocytes, the fibrinogen on the erythrocytes reacts with the clumping factor and is converted to insoluble fibrin; this in turn causes the erythrocytes to agglutinate.

15.3.12.5. Quellung Reaction

Spread a loopful of broth culture on a clean slide, and allow it to air dry without heating. (If a similar smear stained with crystal violet shows more than 15 to 25 organisms per oil immersion field, dilute the culture, since too high a concentration of cells may obscure the results.) Rehydrate the antiserum, and place a large loopful of it on a coverslip. Mix a loopful of 1% aqueous methylene blue into the antiserum, and then place the coverslip, fluid side down, onto the dried smear on the slide. Examine by oil immersion. If the test is positive, the capsules around the blue-stained cells will appear distinctly outlined. A comparison should always be made with a control slide prepared with normal rabbit serum instead of antiserum. If the results of the test are negative, reexamine after 1 h; prevent the preparation from drying by incubating it in a moist chamber or by sealing the edges with melted Vaspar (a mixture of equal volumes of petrolatum and mineral oil).

15.3.12.6. Fluorescent-Antibody Tests

Rehydrate the fluorescent antiserum (termed the conjugate) and fluorescent normal serum. Prepare a series of twofold dilutions of the fluorescent antiserum (e.g., 1:5, 1:10, 1:20, 1:40, and 1:80) in PBS (see section 15.5.16). Using the staining procedure given below with a known culture of the organism, and using a suitable fluorescence microscope system (see chapter 1.6), determine the degree of fluorescence of the bacterial cells as obtained with each antiserum dilution. Rate the fluorescence on a scale of 0, 1+, 2+, 3+, and 4+ (maximum). The working dilution of the antiserum is one-half of the highest dilution that gives 4+ fluorescence. For example, if a 1:20 dilution of the serum is the highest dilution that gives 4+ fluorescence, the working dilution would be 1:10. An equivalent dilution of the fluorescent normal serum should give no fluorescence.

Centrifuge a young broth culture of the organism to be tested, and decant the supernatant fluid. Suspend the cells in PBS, and recentrifuge. Suspend the cells to a dense concentration in a small amount of PBS. With a diamond or tungsten carbide pencil, circumscribe a circle about the diameter of a quarter (ca. 20 to 25 mm in diameter) on a glass slide. Clean the slide thoroughly so that it is free from grease, and smear a loopful of bacterial suspension within the circle. Allow the smear to air dry without heat. Place the slide in a jar of 95% ethanol for 1 min, remove, and allow it to air dry.

Add several drops of the working dilution of fluorescent antiserum to the smear. Distribute the antiserum over the entire smear with an applicator stick, but do not touch the smear itself with the stick. Place the slide in a petri dish in which a moistened disk of filter paper has been affixed to the lid; this serves as a humid chamber to prevent evaporation of the antiserum. Incubate at 37°C for 15 to 20 min.

Drain off the excess antiserum by tilting the slide onto a piece of absorbent paper. Place the slide in a jar of PBS for 10 min, occasionally moving the slide up and down. Repeat this step two more times, using a fresh jar of PBS each time. After the third washing in PBS, dip the slide only once into a jar of distilled water to remove the PBS and then immediately blot the slide (without rubbing) with absorbent paper. Allow the slide to dry. Place a small drop of buffered glycerol mounting fluid (section 15.4.2) on the smear, and apply a coverslip. Examine the smear under the high-power dry lens or oil immersion lens by using a fluorescence microscope (see chapter 1.6). When using oil immersion, be sure to use only nonfluorescing oil.

At the same time that the test organism is being stained, prepare two controls by using a similar procedure. Make one control with the working dilution of the fluorescent antiserum and a known culture of the organism. Make the second control with the organism being tested, but use a dilution of fluorescent normal serum. The first control should give 4+ fluorescence, and the second should give no fluorescence.

Note: when examining smears under the fluorescence microscope, be sure to move the field often, because the fluorescence fades quickly (in ca. 15 s) when any particular field is being observed. Fluorescence is always brightest when a fresh field is moved into position.

Note: the working dilution of the fluorescent antiserum should be prepared fresh daily. The undiluted stock antiserum is stable when stored in a refrigerator below 8°C.

15.3.12.7. Precipitin Test (for Identification of Streptococci)

Culture the organism in 40 ml of Todd-Hewitt broth (section 15.4.37) contained in 50-ml plastic (polycarbonate or polypropylene) screw-cap centrifuge tubes. After incubation at 37°C for 18 to 24 h, centrifuge the culture at top speed in a clinical-type centrifuge with a swinging-bucket type of rotor. Decant the supernatant fluid carefully into an Erlenmeyer flask containing disinfectant, and wipe the lip of the centrifuge tube with a towel moistened with disinfectant. Be sure to decant all of the supernatant fluid without suspending the sediment (cells) at the bottom of the tube. Add 0.5 ml of saline (0.85% NaCl) to the sediment, and suspend the organisms by gently agitating the tube from side to side. Replace the screw cap tightly, and autoclave the tube for 15 min at 121°C. Centrifuge again at top speed to sediment the autoclaved cells. Taking care not to disturb the packed cells, withdraw most of the supernatant fluid with a Pasteur pipette into a small tube. The fluid must be perfectly clear; if it is cloudy, it must be recentrifuged. The supernatant fluid contains the soluble cell wall antigens which have been extracted from the streptococcal cell walls.

Rehydrate the antiserum as specified by the manufacturer. Dip a capillary tube (75 to 90 mm by 0.7 to 1.0 mm [inside diameter]) in a tilted position into the antiserum, so that 2 to 3 cm of serum rises into the capillary. Place the forefinger over the opposite end of the capillary, remove the capillary from the serum bottle, and wipe off excess serum from the outside of the capillary. Dip the end of the capillary tube into the antigen extract, and remove the forefinger to allow a rise of 2 to 3 cm of antigen. There must be no air space between the antiserum and the extract in the capillary as the extract rises. Replace the forefinger, and withdraw the capillary. Tilt the capillary to a nearly horizontal position, and with the finger removed, allow the column of fluid to move to the center part of the capillary. Leave an air space of at least 1 cm at each end. Plunge the end of the capillary into a small lump of modeling clay. Wipe the surface of the capillary free from fingerprints, and incubate it for 10 to 15 min. Examine by holding the capillary near a strong light and viewing it against a dark background with a hand lens. A positive test is indicated by the development of a milky haze near the center of the capillary at the junction of the antiserum and the extract. With further incubation, this haze will become a precipitate that settles to the bottom of the column of fluid.

15.3.13. RAPD-PCR

RAPD-PCR is a high-resolution genetic fingerprinting method for closely related strains, generally members of a single species (290, 297). Motivation for performing a RAPD-PCR may come from the need for a strain similarity or population genetic characterization. RAPD-PCR is easy and cheap to perform. There are no specific technical requirements, and no sequence knowledge of the test organism is required. Experiments can be performed with standard PCR machines, *Taq* polymerases, and agarose gel electrophoresis. The major drawback, which led to frequent criticism of this method, is problems with reproducibility. However, when conditions are kept strictly constant, reasonable reproducibility can be achieved within labs. Therefore, verification of reproducibility prior to strain collection fingerprinting is crucial.

RAPD-PCR utilizes a single primer under nonstringent PCR conditions. The primer hybridizes as both a forward and a backward primer and thereby generates a PCR product. Hybridization may occur simultaneously at several parts of the genome, and thereby a characteristic banding pattern is generated. Strains identical in genomic sequence will yield the same banding pattern, whereas strains with increasing sequence dissimilarities will yield increasingly different banding patterns. Detailed reports on RAPD-PCR performance and reproducibility and comparisons to other fingerprinting methods have been published elsewhere (34, 39, 70, 108, 127, 165, 179, 183, 199, 227, 277, 279, 286).

In the following sections, the strategic outline for planning, performing, and analyzing a RAPD-PCR study is presented.

15.3.13.1. PCR Protocol

The following protocol combines the most frequent features of previously published protocols. It requires primers of standard sizes (18 to 25 bases) generally present in the lab for other PCR purposes. RAPD reaction mix (25-μl volume) may contain 1× buffer, 1.5 mM MgCl$_2$, 0.2 μM primer, 0.75 U of *Taq* polymerase, and 100 μM (each) deoxynucleoside triphosphates. The PCR amplification protocol may be as follows: four cycles at 94°C, 40°C, and 70°C for 5 min each followed by 30 cycles at 94°C and 55°C for 1 min each and at 70°C for 2 min, with a final primer extension at 70°C for 5 min. The mixture for all steps should be pipetted on ice.

15.3.13.2. PCR Machines and DNA Polymerases

Generally, any PCR machine that allows for ramping time between denaturation, annealing, and extension steps can be used. PCR machines with an immediate switch between the PCR steps (e.g., Robocycler; Stratagene) are not recommended, since abrupt temperature changes do not allow sufficient time for the desired semispecific binding. Standard *Taq* polymerases may be used. There is no need for a *Taq* polymerase specifically designed for RAPD-PCR.

15.3.13.3. Choice of Primers

RAPD-PCR primers have to be evaluated empirically for suitability. Primer choice should start with 20 to 30 different primers. Each primer should be tested with two to four strains from the collection under investigation. Use high-quality DNA (10 ng per reaction mixture of 25-μl volume) for the RAPD-PCR at the initial stage. Approximately 10 to 30% of the primers might yield suitable banding patterns for analyses. These should ideally comprise 7 to 12 bands in the size

range from 100 to 3,000 bp. The bands should be well spaced and of sufficient intensity. The final number of primers is dependent on the desired quality of resolution of strain analyses and on the effort necessary to obtain this resolution. However, at least two different primers should be utilized.

15.3.13.4. Template DNA Concentration and Preparation

RAPD-PCR may perform well with template DNA concentrations ranging from 0.2 to 20 ng per reaction. The best concentration has to be determined empirically. The method used for template DNA preparation is dependent on the type and number of test organisms and has to be chosen empirically. For a few strains, it might be feasible to extract high-quality DNA, but for larger numbers of strains (>100), more rapid and easy-to-perform methods, such as those using enhanced but quick preparations (e.g., GeneReleaser®; Bioventures, Inc.) or simply crude extracts (fresh cell material is heated for 10 min at 99°C) should be explored. Take one to three suitable primers, identified as described in section 15.3.13.3, and explore the qualities of RAPD-PCR banding patterns derived from differently prepared template DNAs from two to four strains.

15.3.13.5. Ensuring Reproducibility of the RAPD-PCR Protocol

A reproducible RAPD banding pattern is strongly dependent on the optimal performance of all parameters involved in the PCR. This performance has to be verified empirically for every application. As a basic rule, during the investigation of a strain collection never change a working RAPD-PCR protocol. Never change the PCR machine or *Taq* polymerase. Keep all steps strictly constant.

Up to this point, suitable primers and a template DNA preparation method will have been identified (section 15.3.13.3 and 15.3.13.4). Next, the robustness of reproducibility has to be verified when experimental conditions are altered (intentionally), which should be avoided but still may occur. The following initial control experiments are recommended. (i) Perform separate RAPD-PCR runs with the same template DNA preparation by using the same master mix. This will elucidate the reproducibility of the PCR itself. (ii) Prepare several template DNA samples from the same strain by using the same DNA preparation method, and run a RAPD-PCR with these samples by using the same master mix. This will elucidate the impact of variation in the quality of the template DNA preparation on the banding pattern. (iii) Vary the DNA template concentration by at least fourfold up and down using the same DNA template preparation and the same master mix. This will elucidate the impact of the DNA template concentration. (iv) Vary, in separate experiments, the amount of *Taq* polymerase, primer, and MgCl$_2$ by at least twofold up and down and explore the impact of their variation on the banding pattern. Those primers should be selected which yield the most reproducible banding pattern even when conditions are changed intentionally. However, to exclude chances for experimental variation, prepare for each chosen primer a single master mix containing all components excluding the template and *Taq* polymerase. Prepare at least 1.3-fold more than the amount needed for the expected number of reactions, in order to have reserve material. Be sure to have enough *Taq* polymerase in stock. If necessary, mix several lots of enzyme to ensure that all template DNAs face the same *Taq* polymerase. Good experimental design pays off later.

15.3.13.6. Agarose Gel Electrophoresis

The agarose gel concentration should be between 1.2 and 2.0% and may depend on the distribution of sizes of the predominant bands obtained. For optimal band resolution, generally, the smaller the PCR products are, the higher the agarose concentration should be. Voltage should be approximately 5 V/cm but may be empirically modified. Standard agarose may be used; there is no need for specific agaroses. Use a molecular marker with preferably a band every 100 bp between 100 bp and 1,000 bp and then a band at least every 500 bp between 1,000 bp and 4,000 bp. The outermost left and right lanes of the gel should contain the marker, and not more than six sample lanes should be between two marker lanes. This setup will allow for the correction of "smiling" effects during the gel run. Load between 5 and 10 μl of the 25-μl PCR product onto the gel. To ensure optimal analyzable bands, use a gel comb with teeth at least 6 to 7 mm wide and at least 1.5 mm of space between teeth. Avoid using narrower teeth (3 to 5 mm) for the sake of adding more sample lanes to the gel, since band resolution quality will drop dramatically with narrower teeth. Explore prior to sample electrophoresis the voltage and duration of the gel run in order to ensure optimal gel length exploitation. Attention should be given to whether or not to add the DNA stain ethidium bromide (ETBR) during the gel run. ETBR is intercalated between the bases of DNA and thereby retards DNA mobility. However, the effect of mobility retardation is dependent on the amount of DNA. Larger amounts of DNA intercalate ETBR less rapidly; therefore, the retarding effect is delayed. Thus, when ETBR is present during electrophoresis, a sample with a larger amount of DNA will migrate faster than a sample of the same size with a smaller amount of DNA, (falsely) indicating a different size. Either this effect should be taken into account or the gel should be stained after electrophoresis.

15.3.13.7. Scoring of Bands for Raw-Data Extraction

The presence of a band of a given size is scored as "1," its absence as "0." Generally, bands of the same size are scored as identical. However, this scoring may be subject to interpretation when bands of the same size differ in intensity. If they differ slightly, then this may be due to slight and unintentional variations in the experimental procedures, and thus the bands may be scored as identical. If they differ dramatically in intensity, then this may be due to a difference in amplification efficiencies due to slight sequence variations in the primer hybridization site or to very different but comigrating amplicons. In such cases, the bands should be scored as different.

Band patterns may be scored manually or with the aid of computer software. Taking into account the expenditure for manpower, one to two gels with 15 to 20 sample lanes each may still be analyzed manually; for larger numbers of lanes, the aid of specific software is recommended (e.g., Gene-Profiler [Scanalytics] and GelCompar [Applied Maths]). However, despite the enormous sophistication of such software packages, the user should be aware that they are mainly data registration, storage, and processing devices, which are not able to, with 100% accuracy, replace the experienced eye of a researcher judging the band pattern. Therefore, the outcome from software-aided analyses should not be trusted blindly but should be at least randomly checked by visualizing the original gel picture. In the case of incongruencies between software evaluation and user evaluation, the user has to decide whether and how to intervene and to apply manual corrections during the software processing.

15.3.13.8. Data Analyses

Data from RAPD-PCR may be utilized for a variety of approaches. Appropriate software packages for genetic analyses are TREECON (281), MEGA 2.1 (139), and PAUP (http://paup.csit.fsu.edu/index.html). Appropriate software packages for population genetic analyses are TPFGA (http://bioweb.usu.edu/mpmbio/), PopGene (http://www.ualberta.ca/~fyeh/), and Arlequin 2.000 (231).

15.3.14. Gene Probes

The advent of molecular genetic methods coupled with the use of fluorescently labeled nucleic acid oligomers has meant that it is now possible to use such methods to detect particular gene sequences in prokaryotic cells in situ (see chapters 26 and 39). The underlying methodology is based on extensive studies of nucleic acid binding carried out in the 1960s and 1970s. Although the potential of the gene probe system remains to be fully explored, it should be remembered that gene probes detect primarily either RNA or DNA oligomers. The specificity of these gene probes depends on a number of factors, including how comprehensive the data set is on which the probe is based, as well as accurate information regarding the functions of the genes being detected (whether they encode pathogenicity factors, enzymes, etc.). In the case of the ever-growing 16S rRNA gene (and perhaps whole-genome) database, it is prudent to constantly check whether the specificity of a probe has changed as new data have been added. Most important is the fact that the taxon specificity of a gene probe (species, genus, or family specific) is only as good as the underlying taxonomy. The more comprehensive and broad based the characterization is, the greater is the chance that the classification will remain stable, and the more reliable will be the gene probes which result.

15.4. MEDIA

15.4.1. "A" Medium of King et al. (129)

Peptone	20.0 g
Glycerol	10.0 ml
K_2SO_4 (anhydrous)	10.0 g
$MgCl_2$ (anhydrous)	1.4 g
Agar (dried)	13.6 g
Distilled water	1,000 ml

Combine ingredients, and adjust the pH to 7.2. Boil to dissolve the agar, and sterilize by autoclaving at 121°C for 15 min. Cool in a slanted position.

15.4.2. Acetamide Agar (90)

Acetamide	10.0 g
NaCl	5.0 g
K_2HPO_4	1.39 g
KH_2PO_4	0.73 g
$MgSO_4 \cdot 7H_2O$	0.5 g
Phenol red	0.012 g
Agar	15.0 g
Distilled water	1,000 ml

Combine ingredients, and adjust the pH to 6.9 to 7.2. Boil to dissolve the agar, and sterilize by autoclaving at 121°C for 15 min. Cool in a slanted position.

15.4.3. Arginine Broth of Niven et al. (196)

Yeast extract	5.0 g
Tryptone	5.0 g
K_2HPO_4	2.0 g
Glucose (dextrose)	0.5 g
L-Arginine · HCl	3.0 g
Distilled water	1,000 ml

Note that although the original formula designated the D form of arginine, the L form is satisfactory (53).

Combine the ingredients, and dissolve with heating. Adjust the pH to 7.0. Boil, filter, and sterilize by autoclaving.

15.4.4. Arginine Broth of Evans and Niven (71)

Tryptone	10.0 g
Yeast extract	5.0 g
NaCl	5.0 g
L-Arginine · HCl	3.0 g
Distilled water	1,000 ml

Combine the ingredients, and dissolve with heating. Adjust the pH to 7.0. Boil, filter, and sterilize by autoclaving.

15.4.5. Autotrophic Growth on Hydrogen (159)

KH_2PO_4	2.3 g
$Na_2HPO_4 · 2H_2O$	2.9 g
NH_4Cl	1 g
$MgSO_4 · 7H_2O$	0.5 g
$NaHCO_3$	0.5 g
$CaCl_2 · 2H_2O$	0.01 g
$Fe(NH_4)$ citrate	0.05 g
Trace element solution SL-6	5 ml
Distilled water	980 ml
Agar (if necessary)	15g

Adjust the pH to 6.8.

Trace Element Solution SL-6

$ZnSO_4 · 7H_2O$	0.1 g
$MnCl_2 · 4H_2O$	0.03 g
H_3BO_3	0.3 g
$CoCl_2 · 6H_2O$	0.2 g
$CuCl_2 · 2H_2O$	0.01 g
$NiCl_2 · 6H_2O$	0.02 g
$Na_2MoO_4 · 2H_2O$	0.03 g
Distilled water	1,000 ml

Autoclave for 15 min at 121°C. Sterilize $Fe(NH_4)$ citrate (0.05 g in 20 ml of H_2O) separately, and then add it to the medium. For chemolithotrophic growth, incubate the culture under an atmosphere of 10% (vol/vol) O_2, 10% CO_2, 60% H_2, and 20% N_2. For heterotrophic growth, supplement the mineral medium with an appropriate carbon source (0.2% carbohydrate or 0.1% organic acid). For growth on nitrogen-free medium, omit NH_4Cl, and incubate the culture under an atmosphere of 2% (vol/vol) O_2, 10% CO_2, 10% H_2, and 78% N_2 or heterotrophically under 2% (vol/vol) O_2 and 98% N_2 (160).

15.4.6. *Azospirillum* Semisolid Nitrogen-Free Malate Medium (64)

KH_2PO_4	0.4 g
K_2HPO_4	0.1 g
$MgSO_4 · 7H_2O$	0.2 g
NaCl	0.1 g
$CaCl_2 · 2H_2O$	0.026 g
$FeCl_3 · 6H_2O$	0.017 g
$Na_2MoO_4 · 2H_2O$	0.002 g
L-Malic acid (or sodium malate)	3.58 (5.00) g
Bromothymol blue, 0.5% alcoholic solution	5.0g
Agar	1.75g
Distilled water	1,000 ml

Combine the ingredients, and adjust the pH to 6.8 with KOH. Sterilize by autoclaving. Note: some *Azospirillum* strains require biotin. Add 10 mg of the vitamin to 100 ml of distilled water, heat until dissolved, and add 1 ml of this stock solution per liter of nitrogen-free malate medium.

15.4.7. "B" Medium of King et al. (129)

Pancreatic digest of casein USP	10.0 g
Peptic digest of animal tissue USP	10.0 g
K_2HPO_4	1.5 g
$MgSO_4 · 7H_2O$	1.5 g
Agar (dried)	14.0 g
Distilled water	1,000 ml

Combine the ingredients, and adjust the pH to 7.2. Boil to dissolve the agar, and sterilize by autoclaving at 121°C for 15 min. Cool in a slanted position.

15.4.8. Bile Agar (53)

Beef extract	10.0 g
Peptone	10.0 g
NaCl	5.0 g
Distilled water	1,000 ml

Combine the ingredients, and dissolve with heating. Adjust the pH to 8.0 to 8.4 with 10 N NaOH. Boil for 10 min. Filter, adjust the pH to 7.2 to 7.4, and add 10.0 g of dehydrated ox gall. Readjust the pH if necessary. Add 15.0 g of agar, boil, and sterilize by autoclaving. Cool to 55°C, and aseptically add 50 ml of sterile blood serum. Mix, and dispense into plates. Note: 1% dehydrated ox gall equals 10% bile (vol/vol); to make 40% bile agar, use 4% dehydrated ox gall.

15.4.9. Blood Agar

To sterile blood agar base (see section 15.4.35) which has been melted and cooled to 45 to 50°C, add 5% (vol/vol) sterile defibrinated blood, mix well, and dispense into plates. Avoid getting plates with bubbles or froth on the surface. Defibrinated sheep or rabbit blood is preferred. Horse blood may give incorrect hemolytic reactions. Human blood-bank blood contains citrate and glucose; the citrate may be inhibitory to some bacteria, and the glucose may cause false greening rather than true alpha-hemolysis. Antibiotics or chemotherapeutic agents should not be present in blood used for making blood agar.

Prepare defibrinated blood by drawing whole blood aseptically with a syringe equipped with an 18-gauge needle (to prevent hemolysis caused by mechanical damage) and dispensing the blood immediately into a sterile flask containing a layer of sterile glass beads (ca. 3 mm). Shake the flask from side to side for 10 min. The fibrin formed during clotting will be deposited on the beads. Decant the supernatant containing blood cells and serum into a sterile container, and store it in a refrigerator.

15.4.10. Board and Holding Medium (33)

$(NH_4)H_2PO_4$	0.5 g
K_2HPO_4	0.5 g
Yeast extract	0.5 g
Bromothymol blue	0.03 g
Mineral solution (see section 15.4.20)	20 ml
Agar	5.0 g
Distilled water	880 ml

Dissolve the first three ingredients in the distilled water; add the indicator and the mineral solution. Adjust the pH to 7.2. Add the agar, and boil to dissolve. Dispense 9.0-ml volumes into tubes, and sterilize by holding momentarily at 22 lb/in.[2]. Cool to 45 to 50°C. A precipitate may form during autoclaving but should redissolve as the solution cools. To each tube, add 1.0 ml of a 5% solution of carbohydrate (usually glucose), which has previously been sterilized by filtration. Mix, and allow the medium to cool.

15.4.11. Burk Medium, Modified (206)

Solution A

$MgSO_4 \cdot 7H_2O$	0.2 g
$CaSO_4 \cdot 2H_2O$	0.1 g
$FeSO_4 \cdot 7H_2O$	0.005 g
$Na_2MoO_4 \cdot 2H_2O$	0.00024 g
Glucose	10.0 g
Distilled water	500 ml

Solution B

Potassium phosphate buffer, 0.005 M (pH 7.1)	500 ml

Autoclave solutions A and B separately, and combine equal volumes when cool.

15.4.12. Chopped-Meat Broth, Prereduced (7)

Ground beef, fat-free; use lean beef (cutter-canner grade is suitable) or horsemeat	500 g
Distilled water	1,000 ml
NaOH, 1 N	25 ml

Remove fat and connective tissue before grinding meat. Mix meat, water, and NaOH; bring to a boil while stirring. Cool to room temperature, skim fat off the surface, and filter, retaining both meat particles and filtrate. To the filtrate, add sufficient distilled water to restore the original 1-liter volume. To this filtrate add the following.

Trypticase	30.0 g
Yeast extract	5.0 g
K_2HPO_4	5.0 g
Resazurin solution, 0.025%	4.0 ml

Boil the mixture in a flask equipped with a chimney (to prevent boiling over). For a 1-liter flask, use 750 ml of broth; for a 750-ml flask, use 500 ml of broth. Boil until the resazurin changes from pink to colorless. Remove the flask from the heat, and replace the chimney with a two-hole stopper that has a gas cannula in one hole. Bubble oxygen-free CO_2 through the medium while cooling it to room temperature in an ice bath. Add the following ingredients (per liter).

Cysteine	0.5 g
Hemin solution (see section 15.4.31)	10 ml
Vitamin K_1 solution (see section 15.4.31)	0.2 ml

Adjust the pH to 7.2 with 8 N NaOH. Continue bubbling CO_2 until the pH is lowered to 7.0. Bubble oxygen-free nitrogen through the medium while dispensing it into 18-by-142-mm tubes (Bellco Glass, Inc., Vineland, NJ) that are being flushed with oxygen-free nitrogen. The tubes should already contain the meat particles prepared as described above so as to give 1 part of meat particles per 4 to 5 parts of added broth. Stopper the tubes with black rubber stoppers (size no. 1) as the gas cannula is being removed from the tubes. Autoclave the tubes in a press (to prevent the stoppers from being blown off) for 30 min. Cool. Discard any tubes that develop a light pink color on standing (indicating oxidation of the medium). *Note:* crimp-top tubes or screw-cap tubes fitted with the appropriate stoppers and caps with a hole may be substituted. The description of this method refers to roll tubes originally supplied by Bellco Glass (model no. 2044-18150), which are no longer routinely supplied but may be made to order. Alternatives are the Hungate tubes or the aluminum seal type tubes.

15.4.13. Christensen Citrate Agar (47)

Sodium citrate	3.0 g
Glucose	0.2 g
Yeast extract	0.5 g
Cysteine hydrochloride	0.1 g
Ferric ammonium citrate	0.4 g
KH_2PO_4	1.0 g
NaCl	5.0 g
Sodium thiosulfate	0.08 g
Phenol red	0.012 g
Agar	15.0 g
Distilled water	1,000 ml

Combine ingredients, and adjust the pH to 6.7. Sterilize by autoclaving. Cool in a slanted position for a 2.5-cm butt and a 3.8-cm slant.

15.4.14. Christensen Urea Agar (46)

Peptone	1.0 g
Glucose	1.0 g
NaCl	5.0 g
KH_2PO_4	2.0 g
Phenol red	0.012 g
Agar	20.0 g
Distilled water	1,000 ml
Yeast extract (optional)	0.1 g

Combine ingredients, and adjust the pH to 6.8 to 6.9. Boil to dissolve the agar, and sterilize by autoclaving. Cool to 50°C. Aseptically add sufficient 20% urea solution (sterilized by filtration) to give a final concentration of 2% urea. Mix, and aseptically dispense 2- to 3-ml volumes into sterile small tubes. Cool in a slanted position to give a 1.3-cm butt and a 2.5-cm slant.

15.4.15. *Clostridium pasteurianum* Nitrogen-Free Medium (55)

$MgSO_4 \cdot 7H_2O$	0.0493 g
$FeCl_3 \cdot 6H_2O$	0.0541 g
$MnSO_4 \cdot H_2O$	0.0034 g

$Na_2MoO_4 \cdot 2H_2O$	0.0048 g
$ZnSO_4 \cdot 7H_2O$	0.00058 g
$CuSO_4 \cdot 5H_2O$	0.0005 g
$CoCl_2 \cdot 6H_2O$	0.00048 g
K_2HPO_4	1.132 g
Biotin	trace
Sucrose	20.0 g
$CaCO_3$	3.0 g
Distilled water	1,000 ml

Note: the amount of $CaCO_3$ (which does not dissolve in the medium) can be increased to 30 g for increased buffering.

Combine ingredients, and autoclave. The medium must be made anaerobic before inoculation by bubbling oxygen-free nitrogen through it.

15.4.16. Egg Yolk Agar (7)

Peptone	20.0 g
Na_2HPO_4	2.5 g
NaCl	1.0 g
$MgSO_4$ 0.5% (wt/vol) solution	0.1 ml
Glucose	1.0 g
Agar	12.5 g
Distilled water	500 ml

Combine ingredients, and adjust the pH to 7.3 to 7.4. Boil to dissolve the agar, and sterilize by autoclaving. Cool to 60°C in a water bath. Disinfect the surface of an egg with alcohol, and allow the egg to dry. Crack the egg, and separate the yolk from the white. Add the yolk aseptically to the melted agar, and mix to obtain a homogeneous suspension. Dispense into plates, and allow to solidify. For anaerobes, use plates within 4 h or store them in an anaerobe jar until needed. Beware that eggs bought from shops may contain antibiotics, which may inhibit the strains being tested.

15.4.17. Haynes' Broth (98)

Tryptone	1.5 g
Yeast extract	1.0 g
K_2HPO_4	1.0 g
Potassium gluconate	40.0 g

Combine ingredients, and adjust the pH to 7.0. Sterilize by filtration.

15.4.18. Hino and Wilson Nitrogen-Free Medium for *Bacillus* Species (105)

Solution A

Sucrose	20.0 g
$MgSO_4 \cdot 7H_2O$	0.5 g
NaCl	0.01 g
$FeSO_4 \cdot 7H_2O$	0.015 g
$Na_2MoO_4 \cdot 2H_2O$	0.005 g
$CaCO_3$	10.0 g
Distilled water	500 ml

Solution B

p-Aminobenzoic acid	10 μg
Biotin	5 μg
K_2HPO_4-KH_2PO_4 buffer, 0.1 M (pH 7.7)	500 ml

Autoclave solutions A and B separately. When cool, combine equal volumes aseptically.

15.4.19. Hugh and Leifson Oxidation-Fermentation Medium (110)

Peptone (pancreatic digest of casein)	2.0 g
NaCl	5.0 g
K_2HPO_4	0.3 g
Bromothymol blue	0.03 g
Agar	3.0 g
Distilled water	1,000 ml

Combine ingredients, and adjust the pH to 7.1. Boil to dissolve the agar, and dispense 3- to 4-ml volumes into 13-by-100-mm tubes. Sterilize by autoclaving. Cool to 45 to 50°C. Add sufficient 10% carbohydrate solution (usually glucose) sterilized by filtration to each tube to give a final concentration of 1.0%. Mix, and allow the tubes to cool.

15.4.20. Hutner's Mineral Base (48)

Mineral Solution

Nitrilotriacetic acid	10.0 g
$MgSO_4$	14.45 g
$CaCl_2 \cdot 2H_2O$	3.335 g
$(NH_4)_6Mo_7O_{24} \cdot 4H_2O$	0.00925 g
$FeSO_4 \cdot 7H_2O$	0.099 g
Stock salts solution (see below)	50 ml
Distilled water	950 ml

Stock Salts Solution

EDTA	2.5 g
$ZnSO_4 \cdot 7H_2O$	10.95 g
$FeSO_4 \cdot 7H_2O$	5.0 g
$MnSO_4 \cdot H_2O$	1.54 g
$CuSO_4 \cdot 5H_2O$	0.392 g
$Co(NO3)_2 \cdot 6H_2O$	0.248 g
$Na_2B_4O_7 \cdot 10H_2O$	0.177 g
Distilled water	1,000 ml

Add a few drops of H_2SO_4 to the stock salts solution to decrease precipitation. To prepare the mineral solution, dissolve the nitrilotriacetic acid in the distilled water, and neutralize with approximately 7.3 g of KOH. Add the remaining ingredients, and adjust the pH to 6.8.

15.4.21. Leifson Modified Oxidation-Fermentation Medium (153)

Casitone (Difco)	1.0 g
Yeast extract	0.1 g
$(NH_4)_2SO_4$	0.5 g
Tris buffer	0.5 g
Phenol red	0.01 g
Agar	3.0 g
Artificial seawater (see section 15.4.33)	500 ml
Distilled water	500 ml

Dissolve ingredients in the distilled water, adjust the pH to 7.5, and sterilize by autoclaving. Autoclave the seawater separately. After cooling to 45 to 50°C, mix the two solutions aseptically. Add a sufficient volume of a 10 or 20% solution of the desired carbohydrate (usually glucose) sterilized by filtration to give a final concentration of 0.5 or

1.0%. Dispense 2.5- to 3.0-ml volumes of the medium aseptically into 13-by-100-mm tubes, and allow them to cool.

15.4.22. Litmus Milk (15)

Alcoholic Litmus Solution

Grind 50 g of litmus in a mortar with 150 ml of 40% ethanol. Transfer to a flask, and boil gently on a steam bath for 1 min. Decant the fluid, and save it. Add another 150 ml of 40% ethanol to the residue, and boil again for 1 min. Decant, and combine the two supernatants. Allow to settle overnight, and dilute up to 300 ml with 40% ethanol. Add 1 N HCl drop by drop to adjust the pH to 7.0.

Preparation of Medium

Add sufficient litmus solution to skim milk to give a bluish purple color (usually 40 ml/liter). Adjust the pH to 7.0 with 1 N NaOH. Sterilize by autoclaving at 115°C for 10 min. Overheating will result in caramelization.

15.4.23. Lysozyme Resistance Test Medium (1)

Peptone (Difco)	5.0 g
Beef extract (Difco)	3.0 g
Glycerol	70 ml
Distilled water	1,000 ml

Dispense 500 ml of this medium into 16-by-125-mm screw-cap glass test tubes, 5 ml per tube. Dispense the remaining medium into screw-cap dilution bottles, 95 ml per bottle. Autoclave the tubes and bottles for 15 min at 121°C. After cooling, tighten the screw caps and store at 4°C for a maximum of 2 months. To prepare lysozyme-containing medium, add 5 ml of sterile lysozyme solution (see below) to a 95-ml portion of the sterile medium and then dispense 5.0-ml portions aseptically into sterile 16-by-125-mm screw-cap test tubes. The lysozyme-containing medium can be stored at 4°C for a maximum of 2 weeks.

Lysozyme Solution

Lysozyme (Sigma)	100 mg
HCl, 0.01 N	100 ml

Sterilize by filtration through a membrane filter (pore size, 0.45 μm).

15.4.24. Malonate Broth (72)

Yeast extract	1.0 g
$(NH_4)_2SO_4$	2.0 g
K_2HPO_4	0.6 g
KH_2PO_4	0.4 g
NaCl	2.0 g
Sodium malonate	3.0 g
Glucose	0.25 g
Bromothymol blue	0.025 g
Distilled water	1,000 ml

Combine ingredients, and adjust the pH to 7.0. Sterilize by autoclaving.

15.4.25. Milk Medium, Prereduced (7)

Fresh skim milk	100 ml
Resazurin solution, 0.025%	0.4 ml
Hemin solution (section 15.4.31)	1.0 ml
Vitamin K_1 solution (section 15.4.31)	0.02 ml

Add the milk and resazurin to a flask, and prepare as described for PY broth (section 15.4.31). Add the hemin and vitamin K_1 after boiling and cooling. The pH after bubbling with CO_2 should be 7.1. Dispense the medium into tubes being purged with nitrogen, stopper the tubes, and autoclave them in a press for 12 min at 121°C.

15.4.26. Møller's Broth Base (175)

Peptide digest of animal tissue USP (see note)	5.0 g
Beef extract	5.0 g
Bromocresol purple solution, 1.6%	0.625 ml
Cresol red solution, 0.2%	0.005 g
Glucose	0.5 g
Pyridoxal	0.005 g
Distilled water	1,000 ml

Note: the original formula designates Orthana special peptone; however, Thiotone (BBL; peptic digest of animal tissue) or Difco Proteose Peptone 3 is satisfactory. Different editions of the *Manual of Clinical Microbiology* (12) give different information.

Combine ingredients, and adjust the pH to 6.0 or 6.5. Dispense into 13-by-100-mm tubes, and sterilize by autoclaving.

15.4.27. Methyl Red–Voges-Proskauer Broth

Polypeptone or buffered peptone	7.0 g
K_2HPO_4 (see note)	5.0 g
Glucose	5.0 g
Distilled water	1,000 ml

Note: for testing the Voges-Proskauer reaction of members of the genus *Bacillus*, replace the phosphate with 5.0 g of NaCl (7).

Combine ingredients, and adjust the pH so that after autoclaving and cooling the pH is 6.9 (6.5 when used for *Bacillus* species). Dispense in 5-ml volumes, and sterilize at 121°C for 10 min.

15.4.28. Nutrient Agar

Beef extract	3.0 g
Peptone	5.0 g
Agar	15.0 g
Distilled water	1,000 ml

Combine ingredients, and adjust the pH to 6.8. Sterilize by autoclaving.

15.4.29. Phenylalanine Agar

Yeast extract	3.0 g
L-Phenylalanine	1.0 g
Na_2HPO_4	1.0 g
NaCl	5.0 g
Agar	12.0 g
Distilled water	1,000 ml

Combine ingredients, and adjust the pH to 7.3. Boil to dissolve the agar, dispense into tubes, and sterilize by autoclaving. Cool the tubes in a slanted position.

15.4.30. PY Agar, Prereduced (7)

Prepare PY broth (section 15.4.31), and dispense 10-ml volumes into tubes containing 0.2 g of agar while purging the tubes with oxygen-free nitrogen. Stopper the tubes while removing the gas cannula, and autoclave them in a press for 15 min, using fast exhaust. After autoclaving, mix by inverting the tubes in the press several times. The description of this method refers to roll tubes originally supplied by Bellco Glass (model no. 2044-18150), which are no longer routinely supplied but may be made to order. Alternatives are the Hungate tubes or the aluminum seal type tubes.

15.4.31. PY Broth, Prereduced (7)

Peptone	5.0 g
Trypticase	5.0 g
Yeast extract	10.0 g
Resazurin solution, 0.025%	4.0 ml
Salts solution (see below)	40.0 ml
Hemin solution (see below)	10.0 ml
Vitamin K_1 solution (see below)	0.2 ml
Cysteine · HCl	0.5 g
Distilled water	1,000 ml

Salts Solution

$CaCl_2$ (anhydrous)	0.2 g
$MgSO_4 \cdot 7H_2O$	0.48 g
K_2HPO_4	1.0 g
KH_2PO_4	1.0 g
$NaHCO_3$	10.0 g
NaCl	2.0 g

Mix the $CaCl_2$ and $MgSO_4$ in 300 ml of distilled water until dissolved. Add 500 ml of distilled water, and while swirling, add the remaining salts. After the salts are dissolved, add 200 ml of distilled water.

Hemin Solution
Dissolve 0.05 g of hemin in 1 ml of 1 N NaOH. Dilute to 100 ml with distilled water. Autoclave at 121°C for 15 min.

Vitamin K_1 Solution
Dissolve 0.15 ml of vitamin K_1 in 30 ml of 95% ethanol.

To prepare PY broth, place dry ingredients (except cysteine) into a flask equipped with a chimney to prevent boiling over. For a 1-liter flask, use ingredients sufficient to prepare 750 ml of medium; for a 750-ml flask, prepare 500 ml of medium. Add water-salts solution and resazurin. Boil until the resazurin changes from pink to colorless. Remove the flask from the heat, and replace the chimney with a two-hole stopper that has a gas cannula in one hole. Bubble oxygen-free CO_2 through the medium while cooling to room temperature in an ice bath. Add hemin solution, vitamin K_1 solution, and cysteine. Adjust the pH to 7.1 with 8 N NaOH. Continue bubbling CO_2 until the pH is lowered to 6.9. Then bubble oxygen-free nitrogen through the medium while dispensing into 18-by-142-mm tubes (Bellco Glass, Inc.) that are being flushed with nitrogen. Stopper with black rubber stoppers (size no. 1) as the gas cannula is being removed from the tubes. Autoclave the tubes in a press (to keep the stoppers from being blown off). The description of this method refers to roll tubes originally supplied by Bellco Glass (model no. 2044-18150), which are no longer routinely supplied but may be made to order. Alternatives are the Hungate tubes or the aluminum seal type tubes.

15.4.32. Pyrazinamide Medium
For Method 1

Dubos broth base, dehydrated (Difco)	6.5 g
Pyrazinamide	0.1 g
Sodium pyruvate	2.0 g
Agar	15.0 g
Distilled water	1,000 ml

Combine ingredients, boil to dissolve the agar, and dispense 5-ml volumes into 16-by-125-mm screw-cap tubes. Sterilize by autoclaving. Allow the medium to solidify upright to form a butt.

For Method 2

Tryptic soy agar, dehydrated (Difco)	30.0 g
Yeast extract	3.0 g
Pyrazinamide	1.0 g
Tris-maleate (Sigma)	47.4 g
NaOH solution, 1 N	ca. 120 ml
Distilled water	to 1,000 ml

Add the solid ingredients to ca. 800 ml of distilled water, and dissolve. Adjust the pH to 6.0 with the 1 N NaOH solution, and add water to bring the final volume to 1,000 ml. Boil to dissolve the agar. Dispense 5-ml volumes into 16-by-160-mm screw-cap tubes, and sterilize by autoclaving. Slant the tubes before the medium solidifies.

15.4.33. Seawater, Artificial
Formula 1 (214)

NaCl	27.5 g
$MgCl_2$	5.0 g
$MgSO_4$	2.0 g
$CaCl_2$	0.5 g
KCl	1.0 g
$FeSO_4$	0.001 g
Distilled water	1,000 ml

Formula 2 (307)

NH_4NO_3	0.002 g
H_3BO_3	0.027 g
$CaCl_2$	1.140 g
$FeSO_4$	0.001 g
$MgCl_2$	5.143 g
KBr	0.100 g
KCl	0.690 g
$NaHCO_3$	0.200 g
NaCl	24.320 g
NaF	0.003 g
Na_2SiO_3	0.002 g
Na_2SO_4	4.060 g
SrCl	0.026 g
Distilled water	1,000 ml

15.4.34. Simmons' Citrate Agar (238)

Sodium citrate	2.0 g
NaCl	5.0 g
$MgSO_4$	0.2 g
$(NH_4)H_2PO_4$	1.0 g
K_2HPO_4	1.0 g

Bromothymol blue . . . . 0.08 g
Agar . . . . 15.0 g
Distilled water . . . . 1,000 ml

Combine the ingredients, and adjust the pH to 6.9. Boil to dissolve the agar, dispense into tubes, and sterilize by autoclaving. Cool in a slanted position.

15.4.35. Soybean-Casein Digest Agar
See note.

Pancreatic digest of casein USP . . . . 15.0 g
Papaic digest of soy meal USP . . . . 5.0 g
NaCl . . . . 5.0 g
Agar . . . . 15.0 g
Distilled water . . . . 1,000 ml

Combine ingredients, and adjust the pH to 7.3. Boil to dissolve the agar, and sterilize by autoclaving. *Note*: this medium is also known as tryptic soy agar or Trypticase soy agar; it can be used as a blood agar base.

15.4.36. Thornley's Semisolid Arginine Medium (269)

Peptone . . . . 1.0 g
NaCl . . . . 5.0 g
K_2HPO_4 . . . . 0.3 g
Phenol red . . . . 0.01 g
L-Arginine · HCl . . . . 10.0 g
Agar . . . . 3.0 g
Distilled water . . . . 1,000 ml

Combine ingredients, and adjust the pH to 7.2. Boil to dissolve the agar, and sterilize by autoclaving.

15.4.37. Todd-Hewitt Broth, Modified (76)

Beef heart, infusion from . . . . 1,000 ml
Neopeptone . . . . 20.0 g

Combine, and adjust the pH to 7.0 with 1 N NaOH. Then add the following.

NaCl . . . . 2.0 g
$NaHCO_3$ . . . . 2.0 g
Na_2HPO_4 . . . . 0.4 g
Glucose . . . . 2.0 g

Adjust the pH to 7.8. Boil for 15 min, filter through paper, and dispense into tubes. Sterilize by autoclaving at 121°C for 10 min.

15.4.38. Triple Sugar Iron Agar

Pancreatic digest of casein USP (see note) . . . 10.0 g
Peptic digest of animal tissue USP
   (see note) . . . . 10.0 g
Glucose . . . . 1.0 g
Lactose . . . . 10.0 g
Sucrose . . . . 10.0 g
Ferrous sulfate or ferrous ammonium sulfate . . 0.2 g
NaCl . . . . 5.0 g
Sodium thiosulfate . . . . 0.3 g
Phenol red . . . . 0.024 g
Agar . . . . 13.0 g
Distilled water . . . . 1,000 ml

Note: the following combination of ingredients can substitute for the first two components listed: beef extract, 3.0 g; yeast extract, 3.0 g; and peptone, 20.0 g.

Combine ingredients, and adjust the pH to 7.3. Boil to dissolve the agar, and dispense into tubes. Sterilize by autoclaving at 121°C for 15 min. Cool in a slanted position to give a 2.5-cm butt and a 3.8-cm slant.

15.4.39. Yoch and Pengra Nitrogen-Free Medium for *Klebsiella* Species (305)

Solution A

Na_2HPO_4 . . . . 6.25 g
KH_2PO_4 . . . . 0.75 g
Distilled water . . . . 500 ml

Solution B

$MgSO_4 · 7H_2O$ . . . . 6.25 g
$FeSO_4 · 7H_2O$ . . . . 0.04 g
Na_2MoO_4 (anhydrous) . . . . 0.005 g
Sucrose . . . . 20.0 g
NaCl . . . . 8.5 g
Distilled water . . . . 500 ml

Autoclave solutions A and B separately, and aseptically combine equal volumes when cool.

15.5. REAGENTS

15.5.1. Benedict's Reagent

Sodium citrate . . . . 17.3 g
Na_2CO_3 . . . . 10.0 g
$CuSO_4 · 5H_2O$ . . . . 1.73 g

Dissolve the first two ingredients in 80 ml of water by heating. Filter. Dilute to 85 ml. Dissolve the $CuSO_4$ in 10 ml of water, and add to the carbonate-citrate solution with stirring. Dilute to 100 ml with distilled water.

15.5.2. Buffered Glycerol Mounting Fluid

Glycerol . . . . 90 ml
PBS (section 15.5.16) . . . . 10 ml

Combine ingredients, and adjust the pH to 8.0 with NaOH. The solution should be tightly stoppered to prevent absorption of carbon dioxide. Fresh solutions should be prepared monthly.

15.5.3. Charcoal-Gelatin Disks (133)
Mix 15 g of dehydrated nutrient gelatin (Difco) with 100 ml of tap water. Heat to dissolve. Add 3 to 5 g of finely powdered charcoal, shake thoroughly, and pour into petri dishes to form a layer about 3 mm thick. India ink may be substituted for charcoal. Pour the mixture when quite cool so that it sets quickly before the charcoal can sediment. It is advisable to smear the bottom of the dish very thinly with petrolatum before pouring the gelatin. After the mixture has set, lift off the whole sheet from the dish and place it in 10% Formalin for 24 h. Then punch the formalinized sheet into disks (about 1 cm in diameter) or cut it into strips (about 5 to 8 mm by 20 mm). Wrap the pieces in gauze, and place them in a basin under running tap water for 24 h. Then

place the pieces into screw-cap bottles, and cover them with water. Sterilize by steaming for 30 min or by repeated heating in a water bath at 90 to 100°C for 20 min each time. Do not autoclave. After sterilization, decant the water and add the pieces aseptically to sterile broth (one piece per tube). Incubate the tubes containing the broth to test for sterility. The prepared media are then ready for use.

15.5.4. Developing Solution for Arginine (224)

α-Naphthol, 25% (wt/vol) solution in
    n-propanol . . . . 20 ml
Diacetyl, 1% (wt/vol) in n-propanol . . . . 2.5 ml

Combine ingredients, and dilute to 100 ml with n-propanol.

15.5.5. Ehrlich's Reagent

p-Dimethylaminobenzaldehyde . . . . 1.0 g
Ethanol, 95% . . . . 95 ml
HCl, concentrated . . . . 20 ml

Dissolve the aldehyde in the ethanol, and then add the acid. Protect from the light, and store in a refrigerator.

15.5.6. Formate-Fumarate Solution (7)

Sodium formate . . . . 3.0 g
Fumaric acid . . . . 3.0 g
Distilled water . . . . 50 ml

Combine ingredients. Add 20 pellets of NaOH with stirring until the pellets are dissolved and the fumaric acid is in a solution. Adjust the pH to 7.0 with about 15 drops of 4 N NaOH. Sterilize by filtration.

15.5.7. Kovács' Reagent

p-Dimethylaminobenzaldehyde . . . . 3.0 g
Pentanol or butanol (amyl or butyl alcohol) . . . 75 ml
HCl, concentrated . . . . 25 ml

Dissolve the aldehyde in the alcohol at 50 to 55°C. Cool, and add the acid. Protect from the light, and store in a refrigerator.

15.5.8. Lactic Acid Reagents for Optical Rotation (7)

Perchloric Acid Solution

Perchloric acid, 70% . . . . 5.8 ml

Make up to 100 ml with distilled water.

Buffer Solution

Glycine . . . . 3.75 g
Hydrazine sulfate . . . . 5.2 g

Add ingredients to 60 ml of distilled water. Adjust the pH to 9.0 with 8 N NaOH, and make the volume up to 100 ml with distilled water.

Lactic Dehydrogenase Solution

Lactic dehydrogenase, rabbit muscle
    (no. L2375; Sigma) . . . . 0.8 ml
Distilled water . . . . 0.8 ml

Make up the solution in an ice bath, and store in ice. The solution should be prepared on the same day it is used.

NAD Solution

NAD (no. N7004; Sigma) . . . . 0.16 ml
Distilled water . . . . 8.0 ml

Dissolve the NAD in the water. Store in a refrigerator. The solution should be freshly prepared.

15.5.9. McFarland Turbidity Standards (166)

Prepare a 1% solution of anhydrous $BaCl_2$ and a 1% solution of H_2SO_4. The McFarland standards consist of mixtures of the two solutions in various proportions (Table 7) to form a turbid suspension of $BaSO_4$.

Tightly seal the tubes containing these mixtures, and store them at room temperature in the dark. They will remain stable for at least 6 months. Before comparing the turbidity of a bacterial suspension with that of the standards, be sure to invert the standards several times to suspend the $BaSO_4$ precipitate evenly. Comparison of a bacterial suspension with the standards is best done by placing a white card with thick, horizontal, black lines behind the tube of bacterial suspension and the tube of standard and viewing the black lines through the tubes.

15.5.10. Methyl Green for DNase Test (243)

Methyl green (no. M295, certified biological
    stain, C.I. no. 43590; Fisher Scientific Co.,
    Pittsburgh, PA) . . . . 0.5 g
Distilled water . . . . 100 ml

Dissolve the dye in the water. Extract the solution with approximately equal volumes of chloroform until the chloroform is colorless (usually six to eight extractions). Store the resulting aqueous solution of methyl green at 4°C. The dye may be added to DNase medium before autoclaving, or it may be sterilized by filtration and added aseptically to the autoclaved medium at 45 to 50°C.

15.5.11. Nessler's Solution

KI . . . . 5.0 g
Distilled water, ammonia free . . . . 5.0 ml

TABLE 7 Preparation of McFarland turbidity standards

McFarland scale no.	Amt of 1% $BaCl_2$ (ml)	Amt of 1% H_2SO_4 (ml)
0.5	0.05	9.95
1	0.1	9.9
2	0.2	9.8
3	0.3	9.7
4	0.4	9.6
5	0.5	9.5
6	0.6	9.4
7	0.7	9.3
8	0.8	9.2
9	0.9	9.1
10	1	9

Dissolve the KI in the water, and add a saturated solution of $HgCl_2$ (about 2.0 g in 35 ml) until an excess is indicated by the formation of a precipitate. Then add 20 ml of 5 N NaOH, and dilute to 100 ml. Let any precipitate settle out, and draw off the clear supernatant.

15.5.12. Nile Blue Sulfate Fat (15)

Nile Blue Sulfate Solution

Prepare a saturated aqueous solution of Nile blue sulfate. Add 1 N NaOH drop by drop until precipitation is complete. Filter, and wash the precipitate with distilled water at pH 7.5. Dry, and store for use.

Staining of Fat

Prepare a saturated solution of the Nile blue sulfate oxazine base by using the dried precipitate prepared as described above. Mix 1 ml of the solution with 10 ml of the fat (tripropionine, tributyrin, tricaproine, tricaprylin, triolein, beef tallow, butterfat, coconut oil, corn oil, cottonseed oil, lard, linseed oil, or olive oil, as desired). If necessary, work in a heated water bath to liquefy the fat and maintain it in a liquid state throughout the subsequent washing procedure.

Washing Procedure

With fats that are liquid at room temperature, add 2 volumes of diethyl ether to the dye-fat mixture in a separating funnel; separate the red ether-soluble layer from the water layer, and wash it several times with water. (CAUTION: ether is flammable! For all operations involving ether, use a fume hood and avoid any flames or sparks that could cause an explosion. *tert*-Butylmethyl ether may be substituted for ether.) Finally, separate the ether-fat layer and evaporate off the ether. Separate the fat from the residual water, and sterilize by autoclaving. Store in a refrigerator. For use, add 1 ml of the dyed fat to 10 ml of sterile melted basal medium, mix well, and dispense into plates.

With fats that are solid at room temperature, wash the dyed fat with several changes of hot water, and then disperse the fat in a melted, neutral 0.5% agar solution by using 10 ml of fat per 90 ml of agar. Sterilize by autoclaving. For use, melt the fat emulsion and add 1 ml of fat emulsion to 20 ml of the sterile melted basal medium. Mix well, and dispense into plates.

15.5.13. Nitrite Test Reagents

For Method 1

Solution A
See cautionary note.

N-(1-Naphthyl)ethylencdiamine
  dihydrochloride (Sigma) . . . . 0.02 g
HCl, 1.5 N solution . . . . 100 ml

Dissolve by gentle heating in a fume hood.

Solution B

Sulfanilic acid . . . . 1.0 g
HCl, 1.5 N solution . . . . 100 ml

Dissolve by gentle heating.
Mix equal volumes of solutions A and B just before use to make the test reagent.

For Method 2

Solution A
See cautionary note.

N,N-Dimethyl-1-naphthylaminedihydrochloride
  (Sigma; or use α-naphthylamine) . . . . . 0.6 (0.5) g
Acetic acid, 5 N solution . . . . 100 ml

Dissolve by gentle heating in a fume hood.

Solution B

Sulfanilic acid . . . . 0.8 g
Acetic acid, 5 N solution . . . . 100 ml

Dissolve by gentle heating in a fume hood.
CAUTION: α-naphthylamine is a carcinogen. Although N-(1-naphthyl)ethylenediamine and N,N-dimethyl-1-naphthylamine have not been listed as carcinogenic, one would suspect on the basis of structural similarity that they might also be dangerous. All precautions pertinent to the use, handling, and subsequent disposal of carcinogens should be observed for these compounds.

15.5.14. Peptidase-Developing Reagents (56)

Reagent A

Tris . . . . 250 g
HCl, 37% . . . . 110 ml
SDS . . . . 100 g
Distilled water . . . . to 1,000 ml

The final pH should be 7.6.

Reagent B

Fast Blue BB . . . . 3.0 g
2-Methoxyethanol . . . . 1,000 ml

15.5.15. pH Indicators for Culture Media

Some common pH indicators for use in culture media are listed in Table 8. Also see chapter 14.1.1.

15.5.16. PBS

10× Stock Solution

Na_2HPO_4, anhydrous, reagent grade . . . . 12.36 g
$NaH_2PO_4 \cdot H_2O$, reagent grade . . . . 1.80 g
NaCl, reagent grade . . . . 85.00 ml

Dissolve ingredients in distilled water to a final volume of 1,000 ml.

Working Solution (0.01 M Phosphate, pH 7.6)

Stock solution . . . . 100 ml
Distilled water . . . . 900 ml

15.5.17. PHB Granule Suspension (60)

Bacillus megaterium KM (ATCC 13632) was originally used as the source of the granules, but *Pseudomonas* strains may be used alternatively and may give higher yields of PHB (209).

Culture the bacterial strain under conditions conducive to PHB production. For *Bacillus megaterium*, these conditions are described by Macrae and Wilkinson (157); for pseudomonads, see reference 258. When the level of PHB is sufficiently high, as judged by staining with Nile blue A (see section 15.2.70), harvest the cells by centrifugation and suspend them to a concentration of 15% (wt/vol) in

TABLE 8 pH indicators for culture media

Name	pK'	pH range and colors[a]	Concn[b] (g/liter) usually used in media	Amt (ml) of 1 N NaOH needed to dissolve 0.1 g
Phenol red	7.8	6.9(Y)–8.5(R)	0.010–0.030	28.2
Bromothymol blue	7.1	6.1(Y)–7.7(B)	0.010–0.032	16
Bromocresol purple	6.2	5.4(Y)–7.0(P)	0.010–0.032	18.5
Chlorophenol red	6	5.1(Y)–6.7(R)	0.015	23.6
Bromocresol green	4.7	3.8(Y)–5.4(B)	0.02	14.3

[a]Abbreviations: Y, yellow; B, blue; P, purple; R, red.
[b]For addition to culture media, dissolve in alcohol or prepare an aqueous solution with 0.01 NaOH.

ice-cold 0.02 M phosphate buffer, pH 7.2 (see section 15.5.18). Disrupt the cells by sonication (see chapter 7.1.1.2), and centrifuge at high speed. Suspend the pellet in the alkaline hypochlorite reagent of Williamson and Wilkinson (see section 15.5.20) to a density of ca. 8 mg (dry weight)/ml or less (299). Alternatively, commercial 5% hypochlorite (Clorox bleach) can be used (209).

Incubate at room temperature for 2 days with stirring. Either allow the suspension to settle and carefully remove and discard the supernatant fluid, or collect the granules by centrifugation. Suspend the granules in water, and dialyze against running tap water until free from chloride (use AgNO$_3$ on a small portion; there should be no precipitate of AgCl formed). Dialyze further against distilled water.

In a series of centrifugations, wash the granules several times with acetone. Place the granules in a Soxhlet apparatus, and continuously extract them with acetone-ether (2:1, vol/vol) for 3 days to remove non-PHB lipids. Further extract the granules with hot diethyl ether, pulverize them, and dry them under a vacuum. Determine the dry weight of the granules. After drying, disperse the granules in 0.02 M phosphate buffer by means of a tissue grinder and sonic oscillator to form a stable, concentrated suspension. Sterilize the suspension by autoclaving, and store it in a refrigerator. Calculate how much of the concentrated suspension to add to molten mineral agar to give a final concentration of 0.25% (wt/vol) in the overlay described in section 15.2.71.

15.5.18. Potassium Phosphate Buffer

Prepare 0.2 M solutions of K$_2$HPO$_4$ and KH$_2$PO$_4$. Mix the solutions in the ratios shown in Table 9 to obtain the appropriate pH. The mixture (0.2 M phosphate buffer) can be further diluted to give the desired molarity. Also see chapter 14.1.2.

TABLE 9 Numbers of parts of K$_2$HPO$_4$ solution and KH$_2$PO$_4$ solution required to obtain potassium phosphate buffer of various pHs

pH	No. of parts of:	
	K$_2$HPO$_4$	KH$_2$PO$_4$
6.8	49	51
6.9	55	45
7.0	61	39
7.1	67	33
7.2	72	28
7.3	77	23

15.5.19. Standard Solutions for Gas Chromatography (7)

Standard Mixture of Volatile Fatty Acids (ca. 1 meq/100 ml)

Formic acid	0.037 ml
Acetic acid, glacial	0.057 ml
Propionic acid	0.075 ml
Isobutyric acid	0.092 ml
Butyric acid	0.091 ml
Isovaleric acid	0.109 ml
Valeric acid	0.109 ml
Isocaproic acid	0.126 ml
Caproic acid	0.126 ml
Heptanoic acid	0.142 ml
Distilled water	100 ml

When using the standard mixture, acidify, and extract 1 ml of the solution with ether for gas chromatography.

Standard Mixture of Alcohols
The following list shows the amount of each component to be added to 100 ml of distilled water to obtain the final concentration shown in parentheses.

Ethanol	0.1 ml (1.7 mM)
Propanol	0.035 ml (0.5 mM)
Isobutanol	0.005 ml (0.05 mM)
Butanol	0.01 ml (0.1 mM)
Isopentanol	0.005 ml (0.05 mM)
Pentanol	0.005 ml (0.05 mM)

Extract 1 ml of the mixture with ether for gas chromatography.

Standard Mixture of Nonvolatile Acids (ca. 1 meq/100 ml)

Pyruvic acid	0.068 ml
Lactic acid, 85% syrup	0.084 ml
Oxalacetic acid	0.06 g
Oxalic acid	0.06 g
Methylmalonic acid	0.060 g
Malonic acid	0.06 g
Fumaric acid	0.06 g
Succinic acid	0.06 g
Distilled water	100 ml

Methylate 1 ml of the mixture, and extract with chloroform.

15.5.20. Williamson and Wilkinson Hypochlorite Reagent (299)

Triturate 200 g of fresh bleaching powder (chlorinated lime) with a little distilled water, and make the volume up to 1 liter. With stirring, add 1 liter of 30% (wt/vol) Na_2CO_3. Allow the mixture to stand for 2 to 3 h with shaking at intervals. Filter through paper. Adjust the pH of the filtrate to 9.8 with concentrated HCl. Warm the solution to 37°C, and remove the precipitate by filtration. Store the clear filtrate in a stoppered bottle in a refrigerator. The solution is stable for several months.

15.6. SPRAY REAGENTS FOR THIN-LAYER CHROMATOGRAPHY

CAUTION: all spray reagents should be considered to be toxic or corrosive. Prepare spray reagents in a fume cupboard. Solutions will heat up when concentrated sulfuric acid is added.

All plates must be sprayed in a fume cupboard, and using a properly constructed spraying chamber minimizes risks.

When plates are heated, any solvents should be allowed to evaporate before heating. Thin-layer chromatography plates may be heated in a properly ventilated oven or on a heating plate (Thermoplate S, Desaga) placed in the fume cupboard.

A comprehensive listing of spray reagents may be found in a number of good reference works (9, 256, 309).

15.6.1. Sugars

15.6.1.1. Anisaldehyde-Sulfuric Acid

Reagent 1

Add 1 ml of concentrated sulfuric acids to 50 ml of acetic acid, followed by 0.5 ml of anisaldehyde (4-methoxybenzaldehyde). Spray the plates evenly.

Heat the plates at 100 to 105°C until the spots have reached maximum color intensity. This reagent will detect terpenes (including straight chain isoprenoids), fatty acids (purple), and sugars (green, although glucose will stain blue).

Reagent 2

Mix 9.0 ml of ethanol, 0.5 ml of anisaldehyde (4-methoxybenzaldehyde), 0.5 ml of ethanol, and 0.1 ml of acetic acid. Spray the plates evenly.

Heat the plates at 90 to 100°C.

15.6.1.2. α-Naphthol-Sulfuric Acid

Mix 1.5 g of α-naphthol with 50 ml of ethanol, and then add 4 ml of distilled water and 6.5 ml of sulfuric acids. Spray the plates evenly.

Heat the plates for about 5 min at 100°C. Sugar-containing lipids produce a red-purple spot.

15.6.2. Ninhydrin (for Lipids Containing Free Amino Groups)

Dissolve 0.2 g of ninhydrin in 100 ml of ethanol. Spray the plates evenly.

Heat the plates for about 15 min at 100°C. Most lipids with free amino groups will produce red-purple spots. However, there may be slight differences in the shade of the color and some compounds may produce yellow to orange spots. It is possible to overspray the ninhydrin plates with the phosphate reagent. The red-purple spots decolorize, and the blue, phosphate-positive spots appear after about 15 min (see section 15.6.7).

15.6.3. Secondary Amines

Either generate chlorine gas from a sodium hypochlorite solution or use a lecture bottle of chlorine gas. Place the damp thin-layer chromatography plates in a small thin-layer chromatography tank or sealable plastic box which is flooded with chlorine. Leave the thin-layer chromatography plates in the chlorine gas for about 5 min. Remove the thin-layer chromatography plates, and allow them to stand in the fume cupboard until no chlorine is left in the silica gel (about 2 h). To test for the presence of chlorine, spray a corner of the plate or a suitable blank plate with either of the two reagents listed below.

Reagent 1

Mix 10 ml of a saturated o-toluidene solution in 2% acetic acid and 0.85% potassium iodide. Spray the plates evenly, and leave at room temperature.

Secondary amines (which bind chlorine) will give blue-black spots.

Reagent 2

Mix 10 ml of 2% potassium iodide solution and 10 ml of 2% soluble starch. Spray the plates evenly.

Secondary amines (which bind chlorine) will give blue-black spots.

15.6.4. Periodate-Schiff

Solution 1

Prepare 0.5% sodium periodate in water.

Solution 2

Prepare an aqueous solution of 0.5% pararosanaline base. Decolorize the solution with SO_2. SO_2 may be obtained either from a lecture bottle or from the acidification of sodium metabisulfite.

Spray the plates evenly with solution 1. Place the plates in a sealed plastic box, and leave for 15 minutes. Treat the plates (while still damp) with SO_2. The plates will initially turn brown-yellow, due to the release of iodine, but should be colorless at the end of the treatment. Spray the plates lightly with solution 2, and treat with SO_2. If possible, leave the plates in an atmosphere of SO_2. Purple spots, which appear almost immediately (rapid periodate-Schiff positive), are due to the presence of formaldehyde, which is released from the terminal CH_2OH group, such as in glycerol. Sugars will slowly develop a purple color, turning blue overnight. Lipids containing only inositol will appear as yellow-brown spots.

15.6.5. Dragendorff's Reagent

Stock Solution

Prepare a solution of 8.0 g of basic bismuth nitrate in 20 to 25 ml of 25% nitric acid. Prepare a slurry of 20 g of potassium iodide in 1 ml of 6 N HCl and 5 ml of water. Add the acidic bismuth nitrate solution to the potassium iodide slurry. Slowly add water to this mixture until an orange-red solution is obtained. The final volume should be about 95 ml. Remove any insoluble residue by filtration, and make up to 100 ml with water. This stock solution keeps for several weeks in the dark in the refrigerator.

Spray Reagent

Mix the spray reagent in the following order.

20 ml of water
6 ml of 6 N HCl
2 ml of the stock solution
6 ml of 6 N freshly prepared NaOH

If all of the bismuth hydroxide does not dissolve on shaking, then a few drops of 6 N HCl should be added. Spray the plates evenly.

Quaternary nitrogen compounds (such as phosphatidylcholine) produce a red-orange color.

15.6.6. Molybdophosphoric Acid

Dissolve 5% molybdophosphoric acid in ethanol. The solution is initially yellow-green but turns green with storage. Spray the plates evenly.

Heat at 150°C for 15 min. All lipids (with fatty acid, isoprenoid, or terpenoid side chains) stain blue-black. The background is yellow-green for the freshly prepared solution but green for the stored solution. Note that lipids containing no, or only a small proportion of, unsaturated side chains will either take longer to stain or be stained more quickly by increasing the temperature.

15.6.7. Phosphate Reagents

CAUTION: both reagents present a risk to health during their preparation, and they are only presented for completeness. A suitable phosphate spray reagent is available commercially from Sigma.

Reagent 1, Dittmer and Lester

Solution 1
Add 40.11 g of MoO_3 to 1 liter of 25 N H_2SO_4. The mixture is boiled until the MoO_3 is dissolved.

Solution 2
To 500 ml of solution 1, add 1.78 g of powdered molybdenum. Boil gently for 15 min. Cool, and decant the solution from any residue, which may remain.

Spray reagent
Mix equal volumes of solution 1 and solution 2, and then dilute with 2 volumes of water. Spray the plates evenly.

Phosphate-containing lipids appear within about 15 min as blue spots. Plates may be previously sprayed with the ninhydrin reagent (the outline of the positive spots being marked with a soft pencil), and then sprayed with the phosphate reagent (see section 15.6.2). In order to visualize all lipids, the plates may also be heated at 100°C for 15 min after the presence of phosphate-positive spots has been noted.

Reagent 2, Vaskovsky and Kostetsky

Solution 1
Sixteen grams of ammonium molybdate are dissolved in 120 ml of water.

Solution 2
Shake 80 ml of solution 1 with 40 ml of concentrated HCl and 10 ml of mercury for 30 min. Filter to recover solution 2. Combine all of solution 2 with the remainder of solution 1. Cool the mixture and dilute to 1 liter. Spray the plates evenly.

Phosphate-containing lipids appear within about 15 min as blue spots. Plates may be previously sprayed with the ninhydrin reagent (the outline of the positive spots being marked with a soft pencil), and then sprayed with the phosphate reagent (see section 15.6.2).

The residue (containing mercury) may be diluted with water and strips of aluminum foil in order to recover the mercury (144).

15.7. REFERENCES

15.7.1. General References

1. **Baron, E. J., and S. M. Finegold.** 1990. *Bailey & Scott's Diagnostic Microbiology,* 8th ed. The C. V. Mosby Co., St. Louis, MO.
   Contains a wealth of information about clinical characterization methods as well as recipes for reagents and culture media.
2. **Bascomb, S.** 1987. Enzyme tests in bacterial identification, p. 105–160. *In* R. R. Colwell and R. Grigorova (ed.), *Methods in Microbiology,* vol. 19. Academic Press, Inc., New York, NY.
   A comprehensive review of the various substrates and testing methods for detecting the activities of enzymes such as esterases, glycosidases, oxidases, nucleases, peptidases, and phosphatases.
3. **Blazevic, D. J., and G. M. Ederer.** 1975. *Principles of Biochemical Tests in Diagnostic Microbiology.* John Wiley & Sons, New York, NY.
   Contains a good overview of the bases of many biochemical tests.
4. **Garrity, G. M. (ed.).** 2001. *Bergey's Manual of Systematic Bacteriology,* 2nd ed., vol. 1. Springer-Verlag, New York, NY.
   Describes the genera and species of the archaea, the "deep rooting Bacteria," and the cyanobacteria.
5. **Goodfellow, M., and D. E. Minnikin.** 1985. *Chemical Methods in Bacterial Systematics.* SAB technical series 20. Academic Press, London, United Kingdom.
   An overview of chemotaxonomic methods.
6. **Goodfellow, M., and A. G. O'Donnell.** 1994. *Chemical Methods in Prokaryotic Systematics.* John Wiley & Sons, Chichester, United Kingdom.
   An overview of chemotaxonomic methods.
7. **Holdeman, L. V., E. P. Cato, and W. E. C. Moore (ed.).** 1977. *Anaerobe Laboratory Manual,* 4th ed. Virginia Polytechnic Institute and State University, Blacksburg.
   Identification charts, characterization methods, and media for anaerobic bacteria; describes in detail the preparation and use of prereduced media and culture techniques.
8. **Holding, A. J., and J. G. Colee.** 1971. Routine biochemical tests, p. 2–32. *In* J. R. Norris and D. W. Ribbons (ed.), *Methods in Microbiology,* vol. 6A. Academic Press, Inc., New York, NY.
   Principles and procedures for a wide variety of routine biochemical characterization tests.
9. **Kates, M.** 1986. *Techniques of Lipidology: Isolation, Analysis and Identification of Lipids,* 2nd ed. Elsevier, Amsterdam, The Netherlands.
   A collection of methods for the analysis of lipids.
10. **Krieg, N. R., and J. G. Holt (ed.).** 1984. *Bergey's Manual of Systematic Bacteriology,* vol. 1. The Williams & Wilkins Co., Baltimore, MD.
    Describes gram-negative chemoheterotrophic eubacteria that have clinical, industrial, or agricultural importance. Includes many identification tables and keys.
11. **Lanyi, B.** 1987. *Classical and Rapid Identification Methods for Medically Important Bacteria,* p. 1–67. *In* R. R. Colwell and R. Grigorova (ed.), *Methods in Microbiology,* vol. 19. Academic Press, Inc., New York, NY.
    An extensive compilation of morphological, cultural, and biochemical methods that are commonly used for identifying bacteria in the clinical microbiology laboratory.
12. **Murray, P. R., E. J. Baron, J. H. Jorgensen, M. L. Landry, and M. A. Pfaller.** 2007. *Manual of Clinical Microbiology,* 9th ed. ASM Press, Washington, DC.
    Contains microscopic, cultural, biochemical, and serological methods for the identification of pathogenic microorganisms.
13. **Ratledge, C., and S. G. Wilkinson.** 1988. *Microbial Lipids,* vol. 1. Academic Press, London, United Kingdom.

A collection of review articles dealing with the lipid compositions of microorganisms.

14. **Ratledge, C., and S. G. Wilkinson.** 1988. *Microbial Lipids,* vol. 2. Academic Press, London, United Kingdom.
A collection of articles covering the biosynthesis of the lipid compositions of microorganisms.

15. **Skerman, V. B. D.** 1967. *A Guide to the Identification of the Genera of Bacteria.* The Williams & Wilkins Co., Baltimore, MD.
Based on the 7th edition of Bergey's Manual of Determinative Bacteriology. The methods section is a gold mine of information on useful procedures and media for many kinds of bacteria.

16. **Skerman, V. B. D.** 1969. *Abstracts of Microbiological Methods.* Wiley-Interscience, New York, NY.
Descriptions of a wide variety of routine physiological and biochemical characterization methods, abstracted from original articles.

17. **Sneath, P. H. A., N. S. Mair, M. E. Sharpe, and J. G. Holt (ed.).** 1986. *Bergey's Manual of Systematic Bacteriology,* vol. 2. The Williams & Wilkins Co., Baltimore, MD.
Describes gram-positive chemoheterotrophic eubacteria that have clinical, industrial, or agricultural importance. Includes many identification tables and keys.

18. **Staley, J. T., M. P. Bryant, N. Pfennig, and J. G. Holt (ed.).** 1989. *Bergey's Manual of Systematic Bacteriology,* vol. 3. The Williams & Wilkins Co., Baltimore, MD.
Describes phototrophs, autotrophs, sheathed and budding bacteria, myxobacteria, archaeobacteria, and other aquatic and soil bacteria. Includes many identification tables and keys.

19. **Tilton, R. C. (ed.).** 1982. *Rapid Methods and Automation in Microbiology. Proceedings of the Third International Symposium on Rapid Methods and Automation in Microbiology.* American Society for Microbiology, Washington, DC.
Rapid methods, automation, computers, and miniaturized techniques relating to the characterization and identification of bacteria.

20. **Williams, S. T., M. E. Sharpe, and J. G. Holt (ed.).** 1989. *Bergey's Manual of Systematic Bacteriology,* vol. 4. The Williams & Wilkins Co., Baltimore, MD.
Describes the genera and species of actinomycetes. Includes many identification tables and keys.

15.7.2. Specific References

20. **Adachl, T., Y. Murooka, and T. Harada.** 1973. Depression of arylsulfatase synthesis in *Aerobacter aerogenes* by tyramine. *J. Bacteriol.* **116:**19–24.

21. **Adler, J.** 1966. Chemotaxis in bacteria. *Science* **153:**708–716.

22. **Ahlborn, F.** 1895. Über die Wasserblüte Byssus Flos-aquae und ihr Verhalten gegen Druck. *Verh. Naturwiss. Ver. Hamburg III* **2:**25.

23. **Anderson, R.** 1983. Alkylamines: novel lipid constituents in *Deinococcus radiodurans. Biochem. Biophys. Acta* **753:**266–268.

24. **Anderson, R., and K. Hansen.** 1985. Structure of a novel phosphoglycolipid from *Deinococcus radiodurans. J. Biol. Chem.* **260:**12219–12223.

25. **Badger, J. H., J. A. Eisen, and N. L. Ward.** 2005. Genomic analysis of *Hyphomonas neptunium* contradicts 16S rRNA gene-based phylogenetic analysis: implications for the taxonomy of the orders "Rhodobacterales" and Caulobacterales. *Int. J. Syst. Evol. Microbiol.* **55:**1021–1026.

26. **Balderston, W. L., B. Sherr, and W. J. Payne.** 1976. Blockage by acetylene of nitrous oxide reduction in *Pseudomonas perfectomarinus. Appl. Environ. Microbiol.* **31:**504–508.

27. **Bapteste, E., and C. Brochier.** 2004. On the conceptual difficulties in rooting the tree of life. *Trends Microbiol.* **12:**9–13.

28. **Baumann, L., R. D. Bowditch, and P. Baumann.** 1983. Description of *Deleya* gen. nov. created to accommodate the marine species *Alcaligenes aestus, A. pacificus, A. cupidus, A. venustus,* and *Pseudomonas marina. Int. J. Syst. Bacteriol.* **33:**793–802.

29. **Bernaerts, M. J., and J. De Ley.** 1963. A biochemical test for crown gall bacteria. *Nature* (London) **187:**406–407.

30. **Beveridge, T. J., and J. A. Davies.** 1983. Cellular responses of *Bacillus subtilis* and *Escherichia coli* to the Gram stain. *J. Bacteriol.* **156:**846–858.

31. **Biebl, H., and G. Drews.** 1969. Das *in vivo* Spektrum als taxonomische Merkmal bei Unteruschungen zur Verbreitung der Athiorhodaceae. *Zentbl. Bakteriol. Parsitenkd. Infektkrankh. Hyg. Abt. 2* **123:**425–452.

32. **Bille, J., and M. P. Doyle.** 1991. *Listeria* and *Erysipelothrix,* p. 287–295. *In* A. Balows, W. J. Hausler, Jr., K. L. Herrmann, H. D. Isenberg, and H. J. Shadomy (ed.), *Manual of Clinical Microbiology,* 5th ed. American Society for Microbiology, Washington, DC.

33. **Board, R. G., and A. J. Holding.** 1960. The utilization of glucose by aerobic gram-negative bacteria. *J. Appl. Bacteriol.* **23:**xi.

34. **Boerlin, P., E. Bannerman, F. Ischer, J. Rocourt, and J. Bille.** 1995. Typing *Listeria monocytogenes:* a comparison of random amplification of polymorphic DNA with 5 other methods. *Res. Microbiol.* **146:**35–49.

35. **Botsford, J. L., and R. D. DeMoss.** 1971. Catabolic repression of tryptophanase in *Escherichia coli. J. Bacteriol.* **105:**303–312.

36. **Brenner, D. J., J. T. Staley, and N. R. Krieg.** 2001. Classification of prokaryotic organisms and the concept of bacterial speciation, p. 27–31. *In* G. M. Garrity (ed.), *Bergey's Manual of Systematic Bacteriology,* 2nd ed., vol. 1. Springer-Verlag, New York, NY.

37. **Bruns, A., H. Cypionka, and J. Overmann.** 2002. Cyclic AMP and acyl homoserine lactones increase the cultivation efficiency of heterotrophic bacteria from the central Baltic Sea. *Appl. Environ. Microbiol.* **68:**3978–3987.

38. **Bürger, H.** 1967. Biochemische Leistungen nicht proliferierender Mikroorganismen. II. Nachweis von Glycosid-Hydrolasen, Phosphatasen, Esterasen und Lipasen. *Zentbl. Bakteriol. Parasitenkd. Infektkrankh. Hyg. Abt. 1 Orig.* **202:**97–109.

39. **Busse, H. J., E. B. M. Denner, and W. Lubitz.** 1996. Classification and identification of bacteria: current approaches to an old problem. Overview of methods used in bacterial systematics. *J. Biotechnol.* **47:**3–38.

40. **Busse, H.-J., and G. A. Auling.** 1988. Polyamine patterns as a chemotaxonomic marker within the Proteobacteria. *Syst. Appl. Microbiol.* **11:**1–8.

41. **Cain, A. J., and G. A. Harrison.** 1960. Phyletic weighting. *Proc. Zool. Soc. Lond.* **135:**1–31.

42. **Campbell, N. E. R., and H. J. Evans.** 1969. Use of Pankhurst tubes to assay acetylene reduction by facultative and anaerobic nitrogen-fixing bacteria. *Can. J. Microbiol.* **15:**1342–1343.

43. **Cato, E. P., and W. E. C. Moore.** 1965. A routine determination of the optically active isomers of lactic acid for bacterial identification. *Can. J. Microbiol.* **11:**319–324.

44. **Cato, E. P., D. E. Hash, L. V. Holdeman, and W. E. C. Moore.** 1982. Electrophoretic study of *Clostridium* species. *J. Clin. Microbiol.* **15:**688–702.

45. **Cerny, G.** 1976. Method for the distinction of Gram-negative from Gram-postive bacteria. *Eur. J. Appl. Microbiol.* **3:**223–225.

46. **Christensen, W. B.** 1946. Urea decomposition as a means of differentiating *Proteus* and paracolon cultures from each

other and from *Salmonella* and *Shigella* types. *J. Bacteriol.* **52**:461–466.

47. **Christensen, W. B.** 1949. *Hydrogen Sulfide Production and Citrate Utilization in the Differentiation of Enteric Pathogens and Coliform Bacteria.* Research Bulletin no. 1. Weld County Health Department, Greeley, CO.

48. **Cohen-Bazire, G., W. R. Sistrom, and R. Y. Stanier.** 1957. Kinetic studies of pigment synthesis by nonsulfur purple bacteria. *J. Cell. Comp. Physiol.* **49**:25-68.

49. **Collins, M. D.** 1985. Isoprenoid quinone analyses in bacterial classification and identification, p. 267-287. *In* M. Goodfellow and D. E. Minnikin (ed.), *Chemical Methods in Bacterial Systematics.* SAB technical series 20. Academic Press, London, United Kingdom.

50. **Collins, M. D.** 1994. Isoprenoid quinones, p. 345–401. *In* M. Goodfellow and A. G. O'Donnell (ed.), *Chemical Methods in Prokaryotic Systematics.* John Wiley & Sons, Chichester, United Kingdom.

51. **Constance, L.** 1964. Systematic botany—an unending synthesis. *Taxon* **13**:257–273.

52. **Corbel, M. J., and W. J. Brinley-Morgan.** 1984. Genus *Brucella*, p. 377–388. *In* N. R. Krieg and J. G. Holt (ed.), *Bergey's Manual of Systematic Bacteriology*, vol. 1. The Williams & Wilkins Co., Baltimore, MD.

53. **Cowan, S. T.** 1974. *Cowan and Steel's Manual for the Identification of Medical Bacteria*, 2nd ed. Cambridge University Press, London, United Kingdom.

54. **Cowan, S. T.** 1978. A *Dictionary of Microbial Taxonomy.* Cambridge University Press, Cambridge, United Kingdom.

55. **Daescb, G., and L. E. Mortensen.** 1972. Effect of ammonia on the synthesis and function of the N_2-fixing enzyme system in *Clostridium pasteurianum. J. Bacteriol.* **110**:103–109.

56. **D'Amato, R. F., L. A. Enriquez, K. M. Tomfohrde, and E. Singerman.** 1978. Rapid identification of *Neisseria gonorrhoeae* and *Neisseria meningitidis* by using enzymatic profiles. *J. Clin. Microbiol.* **7**:77–81.

57. **Davies, J. A., G. K. Anderson, T. J. Beveridge, and H.C. Clark.** 1983. Chemical mechanism of the Gram stain and synthesis of a new electron-opaque marker for electron microscopy which replaces the iodine mordant. *J. Bacteriol.* **156**:837-845.

58. **Dawson, R. M. C., D. C. Elliott, W. H. Elliott, and K. M. Jones.** 1986. *Data for Biochemical Research*, 3rd ed. Oxford Science Publications, Oxford University Press, Oxford, United Kingdom.

59. **Dees, S. B., C. W. Moss, D. G. Hollis, and R. E. Weaver.** 1986. Chemical characterization of *Flavobacterium odoratum, Flavobacterium breve*, and *Flavobacterium*-like groups IIe, IIb, and IIf. *J. Clin. Microbiol.* **23**:267–273.

60. **Delafield, F. P., M. Doudoroff, N. J. Palleront, C. J. Lusty, and R. Contopoulos.** 1965. Decomposition of poly-β-hydroxybutyrate by pseudomonads. *J. Bacteriol.* **90**:1455–1466.

61. **Delbel, R. H., and J. B. Evans.** 1960. Modified benzidine test for the detection of cytochrome-containing respiratory systems in microorganisms. *J. Bacteriol.* **79**:356–360.

62. **De Ley, J., and J. Swings.** 1984. Genus *Gluconobacter*, p. 275–278. *In* N. R. Krieg and J. G. Holt (ed.), *Bergey's Manual of Systematic Bacteriology*, vol. 1. The Williams & Wilkins Co., Baltimore, MD.

63. **den Dooren de Jong, L. E.** 1926. *Bijdrage tot de kennis van het Mineralisatieprocess.* Nijgh and Van Ditmar, Rotterdam, The Netherlands.

64. **Döbereiner, J., and J. M. Day.** 1976. Associative symbioses in tropical grasses: characterization of microorganisms and dinitrogen fixing sites, p. 518–538. *In* W. E. Newton and C. J. Nymans (ed.), *Symposium on Nitrogen Fixation.* Washington State University Press, Pullman, WA.

65. **Doolittle, W. F., Y. Boucher, C. L. Nesbo, C. J. Douady, J. D. Anderson, and A. J. Roger.** 2002. How big is the iceberg of which organellar genes in nuclear genomes are but the tip. *Phil. Trans. R. Soc. Lond. B* **358**:39–57.

66. **Eisen, J.** 1998. Phylogenomics: improving functional predictions for uncharacterised genes by evolutionary analysis. *Genome Res.* **8**:163–167.

67. **Ekiel, I., and G. D. Sprott.** 1992. Identification of degradation artifacts formed upon treatment of hydroxydiether lipids from methanogens with methanolic HCl. *Can. J. Microbiol.* **38**:764–768.

68. **Elder, B. L., I. Trujillo, and D. J. Blazevic.** 1977. Rapid deoxyribonuclease test with methyl green. *J. Clin. Microbiol.* **6**:312–313.

69. **Elsgaard, L., and L. W. Jørgensen.** 2002. A sandwich-designed temperature-gradient incubator for studies of microbial temperature responses. *J. Microbiol. Methods* **49**:19–29.

70. **Espinasa, L., and R. Borowsky.** 1998. Evolutionary divergence of AP-PCR (RAPD) patterns. *Mol. Biol. Evol.* **15**: 408–414.

71. **Evans, J. B., and C. F. Niven.** 1950. A comparative study of known food-poisoning staphylococci and related varieties. *J. Bacteriol.* **59**:545-550.

72. **Ewing, W. H., B. R. Davis, and R. W. Reavis.** 1957. Phenylalanine and malonate media and their use in enteric bacteriology. *Public Health Lab.* **15**:153.

73. **Farris, J. S.** 1979. On the naturalness of phylogenetic classification. *Syst. Zool.* **28**:200-214.

74. **Farris, J. S.** 1979. The information content of the phylogenetic system. *Syst. Zool.* **28**:483–519.

75. **Fenselau, C., and P. A. Demitrev.** 2001. Characterisation of intact microorganisms by MALDI mass spectrometry. *Mass Spectrom. Rev.* **20**:157–171.

76. **Finegold, S. M., W. J. Martin, and E. G. Scott.** 1978. *Bailey and Scott's Diagnostic Microbiology*, 5th ed. The C. V. Mosby Co., St. Louis, MO.

77. **Fitch, W. M.** 2000. Homology: a personal view of some of the problems. *Trends Genet.* **16**:227–231.

78. **Franzmann, P. D., and V. B. D. Skerman.** 1989. Genus *Agitococcus*, p. 2133–2135. *In* J. T. Staley, M. P. Bryant, N. Pfennig, and J. G. Holt (ed.), *Bergey's Manual of Systematic Bacteriology*, vol. 3. The Williams & Wilkins Co., Baltimore, MD.

79. **Franzmann, P. D., P. Höpfl, N. Weiss, and B. J. Tindall.** 1991. Psychrotrophic, lactic acid-producing bacteria from anoxic waters in Ace Lake, Antarctica: *Carnobacterium funditum* sp. nov. and *Carnobacterium alterfunfitum* sp. nov. *Arch. Microbiol.* **156**:255–262.

80. **Fung, D. Y. C., and P. A. Hartman.** 1972. Rapid characterization of bacteria, with emphasis on *Staphylococcus aureus. Can. J. Microbiol.* **18**:1623–1627.

81. **Fung, D. Y. C., and P. A. Hartman.** 1975. Miniaturized microbiological techniques for rapid characterization of bacteria, p. 347–370. *In* C. G. Hedén and T. Illéni (ed.), *New Approaches to the Identification of Microorganisms.* John Wiley & Sons, Inc., New York, NY.

82. **Fung, D. Y. C., and R. D. Miller.** 1970. Rapid procedure for the detection of acid and gas production by bacterial cultures. *Appl. Microbiol.* **20**:527–528.

83. **Gibbons, N. E., P. H. A. Sneath, and S. P. Lapage.** 1989. Reference collections of bacteria: the need and requirement for type strains, p. 2325–2328. *In* S. T. Williams, M. E. Sharpe, and J. G. Holt (ed.), *Bergey's Manual of Systematic Bacteriology*, vol. 4. The Williams & Wilkins Co., Baltimore, MD.

84. **Goebel, E. M., and N. R. Krieg.** 1984. Fructose catabolism in *Azospirillum brasilense* and *Azospirillum lipoferum. J. Bacteriol.* **159**:86–92.

85. **Gogarten, J. P.** 1994. Which is the most conserved group of proteins? Homology-orthology, paralogy, xenology and

the fusion of independent lineages. *J. Mol. Biol.* **39:**541–543.

86. **Goodacre, R., E. M. Timmins, P. J. Rooney, J. J. Rowland, and D. B. Kell.** 1996. Rapid identification of *Streptococcus* and *Enterococcus* species using diffuse reflectance-absorbance Fourier transform infrared spectroscopy and artificial neural networks. *FEMS Microbiol. Lett.* **140:**233–239 .

87. **Goodacre, R., E. M. Timmins, R. Burton, N. Kaderbhai, A. M. Woodward, D. B. Kell, and P. J. Rooney.** 1998. Rapid identification of urinary tract infection bacteria using hyperspectral whole-organism fingerprinting and artificial neural networks. *Microbiology* **144:**1157–1170.

88. **Gram, C.** 1884. *Forschitte der Medicin,* vol. 2. p. 185–189.

89. **Grange, J. M.** 1978. Fluorimetric assay of mycobacterial group-specific hydrolase enzymes. *J. Clin. Pathol.* **31:**378–381.

90. **Greenberg, A. E., R. R. Trussel, and L. C. Clesceri (ed.).** 1985. *Standard Methods for the Examination of Water and Wastewater,* 16th ed. American Public Health Association, Washington, DC.

91. **Gregersen, T.** 1978. Rapid method for distinction of Gram-negative from Gram-positive bacteria. *Eur. J. Appl. Microbiol. Biotechnol.* **5:**123–127.

92. **Hamana, K., and S. Matsuzaki.** 1992. Polyamines as a chemotaxonomic marker in bacterial systematics. *Crit. Rev. Microbiol.* **18:**261–283.

93. **Han, Y.-H., R. M. Smibert, and N. R. Krieg.** 1991. *Wolinella recta, Wolinella curva, Bacteroides ureolyticus,* and *Bacteroides gracilis* are microaerophiles, not anaerobes. *Int. J. Syst. Bacteriol.* **41:**218–222.

94. **Hardie, J. M.** 1986. Genus *Streptococcus,* p. 1043–1047. *In* P. H. A. Sneath, N. S. Mair, M. E. Sharpe, and J. G. Holt (ed.), *Bergey's Manual of Systematic Bacteriology,* vol. 2. The Williams & Wilkins Co., Baltimore, MD.

95. **Hardy, R. W. F., R. C. Burns, and R. D. Holsten.** 1973. Applications of the acetylene-ethylene assay for measurement of nitrogen fixation. *Soil Biol. Biochem.* **5:**47–81.

96. **Hardy, R. W. F., R. D. Holsten, E. K. Jackson, and R. C. Burns.** 1968. The acetylene-ethylene assay for N_2 fixation: laboratory and field evaluation. *Plant Physiol.* **43:**1185–1207.

97. **Harrison, G. A.** 1964. [Referenced], p. 161. *In* V. H. Heywood and J. McNeill (ed.), *Phenetic and Phylogenetic Classification.* Systematics Association publication 6. The Systematics Association, London, United Kingdom.

98. **Haynes, W. C.** 1951. *Pseudomonas aeruginosa*—its characterization and identification. *J. Gen. Microbiol.* **5:**939–950.

99. **Helander, I. M., and A. Haikara.** 1995. Cellular fatty acyl and alkenyl residues in *Megasphaera* and *Pectinatus* species: contrasting profiles and detection of beer spoilage. *Microbiology* **141:**1131–1137.

100. **Helm, D., H. Labischinski, and D. Naumann.** 1991. Elaboration of a procedure for identification of bacteria using Fourier-transform IR libraries: a step wise correlation approach. *J. Microbiol. Methods* **14:**127–142.

101. **Helm, D., H. Labischinski, G. Schallehn, and D. Naumann.** 1991. Classification and identification of bacteria by Fourier-transform infrared spectroscopy. *J. Gen. Microbiology* **137:**69–79.

102. **Hennig, W.** 1982. *Phylogenetische Systematik.* Pareys Studientexte 34. Verlag Paul Parey, Berlin, Germany.

103. **Henrichsen, J.** 1972. Bacterial surface translocation: a survey and a classification. *Bacteriol. Rev.* **36:**478–503.

104. **Hillis, D. M., M. W. Allard, and M. Miyamoto.** 1993. Analysis of DNA sequence data: phylogenetic inferences. *Methods Enzymol.* **224:**456–487.

105. **Hino, S., and P. W. Wilson.** 1958. Nitrogen fixation by a facultative bacillus. *J. Bacteriol.* **75:**403–408.

106. **Hippe, H. H.** 1991. Maintenance of methanogenic bacteria, p. 101–113. *In* B. E. Kirsop and A. Doyle (ed.), *Maintenance of Microorganisms.* Academic Press, London, United Kingdom.

107. **Hirao, T., M. Sato, A. Shirahata, and Y. Kamio.** 2000. Covalent linkage of polyamines to peptidoglycan in *Anaerovibrio lipolytica. J. Bacteriol.* **182:**1154–1157.

108. **Hoi, L., A. Dalsgaard, J. L. Larsen, J. M. Warner, and J. D. Oliver.** 1997. Comparison of ribotyping and randomly amplified polymorphic DNA PCR for characterization of *Vibrio vulnificus. Appl. Environ. Microbiol.* **63:**1674–1678.

109. **Holt, S. C., and S. A. Kinder.** 1989. Genus *Capnocytophaga,* p. 2050–2058. *In* J. T. Staley, M. P. Bryant, N. Pfennig, and J. G. Holt (ed.), *Bergey's Manual of Systematic Bacteriology,* vol. 3. The Williams & Wilkins Co., Baltimore, MD.

110. **Hugh, R., and E. Leifson.** 1953. The taxonomic significance of fermentative versus oxidative metabolism of carbohydrates by various gram-negative bacteria. *J. Bacteriol.* **66:**22–26.

111. **Hungate, R. E.** 1966. *The Rumen and Its Microbes.* Academic Press Inc., New York, NY.

112. **Huxley, T. S.** 1958. Evolutionary processes and taxonomy with special reference to grades and clades. *Uppsala Univ. Arsskrift* **2:**21–39.

113. **Hwang, M., and G. M. Ederer.** 1975. Rapid hippurate hydrolysis method for presumptive identification of group B streptococci. *J. Clin. Microbiol.* **1:**114–115.

114. **Hylemon, P. B., J. S. Wells, Jr., N. R. Krieg, and H. W. Jannasch.** 1973. The genus *Spirillum:* a taxonomic study. *Int. J. Syst. Bacteriol.* **23:**340–380.

115. **Jackman, P. J. H.** 1985. Bacterial taxonomy based on electrophoretic whole-organism protein patterns, p. 115–129. *In* M. Goodfellow and D. E. Minnikin (ed.), *Chemical Methods in Bacterial Systematics.* SAB technical series 20. Academic Press, London, United Kingdom.

116. **Jensen, M. T., J. Knudsen, and J. M. Olson.** 1991. A novel aminoglycosphingolipid found in *Chlorobium limicola* f. *thiosulfatophilum* 6230. *Arch. Microbiol.* **156:**248–254.

117. **Johnson, R., and P. H. A. Sneath.** 1973. Taxonomy of *Bordetella* and related organisms of the families *Achromobacteraceae, Brucellaceae,* and *Neisseriaceae. Int. J. Syst. Bacteriol.* **23:**381–404.

118. **Jones, D., and N. R. Krieg.** 1989. Serology and chemotaxonomy, p. 2313–2316. *In* S. T. Williams, M. E. Sharpe, and J. G. Holt (ed.), *Bergey's Manual of Systematic Bacteriology,* vol. 4. The Williams & Wilkins Co., Baltimore, MD.

119. **Kamio, Y.** 1987. Structural specificity of diamines covalently linked to peptidoglycan for cell growth of *Veillonella alcalescens* and *Selenomonas ruminantium. J. Bacteriol.* **169:**4837–4840.

120. **Kamio, Y., Y. Itoh, and Y. Terawaki.** 1981. Chemical structure of peptidoglycan in *Selenomonas ruminantium:* cadaverine links covalently to the D-glutamic acid residue of peptidoglycan. *J. Bacteriol.* **146:**49–53.

121. **Kamio, Y., Y. Itoh, Y. Terawaki, and T. Kusano.** 1981. Cadaverine is covalently linked to peptidoglycan in *Selenomonas ruminantium. J. Bacteriol.* **145:**122–128.

122. **Kamio, Y., and K. Nakamura.** 1987. Putrescine and cadaverine are constituents of peptidoglycan in *Veillonella alcalescens* and *Veillonella parvula. J. Bacteriol.* **169:**2881–2884.

123. **Kamio, Y., H. Poso, Y. Terawaki, and L. Paulin.** 1986. Cadaverine covalently linked to a peptidoglycan is an essential constituent of the peptidoglycan necessary for the

normal growth in *Selenomonas ruminantium*. *J. Biol. Chem.* **261**:6585–6589.

124. **Kamio, Y., Y. Terawaki, and K. Izaki.** 1982. Biosynthesis of cadaverine-containing peptidoglycan in *Selenomonas ruminantium*. *J. Biol. Chem.* **257**:3326–3333.

125. **Kandolo, K., and G. Wauters.** 1985. Pyrazinamidase activity in *Yersinia enterocolitica* and related organisms. *J. Clin. Microbiol.* **21**:980–982.

126. **Kersters, K., and J. De Ley.** 1975. Identification and grouping of bacteria by numerical analysis of their electrophoretic patterns. *J. Gen. Microbiol.* **87**:333–342.

127. **Khandka, D. K., M. Tuna, M. Tal, A. Nejidat, and A. Golan-Goldhirsh.** 1997. Variability in the pattern of random amplified polymorphic DNA. *Electrophoresis* **18**: 2852–2856.

128. **Killinger, A. H.** 1974. *Listeria monocytogenes*, p. 135–139. *In* E. H. Lennette, E. H. Spaulding, and J. P. Truant (ed.), *Manual of Clinical Microbiology*, 2nd ed. American Society for Microbiology, Washington, DC.

129. **King, E. O., M. K. Ward, and D. E. Raney.** 1954. Two simple media for the demonstration of pyocyanin and fluorescein. *J. Lab. Clin. Med.* **44**:301–307.

130. **Klebahn, H.** 1895. Gasvakuolen, ein Bestandteil der zellen der Wasserbluitebildenden Phycochromaceen. *Flora* (Jena) **80**:241–282.

131. **Klebahn, H.** 1922. Neue Untersuchungen über die Gasvakuolen. *Jahrb. Wiss. Bot.* **61**:535–589.

132. **Knapp, J. S.** 1988. Historical perspectives and identification of *Neisseria* and related species. *Clin. Microbiol. Rev.* **1**:415–431.

133. **Kohn, J.** 1953. A preliminary report of a new gelatin liquefaction method. *J. Clin. Microbiol.* **6**:249.

134. **Konstantinidis, K.T., and J. M. Tiedje.** 2004. Trends between gene content and genome size in prokaryotic species with larger genomes. *Proc. Natl. Acad. Sci. USA* **101**:3160–3165.

135. **Krader, P., and D. Emerson.** 2004. Identification of archaea and some extremophilic bacteria using matrix-assisted laser desorption/ionisation time-of-flight spectrometry (MALDI-TOF) mass spectrometry. *Extremophiles* **8**:259–268.

136. **Kramer, J. K. G., F. D. Sauer, and B. A. Blackwell.** 1987. Structure of two new aminophospholipids from *Methanobacterium thermoautotrophicum*. *Biochem. J.* **245**:139–143.

137. **Krieg, N. R.** 2001. Identification of procaryotes, p. 33–38. *In* G. M. Garrity (ed.), *Bergey's Manual of Systematic Bacteriology*, 2nd ed., vol. 1. Springer-Verlag, New York, NY.

138. **Kumar, R., S. T. Weintraub, and D. J. Hanaham.** 1983. Differential susceptibility of mono- and di-O-alkyl ether phosphoglycerides to acetolysis. *J. Lipid Res.* **24**:930–937.

139. **Kumar, S., K. Tamura, I. B. Jakobsen, and M. Nei.** 2001. MEGA2: molecular evolutionary genetics analysis software. *Bioinformatics* **17**:1244–1245.

140. **Kurland, C. G., B. Canback, and O. G. Berg.** 2003. Horizontal gene transfer: a critical review. *Proc. Natl. Acad. Sci. USA* **100**:9658–9662.

141. **Labeda, D. P.** 2001. Culture collections: an essential resource for microbiology, p. 111–113. *In* G. M. Garrity (ed.), *Bergey's Manual of Systematic Bacteriology*, 2nd ed., vol. 1. Springer-Verlag, New York, NY.

142. **Lambert, M. A., and C. W. Moss.** 1983. Comparison of the effects of acid and base hydrolyses on hydroxy and cyclopropane fatty acids in bacteria. *J. Clin. Microbiol.* **18**:1370–1377.

143. **Lamm, R. B., and C. A. Neyra.** 1981. Characterization and cyst production of azospirilla isolated from selected grasses growing in New Jersey and New York. *Can. J. Microbiol.* **27**:1320–1325.

144. **Langworthy, T. A.** Personal communication.

145. **Langworthy, T. A., G. Holzer, J. G. Zeikus, and T. G. Tornabene.** 1983. Iso- and anteiso-branched glycerol diethers of the thermophilic anaerobes *Thermodesulfobacterium commune*. *Syst. Appl. Microbiol.* **4**:1–17.

146. **Langworthy, T. A., W. M. Mayberry, and P. F. Smith.** 1974. Long-chain glycerol diether and polyol dialkyl glycerol triether lipids of *Sulfolobus acidocaldarius*. *J. Bacteriol.* **119**:106–116.

147. **Lapage, S. P., P. H. A. Sneath, E. F. Lessel, V. B. D. Skerman, H. P. R. Seeliger, and W. A. Clark (ed.).** 1992. *International Code of Nomenclature of Bacteria (1990 Revision). Bacteriological Code*. American Society for Microbiology, Washington, DC.

148. **Laung, K. T., Y. J. Chang, Y. D. Gan, A. Peacock, S. J. MacNaughton, J. R. Stephen, R. S. Burkhalter, C. A. Flemming, and D. C. White.** 1999. Detection of Sphingomonas spp. in soil by PCR and sphingolipid biomarker analysis. *J. Ind. Microbiol. Biotechnol.* **23**:252–260.

149. **Law, J. H., and R. A. Slepecky.** 1961. Assay of poly-β-hydroxybutyric acid. *J. Bacteriol.* **82**:33–36.

150. **Lay, J. O.** 2001. MALDI-TOF-MS of bacteria. *Mass Spectrom. Rev.* **20**:172–194.

151. **Lederberg, J., and E. M. Lederberg.** 1962. Replica plating and indirect selection of bacterial mutants. *J. Bacteriol.* **63**:399–406.

152. **Lee, K.-B., C.-T. Liu, Y. Anzai, H. Kim, T. Aono, and H. Oyaizu.** 6 May 2005, posting date. The hierarchical system of the 'Alphaproteobacteria:' description of *Hyphomonadaceae* fam. nov., *Xanthobacteracea* fam. nov. and *Erythrobacteraceae* fam. nov. *Int. J. Syst. Evol. Microbiol.* **55**:1907–1919. doi:10.1099/ijs.0.63663-0.

153. **Leifson, E.** 1963. Determination of carbohydrate metabolism of marine bacteria. *J. Bacteriol.* **85**:1183–1184.

154. **Lerat, E., V. Daubin, H. Ochman, and N. A. Moran.** 2005. Evolutionary origins of genomic repertoires in bacteria. *PLoS Biol.* **3**:e130.

155. **Lovelace, T. E., and R. R. Colwell.** 1968. A multipoint inoculator for petri dishes. *Appl. Microbiol.* **16**:944–945.

156. **Ludwig, W., and H.-P. Klenk.** 2001. Overview: a phylogenetic backbone and taxonomic framework for prokaryote systematics, p. 49–65. *In* G. M. Garrity (ed.), *Bergey's Manual of Systematic Bacteriology*, 2nd ed., vol. 1. Springer-Verlag, New York, NY.

157. **Macrae, R. M., and J. F. Wilkinson.** 1958. Poly-β-hydroxybutyrate metabolism in washed suspensions of *Bacillus cereus* and *Bacillus megaterium*. *J. Gen. Microbiol.* **19**:210–222.

158. **Maddocks, J. L., and M. J. Greenan.** 1975. Technical method. A rapid method for identifying bacterial enzymes. *J. Clin. Pathol.* **28**:686–687.

159. **Malik, K. A., and D. Claus.** 1979. *Xanthobacter flavus*, a new species of nitrogen-fixing hydrogen bacteria. *Int. J. Syst. Bacteriol.* **29**:283–287.

160. **Malik, K. A., and H. G. Schlegel.** 1981. Chemolithoautotrophic growth of bacteria able to grow under N$_2$-fixing conditions. *FEMS Microbiol. Lett.* **11**:63–67.

161. **Margoliash, E., W. M. Fitch, and R. E. Dickerson.** 1968. Molecular expression of evolutionary phenomena in the primary and tertiary structures of cytochrome c. *Brookhaven Symp. Biol.* **21**:259–305.

162. **Matsuzaki, S., K. Hamana, K. Imai, and K. Matsuura.** 1982. Occurrence in high concentrations of N'-acetylspermidine and *sym*-homospermidine in the hamster epididymus. *Biochem. Biophys. Res. Commun.* **107**: 307–313.

163. **Mayberry, W. R.** 1980. Hydroxy fatty acids in *Bacteroides* species: D-[-]-3-hydroxy-15-methylhexadecanoate and its homologs. *J. Bacteriol.* **143**:582–587.

164. **Mayr, E.** 1981. Biological classification: toward a synthesis of opposing methodologies. *Science* **214:**510–516.

165. **McEwan, N. R., J. Bakht, E. Cecchini, A. Coultart, C. Geri, F. M. McDonald, M. S. McDonald, N. N. A. Rahman, and G. Williamson.** 1998. Examination of the pipetting variation between individuals performing random amplified polymorphic DNA (RAPD) analysis. *J. Microbiol. Methods* **32:**213–215.

166. **McFarland, J.** 1907. Nephelometer: an instrument for estimating the number of bacteria in suspensions used for calculating the opsonic index and for vaccines. *JAMA* **14:**1176–1178.

167. **McGinnis, M. R., R. F. D'Amato, and G. A. Land.** 1982. *Pictorial Handbook of Medically Important Fungi and Aerobic Actinomycetes.* Praeger Publishers, New York, NY.

168. **Meynell, G. G., and E. Meynell.** 1965. *Theory and Practice in Experimental Bacteriology.* Cambridge University Press, London, United Kingdom.

169. **Meynell, G. G., and E. Meynell.** 1970. *Theory and Practice in Experimental Bacteriology,* 2nd ed. Cambridge University Press, London, United Kingdom.

170. **Milazzo, F. H., and J. W. Fitzgerald.** 1967. The effect of some culture conditions on the arylsulfatase of *Proteus rettgeri. Can. J. Microbiol.* **13:**659–664.

171. **Miller, C. G., and K. MacKinnon.** 1974. Peptidase mutants of *Salmonella typhimurium. J. Bacteriol.* **120:**355–363.

172. **Miller, L. T.** 1982. Single derivatization method for routine analysis of bacterial whole-cell fatty acid methyl esters, including hydoxy fatty acids. *J. Clin. Microbiol.* **16:**584–586.

173. **Mills, C. K., and R. L. Gherna.** 1987. Hydrolysis of indoxyl acetate by *Campylobacter* species. *J. Clin. Microbiol.* **25:**1560–1561.

174. **Molisch, H.** 1907. *Die Purpurbakterien nach neuen Untersuchungen.* Gustav Fischer, Jena, Germany.

175. **Møller, V.** 1955. Simplified test for some amino acid decarboxylases in Enterobacteriaceae. *Acta Pathol. Microbiol. Scand.* **36:**158–172.

176. **Moore, L. V. H., D. M. Bourne, and W. E. C. Moore.** 1994. Comparative distribution and taxonomic value of cellular fatty acids in thirty-three genera of anaerobic Gram-negative bacilli. *Int. J. Syst. Bacteriol.* **44:**338–347.

177. **Moore, P. B., and T. A. Steitz.** 2002. The involvement of RNA in ribosome function. *Nature* **418:**229–235.

178. **Moore, W. E. C., D. E. Hash, L. V. Holdeman, and E. P. Cato.** 1980. Polyacrylamide slab electrophoresis of soluble proteins for studies of bacterial floras. *Appl. Environ. Microbiol.* **39:**900–907.

179. **Mori, E., P. Lio, S. Daly, G. Damiani, B. Perito, and R. Fani.** 1999. Molecular nature of RAPD markers from *Haemophilus influenzae* Rd genome. *Res. Microbiol.* **150:**83–93.

180. **Morii, H., M. Nishihara, M. Ohga, and Y. Koga.** 1986. A diphytanoyl ether analog of phosphatidylserine from a methanogenic bacterium, *Methanobrevibacter arboriphilus. J. Lipid Res.* **27:**724–730.

181. **Moss, C. W., P. L. Wallace, D. G. Hollis, and R. E. Weaver.** 1988. Cultural and chemical characterization of CDC groups EO-2, M-5, and M-6, *Moraxella* (*Moraxella*) species, *Oligella urethralis, Acinetobacter* species, and *Psychrobacter immobilis. J. Clin. Microbiol.* **26:**484–492.

182. **Moss, C. W., M. A. Lambert, and W. H. Merwin.** 1974. Comparison of methods for analysis of bacterial fatty acids. *Appl. Microbiol.* **28:**80–85.

183. **Mueller, U. G., and L. L. Wolfenbarger.** 1999. AFLP genotyping and fingerprinting. *Trends Ecol. Evol.* **14:**389–394.

184. **Mundt, J. O.** 1986. Enterococci, p. 1063–1065. *In* P. H. A. Sneath, N. S. Mair, M. E. Sharpe, and J. G. Holt (ed.), *Bergey's Manual of Systematic Bacteriology,* vol. 2. The Williams & Wilkins Co., Baltimore, MD.

185. **Mundt, J. O.** 1986. Lactic acid streptococci, p. 1065–1066. *In* P. H. A. Sneath, N. S. Mair, M. E. Sharpe, and J. G. Holt (ed.), *Bergey's Manual of Systematic Bacteriology,* vol. 2. The Williams & Wilkins Co., Baltimore, MD.

186. **Murray, R. G. E., D. J. Brenner, R. R. Colwell, P. De Vos, M. Goodfellow, P. A. D. Grimont, N. Pfenning, E. Stackebrandt, and G. A. Zavarzin.** 1990. Report of the ad hoc committee on approaches to taxonomy within the Proteobacteria. *Int. J. Syst. Bacteriol.* **40:**213–215.

187. **Naka, T., N. Fujiwara, I. Yano, S. Maeda, M. Doe, M. Minamino, N. Ikeda, Y. Kato, K. Watabe, Y. Kumazawa, I. Tomiyasu, and K. Kobayashi.** 2003. Structural analysis of sphingophospholipids derived from *Sphingobacterium spiritivorum,* the type species of the genus *Sphingobacterium. Biochem. Biophys. Acta* **1635:**83–92.

188. **Naumann, D., D. Helm, and H. Labischinski.** 1991. Microbial characterisation by FT-IR spectroscopy. *Nature* **351:**81–82.

189. **Neal, J. L., Jr., K. C. Lu, W. B. Bollen, and J. M. Trappe.** 1966. Apparatus for rapid replica plating in rhizosphere studies (soil surrounding and influenced by roots of plants). *Appl. Microbiol.* **14:**695–696.

190. **Neyra, C. A., J. Döbereiner, R. LaLande, and R. Knowles.** 1977. Denitrification by N_2 fixing *Spirillum lipoferum. Can. J. Microbiol.* **23:**300–305.

191. **Nichols, P. D., P. M. Shaw, C. A. Mancuso, and P. D. Franzmann.** 1993. Analysis of archaeol phospholipids-derived di- and tetraether lipids by high temperature capillary gas chromatography. *J. Microbiol. Methods* **18:**1–9.

192. **Nicolet, J., P. Paroz, and M. Krawinkler.** 1980. Polyacrylamide gel electrophoresis of whole-cell proteins of porcine strains of *Haemophilus. Int. J. Syst. Bacteriol.* **30:**69–76.

193. **Nishihara, M., and Y. Koga.** 1987. Extraction and composition of polar lipids from the archaebacterium *Methanobacterium thermoautotrophicum:* effective extraction of tetraether lipids by an acidified solvent. *J. Biochem.* **101:**997–1005.

194. **Nishihara, M., H. Morii, K. Matsuno, M. Ohga, K. Stetter, and Y. Koga.** 2002. Structural analysis by reductive cleavage with $LiAlH_4$ of an allyl ether choline-phospholipid, achaetidylcholine, from the thermophilic methanoarchaeon *Methanopyrus kandleri. Archaea* **1:**123–131.

195. **Nitsu, M., K. Samejima, S. Matsuzaki, and K. Hamana.** 1993. Systematic analysis of naturally occurring linear and branched polyamines by gas chromatography and gas-chromatography-mass spectrometry. *J. Chromatogr.* **641:**115–123.

196. **Niven, C. F., Jr., K. L. Smiley, and J. M. Sherman.** 1942. The hydrolysis of arginine by streptococci. *J. Bacteriol.* **43:**651–660.

197. **Noel, K. D., and W. J. Brill.** 1980. Diversity and dynamics of indigenous *Rhizobium japonicum* populations. *Appl. Environ. Microbiol.* **40:**931–938.

198. **Nohynek, L.J., E. L. Suhonen, E.-L. Nurmiaho-Lassila, J. Hantula, and M. Salkinoja-Salonen.** 1995. Description of four pentachlorophenol-degrading bacterial strains as *Sphingomonas chlorophenolica* sp. nov. *Syst. Appl. Microbiol.* **18:**527–538.

199. **Oakey, H. J., L. F. Gibson, and A. M. George.** 1998. Co-migration of RAPD-PCR amplicons from *Aeromonas hydrophila. FEMS Microbiol. Lett.* **164:**35–38.

200. **O'Callaghan, C. H., A. Morris, S. M. Kirby, and A. H. Shingler.** 1972. Novel method for detection of β-lactamases by using a chromogenic cephalosporin substrate. *Antimicrob. Agents Chemother.* **1:**283–288.

201. **Ohta, H., and J. C. Gottschal.** 1988. Microaerophilic growth of *Wolinella recta. FEMS Microbiol. Ecol.* **53:**79–96.

202. **Okon, Y., S. L. Albrecht, and R. H. Burris.** 1976. Carbon and ammonia metabolism of *Spirillum lipoferum*. *J. Bacteriol.* **128:**592–597.

203. **Orla-Jensen, S.** 1909. Die Hauptlinie des Natürlichen Bakteriensystem. *Zentbl. Bakteriol. Parasitenkd. Infektkrankh. Hyg. Abt. II* **22:**97–98, 305–346.

204. **Orla-Jensen, S.** 1921. The main lines of the natural bacterial system. *J. Bacteriol.* **6:**263–273.

205. **Ostle, A. G., and J. G. Holt.** 1982. Nile blue A as a fluorescent stain for poly-β-hydroxybutyrate. *Appl. Environ. Microbiol.* **44:**238–241.

206. **Page, W. J., and H. L. Sadoff.** 1976. Physiological factors affecting transformation of *Azotobacter vinelandii*. *J. Bacteriol.* **125:**1080–1087.

207. **Palleroni, N. J.** 1984. Genus *Pseudomonas*, p. 141–199. *In* N. R. Krieg and J. G. Holt (ed.), *Bergey's Manual of Systematic Bacteriology*, vol. 1. The Williams & Wilkins Co., Baltimore, MD.

208. **Palleroni, N. J.** 2003. Prokaryote taxonomy of the 20th century and the impact of studies on the genus *Pseudomonas*: a personal view. *Microbiology* **149:**1–7.

209. **Palleroni, N. J.** Personal communication.

210. **Palleroni, N. J.** 1997. Prokaryotic diversity and the importance of culturing. *Antonie Leeuwenhoek* **72:**3–19.

211. **Paściak, M., O. Holst, B. Lindner, H. Mordarska, and A. Gamian.** 2003. Novel bacterial polar lipids containing ether-linked alkyl chains, the structures and biological properties of the four major glycolipids from *Propionibacterium propionicum* PCN 2431 (ATCC 14157T). *J. Biol. Chem.* **278:**3948–3956.

212. **Patterson, C.** 1987. Introduction, p. 1–22. *In* C. Patterson (ed.), *Molecules and Morphology in Evolution: Conflict or Compromise?* Cambridge University Press, Cambridge, United Kingdom.

213. **Patterson, C.** 1988. Homology in classical and molecular biology. *Mol. Biol. Evol.* **5:**603–625.

214. **Pelczar, M. J., Jr. (ed.).** 1957. *Manual of Microbiological Methods.* McGraw-Hill Book Co., New York, NY.

215. **Pond, J. L., and T. A. Langworthy.** 1987. Effect of temperature on the long-chain diols and fatty acids of *Thermomicrobium roseum*. **169:**1328–1330.

216. **Pond, J. L., T. A. Langworthy, and G. Holzer.** 1986. Long-chain diols: a new class of membrane lipids from a thermophilic bacterium. *Science* **231:**1134–1136.

217. **Porschen, R. K., and S. Sonntag.** 1974. Extracellular deoxyribonuclease production by anaerobic bacteria. *Appl. Microbiol.* **27:**1031–1033.

218. **Postgate, J. R.** 1972. The acetylene reduction test for nitrogen fixation, p. 343–356. *In* J. R. Norris and D. W. Ribbons (ed.), *Methods in Microbiology*, vol. 6B. Academic Press, Inc., New York, NY.

219. **Pot, B., P. Vandamme, and K. Kersters.** 1994. Analysis of electrophoretic whole-organism protein fingerprints, p. 493–521. *In* M. Goodfellow and A. G. O'Donnell (ed.), *Chemical Methods in Prokaryotic Systematics.* John Wiley & Sons, Chichester, United Kingdom.

220. **Rappé, M. S., S. A. Common, K. L. Vergin, and S. Giovannoni.** 2002. Cultivation of the ubiquitous SAR11 marine bacterioplankton clade. *Nature* **418:**630–633.

221. **Razin, S., and S. Rottem.** 1967. Identification of *Mycoplasma* and other microorganisms by polyacrylamide gel electrophoresis of cell proteins. *J. Bacteriol.* **94:**1807–1810.

222. **Richards, W. R.** 1994. Analysis of pigments: bacteriochlorophylls, p. 345–401. *In* M. Goodfellow and A. G. O'Donnell (eds.), *Chemical Methods in Prokaryotic Systematics.* John Wiley & Sons, Chichester, United Kingdom.

223. **Roop, R. M., H. R. M. Smibert, J. L. Johnson, and N. R. Krieg.** 1984. Differential characteristics of catalase-positive campylobacters correlated with DNA homology groups. *Can. J. Microbiol.* **30:**938–951.

224. **Rosenberg, H., A. H. Ennor, and V. F. Morrison.** 1956. The estimation of arginine. *Biochem. J.* **63:**153–159.

225. **Rütters, H., H. Sass, H. Cypionka, and J. Rullkötter.** 2001. Monalkylether phospholipids in the sulfate-reducing bacteria *Desulfosarcina vaiabilis* and *Desulforhabdus aminigenus*. *Arch. Microbiol.* **176:**435–442.

226. **Sadasivan, L., and C. A. Neyra.** 1987. Cyst production and brown pigment formation in aging cultures of *Azospirillum brasilense* ATCC 29145. *J. Bacteriol.* **169:**1670–1677.

227. **Sakallah, S. A., R. W. Lanning, and D. L. Cooper.** 1995. DNA fingerprinting of crude bacterial lysates using degenerate RAPD primers. *PCR Methods Appl.* **4:**265–268.

228. **Scherer, P., and H. Kneifel.** 1983. Distribution of polyamines in methanogenic bacteria. *J. Bacteriol.* **154:**1315–1322.

229. **Schleifer, K. H., M. Leuteritz, N. Weiss, W. Ludwig, G. Kirchhof, and H. Seidel-Rüfer.** 1990. Taxonomic study of anaerobic, Gram-negative, rod-shaped bacteria from breweries: emended description of *Pectinatus cerevisiiphilus* and description of *Pectinatus frisingensis* sp. nov., *Selenomonas lacticifex* sp. nov., *Zymophilus raffinosivorans* gen. nov., sp. nov., and *Zymophilus paucivorans* sp. nov. *Int. J. Syst. Bacteriol.* **40:**19–27.

230. **Schmidt, K., A. Connon, and G. Britton.** 1994. Analysis of pigments: carotenoids and related polyenes, p. 403–461. *In* M. Goodfellow and A. G. O'Donnell (ed.), *Chemical Methods in Prokaryotic Systematics.* John Wiley & Sons, Chichester, United Kingdom.

231. **Schneider, S., J.-M. Kueffer, D. Roessli, and L. Excoffier.** 2000. *Arlequin ver. 2.000: A Software for Population Genetic Data Analysis.* Genetics and Biometry Laboratory, University of Geneva, Switzerland.

232. **Schreier, J. B.** 1969. Modification of deoxyribonuclease test medium for rapid identification of *Serratia marcescens*. *Am. J. Clin. Pathol.* **51:**711–716.

233. **Schweickert, B., A. Moter, M. Lefmann, and U. B. Göbel.** 2004. Let them fly or light them up: matrix-assisted laser desorption/ionisation time-of-flight spectrometry (MALDI-TOF) mass spectrometry and fluorescence *in situ* hybridisation. *APMIS* **112:**856–885.

234. **Shah, H. N., C. J. Keys, O. Schmid, and S. E. Gharbia.** 2002. Matrix-assisted laser desorption/ionisation time-of-flight spectrometry and proteomics: a new era in anaerobic microbiology. *Clin. Infect. Dis.* **35**(Suppl. 1):S58–S64.

235. **Shinoda, S., and K. Okamoto.** 1977. Formation and function of *Vibrio parahaemolyticus* lateral flagella. *J. Bacteriol.* **129:**1266–1271.

236. **Siefert, J. L., and G. E. Fox.** 1998. Phylogenetic mapping of bacterial morphology. *Microbiology* **144:**2803–2808.

237. **Sierra, G.** 1957. A simple method for the detection of lipolytic activity of microorganisms and some observations on the influence of the contact between cells and fatty substrates. *Antonie Leeuwenhoek J. Microbiol. Serol.* **23:**15–22.

238. **Simmons, J. S.** 1926. A culture medium for differentiating organisms of the typhoid-colon aerogenes groups and for isolation of certain fungi. *J. Infect. Dis.* **39:**201–214.

239. **Simpson, G. G.** 1964. Organisms and molecules in evolution. *Science* **146:**1535–1538.

240. **Skerman, V. D. B., V. McGowan, and P. H. A. Sneath.** 1989. The approved lists of bacterial names. *Int. J. Sys. Bacteriol.* **30:**225–420.

241. **Smibert, R. M.** 1984. Genus *Campylobacter*, p. 111–118. *In* N. R. Krieg and J. G. Holt (ed.), *Bergey's Manual of Systematic Bacteriology*, vol. 1. The Williams & Wilkins Co., Baltimore, MD.

242. **Smibert, R. M., J. L. Johnson, and R. R. Ranney.** 1984. *Treponema socranskii* sp. nov., *Treponema socranskii* subsp. *socranskii* subsp. nov., *Treponema socranskii* subsp. *buccale* subsp. nov., and *Treponema socranskii* subsp. *paredis* subsp. nov. isolated from the human periodontia. *Int. J. Syst. Bacteriol.* **34:**457–462.

243. **Smith, P. B., G. A. Hancock, and D. L. Rhoden.** 1969. Improved medium for detecting deoxyribonuclease-producing bacteria. *Appl. Microbiol.* **18:**991–993.

244. **Smith, R. F., R. R. Rogers, and C. L. Bettge.** 1972. Inhibition of the indole test reaction by sodium nitrite. *Appl. Microbiol.* **23:**423–424.

245. **Sneath, P. H. A.** 1989. Analysis and interpretation of sequence data for bacterial systematics: the view of a numerical taxonomist. *Syst. Appl. Microbiol.* **12:**15–31.

246. **Sneath, P. H. A.** 1989. Numerical taxonomy, p. 2303–2305. *In* S. T. Williams, M. E. Sharpe, and J. G. Holt (ed.), *Bergey's Manual of Systematic Bacteriology*, vol. 4. The Williams & Wilkins Co., Baltimore, MD.

247. **Sneath, P. H. A.** 1989. Bacterial nomenclature, p. 2317–2321. *In* S. T. Williams, M. E. Sharpe, and J. G. Holt (ed.), *Bergey's Manual of Systematic Bacteriology*, vol. 4. The Williams & Wilkins Co., Baltimore, MD.

248. **Sneath, P. H. A.** 2001. Bacterial nomenclature, p. 83–88. *In* G. M. Garrity (ed.), *Bergey's Manual of Systematic Bacteriology*, 2nd ed., vol. 1. Springer-Verlag, New York, NY.

249. **Sneath, P. H. A.** 2001. Numerical taxonomy, p. 39–42. *In* G. M. Garrity (ed.), *Bergey's Manual of Systematic Bacteriology*, 2nd ed., vol. 1. Springer-Verlag, New York, NY.

250. **Sneath, P. H. A., and M. Stevens.** 1967. A divided petri dish for use with multipoint inoculators. *J. Appl. Bacteriol.* **30:**495–497.

251. **Sneath, P. H. A., and R. R. Sokal.** 1973. *Numerical Taxonomy: the Principle and Practice of Numerical Classification.* W. H. Freeman and Company, San Francisco, CA.

252. **Snell, J. J. S.** 1984. Genus *Kingella*, p. 307–309. *In* N. R. Krieg and J. G. Holt (ed.), *Bergey's Manual of Systematic Bacteriology*, vol. 1. The Williams & Wilkins Co., Baltimore, MD.

253. **Sorokin D. Y.** 15 April 2005, posting date. Is there a limit for high-pH life? *Int. J. Syst. Evol. Microbiol.* **55:**1405–1406. doi:10.1099/ijs.0.63737-0.

254. **Sprott, G. D., I. Ekiel, and C. Dicaire.** 1990. Novel, acid-labile, hydroxydiether lipid cores in methanogenic bacteria. *J. Biol. Chem.* **265:**13735–13740.

255. **Stackebrandt, E., O. Päuker, and M. Erhard.** 2005. Grouping myxococci (*Corallococcus*) strains by matrix-assisted laser desorption/ionisation time-of-flight spectrometry (MALDI-TOF) mass spectrometry: comparisons with gene sequence phylogenies. *Curr. Microbiol.* **50:**71–77.

256. **Stahl, E.** 1969. *Thin Layer Chromatography: a Laboratory Manual.* 2nd ed. Academic Press Inc., New York, NY.

257. **Staley, J. T., and N. R. Krieg.** 1989. Bacterial classification of procaryotic organisms: an overview, p. 2299–2302. *In* S. T. Williams, M. E. Sharpe, and J. G. Holt (ed.), *Bergey's Manual of Systematic Bacteriology*, vol. 4. The Williams & Wilkins Co., Baltimore, MD.

258. **Stanier, R. Y., N. J. Palleroni, and M. Doudoroff.** 1966. The aerobic pseudomonads: a taxonomic study. *J. Gen. Microbiol.* **43:**159–271.

259. **Stockdale, H., D. W. Ribbons, and E. A. Dawes.** 1968. Occurrence of poly-β-hydroxybutyrate in the Azotobacteraceae. *J. Bacteriol.* **95:**1798–1803.

260. **Strodtmann, S.** 1895. Die Ursache des Schwebevermæogens bei den Cyanophyceen. *Biol. Zentbl.* **15:**113–115.

261. **Strömpl, C., B. J. Tindall, G. N. Jarvis, H. Lünsdorf, E. R. B. Moore, and H. Hippe.** 1999. A re-evaluation of the taxonomy of the genus *Anaerovibrio*, with the reclassification of *A. glycerini* as *Anaerosinus glycerini* gen. nov., comb. nov., and *A. burkinabensis* as *Anaeroarcus burkinabensis* gen. nov., comb. nov. *Int. J. Syst. Bacteriol.* **49:**1861–1872.

262. **Strömpl, C., B. J. Tindall, N. Lünsdorf, T.-Y. Wong, E. R. B. Moore, and H. Hippe.** 2000. Reclassification of *Clostridium quercicolum* as *Dendrospora quercicola* gen. nov., comb. nov. *Int. J. Syst. Bacteriol.* **50:**101–106.

263. **Sulea, I. T., M. C. Pollice, and L. Barksdale.** 1980. Pyrazine carboxylamidase activity in *Corynebacterium*. *Int. J. Syst. Bacteriol.* **30:**466–472.

264. **Takatsuka, Y., and Y. Kamio.** 2004. Molecular dissection of the *Selenomonas ruminantium* cell envelope and lysine decarboxylase involved in the biosynthesis of a polyamine covalently linked to the cell wall peptidoglycan layer. *Biosci. Biotechnol. Biochem.* **68:**1–19 .

265. **Takatsuka Y., M. Onoda, T. Sugiyama, K. Muramoto, T. Tomita, and Y. Kamio.** 1999. Novel characteristics of *Selenomonas ruminantium* lysine decarboxylase capable of decarboxylating both L-lysine and L-ornithine. *Biosci. Biotechnol. Biochem.* **63:**1063–1069.

266. **Tanner, A. C. R., and S. S. Socransky.** 1984. Genus *Wolinella*, p. 646–650. *In* N. R. Krieg and J. G. Holt (ed.), *Bergey's Manual of Systematic Bacteriology*, vol. 1. The Williams & Wilkins Co., Baltimore, MD.

267. **Tanner, A. C. R., M. A. Listgarten, and J. L. Ebersole.** 1984. *Wolinella curva* sp. nov.: "*Vibrio succinogenes*" of human origin. *Int. J. Syst. Bacteriol.* **34:**275–282.

268. **Tarrand, J. J., and D. H. M. Gröschel.** 1982. Rapid, modified oxidase test for oxidase-variable bacterial isolates. *J. Clin. Microbiol.* **16:**772–774.

269. **Thornley, M. J.** 1960. The differentiation of *Pseudomonas* from other gram-negative bacteria on the basis of arginine metabolism. *J. Appl. Bacteriol.* **23:**37–52.

270. **Tindall, B. J.** 1990. A comparative study of the lipid composition of *Halobacterium saccharovorum* from various sources. *Syst. Appl. Microbiol.* **13:**128–130.

271. **Tindall, B. J.** 1990. Lipid composition of *Halobacterium lacusprofundi*. *FEMS Microbiol. Lett.* **66:**199–202.

272. **Tindall, B. J.** 1996. Respiratory lipoquinones as biomarkers, section 4.1.5., suppl. 1. *In* A. Akkermans, F. de Bruijn, and D. van Elsas (ed.), *Molecular Microbial Ecology Manual*, Kluwer Publishers, Dordrecht, The Netherlands.

273. **Tindall, B. J.** 2002. *Encylopedia of the Life Sciences*, vol. 15, p. 244–251. Nature Publishing Group, London, United Kingdom.

274. **Tindall, B. J., E. Brambilla, M. Steffen, R. Neumann, R. Pukall, R. M. Kroppenstedt, and E. Stackebrandt.** 2000. Cultivatable microbial biodiversity: gnawing at the gordian knot. *Environ. Microbiol.* **2:**310–318.

275. **Tinghitella, T. J., and S. C. Edberg.** 1991. Agglutination tests and *Limulus* assay for the diagnosis of infectious disease, p. 61–72. *In* A. Balows, W. J. Hausler, Jr., K. L. Herrmann, H. D. Isenberg, and H. J. Shadomy (ed.), *Manual of Clinical Microbiology*, 5th ed. American Society for Microbiology, Washington, DC.

276. **Valentine, N., S. Wunschel, D. Wunschel, C. Petersen, and K. Wahl.** 2005. Effect of culture conditions on microorganism identification by matrix-assisted laser desorption ionisation mass spectrometry. *Appl. Env. Microbiol.* **71:**58–64.

277. **Valentini, A., A. M. Timperio, I. Cappuccio, and L. Zolla.** 1996. Random amplified polymorphic DNA (RAPD) interpretation requires a sensitive method for the detection of amplified DNA. *Electrophoresis* **17:**1553–1554.

278. **Van Baar, B. L. M.** 2000. Characterisation of bacteria by matrix-assisted laser desorption/ionisation and electrospray mass spectrometry. *FEMS Microbiol. Rev.* **24:**193–219.

279. **Vandamme, P., B. Pot, M. Gillis, P. De Vos, K. Kersters, and J. Swings.** 1996. Polyphasic taxonomy, a consensus approach to bacterial systematics. *Microbiol. Rev.* **60:**407–438.

280. **Vandamme, P., M. I. Daneshvar, F. E. Dewhirst, B. J. Paster, K. Kersters, H. Goosens, and C. W. Moss.** 1995. Chemotaxonomic analyses of *Bacteroides gracilis* and *Bacteroides ureolyticus* and reclassification of *B. gracilis* as *Campylobacter gracilis* comb. nov. *Int. J. Syst. Bacteriol.* **45:**145–152.

281. **Van de Peer, Y., and R. De Wachter.** 1994. TREECON for Windows: a software package for the construction and drawing of evolutionary trees for the Microsoft Windows environment. *Comput. Appl. Biosci.* **10:**569–570.

282. **Vaneechoutte, M., G. Verschraegen, G. Claeys, and P. Flamen.** 1988. Rapid identification of *Branhamella catarrhalis* with 4-methylumbelliferyl butyrate. *J. Clin. Microbiol.* **26:**1227–1228.

283. **Vedros, N. A.** 1984. Genus *Neisseria*, p. 290–296. *In* N. R. Krieg and J. G. Holt (ed.), *Bergey's Manual of Systematic Bacteriology*, vol. 1. The Williams & Wilkins Co., Baltimore, MD.

284. **von Wintzingerode, F., S. Börker, C. Schlötelburg, N. H. L. Chiu, N. Storm, C. Jurinke, C. R. Cantor, U. B. Göbel, and D. van der Boom.** 2002. Base-specific fragmentation of amplified 16S rDNA genes analysed by mass spectrometry: a tool for rapid bacterial identification. *Proc. Natl. Acad. Sci. USA* **99:**7039–7044.

285. **Walsby, A. E.** 1972. Structure and function of gas vacuoles. *Bacteriol. Rev.* **36:**1–32.

286. **Wang, G., T. S. Whittam, C. M. Berg, and D. E. Berg.** 1993. RAPD (arbitrary primer) PCR is more sensitive than multilocus enzyme electrophoresis for distinguishing related bacterial strains. *Nucleic Acids Res.* **21:**5930–5933.

287. **Watson, R. R.** 1976. Substrate specificities of aminopeptidases: a specific method of microbial differentiation, p. 1–14. *In* J. R. Norris (ed.), *Methods in Microbiology*, vol. 9. Academic Press Ltd., London, United Kingdom.

288. **Wayne, L. G., D. J. Brenner, R. R. Colwell, P. A. D. Grimont, O. Kandler, M. I. Krichevsky, L. H. Moore, W. E. C. Moore, R. G. E. Murray, E. Stackebrandt, M. P. Starr, and H. G. Trüper.** 1987. Report of the ad hoc committee on the reconciliation of approaches to bacterial systematics. *Int. J. Syst. Bacteriol.* **37:**463–464.

289. **Weilenmann, H.-U., B. Engeli, M. Bucheli-Witschel, and T. Egli.** 2004. Isolation and growth characteristics of an EDTA-degrading member of the α-subclass of Proteobacteria. *Biodegradation* **15:**289–301.

290. **Welsh, J., and M. McClelland.** 1990. Fingerprinting genomes using PCR with arbitrary primers. *Nucleic Acids Res.* **18:**7213–7218.

291. **Westley, J. W., P. J. Anderson, V. A. Close, B. Halpern, and E. M. Lederberg.** 1967. Aminopeptidase profiles of various bacteria. *Appl. Microbiol.* **15:**822–825.

292. **Wheater, D. M.** 1955. The characteristics of *Lactobacillus acidophilus* and *Lactobacillus bulgaricus*. *J. Gen. Microbiol.* **12:**123–132.

293. **Whittenbury, R.** 1964. Hydrogen peroxide formation and catalase activity in the lactic acid bacteria. *J. Gen. Microbiol.* **35:**13–26.

294. **Wiegel, J.** 1981. Distinction between the Gram reaction and the Gram type of bacteria. *Int. J. Syst. Bacteriol.* **31:**88.

295. **Wilkins, T. D., and C. B. Walker.** 1975. Development of a micromethod for identification of anaerobic bacteria. *Appl. Microbiol.* **30:**825-830.

296. **Wilkins, T. D., C. B. Walker, and W. E. C. Moore.** 1975. Micromethod for identification of anaerobic bacteria: design and operation of apparatus. *Appl. Microbiol.* **30:**831–837.

297. **Williams, J. G., A. R. Kubelik, K. J. Livak, J. A. Rafalski, and S. V. Tingey.** 1990. DNA polymorphisms amplified by arbitrary primers are useful as genetic markers. *Nucleic Acids Res.* **18:**6531–6535.

298. **Williams, M. A.** 1960. Flagellation in six species of *Spirillum*—a correction. *Int. Bull. Bacteriol. Nomencl. Taxon.* **10:**193–196.

299. **Williamson, D. H., and J. F. Wilkinson.** 1958. The isolation and estimation of poly-β-hydroxybutyrate inclusions in *Bacillus* species. *J. Gen. Microbiol.* **19:**198–209.

300. **Wilson, A. C., S. S. Carlso, and T. J. White.** 1997. Biochemical evolution. *Annu. Rev. Biochem.* **46:**573–639.

301. **Woese, C. R.** 1987. Macroevolution in the microscopic world, p. 177–202. *In* C. Patterson (ed.), *Molecules and Morphology in Evolution: Conflict or Compromise?* Cambridge University Press, Cambridge, United Kingdom.

302. **Woese, C. R.** 1987. Bacterial evolution. *Microbiol. Rev.* **51:**221–271.

303. **Yamamoto, A., I. Yano, M. Masui, and E. Yabuuchi.** 1978. Isolation of a novel sphingoglycolipid containing glucuronic acid and 2-hydroxy fatty acid from *Flavobacterium devorans* ATCC 10829. *J. Biochem.* **83:**1213–1216.

304. **Yano, I., J. Tomiyasu, and E. Yabuuchi.** 1982. Long chain bases of strains of three species of *Sphingobacterium* gen. nov. *FEMS Microbiol. Lett.* **15:**303–307.

305. **Yoch, D. C., and R. M. Pengra.** 1966. Effect of amino acids on the nitrogenase system of *Klebsiella pneumoniae*. *J. Bacteriol.* **92:**618–622.

306. **Zellner, G., and H. Kneifel.** 1993. Caldopentamine and caldohexamine in cells of *Thermotoga* species, a possible adaptation to the growth at high temperatures. *Arch. Microbiol.* **159:**472–476.

307. **Zobell, C. E.** 1946. *Marine Microbiology.* Chronica Botanica Co., Waltham, MA.

308. **Zuckerkandl, E., and L. Pauling.** 1965. Molecules as documents of evolutionary history. *J. Theor. Biol.* **8:**357–366.

309. **Zweig, G., and J. Sherman.** 1972. *CRC Handbook of Chromatography*, vol. 2. *Chromatographic Data and Techniques.* CRC Press, Cleveland, OH.

16

General Methods To Investigate Microbial Symbioses

TODD A. CICHE AND SHANA K. GOFFREDI

16.1. INTRODUCTION

16.1.1. Symbiosis

Symbiosis, defined by de Bary as the nontransient interaction between dissimilar organisms (38), has dramatically affected life on Earth. For example, eukaryotic cells in our own bodies have resulted from symbiotic mergers between microbes millions of years ago (111). Entire biomes, including forests and grasslands, rely upon symbiosis for stability and efficiency; for example, roots require infection by fungal mycorrhizae and rhizobia for the procurement of phosphate and reduced nitrogen, respectively. Symbiosis, in sum, is a powerful evolutionary mechanism that has resulted in novel characteristics and allowed for the exploitation of diverse environmental niches.

All macroorganisms and microorganisms are surrounded by a diverse microbiota, with which they have evolved to interact for part or all of their lives. Organisms tend to resist pathogenic or parasitic interactions (i.e., immunity) while fostering beneficial interactions. It has become increasingly apparent that components of the innate immune system are required for animals and plants to interact both in mutualism and pathogenesis (125). How then does the host differentiate between benign or beneficial microorganisms and those that are pathogenic? Understanding the molecular mechanisms of the establishment and maintenance of symbiosis will shed light on the factors affecting the stability of beneficial microbial interactions that are important for most plant and animal life.

Unlike parasitic and pathogenic interactions that cause disease, symbiotic associations are usually operating incognito in healthy hosts and only become apparent upon disruption of the association and close examination by the researcher. Although different in effect on host fitness, mutualism and pathogenesis share many characteristics that have recently been revealed by genomics, genetics, and molecular biology (78). We use the general definition of symbiosis by de Bary, acknowledging similarities between pathogenic and mutual host-bacterial interactions, as long as they are nontransient, which result in a great range of fitness to the host. We also use similarities between pathogenesis and symbiosis to develop a suite of criteria, described below, that one can use to infer symbiotic relationships involving bacteria.

A common conception of symbiosis is of two organisms cooperating for their common benefit. However, the fitness of each organism resulting from symbiosis is difficult to quantify and is usually dynamic, influenced by environmental and genetic variables. Thus, de Bary defined symbiosis as a gradient of interactions ranging from parasitism to commensalism and mutualism whereby the host fitness is decreased, unaffected, or increased, respectively, as a result of the interaction. It is also often assumed that the host is a eukaryotic organism, but the definition from de Bary does not exclude bacterial-bacterial or bacterial-phage interactions as being symbiotic (examples of these are shown in Table 1).

In this chapter, methods are presented that highlight two general areas of symbiosis research: (i) the detection and characterization of environmental symbioses and (ii) elucidation of the molecular and cellular mechanisms of symbiosis in model systems. We describe here methods to demonstrate that an organism is symbiotic and examples of techniques used to elucidate the molecular, physiological, cellular, and evolutionary basis of symbiosis. Research in these areas is in its infancy and should provide many more interesting discoveries in the future, taking advantage of some of the techniques described here. These studies should result in an increasing knowledge about the role of symbiotic microbes to host biology and ecology and should indicate that symbiotic associations involve sophisticated mechanisms for host-bacterial interactions, somewhat comparable to those of host-pathogen interactions.

16.1.2. Koch's Postulates Applied to Symbiosis

If we want to identify a potential symbiosis involving two or more organisms, we must first demonstrate a nontransient interaction, distinguish the residents (i.e., the symbionts)

TABLE 1 Examples of microbial symbioses

Symbiosis	Type	Reference(s)
Bacteria-phage		
Vibrio cholerae–CTXφ phage	Broadened host range	169
Shigella flexneri–Sf(I-V) phage	Broadened host range	1
Microbe-microbe		
"*Chlorochromatium*"	Phototrophic consortia	60, 70
Archaea-bacteria		
Anaerobic methane-oxidizing–sulfate-reducing consortium	Syntrophy	13, 129; Color Plate 9A
Archaea-archaea		
Nanoarchaeum equitans–*Ignicoccus*	Parasitism	173
Protozoa-bacteria		
Staurojoenina–*Vestibaculum*	Unknown	158
Amoeba proteus D strain–X symbiont	S-adenosylmethionine dependence	87
Plant-bacteria		
Legume–rhizobia	Diazotrophy (N_2 fixation)	166; Color Plate 9F
Plant-growth-promoting bacteria	Nutrition, defense	142
Animal-bacteria		
Aphid–*Buchnera*	Nutrition, essential amino acids	9, 42
Euprymna scolopes–*Vibrio fischeri*	Bioluminescence	125; Color Plate 9E
Riftia pachyptila–sulfur-oxidizing autotrophic bacteria	Nutrition-carbon energy	26; Color Plate 9C
Nematodes/polychaetes/shrimp–sulfide-oxidizing ε-Proteobacteria	Nutrition-epibionts	76, 140, 141; Color Plate 9B
Heterorhabditis bacteriophora–*Photorhabdus luminescens*	Insect pathogenicity-nutrition	33, 54; Color Plate 9G
Animal-polymicrobial		
Termite–bacteria, archea, protozoa, fungi	Cellulose degradation-nitrogen fixation	18; Color Plate 9D
Phallodriline worms–α-, γ-, δ-Proteobacteria	Nutrition, unknown	12, 45, 49, 94; Color Plates 9H, 10
Human gastrointestinal tract–bacteria, archaea	Nutrition, other unknown	80

from the tourists (i.e., free-living "contaminants"), and assess the effect of the symbiosis on the fitness of the host. Since the broad definition of symbiosis includes parasitism (i.e., pathogenesis), criteria for determining the causative agent of disease (Koch's postulates) can be applied to symbiosis where the characteristics of the symbiosis (e.g., colonization, tissue structure, and fitness) can be substituted for the symptoms of disease.

The contributions of many scientists, including Koch and Pasteur in the late 19th century, on the ability of bacteria to cause disease are useful for our understanding of host-bacterial interactions in general. Robert Koch presented four postulates that, when fulfilled, establish causal evidence for the microbe causing a given disease, although failure to meet these criteria does not prove otherwise. Koch's postulates include the following.

1. The pathogen must be present in each diseased host.
2. The causative pathogen must be isolated and grown in pure culture.
3. The pure culture must cause disease in the new host.
4. The causative pathogen must be reisolated from the new host in pure culture and identified.

Symbiotic microorganisms do not fit easily into Koch's postulates, because the resulting change in observable fitness of the host is usually less obvious than acute disease caused by a pathogenic bacterium. Because symbiosis encompasses a wide range of interactions, the effect of the association on host fitness will vary from unobservable (commensalisms) to obvious or even essential to host and/or symbiont fitness (mutualism). However, symbiosis is often accompanied by novel characteristics not observed in apo- (or non-) symbiotic organisms. Even for commensal relationships, a nontransient presence (i.e., colonization) of symbionts on or in host tissues can be reliably observed. These characteristics can be substituted for disease to adapt Koch's postulates to symbiosis as given below.

Koch's postulates have been adopted for symbiosis, and these include the following.

1. Identify symbiont from the host.
2. Eliminate the symbiont from the host.
3. Observe difference(s) between symbiotic and aposymbiotic host.
4. Add the symbiont to a new aposymbiotic host and observe symbiotic state.
5. Identify the symbiont from the new host.

An example of the fulfillment of these modified postulates is the symbiosis between the Hawaiian bobtail squid, *Euprymna scolopes*, and the bioluminescent bacterium, *Vibrio fischeri*, which resides in the squid's light organ (Color Plate 9, described in section 16.3): symbiotic bacteria are readily isolated and identified from the light organs of symbiotic hosts (postulate 1); hatchling squid are aposymbiotic in the absence of symbiotic bacteria, and symbiotic bacteria can be eliminated from adult squid by antibiotic treatment (postulate 2); aposymbiotic hosts are not bioluminescent (postulate 3); addition of symbiotic bacteria to aposymbiotic hosts reestablishes the symbiotic function (i.e., bioluminescence of the light organ) (postulate 4); and symbiotic *V. fischeri* can be reisolated from symbiotic hosts (postulate 5). These postulates have been met for the majority of symbioses described in the introduction to model symbioses (section 16.3.1 below).

Many symbioses, however, present additional challenges to fulfilling Koch's postulates. These include the inability to isolate and cultivate symbiotic bacteria and the inability to selectively eliminate the symbiont or generate an aposymbiotic host. These challenges are similar to those encountered when investigating pathogenesis caused by uncultured bacteria, viruses, or chronic infections. In 1996, Relman adapted Koch's postulates for criteria to determine the causative agent of disease by use of sequence-based detection methods for pathogens (58). Similar sequence-based criteria can be adapted to symbiosis to provide evidence that an organism is symbiotic (listed below). These criteria are generally listed from strongest (i.e., by definition, a nontransient association between symbiont nucleic acid and host is required for symbiosis) to weakest (the specific nature of the association may be unknown). Even though many symbioses only partially meet these criteria, they are presented here to aid the researcher when examining symbiosis from a molecular perspective.

Biochemical and Molecular Criteria for Symbiosis

1. A nucleic acid sequence (or ribotype) belonging to the putative symbiont should be present in most cases of symbiosis and found preferentially in organs, tissues, or cells known to be involved in symbiosis (based on other anatomic, histologic, or chemical evidence of symbiosis).
2. Fewer or no copy numbers of symbiont-associated nucleic acid sequences should occur in aposymbiotic hosts or tissues not involved in housing symbionts.
3. Efforts should be made to demonstrate the microbial ribotype in situ (via fluorescence in situ hybridization [FISH] microscopy on tissues or cellular level; see section 16.2.3.1 below).
4. These sequence-based forms of evidence for microbial causation should be reproducible.
5. After elimination of symbionts (i.e., by antibiotic treatment), the symbiont-associated ribotype should decrease or become nondetectable, manifested in loss of symbiont-derived characteristics.
6. When nucleic acid sequence detection predates symbiosis (e.g., during ontogenetic development), or sequence copy number correlates with symbiosis (e.g., health of the host for mutualism), the sequence-symbiosis association is more likely to be a causal relationship. The term "ontogenetic" refers to the origin and development of an individual organism from embryo to adult.
7. The nature of the microorganism inferred from the available sequence should be consistent with the known characteristics for that group of organisms. When phenotype (e.g., microbial morphology and physiology) is predicted by sequence-based phylogenetic relationships, the meaningfulness of the sequence is enhanced but should be supported by other evidence (e.g., molecular or physiological; see section 16.2).

These criteria have been met for many of the symbioses discussed in section 16.2, including the chemoautotrophic symbiosis involving the giant tube worm *Riftia pachyptila*, which inhabits hydrothermal vents. For example, the symbiont nucleic acid (or ribotype) and bacterial-specific enzymes are always found in symbiotic tube worms (criterion 1), the nonsymbiotic tissues of *R. pachyptila* are devoid of symbionts (criterion 2), the direct identification and localization of the symbiont has been observed via fluorescence in situ microscopy of nucleic acids and immunohistochemistry of intracellular bacteria-specific enzymes (criterion 3), these results are reproducible from worm to worm (criterion 4), symbiont nucleic acid is absent from *R. pachyptila* larvae (criterion 6), and sulfide oxidation and CO_2 fixation are metabolic properties consistent with certain members of the γ-Proteobacteria, with which the symbiont is phylogenetically related (criterion 7).

16.1.3. Diversity and Significance of Symbiosis

Symbiosis is prevalent in almost all types of organisms, from bacteria to protists, plants, and animals (Color Plate 9). In fact, among plants and animals, symbioses with microorganisms are the norm, rather than the exception. Historically, the important and often essential role of symbiotic organisms to the lives of their hosts has been overlooked. For example, biologists studied for decades the symbiosis between leaf-cutting Attine ants and their fungal symbionts without realizing that a conspicuous patch of "waxy bloom" on the ant's thorax was an actinomycete symbiont (36). The symbiotic actinomycete produces antibiotics specific to the parasitic fungus Escovopsis, which can infect the symbiotic fungal garden, and is preferentially found in ant castes that groom the primary fungal symbiont. Thus, even symbiosis researchers can overlook symbiosis.

There is an enormous diversity in the number of organisms that function as hosts, as well as the types of microbes that become symbionts. Examples of the range of bacterial-host interactions are listed in Table 1 (see also Color Plate 9). We mention only a few examples here to illustrate the breadth and diversity of symbioses, focusing on bacterial symbioses despite the abundance of many other host-eukaryotic associations (i.e., algae, protists, and fungi).

16.1.3.1. Bacteria-Phage

Viruses and phages, which by de Bary's definition can be considered symbiotic, affect their hosts in a variety of ways, including acute virulence, nonvirulence, or even benefit. Microbial genomes are littered with genes that confer added capabilities to the host organism and in many cases have been shown to be transmitted by phages. This is most evident in pathogens where toxins and/or virulence factors are encoded on a lysogenic phage (19), i.e., the cholera toxin genes, ctxAB, are carried and transmitted by CTXphi, which itself utilizes the colonization factor toxin coregulated pilus (TCP), encoded on another phage-like element, for its infection (50). Another type of bacteria-phage symbiosis is the gene transfer agents found in several diverse prokaryotes, including Rhodobacter capsulatus (99) and Brachyspira hyodysenteriae (114). These phages appear to be obligate symbionts of the host cell and function mainly as gene transfer agents by indiscriminately packaging host DNA. The study of bacteria-phage symbioses is important for our understanding of bacterial evolution and emergence of disease as well as providing powerful experimental models for the study of symbiosis in a test tube.

16.1.3.2. Microbe-Microbe

Bacteria-bacteria syntrophies are one type of symbiosis in which energetically unfavorable reactions are made more favorable by the close association of two or more microorganisms. In marine cold seeps and other subseafloor diffusion-limited environments, for example, there are cell aggregations of Archaea belonging to the Methanosarcinales, surrounded by, and often in direct physical contact with, sulfate-reducing bacteria related to the Desulfosarcina genus (13). These organisms work together to achieve the thermodynamically unfavorable anaerobic oxidation of methane. In this association, the Archaea are commonly located in the central core of each microbial aggregate and are surrounded by a shell of Desulfosarcina cells. The Archaea are thought to mediate methane oxidation, while the bacterial partner is presumed to consume some reduced intermediate for the reduction of sulfate, thereby creating a favorable energy gradient for the conversion of methane to carbon dioxide. Similarly, a likely syntrophy involves the phototrophic consortium "Chlorochromatium aggregatum," originally thought to be a single organism. This consortium can be very abundant in freshwater lakes and is thought to be important in sulfur and carbon cycles. This consortium was recently isolated for the first time (59) and found to consist of a single motile nonpigmented central β-proteobacterium surrounded by green sulfur photosynthetic epibionts (60). It is hypothesized that the consortium utilizes a closed sulfur cycle in which the epibionts oxidize sulfide during anaerobic photolithotrophy and the central symbionts reduce sulfur while oxidizing carbon compounds from the host. The consortium may also interact during phototaxis (60) and coordinate cell division. A draft genome sequence of this consortium is available (htpp://genome.jgi-psf.org/finished_microbes/chlag/chlag.home .html), further adding to the utility of this model phototrophic consortium for the study of bacteria-bacteria symbiosis.

16.1.3.3. Bacteria-Protists

Many diverse protists, including at least three types of flagellates that occur as symbionts within the termite gut, are associated with symbiotic bacteria. One recent example of this is a cellulose-degrading hypermastigote, Staurojoenina sp., found within the gut of the termite Neotermes cubanus (158). Electron microscopy revealed a dense coat of epibiotic rod-shaped bacteria (up to 3,500 per flagellate) that interact with the cuticle via special attachment sites on the cell envelope of the host. The nature and function of this relationship is unknown but is thought to be different from the motility-conferring epibionts of other gut flagellates (e.g., Treponema [spirochetes] found on Mixotricha sp.) (34, 162, 174).

16.1.3.4. Rhizobium-Legume Symbiosis

The symbioses between plants in the family Leguminosae and the α-Proteobacteria Rhizobium, Mesorhizobium, Sinorhizobium, Bradyrhizobium, and Azorhizobium, collectively called rhizobia, are among the best understood symbioses (Color Plate 9F). Only a general introduction to the detailed knowledge of this complex symbiosis is given here. The rhizobium-legume symbiosis is one in which the symbiont fixes dinitrogen in return for photosynthate and a protected niche (root nodule) from the host. The establishment of this symbiosis is well understood and involves a complex developmental program of binding, signaling, and growth between symbiont and plant (reviewed in reference 62). The rhizobia likely bind to root hairs, at first weakly, in a Ca^{2+}-dependent manner partially through rhicadhesin. The rhizobia sense plant flavonoids and produce signaling molecules called nod factors or lipo-chito-oligosaccharides that trigger root hair curling and infection thread formation. A further dialogue ensues when the rhizobia migrate through the infection thread, which in Sinorhizobium meliloti requires exopolysaccharide production (reviewed in reference 57) as bacteria grow from a few colonizing cells in the infection thread (61). Several lines of evidence suggest that the plants control the infection by producing reactive oxygen and nitrogen species, phenolics, etc., factors similar to a hypersensitive response (reviewed in reference 7). After passing through the gauntlet, the rhizobia then differentiate into bacterioids, capable of nitrogen fixation, while the plant develops a mature nodule.

The legume maintains a suitable environment for nitrogen fixation by producing leghemoglobin to protect the symbiont nitrogenase that is irreversibly inactivated by oxygen and by providing photosynthate to fuel the high energy demands of symbiont nitrogen fixation.

16.1.3.5. *Buchnera*-Aphids

The aphid symbiosis with *Buchnera aphidicola* is one of the most well-understood bacterial-animal endosymbioses. The symbionts provide the insect host (e.g., *Schizaphis graminum*) with essential amino acids which are lacking in the plant phloem on which it feeds. Aphids are not able to meet their nutritional requirements without *Buchnera*, and *Buchnera* has so far not been found outside the aphid hosts. The capacity for synthesizing certain essential amino acids, such as tryptophan and leucine, has been relegated to symbiont plasmids, thus allowing for independent regulation of these genes in relation to the rest of the genome. *Buchnera* organisms inhabit vesicles inside specialized aphid cells called bacteriocytes and are vertically transmitted by infection of aphid gametes during ovulation (42). The strict vertical transmission of symbionts is manifested in the coevolution of *Buchnera* symbiont and aphid host (113), which has significant consequences to both partners, including dramatic genome reduction of the symbiont (123).

16.1.3.6. Termites

Termites are now regarded as one of the classic examples of an obligate mutualistic symbiosis (Color Plate 9D). Early observations suggested that nearly all of the termite species lived almost exclusively on cellulose-rich materials, a difficult task for most animals. The secret to their success is the presence of a diverse microbiota located in the hindgut region of the alimentary tract, the major site of nutrient absorption by the host (85). Over the past century, much has been learned about the obligate dependence of termites on protozoans and bacteria for the supply of carbon compounds resulting from cellulose digestion and on bacteria for nitrogen fixation and recycling. The tremendous diversity of termite symbionts has been elucidated by molecular surveys as well as cultivation studies. In fact, this is one instance where cultivation has been reasonably successful (see section 16.2.4.2), even though many of the potentially hundreds of symbiont species within the gut have not been isolated in the lab. Bacteria associated with termites include methanogens, acetogens, fermenters, sulfate reducers, and nitrogen fixers (95, 127). A portion of the dinitrogenase reductase gene (*nifH*) was directly amplified from DNA extracted from the mixed microbial population in the termite gut (128). Sequences from the termites included four groups of microbial *nifH* sequences, some distinct from those previously recognized in studies using classical microbiological techniques, indicating the presence of diverse nitrogen-fixing microbial assemblages in the guts of termites. Some of the termite gut microbes are associated directly with the termite, while others are symbionts on protists within termites, and some are involved in specific bacterial-bacterial interactions. Many of the termite symbionts are endemic to termite guts.

16.1.3.7. *Riftia pachyptila*

R. pachyptila, the gutless giant hydrothermal vent tube worm, was the first demonstration of a chemoautotrophic bacteria-marine invertebrate association (Color Plate 9C). Since its discovery in 1977, a number of other, similar associations have been discovered in a variety of habitats including vents, cold seeps, sewage outfalls, eelgrass beds, and anoxic basins. *R. pachyptila* houses symbiotic γ-proteobacteria in a unique internal organ known as the trophosome, which can account for up to 50% of the host volume and is highly vascularized (23, 97). First observed via transmission electron microscopy (23, 26), these intracellular bacteria were discovered to be sulfide-oxidizing and carbon dioxide-fixing symbionts solely responsible for the nutrition of the worm (51). Many important studies have been conducted on the metabolic needs of both partners as well as the nutritional dialogue between them (see section 16.2.4). For example, it is now known that the host has evolved many biochemical adaptations for symbiont accommodation, including effective mechanisms to concentrate inorganic carbon internally, sulfide acquisition from the environment using specialized hemoglobin molecules, as well as unprecedented pH and ion regulation. The symbiont, in turn, provides the host with organic carbon, which enables the worm to grow rapidly, quickly dominating communities around newly established hydrothermal vents.

16.1.3.8. Epibiotic Associations

There are many examples of commensal or mutual associations between host organisms and symbionts on the external surface of the host ("epibiont"). For instance, a diverse number of marine hosts, including the Stilbonematinae (nematodes [Color Plate 9B]), the Alvinellidae (polychaete worms), and the Bresiliidae (shrimp) have dense growth of epibionts often consisting of highly ordered sulfide-oxidizing, ε-proteobacteria (39, 132, 165). In all cases, the hosts physically bridge the gap between oxic and anoxic waters via movement, thereby exposing the symbionts to all necessary metabolites. The hosts are thought to graze on the symbionts as a source of nutrition. Epibiotic associations are also abundant on nonmarine organisms and can affect the host in a variety of ways, for example in disease suppression in plants (142).

16.1.4. Methodological Dissection of Symbiosis

Within the last decade, two main areas of symbiosis research have emerged, mostly as a result of the development of molecular biological tools: (i) discovering and elucidating the important role of symbiotic microorganisms to life and (ii) dissecting the dialogue required to establish and maintain a symbiosis. Because there are increasing numbers of described symbioses, we have adapted Koch's postulates and molecular Koch's postulates as criteria in which to judge the degree of evidence for a particular symbiosis and a causal relationship of the symbiont to one or many host attributes. In section 16.2, we describe some examples of the techniques used to identify and characterize novel or poorly understood symbioses from the environment, from insects to deep-sea worms. Most of the techniques described are independent of the ability to cultivate the symbiont or host. We present a case study involving the dual symbionts of phallodrine oligochaetes, for which much has been learned despite the inability to specifically manipulate either partner. In section 16.3, we describe established and emerging model systems and genetic and molecular tools used to elucidate symbiotic factors in both the symbiont and host at the cellular, molecular, and evolutionary level.

16.2. TECHNIQUES USED TO STUDY ENVIRONMENTAL SYMBIOSES

16.2.1. Environmental Symbioses

The study of presently uncultured symbioses, for which either cultivation of the symbiont or live maintenance of the host is intractable or has not been attempted, can be accomplished using many recent advances in molecular biology and biochemistry, as well as more traditional methodologies. We now know a great deal about hundreds of these uncultured symbioses, most of which have been discovered only within the past 40 years. The following section summarizes many of the strategies that have helped to understand some of these remarkable host-microbial systems and will undoubtedly be useful for the continued investigation of potential symbioses.

16.2.1.1. The Phallodrine Oligochaete-Microbial Symbiosis

A fascinating host-microbe symbiotic system is that found in many members of the annelid subfamily Phallodrilinae, marine relatives of earthworms (Color Plate 9H). First discovered and identified in 1979 (64), there are now over 100 species known worldwide today, all belonging to the genera *Inanidrilus* and *Olavius* (49). All known species of these genera lack digestive and excretory systems, and at least 20 species possess multiple symbionts belonging to a variety of bacterial groups (63, 147). The worms are thought to depend entirely on bacterial symbionts for nutrition, as well as potential recycling and elimination of nitrogenous compounds. Many of the techniques described in this section have been carried out with phallodriline worms, illustrating a successful attempt to characterize an uncultured symbiosis.

16.2.2. Microscopy

Microscopy is a traditional, and very reliable, method for examining the ultrastructure of host tissues to look for evidence of bacteria-like cells within (intracellular) or between (extracellular) host cells or on external surfaces (epibiotic). Light and electron (both transmission and scanning) microscopy have played very significant roles in the initial characterization of many symbioses known to date. In some of the earliest microscopic observations of symbiotic systems, "parasites" were discovered in the intestine of the termite by Leidy and Hungate in the late 1800s and early 1900s using light microscopy (85, 104). Buchner, in the 1960s, continued to pioneer the use of microscopy and documented a surprising array of microbial symbionts in insects (20).

Light microscopy (detailed below) can reveal much about the morphology of cells and tissues. It is probably the most commonly used research tool in microbiology, with continuing advances in techniques and staining protocols allowing for more precise observations. Embedding of samples and staining of eukaryotic tissue allows for better orientation within complex samples. Both are described below, including toluidine blue and hematoxylin, which allow for enhanced visualization of bacteria.

Light Microscopy

In almost all cases, depending on the level of integration between the two partners, symbionts are restricted to a specific region or tissue within the host. As evidence for symbiosis, one should generally look for (i) the presence of bacterial cells, (ii) unusual or novel structures, (iii) lack of typical structures or organs, (iv) evidence for extreme vascularization, (v) precipitation of minerals or other substances, or (vi) unusual color patterns within the tissues. All of these features are common anatomical characteristics of many hosts involved in mutual symbiotic arrangements.

1. Specimens for routine histology should be preserved (i.e., fixed) quickly, preferably in 2 to 4% paraformaldehyde (made fresh) or 4% formalin (both in 1× phosphate-buffered saline [PBS]; 1× PBS = 130 mM NaCl, 10 mM Na_2HPO_4, pH 7.2, 0.2-μm-pore-size filter sterilized), but specimens preserved in 70 to 100% ethanol may be sufficient and can even be postfixed in 4% formalin.

2. Preserved specimens for light microscopy should be rinsed and dehydrated before embedding. The benefit of sectioning is the retention of tissue and cell morphology and a better understanding of symbiont localization within the host.

3. Embedding in Steedman's polyester wax is described; however, a number of embedding resins are available. Steedman's wax has a lower melting point than other resins and, thus, reduces heat-induced artifacts such as tissue shrinkage. Melt 90 g of polyethylene glycol distearate at 60°C. Add 10 g of 1-hexadecanol (cetyl alcohol) and shake until dissolved. All tissue incubation steps are performed at 37°C for 1 h; 3:1 EtOH (100%):wax; then, 2:1 EtOH:wax; then, 1:1 EtOH:wax. Incubate tissue three times in 100% wax and place in Peel-A-Way mold with fresh 100% wax. Let harden at room temperature for 3 to 4 h. Store at −20°C until sectioning.

4. Sections should be cut (6 to 8 μm thick with a microtome) from wax-embedded specimens and mounted on glass slides. After wax removal via ethanol (three rinses in 100% EtOH for 5 min each), sections can be stained with either hematoxylin-eosin (suitable for tissues fixed in formalin or alcohol) or toluidine blue.

5a. Hematoxylin: dissolve ethylene glycol (25% final concentration) in distilled water. Add hematoxylin (0.6%, wt/vol), then sodium iodate (0.06%, wt/vol). Add aluminum sulfate (8%, wt/vol) and glacial acetic acid. Let sit 1 h at room temperature. Eosin: add eosin Y (0.5%, wt/vol) to 96% ethanol with several drops of glacial acetic acid. Procedure: stain in hematoxylin for 5 to 15 min. Wash in tap water for 10 min. Rinse with acid alcohol (1% HCl, 70% EtOH) followed by tap water rinse. Soak twice in distilled water 2 min. Counterstain with eosin for 2 to 5 min. Dehydrate using 96 and 100% ethanol.

5b. Toluidine blue: place a drop of toluidine blue on the section for 30 to 60 s. Drain and wash in distilled water for 5 s. Drain and dry thoroughly on a hot plate at 80°C for 2 min before using a coverslip with a permanent mounting medium.

6. View with a compound microscope at ×100 magnification to look for bacteria, including cells smaller (1 to 10 μm) than eukaryotic cells (10 to 500 μm), sometimes enclosed within secondary vacuoles.

Note: electron microscopy (not described here) allows for magnification greater than ×1,000 and resolution greater than 0.2 μm, the approximate limits of light microscopy. Electron microscopy can be used to examine objects on a very fine scale to learn more about tissue, cellular, and subcellular morphology and even elemental composition. For scanning electron microscopy (4), whole samples are dried and sputter coated and the electrons that are produced as a beam hits the sample reveal information about the three-dimensional external structure and surface features. For transmission electron microscopy (TEM), tissues are infiltrated with embedding medium (plastics such as Spurr's epoxy resin, Embed 812, or araldite resin) and sliced into ultrathin sections (50 to 100 nm). Electrons that pass through the sample reveal information about the internal structure of the specimen.

Microscopic investigation (including light microscopy, TEM, and scanning electron microscopy) of phallodrine oligochaetes provided evidence about the nature of these associations, including the presence of multiple symbionts and the mode of transmission from one generation to the next. Bacteria-like cells were observed just below the cuticle between extensions of epidermal cells (Color Plate 10A), suggesting that up to 25% of the host volume was comprised of symbionts, for which there was significant difference in appearance (including cytoplasm and cell membrane) and size between two distinct morphotypes (66, 67). Via microscopy, Giere and Langheld (66) discovered that the symbionts could penetrate a developing egg while it was still adhered to the adult female. This "external intrusion" was observed via TEM on cultured eggs and developing embryos in various ontogenetic stages, suggesting that there is an initial infection by the symbiont.

16.2.3. Molecular Dissection of Symbiosis

16.2.3.1. PCR Fingerprinting

The enormous diversity of bacterial symbionts has only recently been appreciated, in large part due to the introduction of molecular techniques and the ability to construct molecular phylogenies of microorganisms. Prior to this, little was understood of the identity and function of bacterial symbionts, since limited morphology did not allow for meaningful interpretations of relatedness or function, and most, to this day, remain uncultured (although some advances in that area are described in section 16.2.4.3). Nucleic acid techniques not only allow for the identification of microbes potentially associated with host organisms (fulfilling biochemical and molecular criterion for symbiosis no. 1) but also may reveal insight into their function.

There are many studies in which the phylogenetic identity of symbionts within a host has been determined via 16S ribosomal sequences, long recognized as a suitable molecular marker for identifying organisms (44, 72, 77, 106, 127, 134). In some cases, including many monospecific (monoxenic) chemoautotrophic symbioses, direct bacteria-specific PCR amplification from symbiont-containing tissues of the host results in only one unambiguous symbiont sequence. For example, Distel and Felbeck (40) used direct sequencing of bacterial 16S rRNA to characterize the symbionts from the tissues of six sulfur-oxidizing polychaete and bivalve hosts. All of the sulfur-oxidizing symbionts examined clustered within two groups of the γ-Proteobacteria, with *Thiomicrospira* as the closest free-living genera. In other polyxenic cases, especially hosts with dual symbionts (for example, many insects, oligochaetes, and deep-sea mussels), microbial consortia (e.g., termites), or in systems with high potential for contamination by external bacteria (for example, bivalve gills, sponges, or epibiotic associations), it is necessary to construct bacterial clone libraries for adequate sampling of possible microbial variability (see surface sterilization considerations during the DNA extraction protocol below). In all of these cases, additional FISH assays are necessary to show definitively which recovered ribotype belongs to an actual symbiont (see the section on FISH microscopy below).

For DNA extraction, many standard kits (Qiagen DNAeasy, MoBio, Isoquick, etc.) or protocols (CTAB [hexadecyltrimethyl ammonium bromide], GITC [guanidine thiocyanate], traditional phenol:chloroform, etc.) are available and work well, even with very small amounts of tissue. The CTAB protocol for DNA extraction is particularly effective at removing exopolysaccharides from cells and is detailed below. If it is not possible to process fresh tissue di-

rectly, tissues can be preserved using any number of DNA-stabilizing methods. Tissue frozen at −20°C is best, but preservation in ethanol, RNALater (Ambion), or guanidine-containing solutions like DNAzol (GibCo) may also be sufficient. Once DNA extraction is complete, DNA fragments can be PCR amplified using bacteria-specific 16S rRNA primers (usually 27F and 1492R [98]). Provided that caution has been taken to avoid contamination during tissue dissection, a positive bacterial amplification with general PCR primers may be evidence of the presence of a potential symbiont. Usually, however, subsequent PCR amplification is undertaken using symbiont-specific primer sets (see, for example, the extracellular endosymbionts of *Bugula* sp. [bryzoan] larvae [77, 106]), particularly in order to screen additional individuals for the presence of the symbiont (biochemical and molecular criterion for symbiosis no. 4).

Whether one (monoxenic) or many (polyxenic) symbionts are present can be determined by digesting an aliquot of the ~1.5-kb PCR product with any number of restriction endonucleases (e.g., RsaI, HaeIII) and analysis of the digested products by agarose gel electrophoresis. If the restriction fragment lengths add to multiples of 1.5 kb, this indicates the likelihood of a polyxenic symbiosis. To analyze polyxenic symbioses, the 16S sequences are usually isolated by denaturing gradient gel electrophoresis (DGGE) or by cloning. In DGGE, the two strands of a DNA molecule separate upon exposure to heat and/or a chemical denaturant. When separated by electrophoresis through a gradient of increasing chemical denaturant (usually formamide and urea), the mobility of the molecule is affected by the number of GC-rich regions, which require higher temperatures or denaturant concentrations to separate than AT-rich regions. During DGGE, an artificial GC clamp is added to one end of the molecule (i.e., PCR amplification using a PCR primer with a 5′ tail consisting of a sequence of 40 GC bp); thus, the molecule becomes entangled in the gel as the branched structure separates into a semisingle stranded moiety. In this way, even DNA molecules which differ by only one nucleotide can be distinguished (12).

DNA Extraction: Dissection

1. Careful dissection is critical in order to reduce the risk of recovering bacterial contaminants by mistake. Usually symbiont-containing tissues are separated from symbiont-free tissues prior to analysis or preservation.

2. Tissues should be rinsed with 90% ethanol or dilute bleach (0.1% sodium hypochlorite) solution to remove surface contaminants, especially with tissues that are typically in contact with the environment (e.g., bivalve gills).

DNA Extraction Using CTAB
(Modified from references 28 and 176.)

3. Add ~300 mg of sample to 567 µl of TE buffer (10 mM Tris, 1 mM EDTA, pH 8.0)—gently homogenize or vortex.

4. Add 30 µl of 10% SDS (sodium dodecyl sulfate) and 3 µl of 20 mg/ml pronase E, a mixture of endo- and exoproteinases (final concentration of 100 µg/ml in 0.5% SDS). Mix thoroughly and incubate 1 to 4 h (or overnight when required) at 56°C.

5. Add 100 µl of 5 M NaCl and mix thoroughly.

6. Add 80 µl of CTAB-NaCl solution (0.7 M NaCl, 10% CTAB; first dissolve NaCl in water, then slowly add CTAB, while heating). Mix thoroughly and incubate 10 min at 65°C.

7. Add an equal volume (0.7 to 0.8 ml) of chloroform/isoamyl alcohol, vortex.

8. Centrifuge at 16,000 × g for 5 min.

9. Remove aqueous, viscous supernatant to a fresh microcentrifuge tube, leaving the interface behind. Add an equal volume of phenol/chloroform/isoamyl alcohol, vortex.

10. Centrifuge at 16,000 × g for 5 min.

11. Transfer the supernatant to a fresh tube. Add 0.6 volume isopropanol to precipitate the nucleic acids. Gently vortex.

12. Pellet the precipitate by brief centrifugation at 16,000 × g for 1 min.

13. Wash the DNA with 70% ethanol to remove residual CTAB.

14. Centrifuge at 16,000 × g for 5 min at room temperature. Air dry pellet.

15. Suspend pellet in 100 μl of TE buffer (up to 1 h). *Alternatively, the supernatant from step no. 11 can be added to a Centricon® centrifugal filter unit (Millipore) and spun to a volume of 100 μl (in place of steps 12 to 15).

Once sequences are generated, the molecular identity of a potential symbiont can then be assigned via comparison of new sequences to those in existing databases (including GenBank and RDPII). When examining symbionts, a problem occasionally arises when appropriately similar sequences (from close relatives) do not exist in public sequence databases. At the very least, a bacterial subdivision (and likely family) can be inferred by immediate comparison to known databases.

One can also take a nucleic acid approach to infer metabolic pathways that may be utilized by a symbiont. The specific physiological capabilities of the symbiont, whether isolated or not, can be explored through methods such as DNA amplification and Southern blotting for specific coding regions, RNA amplification (via reverse transcription-PCR and Northern blotting), and simple PCR amplification for genes that encode proteins with certain functions. For example, Millikan and colleagues cloned and characterized a flagellin gene, one of the subunits that form the flagellum in a variety of bacterial species, from the *Riftia pachyptila* symbiont (120). Degenerate primers were designed by aligning conserved sequences of known enteric *fliC* genes. The symbiont protein was expressed in nonmotile *Escherichia coli*, and flagella were observed by electron microscopy (120), suggesting the retention of motility by the symbiont. Similarly, many genes involved in essential amino acid biosynthesis (e.g., tryptophan and leucine) have been detected within the *Buchnera* symbiont genome (8). In phallodrine oligochaetes, the gene that codes for dissimilatory sulfite reductase, an essential and key enzyme of sulfate reduction in free-living microbes, was amplified from the symbiont-containing tissues of *Olavius algarvensis* (45). The dissimilatory sulfite reductase sequence of the *O. algarvensis* symbiont was most closely related to that of the sulfate-reducing δ-proteobacterium *Desulfosarcina*, the same closest relative to the secondary symbiont observed during 16S ribosomal surveys (45).

16.2.3.2. FISH

Many symbioses have been characterized by a combination of the above molecular methods followed by FISH microscopy. FISH microscopy was developed in the late 1980s for environmental bacterial applications, and for symbiosis research it complements other types of microscopy by po-

tentially providing simultaneous symbiont identity and host ultrastructure. The detection of specific rRNA sequences is performed by first hybridizing a fluorescently labeled probe with the target rRNA, in this case that of the putative symbiont. Usually one starts with probes that target particular bacterial subdivisions (for example, group-specific probes include Gam42a [5′- gCCTTCCCACATCgTTT-3′] [110]; alf968 [5′-ggTAAggTTCTgCgCgTT-3′] [124], and Delta495a [5′-AgTTAgCCggTgCTTCCT-3′] [109] for the γ-, α- and δ-*Proteobacteria*, respectively). In addition to confirming the presence of a particular ribotype in the host (biochemical and molecular criterion for symbiosis no. 3), FISH can also provide specific ultrastructural integration between the bacteria and host (e.g., intra- versus extracellular, nonrandom distribution within tissues, etc.). For example, FISH examination of *Alvinella pompejana*, a hydrothermal vent polychaete annelid, revealed the identity and localization of two filamentous bacterial phylotypes attached to the dorsal integument of the host (22). These ε-proteobacteria are thought to play a role in detoxification of high levels of poisonous sulfide and metals within the host's environment.

Similarly, Sipe et al. (154) investigated the mode of transmission of symbionts within two shipworm species, *Lyrodus pedicelatus* and *Bankia setacea*, using specifically designed FISH probes, citing evidence for vertical transmission based on the microscopic presence of the symbiont ribotype in eggs and ovarial tissue, as well as gill tissue (biochemical and molecular criteria for symbiosis no. 2 and 3). There are many other examples, too numerous to include in detail, so the reader is encouraged to explore recent examples including *Thiothrix* epibionts on amphipods (68), *Bacteroides* epibionts on flagellate protists (158), *Mollicutes* gut symbionts in isopods (170), *Acidovorax* symbionts in the nephridia of earthworms (151), and extracellular symbionts of bryzoan larvae (106), to name a few. Two very good references for general FISH protocols are Pernthaler et al. (136, 138).

The primary considerations for performing FISH assays, besides those detailed in the FISH protocol below, include probe design, probe labeling, and tissue preparation. Using variable regions of the 16S rRNA sequence, symbiont-specific probes can be designed to selectively label a symbiont even in complex microbial communities. Probe development should be based on the following criteria: desired specificity for a particular target bacterium or group of bacteria and adequate in situ accessibility. For maximum penetration into the highly structured ribosome during hybridization, it is best to target regions of the rRNA that have been empirically demonstrated to be accessible by oligonucleotide probes (10). Probes can be labeled (commercially or in-house) with any number of fluorescent dyes. Fluorescein is commonly used; however, cyanine dyes (Cy3 and Cy5) are brighter, pH-independent alternatives that avoid shorter-wavelength tissue autofluorescence. Probe modifications (e.g., labeling with heat-stabilized horseradish peroxidase) such as those for CARD (catalyzed reporter deposition)-FISH can also increase the sensitivity of the assay (137).

Protocol for FISH

1. Preserve tissue for 3 h in 4% paraformaldehyde (made fresh in 1× PBS) at 4°C. Rinse twice with 1× PBS. Transfer to 1:1 1× PBS:100% ethanol, store at 4°C.

2. Homogenize and gently sonicate sample (in 1× PBS:ethanol). *Whole tissue can also be embedded in

Steedman's wax (see section 16.2.2, Light Microscopy, step 3). If so, proceed to step 4.

3. Spot ~50 μl of homogenate onto slide (e.g., coated SuperFrost Plus slides; Fisher Scientific). Let dry.

4. Take slide through ethanol series (50, 75, and 100% EtOH; 5 min for each). Let dry (can store at −20°C).

5. Remove tissue autofluorescence with 10-min rinse in 0.1 M triethanolamine (pH 8.0) with 0.5% acetic anhydride (vol/vol). Repeat three times. *Alternatively, a 30-min dark rinse in 50 mM sodium borohydride (made in 100mM Tris-HCl, pH 8.0) is also effective at removing tissue autofluorescence.

6. Make hybridization buffer (0.9 M NaCl, 20 mM Tris-HCl, 0.01% SDS, 35% formamide).

7. Add 3 μl of Cy3-labeled (or other fluorescently labeled) oligonucleotide probe (50 μg/ml, specific to the endosymbiont; see text for design tips) to 30 μl of hybridization buffer. Add to sample on slide.

8. Hybridize at 46°C for 2 h to overnight in a chamber to prevent drying out (such as a capped 50-ml Falcon tube).

9. Make wash buffer (e.g., 50 ml per slide; 70 mM NaCl, 20 mM Tris-HCl, 0.01% SDS, 5 mM EDTA). Heat to 48°C before proceeding to step 10.

10. Remove slide from hybridization chamber and place in wash buffer for 15 min at 48°C.

11. Remove slide from wash buffer and rinse in distilled water.

12. Stain with a dilute 4′6′-diamidino-2-phenylindole (DAPI) solution (5 μg ml^{-1}) for 1 min.

13. Rinse in distilled water, let slide dry completely, add antifade mounting compound (such as Citifluor [Ted Pella, Inc.] or VectaShield® [Vector Laboratories]), add coverslip, and examine under epifluorescence microscopy.

The biggest challenge when using FISH for investigation of potential symbioses is background interference caused by the inherent autofluorescence of many animal and plant tissues. Certain reagents, such as glutaraldehyde fixative, are additional sources of autofluorescence. Methods to reduce autofluorescence, before hybridization, include acetylation or sodium borohydride treatment (described above in step 5).

Confocal laser scanning microscopy (and the alternative, deconvolution microscopy) has also been established as a valuable tool for the study of symbiosis. Confocal microscopy, in contrast to conventional epifluorescence microscopy, allows for the possibility to distinguish interior detail, by obtaining high-resolution images and three-dimensional reconstructions via serial optical sectioning. Confocal imaging and deconvolution can be applied to all fluorescence microscopy applications and thus proves to be an invaluable tool with which to study the structural intricacies between host and symbiont.

Many other modifications to the FISH methodology have great potential for application to symbiotic systems. For example, a method known as mRNA-FISH allows for the simultaneous detection of functional gene transcripts and traditional rRNA FISH cell identification (136). This technique has been used extensively with eukaryotes; however, decreased mRNA stability and copy number and decreased permeability of bacterial cell membranes has made this technique underutilized in terms of symbiosis. Recently, these issues have been overcome by increased amplification of the mRNA signal and rigorous permeabilization steps, allowing for the investigation of symbiont function without the requirement of symbiont isolation. In the vent mussel *Bathymodiolus puteoserpentis*, for example, mRNA-FISH re-

vealed the presence of particulate methane monooxygenase transcript (*pmoA*), encoding a bacterial enzyme responsible for conversion of methane to methanol, suggesting this capability in one of the two mussel symbionts (A. Pernthaler, Max Planck Institute, personal communication [Color Plate 11]). The successful visualization of other expressed genes within symbiotic arrangements is a very real and exciting possibility to link symbiont phylogeny and location within the host with metabolism and physiology.

Molecular characterization of the phallodrine oligochaete *Inanidrilus leukodermatus* symbiosis, using 16S rRNA (PCR) characterization and FISH microscopy, determined that the primary symbiont was a unique member of the γ-*Proteobacteria* and clustered with other known chemoautotrophic symbionts (94). FISH microscopy showed a fluorescent signal in a region between the cuticle and epidermis, confirming that the recovered 16S rRNA originated from an endosymbiont.

The possibility of multiple symbionts in these oligochaetes was explored again in 1999 when Dubilier and colleagues used TEM and 16S rRNA characterization, including PCR fingerprinting, FISH microscopy, and DGGE to compare the symbionts of *Olavius loisae* to known bacteria (12). PCR fingerprinting resulted in a heterogeneous pattern, with two dominant "families" detected as a result of cloning efforts. Again one symbiont clustered with known γ-proteobacterial symbionts; however, the other symbiont clustered within the α-*Proteobacteria*. FISH probes designed for each of the clone families confirmed the presence of two ribotypes within the worm host, the larger γ-proteobacteria and the generally smaller α-proteobacteria distributed evenly between the cuticle and epidermis. DGGE patterns from individual worms showed the presence of both symbiont types, resulting in a fast method for screening other oligochaetes for the presence of these symbionts as well as the distribution of symbionts within different tissue regions within one host. Phylogenetic screening (PCR fingerprinting and FISH microscopy) of *O. algarvensis* and *O. crassitunicatus* (Color Plate 10B), on the other hand, revealed one dominant clone group related to known sulfide-oxidizing γ-proteobacteria and the other, surprisingly, a δ-proteobacterium most closely related to sulfate-reducing bacteria within the *Desulfosarcina* cluster (12, 45). Thus, it appears that most phallodrine oligochaetes possess a common vertically transmitted primary symbiont (a γ-proteobacterium), yet they form interactions with a diverse group of potentially horizontally transmitted secondary symbionts (including δ- and α-proteobacteria), with some reports of additional symbionts as well, including spirochetes (N. Dubilier, Max Planck Institute, personal communication).

16.2.4. Physiological Dissection of Symbiosis

16.2.4.1. Direct Measures of Enzyme Activity

The presence and physiological capabilities of symbionts can be inferred from rRNA similarity as described above; however, it is not necessarily proof that they share similar function with close relatives. Direct evidence for potential pathways utilized by a symbiont involves the measure of actual enzyme activity. In many cases, the presence of a particular prokaryote enzyme within animal, plant, or protozoan tissues can provide evidence for a bacterial symbiont-host arrangement and fulfill biochemical and molecular criteria no. 2 and 3 (see section 16.1.2).

For example, microbes can potentially use at least four major pathways for CO_2 fixation, including the Calvin

Benson cycle, the reverse tricarboxylic acid cycle, the acetyl-coenzyme A CoA pathway, and the 3-hydroxypropionate cycle. All of these involve specific enzymes, which have been demonstrated by a number of studies investigating the function of symbionts. In particular, one of the first demonstrations of a chemoautotrophic bacteria-invertebrate symbiosis was based on an enzyme assay for ribulose 1,5-bisphosphate carboxylase/oxygenase (RuBisCO), a key enzyme in the Calvin Benson cycle (51). This enzyme, which has been measured enzymatically (described below), immunochemically, and molecularly (via PCR, sequencing, and expression), is now known to be critical to the interaction between hosts and symbionts in many nutritional associations (25, 94, 152). In phallodrine oligochaetes, the presence of RuBisCO was enzymatically determined in whole animal homogenates from two host species, suggesting production of organic carbon by the symbionts via the Calvin Benson cycle (53). Similarly, the possibility of autotrophic CO_2 fixation via RuBisCO in the oligochaete *I. leukodermatus* was measured, at the ultrastructural level, using an antibody against a specific form of RuBisCO found in many symbiotic bacteria (94).

Enzymatic Technique for Measuring RuBisCO (EC 4.1.1.39)

RuBisCO catalyzes the irreversible carboxylation of ribulose 1,5'-diphosphate to two molecules of 3-phosphoglycerate. It is a good indicator for the presence of bacteria and eukaryotic algae in animal tissue as it has been found only in autotrophic organisms. RuBisCO enzyme activity can be determined by measuring incorporation of ^{14}C-labeled CO_2 (usually in the form of HCO_3^-) into phosphoglycerate, or following the rate of CO_2-dependent NADH oxidation in a nonradioactive enzymatic assay that couples production of 3-phosphoglycerate to its phosphorylation and reduction to glyceraldehyde-3-phosphate.

Protocol for radioactive isotopic method for measuring RuBisCO activity (modified from reference 171)

1. Make homogenization buffer: 10mM Tris-HCl (pH 7.8 at 25°C), 0.1 mM EDTA, 10 mM β-mercaptoethanol.
2. Gently homogenize tissue in ground glass homogenizer and sonicate (5 × 30 s on ice). Sonication is reported to cause a threefold increase in RuBisCO activity (24).
3. To remove cell debris, centrifuge homogenate at 27,000 × g for 10 to 30 min (at 4°C).
4. Make reaction mixture (usually 0.5 ml) containing 100 μmol Tris-HCl (pH 7.6), 5 μmol of $MgCl_2$, 25 μmol of $NaCO_3$, 0.03 μmol of EDTA (pH 6.5), 3 μmol of glutathione, 1 μmol of D-ribulose 1,5-bisphosphate, and ~2 μCi of $NaH^{14}CO_3$. *Alternatively, $KH^{14}CO_3$ can be added at the same specific activity.
5. Add sample (20 to 100 μl) to the reaction mixture and incubate for 5 to 10 min at 20 to 30°C.
6. Stop reaction with 0.2 ml of 6N HCl and heat at 90°C for 60 min.
7. Aliquot 0.1 ml into glass vial. Dry for 1 h at 95°C. Add 0.3 ml of water followed by 3 ml of liquid scintillant (many are available on the market, e.g., Kodak or Packard Instruments).
8. Measure incorporation of radioactivity into the acidified samples by using a liquid scintillation counter.

Units: one unit of RuBisCO is defined as that amount of enzyme which catalyzes the carboxylation of 1.0 μmol of ribulose diphosphate per min.

Coupled enzyme method for measuring RuBisCO activity (modified from reference 143)

1–3. Follow steps as described in the protocol above.
4. Make reaction mixture (usually 0.5 ml) containing 50 μmol of Tris-HCl (pH 7.8), 12 μmol of ATP, 10 μmol of $MgCl_2$, 75 μmol of $NaHCO_3$ (or $KHCO_3$), 0.03 μmol of EDTA (pH 6.5), 10 μmol of glutathione, 1.0 μmol of D-ribulose 1,5-bisphosphate, 0.15 μmol of reduced NADH, 30,000 U of glyceraldehyde 3-phosphate dehydrogenase, and 30,000 U of 3-phosphoglycerate kinase.
5. Add sample (20 to 100 μl) to reaction mixture and measure change in absorbance at 340 nm, after an initial lag of several minutes.

There are a number of other enzymes and biochemical pathways that may be diagnostic for the presence of bacterial symbionts. For example, evidence of the reverse tricarboxylic acid cycle has recently been demonstrated in epibionts of alvinellid polychaetes (21) and the ε-proteobacterial endosymbiont of a snail, *Alvinoconcha* sp. (161; S. K. Goffredi, unpublished results), by the positive PCR amplification of ATP-dependent citrate lyase in certain tissues. Methodologies for measuring many prokaryotic enzymes in animal (and plant) tissues are well established. General concerns when applying these techniques to the study of potential symbioses include appropriate negative controls (e.g., boiled tissues) and positive controls (or standards).

16.2.4.2. Measures of Bacterial-Specific Biomarkers and Stable Isotopes

There are many bacteria-specific compounds that, if positively detected within eukaryote tissues, may indicate that symbiotic bacteria are responsible for their production. These include vaccenic acid, a fatty acid produced only by bacteria (86, 149); poly-β-hydroxybutyrate (93, 115), a bacteria-specific polyester storage product; elemental sulfur or polysulfides (135, 150, 167), energy storage products not known to be produced by metazoans and likely too difficult to take up across membranes from the environment; lipopolysaccharide (26); and cell components (e.g., N-acetylglucosamine and N-acetyl muramic acid) and the *Vibrio fischeri*-derived tracheal cytotoxin component of petidoglycan required for morphogenesis of the *Euprymna scolopes* light organ (see section 16.3). In phallodrine oligochaetes, structures resembling poly-β-hydroxybutyrate were observed microscopically and later confirmed, via biochemical assays, to account for up to ~10% (dry weight) of worm tissues. Additionally, energy-dispersive X-ray analysis indicated the presence of elemental sulfur in the symbiont-containing tissues, suggesting the presence of chemosynthetic sulfide-oxidizing bacterial symbionts (66). Certain body segments of the worm even appeared white, presumably due to sulfur inclusions produced by the symbionts themselves, and reflected the distribution of bacteria on the oligochaete body.

Additionally, natural stable isotope measurements can be useful in determining the presence of bacteria within eukaryotic tissues. Many factors can affect the stable isotope value within an organism; however, many bacteria possess a distinct isotopic "signature" that gives clues to their metabolic pathways. Isotopic compositions are typically reported as delta values, defined as follows: δ (per mil) $= 1,000 \times [(R_{sample}/R_{standard}) - 1]$ where R is the ratio between two isotopes (e.g., $^{13}C/^{12}C$) and the $R_{standard}$ used is the Vienna PeeDee Belemnite (VPDB). Thus, the abundance of the rare, heavy isotope is always expressed relative to the more common light isotope. For instance, bacteria that consume methane have $\delta^{13}C$ isotopic values that are very "light" (-40 to $-60‰$), whereas chemoautotrophic bacteria that fix CO_2 via the Calvin Benson cycle have $\delta^{13}C$ values typically in the -20 to $-30‰$ range (with some exceptions, of course). Giere et al. measured stable carbon isotopes in phallodrine oligochaetes and determined the symbiotic bacteria to be chemoautotrophic, based on similarity of the $\delta^{13}C$ value ($-26‰$) with free-living chemoautotrophic microbes (65).

Stable isotope ratios provide not only clues as to the functioning of the bacterial symbiont but also the potential nutritional integration between the symbiont and host. Recently several innovative techniques have been used to measure stable isotope signatures associated with symbiotic associations. One such method combines the techniques of FISH microscopy with secondary ion mass spectrometry (130, 131) to measure simultaneously the metabolic function and the identity of individual microbes. This technique was used to investigate the utilization and alteration of specific compounds by syntrophic cooperation of *Archaea* belonging to the *Methanosarcinales*, and sulfate-reducing bacteria related to the *Desulfosarcina* genus, described in section 16.1.3. Both members of the consortia had isotopic signatures that reflected the initial assimilation of isotopically light methane by the archaeal partner and transfer of this carbon to the sulfate-reducing partner (131).

16.2.4.3. Physiological Studies

Physiological investigation of currently uncultured symbionts is often difficult due to poorly understood requirements for survival and a fragile balance between symbiotic partners. The physiological dissection of these symbioses has been successful, nonetheless, and can take many forms including investigations of symbiont enrichments, pure symbiont cultures (which is rarely achieved), and whole intact associations.

16.2.4.3.1. Symbiont Enrichments

In the absence of pure symbiont cultures, much can still be learned about the physiology, metabolism, and biochemistry of a symbiont if a bacterial enrichment, separated from host tissues, cellular debris, etc., is achieved. The procedure for initial separation of symbionts from host tissue is described below. Bacterial symbionts have been successfully purified from the vent tube worm *Riftia pachyptila* and were observed to excrete succinate and glutamate into the culture medium (52). Researchers attributed this microbial release of carbon as the likely source of nutrition for the worm host. Likewise, Wilmot and Vetter (175) incubated a purified bacterial fraction from *R. pachyptila* trophosome with radiolabeled sulfur compounds and discovered the strict utilization of ^{35}S-sulfide, and not other sulfur compounds, by the symbionts. In some cases, metabolic activity of purified bacterial fractions was observed to be higher than crude homogenates (11); thus, the additional step of removing host cellular debris may be advantageous.

Protocol for Symbiont Enrichments

1. To achieve a preparation from host tissue that is enriched in symbiont cells, gently homogenize tissue in $1 \times$ PBS in a ground glass homogenizer (keep cold), prefilter through coarse (i.e., >10-μm mesh) Nitex® (optional), and separate by:

    A. Differential centrifugation: spin at $3,000 \times g$ (twice) to remove large cellular debris, followed by a spin at $15,000 \times g$ to pellet the symbiont cells (11).

B. Density gradients:

    i. Percoll — 60% (in $1\times$ PBS). Establish a gradient by centrifugation at $13,000 \times g$ (1 h). Overlay the homogenate and centrifuge at $13,000 \times g$ (1 h). Remove layers and wash by low-speed ($750 \times g$) centrifugation (1 h) in $1\times$ PBS, or a buffer solution that simulates host fluids, for example, imidazole buffer solution (50 mM imidazole, 500 mM NaCl, 20 mM $MgSO_4$, 10 mM KCl, 10 mM $CaCl_2$, pH 7.1) (40).

    ii. Nycodenz (also known as Histodenz), 16 to 28% (usually made up in $1\times$ PBS). Establish the gradient at 4°C overnight. Overlay homogenate and centrifuge at $10,000 \times g$ (1 h). Remove layers and bring each to ~2 ml with $1\times$ PBS. Wash, via centrifugation at $10,000 \times g$ for 30 min, and collect pellet, if present (72).

2. To confirm successful separation of symbionts from host cells, bacterial cell counts can be made via DAPI staining and epifluorescence microscopy, by acridine orange staining, or by hemocytometer and/or light microscopy counting.

3. Suitability of the enriched bacteria for physiological studies can be assessed via trypan blue or Hoechst staining to observe cell membrane integrity. For trypan blue, mix cells 1:1 with trypan blue solution (0.8 mM in $1 \times$ PBS) for less than 30 min. Dead cells stain blue due to trypan blue uptake. For Hoechst stain, pellet cells briefly (8 to 10 s) at high speed (approximately $16,000 \times g$) in a microcentrifuge. Pour off supernatant and resuspend cells in 50 μl of freshly made Hoescht-33342 (100 μg/ml in $1\times$ PBS). Incubate on ice for 5 min. The Hoescht-33342 stain (which is permeable and can stain both fixed and nonfixed cells) is prepared from a 10 mg/ml solution. Cells can be examined immediately by fluorescence microscopy or fixed and stored for later examination. If cells are fixed in 10% formalin, they can be stored for up to 1 week at 4°C without significant loss of staining. Live cells have evenly stained nucleosomes. Dead cells show fragmented nuclei.

4. Physiological experiments should be initiated immediately following isolation, if not fixed.

16.2.4.3.2. Pure Symbiont Cultures

Improvements in culture methods over the last few decades have made the isolation of symbiotic microbes possible, especially for those that are heterotrophic in nature (e.g., shipworm symbionts, described below). Many physiological properties of symbionts, including favorable environmental conditions, can be observed only through the study of cultured isolates. Although it is an important method for the investigation of symbiotic host-bacterial interactions, we provide only a brief discussion on the cultivation of symbionts, with emphasis on two cases in which symbiont cultures have been obtained.

In 1983, Waterbury and colleagues successfully cultured the symbionts from six species of bivalve shipworm (including *Lyrodus pedicellatus*) by serial dilution (10^7 to 10^8) of homogenate from the symbiont-containing tissue, the gland of Deshayes (172). A single species of bacterium was isolated from all species with the ability to grow on cellulose as a sole carbon source without a fixed nitrogen source. Semisolid media (0.2 to 0.9% agar in tubes and plates) were used to create a gradient of both oxygen and cellulose, allowing the microbe to "choose" the most suitable environment within the isolation chamber, a technique that alleviates the need for complete knowledge of the in vivo habitat.

Similarly, many microbial symbionts, including spirochetes, methanogens, and other microbial representatives within termite guts, have been cultivated. In the 1990s, Breznak and colleagues achieved pure cultures of termite methanogens by using long incubations (>8 weeks), dilution-to-extinction methods, and antibiotics to which methanogens were naturally resistant (100, 101). They examined oxygen tolerance, pH, and temperature optima, as well as energy sources and nutritional requirements of the symbiotic methanogens. Acetogenic, nitrogen-fixing spirochetes from the termite gut were isolated, for the first time, shortly thereafter (74, 102, 105). This isolation of spirochetes, in an anoxic dilution series that contained antimicrobial compounds (102), facilitated the observation of metabolic capabilities that were previously unknown for spirochetes, including H_2/CO_2 acetogenesis and nitrogen fixation (74, 102, 105).

16.2.4.3.3. Intact Associations

The most direct way to investigate physiological or biochemical interactions between symbiotic partners is to monitor intact associations. This, in the case of many newly discovered symbioses, is a difficult prospect when one considers the lack of knowledge regarding environmental requirements for both partners. Nevertheless, many attempts to manipulate and/or characterize intact symbiotic associations have resulted in a wealth of information, including organismal studies of metabolite production and transfer.

Methods for physiological experiments on intact symbioses are too diverse to cover in detail, and creativity is encouraged as most nonmodel associations are somewhat difficult with which to work. Many physiological experiments using respiration chambers and techniques to measure fluxes of gaseous or ionic metabolites have also contributed to our understanding of symbiont and host metabolism in many symbiotic associations. For example, Girguis et al. (69) conducted the first live animal experiments detailing proton exchange in the deep-sea tube worm host, *Riftia pachyptila*. This symbiosis relies on the ability of the worm to rapidly remove symbiont waste products internally, the absence of which would result in unsuitable conditions for both symbiont survival and host protein function. By inhibiting specific biochemical properties of both partners, they observed an increase in host elimination of protons as a result of increased symbiont productivity. Exposure to specific inhibitors of symbiont and/or host function is another effective way to tease apart the functioning of either partner in whole association experiments. Examples of these include inhibitors of host ion elimination (e.g., N-methylmaleimide and vanadate) (71) and inhibitors of autotrophic CO_2 fixation via RuBisCO (e.g., D,L-glyceraldehyde) (30). Successful organismal studies rely on good controls, usually an unmanipulated association (positive control) or incubation chambers without the organisms present (negative control). Similarly, it is important to have good a priori knowledge of the in situ environmental conditions, such that environmentally relevant conditions are simulated and healthy individuals are maintained.

One common physiological examination of intact symbioses includes radiolabeling/autoradiography experiments (detailed below). These studies mostly examined partitioning and movement of radiolabeled tracers and were able to quantify the nutritional importance of bacterially produced compounds (160). Odelson and Breznak (126), for example, used ^{14}C-labeled substrates to show that acetate results from the breakdown of cellulose in the intact termite gut and contributes to the nutrition of the host. Douglas et al. (43) demonstrated that only 8% of aphids experimentally stripped of *Buchnera* by antibiotic treatment survived (compared to 72% in nonmanipulated aphids). Aphids cannot synthesize essential amino acids and developed normally without symbionts only if the diet was supplemented with specific essential amino acids, based on incorporation of ^{14}C-labeled compounds. This provides evidence for the utilization of symbiont-produced essential amino acids by the host (43), thus fulfilling biochemical and molecular criterion for symbiosis no. 5.

For successful radiolabeling experiments, one must first choose a labeled substrate that is likely to be utilized by the symbiont (i.e., CO_2 and CH_4 for organic carbon production, acetate incorporation into fatty acids, etc.). The specific activity often has to be determined empirically. Exposure of the intact association, maintained in a healthy state, to the label is usually for a short time period (0 to 6 h). Samples are taken at intervals to monitor the movement of the label throughout the symbiont and/or host. Samples, including symbiont/host tissues or the surrounding medium, can be taken and separated into specific carbon (or other) compounds via liquid or thin-layer chromatography, followed by liquid scintillation counting or autoradiography.

This technique can also be performed on purified symbionts. For example, when incubated with ^{14}C-acetate and ^{14}C-glucose, purified symbionts of termites were found to excrete amino acids that were eventually absorbed by the host (14). Similarly, Felbeck et al. (53) used autoradiography with labeled substrates to demonstrate the uptake of CO_2 and glucose and production of organic acids by symbiont enrichments from two phallodrine oligochaete host species, further confirming the potentially autotrophic nature of the symbionts.

16.2.4.4. Tissue Autoradiography

This procedure involves the microscopic examination of visible reduction of silver halide by incorporation of a radioactive substance, most commonly a ^{14}C-, ^3H-, or ^{35}S-labeled compound, into living tissue through the use of a photographic emulsion (modified from reference 15).

1. Incubate living (freshly collected) tissue for 1 h in 0.2-μm-pore-size filtered sea water (for marine animals) containing 40 μCi ml^{-1} of NaH^{14}CO$_3$.
2. Fix tissue immediately in phosphate-buffered 3% glutaraldehyde (0.1 M, pH 7.4, made fresh in 1× PBS), for 1 to 2 h.
3. Cut semithin (1.5-μm) sections and mount on slides. *It may be necessary to embed in paraffin, with a subsequent procedure to deparaffinize, and rehydrate as described above in section 16.2.3.1.
4. For autoradiography, dip in liquid emulsion (e.g., 50% Kodak Autoradiography Emulsion Type NTB, melted in a water bath at 42°C) in darkness, allow to dry.
5. Expose for 5 to 14 days in darkness, desiccate at 4°C. During exposure, trial slides can be developed. Examine the slide at a magnification of ×10 to ×40 with a bright-field microscope and look for the presence of black silver grains located over the cells.
6. Develop in, for example, Kodak D-19 Developer at 20°C for 3 min.
7. Wash for 10 s in distilled water.
8. Fix with Kodak Fixer for 3 min.
9. Wash in running water for 15 min.

406 ■ GROWTH

10. Dehydrate by taking through an ethanol series (50-70-100%) and final xylene (100%) rinse (3 min each).

11. Stain slides with 0.25% azure II, 0.25% toluidine blue, and 0.25% methylene blue in 0.25% sodium borate.

12. Mount coverslip.

13. Visualize silver grains using light microscope with dark-field illumination.

A recent physiological characterization of symbiont function in *O. algarvensis* involved the measurement of radiotracers in the intact association (45). Worms were incubated in radiolabeled sulfate ($^{35}SO_4$). Thin silver needles were then inserted into the worms to collect ^{35}S-labeled sulfide that was produced via the reduction of sulfate. Because sulfate reduction can proceed biologically or chemically, sulfide production of live worms was compared with that of heat-killed and formalin-fixed worms. For the live worms only, autoradiographic analysis demonstrated that sulfide precipitated on the needles. The reduction of radioactive sulfate to sulfide occurred only under microaerobic conditions, a behavior common to almost all free-living, nonsymbiotic sulfate reducers, suggesting their presence within the worm tissues.

16.3. METHODS TO STUDY SYMBIOSIS IN MODEL SYSTEMS

16.3.1. Model Systems

Model symbioses are considered here as those that are easily maintained in the laboratory, where the symbiont and host can be separated and reassociated, development of the symbiosis is rapid, and one or both partners can be genetically manipulated. The tractability of model systems allows fundamental processes involved in symbiosis (e.g., the establishment and maintenance of symbiosis, nutrient exchange, and signaling) to be studied in great detail. The few established symbiotic model systems involve monoxenic (i.e., one symbiont type) associations where effects on the host are known (e.g., bioluminescence). Other characteristics of value for model symbioses are that (i) they are *nonobligate*; both the symbiont and host can be maintained or obtained in an aposymbiotic state; (ii) symbionts are *horizontally transmitted* (e.g., acquired from the environment); (iii) one or both partners have a *complete genome sequence*; and (iv) the symbiont is mechanistically or phylogenetically *related to pathogens*, allowing functional and evolutionary studies of virulence determinants in a symbiosis context. Two established models for symbioses are the legume-rhizobia and *E. scolopes* light organ-*V. fischeri*, which share all of the above attributes. Additional models that also share many of the attributes listed above are emerging. Utilization of models (both established and emerging) to study the diverse molecular mechanisms involved in symbiotic host-bacterial interactions can reveal unique attributes and common themes of symbiotic interactions. These model systems help address fundamental questions in symbiosis research: (i) What are the mechanisms whereby symbiosis is initiated and maintained? (ii) Are there common themes unifying all mutualistic symbioses? (iii) Are there similarities between mutualistic symbioses and disease? Specifically, are "virulence" factors also involved in mutualistic interactions and do any symbiotic microorganisms interact pathogenically in atypical hosts? Understanding symbiotic host-bacterial interactions is essential not only for full understanding of organismal and ecosystem health but also probably for the evolution of virulence.

16.3.1.1. Monoxenic or Binary Symbioses

We describe here only a few of the many potential monoxenic symbiotic models that contribute greatly to our understanding of the conserved and diverse mechanisms in which symbioses are initiated, are maintained, and evolve. For example, the legume-rhizobial symbiosis is arguably the most mature model system, as a result of increased attention due to its agricultural value (i.e., biological fixation of nitrogen by symbiotic legumes), the ability to associate and experimentally manipulate the symbiosis in the laboratory and to utilize genetic and molecular biological tools in both the symbiont and host (see section 16.1.4). In this model system comprising several legume-rhizobial pairs, the symbiont can be genetically manipulated (e.g., via transposon mutagenesis, allelic exchange, and green fluorescent protein [GFP] labeling; see below), and several completed genome sequences are available. The legume host also can be genetically manipulated, and symbiotic mutants have been isolated and symbiotic genes have been identified (3, 107, 121, 122).

In particular, the utilization of model systems increases our ability to determine the molecular dialogue between partners involved in the initiation, establishment, and maintenance of symbiosis. For example, a dual organism microarray containing both rhizobial and legume genes has been recently used to analyze transcription during symbiosis (6). This and other techniques, including gene silencing using RNA interference in the legume host (133), enable a better understanding of the reciprocal signaling at key steps during the initiation and maintenance of mutualistic host-bacterial interactions.

The legume-rhizobia mutualism has also revealed examples and parallels with animal-bacterial associations; for example, type III secretion system effectors in some *Rhizobia* spp. (168), and the type IV secretion system in *Mesorhizobium loti* (84) are systems known to be important in a variety of pathogenic bacteria. An emerging picture of these host-bacterial interactions is that strong selection for the symbiont is employed by the legume, involving components of the plant innate defenses and production of lipopolysaccharide and signaling molecules (e.g., nod factors) to facilitate symbiont-specific association while excluding nonsymbiotic or pathogenic bacteria from the symbiosis.

The *E. scolopes*-*V. fischeri* symbiosis is a well-developed animal-bacterial symbiosis. The Hawaiian bobtail squid, *E. scolopes*, inhabits shallow sand flats in the Hawaiian archipelago, where it is uses a ventrally located light organ for counterillumination against the night sky during feeding (88). The bioluminescence of the light organ is due to the presence of *V. fischeri* at high cell densities inside the light organ. Hatchling squid have light organ primordia and ciliated epithelia appendages, which are used to acquire symbiotic bacteria from the surrounding environment (i.e., horizontal transmission) (118). As elegantly summarized by Nyholm and McFall-Ngai, a winnowing occurs to select for symbiotic *V. fischeri*, which is vastly outnumbered by other marine planktonic microorganisms (125). After *V. fischeri* colonizes the light organ, more than 90% of *V. fischeri* organisms are vented from the light organ before dawn. The remaining *V. fischeri* then repopulates the light organ during the day while the squid remains burrowed in sand on the seafloor. The diurnal venting of *V. fischeri* reinoculates the environment with symbiotic bacteria (103).

Development of the mature light organ is coincident with colonization by symbiotic bacteria and requires reciprocal and sequential host-bacterial interactions (125). Colonization involves mucus secretion, quorum sensing, biofilm formation, oxygen radical physiology, lipopolysaccharide and peptidoglycan signaling, apoptosis, and actin remodeling, many processes relevant to human health and disease. In fact, analysis of the completed genome sequences reveals striking similarities between V. fischeri and V. cholerae (148). These include shared virulence factors and components of pathogenicity islands. One such island, TCP, is lacking certain phage sequences flanking it, and some genes found in TCP in V. cholerae are distributed in the V. fischeri genome that may indicate a more recent ancient acquisition of TCP in V. fischeri. It will be of great interest to determine the role of these virulence factors in V. fischeri, e.g., homologs of the RTX proteins in V. fischeri might affect the actin cytoskeleton of host cells (see phalloidin staining protocol below).

Additional models are also emerging for the study of monoxenic or relatively simple polyxenic associations. One of these involves the gut symbiosis between the medicinal leech, *Hirudo medicinalis*, and the predominant symbiont *Aeromonas veronii*, which facilitates the utilization of the blood diet for the leech (75). Other emerging models are the symbioses between two types of entomopathogenic (insect pathogenic) nematodes, e.g., *Heterorhabditis bacteriophora* (Color Plate 9G) and *Steinernema carpocapsae*, which specifically transmits the symbionts, *Photorhabdus luminescens* and *Xenorhabdus nematophila*, respectively, to the blood cavity (hemocoel) of insect hosts (33, 54, 112). These nematodes require the bacteria for their entomopathogenic lifestyle and are able to distinguish the symbiont from other bacteria for colonization of the nonfeeding infective stage (33, 112). The nematodes also depend on symbiotic bacteria to establish a protected niche inside the insect cadaver (by producing secondary metabolites that are antibiotic and antihelminthic and inhibit scavenging insects) and for growth and reproduction (32). Like the legume-rhizobia and squid light organ-V. fischeri symbioses, colonization of the infective juvenile host is highly selective, suggesting a strong pressure to inhibit nonsymbiotic or pathogenic bacteria. The nematode-bacterial complexes can be separated in the laboratory, propagated, and associated on agar-based media. Additionally, genome sequencing is completed or in progress for several of the symbionts and for *Heterorhabditis bacteriophora* nematode. The symbionts are genetically tractable, and genetics is being developed for the nematode hosts. *Photorhabdus* and *Xenorhabdus* possess an impressive arsenal of virulence factors (46); thus, in addition to being excellent models for colonization of specific bacteria in animal guts, they also provide good comparisons between mutualism and pathogenesis.

16.3.1.2. Polyxenic or Consortial Symbioses

Polyxenic symbioses (i.e., many symbiont types) can be very important or even essential to host biology. They are ubiquitous in animals and plants but they present challenges to the researcher due to high symbiont diversity and variability with respect to the individual or environment (e.g., influence of diet on intestinal microbiota). A few examples of polyxenic model systems, most involving gut microbiota, are briefly described below. Ruminants and xylophagous insects have complex microbiotas that are required for converting complex plant polymers (e.g., lignocellulose) into a utilizable substrate (e.g., acetate) for the animal hosts (14). The mammalian gastrointestinal tract harbors a complex microbial community (4, 80), of which certain components have been successfully investigated using gnotobiotic (i.e., germ-free) animals (81). One symbiont of the distal small intestine, *Bacteroides thetaio-*

taomicron stimulates production of fucosylated-glycans on the gut epithelium (82), angiogenesis (157), and beneficial antimicrobial peptide synthesis (81). Similarly, it was determined using the gnotobiotic mouse model that commensal gut microbiotas signal through the production of lipopolysaccharide and lipoteichoic acid via Toll-like receptors TLR4 and TLR2 (145). This signaling is normally associated with the activation of inflammation and innate immunity, but surprisingly, signaling by commensal microbiota is required for intestinal homeostasis (145). A bacterial polysaccharide from the gut commensal *Bacteroides fragilis* has immunomodulatory function, including the correction of systemic T-cell deficiencies, lymphoid organogenesis, and correcting T_H1/T_H2 imbalances related to tolerance (117). In other model animals, gut microbiotas were found to increase the longevity of *Drosophila melanogaster* when present in early adult life (17) and for normal gut development in the zebra fish *Danio rerio* (146). Knowledge of symbiotic microbiotas associated with insect pests (for example, the gypsy moth is an invasive insect pest of North American forests [16]) may enable novel approaches for insect control. These examples illustrate the importance of microbial gut communities to animal health and allow for further dissection, using model systems, of the molecular interactions involved in establishing and maintaining these symbioses.

16.3.2. Colonization Assays

The ability to generate and colonize aposymbiotic hosts is important for fulfillment of Koch's postulates adapted to symbiosis (postulates 2 to 4) and for the experimental manipulation of symbioses (e.g., through biochemical or genetic means). To perform colonization assays, symbiont-free (aposymbiotic) hosts and a source of the symbiont, preferably a pure culture, should be available. Aposymbiotic hosts can usually be obtained for horizontally transmitted symbioses by surface sterilizing and isolating embryos from symbiotic bacteria, since embryos are usually aposymbiotic. If aposymbiotic hosts cannot be naturally obtained, symbiont bacteria can be eliminated, or nearly so, by antibiotic treatments. Although embryos are typically permissive to symbiont colonization, this is not always the case and developmental stages of the host that are competent for symbiont colonization should be used. In addition, the symbionts should be prepared under conditions, and added to aposymbiotic hosts at densities, suitable for colonization. For insect eggs, plant seeds, nematode eggs, and infective juveniles, which are resistant to chemical treatment, treatment with 0.1% sodium hypochlorite is effective for disinfecting the surfaces of these organisms. For squid embryos or other more sensitive organisms, freshly laid embryos are washed extensively and incubated under symbiont-free conditions, usually accompanied with antibiotic treatments.

Surface Sterilization of Aposymbiotic Embryos (with Resistant Surfaces)

1. Obtain fresh embryos, purified from other tissues.
2. Wash embryos three times in sterile water, saline (0.85% [wt/vol] NaCl), or buffer (e.g., PBS).
3. Incubate embryos in 2% commercial bleach (Clorox® is 5.75% sodium hypochlorite) for 5 min.
4. Wash three times in sterile saline.

Antibiotic Treatment To Eliminate Bacteria Present in Whole Embryos or Animals (e.g., Insects)

1. Prepare a 100× antibiotic cocktail consisting of 5 mg/ml tetracycline, 20 mg/ml rifamycin, 10 mg/ml streptomycin in 50% ethanol/water.

2. Add 1 ml of antibiotic cocktail to 100 ml of animal food. Also, add 1 ml of 50% ethanol/water to 100 ml of animal food to control for the effect of 0.5% ethanol.

3. Incubate 72 h; repeat if necessary.

4. Test sterilization by performing PCR for bacterial 16S rRNA using general forward and reverse primers such as 27f or 1492r primers (see section 16.3.1), and/or by performing plate counts on homogenized animals.

5. Transfer to sterile food (72 h).

6. Colonize with symbiotic bacteria or compare aposymbiotic with symbiotic animals.

After the hosts are found to be axenic, they can be colonized by symbiotic bacteria and the symbionts can be shown to be causal to a symbiotic phenotype (e.g., nitrogen fixation or bioluminescence), thus allowing the fulfillment of symbiotic Koch's postulate no. 4. In addition, axenic hosts can be used for the biochemical or genetic analysis of the symbiosis. For biochemical analyses, purified components of the symbiont are typically added to axenic hosts and assayed for host response, e.g., lipopolysaccharide (55, 117), peptidoglycan (92), nod factors, and other infection factors produced by symbiotic bacteria (156). Genetics is often employed, usually in the bacterial symbiont, to determine genes that are involved in symbiosis, usually performed by first mutating the symbiont and then testing for a symbiotic defect by testing colonization competence of an aposymbiotic host. Protocols for forward and reverse genetics and reporter methods to aid in symbiont detection are presented below.

16.3.3. Molecular Genetic Analysis of Symbiotic Bacteria

Great strides have been made during the last decade for determining the molecular basis for pathogenic host-bacterial interactions, due to the application of powerful tools of molecular genetics and whole-genome sequencing. These strategies are positively impacting symbiosis research and the potential for the genetic analysis of symbiosis in general. With the ability to generate colonization-competent aposymbiotic hosts, many of Koch's postulates adapted to symbiosis can be fulfilled and genetic analysis of the symbiosis can be performed. By having an assay for symbiosis, e.g., nodule formation or bioluminescence for the legume-rhizobium and *E. scolopes-V. fischeri* light organ symbioses, respectively, bacterial or host mutants that are defective in symbiosis can be isolated. However, it is usually advantageous to label the symbionts with reporters (e.g., by expressing the GFP) to facilitate their detection in host tissues and during symbiosis assays. In this section, we present molecular genetic methods to label symbionts with GFP, forward genetics of symbionts using transposon mutagenesis, and reverse genetics by targeted gene disruption by allelic exchange. Although these techniques have been applied to only select members of the *Proteobacteria*, they illustrate general strategies that may be applicable to more diverse systems.

16.3.3.1. Use of Reporters for Symbiont Detection

Even for symbioses that result in obvious characteristics (e.g., bioluminescence in *E. scolopes-V. fischeri* symbiosis or nodule formation in the legume-rhizobia symbiosis), it is advantageous to label the symbiotic bacteria with a reporter to facilitate symbiont detection and enumeration. Several reporters have been used for this purpose: (i) β-galactosidase encoded by *lacZ*, which uses galactopyranosides as substrates and can be detected by colorimetric, fluorescent, chemiluminescent, and electrochemical methods (153); (ii) *luxAB*, minescent, and electrochemical methods (153); (ii) *luxAB*,

a bacterial luciferase that emits light during reduction of a long-chain aldehyde that can be supplied externally (48); (iii) glucuronidase encoded by *gusA*, which utilizes β-glucuronides and can be detected by colorimetric and fluorescent methods (177); (iv) ice-nucleation protein encoded by *ice-1*, which is detected by ice-nucleation activity of supercooled water (108); and (v) autofluorescent proteins, e.g., GFP, that are detected by fluorescence when excited by the appropriate wavelength of light (27).

Reporters are often placed on the chromosome because antibiotic selection of plasmid constructs can be difficult to maintain in vivo. This is usually performed by placing the reporter in a minitransposon where the transposase is separated from the transposon (i.e., mini-Tn5 or Tn10). The method described below involves inserting a mini-Tn7KSGFP (Fig. 1A) functionally similar to mini-Tn7*gfp*3 (91) into a proteobacterial symbiont genome by triparental mating. Other variants of this system are also available (e.g., other autofluorescent proteins) (91, 96) but are contained in mobilizable pUC-based plasmids that can replicate in some enterics, unlike the R6K vectors described here (a similar system was recently described [31]). The triparental mating system described here utilizes *E. coli* BW29427 {*thrB1004 pro thi rpsL hsdS lacZ*ΔM15 RP4-1360 Δ(*araBAD*)567 Δ*dapA*1341 ::[*erm pir*(wt)]; K. A. Datsenko and B. L. Wanner, Purdue University, personal communication} as a donor. BW29427 contains an RP4 *mob* for conjugation and the *pir* gene for replicating R6K *ori*-based replicons and is a diaminopimelic acid (DAP) auxotroph unable to cross-link the peptidoglycan cell wall without DAP added to the media. BW29427 is transformed individually with the mini-Tn7KSGFP-containing plasmid (pURR25) and a plasmid containing a Tn7 transposase (pUX-BF13) (5) (Fig. 1B); both plasmids contain only an R6K *ori* and are unable to replicate in bacteria lacking the *pir* gene. Triparental mating is then performed to introduce these plasmids into recipient bacteria (Fig. 1C). Transconjugants containing the mini-Tn7KSGFP are selected for by plating on media without DAP and transposon insertions by resistance to kanamycin and streptomycin. The mini-Tn7 usually inserts in the neutral *att* site located downstream of the highly conserved *glmS* genes present in proteobacterial genomes (35, 139). If mini-Tn7 transposition is not feasible, the reporter can be inserted randomly into the genome by using another transposon appropriate for the bacterial symbiont of interest or by electroporating transposomes (73) containing the reporter. Additionally, other promoters or reporters can be used. For example, a promoter of interest can be fused to an unstable *gfp* variant (AAV*gfp* [2]) to determine the transcription of this promoter in vivo (89).

Below is a protocol that has been successful for *Photorhabdus luminescens* and other *Proteobacteria* (e.g., *Pseudomonas aeruginosa*, *Serratia* spp.; T. Ciche, unpublished data) and adapted from that developed for *Shewanella oneidensis* (D. Lies and D. K. Newman, Caltech, personal communication). Appropriate growth conditions, cell number, and incubation times for the triparental matings should be experimentally determined for the bacterial symbiont being conjugated.

16.3.3.2. Labeling of Symbiotic Bacteria with GFP

The following bacterial strains are used in this subsection: BW29427 pURR25 (donor + pURR25 containing the miniTn7KSGFP transposon) (Fig. 1A) and BW29427 pUX-BF13 (donor strain + pUX-BF14 encoding the Tn7 transposase) (Fig. 1B)

1. Inoculate a 3-ml starter culture of recipient cells in appropriate media and donor cells in 3 ml of Luria-Bertani

A.

mini-Tn7KSGFP

Tn7L   Km   ΩStrep/Spec Plac::*gfpmut3**   Cm Tn7R

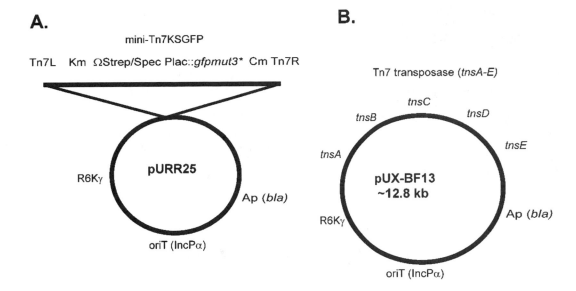

B.

Tn7 transposase (*tnsA-E*)

C.

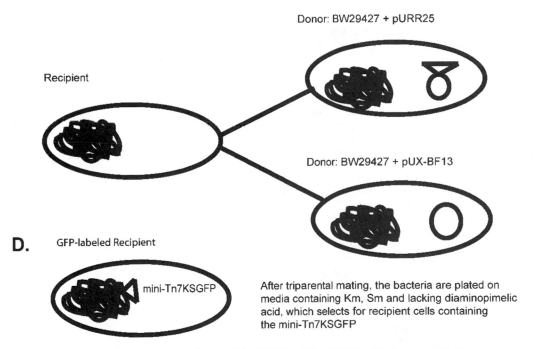

Donor: BW29427 + pURR25

Recipient

Donor: BW29427 + pUX-BF13

D.   GFP-labeled Recipient

mini-Tn7KSGFP

After triparental mating, the bacteria are plated on media containing Km, Sm and lacking diaminopimelic acid, which selects for recipient cells containing the mini-Tn7KSGFP

FIGURE 1   Labeling bacteria with mini-Tn7KSGFP. (A) pURR25 (D. Lies and D. Newman, Caltech) containing the mini-Tn7KSGFP where GFP expression is driven from the constitutive P_{lac} promoter ($P_{A1/04/03}$) on a mobilizable ($oriT_{IncP\alpha}$) suicide plasmid ($oriR_{R6K\gamma}$). (B) pUX-BF13 containing the genes (*tnsA-E*) encoding the Tn7 transposase on a mobilizable suicide plasmid similar to pURR25. (C) Triparental mating introduces pURR25 (mini-Tn7KSGFP) and pUX-BF13 (Tn7 transposase) into the recipient cells, which allows the mini-Tn7KSGFP to transpose, usually in an *att* site, downstream of *glmS*, of the recipient's genome. (D) Because pURR25 and pUX-BF13 cannot replicate in recipient cells and BW29427 donor cells require DAP for growth, plating on kanamycin- and streptomycin-containing media selects for recipient cells containing the mini-Tn7KSGFP.

medium (LB) (prepared by adding 10 g of Bacto tryptone, 5 g of Bacto yeast extract and 5 g of NaCl to 1 liter, adjust to pH 7.5 by adding 5 M NaOH, sterilize by autoclaving) containing 300 μg/ml DAP, 100 μg/ml ampicillin (Ap). Incubate recipient cells at the appropriate temperature, donor cells at 37°C.

2. Transfer 1/100 volume of starter cultures to 10 ml of LB for recipient cells, 10 ml of LB with DAP, Ap for donor cells. Grow to an optical density at 600 nm (OD$_{600}$) (1-cm path length) of 0.6.

3. Wash recipient cells twice with 1.5 ml of LB, donor cells with LB + DAP by centrifuging 30 s at 10,000 × g. Resuspend in 0.5 ml of LB (recipient cells), 0.5 ml of LB DAP (donor cells).

4. Mix 0.5 ml of each of the donor cells with 0.5 ml of recipient. Centrifuge 30 s at 10,000 × g, decant supernatant, and resuspend remaining cells in the approximately 50 μl of remaining fluid.

5. Add donor and recipient cell mixture to a dry LB DAP plate. Incubate at 28°C (for *P. luminescens*) for 6 h (4- to 24-h mating times are used depending on recipient bacteria).

6. Wash cells off plate with 1.5 ml of LB. Then, wash the cells by centrifuging for 30 s at 10,000 × g, decanting the supernatant and resuspending cells with 1.5 ml of LB. Repeat wash.

7. Plate cells on LB kanamycin (usually 30 μg/ml, 3.75 μg/ml for *P. luminescens*) and streptomycin or spectinomycin (20 to 100 μg/ml).

8. Observe cells for fluorescence using a stereofluorescent microscope (a handheld UV lamp may be adequate) and by epifluorescence.

16.3.3.3. Transposon Mutagenesis of Symbiotic Bacteria

Genetics can be a useful tool to identify genes and corresponding functions that are involved in symbiosis. Typically, the symbiont is mutated and subsequently screened for an ability to initiate or reestablish interactions with the host (i.e., those mutants defective in forming a symbiosis are elucidated). In addition to the application of the mini-Tn7 described above, transposons are often used for bacterial mutagenesis because (i) they typically contain a selectable marker for the selection of mutants and (ii) mini-Tn5 and mini-Tn10 (and others) have been engineered so that the transposase gene(s) are *cis* to the transposon (e.g., the sequence will be transposed outside the transposon, creating more stable insertions). Additionally, transposons are available that can create transcriptional or translational fusions (i.e., promoterless *lacZ*), that can also detect intracellular (*lacZ* transcriptional fusions) or secreted proteins (e.g., *phoA*), and that allow direct cloning of DNA flanking the transposon (by containing a low-copy-number origin of replication (e.g., p15A [32]) or a conditionally replicative *ori* (e.g., R6K). Lastly, transposons that contain a unique sequence tag, called signature-tagged mutagenesis, can be constructed. In signature-tagged mutagenesis, probes are constructed from pools of mutants and hybridized to DNA from individual target clones. Typically, one screens for mutants that fail to hybridize after symbiosis, which indicates that the mutant and its corresponding DNA were lost during symbiosis (79). Concerns regarding transposons are the creation of polar mutations, e.g., they can disrupt transcription of genes downstream of the insertion, and nonrandom insertions (i.e., "hot and cold" spots). A method is described below that utilizes a hyperactive mariner transposon (178) containing a gentamicin resistance gene and an R6K

origin of replication initiation (Fig. 2A). This transposon is present on a suicide plasmid and introduced into symbiotic bacteria by conjugation (Fig. 2B).

Protocol for Transposon Mutagenesis of Symbiotic Bacteria

Bacterial strains: BW29427 pURE10 (kindly provided by D. Lies and D. Newman, Caltech); and symbiont (optionally GFP-labeled), DH5α *pir*.

1. Inoculate recipient cells (i.e., symbiont) in 3 ml of LB, donor cells in 3 ml of LB + 300 μg/ml DAP and 100 μg/ml Ap and grow overnight (at temperature appropriate for the symbiont; 37°C for BW29427).

2. Transfer 1/100 volume of overnight culture to 10 ml LB for recipient cells and 10 ml of LB-DAP-Ap for donor cells. Grow to an OD$_{600}$ of 0.6.

3. Pellet cells by centrifugation, 30 s at 10,000 × g, and wash three times in LB-DAP. Decant supernatant and resuspend in 0.5 ml of LB-DAP.

4. Combine donor and recipient cells and centrifuge as above; resuspend cells in remaining ~50 μl and pour the mixture on an LB-DAP plate. Incubate 8 h.

5. Wash cells off plate with 1.5 ml of LB and an additional two times by centrifugation. Resuspend in 1.5 ml of LB.

6. Plate 0.1 ml (depending on the level of conjugation efficiency of the recipient cells) on LB + transposon antibiotic (e.g., kanamycin, streptomycin, or Ap).

7. Isolate putative mutants by transferring to grids on LB plates or into 96-well titer dishes.

8. Assay mutants for symbiotic competence by adding mutants to aposymbiotic hosts (or other assays potentially

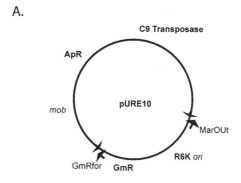

A.

B.

Recipient cells containing the HimarGm are selected for on media containing gentamicin and lacking diaminopimelic acid.

FIGURE 2 Transposon mutagenesis of symbiotic bacteria. (A) pURE10 is a mobilizable suicide plasmid as described in Fig. 1 and contains a hyperactive *mariner* transposon (HimarGm) similar to pSC189 (29) with *gm* replacing *kan* of pSC189. (B) BW29427 containing pURE10 is mated with symbiont recipient cells. Transconjugants containing the HimarGm are selected for on media containing Gm and lacking DAP.

involved in symbiosis, i.e., motility and siderophore production, and then assay for symbiosis competency).

9. Transposon retrieval. First, purify the DNA (use CTAB method described above) from 3 ml of overnight culture of the putative symbiotic mutants and digest 10 μg of genomic DNA overnight with an enzyme (10 U) that does not cleave the transposon (e.g., for *Himar*Gm, SphI, SpeI, or NsiI). Heat inactivate the restriction endonuclease and ligate in 500 μl containing 1× ligase buffer and 10 U of T4 ligase (this favors intramolecular over intermolecular ligation), ethanol precipitate by adding 1/10 volume of 10 mM sodium acetate and 0.6 volume of isopropanol, resuspend in 10 μl of sterile distilled water, and transform into DH5α *pir* (or other strains containing the *pir* gene).

10. Perform a plasmid prep on several of the putative transposon-containing clones. Digest with the restriction endonuclease used to digest the mutant genomic DNA. A single band indicates that only an intramolecular ligation occurred.

11. Sequence the DNA flanking the insertion (MAROUT, 5′CCGGGGGACTTATCAGCCAACC-3′; GMRFor, 5′CGGTAAATTGTCACAACGCC-3′ [Ciche, unpublished]).

12. Determine causality of the insertion to the mutant phenotype. This can be performed by cloning the retrieved DNA (or ~1.5 kb of DNA upstream and downstream of Tn) into a mobilizable suicide plasmid, e.g., pWM91 (119). This plasmid is conjugated into wild-type (wt) bacteria, and single recombinants are selected by the antibiotic resistance on the plasmid (e.g., ApR for pWM91, CmR for pEVS114). The bacteria containing the single recombinants are grown overnight in the absence of selection and diluted and plated on media containing the antibiotic to which resistance is conferred by the transposon. These are then transferred in replica to plates containing the transposon and antibiotic resistance plasmids (see allelic exchange protocol in section 16.3.3.4). Clones containing the transposon but not the plasmid antibiotic resistances have the mutated gene replacing the wt gene. This is confirmed by PCR of the DNA containing the insertion (e.g., by amplification of a PCR product that equals the length of the wt gene plus the transposon). Alternatively, a suicide plasmid that contains the *sacBR* cassette that confers sensitivity to 5% sucrose to many *Proteobacteria* can be used to select for recombinants that excised the plasmid, e.g., pWM91 (119).

16.3.3.4. Targeted Gene Disruption by Allelic Exchange

With the abundance of bacterial genome sequences and the increasing number of completed symbiotic bacterial genomes available, it may be desirable to perform gene disruption to determine functions for genes of interest (e.g., reference 164). There are a variety of strategies to construct and deliver disrupted genes. A protocol is described below for construction of unmarked gene deletions of interest. First, a deletion is constructed using strand overlap extension (SOEing) PCR (83). In this method ~1.5 kb flanking the 5′ and 3′ ends of the gene are amplified where the two internal primers contain DNA sequence complementary to the other 1.5-kb sequence (Fig. 3A). The 1.5-kb fragments are first individually amplified by PCR and purified by gel extraction. The PCR products are then amplified using primers UPfor and DNrev and the 1.5-kb fragments as templates. Because the internal primers were removed by gel extraction, PCR amplification using the end primers amplifies only the fused template (Fig. 3B). This is then cloned into a mobilizable suicide vector containing the *sacBR* gene,

e.g., pWM91 (119), and transformed into BW29427 (Fig. 3C). Conjugation of BW29427 containing the deletion construct results in a single recombination in which the deletion and suicide vector are recombined in the gene of interest. The recombinants are grown overnight without selection and then plated on 5% sucrose to select for bacteria that have excised the plasmid and either the mutant or wt gene of interest. PCR is used to screen the recombinants for those that contain only the deletion (119). The deleted strain is then characterized for symbiosis directly or labeled with GFP prior to characterization (see section 16.3.3.1).

16.3.3.5. Strand-Overlap Extension (SOE) PCR and Allelic Exchange for Targeted Gene Disruption

1. Design oligonucleotide primers for ~1.5 kb flanking each end of the gene of interest which contain ~27 bp complementary on the 5′ ends of the internal primers to the internal end of the other 1.5-kb fragment. Add EagI sites to the 5′ ends of the distal primers for cloning of the deletion into the NotI site of a suicide vector.

UPfor = 5′ gtac (noncomplementary 4-bp sequence) CGGCCG(EagI site) ~26 bp 5′ complementary to the sense strand of gene X (of upstream DNA flanking gene X) 3′

UPrev = 5′~26 bp complementary to 5′ DNfor sequence, then ~26 bp 1.5 kb downstream of Afor and complementary to the antisense strand of gene X

DNfor = 5′ ~26 bp complementary to 3′ of UPrev sequence, then ~26 bp complementary to sense strand (of downstream DNA flanking gene X)

DNrev = 5′ tcga (noncomplementary 4-bp sequence) CGGCCG (reverse complementary of EagI site, in this case the same) ~26 bp complementary to the antisense strand of gene X (of downstream DNA flanking gene X)

2. In separate PCR amplify 1.5 kb of DNA upstream and downstream of gene X using UPfor/UPrev and DNfor/DNrev, respectively.

3. Electrophorese on a 1.5% agarose gel. If a single 1.5-kb band is present, clean 5 μl of the PCR product with ExoSAP according to the manufacturer's instructions (ExoSap-IT; USB Corporation, Cleveland, OH), then dilute the PCR product 1:1,000 in sterile distilled water. If multiple bands are present, extract the 1.5-kb fragment (use QIAquick Gel Extraction Kit [QIAGEN Sciences] or similar kit) and dilute extracted fragment 1:500 (ExoSap treatment is unnecessary for gel extract PCR products).

4. PCR amplify using 1 μl of each diluted 1.5-kb fragment using UPfor and DNrev primers, then digest with 10 U of EagI for 3 h. Electrophorese, and if a single 3.0-kb band is present proceed to step 5 or gel extract 3.0-kb fragment (also with above kit).

5. Digest suicide vector (pWM91 or similar) with NotI, alkaline phosphatase (e.g., Antarctic phosphatase, New England Biolabs, Beverly, MA), heat inactivate at 68°C for 15 min.

6. Ligate SOEed (i.e., fused) 3.0-kb fragment to NotI-digested pWM91 (119).

7. Transform into DH5α *pir*, plate transformation mixture on LB + Ap X-Gal 5-bromo-4-chloro-3-indolyl-β-D-glucuronic acid plasmid prep white colonies, verify insert by restriction enzyme digests and gel electrophoresis.

8. Transform into BW29427, select transformants on LB Ap DAP.

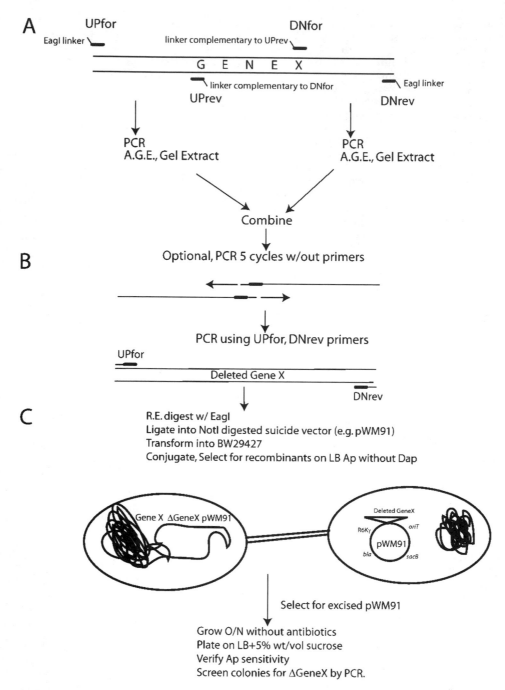

FIGURE 3 Creating unmarked deletions using SOE PCR and allelic exchange. (A) Two fragments (~0.75 to 1.5 kb) are amplified by PCR using primers that are complementary (Arev/Bfor) and that have EagI sequences attached (Afor/Brev). The PCR fragments are purified by agarose gel electrophoresis (AGE). (B) Each fragment is then combined and fused by SOEing (83). (C) The fused PCR products are then digested with EagI and then cloned into the NotI site of the suicide plasmid pWM91 creating pWM91Δ*geneX*, which is transformed into BW29427 and mated with symbiotic bacteria. Transconjugants containing pWM91Δ*geneX* inserted into *geneX* are selected based on their resistance to ampicillin, creating a merodiploid strain. Strains that have excised pWM91 are selected for by plating on media containing 5% sucrose. A portion of the sucrose-resistant strains have only the deleted *geneX*. These are screened by PCR. The presence of the wt *geneX* can be determined by using one primer for the deleted sequence and another flanking it. Afor and Bfor can also be used to verify the presence of the deletion.

9. Conjugate BW29427 to symbiont bacteria as described above (section 16.3.3.2). Select for plasmid integrants on LB-Ap.

10. Grow integrants overnight without selection, and then plate on LB + 5% sucrose (wt/vol).

11. Screen for strains containing the deletion construct by PCR using primer pairs where one primer is specific for the deleted region and therefore results in a product in the wt strain and no product in the mutant strain, or primer pairs designed to amplify small products in the deleted strain and large or no products in the wt.

12. Assay deletion mutants for symbiosis or label mutants with GFP (section 16.3.3.1).

Note the SOEing protocol described above can be applied for the construction of transcriptional and translational fusions for analysis of in situ gene expression, perhaps for symbiosis genes identified by random (see section 16.3.3.3) or targeted (section 16.3.3.4) mutagenesis.

16.3.4. Genetic, Cellular, and Molecular Analysis of Symbiosis in the Host

Genetic analysis of symbiosis has predominantly focused on the microbial symbiont, due to the relative ease of doing this. However, with the application of transcriptional profiling of large gene sets (e.g., microarray analysis) and powerful reverse genetic tools (e.g., gene silencing by RNAi), genetic analysis of the host biology related to symbiosis is now more tractable. One example is the pea, *Lotus japonicus*, where forward and reverse genetics and genomics are available (reviewed in reference 159). These tools have been applied for the identification of several plant genes involved in symbiosis, including a lipochito-oligosaccharide receptor required for pea response to nod factors (144). In addition, symbiotic interactions are now being studied in animal models (e.g., *Danio rario* and *D. melanogaster*) and animal hosts with emerging genetics (e.g., the nematode *H. bacteriophora*).

16.3.4.1. Cell Biological Analysis

Symbiotic host-bacterial interactions usually involve intimate cell-cell interactions and changes in host cell biology as a consequence. These changes can be studied using cell biological methods developed to detect and manipulate a variety of cell processes and structures. For example, injection of calcium indicators (Calcium Green or Fura-2-dextrans) into alfalfa roots revealed oscillations in cytoplasmic calcium in response to nod factor (47). Using NADPH-diaphorase staining and immunocytochemistry to detect nitric oxide synthase and diaminofluorescein to detect its product, nitric oxide, both were detected at high levels in the light organ of *E. scolopes* and found to be attenuated after colonization by *V. fischeri* (37). Furthermore, using several techniques, including terminal deoxynucleotidyl-transferase-mediated dUTP-biotin nick end labeling (TUNEL), lipopolysaccharide was shown to induce apoptosis in the ciliated epithelia cells of the squid light organ (55, 56). These examples illustrate the variety of cell biological techniques that can be used to detect changes in host cells in response to symbiosis.

Bacteria that live in close association with eukaryotic host cells often affect the cytoskeleton of the host cells, which can be visualized by staining actin. The protocol used to stain polymerized actin using fluorescently labeled phalloidin described below illustrates cell-staining procedures common in cell biology and is generally applicable to host-bacterial interactions (see microscopy protocols above).

Phallotoxins from the deadly *Amanita phalloides* mushroom bind to the filamentous form of actin (F-actin) and several fluorescently conjugated phallotoxins (i.e., phalloidins) are available commercially. Rhodamine-conjugated phalloidin (R415; Molecular Probes, Eugene, OR) binds to filamentous actin (F-actin) and is used to observe the cell cytoskeleton using fluorescence microscopy. Rhodamine-conjugated phalloidin is highly water soluble but diffuses poorly across most cell membranes; therefore, fixation and permeabilization of cells is usually necessary for optimal labeling. In animal-bacterial symbiosis, bacteria often interact with the apical side of polarized epithelial cells. The apical side of polarized epithelia cells (i.e., in the intestine) typically contains microvilli that contain actin filaments. Underlying the microvilli is an actin lattice termed the terminal web. Changes in the actin cytoskeleton such as microvilli density, cell shape, and endocytosis can be visualized by phalloidin staining and contrasted between symbiotic and aposymbiotic animals.

Phalloidin Staining of Polymerized Actin in the Squid Light Organ

1. Anesthetize animals (i.e., squid) in artificial seawater containing 2% ethanol.

2. Fix with 4% formaldehyde in marine PBS (as described in section 16.2.1.1., except prepared in three-quarter strength artificial seawater such as Instant Ocean®) for 1 h.

3. Dissect out light organs and wash twice for 10 min in marine PBS.

4. Permeabilize with 1% Triton X-100 for 20 min in marine PBS.

5. Stain samples overnight with 2 mg/ml rhodamine phalloidin in 1% Triton X-100 in marine PBS.

6. Rinse three times for 10 min in marine PBS.

7. View using confocal laser scanning microscopy (epifluorescence might be adequate for some thin tissues).

Note: this technique can be used in conjunction with GFP-labeled symbiotic bacteria (see above). Adapted from reference 90.

16.3.4.2. Molecular and Proteomic Analysis

In addition to cellular biology, genomic and proteomic approaches have also been employed to identify genes and proteins involved in the development and maintenance of host tissues involved in symbiosis (e.g., the light organ). A common method used to identify specific host genes involved in symbiosis is the construction and sequence analysis of a cDNA library (i.e., single-pass DNA sequencing of a subset of the cDNA library to create an expressed sequence tag database [EST]). There are a variety of methods to isolate mRNA, reverse transcribe mRNA into cDNA, and normalize and enrich differentially expressed sequences (e.g., symbiotic versus aposymbiotic squid). In general, an EST library is constructed by (i) isolating mRNA from the tissues and/or whole animals of interest, (ii) synthesizing antisense DNA using reverse transcriptase and a polythymidine primer, and (iii) use of strand switching to synthesize sense DNA (116) and cloning of resulting cDNA. We recommend the Trizol reagent (Invitrogen, Carlsbad, CA) for the isolation of mRNA, SMART technology that utilizes terminal transferase and strand-switching activity of MMLV reverse transcriptase to attach a primer on the 5' end of the mRNA, which also allows cDNA amplification (SMART technology; BD Biosciences, Palo Alto, CA), and suppressive subtraction hybridization to enrich the library for rare and differentially expressed transcripts (SMART-PCR; BD Biosciences). By analyzing a cDNA library constructed from mRNA isolated from the squid light organ, a highly abundant mRNA with homology to halide peroxidases was identified (163), which likely selects for the symbiont's ability to eliminate oxygen radicals, possibly through the production of catalase and luciferase (155). A large-scale EST project is underway for *E. scolopes*, which will greatly facilitate the analysis of gene expression in the symbiotic light organ (McFall-Ngai, personal communication). Transcriptional analysis of symbiotic tissue can also be performed using a microarray containing both host and symbiont genes (6). This was recently performed for the legume symbiosis and should be applicable to the *E. scolopes-V. fischeri* light organ symbiosis or to other symbioses where large gene sets or complete genome sequences of symbiont and host are known.

Similarly, important changes at the protein level during symbiosis have also been investigated. Proteins produced in the tissue of interest can be identified by separation in two dimensions, by charge and size. This technique was used to assess changes in the proteome between early and mature light organ tissue in *E. scolopes* (41). Proteins of interest can then be extracted and identified by mass spectrometry, N-terminal sequencing, and/or immunological methods, often in conjunction with analysis of a cDNA library, i.e., a nucleic acid probe designed using the N-terminal sequence or an antibody can be used to probe a cDNA encoding the protein of interest.

16.4 REFERENCES

1. **Allison, G. E., and N. K. Verma.** 2000. Serotype-converting bacteriophages and O-antigen modification in *Shigella flexneri. Trends Microbiol.* **8:**17–23.

2. **Andersen, J. B., C. Sternberg, L. K. Poulsen, S. P. Bjorn, M. Givskov, and S. Molin.** 1998. New unstable variants of green fluorescent protein for studies of transient gene expression in bacteria. *Appl. Environ. Microbiol.* **64:**2240–2246.

3. **Ane, J. M., G. B. Kiss, B. K. Riely, R. V. Penmetsa, G. E. Oldroyd, C. Ayax, J. Levy, F. Debelle, J. M. Baek, P. Kalo, C. Rosenberg, B. A. Roe, S. R. Long, J. Denarie, and D. R. Cook.** 2004. *Medicago truncatula* DMI1 required for bacterial and fungal symbioses in legumes. *Science* **303:**1364–1367.

4. **Backhed, F., H. Ding, T. Wang, L. V. Hooper, G. Y. Koh, A. Nagy, C. F. Semenkovich, and J. I. Gordon.** 2004. The gut microbiota as an environmental factor that regulates fat storage. *Proc. Natl. Acad. Sci. USA* **101:**15718–15723.

5. **Bao, Y., D. P. Lies, H. Fu, and G. P. Roberts.** 1991. An improved Tn7-based system for the single-copy insertion of cloned genes into chromosomes of Gram-negative bacteria. *Gene* **109:**167–168.

6. **Barnett, M. J., C. J. Toman, R. F. Fisher, and S. R. Long.** 2004. A dual-genome symbiosis chip for coordinate study of signal exchange and development in a prokaryote-host interaction. *Proc. Natl. Acad. Sci. USA* **101:**16636–16641.

7. **Baron, C., and P. C. Zambryski.** 1995. The plant response in pathogenesis, symbiosis, and wounding: variations on a common theme? *Annu. Rev. Genet.* **29:**107–129.

8. **Baumann, L., P. Baumann, and M. L. Thao.** 1999. Detection of messenger RNA transcribed from genes encoding enzymes of amino acid biosynthesis in *Buchnera aphidicola* (endosymbiont of aphids). *Curr. Microbiol.* **38:**135–136.

9. **Baumann, P., L. Baumann, C. Y. Lai, D. Rouhbakhsh, N. A. Moran, and M. A. Clark.** 1995. Genetics, physiology, and evolutionary relationships of the genus *Buchnera:* intracellular symbionts of aphids. *Annu. Rev. Microbiol.* **49:**55–94.

10. **Behrens, S., B. M. Fuchs, F. Mueller, and R. Amann.** 2003. Is the in situ accessibility of the 16S rRNA of *Escherichia coli* for Cy3-labeled oligonucleotide probes predicted by a three-dimensional structure model of the 30S ribosomal subunit? *Appl. Environ. Microbiol.* **69:**4935–4941.

11. **Belkin, S., D. C. Nelson, and H. W. Jannasch.** 1986. Symbiotic assimilation of CO_2 in two hydrothermal vent animals, the mussel *Bathymodiolus thermophilus* and the tube worm *Riftia pachyptila. Biol. Bull.* **170:**110–121.

12. **Blazejak, A., C. Erseus, R. Amann, and N. Dubilier.** 2005. Coexistence of bacterial sulfide oxidizers, sulfate reducers, and spirochetes in a gutless worm (Oligochaeta) from the Peru margin. *Appl. Environ. Microbiol.* **71:**1553–1561.

13. **Boetius, A., K. Ravenschlag, C. J. Schubert, D. Rickert, F. Widdel, A. Gieseke, R. Amann, B. B. Jorgensen, U. Witte, and O. Pfannkuche.** 2000. A marine microbial consortium apparently mediating anaerobic oxidation of methane. *Nature* **407:**623–626.

14. **Breznak, J. A.** 1982. Intestinal microbiota of termites and other xylophagous insects. *Annu. Rev. Microbiol.* **36:**323–343.

15. **Bright, M., H. Keckeis, and C. R. Fisher.** 2000. An autoradiographic examination of carbon fixation, transfer and utilization in the *Riftia pachyptila* symbiosis. *Mar. Biol.* **136:**621–632.

16. **Broderick, N. A., K. F. Raffa, R. M. Goodman, and J. Handelsman.** 2004. Census of the bacterial community of the gypsy moth larval midgut by using culturing and culture-independent methods. *Appl. Environ. Microbiol.* **70:**293–300.

17. **Brummel, T., A. Ching, L. Seroude, A. F. Simon, and S. Benzer.** 2004. Drosophila lifespan enhancement by exogenous bacteria. *Proc. Natl. Acad. Sci. USA* **101:**12974–12979.

18. **Brune, A., and M. Friedrich.** 2000. Microecology of the termite gut: structure and function on a microscale. *Curr. Opin. Microbiol.* **3:**263–269.

19. **Brussow, H., C. Canchaya, and W.-D. Hardt.** 2004. Phages and the evolution of bacterial pathogens: from genomic rearrangements to lysogenic conversion. *Microbiol. Mol. Biol. Rev.* **68:**560–602.

20. **Buchner, P.** 1965. *Endosymbiosis of Animals with Plant Microorganisms.* Interscience, New York, NY.

21. **Campbell, B. J., J. L. Stein, and S. C. Cary.** 2003. Evidence of chemolithoautotrophy in the bacterial community associated with *Alvinella pompejana*, a hydrothermal vent polychaete. *Appl. Environ. Microbiol.* **69:**5070–5078.

22. **Cary, S. C., M. T. Cottrell, J. L. Stein, F. Camacho, and D. Desbruyeres.** 1997. Molecular identification and localization of filamentous symbiotic bacteria associated with the hydrothermal vent annelid *Alvinella pompejana. Appl. Environ. Microbiol.* **63:**1124–1130.

23. **Cavanaugh, C. M.** 1985. Symbiosis of chemoautotrophic bacteria and marine invertebrates from hydrothermal vents and reducing sediments. *Bull. Biol. Soc. Wash.* **6:**373–388.

24. **Cavanaugh, C. M.** 1983. Symbiotic chemoautotrophic bacteria in marine invertebrates from sulphide-rich habitats. *Nature* **302:**58–61.

25. **Cavanaugh, C. M., M. Abbott, and M. Veenhuis.** 1988. Immunochemical localization of ribulose-1,5-bisphos-

phate-carboxylase in the symbiont-containing gills of *Solemya velum* (Bivalvia: Mollusca). *Proc. Natl. Acad. Sci. USA* **85**:7786–7789.

26. **Cavanaugh, C. M., S. L. Gardiner, M. L. Jones, H. W. Jannasch, and J. B. Waterbury.** 1981. Procaryotic cells in the hydrothermal vent tube worm *Riftia pachyptila* Jones: possible chemoautotrophic symbionts. *Science* **213**:340–342.

27. **Chalfie, M., Y. Tu, G. Euskirchen, W. W. Ward, and D. C. Prasher.** 1994. Green fluorescent protein as a marker for gene expression. *Science* **263**:802–805.

28. **Chan, J. W. Y. F., and P. H. Goodwin.** 1995. Extraction of genomic DNA from extracellular polysaccharide synthesizing Gram-negative bacteria. *BioTechniques* **18**:419–422.

29. **Chiang, S. L., and E. J. Rubin.** 2002. Construction of a mariner-based transposon for epitope-tagging and genomic targeting. *Gene* **296**:179–185.

30. **Childress, J. J., C. R. Fisher, J. A. Favuzzi, R. E. Kochevar, N. K. Sanders, and A. M. Alayse.** 1991. Sulfide-driven autotrophic balance in the bacterial symbiont-containing hydrothermal vent tubeworm *Riftia pachyptila* Jones. *Biol. Bull.* **180**:135–153.

31. **Choi, K. H., J. B. Gaynor, K. G. White, C. Lopez, C. M. Bosio, R. R. Karkhoff-Schweizer, and H. P. Schweizer.** 2005. A Tn7-based broad-range bacterial cloning and expression system. *Nat. Methods* **2**:443–448.

32. **Ciche, T. A., S. B. Bintrim, A. R. Horswill, and J. C. Ensign.** 2001. A phosphopantetheinyl transferase homolog is essential for *Photorhabdus luminescens* to support growth and reproduction of the entomopathogenic nematode *Heterorhabditis bacteriophora*. *J. Bacteriol.* **183**:3117–3126.

33. **Ciche, T. A., and J. C. Ensign.** 2003. For the insect pathogen *Photorhabdus luminescens*, which end of a nematode is out? *Appl. Environ. Microbiol.* **69**:1890–1897.

34. **Cleveland, L. R., and A. V. Grimstone.** 1964. The fine structure of the flagellate *Mixotricha paradoxa* and its associate micro-organisms. *Proc. R. Soc. Lond. B* **159**:668–686.

35. **Craig, N. L.** 1991. Tn7: a target site-specific transposon. *Mol. Microbiol.* **5**:2569–2573.

36. **Currie, C. R., J. A. Scott, R. C. Summerbell, and D. Malloch.** 1999. Fungus-growing ants use antibiotic-producing bacteria to control garden parasites. *Nature* **398**:701–704.

37. **Davidson, S. K., T. A. Koropatnick, R. Kossmehl, L. Sycuro, and M. J. McFall-Ngai.** 2004. NO means 'yes' in the squid-vibrio symbiosis: nitric oxide (NO) during the initial stages of a beneficial association. *Cell. Microbiol.* **6**:1139–1151.

38. **de Bary, A.** 1879. *Die Erscheinung der Symbiose.* Trubner, Strasbourg, France.

39. **Desbruyères, D., F. Gaill, L. Laubier, and Y. Fouquet.** 1985. Polychaetous annelids from hydrothermal vent ecosystems: an ecological overview. *Bull. Biol. Soc. Wash.* **6**:103–116.

40. **Distel, D. L., and H. Felbeck.** 1988. Pathways of inorganic carbon fixation in the endosymbiont bearing lucinid clam *Lucinoma aequizonata*. Part 1. Purification and characterization of the endosymbiotic bacteria. *J. Exp. Zool.* **247**:11–22.

41. **Doino Lemus, J., and M. J. McFall-Ngai.** 2000. Alterations in the proteome of the *Euprymna scolopes* light organ in response to symbiotic *Vibrio fischeri*. *Appl. Environ. Microbiol.* **66**:4091–4097.

42. **Douglas, A. E.** 1998. Nutritional interactions in insect-microbial symbioses: aphids and their symbiotic bacteria Buchnera. *Annu. Rev. Entomol.* **43**:17–37.

43. **Douglas, A. E., L. B. Minto, and T. L. Wilkinson.** 2001. Quantifying nutrient production by the microbial symbionts in an aphid. *J. Exp. Biol.* **204**:349–358.

44. **Dubilier, N., O. Giere, D. L. Distel, and C. M. Cavanaugh.** 1995. Characterization of chemoautotrophic bacterial symbionts in a gutless marine worm (Oligochaeta, Annelida) by phylogenetic 16S rRNA sequence analysis and in situ hybridization. *Appl. Environ. Microbiol.* **61**:2346–2350.

45. **Dubilier, N., C. Mulders, T. Ferdelman, D. de Beer, A. Pernthaler, M. Klein, M. Wagner, C. Erséus, F. Thiermann, J. Krieger, O. Giere, and R. Amann.** 2001. Endosymbiotic sulphate-reducing and sulphide-oxidizing bacteria in an oligochaete worm. *Nature* **411**:298–302.

46. **Duchaud, E., C. Rusniok, L. Frangeul, C. Buchrieser, A. Givaudan, S. Taourit, S. Bocs, C. Boursaux-Eude, M. Chandler, J. F. Charles, E. Dassa, R. Derose, S. Derzelle, G. Freyssinet, S. Gaudriault, C. Medigue, A. Lanois, K. Powell, P. Siguier, R. Vincent, V. Wingate, M. Zouine, P. Glaser, N. Boemare, A. Danchin, and F. Kunst.** 2003. The genome sequence of the entomopathogenic bacterium *Photorhabdus luminescens*. *Nat. Biotechnol.* **21**:1307–1313.

47. **Ehrhardt, D. W., R. Wais, and S. R. Long.** 1996. Calcium spiking in plant root hairs responding to Rhizobium nodulation signals. *Cell* **85**:673–681.

48. **Engebrecht, J., M. Simon, and M. Silverman.** 1985. Measuring gene expression with light. *Science* **227**:1345–1347.

49. **Erséus, C.** 1984. Taxonomy and phylogeny of the gutless phallodrilinae (Oligochaeta, Tubificidae), with descriptions of one new Genus and twenty-two new species. *Zool. Scripta* **13**:239–272.

50. **Faruque, S. M., and J. J. Mekalanos.** 2003. Pathogenicity islands and phages in *Vibrio cholerae* evolution. *Trends Microbiol.* **11**:505–510.

51. **Felbeck, H., J. J. Childress, and G. N. Somero.** 1981. Calvin-Benson cycle and sulphide oxidation enzymes in animals from sulphide-rich habitats. *Nature* **293**:291–293.

52. **Felbeck, H., and J. Jarchow.** 1998. Carbon release from purified chemoautotrophic bacterial symbionts of the hydrothermal vent tubeworm *Riftia pachyptila*. *Physiol. Zool.* **71**:294–302.

53. **Felbeck, H., G. Liebezeit, R. Dawson, and O. Giere.** 1983. CO_2 fixation in tissues of marine oligochaetes (*Phallodrilus leukodermatus* and *P. planus*) containing symbiotic, chemoautotrophic bacteria. *Mar. Biol.* **75**:187–191.

54. **Forst, S., B. Dowds, N. Boemare, and E. Stackebrandt.** 1997. *Xenorhabdus* and *Photorhabdus* spp.: bugs that kill bugs. *Annu. Rev. Microbiol.* **51**:47–72.

55. **Foster, J. S., M. A. Apicella, and M. J. McFall-Ngai.** 2000. *Vibrio fischeri* lipopolysaccharide induces developmental apoptosis, but not complete morphogenesis, of the *Euprymna scolopes* symbiotic light organ. *Dev. Biol.* **226**:242–254.

56. **Foster, J. S., and M. J. McFall-Ngai.** 1998. Induction of apoptosis by cooperative bacteria in the morphogenesis of host epithelial tissues. *Dev. Genes Evol.* **208**:295–303.

57. **Fraysse, N., F. Couderc, and V. Poinsot.** 2003. Surface polysaccharide involvement in establishing the rhizobium-legume symbiosis. *Eur. J. Biochem.* **270**:1365–1380.

58. **Fredericks, D., and D. Relman.** 1996. Sequence-based identification of microbial pathogens: a reconsideration of Koch's postulates. *Clin. Microbiol. Rev.* **9**:18–33.

59. **Frostl, J. M., and J. Overmann.** 2002. Phylogenetic affiliation of the bacteria that constitute phototrophic consortia. *Arch. Microbiol.* **174**:50–58.

60. **Frostl, J. M., and J. Overmann.** 1998. Physiology and tactic response of the phototrophic consortium "*Chlorochromatium aggregatum.*" *Arch. Microbiol.* **169:**129–135.

61. **Gage, D. J.** 2002. Analysis of infection thread development using Gfp- and DsRed-expressing *Sinorhizobium meliloti. J. Bacteriol.* **184:**7042–7046.

62. **Gage, D. J.** 2004. Infection and invasion of roots by symbiotic, nitrogen-fixing rhizobia during nodulation of temperate legumes. *Microbiol. Mol. Biol. Rev.* **68:**280–300.

63. **Giere, O.** 1981. The gutless marine oligochaete *Phallodrilus leukodermatus.* Structural studies on an aberrant tubificid associated with bacteria. *Mar. Ecol. Prog. Ser.* **5:**353–357.

64. **Giere, O.** 1979. Studies on marine oligochaeta from Bermuda, with emphasis on new Phallodrilus-species (Tubificidae). *Cah. Biol. Mar.* **20:**301–314.

65. **Giere, O., N. M. Conway, G. Gastrock, and C. Schmidt.** 1991. "Regulation" of gutless annelid ecology by endosymbiotic bacteria. *Mar. Ecol. Prog. Ser.* **68:**287–299.

66. **Giere, O., and C. Langheld.** 1987. Structural organization, transfer and biological fate of endosymbiotic bacteria in gutless oligochaetes. *Mar. Biol.* **93:**641–650.

67. **Giere, O., R. Windoffer, and E. C. Southward.** 1995. The bacterial endosymbiosis of the gutless nematode, *Astomonema southwardorum*: ultrastructural aspects. *J. Mar. Biol. Assoc.* **75:**153–164.

68. **Gillan, D. C., and N. Dubilier.** 2004. Novel epibiotic thiothrix bacterium on a marine amphipod. *Appl. Environ. Microbiol.* **70:**3772–3775.

69. **Girguis, P. R., J. J. Childress, J. K. Freytag, K. Klose, and R. Stuber.** 2002. Effects of metabolite uptake on proton-equivalent elimination by two species of deep-sea vestimentiferan tubeworm, *Riftia pachyptila* and *Lamellibrachia* cf. *luymesi*: proton elimination is a necessary adaptation to sulfide-oxidizing chemoautotrophic symbionts. *J. Exp. Biol.* **205:**3055–3066.

70. **Glaeser, J., and J. Overmann.** 2004. Biogeography, evolution, and diversity of epibionts in phototrophic consortia. *Appl. Environ. Microbiol.* **70:**4821–4830.

71. **Goffredi, S. K., J. J. Childress, F. H. Lallier, and N. T. Desaulniers.** 1999. The internal ion composition of the hydrothermal vent tubeworm *Riftia pachyptila*; evidence for the elimination of SO_4^{2-} and H^+ and for a Cl^-/HCO_3^- shift. *Phys. Biochem. Zool.* **72:**296–306.

72. **Goffredi, S. K., V. J. Orphan, G. W. Rouse, L. Jahnke, T. Embaye, K. Turk, R. Lee, and R. C. Vrijenhoek.** 2005. Evolutionary innovation: a bone-eating marine symbiosis. *Environ. Microbiol.* **7:**1369–1378.

73. **Goryshin, I. Y., J. Jendrisak, L. M. Hoffman, R. Meis, and W. S. Reznikoff.** 2000. Insertional transposon mutagenesis by electroporation of released Tn5 transposition complexes. *Nat. Biotechnol.* **18:**97–100.

74. **Graber, J. R., J. R. Leadbetter, and J. A. Breznak.** 2004. Description of *Treponema azotonutricium* sp. nov. and *Treponema primitia* sp. nov., the first spirochetes isolated from termite guts. *Appl. Environ. Microbiol.* **70:**1315–1320.

75. **Graf, J.** 1999. Symbiosis of *Aeromonas veronii* biovar sobria and *Hirudo medicinalis*, the medicinal leech: a novel model for digestive tract associations. *Infect. Immun.* **67:**1–7.

76. **Haddad, A., F. Camacho, P. Durand, and S. C. Cary.** 1995. Phylogenetic characterization of the epibiotic bacteria associated with the hydrothermal vent polychaete *Alvinella pompejana. Appl. Environ. Microbiol.* **61:**1679–1687.

77. **Haygood, M. G., and S. K. Davidson.** 1997. Small-subunit rRNA genes and in situ hybridization with oligonucleotides specific for the bacterial symbionts in the larvae of the bryozoan *Bugula nertitina* and proposal of "*Candidatus* Endobugula sertula." *Appl. Environ. Microbiol.* **63:**4612–4616.

78. **Hentschel, U., and M. Steinert.** 2001. Symbiosis and pathogenesis: common themes, different outcomes. *Trends Microbiol.* **9:**585.

79. **Heungens, K., C. E. Cowles, and H. Goodrich-Blair.** 2002. Identification of *Xenorhabdus nematophila* genes required for mutualistic colonization of *Steinernema carpocapsae* nematodes. *Mol. Microbiol.* **45:**1337–1353.

80. **Hooper, L. V., and J. I. Gordon.** 2001. Commensal host-bacterial relationships in the gut. *Science* **292:**1115–1118.

81. **Hooper, L. V., M. H. Wong, A. Thelin, L. Hansson, P. G. Falk, and J. I. Gordon.** 2001. Molecular analysis of commensal host-microbial relationships in the intestine. *Science* **291:**881–884.

82. **Hooper, L. V., J. Xu, P. G. Falk, T. Midtvedt, and J. I. Gordon.** 1999. A molecular sensor that allows a gut commensal to control its nutrient foundation in a competitive ecosystem. *Proc. Natl. Acad. Sci. USA* **96:**9833–9838.

83. **Horton, R. M., Z. L. Cai, S. N. Ho, and L. R. Pease.** 1990. Gene splicing by overlap extension: tailor-made genes using the polymerase chain reaction. *BioTechniques* **8:**528–535.

84. **Hubber, A., A. C. Vergunst, J. T. Sullivan, P. J. J. Hooykaas, and C. W. Ronson.** 2004. Symbiotic phenotypes and translocated effector proteins of the *Mesorhizobium loti* strain R7A VirB/D4 type IV secretion system. *Mol. Microbiol.* **54:**561–574.

85. **Hungate, R. E.** 1939. Experiments on the nutrition of *Zootermopsis*. III. The anaerobic carbohydrate dissimilation by the intestinal protozoa. *Ecology* **20:**230–245.

86. **Jahnke, L. L., R. E. Summons, L. M. Dowling, and K. D. Zahiralis.** 1995. Identification of methanotrophic lipid biomarkers in cold-seep mussel gills: chemical and isotopic analysis. *Appl. Environ. Microbiol.* **61:**576–582.

87. **Jeon, K. W.** 2004. Genetic and physiological interactions in the amoeba-bacteria symbiosis. *J. Eukaryot. Microbiol.* **51:**502–508.

88. **Jones, B. W., and M. K. Nishiguchi.** 2004. Counter-illumination in the Hawaiian bobtail squid, *Euprymna scolopes* Berry (Mollusca: Cephalopoda). *Mar. Biol.* **144:**1151–1155.

89. **Karunakaran, R., T. H. Mauchline, A. H. Hosie, and P. S. Poole.** 2005. A family of promoter probe vectors incorporating autofluorescent and chromogenic reporter proteins for studying gene expression in Gram-negative bacteria. *Microbiology* **151:**3249–3256.

90. **Kimbell, J. R., and M. J. McFall-Ngai.** 2004. Symbiont-induced changes in host actin during the onset of a beneficial animal-bacterial association. *Appl. Environ. Microbiol.* **70:**1434–1441.

91. **Koch, B., L. E. Jensen, and O. Nybroe.** 2001. A panel of Tn7-based vectors for insertion of the gfp marker gene or for delivery of cloned DNA into Gram-negative bacteria at a neutral chromosomal site. *J. Microbiol. Methods* **45:**187–195.

92. **Koropatnick, T. A., J. T. Engle, M. A. Apicella, E. V. Stabb, W. E. Goldman, and M. J. McFall-Ngai.** 2004. Microbial factor-mediated development in a host-bacterial mutualism. *Science* **306:**1186–1188.

93. **Kranz, R. G., K. K. Gabbert, and M. T. Madigan.** 1997. Positive selection systems for discovery of novel polyester biosynthesis genes based on fatty acid detoxification. *Appl. Environ. Microbiol.* **63:**3010–3013.

94. **Krieger, J., O. Giere, and N. Dublier.** 2000. Localization of RubisCO and sulfur in endosymbiotic bacteria of the gutless marine oligochaete *Inanidrilus leukodermatus* (Annelida). *Mar. Biol.* **137:**239–244.

95. **Kudo, T., M. Ohkuma, S. Moriya, S. Noda, and K. Ohtoko.** 1998. Molecular phylogenetic identification of the intestinal anaerobic microbial community in the hindgut of the termite, *Reticulitermes speratus*, without cultivation. *Extremophiles* **2:**155–161.

96. **Lambertsen, L., C. Sternberg, and S. Molin.** 2004. Mini-Tn7 transposons for site-specific tagging of bacteria with fluorescent proteins. *Environ. Microbiol.* **6:**726–732.

97. **Land, J. v. d., and A. Nørrevang.** 1975. The systematic position of *Lamellibrachia* [sic] (Annelida, Vestimentifera). *Zeit. Zool. Syst. Evol.* **1:**86–101.

98. **Lane, D. J., B. Pace, G. J. Olsen, D. A. Stahl, M. L. Sogin, and N. R. Pace.** 1985. Rapid determination of 16S ribosomal RNA sequences for phylogenetic analyses. *Proc. Natl. Acad. Sci. USA* **82:**6955–6959.

99. **Lang, A. S., and J. T. Beatty.** 2001. The gene transfer agent of *Rhodobacter capsulatus* and "constitutive transduction" in prokaryotes. *Arch. Microbiol.* **175:**241–249.

100. **Leadbetter, J. R., and J. A. Breznak.** 1996. Physiological ecology of *Methanobrevibacter cuticularis* sp. nov. and *Methanobrevibacter curvatus* sp. nov., isolated from the hindgut of the termite *Reticulitermes flavipes*. *Appl. Environ. Microbiol.* **62:**3620–3631.

101. **Leadbetter, J. R., L. D. Crosby, and J. A. Breznak.** 1998. *Methanobrevibacter filiformis* sp. nov., a filamentous methanogen from termite hindguts. *Arch. Microbiol.* **169:**287–292.

102. **Leadbetter, J. R., T. M. Schmidt, J. R. Graber, and J. A. Breznak.** 1999. Acetogenesis from H$_2$ plus CO$_2$ by spirochetes from termite guts. *Science* **283:**686–689.

103. **Lee, K.-H., and E. G. Ruby.** 1994. Effects of the squid host on the abundance and distribution of symbiotic *Vibrio fischeri* in nature. *Appl. Environ. Microbiol.* **60:**1565–1571.

104. **Leidy, J.** 1881. The parasites of the termites. *J. Nat. Acad. Sci.* (Philadelphia) **8:**425–447.

105. **Lilburn, T. G., K. S. Kim, N. E. Ostrom, K. R. Byzek, J. R. Leadbetter, and J. A. Breznak.** 2001. Nitrogen fixation by symbiotic and free-living spirochetes. *Science* **292:**2495–2498.

106. **Lim, G. E., and M. G. Haygood.** 2004. "*Candidatus* Endobugula glebosa," a specific bacterial symbiont of the marine bryozoan *Bugula simplex*. *Appl. Environ. Microbiol.* **70:**4921–4929.

107. **Limpens, E., C. Franken, P. Smit, J. Willemse, T. Bisseling, and R. Geurts.** 2003. LysM domain receptor kinases regulating rhizobial nod factor-induced infection. *Science* **302:**630–633.

108. **Loper, J. L., and S. E. Lindow.** 1994. A biological sensor for iron available to bacteria in their habitats on plant surfaces. *Appl. Environ. Microbiol.* **60:**1934–1941.

109. **Loy, A., A. Lehner, N. Lee, J. Adamczyk, H. Meier, J. Ernst, K.-H. Schleifer, and M. Wagner.** 2002. Oligonucleotide microarray for 16S rRNA gene-based detection of all recognized lineages of sulfate-reducing prokaryotes in the environment. *Appl. Environ. Microbiol.* **68:**5064–5081.

110. **Manz, W., R. Amann, W. Ludwig, M. Wagner, and K.-H. Schleifer.** 1992. Phylogenetic oligodeoxynucleotide probes for the major subclasses of Proteobacteria: problems and solutions. *Syst. Appl. Microbiol.* **15:**593–600.

111. **Margulis, L.** 1970. *Origin of Eukaryotic Cells: Evidence and Research Implications for a Theory of the Origin and Evolution of Microbial, Plant, and Animal Cells on the Precambrian Earth.* Yale University Press, New Haven, CT.

112. **Martens, E. C., K. Heungens, and H. Goodrich-Blair.** 2003. Early colonization events in the mutualistic association between *Steinernema carpocapsae* nematodes and *Xenorhabdus nematophila* bacteria. *J. Bacteriol.* **185:**3147–3154.

113. **Martinez-Torres, D., C. Buades, A. Latorre, and A. Moya.** 2001. Molecular systematics of aphids and their primary endosymbionts. *Mol. Phylogen. Evol.* **20:**437–449.

114. **Matson, E. G., M. G. Thompson, S. B. Humphrey, R. L. Zuerner, and T. B. Stanton.** 2005. Identification of genes of VSH-1, a prophage-like gene transfer agent of *Brachyspira hyodysenteriae*. *J. Bacteriol.* **187:**5885–5892.

115. **Matsusaki, H., S. Manji, K. Taguchi, M. Kato, T. Fukui, and Y. Doi.** 1998. Cloning and molecular analysis of the poly(3-hydroxybutyrate) and poly(3-hydroxybutyrate-co-3-hydroxyalkanoate) biosynthesis genes in *Pseudomonas* sp. strain 61–3. *J. Bacteriol.* **180:**6459–6467.

116. **Matz, M., D. Shagin, E. Bogdanova, O. Britanova, S. Lukyanov, L. Diatchenko, and A. Chenchik.** 1999. Amplification of cDNA ends based on template-switching effect and step-out PCR. *Nucleic. Acids Res.* **27:**1558–1560.

117. **Mazmanian, S. K., C. H. Liu, A. O. Tzianabos, and D. L. Kasper.** 2005. An immunomodulatory molecule of symbiotic bacteria directs maturation of the host immune system. *Cell* **122:**107–118.

118. **McFall-Ngai, M. J., and M. K. Montgomery.** 1990. The anatomy and morphology of the adult bacterial light organ of *Euprymna scolopes* Berry (Cephalopoda: Sepiolidae). *Biol. Bull.* **179:**332–339.

119. **Metcalf, W. W., W. Jiang, L. L. Daniels, S. K. Kim, A. Haldimann, and B. L. Wanner.** 1996. Conditionally replicative and conjugative plasmids carrying lacZ alpha for cloning, mutagenesis, and allele replacement in bacteria. *Plasmid* **35:**1–13.

120. **Millikan, D. S., H. Felbeck, and J. L. Stein.** 1999. Identification and characterization of a flagellin gene from the endosymbiont of the hydrothermal vent tubeworm *Riftia pachyptila*. *Appl. Environ. Microbiol.* **65:**3129–3133.

121. **Mitra, R. M., and S. R. Long.** 2004. Plant and bacterial symbiotic mutants define three transcriptionally distinct stages in the development of the *Medicago truncatula/Sinorhizobium meliloti* symbiosis. *Plant Physiol.* **134:**595–604.

122. **Mitra, R. M., S. L. Shaw, and S. R. Long.** 2004. Six nonnodulating plant mutants defective for Nod factor-induced transcriptional changes associated with the legume-rhizobia symbiosis. *Proc. Natl. Acad. Sci. USA* **101:**10217–10222.

123. **Moran, N. A., and A. Mira.** 2001. The process of genome shrinkage in the obligate symbiont *Buchnera aphidicola*. *Genome. Biol.* **2:**research0054.1–0054.12.

124. **Neef, A.** 1997. *Anwendung der in situ Einzelzell-Identifizierung von Bakterien zur Populationsanalyse in komplexen mikrobiellen Biozönosen.* Technische Universität München, Munich, Germany.

125. **Nyholm, S. V., and M. J. McFall-Ngai.** 2004. The winnowing: establishing the squid-vibrio symbiosis. *Nat. Rev. Microbiol.* **2:**632–642.

126. **Odelson, D. A., and J. A. Breznak.** 1983. Volatile fatty acid production by the hind-gut microbiota of xylophagous termites. *Appl. Environ. Microbiol.* **45:**1602–1613.

127. **Ohkuma, M., and T. Kudo.** 1996. Phylogenetic diversity of the intestinal bacterial community in the termite *Reticulitermes speratus*. *Appl. Environ. Microbiol.* **62:**461–468.

128. **Ohkuma, M., S. Noda, R. Usami, K. Horikoshi, and T. Kudo.** 1996. Diversity of nitrogen fixation genes in the symbiotic intestinal microflora of the termite *Reticulitermes speratus*. *Appl. Environ. Microbiol.* **62:**2747–2752.

129. **Orphan, V. J., K. U. Hinrichs, W. Ussler III, C. K. Paull, L. T. Taylor, S. P. Sylva, J. M. Hayes, and E. F. Delong.** 2001. Comparative analysis of methane-oxidizing archaea and sulfate-reducing bacteria in anoxic marine sediments. *Appl. Environ. Microbiol.* **67:**1922–1934.

130. **Orphan, V. J., C. H. House, K.-U. Hinrichs, K. D. McKeegan, and E. F. DeLong.** 2001. Methane-consuming Archaea revealed by directly coupled isotopic and phylogenetic analysis. *Science* **293**:484–487.

131. **Orphan, V. J., C. H. House, K.-U. Hinrichs, K. D. McKeegan, and E. F. DeLong.** 2002. Multiple archaeal groups mediate methane oxidation in anoxic cold seep sediments. *Proc. Natl. Acad. Sci. USA* **99**:7663–7668.

132. **Ott, J.** 1995. Sulfide symbioses in shallow sands, p. 143–147. *In* A. Eleftheriou, A. Ansell, and C. Smith (ed.), *Biology and Ecology of Shallow Coastal Waters.* Olsen & Olsen, Fredensborg, Denmark.

133. **Ott, T., J. T. van Dongen, C. Gunther, L. Krusell, G. Desbrosses, H. Vigeolas, V. Bock, T. Czechowski, P. Geigenberger, and M. K. Udvardi.** 2005. Symbiotic leghemoglobins are crucial for nitrogen fixation in legume root nodules but not for general plant growth and development. *Curr. Biol.* **15**:531–535.

134. **Pace, N. R.** 1997. A molecular view of microbial diversity and the biosphere. *Science* **276**:734–740.

135. **Pasteris, J., J. Freeman, S. Goffredi, and K. Buck.** 2001. Raman spectroscopic and laser scanning confocal microscopic analysis of sulfur in living sulfur-precipitating marine bacteria. *Chem. Geol.* **180**:3–18.

136. **Pernthaler, A., and R. Amann.** 2004. Simultaneous fluorescence in situ hybridization of mRNA and rRNA in environmental bacteria. *Appl. Environ. Microbiol.* **70:** 5426–5433.

137. **Pernthaler, A., J. Pernthaler, and R. Amann.** 2004. Sensitive multi-color fluorescence in situ hybridization for the identification of environmental microorganisms. *In* G. Kowalchuk, F. J. de Bruijn, I. M. Head, A. D. L. Akkermans, and J. D. van Elsas (ed.), *Molecular Microbial Ecology Manual.* Kluwer Academic Press, Boston, MA.

138. **Pernthaler, J., F. O. Glöckner, W. Schönhuber, and R. Amann.** 2001. Fluorescence in situ hybridization (FISH) with rRNA-targeted oligonucleotide probes, p. 207–226. *In* J. Paul (ed.), *Methods in Microbiology,* vol. 30. Academic Press, San Diego, CA.

139. **Peters, J. E., and N. L. Craig.** 2000. Tn7 transposes proximal to DNA double-strand breaks and into regions where chromosomal DNA replication terminates. *Mol. Cell* **6**:573–582.

140. **Polz, M. F., and C. M. Cavanaugh.** 1995. Dominance of one bacterial phylotype at a Mid-Atlantic Ridge hydrothermal vent site. *Proc. Natl. Acad. Sci. USA* **92:** 7232–7236.

141. **Polz, M. F., D. L. Distel, B. Zarda, R. Amann, H. Felbeck, J. A. Ott, and C. M. Cavanaugh.** 1994. Phylogenetic analysis of a highly specific association between ectosymbiotic, sulfur-oxidizing bacteria and a marine nematode. *Appl. Environ. Microbiol.* **60**:4461–4467.

142. **Preston, G. M.** 2004. Plant perceptions of plant growth-promoting Pseudomonas. *Philos. Trans. R. Soc. Lond. B* **359**:907–918.

143. **Racker, E.** 1957. The reductive pentose phosphate cycle. I. Phosphoribulokinase and ribulose diphosphate carboxylase. *Arch. Biochem. Biophys.* **69**:300–310.

144. **Radutoiu, S., L. H. Madsen, E. B. Madsen, H. H. Felle, Y. Umehara, M. Gronlund, S. Sato, Y. Nakamura, S. Tabata, N. Sandal, and J. Stougaard.** 2003. Plant recognition of symbiotic bacteria requires two LysM receptor-like kinases. *Nature* **425**:585–592.

145. **Rakoff-Nahoum, S., J. Paglino, F. Eslami-Varzaneh, and S. Edberg.** 2004. Recognition of commensal microflora by toll-like receptors is required for intestinal homeostasis. *Cell* **118**:229–241.

146. **Rawls, J. F., B. S. Samuel, and J. I. Gordon.** 2004. Gnotobiotic zebrafish reveal evolutionarily conserved responses to the gut microbiota. *Proc. Natl. Acad. Sci. USA* **101**:4596–4601.

147. **Richards, K. S., T. P. Fleming, and B. G. M. Jamieson.** 1982. An ultrastructural study of the distal epidermis and the occurrence of subcuticular bacteria in the gutless tubificid *Phallodrilus albidus* (Oligochaeta : Annelida). *Aust. J. Zool.* **30**:327–336.

148. **Ruby, E. G., M. Urbanowski, J. Campbell, A. Dunn, M. Faini, R. Gunsalus, P. Lostroh, C. Lupp, J. McCann, D. Millikan, A. Schaefer, E. Stabb, A. Stevens, K. Visick, C. Whistler, and E. P. Greenberg.** 2005. Complete genome sequence of *Vibrio fischeri:* a symbiotic bacterium with pathogenic congeners. *Proc. Natl. Acad. Sci. USA* **102**:3004–3009.

149. **Russel, N. J., and D. S. Nichols.** 1999. Polyunsaturated fatty acids in marine bacteria—a dogma rewritten. *Microbiology* **145**:767–779.

150. **Schedel, J., and H. Truper.** 1980. Anaerobic oxidation of thiosulfate and elemental sulfur in *Thiobacillus denitrificans. Arch. Microbiol.* **124**:205–210.

151. **Schramm, A., S. K. Davidson, J. A. Dodsworth, H. L. Drake, D. A. Stahl, and N. Dubilier.** 2003. *Acidovorax*-like symbionts in the nephridia of earthworms. *Environ. Microbiol.* **5**:804–809.

152. **Schwedock, J., T. L. Harmer, K. M. Scott, H. J. Hektor, A. P. Seitz, M. C. Fontana, D. L. Distel, and C. M. Cavanaugh.** 2004. Characterization and expression of genes from the RubisCO gene cluster of the chemoautotrophic symbiont of *Solemya velum:* cbbLSQO. *Arch. Microbiol.* **182**:18–29.

153. **Silhavy, T. J., M. J. Casadaban, H. A. Shuman, and J. R. Beckwith.** 1976. Conversion of beta-galactosidase to a membrane-bound state by gene fusion. *Proc. Natl. Acad. Sci. USA* **73**:3423–3427.

154. **Sipe, A. S., A. E. Wilbur, and S. C. Cary.** 2000. Bacterial symbiont transmission in the wood-boring shipworm *Bankia setacea* (Bivalvia: Teredinidae). *Appl. Environ. Microbiol.* **66**:1685–1691.

155. **Small, A. L., and M. J. McFall-Ngai.** 1999. Halide peroxidase in tissues that interact with bacteria in the host squid *Euprymna scolopes. J. Cell Biochem.* **72**:445–457.

156. **Spaink, H. P.** 2000. Root nodulation and infection factors produced by rhizobial bacteria. *Annu. Rev. Microbiol.* **54**:257–288.

157. **Stappenbeck, T. S., L. V. Hooper, and J. I. Gordon.** 2002. Developmental regulation of intestinal angiogenesis by indigenous microbes via Paneth cells. *Proc. Natl. Acad. Sci. USA* **99**:15451–15455.

158. **Stingl, U., A. Maass, R. Radek, and A. Brune.** 2004. Symbionts of the gut flagellate *Staurojoenina* sp. from *Neotermes cubanus* represent a novel, termite-associated lineage of Bacteroidales: description of 'Candidatus Vestibaculum illigatum.' *Microbiology* **150**:2229–2235.

159. **Stougaard, J.** 2001. Genetics and genomics of root symbiosis. *Curr. Opin. Plant Biol.* **4**:328–335.

160. **Streams, M. E., C. R. Fisher, and A. Fiala-Médioni.** 1997. Methanotrophic symbiont location and fate of carbon incorporated from methane in a hydrocarbon seep mussel. *Mar. Biol* **129**:465–476.

161. **Suzuki, Y., T. Sasaki, M. Suzuki, Y. Nogi, T. Miwa, K. Takai, K. H. Nealson, and K. Horikoshi.** 2005. Novel chemoautotrophic endosymbiosis between a member of the Epsilonproteobacteria and the hydrothermal-vent gastropod *Alviniconcha* aff. *hessleri* (Gastropoda: Provannidae) from the Indian Ocean. *Appl. Environ. Microbiol.* **71**:5440–5450.

162. **Tamm, S. L.** 1982. Flagellated ectosymbiotic bacteria propel a eucaryotic cell. *J. Cell Biol.* **94**:697–709.

163. **Tomarev, S. I., R. D. Zinovieva, V. M. Weis, A. B. Chepelinsky, J. Piatigorsky, and M. J. McFall-Ngai.** 1993. Abundant mRNAs in the squid light organ encode proteins with a high similarity to mammalian peroxidases. *Gene* **132:**219–226.

164. **Uchiumi, T., T. Ohwada, M. Itakura, H. Mitsui, N. Nukui, P. Dawadi, T. Kaneko, S. Tabata, T. Yokoyama, K. Tejima, K. Saeki, H. Omori, M. Hayashi, T. Maekawa, R. Sriprang, Y. Murooka, S. Tajima, K. Simomura, M. Nomura, A. Suzuki, Y. Shimoda, K. Sioya, M. Abe, and K. Minamisawa.** 2004. Expression islands clustered on the symbiosis island of the *Mesorhizobium loti* genome. *J. Bacteriol.* **186:**2439–2448.

165. **Van Dover, C. L., B. Fry, J. F. Grassle, S. Humphris, and P. A. Rona.** 1988. Feeding biology of the shrimp *Rimicaris exoculata* at hydrothermal vents on the Mid-Atlantic Ridge. *Mar. Biol.* **98:**209–216.

166. **van Rhijn, P., and J. Vanderleyden.** 1995. The Rhizobium-plant symbiosis. *Microbiol. Rev.* **59:**124–142.

167. **Vetter, R. D.** 1985. Elemental sulfur in the gills of three species of clams containing chemoautotrophic symbiotic bacteria: a possible inorganic energy storage compound. *Mar. Biol.* **88:**33–42.

168. **Viprey, V., A. Del Greco, W. Golinowski, W. J. Broughton, and X. Perret.** 1998. Symbiotic implications of type III protein secretion machinery in *Rhizobium*. *Mol. Microbiol.* **28:**1381–1389.

169. **Walder, M. K., and J. J. Mekalanos.** 1996. Lysogenic conversion by a filamentous bacteriophage encoding cholera toxin. *Science* **272:**1910–1914.

170. **Wang, Y., U. Stingl, F. Anton-Erxleben, S. Geisler, A. Brune, and M. Zimmer.** 2004. "*Candidatus* Hepatoplasma crinochetorum," a new, stalk-forming lineage of *Mollicutes* colonizing the midgut glands of a terrestrial isopod. *Appl. Environ. Microbiol.* **70:**6166–6172.

171. **Washnick, M., and D. M. Lane.** 1971. Ribulose biphosphate carboxylase from spinach leaves. *Methods Enzymol.* **23:**570–577.

172. **Waterbury, J. B., C. B. Calloway, and R. D. Turner.** 1983. A cellulolytic-nitrogen fixing bacterium cultured from the gland of Deshayes in shipworms (Bivalvia: Teredinidae). *Science* **221:**1401–1403.

173. **Waters, E., M. J. Hohn, I. Ahel, D. E. Graham, M. D. Adams, M. Barnstead, K. Y. Beeson, L. Bibbs, R. Bolanos, M. Keller, K. Kretz, X. Lin, E. Mathur, J. Ni, M. Podar, T. Richardson, G. G. Sutton, M. Simon, D. Soll, K. O. Stetter, J. M. Short, and M. Noordewier.** 2003. The genome of *Nanoarchaeum equitans*: insights into early archaeal evolution and derived parasitism. *Proc. Natl. Acad. Sci. USA* **100:**12984–12988.

174. **Wenzel, M., R. Radek, G. Brugerolle, and H. König.** 2003. Identification of the ectosymbiotic bacteria of *Mixotricha paradoxa* involved in movement symbiosis. *Eur. J. Protistol.* **39:**11–24.

175. **Wilmot, D. B. J., and R. D. Vetter.** 1990. The bacterial symbiont from the hydrothermal vent tubeworm *Riftia pachyptila* is a sulfide specialist. *Mar. Biol.* **106:**273–283.

176. **Wilson, K.** 1990. Preparation of genomic DNA from bacteria, p. 2.4.1–2.4.5. *In* F. M. Ausubel, R. Brent, R. E. Kingston, D. D. Moore, J. G. Seidman, J. A. Smith, and K. Struhl (ed.), *Current Protocols in Molecular Biology.* Greene Publishing Associates and Wiley Intersciences, New York, NY.

177. **Wilson, K. J., A. Sessitsch, J. C. Corbo, K. E. Giller, A. D. Akkermans, and R. A. Jefferson.** 1995. beta-Glucuronidase (GUS) transposons for ecological and genetic studies of rhizobia and other gram-negative bacteria. *Microbiology* **141:**1691–1705.

178. **Wong, S. M., and J. J. Mekalanos.** 2000. Genetic footprinting with mariner-based transposition in Pseudomonas aeruginosa. *Proc. Natl. Acad. Sci. USA* **97:**10191–10196.

METABOLISM

Introduction to Metabolism

C. A. REDDY

Metabolism is central to the question of how microbes make a living. The metabolic systems that are involved in generating energy for growth, motion, maintenance, and cell division are studied through physical, chemical, and enzymatic methods. Methods are also needed to study oxidative and fermentation mechanisms for producing ATP and biosynthetic intermediates needed by the cell. Furthermore, there is a growing interest in studying plant biopolymers, as exemplified by cellulose, hemicellulose, and lignin, the most abundant sources of organic carbon and energy in the biosphere. There is a great deal of current interest in the economic conversion of these plant polymers as sources of food, feed, fuels, and chemicals. Methods to study the above-mentioned topics in metabolism constitute the main focus of this section, which has been significantly expanded with nine chapters, compared to four chapters in the previous edition.

Chapters 17 to 20, dealing with physical, chemical, and enzymatic analysis and with permeability and transport, are retained from previous edition of this manual, *Methods for General and Molecular Bacteriology* (MGMB). Mulrooney revised and updated chapter 17, previously authored by Wood and Paterek. This chapter deals with the basic physical analytical methods and separation procedures such as spectrophotometry, electrodes, chromatography, radioactivity, and electrophoresis. Chapter 18 on chemical analysis, authored by Daniels et al. as chapter 22 in the previous edition, required relatively little revision and is largely intact in this edition. Chapter 19 on enzymatic activity, authored by Phillips as chapter 23 in the previous edition, was revised,

rearranged, and updated substantially by Robert P. Hausinger and Phillips for this edition. The chapter authored by Marquis on permeability and transport (chapter 20) is essentially intact from the previous edition, with minor revisions.

Chapters 21, 22, and 23, "Bacterial Respiration," "Carbohydrate Fermentations," and "Metabolism of Aromatic Compounds," are completely new to this edition. An understanding of the oxidation and fermentation pathways as sources of energy to a microbial cell is of fundamental interest to microbiologists. The chapter on bacterial respiration, authored by Gunsalus et al., presents methods for studying localization of respiratory enzymes and related assays as well as basic methods used for studying aerobic and anaerobic respiratory enzymes. "Carbohydrate Fermentations," by Meganathan et al., is one of the most comprehensive presentations of its kind, with detailed schematics depicting most of the major fermentation pathways, including details on relative ATP yields and key oxidation/reduction steps in each of the pathways and details of assay procedures for various enzymes involved in these pathways. In chapter 23, Kukor et al. have done an excellent job of presenting succinct and current information on methods for studying metabolism of aromatic compounds.

Chapters 24 and 25 focus on methods for studying key enzymes involved in the degradation of plant polymers. Himmel et al. present useful information related to the study of cellulases, hemicellulases, and pectinases, while Dosoretz and Reddy discuss current approaches for studying fungal lignin-modifying enzymes.

17

Physical Analysis and Purification Methods

SCOTT B. MULROONEY, WILLIS A. WOOD, AND J. R. PATEREK

This chapter deals with the basic physical analytical methods and separation procedures of spectrophotometry, electrodes, chromatography, radioactivity, and electrophoresis. Application of these methods in microbiology has become routine. The approach in this manual is to give a general description along with strengths and weaknesses of a method, specific techniques and preparations where appropriate, and examples of application to investigations with bacteria. The purification of enzyme proteins is particularly exemplified.

In this chapter, especially, it has been necessary to confine the scope to the simple procedures which are appropriate for laboratory classes or are routinely used in a research laboratory. In-depth treatment of the methods presented here can be found in books specifically dedicated to individual techniques, some of which are listed in section 17.8. In addition, catalogs and commercial literature from manufacturers and suppliers contain a wealth of information that is highly valuable to the investigator. Some specialized techniques such as thin-layer chromatography and capillary electrophoresis are mentioned only briefly.

17.1. SPECTROPHOTOMETRY

A wide variety of photometric methods have found important uses in bacteriological research and applications. Absorbance spectroscopy and spectrofluorimetry (1–6) have been particularly useful. These have facilitated identification of compounds, determination of structure, estimation of concentrations, and measurement of enzymatic reaction rates. For details of photometric measurement of bacterial growth, see chapter 9.6.

17.1.1. Absorbance Spectroscopy

UV-visible spectrophotometers can be classified into two broad categories: those containing a wavelength-dispersive monochromator that sends monochromatic light through a sample, and diode-array instruments that pass a broad spectrum through the sample and subsequently measure light at individual wavelengths. No matter what the design, the basic principles of light absorption are common to all spectrophotometers. Light from a radiant source is directly or indirectly (through a monochromator) passed through the sample to a detector, which quantitates the energy received and expresses it as the ratio of the transmitted light (I_T) to the incident light (I_0). When I_0 is set at 100, the instrument response to light passing through the sample is percent transmittance (%T). This can be related to the molar concentration (C) of an absorbing solute through the Beer-Lambert law:

$$-\log \%T = \varepsilon l C = A$$

where ε is the molar extinction coefficient of the absorbing species at a specified wavelength and l is the length of the light path in centimeters. Since both ε and l are constant in practice, C is proportional to the negative logarithm of %T or directly proportional to absorbance (A). For this reason, newer spectrophotometers reading linearly in A are more convenient and more accurate for samples of moderate to high absorbance.

Other spectroscopic methods that are not discussed in detail here are measurement of the ability of an optically active chromophore to rotate polarized light (optical rotatory dispersion) or the differential absorption of right- and left-polarized light (circular dichroism). The spectropolarimeter instrumentation is complex and expensive, and is not routinely found in the microbiology laboratory. Of these methods, circular dichroism is the most popular; it can be used to examine structural features of nucleic acids and proteins as well as observing stereoisomer preference of an enzymatic reaction by substrate-monitored turnover. More details can be found in reference 3.

17.1.1.1. Spectrophotometer Components: Dispersive and Diode Array Instruments

All absorbance spectrophotometers have several common components. These are described briefly below along with how they fit into the overall instrument design. The first element, the light source, is typically a tungsten-halogen lamp for visible-light wavelengths of 350 to 800 nm, and a deuterium arc lamp for UV wavelengths of 190 to 350 nm. Another type of light source is a xenon lamp that emits a broad range of 200 to 900 nm; thus, a single lamp can be used instead of deuterium and tungsten lamps. Xenon lamps have the added advantage of providing higher-energy emissions in the UV range than deuterium. Furthermore, new designs incorporate a flash xenon lamp that is energized only when a reading is being taken.

Energy emitted from the light source then passes through a very narrow entrance slit. The light passing through the slit is usually collimated by a mirror or lens with the overall goal of presenting nearly parallel light rays to the sample. (The location of the slit differs in typical diode-array instruments, where the light is passed first through the sample and then through the slit.) In dispersive spectrophotometers, the light is next directed to an element that separates different wavelengths. This element, typically a diffraction grating (but some instruments use a prism), is mounted on an assembly connected to a step motor that allows the spectrum to be scanned at various rates.

The light is next directed through the sample compartment which contains a cuvette holder. Dispersive spectrophotometers can pass the light through a single sample (single beam), or the light is first split into two beams (double beam) where one passes through a reference cell and the second passes through the cell being measured. The double beam design allows for correction of any fluctuation in lamp energy by always measuring the difference between the reference and sample beams. After leaving the sample compartment, the light is quantitated by a detector that converts the energy to a voltage. Typical dispersive spectrophotometers use photomultiplier tubes while the photodiode array instrument projects the light off a zero-order grating onto the array, where each diode responds to a discrete wavelength. The instrument contains circuitry that polls the voltage of each diode. One advantage of the diode-array design is that all of the diodes can be scanned in as little as a few milliseconds, making data collection less time-consuming.

Finally, most moderate to high-end spectrophotometers produced today have some type of computer interface. This allows the researcher complete control of the instrument from the computer and allows more rapid saving, display, and preparation of presentation quality results.

17.1.1.2. Performance Factors for Dispersive Spectrophotometers

For laboratory spectrophotometers, quality is determined by the purity of the light presented to the sample, the intensity and area of the beam (numerical aperture), wavelength accuracy, the accuracy of photometer calibration, the linearity of response, signal-to-noise ratio, and short- and long-term drift. Manufacturers of some instruments go to great lengths and expense to achieve high quality.

Effect of spectral bandwidth. Spectral purity of light from a monochromator, or spectral bandwidth, is the width in nanometers at one-half the height of the emission energy peak at a specified nominal wavelength (Fig. 1B). This bandwidth is specified for the instrument and is not readily determined in the laboratory. Bandwidth is reasonably constant across the wavelength range in monochromators in which a grating is the dispersing element but varies with wavelength for prism monochromators. It varies with slit width in both cases. The accuracy of a measured absorbance, assuming no stray light (see below), depends on the ratio of the spectral bandwidth (a property of the instrument) to the natural bandwidth (the width in nanometers at one-half the peak height) of the peak of the solute in the light beam (a property of the chromophore in solution) (Fig. 1A). Therefore, in an inexpensive spectrophotometer with a spectral bandwidth at 340 nm of 8 nm, a peak with an 80-nm natural bandwidth (giving a ratio of 0.1) allows absorbance measurements with 99.5% accuracy (9). Similarly, with an 8-nm bandwidth, 99% accuracy is achieved with a chromophore with a 50-nm natural bandwidth. The most commonly measured material at 340 nm, NADH, has a natural bandwidth of 58 nm at that wavelength. Therefore, it is readily apparent that absorbing compounds such as NADH can be measured at better than 99% accuracy in this inexpensive spectrophotometer (4, 9).

Effect of stray light. Another consequence of wide spectral bandwidth, or low spectral purity, is an increased content of stray light (i.e., energies at wavelengths removed from the nominal value and not part of the spectral bandwidth wavelengths). Since spectral bandwidth for prism monochromators is related to both slit width and wavelength, so also is stray-light content. Stray light derives from incomplete removal of unwanted wavelengths owing to faulty design or dirty optics. The stray-light content of the exit beam diminishes the linear range of response to chromophore concentration, i.e., the range where the Beer-Lambert law applies. The operator should determine the useful linear range in the wavelength region at which determinations are to be made.

Newer spectrophotometers in wide use can make accurate absorbance measurements with an upper range of 1.5 A, although some designs can go as high as 2.0 to 5.0 A. The ability to maintain linearity at such high absorbance is the result of newer designs that incorporate a premonochromator in the light path between the light source and the monochromator. This results in decreased stray light and better sensitivity to the minute amounts of light impinging on the detector. Furthermore, many spectrophotometers are now designed to be operated with sample compartments open while eliminating most of the potential for room light that would otherwise interfere with the optical measurements.

Photometric accuracy and linearity. For certain spectrophotometers, electronic adjustment of photometric calibration is not available and linearity of response is assumed. It is a good policy to routinely check both absorbance accuracy and linearity of absorbance measurement. This allows the researcher to have confidence in measurements for a particular series of experiments and can serve as an early indicator that some component of the instrument is failing. For instruments for which absorbance calibration is possible, neutral density filter standards of known absorbance at specified wavelengths are supplied and are used to adjust the photometer response to give the same absorbance value as the standards at one or a few absorbance values. With linearity between these points assumed, the total absorbance range is calibrated. Most spectrophotometer manufacturers provide protocols to check for instrument performance and linearity.

Effect of slit width. For prism monochromators working at a fixed slit setting, there is a very marked nonlinear dependency of band pass on wavelength. Operating manuals from spectrophotometer manufacturers and books on spectrophotometry contain graphs of this function. Thus, to make measurements at constant band pass with prism optics, take the slit setting at the wavelength of measurement from the graph. To be rigorous, determine the solute concentration at the slit width setting used for the determination of its extinction coefficient. Conversely, when reporting an extinction coefficient, state the band pass used. For prism instruments this involves both the slit width and the graph of band pass versus wavelength for that instrument.

Similarly, spectra should be acquired at constant band pass, not constant slit width. The absorbance values at fixed wavelength or the absorbance spectra determined at two different slit widths seldom are identical. The error is magnified when the absorbance peaks of the solutes are sharp and may be negligible when peaks are broad. Therefore, it is prudent to use the minimum slit width. Slit width also is a function of the sensitivity of the photometer, the spectral characteristics of the photodetector, and the source. More importantly, the intensity of the lamp, cleanliness of optical surfaces, and critical focusing of the source on the slit greatly affect slit width. As a rule of thumb, the average spectrophotometer should be capable of nulling by opening the slit to 2 mm at 210 nm or lower. If this is not possible, one or more of the above factors may require attention.

Single-beam versus double-beam spectrophotometers. Double-beam instruments are constructed so that there are two

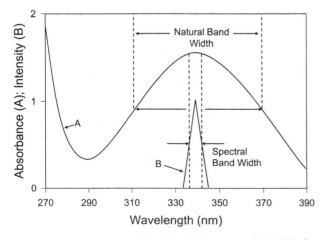

FIGURE 1 (A) Natural bandwidth diagram of NADH; the bandwidth is 58 nm. (B) Spectral bandwidth diagram (schematic) of spectrophotometer exit beam; the spectral bandwidth is 8 nm.

light beams giving a simultaneous or parallel comparison between I_0 and I_T. Single-beam units make these two measurements serially, whereas double-beam spectrophotometers split the light beam so that part of the beam passes through the solution used to set the instrument at $I_0 = \%T = 100$ or $A_0 = 0$, while the second part passes through the sample to measure $\%T$ or A_1. For this approach to be valid, the parallel optical systems and their associated electronics must perform identically. In single-beam units, the same end is achieved by first setting $\%T = 100$ or $A_0 = 0$ with the blank and then reading $\%T$ or A_1 when the sample is introduced into the light beam.

17.1.1.3. Performance Factors for Diode-Array Spectrophotometers

Many of the performance factors mentioned above also apply to diode-array instruments. One disadvantage of this design is that the photodiode array has lower sensitivity to light compared to photomultipliers. Instrument manufacturers compensate for this by increasing the intensity of light that is presented to the sample; however, this can have a deleterious effect on samples that are light sensitive. This shortcoming is to some extent ameliorated by the ability to scan the entire spectrum rapidly, minimizing the exposure time of the sample to the intense light. Most diode-array spectrophotometers are single-beam instruments and usually have motor-driven cuvette holders that place the desired sample in the light path. Instruments should be checked at regular intervals for precise alignment of the cell holders. In some cases, black-masked cuvettes provide improved spectral data by reducing the amount of stray light that enters the detector.

17.1.1.4. Determination of Solute Concentration

Direct concentration determination of a chromophore is straightforward as long as there is no significant interference by other absorbing species. With a properly calibrated spectrophotometer and standard 1-cm light path cuvettes, absorbance at the wavelength specified for the molar extinction coefficient (ε) of a chromophore is measured. Calculation of A/ε yields molar concentration. If the extinction coefficient for a particular chromophore is not known, it can be easily measured. In that case, prepare a standard regression line which relates absorbance to concentration for a set of concentration standards that are measured directly. The plot will show the linear region and the variability of individual determinations. Standard statistical considerations apply, i.e., replication, least-square fits of the data to straight lines, determination of standard deviation from the mean, and standard error. If a particular substance to be measured has no significant absorbance, then secondary color-development reactions will be needed. In all cases, care must be taken to ensure that the absorbance readings are within the range of linearity for the instrument.

17.1.1.5. Determination of Particle Concentration

Spectrophotometers and colorimeters can also be used to determine the concentration of microbial cells. The method depends on attenuation of the transmitted beam as a result of light scattering rather than absorbance. The photometric response does not follow the Beer-Lambert law; however, a calibration curve relating instrument response to bacterial cell concentration, for instance, gives satisfactory results (see chapter 9.6).

17.1.1.6. Determination of Reaction Rates

When either the product or the reactant of a reaction is a chromophore, the rate of chromophore production or utilization can be determined at fixed wavelength by reading the absorbance at time intervals or by monitoring the absorbance change versus time (2, 8). A plot of absorbance versus time permits determination of velocity from the slope of the line. Specialized spectrophotometer attachments that allow for temperature control and/or stirring of the sample cuvettes have been developed for this purpose. Automatic sample changers can be used to assay several simultaneous reactions and record readings at intervals as short as 10 to 20 s, and spectrophotometers are available to record kinetic results from 96-well plates. Spectrophotometers equipped with a computer interface usually have software available to rapidly calculate and plot reaction parameters.

17.1.2. Fluorescence Spectroscopy

Fluorescence is the absorbance of light by a chromophore (fluorophore) with concomitant emission of light at a longer wavelength (lower energy) (2, 5). Since different wavelengths are involved in excitation and fluorescent emission, the instrument must be designed to prevent excitation energy from reaching the photodetector. This is usually accomplished by placing the detector at 90° to the excitation beam. Thus, the measurement involves the appearance of radiant energy against the very low level of background light from the blank and the signal-to-noise ratio is very high, so fluorimetric methods are inherently more sensitive than photometric methods that are based on absorption of light. For example, the fluorescent method for determination of NADH is as much as 1,000 times more sensitive than the spectrophotometric method. Fluorescence intensity depends on the concentration of fluorophore and, unlike absorption, also depends on the intensity of the excitation beam. Stray light arising from scattering both in the solution and from surfaces can be a major problem, but this can be minimized by placing before the photodetector a sharp cutoff filter, which passes light at the wavelength of fluorescent emission but blocks excitation energy.

Since fluorescence intensity is dependent on excitation intensity, lamp emission characteristics and stability are of major importance. Some instruments monitor beam intensity through a second photodetector and display the ratio of fluorescence to excitation intensities. In other spectrofluorimeters, the excitation beam is split, with one going to the sample and the other going to a cuvette containing a reference dye such as rhodamine B, resulting in a ratio spectrum (see below). At low emission intensities, photodetector output voltage is nearly linear with intensity; hence, concentration can be determined by measuring fluorescence at wavelengths characteristic for a particular fluorophore. The process basically involves preparation of a solvent blank and a set of standards, construction of a standard curve, and comparison with unknowns.

17.1.2.1. Types of Fluorescence Spectra

Excitation and emission spectra. Most spectrofluorimeters possess two wavelength dispersive elements; an excitation monochromator that directs single-wavelength light to the sample cuvette, and an emission monochromator positioned at a right angle relative to the excitation beam that scans the spectrum of light emitted by the sample. One monochromator can be used to scan a spectrum while the second monochromator is held at a fixed wavelength. An excitation spectrum is taken by scanning the excitation monochromator and recording the emitted light at a wavelength that is maximal for that particular sample. For example, the excitation spectrum of NADH is taken by scanning the excitation monochromator while monitoring the emission at a fixed wavelength of 470 nm. An emission spectrum is taken by holding the excitation monochromator wavelength of peak absorbance (340 nm for NADH) while scanning the emission monochromator.

Corrected and uncorrected spectra. Spectrofluorimeters produce spectra that are inherently distorted due to variation of lamp intensity across the spectrum, variation in detector sensitivity at different wavelengths, and other optical and electronic artifacts. For many purposes such as single-wavelength reaction assays or a series of experiments all done on the same instrument, direct, or *uncorrected*, spectral data are sufficient. If, however, the spectra are to be compared with those obtained on different instruments or compared with absorbance spectra as shown in Fig. 2, then some correction must be made for the instrument artifacts. A *corrected* spectrum is recorded on a double-excitation beam spectrofluorimeter where the instrument aberrations are subtracted out, usually by multiplying the data by a calibration spectrum file. This calibration is typically performed by placing two rhodamine B-containing cuvettes in the reference and sample compartments and simultaneously scanning both the excitation and emission monochromators. An

example of corrected excitation and emission fluorescence spectra compared to an absorbance spectrum is shown in Fig. 2, where panel A is an absorbance spectrum of the flavin adenine dinucleotide-containing thioredoxin reductase (expressed as extinction coefficient) and panel B shows the corrected fluorescence excitation spectrum on the left and the fluorescence emission spectrum on the right. Note how the shape and peak locations of the corrected fluorescence excitation and absorbance spectra are nearly identical. Computer interfaces available on many instrument models make this procedure simple. It is important that researchers state in published spectra whether they are corrected or uncorrected and what instrument model and settings were used.

17.1.2.2. Common Applications of Fluorescence

Fluorescence spectroscopy can be a useful tool for a wide variety of analytical situations. In kinetic assays such as enzyme reactions utilizing NAD(P)H, there is much greater sensitivity than in absorbance-based monitoring. Intrinsic fluorescence of proteins or enzymes containing tryptophan can be observed to obtain data on unfolding, substrate binding, and conformational changes because of the extreme sensitivity of this fluorophore to small changes in its environment (5, 10). Various fluorescent dyes (fluorescent donor), used in conjunction with dyes that absorb light emitted by the donor (acceptor), can be used to estimate intermolecular and intramolecular distances by resonance energy transfer (5). Green fluorescent protein, a commonly used reporter for gene expression studies, is quantitated using fluorescence-based methods like UV microscopy, multiwell high-throughput screening, and cell sorting (7).

17.1.2.3. Factors Affecting Accuracy

Instrument variation and drift. Care should be taken to correct for or prevent fluctuations in excitation intensity as a result of variations in lamp intensity. It may be necessary to introduce the solvent blank and standards frequently. It is best to use a known standard that has fluorescence properties similar to what is being measured and to check that an identical reading is obtained throughout the length of the experiment. If readings begin to drift, then signal gain can be adjusted to maintain accuracy.

Absorbance of solutions. High absorbance of solutions produces two kinds of errors: (i) reabsorption of fluorescent energy to give an apparently lower fluorescence yield, and (ii) excessive absorbance of excitation light due to high absorbance by the sample at this wavelength. For this reason, absorbance should not exceed 0.2 Å, and independent absorbance measurements on standards and unknowns should be made at the same excitation and emission wavelengths.

Fluorescence of impurities. The limiting sensitivity is determined by background fluorescence. This is minimized by use of pure reagents. In addition, errors can result from nonuniform contamination of samples and glassware by a variety of fluorescent materials, including rubber, grease, fingerprints, and extracts of filter paper or dialysis tubing. Turbidity contributes to error through scattering, which, unless an excitation energy-blocking filter is present, contributes to the measured fluorescence intensity. Therefore, maintenance of scrupulously clean glassware for measurement and storage of solutions is necessary.

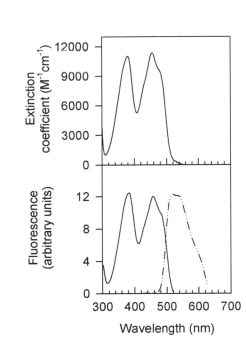

FIGURE 2 Comparison of absorbance and corrected fluorescence spectra. (A) Absorbance spectrum of the flavin-containing enzyme thioredoxin reductase. (B) Corrected fluorescence spectra: the excitation spectrum (excitation with a scanned spectrum and recorded at 530 nm emission) is on the left (solid line) and the emission spectrum (recorded by exciting at 455 nm and recording the spectrum of emitted light) is on the right (dashed line). Note the similar shapes and peak locations of the absorbance and fluorescence excitation spectra.

Many of the above effects can be minimized by the use of blocking filters at the photomultiplier.

Effects of temperature and ionic strength. Since fluorescent phenomena are temperature dependent, all measurements should be made at constant temperature, preferably in a thermostated cell. Fluorescence is also extremely sensitive to ionic strength, and care must be taken when performing experiments that will result in changes in the ion concentration of the sample.

17.2. ELECTRODES

The Nernst equation,

$$E = E_0 + \frac{RT}{nF} \ln \frac{[\text{ox}]}{[\text{red}]}$$

can be used to measure the concentration (activity), or the ratio of concentrations, of certain molecules or ions. E is the measured potential, E_0 is the standard electrode potential, R is 8.315 J/°C, T is the absolute temperature, n is the number of electrons, F (Faraday) is 96,500 C, and [ox] and [red] are the concentrations of oxidized and reduced members of the half-cell pair, respectively.

This basic behavior has made possible the determination of a number of biologically important molecules, e.g., protons, H_2, O_2, ammonium, and chloride. Incorporation of these principles into a usable instrument involves (i) the availability of a measurement electrode capable of sensing the potential generated by the internal oxidant-reductant system, (ii) a standard half-cell of known potential, and (iii) an electrometer or potentiometer which measures the voltage between the two half-cells (11–13). Solution of the Nernst equation for the measuring half-cell results from the relationship

$$E_{\text{(measuring half-cell)}} = E + E_0$$

The ability to measure a specific molecule is dependent on the construction of the measuring electrode. For instance, a glass electrode develops a potential proportional to hydrogen ion concentration, and several ion-selective electrodes develop potentials for specific ions. An ion-specific oxidation-reduction occurs in the coating material that isolates the electrode from the medium. There are, in addition, many other strategies for determination of concentration of various ions on the basis of electrode behavior. The oxygen electrode device, for instance, is a form of polarograph, because there must be a polarizing voltage to ionize the oxygen.

17.2.1. Hydrogen Ion Electrode

The glass electrode is the only practical means of measuring hydrogen-ion activity (pH). A thin membrane of special glass isolates a silver-silver chloride half-cell from the measured solution (12–14). The electrical potential of the glass electrode varies linearly with pH over a wide range, pH 1 to 11. In the region of pH 11 and above, there is nonlinearity unless a special alkali-resistant glass is used. The reference is a standard calomel (Hg/Hg_2Cl_2) or Ag/AgCl half-cell, which is connected to the measured solution by a salt bridge of KCl. Both the glass electrode and the standard electrode are immersed in the sample or assembled into a single, combination electrode. The latter can be constructed to enter long test tubes or to accommodate very small samples.

In addition to the static determination of the pH of samples, the pH meter or an automated version can be used to monitor over time the reactions in which a hydrogen ion is evolved or consumed. In this application, alkali or acid is added to maintain constant pH; a plot of acid or alkali consumption versus time expresses the reaction progress. For instance, the oxidation of glucose and glucose 6-phosphate either enzymatically or chemically, and the phosphorylation of substrates with adenosine triphosphate by kinases, are reactions that can be tracked with the glass electrode.

17.2.1.1. Calibration of the pH Meter

The system must be calibrated by using buffers of known pH. A popular but expensive approach to calibration is to buy prepackaged liquid or solid buffers of specified pH, but this is by no means necessary or superior. Table 1 lists some standard buffers (13) which have dependable pH values. With linear electrode behavior, a single standard buffer should serve to calibrate across the useful range. However, electrode behavior is seldom ideal; hence, it is wise to use two standard buffers (for instance, pH 4 and 6.88) and to bracket the pH region of the unknown.

17.2.1.2. Problems

The major problems in measurement of pH come from nonlinearity in the alkaline region, absorption of CO_2 in neutral and alkaline samples and in standards, and slow evaporation of standards. Additional problems of unsatisfactory instrument behavior commonly arise from poor maintenance of electrodes (16). The glass electrode needs an equilibrium-hydrated silica gel layer for accurate and constant pH measurement. Electrode manufacturers recommend that a new or cleaned electrode be soaked in water, or in buffer followed by water, for up to several days to establish the gel layer. They also recommend that the electrodes be maintained in distilled water between measurements. Electrodes must be cleaned after use by washing and careful wiping to remove adhering materials.

Unfortunately, incomplete cleaning plus storage in water can result in a buildup of contaminants and fouling of the glass surface. Lipid and protein contribute to the nonlinearity of electrode behavior and produce a slow response. In addition, bacterial growth may occur, with both fouling of the glass surface and plugging of the asbestos fiber which functions as a salt bridge between the standard electrode and the solution being measured. The KCl level in the calomel electrode should be maintained, and the cap should be removed during measurement so that KCl drainage through the fiber prevents impurities from lodging in the fiber.

Once the glass electrode is fouled, the temptation is to replace it with a new one. However, it can often be cleaned

TABLE 1 Standard buffers for calibration of pH meters at 25°C[a]

pH	Buffer
1.10	0.1 M HCl
2.16	0.1 M potassium tetroxalate
3.56	Saturated potassium hydrogen tartrate
4.00	0.05 M potassium hydrogen phthalate
6.88	0.025 M KH_2PO_4 + 0.025 M Na_2HPO_4
7.41	0.008695 M KH_2PO_4 + 0.03043 M Na_2HPO_4
9.22	0.01 M sodium borate decahydrate (borax)
12.45	Saturated calcium hydroxide

[a]Data from reference 1.

effectively by soaking alternately in 0.1 N HCl and 0.1 N NaOH at 50°C for several cycles, followed by reestablishment of the gel layer by soaking. In difficult cases, specific methods recommended by the manufacturers should be employed for more aggressive cleaning.

17.2.1.3. Effect of Salt

A high ionic strength of the solution being measured results in errors in pH measurement by as much as 0.5 pH unit. This problem is commonly encountered in preparing high molarity ammonium sulfate solutions in the neutral and alkaline pH region. It is common practice to dilute such samples 1:20 before pH measurement. Dilution of solutions of completely dissociated salts has little effect on pH because the ratio of ions remains nearly constant. The magnitude of the salt effect for specific circumstances is ascertained by making up standards in the salt solutions involved.

17.2.2. Other Ion-Specific Electrodes

Ion-specific electrodes are available for electrometric determination of a large number of ions and gases (Table 2) (12). Three examples of commonly used electrodes, ammonium, fluoride, and oxygen, are briefly described below. Although most of these have applications in industrial processes, several measure biologically important ions with sufficient sensitivity and selectivity to be of use in bacteriological investigations. Ion-specific membrane probes usually have a solid crystalline or compressed crystalline pellet, or a plastic polymer impregnated with ion-carrier organic molecules. In addition, there are now many proprietary designs in use. The majority of these electrodes are constructed of polyvinyl chloride, glass, or epoxy. Electrode response is, broadly speaking, the result of an ion-exchange process, and potentials follow the Nernst equation or one of its expanded forms. The response times for many ion-specific electrodes are much slower than for pH electrodes; it can take from 30 s to 2 min for a stable reading to be obtained, especially at very low concentration ranges.

Calibration

Since there is no system for absolute calibration of probes, as there is for pH with the hydrogen electrode, it has been common practice to use dilutions of stock solution standards made up of a completely dissociating salt (18). Many concentration standard solutions are commercially available for a wide variety of ions. These stock solutions are tested by more expensive methods to verify the specific ion in each production lot.

Ion-specific electrodes generate a potential in response to the activity of a particular ion, not necessarily its concentration. Thus, for many types of electrodes when measuring within certain concentration ranges, it is necessary to adjust the total ionic strength of standard and unknown so-lutions to provide a stable activity coefficient for the particular ion to be measured. Reagents such as Total Ionic Strength Adjustment Buffers can be used to bring the ionic strength within proper parameters for an ion-specific electrode (18).

Sensitivity

With respect to sensitivity, various manufacturers' specifications show that measurements can be made over the range of 10^{-6} to 10^{-8} M to 1 M for certain ions. However, at lower concentrations some time is required to establish equilibrium with the electrode. Seldom is the sensor highly specific. In fact, interference with other ions is quite common. For instance, a calcium electrode gives 10% response to equivalent amounts of Pb^{2+}, Zn^{2+}, Cu^{2+}, and Fe^{2+} in the 10^{-5} to 10^{-6} M range. Chloride electrodes respond similarly to iodide and bromide, i.e., 10% response at 10^{-5} to 10^{-6} M. On the other hand, the K^+ electrode is quite insensitive to Na^+, Li^+, Cs^+, and H^+. The specific details of sensitivity and selectivity depend on the particular electrode being used. Refer to the manufacturer's literature for detailed information.

17.2.2.1. Ammonia Electrode

The ammonia electrode uses a gas-permeable membrane (12, 18, 19) through which ammonia diffuses to react with the filling solution:

$$NH_3 + H_2O = NH_4 + OH^-$$

The hydroxide level of the filling solution is then measured against a reference electrode. The electrode is specific for ammonia; however, the ammonium ion concentration can be measured after deprotonation by addition of NaOH to pH 11. Only volatile amines interfere, in the range from 1 $\times 10^{-7}$ to 5 $\times 10^{-7}$ M. Nitrate and nitrite can also be measured after reduction to ammonia, and urea-N can also be measured in a special adaptation of the ammonia electrode in which the enzyme urease is immobilized on the membrane (17). Alternatively, the samples may be pretreated with urease and the normal ammonia electrode may be used for measurement.

17.2.2.2. Fluoride Electrode

The fluoride electrode is typical of a solid-state type of sensor. The membrane consists of a single lanthanum fluoride crystal that is doped with europium fluoride to increase the conductivity. Some interference from hydroxide anion occurs, but this can be minimized by maintaining the pH within the 5 to 8 range and using Ionic Strength Adjustment Buffers where appropriate. When measuring F^- at very low concentrations, it may be necessary to reduce interference by taking measurements in plastic or Teflon containers instead of glass.

TABLE 2 Ion-selective electrodes[a]

Type	Examples of ions and gases detected
Glass membrane	H^+, Na^+
Solid state	Ag^+, Br^-, Cd^{2+}, Cl^-, Cu^{2+}, CN^-, F^-, I^-, Pb^{2+}, S^{2-}
Gas sensing	NH_3, CO_2, SO_2
Polymer membrane	NH_4, Ca^{2+}, BF^{4-}, Li^+, NO_3^-, K^+

[a] Data compiled from manufacturers' specifications.

17.2.3. Oxygen Polarograph

A platinum electrode at −0.5 to −0.8 V consumes oxygen according to the equation

$$O_2 + 2H_2O + 4e^- \rightarrow 4OH^-$$

The equation shows that the process occurs with the flow of electrons or current toward the electrode. This current flow through a resistor appears as a voltage across the resistor. The "potential" of such a half-cell, when used with a standard Hg/Hg_2Cl_2 or $Ag/AgCl$ (silver wire in chloride-containing solution) half-cell, can be measured if the two are connected by a salt bridge. Alternatively, the current may be measured with a galvanometer or the voltage can be measured after amplification with a potentiometer, and these do not require a reference electrode.

The platinum electrode may be stationary, rotated, or oscillated to minimize diffusion gradients. However, such electrodes tend to be poisoned in biological measurements. A popular solution to this problem is the Clark oxygen electrode (20), in which the platinum electrode is covered with a gas-permeable membrane. Whereas the platinum electrode measures the number of oxygen molecules present at its surface as governed mostly by collision theory, the Clark electrode response is proportional to the rate of diffusion across the membrane isolating the electrode. In this arrangement a gradient across the membrane results from the zero oxygen concentration at the electrode surface where oxygen consumption occurs. Fortunately, the rate of diffusion is linearly dependent on oxygen concentration or partial pressure. The current output is also dependent on the platinum cathode area.

A plot of current versus polarizing voltage shows a flat or nearly constant response in the region of −0.5 to −0.8 V. Therefore, there should be a polarizing voltage adjusted to the center of the constant-current region. The behavior in response to polarizing voltage is the major distinction of the oxygen electrode.

17.2.3.1. Effects of Temperature

The sensor current is quite sensitive to temperature because the membrane material permeability has a high temperature coefficient (about 3 to 59%/°C). In addition, oxygen solubility varies with temperature. For this reason, good temperature control and adequate equilibration are required.

17.2.3.2. Calibration and Calculations

Calibration is based on the known oxygen solubility of a solution under defined conditions (20). Zero oxygen level or "bleed current" can be defined as the response in the presence of added sodium dithionite. Full response is set at 100% with an oxygen-saturated solution. The oxygen concentration at the operating temperature is obtained from oxygen solubility tables in the *Handbook of Chemistry and Physics* (15) or a similar reference publication. The oxygen content of solutions to be used as working standards can then be measured. For most work the actual oxygen concentration of standards is not needed; rather, a reproducible concentration is necessary to set the instrument. The measurements needed are the relative differences between experimental samples, or assays can be monitored over time in a continuous manner. For example, if Ringer's solution contains 5 µl of O_2/ml when air saturated, 3 ml of solution giving an instrument response of 92% saturation contains 13.8 µl of O_2 (3 × 5.0 × 0.92). Rates of oxygen consumption or evolution are either calculated by taking measurements at two times or obtained from the slope of plots of percent full scale versus time involving several readings or continuously recorded in a real-time assay.

Unfortunately, it is unlikely that reaction mixtures under study have the same composition as the oxygen standard, and hence, there will be some error. However, the error can be minimized by maintaining a low total solute concentration. When solutions of greater oxygen concentration are needed, increased partial pressures can be generated by saturation of standards and working solutions with oxygen or oxygen-nitrogen mixtures.

The oxygen content of standards and reagents is influenced by temperature and altitude. Barometric pressure fluctuations seldom change oxygen concentration by more than 3%. Correction of the oxygen solubility coefficient for barometric change involves the relationship

$$\frac{\text{observed barometric pressure}}{760}$$
$$\times \text{ solubility at 760 mm Hg (101.3 kPa)}$$

17.2.3.3. Operating Information

At the beginning of an experiment, the probe should be tested for (i) plateau setting of polarizing voltage, (ii) response time for a step change in polarizing voltage, (iii) noise, and (iv) drift. The lack of horizontal plateau indicates a dirty electrode; however, some slope can be tolerated in many instances. The response time for a step change in polarizing voltage should be 1 to 1.5 min. Noise may result from poor connections and grounding, a damaged membrane (folds, holes, KCl crystals, or drying), or poor contact with the silver electrode, in which case cleaning with ammonia is recommended.

A properly functioning probe should have less than 2% drift per h. Drift may be caused by atmospheric changes or by contamination with organic material. Thus, proper care and cleaning of the probe are essential. In one study, reproducible results were obtained when the probe and receptacle were washed with reagents used in the test.

17.2.3.4. Assembly of Electrode Membrane

The demanding aspect of the Clark electrode is attaching the membrane. Directions and spare membranes are furnished by manufacturers of the probe. Care must be taken in assembly to ensure that the membrane is smoothly stretched over the probe, that there is a true seal at all points, and that no air bubbles are trapped on the electrode side of the membrane. To obtain good wetting in assembly, add 3 or 4 drops of Kodak Photoflow per eyedropper bottle of KCl used in assembly.

17.3. CHROMATOGRAPHY

Chromatography and related separation techniques depend on many kinds of distribution processes, which take place repeatedly in a small local environment (22, 33, 34). Therefore, the operations of ion-exchange, solvent partition (including reversed-phase), adsorption-desorption, hydrophobic-interaction, ion pair and hydrogen-bonding, and affinity (association-dissociation) chromatography, as well as the operationally related gel permeation (size exclusion) chromatography based on a sieving principle, can all be carried out with the same laboratory equipment to make separations based on widely different properties of the solute molecules.

17.3.1. Ion-Exchange Chromatography

Sorption of solutes is based on ionic charge, although other types of bonding may have some influence on separation. Charged groups $(-X)$ on a permeable matrix are able to exchange sorbed ions with those in solution.

$$-(X)SO_3Na^+ + K^+Cl^- \quad -(X)SO_3^- K^+ + Na^+Cl^-$$

Matrix materials include cross-linked, substituted, or derivatized resins such as polystyrene or polymethacrylate, as well as derivatized carbohydrate containing natural polymers such as cellulose, dextran, or agarose.

Exchangers are available with a great variety of functional groups, support materials, "purity," porosity, capacity, size, shape, selectivity, and stability (34–36). It is therefore necessary to consult technical literature from manufacturers for the specific information needed, for instance, Dow Chemical Co. for Dowex resins, Rohm and Haas Co. for Amberlites, Bio-Rad Laboratories for a wide variety of exchangers, and Amersham Biosciences for Sephadexes and Sepharoses. A more extensive listing can be found in reference 35.

17.3.1.1. Resin Exchangers

Many synthetic resins are available with strong and weak acidic and basic exchange groups as well as with different degrees of cross-linking. Lower cross-linking results in higher permeability and water uptake and an ability to interact with larger molecules. For instance, for Bio-Rad (Hercules, CA) resins of the AG 1 series, 2% cross-linking produces an exclusion limit of 2,700 Da whereas 8% cross-linked resin excludes 1,000-Da molecules. Thus, high cross-linking favors sorption of substrates and products of enzyme reactions while discriminating against sorption of the enzyme involved. Ion-exchange wet-volume capacity and selectivity increase with higher cross-linking and also increase as the particle size decreases. Thus, fine-mesh exchangers have increased resolution per bed length simply because of the increased number of exchange cycles taking place. However, the increased packing of small particles decreases the flow rate, which in turn requires increased operating pressures. Newer, more rigid polymer resins, ceramic encapsulation of hydrophilic gels, polyacrylamide-based gels, or other proprietary designs can help to overcome this flow restriction by making the column bed less compressible. In addition, strong anion, cation, and chelating resin beads have been cast into membranes with a pore size of 0.45 μm. Such membranes, when fabricated into convenient plastic housings, offer the advantages of speed and ease of use for certain applications.

Sorption, purification, and fractionation are major applications of exchangers. In sorption, a desired ion is selectively removed from extraneous material by ion exchange. In purification, the ion desired is selectively sorbed and eluted while other ions are not sorbed or are retained. Fractionation is the process of separating several similarly charged ions by introduction of solvent-solute systems which progressively elute sorbed ions of increasing charge. The charge state is often a function of the pK of dissociating groups and the pH of the environment.

As a rule of thumb, resolution is related to bed length, whereas for a given resolution, capacity is related to bed cross-section. Since binding of soluble ions is stoichiometric and reversible, the amount of resin needed can be calculated from its published capacity. In general, the strength of binding of inorganic ions is directly related to valence (number of charges) and inversely related to atomic number. For weak acids and bases, strength of binding is related to the pK of the ionizing group if the pH of operation is in the dissociation range.

Exchange of counterions and deionization. Since ion exchangers are charged with a counterion (for instance, for Dowex-1 this might be OH^-, Cl^-, or formate), it is possible to exchange virtually completely an ion in the solvent with the charged counterion. For instance, chloride can be exchanged for OH^- with Dowex-1 and H^+ can be exchanged for Na^+ with Dowex-50. Therefore, a major application of exchangers is the deionization of uncharged solutes such as urea and polyols. This can be accomplished by serial passages of the sample through cationic and anionic exchangers or through a mixed bed. However, the mixed bed is more difficult to regenerate because the cationic and anionic exchangers must first be separated on the basis of density difference. Alternatively, the spent exchanger is discarded.

Selective removal of divalent and trivalent cations is accomplished by specialized resins which have EDTA, iminodiacetate, or related groups to chelate the metal ion. Chelex (Bio-Rad), Prosep chelating columns (Millipore, Bedford, MA), and other products perform this function.

Concentration. As a consequence of the high affinity for ions plus the fact that fresh resin is encountered with movement of the sample through the column, it is possible to bind ions from a large volume of dilute solution. The sorbed ions can then be recovered in a small volume of eluant containing an ion of higher affinity or of higher concentration. In this way, any number of common and trace inorganic elements as well as radionuclides and trace organic compounds have been concentrated for analysis.

Purification and fractionation. If two ions have different affinities for an exchanger at a given pH, they can be separated by finding conditions which sorb and/or elute one ion but not the other. In such extreme cases, there is a "frontal" elution of each ion. In cases when the affinity is similar, separation is still possible by finding conditions which selectively elute the ion. Development of the column with eluants then leads to a gradual separation. If the column length is sufficient, separation can be attained. However, diffusion of ions limits the ability of increased column length to resolve the components. Changing pH or ionic strength elutes sorbed ions of different pK values or of increasing affinities.

Gradient elution. Elution of sorbed molecules by continuous change of eluant composition is a useful way to maximize the separation potential of a chromatographic system. Rather than making step changes in concentration by application of different eluants, the change is made to be continuous within given limits. Devices and procedures range from the very complicated using expensive microprocessor-controlled pumps that produce complex gradients, to the relatively simple linear concentration gradient involving mixing of two components. In recent years the simple linear gradient has been used most frequently. For this purpose, two reservoirs are connected so that the outlet of one is fed into the other. Another outlet of the second reservoir feeds the column. A mixer in the second vessel keeps the eluant composition homogeneous. When both vessels are open to the atmosphere and each vessel has the same cross-sectional area, the eluant composition will change linearly from the limit in the second (mixing) reservoir to the limit in the first. The steepness of the gradient is determined by the volume in the two reservoirs relative to the column bed. Many varieties of specially designed simple gradient formers are available and consist of two reservoirs connected through a valve-containing tube or orifice with an outlet from one side going to the column.

17.3.1.2. Cellulose, Dextran, and Agarose Exchangers

For separation of macromolecules, natural polymers have been modified to contain weak acid or base functional groups (22, 27). These matrices also have capacities and other properties generally more suitable than resins for fractionation of biomolecules. With dextrans and agaroses, cross-linking and particle size distribution can be controlled, and this minimizes shrinkage and swelling and produces exchangers with high capacity for macromolecules as well as high and uniform flow rates. Generally, agaroses better maintain bed volume when ionic strength increases and when the pH is outside the neutral region.

Since proteins are amphoteric and have a large number of charges, often in clusters, it is possible to treat them as cations or anions by choosing a pH for ion-exchange separations that is at least 1 unit above or below the isoionic point or the pI value. Cycles of binding and elution make use of a pH change which can cause disappearance of ionic groups, e.g., carboxyl or amino groups on the protein or on the exchanger. Alternatively, elution is effected by displacement with another ion.

The DEAE derivative of cellulose, dextran, or agarose is a weak anion exchanger (pK of about 9.5) (Table 3). Hence, in the neutral or lower pH region, the amino group is protonated and binds proteins and peptides if the pH is near or above their pI values. The carboxymethyl derivatives of the same supports are cationic exchangers with a pK of about 3.5. Separations are generally carried out above pH 4. When the pH of the separation is below the isoelectric point of the protein and above pH 4, sorption is efficient. Other cellulose and dextran derivatives used in ion-exchange chromatography are listed in Table 3.

Practical Considerations

Successful use of exchangers rests on the judicious choice of a number of parameters. For the exchanger these include the type of charged group, matrix, particle size and shape, and porosity; for the eluant they include the buffer type and pH; for the system they include the column dimensions, flow rate, and elution program. In addition, attention must be paid to exchanger packing, sample addition, maintenance of low sample viscosity, and column design, since these also determine the degree of resolution. Treatment of these aspects in adequate detail is beyond the space available here, and hence, only general guides for each are given. Publications by Amersham Biosciences (22) for exchanger column media provide more details.

In general, uniform spherical dextran beads (e.g., Sephadex and Sepharose) give higher flow rates, although the cellulose derivatives now available also perform well. Smaller bead or particle sizes give lower flow rates and increased resolution. Greater porosity increases the capacity for a given molecule and is particularly important for work with macromolecules. The choice of exchanger is based on the ionic forms to be separated as described above. The pH chosen should be at least 1 pH unit removed from the isoelectric point or the pK of the dissociating group. When possible, cationic buffers (alkylamines, ammonia, barbital, imidazole, and Tris) should be used with anionic exchangers; anionic buffers (acetate, citrate, glycine, and phosphate) are used with cationic exchangers. The recommended pH range for DEAE-Sephadex is 2 to 9, and that for carboxymethyl-Sephadex is 6 to 10. Ionic strength influences the binding and therefore the capacity for ions. A relatively high ionic strength of ca. 0.1 M is recommended, but the concentration should be somewhat lower than is needed to elute the ions desired.

Column height determines the resolution achieved. When the zone of absorbed ions is 1 to 2 cm, a 20-cm bed height is recommended as a starting point. At fixed height, the column capacity is linearly related to the cross-sectional area.

Elution is effected either by changing the pH to cause the ionic charge to diminish or disappear or by increasing the ionic strength to the point at which the buffer ion displaces the ion of interest. These can be done by either stepwise or gradient change. For anion exchangers the pH is decreased or the ionic strength is increased; for cation exchangers the elution is effected by increasing the pH or the ionic strength.

Either the sample total ionic content or column size must be determined first, because available capacity is the property of a specified volume of exchanger for each exchanger type. The ionic composition of the sample ideally should be that of the starting buffer. When necessary, the ionic composition is changed to that desired by means of gel filtration on Sephadex G-25, dialysis, or dilution. The sample volume is of minor importance for all elution programs except when using only the starting buffer for development. The sample must be added without disturbing the bed and without becoming distributed in the oncoming developing buffer.

As with other types of chromatography, it is essential to use glassware designed for ease of packing, addition of eluant and sample, application of pressure (if needed), regulation of flow rate, prevention from running dry, and low holdup volume above and below the exchanger bed. It is also necessary to establish criteria for achieving the objective in mind and to analyze the effluent to determine the performance of the column.

17.3.1.3. Chromatofocusing

Isoelectric focusing (IEF) by ion exchange, or chromatofocusing, is a specialized type of ion-exchange chromatography in which a pH gradient is used to elute bound proteins instead of a salt gradient (52). A pH gradient along the length

TABLE 3 Chemical groups used in ion-exchange chromatography[a]

Ion exchanger	Chemical name	pK
TMAE	Trimethylammoniumethyl	>13
QAE	Diethyl-2-hydroxypropyl aminoethyl	12
DEAE	Diethylaminoethyl	9.5
TEAE	Triethylaminoethyl	9.5
DMAE	Dimethylaminoethyl	8–9
CM	Carboxymethyl	3.5–4
P	Orthophosphate	3, 6
SP	Sulfopropyl	2–2.5

[a]Data from various manufacturers' specifications.

of the column can be made by gradually mixing a buffer with one pH with a buffer with a different pH, or the gradient can be established when a column that is equilibrated at one pH is switched to a buffer of a different pH. When the pH at a particular cross section of the column is equal to the pI of a protein, its charges are neutralized and it elutes from the column matrix. As the pH gradient moves down the column, the bound protein continues to elute, resulting in a focusing effect or banding for each particular protein. The net result is that proteins elute from the column according to their pI values. Chromatofocusing is especially suited to high-resolution methods such as fast protein liquid chromatography using a Mono-P column (Amersham Biosciences) or by ion-exchange high-performance liquid chromatography (HPLC) (79).

17.3.2. Adsorption Chromatography

Adsorption involves van der Waals interactions as well as electrostatic, hydrophobic, and steric factors. Adsorption of a single component is typically described by Langmuir's adsorption isotherm, a rectangular hyperbola in a plot of solute concentration versus amount adsorbed. The same curve describes the saturation behavior of substrate for an enzyme and the "law of diminishing returns" among others. Normal phase surface sorption and elution have been successful with silicic (Celite) acid and magnesium silicate (Florosil), kieselguhr (diatomaceous earth), aluminas, and charcoal. Proteins and nucleic acids have been purified by either batch treatment or column chromatography with alumina Cγ or various forms of calcium phosphate including hydroxyapatite and calcium phosphate gel (53, 66).

17.3.2.1. Types of Adsorbant

Silicic acid has been used as adsorbant for purification of simple and complex lipids, fatty acids, sterols, phenols, and hydrocarbons. Elution is effected by changing the solvent composition from partially nonpolar toward polar character by use of a concentration gradient. Neutral alumina (pH 6.9 to 7.1) separates steroids, alkaloids, hydrocarbons, esters, organic acids and bases, and other organic compounds in nonaqueous solution. Basic alumina (pH 10 to 10.5) has similar properties in adsorbing compounds from organic solvents. However, with aqueous media it acts as a strong cation exchanger, sorbing basic substances such as amino acids and amines. Acid-washed alumina (pH 3.5 to 4.5) acts as an anion exchanger in aqueous medium.

Adsorbents for purification of macromolecules in aqueous solution are alumina Cγ, calcium phosphate gel, and hydroxyapatite. Some of these do not lend themselves well to column chromatography and hence are used with batch methods. Column methods for hydroxyapatite are well known (53). Hydroxyapatite columns are typically equilibrated with low-ionic-strength phosphate buffers for separation of basic proteins, NaCl for separating acidic proteins, or $MgCl_2$ or $CaCl_2$ for acidic proteins that do not bind to NaCl-equilibrated columns. Bound proteins are eluted with gradients of NaCl, phosphate, or a combination of both. For example, commercially purchased papaya lysozyme can be purified on hydroxyapatite columns using NaCl or phosphate gradients (53).

17.3.3. Reversed-Phase and Hydrophobic Chromatography

Both reversed-phase and hydrophobic interaction chromatography separate molecules on the basis of hydrophobicity (28). For reversed-phase work, an alkyl or phenyl arm is bonded to silica and organic solvents are used as eluting agents. This method is useful for solutes that are not adversely affected by an organic solvent, usually small molecules. However, some enzymes and bioactive peptides have been purified by elution from the matrix with a water-acetonitrile gradient containing trifluoroacetic acid.

In hydrophobic chromatography, a phenyl or alkyl group is attached to a polar matrix such as Sepharose. Hydrophobic interaction between the matrix ligands and hydrophobic regions of a protein is enhanced by raising the hydrophilicity of the aqueous mobile phase with salt. Decreasing the salt concentration, and when necessary followed by increasing the hydrophobicity of the eluant with an alcohol or glycol, elutes the sorbed proteins. This procedure is especially useful following an ammonium sulfate precipitation in which the dissolved precipitate contains a relatively high concentration of ammonium sulfate. The salt is useful in enhancing binding of proteins, but it must also be removed before adsorption or ion-exchange chromatography can be used as a next purification step.

17.3.4. Liquid-Liquid Partition Chromatography

When a two-phase solvent system is set up, a large number of solutes, partially soluble in each phase, reach an equilibrium distribution between the two phases, which reflects the relative solubility in each phase. This has led to useful solvent extraction procedures, both batch and continuous. The commercial Craig and Podbilniak countercurrent extractors are examples of these techniques at their best. However, the ability to separate similar materials is limited by the number of distribution cycles one can make. Fortunately, when one phase of a two-phase system is immobilized as a coating on a small particle, the distribution process can occur repeatedly in a column or paper chromatographic arrangement. Thus, paper and cellulose column chromatography separate by partition between an aqueous phase coating on the cellulose fiber and a less polar phase which flows through the interstitial spaces. Silicic acid is another support which has been widely used in partition chromatography, both in column and in thin layers.

Gas-liquid chromatography and some forms of HPLC also separate by solute partition. Both gas-liquid chromatography and HPLC have been very highly developed in terms of equipment, instrumentation, column packing matrices, and procedures, originally for analytical purposes, for which high resolution is desired. These specialized forms of partition chromatography warrant separate treatment (see sections 17.3.5 and 17.3.6, respectively).

17.3.5. Gas-Liquid Chromatography

Gas-liquid chromatography (gas chromatography [GC]) has a long history in the separation, identification, and quantitation of compounds that are volatile or can be made volatile at temperatures up to 450°C. Many compounds of biological interest exhibit these characteristics. One of the first reports by James and Martin in 1952 (58) involved volatile fatty acids. Separation was improved with the introduction of capillary columns in 1957. GC is widely used because of its ability to (i) separate diverse mixtures, (ii) separate chemically and physically similar components, (iii) analyze mixtures with a wide variety of components, and (iv) achieve rapid analysis.

17.3.5.1. Principles

Separation by GC is defined as the partitioning of a solute between a mobile phase of gas and a stationary phase of polar or nonpolar material that is liquid at the operating temperature of the system. The material is coated on inert particles packed into a column or bound to the interior surface of a capillary column. The components have various affinities for the liquid phase and hence move or are retained to various degrees. This retention of each component or analyte is a function of the chemical nature of the analyte, column-packing material (liquid phase), temperature, and flow rate of the mobile phase. Chemical reaction between the analyte and carrier is minimal and is usually disregarded. Each solute moves through the column as a band with a specific speed, and the bands exit the column and enter the detector distributed as symmetrical peaks.

Four parameters, resolution, efficiency, selectivity, and capacity, are considered in the effective separation of the sample of interest. Under specific operating conditions, each substance has a net retention time, $t_{R'}$, in the stationary phase. For a given, constant flow rate, the total retention time, t_R, is the sum of the net retention time and time spent in the mobile phase, (t_0); thus,

$$t_R = t_0 + t_{R'}$$

All analytes should have equal t_0 values under the same operating conditions. The term t_0 is also known as dead time. Its value is obtained by injecting an inert gas such as ethane and recording the elution time under operating conditions. Because the flow rate of the mobile phase is a factor, retention volume, V_R, is often another term calculated. The liquid phase is the major factor in the behavior of the analytes being analyzed, so choosing the proper column for a specific sample is critical.

Because of the separation technique involved, GC is readily described as analytical distillation, and the performance of the column with the particular sample can be measured by using the concept of theoretical plates. Separation or resolution (R) of the components in the sample is calculated by the following equation:

$$R = \frac{n^{1/2}(\alpha - 1)\kappa}{4\alpha(\kappa + 1)}$$

where n is the number of theoretical plates, α is the relative retention time, and κ is the capacity (or partition) ratio. The number of theoretical plates represents efficiency and is determined by

$$n = 5.545\, t_R^2/W_h$$

where t_R is the retention time of the peak of interest and W_h is the peak width at half-height of peak of interest.

The relative retention time, α, is the ratio of retention times for two solutes for which separation is desirable. The capacity or partition ratio, κ is derived from the ratio of the corrected retention time, t_R, of the solute and the column dead time.

Efficiency is a measure of the broadening of the solute plug or front as it moves through the column and generates a peak at the detector. A number of methods are available to calculate the efficiency of a column separation. A common method is to assume a Gaussian peak shape and to calculate efficiency in terms of a theoretical plate number, n (above). Other efficiency methods that are beyond the scope of this manual include relating efficiency to the average linear velocity of the mobile or gas phase or, more accurately, relating retention to the volume of the gas phase. These topics are examined in depth in specialized texts on chromatography in general and GC in particular (44, 78).

Selectivity is the third criterion of the separation, and this refers to the interactions between the analyte(s) and the chromatographic system. In GC this factor is controlled by the liquid component of the stationary phase. Components controlling selectivity include a mixture of physicochemical interactions that may be dispersive and dipolar. The degree of polarity of the stationary phase is given by the McReynolds constant. This constant is defined by using a set of reference compounds and comparing the retention of the reference compounds on a "standard" liquid phase with that in the liquid phase in question. The relative retention of an analyte pair, expressed as α (above), is a common expression of selectivity. If α is greater than 1, separation is achievable. Selectivity is a major consideration in analyses involving packed columns which are usually rated at low efficiency.

The final parameter of separation is capacity, and this is affected by the partitioning of the solute between the mobile and stationary phases. The liquid phase and the supporting solid phase or carrier both contribute to the separation of the analyte and to the total sample-carrying capacity of the system. Manipulation of the liquid- and solid-phase polarity can be used to optimize analysis. Absorption to the solid phase can be used to enhance detection of trace components in gas analysis (59). This is a dynamic equilibrium defined by the distribution coefficient, K. A complete discussion of this factor is available in texts on chromatography (44, 78).

17.3.5.2. Instrumentation

There is a great variety of instruments available, but all consist of a few major components: injectors and inlets, column ovens for temperature control and programming, detector, data collection and handling, and a source of gas for the mobile phase. The choice of column is the first step in setting up a GC system. A survey of published methods shows that most bacteriologists and chromatographers use commercial columns rather than preparing their own. Choosing the proper column from the vast array of those available is very difficult. The two types of columns used differ in the attachment sites of the liquid phase. Packed columns contain an inert carrier, such as silica, which is coated with the liquid-phase material. Capillary and large-bore columns lack a carrier and have the liquid phase sheathing the interior walls of the column.

The advantages of packed columns are lower expense, the ability to accept larger injection volumes (1 to 10 μl), the capacity to analyze samples with high concentrations of analytes, the ability to perform in the presence of interfering compounds such as salts, and the variety of liquid phases and inert carriers available. The shortcomings of packed columns include lower selectivity, especially for trace analysis, difficulty in separating complex mixtures, and requirements for higher volumes of the carrier phase (gas).

Capillary and large-bore (megabore) coated columns limit or eliminate these problems. The megabore columns have a greater internal diameter than the capillary columns but perform in a similar manner. Capillary columns are used with complex mixtures at low concentrations and when similar compounds are present (70). The greatest disadvantage is their expense.

The selection of the liquid phase in the column is not controlled by any specific guidelines, but a common rule of thumb is to separate analytes with liquid phases that are chemically similar. For instance, use a polar liquid phase for polar analytes. The suppliers of packed and capillary columns furnish a variety of systems for most compounds of interest to bacteriologists.

The most commonly used detector system is flame ionization detection (FID). The sample enters a hydrogen-air flame and is burned to its component ions. All the ions are collected and measured at a collector electrode that surrounds the flame. The collector has an applied potential of -150 V or greater to ensure efficient ion collection. This detector responds to virtually all organic compounds. It does not respond to the rare gases, oxygen, nitrogen, carbon monoxide, carbon dioxide, hydrogen, hydrogen sulfide, halogen, hydrogen halides, ammonia, or oxides of nitrogen. Sensitivity depends on the gas flow rates, for both the flame gases of air and hydrogen and the carrier gas. The detector is destructive and mass sensitive (its response depends on the sample component entry rate). This allows a broad range of sensitivities. An example is the analysis of hydrocarbons with a detection limit of 10^{-12} g of carbon per s. The concentration range in a sample can be 0.1 ppm to 100% molar (78).

The thermal conductivity detector is one of the oldest and simplest detection systems available. The sample exits the column and passes over an electrically heated element in a thermostat-controlled metal block. The changes in the thermal conductivity of the gas stream as a result of the sample components cause changes in the heat loss of the element, and these changes are recorded as the signal (peaks). Since the signal results from a minute change in temperature, dual elements are used to compare changes in signal as a result of environmental temperature, changes in block temperature, and changes in the temperature, pressure, and flow rate of the carrier gas. This detector has a broad sensitivity range because the component in the carrier gas changes its thermal conductivity. The detection limits are not as low as for FID; for example, hydrocarbons can be determined in concentrations ranging from approximately 50 ppm to 100% molar. Analysis of permanent gases is a common use of thermal conductivity detectors.

The electron capture detector reports changes in a standing current generated by a β-emitting source such as ^{63}Ni in an ionization chamber creating a standing current.

$$\beta + N_2 \rightarrow N_2^+ + e^-$$

The analytes capture the electrons to form stable ions, and some ions can further react with charged carrier gas.

$$\text{Solute (X)} + e^- \leftrightarrows X^-$$

$$X^- + N_2^+ \rightarrow \text{uncharged products}$$

The changes in the standing current so created generate the signals. Electron capture detectors are now available without radionuclides to generate electrons. Polyatomic compounds such as alkyl halides and conjugated carbonyls capture electrons. Many compounds of interest to microbiologists, such as hydrocarbons, alcohols, and aldehydes, do not capture electrons and hence are not detected.

Mass-selective detectors (MSD) and dedicated mass spectrometers have also been coupled to the gas chromatograph (47, 57, 83). The MSD can identify and quantify analytes on the basis of mass, charge, and abundance of the analyte ion and its fragmentation ions. The most common of the "small" (benchtop) MSDs is the negative-ion quadrupole system (model 5970; Hewlett Packard Co.). This system generates a stream of ions derived from the analytes, with the mass analysis being accomplished at a detector while changing the voltage applied to the quadrupoles. The masses of the ions in the gas stream are scanned, and signal or peaks are generated on the basis of the abundance of each mass level (in atomic mass units). Because of the availability of computer data-handling systems and pattern recognition software, the compounds in the sample can be identified by comparison of the mass spectrum of the analyte(s) and its fragments can be compared to scans of compounds in a computer library. The major limitation to the systems is the requirement for a strong vacuum in the detector, so only capillary columns with a lower carrier gas flow rate can be used.

17.3.5.3. Sample Preparation and Derivatization

Some samples, e.g., culture supernatants, cell extracts, and biological fluids, may contain interfering substances. These include highly polar and/or highly reactive components. Preliminary separations or preparations of the samples may be required. The compounds of interest can be extracted with a nonpolar solvent, or solid-phase extraction systems can be used to clean up the sample (84). An internal standard should be added to the sample prior to any treatment, to permit calculation of the recovery efficiency of the pretreatment. The internal standard consists of a compound chemically similar to the analytes of interest but not present in the sample. This material is added at a known concentration estimated to be near that of the analytes. The recovery percentage of this internal standard can be used to correct for any sample component loss. The procedures for the pretreatment or extraction can be found in general references and from the vendors of the extraction material, as for the solid-phase extraction.

Some analytes are of insufficient volatility and hence must be converted to a more volatile derivative, e.g., methyl esters of fatty acids, trimethylsilyl ethers of sugars, and methyl esters and N-acyl forms of amino acids and peptides. These methods and others are available from references 50 and 74 and from vendors supplying the derivatization compounds.

17.3.5.4. Metabolic Products

Bacteria generate end products during their growth on various substrates. Aerobic bacteria generate only carbon dioxide and water as products, but fermenting anaerobic bacteria produce end products including alcohols (C_1 to C_5), volatile and nonvolatile acids (C_2 to C_6), aldehydes, and ketones. Published procedures (82) allow the determination of end products (alcohols, volatile fatty acids [VFAs], lactic acid, and acetone) to assess the extent and nature of fermentation and aid in the identification of anaerobic bacteria involved. There are now completely automated systems to identify microbes, and at least one of these uses VFA analysis. The automated systems are discussed in more detail in section 17.3.5.6.

17.3.5.5. Gas Production and Consumption

Especially during anaerobic growth, bacteria can consume or produce various gases. The presence and concentration of these gases can be determined readily by GC. One method involves the use of the thermal conductivity detector and a column packed with a solid-phase carrier material, such as 100/120 Carbosieve S-II (Supelco Inc.). Because there is no liquid phase, the chromatography is a form of adsorption separation rather than partitioning. Pack the material into a 6- to 10-ft. (183- to 305-cm) stainless-steel column (diameter, 1/8 in. [0.32 cm]). Hold the detector at 250°C, and program the column oven to hold at 35°C for 7 min and then to increase the temperature to 225°C at a rate of 32°C/min. When analyzing gases other than hydrogen, use helium at a flow rate of 30 ml/min as the carrier gas. For determination of hydrogen, the carrier is nitrogen at the same settings. Collect and inject 500-μl samples by using a Pressure-Lok gas syringe. This protocol is based on our experience in measuring production of hydrogen, methane, and carbon monoxide by pure-culture and mixed-culture fermentations from marine and hypersaline ecosystems, as well as the consumption of hydrogen, methane, ethylene, carbon monoxide, and ethane in other experiments.

17.3.5.6. Membrane Lipid Analysis for Identification of Bacteria

The membranes of bacteria contain lipids as a major component. The fatty acid composition of these lipids is characteristic of the different taxa (also see chapter 15.3.4). Specific fatty acids of these lipids can be used as biomarkers, and the quantitative recovery and analysis of these fatty acids can be used to define the members of a microbial community (83, 87).

A number of methods for the analysis of whole-cell lipid fatty acids have been developed (48, 49, 63, 70, 74, 77, 87). The lipids are extracted from the cells with organic solvent, and specific classes of lipids such as the phospholipids are isolated by silicic acid chromatography (82). Following saponification of the phospholipids, the fatty acids are made volatile by an esterification by methanolysis. The methyl esters of the fatty acids are separated and detected by capillary GC with FID. The sensitivity of this method, based on palmitic acid, is approximately 2×10^{-12} mol (74).

Samples with low biomass, such as biofilms, ground water, and subsurface soils, usually require greater sensitivity. Picomolar concentrations of phospholipid fatty acids can be measured by using GC and chemical ionization mass spectrometry (83). A combined hardware (Agilent Technologies gas chromatograph) and software (MIDI Inc., Newark, DE) system identified as the Microbial Identification System (MIS) is commercially available. The MIS analysis is based on fatty acids of 9 to 20 carbons that have straight and branched chains, saturated or unsaturated, and contain hydroxy and cyclopropane groups. MIS uses computerized high-resolution GC to determine the fatty acid composition of isolates of interest and then searches against libraries of known composition for species identification. Libraries include those available from MIDI and those generated by the user or other researchers (44). The sample must be processed by the operator before being introduced into the system. This involves (i) saponification with methanol, water, and NaOH, (ii) methylation with HCl and methanol, (iii) extraction with hexane and methyl t-butyl ether, and (iv) sample cleanup with dilute NaOH. The software included with the system allows the generation of dendrograms and two-dimensional association plots from the fatty acid data from the unknown organism compared with the library of known compositions of selected bacterial groups.

Whole-cell fatty acid analysis has been applied to the identification of nonfermenting gram-negative rods in clinical specimens (56, 75, 77), soil microbial communities (63), and members of the *Archaea* (70).

17.3.5.7. Biomineralization of Xenobiotic Organic Compounds

Microorganisms are able to mineralize most, if not all, known naturally occurring organic compounds (the doctrine of microbial metabolic infallibility), and this sweeping ability is being put to use by biotechnology. Manmade and naturally occurring toxic organic compounds contaminate soil, sediments, and waters at many locations. The use of microorganisms to convert these polluting xenobiotics to harmless end products is the goal of a number of microbiology laboratories. GC can be used to monitor the concentration of these contaminating compounds (41, 60).

Phenol is on the Environmental Protection Agency priority pollutant list and is the basic aromatic structure of many synthetic organic compounds. The concentration of phenol and possible degradation intermediates are detectable by GC with a 5% phenyl-methyl-silicone capillary column (25 m by 0.2 mm [inner diameter]) (41). For this determination, use nitrogen as the carrier gas and detect components by FID. Hold the temperature of the column at 50°C, inject 2 ml of sample, and increase the temperature to 230°C at 11°C/ min. Set the injector and detector temperatures at 250 and 300°C, respectively. Prepare the sample (1.0 ml) for injection by addition of 0.2 ml of 9 M H_2SO_4. Then, add 0.4 g of NaCl and 1 ml of ethyl ether. After agitation and centrifugation, remove the ether layer. Perform extraction for the possible intermediates, cyclohexanol and cyclohexanone, at pH 7.0. The ether phase (0.5 ml) is evaporated under nitrogen and derivatized with N,O-bis-(trimethylsilyl)trifluoroacetamide in acetonitrile (1:4). Add an internal standard, m-cresol, at a concentration of 150 mg/liter, and incubate the mixture at 70°C prior to injection. Run standards for tentative identification of peaks by retention time and for quantification. Verify the identification of peaks by GC-mass spectrometry. In this study, GC is also used to monitor production of methane gas, an end product of the consortium growing on the phenol.

GC has also been used to monitor numerous compounds that contaminate ecosystems such as aromatic and saturated compounds in crude oil (60), chlorinated and brominated phenols and biphenyls (39), and organic sulfur compounds such as methanethiol, dimethyl sulfide, and dimethyl disulfide (45).

17.3.6. High-Performance Liquid Chromatography

HPLC (also known as high-pressure liquid chromatography and high-performance liquid chromatography) is a chromatographic separation in which the mobile phase is not a gas (as in GC) but a liquid. Most of the theory described in the previous section on GC applies to HPLC. The compounds of biological interest that can be analyzed by HPLC are more varied than those amenable to GC investigation. This advantage over GC is due to the removal of the volatility requirement. The lower-molecular-mass compounds (<800 to 1,000 Da) must compete with the mobile solvent phase that is present in relative excess for sorption sites on the stationary phase. The larger molecules (>2,000 to 3,000 Da) are affected by exclusion mechanisms, so they are best analyzed by gel permeation chromatography (24) (see also section 17.3.8).

Examples of compounds that can be resolved well by HPLC include VFAs such as formic, acetic, propionic, and butyric acids; nonvolatile fatty acids such as pyruvic, lactic, fumaric, and oxalic acids; carbohydrates such as cellobiose, maltose, lactose, and glucose; amino acids such as glycine, alanine, leucine, and lysine (30); nucleotides and nucleosides (61, 72, 73, 85); secondary products of metabolism (82); and pigments such as carotenoids, chlorophyll, and bacteriochlorophylls (64, 65, 86). Examples for two of these are described in greater detail in sections 17.3.6.2 and 17.3.6.3.

Since the general introduction of modern liquid column chromatography in 1969, the method has undergone spectacular growth (24). The development of sensitive detectors, the transfer of knowledge and techniques from GC, the advances in pump and metering technology, and the availability of narrow columns with small particles (50 μm or less) have led to this rapid increase in use. High-efficiency columns are available (Supelco, Bellefonte, PA) with particles of 3 to 5 μm, which deliver >120,000 to 80,000 theoretical plates per meter, respectively. To maintain a reasonable flow rate of 0.1 to 5.0 cm/s or greater, the mobile liquid phase is operated at elevated pressures from 1,000 to 40,500 kPa (10 to 400 atm).

17.3.6.1. Instrumentation

HPLC systems are available either as a complete unit such as model 1098 (Hewlett-Packard, now Agilent Technologies, Palo Alto, CA) or in a modular form in which various components are interconnected. The two systems have the same major components. These include high-pressure liquid pumps (one pump for isocratic analysis, or two or more pumps for gradients of two or three solvents), pressure transducer and recorder, high-pressure sample addition valve, column, and detector system. Columns and detector systems are the components that are variable and are the areas of the fastest growth and technical improvements.

The column form dictates the mode of chromatography. An uncoated normal-phase packing consists of silica particles with pore diameter ranging from 3 to 20 μm and is used for adsorption chromatography. Bonded-phase columns have particles with a bonded liquid phase that incorporate one of the following groups: cyanopropylmethylsilyl, aminopropylsilyl, and 3-glycerylpropylsilyl for normal phase or trimethylsilyl (C_1), butyldimethylsilyl (C_4), octyldimethylsilyl (C_8), and octadecyldimethylsilyl (C_{18}) for reverse-phase separations. A number of resin-based columns are available that separate analytes by affinity, hydrophobic interaction, gel filtration, and ion-exchange chromatography. Microbore columns with a diameter of 1 mm are also commercially available. These columns have the advantage of lower sample volume requirement and enhanced sensitivity. The decreased diameter of microbore columns decreases the dilution within the column (at the same solvent velocity), thus increasing sensitivity.

No universal liquid chromatography detector exists, so most instruments require two or more detector types for broad applicability. A good approach is to have a "specific" detector such as a UV or fluorescence photometer, conductivity, or polarographic detector, as well as a "differential" detector, such as a refractometer. The introduction of the microbore column with decreased total flow has facilitated the direct connection to a mass spectrometry detector, which is a very powerful analytical tool.

The UV detector has a low sensitivity to temperature and flow rate fluctuations. It can use single, adjustable, or multiple-wavelength monitoring, the last by means of photodiode array detectors (see section 17.1.1). Spectrophotometers can be used with simple HPLC systems in which a fraction collector receives the column effluent. The major disadvantage of UV detectors lies in their specificity, but this limitation can be minimized with photodiode array monitoring of multiple wavelengths.

The fluorescence detector is useful because many compounds of biological interest are fluorescent or can be derivatized to a fluorescent form. Amino acids, alkaloids, and some amines can be converted to dansyl derivatives (treatment with 1-dimethylaminonaphthalene-8-sulfonic acid) for ready separation and detection at low concentrations. Methods are available in the literature and in the catalogs of reagent suppliers to create dansyl and other fluorescent derivatives of compounds that may be of interest to bacteriologists.

17.3.6.2. Guanine-Plus-Cytosine Content of DNA

The G+C content of the bacterial chromosome (expressed as the percent G+C of the total base content) is a commonly used taxonomic marker in characterization of an organism. A number of indirect physical methods and a direct chemical method are used to determine this value. The DNA thermal denaturation temperature, or melting temperature (T_m), is related to the G+C content. This value is obtained by using isolated and purified DNA (71). Alternatively, it can be obtained by using intact cells, in which case intracellular melting of DNA is observed by differential scanning calorimetry (69). All of these methods require calibration with DNAs or cells of known G+C content.

Direct HPLC measurement of the bases following hydrolysis of DNA is a precise and rapid alternative to the indirect methods above. Nuclease P1 can be used to hydrolyze DNA isolated from the microorganism to 5′-deoxynucleoside monophosphates (dNMP) which can readily be identified and quantitated by HPLC (61). Isolate and dissolve the DNA in 0.15 M NaCl–15 mM trisodium citrate (pH 7.0) at about 2 mg/ml. Treat the solution with 10 μg of proteinase K per ml for 1 h at 37°C, extract twice with phenol, and keep the aqueous layer. Precipitate the nucleic acids with cold ethanol, and recover the precipitate by centrifugation. Two nuclease P1 treatments are used; in the first, to digest single-stranded DNA and RNA, add 2 ng of nuclease P1 and 1 mg of DNA in 2 ml of 10 mM acetate buffer (pH 5.3), and incubate for 30 min at 40°C. Dialyze overnight against two changes of the above buffer, and then heat the mixture to 100°C for 10 min and rapidly cool in an ice bath. For the second treatment, add 2.0 μg of nuclease P1 to the mixture and incubate at 40°C for 15 min. Heat the mixture to 100°C for 10 min, and centrifuge at 15,000 × g for 25 min. Analyze 2 to 5 μl of supernatant (10 to 15 μg of dNMP) on an anion-exchange column of Zorbax NH_2 (25 cm by 4.6 mm [inner diameter]) (Agilent Technologies). Use a mobile phase of 20 mM KH_2PO_4 (pH 2.8) at a flow rate of 1 ml/min. Set the detector at 258 nm. Determine the concentration of each deoxynucleotide by using standards.

An alternate HPLC procedure for determination of G+C content has been reported (72, 73); in this method, the DNA is digested by the same proteinase and nuclease as described above but with different conditions and extraction protocols. To quantify the dNMPs, use a C_{18} reversed-phase column (Econosphere; Alltech Associates, Inc., Deerfield, IL), a mobile phase of 12% methanol–20 mM triethylamine phosphate (pH 5.1), and a flow rate of 1.0 ml/min. Detect the hydrolysis products by measuring the A_{254}. Also, see references 72 and 73 for a procedure for isolation and analysis of DNA from single colonies.

Two shortcomings of the method have been noted in practice. One is the difficulty of obtaining pure DNA by the published methods. For example, the traditional lysis and isolation procedures may not perform as expected for bacteria isolated from an extreme environment such as brine. In that event, the intact-cell method (69) should be tried, or a new DNA isolation procedure must be developed. The second shortcoming is the appearance of many more peaks in the column effluent than expected, as a result of incomplete digestion or the presence in the DNA of methylated bases such as 5-methylcytosine. Incomplete digestion can be identified and overcome by using less DNA, by adding more of the hydrolytic enzymes, or by using longer incubation times. The presence of methylated bases can be ascertained with standards. Following quantitation, the content of methylated base(s) is added to the content of the related nonmethylated base.

17.3.6.3. Separation and Quantitation of Bacterial Photosynthetic Pigments

Anaerobic photosynthetic (phototrophic) bacteria can be studied in pure culture, or in natural samples, by analysis of the bacteriochlorophyll present. Bacteriochlorophylls a, b, c, and e, as well as plant or cyanobacterial chlorophyll a, can be separated by HPLC. Pigment analysis can be used as a signature to aid in the taxonomic characterization of pure cultures of phototrophic bacteria and in the identification of species in natural samples. Yacobi et al. (86) described a method for comparing pigment extracts from isolated and identified species with those from lake water samples as follows. Filter water samples and culture suspensions of phototropic bacteria onto Whatman GF/C filters. Use 2 ml of pure culture or various volumes of lake water. Extract the pigments from the filters in 3 ml of HPLC grade methanol for 30 min in the dark at 2 to 4°C. Filter the extract through Whatman GF/F filters, and analyze by reversed-phase chromatography with a Spherisorb C_{18} column (25 cm by 4.6 mm [inner diameter]) (Supelco). For the mobile phase, use a gradient-producing pump to make a gradient by mixing 30% 1 M ammonium acetate in double-distilled water–70% HPLC grade methanol (component 1) with 30% ethyl acetate–70% methanol (component 2). Start the gradient with 80% component 1 and 20% component 2. Program the pump to produce a linear increase from 20 to 60% component 2 in 7 min, hold at 60% for 5 min, increase linearly from 60 to 100% in 12 to 20 min, and hold at 100% component 2 for 20 min. Detect and scan eluting components with a photodiode array spectrophotometer that covers the range from 300 to 800 nm. This method is effective in developing pigment profiles that are detectable and quantifiable in the raw lake water samples.

17.3.7. Ion Chromatography

HPLC with an ion-exchange column provides excellent separations of anions and cations, both inorganic and organic, on the basis of their pK values and relative affinities for the exchanger (43, 51). In practice, however, application to most ions has suffered from the lack of chromophoric behavior in the visible or UV spectral region above 220 nm. Therefore, it has been necessary either to resort to derivatization by a chromophoric reagent or to monitor the column effluent by conductivity. To decrease the background conductivity of the eluant buffer and thereby to increase sensitivity, postcolumn conductivity suppression is required. Although cumbersome, conductivity monitoring has proven to be useful, especially because of improvements in suppression methods (67).

Of more use to many investigators is column effluent monitoring by indirect photometric chromatography as developed by Small and Miller (80), in which a chromophoric ion buffer is used as the eluant. Proper choice of eluant buffer and its concentration, matched to the appropriate column packing, allows detection and quantitation of a large number of ions of interest to the microbiologist using HPLC equipped with a photometer designed for chromatographic work.

This strategy makes use of low-capacity and low-affinity ion exchangers, which in turn require low concentrations of eluant to displace the sorbed ions. The wavelength for detection is selected to give the best signal-to-noise ratio (55). The broad absorbance peaks of the eluant ions allow use of broad bandwidth conditions (see section 17.1.1.1), which allows low concentrations of eluant and increased sensitivity of the method (55).

The light-absorbing eluant ion passes continuously through the photometer, giving a high and constant absorbance. When an ion elutes, a corresponding amount of displacing eluant is bound to the exchanger, resulting in a decrease in absorbance. The area of the negative "peak" is proportional to the concentration of ion being eluted. Thus, conditions are adjusted such that the concentration of eluant ion both is able to displace from the exchanger the ions of interest in a reasonable amount of time and is only moderately higher than that of the eluting ions.

17.3.7.1. Instrumentation

The standard components of HPLC, including the photometer plus columns specific to ion chromatography, are all that is required. A variable-wavelength photometer either with substantial electronic zero baseline offset or with double-beam optics and reference cell, although not absolutely required, is highly desirable. Such photometers have very low volume flow cells designed for column work. Both low-capacity silica- and resin-based exchangers have been used successfully (22, 24, 40, 43). Silica exchangers are said to be more efficient, whereas resin exchangers have more alkali resistance.

17.3.7.2. Ions Determined

The following selected inorganic and organic anions and cations can be quantitated by ion chromatography: halogen, carbonate, chloride, perchlorate, nitrate, nitrite, cyanate, thiocyanate, selenite, selenate, sulfate, sulfite, P_i, PP_i, acetate, citrate, glycolate, propionate, succinate, sodium, ammonium, amines, potassium, and many others (68). Joint determination of anions and cations is possible under some circumstances.

17.3.7.3. Eluants

Anions are most commonly determined by this method, primarily because of a wider variety of chromophoric anion buffers. The order of eluting power for anions is 1,2-sulfobenzoate > 1,3,5-benzenetricarboxylate > phthalate > iodide. The most common eluants for cations are copper sulfate and copper nitrate.

17.3.7.4. Sensitivity

Detection limits are well below 50 ppb for several common anions in routine laboratory settings (54) when careful attention is given to (i) controlling the temperatures of both the column and the eluant, (ii) eliminating pump noise with a pump-damping accessory, (iii) minimizing the dead volume of the entire system, and (iv) using the maximum sensitivity of the photometer-recorder system (0.005 to 0.01 absorbance units). Less-stringent observance of these requirements still leads to determinations of less than 1 ppm.

17.3.7.5. Ion Chromatography Coupled with Atomic Spectrometry

There is a trend to couple ion chromatography with extremely sensitive atomic spectroscopic methods for detection, especially for metal ions. Atomic absorption spectrometry can be carried out with flame (FAAS), graphite furnace (GFAAS), or electrothermal vaporization (ETVAAS). These instruments are dedicated to analyzing only one ion at a time. Inductively coupled plasma atomic emission spectrometry (ICP-AES) is capable of detecting many elements simultaneously at sensitivities at least 1 order of magnitude greater than that of atomic absorption (68). However, this large and expensive instrument is not found in the routinely equipped laboratory and is usually located at a larger centralized analytical facility.

17.3.8. Affinity Chromatography

Affinity chromatography makes use of highly specific binding equilibria in which at least one member is a complex macromolecule to effect separations that are difficult to accomplish by other means (21, 23, 26, 29, 32, 35). The principle simply involves chemical linking of one component of a reversible binding system on a solid support followed by mixing with the other component(s) of the equilibrium in solution. Following binding between members of the complementary pair, separation of the solid phase leaves the contaminants in solution. Thus, the basis of separation is the specificity of the binding system conferred by the three-dimensional location of the complementary binding groups rather than simple differences in physical and chemical properties.

Affinity chromatography techniques have been highly developed so that they make use of (i) small molecule-biopolymer interactions such as hormone-binding protein and enzyme-ligand interactions, (ii) the group-specific interactions among proteins and nucleic acids, (iii) the exceptionally high specificity of interaction between proteins such as the binding between antibody and antigen, and (iv) special affinity matrices that are designed to bind certain polypeptide regions joined to either the amino or carboxy terminus of genes encoding recombinant proteins. The major factors involved are the nature and chemistry of the matrix, the chemistry of ligand attachment, the requirement for a spacer arm, and conditions for adsorption and elution of the mobile binding member of the system. General information is available in references 23 and 35 and in manufacturers' literature (21).

Many types of affinity materials are commercially available, allowing the researcher to make a column to fit specific needs, with choices of (i) the type of matrix or support, (ii) the presence and kind of spacer arm and terminating functional groups, and (iii) bonding to certain biopolymers which act as group-specific binding agents. Affinity chromatographic media can be custom-made to suit particular requirements by attaching a ligand to a chemically activated matrix, or there are many ready-made matrices that can be purchased.

17.3.8.1. Preparation of a Custom Affinity Matrix: Coupling of Ligand

Although small ligands can attach to an activated matrix directly, large ligands such as proteins are often hindered sterically, and the reaction is slow and incomplete. When proteins do react, it is likely that several covalent bonds will be formed with the matrix. For this reason, the specific ligand is often attached to the matrix through a spacer arm. This arm generally is a linear saturated aliphatic chain of 6 to 12 carbon atoms containing reactive groups at each end to facilitate bonding to both the matrix and the ligand conferring specificity. The matrix material is first activated by derivatization, for example with N-hydroxysuccinimide or

cyanogen bromide or by conversion to the hydrazide. The first two activated supports react with amino groups of spacer arms or ligands, and the last activated matrix reacts with carbohydrate moieties such as that of the Fc region of immunoglobulin G. The latter special case provides oriented coupling to IgG as opposed to random coupling through amino groups as occurs with the other activated supports. Spacer arms can terminate in amino, carboxyl, thiol, or other functional groups. These can act as affinity ligands as such or, alternatively, can undergo further coupling through these functional groups to biospecific ligands such as proteins and nucleic acids.

Matrix material can be derivatized by the investigator (26) or purchased with an arm which is terminated with a reactive group such as N-hydroxysuccinimide (Affi-gel 10 and Affi-gel 15; Bio-Rad). This group reacts with alkyl and aryl amines. With CNBr-activated Sepharose, the most common arms are produced by reaction with 1,6-diaminohexane and 6-aminohexanoic acid. Ligand bonding then involves coupling with a water-soluble carbodiimide to form amides. For this reaction, 1-ethyl-3-(3-dimethylaminopropyl)carbodiimide hydrochloride or 1-cyclohexyl-3-(2-morpholinoethyl)carbodiimide metho-p-toluene sulfonate gives good yields in aqueous and nonaqueous media at pH 4.5 to 6.0. Exact protocols for attachment of ligands to activated chromatography media can be found in the manufacturer's instructions.

17.3.8.2. Special Considerations for Coupling Reactions

For adsorption, choose the starting buffer to maximize binding and minimize nonspecific protein-protein interactions. Therefore, consider the pH and presence of specific ions or factors, and use a buffer of high ionic strength, e.g., 0.5 M NaCl. Add the sample in the starting buffer, and if necessary, exchange the buffer ion in the sample either by dialysis or by membrane or gel filtration.

Elution may involve one or a combination of pH shift, change in ionic strength, or displacement by a solute related or identical to the immobilized ligand. For the first two methods, the specificity characteristic of affinity chromatography lies only in the adsorption phase. In elution by specific displacement of immobilized ligand, specificity is exhibited in both the adsorption and elution phases.

In instances when a highly reactive support material is to be coupled to macromolecules that are highly sensitive to excessive multipoint coupling, it may be necessary to reduce the number of coupling groups in the matrix. This is accomplished for CNBr-Sepharose and some other materials by controlled hydrolysis of the reactive group and for other activated matrices by controlled pretreatment with a small ligand. Even without such sensitivity, it is likely that some distortion of macromolecular structure will occur. For enzymes, this is displayed by a lower specific activity and other changes in catalytic properties.

After completion of the procedure for binding a ligand to a matrix, any remaining reactive sites on the matrix should be destroyed by adding an excess of small molecules containing the same reactive group; e.g., for CNBr-activated Sepharose, add ethanolamine.

17.3.8.3. Affinity Chromatography of Tagged Recombinant Proteins

The ease and widespread use of recombinant DNA methods have led to a myriad of new ways to purify proteins by linking the respective gene to a segment of DNA that encodes a short polypeptide that has high affinity for a cognate group that is linked to a column matrix. A summary of the most popular strategies is shown in Table 4. The major advantage of this approach is that purification can often be achieved by using the affinity column on crude cell extracts without any additional downstream purification steps. The use of affinity-tagged expression systems lends itself to high-throughput purification schemes (46, 88), and to situations where recombinant proteins are expressed at low amounts or are especially difficult to purify using conventional techniques. In some cases the recombinant chimeric protein resulting from the fusion with an affinity tag has greater stability and solubility than the native form of the protein alone. The major disadvantage is that the purified protein, because it has an affinity tag attached, will be different from the native form. Some affinity tag expression systems are designed to produce a specific protease site to remove most or all of the affinity tag, but this processing results in additional steps and the use of expensive proteases.

The most popular scheme used for affinity purification of recombinant proteins is a polyhistidine tag added to the amino or carboxy terminus of a cloned gene (76). The purification method, generally known as immobilized metal affinity chromatography, or chelation chromatography, uses a column support matrix which has metal chelating groups attached. The column is "charged" with divalent metal ions which bind to the chelating groups on the column. Most often, the metal ion is Ni^{2+}, but Zn^{2+}, Cu^{2+}, or Co^{2+} can be used. The crude preparation is then passed over the column and washed to remove contaminating proteins. The bound His-tagged protein is eluted with buffers containing high concentrations of a competing ligand such as imidazole (76).

17.3.9. Gel Permeation Chromatography

The availability of spherical gel beads of relatively homogeneous pore size distribution makes possible a number of kinds of molecular separations which are based to a great degree on molecular size (25, 31, 37, 81). This method has been given various names including gel filtration and size exclusion chromatography. When a sample is introduced on a gel permeation column, molecules much larger than the pore size of the beads proceed through the column with the displacing solvent without penetrating the beads and exit as a peak after the void or interstitial volume, V_0, has passed through. Small molecules, which freely equilibrate with the internal space of the beads, V_i, theoretically proceed down the column by eluant displacement as though beads were not present and exit after passage of the total column volume minus the volume of the matrix. Thus, the total column volume, V_t, is the sum of the excluded volume, V_0, the internal volume, V_i, and the matrix or bead material volume, V_g.

$$V_t = V_0 + V_i + V_g$$

Molecules intermediate between these extremes "partition" between interstitial and internal spaces largely on the basis of molecular size. Since a wide variety of pore sizes, and hence fractionation ranges, of gel filtration media are available, it is possible to separate molecules over a molecular mass range of 1×10^2 to 4×10^7 Da. Although many molecules behave "ideally" as described above, many others are retarded more than expected. This is due to adsorption, ion exchange with a small number of carboxyl groups, or interaction between aromatic rings and the matrix. The behavior of molecules as they pass through the column is usually deduced by analyzing the column effluent.

17.3.9.1. Gel Permeation Media

Many kinds of gel beads are available: cross-linked dextrans and derivatives, polyacrylamides, agaroses, and cross-linked agaroses. In addition, there are a wide variety of proprietary polymer gels that allow for small and uniform bead size (for improved resolution) and greater physical strength (for higher pressure operation). Gel filtration media are available in many bead sizes and porosities from many manufacturers. It is therefore essential to consult the manufacturers' literature for details of their properties.

17.3.9.2. Multicomponent Analysis

By proper choice of bead porosity and carefully worked-out conditions, it is possible to separate closely sized molecules or, conversely, to separate classes of widely different molecules. Analysis of mixtures places the greatest demand on separation, but under these conditions a high flow rate is not as important and is deleterious for good resolution. Since quantitative yields are usually obtained on small columns, separation of components into nonoverlapping bands followed by quantitative analysis by nonspecific methods is possible. Usually, such separations are based on molecular weight. However, such separations have also been made because particular molecules are retarded for other reasons, such as adsorption of a polar solute from a nonpolar solvent. Gel permeation chromatography has also been used to study protein-ligand binding (42, 62).

17.3.9.3. Fractionation of Proteins and Determination of Native Molecular Weight

For gel permeation chromatography on a preparative scale, such as in the purification of enzymes and other proteins, it is often desirable to compromise between resolution and flow rate. Thus, some resolution is sacrificed by using a larger gel particle size to accommodate the volume involved. The porosity is chosen (often empirically) to resolve as many components as possible and to place the desired component intermediate in the elution diagram between V_0 and $V_0 + V_i$. Successive gel filtrations or a gel

TABLE 4 Affinity tags used for purification of recombinant proteins[a]

Affinity tag	Binding partner on column matrix	Elution strategy
Polyhistidine	Metal chelating group	Imidazole displaces histidines
Glutathione S-transferase	Glutathione	Reduced glutathione
Maltose binding domain	Amylose	Maltose
S-Tag (small subtilisin fragment of RNase A)	S-protein (large subtilisin fragment of RNase A)	In situ protease cleavage
Thioredoxin containing metal-binding histidines	Metal chelating group	Imidazole displaces histidines

[a] Data from reference 32.

filtration step in combination with other separation procedures has been used effectively. The native molecular weight of soluble proteins (or multisubunit proteins or protein complexes) can be estimated by gel filtration on Superose 6 or Superose 12 (Amersham Biosciences) and Ultrogel (26, 31). A regression line of the elution position V_x/V_0, where V_x is the volume to the peak of the unknown and V_0 is the void volume, versus the log of molecular weight is established with a group of enzymes and proteins of known molecular weight. The molecular weight of the unknown is taken from the graph.

17.3.9.4. Practical Considerations

Choice of Gel Filtration Medium

It is necessary to choose a medium with a fractionation range to suit the application in mind. For total exclusion of a macromolecule, such as desalting, use a gel with a molecular mass fractionation range well below that of the material, i.e., Sephadex G-25 (fractionation range, 100- to 5,000-kDa molecular mass) or Bio-Gel P-6 (Bio-Rad), for exclusion of most proteins. For total equilibration with the gel interior, V_i, and elution with the "salt" volume, $V_0 + V_i$, select a gel with a fractionation range whose lower limit is above the molecular mass of the molecule of interest. For fractionation rather than desalting, the gel should have a molecular weight fractionation range such that the molecule(s) of interest will be within the range for separation, i.e., Sephacryl S-200 (fractionation range, 5 to 600 kDa) or Bio-Gel A 0.5 M (fractionation range, 10 to 500 kDa). If a protein has a native molecular mass that is near the extremes of the fractionation range for a particular gel permeation medium, then the experimenter should consider changing resins. For example, Superdex-200 (fractionation range, 10 to 600 kDa) would theoretically work as a polishing step for purifying a 34-kDa protein, but Superdex-75 (fractionation range, 3 to 70 kDa) would be more suitable.

It is important to note that the above values for molecular weight range are determined for globular proteins that have a spherical shape. Polymeric substances such as dextrans and DNA elute at an apparently lower molecular weight. For example, the upper limit of Superdex-200 for separating globular proteins is 600 kDa, the limit for fractionating DNA fragments is 200 bp (~130 kDa).

The efficiency of resolution is a function primarily of the flow rate, which in turn is related to bead size and uniformity. Thus, very small beads and slow flow give the highest resolution. For Bio-Gels P and A, flow rates of 2 to 10 ml/h/cm^2 of bed cross section give the highest resolution, and this is much slower than the maximum gravity flow rate for some bead sizes and porosities. With the most porous gels, for instance, Bio-Gels A 50 and 150 M (200 to 400 mesh), increased hydrostatic head is needed. With Bio-Gels P-2, P-4, and P-6, elevated pressures of several atmospheres may be used with 400-mesh beads capable of high-resolution separation of small molecules. When pressure is used, measures must be taken to ensure use of appropriate vessels and columns. The maximum recommended pressure should be ascertained from the manufacturer's literature. With some highly porous materials (Sephadex-200 and Sephacryl-500), pressures above that recommended by the manufacturer lead to collapse of the bead structure and very low flow rates. Some newer gel permeation media such as Superdex-200, which is made of a more rigid cross-linked agarose/dextran composite matrix can be used at pressures over 200 lb/in^2.

Preparation of biological materials rarely involves chemical conditions that would threaten the integrity of the gel filtration medium. However, the general stability characteristics are included here as a guide. Dextrans such as Sephadex are stable in salt solution, organic solvents, alkaline solutions, and weakly acidic solvents but are hydrolyzed in strong acid. Dextrans are stable to 30 min at 121°C, wet or dry, at neutral pH. Agaroses such as Sepharoses do not contain cross-links and hence should be maintained between pH 4 and 9 and between 2 and 30°C. They are stable to 1 M NaCl and 2 M urea. Cross-linked agarose, i.e., Sepharose CL, is stable between pH 3 and 9, but oxidizing conditions should be avoided. It is highly stable in 3 M potassium thiocyanate, a strong chaotropic agent. It can be transferred from water to a number of organic solvents with little change in pore size. It can be sterilized repeatedly at neutral pH at 110 to 120°C. Polyacrylamides such as Bio-Gel P are reasonably stable in dilute organic acids, 8 M urea, 6 M guanidine HCl, chaotropic agents, and detergents. Buffers with 0.5 M ionic strength are recommended for protein separations. Miscible solvents, e.g., alcohol up to 20%, do not materially change the pore size. Bio-Gel P is stable between pH 2 and 10 at room temperature and to autoclaving (121°C) at pH 6 to 8. Composite media such as Superdex can withstand pH of 1 to 14 for short periods and 3 to 12 for long exposure. Although it is not routine to use any of these column materials at extreme pH, the ability to use strong acids or bases for short periods allows for efficient cleaning of columns without repacking.

Column Design and Size

Many gel filtration applications such as desalting do not require high resolution and hence can be carried out on homemade or disposable columns of ordinary design. For desalting samples in the volume range of 50 to 100 μl, centrifugal columns that allow for efficient recovery of the desalted material can be purchased. However, success in critical applications requires properly designed columns and associated equipment. Areas of buffer volume above and below the gel bed must be avoided, or the peaks will become distorted by passing through these reservoirs. Commercial columns which have virtually no eluant retention below the gel bed are available. Others have fittings both above and below the gel bed to add eluant and collect effluent in the absence of reservoirs. Sintered-glass supports are to be avoided, particularly for low cross-linked and high-porosity gels.

The column length affects resolution, and the column cross-sectional area is proportional to capacity. On the other hand, finer particle size increases resolution and requires shorter gel beds. Since the sample size also dictates the initial bandwidth, the length and diameter become functions of the degree of separation needed for a given sample volume, longer for greater separation and wider for a narrower starting band. For example, a 51-cm bed height will easily desalt a protein in a small volume, whereas very long columns (300 cm) are needed to partially separate a set of peptides derived from a protein by treatment with cyanogen bromide, or as a final polishing step in a protein purification. An internal diameter of 2.5 cm may be a good starting point, although a somewhat narrower column can also be used. With very narrow columns, there are wall effects caused by a greater proportion of laminar flow, which broadens the bands. With excessive diameter, there is unnecessary dilution and difficulty in maintaining horizontal zones.

Column Packing

Calculate the amount of dry gel needed from the swollen volume furnished in the manufacturers' literature (for preswollen gels, this step is unnecessary). Swell the gel in twice the expected bed volume of eluant. The length of time varies with volume regain. Highly cross-linked and lower-regain gels such as Sephadexes G-10, G-15, G-25, G-50, and LH-50 and Bio-Gel P require 3 to 4 h at room temperature, whereas Sephadexes G-75 to G-200 require 1 to 3 days. At 90 to 100°C, these times are reduced to 1 to 5 h. Swelling in boiling water aids in removing bubbles, which at room temperature must be removed by vacuum deaeration. Dilute preswollen gels enough to allow bubbles to escape, but avoid stirring. Allow the swollen gel to stand, and remove excess eluant to give a thick slurry of a consistency which will still allow air bubbles to escape. Dilute the preswollen gels if necessary to give a slurry of the same consistency.

Mount and align the column vertically. Fill the column with eluant, preferably from the bottom, until the dead space below the gel bed is occupied and all air bubbles are removed. With the outlet closed, pour the gel slurry smoothly down the column wall or down a rod all at one time. Add eluant to fill the column completely, connect an eluant reservoir, and remove all bubbles through an air vent in the column top piece. If the total gel suspension cannot be added at one time, use a column extension. Start the flow immediately. Bed formation from a thin suspension, addition of a slurry to a column containing eluate, or packing in stages is not recommended. Two or three column volumes of eluant are passed through at about the flow rate to be used. The column must not be allowed to go dry. This can be prevented by making a loop in the outlet tube which is higher than the gel surface.

Sample Preparation and Addition

Gel permeation chromatography is most effective when the sample volume is kept to a minimum, ideally 1 to 2% of the column bed volume. However, if sample concentration becomes excessive, viscosity can have detrimental effects on separation. In such cases, several chromatographic runs may be required rather than overloading the column in a single run. If a particular sample is too dilute, then concentration can be achieved by using ultrafiltration membranes with a molecular weight cutoff that allows solvent and salt molecules to pass through while retaining the desired macromolecules. Smaller molecules are forced through the filtration membrane with pressurized gas (such as the Amicon stirred cells [Millipore]) or by centrifugation with manufactured single-use units such as Centricon or Centriprep (Millipore). These concentration systems are available with ultrafiltration membranes of various molecular weight exclusion limits fabricated of polymers with minimal protein binding properties.

A flow adapter is adjusted so that it just touches the top of the column bed to minimize dilution of the sample as it enters the column. The adapters consist of plug-like devices attached to threaded tubes which pass through the column ends and terminate in capillary tubing. Two of the adapters with tight fits to the internal walls can be adjusted via the threaded rods to fit snugly at the bottom and top of the gel bed. With this arrangement, sample followed by eluate is added directly to the gel bed and eluate is removed from the bed bottom.

If no flow adapter is available, then other methods must be used to carefully introduce the sample without disturbing the top of the column bed. Before sample addition, place a disk of filter paper or perforated plastic, or a sample applicator cup, on the gel surface. The cup is a plastic cylinder with a nylon net bottom. Remove excess eluant, layer the sample on the gel surface, and open the outlet until the sample has moved into the bed. Wash the column wall in the region of the surface with a small amount of eluant. Carefully fill the column with eluant, and connect the column to a reservoir of eluant to eliminate any bubbles in the column or tubing. It is also possible to layer the sample on the gel bed and beneath the eluant. The sample must be more dense than the eluant; if necessary, sucrose or salt is added. Layering is done with a pipette or capillary tubing attached to a syringe.

If desired, the homogeneity of packing, as well as the void volume (V_0) and fluid volume ($V_0 + V_i$) can be determined by passage of a mixture of blue dextran (2 mg/ml) and potassium ferricyanide or other colored small molecule through the column.

Elution

The flow rate can be regulated by hydrostatic pressure, i.e., by the difference between eluant surface and outlet, as long as the maximum pressure permitted for a particular gel is not exceeded. The effluent reservoir (Mariotte flask), any other devices, and the outlet tube end are then placed so that the difference in fluid level in the adjustable vent tube of the Mariotte flask and the column outlet end does not exceed the permissible difference in height (hydrostatic pressure). The Mariotte flask is a vessel with an outlet at the bottom to connect to the column via tubing and closed at the top with a stopper or plug through which passes an adjustable-height glass tube; the tube extends below the eluate surface and maintains a constant head. A pump can also be used under the same limitations. For more highly cross-linked gels, the maximum gravity flow rates greatly exceed the flow rate for optimal separation or peak sharpness. With less highly cross-linked gels, the flow rate may be limited by the permissible hydrostatic pressure.

Factors Affecting Peak Broadening

Gel permeation is a diffusion-controlled process; hence, peak sharpness and resolution require a very low flow rate and bead size uniformity. In other words, the highest resolution is obtained when the void and internal volumes of the beads are in equilibrium. Band broadening can be caused by inefficient mass transfer, diffusion along the column, and eddy diffusion. Selecting an optimal flow rate and using gel permeation media with more uniform particle size can minimize these effects. In analytical work, resolution is primary; in preparative-scale work, resolution is usually secondary to the speed of separation.

Elution patterns, i.e., peak shapes, are highly sensitive to the viscosity of the sample. Therefore, for maintenance of peak shape, samples containing sucrose or very high protein concentrations must be diluted so that the relative viscosity is not more than twice that of the eluant.

Storage

If the column is to be inactive for more than 1 week at a time, it is necessary to prevent microbial growth by adding either sodium azide (0.02%), Cloretone (0.05%), Merthiolate (0.005%), or 10 to 20% ethanol. Chloroform, butanol, and toluene should not be used. Follow the specific recommendations of the manufacturers. Before the column is returned to use, the above agents can be removed by passing eluant through the column.

17.4. RADIOACTIVITY

In most bacteriological studies, radioisotopes can be used to advantage in the experimental program (89–91). The choice of radionuclide depends on the half-life, the kind of decay process and its energy, and a number of other factors peculiar to the experiment. The half-life, short or long, determines the practicality of a given strategy as well as the necessity to correct for decay during the experiment. The type of decay and its energy determine the kind of counting instrument to be used as well as the resolution on photographic emulsions. Table 5 gives information on the isotopes most likely to be used in bacteriological work. Unfortunately, there are no useful radioactive isotopes of nitrogen or oxygen.

Most measurements of radioactivity in biology are of particles (electrons) and gamma rays. Beta radiation, for instance, from tritium and ^{14}C, can be measured in a gas counting chamber attached to an electrometer, by Geiger counter, by proportional methods, or by scintillation techniques. In recent years the liquid scintillation spectrometer has replaced all other methods and is the only system described here for quantifying beta radiation. Gamma counters are used primarily to measure ^{131}I and ^{125}I.

17.4.1. Safety Considerations

The conduct of experiments involving radioisotopes can involve risks to human health. Therefore, it is essential that safety practices and procedures be strictly followed. Such procedures not only minimize the risk to investigators but also improve the conduct of the experiment by requiring a range of good laboratory practices. The use of radioisotopes for any purpose is strictly regulated in many countries. In the United States, regulation is carried out by the Nuclear Regulatory Commission of the Federal Government as well as by regulatory agencies in many states. Furthermore, all universities, corporations, and other research facilities have their own radiation safety and disposal organizations that oversee required regulations and, in many cases, implement their own regulatory protocols that exceed government regulations. Licensing and periodic inspection are usually required for the parent institution under which the research is conducted. In turn, the institution, through a radiation safety officer, ensures that safe practices and records are maintained, regulates access control measures, and supplies working rules for the laboratory.

1. Conduct work in a laboratory area that is appropriate for the kind of radioisotope compound involved. Often, this is a separate area with a hood if volatile radioactive materials (e.g., ^3H_2O) are involved. Control the amounts used and their disposition. Work in a setup that has specific waste and cleanup capabilities available and ready for use.

2. Do not eat, drink, or smoke in this area.

3. Do not pipette by mouth.

4. Wear a laboratory coat, gloves, and safety glasses.

5. Label all containers with specific radioisotope labels, and give specific data.

6. Use disposable laboratory items when possible, and deposit them in designated solid-waste containers.

7. Monitor and decontaminate all laboratory glassware before returning it to general use.

8. Maintain an isotope inventory and record of isotope use.

9. Monitor the work area, as well as clothes and hands, before leaving the laboratory.

10. Report spills both to others in the laboratory and to the radiation safety officer.

A thin-window Geiger counter can monitor all isotopes except tritium. For tritium, working areas are sampled after experiments and periodically with absorbent wipes which are then evaluated by scintillation counting. Film badges are required for persons using gamma-emitters and for persons using ^{32}P and other strong beta-emitters. Persons using ^{14}C, ^3H, and ^{35}S do not need film badges. However, specific regulations at some institutions may require additional monitoring.

Protect counting equipment against contamination. This means having a separate location and paying special attention to cleanliness of the *outside* of counting vials, using leak-free closures tightly sealed, and maintaining the sample changer to prevent mishaps with vials. Distinguish between, and separate, high-level and low-level work. Clean up. Do not leave contaminated glassware and equipment where others can handle it.

Tritium and ^{14}C in the quantities used in most experiments present no radiation danger. However, ^3H as in ^3H_2O is volatile, and $^3H_2$ gas as used in compound labeling is of very high specific activity; these require special precautions. ^{14}C, although a weak beta-emitter, has a very long half-life (5,000 years). Hence, its incorporation into biological material (DNA and bone) presents a potential long-term hazard. ^{35}S and ^{32}P are strong beta-emitters with relatively short half-lives (87.4 and 14.3 days, respectively). High levels of these should be handled behind a radiation shield of acrylic or lead bricks (for high levels) and the radiation level at the worker location should be monitored.

17.4.2. Purity of Isotopically Labeled Materials

Isotopic materials supplied by commercial sources are generally of high quality, and analysis is performed for each lot. However, one should keep in mind that both radiochemical purity and chemical purity may be critical in the proposed experiment. In cases where a radiochemical purity is suspect, or for one that is synthetic or enzymatically produced in the lab, it may be necessary for the researcher to determine purity using one of the methods described below. At least, it should be known that impurities, if present, will have no effect. A compound that is radiochemically pure contains one kind of radioactive molecule, but determinations of radiochemical purity give no indication of nonradioactive contamination. Radiochemical purity may be determined chromatographically and by reverse isotope dilution (see below), that is, by addition of the pure nonradioactive compound (carrier) followed by isolation and determination of specific activity. If the compound is radiochemically pure, the new specific activity is that expected by simple dilution with nonradioactive compound. If radioactive impurities are present, during reisolation with carrier

TABLE 5 Physical properties of selected radionuclides[a]

Radionuclide	Half-life	Type	Energy (MeV)
Tritium (^3H)	12.26 yr	β	0.018
Carbon-14 (^{14}C)	5,730 yr	β	0.159
Sulfur-35 (^{35}S)	87.4 days	β	0.167
Phosphorus-32 (^{32}P)	14.3 days	β	1.71
Iodine-131 (^{131}I)	8.06 days	β	0.81
		γ	0.08–0.72
Iodine-125 (^{125}I)	60 days	γ	0.035

[a]Data from reference 15.

these would be expected to be lost and the specific activity would be lower than expected. If the impurity is not radioactive, the specific activity after reisolation is higher than expected after dilution with carrier. If the impurity is known, its concentration can be determined by isotope dilution, i.e., by addition of a pure nonradioactive contaminant, isolation, and determination of specific activity. Chemical purity is analyzed in the usual ways, e.g., spectrophotometry, optical rotation, or GC.

Substantial changes in the radiochemical and chemical purities of a compound may occur during storage, primarily as a result of radiation damage from high specific activities. This process can be minimized by adherence to the specific conditions recommended by the supplier or more generally to one of the following: (i) storing at the lowest practical specific activity; (ii) subdividing into smaller lots; (iii) keeping dry if solid; (iv) storing in vacuo or under inert gas; (v) storing in pure benzene at 5 to 10°C; (vi) for water-soluble, benzene-insoluble compounds, adding 2 to 10% ethyl alcohol; and (vii) storing at the lowest possible temperature. Storage of solids often does not require a vacuum or inert gas, and hence, solids can be stored as dry as possible in screw-cap vials. Volatile materials and liquids should be resealed in ampoules under vacuum or inert gas.

17.4.3. Isotope Dilution Techniques

Radioactive compounds can be used to determine the amount of the same unlabeled compound in a mixture. It is necessary to know the specific activity and weight of the radioactive compound added and to isolate some of the pure compound for determination of its specific activity. The amount of unknown is derived from the equation

$$\text{weight of unknown (g)} =$$
$$\text{weight of added radioactive compound (g)} \times \frac{\text{SA}_{\text{final}}}{\text{SA}_{\text{initial}}}$$

where SA is specific activity. Purity is attained when the specific activity after several reisolations (or recrystallizations) is constant.

A reverse approach can measure the total amount of a radioactive compound in a mixture. In this method, pure nonradioactive compound is added and the compound is reisolated. The initial specific activity of the compound and the amount of nonradioactive carrier must be known, and the final specific activity (after reisolation) is determined. The weight of isotopic material is calculated from

$$\text{weight of unknown (radioactive) (g)} =$$
$$\text{total weight (carrier plus unknown) (g)} \times \frac{\text{SA}_{\text{final}}}{\text{SA}_{\text{initial}}}$$

A double-isotope dilution technique is valuable for isolation of very small quantities of, say, a sterol from a homogenate by a procedure involving many steps and partial recoveries. For instance, a ^{14}C-labeled sterol of known total activity can be added. The total ^{14}C radioactivity of the pure isolated compound is determined by measuring the recovery of the sterol in the procedure. In another procedure, the sterol in the sample is derivatized either in the crude extract if practical or at the earliest possible purification stage. For instance, (^{3}H)acetic anhydride can be used to form the acetyl ester of the sterol. Then, determination of ^{3}H in the isolated material gives the amount of sterol originally present. For this, only the specific activity of the acetic anhydride is needed; i.e.,

$$\text{Sterol recovery} = \frac{\text{cpm recovered}}{\text{cpm added}}$$

$$\text{Amount of sterol (}\mu\text{mol)} = \frac{\mu\text{mol of }^{3}\text{H-compound}}{\text{recovery}}$$

$$= \frac{\dfrac{\text{cpm of }^{3}\text{H/counting efficiency}}{\text{SA of acetic anhydride (}\mu\text{Ci/}\mu\text{mol)}}}{\text{recovery}}$$

where $1 \ \mu\text{Ci} = 2.22 \times 10^{6}$ dpm.

17.4.4. Calculation of Amount of Isotope Required

In general, the total amount of activity should be at least threefold above the minimum needed to make the final or lowest activity measurement. In this way, radiation effects and laboratory contamination are avoided and radiation safety is maximized. The easiest method, of course, is to use activities already reported for similar experiments, making adjustments for changes in volumes and instrument characteristics. Before starting the experiment, first decide on the minimum count rate which will be satisfactory (section 17.4.8), and then calculate back to the start of the experiment to obtain the total amount (disintegrations per minute) that is required. To do this, it is necessary to bring into the calculations many of the following factors.

Half-life of radioactive compound
Duration of experiment
Initial sample size
Dilution factor
Counting efficiency
Molar specific activity
Size of sample counted
Biological half-life
Losses in isolation
Purity and stability of labeled compound
Metabolism of labeled compound

As an example, assume that it is desired to isolate on a column a tritium-labeled metabolite derived from the incubation of glucose in ^{3}H$_2$O. The material will be distributed in five 10-ml fractions from the column and will contain 1 atom of ^{3}H per molecule of metabolite. The yield from glucose added after the column chromatography is 5%, taking into account all factors of metabolite pools, side reactions, and losses in isolation. One-half of the total incubation mixture is used for this isolation.

In this experiment, ^{3}H from water is incorporated into the material isolated. In the incorporation process, protons compete very effectively with ^{3}H^{+} both because there are many more and because the mass and the chemical characteristics of ^{3}H^{+} create an isotope discrimination effect, which, for this example, is assumed to be 4:1 against ^{3}H^{+}.

The counting solution to be used functions well with a maximum of 0.5 ml of aqueous sample; 0.2 ml is used in this case. The counting efficiency for ^{3}H is 43%, and there is 20% quenching. Sometimes these two together are called counting efficiency. For ^{14}C and ^{3}H, no calculations are needed to correct for decay of radioactivity.

From the standpoint of counting-error statistics (see below), 5,000 count events are desired. Since the scintillation counter cannot be tied up for more than 1 to 2 h, this count should be obtained in 10 min per sample. Therefore, a 0.2-ml sample should contain radioactivity giving 500 cpm for *the lowest-activity fraction*. The radioactivity is estimated to be distributed into five fractions as 5, 20, 50, 20, and 5% of the total; hence, the count rate in each 0.2-ml sample should be 500, 2,000, 5,000, 2,000, and 500 cpm, or 8,000 cpm for the five samples counted out of a total of 50 ml of eluate. To continue: 8,000 cpm × 50 (total sample) × 2 (total reaction mixture) × 4 (isotope effect)/0.43 (counting efficiency)/0.8 (quench)/0.05 (yield) = 1.86×10^8 total dpm of ^3H_2O to be added to the reaction mixture. The ^3H_2O purchased contains 25 mCi/ml or 25 mCi × 2.22 × 10^9 (dpm/mCi) = 5.55×10^{10} dpm/ml in stock solution. Thus, $1.86 \times 10^8/5.55 \times 10^{10} = 0.00335$ μl, or, better, 3.35 ml of 1:1,000 dilution of ^3H_2O solution.

Total activity calculations thus provide information about quantities to be used. However, it is necessary to make specific activity calculations as well, i.e., disintegrations per minute per mole or disintegrations per minute per micromole, to determine how many atoms of tritium have exchanged or whether other processes equilibrating with hydrogen atoms are involved. In the above example, it was calculated that 5.55×10^{10} dpm was present in 1 ml or 55.56 mmol of water or 111 meq of hydrogen per ml. The specific activity is $5.55 \times 10^{10}/1.11 \times 10^2$ or 5×10^8 dpm/meq of H.

In the 10-ml incubation, the dilution of specific activity is 10 ml/0.00335 ml of ^3H = 2,985-fold, giving a specific activity of $5 \times 10^8/2,985 = 1.67 \times 10^5$ dpm/meq of H. If the metabolite isolated had the same disintegrations per minute per millimole, there was 1 eq of $^3H^+$ incorporated with the assumed isotope effect of 4. If the specific activity is higher or lower than this value, the magnitude of the isotope effect may be different, or there may be exchanges with hydrogen atoms not derived from ^3H_2O (lower specific activity), or more than one proton from ^3H_2O has been incorporated (higher specific activity). The isotope effect is significant with tritium but is small enough with ^{14}C, ^{32}P, and ^{35}S to be ignored. It is often necessary, as in this example, to make an independent determination of the isotope effect by measuring the rate with H_2O or ^2H_2O and the rate with ^3H_2O.

17.4.5. Enzyme Assays with Isotopes

A wide range of enzyme assays which depend on radioisotopes have been developed (see also chapter 19). In many cases there are no other convenient methods. Further, these have intrinsically high sensitivity, which is demanded in many kinds of work. The amount of enzyme is proportional to the rate of substrate utilization or product appearance. Hence, determination of the concentration of either substrate or product with time is needed. For an assay to be valid, the initial rate must be constant for a short period and must also be proportional to the level of enzyme in the assay. In some instruments such as spectrophotometers and pH meters, rates can be monitored continuously on a single assay mixture, often with the progress of the reaction being displayed as a strip chart recording. In other cases, however, such as with radioisotope enzyme assays, the rate must be established by measuring radioactivities on discrete aliquots of a reaction mixture taken at intervals.

One of the major problems characteristic of such assays is the necessity of separating the radioactive substrate from the radioactive products. Usually, the labeled substrate is present in an amount required to saturate the enzyme and produce pseudo-zero-order kinetics in which the rate of reaction is constant with time for an appreciable period. In sensitive and valid assays, the amount of product formed may be only 1% of the substrate present. It is therefore necessary to separate a small amount of radioactive product from a very large amount of radioactive substrate. Hence, the success of the method depends on finding an efficient separation that will give 0.1% or less cross-contamination. These procedures are often specialized and hence are outside the scope of this manual. The general types are summarized below.

The separation methods include (i) precipitation of macromolecules synthesized or utilized, i.e., polynucleotides, polypeptides, and polysaccharides; (ii) release or uptake of a volatile radionuclide which can be separated by distillation or an equivalent process, i.e., 3^3H_2O, $^{14}CO_2$, or VFAs; (iii) solvent extraction; (iv) ion exchange; (v) paper and thin-layer chromatography or electrophoresis; (vi) adsorption and elution; (vii) isolation of derivatives; (viii) conversion of unlabeled product to labeled derivative; (ix) dialysis or gel filtration; and (x) reverse isotope dilution.

The degree of cross-contamination of substrate into product in the isolation procedure limits the sensitivity of the method. For instance, assume that there is a 0.1% cross-contamination of substrate into product in the separation used. Further assume that the zero-time (i.e., with no product formed) count rate due to cross-contamination is 100 cpm. It is desirable to have the count rate due to product be at least equal to the rate at zero time or 100 cpm for 0.5% conversion of substrates. In this case the signal-to-noise ratio is 1:1, or the minimum sensitivity without very long count rates (see section 17.4.8). To attain these minimum count rates, 100,000 cpm in the substrate is the minimum to be used if the total sample is counted. The use of larger amounts of radioactivity will not improve the sensitivity, because the zero-time value increases correspondingly. Of course, larger amounts of radioactivity in the product above the zero-time count rate decrease the error of the assay. However, a very substantial amount of product formation causes loss of the pseudo-zero-order condition needed for a valid assay.

From the above example, it is clear that although it is necessary to add enough substrate to saturate the enzyme, its radioactivity need not be high. On the other hand, if separation of substrate and product is complete, the sensitivity is proportional to the specific activity of the substrate and is limited only by cost, practicality, and availability. Under these circumstances, investigators may prefer to use ^3H-rather than ^{14}C-labeled substrates because of the availability of 100-fold-higher specific activities at similar costs (see section 17.4.6 for a discussion of problems with ^3H substrates).

Dilution of radioactive substrate with unlabeled substrate affords a number of advantages in addition to facilitating saturation of the enzyme with substrate. It reduces (i) cost by using less isotope, (ii) effects of contaminants and radiation degradation products in radioactive substrates, (iii) the effect of an unwanted stereoisomer in radioactive substrate containing a mixture of isomers by adding carrier substrate of the desired stereoconfiguration, and (iv) radiation decomposition of the substrate. Such dilution also increases the accuracy of determinations.

Since enzyme activity is expressed in international units (IU), where 1 IU is 1 μmol of substrate converted per min under standard conditions, it is necessary to convert count rates to micromoles per minute by using the specific activity of the substrate (which will be converted to product presumably of the same specific activity). This value is subject to errors in measurement of both radioactivity and concentration. Although the specific activity of the radionuclide as applied is usually known to ±5%, it is seldom used as such. Rather, the specific activity is diluted with large amounts of the nonradioactive form, with attendant errors of measurement and purity. It is wise, therefore, to carefully determine the specific activity of the substrate as used in the assay. It may also be necessary to demonstrate radiochemical purity of the substrate or to purify it prior to using it to remove decomposition products and other impurities.

17.4.6. Special Problems with Tritiated Substrates

Tritium shows isotope effects relative to hydrogen because of the large change in mass involved. The effect is most pronounced when making or breaking the bond between tritium or hydrogen and another atom is a rate-limiting step in the reaction. This effect may amount to virtual exclusion of reaction with the tritiated species, as has been observed in the enzymatic reductive carboxylation of L-ribulose-5-phosphate by [^{3}H]NADPH and CO_2. The isotope effect may be displayed as a change in V_{max} of the enzyme when using a tritiated substrate or as a change in K_m.

When tritium is used as a label only to monitor the conversion of substrate to product, it must be remembered that (unlike ^{14}C) ^{3}H often is released as ^{3}H$^+$ into the medium in unrelated spontaneous reactions, giving substrate and products a lower specific activity and hence erroneously low enzyme activities. In some cases the side reaction is enzyme catalyzed, not necessarily accompanied by the complete reaction. That is, there is a proton-tritium exchange reaction which is independent of, and often more rapid than, the total reaction. In addition, it is characteristic of tritiated compounds to "leak" ^{3}H$^+$ by chemical reaction. This varies with the location of ^{3}H in the molecule. For example, α-^{3}H-amino acids lose ^{3}H in many enzyme-catalyzed conversions, whereas ^{3}H in other positions, especially if remote from functional groups, is relatively stable. ^{3}H may also be lost by spontaneous exchange reactions during isolation of the product.

Some enzyme assays are based on release of ^{3}H from a specific position in the substrate. Errors result if ^{3}H in the substrate is not exclusively located in the position assumed and/or in the correct stereo location, because it has been demonstrated many times that enzymes selectively or specifically labilize tritium in one of two or more bonding positions occupied by ^{3}H or H.

The low energy of ^{3}H radiation causes low counting efficiency and losses due to quenching and absorption by solid materials in the counting vial. These in turn lead to greater inaccuracies or even lower apparent enzyme velocities. Good efficiency and quench corrections made on a sample-by-sample basis often are required.

17.4.7. Liquid Scintillation Counting

Liquid scintillation counting (LSC) has now supplanted Geiger, proportional, and gas counting methods for ^{3}H, ^{14}C, ^{35}S, and ^{32}P, even though the instruments are quite expensive. This preference is due to several advantages of liquid scintillation counters such as elimination of self-absorption, more favorable counting efficiency, and greater convenience in sample preparation.

In LSC, the radiation event excites the emission of a light pulse from a fluorescent material, which is then "seen" by a pair of photomultiplier tubes and registered as a radiation event or count. The sophistication of this instrument lies in the fact that both photomultiplier tubes must see the light flash simultaneously before a "count" is registered.

This eliminates a large number of spurious, electronically derived pulses. In addition, the photomultiplier produces voltage pulses whose amplitude or height is proportional to the energy of the radiation event. By use of pulse height analyzers, with adjustable discriminator settings, the instrument can be set to count only pulses appearing in the "window" corresponding to desired minimum and maximum radiation energies. Therefore, the instrument can be set to (i) eliminate many low background pulses, (ii) selectively count one isotope (one class of pulse heights), or (iii) count all isotopes (all pulse heights); it usually does these simultaneously via several parallel analyzer channels.

The radiation event is converted into a scintillation via solvent molecules that become activated by the decay event. The activated solvent then transfers the excitation energy to a compound which fluoresces (fluor). The excited fluor either emits a photon of light at a characteristic wavelength or transfers the excitation energy to a secondary fluor, which in turn emits a photon of light at a more advantageous wavelength with respect to the spectral sensitivity of the photomultiplier. PPO (2,5-diphenyloxazole) and POPOP [1,4-bis-2-(5-phenyloxazolyl)benzene] and its dimethyl derivative were traditionally the most popular primary and secondary fluors. These are highly nonpolar substances which are dissolved in a nonpolar solvent such as toluene to make up the scintillation solution. The principal drawback in scintillation counting using organic solvent cocktails is the difficulty of introducing aqueous samples containing polar solutes into the nonpolar counting solution. Further, water and solvents of increasing polarity tend to diminish the counting efficiency; that is, they, along with colored compounds, quench the normal chain of events leading to light emission.

Because of the problems described above, a variety of scintillation mixtures have been developed for polar and nonpolar materials in solution, for solid materials, and for other specific purposes (see below). In addition, high background counting rates can occur as a result of fluorescent light or sunlight-induced phosphorescence, which may require hours or days to decay. Also, chemiluminescence of samples in counting medium may result in initially high count rates which decay slowly in the dark. Glass vials may contain ^{40}K, which raises the background level.

17.4.7.1. Solid Materials

Small pieces of paper or removed sections of a thin-layer plate may be counted in scintillation medium. The main problems are associated with self-absorption by the particles, uniformity of orientation, uniformity of solubilization of radioactive material, and difficulties in using the external standard method for determining counting efficiency (see section 17.4.7.4). Since there is no direct way of determining self-absorption, only relative radioactivity of a series of samples counted under the same conditions can be obtained. Variability is greatly decreased if the material to be counted is either completely dissolved or is completely insoluble in the scintillant. For instance, hyamine hydroxide 10-X in methanol has been used to remove amino acids and sugars from paper, followed by addition of a toluene scintillant. Some scintillation fluids are designed to form a gel under certain conditions, thus allowing particulates to remain suspended in the counting vial rather than settling to the bottom. In some extreme cases, it may be necessary to burn the paper, trap the CO_2 in an organic base or as sodium or barium carbonate, and count these materials. Additional options for filters are offered by scintillation cocktails that dissolve cellulose acetate, nitrocellulose, and other types of filter membranes (e.g., Filtron-X; National Diagnostics, Atlanta, GA), aiding reproducibility and counting efficiency. There are also several methods for solubilization of cells and tissue samples, using scintillation fluid additives (96–98).

17.4.7.2. Polyacrylamide Gels

Radioactive samples separated on polyacrylamide gels can be quantitated by fluorography (gels soaked with a scintillator solution or intensifier screen and exposed to film [93]) or commercially available high-resolution phosphorimaging systems such as the Storm, PhosphorImager (Amersham Biosciences, Piscataway, NJ) and Bio-Rad FX (Bio-Rad), or by LSC methods described below. Phosphorimaging systems, while not a typical component of most labs, are often found in centralized shared equipment areas and are becoming more widely available. The gel containing radioactivity (in some cases, dried or blotted onto a membrane) is exposed to a storage phosphor screen. The screen contains crystals of $BaFBrEu^{2+}$, which upon excitation by beta or gamma particles, becomes oxidized to $BaFBrEu^{3+}$. Exposure times can be over 10-fold faster than for conventional film methods, and the linearity of response is much greater. The captured image is then scanned by a laser which releases an electron and a photon and returns the crystals to the $BaFBrEu^{2+}$ ground state. In the phosphorimaging system the photons are quantitated using the supplied software tools. The phosphor screen is erased using light in a specific wavelength range and reused.

LSC methods can be used on excised gel bands or evenly spaced slices along the length of the gel. Many tissue and/or gel solubilizers are commercially available that allow efficient counting of samples separated on polyacrylamide gels. These solubilizers can be acidic (e.g., perchloric acid) or, more commonly, alkaline (e.g., Soluene-350; PerkinElmer Biosciences; Solusol; National Diagnostics). Preparation of the sample can take place starting with a small volume of solubilizer with a gel fragment, or in many cases, the scintillation cocktail, solubilizer, and gel can be mixed together, incubated for a length of time specified by the manufacturer, and then counted. An additional option for counting polyacrylamide gels is available by using reversible cross-linkers in the gel (92).

17.4.7.3. Carbon Dioxide

Measurement of ^{14}C radioactivity as CO_2 is of major importance in metabolic studies and in the large number of applications that require combustion or digestion of carbonaceous material. For scintillation counting, the CO_2 is trapped in an organic base, for instance, hyamine hydroxide Carbo-Sorb (Packard BioScience BV, Groningen, The Netherlands), ethanolamine, KOH, or NaOH. With all of these, the CO_2 must be released with acid and swept through a train of traps containing the base. The base with trapped CO_2 is then added to a scintillant. The optimal scintillation cocktail depends on the trapping method used. Alternatively, scintillation cocktails that allow direct extraction of the $^{14}CO_2$ gas are available. For example, Oxosol C^{14} (National Diagnostics) can be used by sparging gas containing the $^{14}CO_2$ through the solution, or the gas can be shaken in a vial containing the cocktail (95).

17.4.7.4. LSC Instrument Operation

Single-Isotope Counting

The goal in single-isotope counting is to set the scintillation counter for the highest efficiency and the lowest background (91). Two types of adjustment are provided for each channel: (i) high and low discriminator settings and (ii) photomultiplier voltage and amplifier gain. Higher settings of the latter pair increase the pulse height. The high and low discriminator settings determine the pulse height range that will be accepted. The pulse height or energy spectrum of beta-emission is determined by setting the upper and lower discriminator with a fixed differential and then counting a radioisotope at increasing positions of the discriminator band, e.g., 0 to 10, 10 to 20, and 20 to 30. A plot of the count versus discriminator range midpoint gives the spectrum. Because beta-emitters have different energies of emission, each isotope has a characteristic peak height distribution, and hence the discriminators in various channels may be set to include or exclude various parts of the isotope energy spectrum. Once this has been determined, the lower discriminator is set to exclude most of the background (i.e., set at 10) and the upper discriminator is set to just include the maximum pulse height of the isotope being measured. In this way, many of the low-energy and high-energy background events are eliminated.

Determination of the peak height spectrum at various photomultiplier voltage settings also gives a family of curves (Fig. 3) which are especially useful in separating the radiation of each isotope in dual-labeled samples. Note that lower voltage settings narrow the discriminator settings necessary to produce the window.

In all cases, the background count rate must be determined with a vial of scintillation solution with no radioisotope and subtracted from each sample count. It is assumed that quench of the background measured as above is nonexistent when high sample count rates are measured. However, at very low count rates it may be necessary to determine the effect of sample quench as described below.

Dual-Label Counting

Dual-label counting can be accomplished when the energies of the isotope pair are sufficiently different that there is resolution of the two pulse height spectra (Fig. 3). This is practical for the ^3H and ^{14}C, ^3H and ^{35}S, ^3H and ^{32}P, and ^{14}C and ^{32}P pairs. The more energetic isotope is counted exclusively in channel 1. However, the count for the weaker

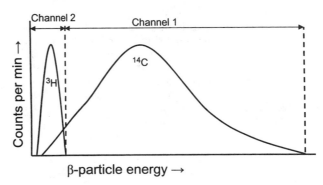

FIGURE 3 Pulse height spectrum showing counts per minute as a function of β-particle energy. Discriminators are set so that channel 1 counts exclusively the higher-energy ^{14}C and channel 2 counts all of the lower-energy ^3H with a small amount of ^{14}C overlap. After determining the counting efficiency of each isotope in each channel, the equations in section 17.4.7.4 are used to calculate the disintegrations per minute.

isotope in channel 2 will always have counts of the more energetic member of the pair. It is necessary to determine counting efficiency (instrument efficiency plus quench) for each isotope separately, preferably by internal or external standard methods, and then to calculate the count for the second channel from simultaneous equations:

$$\text{dpm of } {}^{14}\text{C} = \frac{\text{net count in channel 1}}{\text{efficiency of } {}^{14}\text{C in channel 1}}$$

$$\text{dpm of } {}^{3}\text{H} = \frac{(\text{net counts in channel 2}) - (\text{dpm of } {}^{14}\text{C} \times \text{counting efficiency of } {}^{14}\text{C in channel 2})}{\text{tritium efficiency in channel 2}}$$

The counting efficiency of channel 2 for the lower-energy isotope (^{3}H in this case) can be optimized first by adjustment of discriminator settings with a tritium-containing standard. Then, the lower discriminator of channel 1 is set to exclude counts from ^{3}H and the upper discriminator is set to give maximum efficiency for ^{14}C. The above procedure applies if the quench and efficiency of samples correspond to those of standards used. When quench is encountered, more of the higher-energy counts appear in the low-energy channel. The original relationship of channels can be reestablished by changing the discriminator settings if the needed information is available from similarly quenched standards. Alternatively, the gain of the photomultiplier or amplifier can be advanced to reestablish the pulse height pattern of the original window settings. Dual-label counting is fraught with difficulties and errors when low count rates are involved.

Quench Correction and Counting Efficiency

Under ideal conditions, not all radiation events are detected even in the best scintillation counters, although efficiencies of above 90% are obtained with intermediate and strong beta-emitters, e.g., ^{14}C and ^{32}P. Instrument efficiency can be determined by several means (described below) with a nonpolar counting medium which has neither chemical nor physical quenchers. More commonly, however, a combined instrument efficiency and sample quench correction factor is produced by one of the following three methods, each of which has advantages and disadvantages.

Internal-Standard Method

The original method used, the internal-standard method, is relatively simple and easy to understand. It is highly reliable and potentially very accurate for counting single isotopes. The correction includes both chemical and color quenching. It consists of first counting a sample (C) and then adding a small volume containing a known number of disintegrations per minute (D_s) of the same isotope: e.g., for ^{14}C counting, [^{14}C]benzoic acid, and for ^{3}H counting, [^{3}H]toluene. Then, the sample plus standard (C_t) is recounted. From this, the overall efficiency is (C_t - C)/D_s and the radioactivity of the sample is

$$\text{Sample dpm} = C \times D_s/(C_t - C)$$

The main disadvantages are that the vials must be removed from the counter to add the standard and then be recounted. The sample alone cannot be recounted as an afterthought. Addition of the standard to the vial must be accurate. For large numbers of vials, this is now accomplished with improved accuracy by using various kinds of automatic pipettes or syringes.

Channel Ratio Method

The channel ratio method is based on the observation that the spectrum of pulse heights in a given scintillant-isotope system shifts with the efficiency. In an instrument with two counting channels, the discriminators in each channel are set differently and the count in each channel is determined simultaneously and expressed as a ratio. This method can also be used in a single-channel instrument; however, it is necessary to make two counts, each at a different discriminator setting.

Using a set of standards prepared to have different efficiencies, a standard curve of efficiency versus channel ratio is constructed and used to correct the unknowns. In both channels the lower discriminator is set above the pulse height of background noise. One discriminator setting, either upper or lower, may be the same in both channels, with the other different. Alternatively, both upper and lower discriminator settings may be different for the two channels. Since quenching may be due to colorless chemicals such as water and CC1$_4$ and also to colored compounds, it may be difficult to prepare standards of the same quencher content as the unknowns. Therefore, there will be some error if colored quenchers are present in the unknowns, especially at high concentrations. However, this method applies over a wide range of conditions (94). Dual-channel systems require less time unless low counts are involved. In that case, much more counting time is required to obtain a ratio of given precision than for the internal-standard method. In this method, sample counting can be repeated.

External-Standard Method

Newer instruments are sometimes equipped with a gamma-source of known disintegrations per minute, which, on call by the instrument program, is positioned directly below the counting vial. Counts are made with and without the gamma-source. A standard curve of external standard counts versus counting efficiency (produced by introducing quenchers) is constructed and used to correct counts in unknowns. The method is susceptible to variations in volume of scintillant, thickness of the glass vial, and different scintillants. In addition, heterogeneity of solutions or counting of paper strips produces errors. Long counts are not needed at low radioactivity levels as for the channel ratio method. It is possible to improve the external-standard method described above by making a channel ratio approach with and without the gamma-source. In this case the external standard is counted in two channels with discriminators set above the pulse height of the isotope being counted to obtain counting efficiency. This method requires the availability of four channels or two counting runs at different discriminator settings.

17.4.7.5. Counting Solutions

Classical scintillation solutions were made using organic solvents such as toluene, pseudocumene, or xylene and containing a primary fluor (e.g., PPO) and a secondary fluor (e.g., POPOP). Other additives often include detergents and tissue- or gel-solubilizing agents. These scintillation cocktails can be easily made, or premixed versions can be purchased commercially (90). They have the advantage of low cost and generally high counting efficiency. A significant drawback is that these solvent-based scintillation cocktails are considered hazardous, flammable compounds and the added precautions and cost of disposal can be problematic. These can be lessened to some extent by using scintillation vials that have smaller geometry and use less volume.

The trend in recent years has been to switch to commercially available safer scintillation cocktails that are less hazardous, possess higher flash points, and in some cases, are nonflammable and biodegradable. There are many proprietary formulations, but most are 70 to 80% phenyl alkanes (e.g., Econoscint A; National Diagnostics) or naphthalenes (e.g., Ultima Gold; PerkinElmer Lifesciences, Boston, MA) with the remainder being composed of other proprietary compounds. Various other formulations are designed to optimize scintillation cocktails for the ability to accommodate samples of high water content, provide for negligible diffusion through plastic scintillation vials, and be biodegradable. Note that even though these formulations are classified as biodegradable, nonradioactive aliquots still need to be disposed of as organic hazardous waste by the proper chemical or radiation safety authorities.

17.4.8. Statistics of Counting

Determinations of radioactivity from count and time measurements under ideal conditions contain random errors which cause the measured count rate to depart from the true value (91). For this reason, one should have at hand statistical tools which aid in estimating error. The intelligent use of these methods requires that one be informed of the assumptions and terminology involved, if not how the final useful expressions are derived. It is not within the scope of this manual to present the application of these tools with the appropriate background, because various tables and graphs are needed to perform the operations. Rather, it is recommended that a general text such as reference 91 be consulted.

Nevertheless, several important statistical tools are listed and their equations are given below. These include average of several counts, standard deviation estimates of single and multiple counts, percent standard error (%SE) of counts and count rates, and time required for a desired %SE of count rates of samples with and without background for equal and unequal times for counting background. Additional expressions and tables are available for determination of times required for counting background and for total count with a given standard error, limits of detection of radioactivity, and the correct way to eliminate a count value which varies widely from the mean.

There are a number of different kinds of error: probable error, for which the true value is likely to exceed the error expressed 50% of the time; standard error, 68% of the time; and 99/100 error, 99% of the time. Since SE has a coefficient of 1 in many statistical derivations, and since this expression is most commonly used, only "standard" forms of equations will be given here. This means that, among members of a set of values, one value will not vary from the mean by (the standard deviation) more than 32% of the time. Since count errors are independent of time, one can determine the number of counts required for a given %SE by using the expressions below. For instance, for 1% SE, 10,000 counts are required; for 2.5% SE, 1,600; for 5% SE, 400; and for 10% SE, 100. From the corresponding expression below, the times required to determine both background and total count rates at 1% SE when both rates are equal (see calculations of the amount of radionuclide to use) are 804 min for background and 1,140 min for sample plus background; for 10% SE, 8 min and 11 min are required, respectively.

Equations for counting statistics are as follows.

1. Average of several counts

$$X = \frac{\Sigma \text{ individual counts (X)}}{\text{number of determinations } (n)}$$

2. Standard deviation from a number of counting cycles

$$SD = \left(\frac{(\text{deviation of } X \text{ from } \overline{X})^2}{n-1} \right)$$

3. SE (%) in count rate

$$SE \text{ of rate (\%)} = \frac{100}{\text{rate } (R) \times \text{time } (t)}$$

4. Time required to give the desired SE (%)

$$t = \frac{10^4}{R(\%SE)^2}$$

5. %SE of net sample count when the times of counting sample and background are equal

$$\%SE = \frac{100(X_{\text{total}} - X_{\text{background}})}{X_{\text{total}} - X_{\text{background}}}$$

6. %SE of sample count without background when there are unequal counting times for background and sample

$$\%SE = \frac{200(R_t/t_t + R_b/t_b)}{R_t - R_b}$$

where R_t and R_b are the total and background rates, respectively, and t_t and t_b are the times for R_t and R_b.

17.5. ELECTROPHORESIS

A number of separation methods make use of the electrophoretic force that propels charged molecules through various media. Perhaps the most popular electrophoretic method used in research laboratories is gel electrophoresis. Gels used for separations of proteins and short oligonucleotides are usually made of polymerized and cross-linked acrylamide and are discussed in the following section. Larger DNA segments and plasmids are usually separated on agarose gels and are presented elsewhere (chapter 30.3.2 and 30.4.1). Another method of electrophoresis that is being more commonly used is capillary electrophoresis, which is described in section 17.5.3. For these methods, the same principles apply: charged molecules are exposed to an electric field and move towards the electrode of opposite polarity. This migration occurs through a gel matrix or through a capillary column that is lined with a charged silica wall or other types of support.

17.5.1. Gel Electrophoresis

Separation of macromolecules by electrophoresis through a sieving gel has become a very important tool in biological and medical research as well as in clinical diagnosis and monitoring of industrial processes. Since the inception of the discontinuous disk gel method in 1964 by Ornstein (104) and Davis (101), for separation of blood proteins, the most popular method used is the modified technique developed by Laemmli (114) using denaturing polyacrylamide gels. Discontinuous gels are made of two stacked but contiguous gels that have differing porosities and pH: a lower resolving gel below a shorter stacking gel. Two different ions are present, one with greater mobility and the other with lesser mobility than the mobilities of the macroions of the biological material to be separated. Conditions are such that the sample macroions are sandwiched between the fast- and slow-moving buffer ions. When electrophoresis starts, free movement in the stacking gel concentrates the macroions into a series of very narrow bands at the top of the resolving gel. Applications have expanded beyond visualization of the number of components in a sample to include determination of apparent molecular weights of protein monomers and oligomers, estimation of isoelectric point (pI) by IEF, and visualization and sizing of intact and fragmented DNA molecules. Intensive investigations have led to a large body of theory, detailed information on conditions, and many variants of the buffer and gel system (99, 103, 107, 108).

It is the intent here to describe only the Laemmli method of polyacrylamide gel electrophoresis using sodium dodecyl sulfate (SDS)-denatured proteins, as well as a few commonly used variations of this technique. Refer to references 99 and 103 for more-comprehensive treatments of current general- and special-purpose methods, which include buffers for operation at constant and higher pH as well as a conductivity shift to aid in concentrating the sample and polyacrylamide gel separation of oligonucleotides. For pulsed-field gel electrophoresis, see chapter 30.3.2.2.

17.5.1.2. SDS-Polyacrylamide Gel Electrophoresis of Proteins

Electrophoresis of native proteins through polyacrylamide gels results in separation according to size, hydrodynamic shape, and charge. With all of these factors in play, it is possible that proteins of different molecular weights will migrate at the same rate. In addition, proteins that maintain tight associations will migrate as a single band, further complicating interpretation of the gel. Because of these concerns, electrophoretic estimation of the native mass of a protein or protein complex is to be discouraged. By running samples and controls in several gels with varying extents of polymerization, or by using pore gradient gels, some of these concerns can be ameliorated (113, 121). In contrast, excellent results can be obtained by analysis of proteins denatured with a detergent such as SDS before electrophoresis. The anionic detergent binds to typical proteins at a relatively constant ratio yielding an unfolded protein possessing an overall negative charge that is proportional to its molecular weight (108).

SDS-polyacrylamide gel electrophoresis is usually carried out in a vertical gel slab. Polyacrylamide gels are typically made vertically between two glass plates. Many commercially available devices are designed for this purpose to allow easy production of one to 10 or 20 gels simultaneously. The most popular glass plates are usually about 10 by 10 cm separated by 0.75- to 1.5-mm spacers at each side. The entire assembly is clamped together and placed on an apparatus to seal the bottom. The resolving gel is mixed, poured to a height that almost fills the space between the plates, and allowed to polymerize. The 30% (wt/vol) acrylamide stock solution used for making the gels is usually a mixture of acrylamide and a smaller amount of a crosslinker such as N,N′-methylene-bis-acrylamide (BIS). These two ingredients are present at a specific ratio, usually 29:1 or 29.2:0.8 acrylamide:BIS (expressed as % wt/vol), and are diluted into a solution containing the resolving gel buffer (at pH 8.8), water, and SDS, along with the ammonium persulfate and N,N,N′,N′-tetramethylethylenediamine (TEMED) required for the polymerization reaction. After filling the gel apparatus to the desired level with resolving gel solution, it can be overlaid with a nonmiscible solvent such as N-butanol to exclude air (which hinders the polymerization) and to create a uniform surface at the interface with the stacking gel. After the resolving gel has polymerized and the solvent is removed, a smaller 0.5- to 1-cm-high stacking gel of low porosity (in a pH 6.8 buffer) is poured and allowed to polymerize. Protein samples are boiled with buffer containing SDS and 2-mercaptoethanol to ensure complete denaturation before loading onto the gel.

It is important to note that acrylamide is a potent neurotoxin and care must be taken when handling solutions used for gel preparation. Acrylamide should be weighed in a well-ventilated area by using appropriate gloves, and any unused solutions and contaminated items must be disposed of as hazardous waste. As an alternative to preparation of gel solutions in the lab, premade acrylamide solutions, as well as precast gels designed to fit many popular formats, are available from several commercial sources.

Once an electric field is applied, the protein mixture is concentrated before entering the resolving gel by using an appropriate pair of buffer ions and pH such that the mobility multiplied by the degree of dissociation for one ion (Cl^-) is greater than the same value for protein, which in turn is greater than the value for the second ion (glycinate) (99, 109). The separation gel is more alkaline (pH 8.8), whereas the stacking gel and sample are in a different buffer at pH 6.8. When a potential is applied, the proteins quickly traverse the stacking gel and form a very thin band at the top of the separation gel with the proteins sandwiched between Cl^- and glycinate ions. The denatured, SDS-bound proteins then slowly penetrate the separation gel. The migration rate in the separation gel is a result of charge (dictated by the amount of bound SDS), which is proportional to molecular size. Since the bands were originally very thin, resolution is exceptionally high.

Separation of proteins in the discontinuous system is the result of two major factors: differences in migration as a result of the sieving effect of the cross-linked gel, causing separations based on size, and differences due to charge. In addition, unusually sharp bands exist after electrophoreses as a result of concentration of the proteins into very thin bands or disks before they enter the separation gel. Thus, molecular sieving is controlled by the degree of cross-linking, which is in turn dictated by the concentration of cross-linking acrylamide. Concentrating proteins into a narrow band before they enter the separation gel is characteristic of the discontinuous system and results from an ingenious manipulation of discontinuities of pH, buffer ion composition, and gel pore size of the sieving gel. Efficiency of the separation gel is increased by using pore gradient gels (121), in which the concentration of polyacrylamide uniformly increases throughout the length of the gel.

Analysis of Gels

Several methods can be used to directly detect protein bands in gels. Gels can be cut and scanned using specially equipped UV spectrophotometers. The proteins can be transferred onto membranes for Western blot analysis using immunological techniques (chapter 8.5.5). Gels of radioactive samples can be exposed to film for autoradiography or quantitated using phosphorimaging systems. More commonly, gels are stained with one of a variety of methods that allow qualitative visual analysis of results as well as quantitative measurements of band intensities (109, 120). The methods discussed here are staining with organic dyes and metal ions,

Staining using colored or fluorescent organic dyes

The most commonly used method of detecting proteins in gels is by staining with organic dyes. These include Ponceau S, amido black 10B, Coomassie brilliant blue R-250 (R, reddish hue), and Coomassie brilliant blue G-250 (G, greenish hue) (120). Coomassie brilliant blue dye staining is usually performed under acidic conditions which serve to fix the protein within the gel and prevent diffusion and to enhance binding of the dye to basic side chains on the protein. Staining solutions contain 7 to 10% (vol/vol) acetic acid, 10 to 45% (vol/vol) methanol, and 0.1 to 0.2% (wt/vol) dye. The amount of time required for complete staining is determined mainly by the thickness of the gel and can vary from several hours to overnight. The excess stain is then removed by successive washes in 10% (vol/vol) methanol, 7 to 10% acetic acid. Proprietary formulations of the Coomassie brilliant blue method are commercially available (e.g., BluePrint Fast Coomassie Stain from Promega, Madison, WI) that accelerate the staining and destaining process. Another type of organic dye used for protein detection is fluorescent stains. For example, new SYPRO stains (Molecular Probes, Eugene, OR) have the same sensitivity as silver-based methods (122). The SYPRO-stained gels are visualized by irradiation on a UV light box, and digital imaging can yield accurate quantitation of protein bands.

Silver staining

For the majority of circumstances, staining protein gels with organic dyes has sufficient sensitivity to provide the desired information. However, there are times when the absolute purity of a protein must be verified, or when the protein of interest is present in low abundance. Silver staining provides much higher sensitivity than that obtained by Coomassie brilliant blue, but multistep protocols are necessary. Three general types of silver stains are used: (i) diamine or ammoniacal, (ii) nondiamine chemically developed, and (iii) photoreduction (109, 120). Silver provides an excellent stain because it reacts with free amines and sulfur groups on proteins (117). Diamine silver stains utilize ammonium hydroxide to form soluble and stable diamine complexes. After washing the gel to remove SDS, silver diamine solution is added and acidified with citric acid in the presence of formaldehyde. The citric acid decreases the free ammonium ion concentration, and the released silver ions are reduced to metallic silver by the formaldehyde. When done carefully, the deposition of metallic silver occurs on the proteins without nonselective staining. The nondiamine chemical development methods rely on reaction of dissolved silver nitrate with protein under acidic conditions with subsequent reduction to metallic silver by formaldehyde under alkaline conditions. Alkaline buffers such as sodium carbonate can be used for the development step so that the formic acid that is formed from the oxidized formaldehyde does not alter the pH significantly. The photoreduction method for silver staining is accomplished by using light energy to reduce ionic silver to its elemental state. It is fast and can provide visualization of protein bands in as little as 10 min; however, it is less sensitive than the other silver methods.

Reverse staining using copper or zinc

The organic dye and silver staining methods described above are positive stains whereby black or colored bands appear in an unstained background. Several methods have been developed for "reverse" or negative staining. These methods are used almost exclusively on SDS-polyacrylamide gels and result in an opaque background formed of insoluble metal with translucent or clear protein bands. The lighting arrangement required for photographic documentation is different from what is used for standard backlit positively stained gels. The method has greater sensitivity than Coomassie brilliant blue, but less than silver staining, and is

quite fast since the gel does not require prior fixation (109). Since the protein bands themselves are unstained, they can be eluted and used for a variety of subsequent procedures such as amino acid analysis and sequencing.

Recording results and quantitation

Stained gels can be photographed, dried on filter paper, or dried onto cellulose membranes to yield a clear, flat gel suitable for photography or direct overhead projection. Quantitative measurements are obtained from (i) digital photography and quantitation of stain intensity by using one of several commercially available systems, (ii) similar methods using gel stained with fluorescent dyes and illuminated with UV light, or (iii) analysis of radioactivity-containing gels as described in section 17.4.7.2.

17.5.2. Other Types of Polyacrylamide Gel Analysis

In addition to the SDS-polyacrylamide gels described above, several variations are commonly used as summarized below. Other methods involving immunological techniques are presented in chapter 8.5.3.2.

Native Gel Electrophoresis

It is possible to separate proteins in different ways by selective use of denaturants. For example, omission of SDS, 2-mercaptoethanol, or both from the gel and sample buffer and elimination of boiling of the sample during its preparation can be used for gel electrophoresis. For example, comparison of SDS gels run with and without the 2-mercaptoethanol disulfide reducing agent can give information about inter- or intrasubunit disulfide bonds. Electrophoresis of proteins omitting both denaturing agents is the classic method of Ornstein (104) for separation of native proteins.

Location of enzymes

An enzyme in a nondenaturing gel can be located by immersing the gel or gel sections in a catalytic reaction mixture in which conversion of the substrate leads to a colored product (116). After incubation for a few minutes, color develops in a disk where the diffusing reactants contact the enzyme. Alternatively, when reagents do not diffuse into the gel, the reaction produces a ring around the gel where the band of enzyme is exposed.

Two-Dimensional Electrophoresis

A single band on a gel can contain more than one protein, and incomplete separation of complex mixtures can occur. The most effective approach to achieve separation is to base the separation on two strategies, pI and size. After initial separation by IEF, SDS gel electrophoresis is carried out and the slab is stained. A particularly effective example of the original method is described by O'Farrell (118), who separated 1,000 proteins from an *Escherichia coli* extract. Many improvements have been made since the mid-1970s, and the procedure has of necessity become very complex. This method is capable of separating up to several thousands of proteins on a single gel (100, 103, 111, 112). Combining two-dimensional gels with extremely accurate mass spectroscopic analysis of protein spots has given rise to the field of proteomics (119).

Samples are usually prepared from disrupted cells or tissues, and the proteins are then solubilized with denaturants such as urea or detergents. Gels consisting of an immobilized pH gradient (IPG) in a low-porosity polyacrylamide matrix can be prepared in the lab (111) or, more commonly, purchased commercially (112). These commercially available precast IPG gels are usually supplied in a dried form that must be rehydrated before use. Separation is achieved using dedicated electrophoresis systems designed to allow for easy removal of the IPG gel in a format that is suitable for placing horizontally atop a subsequent SDS-polyacrylamide gel. After electrophoresis in the second dimension, proteins are visualized by staining or blotted onto membranes for subsequent analysis.

Postelectrophoresis Analysis

In many cases, separation of proteins by gel electrophoresis is only a first step in analysis of a sample. Stained protein bands or spots can be excised from a gel for extraction, proteolytic cleavage, and/or mass spectrometry. Gels can also be blotted onto protein-binding membranes for subsequent analysis using immunological methods, microsequencing, amino acid composition, or mass spectrometry. Details of these methods are presented in chapters 8, 27, and 30 and in appropriate chapters in reference 119.

17.5.3. Capillary Electrophoresis

Capillary electrophoresis (CE) separates molecules through a capillary tube driven by an electric field (106, 115). The columns are often made of fused silica, are 20 to 100 μm in diameter and 10 to 100 cm long, and contain an electrolyte solution. When an electric field is applied, the sample is introduced at the anode end of the capillary and the charge on the inner wall of the capillary results in flow (electro-osmotic flow) from the anode to the cathode, which drives analyte molecules through the system. At the cathode end of the capillary is a detector to record the chromatographic process. In addition to electro-osmotic flow, electrophoresis separates the molecules on the basis of charge. Molecules with a positive charge flow in conjunction with the flow while negatively charged molecules move in opposition, resulting in an enhanced separation. Furthermore, the pH of the electrolyte can be adjusted to alter the electro-osmotic flow, or special coated capillaries that have reduced or no flow can be used. CE is useful for separation of a wide variety of molecules including metal ions, carbohydrates, chiral separation of small molecules, proteins and polypeptides, and oligonucleotides. Detection for most commercially available systems is by UV absorbance, although diode array and fluorescence detection are also available. CE instruments can also be coupled with mass spectroscopy to yield molecular weight and structural information. Other modifications of capillaries used in CE are packing with silica or ion exchange resins (capillary electrochromatography) and packing with affinity matrix (affinity capillary electrophoresis). The major advantages of CE include the ability of a single instrument to be used for a wide variety of column types, short separation times, and the small amount of material required. However, the necessity for a very high voltage power supply and instrumentation for detection using short light paths results in a significant cost for the entire integrated system (115).

17.5.4. Isoelectric Focusing

IEF is a special kind of electrophoresis that involves migration of charged molecules in an electric field across a liquid column with a stable pH gradient (99, 102, 103, 105). Amphoteric molecules such as proteins change charge as they traverse the pH gradient and come to a region where the pH is the same as the pI of the protein. At this point of equal plus and minus charges, the protein no longer migrates and the band is maintained in an equilibrium between the forces of diffusion and electrophoretic movement. Thus, separation is based on different pI values. Early

applications of this method stabilized the column against convection by use of either a sucrose density gradient or Sephadex. Most IEF is done using horizontal or tube polyacrylamide gels of low porosity to allow migration of the proteins. A mixture of proteins is introduced and separates into zones at the pI values of the components of the mixture. The procedure can separate components that differ by 0.005 pH unit or less. The ampholytes used to develop the stable pH gradient are available that cover wide and narrow pH ranges. Both microanalytical (102) and preparative (105) methods have been developed.

17.5.5. Immunoelectrophoresis and Western Blotting

See chapter 8.5.3 and 8.5.5.

17.6. ENZYME PURIFICATION

Investigation of the function of enzymes and other biologically active proteins often requires purification to remove the influence of extraneous factors and facilitate measurements. The purification process involves the application of a series of treatments which differentially separate components to purify and concentrate the target protein. Usually, preservation of biological activity is required, in which case an acceptable proportion of the original activity (units) must be recovered and the specific activity of the purified fraction (units per milligram of protein) should be greatly increased as well.

A detailed presentation of many aspects of this field is beyond the scope of this volume. The investigator is referred to specialized books (109, 124, 127), as well as to the series *Methods in Enzymology* (124, 125), for comprehensive treatment of, and procedures for, specific enzymes. In most cases, one or more purification procedures will have been published for a protein of interest. The general experience has been, however, that initial attempts to follow a procedure, even by experienced workers, often give disappointing results. There are many reasons for this, including an inability to closely follow the directions, the omission of details not recognized as important by the author(s), and variation in chemicals, apparatus, and starting material. Further, performing the procedure "blind," that is, without monitoring for biological activity and protein content, almost always results in failure. Nevertheless, careful work accompanied by appropriate assays usually yields the desired material.

A broad strategy for enzyme purification often includes the following steps: (i) optimization of starting material to achieve the highest possible fractional abundance of the desired enzyme by varying growth conditions, strain, and recombinant plasmid; (ii) bulk precipitation methods using either cell extracts or extracellular culture fluid to remove contaminating material and further increasing fractional abundance; and (iii) different chromatographic, electrophoretic, or other procedures as needed for purification. Maintaining enzyme stability and appropriate enzyme assay procedures are important prerequisites. Details of these general methods are presented below.

The design and testing of a new purification procedure is a challenging but rewarding venture. There are numerous individual operations that can be applied to enrich the target protein; some of these are described in this and other chapters. A primary but often unrecognized purification strategy can be effected before the extract is prepared, by simply enhancing the activity present in the cells.

Understanding growth conditions for maximizing enzyme production, including enzyme induction and recombinant DNA manipulation for increased expression, has the effect of raising the starting specific activity and hence amounts to a purification step.

Steps must be taken to protect the activity of enzymes, especially in crude extracts. Losses are caused by proteolytic and other enzyme activities, by oxidation of sulfhydryl groups, and by dissociation of coenzyme and metal cofactors. At a minimum, the preparation of crude extracts and manipulation of various fractions should be performed in an ice bath and using a minimum of delay. In addition, proteinase inhibitors, sulfhydryl-reducing agents, and coenzyme/cofactor and other treatments may help stabilize activity.

To produce an efficient procedure, the operational parameters (pH, temperature, salt concentration, etc.) of each step must be optimized, usually on the basis of a series of pilot tests, as follows. The use of sequence information can help in decisions about what types of purification steps to use. For example, many NAD(P)/NAD(P)H-utilizing oxidoreductases bind to a 2',5'-ADP Sepharose affinity resin, and the calculated pI may be useful for deciding what pH range to test. Select a separation process, and apply it to a series of small aliquots of the preparation to be purified while varying the level of one parameter to be optimized (e.g., the quantity of adsorbant). Following separation (often into a series of fractions), assay for activity and protein. Construct graphs of activity recovered and specific activity versus levels of the parameter being studied to establish the optimum for that parameter. Repeat this procedure for each parameter under study. When this is finished, run a second pilot test using the optimal value established for each parameter. If there is purification (i.e., a satisfactory increase in specific activity) and good recovery of the total units treated, process a larger amount of the preparation by using the best conditions found for that step to obtain material for pilot tests performed by a different separation method. Proceed in this fashion with additional steps until a satisfactory level of purification is achieved and there is sufficient activity remaining for the experiments planned. This iterative exercise consumes considerable material and may require starting over several times. It is useful to develop an accounting spreadsheet in which vertical columns record such important information as volumes, assay results, units per milliliter, total units, protein content, specific activity, and purification factor. Horizontal rows are assigned to different steps in the process. From this assembly of data into a purification table, it is easy to monitor success or failure as the procedure is applied.

In developing a procedure, a twofold improvement or less may be considered sufficient when starting with a crude extract, because of the large volume involved and the large amount of protein and other solubles present. However, a greater purification must be achieved in subsequent steps, otherwise too many steps (with unacceptably high losses in activity) would be needed to reach reasonable purity. For many steps a 5- to 10-fold purification or more is to be expected. Similarly, up to a 50% loss in total activity may be acceptable in the first step but not in the following steps.

This section describes a few of the popular and traditional procedures that were not covered earlier in this chapter. Many types of chromatography, treated above in a general way, adequately describe application to enzyme purification and are only referenced here. Thus, ion-exchange (17.3.1), adsorption (17.3.2), reversed-phase and hydrophobic (17.3.3), high-performance (17.3.6), affinity (17.3.8), and gel permeation (17.3.9) chromatographic methods have been successfully used in enzyme purification. The following additional techniques are specific to enzyme purification.

17.6.1. Preparation of Cell Extract

Preparation of a cell extract, which must precede the fractionation operations, is essentially a purification step. The goal is to achieve complete breakage with minimal loss in activity. Following centrifugation to recover the extract, considerable inactive material is removed. For a discussion of the various breakage methods, see chapters 7 and 21 and also reference 128.

17.6.2. Removal of Nucleic Acids

In general, bacterial extracts differ from eukaryotic extracts in containing high content of nucleic acids (mostly DNA), which interfere with the differential separation usually obtained with many methods of purification. Therefore, the first step is to dissociate nucleic acid-protein complexes and precipitate nucleic acids. This situation holds true whether the DNA is highly polymerized, fragmented, or hydrolyzed. In some circumstances where the enzyme of interest is of high fractional abundance (when expressed from a recombinant plasmid), the amount of nucleic acids may not be a problem for the first purification step. A centrifugation at $100,000 \times g$ often suffices for removal of cellular debris. Treatment of extracts with nucleases such as DNase I, while not removing the DNA, decreases viscosity and allows for easier manipulation. Much of the resulting fragmented DNA is lost during subsequent dialysis and chromatographic steps.

When it is necessary to remove nucleic acids, two different strategies are used, depending on whether there is substantial polysaccharide in the crude extract. In the absence of polysaccharides, add solid ammonium sulfate to a final concentration of 0.2 M and then add protamine sulfate (2%, pH 5) to 20% of the crude extract volume and centrifuge. When polysaccharide is present, totally precipitate the protein with ammonium sulfate (2.8 to 3.2 M) or with about 1.3 volumes of acetone (section 17.7.3). Recover the precipitate by centrifugation, and dissolve the protein in a buffer (pH 6 to 7). Add protamine sulfate as above, and centrifuge to remove the protamine-DNA complex. A rough evaluation of the success of this step is made by determining the ratio of A_{260}/A_{280}. The ratio of nucleic acid to protein can then be estimated (127); the ratio for a crude bacterial extract is usually about 0.5. Following treatment with protamine, the value usually increases to an unimpressive 0.6 to 0.8, and hence the success of the step is in doubt. However, the general experience is that in subsequent steps, the ratio continues to rise and approaches the value for pure protein (ca. 1.75).

17.6.3. Fractionation by Precipitation

Proteins are precipitated either by high salt concentrations, by solvents, or by approaching the pI of the target protein (129). At best, fractionation by precipitation gives a modest purification increase as a result of coprecipitation of other proteins. However, such a step early in the procedure may be advantageous as a means of handling a large volume and removing a large amount of nonproteinaceous material. Pilot tests are used to determine the cuts to make. Each fraction is recovered by centrifugation and dissolved in buffer at a higher protein concentration. Ammonium sulfate is commonly used for salt fractionation. Enzyme grade salt is added slowly with stirring to the desired concentration. Alternatively, a saturated solution (with or without pH adjustment) may be used. In the past "percent saturation" has been used to express the concentration of salt involved. This has proven to be a crude and unreliable term, largely because the concentration at saturation varies, especially with temperature. This term has been largely supplanted by the use of molarity. Consult references 129 or 132 for amounts and volumes to use to achieve desired ammonium sulfate concentrations. The protein precipitates are removed by centrifugation and dissolved in buffer.

Additional precautions are necessary when using acetone or ethanol for purification. Close temperature control is required because there is a substantial temperature rise in diluting acetone or ethanol. In addition, proteins may denature unless the temperature is maintained near 0°C or below.

A general solvent precipitation procedure is as follows. Dilute the acetone or ethanol to 90%, and chill to −20°C. Set up a stirrer, a thermometer, and a separatory funnel to serve the contents of a beaker in a salt-ice bath. Arrange the funnel so that the solvent is introduced to the wall of the beaker. Add the preparation to the beaker, and start rapid stirring. As the temperature approaches 0°C, start the slow addition of cold solvent. The temperature should continue to drop as solvent is added, but the solution should not freeze. When solvent addition is complete, rapidly centrifuge the preparation in prechilled vessels at −10 to −15°C. Rapidly dissolve the precipitate in buffer, and, if necessary, remove insoluble material by centrifugation. Some solvent will remain in the recovered fraction, and this could lead to inactivation with time. If a subsequent fractionation step can follow immediately, then proceed. Otherwise, completely precipitate the protein with ammonium sulfate and dissolve the precipitate in buffer.

17.6.4. Fractionation by Denaturation

Some enzymes are remarkably stable under conditions that denature and precipitate most other proteins. If such a situation can be defined by pilot tests, the target protein may be substantially purified merely by removing the denatured protein. For this purpose, heat, acid, and solvents are commonly used.

Heat denaturation is pH and time dependent; therefore, run the pilot tests over a temperature range at several pH values. Prepare graphs to visualize the profiles obtained. Take care in scaling up from pilot studies because the rate of temperature rise and fall will be quite difficult for the larger volume. Therefore, place the preparation in a stainless-steel beaker, and immerse it in a water bath with a temperature substantially above the desired holding temperature. When the desired temperature is reached, transfer the beaker to a water bath at that temperature hold for the specified period and rapidly cool the beaker in an ice bath with stirring. The time-temperature relationship should be held as closely as possible to that of the selected pilot condition. For acid denaturation, the pI and the pH of maximal activity usually are not in the region of maximal stability. Hence, pilot tests are required to find the pH region for maximal stability. Either HCl or acetic acid is added, depending on the pH to be achieved.

A dramatic example of acid denaturation is seen in the purification of 2-keto-3-deoxy-6-phosphogluconate aldolase from *Pseudomonas putida* (130). This is carried out as follows. Add 4 N HCl to the crude extract at 1 to 4°C to give a final concentration of 0.2 N HCl. After 15 min, re-

move the copious precipitate by centrifugation and, without prior neutralization, perform an ammonium sulfate fractionation on the supernatant. A total of 70% of the activity is recovered, and a 10-fold purification is achieved in the acid step.

Protein denaturation by solvent takes place at a higher temperature than protein fractionation by solvent. Pilot studies should establish the pH, salt concentration, temperature, and solvent type and concentration for optimal results.

17.6.5. Purification by Chromatography

Separations by chromatography and gel filtration make use of charge/pH, hydrophobic/hydrophilic, ligand-binding, and size properties of proteins. Specialized supports and matrices which optimize the binding of macromolecules have been developed. Thus, ion-exchange purification of enzymes uses derivatized celluloses, dextrans, and agaroses which have better operating characteristics than do the corresponding resins. See section 17.3.1.2 for details. Similarly, media for adsorption and elution (section 17.3.2) in both batch and column mode and hydrophobic (section 17.3.3) and affinity (section 17.3.8) strategies are well developed and are especially suited to protein purification in standard laboratory equipment. HPLC (section 17.3.6), while developed as an analytical tool, has been adapted to larger-scale separations by using specifically developed equipment. Gel permeation chromatography (section 17.3.9), using media with highly controlled pore sizes, also provides useful separations.

17.6.6. Integrated Purification Systems

A variety of totally integrated purification systems are commercially available that automate sample introduction, column pumping and gradients, effluent monitoring, and fraction collecting. These systems can generally be classified into two categories: low pressure and medium to high pressure. Low-pressure systems operate at a maximum of around 30 lb/in^2 and consist of a pump, injection valve, column fittings, online absorbance monitor, and fraction collector. All of the modular components are orchestrated by a computer interface (e.g., Bio-Rad Biologic LP system) or by a programmable fraction collector (ProTeam LC Gradient System; Isco, Lincoln, NE). Columns can be purchased premade, but in many cases adequate purification results can be achieved from columns prepared in the lab using medium pressure columns (e.g., XK columns; Amersham Biosciences). Computer or microprocessor control allows for the purification step, after optimization in pilot studies, to be fully automated with minimal time required by the operator after sample injection. Higher-pressure systems (e.g., ÄKTA System from Amersham Biosciences or BioLogic DuoFlow System from Bio-Rad) have similar features but have pumps that go to pressures of 1,000 to over 3,000 lb/in^2 and HPLC-type tubing and fittings. These systems have designs similar to those of HPLC instruments, but all surfaces, pump heads, valves, and tubing are biocompatible and consist of plastic or noncorroding and nonleaching metal. Columns are usually purchased ready-made. For example, a Mono-Q ion-exchange column from Amersham Biosciences is made of resin with a very uniform bead size of 10 μm. This results in very high resolution of peaks, but the small bead size restricts flow and thus the higher pressures are necessary. All of these integrated systems can save operator time and possibly provide material of higher purity than conventional gravity chromatography. However, the down side is the high cost, and careful consideration must be given to the time savings versus the price. In many cases, early fractionation steps can be performed by precipitation methods or gravity flow low-resolution chromatography, and the purification system used for the final steps.

17.6.7. Affinity Elution with Substrate

In a few but striking instances, sorbed enzyme can be specifically eluted from an ion exchanger with substrate or competitive inhibitors (and presumably also with effectors). Because of the specific binding of these ligands to the enzyme, this process is a form of affinity elution chromatography. In many cases the binding to substrate imparts an elution behavior best explained by the cancellation of charges on the enzyme by the substrate. In others, a change in tertiary structure in response to substrate binding may be involved. Therefore, the adsorption of positively charged enzymes to CM-cellulose (negatively charged) can be overcome by binding the substrate (negatively charged) to the enzyme, which effectively displaces the CM-cellulose and elutes the enzyme. See reference 127 for detailed information.

17.6.8. Dye Ligand Chromatography

A number of dyes bind to proteins, especially enzymes. When the dye is bound to a porous support such as dextran or agarose, the enzyme is selectively picked up from solution. Following a wash to remove unbound material, the enzyme is eluted, with usually substantial purification. Originally, Cibachron Blue F3GA was found to bind a number of enzymes, especially those that had nucleotide phosphate coenzyme or substrate-binding sites. Recently, this method has been highly developed (123). A large number of dyes are available, and procedures for attachment to solid supports, selection of dye ligand for adsorption, and conditions for elution have been published.

The preferred strategy is to select a pair of dye-liganded supports such that one binds a substantial fraction of contaminating proteins but not the target enzyme and the second binds the target enzyme preferentially. A series of dye supports are screened to find the best pair for the purpose. A comprehensive dye ligand kit is available from Sigma (St. Louis, MO). Dye-liganded supports are also available from Amicon Co., Danvers, MA. Elution may be nonspecifically effected by high salt (up to 1 M NaCl), high pH, incorporation of solvent (ethylene glycol), or chaotropic agents. In addition, affinity elution can be effected with a substrate or competitive inhibitor, often with improved purification.

17.6.9. Crystallization

Crystallization of enzymes is often thought of as a process of controlled crystal growth for the purposes of X-ray structure determination. Crystallization is also used as a method of purification. However, crystallization from impure preparations is difficult, and hence this step is usually the last in a purification procedure. Considerable experimentation can be performed quite satisfactorily with noncrystalline but substantially purified material. In addition, crystallization has proven to be a dubious criterion of high purity of the protein. However, chemical and physical analyses of enzyme composition and structure require a purity level obtained by crystallization, and of course, the crystals themselves, generally grown under slower, more controlled conditions, are needed for X-ray crystallography.

Crystallization occurs when a gradual change in conditions causes a transition from the solution phase to the solid phase. When this occurs rapidly, disordered precipitation results, yielding amorphous material; when it occurs slowly, ordered deposition-yielding crystals are obtained. Dialysis is a preferred means of developing a time-dependent decrease or increase in salt or solvent (polyethylene glycol) concentration. An example in the use of crystallization as a final purification step is with recombinant lactate monooxygenase, where crystallization is obtained by overnight dialysis in pH 5.4 buffer (131). In some cases, however, gradual addition of a salting-out agent is effective. Microdialysis/diffusion devices which use 100 μl or less of material and which provide for regulation of the rate of concentration increase or decrease have been developed.

17.7. CONCLUDING COMMENTS AND FUTURE TRENDS

This chapter attempts to describe analytical and purification methods most commonly used by the microbiologist. There are many other methods that rely on instrumentation that is too expensive to be found in all but a few individual labs. These methods are therefore usually performed in a centralized dedicated facility located within a university or corporate setting. While it is beyond the scope of this chapter to provide complete details of these methods and instrumentation, it is useful to provide a brief summary along with some specific examples of recent innovative applications to microbiology.

Calorimetry. There are two main types of calorimetric techniques used to examine biological molecules: differential scanning calorimetry (DSC) and isothermal titration calorimetry (ITC) (133, 134). DSC measures change in partial heat capacity and is usually used to examine thermodynamics of temperature-induced conformational changes such as protein unfolding or double-stranded to single-stranded DNA or RNA transitions. For example, a common experiment involves observing thermal denaturation of proteins or protein-ligand complexes as the sample temperature is increased. ITC measures the amount of energy that is required to maintain a constant temperature of a sample upon addition of a ligand into a solution containing its cognate binding partner. In the case of an enzyme binding to and reacting with its substrate(s), this sensitive measurement can be used as a nearly universal assay method (136). DSC and ITC can yield important thermodynamic parameters of the particular process being examined.

Flow cytometry. Flow cytometry is a method of analyzing populations of cells (135). The technique relies on a continuous flow in which individual suspended cells are exposed to a light source or laser beam and the resulting absorbance, light scattering, or fluorescence is recorded. Instrumentation consists of a fluidic injection system, flow cell, light source, a light measurement subsystem (photomultiplier with associated filters, or a photodiode array), and computer interface. The internal shape of the flow cell places the passing cells in a very narrow stream such that up to 10^4 cells/s can be analyzed. Light scattering and absorbance yields information about cell size distribution. The most common method of monitoring cell populations is by fluorescence, either intrinsic, or extrinsic using added dyes that bind to specific cell components. For example, bacterial viability can be determined by using certain dyes that specifically bind to DNA with a concomitant increase in fluorescence quantum yield (143). Another area in which flow cytometry has proven useful is in analysis of complex microbial communities (140, 144).

In addition to the above analytical uses, flow cytometers can be augmented to perform cell sorting. The flowing suspension is formed into droplets that are given an electrical charge immediately after scanning by the light source. The droplets are then deflected by electric fields into collection tubes (135). Thus, it is possible to use cell sorting in combination with simultaneous fluorescence and light-scattering measurements to separate members of an aquatic planktonic bacterial community (142). The major drawbacks of flow cytometry are the high cost of the instrumentation and the requirement for dedicated trained personnel to operate the system.

Mass Spectrometry. Mass spectrometry is an extremely accurate method of determining the molecular weight of small and large molecules (137). The instrumentation involves ionization of the sample, separation of the resulting ions according to their mass to charge ratio by mass analyzers, and detection of the separated ions. Two of the most popular methods for ionization are matrix-assisted laser desorption ionization (MALDI) and electrospray ionization (ESI). MALDI involves mixture of a sample with an organic matrix followed by exposure to laser light and can be used for analysis of mixtures of proteins, peptides, oligonucleotides, glycoproteins, and polysaccharides. In ESI, the sample is introduced as fine droplets that become charged in an electric field. As the droplet decreases in size, ions eventually begin to dissipate and are then directed into the mass analyzer. ESI does not work as well as MALDI for analysis of complex mixtures. Thus, this method is ideal for analyzing biomolecules after some type of separation has occurred. Examples such as coupling ESI to effluents of analytical chromatographic methods are mentioned in section 17.3. Small molecules can be singly or doubly charged, while molecules >2,000 Da lead to a series of charged forms.

Mass spectrometry has numerous applications in the field of microbiology. One of the most widely used is proteome analysis (138). Mass spectrometry is also commonly used for analysis of isolated proteins, lipids, carbohydrates, and other cellular components. MALDI analysis is often applied to protease-treated proteins for mapping disulfide bonds and investigation of protein-protein interactions in which the associating proteins have been chemically cross-linked (137). An emerging area in the application of mass spectrometry to microbiology is analysis of whole cells and crude cell extracts that have not undergone any previous separation step. This approach can be also used to monitor recombinant protein expression, analyze cellular RNA and DNA, and detect specific virulence markers. Furthermore, protein profiles of cells, cell lysates, and cell-free extracts can be used for very rapid taxonomic identification of microbes (138, 139, 141).

17.8. REFERENCES

17.8.1. Photometry

17.8.1.1. General References

1. **Christian, G. D.** 1994. *Analytical Chemistry*, p. 323. John Wiley & Sons, New York, NY.
2. **Clark, B. J., T. Frost, and M. A. Russell (ed.).** 1993. *UV Spectroscopy: Techniques, Instrumentation, Data Handling.* Chapman & Hall, London, United Kingdom.
3. **Gore, M. (ed.).** 2000. *Spectrophotometry and Spectrofluorimetry : a Practical Approach.* Oxford University Press, Oxford, United Kingdom.
4. **Knowles, A., and C. Burgess.** 1984. *Techniques in Visible and Ultraviolet Spectrometry*, vol. 3. *Practical Absorption Spectrometry.* Chapman and Hall, New York, NY.
5. **Lakowicz, J. R.** 1999. *Principles of Fluorescence Spectroscopy.* Kluwer Academic/Plenum, New York, NY.
6. **Sommer, L.** 1989. *Analytical Absorption Spectrophotometry in the Visible and Ultraviolet: the Principles.* Elsevier, Amsterdam, The Netherlands.

17.8.1.2. Specific References

7. **Bongaerts, R. J. M., I. Hautefort, J. M. Sidebotham, and J. C. D. Hinton.** 2002. Green fluorescent protein as a marker for conditional gene expression in bacterial cells. *Methods Enzymol.* **358:**43–66.
8. **Eisenthal, R., and M. J. Danson (ed.).** 1992. *Enzyme Assays: a Practical Approach.* Oxford University Press, Oxford, United Kingdom.
9. **Gil, M., D. Escolar, N. Iza, and J. L. Montero.** 1986. Accuracy and linearity in UV spectrophotometry with a liquid absorbency standard. *Appl. Spectrosc.* **40:**1156–1161.
10. **Mulrooney, S. B., and C. H. Williams, Jr.** 1997. Evidence for two conformational states of thioredoxin reductase from Escherichia coli: use of intrinsic and extrinsic quenchers of flavin fluorescence as probes to observe domain rotation. *Prot. Sci.* **6:**2188–2195.

17.8.2. Ion Electrodes

17.8.2.1. General References

11. **Bassey, E. J. S., and E. E. Edouk.** 2000. *Basic Calculations for Chemical and Biological Analysis.* A.O.A.C. International, Gaithersburg, MD.
12. **Fry, C., and S. E. M. Langley.** 2000. *Ion-Selective Electrodes for Biological Systems.* Gorden & Breach Publishing Group, New York, NY.
13. **Glaster, H.** 1991. *pH Measurement: Fundamentals, Methods, Applications, Instrumentation.* John Wiley & Sons, Somerset, NJ.
14. **Prichard, E.** 2003. *Measurement of pH.* The Royal Chemical Society, Cambridge, United Kingdom.
15. **Weast, R. C. (ed.).** 1980. *CRC Handbook of Chemistry and Physics*, 60th ed. CRC Press, Inc., Boca Raton, FL.
16. **Westcott, C. C.** 1978. Selection and care of pH electrodes. *Am. Lab.* **10(8):**71–73.

17.8.2.2. Specific references

17. **Bertocchi, P., D. Compagnone, and G. Palleschi.** 1996. Amperometric ammonium ion and urea determination with enzyme-based probes. *Biosens. Bioelectron.* **11:**1–10.
18. **Buck, R. P., and V. V. Cosofret.** 1993. Recommended procedures for calibration of ion-selective electrodes. *Pure Appl. Chem.* **65:**1849–1858.
19. **Cabral, J. P. S.** 1994. Comparison of methods to assay ammonia in bacterial suspensions. *J. Microbiol. Methods* **19:**207–213.
20. **Wise, R. R., and A. W. Naylor.** 1985. Calibration and use of a Clark-type oxygen-electrode from 5 to 45-degrees-C. *Anal. Biochem.* **146:**260–264.

17.8.3. Chromatography

17.8.3.1. General References

21. **Amersham Biosciences.** 2003. *Affinity Chromatography, Principles and Methods.* Amersham Biosciences, Piscataway, NJ.

22. **Amersham Biosciences.** 2003. *Ion Exchange Chromatography, Principles and Methods.* Amersham Biosciences, Piscataway, NJ.

23. **Bailon, P., G. K. Ehrlich, and W. Fung.** 2000. *Affinity Chromatography: Methods and Protocols.* Humana Press, Totowa, NJ.

24. **Gooding, K. E., and F. E. Regnier.** 2002. *HPLC of Biological Macromolecules.* Marcel Dekker, New York, NY.

25. **Hagel, L.** 1998. *Gel Filtration,* p. 79–143. *In* J. C. Janson and L. Ryden (ed.), *Protein Purification, Principles, High Resolution Methods, and Applications,* John Wiley & Sons, Hoboken, NJ.

26. **Hermanson, G. T., A. Krishna Mallia, and P. K. Smith.** 1992. *Immobilized Affinity Lligand Techniques.* Academic Press, San Diego, CA.

27. **Kastner, M. (ed).** 1999. *Protein Liquid Chromatography.* Elsevier, New York, NY.

28. **Kennedy, R. M.** 1990. Hydrophobic chromatography. *Methods Enzymol.* **182:**339–343.

29. **Kline, T. (ed.).** 1993. *Handbook of Affinity Chromatography.* M. Dekker, New York, NY.

30. **Lucarelli, C., L. Radin, R. Corlo, and C. Eftimiadi.** 1990. Applications of high-performance chromatography in bacteriology. *J. Chromatogr.* **515:**415–434.

31. **Mori, S., and H. G. Barth.** 1999. *Size Exclusion Chromatography.* Springer Verlag, New York, NY.

32. **Ostrove, S.** 1990. Affinity chromatography: general methods. *Methods Enzymol.* **182:**357–371.

33. **Scott, R. P. W.** 1995. *Techniques and Practice of Chromatography.* M. Dekker, New York, NY.

34. **Sheehan, D.** 2003. *Physical Biochemistry, Principles and Applications.* John Wiley & Sons, Hoboken, NJ.

35. **Smith, C.** 1998. Liquid chromatography: products in the protein chemist's tool chest. *Scientist,* **12:**14.

36. **Weiss, J.** 2003. *Handbook of Ion Chromatography.* John Wiley & Sons, Hoboken, NJ.

37. **Wu, C.** 1995. *Handbook of Size Exclusion Chromatography.* Marcel Dekker, Inc., New York, NY.

17.8.3.2. Specific References

38. **Andrea, J. M.** 1985. Measurement of protein-ligand interactions by gel chromatography. *Methods Enzymol.* **117:**346–354.

39. **Aronstein, B. N., J. R. Paterek, R. L. Kelley, and L. E. Rice.** 1995. The effect of chemical pretreatment on the aerobic microbial-degradation of PCB congeners in aqueous systems. *J. Ind. Microbiol.* **15:**55–59.

40. **Ausserer, W. A., and M. L. Biros.** 1995. High-resolution analysis and purification of synthetic oligonucleotides with strong anion-exchange HPLC. *BioTechniques* **19:**136–139.

41. **Bechard, G., J.-G. Bisaillon, and R. Beaudet.** 1990. Degradation of phenol by a bacterial consortium under methanogenic conditions. *Can. J. Microbiol.* **36:**573–578.

42. **Beeckmans, S.** 1999. Chromatographic methods to study protein-protein interactions. *Methods* **19:**278–305.

43. **Benning, M.** 1988. Single ion chromatography. *Am. Lab.* **20:**74–79.

44. **Berezkin, V. G., and J. de Zeeuw.** 1998. *Capillary Gas Adsorption Chromatography.* John Wiley & Sons, Hoboken, NJ.

45. **Cho, K. S., L. Zhang, M. Harm, and M. Shoda.** 1991. Removal characteristics of hydrogen sulfide, and methanethiol by *Thiobacillus sp.* isolated from peat in biological deodorization. *J. Ferment. Bioeng.* **71:**44–49.

46. **Christendat. D., A. Yee, A. Dharamsi, Y. Kluger, A. Savchenko, J. R. Cort, V. Booth, C. D. Mackereth, V. Saridakis, I. Ekiel, G. Kozlov, K. L. Maxwell, N. Wu, L. P. McIntosh, K. Gehring, M. A. Kennedy, A. R. Davidson, E. F. Pai, M. Gerstein, A. M. Edwards, and C. H. Arrowsmith.** 2000. Structural proteomics of an archaeon. *Nat. Struct. Biol.* **7:**903–909.

47. **Cserháti, T.** 2002. Mass spectrometric detection in chromatography. Trends and perspectives. *Biomed. Chromatogr.* **16:**303–310.

48. **Dworzanski, J. P., L. Berwald, and H. L. C. Meuzelaac.** 1990. Pyrolytic methylation-gas chromatography of whole bacterial cells for rapid profiling of cellular fatty acids. *Appl. Environ. Microbiol.* **56:**1717–1724.

49. **Eder, K.** 1995. Gas chromatographic analysis of fatty acid methyl esters. *J. Chromatogr. B* **671:**113–131.

50. **Fox, A.** 1999. Carbohydrate profiling of bacteria by gas chromatography-mass spectrometry and their trace detection in complex matrices by gas chromatography-tandem mass spectrometry. *J. Chromatogr. A* **843:**287–300.

51. **Fritz, J. S., and D. T. Gjerde.** 2000. *Ion Chromatography.* John Wiley & Sons, Hoboken, NJ.

52. **Giri, L.** 1990. Chromatofocusing. *Methods Enzymol.* **182:**380–392.

53. **Gorbunoff, M. J.** 1990. Protein chromatography on hydroxyapatite columns. *Methods Enzymol.* **182:**329–339.

54. **Heckenberg, A. L., and P R. Haddad.** 1984. Determination of inorganic ions at parts per billion levels using single column ion chromatography without preconcentration. *J. Chromatogr.* **299:**301–305.

55. **Hordijk, C. A., C. P C. M. Hagenaars, and T. E. Cappenberg.** 1984. Analysis of sulfate at the mud-water interface of freshwater lake sediments using indirect photometric chromatography. *J. Microbiol. Methods* **2:**49–56.

56. **Hsueh, P. R., T. L. Jene, P. H. Ju, C. Y. Chi, S. C. Chuan, and H. S. Wu.** 1998. Outbreak of *Pseudomonas fluorescens* bacteremia among oncology patients. *J. Clin. Microbiol.* **36:**2914–2917.

57. **Hubschmann, H.-J.** 2001. *Handbook of GC/MS: Fundamentals and Applications.* John Wiley & Sons, Hoboken, NJ.

58. **James, A. T., and A. J. P Martin.** 1952. Gas-liquid partition chromatography: the separation and microestimation of volatile fatty acids from formic to dodecanoic acid. *Biochem. J.* **50:**679–690.

59. **Janak, J.** 1990. The role of adsorption in gas-liquid systems and its use for enrichment of trace amounts. *Chromatographia* **30:**489–492.

60. **Jensen, T. S., E. Arvin, B. Svensmark, and P. Wrang.** 2000. Quantification of compositional changes of petroleum hydrocarbons by GC/FID and CC/MS during a long-term bioremediation experiment. *Soil Sediment Contam.* **9:**549–577.

61. **Katayama-Fujimura, Y., Y. Komatsu, H. Kuraishi, and T. Kanerko.** 1984. Estimation of DNA base composition by high performance liquid chromatography of its nuclease PI hydrolysate. *Agric. Biol. Chem.* **48:**3169–3172.

62. **Kido, H., A. Vita, and B. L. Horecker.** 1985. Ligand binding to proteins by equilibrium gel penetration. *Methods Enzymol.* **117:**342–346.

63. **Kim, J. S., J. B. Joo, H. Y. Weon, C. S. Kang, S. K. Lee, and C. S. Yahng.** 2002. FAME analysis to monitor impact of organic matter on soil bacterial populations. *J. Microbiol. Biotechnol.* **12:**382–388.

64. **Korthals, H. J., and C. L. M. Steenbergen.** 1985. Separation and quantification of pigments from natural phototrophic microbial populations. *FEMS Microbiol. Ecol.* **31:**177–185.

65. **Le Bris, S., M. R. Plante-Cuny, and E. Vacelet.** 1998. Characterisation of bacterial and algal pigments and breakdown products by HPLC in mixed freshwater planktonic populations. *Arch. Hydrobiol.* **143:**409–434.

66. **Levin, O.** 1962. Column chromatography of proteins: calcium phosphate. *Methods Enzymol.* **5:**27–33.

67. **Linhardt, R. J., K. N. Gu, D. Loganathan, and S. R. Carter.** 1989. Analysis of glycosaminoglycan-derived oligosaccharides using reversed-phase ion-pairing and ion-

exchange chromatography with suppressed conductivity detection. *Anal. Biochem.* **181:**288–296.

68. **López-Ruiz, B.** 2000. Advances in the determination of inorganic anions by ion chromatography. *J. Chromatogr. A* **881:**607–627.

69. **Mackey, B. M., S. E. Parsons, C. A. Miles, and R. J. Owen.** 1988. The relationship between the base composition of bacterial DNA and its intracellular melting temperature as determined by differential scanning calorimetry. *J. Gen. Microbiol.* **134:**1185–1195.

70. **Mancuso, C. A., P D. Nichols, and D. C. White.** 1986. A method for the separation and characterization of archaebacterial signature ether lipids. *J. Lipid Res.* **27:**49–56.

71. **Marmur, J., and P. Doty.** 1962. Determination of the base composition of deoxyribonucleic acid from its thermal denaturation temperature. *J. Mol. Biol.* **5:**109–118.

72. **Mesbah, M., and W. B. Whitman.** 1989. Measurement of deoxyguanosine/thymidine ratios in complex mixtures by high-performance liquid chromatography for determination of the mole percentage guanine + cytosine of DNA. *J. Chromatogr.* **479:**297–306.

73. **Mesbah, M., U. Premachandran, and W. B. Whitman.** 1989. Precise measurement of the G+C content of deoxyribonucleic acid by high-performance liquid chromatography. *Int. J. Syst. Bacteriol.* **39:**159–167.

74. **Miller, L., and T. Berger.** 1985. *Bacterial Identification by Gas Chromatography of Whole Cell Fatty Acids.* Application note 228-41. Hewlett-Packard Co., Palo Alto, CA.

75. **Odumeru, J. A., M. Steele, L. Fruhner, C. Larkin, J. Jiang, E. Mann, and W. B. McNab.** 1999. Evaluation of accuracy and repeatability of identification of food-borne pathogens by automated bacterial identification systems. *J. Clin. Microbiol.* **37:**944–949.

76. **Ostrove, S., and S. Weiss.** 1990. Affinity chromatography: specialized techniques. *Methods Enzymol.* **182:**371–379.

77. **Peltroche-Llacsahuanga, H., S. Schmidt, R. Lutticken, and G. Haase.** 2000. Discriminative power of fatty acid methyl ester (FAME) analysis using the Microbial Identification System (MIS) for *Candida* (Torulopsis) *glabrata* and *Saccharomyces cerevisiae*. *Diagn. Microbiol. Infect. Dis.* **38:**213–221.

78. **Scott, R. P. W.** 1998. *Introduction to Analytical Gas Chromatography.* Marcel Dekker, New York, NY.

79. **Shan, L., and D. J. Anderson.** 2002. Gradient chromatofocusing. Versatile pH gradient separation of proteins in ion-exchange HPLC: characterization studies. *Anal. Chem.* **74:**5641–5649.

80. **Small, H., and T. E. Miller.** 1982. Indirect photometric chromatography. *Anal. Chem.* **54:**262–269.

81. **Stellwagen, E.** 1990. Gel filtration. *Methods Enzymol.* **182:**317–328.

82. **Teunissen, M. J., S. A. E. Marras, H. J. M. Op den Camp, and G. D. Vogels.** 1989. Improved method for simultaneous determination of alcohols, volatile fatty acids, lactic acid or 2,3-butanediol in biological samples. *J. Microbiol. Methods* **10:**247–254.

83. **Tunlid, A., D. Ringelberg, T. J. Phelps, C. Low, and D. C. White.** 1989. Measurement of phospholipid fatty acids at picomolar concentrations in biofilms and deep subsurface sediments using gas chromatography and chemical ionization mass spectrometry. *J. Microbiol. Methods* **10:**139–153.

84. **Ulrich, S.** 2000. Solid-phase microextraction in biomedical analysis. *J. Chromatogr. A* **902:**167–194.

85. **Villegas, S., and L. L. Brunton.** 1996. Separation of cyclic GMP and cyclic AMP. *Anal. Biochem.* **235:**102–103.

86. **Yacobi, Y. Z., W. Eckert, H. G. Truper, and T. Berman.** 1990. High performance liquid chromatography detection of phototrophic bacterial pigments in aquatic environments. *Microb. Ecol.* **19:**127–136.

87. **Zelles, L.** 1999. Fatty acid patterns of phospholipids and lipopolysaccharides in the characterisation of microbial communities in soil: a review. *Biol. Fert. Soils* **29:**111–129.

88. **Zou, H., Q. Luo, and D. Zhou.** 2001. Affinity membrane chromatography for the analysis and purification of proteins. *J. Biochem. Biophys. Methods* **49:**199–240.

17.8.4. Radioactivity

17.8.4.1. General References

89. **Klein, R. C., and E. L. Gershey.** 1990. Biodegradable liquid scintillation-counting cocktails. *Health Phys.* **59:**461–470.

90. **Ross, H. (ed.).** 1991. *Liquid Scintillation Counting and Organic Scintillators.* CRC Press, Boca Raton, FL.

91. **Wang, C., D. L. Willis, and W. D. Loveland.** 1975. *Radiotracer Methodology in the Biological, Environmental, and Physical Sciences.* Prentice-Hall, Inc., Englewood Cliffs, NJ.

17.8.4.2. Specific References

92. **Anderson, L. E., and W. O. McClure.** 1973. An improved scintillation cocktail of high solubilizing power. *Anal. Biochem.* **51:**173–179.

93. **Bonner, W. M.** 1983. Use of fluorography for sensitive isotope detection in polyacrylamide-gel electrophoresis and related techniques. *Methods Enzymol.* **96:**215–222.

94. **Thomson, J.** 2002. *Use and Preparation of Quench Curves.* Liquid Scintillation Application Note P11399. PerkinElmer Lifesciences, Boston, MA.

95. **Thomson, J., and D. A Burns.** 1995. *Radio-Carbon Dioxide ($^{14}CO_2$) Trapping and Counting.* Liquid Scintillation Application Note CS-001. PerkinElmer Lifesciences, Boston, MA.

96. **Thomson, J., and D. A. Burns.** 1996. *LSC Sample Preparation by Solubilization.* Liquid Scintillation Application Note CS-003. PerkinElmer Lifesciences, Boston, MA.

97. **Turner, J. C.** 1967. *Sample Preparation for Liquid Scintillation Counting.* The Radiochemical Centre, Amersham, England.

98. **Woo, H. J., S. K. Chun, S. Y. Cho, Y. S. Kim, D. W. Kang, and E. H. Kim.** 1999. Optimization of liquid scintillation counting techniques for the determination of carbon-14 in environmental samples. *J. Radioanal. Nucl. Chem.* **239:**649–655.

17.8.5. Gel Electrophoresis

17.8.5.1. General References

99. **Allen, R. C.** 1994. *Gel Electrophoresis of Proteins and Nucleic Acids.* Walter DeGruyter, New York, NY.

100. **Bravo, R.** 1984. *Two-Dimensional Gel Electrophoresis of Proteins.* Academic Press, Inc., New York, NY.

101. **Davis, M. B. J.** 1964. Disc electrophoresis. II. Method and application to human serum proteins. *Ann. N. Y. Acad. Sci.* **121:**404–427.

102. **Garfin, D. E.** 1990. Isoelectric focusing. *Methods Enzymol.* **182:**459–477.

103. **Hames, B. D.** 1998. *Gel Electrophoresis of Proteins: A Practical Approach.* Oxford University Press, New York, NY.

104. **Ornstein, L.** 1964. Disc electrophoresis. I. Background and theory. *Ann. N. Y. Acad. Sci.* **121:**321–403.

105. **Richetti, P. G.** 1984. *Isoelectric Focusing Theory, Methodology and Applications.* Elsevier Biochemical Press, New York, NY.

106. **Weinberg, R.** 2000. *Practical Capillary Electrophoresis.* Academic Press, San Diego, CA.

17.8.5.2. Specific References

107. **Anderson, D., and C. Peterson.** 1981. High resolution electrophoresis of proteins in SDS polyacrylamide gels, p. 41. *In* R. C. Allen and P. Arnaud (ed.), *Electrophoresis.* Walter deGruyter, Berlin, Germany.

108. **Blackshear, P. J.** 1984. Systems for polyacrylamide gel electrophoresis. *Methods Enzymol.* **104:**237–255.

109. **Bollag, D. M., M. D. Rozyski, and S. J. Edelstein.** 1996. *Protein Methods.* John Wiley & Sons, Hoboken, NJ.

110. **Castellanos-Serra, L., and E. Hardy.** 2001. Detection of biomolecules in electrophoresis gels with salts of imidazole and zinc II: a decade of research. *Electrophoresis* **22:**864–873.

111. **Dunbar, B. S., H. Kimura, and T. M. Timmons.** 1990. Protein analysis using high-resolution two-dimensional polyacrylamide gel electrophoresis. *Methods Enzymol.* **182:**441–459.

112. **Görg, A., C. Obermaier, G. Boguth, A. Harder, B. Scheibe, R. Wildgruber, and W. Weiss.** 2000. The current state of two-dimensional electrophoresis with immobilized pH gradients. *Electrophoresis* **21:**1037–1053.

113. **Hendrick, J. L., and A. J. Smith.** 1968. Size and charge isomer separation and estimation of molecular weight of proteins by disc gel electrophoresis. *Arch. Biochem. Biophys.* **126:**155–164.

114. **Laemmli, U. K.** 1970. Cleavage of structural proteins during the assembly of the head of bacteriophage T4. *Nature* **227:**680–685.

115. **Linhardt, R. J., and T. Toida.** 2002. Ultra-high resolution separation comes of age. *Science* **298:**1441–1442.

116. **Manchenko, G. P.** 2002. *Handbook of Detection of Enzymes on Electrophoretic Gels.* CRC Press, Boca Raton, FL.

117. **Merril, C. R., and M. E. Pratt.** 1986. A silver stain for the rapid quantitative detection of proteins or nucleic-acids on membranes or thin-layer plates. *Anal. Biochem.* **156:**96–110.

118. **O'Farrell, P.** 1975. High resolution two-dimensional electrophoresis of proteins. *J. Biol. Chem.* **250:**4007–4021.

119. **Rabilloud, T.** 1999. *Proteome Research: Two-Dimensional Gel Electrophoresis and Identification Methods.* Springer-Verlag, New York, NY.

120. **Wirth, P. J., and A. Romano.** 1995. Staining methods in gel-electrophoresis, including the use of multiple detection methods. *J. Chromatogr. A* **698:**123–143.

121. **Yamaoka, T.** 1998. Pore gradient gel electrophoresis: theory, practice, and applications. *Anal. Chim. Acta* **372:**91–98.

122. **Yan, J. X., R. A. Harry, C. Spibey, and M. J. Dunn.** 2000. Postelectrophoretic staining of proteins separated by two-dimensional gel electrophoresis using SYPRO dyes. *Electrophoresis* **21:**3657–3665.

17.8.6. Enzyme Purification

17.8.6.1. General References

123. **Amicon Corp.** 1990. *Dye-Ligand Chromatography.* Amicon Corp, Danvers, MA.

124. **Deutscher, M. P. (ed.).** 1990. Guide to protein purification. *Methods Enzymol.* **182:**1–894.

125. **Jakoby, W. (ed.).** 1984. Enzyme purification and related techniques, part C. *Methods Enzymol.* **104:**1–528.

126. **Roe, S.** 2001. *Protein Purification: A Practical Approach.* Oxford University Press, New York, NY.

127. **Scopes, R. P.** 1994. *Protein Purification: Principles and Practice.* Springer-Verlag, New York, NY.

17.8.6.2. Specific References

128. **Cumming, R. H., and G. Iseton.** 2001. Cell disintegration and extraction techniques, p. 111-155. *In* S. Roe (ed.), *Protein Purification Techniques: A Practical Approach.* Oxford University Press, New York, NY.

129. **England, S., and S. Seifter.** 1990. Precipitation techniques. *Methods Enzymol.* **182:**295–300.

130. **Hammerstedt, R. H., H. Mohler, K. A. Decker, and W. A. Wood.** 1971. Structure of 2-keto-3-deoxy-6-phosphogluconate aldolase. I. Evidence for a three-subunit molecule. *J. Biol. Chem.* **246:**2069–2074.

131. **Muh, U., V. Massey, and C. H. Williams.** 1994. Lactate monooxygenase. 1. Expression of the mycobacterial gene in *Escherichia coli* and site-directed mutagenesis of lysine-266. *J. Biol. Chem.* **269:**7982–7988.

132. **Wood, W. I.** 1976. Tables for the preparation of ammonium sulfate solutions. *Anal. Biochem.* **73:**250–257.

17.8.7. Concluding Comments and Future Trends

17.8.7.1. General References

133. **Jelesarov, I., and H. R. Bosshard.** 1999. Isothermal titration calorimetry and differential scanning calorimetry as complementary tools to investigate the energetics of biomolecular recognition. *J. Mol. Recognit.* **12:**3–18.

134. **Ladbury, J. E., and B. Z. Chowdhry (ed.).** 1998. *Biocalorimetry: Applications of Calorimetry in the Biological Sciences.* John Wiley & Sons, Hoboken, NJ.

135. **Rieseberg, M., C. Kasper, K. F. Reardon, and T. Scheper.** 2001. Flow cytometry in biotechnology. *Appl. Microbiol. Biotechnol.* **56:**350–360.

136. **Todd, M. J., and J. Gomez.** 2001. Enzyme kinetics determined using calorimetry: a general assay for enzyme activity? *Anal. Chem.* **296:**179–187.

137. **Watson, J. T.** 1997. *Introduction to Mass Spectrometry.* Lippincott Williams & Wilkins, Philadelphia, PA.

17.8.7.2. Specific References

138. **Dalluge, J. J.** 2000. Mass spectrometry for direct determination of proteins in cells: applications in biotechnology and microbiology. *Fresenius J. Anal. Chem.* **366:**701–711.

139. **Fenselau, C., and P. A. Demirev.** 2001. Characterization of intact microorganisms by MALDI mass spectrometry. *Mass Spectrom. Rev.* **20:**157–171.

140. **Katsuragi, T., and Y. Tani.** 2000. Screening for microorganisms with specific characteristics by flow cytometry and single-cell sorting. *J. Biosci. Bioeng.* **89:**217–222.

141. **Lay, J. O.** 2001. MALDI-TOF mass spectrometry of bacteria. *Mass Spectrom. Rev.* **20:**172–194.

142. **Lebaron, P., P. Servais, A. C. Baudoux, M. Bourrain, C. Courties, and N. Parthuisot.** 2002. Variations of bacterial-specific activity with cell size and nucleic acid content assessed by flow cytometry. *Aquat. Microb. Ecol.* **28:**131–140.

143. **Roth, B. L., M. Poot, S. T. Yue, and P. J. Millard.** 1997. Bacterial viability and antibiotic susceptibility testing with SYTOX Green nucleic acid stain. *Appl. Environ. Microbiol.* **63:**2421–2431.

144. **Vives-Rego, J., P. Lebaron, and G. Nebe-von Caron.** 2000. Current and future applications of flow cytometry in aquatic microbiology. *FEMS Microbiol. Rev.* **24:**429–448.

18

Chemical Analysis

LACY DANIELS, RICHARD S. HANSON, AND JANE A. PHILLIPS

Analysis to estimate amounts of the major and minor components of microorganisms and their metabolites is often required to address certain research questions. In other instances, a microbial component may need to be purified to study its properties, e.g., a labeling pattern to ascertain a metabolic pathway. The methods selected for this chapter are most commonly used in research relating to microbial components and metabolism, but most are also suited to advanced undergraduate and graduate teaching laboratories. Many other methods and more detailed descriptions of those selected are presented in publications listed at the end of the chapter. Techniques that require costly instrumentation or significant training are not described in detail. Specialized methods (which vary with the nature of the samples, equipment, and expertise) are treated by providing references to other sources of information rather than by describing experimental details. As with virtually all laboratory methods, some steps in the described methods require the use of dangerous chemicals or equipment, and appropriate attention to safety issues is essential.

18.1. FRACTIONATION OF MAIN CHEMICAL COMPONENTS

It is sometimes useful to know the amounts of major bacterial cell components present at a given time and to know how rapidly such components are synthesized. The incorporation of radioactively labeled compounds into bacterial cells is used as a means of estimating the rates and amounts of synthesis as well as the distribution of the main small-molecule and macromolecule fractions of the cells (2). After the cells have been incubated with the labeled substrate and the various molecular components have been fractionated, the radioactivity in each fraction can be determined.

Macromolecules are precipitated by cold dilute solutions of trichloroacetic acid. The (denatured) macromolecules are separated from the small molecules by centrifugation. The supernatant fraction of small molecules contains the pooled metabolites of sugars, sugar derivatives, organic acids, amino acids, nucleotides, and coenzymes; oligonucleotides and basic proteins with fewer than 20 nucleotide or amino acid residues are also contained in this fraction (2). (Note that some coenzymes are destroyed by acid extraction; see section 18.2 for alternate methods of coenzyme measurements.) The precipitate fraction of macromolecules is sequentially extracted with organic solvents for lipids, alkali for RNAs, and hot trichloroacetic acid for DNAs; the precipitate remaining after these extractions contains the cell proteins and peptidoglycan. A flow diagram of this fractionation procedure is shown in Fig. 1.

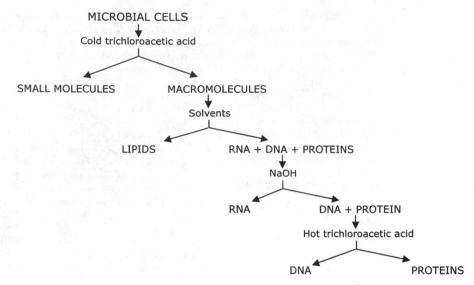

FIGURE 1 Flow diagram of fractionation procedure.

No single procedure gives complete separation of all of the chemical components and total recovery of a given type of macromolecule in a single fraction. The procedures described by Kennell (1) are presented below. They are modifications of the methods of Roberts et al. (2).

18.1.1. Reagents and Materials

Saline solution: 0.85% (wt/vol) NaCl in distilled water

Trichloroacetic acid solutions (*CAUTION*: See section 18.10): 50, 20, 10, and 5% (wt/vol) trichloroacetic acid in distilled water

Ethanol (*CAUTION*: See section 18.10): 70% (vol/vol) in distilled water

Diethyl ether (*CAUTION*: See section 18.10)

Sodium hydroxide (0.5 N) (*CAUTION*: See section 18.10). Dissolve 2.0 g in 100 ml of distilled water.

Bovine serum albumin (BSA). Dissolve 0.1 g in 10 ml of distilled water. Store at 4°C.

Scintillation fluid (available from a variety of vendors)

Glassware: 20-ml Pyrex glass beakers, 10-ml Erlenmeyer flasks, 25-ml volumetric flasks, capped 15- and 30-ml centrifuge tubes

Filters: fiberglass or synthetic membrane filters (pore size, 0.45 μm), usually 25 mm in diameter. The filters must be acid and solvent resistant. Pyrex filter holders with fritted-glass filter supports.

Vacuum filtering flask: 125-ml flask and test tubes cut to fit below the filter holder inside the flask. The test tubes are used to collect filtrates.

Heating lamp: 250-W infrared heat lamp, and a suitable reflector for sample drying; alternatively, an oven heated at 50 to 80°C is suitable but should not be used with significant amounts (>1 ml) of ether.

Ice-water bath

Water bath, adjustable to 80°C

Refrigerated centrifuge, capable of achieving 10,000 × g, with a centrifuge head that holds 30-ml centrifuge tubes

Scintillation counter

Vortex mixer

18.1.2. Radioactive Labeling

(*CAUTION*: See section 18.10.) The use of radioactive isotopes can be hazardous and is thus strictly regulated. All personnel involved in their use must be familiar with safety issues and regulations regarding their use (see also chapter 17.4.13).

The culture volume to be used will vary with the organism and growth conditions. For example, 30 ml of a mid-exponential-phase culture of *Escherichia coli* is adequate to provide the 10 mg of cells required.

If the purpose of the experiment is to determine cell composition, a uniformly labeled substrate such as [^{14}C]glucose (5 μCi/30 ml of a culture containing 1% glucose as the only carbon source) is suitable (1 μCi = 37,000 Bq). The incorporation of an amino acid into protein can be measured by adding 2 to 5 μCi of a labeled amino acid to 30 ml of a culture growing on a mineral salts medium plus glucose. Ideally, an auxotrophic mutant that is unable to synthesize the amino acid should be used when incorporation into protein is measured. In this case, 10 μg of the unlabeled amino acid per ml and 2 to 5 μCi of the labeled amino acid should be added to the medium. It is important to use a nonmetabolizable amino acid as a labeled protein precursor. Leucine, lysine, phenylalanine, and tyrosine are not significantly metabolized by most bacteria. When labeled thymidine is used as a precursor of DNA, and when uracil or uridine is used as a precursor of RNA, auxotrophs should also be used to obtain incorporation of a larger fraction of the label added to the medium. The cultures (inoculated with healthy, mid-exponential-phase cultures) should be harvested during exponential growth when cell composition studies are performed. The cell composition changes during the lag and stationary phases of growth.

18.1.3. Small Molecules

1. Cool the culture rapidly by pouring it over ice. Centrifuge the radioactively labeled cells (approximately

10 mg [dry weight]) from the growth medium at 5,000 × g for 10 min at 5°C, and wash the sedimented cells twice by centrifugation with ice-cold (5°C) 0.85% saline (one-fifth of the culture volume).

2. Suspend the cells with 10 ml of distilled water in the centrifuge tube. Place the suspension in an ice bath until cold, and add 10 ml of ice-cold 20% trichloroacetic acid. Cap the centrifuge tube, and mix the contents by inversion. Place the tube in the ice bath for 5 min to allow complete precipitation.

3. Centrifuge (5,000 to 10,000 × g for 15 min) at 0 to 4°C. Carefully decant the supernatant fraction, without disturbing the precipitate, into a 25-ml volumetric flask.

4. Add 4 ml of ice-cold 10% trichloroacetic acid to the precipitate, suspend it, and centrifuge again. Decant the supernatant fraction into that obtained after the first centrifugation. Adjust the volume of the combined supernatant fractions to 25 ml with distilled water.

5. Determine the radioactivity of a sample of the combined supernatant fraction containing the small molecules (metabolite pools) directly in a solution used for determining radioactivity in aqueous samples, as described elsewhere in this book (see chapter 17.4). Internal or external standards should be used to correct for quenching by trichloroacetic acid. Alternatively, trichloroacetic acid can be removed by three extractions with equal volumes of ether. However, ether extraction will remove organic acids from the small-molecule fraction. Samples of the ether-extracted portion can be counted separately in a counting solution used for aqueous solvents or can be dried in vials with the assistance of a heat lamp. Large volumes (>1 ml) of ether solutions should not be dried in an oven because of the explosion hazard, but it is possible to evaporate most of the ether at room temperature, preferably in a fume hood, and then place the container in the oven. After drying, the residue can be counted in a suitable scintillation cocktail (see chapter 17.4 and section 18.2).

18.1.4. Lipids

1. Separate lipids from the precipitated fraction of macromolecules in the centrifuge tube (see above). Suspend the sediment in 10 ml of ice-cold 10% trichloroacetic acid with the aid of a vortex mixer. (CAUTION: Cap the tube to avoid trichloroacetic acid and radioisotope aerosols.)

2. Filter 1 ml of the homogeneous suspension, and wash the filter twice with 2 ml of ice-cold 10% trichloroacetic acid. Wash the filter with 3 ml of ice-cold 70% ethanol to remove residual trichloroacetic acid, which may cause solubilization of some protein in the next step. Discard the filtrate.

3. Place the filter holder (with the filter and suction flask) and a test tube for collecting the filtrate in an incubator at 45°C, and allow the assembly to warm to temperature. Then, remove the filter flask from the incubator, quickly add 10 ml of 70% ethanol (warmed in a water bath to 45°C), and allow the solvent to pass slowly through the filter. Collect the filtrate in a test tube. The filtration should take approximately 10 min to effect the extraction.

4. Wash the filter in a hood with 2- to 5-ml volumes of ethanol-diethyl ether (1:1, vol/vol) warmed to 45°C in a water bath by allowing the solvents to pass slowly through the filter (10 min). Collect the filtrate in a test tube, and combine it with the 45°C ethanol wash. Retain the filter containing the residual precipitated fraction of macromolecules for further fractionation.

5. Adjust the volume of the combined filtrates (which represents the lipid fraction) to 25 ml, and dry a sample in

a scintillation vial under a stream of filtered air or nitrogen at 40°C. Gas filtration (to remove oil and particulates) can be accomplished by passing compressed gas through glass wool or an in-line 0.45-μm-pore-size bacterial filter.

6. Determine the radioactivity of the dried lipid sample by use of a suitable scintillation cocktail (see chapter 17.4).

18.1.5. RNA

1. Separate RNA from the precipitated fraction of macromolecules, with the lipids now removed. Remove the filter containing the precipitate from the filter holder (step 4 above), and place upside down in a 20-ml beaker. Allow the filter to air dry in a hood (about 10 min).

2. Add 2 ml of 0.5 N NaOH (warmed to 37°C) to the beaker, and incubate with occasional shaking at 37°C for precisely 40 min. Make sure the filter is completely covered with the NaOH solution. Longer incubation causes hydrolysis of some proteins (1, 3).

3. Transfer the beaker to an ice bath, and cool rapidly. Add 0.6 ml of ice-cold 50% trichloroacetic acid, mix, and leave the beaker in the ice bath for about 30 min.

4. Transfer the liquid in the beaker to a new filter held in a filter holder, apply a vacuum, and collect the acid-soluble products in a test tube. Rinse the beaker and original filter with 3 ml of ice-cold 5% trichloroacetic acid, transfer the liquid to the same filter, and collect the filtrate in the same test tube. Adjust the volume of the combined filtrates to 15 ml in a volumetric flask with distilled water. The filtrate contains about 95% of the ribonucleotides in RNA.

5. Determine the radioactivity of the hydrolyzed RNA fraction in the same way as for the small-molecule fraction (see above and chapter 17.4).

18.1.6. DNA

1. After removing lipids and RNA from the macromolecular fraction, separate the DNA from the precipitated fraction of macromolecules by treatment with hot trichloroacetic acid. Use a duplicate sample from the suspended sediment in cold trichloroacetic acid, as described in the first steps of the lipid fractionation (section 18.1.4). Repeat steps 1 through 4 for removal of lipids. After washing the filter with ethanol and ethanol-ether to remove lipids, place the filter in a beaker, add 2 ml of 0.5 N NaOH, mix, and incubate for 90 min at 37°C. The incubation time with NaOH is increased to give total hydrolysis of RNA. DNA is present in smaller amounts than RNA, and traces of unhydrolyzed RNA can significantly increase estimates of radioactivity in DNA. The solubilization of some proteins will not affect the estimations of DNA.

2. Cool the sample on ice, add 0.6 ml of cold 50% (wt/vol) trichloroacetic acid, and keep it on ice for 30 min; then, pour the liquid in the beaker onto a new filter that has been presoaked in cold 5% trichloroacetic acid. Apply a vacuum to facilitate filtration. Wash the original filter twice with 3 ml of ice-cold 5% trichloroacetic acid, and discard the filtrate.

3. Place the two filters in a 10-ml Erlenmeyer flask, add 3 ml of 5% trichloroacetic acid, and heat for 30 min in a water bath at 80°C; this step hydrolyzes purines from the DNA. Make sure the filters are entirely immersed in the liquid.

4. Add 0.1 ml of a solution (1.0 mg/ml) of BSA, mix rapidly, cool the flask for 30 min in an ice bath, and filter the entire contents of the Erlenmeyer flask through a new filter. The BSA ensures precipitation of minute amounts of macromolecules that may otherwise remain in suspension. Use a slight vacuum to increase the filtration rate. Collect the

filtrate in a test tube placed inside the vacuum flask. Wash the Erlenmeyer flask and filters twice with 3 ml of cold 5% trichloroacetic acid. Collect the filtrates in the same test tube. Adjust the volume of the combined filtrates to 15 ml with water in a volumetric flask. Discard the filters.

5. Determine the radioactivity of the hydrolyzed DNA fraction in the same way as for the small-molecule fraction (see above and chapter 17.4).

18.1.7. Proteins

1. Separate proteins from the precipitated fraction of macromolecules. Use a duplicate sample from the suspended sediment in cold trichloroacetic acid, as described in the first step of the lipid fractionation (section 18.1.4). Transfer 1 ml into a test tube (15 by 100 mm). Dilute with 1 ml of water to give a trichloroacetic acid concentration of 5% (wt/vol).

2. Heat the diluted sample for 30 min at 80°C to hydrolyze both RNA and DNA.

3. Cool the hydrolyzed sample on ice for 30 min to precipitate proteins. Collect the precipitate on a filter, and discard the filtrate. Wash the test tube twice with 2 ml of ice-cold 10% trichloroacetic acid, and apply both washes to a filter; wash the filter once with 3 ml of ice-cold 70% ethanol to remove residual trichloroacetic acid that would otherwise solubilize some protein in the next step.

4. Wash the filter twice with 5 ml of 70% ethanol (45°C), twice with 5.0 ml of ethanol-diethyl ether (1:1) at 45°C, and once with 5 ml of diethyl ether.

5. Air dry the filter, place it in a scintillation vial, and dry it under a heat lamp.

6. Determine the radioactivity of the dried protein by use of a suitable scintillation cocktail (see chapter 17.4).

18.1.8. Applications

The techniques described above are usually applied to the fractionation of bacterial cells labeled with a medium component that is nonspecifically incorporated into all of the cell molecules. When bacterial cells are incubated in the presence of a specific precursor of a macromolecule to measure its rate of synthesis, it is important to determine that the macromolecule of interest contains all of the radioactivity. This can be determined by fractionating the cells and estimating the radioactivity of each fraction by the techniques described above. If all the radioactivity occurs in the fraction of interest, subsequent experiments can be simplified. For example, if radioactive thymidine is used to measure the amount of DNA synthesis and preliminary experiments indicate that it is not found in other fractions, the cells in their culture medium can be cooled and treated with ice-cold 10% trichloroacetic acid. The precipitate containing the DNA can then be collected on filters, washed with cold ethanol, and dried, and the radioactivity can be counted. This procedure is suitable for measuring the amount of DNA synthesis. The same technique can be applied to measure the amount of protein or RNA synthesis if the precursor appears only in the one fraction. Other factors must be considered in calculating the rates of synthesis. A primary concern is the dilution of exogenously added substrates by unlabeled precursors in the cell pools. In kinetic experiments, it is necessary to determine the rate of saturation of the pool by radioactive precursors.

Although the methods described above provide a means of obtaining an estimate of the distribution of carbon from a ^{14}C-labeled compound in bacteria, several cell components are not accounted for. Products of hydrolysis of teichoic acids and polyphosphates will appear in the RNA fraction, and peptidoglycan will be precipitated with protein.

Poly-β-hydroxybutyrate, if present, can be extracted with chloroform-methanol (2:1, vol/vol) at 60°C after the ether-ethanol washes (see also section 18.4.9). If this solvent is used, care should be taken to use filters that are stable in methanol and chloroform. (CAUTION: The extraction should be performed in a hood [see section 18.10].)

Polysaccharides, if present, are partially extracted by cold trichloroacetic acid and can significantly affect estimates of the small molecules. When present in large amounts, polysaccharides present a variety of troublesome problems if nonspecific precursors of macromolecules such as radioactive glucose are used. When these problems occur, these procedures must be modified (3).

Some bacterial cells leak part of their pools of small molecules when chilled during centrifugation and washing. Some cells have high activities of nucleases or proteases that may be activated during harvesting and washing. If cells are frozen and then thawed, even more small molecules and enzymes will leak.

Significant losses of macromolecules occur when smaller amounts of material than indicated are used in the first steps. It is possible to add unlabeled bacterial cells as a carrier to facilitate the precipitation of macromolecules. Precipitation of DNA and protein can also be facilitated by addition of serum albumin to the solutions.

Each organism potentially presents a unique problem in fractionation. Although the procedures described above have been widely used, the data obtained should be interpreted with proper regard for the problems of obtaining total separation of the small molecules and macromolecules (3).

Sequential methods for treating a single sample have been described (2, 3). However, they are likely to result in greater cross-contamination of fractions than the procedure described here.

18.2. SMALL-MOLECULE EXTRACTION AND ANALYSIS

In some cases the small molecules of interest are unstable to acid treatment or require further fractionation or analysis for quantitation. Also, coenzyme analysis may be needed without the need for any of the other analyses described above for macromolecules. For example, coenzymes could be extracted under quite different conditions from those described above, from either labeled or unlabeled cells, and analyzed further. Detailed extraction procedures for a wide range of coenzymes has been provided for a wide range of coenzymes in *Methods in Enzymology* (4, 11–14). In some cases, anaerobic methods are required to maintain the stability of oxygen-sensitive components, e.g., by conducting extraction under nitrogen or argon gas as is done for tetrahydromethanopterin from methanogens (7). In other cases, some coenzymes (e.g., B_{12}) are sensitive to light, and procedures should be conducted under low-light conditions. When a small molecule is stable to heat, boiling or briefly autoclaving cells is generally one of the easiest and most effective extraction methods. One type of more gentle extraction is the use of 70% aqueous ethanol or 50% aqueous acetone at 4°C to resuspend the original cell pellet (15, 16). (CAUTION: See section 18.10.) The suspension is stirred at 4°C for 5 to 15 h and centrifuged, and the supernatant is saved; the pellet is then extracted one more time with fresh solvent. The combined supernatants contain virtually all the water-soluble, free coenzymes. Extraction and analysis of coenzyme F_{420} illustrates the importance of understanding the physiology of the organism being examined for coenzyme content. F_{420} is found in methanogenic archaea and some aerobic actinomycetes but is absent from most other microbes (5, 6, 8, 10). F_{420} is chemically stable to oxygen, i.e., pure F_{420} is unaffected by exposure to O_2. However, if methanogens are extracted under conditions that allow enzyme activity to proceed in the presence of air while cells are being heated relatively slowly to lyse them to release F_{420}, then an oxygen-dependent enzyme called F_{390} synthase will convert F_{420} into F_{390}, resulting in significant loss of F_{420} (9, 15). In contrast, aerobic actinomycetes do not have this enzyme, and extracting by heating in the air causes no loss of F_{420} (10).

Some coenzymes are not very water soluble, and thus, in these cases extraction with a suitable nonaqueous solvent is required, e.g., menaquinones can be extracted with chloroform-methanol, acetone:methanol, or ethanol (14).

After extraction, analysis must be conducted to measure the coenzyme content. In some cases, the extract (diluted or evaporated to remove possibly inhibitory solvents in some cases) can be examined by using bioassay techniques. However, more commonly the supernatant is fractionated by chromatography on suitable columns, e.g., DEAE-Sephadex, or C_{18} reverse-phase high-pressure liquid chromatography (HPLC). Coenzyme elution can be monitored by absorbance at an appropriate UV or visible wavelength or by fluorescence. In some cases, if the material is sufficiently pure after one chromatographic step, the coenzyme can be quantitated by measuring the absorbance or the fluorescence of the pooled column fractions. Alternatively, the area under the HPLC chromatographic profile can be measured; care should be taken to ensure that the proper extinction coefficients are used for the pH and solvent in which the material is dissolved as it passes through the detector, since organic solvents and pH can greatly affect the spectra of some small molecules (see chapter 17.1.1.4). A variety of coenzymes can be quantitated at one time by HPLC methods if retention times are known and if there are no interfering peaks; the use of a rapid spectral scan of HPLC peaks while the material passes through the detector is an excellent tool to examine the purity of peaks and thus confirm the validity of the quantitation and the purity of the material. Examples of HPLC methods for quantitation of ascorbic acid, thiamine, biotin, NADH, flavins, deazaflavins such as F_{420}, folates, and methanopterin species have been described (8).

If labeled material has been used to feed the cells prior to coenzyme analysis, the specific radioactivity of the pure coenzymes can be determined by comparing the molar quantity of coenzyme and the radioactivity measured by scintillation counting. This approach may be useful in the study of precursors of coenzyme production.

In addition to coenzymes, other small molecules of interest, including amino acids and sugars, can be extracted by this general method and analyzed as described below (sections 18.3 and 18.7).

18.3. CARBOHYDRATES

A large number of different carbohydrates can be found in bacteria and archaea: free sugars, sugar derivatives, simple polysaccharides (polymers of a single sugar or sugar derivative, such as glycogen), and complex polysaccharides (polymers of more than one different sugar, amino sugars, uronic acids, etc.). Some macromolecules contain both carbohydrate and noncarbohydrate components (lipopolysaccharides, peptidoglycans, teichoic acids, lipoteichoic acids, teichuronic acids, nucleic acids, glycoproteins, etc.).

Colorimetric methods based on the Molisch test for carbohydrates are useful for estimating the amount of simple sugars and their polymers (17, 29, 30, 46). Specific assays for pentoses in nucleic acids (ribose and deoxyribose) are described below in the sections that deal with nucleic acids (sections 18.6.1 and 18.6.2). Specific enzymatic assay examples are given for glucose, galactose, lactose, and sucrose. More complex descriptions of analytical methods for the analysis of carbohydrates have been presented (17–20, 28, 33, 35, 38, 41). Details of automated enzymatic methods can be obtained from the suppliers of enzyme kits or from instrument manufacturers.

18.3.1. Total Carbohydrates by Anthrone Reaction

The most convenient assays for total carbohydrates involve heating media, cells, or isolated carbohydrates with sulfuric acid to hydrolyze the polysaccharides and dehydrate the monosaccharides to form furfural from pentoses and hydroxymethylfurfural from hexoses. The solutions of furfural and hydroxymethylfurfural are then treated with a reagent (an aromatic compound or phenol) to produce a colored compound, which is measured in a spectrophotometer. Many variations of this Molisch test have been described (17, 29, 30, 46). Pentoses, hexoses, heptoses, and their derivatives (except the amino sugars) yield colored products, but trioses and tetroses do not.

No method accurately measures all the carbohydrate in bacteria, because different sugars give different color intensities at any selected wavelength (19). The anthrone and the phenol methods (described below) are presented because of their simplicity, relative insensitivity to interference by other cell components, and similar response to most common hexoses in cells, thus providing a reliable index of "total carbohydrates."

Either the anthrone or the phenol procedure is suitable for estimation of total carbohydrate in bacteria that contain 10% or more hexose polymers. The pentoses in nucleic acid interfere with these essays when the hexose content is low. Two notable disadvantages of these methods are that individual sugars are not identified and that working with concentrated sulfuric acid and phenol is dangerous.

Reagents and Equipment

Saline solution: 0.85% (wt/vol) NaCl in distilled water

Stock glucose solutions. Dissolve 50 mg of glucose in 50 ml of 0.15% (wt/vol) benzoic acid, which is added as a preservative. Store at 5°C. This solution is stable for several months. Dilute 1:10 in distilled water just before use to give a solution containing 100 μg/ml. Prepare standard solutions (10 to 100 μg/ml) from the diluted solution.

Sulfuric acid solution (75%, vol/vol). Add 150 ml of reagent-grade concentrated sulfuric acid to 50 ml of distilled water (CAUTION: Sulfuric acid is very dangerous [see section 18.10]). Also, avoid contact of paper with either of the sulfuric acid-containing reagents, since cellulose will be hydrolyzed to sugars, resulting in a high background; one way in which this contamination can occur is to wipe acid off a pipette tip with a paper towel and then to continue using the pipette. Also, the concentrated sulfuric acid should be of good quality and free of excess water; excess water in the assay causes turbidity.

Anthrone reagent. Add 100 mg of anthrone to 2.5 ml of absolute ethanol, and make up to 50 ml with 75% H_2SO_4 (*CAUTION*: See section 18.10). Stir until dissolved. Prepare fresh daily, and store in a refrigerator.

Thick-walled Pyrex boiling tubes

Glass marbles to fit boiling tubes (to avoid water dropping in, and to avoid evaporation; a drop of water can cause turbidity)

Spectrophotometer or colorimeter

Glass cuvettes or tubes

Goggles

Procedure

Wash samples of bacterial cells free from medium components by centrifugation at 5,000 × *g* for 10 min. Resuspend in saline solution, and centrifuge again. Resuspend the sedimented cells in distilled water, and pipette samples (0.06 to 0.6 ml) of the cell suspension into the boiling tubes. (Alternatively, culture medium can be assayed after its separation from cells by centrifugation.) Adjust the volume of all samples to 0.6 ml with distilled water. Prepare a reagent blank of distilled water (0.6 ml) and a standard curve by adding 0.6 ml of standard solutions containing 10 to 100 μg of glucose per ml. Chill the tubes and the anthrone reagent in an ice-water bath until cold. Add 3.0 ml of cold anthrone reagent, mix rapidly by swirling the tubes in ice water, and continue mixing in the ice-water bath for 5 min. Then, transfer the tubes to a bath of boiling water for precisely 10 min. Return the tubes to the ice water. Measure the A_{625} of each solution. Determine the concentration of sugars in the samples from a standard curve prepared by plotting the absorbances of the standards versus the concentration of glucose.

18.3.2. Total Carbohydrates by Phenol Reaction

Reagents and Equipment

Phenol reagent. Dissolve 5 g of reagent-grade phenol in 100 ml of distilled water (*CAUTION*: See section 18.10)

Reagent-grade concentrated sulfuric acid (*CAUTION*: See section 18.10)

Glucose standard (see anthrone procedure above)

Thick-walled Pyrex boiling tubes (15 by 2.5 cm)

Spectrophotometer or colorimeter

Glass cuvettes or tubes

Goggles

Procedure

Pipette the samples to be analyzed into thick-walled Pyrex tubes. Adjust the volume of all samples to 0.5 ml with distilled water. Prepare a reagent blank by adding 0.5 ml of distilled water, and prepare a standard curve by adding 0.5 ml of solutions containing 10 to 100 μg of glucose per ml. Add 0.5 ml of the phenol reagent; mix rapidly and thoroughly. Add 2.5 ml of concentrated sulfuric acid, mix rapidly, and let stand for 10 min. (*CAUTION*: Addition of sulfuric acid to aqueous solutions causes rapid heating and occasionally boiling; see also section 18.10.) Place the tubes in a water bath at 25°C for 15 min. Read the A_{488} of each tube against the blank prepared without glucose. Determine the concentration of sugars in the sample from a standard curve prepared by plotting the absorbances of standards versus the concentration of glucose. (See also section 18.3.1, on the

anthrone reaction, for a discussion of the disadvantages of these two assays, and see below for more-specific assays.)

18.3.3. D-Glucose, Galactose, Lactose, and Sucrose

A wide variety of monosaccharides are liberated by the hydrolysis of simple and complex polysaccharides in bacterial cells. Because the polymers exist as mixtures, except in homopolymers such as glycogen, the estimation of specific sugars sometimes presents difficult analytical problems that can be best solved with HPLC, ion chromatography, or gas-liquid chromatography procedures (18, 24, 33, 35, 52, 55, 62). However, some sugars can be estimated by specific colorimetric reactions or enzymatic methods (17, 32, 47, 57, 58); two good examples are the hexoses glucose and galactose, for which a variety of enzymatic methods are available. A specific example is the use of glucose oxidase to assay glucose, but a very similar system works for galactose. Glucose oxidase catalyzes the following reaction:

$$\beta\text{-D-glucose} + O_2 \rightarrow \text{D-glucono-}\delta\text{-lactone} + H_2O_2$$

The hydrogen peroxide, in the presence of peroxidase and a reduced chromogen, causes the production of a brown or blue oxidized chromogen that can be measured spectrophotometrically. Different chromogenic compounds are used in commercially available kits, which also contain the needed enzymes and sometimes glucose standards. The glucono-δ-lactone can be converted into gluconic acid under basic conditions, but this is not needed for the assay. The glucose oxidase assay is specific, sensitive, and easily performed. Only glucose and 2-deoxyglucose are known to produce colored products.

Galactose oxidase assay kits are also available from commercial sources. The kits contain galactose oxidase, chromogen, and standards. Since galactose oxidase oxidizes the C-6-hydroxymethyl group of galactose to produce galacturonic acid, the enzyme will also oxidize some galactosides.

Commercial kits and/or the constituent enzymes are available from many sources, but a few examples are Sigma Chemical Co., St. Louis, MO; Worthington Biochemical Co., Freehold, NJ; and Boehringer Mannheim Biochemicals, Indianapolis, IN.

A variation on the glucose oxidase assays has been developed for use in an automated system, in which hydrogen peroxide is detected electrochemically (44, 49, 50, 63; Yellow Springs Instrument Co. [YSI], *Immobilized Enzyme Biosensor*, 2001 [http://www.ysi.com/lifesciences.html]). This method is particularly attractive in classroom settings, where a large number of students have to do assays in a limited period and with a higher degree of safety than is possible with the anthrone and phenol assays; it is also cost-effective in an industrial setting, where many assays are done routinely.

A convenient method for glucose analysis of some types of samples (e.g., liquid medium containing growing microbes) is the use of strips that are sold for glucose determination in urine, e.g., for diabetic patients. These are readily available from drug stores, are inexpensive, and require no additional equipment, making them ideal for many classroom and research settings. Proper dilution of samples to an appropriate glucose concentration and appropriate controls with no added glucose are required.

Several other sugars can also be assayed by a system very similar to those above with an enzyme that generates hydrogen peroxide, sometimes assisted by other immobilized enzymes. Lactose, galactose, and sucrose are three examples of additional sugars that can be determined with the automated analyzer from Yellow Springs Instrument Co. (YSI),

Yellow Springs, OH (see above). The same system can also estimate ethanol, methanol, L-glutamate, L-glutamine, and L-lactate by using an enzymatic approach.

An alternative to the automated glucose oxidase approach is an automated glucose dehydrogenase reaction, as used in the HemoCue Betaglucose analyzer (61).

An example of the use of glucose oxidase in a manual spectroscopic system is as follows. The assay for galactose would be very similar.

Reagents

Glucose stock solution. Dissolve 300 mg of D-glucose in 100 ml of 0.15% (wt/vol) benzoic acid. Store in a refrigerator. Dilute 1:10 with distilled water, and adjust to pH 7.0 before use.

Commercial glucose oxidase-peroxidase preparation. Prepare as specified by the manufacturer.

4 M HCl (CAUTION: See section 18.10)

Chromogen: available as part of the commercial kits. Prepare as specified by the manufacturer.

Procedure

Dilute samples with distilled water so that they contain 50 to 300 μg of glucose per ml. Adjust the pH to between 6.8 and 7.2 with 0.1 M KOH or 0.1 M HCl. A pH indicator paper is adequate. Prepare a blank and a set of standards containing 50 to 300 μg of glucose in 1.0 ml of distilled water. Add the enzyme and chromogen, and incubate the mixture as specified by the manufacturer. Add 1 drop of 4 M HCl with a Pasteur pipette, and read the A_{400}. Determine the concentration of glucose in the samples from a standard curve prepared by plotting the absorbances of the standards versus the concentration of glucose.

Applications

The oxidase tests for glucose, galactose, lactose, and sucrose are suitable for estimating levels of each in growth media, in neutralized hydrolysates of polysaccharides, and in enzyme reaction mixtures. For example, the enzymatic hydrolysis of cellulose can be monitored by measurement of glucose released, and the metabolism of sucrose in a fermentor containing molasses can be monitored by measurement of sucrose disappearance. A particularly attractive monitoring system is found in the YSI analyzer equipped for simultaneous measurement of both glucose consumption and either ethanol or lactate production due to metabolizing microorganisms, e.g., yeasts or lactic acid bacteria, respectively (see YSI website above).

One of the major attractions of enzymatic methods for sugar analyses is their relative specificity. In general, there are many fewer interferences than with chemical methods. However, caution is always a good practice in analysis, and appropriate controls should always be used. It is wise to add a standard to each type of unknown solution to check for interference by salt and other possible inhibitors of oxidases.

18.3.4. Glycogen

Glycogen, in contrast with many bacterially produced polysaccharides, is easy to quantify by unsophisticated methods. The method for isolation of glycogen is presented because this polymer functions as a reserve material in many bacteria and can account for a significant fraction of the biomass in cultures. Methods for isolation of other commonly occurring complex polysaccharides (capsules, lipopolysaccharides, and peptidoglycan) are presented in separate sections.

Glycogen, like other polysaccharides, is resistant to hydrolysis by alkali, but it is readily soluble in water and insoluble in ethanol. The quantitative isolation of the polymer takes advantage of these properties. After isolation, and after hydrolysis for 6 h in 4 N HCl at 100°C in a sealed ampoule, the amount of glucose can be estimated by the anthrone or phenol reaction or by enzymatic assay.

Reagents and Equipment

Absolute ethanol (CAUTION: See section 18.10)

Potassium hydroxide (30%, wt/vol) in distilled water (CAUTION: See section 18.10)

Refrigerated centrifuge, capable of achieving 10,000 × g

30-ml glass centrifuge tubes

Vacuum desiccator or lyophilizer

Procedure

Harvest bacterial cells by centrifugation at 5,000 to 10,000 × g for 15 min, resuspend the cells in an equal volume of 0.85% (wt/vol) NaCl, and centrifuge again. Wash the cells by centrifugation once more, and then lyophilize (see chapter 47.2.1). Treat 100 mg of dry cells in a Pyrex screw-cap test tube with a Teflon-lined cap with 1 ml of 30% KOH at 100°C in a steamer or a boiling-water bath for 3 h. If lyophilization is not practical, suspend 500 mg of packed wet cells in 0.6 ml of 50% (wt/vol) KOH and heat for 3 h at 100°C to solubilize the glycogen. Cool without opening the tube, and then add 3 ml of distilled water and 8 ml of ethanol to precipitate the glycogen. Centrifuge at 10,000 × g for 15 min, and wash the precipitate with 8 ml of 60% (vol/vol in distilled water) ice-cold ethanol by centrifugation. Dry the precipitate in a vacuum desiccator or lyophilize. The washed precipitate can be analyzed for glucose by the anthrone or phenol method (see section 18.3.1 or 18.3.2) or by enzymatic methods (see section 18.3.3).

Mechanical disruption of gram-positive cells is sometimes necessary to obtain complete extraction of glycogen (31). If cells have a polysaccharide capsule, the capsular material will contaminate the glycogen preparation since, in most cases, it has similar solubility properties in ethanol.

18.3.5. Starch

Starch is not a common component of microbial cells, but it is used as a substrate for a variety of bacteria and yeasts. It is therefore sometimes of value to know the level of starch in a microbial culture. Soluble starch can be obtained free of cells by use of selective centrifugation, and some organic solvents can be of value in its precipitation and purification. It is easily hydrolyzed by acid and heat, or by using enzymes (e.g., glucoamylase), and the resultant glucose can be estimated by any of the methods above in sections 18.3.1, 18.3.2, and 18.3.3. An automated enzymatic method is also available in the YSI analyzer discussed above. Insoluble crude starch is more problematic, because of the difficulty of purification separate from the growing cells.

18.3.6. Hexosamines in Purified Peptidoglycans

Hexosamines are found in lipopolysaccharides, peptidoglycans, and teichoic acids of bacteria (19, 20, 22, 30, 31, 37, 40, 56). The most common hexosamines are glucosamine, galactosamine, and muramic acid. All N-acetylated hexosamines react with p-dimethylaminobenzaldehyde to produce a red product. The test is very sensitive and specific for hexosamines but does not distinguish among glucosamine, muramic acid, and galactosamine. Muramic acid can be estimated separately by other procedures (section 18.3.8).

The following procedure modified by Ghuysen et al. (37) is more precise and convenient than the Morgan-Elson procedure as originally described and is applicable to the estimation of total hexosamines in purified peptidoglycans. An alternative set of procedures (not described here) which can distinguish between glucosamine and galactosamine, and which can also estimate total hexosamine, has been described by Blumenkrantz and Asboe-Hansen (22). However, estimation of hexosamines in complex polymers other than peptidoglycan or bacterial cells requires separation of hexosamines from neutral sugars by ion-exchange chromatography, as described in the method in the next section ("Hexosamines in Complex Solutions").

Reagents, Equipment, and Supplies

Morgan-Elson reagents. Dissolve 16 g of p-dimethyl-aminobenzaldehyde in sufficient acetic acid to give a volume of 95 ml. Add 5 ml of concentrated HCl. Dilute 1:5 with acetic acid to prepare the color reagent. (CAUTION: See section 18.10.)

3 N HCl (CAUTION: See section 18.10.)

3 N NaOH (CAUTION: See section 18.10.)

Saturated solution of sodium bicarbonate. Add ca. 15 g of NaHCO$_3$ to 100 ml of distilled water, and stir at room temperature for 20 min.

5% (vol/vol) acetic anhydride in distilled water. Prepare just before use, and keep on ice until use. (CAUTION: See section 18.10.)

5% Potassium tetraborate in distilled water

Glucosamine standard. Dissolve 17.92 mg of glucosamine in 10 ml of 3 N HCl; store this stock solution in the refrigerator (for no longer than 1 week). This standard contains 0.2 μmol in 20 μl. Dilute with 3 N HCl to obtain standards containing 0.01 to 0.2 μmol/20 μl.

Ampoules: 1-ml lyophilization ampoules or l-ml screw-cap test tubes with Teflon-lined caps

Pyrex test tubes, 3 ml

Spectrophotometer or colorimeter

Glass cuvettes or tubes for spectrophotometry, 1 ml

Lyophilizer or vacuum desiccator

Procedure

Lyophilize, or dry in a vacuum desiccator, samples containing between 0.01 and 0.1 μmol of hexosamine. Add 20 μl of 3 N HCl to each sample. Prepare standards of glucosamine (0.01 to 0.2 μmol/20 μl). Heat the samples and standards at 95°C for 4 h in a sealed ampoule or tube. Cool and neutralize the hydrolysates with 20 μl of 3 M NaOH. Transfer a 30-μl sample of each hydrolysate and standard to separate 3-ml test tubes. Add 10 μl of a saturated solution of NaHCO$_3$ and 10 μl of freshly prepared 5% acetic anhydride. Allow the tubes to stand at room temperature for 10 min, during which N acetylation of the hexosamines occurs. Then, immerse them in boiling water for precisely 3 min to destroy the unreacted acetic anhydride. Add 50 μl of 5% K$_2$B$_4$O$_7$, mix, and heat in a boiling-water bath for 7 min. Cool, and add 0.7 ml of the color reagent (a 1:5 dilution of the Morgan-Elson reagent). Mix again, and incubate at 37°C for 20 min. Read the A$_{585}$ of each tube.

Determine the amount of total hexosamine in each sample from a standard curve prepared by plotting the A$_{585}$ versus the amount of hexosamine in each standard. Reaction products of glucosamine, galactosamine, and muramic acids have the same extinction coefficients.

Applications

As mentioned above, this procedure is not suited for analyses of hexosamines in complex mixtures of sugars and sugar derivatives. Hexosamine derivatives such as N-acetylmuramic acid substituted with peptides produced by the action of lysozyme on cell walls yield the same colored products.

The precise neutralization of the HCl after hydrolysis of samples is critical to the reaction. Therefore, it is important to prepare the HCl and NaOH carefully and to be sure that the ampoules or tubes are tightly sealed during hydrolysis.

Glucosamine, galactosamine, and muramic acid can be separately estimated by other methods (26, 60). An enzymatic assay for galactosamine has been described previously (58). Other assays for muramic acid are described in section 18.3.8.

18.3.7. Hexosamines in Complex Solutions

Hexosamines in complex solutions and those produced by the hydrolysis of mixed polymers can be separated from neutral sugars that interfere with the Morgan-Elson reaction by ion-exchange chromatography. The purified hexosamines can then be quantified by the modified reaction described above. Alternative procedures (not described here) that can distinguish between glucosamine and galactosamine and can also estimate total hexosamine have been described by Blumenkrantz and Asboe-Hansen (22).

Reagents, Equipment, and Supplies

4 N HCl (CAUTION: See section 18.10)

1 N NaOH (CAUTION: See section 18.10)

Lyophilization ampoules or Pyrex screw-cap test tubes with Teflon-lined caps

Dowex-50 (H$^+$) columns:

Preparation (precycling) procedure

Add Dowex-50 resin (1 g for each sample to be analyzed) to 2 N NaOH (20 ml/g of resin). Mix for ca. 8 min, and filter onto a Buchner funnel. Wash the resin with distilled water (50 ml/g), and slowly stir the resin with 3 N HCl (20 ml/g). Remove the HCl by filtration on a Buchner funnel, and wash the resin on the funnel with ca. 2 liters of distilled water to remove the acid. Remove the water by applying suction, and suspend the resin in distilled water (1 g/5 ml). This preparation procedure is important for removal of impurities in the Dowex resin.

Use

Transfer 5 ml of the suspended resin to a Pasteur pipette or a 3-ml disposable syringe with a ca. 18-gauge needle, plugged at the bottom with glass wool, or to a commercially prepared minicolumn. Attach rubber or silicon tubing and a clamp at the bottom to control effluent flow. Do not allow the liquid surface to pass below the top of the resin bed.

Vacuum desiccator containing NaOH pellets

Nitrogen gas supply (CAUTION: See section 18.10)

Sintered-glass funnel, fine porosity

Buchner funnel

Procedure

Mix the sample to be assayed with 4 N HCl in a small lyophilization ampoule. The amount of material and volume of acid will vary with the material used. Samples should contain 50 to 100 mg (dry weight) of bacterial cells and lesser amounts of purified material in 3 ml of 4 N HCl. Each sample should contain 0.15 to 0.50 mg of hexosamine in 3 ml. Hydrolyze the preparation for 6 h at 100°C in a vial which has been flushed with nitrogen gas and sealed.

Hydrolysis times should be optimized by using several subsamples treated for different times. The amount of hexosamine will increase to a maximum and then decrease with time of hydrolysis because of hexosamine destruction. Corrections for losses can be made by adding a known amount of hexosamine to one sample.

Dry the hydrolysate in vacuo over NaOH to remove HCl. Drying can be accomplished in a vacuum desiccator containing NaOH pellets; add water and redry. Dissolve the hydrolysate in water, and filter to remove insoluble material. Add the filtered hydrolysate to a small column containing 1 g of Dowex-50 resin. Wash the column with 15 ml of water, and discard the eluate. Elute the hexosamines with 10 ml of 2 N HCl, and dry the eluate in a vacuum desiccator containing NaOH pellets. Add 2 ml of water, and redry.

Drying of the hydrolysates and column eluates causes considerable destruction of muramic acid as a result of concentration of HCl. The losses of this compound are variable; this procedure tends to underestimate total hexosamines.

Dissolve the dried hexosamine fraction in 0.1 ml of water, and proceed with the estimation of hexosamines as described above (section 18.3.6). Because hydrolysis has been completed, treatment with 3 N HCl and subsequent neutralization can be eliminated.

Applications
This procedure for purification of hexosamines is necessary when complex polymers other than peptidoglycan (e.g., lipopolysaccharides) are examined. It can be applied to whole bacterial cells with the knowledge that muramic acid recoveries will be variable. It is also applicable to estimation of hexosamines in culture media. For galactosamine analysis, an alternative to separation of the components of complex polymers is possible when galactose oxidase is used in an enzymatic assay (58).

18.3.8. Muramic Acid
Muramic acid is found only as a component of the cell wall of bacteria and is not found in any of the archaea (34, 37, 119). The amount of muramic acid in a bacterial cell suspension is a measure of the presence and the amount of peptidoglycan.

The most convenient and specific assays for muramic acid involve treatment of acid hydrolysates of cell walls with alkali to remove D-lactate from the 3-position of muramic acid. The D-lactate is then estimated by an enzymatic method (64) involving lactate dehydrogenase or, as mentioned above (section 18.3.3), in an automatic enzymatic method, or by the colorimetric method of Hadzija (40) (see below). (Some archaea contain a peptidoglycan cell wall material called pseudomurein, but although it contains hexosamines, it does not contain D-lactate or muramic acid and thus would be negative in the assay described here [119].)

Recently there has been an increased interest in a structural variation in muramic acid, where O acetylation occurs specifically at the C-6 hydroxyl of muramoyl residues. This variation occurs in a least 10 genera of gram-positive and gram-negative bacteria (54, 65). It is thought that O acetylation could help some bacteria resist lysozyme action and thus could be important for virulence (21, 54). Thus, esti-

mation of the O acetylation of muramic acid may be of interest. This can be accomplished using a modified peptidoglycan protocol, NaOH to hydrolyze the ether-linked acetate, and HPLC or enzymatic analysis of the acetate (54).

Reagents, Equipment, and Glassware
Copper sulfate solution. Dissolve 4 g of $CuSO_4$ in 100 ml of distilled water.

p-Hydroxydiphenyl reagent. Dissolve 1.5 g of p-hydroxydiphenyl in 100 ml of 95% (vol/vol) ethanol (CAUTION: See section 18.10).

Muramic acid stock solution. Dissolve muramic acid (Sigma Chemical Co.) in water to give 0.25 mg/ml. Muramic acid contains 30% by weight of O-ether-linked lactic acid.

Screw-cap test tubes, 20 ml with Teflon-lined screw caps

Spectrophotometer and glass cuvettes or tubes. All glassware must be carefully cleaned, rinsed with sulfuric acid, and then rinsed very well with high-quality distilled water. After cleaning, contamination with dust and fingerprints must be avoided.

Concentrated sulfuric acid (CAUTION: See section 18.10)

Procedure
Neutralize samples of hydrolyzed peptidoglycan in 3 N HCl (see section 18.3.6) containing 5 to 50 μg of muramic acid in screw-cap test tubes with Teflon-lined caps by adding an equal volume of 3 N NaOH. Prepare standards containing 1 to 100 μg of lactic acid from the muramic acid stock solution. Also prepare a reagent blank. Adjust the volume of the samples, standards, and the reagent blank to 1.0 ml with distilled water. Add an additional 0.5 ml of 1.0 N NaOH, and incubate the solution for 30 min at 37°C. Add 10 ml of concentrated sulfuric acid, seal the tubes, and heat them in boiling water for 10 min. After cooling, add 0.1 ml of the copper sulfate reagent and 0.2 ml of the p-hydroxydiphenyl reagent and mix rapidly. Incubate the tubes at 30°C for 30 min in a water bath, and read the A_{560}. Determine the concentration of muramic acid in the samples from a standard curve prepared by plotting the absorbance of each standard versus the concentration of muramic acid in the standard.

Applications
The method described above is a simple, rapid, inexpensive procedure for detection of muramic acid when metabolic lactic acid is not present. Other components of the muramic acid do not interfere, nor do uronic acids and neutral sugars up to 250 μg per sample.

18.3.9. 2-Keto-3-Deoxyoctanoic Acid
The 2-keto-3-deoxyoctanoic acid (KDO) in lipopolysaccharide can be determined by the method of Weisbach and Hurwitz (66) as modified by Osborn (53). When oxidized with periodate, KDO (as well as other 2-keto-3-deoxy carbohydrates) yields formylpyruvic acid, which reacts with thiobarbituric acid to yield a chromogen with an absorption maximum at 550 nm. Deoxy sugars other than 2-keto-3-deoxy sugar acids produce different products, which form chromogens that absorb light maximally outside the 545- to 550-nm spectral range.

However, some researchers choose to use more than one type of KDO assay and average the results in an effort to increase the accuracy, since KDO is generally present at low levels (<2% of lipopolysaccharide [dry weight]) compared to other sugars. For example, Gillespie et al. (38) used an average of three different KDO assays (23, 26, 43) to demonstrate that KDO accounted for 0.6% of the dry weight of *Wolinella recta* lipopolysaccharide. More recently a fluorometric HPLC method has been described which estimates KDO in sialic acid, which should be adaptable to lipopolysaccharide analysis (41).

Reagents

Periodate reagent. Dissolve HIO_4 in 0.125 N H_2SO_4 to give a concentration of 0.025 N. (*CAUTION*: Iodic acid and periodate are dangerous acids and should be handled in the hood, using precautions as described for other volatile acids in section 18.10)

Dilute sulfuric acid, 0.02 N (*CAUTION*: See section 18.10)

Sodium arsenite reagent: 0.2 N sodium arsenite dissolved in 0.5 N HCl (*CAUTION*: See section 18.10)

Thiobarbituric acid reagent: 0.3% thiobarbituric acid in high-quality distilled water; adjust to pH 2

Procedure

Dissolve lyophilized polysaccharide containing 0.01 to 0.25 µmol of KDO in 0.1 ml of 0.02 N H_2SO_4, and heat at 100°C for 20 min to release KDO from the polymer.

Adjust samples of the hydrolysate to 0.2 ml with 0.02 N H_2SO_4. Add 0.25 ml of the periodate reagent, mix, and let stand for 20 min at room temperature. Add 0.5 ml of 0.2 N sodium arsenite solution. Incubate for 2 min at room temperature, and add 2 ml of the thiobarbituric acid reagent. Mix, and heat at 100°C for 20 min. Cool, and read the A_{548}. One micromole of KDO per ml gives an A_{548} of 19.0 in a 1-cm cuvette.

Applications

The procedure described above is suitable for the determination of KDO in purified lipopolysaccharide or cell wall preparations.

18.3.10. Heptoses

The product formed when a heptose is heated with sulfuric acid reacts with cysteine hydrochloride to produce a chromogen with an intense orange color. This chromogen changes to a pink compound which absorbs at 505 nm (53).

Reagents, Supplies, and Equipment

Sulfuric acid reagent. Add 6 volumes of concentrated H_2SO_4 to 1 volume of water (*CAUTION*: See section 18.10).

Cysteine hydrochloride reagent: 0.3 g of cysteine hydrochloride in 10 ml of distilled water

Water bath at 20°C

Ice-water bath

Boiling-water bath

Spectrophotometer

Cuvettes or tubes for spectrophotometer

Procedure

Adjust sample volume to 0.5 ml. Place in an ice bath. Add 2.25 ml of the sulfuric acid reagent to samples. Mix the tubes in the ice-water bath for 3 min. Transfer to the 20°C water bath for 3 min, and then heat in boiling water for 10 min. Cool, add 0.05 ml of the cysteine reagent, and mix

well. Read the A_{505} and A_{545} 2 h after adding cysteine reagent. Use a blank containing the same reagents described above, but lacking the cysteine, to zero the spectrophotometer at each wavelength.

Subtract the A_{545} from the A_{505}. After this subtraction, 1 µmol of L-glycero-D-mannoheptose in 1 ml gives an absorbance of 1.07 in a 1-cm cuvette.

18.3.11. Polysaccharides and Complex Carbohydrates

In some cases, experiments require analysis of multicomponent carbohydrates containing a variety of possible sugars and sometimes organic acids, proteins, and lipids. Analysis of this type of material is considerably more complex than that of simple polymers of monosaccharides, and often the structural information required may involve the use of nuclear magnetic resonance (NMR) and mass spectral data, both of which are beyond the scope of this chapter. However, a variety of capsule or slime polysaccharides provide examples of the utility of techniques available in a large number of well-equipped laboratories and departments. In some cases, noncarbohydrate components, if they are limited in number, can be analyzed by chromatography systems identical to or not too dissimilar from those already being used for sugar analysis.

Hydrolysis needs will vary somewhat, depending on the material, and thus, it is advisable to conduct hydrolysis under several conditions (in some cases with internal sugar controls to estimate sugar destruction) and to examine the products by relevant and informative methods, e.g., HPLC or ion chromatography. Both of these chromatographic techniques allow the separation of a number of compounds and the use of variable detection methods, e.g., refractive index, pulsed amperometric detection, or absorbance at various wavelengths. Evaluation of polysaccharides and complex carbohydrates will be greatly strengthened by additional confirmatory work, e.g., by enzymatic assays for glucose or galactose, so that chromatographic information can be verified when multiple and sometimes overlapping peaks are present in elution profiles. Likewise, thin-layer chromatography can play a valuable confirmatory role in identification.

Although exocellular polysaccharides vary considerably in their sugar content, most can be harvested and partly purified by solvent precipitation, and the component sugars can be determined by HPLC or ion chromatography. Described below is an example of a procedure that would be suitable for recovery and analysis of exopolysaccharides from *Acinetobacter calcoaceticus* (24, 42), *Klebsiella* spp. (24), *Rhodotorula glutinis* (36), and (possibly with a slight modification to use ethanol or isopropanol instead of acetone) *Xanthomonas campestris* (25), *Pseudomonas mendocina* (39), and *Azotobacter vinelandii* (27). The general problem of microbial polysaccharide recovery has been discussed by Smith and Pace (59).

Reagents, Materials, and Equipment

Sodium chloride

Acetone (*CAUTION*: See section 18.10)

Sodium azide (*CAUTION*: See section 18.10)

Incubator at 4°C

Dialysis tubing

Screw-cap tubes with Teflon-lined caps

Trifluoroacetic acid (8 N aqueous solution) (*CAUTION:* See section 18.10)

Boiling-water bath

Vacuum desiccator or rotary evaporator

Suitable analysis instrument (e.g., HPLC or ion chromatograph) and supplies for its use

Centrifuge, refrigerated

Magnetic stirrer

Lyophilizer

Procedure

Grow cells in 50-ml volumes in a 250-ml flask with an appropriate medium for polysaccharide production (for examples, see above references). Remove cells by centrifugation at 5,000 × g for 20 min. To the supernatant of ca. 45 ml add 0.05 g of NaCL, and mix. Pour this mixture into 200 ml of acetone, and stir well (acetone can be replaced with 95% ethanol or neat isopropanol in most cases). Allow this to sit without stirring overnight at 4°C, and the following day centrifuge the material at 10,000 × g for 20 min (at 4°C) and decant most of the supernatant, leaving the precipitate undisturbed. Resuspend the precipitate in a minimum of the remaining solution, and place the slurry in a dialysis bag. (If necessary, the polysaccharide could be washed one or more times with ethanol or another solvent, as described above, before starting dialysis.) Dialyze overnight at 4°C in distilled water containing 0.01% sodium azide. Replace the water with pure distilled water, and dialyze for an additional 3 to 4 h. Lyophilize the remaining slurry to make a dry polysaccharide powder.

Place 4 mg of polysaccharide powder in a Teflon-lined screw-top glass tube, add 1 ml of 8 N trifluoroacetic acid, and tightly cap the tube. Heat the tube at 100°C for 2 h. Cool the tube to room temperature, uncap, and dry under reduced pressure (e.g., in a vacuum desiccator containing NaOH pellets, under continuous vacuum) for ca. 48 h until only a residue remains (24, 52).

The residue can then be examined for sugar content by any of a variety of methods including HPLC (18; also see informational literature from HPLC suppliers), gas-liquid chromatography (51, 52) (see chapter 17.3.5), or ion chromatography (24, 45, 55) (see chapter 17.3.7; e.g., with a Dionex Q1C analysis system equipped with an HPIC-AS6 anion-exchange column). In some cases gas-liquid chromatography can be useful when coupled with mass spectrometry (28, 48) The enzymatic methods described above (section 18.3) are also excellent and specific methods for some possible component sugars. See also chapter 17.3 for a description of types of chromatographic analysis.

18.4. LIPIDS

Lipids are a chemically diverse group of biological molecules that are insoluble in water but soluble in organic solvents and that commonly contain long-chain hydrocarbon groups. Lipids include long-chain hydrocarbons, alcohols, aldehydes, fatty acids, and their derivatives including glycerides, phospholipids, glycolipids, and sulfolipids. Sterols, fatty acid esters of sterols, vitamins A, D, E, and K, carotenoids, and other polyisoprenoids often are considered in this class because they, like lipids, are associated with membranes and can be extracted by lipid solvents. Surveys of the types and composition of microbial and archaeal lipids have been compiled (69, 81, 85, 96, 97).

Cytoplasmic membranes of all living cells contain lipids. Bacterial membranes are much less complex than those of eukaryotic cells. However, bacteria contain a wide variety of lipids in addition to phospholipids. These include sphingolipids, neutral lipids, glycolipids, and a variety of unusual lipids associated with organisms of specific genera such as *Corynebacterium*, *Nocardia*, and *Mycobacterium* (69). Prokaryotes were believed not to contain sterols; however, it is now known that some bacteria can contain squalene and a variety of sterols (70, 91). Gram-negative bacteria contain unique lipid constituents in lipopolysaccharide complexes as part of their outer membrane (69). Poly-β-hydroxybutyrate is accumulated in large amounts in the cytoplasm of several bacteria and serves as a carbon and energy reserve (88, 89).

Archaeal lipids are quite different from those found in bacteria (81, 83–87, 96, 97). Archaeal lipids have an ether bond linking the hydrophobic polyisoprenoid group to the hydrophilic glycerol, whereas bacterial lipids have an ester bond between the glycerol and an alkane chain. A diversity of novel variations on this ether-linkage and polyisoprenoid structural motif can be found among the archaea. Since no archaea have yet been demonstrated to have an outer membrane, the major location of these lipids is the cytoplasmic membrane. In general, total archaeal lipids account for about 2 to 5% of the cell (dry weight).

Because of the complexity of microbial lipids, a description of procedures that would provide a total analysis of each class is impractical here. For variations of the procedures described below, and for detailed techniques required for a more complete analysis of bacterial lipids, the reader should consult one of the specialized articles on the subject (67, 68, 71, 76–87, 90–101).

18.4.1. Crude Total Lipids

The procedure for extraction of total crude lipids described by Bligh and Dyer (71) is less time-consuming and requires less sample manipulation than other methods. The procedure has been widely used with bacteria. The bacterial cells are mixed with chloroform and methanol in amounts that yield a monophasic solution when combined with water in the tissue (75). Upon dilution with water or chloroform, a biphasic solution results. The lipids are retained in the chloroform phase, and nonlipid materials are retained in the methanol-water phase. The chloroform layer is isolated, and the lipids can be purified and separated into classes as described below. The recovery of total crude lipids after separation from the nonlipid contaminants should exceed 95%.

Hara and Radin (76) have described an alternative to this procedure which does not involve chloroform use and which has the advantage that centrifugation can be used instead of the filtration steps in the Bligh and Dyer procedure. A hexane-isopropanol mixture is used to extract lipids, and nonlipid contaminants are removed by washing the extract with aqueous sodium sulfate. The lipid extract can be fractionated on silica gel columns by use of hexane, isopropanol, and water mixtures.

Reagents, Materials, and Equipment

Methanol (*CAUTION:* See section 18.10)

Chloroform (*CAUTION:* See section 18.10)

Filter paper, Whatman no. 1

Porcelain Büchner funnel, ca. 43-mm diameter

Suction flask: glass, 125 ml

Fume hood
Nitrogen gas supply (CAUTION: See section 18.10)

Procedure

Mix a thick cell paste prepared from 5 g of wet packed cells (or 1 g of lyophilized cells, in 4 ml of distilled water) in a glass-stoppered tube with 5 ml of chloroform and 10 ml of methanol. Shake the mixture for 5 min, and leave it at room temperature with intermittent shaking for 3 to 4 h. Add another 5 ml of chloroform and 5 ml of distilled water, and shake the mixture at room temperature for another 30 min.

Rapidly filter the extract through Whatman no. 1 filter paper with slight suction. Transfer the filtrate to a graduated glass cylinder to allow separation of the layers. Completely remove the methanol-water layer by aspiration.

Reextract the residue of cell material on the filter paper by shaking with 5 ml of chloroform. Filter the mixture again, and rinse the filter with 12.5 ml of chloroform. Add the combined filtrates to the chloroform layer from the first extraction.

Add the chloroform extracts to a tared glass or porcelain container, and evaporate to dryness under a stream of nitrogen in a water bath at 35 to 40°C to give a dry crude total-lipid preparation. Work in the fume hood. Measure the amount of crude total lipid with an analytical balance.

(Note that the procedure of Hara and Radin [76] can be used as an alternative, to avoid the danger of chloroform [see section 18.10]. This is particularly attractive for larger-scale preparations, in which large volumes of chloroform must be handled. Nonetheless, hexane and isopropanol present hazards of their own, e.g., flammability.)

Applications

This procedure is the starting point for most types of lipid analysis. It can be used to obtain the total dry weight of lipids in cells, as the starting point for the purification of many types of specific lipids, and to determine the structures of specific lipids.

18.4.2. Purified Total Lipids

Molecular sieve chromatography of lipids in an appropriate solvent can be used to provide material that is more pure than that obtained by extraction alone (101).

Reagents and Materials

Solvents. Mix 200 parts of redistilled or fresh chloroform and 100 parts of methanol with 75 parts by volume of water. Shake vigorously in a separatory funnel, and allow the mixture to separate into an upper phase (solvent I) and a lower phase (solvent II). Collect the phases separately, and store them at room temperature. (CAUTION: See section 18.10.)

Sephadex G-25, fine bead form (Pharmacia Fine Chemicals, Inc., Piscataway, NJ)

Chromatography tube: ca. 1 by 15 cm with sintered-glass bottom and Teflon stopcock or other flow control device

Paraffin shavings

Procedure

(See Chapter 17.3.9 for principles of gel permeation chromatography.)

Soak 25 g of Sephadex G-25 beads overnight in 100 ml of solvent I. Allow the beads to settle, decant the solvent, and rinse the beads four times with solvent I. Avoid rough treatment of beads; e.g., do not use stirring bars to mix the solution.

Prepare a thick slurry of beads (total volume, 50 ml) in solvent I. Degas the slurry by placing it in a vacuum flask, and draw a vacuum with an aspirator for ca. 10 min. Swirl occasionally. Pour the slurry of beads into a column containing a few milliliters of solvent I. Continue adding slurry until a bed height of 10 cm is obtained. Apply a pressure of ca. 0.1 atm (1 atm = 1.013×10^5 Pa) from a nitrogen tank to settle the resin bed, and place a fiberglass filter disk on top of the resin bed. Rinse the column with 10 ml of solvent I and 10 ml of solvent II.

Dissolve ca. 175 mg of the dry crude total lipids (see the preceding procedure) in ca. 4 ml of solvent II. Filter the dissolved lipid fraction through a medium-porosity sintered-glass funnel directly onto the Sephadex column. Rinse the funnel with 2 ml of solvent II. Apply a pressure of 0.15 to 0.2 atm with a nitrogen tank to obtain a flow rate of 0.5 to 1.0 ml/min. Add solvent II, and elute the column until 25 ml of solvent is collected.

Concentrate the eluate under reduced pressure at room temperature until a cloudy aqueous emulsion forms. Lyophilize this emulsion to dryness. Dissolve the dried lipid in chloroform-methanol (2:1, vol/vol), transfer it to tared glass-stoppered vials, and evaporate the solvents under a stream of nitrogen at ca. 37°C. Complete the drying in an evacuated vacuum desiccator containing paraffin shavings. After drying, weigh the vial containing the dried lipid extract.

Recovery of lipids can be checked by using known amounts of purified total lipids in the extraction and purification steps.

18.4.3. Lipid Classes

Lipids can be bound to a chromatographic packing in a column and can be eluted by solvent solutions; the adsorbent materials most often used are silicic acid, alumina, and magnesium oxide (see also chapter 17.3). The lipids are bound by polar, ionic, and van der Waals forces. Separation is effected by using solvents of increasing polarities, which separates lipids into classes determined largely by the number and types of polar groups in the molecules of each class (67, 68).

In the method described below, a chromatography grade of silicic acid is used. The solvents used to elute the adsorbed lipids are chloroform, a chloroform-acetone mixture, and methanol. These have been selected because of the relative ease and reproducibility of the technique and because all of the phospholipids are eluted in a single fraction.

Reagents and Materials

Silicic acid for chromatography: 100 g of Bio-Sil A (100 to 200 mesh, activated at 125°C for 4 h; Bio-Rad Laboratories, Richmond, CA)

Chloroform (CAUTION: See section 18.10)

Acetone (CAUTION: See section 18.10)

Methanol (CAUTION: See section 18.10)

Glass chromatographic tube, approximately 1 by 30 cm, fitted with a solvent reservoir (separatory funnel with a Teflon stopcock) coupled to the top of the column by a 24/40 or standard taper greaseless glass joint. The column should be fitted with a fritted-glass disc or a glass-wool plug and greaseless Teflon stopcock at the bottom.

Procedure

Make a slurry of 12 g of Bio-Sil A in redistilled or fresh chloroform. Place the slurry in a vacuum flask or desiccator, and apply a vacuum for 10 min to degas it. Pour the degassed slurry into the chromatography tube. Allow the chloroform to drain through the column until the meniscus is about 1 mm above the resin.

Dissolve 150 to 200 mg of the purified total lipids (see the preceding procedure) in 3 ml of chloroform. Add the solution to the column, and drain the column until the chloroform just enters the resin layer. Add an additional 5 ml of chloroform, and drain the column again until all the lipid just enters the resin bed. Carefully add 50 ml of chloroform to the column and reservoir without disturbing the resin bed surface. Allow the solvent to flow through the column into a glass collection vessel at a rate not exceeding 1 ml/min. When the chloroform just enters the resin bed, change collection vessels and add 60 ml of chloroform-acetone to the column and reservoir. Collect the effluent at a rate not exceeding 1 ml/min in the second collection flask. When this solvent mixture just enters the resin bed, add 60 ml of acetone and collect the eluate in the same vessel; then, add 60 ml of methanol to the column and collect the eluate in the third vessel.

The first collection vessel will contain neutral lipids (hydrocarbons, carotenoids, chlorophyll, sterols and sterol esters, glycerides, waxes, long-chain alcohols, and aldehydes) and fatty acids. The second reservoir will contain glycolipids, sulfolipids, and occasionally some phosphatides. The third reservoir will contain phospholipids. These can be detected and quantified by assaying for organic phosphate (see section 18.4.5).

18.4.4. Phospholipid Fractionation

Thin-layer chromatography with silica gel is commonly used for phospholipids fractionation. There are many different solvent systems and choices of developing chambers for thin-layer chromatography plates to be used in the analysis of phosphopholipids (67, 68, 94).

Reagents and Materials

Silica gel glass plates, 20 by 20 cm, 250 μm thick. Prepared plates are available from several manufacturers, or they can be prepared by coating silica gel onto glass plates (68). Develop the plates to the top in a chromatography chamber with redistilled or fresh acetone. Air dry in a fume hood, and heat for 30 min at 110°C. Cool in a desiccator, and use on the same day.

Glass chromatography tanks. Thin-layer chromatography tanks or sandwich developing tanks can be purchased from several sources. Add freshly mixed solvent (60 ml) to multiplate tanks 2 h before developing the chromatograms. Line the tanks with sheets of filter paper saturated with solvent, and seal the tanks to equilibrate the atmosphere with the solvent.

Solvents (*CAUTION:* See section 18.10)

Solvent I: Chloroform-methanol-water, 65:25:4 (vol/vol/vol), prepared from fresh solvents

Solvent II: Chloroform-methanol-7 N ammonium hydroxide, 60:35:5 (vol/vol/vol)

Nitrogen gas: in a compressed gas cylinder; use carefully (*CAUTION:* See section 18.10)

Micropipettes or microsyringes, 50 μl, graduated at intervals

Iodine (I₂) crystals (*CAUTION:* See section 18.10)

Procedure

Dissolve the dry phospholipid fraction from the Bio-Sil column in chloroform-methanol (1:1, vol/vol [about 100 mg/ml]). Record the volume of the solution. Apply a measured amount of each solution as a spot, <10 mm in diameter, 2.5 cm from the bottom and 2.5 cm from the right side of a separate chromatography plate. Use a graduated 50-μl disposable pipette, micropipettor, or microsyringe. Dry between applications with a stream of nitrogen gas or a hair dryer (with gentle heat on, the hair dryer is faster). Apply up to 0.1 ml of a solution to a plate. Dry the spot, and place in a developing tank containing solvent I. Seal the tank, and allow the chromatogram to develop until the solvent front reaches the top of the plate (1 to 2 h). Remove and air dry the plate. Place the plate, rotated 90° clockwise, relative to the first solvent flow direction in another tank containing solvent II, and develop in the second direction. Air dry the developed plate.

Stain the developed chromatogram with iodine vapors as described below to visualize the phospholipid spots. Wet the chromatogram slightly with a water mist, and scrape each spot off with a stainless-steel spatula or a razor blade, collecting the powder on weighing paper. Transfer all the silica gel from each spot to a commercially available minicolumn (or into a Pasteur pipette plugged with glass wool in the constricted end). Tap the minicolumn until the silica gel forms a layer on the frit. Rinse the minicolumn and silica gel with 2 ml of chloroform-methanol (1:1, vol/vol) followed by 2 ml of methanol to elute the phospholipids. Collect the eluates in glass-stoppered tubes. Distribute each phospholipid into three equal portions in two glass-stoppered 15-ml test tubes and a lyophile ampoule for further analysis. Evaporate the solvents at 37°C under a stream of nitrogen. One tube containing each of the separated phospholipids is to be assayed for total organic phosphorus to determine the amount of phospholipid present. Other samples of each phospholipid can be subjected to acid and alkaline hydrolysis followed by identification of the products to establish their probable structures.

Iodine stain for detection of lipids on chromatograms (51, 52). Working in a hood, add crystals of iodine to a chromatographic tank, cover the tank, and warm it in a water bath at 60°C (or with a heated hair dryer) to vaporize the iodine (*CAUTION:* Iodine vapors are dangerous; see section 18.10). Place the chromatograms in the tank on glass supports to hold them 2.5 cm off the bottom. Cover the tank for 1 min, remove the chromatogram, and view it in daylight and with a UV lamp at 365 nm (*CAUTION:* UV light is dangerous; see section 18.10). The lipids appear as brown spots in daylight and as very dark spots under UV light. Mark the outline of the spots with a pencil before the color fades.

18.4.5. Phosphate in Phospholipids

Phosphate levels in each phospholipid fraction obtained by thin-layer chromatography can be estimated by the procedure of Bartlett, as described by Kates (68) and as follows.

Reagents and Materials

Clear Pyrex test tubes, 13 by 100 mm. Clean in hot 1 N nitric acid for 1 h, and rinse with distilled water. All glassware must be carefully cleaned to remove traces of detergent and other sources of phosphorus contamination. (*CAUTION:* Nitric acid is dangerous; see section 18.10.)

Acid-washed glass beads (ca. 1 to 2 mm in diameter) and marbles (sized to fit top of the Pyrex tubes above). Clean with the test tubes above.

Perchloric acid (72%, vol/vol) in distilled water (CAUTION: Perchloric acid is dangerous; see section 18.10)

Ammonium molybdate solution (5%, wt/vol) in distilled water

Amidol reagent. Add 0.5 g of 2,4-diaminophenol dihydrochloride to 50 ml of 20% (wt/vol) sodium bisulfite solution. Filter.

Phosphorus standard (10 μg of phosphorus per ml). Dissolve 1.097 g of KH_2PO_4 in 250 ml of distilled water. Dilute 1:10 with distilled water to give 10 μg/ml.

Spectrophotometer to read at 830 nm (near-infrared)

Procedure

Place a measured volume of lipid sample in test tubes, and evaporate the solvent under a stream of filtered air or nitrogen. Dry the phosphate standards containing 1 to 10 μg of phosphorus per tube (0.1 to 1.0 ml of standard) in similar fashion. Add 0.4 ml of 72% perchloric acid and an acid-washed glass bead. Heat at 100°C until the solution is clear and colorless. Work in a fume hood that has been approved for use with perchloric acid. Add 4.2 ml of distilled water, 0.2 ml of ammonium molybdate solution, and 0.2 ml of the amidol reagent. Mix, and cover the tubes with acid-washed glass marbles. Heat in a boiling-water bath for 7 min. Cool rapidly, and read the A_{830} after 15 min. A blank without phosphate is used to zero the spectrophotometer. Determine the concentration of phosphate in each sample from a standard curve prepared by plotting the amount of phosphorus in each standard tube versus the absorbance of the standard. The absorbance is linear up to 10 μg of phosphorus per ml. The total phospholipid in each fraction can be converted into micromoles of each phospholipid by assuming that each phospholipid contains one phosphate group.

18.4.6. Phospholipid Identification

Phospholipids can be deacylated by mild alkaline hydrolysis to yield water-soluble glycerol phosphate esters and fatty acids, which are soluble in organic solvents. The glycerol phosphate esters can be identified by specific staining reactions and their R_f values as determined by paper chromatography (70) or thin-layer chromatography. The fatty acids can best be identified by gas-liquid chromatography of methyl esters (81, 96, 101) (see chapter 17.3.5).

The identity of the phospholipids can be further confirmed by analysis of the free bases liberated by acid hydrolysis. Free bases are identified by R_f values and color reactions after thin-layer or paper chromatography (67, 94).

Reagents and Materials

Methanolic sodium hydroxide (0.2 N). Dissolve 0.4 g of NaOH in 50 ml of methanol. Prepare fresh (CAUTION: See section 18.10).

Methanolic ammonium hydroxide (1.5 N). Dilute 10 ml of concentrated ammonium hydroxide to 100 ml with methanol (CAUTION: Ammonium hydroxide is dangerous; see section 18.10).

Sodium hydroxide (1.25 N) (CAUTION: See section 18.10)

Chloroform (CAUTION: See section 18.10)

Methanol (CAUTION: See section 18.10)

Cation-exchange resin: Bio-Rad AG 50W x 8, Dowex-50 (H^+), Amberlite IR-100 (H^+), or a similar resin

Hydrochloric acid, 1 N (CAUTION: See section 18.10)

Water-saturated phenol. Place equal volumes of fresh, high-quality liquid phenol and water in a well-capped bottle,

shake well, and remove the phenol layer. Store in a brown bottle at 4°C. Prepare fresh for each use from redistilled phenol stored at −20°C. (CAUTION: Phenol is dangerous; see section 18.10.)

Methanol-water (10:9, vol/vol) (CAUTION: See section 18.10)

Ethanol, reagent grade, absolute (CAUTION: See section 18.10)

Acetic acid, reagent grade, absolute (CAUTION: See section 18.10)

Hexane, spectroscopic grade (CAUTION: See section 18.10)

Glass-stoppered 15-ml test tubes

Whatman no. 1 chromatography paper. The size of the paper depends on the developing tank available; alternatively, use cellulose thin-layer chromatography plates, available from many commercial sources.

Tank for descending chromatography, or for thin-layer chromatography

Lyophile ampoules, 10 ml

Lyophilizer or flash evaporator

Phospholipid standards (available from commercial sources)

Phosphatidylcholine, distearoyl ester

Phosphatidylserine from bovine brain

Phosphatidylethanolamine from E. coli

Phosphatidyl-N,N-dimethylethanolamine, dipalmitoyl ester

Diphosphatidylglycerol (cardiolipin) from bovine heart

Dragendorf spray reagent. Dissolve 1.7 g of bismuth nitrate in 100 ml of 20% acetic acid. Dissolve 10 g of potassium iodide in 25 ml of water. Just before use, mix 5 ml of the potassium iodide solution, 20 ml of the bismuth nitrate solution, and 70 ml of water to use as a spray reagent.

Ninhydrin reagent. Dissolve 0.25 g of reagent-grade ninhydrin in 100 ml of acetone-lutidine (2,6-dimethyl pyridine) (9:1, vol/vol). Prepare fresh for each use. (CAUTION: Some individuals are allergic to ninhydrin.)

Salicylsulfonic acid-ferric chloride reagent (67). Dissolve 1.5 g of $FeCl_3 \cdot 6H_2O$ in 30 ml of 0.3 N HCl. Dilute with 90 ml of acetone. Dip paper strips from chromatograms in this reagent, dry in a fume hood, and dip in a solution prepared by dissolving 1.5 g of salicylsulfonic acid in 100 ml of acetone.

Procedures

Acid hydrolysis of phospholipids and identification of free bases

Add 2 ml of 1 N HCl to the lyophile ampoule containing the purified individual phospholipids from the thin-layer chromatographic plates (section 18.4.4). Seal the ampoule, and heat to 100°C for 4 h. Cool, open the ampoule, transfer the contents to a 10-ml screw-cap tube, and add 2 ml of redistilled hexane. Mix on a Vortex mixer, and let stand to allow the phases to separate. Remove the aqueous phase, and lyophilize to dryness or dry in a rotary evaporator.

Dissolve the residue in 0.1 ml of distilled water. Spot three samples of each aqueous phase 10 mm apart on Whatman no. 1 chromatographic paper, and develop the chromatogram by descending chromatography by using the solvent system phenol-ethanol-acetic acid (50:5:6, vol/vol/vol). Known phospholipids treated identically, or choline, ethanolamine, serine, and *N,N*-dimethylethanolamine, can be used as standards for chromatography and identification of the free bases (67). With minor modifications, thin-layer chromatography can be used for this separation.

After the chromatogram is developed, dry it overnight in a fume hood and cut it into strips. Spray one strip, containing a standard or an unknown, with ninhydrin reagent, and spray another strip, with the same unknown or standard, with the Dragendorf reagent. Serine and ethanolamine give blue spots with R_f values of approximately 0.27 and 0.51, respectively, after treatment with ninhydrin. Choline and *N,N*-dimethylethanolamine do not react with the ninhydrin spray. Choline gives an orange spot, and dimethylethanolamine gives a weak orange spot with the Dragendorf spray reagent. The R_f values for choline and *N,N*-dimethylethanolamine are approximately 0.93 and 0.86, respectively.

Alkaline hydrolysis of phospholipids and identification of glycerol phosphate esters (67, 68)

Add 0.2 ml of chloroform, 0.3 ml of methanol, and 0.5 ml of methanolic sodium hydroxide to 1 to 5 mg of phospholipid prepared by chromatography on Bio-Sil A (section 18.4.3). Mix in a glass-stoppered test tube on a Vortex mixer, and leave at room temperature for 15 min.

Add 1.9 ml of chloroform-methanol–1.25 N NaOH (1:4:4.5, vol/vol/vol). Transfer to a conical centrifuge tube, mix on a Vortex mixer, and centrifuge for 1 min at 600 × g. Remove the upper methanol-water phase with a Pasteur pipette. Add cation-exchange resin to the methanol-water phase with vigorous mixing until the pH, measured with indicator paper, is slightly acidic (pH 5 to 6). Centrifuge, and remove the liquid from the resin with a Pasteur pipette. Add 1 or 2 drops of methanolic ammonium hydroxide to make the supernatant slightly alkaline (pH 7.5 to 9.0). Check the pH with indicator paper. Concentrate to dryness under a stream of nitrogen at 30°C, and dissolve the residue in 0.1 ml of methanol-water (10:9, vol/vol).

The dissolved glycerol phosphate esters can be identified by cochromatography with products of known phospholipids treated in an identical fashion (67). Three 25-μl aliquots of each sample should be spotted 10 cm apart on Whatman no. 1 paper and developed by descending chromatography with a solvent system of phenol-ethanol-acetic acid (50:5:6). After the chromatogram is developed, dry it overnight in a fume hood and cut into strips. Spray one of the strips containing the glycerol-phosphate ester sample with the Dragendorf spray reagent, one with the ninhydrin reagent, and another with the salicylsulfonic acid-ferric chloride spray reagent of Vorbeck and Marinetti (99). Phosphatidylethanolamine (R_f = 0.66) and phosphatidylserine (R_f = 0.28) give mauve spots when sprayed with the ninhydrin spray and left for a few hours at room temperature. Phosphatidylcholine (R_f = 0.86) and phosphatidyldimethylethanolamine (R_f = 0.80) give orange spots and weak orange spots, respectively, with the Dragendorf reagent and no color with the ninhydrin spray reagent. All react with the salicylsulfonic acid-ferric chloride reagent; glycerol phosphate esters appear as white spots on a mauve background. Note that thin-layer chromatography can be used as a substitute for the paper chromatography in all of the above procedures, with appropriate modifications.

18.4.7. Archaeal Lipids

The lipids of archaea are structurally distinct from lipids of bacteria: the glycerol moiety is bound to the hydrophobic portion by an ether bond, not an ester bond; the hydrophobic portion is composed of a polyisoprenoid chain, not an alkane; in some cases a tetraether structure occurs, where two glycerol moieties are bound to an intervening double-length polyisoprenoid (77, 81–87, 96, 97). Despite these major differences, these ether lipids can be extracted and examined by using methods very similar to those used for bacteria such as *E. coli* or *Bacillus* species. Since variation in lipid content between genera or species may be of taxonomic value, lipid analysis of archaea is in some cases useful for identification and for classification of new isolates. Provided here is a general overview of how archaeal lipids can be extracted, separated, and analyzed.

18.4.7.1. Extraction

The Bligh and Dyer extraction method (71) described above (section 18.4.1) is appropriate for most archaea, but modified procedures have also been described (78, 83, 96). Extraction with dichloromethane after reflux in 4 N HCl has been described by Hopmans et al. (77) The nonchloroform procedure described by Hara and Radin (76) should work, but its use has not yet been described. One gram of cells (dry weight) should yield 20 to 50 mg of total lipids.

18.4.7.2. Separation

Reagents and Materials

Silicic acid for chromatography: 10 g of Bio-Sil A, 100 to 200 mesh, activated at 125°C for 4 h (Bio-Rad)

Glass chromatography column (ca. 1 by 30 cm or 2 by 10 cm)

Chloroform (*CAUTION: See section 18.10*)

Acetone (*CAUTION: See section 18.10*)

Methanol (*CAUTION: See section 18.10*)

Procedures

Make a slurry of 10 g of Bio-Sil A in chloroform. Place the slurry in a vacuum flask or desiccator, and apply a vacuum for 10 min to degas it. Pour the degassed slurry into the chromatography column. Drain the column until the meniscus is about 1 mm above the resin. (*CAUTION: Chloroform is hepatotoxic and a carcinogen. Avoid exposure to either liquid or vapor, and use a fume hood. See section 18.10.*)

Dissolve 50 to 200 mg of the total lipid in 3 ml of chloroform. (Alternatively, lipid treated to yield "purified total lipid" as described in section 18.4.2 can be used.) Add the solution to the column, drain until the liquid is about 1 mm above the resin surface, carefully add another 5 ml of chloroform, and drain in the same way. Carefully add 5 ml of chloroform, and elute with another 45 ml of chloroform; collect this fraction 1 in a glass vessel (contains neutral lipids). Elute the column with 50 ml of acetone by using a similar technique; this fraction 2 should contain principally glycolipids and phosphatidyl glycerol. Elute the column with 65 ml of chloroform-methanol (3:2, vol/vol); this fraction 3 should contain principally phosphoglycolipids. Elute the column with 200 ml of methanol; fraction 4 should contain principally phosphoglycolipids.

18.4.7.3. Analysis

Polar lipids (fractions 2, 3, and 4) and neutral lipids (fraction 1) can be analyzed by a variety of methods. The general scheme for certain identification of the structures of lipids is to further purify the polar lipids by thin-layer chromatography, with and without methanolic-HCl hydrolysis, to degrade the lipid products by cleavage and reduction (e.g., with HI and LiAlH$_4$), and to analyze the resultant products by gas chromatography coupled with mass spectrometry and/or with HPLC-mass spectroscopy. Structural identity is further strengthened with infrared and NMR data, and information on hydrophilic groups is obtained by sugar assays, HPLC, liquid chromatography-mass spectrometry, and NMR. These methods and strategies are described in a variety of references (68, 77, 78, 80–87, 96–98) and are beyond the scope of this chapter. However, we do provide here a method by which the distribution of diethers and tetraethers in polar lipids can be determined by thin-layer chromatography.

Reagents and Materials

Silica gel glass thin-layer chromatography plates and chromatography tank (as described in section 18.4.4)

Thin-layer chromatography solvent: n-hexane, diethyl ether, and acetic acid (80:20:1, vol/vol/vol) (CAUTION: See section 18.10)

Hydrolysis reagent: 2.5% anhydrous methanolic HCl (made by bubbling anhydrous methanol with gaseous HCl). (CAUTION: This procedure is extremely dangerous and requires the use of a fume hood for the containment of a gaseous HCl tank and apparatus for making the reagent. Gaseous HCl is a very corrosive and toxic gas. If unfamiliar with this type of procedure, consultation with staff familiar with the techniques is essential.)

Iodine (I_2) crystals (CAUTION: See section 18.10)

Petroleum ether (CAUTION: See section 18.10)

Procedure

Hydrolyze each polar lipid fraction by addition to anhydrous 2.5% methanolic HCl, and reflux for 1 to 2 h, using a $CaCl_2$ tube on the reflux exit. Extract the products with petroleum ether. Apply the extracts to thin-layer chromatography plates, and develop until the solvent front reaches the top of the plates. Remove the plates, and dry with a hair dryer. Use the iodine stain (see section 18.4.4) to detect the lipids. Diphytanyl glycerol diether has an R_f value of ca. 0.12, and dibiphytanyl diglycerol tetraether has an R_f of ca. 0.46.

A variety of methanogens, including *Methanobacterium* and *Methanospirillum* species, have both diethers and tetraethers, whereas *Methanococcus vannielii* and *Methanosarcina barkeri* have only the diether version (81, 96). This technique is valuable for characterization and identification of newly isolated archaea of any group and may also be valuable in monitoring lipid changes in response to growth conditions.

18.4.8. Fatty Acids

A convenient method for identification of fatty acids in bacterial cells is gas-liquid chromatography of their methyl esters prepared from phospholipids, total lipids, or other lipid fractions. Methanolysis of lipids is accomplished by heating the lipids in sulfuric acid and methanol. The methyl esters can be extracted into hexane, and quantitative measurements of each fatty acid can be made with a gas chromatograph equipped with appropriate columns (see chapter 17.3.5).

Several procedures are available for preparation of methyl esters of fatty acids. One method selected for its simplicity and convenience is presented below. Alternative methods have been described (80, 95, 100).

Reagents and Materials

Gas chromatograph, equipped with a flame ionization detector and suitable recorder. Several choices for columns and temperature programs are available. Investigators should consult one of the references to select a gas-liquid chromatography procedure suitable for the instrument available and the material to be analyzed (52, 60, 68, 70, 76). In most cases, catalogs available from commercial gas chromatography suppliers are very helpful in selection of the proper column, especially when samples of chromatographic elution profiles are provided.

n-Hexane, spectroscopic grade (CAUTION: See section 18.10).

Sulfuric acid-methanol (5%). Add 5 ml of reagent-grade concentrated sulfuric acid to 95 ml of water-free methanol. Prepare fresh before each use. (CAUTION: See section 18.10.)

Standards: methyl esters of straight- and branched-chain C-13 to C-20 fatty acids (available from a variety of gas chromatography supply companies)

Glass-stoppered 15-ml Pyrex test tubes

Procedure

Transfer 0.5 to 1 mg of the purified total lipids (section 18.4.2) to glass-stoppered vials, and evaporate the solvent under a stream of nitrogen at 30 to 37°C. Add 4 ml of 5% (wt/vol) sulfuric acid in methanol to the dried lipid, and heat for 2 h at 70°C. After cooling, add 4 ml of n-hexane and shake the tube for 10 min. Remove and save the hexane phase. Reextract the methanol phase with 4 ml of hexane. Use the pooled hexane phases for gas-chromatographic analysis (see chapter 17.3.5). The fatty acid methyl esters can be identified by comparing retention times with those of known fatty acids. Quantitation is achieved by comparing peak areas with those for an internal standard added to each sample (80).

18.4.9. Poly-β-Hydroxybutyrate

Poly-β-hydroxybutyrate (PHB), although a lipid, is not extracted from bacterial cells by the usual solvents used for lipid extraction (88). In the procedure described by Law and Slepecky (88), PHB is extracted into hot chloroform from cell material that is obtained by first treating bacterial cells with hypochlorite to remove interfering substances. The extract is hydrolyzed to yield β-hydroxybutyrate, which undergoes dehydration to crotonic acid in concentrated sulfuric acid. Crotonic acid absorbs UV light with an absorption maximum at 235 nm. (Although butyrate-based polymers are the most common naturally occurring polyesters, in many cases polymers containing valerate and other longer chains may be present, as discussed below in section 18.4.10.)

Alternative procedures not described here include conversion of PHB constituents into their methyl esters followed by gas chromatography (74, 79) and the use of an automated Fourier transform infrared spectroscopy method (79). However, the crotonic acid method described here is less expensive and generally more convenient than these methods.

Reagents and Materials

Sodium hypochlorite solution: a commercial bleach solution such as Clorox (CAUTION: See section 18.10)

Cuvettes: 1 cm, made of quartz or other material transparent to UV light

Spectrophotometer, capable of UV absorbance measurements

Polypropylene or glass centrifuge tubes, 12 by 100 mm, or 30 ml. The plastic centrifuge tubes should be washed with hot chloroform and ethanol to remove plasticizers. (*CAUTION:* See section 18.10.)

Centrifuge: a refrigerated centrifuge capable of achieving $10,000 \times g$

Pyrex glass boiling tubes

Glass marbles to fit boiling tubes

Volumetric flask, 10 ml

Boiling-water bath

Chloroform (*CAUTION:* See section 18.10)

Acetone (*CAUTION:* See section 18.10)

Reagent-grade sulfuric acid. The quality of the sulfuric acid is important. Low grades contain materials that absorb UV light and cause high absorbance in the blanks used to zero the spectrophotometer. (*CAUTION:* See section 18.10.)

Standard. Dissolve 3-hydroxybutyric acid (0.5 mg/ml) in ethanol. Dilute 1:10 in ethanol to obtain a solution of 50 μg/ml. (*CAUTION:* See section 18.10.)

Procedure

Centrifuge bacterial cells from a volume of culture containing 10 to 20 mg (dry weight) for 15 min at $10,000 \times g$ and 4°C. Determine the dry weight of cells from an equal volume of culture. Add a volume of sodium hypochlorite equal to the original culture volume, mix, and transfer the suspension to a centrifuge tube. Incubate the suspension at 37°C for 1 h. Centrifuge the suspension at $10,000 \times g$ for 15 min to sediment the lipid granules. The temperature during centrifugation is not important. Suspend the sediment in water, and centrifuge at $10,000 \times g$ for 15 min. Discard the supernatant fraction, and wash lipid granules adhering to the centrifuge tube walls with 5 ml of acetone followed by 5 ml of ethanol. The washings, which remove water, are accomplished by suspending the granules in 5 ml of acetone with a Vortex mixer followed by centrifugation at $10,000 \times g$ for 15 min. Suspend the sediment in 5 ml ethanol by using a Vortex mixer, and centrifuge again at $10,000 \times g$ for 15 min.

Dissolve the washed granules with boiling chloroform by heating with 3 ml of chloroform in a boiling-water bath for 2 min. Cool the solution, and centrifuge it at $10,000 \times g$ for 15 min at room temperature. Save the supernatant fraction in a 10-ml volumetric flask, and reextract the sediment twice with 3 ml of chloroform at 100°C for 2 min. If the solution is not clear, filter while hot through Whatman no. 1 filter paper (previously washed with hot chloroform). The chloroform extracts are pooled and made to 10 ml with chloroform. Work in a fume hood while extracting with chloroform. (*CAUTION:* See section 18.10 for precautions involved in the use of chloroform.)

Add samples of the chloroform extract containing 5 to 50 μg of the polymer and 5 to 8 standards containing 5 to 50 μg of 3-hydroxybutyric acid in ethanol to glass boiling tubes. Immerse the tubes in a boiling-water bath until all the solvents are evaporated. Add 10 ml of sulfuric acid, cap the tubes with a glass marble, and heat them in a boiling-water bath for 10 min. Cool the solutions, and mix them thoroughly.

Read the A_{235} of the solutions against a sulfuric acid blank. Calculate the amount of β-hydroxybutyrate from the standard curve. The amount of PHB per gram (dry weight) of bacterial cells can be determined from the dry-weight measurements of the cells (see chapter 9.5.2).

Applications

The technique described above is suitable for measuring the amount of PHB in freeze-dried cells or cell pastes obtained by centrifugation. Other β-hydroxy acids and some sugars interfere because they are converted to products that absorb at 235 nm. Sugars are removed by hypochlorite treatment followed by extraction of PHB into chloroform. Other substances in cells occasionally produce products that absorb UV light. The absorption spectrum of the product from bacterial cells should be compared with that of the standard to ensure that the absorbance is due to crotonic acid. (See chapter 21.1.1.3 for determining absorption spectra.)

18.4.10. Poly-β-Hydroxyalkanoate Polymers

A wide variety of organisms can make poly-β-hydroxyalkanoate (PHA) polymers that contain repeating units with chain lengths longer than butyrate in the polymer (72, 73, 89, 93). The makeup of the major carbon source can, in some cases, determine the chain length of the monomers; e.g., with *Pseudomonas oleovorans*, alkanes with chain lengths of 6 to 10 result in monomers of different lengths, generally with most monomers of the same length as the alkanoic substrates (72). In some cases, the polyester makes up more than 50% of the dry weight of the culture, and these polymers are being examined as the basis for renewable and biodegradable plastic production. However, thus far the low cost of petroleum has resulted in these plastics not being able to compete with petroleum-based plastics.

Analysis by an assay that is effective at detecting β-hydroxybutyrate only by its conversion to crotonic acid would obviously be inaccurate for some polymers but may be acceptable for many natural samples, because of the predominance of butyrate in the naturally occurring polymers. If nonbutyrate polymers are to be examined, a gas-chromatographic method is suitable. Described here is a method by which monomers containing 4 to 12 carbons can be estimated.

Reagents and Materials

Sodium hypochlorite solution: a commercial bleach solution such as Clorox (*CAUTION:* See section 18.10)

Polypropylene or glass centrifuge tubes: 12 by 100 mm, or 30 ml. The plastic centrifuge tubes should be washed with hot chloroform and ethanol to remove plasticizers (*CAUTION:* See section 18.10).

Centrifuge: a refrigerated centrifuge capable of achieving $10,000 \times g$

Pyrex glass boiling tubes with glass marbles to fit tubes, or screw-cap Pyrex tubes with Teflon-coated caps

Volumetric flask, 10 ml

Boiling-water bath

Chloroform (*CAUTION:* See section 18.10)

Methanol (*CAUTION:* See section 18.10)

Sulfuric acid (*CAUTION:* See section 18.10)

Gas chromatograph with flame ionization detector; column of Durabond Carbowax M15 (72) or its equivalent (e.g., Shimadzu Carbowax column PEGM20 [93]); injector, 230°C; detector, 275°C; column, starting at 80°C for 4 min, then with a temperature ramp of 8°C/min to reach 160°C, then 160°C for 6 min

Procedure

The dry weight of the PHA polymers can be determined by using the extraction method described above for PHB, with sufficient cells to produce an accurate measurement by weighing the extracted polymer. After the extraction, place the solution in tared weighing boats and dry it to a constant weight in an oven at ca. 100°C. (Alternatively, a Soxhlet extraction as described by Brandl et al. [72] can be used to extract the PHA.) This weight can be compared with the dry cell weight, which is the sum of the weights of the cell components and PHA.

After the PHA is extracted by either method, hydrolyze and methylate the polymer and perform gas-chromatographic analysis. (Alternatively, whole dried cells can be treated in the same way; if PHA makes up a substantial percentage of the dry cell weight, this is a good method, but if not, cellular lipids may contribute to a background of peaks which reduce the accuracy of the analysis.)

For hydrolysis, derivatization, and analysis, use the following procedure (72, 93). Place the PHA fraction (ca. 2 mg) or dried cells (ca. 4 mg) in a small screw-cap test tube, and react for 140 min at 100°C with a solution of 1.0 ml of chloroform, 0.85 ml of methanol, and 0.15 ml of sulfuric acid. This degrades the PHA to its constituent β-hydroxyalkanoic methyl esters. After the reaction, add 0.5 ml of distilled water and shake the tube vigorously for 1 min. The solution then resolves into two phases; transfer the bottom, organic phase to another vial, and either store it frozen or analyze it. Inject a 2-μl volume into the gas chromatograph system as described above, using temperature programming. Use standards of C-4 to C-12 β-hydroxyalkanoic methyl esters to determine the retention times of the compounds being separated. Peak areas of different concentrations of the compounds can be used to establish a standard curve.

18.5. CELL WALL POLYMERS

Cell wall polymers constitute a major portion of the prokaryotic cell. The term peptidoglycan refers to a large structural component, composed of both sugars and peptides, found in the cell wall structure of most bacteria. It most commonly refers to the murein found in bacteria but also as a general term refers to the pseudomurein found in archaea (119). The specific type of peptidoglycan called murein (containing N-acetylmuramic acid) may make up 40 to 90% of the weight of isolated cell walls of gram-positive bacterial cells (15 to 20% of the cell [dry weight]), usually less than 10% of the gram-negative bacterial cell walls, and as little as 1.2% of the cell wall of some pseudomonads (102–105, 111, 113, 129). Murein controls the shape of prokaryotes except mycoplasmas and some species of prokaryotes belonging to the taxonomic grouping of archaea (103, 113, 117–119). The archaea have a variety of structural alternatives to murein, some of which are described in section 18.5.4.

The peptidoglycan (murein) of most bacteria has a backbone containing alternate residues of N-acetylglucosamine and N-acetylmuramic acid linked by β(1-4)-glycosidic bonds (124). Minor exceptions to this backbone structure are found in a few Streptomyces species (105) in which N-glycolylmuramic acid exists, and a more modified peptidoglycan containing a muramyl lactam is found in the cortex of bacterial spores (126). Some bacteria, especially virulent strains, can modify their peptidoglycan by O acetylation (21, 54, 65). (The estimation of muramic acid as an individual component has been separately dealt with in section 18.3.8 above.)

A tetrapeptide is attached to the carboxyl group of the 3-O-D-lactic acid linked to many of the N-acetylmuramic acid residues. The tetrapeptide most frequently contains amino acids in the order L-alanine, L-glutamic acid, a diamino acid, and a terminal D-alanine. The diamino acids found in the tetrapeptide include meso- and LL-diaminopimelic acid, L-lysine, L-ornithine, and L-diaminobutyric acid. Descriptions of these and several variations of the tetrapeptide sequences can be found in reviews by Ghuysen (116), Rogers et al. (112), and Cummins (105).

Many of the tetrapeptide chains are cross-linked to one another. The extent of cross-linking and the nature of the linkages vary from organism to organism. The linkage may be a direct peptide bond from a dibasic amino acid on one tetrapeptide to a D-alanine on another tetrapeptide, or it may involve one or several amino acids as bridges between one tetrapeptide and another (103, 105, 112, 116).

The unique occurrence of N-acetylmuramic acid in murein provides a means of determining the presence of peptidoglycan in bacterial cells and can be used to estimate bacterial biomass in soil and water samples. Diaminopimelic acid is found in macromolecules only as a component of peptidoglycan, and positive assays for this compound indicate the presence of peptidoglycan. However, not all peptidoglycan polymers contain this diamino acid (105, 119).

Cell walls of gram-positive bacteria often also contain teichoic acids, teichuronic acids, proteins, and lipoteichoic acids. Teichoic acids are complex polymers of polyols (ribitol and glycerol) linked together in a backbone by phosphodiester bonds. Several variations in the backbone structure have been described previously (105, 115). These polymers, which make up 50% of the cell wall of some gram-positive bacteria, are good antigens, and their serological reactivity has been used to define antigenic groups of several genera (115). Cell wall teichoic acids are covalently linked to peptidoglycan. Teichoic acids of the glycerol phosphate polymer type also occur as lipoteichoic acids, which contain covalently linked glycolipid. They are found in the plasma membrane of cells. The most common antigenic determinants are sugars or amino sugars attached through the hydroxyl groups of the polyols of the backbone. Alanine is often found, attached to the polyols as well.

Teichuronic acids are acidic polysaccharides which contain uronic acids but no phosphate. Polymers of glucuronic acid with N-acetylglucosamine and aminomannuronic acid and glucose occur in Bacillus licheniformis and Micrococcus lysodeikticus, respectively. Other uronic acid polymers have been observed as cell wall components of other gram-positive bacteria (105).

Cell wall polysaccharides and lipoteichoic acids of gram-positive bacteria, particularly the streptococci, are responsible for antigenic differences that have enabled the separation of these bacteria into serological groups (105). Cell wall polysaccharides of other gram-positive bacteria form a heterogeneous group of macromolecules that usually contain neutral sugars and occasionally contain amino sugars. These polysaccharides are of value in distinguishing between strains and species, especially when unusual or uncommon sugars are present.

Proteins occur widely as surface components of gram-positive bacteria. Surface antigens of streptococci include numerous acid-soluble proteins (105).

The cell walls of gram-negative bacteria are more complex than those of gram-positive cells. They have a comparatively low peptidoglycan content, and in enteric bacteria the peptidoglycan is covalently bonded to lipoproteins, which probably form a bond between the peptidoglycan and the outer membranous layer of the cell wall (102, 122, 123). The outer membranous layer is rich in lipopolysaccharide.

Lipopolysaccharides are extremely complex molecules with molecular weights higher than 10,000 (102, 103, 106, 108, 111, 114, 120, 122, 123, 125, 128). The lipopolysaccharide from *Salmonella* spp. is made up of three different parts: the lipid moiety, a core polysaccharide, and the O-polysaccharide. The lipid moiety, termed lipid A, contains a phosphorylated glucosamine disaccharide heavily esterified through hydroxyl and amino groups with fatty acids that are 10 to 22 carbon atoms in length. A major fatty acid component of the lipid A fraction is β-hydroxymyristic acid, which is unique to the lipid A portion of the lipopolysaccharide molecule.

The core polysaccharide serves as a link between lipid A and the O-polysaccharide. It is usually composed of four basal sugars, KDO, phosphate, and ethanolamine. KDO is unique to this part of the lipopolysaccharide molecule.

In *Salmonella enterica* serovar Typhimurium the sugars of the core polysaccharide are glucose, galactose, heptose, and glucosamine (102, 128). The core polysaccharide is covalently bonded through C-2 of KDO to C-2 of a glucosamine residue in lipid A. The core polysaccharide is also covalently linked to the variable O-polysaccharide chains. Several lipopolysaccharide subunits may be bonded together by pyrophosphate bonds in the lipid A portion of the molecule. The lipid A portion is embedded in the outer membrane with the O-polysaccharides extending outward (102, 103). The O-polysaccharides are the major antigenic determinants (the O-antigens) of the surface of gram-negative bacteria. They often serve as receptors of bacteriophage attachment.

The fine structure and composition of the outer membrane of gram-negative bacteria suggest that it is a bilayer containing phospholipids and proteins with variable amounts of lipopolysaccharide. The outer membrane encloses the periplasmic space, an enzyme-containing compartment bounded by the plasma membrane on the inside (102, 103). Several gram-negative cells possess proteinaceous outer coats or thick carbohydrate layers outside the lipopolysaccharide zone extending from the outer membrane.

It is impractical to present here all the methods used for isolating and characterizing structural components of cell walls. The methods used depend on the organism, the facilities available, and the objective of the investigation. For some investigations, intact highly purified polymers are not required; for others, macromolecular integrity and purity are required. Persons who wish to thoroughly characterize one of these polymers from a bacterium should first consult the literature describing the different isolation techniques and the advantages and limitations of each (102, 103, 114, 116–118, 122, 127–129). The simplest methods for isolation and preliminary characterization of murein peptidoglycans and lipopolysaccharides from the more ordinary laboratory strains of bacteria are presented here. As mentioned below, these methods are generally applicable to isolation of the pseudomureins found in some archaea, but their characterization requires the examination of different components (section 18.5.4).

18.5.1. Peptidoglycan

Reagents and Equipment

Saline solution: 0.9% (wt/vol) NaCl in distilled water

Trypsin: crystalline enzyme preparation

Tris buffer. Dissolve 1.21 g of Tris in 90 ml of distilled water. Adjust the pH to 7.0 with 5 N HCl, and add additional water to 100 ml. (*CAUTION*: HCl is dangerous; see section 18.10.)

DNase. Dissolve 5 mg of dry powdered enzyme preparation in 5.0 ml of Tris buffer.

RNase. Dissolve 5 mg of dry powdered enzyme preparation in 1.0 ml of Tris buffer.

Refrigerated centrifuge, capable of achieving 22,000 × *g*. Heavy-walled glass or Pyrex 30-ml centrifuge tubes, centrifuge bottles (100 to 250 ml), and rotors to accommodate each. Plastic centrifuge tubes and bottles can be used if they are stable to chloroform. (*CAUTION*: Thin-walled Pyrex glass tubes will not withstand 22,000 × *g*.)

Sonicator, French pressure cell, colloid mill, or other suitable cell breakage equipment.

Procedures

The following procedure is appropriate for ca. 200 to 1,000 ml of culture (corresponding to ca. 0.4 to 4 g of cells [dry weight]) and can be scaled up or down as needed. Cool the bacterial cells in their growth medium to 4°C with ice, and harvest by centrifugation in 100- to 250-ml centrifuge bottles (5,000 × *g* for 15 min). Resuspend the cells in 20 ml of cold saline solution, and centrifuge again. Use the packed cells immediately, or lyophilize them and store them in a freezer to prevent autolysis. Suspend the cells homogeneously in 20 ml of cold water. If the cells have been lyophilized, suspend them in stages by first wetting the cells with cold water to form a paste and subsequently adding liquid slowly while stirring the suspension. Mix the cells mechanically or with a hand-operated homogenizer to obtain complete dispersal of the cells. Resuspension by use of a Pasteur pipette is adequate if mechanical homogenizers are unavailable. Keep the suspension cold to prevent autolysis of cell walls during all manipulations. Disintegrate the cells by using a sonicator, mechanical grinding device, French pressure cell, or other device.

As soon as possible after disruption of the cells, quickly heat the suspension to 75°C. Keep the suspension at this temperature for 15 min to inactivate autolytic enzymes. (For mesophiles, this is effective. However, thermophiles may have to be heated at higher temperatures, ca. 30°C above their maximal growth temperature. Alternatively, autolytic activity may best be controlled by keeping the extract at 4°C. Experience with a given organism is needed to choose the best approach.) Centrifuge at a low speed (1,000 × *g* for 10 min) to remove unbroken cells. Discard the sediment, and centrifuge the supernatant fraction in 30-ml centrifuge tubes at 22,000 × *g* for 15 min. A small opaque layer of unbroken cells under a translucent layer of cell walls should be obtained. Remove the upper layer carefully by gently washing the sediment with a stream of ice-cold 0.1 M Tris buffer (pH 7) delivered from a Pasteur pipette. When the upper layer is totally suspended in 15 ml of buffer, centrifuge again at 1,000 × *g* for 10 min, discard the sediment, and centrifuge the supernatant fraction at 22,000 × *g* for 15 min. Resuspend the upper layer in the Tris buffer. Repeat these steps until a homogeneous cell wall fraction, free from intact cells, is obtained.

Suspend the cell wall fraction in 100 ml of Tris buffer that has been saturated with chloroform. Add DNase (0.5 ml of a 1-mg/ml solution) and RNase (1.0 ml of a 5-mg/ml solution), and incubate the suspension at 37°C for 30 min. Add 100 μg of crystalline trypsin, and continue the incubation for 6 h at 37°C. Trypsin is added to remove denatured proteins. This enzyme causes lysis of the peptidoglycan of some bacteria. If this occurs, centrifugation will not yield a precipitate, and the trypsin treatment must be omitted. Sediment the cell walls by centrifugation, and wash them as described above. Continue the sedimentation and washing until a homogeneous layer is obtained.

Purification of peptidoglycan from most gram-negative cells involves the use of much larger amounts of cell material (20 to 30 g) because of the small amount of peptidoglycan per unit weight of cells and the greater susceptibility of the peptidoglycan of these bacteria to autolysis. A procedure for preparation of 150 mg of peptidoglycan from *E. coli* has been described by Weidel et al. (127).

The amounts of hexosamine and muramic acid can be estimated after acid hydrolysis of the purified cell walls (see sections 18.3.7 and 18.3.8). The amino acid content can be determined with an amino acid analyzer, by HPLC, or by thin-layer chromatography (129). The presence of many amino acids indicates contamination by membrane, cell wall, or proteins.

Good examples of approaches for a more complete characterization of cell walls involving a variety of chemical assays to determine reducing sugars, unsubstituted amino groups of amino acids, etc., have been provided by Kottell et al. (121), Ghuysen et al. (107, 116), and Kandler et al. (117–119).

18.5.2. Total Lipopolysaccharides (109)

Reagents, Materials, and Equipment

Saline solution: 0.9% (wt/vol) NaCl in distilled water

Phenol solution. Using a fume hood, add 10 ml of distilled water to 90 ml of warm (60°C) high-quality phenol. (*CAUTION*: Phenol is dangerous; see section 18.10.)

Centrifuges, rotors, and tubes: a centrifuge capable of achieving 10,000 × *g* and an ultracentrifuge capable of achieving 100,000 × *g*. Centrifuge bottles (100 to 300 ml) and a rotor for them to be used for harvesting cells from the medium, and ca. 30-ml centrifuge tubes and rotors to hold them for both centrifuges.

Dialysis tubing

Rotary evaporator

Lyophilizer or vacuum desiccator and a vacuum source

Procedures

Centrifuge the bacteria from ca. 3 liters of growth medium at 10,000 × *g* for 15 min. Suspend the sedimented cells in 300 ml of ice-cold saline solution, and centrifuge again. Repeat the resuspension and centrifugation to remove medium constituents. Lyophilize the bacteria in the final sediment, or store the packed cells in a freezer until used.

Thoroughly suspend 20 g (dry weight) of bacteria in 350 ml of water at 65 to 68°C. Heat 350 ml of phenol solution to 65 to 68°C, and add it to the bacterial cell suspension (*CAUTION*: Phenol is dangerous; see section 18.10). Stir vigorously, and keep the mixture at 65°C for 15 min; then, cool it for 2 min in room temperature water, and finally cool it to 10°C in an ice bath. Centrifuge the emulsion at 10,000 × *g* for 30 min. Three layers result: a water layer, a phenol layer, and a sediment. Sometimes a layer of denatured protein is formed at the phenol-water interface. Remove the upper, aqueous layer, and save it. Remove and discard the material at the interface. Add another 350 ml of 65°C water to the phenol layer and sediment, and mix thoroughly while the temperature is held at 65 to 68°C for 15 min. Cool and centrifuge the mixture again, and combine the upper, water layer with that obtained from the first centrifugation step. Dialyze the aqueous suspension for 3 to 4 days at 4°C

against distilled water to remove the phenol and low-molecular-weight contaminants. Concentrate the dialyzed solution to 5 ml at 37°C under a vacuum. A rotary evaporator is useful for this purpose if one is available. Centrifuge the concentrate at 3,000 × *g* for 10 min to remove insoluble material. Discard the sediment. One-half of the suspended organic material in the supernatant fraction should be lipopolysaccharide. The major contaminant is RNA. Centrifuge the supernatant fraction at 100,000 × *g* for 1 h. Thoroughly resuspend the sediment in saline solution, and centrifuge again at 100,000 × *g* for 1 h. Repeat the centrifugation and resuspension of the sediment five times to wash it free from contaminating material. Finally, dissolve the sediment in distilled water, and lyophilize the material until dry. If a lyophilizer is unavailable, the suspension can be dried in a vacuum desiccator. A yield of 100 to 500 mg of lipopolysaccharide should be obtained from 20 g (dry weight) of species of the family *Enterobacteriaceae*. This lipopolysaccharide is a potent endotoxin (102, 106, 123) and contains firmly bound lipid A (*CAUTION*: See section 18.10). It should be handled carefully to avoid contact that could lead to ingestion or inhalation.

18.5.3. Lipid A and Polysaccharide from Lipopolysaccharides

The lipid A moiety can be obtained as a water-insoluble product after mild-acid hydrolysis of purified lipopolysaccharide, as described below. However, alternative methods using more-severe-acid hydrolysis have also proven useful to prepare core oligosaccharides (114). The polysaccharide portion of the polymer is soluble in water.

Reagents, Materials, and Equipment

Acetic acid solution (1%, vol/vol). Add 1 ml of acetic acid to 99 ml of distilled water. (*CAUTION*: See section 18.10.)

Lyophilized lipopolysaccharide preparation (*CAUTION*: Lipopolysaccharide is toxic; see section 18.10)

Lyophilization ampoules, 15 ml

Dialysis tubing

Tank of compressed nitrogen gas (*CAUTION*: See section 18.10)

Vacuum desiccator or lyophilizer

Centrifuge capable of achieving 5,000 × *g*, 15-ml centrifuge tubes, and an appropriate rotor

Temperature control is not important

Procedure

Suspend 0.1 g of the dried lipopolysaccharide, obtained as described above, in 10 ml of a 1% acetic acid solution that has been gassed with nitrogen for 10 min.

Seal the suspension in an ampoule under N_2, and heat it in a steamer for 4 h. Cool and open the vial, transfer the contents to a centrifuge tube, and centrifuge at 5,000 × *g* for 10 min. Carefully remove the supernatant from the insoluble lipid A. Save the supernatant fraction which contains the polysaccharide portion of the lipopolysaccharide.

Resuspend the sediment in distilled water, and centrifuge again. Repeat the resuspension and centrifugation five times. Discard the supernatant fractions. Dry the sediment in a vacuum desiccator or by lyophilization.

Dialyze the polysaccharide remaining in solution after acid hydrolysis against 200 volumes of water, and dry the contents of the bag by lyophilization or in a vacuum desiccator. Between 40 and 100 mg of lipid-free polysaccharide and 30 to 150 mg of lipid A should be obtained from 100 to 500 mg of lipopolysaccharide.

Assays for KDO and heptose can be used to estimate the amount of lipopolysaccharide or lipid-free polysaccharide (section 18.3.9). Hexosamine assays (section 18.3.6) can be used to estimate the amount of lipid A in purified preparations. Alternatively, in some cases more-detailed information may be obtained by using gas-liquid chromatography with mass spectrometry supplemented when necessary with NMR data (48, 114).

18.5.4. Archaeal Cell Wall Components

The archaea as a group are different in several ways from bacteria. These differences include the lack of murein-containing peptidoglycan, D-alanine, muramic acid, and ester-linked alkane lipids (119). Most archaea do have a cell wall and can be characterized by the color of staining as gram positive or gram negative; however, no outer membrane or lipopolysaccharide is present in the gram-negative types and no common cell wall component is known within the archaea. Instead, a variety of cell wall structures are present. The most common gram-positive cell walls include the pseudomurein found in *Methanobacterium* and *Methanobrevibacter* species and the heteropolysaccharides found in *Methanosarcina* and *Halococcus* species. Several types of protein and glycoprotein cell walls have been found in the gram-negative archaea.

Both pseudomurein and the heteropolysaccharide can be isolated by methods similar to those used with bacteria, as described above in section 18.5.1, i.e., physical disruption and trypsin degradation (119). Amino acid components can be determined after acid hydrolysis by using methods similar to those in section 18.7.13, and carbohydrates and aminosugars can be analyzed by methods similar to those in section 18.3. The pseudomurein is characterized by the presence of *N*-acetylglucosamine and of *N*-acetyltalosaminuronic acids as a replacement for *N*-acetylmuramic acid. No known archaeal cell walls contain D-alanine or diaminopimelic acid. The heteropolysaccharide cell walls principally contain glucose, galactosamine, and uronic acids.

18.6. NUCLEIC ACIDS (also see chapter 26)

For most molecular biology applications, DNA or RNA is purified following standard techniques described in chapter 26.2 and 26.3 and by Sambrook and Russell (138), and its concentration is determined by A_{260}. This approach is suitable for chromosomal or plasmid DNA that will be used for cloning and for RNA that may be used for other molecular biology applications, e.g., primer extension analysis. However, for other applications, it may be useful to know the concentration of classes of nucleic acids in cells without purifying the nucleic acids. Alternatively, using purified DNA, it may be useful to have an assay that is more sensitive than the A_{260} approach. This section deals with these alternative methods.

A widely used method for estimation and identification of DNA and RNA in cell extracts involves extraction and hydrolysis with a hot solution of trichloroacetic (or perchloric) acid followed by specific colorimetric assays for pentose components of nucleic acids (131). A more convenient means of measuring the amounts of DNA and RNA in pure solutions takes advantage of their A_{260}. Rapid and reliable quantitation of nanogram to microgram amounts of DNA can be achieved by fluorescence techniques (132, 134, 135, 137). These techniques are much more sensitive than either of the spectrophotometric methods, and one fluorescent method using bisbenzimide is less sensitive to interference by small amounts of RNA, protein, and other macromolecules. Other procedures less applicable to the purposes of this manual are described in other reference books (130, 138, 139).

The estimation of nucleic acids in bacterial cultures by the techniques described below requires that the cells be present in sufficient numbers in a noninterfering medium or that they can be harvested from the growth medium, washed free from interfering medium components, and resuspended at an appropriate concentration without lysis of the cells or hydrolysis of the nucleic acids by nucleases. Total extraction of nucleic acids from the cells is essential for accurate determination of their nucleic acid content.

18.6.1. Total DNA by Diphenylamine Reaction

Reaction of DNA with diphenylamine in a mixture of acetic and sulfuric acids is an appropriate colorimetric procedure for the determination and identification of this polymer in complex mixtures. The chemistry of the reaction and the nature of the chromophore formed remain unknown. The Burton method (133), described below, has the advantages that it is sensitive and that few substances in bacterial cells interfere with its accurate determination of DNA.

Reagents, Materials, and Equipment

Perchloric acid: 0.5, 1.0, and 2.5 N solutions (*CAUTION*: Perchloric acid is very dangerous; see section 18.10)

Diphenylamine reagent. Dissolve 1.5 g of purified high-quality diphenylamine (white crystals; melting point, 52.9°C) in 100 ml of acetic acid, and add 1.5 ml of reagent-grade concentrated sulfuric acid (*CAUTION*: Acetic and sulfuric acids are dangerous; see section 18.10). On the day it is used, add 0.1 ml of a 16-mg/ml aqueous solution of acetaldehyde (1 ml of acetaldehyde in 50 ml of water) to each 20 ml of the diphenylamine reagent (*CAUTION*: Acetaldehyde is very volatile and toxic and boils at 21°C; see section 18.10).

Saline-citrate solution (SSC): 0.15 M NaCl, 0.015 N trisodium citrate (pH 7.0)

Standards: dissolve crystalline DNA (0.4 mg/ml), from a commercial source, in 5 mM NaOH. Prepare standards every 2 weeks by mixing measured volumes of the stock DNA solution with an equal volume of 1 N perchloric acid, and heat at 70°C for 15 min. Prepare dilutions in 0.5 N perchloric acid. (*CAUTION*: See section 18.10.) See note below.

Spectrophotometer or colorimeter and glass cuvettes

Centrifuge, capable of achieving 10,000 × *g*, which can be refrigerated or placed in a cold room

Centrifuge tubes, 15 or 30 ml

Water baths at 70°C and an incubator or water bath at 30°C

Test tubes, Pyrex glass, 15 by 125 mm

Note: Most DNA preparations contain some salts and water, and it is difficult to obtain or prepare DNA totally free from water to use in the preparation of standards. One should be aware of the inaccuracies resulting from the use of DNA as a standard and should account for the salts and an estimate of the water, both of which can be obtained from the supplier. One alternative is to calibrate the DNA by using the UV method described below in section 18.6.4. As an alternative, a DNA standard can be hydrolyzed, and its phosphate content per milligram of DNA can be determined (136). The results can then be reported as micrograms (or milligrams) of DNA-phosphorus rather than micrograms of DNA.

Deoxyribose prepared as an aqueous solution (1.0 mM) is also a convenient standard. Solutions are stable for 3 months in a refrigerator. Dilutions made in 0.5 N perchloric acid can be used to construct a standard curve for the diphenylamine assay. Results are reported as deoxyribose equivalents of DNA if this standard is used.

Procedure

A suspension of 5 to 10 mg (dry weight) of bacterial cells per ml (about 5×10^{10} to 1×10^{11} cells per ml for bacteria the size of E. coli) is required. Harvest the cells from the medium if the culture does not contain sufficient cells or if the medium contains interfering compounds. Complex media contain nucleic acid components and sugars that must be eliminated by harvesting the cells from the media and washing them by resuspension and centrifugation. Most bacteria can be harvested by centrifugation at $5,000 \times g$ for 15 min at 4 to 5°C. Suspend the sedimented cells in 1/10 of the culture volume of SSC. Centrifuge the cells at $5,000 \times g$ for 15 min at 5°C, and resuspend them in ice-cold SSC. If the cells are grown in mineral salts medium and the cell concentration is sufficient, samples of the culture can be used directly for extraction of DNA.

Acidify each sample of bacterial culture in a suspension of bacteria in SSC with sufficient 2.5 N perchloric acid to give a final concentration of 0.25 N perchlorate. Cool the acidified sample on ice for 30 min, and centrifuge at $10,000 \times g$ for 10 min. Using a glass rod, suspend the precipitate containing the nucleic acids in 0.5 ml of 0.5 N perchloric acid. Add an additional 3.5 ml of 0.5 N perchloric acid, and heat the suspension at 70°C for 15 min. It is important to stir or mix the suspension occasionally during this step. Centrifuge the heated suspension at room temperature for 10 min at $5,000 \times g$ and carefully decant the supernatant fraction into a 10-ml graduated test tube. Reextract the precipitate in an identical fashion. Combine the two extracts, and mix. Record the volume of the combined extracts.

For the diphenylamine reaction, mix samples of various volumes from the combined extracts with sufficient 0.5 N perchloric acid to bring the final volume to 2 ml. It is usually wise to use a 10-fold range of sample sizes from each extract (e.g., 0.2, 1.0, and 2.0 ml) to ensure that the amount of color developed in one sample falls on the standard curve. After the volumes of each sample have been adjusted to 2.0 ml, add 4.0 ml of the diphenylamine reagent to each tube, and mix thoroughly. Adjust the volume of the liquid in tubes containing dilutions of the standard (5 to 100 μg/ml) to 2.0 ml with 0.5 N perchloric acid, and mix the contents with 4.0 ml of the diphenylamine reagent. Prepare a blank for the spectrophotometer by adding 4 ml of the diphenylamine reagent to 2.0 ml of 0.5 N perchloric acid. Incubate the tubes at 30°C for 16 to 20 h (or overnight). If the contents of the tubes are cloudy, centrifuge the tubes at $10,000 \times g$ for 15 min to remove unhydrolyzed protein. Determine the A_{600} of the supernatant fractions against the blank. Estimate the concentration of DNA in each tube from the absorbance values by using a standard curve prepared by plotting the absorbance of standard solutions against the DNA, deoxyribose, or phosphate concentrations of the standards. (If small [1-ml]

cuvettes are available, the assay volumes can be appropriately reduced.)

It is not important to control the temperature carefully. If all tubes, including the standards, are incubated at the same temperature, the results should not be affected. A standard curve should be prepared with each set of determinations.

18.6.2. Total DNA by Fluorescence Spectroscopy

Two dyes that bind to DNA, bisbenzimide and ethidium bromide, have commonly been used for fluorescence measurements (132, 134, 137). Bisbenzimide (also called Hoechst 33258 dye) has little affinity for RNA and preferentially binds to regions of DNA rich in adenosine and thymidine residues. The fluorescence of the dye is enhanced when bound to DNA, and the excitation maximum shifts from 356 nm for the unbound dye to 365 nm when the dye is bound to DNA. The emission spectrum of the free dye peaks at 492 nm, and the emission maximum in the presence of DNA is 458 nm.

Ethidium bromide intercalates into the DNA double helix, and its fluorescence is enhanced. This reaction is often used to detect DNA in agarose gels. Detection of DNA with ethidium bromide is not as sensitive as with Hoechst 33258. However, the binding of ethidium bromide to DNA is not affected by the base composition. DNA samples with high guanosine and cytosine contents give less fluorescence than do DNA samples rich in adenosine and thymidine when Hoechst 33258 is used. Ethidium bromide also binds to RNA and thus should not be used to quantitate DNA if significant amounts of RNA are also present.

18.6.2.1. Bisbenzimide (Hoechst 33258 Dye) Binding

Reagents and Equipment

Concentrated dye solution: 10 mg of Hoechst 33258 dissolved in 10 ml of distilled water. The solution should be stored refrigerated and protected from light. It is stable for 6 months.

10× TNE buffer: 100 mM Tris buffer, 1 M NaCl, 100 mM EDTA (pH 7.4).

Working dye solution A: suitable for measuring solutions containing 10 to 400 μg of DNA per ml. Add 10 ml of 10× TNE buffer to 90 ml of distilled water; mix and filter through a nitrocellulose filter (pore size, 0.45 μm). Add 10 μl of the concentrated dye solution to 100 ml of the filtered 1× TNE buffer.

Working dye solution B: suitable for measuring DNA in solution at concentrations between 0.10 and 15 μg/ml. Prepare and filter 1× TNE buffer as described for working dye solution A. Add 100 μl of the concentrated dye solution to 99.9 ml of the 1× TNE buffer.

DNA standard solutions: a DNA with approximately the same base composition as the DNA sample should be used as a standard. Supercoiled plasmid DNA has different dye-binding characteristics from those of linear DNA. Therefore, a plasmid DNA such as pBR322 should be used as a standard when the sample to be assayed is plasmid DNA. Prepare solutions of DNA from 25 to 250 mg/ml in 1× TNE buffer. The range of standards to be used will depend on the concentration of DNA in the samples to be assayed. A 10-fold range of DNA concentrations in the standard should be used to establish a standard curve.

Fluorimeter. Dedicated instruments with preset activation and emission wavelengths are relatively inexpensive (e.g., Hoeffer model TKO 100). A scanning spectrofluorimeter is also suitable but is more expensive.

Fluorimetric cuvettes: glass cuvettes with 1-cm sides. Liquid measurement devices to accurately measure 2.0 ml and 2.0 μl.

Procedure

Add 2.0 ml of the working dye solution A or B to a cuvette, depending on the amount of DNA in the standard or sample solutions. Set the excitation wavelength at 365 nm and the emission wavelength at 458 nm if a scanning spectrofluorimeter is used. Set the fluorimeter to read zero with the working dye solution as a blank. Add 2 μl of a standard DNA solution and mix with a Teflon stirring rod or by capping and inverting the cuvette. Record the emission in relative fluorescence units. Repeat with fresh working dye solutions and the other standards, and plot a standard curve of relative fluorescence versus DNA concentrations. If a concentration readout is available, the readout can be set to equal that of a DNA standard. Add 2.0 μl of the sample solutions, and read the relative fluorescence. Estimate the concentration of DNA in the samples by comparison with the standard curve or by using the concentration readout.

18.6.2.2. Ethidium Bromide Binding

Reagents and Equipment

10× TNE buffer: see above

Ethidium bromide assay solution. Add 10 ml of 10× TNE buffer to 89.5 ml of distilled water. Filter through a 0.45 μm-pore-size nitrocellulose filter, and add 0.5 ml of a solution containing 1 mg of ethidium bromide per ml in distilled water. (CAUTION: Ethidium bromide is a mutagen; see section 18.10.) Commercial solutions are available (e.g., from Bio-Rad) and are safer to use than weighing dry ethidium bromide in the lab.

Fluorimetric cuvettes: glass cuvettes with 1-cm sides

Scanning spectrofluorimeter

Liquid measurement devices to accurately measure 2.0 ml and 2.0 μl

Procedure

Add 2.0 ml of the ethidium bromide assay solution into a cuvette. Set the excitation wavelength at 546 nm and the emission wavelength at 590 nm. The remainder of the procedure is identical to that described for assaying DNA samples by using Hoechst 33258 dye.

18.6.3. Total RNA by Orcinol Reaction

The reaction of aldopentoses with acidified orcinol to produce a green chromogen is a classical reaction of sugar chemistry. This reaction detects ribose moieties present in RNA. Deoxyribose gives 20% of the color of ribose. The method is suitable for the measurement of RNA or ribonucleotides in solution as well as in tissue or cell extracts (131, 136). Since the quantity of RNA in bacteria exceeds the total DNA about sixfold, the contribution by DNA to the assay of unpurified extracts should be less than 4% of the total signal. However, if intact RNA is needed for further molecular biology work, more appropriate methods are described by Sambrook and Russell (138).

Reagents and Equipment

Ferric chloride reagent. Dissolve 100 mg of $FeCl_3 \cdot 6H_2O$ in 100 ml of reagent-grade concentrated HCl. (CAUTION: HCl is a strong and volatile acid. Use in a fume hood. See section 18.10.)

Ethanolic orcinol solution. Dissolve 6.0 g of orcinol in 100 ml of ethanol (CAUTION: See section 18.10). Store in a dark bottle. If the crystals are yellow or brown, use a new bottle or seek advice from a chemist on recrystallization.

Orcinol reagent. On the day the reagent is to be used, mix equal volumes of the ferric chloride reagent with the ethanolic orcinol solution. The complete reagent solution should be bright yellow when prepared (CAUTION: This is a strong HCl solution; see section 18.10.)

Perchloric acid (CAUTION: This is a dangerous acid; see section 18.10).

Standard solutions for RNA analysis.

Heat a solution containing 200 μg of RNA per ml with 0.25 N perchloric acid at 70°C for 30 min. Add an equal volume of 0.1 N HCl to give a final RNA concentration of 100 μ/ml.

AMP is a preferred standard because its concentration can be accurately measured. Prepare a solution containing 100 μg/ml by heating 200 μg/ml in 0.25 N perchloric acid. Dilute the solution with an equal volume of 0.1 N HCl.

Spectrophotometer or colorimeter and cuvettes or tubes

Water bath, set at 90°C

Test tubes, Pyrex glass, 18 by 125 mm

Clear glass marbles, approximately 20 mm in diameter

Procedure

Prepare hydrolysates from suspensions of bacterial cells by treatment with perchloric acid exactly as described for the extraction of DNA (section 18.6.1). Dilute the cell hydrolysates with 0.1 N HCl so that they contain 10 to 100 μg of RNA per ml. (Cell hydrolysates from 0.8 to 1 mg [dry weight] of cells contain ca. 100 μg of RNA.) Mix the diluted perchlorate hydrolysates of the cells or standards with 2 volumes of the orcinol reagent, cap the tubes with marbles, and heat them at 90°C for 30 min. (The volumes used can vary depending on the spectrophotometer or colorimeter available. For most spectrophotometers and colorimeters, mix 1.0 ml of the extracts or standards with 2 ml of the orcinol reagent.) Cool the tubes under running tap water, and determine the A_{665} of each tube. Determine the concentration of RNA or nucleotides from a standard curve prepared with known amounts of AMP or RNA. The standard curve should be repeated each time a sample or series of samples is assayed. The assay can reliably measure 15 to 100 μg of RNA in an extract of bacterial cells. When AMP is used as a standard, the results should be reported as purine riboside equivalents of RNA.

18.6.4. DNA and RNA by Other Methods

Pure DNA or RNA solutions, free from contaminating protein and carbohydrate, can be conveniently measured by UV spectroscopy. A solution containing 1 mg of pure nucleic acid per ml has an A_{260} of ca. 22 to 23. In most molecular biology labs, a similar UV method is used to measure purified nucleic acids, as described extensively by Sambrook and Russell (138). The methods described in this section differ mainly in how the DNA is purified and prepared.

Applications

Perchloric acid extracts approximately 95% of the nucleic acid from bacterial cells and endospores. Alternative extraction procedures can be used if incomplete extraction is indicated by the presence of DNA in the soluble fraction after a third extraction of the precipitate from the second hot perchloric acid treatment. Cells can also be extracted conveniently with 5% (wt/vol in water) trichloroacetic acid by heating at 90°C for 15 min. The extract can be diluted to the desired volume with 5% trichloroacetic acid and assayed for ribose by the orcinol procedure described above (18.6.3).

When small amounts of DNA are available, the fluorescence assays described in section 18.6.3 are useful because very small amounts of sample are consumed and the assays are much more sensitive than other methods. These methods are particularly useful for measuring PCR products, plasmids, and other DNA preparations prior to ligation reactions and transformation of cells during cloning of recombinant DNA molecules. These reactions are not subject to interference by small amounts of protein, and they have been used to measure DNA levels in cellular homogenates (132). However, direct A_{260}/A_{280} measurements with very small cuvettes that accommodate as little as 25 μl (microcuvettes) also use very little DNA, and in many cases this method is more convenient (138). Furthermore, UV methods are nondestructive.

If an estimate of the concentration of purified DNA in solution is to be made, the sample of the solution can be treated once in 0.5 N $HClO_4$ at 70°C for 15 min. Portions of the treated DNA solution can be assayed by the diphenylamine reaction described above. The precipitation of the nucleic acid with cold perchloric acid causes losses when small amounts of DNA or other macromolecules are present. Therefore, this step should be omitted when pure DNA solutions are assayed. If it is necessary to precipitate the DNA from dilute solutions to remove interfering compounds such as nucleotides, the addition to each tube of 100 μg of a protein such as BSA facilitates complete precipitation of the DNA (except when UV methods are to be used, since protein also absorbs in the UV).

The colorimetric and UV absorption assays are applicable to any class of RNA. All RNA classes have the same extinction coefficient when the colorimetric assay is used. Proteins, polysaccharides, and many other UV-absorbing compounds (including phenol, which is often used in the preparation of nucleic acids) compromise the accuracy of the UV absorption assay. Deviations from a normal spectrum indicate contamination with protein, carbohydrate, or other compounds. The orcinol procedure for measuring RNA is subject to interference by hexoses, which produce a brown color. Therefore, the UV absorbance assay is preferred for the measurement of RNA in sucrose gradients or solutions containing significant levels of hexoses or hexose polymers.

It is possible to calculate the DNA content per cell from the DNA content of a suspension if the number of cells per unit volume is known. The nucleotide content per cell can be determined, and these data can be used to calculate the genome size of an organism (130).

18.7. NITROGEN COMPOUNDS

Choice of the most suitable method for analysis of nitrogen in a compound depends on the form of nitrogen analyzed, the concentration of nitrogen, and the presence of interfering compounds. For nitrite analysis, the most widely accepted method involves modification of the diazotization and coupling reactions. For nitrate nitrogen, ammonia nitrogen, and organic nitrogen, many other methods have been used. A comparison of the sensitivities and uses of some of these analyses is given in Tables 1 and 2. For purposes of this manual, the simplest widely applicable methods for each nitrogen compound are described. References to other methods are given in Tables 1 and 2. Because of the diversity of substances that may be analyzed for nitrogen (such as milk, fertilizer, soil, and plant and animal materials), many of the methods in Table 1 must be modified or the sample must be pretreated to fit the method. Many variations of the procedures listed or described here have been published, and the reader should refer to the literature for procedures directly applicable to a particular kind of sample.

18.7.1. Nitrite

The analysis of nitrite nitrogen is based on the reaction of NO_2^- with sulfanilamide in an acid solution to form a diazo compound. This compound reacts with N-(1-naphthyl)ethylenediamine to form a colored dye whose A_{543} is measured in a spectrophotometer (150, 151). Sulfanilic acid and α-naphthylamine can be substituted for the two analytical chemicals; however, α-naphthylamine has been identified as a carcinogen and should be used only under carefully controlled conditions.

This reaction can detect as little as 1 μg of NO_2^- per liter when cuvettes with a 10-cm light path are used. The absorbance of the colored compound follows Beer's law up to 180 μg of NO_2 per liter (145). Ions that interfere by causing precipitation are listed in Table 1. The presence of any of these may require the use of an alternate method of NO_2^- detection, such as reduction and subsequent analysis as NH_4^+. As with any colorimetric procedure, colored or particulate substances in the sample may interfere; this is particularly important with 10-cm cuvettes. Interference by colored substances may be lessened by dilution prior to the reaction if the NO_2^- concentration is high enough to permit this. Alternatively, treatment with activated charcoal often is effective for removal of interfering compounds. Particulate substances (such as bacterial cells) can be removed by centrifugation or filtration through a 0.45-μm-pore-size filter.

An ion-specific electrode is also available for estimation of nitrite (see chapter 17.2.2).

Sampling

Samples should be assayed immediately or frozen at −20°C until the assay can be run. Samples for NO_2^- assays should not be preserved with acid. Neutralization of all samples to pH 7.0 is suggested before analysis.

Reagents, Materials, and Equipment

NO_2^- free water. Most sources of distilled or demineralized water are free from NO_2^-. If not, water may be prepared as follows. Add 1 ml of concentrated H_2SO_4 (CAUTION: This acid is corrosive and dangerous; see section 18.10) and 0.2 ml of $MnSO_4$ solution (36.4 g of $MnSO_4 \cdot H_2O$ per 100 ml of distilled water) to each liter and make pink with 1 to 3 ml of $KMnO_4$ solution (400 mg of $KMnO_4$ per liter).

Sulfanilamide reagent. Add 5 g of sulfanilamide to 300 ml of distilled water containing 50 ml of concentrated HCl (CAUTION: Concentrated HCl is corrosive and dangerous; see section 18.10). Adjust the volume to 500 ml

TABLE 1 Comparison of analytical techniques for analysis of nonprotein nitrogen compounds

Compound	Method	Sensitivity (μg of N/liter in sample)	Interference	Advantages	Disadvantages	Reference(s)
NO_2^-	Diazotization	1	Cl^-, colored samples, particulates; Sb, Fe(III), Pb, Hg, Ag, Cu(III) ions	Simple	Cl^- and Fe^{3+} interfere	140, 145, 150, 160, 176
	Reduction to NH_4^+, further analysis	>2,000	NO_3^-, NH_4^+	Good for high nitrite	Hot alkali, distillation, nonspecific; low sensitivity	145, 147
NO_3^-	Reduction to NO_2^-, further analysis[a]					145, 150, 162, 176
	Measurement of A_{220}	40	Many organics, NO_2^-, surfactants, chromate, colored samples	Simple	Requires UV spectrophotometer; organic interference	145, 152, 153
	Brucine	100	None	None	Brucine toxicity, low sensitivity	145
	Chromotropic acid	100	Colored samples; Ba, Pb, I Se, Cr ions	Color stable for 24 h	Low sensitivity	145
	Nitrophenoldisulfonic	20	Organics, Cl^-	Simple	Requires drying	140, 141, 178, 192
	Reduction to NH_4^+, further analysis	>2,000	NO_2^-, NH_4^+	Good for high nitrate	Hot alkali, distillation, nonspecific; low sensitivity	140, 141, 145, 165
NH_4^+	Oxidation to NO_2^-, further analysis	Varies	Amino acids, atmospheric NH_3	No distillation	Some amino acids yield nitrite	150
	Indophenol blue (phenate)	10	Particulates, colored samples, excess acid or base, phenol, small N compounds, Cl^-	Simple; preferred for Kjeldahl method	Phenol; may need distillation	140, 145, 189
	Nessler reagent	400	Particulates, colored samples; Mg, Fe, Ca, S, Mn, Cl ions; small organics	Good for pure water samples	Hg; may need distillation; low sensitivity	140, 145
	Acidimetric	2,000	Cl^-	Good for high ammonium; after distillation, simple; preferred for Kjeldahl method	Distillation; low sensitivity	140, 145, 147
	Pyridine-pyrazole	50	Fe, Zn, Ag, Cu ions; cyanate	Specific for NH_4^+	Pyridine	150, 171
Total N	Kjeldahl digestion, reduction to NH_4^+	Variable, insensitive	Few	Quantitative	Distillation; sulfuric acid; low sensitivity	140–142, 145, 147, 148, 159, 165

[a] For further information and sensitivity, see analysis listed under appropriate nitrogen class.

TABLE 2 Comparison of analytical techniques available for protein assay

Method	Sensitivity[a] (μg/ml in assay solution)	Interference	Advantages	Disadvantages	Reference(s)
Lowry	ca. 8	Ammonium sulfate, amino acid and peptide buffers, mercaptans	Very sensitive	Variation with amino acid content; nonlinear with Beer's law; multistep and timed	163, 173, 174
Modified Lowry					
Pierce	2–3	As above for Lowry	Very sensitive	As above for Lowry	183
DC (Bio-Rad)	5–10	Ammonium sulfate, amino acid and peptide buffers	Detergent compatible, very sensitive	As above for Lowry	155
RC DC (Bio-Rad)	300–400	As above for DC	Detergent, mercaptoethanol, and DTT[b] compatible	As above for Lowry, not sensitive	156
Biuret	ca. 200	Amino acids and peptide buffers	Rapid and simple	Not highly accurate or sensitive; variation with amino acid content	143, 147, 172
Coomassie blue	1–4	Some detergents	More uniform response than Lowry to proteins; mercaptoethanol and DTT compatible; rapid, simple; very sensitive	Use of phosphoric acid; nonlinear with Beer's law	157, 158, 185, 186
Bicinchoninic acid/copper	1–5	Thiols and mercaptans, uric acid, others[c]	Very sensitive	Timed reaction	167, 182, 191
UV absorbance	ca. 0.5	Nucleic acid, any aromatic or phenolic compound, many coenzymes, some buffers	Very sensitive, simple, and rapid	Serious interference; requires UV spectrophotometer	169, 170, 172, 194
Fluorometric	ca. 0.0003	Amino acid or peptide buffers, some coenzymes	Very sensitive, simple, and rapid	Variable protein response; requires fluorometer and special cuvette; dioxane use	181

[a]Sensitivity is given in micrograms per ml of solution in the assay. This can be converted easily to micrograms per ml of original solution by accounting for the dilution factors. When a range is given, both the extrasensitive microassay and the routine, less sensitive assays are described. Interfering substances may reduce this sensitivity but still allow accurate assays in an appropriate range.

[b]DTT, dithiothreitol.

[c]Glucose, Tris buffer, and ammonium sulfate can cause interference but can be corrected with proper blanks or pH adjustment.

with distilled water. The solution is stable at room temperature for many months.

Standard NO_2^- solutions (120). Prepare a stock solution by dissolving 0.13 g of high-purity $NaNO_2$ in 500 ml of distilled water. Use a new bottle of $NaNO_2$ since NO_2^- is readily oxidized in the presence of water. Keep the stock bottle tightly stoppered. One milliliter contains ca. 50 μg of NO_2^--N. Dilute the stock to prepare standards for the standard curve.

Spectrophotometer and cuvettes

Procedure

Add 0.1 ml of the sulfanilamide solution to 5.0 ml of the sample. Allow it to react for a minimum of 2 min but not longer than 8 min. Undesirable side reactions and decomposition are significant after 10 min (150). Add 0.1 ml of N-(1-naphthyl)ethylenediamine solution, and mix immediately. Allow color development for at least 10 min, and then measure the A_{543}.

Prepare a set of standards from the 50-μg/ml NO_2^- stock solution to give 0 to 25 μg of NO_2^--N per ml. Plot $[NO_2^-]$ versus A_{543}. Using the standard curve, determine the concentration of NO_2^--N in the sample by reading the concentration from the curve or by determining the extinction coefficient (ε) (the slope of the line) from linear regression analysis. This extinction coefficient can be used in successive determinations with the same reagents. A new standard curve should be prepared when new reagents are prepared.

18.7.2. Nitrate

The method for analysis of NO_3^- outlined here was chosen because of its relative simplicity, and the reagents prepared for the NO_2^- determinations are also used in this assay. The basis for this method is the reduction of NO_3 to NO_2^-, followed by the analysis of NO_2^-. Two procedures have been commonly used and are described here: reduction by zinc (162, 176) and reduction by cadmium (145).

The reduction of NO_3^- to NO_2^- with Zn is a simple procedure involving the reaction of the sample with Zn dust for 30 min, followed by filtration. The disadvantage of this procedure is the manipulation time required for each sample.

The reduction of NO_3^- with Cd amalgamated with Cu is performed by passing the sample treated with NH_4Cl (to prevent deactivation of the column) through a previously prepared column of Cd-Cu^{2+} filings. This procedure requires little effort at the time of analysis but requires prior preparation of the reduction column.

After either of the reduction procedures described here, the analysis for NO_2^- is performed as outlined above in the method for analysis of nitrite. Since the analysis is for total nitrite, part of the sample must be analyzed for NO_2^- before reduction, and this amount must be subtracted from the total NO_2^- to determine the NO_3^- concentration.

An ion-specific electrode is also available for measurement of nitrate (see chapter 17.2.2).

18.7.2.1. Sampling

Samples should be treated as described above for nitrite.

18.7.2.2. Zinc Reduction Procedure

Reagents

Use of NO_2^--free distilled water (section 18.6.1) is recommended for preparation of all reagents.

Buffer. Mix 50 ml of 0.2 M KCl with 34 ml of 0.2 M HCl, and dilute to 400 ml. Adjust the pH to 1.5 with HCl. (*CAUTION:* See section 18.10.)

NH_4Cl (2%, wt/vol)

Zinc powder

Procedure

Add 1.0 ml of 2% NH_4Cl and 2 ml of buffer to 25 ml of each sample in 50-ml stoppered flasks, and bring to 20 to 25°C. Add 0.2 g of zinc powder, and let stand for 30 min with occasional shaking. Filter out the zinc particles by using a low-porosity filter paper.

18.7.2.3. Cadmium Reduction Procedure

Reagents

Copper sulfate reagent: 2% (wt/vol) $CuSO_4 \cdot 5H_2O$

Concentrated NH_4Cl. Add 175 g of NH_4Cl to 500 ml of distilled water.

Dilute NH_4Cl. Dilute 1 volume of concentrated NH_4Cl with 40 volumes of NO_2^--free water to give 4.38 g of NH_4Cl in 500 ml of water.

Reagent grade cadmium sticks

Preparation of Columns

Columns should be ca. 1 by 30 cm with an appropriate opening (e.g., thistle tube or funnel) at the top so that the total sample can be placed on the column at once. The bottom of the column should have a burette-type stopcock or be attached to rubber tubing with a pinch clamp to allow the flow to be stopped before the column runs dry. The bottom of the column should be packed with glass wool or fine Cu "wool" turnings. A small amount of glass wool or copper filings is often added to the top of the column to prevent disturbance of the amalgamated cadmium-copper filings when adding the sample.

Cadmium filings can be prepared by filing sticks of reagent-grade cadmium metal with a coarse metal hand file (about second cut). Collect the fraction that passes a 2-mm sieve but is retained on a 0.5-mm sieve. Stir approximately 50 g of filings with 250 ml of 2% $CuSO_4 \cdot 5H_2O$ until the blue color disappears from solution and semicolloidal Cu particles begin to enter the liquid above the sediment. Allow the suspension to settle for 10 min. Fill the column with dilute NH_4Cl solution or with the supernatant fraction from the Cd-Cu amalgamation. Then, slowly pour in the Cd-Cu filings, and let them settle until a column height of about 30 cm is achieved. Wash the column three times with 100 ml of the dilute NH_4Cl solution with a flow rate of approximately 10 ml/min.

Procedure

Add 2 ml of concentrated NH_4Cl reagent to 100 ml of the sample. Add 5 ml of this solution to the column, and allow the liquid to drain to the top of the packing. This small addition ensures that when the bulk of the sample is added to the column, no error results from mixing with the previously run sample.

Add the remainder of the sample NH_4Cl solution. Collect and discard the first 25 to 30 ml of the eluting liquid. Then, collect 50 ml for the analysis. Discard the remainder of the eluting liquid. A column can be reused up to 100 times and usually need not be rinsed between samples. However, if the column will not be used for a few hours, rinse it with dilute ammonium chloride.

The 50 ml of eluate collected for analysis is assayed for NO_2^- (section 18.6.1) as soon as possible after reduction.

18.7.3. Ammonium by Indophenol Blue Reaction

Many methods are available to determine the concentration of NH_4^+ in samples (Table 1). An easy determination of NH_4^+ is based on the reaction of ammonia, hypochlorite, and phenol to form indophenol blue. The reaction is catalyzed by Mn^{2+}. The color is measured in a colorimeter or spectrophotometer at 630 nm (140, 145, 189, 192). The major disadvantages of this method are the use of phenol and the instability of the reagents.

Another easy determination of NH_4^+ is Nessler's method, in which the NH_4^+ reacts with mercury and iodide in an alkaline solution to yield a colored complex which is measured in a colorimeter or spectrophotometer (140, 145). The major disadvantage of this method is the use of mercury compounds. Nessler's method is given in section 18.7.4.

An alternative, and almost as simple, method, which can use the reagents prepared for NO_2^- determination, involves the oxidation of NH_4^+ to NO_2^- (150) with subsequent analysis of NO_2^-. However, the oxidation requires several reagents, as well as 3.5 h for the oxidation; therefore, the indophenol blue method or Nessler's method is recommended. If a technique very specific for NH_4^+ is required, the pyridine-pyrazole method (150, 171) may be more satisfactory than the other techniques. Also see chapter 17.2.2.1 for information on commercially available NH_3 electrodes. Some applications may benefit from the ammonium ion-specific electrodes, but they are generally not as accurate or as sensitive as the methods described here; however, they can provide a continuous readout, which can be very useful for some applications, e.g., monitoring the contents of a fermentor.

Samples

Samples should be treated as described in section 18.7.1. Samples high in interfering compounds (Table 1) may require distillation before the analysis is performed (145).

Reagents

Ammonium-free water. Add 0.1 ml of concentrated H_2SO_4 (CAUTION: This is a dangerous acid; see section 18.10) to each liter of distilled water, and redistill. Prepare fresh for each sample group, since NH_4^+ fumes from the laboratory will rapidly contaminate the water. Alternatively, an ion-exchange column may be used. Resins should be selected for maximum binding of NH_4^+ and other interfering organic compounds.

0.003 M $MnSO_4$. Dissolve 50 mg of $MnSO_4 \cdot H_2O$ in 100 ml of distilled NH_4^+-free water.

Hypochlorite reagent. Add 10 ml of commercial bleach (5% sodium hypochlorite solution) to 40 ml of NH_4^+-free distilled water. Adjust the pH to 6.5 to 7.0. Prepare fresh weekly. (CAUTION: See section 18.10.)

Phenate reagent. Dissolve 2.5 g of NaOH and 10 g of phenol in 100 ml of NH_4^+-free distilled water. Prepare fresh weekly. (CAUTION: Phenol is dangerous [see section 18.10], and NaOH is a dangerous base [see section 18.10].)

Standard NH_4^+ solution. Prepare a stock NH_4^+ solution by dissolving 381.9 mg of anhydrous NH_4Cl (dried at 100°C) in NH_4^+-free water, and adjust the volume to 1

liter. One ml contains 100 μg of N and 122 μg of NH_3. To prepare standards, dilute the stock solution to obtain a range of 0.1 to 5 μg of NH_4^+-N per 10 ml.

Procedure

Place 10 ml of each sample and 0.05 ml of the $MnSO_4$ solution in a small beaker or flask. With constant and vigorous stirring, preferably with a magnetic stirrer, add 0.5 ml of the hypochlorite reagent. Immediately add 0.6 ml of phenate reagent (alkaline phenol) dropwise.

The reaction is complete in 10 min, and the color is stable for 24 h. Measure the color in a colorimeter or spectrophotometer at 630 nm. Prepare a blank and a standard curve each day, since the color intensity may change with the age of reagents.

18.7.4. Ammonium by Nessler Reaction

Samples

Samples should be treated as described in section 18.7.1. Those high in interfering compounds (Table 1) may require distillation before analysis (145).

Reagents

Ammonium-free water. Prepare as described in section 18.7.3.

Reagents for removing large amounts of interfering cations. Prepare $ZnSO_4$ solution by dissolving 100 g of $ZnSO_4 \cdot 7H_2O$ in ammonium-free water to a final volume of 1 liter. Prepare 6 N NaOH by dissolving 240 g of NaOH in ammonium-free water to a final volume of 1 liter. (CAUTION: NaOH is a dangerous base. See section 18.10.)

Reagent for removing trace amounts of interfering cations. Dissolve 50 g of disodium EDTA in 60 ml of ammonium-free water containing 10 g of NaOH. (Heat may be required to dissolve the chemicals.) Adjust the volume of the cooled solution to 100 ml with ammonium-free water.

Nessler's reagent. Dissolve 100 g of HgI_2 and 70 g of KI in a minimal amount of water (100 to 200 ml). Add this slowly and with stirring to 500 ml of 8 N NaOH at 20 to 25°C. Adjust the volume of this mixture to 1 liter. Store in a rubber-stoppered glass vessel in the dark. The reagent is stable for up to 1 year. (CAUTION: This reagent is toxic; see section 18.10.) Check the reagent monthly by comparing a standard curve of NH_4^+-N using the reagent with the standard curve prepared when the reagent was new. When the curves deviate significantly, discard the reagent.

Standard NH_4^+ solution. Prepare a stock NH_4^+ solution as outlined in section 18.7.3. To prepare the standards, dilute the stock solution to obtain a range of 20 to 500 μg of NH_4^+-N per 50 ml.

Procedure

To remove interfering ions of Ca, Mg, S, and Fe, two methods are used. For large quantities of interfering ions, add 1 ml of the $ZnSO_4$ reagent to 100 ml of sample. Adjust the pH to 10.5 with 0.4 to 0.5 ml of 6 N NaOH by using a pH meter. The interfering ions precipitate out of solution within 15 min. The precipitate usually sediments other suspended matter as well. Remove the precipitate by filtration (use NH_4^+-free filter paper) or by centrifugation. If experience with the samples indicates that only trace amounts of the interfering ions are present, add 0.05 ml of the EDTA solution, with thorough mixing, to 50 ml of sample. Samples pretreated with $ZnSO_4$ and alkali are often also treated with EDTA. Samples with no interfering cations may be treated with Nessler's reagent without pretreatment.

The Nessler reaction is performed by adding 1 ml of Nessler's reagent to 50 ml of sample. If EDTA has been added to the sample, 2 ml of Nessler's reagent should be used. Mix the sample-reagent solution well, and let it stand for at least 10 min at room temperature. (For low NH_4^+ levels, 30 min is preferred.) Keep the temperature and reaction time constant for all samples, standards, and blanks. Read the absorbance of the resulting colored compound at a single wavelength between 400 and 425 nm for samples and standards with low NH_4^+ (20 to 250 μg of NH_4^+-N per 50 ml) and at a wavelength between 450 and 500 nm for those with high NH_4^+ (250 to 500 μg/50 ml). This reaction can be used with proportionally smaller volumes of sample, standards, and reagents as required by the investigator. The sensitivity of the method can be increased if a cuvette with a 5-cm light path is available. Calculate the amount of ammonia in each sample from a standard curve prepared by plotting the absorbance of each standard versus the concentration of ammonia in the standard solution.

18.7.5. The Challenge of Protein Analysis

A wide variety of methods are available to estimate protein concentrations in biological samples, several of which are described below. Most sensitivities vary from ca. 1 to 200 μg/ml, and this factor may determine which assay is preferable. Kjeldahl analysis of crude protein is also possible (188), but is seldom used for bacteria. Several problems prevent any one method from being the best for all purposes. First, a variety of substances interfere to different degrees with the different assays, e.g., commonly used detergents, reducing agents, or buffers (Table 2). Second, different proteins often give different responses, including the proteins used for standardization. Third, some assays are more time-consuming to perform or require careful timing for proper quantitation. From ca. 1960 until 1980, the Folin reaction (Lowry assay) (173) was the most commonly used, but in the past 2 decades, the Coomassie blue G method (Bradford assay) has become increasingly popular (157, 158, 184, 190). The newest method gaining popularity is the bicinchoninic acid-copper (BCA) assay (182, 191). Currently, the majority of labs use the Coomassie blue method as the preferred protein assay, with a modified Lowry (155–157, 183) or the BCA assay commonly used when an alternative is needed. Nonetheless, some labs use either of the last two assays for virtually all their protein assays.

When conducting protein assays, the practitioner is well advised to examine the literature for known interferences and to examine blanks experimentally, using the buffer and other components minus protein, for possible problematic reactions. Detailed information on interfering compounds is generally available at websites for the major manufacturers of protein assay solutions (e.g., Pierce and Bio-Rad). In most cases, a high level of precision is not required, e.g., when measuring protein eluting from a chromatography column or even when measuring specific activities during a purification procedure. When a fairly precise quantification is needed, the use of buffer controls minus protein is mandatory, and all assays should be performed in duplicate. For high accuracy, the use of two types of assays is recommended to counter the variation in the response of individual proteins; two such good pairs are the Bradford and BCA or the Bradford and Lowry assays. If samples of a specific protein of interest are available, it would be appropriate to use this protein as a standard in some cases. However, to ensure that the assays are comparable to assays done with other samples, and by other researchers, it is often preferable for a lab to use one of the two most common standards, BSA or bovine gamma globulin.

Protein assay is also used in the development of gel electrophoresis experiments. The Coomassie blue R dye method is commonly used, but the silver stain method is gaining popularity, especially because of its high sensitivity. However, the gel-staining methods are not described in this section; see chapter 17.5.1.

18.7.6. Protein by the Lowry Assay (Folin Reaction)

From ca. 1960 until 1980, the most widely used colorimetric method for estimation of proteins was that of Lowry et al. (173), using the Folin reaction. The blue color appearing in the assay is due to the reaction of protein with copper ion in alkaline solution (the biuret reaction) and the reduction of the phosphomolybdate phosphotungstic acid in the Folin reagent by the aromatic amino acids in the treated protein. The following procedure is useful for proteins that are already in solution or that are soluble in dilute alkali. Commercial kits have made this assay much more convenient than if all reagents were prepared from scratch. The Pierce modified-Lowry and the Bio-Rad DC modified-Lowry assays are compatible with many detergents, and the Bio-Rad RC DC modified-Lowry is in addition compatible with some common reducing agents (Table 2).

Reagents

Reagent A. Dissolve 20 g of Na_2CO_3 in 1 liter of 0.1 N NaOH (*CAUTION*: See section 18.10).

Reagent B. Dissolve 0.5 g of $CuSO_4 \cdot 5H_2O$ in 100 ml of a 1% (wt/vol) aqueous solution of sodium tartrate.

Reagent C. Just before use, mix 50 ml of reagent A and 1 ml of reagent B. Discard after 1 day.

Reagent D. Diluted Folin reagent. Concentrated Folin-Ciocalteu phenol reagent can be purchased commercially from many suppliers and is diluted as specified by the manufacturer; normally, it is provided as a 2 N acid solution by suppliers and should be diluted to 1 N with distilled water on the day of use. This reagent is usually purchased premade and is not made in individual laboratories. (*CAUTION*: This reagent contains a strong volatile acid [HCl] and phosphoric acid; see also section 18.10.)

Standard protein solution. Any one of several commercially available proteins or standards is suitable. Crystalline BSA or prepared solutions of BSA that can be purchased from biological supply houses produce suitable linear standard curves. The stock standard solution should contain 0.5 mg of protein per ml in distilled water. Different proteins give different standard curves. The protein standard used should be specified when results are reported.

Procedure

Add up to 0.5 ml of a sample containing 25 to 125 μg of protein to a clean 10-ml test tube. Adjust the volumes of each sample to 0.5 ml with distilled water. Add 2.5 ml of reagent C, and mix well. Allow this mixture to stand for at least 10 min at room temperature. Add 0.25 ml of reagent D, and mix immediately. The reactivity of this reagent lasts only seconds after addition to the alkaline, protein-containing solution. After 30 min or longer at room temperature, measure the A_{500} in a spectrophotometer or colorimeter. Greater sensitivity can be achieved by measuring the A_{750} if an appropriate spectrophotometer is available.

If 1-ml cuvettes and a suitable instrument are available, all volumes can be reduced by a factor of 2.5 and reagents and sample can be conserved. The samples and standards can be adjusted to 0.2 ml, and 1.0 ml of reagent C and 0.1 ml of reagent D are then added at the time intervals indicated above.

When protein is to be measured in bacterial-cell suspensions or when proteins are insoluble in dilute alkali, it is necessary to heat the suspension at 90°C in 1 N NaOH for 10 min to obtain complete solubilization. After solubilization, the suspension should generally be centrifuged to remove particulates prior to assay, and it may be advisable to wash the pellet with 1 N NaOH to obtain good recovery of the proteins. The standard should be treated in the same way, because this treatment reduces the intensity of the color. When this procedure is used to dissolve or extract proteins, a reagent containing 20 g of Na_2CO_3 in 1 liter of water (no NaOH) is substituted for reagent A. The assay procedure is otherwise identical.

Ammonium sulfate in amounts greater than 0.15% reduces color development. Investigators should be aware of this when this assay is used for proteins purified by fractionation procedures based on differential solubilities in ammonium sulfate solutions. High concentrations (1 to 5 M) of many other salts, some sugars, glycine, Tris buffer, some chelating agents, and compounds with sulfhydryl groups also interfere with the assay. Aromatic amino acids produce color with the phenol reagent, and glycine interferes with color development.

A modification of the Lowry assay that permits its use with samples containing sucrose or EDTA and with membrane and lipoprotein preparations without prior solubilization of the proteins has been described by Markwell et al. (174). The procedure employs a detergent (sodium dodecyl sulfate) in the alkaline carbonate reagent to improve solubilization of samples containing lipoidal material and an increase in the amount of the copper tartrate reagent to overcome interference by sucrose and EDTA. This procedure has several applications, including assays of proteins in fractions from sucrose gradients.

Commercially available modified-Lowry kits are available (155–157, 183). Advantages include a more convenient Lowry assay and one which is insensitive to some types of interference.

18.7.7. Protein by Biuret Reaction

Peptide bonds of proteins form blue chelates with copper ions in alkaline solutions (145, 164). This reaction is much less sensitive than the Lowry procedure but has the advantage that it is less time-consuming and can be used in the presence of ammonium salts and other materials that interfere with the Lowry procedure. There is also less variability among proteins in the amount of color formed per unit weight of protein.

The modification of the biuret procedure (164) described below involves the use of tartrate to form a copper complex that is soluble in NaOH. The protein displaces the copper ion from the complex to form a copper-protein complex with a different color and absorption intensity.

Reagent

Dissolve 1.5 g of cupric sulfate · $5H_2O$ and 6 g of sodium potassium tartrate · $4H_2O$ in 500 ml of water. Add, with stirring, 300 ml of a 10% (wt/vol) solution of sodium hydroxide in water (*CAUTION*: NaOH is a dangerous base; see section 18.10). Adjust the volume of the mixture to 1 liter with distilled water. Discard if a reddish or black precipitate is evident.

Procedure

Add 2.0 ml of the above biuret reagent to a neutralized sample (0.5 ml) containing 0.5 to 5 mg of protein. Mix thoroughly, and incubate at room temperature for 30 min. The time and temperature for all samples, standards, and blanks should be the same. Measure the A_{500}. Prepare a blank with 0.5 ml of water or aqueous solution to dissolve proteins and 2.0 ml of biuret reagent. A standard curve can be prepared by using any completely soluble protein such as BSA. The response is not linear with protein concentration because of competition between tartrate and protein for copper ions.

Applications

This is a robust and generally effective assay suitable for use with a variety of sample types, so long as high sensitivity is not required. A modification of this procedure described by Stickland (193) is suitable for the analysis of total protein in bacterial cells. In this procedure the cellular protein is dissolved in 1.0 N NaOH, and then $CuSO_4$ without tartrate is added. The insoluble cellular material and insoluble $Cu(OH)_2$ are removed by centrifugation, leaving the colored Cu-protein complex in solution.

18.7.8. Protein by Coomassie Blue Reaction (Bradford Assay)

Proteins bind several dyes, causing a shift in the absorption spectrum of the dye. In the procedure described below, Coomassie brilliant blue G is used as the dye. This dye exists in two forms: the red anionic form is converted to a blue form when the dye binds to amino groups of proteins (161). This procedure, described in detail by Bradford, is often more sensitive and less subject to interference by many compounds that restrict the use of other assays (158, 190). The dye assay can be performed more rapidly than those described above. The sensitivity of the dye assay is a little higher than that of the Lowry procedure (Table 2). The response to different proteins is less variable than the color reactions in the other assays described above.

Reagent

Dissolve 100 mg of Coomassie brilliant blue G250 in 50 ml of 95% ethanol (*CAUTION*: See section 18.10). Add 100 ml of 85% (wt/vol) phosphoric acid (*CAUTION*: This is a strong acid; see section 18.10). Adjust the volume to 1 liter with distilled water. (Commercially prepared solutions are available, e.g., from Bio-Rad Laboratories or Pierce Chemical Co., and are more convenient.)

Procedure

Add a protein solution containing 5 to 50 μg of protein in 0.05 ml to a test tube. Add 2.5 ml of the dye reagent, and thoroughly mix the contents (or follow instructions for the commercially prepared reagents). Measure the A_{595} after 2 min and before 1 h against a blank containing 0.05 ml of the buffer or salt solution and 2.5 ml of the dye reagent. Relatively inexpensive plastic cuvettes can be used multiple times for assays. Moderately expensive glass cuvettes can also be used; however, quartz cuvettes bind to the dye more firmly and, if used, must be rinsed with ethanol or acetone at appropriate intervals.

The protein content of the sample is determined from a standard curve obtained by plotting the absorbance of standard solutions containing 10 to 60 μg of protein in the same

2.55-ml assay volume as the samples, versus the concentration of protein in each standard solution. The same buffer used to dissolve the unknown samples is used with the standards.

Applications

The Coomassie blue procedure is applicable to the assay of many soluble proteins and is currently the most commonly used protein assay in the literature. Detergents interfere with color development; therefore, glassware should be rinsed well. The assay is linear from 25 to 75 μg of protein per sample. Other limitations of the assay are described by Pierce and Suelter (185). A microassay procedure (158) and another variation that has high reproducibility and can detect 1.0 μg of protein in an assay volume of 1 ml have also been described (190).

18.7.9. Protein by Bicinchoninic Acid-Copper Reaction (BCA Assay)

Proteins in an alkaline solution react with Cu^{2+} to produce Cu^+, as in the biuret assay. Bicinchoninic acid (BCA) reacts with the Cu^+, forming a purple, water-soluble product which can be measured by its A_{562} (191). The sensitivity of the BCA method is about equal to that of the Bradford method and a little higher than the Lowry method (Table 2). It is compatible with most detergents used in the laboratory and with many buffer components. It may be the preferred assay for laboratories in which detergent is commonly used in protein solutions (191). The assay is interfered with by high levels of ammonium sulfate, EDTA, and sulfhydryl reagents; the sulfhydryl types of interference can be remedied by their modification with iodoacetamide (167). The assay requires mixing two or three reagents on the day of the assay, but the individual solutions are stable for many days. The response variation between different proteins is similar to that of the Lowry and Bradford methods and can vary a maximum of 20 to 40%, depending on the protein standard used. The assay time is slightly shorter than required for the Lowry method and involves incubation for a set time and at a specific temperature. The assay is less complex than the Lowry method but more so than the Bradford method.

Reagents

A commercially prepared reagent is available from Pierce Chemical Co.; it is essentially the same as described here and as originally described by Smith et al. (182, 191).

The regular working reagent is prepared by mixing two stock solutions (reagent A and reagent B). The stock solutions can be stored indefinitely at room temperature, but the mixture of the two can be stored for only 1 day.

Reagent A. Dissolve 10 g of sodium bicinchoninate in a 1-liter aqueous solution containing 20 g of $Na_2CO_3 \cdot H_2O$, 1.6 g of sodium tartrate, 4.0 g of NaOH, and 9.5 g of $NaHCO_3$. The pH may be adjusted, if necessary, to pH 11.25 by the addition of concentrated NaOH or solid $NaHCO_3$. (*CAUTION*: This reagent is a strong base and contains both sodium hydroxide and sodium carbonate; see section 18.10.)

Reagent B. Dissolve 4 g of $CuSO_4 \cdot 5H_2O$ in 100 ml of deionized or distilled water.

Procedure

A working reagent is made by mixing 100 volumes of reagent A with 2 volumes of reagent B, resulting in an apple-green solution (182, 191). Add 1 volume of protein sample or buffer blank to 20 volumes of the working reagent in an appropriate tube (e.g., 0.1 ml of sample plus 2.0 ml of reagent), and mix gently. Incubate the tubes at 37°C for 30 min, and then cool them to room temperature; the protein-containing tubes will turn purple. (Alternatively, assays can be incubated at room temperature for 2 h or at 60°C for 30 min; the latter method enhances the sensitivity ca. four-fold.) Read the A_{562}, using a water blank. Subtract the absorbance of the buffer blank from all sample absorbances, and compare with a standard curve of protein concentration versus absorbances. BSA is a suitable standard protein and should be diluted in the same buffer as used for the samples to be assayed and at concentrations similar to those of the protein being assayed, generally in the range of 10 to 120 μg/ml of assay working reagent.

Applications

The BCA method is a good method for the determination of proteins for a variety of purposes, but it is particularly appropriate for solutions containing most detergents, since most of those used as solubilizing agents do not interfere with the assay. Although thiol-containing reagents interfere with this method, suitable methods involving iodoacetamide have been developed to circumvent some of these problems (167); trichloroacetic acid precipitation or other methods may also be of value (177).

18.7.10. Protein by UV Absorption

Proteins have an absorption maximum at 280 nm due to the presence of the aromatic amino acids tyrosine and tryptophan (with a very small contribution from phenylalanine). These two amino acids are present in nearly all proteins, and their proportions relative to other amino acids usually vary over a narrow range. The absorption of UV light is a suitable means of estimating proteins in solution if the protein solution does not contain more than 20% by weight of other UV light-absorbing compounds, such as nucleic acids or phenols, and if the solution is not turbid. Correction for the absorbance of smaller amounts of nucleic acids in protein can be made by the method of Warburg and Christian (194), of Kalckar (170), or of Kalb and Bernlohr (169).

Equipment

UV spectrophotometer, calibrated with an absorbance standard

Cuvettes, made from quartz or silica, which do not prevent transmittance of UV light

Procedure

The A_{280} and A_{260} are measured with the protein dissolved in a suitable buffer. A solvent blank containing the buffer is used to zero the instrument at each wavelength. If the ratio of A_{280}/A_{260} is not greater than 1.70, except with solutions known to contain pure protein, the following equation (170) should be used to calculate the concentration of protein, or an appropriate table (172) should be consulted. The equation is used to subtract the contribution of nucleic acids to the UV absorbance of the solution.

$$\text{protein concentration (mg/ml)} = 1.45A_{280} - 0.74A_{260}$$

Applications

When preservation of a protein sample is important because of its limited availability, this nondestructive assay procedure is the method of choice, provided that the protein is sufficiently pure to permit its application. Different proteins can yield different extinction coefficients, and a standard prepared from an identical protein (if available) should be used if precise measurements are important.

The use of absorbance to estimate protein is common during protein purification involving column chromatography, but the correction for A_{260} is seldom used. This method should be used with caution, since coenzymes present in early fractions can contribute significant nonprotein A_{280} values.

18.7.11. Total Nitrogen (Kjeldahl Digestion)

A sample may be analyzed for total nitrogen by adding the results of analyses for the separate nitrogenous components (NO_2^-, NO_3^-, NH_4^+, and organic nitrogen). However, if one is interested in only the total nitrogen, performing all these analyses is time-consuming. The most commonly used analysis for total nitrogen is the Kjeldahl digestion of organic nitrogen compounds to yield NH_4^+, which is then assayed by one of the techniques for NH_4^+ analysis. NO_3^- and NO_2^- can be included in this analysis when desired by reduction to NH_4^+ with hot alkali (145) or salicylic acid and zinc (141). An alternative method, which allows use of reagents already prepared for the NO_2^- analysis, is based on the oxidation of the NH_4^+ from digestion to nitrite and subsequent analysis of nitrite.

The procedure used to digest the organic material depends on the state of the sample (solid or liquid). Because procedures for the Kjeldahl digestions are variable, complex, and easily available in various texts, the reader is referred to the literature that treats the specific procedure. Kjeldahl digestion techniques developed for liquid samples include those for seawater (150), lake water (140, 145), and wastewater (140, 145). Techniques for solids include those for soil (140–142, 148), sediment (145), plant materials (140, 147), fertilizers (147), animal materials (147), and microorganisms (145).

18.7.12. Molecular Nitrogen

The need to detect molecular nitrogen transformation in a biological system is usually limited to nitrogen fixation studies in which nitrogen (dinitrogen) is reduced by an enzyme system (called nitrogenase) to ammonia, which can then be assimilated by the cell. Nitrogen fixation can be estimated or monitored in two very different ways: by direct measurement of N_2 assimilated, or by assay for the enzyme nitrogenase (166).

Direct measurement could be made of the disappearance of N_2 (determined by gas chromatography or other physical analysis), but the abundance of N_2 in the atmosphere makes such analysis difficult. Direct measurement of ammonia formation is also possible, using analysis described in the above sections (sections 18.7.3 and 18.7.4).

In practice, neither of these direct measurements is made. Instead, the stable isotope, $^{15}N_2$, is used. This isotope is provided to cells or plants, enriched significantly above the natural abundance of 0.4%. In one variation of this approach, after time has been allowed for dinitrogen metabolism, the total organic nitrogen, as described in the section above, is obtained and then analyzed by mass spectrometry (151, 186). A more contemporary alternative is the use of ^{15}N-NMR, in which cells are allowed to metabolize the isotope and either extracts or whole cells are examined for ^{15}N enrichment in the proteins (154). Use of either the mass-spectrometric or NMR approach may be the best way to monitor N_2 metabolism, as well as metabolism of other nitrogenous substrates (e.g., nitrate, nitrite, and ammonium), and even some specific intermediates may be identified by the NMR approach (154). However, these approaches require the use of facilities dedicated to these types of analyses and equipment operated by well-trained staff. Despite this, most major universities and biotech companies have such facilities, and these methods are not particularly expensive by modern research standards.

However, the nitrogen fixation capability of cells is commonly measured by the ability of nitrogenase to reduce acetylene to ethylene. This requires less-expensive equipment than other methods and can be conducted in most microbiology and biochemistry laboratories. Acetylene and ethylene can be detected and quantified by use of a gas chromatograph (see chapter 17.3.5). For qualitative determination of nitrogenase in systematic bacteriology, and for further discussion, see chapter 15.3.1.1.

18.7.13. Specific Nitrogenous Compounds

A variety of specific nitrogen-containing compounds may be of interest to the investigator. These compounds include amino acids and nitrogen-containing cellular components not discussed above. Many of these components can be most conveniently measured by amino acid analyzers (sometimes with alternative detection methods) normally present in central facilities, but they can also be measured by the individual scientist using a variety of methods. Broadly applicable analysis methods include ion-exchange chromatography or HPLC, coupled with ninhydrin assay of eluates or coupled with spectrophotometric detectors of several sorts. Gas chromatography and mass spectrometry may also be of value (146, 149, 151). Discussed below are three specific examples of nitrogenous-compound analysis particularly applicable to the bacterial physiologist.

18.7.14. Total Amino Acids

It is often necessary to determine the amino acid composition of proteins, either from whole cells or from those specifically purified. Most research institutions have specific facilities for this analytical task. Because of the large number of amino acids normally found in proteins (20 amino acids) and a smaller number of unusual or atypical amino acids, such analysis is not feasible in most individual laboratories. However, HPLC methods coupled with sophisticated detectors can be used by some laboratories to supplant the more traditional "amino acid analyzer." Information on the variety of equipment for HPLC amino acid analysis can be obtained from a variety of suppliers (e.g., Waters, Beckman, Varian, and Ranin) and from institutional facilities. In general, for work needed by most bacteriologists, protein hydrolysis and submission to a central facility is the best route; this is also likely to be the fastest and least expensive.

For analysis by central facilities, a wide array of methods are available for sample hydrolysis to convert the protein to individual amino acids. In general, treatment involves heating solutions of the protein in a strong acid in a sealed glass tube. Most methods create difficulties for the stability and quantitation of at least one amino acid, and thus, two parallel methods must often be used for measurement of all the desired amino acids. The resulting amino acid mixture can also be separated by a variety of methods, usually either ion-exchange or C_{18} reverse-phase chromatography. Detection can be carried out by a postcolumn assay (e.g., the ninhydrin reaction) or by spectrophotometric or fluorometric methods involving either pre- or postcolumn derivatization (e.g., with dimethylaminonaphthalene sulfonyl chloride, phenylisothiocyanate, or orthophthalaldehyde). These methods are very sensitive and require very small amounts of protein; a corollary to this is that the material must be very pure and contamination must be avoided during sample preparation (for example, amino acids left by a fingerprint can be significant).

18.7.15. *meso*-Diaminopimelic Acid

meso-Diaminopimelic acid (DAP) is found in most gram-negative peptidoglycans and also serves as the precursor of dipicolinic acid in spores (144, 179, 180, 195). DAP is not a normal component of proteins, but its measurement in cell walls or spores is sometimes of interest. One analytical strategy is to submit a hydrolyzed sample for amino acid analysis as described above, because this can provide a good quantitative method. Another approach is to use HPLC methods, as indicated above for amino acids.

Described here is an alternative colorimetric procedure for the estimation of DAP, in which the ninhydrin reaction is used as described by Work (196). In this method, the reaction is carried out under acidic conditions, which makes it more selective for diamino acids.

Reagents, Supplies, and Equipment

Acetic acid (*CAUTION:* See section 18.10)

Ninhydrin reagent: 250 mg of ninhydrin dissolved in 6 ml of acetic acid and 4 ml of aqueous 0.6 M phosphoric acid (*CAUTION:* See section 18.10)

Spectrophotometer

Glass or quartz cuvettes or spectrophotometer tubes (plastic can also be used if it is stable to concentrated acetic acid)

Glass test tubes capable of holding 10 ml

DAP standard

Samples, properly prepared for analysis

Water bath at 37°C

Procedures

Samples containing DAP can be prepared in any of a variety of ways. A processed portion of cells (e.g., peptidoglycan as described above in section 18.5.1) can be treated as described for protein hydrolysis in section 18.7.14, or alternative hydrolysis methods can be used. In most cases samples contain low levels of a variety of amino acids, but contamination with significant quantities of protein, and the accompanying amino acids, creates difficulties with the assay unless some purification step is performed. Prior to hydrolysis, many protein-containing solutions can be deproteinized by addition of 1 volume of acetic acid to 1 volume of sample, mixing, and centrifugation to remove the protein as a pellet. The supernatant is used as the sample.

The following procedure can be conducted with samples containing lysine, ornithine, proline, cysteine, or tryptophan at levels similar to the DAP. Mix a 0.25-ml sample with 2.0 ml of acetic acid, and add 0.25 ml of the ninhydrin reagent. Heat the solution at 37°C for 1.5 h, and then read the A_{440} of the standard and sample by using a blank containing water instead of the sample added to the acetic acid and ninhydrin reagent. Construct a standard curve by using sample solutions containing ca. 10 to 450 μg of DAP per ml. (Hydrolysis of the original sample should not be done with HCl.)

This assay would also be appropriate for examination of fractions eluted during column chromatography of samples; e.g., acid-hydrolyzed samples could be applied to an AG 50W-X8 (H⁺ form) column and eluted with a gradient of acid, with or without additional salts. The resulting fractions

could be assayed with this procedure. (See also chapter 17.3.1 for information on ion-exchange chromatography.)

18.7.16. Dipicolinic Acid

Dipicolinic acid (DPA) is found almost entirely in bacterial endospores (spores made by *Bacillus* and *Clostridium*), where it is a major and unique component, contributing 5 to 14% (dry weight) of the spore (179, 180). Following its extraction from spores, it can be estimated spectrophotometrically.

Reagents, Supplies, and Equipment

Culture of sporeforming bacterium grown on a medium appropriate for sporulation (alternatively, a pure spore preparation)

Methylene chloride (*CAUTION:* See section 18.10)

Toluene (*CAUTION:* Toluene is flammable and toxic; use a hood, and keep away from flames)

Centrifuge and centrifuge tubes

Refrigerator at ca. 4°C

Boiling-water bath

Sulfuric acid (0.1 and 1.0 M solutions) (*CAUTION:* See section 18.10)

KOH (1.0 M solution) (*CAUTION:* See section 18.10)

Diethyl ether (*CAUTION:* See section 18.10)

pH meter

DPA standard

Spectrophotometer capable of use in UV- and visible-light ranges

Quartz cuvettes (or glass cuvettes for alternative assay)

Fe(NH₄)₂(SO₄)₂ · 6H₂O

Ascorbic acid

Acetic acid (*CAUTION:* See section 18.10)

Procedures

Preparation of spores

Grow any *Bacillus* or *Clostridium* species under sporeforming conditions (see chapter 7.4.10). Harvest cells and spores by centrifugation (ca. 5,000 × g for 15 min), and wash them once with water. In many cases, virtually all cells have undergone sporulation, but cell debris and vegetative cells contaminate the spores, so some cleanup of the preparation is needed for some purposes; for other purposes, further cleanup is not needed. Either of two general approaches can be used to purify spores: degradation of nonspore material, or density gradient centrifugation (see chapter 7.3.3).

In the degradation approach, residual vegetative cells are removed by treatment of a water suspension of cells plus spores with a protease (1 mg of papain per ml) at 4°C for 5 to 20 days in the presence of 0.1% of a 1:1 methylene chloride-toluene mixture (adapted from reference 196 with modifications). After this treatment, spores should be washed six times with a 20-fold volume excess of high-purity distilled water and either used directly, stored frozen, or dried at 80°C overnight and then stored dry at 4°C in a dry closed container.

Extraction of DPA

Since chelating metals should be avoided for analysis, all water should be high-quality distilled water. Resuspend spores (ca. 0.05 g [dry weight] or 0.15 g [frozen wet weight]) in 10 ml of 0.1 M H₂SO₄, heat in boiling water for 5 min, and then neutralize with 1.0 M KOH. Extract the suspension with 30 ml of alkali-washed diethyl ether, and, after phase separation, discard the ether and adjust the aqueous phase to pH 1.0 with 1.0 M H_2SO_4. Extract this solution with 30 ml of alkali-washed diethyl ether, and discard the aqueous phase. Add 5.0 ml of water to the ether, and evaporate the ether. This aqueous extract will contain the DPA. (Adopted from reference 175 with modifications.)

Estimation of DPA

Dilute the aqueous extract and prepared DPA standards in water as needed, to yield concentrations in the range of 30 to 150 μg/ml. Add 0.2 ml of these solutions to 1.8 ml of 0.1 M H_2SO_4 in a quartz spectrophotometer cuvette, and read against a water blank at 273 nm. Compare the samples with a DPA standard curve. A DPA concentration in the assay cuvette of 10 μg/ml will give an A_{273} of ca. 0.4. (Adopted from references 175 and 195 with modifications.)

An alternative assay method that is not sensitive to contamination by UV-absorbing material is available (168) and may not require ether extraction prior to assay. The sample from the ether extraction above (or in some cases from a direct dilution of the spore hydrolysis solution) is diluted in water to the same DPA concentrations as above. To 4.0 ml of this dilution is added 1.0 ml of the freshly prepared reagent containing 1.0% $Fe(NH_4)_2(SO_4)_2 \cdot 6H_2O$ and 1.0% ascorbic acid, both dissolved in 0.5 M potassium acetate buffer adjusted to pH 5.5 after their addition. After the sample and reagent are mixed, the A_{440} is measured within 2 h, and the DPA concentration is calculated from a standard curve. This assay has the advantage that quartz cuvettes are not required and a spectrophotometer capable of use below 320 nm is not required; as well, the assay is more specific, and with proper caution it can be used without ether extraction to purify the DPA prior to analysis.

18.8. ORGANIC ACIDS AND ALCOHOLS

A number of short-chain organic acids and alcohols are formed in bacterial cells as intermediates or end products of the citric acid, glycolytic, and other metabolic pathways. These compounds may be accumulated in the cytoplasm or excreted into the medium. In bacterial cell fractions, these compounds occur in the small-molecule, supernatant fluid fraction after precipitation of the macromolecules with cold trichloroacetic acid (Fig. 1) or after boiling-water extraction. The small-molecule fraction may include acetate, acetoacetate, citrate, α-ketoglutarate, formate, lactate, succinate, pyruvate, malate, fumarate, ethanol, and many other compounds.

These small molecules may be analyzed by ion-exchange chromatography (chapter 17.3.1), absorption chromatography (chapter 17.3.2), HPLC with a variety of column materials, or gas-liquid partition chromatography (chapter 17.3.5). Lactate, ethanol, acetate, and succinate may be also analyzed by enzymatic or colorimetric methods (see also sections 18.2 and 18.3.8).

18.9. MINERALS

The amounts of different elements in bacterial cells and the biological roles of trace elements are described in references 198 and 200; references 198 and 202 describe sample preparation and storage. Metal analysis in samples of interest to bacteriologists can be accomplished by colorimetric methods, atomic absorption, photometry, spectrophotometry, flame emission photometry, nuclear activation analysis, the use of ion-specific electrodes, and other means; these and other methods for determining metals are described in references 197 to 200.

A rapid, accurate, and convenient titrimetric procedure that requires no specialized equipment can be applied to the analysis of some elements, such as calcium. Interfering elements and compounds present in small amounts can sometimes be removed by treatment of the samples. These procedures are described in references 200 and 201.

Flame emission photometry (198, 199) is ideally suited to the analysis of alkali and alkaline earth metals. **Atomic absorption analysis** (197, 199) is also an accurate and relatively convenient method for the analysis of several metals. Both of these methods require experienced personnel and a specialized spectrophotometer. The inaccessibility of proper advice or of a suitable instrument may limit the use of the technique. Several elements and compounds interfere with atomic absorption analysis of an element, but proper treatment of samples can reduce or eliminate interferences.

Nuclear activation analysis (199) involves the bombardment of a sample to make one or more elements radioactive. The radioactive species are identified and measured quantitatively. This method of elemental analysis has the advantage of sensitivity, and many elements can be assayed in a single small sample. The use of this technique requires access to a facility that has a high-energy neutron source (a nuclear reactor or accelerator), sophisticated isotope detection equipment, and expert personnel to operate the facility. When available, this method is relatively inexpensive in comparison with other methods if time and other costs are taken into account and if several elements are determined simultaneously in a single sample. This is particularly true when the availability of sample material for analysis is limited.

X-ray emission spectroscopy (199) is a nondestructive method of analysis applicable to the estimation and identification of some elements that are present in sufficient abundance in biological material.

Electron probe X-ray microanalysis can be applied to the estimation and location of some elements in situ. The resolution of the technique is suitable for use in individual bacterial cells or spores (203), and even in prokaryotic cells some estimation of the location of the elements within the cell can be made. This technique also requires sophisticated equipment and expertise. Thus, its application is generally limited to specialized research laboratories, but many universities and industrial companies have access to such instrumentation via their electron microscopy central facilities.

Methods for **colorimetric and turbidimetric analysis** of metals and inorganic compounds, including automated analysis, have been described by Snell and Snell (200) and in publications from the U.S. Environmental Protection Agency (201).

Commercially available **selective ion electrodes** also provide a very convenient and inexpensive means of measuring some elements. Their application is limited to a few types of samples and elements, including sodium, chloride, bromide, fluoride, copper, silver, and calcium (see chapter 17.2.2).

18.10. HAZARDS AND PRECAUTIONS

The methods described above almost always require the use of chemicals, supplies, or equipment that are hazardous when handled or applied improperly. Provided here is a listing of chemical precautions needed to safely carry out most of the methods in this chapter. However, these warnings and precautions should be used in an intelligent way by suitably trained personnel and are meant to supplement their awareness of laboratory hazards. A more complete coverage of these and related topics is found in chapter 46. In addition, any unfamiliar laboratory work should be attempted only after consultation with more experienced personnel. The laboratories to be used should meet minimum standards for safety, including adequate fire safety, plans for spills, and availability of safety devices and fume hoods. Common sense should be used in all laboratory work.

If any serious exposure or injury occurs, seek medical attention promptly, after immediately initiating appropriate decontamination or first-aid procedures.

Hazards of using chemicals are often dependent on the scale of chemical use. In many cases an experiment can be adjusted in size to require less material and thus to create less hazard in the conduct of the work, as well as to generate a smaller amount of dangerous waste to process. Conducting experiments at the smallest scale that will provide good data is both safer and cheaper.

Several general rules apply to all chemicals as follows.

- Never mouth pipette anything in the laboratory, but use a pipetting device instead.
- Avoid contact with chemicals by using proper laboratory practices.
- Learn about the chemicals used before an accident occurs with them.

Listed here in alphabetical order are a variety of chemical and physical hazards that are encountered in the work described above. Each of these is denoted by "*CAUTION:*" in the relevant section.

Acetaldehyde   Vapors and liquid are toxic. Flammable. Chill bottle before opening, since the boiling point is 21°C. Use only in a fume hood.

Acetic acid   A strong and volatile acid. Avoid exposure to both liquid and vapors. Use in a fume hood. Wear gloves and eye protection.

Acetic anhydride   Flammable; avoid contact with flames or sparks. Irritates tissue; avoid contact with skin and eyes. Use in a fume hood.

Acetone   Highly flammable. Avoid contact of liquid or vapors with flame or sparks. Skin and respiratory tract irritant. Use in a fume hood when large quantities are handled.

Ammonium hydroxide   Fumes and liquid are both caustic bases. Avoid contact with liquid and vapor. Use in a fume hood. Wear gloves and eye protection.

Chloroform   Chloroform is hepatotoxic, is a carcinogen, and acts as an anesthetic. Contact with liquid or vapor should be strictly avoided. Use in a fume hood.

Diethyl ether   Ethers are very flammable and may explode when vapors contact flames or electrical sparks. Keep the minimum amount necessary in a laboratory. Warm all solutions in water baths rather than incubators. Avoid inhalation of vapors by use of a fume hood. Ethers cause irritation of mucous tissues and can cause fainting. Ethyl ether can form explosive peroxides when evaporated to dryness; use only fresh containers, and properly dispose of older containers. Recycling of ethers should be done with great caution; if one container is used for several evaporations without being cleaned to remove peroxides, a spontaneous and life-threatening explosion may occur.

Ethanol   Flammable. Avoid using near open flame and electrical appliances that cause sparks.

Ethidium bromide   Mutagen and possible carcinogen. Use gloves. Avoid contact with skin. Wash thoroughly with water if exposure occurs. Purchase premade solutions if possible, to avoid exposure to dust during weighing.

Hexane   Very flammable. Avoid use near open flames and electrical appliances that cause sparks. Respiratory irritant. Use in a fume hood.

Hydrochloric acid   A strong and volatile acid. Avoid exposure to both liquid and vapors. Use in a fume hood. Wear gloves and eye protection.

Iodine   Iodine vapors are a corrosive and dangerous eye, respiratory, and skin irritant. Avoid vapors by using in a fume hood.

Lipopolysaccharide (LPS)   Toxic and pyogenic. Avoid all contact. Wear gloves.

Mercuric iodide   Poisonous. Wash hands after use. Avoid dust. Dispose of properly.

Methanol   Very flammable; avoid contact of liquid or vapor with flames or sparks. Toxic when swallowed or inhaled, and causes blindness. Use in a fume hood when dealing with large quantities.

Methylene chloride   Narcotic in high concentrations.

Nitric acid   Strong acid and oxidizing agent. Avoid contact with liquid or vapor. Use in a fume hood. Wear eye protection and gloves. Concentrated nitric acid can react violently with organics, including some alcohols. If skin or eyes are exposed, wash with abundant quantities of tap water.

Nitrogen gas cylinder   Nitrogen gas is nontoxic and not flammable. However, compressed gas cylinders are always dangerous. Use a proper regulator. Secure the cylinder carefully to prevent it from falling. Make sure that the pressure used does not exceed the strength of any hosing attached to carry gas from the tank. Use a proper cart to move the cylinder, and never move a cylinder that does not have a cap. If the top of a gas cylinder is damaged by a fall, the entire container can become an extremely dangerous missile and is capable of penetrating walls.

Perchloric acid   A very strong acid and oxidizer. Liquid and vapors are corrosive and cause severe burns. Mixture with water is exothermic, and exposure of the concentrated reagent to organic compounds may cause a fire or explosion. Use only in a fume hood. Wear eye protection and gloves. If skin is exposed, wash with abundant quantities of cold water. Purchase and use the minimum quantity needed.

Petroleum ether   Highly flammable; use only in absence of flames and sparks. Vapors may explode if lighted. Use in fume hood.

Phenol   Poisonous and caustic. Exposure of skin, eyes, or the respiratory tract to phenol or its vapors can cause tissue damage. Wear gloves and eye protection. Use a fume hood if the phenol is hot. If skin or eyes are exposed, wash with abundant quantities of tap water.

Phosphoric acid   Strong acid. Avoid contact with skin and eyes. Wear eye protection and gloves. If skin or eyes are exposed, wash with abundant quantities of tap water.

Potassium hydroxide   A strong base. Avoid contact with skin and eyes; wash with abundant quantities of water if contact occurs. Wear eye protection and gloves. Be aware when making strong solutions that the dissolution of KOH in water or alcohol is exothermic, and the solution can get very hot (so hot that it may crack glass containers).

Radioactive chemicals Use of radioactive material is strictly regulated. Materials may be purchased only by licensed laboratories and must be used under careful supervision. The hazards of different materials vary considerably. Consult appropriate references and local supervising offices for details.

Sodium arsenite Very poisonous; do not breathe dust. Wash hands well after use.

Sodium azide Poisonous. Avoid contact. Wash hands well after use.

Sodium carbonate Strong alkali. Avoid contact with aqueous solution. If skin or eye exposure occurs, wash with abundant quantities of tap water. Wear eye protection and gloves.

Sodium hydroxide A strong base. Avoid contact with skin, wash with abundant quantities of tap water if skin is exposed to NaOH solutions. Wear eye protection and gloves. Be aware when making strong solutions that the dissolution of NaOH in water is exothermic and may make the solution very hot (hot enough to crack glass containers).

Sodium hypochlorite Oxidant. Ingestion of the liquid is dangerous and causes severe tissue destruction. Avoid vapors and contact with skin.

Sulfuric acid A very strong acid. Avoid contact with skin, and wash with abundant quantities of cold tap water if skin or eyes are exposed to this acid. Wear eye protection and gloves. Be aware that the dissolution of sulfuric acid in water is very exothermic and can make the solution very hot; the heat generated may be sufficient to burn skin and to break non-Pyrex containers. Do not mix by adding water to the acid, since it can spatter; instead, add acid to the water. For preparation of fairly concentrated solutions, cooling the mixing container with cold water or ice water may be appropriate, but do not place an already very hot glass container in cold water, since it may break. Be aware of where the nearest safety shower and sink is when handling bottles of concentrated acid. Purchase the smallest appropriate bottle.

Trichloroacetic acid Strong acid. Very corrosive. Causes skin irritation and is dissolved in lipids of the skin. Avoid contact by wearing protective gloves, appropriate laboratory clothing, and eye protection. If skin contact occurs, wash well with abundant quantities of tap water.

Trifluoroacetic acid Strong acid. Avoid contact with liquid and vapor. Wear eye protection and gloves. Use in a fume hood. If exposure to skin occurs, wash with abundant quantities of tap water.

UV light UV light can cause severe retinal burns. Chromatograms and other material should be observed in UV light only when plastic goggles are used to cover the eyes, although normal glass eyeglasses decrease the UV somewhat. A variety of suppliers provide special goggles that reduce the UV exposure even more.

18.11. REFERENCES

18.11.1. Fractionation and Radioactivity

1. **Kennell, D.** 1967. Use of filters to separate radioactivity in RNA, DNA, and protein. *Methods Enzymol.: Nucleic Acids* **12A:**686–692.

2. **Roberts, R. B., D. B. Cowie, E. T. Bolton, P. H. Abelson, and R. J. Britten.** 1955. Studies of biosynthesis in *Escherichia coli.* Publ. 607. Carnegie Institute of Washington, Washington, DC.

3. **Sutherland, I. W., and J. F. Wilkinson.** 1971. Chemical extraction methods of microbial cells, p. 345–383. *In* J. R. Norris and D. W. Ribbons (ed.), *Methods in Microbiology,* vol. 5B. Academic Press, Inc., New York, NY.

18.11.2. Small Molecules

4. **Chytil, F., and D. B. McCormick (ed.).** 1986. Vitamins and coenzymes. *Methods Enzymol.* **122:**3–425.

5. **Daniels, L., N. Bakhiet, and K. Harmon.** 1985. Widespread distribution of a 5-deazaflavin cofactor in Actinomycetes and related bacteria. *Syst. Appl. Microbiol.* **6:**12–17.

6. **Eirich, L. D., G. D. Vogels, and R. S. Wolfe.** 1978. Proposed structure for coenzyme F_{420} from *Methanobacterium. Biochemistry* **17:**4583–4593.

7. **Escalente-Semerena, J. C., K. L. Rinehart, Jr., and R. S. Wolfe.** 1984. Tetrahydromethanopterin, a carbon-carrier in methanogenesis. *J. Biol. Chem.* **259:**9447–9455.

8. **Gorris, L. G., and C. van der Drift.** 1994. Cofactor contents of methanogenic bacteria reviewed. *Biofactors* **4:**139–145.

9. **Hausinger, R. P., W. H. Orme-Johnson, and C. Walsh.** 1985. Factor 390 chromophores: phosphodiester between AMP or GMP and methanogen factor 420. *Biochemistry* **24:**1629–1633.

10. **Isabelle, D. W., D. R. Simpson, and L. Daniels.** 2002. Large-scale production of coenzyme F_{420} by using *Mycobacterium smegmatis. Appl. Environ. Microbiol.* **68:**5750–5755.

11. **McCormick, D. B., J. W. Suttie, and C. Wagner (ed.).** 1997. *Methods in Enzymology, Part I.* **279:**3–502.

12. **McCormick, D. B., J. W. Suttie, and C. Wagner (ed.).** 1997. *Methods in Enzymology, Part J.* **280:**3–496.

13. **McCormick, D. B., J. W. Suttie, and C. Wagner (ed.).** 1997. *Methods in Enzymology, Part K.* **281:**3–469.

14. **McCormick, D. B., J. W. Suttie, and C. Wagner (ed.).** 1997. *Methods in Enzymology, Part L.* **282:**3–505.

15. **Purwantini, E.** 1991. Coenzyme F_{420}: factors affecting its purification from *Methanobacterium thermoautotrophicum.* M.S. Thesis. University of Iowa, Iowa City.

16. **Schonheit, P., H. Keweloh, and R. K. Thauer.** 1981. Factor F_{420} degradation in *Methanobacterium thermoautotrophicum* during exposure to oxygen. *FEMS Microbiol. Lett.* **12:**347–349.

18.11.3. Carbohydrates

18.11.3.1. General References

17. **Colowick, J., and N. Kaplan (ed.).** 1966. Complex carbohydrate. *Methods Enzymol.* **8:**1–759.

18. **Ginsburg, V. (ed.).** 1989. Complex carbohydrates. *Methods Enzymol.* **179:**3–287.

19. **Herbert, D., P. J. Phipps, and R. E. Strange.** 1971. Chemical analysis of microbial cells, p. 209–344. *In* J. R. Norris and D. W. Ribbons (ed.), *Methods in Microbiology,* vol. 5B. Academic Press, Inc., New York, NY.

20. **Work, E.** 1971. Cell walls, p. 361–418. *In* J. R. Norris and D. W. Ribbons (ed.), *Methods in Microbiology,* vol. 5A. Academic Press, Inc., New York, NY.

18.11.3.2. Specific References

21. **Bera, A., R. Biswas, S. Herbert, and F. Gotz.** 2006. The presence of peptidoglycan O-acetyltransferase in various staphylococcal species correlates with lysozyme resistance and pathogenicity. *Infect. Immun.* **74:**4598–4604.

22. **Blumenkrantz, N., and G. Asboe-Hansen.** 1976. An assay for total hexosamine and a differential assay for glucosamine. *Clin. Biochem.* **9:**269–274.

23. **Brade, H., C. Galanos, and O. Luderitz.** 1983. Differential determination of the 3-deoxy-d-mannooctulosonic acid residues in lipopolysaccharides of *Salmonella minnesota* rough mutants. *Eur. J. Biochem.* **131:**195–200.

24. **Bryan, B. A., R. J. Linhardt, and L. Daniels.** 1986. Variation in composition and yield of exopolysaccharides produced by *Klebsiella* sp. strain K32 and *Acinetobacter calcoaceticus* BD4. *Appl. Environ. Microbiol.* **51:**1304–1308.

25. **Cadmus, M. C., C. A. Knutson, A. A. Lagoda, J. E. Pittsley, and K. A. Burton.** 1978. Synthetic media for production of quality xanthan gum in 20 liter fermentors. *Biotechnol. Bioeng.* **20:**1003–1014.

26. **Caroff, M., S. Lebbar, and L. Szabo.** 1987. Detection of 3-deoxy-2-octulosonic in thiobarbiturate-negative endotoxins. *Carbohydr. Res.* **161:**c4–c7.

27. **Chen, W.-P., J.-Y. Chen, S.-C. Chang, and C.-L. Su.** 1985. Bacterial alginate produced by a mutant of *Azotobacter vinelandii*. *Appl. Environ. Microbiol.* **49:**543–546.

28. **Cox, A. D., J. C. Wright, M. A. J. Gidney, S. Lacelle, J. S. Plested, A. Martin, E. R. Moxton, and J. C. Richards.** 2003. Identification of a novel inner-core oligosaccharide structure in *Neisseria meningitidis* lipopolysaccharide. *Eur. J. Biochem.* **270:**1759–1766.

29. **Dische, Z.** 1953. Qualitative and quantitative colorimetric determination of heptoses. *J. Biol. Chem.* **204:**983–997.

30. **Dische, Z.** 1962. Color reactions of carbohydrates. *Methods Carbohydr. Chem.* **1:**477–514.

31. **Elson, L. A., and W. T. Morgan.** 1933. A colorimetric method for the determination of glucosamine and chondrosamine. *Biochem. J.* **27:**1824–1828.

32. **Ford, J. D., and J. C. Haworth.** 1964. The estimation of galactose in plasma using galactose oxidase. *Clin. Chem.* **10:**1002–1006.

33. **Fox, A., and G. Black.** 1994. Identification and detection of carbohydrate markers for bacteria: derivatization and gas chromatography-mass spectrometry, p. 107–131. *In* C. Fenselau (ed.), *Mass Spectrometry for the Characterization of Microorganisms.* American Chemical Society, Washington, DC.

34. **Fox, G. E., L. J. Magrum, W. E. Balch, R. S. Wolfe, and C. R. Woese.** 1977. Classification of methanogenic bacteria by 16S ribosomal RNA. *Proc. Natl. Acad. Sci. USA* **74:**4537–4541.

35. **Fox, K. F., A. Fox, M. Nagpal, P. Steinberg, and K. Heroux.** 1998. Identification of *Brucella* by ribosomal-spacer-region PCR and differentiation of *Brucella canis* from other *Brucella* spp. pathogenic for humans by carbohydrate profiles. *J. Clin. Microbiol.* **36:**3217–3222.

36. **Fukagawa, K., H. Yamaguchi, D. Yonezawa, and S. Murao.** 1974. Isolation and characterization of polysaccharide produced by *Rhodotorula glutinis* K-24. *Agr. Biol. Chem.* **38:**29–35.

37. **Ghuysen, J.-M., D. J. Tipper, and J. L. Strominger.** 1966. Enzymes that degrade bacterial cell walls. *Complex Carbohydr.* **8:**685–699.

38. **Gillespie, J., S. T. Weintraub, G. G. Wong, and S. C. Holt.** 1988. Chemical and biological characterization of the lipopolysaccharide of the oral pathogen *Wolinella recta* ATCC 33238. *Infect. Immun.* **56:**2028–2035.

39. **Hacking, A. J., I. W. F. Taylor, T. R. Jarman, and J. R. W. Govan.** 1983. Alginate biosynthesis by *Pseudomonas mendocina*. *J. Gen. Microbiol.* **129:**3473–3480.

40. **Hadzija, O.** 1974. A simple method for the quantitative determination of muramic acid. *Anal. Biochem.* **60:**512–517.

41. **Ito, M., K. Ikeda, Y. Suzuki, K. Tanaka, and M. Saito.** 2002. An improved fluorometric high-performance liquid chromatography method for sialic acid determination: an internal standard method and its application to sialic acid analysis of human apolipoprotein E. *Anal. Biochem.* **300:**260–266.

42. **Kaplan, N., E. Rosenberg, B. Jann, and K. Jann.** 1985. Structural studies of the capsular polysaccharide of *Acinetobacter calcoaceticus* BD4. *Eur. J. Biochem.* **152:**453–458.

43. **Karkhanis, Y. D., J. Y. Zeltner, J. J. Jackson, and D. J. Carlo.** 1978. A new and improved microassay to determine 2-keto-3-deoxyoctonate in lipopolysaccharide of gram-negative bacteria. *Anal. Biochem.* **85:**595–601.

44. **Kosinski, E. D.** 1981. The indirect determination of lactose using a glucose analyzer with particular reference to solutions containing low levels of lactose in high levels of glucose and galactose. *J. Soc. Dairy Technol.* **43:**28–31.

45. **Lee, Y. C.** 1990. High-performance anion-exchange chromatography for carbohydrate analysis. *Anal. Biochem.* **189:**151–162.

46. **Linhardt, R. J., R. Bakhit, L. Daniels, F. Mayerl, and W. Pickenhagen.** 1989. Microbially produced rhamnolipid as a source of rhamnose. *Biotechnol. Bioeng.* **33:**365–368.

47. **Lott, J. A., and K. Turner.** 1975. Evaluation of Trinder's glucose oxidase method for measuring glucose in serum and urine. *Clin. Chem.* **21:**1754–1760.

48. **Lysenko, E., J. C. Richards, A. D. Cox, A. Stewart, A. Martin, M. Kapoor, and J. N. Weiser.** 2000. The position of phosphoryl-choline on the lipopolysaccharide of Haemophilus influenzae affects binding and sensitivity to C-reactive protein-mediated killing. *Mol. Microbiol.* **35:**234–245.

49. **Mason, M.** 1982. A new method for ethanol measurement utilizing an immobilized enzyme. *J. Am. Soc. Brew. Chem.* **40:**78–79.

50. **Mason, M.** 1983. Determination of glucose, sucrose, lactose, and ethanol in foods and beverages, using immobilized enzyme electrodes. *J. Assoc. Off. Anal. Chem.* **66:**981–984.

51. **Mukumoto, T., and H. Yamaguchi.** 1977. The chemical structure of a mannofucogalactan from the fruit bodies of *Flammulina velutipes* (Fr.) Sing. *Carbohydr. Res.* **59:**614–621.

52. **Nesser, J.-R., and T. F. Schweizer.** 1984. A quantitative determination by gas-liquid chromatography of neutral and amino sugars (as O-methyloxime acetates), and a study on hydrolytic conditions for glycoproteins and polysaccharides in order to increase sugar recoveries. *Anal. Biochem.* **142:**58–67.

53. **Osborn, M. J.** 1963. Studies on the Gram-negative cell wall. I. Evidence for the role of 2-keto-3-deoxyoctanoate in the lipopolysaccharide of *Salmonella typhimurium*. *Proc. Natl. Acad. Sci. USA* **50:**499–506.

54. **Pfeffer, J. M., H. Strating, J. T. Weadge, and A. J. Clarke.** 2006. Peptidoglycan O acetylation and autolysin profile of *Enterococcus faecalis* in the viable but nonculturable state. *J. Bacteriol.* **188:**902–908.

55. **Rocklin, R. D., and C. A. Pohl.** 1983. Determination of carbohydrates by anion exchange chromatography with pulsed amperometric detection. *J. Liquid Chromatogr.* **6:**1577–1590.

56. **Rondle, C. J. M., and W. T. J. Morgan.** 1955. The determination of glucosamine and galactosamine. *Biochem. J.* **61:**586–590.

57. **Roth, H., S. Segal, and D. Bertoli.** 1965. The quantitative determination of galactose—an enzymic method using galactose oxidase, with applications to blood and other biological fluids. *Anal. Biochem.* **10:**32–52.

58. **Sempre, J. M., C. Gancedo, and C. Asensio.** 1965. Determination of galactosamine and N-acetylgalactosamine in the presence of other hexosamines with galactose oxidase. *Anal. Biochem.* **12:**509–515.

59. **Smith, I. H., and G. W. Pace.** 1982. Recovery of microbial polysaccharides. *J. Chem. Technol. Biotechnol.* **32:** 119–129.

60. **Stewart-Tull, D. E. S.** 1968. Determination of amino sugars in mixtures containing glucosamine, galactosamine, and muramic acid. *Biochem. J.* **109:**13–18.

61. **Stork, A. D. M., H. Kemperman, D. W. Erkelens, and T. F. Veneman.** 2005. Comparison of the accuracy of the HemoCue Glucose analyzer with the Yellow Springs Instrument glucose oxidase analyzer, particularly in hypoglycemia. *Eur. J. Endocrinol.* **153:**275–281.

62. **Sweeley, C. C., W. W. Wells, and R. Bentley.** 1966. Gas chromatography of carbohydrates. *Methods Enzymol.: Complex Carbohydr.* **8:**95–107.

63. **Taylor, P. J., E. Kmetec, and J. M. Johnson.** 1977. Design, construction, and applications of a galactose selective electrode. *Anal. Chem.* **49:**789–794.

64. **Tipper, D. J.** 1968. Alkali-catalyzed elimination of D-lactic acid from muramic acid and its derivatives, and the determination of lactic acid. *Biochemistry* **7:**1441–1449.

65. **Weadge, T. J., J. M. Pfeffer, and A. J. Clarke.** 2005. Identification of a new family of enzymes with potential O-acetylpeptidoglycan esterase activity in both gram-positive and gram-negative bacteria. *BMC Microbiol.* **5:**49.

66. **Weisbach, A., and J. Hurwitz.** 1959. The formation of 2-keto-3-deoxyheptanoic acid in extracts of *E. coli* B. *J. Biol. Chem.* **234:**705–712.

18.11.4. Lipids

18.11.4.1. General References

67. **Dittmer, J. C., and M. A. Wells.** 1969. Quantitative and qualitative analysis of lipids and lipid components. *Methods Enzymol.: Lipids* **14:**482–530.

68. **Kates, M.** 1972. Techniques of lipidology isolation, analysis, and identification of lipids, p. 275–327. *In* T. S. Work and E. Work (ed.), *Laboratory Techniques in Biochemistry and Molecular Biology*, vol. 3. American Elsevier Publishing Co., Inc., New York, NY.

69. **O'Leary, W.** 1974. Chemical and physical characterization of fatty acids, p. 275–327. *In* A. I. Laskin and H. A. Lechevalier (ed.), *Handbook of Microbiology*, vol. 2. Chemical Rubber Co., Cleveland, OH.

18.11.4.2. Specific References

70. **Bird, C. W., J. M. Lynch, S. J. Pirt, W. W. Reid, C. J. Brooks, and B. S. Middleditch.** 1971. Steroids and squalene in *Methylococcus capsulatus* grown on methane. *Nature* (London) **230:**473–474.

71. **Bligh, E. G., and W. J. Dyer.** 1959. A rapid method of total lipid extraction and purification. *Can. J. Biochem. Physiol.* **37:**911–917.

72. **Brandl, H., R. A. Gross, R. W. Lenz, and R. C. Fuller.** 1988. *Pseudomonas oleovorans* as a source of poly-β-hydroxyalkanoates for potential applications as biodegradable polyesters. *Appl. Environ. Microbiol.* **54:**1977–1982.

73. **Brandl, H., R. A. Gross, R. W. Lenz, and R. C. Fuller.** 1990. Plastics from bacteria and for bacteria: poly-β-hydroxy-alkanoates as natural, biocompatible, and biodegradable polyesters. *Adv. Biochem. Eng. Biotechnol.* **41:** 77–93.

74. **Braunegg, G., B. Sonnleitner, and R. M. Lafferty.** 1978. A rapid gas chromatographic method for the determination of poly-beta-hydroxybutyric acid in microbial biomass. *Appl. Microbiol. Biotechnol.* **6:**29–37.

75. **Folch, J.** 1957. A simple method for the isolation and purification of total lipids from animal tissue. *J. Biol. Chem.* **226:**497–509.

76. **Hara, A., and N. S. Radin.** 1978. Lipid extraction with a low-toxicity solvent. *Anal. Biochem.* **90:**420–426.

77. **Hopmans, E. C., S. Schouten, R. D. Pancost, M. T. J. van der Meer, and J. S. S. Damste.** 2000. Analysis of intact tetraether lipids in archaeal cell material and sediments by high performance liquid chromatography/atmospheric pressure chemical ionization mass spectrometry. *Rapid Commun. Mass Spectrom.* **14:**585–589.

78. **Jahn, U., R. Simmons, H. Sturt, E. Grosjean, and H. Huber.** 2004. Composition of the lipids of *Nanoarchaeum equitans* and their origin from its host *Ignicoccus* sp. Strain KIN4/I. *Arch. Microbiol.* **182:**404–413.

79. **Jarute, G., A. Kainz, G. Schroll, J. R. Baena, and B. Lendl.** 2004. On-line determination of the intracellular poly(β-hydroxybutyric acid) content in transformed *Escherichia coli* and glucose during PHB production using stopped-flow attenuated total reflection FT-IR spectrometry. *Anal. Chem.* **76:**6353–6358.

80. **Johnson, A. R., and R. B. Stocks.** 1971. Gas chromatography of lipids, p. 195–218. *In* A. R. Johnson and J. B. Davenport (ed.), *Biochemistry and Methodology of Lipids*. Wiley-Interscience, New York, NY.

81. **Kates, M.** 1993. Membrane lipids of archaea, p. 261–295. *In* M. Kates, D. J. Kushner and A. T. Matheson (ed.), *The Biochemistry of Archaea*. Elsevier, Amsterdam, The Netherlands.

82. **Kates, M.** 1964. Simplified procedure for hydrolysis or methanolysis of lipids. *J. Lipid Res.* **5:**132–135.

83. **Kates, M., S. C. Kushwaha, and G. D. Sprott.** 1982. Lipids of purple membrane from extreme halophiles and of methanogenic bacteria. *Methods Enzymol.* **88:**98–111.

84. **Kates, M., L. S. Yengoyan, and P. S. Sastry.** 1965. A diether analog of phosphytidyl glycerophosphate in *Halobacterium cutirubrum*. *Biochim. Biophys. Acta* **98:**252–268.

85. **Koga, Y., and H. Morii.** 2005. Recent advances in structural research on ether lipids from Archaea including comparative and physiological aspects. *Biosci. Biotechnol. Biochem.* **69:**2019–2034.

86. **Kushwaha, S. C., M. Kates, G. D. Sprott, and I. C. P. Smith.** 1981. Novel complex polar lipids from the methanogenic archaebacterium *Methanospirillum hungatei*. *Science* **211:**1163–1164.

87. **Langworthy, T. A.** 1977. Comparative lipid composition of heterotrophically and autotrophically grown *Sulfolobus acidocaldarius*. *J. Bacteriol.* **130:**1326–1332.

88. **Law, J. H., and R. A. Slepecky.** 1961. Assay of poly-β-hydroxybutyric acid. *J. Bacteriol.* **82:**33–36.

89. **Lenz, R. W., B. W. Kim, H. W. Ulmer, K. Fritzche, E. Knee, and R. C. Fuller.** 1990. Functionalized poly-β-hydroxyalkanoates produced by bacteria, p. 23–35. *In* E. A. Dawes (ed.), *Novel Biodegradable Microbial Polymers*. Kluwer Academic Publishers, Dordrecht, The Netherlands.

90. **Paton, J. C., E. J. McMurchie, B. K. May, and W. H. Elliot.** 1978. Effect of growth temperature on membrane fatty acid composition and susceptibility to cold shock of *Bacillus amyloliquefaciens*. *J. Bacteriol.* **135:**754–759.

91. **Patt, T. E., and R. S. Hanson.** 1978. Intracytoplasmic membrane, phospholipid, and sterol content of *Methylobacterium organophilum* cells grown under different conditions. *J. Bacteriol.* **134:**636–644.

92. **Rilfors, L., A. Wieslander, and S. Stahl.** 1978. Lipid and protein composition of membranes of *Bacillus megaterium* variants in the temperature range of 5 to 70°C. *J. Bacteriol.* **135:**1043–1052.

93. **Shamala, T. R., A. Chandrashekar, S. V. N. Vijayendra, and L. Kshama.** 2003. Identification of polyhydroxyalka-

noate (PHA)-producing *Bacillus* spp. using the polymerase chain reaction (PCR). *J. Appl. Microbiol.* **94:**369–374.

94. **Skipski, V. P., and M. Barlday.** 1969. Detection of lipids on thin-layer chromatograms. *Methods Enzymol.: Lipids* **14:**541–548.

95. **Stein, R. A., V. Slawson, and J. F. Mead.** 1976. Gas-liquid chromatography of fatty acids and derivatives, p. 857–896. *In* G. V. Marinetti (ed.), *Lipid Chromatographic Analysis*, 2nd ed. Marcel Dekker, Inc., New York, NY.

96. **Tornabene, T. G., and T. A. Langworthy.** 1979. Diphytanyl and dibiphytanyl glycerol ether lipids of methanogenic archaebacteria. *Science* **203:**51–53.

97. **Tornabene, T. G., R. S. Wolfe, W. E. Balch, C. Holzer, G. E. Fox, and J. Oro.** 1978. Phytanyl-glycerol ethers and squalenes in the archaebacterium *Methanobacterium thermoautotrophicum*. *J. Mol. Evol.* **11:**259–266.

98. **Virtue, P., P. D. Nichols, and P. I. Boon.** 1996. Simultaneous estimation of microbial phospholipids fatty acids and diether lipids by capillary gas chromatography. *J. Microbiol. Methods* **25:**177–185.

99. **Vorbeck, M. L., and G. V. Marinetti.** 1965. Intracellular distribution and characterization of the lipids of *Streptococcus faecalis*. *Biochemistry* **4:**296–305.

100. **Welch, D. F.** 1991. Applications of fatty acid analysis. *Clin. Microbiol. Rev.* **4:**422–438.

101. **Wuthier, R. E.** 1966. Purification of lipids from non-lipid contaminants on Sephadex bead columns. *J. Lipid Res.* **7:**558–561.

18.11.5. Cell Wall Polymers

18.11.5.1. General References

102. **Braun, V.** 1978. Structure-function relationship of the Gram-negative bacterial cell envelope. *Symp. Soc. Gen. Microbiol.* **28:**111–138.

103. **Braun, V., and K. Hantke.** 1974. Biochemistry of bacterial cell envelopes. *Annu. Rev. Biochem.* **43:**89–121.

104. **Costerton, J. W., J. M. Ingram, and K. J. Cheng.** 1974. Structure and function of the cell envelope of gram-negative bacteria. *Bacteriol. Rev.* **38:**87–110.

105. **Cummins, C. S.** 1974. Bacterial cell wall structures, p. 251–284. *In* A. I. Laskin and H. A. Lechevalier (ed.), *Handbook of Microbiology*, vol. 2. *Microbial Composition.* Chemical Rubber Co., Cleveland, OH.

106. **Elin, R. J., and S. M. Wolff.** 1974. Bacterial endotoxin, p. 674–731. *In* A. I. Laskin and H. A. Lechevalier (ed), *Handbook of Microbiology*, vol. 2. *Microbial Composition.* Chemical Rubber Co., Cleveland, OH.

107. **Ghuysen, J. M.** 1968. Use of bacteriolytic enzymes in determination of wall structure and their role in cell metabolism. *Bacteriol. Rev.* **32:**425–464.

108. **Leive, L. (ed.).** 1973. *Bacterial Membranes and Walls.* Marcel Dekker, Inc., New York, NY.

109. **Luderitz, O., A. M. Staub, and O. Westphal.** 1966. Immunochemistry of O and R antigens of *Salmonella* and related *Enterobacteriacae*. *Bacteriol. Rev.* **30:**192–255.

110. **Milner, K. C., J. A. Rudbach, and E. Ribi.** 1971. General characteristics, p. 1–65. *In* G. Weinbaum, S. Kadis, and S. Ajl (ed.), *Microbial Toxins: Bacterial Endotoxins*, vol. 4. Academic Press, Inc., New York, NY.

111. **Reaveley, D. A., and R. E. Burge.** 1972. Walls and membranes in bacteria. *Adv. Microb. Physiol.* **7:**1–81.

112. **Rogers, H. J., J. B. Ward, and I. D. J. Burdett.** 1978. Structure and growth of the walls of Gram-positive bacteria. *Symp. Soc. Gen. Microbiol.* **28:**139–176.

113. **Schleifer, K. H., and O. Kandler.** 1972. Peptidoglycan types of bacterial cell walls and their taxonomic implications. *Bacteriol. Rev.* **36:**407–477.

18.11.5.2. Specific References

114. **Cox, A. D., J. C. Wright, M. A. J. Gidney, S. Lacelle, J. S. Plested, A. Martin, E. R. Moxton, and J. C. Richards.** 2003. Identification of a novel inner-core oligosaccharide structure in *Neisseria meningitidis* lipopolysaccharide. *Eur. J. Biochem.* **270:**1759–1766.

115. **Davidson, A. L., and J. Badiley.** 1964. Glycerol teichoic acids in walls of *Staphylococcus epidermidis*. *Nature* (London) **202:**874.

116. **Ghuysen, J. M., D. Tipper, and J. L. Strominger.** 1965. Structure of the cell wall of *Staphylococcus epidermidis* strain Copenhagen. IV. The soluble glycopeptide and its sequential degradation by peptidase. *Biochemistry* **4:** 2245–2256.

117. **Kandler, O., and H. Hippe.** 1977. Lack of peptidoglycan in the cell walls of *Methanosarcina barkeri*. *Arch. Microbiol.* **113:**57–60.

118. **Kandler, O., and H. Konig.** 1978. Chemical composition of the peptidoglycan-free cell walls of methanogenic bacteria. *Arch. Microbiol.* **118:**141–152.

119. **Kandler, O., and H. Konig.** 1985. Cell envelopes of archaebacteria, p. 413–457. *In* C. R. Woese and R. S. Wolfe (ed.), *The Bacteria*, vol. 8. Academic Press, Orlando, FL.

120. **Knox, K. W.** 1966. The relation of 3-deoxy-2-oxo-octanoate to the serological and physical properties of a lipopolysaccharide from a rough strain of *Escherichia coli*. *Biochem. J.* **100:**73–78.

121. **Kottel, R. H., K. Bacon, D. Clutter, and D. White.** 1975. Coats from *Myxococcus xanthus*: characterization and synthesis during myxospore differentiation. *J. Bacteriol.* **124:**550–557.

122. **Osborn, M. J.** 1963. Studies on the Gram-negative cell wall. *Proc. Natl. Acad. Sci. USA* **50:**499–514.

123. **Raetz, C. R. H.** 1990. Biochemistry of endotoxins. *Annu. Rev. Biochem.* **59:**129–170.

124. **Strominger, J. L., and D. J. Tipper.** 1974. Structure of bacterial cell walls: the lysozyme substrate, p. 169–184. *In* E. F. Osserman, R. E. Canfield, and S. Beychok (ed.), *Lysozyme*. Academic Press, Inc., New York, NY.

125. **Taylor, A., K. W. Knox, and E. Work.** 1966. Chemical and biological properties of an extracellular liposaccharide from *Escherichia coli* grown under lysine-limiting conditions. *Biochem. J.* **99:**53–61.

126. **Warth, A. W., and J. L. Strominger.** 1969. Structure of the peptidoglycan of bacterial spores: occurrence of the lactam of muramic acid. *Proc. Natl. Acad. Sci. USA* **64:** 528–535.

127. **Weidel, W., H. Frank, and H. H. Martin.** 1960. The rigid layer of the cell wall of *Escherichia coli* strain B. *J. Gen. Microbiol.* **22:**158–166.

128. **Westphal, O., and K. Jan.** 1965. Bacterial lipopolysaccharide extraction with phenol-water and further applications of the procedure. *Methods Carbohydr. Chem.* **5:** 83–91.

129. **Work, E.** 1971. Cell walls, p. 361–418. *In* J. R. Norris and D. W. Ribbons (ed.), *Methods in Microbiology*, vol. 5A. Academic Press, Inc., New York, NY.

18.11.6. Nucleic Acids

18.11.6.1. General References

130. **DeLey, J.** 1971. The determination of the molecular weight of DNA per bacterial nucleoid, p. 301–311. *In* J. R. Norris and D. W. Ribbons (ed.), *Methods in Microbiology*, vol. 5A. Academic Press, Inc., New York, NY.

131. **Parish, J. H.** 1972. *Principles and Practice of Experiments with Nucleic Acids.* Longman Group, Ltd., London, United Kingdom.

18.11.6.2. Specific References

132. **Brunk, C. F., K. C. Jones, and T. W. James.** 1979. Assay for nanogram quantities of DNA in cellular homogenates. *Anal. Biochem.* **92:**497–500.

133. **Burton, K.** 1957. A study of the conditions and mechanism of the diphenylamine reaction for the calorimetric estimation of deoxyribonucleic acid. *Biochem. J.* **62:**315–323.

134. **Cesarome, C., C. Bolognesi, and L. Santi.** 1979. Improved microfluorometric DNA determination in biological material using 33258 Hoechst. *Anal. Biochem.* **179:**401–403.

135. **Gallagher, S.** 1989. Quantitation of DNA and RNA with absorption and fluorescence spectroscopy, p. A.3.9–A.3.15. *In* F. A. Ausubel, R. Brent, R. E. Kingston, D. D. Moore, J. M. Seidman, J. A. Smith, and K. Struhl (ed.), *Current Protocols in Molecular Biology,* supplement 8. Greene Publishing Associates and John Wiley & Sons, Inc., New York, NY.

136. **Griswold, B. L., E. L. Humoller, and A. R. McIntyre.** 1951. Inorganic phosphates and phosphate esters in tissue extracts. *Anal. Chem.* **23:**192–194.

137. **Labarca, C., and K. Paigen.** 1980. A simple, rapid, and sensitive DNA assay procedure. *Anal. Biochem.* **102:**344–352.

138. **Sambrook, J., and D. W. Russell.** 2001. *Molecular Cloning: a Laboratory Manual,* 3rd ed. Cold Spring Harbor Laboratory Press, Cold Spring Harbor, NY.

139. **Tilzer, L., S. Thomas, and R. F. Moreno.** 1989. Use of silica gel polymer for DNA extraction with organic solvents. *Anal. Biochem.* **183:**13–15.

18.11.7. Nitrogen Components

18.11.7.1. General References

140. **Allen, S. E., H. M. Grimshaw, J. A. Parkinson, and C. Quarmby.** 1974. Inorganic constituents: nitrogen, p. 184–206. *In* S. E. Allen (ed.), *Chemical Analysis of Ecological Materials.* Blackwell Scientific Publications, London, United Kingdom.

141. **Black, C. A. (ed.).** 1965. *Methods of Soil Analysis,* part 2. *Chemical and Microbiological Properties.* American Society of Agronomy, Inc., Madison, WI.

142. **Bremner, J. M., and D. R. Keeney.** 1965. Steam distillation methods for determination of ammonium, nitrate, and nitrite. *Anal. Chim. Acta* **32:**485–495.

143. **Cooper, T. C.** 1977. *The Tools of Biochemistry.* John Wiley & Sons, Inc., New York, NY.

144. **Davis, B. D., R. Dulbecco, H. N. Eisen, and H. S. Ginsberg.** 1990. *Microbiology,* p. 45–49. J. B. Lippincott, Philadelphia, PA.

145. **Franson, M. A. (ed.).** 1976. *Standard Methods for the Examination of Water and Wastewater,* 14th ed. American Public Health Association, Washington, DC.

146. **Gudinowlcz, B. J., M. J. Gudlnowlcz, and H. F. Martin.** 1976. *Fundamentals of Integrated GC-MS.* Part II. *Mass Spectrometry.* Marcel Dekker, Inc., New York, NY.

147. **Horowitz, W. (ed.).** 1975. *Official Methods of Analyses of the Association of Official Analytical Chemists,* 12th ed. Association of Official Analytical Chemists, Washington, DC.

148. **Jackson, M. L.** 1958. *Soil Chemical Analysis.* Prentice-Hall, Inc., Englewood Cliffs, NJ.

149. **Safe, S., and O. Hutzinger.** 1973. *Mass Spectrometry of Pesticides and Pollutants.* CRC Press, Inc., Cleveland, OH.

150. **Strickland, J. D. H., and T. R. Parsons.** 1960. *A Practical Handbook of Seawater Analysis.* Fisheries Research Board of Canada, Ottawa, Canada.

151. **Waller, G. R. (ed.).** 1972. *Biochemical Applications of Mass Spectrometry.* John Wiley & Sons, Inc., New York, NY.

18.11.7.2. Specific References

152. **Armstrong, F. A. J.** 1963. Determination of nitrate in water by ultraviolet spectrophotometry. *Anal. Chem.* **35:**1292.

153. **Baca, P., and H. Freiser.** 1977. Determination of trace levels of nitrates by an extraction-photometric method. *Anal. Chem.* **49:**2249–2250.

154. **Belay, N., R. Sparling, B.-S. Choi, M. Roberts, J. E. Roberts, and L. Daniels.** 1988. Physiological and ^{15}N-NMR analysis of molecular nitrogen fixation by *Methanococcus thermolithotrophicus, Methanobacterium bryantii,* and *Methanospirillum hungatei. Biochim. Biophys. Acta* **971:**233–245.

155. **Bio-Rad Laboratories.** 1994. *Bio-Rad DC Protein Assay.* Bulletin 1731. Bio-Rad Laboratories, Richmond, CA.

156. **Bio-Rad Laboratories.** 2005. *RC DC Protein Assay.* Bulletin 2610. Bio-Rad Laboratories, Richmond, CA.

157. **Bio-Rad Laboratories.** 2003. *Colorimetric Protein Assays.* Tech note 1069. Bio-Rad Laboratories, Richmond, CA.

158. **Bradford, M. M.** 1976. A rapid and sensitive method for the quantitation of microgram quantities of protein utilizing the principle of protein-dye binding. *Anal. Biochem.* **72:**248–254.

159. **Bremner, J. M.** 1960. Determination of nitrogen in soil by the Kjeldahl method. *J. Agric. Sci.* **55:**11–33.

160. **Canney, P. J., D. E. Armstrong, and J. H. Wiersma.** 1974. *Determination of Nitrite and Nitrate Ions in Natural Waters Using Aromatics or Diamines as Reagents.* Technical Report. University of Wisconsin Water Resources Center, Madison.

161. **Chial, H. J., H. B. Thompson, and A. G. Splittgerber.** 1993. A spectral study of the charge forms of Coomassie blue G. *Anal. Biochem.* **209:**258–266.

162. **Chow, T. J., and M. S. Johnstone.** 1962. Determination of nitrate in sea water. *Anal. Chim. Acta* **27:**441–446.

163. **Coakley, W. T., and C. J. James.** 1978. A simple linear transformation for the Folin-Lowry protein calibration curve to 1.0 mg/ml. *Anal. Biochem.* **85:**90–97.

164. **Gornall, A. G., C. S. Bardawill, and M. M. David.** 1949. Determination of serum protein by means of the Biuret reaction. *J. Biol. Chem.* **177:**751–766.

165. **Guiraud, G., J. C. Fardeau, G. Llimous, and M. A. Barral.** 1977. Determination of nitrate in soils and plants by the Kjeldahl method. *Ann. Agron.* **28:**329–333.

166. **Hardy, R. W. F., and A. H. Gibson (ed.).** 1977. *A Treatise on Dinitrogen Fixation.* Section IV. *Agronomy and Ecology.* John Wiley & Sons, Inc., New York, NY.

167. **Hill, H. D., and J. G. Straka.** 1988. Protein determination using bicinchoninic acid in the presence of sulfhydryl reagents. *Anal. Biochem.* **170:**203–208.

168. **Janssen, F. W., A. J. Lund, and L. E. Anderson.** 1957. Colorimetric assay for dipicolinic acid in bacterial spores. *Science* **127:**26–27.

169. **Kalb, V. F., Jr., and R. W. Bernlohr.** 1977. A new spectrophotometric assay for protein in cell extracts. *Anal. Biochem.* **82:**362–371.

170. **Kalckar, H. M.** 1947. Differential spectrophotometry of purine compounds by means of specific enzymes. I. Determination of hydroxypurine compounds. *J. Biol. Chem.* **167:**429–475.

171. **Kruse, J. M., and M. G. Mellon.** 1953. Colorimetric determination of ammonia and cyanate. *Anal. Chem.* **25:**1188–1192.

172. **Layne, E.** 1957. Spectrophotometric and turbidometric methods for measuring proteins. *Methods Enzymol.* **3:**447–454.

173. **Lowry, O. H., N. J. Rosebrough, A. L. Farr, and R. J. Randall.** 1951. Protein measurement with the Folin phenol reagent. *J. Biol. Chem.* **193:**265–275.

174. **Markwell, M. A., S. M. Haas, L. L. Bieber, and N. E. Tolbert.** 1978. A modification of the Lowry procedure to

simplify protein determination in membrane and lipoprotein samples. *Anal. Biochem.* **87**:206–210.

175. **Martin, H. H., and J. W. Foster.** 1958. On the chromatographic behavior of dipicolinic acid. *Arch. Mikrobiol.* **31**:171–178.

176. **Matsunaga, K., and M. Nishimura.** 1969. Determination of nitrate in sea water. *Anal. Chim. Acta* **45**:350–353.

177. **McClard, R. W.** 1981. Removal of sulfhydryl groups with 1,3,4,6-tetrachloro-3a-6a-diphenylglycoluril: application to the assay of protein in the presence of thiol reagents. *Anal. Biochem.* **112**:278–281.

178. **Mubarak, A., R. A. Howald, and R. Woodriff.** 1977. Elimination of chloride interferences with mercuric ions in the determination of nitrates by the phenol disulfonic acid method. *Anal. Chem.* **49**:857–860.

179. **Murrell, W. G.** 1969. Chemical composition, p. 249–251. *In* G. W. Gould and A. Hurst (ed.), *The Bacterial Spore*. Academic Press, Inc., New York, NY.

180. **Murrell, W. G., and A. D. Warth.** 1965. Composition and heat resistance of bacterial spores, p. 1–24. *In* L. L. Campbell and H. O. Halvorson (ed.), *Spores III*. American Society for Microbiology, Washington, DC.

181. **Nakamura, H., and J. J. Pisano.** 1976. Sensitive fluorometric assay for proteins. Use of fluorescamine and membrane filters. *Arch. Biochem. Biophys.* **172**:102–105.

182. **Pierce Chemical Co.** 2003. *Bicinchoninic Acid (BCA) Method.* Pierce Chemical Co., Rockford, IL.

183. **Pierce Chemical Co.** 2003. *The Lowry Method.* Pierce Chemical Co., Rockford, IL.

184. **Pierce Chemical Co.** 2003. *Coomassie Dye Binding Method.* Pierce Chemical Co., Rockford, IL.

185. **Pierce, J., and C. H. Suelter.** 1977. An evaluation of the Coomassie brilliant blue G-250 dye-binding method for quantitative protein determination. *Anal. Biochem.* **81:** 478–480.

186. **Postgate, J. R. (ed.).** 1971. *The Chemistry and Biochemistry of Nitrogen Fixation.* Plenum Press, New York, NY.

187. **Raganowicz, E., and A. Niewiadomy.** 1976. Colorimetric determination of the nitrites content in water by means of 2-sulfanilamidothiazole method. *Pol. Arch. Hydrobiol.* **23**:1–4.

188. **Rexroad, P. R., and R. D. Cathey.** 1976. Pollution-reduced Kjeldahl method for a crude protein. *J. Assoc. Off. Anal. Chem.* **59**:1213–1217.

189. **Russel, J. A.** 1944. The colorimetric estimation of small amounts of ammonia by the phenol-hypochlorite reaction. *J. Biol. Chem.* **156**:457–461.

190. **Sedmak, J. J., and S. E. Grossberg.** 1977. A rapid, sensitive, and versatile assay for protein using Coomassie brilliant blue G-250. *Anal. Biochem.* **79**:544–552.

191. **Smith, P. K., R. I. Krohn, G. T. Hermanson, A. K. Mallia, F. H. Gartner, M. D. Provenzano, E. K. Fujimoto, N. M. Goeke, B. J. Olson, and D. C. Klenk.** 1985. Measurement of protein using bicinchoninic acid. *Anal. Biochem.* **150**:76–85.

192. **Snell, F. D., and C. T. Snell.** 1945. *Colorimetric Methods of Analysis.* D. Van Nostrand Co., Inc., New York, NY.

193. **Stickland, H. L.** 1951. The determination of small quantities of bacteria by means of the Biuret reaction. *J. Gen. Microbiol.* **5**:698–703.

194. **Warburg, O., and W. Christian.** 1942. Isolierung und Kristallisation des Garungsferments enolase. *Biochem. Z.* **310**:384–421.

195. **Warth, A. D., D. F Ohye, and W. C. Murrell.** 1963. The composition and structure of bacterial spores. *J. Cell Biol.* **16**:579–592.

196. **Work, E.** 1963. α,ε-Diaminopimelic acid. *Methods Enzymol.* **6**:624–634.

18.11.8. Minerals

18.11.8.1. General References

197. **Christian, G. D., and F. J. Feldman.** 1970. *Atomic Absorption Spectroscopy. Applications in Agriculture, Biology and Medicine.* Wiley-Interscience, New York, NY.

198. **Heirman, R., and C. T. J. Allkemade.** 1963. *Chemical Analysis by Flame Photometry,* 2nd ed. Translated by P. T. Gilbert, Jr. Interscience Publishers, Inc., New York, NY.

199. **Morrison, G. H. (ed.).** 1956. *Trace Analysis.* Interscience Publishers, Inc., New York, NY.

200. **Snell, F. D., and C. T. Snell.** 1963. *Colorimetric Methods of Analysis,* vol. 2, 3rd ed. D. Van Nostrand Co., Inc., Princeton, NJ.

201. **U.S. Environmental Protection Agency.** 1974. *Methods for Chemical Analysis of Water and Wastes.* Office of Technology Transfer, Washington, DC.

202. **Weinberg, E. D. (ed.).** 1977. *Microorganisms and Minerals.* Marcel Dekker, Inc. New York, NY.

18.11.8.2. Specific References

203. **Stewart, M., A. P. Somlyo, A. V. Somlyo, H. Shuman, J. A. Lindsay, and W. C. Murrell.** 1980. Distribution of calcium and other elements in cryosectioned *Bacillus cereus* T spores, determined by high-resolution scanning electron probe X-ray microanalysis. *J. Bacteriol.* **143**:481–491.

19

Enzymatic Activity

ROBERT P. HAUSINGER AND ALLEN T. PHILLIPS

The study of enzymes and enzyme-catalyzed reactions has contributed greatly to our understanding of microbial metabolism, just as current advances in protein structure promise new insights into the detailed workings of enzymes at the molecular level. Nevertheless, there remain enormous gaps in our knowledge regarding the biochemistry and physiology of all but the best-studied bacteria, and thus, a thorough analysis of enzymes, their action, and their properties continues to be an important effort for bacteriologists. This chapter is intended to provide a brief description of the most important principles and approaches used in enzyme activity measurements. In addition, the discussion presents practical information useful to investigators who seek to design enzyme assays for performing meaningful and accurate evaluations of catalytic activities.

19.1. PRINCIPLES

19.1.1. Units and Specific Activity

Although most enzymes are proteins (with the exception of catalytic RNA species) and can be quantified by protein-specific methods, it is almost invariably useful to express the quantity of an enzyme present in terms of its catalytic activity rather than the content of protein. This has the advantage of relative specificity, assuming no interfering activities are present, and measurements can usually be done in impure systems containing numerous other protein species as well as on purified preparations. The most common definition of one **unit (U) of enzyme activity** is the amount of enzyme which catalyzes the conversion of 1 μmol of substrate to product in 1 min under a prescribed set of assay conditions, i.e., at a stated pH, temperature, buffer type and concentration, and substrate and cofactor concentrations. This definition is widely used and is applicable to most situations. When the resulting numbers are inconveniently large or small, multipliers such as kilounits or milliunits are preferable to defining a nonstandard unit. An exception to this general practice exists for the case when the molecular weight of a substrate is unknown, and thus, a micromole quantity cannot be specified. In this instance it is acceptable to use some convenient mass of substrate converted to product per minute as the basis for the unit definition. (Another unit occasionally encountered is the katal, the amount of enzyme that converts 1 mol of substrate to product in 1 s. One enzyme unit as described above is equal to 16.67 nkat.)

Whereas an enzyme unit is a measure of quantity expressed in catalytic terms, the relative purity of an enzyme is specified by its **specific activity.** Specific activity typically is indicated as the number of enzyme units present per milligram of protein. Since estimates of protein content are highly influenced by the method used for their determination, values of specific activity should always be accompanied by information about how the protein is quantified (both the method and the reference protein used) as well as how an enzyme unit is defined. For cell extracts and other impure preparations, simple colorimetric protein assays (e.g., Lowry, Bradford, and bicinchoninic acid methods; see reference 4) are suitable. These methods yield widely variable extinction coefficients for different purified protein samples; hence, it is necessary to determine the appropriate correction factor for use with the isolated protein of interest. Precise measurements of purified protein concentrations are obtained by quantitative amino acid analysis or by using the molar absorption at 280 nm calculated on the basis of the numbers of tryptophan, tyrosine, and cystine residues in the unfolded protein (26). Comparisons of enzyme specific activities among cells grown under different medium conditions can provide important information about the relative levels of biosynthesis of the enzyme of interest.

19.1.2. Purification Tables and k_{cat}

For a given enzyme, the specific activity increases during purification up to a limit value at which all protein present is the active enzyme. For reporting the results of enzyme isolation, data should be presented in a **purification table** that includes total units and specific activities at each stage of purification. Typically, such a table also includes values for enzyme recovery and increase in purification ("fold" purification) for each step. The percent recovery is usually calculated on the basis of remaining total units versus the total units measured in crude cell extracts, whereas the fold purification is calculated by dividing the specific activity at each step by the specific activity of the initial sample. These values are useful in identifying, and overcoming, steps that cause substantial losses of enzyme activity (e.g., due to protein instability, loss of a cofactor, or other reason). In addition, the final fold purification can give insight into the cellular abundance of the particular enzyme of interest. For example, a highly overproduced enzyme comprising 10% of the total cell protein correspondingly would have a maximum of 10-fold purification. By contrast, most enzymes are present at much lower abundance and may require several hundred fold, or higher, purification.

An important term for describing the activity of an isolated enzyme is k_{cat}, the **molecular activity** of the enzyme (also referred to as its **turnover number**). This number describes the molecules of substrate(s) consumed per minute per molecule of enzyme and represents a composite of rate constants that limit reaction velocity. It has units of reciprocal minutes and is calculated from the specific activity at saturating substrate concentration(s) divided by the molecular weight of the enzyme. The reciprocal of k_{cat} is the time required for one catalytic cycle. For example, an enzyme with an M_r of 30,000 (30 mg/μmol) and a measured specific activity of 8,000 U/mg has a k_{cat} of 2.4×10^5 min^{-1} and thus completes each catalytic cycle in 250 μs.

19.1.3. Reaction Classification and Nomenclature

The current system of enzyme nomenclature recognizes six classes of enzymes and defines the basis for a systematic name for each enzyme. A listing of enzymes along with their recommended names and numerical **enzyme codes (EC numbers)** is periodically updated by the International Union of Biochemistry and Molecular Biology (http://www.chem.qmul.ac.uk/iubmb/enzyme/) and should be consulted when preparing results of enzyme studies for publication. The EC number (e.g., 3.2.1.23) is made up of four parts: the first decimal indicates the enzyme's class; the next two decimals classify the enzyme according to type of bonds broken, coenzymes used, and other important characteristics; and the final decimal is a serial number used to distinguish among enzymes otherwise classified together. Because EC numbers provide generally unambiguous identification of enzymes, they are of increasing importance in assisting the cataloging of sequence data for computer searches. A brief explanation of the six enzyme classes follows.

Class 1: Oxidoreductases. Enzymes in class 1 catalyze oxidation-reduction reactions and are assigned systematic names based on the form **hydrogen donor:acceptor oxidoreductase.** The name **dehydrogenase** is recommended for normal use, but **reductase** is acceptable in cases where this more correctly describes the usual reaction direction; **oxygenase** or **oxidase** is used when O_2 is the acceptor species. Thus, the NAD$^+$-dependent isocitrate dehydrogenase of the tricarboxylic acid cycle has the assigned systematic name isocitrate:NAD$^+$ oxidoreductase (decarboxylating), EC 1.1.1.41.

Class 2: Transferases. Class 2 enzymes catalyze group transfer from a donor to an acceptor. Accordingly, the systematic designation for the enzyme catalyzing the ATP-dependent phosphorylation of adenosine to yield AMP + ADP is ATP:adenosine 5′-phosphotransferase, EC 2.7.1.20, more commonly known as adenosine kinase. Enzymes catalyzing transamination (aminotransferases) and hexose transfer (e.g., glycogen synthase, with its systematic name UDP-glucose:glycogen 4-α-D-glucosyltransferase, EC 2.4.1.11) are also included in this class.

Class 3: Hydrolases. Class 3 is a broad category that covers enzymes catalyzing the hydrolytic cleavage of a substrate. These enzymes comprise a special group of transferases in which the acceptor group is water. In cases such as proteolytic enzymes (carboxypeptidase or collagenase) or phosphatases, specificity is usually directed toward a family of related substrates containing common structural features rather than specific compounds. Although their systematic names always include the term hydrolase, as in urea amidohydrolase (EC 3.5.1.5), members of this class are frequently referred to by the name of their substrate with the suffix -ase (e.g., urease, penicillinase, and β-galactosidase). Also found in this class are proteinases (proteases), many with nonstandard but still widely used names, such as trypsin and papain.

Class 4: Lyases. Class 4 enzymes catalyze cleavage of C-C, C-O, and C-N bonds (plus a few others) with elimination of a group to leave an unsaturated residue; the converse addition of groups to double bonds is included as well. Most lyase reactions are referred to as uni-bi type, meaning that they involve conversion of a single substrate to two products. Their systematic names are based on the form **substrate group-lyase**, as in *S*-adenosylmethionine carboxy-lyase, EC 4.1.1.50 (also known as a **decarboxylase**), and L-malate hydro-lyase, commonly called fumarase or fumarate hydratase. Several enzymes involving coenzyme A (CoA)-dependent cleavages, as in citrate synthase, or addition of another element concomitant with substrate cleavage (e.g., anthranilate synthase) are less obvious members of the lyase family.

Class 5: Isomerases. Class 5 enzymes catalyze geometric or structural changes within one molecule. Class 5 includes enzymes variously known as **racemases**, **epimerases**, **isomerases**, **tautomerases**, and **mutases**. There is no single form for the systematic name, and thus, one finds alanine racemase (interconverting D- and L-alanine, EC 5.1.1.1) as well as D-ribulose-5-phosphate 3-epimerase (EC 5.1.3.1) and D-glyceraldehyde-3-phosphate ketolisomerase (EC 5.3.1.1) (better recognized as triosephosphate isomerase).

Class 6: Ligases (Synthetases). Ligases catalyze the joining of two molecules, coupled with the hydrolysis of a pyrophosphate bond from ATP or similar triphosphate. Their systematic names are based on the convention substrate 1:substrate 2 ligase (ADP-forming). The recommended class name is **synthetase**, not to be confused with **synthase**, which simply describes an enzyme that joins parts of two molecules without involvement of pyrophosphate bond cleavage. Examples of synthetases are pyruvate carboxylase (pyruvate:carbon-dioxide ligase [ADP-forming], EC 6.4.1.1), glutamine synthetase [(L-glutamate:ammonia ligase [ADP-forming], EC 6.3.1.2), and leucyl-tRNA synthetase (L-leucine:tRNA^{Leu} ligase [AMP-forming], EC 6.1.1.4).

19.1.4. Cofactors, Coenzymes, and Metal Ions

Many enzyme reactions require substances other than the protein and substrate for optimal activity. These accessory materials, termed **cofactors**, are in most cases unchanged at the end of a catalytic cycle and are regarded as important (often essential) parts of the reaction mechanism. One important class of cofactors involves **coenzymes**, organic molecules that have a precisely defined role in the reaction. In some cases, these coenzymes are tightly bound (e.g., flavin and cobalamin) or even covalently bonded (e.g., biotin and lipoic acid) to the enzyme protein and are known as **prosthetic groups**. Alternatively, the coenzyme may be more loosely associated with the enzyme of interest (e.g., pyridoxal phosphate) and can be lost during purification. Many investigators refer to compounds such as ATP, NAD$^+$, and CoA as cellular coenzymes, since they are regenerated by other enzyme systems within the cell, even though they are modified during the specific enzyme reaction. When working with the isolated enzymes, such compounds must be treated as substrates unless a regenerating system is provided. Metal ions comprise a second large group of cofactors. It has been estimated that more than one-third of all enzymes possess bound metal ions, often with the metals playing a critical role in substrate binding, electron transfer, or protein stabilization. Although substitution with related metal ions often is possible, metal-containing enzymes typically are highly specific to a particular metal ion due to structural constraints on the size or allowed coordination geometry. Metalloenzymes include examples with a wide variety of mononuclear, dinuclear, and polynuclear metallocenters (10). Additional cellular substances also can act as enzyme activators by enhancing catalysis or stabilizing enzyme structure. Cellular control of the levels of these substances may allow for allosteric regulation of selected enzymes (see section 19.4.3.1).

19.1.5. Unknown Substrates

Microbiologists increasingly are faced with the challenge of characterizing gene products suspected of being enzymes, but for which the reactions catalyzed are unknown. Annotations derived from genomics studies often are flawed or misleading, and even protein structures obtained from proteomics studies can fail to assign proper function. Nevertheless, the results of such studies should be carefully scrutinized as a first step to identifying the true substrate. A target gene sequence may be related to those of family members that catalyze mechanistically similar reactions or that use structurally similar substrates (7). Examination of flanking DNA regions may reveal sequences encoding enzymes catalyzing other steps in a pathway and thus help to define the role of the gene of interest. Structural information about a gene product should be compared to known three-dimensional structures, again with the goal of seeking features that provide mechanistic insight or substrate preferences. Gene array or other approaches to assess the effects of varied growth conditions on gene expression can yield additional hints about function. After compiling such information, reasonable guesses of possible substrates will need to be tested by appropriately designed assays. If transformation of any compound is detected, structural variations of that compound can be tested. Comparison of possible substrates typically involves analysis of the **catalytic efficiency** (V_{max}/K_m), a ratio of kinetic constants described in section 19.4, that accounts for both the rate of the reaction and the affinity of the enzyme for the test substrate. This value is also known as the **specificity constant** and is the second-order rate constant for a reaction at low substrate concentrations. The enzyme often, but not always, exhibits the highest catalytic efficiency/specificity constant with the true substrate of a reaction.

A different situation is encountered in many oxidation-reduction reactions in which the cellular electron carrier is unknown. Common electron carriers include a variety of cytochromes, ferredoxins, quinones, and other components, many of which are distinct for different microorganisms. In some cases, a commercially available, easily purified, or more soluble surrogate electron carrier can be used, such as horse heart cytochrome c, spinach ferredoxin, or menaquinone; however, caution must be used in the interpretation of kinetic studies in which the natural electron carrier is replaced by these reagents. In addition, some enzyme activities are monitored by using artificial electron acceptors such as synthetic dyes (e.g., methyl or benzyl viologen). These compounds undergo a large change in color intensity depending on their oxidation-reduction state, allowing for convenient and sensitive assays, but again the relevance to the authentic biological electron carrier can be questioned.

19.2. TYPES OF ENZYME PREPARATIONS

19.2.1. Secreted Enzymes

Particular microorganisms are adept at degrading various biopolymers by secreting the corresponding biodegradative enzymes, including proteases, glycosidases, lipases, and nucleases. In addition, a secreted enzyme can result from use of any of several commercially available cloning vectors that are designed to export the product of any selected gene. Analysis of such secreted enzymes simply requires removal of the intact cells by standard centrifugation or filtration methods and use of an appropriate assay. If desired, additional enrichment steps can be carried out by traditional purification methods prior to characterization of the enzyme activity.

19.2.2. Intact and Permeabilized Cells

Intact cells are often used to provide qualitative information related to an enzyme activity. For example, whole-cell assays are very useful for identifying particular isolates possessing high levels of a desired activity or for screening recombinants to identify those that express a desired gene. In contrast, in only a few cases is it possible to obtain reasonably quantitative enzyme activity results by using intact cells. For example, the enzyme nitrogenase typically is assayed by monitoring conversion of acetylene (an alternate substrate) to ethylene using gas chromatography (see section 19.3.6.4). Both of these gases readily diffuse through the cellular membrane so the substrate can be delivered in known and nonlimiting quantities while the product does not collect within the cell. Other gases (e.g., O_2, CO_2, and H_2) and some nongaseous substrates also diffuse readily through membranes and can be delivered in a controlled fashion; however, assays involving these compounds must also take into consideration whether cosubstrates or products also have the ability to cross the membrane. In the assay of a specific oxygenase, an oxygen electrode can measure oxygen consumption activity in a suspension of whole cells (see section 19.3.6.3), but the second substrate undergoing oxidation may be rate limiting because it has not

reached high intracellular concentrations, or alternatively, the product formed may be consumed by subsequent reactions catalyzed by other oxygenases, thus producing an erroneous rate of oxygen consumption. Nuclear magnetic resonance (NMR) analysis is an approach that partially overcomes these limitations in studies of in vivo activities. For example, NMR using ^{31}P- and ^{13}C-containing substrates enables whole-cell measurements of the rates of specific enzymes in cells maintained under various conditions (2, 34). The low sensitivity of NMR necessitates high cell densities in order to have sufficient concentrations of metabolic intermediates; for many analyses, 10^{10} cells per ml are required, and this presents some difficulty if oxygenation is important. Due to the above concerns, quantitative assays are not generally recommended with whole cells.

One approach to overcome problems involving substrate or product transport across the membrane is to permeabilize the cells by chemical or enzymatic treatment. Among the gentlest permeabilization methods is Tris-EDTA treatment. Solutions containing moderate cell densities (1 mg [dry weight]/ml or less) are adjusted to 0.1 M Tris-HCl (pH 7.5) and 1 mM EDTA, which alters the permeability of the lipopolysaccharide-containing outer membrane of gram-negative bacteria by a combination of chelating Mg^{2+} ions and some incompletely understood effects of Tris buffer (11). Toluene is another widely used permeabilization agent that probably disrupts membrane lipid regions, permitting a wide variety of low-molecular-weight materials access to the cell interior while also allowing many metabolites to diffuse out. For example, 1 drop (50 μl) in a 1-ml assay volume is sufficient to render *Escherichia coli* cells permeable to various amino acids and carbohydrates. Chloroform with detergent added (e.g., 15 μl of chloroform and 50 μl of a 1% [wt/vol] sodium deoxycholate solution per ml of cell suspension) also is effective for permeabilizing enteric organisms. Detergents alone are used as permeabilization agents, although in many instances it is unclear whether the cells remain intact during the subsequent assay. The cationic detergent cetyltrimethylammonium bromide (CTAB) (used at a final concentration of 0.05 to 0.1 mg/ml) is representative of these compounds. In all applications with permeabilized cells, there may be some variation in results as a function of the concentration of cells used or the time allowed for the permeabilizer to function. Standardization of cell density in the assay is generally desirable for maximum reproducibility. Finally, it is important to include controls with no permeabilizing agent to ascertain that the treatment is effective.

19.2.3. Soluble Extracts

The great majority of enzyme assays are conducted on cell-free preparations. This approach allows for careful control of substrate and cofactor concentrations, reaction conditions such as pH and ionic strength, and the use of coupled assays where desired. Selective enrichment of **periplasmic enzymes** is possible by use of methods that lyse the outer membrane while keeping the inner membrane intact. The remaining cellular material is simply removed by centrifugation, and the soluble periplasmic material is further characterized. This approach is especially useful for analysis of the products of genes that are cloned into commercially available plasmid vectors designed for recombinant protein export to the periplasmic region. For characterization of **cytoplasmic enzymes**, a wide variety of physical methods is available to disrupt cells including those based on hydrodynamic shear, ultrasonic disruption, and homogenization with glass beads. In addition, release of cytoplasmic proteins can be achieved using enzymatic lysis and commercially available cell lysis reagents. Membranes and other cellular debris are readily removed by ultracentrifugation, leaving a soluble solution of cell extracts. Further information on cell disruption techniques is found in chapter 7.

19.2.4. Membrane Fractions

Some enzymes are membrane associated, leading to numerous complications in their characterization. The preparation of a membrane fraction begins with one of the cell disruption methods mentioned above. Various centrifugation methods are available (e.g., standard ultracentrifugation as well as sucrose gradient and isopycnic methods) to further enrich a general membrane fraction or to isolate a particular type of membrane, as described in chapter 7. The resulting preparation generally contains membrane vesicles, which are problematic for enzyme studies. Light refraction by such vesicles interferes with many common enzyme assays. Also, the enzyme of interest may be positioned randomly toward each face or selectively facing either the inner or outer surface depending on the physical method used to disrupt the cells; the orientation can impact substrate delivery to, and product release from, the enzyme. Finally, the presence of other membrane-associated proteins complicates efforts at enzyme isolation. In particular situations, enzyme characterization studies can be carried out with such vesicles (e.g., examining cytochrome *c* oxidase or other enzymes involved in bioenergetics where an intact membrane is needed), but in general further solubilization is desired. For loosely associated proteins, the membrane interaction often can be disrupted by high ionic strength (e.g., inclusion of 0.5 to 5 M KCl or other salt). In other cases, it is possible by careful use of proteases to separate a soluble domain associated with activity from its attachment to a hydrophobic domain inserted into the membrane. More typically, solubilization requires the selective use of low concentrations (e.g., 0.1 to 3%, wt/vol) of a mild ionic or nonionic detergent such as sodium deoxycholate or Triton (see chapter 7). In each of these cases, the solubilized protein can be further purified by most routine methods (although it may be necessary to retain detergent in all buffers). If necessary, purified and solubilized membrane proteins can be reconstituted into artificially created lipid vesicles in order to assay reactions that require intact membranes.

19.2.5. Purified Enzymes

Serious kinetic studies and documentation of most enzyme properties are best conducted with purified preparations, but there are situations in which quantitative assays of enzymes in unpurified extracts are desirable and even preferable. For example, some important enzymatic properties can be altered by purification, such as the removal of a key allosteric effector (see section 19.4.3.1). More commonly, assays are performed with crude extracts simply because only comparative activities are sought among samples in a group and pure enzymes are not needed for the desired comparisons. This might be the case when conducting a physiological study, such as the effect of a growth condition on enzyme production, or when assaying mutants.

The principal limitations on the use of crude extracts for assays are the presence of competing activities, endogenous inhibitors, endogenous substrate, and proteolytic enzymes that might reduce the assayable activity. Frequently, dialysis or gel filtration can be used to remove contaminating low-molecular-weight components. A reasonable selection of protease inhibitors is available, including metal chelators for metalloproteases, serine protease inhibitors such as phenylmethylsulfonyl fluoride, and inhibitory peptides such as leupeptin. These inhibitors may be unnecessary for the high protein concentration of most crude extracts unless extracts are to be stored prior to assay, an inadvisable option. The problem of competing activities, e.g., contaminating ATPase activity in assays which measure kinase-dependent production of ADP from ATP, is clearly a major limitation on the conduct of assays and one for which there is no general solution. Fortunately, innovative assay methods or a modified design can help in some instances, but when that is not possible, partial or complete purification is the only recourse. Summation of the numerous detailed procedures available for enzyme purification (for an example, see reference 4) is beyond the scope of this chapter.

19.3. PRACTICAL ASSAY CONSIDERATIONS

19.3.1. General Comments

Depending on the needs of the investigator, assays are performed to ascertain whether a particular activity is present, assess the activity level under precisely defined conditions, or monitor the effects of various conditions on enzymatic rates. In the most simple situation, samples are incubated with a given amount (usually a near saturating level) of substrate(s) and any additional required components (cofactors, metal ions, stabilizing agents, or allosteric effectors) in a specific buffer (and pH) at a chosen temperature and for a particular time, and then the extent of product formation or substrate consumption is monitored. The amount of transformation is compared to the extent of nonenzymatic conversion to assess activities (e.g., in cell extracts derived from various isolates or from a single strain grown under a variety of growth conditions). The activity values obtained from these measurements allow reasonable comparisons to be made among the various samples for the given experimental conditions. For more detailed characterization of an enzyme activity, the approach usually involves precise determination of initial rates of the enzyme activity in an optimized assay system, often using varied substrate concentration(s). Key criteria in developing an assay for an enzyme of interest include its specificity, sensitivity, precision, and convenience.

Two choices in developing an assay involve use of a continuous or discontinuous method and whether to measure product formation or substrate disappearance. Only certain assay procedures (e.g., spectrophotometry, fluorescence methods, and various types of electrodes) allow continuous data collection, whereas others require the taking of time points. The former approach generally requires little effort by an investigator during a single assay and provides immediate information; however, instrument design may limit monitoring to only one or a few samples at a time. Although discontinuous methods do not provide immediate feedback, automated sampling systems have increased the ease of analysis for multiple samples and allow extensive sets of data to be collected on previously quenched samples (in which the reaction has been stopped). Thus, both methods are commonly employed in enzyme assays. For either assay approach, the product concentrations typically undergo dramatic increases during the initial stages of an enzyme assay while substrate concentration changes may be minimal; thus, analysis of product formation often offers the greatest sensitivity for measurements of initial rates. Nevertheless, it is possible to obtain initial rates by following changes in substrate concentration if the detection limits allow precise measurements in samples that retain saturating substrate levels.

The following sections provide additional guidelines for development and use of quantitative assays. As described in sections 19.3.2 and 19.3.3, it is important that the enzymatic rate (also known as velocity) be linear for an acceptable time period and is directly proportional to the amount of added enzyme. Substrate concentrations and other assay conditions must be carefully controlled, as detailed in sections 19.3.4 and 19.3.7. Furthermore, one must choose the best assay procedure (several specific examples are described in section 19.3.6) based on its sensitivity, ease of use, equipment available, or other factors related to an enzyme of interest. In some assays, the immediate product of the target enzyme is converted to another species for ease of detection; use of such coupling enzymes can introduce additional concerns as described in section 19.3.5. Regardless of the assay method, it is critical to calibrate the procedure with standards of known purity. For example, in assays that monitor an absorbance change it may be necessary to determine the extinction coefficient of a product or substrate. More generally, it is advisable to perform mock assays that contain all ingredients but one (e.g., no substrate or no enzyme) and then add known amounts of the material to be analyzed. It is also important to emphasize that development of an assay is an iterative procedure in which changes made in one parameter may affect the enzyme behavior toward other parameters. For a much more complete description of specific assays, multivolume collective works as *Methods in Enzymology* and *Methods of Enzymatic Analysis* should be consulted.

19.3.2. Linearity

An important criterion in any enzyme assay is establishing that the velocity is constant throughout the incubation period. To evaluate the constancy of velocity, a series of identical assays are set up (with the same enzyme concentration throughout) and terminated at different times. The amount of product formed (or substrate consumed) should yield, in a favorable situation, a linear plot as shown by the circles in Fig. 1. If the data are linear over a reasonable time period, a single-time-point assay taken within the linear range may suffice for further enzyme characterization.

Nonlinearity of the data can take two forms: a lag phase followed by a linear plot (Fig. 1, triangles) or the more common situation of decreasing velocity with time (Fig. 1, squares). A lag phase observed during the assay may be attributed to a change in the state of aggregation of the enzyme upon dilution into the assay mixture (for example, a monomeric species may exhibit higher activity than a polymer), dissociation of a tight-binding inhibitor (present within the cell or added during a stage of purification; see section 19.4.1.5), slow activation by some component in the assay mix, stimulation of activity by the product, or any of several other reasons. In addition, this behavior is often observed in coupled enzyme systems as described in section 19.3.5. Simply measuring the enzymatic rate during the linear phase generally provides suitable values for maximal rates; however, this approach may be inadequate for more complete kinetic analyses. It may be worthwhile to examine modifications of the assay in an effort to eliminate the lag phase. To reduce the possible complications of enzyme dis-

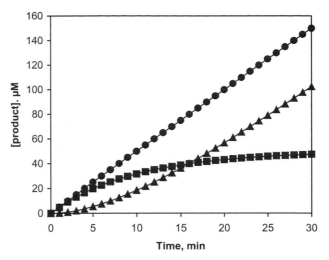

FIGURE 1 Comparison of progress curves for enzymes that are stable, activated, or inactivated during the assay. For the favorable situation of a stable enzyme (●), the concentration of product formed per unit of time increases in a linear fashion (shown here as $v = 5$ µM/min). For a particular enzyme, the activity may increase during the assay to reach a steady-state level (▲). This situation is observed for some coupled enzyme systems, or it may be associated with enzyme disaggregation, dissociation of a slow-binding inhibitor, or other type of activation that occurs under assay conditions. Progress curve data for an enzyme exhibiting activation can be fit to equation 12 (with $v_0 = 0$) to obtain the final rate (v_f) and the apparent first-order rate constant for establishment of equilibrium (k_{app}). In the opposite situation (■), the rate of product formation may decrease over time and eventually result in complete enzyme inactivation. Progress curve data for such a labile enzyme can be fit to equation 1 to obtain the initial rate and the inactivation rate constant ($v_0 = 5$ µM/min and $k_{inact} = 0.1$ min^{-1} for the situation shown). Attempts to estimate v_0 by taking the tangent to the first few points of such data generally result in an underestimation of the true value. By proper choice of assay conditions (inclusion of reductant, addition or removal of metal chelators, change in ionic strength, provision of glycerol or other additives, etc.), it may be possible to sufficiently enhance the stability of an enzyme so that a linear response is obtained, thus facilitating further characterization.

aggregation during the assay or the slow binding of an activator in the mixture, the enzyme can be diluted into the assay mix and allowed to equilibrate prior to adding a small volume of concentrated substrate to initiate the reaction. Problems related to dissociation of a tightly bound inhibitor also may be alleviated by extensive dialysis of the sample prior to doing the assay.

The opposite situation, a loss of activity over time, is depicted in Fig. 1 (squares) and can arise from many possible types of inactivation reactions. One situation giving rise to this type of behavior involves substrate depletion; the use of higher initial substrate concentrations may eliminate this concern, provided substrate inhibition is not a problem. In other cases, the enzyme can be stabilized, thus increasing the assay linearity, by suitable adjustment of pH, decreasing the temperature, adding stabilizing agents, or removing oxygen, as described in section 19.3.7. It is generally advisable to expend effort at an early stage of investigation to eliminate or minimize nonlinearity of an assay in order to ensure that meaningful results are obtained during enzyme characterization.

Nonlinear assays cannot be eliminated for some enzymes due to various types of autoinactivation events. For example, some nonheme iron oxygenases react with oxygen to yield an activated oxygenating intermediate that partitions between two pathways: catalysis and enzyme inactivation. The latter pathway may simply involve metal oxidation (a reversible process) or reaction with an active-site protein side chain, thus inactivating the enzyme (28). In addition, some enzymes must be chemically derivatized in order to achieve an activated state (e.g., citrate lyase is acetylated) and this modification may be lost during the assay. In such cases, the plot of product formation as a function of time (also known as a **progress curve**) can be analyzed by use of equation 1.

$$P_t = v_0 \cdot (1 - e^{-k(\text{inact})t})/k(\text{inact}) \qquad (1)$$

In this equation, P_t is the accumulated product at time t, v_0 is the initial velocity, and $k(\text{inact})$ is the inactivation rate constant. As illustrated in Fig. 1 (squares) the product concentration increases over time but levels off at longer time points. Data should be fit (e.g., using a commercially available fitting program) to equation 1 to obtain an inactivation rate constant and the initial rates that are needed for kinetic analysis. By contrast, efforts to obtain v_0 by estimation of a tangent to the initial points will almost certainly yield a low value.

19.3.3. Enzyme Concentration

Fundamental to any assay is demonstrating a linear dependence of the observed reaction velocity on the concentration of enzyme present. To assess whether the chosen assay conditions meet this criterion, varied amounts of enzyme are added to otherwise identical assay solutions and the reaction velocities are monitored. Typically, one observes a region showing direct proportionality between the measured velocity and enzyme supplied, followed by a region of declining increments in velocity with further enzyme increases. The nonlinearity may arise from substrate depletion, product inhibition, pH shifts, or other changes during the assay. Routine assays must be carried out in the linear region; thus, the enzyme solution must be diluted if this linear region is exceeded and a reassay must be performed.

19.3.4. Substrate Concentration

For normal assessment of enzyme activity and for calculation of related kinetic constants, initial velocity conditions are essential and there should be a negligible decrease in the starting substrate concentration. In practice, substrate consumption should not exceed 5% over the course of the measured assay period. This means that short time points coupled to exquisite detection sensitivity may be needed for assays involving low substrate concentrations. High concentrations of substrate are utilized to measure the maximum velocities of the desired enzyme; however, inhibition by excess substrate concentration is not uncommon (see section 19.4.1.3) so it is important to monitor the effects of a wide range of substrate concentrations. Examination of the effect of substrate concentration also is required to determine the affinity of an enzyme for a substrate. Preliminary studies can define the approximate substrate concentration leading to approximately half-maximal rates; a series of assays should then be carried out in which the substrate concentrations range from about 20% of this value to at least fivefold this amount. Further discussion of the methods used to abstract kinetic constants by studies involving variation of substrate concentration is found in section 19.4.

19.3.5. Coupling Enzymes

Measurement of reaction products is often made easier by adding an auxiliary enzyme that converts the primary product to a new material while simultaneously providing a useful detection system, such as a convenient absorbance change. The assay for phosphoglucomutase nicely illustrates this concept. Phosphoglucomutase catalyzes the interconversion of glucose-1-phosphate (glucose-1-P) and glucose-6-phosphate (glucose-6-P), two compounds that are not easily distinguished by typical assay methods.

$$\text{Glucose-1-P} \rightleftarrows \text{glucose-6-P}$$

The enzyme is assayed in a reaction that initially contains only glucose-1-P, with conversion to glucose-6-P followed spectrophotometrically by use of purified glucose-6-P dehydrogenase and NADP$^+$. Glucose-6-P generated by phosphoglucomutase catalysis is immediately oxidized by the dehydrogenase concomitant with NADP$^+$ reduction to NADPH, which is easily monitored in a continuous manner as a change in A_{340} (see section 19.3.6.1).

Several conditions must be met for such coupling reactions to provide valid measurements of the rate for the first enzyme. These are as follows: (i) the substrate(s) for the first reaction must be in large excess (known as saturating conditions and associated with zero-order reactions); (ii) the velocity of the second reaction must be linear with the concentration of its substrate that is provided by the first reaction (resulting in what is termed a first-order reaction); and (iii) neither reaction should be significantly reversible under the assay conditions. In the example cited above, this would mean that excess glucose-1-P would have to be supplied for the phosphoglucomutase reaction, that excess NADP$^+$ is provided for the coupling reaction, and that sufficient glucose-6-P dehydrogenase is available so that the velocity of this step is a direct function only of the amount of glucose-6-P being produced by phosphoglucomutase. Functional irreversibility of the two reactions is achieved by having the dehydrogenase remove glucose-6-P as it is formed and by monitoring the overall velocity before the final products have accumulated appreciably. The requirement that the coupling enzyme be in excess can be met by a trial-and-error process in which its concentration is increased until further additions have no effect on the measured velocity of the overall reaction. If sufficient coupling enzyme is not provided in the assay, especially if it exhibits low affinity for the substrate produced by the first reaction, long lag phases may obscure the true enzymatic rate (see Fig. 1, triangles). On the other hand, a large excess of coupling enzyme may be disadvantageous due to its expense and to possible interference with the reaction of interest (e.g., by introducing an inhibitory substance along with the coupling enzyme). A detailed discussion of the kinetic consequences of multiply coupled enzymes is found in references 6 and 20.

19.3.6. Choice of Assay Method

19.3.6.1. Spectrophotometry

The sensitivity, wide availability, and ease of use of UV-visible spectrophotometers have resulted in their prominence as a versatile tool for enzymatic analysis. As a general precaution, assays must be performed under conditions where the light absorption is directly proportional to the concentration of absorbing species. Optimal sensitivity and reliability typically are achieved by designing assays to measure values in the range of 0.1 to 1.5 absorption units, although some instruments can extend this range. Several common types of assays based on spectrophotometry are illustrated below, and additional details about the use of absorption spectroscopy are found in chapter 17

The reaction catalyzed by glucose-6-P dehydrogenase (EC 1.1.1.49 ref. [19]) shows why oxidoreductases are among the easiest enzymes to assay by continuous spectrophotometric methods. With only minor variations, the following comments also apply to almost any dehydrogenase using $NAD(P)^+$ as a substrate. The reaction measured is

$$\beta\text{-D-Glucose-6-P} + NAD(P)^+ \rightleftarrows$$
$$\text{D-gluconolactone-6-P} + NAD(P)H + H^+$$

When examined in the forward $[NAD(P)^+$ reduction] direction, assays are usually run at relatively high pH to consume protons and shift the equilibrium to the right. By contrast, the oxidation of $NAD(P)H$ is normally run at a lower pH since H^+ is a substrate in that reaction. This adjustment of the assay pH with reaction direction is particularly necessary when assays reveal a very short reaction period of linear product formation, indicative of an unfavorable equilibrium; however, the choice of pH must also be dictated by the stability of the enzyme (section 19.3.7.1). Assays run in the forward direction typically use a quartz spectrophotometer cell (preferably using a thermojacketed holder for precise temperature control) with a 1-cm light path and involve the monitoring of absorption increases at 340 nm where the extinction coefficient of NADH or NADPH is $6.22 \text{ mM}^{-1} \text{ cm}^{-1}$. Using this approach, a change of 0.10 optical density (OD) unit/min corresponds to the production of 16 μM NAD(P)H per min. Analysis in the reverse direction is facilitated by use of a 0.2-cm-path-length cell to increase the allowable concentration of this substrate. For example, a 1 mM solution of NADH possesses an out-of-range absorption of 6.22 in a 1-cm cell, but can be monitored in a 0.2-cm cell yielding an absorption of 1.24. When examined in this direction, a substrate-free control should be included to detect background NAD(P)H oxidase activity. Solutions of NAD^+ and $NADP^+$ should not be stored at pH values over 7, since they are unstable under this condition. Stock NAD(P)H solutions should be stored in alkaline solutions for enhanced stability.

Oxidoreductases are often used in coupled assays, as already indicated in section 19.3.5, due to the ease of monitoring the associated spectroscopic changes. To further reinforce this point, two coupled-enzyme methods are described to assay phosphofructokinase (EC 2.7.1.11), which catalyzes the ATP-dependent phosphorylation of fructose-6-phosphate to produce fructose-1,6-bisphosphate and ADP. In one approach, fructose-1,6-bisphosphate formation is coupled to NADH oxidation via three linking enzymes: fructose bisphosphate aldolase, triosephosphate isomerase, and α-glycerolphosphate dehydrogenase (18). Calculation of the units of phosphofructokinase present in the assay is based on the decrease in 340 nm absorption of NADH and must take into account that for each mole of hexose-phosphate consumed, 2 mol of triosephosphate are formed in the assay and thus 2 mol of NADH are oxidized. A second system to monitor phosphofructokinase activity requires the two substrates of the target enzyme as well as

NADH, phosphoenolpyruvate, and two coupling enzymes: pyruvate kinase and lactate dehydrogenase. The ADP generated by phosphofructokinase is cycled back to ATP by pyruvate kinase, and the pyruvate formed in this reaction is reduced by lactate dehydrogenase. Contamination with ATPase or NADH oxidase interferes with this spectrophotometric analysis.

Spectrophotometric assays also are useful for many non-oxidoreductase reactions, as exemplified by an assay for the simple sugar hydrolase β-galactosidase (EC 3.2.1.23 [22]). This is one of the most frequently assayed enzymes, albeit often qualitatively, because of the many *lacZ* gene fusions constructed. However, it is also easily assayed by a quite precise spectrophotometric procedure using the chromogenic substrate o-nitrophenyl-β-D-galactoside (ONPG) that is converted by the enzyme to galactose and o-nitrophenol. The latter compound is faintly yellow, and at high pH, at which the nitrophenolate ion dominates, a more intense yellow is produced and can be measured at 420 nm. Quartz cells are not required to monitor absorption changes at this wavelength, so glass or plastic cells can be utilized. An enzyme solution is incubated with a buffered solution containing ONPG and other additives (e.g., 0.1 M sodium phosphate [pH 7.0], 10 mM KCl, 1 mM $MgSO_4$, and 50 mM mercaptoethanol), aliquots are removed at timed intervals, and 0.5 M Na_2CO_3 is added to adjust the pH to ~11 and inactivate the β-galactosidase. The activity is determined by monitoring absorption changes at 420 nm and by using the extinction coefficient of $4.5 \text{ mM}^{-1}\text{cm}^{-1}$ for o-nitrophenol at alkaline pH. It is important to include control reactions lacking ONPG (to correct for color due to the sample) and without enzyme (to correct for the spontaneous hydrolysis of ONPG). The ONPG assay can be adapted for use in permeabilized whole cells by adding 2 drops of $CHCl_3$ and 1 drop of 0.1% sodium dodecyl sulfate per ml of cell suspension. It is important to note that ONPG and related chromogenic compounds are not the natural substrates of the enzymes; however, data obtained with these substances are generally useful for routine analyses during protein purification and, depending on the substrate specificity of the enzyme, may be relevant to catalysis using the authentic substrates.

Although artificial chromogenic substrates that yield easily monitored colored products are very conveniently used (ONPG is just one of many examples), assays can also be based on spectroscopic changes accompanying enzymatic conversion of a substrate to product in reactions that do not involve nicotinamide coenzymes. One such example is an assay for histidine ammonia-lyase (EC 4.3.1.3; usually called histidase), an enzyme that catalyzes the nonoxidative deamination of L-histidine to form urocanate (imidazole acrylate) and ammonia (27). Because it is unsaturated, urocanate production can be followed on the basis of its absorption at 277 nm, using an extinction coefficient of $18.8 \text{ mM}^{-1}\text{cm}^{-1}$. Thus, an OD change of 0.10/min corresponds to a urocanate formation rate of 5.3 μM/min. A similar approach can be used with other enzymes producing unsaturated products (e.g., fumarate or cinnamate) or consuming unsaturated substrates (e.g., protocatechuate and other aromatic compounds).

19.3.6.2. Fluorescence

Fluorescence spectroscopy can be used to monitor an enzyme reaction, either continuously or discontinuously, as long as the substrate and product exhibit distinct fluorescence properties. For example, tryptophan hydroxylase converts tryptophan to hydroxytryptophan. Hydroxytryptophan's fluorescence excitation extends out to 310 nm, whereas that of tryptophan is only at a wavelength of 280 nm. By monitoring the 330-nm fluorescence emission properties using 300-nm light, the enzyme's formation of hydroxytryptophan can be monitored (24). Direct detection of nicotinamide coenzyme reduction also is possible on the basis of its blue fluorescence. Similarly, the decrease in yellow-green fluorescence of the methanogen coenzyme F_{420}, a 5-deazaflavin, can be used to monitor its reduction. Analogous to the chromogenic assay approach described above involving such compounds as ONPG, fluorogenic substrates can be utilized. For example, 4-methylumbelliferyl-linked substrates may be used to assay the cognate enzymes on the basis of the released fluorescent compound. Four drawbacks to the use of fluorescence include the more limited availability of the necessary instrumentation, interference by other compounds that absorb the incident or fluorescently emitted light, complications introduced by membranes or denatured proteins that scatter the incident radiation, and lack of a quantitative unit of fluorescence comparable to an absorbance unit. Nevertheless, fluorescence assays are widely used because of their ease and sensitivity, and they provide useful quantitative values if directly compared to an appropriate standard curve.

19.3.6.3. Selective Electrodes

A variety of commercially available electrodes exhibit selective responses to particular inorganic ions or gases and can be utilized in enzyme assays. For example, chloride ion-selective electrodes allow ease of monitoring dehalogenase activities and an ammonia gas-sensing electrode facilitates assays involving deamination reactions. Electrodes are available specific for many compounds of interest (including protons, oxygen, hydrogen, carbon dioxide, nitrate, nitrite, potassium, sulfide, and sodium). Each has a particular working concentration range and interferences; thus, it is imperative to carefully study the literature accompanying the electrode of interest. One important consideration is response time, i.e., whether the electrode senses the rapid changes in target compound concentration during the reaction (allowing use in a continuous assay) or whether a slow response dictates use in a discontinuous fashion. Additional comments about the use of selective electrodes are found in chapter 17. Two widely used electrode-based assays are described in the following paragraphs.

Many biochemical reactions result in net proton uptake or release; thus, potentiometric measurement of pH is a valuable assay approach. High-quality instruments permit a resolution of 0.005 pH units, so the pH of the assay solution need not shift appreciably during the assay; however, any significant change in pH can alter the activity of an enzyme. Another concern is that the observed pH changes will be highly dependent on the concentration of buffer present. Yet another disadvantage of this assay is that the readout of a pH meter is not linear with proton concentration (recall that pH is the negative log of the hydrogen ion concentration), so the data must be converted prior to initial rate determination. An approach to at least partially overcome these concerns involves continuous titration while maintaining constant pH. In such a pH-stat system, the rate of added acid or base is monitored and reflects the concentration of protons consumed or released. An example of this approach is found in the assay of urease (EC 3.5.1.5 [1]), an enzyme that hydrolyzes urea to yield two molecules of ammonia and one of carbonate, resulting in alkalinization of the solution.

For oxidoreductases that catalyze oxygen uptake, polarographic procedures are an excellent approach to measure activity. As an example, protocatechuate-3,4-dioxygenase (EC 1.13.11.3 [35]) can be assayed using a Clark-type oxygen electrode. This Fe-dependent enzyme catalyzes intradiol cleavage of protocatechuate to form β-carboxy-cis,cis-muconate. The assay simply involves filling a sample chamber of a polarograph (containing a Clark electrode) with buffer that has been well equilibrated with air, addition of the substrate protocatechuate, recording of the O_2 uptake until a stable baseline reading is obtained, addition of a precise volume of the dioxygenase, and measurement of the initial rate of oxygen uptake. The rate (measured as recorder units per minute) is converted to micromoles of O_2 consumed per minute by conducting a calibration reaction in a similar fashion, but with a limiting and known amount of protocatechuate. Variations of this procedure can be used with a wide variety of oxygenases; however, it is important to note that uncoupling reactions are observed in some enzymes. Thus, the amount of oxygen consumed may be greater than the amount of primary substrate oxidized. Control experiments examining oxygen consumption in the absence of substrate are important to perform, but these may not account for all uncoupling activity since this side reaction is sometimes stimulated by substrate.

19.3.6.4. Chromatography

Many chromatographic approaches are available to allow the separation of product(s) from initial reactant(s) for use in enzyme assays. Gas chromatography (GC) is the method of choice for volatile substances or compounds that are readily converted to volatile compounds by simple chemical steps. The already mentioned assay for nitrogenase utilizes this approach to separate the two gases acetylene and ethylene from one another. Because high temperatures can be used during the separation, this technique also can be applied to monitoring the transformations of less volatile substances such polyaromatic hydrocarbons or polychlorinated biphenyls during biodegradation studies. Quantification of substances separated by GC can occur by flame ionization detectors or other means. More details about the principles of the method, instrumentation, and sample preparation and derivatization are found in chapter 17.

In contrast to the gas-liquid partitioning of substances associated with GC methods, thin-layer chromatography and various types of column chromatography separate compounds on the basis of their differential interactions between solid and liquid (either constant [isocratic] or variable [gradient]) phases. The solid support can encompass a wide range of compounds, with a corresponding wide range of properties. Similarly, the liquid phases are carefully chosen to provide the best resolution. For example, the substrate(s) and product(s) may be separated on the basis of differential hydrophobicity by using a highly nonpolar C_{18} column (a resin with bound alkyl groups) and eluting with solvent of decreasing polarity, a process known as reverse-phase chromatography. Alternatively, separation may focus on differences in charge (ion-exchange chromatography), size (gel permeation chromatography), or more specific interactions (affinity chromatography), with appropriate changes in solvent as needed to enhance separation while allowing timely elution. Visualization of components separated by thin-layer chromatography often utilizes quenching of a fluorophore that is impregnated in the support; however, many alternative procedures are available such as those that specifically stain the samples based on selective chemistry of their reactive groups. Recent developments in scanning instruments have facilitated ease of quantification of such results. Detection of components eluting from a column may be based on UV or visible light absorption, changes in refractive index, fluorescence, or reactivity with chemical reagents. Additional comments about the instrumentation available for these approaches are provided in chapter 17, and examples of methods for chemical analysis are found in chapter 18.

19.3.6.5. Radioactivity

Some assays utilize radiolabeled substrates that contain radioisotopes such as ^{14}C, ^{3}H, or ^{32}P. Each of these isotopes undergoes radioactive decomposition to release a β particle (alternative types of radioactive decay occur for some other isotopes), and this event can be very precisely quantified. By mixing aliquots of such radiolabeled samples with a "counting cocktail" containing scintillants (chemicals that emit light after colliding with a β particle), the number of counts per minute (cpm) can be monitored by use of a scintillation counter. On the basis of the instrument's counting efficiencies for each radioisotope, the number of disintegrations per minute (dpm) can be calculated. Investigators using radioactive material generally undergo rigorous training in the special procedures needed to ensure their safety and proper handling of materials and equipment during the experiment and to learn proper disposal methods. Additional details about the types of scintillants available, instrument operation, statistics, problems involving proton exchange with solvent, and other complications are given in chapter 17.

The simplest types of assays using radioisotopes involve selective extraction or precipitation of a radiolabeled substrate or product. Such a procedure is commonly used to assay aminoacyl-tRNA synthetases, enzymes catalyzing essential ligase reactions that activate amino acids for protein synthesis. For example, the assay for glutaminyl-tRNA synthetase (EC 6.1.1.18 [9]) involves incubation of the ^{14}C-labeled amino acid with tRNA (a mixture or the specific species), MgATP, buffer, and sample containing the enzyme for varied time periods. The mixtures are treated with 5% trichloroacetic acid, resulting in precipitation of the tRNA species (both free and containing covalently bound glutamine), and the precipitate is collected by filtration, washed, dried, and counted. A time course should exhibit linear increases in the amount of radioactivity collected on the filters.

Use of radioisotopes also can increase dramatically the sensitivity of some chromatography-based assays. Typically, tracer levels of the specifically labeled substrate (commercially available or synthesized to contain a known amount of radioisotope per known amount of sample) are mixed with unlabeled material prior to use. Labeled substrate is incubated with the enzyme of interest, aliquots are removed at timed intervals, the product is separated from the remaining substrate (e.g., using one of the chromatographic methods described above), and the radioactivity associated with one or both of the components is measured. An assay for chloramphenicol acetyltransferase (CAT) (EC 2.3.1.28) provides an example of this approach. This enzyme, responsible for the resistance of many bacteria to chloramphenicol, acetylates the substrate by using acetyl-CoA to form chloramphenicol-3-acetate plus free CoA thiol. Thin-layer silica gel chromatography separates $[^{14}C]$chloramphenicol from the labeled acetylated product. The thin-layer support can be exposed to X-ray film for autoradiographic analysis, using a densitometer to quantify the results by scanning the

autoradiogram. It is also possible to scrape the appropriate band of chloramphenicol-3-acetate from the plate and quantify the product by scintillation counting. This assay is often used for studying on/off regulation of a promoterless CAT gene to detect the function of a suspected promoter sequence.

Another method to measure CAT activity illustrates how radiolabeled samples can be used in coupled assays. In this case, the assay uses $[^{14}C]$acetyl-CoA and chloramphenicol. Because $[^{14}C]$acetyl-CoA is unstable, a cycling system composed of ATP, CoA, phosphoenolpyruvate, acetate kinase, phosphotransacetylase, and pyruvate kinase is necessary to maintain its excess throughout the reaction (30). Thus, acetate released by hydrolysis is activated by phosphorylation and transferred onto CoA in a process driven by ATP, which is replenished by pyruvate kinase using phosphoenolpyruvate. The labeled chloramphenicol-3-acetate is monitored after extraction into an organic solvent (toluene) under conditions where $[^{14}C]$acetyl-CoA or $[^{14}C]$acetate is not extracted.

Radiolabeling studies are particularly well suited to assays involving CO_2 fixation by enzymes such as ribulose-1,5-bisphosphate carboxylase (EC 4.1.1.39 [15]), which carboxylates this substrate to form two molecules of 3-phosphoglycerate. A buffered ribulose bisphosphate solution and radiolabeled bicarbonate are sealed within stoppered vials, enzyme is injected through the stoppers, and the reactions are allowed to proceed for precisely measured time periods. The reactions are quenched by addition of a 10% volume of 6 M acetic acid, which inactivates the enzyme and converts the bicarbonate to gaseous CO_2. The vial caps are removed in a chemical hood, and the vials are allowed to stand open for at least 30 min to remove unreacted radiolabeled substrate. The remaining radioactivity, associated with product, is measured by addition of scintillation fluid and counting.

Complementary procedures are available to monitor CO_2 release by using radiolabeled substrate. In this case, the substrate must be purchased or synthesized with ^{14}C incorporated into the carbon that is lost as CO_2. Assays are run using multiple enclosed vials, each of which contains a separate chamber containing a piece of filter paper soaked in basic solution. The reactions are quenched at predetermined time points by injection of acid; acidification of the assay mixture causes the generated $^{14}CO_2$ to transfer to the basic filter paper that subsequently is mixed with scintillation fluid and counted. This approach is illustrated in chapter 17.

19.3.6.6. Chemical Derivatization

Discontinuous colorimetric assays are widely used to monitor enzyme reactions in microbiology. In this approach, aliquots of a reaction mixture are removed at timed intervals and the product (or remaining substrate) is chemically derivatized to form a colored species. Specific methods to quantify inorganic phosphate, ammonia, nitrate, nitrite, and several other compounds are described in chapter 18 and can be incorporated into enzyme assays. Two specific examples involving chemical derivatization are described below, one that strictly requires a discontinuous assay and a second that can be used in a continuous assay.

An assay for ribosephosphate isomerase (EC 5.3.1.6), a key enzyme of the pentose phosphate pathway and a component of the Calvin carbon reduction cycle, makes use of the reddish purple color derived by reaction of ribulose-5-phosphate with a cysteine-carbazole reagent (17). Enzyme is added to a series of vials containing buffered solutions of ribose-5-phosphate and other additives (e.g., 1 mM EDTA and 1 mM mercaptoethanol) to a final volume of 0.5 ml. At various time points the reactions are quenched by addition of equal volumes of 0.03 M cysteine-HCl, and five ml of 70% (vol/vol) H_2SO_4 are immediately added to each sample along with 0.2 ml of 0.1% carbazole (recrystallized from xylene) in absolute ethanol. After 30 min at 37°C, the absorption at 540 nm is determined. An A_{540} of 0.14 corresponds to the formation of 0.1 μmol of ribulose-5-phosphate. Linearity of the assay is limited to approximately 0.5 OD units or roughly 0.3 μmol of ribulose-5-phosphate formed.

An assay based on specific quantification of the amount of thiol released from a substrate is widely applicable, including its use as an alternative method to analyze CAT activity. In this case, the enzymatic release of CoA-SH from acetyl-CoA is followed by thiol exchange with 5,5′-dithiobis(2-nitrobenzoate) (DTNB, also known as Ellman's reagent) to yield 2-nitro-5-thiobenzoate. The anion of the latter compound possesses an extinction coefficient of 13.6 $mM^{-1}cm^{-1}$ at 412 nm. This assay is subject to serious interferences by other thiol compounds and by contaminating thiolesterases; therefore, it is recommended only for partially purified extracts or when CAT levels are moderately high.

19.3.6.7. Other Assay Approaches

Methods to assay enzyme activities are limited only by the imagination of the researcher. Perhaps the enzyme of interest causes a change in optical rotation of the solution, as in the case of sucrase, so that optical rotary dispersion or circular dichroism methods can be used to monitor activity. NMR spectroscopy can be used to monitor transformations of a wide range of molecules on the basis of the changes in chemical shift of particular proton, carbon, or phosphorus atoms. Additionally, an assay may involve a highly specific protease that recognizes a unique protein substrate. In this case, an assay may be developed using denaturing gel electrophoresis or light-scattering methods, with the purified protein as a substrate. The literature is replete with examples of various types of assays that can be used as precedents for developing a procedure for any enzyme under study. Two final examples of assay methods are provided below. These methods are representative of procedures used to assay the important class of enzymes known as restriction endonucleases.

EcoRI endonuclease (EC 3.1.21.4) catalyzes the hydrolytic cleavage of double-stranded DNA containing the following hexanucleotide sequence:

deoxy-5′. . pG↓pApApTpTpCp. .

deoxy-3′. . pCpTpTpApAp↓Gp. .

The sites of cleavage in the two strands are indicated by arrows, leaving a staggered cut near the 5′ ends of the twofold symmetrical hexanucleotide sequence. One approach to assay for EcoRI endonuclease activity involves measuring the amount of enzyme needed to digest 1 μg of lambda DNA in 1 h, with digestion completeness being determined by agarose gel electrophoresis. The sizes of the five resulting EcoRI-generated fragments are 4.878, 5.643, 5.804, 7.421, and 24.756 kbp. An alternative, more quantitative, assay has been described by Modrich and Zabel (23). This method is based on the conversion of radioactively labeled circular colicin E1 (ColE1) DNA containing one EcoRI site

to a linear form, which then is hydrolyzed by exonuclease V (recBC DNase) to release acid-soluble nucleotides. The ^3H-labeled covalently closed circular ColE1 DNA is prepared by growing E. coli JC411 thy (ColE1) in the presence of [^3H]thymidine and then isolating the plasmid. Assay mixtures contain the radiolabeled DNA in buffered solution with appropriate additives (e.g., NaCl, EDTA, ATP, and dithiothreitol), exonuclease V, and the sample of EcoRI endonuclease. The reaction is terminated at selected time points by adding salmon sperm DNA immediately followed by 30% (wt/vol) trichloroacetic acid. Intact DNA is removed by centrifugation, and the supernatant solution is mixed with scintillant and analyzed in a scintillation counter. One unit of EcoRI endonuclease activity in this assay typically is based on the conversion of 1 pmol of ColE1 circular DNA to an exonuclease V-sensitive form per min under the stated conditions. Although somewhat convoluted, this procedure has been used in kinetic studies to provide useful binding constants.

19.3.7. Optimization

Whether newly developed or simply adapted from a method used for the same enzyme from another source, an assay should be optimized for the specific enzyme being studied. Kinetic constants, specific activity in the homogeneous state, and pH optimum can be greatly influenced by assay conditions (e.g., temperature, buffer type, ionic strength, and presence of various additives), and it is therefore desirable to determine these properties under conditions that seem most favorable for catalytic activity. Although it can be argued that these are not "physiological" conditions, in the absence of sufficient information to describe the intracellular environment it seems reasonable to study the enzyme under otherwise optimal conditions.

Most of the steps for accomplishing optimization are relatively obvious (see below); however, it is important to recognize that some of the results may be interactive. For example, it is possible that the pH optimum varies with buffer type; in such an instance, the buffer and pH condition that gives the highest relative catalytic activity should be selected. On the other hand, one should avoid impractical situations even if they appear to be optimum, such as temperatures or pH conditions that give high activity but under which the enzyme is detectably unstable. It is sometimes necessary to adopt an assay condition that is suboptimal in order to improve day-to-day reproducibility. Similarly, if using kinetically saturating substrate or cofactors is not feasible or appropriate, simply be aware of the limitations imposed by assaying at less than optimal reactant concentrations.

19.3.7.1. Effects of pH and Buffer

Both the stability and the catalytic activity of enzymes are affected by pH, and very different experimental approaches must be used to define these effects. In both cases, the type of buffer used can markedly affect the results due to enzyme inactivation or to inhibitory (or occasionally stimulatory) effects on the enzyme. For example, buffers such as citrate and pyrophosphate should be avoided because of their strong metal-chelating tendencies, especially when dealing with metalloenzymes. Zwitterionic buffers such as N-2-hydroxyethylpiperazine-N′-2-ethanesulfonic acid (HEPES), 3-(N-morpholino)propanesulfonic acid (MOPS), and 2-(4-morpholino)ethanesulfonic acid (MES) are recommended in these instances. Furthermore, many enzymes acting on phosphorylated substrates are inhibited by inorganic phosphate in one or more of its protonation states. Another complication is that some buffers undergo large temperature-dependent changes in pH (4). In general, several buffers should be examined in any studies to define the effects of pH on enzyme stability or activity.

Determining the effect of pH on stability is quite straightforward, although the results obtained may vary with the conditions under which the study is conducted. Aliquots from a concentrated stock solution of enzyme are diluted into a selection of buffers adjusted to desired pH values at the same ionic strength (μ, defined as $\frac{1}{2} \Sigma c_i z_i^2$ where c_i is the concentration of each ionic species [i] and z is the charge on each ion). Because buffers usually change ionic strength as a function of pH, buffer concentrations may have to be varied to hold ionic strength at a fixed value as pH changes. Alternatively, an innocuous salt such as KCl can be added to adjust μ without affecting the pH. The diluted enzyme samples are incubated at a set temperature for a predetermined time (e.g., 30 min or 12 h), and a portion of each sample is assayed under standard assay conditions. It is important that the addition of buffered enzyme to the assay mixture does not change the assay pH. From the activity results, a plot of enzyme remaining (measured as units) versus pH can be constructed for the chosen incubation conditions; thus, a region of relative stability can be identified. For proteins that are inherently stable, it may be necessary to increase the time and temperature of incubation to induce moderate instability and allow greater ease of observation. The resulting **pH of optimum stability** should be appropriate for purification studies or for storage of enzyme. It is important to realize, however, that the pH of greatest stability may differ widely from the pH of maximum activity.

The effect of pH on enzyme catalytic properties is much more complicated to analyze, although the information return is potentially greater. To analyze the effect of pH on enzyme activity, assay solutions should be generated using a number of different buffers (each including several pH values within 1.0 pH unit of the buffer's pK_a) and incorporating overlapped pH ranges. The ionic strength should be constant across all buffers and pH values tested. Mixed buffers are sometimes used, such as Tris-malate (pH 5.5 to 8.5) and citrate-phosphate (pH 3.5 to 7.5), but these are not ideal in terms of overlapping pK_a values for each buffer pair. The goal of a pH dependence study typically is to establish an **optimum assay pH**. Definition of the pH optimum requires that excess substrate(s) be present in all assay conditions so that maximum rates are measured; however, many pH curves found in the literature fail to fulfill this criterion because the investigators did not account for pH-induced perturbation of the substrate binding affinity or enzyme stability. The maximum velocity pH profile allows identification of an optimum pH for an enzyme that should be used in standard assays; this optimum often can vary by ±0.1 pH unit without appreciable activity change so that small variations in buffer preparation will not lead to large changes in activity. Experiments assessing the effects of pH on enzyme velocity at varied substrate concentrations are also commonly performed and may provide mechanistically useful information regarding the influence of pH on ionization of groups in the enzyme and enzyme-substrate complex. Procedures for carrying out such studies and interpreting the data are well described (3, 5) and are often coupled with mutagenesis studies to identify specific groups involved in catalysis or substrate binding. Investigators interested in this approach are encouraged to examine a review by Knowles (12), which highlights the limitations of these studies and discusses common misinterpretations encountered in the literature.

19.3.7.2. Effects of Temperature

The term "optimum temperature" is a misnomer when applied to an enzyme and should not be used. Enzyme temperature profiles arise from two distinct processes: increasing activity with increasing temperature (as found with all chemical reactions) and inactivation at the higher temperature conditions (due to protein denaturation). The precise maximum in such a temperature profile has no physical meaning because it may vary with the buffer, pH, ionic strength, substrate and cofactor concentrations, presence of additives, and length of time used in the assay. For example, a fixed-time enzyme assay conducted at 60°C for 30 min might give less total product formation than one done at 40°C for the same period, thereby indicating that 60°C was above the "temperature optimum." However, conducting the same assays for only 10 min could reveal greater activity at 60°C than at 40°C since there is less denaturation during the shorter assay. In contrast, a more meaningful parameter is the highest temperature that permits a constant level of activity over the entire assay time. This value can be determined by incubating a fixed amount of enzyme in the standard assay mixture lacking substrate at different temperatures for the period of the assay and then assaying the remaining active enzyme by using a lower temperature at which the enzyme is known to be stable. The highest temperature that retains full enzyme activity should be used for routine assays in order to ensure the highest meaningful rates.

Careful examination of the effect of temperature on enzyme activity over the range where the enzyme is stable can provide a measurement of the **Arrhenius energy of activation**, E_a. The Arrhenius activation energy (i.e., the threshold energy needed for a substrate to undergo reaction) is central to transition state theory and is a term found in the empirical relationship describing the effect of temperature on a rate constant (k), namely,

$$d(\ln k)/dT = E_a/RT^2 \qquad (2)$$

where R is the gas constant (1.98 cal mol^{-1} K^{-1}) and T is the temperature in kelvins. Integration of this equation provides a form that is more easily used for estimation of E_a from rate measurements at different temperatures. The result is shown in equation 3.

$$\ln k = -E_a/RT + \ln A \qquad (3)$$

In this equation, A is a constant of integration. The plot of ln k versus $1/T$ yields a linear relationship with a slope of $-E_a/R$ (or $-E_a/2.303\ R$ if log k is used). For most enzymes, E_a is easily calculated by this type of analysis using standard assays with saturating concentrations of substrate; i.e., plotting ln V_{max} versus inverse temperature. In some cases such plots are biphasic, leading to two slopes and thus two values of E_a. This break in the plot may indicate a change in the rate-limiting step with temperature or can be due to a conformational alteration in the enzyme. In these experiments it is important to recall that some buffers undergo temperature-dependent changes in pH; thus, care must be exercised in carrying out these studies and in making interpretations.

19.3.7.3. Stabilizing Agents

A variety of approaches can be used to enhance the linearity of assays as well as to stabilize enzyme activity during storage. As already mentioned (section 19.3.2.), the initial time and efforts expended to obtain a linear assay often will reward the investigator by allowing far greater ease in obtaining meaningful and reproducible results from subsequent assays. For proteins that are unstable when diluted to assay concentration, the inclusion of another innocuous protein (e.g., bovine serum albumin) in the assay mixture may provide stabilization. In addition, a change in the type of assay vial should be examined in case the enzyme of interest (when highly diluted) binds to the glass or plastic container. Oxidation of surface cysteines to their disulfide forms can lead to undesirable aggregation (and precipitation) of a protein, while oxidation of an active-site cysteine can inactivate an enzyme. These processes can be slowed and sometimes reversed by addition of thiol-containing reagents. Dithiothreitol is generally the thiol reagent of choice due to its favorable thermodynamics; its two thiols are in close proximity so that generation of a mixed disulfide (between dithiothreitol and enzyme cysteine) is rapidly followed by formation of the oxidized form of this compound (an intramolecular disulfide in a stable six-membered ring) and a corresponding reduced cysteine in the enzyme. Ascorbic acid is another reducing agent that has wide application in assays, especially for enzymes requiring a reduced metal ion or cofactor. Because thiols and ascorbic acid undergo spontaneous oxidation in aerobic environments, fresh solutions always should be prepared. Additional compounds known to stabilize selected proteins and enhance the linearity of certain assays include high salt, calcium, or other metal ions, various detergents at low concentration, and polyols such as sucrose or glycerol.

The same types of compounds that enhance assay linearity may also stabilize enzymes for long-term storage. In addition, a very important approach makes use of decreased temperatures. Many enzymes can retain their activity indefinitely when frozen at $-80°C$ (and some even at $-20°C$ or on ice). On the other hand, the thawing process can wreak havoc on the activity of many enzymes. In particular, eutectic salts such as sodium phosphate should not be used as buffers when freezing samples because thawing leads to microenvironments of high and low pH that can denature the protein (by contrast, potassium phosphate does not lead to this problem). Enzyme samples should be stored in small aliquots, rather than in a single fraction, in order to minimize the number of freeze-thaw cycles to which any one sample is exposed. As a counterpoint to the above comments, some enzymes are denatured by cold storage and must be maintained at higher temperatures.

19.3.7.4. Oxygen Lability

The characterization of some enzymes is confounded by their extreme oxygen lability. Within the cell, these enzymes (e.g., nitrogenase, pyruvate-formate lyase, and acetyl-CoA decarbonylase/synthase) retain their activity because they are derived from strictly anaerobic microorganisms, produced only during anaerobic growth, stabilized by protective proteins, or constantly repaired by appropriate regenerating enzymes. The detailed inactivation mechanisms of these enzymes generally are not well understood, but most investigations have shown that metallocenters or side chain radicals in the protein react with oxygen to initiate these damaging processes. Two general approaches can be used to characterize the enzymatic properties of such labile systems. In the first option, all purification steps as well as assays are carried out in the complete absence of oxygen. This approach often requires elaborate modifications to

general laboratory equipment (and typically the use of anaerobic chambers or glove boxes) as well as exquisite care at each stage of analysis. The inclusion of reductants such as sodium dithionite in all buffers helps to ensure maintenance of anaerobicity; however, such compounds may inhibit or inactivate the enzyme of interest. The second option is possible only in selected cases and involves purification of an inactive form of the protein that subsequently is activated to yield functional enzyme. Measurement of enzyme activity typically requires anaerobic procedures even when using the second approach. Fortunately, most assay methods are readily adapted for use with anaerobic conditions. For example, spectrophotometer cuvettes are available with septum enclosures. These or other assay vials can be purged through needles with nitrogen or argon gas to remove oxygen, often via a process that involves multiple cycles of vacuum and gas addition. For exploratory studies to assess whether anaerobic conditions improve assay linearity, cells can be disrupted (e.g., by sonication) while under a constant stream of nitrogen gas and assayed immediately by injection into an anaerobic assay vial.

19.4. BASIC ENZYME KINETICS

The following discussion covers only the most basic elements of enzyme kinetics; yet, this information should be adequate for the characterization of most enzymes of microbiological interest. Single-substrate reactions are used to exemplify the key features of kinetic studies, and the discussion is then extended to include bisubstrate reactions. For more complex situations, the reader is referred to one of the many excellent texts covering this topic (for example, references 3, 5, and 29) and to a very useful website prepared by the International Union of Biochemistry and Molecular Biology (http://www.chem.qmul.ac.uk/iubmb/kinetics/).

19.4.1. Single Substrate

In a single-substrate reaction, the free enzyme (E) combines with substrate (S) to produce an intermediate complex (ES) with a rate constant k_1; this complex decomposes either with first-order rate constant k_2 to produce product(s) (P) and release free enzyme or with k_{-1} to regenerate the starting materials. Reversibility from P to ES can be ignored if only initial velocities in the forward direction are considered and no P has accumulated or is present.

$$E + S \underset{k_{-1}}{\overset{k_1}{\rightleftharpoons}} ES \overset{k_2}{\rightleftharpoons} P + E \qquad (4)$$

The observed forward enzyme velocity or initial rate (v) is proportional to $k_2[ES]$ (i.e., k_2 times the concentration of the ES complex). When [S] (the substrate concentration) is sufficiently large to tie up all enzyme as the ES intermediate, [ES] will equal $[E]_T$, the total enzyme concentration, and the reaction velocity is at its maximum value. Restated, at large [S] the $v = V_{max} = k_2[E]_T$. In more complex reaction schemes, k_2 may not be the only rate constant relating velocity to the amount of ES complex present. The inability to identify which constants are involved in the limiting-rate step often prompts the use of a less committed designation, k_{cat}, a composite rate constant for catalysis.

19.4.1.1. Michaelis-Menten Kinetics

For many single-substrate enzymes, the relationship between v and [S] is given by the Michaelis-Menten equation.

$$v = \frac{V_{max}[S]}{K_m + [S]} \quad \text{or} \quad \frac{k_{cat}[E]_T[S]}{K_m + [S]} \quad (5)$$

In this equation, K_m (the Michaelis constant) is an equilibrium constant defined as $(k_{-1} + k_2)/k_1$. When k_2 is small relative to k_{-1}, K_m approximates the dissociation constant (K_d) for the dissociation of ES to S plus E. Although this assumption is frequently invalid, K_m values usually reflect in some manner the affinity of an enzyme for a substrate. For an enzyme that provides data fitting equation 5, K_m is equal to the substrate concentration at which v is one-half of V_{max}. Accurate determination of K_m generally requires that velocity data be obtained for [S] ranging from about 0.2 to 5 times the expected K_m value. Routine assays of the enzyme are facilitated by use of [S] that is at least fivefold the K_m to approach nonlimiting substrate conditions. With this much substrate, the enzyme usually gives the highest practically attainable velocity (provided there is no inhibition by excess substrate) and the reaction rate is influenced only by $[E]_T$, a condition known as zero-order kinetics. It is useful to keep in mind, however, that even when [S] is equal to 10 times K_m, v is only 0.91 of V_{max}. Under conditions of substrate scarcity, high cost, insolubility, or inhibition (and when a high-specific-radioactivity substrate is being used in an enzyme assay), reliable assays can be readily achieved at lower [S] provided the initial substrate concentration is constant from assay to assay and is not sufficiently changed during the course of an assay so as to affect the observed velocity.

19.4.1.2. Graphical Analysis Methods

With the assistance of modern computational tools, the preferred graphical method for analysis of single-substrate kinetic data is simply to plot v versus [S], as shown in Fig. 2. Representation of data by this approach has two critical advantages over alternative plotting methods. First, the graph generally permits a decision as to whether or not Michaelis-Menten kinetics are being followed. Indeed, if a sufficient number of data points are taken (including substrate concentrations that approach saturating conditions) it allows gross estimation of K_m and V_{max} directly, but these are mainly useful for purposes of providing starting estimates in subsequent curve-fitting routines to obtain more accurate values. Secondly, fits of the data using commercially available curve-fitting programs to the hyperbolic mathematical relationship demanded by the Michaelis-Menten equation provide robust values of K_m and V_{max} as well as meaningful standard deviations. By contrast, estimation of V_{max} "by eye" often leads to a low value (e.g., substrate concentrations that are fivefold the K_m exhibit rates that are only 83% of the V_{max}), which then results in an underestimation of K_m. Other plotting methods present the data in a linear form (rather than as a rectangular hyperbola) and frequently are used. A major drawback of these plots is that the individual data points often are not weighted appropriately during calculation of the best-fit line used to derive kinetic constants (14); thus, one or a few points can unduly bias the results. Furthermore, and of greater concern to a novice kineticist, some of the plots are difficult to interpret by casual inspection. A brief description of three of these methods is provided below, but readers are encouraged to use the direct v-versus-[S] plot as in Fig. 2 for routine analysis of single-substrate reactions.

The Lineweaver-Burk double-reciprocal plot is possibly the most common form for graphing kinetic data to esti-

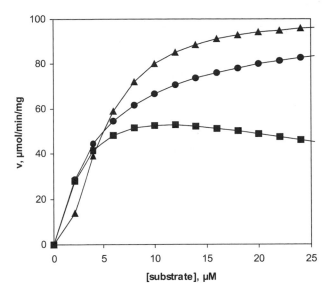

FIGURE 2 Comparison of v-versus-[S] plots for enzymes exhibiting Michaelis-Menten kinetics, substrate inhibition, or cooperative behavior. A single-substrate enzyme following Michaelis-Menten kinetics will yield data forming a rectangular hyperbola that can be fit to equation 5 (●). For the example shown, $V_{max} = 100$ μmol/min/mg and $K_m = 5$ μM. An enzyme exhibiting the same values of V_{max} and K_m, but inhibited by excess substrate (with $K_i = 25$ μM), gives rise to a plot (■) that shows an increase in rate at low concentrations of increasing substrate followed by a decreasing rate at higher [S]. Data such as these should be fit to equation 9 (in particular, they should not be truncated to include only the initial phase). Significantly, the largest observed v (53 μmol/min/mg for the case shown) can be significantly smaller than V_{max}. Positive cooperativity typically results in sigmoidal plots (▲); however, a nonsigmoidal plot with a sharper bend alternatively can arise from negative cooperativity. The data associated with positive cooperativity can be fit to equation 10 to calculate V_{max}, $[S]_{0.5}$ (the substrate concentration when $v = 1/2 V_{max}$), and the Hill coefficient, h (100 μmol/min/mg, 5 μM, and 2, respectively, for the data shown).

mate K_m and V_{max}. This method makes use of the fact that the rectangular hyperbolic function of equation 5 is readily transformed to a linear function in equation 6.

$$\frac{1}{v} = \frac{K_m}{V_{max}[S]} + \frac{1}{V_{max}} \quad (6)$$

Thus, a plot of $1/v$ versus $1/[S]$ yields a straight line of slope K_m/V_{max} and a y-axis intercept of $1/V_{max}$. Also, by setting $1/v$ equal to 0, equation 6 reveals that the x-axis intercept will be $-1/K_m$. In such a plot, the data obtained at low substrate concentrations generally have larger standard deviations than those obtained at high [S]; thus, it is inappropriate to carry out unweighted least-squares analysis to determine the best-fit line associated with the transformed data. Nevertheless, this graphical procedure is useful for preliminary analysis of inhibition patterns (see section 19.4.3) and initial characterization of bisubstrate reactions (section 19.4.2).

Two other linear plotting methods are sometimes encountered. One is the Hanes transformation of the

Michaelis-Menten equation. In this method, equation 5 is rewritten as equation 7.

$$\frac{[S]}{v} = \frac{[S]}{V_{max}} + \frac{K_m}{V_{max}} \qquad (7)$$

If $[S]/v$ is plotted versus $[S]$, a line of slope $1/V_{max}$ is obtained and the y-intercept will be equal to K_m/V_{max}. Extending the line to its x-intercept (i.e., $[S]/v = 0$) gives the value $-K_m$. The second method, generally preferred by those who dislike the double-reciprocal method, is the Eadie-Hofstee plot. This is based on equation 8.

$$v = \frac{-K_m v}{[S]} + V_{max} \qquad (8)$$

A plot of v versus $v/[S]$ normally has a negative slope (i.e., $-K_m$) and a y-intercept of V_{max}. Extrapolation of the line to the x-intercept (where $v = 0$) will give V_{max}/K_m. This plot has the advantage that in most instances K_m and V_{max} can be determined graphically without resorting to the use of a negative axis. The Eadie-Hofstee plot is also more sensitive to deviations from Michaelis-Menten behavior than are the other plot forms and thus may be useful in detecting cooperative substrate binding as described in section 19.4.1.4. Caution is needed, however, since nonlinearity in these plots is more often due to poor data than to non-Michaelis behavior.

19.4.1.3. Substrate Inhibition

The use of high substrate concentrations can in some instances lead to an inhibition of enzyme activity, with a resulting poor fit of single-substrate data to equation 5. As illustrated in Fig. 2 (squares), data analyzed by a direct v-versus-$[S]$ plot show an increase in rate as low concentrations of substrate increase followed by a decreasing rate at higher substrate concentrations. Similarly, none of the various linear graphing procedures yield straight lines using these data. It is inappropriate to treat such data by truncation so that only the low substrate concentration range is used. Rather, an effort should be made to fit these data to an alternative equation. In many cases, the observed behavior may be attributed to binding of a second substrate molecule to the ES complex (3, 5). For this situation, the data are fit to equation 9, which provides values and standard deviations of K_m, V_{max}, and the inhibition constant K_i. It is important to note that the highest rate measured may fall substantially below the true V_{max}; e.g., for the situation where K_i is fivefold the K_m (as shown in Fig. 2), the highest measured rate is only 53% of V_{max}. The three calculated values and their standard deviations should be reported.

$$v = \frac{V_{max}[S]}{K_m + [S] + [S]^2/K_i} \qquad (9)$$

Decreasing activity at high substrate concentrations also may occur in at least three other situations. First, competitive substrate inhibition involves one molecule of substrate binding to the enzyme in an ineffective mode (ES′) that then binds a second substrate molecule producing the noncatalytic ES′S. The equation describing this behavior resembles equation 9 with the additional term $K_m[S]/K_i$ in the denominator (5). Second, apparent substrate inhibition may be observed when the substrate is contaminated with an inhibitory molecule. An equation describing this situation also resembles equation 9, but with $[S]^2/K_i$ replaced

with $K_m[S]/K_i$ (5). Finally, this general type of behavior may arise when an activator molecule (A) is required for catalysis, where substrate binding in the absence of activator leads to an ineffective state (i.e., A binds to E concomitant with or before S, and the ES complex cannot bind A). An example of this situation would be a phosphatase that acts on the Mg-substrate complex, whereas the substrate without associated metal is an inhibitor of the enzyme. Mathematic treatment of several versions of this behavior (see reference 29) is beyond the scope of this chapter.

19.4.1.4. Cooperative Behavior

Another reason that single-substrate data may exhibit a poor fit to equation 5 involves cooperative effects; i.e., binding of substrate to one monomer in a multisite enzyme influences the catalytic behavior of other monomers. In plots of initial velocity versus substrate concentration, two types of cooperativity-based deviations from hyperbolic behavior are observed. Most commonly, such plots exhibit a pronounced sigmoidal nature that is referred to as **positive cooperativity** (depicted in Fig. 2, triangles). This phenomenon arises either from an increase in affinity of unoccupied sites for substrate as a consequence of their interaction with occupied sites (i.e., a decrease in K_m) or to an increase in the catalytic efficiency of one active site as others become saturated with substrate (i.e., an increase in k_{cat}). The net result is to exacerbate the change in rate over a smaller substrate range. An alternative, less common, type of cooperativity yields a sharper bend in the v-versus-$[S]$ curve and is referred to as **negative cooperativity**. This situation arises from the opposite effects of those just described and reduces the sensitivity of the enzyme to small changes in the substrate concentration. The existence of non-Michaelis kinetics is easily discerned by examination of the substrate concentrations that correspond to velocities of $0.9\,V_{max}$ and $0.1\,V_{max}$. This ratio of substrate concentrations (i.e., $[S]_{0.9}/[S]_{0.1}$) is 81 for a true hyperbolic condition but significantly less (usually below 40) for positive cooperativity and higher (often 250 or more) for negative cooperativity.

An expression to describe the behavior of many cooperative enzymes is shown in equation 10.

$$v = \frac{V_{max}[S]^h}{[S]_{0.5}^h + [S]^h} \qquad (10)$$

In this equation, the exponent h is the Hill coefficient, an empirical parameter without physical significance except that it is always less than or equal to the number of interacting sites on the enzyme, and $[S]_{0.5}$ is the substrate concentration when $v = \frac{1}{2} V_{max}$. In the situation where $h = 1.0$, equation 10 becomes equivalent to the Michaelis-Menten relationship for a single-substrate enzyme with independently saturating sites (i.e., no cooperativity). If V_{max} is known, h and $[S]_{0.5}$ can be calculated from the relationship shown in equation 11.

$$\log\left(\frac{v}{V_{max} - v}\right) = h\log[S] - h\log[S]_{0.5} \qquad (11)$$

A plot of $\log[v/(V_{max} - v)]$ versus $\log[S]$ yields a line of slope h. $[S]_{0.5}$ is equal to $[S]$ when $\log[v/V_{max} - v)] = 0$. Equation 11 must be used with care in the calculation of h, however, since the slope at very low and very high substrate concentrations approaches 1.0 even for interacting multisite cooperative enzymes. In practice, one should attempt to estimate h only from data over an intermediate substrate concentration range. Frequently h is determined as the slope at $v = \frac{1}{2} V_{max}$ to avoid uncertainty about where the nonlinear regions might be encountered. Values of h in excess of 1.0 are indicative of positive cooperativity, although errors in its estimation are such that one should be cautious about interpretation in the range between 1.0 and 1.5. Values less than 1.0 suggest negative cooperativity.

19.4.1.5. Reversible Inhibitors

Inhibitors of an enzyme can be used to probe the chemical mechanism of the catalyzed reaction and sometimes can be developed into an antibiotic for pharmaceutical use. This discussion begins with a brief introduction to slow-binding inhibitors and then describes in more detail the basic aspects of rapid-equilibrium inhibitor kinetic analysis along with methods for their analyses. Irreversible inhibitors, better termed inactivators, are dealt with separately in section 19.4.1.6.

Reversible inhibitors fall into two general classes: **slow-binding** and **rapid-equilibrium** inhibitors. In both cases the compounds are specifically recognized and bound by the enzyme, but the distinction refers to the rates of inhibitor binding and dissociation. As illustrated by the progress curves shown in Fig. 3, a slow-binding reversible inhibitor results in a time-dependent loss of enzyme activity (squares) compared to a noninhibited control (circles), whereas a rapid-equilibrium reversible inhibitor (triangles) has an immediate effect on enzyme catalysis. In the slow-binding inhibitor example shown, the initial enzyme rate of 5 µM/min slowly decays over the first few minutes to a final enzyme rate of 1 µM/min. Progress curves for slow-binding inhibition can be analyzed by fitting to equation 12.

$$P_t = v_f t + (v_o - v_f)(1 - e^{-k(app)t})/k(app) \qquad (12)$$

In this equation, P_t, t, and v_o are equivalent to the same values defined for equation 1 and refer to accumulated product at time t and initial velocity (5 µM/min in this example). The term v_f is the final steady-state velocity (1 µM/min here), and k_{app} is the apparent first-order rate constant for establishment of the final equilibrium (0.2 min^{-1}). The latter value comprises several reaction parameters and provides important information about the interaction of enzyme with inhibitor. Slow-binding inhibitors typically exhibit slow dissociation of the enzyme-inhibitor complex. Dilution of such a complex into the assay mixture can result in a curve similar to that shown by the triangles in Fig. 1. Equation 12 can be used to fit this progress-curve with v_o set to zero and k_{app} representing the apparent first-order rate of dissociation to result in establishment of v_f. Slow-binding inhibitors may require very extensive dialysis for dissociation, whereas rapid-equilibrium inhibitors typically are readily reversed by dilution or dialysis. Further analysis of slow-binding inhibitors to extract inhibition constants and other information is beyond the scope of this chapter, but procedures are available in some kinetic textbooks and in a superb review article (25). Significantly, many clinically relevant compounds belong to this group of inhibitors. Further discussion of reversible inhibitors below focuses solely on those exhibiting rapid-equilibrium kinetics.

Rapid-equilibrium reversible inhibitors are usually classified as **competitive**, **noncompetitive**, or **uncompetitive**, depending on how they interact with the enzyme. More complex situations, such as **partial-inhibition** and **mixed-type inhibition**, also occur and are described only briefly below. At the mechanistic level, a competitive inhibitor (I) interacts reversibly only with free enzyme to form an EI complex;

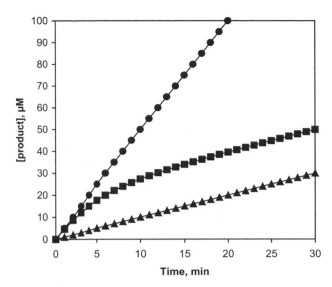

FIGURE 3 Progress curve comparison of slow-binding versus rapid-equilibrium inhibitors. An uninhibited enzyme is shown to form product at a rate of 5 µM/min (●). In the presence of a slow-binding inhibitor (■), the rate decreases from v_o to v_f with an apparent first-order rate constant k_{app} (5 µM/min, 1 µM/min, and 0.2 min^{-1}, respectively, for the data shown). These data can be analyzed by using equation 12. For comparison, the plot associated with a rapid-equilibrium inhibitor (▲) exhibits no time-dependent decrease in enzyme rate (here shown as 1 µM/min).

there can be no ESI complex since substrate and inhibitor are mutually exclusive. A noncompetitive inhibitor, on the other hand, can interact with either free enzyme or enzyme-substrate complex to form EI or ESI, respectively; substrate and inhibitor do not exclude one another, and the ESI complex does not proceed to form product (except when I is a partial-type inhibitor where complete inhibition is never observed). For mixed-type inhibitors, the affinity of I for E and ES differ, somewhat complicating the analysis. An uncompetitive inhibitor in the case of a single-substrate reaction interacts only with ES and not E to form an ESI complex. Although common in multisubstrate reactions, this type of inhibition is seldom encountered for a single-substrate situation. Indeed, in cases where a single-substrate reaction appears to result in uncompetitive inhibition an investigator should consider an underappreciated, alternative model to explain the behavior. Specifically, pseudouncompetitive inhibition can result from a situation where I binds to a form of the enzyme generated by catalysis that is distinct from the resting enzyme species (equation 13) (for an example, see reference 33).

$$E + S \rightleftharpoons ES \rightleftharpoons P + E' \rightleftharpoons E \qquad (13)$$
$$\Updownarrow$$
$$IE'$$

These inhibitors are easily distinguished by differences in their effects on the kinetics of enzyme assays. The assays are run with varied concentrations of substrate at each of several concentrations of inhibitor (including an inhibitor-free control to establish the uninhibited K_m and V_{max}). The data from each inhibitor concentration are analyzed by use of equation 5, resulting in apparent K_m (K_m^{app}) and apparent V_{max} (V_{max}^{app}) values. Competitive inhibitors exhibit changes in K_m^{app} for each [I] while not affecting the V_{max}^{app}, which is equivalent to the true V_{max}. A plot of K_m^{app}/V_{max} versus [I] yields a straight line that intercepts the x axis at $-K_{ic}$, the **inhibition constant** for competitive inhibition. The competitive inhibition data are described by equation 14.

$$v = \frac{V_{max}[S]}{K_m(1 + [I]/K_{ic}) + [S]} \tag{14}$$

Noncompetitive inhibitors yield changes in V_{max}^{app} and no change in K_m^{app}, which is equivalent to the true K_m. The K_i is obtained from the negative of the x-intercept in a plot of K_m/V_{max}^{app} or $1/V_{max}^{app}$ versus [I]. For mixed inhibition, the first plot is used to discern the K_i associated with I binding to E (abbreviated K_i^E) and the second plot is used to obtain the K_i for I binding to ES (i.e., K_i^{ES}). Mixed inhibition data are described by equation 15, and simple noncompetitive inhibition uses the same equation with K_i^E equal to K_i^{ES}.

$$v = \frac{V_{max}[S]}{K_m(1 + [I]/K_i^E) + [S](1 + [I]/K_i^{ES})} \tag{15}$$

Uncompetitive inhibitors exhibit changes in both V_{max}^{app} and K_m^{app}, but with constant V_{max}^{app}/K_m^{app}. The uncompetitive inhibition constant, K_{iu}, is readily calculated as the x-intercept in a plot of $1/V_{max}^{app}$ versus [I]. The equation describing these data is shown below.

$$v = \frac{V_{max}[S]}{K_m + [S](1 + [I]/K_{iu})} \tag{16}$$

Note that each of the equations reverts to equation 5 when [I] is zero.

Despite their drawbacks (see section 19.4.1.2.), double-reciprocal plots are often used to evaluate inhibitor data because of the graphic differences in appearance of the different inhibition types. Namely, competitive inhibitors yield lines that merge on the y axis (Fig. 4A), pure noncompetitive inhibitors result in plots with lines merging on the x axis (Fig. 4B), and uncompetitive inhibitors lead to a series of parallel lines in this type of plot (Fig. 4C). Mixed inhibition leads to convergence points that do not lie on an axis. The double-reciprocal plots of competitive, noncompetitive, and uncompetitive inhibition (Fig. 4) can be fit to

equations 17 through 19, and the V_{max}, K_m, and appropriate K_i values can be determined.

$$\frac{1}{v} = \frac{K_m(1 + [I]/K_{ic})}{V_{max}[S]} + \frac{1}{V_{max}} \tag{17}$$

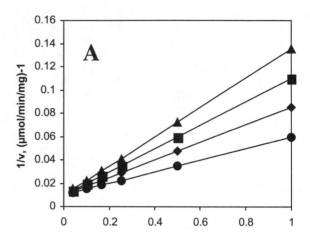

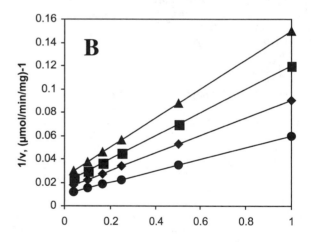

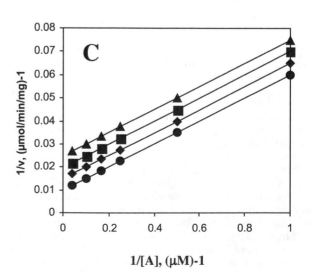

FIGURE 4 Double-reciprocal plots illustrating three patterns of inhibition. (A) Kinetic analysis of a single-substrate reaction in the presence of a rapid-equilibrium competitive inhibitor yields a series of lines intersecting on the y axis. These data can be fit to equation 17. (B) Noncompetitive inhibitors result in lines intersecting to the left of the y axis at a point above, below, or on the x axis (as shown), depending on the relative values of K_i^E and K_i^{ES}. These data can be fit to equation 18. (C) Uncompetitive inhibition can give rise to a series of parallel lines that can be fit to equation 19. This type of inhibition is seldom encountered with single-substrate enzymes, and an alternative model involving pseudouncompetitive inhibition (33) should be considered. For the plots shown, $V_{max} = 100$ μmol/min/mg, $K_m = 5$ μM, K_{ic}, K_i^E, K_i^{ES}, and K_{iu} were each set to 2 μM, and [I] = 0 (●), 1 (◆), 2 (■), and 3 (▲) μM.

$$\frac{1}{v} = \frac{K_m (1 + [I]/K_i^E)}{V_{max}[S]} + \frac{1 + [I]/K_i^{ES}}{V_{max}} \qquad (18)$$

$$\frac{1}{v} = \frac{K_m}{V_{max}[S]} + \frac{1 + [I]/K_{iu}}{V_{max}} \qquad (19)$$

19.4.1.6. Inactivators

Compounds that irreversibly inhibit an enzyme (e.g., by chelating a required metal ion or forming a covalent bond with targeted side chains) should be termed inactivators, rather than inhibitors. This is true even in cases where the inactivation might be reversible under selective conditions, such as the reaction of Ellman's reagent (DTNB) with protein cysteinyl groups that can be reversed by addition of thiol reagent. Careful analysis of inactivator reactivity with an enzyme can help to identify functional groups at the active site; a treatise summarizing the amino acid side chain specificity of various reagents is available (16). It is important to emphasize, however, that kinetic characterization of these reactions is distinct from that used to characterize reversible inhibitors.

Kinetic analysis of an inactivator involves measuring remaining enzyme activity as a function of time after initiation of inhibitor treatment, at a variety of inactivator concentrations. Time course incubations are best conducted at sufficient enzyme concentrations that samples can be diluted to negligible inactivator concentration during the assays. Since the concentration of the inhibitor generally is much greater than that of the enzyme, its concentration will not significantly change during inactivation and the process will be pseudo-first order. The slope of a plot of log(activity remaining) versus time gives an observed first-order rate constant in units of reciprocal time (time^{-1}) for the concentration of inactivator used. If the reaction is of the simple type E + I → EI*, where EI* is inactive, a plot of observed rate constant versus [I] will be linear with a slope reflecting the influence of [I] on the inactivation rate. This result has the units of a second-order rate constant, molar^{-1} time^{-1}. This value should be reported as the measure of the effectiveness of the inactivator, rather than terms such as the inactivator concentration necessary to give 50% inhibition in a given time. When the inactivation proceeds in two steps, initially forming an EI intermediate that then is converted to EI* (E + I ⇌ EI → EI*), the plot of observed rate constant versus [I] will be hyperbolic rather than linear. Fitting of these data to a hyperbolic equation analogous to equation 5 {or linearization by plotting 1/(observed rate) versus 1/[I] and computing the inverse of the slope} provides both a value reflecting the binding affinity of the inactivator and a second-order rate constant that reflects the maximum rate constant of enzyme inactivation.

19.4.2. Bisubstrate Reactions

Most biological reactions involve two (or more) substrates, and the observed rate depends on the concentrations and binding affinities of each. Kinetic analysis of bisubstrate reactions is often much more complicated than that of single-substrate reactions (and of further increasing complexity for each additional substrate), but the same general principles as described above apply (i.e., increasing activity with increasing substrate, inhibition by excess substrate, cooperativity in multisite enzymes, reversible inhibitors acting by various distinct mechanisms, and covalent modification of critical residues resulting in enzyme inactivation). It is important to understand that the apparent K_m for any particular substrate may be influenced by the concentrations of all other substrates; thus, the "true" K_m of a substrate should be estimated only when all cosubstrates are present at saturating concentrations. Apparent K_m values are of little use and generally should not be reported. Likewise, a true V_{max} value is measured only when all substrates are at saturating concentrations. Questions concerning the order of substrate binding and product release, whether new intermediates form before all substrates are bound, and how inhibitors might affect overall kinetics while interacting with only one form of the enzyme are all of interest and can be approached through kinetic analysis. In the following sections, a general experimental design is provided along with descriptions of different types of data analysis for enzymes that utilize different kinetic mechanisms.

19.4.2.1. Experimental Design

In favorable cases, one can determine K_m for one substrate by varying its concentration while maintaining the other(s) at a kinetically saturating level. Each set of v-versus-[S] data is analyzed as described for the single-substrate case to yield the kinetic constants. Unfortunately, this simple approach often cannot be used. For example, a very large excess of the cosubstrate may not be possible due to high expense, limited availability, or solubility concerns. Also, excess concentrations of a substrate can be inhibitory, thereby preventing its inclusion in a reaction mixture at a high level. These situations mandate a more thorough and general approach, but one that also can provide additional information about the enzyme and its action.

A general experimental design is to vary each substrate in a systemic manner that allows estimation of kinetic saturation by extrapolation to infinite concentration. Depending on the complexity of the assay, expense involved, and time required, between four and eight concentrations of each substrate should be used, ranging three- to fourfold above and below the preliminary K_m estimate. The total number of assays generally should be between 16 (four concentrations of the varied substrate at each of four concentrations of the unvaried substrate) and 64 (eight concentrations of each substrate). It is advisable to conduct an initial experiment of only 16 assays to gain a rough idea of the best choice of substrate concentrations so as to span a satisfactory range. As in the single-substrate analysis, enzyme concentration should be the same for all assays, so special attention must be given to stabilization of the enzyme sample if all assays are not done concurrently.

The data from these assays are analyzed to discern the effects of each substrate on the K_m^{app} and V_{max}^{app} of the other substrate. As described earlier, direct v-versus-[S] plots can be used for precise calculation of the kinetic constants; however, double-reciprocal plots of data for bisubstrate reactions have great value in distinguishing between two general types of enzyme mechanisms. In particular, plots of 1/v versus 1/[S] that exhibit a set of parallel lines (reminiscent of the situation shown in Fig. 4C for double-reciprocal plots of uncompetitive inhibition) are diagnostic of a **substituted enzyme** (or **ping-pong**) mechanism (Fig. 5), in which at least one product must be released before all substrates bind. In contrast, double-inverse plots with lines merging to the left of the y axis indicate a **sequential mechanism** (Fig. 6) in which all substrates must bind prior to release of any product. Care must be used in deciding between these two options since lines may appear to be parallel but in fact actually converge upon extrapolation. The straightforward analysis of substituted-enzyme data to extract K_m and V_{max} information

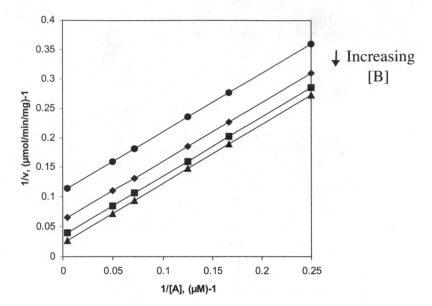

FIGURE 5 Double-reciprocal plot of kinetic data for a substituted enzyme mechanism. A plot of $1/v$-versus-$1/[A]$ is illustrated for a bisubstrate enzyme reaction ($V_{max} = 100$ μmol/min/mg, $K_{mA} = 100$ μM, and $K_{mB} = 5$ μM) at [B] = 0.5 (●), 1 (♦), 2 (■), and 4 (▲) μM. The data are fit to equation 23.

is described further in section 19.4.2.2. The analysis of sequential mechanism enzymes is more complex because the substrates may bind to the enzyme in either an ordered or random fashion. Section 19.4.2.3 describes how product inhibition patterns can be used to distinguish these possibilities and thereby allow definition of the kinetic constants.

19.4.2.2. Substituted Enzymes

Double-reciprocal plots giving rise to parallel lines (as in Fig. 5) are common among flavin-containing enzymes, phosphotransferases, transaminases, and other types of enzyme transferases. These enzymes all catalyze a two-step reaction (equations 20 and 21) in which the first step transforms one substrate (A) to product (P), yielding a modified form of the enzyme (E'). After product release, the modified enzyme reacts with the second substrate (B) to yield the final product (Q) and restore the initial form of the enzyme.

$$E + A \rightleftarrows (EA \rightleftarrows E'P) \rightleftarrows E' + P \quad (20)$$

$$E' + B \rightleftarrows (E'B \rightleftarrows EQ) \rightleftarrows E + Q \quad (21)$$

For the examples mentioned above, the difference between E and E' relates to the flavin reduction state, the presence of enzyme phosphorylation, or whether ammonia is bound to the associated pyridoxal phosphate cofactor, but other possibilities include changes in protein acetylation, methylation, carboxylation, etc.

The steady-state velocity equation for this type of reaction in the absence of significant P and Q is illustrated in equation 22.

$$v = \frac{V_{max}[A][B]}{K_{mB}[A] + K_{mA}[B] + [A][B]} \quad (22)$$

The terms K_{mA} and K_{mB} denote the individual Michaelis constants of the two substrates. Equation 22 can be rewritten to illustrate the effect of varying [A] or [B] in reciprocal plots, as illustrated in equations 23 and 24.

$$\frac{1}{v} = \frac{K_{mA}}{V_{max}[A]} + \frac{1 + K_{mB}/[B]}{V_{max}} \quad (23)$$

$$\frac{1}{v} = \frac{K_{mB}}{V_{max}[B]} + \frac{1 + K_{mA}/[A]}{V_{max}} \quad (24)$$

Thus, for each [B] the apparent K_{mA} and V_{max}^{app} are calculated on the basis of values for v at varied [A]; similarly for each [A] the apparent K_{mB} and V_{max}^{app} are determined for varied [B]. A replot of $1/V_{max}^{app}$ versus $1/[B]$ yields a line with y intercept of $1/V_{max}$ and x-intercept of $-1/K_{mB}$. In the same manner, a replot of $1/V_{max}^{app}$ versus $1/[A]$ provides a y-intercept of $1/V_{max}$ and x-intercept of $-1/K_{mA}$. These kinetic methods to define the associated constants are often coupled with additional methods to confirm a substituted enzyme mechanism. For example, changes in flavin absorption, incorporation of radiolabeled phosphate into the enzyme, and other approaches can demonstrate the presence of two enzyme forms related to catalysis.

19.4.2.3. Sequential Enzymes

Double-reciprocal plots showing lines that intersect to the left of the y axis (as illustrated in Fig. 6) can arise from **ordered binding** or **random binding** (as well as other less common schemes that are not considered here).

A typical ordered-binding mechanism is illustrated in equation 25. Free E binds only one of the two substrates, the second substrate then binds, catalysis occurs, and the products are released in an ordered fashion.

$$E + A \underset{k_{-1}}{\overset{k_1}{\rightleftarrows}} \underset{+}{\overset{}{EA}} \qquad \underset{+}{\overset{}{EQ}} \underset{k_{-4}}{\overset{k_4}{\rightleftarrows}} Q + E$$
$$B \qquad \qquad P$$
$$k_{-2} \left\| k_2 \qquad k_3 \right\| k_{-3}$$
$$\left(EAB \underset{k_{-c}}{\overset{k_c}{\rightleftarrows}} EPQ \right) \qquad (25)$$

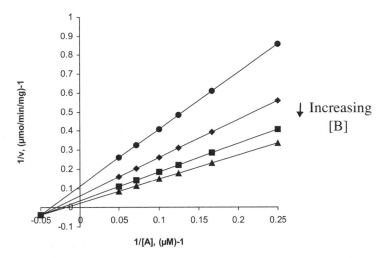

FIGURE 6 Double-reciprocal plot of kinetic data for a sequential enzyme mechanism. A plot of $1/v$-versus-$1/[A]$ is illustrated for a bisubstrate nonsubstituted enzyme reaction. Inspection of such a plot cannot distinguish between ordered-binding and random-binding reactions, because either mechanism yields lines that intersect to the left of the y axis (this point may lie above, on, or below the x axis, depending on the specific kinetic parameters). Product inhibition studies should be carried out (Table 1) to discern whether the data fit to equation 29 or 31. The data shown are for an enzyme operating by an ordered-binding mechanism with $V_{max} = 100$ μmol/min/mg, $K_{mA} = 100$ μM, $K_{mB} = 5$ μM, and $K_{iA} = 20$ μM, at $[B] = 0.5$ (●), 1 (◆), 2 (■), and 4 (▲) μM.

The corresponding velocity equation when no products are present is given in equation 26, where the Michaelis constants for the two substrates are comprised of groups of rate constants and K_{iA} is an inhibition constant (the details of which are not described here but are discussed in most kinetic texts).

$$v = \frac{V_{max}[A][B]}{K_{iA}K_{mB} + K_{mB}[A] + K_{mA}[B] + [A][B]} \quad (26)$$

The opposite situation, a random mechanism, is illustrated in equation 27; either substrate can bind to the free enzyme, and either product can depart first.

$$ \begin{array}{cccc} & K_A & & \\ E + A & \rightleftharpoons EA & EQ \rightleftharpoons Q + E & \\ + & + & + & + \\ B & B & P & P \\ \Big\| K_B & \Big\| \alpha K_B & \Big\| & \Big\| \\ \alpha K_A & k_{cat} & & \\ EB & \rightleftharpoons EAB \rightleftharpoons EPQ \rightleftharpoons Q + EP & \end{array} \quad (27)$$

The velocity equation representing this mechanism (equation 28) looks very similar to that shown in equation 26; however, each substrate has two dissociation constants (not necessarily identical to the Michaelis constants) that are related by the binding interaction factor, α. When $\alpha = 1$, the binding of either substrate to the enzyme is independent of the other. If $\alpha > 1$, the binding of one substrate decreases the affinity of the enzyme for the other substrate. Conversely, $\alpha < 1$ indicates that binding of one substrate increases the affinity of the enzyme for the other substrate.

$$v = \frac{V_{max}[A][B]}{\alpha K_A K_B + \alpha K_B[A] + \alpha K_A[B] + [A][B]} \quad (28)$$

In order to determine the values of the kinetic constants for a bisubstrate enzyme of interest, an investigator must first identify which kinetic scheme applies: ordered-binding (equation 25) or random (equation 27). This analysis is most easily done by examination of product inhibition profiles. (Note that the addition of one product as an inhibitor does not violate initial velocity conditions because the reverse reaction can proceed only if both products are present.) As briefly summarized in Table 1, inhibition of the forward reaction by individual products yields different patterns of inhibition for the two mechanisms. Definitions

TABLE 1 Product inhibition patterns for two common types of bisubstrate reactions[a]

Mechanism	Variable	Fixed (conc)[b]	Product	Pattern[c]
Ordered	A	B (U or S)	Q	C
Ordered	B	A (U)	Q	NC[d]
Ordered	B	A (S)	Q	None
Ordered	A	B (U)	P	NC[d]
Ordered	A	B (S)	P	UC
Ordered	B	A (U or S)	P	NC[d]
Random[e]	A	B (U)	Q	C
Random[e]	A	B (S)	Q	None
Random[e]	A	B (U)	P	C
Random[e]	A	B (S)	P	None

[a]Patterns are described for double-reciprocal plots of $1/v$ versus $1/[variable substrate]$ for varied concentrations of product at the stated [fixed substrate].

[b]Fixed substrate concentration range: U, unsaturated; S, saturated.

[c]C, competitive; NC, noncompetitive; U, uncompetitive; none, not inhibitory.

[d]Convergence point is not necessarily on the horizontal axis but is always to the left of the vertical $1/v$ axis.

[e]In this mechanism, assignments of A and B are arbitrary, as are P and Q. The patterns stated apply only if there are no dead-end complexes, such as EAP or EBQ.

of these inhibition patterns are described in section 19.4.1.5, and the various types (competitive, noncompetitive, and uncompetitive) are easily distinguished by graphical methods akin to those shown in Fig. 4.

Lactate dehydrogenase is used as an example to illustrate the power of product inhibition analysis. This enzyme follows an ordered reaction, with NADH (substrate A) binding before pyruvate (substrate B) and products released in the following order: lactate (product P), then NAD$^+$ (product Q). According to Table 1, NAD$^+$ should be a competitive inhibitor toward NADH regardless of whether the amount of pyruvate is rate limiting or saturating (whereas this is not the case for a random mechanism). Table 1 further indicates that pyruvate should be noncompetitively inhibited by lactate (whereas, again, this is not the case in a random mechanism). Several other patterns of inhibition are indicated in Table 1 and should be examined before accepting the identity of which substrate is A versus B. In general, examination of the patterns of inhibition resulting from the presence of each product can help to identify which substrate binds first and which mechanism is utilized. Once the type of bisubstrate mechanism is identified, the kinetic constants can be determined.

19.4.2.4. Analysis for Ordered Binding

Further analysis of ordered-binding reactions focuses on methods to calculate the kinetic constants. Equation 26 can be transformed into equations 29 and 30 to illustrate the effect on $1/v$ for changing $1/[A]$ while holding [B] fixed, and vice versa. These equations are not symmetrical because the two substrates bind in a defined order.

$$\frac{1}{v} =$$

$$\frac{K_{mA}}{V_{max}}\left(1 + \frac{K_{iA}K_{mB}}{K_{mA}[B]}\right)\frac{1}{[A]} + \frac{1}{V_{max}}\left(1 + \frac{K_{mB}}{[B]}\right) \quad (29)$$

$$\frac{1}{v} = \frac{K_{mB}}{V_{max}}\left(1 + \frac{K_{iA}}{[A]}\right)\frac{1}{[B]} + \frac{1}{V_{max}}\left(1 + \frac{K_{mA}}{[A]}\right) \quad (30)$$

Reciprocal plots of data for an ordered bisubstrate situation result in two families of lines (one for each 1/[substrate] shown in the x axis) that intersect at a common point to the left of the y axis. For the $1/v$ versus $1/[A]$ plot associated with equation 29, the x axis value corresponding to the point of intersection equals $-1/K_{iA}$. The intersection point may be above, on, or below the x axis, depending on the K_{iA} and K_m values. Extraction of the y-intercept values of this same plot (i.e., $1/V_{max}^{app}$) and comparison to $1/[B]$ yield a line of slope K_{mB}/V_{max}, intercept equal to $1/V_{max}$, and x-intercept of $-1/K_{mB}$. Similarly, for $1/v$ versus $1/[B]$ data plotted according to equation 30, a replot of $1/V_{max}^{app}$ versus $1/[A]$ results in a line of slope K_{mA}/V_{max}, intercept equal to $1/V_{max}$, and x-intercept of $-1/K_{mA}$. Also, the slopes of the lines in this reciprocal plot can be replotted versus $1/[A]$ to provide a line with slope of $K_{mB}K_{iA}/V_{max}$ and x-axis intercept of $-1/K_{iA}$. Thus, after establishing that a reaction proceeds by an ordered mechanism and identifying which substrate binds first, the determination of the kinetic constants is straightforward.

19.4.2.5. Analysis for Random Binding

If product inhibition studies reveal that the bisubstrate reaction of interest involves a random order of binding sub-

strates (equation 27), an alternative approach is used to calculate the kinetic constants. By appropriate transformation of equation 28, equations 31 and 32 are derived to illustrate the effect on $1/v$ of changing $1/[A]$ while holding [B] fixed, and vice versa. These equations are symmetrical, unlike the case of ordered binding.

$$\frac{1}{v} = \frac{\alpha K_A}{V_{max}}\left(1 + \frac{K_B}{[B]}\right)\frac{1}{[A]} + \frac{1}{V_{max}}\left(1 + \frac{\alpha K_B}{[B]}\right) \quad (31)$$

$$\frac{1}{v} = \frac{\alpha K_B}{V_{max}}\left(1 + \frac{K_A}{[A]}\right)\frac{1}{[B]} + \frac{1}{V_{max}}\left(1 + \frac{\alpha K_A}{[A]}\right) \quad (32)$$

The reciprocal plots derived from a random order sequential mechanism exhibit an intersection point above the x axis if $\alpha < 1$, below the x axis if $\alpha > 1$, and on the x axis if $\alpha = 1$. The intercept occurs at an x-axis value of $-1/K_A$ for plots of different fixed levels of [B] and at $-1/K_B$ for plots of different fixed levels of [A]. Furthermore, as shown by equations 33 and 34 the values for K_A, K_B, αK_A, αK_B, and V_{max} are easily extracted from secondary plots of $1/V_{max}^{app}$ versus $1/[B]$ along with plots of slope values (from the $1/[A]$ lines) versus $1/[B]$. Parallel equations exist for lines obtained for v-versus-$1/[B]$ plots.

$$\frac{1}{V_{max}^{app}} = \frac{\alpha K_B}{V_{max}}\frac{1}{[B]} + \frac{1}{V_{max}} \quad (33)$$

$$\text{Slope of } \frac{1}{1/[A] \text{ plot}} = \frac{\alpha K_A K_B}{V_{max}}\frac{1}{[B]} + \frac{\alpha K_A}{V_{max}} \quad (34)$$

19.4.3. Regulation of Enzyme Activity

19.4.3.1. Allosteric Regulators

Allosteric regulators are cellular effector molecules that bind noncovalently to noncatalytic sites on enzymes and influence their catalytic activity (8). These effectors are not acted on by the enzyme and may be structurally unrelated to the substrate of the enzyme that they affect. Significantly, they are nearly always related to the metabolic function of the enzyme. For example, an allosteric regulator may be the end product in a pathway for which the enzyme serves as a major control point. The presence of the effector molecule may be stimulatory to catalysis (activation) or inhibitory. Many allosteric enzymes also happen to exhibit cooperativity, and cooperative binding may be noted for effector molecules. In some cases, the allosteric regulators lead to an exaggeration (or reduction) of the sigmoidal response of velocity to substrate concentration.

When a sigmoidal response is noted in a plot of v-versus-[I] (here [I] refers to the allosteric regulator concentration) at a fixed substrate concentration, the relationship can be analyzed as was done for analysis of cooperative interactions (see section 19.4.1.4). For an effector serving as an allosteric inhibitor, the velocity (v_i) measured at some concentration of effector, [I], is compared with the uninhibited velocity, v_o. If the inhibition is not complete when [I] is in great excess, the quantity v_{lim} should be determined; v_{lim} is the rate observed when [I] is infinite. Equation 35 then provides a means for calculating both the Hill coefficient (h) associated with I and [I]$_{0.5}$, the concentration at which ($v_o - v_i$) equals $\frac{1}{2}$ ($v_o - v_{lim}$). A Hill-type plot of $\log[(v_o - v_i)/(v_i - v_{lim})]$ versus $\log[I]$ has a slope of h.

$$\log[(v_o - v_i)/(v_i - v_{lim})] = h \log[I] - h \log[I]_{0.5} \quad (35)$$

Kurganov (13) cautions against assuming that v_{lim} is zero and cites examples where this error has occurred. If the effector is an activator rather than an inhibitor, a related equation can be used to calculate h (8, 13).

19.4.3.2. Covalent Regulation

Reversible covalent changes in bacterial proteins also can have regulatory significance. For example, the best-documented instance of this type of regulatory behavior involves glutamine synthetase of *E. coli* (32). This enzyme is inactivated upon adenylylation of a specific tyrosyl side chain in the protein, and specific enzymes are present to control the extent of covalent modification. Additional proteins are known to undergo phosphorylation (e.g., the *E. coli dnaK* protein), methylation (e.g., chemotaxis transducer proteins), ADP-ribosylation (nitrogenase reductase of *Rhodospirillum rubrum*), and other types of covalent modification as a means of regulating activity. The changes are not intended to be permanent; thus, enzymes exist for carrying out the modifications (i.e., adenylyltransferase, specific protein kinases, or methyltransferases) and for their removal (i.e., adenylyl-removing enzyme, phosphatases, and methylesterases). The keys to understanding these systems are to detect the existence of a covalent modification (e.g., by use of radiolabeling studies, mass spectrometry, or shift in electrophoretic or chromatographic behavior) and determine the chemical structure of the added moiety, establish the catalytic consequences of the derivatization, and identify effectors or conditions that influence the relative activities of the modification and demodification enzymes.

19.5. CONCLUDING COMMENTS

Any microbial enzyme activity of interest can be investigated by using the information provided in sections 19.2 to 19.4, which summarize various types of enzyme preparations, highlight practical considerations for the design and optimization of an enzyme assay, and detail basic approaches for obtaining and analyzing of enzyme kinetic data. The results obtained by such analysis for an enzyme from one particular source should be compared to the wealth of information already obtained for analogous enzymes from other sources. A useful resource to search this vast information landscape is an online website operated by the Cologne University BioInformatics Center (Brenda: The Comprehensive Enzyme Information System; http://brenda.bc.uni-koeln.de/). This site summarizes the substrates and products associated with a target reaction, specifies whether cofactors or metal ions are required, describes sources of the purified enzyme, and highlights features related to enzyme purification and storage, temperature characteristics, pH optimum and ranges, kinetic parameters, inhibitors, activators, presence of covalent modifications, and protein structures. Such comparison can provide useful insights into the metabolic differences among microorganisms and may reveal a unique aspect of the target enzyme under study. More significant is the possibility that no comparable enzyme has been characterized previously. In such a situation, the enzymatic information gained may be especially useful; e.g., it could lead to the search for an antibiotic with new target specificity.

Information garnered from the detailed characterization of individual enzymes can be applied, with caution, to broader questions of intermediary metabolism. For example, the sum of all known *E. coli* reactions has been assembled into a set of interconnected pathways representing the total metabolism of that microorganism (Ecocyc: Encyclopedia of *Escherichia coli* genes and metabolism; http://www.ecocyc.org/). This website also includes partial metabolic reaction schemes for over 190 other microorganisms. Another metabolism-related resource focuses on biocatalysis/biodegradation pathways among a wider group of microorganisms (The University of Minnesota Biocatalysis/Biodegradation Database; http://umbbd.ahc.umn.edu/). At the time of this writing, over 1,100 reactions representing about 165 degradation pathways are available with information on the microbial species capable of performing the reactions, the associated enzymes and their reactions, and the structures and properties of the chemicals making up the pathway, along with key references. Elucidation of individual pathways such as these is obtained by a combination of experimental methods including use of metabolic inhibitors, examination of mutants that are defective in the pathway, isolation and characterization of intermediates, and analysis of upstream and downstream reactions. It is important to stress, however, that the pathways shown in these resources only indicate which chemical reactions exist; i.e., one cannot infer details about chemical fluxes through the associated enzymes.

A critical point of caution when drawing in vivo conclusions based on in vitro activity measurements is that the cellular environment is clearly distinct from that of an isolated enzyme. For example, the enzyme activity may be affected by the presence of allosteric regulators (as described in section 19.4.3.1), by association with membranes, and by interaction with other proteins. To illustrate the latter point, there is growing evidence for associations of enzymes within some pathways leading to a channeling of metabolites from one enzyme to another without release of a free intermediate into the cellular milieu (31). This means that the apparent K_m measured for a downstream enzyme does not reflect its ability to utilize the substrate that is directly provided by an earlier enzyme in the pathway. In some cases, one enzyme may sterically block access to the active site of the second enzyme and prevent binding by exogenous substrate. Detailed molecular-level understanding of substrate channeling is now available in selected examples from X-ray crystallographic structures (21). Another important point to note is that enzyme activity measurements are typically carried out under very-dilute-protein conditions, whereas the intracellular concentration may be much higher for particular enzymes. As one example, ribulosebisphosphate carboxylase concentrations typically exceed 10 mg/ml in chloroplasts so that the concentration of enzyme active sites may approach, or even be larger than, the substrate concentration; similar assay conditions generally would not be examined during in vitro enzyme activity measurements. In other cases, the aggregation state and kinetic properties of the enzyme may be influenced by the protein concentration; thus, the activities measured with the diluted enzyme may not pertain to the in vivo activities.

Under the appropriate circumstances it is possible to overcome some of the limitations mentioned above by monitoring enzyme activities within the cell. As an example, Walsh and Koshland (34) evaluated the fluxes through the tricarboxylic acid cycle and the glyoxylate bypass for *E. coli* growing on acetate. Their approach included the use of radioactive acetate that was released as $^{14}CO_2$ or became incorporated into various cell constituents (e.g., saponifiable lipids) and intercellular metabolites, which were identified and quantified by liquid chromatography. These labeling data were supplemented with ^{13}C-NMR results from metabolites formed during growth on [^{13}C]acetate, most notably the labeling pattern in glutamate, and by kinetic values determined for isocitrate dehydrogenase and isocitrate lyase. The fluxes estimated from the in vivo-labeling measurements were in general agreement with fluxes calculated from kinetic data on the enzymes, a satisfying situation but one that may not hold in other instances.

Taking the above free-enzyme characterization and metabolic-flux approaches a step further, efforts are underway to develop an accurate simulation of all activities in an *E. coli* cell that can respond to changes in its environment (Project Cybercell; http://cybercell.biochem.ualberta.ca/). The virtual cell is being designed on the basis of knowledge at the whole-cell level concerning the fluxes (diffusion rates in the cytoplasm and across the membrane), concentrations, and transformations of small-molecule metabolites with time in three dimensions. It incorporates the influence of molecular complexes on activity and allows for monitoring dynamic responses to altered input and examination of the effects of mutations. While the development of a virtual cell clearly demonstrates that a cell is much more than simply "a bag of enzymes," it is important to keep in mind that an integrated understanding of cellular metabolism requires a foundation that begins with analyses of enzyme activities as described in this chapter.

19.6. REFERENCES

1. **Blakeley, R. L., E. C. Webb, and B. Zerner.** 1969. Jack bean urease (EC 3.5.1.5). A new purification and reliable rate assay. *Biochemistry* **8:**1984–1990.
2. **Campbell-Burk, S. L., and R. G. Shulman.** 1987. High-resolution NMR studies of *Saccharomyces cerevisiae. Annu. Rev. Microbiol.* **41:**595–616.
3. **Cornish-Bowden, A.** 1995. *Fundamentals of Enzyme Kinetics.* Portland Press Ltd., London, United Kingdom.
4. **Deutscher, M. P.** 1990. *Guide to Protein Purification,* Academic Press, Inc., San Diego, CA.
5. **Dixon, M., and E. C. Webb.** 1979. *Enzymes,* 3rd ed. Academic Press, Inc., New York, NY.
6. **Easterby, J. S.** 1984. The kinetics of consecutive enzyme reactions. *Biochem. J.* **219:**843–847.
7. **Gerlt, J. A., and P. C. Babbitt.** 2001. Divergent evolution of enzymatic function: mechanistically diverse superfamilies and functionally distinct suprafamilies. *Annu. Rev. Biochem.* **70:**209–246.
8. **Herve, G.** 1989. *Allosteric Enzymes.* CRC Press, Inc., Boca Raton, FL.
9. **Hoben, P., N. Royal, A. Cheung, F. Yamao, K. Biemann, and D. Soll.** 1982. *Escherichia coli* glutaminyl-tRNA synthetase. II. Characterization of the *glnS* gene product. *J. Biol. Chem.* **257:**11644–11650.
10. **Holm, R. H., P. Kennepohl, and E. I. Solomon.** 1996. Structural and functional aspects of metal sites in biology. *Chem. Rev.* **96:**2239–2314.
11. **Irvin, R. T., T. J. MacAlister, and J. W. Costerton.** 1981. Tris (hydroxymethyl)aminomethane buffer modification of *Escherichia coli* outer membrane permeability. *J. Bacteriol.* **145:**1397–1403.
12. **Knowles, J. R.** 1976. The intrinsic pK_a-values of functional groups in enzymes: improper deductions from the pH-dependence of steady-state parameters. *CRC Crit. Rev. Biochem.* **4:**165–173.
13. **Kurganov, B. I.** 1982. *Allosteric Enzymes. Kinetic Behavior,* p. 56–64. John Wiley & Sons, Inc., New York, NY.
14. **Leatherbarrow, R. J.** 1990. Using linear and non-linear regression to fit biochemical data. *Trends Biochem. Sci.* **19:**455–458.
15. **Lorimer, G. H., M. R. Badger, and T. J. Andrews.** 1977. D-Ribulose-1,5-bisphosphate carboxylase-oxygenase. Improved methods for the activation and assay of catalytic activities. *Anal. Biochem.* **78:**66–73.
16. **Lundblad, R. L.** 1995. *Techniques in Protein Modification.* CRC Press, Boca Raton, FL.
17. **MacElroy, R. D., and C. R. Middaugh.** 1982. Bacterial ribosephosphate isomerase. *Methods Enzymol.* **89:**571–579.
18. **Marschke, C. K., and R. W. Bernlohr.** 1973. Purification and characterization of phosphofructokinase of *Bacillus licheniformis. Arch. Biochem. Biophys.* **156:**1–16.
19. **Maurer, P., D. Lessmann, and G. Kurz.** 1982. D-Glucose-6-phosphate dehydrogenases from *Pseudomonas fluorescens. Methods Enzymol.* **89:**261–270.
20. **McClure, W. R.** 1969. A kinetic analysis of coupled enzyme assays. *Biochemistry* **8:**2782–2786.
21. **Miles, E. W., S. Rhee, and D. R. Davies.** 1999. The molecular basis of substrate channeling. *J. Biol. Chem.* **274:**12193–12196.
22. **Miller, J. H.** 1972. *Experiments in Molecular Genetics,* p. 352–355. Cold Spring Harbor Laboratory, Cold Spring Harbor, NY.
23. **Modrich, P., and D. Zabel.** 1976. *Eco*RI endonuclease. Physical and catalytic properties of the homogeneous enzyme. *J. Biol. Chem.* **251:**5866–5874.
24. **Moran, G. R., and P. F. Fitzpatrick.** 1999. A continuous fluorescence assay for tryptophan hydroxylase. *Anal. Biochem.* **266:**148–192.
25. **Morrison, J. F., and C. T. Walsh.** 1988. The behavior and significance of slow-binding enzyme inhibitors. *Adv. Enzymol.* **61:**201–301.
26. **Pace, C. N., F. Vajdos, L. Fee, G. Grimsley, and T. Gray.** 1995. How to measure and predict the molar absorption coefficient of a protein. *Protein Sci.* **4:**2411–2423.
27. **Rechler, M. M., and H. Tabor.** 1971. Histidine ammonia-lyase (Pseudomonas). *Methods Enzymol.* **17B:**63–69.
28. **Saari, R. E., and R. P. Hausinger.** 1998. Ascorbic acid-dependent turnover and reactivation of 2,4-dichlorophenoxyacetic acid/α-ketoglutarate dioxygenase using thiophenoxyacetic acid. *Biochemistry* **37:**3035–3042.
29. **Segel, I. H.** 1975. *Enzyme Kinetics.* John Wiley & Sons, Inc., New York, NY.
30. **Shaw, W. V.** 1975. Chloramphenicol acetyltransferase from chloramphenicol-resistant bacteria. *Methods Enzymol.* **43:**737–755.
31. **Srere, P. A.** 1987. Complexes of sequential metabolic enzymes. *Annu. Rev. Biochem.* **56:**89–124.
32. **Stadtman, E. R., P. Z. Smyrniotis, J. N. Davis, and M. E. Wittenberger.** 1979. Enzymic procedures for determining the average state of adenylylation of *Escherichia coli* glutamine synthetase. *Anal. Biochem.* **95:**275–285.
33. **Todd, M. J., and R. P. Hausinger.** 2000. Fluoride inhibition of *Klebsiella aerogenes* urease: mechanistic implications of a pseudo-uncompetitive, slow-binding inhibitor. *Biochemistry* **39:**5389–5396.
34. **Walsh, K., and D. E. Koshland, Jr.** 1984. Determination of flux through the branch point of two metabolic cycles: the tricarboxylic acid cycle and the glyoxylate shunt. *J. Biol. Chem.* **259:**9646–9654.
35. **Whittaker, J. W., A. M. Orville, and J. D. Lipscomb.** 1990. Protocatechuate 3,4-dioxygenase from *Brevibacterium fuscum. Methods Enzymol.* **188:**82–88.

20

Permeability and Transport

ROBERT E. MARQUIS

Two general types of processes for entry and exit of solutes across cell membranes can be distinguished. **Permeation** denotes passive movement of a solute into (uptake) or out of (efflux) a cell by diffusion to diminish the electrochemical potential of the specific solute across the cell membrane. Solute passage may be facilitated by chemical interactions with cell structures, say, within the cell membrane. Permeability is a property of the cell and not of the solute. The Fick diffusion equation can be applied to permeation of a solute through the membrane of a cell:

$$ds/dt = PA(\Delta C)$$

where ds/dt is the change in amount of internal solute per unit of time, P is the permeability coefficient (which has units of distance divided by time), A is the surface area of the membrane or other structure across which diffusion occurs, and ΔC is the difference in solute concentration (or electrochemical potential if the solute is ionized) between the interior and the exterior of the cell or membrane-enclosed organelle. If the internal concentration (C_{in}) is plotted as a function of time after exposure of the cell to a higher environmental level of the solute, the resulting curve is parabolic with a plateau at $C_{in} = C_{out}$, i.e., when the internal concentration equals the external concentration or, more correctly, when the electrochemical activity of the solute is the same on both sides of the membrane. For such a curve, $C_{in} = (\text{maximum } C_{in}) (1 - e^{-kt})$, where k is a constant and t is the time. Diffusion can often be speeded up by interaction of solutes with membrane structures, and the process is then called **facilitated diffusion**. However, even for facilitated diffusion the electrochemical potentials inside and outside become equal at equilibrium.

Transport denotes active movement of a solute into or out of a cell or organelle by means of specific transport systems which are coupled to metabolism and involves energized carriers. Two well-studied general types of microbial transport systems are the phosphenolpyruvate:sugar phosphotransferase system (PTS) (53) and various permease systems (17, 43). The PTS functions in facultatively anaerobic bacteria for transport of sugars, while essentially all bacteria have permease systems for transport of mineral cations, inorganic anions, amino acids, inorganic acids, nucleic acid precursors, sugars, and a wide variety of other solutes. The general regulatory roles of the PTS have become more and more appreciated, and in some organisms its functions are more for regulation of metabolism than for transport. Permease systems may be energized by ATP, for example, ATP binding cassette systems, or by the proton motive force across the cell membrane. In addition, P-ATPases may act as solute transporters, for example, Ca-ATPase, and some F-ATPases may transport Na^+ rather than H^+.

A curve relating solute uptake to time for a specific transport process does not differ qualitatively from one for a general permeation process. However, active or energized transport does not stop when $C_{in} = C_{out}$ but continues often until a substantial concentration gradient has developed and the process has become one of concentrative transport. Maximum uptake is then determined by the rates of the inward and outward flows, but any maintained gradient of electrochemical potential requires continued energy coupling.

As indicated, transport results in net movement of solutes into or out of cells, that is, uptake or secretion. For many solutes, a transport process of so-called **exchange diffusion** occurs in which there is a one-for-one exchange of internal and external solutes across the membrane. This exchange is generally assessed by use of radioactively labeled solutes; details of the technique are presented by Maloney et al. (35). Exchange may involve antiporters and heterologous exchange. For example, Driessen et al. (16) have characterized the arginine/ornithine antiporter for the arginine deiminase system of *Lactococcus lactis*. The antiporter catalyzes exchange of ornithine, the product of the system for arginine, the substrate of the system, without need for expenditure of ATP. Concentrative solute uptake does not necessarily require a membrane with transport catalysts. Ion-exchange resins, for example, can take up and concentrate ions selectively. Solutes, then, can be taken up and retained as they would be by an ion-exchange resin. In fact, most of the mineral ions in cells are counterions for charged groups of macromolecules in solid phases, e.g., cell walls, membranes, ribosomes, nuclear bodies, organelles, etc.

20.1. PERMEABILITY

20.1.1. Assay Procedures

A standard technique for assessing solute permeation into bacterial cells or cell structures is the **space** (or thick-suspension) **technique** described by Conway and Downey (10) and modified by Mitchell and Moyle (40) and MacDonald and Gerhardt (33) and others. Large masses of cells are used for the assay—generally 3 g, wet weight, or more per assay tube. Obtaining this quantity of cells can be difficult but not when common, easily grown organisms are used. Techniques that require less biomass and are based on use of membrane filters or a high-speed microcentrifuge are described in the section on transport (section 20.2). For any procedure, careful attention should be paid to details of growth and harvest because permeability may change during the culture cycle in batch cultures or as a result of washing procedures used to rid cells of medium constituents. There can be an advantage in using cells obtained from continuous cultures. It is usually best to take large harvests from the main vessel rather than using so-called runoff cells, which may have altered permeability because of storage on ice during collection. Changes in growth medium, growth temperature, pH, etc., can also be expected to alter permeability, although alterations will generally be less severe

than for energized transport coupled to ATP hydrolysis or Δp across the cell membrane.

The space technique is designed specifically to give an estimate of the fractional space or volume of cells penetrated by the test solute in the absence of energized transport. The technique is based simply on determination of the degree of dilution of a solution containing the solute as a result of mixing with a pellet of centrifuged cells. The following information is needed: the initial volume and concentration of the test solution, the volume of the cell pellet, and the final concentration of test solute in the extracellular phase of the mixture. Cell pellets contain not only cells but also interstitial (intercellular) space. The fractional volume of the interstitial space is generally estimated by use of high-molecular-weight polymeric solutes, which can penetrate the interstitium but not the cell wall.

The steps involved in the space technique are described below.

20.1.1.1. Washing of Cells

After harvest, cells generally need to be washed before use for permeability assays. A single wash with a large volume of fluid is generally sufficient to remove medium constituents without depleting most types of bacteria, archaea, or fungi of internal solutes. Two washings may be done, but normally not three. Gram-positive bacteria are osmotically robust and can be washed with water or dilute buffers without harm, especially when they are chilled. Other organisms, particularly gram-negative bacteria and many archaea, are more sensitive and can be damaged by water washing. Commonly used wash solutions for gram-negative bacteria contain magnesium ions (1 to 10 mM), a buffer, sometimes an osmotic stabilizer (e.g., 0.4 M KCl or 0.5 M sucrose), and sometimes an agent protective against denaturation, such as glycerol or another compatible solute. Maloney et al. (35) recommended using growth medium 63 with deletion of the carbon source for washing *Escherichia coli* cells. With any previously untested organism, it is worthwhile to test a number of washing solutions to determine the best one. The criteria for efficacy are that the cells should be freed of medium constituents but retain internal pools of potassium, amino acids, or P_i. No wash solution is perfect, and some compromise must be accepted. When dense suspensions or biofilms are used, the composition of the wash fluid is less critical than when dilute suspensions of cells are used.

20.1.1.2. Centrifugation of Cells

Harvested cells need to be centrifuged to obtain a tight pellet, and the details of centrifugation depend on the particular organism. A packing curve of the type shown in Fig. 1 can be prepared by plotting pellet weight against centrifugation time, using a series of cell suspensions to obtain the curve. The centrifugation time chosen should be well out in the plateau region so that minor variations in centrifugation time do not affect pellet weight. The final centrifugation before solute addition and mixing is generally done with a tared tube so that pellet wet weight can be obtained by assessing change in weight. Prior to weighing, supernatant fluids should be decanted carefully, the tubes inverted to allow drainage before careful wiping with absorbent tissue. Attempts to blot the pellet surface are best avoided because cells can be removed with the liquid.

20.1.1.3. Measurement of Pellet Weight and Volume

Pellet wet weight is usually an adequate indicator of pellet volume and may be used instead of volume in calculations

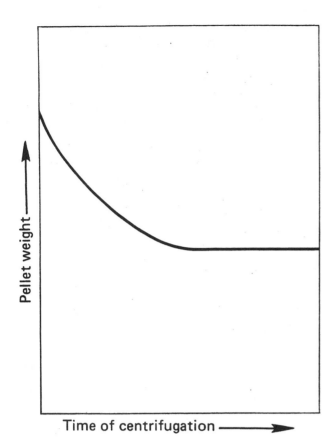

FIGURE 1 Typical packing curve for a cell pellet.

because wet densities of vegetative cells are generally about 1.02 to 1.04 g/ml. However, bacterial spores have significantly higher wet densities, as high as 1.22 g/ml for *Geobacillus stearothermophilus* spores, and pellet weight is not a good substitute for pellet volume. Certain inclusion granules, e.g., poly-β-hydroxybutyrate, of vegetative cells also may have high wet densities and may increase the difference between pellet weight and volume.

Pellet volumes can be estimated also with the use of a standard pycnometer or a volumetric flask. For example, to assess the volume of a 3-g wet-weight pellet, one can mix the pellet with water, transfer it to a 50-ml, tared pycnometer vessel or volumetric flask, and add water until the mixture comes up to the mark. Suppose the weight of the suspension determined with an analytical balance is 50.110 g (weight mixture plus pycnometer minus weight of pycnometer) and the pycnometer has a calibrated volume of exactly 50.000 ml at 25°C, the temperature of weighing. Suppose that the pellet was found to weigh exactly 3.000 g. Then, the weight of water added to the pellet must be 50.110 −3.000, or 47.110 g. At 25°C, this weight of water would have a volume of 47.249 ml based on water density at that temperature obtained from tables in a standard chemical handbook. Therefore, the pellet volume must be 50.000 −47.249, or 2.752 ml. The density of the pellet is then 3.000 g/2.752 ml, or 1.090 g/ml. Once the density is known, and if standard centrifugation conditions are used, the pellet volume can be readily calculated from the pellet weight.

Pellet volumes also have been estimated directly by use of calibrated centrifuge tubes, which yield cytocrit data similar to the hematocrit data obtained routinely in hematol-

ogy. The cytocrit method gives useful volume estimates, but it has difficulties in practice and requires use of a swinging-bucket centrifuge head.

20.1.1.4. Mixing and Incubation of Cells with Solute
The cell pellet to be used for a permeability assay is brought to the chosen experimental temperature and mixed with test solute in solution. Mixing can be enhanced by vortexing the material. However, care must be taken when working with samples of sensitive organisms, such as anaerobes or spheroplasts. The volume of solution should, in general, be about equal to the pellet volume because there should be sufficient dilution of the solution to allow changes in concentration to be readily assessed. After mixing, the suspension should be incubated for a time sufficient to allow diffusion equilibrium to be reached even in clumps of cells, say, 5 to 15 min depending on the tendency of the cells to clump. Vortex mixing may help to reduce clumping.

20.1.1.5. Recentrifugation
After appropriate incubation, the mixture should be centrifuged sufficiently to obtain a cell-free supernatant fluid for solute assay. The supernatant fluid may require an additional centrifugation in the absence of the main pellet.

20.1.1.6. Determination of Solute Concentration
The concentration of solutes should be assayed for the initial solution and for the supernatant fluid after recentrifugation. Sometimes the assay can be a gravimetric one based simply on weighing the material dried from known volumes of the solutions or, alternatively, by means of refractometry. Unfortunately, these techniques are not chemically specific and material leached from the cells may interfere. Therefore, it is usually necessary to use specific chemical or radiochemical assays to assess solute concentration.

20.1.1.7. Determination of Interstitial Space
Since cell pellets contain intercellular or interstitial space into which solutes can diffuse readily, there is a need to assess the volume of this space. The assessment is usually made with solutes of high molecular weight, which are able to penetrate the interstitium but cannot pass through the cell wall or outer membrane. Dextrans with average molecular weights of 2,000,000 (Dextran T2000; Pharmacia-Pfizer Fine Chemicals, Piscataway, NJ) serve the purpose. Dextrans are polydispersed, but even the smallest molecules in the distribution of this large dextran preparation do not penetrate even relatively porous cell walls. Dextran concentrations can be determined gravimetrically, by means of refractometry, or chemically by the phenol-sulfuric acid method (18).

A number of other polymeric materials have been used instead of dextrans. Proteins, such as serum albumin, have an advantage in being nearly monodispersed and easy to assay. However, they have a disadvantage in that they may bind to cells. Inulin has also been used often but may partially penetrate bacterial cell walls. The major need is for an easily assayed polymer that does not bind to cells.

The interstitial volume fraction may be an appreciable part of the total pellet volume. For closely packed, nondeformable spheres, the interstitium makes up some 27% of the total volume. For bacterial cells packed by centrifugation to essentially constant volume, the interstitium may represent only some 10 to 15% of the pellet volume because of irregular shapes and sizes of the cells and their deformability. If solute uptake is not great, uptake into the interstitium may be a large part of the total uptake. However, for solutes that are concentrated by cells, uptake into the interstitium will be a smaller part of the total uptake.

20.1.1.8. Expression of Results

A commonly used permeability index is the space value, or S value, which can be calculated by use of the formula.

$$S = [V_s/V_p][(C_o/C_f) - 1]$$

where V_s is the volume of solution added to the pellet, V_p is the volume of the pellet or, commonly, the pellet wet weight (W_p), C_o is the initial concentration of solute, and C_f is the final concentration. S values indicate the fraction of the cell pellet penetrated by the solute. However, the desired information is the fraction of the cell volume penetrated. The cell volume fraction penetrated is called the R value. It is calculated by use of the formula

$$R = (S_{sol} - S_{int})/(1 - S_{int})$$

where S_{sol} is the corrected space value for the solute being tested and S_{int} is the volume of the interstitial space; if dextran is used to measure the volume, $S_{dextran}$ is used instead of S_{int}. A zero value for R indicates that the cells are impermeable to the solute. Values of zero to one indicate various degrees of penetration, and values greater than one indicate that the cells actually concentrate the solute. It is often desirable to express results in terms of cell dry weight or cell water, e.g., sucrose-permeable space per gram of cell dry weight or per milliliter of cell water. The pellet volume impermeable to small solutes such as sucrose can also be readily estimated, and if the cell membrane is impermeable to the small solute, then the, say, sucrose-impermeable volume is approximately equal to the protoplast volume.

20.1.1.9. Cell Volume Determination

Useful cytological information can be gained from space values. For example, the technique offers the most accurate and direct means available for determining average cell size. The pellet volume minus the interstitial volume is equal to the volume of cells. If the pellet is resuspended in water to some known volume, say, 50 ml, in a volumetric flask, the number of cells can be assessed with use of a Petroff-Hausser counter, and the average volume of a cell can then be calculated. The technique can also be used to estimate the volume of the cell protoplast if the cell wall volume or the space impermeable to a small solute that cannot cross the cell membrane is known.

Cell counting with the Petroff-Hausser chamber has high levels of inherent error, as does plate counting, because of the small numbers of cells or colonies counted and the Poisson distribution of possible counts (28). Electronic counters capable of counting large numbers of cells per sample can be used to obtain more accurate estimates of cell numbers. Cells are counted by electronic counters on the basis of differences in electrical conductivity between cells and their suspending medium. Manufacturers of electronic cell counters include Beckman Coulter (http://www .beckman.com/).

20.1.1.10. Cell Water Content Determinations

The resuspended cells can also be used for dry-weight determinations from which one can estimate the dry weight per cell. The weight of cell water is then equal to the difference between the wet weight of the cell and its dry weight, and the weight can be converted to volume simply from knowledge of the specific volume of water at the temperature of the measurements, generally, about 1 ml/g. Since the volume of interstitial water in centrifuged pellets can be estimated in terms, say, of dextran-permeable volume, the amount of cell water in the pellet can also be estimated.

Sometimes, it is convenient to estimate the amount of cell water directly from the R value for solutes such as ^3H_2O or $[^{14}C]$urea or glycerol, which readily permeate but are not normally concentrated by resting cells, although many cells do have energized transport systems for urea or glycerol. Tritiated water has an advantage in terms of its speed of entry into cells, but during prolonged incubations, tritium exchange can occur between labeled water and cell components. There is value for many studies in estimating the volume of cell wall water, which is outside the cell membrane but within what would ordinarily be considered the outer wall boundary of the cell. Small solutes penetrate readily into the heteroporous cell walls of most gram-positive organisms (51) but may not penetrate the outer, cell wall membrane of gram-negative bacteria. Cell wall water is generally assessed in terms of the cell volume or space permeable to small solutes not transported across the cell membrane of the particular cells used minus the interstitial space. Sucrose and raffinose have been used for this purpose for cells without sucrose or raffinose transport systems or with nonmetabolizing, chilled cells. Radioactively labeled taurine also has often been used, for example, by Tran and Unden (56), for assessing the cell wall water volumes of cells of E. coli. The only requirements for a useful probe are that it should be small and able to penetrate the wall water volume, that it is not strongly bound by wall components, and that it is not transported across the cell membrane.

20.1.1.11. Live-Dead Stains and Permeability

The use of live-dead stains has become increasingly popular, especially for microscopic estimations of cell viability under stress conditions or in biofilms. The most commonly used stain is propidium iodide, which can be obtained from Molecular Probes (Eugene, OR). The differentiation between live and dead cells is based on live cells with intact membranes being impermeable to propidium iodide, while dead cells with damaged membranes are permeable to the dye. There are questions regarding the relationship of increased cell permeability and capacity to, say, grow on agar plates to yield a colony. However, live-dead staining methods do offer a convenient way for obtaining microscopic assessments of cell viability.

20.2. TRANSPORT

20.2.1. Assay Procedures

20.2.1.1. Preparation of Cells

For studies of rates of solute uptake, some means of rapidly separating cells from their suspending medium is required. Separation is often accomplished by filtration through membrane filters, and therefore, it is necessary to use dilute cell suspensions, generally with less than 200 μg (dry weight) per ml, if 1-ml samples are to be filtered through standard 22-mm-diameter filters. The use of large filters allows for use of more concentrated cell suspensions but can present problems in terms of rapid manipulations.

As with permeability assays, care must be taken in transport assays to control closely the conditions of cell growth and harvest. Wash and resuspension media tend to be more important in preparing cell suspensions for transport assays because cells in dilute versus concentrated suspensions are more liable to be harmed by unfavorable environmental conditions, including those leading to leaching of internal solutes from the cells.

For transport studies, cells often have to be preconditioned before addition of the substrate to be transported. For example, they may be starved to deplete endogenous pools of substrate, which may be of advantage when radioactive substrates are used to reduce levels of endogenous unlabeled substrate and the need to make major corrections in estimated specific radioactivity. Alternatively, they may be fed to allow for synthesis of ATP and establishment of Δp (the proton motive force) across the cell membrane. The physiology of many microbial cells is set to maintain pools of energy transfer substrates such as ATP or phosphoenolpyruvate.

20.2.1.2. Mixing and Incubating Cells with Solute

The cell suspension and test solution are brought to the experimental temperature separately and then mixed at zero time. Care has to be taken to avoid cells being severely shocked during mixing. For example, there should not be major changes in osmolality or ionic strength associated with mixing arising from differences between the suspending medium for cells and the solution containing the test solute. Rapid mixing is often required because of the rapidity of transport processes, and incubation of the mixtures is generally of short duration, seconds or minutes rather than hours.

20.2.1.3. Sampling and Separation

Samples of the reaction mixture are withdrawn shortly after mixing. The cells are rapidly separated from the suspending medium, for example, by filtration through membrane filters with 0.45-μm-diameter pores, previously wetted with wash solution. Lower-porosity filters may be used, but they have more resistance to fluid flow with resultant slower filtration. The 0.45-μm-pore-size filters are effective for most bacterial cells. Vacuum from laboratory lines is often sufficient for rapid filtration, although vacuum pumps offer an advantage for developing higher levels of vacuum. The common, clip-on, vacuum-funnel attachment can be replaced by custom-made, heavy brass rings to hold the filter membrane in place and allow for more rapid processing of samples. Also, convenient manifold filter units can be purchased if many samples have to be processed in a short time.

There may be an advantage to mixing assay components in a syringe with a Swinney filter attachment from which samples of cell-free suspending medium can be expressed. The obvious disadvantage of the method is that calculation of solute uptake depends on estimation of changes in solute concentration in the suspending fluid rather than changes in solute levels in the cells. Moreover, repeated removal of suspending fluid changes the ratio of cells to suspending medium in the remaining mixture.

Often, rapid centrifugation can be used to separate cells and suspending medium, even when rate data are required. Microcentrifuges, which accommodate small tubes of 250- to 1,500-μl capacity, can attain centrifugal forces of 12,000 $\times g$ in only a few seconds. Another commonly used means to separate cells from their suspending medium is to centrifuge them through silicone oils (22). An advantage of this method is that very little interstitial aqueous solution is pelleted with the cells and washing is less of a problem.

It is often possible to stop transport processes by adding samples directly to iced or chilled wash medium. Centrifugation can then be carried out in the cold. For this method to be useful, chilling must stop influx but not induce efflux. Satisfactory results are more likely to be obtained with gram-positive than with gram-negative organisms.

Sometimes, it is not necessary to separate the cells from the test solution. For example, ion-specific or solute-specific electrodes can allow for continuous monitoring of pH, partial pressure of oxygen (pO_2), potassium, or other metal ions or organic solutes. In studies of salt uptake, it may be possible to assess uptake simply by assessing changes in conductivity of the suspension measured with a conventional conductivity cell and meter.

20.2.1.4. Washing of Cells

Cells retained on filters or pelleted must generally be washed to remove adherent or interstitial suspending medium. For gram-negative organisms particularly, the wash liquid should be osmotically buffered to avoid leaching of solutes from the cytoplasm. Concentrated sugar solutions, e.g., 0.5 M sucrose solution, are often used for washing. Magnesium ions, usually 1 to 10 mM, may also be added as chloride or sulfate salts. An ideal wash solution should remove test solute from interstitial and adherent suspending medium during the first wash but not appreciably in subsequent washes. Some compromise generally has to be accepted. As indicated above, washing can be avoided by centrifuging cells through silicone oils.

20.2.1.5. Assay of Cells for Content of Transported Solute

Cells may be eluted from membrane filters, resuspended in a known volume of fluid, and assayed for the test solute by chemical, radiometric, or other methods. When radioactive solutes are used, the membrane filters can be placed directly into scintillation fluid for counting. The amount of solute incorporated into polymeric materials can be estimated by treating the cells on the filter with cold, 5% (wt/vol) trichloroacetic acid solution and then washing them with water before counting. The acid treatment damages the cell membrane so that small solutes but not macromolecules are released from the cells. The acidification results also in protonation of carboxyl and phosphate groups with release of bound cations.

Membrane filters may bind test solutes, especially ionized ones, and corrections to uptake values may be needed because of this binding. An estimate of the extent of binding can be made by passing some of the test solution through a membrane filter, washing the filter in the same way the cells are washed, and then assaying the filter for the retention of the test solute.

20.2.1.6. Expression of Results

Transport is usually expressed in terms of uptake or excretion in molar amounts per unit of biomass, e.g., cell dry weight. It may also be expressed per unit of cell water, especially if there is a desire to determine if transport is concentrative and, therefore, requires some type of energetic coupling. Results may also be expressed per cell. However, there is then the problem reviewed earlier of obtaining an accurate cell count. If cell numbers are determined by plate counting, transport can be expressed in terms of CFU, which may or may not be the same as total cell count. When using radioactive solutes, data should be converted from counts per minute, curies, or becquerels to molar equivalents.

Transport processes are generally viewed as a subclass of enzyme-catalyzed reactions, and the principles and practices of enzyme kinetics can be applied to transport kinetics. For example, Lineweaver-Burk, double-reciprocal plots of the inverse of transport rate versus the inverse of solute concentration often yield straight lines from which one can calculate values for K_m (Michaelis constant) and V_{max} (maximum velocity). Alternatively, Eadie-Hofstee plots of velocity of transport versus the velocity divided by substrate concentration can be used for estimating kinetic parameters and substrate affinities. The complications of kinetics that arise in the study of enzyme reactions arise also in studies of transport.

It is not uncommon for cells to have more than one transport system for a particular solute. For example, many bacteria have a high-affinity system that is highly specific for an individual aromatic amino acid, in addition to a low-affinity system that can catalyze transport of a variety of amino acids. Methods for assessing these multiple systems are described by Ames (2).

In addition, many solutes can enter cells passively by diffusion (permeation) as well as actively by energized transport. The diffusion component generally does not show saturation at high substrate concentrations. Its extent can be estimated from plots such as the one shown in Fig. 2. With intact bacteria, the diffusion component may include diffusion of solute into the water of the cell wall (36), which can account for some 50% of the cell volume of organisms such as *Micrococcus luteus*.

20.2.1.7. General Considerations

A major advantage of the use of dilute compared with dense suspensions of cells is the ease with which kinetic data can be obtained. Also, with dilute suspensions, closer control of solute concentration, pO$_2$, pH, ionic strength, etc., can be maintained. The disadvantages include problems associated with solute leaching from cells during initial suspension or washing, the general need to assess solute uptake or loss from changes in intracellular rather than extracellular concentrations, and the inherent vagaries in the use of dilute

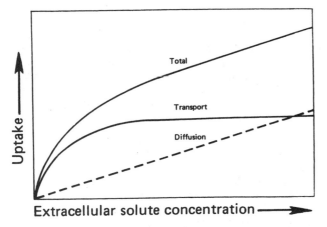

FIGURE 2 Graphic estimation of the diffusion component in the total uptake of a solute by microbial cells. The transport uptake is the total uptake minus the uptake due to passive diffusion. The uptake measure can be either the rate or extent of uptake.

suspensions in which cells may be highly sensitive to changes in environmental parameters.

20.3. SPECIAL METHODS

20.3.1. Proton Motive Force, Intracellular pH, and Proton Permeability

The **proton motive force** is considered to be a major parameter related to energy transfer in biological systems. It is defined as

$$\Delta p = \Delta \psi - 2.3(RT/F)\,(\Delta pH)$$

where Δp is the proton motive force, $\Delta \psi$ is the membrane potential, R is the gas constant (8.318 volt-coulomb/°K charge equivalent), T is the temperature (kelvin), F is the faraday (96,500 kcal/charge equivalent), and ΔpH is the difference in pH between the interior and the exterior of the cell. At 25°C, the value of 2.3 RT/F (which is often given the symbol z) is approximately 59 mV. Thus, to estimate Δp, it is necessary to estimate $\Delta \psi$ and ΔpH separately. Techniques for estimating these parameters have been reviewed critically by Kashket (26).

$\Delta \psi$ for giant cells of *E. coli* has been assessed directly by means of microelectrodes (19), but the method is not applicable to ordinary bacterial cells. $\Delta \psi$ is usually assessed indirectly from determinations of the distribution of permeant cations between the interior and the exterior of cells. Harold and Altendorf (21) indicated that a useful indicator of $\Delta \psi$ must "diffuse rapidly across the membrane . . . be fully dissociated by physiological pH, metabolically innocuous and not subject to translocation by a biological transport system." These criteria can be met by K^+ or Rb^+ in cells treated with 1 to 10 μM valinomycin to render the cytoplasmic membrane permeable to the cations. The measured membrane potential of bacterial cells is generally about 180 mV with the interior negative, so the K^+ concentration in the cytoplasm would be some 20 times that in the suspending medium. Potassium concentrations can be determined by flame photometry or with ion-specific electrodes. The uptake of K^+ can be assessed by use of the space-value method, and the concentration can be expressed in terms of moles of K^+ taken up per unit of cell water. $^{86}Rb^+$ can be used for estimation of Rb^+ uptake.

$\Delta \psi$ has often been assessed by determining the distribution of permeant, lipophilic organic cations or fluorescent dyes. Kashket (26) recommended tetraphenylphosphonium (TPP$^+$) for assessing $\Delta \psi$ and describes basic procedures for its use. There have been concerns about the use of TPP$^+$ and similar compounds in many organisms since Midgley (39) pointed out that *E. coli* has an efflux system with broad specificity for complex organic cations, including TPP$^+$. In fact, these sorts of multidrug transporters occur widely among microbes and cause concerns about the uses of dyes for $\Delta \psi$ estimates. The use of K^+ or Rb^+ with valinomycin, then, has an advantage for studies of organisms with multidrug transporters. Improved methods to determine both ΔpH and $\Delta \psi$ for *Methanobacterium thermoautotrophicum* have been developed (11) based on the pH-dependent fluorescence of coenzyme F$_{420}$ to assess ΔpH and the fluorescent probe bis-(1,3-dibutylbarbituric acid)trimethine oxonol to assess $\Delta \psi$. Fluorescent dyes from Molecular Probes have also been used (9) for assessing $\Delta \psi$, but again, there is a need to know if they can be transported by multidrug transporters.

For determinations of ΔpH, a weak acid or base to which the cell is permeable is used. The compound chosen must dissociate at physiologic pH values so that its distribution between the cell and its environment will depend on the pH difference between the two phases. When a weak acid is used, the membrane of the cell should be permeable to the undissociated form but impermeable to the ionized or dissociated form. Compounds that are commonly used for ΔpH estimates include acetate, benzoate, 5,5-dimethyl-2,4-oxazolidinedione, or salicylate when the cytoplasm is alkaline relative to the exterior pH. When the cytoplasm is more acidic than the environment, weak bases such as methylamine, hexylamine, or benzylamine are used. As Kashket (26), emphasized, the ΔpH calculation is valid only when the pK of the probe is one pH unit or more different than the exterior pH value. In other words, one probe does not do for all cells or for any one cell type under different conditions. Fluorescent probes such as 2′,7′-bis(carboxyethyl)-4- or 5-carboxyfluorescein have been used extensively, for example, by Iwami et al. (23) to assess intracellular pH. In the unhydrolyzed form, they can cross the cell membrane to be hydrolyzed by esterases in the cytoplasm to yield the actual probe.

The Henderson-Hasselbalch equation is used for estimating internal pH values. In the equation, A stands for the anionic form of the weak acid used, HA stands for the protonated form, and subscripts i and o refer to concentrations inside and outside the protoplast, or outside the cell if no correction has been made for the cell wall water volume. Therefore,

$$pH_i = pK_i + \log [A_i]/[HA_i]$$
$$pH_o = pK_o + \log [A_o]/[HA_o]$$

It is generally assumed that $pK_i = pK_o$, although they are not exactly equal because of differences in ionic strength on the two sides of the membrane but are sufficiently close for most purposes. Since the membrane is considered to be highly permeable to the undissociated form, $[HA_i] = [HA_o]$. Therefore,

$$pH_i = pK_i + \log [A_i] - \log [HA_i]$$
$$pH_o = pK_o + \log [A_o] - \log [HA_o]$$

Since

$$-\log [HA_i] = -\log [HA_o]$$

and

$$pK_i = pK_o$$

then

$$pH_i - \log [A_i] = -\log [A_o]$$

and

$$pH_i - pH_o = \Delta pH = \log [A_i]/[A_o]$$

ΔpH can then be assessed from estimates of the distribution of, say, benzoate between cells and suspending medium by the methods described above for permeability or transport assays. From the external benzoate concentration and the measured external pH, one can calculate the external concentration of HA or, here, protonated benzoate, which is equal to the internal concentration. The internal concentration of A, here the unprotonated form of benzoate, is equal to the total internal concentration minus the concentration of HA. From the ratio $[A_i]/[HA_i]$ and a knowledge of pK_i (assumed to be equal to pK_o) the internal pH value can be estimated by use of the Henderson-Hasselbalch equation.

A flow dialysis method that does not require separation of cells and suspending medium has been described in detail by Ramos et al. (45, 46) and applied in studies of membrane vesicles. Methods have been described by Avison et al. (3) for estimation of intracellular pH for bacterial cells by use of magnetic resonance techniques. pH_i has been assessed by using fluorescent dyes such as 9-aminoacridine (26). When the value of pH_i is known, ΔpH can be calculated directly from knowledge of the pH of the suspending medium.

Internal pH values can also be determined by a so-called null-point method. For this method, cells are suspended in thick suspensions in a set of dilute buffers at various pH values. A membrane-damaging agent, such as butanol, toluene or, say, nigericin, is added to render the cells fully proton permeable, and changes in suspension pH as a result of damage are recorded. The changes are then plotted against the initial pH values. The initial pH value at which there was no change in pH induced by membrane damage, obtained by extrapolation, can be taken as the internal pH of the cells. The method is labor-intensive and requires large quantities of cells.

Assessments of Δp, $\Delta\psi$, and ΔpH are subject to many inaccuracies, as reviewed in some detail by Kashket (26). There is also the problem of pH_i and ΔpH being very sensitive to manipulations of the cells, for example, centrifuging or filtering them. Moreover, there are questions about the capacities of bacterial cells to maintain internal pH. Bacteria such as *E. coli* are considered to have homeostatic mechanisms for maintaining internal pH, and a set point pH for maintenance. Other bacteria, such as the streptococci, do not appear to have a set point pH for homeostasis but appear to do the best they can to maintain ΔpH across the cell membrane of about one pH unit. As the environmental pH drops due to, say, acid production by the glycolytic system, the internal pH falls as well. In fact, it appears that the fall in internal pH may have a desirable function in slowing metabolism under unfavorable conditions (50).

Because of the many problems in estimating internal pH values and ΔpH, in the past few years, we have more and more chosen to assess permeabilities of cells to protons rather than ΔpH (42). In fact, when studying membrane-active agents, changes in proton permeabilities induced by the agents are the most pertinent parameters. Methods for assessing proton permeability are described by Bender et al. (5) based on procedures described earlier by Scholes and Mitchell (52) and Maloney (34). Total flux of protons across the membrane per cell or per unit of biomass is readily calculated, usually from initial slopes of pH-versus-time curves of the types shown in Fig. 3. To obtain the data, proton permeabilities of cells in suspensions or in biofilms were assessed by the procedures described previously (32). Basically, the suspensions or biofilms immersed in liquid medium were brought to constant pH values, and then HCl was added to drop the pH value, usually by 0.1 to 0.2 unit. The subsequent rise in pH value as protons moved into the cells was recorded with a glass electrode in the suspending medium. Initial rates of proton entry were estimated from pH changes during the first few minutes after acidification. The cells used had buffer capacities of approximately 330 μmol H$^+$ per pH unit change per gram of cell dry weight. (8). However, cell buffering would not have much effect on the calculation of proton flux into the cells because the pH changes were small initially. Internal buffering by cell constituents would mean that the change in pH in the cytoplasm would not necessarily be the same as the change in

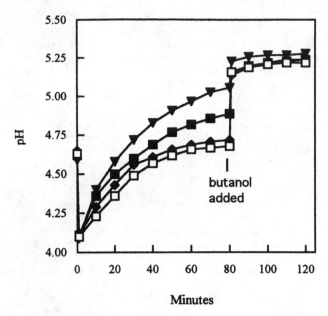

FIGURE 3 Proton permeability of biofilms of *Actinomyces naeslundii* in the presence of 0 (open squares), 0.5 (closed triangles), 1.0 (closed squares), or 5.0 (closed inverted triangles) mM NaF, a weak acid enhancer of proton movements across cell membranes. Butanol was added (5% [vol/vol]) at the indicated time to damage the cell membrane and render it totally permeable to protons.

pH in the environment. Rates of proton movement decreased with time, but this decrease is expected because of reductions in ΔpH across the cell membrane with time. However, the cells continued to have some capacity to maintain ΔpH as indicated by the rise in pH that occurred when cells nearly equilibrated in terms of pH change with the suspending medium were treated with butanol, which opens the membrane to proton movements. Still, it seems that changes in initial rates of proton uptake give a good indication of the extent of effects of membrane-active agents on proton permeability of membranes and on their capacities to maintain ΔpH between the interior and exterior of the cell protoplast.

20.3.2. Buffer Capacities of Cells

Buffer capacities of cells can be estimated by the methods described by Krulwich et al. (31) for assessing external buffering (B_o), total buffering (B_t), and internal buffering (B_i) on the basis of acid-base titrations of thick suspensions of intact or permeabilized cells and the differences between the two. As expected, buffering capacities vary with pH and tend to be dominated by carboxyl, phosphate, and amino groups. Capacity is generally expressed in nanomoles of H^+ per pH unit per unit of cell weight or cell protein, in other words, the numbers of protons required to change the pH of a suspension containing a unit of cells by one pH unit.

20.4. OSMOTICALLY SENSITIVE CELLS

When a solute is taken up by a cell, the process results in a lowering of water activity in the interior and a consequent influx of water with resultant swelling of the cell. Biological membranes are highly permeable to water, and the swelling

reaction is extremely rapid. Matts and Knowles (38) used a stopped-flow spectrophotometer to show that the osmotic movement of water from *E. coli* cells transferred into a 0.4 M $MgCl_2$ solution was complete within 50 ms at 37°C. Since water movement is so rapid, the rate of swelling accompanying solute uptake, indicated by decreased cell refractive index and light scattering, can be used as an indicator of transport in studies of protoplasts or spheroplasts with damaged cell walls. However, the walls of intact bacterial cells have sufficient elasticity to resist swelling (14). This elasticity of the wall results in buildup of tension in the wall and hydrostatic pressure within the cell, called turgor pressure. The turgor pressure ($P^{in} - P^{out}$) can be calculated by use of the equation

$$P^{in} - P^{out} = (2.3\ RT/V)(\log_{10} a^{out}/a^{in})$$

where R is the gas constant, T is the temperature (kelvin), V is the molar volume of water, a^{out} is the water activity outside the cell, and a^{in} is the water activity within the cell. At a temperature of 25°C, turgor pressure is equal to (311 MPa) × (log [a^{out}/a^{in}]). The symbol π is generally used for turgor pressure.

The swelling and shrinking of an ideal osmometer can be predicted from the van't Hoff-Boyle equation,

$$V - b = a/\pi$$

where V is the volume of the cell, b is the so-called osmotically dead space, which generally is approximately equal to the calculated volume of dry matter in the cell, and π is the osmolality of the suspending medium. For an ideal osmometer, a plot of V versus $1/\pi$ yields a straight line with intercept b and slope a. The actual responses of bacterial cells lead to curved lines (Fig. 4). Clearly, bacterial cells are not ideal osmometers.

Osmotic responsiveness is markedly muted at high osmolality, and a value for b can be obtained only by extrapolation (Fig. 4). It is not clear why the cell would be so poorly

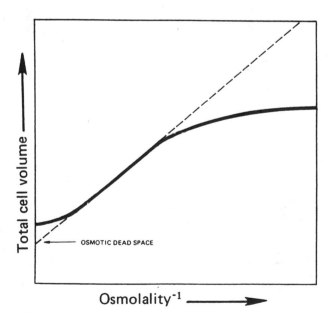

FIGURE 4 Osmotic volume changes for microbial cells. The dashed line indicates the ideal behavior predicted by the van't Hoff-Boyle equation. The solid curve indicates the actual behavior.

responsive in media of high osmolality. The behavior may have to do with bound water or with gel-like states of the cytoplasm, or, possibly, with reduced permeability of the collapsed membrane to water.

When placed in media of low osmolality, bacterial cells are resistant to osmotic swelling primarily because they have elastic cell walls. When placed in concentrated media, they become plasmolyzed; that is, the protoplast shrinks so that the membrane is pulled away from the wall, and plasmolysis vacuoles are formed between the wall and the protoplast membrane. These vacuoles can often be seen microscopically, especially in gram-negative bacteria. However, even if they cannot be seen, their presence can be detected by permeability determinations. Thus, one can assess the total cell volume in terms of the dextran-impermeable space. The protoplast volume is essentially the raffinose-impermeable volume under conditions in which raffinose cannot cross the cell membrane. The aqueous volume of the cell wall and the plasmolysis vacuoles together is then equal to the dextran-impermeable volume minus the raffinose-impermeable volume. A reasonable estimate of the aqueous wall volume alone can be made from the raffinose-impermeable volume of the unplasmolyzed cell.

If plasmolyzed cells are returned to more dilute media, water flows into the protoplast across the water-permeable cell membrane, the protoplast swells to occupy the volume previously occupied by the plasmolysis vacuoles, and the optical density of the suspension increases. There is a direct relationship between the protoplast volume and the inverse of the optical density (36). Initially when the protoplast swells, there is little resistance to swelling, and the process is nearly ideal; i.e., it follows the van't Hoff-Boyle relationship. However, when the protoplast abuts against the cell wall, the elasticity of the wall resists further swelling, and major deviation from the ideal behavior occurs. Now water uptake involves an increase not only in the volume of the protoplast but also in that of the whole cell, with stretching of the wall. A break in the curve for V versus $1/\pi$ occurs at what is called the point of incipient plasmolysis. Further reductions in medium osmolality result in the buildup of turgor pressure. It is considered that the medium osmolality at the point of incipient plasmolysis is essentially equal to the internal osmolality of the cell, as discussed primarily for plant cells by Stadelmann (54). One can obtain another estimate of internal osmolality from the value of the constant a (the number of ideal osmomolar equivalents within the cell) and a knowledge of the protoplast volume.

The elasticity of the cell wall does not affect swelling of isolated protoplasts, which can be used for transport studies. If protoplasts or spheroplasts are made to swell very slowly, their membranes do not undergo the sort of brittle fracture that occurs when they are rapidly shocked osmotically. However, the membrane can then become taut, and the magnitude of swelling is not predicted by the van't Hoff-Boyle equation (36).

In a study of regulation of turgor pressure by *Bacillus subtilis*, Whatmore and Reed (59) used a Coulter Counter (Beckmans Coulter) to assess changes in cell volume, and these changes were used to construct a van't Hoff-Boyle plot. The authors found that spectrophotometric analyses of volume changes gave reasonable but somewhat low estimates for turgor. For bacteria with internal gas vesicles, changes in light scattering as a result of collapse of the vesicles under pressure can be used to estimate cell turgor, as described by Reed and Walsby (47).

Bacterial cells require turgor for growth, cell division, and certain other functions. How membrane proteins sense water stress has been reviewed by Poolman et al. (44). Bacterial cells have well developed means to maintain turgor when placed in hypertonic media through pooling of solutes, including K^+, glycine betaine, proline, and so-called compatible solutes. Osmosensing and osmoregulation have been reviewed in papers by Wood (61), Roessler and Müller (49), and Morbach and Krämer (41).

20.5. SOLUTE UPTAKE BY BIOFILMS

Over the past few years, a greater appreciation for the communal lives of most microbes in nature has developed. In fact, most microorganisms are members of biofilm communities, especially at solid-liquid interfaces but also at liquid-gas interfaces. The growing view is that microbiologists who study pure cultures are studying peculiar cells without the normal range of intercellular interactions of normal cells living in association with a diverse community. Pure cultures can show intercellular phenomena such as quorum sensing but do not have to deal with interactions in a diverse community. Biofilms allow for a wide range of biodiversity. For example, anaerobes and aerobic or facultative organisms can live together in biofilms because the aerobes or facultative organisms can protect anaerobes from oxidative damage. Because of the usual slow growth in biofilm communities, slowly growing organisms can exist in association with organisms capable of very rapid growth. The result is that slowly growing organisms are not eliminated from biofilms, as they are from suspension cultures. Moreover, slow growth can also protect organisms from damage by antibiotics, such as the β-lactams, which require that cells grow rapidly for maximum effect.

Methods for producing biofilms in the laboratory have been described in multiple publications and involve a wide variety of variations and specific designs for biofilm reactors. Good sources for information of growth and manipulation of biofilms are the three volumes of *Methods in Enzymology*, volumes 310, 336, and 337, edited by R. J. Doyle (12, 13), as well as multiple other books on the subject of biofilms. Mixed populations with multiple organisms can be maintained in biofilms under conditions of continuous culture. For example, with oral microbes, it has been possible to develop systems for mixed, continuous cultures with as many as 10 separate species in the biofilm consortium (6). Growth of biofilms is often carried out with continuous feeding, although batch, or more commonly, fed-batch culture is widely used. Also, the films can be grown in constant-depth fermentors or without control of film depth, which allows for sloughing of cells or parts of the film and recruitment of newer organisms from the planktonic phase. In general, biofilms have associated planktonic organisms in the suspension around the film. Although the planktonic cells are not technically part of the biofilm, they can interact with it and may even become incorporated into the film. A need in working with biofilms is to take account of the age of the films used. Biofilms do age, even those grown in the laboratory with a continuous supply of nutrients. The transport and other physiological capacities of the films can change significantly with age, so there is need for fairly rigid standardization of growth and processing procedures if fairly uniform results are needed. Still, the usual finding is that biofilm cultures are more variable than suspension cultures, and greater numbers of replicates of experiments are required to obtain data with acceptable levels of statistical significance.

Solute transport into biofilms has a number of peculiarities, which currently are not fully understood. It is generally felt that growth in biofilms is often diffusion limited and that solute diffusion into biofilms is slow. The topic of diffusion in biofilms has been reviewed by Stewart (55). The general finding is that life is at a slow pace in biofilms. Certainly, growth is slow, and generation times of a day or more are common even for organisms that are able to grow with generation times of less than an hour in suspension cultures. Catabolism as well is generally slowed in biofilms, say, to about one-third the rate for cells in suspensions. Moreover, there is ample evidence that cells can undergo physiological adaptations peculiar to growth in biofilms. Much of the current work on biofilm biology involves use of single-organism biofilms. However, ultimately, the need is for information with multiorganism biofilms and natural biofilms in the human or animal body or in the environment. The perspective will then be one of study of permeability and solute uptake of a community rather than a single species of organism.

A major orientation currently with biofilms is to do in situ measurements, for example, of pH. One seemingly simple way to measure pH in biofilms is to use tiny pH electrodes, for example, the Beetrode pH electrode from World Precision Instruments (New Haven, CT), and to place the tip directly into the biofilm, as described by Burne and Marquis (7). More elaborate methods involve two-photon excitation microscopy and use of fluorescent dyes such as carboxyfluorescein, as described by Vroom et al. (58).

Biofilms are actually thin films, at least as viewed by a chemical engineer, and diffusion of small molecules into the films can be rapid. Diffusion of larger molecules can be greatly slowed, and for organisms dining on, say, proteins, access to substrate within the biofilm can be highly restricted. Of course, cells at the surface of the biofilm can secrete enzymes to hydrolyze large molecules outside the community so that the digested pieces can diffuse into the biofilm community. Biofilms commonly have channels that penetrate the films, and these channels can be conduits for solutes to reach the bottom parts of the film. They have been visualized by means of confocal laser scanning microscopy (60), especially in Pseudomonas biofilms, but also for complex dental-plaque biofilms grown in the human mouth on enamel pieces (62). However, most biofilms are stratified. For example, the more aerobic organisms may be positioned at the liquid interface, where O_2 is more abundant, and obligate anaerobes may be in the depth of the film, where very little O_2 can penetrate without being metabolized. Often studies of solute transport involve use of entire biofilms, so that an average value is obtained for what may be a very heterogeneous set of uptakes by individual cells. However, more and more, techniques such as confocal microscopy and use of fluorescent tags are being exploited to analyze particular solutes in particular parts of the biofilm. In a very direct approach, Robinson et al. (48) developed a system in which they froze intact biofilms, sectioned them with a microtome, and then measured previously added metabolites or chemicals in the slices as a function of depth into the biofilm.

There can be a problem in comparing biofilms to cells from suspension cultures in relation to the base for comparison. Often, biomass is used as a base. However, biofilms are generally very rich in polysaccharides because the biofilm matrix is mainly polysaccharide. Protein content can be used as a base, but again, there can be problems in that biofilms are relatively low in protein content, although the actual biofilm cells may not be. For solute uptake studies, cell water or protoplast water is probably the best base to use. There could be an advantage in using cell or cytoplasmic water as a base for considering solute uptake or excretion. The need, then, for estimating cytoplasmic water in biofilms is for a solute that can penetrate readily the polymer matrix of the biofilm and the cell wall but cannot penetrate the inner cell membrane. The best candidates are again the ones used with suspension cells, small solutes such as sucrose or raffinose for which the cells under study do not have significant permeability or transport systems, or nontransported analogues of known substrates. The total water of the biofilm can be assessed with tritiated water or with solutes such as labeled glycerol or urea, although many cells actually do have transport systems for urea or glycerol.

20.6. MEMBRANE VESICLES AND LIPOSOMES

Membrane vesicles have been useful for physiological studies since the early use by Kaback and Stadtman (25) of cytoplasmic membrane preparations from E. coli cells to study proline uptake. Generally, the first step in preparing vesicles is to convert the test organism into an osmotically sensitive form, which is then easily disrupted osmotically or mechanically under conditions that allow for resealing of membrane fragments to form vesicles. Spheroplasts of gram-negative bacteria such as E. coli can be prepared by the lysozyme-EDTA method (24), which leaves the outer membrane of the cell wall associated with the spheroplast. Protoplasts of gram-positive bacteria can often be completely freed from cell wall structures by treatment with wall-lytic enzymes, such as lysozyme for cells of Bacillus megaterium, B. subtilis, Enterococcus hirae (Streptococcus faecalis), or Micrococcus luteus; lysostaphin for Staphylococcus aureus; or mutanolysin for Streptococcus mutans (29).

Kobayashi et al. (27) isolated vesicles from E. hirae ATCC 9790, and their procedure is illustrative. They first suspended cells in an osmotic stabilizing solution of 500 mM glycylglycine plus 2 mM $MgSO_4$ (pH 7.2), added 0.5 mg of lysozyme per ml, and incubated the mixture at 37°C for 1 h. (Protoplasts can also be stabilized with sucrose instead of glycylglycine.) The protoplast suspension was centrifuged; resuspended in a hypotonic solution of 50 mM Tris-maleate (pH 7.4), 250 mM sucrose, and 5 mM $MgSO_4$; and then treated with DNase I, which requires magnesium ions for activity. The suspension was chilled, homogenized with a tissue homogenizer, centrifuged at 5,000 $\times$ g for 10 min to remove debris, and then centrifuged at 27,000 $\times$ g for 10 min to pellet the membranes. The membranes were resuspended in the buffer solution of Tris-maleate plus sucrose plus $MgSO_4$ and then passed through a French press (10,000 lb/in^2, or 68 MPa). The resulting suspension was centrifuged twice, once at 27,000 $\times$ g to remove large particles and once at 48,000 $\times$ g for 90 min to pellet the vesicles.

There are many possible variants of this procedure, and some preliminary experimentation is needed to determine optimal procedures for a particular organism. A shortened, one-step procedure for use with gram-positive bacteria, notably B. subtilis, was described by Konings et al. (30).

The vesicles in these preparations tend to be heterogeneous in size, and the preparations by Kobayashi et al. (27) contained both right-side-out and wrong-side-out vesicles, as indicated by their appearance in electron micrographs after freeze-fracturing. With E. coli, it is possible to prepare populations of vesicles all in the right-side-out conformation. For example, Kaback (24) described a procedure involving osmotic lysis without passage through a French pressure cell. *E. coli* vesicles prepared with use of the French pressure cell are mainly wrong-side-out and, for example, will not extrude Ca^{2+} or take up proline (1). Both right-side-out and wrong-side-out vesicles can be useful for physiological studies. However, it is best to have preparations nearly all in one conformation or the other. Membranes of many organisms are very resistant to resealing after rupture, so it is nearly impossible to prepare vesicles from organisms such as *S. mutans* and only very small numbers of sealed vesicles from large amounts of isolated membrane.

The membrane continuity of vesicles is best tested by assessing the extent of swelling or shrinking in response to changes in osmolality of the suspending medium. The procedures described in the section on osmotically sensitive cells can be adapted for use with vesicles, especially those involving measurements of changes in light scattering (turbidity) associated with swelling (decreased optical density) and shrinking (increased optical density) of vesicles. The functionality of vesicles is determined in terms of their capacities to transport solutes.

The workings of individual transport components, such as permease proteins, antiporters, symporters, or ion-translocating ATPases, can often best be studied by isolating the catalysts from cell membranes and incorporating them into liposomes or proteoliposomes. The test system is then somewhat simpler than the intact cell, but there commonly is a need to set up some sort of membrane potential or ΔpH or to incorporate ATP or phosphoenolpyruvate into the liposomes or proteoliposomes to energize the translocators. A thorough description of the use of proteoliposomes containing *E. coli* phospholipids and Lac permease is given by Viitanen et al. (57). In a more elaborate system, other proteins can be incorporated into proteoliposomes so that Δp can be generated by the structures themselves. For example, Fristedt et al. (20) used proteoliposomes with incorporated purified cytochrome *c* oxidase from beef heart mitochondria to study the functioning of the proton-coupled Pho84 phosphate permease of *Saccharomyces cerevisiae*. The cytochrome oxidase acted to generate Δp across the vesicle membrane to drive uptake of phosphate. Fusion of the components of the proteolysosomes was accomplished by a freeze-thaw-sonication method (15).

20.7. PERMEABILIZED CELLS

Methods to render cells permeable to solutes by chemical or enzymatic treatment are used widely. Perhaps the most common is that involving treatment of cells with toluene in standard assays of β-galactosidase in bacteria such as *E. coli*. Gram-positive bacteria may require more rigorous procedures; for example, permeabilization of cells of *S. mutans* requires treatment with 10% (vol/vol) toluene (which exceeds the solubility of toluene in water) followed by two rounds of freezing and thawing (4).

20.8. REFERENCES

1. **Altendorf, K. H., and L. A. Staehelin.** 1974. Orientation of membrane vesicles from *Escherichia coli* as detected by freeze-cleave electron microscopy. *J. Bacteriol.* **117:**888–899.
2. **Ames, G. F.-L.** 1974. Two methods for the assay of amino acid transport in bacteria. *Methods Enzymol.* **32:**843–849.
3. **Avison, M. J., H. P. Hetherington, and R. G. Shulman.** 1984. Applications of NMR to studies of tissue metabolism. *Annu. Rev. Biophys. Biophys. Chem.* **15:**377–402.
4. **Belli, W. A., and R. E. Marquis.** 1991. Adaptation of *Streptococcus mutans* and *Enterococcus hirae* to acid stress in continuous culture. *Appl. Environ. Microbiol.* **57:**1134–1138.
5. **Bender, G. R., S. V. W. Sutton, and R. E. Marquis.** 1986. Acid tolerance, proton permeabilities, and membrane ATPases of oral streptococci. *Infect. Immun.* **53:**331–338.
6. **Bradshaw, D. J., P. D. Marsh, K. M. Schilling, and D. Cummins.** 1996. A modified chemostat system to study the ecology of oral biofilms. *J. Appl. Bacteriol.* **80:**124–130.
7. **Burne, R. A., and R. E. Marquis.** 2001. Biofilm acid/base physiology and gene expression in oral bacteria. *Methods Enzymol.* **337:**403–415.
8. **Burne, R. A., R. G. Quivey, Jr., and R. E. Marquis.** 1999. Physiologic homeostasis and stress response in oral biofilms. *Methods Enzymol.* **310:**441–460.
9. **Chung, H-J., T. J. Montville, and M. L. Chikindas.** 2000. Nisin depletes ATP and proton motive force in mycobacteria. *Lett. Appl. Microbiol.* **31:**416–420.
10. **Conway, E. J., and M. Downey.** 1950. An outer metabolic region of the yeast cell. *Biochem. J.* **47:**347–355.
11. **de Poorter, L. M. I., and J. T. Keltjens.** 2001. Convenient fluorescence-based methods to measure membrane potential and intracellular pH in the Archaeon *Methanobacterium thermoautotrophicum*. *J. Microbiol. Methods* **47:**233–241.
12. **Doyle, R. J. (ed.).** 1999. *Methods in Enzymology*, vol. 310. *Biofilms*. Academic Press, Inc., San Diego, CA.
13. **Doyle, R. J. (ed.).** 2001. *Methods in Enzymology*, vol. 336/337. *Microbial Growth in Biofilms. Part A, Developmental and Molecular Aspects. Part B, Special Environments and Physicochemical Aspects.* Academic Press, Inc., San Diego, CA.
14. **Doyle, R. J., and R. E. Marquis.** 1994. Elastic, flexible peptidoglycan and bacterial cell wall properties. *Trends Microbiol.* **2:**57–60.
15. **Driessen, A. J. M., and W. N. Konings.** 1993. Insertion of lipids and proteins into bacterial membranes by fusion with liposomes. *Methods Enzymol.* **221:**394–408.
16. **Driessen, A. J. M., D. Molenaar, and W. N. Konings.** 1989. Kinetic mechanism and specificity of the arginine-ornithine antiporter of *Lactococcus lactis*. *J. Biol. Chem.* **264:**10361–10370.
17. **Driessen, A. J. M., B. P. Rosen, and W. N. Konings.** 2000. Diversity of transport mechanisms: common structural principles. *Trends Biochem. Sci.* **25:**397–401.
18. **Dubois, M., K. A. Gilles, J. K. Hamilton, P. A. Rebers, and F. Smith.** 1956. Colorimetric method for determination of sugars and related substances. *Anal. Chem.* **28:**350–356.
19. **Felle, H., J. S. Porter, C. L. Slayman, and H. R. Kaback.** 1980. Quantitative measurements of membrane potential in *Escherichia coli*. *Biochemistry* **19:**3585–3590.
20. **Fristedt, U., M. van der Rest, B. Poolman, W. N. Konings, and B. L. Persson.** 1999. Studies of cytochrome c oxidase-driven H^+-coupled phosphate transport catalyzed by the *Saccharomyces cerevisiae* Pho84 permease in coreconstituted vesicles. *Biochemistry* **38:**16010–16015.
21. **Harold, F. M., and K. Altendorf.** 1974. Cation transport in bacteria: K^+, Na^+, and H^+. *Curr. Top. Membr. Transp.* **5:**1–50.
22. **Hurwitz, C., C. B. Braun, and R. A. Peabody.** 1965. Washing bacteria by centrifugation through a water-immiscible layer of silicone. *J. Bacteriol.* **90:**1692–1695.
23. **Iwami, Y., S. Hata, C. F. Schachtele, and T. Yamada.** 1995. Simultaneous monitoring of intracellular pH and proton excretion during glycolysis by *Streptococcus mutans* and *Streptococcus sanguis*: effect of low pH and fluoride. *Oral Microbiol. Immunol.* **10:**355–359.

24. **Kaback, H. R.** 1971. Bacterial membranes. *Methods Enzymol.* **22:**99–120.

25. **Kaback, H. R., and E. R. Stadtman.** 1966. Proline uptake by an isolated cytoplasmic membrane preparation of *Escherichia coli. Proc. Natl. Acad. Sci. USA* **55:**920–927.

26. **Kashket, E. R.** 1985. The proton motive force in bacteria: a critical assessment of methods. *Annu. Rev. Microbiol.* **39:**219–242.

27. **Kobayashi, H., J. Van Brunt, and F. M. Harold.** 1978. ATP-linked calcium transport in cells and membrane vesicles of *Streptococcus faecalis. J. Biol. Chem.* **253:**2085–2092.

28. **Koch, A. L.** 1994. Growth measurement, p. 248–277. *In* P. Gerhardt (ed.), *Methods for General and Molecular Bacteriology.* ASM Press, Washington, DC.

29. **Konings, W. N.** 1977. Active transport of solutes in bacterial membrane vesicles. *Adv. Microb. Physiol.* **15:**175–251.

30. **Konings, W. N., A. Bisschop, M. Veenhuis, and C. A. Vermeulen.** 1973. New procedure for the isolation of membrane vesicles of *Bacillus subtilis* and an electron microscopic study of their ultrastructure. *J. Bacteriol.* **116:**1456–1465.

31. **Krulwich, T. A., R. Agus, M. Schneier, and A. A. Guffanti.** 1985. Buffering capacity of bacilli that grow at different pH ranges. *J. Bacteriol.* **162:**768–772.

32. **Ma, Y., T. M. Curran, and R. E. Marquis.** 1997. Rapid procedures for acid adaptation of oral lactic-acid bacteria and further characterization of the response. *Can. J. Microbiol.* **43:**143–148.

33. **MacDonald, R. E., and P. Gerhardt.** 1958. Bacterial permeability: the uptake and oxidation of citrate by *Escherichia coli. Can. J. Microbiol.* **4:**109–124.

34. **Maloney, P. C.** 1979. Membrane H^+ conductance of *Streptococcus lactis. J. Bacteriol.* **140:**197–205.

35. **Maloney, P. C., E. R. Kashket, and T. H. Wilson.** 1975. Methods for studying transport in bacteria. *Methods Membr. Biol.* **5:**1–49.

36. **Marquis, R. E.** 1967. Osmotic sensitivity of bacterial protoplasts and the response of their limiting membrane to stretching. *Arch. Biochem. Biophys.* **118:**323–331.

37. **Marquis, R. E., and P. Gerhardt.** 1964. Respiration-coupled and passive uptake of α-aminoisobutyric acid, a metabolically inert transport analogue, by *Bacillus megaterium. J. Biol. Chem.* **239:**3361–3371.

38. **Matts, T. C., and C. J. Knowles.** 1971. Stopped-flow studies of salt-induced turbidity changes of *Escherichia coli. Biochim. Biophys. Acta* **249:**583–587.

39. **Midgley, M.** 1987. An efflux system for cationic dyes and related compounds in *Escherichia coli. Microbiol. Sci.* **4:**125–127.

40. **Mitchell, P., and J. Moyle.** 1959. Permeability of the envelopes of *Staphylococcus aureus* to some salts, amino acids, and non-electrolytes. *J. Gen. Microbiol.* **20:**434–441.

41. **Morbach, S., and R. Krämer.** 2002. Body shaping under water stress: osmosensing and osmoregulation of solute transport in bacteria. *ChemBioChemistry* **3:**384–397.

42. **Phan, T.-N., J. S. Reidmiller, and R. E. Marquis.** 2000. Sensitization of *Actinomyces naeslundii* and *Streptococcus sanguis* in biofilms and suspensions to acid damage by fluoride and other weak acids. *Arch. Microbiol.* **174:**248–255.

43. **Poolman, B.** 2002. Transporters and their roles in LAB cell physiology. *Antonie Leeuwenhoek* **82:**147–164.

44. **Poolman, B., P. Blout, J. H. A. Folgering, R. H. E. Friesen, P. Moe, and T. der Heide.** 2002. How do membrane proteins sense water stress? *Mol. Microbiol.* **44:**889–902.

45. **Ramos, S., and H. R. Kaback.** 1977. The electrochemical proton gradient in *Escherichia coli* membrane vesicles. *Biochemistry* **16:**848–854.

46. **Ramos, S., S. Schuldiner, and H. R. Kaback.** 1976. The electrochemical gradient of protons and its relationship to active transport in *Escherichia coli* membrane vesicles. *Proc. Natl. Acad. Sci. USA* **73:**1892–1896.

47. **Reed, R. H., and A. E. Walsby.** 1985. Changes in turgor pressure in response to increases in external NaCl concentration in the gas-vacuolate cyanobacterium *Microcystis* sp. *Arch. Microbiol.* **143:**290–296.

48. **Robinson, C., J. Kirkham, R. Percival, R. C. Shore, W. A. Bonass, S. J. Brookes, L. Kusa, H. Nakagaki, K. Kato, and B. Nattress.** 1997. A method for the quantitative, site-specific study of the biochemistry within dental plaque biofilms in vivo. *Caries Res.* **31:**194–200.

49. **Roessler, M., and V. Müller.** 2001. Osmoadaptation in bacteria and archaea: common principles and differences. *Environ. Microbiol.* **3:**743–754.

50. **Russell, J. B., and F. Diez-Gonzales.** 1998. The effects of fermentation acids on bacterial growth. *Adv. Microb. Physiol.* **39:**205–234.

51. **Scherrer, R., and P. Gerhardt.** 1971. Molecular sieving by the *Bacillus megaterium* cell wall and protoplast. *J. Bacteriol.* **107:**718–735.

52. **Scholes, P., and P. Mitchell.** 1970. Acid-base titration across the plasma membrane of *Micrococcus denitrificans:* factors affecting the effective proton conductance and the respiratory rate. *J. Bioenerg.* **1:**61–72.

53. **Siebold, C., K. Flükiger, R. Beutler, and B. Erni.** 2001. Carbohydrate transporters of the bacterial phosphoenolpyruvate:sugar phosphotransferase system (PTS). *FEBS Lett.* **504:**104–111.

54. **Stadelmann, E. J.** 1966. Evaluation of turgidity, plasmolysis and deplasmolysis of plant cells. *Methods Cell Physiol.* **2:**143–216.

55. **Stewart, P. S.** 2003. Diffusion in biofilms. *J. Bacteriol.* **185:**1485–1491.

56. **Tran, Q. H., and G. Unden.** 1998. Changes in the proton potential and the cellular energetics of *Escherichia coli* during growth by aerobic and anaerobic respiration or by fermentation. *Eur. J. Biochem.* **251:**538–543.

57. **Viitanen, P. M., J. Newman, D. L. Foster, T. H. Wilson, and H. R. Kaback.** 1986. Purification, reconstitution, and characterization of the *lac* permease of *Escherichia coli. Methods Enzymol.* **125:**429–452.

58. **Vroom, J. M., K. J. de Grauw, H. C. Gerritsen, D. J. Bradshaw, P. D. Marsh, G. K. Watson, J. J. Birmingham, and C. Allison.** 1999. Depth penetration and detection of pH gradients in biofilms by two-photon excitation microscopy. *Appl. Environ. Microbiol.* **65:**3502–3511.

59. **Whatmore, A. M., and R. H. Reed.** 1990. Determination of turgor pressure in *Bacillus subtilis:* a possible role for K^+ in turgor regulation. *J. Gen. Microbiol.* **136:**2521–2526.

60. **White, N. S.** 1995. Visualization systems for multidimensional CLSM images, p. 211–254. *In* J. B. Pawley (ed.), *Handbook of Biological Confocal Microscopy.* Plenum Press, New York, NY.

61. **Wood, J. M.** 1999. Osmosensing by bacteria. Signals and membrane-based sensors. *Microbiol. Mol. Biol. Rev.* **63:**230–262.

62. **Wood, S. R., J. Kirkham, P. D. Marsh, R. C. Shore, B. Nattress, and C. Robinson.** 2000. Architecture of intact natural human plaque biofilms studied by confocal laser scanning microscopy. *J. Dent. Res.* **79:**21–27.

21

Bacterial Respiration

ROBERT P. GUNSALUS, GARY CECCHINI, AND IMKE SCHRÖDER

Prokaryotes as a group exhibit vast respiratory diversity. This is manifested in the large range of electron donors and electron acceptors that may be redox coupled for energy generation. Respiring microbes include obligate aerobes, obligate anaerobes, microaerophiles, and facultative anaerobes. Only some types of fermentative bacteria are incapable of any respiration. Depending on the species and strain, a microbe may possess a relatively simple electron transport pathway (e.g., oxidation of NADH coupled to di-oxygen reduction to water) as in the mitochondria of yeasts, while other microbes possess multiple branching respiratory pathways for energy generation from a large variety of electron donors and electron acceptors. For example, *Escherichia coli* can utilize over 20 combinations of electron donors and electron acceptors to respire aerobically or

anaerobically (8, 26). This respiratory diversity among all life forms thus allows many different strategies to obtain energy and to sustain cell growth and survival. The respiratory ability is predicated by the genetic makeup of the organism. Associated with this prokaryotic respiratory diversity is a general ability to use the energetically more favorable electron acceptor-donor combinations in lieu of lower-energy releasing reactions (26, 34, 91).

For all respiring microbes, the paired oxidation-reduction reactions are coupled to ion transport and to ATP synthesis. The maximal energy released depends on the midpoint potential difference between the electron donor and electron acceptor pair (Table 1), while the energy yield depends on the efficiency of the specific types of respiratory complexes made by the organism. Thus, the overall energy yield achieved for a given electron donor and electron acceptor pair may differ between species.

The oxidation-reduction midpoint potentials for many aerobic and anaerobic respiratory substrates are presented in Table 1. Common bacterial electron acceptors include oxygen, nitrate, nitrite, nitrous oxide, nitric oxide, methyl sulfoxides, methyl amine-N-oxides, fumarate, sulfate, sulfite, elemental sulfur, ferric ion, and carbon dioxide plus less well-studied acceptors like selenate, arsenate, tellurate, manganous ion, and uranium(VI). Commonly used bacterial electron donors include hydrogen gas, formate, glycerol, ethanol, acetate, lactate, and succinate (7, 26, 40, 91, 93). Several examples of aerobic and anaerobic oxidation-reduction reactions are presented below along with the energy released under standard reaction conditions.

General scheme:

Acceptor + donor → products free energy released
$$(\Delta G'^0 = -nF \, \Delta E'^0)$$

Aerobic respiration

$$O_2 + 2NADH + 2H^+ \rightarrow 2H_2O + 2NAD^+$$
$$\Delta G'^0 = -438 \text{ kJ}$$

$$O_2 + 2H_2 \rightarrow 2H_2O + 2NAD^+$$
$$\Delta G'^0 = -474 \text{ kJ}$$

TABLE 1 Standard redox potentials of selected electron donors and acceptors involved in bacterial oxidation-reduction reactions

Redox pair (ox/red)	E'^0 (mV)[a]
SO_4^{2-}/HSO_3^-	−516
CO_2/formate	−432
H^+/H_2	−414
$S_2O_3^{2-}/HS^- + HSO_3^-$	−402
$NAD^+/NADH + H^+$	−320
CO_2/acetate	−290
S^0/HS^-	−270
CO_2/CH_4	−244
Acetaldehyde/ethanol	−197
Pyruvate/lactate	−190
Dihydroxyacetone-P/glycerol-P	−190
$HSO_3^-/S_3O_6^{2-}$	−173
HSO_3^-/HS^-	−116
MQ ox/red	−74
APS/AMP + HSO_3^-	−60
Fumarate/succinate	+33
Ubiquinone ox/red	+110
TMAO/TMA	+130
DMSO/DMS	+160
NO_2^-/NO	+350
NO_3^-/NO_2^-	+433
SeO_4^{2-}/SeO_3^{2-}	+475
Fe^{3+}/Fe^{2+}	+772
O_2/H_2O	+818
NO/N_2O	+1,175
N_2O/N_2	+1,355

[a] Values derived from reference 91.

Anaerobic respiration

$$NO_3^- + NADH + H^+ \rightarrow NO_2^- + NAD^+ + H_2O$$
$$\Delta G'^0 = -143 \text{ kJ/mol}$$

$$TMAO + NADH + H^+ \rightarrow TMA + NAD^+ + H_2O$$
$$\Delta G'^0 = -86 \text{ kJ/mol}$$

$$Fumarate + NADH + H^+ \rightarrow succinate + NAD^+$$
$$\Delta G'^0 = -67 \text{ kJ/mol}$$

$$NO_3^- + formate \rightarrow NO_2^- + CO_2 + H_2O$$
$$\Delta G'^0 = -167 \text{ kJ/mol}$$

The **Gibbs free energy** ($\Delta G'^0$) is determined under standard biochemical conditions.

$$\Delta G'^0 = -nF\Delta E'^0$$

where $\Delta E'^0 = [E'^0$ (for A/AH) $-E'^0$ (for D/DH)] (E is in volts [the redox values are in Table 1]), $\Delta E'^0$ is the difference in redox potentials between the acceptor and donor pair, n is the number of electrons transferred, and F is the Faraday constant (96.5 kJ/mol/V)

Depending on the genetic blueprint of a microbe, it may only produce a few respiratory enzymes or it may be capable of synthesizing many alternative respiratory complexes. For example, *E. coli* possesses two distinct cytochrome oxidases, three nitrate reductases, two nitrite reductases, two trimethylamine-N-oxide (TMAO) reductases, a dimethyl sulfoxide (DMSO) reductase, and a fumarate reductase. Most enzymes are differentially synthesized in response to oxygen and nitrate availability and, for some species, in response to the presence of TMAO or fumarate. The level of a given enzyme in an organism may also vary depending on the conditions used for cell growth. Thus, prior to cell harvest and enzyme assay, consideration should be given to the composition of the culture medium used (oxygen, nitrate, formate, etc.). One may also consider other environmental parameters for cell growth (e.g., pH, salinity, and temperature). Finally, if the genomic DNA sequence of the organism is known, inspection of the predicted gene and thus protein content may provide additional insight to the potential for cell respiration.

The goal of this chapter is to provide a brief introduction to assay the activity of respiratory enzymes. Of the large range of organic and/or inorganic electron donors and acceptors used by microbes (91), a subset of these respiratory substrates is considered here. For more in-depth information or specific enzyme assays, the reader is directed to the literature.

21.1. ENZYME ASSAYS

Assays for many commonly encountered respiratory enzymes are presented below, while literature citations are presented for others. Assays of several anaerobic respiratory enzymes associated with fermenting conditions (e.g., lactate dehydrogenase, hydrogenase, and formate dehydrogenase) are described in a companion chapter in this book (chapter 22).

21.1.1. Reaction Classification and Nomenclature

Enzyme nomenclature as defined by the International Union of Biochemistry and Molecular Biology is not discussed here but may be explored at http://www.chem.qmul

.ac.uk/iubmb/enzyme/. Additional enzyme information is available at BRENDA, the Comprehensive Enzyme Information System at http://www.brenda.uni-koeln.de/, and chapter 19 of this volume.

21.1.2. Units and Specific Activity

Unless indicated otherwise, one unit of enzyme activity is defined as micromoles of substrate reduced or oxidized per minute. The specific activity refers to units per milligram of protein (see also chapter 19 of this volume and reference 69).

21.1.3. Types of Activity Assays

In general, quantitative enzyme assays are routinely performed to calculate an enzyme's specific activity and other kinetic properties. However, qualitative enzyme assays may be conveniently applied for some studies to screen for the presence of a desired enzyme. These methods include in situ activity stains of native polyacrylamide gels (i.e., "zymograms"), or qualitative or semiquantitative "spot plate" assays. The investigator is encouraged to improvise specific applications for different enzymes and/or strains based on the examples described below.

21.1.3.1. Qualitative Assays

In exploratory studies to screen for the presence of a given enzyme activity, it is often convenient to use qualitative screening methods. Applications employ the enzyme's specific substrate and a redox dye that changes color as a result of the reaction catalyzed by the enzyme. Examples follow for detection of nitrate reductase.

21.1.3.1.1. Use of Spot Assays or Color Indicator Assays

Spot assays can facilitate the rapid detection of enzyme activities in whole or permeabilized cells, in subcellular samples, or in chromatography fractions to follow protein purification. Compared to quantitative cuvette assays that may need to be performed under controlled conditions, spot assays may be done in microtiter dishes or in microcentrifuge tubes under less rigorous conditions. The tube or microtiter dish containing the assay mixture and enzyme sample is mixed with a colorimetric indicator reagent. Typically, an increase or decrease in color change indicates the presence and relative amount of enzyme activity present. The degree of color change can be visually estimated, or semiquantified by spectral methods to determine what fraction(s) contain the most abundant enzyme activity. However, since no clear attempt is made to validate experimental conditions with respect to assay linearity, substrate limitation, or presence of interfering materials, this approach is not suitable for enzyme rate determinations. Rate determinations require the measurement of a color-based activity change with time under carefully established assay conditions.

21.1.3.1.1.1. Nitrate reductase spot assay. Nitrate reductase catalyzes the reduction of nitrate to nitrite and can be assayed by following the color change of a redox dye or by detection of end product formed (i.e., nitrite).

Method A. Nitrate reduction in this assay is monitored by following the oxidation of the artificial electron donor, reduced benzyl viologen (BV) or methyl viologen (MV) in 1.5-ml microcentrifuge tubes. The tubes are filled with an aerobic solution of 0.1 M potassium phosphate (pH 7.2), 2 mM BV, plus the protein sample to be assayed. After equilibrating the tubes at the desired temperature, 5 mM sodium dithionite or less ($Na_2S_2O_4$ dissolved in 0.1 M $NaHCO_3$, stored on ice) is added to reduce the viologen dye. The appearance of a blue-purple color indicates the presence of reduced viologen. The amount of dithionite used in the spot assay is typically sufficient to reduce any oxygen present in the tube since reduced viologen is readily autoxidized by oxygen. However, excess dithionite will also inhibit nitrate reductase activity, so care should be taken in this step of the procedure. The nitrate reductase assay is then initiated by the addition of 10 mM sodium nitrate (i.e., from a concentrated aerobic solution), and the microcentrifuge tube containing the assay mixture is incubated at the desired temperature for a fixed time period (e.g., about 10 min) depending on the level of nitrate reductase present in the sample. The disappearance of purple color indicates the nitrate-dependent oxidation of reduced BV dye where the degree of color change is roughly correlated to the amount of enzyme activity present. It is important to include a nonenzyme control tube to monitor autoxidation of reduced viologen dye. Activity is scored by the relative difference in color across the range of assay tubes. Following completion of the assay as described above, the protein samples are either evaluated immediately for nitrite formation or frozen at −20°C. In the latter case, the samples should have a neutral pH (e.g., not preserved with acid).

Method B. Since it is often problematic to rely on decoloration of reduced viologen dye, the detection of nitrite presents a more reliable method. Formation of nitrite is based on the reaction of nitrite with sulfanilamide under acidic conditions to form a diazo compound (32; see also chapter 18 in this volume). N-(1-Naphtyl)-ethylenediamine is a colorless compound which in the presence of nitrite forms a purple compound with absorbance at 543 nm. The sensitivity of this assay ranges from 1 to 180 nmol of NO_2^-/ml. Certain substances interfere with the assay, including colored or particulate materials that may be present in the protein sample, or the presence of Cl^-, Fe^{3+}, or Cu^{3+} ions. To prepare the color reagent, 1 part 0.08% N-(1-naphtyl)-ethylenediamine, dissolved in H_2O, is mixed with 2 parts of 4% sulfanilamide (p-aminobenzene sulfonamide, dissolved in 25% HCl). Note that the 0.08% N-(1-naphtyl)-ethylenediamine solution should be stored in the dark and discarded when it turns brown. To initiate the nitrite detection reaction, 0.3 ml of color reagent is added to a 1-ml nitrate reductase spot assay from method A. A solution of the diluted enzyme sample is mixed and incubated for 10 min at room temperature. Formation of a purple color indicates the formation of nitrite ions where the color intensity is proportional to the amount of nitrite present. Remaining reduced viologen dye typically oxidizes during this procedure and does not interfere with the nitrite detection assay. Note: the presence of any nitrite reductase activity in the protein sample may interfere with the assay.

21.1.3.1.2. Zymograms, or "In Gel" Polyacrylamide Activity Stains

In-gel activity stains, or zymograms, offer an alternative approach to detect oxidoreductase enzymes, especially when the level of enzyme activity is at or below the sensitivity limit of a standardized quantitative enzyme assay. Low enzyme activity may be due to a low cellular abundance of the enzyme in the sample, or due to suboptimal conditions used for the enzyme assay. Zymograms can also be used to detect isozymes that differ in molecular mass when multiple enzymes with similar activities are present in a cell extract. For extensive information regarding the principles and applications of enzyme analysis using zymograms, see reference 59. One specific application to detect soluble nitrate reductase in cell extracts is provided below.

21.1.3.1.2.1. Nitrate reducase zymogram assay. The (halo)alkaliphilic sulfur-oxidizing bacterium *Thioalkalivibrio nitratireducens* contains a soluble nitrate reductase (5). Following cell harvest, cells are broken by sonication (see section 21.2). The soluble protein fraction is isolated via centrifugation and separated by native polyacrylamide electrophoresis. The polyacrylamide gel lanes are then cut into vertical strips and placed in a shallow tray containing the enzyme assay solution (0.2 M sodium phosphate buffer [pH 7.0], 5 mM MV, and 10 mM KNO_3). A sodium dithionite solution (10 mM) is then slowly added to reduce the MV, which in turn provides electrons for the nitrate reduction reaction. The immersed gels strips are then incubated with gentle shaking at the preferred temperature (e.g., room temperature) until transparent bands appear in the gel strip against the blue background of reduced viologen dye. The transparent band(s) in the gel correspond to the location(s) of the nitrate reductase activity. The gel is subsequently washed with 0.05% (wt/vol) triphenyltetrazolium chloride and fixed in a 5% (vol/vol) acetic acid solution. The gel is then washed in water and stored in 50% ethanol before drying for archival purposes.

21.1.3.1.2.2. Detection of other respiratory enzymes by zymogram assay. A variety of other substrate-specific oxidoreductase enzymes may be detected by using the above zymogram approach where nitrate is replaced by the desired alternative substrate. For example, nitrate is replaced by 1 mM $NaNO_2$ to detect nitrite reductase, by 2 mM Na_2SeO_4 for selenate reductase, by 2 mM $AsHNa_2O_4$ for arsenate reductase, or by 2 mM Na_2SO_3 for sulfite reductase. Other substrates can also be tested in place of nitrate at concentrations in the range of 2 to 10 mM to screen for the presence of the corresponding oxidoreductase activity.

Membrane-bound oxidoreductases can be identified in an analogous fashion. However, the enzyme must first be extracted from the cytoplasmic membranes using detergents prior to protein separation by native polyacrylamide gel electrophoresis. Typically, the cytoplasmic membrane fraction (section 21.2) is incubated with 2% (or higher) Triton X-100 for 60 min. The sample is then subjected to ultracentrifugation to remove the lipid fraction prior to separation of the solubilized proteins by polyacrylamide gel electrophoresis. The extraction efficiency of an enzyme is dependent on its hydrophobicity and charge; therefore, the extraction protocol for each protein must be addressed empirically (99). For improved separation performance a detergent can be included in the polyacrylamide gel (99).

The zymogram-based enzyme assay can also be performed under oxygen-limiting conditions to improve the assay efficiency since reduced MV is readily oxidized by any O_2 that has diffused into the assay chamber during incubation. To reduce autoxidation of reduced MV, it is possible to incubate the gel slices within a closed plastic container that has been perforated with a gas inlet and outlet port to allow a constant flow of oxygen-free N_2 or argon gas over the gel slices during the assay. To further reduce dye autoxidation, careful titration of dithionite into the assay mixture may be performed prior to gassing. The investigator is encouraged to improvise on and optimize assay conditions for his/her specific application.

21.1.3.2. Quantitative Assays

Quantitative assay methods are applied to first detect enzyme activity in cells or cell fractions and then, more importantly, to obtain a detailed understanding of the enzyme's kinetic properties. As noted above, both require optimization of the enzyme assay conditions and requirements of the specific enzyme being studied. The enzyme assays described in the following subsections are based on specific examples from the literature. Therefore, the investigator may need to optimize the assay conditions depending on the source of the enzyme (i.e., the microorganism that makes it) and the enzyme's requirements. The investigator is directed to the microbiological and biochemical literature for additional details. Also, see chapters 19 and 20 in this volume for further details on enzymatic activity, bacterial transport, and membrane permeability.

The choices of electron donors or acceptors include the physiological electron donors or acceptors such as ubiquinone, menaquinone (MQ), NAD(P)H, c-type cytochromes, or iron-sulfur proteins that provide or accept reducing potential to the oxidoreductase complex. In addition, many respiratory enzymes are commonly assayed with artificial electron donors or acceptors that are coupled with the reduction or oxidation of their specific substrate. This is because many membrane-associated enzymes interact with quinones that are difficult to monitor by UV-visible spectroscopy because of protein absorption at the same wavelength. Properties of commonly used artificial redox dyes are listed in Table 2. Assays employing membrane-impermeative artificial dyes may be performed to determine the orientation of the oxidoreductase enzyme active site on the inside or outside of the cell membrane (see section 21.2.6 below). Finally, since anaerobic assays require specialized techniques, not all respiratory enzyme assays are equally easy to master.

21.1.3.2.1. General Consideration of Enzyme Stability

The general assay methods described below apply to most neutrophilic gram-negative bacteria such as *E. coli*. The assays may require adjustment of the pH, temperature, salinity, and other parameters for application to different microbes. For general methods of cell growth, harvest, breakage, and fractionation, see reference 69. Note that, while many respiratory enzymes are relatively stable, other enzymes can be extremely labile. Therefore, before embarking on a major study it is useful to compare enzyme activities in cells grown to stationary versus exponential phase, to evaluate methods used for cell breakage and sensitivity to oxygen. Furthermore, if stability varies, compare enzyme activities in freshly harvested cells versus frozen cell material.

For certain detergent-solubilized preparations of membrane enzymes, lipids such as soybean asolectin or native *E. coli* lipids can be added to the assay mixtures to stimulate activity (29, 80). The increase in activity can be significant (up to eightfold). Therefore, conditions of solubilization and lipid addition to the assay must be considered for accurate activity determinations. General methods for cell disruption by French press or cell sonication are described in chapter 7 in this volume.

TABLE 2 Standard redox potentials of artificial and physiologicial redox substrates commonly used to assay respiratory enzymes

Substrate	$E^{0'}$ (mV)	Wavelength (nm)	ε^a (mM^{-1}cm^{-1})	Reference
MV	−446	603	13.7	101
BV	−356	550	7.8	43
Morfamquat	−374	600	9.0	43
Diquat	−349	460	2.7	43
NAD$^+$	−320	340	6.22	16
Phenosafranin	−252	520	40	97
Lapachol	−157	481	2.66	76
MQ	−80	263	11.3	50, 106
Q$_1$	−70	277		106
DMN	−80	260	8.0	108
Plumbagin	−40	419	3.95	76
Methylene blue	+ 11	578	17.1	16
PES	+ 55	366	5.1	16
Ascorbic acid	+ 58			106
PMS	+ 80	366	5.1	36
Ubiquinone	+ 90	292	13	20, 31
DCIP	+217	600	21.8	16
TMPD	+260	606	11.6	71
Ferricyanide	+408	420	1.09	16
Cytochrome c	+270	550	19.7	16

^{a}Extinction coefficient expressed at the indicated wavelength.

21.2. ENZYME LOCALIZATION

Respiratory enzymes are not all equally positioned within the cell. Some are membrane associated, while others are soluble. If membrane associated, the enzyme may have the active site for interaction with a substrate exposed to the cytoplasm or, alternatively, facing the periplasmic space of gram-negative bacteria (Fig. 1). Soluble enzymes may be located within the cytoplasm or in the periplasmic space. An understanding of the cellular location(s) for substrate utilization and product formation is useful in deciding what type of diagnostic enzyme assay(s) to perform. Conversely, if the cellular location of the enzyme is unknown, it can be determined experimentally. Finally, knowledge of the enzyme location can be useful in understanding the physiological role(s) of the enzyme in cell metabolism and in energy conservation.

This section outlines experimental approaches to determine the cellular location of a redox enzyme following cell fractionation. While this may reveal the presence of isoenzymes located in distinct cellular fractions, it will, in general, not distinguish between two or more isoenzymes located in the same cellular compartment. The latter issue may be resolved only after separating the isoenzymes chromatographically or by applying genetic methods to delete specific genes before analyzing their individual kinetic properties by biochemical methods.

To identify the cellular location of the enzyme, assays should be performed on intact cells and/or permeabilized cells. This is followed by assays with cell spheroplasts, the periplasmic fraction, the cytoplasmic membrane fraction, and the insoluble cytoplasmic fraction. Enzyme activity is evaluated in total units for each cellular compartment to identify the compartment with the highest percentage of activity. The specific activity (units per milligram of pro-

tein) is an indication of an enzyme's purity. It is generally the highest in the compartment where the enzyme is located. However, activity levels may be lower if the cell fractionation method used leads to enzyme complex disintegration or loss of essential cofactors (also see chapters 7 and 19 of this volume).

21.2.1. Intact Cells

Enzyme assays are performed on intact cells to provide a starting point to define the overall level of enzyme activity. It also serves as a control for any leakage of enzyme from the cellular preparation. Cells are harvested and washed (e.g., with 50 mM Tris-HCl [pH 8]–10 mM EDTA–0.3 M sucrose). After recentrifugation at 30,000 × g for 20 min at 4°C, the cells are resuspended in the same buffer and stored on ice prior to enzyme assay. Intact cells, when assayed with a membrane-permeative redox dye (Table 3), exhibit high activity (i.e., defined to be 100% activity). If little or no activity is detected using an impermeative redox dye as cosubstrate (Table 4), the enzyme is likely to be located inside the cell (e.g., soluble within the cytoplasm or membrane associated with the active site facing the cytoplasm [Fig. 1]). If high enzyme activity is detected, the enzyme is likely to be localized on the outer surface of the cytoplasmic membrane or located in the periplasmic space. Examples of the latter enzymes are the *E. coli* TorC TMAO reductase and DmsABC DMSO reductases, and an enzyme for ferrihydroxide reduction (55).

21.2.2. Permeabilized Cells

Cells are readily permeabilized by the addition of a detergent such as 0.1% Triton X-100 or 0.1% dodecyl maltoside. This provides an important control for enzymes that may

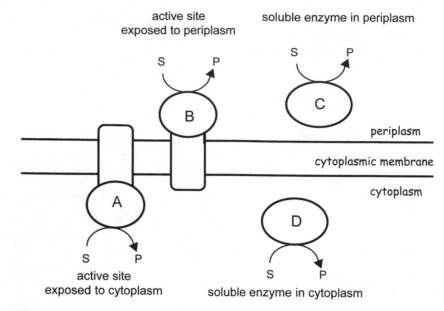

FIGURE 1 Location of redox respiratory enzymes in the cell. The active site of a membrane-bound oxidoreductase may face the cytoplasm (A) or the periplasm (B). A soluble-type oxidoreductase may be located either in the periplasmic space (C) or in the cytoplasm (D). The electron acceptor substrate (S) is converted to the reduced product (P).

lose activity upon cell breakage. However, this activity measurement may reveal limited information as to the enzyme's cellular compartment.

21.2.3. Spheroplasts and the Periplasmic Fraction

To separate the periplasmic proteins, gram-negative cells are incubated in 50 mM Tris-HCl (pH 8.0)–10 mM EDTA–0.3 M sucrose for 10 min at room temperature. Lysozyme is then added to a final concentration of 20 mg/liter, and the cell suspension is gently stirred for 1 h at 4°C. Following centrifugation for 5 min at 3,000 × g and 4°C, the supernatant containing the periplasmic protein complement is concentrated by filter centrifugation or by other methods and stored on ice (4°C). The pelleted spheroplasts are resuspended (50 mM Tris-HCl [pH 8.0], 10 mM EDTA, 0.3 M sucrose) and stored on ice. It is essential to measure the periplasmic fraction for the presence of marker enzymes for the cytoplasmic membrane and cell cytoplasm, e.g., succinate dehydrogenase (membrane bound) and malate dehydrogenase (soluble), to monitor cross-contamination due to cell lysis.

21.2.4. Cell Extracts

Pelleted spheroplasts are resuspended in 10 mM phosphate or other suitable buffer, pH 7.5, with 2 mM MgCl₂, and 4 mg/liter DNase, and stirred for 30 min under a stream of N₂ at 4°C. Spheroplast breakage can be facilitated by freeze-thawing the suspension.

21.2.5. Membrane and Cytoplasmic Protein Fractions

To separate membranes from the cytoplasmic fraction the cell extract is centrifuged for 1 h at 100,000 × g at 4°C. The supernatant fraction containing the cytoplasmic proteins can be further concentrated as necessary and stored on ice. The pellet containing the membranes is washed once in 50 mM phosphate or other suitable buffer, pH 7.5, resuspended in the same buffer, and also stored on ice.

21.2.6. Location(s) of Enzyme Activity in Cell Compartments

Enzyme assays are performed on the above cells and cell fractions to (i) establish total activity and (ii) determine the cell compartment(s) that contains the activity. If significant activity is detected in more than one compartment (assuming minimal cross-contamination of fractions during

TABLE 3 Artificial redox dye substrates that are able to permeate to the cytoplasmic membrane

Dye	Form	Reference
DCIP	Reduced and oxidized	51
MV	Oxidized	43
PES/PMS	Reduced	51
PMS	Reduced	36
Q, MQ, and analogs	Reduced and oxidized	51
Methylene blue	Reduced and oxidized	51

TABLE 4 Artificial redox dye substrates that are unable to permeate to the cytoplasmic membrane

Dye	Form	Reference
BV	Reduced and oxidized	43
MV	Reduced	43
Diquat	Reduced and oxidized	43
PMS	Oxidized	36
Morfamquat	Reduced and oxidized	43
Morfamquat	Reduced and oxidized	16
Dithionite		51
Ferricyanide/ferrocyanide		52

preparation), the cell may contain two or more distinct enzyme species. For example, some bacteria possess a cytoplasmic hydrogenase, a periplasmic hydrogenase, and/or a distinct membrane-bound hydrogenase. Finally, detection of enzyme activity in a single compartment does not rule out the presence of several isoenzymes in that compartment (e.g., distinct hydrogenases).

21.2.7. Determining the Orientation of the Enzyme's Substrate Site

Knowing the orientation of the active site of a membrane-bound enzyme with respect to the cytoplasmic membrane is useful in evaluating its potential bioenergetic role, and in choosing appropriate enzyme assays for further investigations. The active site may be facing the periplasm or facing the cytoplasm. Active-site orientation is determined using redox dyes as electron donors or acceptors that are not able to permeate the cytoplasmic membrane (Table 4). Whole cells exhibit little activity if the enzyme active site is located on the cytoplasmic side of the membrane or if it is soluble within the cytoplasm. Easily assayed cytoplasmic enzymes, such as the tricarboxylic acid (TCA) cycle enzymes malate and succinate dehydrogenase, are used as a control to evaluate cell intactness. When cells are permeabilized by the addition of detergent, the redox dyes gain access to the active site in the cytoplasm and full activity is observed. Periplasmic active sites or enzymes are usually not affected by detergent addition and exhibit full activity like that seen for intact cells. Even though most natural enzyme substrates are charged, they can cross the cytoplasmic membrane via their cognate transporters and are thus rapidly delivered to the enzyme's active site. A number of artificial electron donors and acceptors unable to permeate to the cytoplasmic membrane are listed in Table 3 (51). Note: for certain redox dyes, only the reduced or oxidized form is not membrane permeative. For comparison, redox dyes that are membrane permeative are listed in Table 3.

An example of the cellular orientation of fumarate reductase from the rumen bacterium *Wolinella succinogenes* is described by Kröger and coworkers (51). The fumarate reductase assays are performed anaerobically as described below on intact and lysed cells. Cell lysis is easily accomplished by either sonication or osmotic shock. Alternatively, the cells can be permeabilized with a detergent as follows. After initiation of the fumarate reductase assay by the addition of intact cells, the initial rate is recorded. A small volume of an aerobic detergent stock solution (e.g., 10 μl of 10% Triton X-100 or 10% lauryl maltoside) is then added to the assay mixture and the rate is continuously monitored. An increase in rate upon detergent addition suggests that the enzyme's active site is oriented to the cytoplasm. Note: the orientation of an enzyme active site may not be clearly resolved if the organism contains isoenzymes located in both the periplasmic and cytoplasmic compartments.

21.3. AEROBIC RESPIRATORY ENZYMES

The following enzymes participate in aerobic electron transport pathways in many types of aerobic and facultative microorganisms. NADH:Q oxidoreductase is also present in many anaerobic respiring microbes.

21.3.1. NADH:Q Oxidoreductase

Many bacteria contain either one or two types of NADH-ubiquinone oxidoreductases. For *E. coli* these have been termed **NDH-1** and **NDH-2**.

Physiological reaction: $NADH + H^+ + Q \rightarrow NAD^+ + QH_2$

where Q is ubiquinone and QH_2 is ubiquinol. The NDH-1 enzyme is energy conserving, is located on the inner face of the cytoplasmic membrane, and contains 13 to 14 subunits with multiple prosthetic groups including flavin mononucleotide (FMN) and up to 9 iron-sulfur centers. It is similar to the much larger mitochondrial complex I of eukaryotes. The NDH-1 enzymes in many bacteria and archaea are like that found in the well-studied microbe *Escherichia coli* and, by inference, in *Paracoccus denitrificans*, *Rhodobacter capsulatus*, *Thermus thermophilus*, *Klebsiella pneumoniae*, and *Synechocystis* (23). In contrast, the NDH-2 class of NADH dehydrogenase is composed of a single polypeptide with flavin adenine dinucleotide (FAD) as the only redox group. It is not energy conserving (60). As both NDH-1 and NDH-2 utilize NADH as a substrate, specific assays have been developed to discriminate between the two enzymes (60). NDH-1 utilizes both NADH and deamino (d)-NADH as the electron donor, and enzyme turnover generates a membrane potential in the presence of KCN and membrane-intrinsic ubiquinone (60).

In contrast, the single-subunit NDH-2 enzyme does not generate a membrane potential when using NADH to reduce ubiquinone. Thus, use of d-NADH and appropriate electron acceptors can measure NDH-1 activity even in membrane preparations containing NDH-2. It should be noted that both NADH and d-NADH are unable to permeate the inner membrane (i.e., cytoplasmic membrane), and thus it is essential that either inverted membrane vesicles or permeabilized cells be used in order to assay maximal NDH-1 activity.

The *d-NADH–ferricyanide reductase* assay is useful for determining the content of functional NDH-1 in membrane preparations. Various buffers can be used where a typical assay buffer contains 50 mM potassium phosphate (pH 7.5), with 0.2 mM EDTA, 2 mM KCN, and 1 mM potassium ferricyanide as the electron acceptor. Addition of the inhibitor KCN prevents the reoxidation of ferrocyanide by cytochrome oxidase. Membranes in this buffer are equilibrated in a temperature-controlled cuvette (37°C). Reactions are initiated by the addition of 125 to 150 μM d-NADH, and the rapid reduction of ferricyanide is monitored in a spectrophotometer at 420 nm. Activity is calculated by using the millimolar extinction coefficient of 1.0 for potassium ferricyanide ($\varepsilon_{420} = 1.09$ mM^{-1} cm^{-1} [Table 2]). The site in the enzyme that donates electrons to ferricyanide is unknown, although it is believed to involve an FMN molecule at the enzyme active site since subcomplexes of NDH-1 with FMN still retain NADH-ferricyanide reductase activity (98).

Physiologically, NDH-1 reduces ubiquinone as the electron acceptor: the *NADH-Q reductase* assay is commonly used to monitor the fully intact NDH-1 activity. The same buffer and incubation conditions as described above can be used. However, ubiquinone analogs are used as the electron acceptor in place of the natural substrate, ubiquinone-8. One common analog is ubiquinone-1 or Q_1, a ubiquinone analog with one isoprenoid unit at position 6 of the benzoquinone ring, available from Eisai Co. Ltd. (Tokyo, Japan). Other quinone analogs are decylubiquinone and undecylubiquinone (available from Sigma-Aldrich). However, the quinone analogs give different rates of activity, so caution should be exercised in comparing with NDH-1 activity among different electron acceptors. Because quinone analogs have limited solubility, a small amount of detergent (0.03 to 0.1% dodecyl maltoside) is added to the assay mixture prior to addition of Q_1 in order to increase solubility. Following the addition of 20 to 70 μM Q_1 (or another quinone analog) to the assay mixture, the reaction is initiated by adding 125 to 150 μM d-NADH and the reaction rate is followed by monitoring the decrease of NADH absorbance at 340 nm (ε_{340} = 6.22 mM^{-1} cm^{-1} [Table 2]). It is known that when NDH-1 from either bacterial or mitochondrial preparations is solubilized with detergents, added lipids can stimulate NADH-quinone reductase activity. Thus, for solubilized preparations of NDH-1, lipids such as soybean asolectin or native *E. coli* lipids are added to the assay mixtures to stimulate activity (29, 80). There is a significant (up to eightfold) stimulation of activity, so conditions of solubilization and addition of lipids must be considered to accurately measure complex I activity for an isolated enzyme.

21.3.2. Succinate Dehydrogenase

Succinate dehydrogenase (**SQR**, succinate:ubiquinone oxidoreductase) is the only membrane-bound enzyme of the TCA cycle. It provides a dual role in generating the TCA cycle intermediate, fumarate, and in providing electrons derived from succinate oxidation for the reduction of ubiquinone to ubiquinol. Reducing equivalents are then passed along the electron transport chain of the cytoplasmic membrane to oxygen in bacteria and archaea (14, 41).

Physiological reaction: succinate + Q → fumarate + QH$_2$

21.3.2.1. Activation of Succinate Dehydrogenase

Since succinate dehydrogenase may be partially deactivated by tightly bound oxaloacetate at its active site, the enzyme requires activation prior to enzyme assay by incubation with either malate or succinate. To accomplish this, the succinate dehydrogenase preparation is diluted to 0.5 to 1.0 mg of protein per ml in buffer (30 mM potassium phosphate [pH 7.4], 0.2 mM EDTA, 0.1 % [wt/vol] Thesit [C$_{12}$E$_9$; polyoxyethylene dodecyl ether], and either 10 mM malonate or succinate) and incubated for 10 to 20 min at 37°C . The "activated" enzyme solution can then be stored at 4°C for subsequent assays.

21.3.2.2. Succinate Oxidation with PES or PMS

Succinate oxidation by succinate dehydrogenase is routinely assayed spectrophotometrically using either phenazine ethosulfate (PES; *N*-ethylphenazonium ethyl sulfate salt) or phenazine methosulfate (PMS; *N*-methylphenazonium methyl sulfate) as the electron acceptor (Table 2). When using PES or PMS, the electrons are chemically mediated to 2,6-dichlorophenol indophenol (DCIP) as the final electron acceptor. The coupled reactions using PES are shown below.

Reaction 1: succinate + PES$_{ox}$ → fumarate + PES$_{red}$

Reaction 2: PES$_{red}$ + DCIP$_{ox}$ → PES$_{ox}$ + DCIP$_{red}$

The succinate:PES (or succinate:PMS) reaction is measured aerobically in 1-cm-path-length cuvettes at an assay temperature of 37°C. The cuvette contains 50 mM potassium phosphate (pH 7.6) (or another appropriate buffer), 0.006% (wt/vol) Anapoe C$_{12}$E$_9$ (Anatrace, Maumee, OH) to maintain enzyme solubility, and 0.2 mM EDTA. Then 10 mM succinate, 1.5 mM PES, and 50 μM DCIP are added to the cuvette and the reaction is initiated by addition of en-

zyme. The activity is monitored spectrophotometrically by following the reduction of DCIP at 600 nm (ε_{600} = 21.8 mM^{-1} cm^{-1} [Table 2]).

21.3.2.3. Succinate Oxidation with Ferricyanide

Ferricyanide is another electron acceptor commonly used to assay succinate dehydrogenase activity.

Nonphysiological reaction: succinate + ferricyanide → fumarate + 2H$^+$ + ferrocyanide

Succinate-ferricyanide oxidoreductase activity is determined in the buffer described above but containing 0.45 mM potassium ferricyanide (ε_{420} = 1.09 mM^{-1} cm^{-1} [Table 2]) in place of PES and DCIP. The overall assay protocol is otherwise the same as described above.

21.3.2.4. Succinate Oxidation with Ubiquinone Analogs

Either soluble or membrane-bound forms of succinate dehydrogenase can by assayed by the PES-DCIP method (58). However, reduction of ubiquinone, or various ubiquinone analogs, can only be accomplished by membrane-bound forms of succinate dehydrogenase since the membrane anchor peptides (e.g., SdhC and SdhD) are required to form a binding site for quinones. The latter assay is monitored by the following coupled reaction.

Reaction 1: succinate + Q → fumarate + QH$_2$

Reaction 2: QH$_2$ + DCIP$_{ox}$ → Q + DCIP$_{red}$

The succinate:quinone oxidoreductase assay is performed essentially as described above for the succinate:PES assay. However, a ubiquinone analog is used in place of PES. Low solubility of the quinone limits its accessibility to succinate dehydrogenase. Quinone reduction is therefore assayed by determination of the V_{max} using increasing quinone concentrations. Quinones that are typically used to assay succinate:quinone oxidoreductase activity include decylubiquinone or DB (2,3-dimethoxy-5-methyl-6-decyl-1,4 benzoquinone), DPB (2,3-dimethoxy-5-methyl-6-pentyl-1,4-benzoquinone), or Q$_1$ (Table 2). DB and DPB can be obtained from Sigma-Aldrich, and Q$_1$ may be obtained from Eisai Chemical Company. The hydrophobic nature of quinones usually precludes using quinone analogs that contain greater than two isoprenoid side chains in length. The quinone analogs are dissolved in ethanol and used in the assay at 1 to 40 μM concentrations in order to obtain V_{max}. DCIP reduction is stimulated by the addition of elevated levels of the quinone analog. Usually there is a short lag phase after the addition of quinone that precedes the linear decrease in DCIP absorbance. The lag phase is dependent upon the solubility of the quinone. Succinate-quinone oxidoreductase activity is inhibited by quinone site inhibitors, with pentachlorophenol usually effective for most bacterial enzymes at 20 μM. Detailed descriptions of assays for both succinate oxidase and succinate-quinone oxidoreductase activities are provided in references 1 and 58 and references therein.

21.3.3. Cytochrome Oxidase

Prokaryotes exhibit considerable enzymatic diversity with respect to the number and types of cytochrome oxidase enzymes present for reduction of molecular oxygen to water. Some microbes contain only a single type of oxidase, while others are capable of synthesizing two, three, or more distinct enzyme types. They may differ in type of electron donor used (quinol or cytochrome *c*), in number and types of cofactors present, and in efficiency for energy conservation. This diversity may be explained in part by the ability of bacteria to respond to environmental stress and to changing O_2 conditions (70). For general reviews of cytochrome oxidase enzymes, see references 4, 25, and 105.

General reaction: electron donor + O_2
$$\rightarrow \text{oxidized electron donor} + 2H_2O$$

Based on the immediate electron donor, cytochrome oxidases are divided into two classes: (i) quinol oxidases that receive electrons from the membrane-intrinsic reduced quinone pool (Q), i.e., **cytochrome *bo₃*, *ba₃*, and *bd* oxidases**, and (ii) cytochrome *c* oxidases that obtain electrons from a small periplasmically located *c*-type cytochrome (i.e., **cytochrome *aa₃* and *cbb₃* oxidases**). With the exception of the *bd*-type oxidase, all cytochrome oxidases belong to the heme-copper superfamily (12, 13). Both heme and copper are important to transfer electrons to catalyze oxygen reduction within the enzyme complex. Various hemes (tetrapyrrole ring structures with a central iron atom) are present in the oxidases such as *a*, *b*, *d*, and *o* heme. The hemes differ by their tetrapyrrole ring substituents and saturation, which cause the heme-specific light absorbance characteristics that can be determined by optical spectroscopy. Of all oxidases, only cytochrome *aa₃* oxidase is also present in eukaryotes, where it is located in the mitochondria (68). Several well-established assays are described below as a starting point to evaluate cellular cytochrome oxidase levels.

21.3.3.1. Oxygen Reduction with Ubiquinol as Electron Donor

The quinol oxidases include the cytochrome *bo₃*, *ba₃*, and *bd* oxidases that receive electrons from the membrane-intrinsic reduced quinone pool. They are therefore assayed with the ubiquinol-1 analog, Q_1, or the artificial electron donor TMPD (tetramethylphenylene-diamine) as the electron donor. For both assays, oxygen consumption is monitored using a Clark-type electrode (Yellow Springs Instrument Co. [http://www.ysi.com]). See also chapter 6 of *Methods for General and Molecular Bacteriology* (MGMB) (8a). If the cell material being evaluated contains more than one quinol-type oxidase, the enzyme measurement will yield a combination of the activities. In this event, further analysis using purified enzyme preparations is needed to establish the contributions of each complex.

Physiological reaction: $2QH_2 + O_2 \rightarrow 2Q + 2H_2O$

21.3.3.1.1. Oxygen Reduction with Q_1

The following procedure is used to assay cytochrome *bd* oxidase of *Escherichia coli* and related organisms (109). Membranes are prepared (section 21.2.5) and then resuspended by homogenizing in 30 mM Tris-HCl (pH 7.5), 1 mM EDTA. Appropriately diluted membranes are then introduced into the same buffer but containing 2.5 mM dithiothreitol (DTT) or 4 mM ascorbate equilibrated at 37°C. Once a stable oxygen electrode baseline signal is obtained, the reaction is initiated by addition of Q_1 at a final concentration of 250 μM (see section 21.3.2.4). A value of 237 μM O_2 is calculated assuming 100% air saturation at 37°C (109). Details of oxygen measurement by electrode are addressed in reference 107 and chapter 17 of this volume. The assay volume may vary depending on the sample chamber size selected, where a 1.9-ml liquid volume is convenient. Oxygen consumption is then monitored using a Clark-type electrode (see above). Specific enzyme activity is calculated in micromoles of oxygen consumed per minute per milligram of protein. If purified enzyme preparations are used, activity may be calculated as electrons consumed per second per complex (i.e., the "turnover" number).

21.3.3.1.2. Oxygen Reduction with TMPD

As for the electron donor for oxygen reduction, Q_1 can also be replaced with TMPD at a final concentration of 1 mM (37, 109). Oxygen consumption is monitored using a Clark-type electrode as described above, and enzyme activity is calculated in micromoles of oxygen consumed per minute per milligram of protein.

21.3.3.2. Cytochrome *c* as Electron Donor

The cytochrome *c*-type oxidases (e.g., **cytochrome *aa₃* and *cbb₃* oxidases**) obtain electrons for oxygen reduction from a small soluble periplasmic *c*-type cytochrome in the gram-negative bacteria, or from a membrane-bound *c*-type cytochrome located at the outer surface of the cytoplasmic membrane in gram-positive bacteria and in archaea.

Physiological reaction: $4\text{cyt } c_{red} + 4H^+ + O_2$
$$\rightarrow 4 \text{ cyt } c_{ox} + 2H_2O$$

where cyt c_{red} is the cytochrome *c* reduced form and cyt c_{ox}, is the cytochrome *c* oxidized form.

The following procedure is used to assay the cytochrome *c* oxidase of *Rhodobacter sphaeroides* and related organisms (67). Membranes are prepared and resuspended by homogenization in 50 mM phosphate (pH 7.4) and detergent (e.g., 1% dodecyl-maltoside). Appropriately diluted membranes are then added to the same buffer but containing 1.1 mg/ml asolectin (Sigma-Aldrich), and 2.8 mM ascorbate contained in the oxygen electrode chamber, and are equilibrated to the desired assay temperature. Cytochrome *c* is prereduced using ascorbate, and the reaction is initiated by addition of horse heart cytochrome *c* (Sigma-Aldrich) at a final concentration of 40 μM. As described above for the ubiquinol-1 and TMPD assays, oxygen consumption is monitored using a Clark-type electrode where a value of 237 μM O_2 is used for the buffer, assuming air saturation at 37°C (109). Assays of the cytochrome *c*-type oxidase enzymes (e.g., cytochrome oxidase *caa₃*) in the gram-positive bacteria *Paracoccus denitrificans* and *Bacillus subtilis* are similar (37, 73). Note: if a different assay temperature is used, the oxygen saturation value will differ accordingly (see chapter 19.3 in this volume).

21.4. ANAEROBIC RESPIRATORY ENZYMES

The following section describes a number of commonly used assays for anaerobic respiratory enzymes that act on terminal electron acceptors, including nitrate, nitrite, nitric oxide, nitrous oxide, fumarate, TMAO, DMSO, and metal oxides. Many of these respiratory enzymes are assayed by following the substrate-dependent oxidation of a reduced nonphysiological electron donor such as BV, MV, or phenosafranin (Table 2).

For example: $2 \text{ BV}_{red} + 2H^+ + \text{natural substrate}$
$$\rightarrow 2\text{BV}_{ox} + \text{natural product}$$

Alternatively, Q_1, MQ, or lapachol (Table 2) can be used in place of the above nonphysiological dyes if the enzyme is membrane bound and interacts with quinol.

General reaction: MQH_2 + natural substrate

$$\rightarrow MQ + \text{product}$$

where MQH_2 is menaquinol.

The anaerobic assays described in detail below for fumarate reductase can also be applied to assay many other anaerobic respiratory reductases (e.g., nitrate reductase, DMSO reductase, TMAO reductase, and selenate reductase). Ferric reductase and sulfur reductase activities are determined using alternative assays since reduced BV reacts directly with Fe^{3+} or S^0 and is therefore not a suitable electron donor.

21.4.1. Anaerobic Enzyme Assays

Since the reduced forms of many artificial electron donors have a low redox potential (Table 2) they are thus easily autoxidized by molecular oxygen. It is therefore necessary to assay enzyme activities under strictly anaerobic conditions. To ensure fully anaerobic conditions, the assay should be performed in anaerobic 1-ml glass cuvettes filled with an anaerobic assay solution and sealed using butyl rubber or silicone stoppers (Fig. 2). The assay buffer solutions can be prepared in an anaerobic glove box and then introduced into the cuvettes once they are oxygen free. The latter is accomplished by placing the cuvettes and stoppers into the glove box for several hours before use to allow outgassing. Alternatively, anaerobic buffer solutions can be prepared in serum flasks using the Hungate technique (57, 63) and subsequently filled into stoppered cuvettes that have been preflushed with oxygen-free nitrogen or argon gas using an N_2-flushed needle and syringe. A third approach involves the alternating evacuation and gassing of the stoppered cuvette containing the assay solution using an oxygen-free inert gas (i.e., N_2 or Ar). When using this approach it is important to gently tap the cuvette during the gas evacuation cycle to more efficiently remove residual O_2. To maintain anaerobicity by scavenging trace oxygen, a solution of glucose/glucose oxidase/catalase can be added to the anaerobically prepared assay solution as an optional approach (see below for fumarate reduction). Small additions (up to 50 μl) to the stoppered anaerobic cuvettes are made from anaerobic stock solutions using a gastight Hamilton syringe.

21.4.1.1. Fumarate Reduction

The ability to respire with fumarate as a terminal electron acceptor is a widely distributed feature among the facultative anaerobic bacteria. Most fumarate reductase enzymes are membrane bound and have similar subunit compositions and structures (for reviews see references 14 and 41).

Physiological reaction: fumarate + MQH_2

$$\rightarrow \text{succinate} + MQ$$

The enzyme complex is usually present as a membrane-bound menaquinol:fumarate oxidoreductase (QFR) complex with significant structural and functional homology to succinate:ubiquinone oxidoreductase (SQR). It should be noted that succinate dehydrogenases can be assayed in the reverse direction as fumarate reductase and vice versa. These enzymes have their catalytic sites inward-facing towards the cytoplasm (Fig. 1A).

Some bacteria that are versatile with regard to respiration, such as *Shewanella*, contain a soluble fumarate reductase in the cell periplasm (Fig. 1C) (30). These are single-subunit flavocytochromes that are homologous to the FrdA flavoprotein subunit of the membrane-bound fumarate reductases. Both the soluble and membrane-bound fumarate reductase enzymes can be assayed using reduced BV as an electron donor as described below. Unlike succinate dehydrogenase, fumarate reductase does not require preactivation to remove oxaloacetate since binding of this inhibitor is considerably weakened upon reduction of the enzyme by the artificial electron donor.

21.4.1.1.1. Fumarate Reduction by Reduced MV, BV, or Phenosafranin

To measure the steady-state kinetics of soluble and membrane-bound fumarate reductase, the assay solution includes 30 mM bis-Tris-propane buffer (pH 7.0), 200 μM MV, 0.1 mM EDTA, and 10 mM glucose in a 1-ml volume. Fumarate concentrations can be varied from 10 μM to 10 mM to obtain K_m and V_{max} values. After establishing anaerobic conditions, glucose oxidase (40- to 100-μg/ml final concentration) and 5 μl of catalase (diluted 50-fold from the original crystalline suspension obtained from Sigma-Aldrich) are added. The cuvettes are then incubated for 7 to 10 min at the temperature of the assay (usually 37°C) to remove residual oxygen. MV is reduced by the addition of sodium dithionite from a stock solution (10 mg of $Na_2S_2O_4$/ml, 0.1 M $NaHCO_3$, stored on ice) using a gastight Hamilton syringe until an absorbance reading at 603 nm of about 2.0 is obtained. It is important to note that excess dithionite will inhibit most reductase enzymes. The absorbance is monitored for 1 to 2 min to establish a stable baseline and to determine if all residual oxygen has been removed from the system; no absorbance change should be observed if oxygen is absent. The enzymatic reaction is started by the addition of a known amount of enzyme using a gastight Hamilton syringe (usually from a 0.1- to 1.0-mg protein/ml stock solution). The reaction is monitored by recording the decrease of absorbance of MV ($\varepsilon_{603} = 9.2$ mM^{-1} cm^{-1} [Table 2]). The assay with reduced BV is identical to the MV assay except that absorbance is monitored at 550 nm ($\varepsilon_{550} = 7.8$ mM^{-1} cm^{-1} [Table 2]). As both reduced BV and MV are 1 e^- donors and fumarate is a 2 e^- acceptor, by convention the rate of the reaction is divided by 2 to express rates of fumarate reduction. Phenosafranin, by contrast, is a 2 e^- donor. The assay with reduced phenosafranin as the electron donor is performed as described above; however, the reaction is monitored at 520 nm ($\varepsilon_{520} = 5.1$ mM^{-1} cm^{-1} [Table 2]).

Assays with artificial electron donors:

$$\text{fumarate} + 2H^+ + 2MV_{red} \rightarrow \text{succinate} + 2MV_{ox}$$

$$\text{fumarate} + 2H^+ + 2BV_{red} \rightarrow \text{succinate} + 2BV_{ox}$$

fumarate + $2H^+$ + phenosafranin$_{red}$

$$\rightarrow \text{succinate} + \text{phenosafranin}_{ox}$$

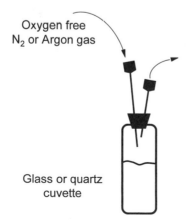

Oxygen free
N_2 or Argon gas

Glass or quartz
cuvette

FIGURE 2 Anaerobic cuvette gassing arrangement for performing anaerobic dye-dependent enzyme assays.

21.4.1.1.2. Fumarate Reduction by Quinol

Various methods have been developed to measure the quinol:fumarate oxidoreductase activity of membrane-bound fumarate reductase. As quinones have a maximum absorbance at 277 nm, protein absorbance significantly interferes with most attempts to directly monitor the formation of quinone from the reduced quinol. Thus, quinol oxidation can be measured effectively only with a dual-wavelength spectrophotometer. However, alternative assays have been developed that require less sophisticated instrumentation.

A recently developed assay system uses hydroxylated naphthoquinones as electron donor substrates for anaerobic menaquinol:oxidant oxidoreductases (76). The naphthoquinone analogs lapachol and plumbagin can be used as an artificial electron donor (Table 2). Both compounds can be obtained from Sigma-Aldrich. The reduction of quinone is accomplished by exposing a 20 mM stock solution of either compound (in ethanol containing 0.18 M HCl) to metallic zinc powder in a small glass vial sealed with a septum from which quinol can be removed using a gastight Hamilton syringe. Anaerobic assay conditions as described above are used for the quinol:fumarate oxidoreductase assay. The cuvettes contain the glucose oxidase/catalase O_2 salvaging system (section 21.4.1.1.1), EDTA, and fumarate. After preincubation of the cuvettes at 37°C, reduced lapachol or plumbagin is added to approximately 0.3 mM. The reaction is initiated by addition of enzyme. For reduced plumbagin the absorbance is followed at 419 nm (ε_{419} = 3.95 mM^{-1} cm^{-1} [Table 2]), whereas for reduced lapachol the absorbance is monitored at 481 nm (ε_{481} = 2.66 mM^{-1} cm^{-1} [Table 2]). Note that the general quinol:fumarate oxidoreductase assay is useful for assay of nitrate reductase and DMSO reductase activities (76) simply by supplying the appropriate substrate as the electron acceptor in place of fumarate.

Hydroxylated naphthoquinone derivatives are highly soluble; however, they generate high K_m values. Thus, some reductase enzymes may need to be assayed with the more hydrophobic quinones, for which they exhibit a higher affinity. Thus, a simple coupled enzyme assay has been developed using NADH, NAD(P)H-quinone oxidoreductase (DT-diaphorase) and MQ to measure the quinol:fumarate oxidoreductase reaction (33, 58). MQ-1 is available from Eisai Co. Ltd. and is the preferred substrate for the reaction due to its high reactivity with fumarate reductase. The assay mixture contains 120 μM NADH, fumarate reductase, and 20 to 80 μg of protein/ml of DT-diaphorase (either rat or human enzyme obtained from Sigma-Aldrich). The assay is carried out in anaerobic cuvettes and the reaction is initiated by the addition of 10 mM fumarate. The decrease in NADH absorbance at 340 nm (ε_{340} = 6.22 mM^{-1} cm^{-1} [Table 2]) is monitored following the addition of fumarate. This assay is very sensitive and can be done with a variety of quinone analogs that interact with fumarate reductase. The coupled reactions are indicated below.

Reaction 1: NADH + H$^+$ + MQ → NAD$^+$ + MQH$_2$

Reaction 2: fumarate + MQH$_2$ → succinate + MQ

21.4.1.2. Nitrate Reduction (Denitrification, Ammonification, and Assimilation)

Nitrate reduction serves one of two general purposes (64): (i) it is coupled to the generation of cellular energy, a process that is termed dissimilatory nitrate reduction (DNR), or (ii) it functions to provide the cell with fixed nitrogen for amino acid, nucleoside, and cofactor biosynthesis. In this case, the end product of the nitrate reduction pathway is assimilated into the cell mass, and the pathway is labeled assimilatory nitrate reduction (ANR). The assimilatory pathway is common to many microbes, fungi, and plants. The enzymes involved are soluble, localized in the cytoplasm, and NAD(P)H dependent. For assays to measure the activities of the assimilatory pathway, please refer to the corresponding literature (11, 21).

DNR is performed exclusively by certain bacteria, archaea, and fungi and occurs primarily under anaerobic or microaerophilic conditions (72, 90). Depending on the genetic capability of the microorganism, nitrate reduction can proceed by the **denitrification** pathway to produce dinitrogen gas or by the **ammonification** pathway to generate ammonium. Both types of DNR pathways release the end product to the environment, in contrast to the assimilatory pathway. Not all enzymes of the ammonification or the denitrification pathway participate in the generation of a proton gradient to generate energy. They may rather serve to detoxify the harmful pathway intermediates. The following section provides a short summary of the denitrification and ammonification pathways as they occur in prokaryotes.

Denitrification proceeds via four enzymatic steps whereby a total of five electrons are transferred from a suitable electron donor to the substrates (nitrate and the reduced intermediates).

$$NO_3^- \longrightarrow NO_2^- \longrightarrow NO \longrightarrow N_2O \longrightarrow N_2$$

nitrate reductase   nitrite reductase   nitric oxide reductase   nitrous oxide reductase

Ammonification involves only two enzymes that catalyze the following reactions.

$$NO_3^- \longrightarrow NO_2^- \longrightarrow NH_4^+$$

nitrate reductase   nitrite reductase

A number of well-characterized enzyme assays are described for the various reduction reactions. Several are provided below, and the reader is directed to more advanced descriptive materials for additional assays and details.

21.4.1.2.1. Nitrate Reductase

Two types of respiratory nitrate reductase enzymes have been described based on their location with respect to the cytoplasmic membrane: these include a membrane-associated NAR-type enzyme with the active site exposed in the cytoplasm in bacteria (Fig. 1A) and a periplasmic NAP-type enzyme (Fig. 1C). For further details, see reference 72. Both types of enzymes catalyze the quinol-dependent reduction of nitrate to nitrite and can be a component of either the denitrification or ammonification pathway. The membrane-bound NAR complex interacts directly with the quinone pool, while the periplasmic NAP receives its electrons via a membrane-bound and/or a periplasmic, soluble cytochrome c.

$$NO_3^- + MQH_2 \rightarrow NO_2^- + MQ + H_2O$$

Both NAR and NAP activities are measured essentially as described for the fumarate reductase assay in section 21.4.1.1.1, with reduced BV (or MV) as the electron donor and with 10 mM KNO$_3$ as the electron acceptor.

$$NO_3^- + 2H^+ + 2BV_{red} \rightarrow NO_2^- + 2BV_{ox} + H_2O$$

21.4.1.2.2. Nitrite Reductase

The ammonification and denitrification pathways employ distinct nitrite reductases. In the ammonification pathway a pentaheme cytochrome *c*-type nitrite reductase reduces nitrite to ammonium. This nitrite reductase occurs in two types, a membrane-associated Nrf-type nitrite reductase that receives its electrons from the quinol pool and plays a respiratory role, and a soluble cytoplasmic, NADH-dependent NirB-type nitrite reductase enzyme that plays a role in nitrite detoxification.

The membrane-associated Nrf nitrite reductase:

$$NO_2^- + 3MQH_2 + 2H^+ \rightarrow NH_4^+ + 3MQ + 2H_2O$$

The soluble NirB nitrite reductase:

$$NO_2^- + 3NADH + 5H^+ \rightarrow NH_4^+ + 3NAD^+ + 2H_2O$$

The denitrification pathway employs two distinctly different types of nitrite reductases, a cytochrome cd_1 (NirS) and a copper-containing enzyme (NirK) (see references 100 and 110 for reviews). Both are located in the periplasmic space and are rarely membrane bound. They receive their electrons from small periplasmic proteins, cytochrome *c*, or small blue copper proteins (azurin and pseudoazurin) and reduce nitrite to nitric oxide (NO).

The NirS and NirK denitrification nitrite reductases:

$$NO_2^- + cyt\ c_{red} + 2\ H^+ \rightarrow NO + cyt\ c_{ox} + H_2O$$

All nitrite reductases can be assayed using a BV or MV assay as described above for fumarate reductase (section 21.4.1.1.1) The activity of NADH-dependent NirB-type enzymes is assayed by monitoring the oxidation of NADH. NirS and NirK enzyme activities can be measured using 10 μM reduced cytochrome *c* as their physiological electron donor. For this type of assay, yeast or horse heart ferricytochrome *c* can be used as an electron donor (obtained from Sigma-Aldrich). Samples are removed periodically and assayed by colorimetric determination of the residual nitrite (74). Alternatively, oxidation of cytochrome *c* can be followed at 550 nm ($\varepsilon_{550} = 19.7$ mM^{-1} cm^{-1} [Table 2]).

21.4.1.2.3. Nitric Oxide Reductase (NO to N$_2$O)

Three types of nitric oxide reductase (NOR) enzymes have been identified that participate in the denitrification pathway (see references 100 and 110 for reviews). All NORs share a membrane localization and an outward-facing orientation of their catalytic site (Fig. 1B). They include the cytochrome *bc* complex-type cNOR that uses membrane or soluble *c*-type cytochromes or small blue copper proteins (azurin and pseudoazurin) as physiological electron donors. Another type of NOR, the cytochrome *b*-containing qNOR, uses ubiquinol or menaquinol as the electron donor. The recently discovered copper-containing qCuANOR uses menaquinol and cytochrome *c* as electron donors. Thus, two types of electron donors should be evaluated.

Donor type 1: $2NO + 2\ cyt\ c_{red} + 2H^+$
$$\rightarrow N_2O + H_2O + 2cyt\ c_{ox}$$
Donor type 2: $2NO + MQH_2 \rightarrow N_2O + H_2O + MQ$

NOR activity is usually assayed polarographically using a modified Clark electrode (28, 89). Rates are determined from the steepest slope of the curved activity trace. Briefly, the reaction is performed in a closed 1.4-ml chamber at a temperature optimal for the enzyme. The assay chamber contains an anaerobic buffer of 20 mM potassium phosphate, pH 7, which is saturated with a 5% NO–95% N$_2$ gas

mixture yielding approximately 100 μM dissolved NO at room temperature. Following the addition of membranes or pure NOR, the reaction is initiated with ascorbate (10 mM)-phenazine methosulfate (100 μM).

The activity of quinol-reactive NOR is measured with MQH$_2$ (65 to 300 μM). Menaquinol is prepared by reducing 100 μM MQ with 200 μM NADH using 0.2 U of diaphorase (Sigma-Aldrich). It is important to choose NADH and diaphorase concentrations that are not rate limiting to NO reduction. This may have to be adjusted empirically. Alternatively, menaquinol can be prepared by reducing MQ with a 10% molar excess of sodium borohydride under a nitrogen atmosphere. Cytochrome *c*-dependent NORs can be assayed with 20 μM horse heart cytochrome *c* as the electron donor (Table 2) in the presence or absence of electron mediators such as PES.

21.4.1.2.4. Nitrous Oxide Reductase (N$_2$O to N$_2$)

Nitrous oxide reductases (**NOS** or **N$_2$OR**) are copper-containing periplasmic soluble enzymes (Fig. 1C) that catalyze the reduction of N$_2$O to N$_2$ using periplasmic, soluble *c*-type cytochromes, azurin or pseudoazurin, as their physiological electron donor.

$$N_2O + 2cyt\ c_{red} + 2H^+ \rightarrow N_2 + H_2O + 2cyt\ c_{ox}$$

N$_2$OR activity is assayed anaerobically in stoppered cuvettes with reduced BV as the electron donor (see above for fumarate reductase) (18, 49, 78). The assay mixture (1 ml) contains 50 mM Tris-HCl buffer, pH 7, and 0.4 mM BV and is prereduced with 0.48 mM sodium dithionite. The enzyme is then added to the assay and incubated at 25.0°C for 90 min for complete activation of N$_2$OR. The enzyme reaction is initiated by the injection of a saturated nitrous oxide solution (final concentration, 1.5 mM). The oxidation of reduced BV is monitored as the decrease in absorbance at 550 nm ($\varepsilon_{550} = 7.8$ mM^{-1} cm^{-1} [Table 2]) as described above for fumarate reductase, section 21.4.1.1.1.

21.4.1.3. Anaerobic Reduction of Amine *N*-Oxides and Sulfoxides

A variety of amine *N*-oxides and organic sulfoxides are reduced as terminal electron acceptors during anaerobic cell growth by many enteric bacteria, by certain marine bacteria, and by nonsulfur purple bacteria (61). While some enzymes are specific for either DMSO or TMAO, others exhibit broad substrate specificity for these plus other related amine *N*-oxides and organic sulfoxides. For example, the *E. coli* enzyme can reduce methionine sulfoxide, nicotinic acid-*N*-oxide, hydroxylamine, and 2-hydroxy pyridine-*N*-oxide (56, 104).

DMSO and TMAO reduction occurs by two types of reductase enzymes. The soluble, single-subunit DorA and TorA enzymes are located in the periplasmic space of the cells (Fig. 1C) and receive electrons from a pentaheme c-type cytochrome (called DorC and TorC, respectively) via the cytoplasmic membrane quinone pool (61). While DorA catalyzes the reduction of both DMSO and TMAO, TorA exhibits high specificity only for TMAO. In contrast, the DmsABC-type enzyme is membrane bound by a hydrophobic anchor polypeptide, very similar to the membrane-bound nitrate reductase (NarGHI) enzyme, although the DmsA catalytic site is exposed to the periplasm (Fig. 1B).

Physiological reactions:

TMAO (or DMSO) + MQH$_2$
$$\rightarrow TMA\ (or\ DMS) + H_2O + MQ$$

TMAO (or DMSO) + 2cyt c_{red} + 2H$^+$
$$\rightarrow \text{TMA (or DMS)} + H_2O + 2\text{cyt } c_{ox}$$

Reduced BV (BV$_{red}$) is used as the electron donor in the anaerobic assay of either the membrane-bound or the periplasmic form of the enzyme. For details, see the fumarate reductase assay (section 21.4.1.1.1) where 5 mM DMSO or 5 mM TMAO is used in place of fumarate.

TMAO (or DMSO) + 2BV$_{red}$ + 2H$^+$
$$\rightarrow \text{TMA (or DMS)} + 2BV_{ox} + H_2O$$

Steady-state kinetics of the DMSO enzymes can be determined using the dithionite-reduced MV:DMSO oxidoreductase activity. The PES-dependent dimethylsulfide (DMS):DCIP oxidoreductase activity, as described by McEwan and coworkers, is in principle similar to that detailed above for fumarate reductase (35, 62).

21.4.1.4. Reduction of Metal Oxides (Selenate, Tellurate, Arsenate, and Iron)

A variety of metal oxides are reduced by various prokaryotes. In some instances the reduction is coupled to an electron transport chain, where energy is conserved. In other instances, the reduction is driven within the cytoplasm, where energy is expended.

21.4.1.4.1. Selenate Reduction

The selenate reductase enzymes belong to the nitrate reductase, formate dehydrogenase family of molybdopterin oxidoreductase enzymes. There are two distinct types that vary in cellular location (103). The *Thauera selenatis* selenate reductase enzyme is located in the periplasm and resembles the NAP-type nitrate reductase enzyme (46, 102). It contains three subunits with molybdopterin, iron-sulfur, and *b*-type heme prosthetic groups. Assay of selenate reductase is performed in anaerobic cuvettes using reduced BV dye as the electron donor (82). The assay is performed essentially as described for fumarate reductase (section 21.4.1.1) except that selenate (5 mM) is used in place of fumarate.

$$SeO_4^{2-} + 2H^+ + 2BV_{red} \rightarrow SeO_3^{2-} + 2BV_{ox} + H_2O$$

The second type of selenate reductase is membrane bound and resembles the NAR-type nitrate reductase that interacts with the reduced menaquinol pool (103). Enzyme assays are performed as for the fumarate reductase enzyme described above (section 21.4.1.1.1) where Q$_1$ (Eisai Co. Ltd.) is used as the electron donor. Membrane preparations are used in place of soluble protein, and 5 mM selenate is used in place of fumarate.

$$SeO_4^{2-} + MQH_2 \rightarrow SeO_3^{2-} + MQ + H_2O$$

21.4.1.4.2. Tellurate Reduction

Tellurate is reduced by a variety of gram-negative and gram-positive bacteria. However, the enzymology of tellurite reduction is not as well understood as that of selenate, fumarate, or nitrate (45, 65, 77). Some enzyme types are respiratory, while others are involved in detoxification where elemental tellurium is formed (92). Reduced BV may be used as the electron donor as in the anaerobic fumarate reductase assay (section 21.4.1.1.1) where 5 mM tellurate is used in place of fumarate.

$$TeO_4^{2-} + 2H^+ + 2BV_{red} \rightarrow TeO_3^{2-} + 2BV_{ox} + H_2O$$

21.4.1.4.3. Arsenate Reduction

Arsenate reduction also occurs in a variety of gram-negative and gram-positive bacteria. Some enzyme types are respiratory (47, 66). In many microbes, the reduction is involved in detoxification (42, 75). Reduced BV may be used as the electron donor as in the anaerobic fumarate reductase assay (section 21.4.1.1.1) where 5 mM arsenate is used in place of fumarate.

$$AsO_4^{2-} + 2H^+ + 2BV_{red} \rightarrow AsO_3^{2-} + 2BV_{ox} + H_2O$$

21.4.1.4.4. Ferric Iron Reduction

Ferric reductase enzymes (FER) constitute a heterogeneous group of oxidoreductases that catalyze the reduction of ferric iron (Fe^{3+}) to ferrous iron (Fe^{2+}) (55, 83). Dissimilatory ferric reductases function in the terminal step of the iron respiratory pathway in iron-reducing bacteria and archaea. Assimilatory ferric reductases function to generate soluble ferrous iron that is incorporated in the prosthetic groups (hemes, Fe-S centers, etc.) of metallo-redox enzymes. Both types of ferric reductases can reduce ferric iron in the form of Fe^{3+}-citrate or Fe^{3+}-EDTA as substrate. Alternatively, an entirely different class of ferric iron enzymes can interact with insoluble ferric iron, although the nature of these enzymes is not well understood (for reviews see references 55 and 83).

21.4.1.4.4.1. Ferric reductase assay. Many ferric reductases are assayed using NADH or NADPH as the electron donor for iron reduction. Catalytic amounts of FMN and FAD can be added to stimulate the activity.

NADH + H$^+$ + 2Fe^{3+}-EDTA $\rightarrow$
$$NAD^+ + 2Fe^{2+}\text{-EDTA} + 2H^+$$

Ferric reductase is assayed spectrophotometrically by following Fe^{3+}-dependent oxidation of NADH or NADPH at 340 nm (94). The assay is performed anaerobically in stoppered quartz cuvettes to prevent oxidation of the reduced flavin by O$_2$. The assay mixture contains 50 mM sodium phosphate (pH 7.0), 0.25 mM Fe^{3+}-EDTA, 5 μM FMN or FAD, and enzyme sample. The reaction is initiated by the addition of NADH or NADPH to a final concentration of 0.1 mM using a gastight Hamilton syringe, and the oxidation of NADH or NADPH is monitored at 340 nm (ε_{340} = 6.22 mM^{-1} cm^{-1} [Table 2]).

Ferric reductase activity can also be assayed anaerobically by photometric monitoring of the appearance of Fe^{2+} by the method of Lascelles and Burke (24, 54). Either 1 mM NADH or 4 mM dithionite-reduced horse heart cytochrome *c* is used as the electron donor, while 0.15 mM Fe^{3+}-nitriloacetic acid (NTA), Fe^{3+}-citrate, or Fe^{3+}-EDTA is used as the electron acceptor (purchased commercially). The assay mixture contains N$_2$-saturated 50 mM HEPES buffer (pH 7) with 0.5 mM ferrozine as the Fe^{2+}-chelating agent. Horse heart cytochrome *c* is added from a stock solution that is prepared by adding 50 ml of a 4.6 mM sodium dithionite solution to 1 ml of 50 μM HEPES buffer, pH 7, containing 2 mg of horse heart cytochrome *c*. The concentration of reduced horse heart cytochrome *c* in the sample cuvette can be calculated by measuring the absorbance at 552 nm (ε_{552} = 29.5 mM^{-1} cm^{-1}). The reaction is initiated by addition of whole cells or cell extracts (0.03 to 0.5 mg of protein) to both the sample and the reference cuvette. The formation of an Fe^{2+}-ferrozine complex is monitored by following the increase in absorbance at 562 nm (ε_{562} = 28.0 mM^{-1} cm^{-1}) (88). The reaction rate should be compared to a reference cuvette where no electron donor was added. In the absence of ferrozine, the enzyme activity may also be monitored at 552 nm, following the oxidation of the reduced horse heart cytochrome *c* (39).

21.4.1.5. Sulfur Oxide Reduction

Dissimilatory sulfate reduction leads to hydrogen sulfide production in a variety of anaerobic habitats. Sulfate-reducing bacteria include species of the gram-negative genera *Desulfovibrio*, *Desulfuromonas*, and *Desulfobacterium*; the gram-positive genus *Desulfotomaculum*; and the archaeon *Archaeoglobus fulgidus*. Cell growth is supported by a variety of organic compounds that serve as electron donors either directly or indirectly. Typical substrates include lactate, acetate, propionate, fatty acids, methanol, ethanol, formate, hydrogen, and a variety of aromatic compounds.

All enzymes involved in sulfate reduction are soluble cytoplasmic enzymes, and the coupling of sulfate reduction to the generation of a proton motive force across the cytoplasmic membrane is not well understood. Commonly assayed reactions of the dissimilatory sulfate reduction pathway include the four enzymatic steps shown in Fig. 3.

Assays for the sulfate reduction enzymes are detailed elsewhere (22, 48) and described briefly here. Cell extracts are prepared as outlined in section 21.2.4 above using an N_2 gassed buffer containing the reducing agent dithiothreitol (DTT) (2 mM). Cell extracts can be assayed directly or stored under anaerobic conditions at -20 or $-70°C$ (22, 48). If cells are to be evaluated for hydrogenase and/or formate dehydrogenase activity, the investigator should also consider the cellular location(s) of these enzymes (e.g., membrane associated, cytoplasmic, or periplasmic). Please consult the section above (21.2.6) for a discussion of enzyme localization.

21.4.1.5.1. ATP Sulfurylase

ATP sulfurylase catalyzes the first step in the dissimilatory pathway for sulfate reduction. It is an energy-consuming reaction that requires ATP to form adenosine-5-phosphosulfate (APS) and PP_i. Due to the very negative redox potential of the sulfate/sulfite couple (about -515 mV), sulfate is not a suitable electron acceptor when NADH is the reductant (-320 mM). Rather, sulfate must first be activated prior to its reduction. APS has a much higher redox potential, about -60 mV, and is then easily reduced with NADH-mediated APS reductase.

$$ATP + SO_4^{2-} \xrightarrow{\text{ATP sulfurylase}} APS + PP_i$$

ATP sulfurylase activity is assayed in the reverse direction by following the pyrophosphate-dependent formation of ATP in a coupled assay. $NADP^+$ reduction is followed spectrophotometrically at 320 nm using glucokinase to phos-phorylate glucose. Glucose-6-phosphate is then consumed by glucose-6-phosphate dehydrogenase to reduce $NADP^+$ (22, 48).

Reaction 1: $APS + PP_i \rightarrow ATP + SO_4^{2-}$

Reaction 2: $ATP + glucose \rightarrow glucose\text{-}6\text{-}phosphate$

Reaction 3: glucose-6-phosphate + $NADP^+$
$\rightarrow$ glucose + $NADPH^+$

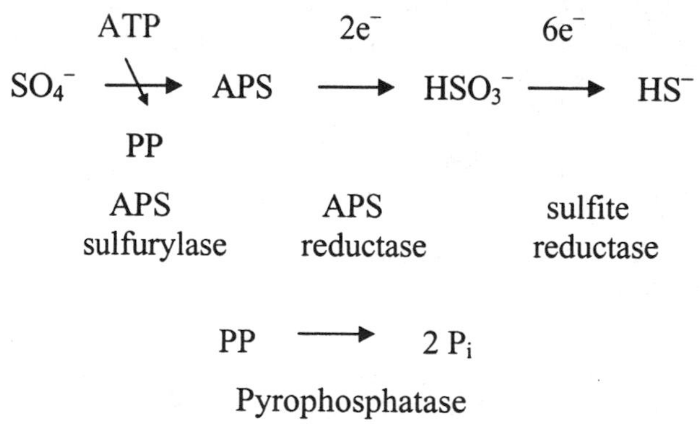

21.4.1.5.2. APS Reductase

The APS reductase enzyme assay is based on the AMP-dependent reduction of the artificial electron acceptor, ferricyanide ($\varepsilon_{420} = 1.09$ mM^{-1} cm^{-1} [Table 2]), using APS as the electron donor (22, 48).

$$HSO_3^- + AMP + 2\text{ferricyanide} \xrightarrow[\text{reductase}]{\text{APS}} APS + 2\text{ferrocyanide}$$

21.4.1.5.3. Sulfite Reductase

Sulfite reductase, a siroheme enzyme, reduces sulfite to sulfide and is present in most sulfate-reducing bacteria. The assay is based on the sulfite-dependent oxidation of reduced MV dye under strict anaerobic conditions (19, 22), and is performed essentially as described for the fumarate reductase assay (section 21.4.1.1.1). The MV dye is first reduced with limiting amounts of dithionite, and the reaction is initiated by addition of cell extract where the decrease in absorption is followed at 600 nm ($\varepsilon_{600} = 13.0$ mM^{-1} cm^{-1}).

$$6BV_{red} + 6H^+ + HSO_3^- \xrightarrow[\text{reductase}]{\text{Sulfite}} 6BV_{ox}^+ + HS^- + 3H_2O$$

21.4.1.5.4. Pyrophosphatase

The diphosphate produced in the formation of APS in the ATP sulfurylase reaction above is subsequently hydrolyzed by the enzyme diphosphatase (pyrophosphatase). Two moles of ATP is then required to regenerate ATP from $AMP + 2 P_i$. This initial reaction of dissimilatory sulfate reduction is identical to the assimilatory process that provides the cell with sulfide for the biosynthesis of the sulfur containing amino acids (i.e., cysteine and methionine), and coenzymes (lipoate and coenzyme A).

Pyrophosphatase is assayed anaerobically by following the formation of orthophosphate from pyrophosphate (15, 48).

$$PP_i + H_2O \longrightarrow 2P_i$$

21.4.1.6. Formate Dehydrogenase

Formate dehydrogenase (FDH) catalyzes the oxidation of formate to CO_2 and H^+. Formate serves as a major electron donor to a variety of respiratory terminal acceptors. The location of the enzyme can vary depending on the physiological role of the enzyme. For example, *E. coli* contains three types of formate dehydrogenases. The one-subunit FDH-H is part of the anaerobic formate hydrogen-lyase complex and is located in the cytoplasmic fraction of the cell (79). In contrast, the three-subunit FDH-N is membrane bound, with a periplasmic orientation of its active site (44). The latter enzyme serves as an electron donor to nitrate respiration. A third enzyme, FDH-O, is similar to FDH-N but is produced in transition to anaerobic cell growth. Multimeric periplasmic formate dehydrogenases have been isolated from *Desulfovibrio* species (17, 85). In many bacteria the expression of formate dehydrogenase is influenced by envi-

FIGURE 3 Reactions of the dissimilatory sulfate reduction pathway.

ronmental conditions. Therefore, cell growth parameters that are optimal for enzyme production should be considered prior to assay, and for determining cellular enzyme location(s). The reaction catalyzed by the respiratory-type FDH-N enzyme follows.

Physiological reaction: $HCO_3^- + H^+ + MQ$
$$\rightarrow CO_2 + MQH_2$$

A general assay capable of detecting each activity is presented below. Formate dehydrogenase activity is measured under anaerobic conditions by following the reduction of BV, methylene blue (52), or ferricyanide (Table 2). The assay mixture contains 50 mM phosphate buffer (pH 7.0), 0.5 mM BV, and 0.2 mM methylene blue or 1.0 mM ferricyanide. After the assay mixture has been equilibrated to the enzyme's optimal temperature, enzyme is added with a Hamilton syringe and the assay is started by the addition of 10 mM sodium formate. Certain formate dehydrogenases, especially tungsten-containing enzymes, are sensitive to oxygen exposure and may require reduction for activation. In this case, enzyme addition is followed by approximately 0.05 mM sodium dithionite. Dithionite will reduce some of the electron acceptor in the absence of formate. A lag phase of several minutes may precede the formate-dependent reduction of the electron acceptor caused by formate dehydrogenase.

21.4.1.7. Hydrogenase

Hydrogenase (H_2ase), the enzyme of hydrogen metabolism, can function in bacteria for either the physiological uptake or the release of hydrogen gas (84). Hydrogen production arises mainly as a result of fermentation of decomposing organic matter by bacteria that reduce protons to H_2. Hydrogen consumption, on the other hand, occurs when hydrogen gas serves as an electron donor during anaerobic or aerobic respiration.

The number and location of hydrogenase enzymes vary in bacteria. For example, *Ralstonia eutropha* contains three distinct hydrogenases involved in hydrogen uptake (10). One, a membrane-bound enzyme with a periplasmically oriented active site, serves as an electron donor to the respiratory chain. A second is a soluble cytoplasmic enzyme that reduces $NAD(P)^+$ with H_2 as the electron donor to provide the cell with reducing power for other cellular reactions. The third is a low-activity H_2-sensing regulatory hydrogenase that mediates gene expression in response to H_2 availability. The facultative bacterium *E. coli* has four distinct hydrogenases that are membrane bound and interact with the quinone pool, where at least two are involved in hydrogen production and one in hydrogen uptake. Several physiological reactions are as follows.

Hydrogen uptake reaction

$$H_2 + MQ \rightarrow MQH_2$$

$$H_2 + NAD(P)^+ \rightarrow NADPH + H^+$$

Hydrogen release reaction

$$2 H^+ + 2 Fd_{red} \rightarrow H_2 + 2 Fd_{ox}$$

where Fd_{red} is reduced ferredoxin and Fd_{ox} is oxidized ferredoxin.

Hydrogenases that catalyze H_2 reduction and/or H_2 evolution are divided into three classes based on the metal content at their active site: [NiFe] hydrogenase, [FeFe] hydrogenase, and [Fe] hydrogenase (6, 96). [NiFe] hydrogenases generally oxidize H_2, while [FeFe] hydrogenases usually re-

duce protons and thus function to generate H_2. The enzymes also differ in subunit composition, electron carrier specificity, cofactor content, and sensitivity to inactivation by O_2 (2, 3). Depending on the nature of the enzyme's O_2 sensitivity, either the protein must be maintained under strict anoxic conditions or an enzyme activation step must precede activity measurement.

Suitable artificial electron acceptors that are commonly used to measure hydrogenase activity include MV, BV, methylene blue, or ferricyanide (Table 2) (81). As noted previously, the investigator should consider conditions of cell growth optimal for enzyme production, and what cell compartment the enzyme is located in (i.e., active site on either the inner or outer surface of the cytoplasmic membrane, soluble within the cytoplasm, or within the periplasmic space).

21.4.1.7.1. Assays and Activation Procedures

H_2 oxidation activity with redox dyes as electron acceptors is determined by following the reduction of MV or BV at 578 nm (86). The anaerobic assay mixture (e.g., 0.8 ml) containing 50 mM morpholinepropanesulfonic acid (MOPS)/KOH buffer (pH 7.0), 2 mM DTT, and either 0.5 mM MV or BV is equilibrated with 100% H_2 headspace. The assay temperature is adjusted to the optimum growth temperature of the microorganism. The reaction is then initiated by addition of enzyme, and the change in absorbance is followed at the appropriate wavelength spectrophotometrically (Table 2).

$$H_2 + 2BV_{ox} \rightarrow 2BV_{red} + 2H^+$$

H_2 formation (reduction of protons) activity is measured by following the oxidation of reduced MV at 578 nm. The standard assay is performed essentially as described for fumarate reductase (section 21.4.1.1.1) in anaerobic buffer containing 50 mM MOPS/KOH (pH 7.0), 2 mM DTT, and 0.5 mM MV with an N_2 gas phase. MV is prereduced with sodium dithionite to an absorbance (A_{578}) of approximately 2.0, and the reaction is initiated by addition of enzyme.

$$2H^+ + 2MV_{red} \rightarrow H_2 + 2MV_{ox}$$

NAD(P)-dependent hydrogenase activity is monitored by following the oxidation of NAD(P)H with $K_3Fe(CN)_6$ as an electron acceptor, as described by Soboh et al., at 420 nm (86). This assay can be performed aerobically using a reaction mixture containing 50 mM Tris-HCl (pH 8.0), 1.25 mM NAD(P)H, and enzyme. The reaction is started by the addition of 1 mM $K_3Fe(CN)_6$. The assay is performed at room temperature since $K_3Fe(CN)_6$ is chemically reduced by NAD(P)H at a low rate at higher temperatures. Assays without added enzyme are used to correct for the rate of chemical $K_3Fe(CN)_6$ reduction.

$NAD(P)H + H^+ +$ ferricyanide
$$\rightarrow NAD(P)^+ + \text{ferrocyanide}$$

Hydrogenase activity can also be assayed by directly following the decrease of H_2 concentration amperometrically in a closed, stirred 2.1-ml cell equipped with a Clark-type electrode as described by van der Linden et al. (95). This assay can be performed aerobically (or anaerobically for oxygen-sensitive enzyme) using 50 mM Tris-HCl, pH 8.0, to which 1 μM FMN is added as an electron mediator. For the aerobic assay, H_2 is added as H_2-saturated water to a final concentration of 36 to 90 μM. Enzyme and the desired electron acceptor, such as BV (2 mM), MV (2 mM), or NAD^+ (5 mM), is added (10). To perform the hydrogenase assay anaerobically all solutions are flushed with argon, and glucose (50 mM) and glucose oxidase (9 U/ml) (see section 21.4.1.1.1) are added to the assay solution to remove residual oxygen (95). The solution is then incubated for 3 min prior to the addition of NADH for enzyme activation. Before use, H_2 must be passed over a palladium catalyst (type E236P; Degussa, Düsseldorf, Germany), or alternatively, argon is passed through an Oxisorb cartridge (MG Industries) to remove trace amounts of O_2 (9).

As noted above, most hydrogenase enzymes are inhibited by molecular oxygen. The O_2 inhibition of [FeFe] hydrogenases is generally irreversible, while [NiFe] hydrogenases can be reactivated. In [NiFe] hydrogenases O_2 forms an O or OH bridge between the Ni and the Fe atoms. This bridging ligand can be removed upon reduction of the enzyme prior to the activity assay (38, 53). Hydrogenase activation can be accomplished by incubating the enzyme with a small amount of a reduced electron donor under an H_2 gas phase prior to the addition of the oxidized electron acceptor. For example, NAD-dependent hydrogenases are activated with 5 μM NADH followed by the addition of BV, MV, or NAD^+ as an electron acceptor (5 mM) to assay hydrogen oxidation. When hydrogenase activity is assayed with the amperometric method using methylene blue or another electron acceptor, a small amount of dithionite is added from a 2 mM stock solution until a slight reduction of the methylene blue is observed. If desired, additional methylene blue can be added to restore the original level of oxidized methylene blue. The reaction is initiated and the activity is followed spectrophotometrically at the desired wavelength (Table 2) by adding H_2-saturated buffer.

21.5. REFERENCES

1. **Ackrell, B. A. C., E. B. Kearney, and T. P. Singer.** 1978. Mammalian succinate dehydrogenase. *Methods Enzymol.* **53:**466–483.
2. **Adams, M. W. W.** 1990. The structure and mechanism of iron hydrogenases. *Biochim. Biophys. Acta* **1020:**115–145.
3. **Adams, M. W. W., L. E. Mortenson, and J. S. Chen.** 1981. Hydrogenases. *Biochim. Biophys. Acta* **594:**105–176.
4. **Anraku, Y., and R. B. Gennis.** 1987. The aerobic respiratory chains of *Escherichia coli*. *Trends Biochem. Sci.* **12:**262–266.
5. **Antipov, A. N., D. Y. Sorkin, N. P. L'vov, and H. G. Kuenen.** 2003. New enzyme belonging to the family of molybdenum-free nitrate reductases. *Biochem. J.* **369:**185–189.
6. **Armstrong, F. A., and S. P. Albracht.** 2005. [NiFe]-hydrogenases: spectroscopic and electrochemical definition of reactions and intermediates. *Phil. Trans. R. Soc.* **363:**937–954.
7. **Baker, S. C., S. J. Ferguson, B. Ludwig, M. D. Page, O. M. Richer, and R. J. van Spanning.** 1998. Molecular genetics of the genus *Paracoccus*: metabolically versatile bacteria with bioenergetic flexibility. *Microbiol. Mol. Biol. Rev.* **62:**1046–1078.
8. **Böck, A., and G. Sawers.** 1996. Fermentation, p. 262–282. *In* F. C. Neidhardt, R. Curtiss III, J. L. Ingraham, E. C. C. Lin, K. B. Low, B. Magasanik, W. S. Reznikoff, M. Riley, M. Schaechter, and H. E. Umbarger (ed.), *Escherichia coli and Salmonella: Cellular and Molecular Biology*, 2nd ed. ASM Press, Washington, DC.
8a. **Breznak, J. A., and R. N. Costilow.** 1994. Physicochemical factors in growth, p. 135–154. *In* P. Gerhard, R. G. E. Murray, W. A. Wood, and N. R. Krieg (ed.), *Methods for General and Molecular Bacteriology*. American Society for Microbiology, Washington, DC.
9. **Burgdorf, T., O. Lenz, T. Buhrke, E. van der Linden, A. K. Jones, S. P. Albracht, and B. Friedrich.** 2005. [NiFe]-hydrogenases of *Ralstonia eutropha* H16: modular enzymes for oxygen-tolerant biological hydrogen oxidation. *J. Mol. Microbiol. Biotechnol.* **10:**181–196.
10. **Burgdorf, T., E. van der Linden, M. Bernhard, Q. Y. Yin, J. W. Back, A. F. Hartog, A. O. Muijsers, C. G. de Koster, S. P. Albracht, and B. Friedrich.** 2005. The soluble NAD+-reducing [NiFe]-hydrogenase from *Ralstonia eutropha* H16 consists of six subunits and can be specifically activated by NADPH. *J. Bacteriol.* **187:**3122–3132.
11. **Cabello, P., M. D. Rolan, and C. Moreno-Vivian.** 2004. Nitrate reduction and the nitrogen cycle in archaea. *Microbiology* **150:**3527–3546.
12. **Castresana, J., M. Lubben, and M. Saraste.** 1995. New archaebacterial genes coding for redox proteins: implications for the evolution of aerobic metabolism. *J. Mol. Biol.* **250:**202–210.
13. **Castresana, J., M. Lubben, M. Saraste, and D. G. Higgens.** 1994. Evolution of cytochrome oxidase, an enzyme older than atmospheric oxygen. *EMBO J.* **13:**2516–2525.
14. **Cecchini, G., I. Schröder, R. P. Gunsalus, and E. Maklashina.** 2002. Succinate dehydrogenase and fumarate reductase from *Escherichia coli*. *Biochim. Biophys. Acta* **1553:**140–157.
15. **Chen, P. S., T. Y. Toribara, and H. Warner.** 1956. Microdetermination of phosphorus. *Anal. Chem.* **28:**1756–1758.
16. **Clark, W. M.** 1972. *Oxidation-Reduction Potentials of Organic Systems.* Robert E. Krieger, Huntington, NY.
17. **Costa, C., M. Teixeira, J. LeGall, J. J. G. Moura, and I. Moura.** 1997. Formate dehydrogenase from *Desulfovibrio desulfuricans* ATCC 27774: isolation and spectroscopic characterization of the active sites (heme, iron-sulfur centers and molybdenum). *J. Biol. Inorg. Chem.* **2:**198–208.
18. **Coyle, C. L., W. G. Zumft, P. M. Kroeneck, H. Korner, and W. Jakob.** 1985. Nitrous oxide reductase from denitrifying *Pseudomonas perfectomarina*. Purification and properties of a novel multicopper enzyme. *Eur. J. Biochem.* **153:**459–467.
19. **Dahl, C., N. Speich, and H. G. Truper.** 1994. Enzymology and molecular biology of sulfate reduction in extremely thermophilic archaeon *Archaeoglobus fulgidus*. *Methods Enzymol.* **243:**331–349.
20. **Ding, H., C. C. Moser, D. E. Robertson, M. K. Tokito, F. Daldal, and P. L. Dutton.** 1995. Ubiquinone pair in the Q_o site central to the primary energy conversion reactions of cytochrome bc1 complex. *Biochemistry* **34:**15979–15996.
21. **Flores, E., J. E. Frias, L. M. Rubio, and A. Herrero.** 2005. Photosynthetic nitrate assimilation in cyanobacteria. *Photosynth. Res.* **83:**117–133.
22. **Frederiksen, T. M., and K. Finster.** 2003. Sulfite-oxidoreductase is involved in the oxidation of sulfite in *Desulfocapsa sulfoexigens* [sic] during disproportionation of thiosulfate and elemental sulfur. *Biodegradation* **14:**189–198.
23. **Friedrich, T.** 1998. The NADH:ubiquinone oxidoreductase (complex I) from *Escherichia coli*. *Biochim. Biophys. Acta* **1364:**134–146.
24. **Gaspard, S., F. Vazquez, and C. Holliger.** 1998. Localization and solubilization of the iron(III) reductase of *Geobacter sulfurreducens*. *Appl. Environ. Microbiol.* **64:**3188–3194.
25. **Gennis, R. B.** 1987. Cytochromes of *Escherichia coli*. *FEMS Microbiol. Rev.* **46:**387–399.
26. **Gennis, R. B., and V. S. Stewart.** 1996. Respiration, p. 217–261. *In* F. C. Neidhardt et al. (ed.), *Escherichia coli and Salmonella typhimurium: Cellular and Molecular Biology*, 2nd ed., vol. 1. American Society for Microbiology, Washington, DC.
27. [Reference deleted.]

28. **Girsch, P., and S. de Vries.** 1997. Purification and initial kinetic and spectroscopic characterization of NO reductase from *Paracoccus denitrificans. Biochim. Biophys. Acta* **1318**:202–216.

29. **Gong, X., T. Xie, L. Yu, M. Hesterberg, D. Scheide, T. Friedrich, and C.-A. Yu.** 2003. The ubiquinone-binding site in NADH:ubiquinone oxidoreductase from *Escherichia coli. J. Biol. Chem.* **278**:25731–25737.

30. **Gordon, E. H., S. L. Pealing, S. K. Chapman, F. B. Ward, and G. A. Reid.** 1998. Physiological function and regulation of flavocytochrome c3, the soluble fumarate reductase from *Shewanella putrefaciens* NCIMB 400. *Microbiology* **144**:937–945.

31. **Green, G. N., and R. B. Gennis.** 1983. Isolation and characterization of an *Escherichia coli* mutant lacking cytochrome d terminal oxidase. *J. Bacteriol.* **154**:1269–1275.

32. **Greenberg, A. E., L. S. Clesceri, and A. D. Eaton.** 1992. *Standard Methods for the Examination of Water and Wastewater*, 18th ed. American Public Health Association, Washington, DC.

33. **Grivennikova, V. G., E. V. Gavrikova, A. A. Timoshin, and A. D. Vinogradov.** 1993. Fumarate reductase activity of bovine heart succinate-ubiquinone reductase. New assay system and overall properties of the reaction. *Biochim. Biophys. Acta* **1140**:282–292.

34. **Gunsalus, R. P.** 1992. Control of electron flow in *Escherichia coli*: coordinated transcription of respiratory pathway genes. *J. Bacteriol.* **174**:7069–7074.

35. **Halnon, S. P., T. H. Toh, P. S. Solomon, R. A. Holt, and A. G. McEwan.** 1996. Dimethylsulfide:acceptor oxidoreductase from *Rhodobacter sulfidophilus*. The purified enzyme contains b-type haem and a pterin molybdenum cofactor. *Eur. J. Biochem.* **239**:391–396.

36. **Hauska, G., A. Trebst, and W. Draber.** 1973. Lipophilicity and catalysis of photophosphorylation. II. Quinoid compounds as artificial carriers in cyclic photophosphorylation and photoreductions by photosystem I. *Biochim. Biophys. Acta* **5**:222–232.

37. **Henning, W., L. Vo, V. Albanese, and B. C. Hill.** 1995. High-yield purification of cytochrome aa3 and cytochrome caa3 oxidases from *Bacillus subtilis* plasma membranes. *Biochem. J.* **309**:279–283.

38. **Higuchi, Y., H. Ogata, K. Miki, N. Yasuoka, and T. Yagi.** 1999. Removal of the bridging ligand atom at the Ni-Fe active site of [NiFe] hydrogenase upon reduction with H2, as revealed by X-ray structure analysis at 1.4 A resolution. *Structure* **7**:549–556.

39. **Hulse, C. L., J. M. Tiedje, and B. A. Averill.** 1988. A spectrophotometric assay for dissimilatory nitrite reductases. *Anal. Biochem.* **172**:420–426.

40. **Ingledew, W. J., and R. K. Poole.** 1984. The respiratory chains of *Escherichia coli. Microbiol. Rev.* **48**:222–271.

41. **Iverson, T. M., C. Luna-Chavez, I. Schröder, G. Cecchini, and D. C. Rees.** 2000. Analyzing your complexes: structure of the quinol-fumarate reductase respiratory complex. *Curr. Opin. Struct. Biol.* **10**:448–455.

42. **Ji, G., and S. Silver.** 1995. Bacterial resistance mechanisms for heavy metals of environmental concern. *J. Ind. Microbiol.* **14**:61–75.

43. **Jones, R. W., and P. B. Garland.** 1977. Sites and specificity of the reaction of bipyridylium compounds with anaerobic respiratory enzymes of *Escherichia coli*. Effects of permeability barriers imposed by the cytoplasmic membrane. *Biochem. J.* **164**:199–211.

44. **Jormakka, M., B. Bryne, and S. Iwata.** 2003. Formate dehydrogenase—versatile enzyme in changing environments. *Curr. Opin. Struct. Biol.* **13**:418–423.

45. **Klonowska, A., T. Heulin, and A. Vermeglio.** 2005. Selenite and tellurite reduction by *Shewanella oneidensis. Appl. Environ. Microbiol.* **71**:5607–5609.

46. **Krafft, T., A. Bowen, F. Theis, and J. M. Macy.** 2000. Cloning and sequencing of the genes encoding the periplasmic-cytochrome b-containing selenate reductase of *Thauera selenatis. DNA Sequence* **10**:365–377.

47. **Krafft, T., and J. M. Macy.** 1988. Purification and characterization of the respiratory arsenate reductase of *Chrysiogenes arsenatis. Eur. J. Biochem.* **255**:647–653.

48. **Kramer, M., and H. Cypionka.** 1989. Sulfate formation via ATP-sulfurylase in thiosulfate-disproportionating and sulfide-disproportionating bacteria. *Arch. Microbiol.* **151**:232–237.

49. **Kristjansson, J. K., and T. C. Hollocher.** 1980. First practical assay for soluble nitrous oxide reductase of denitrifying bacteria and a partial kinetic characterization. *J. Biol. Chem.* **255**:704–707.

50. **Kröeger, A.** 1978. Determination of contents and redox states of ubiquinone and menaquinone. *Methods Enzymol.* **53**:579–591.

51. **Kröger, A., E. Dorrer, and E. Winkler.** 1980. The orientation of the substrate sites of formate dehydrogenase and fumarate reductase in the membrane of *Vibrio succinogenes. Biochim. Biophys. Acta* **589**:118–136.

52. **Kröger, A., E. Winkler, A. Innerhofer, H. Hackenberg, and H. Schagger.** 1979. The formate dehydrogenase involved in electron transport from formate to fumarate in *Vibrio succinogenes. Eur. J. Biochem.* **94**:465–475.

53. **Lamle, S. E., S. P. Albracht, and F. A. Armstrong.** 2005. The mechanism of activation of a [NiFe]-hydrogenase by electrons, hydrogen, and carbon monoxide. *J. Am. Chem. Soc.* **127**:6595–6604.

54. **Lascelles, J., and K. A. Burke.** 1978. Reduction of ferric iron by L-lactate and DL-glycerol-3-phosphate in membrane preparations from *Staphylococcus aureus* and interactions with the nitrate reductase system. *J. Bacteriol.* **134**:585–559.

55. **Lovley, D. R., D. E. Holmes, and K. P. Nevin.** 2004. Dissimilatory Fe(III) and Mn(IV) reduction. *Adv. Microb. Physiol.* **49**:219–286.

56. **Lubitz, S. P., and J. H. Weiner.** 2003. The *Escherichia coli ynfEFGHI* operon encodes polypeptides which are paralogues of dimethyl sulfoxide reductase (DmsABC). *Arch. Biochem. Biophys.* **418**:205–216.

57. **Macy, J. M., J. E. Snellen, and R. E. Hungate.** 1972. Use of syringe methods for anaerobiosis. *Am. J. Clin. Nutr.* **25**:1318–1323.

58. **Maklashina, E., and G. Cecchini.** 1999. Comparison of catalytic activity and inhibitors of quinone reactions of succinate dehydrogenase (succinate-ubiquinone oxidoreductase) and fumarate reductase (menaquinol-fumarate oxidoreductase) from *Escherichia coli. Arch. Biochem. Biophys.* **369**:223–232.

59. **Manchenko, G. P.** 1994. *Handbook of Detection of Enzymes on Electrophoretic Gels.* CRC Press, Boca Raton, FL.

60. **Matsushita, K., T. Ohnishi, and H. R. Kaback.** 1987. NADH:ubiquinone oxidoreductases of the *Escherichia coli* aerobic respiratory chain. *Biochemistry* **26**:7732–7737.

61. **McCrindle, S. L., U. Kappler, and A. G. McEwan.** 2005. Microbial dimethylsulfoxide and trimethylamine-N-oxide respiration. *Adv. Microb. Physiol.* **50**:147–198.

62. **McEwan, A. G., S. J. Ferguson, and J. B. Jackson.** 1991. Purification and properties of dimethyl sulphoxide reductase from *Rhodobacter capsulatus*. A periplasmic molybdoenzyme. *Biochem. J.* **274**:305–307.

63. **Miller, T. L., and M. J. Wolin.** 1974. A serum bottle modification of the Hungate technique for cultivating obligate anaerobes. *Appl. Microbiol.* **27**:985–987.

64. **Moreno-Vivian, C., and S. J. Ferguson.** 1998. Definition and distinction between assimilatory, dissimilatory and respiratory pathways. *Mol. Microbiol.* **29**:664–666.

65. **Moscoso, H., C. Saavedra, C. Loyola, S. Pichuantes, and C. Vasquez.** 1998. Biochemical characterization of the

tellurite-reducing activities of *Bacillus stearothermophilus* V. *Res. Microbiol.* **149:**389–397.

66. **Newman, D. K., E. K. Kenney, J. D. Coates, D. Ahmann, D. J. Ellis, D. R. Lovely, and F. M. M. Morel.** 1977. Dissimilatory arsenate and sulfate reduction in *Desulfotomaculum auripigmentum* sp. nov. *Arch. Microbiol.* **168:**380–388.

67. **Pawate, A. S., J. Morgan, A. Namslauer, D. Mills, P. Brzezinski, S. Ferguson-Miller, and R. B. Gennis.** 2002. A mutation in subunit I of cytochrome oxidase from *Rhodobacter sphaeroides* results in an increase in steady-state activity but completely eliminates proton pumping. *Biochemistry* **41:**13417–13423.

68. **Pereira, M. M., M. Santana, and M. Teixeira.** 2001. A novel scenario for the evolution of haem-copper oxygen reductases. *Biochim. Biophys. Acta* **1505:**185–208.

69. **Phillips, A. T.** 1994. Enzymatic activity, p. 555–586. *In* P. Gerhardt, R. G. E. Murray, W. A. Wood, and N. R. Krieg (ed.), *Methods for General and Molecular Bacteriology.* American Society for Microbiology, Washington, DC.

70. **Poole, R. K., and G. M. Cook.** 2000. Redundancy of aerobic respiratory chains in bacteria? Routes, reasons and regulation. *Adv. Microb. Physiol.* **43:**165–224.

71. **Prince, R. C., S. J. Linkletter, and P. L. Dutton.** 1981. The thermodynamic properties of some commonly used oxidation-reduction mediators, inhibitors and dyes, as determined by polarography. *Biochim. Biophys. Acta* **635:**132–148.

72. **Richardson, D. J., B. C. Berks, D. A. Russell, S. Sprio, and C. J. Taylor.** 2001. Functional, biochemical and genetic diversity of prokaryotic nitrate reductases. *Cell. Mol. Life Sci.* **58:**165–178.

73. **Riistama, S., A. Puustinen, M. I. Verkhovsky, J. E. Morgan, and M. Wikstrom.** 2000. Binding of O_2 and its reduction are both retarded by replacement of valine 279 by isoleucine in cytochrome c oxidase from *Paracoccus denitrificans.* *Biochemistry* **39:**6365–6372.

74. **Robinson, M. K., K. Martinkus, P. J. Kennelly, and R. Timkovich.** 1979. Implications of the integrated rate law for the reactions of *Paracoccus denitrificans* nitrite reductase. *Biochemistry* **18:**3921–3926.

75. **Rosen, B. P., H. Bhattacharjee, and W. Shi.** 1995. Mechanisms of metalloregulation of anion-translocating ATPase. *J. Bioenerg. Biomembr.* **27:**85–91.

76. **Rothery, R. A., I. Chatterjee, G. Kiema, M. T. McDermott, and J. H. Weiner.** 1998. Hydroxylated naphthoquinones as substrates for *Escherichia coli* anaerobic reductases. *Biochem. J.* **332:**35–41.

77. **Sabaty, M., C. Avazeri, D. Pignol, and A. Vermglio.** 2001. Characterization of the reduction of selenate and tellurite by nitrate reductases. *Appl. Environ. Microbiol.* **67:**5122–5126.

78. **Sato, K., A. Okubo, and S. Yamazaki.** 1999. Anaerobic purification and characterization of nitrous oxide reductase from *Rhodobacter sphaeroides* f. sp. *denitrificans* IL106. *J. Biochem.* (Tokyo) **125:**864–868.

79. **Sawers, G.** 1994. The hydrogenases and formate dehydrogenases of *Escherichia coli.* *Antonie Leeuwenhoek* **66:**57–88.

80. **Sazanov, L. A., J. Carroll, P. Holt, L. Toime, and I. M. Fearnley.** 2003. A role for native lipids in the stabilization and two-dimensional crystallization of the *Escherichia coli* NADH-ubiquinone oxidoreductase (complex I). *J. Biol. Chem.* **278:**19483–19491.

81. **Schneider, K., H. G. Schlegel, and K. Jochim.** 1984. Effect of nickel on activity and subunit composition of purified hydrogenase from *Nocardia opaca* 1b. *Eur. J. Biochem.* **138:**533–541.

82. **Schröder, I., S. Rech, T. Krafft, and J. M. Macy.** 1997. Purification and characterization of the selenate reductase from *Thauera selenatis.* *J. Biol. Chem.* **272:**23765–23768.

83. **Schroeder, I., E. Johnson, and S. DeVries.** 2003. Microbial ferric iron reductases. *FEMS Microbiol. Rev.* **27:**427–447.

84. **Schwartz, E., and B. Friedrich.** 2006. The H2-metabolizing prokaryotes. *In* M. Dworkin, S. Falkow, E. Rosenberg, K.-H. Schleifer, and E. Stackebrandt (ed.), *The Prokaryotes,* 3rd ed., vol. 3. Springer, New York, NY.

85. **Sebban, C., L. Blanchard, M. Bruschi, and F. Guerlesquin.** 1995. Purification and characterization of the formate dehydrogenase from *Desulfovibrio vulgaris* Hildenborough. *FEMS Microbiol. Lett.* **133:**143–149.

86. **Soboh, B., D. Linder, and R. Hedderich.** 2004. A multisubunit membrane-bound [NiFe] hydrogenase and an NADH-dependent Fe-only hydrogenase in the fermenting bacterium *Thermoanaerobacter tengcongensis.* *Microbiology* **150:**2451–2463.

87. **Sprott, G. D., S. F. Koval, and C. A. Schnaitman.** 1994. Cell fractionation, p. 72–103. *In* P. Gerhardt, R. G. E. Murray, W. A. Wood, and N. R. Krieg (ed.), *Methods for General and Molecular Bacteriology.* American Society for Microbiology, Washington, DC.

88. **Stookey, L. L.** 1970. Ferrozine: a new spectrophotometric reagent for iron. *Anal. Chem.* **42:**779–781.

89. **Suharti, S., M. J. Strampraad, I. Schröder, and S. de Vries.** 2001. A novel copper A containing menaquinol NO reductase from *Bacillus azotoformans.* *Biochemistry* **40:**2632–2639.

90. **Takaya, N.** 2002. Dissimilatory nitrate reduction metabolisms and their control in fungi. *J. Biosci. Bioeng.* **94:**506–510.

91. **Thauer, R. K., K. Jungermann, and K. Decker.** 1977. Energy conservation in chemotrophic anaerobic bacteria. *Bacteriol. Rev.* **41:**100–180.

92. **Toptchieva, A., G. Sisson, L. J. Bryden, D. E. Taylor, and P. S. Hoffman.** 2003. An inducible tellurite-resistance operon in *Proteus mirabilis.* *Microbiology* **149:**1285–1295.

93. **Unden, G., and J. Bongaerts.** 1997. Alternative respiratory pathways of *Escherichia coli*: energetics and transcriptional regulation in response to electron acceptors. *Biochim. Biophys. Acta* **1320:**217–234.

94. **Vadas, A., H. G. Monboquette, E. Johnson, and I. Schröder.** 1999. Identification and characterization of a novel ferric reductase from the hyperthermophilic archaeon *Archaeoglobus fulgidus.* *J. Biol. Chem.* **274:**36715–36721.

95. **van der Linden, E., T. Burgdorf, A. L. de Lacey, T. Buhrke, M. Scholte, V. M. Fernandez, B. Friedrich, and S. P. J. Albracht.** 2006. An improved purification procedure for the soluble [NiFe]-hydrogenase of *Ralstonia eutropha*: new insights into its (in)stability and spectroscopic properties. *J. Biol. Inorg. Chem.* **11:**247–260.

96. **Vignais, P. M., and A. Colbeau.** 2004. Molecular biology of microbial hydrogenases. *Curr. Issues Mol. Biol.* **6:**159–188.

97. **Vik, S. B., and Y. Hatefi.** 1981. Possible occurrence and role of an essential histidyl residue in succinate dehydrogenase. *Proc. Natl. Acad. Sci. USA* **78:**6749–6753.

98. **Vinogradov, A. D.** 1998. Catalytic properties of the mitochondrial NADH-ubiquinone oxidoreductase (complex I) and the pseudo-reversible active/inactive enzyme transition. *Biochim. Biophys. Acta* **1364:**169–185.

99. **Von Jagow, G., and H. Schäger.** 1994. *A Practical Guide to Membrane Protein Purification,* 2nd ed. Academic Press, San Diego, CA.

100. **Wasser, I. M., S. de Vries, P. Moenne-Loccoz, I. Schröder, and K. D. Karlin.** 2002. Nitric oxide in biological denitrification: Fe/Cu metalloenzyme and metal complex NO(x) redox chemistry. *Chem. Rev.* **102:**1201–1234.

101. **Watanabe, T., and K. Honda.** 1982. Measurement of extension coefficient of the methyl viologen cation radical

and the efficiency of its formation by semiconductor photocatalysis. *J. Phys. Chem.* **86:**2617–2619.

102. **Watts, C. A., H. Ridley, K. L. Condie, J. T. Leaver, D. J. Richardson, and C. S. Butler.** 2003. Selenate reduction by *Enterobacter cloacae* SLD1a-1 is catalysed by a molybdenum-dependent membrane-bound enzyme that is distinct from the membrane-bound nitrate reductase. *FEMS Microbiol. Lett.* **228:**273–379.

103. **Watts, C. A., H. Ridley, E. J. Dridge, J. T. Leaver, A. J. Reilly, D. J. Richardson, and C. S. Butler.** 2005. Microbial reduction of selenate and nitrate: common themes and variations. *Biochem. Soc. Trans.* **33:**173–175.

104. **Weiner, J. H., D. P. Maclsaac, R. E. Bishop, and P. T. Bilous.** 1988. Purification and properties of *Escherichia coli* dimethyl sulfoxide reductase, an iron-sulfur molybdoenzyme with broad substrate specificity. *J. Bacteriol.* **170:**1505–1511.

105. **Wikström, M.** 2004. Cytochrome c oxidase: 25 years of the elusive proton pump. *Biochim. Biophys. Acta* **1655:**241–247.

106. **Wilson, G. S.** 1978. Determination of oxidation-reduction potentials. *Methods Enzymol.* **54:**396–410.

107. **Wood, W. A., and J. R. Paterek.** 1994. Physical analysis, p. 465–511. *In* P. Gerhardt, R. G. E. Murray, W. A. Wood, and N. R. Krieg (ed.), *Methods for General and Molecular Bacteriology.* American Society for Microbiology, Washington, DC.

108. **Yamamoto, I., N. Okubo, and M. Ishimoto.** 1986. Further characterization of trimethylamine N-oxide reductase from *Escherichia coli*, a molybdoprotein. *J. Biochem.* **99:**1773–1779.

109. **Zhang, J., P. Hellwig, J. P. Osborne, H. W. Huang, P. Moenne-Loccoz, A. A. Konstantinov, and R. B. Gennis.** 2001. Site-directed mutation of the highly conserved region near the Q-loop of the cytochrome bd quinol oxidase from *Escherichia coli* specifically perturbs heme b595. *Biochemistry* **40:**8548–8556.

110. **Zumft, W. G.** 1997. Cell biology and molecular basis of denitrification. *Microbiol. Mol. Biol. Rev.* **61:**533–616.

22

Carbohydrate Fermentations

R. MEGANATHAN, YAMINI RANGANATHAN, AND C. A. REDDY

Fermentation was defined by Pasteur as "la vie sans air" (life without air) during the middle of the 19th century. Even though in common usage now, the term fermentation is used for an aerobic process (as in antibiotic production) or an anaerobic process (as in lactic acid production), strictly speaking, "fermentation" is defined as an energy-yielding metabolic process (in the dark) in which an organic compound(s) serves as both an electron donor and an electron acceptor. Fermentations are carried out by obligate anaerobes or facultative anaerobes, and ATP is formed by substrate level phosphorylation. Facultative anaerobes such as enterobacteria grow by fermentative metabolism under anaerobic conditions, while under aerobic conditions they grow as aerobic heterotrophs. Obligate anaerobes, on the other hand, obtain energy for growth mostly by fermentative metabolism. However, a few anaerobes obtain at least part of their ATP for growth by anaerobic respiration, which is defined as an energy-yielding metabolism in which either an organic compound (such as formate) or inorganic compound (such as H_2) serves as an electron donor and an organic compound (such as fumarate) or inorganic compound (such as nitrate, sulfate, or CO_2) serves as an electron acceptor.

Historically, studies on ethanolic fermentation of glucose via the glycolysis or the Embden-Meyerhof-Parnas (EMP) pathway were very important because these led to profound fundamental discoveries that laid the foundations for modern physiology and biochemistry as well as for the fermentation industry. More than one hundred years after these initial investigations, important new discoveries continue to be made in the fermentation field. Recently, some important discoveries were made in the study of glycolysis.

These include for the first time the rather dramatic and startling demonstration that ADP, rather than ATP, is the phosphoryl donor for the phosphorylation of glucose and fructose-6-phosphate in thermophilic archaea. Another surprising finding is the reported presence of a bifunctional kinase capable of phosphorylating both glucose and fructose-6-phosphate using ADP as the phosphoryl donor in the hyperthermophilic archaeon *Methanococcus jannaschii*.

This chapter provides a brief review of some of the more common carbohydrate fermentation pathways, schematics of the flow of carbon in these pathways, characteristic products produced, typical energy yields (in the form of ATP), and assay procedures for the key enzymes. The reader is referred to a wide array of books and reviews (18, 32, 60, 68, 76, 103, 118, 123) that give more details on the topics covered here. Anaerobic respiratory processes such as sulfate reduction are not covered in this chapter, but some aspects of bacterial respiration are covered in chapter 21 of this volume.

22.1. FERMENTATION PRODUCTS FROM PYRUVATE

Fermentations are classified according to the key fermentation end products of each as exemplified by bacterial ethanolic, homolactic, heterolactic, propionic, mixed-acid, butyrate-butanol, homoacetogenic, and other fermentations (see below). Pyruvate is a key intermediate in many of these fermentations; i.e., substrate carbon flows via pyruvate to produce an end product(s) characteristic of a given fermentation. The various fermentation products derived from pyruvate are shown in Fig. 1. Microorganisms catabolize sugars to pyruvate and obtain energy for growth by substrate level phosphorylation. Since by definition, a fermentative organism is unable to use O_2 as an electron acceptor, it has to generate its own alternate electron acceptor. This alternate electron acceptor to oxygen is usually pyruvate or

a compound derived from pyruvate (Fig. 1). The type of electron acceptor compound formed leading to the production of a given end product of fermentation is determined by the substrate fermented, the organism responsible for the fermentation, and various physiological factors.

An examination of Fig. 1 also shows that formation of acetyl coenzyme A (acetyl-CoA) from pyruvate cleavage or from acetate activation is a key reaction in the pathways presented. Acetyl-CoA is converted to acetate via acetyl-PO_4 or undergoes condensation with another molecule of acetyl-CoA to eventually form butyryl-CoA, or undergoes reduction to form ethanol. Furthermore, it is worthy of note that pyruvate undergoes two types of clastic reactions: one yields acetyl-CoA and formate (coli-aerogenes type), while the other yields acetyl-CoA, CO_2, and H_2 (clostridial type). Fermentative organisms are constrained to grow and survive mostly on substrate level phosphorylation that results in relatively poor energy yield per mole of substrate fermented, in contrast to aerobic respiring organisms that produce large amounts of ATP by oxidative phosphorylation (in addition to substrate level phosphorylation). Thus, the inherent limitations in energy yield in fermentation necessitate catabolism of large amounts of substrate, resulting in the accumulation of large quantities of end products. Efficient and economic production of fermentation products such as ethanol is currently a multibillion-dollar industry and is expected to grow in the future.

22.2. FERMENTATION BALANCE

22.2.1. Construction of Fermentation Balance

The basic requirement for studying fermentation is to determine the optimum conditions for the growth of the organism and determine the fermentation products quantitatively. A variety of media used for growing of bacteria is

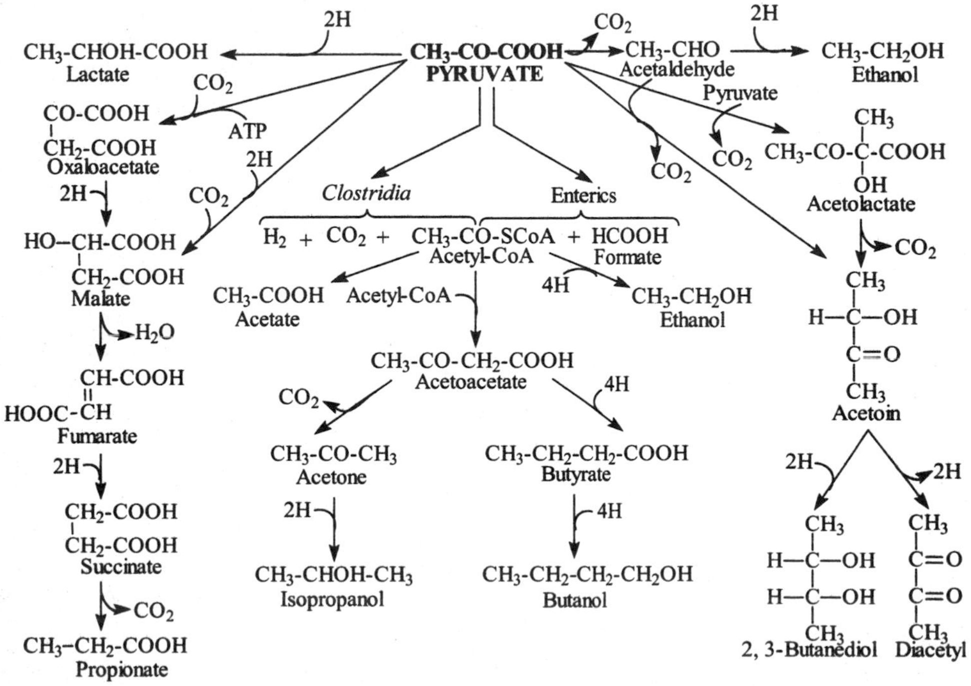

FIGURE 1 Fermentation products formed from pyruvate by various organisms.

listed in chapter 10 in this volume. A complete description of methods used for the determination of fermentation products is beyond the scope of this chapter, but an excellent review on the chemical methods for the analysis of fermentation products is that by Dawes et al. (24). Also see chapter 18 in this manual. With the recent advances in technology, chemical methods have been superseded by techniques such as high-performance liquid chromatography and liquid chromatography/mass spectrometry. High-performance liquid chromatography is now the most common method used for the analysis of organic acid and solvents produced during fermentation (42, 108, 113). Fermentatively produced gases can be determined with an acoustic gas monitor (17, 42). Also see chapters 17 and 18 in this volume for selected analytical procedures.

Since fermentation often involves reactants, the substrate, and water only, stoichiometry of the substrate consumed and products formed in fermentation can be determined accurately and this bookkeeping serves as an invaluable tool that will aid in deducing the presumed pathway employed by the organism. In the construction of a fermentation balance, usually the carbon recovery, oxidation/reduction (O/R) balance, and a balance of available hydrogens are determined. Data on butanol-isopropanol fermentation carried out by *Clostridium beijerinckii* (syn. *Clostridium butylicum*) were presented in the classic review by Wood (123). A typical fermentation balance is illustrated in Table 1.

22.2.2. Carbon Recovery

The carbon recovered in the fermentation shown in Table 1 is about 96%, which is in agreement with the expected value. It is not uncommon in certain fermentations to recover significantly more carbon than expected. Such an increase in recovered carbon indicates CO_2 fixation (32, 123). In anticipation of this, it is customary to include bicarbonate in the fermentation media (see chapter 10).

22.2.3. Oxidation/Reduction Balance

For the calculation of O/R balance, it is usually customary to assign an arbitrary O/R value of 0 for formaldehyde. Each excess 2H is counted as -1, and deficiency of 2H is counted as $+1$ (32, 118). However, in a simpler alternative procedure, hydrogen is assigned an arbitrary value of -0.5 and oxygen is assigned a value of $+1$. The two methods yield the same results.

22.2.4. Available Hydrogen Balance

For the calculation of the available hydrogen balance, it is customary to determine the number of available hydrogens in the substrate and the products of fermentation. During fermentation, the products derive their hydrogens from the substrate and water. Hence, the substrate and products are completely oxidized to CO_2 with water and the numbers of available hydrogens are determined.

From the nature and quantity of the fermentation products formed, it is often possible to deduce the pathway. This information can be verified by radiorespirometry (see section 22.3.3 below) and other techniques (see chapters 17 and 18), and the validity of the results can be further confirmed by the assay of key enzymes of the pathways.

22.3. ALCOHOLIC FERMENTATION

Understanding of microorganisms as agents of fermentation largely came from detailed studies on alcoholic fermentation in which breakdown of glucose to pyruvate occurs by one of two main pathways: the EMP pathway and the Entner-Duodoroff (ED) pathway. Other six-carbon sugars are generally interconverted to glucose by the action of accessory cellular enzymes.

22.3.1. Embden-Meyerhof-Parnas Pathway

Much of our understanding of early fermentations came from alcoholic fermentation of sugar by the yeast *Saccha-*

TABLE 1 Glucose fermentation by *Clostridium butylicum*: calculation of carbon recovery, O/R balance, and available hydrogen balance

Substrate or product [1]	mol/100 mol of substrate[a] [2]	mol of carbon[b] [3]	O/R value [4]	O/R value mol/100 ml[c] [5] Reduced	O/R value mol/100 ml[c] [5] Oxidized	Available H[d] [6]	Available H[e] (mol/100 mol) [7]
Substrate							
Glucose	100	600	0	—		24	2,400
Products							
Butyrate	17	68	-2	-34		20	340
Acetate	17	34	0	—		8	136
CO_2	204	204	$+2$		$+408$	0	—
H_2	78	—	-1	-78		2	156
Butanol	59	236	-4	-236		24	1,416
Isopropanol	12	36	-3	-36		18	216
Total		578		-384	$+408$		2,264

[a]Data in this column are taken from reference 123.

[b]This value is the product of the number of carbons × the number in column [2]. Carbon recovered = 578/600 × 100 = 96.3%; O/R balance = 408/384 = 1.06; balance of available H = 2,400/2,264 = 1.06.

[c]This value is the product of column [2] × column [4].

[d]Calculation of available [H] is based on the following equations: $C_6H_{12}O_6 + 6H_2O \rightarrow 24H + 6H_2O$; $C_4H_8O_2 + 6H_2O \rightarrow 20H + 4CO_2$; $C_2H_4O_2 + 2H_2O \rightarrow 8H + 2CO_2$; $C_4H_{10}O + 7H_2O \rightarrow 24H + 4CO_2$; $C_3H_8O + 5H_2O \rightarrow 18H + 3CO_2$.

[e]This value is the product of column [2] × column [6].

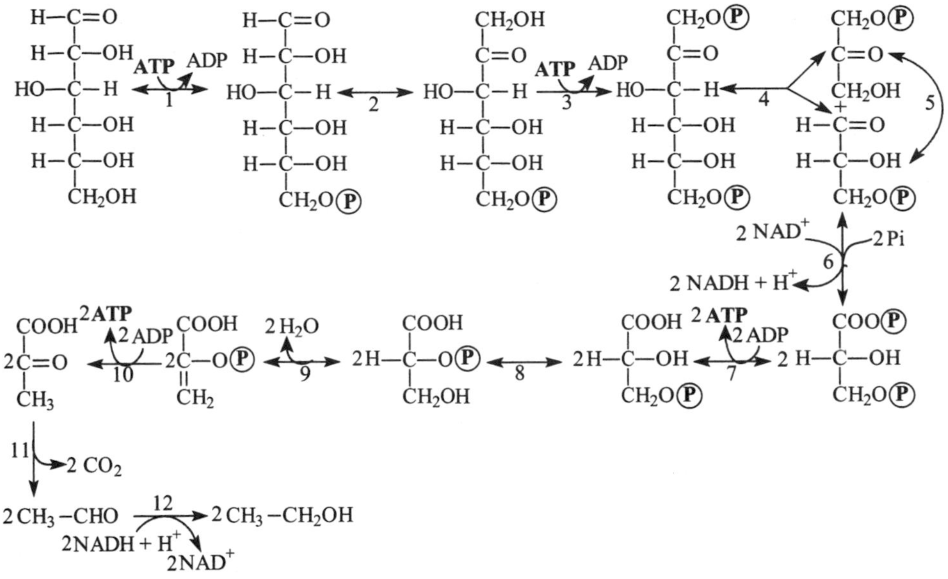

FIGURE 2 Ethanolic fermentation of glucose by the EMP pathway. 1, hexokinase; 2, glucose-6-phosphate isomerase; 3, PFK; 4, fructose bisphosphate aldolase; 5, triose phosphate isomerase; 6, GAPDH; 7, 3-phosphoglycerate kinase; 8, phosphoglycerate mutase; 9, enolase; 10, pyruvate kinase; 11, pyruvate decarboxylase; 12, ADH.

romyces cerevisiae. This is the classical fermentation employed in the manufacture of wine and other popular alcoholic drinks. In this fermentation, glucose is degraded via glycolysis or the EMP pathway to produce 2 mol of ethanol for each mole of glucose metabolized (Fig. 2). Yeast cells produce three enzymes for the phosphorylation of glucose and other sugars: hexokinases 1 and 2 and a glucokinase (31, 37). Hexokinase 2 is thought to be the major enzyme during growth on glucose since it is the isozyme that predominates. The other two isozymes present are individually sufficient to support growth on glucose, but their expression is higher when the cells are grown on other carbon sources (37, 61). Further, hexokinase 2 plays an important role in the regulation of glucose repression (29, 30).

In alcoholic fermentation, 2 mol of pyruvate is produced for each mole of glucose degraded. The key enzyme that is unique to the EMP pathway in the formation of pyruvate is 6-phosphofructokinase. All the other enzymes are involved in other pathways or reactions of intermediary metabolism. The unique enzyme that is involved in the conversion of pyruvate to ethanol is pyruvate decarboxylase. The electron acceptor is the acetaldehyde. The 2 mol of NADH produced in the glyceraldehyde-3-phosphate dehydrogenase (GAPDH) reaction are consumed in the reduction of acetaldehyde to ethanol, resulting in an even hydrogen balance. The net energy yield of yeast ethanolic fermentation is 2 mol of ATP/mol of glucose consumed (Fig. 2). The assay methods for the various enzymes of the pathway are shown in Table 2. (See assay tables in section 22.11.)

22.3.2. Entner-Doudoroff Pathway

Zymomonas mobilis is the organism used in the fermentation of agave juice to Mexican pulque and tequila (111); it ferments glucose via the ED pathway, resulting in the production of 2 mol each of ethanol and CO_2 from each mole of glucose (Fig. 3). *Z. mobilis* contains pyruvate decarboxylase, which is unusual for a bacterium. The key enzymes of the

pathway are 6-phosphogluconate dehydratase and 2-keto-3-deoxy-6-phosphogluconate (KDPG) aldolase. In contrast to the EMP pathway, which yields 2 mol of ATP per mol of glucose fermented, the ED pathway yields only 1 mol of ATP/mol of glucose. The low ATP yield might explain why the presence of the ED pathway in anaerobes is rare in spite of its widespread occurrence in aerobes.

For demonstrating the presence of the ED pathway in an organism, extracts from glucose-grown cells are assayed for the key enzymes, dehydratase and KDPG aldolase. The presence of these enzymes and the absence of phosphofructokinase (PFK) (the key enzyme of the EMP pathway) are clear indicators for the operation of the ED pathway (Fig. 3). The two molecules of pyruvate formed are decarboxylated to acetaldehyde and then reduced to ethanol by alcohol dehydrogenase (ADH). It should be mentioned that there are two ADHs involved in the conversion of acetaldehyde to ethanol. These enzymes are designated as ADH I and ADH II. These enzymes constitute approximately 20 and 80% of the total ADH activity. ADH I contains zinc, while ADH II contains ferrous iron as a cofactor. For further information, the reader should consult the original publications (5, 22, 52, 62, 72). The assay methods for the various enzymes of the pathway are shown in Table 3 (see section 22.11).

22.3.3. Radiorespirometry

From the nature and quantity of the fermentation products formed, it is possible to deduce the likely pathway. This information can be confirmed by radiorespirometry. For example, radiorespirometry is a useful method for distinguishing between the EMP pathway and the ED pathway. The expected labeling patterns during fermentation of glucose by the EMP and ED pathways are shown in Fig. 4. Each pathway produces 2 mol of pyruvate for each mole of glucose consumed. However, the pyruvate originating from the first three carbon atoms of glucose is different. In the

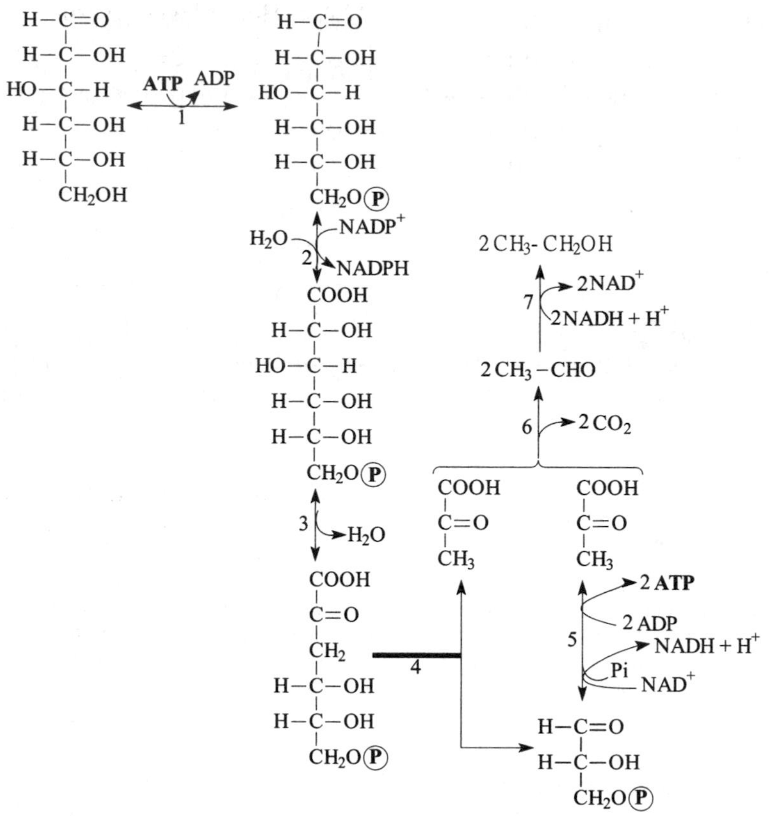

FIGURE 3 Ethanolic fermentation of glucose by the ED pathway. 1, glucokinase; 2, glucose-6-phosphate dehydrogenase; 3, 6-phosphogluconate dehydratase; 4, KDPG aldolase; 5, GAPDH; 6, 3-phosphoglycerate kinase; 7, phosphoglycerate mutase; 8, enolase; 9, pyruvate kinase; 10, pyruvate decarboxylase; 11, ADH.

ED pathway, the carbon one (aldehyde group) of glucose becomes the carboxyl of pyruvate, while in the EMP pathway, the aldehyde group becomes the methyl group of pyruvate. The converse is true in the case of carbon 3 of glucose in the EMP pathway. If glucose ^{14}C labeled in carbons 1, 3, and 4 is fermented, and if the organism uses the EMP pathway, carbons 3 and 4 will appear as CO_2. On the other hand, if the ED pathway is used, then carbons 1 and 4 will appear as CO_2 (Fig. 4).

22.4. LACTIC ACID FERMENTATION

A group of bacteria commonly referred to as lactic acid bacteria ferment glucose and produce lactic acid. If a bacterium ferments each mole of glucose to 2 mol of lactate (lactate is the sole product), then the pathway is referred to as homolactate fermentation. If glucose is fermented to lactate, ethanol, and CO_2, then it is referred to as heterolactate fermentation. Depending on the organism, the lactic acid bacteria produce either L-(+), D-(−) or, DL-lactic acid (123). The lactic acid produced by muscle, on the other hand, is L-(+).

22.4.1. Homolactate Fermentation

The homolactate fermentation pathway is shown in Fig. 5. The lactic acid bacteria usually phosphorylate glucose by the phosphoenolpyruvate phosphotransferase system (PEP-PTS). (For an excellent review on the genomic analysis of PEP-PTS, see reference 6a). But there are instances

where hexokinase is used (76, 78). Glucose-6-phosphate is fermented by the EMP pathway to pyruvate. The 2 mol of pyruvate produced is reduced using the 2 mol of NADH derived from the GAPDH reaction. This fermentation yields 2 mol of ATP/mol of glucose fermented. The assay methods for the various enzymes of the pathway are shown in Table 4 (section 22.11).

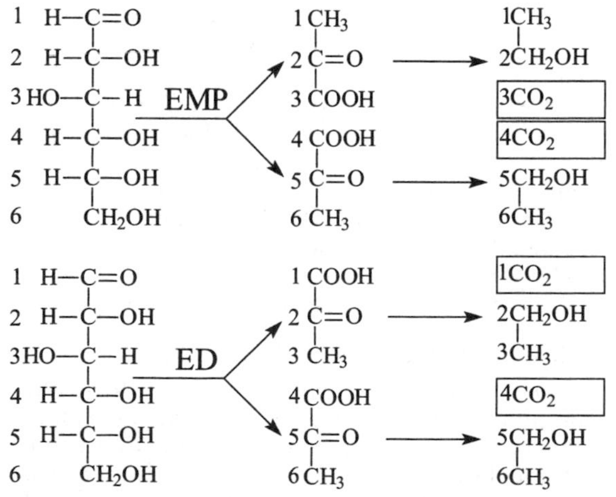

FIGURE 4 Flow of carbon when ^{14}[C]glucose is fermented to ethanol and CO_2 via the EMP and ED pathways.

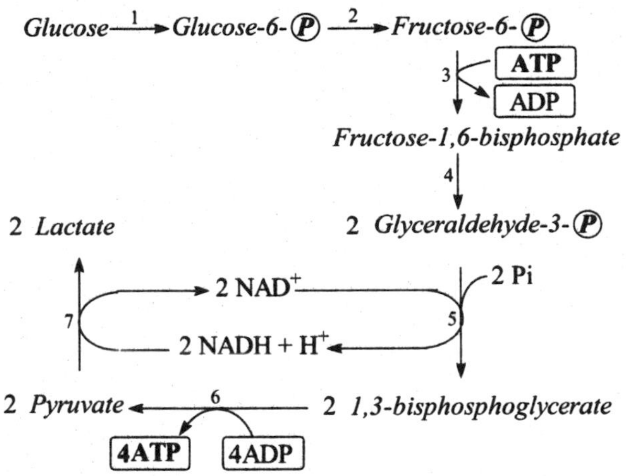

FIGURE 5 Homolactate fermentation. 1, PEP-PTS; 2 to 6, conversion of glucose-6-phosphate to pyruvate, which is accomplished by the same enzymes as in Fig. 2; 7, lactate dehydrogenase.

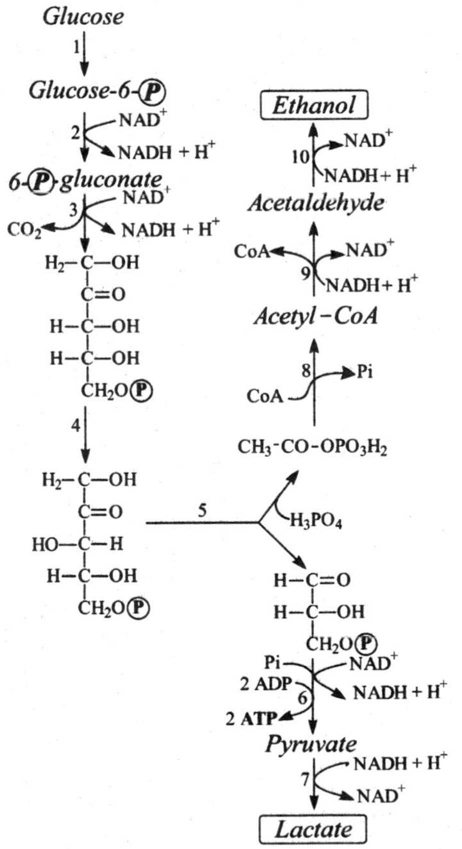

FIGURE 6 Heterolactate fermentation. 1, PEP-PTS; 2, glucose-6-phosphate dehydrogenase; 3, 6-phosphogluconate dehydrogenase; 4, ribulose 5-phosphate 3-epimerase; 5, phosphoketolase; 6, conversion of GAP to pyruvate, which is accomplished by the same enzymes as in Fig. 2 and 3; 7, lactate dehydrogenase; 8, phosphotransacetylase; 9, acetaldehyde dehydrogenase; 10, ADH.

22.4.2. Heterolactate Fermentation

The initial transport and phosphorylation of glucose in heterolactate fermentation (Fig. 6) occur the same way as in homolactate fermentation. The glucose-6-phosphate is converted to gluconate 6-phosphate by the same reactions as in the ED pathway. In the subsequent reactions, the gluconate 6-phosphate is decarboxylated to ribulose 5-phosphate and epimerized to xylulose 5-phosphate. A thiamine pyrophosphate-dependent phosphoketolase cleaves the xylulose 5-phosphate to GAP and acetylphosphate. The GAP is converted to lactate by the same reactions as in the homolactate pathway. The acetylphosphate is first converted to acetyl-CoA by a phosphotransacetylase and reduced to acetaldehyde and, finally, to ethanol. The ATP yield is 1 mol/mol of glucose fermented. The assay methods for the various enzymes of the pathway are shown in Table 5 (section 22.11).

22.4.3. Heterofermentative Bifidum Pathway

The transport and phosphorylation of glucose in the bifidobacteria are similar to that described above for homo- and heterolactic bacteria. In the bifidum pathway, *Bifidobacterium bifidum* ferments glucose to lactate and acetate (Fig. 7). There are two mechanistically identical phosphoclastic reactions that cleave fructose-6-phosphate and xylulose 5-phosphate with inorganic phosphate. In human-derived species like *Bifidobacterium dentium*, there is a separate enzyme for each substrate, namely, a fructose-6-phosphate phosphoketolase and a xylose 5-phosphate phosphoketolase. In contrast, in animal-derived species such as *Bifidobacterium globosum* and *Bifidobacterium lactis*, a single enzyme cleaves both substrates (66, 77).

There is no reduction or oxidation of NAD$^+$ in this pathway until GAP is formed. The net result is the formation of 1 mol of lactate and 1.5 mol of acetate per mol of glucose fermented. The ATP yield of the pathway is higher than that of other fermentative pathways: 2.5 mol ATP/mol of glucose fermented. The assay methods for the various enzymes of the pathway are shown in Table 6 (section 22.11).

22.5. BUTYRIC ACID-BUTANOL AND ACETONE-ISOPROPANOL FERMENTATIONS

Butyrate as a major fermentation product of glucose fermentation is primarily seen in anaerobes such as clostridia and in strains of genera *Butyrivibrio*, *Fusobacterium*, and *Eubacterium*. The clostridia employ PEP-PTS for the uptake and phosphorylation of glucose to glucose-6-phosphate (Fig. 8). A full description of phosphotransferase is beyond the scope of this chapter, but recent reviews (28, 68) provide more details. The glucose-6-phosphate formed is converted to 2 mol of pyruvate by reactions of the EMP pathway (as shown in Fig. 2). Pyruvate is then oxidized to acetyl-CoA, H$_2$, and CO$_2$ (Fig. 8) mediated by pyruvate-ferredoxin oxidoreductase (reaction 1 below) and ferredoxin-linked hydrogenase (reaction 2 below).

Reaction 1: pyruvate + ferredoxin + CoA
$$\rightarrow \text{acetyl-CoA} + \text{ferredoxin} \cdot \text{H}_2$$

Reaction 2: ferredoxin · H$_2$ → ferredoxin + H$_2$

Evolution of H$_2$ is an advantage to the organism, as it is an effective way to dispose of the electrons and protons released during pyruvate oxidation in that, even in an environment containing high levels of hydrogen gas, reduced ferredoxin can transfer electrons to hydrogenase to produce H$_2$. Some of the clostridia also contain a pyridine nucleotide-linked

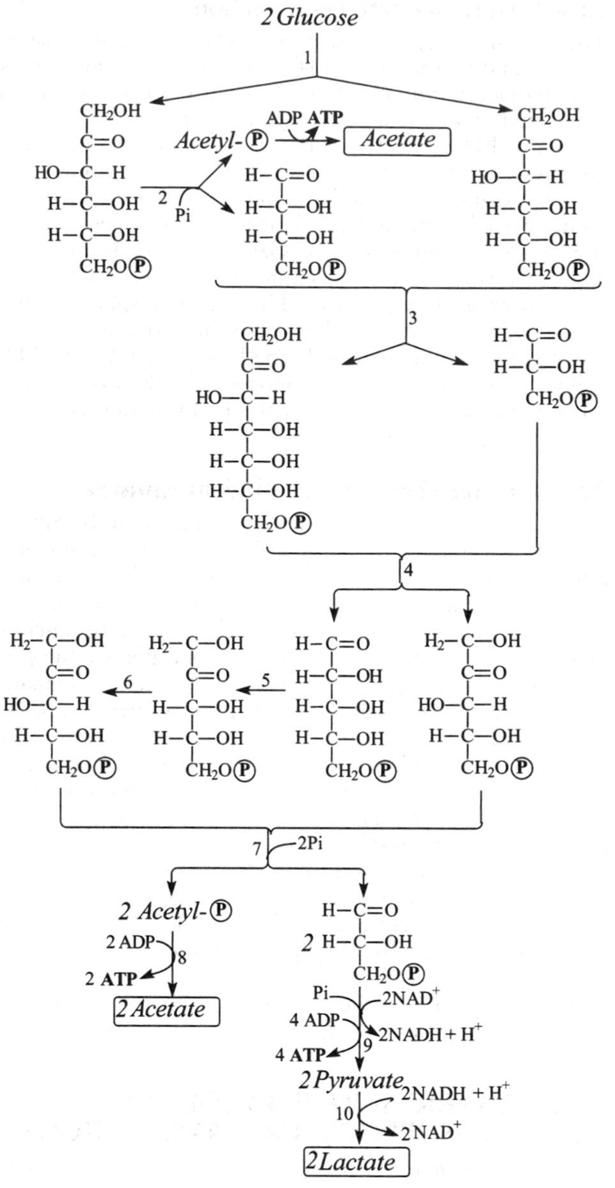

FIGURE 7 Formation of acetate and lactate from glucose by the heterofermentative bifidum pathway. 1, PEP-PTS; 2, fructose-6-phosphate phosphoketolase; 3, transaldolase; 4, transketolase; 5, ribose-5-phosphate isomerase; 6, ribulose 5-phosphate 3-epimerase; 7, xylulose 5-phosphate phosphoketolase; 8, acetate kinase; 9, conversion of GAP to pyruvate, which is accomplished by the same enzymes as in Fig. 2 and 3; 10, lactate dehydrogenase.

hydrogenase for the oxidation of NADH via ferredoxin to yield H_2 gas. When the levels of hydrogenase and ferredoxin synthesis decrease due to iron limitation or when the hydrogenase activity is inhibited by carbon monoxide, some clostridia use pyruvate reduction to lactate as an alternate sink for disposing of the electrons and protons (46).

In butyrate production in clostridia, 2 mol of acetyl-CoA (derived from pyruvate produced by the EMP pathway) is converted to 1 mol of acetoacetyl-CoA. The acetoacetyl-CoA is reduced using 1 mol of NADH derived from the GAPDH reaction and then undergoes dehydration to yield crotonyl-CoA. The crotonyl-CoA is reduced to butyryl-

CoA by another mole of NADH. The butyryl-CoA undergoes a phosphorylytic cleavage resulting in the formation of the high-energy phosphorylated intermediate butyryl phosphate, which in turn is converted to butyrate with the concomitant phosphorylation of ADP to ATP by substrate level phosphorylation. The reactions involved in the conversion of a mole of glucose to 1 mol of butyrate result in the net production of 3 ATPs/mol of glucose fermented (Fig. 8).

During butyrate fermentation by *Clostridium acetobutylicum* and certain other clostridia, the production of acids ceases when the pH of the medium drops below pH 5.0 and two neutral products, butanol and acetone, are produced with the consumption of butyrate and a concomitant rise in pH. This observation was made very early during the history of fermentation research (23, 46). *C. acetobutylicum* produces two butanol dehydrogenase isozymes that are NADH dependent; however, the number of isozymes varies depending on the strain. Similarly, *Clostridium beijerinckii* contains three isozymes of NADH/NADPH-dependent primary ADHs. Again, the number of isozymes is dependent on the strain examined (16).

Certain clostridia such as *C. beijerinckii* possess the ability to further reduce acetone to isopropanol. The regulation of the ratio of acid and solvent products has been dealt with in excellent reviews by Thauer et al. (109) and Jones and Woods (46). The assay methods for the various enzymes of the pathway are shown in Table 7 (section 22.11).

22.6. ETHANOL-ACETATE FERMENTATION

One of the most interesting and remarkable of the fermentations is that by *Clostridium kluyveri*. The organism requires a mixture of ethanol and acetate for growth. However, the acetate can be replaced by propionate. The primary products of this fermentation are butyrate and caproate in addition to hydrogen. The ratio of the two fatty acids is determined by the concentration of ethanol in the medium. Thus, an increase in ethanol concentration favors an increase in the production of caproate and vice versa. The pathway for the formation of butyrate is presented in Fig. 9. As seen from the figure, the oxidation of ethanol is in balance with the reduction of acetyl-CoA. Thus, the 2 mol of NAD(P)H generated in the formation of acetyl-CoA is oxidized during the reduction of acetoacetyl-CoA to butyrate. Evolution of hydrogen from the reduced NAD(P)H formed in the first and second reactions of the pathway plays a central role in the generation of ATP. For each mole of H_2 evolved, half a mole of acetyl-CoA is not required for the butyric acid cycle and the surplus can be used for the generation of ATP. The stoichiometry between H_2 evolution and ATP generation can be calculated (32, 103). For convenience, in Fig. 9, the concentrations of various intermediates produced during the production of 1 mol of ATP are shown. The assay methods for the various enzymes of the pathway are shown in Table 8.

22.7. MIXED ACID AND BUTANEDIOL FERMENTATION

The enterobacteria carry out mixed acid and butanediol fermentation, and this is the basis for the methyl red/Voges-Proskauer test (see chapter 18) that is used to distinguish the genera. Members of the genera *Escherichia*, *Salmonella*, and *Shigella* ferment glucose to lactate, acetate, succinate, and formate (Fig. 10). They also form CO_2, H_2, and ethanol. In contrast, *Enterobacter*, *Erwinia*, and *Serratia* produce CO_2,

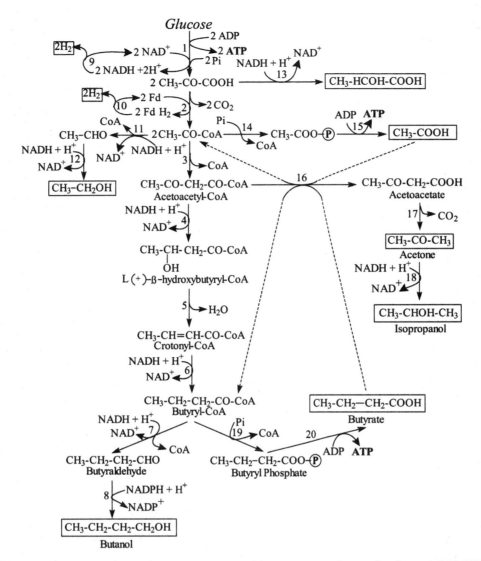

FIGURE 8 Butyrate-butanol-acetone-isopropanol fermentation in C. *acetobutylicum*. 1, PEP-PTS and EMP pathway enzymes; 2, pyruvate-ferredoxin oxidoreductase; 3, acetyl-CoA acetyltransferase (thiolase); 4, L-(+)-β-hydroxybutyryl-CoA dehydrogenase; 5, crotonase; 6, butyryl-CoA dehydrogenase; 7, butyraldehyde dehydrogenase; 8, butanol dehydrogenase; 9, NADH-ferredoxin oxidoreductase and hydrogenase; 10, hydrogenase; 11, acetaldehyde dehydrogenase; 12, ethanol dehydrogenase; 13, lactate dehydrogenase; 14, phosphotransacetylase; 15, acetate kinase; 16, acetoacetyl-CoA:acetate/butyrate:CoA transferase; 17, acetoacetate decarboxylase; 18, isopropanol dehydrogenase; 19, phosphotransbutyrylase; 20, butyrate kinase.

ethanol, and large amounts of 2,3-butanediol and produce less acid. The enterobacteria employ PEP-PTS for the transport and phosphorylation of glucose. The glucose-6-phosphate is converted to pyruvate by the EMP pathway. The formation of the various fermentation products from PEP and pyruvate and the enzymes involved are shown in Fig. 10. A branch of the pathway leading from PEP forms succinic acid. The rest of the products are formed from pyruvate. Large amounts of lactate are formed in the mixed acid fermentation of *Escherichia coli*. There are three lactate dehydrogenases in this organism. One of these enzymes is a soluble NAD-linked fermentative lactate dehydrogenase which produces D-lactate from pyruvate (12, 18). The other two are membrane-bound flavoproteins, one of which is

specific for the D and the other for the L isomer. These two unidirectional aerobic enzymes are not produced under fermentative conditions and should properly be described as oxidases rather than as dehydrogenases (18). In butanediol fermentation, lactate formation is reduced and the formation of other products of the pathway is determined by three enzyme activities: (i) pyruvate-formate lyase (PFL), (ii) α-acetolactate synthase, and (iii) formate hydrogen lyase (FHL). Assay procedures for the enzymes involved in these fermentations are given in Table 9 (section 22.11).

22.7.1. Pyruvate-Formate Lyase

PFL plays a key role in the fermentation of glucose by a number of bacteria. The enzyme catalyzes the CoASH-

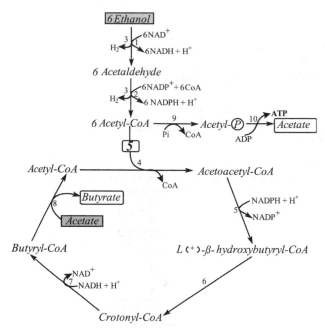

FIGURE 9 Ethanol-acetate fermentation by *C. kluyveri*. 1, ADH; 2, acetaldehyde dehydrogenase; 3, H_2-evolving enzyme system; 4, thiolase; 5, L-(+)-β-hydroxybutyryl-CoA dehydrogenase; 6, crotonase; 7, butyryl-CoA dehydrogenase; 8, CoA transferase; 9, phosphotransacetylase; 10, acetate kinase. The substrates, ethanol and acetate, are enclosed in shaded rectangles, while the major product, butyrate, and the minor product, acetate, are enclosed in rectangles. Formation of acetate primarily yields the energy required for growth of the organism.

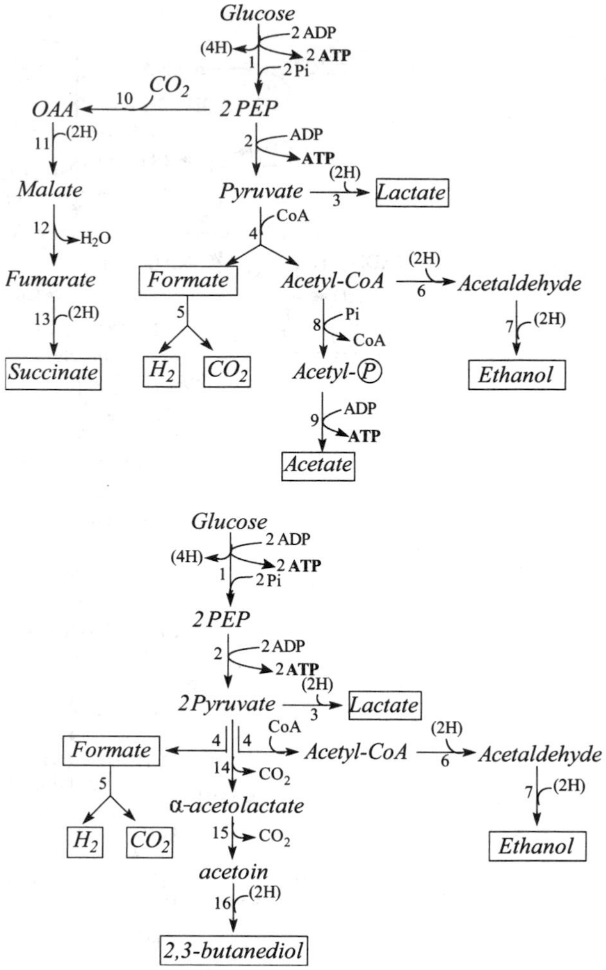

FIGURE 10 Mixed acid (top) and butanediol (bottom) fermentation. 1, PEP-PTS; 2, pyruvate kinase; 3, lactate dehydrogenase; 4, PFL; 5, FHL; 6, acetaldehyde dehydrogenase; 7, ADH; 8, phosphotransacetylase; 9, acetate kinase; 10, PEP carboxylase; 11, malate dehydrogenase; 12, fumarase; 13, fumarate reductase; 14, α-acetolactate synthase; 15, α-acetolacate decarboxylase; 16, acetoin reductase.

dependent cleavage of pyruvate, resulting in the formation of formate and acetyl-CoA. The enzyme is induced only under anaerobic conditions when the aerobic pyruvate dehydrogenase complex is completely repressed. The advantage of this switch is obvious in that the PFL forms acetyl-CoA anaerobically without the requirement for the reduction of NAD. For the catalysis of the reaction, the enzyme has to be activated by PFL-activating enzyme, which requires *S*-adenosylmethionine and dihydroflavodoxin as cosubstrates (see reference 54 for discussion of the activation process).

22.7.2. α-Acetolactate Synthase

As the name suggests, α-acetolactate synthase synthesizes acetolactate from pyruvate. The enzyme decarboxylates pyruvate by a thiamine diphosphate (TPP)-dependent reaction, with the formation of hydroxyethyl-TPP and CO_2. The enzyme-bound active acetaldehyde is subsequently transferred to a second molecule of pyruvate, resulting in the formation of α-acetolactate. Consistent with its role in shifting the fermentation from the formation of acid end products to that of neutral products, the enzyme is synthesized and active under acidic conditions (~pH 6).

22.7.3. Formate Hydrogen Lyase

FHL is synthesized by *E. coli* and *Enterobacter* during fermentative growth. The enzyme cleaves formate to CO_2 and H_2. In contrast, *Erwinia* and *Shigella* species accumulate formate due to the absence of this enzyme.

E. coli produces three formate dehydrogenases. Formate dehydrogenase N (Fdh-N) is synthesized under anaerobic

conditions when nitrate is present (also see chapter 21). The second enzyme, formate dehydrogenase O (Fdh-O), is synthesized at low levels independent of the availability of either oxygen or nitrate. The third enzyme, formate dehydrogenase H (Fdh-H), is the enzyme that in complex with hydrogenase is synthesized under anaerobiosis in the presence of formate (6, 117).

22.7.4. Acetaldehyde Dehydrogenase and Alcohol Dehydrogenase

As shown in Fig. 10, acetaldehyde dehydrogenase converts acetyl-CoA to acetaldehyde, which is further reduced to ethanol by ADH. In *E. coli*, these two reactions are carried out by a single bifunctional protein encoded by the *adhE* gene (41, 59).

22.7.5. Fumarase

E. coli produces three distinct fumarases known as FumA, FumB, and FumC. The major enzyme produced during fermentative growth is FumB (124). The assay methods for the various enzymes of the pathway are shown in Table 9 (section 22.11).

22.8. PROPIONATE AND SUCCINATE FERMENTATION

Propionate is an end product of many fermentative bacteria. These organisms ferment glucose to acetate, propionate, and CO_2. The most common substrate for propionate-producing bacteria is lactate, the end product of lactate fermentations (see above). Propionic acid-producing bacteria contain a lactate racemase because of which these organisms are capable of fermenting D-, L-, or DL-lactate. There are two different pathways for the formation of propionate. Lactate is reduced to propionate in the acrylate pathway, which is of limited distribution in bacteria. In contrast, in the propionate-succinate pathway (randomizing pathway), the formation of propionate involves pyruvate and succinate as intermediates and appears to be much more widespread.

22.8.1. Acrylate Pathway

The acrylate pathway is present in *Clostridium propionicum*, *Bacteroides ruminicola*, *Megasphaera elsdenii*, and a few other organisms. These organisms contain lactate racemase for the interconversion of L- and D-lactate. In the oxidative branch of the pathway, D-lactate is converted to pyruvate by the enzyme lactate dehydrogenase. There are two types of lactate dehydrogenases present in organisms that use the acrylate pathway. For example, *M. elsdenii* has an NAD^+-independent D-lactate dehydrogenase (8), while *C. propionicum* has an NAD^+-dependent enzyme (93). In organisms containing the NAD^+-independent enzyme, an electron-transferring flavoprotein is involved in the transfer of electrons.

The pyruvate formed by the lactate dehydrogenase is converted to acetyl-CoA by pyruvate-ferredoxin oxidoreductase (see section 22.5). The enzymes phosphotransacetylase and acetate kinase are responsible for the formation of acetate from acetyl-CoA. In this pathway, four [H] released during the oxidation of one molecule of lactate to acetate are used to reduce the two molecules of acrylyl-CoA arising from dehydration of two molecules of lactate. This pathway yields 1 mol of ATP/3 mol of lactate metabolized (Fig. 11). The assay methods for the various enzymes of the pathway are shown in Table 10 (section 22.11).

22.8.2. Succinate-Propionate Pathway

A majority of the propionate-producing organisms utilize the succinate-propionate pathway, and many of the classical early investigations that led to the characterization of this pathway were done using *Propionibacterium* (Fig. 12). This pathway is commercially important in the manufacture of Swiss cheese. The characteristic flavor of Swiss cheese is due to the propionic acid, and the holes are due to CO_2 production. As seen from Fig. 12, succinate is an intermediate in the pathway and in addition, depending on the organism, it also accumulates in various amounts as an end product, whereas succinate is not an end product in the acrylate pathway described above.

In this pathway, three molecules of lactate are oxidized to pyruvate by lactate dehydrogenase, resulting in the removal of six electrons. One of the pyruvates formed undergoes oxidative decarboxylation, resulting in the formation of acetyl-CoA and release of two additional electrons (not shown). The acetyl-CoA is converted to acetate via acetylphosphate, generating an ATP in the process.

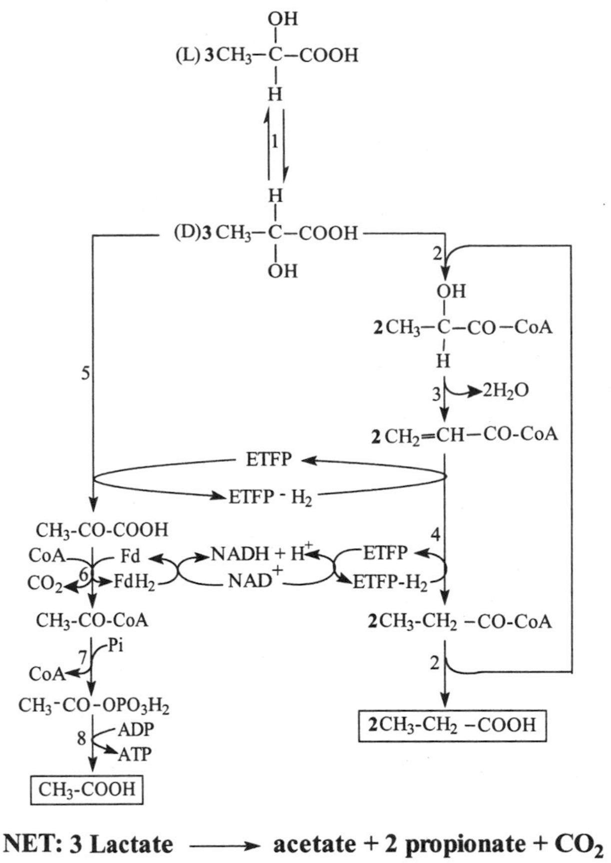

NET: 3 Lactate ⟶ acetate + 2 propionate + CO_2

FIGURE 11 Fermentation of lactate to propionate by the acrylate pathway. 1, lactate racemase; 2, propionyl-CoA transferase; 3, lactyl-CoA dehydratase; 4, acrylyl-CoA reductase; 5, D-lactate dehydrogenase; 6, pyruvate-ferredoxin oxidoreductase; 7, phosphotransacetylase; 8, acetate kinase. ETFP, electron-transferring flavoprotein.

The remaining two pyruvates are carboxylated to yield oxaloacetate (OAA). It should be pointed out that in other anaplerotic reactions employed by bacteria the carboxylation of pyruvate to OAA is an energy-consuming reaction requiring ATP. However, *Propionibacterium* carboxylates pyruvate and forms OAA by methylmalonyl-CoA-pyruvate transcarboxylase, without consuming ATP. The two OAAs are reduced to two molecules of malate, and four electrons are consumed for this reaction. The two malates are dehydrated to two fumarates which are in turn reduced to two succinates, consuming four more electrons. Thus, the eight electrons generated are consumed and the pathway is balanced. The reduction of two fumarates to two succinates also yields two ATPs by substrate level phosphorylation. The succinates formed are esterified to succinyl-CoA by a CoA transferase. Thus, the thioesterification of succinate also spares the consumption of energy. Isomerization of the two molecules of succinyl-CoA results in the formation of two molecules of methylmalonyl-CoA. This is an unusual reaction where the CoASH molecule moves from the α-carbon to the β-carbon of succinyl-CoA, resulting in the formation of (R)-methylmalonyl-CoA in a vitamin B_{12}-dependent reaction. The (R)-methylmalonyl-CoA molecules are first racemized to (S)-methylmalonyl-CoA before being transcarboxylated, resulting in the formation of propionyl-CoA. The CoA from

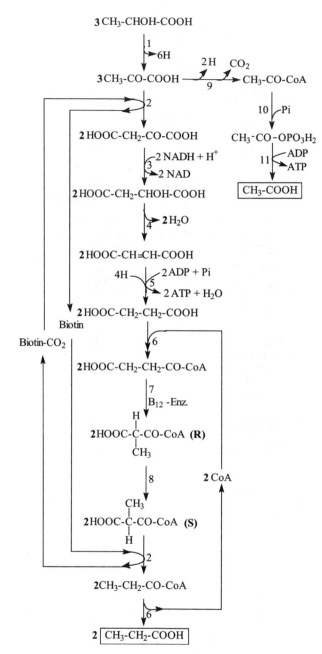

FIGURE 12 Fermentation of lactate by the succinate-propionate pathway. 1, lactate dehydrogenase (H acceptor is flavoprotein); 2, (S)-methylmalonyl-CoA-pyruvate transcarboxylase; 3, malate dehydrogenase; 4, fumarase; 5, fumarate reductase; 6, CoA transferase; 7, (R)-methylmalonyl-CoA mutase; 8, methylmalonyl-CoA racemase; 9, pyruvate-ferredoxin oxidoreductase; 10, phosphotransacetylase; 11, acetate kinase.

the two molecules of propionyl-CoA are transferred to succinate, resulting in the formation of two propionate.

In this connection, it is worth mentioning that the anaerobes *Bacteroides fragilis* and *Selenomonas ruminantium* ferment glucose to propionate (63, 67). However, these organisms lack methylmalonyl-CoA-pyruvate transcarboxylase, a key enzyme of the pathway which normally carboxy-

lates pyruvate, resulting in the formation of OAA. Instead, *B. fragilis* and *S. ruminantium* form OAA by carboxylating the glycolytic high-energy intermediate PEP by the enzyme PEP-carboxykinase. The subsequent reactions of the pathway from OAA to (S)-methylmalonyl-CoA are identical. (S)-methylmalonyl-CoA is decarboxylated to propionyl-CoA. The enzyme involved in this reaction, methylmalonyl-CoA decarboxylase, is membrane bound and acts as a primary Na^+ pump. The Na^+ ion gradient established by the decarboxylase is used to drive ATP synthesis by an Na^+- translocating ATP synthase (for a review on energy conservation by decarboxylation of decarboxylic acids, see reference 25).

As the name of the pathway indicates, propionibacteria can produce succinate in addition to propionate as the end product. When the organism is growing on glucose or other glycolytic substrates, it carboxylates the C_3 intermediate PEP to OAA, which is then reduced to succinate. The enzyme that carboxylates PEP is PEP carboxytransphosphorylase. It carries out the following unusual anaplerotic reaction:

$$PEP + CO_2 + P_i \rightarrow OAA + PP_i$$

In this reaction, the phosphate is transferred from PEP to P_i, resulting in the formation of PP_i. In this connection, it is worth mentioning that propionibacteria use PP_i to phosphorylate fructose-6-phosphate to fructose-1,6-diphosphate, originally discovered in *Entamoeba histolytica* (82). The assay methods for the various enzymes of the pathway are shown in Table 11 (section 22.11).

22.9. ACETATE FERMENTATION BY BACTERIA

A number of bacteria produce acetate as the sole product during fermentation and are called homoacetogens. Many genera and species of bacteria are able to obtain energy and grow solely on the reactions involving the formation of acetate from CO_2 and H_2. However, in this chapter, only the acetate fermentative pathway of hexoses is considered. The gram-positive, anaerobic, spore-forming bacterium *Moorella thermoacetica* (previously *C. thermaceticum*) ferments hexose to 3 mol of acetate (Fig. 13). The organism uses the EMP pathway for the catabolism of glucose to two pyruvates. The two pyruvates are degraded by the action of pyruvate-ferredoxin oxidoreductase, phosphotransacetylase, and acetate kinase, resulting in the formation of acetate and CO_2. Thus, from each molecule of glucose, two molecules each of acetate and CO_2 are formed. The synthesis of the third molecule of acetate is initiated from the two molecules of CO_2. One of the CO_2 molecules is first reduced to formate. In *M. thermoacetica*, the electron donor for the reaction is NADPH. The formate combines with tetrahydrofolate (THF) by an ATP-requiring reaction, resulting in the formation of formyl-THF. It, in turn, undergoes a dehydration reaction followed by two reductions resulting in the formation of CH_3-THF (Fig. 13). The methyl group is transferred to a corrinoid iron-sulfur protein (CFeSP). The CO_2 from the second molecule of pyruvate is not released but becomes enzyme bound with carbon monoxide dehydrogenase/acetyl-CoA synthetase (CODH/ACS). The CO_2 is now reduced to CO and finally becomes the carboxyl group of acetate. CODH/ACS accepts the methyl group from CFeSP. CODH also contains a binding site for CoASH and is responsible for the synthesis of acetyl-CoA (60, 70, 79, 121). The typical enzymes acting on acetyl-CoA form the third acetate and ATP. This pathway is unique in that all six car-

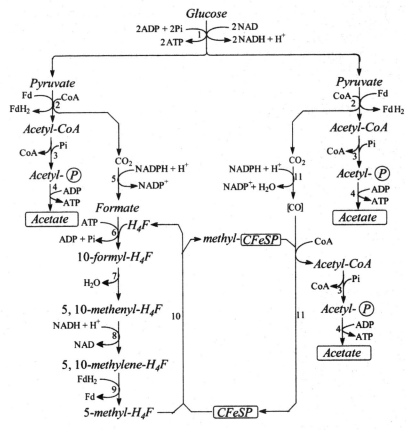

FIGURE 13 Acetate fermentation of glucose by bacteria. 1, enzymes of the EMP pathway; 2, pyruvate-ferredoxin oxidoreductase; 3, phosphotransacetylase; 4, acetate kinase; 5, formate dehydrogenase; 6, formyl-THF synthetase; 7, methenyl-THF cyclohydrolase; 8, 5,10-methylene-THF dehydrogenase; 9, 5,10-methylene-THF reductase; 10, THF:B$_{12}$ methyltransferase; 11, CO dehydrogenase/acetyl-CoA synthase.

bons of hexose are converted to three molecules of two carbon compounds. The assay methods for the various enzymes of the pathway are shown in Table 12 (section 22.11).

22.10. ACETATE FERMENTATION IN HYPERTHERMOPHILIC ARCHAEA

Hyperthermophiles are those organisms that have the remarkable ability to grow at, near, or above the boiling point of water. By definition, according to Stetter, any organism with an optimum growth temperature of between 80 and 106°C is a hyperthermophile (104). It is worth noting that the upper temperature limit of life (121°C) is held by an unidentified chemolithoautotrophic archaeon related to *Pyrodictium occultum* and *Pyrobaculum aerophilum* (48). Further, it has been demonstrated that vegetative cells of *Pyrolobus* and *Pyrodictium* are able to survive autoclaving at 121°C for 1 h (105)!

Many of the organotrophic hyperthermophiles ferment sugars or peptides. Carbohydrates fermented include the polysaccharides, such as starch and glycogen; the disaccharides maltose, cellulose, and lactose; and the monosaccharides, such as glucose, galactose, pyruvate, and lactate. The fermentation products formed include acetate, alanine, propionate, isovalerate, isobutyrate, and butanol (for a list of substrates fermented and products formed, see reference

90). However, quantitative determination of substrate utilization and product formation and construction of fermentation balances have only been described for a very few organisms.

Sugars are catabolized via the modified versions of the EMP pathway (Fig. 14). The modifications usually occur in (i) the phosphorylation of glucose and fructose-6-phosphate, (ii) the oxidation of GAP, and (iii) the conversion of acetyl-CoA to acetate.

22.10.1. Kinases

22.10.1.1. Phosphorylation of Glucose

The phosphorylation of glucose is carried out by either glucokinase or hexokinase. It appears that among the archaea so far examined, the organisms that contain glucokinase appear to use ADP as the phosphoryl donor for the phosphorylation of glucose. In contrast, the few archaea that have been reported to contain the enzyme hexokinase appear to use ATP as the phosphoryl donor (Table 13).

22.10.1.2. Phosphorylation of Fructose-6-Phosphate

The phosphorylation of fructose-6-phosphate is carried out by the enzyme PFK. The archaea contain three different PFKs which use different phosphoryl donors to phosphorylate fructose-6-phosphate. They are referred to as ATP-PFK,

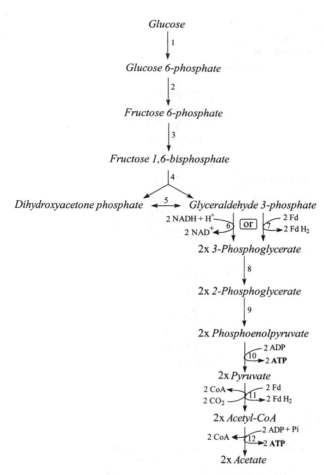

Glucose

↓ 1

Glucose 6-phosphate

↓ 2

Fructose 6-phosphate

↓ 3

Fructose 1,6-bisphosphate

↓ 4

Dihydroxyacetone phosphate ⇄ 5 *Glyceraldehyde 3-phosphate*

2 NADH + H⁺ ↘ 6 or 7 ↗ 2 Fd
2 NAD⁺ ↗   ↘ 2 Fd H₂

2x 3-Phosphoglycerate

↓ 8

2x 2-Phosphoglycerate

↓ 9

2x Phosphoenolpyruvate

10 ← 2 ADP
   → **2 ATP**

2x Pyruvate

2 CoA ← 11 ← 2 Fd
2 CO₂ ↗   → 2 Fd H₂

2x Acetyl-CoA

2 CoA ← 12 ← 2 ADP + Pi
           → **2 ATP**

2x Acetate

FIGURE 14 Acetate fermentation in archaea. 1, kinase (glucokinase/hexokinase); 2, glucose-6-phosphate isomerase; 3, PFK; 4, fructose bisphosphate aldolase; 5, triose phosphate isomerase; 6, GAPDH; 7, GAP ferredoxin oxidoreductase; 8, phosphoglycerate mutase; 9, enolase; 10, pyruvate kinase; 11, pyruvate-ferredoxin oxidoreductase; 12, acetyl-CoA synthetase (ADP forming).

which uses ATP as the phosphoryl donor; ADP-PFK, which uses ADP as the phosphoryl donor; and PPᵢ-PFK, which uses pyrophosphate as the phosphoryl donor (Table 13).

22.10.1.3. Bifunctional Kinase

Recently, an unusual kinase has been described for *Methanococcus jannaschii*. The enzyme is capable of phophorylating both glucose and fructose-6-phosphate. The phosphoryl donor for the reaction is ADP. Hence, the enzyme is referred to as ADP-dependent glucokinase/PFK.

22.10.2. Oxidation of Glyceraldehyde-3-Phosphate

In many of the anaerobic species of archaea examined so far (*Pyrococcus, Thermococcus, Thermoproteus,* and *Desulfurococcus*), GAP is directly oxidized to 3-phosphoglyceric acid. Depending on the organism, this oxidation is carried out by either glyceraldehyde:ferredoxin oxidoreductase or the NAD⁺-dependent GAPDH. Both the enzymes carry out irreversible conversion of GAP to 3-phosphoglyceric acid. In the conventional EMP pathway, the conversion of GAP to 3-phosphoglycerate is carried out by two enzymes: a Pᵢ-dependent, NADP-requiring GAPDH and a phosphoglycerate kinase. In this reaction sequence, 1,3-bisphosphoglycerate is an intermediate and ADP is phosphorylated to ATP during its conversion to 3-phosphoglycerate. The anaerobic archaea, by eliminating the formation of 1,3-bisphosphoglycerate as an intermediate, also dispense with the formation of 2 mol of ATP/mol of glucose oxidized.

22.10.3. Conversion of Acetyl Coenzyme A to Acetate

During the fermentation of sugars or lactate, the acetate-forming archaea oxidize pyruvate by the enzyme pyruvate-ferredoxin oxidoreductase (see discussion of clostridia above) to acetyl-CoA and CO₂. The hyperthermophilic acetate-forming archaea convert acetyl-CoA to acetate by the action of a unique enzyme, acetyl-CoA synthetase (ADP forming). This novel enzyme couples substrate level phosphorylation of ADP to the conversion of acetyl-CoA to acetate. In this connection, it is worth mentioning that the conversion of acetyl-CoA to acetate in bacteria requires two enzymes, phosphate acetyltransferase and acetate kinase.

The conversion of acetyl-CoA to acetate and ATP by acetyl-CoA synthetase is the most important energy-conserving reaction during sugar fermentation in archaea. The assay methods for the various enzymes of the pathway are shown in Table 14 (section 22.11).

TABLE 13 Phosphoryl donors for kinases of archaea[a]

Organism(s)	ADP-GK	ATP-HK	ATP-PFK	ADP-PFK	PPᵢ-PFK	Reference(s)
Aeropyrum pernix			+	+		27, 35
Archaeoglobus fulgidus	+			+		57, 125
Desulfurococcus amylolyticus		+	+			34, 97
Methanococcus jannaschii[b]	+		+			87
Many methanogens				+		116
Pyrococcus furiosus	+			+		27, 51, 115
Thermococcus celer	+			+		97
Thermococcus litoralis	+			+		54, 97
Thermococcus zilligii	+			+		84, 125
Thermoproteus tenax		+			+	97, 100

[a]ADP-GK, ADP-glucokinase; ATP-HK, ATP-hexokinase.
[b]In M. *jannaschii*, ADP-GK and ADP-PFK activities are carried out by a single bifunctional enzyme.

TABLE 2 Assay procedures for enzymes of the EMP pathway [a]

Enzyme	Assay mixture	ΔA (nm)	ε (M^{-1}cm^{-1})
1. Hexokinase (65) (EC 2.7.1.1)	50 mM triethanolamine hydrochloride (pH 7.4), 10 mM MgCl$_2$, 5 mM glucose, 1 mM ATP, 0.05 mM NADP$^+$, 0.3 U of glucose-6-phosphate dehydrogenase, and cell extract	340	6.23×10^3
2. Phosphoglucoisomerase (65) (EC 5.3.1.9)	50 mM triethanolamine hydrochloride (pH 7.4), 10 mM MgCl$_2$, 1 mM fructose-6-phosphate-free glucose-6-phosphate, 0.05 mM NADP$^+$, 0.3 U of glucose-6-phosphate dehydrogenase, and cell extract	340	6.23×10^3
3. PFK (65) (EC 2.7.1.11)	50 mM triethanolamine hydrochloride (pH 7.4), 10 mM MgCl$_2$, 5 mM glucose-6-phosphate, 2 U of phosphoglucoisomerase, 0.03 mM NADH, 1 mM ATP, 0.3 mM ADP, 1 U each of aldolase and α-glyceraldehyde phosphate dehydrogenase, 10 U of triose-phosphate isomerase, and cell extract. The reaction velocity should be divided by 2.	340	6.23×10^3
4. Fructose bisphosphate aldolase (65) (EC 4.1.2.13)	50 mM triethanolamine hydrochloride (pH 7.4), 10 mM MgCl$_2$, 0.03 mM NADH, 1 mM fructose 1,6-bisphosphate, 1 U α-glyceraldehyde-phosphate dehydrogenase, 10 U triose-phosphate isomerase, and cell extract. The reaction velocity should be divided by 2.	340	6.23×10^3
5. Triose phosphate isomerase (65) (EC 5.3.1.1)	50 mM triethanolamine hydrochloride (pH 7.4), 10 mM MgCl$_2$, 0.03 mM NADH, 0.4 mM GAP, 1 U of α-glyceraldehyde phosphate dehydrogenase, and cell extract	340	6.23×10^3
6. GAPDH (65) (EC 1.2.1.12)	50 mM triethanolamine hydrochloride (pH 7.4), 10 mM MgCl$_2$, 0.03 mM NADH, 5 mM cysteine, 1 mM 3-phosphoglycerate, 1 mM ATP, 1 U of phosphoglycerate kinase, and cell extract	340	6.23×10^3
7. 3-Phosphoglycerate kinase (65) (EC 2.7.2.3)	50 mM triethanolamine hydrochloride (pH 7.4), 10 mM MgCl$_2$, 0.03 mM NADH, 5 mM cysteine, 1 mM 3-phosphoglycerate, 1 mM ATP, 1 U of GAPDH, and cell extract	340	6.23×10^3
8. Phosphoglycerate mutase (65) (EC 5.4.2.1)	50 mM triethanolamine hydrochloride (pH 7.4), 10 mM MgCl$_2$, 0.03 mM NADH, 1 mM 3-phosphoglycerate, 0.5 U enolase, 1 mM ADP, 1 U each of pyruvate kinase and muscle lactate dehydrogenase, and cell extract	340	6.23×10^3
9. Enolase (65) (EC 4.2.1.11)	50 mM triethanolamine hydrochloride (pH 7.4), 10 mM MgCl$_2$, 0.03 mM NADH, 1 mM 3-phosphoglycerate, 0.5 U of phosphoglycerate mutase, 1 mM ADP, 1 U each of pyruvate kinase and muscle lactate dehydrogenase, and cell extract	340	6.23×10^3
10. Pyruvate kinase (65) (EC 2.7.1.40)	50 mM triethanolamine hydrochloride (pH 7.4); 10 mM MgCl$_2$; a 1 mM concn each of PEP, ADP, and fructose-1,6-diphosphate; 0.03 mM NADH; 1 U of muscle lactate dehydrogenase; and cell extract	340	6.23×10^3
11. Pyruvate decarboxylase (65) (EC 4.1.1.1)	50 mM triethanolamine hydrochloride (pH 7.4), 10 mM MgCl$_2$, 50 mM pyruvate, 5 mM cysteine, 0.03 mM NADH, 0.25 mM TPP, 1 U of ADH, and cell extract	340	6.23×10^3
12. ADH (26) (EC 1.1.1.1)	21 mM glycine, 75 mM semicarbazide HCl, 75 mM sodium pyrophosphate buffer (pH 8.7), 150 mM ethyl alcohol, 1 mM NAD$^+$, and cell extract	340	6.23×10^3

[a] Most reagents and enzymes described are readily available from various commercial sources, such as Sigma-Aldrich.

TABLE 3 Assay procedures for the enzymes of the ED pathway

Enzyme	Assay mixture	ΔA (nm)	ε ($M^{-1}cm^{-1}$)
1. Glucokinase (96) (EC 2.7.1.2)	20 mM imidazole, 20 mM KH_2PO_4 buffer (pH 6.8), 5 mM $MgCl_2$, 1 mM ATP, 10 mM glucose, 1 mM NAD^+, 2 U of glucose-6-phosphate dehydrogenase, and cell extract	340	6.23×10^3
2. Glucose-6-phosphate dehydrogenase (96) (EC 1.1.1.49)	30 mM Tris–30 mM KCl–2 mM $MgSO_4$ buffer (pH 6.8), 1 mM glucose-6-phosphate, 1 mM NAD^+, 0.2 mg of BSA,[a] and cell extract	340	6.23×10^3
3. 6-Phosphogluconate dehydratase (95) (EC 4.2.1.12)	20 mM K-MES buffer (pH 6.5) containing 2 mM $MgCl_2$ and 50 mM NaCl, 1 mM 6-phosphogluconate, 0.15 mM NADH, 5 U each of lactate dehydrogenase and KDPG aldolase, and cell extract	340	6.23×10^3
4. KDPG aldolase (94) (EC 4.1.2.14)	20 mM K-MES buffer (pH 6.5) containing 50 mM NaCl and 2 mM $MgCl_2$, 0.15 mM NADH, 5 U of rabbit muscle lactate dehydrogenase, 1 mM KDPG, and cell extract	340	6.23×10^3
5. GAPDH (75) (EC 1.2.1.12)	50 mM K-MES buffer (pH 6.5), 30 mM KCl, 3 mM $MgCl_2$, 5 mM 3-phosphoglycerate, 1 mM ATP, 0.15 mM NADH, 10 mM β-mercaptoethanol, 10 U of phosphoglycerate kinase, and cell extract	340	6.23×10^3
6. 3-Phosphoglycerate kinase (75) (EC 2.7.2.3)	Same as that described for GAPDH except for 10 U of GAPDH as the coupling enzyme		
7. Phosphoglycerate mutase (75) (EC 5.4.2.1)	50 mM K-MES buffer (pH 6.5) containing 30 mM KCl and 3 mM $MgCl_2$; 2 mM 3-phosphoglycerate; 0.5 mM ADP; 0.15 mM NADH; 10 mM β-mercaptoethanol; 10 U each of enolase, pyruvate kinase, and lactate dehydrogenase; and cell extract	340	6.23×10^3
8. Enolase (75) (EC 4.2.1.11)	30 mM triethanolamine buffer (pH 7.5) containing 0.5 mM ADP, 30 mM KCl, and 3 mM $MgCl_2$; 10 U each of pyruvate kinase and lactate dehydrogenase; 0.15 mM NADH; 0.4 mM 2-phosphoglycerate; and cell extract	340	6.23×10^3
9. Pyruvate kinase (75) (EC 2.7.1.40)	50 mM MES buffer (pH 6.5) containing 100 mM KCl and 0.5 mM ADP, 0.15 mM NADH, 0.5 mM PEP, 10 U of lactate dehydrogenase, and cell extract	340	6.23×10^3
10. Pyruvate decarboxylase (71) (EC 4.1.1.1)	50 mM K-MES buffer (pH 6.5), 5 mM $MgCl_2$, 0.1 mM TPP, 0.1 mg of BSA, 0.15 mM NADH, 10 U of yeast alcohol dehydrogenase, 5 mM pyruvate, and cell extract	340	6.23×10^3
11. ADH (5, 72) (EC 1.1.1.1)	30 mM Tris-HCl buffer (pH 8.5), 1 M ethanol, 1 mM NAD^+, and cell extract	340	6.23×10^3

[a]BSA, bovine serum albumin; MES, morpholineethanesulfonic acid.

TABLE 4 Assay procedures for enzymes of the homolactate fermentation pathway

Enzyme	Assay mixture	ΔA (nm)	ε ($M^{-1}cm^{-1}$)
1. PEP-PTS (15, 56) (EC 2.7.3.9)	2.5, 5, and 10 μl of permeabilized cells[a] and 100 μl of 50 mM KPO_4 buffer (pH 6.5) containing 12.5 mM NaF, 5 mM $MgCl_2$, 2.5 mM DTT, 10 mM PEP, and 10 mM ^{14}C-labeled carbohydrate (specific activity, 200 dpm · $nmol^{-1}$) are incubated for 15–30 min at 37°C; the phosphorylated carbohydrates are separated[b] by columns and the radioactivity is determined by liquid scintillation counting (see chapter 17)		
2–6. Glucose-6-phosphate to pyruvate	Assays as described in Table 2		
7. Lactate dehydrogenase (4) (EC 1.1.1.27)	50 mM triethanolamine (pH 7.5), 1 mM fructose-1,6-bisphosphate, 0.2 mM NADH, and 10 mM sodium pyruvate are used to initiate the reaction with permeabilized cells at 28°C	340	6.23×10^3

[a]**Preparation of permeabilized cells:** Bacterial culture is washed twice with ice-cold 50 mM KPO_4 buffer (pH 6.5) containing 2 mM $MgSO_4$ (KPM buffer), resuspended in 1/100 of culture volume of KPM buffer containing 20% (wt/vol) glycerol, and rapidly frozen in liquid nitrogen and kept at −80°C until use. After thawing, cells are washed once with KPM buffer; resuspended in 50 mM KPO_4 buffer (pH 6.5) containing 12.5 mM NaF, 5 mM $MgCl_2$, and 2.5 mM dithiothreitol (DTT); and permeabilized. First, 2.5 μl of toluene/acetone (1:9, vol/vol) is added per 250 μl of cell suspension and vortexed for 5 min at 4°C. The cells are then centrifuged (150 × g at 4°C for 2 min) and the cell pellet is resuspended in the same buffer (optical density at 600 nm of 50) and again treated with toluene/acetone as described above (50). After 5 min of vortexing, permeabilized cells are kept on ice.
[b]**Separation method:** The product, [^{14}C]hexose-phosphate, is separated from excess ^{14}C-substrate by ion-exchange chromatography. The incubation mixtures are diluted with 0.5 ml of water and transferred to columns, 0.8 by 9 cm, of analytical-grade, chloride form, ion-exchange resin (AG 1-X2, 50 to 100 mesh; Bio-Rad); the excess ^{14}C-substrate is washed from each column with 15 ml of water; and the labeled product is eluted from the column with 6 ml of 1.0 M LiCl. Each elute is collected in a liquid scintillation spectrometer vial, and the ^{14}C is counted after the addition of 15 ml of a mixture containing 333 ml of Triton X-100 (Packard Instrument Company), 666 ml of toluene, 5.5 g of 2,5-diphenyloxazole, and 125 mg of dimethyl-1,4-bis[2-(5-phenyloxazolyl)] benzene (dimethyl POPOP). The efficiency of the counting system is determined with ^{14}C-standards (see chapter 17).

TABLE 5 Assay procedures for enzymes of the heterolactate fermentation pathway

Enzyme	Assay mixture[a]	ΔA (nm)	ε (M⁻¹cm⁻¹)
1. PEP-PTS (EC 2.7.3.9)	Assay as described in Table 4		
2. Glucose-6-phosphate dehydrogenase (EC 1.1.1.49)	Assay as described in Table 3		
3. 6-Phosphogluconate dehydrogenase (EC 1.1.1.44)	Assay as described in Table 3		
4. Ribulose-5-phosphate-3-epimerase (122) (EC 5.1.3.1)	100 mM triethanolamine buffer (pH 7.4), 10 mM D-ribose-5-phosphate, 2.2 U of heat-treated ribose-5-phosphate isomerase, and cell extract	290	72
5. Phosphoketolase (66) (EC 4.1.2.9)	33.3 mM KPO₄ (pH 6.5), 1.9 mM L-cysteine hydrochloride, 23 mM sodium fluoride, 8 mM sodium iodoacetate, 27 mM D-fructose-6-phosphate or D-xylulose-5-phosphate, and cell extract. Incubate at 37°C for 30 min, and then add 0.075 ml of 2 M hydroxylamine hydrocholoride (pH 6.5) at room temp. After 10 min, add 0.05 ml of 15% (wt/vol) trichloroacetic acid, 4 M HCl, and 5% (wt/vol) FeCl₃ · 6H₂O in 0.1 M HCl. Acetylphosphate is used as the standard.	505	
6. Conversion of GAP to pyruvate	Assays as described in Tables 2 and 3		
7. Lactate dehydrogenase	Assay as described in Table 4		
8. Phosphotransacetylase (9) (EC 2.3.1.8)	100 mM Tris-HCl buffer (pH 8), 5 mM MgCl₂, 0.5 mM NAD⁺, 0.5 mM CoA, 5 mM L-malate, 12.5 μg of crystalline malate dehydrogenase, 25 μg of crystalline citrate synthase, 10 mM lithium acetyl phosphate, and cell extract	340	6.23 × 10³
9. Acetaldehyde dehydrogenase (19) (EC 1.2.1.4)	50 mM CHES buffer (pH 9.5), 100 μM CoA, 1.0 mM DTT, 75 μM NAD⁺, 10 mM acetaldehyde, and cell extract	340	6.23 × 10³
10. ADH (19) (EC 1.1.1.1)	12 mM sodium pyrophosphate (pH 8.5), 75 μM NAD⁺, 20 μl of ethanol, and cell extract	340	6.23 × 10³

[a]CHES, 2(N-cyclohexylamino) ethanesulfonate; DTT, dithiothreitol.

TABLE 6 Assay procedures for enzymes of the heterofermentative bifidum pathway

Enzyme	Assay mixture	ΔA (nm)	ε (M⁻¹cm⁻¹)
1. PEP-PTS (EC 2.7.3.9)	Assay as described in Table 4		
2. Fructose-6-phosphate phosphoketolase (EC 4.1.2.9)	Assay as described in Table 5		
3. Transaldolase (114) (EC 2.2.1.2)	40 mM triethanolamine buffer (pH 7.6), 100 mM EDTA, 3.2 mM fructose-6-phosphate, 0.2 mM erythrose-4-phosphate, 0.1 mM NADH, 10 μg α-glycerophosphate dehydrogenase-triose phosphate isomerase mixture, and cell extract	340	6.23 × 10³
4. Transketolase (101) (EC 2.2.1.1)	50 mM glycylglycine buffer (pH 8.5), 0.2 mM NADH, 2 mM xylulose 5-phosphate, 5 mM MgCl₂, 1 mM TPP, and cell extract	340	6.23 × 10³
5. Ribose-5-phosphate isomerase (122) (EC 5.3.1.6)	100 mM triethanolamine buffer (pH 7.4), 10 mM D-ribose-5-phosphate, 2.2 U of ribose-5-phosphate epimerase, and cell extract	290	72
6. Ribulose-5-phosphate-3-epimerase (EC 5.1.3.1)	Same as above except 2.2 U of heat-treated ribose-5-phosphate isomerase was added		
7. Xylulose-5-phosphate phosphoketolase (EC 4.1.2.9)	GAP formed is measured by coupling it to the GAP assay as described in Table 2		
8. Acetate kinase (102) (EC 2.7.2.1)	100 mM Tris-HCl buffer (pH 6.5), 3 mM MgCl₂, 2 mM glucose, 0.5 mM NADP⁺, 1 U of hexokinase, 1 U of glucose-6-phosphate dehydrogenase, 4 mM acetylphosphate, 1 mM ADP, and cell extract	340	6.23 × 10³
9. Conversion of GAP to pyruvate	Assays as described in Tables 2 and 3		
10. Lactate dehydrogenase	Assay as described in Table 4		

TABLE 7 Assay procedures for the enzymes of the butyrate-butanol-acetone-isopropanol fermentation in *C. acetobutylicum*

Enzyme	Assay mixture[a]	ΔA (nm)	ε (M^{-1}cm^{-1})
1. PEP-PTS and EMP pathway enzymes			
PEP-PTS (69)	10 mM Tris-HCl (pH 7.5), 0.5 mM PEP, 2 mM DTT, 5 mM MgCl$_2$, 12 mM KF, and 0.1 mM radiolabeled sugar (1.05 Ci mol^{-1}), with 25 mM potassium phosphate (pH 7.0) added to ensure adequate precipitation of sugar phosphate. At intervals, 0.15-ml samples were withdrawn and added to 2 ml of 1% barium bromide in 80% ethanol. Precipitates were removed by filtration on fiberglass disks (Whatman GF/F), washed with 5 ml of 80% ethanol, and dried under a heat lamp. The disks were added to 4 ml of scintillation cocktail and the radioactivity was determined by scintillation counting.		
EMP pathway enzymes	Assays as described in Table 2		
2. Pyruvate ferredoxin oxidoreductase (58, 102) (EC 1.2.7.1)	50 mM K$_2$HPO$_4$ buffer (pH 7.0), 0.1 mM CoA, 20 mM DTT, 2 mM benzyl viologen, 5 mM pyruvate, and cell extract	578	8.65 × 10^3
3. Acetyl-CoA acetyltransferase (36) (EC 2.3.1.9)	100 mM Tris-HCl (pH 8.0), 100 mM KCl, 0.6 mM CoA, 10 mM lithium acyl phosphate, and cell extract	233	4.44 × 10^3
4. L-(+)-β-Hydroxybutyryl-CoA-dehydrogenase (36) (EC 1.1.1.157)	50 mM 2-(N-morpholine)propanesulfonic acid-KOH buffer (pH 7.0), 1 mM DTT, 0.2 mM NADH, 75 μM acetoacetyl-CoA, and cell extract	340	6.23 × 10^3
5. Crotonase (36) (EC 4.2.1.17)	100 mM Tris-HCl buffer (pH 7.6), 0.15 mM crotonyl-CoA, and cell extract	263	6.7 × 10^3
6. Butyryl-CoA dehydrogenase (36) (EC 1.3.99.2)	100 mM KPO$_4$ buffer (pH 7.0), 0.15% Triton X-100, 0.2 mM meldola-blau (ZnCl salt of 8-dimethyl-amino-2,3-benzophenoxazine), 0.3 mM iodonitrotetrazolium, and cell extract. Butyryl-CoA (0.1 mM) is added to initiate the reaction after flushing with H$_2$ gas for 2 min.	492	19.4 × 10^3
7. Butyraldehyde dehydrogenase (3) (EC 1.2.1.57)	80 mM Tris-maleate buffer (pH 6.0), 0.2 mM butyryl-CoA, 0.4 mM NADH, 72 mM semicarbazide hydrochloride, and cell extract	340	6.23 × 10^3
8. Butanol dehydrogenase (3)	70 mM Tris-HCl buffer (pH 7.8), 20 mM butanol, 0.4 mM NAD, 72 mM semicarbazide-HCl, and cell extract	340	6.23 × 10^3
9. NADH-ferredoxin oxidoreductase (7) (EC 1.18.1.3)	100 mM Tris-HCl (pH 7.6), NADH-regenerating system (250 μM NADH, 30 μl of 96% ethanol, 45 U of ADH from yeast), and acetyl-CoA-generating system [20 mM acetylphosphate, 2 U of phosphotransacetylase, 1 mM CoA, 30 mM (NH$_4$)$_2$SO$_4$, and 2 mM GSH to activate phosphotransacetylase], 10 μM FAD, 0.1 mM metronidazole, 0.5–8 μM ferredoxin, and cell extract. The reaction is carried out under an atmosphere of CO in anaerobic cuvettes fitted with serum stoppers.	320	9.31 × 10^3
10. Hydrogenase (102) (EC 1.12.7.2)	100 mM Tris-HCl buffer (pH 6.5), 2 mM benzyl viologen, 2 mM DTT, 1 atm of H$_2$ gas, and cell extract	578	8.65 × 10^3
11. Acetaldehyde dehydrogenase (58, 102) (EC 1.2.1.4)	100 mM Tris-HCl (pH 6.5), 1 mM NAD(P), 1 mM DTT, 0.1 mM CoA, 7 mM sodium arsenate, 0.01 mM acetaldehyde, 0.5 U of phosphotransacetylase, and cell extract	340	6.23 × 10^3
12. Ethanol dehydrogenase (102) (EC 1.1.1.1)	100 mM Tris-HCl (pH 6.5), 10 mM DTT, 0.3 mM NAD(P)H, 10 mM acetaldehyde, and cell extract	340	6.23 × 10^3
13. Lactate dehydrogenase (12) (EC 1.1.1.27)	20 mM MOPS buffer (pH 7.0), 30 mM sodium pyruvate (pH 7.5), 0.2 mM NADH, and cell extract	340	6.23 × 10^3
14. Phosphotransacetylase (67, 102) (EC 2.3.1.8)	100 mM Tris-HCl buffer (pH 6.5), 1 mM CoA, 30 mM NH$_4$Cl, 10 mM DTT, 2 mM acetylphosphate, and cell extract	233	4.44 × 10^3
15. Acetate kinase (102) (EC 2.7.2.1)	100 mM Tris-HCl buffer (pH 6.5), 3 mM MgCl$_2$, 2 mM glucose, 0.5 mM NADP$^+$, 1 U of hexokinase, 1 U of glucose-6-phosphate dehydrogenase, 4 mM acetylphosphate, 1 mM ADP, and cell extract	340	6.23 × 10^3
16. Acetoacetyl-CoA:acetate/butyrate:CoA transferase (13) (EC 2.8.3.9)	100 mM Tris-HCl (pH 7.5), 150 mM carboxylic acid (potassium salt of acetate, butyrate, or propionate) (pH 7.5), 40 mM MgCl$_2$, 0.1 mM acetoacetyl-CoA, 5% (vol/vol) glycerol, and cell extract	310	8.0 × 10^3

(Continued on next page)

TABLE 7 *(Continued)*

Enzyme	Assay mixture[a]	ΔA (nm)	ε (M^{-1}cm^{-1})
17. Acetoacetate decarboxylase (3) (EC 4.1.1.4)	40 mM sodium acetate buffer (pH 5.0), 80 mM lithium acetoacetate, and cell extract in a 3-ml total reaction volume. CO$_2$ is measured in a Warburg manometer.	270	55.0
18. Isopropanol dehydrogenase (44) (EC 1.1.1.80)	50 mM Tris-HCl buffer (pH 7.5), 1 mM DTT, 0.2 mM NADPH, 6.7 mM acetone, and cell extract	340	6.23×10^3
19. Phosphotransbutyrylase (36) (EC 2.3.1.19)	100 mM KPO$_4$ buffer (pH 7.0), 0.1 mM butyryl-CoA, and cell extract	233	4.44×10^3
20. Butyrate kinase (36) (EC 2.7.2.7)	75 mM Tris-HCl (pH 8.0), 6 mM ADP, 10 mM MgCl$_2$, 2 mM glucose, 1 U each of hexokinase and glucose-6-phosphate dehydrogenase, 0.2 mM NADP$^+$, 7 mM butyryl phosphate, and cell extract	340	6.23×10^3

[a]DTT, dithiothreitol; KF, potassium fluoride; GSH, glutathione (reduced); FAD, flavin adenine dinucleotide; MOPS, morpholinepropanesulfonic acid.

TABLE 8 Assay procedures for the enzymes involved in ethanol and acetate fermentation by *C. kluyveri*

Enzyme	Assay mixture[a]	ΔA (nm)	ε (M^{-1}cm^{-1})
1. ADH (EC 1.1.1.1)	Assay as described in Table 7		
2. Acetaldehyde dehydrogenase (EC 1.2.1.4)	Assays as described in Table 7		
3. H$_2$-evolving enzymes (EC 1.12.2.1)			
NADH (110)	75 mM Tris-HCl (pH 7.4), 2 mM glutathione red, 12 μM FAD, 1 U of phosphotransacetylase, 25 mM acetylphosphate (potassium-lithium salt), 0.75 mM CoA, 0.9 mg of *C. kluyveri* ferredoxin (Fdkl), NADH-RS, and 5 mg of DEAE-cellulose-treated extract. NADH-RS is 40 mM galactose, 2.5 mM NADH, and 0.5 U of galactose dehydrogenase. Assays are carried out in a total volume of 1 ml in 17.5-ml Thunberg tubes at 37°C with shaking; gas phase, argon. H$_2$ is detected by gas chromatography.[b]		
NADPH (47)	70 mM Tris-HCl buffer (pH 8.0), 2 mM glutathione red, 12 μM FAD, 1 mg of ferredoxin (Fdkl), NADPH-RS, NAD$^+$-RS, and 5 mg of DEAE-cellulose-treated extract. NADPH-RS is 16 mM glucose-6-phosphate, 3.5 U of glucose-6-phosphate dehydrogenase, and 0.5 mM NADP$^+$. NAD-RS is 2 mM fructose-1,6-diphosphate, 0.5 U of aldolase, 2.5 U of triose phosphate isomerase, 0.7 U of glycerol-l-phosphate dehydrogenase, and 0.5 mM NAD. Assays are carried out in a total volume of 2.5 ml in 17.5-ml Thunberg tubes at 37°C using an argon gas phase. H$_2$ is detected by gas chromatography.[b]		
4. Thiolase (120) (EC 2.3.1.9)	67 mM Tris-HCl buffer (pH 8.1), 0.2 mM acetoacetyl-CoA, 25 mM potassium arsenate (pH 8.1), 2 U of phosphotransacetylase, 0.2 mM CoA, and cell extract	233	4.4×10^3
5. L-(+)-β-Hydroxybutyryl-CoA dehydrogenase (EC 1.1.1.157)	Assay as described in Table 7		
6. Crotonase (EC 4.2.1.17)	Assay as described in Table 7		
7. Butyryl-CoA dehydrogenase (EC 1.3.99.2)	Assay as described in Table 7		
8. CoA transferase (119) (EC 2.8.3.8)	100 mM Tris-HCl buffer (pH 7.5), 100 mM potassium butyrate (pH 7.5), 20 mM MgCl$_2$, 0.10 mM acetoacetyl-CoA, 5% (vol/vol) glycerol, 0.05 U of CoA transferase, and cell extract	310	8.0×10^3
9. Phosphotransacetylase (EC 2.3.1.8)	Assay as described in Table 7		
10. Acetate kinase (EC 2.7.2.1)	Assay as described in Table 7		

[a]FAD, flavin adenine dinucleotide; RS, regenerating system.
[b]H$_2$ is quantified by gas chromatography. Sample column: Length, 4 m; inner diameter, 2 mm; material, steel; temperatures, 50°C for both injection port and column; carrier gas, argon; detection, thermal conductivity detector; gas samples, 2 ml of 15-ml gas phase is injected with a gastight syringe.

TABLE 9 Assay procedures for the enzymes of the mixed acid and butanediol fermentation[a]

Enzyme	Assay mixture	ΔA (nm)	ε (M^{-1}cm^{-1})
1. PEP-PTS (1) (EC 2.7.3.9)	50 mM KPO$_4$ buffer (pH 7.4), 10 μM ^{14}C-sugar, 5 mM PEP, 12.5 mM MgCl$_2$, 25 mM potassium fluoride, 2.5 mM DTT, plus excess quantities of the soluble PTS enzymes for PEP-dependent reaction. For the sugar-P-dependent reaction (transphosphorylation), PEP is replaced by 10 mM sugar, the pH of the KPO$_4$ buffer is 6.0, and the soluble enzymes are omitted.[b]		
2. Pyruvate kinase	Assay as described in Tables 2 and 3		
3. Lactate dehydrogenase (EC 1.1.1.27)	Assay as described in Table 7		
4. PFL (53) (EC 2.3.1.54)			
Activation of PFL	A two-armed test tube equipped with a ground-glass stopper with an outlet for gas exchange is used for the anaerobic activation of PFL. To one arm, enzyme II in MOPS buffer and 9 mM DTT are added to a volume of 0.25 ml. To the second arm, 0.1 M MOPS buffer, 9 mM DTT, 0.4 mM Fe(NH$_4$)$_2$(SO$_4$)$_2$, 20 μM dichloro-phenol-indophenol, 10 μM 3(3,4-dichlorophenyl)-1,1-dimethylurea, 4 mM potassium oxamate, 0.1 mM adenosylmethionine, 0.02 mg of flavodoxin, chloroplast fragments (with 10 μg of chlorophyll), and 0.2 mg of inactive PFL are added. Volume, 0.25 ml; pH, 7.7; temp, 30°C.[c]		
Coupled assays for PFL			
Forward reaction	Tris-HCl (0.1 M) buffer (pH 8.1), 10 mM DTT, 0.1 mM Fe(NH$_4$)$_2$(SO$_4$)$_2$, 5 mM DL-malate, 1 mM NAD$^+$, 10 mM sodium pyruvate, 0.055 mM CoASH, 20 μg of citrate synthase, 25 μg of malate dehydrogenase, 0.1 mg of BSA, and activated enzyme (in a total volume of 1 ml) are added to the anaerobic cuvette.	340	6.24×10^3
Reverse reaction	Tris-HCl (0.1 M) buffer (pH 8.1), 10 mM DTT, 0.1 mM Fe(NH$_4$)$_2$(SO$_4$)$_2$, 10 mM lithium acetylphosphate, 0.3 mM NADH, 0.055 mM CoA, 53 mM potassium formate, 25 μg of lactate dehydrogenase, 5 μg of phosphate acetyltransferase, 0.1 mg of BSA, and activated enzyme are added to an anaerobic cuvette in a final volume of 1 ml.	340	6.24×10^3
5. FHL (6) (EC 1.2.1.2)	50 mM Tris-HCl buffer (pH 7.5), 20 mM sodium formate, 2 mM benzyl viologen (dichloride), and cell extract	578	8.65×10^3
6. Acetaldehyde dehydrogenase (EC 1.2.1.4)	Assay as described in Table 5		
7. ADH (EC 1.1.1.1)	Assay as described in Table 5		
8. Phosphotransacetylase (EC 2.3.1.8)	Assay as described in Table 5		
9. Acetate kinase (EC 2.7.2.1)	Assay as described in Table 7		
10. PEP carboxylase (67, 102) (EC 4.1.1.31)	100 mM Tris-HCl (pH 6.5), 10 mM DTT, 0.15 mM NADH, 10 mM MgCl$_2$, 25 mM NaHCO$_3$, 1 U of malate dehydrogenase, 5 mM PEP, and cell extract	340	6.24×10^3
11. Malate dehydrogenase (107) (EC 1.1.1.37)	50 mM Tris (pH 7.6), 50 mM L-malate (pH 7.0), 2 mM NAD$^+$, and cell extract	340	6.24×10^3
12. Fumarase (39) (EC 4.2.1.2)	50 mM L-malate, 50 mM sodium phosphate buffer (pH 7.3), and cell extract	240	2.53×10^3
13. Fumarate reductase (14) (EC 1.3.99.1)	50 mM Tris-SO$_4$, 0.1 mM EDTA buffer, 250 μM benzyl viologen, 20 mM fumarate, and 20 mM glucose. The cuvette is evacuated and flushed with O$_2$-free argon before the addition of glucose oxidase and catalase to ensure complete removal of residual oxygen. Sodium dithionite (0.2 mM) is then added to reduce the benzyl viologen and incubated at 38°C for 3 min prior to the addition of enzyme to start the reaction.	578	8.65×10^3
14. α-Acetolactate synthase (106) (EC 2.2.1.6)	50 mM sodium acetate (pH 5.8), 40 mM sodium pyruvate, 0.87 mM TPP, 0.5 mM MnCl$_2$, and cell extract. After 25 min of incubation at 37°C, 0.9 ml of 2.5 N NaOH is added and 0.5 ml of this solution is mixed with 2 ml of solution containing creatine and α-naphthol (1:1). This reaction mixture is incubated for 20 min with intermittent shaking, and the reaction is terminated by adding 2.5 ml of 2.5 N NaOH.	540	

(Continued on next page)

TABLE 9 (Continued)

Enzyme	Assay mixture	ΔA (nm)	ε ($M^{-1}cm^{-1}$)
15. α-Acetolactate decarboxylase (106) (EC 4.1.1.5)	200 mM KPO$_4$ (pH 6.2), 10 mM DL-α-acetolactate, and cell extract. After 25 min of incubation at 37°C, 0.9 ml of 2.5 N NaOH is added and 0.5 ml of this solution is mixed with 2 ml of solution containing creatine and α-naphthol (1:1). This reaction mixture is incubated for 20 min with intermittent shaking. The reaction is terminated by adding 2.5 ml of 2.5 N NaOH.	540	
16. Acetoin reductase (106) (EC 1.1.1.4)		340	6.23 × 10^3
Reduction of acetoin	50 mM KPO$_4$ (pH 7.0), 5 mM acetoin, 0.1 mM NADH, and cell extract	340	6.23 × 10^3
Oxidation of 2,3-butanediol	50 mM KPO$_4$ (pH 7.0), 100 mM 2,3-butanediol, 0.1 mM NAD, and cell extract		

^{a}DTT, dithiothreitol; MOPS, morpholinepropanesulfonic acid; BSA, bovine serum albumin.

^{b}The sugar phosphates used are glucose-6-phosphate with [^{14}C]glucose, fructose-1-phosphate with [^{14}C]fructose, and mannitol-1-phosphate with [^{14}C]mannitol. The resin used for separating ^{14}C-sugar from ^{14}C-sugar-phosphate is AG1-X2, 50- to 100-mesh size, chloride form.

^{c}The tube is evacuated and filled with O$_2$-free argon, and 10 µl of 20 mM Fe(NH$_4$)$_2$(SO$_4$)$_2$ is then added to the first arm. The tube is preincubated under argon for 30 min. The contents of the two arms are mixed and the tube is illuminated by a daylight lamp. The enzyme is fully activated usually within 30 min. The larger chloroplast fragments are centrifuged and the mixture is stored at 0°C. The low-molecular-weight components of the activated enzyme solution are removed by gel filtration under argon with a Sephadex G-25 column (1.6 by 20 cm) with an anaerobic buffer containing 50 mM MOPS (pH 7.6), 9 mM DTT, and 0.2 mM Fe(NH$_4$)$_2$(SO$_4$)$_2$. Temperature, 4°C; flow rate, 2 ml/min. The protein fraction is collected into argon-flushed tubes and stored at 0°C.

TABLE 10 Assay procedures for the enzymes of the acrylate pathway for propionate production

Enzyme	Assay mixture	ΔA (nm)	ε ($M^{-1}cm^{-1}$)
1. Lactate racemase (40) (EC 5.1.2.1)	100 mM KPO$_4$ (pH 7.2), 0.4 mM NAD$^+$, 125 mM L-lactate, 2 U of D-lactate dehydrogenase, and cell extract	340	6.23 × 10^3
2. Propionyl-CoA transferase (10, 93) (EC 2.8.3.1)	111 mM KPO$_4$ (pH 7.0), 11 mM sodium acetate, 0.44 mM 5,5′-dithiobis(2-nitrobenzoate), 1.1 mM OAA, 0.2 mM propionyl-CoA, 2.2 U of citrate synthase, 11 mM D-lactate, and cell extract	412	1.36 × 10^4
3. Lactyl-CoA dehydratase (10) (EC 4.2.1.54)	50 mM KPO$_4$ (pH 7.5), 1 mM acrylyl-CoA, 10 mM sodium dithionite, 0.4 mM ATP, 10 mM MgSO$_4$, and cell extract. The reaction is terminated after 5 min by adding 20 µl of concentrated perchlorate, the product is centrifuged, and lactate in the supernatant is measured.		
4. Acrylyl-CoA reductase (38)	50 mM Tris-HCl (pH 7.0), 0.1 mM NADH, and cell extract. The reaction is started by the addition of 0.03 mM acrylyl-CoA under anaerobic conditions.	340	6.23 × 10^3
5. D-Lactate dehydrogenase (40, 63) (EC 1.1.1.28)			
NAD$^+$ independent	50 mM KPO$_4$ (pH 7.6), 0.15 mM 2,6-dichlorophenol indophenol, and 1 mM D-lactate or L-lactate, and cell extract or 100 mM KPO$_4$ (pH 7.0), 1 mM K$_3$Fe(CN)$_6$, 0.2 mg of BSA,^{a} 125 mM sodium-D-lactate, and cell extract	600 / 420	17.3 × 10^3
NAD dependent	13.2 mM KPO$_4$ (pH 7.5), 6.6 mM pyruvate, 0.12 mM NADH, and cell extract	340	6.23 × 10^3
6. Pyruvate-ferredoxin oxidoreductase (EC 1.2.7.1)	Assay as described in Table 7		
7. Phosphotransacetylase (EC 2.3.1.8)	Assay as described in Table 7		
8. Acetate kinase (EC 2.7.2.1)	Assay as described in Table 7		

^{a}BSA, bovine serum albumin.

TABLE 11 Assay procedures for the enzymes of the succinate pathway for propionate production

Enzyme	Assay mixture	ΔA (nm)	ε (M^{-1}cm^{-1})
1. Lactate dehydrogenase (EC 1.1.1.28)	Assay as described in Table 10		
2. Methylmalonyl-CoA-pyruvate trans-carboxylase (112) (EC 2.1.3.1)	350 mM KPO$_4$ (pH 6.8), 10 mM sodium pyruvate, 0.4 mM methylmalonyl-CoA, 0.6 mM NADH, and cell extract	340	6.23×10^3
3. Malate dehydrogenase (EC 1.1.1.37)	Assay as described in Table 9		
4. Fumarase (EC 4.2.1.2)	Assay as described in Table 9		
5. Fumarate reductase (67) (EC 1.3.1.6)	100 mM KPO$_4$ (pH 7.2), 5 mM fumarate, 0.15 mM NADH, and cell extract	340	6.23×10^3
6. CoA Transferase (119) (EC 2.8.3.8)	100 mM Tris-HCl (pH 7.5), 100 mM potassium butyrate (pH 7.5), 20 mM MgCl$_2$, 0.10 mM acetoacetyl-CoA, 5% (vol/vol) glycerol, and cell extract	233	4.44
7. (R)-Methylmalonyl-CoA mutase (49) (EC 5.4.99.2)	12.5 mM KPO$_4$ (pH 7.5), 3 mM glutathione (reduced), 6.25 mM sodium pyruvate, 6.25 mM sodium succinate, 0.05 U of malate dehydrogenase (diluted in 1% bovine albumin), 0.1 U of OAA transcarboxylase, 0.1 U of methylmalonyl-CoA racemase, 0.1 U of CoA transferase (enzymes are diluted in 50 mM phosphate buffer, pH 7.4), 0.125 mM NADH, 1.5 mM acetyl-CoA, 0.003 mM DBC,[a] and cell extract. All the reagents except DBC are mixed by inversion in the cuvette and incubated in a water bath with same temp as the spectrophotometric cell chamber for 5 min. DBC is then added under dim light.	340	6.23×10^3
8. Methylmalonyl-CoA racemase (2) (EC 5.1.99.1)	12.5 mM Tris-HCl (pH 7.8), 3 mM glutathione (reduced), 6.25 mM sodium pyruvate, 6.25 mM sodium succinate, 0.05 U of malate dehydrogenase, 0.125 mM NADH, 0.1 U of OAA transcarboxylase, 0.005 U of methylmalonyl-CoA mutase, 1.5 mM acetyl-CoA, 0.1 U of CoA transferase, 0.01 μM DBC, and cell extract. The reaction conditions are as described above for (R)-methylmalonyl-CoA mutase.	233	4.44
9. Pyruvate-ferredoxin oxidoreductase (EC 1.2.7.1)	Assay as described in Table 7		
10. Phosphotransacetylase (EC 2.3.1.8)	Assay as described in Table 7		
11. Acetate kinase (EC 2.7.2.1)	Assay as described in Table 7		

[a]DBC [(5,6-dimethyl benzimidazolyl) Co-5′-deoxyadenosine cobamide] should be stored in a light-proof container. Benzimidazolylcobamide or adenylcobamide can also be used as coenzymes.

TABLE 12 Assay procedures for the enzymes of the bacterial acetate fermentation pathway

Enzyme	Assay mixture	ΔA (nm)	ε (M^{-1}cm^{-1})
1. EMP pathway enzymes	Assay as described in Table 2		
2. Pyruvate-ferredoxin oxidoreductase (EC 1.2.7.1)	Assay as described in Table 7		
3. Phosphotransacetylase (EC 2.3.1.8)	Assay as described in Table 7		
4. Acetate kinase (EC 2.7.2.1)	Assay as described in Table 7		
5. Formate dehydrogenase (74) (EC 1.2.1.2)	100 mM triethanolamine hydrochloride (pH 7.5), 20 mM sodium formate, 3 mM DTT,[a] 1 mM NAD, or 1.4 mM methyl viologen and cell extract.[b]	340 / 600	6.23×10^3 / 1.13×10^4
6. Formyl-THF synthetase (74) (EC 6.3.4.3)	100 mM triethanolamine-hydrochloride (pH 8.0), 40 mM sodium formate, 5 mM (ATP), 10 mM magnesium chloride, 100 mM 2-mercaptoethanol, 40 mM Tris-HCl (pH 7.4), 1 mM THF (stored in the Tris-HCl buffer supplemented with mercaptoethanol), and cell extract. After 10 min of incubation, 1 ml of 0.36 N HCl is added to stop the reaction. The reaction conditions are as described for formate dehydrogenase.	350	2.49×10^4

(Continued on next page)

TABLE 12 (Continued)

Enzyme	Assay mixture	ΔA (nm)	ε ($M^{-1}cm^{-1}$)
7. Methenyl-THF cyclo-hydrolase (74) (EC 3.5.4.9)	200 mM potassium maleate, (pH 7.0), 80 mM 2-mercaptoethanol, 40 mM KOH, 0.2 mM 5,10-methenyl-THF, and cell extract. The reaction conditions are as described for formate dehydrogenase above.	356	24.9×10^3
8. 5,10-Methylene-THF dehydrogenase (74) (EC 1.5.1.5)	200 mM potassium maleate (pH 7.0), 2.4 mM NAD$^+$, 0.96 mM 5,10-methylene-THF, 240 mM 2-mercaptoethanol, and cell extract. The reaction measures the formation of NADH and 5,10-methylenetetra-hydrofolate. The reaction conditions are as described for formate de-hydrogenase above.	356	2.94×10^4
9. 5,10-Methylene-THF reductase (20) (EC 1.5.1.20)	100 mM KPO$_4$ (pH 7.2), 20 mM sodium ascorbate, 20 mM benzyl viologen, 0.13 mM methyl-THF, and cell extract. The assay is per-formed in argon-filled, serum-stoppered anaerobic cuvettes.	578/555	8.65×10^3
10. THF:B$_{12}$ methyltrans-ferase (83) (EC 2.1.1.13)	0.066 mM methyl-B$_{12}$, 0.3 mM H$_4$ folate, 5 mM DTT, and cell extract in a total volume of 0.8 ml	525	8.6 mM
11. CO dehydrogenase/ acetyl-CoA synthase (21, 80, 81) (EC 1.2.99.2)			
CO-MV reductase	100 mM KPO$_4$ buffer (pH 7.0), 30 mM methyl viologen, 3.2 mM DTT, and cell extract. O$_2$-free hydrogen gas or CO is bubbled into the assay mixture in a serum-stoppered cuvette at 60°C for 5 min. The nonspe-cific reduction of the dye is corrected by using nitrogen as the gas phase.	600	11.4×10^3
Exchange between 1-^{14}C-acetyl-CoA and CO	150 mM/5mM KPO$_4$/DTT (pH 6.0), 200 nmol/liter ^{14}C-acetyl-CoA, and 10 μg of ferredoxin II. The amount of CO exchanged into the C-1 of acetyl-CoA is determined by bubbling the acidified reaction mixture with nitrogen to remove ^{14}CO from the solution, and the ^{14}C in an aliquot of the reaction mixture is determined. The nanomoles of [1-^{14}C]acetyl-CoA exchanged with CO are calculated by subtracting the counts per minute left in the solution from the counts per minute in the acetyl-CoA at time zero.		
Exchange between ^{14}CO and C-1 of acetyl-CoA	100 mM KPO$_4$ (pH 6), 0.266 mM acetyl-CoA, 4 mM ATP, 1.2 mM Fe(NH$_4$)$_2$(SO$_4$)$_2$, 18.4 mM DTT, and cell extract. The assay mixture is bubbled with CO for 5 min, and then 200 μl of CO is replaced with 200 μl of ^{14}CO (1.4 mM, 4.7×10^6 cpm) and cell extract. Incubation is at 55°C for 3 min.^{c}		

^{a}DTT, dithiothreitol.

^{b}All reagents are prepared anaerobically by boiling the water containing reagents and storing them under nitrogen. The assay is performed in nitrogen-filled, serum-stoppered anaerobic cuvettes and the enzyme is added via a syringe. NAD$^+$ reduction is monitored by measuring the absorbance at 340 nm, and methyl viologen re-duction is measured at 600 nm.

^{c}The reaction is initiated with the addition of acetyl-CoA and is allowed to proceed for 15 min at 55°C. The reaction is stopped by adding 0.04 ml of 4 M acetic acid, which drops the pH to 3.5. The solution is then made alkaline with the addition of 0.2 ml of 2 N NaOH to hydrolyze acetyl-CoA over a period of 6 h at room temperature. Carrier acetic acid (0.0032 mM) is added and the acetate is then isolated using chromatography on Celite before assaying for ^{14}C.

TABLE 14 Assay procedures for the enzymes of the acetate fermentation pathway in archaea

Enzyme	Assay mixture	ΔA (nm)	ε ($M^{-1}cm^{-1}$)
1. Kinases			
Glucokinase (ADP) (50, 51) (EC 2.7.1.2)	100 mM Tris-HCl (pH 7.8), 10 mM Mg Cl$_2$, 0.5 mM NADP$^+$, 15 mM D-glucose, 2 mM ADP, 0.35 U of D-glucose-6-phosphate dehydrogenase, and cell extract	340	6.23×10^3
Glucokinase (ATP) (86) (EC 2.7.1.2)	50 mM Tris-HCl buffer (pH 8.0), 10 mM glucose, 2.5 mM ATP, 10 mM MgCl$_2$ · 6H$_2$O, 1 U of glucose-6-phosphate dehydrogenase, 1 mM NADP, and cell extract	340	6.23×10^3
Hexokinase (ATP) (27) (EC 2.7.1.1)	100 mM Tris-HCl (pH 7.5), 4 mM ATP, 4 mM MgCl$_2$, 1 mM NADP$^+$, 3 U of yeast glucose-6-phosphate dehydrogenase, 10 mM glucose, and cell extract. The reaction is carried out at 50°C.	340	6.23×10^3
2. Glucose-6-phosphate isomerase			
Formation of fructose-6-phosphate from glucose-6-phosphate (33) (EC 2.7.1.11)	100 mM Tris-HCl (pH 7.0), 40 mM glucose-6-phosphate, 3 mM ATP, 5 mM MgCl$_2$, 0.5 mM NADH, 1 U of PFK, 1 U of fructose bisphosphate aldolase, 50 U of triose phosphate isomerase, 9 U of glycerol-3-phosphate dehydrogenase, and cell extract. The reaction is carried out at 50°C.	340	6.23×10^3
Formation of glucose-6-phosphate from fructose-6-phosphate (33) (EC 2.7.1.11)	100 mM Tris-HCl (pH 7.0), 10 mM fructose-6-phosphate, 0.5 mM NADP$^+$, 0.3 U of glucose-6-phosphate dehydrogenase, and cell extract. The reaction is carried out at 50°C.	365/340	6.23×10^3
3. PFK (ATP) (34) (EC 2.7.1.11)			
Forward reaction (discontinuous)	100 mM Tris-HCl (pH 6.0), 10 mM fructose-6-phosphate, 2.5 mM ATP, and 12.5 mM MgCl$_2$. After preincubation at 85°C the reaction is started with an aliquot of cell extract (1 mg of enzyme), incubated for 1–20 min, and stopped by rapid addition of EDTA to a final concn of 50 mM. Fructose-1,6-bisphosphate formation is quantified by adding 0.6 mM NADH, 0.2 U of aldolase, 1 U of triose phosphate isomerase, and 0.3 U of glycerol-3-phosphate dehydrogenase.	365/340	6.23×10^3
Reverse reaction (continuous)	100 mM Tris-HCl (pH 6.0,), 20 mM fructose-1,6-bisphosphate, 5 mM ADP, and 25 mM MgCl$_2$. After preincubation at 85°C, the reaction is started with an aliquot of cell extract (3 mg), incubated for 1–30 min, and stopped by rapid addition of EDTA to a final concn of 50 mM. Fructose-6-phosphate formation is quantified by adding 0.6 mM NADP, 0.2 U of glucose-6-phosphate isomerase, and 0.2 U of glucose-6-phosphate dehydrogenase.	340	6.23×10^3
PFK (ADP) (50, 115) (EC 2.7.1.146)	100 mM MES buffer (pH 6.5), 10 mM MgCl$_2$, 10 mM fructose-6-phosphate, 0.2 mM NADH, 2.5 mM ADP, 3.9 U of glycerol-3-phosphate dehydrogenase, 11 U of triose phosphate isomerase, 0.23 U of aldolase, and cell extract	340	6.23×10^3
PFK (PP$_i$) (EC 2.7.1.90)			
Fructose-6-phosphate as substrate (99)	100 mM Tris-HCl (pH 7.0), 0.4 mM NADH, 1 mM fructose-6-phosphate, 5 mM potassium pyrophosphate, 2 U of fructose bisphosphate aldolase, 5 U of glycerol-3-phosphate dehydrogenase, 5 U of triose phosphate isomerase, and cell extract	340	6.23×10^3
Fructose-1,6-bisphosphate as substrate (99)	100 mM Tris-HCl (pH 7.0), 5 mM NADP$^+$, 10 mM fructose-1,6-bisphosphate, 5 mM potassium pyrophosphate, 3 U of glucose-6-phosphate dehydrogenase, 2 U of glucose-6-phosphate isomerase, and cell extract	340	6.23×10^3
4. Fructose bisphosphate aldolase (98) (EC 4.1.2.13)	50 mM Tris-HCl (pH 7.0), 0.2 mM NADH, 2.5 mM fructose bisphosphate, 4 U glycerol-3-phosphate dehydrogenase, 11 U of triose phosphate isomerase, and cell extract. The reaction is carried out at 50°C.	340	6.23×10^3
5. Triose phosphate isomerase (EC 5.3.1.1)			
Dihydroxyacetone phosphate as substrate (91)	0.1 M Tris-HCl (pH 7.3), 5 mM potassium arsenate, 8 mM NAD$^+$, 4 mM DHAP, 4 U of GAPDH, and cell extract	340	6.23×10^3

(Continued on next page)

TABLE 14 *(Continued)*

Enzyme	Assay mixture	ΔA (nm)	ε ($M^{-1}cm^{-1}$)
GAP as substrate (91)	100 mM HEPES (pH 7.3), 0.5 mM NADH, 4 mM GAP, 4 U of glycerol-3-phosphate dehydrogenase, and cell extract	340	6.23×10^3
6. GAPDH (NAD) (EC 1.2.1.12)			
Forward reaction (11)	100 mM HEPES-KOH buffer (pH 7), 200 mM KCl, 10 mM NAD^+, 2 mM DL-glyceraldehyde-3-phosphate, and cell extract	340	6.23×10^3
Reverse reaction (11)	100 mM HEPES-KOH buffer (pH 7), 200 mM KCl, 1 mM NADH, 0.5–10 mM 3-phosphoglycerate, 100–500 μM 1,3-bisphosphoglycerate, and cell extract	340	6.23×10^3
7. GAP:ferredoxin oxidoreductase (85) (EC 1.2.7.6)	2 ml of 100 mM EPPS buffer (pH 8.4), 50 μl of 100 mM benzyl viologen, a few microliters of 100 mM sodium dithionite, 250 μM GAP, and cell extract	578/600	7.4×10^3
8. Phosphoglycerate mutase (88) (EC 5.4.2.1)	100 mM Tris-HCl (pH 8.0), 5 mM 2-phosphoglycerate, 5 mM PEP, 0.5 mM ATP, 10 mM $MgCl_2$, 0.4 mM NADH, 4.5 U of phosphoglycerate kinase, 2.5 U of pyruvate kinase, 7.5 U of lactate dehydrogenase, and cell extract	340	6.23×10^3
9. Enolase (89) (EC 4.2.1.11)			
PEP formation	100 mM Tris-HCl (pH 8.0), 4 mM 2-phosphoglycerate, 3 mM $MgSO_4$, and cell extract (protein prepared in 50 mM glycylglycine/NaOH buffer, pH 8.0)	240	1.50?
PEP formation with NADH oxidation	100 mM Tris-HCl (pH 8.0), 2.5 mM 2-phosphoglycerate, 0.3 mM NADH, 3 mM $MgSO_4$, 5 mM ADP, 10 U of pyruvate kinase, 7.5 U of lactate dehydrogenase, and cell extract. The reaction is carried out at 60°C.	340	6.23×10^3
10. Pyruvate kinase (45) (EC 2.7.1.40)			
Discontinuous assay	100 mM Tris-HCl, 1–5 mM PEP, 2 mM ADP, 5–10 mM $MgCl_2$, and cell extract are incubated for 15–20 s, and the reaction is stopped by rapid addition of 750 μl of ice-cold buffer (100 mM Tris-HCl, pH 7.0), 0.6 mM NADH, and 0.5 U of lactate dehydrogenase	340	6.23×10^3
Continuous assay	100 mM triethanolamine (pH 7.0), 1 mM PEP, 2 mM ADP, 5 mM $MgCl_2$, 0.3 mM NADH, 1 U of lactate dehydrogenase, and cell extract. The reaction is carried out at 60°C.	340	6.23×10^3
11. Pyruvate-ferredoxin oxidoreductase (92) (EC 1.2.7.1)	50 mM EPPS buffer (pH 8.4), 1 mM methyl viologen, 2 mM $MgCl_2$, 0.4 mM thiamine PP_i, 0.1 mM CoA, 500 mM sodium pyruvate, and cell extract	578	8.65×10^3
12. ACS (ADP forming) (43, 64) (EC 6.2.1.13): coupled assay for acid production from CoA derivative			
ACS I	10 mM pyruvate, 5 mM $MgCl_2$, 0.4 mM thiamine PP_i, 0.025 mM CoA, 5 mM methyl viologen, 1 mM ADP, 10 mM K_2HPO_4, and 40 mg of pyruvate-ferredoxin oxidoreductase in 50 mM EPPS buffer (pH 8.4). Assays are carried out in serum-stoppered cuvettes under Ar.	600	9.78×10^3
ACS II	Same as above except that the assay mixture contained 40 mg of indole pyruvate-ferredoxin oxidoreductase instead of pyruvate-ferredoxin oxidoreductase and 10 mM of indole pyruvate instead of pyruvate	600	9.78×10^3

[a]DHAP, dihydroxyacetone phosphate; EPPS, N-(2-hydroxyethyl) piperazine-N9-3-propanesulfonic acid; ACS, acetyl-CoA synthetase.

22.12. REFERENCES

1. **Aboulwafa, M., and M. H. Saier, Jr.** 2002. Dependency of sugar transport and phosphorylation by the phosphoenolpyruvate-dependent phosphotransferase system on membranous phosphatidyl glycerol in *Escherichia coli*: studies with a *pgsA* mutant lacking phosphatidyl glycerophosphate synthase. *Res. Microbiol.* **153:**667–677.

2. **Allen, S. H. G., R. W. Kellermeyer, and H. G. Wood.** 1969. Methylmalonyl CoA racemase from *Propionibacterium shermanii. Methods Enzymol.* **13:**194–198.

3. **Andersch, W., H. Bahl, and G. Gottschalk.** 1983. Level of enzymes involved in acetate, butyrate, acetone and butanol formation by *Clostridium acetobutylicum. Eur. J. Appl. Microbiol. Biotechnol.* **18:**327–332.

4. **Andersen, H. W., M. B. Pedersen, K. Hammer, and P. R. Jensen.** 2001. Lactate dehydrogenase has no control on lactate production but has a strong negative control on formate production in *Lactococcus lactis. Eur. J. Biochem.* **268:**6379–6389.

5. **Arfman, N., V. Worrell, and L. O. Ingram.** 1992. Use of the *tac* promoter and *lacI*^{q} for the controlled expression of *Zymomonas mobilis* fermentative genes in *Escherichia coli* and *Zymomonas mobilis. J. Bacteriol.* **174:**7370–7378.

6. **Axley, M. J., D. A. Grahame, and T. C. Stadtman.** 1990. *Escherichia coli* formate-hydrogen lyase. Purification and properties of the selenium-dependent formate dehydrogenase component. *J. Biol. Chem.* **265:**18213–18218.

6a. **Barabote, R. D., and M. H. Saier, Jr.** 2005. Comparative genomic analyses of the bacterial phosphotransferase system. *Microbiol. Mol. Biol. Rev.* **69:**608–634.

7. **Blusson, H., H. Petitdemange, and R. Gay.** 1981. A new, fast and sensitive assay for NADH-ferredoxin oxidoreductase detection in *Clostridia. Anal. Biochem.* **110:**176–181.

8. **Brockman, H. L., and W. A. Wood.** 1975. D-Lactate dehydrogenase of *Peptostreptococcus elsdenii. J. Bacteriol.* **124:**1454–1461.

9. **Brown, T. D., M. C. Jones–Mortimer, and H. L. Kornberg.** 1977. The enzymic interconversion of acetate and acetyl-coenzyme A in *Escherichia coli. J. Gen. Microbiol.* **102:**327–336.

10. **Brunelle, S. L., and R. H. Abeles.** 1993. The stereochemistry of hydration of acryly-CoA catalyzed by lactyl-CoA dehydratase. *Bioorg. Chem.* **21:**118–126.

11. **Brunner, N. A., and R. Hensel.** 2001. Nonphosphorylating glyceraldehyde-3-phosphate dehydrogenase from *Thermoproteus tenax. Methods Enzymol.* **331:**117–131.

12. **Bunch, P. K., F. Mat-Jan, N. Lee, and D. P. Clark.** 1997. The ldhA gene encoding the fermentative lactate dehydrogenase of *Escherichia coli. Microbiology* **143:**187–195.

13. **Cary, J. W., D. J. Petersen, E. T. Papoutsakis, and G. N. Bennett.** 1990. Cloning and expression of *Clostridium acetobutylicum* ATCC 824 acetoacetyl-coenzyme A:acetate/butyrate:coenzyme A-transferase in *Escherichia coli. Appl. Environ. Microbiol.* **56:**1576–1583.

14. **Cecchini, G., B. A. Ackrell, J. O. Deshler, and R. P. Gunsalus.** 1986. Reconstitution of quinone reduction and characterization of *Escherichia coli* fumarate reductase activity. *J. Biol. Chem.* **261:**1808–1814.

15. **Chaillou, S., P. W. Postma, and P. H. Powels.** 2001. Contribution of the phosphoenolpyruvate:mannose phosphotransferase system to carbon catabolite repression in *Lactobacillus pentosus. Microbiology* **147:**671–679.

16. **Chen, J. S.** 1995. Alcohol dehydrogenase: multiplicity and relatedness in the solvent-producing clostridia. *FEMS Microbiol. Rev.* **17:**263–273.

17. **Christensen, L. H., U. Schulze, J. Nielsen, and J. Villadsen.** 1995. Acoustic off-gas analyser for bioreactors: precision, accuracy and dynamics of detection. *Chem. Eng. Sci.* **50:**2601–2610.

18. **Clark, D. P.** 1989. The fermentation pathways of *Escherichia coli. FEMS Microbiol. Rev.* **5:**223–234.

19. **Clark, D. P., and J. E. Cronan, Jr.** 1980. Acetaldehyde coenzyme A dehydrogenase of *Escherichia coli. J. Bacteriol.* **144:**179–184.

20. **Clark, J. E., and L. G. Ljungdahl.** 1984. Purification and properties of 5,10-methylenetetrahydrofolate reductase, an iron-sulfur flavoprotein from *Clostridium formicoaceticum. J. Biol. Chem.* **259:**10845–10849.

21. **Clark, J. E., S. W. Ragsdale, L. G. Ljungdahl, and J. Wiegel.** 1982. Levels of enzymes involved in the synthesis of acetate from CO_2 in *Clostridium thermoautotrophicum. J. Bacteriol.* **151:**507–509.

22. **Conway, T., G. W. Sewell, Y. A. Osman, and L. O. Ingram.** 1987. Cloning and sequencing of the alcohol dehydrogenase II gene from *Zymomonas mobilis. J. Bacteriol.* **169:**2591–2597.

23. **Davies, R., and M. Stephenson.** 1941. Studies on the acetone butyl alcohol fermentation. 1. Nutritional and other factors involved in the preparation of active suspension of *Cl. acetobutylicum* (Wiezmann). *Biochem. J.* **35:**1320–1331.

24. **Dawes, E. A., D. J. McGill, and M. Midgley.** 1971. Analysis of fermentation products. *Methods Microbiol.* **6A:**63–215.

25. **Dimroth, P., and B. Schink.** 1988. Energy conservation in the decarboxylation of dicarboxylic acids by fermenting bacteria. *Arch. Microbiol.* **170:**69–77.

26. **Dombek, K. M., and L. O. Ingram.** 1987. Ethanol production during batch fermentation with *Saccharomyces cerevisiae*: changes in glycolytic enzymes and internal pH. *Appl. Environ. Microbiol.* **53:**1286–1291.

27. **Dorr, C., M. Zaparty, B. Tjaden, H. Brinkmann, and B. Siebers.** 2003. The hexokinase of the hyperthermophile *Thermoproteus tenax.* ATP-dependent hexokinases and ADP-dependent glucokinases, two alternatives for glucose phosphorylation in Archaea. *J. Biol. Chem.* **278:**18744–18753.

28. **Duree, P., and H. Bahl.** 1996. Microbial production of acetone/butanol/isopropanol, p. 230–267. *In* H. G. Rehm and G. Reed (ed.), *Biotechnology*, 2nd ed, vol. 6. VCH, New York, NY.

29. **Entian, K. D.** 1980. Genetic and biochemical evidence for hexokinase PII as a key enzyme involved in carbon catabolite repression in yeast. *Mol. Gen. Genet.* **178:**633–637.

30. **Fernandez, R., P. Herrero, M. T. Fernandez, and F. Moreno.** 1986. Mechanism of inactivation of hexokinase PII of *Saccharomyces cerevisiae* by D-xylose. *J. Gen. Microbiol.* **132:**3467–3472.

31. **Gancedo, J. M., D. Clifton, and D. G. Fraenkel.** 1977. Yeast hexokinase mutants. *J. Biol. Chem.* **252:**4443–4444.

32. **Gottschalk, G.** 1985. *Bacterial Metabolism.* Springer-Verlag, New York, NY.

33. **Hansen, T., M. Oehlmann, and P. Schönheit.** 2001. Novel type of glucose-6-phosphate isomerase in the hyperthermophilic archaeon *Pyrococcus furiosus. J. Bacteriol.* **183:**3428–3435.

34. **Hansen, T., and P. Schonheit.** 2000. Purification and properties of the first-identified archaeal, ATP-dependent 6-phosphofructokinase, an extremely thermophilic nonallosteric enzyme, from the hyperthermophile *Desulfurococcus amylolyticus. Arch. Microbiol.* **173:**103–109.

35. **Hansen, T., and P. Schonheit.** 2001. Sequence, expression, and characterization of the first archaeal ATP-dependent 6-phosphofructokinase, a non-allosteric enzyme related to the phosphofructokinase-B sugar kinase family, from the hyperthermophilic crenarchaeote *Aeropyrum pernix. Arch. Microbiol.* **177:**62–69.

36. **Hartmanis, M. G., and S. Gatenback.** 1984. Intermediary metabolism in *Clostridium acetobutylicum*: levels of enzymes involved in the formation of acetate and butyrate. *Appl. Environ. Microbiol.* **47**:1277–1283.

37. **Herrero, P., J. Galindez, N. Ruiz, C. Martinez-Campa, and F. Moreno.** 1995. Transcriptional regulation of the *Saccharomyces cerevisiae* HXK1, HXK2 and GLK1 genes. *Yeast* **11**:137–144.

38. **Hetzel, M., M. Brock, T. Selmer, A. J. Pierik, B. T. Golding, and W. Buckel.** 2003. Acryloyl-CoA reductase from *Clostridium propionicum*. An enzyme complex of propionyl-CoA dehydrogenase and electron-transferring flavoprotein. *Eur. J. Biochem.* **270**:902–910.

39. **Hill, R. L., and R. H. Bradshaw.** 1969. Fumarase. *Methods Enzymol.* **13**:91–99.

40. **Hino, T., and S. Kuroda.** 1993. Presence of lactate dehydrogenase and lactate racemase in *Megasphaera elsdenii* grown on glucose or lactate. *Appl. Environ. Microbiol.* **59**:255–259.

41. **Holland-Staley, C. A., K. Lee, D. P. Clark, and P. R. Cunningham.** 2000. Aerobic activity of *Escherichia coli* alcohol dehydrogenase is determined by a single amino acid. *J. Bacteriol.* **182**:6049–6054.

42. **Horvath, S. I., C. J. Franzen, M. J. Taherzadeh, C. Niklasson, and G. Liden.** 2003. Effects of furfural on the respiratory metabolism of *Saccharomyces cerevisiae* in glucose-limited chemostats. *Appl. Environ. Microbiol.* **69**:4076–4086.

43. **Hutchins, A. M., X. Mai, and M. W. W. Adams.** 2001. Acetyl-CoA synthetases I and II from *Pyrococcus furiosus*. *Methods Enzymol.* **331**:158–167.

44. **Ismaiel, A. A., C. X. Zhu, G. D. Colby, and J. S. Chen.** 1993. Purification and characterization of a primary secondary alcohol dehydrogenase from two strains of *Clostridium beijerinckii*. *J. Bacteriol.* **175**:5097–5105.

45. **Johnsen, U., T. Hansen, and P. Schonheit.** 2003. Comparative analysis of pyruvate kinases from the hyperthermophilic archaea *Archaeglobus fulgidus, Aeropyrum pernix*, and *Pyrobaculum aerophilum* and the hyperthermophilic bacterium *Thermotoga maritima*: unusual regulatory properties in hyperthermophilic archaea. *J. Biol. Chem.* **278**:25417–25427.

46. **Jones, D. T., and D. R. Woods.** 1986. Acetone-butanol fermentation revisited. *Microbiol. Rev.* **50**:484–524.

47. **Jungermann, K., R. K. Thauer, E. Rupprecht, C. Ohrloff, and K. Decker.** 1969. Ferredoxin mediated hydrogen formation from NADPH in a cell-free system of *Clostridium kluyveri*. *FEBS Lett.* **3**:144–146.

48. **Kashefi, K., and D. R. Lovley.** 2003. Extending the upper temperature limit of life. *Science* **301**:934.

49. **Kellermeyer, R. W., and H. G. Wood.** 1969. 2-Methylmalonyl CoA mutase from *Propionibacterium shermanii*. *Methods Enzymol.* **13**:207–215.

50. **Kengen, S. W., J. E. Tuininga, C. H. Verhees, J. van der Oost, A. J. Stams, and W. M. de Vos.** 2001. ADP-dependent glucokinase and phosphofructokinase from *Pyrococcus furiosus*. *Methods Enzymol.* **331**:41–53.

51. **Kengen, S. W. M., J. E. Tuininga, F. A. M. de Bok, A. J. M. Stams, and W. M. de Vos.** 1995. Purification and characterization of a novel ADP-dependent glucokinase from the hyperthermophilic archaeon *Pyrococcus furiosus*. *J. Biol. Chem.* **270**:30453–30457.

52. **Keshav, K. F., L. P. Yomano, H. J. An, and L O. Ingram.** 1990. Cloning of the *Zymomonas mobilis* structural gene encoding alcohol dehydrogenase I (*adhA*): sequence comparison and expression in *Escherichia coli*. *J. Bacteriol.* **172**:2491–2497.

53. **Knappe, J., H. P. Blaschkowski, P. Grobner, and T. Schmitt.** 1974. Pyruvate formate lyase of *Escherichia coli*: the acetyl-enzyme intermediate. *Eur. J. Biochem.* **50**:253–263.

54. **Koga, S., I. Yoshioka, H. Sakuraba, M. Takahashi, S. Sakasegawa, S. Shimizu, and T. Ohshima.** 2000. Biochemical characterization, cloning, and sequencing of ADP-dependent (AMP-forming) glucokinase from two hyperthermophilic archaea, *Pyrococcus furiosus* and *Thermococcus litoralis*. *J. Biochem.* (Tokyo) **128**:1079–1085.

55. **Kundig, W., S. Ghosh, and S. Roseman.** 1964. Phosphate bound to histidine in a protein as an intermediate in a novel phosphotransferase system. *Proc. Natl. Acad. Sci. USA* **52**:1067–1074.

56. **Kundig, W., and S. Roseman.** 1971. Sugar transport. II. Characterization of constitutive membrane-bound enzymes II of the *Escherichia coli* phosphotransferase system. *J. Biol. Chem.* **246**:1407–1418.

57. **Labes, A., and P. Schonheit.** 2003. ADP-dependent glucokinase from the hyperthermophilic sulfate-reducing archaeon *Archaeoglobus fulgidus* strain 7324. *Arch. Microbiol.* **180**:69–75.

58. **Lamed, R., and J. G. Zeikus.** 1980. Glucose fermentation pathway of *Thermoanaerobacterium brockii*. *J. Bacteriol.* **141**:1251–1257.

59. **Leonardo, M. R., P. R. Cunningham, and D. P. Clark.** 1993. Anaerobic regulation of the *adhE* gene, encoding the fermentative alcohol dehydrogenase of *Escherichia coli*. *J. Bacteriol.* **175**:870–878.

60. **Ljungdahl, L. G.** 1986. The autotrophic pathway of acetate synthesis in acetogenic bacteria. *Annu. Rev. Microbiol.* **40**:415–450.

61. **Lobo, Z., and P. K. Maitra.** 1977. Physiological role of glucose-phosphorylating enzymes in *Saccharomyces cerevisiae*. *Arch. Biochem. Biophys.* **182**:639–645.

62. **Mackenzie, K. F., C. K. Eddy, and L. O. Ingram.** 1989. Modulation of alcohol dehydrogenase isoenzyme levels in *Zymomonas mobilis* by iron and zinc. *J. Bacteriol.* **171**:1063–1067.

63. **Macy, J. M., L. G. Ljungdahl, and G. Gottschalk.** 1978. Pathway of succinate and propionate formation in *Bacteroides fragilis*. *J. Bacteriol.* **134**:84–91.

64. **Mai, X., and M. W. Adams.** 1996. Purification and characterization of two reversible and ADP-dependent acetyl coenzyme A synthetases from the hyperthermophilic archaeon *Pyrococcus furiosus*. *J. Bacteriol.* **178**:5897–5903.

65. **Maitra, P. K., and Z. Lobo.** 1971. A kinetic study of glycolytic enzyme synthesis in yeast. *J. Biol. Chem.* **246**:475–488.

66. **Meile, L., L. M. Rohr, T. A. Geissmann, M. Herensperger, and M. Teuber.** 2001. Characterization of the D-xylulose 5-phosphate/D-fructose 6-phosphate phosphoketolase gene (*xfp*) from *Bifidobacterium lactis*. *J. Bacteriol.* **183**:2929–2936.

67. **Melville, S. B., T. A. Michel, and J. M. Macy.** 1988. Pathway and sites for energy conservation in the metabolism of glucose by *Selenomonas ruminantium*. *J. Bacteriol.* **170**:5298–5304.

68. **Mitchell, W. J.** 1998. Physiology of carbohydrate to solvent conversion by clostridia. *Adv. Microb. Physiol.* **39**:31–130.

69. **Mitchell, W. J., J. E. Shaw, and L. Andrews.** 1991. Properties of the glucose phosphotransferase system of *Clostridium acetobutylicum* NCIB 8052. *Appl. Environ. Microbiol.* **57**:2534–2539.

70. **Muller, V.** 2003. Energy conservation in acetogenic bacteria. *Appl. Environ. Microbiol.* **69**:6345–6353.

71. **Neale, A. D., R. K. Scopes, J. M. Kelly, and R. E. H. Wettenhall.** 1986. The two alcohol dehydrogenases of *Zymomonas mobilis*. *Eur. J. Biochem.* **154**:119–124.

72. **Neale, A. D., R. K. Scopes, R. E. H. Wettenhall, and N. J. Hoogenraad.** 1987. Pyruvate decarboxylase of

Zymomonas mobilis: isolation, properties, and genetic expression in *Escherichia coli*. *J. Bacteriol.* **169**:1024–1028.

73. **O'Brien, W. E., S. Bowien, and H. G. Wood.** 1975. Isolation and characterization of a pyrophosphate-dependent phosphofructokinase from *Propionibacterium shermanii*. *J. Biol. Chem.* **250**:8690–8695.

74. **O'Brien, W. E., and L. G. Ljungdahl.** 1972. Fermentation of fructose and synthesis of acetate from carbon dioxide by *Clostridium formicoaceticum*. *J. Bacteriol.* **109**:626–632.

75. **Pawluk, A., R. K. Scopes, and K. Griffiths-Smith.** 1986. Isolation and properties of the glycolytic enzymes from *Zymomonas mobilis*. *Biochem. J.* **238**:275–281.

76. **Poolman, B.** 1993. Energy transduction in lactic acid bacteria. *FEMS Microbiol. Rev.* **12**:125–147.

77. **Posthuma, C. C., R. Bader, R. Engelmann, P. W. Postma, W. Hengstenberg, and P. H. Pouwels.** 2002. Expression of the xylulose 5-phosphate phosphoketolase gene, *xpkA*, from *Lactobacillus pentosus* MD363 is induced by sugars that are fermented via the phosphoketolase pathway and is repressed by glucose mediated by CcpA and the mannose phosphoenolpyruvate phosphotransferase system. *Appl. Environ. Microbiol.* **68**:831–837.

78. **Postma, P. W., J. W. Lengeler, and G. R. Jacobson.** 1993. Phosphoenolpyruvate:carbohydrate phosphotransferase system of bacteria. *Microbiol. Rev.* **57**:543–594.

79. **Ragsdale, S. W.** 1992. Enzymology of the acetyl-CoA pathway of autotrophic CO_2 fixation. *Crit. Rev. Biochem. Mol. Biol.* **26**:261–300.

80. **Ragsdale, S. W., J. E. Clark, L. G. Ljungdahl, L. L. Lundie, and H. L. Drake.** 1983. Properties of purified carbon monoxide dehydrogenase from *Clostridium thermoaceticum*, a nickel, iron-sulfur protein. *J. Biol. Chem.* **258**:2364–2369.

81. **Ragsdale, S. W., and H. G. Wood.** 1985. Acetate biosynthesis by acetogenic bacteria. Evidence that carbon monoxide dehydrogenase is the condensing enzyme that catalyzes the final steps of the synthesis. *J. Biol. Chem.* **260**:3970–3977.

82. **Reeves, R. E., D. J. South, H. J. Blytt, and L. G. Warren.** 1974. Pyrophosphate:D-fructose 6-phosphate 1-phosphotransferase. A new enzyme with the glycolytic function of 6-phosphofructokinase. *J. Biol. Chem.* **249**:7737–7741.

83. **Roberts, D. L., S. Zhao, T. Doukov, and S. W. Ragsdale.** 1994. The reductive acetyl coenzyme A pathway: sequence and heterologous expression of active methyltetrahydrofolate:corrinoid/iron-sulfur protein methyltransferase from *Clostridium thermoaceticum*. *J. Bacteriol.* **176**:6127–6130.

84. **Ronimus, R. S., E. de Heus, and H. W. Morgan.** 2001. Sequencing, expression, characterisation and phylogeny of the ADP-dependent phosphofructokinase from the hyperthermophilic, euryarchaeal *Thermococcus zilligii*. *Biochim. Biophys. Acta* **1517**:384–391.

85. **Roy, R., A. L. Menon, and M. W. W. Adams.** 2001. Aldehyde oxidoreductases from *Pyrococcus furiosus*. *Methods Enzymol.* **331**:132–143.

86. **Sakuraba, H., Y. Mitani, S. Goda, Y. Kawarabayasi, and T. Ohshima.** 2003. Cloning, expression, and characterization of the first archaeal ATP-dependent glucokinase from aerobic hyperthermophilic archaeon *Aeropyrum pernix*. *J. Biochem.* (Tokyo) **133**:219–224.

87. **Sakuraba, H., I. Yoshioka, S. Koga, M. Takahashi, Y. Kitahama, T. Satomura, R. Kawakami, and T. Ohshima.** 2002. ADP-dependent glucokinase/phosphofructokinase, a novel bifunctional enzyme from the hyperthermophilic archaeon *Methanococcus jannaschii*. *J. Biol. Chem.* **277**:12495–12498.

88. **Schafer, T., and P. Schonheit.** 1992. Maltose fermentation to acetate, CO_2 and H_2 in the anaerobic hyper-
therophilic archaeon *Pyrococcus furiosus*: evidence for the operation of a novel sugar fermentation pathway. *Arch. Microbiol.* **158**:188–202.

89. **Schafer, T., and P. Schonheit.** 1993. Gluconeogenesis from pyruvate in the hyperthermophilic archaeon *Pyrococcus furiosus*: involvement of reactions of the Embden-Meyerhof pathway. *Arch. Microbiol.* **159**:354–363.

90. **Schonheit, P., and T. Schafer.** 1995. Metabolism of hyperthermophiles. *World J. Microbiol. Biotechnol.* **11**:26–57.

91. **Schramm, A., M. Kohlhoff, and R. Hensel.** 2001. Triose-phosphate isomerase from *Pyrococcus woesei* and *Methanothermus fervidus*. *Methods Enzymol.* **331**:62–77.

92. **Schut, G. J., A. L. Menon, and M. W. Adams.** 2001. 2-Keto acid oxidoreductases from *Pyrococcus furiosus* and *Thermococcus litoralis*. *Methods Enzymol.* **331**:144–158.

93. **Schweiger, G., and W. Buckel.** 1984. On the dehydration of (R)-lactate in the fermentation of alanine to propionate by *Clostridium propionicum*. *FEBS Lett.* **171**:79–84.

94. **Scopes, R. K.** 1984. Use of differential dye-ligand chromatography with affinity elution for enzyme purification: 2-keto-3-deoxy-6-phosphogluconate aldolase from *Zymomonas mobilis*. *Anal. Biochem.* **136**:525–529.

95. **Scopes, R. K., and K. Griffiths-Smith.** 1984. Use of differential dye-ligand chromatography with affinity elution for enzyme purification: 6-phosphogluconate dehydratase from *Zymomonas mobilis*. *Anal. Biochem.* **136**:530–534.

96. **Scopes, R. K., V. Testolin, A. Stoter, K. Griffiths-Smith, and E. M. Algar.** 1985. Simultaneous purification and characterization of glucokinase, fructokinase, and glucose-6-phosphate dehydrogenase from *Zymomonas mobilis*. *Biochem. J.* **228**:627–634.

97. **Selig, M., K. B. Xavier, H. Santos, and P. Schonheit.** 1997. Comparative analysis of Embden-Meyerhof and Entner-Doudoroff glycolytic pathways in hyperthermophilic archaea and the bacterium *Thermotoga*. *Arch. Microbiol.* **167**:217–232.

98. **Siebers, B., H. Brinkmann, C. Dörr, B. Tjaden, H. Lilie, J. van der Oost, and C. H. Verhees.** 2001. Archaeal fructose-1,6-bisphosphate aldolases constitute a new family of archaeal type class I aldolase. *J. Biol. Chem.* **276**:28710–28718.

99. **Siebers, B., and R. Hensel.** 2001. Pyrophosphate-dependent phosphofructokinase from *Thermoproteus tenax*. *Methods Enzymol.* **331**:54–62.

100. **Siebers, B., H. P. Klenk, and R. Hensel.** 1998. PPi-dependent phosphofructokinase from *Thermoproteus tenax*, an archaeal descendant of an ancient line in phosphofructokinase evolution. *J. Bacteriol.* **180**:2137–2143.

101. **Sprenger, G. A., U. Schorken, G. Sprenger, and H. Sahm.** 1995. Transketolase A of *Escherichia coli* K12. *Eur. J. Biochem.* **230**:525–532.

102. **Sridhar, J., M. A. Eiteman, and J. W. Wiegel.** 2000. Elucidation of enzymes in fermentation pathways used by *Clostridium thermosuccinogenes* growing on inulin. *Appl. Environ. Microbiol.* **66**:246–251.

103. **Stanier, R. Y., J. L. Ingraham, M. L. Wheelis, and P. R. Painterm.** 1986. *The Microbial World.* Prentice Hall, NJ.

104. **Stetter, K. O.** 1988. Hyperthermophiles—physiology and enzymes. *J. Chem. Technol. Biotechnol.* **42**:315–317.

105. **Stetter, K. O.** 1999. Extremophiles and their adaptation to hot environments. *FEBS Lett.* **452**:22–25.

106. **Stormer, F. C.** 1975. 2,3-Butanediol biosynthetic system in *Aerobacter aerogenes*. *Methods Enzymol.* **41**:518–532.

107. **Sutherland, P., and L. McAlister-Henn.** 1985. Isolation and expression of the *Escherichia coli* gene encoding malate dehydrogenase. *J. Bacteriol.* **163**:1074–1079.

108. **Taherzadeh, M. J., L. Gustafsson, C. Niklasson, and G. Liden.** 2000. Inhibition effects of furfural on aerobic batch cultivation of *Saccharomyces cerevisiae* growing on ethanol and/or acetic acid. *J. Biosci. Bioeng.* **90:**374–380.

109. **Thauer, R. K., K. Jungermann, and K. Decker.** 1977. Energy conservation in chemotrophic anaerobic bacteria. *Bacteriol. Rev.* **41:**100–180.

110. **Thauer, R. K., K. Jungermann, E. Rupprecht, and K. Decker.** 1970. Hydrogen formation from NADH in cell-free extracts of *Clostridium kluyveri.* Acetyl coenzyme A requirement and ferredoxin dependence. *FEBS Lett.* **4:**108–112.

111. **Thimann, K. V.** 1963. *The Life of Bacteria.* The Macmillan Company, New York, NY.

112. **Thompson, T. E., and J. G. Zeikus.** 1988. Regulation of carbon and electron flow in *Propionispira arboris:* relationship of catabolic enzyme levels to carbon substrates fermented during propionate formation via the methyl-malonyl coenzyme A pathway. *J. Bacteriol.* **170:**5298–5304.

113. **Toh, H., and H. Doelle.** 1997. Changes in the growth and enzyme level of *Zymomonas mobilis* under oxygen-limited conditions at low glucose concentration. *Arch. Microbiol.* **168:**46–52.

114. **Tsolas, O., and B. L. Horecker.** 1970. Isoenzymes of transaldolase in *Candida utilis. Arch. Biochem. Biophys.* **136:**287–302.

115. **Tuininga, J. E., C. H. Verhees, J. van der Oost, and W. M. de Vos.** 1999. Molecular and biochemical characterization of the ADP-dependent phosphofructokinase from the hyperthermophilic archaeon *Pyrococcus furiosus. J. Biol. Chem.* **274:**21023–21028.

116. **Verhees, C. H., J. E. Tuininga, S. W. Kengen, A. J. Stams, J. van der Oost, and W. M. de Vos.** 2001. ADP-dependent phosphofructokinases in mesophilic and thermophilic methanogenic archaea. *J. Bacteriol.* **183:**7145–7153.

117. **Wang, H., and R. P. Gunsalus.** 2003. Coordinate regulation of the *Escherichia coli* formate dehydrogenase *fdnGHI* and *fdhF* genes in response to nitrate, nitrite, and formate: roles for NarL and NarP. *J. Bacteriol.* **185:**5076–5085.

118. **White, D.** 2000. *The Physiology and Biochemistry of Prokaryotes,* 2nd ed. Oxford University Press, New York, NY.

119. **Wiesenborn, D. P., F. B. Rudolph, and E. T. Papoutsakis.** 1988. Thiolase from *Clostridium acetobutylicum* ATCC 824 and its role in the synthesis of acids and solvents. *Appl. Environ. Microbiol.* **54:**2717–2722.

120. **Wiesenborn, D. P., F. B. Rudolph, and E. T. Papoutsakis.** 1989. Coenzyme A transferase from *Clostridium acetobutylicum* ATCC 824 and its role in the uptake of acids. *Appl. Environ. Microbiol.* **55:**323–329.

121. **Wood, H. G., S. W. Ragsdale, and E. Pezacka.** 1986. The acetyl-CoA pathway of autotrophic growth. *FEMS Microbiol. Rev.* **39:**345–362.

122. **Wood, T.** 1970. Spectrophotometric assay for D-ribose-5-phosphate ketol isomerase and for D-ribulose-5-P 3-epimerase. *Anal. Biochem.* **33:**297–306.

123. **Wood, W. A.** 1961. Fermentation of carbohydrates and related compounds, p. 59–149. *In* I. C. Gunsalus and R. Y. Stanier (ed.), *The Bacteria: a Treatise of Structure and Function,* vol. 2. Academic Press, New York, NY.

124. **Woods, S. A., S. D. Schwartzbach, and J. R. Guest.** 1988. Two biochemically distinct classes of fumarase in *Escherichia coli. Biochim. Biophys. Acta* **954:**14–26.

125. **Xavier, K. B., M. S. da Costa, and H. Santos.** 2000. Demonstration of a novel glycolytic pathway in the hyperthermophilic archaeon *Thermococcus zilligii* by (13)C-labeling experiments and nuclear magnetic resonance analysis. *J. Bacteriol.* **182:**4632–4636.

23

Metabolism of Aromatic Compounds

JEROME J. KUKOR, BORIS WAWRIK, AND GERBEN J. ZYLSTRA

There has been a long-standing fundamental and practical interest in microbial metabolism of aromatic compounds. The topic is of interest to ecologists and soil scientists who are concerned with organic matter sequestration and turnover as part of the global carbon cycle; to environmental engineers who want to design treatment systems to remediate aquifers contaminated with aromatic hydrocarbon-containing fuels or solvents; to geneticists and evolutionary biologists who use genes encoding aromatic biodegradative pathways as models to understand transcriptional control, lateral gene transfer, and patchwork assembly of catabolic pathways; to biochemists who seek to probe the mechanisms by which the stable aromatic nucleus can be broken; and to biotechnologists in their quest to bioprospect for—or construct via molecular rearrangement—novel enzymatic catalysts that can be used to specifically modify chemical structures.

Given this long-standing interest in the metabolism of aromatic compounds from such a diverse array of scientists, it is not surprising that the publications focused on the topic are legion. A perusal of the Biocatalysis and Biodegradation Database (http://umbbd.msi.umn.edu/) allows one to readily access several thousand publications in PubMed that deal with specific enzymes or pathway intermediates related to aromatic compound biodegradation. A single chapter that would cover all methods for all aromatic compounds is an impossibility. Instead, the goal of this chapter is to provide an overview of the generally used, well-established methods that are the hallmark of aromatic metabolism studies. To do this, we have chosen to focus on toluene as a model aromatic substrate. In addition, we restrict our focus to methods developed to study metabolism of aromatics under aerobic conditions.

23.1. GROWTH OF BACTERIA WITH VOLATILE AROMATIC SUBSTRATES

Bacteria variably tolerate the presence of aromatic compounds in culture medium, where the toxicity of the aromatic compound is somewhat dependent on the logP value (the logarithm of the partition coefficient in an n-octanol–water two-phase system) and the propensity of the substrate to disrupt the cell membrane and cause lysis. In fact, many different types of bacteria have evolved mechanisms to tolerate aromatic compounds and other solvents, mostly involving metabolic changes resulting in membrane strengthening and efflux pumps to remove the compound from the membrane (25, 30, 46). Thus, it is not unexpected that direct addition of liquid or solid aromatic compounds to growth media can cause erratic or unpredictable bacterial growth due to toxicity of high concentrations of the substrate. The most obvious way to overcome this problem is to initially add a small concentration of the substrate to the culture medium and then continually feed the substrate to the culture over time or to spike the culture at appropriate time intervals with new substrate. Continually feeding aromatic substrates to the culture can be accomplished in a variety of ways, as described in the following paragraphs.

The most efficient method of providing an aromatic compound as a carbon and energy substrate for growth is through the vapor phase, provided that the compound is volatile. This methodology can be applied to cultures grown in either liquid or solid media as shown in Fig. 1 (49). In the case of a liquid culture in a flask, a glass bulb containing the liquid volatile aromatic substrate is suspended above the medium (Fig. 1A). As the substrate evaporates, the atmosphere inside the flask becomes saturated and the substrate diffuses into the culture medium. For most substrates it is important to vent the flask to the outside atmosphere (as shown in Fig. 1) in order to prevent the substrate concentration from becoming too high. In the case of highly toxic compounds it is sometimes advantageous to insert a second glass tube through the stopper to allow more outside air into the flask to keep the vapor concentration of the substrate low. The upper part of the glass bulb should be stoppered with nonabsorbent cotton to maintain sterility. Since the stoppered flasks are open to the atmosphere, cultures should be grown in incubators

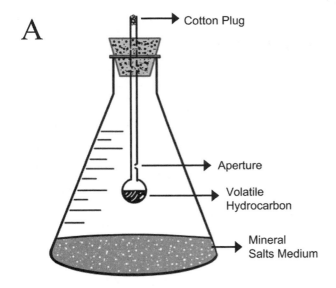

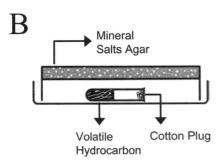

FIGURE 1 Method for growing bacteria in the presence of volatile liquid aromatic hydrocarbons either in a shake flask (A) or on a solid medium (B).

with appropriate ventilation, for instance, with a shaker in a hood. In cases where the growth chamber is not adequately vented, an alternative is to place activated charcoal in the glass tubing between two cotton plugs to absorb the substrate and prevent it from diffusing into the atmosphere of the incubator. As most rubber stoppers absorb hydrocarbons, it is important to label each stopper with the substrate chemical used and, after cleaning and resterilization, use the stopper-bulb pair for the same substrate in subsequent experiments. An alternative to the suspended bulb technique is to utilize a flask with an enclosed side arm, such as a biometer flask or a Klett flask. The volatile hydrocarbon can be placed in the side arm in a fashion similar to that of placing it in a suspended bulb.

A similar approach can be used for growth of bacteria on solid medium. A small (6- by 50-mm) tube containing the volatile liquid hydrocarbon can be placed in the lid of a petri dish (Fig. 1B). Evaporated liquid will saturate the atmosphere, diffuse into the solid medium, and be available for growth of the bacteria. When stoppering these tubes with nonabsorbent cotton, it is important to make sure that the cotton is fully inserted into the tube. If any part of the cotton extends out of the tube, it is possible that this will act as a wick and the liquid hydrocarbon substrate will escape from the tube and perhaps melt the petri dish lid. When testing for growth on different volatile substrates, it is important to place petri dishes prepared in this fashion at

a distance from each other in an appropriately vented incubator. Alternatively, petri dishes can be placed within Nalgene screw-cap containers to contain the vapors. In cases where a particularly volatile substrate is being used as a growth substrate, several plates can be stacked up together in one container, with only one or two tubes of substrate. When using a less volatile substrate, it is important to include one tube of substrate per plate. An alternative to this approach is to place several petri plates in a glass desiccator with a small beaker containing the growth substrate. It is especially important in these cases that the substrate concentration not rise above the toxicity level for the organisms. Since glass desiccators are completely sealed systems, it is possible to add a substrate, let it evaporate completely, and saturate the atmosphere at a target level. Thus, the calculated volume of the desiccator determines the amount of liquid substrate added.

Although adding a volatile liquid substrate through the vapor phase, as described above, works well, these methods do not work very well with volatile solid substrates such as naphthalene and biphenyl. These substrates can often be added directly to liquid culture medium for growth, provided that the toxicity levels are low (see alternative methods below). Growth on solid medium can be accomplished by placing crystals of the volatile solid substrate in the petri dish lid. One disadvantage to this technique is that although the entire atmosphere will become saturated with vapors from the substrate, the area of the medium closest to the crystals is exposed to a higher concentration. This can often result in erratic growth or even no growth of the organism in areas on the petri dish where there are higher concentrations of the substrate. To evenly distribute the substrate across the entire plate, a filter method can be employed. In this case, a sterile Whatman no. 1 filter is placed in the petri dish lid. A solution (typically 100 to 300 μl) of substrate in methanol (or other appropriate volatile carrier) is applied to the filter. After the carrier liquid evaporates, the filter is evenly saturated with the solid volatile compound. The concentration of the solid substrate added to the filter can be empirically adjusted based on the growth rate of the strain and tolerance for the hydrocarbon.

Less volatile and insoluble aromatic hydrocarbons, such as phenanthrene, carbofuran, atrazine, etc., are more difficult to apply to solid culture medium. Three techniques are commonly used in these cases: the use of top agar, spraying the surface of the plate, or deposition by sublimation. In the case of top agar, an appropriate amount of substrate (finely ground) can be added to 5 ml of 0.5% Noble agar in media salts. Even though the substrate does not dissolve, the solution can be mixed thoroughly and immediately poured on top of prepared medium in a petri dish. After the medium sets, cells can be applied to the surface and growth will occur as the substrate diffuses slowly through the agar from the crystals. Alternatively, the substrate can be added to a low-melting-point agar, the solution is cooled to just above gelling temperature, and then the cells are added directly to the solution before it is poured onto base agar in a petri dish. This way, colonies will form within the top agar in close proximity to crystals of the growth substrate. The spray plate technique can be utilized to apply a solid aromatic compound directly to the surface of the medium (32). In this case the cells are commonly plated out on the agar first (and allowed to dry if cells were applied in solution). A 0.1% ether solution of the substrate is then carefully sprayed on the plate in a chemical hood using a sprayer originally designed for spraying thin-layer chromatography (TLC) plates. Care must be taken to cover the entire plate in an even fashion. Varying the thickness of the applied layer controls the amount of substrate available for growth. More-controlled deposition can be achieved by the sublimation technique (1). A visible layer of a water-insoluble compound can be deposited onto plates by chilling and inverting them onto a heated aluminum dish containing the substrate. As the bacteria grow and form colonies, clear zones should develop, indicating that the substrate is being utilized (Fig. 2). Although ether is commonly used as a base to apply the substrate, any sufficiently volatile liquid can be utilized. Care must be taken that the solvent used is not harmful to the bacteria and does not itself serve as a growth substrate for the bacteria.

Another way to reduce the toxicity of volatile aromatic substrates to cells in batch cultures is to provide the volatile substrate in a reservoir of a less-volatile organic solvent. For example, Hartmans et al. (24) were able to successfully obtain and grow styrene-degrading bacteria by providing 25 μl of styrene in a small tube containing 5 ml of dibutylphthalate. This tube was placed inside a larger flask in which the bacteria were to be grown. Using an organic solvent as a reservoir for potentially toxic volatile aromatic compounds in enrichment and growth experiments allows for the slow release of the substrate into the vapor phase. One can then achieve high biomass concentrations at low aqueous substrate concentrations in the growth medium without the need for continuous monitoring and addition of new substrate. Care should be taken to ensure that a culture is not, in fact, growing on the less volatile solvent.

Amberlite XAD-7 resin can also be added to culture media as a way to provide continuous but gradual release of potentially toxic aromatic compounds. Nishino et al. (42) used this approach to grow dinitrotoluene degraders. The protocol involved addition of an amount of dinitrotoluene

(dissolved in acetone) to an empty culture flask to give a final effective concentration of 1 to 4 mM. The acetone was evaporated under a stream of air, leaving a coat of fine dinitrotoluene crystals on the bottom of the culture flask. An appropriate amount of basal salts medium and XAD-7 resin (3.5 g [dry weight] per liter; washed three times with methanol) were added to the dinitrotoluene-coated flask prior to autoclaving. A final dissolved dinitrotoluene concentration of 20 to 200 mM was achieved in this manner after autoclaving and cooling. Following growth of a bacterial culture in this medium, the culture was filtered through glass wool to remove the XAD-7 resin prior to centrifugation to harvest the cells. A similar procedure was developed by the same authors (42) to provide continuous but gradual release of dinitrotoluene to bacteria in solidified media. For this, the agar plates were prepared by grinding the XAD-7 resin to a paste with a mortar and pestle. The ground resin was added to media at a concentration of 7 g (hydrated weight) per liter.

Growing cells in large liquid volumes such as in a fermentor or chemostat is a challenge as well. The high aeration rate of these cultures strips volatile aromatic substrates out of the medium. This necessitates addition of the volatile growth substrate through the culture air feed. One way of doing this is to pass the air that is being used to sparge the culture through a column or flask containing substrate crystals or liquid. Although a single air feed set up in this way provides for efficient growth, two air feeds, one with air only and one with air plus substrate, allow for more control over the final substrate concentration in the growth medium. Flowmeters should be used to control the influent air stream(s) and hence the concentration of influent substrate. Substrate concentrations can be measured in the influent and effluent air by gas chromatography (24, 47, 53).

Several methods for the enumeration of hydrocarbon-degrading microorganisms are available. These methods are generally based either on direct counting on agar plates or on a type of most probable number (MPN) assay performed in microtiter plates. Direct plate counting of hydrocarbon-degrading bacteria can be performed by spraying agar plates of serial dilutions of cells (26, 32) and counting the resulting clearing zones. Alternatively the substrate can be incorporated into a soft-agar overlay (7) to obtain clearing zones. Many different variations of MPN methods are available. Oil-degrading microorganisms, for example, can be enumerated by the "Sheen Screen" method (8). This procedure estimates the number of oil-degrading bacteria by recording their ability to emulsify when grown on oil as a sole carbon source. Other methods consider the fact that complex environmental samples often contain both aliphatic and aromatic constituents. Aliphatic- and aromatic-hydrocarbon-degrading bacteria can be enumerated separately by using the MPN approach (see chapter 9) in microtiter plates containing different substrates (62). In this method, alkane degradation is indicated after 2 weeks of growth by the reduction of iodonitrotetrazolium violet, which was added to the medium. Similarly, aromatic-hydrocarbon degradation is indicated by the accumulation of yellow to greenish-brown partially oxidized substrate.

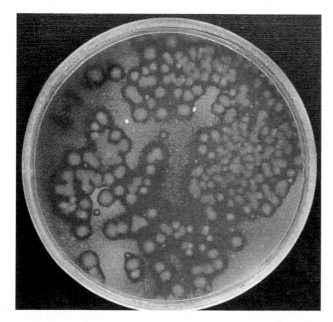

FIGURE 2 Photograph of bacterial colonies on a plate that has phenanthrene applied to it by the spray plate technique. As the colonies grow, they form clear zones, indicating that the substrate is being utilized. This photograph was kindly provided by J. Philp, Napier University, Edinburgh, United Kingdom.

23.2. APPROACHES TO THE STUDY OF AROMATIC HYDROCARBON DEGRADATION
Aromatic hydrocarbon degradation is as complex as there are potential substrates. There are many ways by which microorganisms can metabolize aromatic hydrocarbons, but

while the enzymes and their substrates and products vary, the mechanisms utilized and the catabolic trends observed are the same for all. Toluene and related aromatic compounds have long been used as models for the analysis of aromatic hydrocarbon degradation due to the fact that there are several catabolic pathways known for its degradation, illustrating the many possible ways by which aromatic hydrocarbons can be metabolized (for reviews see references 67 and 68). As shown in Fig. 3, toluene degradation under aerobic conditions generally proceeds by oxidation to dihydroxylated intermediates which undergo ring cleavage and further metabolism to tricarboxylic acid (TCA) cycle intermediates. The initial oxidation reactions are catalyzed by mono- or dioxygenases which add one atom or two atoms, respectively, of molecular oxygen to the substrate. In the case of toluene, initial monooxygenases are known that can oxidize the aromatic ring at any of the three possible ring positions, resulting in phenolic compounds. Also, dioxygenases that oxidize the aromatic ring adjacent to the methyl group (2,3-oxidation), resulting in *cis*-dihydrodiol compounds and monooxygenases that oxidize the methyl group to a benzylic alcohol, have been described (see references 67 and 68). These initial catabolic enzymes act to channel growth substrates into their respective catabolic pathways, and thus, their substrate range (and identity of the product formed) determines the ability of the catabolic pathway to metabolize a particular substrate. For instance, in general, the different toluene catabolic pathways can be distinguished based on the substrates upon which a microorganism can grow. Organisms with the methyl-oxidizing pathway, exemplified by the TOL plasmid (60), can grow

on toluene, *m*- and *p*-xylene, and many other methyl-substituted aromatic compounds but not on benzene or ethylbenzene. Organisms with the ring-oxidizing dioxygenase pathway can grow on toluene, benzene, and ethylbenzene but normally not on *m*- or *p*-xylene. Organisms with the ring-oxidizing monooxygenase pathway can often grow on toluene, benzene, phenol, and cresol. Of course, although the substrate range for growth falls into distinct patterns, the exact substrate range for growth within each group is variable based on the ability of the initial oxygenase to oxidize the substrate, the ability of subsequent enzymes in the pathway to continue metabolism, and the ability of the strain to tolerate the toxic effects of a given aromatic compound and its metabolic products.

Although determining the substrate range for growth of a particular strain points the investigator toward the particular class of enzyme and type of pathway involved in the degradation of a particular hydrocarbon, further in-depth chemical and biochemical investigations must be performed to positively identify the enzymes and pathways involved. Elucidation and confirmation of the catabolic pathway and the enzymes involved come from a combination of experiments involving analysis of wild-type and mutant culture filtrates for catabolic intermediates, $^{18}O_2$ experiments (to distinguish mono- from dioxygenases), oxygen uptake assays and biotransformation assays with whole cells, and enzyme assays using crude cell extracts. The exact combination of approaches utilized depends on the goal(s) of the investigator, the resources available, and the vagaries of the organism (ease of growth, stability and activity of the enzymes, etc.). Each of these approaches is outlined in depth in the sections below.

23.3. EXTRACTION, DERIVATIZATION, AND ANALYSIS OF METABOLITES

Bacterial metabolism of aromatic hydrocarbons can result in a variety of chemically different compounds, resulting from the ring oxidation and the later ring cleavage steps. These metabolites can be either neutral or charged and either chemically stable or unstable, and thus, care must be taken in the extraction process to stabilize and competently extract the chemicals. These metabolites can be directly identified by high-performance liquid chromatography (HPLC) in culture supernatants if in high concentration or identified by HPLC or gas chromatography (GC) following liquid:liquid extraction of the culture. Extraction of neutral metabolites is commonly performed with the addition of an equal volume of ethyl acetate (or other solvent) to the culture broth or enzyme reaction mixture. After mixing and phase separation, the organic layer is removed and the extraction is repeated two more times with equal volumes of solvent. The aqueous phase is then acidified to pH 1.8 with hydrochloric or sulfuric acid. Acidic metabolites are then extracted from the medium by three additional extractions with equal volumes of ethyl acetate. The ethyl acetate extracts from the neutral and acidic extractions are then separately dried over sodium sulfate, and the ethyl acetate is removed by evaporation under vacuum with a rotary evaporator. Because several of the metabolites are unstable at elevated temperatures, care should be taken during rotary evaporation not to heat the solvent above the temperature at which the culture was grown. Following ethyl acetate removal, residues are taken up in a small volume of methanol for further chromatographic analysis. These methanolic extracts can be stored at $-20°C$ until analyzed.

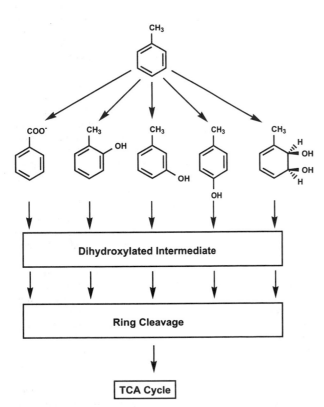

FIGURE 3 Representative known pathways for the degradation of toluene illustrating typical reactions catalyzed by enzymes that act on aromatic compounds.

The extracted compounds can directly be examined by HPLC, and their retention times can be compared to authentic compounds as standards. Typical HPLC conditions involve use of different solvent gradients under either neutral or acidic conditions. The exact solvent gradient utilized depends on the polarity and complexity of the suspected metabolites. More-polar compounds elute early, whereas compounds with multiple rings elute later. Methanol or acetonitrile are often utilized as the mobile phase, with acetonitrile commonly giving shorter retention times. Acidic HPLC effluents (resulting from the use of 0.1% acetic acid or trifluoroacetic acid as the mobile phase) are commonly monitored based on absorbance at 254 or 280 nm for aromatic compounds and 190 or 210 nm for products resulting from ring cleavage.

Although neutral extracted compounds can often be directly examined by GC-mass spectrometry (GC-MS), it is often advantageous to derivatize the compounds to produce sharper GC peaks and to aid in MS identification. Common methods involve acetylation or silylation of the hydroxyl groups and esterification of acidic groups. There are many different reagents and methods for acetylation and silylation available commercially from companies such as Alltech (Deerfield, IL) or Pierce (Rockford, IL). A common method for acetylation is to evaporate the solvent from a portion of the extracted metabolite with a gentle stream of nitrogen in a test tube. Two milliliters of anhydrous pyridine and 1 ml of anhydrous acetic anhydride are added, and the solution is heated to 37°C for 15 min. One milliliter of water is then added, and the derivatized metabolite is extracted from the reaction mixture with ether. The ether layer is removed, dried with sodium sulfate, and evaporated under a gentle stream of nitrogen, and the derivatized metabolite is suspended in methanol or another suitable solvent for analysis. There is an amazing array of silylation reagents and procedures available for derivatization of acidic and hydroxylation metabolites, many of which are explained in detail elsewhere (6, 33).

Analysis of *cis*-dihydrodiol compounds, the products of dioxygenase-catalyzed addition of molecular oxygen to the aromatic ring, can be problematic as these are not very stable and readily dehydrate to the corresponding phenols in a predictable fashion. For example, toluene *cis*-dihydrodiol dehydrates under acidic and/or elevated temperature conditions to *o*-cresol (16) while biphenyl *cis*-dihydrodiol dehydrates to a 1:2 mixture of 2-hydroxy- and 3-hydroxy-biphenyl (17). Thus, *cis*-dihydrodiols may dehydrate when accumulating in culture medium (especially if the pH is less than neutral), during the extraction process, and even during injection onto a GC or in the GC column. It is important therefore that these compounds be derivatized for appropriate analysis. Derivatization reactions have been designed that not only derivatize the *cis*-dihydrodiols but also, due to steric effects, demonstrate unequivocally that the hydroxyl groups are in the *cis* configuration. Two widely used methods are the synthesis of phenylboranate and acetonide derivatives. Acetonide derivatives (16, 17) are synthesized by drying the metabolite extract under a gentle stream of nitrogen and suspending the metabolites in 2 ml of dimethoxypropane. *p*-Toluenesulfonic acid (100 μl of a 10 mg ml^{-1} solution in dimethoxypropane) is added, and the mixture is incubated for 1 h on ice. Phenylboronate derivatives (48) are prepared in a similar fashion, suspending the dried metabolite extract in 2 ml of ethyl acetate, adding 20 μl of a 10 mg ml^{-1} phenylboronic acid solution (in ethyl acetate), and incubating at 75°C for 15 min or at room tem-

perature for several hours. Following derivatization, the reaction mixture is evaporated at room temperature with a gentle stream of nitrogen and the dried derivatized metabolites are suspended in methanol or another suitable solvent. If any *cis*-dihydrodiols are present in the extract, then either of these methodologies will result in a stable derivatized product with a corresponding identifiable peak on the GC.

Although late-log-phase cultures actively growing on the aromatic compound of interest or stationary-phase cultures are the preferred starting points for metabolite analysis, they often do not yield a large enough quantity of metabolites in the culture broth for full analysis and identification. Mutant strains blocked in metabolism of the aromatic hydrocarbon (16, 17, 48) or modular units of genes cloned and expressed in *Escherichia coli* (48, 69) provide for the accumulation of large amounts of pathway intermediates for chemical analysis. A concentrated suspension of bacterial cells that have been grown on a particular aromatic substrate can be used in a catalytic, resting-cell mode to test the ability of these cells to transform the original substrate as well as an array of other substrates (18). Resting-cell analyses involving highly concentrated induced bacterial cells can also be utilized to increase the possibility of obtaining identifiable intermediates from the culture medium due to differential rates of catalysis of the individual enzymes in the catabolic pathway.

Resting-cell assays can be performed on wild-type strains, mutant strains, or induced clones in heterologous hosts. A typical resting-cell experiment consists of growing a wild-type bacterial culture in an appropriate basal salts medium with an aromatic substrate as the sole carbon source or growing a mutant bacterial strain on a suitable non-catabolite-repressing substrate in the presence of the pathway-inducing compound to an A_{600} of 0.6 to 1.0. In the case of heterologous expression of genes in *E. coli*, recombinant cells are grown to an A_{600} of 0.5 to 0.7, induced with the appropriate method (depending on the expression vector), and allowed to grow for one or two more hours. In all cases, cells are harvested by centrifugation, washed at least once in a suitable buffer (e.g., 50 mM sodium phosphate, pH 7.5) to remove the growth medium, and resuspended in one-half to 1/20 of the original culture volume of the same buffer. This concentrated suspension of cells is no longer capable of significant cell division since essential macronutrients such as nitrogen are lacking, hence the designation "resting cells." Resting cells can be used to test for the ability of the strain to transform other carbon sources, to determine the substrate range of a particular catabolic enzyme (where cell-free assays are not possible), to determine potential pathway intermediates, and to obtain transformation products that can be subsequently extracted and chemically characterized.

Examples showing the utility of resting cells for the production and analysis of *cis*-dihydrodiols from a wide range of monocyclic and polycyclic compounds using various mutant strains (16, 17, 48) and cloned genes (48, 69) have been elegantly described by Gibson and his coworkers. Seeger et al. (51, 52) also used resting cells of *E. coli* carrying various cloned DNA fragments from *Burkholderia* sp. strain LB400 or from *Rhodococcus globerulus* P6. The clones expressed enzymes in a pathway for degradation of substituted biphenyls. Use of individual recombinants allowed the investigators to produce pathway intermediates that were extracted, derivatized, and analyzed by GC-MS. Similar approaches to obtain and identify products of aromatic substrate transformation using cloned genes in resting cells are given by Habe et al. (20) for transformation products of dibenzofurans and dibenzo-*p*-dioxins, by Pollmann et al. (45) for products of chlorotoluenes, and by Lessner et al. (36) for products of nitrobenzene.

23.4. ARENE MONOOXYGENASES VERSUS DIOXYGENASES

As mentioned above, under oxic and suboxic conditions, bacteria utilize a variety of oxygenases to initiate attack on aromatic compounds. Oxygen incorporation into the aromatic nucleus mediated by arene oxygenases results in the production of dihydrodiols when the initial oxygenase is a dioxygenase and in the production of phenolic intermediates when the initial oxygenase is a monooxygenase. When analyzing new bacterial strains with novel catabolic capabilities, it is often necessary to demonstrate that an arene oxygenase is involved in a particular catabolic pathway and that the detected hydroxylated intermediates have arisen from incorporation of molecular oxygen into their structures. If one can identify a *cis*-dihydrodiol intermediate in culture supernatants, then a priori the pathway proceeds via an initial dioxygenase attack of the aromatic nucleus. However, if one detects only phenolic or dihydroxylated products in the culture medium then it is uncertain whether the catabolic pathway proceeds by an initial dioxygenase or by two initial monooxygenases. Identifying which class of oxygenase initiates the catabolic pathway can be accomplished through the use of an experimental system in which the atmosphere is enriched in $^{18}O_2$ (15). For example, determination of whether a nonhydroxylated aromatic compound is being converted to a dihydroxylated intermediate (e.g., a catecholic intermediate) via the action of an initial dioxygenase or via the sequential action of two monooxygenations can be done in an atmosphere containing a 50:50 mixture of $^{16}O_2$ and $^{18}O_2$. From analysis of the dihydroxylated intermediates, one would expect a 50:50 mixture of a dihydroxylated intermediate containing two atoms of ^{16}O or two atoms of ^{18}O from an initial dioxygenase attack, whereas the product distribution of the dihydroxylated intermediate would be 25% containing two atoms of ^{16}O, 25% containing two atoms of ^{18}O, and 50% containing one atom of ^{16}O and one atom of ^{18}O from sequential monooxygenase attacks. The explanation for this type of product distribution is that a dioxygenase incorporates both atoms of molecular oxygen yielding a *cis*-dihydrodiol intermediate which is subsequently dehydrogenated to yield a catechol, whereas a monooxygenase incorporates only one atom of molecular oxygen at a time (the other atom being converted to water).

A typical experimental setup (23, 54, 56) consists of induced cells harvested by centrifugation and resuspended in a suitable buffer (e.g., 20 mM sodium phosphate at pH 7.0) to an A_{600} of 0.5 to 1.0. The suspension is transferred to a flask that is sealed with a stopcock and placed on a magnetic stirrer. The air in the headspace of the flask is removed under vacuum and is replaced with nitrogen four times. The headspace is then evacuated and refilled with air containing a 50:50 mixture of $^{16}O_2$ and $^{18}O_2$. The gas in the headspace is analyzed by MS to determine the final composition of $^{16}O_2$ and $^{18}O_2$. The experimental system is allowed to equilibrate for 15 min, and then the aromatic substrate to be tested is added, typically to a final concentration of 10 to 100 μM. The suspension is incubated with stirring for about 1 h and sparged with nitrogen for 5 min, and the cells are then removed by centrifugation. Metabolites in the supernatant can be extracted (see above) and analyzed by GC-MS.

Caution should be exercised in the use of $^{18}O_2$ incorporation experiments to determine whether a dioxygenase or a monooxygenase is involved in a new catabolic pathway in an incompletely characterized isolate. Spain et al. (54) have shown that toluene dioxygenase of *P. putida* F1 appears to function as a monooxygenase in the transformation of 2,5-dichlorophenol to 3,6-dichlorocatechol. Therefore, $^{18}O_2$ incorporation studies alone are not always sufficient to unambiguously determine the type of oxygenase that is involved in an aromatic catabolic pathway.

23.5. OXYGEN CONSUMPTION ASSAYS

Aerobic metabolism of aromatic compounds by bacteria involves consumption of oxygen, either as a cosubstrate by bacterial aromatic oxygenases or as a terminal respiratory electron acceptor. Oxygen consumption can be used as a measure of the ability of whole cells or cell extracts to utilize an aromatic substrate. For whole cells, oxygen uptake can be measured polarographically with a Clark-type oxygen electrode (obtainable from Rank Brothers Ltd., Cambridge, England, or Yellow Springs Instrument Co., Yellow Springs, OH). A typical reaction mixture (53, 56) would contain a cell suspension (typically containing 0.25 and 0.5 mg of protein) that has been washed free of growth medium by using aerated 20 to 50 mM sodium phosphate buffer (pH 7.0). After equilibration, background oxygen consumption is measured and substrate is added to a final concentration of 500 μM (typically from a 10 mM stock). Substrates can be added from aqueous stock solutions or, for substrates that are poorly water soluble, from stocks made in carriers such as dimethylformamide or dimethylsulfoxide. It is important to ascertain the effect of the substrate carrier on the basal oxygen consumption rate. The rate of substrate-dependent oxygen consumption must be corrected by subtracting the rate of endogenous oxygen consumption. The rates of oxygen consumption without a substrate (i.e., endogenous consumption) and with an aromatic substrate are measured for at least 5 min.

It should be noted that oxygen uptake experiments measure all processes within the cell that consume oxygen. Thus, addition of any substrate that enters the aromatic hydrocarbon catabolic pathway will result in whole-cell oxygen consumption provided that it is metabolized through one of the three routes requiring oxygen (initial oxygenase, ring-cleavage oxygenase, or tricarboxylic acid cycle). It is thus possible in the analysis of toluene degradation, for instance, to add *cis*-toluene dihydrodiol to toluene-induced cells and observe oxygen uptake for strains metabolizing toluene by the *cis*-dihydrodiol pathway, even though the nearest oxygen-consuming enzyme is the second metabolic step downstream. Care should be taken in interpreting oxygen uptake experiments, as many aromatic hydrocarbon catabolic enzymes have broad substrate ranges. Thus, the ability of toluene-grown cells to oxidize cresols is not necessarily due to a monooxygenase but could also be due to a dioxygenase (54). Certain bacterial strains have more than one catabolic pathway for a particular aromatic hydrocarbon (27), and this must also be taken into account when interpreting the data. One drawback to whole-cell oxygen uptake experiments is the ability of the added substrate to cross the cell membrane. While many aromatic hydrocarbons partition into and diffuse across the membrane without difficulty, many aromatic and aliphatic compounds, especially polar or charged compounds, have difficulty in diffusing across the membrane. This, however, is a problem only for testing compounds related to the growth substrate or testing potential intermediates in the catabolic pathway, since by definition the cells should be induced for transport of the growth substrate (if a transport mechanism is needed).

23.6. CELL-FREE ASSAYS: ARENE OXIDATION

In order to get around the problem with transport of the enzyme substrate into the cell and to more precisely assay the enzymes for aromatic-hydrocarbon degradation, cell-free assays should be performed. This is not an easy task, as many of the arene monooxygenases and dioxygenases consist of multiple components including an electron transport chain that mediates transfer of electrons from NAD(P)H to the oxygenase component, which actually performs the catalytic reaction. Due to this multicomponent nature of many oxygenases, cell-free enzymatic activity is quite low and does not show linearity with protein concentration. Phthalate dioxygenase, one of the simplest multicomponent enzymes, consisting of a reductase and an oxygenase component, can be assayed spectrophotometrically by following the phthalate-dependent oxidation of NADH at 340 nm (3). In a typical reaction, crude cell extract, ferrous ammonium sulfate (10 μM final concentration from a freshly prepared stock solution), and NADH (100 μM final concentration) are added to 100 mM HEPES, pH 8.0. After the background rate of NADH oxidation is measured, phthalate is added at a final concentration of 2 mM. The rate of oxidation of NADH is again monitored, and the difference in the rates before and after phthalate addition is the enzymatic activity of phthalate dioxygenase.

Measuring the rate of oxygenases consisting of more than two components in crude cell extracts is problematic due to the very low levels of measurable activity. Purified electron transfer components, if available, can be added in excess to the crude cell extract, and the rate of substrate-induced NAD(P)H oxidation, which is equivalent to the amount of oxygenase component present, is monitored. In many cases, purified electron transfer components are not readily available and enzyme assays must be performed using the native concentration of oxygenase. Given these conditions, a radiometric assay is valuable due its extreme sensitivity. In general terms, the oxygenase reaction can be performed using a radiolabeled substrate, the volatile substrate can be removed at a defined endpoint, and the polar nonvolatile radioactive products can be measured in a scintillation counter. In a typical 400-μl assay for toluene dioxygenase (55, 64), crude cell extract is added to 50 mM Tris-HCl buffer, pH 7.5, along with 400 μM freshly prepared ferrous sulfate. The mixture is incubated for 2 min to allow the ferrous iron to reconstitute any enzyme that has lost its original ferrous iron. NADH (1.2 mM final concentration) and ^{14}C-radiolabeled toluene (80 μM final concentration from an N,N-dimethylformamide stock) are added, the reactants are mixed, and the incubation is continued for 5 min. A 10-μl aliquot is then removed and spotted onto a 1- by 1-cm plastic-backed silica gel square (Kodak chromagram sheet 13181). The TLC square is placed in a chemical hood to allow the volatile toluene to evaporate from the spotted reaction mixture. Nonvolatile polar products remain on the TLC square. Following an adequate drying period (30 min), the TLC square is added to a scintillation vial along with scintillation cocktail and the radioactivity is measured in a

scintillation counter. A negative control should be performed to determine the background counts, especially as commercial radiolabeled toluene is often contaminated with minor amounts of radiolabeled polar compounds. A similar assay has been developed for measuring the rate of oxidation of compounds less volatile than toluene. In the case of naphthalene (11, 12) or biphenyl (21, 22) dioxygenase a similar reaction can be performed. In these cases, however, the less volatile substrate must be removed under a stream of warm air for 30 min in a chemical hood prior to scintillation counting of the products.

Simple single-component aromatic oxygenases can easily be assayed by measuring the disappearance of a specific cosubstrate. For example, the activity of phenol hydroxylase from the yeast *Trichosporon cutaneum* (39) or chlorophenol hydroxylase from *Ralstonia eutropha* JMP134(pJP4) (37) can be measured by monitoring the disappearance at 340 nm of the cosubstrates NADPH or NADH (as appropriate for the individual enzyme). A typical assay mixture would contain, in a 1-cm quartz cuvette, the following: 100 μmol of an appropriate buffer (e.g., potassium phosphate at pH 7.6), 0.5 μmol of NAD(P)H, suitably diluted enzyme or cell extract (diluted so that the reaction rate will be linear and proportional to the enzyme concentration), 0.5 μmol of phenolic substrate, and water to 3 ml. The reaction is initiated by addition of the substrate, and its rate is recorded for 1 to 4 min. When using crude extracts, endogenous oxidation of NAD(P)H must be taken into account. However, if this presents a problem, endogenous NAD(P)H oxidase activity can be removed by centrifugation of the extract at 100,000 × g and using the supernatant for the assay (see, for example, Chapman and Ribbons [9]). One enzyme unit is defined as the amount of enzyme which in the presence of a phenolic substrate causes the oxidation of 1 μmol of NAD(P)H per min.

Some aromatic oxygenases require the addition of other cofactors besides NADH for optimal activity. For instance, p-hydroxybenzoate hydroxylase, an enzyme involved in the degradation of toluene via p-cresol (57, 58, 63, 65, 66), requires the addition of flavin adenine dinucleotide (FAD) for maximal activity. p-Hydroxybenzoate hydroxylase acts as a monooxygenase in the conversion of p-hydroxybenzoate to the dihydroxylated compound protocatechuate. p-Hydroxybenzoate hydroxylase activity can be measured spectrophotometrically by following the rate of p-hydroxybenzoate-dependent oxidation of NADPH at 340 nm. The assay mixture (13) contains, in a 1-cm cuvette in a final volume of 3 ml, the following: 33 mM Tris-sulfate (pH 8.0), 0.33 mM sodium EDTA, 0.33 mM sodium p-hydroxybenzoate, 0.23 mM NADPH, and 3.3 μM FAD. The reaction is initiated by adding suitably diluted enzyme or cell extract. When using cell extracts, endogenous oxidation of NADPH should first be determined. The FAD in this case is needed to reconstitute those enzymes which have lost their FAD.

Dihydrodiols are products of the action of dioxygenases on aromatic substrates, as illustrated in Fig. 3. Dihydrodiol dehydrogenases are assayed using an NAD+/NADH-linked assay. Dihydrodiol dehydrogenase activity can be measured spectrophotometrically by following the increase in absorbance at 340 nm concomitant with the conversion of NAD+ to NADH. A typical reaction mixture (53) would contain 0.5 μmol of NAD+, 0.1 μmol of the dihydrodiol substrate, 18 μmol of an appropriate buffer (e.g., Tris hydrochloride at pH 8.0), and suitably diluted cell extract (typically 0.02 to 0.06 mg of protein) in a final volume of

1.0 ml. When measuring extremely low rates of dihydrodiol dehydrogenase in crude cell extracts, it is important to use a sealable cuvette with gas ports so that the reaction mixture can be sparged with nitrogen. This prevents the oxidation of NADH by NADH oxidase in the cell extract.

23.7 CELL-FREE ASSAYS: RING CLEAVAGE AND SUBSEQUENT REACTIONS

As illustrated in Fig. 3, metabolism of toluene yields dihydroxylated intermediates such as catechol or methylcatechols, which are further converted into intermediates that ultimately enter the TCA cycle. Metabolism of dihydroxylated compounds proceeds by cleavage of the aromatic ring either between (*ortho*-cleavage) or adjacent to (*meta*-cleavage) the hydroxyl groups. Aromatic ring cleavage in the latter case is mediated by an extradiol catechol dioxygenase, commonly designated (methyl)catechol 2,3-dioxygenase (C23O) or metapyrocatechase. The activity of C23O is determined spectrophotometrically from the rate of product accumulation. A typical assay mixture (4, 5, 29, 31, 34, 43) contains in a quartz cuvette (1-cm light path and a final volume of 3 ml) the following: 0.3 ml of 50 mM potassium phosphate buffer (pH 7.5), 0.1 ml of 10 mM catechol or methylcatechol dissolved in water (made fresh for each use as the material rapidly auto-oxidizes in the presence of air), and 0.1 ml of suitably diluted enzyme or cell extract. The rate of increase in absorbance is monitored at 375 nm for catechol, 388 nm for 3-methylcatechol, and 382 nm for 4-methylcatechol. When using cell extracts, heat treatment of the extract at 60°C for 10 min prior to use allows the ring fission product to accumulate by inactivating the enzymes involved in further metabolism of the muconate semialdehydes (53).

Metabolism of chlorobenzene yields chlorocatechol, which is further converted into intermediates that ultimately enter the TCA cycle by cleavage of the aromatic ring between the hydroxylated carbons (i.e., intradiol cleavage) to produce chloro-*cis,cis*-muconic acid. Further metabolism of chloro-*cis,cis*-muconic acid proceeds by a modified form of the β-ketoadipate pathway, with a diene lactone (*trans*-4-carboxymethylene-but-2-en-4-olide) and maleylacetate as key intermediates (10, 28, 35, 44, 47, 50). Chlorocatechol 1,2-dioxygenase (also known as catechol dioxygenase II) has a broad substrate specificity (10, 38) and relatively low k_{cat} values with catechol and chlorocatechols (41). The activity of chlorocatechol 1,2-dioxygenase is determined spectrophotometrically from the rate of accumulation of the chloro-*cis,cis*-muconic acid product at 260 nm. A typical assay mixture (40) contains (final volume of 3 ml in a quartz cuvette with a 1-cm light path) 0.1 mM catechol or chlorocatechol, 1 mM EDTA, 33 mM Tris-HCl buffer (pH 8.0), and suitably diluted enzyme or cell extract. The enzyme that further transforms the *cis,cis*-muconic acid requires Mn^{2+} for its activity, and unlike the chlorocatechol 1,2-dioxygenase, this enzyme is inhibited by EDTA, allowing accumulation of the ring fission product.

Many other aromatic substrates are also metabolized by way of catechol as the key ring fission intermediate (see, for example, p. 157–162 in Gottschalk [19]) and the catecholic substrate is cleaved by an intradiol catechol 1,2-dioxygenase (also known as catechol dioxygenase I, or pyrocatechase). Assay of this enzyme's activity is conducted in the same manner as for chlorocatechol 1,2-dioxygenase.

Besides catechol and methylcatechol, protocatechuate (the dihydroxylated substrate) is a common ring cleavage substrate found in aromatic catabolic pathways. Protocatechuate 3,4-dioxygenase catalyzes intradiol cleavage of protocatechuate to yield β-carboxy-*cis,cis*-muconic acid. This enzyme can be assayed by monitoring the decrease in A_{290} due to the degradation of the substrate (14). The assay mixture contains, in a final volume of 3 ml, 0.3 ml of 50 mM Tris-acetate (pH 7.5), 0.12 ml of 10 mM protocatechuate, and 0.1 ml of suitably diluted enzyme or cell extract. An alternative assay method is to polarographically monitor the consumption of oxygen during the reaction (59, 70). For this, the chamber of the oxygen polarograph is filled with air-saturated MOPS (3-[N-morpholino]propanesulfonic acid) buffer (pH 7.0). Once the temperature has equilibrated, 1 μl of 100 mM protocatechuate is added and a baseline is established on the recording device. A suitably diluted aliquot of enzyme or cell extract is added, and oxygen consumption is monitored. As in the case of (methyl)catechol, ring cleavage of protocatechuate can be via *meta*-cleavage at either the 2,3- or the 4,5- position on the aromatic ring. The activities of these two enzymes can also be measured polarographically as for protocatechuate 3,4-dioxygenase (2, 61).

We thank Kristen Foster for artwork. J.J.K. thanks NIEHS for their support through Superfund basic research program grant P42-ES04911, and G.J.Z. thanks NSF for their support through grants MCB-0078465 and CHE-9810248.

23.8. REFERENCES

1. **Alley, J. F., and L. R. Brown.** 2000. Use of sublimation to prepare solid microbial media with water-insoluble substrates. *Appl. Environ. Microbiol.* **66:**439–442.
2. **Arciero, D. M., A. M. Orville, and J. D. Lipscomb.** 1990. Protocatechuate 4,5-dioxygenase from *Pseudomonas testosteroni. Methods Enzymol.* **188:**89–95.
3. **Batie, C. J., E. LaHaie, and D. P. Ballou.** 1987. Purification and characterization of phthalate oxygenase and phthalate oxygenase reductase from *Pseudomonas cepacia. J. Biol. Chem.* **262:**1510–1518.
4. **Bayly, R. C., S. Dagley, and D. T. Gibson.** 1966. The metabolism of cresols by species of *Pseudomonas. Biochem. J.* **101:**293–301.
5. **Bayly, R. C., and G. J. Wigmore.** 1973. Metabolism of phenol and cresols by mutants of *Pseudomonas putida. J. Bacteriol.* **113:**1112–1120.
6. **Blau, K., and J. Halket.** 1993. *Handbook of Derivatives for Chromatography,* 2nd ed. John Wiley & Sons, New York, NY.
7. **Bogardt, A. H., and B. B. Hemmingsen.** 1992. Enumeration of phenanthrene-degrading bacteria by an overlayer technique and its use in evaluation of petroleum-contaminated sites. *Appl. Environ. Microbiol.* **58:**2579–2582.
8. **Brown, E. J., and J. F. Braddock.** 1990. Sheen Screen, a miniaturized most-probable-number method for enumeration of oil-degrading microorganisms. *Appl. Environ. Microbiol.* **56:**3895–3896.
9. **Chapman, P. J., and D. W. Ribbons.** 1976. Metabolism of resorcinylic compounds by bacteria: orcinol pathway in *Pseudomonas putida. J. Bacteriol.* **125:**975–984.
10. **Dorn, E., and H. J. Knackmuss.** 1978. Chemical structure and biodegradability of halogenated aromatic compounds. Two catechol 1,2-dioxygenases from a 3-chlorobenzoate-grown pseudomonad. *Biochem. J.* **174:**73–84.
11. **Ensley, B. D., and D. T. Gibson.** 1983. Naphthalene dioxygenase: purification and properties of a terminal oxygenase component. *J. Bacteriol.* **155:**505–511.

12. **Ensley, B. D., D. T. Gibson, and A. L. Laborde.** 1982. Oxidation of naphthalene by a multicomponent enzyme system from *Pseudomonas* sp. strain NCIB 9816. *J. Bacteriol.* **149:**948–954.

13. **Entsch, B.** 1990. Hydroxybenzoate hydroxylase. *Methods Enzymol.* **188:**138–147.

14. **Fujisawa, H.** 1970. Protocatechuate-3,4-dioxygenase (*Pseudomonas*). *Methods Enzymol.* **17A:**526–529.

15. **Gibson, D. T., G. E. Cardini, F. C. Maseles, and R. E. Kallio.** 1970. Incorporation of oxygen-18 into benzene by *Pseudomonas putida*. *Biochemistry* **9:**1631–1635.

16. **Gibson, D. T., M. Hensley, H. Yoshioka, and T. J. Mabry.** 1970. Formation of (+)-cis-2,3-dihydroxy-1-methylcyclohexa-4,6-diene from toluene by *Pseudomonas putida*. *Biochemistry* **9:**1626–1630.

17. **Gibson, D. T., R. L. Roberts, M. C. Wells, and V. M. Kobal.** 1973. Oxidation of biphenyl by a *Beijerinckia* species. *Biochem. Biophys. Res. Commun.* **50:**211–219.

18. **Gibson, D. T., G. J. Zylstra, and S. Chauhan.** 1990. Biotransformations catalyzed by toluene dioxygenase from *Pseudomonas putita* F1, p. 121–132. *In* S. Silver, A. M. Chakrabarty, B. Iglewski, and S. Kaplan (ed.), *Pseudomonas: Biotransformations, Pathogenesis, and Evolving Biotechnology*. American Society for Microbiology, Washington, DC.

19. **Gottschalk, G.** 1986. *Bacterial Metabolism*, 2nd ed. Springer-Verlag, New York, NY.

20. **Habe, H., J. S. Chung, J. H. Lee, K. Kasuga, T. Yoshida, H. Nojiri, and T. Omori.** 2001. Degradation of chlorinated dibenzofurans and dibenzo-*p*-dioxins by two types of bacteria having angular dioxygenases with different features. *Appl. Environ. Microbiol.* **67:**3610–3617.

21. **Haddock, J. D., and D. T. Gibson.** 1995. Purification and characterization of the oxygenase component of biphenyl 2,3-dioxygenase from *Pseudomonas* sp. strain LB400. *J. Bacteriol.* **177:**5834–5839.

22. **Haddock, J. D., L. M. Nadim, and D. T. Gibson.** 1993. Oxidation of biphenyl by a multicomponent enzyme system from *Pseudomonas* sp. strain LB400. *J. Bacteriol.* **175:** 395–400.

23. **Haigler, B. E., and J. C. Spain.** 1991. Biotransformation of nitrobenzene by bacteria containing toluene degradative pathways. *Appl. Environ. Microbiol.* **57:**3156–3162.

24. **Hartmans, S., M. J. van der Werf, and J. A. de Bont.** 1990. Bacterial degradation of styrene involving a novel flavin adenine dinucleotide-dependent styrene monooxygenase. *Appl. Environ. Microbiol.* **56:**1347–1351.

25. **Heipieper, J. H., F. J. Weber, J. Sikkema, H. Keweloh, and J. A. M. de Bont.** 1994. Mechanisms of resistance of whole cells to toxic organic solvents. *Trends Biotechnol.* **12:**409–415.

26. **Heitkamp, M. A., and C. E. Cerniglia.** 1987. Effects of chemical structure and exposure on the microbial degradation of polycyclic aromatic hydrocarbons in freshwater and estuarine ecosystems. *Environ. Toxicol. Chem.* **6:**535-546.

27. **Kahng, H. Y., J. C. Malinverni, M. M. Majko, and J. J. Kukor.** 2001. Genetic and functional analysis of the *tbc* operons for catabolism of alkyl- and chloroaromatic compounds in *Burkholderia* sp. strain JS150. *Appl. Environ. Microbiol.* **67:**4805–4816.

28. **Kaphammer, B., J. J. Kukor, and R. H. Olsen.** 1990. Regulation of *tfdCDEF* by *tfdR* of the 2,4-dichlorophenoxyacetic acid degradation plasmid pJP4. *J. Bacteriol.* **172:**2280–2286.

29. **Kataeva, I. A., and L. A. Golovleva.** 1990. Catechol 2,3-dioxygenases from *Pseudomonas aeruginosa* 2x. *Methods Enzymol.* **188:**115–121.

30. **Kieboom, J., J. J. Dennis, J. A. de Bont, and G. J. Zylstra.** 1998. Identification and molecular characterization of an efflux pump involved in *Pseudomonas putida* S12 solvent tolerance. *J. Biol. Chem.* **273:**85–91.

31. **Kim, E., and G. J. Zylstra.** 1995. Molecular and biochemical characterization of two *meta*-cleavage dioxygenases involved in biphenyl and *m*-xylene degradation by *Beijerinckia* sp. strain B1. *J. Bacteriol.* **177:**3095–3103.

32. **Kiyohara, H., K. Nagao, and K. Yana.** 1982. Rapid screen for bacteria degrading water-insoluble, solid hydrocarbons on agar plates. *Appl. Environ. Microbiol.* **43:**454–457.

33. **Knapp, D. R.** 1979. *Handbook of Analytical Derivatization Reactions.* John Wiley & Sons, New York, NY.

34. **Kukor, J. J., and R. H. Olsen.** 1991. Genetic organization and regulation of a *meta* cleavage pathway for catechols produced from catabolism of toluene, benzene, phenol, and cresols by *Pseudomonas pickettii* PKO1. *J. Bacteriol.* **173:**4587–4594.

35. **Kukor, J. J., R. H. Olsen, and J. S. Siak.** 1989. Recruitment of a chromosomally encoded maleylacetate reductase for degradation of 2,4-dichlorophenoxyacetic acid by plasmid pJP4. *J. Bacteriol.* **171:**3385–3390.

36. **Lessner, D. J., G. R. Johnson, R. E. Parales, J. C. Spain, and D. T. Gibson.** 2002. Molecular characterization and substrate specificity of nitrobenzene dioxygenase from *Comamonas* sp. strain JS765. *Appl. Environ. Microbiol.* **68:**634–641.

37. **Liu, T., and P. J. Chapman.** 1984. Purification and properties of a plasmid-encoded 2,4-dichlorophenol hydroxylase. *FEBS Lett.* **173:**314–318.

38. **Nakagawa, H., H. Inoue, and Y. Takeda.** 1963. Characteristics of catechol oxygenase from *Brevibacterium fuscum. J. Biochem.* (Tokyo) **54:**65–74.

39. **Neujahr, H. Y., and A. Gaal.** 1973. Phenol hydroxylase from yeast. Purification and properties of the enzyme from *Trichosporon cutaneum. Eur. J. Biochem.* **35:**386–400.

40. **Ngai, K. L., E. L. Neidle, and L. N. Ornston.** 1990. Catechol and chlorocatechol 1,2-dioxygenases. *Methods Enzymol.* **188:**122–126.

41. **Ngai, K. L., and L. N. Ornston.** 1988. Abundant expression of *Pseudomonas* genes for chlorocatechol metabolism. *J. Bacteriol.* **170:**2412–2413.

42. **Nishino, S. F., G. C. Paoli, and J. C. Spain.** 2000. Aerobic degradation of dinitrotoluenes and pathway for bacterial degradation of 2,6-dinitrotoluene. *Appl. Environ. Microbiol.* **66:**2139–2147.

43. **Nozaki, M.** 1970. Metapyrocatechase. *Methods Enzymol.* **17A:**522–525.

44. **Pieper, D. H., K.-H. Engesser, and H.-J. Knackmuss.** 1989. Regulation of catabolic pathways of phenoxyacetic acids and phenol in *Alcaligenes eutrophus* JMP134. *Arch. Microbiol.* **151:**365–371.

45. **Pollmann, K., S. Beil, and D. H. Pieper.** 2001. Transformation of chlorinated benzenes and toluenes by *Ralstonia* sp. strain PS12 *tecA* (tetrachlorobenzene dioxygenase) and *tecB* (chlorobenzene dihydrodiol dehydrogenase) gene products. *Appl. Environ. Microbiol.* **67:**4057–4063.

46. **Ramos, J. L., E. Duque, J. J. Rodriguez-Herva, P. Godoy, A. Haidour, F. Reyes, and A. Fernandez-Barrero.** 1997. Mechanisms for solvent tolerance in bacteria. *J. Biol. Chem.* **272:**3887–3890.

47. **Reineke, W., and H. J. Knackmuss.** 1984. Microbial metabolism of haloaromatics: isolation and properties of a chlorobenzene-degrading bacterium. *Appl. Environ. Microbiol.* **47:**395–402.

48. **Resnick, S. M., and D. T. Gibson.** 1996. Regio- and stereospecific oxidation of fluorene, dibenzofuran, and dibenzothiophene by naphthalene dioxygenase from *Pseudomonas* sp. strain NCIB 9816-4. *Appl. Environ. Microbiol.* **62:**4073–4080.

49. **Rosenberg, E.** 1992. The hydrocarbon-oxidizing bacteria, p. 446–459. *In* A. Balows, H. G. Truper, M. Dworkin, W. Harder, and K.-H. Schleifer (ed.), *The Prokaryotes*, 2nd ed. Springer-Verlag, New York, NY.

50. **Schmidt, E., and H. J. Knackmuss.** 1980. Chemical structure and biodegradability of halogenated aromatic compounds. Conversion of chlorinated muconic acids into maleoylacetic acid. *Biochem. J.* **192:**339–347.

51. **Seeger, M., B. Camara, and B. Hofer.** 2001. Dehalogenation, denitration, dehydroxylation, and angular attack on substituted biphenyls and related compounds by a biphenyl dioxygenase. *J. Bacteriol.* **183:**3548–3555.

52. **Seeger, M., K. N. Timmis, and B. Hofer.** 1995. Conversion of chlorobiphenyls into phenylhexadienoates and benzoates by the enzymes of the upper pathway for polychlorobiphenyl degradation encoded by the *bph* locus of *Pseudomonas* sp. strain LB400. *Appl. Environ. Microbiol.* **61:**2654–2658.

53. **Spain, J. C., and S. F. Nishino.** 1987. Degradation of 1,4-dichlorobenzene by a *Pseudomonas* sp. *Appl. Environ. Microbiol.* **53:**1010–1019.

54. **Spain, J. C., G. J. Zylstra, C. K. Blake, and D. T. Gibson.** 1989. Monohydroxylation of phenol and 2,5-dichlorophenol by toluene dioxygenase in *Pseudomonas putida* F1. *Appl. Environ. Microbiol.* **55:**2648–2652.

55. **Subramanian, V., T. N. Liu, W. K. Yeh, and D. T. Gibson.** 1979. Toluene dioxygenase: purification of an iron-sulfur protein by affinity chromatography. *Biochem. Biophys. Res. Commun.* **91:**1131–1139.

56. **Warhurst, A. M., K. F. Clarke, R. A. Hill, R. A. Holt, and C. A. Fewson.** 1994. Metabolism of styrene by *Rhodococcus rhodochrous* NCIMB 13259. *Appl. Environ. Microbiol.* **60:**1137–1145.

57. **Whited, G. M., and D. T. Gibson.** 1991. Separation and partial characterization of the enzymes of the toluene-4-monooxygenase catabolic pathway in *Pseudomonas mendocina* KR1. *J. Bacteriol.* **173:**3017–3020.

58. **Whited, G. M., and D. T. Gibson.** 1991. Toluene-4-monooxygenase, a three-component enzyme system that catalyzes the oxidation of toluene to *p*-cresol in *Pseudomonas mendocina* KR1. *J. Bacteriol.* **173:**3010–3016.

59. **Whittaker, J. W., A. M. Orville, and J. D. Lipscomb.** 1990. Protocatechuate 3,4-dioxygenase from *Brevibacterium fuscum. Methods Enzymol.* **188:**82–88.

60. **Williams, P. A., and K. Murray.** 1974. Metabolism of benzoate and the methylbenzoates by *Pseudomonas putida* (*arvilla*) mt-2: evidence for the existence of a TOL plasmid. *J. Bacteriol.* **120:**416–423.

61. **Wolgel, S. A., and J. D. Lipscomb.** 1990. Protocatechuate 2,3-dioxygenase from *Bacillus maceRans. Methods Enzymol.* **188:**95–101.

62. **Wrenn, B. A., and A. D. Venosa.** 1996. Selective enumeration of aromatic and aliphatic hydrocarbon degrading bacteria by a most-probable-number procedure. *Can. J. Microbiol.* **42:**252–258.

63. **Wright, A., and R. H. Olsen.** 1994. Self-mobilization and organization of the genes encoding the toluene metabolic pathway of *Pseudomonas mendocina* KR1. *Appl. Environ. Microbiol.* **60:**235–242.

64. **Yeh, W. K., D. T. Gibson, and T. N. Liu.** 1977. Toluene dioxygenase: a multicomponent enzyme system. *Biochem. Biophys. Res. Commun.* **78:**401–410.

65. **Yen, K. M., and M. R. Karl.** 1992. Identification of a new gene, *tmoF*, in the *Pseudomonas mendocina* KR1 gene cluster encoding toluene-4-monooxygenase. *J. Bacteriol.* **174:**7253–7261.

66. **Yen, K. M., M. R. Karl, L. M. Blatt, M. J. Simon, R. B. Winter, P. R. Fausset, H. S. Lu, A. A. Harcourt, and K. K. Chen.** 1991. Cloning and characterization of a *Pseudomonas mendocina* KR1 gene cluster encoding toluene-4-monooxygenase. *J. Bacteriol.* **173:**5315–5327.

67. **Zylstra, G. J.** 1995. Molecular analysis of aromatic hydrocarbon degradation, p. 83–115. *In* S. J. Garte (ed.), *Molecular Environmental Biology*. Lewis Publishers, Boca Raton, FL.

68. **Zylstra, G. J., and D. T. Gibson.** 1997. Aromatic hydrocarbon degradation: a molecular approach, p. 183–203. *In* J. K. Setlow (ed.), *Genetic Engineering: Principles and Methods*. Plenum Press, New York, NY.

69. **Zylstra, G. J., and D. T. Gibson.** 1989. Toluene degradation by *Pseudomonas putida* F1. Nucleotide sequence of the *todC1C2BADE* genes and their expression in *Escherichia coli. J. Biol. Chem.* **264:**14940–14946.

70. **Zylstra, G. J., R. H. Olsen, and D. P. Ballou.** 1989. Cloning, expression, and regulation of the *Pseudomonas cepacia* protocatechuate 3,4-dioxygenase genes. *J. Bacteriol.* **171:**5907–5914.

24

Cellulases, Hemicellulases, and Pectinases

MICHAEL E. HIMMEL, JOHN O. BAKER, WILLIAM S. ADNEY,
AND STEPHEN R. DECKER

24.1. INTRODUCTION

Fossil fuels, which are finite and nonrenewable, supplied 86% of the energy consumed in the United States in 2002. Biomass is renewable, offers an alternative to conventional energy sources, and provides national energy security, economic growth, and environmental benefits. The potential impact of biofuels is great, but both mid- and long-term research strategies to obtain a better understanding of plant cell wall biosynthesis and degradation are needed.

Plant cell walls are the planet's dominant form of biomass. Plant cell walls and structural tissues are composed of three primary structural polymers: cellulose, a homopolymer of β-(1,4)-linked cellobiose residues; hemicellulose, a branched cross-linking heteropolymer of varied composition; and lignins (see chapter 25 in this volume for a discussion of lignin and ligninolytic enzymes). Plant cell wall is a complex of cellulose microfibrils embedded in a matrix of hemicellulose and lignin. Studies consistently show positive correlation between the enzymatic digestibility of cellulose and the removal of hemicellulose (62, 81), supporting the notion of close spatial relationships between these two polymers. Cellulose is the most abundant renewable organic polymer in the biosphere and is highly crystalline, water insoluble, and relatively resistant to depolymerization. Crystalline cellulose is of particular scientific interest because strands of cellulose make up the core of the structural elements comprising plant cell walls. Not only do these elementary microfibrils provide strength and flexibility to plants, but also the crystalline cores are quite resistant to chemical and biological degradation. As a result, proposed renewable energy schemes that rely on fermentable sugars from plant polysaccharides are impeded partly because of the recalcitrance of one of the major constituents, crystalline cellulose. The study of cellulases is also important from the standpoint of microbial conversion of biomass to feeds and chemical feed stock.

The complex structure of hemicelluloses has dictated a correspondingly diverse array of hemicellulases. Generally, each structural feature in hemicellulose has an associated enzyme that can hydrolyze or chemically modify this feature; however, many enzymes needed to form and deconstruct cell wall structures remain a mystery today. The definitive enzymatic degradation of cellulose to glucose, probably the most desirable fermentation feedstock, is generally accomplished by the synergistic action of three distinct classes of enzymes (Fig. 1): endoglucanases, exoglucanases, and β-glucosidases (cellobiases) (115). The enzymes involved in the deconstruction of various hemicelluloses have significantly more variability, as the bond specificities are much more complex than those in cellulose (Fig. 2 and 3). A recent review by Wyman et al. (163) covers more details on these enzymes and their activity on plant biomass.

Nature has evolved numerous enzymes that effect the depolymerization of cellulose to glucose. Bacterial cellulases allow herbivores to digest cellulose into glucose, and

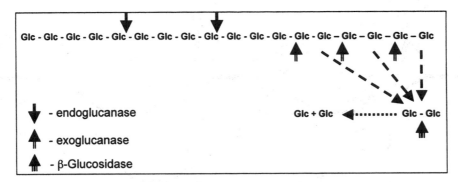

FIGURE 1 Action of the major cellulase enzymes on cellulose. Endoglucanases cleave random internal β-1→4 bonds. The major exoglucanases cleave cellobiose units from the reducing chain ends in a processive manner, whereas minor exoglucanases may cleave glucose units from the non-reducing end of the chain (not shown). β-Glucosidases cleave cellobiose to two glucose units.

fungal celluloses allow organisms to decompose woody materials. Of particular interest in the conversion of biomass to transportation fuels is the cellobiohydrolase I (Cel7A) enzyme isolated from *Trichoderma reesei*. In many ways, this highly active enzyme with specificity for crystalline cellulose is the key component in a complex cellulase system containing many other (lesser) activities. Also of keen interest are the pathways of degradation utilized by microbial plant decay enzyme systems, for hidden in the knowledge of

the natural deconstruction of plants lies the key to improved biomass conversion processes.

24.1.1. Cellulases

1. The "endo-1,4-β-glucanases" or 1,4-β-D-glucan 4-glucanohydrolases (EC 3.2.1.4) act randomly on soluble and insoluble 1,4-β-glucan substrates. These are commonly measured by detecting the decrease in viscosity or reducing groups released from carboxymethylcellulose (CMC) (137).

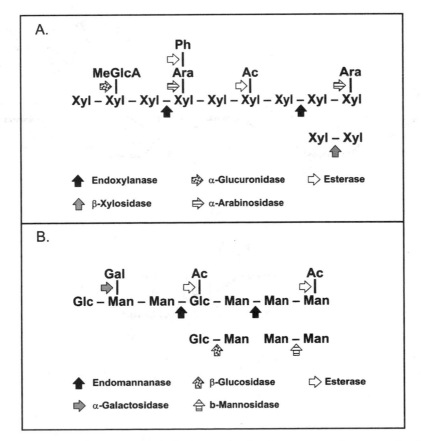

FIGURE 2 Action of major enzymes involved in the depolymerization of generic xylan (A) and glucomannan (B) chains. Additional debranching enzymes are required for the numerous variants found in the different hemicelluloses.

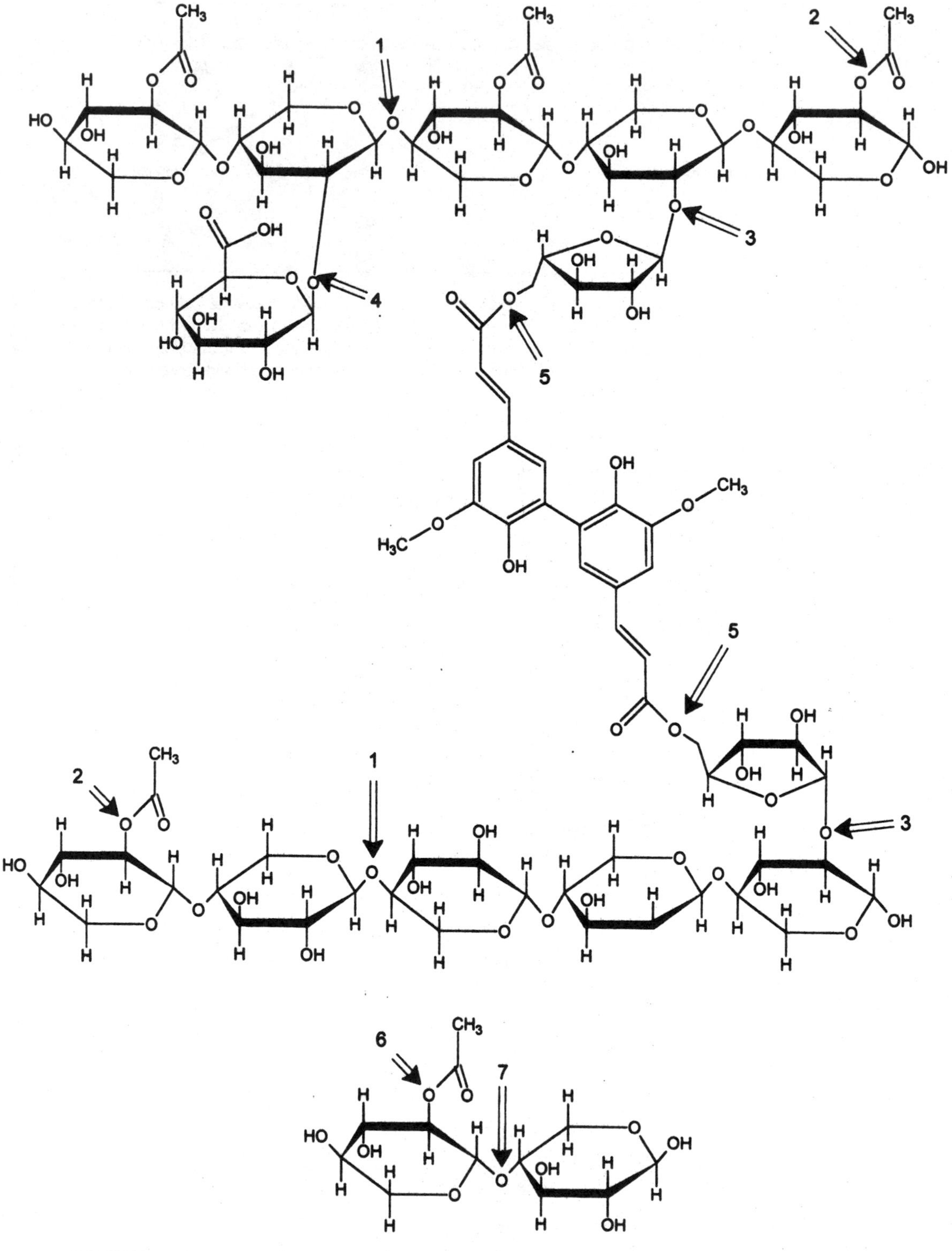

FIGURE 3 More specific action of debranching enzymes involved in the deconstruction of arabinoxylan. The structure is a generalized diagram of arabinoglucuronoxylan. Enzyme activities: 1, endoxylanase; 2, acetyl xylan esterase; 3, α-L-arabinofuranosidase; 4, α-D-glucuronidase; 5, ferulic acid esterase; 6, acetyl esterase; and 7, β-xylosidase.

2. The "exo-1,4-β-D-glucanases" include both the 1,4-β-D-glucan glucohydrolases (EC 3.2.1.74), which liberate D-glucose from 1,4-β-D-glucans and hydrolyze D-cellobiose slowly, and 1,4-β-D-glucan cellobiohydrolase (EC 3.2.1.91), which liberates D-cellobiose from 1,4-β-glucans. Differentiation of these enzyme classes requires analytical techniques to distinguish glucose and cellobiose and is usually carried out by high-performance liquid chromatography

(HPLC) or gas chromatography (GC) (5, 22). These enzymes can be further distinguished by their ability to liberate free sugars from either the reducing or the nonreducing end of the cellulose chain (59, 90). Determination of which specificity a given enzyme has is usually carried out through synergistic studies with enzymes of known orientation (67, 70, 128).

3. The "β-D-glucosidases" or β-D-glucoside glucohydrolases (EC 3.2.1.21) release D-glucose units from cellobiose and soluble cellodextrins, as well as an array of glycosides. Measurement of this activity is carried out on cellobiose or cello-oligomers, with product analysis by HPLC or GC or via direct spectrophotometric or fluorometric analysis of various chromogenic and fluorogenic analogues of cellobiose and cello-oligomers (17, 38, 102, 144).

24.1.2. Hemicellulases

1. The endoacting hemicellulase enzymes attack polysaccharide chains internally, with very little activity on short oligomers; i.e., the degree of polymerization (DP) is <3. Endoxylanases (EC 3.2.1.8) are specific for β-(1→4)-xylopyranose polymers (i.e., the backbone of xylan) while others are specific for other hemicellulose polymers, such as mannan [endo-(1→4)-β-mannosidases; EC 3.2.1.78] or β-glucans [endo-(1→3)-β-D-glucosidase; EC 3.2.1.39]. As with endocellulases, these activities can be measured through viscosimetry or by measuring the production of reducing sugar end groups from a given hemicellulosic polymer.

2. The exoacting enzymes tend to act from either the reducing or the nonreducing termini and are specific to the type and length of the polymer. Some exoacting enzymes act preferentially on short-chain substrates (DP, 2 to 4), and some are more active on larger substrates (DP, >4). Xylan (1→4)-β-xylosidase (EC 3.2.1.37), glucan (1→3)-β-glucosidase (EC 3.2.1.58), and mannan (1→4)-mannobiosidase (EC 3.2.1.100) are exoacting enzymes specific for xylan, β-(1→3)-glucan, and mannan, respectively.

3. The so-called "accessory" enzymes are also required for hydrolysis of hemicellulose in native plant tissue. This category includes a variety of acetyl xylan esterases (EC 3.1.1.72), acetyl esterases (EC 3.1.1.6), and esterases that hydrolyze lignin glycoside bonds, such as ferulic acid esterase (EC 3.1.1.73) (146) as well as enzymes involved in the cleavage of specific hemicellulose sidechains (such as α-L-arabinofuranose), glucuronic acid, and 4-O-methyl-glucuronic acid groups.

24.2. CELLULASE, HEMICELLULASE, AND PECTINASE ASSAYS—OVERVIEW

Literature describing the assay of general cellulase activity (or of individual component enzymes) has broadened considerably since the first reports by Mandels et al. (94) indicating that reducing-sugar release and substrate weight loss could serve as suitable cellulase assay methods. To some extent (for appropriate substrates), these methods are still considered adequate. However, modern assays based on molecular weight (MW) analysis detected by HPLC-size exclusion chromatography (SEC), coupled enzymes, viscometry, hydrolysis of dyed or derivatized insoluble and soluble polymers, and hydrolysis of derivatized or labeled low MW substrates have greatly enhanced the understanding of these complex systems. Cellulose structure and physical disruption of cellulose microfibrils have also been exam-

ined by microscopy, using traditional light microscopy, electron microscopy, and more recently atomic force microscopy (63, 66, 85, 89, 120, 126, 131, 166). This type of observation is critical when the properties of the cellulose fiber are in question, not when the goal is total hydrolysis. There has also been work using gel permeation chromatography to characterize changes to cellulose structure by examining the products of cellulase action on wood fiber (121). This said, workers in the field are reminded that only assays designed to measure the conversion of cellulose from the actual biomass substrates in question are ultimately valid performance measures. As with much of biotechnology today, high-throughput methods have also been developed to increase the speed and accuracy of cellulase assay (41).

For critical measurement of cellulase and hemicellulase specific activity, precise and accurate measurements of protein content are also necessary. A round-robin assay of the protein content in commercial cellulase preparations demonstrated that labs already working in the field were not able to agree within 1 or 2 standard deviations when using the Bradford dye binding, Kjeldahl, and Biuret methods (4). Note that for these experiments the samples were "standardized" using protein determination by amino acid analysis.

24.2.1. IUPAC General Methods for Cellulase and Hemicellulase Mixtures

As a result of significant effort by an international committee of cellulase researchers and the International Union of Pure and Applied Chemists (IUPAC), a procedure was published in 1987 describing the use of microcrystalline cellulose and measurement of reducing sugars by the dinitrosalicylic acid method of Miller (109) in the context of a highly specific assay protocol (56). In fact, the text of this protocol must be followed carefully to achieve comparable results. The rationale developed in this IUPAC method is that to be maximally useful, all assays for cellulase activity must be applied to an identical cellulosic substrate, i.e., Whatman no. 1 filter paper, and that enzyme action on the substrate must be measured during the initial first-order reaction rates. The definitive level of hydrolysis has been set at 4% (wt/wt) of the cellulose from a 1.0- by 6.0-cm (50-mg) Whatman no. 1 test coupon; i.e., 2 mg, converted to glucose after a 60-min incubation at 50°C. After accounting for the addition of water to the glycosidic bond, the actual level of hydrolysis measured is 3.6% of the substrate. The concentration (or actually dilution) of enzyme preparation required to effect this level of depolymerization is converted, through a somewhat indirect procedure, to the cellulase activity in filter paper units (FPU) per milliliter. For example, an undiluted cellulase preparation that yields exactly 2 mg of glucose during the IUPAC assay has 0.37 FPU/ml. This fractional unit is the lowest cellulase activity measurable with the IUPAC assay. Note that because the IUPAC FPU assay is nonlinear due to hydrolysis of an insoluble (and structurally heterogeneous) substrate, the use of traditional international units of enzyme activities, based on initial velocities, is invalid. Here, a single incubation time (60 min), a single temperature (50°C), and, crucially important, a single resulting extent of hydrolysis (2 mg of glucose-equivalents released from 50-mg filter paper) are used for all samples. It should be stressed that this assay is difficult to reproduce consistently. Attention must be paid to all of the details, including assay tube diameter, rolling and positioning of the filter paper strip, and avoidance of large, single-step dilutions.

The IUPAC cellulase assay has many significant limitations, and it merely serves as the best existing method. The IUPAC commission warns, for example, that extrapolation of required glucose release from highly diluted or concentrated solutions of enzyme is not permitted. Indeed, the assays used to confirm the release of 2 mg of glucose must be conducted with enzyme dilutions that closely bracket the actual value. The implication is that cellulase solutions too dilute to release 2 mg of glucose must either be concentrated to an appropriate level or pronounced "not assayable" by the IUPAC method. The authors have also noted that the nonlinearities inherent in the assay render erroneous the estimation of new cellulase activities from simple dilution. For highest accuracy, every working solution made from an enzyme stock must be reanalyzed for activity, a condition that complicates most analytical procedures.

A similar approach was proposed for measuring hemicellulases in 1987 (58). This method relies on meeting a standard level of conversion of the xylan fraction in oat spelt xylan to xylose in a specified period of time under standard conditions.

24.2.1.1. Method Using Automated IUPAC Approach

From the above discussion, it is clear that traditional cellulase assays are tedious and time-consuming. Multiple enzyme components, an insoluble substrate, and generally slow reaction rates have plagued cellulase researchers interested in creating cellulase mixtures with increased activities and/or enhanced biochemical properties. Although the IUPAC standard measure of cellulase activity, the filter paper assay (FPA), can be reproduced in most laboratories with some effort, this method has long been recognized for its complexity and susceptibility to operator error. The automated FPA method reported by Decker and coworkers (41) is based on a Cyberlabs C400 robotics deck. This automated system is equipped with customized incubation, reagent storage, and plate-reading capabilities which allow rapid evaluation of cellulases with a maximum throughput of 84 enzyme samples per day when performing the automated FPA. Although the assay is not directly comparable to the standard IUPAC FPA, the results are reproducible and function well to compare cellulase mixtures and components assayed previously.

24.2.2. General Non-IUPAC Methods for Cellulase Mixtures

Many cellulase enzyme preparations are simply not concentrated enough to cause the required release of 2 mg of glucose from the 50-mg filter paper sample in 60 min as in the IUPAC method described above. If these samples cannot be concentrated accurately (which is often the case), traditional FPU cannot be measured. In such cases, however, the IUPAC committee recommends that the reducing-sugar release per unit time be accepted as a "provisional" measure of enzyme activity. This is similar to the pseudo-initial-rate approach often used in the decade previous to the IUPAC report to measure cellulase activity from a wide variety of substrates. These substrates may include filter paper (95), Avicel (161), a plant-fiber-derived "microcrystalline" cellulose originated by FMC Corporation and presently supplied in 50-μm particle size by Sigma-Aldrich Co. (St. Louis, MO) under the brand name BioChemika, dewaxed cotton (57), or phosphoric-acid-swollen cellulose (136). Methods based on the use of antibiotic disks (135) and turbidity development (72) also predated the IUPAC study.

24.2.2.1. Methods from Empirical Mathematical Models

Contributions made in the assay of cellulase preparations by Sattler et al. (133) describe a relationship between extent of hydrolysis, reaction time, and enzyme concentration. This procedure permits the effectiveness of different enzymes and of different pretreatment methods to be ranked. For this method, cellulose hydrolysis data collected from hyperbolic functions of substrate concentration versus cellulase enzyme concentration at various timed incubations are examined. A double-reciprocal plot based on the relationship

$$(Y/C_o)^{-1} = (KC_o/Y_{max})[E]^{-1} + (Y_{max}/C_o)^{-1}$$

where Y/C_o is the fraction of substrate hydrolyzed; $[E]$ is given in filter paper units per gram of substrate initially added; and Y_{max}/C_o is the fraction of substrate that could be maximally hydrolyzed at an infinite enzyme concentration. The y-axis intercept in the double-reciprocal plot, $(Y_{max}/C_o)^{-1}$, may be used to quantify the quality of the enzyme preparation. Ideally, an enzyme should have a high Y_{max} and a low value for KC_o/Y_{max}. Adney and coworkers (3) used this general method successfully to model the action of commercial *Trichoderma reesei* cellulase preparations on Sigmacell 50, a cotton-linter-derived microparticulate cellulose preparation (Sigma-Aldrich Co.). Results from double-reciprocal plots of enzyme activity, (percent conversion)$^{-1}$, versus loading, (FPU/g cellulose)$^{-1}$, enabled extrapolation to infinite enzyme loading or maximal digestibility.

24.2.2.2. Method Using a Membrane Dialysis Reactor

A new saccharification assay uses a continuous buffer flux through a membrane reactor to remove the solubilized saccharification products, thus allowing high extents of substrate conversion without significant inhibitory effects from the buildup of either cellobiose or glucose (11). This diafiltration saccharification assay (DSA) was originally designed on the assumption that it could be used to obtain direct measurements of the performance of combinations of cellulase and substrate under simulated simultaneous saccharification and fermentation (SSF) conditions, without the saccharification results being complicated by factors that may influence the subsequent fermentation step (5). Comparisons between actual SSF and DSA experiments using the same ranges of cellulase loadings (ratios of cellulase protein to the cellulose content of biomass substrates) have since made it plain, however, that there are special conditions existing in the SSF environment, as opposed to the much simpler DSA environment, that can affect the saccharification step as well as the fermentation step of the process. Rather than being an exact representation of the saccharification process that occurs in SSF, DSA data were useful for comparison with SSF data in efforts to identify the influences of factors other than product inhibition on the performance of cellulases in SSF. The principle value of DSA, however, lies in its ability to measure cellulase saccharification performance substantially independent of end product inhibition over a wide range of substrate conversion.

An additional characteristic of DSA that must be taken into account in data analysis is that the appearance of product is measured in the effluent stream, rather than in the reaction vessel itself. For all practical cases, which involve neither an infinitely small reactor volume nor an infinite rate of flow of buffer solution through the membrane, there is a lag between the generation of product in the reactor vessel and the detection of the product in the effluent stream. This lag is especially pronounced at high enzyme loadings, which result in rapid generation of soluble-sugar products, at low buffer flow rates (low dilution rates), or with a combination of the two. This effect may be accommodated by data-processing approaches that account for the amount of product held up in the reactor vessel at a given instant. After corrections are made for the lag in detection, the very detailed and precise progress curves yielded semiautomatically by this assay (Fig. 4) become intrinsic to the saccharification being studied and are transportable and scalable to cases in which the reaction occurs in different reactor geometries.

24.2.2.3. Use of Multiangle Laser Light Scattering To Estimate MW Reduction of Cellulose
More recently, methods to use solvents to solubilize cellulose that has been treated with purified enzymes and characterize the products using SEC have been developed (154). The high-performance size exclusion chromatography-multiangle laser light scattering (HPSEC-MALLS) technique was used to determine the number-average molecular weight ($\overline{M}_n$) of CMC that has been treated with cellulase plus solvents, thus allowing the quantification of the number of the bonds broken during degradation of CMC. However, reproducibility of the method was low, especially for the high-molecular-weight fragments of CMC at the beginning of hydrolysis. As hydrolysis proceeded to the more advanced stages, the HPSEC-MALLS method gave over-

estimation (compared to the reducing end group analysis) of values for ($\overline{M}_n$), probably due to insufficient sensitivity of the light-scattering detector for the low-molecular-weight products of CMC degradation.

24.2.2.4. Detection of Cellulase Activity in Gels
Overall cellulose-depolymerizing activity can also be detected by differential dye binding at low concentrations (0.002% [wt/vol]) of finely ground microcrystalline cellulose. According to the procedures of Sharrock (134) or Bartley et al. (12), Congo red visualization of degraded insoluble cellulose can be used to estimate the presence of general cellulase activity in polyacrylamide gels. Bartley et al. (12) also utilized reduction of tetrazolium dyes by cello-oligomers to distinguish between endo- and exocellulase activity. Congo red staining has also been used in in vitro assays for cellulase activity (27). Although useful for detecting cellulase components that solubilize microcrystalline cellulose, such as endoglucanases and some exoglucanases, its usefulness is limited with components such as endodependent exoglucanases and β-glucosidases, which do not efficiently solubilize cellulose microfibrils. Cellulases may also be detected in slab gels using either Western blotting or enzyme-linked immunosorbent assay, as reported for enzymes from *T. reesei* (80, 116). Although useful for both quantification and detection of cellulase, both methods are dependent on access to appropriate antibodies of known reactivity. One particular problem noted by the authors is the difficulty of obtaining antibodies against the catalytic domain of *T. reesei* cel7A. With one exception reported in the literature (7), anti-*T. reesei* cel7A antibodies are specific to either the linker or cellulose-binding module, making detection of catalytic domain degradation products difficult.

24.2.3. Assays for Specific Cellulase Enzyme Classes

24.2.3.1. Endoglucanase Analysis
Endoglucanase activity is determined primarily by two (widely different) procedures, both of which release reducing sugars (114, 140, 161) from CMC and reduce fluidity, $\Delta\phi$, of CMC measured by either capillary or rotary-type viscometers (33, 43, 74, 97). In general, however, endoglucanases have the widest range of hydrolytic potential of the specific cellulases, because it is possible to hydrolyze polymeric, substituted substrates, such as ostazin brilliant red-hydroxyethylcellulose (OBR-HEC), azo-dyed cellulose (cellulose azure) and azo-dyed and cross-linked HEC (AZCL-HEC), as well as the low MW fluorogenic substrates, such as 4-methylumbelliferyl-cellobiose (MUC), 4-methylumbelliferyl-lactopyranoside (MUL), or 4-methylumbelliferyl-cellotriose (MU-G₃) (6, 17, 165).

Reducing-Sugar Release from CMC
Reducing-sugar detection methods are traditionally based on initial rate measurements introduced by Wood and McCrae in 1977 (161). The reducing sugars are typically measured by the Somogyi (140) and Nelson (114) procedures, which measure reduction of Cu^{2+} to Cu^+ in alkaline solution, or by the DNS assay, whereby reducing sugars reduce DNS to 3-amino,5-nitrosalicylic acid under alkaline conditions. One unit of enzyme activity is defined as the amount of enzyme needed to liberate reducing sugars equivalent to 5 μg of glucose per h. A variation of the original IUPAC method for filter paper activity uses CMC as a substrate (56). Here, the IUPAC CMC unit of activity is found from the dilution of enzyme necessary to produce 0.5 mg of

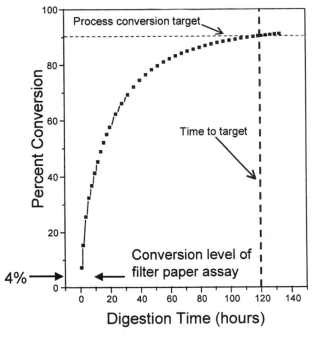

FIGURE 4 DSA progress curve of commercial cellulase preparation (Spezyme from Genencor International) at a loading of 20 FPU/g of cellulose acting on Sigmacell 20 at 38°C. After the actual experimental measurements, empirical fits to the data were done. The expression "time-to-target kinetics" is considered more descriptive of the approach than is the classical "integrated-rate kinetics."

glucose from 0.5 ml of a 2.0% CMC solution after 30 min at 122°F (50°C). The recommended substrate was CMC 7L2 (Aqualon Division of Hercules, Inc., Wilmington, DE; degree of substitution [DS] = 0.7; and approximate DP, 400) in 0.05 M sodium citrate buffer, pH 4.8. For comparison purposes, this method should be used to establish the specific activity of a purified endoglucanase preparation. A more commonly used method is that of Mandels et al. (96), where CMC 4M6F (Hercules, Inc.; DS, 0.38 to 0.48) is used as a substrate and units are expressed as micromoles of reducing sugar released per minute of incubation time. Methods proposed by Shoemaker and Brown (136) and Håkansson et al. (65) also rely on initial-rate reducing-sugar release from various CMCs. In contrast, Rescigno et al. (123) have used ruthidium red to stain CMC and use spectrophotometric absorption of the soluble fragments for determination of endocellulase activity.

The measurement of the initial rate of reducing end group formation from the action of endoglucanases on cellulose can also be accomplished using disodium 2,2′-bicinchoninate (BCA) (73). This reagent was found to be the best choice in a recent comparison of methods for the determination of endoglucanase activity (154). The BCA method was highly sensitive and simple to perform and directly gave the number of bonds broken, thus allowing for expression of endoglucanase activity in international units (micromoles of beta-1,4-glucosidic bonds hydrolyzed in 1 min during the initial period of hydrolysis).

Viscosimetric Assays
Viscometric approaches to cellulase measurement activities are important because other methods measure only the number of glycosidic bonds cleaved in a polymeric substrate, without providing any information about location in the substrate of the bonds cleaved. Viscometric methods measure a substantial change in a physical property of the substrate polymer that is a very sensitive function of both the number and the location of the bonds cleaved (33, 43, 97). For this reason, even though the recommended international units of carboxymethylcellulase are given in terms of glycosidic bonds cleaved, it is important to measure both bond cleavage (most often by measurement of sugar-reducing groups) and the change in solution viscosity as enzymatic hydrolysis proceeds. Vlasenko and coworkers (154) found the viscometric method to be simple to perform and highly sensitive for the internal bonds cleaved. However, this method does not account for the hydrolysis of CMC near the chain end and thus only allows for expression of endoglucanase activity in arbitrary viscometric units.

24.2.3.2. Exoglucanase Analysis
The process of detecting and verifying exoglucanases (cellobiohydrolases [CBHs] in context of the fungal cellulase systems) has long been controversial. If purified proteins are available, careful comparisons of reducing-sugar yields and fluidity values from CMC hydrolysis as a function of enzyme concentration can be used to judge whether an enzyme is more endoglucanase-like or CBH-like. Of course, purified enzymes can also be subjected to product analysis from the hydrolysis of a series of derivatized; i.e., radiolabeled, chromophoric, or fluorophoric cello-oligomers for further verification (153, 155). One class of these derivatives, cellobiosyl fluorides, has been reported to distinguish between CBH I and CBH II from *T. reesei* based on cleavage activity on the alpha and beta conformations of the cellobiosyl fluorides (82). Claeyssens et al. have also reported this type of rigorous analysis for fungal CBH I and CBH II (39). Further

specificities can be determined for proposed exoglucanases from analytical product evaluation by HPLC (13, 76). This is a much more definitive method of distinguishing endo- from exoacting cellulases. In general, exoglucanases such as CBH I can be expected to hydrolyze the aryl substrates MUC and MUL at the agluconic bond, and not the substituted, soluble celluloses such as AZCL-HEC, OBR-HEC, and CMC (152). Phosphoric-acid-swollen cellulose is also used as a substrate for exoglucanases; however, some endoglucanases hydrolyze this cellulose form as well (76). Analysis of higher oligomeric derivatives has proven to be complex, especially with endoglucanases and CBH I (153). Furthermore, as endoglucanases are highly synergistic with exoacting glucanases, the presence of endoglucanases significantly complicates efforts to quantify exoglucanase activity and can be compensated for only by the separate purification and kinetic characterization of the endoglucanase. Attempts persist in trying to link this synergy effect to the cellulose binding module. Some evidence indicates that the cellulose binding module alone can have a synergistic effect on the activity of cellulases, both exo- and endoacting types (87, 88, 93).

24.2.3.3. β-D-Glucosidase Analysis
β-D-Glucosidase and cellobiase activities are usually determined according to the method of Wood (159) as aryl-β-glucosidase activity by the hydrolysis of *p*-nitrophenyl-β-D-glucopyranoside. The concentration of *p*-nitrophenol was determined from the absorbance at A_{410} under alkaline conditions induced by the addition of 2 M Na_2CO_3. One unit of activity was defined as the amount of enzyme that catalyzes the cleavage of 1.0 μmol of substrate per min at 37°C. If necessary, β-D-glucosidases can be distinguished from cellobiases by the relative differences in the initial rates of aryl-β-D-glucosides and cellobiose. Furthermore, the unique and acute sensitivity of β-D-glucosidase to inhibition by gluconolactone provides a method to assess exoglucanase activity in mixed systems of these two enzymes. This approach is necessary because β-D-glucosidase also cleaves the agluconic, as well as the holosidic bond of aryl-glucosides (44).

A similar approach is often used to assay other arylglycosidases (2, 17). This practice has been made possible by the availability of many *o*- and *p*-linked aryl-glycosides including (but not limited to) β-xylosides, β-mannosides, β-galactosides, and L-arabinofuranosides.

24.2.4. General Hemicellulase Assays
In 1991, 20 laboratories participated in a collaborative investigation of assays for endo-1,4-beta-xylanase activity based on production of reducing sugars from polymeric 4-O-methyl glucuronoxylan (9). The standard deviation of the results reported in this analysis was 108% of the mean when the laboratories used their own substrates and methods routinely in each laboratory. Significant reduction in interlaboratory variation was obtained when all the participants used the same substrate for activity determination, each with their own assay procedure. Not surprisingly, the level of agreement was further improved (i.e., 17% of the mean) when both the substrate and the assay procedure were standardized. A subset of these laboratories (9) also participated in preliminary testing of an assay based on the release of dyed fragments from 4-O-methyl glucuronoxylan dyed with Remazol Brilliant Blue (RBB) (16). The relative standard deviations of the results obtained by these laboratories were about 30% for an optimum range of xylanase activity in the reaction mixture. A major distinction in hemicellulase assays arises from the heterogeneous structure of these poly-

saccharides. Ignoring the actual chemical composition of hemicelluloses from various sources, the enzyme activities required are divided into those that debranch the polymeric backbone and those that depolymerize the polysaccharide chain(s).

A general approach to detecting glycosyl hydrolase activity is through the utilization of dyed polysaccharide substrates. Dyed polysaccharides have been utilized to determine activities of cultures on various polysaccharides, both as activity screens and as quantitative measures. These substrates include both soluble and insoluble forms (dependent mainly on the properties of the native polysaccharides) and

include azurine (azo-), RBB, and OBR, among others (14, 19, 55, 106). For azo-, RBB- and OBR-linked substrates, clearing zones on petri plate or acrylamide gel agar overlays indicate active colonies or protein bands (23, 98, 138, 143, 157). The cross-linked version (AZCL-polysaccharides; Megazyme, Inc., Bray, Ireland) has also been used to screen for activity of various glycosyl hydrolases (139, 165). In the case of AZCL substrates, the result of activity is a blue halo surrounding active colonies or dye release into microtiter plate wells. The authors have used this technique extensively to screen both environmental samples and recombinant libraries for glycosyl hydrolase activities (Fig. 5).

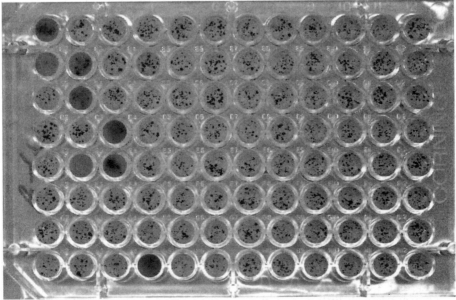

FIGURE 5 AZCL-polysaccharide hydrolysis in a petri plate (top) and a microtiter plate. The dark particulates are the AZCL-β-glucan (top) and AZCL-galactan (bottom). Soluble blue dye is released upon hydrolysis.

24.2.4.1. Hemicellulose-Debranching Enzymes

The debranching of the xylan backbone results in a wide variety of soluble low-molecular-weight compounds as products. Typically, these products are measured either by HPLC or GC. The difficulty in assaying these products is not so much in the detection as in obtaining the correct substrate for the enzyme. Most commercial xylan products are extracted by alkaline treatment, essentially hydrolyzing any ester linkages by saponification; i.e., any acetyl-, coumaroyl-, or feruloyl esters are destroyed. Glycosidic side chains, such as arabinose or glucuronic acid, are left intact; however, the polymer is typically insoluble. Enzyme studies using these substrates must be interpreted with caution, as the native esterified xylan is soluble. Extraction by dimethyl sulfoxide or steam has been used to prepare "native" xylan, in which the esters are still intact and the polymer is soluble in water (37).

Arabinofuranosidases

α-L-Arabinofuranosidase (EC 3.2.1.55) cleaves α-L-arabinofuranosides from the arabinoxylan xylose backbone. It has also been shown to enhance the release of ferulic and coumaric acid from arabinoxylan, presumably through an affinity for hydrolyzing phenolic acid-substituted arabinose side chains (160). Although in the context of hemicellulose hydrolysis the activity most often reported is hydrolysis of the α-(1→2)-glycosidic linkage of the arabinofuranoside to the xylan backbone, some of these enzymes have been shown to cleave linear and/or branched α-(1→5)-linked arabinan side chains found in some pectins, resulting in some confusion regarding the specificity of this enzyme class (99, 105, 129, 164). Although most assays are carried out with extracted arabinoxylan, p-nitrophenyl-arabinofuranoside has also been used as a substrate (42, 60, 86).

Esterases

Acetyl esterase (EC 3.1.1.6) removes acetyl esters from acetylated xylose and short-chain xylo-oligomers. Its polymer-acting counterpart, acetyl xylan esterase (EC 3.1.1.72), has a similar activity but is more active on polymeric xylan (37). In addition to acetate-specific enzyme detection kits, HPLC, or GC analysis of acetate release from native extracted xylan and chemically acetylated xylan, colorimetric substrates such as p-nitrophenol acetate and β-napthyl acetate or the fluorometric substrate 4-methylumbelliferyl acetate are also used (18, 37). The third esterase, ferulic acid esterase (EC 3.1.1.73), hydrolyzes the ester bond between ferulic acid or coumaric acid and the arabinose side chain of arabinoxylan. Assays for this activity are usually carried out using starch-free wheat bran or cellulase-treated graminaceous biomass as a substrate and monitoring ferulic- or coumaric acid released by HPLC or thin-layer chromatography. When preparing enzyme-treated substrates, care must be taken to employ phenolic acid esterase-free cellulases (37). Other substrates include methyl and ethyl esters of the phenolic acids and finely ground plant biomass (20, 30, 104).

Glucuronidases

In hardwood xylans, xylan α-1,2-glucuronosidase (EC 3.2.1.131) and α-glucuronidase (EC 3.2.1.139) are involved in debranching the xylan backbone through removal of α-(1→2)-linked glucurono- and 4-O-methyl-glucuronosides (71, 79, 112, 148). Although relatively little work on these enzymes has been carried out, Tenkanen and Siika-aho reported synergy with endoxylanase utilizing deacetylated birchwood glucuronoxylan. The same report also demonstrated that acetylation interferes with glucuronidase activity and that higher activity was observed on soluble softwood 4-O-methylglucuronoxylan (148). Such synergy has been reported by other workers (35, 45, 148). Paranitrophenyl-αD-glucuronide is used as a substrate for α-glucuronidase (132), whereas xylan α-1,2-glucuronosidase is specific for an α-(1→2)-linked glucuronoside. Some glucuronidases, including membrane-bound enzymes, have been found more active on glucuronoxylo-oligomers as substrates (29, 35, 46, 111). One recent report has demonstrated the specific requirement for the 4-O-methyl group for efficient binding and positioning of the side chain in the enzyme active site (112).

24.2.5. Hemicellulose Depolymerization Enzymes

As noted for cellulases, hemicellulose-depolymerizing enzymes are divided into three classes; the endoacting, exoacting, and oligomer-hydrolyzing. Although research into the mechanisms of hemicellulose hydrolysis has been steady over the years, it has not received the attention given to cellulose hydrolysis. Despite this history, a general pattern of degradation is beginning to emerge.

Xylanases

Depolymerization of the xylan backbone is mediated by endoxylanases, and the oligomers are hydrolyzed by β-xylosidases. Based primarily on structural differences, the endoxylanases are divided into two major categories, the glycosyl hydrolase families 10 and 11. Assays for endoxylanases follow the same general patterns as endocellulase assays. Viscosity reduction, reducing sugar production, dye-release, solubilization, zymogram analysis, and colorimetric/fluorometric analogues are all utilized in determining the endoxylanase activity (10, 25, 49, 61, 91, 92, 108, 125, 142). DNS detection of reducing sugars from xylan is the most cited method. Endoxylanases tend to act preferentially on polymers of a certain DP. *Schizophyllum commune* endoxylanase preferred oligomers with seven subunits (24), while other enzymes exhibited true endo-type activity, with decreasing activity for lower DP oligomers (36, 40, 117). There are numerous reports of β-xylosidases that cleave short-chain xylo-oligomers to xylose. In these cases, product detection was carried out by direct HPLC analysis or hydrolysis of p-nitrophenyl-β-D-xylopyranoside (84, 122, 130, 141).

β-Glucanases

β-Glucan is a glucopyranose polymer containing either β-(1→3) or mixed β-(1→3), β-(1→4) linkages. The ratio of (1→4) to (1→3) linkages varies in different species and gives specific properties to individual β-glucan polymers. Because of the differences in the linkages, different enzymes are required to cleave the two forms of β-glucan (26, 68, 69, 78, 83, 162). Notably, one report utilized an enzyme-linked immunosorbent assay in microtiter plates coated with biotinylated β-glucan to determine activity (156).

Mannanases, Glucomannanases, and Galactomannanases

While mannan is characteristically described as a linear β-(1→4) mannopyranose polymer, galactomannan is composed of a polymeric β-(1→4) mannopyranosyl backbone highly substituted with β-(1→6)-linked galactopyranose residues (32, 107). The degree of substitution varies with the source. Glucomannan, found mainly in the root of the Konjac plant (*Amorphophallus konjac*), consists of a β-(1→4)-linked mannopyranose and glucopyranose backbone in a ratio of 1.6:1 (150). The enzymes involved in depolymerization of the mannans consist of β-mannanase (EC 3.2.1.78), the endoacting enzyme, and β-mannosidase (EC 3.2.1.25), which produces mannose from the nonreducing end of the mannose chain (34, 118, 119). Debranching of galactomannan is primarily carried out by α-D-galactosidase (EC 3.1.2.22) (1, 54). Tenkanen and coworkers have also reported an acetyl glucomannan esterase active on the acetyl sidechains in glucomannan (145, 147, 149). There are few other data on specific debranching enzymes involved in degradation of glucomannan. Assays for mannan hydrolysis have been carried out using extracted polysaccharides as substrates, colorimetric analogues, and dyed polysaccharides (21, 48, 52, 103, 127, 151).

24.2.6. Pectinases

In addition to cellulose and the hemicelluloses, pectins are a third class of polysaccharides found in the cell wall matrix of plant cells. Pectin polymers predominantly consist of 1,4-linked α-D-galacturonic acid residues, the carboxylic groups of which are methoxylated to varying extent. Further information and good structural diagrams can be found in the recent review by Ridley et al. (124). Pectins are major constituents of middle lamellae and primary cell walls of higher plants. Pectins fall into three classes differentiated by their backbone structure and branching patterns (28, 47). Homogalacturonan (xylogalacturonan) is composed of α-(1→4)-linked galacturonic acid chains containing xylose side chains and makes up the smooth region of pectin (101). Homogalacturonan is methylated through ester-linkages to the galacturonic acid residues. Once in place, pectin methyl esterases (EC 3.1.1.11) remove these side chains and allow formation of the gel matrix (158). The rhamnogalacturonans make up the "hairy" region of pectin. In rhamnogalacturonan I (RG I), the backbone chain is composed of the disaccharide (→4)-α-D-galacturonic acid-α-(1→2)-α-L-rhamnopyranose-(1→). The rhamnose is typically substituted at the C-4 position with a branched chain of sugars made up of either galactose or arabinose or a combination of both and other sugars. The arabinose residues can be derivatized with ferulic acid. The galacturonic acid residues in the backbone are usually O-2 or O-3 acetylated and O-6 methylated. The structure and substitution patterns of RG I vary widely across plant species. Where the majority of the side chain is composed of arabinose, the side chains are referred to as arabinans. These arabinans are predominantly α-(1→5)-linked arabinofuranosyl residues substituted at the O-2 and/or O-3 positions (28). Side chains composed of galactose residues are referred to as galactans. When these galactans are further substituted with arabinan chains, they are referred to as arabinogalactans (51). In contrast to RG I, the structure of rhamnogalacturonan II (RG II) is highly conserved across the plant kingdom (100). It comprises 28 glycosyl residues, of which 7 are found in the galacturonic acid backbone. The backbone is specifically branched at four points (designated A through D chains) with some unusual sugars such as 2-O-methly-L-fucose and 3-deoxy-D-manno-2-octulosonic acid, aceric acid, and apiose (100). The structures of each branch are known, although the exact point of attachment of branch D is still unclear.

As is apparent from its complex structure, there is a diverse enzyme suite required to hydrolyze pectin. As in hemicellulase systems, there are depolymerizing and debranching enzymes, mainly esterases, that act synergistically (53, 106). The depolymerizing enzymes include both glyco-syl hydrolases, which cleave glycosidic bonds by an acid-base catalysis mechanism, and polysaccharide lyases, which hydrolyze the glycosidic bond through a β-elimination mechanism, resulting in a double bond between the C-4 and C-5 of the new nonreducing end (106). Recent reviews by Kashyap et al. (77) and Naidu and Panda (113) outline the pectinase enzymes in detail. Assay techniques involve the usual assortment of reducing-sugar production, viscosity reduction, HPLC analysis, and dye release (8, 15, 31, 50, 53, 75, 106, 110). As with other polysaccharide degradation studies, product structure has been determined with nuclear magnetic resonance spectroscopy (110).

Ruthenium red staining in plates and zymograms has also been used for assay of pectinase enzymes (53). Because of its solubility, pectin incorporated into plates can be detected by precipitation with ruthenium red (53) or hexadecyl-trimethyl-ammonium bromide, resulting in clear halos of hydrolysis around active colonies (64). Polygalacturonase (EC 3.2.1.15) is often assayed by measuring the increase in reducing groups resulting from the hydrolysis of polygalacturonic acid to oligo-galacturonates. The latter are measured using the Somogyi-Nelson assay (140). In doing this procedure, citrate buffer should be avoided in the reaction mixture because it interferes with the assay. Hydrolases such as polygalacturonases generally have pH optima at ≤6.0. Pectinlyase (EC 4.2.2.10) cleaves pectin by β-elimination and generates products with 4,5-unsaturated residues which are measured by the increase in absorbance at 232 nm. Lyases generally have a pH optimum around 8.5. Pectin esterase (EC 3.1.1.11) acts on pectin by liberating methanol and pectic acid, and the assay for this enzyme is based on the decrease in pH of the reaction mixture as a result of increase in acid production. For details of these and other assays, the reader is referred to the literature cited above.

This work was funded by the U.S. Department of Energy Office of the Biomass Program.

24.3. REFERENCES

1. **Ademark, P., R. P. de Vries, P. Hagglund, H. Stalbrand, and J. Visser.** 2001. Cloning and characterization of *Aspergillus niger* genes encoding an alpha-galactosidase and a beta-mannosidase involved in galactomannan degradation. *Eur. J. Biochem.* **268:**2982–2990.

2. **Adney, W. S., J. O. Baker, T. B. Vinzant, S. R. Thomas, and M. E. Himmel.** 1995. Kinetic comparison of beta-D-glucosidases of industrial importance. *Abstr. Pap. Am. Chem. Soc.* **209:**119–BTEC.

3. **Adney, W. S., C. I. Ehrman, J. O. Baker, S. R. Thomas, and M. E. Himmel.** 1994. Cellulase assays—methods from empirical mathematical-models, p. 218–235. *In* J. N. Saddler and M. H. Penner (ed.), *Enzymatic Conversion of Biomass for Fuels Production*, vol. 566. American Chemical Society, Washington, DC.

4. **Adney, W. S., A. Mohagheghi, S. R. Thomas, and M. E. Himmel.** 1995. Comparison of protein contents of cellulase preparations in a worldwide round-robin assay, p. 256–271. *In* M. E. Himmel, J. O. Baker, and R. P. Overend (ed.), *Enzymatic Degradation of Insoluble Carbohydrates*, vol. 618. American Chemical Society, Washington, DC.

5. **Agblevor, F. A., A. Murden, and B. R. Hames.** 2004. Improved method of analysis of biomass sugars using high-performance liquid chromatography. *Biotechnol. Lett.* **26:**1207–1210.

6. **Aho, S.** 1991. Structural and functional analysis of *Trichoderma reesei* endoglucanase I expressed in yeast *Saccharomyces cerevisiae. FEBS Lett.* **291:**45–49.

7. **Aho, S., V. Olkkonen, T. Jalava, M. Paloheimo, R. Buhler, M. Niku-Paavola, D. Bamford, and M. Korhola.** 1991. Monoclonal antibodies against core and cellulose-binding domains of *Trichoderma reesei* cellobiohydrolases I and II and endoglucanase I. *Eur. J. Biochem.* **200:**643–649.

8. **Antov, M. G., D. M. Pericin, and G. R. Dimic.** 2001. Cultivation of *Polyporus squamosus* for pectinase production in aqueous two-phase system containing sugar beet extraction waste. *J. Biotechnol.* **91:**83–87.

9. **Bailey, M. J., P. Biely, and K. Poutanen.** 1992. Interlaboratory testing of methods for assay of xylanase activity. *J. Biotechnol.* **23:**257–270.

10. **Bailey, M. J., and K. Poutanen.** 1989. Production of xylanolytic enzymes by strains of *Aspergillus*. *Appl. Microbiol. Biotechnol.* **30:**5–10.

11. **Baker, J. O., T. B. Vinzant, C. I. Ehrman, W. S. Adney, and M. E. Himmel.** 1997. Use of a new membrane-reactor saccharification assay to evaluate the performance of cellulases under simulated SSF conditions—effect on enzyme quality of growing *Trichoderma reesei* in the presence of targeted lignocellulosic substrate. *Appl. Biochem. Biotechnol.* **63–65:**585–595.

12. **Bartley, T. D., K. Murphy-Holland, and D. E. Eveleigh.** 1984. A method for the detection and differentiation of cellulase components in polyacrylamide gels. *Anal. Biochem.* **140:**157–161.

13. **Beldman, G., M. Searle-Van Leeuwen, F. Rombouts, and F. Voragen.** 1985. The cellulase of *Trichoderma viride*; purification, characterization and comparison of all detectable endoglucanases, exoglucanases and beta-glucosidases. *Eur. J. Biochem.* **146:**301–308.

14. **Berens, S., H. Kaspari, and J. H. Klemme.** 1996. Purification and characterization of two different xylanases from the thermophilic actinomycete *Microtetraspora flexuosa* SIIX. *Antonie Leeuwenhoek* **69:**235–241.

15. **Bhattacharya, S., and N. K. Rastogi.** 1998. Rheological properties of enzyme-treated mango pulp. *J. Food Eng.* **36:**249–262.

16. **Biely, P., D. Mislovicova, and R. Toman.** 1988. Remazol brilliant blue xylan—a soluble chromogenic substrate for xylanases. *Methods Enzymol.* **160:**536–541.

17. **Biely, P., D. Mislovicova, and R. Toman.** 1985. Soluble chromogenic substrates for the assay of endo-1,4-beta-xylanases and endo-1,4-beta-glucanases. *Anal. Biochem.* **144:**142–146.

18. **Blum, D. L., X. L. Li, H. Chen, and L. G. Ljungdahl.** 1999. Characterization of an acetyl xylan esterase from the anaerobic fungus *Orpinomyces* sp. strain PC-2. *Appl. Environ. Microbiol.* **65:**3990–3995.

19. **Bolam, D. N., N. Hughes, R. Virden, J. H. Lakey, G. P. Hazlewood, B. Henrissat, K. L. Braithwaite, and H. J. Gilbert.** 1996. Mannanase A from *Pseudomonas fluorescens* ssp. *cellulosa* is a retaining glycosyl hydrolase in which E212 and E320 are the putative catalytic residues. *Biochemistry* **35:**16195–16204.

20. **Borneman, W. S., L. G. Ljungdahl, R. D. Hartley, and D. E. Akin.** 1991. Isolation and characterization of p-coumaroyl esterase from the anaerobic fungus *Neocallimastix* strain MC-2. *Appl. Environ. Microbiol.* **57:**2337–2344.

21. **Bourgault, R., and J. D. Bewley.** 2002. Gel diffusion assays for endo-beta-mannanase and pectin methylesterase can underestimate enzyme activity due to proteolytic degradation: a remedy. *Anal. Biochem.* **300:**87–93.

22. **Boussaid, A., J. Robinson, Y. J. Cai, D. J. Gregg, and J. R. Saddler.** 1999. Fermentability of the hemicellulose-derived sugars from steam-exploded softwood (Douglas fir). *Biotechnol. Bioeng.* **64:**284–289.

23. **Braithwaite, K. L., G. W. Black, G. P. Hazlewood, B. R. Ali, and H. J. Gilbert.** 1995. A non-modular endo-beta-1,4-mannanase from *Pseudomonas fluorescens* subspecies *cellulosa*. *Biochem. J.* **305**(Pt. 3)**:**1005–1010.

24. **Bray, M. R., and A. J. Clarke.** 1992. Action pattern of xylo-oligosaccharide hydrolysis by *Schizophyllum commune* xylanase A. *Eur. J. Biochem.* **204:**191–196.

25. **Bronnenmeier, K., A. Kern, W. Liebl, and W. L. Staudenbauer.** 1995. Purification of *Thermotoga maritima* enzymes for the degradation of cellulosic materials. *Appl. Environ. Microbiol.* **61:**1399–1407.

26. **Brummell, D. A., C. Catala, C. C. Lashbrook, and A. B. Bennett.** 1997. A membrane-anchored E-type endo-1,4-beta-glucanase is localized on golgi and plasma membranes of higher plants. *Proc. Natl. Acad. Sci. USA* **94:**4794–4799.

27. **Carder, J. H.** 1986. Detection and quantitation of cellulase by congo red staining of substrates in a cup-plate diffusion assay. *Anal. Biochem.* **153:**75–79.

28. **Cardoso, S. M., A. M. Silva, and M. A. Coimbra.** 2002. Structural characterisation of the olive pomace pectic polysaccharide arabinan side chains. *Carbohydr. Res.* **337:**917–924.

29. **Castanares, A., A. J. Hay, A. H. Gordon, S. I. McCrae, and T. M. Wood.** 1995. D-xylan-degrading enzyme system from the fungus *Phanerochaete chrysosporium*: isolation and partial characterization of an alpha-(4-O-methyl)-D-glucuronidase. *J. Biotechnol.* **43:**183–194.

30. **Castanares, A., and T. M. Wood.** 1992. Purification and characterization of a feruloyl/p-coumaroyl esterase from solid-state cultures of the aerobic fungus *Penicillium pinophilum*. *Biochem. Soc. Trans.* **20:**275S.

31. **Castilho, L. R., T. L. M. Alves, and R. A. Medronho.** 1999. Recovery of pectolytic enzymes produced by solid state culture of *Aspergillus niger*. *Process Biochem.* **34:**181–186.

32. **Chaubey, M., and V. P. Kapoor.** 2001. Structure of a galactomannan from the seeds of *Cassia angustifolia* Vahl. *Carbohydr. Res.* **332:**439–444.

33. **Chee, K. K.** 1990. Kinetic study of random chain scission by viscometry. *J. Appl. Polym. Sci.* **41:**985–994.

34. **Chhabra, S., K. N. Parker, D. Lam, W. Callen, M. A. Snead, E. J. Mathur, J. M. Short, and R. M. Kelly.** 2001. Beta-mannanases from *Thermotoga* species. *Methods Enzymol.* **330:**224–238.

35. **Choi, I. D., H. Y. Kim, and Y. J. Choi.** 2000. Gene cloning and characterization of alpha-glucuronidase of *Bacillus stearothermophilus* no. 236. *Biosci. Biotechnol. Biochem.* **64:**2530–2537.

36. **Christakopoulos, P., W. Nerinckx, D. Kekos, B. Macris, and M. Claeyssens.** 1997. The alkaline xylanase III from *Fusarium oxysporum* F3 belongs to family F/10. *Carbohydr. Res.* **302:**191–195.

37. **Christov, L. P., and B. A. Prior.** 1993. Esterases of xylan-degrading microorganisms: production, properties, and significance. *Enzyme Microb. Technol.* **15:**460–475.

38. **Claeyssens, M., and G. Aerts.** 1992. Characterization of cellulolytic activities in commercial *Trichoderma reesei* preparations—an approach using small, chromogenic substrates. *Bioresour. Technol.* **39:**143–146.

39. **Claeyssens, M., H. Van Tilbeurgh, P. Tomme, T. M. Wood, and S. I. McRae.** 1989. Fungal cellulase systems. Comparison of the specificities of the cellobiohydrolases isolated from *Penicillium pinophilum* and *Trichoderma reesei*. *Biochem. J.* **261:**819–825.

40. **Debeire, P., B. Priem, G. Strecker, and M. Vignon.** 1990. Purification and properties of an endo-1,4-xylanase excreted by a hydrolytic thermophilic anaerobe, *Clostridium thermolacticum*. A proposal for its action mechanism on

larchwood 4-O-methylglucuronoxylan. *Eur. J. Biochem.* **187**:573–580.

41. **Decker, S. R., W. S. Adney, E. Jennings, T. B. Vinzant, and M. E. Himmel.** 2003. Automated filter paper assay for determination of cellulase activity. *Appl. Biochem. Biotechnol.* **105–108**:689–703.

42. **Degrassi, G., A. Vindigni, and V. Venturi.** 2003. A thermostable alpha-arabinofuranosidase from xylanolytic *Bacillus pumilus*: purification and characterisation. *J. Biotechnol.* **101**:69–79.

43. **Demeester, J., M. Bracke, and A. Lauwers.** 1979. Absolute viscometric method for the determination of endocellulase (Cx) activities based upon light-scattering interpretations of gel chromatographic fractionation data. *Adv. Chem. Ser.* **181**:91–125.

44. **Deshpande, M. V., K.-E. Eriksson, and L. G. Pettersson.** 1984. An assay for selective determination of exo-1,4-β-glucanases in a mixture of cellulolytic enzymes. *Anal. Biochem.* **138**:481–487.

45. **de Vries, R. P., H. C. Kester, C. H. Poulsen, J. A. Benen, and J. Visser.** 2000. Synergy between enzymes from *Aspergillus* involved in the degradation of plant cell wall polysaccharides. *Carbohydr. Res.* **327**:401–410.

46. **de Vries, R. P., C. H. Poulsen, S. Madrid, and J. Visser.** 1998. *aguA*, the gene encoding an extracellular alpha-glucuronidase from *Aspergillus tubingensis*, is specifically induced on xylose and not on glucuronic acid. *J. Bacteriol.* **180**:243–249.

47. **de Vries, R. P., and J. Visser.** 2001. *Aspergillus* enzymes involved in degradation of plant cell wall polysaccharides. *Microbiol. Mol. Biol. Rev.* **65**:497–522.

48. **Downie, B., H. W. M. Hilhorst, and J. D. Bewley.** 1994. New assay for quantifying endo-beta-D-mannanase activity using congo red-dye. *Phytochemistry* **36**:829–835.

49. **Dupont, C., N. Daigneault, F. Shareck, R. Morosoli, and D. Kluepfel.** 1996. Purification and characterization of an acetyl xylan esterase produced by *Streptomyces lividans*. *Biochem. J.* **319(Pt. 3)**:881–886.

50. **Fanta, N., A. Quaas, P. Zulueta, and L. M. Perez.** 1992. Release of reducing sugars from citrus seedlings, leaves and fruits. Effect of treatment with pectinase and cellulase from *Alternaria* and *Trichoderma*. *Phytochemistry* **31**:3359–3364.

51. **Fransen, C. T., S. R. Haseley, M. M. Huisman, H. A. Schols, A. G. Voragen, J. P. Kamerling, and J. F. Vliegenthart.** 2000. Studies on the structure of a lithium-treated soybean pectin: characteristics of the fragments and determination of the carbohydrate substituents of galacturonic acid. *Carbohydr. Res.* **328**:539–547.

52. **Fulop, L., and T. Ponyi.** 1997. Rapid screening for endo-beta-1,4-glucanase and endo-beta-1,4-mannanase activities and specific measurement using soluble dye-labelled substrates. *J. Microbiol. Methods* **29**:15–21.

53. **Gainvors, A., N. Nedjaoum, S. Gognies, M. Muzart, M. Nedjma, and A. Belarbi.** 2000. Purification and characterization of acidic endo-polygalacturonase encoded by the PGL1-1 gene from *Saccharomyces cerevisiae*. *FEMS Microbiol. Lett.* **183**:131–135.

54. **Ganter, J. L., J. C. Sabbi, and W. F. Reed.** 2001. Real-time monitoring of enzymatic hydrolysis of galactomannans. *Biopolymers* **59**:226–242.

55. **Ghangas, G. S., Y. J. Hu, and D. B. Wilson.** 1989. Cloning of a *Thermomonospora fusca* xylanase gene and its expression in *Escherichia coli* and *Streptomyces lividans*. *J. Bacteriol.* **171**:2963–2969.

56. **Ghose, T. K.** 1987. Measurement of cellulase activities. *Pure Appl. Chem.* **59**:257–268.

57. **Ghose, T. K., A. N. Pathak, and V. S. Bisaria.** 1975. Kinetic and dynamic studies of *Trichoderma viride* cellulase production, p. 111. *In* M. Bailey, T.-M. Enari, and M.

58. Linko (ed.), *Proceedings of the Symposium on Enzymatic Hydrolysis of Cellulose.* VTT, Aulanko, Finland.

58. **Ghose, T. K., and V. S. Bisaria.** 1987. Measurement of hemicellulase activities. 1. Xylanases. *Pure Appl. Chem.* **59**:1739–1751.

59. **Gilkes, N. R., E. Kwan, D. G. Kilburn, R. C. Miller, and R. A. J. Warren.** 1997. Attack of carboxymethylcellulose at opposite ends by two cellobiohydrolases from *Cellulomonas fimi*. *J. Biotechnol.* **57**:83–90.

60. **Gomes, J., I. I. Gomes, K. Terler, N. Gubala, G. Ditzelmuller, and W. Steiner.** 2000. Optimisation of culture medium and conditions for alpha-L-arabinofuranosidase production by the extreme thermophilic eubacterium *Rhodothermus marinus*. *Enzyme Microb. Technol.* **27**:414–422.

61. **Green, F., III, C. A. Clausen, and T. L. Highley.** 1989. Adaptation of the Nelson-Somogyi reducing-sugar assay to a microassay using microtiter plates. *Anal. Biochem.* **182**:197–199.

62. **Grohmann, K., R. Torget, and M. E. Himmel.** 1985. Optimization of dilute acid pretreatment of biomass. *Biotechnol. Bioeng. Symp.* **15**:59–80.

63. **Gustafsson, J., L. Ciovica, and J. Peltonen.** 2003. The ultrastructure of spruce kraft pulps studied by atomic force microscopy (AFM) and X-ray photoelectron spectroscopy (XPS). *Polymer* **44**:661–670.

64. **Hadj-Taieb, N., M. Ayadi, S. Trigui, F. Bouabdallah, and A. Gargouri.** 2002. Hyperproduction of pectinase activities by a fully constitutive mutant (CT1) of *Penicillium occitanis*. *Enzyme Microb. Technol.* **30**:662–666.

65. **Håkansson, U., L. Fägerstam, G. Pettersson, and L. Andersson.** 1978. Purification and characterization of a low molecular weight 1,4-beta-glucan glucanohydrolase from the cellulolytic fungus *Trichoderma viride* QM 9414. *Biochim. Biophys. Acta* **524**:385–392.

66. **Helbert, W., J. Sugiyama, M. Ishihara, and S. Yamanaka.** 1997. Characterization of native crystalline cellulose in the cell walls of Oomycota. *J. Biotechnol.* **57**:29–37.

67. **Hoshino, E., M. Shiroishi, Y. Amano, M. Nomura, and T. Kanda.** 1997. Synergistic actions of exo-type cellulases in the hydrolysis of cellulose with different crystallinities. *J. Ferment. Bioeng.* **84**:300–306.

68. **Hrmova, M., and G. B. Fincher.** 2001. Structure-function relationships of beta-D-glucan endo- and exohydrolases from higher plants. *Plant Mol. Biol.* **47**:73–91.

69. **Hu, G., and F. H. Rijkenberg.** 1998. Subcellular localization of beta-1,3-glucanase in *Puccinia recondita* f.sp. tritici-infected wheat leaves. *Planta* **204**:324–334.

70. **Irwin, D. C., M. Spezio, L. P. Walker, and D. B. Wilson.** 1993. Activity studies of 8 purified cellulases—specificity, synergism, and binding domain effects. *Biotechnol. Bioeng.* **42**:1002–1013.

71. **Jeffries, T. W.** 1996. Biochemistry and genetics of microbial xylanases. *Curr. Opin. Biotechnol.* **7**:337–342.

72. **Johnson, E. A., M. Sakajoh, G. Halliwell, A. Madia, and A. L. Demain.** 1982. Saccharification of complex cellulosic substrates by the cellulase system from *Clostridium thermocellum*. *Appl. Environ. Microbiol.* **43**:1125–1132.

73. **Johnston, D. B., S. P. Shoemaker, G. M. Smith, and J. R. Whitaker.** 1998. Kinetic measurements of cellulase activity on insoluble substrates using disodium 2,2′ bicinchoninate. *J. Food Biochem.* **22**:301–319.

74. **Joos, P., W. Sierens, and R. Ruyssen.** 1969. The determination of cellulase activity by viscometry. *J. Pharm. Pharmacol.* **21**:848–853.

75. **Kapoor, M., Q. K. Beg, B. Bhushan, K. S. Dadhich, and G. S. Hoondal.** 2000. Production and partial purification and characterization of a thermo-alkali stable polygalacturonase from *Bacillus* sp. MG-cp-2. *Process Biochem.* **36**:467–473.

76. **Karlsson, J., M. Siika-aho, M. Tenkanen, and F. Tjerneld.** 2002. Enzymatic properties of the low molecular mass endoglucanases Cel12A (EG III) and Cel45A (EG V) of *Trichoderma reesei*. *J. Biotechnol.* **99:**63–78.

77. **Kashyap, D. R., P. K. Vohra, S. Chopra, and R. Tewari.** 2001. Applications of pectinases in the commercial sector: a review. *Bioresour. Technol.* **77:**215–227.

78. **Keitel, T., K. K. Thomsen, and U. Heinemann.** 1993. Crystallization of barley (1-3,1-4)-beta-glucanase, isoenzyme II. *J. Mol. Biol.* **232:**1003–1004.

79. **Khandke, K. M., P. J. Vithayathil, and S. K. Murthy.** 1989. Purification and characterization of an alpha-D-glucuronidase from a thermophilic fungus, *Thermoascus aurantiacus*. *Arch. Biochem. Biophys.* **274:**511–517.

80. **Kolbe, J., and C. P. Kubicek.** 1990. Quantification and identification of the main components of the *Trichoderma* cellulase complex with monoclonal antibodies using an enzyme-linked immunosorbent assay (ELISA). *Appl. Microbiol. Biotechnol.* **34:**26–30.

81. **Kong, F., C. R. Engler, and E. J. Soltes.** 1992. Effects of cell-wall acetate, xylan backbone, and lignin on enzymatic hydrolysis of aspen wood. *Appl. Biochem. Biotechnol.* **34/35:**23–35.

82. **Konstantinidis, A. K., I. Marsden, and M. L. Sinnott.** 1993. Hydrolyses of alpha- and beta-cellobiosyl fluorides by cellobiohydrolases of *Trichoderma reesei*. *Biochem. J.* **291**(Pt. 3):883–888.

83. **Kotake, T., N. Nakagawa, K. Takeda, and N. Sakurai.** 1997. Purification and characterization of wall-bound exo-1,3-beta-D-glucanase from barley (*Hordeum vulgare* L.) seedlings. *Plant Cell Physiol.* **38:**194–200.

84. **La Grange, D. C., I. S. Pretorius, M. Claeyssens, and W. H. van Zyl.** 2001. Degradation of xylan to D-xylose by recombinant *Saccharomyces cerevisiae* coexpressing the *Aspergillus niger* beta-xylosidase (*xlnD*) and the *Trichoderma reesei* xylanase II (*xyn2*) genes. *Appl. Environ. Microbiol.* **67:**5512–5519.

85. **Lee, I., B. R. Evans, and J. Woodward.** 2000. The mechanism of cellulase action on cotton fibers: evidence from atomic force microscopy. *Ultramicroscopy* **82:**213–221.

86. **Lee, R. C., R. A. Burton, M. Hrmova, and G. B. Fincher.** 2001. Barley arabinoxylan arabinofuranohydrolases: purification, characterization and determination of primary structures from cDNA clones. *Biochem. J.* **356:**181–189.

87. **Lemos, M. A., J. A. Teixeira, M. R. M. Domingues, M. Mota, and F. M. Gama.** 2003. The enhancement of the cellulolytic activity of cellobiohydrolase I and endoglucanase by the addition of cellulose binding domains derived from *Trichoderma reesei*. *Enzyme Microb. Technol.* **32:**35–40.

88. **Levy, I., and O. Shoseyov.** 2002. Cellulose-binding domains: biotechnological applications. *Biotechnol. Adv.* **20:**191–213.

89. **Li, H., M. Rief, F. Oesterhelt, H. E. Gaub, X. Zhang, and J. Shen.** 1999. Single-molecule force spectroscopy on polysaccharides by AFM-nanomechanical fingerprint of alpha-(1,4)-linked polysaccharides. *Chem. Phys. Lett.* **305:**197–201.

90. **Limam, F., S. E. Chaabouni, R. Ghrir, and N. Marzouki.** 1995. Two cellobiohydrolases of *Penicillium occitanis* mutant Pol 6: purification and properties. *Enzyme Microb. Technol.* **17:**340–346.

91. **Lin, J., L. M. Ndlovu, S. Singh, and B. Pillay.** 1999. Purification and biochemical characteristics of beta-D-xylanase from a thermophilic fungus, *Thermomyces lanuginosus*-SSBP. *Biotechnol. Appl. Biochem.* **30:**73–79.

92. **Lin, L. L., and J. A. Thomson.** 1991. An analysis of the extracellular xylanases and cellulases of *Butyrivibrio fibrisolvens* H17c. *FEMS Microbiol. Lett.* **68:**197–203.

93. **Linder, M., and T. T. Teeri.** 1997. The roles and function of cellulose-binding domains. *J. Biotechnol.* **57:**15–28.

94. **Mandels, M., G. L. Miller, and R. W. Slater.** 1961. Separation of fungal carbohydrases by starch block zone electrophoresis. *Arch. Biochem. Biophys.* **93:**115–121.

95. **Mandels, M., R. Andreotti, and C. Roche.** 1976. Measurement of saccharifying cellulase. *Biotechnol. Bioeng. Symp.* **6:**21–33.

96. **Mandels, M., D. Sternberg, and R. E. Andreotti.** 1975. Growth and cellulase production by *Trichoderma*, p. 81. *In* M. Bailey, T.-M. Enari, and M. Linko (ed.), *Proceedings of the Symposium on Enzymatic Hydrolysis of Cellulose*. VTT, Aulanko, Finland.

97. **Manning, K.** 1981. Improved viscometric assay for cellulase methods. *J. Biochem. Biotechnol.* **5:**189–202.

98. **Markovic, O., D. Mislovicová, P. Biely, and K. Heinrichová.** 1992. Chromogenic substrate for endopolygalacturonase detection in gels. *J. Chromatogr. A* **603:**243–246.

99. **Matsuo, N., S. Kaneko, A. Kuno, H. Kobayashi, and I. Kusakabe.** 2000. Purification, characterization and gene cloning of two alpha-L-arabinofuranosidases from *Streptomyces chartreusis* GS901. *Biochem. J.* **346**(Pt. 1):9–15.

100. **Mazeau, K., and S. Perez.** 1998. The preferred conformations of the four oligomeric fragments of rhamnogalacturonan II. *Carbohydr. Res.* **311:**203–217.

101. **McCartney, L., A. P. Ormerod, M. J. Gidley, and J. P. Knox.** 2000. Temporal and spatial regulation of pectic (1→4)-beta-D-galactan in cell walls of developing pea cotyledons: implications for mechanical properties. *Plant J.* **22:**105–113.

102. **McCleary, B. V.** 1980. New chromogenic substrates for the assay of alpha-amylase and (1 leads to 4)-beta-D-glucanase. *Carbohydr. Res.* **86:**97–104.

103. **McCleary, B. V.** 1978. Simple assay procedure for beta-D-mannanase. *Carbohydr. Res.* **67:**213–221.

104. **McDermid, K. P., C. W. Forsberg, and C. R. MacKenzie.** 1990. Purification and properties of an acetylxylan esterase from *Fibrobacter succinogenes* S85. *Appl. Environ. Microbiol.* **56:**3805–3810.

105. **McKie, V. A., G. W. Black, S. J. Millward-Sadler, G. P. Hazlewood, J. I. Laurie, and H. J. Gilbert.** 1997. Arabinanase A from *Pseudomonas fluorescens* subsp. cellulosa exhibits both an endo- and an exo- mode of action. *Biochem. J.* **323**(Pt. 2):547–555.

106. **McKie, V. A., J. P. Vincken, A. G. Voragen, L. A. van den Broek, E. Stimson, and H. J. Gilbert.** 2001. A new family of rhamnogalacturonan lyases contains an enzyme that binds to cellulose. *Biochem. J.* **355:**167–177.

107. **Mestechkina, N. M., O. V. Anulov, N. I. Smirnova, and V. D. Shcherbukhin.** 2000. Composition and structure of a galactomannan macromolecule from seeds of *Astragalus lehmannianus* Bunge. *Appl. Biochem. Microbiol.* **36:**502–506.

108. **Milagres, A. M. F., and R. M. Sales.** 2001. Evaluating the basidiomycetes *Poria medula-panis* and *Wolfiporia cocos* for xylanase production. *Enzyme Microb. Technol.* **28:**522–526.

109. **Miller, G. L.** 1959. Dinitrosalicylic acid reagent for determination of reducing sugar. *Anal. Chem.* **31:**426–428.

110. **Mutter, M., C. M. Renard, G. Beldman, H. A. Schols, and A. G. Voragen.** 1998. Mode of action of RG-hydrolase and RG-lyase toward rhamnogalacturonan oligomers: characterization of degradation products using RG-rhamnohydrolase and RG-galacturonohydrolase. *Carbohydr. Res.* **311:**155–164.

111. **Nagy, T., K. Emami, C. M. Fontes, L. M. Ferreira, D. R. Humphry, and H. J. Gilbert.** 2002. The membrane-bound alpha-glucuronidase from *Pseudomonas*

cellulosa hydrolyzes 4-O-methyl-D-glucuronoxylooligosaccharides but not 4-O-methyl-D-glucuronoxylan. *J. Bacteriol.* **184:**4925–4929.

112. **Nagy, T., D. Nurizzo, G. J. Davies, P. Biely, J. H. Lakey, D. N. Bolam, and H. J. Gilbert.** 2003. The alpha-glucuronidase, GlcA67A, of *Cellvibrio japonicus* utilizes the carboxylate and methyl groups of aldobiouronic acid as important substrate recognition determinants. *J. Biol. Chem.* **278:**20286–20292.

113. **Naidu, G. S. N., and T. Panda.** 1998. Production of pectolytic enzymes—a review. *Bioprocess Eng.* **19:**355–361.

114. **Nelson, N.** 1944. A photometric adaptation of the Somogyi method for the determination of glucose. *J. Biol. Chem.* **153:**375–380.

115. **Nidetzky, B., W. Steiner, M. Hayn, and M. Claeyssens.** 1994. Cellulose hydrolysis by the cellulases from *Trichoderma reesei*: a new model for synergistic interaction. *Biochem. J.* **298:**705–710.

116. **Nieves, R. A., Y. C. Chou, M. E. Himmel, and S. R. Thomas.** 1995. Quantitation of *Acidothermus cellulolyticus* E1 endoglucanase and *Thermomonospora fusca* E(3) exoglucanase using enzyme-linked-immunosorbent-assay (ELISA). *Appl. Biochem. Biotechnol.* **51–2:**211–223.

117. **Nishitani, K., and D. J. Nevins.** 1991. Glucuronoxylan xylanohydrolase. A unique xylanase with the requirement for appendant glucuronosyl units. *J. Biol. Chem.* **266:**6539–6543.

118. **Parker, K. N., S. Chhabra, D. Lam, M. A. Snead, E. J. Mathur, and R. M. Kelly.** 2001. beta-Mannosidase from *Thermotoga* species. *Methods Enzymol.* **330:**238–246.

119. **Parker, K. N., S. R. Chhabra, D. Lam, W. Callen, G. D. Duffaud, M. A. Snead, J. M. Short, E. J. Mathur, and R. M. Kelly.** 2001. Galactomannanases man2 and man5 from *Thermotoga* species: growth physiology on galactomannans, gene sequence analysis, and biochemical properties of recombinant enzymes. *Biotechnol. Bioeng.* **75:**322–333.

120. **Pere, J., A. Puolakka, P. Nousiainen, and J. Buchert.** 2001. Action of purified *Trichoderma reesei* cellulases on cotton fibers and yarn. *J. Biotechnol.* **89:**247–255.

121. **Ramos, L. P., A. Zandona Filho, F. C. Deschamps, and J. N. Saddler.** 1999. The effect of Trichoderma cellulases on the fine structure of a bleached softwood kraft pulp. *Enzyme Microb. Technol.* **24:**371–380.

122. **Ratanakhanokchai, K., K. L. Kyu, and M. Tanticharoen.** 1999. Purification and properties of a xylan-binding endoxylanase from alkaliphilic *Bacillus* sp. strain K-1. *Appl. Environ. Microbiol.* **65:**694–697.

123. **Rescigno, A., A. C. Rinaldi, N. Curreli, A. Olianas, and E. Sanjust.** 1994. A dyed substrate for the assay of endo-1,4-beta-glucanases. *J. Biochem. Biophys. Methods* **28:**123–129.

124. **Ridley, B. L., M. A. O'Neill, and D. Mohnen.** 2001. Pectins: structure, biosynthesis, and oligogalacturonide-related signaling. *Phytochemistry* **57:**929–967.

125. **Ruiz-Arribas, A., J. M. Fernandez-Abalos, P. Sanchez, A. L. Garda, and R. I. Santamaria.** 1995. Overproduction, purification, and biochemical characterization of a xylanase (Xys1) from *Streptomyces halstedii* JM8. *Appl. Environ. Microbiol.* **61:**2414–2419.

126. **Rutland, M. W., A. Carambassis, G. A. Willing, and R. D. Neuman.** 1997. Surface force measurements between cellulose surfaces using scanning probe microscopy. *Colloids Surf. A* **123–124:**369–374.

127. **Sachslehner, A., G. Foidl, N. Foidl, G. Gubitz, and D. Haltrich.** 2000. Hydrolysis of isolated coffee mannan and coffee extract by mannanases of Sclerotium rolfsii. *J. Biotechnol.* **80:**127–134.

128. **Sadana, J. C., and R. V. Patil.** 1985. Synergism between enzymes of *Sclerotium rolfsii* involved in the solubiliza-tion of crystalline cellulose. *Carbohydr. Res.* **140:**111–120.

129. **Sakamoto, T., and T. Sakai.** 1995. Analysis of structure of sugar-beet pectin by enzymatic methods. *Phytochemistry* **39:**821–823.

130. **Saluzzi, L., H. J. Flint, and C. S. Stewart.** 2001. Adaptation of *Ruminococcus flavefaciens* resulting in increased degradation of ryegrass cell walls. *FEMS Microbiol. Ecol.* **36:**131–137.

131. **Samejima, M., J. Sugiyama, K. Igarashi, and K.-E. L. Eriksson.** 1997. Enzymatic hydrolysis of bacterial cellulose. *Carbohydr. Res.* **305:**281–288.

132. **Saraswat, V., and V. S. Bisaria.** 1997. Biosynthesis of xylanolytic and xylan-debranching enzymes in *Melanocarpus albomyces* IIS 68. *J. Ferment. Bioeng.* **83:**352–357.

133. **Sattler, W., H. Esterbauer, O. Glatter, and W. Steiner.** 1989. The effect of enzyme concentration on the rate of the hydrolysis of cellulose. *Biotechnol. Bioeng. Symp.* **33:**1221–1234.

134. **Sharrock, K. R.** 1988. Cellulase assay methods: a review. *J. Biochem. Biophys. Methods* **17:**81–106.

135. **Sheir-Neiss, G., and B. S. Montenecourt.** 1984. Characterization of the secreted cellulases of *Trichoderma reesei* wild type and mutants during controlled fermentations. *Appl. Microbiol. Biotechnol.* **20:**46–53.

136. **Shoemaker, S. P., and J. R. D. Brown.** 1978. Characterization of endo-1,4-beta-D-glucanases purified from *Trichoderma viride*. *Biochim. Biophys. Acta* **523:**147–161.

137. **Sieben, A.** 1975. Cellulase and other hydrolytic enzyme assays using an oscillating tube viscometer. *Anal. Biochem.* **63:**214–219.

138. **Sipat, A., K. A. Taylor, R. Y. Lo, C. W. Forsberg, and P. J. Krell.** 1987. Molecular cloning of a xylanase gene from *Bacteroides succinogenes* and its expression in *Escherichia coli*. *Appl. Environ. Microbiol.* **53:**477–481.

139. **Skjot, M., S. Kauppinen, L. V. Kofod, C. Fuglsang, M. Pauly, H. Dalboge, and L. N. Andersen.** 2001. Functional cloning of an endo-arabinanase from *Aspergillus aculeatus* and its heterologous expression in *A. oryzae* and tobacco. *Mol. Genet. Genomics* **265:**913–921.

140. **Somogyi, M.** 1952. Notes on sugar determination. *J. Biol. Chem.* **195:**19–23.

141. **Suzuki, T., E. Kitagawa, F. Sakakibara, K. Ibata, K. Usui, and K. Kawai.** 2001. Cloning, expression, and characterization of a family 52 beta-xylosidase gene (xysB) of a multiple-xylanase-producing bacterium, *Aeromonas caviae* ME-1. *Biosci. Biotechnol. Biochem.* **65:**487–494.

142. **Taguchi, H., T. Hamasaki, T. Akamatsu, and H. Okada.** 1996. A simple assay for xylanase using o-nitrophenyl-beta-D-xylobioside. *Biosci. Biotechnol. Biochem.* **60:**983–985.

143. **Takahashi, R., K. Mizumoto, K. Tajika, and R. Takano.** 1992. Production of oligosaccharides from hemicellulose of woody biomass by enzymatic-hydrolysis. 1. A simple method for isolating beta-D-mannanase-producing microorganisms. *Mokuzai Gakkaishi* **38:**1126–1135.

144. **Ten, L. N., W. T. Im, M. K. Kim, M. S. Kang, and S. T. Lee.** 2004. Development of a plate technique for screening of polysaccharide-degrading microorganisms by using a mixture of insoluble chromogenic substrates. *J. Microbiol. Methods* **56:**375–382.

145. **Tenkanen, M.** 1998. Action of *Trichoderma reesei* and *Aspergillus oryzae* esterases in the deacetylation of hemicelluloses. *Biotechnol. Appl. Biochem.* **27:**19–24.

146. **Tenkanen, M., and K. Poutanen.** 1992. Significance of esterases in the degradation of xylans, p. 203–212. *In* J. G. Visser, M. A. Beldman, K.-V. Someren, and A. G. J. Voragen (ed.), *Xylans and Xylanases*. Elsevier, New York, NY.

147. **Tenkanen, M., J. Puls, M. Ratto, and L. Viikari.** 1993. Enzymatic deacetylation of galactoglucomannans. *Appl. Microbiol. Biotechnol.* **39:**159–165.

148. **Tenkanen, M., and M. Siika-aho.** 2000. An alpha-glucuronidase of *Schizophyllum commune* acting on polymeric xylan. *J. Biotechnol.* **78:**149–161.

149. **Tenkanen, M., J. Thornton, and L. Viikari.** 1995. An acetylglucomannan esterase of *Aspergillus oryzae*—purification, characterization and role in the hydrolysis of O-acetyl-galactoglucomannan. *J. Biotechnol.* **42:**197–206.

150. **Teramoto, A., and M. Fuchigami.** 2000. Changes in temperature, texture, and structure of konnyaku (konjac glucomannan gel) during high-pressure-freezing. *J. Food Sci.* **65:**491–497.

151. **Torto, N., T. Buttler, L. Gorton, G. Markovarga, H. Stalbrand, and F. Tjerneld.** 1995. Monitoring of enzymatic-hydrolysis of ivory nut mannan using online microdialysis sampling and anion-exchange chromatography with integrated pulsed electrochemical detection. *Anal. Chim. Acta* **313:**15–24.

152. **Tuohy, M. G., D. J. Walsh, P. G. Murray, M. Claeyssens, M. M. Cuffe, A. V. Savage, and M. P. Coughlan.** 2002. Kinetic parameters and mode of action of the cellobiohydrolases produced by *Talaromyces emersonii. Biochim. Biophys. Acta/Prot. Struct. Mol. Enzymol.* **1596:**366–380.

153. **van Tilbeurgh, H., M. Claeyssens, and C. K. DeBruyne.** 1982. The use of 4-methylumbelliferyl and other chromophoric glycosides in the study of cellulolytic enzymes. *FEBS Lett.* **149:**152–156.

154. **Vlasenko, E. Y., A. I. Ryan, C. F. Shoemaker, and S. P. Shoemaker.** 1998. The use of capillary viscometry, reducing end-group analysis, and size exclusion chromatography combined with multi-angle laser light scattering to characterize endo-1,4-beta-D-glucanases on carboxymethylcellulose: a comparative evaluation of the three methods. *Enzyme Microb. Technol.* **23:**350–359.

155. **Vrsanska, M., and P. Biely.** 1992. The cellobiohydrolase I from *Trichoderma reesei* QM 9414: action on cello-oligosaccharides. *Carbohydr. Res.* **227:**19–27.

156. **Wang, G., R. R. Marquardt, H. Xiao, and Z. Zhang.** 1999. Development of a 96-well enzyme-linked solid-phase assay for beta-glucanase and xylanase. *J. Agric. Food Chem.* **47:**1262–1267.

157. **Whitehead, T. R., and R. B. Hespell.** 1989. Cloning and expression in *Escherichia coli* of a xylanase gene from *Bacteroides ruminicola* 23. *Appl. Environ. Microbiol.* **55:**893–896.

158. **Willats, W. G., C. Orfila, G. Limberg, H. C. Buchholt, G. J. van Alebeek, A. G. Voragen, S. E. Marcus, T. M. Christensen, J. D. Mikkelsen, B. S. Murray, and J. P. Knox.** 2001. Modulation of the degree and pattern of methyl-esterification of pectic homogalacturonan in plant cell walls. Implications for pectin methyl esterase action, matrix properties, and cell adhesion. *J. Biol. Chem.* **276:**19404–19413.

159. **Wood, T. M.** 1971. The cellulase of *Fusarium solani*: purification and specificity of the beta-(1,4)-glucanase and the beta-D-glucosidase components. *Biochem. J.* **121:**353–362.

160. **Wood, T. M., and S. I. McCrae.** 1996. Arabinoxylan-degrading enzyme system of the fungus *Aspergillus awamori*: purification and properties of an alpha-L-arabinofuranosidase. *Appl. Microbiol. Biotechnol.* **45:**538–545.

161. **Wood, T. M., and S. I. McCrae.** 1977. Cellulase from *Fusarium solani* purification and properties of the C-1 component. *Carbohydr. Res.* **57:**117–133.

162. **Wu, C. T., G. Leubner-Metzger, F. Meins, Jr., and K. J. Bradford.** 2001. Class I beta-1,3-glucanase and chitinase are expressed in the micropylar endosperm of tomato seeds prior to radicle emergence. *Plant Physiol.* **126:**1299–1313.

163. **Wyman, C. E., S. R. Decker, M. E. Himmel, J. W. Brady, C. E. Skopec, and L. Viikari.** 2005. Hydrolysis of cellulose and hemicellulose, p. 995–1034. *In* S. Dimitriu (ed.), *Polysaccharides: Structural Diversity and Functional Versatility.* Marcel Dekker, New York, NY.

164. **Yanai, T., and M. Sato.** 2000. Purification and characterization of a novel alpha-L-arabinofuranosidase from *Pichia capsulata* X91. *Biosci. Biotechnol. Biochem.* **64:**1181–1188.

165. **Zantinge, J. L., H. C. Huang, and K. J. Cheng.** 2002. Microplate diffusion assay for screening of beta-glucanase-producing microorganisms. *BioTechniques* **33:**798, 800, 802.

166. **Zauscher, S., and D. J. Klingenberg.** 2000. Normal forces between cellulose surfaces measured with colloidal probe microscopy. *J. Colloid Interface Sci.* **229:**497–510.

25

Lignin and Lignin-Modifying Enzymes

CARLOS G. DOSORETZ AND C. A. REDDY

25.1. INTRODUCTION

Lignin, next to cellulose, is the most abundant renewable organic polymer in the biosphere. Lignin occurs in plant cell walls in intimate association with the carbohydrate polymers cellulose and hemicellulose, which are, respectively, the first and second most abundant polysaccharide polymers in nature. These three polymers are the major constituents of plant cell walls. Lignin, which is structurally complex and recalcitrant to microbial degradation, and hemicellulose form an amorphous matrix in which lie the cellulose fibrils (23, 49, 51). The sheath-like covering of lignin-hemicellulose makes the cellulose component relatively less accessible to microbial degradation. Thus, lignin biodegradation plays a central role in global carbon cycling, which is of considerable ecological interest. Interest in lignin biodegradation also lies in the fact that lignin removal enhances the efficiency of conversion lignocellulosic biomass as a feed for ruminants, as a source of feedstock chemicals for industry, and for production of fuels such as ethanol and that lignin is itself a potential source of useful industrial chemical feedstock (11, 60, 64, 65, 73, 107). Moreover, there is increasing worldwide interest in the use of ligninolytic fungi for bioremediation purposes and for biopulping applications (11, 12, 32, 53, 78, 87).

Free-radical condensation of lignin precursors, coniferyl, synapyl, and p-coumaryl alcohols by plant cell wall peroxidases results in the formation of a heterogeneous, amorphous, optically inactive, and highly branched aromatic polymer with many different types of intermonomer linkages. It is these unique structural features of lignin that impose unusual restrictions on its degradation by microbes (11, 13, 20, 51, 59, 60). Three families of fungal enzymes, designated lignin-modifying enzymes (LMEs), consist of lignin peroxidases, manganese peroxidases, and laccases, and these play a key role in lignin biotransformation. There is intense worldwide interest in studying LMEs because of their potential biotechnological applications. These include transformation of lignocellulosic biomass into feeds, fuels, and chemicals; for biopulping and biobleaching in pulp and paper production; for decolorizing and detoxifying Kraft bleach plant effluents; and for degrading toxic environmental pollutants such as dioxins, polychlorinated biphenyls, dye effluents, and polyaromatic hydrocarbons (4, 11, 20, 32, 51, 64, 65, 82, 107). A brief review of lignin degradation and the basic methodologies associated with the study of LMEs are presented here. For greater details, the reader is referred to selected books (20, 29, 51, 104) and numerous reviews (3, 4, 9, 11, 13, 14, 47, 49, 50, 52, 58, 59, 62, 69, 77) that have appeared regarding the physiology and biochemistry of lignin degradation and LME. For information on cellulases and hemicellulases, the reader is referred to chapter 24 of this volume.

25.2. FUNGAL LIGNIN DEGRADATION

25.2.1. Primary Degraders of Lignin

Basidiomycete fungi that cause white rot decay of wood are the most efficient lignin degraders in nature and show extensive and complete mineralization of lignin to carbon dioxide and water (13, 20, 29, 51, 59). Two other groups of fungi, designated soft rot fungi and brown rot fungi, are also capable of causing wood decay (8, 35, 29). Brown rot basidiomycetes extensively degrade cellulose and hemicellulose in plant cell walls but are limited in their ability to mineralize lignin (8, 10, 20, 29, 51). Demethoxylation is the most obvious consequence of attack on lignin by these fungi. Soft rot fungi such as *Chaetomium*, *Ceratocystis*, and *Phialophora* degrade cellulose and hemicellulose but cause only limited modification of lignin. Some bacteria such as streptomyces have been shown to partially depolymerize lignin to yield low-molecular-weight products but are not believed to be significant in causing mineralization of lignin to CO_2 (20, 29, 51).

25.2.2. White Rot Fungi

Although white rot fungi are capable of extensive mineralization of lignin to CO_2 and H_2O, lignin is not a significant source of cell carbon or energy for these organisms (13, 51, 59). It is widely believed that the ability to degrade lignin enables white rot fungi to gain access in wood to cellulose and hemicellulose, which serve as the main carbon and energy sources for these fungi. White rot fungi cause more extensive degradation of hardwood from deciduous trees (angiosperms) than of softwood trees such as conifers. In a survey of a total of 65 central European wood-decaying basidiomycetes, 34 were reported to attack angiosperms exclusively and 27 attacked both hardwood and coniferous wood, while only four species exclusively attacked coniferous wood, (8, 9, 86). White rot fungi typically colonize the cell lumen and cause cell wall erosion. Degradation is usually localized to cells colonized by fungal hyphae. Progressive erosion of the cell wall occurs when components are degraded simultaneously during nonselective delignification and eroded zones coalesce as the decay progresses, forming large voids filled with mycelium. During selective delignification, a diffuse attack of lignin occurs and a white pocket or white-mottled type of rot results (8, 10). Electron microscopic studies have revealed that lignin is degraded at some distance from the hyphae and is removed progressively from the lumen towards the middle lamella, which is also degraded (8, 9).

25.2.3. Initial Studies on Lignin Degradation by White Rot Fungi

The main mechanism of lignin degradation by white rot fungi involves one-electron oxidation reactions catalyzed by extracellular oxidases and peroxidases (38, 50, 52, 58, 77). Lignin-degrading ability is commonly studied by growing the organism in media containing relatively low levels of nitrogen (2.4 mM), because higher levels of nitrogen (24 mM) result in cessation of lignin degradation (13, 59, 77). Similarly, low levels of carbon promote lignin degradation, while higher levels inhibit lignin degradation. Lignin degradation in such media is measured by determining the amount of $^{14}CO_2$ produced from ^{14}C-labeled lignin preparations, such as ^{14}C-ring-labeled dehydrogenative polymer (DHP) of lignin (20, 29). To date, measurement of $^{14}CO_2$ evolution from ^{14}C-DHP lignin is reported to be the most reliable method for testing ligninolytic activity of a given organism. Therefore, this method has been extensively used for determining ligninolytic activity of many white rot basidiomycete fungi as well as for elucidating the role of different fungal enzymes and other constituents in lignin degradation (20, 29). Other methods such as nuclear magnetic resonance spectroscopy have also been used to study the degradation of polymeric lignin (33), but these methods are not easily amenable for detailed physiological and biochemical studies on white rot fungi and their enzymes. Decolorization of the polymeric dye poly R has also been used by a number of investigators as a quick screen for ligninolytic organisms

and for isolating lignin-peroxidase-negative and nitrogen-deregulated mutants (see references 11 and 104); however, poly R dye decolorization by a fungus does not necessarily correlate in all cases with its ability to degrade ^{14}C-DHP lignin to $^{14}CO_2$. A simple screening procedure that distinguishes between fungi that cause decay by selectively removing lignin and those that degrade both cellulose and lignin simultaneously has been developed (91). This involves staining of lignin with Astra-blue, which stains cellulose blue only in the absence of lignin. Safranin, on the other hand, stains lignin regardless of the presence of cellulose.

Dimeric model compounds that represent the principal substructures of lignin have been used widely to characterize the ligninolytic systems of white rot fungi (20, 51). Dimeric models also played a large role in revealing the type of lignin substructures that are degraded by fungal lignin peroxidase (LiP) (42, 49–51). For example, oxidation of β-O-4 linkages (which represent 50 to 60% of the bonds in lignin molecule) by LiP indicated that these linkages are cleaved via the cation radical intermediates formed by one-electron oxidation of the aromatic ring of the substrate. Thus, LiP does cleave the predominant linkages in lignin (49, 50, 58) . Dimeric models have also been used to detect LiP activity in situ in fungus-colonized wood, where extraction and conventional assay of the enzyme are technically difficult (90). Degradation of a (β-O-4)-(5-5')-type lignin trimer compound, arylglycerol-β-(dehydrodivanillyl alcohol) ether (substrate I) by LiP showed formation of Cα-Cβ cleavage, β-O-4 bond cleavage, and opening of β-etherified aromatic ring (B-ring) (reviewed in reference 50). The results also showed that the B-ring of substrate (I), which is more difficult to degrade than arylglycerol-β-guaiacyl and -β-(2,6-dimethoxyphenyl) ethers, was oxidized by LiP (50). The disadvantage in the use of dimeric lignin model compounds, however, is the fact that, unlike the lignin polymer, they can be taken up and metabolized intracellularly by microorganisms, which can make it difficult to determine whether the degradation products observed really reflect actual ligninolytic activity. Therefore, ideally, lignin model compounds should be sufficiently macromolecular but at the same time facilitate efficient product analysis. To this end, dimeric lignin compounds attached to a polymer backbone such as polystyrene and polyethylene glycol have been used (57).

25.3. LIGNIN-MODIFYING ENZYMES

The three major families of LMEs produced by white rot fungi that are involved in the oxidative degradation of lignin are LiP (96), manganese-dependent peroxidases or manganese peroxidases (MnP) (36), and laccases (LAC) (4) (Table 1). Both LiP and MnP are heme-containing enzymes that require hydrogen peroxide (H_2O_2) for activity. In addition to LiP and MnP, a versatile peroxidase (VP), which shares some of the properties of both LiP and MnP, has also been described (16, 76, 98). LACs are a group of blue multicopper oxidases that couple substrate oxidation to four electron reduction of molecular oxygen to water. LACs appear to be present in most white rot fungi studied to date (4, 62, 70). Two other groups of enzymes are indirectly involved in lignin degradation. These include (i) enzymes involved in the reduction of aromatic radicals preventing repolymerization such as cellobiose dehydrogenase, cellobiose:quinone oxidoreductase, and pyranose 2-oxidase; and (ii) enzymes involved in the generation of hydrogen

TABLE 1 Key features of LMEs

Enzyme	Electron acceptor/metal	Metal cofactor	Mediator	Bonds cleaved
LiP	H_2O_2/Fe	None	Veratryl alcohol, veratrole, dimethoxycinnamic acid, 1,2-dimethoxybenzene	$C\alpha$-$C\beta$, β-O-4; aryl-$C\alpha$ bonds, β-ring oxidation, demethoxylation; nonphenolic moieties in the presence of mediators (54)
MnP	H_2O_2/Fe	Mn^{2+}	None	$C\alpha$-$C\beta$ cleavage; nonphenolic moieties in the presence of mediators (48, 54)
VP	H_2O_2/Fe	Mn^{2+}	Veratryl alcohol	Presumably similar to MnP (69)
LAC	O_2/Cu	None	3-Hydroxyanthrallic acid, ABTS, hydroxybenzotriazole	$C\alpha$-$C\beta$; nonphenolic moieties in the presence of mediators (54)

peroxide such as glyoxal oxidase, glucose oxidase, and aryl alcohol oxidase (1, 11, 22, 29, 58, 77, 98).

25.3.1. Lignin Peroxidase (LiP)

LiP (EC 1.11.1.14) of *Phanerochaete chrysosporium*, the most extensively studied ligninolytic fungus, is secreted as a series of glycosylated isoenzymes with pIs ranging from 3.2 to 4.0 and molecular masses ranging from 38 to 43 kDa; each isozyme contains 1 mol of heme per mol of protein (14, 59, 77, 83, 105). Up to 15 different isoenzymes have been reported in cell extracts of *P. chrysosporium* which are structurally similar but show some variation in stability, quantity, and catalytic properties (14, 37, 74, 77). LiP possesses a higher redox potential and a lower pH optimum than that of most other isolated peroxidases or oxidases (15, 42, 58). LiP, similar to other peroxidases, is capable of oxidizing most phenolic compounds through the generation of phenoxy radicals; however, due to its exceptionally high redox potential and low pH optimum, it is also able to oxidize nonphenolic aromatic substrates, typically not oxidized by other peroxidases, including the nonphenolic phenylpropanoid units of lignin (22, 34, 45–47, 50, 59). Stable cation centered radicals formed during the oxidation of nonphenolic aromatic nuclei may serve as redox mediators for LiP-catalyzed oxidations, ef-

fectively extending the substrate range. Reactions catalyzed by LiP include benzyl alcohol oxidations, side chain cleavages, ring-opening reactions, demethoxylations, and oxidative dechlorinations (46, 50, 68, 81).

25.3.1.1. Catalytic Cycle

The catalytic cycle of LiP is similar to that of other peroxidases (14, 50, 59, 79). As shown in Fig. 1, reaction of native ferric enzyme (Fe-LiP; Fe^{3+}–P) and H_2O_2 yields LiP-compound I (LiPI), a complex of high valent oxo-iron and porphyrin cation radical ($Fe^{4+}=O$–$P^{\bullet+}$). One-electron oxidation of a phenolic reducing substrate (SH) by LiPI yields a phenoxy radical (S•) and the one-electron-oxidized enzyme intermediate, LiP-compound II (LiPII; $Fe^{4+}=O$–P). A single one-electron oxidation of a second substrate molecule returns the enzyme to Fe-LiP completing the catalytic cycle. LiP is uniquely different from other peroxidases in that it exhibits an unusually high reactivity between LiPII and H_2O_2 (14, 102). In the absence of suitable reducing substrate or at high H_2O_2 concentrations, LiPII is further oxidized by H_2O_2 to LiP-compound III (LiPIII; $Fe^{3+}=O_2^{\bullet-}$–P), a species with limited catalytic activity. LiPIII is inactivated rapidly in the presence of excess H_2O_2.

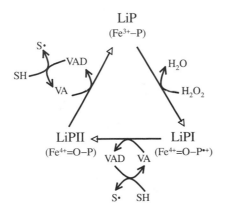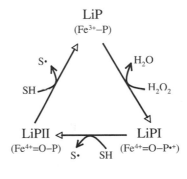

FIGURE 1 Catalytic cycle of lignin peroxidase (LiP) in the presence (left panel) and absence (right panel) of a mediator, as exemplified by the veratryl alcohol in this figure. Symbols: LiP, native lignin peroxidase; LiP I, compound I; LiP II, compound II; VA, veratryl alcohol; VAD, veratraldehyde; SH, reducing substrate; S•, substrate cation radical.

25.3.1.2. Role of Redox Mediators

LiPIII (see above) has been shown to readily return to the native ferric state in the presence of H_2O_2 and veratryl alcohol (3,4-dimethoxybenzyl alcohol) (VA) (6, 14, 49). Ligninolytic cultures of *P. chrysosporium* normally produce VA, one of the physiological roles of which is believed to be to protect LiP from H_2O_2-dependent inactivation by reverting LiPIII to the native state. This is of particular significance since during oxidation of certain chemicals such as phenols, LiPIII has been shown to accumulate, indicating that either they are poor substrates for LiPII or they lack the ability to revert LiPIII to the native state (14, 17, 19, 58). Thus, oxidation of such chemicals is inefficient at high H_2O_2 concentrations. The catalytic cycle of LiP in the presence of VA as mediator is shown in Fig. 1. VA has also been shown to act as a charge-transfer mediator in LiP-catalyzed reactions (41). During the catalytic cycle of LiP, VA is oxidized to VA cation radical ($VA^{+\bullet}$), which in the presence of a suitable reducing substrate is reduced back to VA and then becomes ready for another LiP-catalyzed charge-transfer reaction.

Low-molecular-weight redox mediators such as VA are believed to be important for lignin degradation, since electron microscopic studies have indicated that enzymes as large as peroxidases and LAC cannot have direct contact with lignin, since they appear to be too large for the penetration of the cell wall or the middle lamella (15). Pine decayed by wood-rotting fungi was infiltrated with a concentrated culture filtrate of the white rot fungus *P. chrysosporium* and labeled for LiP by postembedding immunoelectron microscopy. This method demonstrated that enzymatic attack on pine by LiP and presumably also other enzymes of the same size is restricted to the surface of the wood cell wall (92).

Measurements of the stability of $VA^{+\bullet}$ have led to the suggestion that it would be unable to act as a diffusible oxidant without some form of stabilization (17, 22, 50). It was suggested that a LiP-$VA^{+\bullet}$ complex acts as the redox partner, with the cation radical being stabilized by a protein microenvironment of acidic character (see references 22 and 50).

Nonphenolic aromatic compounds other than VA such as 3,4-dimethoxycinnamic acid, 1,2-dimethoxybenzene, and 2-chloro-1,4-dimethoxybenzene have also been shown to be capable of mediating oxidation (94, 100). The mediation phenomenon appears to be driven by the difference in the oxidation potential (OP) and site-binding affinity of the mediators (possessing higher OP values and higher affinity) and the target substrates (possessing lower OP values and lower affinity) (101).

25.3.1.3. LiP Activity Measurement

LiP activity is assayed by measuring the increase in absorbance at 310 nm resulting from the oxidation of VA to veratraldehyde (2,3-dimethoxybenzaldehyde) in the presence of H_2O_2 (14, 96, 104), as shown in Fig. 2.

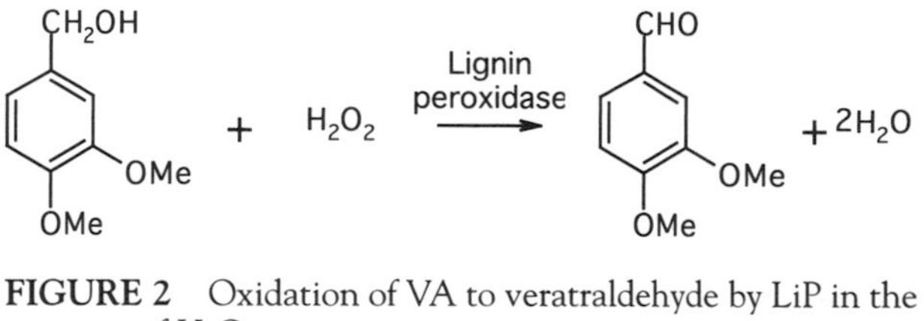

This assay mixture consists of VA, 2mM; H_2O_2, 0.4 mM; *d*-tartaric acid, 50 mM; and the reaction is started by adding a LiP preparation. Optimal conditions for LiP activity are pH 2.5 and 30°C. Care should be taken to minimize preincubation of LiP preparation with buffer or with H_2O_2 (in the absence of VA) because the enzyme is likely to get inactivated and/or show lower activity otherwise.

The oxidation of azure B, involving absorbance measurements in the visible range, was suggested as an alternative method for measuring the activity of LiP. This method is claimed to be less susceptible to interferences due to absorbance of organic material in the near UV range when measuring activity in crude extracellular culture fluid (2); however, this method has not gained much popularity.

25.3.2. Manganese Peroxidase (MnP)

MnP (EC 1.11.1.13) is an extracellular enzyme that is produced as a part of the lignin-degrading enzyme system in some white rot fungi. MnP exists as a series of glycosylated isozymes with pIs ranging from 4.2 to 4.9 and molecular masses ranging from 45 to 47 kDa. Similar to LiP, each isozyme contains 1 mol of iron per mol of protein (22, 39, 48, 51, 52, 58, 75). To date, five major isozymes have been detected in *P. chrysosporium*. MnP catalyzes the following reaction:

$$2Mn(II) + H_2O_2 + 2H^+ \rightarrow 2Mn(III) + 2H_2O$$

The product of MnP reaction is Mn(III), which nonspecifically oxidizes a variety of organic compounds including the key substructures in lignin.

25.3.2.1. Catalytic Cycle

MnP, similar to LiP and several other peroxidases, has a catalytic cycle involving a two-electron oxidation of the heme by H_2O_2, followed by two subsequent one-electron reductions to the native ferric enzyme. The primary reducing substrate in the MnP catalytic cycle is Mn^{2+}, which efficiently reduces MnP I and MnP II (Fig. 3) generating Mn^{3+}, which then serves to oxidize phenols to phenoxy radicals (46–48). Just as cation radicals of aromatic substrates (such as that of VA) maintain the active form of LiP by oxidatively converting compound III to the native enzyme and prevent H_2O_2-dependent inactivation (6), Mn^{3+} has similarly been shown to convert MnP III to native enzyme (50). Additionally, Mn^{2+} also reactivates compound III (although this reaction is slower) and this reaction is also

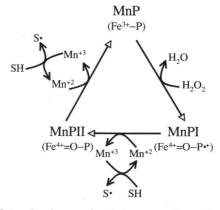

FIGURE 3 Catalytic cycle of manganese-dependent peroxidase (MnP). See the text for details.

FIGURE 2 Oxidation of VA to veratraldehyde by LiP in the presence of H_2O_2.

known to prevent compound III accumulation when excess Mn^{2+} is present.

25.3.2.2. Role of MnP in Lignin Degradation

In many fungi, MnP is thought to play a crucial role in the primary attack on lignin, because it generates Mn^{3+}, a strong diffusible oxidant that is able to penetrate the small "molecular pores" between cellulose microfibrils, which precludes the action of LiP because of steric hindrances (30). Organic acids such as oxalate, fumarate, and malate (52, 67, 84), which are produced by cultures of white rot fungi, chelate Mn^{3+}. These stable complexes are believed to be responsible for the oxidation of the substrate. Although MnP does not oxidize nonphenolic lignin structures during normal turnover, these structures have been shown to be slowly co-oxidized when MnP peroxidatively oxidizes unsaturated fatty acids (54, 55). Bao et al. (5) described the oxidation of a nonphenolic lignin model by a lipid peroxidation system that consisted of *P. chrysosporium* MnP, Mn^{2+}, and unsaturated fatty acid esters. Substrate oxidation occurred via benzylic hydrogen abstraction, and it was suggested that this process might enable the white rot fungi to accomplish the initial delignification of wood. The importance of Mn^{2+} and lipid peroxidation in depolymerization and mineralization of ^{14}C-labeled, polyethylene glycol-linked, β-O-4 lignin model compound by *Ceriporiopsis subvermispora* has been demonstrated in wood block cultures as well as in defined medium (54). Thus, lipid peroxidation has been suggested to be the mechanism involved in the oxidation of the nonphenolic lignin structures by white rot fungi that do not produce LiP (54, 72).

25.3.2.3. MnP Activity Measurement

MnP activity assay is based on the oxidation of a substrate (S) such as ABTS [2,2'-azinobis(3-ethylbenzothiazoline-6-sulfonic acid)] or vanillylacetone [4-(4-hydroxy-3-methoxyphenyl)-3-buten-2-one] in the presence of H_2O_2 and Mn^{2+}, according to the scheme shown in Fig. 4. The use of a buffer having chelating activity such as citrate, lactate, succinate, tartrate, or malonate has been reported to improve the reaction (36, 75). Commonly used substrates for MnP assay are Mn(II), ABTS, 2,6-dimethoxyphenol (DMP), vanillylacetone, phenol red, syringaldazine, and several others (38, 39, 44, 48, 52). Two commonly used assays are described here.

Assay with Mn(II) as the Substrate

This assay measures oxidation of Mn(II) to Mn(III). The reaction mixture (in a total volume of 1 ml) contains 50 μM $MnSO_4$, 50 μM H_2O_2, and 50 mM sodium lactate buffer (pH 4.5). The reaction, performed at room temperature, is initiated by adding the appropriate amount of MnP, and the initial rate of Mn(III)-lactate formation is determined by measuring the increase in absorbance at 240 nm. One unit of MnP is usually defined as the amount of enzyme needed to oxidize 1 μmol of Mn(II) to Mn(III) in 1 min.

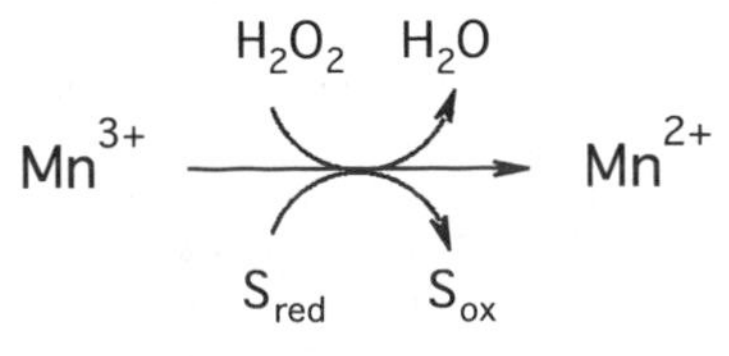

This assay works best with purified enzyme because contaminating metals such as copper and iron in crude enzyme preparations interfere with the activity.

Paszczynski et al. (75) described a slightly different MnP assay procedure in that the reaction mixture contained (in a total volume of 1 ml) 100 mM sodium tartrate (pH 5.0) and 100 μM each $MnSO_4$ and H_2O_2. In this procedure, the reaction is initiated by the addition of H_2O_2 (rather than the enzyme preparation) and increase in absorbance is measured at 238 nm during the first 5 to 30 s of reaction. Sodium malonate buffer (pH 4.5) can be used instead of tartrate, and Mn(II) conversion to Mn(III) is determined by measuring the change in absorbance at 270 nm. Even though the principle is the same, different investigators make some minor changes (to suit their convenience) to the basic procedure.

Assay with ABTS as the Substrate

The reaction mixture (in a total volume of 1 ml) consists of 50 mM each of sodium succinate and sodium lactate buffer (pH 4.5), 100 μM $MnSO_4$, ABTS (40 μg), and 50 μm H_2O_2. The assay is done at room temperature. The reaction is initiated by adding the enzyme preparation. The initial rate of ABTS oxidation is determined spectrophotometrically by measuring the increase in absorbance at 415 nm.

Other Substrates for MnP Assays

Oxidative coupling of 3-methyl-2-benzothiazolinone hydrazone and 3-(dimethylamino) benzoic acid in the presence of H_2O_2 and Mn^{2+} was also reported to be specific for assaying MnP activity (18). This reaction gives a deep purple-blue color with a broad absorption band that peaks at 590 nm. The extinction coefficient is high (53,000 $M^{-1} \cdot cm^{-1}$), and therefore, relatively low MnP activities can be detected. When DMP is used as the substrate, it is oxidized to the dimer 3,3',5,5'-tetramethoxy-p,p'diphenylquinone ($\varepsilon = 49,600$ $M^{-1} \cdot cm^{-1}$). The oxidation of NADH in the presence of H_2O_2 and Mn^{2+} is also employed by some for determining the MnP activity.

25.3.3. LAC

LACs (EC 1.10.3.2) are blue multicopper oxidases that oxidize polyphenols, methoxy-substituted phenols, aromatic diamines, and a range of other aromatic compounds but do not oxidize tyrosine (which is a substrate for tyrosinase) (4, 62). Laccases are found in all the three domains of life: *Eukarya*, *Prokarya*, and *Archaea*. Biochemically, LACs are blue multicopper oxidases that couple four-electron reduction of molecular oxygen to water concomitant with the oxidation of an aromatic substrate (15, 62, 89, 95). LACs are glycosylated enzymes that are generally larger than LiP and MnP, having molecular masses of approximately 60 kDa and above (4, 80). In white rot fungi, they are produced as multiple isozymes (4, 24, 62) and are encoded by gene families (4, 106). A number of excellent reviews on the physiology, biochemistry, and applications of LACs have appeared (4, 25, 50, 62, 70, 89, 92, 95).

In comparison to classical LACs, which are blue, a blue LAC as well as a novel yellow LAC were isolated from cultures of *Pannus tigrinus* (63). The yellow LAC lacks the absorption maxima typically associated with the blue color in classic LACs.

25.3.3.1. Role of LAC in Lignin Degradation

LAC oxidizes phenolic lignin model compounds, but it was widely believed (even a few years ago) that the redox

FIGURE 4 MnP activity assay scheme.

potential of LAC is too low to directly oxidize the nonphenolic components of lignin (4, 50, 60). However, Eggert et al. (26, 27) showed that the white rot fungus *Pycnoporus cinnabarinus*, which does not produce either MnP or LiP but produces LAC, degrades lignin efficiently, suggesting the importance of LAC in lignin degradation (27). Furthermore, *P. cinnabarinus* was shown to produce 3-hydroxyanthranilate (HAA) that apparently mediates the oxidation of nonphenolic substrates by LAC. Based on several lines of evidence, it is now clear that LAC is able to oxidize lignin and a wide range of lignin-related aromatic compounds in the presence of appropriate low-molecular-weight mediators such as ABTS, HAA, and 1-hydroxybenzotriazole (4, 15, 21, 26, 50, 62). The LAX-HAA couple was shown to be more effective in degrading synthetic ^{14}C-lignin than is the LAC-ABTS couple (26). It is widely believed now that low-molecular-weight mediators such as HAA act as diffusible lignin-oxidizing agents and that LACs play a far more important role in lignin degradation than was previously thought (60, 62). The data to date clearly indicate that LACs and MnPs can degrade both phenolic and nonphenolic aromatic moieties of lignin in the presence of selected low-molecular-weight mediator compounds. A LAC-mediated system (Lignozyme^R Process) was reported to be effective in delignifying wood in making paper pulp (15, 25).

25.3.3.2. LAC Activity Measurement

LAC activity assay is based on the oxidation of a phenolic reducing substrate in the presence of O_2. LAC activities are often assayed using ABTS as the substrate (24, 62). Oxidation of ABTS is monitored by determining the increase in A_{420} ($\varepsilon = 36{,}000$ M^{-1}cm^{-1}). LAC activity is also monitored by measuring the oxidation of catechol as evidenced by increase in absorbance at 440 nm or by the oxidation of syringaldazine with the concomitant increase in absorbance at 560 nm (44, 61). Determining LAC activity spectrophotometrically by measuring the conversion of 2 mM DMP to 3,5,3,9,5,9-tetramethoxydiphenoquinone at 468 nm ($\varepsilon = 549.6$ mM^{-1}cm^{-1}) has also been described (4, 62, 66). Furthermore, the use of an oxygen electrode for measuring LAC activity has been reported (66). The optimal conditions for LAC activity are pH 5 and 25°C.

LAC Assay with ABTS as the Substrate

The reaction mixture for this assay (24) contains, in a total volume of 1 ml, 14 μmol of ABTS, the enzyme preparation, and 50 mM glycine-HCl buffer, pH 3.0. The reaction is monitored by determining the increase in absorbance at 420 nm ($\varepsilon = 36{,}000$ M^{-1}cm^{-1}) for 3 to 5 min. The LAC activity is usually expressed as nanokatals (nanomoles/second) per unit amount of enzyme preparation.

When DMP is used as the substrate for LAC, the reaction mixture may contain 1 mM DMP, McIlvane's citrate-phosphate buffer adjusted to pH 5.0, and the MnP enzyme preparation. Oxidation of DMP is monitored by measuring the increase in absorbance at 477 nm ($\varepsilon = 14{,}800$ M^{-1}cm^{-1}).

When syringaldazine is used as the substrate (44, 61), the reaction mixture contains 0.5 mM syringaldazine (dissolved in ethanol) and 50 mM phosphate buffer. Syringaldazine oxidation is followed by measuring the increase in absorbance at 525 nm ($\varepsilon = 65{,}000$ M^{-1} cm^{-1}).

Several other substrates such as guaiacol and 6-hydroxydopamine (2,4,5-trihydroxyphenethyl amine) can also be used as substrates for LAC activity measurements.

25.3.4. Versatile Peroxidase

A heme peroxidase different from other microbial, plant, and animal peroxidases, termed versatile peroxidase (VP), was discovered in *Pleurotus* and *Bjerkandera* species (16, 40, 69, 71, 85, 97). VP shares catalytic properties of both MnP and LiP. The enzyme exhibits high affinity for Mn^{2+}, hydroquinones and dyes, and also oxidizes VA, dimethoxybenzene, and lignin dimers (50, 61, 69, 76, 85). Molecular models and crystal analysis of native and recombinant VP show an Mn^{2+}-binding site formed by three acidic residues near the heme internal propionate accounting for the ability of VP to oxidize Mn^{2+} (69, 76, 85). VP can oxidize Mn^{2+} as well as phenolic and nonphenolic aromatic compounds. With regard to oxidation of aromatic substrates, VP shows a putative long-range electron transfer pathway from an exposed tryptophan to heme, similar to that postulated in LiP (69). Mutagenesis and chemical modification of this tryptophan and the acidic residues forming the Mn^{2+}-binding site confirmed their role in catalysis.

Determination of VP Activity

Since VP shares many catalytic properties of both MnP and LiP, determination of VP activity is essentially similar to the above-mentioned assays for LiP and MnP. In crude preparations, VP is determined as Mn-independent peroxidase by using 0.1 mM DMP in 0.1 M sodium tartrate (pH 4 to 5) in the presence of 0.1 mM H_2O_2 (16) or as LiP in reaction mixtures containing 4 mM VA in 50 mM succinic acid buffer, pH 3.5 (99).

Activity of purified fractions of VP is determined by measuring both the Mn^{+2}-dependent and Mn^{+2}-independent activities (as described above) using purified VP fraction (85, 99).

25.3.5. Role of Reactive Oxygen Species in Lignin Degradation

Reactive oxygen species have been reported to play a direct role in lignin degradation by a number of investigators. The production of $OH^•$ radicals by white rot fungi using a Fenton-type reaction is well documented (7, 31, 93, 103). $OH^•$ radicals are very reactive and can attack the subunits of lignin both by abstracting aliphatic Cα-hydrogens and by adding to aromatic rings (43). Typical reactions of $OH^•$ radical with the major arylglycerol-β-aryl ether structure of lignin can result in demethoxylation, β-O-4 cleavage, hydroxylation, or Cα-oxidation (43). The oxidation of lignin by $OH^•$ radicals therefore results in diverse reactions, some of which are expected to degrade the polymer. Hydroxylation of both phenolic and nonphenolic lignin resulting in new phenolic substructures on the lignin polymer may make it susceptible to attack by LAC or MnP (93).

In considering the role of $OH^•$ radicals in lignin degradation, it is also necessary to consider the possible role of peroxyl ($ROO^•$) and hydroperoxyl ($HOO^•$) radicals on lignin degradation, since both of these radicals are expected as secondary radicals when $OH^•$ radicals oxidize wood polymers (43, 56). MnP of white rot fungi cause peroxidation of unsaturated fatty acids, which results in the formation of $ROO^•$ (72). $ROO^•$ radical-generating systems which appear to oxidize a nonphenolic β-O-4-linked lignin model dimer to products indicative of hydrogen abstraction (56). Since white rot fungi do produce extracellular lipids (28), the formation of $ROO^•$ radicals by an MnP-dependent mechanism and their involvement in lignin degradation seems reasonable.

Both LiP and MnP have been reported to decompose oxalate in the presence of VA and Mn^{2+}, respectively (88), and indeed other organic acids (see references 52 and 98). These reactions account for the observed oxidation of phenol red and kojic acid by MnP in the presence of Mn^{2+}, without exogenous addition of H_2O_2 (see references 14 and 50). The reduction of $VA^{+\bullet}$ or Mn^{3+} by oxalate suggests that as long as oxalate coexists with LiP and MnP, it would inhibit lignin degradation. Indeed, oxalate has been shown to strongly reduce the rate of lignin mineralization in ligninolytic cultures of white rot fungi (88).

C.A.R. acknowledges financial support from U.S. Department of Energy, Biological Sciences Division.

25.4. REFERENCES

1. **Ander, P., C. Mishra, R. Farrel, and K.-E. Eriksson.** 1990. Redox interactions in lignin degradation: interaction between laccase, different peroxidases and cellobiose:quinone oxidoreductase. *J. Biotechnol.* **13:**189–198.

2. **Archibald, F. S.** 1992. A new assay for lignin-type peroxidases employing the dye azure B. *Appl. Environ. Microbiol.* **58:**3110–3116.

3. **Aro, N., T. Pakula, and M. Pentilä.** 2005. Transcriptional regulation of plant cell wall degradation by filamentous fungi. *FEMS. Microbiol. Rev.* **29:**719–739.

4. **Baldrian, P.** 2006. Fungal laccases—occurrence and properties. *FEMS. Microbiol. Rev.* **30:**215–242.

5. **Bao, W. L., Y. Fukushima, K. A. Jensen, M. A Moen, and K. E. Hammel.** 1994. Oxidative-degradation of non-phenolic lignin during lipid-peroxidation by fungal manganese peroxidase. *FEBS. Lett.* **354:**297–300.

6. **Barr, D. P., and S. D. Aust.** 1994. Conversion of lignin peroxidase compound III to active enzyme by cation radicals. *Arch. Biochem. Biophys.* **312:**511–515.

7. **Barr, D. P., M. M. Shah, T. A. Grover, and S. D. Aust.** 1992. Production of hydroxyl radical by lignin peroxidase from *Phanerochaete chrysosporium. Arch. Biochem. Biophys.* **298:**480–485.

8. **Blanchette, R. A.** 1991. Delignification by wood-decay fungi. *Annu. Rev. Phytopathol.* **29:**381–398.

9. **Blanchette, R. A.** 1995. Degradation of the lignocellulose complex in wood. *Can. J. Bot.* **7351:**999–1010.

10. **Blanchette, R. A.** 2000. A review of microbial deterioration found in archaeological wood from different environments. *Int. Biodeterior. Biodegrad.* **46:**189–204.

11. **Boominathan, K., and C. A. Reddy.** 1992. Lignin degradation by fungi: biotechnological applications, p. 763–822. *In* D. K. Arora, K. G. Mukerji, and R. P. Elander (ed.), *Handbook of Applied Mycology*, vol.4. *Biotechnology.* Marcel Dekker Inc, New York, NY.

12. **Breen, A., and F. L. Singleton.** 1999. Fungi in lignocellulose breakdown and biopulping. *Curr. Opin. Biotechnol.* **10:**252–258.

13. **Buswell, J. A., and O. Odier.** 1987. Lignin biodegradation. *CRC Crit. Rev. Biotechnol.* **6:**1–60.

14. **Cai, D., and M. Tien.** 1993. Lignin-degrading peroxidases of *Phanerochaete chrysosporium. J. Biotechnol.* **30:**79–90.

15. **Call, H. P., and I. Mucke.** 1997. History, overview and applications of mediated lignolytic systems, especially laccase-mediator-systems [Lignozym(R)- process]. *J. Biotechnol.* **53:**163–202.

16. **Camarero, S., S. Sarkar, F. J. Ruiz-Duenas, M. J. Martinez, and A. T. Martinez.** 1999. Description of a versatile peroxidase involved in the natural degradation of lignin that has both manganese peroxidase and lignin peroxidase substrate interaction sites. *J. Biol. Chem.* **274:**10324–10330.

17. **Candeias, L. P., and P. J. Harvey.** 1995. Lifetime and reactivity of the veratryl alcohol radical-cation—implications for lignin peroxidase catalysis. *J. Biol. Chem.* **270:**16745–16748.

18. **Castillo, M. D., P. J. Stenstrom, and P. Ander.** 1994. Determination of manganese peroxidase activity with 3-methyl-2-benzothiazolinone hydrazone and 3-(dimethylamino) benzoic acid. *Anal. Biochem.* **218:**399–404.

19. **Chung, N., and S. D. Aust.** 1995. Inactivation of lignin peroxidase by hydrogen-peroxide during the oxidation of phenols. *Arch. Biochem. Biophys.* **316:**851–855.

20. **Crawford, R. L.** 1981. *Lignin Biodegradation and Transformation.* John Wiley & Sons, Inc., New York, NY.

21. **Crestini, C., L. Jurasek, and D. S. Argyropoulos.** 2003. On the mechanism of laccase-mediator system in the oxidation of lignin. *Chem. Eur. J.* **9:**5371–5378.

22. **Cullen, D., and P. J. Kersten.** 2004. Enzymology and molecular biology of lignin degradation, p. 249–273. *In* R. Brambl and G. A. Marzluf (ed.), *The Mycota III. Biochemistry and Molecular Biology*, 2nd ed. Springer-Verlag, Berlin, Germany.

23. **Davin, L. B., and N. G. Lewis.** 2006. Lignin primary structures and dirigent sites. *Curr. Opin. Biotechnol.* **16:**407–415.

24. **D'Souza, T. M., C. S. Merritt, and C. A. Reddy.** 1999. Lignin-modifying enzymes of the white rot basidiomycete *Ganoderma lucidum. Appl. Environ. Microbiol.* **65:**5307–5313.

25. **Durán, N., M. A. Rosa, A. Dannibale, and L. Gianfreda.** 2002. Applications of laccases and tyrosinases (phenol oxidases) immobilized on different supports: a review. *Enzyme Microb. Technol.* **31:**907–931.

26. **Eggert, C., U. Temp, J. F. Dean, and K.-E. Eriksson.** 1997. A fungal metabolite mediates degradation of nonphenolic lignin structures and synthetic lignin by laccase. *FEBS Lett.* **391:**144–148.

27. **Eggert, C., U. Temp, and K.-E. Eriksson.** 1997. Laccase is essential for lignin degradation by the white-rot fungus *Pycnoporus cinnabarinus. FEBS Lett.* **407:**89–92.

28. **Enoki, M., T. Watanabe, S. Nakagame, K. Koller, K. Messner, Y. Honda, and M. Kuwahara.** 1999. Extracellular lipid peroxidation of selective white-rot fungus, *Ceriporiopsis subvermispora. FEMS Microbiol. Lett.* **180:**205–211.

29. **Eriksson, K.-E., R. A. Blanchette, and P. Ander.** 1990. *Microbial and Enzymatic Degradation of Wood and Wood Components.* Springer, New York, NY.

30. **Flournoy, D. S., J. A. Paul, T. K. Kirk, and T. L. Highley.** 1993. Changes in the size and volume of pores in sweetgum wood during simultaneous rot by *Phanerochaete chrysosporium* Burds. *Holzforschung* **47:**297–301.

31. **Forney, L. J., C. A. Reddy, M. Tien, and S. D. Aust.** 1982. The involvement of hydroxyl radical derived from hydrogen peroxide in lignin degradation by the white rot fungus *Phanerochaete chrysosporium. J. Biol. Chem.* **257:**11455–11462.

32. **Gadd, G. M. (ed.).** 2001. *Fungi in Bioremediation.* Cambridge University Press, Cambridge, United Kingdom.

33. **Gamble, G. R., A. Sethuraman, D. E. Akin, and K.-E. Eriksson.** 1994. Biodegradation of lignocellulose in Bermuda grass by white-rot fungi analyzed by solid-state C-13 nuclear magnetic resonance. *Appl. Environ. Microbiol.* **60:**3138–3144.

34. **Geng, X., and K. Li.** 2002. Degradation of non-phenolic lignin by the white-rot fungus *Pycnoporus cinnabarinus. Appl. Microbiol. Biotechnol.* **60:**342–346.

35. **Gilbertson, R. L.** 1980. Wood-rotting fungi of North America. *Mycologia* **72**:1–49.
36. **Glenn, J. K., and M. H. Gold.** 1985. Purification and characterization of an extracellular Mn(II)-dependent peroxidase from the lignin-degrading basidiomycete, *Phanerochaete chrysosporium*. *Arch. Biochem. Biophys.* **242**:329–341.
37. **Glumoff, T., P. J. Harvey, S. Molinari, M. Goble, G. Frank, J. M. Palmer, D. G. Smit, and M. S. A. Leisola.** 1990. Lignin peroxidase from *Phanerochaete chrysosporium*. Molecular and kinetic characterization of isozymes. *Eur. J. Biochem.* **187**:515–520.
38. **Gold, M. H., and M. Alic.** 1993. Molecular biology of the lignin-degrading basidiomycete *Phanerochaete chrysosporium*. *Microbiol. Rev.* **57**:605–622.
39. **Gold, M. H., H. Wariishi, and K. Valli.** 1989. Extracellular peroxidases involved in lignin degradation by the white rot basidiomycete *Phanerochaete chrysosporium*. *ACS Symp. Ser.* **389**:127–140.
40. **Gomez-Toribio, V., A. T. Martinez, M. J. Martinez, and F. Guillen.** 2001. Oxidation of hydroquinones by the versatile ligninolytic peroxidase from *Pleurotus eryngii*. H_2O_2 generation and the influence of Mn^{2+}. *Eur. J. Biochem.* **268**:4787–4793.
41. **Goodwin, D. C., S. D. Aust, and T. A. Grover.** 1995. Evidence for veratryl alcohol as a redox mediator in lignin peroxidase-catalyzed oxidation. *Biochemistry* **34**:5060–5065.
42. **Hammel, K. E., K. A. Jensen, M. D. Mozuch, L. L. Landucci, M. Tien, and E. A. Pease.** 1993. Ligninolysis by a purified lignin peroxidase. *J. Biol. Chem.* **268**:12274–12281.
43. **Hammel, K. E., A. N. Kapich, K. A. Jensen, Jr., and Z. C. Ryan.** 2002. Reactive oxygen species as agents of wood decay by fungi. *Enzyme Microb. Technol.* **30**:445–453.
44. **Harkin, M., and J. R. Obst.** 1973. Syringaldazine, an effective reagent for detecting laccase and peroxidase in fungi. *Experientia* **29**:381–387.
45. **Harvey, P. J., and J. M. Palmer.** 1990. Oxidation of phenolic compounds by ligninase. *J. Biotechnol.* **13**:169–180.
46. **Hatakka, A.** 1994. Lignin-modifying enzymes from selected white-rot fungi—production and role in lignin degradation. *FEMS Microbiol. Rev.* **13**:125–135.
47. **Hatakka, A.** 2001. Biodegradation of lignin, p. 129–184. *In* M. Hofrichter and A. Steinbüchel (ed.), *Biopolymers*, vol. 1. *Lignin, Humic Substances and Coal*. Wiley-VCH, Weinheim, Germany.
48. **Hatakka, A., T. Lundell, M. Hofrichter, and P. Maijala.** 2003. Manganese peroxidase and its role in the degradation of wood lignin, p. 230–243. *In* S. D. Mansfield and J. N. Saddler (ed.), *Applications of Enzymes to Lignocellulosics*. ACS Symposium Series 855. Oxford University Press, Washington, DC.
49. **Higuchi, T.** 1990. Lignin biochemistry: biosynthesis and biodegradation. *Wood Sci. Technol.* **24**:23–63.
50. **Higuchi, T.** 2004. Microbial degradation of lignin: role of lignin peroxidase, manganese peroxidase, and laccase. *Jpn. Acad. Ser. B* **80**:204–214.
51. **Higuchi, T., H.-M. Chang, and T. K. Kirk (ed.).** 1983. *Recent Advances in Lignin Biodegradation Research*. Uni-publishers Co. LTD., Tokyo, Japan.
52. **Hofrichter, M.** 2002. Review: lignin conversion by manganese peroxidase (MnP). *Enzyme Microb. Technol.* **30**:454–466.
53. **Ikehata, K., I. D. Buchanan, and D. W. Smith.** 2004. Recent developments in the production of extracellular fungal peroxidases and laccases for waste treatment. *J. Environ. Eng. Sci.* **3**:1–19.
54. **Jensen, K. A., W. L. Bao, S. Kawai, E. Srebotnik, and K. E. Hammel.** 1996. Manganese-dependent cleavage of

nonphenolic lignin structures by *Ceriporiopsis subvermispora* in the absence of lignin peroxidase. *Appl. Environ. Microbiol.* **62**:3679–3686.
55. **Kapich, A., M. Hofrichter, T. Vares, and A. Hatakka.** 1999. Coupling of manganese peroxidase-mediated lipid peroxidation with destruction of nonphenolic lignin model compounds and C-14-labeled lignins. *Biochem. Biophys. Res. Commun.* **259**:212–219.
56. **Kapich, A. N., K. A. Jensen, and K. E. Hammel.** 1999. Peroxyl radicals are potential agents of lignin biodegradation. *FEBS Lett.* **461**:115–119.
57. **Kawai, S., K. A. Jensen, W. Bao, and K. E. Hammel.** 1995. New polymeric model substrates for the study of microbial ligninolysis. *Appl. Environ. Microbiol.* **61**:3407–3414.
58. **Kirk, T. K., and D. Cullen.** 1998. Enzymology and molecular genetics of wood degradation by white-rot fungi, p. 273–307. *In* A. R. Young and M. Akhtar (ed.), *Environmentally Friendly Technologies for the Pulp and Paper Industry*. John Wiley & Sons, Inc., New York, NY.
59. **Kirk, T. K., and R. L. Farrell.** 1987. Enzymatic "combustion": the microbial degradation of lignin. *Annu. Rev. Microbiol.* **41**:465–505.
60. **Krause, D. O., S. E. Denman, R. I. Mackie, M. Morrison, A. L. Rae, G. T. Attwood, and C. S. McSweeney.** 2003. Opportunities to improve fiber degradation in the rumen: microbiology, ecology and genomics. *FEMS Microbiol. Rev.* **27**:663–693.
61. **Leonowicz, A., and K. Grzywnowicz.** 1981. Quantitative estimation of laccase forms in some white-rot fungi using syringaldazine as a substrate. *Enzyme Microb. Technol.* **3**:55–58.
62. **Leonowicz, A., N. S. Cho, J. Luterek, A. Wilkolazka, M. Wojtaswasilewska, A. Matuszewska, M. Hofrichter, D. Wesenberg, and J. Rogalski.** 2001. Fungal laccases: properties and activity on lignin. *J. Basic Microbiol.* **41**:185–227.
63. **Leontievsky, A., N. Myasoedova, N. Pozdnyakova, and L. Golovleva.** 1997. 'Yellow' laccase of *Pannus tigrinus* oxidizes non-phenolic susbstrates without electron-transfer mediators. *FEBS. Lett.* **413**:446–448.
64. **Lin, Y., and S. Tanaka.** 2006. Ethanol fermentation from biomass resources. *Appl. Microbiol. Biotechnol.* **69**:627–642.
65. **Lynd, H. R., W. H. VanZyl, J. E. Mcbride, and M. Laser.** 2005. Consolidated bioprocessing of cellulosic biomass: an update. *Curr. Opin. Biotechnol.* **16**:577–583.
66. **Mai, C., W. Schorman, O. Milstein, and A. Hutterman.** 2000. Enhanced stability of laccase in the presence of phenolic compounds. *Appl. Microbiol. Biotechnol.* **54**:510–514.
67. **Makela, M., S. Galkin, A. Hatakka, and T. Lundell.** 2002. Production of organic acid and oxalate decarboxylase in lignin-degrading white rot fungi. *Enzyme Microb. Technol.* **30**:542–549.
68. **Martinez, A. T.** 2002. Molecular biology and structure-function of lignin-degrading heme peroxidases. *Enzyme Microb. Technol.* **30**:425–444.
69. **Martinez, A. T., M. Speranza, F. J. Riz-Duenas, P. Ferreira, S. Camarero, F. Guillen, M. J. Martinez, A. Gutierrez, and J. C. del Rio.** 2005. Biodegradation of lignocellulosics: microbial, chemical, and enzymatic aspects of the fungal attack of lignin. *Int. Microbiol.* **8**:195–204.
70. **Mayer, A. M., and R. C. Staples.** 2002. Laccase: new functions for an old enzyme. *Phytochemistry* **60**:551–565.
71. **Mester, T., and J. A. Field.** 1998. Characterization of a novel manganese peroxidase-lignin peroxidase hybrid isozyme produced by *Bjerkandera* species strain BOS55 in the absence of manganese. *J. Biol. Chem.* **273**:15412–15417.

72. Moen, M. A., and K. E. Hammel. 1994. Lipid-peroxidation by the manganese peroxidase of *Phanerochaete chrysosporium* is the basis for phenanthrene oxidation by the intact fungus. *Appl. Environ. Microbiol.* **60:**1956–1961.

73. Mosier, N., C. Wyman, B. Dale, R. Elander, Y. Y. Lee, M. Holtzapple, and M. Ladisch. 2005. Features of promising technologies for treatment of lignocellulosic biomass. *Bioresour. Technol.* **96:**673–686.

74. Ollikka, P., V. M. Leppanen, T. Anttila, and I. Suominen. 1995. Purification of major lignin peroxidase isoenzymes from *Phanerochaete chrysosporium* by chromatofocusing. *Protein Expr. Purif.* **6:**337–342.

75. Paszczynski, A., R. Crawford, and V.-B.Huynh. 1986. Manganese peroxidase of *Phanerochaete chrysosporium:* purification. *Methods Enzymol.* **161:**264–270.

76. Pérez-Boada, M., F. J. Ruiz-Dvenas, R. Pogni, R. Basosi, T. Choinowski, M. J. Martínez, K. Piontek, and A. T. Martínez. 2005. Versatile peroxidase oxidation of high redox potential aromatic compounds: site-directed mutagenesis, spectroscopic and crystallographic investigations of three long-range electron transfer pathways. *J. Mol. Biol.* **354:**385–402.

77. Reddy, C. A., and T. M. D'Souza. 1994. Physiology and molecular biology of the lignin peroxidases of *Phanerochaete chrysosporium. FEMS Microbiol. Rev.* **13:**137–152.

78. Reddy, C. A., and Z. Mathew. 2001. Bioremediation potential of white-rot fungi, p 52–78. *In* G. M. Gadd (ed.), *Fungi in Bioremediation.* Cambridge University Press, Cambridge, United Kingdom.

79. Renganathan, V., and M. H. Gold. 1986. Spectral characterization of the oxidized states of lignin peroxidase, an extracellular heme enzyme from the white rot basidiomycete *Phanerochaete chrysosporium. Biochemistry* **25:**1626–1631.

80. Rheinhammar, B. 1984. Laccase, p. 4–10. *In* R. Lontie (ed.), *Copper Proteins and Copper Enzymes.* CRC Press, Inc., Boca Raton, FL.

81. Rheinhammar, B., and T. K. Kirk. 1990. Comparison of lignin peroxidase, horseradish-peroxidase and laccase in the oxidation of methoxybenzenes. *Biochem. J.* **268:**475–480.

82. Rodriguez, C. S., and H. J. L. Toca. 2006. Industrial and biotechnological applications of laccases: a review. *Biotechnol. Adv.* **24:**500–513.

83. Rothschild, N., C. Novotny, S. Sasek, and C. G. Dosoretz. 2002. Ligninolytic enzymes of the fungus *Irpex lacteus (Polyporus tulipiferae):* isolation and characterization of lignin peroxidase. *Enzyme Microb. Technol.* **31:** 627–633.

84. Roy, B. P., M. G. Paice, F. S. Archibald, S. K. Misra, and L. E. Misiak. 1994. Creation of metal-complexing agents, reduction of manganese-dioxide, and promotion of manganese peroxidase-mediated Mn(III) production by cellobiose-quinone oxidoreductase from *Trametes versicolor. J. Biol. Chem.* **269:**19745–19750.

85. Ruiz-Duenas, F. J., S. Camarero, M. Perez-Boada, M. J. Martinez, and A. T. Martinez. 2001. A new versatile peroxidase from *Pleurotus. Biochem. Soc. Trans.* **29:**116–122.

86. Rypacek, V. 1977. Chemical composition of hemicelluloses as a factor participating in the substrate specificity of wood-destroying fungi. *Wood Sci. Technol.* **11:**59–67.

87. Scott, G. M., M. Akhtar, M. J. Lentz, T. K. Kirk, and R. Swaney. 1998. New technology for papermaking: commercializing biopulping. *Tappi J.* **81:**220–225.

88. Shimada, M., D. B. Ma, Y. Akamatsu, and T. Hattori. 1994. A proposed role of oxalic acid in wood decay systems of wood-rotting basidiomycetes. *FEMS Microbiol. Rev.* **13:** 285–296.

89. Solomon, E. I., U. M. Sundaram, and T. E. Machonkin. 1996. Multicopper oxidases and oxygenases. *Chem. Rev.* **96:**2563–2605.

90. Srebotnik, E., K. A. Jensen, Jr., and K. E. Hammel. 1994. Fungal degradation of recalcitrant nonphenolic lignin structures without lignin peroxidase. *Proc. Natl. Acad. Sci. USA* **91:**12794–12797.

91. Srebotnik, E., and K. Messner. 1994. A simple method that uses differential staining and light microscopy to assess the selectivity of wood delignification by white-rot fungi. *Appl. Environ. Microbiol.* **60:**1383–1386.

92. Srebotnik, E., and K. Messner. 2001. Oxidative mechanisms involved in lignin degradation by white-rot fungi. *Chem. Rev.* **101:**3397–3413.

93. Tanaka, H., S. Itakura, and A. Enoki. 1999. Hydroxyl radical generation by an extracellular low-molecular-weight substance and phenol oxidase activity during wood degradation by the white-rot basidiomycete *Trametes versicolor. J. Biotechnol.* **75:**57–70.

94. Teunissen, P. J., and J. A. Field. 1998. 2-Chloro-1,4-dimethoxybenzene as a mediator of lignin peroxidase catalyzed oxidations. *FEBS Lett.* **439:**219–223.

95. Thurston, C. F. 1994. The structure and function of fungal laccases. *Microbiology* **140:**19–26.

96. Tien, M., and K. Kirk. 1984. Lignin degrading enzyme from *Phanerochaete chrysosporium:* purification, characterization, and catalytic properties of a unique H_2O_2-requiring oxygenase. *Proc. Natl. Acad. Sci. USA* **81:**2280–2284.

97. Tsukihara, T., Y. Honda, T. Watanabe, and T. Watanabe. 2006. Molecular breeding of white rot fungus *Pleurotus ostreatus* by homologous expression of its versatile peroxidase MnP2. *Appl. Microbiol. Biotechnol.* **71:**114–120.

98. Urzua, U., P. J. Kersten, and R. Vicuna. 1998. Manganese peroxidase-dependent oxidation of glyoxylic and oxalic acids synthesized by *Ceriporiopsis subvermispora* produces extracellular hydrogen peroxide. *Appl. Environ. Microbiol.* **64:**68–73.

99. Verdin, J., R. Pogni, A. Baeza, M. C. Baratto, R. Basosi, and R. Vazquez-Duhalt. 2006. Mechanism of versatile peroxidase inactivation by Ca^{2+} depletion. *Biophys. Chem.* **121:**163–170.

100. Ward, G., P. Belinky, Y. Hadar, I. Bilkis, and C. G. Dosoretz. 2002. The influence of non-phenolic mediators and phenolic co-substrates on the oxidation of 4-bromophenol by lignin peroxidase. *Enzyme Microb. Technol.* **30:**490–498.

101. Ward, G., Y. Hadar, I. Bilkis, and C. G. Dosoretz. 2003. Mechanistic features of lignin peroxidase catalyzed oxidation of substituted phenols and 1,2-dimethoxyarenes. *J. Biol. Chem.* **278:**39726–39734.

102. Wariishi, H., K. Valli, and M. H. Gold. 1992. Manganese II oxidation by manganese peroxidase from the basidiomycete *Phanerochaete chrysosporium.* Kinetic mechanism and role of chelators. *J. Biol. Chem.* **267:** 23688–23695.

103. Wood, P. M. 1994. Pathways of production of Fenton reagent by wood-rotting fungi. *FEMS Microbiol. Rev.* **13:**313–320.

104. Wood, W. A., and S. T. Kellogg. 1988. *Lignin, Pectin, and Chitin. Methods in Enzymology,* vol. 161. Academic Press, Inc., San Diego, CA.

105. Wymelenberg, A. V., P. Minges, G. Sabat, D. Martinez, A. Aerts, A. Salamov, I. Grigoriev, H. Shapiro, N. Putnam, P. Belinky, C. Dosoretz, J. Gaskell, P. Kersten, and D. Cullen. 2006. Computational analysis of the *Phanerochaete chrysosporium* v2.0 genome database and mass spectrometry identification of peptides in ligninolytic cultures reveal complex mixtures of secreted proteins. *Fungal Genet. Biol.* **43:**343–356.

106. Yaver, D. S., F. Xu, E. J. Golightly, K. M. Brown, S. H. Brown, M. W. Rey, P. Schneider, T. Halkier, K. Mondorf, and H. Dalbøge. 1996. Purification, characterization,

molecular cloning, and expression of two laccase genes from the white-rot basidiomycete *Trametes villosa. Appl. Environ. Microbiol.* **62:**834–841.

107. **Yoshida, S., A. Chatani, Y. Honda, T. Watanabe, and M. Kuwahara.** 1998. Reaction of manganese peroxidase of *Bjerkandera adusta* with synthetic lignin in acetone solution. *J. Wood Sci.* **44:**486–490.

108. **Zadrazil, F., and P. Reinger (ed.).** 1988. *Treatment of Lignocellulosics with White-Rot Fungi.* Elsevier Applied Science, London, United Kingdom.

MOLECULAR GENETICS

Introduction to Molecular Genetics

C. A. REDDY

Philipp Gerhardt wrote in his Introduction to Molecular Genetics section in *Methods for General and Molecular Bacteriology* (MGMB), "The growth of knowledge about the molecular genetics of bacteria has accelerated like a chain reaction" and that molecular genetics has become a premier part of research in microbiology. These statements are as true today as they were then. The explosive growth in molecular genetics in the past few decades has been so extensive that it has permeated all the subdisciplines of microbiology and in fact practically all areas of modern biology. Many of the major topics in molecular genetics have grown so big that one or more books have been published on the individual topics. Therefore, anything beyond a basic coverage of the methodology in this field is not practical in a book of this kind. However, the methods covered in this section constitute a good start for most of the basic techniques in molecular genetics, and the reader may then seek more detailed and specialized information on advanced procedures from other published sources.

Chapter 26, "Similarity Analysis of DNAs" by Johnson and Whitman, is a new inclusion in this section and is an update of a chapter by the same name from the previous edition (but in a different section) and authored by Johnson (now deceased). This is a comprehensive chapter of basic methods such as DNA isolation and measurement, DNA labeling, and DNA-DNA reassociation. Chapter 27, "Nucleic Acid Analysis," authored by Hendrickson and Walthers, is a revised and updated version of a chapter by the same name in the previous edition and includes several new figures and a table. This is another useful chapter of basic methodology in molecular biology.

"Measuring Spontaneous Mutation Rates" (chapter 28), authored by Foster, is new to this edition. Spontaneous mutations occur in the absence of exogenous causes and can be due to a variety of factors such as errors made by DNA polymerases during replication (or repair), errors made during recombination, and several others. Measuring spontaneous mutation rates is quite challenging, and this chapter should be very useful for researchers in this area.

The chapters "Transposon Mutagenesis" (chapter 29) by Rossbach and de Bruijn, "Plasmids" (chapter 30) by Tolmasky et al., "Gene Transfer in Gram-Negative Bacteria" (chapter 31) by Peters, and "Genetic Exchange in Gram-Positive Bacteria" (chapter 32) by Kristich et al. are retained from the previous edition. Each of these chapters has been revised and updated for this edition and each continues to be a key component in this section. To accommodate the large body of research information that has become available on the molecular biology of Archaea since the last edition of this book, chapter 33, "Genetics of Archaea," authored by Sowers et al., is a welcome new addition to this edition. Another excellent chapter that is being presented for the first time in this edition is chapter 34, "Genetic Manipulations Using Phages," authored by Hatfull et al. In this chapter, the authors focus on the use of bacteriophages to manipulate host bacteria genetically and as sources for the development of genetic tools with wide applications in microbiology.

I thank my coeditor, Loren Snyder, for his assistance in editing the chapters in this section.

26

Similarity Analysis of DNAs

JOHN L. JOHNSON AND WILLIAM B. WHITMAN

Elucidation of the structure and the physical properties of DNA and RNAs has enabled investigators to determine phylogenetic and taxonomic relationships among bacteria by comparing their genomes. A preliminary comparison between two organisms can be made with respect to the overall nucleotide base composition of DNA. This is usually expressed as the moles percent guanine plus cytosine, which is considered a constant feature of that organism. A large difference in the moles percent G+C values from DNAs of two organisms indicates the lack of a close genetic relatedness; however, a similarity in the moles percent G+C does not necessarily mean that the two organisms are closely related with regard to the sequence of nucleotides in their DNAs. In such cases, methods more precise than DNA base composition are required for comparing the genomes, namely **DNA-DNA reassociation** (the formation of duplexes between single-stranded DNA fragments by complementary base pairing), **rRNA-DNA hybridization** (the formation of duplexes between single-stranded DNA and rRNA), and the nucleotide sequencing of specific genes and rRNAs. Methods involving DNA reassociation are covered in this chapter, and those involving rRNA hybridization and nucleotide sequencing are dealt with in a following chapter.

DNA hybridization is especially important because it remains one of the major criteria for the definition of a prokaryotic species (109). In this proposal by an Ad Hoc Committee of the International Committee of Systematic Bacteriology, the criteria for placement of strains in separate species include DNA reassociation values below 70% and a difference in the melting temperature of the DNA hybrids of >5°C. These numerical values for the species boundary represent the values below which phenotypic similarity is difficult to demonstrate (for examples, see references 23 and 54). Thus, when the DNA reassociation is above 70%, the strains possess significant phenotypic similarity. Below 70% DNA reassociation, the phenotypic similarities of strains are often too low to establish taxonomic relationships.

Although sequencing of 16S rRNA genes has proven enormously successful in elucidating the deeper taxonomic relationships at the genus level and below, it has failed to provide reliable information for making species level distinctions (58, 98). Thus, DNA reassociation remains the major method for bridging the modern molecular criteria based on sequencing with the phenotypic descriptions based on numerical taxonomy (97).

The reassociation of denatured DNA fragments or the hybridization between denatured **DNA** and **rRNA** have commonly been referred to as "homologies," with the results reported as relative "percent homology values." In retrospect, the word "homology" was not the best choice to describe these experiments, and it is not used in this chapter because of its long use in a different sense among classical evolutionalists: **homology** (or homologous) refers to phylogenetic development, in which, for example, the arm and the wing are viewed as homologous appendages, with degrees of similarity between the two. Thus, homology refers to a common ancestry and is not used quantitatively. Also, in the literature on DNA reassociation, the relative amounts of DNA from one organism forming duplexes with DNA from the same organism or from another organism have been referred to in the literature as **percent similarity**, **percent identity**, or **percent reassociation**. For consistency, only the term "percent reassociation" is used in this chapter. This terminology is chosen to avoid the misconception that the extent of reassociation between two types of DNA

is directly related to the sequence similarity. In fact, it is the change in the melting temperature of the hybrid DNA complexes that is directly related to sequence similarity (18), while the percent reassociation is the fraction of DNA with a sequence similarity sufficient to form hybrids under the experimental conditions chosen. Thus, the criteria that are used to delineate prokaryotic species, reassociation values below 70% and a difference in the melting temperature of the DNA hybrids of greater than 5°C, suggest that less than 70% of the DNA forms hybrids which have 91 to 92% or greater sequence similarity. Even though the interpretation of DNA reassociation values is complex, these values are still useful to express relatedness between prokaryotes, especially closely related species, subspecies, and strains of a species.

In theory and in practice, DNA reassociation experiments are relatively easy to do. However, there are many technical details or problems associated with the experiments. These include the initial isolation of high-purity DNA, experimental conditions, and the type of experiment. Some problems may be unique for a given group of bacteria, and there are no universal solutions. The methods described in this chapter are, for the most part, those that have been successfully employed in one of the authors' laboratories, and familiarity has contributed to the selection of the specific methods. In some cases specialized instruments are required, such as those found in core facilities (e.g., determination of moles percent G+C in intact cells by flow cytometry or differential scanning colorimetry).

Additional background information on DNA reassociation methods, as well as on DNA base composition methods, can be found in the general references at the end of this chapter (1–6).

26.1. CELL GROWTH AND DISRUPTION

The rapid growth of many bacteria, relative to other organisms, is advantageous for the quick generation of cells from which to isolate DNA. The growth rates and the extent of growth, however, vary greatly and dictate how long cultures must be grown and what volumes are produced.

26.1.1. Growth Conditions (also see chapters 9 to 16)

The group of organisms being investigated determines the culture medium requirements. A peptone-yeast extract-based medium supplemented with a carbohydrate energy source works well for cultivating many chemoheterotrophs. If high levels of acidic by-products are produced, inclusion of a buffer (e.g., 50 mM potassium phosphate [pH 7.0]) may result in larger cell numbers.

The consideration of oxygen requirements is very important. Although reducing agents are usually added to anaerobic media, many anaerobic bacteria grow well in a freshly prepared medium with a nitrogen-filled headspace (26, 43) and a 5% (vol/vol) inoculum of an actively growing culture. For anaerobic bacteria requiring CO_2, sterile sodium bicarbonate can be added at the time of inoculation. Gentle agitation (stirring bar or shaking) may also be important for organisms that tend to settle under static growth conditions. Facultatively anaerobic organisms are best cultured under aerobic conditions, because smaller amounts of acidic end products are produced. Actively growing aerobic cultures rapidly deplete oxygen in the medium; therefore, the rate at which oxygen is dissolved from the atmosphere limits the growth rate. Aeration is optimized by using small volumes of medium per flask (e.g., 250 ml per 2-liter flask), flasks with fluted walls, and rapid shaking (250 to 450 rpm). The oxygen requirements of microaerophilic organisms are the most difficult to fulfill. The medium is first equilibrated by sparging with a mixture of air and CO_2, and the sparging is continued during growth (104).

The preferred time for harvesting a culture is at the early stationary phase, when the amount of growth is near maximum and the death rate is still minimal. The amount of attention that must be given to the growth curve depends on the organism. Many organisms have a long stationary phase with little cell death, so that the exact harvest time is not very critical. Others may start dying quite soon after reaching maximal growth. With sporulating organisms, vegetative-cell DNA can be converted into sporal DNA during the 1- to 3-h sporulation period of the stationary phase.

26.1.2. Cell Disruption (also see chapter 7)

The first step in DNA isolation is the disruption of the bacterial cells. Depending on the nature of the cell wall, bacterial cells can be disrupted (i) with a detergent alone, (ii) with a combination of detergents and hydrolytic enzymes, or (iii) by physical methods such as sonic oscillation, shearing release from a pressure cell, or shaking in the presence of glass beads.

The general protocols are for culture volumes of 250 to 350 ml, and after centrifugation, the cell pellets are suspended in 25 ml of suspending buffer (see below). These volumes can be scaled up or down depending on the amount of growth or the amount of DNA needed. For many organisms, using cell pellets that have been stored frozen greatly reduces the yield of DNA, and these procedures work best with fresh cell material.

26.1.2.1. Gram-Negative Bacteria

Detergents alone can cause many gram-negative bacteria to lyse, but some species may lyse only partially or not at all. An initial incubation of the cells in the presence of hen egg white lysozyme often provides a more uniform lysis on subsequent treatment with the detergent. (Lysozyme is a muramidase, a hydrolytic enzyme that acts on the peptidoglycan of the cell wall. It is inhibited by high salt concentrations and low pH.)

Suspend the cells in a buffer consisting of 10 mM Tris-HCl (pH 8.0), 1 mM sodium EDTA, and 0.35 M sucrose. The sucrose stabilizes the resulting spheroplasts until the addition of the lysing solution. The EDTA binds the divalent cations that stabilize the outer membrane, and Tris buffer also disrupts the outer membrane. Add lysozyme (0.5 to 1.0 mg/ml), and incubate at room temperature for 5 min; then, add the lysing solution (see section 26.2.1). If the cells are not readily lysed by the lysing solution, use a temperature of 37°C for the initial lysozyme digestion. It is also useful to microscopically follow spheroplast formation to ensure successful lysis.

26.1.2.2. Gram-Positive Bacteria

In contrast to gram-negative bacteria, nearly all gram-positive bacteria must be digested with a lytic enzyme before they can be lysed by a detergent. In addition to lysozyme, which is the enzyme most commonly used, several other enzymes are available. These include N-acetylmuramidase, which is isolated from *Streptomyces globisporus* and also cleaves the muramic acid backbone; lysostaphin, an endopeptidase that is isolated from *Staphylococcus* sp. strain K-6-WI and is specific for the cross-linking peptides of

other staphylococci; and achromopeptidase, a peptidase that is isolated from *Achromobacter lyticus* and is also active on the cross-linking peptides.

Lysozyme
Use the same suspending buffer described previously for gram-negative organisms. Add 1 to 3 mg of lysozyme per ml, and incubate at 37°C until the cells are susceptible to lysis by a detergent.

N-Acetylmuramidase
First isolated by Yokogawa et al. (115), N-acetylmuramidase has the same substrate specificity as lysozyme, but the range of organisms that can be lysed by it differs (114). The recommended buffer for this enzyme is 20 mM Tris-HCl (pH 7.0). Add approximately 6 μg of the enzyme per ml to the cell suspension and incubate at 50°C.

Lysostaphin
Lysostaphin is specific for the cross-linking peptides in the cell walls of staphylococci (88). The recommended buffer for the enzyme is 50 mM Tris-HCl–0.145 M NaCl (pH 7.5). Add the enzyme to a concentration of 25 μg/ml and incubate at 37°C.

Achromopeptidase
The *Achromobacter lyticus* strain that produces achromopeptidase was first isolated at Takeda Chemical Industries, Ltd., Osaka, Japan. The enzyme has been found to be useful for plasmid isolation (44) and is also useful for routine DNA isolations. The recommended buffer is 10 mM Tris-HCl (pH 8.2). Add the enzyme to a concentration of 50 μg/ml and incubate at 50°C until the cells become sensitive to lysis.

26.1.2.3. Recalcitrant Bacteria

There are many bacteria whose cell walls are not susceptible to enzymic digestion and detergent lysis. Although this group may include some gram-negative bacteria, most are typical gram-positive bacteria (such as *Actinomyces*, *Streptococcus*, *Peptococcus*, and *Propionibacterium* species) and archaea (such as *Methanosarcina* and *Methanobacterium* species). There are two approaches available for disrupting these recalcitrant bacteria. The first involves making the cells susceptible to one or more lytic enzymes by growing them in the presence of wall-component analogs or antibiotics, and the second involves the use of any of several physical methods for cell disruption.

One can make cells more susceptible to lysozyme by growing them in the presence of glycine, threonine, or lysine (22, 113). However, a problem with using high concentrations of these amino acids is that they may inhibit growth. Another way of making cells susceptible to lysozyme is by adding penicillin at late log phase to inhibit wall synthesis. Here one has to be concerned that the cells do not lyse before or during harvesting. Required penicillin levels may vary from strain to strain. Lysozyme-induced lysis of some gram-negative bacteria can be improved by washing the cells with a mild detergent (0.1% Sarkosyl) and then subjecting them to a mild osmotic shock (89).

Many physical methods have been used to disrupt bacterial cells. The following are some methods that have been routinely used for DNA isolation.

Sonication
The major problems with sonication are that it may not disrupt cells having thick cell walls or small sizes and that it

causes substantial fragmentation of the DNA. The amount of fragmentation increases with time of sonication. DNA preparations from cells disrupted in this manner are not useful for restriction endonuclease digestion and cloning, but if the sonication times can be standardized the DNA preparations can be used in reassociation experiments.

French Pressure Cell
Disruption by passage through a French pressure cell has the same limitations as sonication. The advantage of the French pressure cell is that one obtains more evenly sized DNA fragments.

Glass Bead Disruption
When suspensions of organisms are shaken with glass beads at high speed (4,000 oscillations per min), the cells rupture by being crushed between the colliding beads. The bead size is very important; for bacteria, small beads of about 100 μm in diameter are used. There are several instruments available commercially for shaking cells with glass beads. One that works well is the Braun Cell Homogenizer (Braun Instruments, Melsungen, Germany). The colliding beads produce so much kinetic energy that the sample vessel must be cooled with liquid CO_2. Although it can disrupt all cell types, the homogenizer does not fragment DNA to the same extent as does sonication or passage through a French pressure cell.

Alkaline Hydrolysis
Sodium hydroxide at a concentration of 0.03 N has been used for the mild disruption of gram-negative bacilli and subsequent isolation of native DNA (9) (see section 26.2.1). It has also been used at a concentration of 0.4 N for not only disrupting bacteria but also obtaining denatured (single-stranded) DNA, as in the following procedure (77).

Suspend the bacterial cells in water, and add NaOH from a 10 N stock solution to give a final concentration of 0.4 N. (*CAUTION:* Add the stock solution slowly, and be sure to wear safety goggles.) Incubate in a water bath at 80°C for 30 min. Under these conditions the RNA, proteins, and many of the polysaccharides are hydrolyzed, but the denatured DNA remains intact except for some depurination and strand scission (105). The alkaline lysate can be used directly to immobilize DNA on nylon membranes; alternatively, after neutralization, the denatured DNA can be isolated by using hydroxylapatite (HA) (see section 26.2.3).

26.2. DNA ISOLATION
The pioneering work of Kirby et al. (59, 60) and Marmur (74) has provided the basis for many nucleic acid isolation protocols. Procedures involving cetyltrimethylammonium bromide (CTAB) precipitation have evolved from the work of Jones (53). Britten et al. (15) were the first to use HA adsorption for DNA purification.

26.2.1. Marmur Procedure
The Marmur procedure (74) is widely used for isolation of large amounts of DNA. The bacterial cells are lysed with sodium dodecyl sulfate (SDS) in the presence of sodium chloride, Na-EDTA, and sodium perchlorate. Sodium perchlorate is a chaotropic salt that helps dissociate proteins and polysaccharides from nucleic acids. The proteins are then extracted with chloroform-isopentanol and discarded, and the DNA is precipitated with ethanol. After treatment with RNase and more chloroform-isopentanol extractions, the DNA is again precipitated. The isolated DNA has rea-

sonably large fragments such that, in addition to all of the standard DNA reassociation and RNA hybridization experiments, it can be used in endonuclease restriction digestions and partial digestions for fragment size estimations, fragment size polymorphism studies, and construction of genomic libraries. The major problem with the procedure, depending on the organism, is the coisolation of high-molecular-weight polysaccharides, which have ethanol precipitation properties very much like those of DNA. The current variation of the Marmur procedure that is used in the author's laboratory includes the addition of the reducing agent 2-mercaptoethanol (which was initially used to inactivate nucleases in eucaryotic tissue but also seems to work with bacterial cells) and proteinase K to degrade proteins. The procedure is described for culture volumes of 250 to 500 ml but is amenable to scaling up or down. The protocol below is based on the use of lysozyme to render the cells susceptible to SDS disruption, but the suspending buffer can be modified for use with other lytic enzymes.

Reagents
Cell-suspending buffer: 10 mM Tris-HCl (pH 8.0), 1 mM sodium EDTA, and 0.35 M sucrose.

Na-EDTA, 0.5 M solution (pH 8.0).

SDS, 20% (wt/vol) solution in water.

Proteinase K, 20 mg/ml in Tris-EDTA (TE) buffer. The latter consists of 10 mM Tris-HCl (pH 8.0) containing 1 mM sodium EDTA (pH 8.0).

Lysing solution. The lysing solution consists of 100 mM Tris-HCl buffer (pH 8.0), 0.3 M NaCl, 20 mM EDTA, 2% (wt/vol) SDS, 2% (vol/vol) 2-mercaptoethanol, and 100 μg/ml proteinase K. Add the proteinase K and 2-mercaptoethanol the day of use; do not store the complete solution longer than 1 day. Also, just before use, add 0.5 vol of 5 M sodium perchlorate to the volume of lysing solution needed. A precipitate will form but will quickly dissolve if the solution is warmed in a 50 to 60°C water bath. Add 1.5 vol of this mixture to each volume of cell suspension.

Sodium acetate, 3 M (pH 6.0).

Standard saline-citrate (SSC): 0.15 M NaCl, 0.015 M trisodium citrate (pH 7.0). Other concentrations of SSC are indicated by multipliers; for example, 20× SSC = 20-fold the standard concentration, 0.1× SSC = 1/10 the standard concentration.

Chloroform-isopentanol: chloroform containing 3% (vol/vol) isopentanol.

Phenol-chloroform: water-saturated phenol mixed with an equal volume of chloroform that contains 3% (vol/vol) isopentanol. Add 0.1% (wt/vol) hydroxyquinoline to the mixture.

Ethanol, 80% (vol/vol): 80 parts 100% ethanol plus 20 parts distilled water. Store and use at −20°C.

RNase A: bovine pancreatic RNase A, 1 mg/ml, dissolved in 0.15 M NaCl (pH 5.0). Heat the solution at 80°C for 10 min to inactivate any traces of DNase. Store in tubes containing small portions of the solution at −20°C.

RNase T: dissolve RNase T_1 in 0.02 M Tris-HCl buffer (pH 6) to a concentration of 200 U/ml.

Procedure
1. Harvest the cells by centrifugation. After decanting the spent medium, remove any residual medium with a

Pasteur pipette, and dry the inside of the tube with a tissue. Suspend the cells in 20 ml of suspending buffer, and place them into a ground-glass-stoppered 250-ml Erlenmeyer flask.

2. Add dry lysozyme to the 20-ml cell suspension with a measuring spoon (one-eighth teaspoon gives about 2.5 mg/ml). Incubate the mixture at room temperature or at 37°C until the cells become susceptible to lysis by the addition of the next set of reagents. For gram-negative organisms, only about 5 min may be needed; gram-positive organisms may also become susceptible in a few minutes but sometimes require several hours. Remove small samples (0.1 to 0.2 ml) to a small test tube and mix with 1.5 vol of lysing solution to test for lysis. On lysis, the cell suspension changes from turbid to opalescent and becomes very viscous.

3. Lyse the cells by adding 30 ml of the warm lysing solution. Quickly swirl to uniformly mix the components and obtain a uniform lysis of the cells. Incubate at 50 to 60°C for 1 to 4 h.

4. Add 12 ml of the phenol-chloroform mixture, and swirl and shake to obtain a homogeneous mixture. (Shaking may have to be vigorous for a highly viscous lysate.) Shake on a wrist-action shaker for 20 min at a setting that will provide vigorous mixing; however, if the phenol-chloroform is not first mixed in well by hand, the extraction will be incomplete.

5. Place the extracted lysate into centrifuge tubes (dividing the lysate between two 50-ml straight-wall polypropylene tubes works well), and centrifuge for 10 min at 17,000 × g. Carefully decant and/or pipette the upper, aqueous layer from the centrifuge tubes back into a clean Erlenmeyer flask. An inverted 5-ml pipette placed in a pipetting device or pump works well for removing the viscous lysate. Discard the phenol-chloroform, bottom layer into an organic waste container.

6. Add 10 ml of phenol-chloroform mixture to the lysate, and again shake for 20 min on a wrist-action shaker. Centrifuge, and collect the upper layer as in step 5. Repeat this step until there is no protein layer at the aqueous-chloroform interface. Each time discard the phenol-chloroform, bottom layer into an organic-waste container.

7. Place the aqueous phase into a 125-ml ground-glass-stoppered Erlenmeyer flask, and add 0.6 volume of isopropanol. Swirl the flask so that the precipitated DNA will form a clot. Decant the lysate, leaving the DNA clot in the flask. Add about 20 ml of cold 80% ethanol, swirl, and decant after about 5 min. Repeat this ethanol wash step twice. After the final ethanol wash, allow the clot to stick to the bottom of the flask (this can usually be done by slowly tilting the flask) and then invert the flask on a paper towel to drain. Air dry for 15 to 30 min in a 37°C incubator.

8. Dissolve the DNA by adding 20 ml of 0.1× SSC to the flask. After the DNA is completely dissolved, add 0.25 ml of RNase A solution and 2.5 ml of RNase T_1, solution. Incubate at 37°C for 30 min.

9. Add 5 ml of chloroform-isopentanol to the DNA solution, and shake on a wrist-action shaker for 20 min. Centrifuge at 17,000 × g for 10 min. Remove the aqueous phase with a pipette and save it. If there is a substantial protein layer, repeat the chloroform-isopentanol step one more time. Finally, transfer the aqueous phase to an 80- or 100-ml beaker.

10. Again precipitate the DNA, as follows. Add 2.0 ml of 3 M sodium acetate to the DNA solution, swirl to mix, and then overlay the solution with 2 volumes of 95% ethanol. "Spool" the DNA onto a glass rod (20 cm by 7

mm) by slowly stirring and spinning the rod in the mixture (Fig. 1). Continue the spooling until the two phases are totally mixed. Gently press and turn the rod on the wall of the beaker to squeeze out the remaining ethanol. Wash the rod with 10 to 20 ml of cold 80% ethanol, invert, and stand the rod in a test tube rack to air dry. Dissolve the DNA by placing the rod in screw-cap test tube containing 3 to 5 ml of 0.1× SSC or TE buffer. Usually after about 30 min the DNA can be slipped off the glass rod by gently shaking the rod up and down (be careful not to hit the bottom of the test tube at this point; the rod can easily punch through the bottom). Alternatively, place the tube and rod overnight in a refrigerator to dissolve the DNA.

11. Finally, dissolve the DNA in 3 to 5 ml of 0.1× SSC or TE and store in a freezer.

Comments

There are many DNA isolation procedures, but the Marmur procedure is the best method for most applications. The following are some important variations from the procedure originally described by Marmur. (i) The lower-ionic-strength suspension buffer optimizes the efficiency of cell wall lysis by lysozyme. (ii) The single lysing solution allows for an even distribution of all of the components in what will become a very viscous solution. (iii) The proteinase K digestion helps promotes purification of protein-free DNA. For some organisms, the binding of protein to the DNA can cause the DNA to be extracted with the proteins. (iv) Initial protein extraction with phenol-chloroform seems to be more effective than the use of chloroform-isopentanol alone. (v) The use of 0.6 volume of isopropanol for the initial DNA precipitation promotes less coprecipitation of polysaccharides with the DNA. In the original procedure, this selective precipitation step was near the end. (vi) Letting the DNA form a clot permits efficient washing with

FIGURE 1 Spooling DNA onto a glass rod during precipitation by ethanol. Photo by D. Arbour.

80% ethanol and rapid redissolution in buffer. (vii) In many cases, the use of a single RNase (RNase A) leaves RNA fragments that are still large enough to coprecipitate with the DNA. Using both RNase A and RNase T reduces this problem.

There have been other variations of the Marmur procedure reported in the literature, such as dialyzing DNA preparations to remove small RNA fragments, replacing sodium perchlorate with NaCl (35), and performing an initial alkaline lysis of the cells (8). The major features of the procedure of Beji et al. (9) are as follows.

Harvest late-log-phase cells (500 to 600 ml of culture) and resuspend them in 50 ml of saline-EDTA (150 mM NaCl, 100 mM EDTA [free acid] [pH 8.0]). Centrifuge again and determine the wet weight of the cells. Lyse the cells with 0.03 M NaOH (10 ml/g [wet weight] of cells); leave the cells in contact with NaOH for a maximum of 2 min at room temperature. Add (per gram [wet weight] of cells) 1.5 ml of 25% (wt/vol) SDS and 35 ml of saline-EDTA (pH 7). Mix well, and centrifuge at $17,000 \times g$ for 5 min at 4°C. After treatment with 2.5 mg of RNase at 60°C for 30 min and 0.6 mg of proteinase K at 37°C for 15 min, add 5 M NaCl to increase the NaCl concentration to 1 M. Precipitate the DNA with 1 volume of 95% ethanol and redissolve in 20 ml of $0.1 \times$ SSC. Add 0.45 M NaCl, and again precipitate the DNA by adding an equal volume of isopropanol. Dissolve the final product in $0.1 \times$ SSC, and dialyze it against $0.1 \times$ SSC.

26.2.2. Cetyltrimethylammonium Bromide Lysis

The CTAB procedure has been used extensively for fungal materials, which contain polysaccharide-like contaminants that have plagued DNA isolation procedures (80, 116). The procedure described here also works well for actinomycetes and propionibacteria.

The cetyltrimethylammonium ion is a cationic detergent and cannot be used in conjunction with SDS or phenol, because these mixtures form very insoluble complexes. DNA is soluble in the presence of CTAB if there is also a high concentration of monovalent cations such as Na^+ or NH_4^+.

The protocol given here (52) is for recalcitrant organisms that must be physically disrupted. This disruption results in fragmentation of the DNA, and the DNA works well for reassociation experiments involving only fragmented DNA, i.e., the optical, HA, and S1 nuclease procedures. However, it may not work well if membrane-bound DNA is required for DNA reassociation experiments or rRNA hybridization, and it would not be suitable for restriction endonuclease digestions. The procedure can be used for the isolation of high-molecular-weight DNA if lysozyme-digested cells (or cells digested by some other lytic enzyme) are lysed by the CTAB. The procedure may also be useful when polysaccharide contamination of DNA preparations is a problem.

Reagents

Cell-suspending buffer (TE buffer): 10 mM Tris-HCl buffer (pH 8.0) containing 1 mM Na-EDTA (pH 8.0).

CTAB lysing solution: 50 mM Tris-HCl buffer (pH 8.0), 0.7 M NaCl, 10 mM EDTA, 1% CTAB (wt/vol), 1% (vol/vol) 2-mercaptoethanol. $2 \times$ and $5 \times$ CTAB lysing solutions can be prepared by using two and five times the concentrations of all of the components, respectively.

Proteinase K: see section 26.2.1.

Chloroform-isopentanol: see section 26.2.1.

CTAB dilution-precipitation buffer: 50 mM Tris-HCl, 10 mM EDTA, 1% CTAB (wt/vol).

CTAB wash solution: 0.4 M NaCl.

CTAB-DNA salt dissolving solution: 2.5 M ammonium acetate.

SSC: see section 26.2.1.

RNase A: see section 26.2.1.

RNase T$_1$: see section 26.2.1.

Phenol-chloroform: see section 26.2.1.

Sodium acetate, 3 M (pH 6.0).

80% ethanol: see section 26.2.1.

Procedure

1. Harvest the cells by centrifugation. After decanting the spent medium, remove any residual medium with a Pasteur pipette. Transfer the cell pellet to a tared Braun Disintegrator shaking bottle, and make up total mass to 10 g with TE buffer. Add 10 ml of $2 \times$ CTAB lysing buffer, 0.2 ml of 2-mercaptoethanol, and 20 ml of glass beads (diameter range, ca. 100 μm). Place the bottle in the Braun Disintegrator, and shake for 5 min with CO_2 cooling. The exact shaking time required depends on the particular organism; confirm cell breakage by microscopic observation (phase contrast or Gram staining).

2. Separate the lysate from the glass beads by filtration through a sintered-glass filter (coarse pore). Filtration is facilitated by applying a vacuum (a water aspirator can produce a sufficient vacuum) to a vacuum flask. Use an additional 20 ml of $1 \times$ lysing buffer to wash the remaining lysate away from the glass beads. Add small amounts of buffer at a time, and use a rubber-tipped stirring rod to mix it with the glass beads. Avoid using metal or glass stirring rods, as they will damage the surface of the sintered glass filter.

3. To the pooled lysate, add proteinase K to a final concentration of 50 μg/ml, and incubate the mixture for 1 to 2 h at 60°C.

4. Add 15 ml of chloroform-isopentanol to the lysate, shake on a wrist-action shaker for 20 min, and centrifuge at $17,000 \times g$ for 10 min.

5. Transfer the lysate (upper layer) into a 250-ml centrifuge bottle and add an equal volume of CTAB dilution-precipitation buffer. Mix well, and pellet the CTAB-nucleic acid salts by centrifuging at $8,000 \times g$ for 10 min. Wash the precipitate two or three times with 40-ml volumes of cold 0.4 M NaCl wash solution to remove free CTAB. A glass homogenizer works well to disperse the pellet for more efficient washing.

6. Dissolve the CTAB-nucleic acid pellet in 8 ml of 2.5 M ammonium acetate. Use a 50-ml screw-cap polypropylene centrifuge tube, and warm the material at 50°C to facilitate dissolution. After the nucleic acids are in solution, place the tube on ice for 15 min to precipitate the large RNAs. Remove the RNA by centrifugation at $17,000 \times g$ at 4°C for 10 min. Transfer the supernatant to another centrifuge tube, and precipitate the remaining nucleic acids with 2 volumes of 95% ethanol. Collect the precipitate by centrifugation at $17,000 \times g$ for 10 min.

7. Dissolve the pellet in 10 ml of $0.1 \times$ SSC, and add 0.25 ml of RNase A and 2.5 ml of RNase T$_1$. Incubate at 37°C for 1 h.

8. Extract the digest once with 5 ml of phenol-chloroform. After centrifugation, add 1 ml of 3 M sodium acetate

to the extracted digest, and precipitate by adding 2 volumes of 95% ethanol. Collect the precipitate by centrifugation, and wash the pellet with 10 ml of 80% ethanol.

9. Dissolve the DNA in 3 to 5 ml of $0.1 \times$ SSC or TE and store in a freezer.

Comments

The procedure described above can be used for organisms that are difficult to lyse and have high levels of nuclease activity. A technical difficulty is that the detergent causes foaming and tends to reduce the velocity of the beads, thereby reducing their kinetic energy. Depending on the organism, this may result in inefficient cell breakage. If nuclease activity is not a problem, suspend and disrupt the cells in TE buffer (it may be helpful to increase the EDTA concentration to 10 mM) and add the RNase at this time. Then, remove the glass beads by filtration and add 0.25 volume of $5 \times$ CTAB lysing solution. Then proceed to step 3 of the procedure.

If cell lysis is not difficult and the main reason for using this procedure is to eliminate contaminating polysaccharides, one can start as if doing the Marmur procedure but then lyse cells with $2 \times$ CTAB lysing solution.

26.2.3. Hydroxylapatite Adsorption

Britten et al. (15) were the first to isolate DNA by adsorbing it to HA. Since then, several modifications of the method have been described (2, 49, 73) including the following adaptation.

Reagents

Cell-suspending buffer (TE buffer): see section 26.2.2.

SDS: see section 26.2.1.

RNase A: see section 26.2.1.

RNase T_1: see section 26.2.1.

Phenol: water-saturated phenol containing 0.1% (wt/vol) 8-hydroxyquinoline.

HA, DNA grade Bio-Gel HTP (Bio-Rad Laboratories, Richmond, CA).

Phosphate buffer (PB), 1.0 M (pH 6.8) (prepare by mixing equal volumes of 1 M Na_2HPO_4 and 1 M NaH_2PO_4). This is used for preparing the lower concentrations as indicated in the text.

Procedure

1. Suspend the centrifuged cells in 25 ml of TE buffer, add lysozyme or other lytic enzyme, and incubate until the cells are susceptible to detergent lysis (see section 26.2.1, step 2). Add 1.5 ml of the SDS solution.

2. Add 0.25 ml of RNase A and 2.5 ml of RNase T_1. Swirl the flask and warm in a 50 to 60°C water bath until lysis is complete. For bacteria not disrupted by detergent alone, see Comments, section 26.2.2, for alternate methods.

3. Reduce the viscosity of the lysate by briefly subjecting it to sonic oscillation or by use of a mechanical tissue homogenizer. Proteinase K may be added at this time (50 µg/ml). For some organisms proteinase digestion is not needed, but for others it is essential. If the proteinase is added, incubate it with the lysate at 50°C for 1 h.

4. Add 7 ml of water-saturated phenol, shake by hand to get the two phases well mixed, and then shake the flask on a wrist-action shaker for 20 min.

5. Centrifuge at $17,000 \times g$ in a polypropylene centrifuge tube, using a refrigerated centrifuge at 0 to 4°C.

Carefully draw off the upper (aqueous) layer (see step 5 of section 26.2.1) and return it to the flask.

6. Repeat steps 4 and 5.

7. After again returning the aqueous phase to the flask, add 2.0 ml of 1.0 M PB. Then, add 2 g (1 measuring teaspoon is convenient and sufficiently accurate) of dry HA. Suspend well, and gently shake on a rotary or reciprocal shaker for 1 h at a speed sufficient to keep the HA from settling out.

8. Transfer the suspension to a 50-ml polypropylene centrifuge tube, and centrifuge for 2 to 3 min at $5,000 \times g$ at room temperature. Return the supernatant layer (lysate) to the flask (this can be used for a second DNA adsorption cycle), and use the sedimented HA (to which DNA is now adsorbed) in step 9.

9. To the sedimented HA, add 8 ml of 0.10 M PB. Suspend the HA with the aid of a Vortex mixer. Immediately add an additional 24 ml of the PB (an automatic pipettor works well for these additions). The PB should be added with some force, so that the HA will become evenly mixed for maximum dilution of nucleotides and phenol. Allow the HA to settle for 1 to 2 min, and then centrifuge for 2 to 3 min at $5,000 \times g$. Discard the supernatant.

10. Repeat step 9 six or seven times or until the absorbance at 270 nm (or A_{270}) of the supernatant is less than 0.05 (the wavelength of maximum absorption by phenol).

11. Suspend the HA in 5.0 ml of 0.5 M PB to desorb the DNA. Centrifuge as before, but this time save the DNA-containing supernatant.

12. If most of the DNA has not been removed from the lysate in step 8, wash the HA from step 11 once with distilled water, and then add it to the lysate for a second adsorption cycle. Alternatively, fresh HA may be used. Add the DNA obtained from a second adsorption cycle to that from the first cycle.

13. Filter the DNA preparation through a fiberglass filter (2.4-cm diameter) in a syringe-type filter holder to remove any remaining HA particles.

14. Dialyze the DNA preparation against TE buffer (pH 8.0). Cut 13- to 15-cm lengths of dialysis tubing and wash out UV-absorbing materials by boiling the tubing in a 2 to 5% solution of sodium carbonate for a few minutes, thoroughly rinsing the lengths of tubing under running tap water, and then rinsing them under distilled water. The tubing can be stored in a refrigerator for 2 to 3 days in distilled water. Cellulolytic organisms may hydrolyze the tubing if it is stored for a longer time. Seal one end of the tubing length by tying a knot or using a dialysis tubing clamp, pour in the DNA solution, and seal the other end of the tubing. Attach a string to one end, and wrap a piece of tape around the string for a convenient label. Dialyze in a 400- to 500-volume excess of the buffer for about 3 h, change the buffer, and continue dialyzing overnight. Store the DNA preparation in a freezer, or add a few drops of chloroform and store in a refrigerator.

Comments

If the lysate is digested with proteinase K, a single phenol extraction will usually be sufficient.

An important feature of the HA procedure is that RNA is degraded to such an extent that it will not compete with DNA for adsorption sites on the HA. For some groups of organisms, the RNase is inhibited when added in step 2. In such cases, dialyzing the lysate against saline-EDTA buffer for several hours removes the inhibitory effect.

The major contaminant in the DNA preparations, other than RNA, is polysaccharide. Many polysaccharides do not adsorb to HA and therefore are easily separated from DNA by this procedure. Others, however, can bind to HA and inhibit DNA adsorption.

In some cases a more dilute cell suspension of some organisms may be required to improve the efficiency of this method. Dilution appears to promote more complete lysis and/or RNA degradation. Scale up the volumes four or five times, and follow the above steps until the HA adsorption. After adding the HA to part of the lysate, pour the HA suspension into a 60-ml sintered-glass filter. The HA will quickly settle, and the lysate will flow through the HA as through a column, resulting in more efficient adsorption. Transfer the HA to a centrifuge tube, and complete the procedure as described above.

26.3. DNA MEASUREMENT (also see chapter 18.6)

26.3.1. UV Spectrophotometry

The molar extinction coefficient of native DNA is in the range of 6,650 to 6,700 (51, 68). Conversion of the DNA concentration from molarity to milligrams per milliliter yields the following formula, for cuvettes with a 1-cm light path. For native (double-stranded) DNA, the formula is

$$\text{Concentration (mg/ml)} = A_{260}/20 \qquad (1)$$

For denatured (single-stranded) DNA and for RNA, the formula is

$$\text{Concentration (mg/ml)} = A_{260}/23 \qquad (2)$$

In practice, most stock DNA preparations range from 0.5 to 2 mg of DNA per ml. It is convenient to dilute them 20-fold (0.1 ml of DNA solution added to 1.9 ml of buffer) because the resulting A_{260} can be read directly as milligrams of DNA per milliliter of the stock preparation (see equation 1).

UV spectrophotometry has also been used for detecting contaminating materials (51). (i) Proteins absorb UV light more strongly than nucleic acid polymers at 280 nm. The A_{260}/A_{280} ratio of a nucleic acid solution is used as an indicator of protein contamination. The extinction coefficient for protein, however, is very low compared with that for nucleic acids; consequently, the A_{260}/A_{280} ratio is not a very sensitive indicator. (ii) The absorption spectrum for nucleic acids has maxima at approximately 208 and 260 nm and a minimum at 234 nm. The peak at 208 nm is not specific for nucleic acids, because most compounds absorb light of that wavelength. If contaminating material is present in a nucleic acid preparation and contributes to a large peak at 208 nm, this peak will overlap the 234-nm minimum for the nucleic acid. Therefore, the A_{260}/A_{234} ratio is a more sensitive indicator of contaminating material in a DNA preparation than the A_{260}/A_{280} ratio. (iii) The absorption maximum of phenol is 270 nm; consequently, the A_{260}/A_{270} ratio is quite sensitive for detecting phenol contamination in a nucleic acid preparation.

26.3.2. Hyperchromic Shift

The major contaminant of DNA preparations is usually RNA, and unfortunately, the two nucleic acids cannot be differentiated directly by their UV absorbance. However, the hyperchromic shift (absorbance change) that occurs when DNA is thermally denatured (see section 26.4.1) can be used to estimate the fraction of the preparation that is

DNA. A preparation of pure native DNA increases in A_{260} by approximately 40% (Fig. 2) when denatured. If the secondary structure of the contaminating RNA has been destroyed by RNase, the RNA will have no hyperchromic shift. If the secondary structure is intact, the RNA will have a hyperchromic shift of approximately 30%. The hyperchromic shift for RNA occurs over a lower and wider temperature range than for DNA, so the profiles for RNA and DNA can be easily distinguished. Therefore, to estimate the A_{260} of a nucleic acid preparation that is due only to the DNA, divide the hyperchromic shift due to DNA by 0.4. For example, assume that the A_{260} of a sample (or a dilution of it) is 1.25 and that the change in A_{260} as a result of denaturation is 0.4. The calculated DNA A_{260} would be 0.4/0.4, or 1.00 (i.e., 1/1.25 or 80% of the A_{260} due to DNA).

26.3.3. Colorimetric Method (Diphenylamine Reaction)

Many colorimetric assays have been developed and modified over the years for the measurement of DNA (8, 16, 17, 39, 93, 112). Although they can be used for measuring purified DNA, these procedures were often designed for measuring DNA in tissue material. They have the disadvantage of using high concentrations of mineral acids and of requiring long periods for color development.

The diphenylamine reaction is specific for the deoxyribose in DNA. Although the Burton procedure (16) has

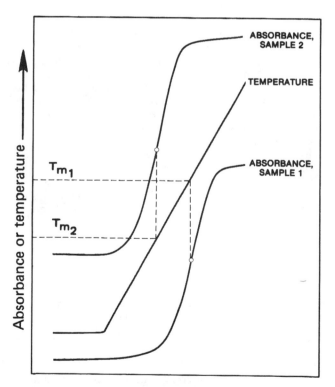

FIGURE 2 Hyperchromic shift tracings for two samples of DNA in an automatic recording spectrophotometer. The points at which any vertical line crosses the absorbance and temperature curves give values for these two parameters at a given time. The inflection points are estimated at one-half of the hyperchromatic shift, and the temperature at that time is the melting temperature (T_m).

often been used, the following variation of the more sensitive Giles and Meyers modification (37) is preferable.

1. Place 1.0 ml of the DNA sample (5 to 50 μg) in a 16-by 125-mm screw-cap tube and add 1.0 ml of 20% (wt/vol) perchloric acid.

2. Add 2.0 ml of glacial acetic acid containing 4% (wt/vol) diphenylamine.

3. Add 0.2 ml of 0.16% (wt/vol) solution of acetaldehyde.

4. Mix and incubate overnight at 30°C.

5. Read the A_{595} and A_{700}, and subtract the A_{700} from the A_{595}. Construct a standard curve, using known amounts (5 to 50 μg) of a pure standard DNA (calf thymus or salmon sperm DNA; Sigma Chemical Co.), and plot the microgram amount of DNA against the difference ($A_{595} - A_{700}$). Determine the concentration of DNA in the sample from the standard curve.

The diphenylamine procedure can also be used for quantifying the DNA bound to nitrocellulose or nylon membranes in hybridization studies. It is particularly useful when one cannot readily measure the amount of DNA being bound to a filter (for an example, see section 26.6.3). After the amounts of probe (radioactive or nonradioactive) bound to the membranes have been measured (use a nonaqueous counting solution so that the membranes will not be dissolved or the DNA eluted), dry the membranes and place each membrane in 2.0 ml of 10% perchloric acid (duplicate membranes can be placed together in the same tube). Heat in a boiling-water bath for 5 min. After cooling to room temperature, proceed with step 2 above.

26.3.4. Fluorimetric Dye-Binding Methods

Fluorescent compounds, usually dyes, have been used for characterizing and quantitating DNA. After interacting with the DNA, the resulting complex causes an enhancement of the fluorescence. The complexes are very specific, in that molecules other than DNA (such as RNA, proteins, and carbohydrates) seldom interfere with assays. A major use of these procedures is the measurement of DNA in tissue homogenates. Certain dyes may complex specifically with high-A+T or high-G+C regions of the DNA, whereas the binding of others is independent of the base composition. As a result, the moles percent G+C content of the DNA may have to be taken into account when constructing standard curves, and in some cases differential binding of two dyes can be used for estimating this value. The following are some of the more commonly used fluorimetric methods for measuring DNA.

26.3.4.1. Mithramycin

The antibiotic mithramycin binds to double-stranded DNA, and the amount of fluorescence is directly proportional to the amount of DNA (42). The antibiotic binds to a lesser extent to denatured DNA and does not bind to DNase-treated DNA or to RNA.

Reagent

Prepare a stock solution containing 200 μg of mithramycin per ml in 300 mM $MgCl_2$.

Procedure

1. Dilute a sample of a DNA standard (a preparation containing no RNA, with the concentration determined by UV absorbance) with distilled water to give amounts ranging from 0.4 to 32 μg in a final volume of 1.9 ml.

2. Dilute the test DNA preparations so that they are within the same range.

3. Add 0.1 ml of the mithramycin stock solution to the standard and test DNA dilutions, mix, and determine the fluorescence (emission) at 540 nm with an excitation wavelength of 440 nm.

4. Plot a standard curve from the values obtained for the DNA standards, and use it to estimate the concentrations of the test DNA samples.

26.3.4.2. 4′,6′-Diamidino-2-Phenylindole·2HCl

DAPI (4′,6′-diamidino-2-phenylindole · 2HCl) forms a specific fluorescent complex with DNA, but the manner by which it binds (i.e., by intercalation, or by a nonintercalating mode) is uncertain (67). DAPI does exhibit a preference for several continuous A-T base pairs (56). The dye is very stable and is specific for DNA. It has been used to measure DNA directly in fixed cells of *Escherichia coli* and might be similarly useful with other bacteria.

Reagents

DAPI stock solution: 3 μg/ml in distilled water. This stock solution is stable at 4°C for months.

Dilution buffer: 10 mM NaCl, 6.6 mM Na_2SO_4, 5 mM N′-2-hydroxyethylpiperazine-N′-2-ethanesulfonic acid (HEPES; pH 7.0).

Toluene-fixed cells. Mix samples of bacteria with toluene (1%, vol/vol), and shake vigorously. Store at 4°C.

Procedure

1. Dilute a sample of a DNA standard in the dilution buffer to give concentrations ranging from 5 ng/ml to 1.0 μg/ml in a final volume of 3.0 ml.

2. Similarly, dilute the fixed bacterial cells in the dilution buffer to a final turbidity of 0.008 to 0.08 at 600 nm (turbidity values may differ from one spectrophotometer to another; the only important requirement is that the fluorescence values for DAPI [see step 3] be linear within the turbidity range used).

3. Add 100 μl of DAPI stock solution to each tube, mix, and measure the fluorescence intensity with excitation at 350 nm and emission at 450 nm. (The mixtures are stable at 4°C for several weeks.) Plot a standard curve from the values obtained for the DNA standard and use it to estimate the concentrations of the test DNA samples.

26.3.4.3. Bisbenzimide (Hoechst 33258)

The Hoechst 33258 dye is probably the most widely used of DNA-binding dyes (20, 29, 62, 82, 83, 99, 100). The dye binds specifically to regions rich in A-T base pairs in the large groove of the DNA helix (99). Consequently, the DNA used for the standard curve should have a moles percent G+C content similar to that of the test DNAs.

Reagents

20× Phosphate-buffered SSC (PB-SSC): 20× SSC (see section 26.2.1) prepared in 200 mM sodium phosphate buffer (pH 7.0).

Hoechst 33258 stock solution: 2 μg/ml in 1× PB-SSC. (This is equivalent to a concentration of 3.7×10^{-6} M; the molecular weight of the dye is 533.89.) The dye concentration can be measured with a spectrophotometer at 338 nm; the molar extinction coefficient is 4.2×10^4 M^{-1} (103).

Procedure

1. Dilute samples of a DNA standard (a preparation containing no RNA, with the concentration determined by UV absorbance) in 1× PB-SSC to give amounts ranging from 0.01 to 10 µg in final volumes of 2.0 ml.

2. Dilute the test DNA preparations so that they are within the same range.

3. Add 2 ml of dye solution to each tube of DNA, mix, and measure the fluorescence intensity with excitation at 360 nm and emission at 450 nm. Plot a standard curve from the values obtained for the DNA standard and use it to estimate the concentrations of the test DNA samples.

26.3.4.4. Ethidium Bromide

The intercalating dye ethidium bromide binds to double-stranded DNA with little dependence on the base composition. Although it binds to a lesser extent to intact RNA, error from contaminating RNA can be eliminated by treating the DNA preparation with RNase (57, 67, 82, 83).

Reagents

20× PB-SSC: see section 26.3.4.3.

Ethidium bromide. Prepare a working solution of 2 µg/ml in 2× PB-SSC.

Procedure

1. Dilute samples of a DNA standard (a preparation containing no RNA and the concentration determined by UV absorbance) in 2× PB-SSC to give amounts ranging from 0.01 to 10 µg in final volumes of 2.0 ml.

2. Dilute the test DNA preparations so that they are within the same range.

3. Add 2 ml of dye solution to each tube, mix, and measure the fluorescence intensity with excitation at 305 nm and emission at 615 nm. Plot a standard curve from the values obtained for the DNA standard, and use it to estimate the concentrations of the test DNA samples.

26.4. GUANINE-PLUS-CYTOSINE CONTENT OF DNA

The moles percent G+C content of DNA can be estimated by several different methods; see reference 1 for a review. The methods described in detail here use thermal denaturation, high-performance liquid chromatography (HPLC), or dye-binding fluorimetry. Although buoyant-density centrifugation has also been used for many moles percent G+C estimations reported in the literature, there are few if any analytical ultracentrifuges still functioning. All of these methods require the prior isolation and purification of DNA.

Estimation of the moles percent G+C directly in intact whole cells is often desirable, especially in characterizing newly isolated bacteria in ecological and systematics studies or bacteria whose DNA is difficult to extract or purify. One of the dye-binding fluorimetric methods described below (section 26.4.3.2) can be used with formalin-fixed whole cells. A calorimetric method involving native intact cells is also described (section 26.4.4); this method requires expensive specialized instrumentation but can be performed quickly.

26.4.1. Thermal Melting Method

The midpoint temperatures of the thermal melting profiles (T_m) were first correlated with the base composition of

DNA preparations by Marmur and Doty (75) (Fig. 2). The method has been reexamined by several investigators, and several equations correlating the T_m with the base composition have been proposed. With the exception of the equation proposed by Mandel et al. (72), which appears to have a different slope (2), all are similar to the one proposed by Marmur and Doty. The T_m values are greatly affected by the ionic strength of the buffer used, and this effect is the same on all DNA preparations. Consequently, one must either know the ionic strength of the buffer that is used or include a reference DNA preparation at the same ionic strength as that used for the unknown samples. The latter is easy to do: one merely dialyzes all of the DNA preparations (including the reference DNA) together in a single batch of buffer. Alternatively, if the DNA preparations are all dissolved in a common buffer, they can be diluted with the same buffer and tested directly.

Procedure

The following procedure is reliable and may be adapted to other buffers or other buffer concentrations than those given.

1. Prepare 2 to 4 liters of 0.5× SSC (section 26.2.1) at pH 7.0. Save some of the buffer for diluting the DNA preparations and for rinsing out cuvettes just prior to putting the DNA samples into them. Divide the rest into two equal parts so that the buffer can be changed once during dialysis (see step 3).

2. Prepare 2- to 5-ml samples (depending on the size of the cuvettes to be used) of DNA at 50 µg/ml, using 0.5× SSC to dilute the stock DNA preparations. Prepare a 5- to 10-ml volume of the reference DNA (i.e., the DNA of *E. coli* b has a G+C content of 51 mol% and a T_m of 90.5°C in 1× SSC), because it will have to be included in each instrument run.

3. Dialyze all of the preparations together overnight in the 0.5× SSC. Change the buffer once after the first few hours of dialysis. After dialysis, return the DNA preparations to screw-cap test tubes.

4. Determine the melting profile with an automatic recording spectrophotometer having a cuvette holder that is heated electronically. Start at 60°C, and increase the temperature linearly at 0.5 or 1.0°C per min over the entire range.

5. Determine the T_m values as in Fig. 2. Calculate the moles percent G+C of the sample DNA by the following equation:

$$\text{mol\% G+C} = (T_m + [90.5 - T_{m(\text{ref})}] - 69.3)/0.41$$

This is the Marmur and Doty equation, modified to normalize the T_m values to those in SSC.

Comments

The thermal stability of double-stranded DNA is highly dependent on the ionic strength of the buffer in which it is dissolved. If the DNA preparations have been spooled out of a sodium acetate-ethanol mixture and the spooled DNA has been rinsed with 80% ethanol (i.e., there is little or no residual salt in the DNA preparations), and if the DNA preparations have been dissolved in 0.1× SSC, they can be diluted in a single batch of 0.1× SSC and not dialyzed. Rinsing the cuvettes can be a cause of error if residual water is allowed to dilute the buffer. If rinse water remains in the cuvettes, give them a final rinse with the T_m buffer prior to use.

26.4.2. High-Performance Liquid Chromatography
(also see chapter 17.3.6)

The only direct approach for measuring the base composition of DNA is to hydrolyze it and quantify the individual nucleosides or nucleotides. Originally this was done by using paper chromatography to separate the nucleosides or nucleotides, which were then eluted and measured with a spectrophotometer. Many other methods have been based on correlations with these early results. However, because of the difficulty in quantifying these chromatographic analyses, few have been done since the early 1960s. Recent advances in HPLC have enabled investigators to determine the base composition of DNA more rapidly and accurately (see reference 2 for a review). The procedure described here is that of Mesbah et al. (78, 79), although similar procedures have been described by others (61, 102). This procedure includes degrading denatured DNA with the single-strand-specific nuclease P_1 from *Penicillium citrinum*. The resulting 5′ nucleotides are then dephosphorylated with alkaline phosphatase, and the resulting nucleosides are separated on a reverse-phase C_{18} column. This method has a number of advantages. One, because of the high resolution of modern columns, ribo- and deoxyribonucleosides can be easily separated, and contamination with up to 50% RNA does not interfere (Fig. 3). Two, by limiting the analyses to deoxyguanosine (dGuo) and thymidine (dThd), the effects of base methylations and other modifications (which are common for deoxycytidine and deoxyadenosine) are negligible. Three, if the moles percent G+C is estimated from the ratio of dGuo and dThd, it is also possible to obtain a very high precision of 0.1 mol% with standard equipment. Lastly, much of the analysis can be performed automatically with an autosampler.

Instrumentation Requirements

The HPLC setup includes a pump that will produce pulse-free isocratic separations of the nucleosides, an injector, a fixed-wavelength (254-nm) UV detector, and an integrator-recorder. The column is a C_{18} reverse-phase column, with 5-μm particle size and column dimensions of 250 by 4.6 mm, or any comparable C_{18} column. Because of the sensitivity of peak shape to temperature, the column temperature should be carefully maintained with a column oven or water jacket.

Reagents

Stock triethylamine PB, 0.5 M. Prepare by adjusting the pH of a solution of 0.5 M triethylamine to 5.1 with 85% phosphoric acid. For preparing this buffer, the triethylamine should be nearly colorless. If yellow, it should be redistilled before use.

Chromatographic buffer: 20 mM triethylamine phosphate buffer (pH 5.1) containing about 8 to 12% (vol/vol) HPLC grade methanol. Before using, filter the buffer through a cellulose triacetate membrane filter with a pore size of 0.45 μm. The actual methanol concentration of the buffer depends greatly on the individual column and the analysis temperature, and it is determined empirically.

P_1 nuclease: the commercial enzyme is diluted to 340 U/ml in 30 mM sodium acetate (pH 5.3). Portions of 50 μl are then stored in microfuge tubes at −20°C for up to 4 months.

Calf intestinal alkaline phosphatase: 20 to 30 U/μl dissolved in a solution consisting of 30 mM triethanolamine, 3 M NaCl, 1 mM MgCl, and 0.1 mM $ZnCl_2$

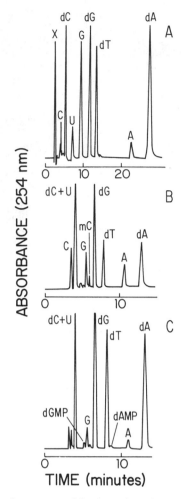

FIGURE 3 Separation of deoxynucleosides from the ribonucleosides by HPLC for determination of the moles percent G+C. (A) Separation of ribonucleosides from deoxyribonucleosides at 24°C in 12% methanol on a C_{18} reverse-phase column. The nucleosides were obtained from enzymatic degradation of *Methanococcus voltae* nucleic acid. (B) Separation of nucleosides under the optimal conditions for resolution of deoxyguanosine (dG) and thymidine (dT), which were 38°C and 12% methanol for this particular column. Under these conditions, guanosine (G) and 5-methyldeoxycytidine (mC) elute as minor peaks just prior to dG and deoxycytidine (dC) and uridine (U) are not resolved. The nucleoside mixture was prepared from standards. (C) Separation of nucleosides from the nucleotide monophosphates. The chromatography was performed as for panel B. The nucleoside mixture contained 2% of the nucleotide monophosphates. Under these conditions, the nucleotides appear as minor peaks near the nucleosides. Possible products of incomplete degradation, the nucleotides can interfer with the determination of dG and dT. Abbreviations: A, adenosine; dA, deoxyadenosine; dAMP, deoxyadenosine monophosphate; C, cytidine; dC, deoxycytidine; mC, 5-methyldeoxycytidine; G, guanosine; dG, deoxyguanosine; dGMP, deoxyguanosine monophosphate; dT, thymidine; U, uridine; X, unidentified compound. From reference 79.

(pH 7.6) or any comparable commercial preparation. Before using, dilute to 0.1 U/μl with alkaline phosphatase buffer (see below).

Sodium acetate buffer: 300 mM, pH adjusted to pH 5.3.

ZnSO₄ solution, 20 mM.
Alkaline phosphatase buffer: 0.1 M glycine–HCl (pH 10.4).
TE buffer: see section 26.2.2.
DNA standard: calf thymus DNA, 1 mg/ml in TE buffer, is an inexpensive standard. Using the method described below, its moles percent G+C has been found to be 44.03% ± 0.05%, and the 5-methyldeoxycytosine content is 1.0 ± 0.1 mol%. Unmethylated lambda phage DNA can also be used. Because it has been sequenced, the exact moles percent G+C is known to be 49.858 mol%.

Degradation Procedure

1. Add standard and unknown DNAs, 2 to 25 μg, to 1.5-ml microfuge tubes in a final volume of 70 μl of water or dilute TE buffer. For highly purified DNA, use 10 μg. Increase the amount of nucleic acid for samples believed to contain large amounts of RNA. Denature the DNA by heating in a boiling-water bath for 2 min, and then rapidly cool it by placing it in an ice bath.

2. Add 13 μl of the P₁ mix. This mix is composed of 50 μl of 0.3 M sodium acetate buffer (pH 5.3), 50 μl of 20 mM ZnSO₄ solution, and 30 μl of diluted P₁ enzyme. Shake gently to mix, centrifuge for 5 s to collect the liquid at the bottom of the tube. Incubate the mixture at 37°C for 2 h.

3. Add 10 μl of alkaline phosphatase diluted in buffer to 0.1 U/μl. Incubate at 37°C for 6 h or overnight.

4. Centrifuge the sample in a microcentrifuge for 5 min, transfer to a clean tube, and store at −20°C until chromatographed. Samples should be stored for no more than 1 week after the degradation and before the analyses.

Chromatographic Procedure

Because the determination of the moles percent G+C is based upon the ratio of deoxyguanosine and thymidine, only these two nucleosides need to be well resolved during the chromatography. Ordinarily, all the major deoxynucleosides are well separated, but coelution of guanosine, 5-methyldeoxycytidine, and deoxyguanosine is frequently a problem, and the chromatography buffer needs to be optimized for the separation of these compounds. In general, these compounds migrate more closely as the methanol concentration and temperature increase (79). However, the effect of temperature is offset somewhat by the smaller peak width at higher temperatures.

In any case, it is necessary to establish the precise chromatographic conditions empirically when using a new type of column. Since calf thymus DNA contains about 1 mol% of 5-methyldeoxycytidine and small amounts of RNA contaminate the commercial preparations, this standard is suitable for testing the chromatography conditions. After the degradation of the calf thymus DNA, perform the chromatography at 35°C and 10% methanol. Look for the appearance of two small peaks eluting just before the deoxyguanosine peak (Fig. 3). These peaks can also be identified by coelution with authentic standards. If both peaks are not observed, decrease the methanol concentration by dilution with the aqueous components of the buffer, increase the temperature, or both. Be sure to allow sufficient time for the column to reequilibrate under the new conditions (at least 60 min) prior to the chromatography.

Because of the sensitivity of the chromatography to small changes in the methanol concentration, large amounts of buffer sufficient to complete the entire analysis are prepared. Typically, the column is equilibrated at a flow rate of 1 ml/min at the beginning of the week. At night or during breaks in the analyses, the flow rate is decreased to 0.1 to 0.2 ml/min. When analyses are complete, the column is washed sequentially with water and storage buffer (such as 70% methanol in water) prior to storage. Because of the salts in the buffer, the column requires a continuous flow of buffer between uses. Moreover, the buffer is always slowly mixed on a magnetic stir plate during the chromatography to prevent the formation of precipitates.

When the chromatographic conditions are known, inject a control sample prepared with the enzymes and buffers but without DNA. Look for minor peaks that elute at the same times as deoxyguanosine and thymidine. This control tests for interference by contaminants that could adversely affect the analyses. Also inject a representative DNA sample and allow the chromatography to run for 60 min. Check the chromatogram to ensure that no minor peaks are eluting after the major deoxynucleosides. Program the auto-injector to load samples after the last peak in the chromatogram, usually every 15 to 30 min.

Typically, the column is equilibrated for several hours at a flow rate of 1 ml/min and the running temperature prior to the analysis. During this time, 10 μl of each sample is analyzed once to establish the best injection volume. The injection volume is then adjusted so that nucleosides from about 1 μg of DNA are injected. If the DNA sample is below about 0.2 μg, the integration will not be accurate. If the DNA sample is too large, the nucleosides will not be well resolved on the column. The injection volume must be between 5 and 40 μl. When the column is equilibrated, samples are run in duplicate or triplicate, with a standard DNA run after every four or five samples.

Calculation

The moles percent G+C is calculated from the ratios of deoxyguanosine and thymidine of the unknown and the standard DNAs:

$$\text{mol\% G+C} = [1 + YT_{app}/G_{app}]^{-1}$$

where T_{app} is the integrator response for thymidine and G_{app} is the integrator response for deoxyguanosine of the unknown DNA. $Y = (G_{app}/T_{app}) \times (T_{act}/G_{act})$ for the standard DNA, where T_{act} is the actual thymidine content and G_{act} is the actual deoxyguanosine content. For calf thymus DNA, T_{act}/G_{act} is 1.2712.

Comments

A common error in this procedure is incomplete degradation of the DNA caused by contaminating salts. Salts change the pH optimum of the P₁ nuclease and greatly reduce the activity at pH 5.3 (78). Moreover, contaminating buffers can maintain the pH near neutrality when the alkaline phosphatase buffer is added and inhibit that reaction as well. These problems can be avoided by dialyzing the DNA against water, 0.01× SSC or TE buffer prior to the degradation reaction. For this reason, it is also important to perform replicate degradation reactions and analyze each digestion in duplicate or triplicate. If the digestion reactions have gone to completion, they will not be significantly different.

The precision of this method is generally better than 0.1 mol%. Most of this precision comes from calculating the moles percent from the deoxyguanosine/thymidine ratio because it allows one to neglect differences in injection volumes and preparation of standards. However, the method is sensitive to systematic errors because of incomplete degradations, contaminants that coelute with the deoxynucleosides, and changes in the chromatography. Even then, the method is likely to be more accurate than many of the values reported in the literature that are based upon indirect measurements, such as thermal denaturation and buoyant density. As with any of these methods it is important to compare the moles percent G+C values obtained by chromatography to values obtained by the same method.

26.4.3. Fluorimetric Dye-Binding Methods

Although fluorimetric procedures have not been used extensively, they are rapid and sensitive, and some can be done directly with fixed cells. They rely on differential binding of fluorescent dyes to G+C and A+T regions of DNA.

26.4.3.1. Purified DNA

This procedure is designed for use with purified preparations of DNA (83). One sample, from a diluted DNA preparation, is assayed with ethidium bromide, which binds independently of the G+C content, whereas another sample is assayed with Hoechst 33258, which binds preferentially to regions of the DNA that are rich in A-T base pairs. There is about a 10-fold range in the relative response ratios (Hoechst 33258/ ethidium bromide) between poly(dA-dT) and poly(dG-dC). Use DNA samples that contain about 1 μg/ml in 1× PB-SSC, which is in the range that is linear for both assays.

Reagents

PB-SSC: prepare 1× and 2× PB-SSC from 20× PB-SSC (see section 26.3.4.3)
Hoechst 33258 solution: 2 μg/ml in 2× PB-SSC
Ethidium bromide solution: 2 μg/ml in 2× PB-SSC

Procedure

1. Mix 1.0 ml of DNA solution (1 μg/ml in 1× PB-SSC), 1.0 ml of 2× PB-SSC, and 2.0 ml of Hoechst 33258 solution. Let stand for at least 30 s before measuring the fluorescence (excitation at 365 nm and emission at 458 nm).

2. Repeat step 1, substituting the ethidium bromide solution for the Hoechst 33258 dye. Measure the fluorescence at an excitation setting of 305 nm and an emission setting of 615 nm.

3. Plot the ratio of fluorescence intensity of Hoechst 33258/ethidium bromide for each known DNA (i.e., known moles percent G+C content) solution, and calculate the moles percent G+C content of the test DNA samples from the standard curve.

26.4.3.2. Intact-Cell DNA

This procedure uses a combination of two dyes, Hoechst 33258 and chromomycin A3. The rationale for using this dye combination is that the Hoechst 33258 binds selectively in regions of the DNA that are rich in A+T, whereas chromomycin A3 selectively binds to regions rich in G+C; thus, the ratio of binding for the two dyes is a function of the moles percent G+C content. The procedures discussed here are for formalin-fixed whole cells analyzed in a conventional fluorimeter (69) or in a dual-beam flow cytometer (82, 87, 107). The protocol given here is a variation of the method of Lutz and Yayanos (69) with a conventional fluorimeter.

Reagents

Hoechst 33258 solution: see section 26.3.4.3. Prepare a 15 μM stock solution in distilled water. The stock solution

can be stored in plastic tubes wrapped in aluminum foil at −20°C for several weeks. Prepare the working dye solution by diluting to 1.5 μM in 1× PB-SSC.
Chromomycin A3 solution. Prepare a 15 μM stock solution in distilled water. Chromomycin A3 has a molecular weight of 1,183.3 and a molar extinction coefficient of 1.86×10^4 M^{-1} cm^{-1} at 281 nm when the dye is dissolved in ethanol. Storage of the stock solution and preparation of the working dye solution are as described for Hoechst 33258.
Formalin

Procedure

1. Grow cultures to 10^7 to 10^{13} cells per ml. Fix the cells by adding 2 to 4% (vol/vol) formalin, mix, and incubate for 20 min. Centrifuge the cells, and suspend them in a similar volume of 1× PB-SSC. The fixed cells are stable for months when stored at 4°C.

2. To 2.0 ml of cell suspension, add 1.0 ml of each dye solution, mix vigorously, and incubate in the dark for no more than 10 min. Measure the fluorescence for each dye: for Hoechst 33258, use emission and excitation wavelengths of 450 and 360 nm, respectively; for chromomycin A3, use 378 and 325 nm, respectively.

26.4.4. Differential Scanning Calorimetry of Intact-Cell DNA

The melting temperature of intracellular DNA, determined by differential scanning calorimetry (DSC), has been correlated with literature values of the moles percent G+C for 58 species of bacteria by Mackey et al. (71). The method is simple and quick, obviating the need to isolate and purify DNA but requiring costly specialized instrumentation. The method has not been tested by the author, but its usefulness has been confirmed by investigators at Michigan State University (55).

A small sample of cells from a centrifuged pellet or an agar colony (50 μl, 10 to 20 mg [dry weight]) is crimp-sealed in a gasket-containing stainless-steel sample pan and heat scanned in a Perkin-Elmer model DSC7 instrument at a rate of 10°C/min from 25 to 130°C, with an empty sealed pan used as reference. The first scan yields a thermogram of endothermic transitions which reflect the denaturation of membranes, ribosomes, cell walls, and DNA (70). The sample pan is then rapidly cooled (200°C/min) to the initial temperature and immediately heat rescanned. The second scan yields only a single major endothermic transition, which reflects DNA reassociated (renatured) during the cooling step. The temperature at which the DNA transition peak is maximum (T_{max}) is characteristic for the melting point of the intracellular DNA of a particular organism. The moles percent G+C content is linearly correlated with T_{max} by the following equation: $T_{max} = 73.8 + 41.0$ (mol% G+C). The entire procedure can be completed within 1 h.

26.5. DNA LABELING

There are many options for preparing labeled probe DNA, either radioactive or nonradioactive. In addition, commercial kits with detailed instructions are available for performing many of the labeling procedures, and the procedures are discussed here only in general terms. Methods for using nucleic acid probes are described in chapter 39 and 40.

In vivo labeling of DNA is limited to radioactive precursors and is dependent on the medium required for culturing the organism and also on the nucleic acid precursors which the organism will take up. The ^{32}P, ^{33}P, ^{3}H, or ^{14}C-labeled nucleic acid precursors most often taken up by bacteria are phosphate, adenine, guanine, hypoxanthine, uracil, thymidine, and deoxyadenosine. For details on the in vivo labeling of DNA, see reference 50.

In vitro labeling can be achieved by direct chemical reactions with the DNA or by DNA polymerase incorporation of nucleotides, using modified nucleoside triphosphates. Examples of direct chemical reactions include iodination, photoreaction (photobiotin), and glutaraldehyde coupling of horseradish peroxidase. DNA polymerases are used in the nick translation and random-primer procedures.

26.5.1. Iodination

Iodine can be bound to cytosine residues in the presence of thallium chloride (pH 4.8) (21, 24, 51, 81, 92, 103). The following is a protocol for labeling 2 to 10 μg of DNA.

Reagents and Components

Buffer-salt solution: 7.2 M sodium perchlorate, 0.25 mM potassium iodide, 100 mM sodium acetate buffer (pH 4.8). Store in refrigerator, and keep on ice when labeling.

TlCl$_3$ solution: 1 mg/ml in 50 mM sodium acetate buffer (pH 4.8). Store in refrigerator, and keep on ice when labeling.

^{125}I: sodium [^{125}I]iodide (pH 10). Usually supplied at 100 μCi/μl. Some examples of sources are Amersham Corp., Arlington Heights, IL, and Dupont New England Nuclear, Boston, MA.

PB-mercaptoethanol: 0.5 M Na$_2$HPO$_4$-NaH$_2$PO$_4$ buffer (pH 6.8) containing 0.1 M 2-mercaptoethanol.

DNA: denatured DNA fragments (i.e., as prepared for S1 nuclease or HA reassociation experiments; see section 26.5.2.1).

Gel filtration columns: Pharmacia NAP 25 columns. These disposable G-25 gel filtration columns work well. Columns from other companies also work well, but be sure to check the void and elution volumes.

HA columns: Bio-Rad DNA grade HA.

HA buffer: 0.14 M Na$_2$HPO$_4$-NaH$_2$PO$_4$ (pH 6.8), 0.5% (wt/vol) SDS.

Radiation safety: check with your radiation safety office to make sure that your labeling facility is in compliance with safety requirements. In general, do all manipulations in a fume hood containing a safety shield and equipped with an auxiliary hood having a charcoal filter. A survey meter that detects ^{125}I is also needed.

Procedure

1. Place 2 to 5 μg of denatured DNA fragments in a reaction tube. Add 3 M sodium acetate (pH 6) to give a final concentration of 0.3 M. Mix, add 2 volumes of 95% ethanol, mix again, and centrifuge for 10 min. Draw off the supernatant, wash the pellet with 100 μl of 80% ethanol, and then dry the pellet in an incubator.

2. Dissolve the DNA in 10 μl of the buffer-salt solution. Deposit 1 or 2 μl of sodium [^{125}I]iodide on one inside wall of the reaction tube, and then deposit 2.0 μl of the TlCl$_3$ solution on the opposite side of the tube. Crimp the serum cap onto the tube, mix the components, and centrifuge the tube briefly at low speed.

3. Place the reaction tube in a 70°C water bath for 20 min. Remove it, let it cool for 1 or 2 min, and inject approximately 0.1 ml of PB-mercaptoethanol buffer into the tube via a disposable tuberculin syringe. Return the tubes to the water bath for an additional 20 min. Remove and cool.

4. Using a 1-ml disposable syringe with a 1.5-in. (3.8-cm) 22-gauge needle, inject approximately 0.15 ml of HA buffer into the reaction tube and then draw the contents of the tube up into the syringe. Load this material onto a NAP 25 gel filtration column that has been equilibrated with HA buffer. Add HA buffer to the column to give a total volume of 2.3 to 2.4 ml (i.e., sample plus buffer; the void volume is 2.5 ml), and discard the eluate. Place the collection tube containing 20 μg of carrier DNA (e.g., sheared salmon sperm DNA) under the column, and add an additional 2.0 ml of HA buffer to the column to collect the labeled DNA. Cap the column, and dispose of it in the laboratory's radioactive-waste container. Holding the tubes up to the survey meter will give you an immediate indication as to how well the iodination reaction worked.

5. Prepare a small HA column in a Pasteur pipette. Cut the tip off so that the pipette will just protrude from the top of a screw-cap culture tube (16 by 125 mm). Place glass wool in the pipette to retain the HA, wash with HA buffer, and add enough HA to make a column that is 0.75 to 1 in. (2 to 2.5 cm) high. Place the tube in a 70°C water bath, and wash the column with heated (70°C) HA buffer. Heat the labeled preparation in a boiling-water bath for 3 to 5 min and then load onto the washed HA column. Transfer the column to a clean tube when the radioactivity begins to elute, and after all of the sample has entered the column, add an additional 0.5 ml of HA buffer to elute the rest of the DNA from the column. Dispose of the HA column as follows. Break the tip by pressing it against the inside wall of an empty culture tube; this shortens the column so that the tube can now be capped. Place the capped tube in the laboratory's radioactive-waste receptacle.

6. Place the labeled DNA (approximately 1.8 to 2.2 ml) on another NAP 25 column, which has been equilibrated with 0.1× SSC. Add 2.5 ml (the void volume) of 0.1× SSC, place a collection tube under the column, and add an additional 3.5 ml of 0.1× SSC to collect the labeled product. Add SDS to the preparation to a final concentration of approximately 0.2%. Store at −20°C.

Comments

This labeling procedure is fast and gives consistent results. Usually the labeled DNA preparations are about 95% digested with S1 nuclease (single-stranded-specific nuclease). Occasionally a DNA preparation will contain contaminating material that will become iodinated, and this will raise the background level of S1-resistant material in the reassociation experiments. There are several advantages to using ^{125}I labeling. It does not affect the DNA fragment lengths because the low energy release during decay does not cause fragmentation of the DNA. If one has a gamma-scintillation counter, there is no need for scintillation fluids, and the 60-day half-life allows one to use the labeled DNA for up to 6 months without any change in relative reassociations. One disadvantage of using ^{125}I is that during the iodination step, when the pH is low, the resulting iodine is volatile and can be a safety hazard if the reaction tubes are not properly cleansed or if proper venting is not used.

26.5.2. Photobiotination

Photobiotin, a photoactive analog of biotin, was developed by Forster et al. (36) and is available in commercial kits, and the directions of the kits are easily followed. In the procedure used by Ezaki et al. (31), the DNA is suspended in a solution of 0.1 mM EDTA (pH 8.0) and heat denatured at 100°C for 5 min. A 3-μl solution containing 0.5 to 5.0 μg of DNA is mixed with an equal volume of the photobiotin reagent (Vector Laboratories, Burlingame, CA) in a microfuge tube and irradiated for 15 min in an ice bath 10 cm below a 500-W sunlamp. The volume is then increased to 100 μl by the addition of 100 mM Tris-HCl (pH 9.0). To remove free photobiotin reagent, 100 μl of 2-butanol is added, the sample is mixed and centrifuged, and the upper layer is discarded. This step is repeated two more times.

26.5.3. Glutaraldehyde Procedure

The glutaraldehyde procedure involves treating horseradish peroxidase to give it a positive charge and then mixing it and glutaraldehyde with the denatured DNA (84). This results in covalent cross-linking between the DNA and the enzyme. The reagents are commercially available as a kit.

26.5.4. Nick Translation Procedure

The nick translation procedure is used for the labeling of high-molecular-weight DNA. It involves the generation of single-strand breaks (nicks) followed by nucleotide removal and replacements by E. coli DNA polymerase 1 (86). Both radioactive analogs and other analogs (biotin or digoxigenin) can be introduced by this procedure. Kits incorporating this technology are available from several companies. The lengths of the labeled fragments are dependent on the reaction time: the longer the time, the shorter will be the fragments. Nicks within the synthesized regions will remove the introduced analogs.

26.5.5. Random Primer Procedure

The random-primer procedure was developed by Feinberg and Vogelstein (33, 34) for radioactively labeling DNA fragments to high specific activity. The procedure can also be used with nucleotide analogs containing biotin or digoxigenin. Labeling kits are available from a number of companies.

26.6. DNA-DNA REASSOCIATION

The reassociation of labeled DNA is performed with an excess of unlabeled DNA either in solution or immobilized on a membrane or surface. The method is used to measure the fraction of the genomes that can form duplexes under specific conditions of ionic strength and temperature. The results are then reported as percent reassociation. In addition, the thermal stability of the duplexes can be measured and is used to estimate the sequence similarity of the DNA participating in the heterologous duplexes (hybrids). There are two general methods for measuring DNA reassociation. The first method, often referred to as the free-solution method, involves the reassociation of the single-stranded nucleic acids in solution. Reassociation of the DNA may be monitored optically by using UV spectrophotometry, or it may be measured by incubating a small amount of denatured DNA having a high specific radioactivity with a large excess of unlabeled, denatured DNA. In the latter instance, the ability of the labeled fragments to form duplexes with unlabeled DNA is assayed by adsorption to HA or by resistance to hydrolysis by S1 nuclease.

The second method, often referred to as the membrane method, involves immobilizing denatured (single-stranded) unlabeled DNA onto a nitrocellulose or nylon membrane filter, treating the filter so that additional DNA will not adsorb to it, and then incubating the filter in the presence of labeled denatured DNA fragments (or RNA, for RNA-DNA hybridization experiments). Duplexes of the labeled DNA that form by reassociation with the DNA fragments bound to the membrane are stable to the subsequent washing of the membrane, and the amount of duplex formation can be estimated by measuring the radioactivity of the membrane (or, for nonradioactive probes, some other unique activity). A variation of this approach is to immobilize the DNA in wells of a microdilution plate.

26.6.1. Special Apparatus

Some of the specially constructed equipment that is useful for these measurements is shown in Fig. 4. The reaction vials can be constructed from 6-mm (outer diameter) Pyrex tubing cut into 7/8-in (2.2-cm) lengths and scaled at one end with a flame. The stoppers for these vials are serum vial stoppers which have had the outside flange cut off. Small polyethylene microcentrifuge tubes also work well and have the advantage of being disposable. When components are being added to the reaction vials, the vials are placed in a Plexiglas holder that has double rows of holes to help the investigator keep track of the additions. After being filled, the vials are stoppered and placed in a stainless-steel holder for incubation in a water bath. This water bath must be large enough to contain the submerged holder and should have a circulatory pump for maintaining an even temperature distribution. If the holder is not totally submerged or the bath not tightly covered, there is danger of distillation and collection of water drops on the vial sides and in the cap, which will greatly alter the ionic strengths and nucleic acid concentrations.

For membrane filter hybridization procedures, denatured DNA can be immobilized on a large filter by using dot blot or slot blot devices; alternatively, the DNA can be immobilized on the entire surface of a large membrane by using a filtration device. In the latter case, small membranes are punched out of the larger membrane for use in individual reaction vials. Commercially available filtering devices can be used for immobilizing DNA on larger membranes, or custom-made ones can be constructed (Fig. 5). A useful one can be constructed from a Plexiglas cylinder and a porous polyethylene filtration surface. The base of the apparatus is a short Plexiglas cylinder with an outside diameter the same as that of the membrane. The porous polyethylene is mounted on the inside and flush with the top. The base has a single exit tube, whose flow rate can be controlled with an adjustable pinch clamp. The top part of the device consists of a Plexiglas cylinder with a neoprene gasket (1/8 in. [3 mm] thick) glued to the bottom rim. The top part is placed on top of the membrane and is held on by spring tension. All of the Plexiglas apparatuses can be made by a local machine shop. Two different designs are illustrated in Fig. 5.

A high surface-to-volume ratio is needed for efficient annealing of probe DNA with the membrane-immobilized DNA. Therefore, the specific reaction vial depends on the paper punch that is used to punch out the membrane. Another device that is useful for working with small membranes is a washing chamber to remove unreacted probe from the membrane surface. For oval membranes, one can use the Plexiglas washing chamber shown in Fig. 4. This chamber contains 120 compartments and consists of a top

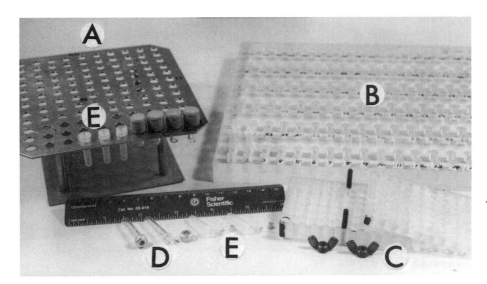

FIGURE 4 Specialized equipment for DNA reassociation experiments. (A) Stainless-steel holder for vials, used during incubation in a water bath. (B) Plexiglas rack for setting up experiments. (C) Plexiglas washing chamber for membrane filters. (D) Autoinjection tubes used for iodinating nucleic acids. (E) Polyethylene tubes (0.2 ml) for use in reassociation experiments.

and bottom piece; the membrane filters are placed in the bottom piece, after which the top piece is bolted on. The assembly shown is small enough to fit into a 400-ml beaker.

The thermal stability of membrane-bound duplexes can be determined with the aid of the devices depicted in Fig. 6 and 7. The membrane skewer makes it easy to transfer the membranes from one tube to the next, while the tube rack fits into a circulating water bath opening and reduces the exposed water surface, which results in more efficient heating at the higher temperatures.

26.6.2. Free-Solution Methods

Free-solution reassociation of denatured DNA fragments can be measured optically by use of a recording spectrophotometer; if radioactive DNA is included, it can be measured by adsorption to HA or by resistance to S1 nuclease. A general scheme for performing these procedures is given in Fig. 8. Both require high-specific-activity labeled DNA (a very small amount) to be incubated with a large excess of unlabeled DNA fragments. The specific activity must be in the range of 5×10^4 to 2.5×10^5 cpm/μg, and the DNA should be at a concentration of less than 5 μg/ml.

26.6.2.1. Preliminary Procedures

DNA Fragmentation

For all of the free-solution reassociation methods, the DNA preparations must be fragmented to similar fragment sizes. The two common methods for fragmenting DNA are passage through a French pressure cell and sonication.

FIGURE 5 Plexiglas filtration devices for immobilizing DNA on membranes.

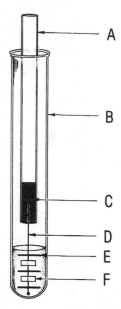

FIGURE 6 Diagram of apparatus for testing the thermostability of hybrid duplexes. A piece of 7-mm-diameter glass tubing (A) is sealed with a rubber plug (C), which is made from a rubber stopper by use of a cork borer. Membrane filters (E) and Teflon washers (F) are impaled on an insect pin (D) close to the head of the pin. The pointed end of the pin is inserted into the rubber plug. This assembly is placed into a 13- by 100-mm glass test tube (B) containing 1.2 ml of 0.5× SSC. The test tube is then placed in a heated-water bath.

French Pressure Cell

1. Dilute the stock DNA to 0.4 mg/ml.
2. Pass the DNA preparation through the French pressure cell three times at a pressure of 16,000 lb/in². The flow rate should be a fast drip, with a constant pressure on the hydraulic pump being maintained. This method results in a Gaussian distribution of fragment sizes, with a mean size of 600 to 700 nucleotides. Check the uniformity of the fragment sizes in the DNA preparations by agarose gel electrophoresis with 5-μg amounts of DNA.

Sonication

1. Dilute the stock DNA to 0.4 mg/ml.
2. Sonicate each preparation for two 30-s periods (e.g., in a Biosonic III sonicator with a 0.375-in. [9.5-mm] probe and an energy setting of 60). If there is a lot of heat buildup, it may be desirable to cool the preparations on ice between sonications. Check for fragment size uniformity by agarose gel electrophoresis as described above.

For optical reassociation experiments, centrifuge each fragmented DNA preparation at 17,000 × g for 10 min to remove any particulate debris.

DNA Denaturation

If the fragmented DNA preparations are to be assayed in reassociation experiments using S1 nuclease or HA, first denature the preparations by heating the tubes in a boiling-water bath for 5 min, cool on ice, and then centrifuge at 17,000 × g for 10 min. Store the DNA preparations at −20°C.

Preliminary denaturation of DNA is not done in the optical reassociation method; the denaturation occurs when the DNA is heated in the spectrophotometer cuvettes.

DNA Labeling

Both the HA and S1 nuclease procedures require a small amount of denatured labeled DNA with a high specific radioactivity to be incubated with a large excess of denatured unlabeled DNA fragments. In the case of iodination, use French pressure cell-fragmented, denatured DNA preparations for labeling; thus, the labeled DNA fragments will have the same size distribution as those in the unlabeled

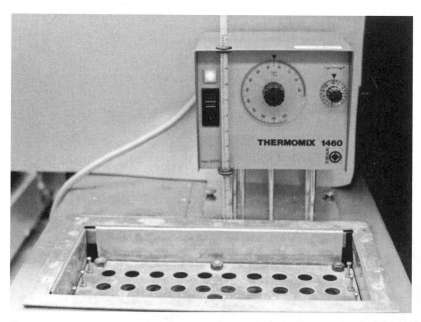

FIGURE 7 Tube rack for thermal stability experiments. This rack fits into the opening of a circulating-water bath. The small amount of exposed surface area cuts down on evaporation.

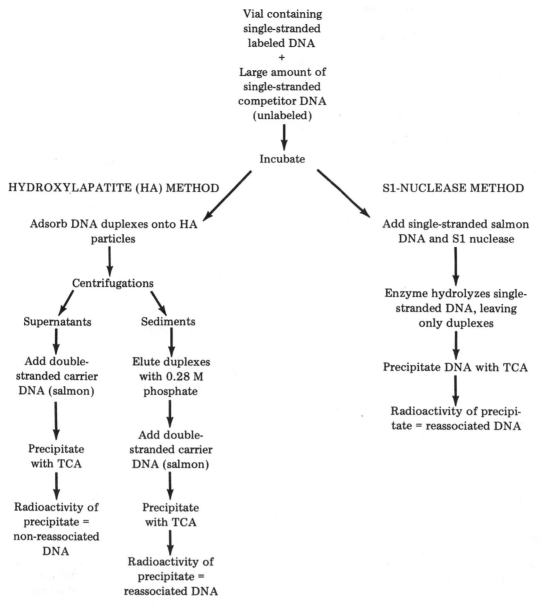

FIGURE 8 Scheme for performing free-solution DNA reassociation experiments, assayed either by the HA or S1 nuclease procedure. TCA, trichloroacetic acid.

DNA preparations. Dilute the labeled DNA preparations to 3×10^6 cpm/ml.

26.6.2.2. Reassociation Measured by Photometry

The protocol presented here is based on those of De Ley et al. (27) and Huss et al. (46). The major variables are the type of UV spectrophotometer and the type of computer interface that may be available. The spectrophotometer must be equipped with a temperature-controlled cuvette chamber or cuvette holding block and must have an automatic cuvette-positioning mechanism. The reassociation rates can be calculated manually or by a microprocessor-controlled spectrophotometer, or the data can be uploaded into a computer system. Procedures presented here are limited to sample preparation and the initial reassociation reactions.

Sample Preparation

Adjust each fragmented DNA sample to a concentration of 40 μg/ml in 2× SSC-1 mM HEPES (pH 7.0) buffer by the following procedure. Prepare a large batch of 2× SSC-1 mM HEPES buffer and a smaller batch containing 4× SSC-2 mM HEPES. Dilute the fragmented DNA by adding 1 volume of the sample to an equal volume of 4× SSC; then, add sufficient 2× SSC-1 mM HEPES to obtain the desired final concentration. Alternatively, each diluted DNA preparation can be dialyzed against 2× SSC-1 mM HEPES buffer. Pass each preparation through a 0.45-μm-pore-size filter to remove any particulate matter.

Procedure

1. Prepare mixtures (M) of each pair of DNA preparations (equal volumes) to be compared. When comparing

two organisms (i.e., a DNA preparation from organism A and a DNA preparation from organism B) by using a spectrophotometer that has a cuvette holder for four cuvettes, fill one cuvette with 2× SSC-1 mM HEPES buffer, one with DNA from organism A, one with DNA from organism B, and the last with the equal-volume mixture of the two DNA preparations. Close the cuvettes tightly so that there is no evaporation during the denaturation and reassociation reaction; be sure to leave an air headspace in each cuvette to allow for the thermal expansion of the solutions. Teflon-stoppered cuvettes seal very well, and they can be filled to about 80 or 90% capacity.

2. Place the cuvettes in the spectrophotometer, heat the samples to 100°C, and hold at that temperature for 10 min. Then, decrease the temperature to the optimal renaturation temperature (T_{OR}), and allow the reassociation to proceed for about 40 min. The optimal renaturation temperature can be calculated by the following formula (27).

$$T_{OR} = 0.51(\text{mol\% G+C}) + 47.0$$

3. Express reassociation rates (v') as the decrease in A_{260} per min, and calculate the percent reassociation by the following formula (27). The percent reassociation was designated %D [degree of binding] in this formula by the original authors.

$$\%D = \frac{100(4v'_M - v'_A - v'_B)}{2(v'_A + v'_B)^{1/2}}$$

26.6.2.3. Reassociation Measured with S1 Nuclease

When radioactive probe DNA is reassociated with an excess of unlabeled DNA, the resulting heteroduplexes can be measured by resistance to S1 nuclease or by adsorption to HA. For the procedures described here, the initial reassociation mixtures are the same for both measurements.

Reassociation Mixtures

When starting an experiment, thaw the DNA preparations and allow them to warm to room temperature. In addition to the test DNA preparations, prepare an unrelated DNA preparation (e.g., 0.4 mg of sheared salmon sperm DNA per ml), for measuring background reassociation. Prepare the reaction mixtures in 6- by 22-mm vials (or 250-μl microcentrifuge tubes) as follows.

Labeled DNA (3×10^6 cpm/ml)	10 μl
13.2× SSC-1 mM HEPES buffer (pH 7.0)	50 μl
Unlabeled DNA, 0.4 mg/ml	50 μl
Total volume	110 μl

Place sheared native salmon sperm DNA (which has no sequence similarity with bacterial DNA) into four control vials to measure the amount of self-renaturation of the labeled fragments. Use four vials for homologous reassociation and two vials for each of the heterologous DNA preparations. Incubate the vials at 25°C below the T_m of the reference DNA (as determined in 6× SSC) for 20 to 24 h. After incubation, quantitatively transfer the reassociation mixture to a large tube for determining the extent of reassociation by either the S1 nuclease or the HA procedure (see section 26.6.2.4).

Comments

The accuracy of the results obtained from these methods is dependent on the reproducibility of the addition of the 10-μl amounts of labeled DNA to the vials. A multiple-dispensing micropipette works well for adding constant vol-

umes to a large number of vials. The 110-μl volumes allow for accurate addition of components but are small enough that vast amounts of DNA are not required for the experiments. If the DNA preparations being compared have a high moles percent G+C value, formamide can be included to lower the T_m (51, 76). For instance, substituting the 50 μl of 13.2× SSC–1 mM HEPES with 25 μl of 23.4× SSC–2 mM HEPES and 25 μl of deionized formamide works well (this will lower the T_m by about 13.6°C).

Procedure

Under carefully controlled conditions, S1 nuclease has little effect on double-stranded DNA but does hydrolyze single-stranded DNA (25). Therefore, the extent of duplex formation between radioactively labeled DNA fragments and an excess of unlabeled DNA fragments can be determined by measuring the amount of S1-resistant (i.e., acid-precipitable) radioactivity.

1. Quantitatively transfer the contents of each reaction vial to a 12- by 75-mm polypropylene test tube containing 1 ml of acetate-zinc buffer (0.05 M sodium acetate, 0.3 M NaCl, 0.5 mM ZnCl$_2$ [pH 4.6]), and 50 μl of denatured sheared salmon sperm DNA (0.4 mg/ml). Add 10 μl (100 U) of diluted S1 nuclease, mix the tube contents on a Vortex mixer, and incubate the tubes in a 50°C water bath for 1 h.

2. Remove the tubes from the water bath, cool to room temperature, add 50 μl of sheared salmon sperm DNA (1.2 mg/ml) and 0.5 ml of HCl-PP$_i$ solution (1 N HCl, 1% Na$_4$P$_2$O$_7$ · 10H$_2$O, 1% NaH$_2$PO$_4$ [86]) to each tube, mix, and incubate at 4°C for 1 h.

3. Collect the precipitates on nitrocellulose membrane filters or on fiberglass filters (GF/F type; Whatman). Dry the filters, place them into scintillation vials, and count the radioactivity. Only the S1-resistant fragments (duplexes) are detected.

Calculate percent similarity values by dividing the counts per minute of the heterologous S1-resistant DNA by the counts per minute of the homologous S1-resistant DNA and multiplying by 100. A hypothetical example of the calculation is given in Table 1.

26.6.2.4. Reassociation Measured with HA

The HA procedure separates double-stranded DNA from single-stranded DNA by selective adsorption to HA. Double-stranded DNA adsorbs to HA in 0.14 M phosphate buffer, whereas single-stranded DNA does not (11, 13, 14). There is variability in the adsorption capacity of different lots of HA, so it is a good idea to determine the binding capacity of each lot. The reassociation mixture in this procedure is the same as that described for the S1 nuclease procedure (section 26.6.2.3).

1. Place 0.5 g of dry HA into a 13- by 100-mm test tube, and add 3.0 ml of 0.14 M PB (equimolar NaH$_2$PO$_4$ and Na$_2$HPO$_4$) containing 0.4% SDS. Suspend the HA with a Vortex mixer, and warm the tube in a 60°C water bath. Centrifuge the tube at 3,000 × g for 3 to 4 min, decant the buffer, and resuspend the HA in 1 ml of PB.

2. Quantitatively transfer the reassociation mixture to the HA-containing tube, using PB for rinsing the reassociation vial. Mix on the Vortex mixer and return to the water bath for 5 min. Add an additional 2 ml of PB, mix on the Vortex mixer, and allow the HA to settle for 1 to 2 min.

3. Centrifuge the tube at 3,000 × g for 3 to 4 min, and decant the buffer into a collection tube. Add another 3 ml

TABLE 1 Hypothetical examples of the calculations of percent similarity values in the free-solution method by the HA and S1 nuclease procedures

Reaction vial	cpm S1 resistant	Net cpm S1 resistant	cpm not adsorbed	cpm adsorbed	% Adsorbed	Net % adsorbed	% Similarity
S1 nuclease procedure							
Labeled DNA fragment only	100						
Homologous reassociation	900	800					100
Heterologous reassociation	600	500					62
HA procedure							
Labeled DNA fragments only			900	100	10		
Homologous reassociation			100	900	90	80	100
Heterologous reassociation			400	600	60	50	62

of 0.14 M PB to the HA. Repeat the mixing, warming, and centrifugation three times, and decant each supernatant into the collection tube.

4. Add 3 ml of 0.28 M PB to the HA, mix, and centrifuge. Decant the supernatant into a second collection tube. Add another 3 ml of buffer to the HA, mix, centrifuge, and decant the supernatant; decant it again into a collection tube. *Note:* When using an [129]I-labeled probe DNA, collect the supernatants in a tube that will fit into a gamma counter and measure the radioactivity directly. If the label is a beta emitter, the fragments must be precipitated and collected on filters before the radioactivity is measured. For this, pool the 0.14 M PB washes for the unbound fragments and the 0.28 M PB elutions before precipitating.

5. To each of the two collection tubes add 60 μg of carrier DNA (50 μl of sheared salmon sperm DNA, 1.2 mg/ml). Mix, and add 0.25 volume of HCl-PP$_i$ solution, and again mix well. Cool in a refrigerator for 1 h.

6. Collect the precipitates on nitrocellulose or fiberglass filters. Dry the filters, place them into scintillation vials, and measure the radioactivity.

The sum of the counts in the two scintillation vials represents the total number of counts per minute in the reassociation vial. The radioactivity in the first scintillation vial represents the nonreassociated DNA, and the radioactivity in the second scintillation vial represents reassociated DNA.

Calculate the percent similarity values by dividing the amount of heterologous reassociation by the amount of homologous reassociation and multiplying by 100. A hypothetical example of the calculation is given in Table 1.

Brenner et al. (11) have described an effective HA procedure in which larger amounts of DNA are used. Lachance (63) has described the use of microcolumns for separating the nonreassociated fragments from the reassociated fragments, and Sibley and Ahlquist (94) have devised an automated instrument for measuring thermal stabilities of duplexes.

26.6.2.5. Thermal Stability Measurement

The extent of base pair mismatching can be estimated by comparing the thermal stability of heterologous duplexes with the stability of homologous duplexes. Thermal-stability measurements can be determined by using either the S1 nuclease or HA assay procedure and can also be measured by using membranes (see section 26.6.3.4). When using the S1

nuclease procedure, the decrease of S1-resistant radioactively labeled DNA is measured as the reassociated DNA is heated stepwise (usually in 2.5 or 5.0°C increments) in a circulating-water bath until denaturing temperatures are reached. When using HA, the nonreassociated DNA fragments are first washed from the HA with 0.14 M PB. Then, with the same 0.14 M PB, the HA is heated stepwise in temperature increments (again, 2.5 or 5.0°C increments) until all of the labeled DNA has denatured and eluted from the HA.

The reassociation mixtures and reassociation conditions are the same as when determining similarity values, except that the probe DNA has double or triple the specific activity (i.e., 10^7 cpm/ml) and the volume is doubled:

Labeled DNA (1 × 10^7 (cpm/ml) . . . . 20 μl
13.2× SSC–1 mM HEPES buffer (pH 7.0) . . . . 100 μl
Unlabeled DNA, 0.4 mg/ml . . . . 100 μl
Total volume . . . . 220 μl

These can be assayed by either the S1 nuclease or HA procedure.

S1 Nuclease Procedure

The S1 nuclease procedure for determining the thermostability of duplexes can be done as follows. After reassociation, dilute the mixture 12-fold by adding 2,420 μl of water to decrease the SSC concentration to 0.5× SSC. Then add 2.5 ml of 0.5× SSC-1 mM HEPES (pH 7.0) buffer. Place 500-μl amounts into 10 tubes (14.5- by 82-mm polypropylene tubes work well).

1. Start the thermal-stability profile by using a circulating-water bath set at 50°C. Maintain the first set of tubes (i.e., one from each reassociation mixture) at that temperature for 5 min, and then remove the tubes and cool them on ice or in a refrigerator. Increase the temperature of the water bath to 55°C, and place the next set of tubes in for 5 min. Continue the cycles through 90°C. Retain one tube that is not heated. *Note:* For low G+C DNAs, the melting profile may have to begin at 45 or 50°C; for higher G+C DNAs, it may have to begin at 60 or 65°C.

2. Add 2.0 ml of S1 nuclease buffer, 50 μl of 0.4-mg/ml denatured sheared salmon sperm DNA, and 10 μl (100 U) of diluted S1 nuclease to each tube. Mix the tube contents on a Vortex mixer, and incubate the tubes in a water bath at 50°C for 1 h.

3. Remove the tubes from the water bath, and cool them to room temperature. Add 50 μl of 1.2-mg/ml sheared

salmon sperm DNA and 0.5 ml of HCl-PP$_i$ solution to each tube, mix, and incubate at 4°C for 1 h.

4. Collect the precipitates on nitrocellulose membrane filters or fiberglass filters. Dry the filters, place them into scintillation vials, and count the radioactivity.

Plot the S1-resistant counts per minute versus the temperatures. The temperature at which 50% of duplexes become S1 sensitive (i.e., have dissociated) is called the $T_{m(i)}$ ("i" meaning "irreversible separation of strands"). The difference between the homologous and heterologous $T_{m(i)}$ values is the $\Delta T_{m(i)}$ value. Estimates for the amount of base mispairing for each 1°C in the $\Delta T_{m(i)}$ range from 1 to 2.2% (18, 106).

HA Procedure

The hydroxylapatite procedure for determining the thermal stability of duplexes can be done by the following protocol.

1. Place 0.5 g of dry HA into a 13- by 100-mm test tube, and add 3.0 ml of 0.14 M PB (equimolar NaH$_2$PO$_4$ and Na$_2$HPO$_4$) containing 0.4% SDS. Suspend the HA with a vortex mixer, and warm the tube in a 50°C water bath. Centrifuge the tube at 3,000 × g for 3 to 4 min, decant the buffer, and resuspend the HA in 1 ml of PB.

2. Add 220 μl of 0.28 M PB buffer to the reassociation mixture and quantitatively transfer it to the HA tube, using 0.14 M PB for rinsing the reassociation vial. Mix on the Vortex mixer, and return to the water bath for 5 min. Add an additional 2 ml of PB, mix on the Vortex mixer, and allow the HA to settle for 1 to 2 min.

3. Centrifuge the tube at 3,000 × g for 3 to 4 min, and decant the buffer into a collection tube. Add another 3 ml of 0.14 M PB to the HA. Repeat the mixing, warming, mixing, settling, and centrifugation a total of three times, decanting each supernatant into a collection tube.

4. Add 3 ml of 0.14 M PB to the HA, mix, and return to the water bath. Increase the temperature of the water bath by 5°C, allow the bath to reach the new temperature, and incubate the tubes for 5 min at that temperature. Mix the tubes on a Vortex mixer, allow the HA to settle, and centrifuge as above. Decant the supernatant into another collection tube. Add another 3 ml of buffer to the HA, and heat the water bath to the next temperature. Continue until the bath reaches 90 or 95°C. *Note:* When using ^{125}I-labeled probe DNA, collect the supernatants in tubes that will fit into a gamma counter, so that the radioactivity can be measured directly. If the label is a beta emitter, the eluate can be mixed with scintillation fluid that is miscible with aqueous samples, or if the volume is too great, the fragments can be precipitated and collected on filters before measuring the radioactivity.

5. For precipitating the eluted fragments, add 60 μg of carrier DNA (50 μl of sheared salmon sperm DNA, 1.2 mg/ml) to each collection tube. Mix, and add 0.25 volume of HCl PP$_i$ solution. Mix well, and cool in a refrigerator for 1 h.

6. Collect the precipitates on nitrocellulose membrane filters or fiberglass. Dry the filters, place them into scintillation vials, and measure the radioactivity.

The radioactivity from the three washes represents nonreassociated DNA fragments, and the radioactivity at each temperature represents the amount of denaturation that occurred at that temperature. Hypothetical data for a thermal stability profile are given in Table 2, and the plotting of the data is illustrated in Fig. 9.

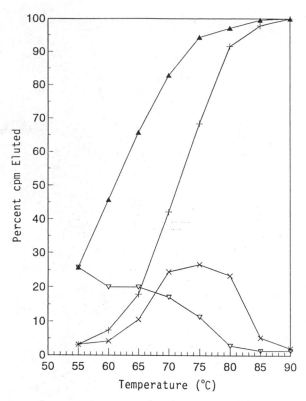

FIGURE 9 Thermal stability profiles obtained with the hypothetical HA procedure data in Table 2. Symbols: ∇, percent counts per minute eluted at each temperature for heterologous duplexes; ▲, sum of percent counts per minute eluted at each temperature for heterologous duplexes; ×, percent counts per minute eluted at each temperature for homologous duplexes; +, sum of percent counts per minute eluted at each temperature for homologous duplexes.

26.6.2.6. Determination of C_0t Values

Reassociation kinetics are most easily studied in solution; consequently, the optical, S1 nuclease, and HA procedures can all be used in these types of experiments. The following is a brief discussion of the equations that have been used for these determinations.

The initial reassociation of DNA in solution follows second-order reaction kinetics (3, 6, 14, 110, 111). The reassociation rate is a function of the genome size and has been used to estimate this value. The usual equation for a second-order reaction (equation 3) is rearranged (equation 4) for plotting the so-called C_0t curves (14):

$$1/C - 1/C_0 = kt \qquad (3)$$

$$C/C_0 = 1/(1 + kC_0t) \qquad (4)$$

where t is the time in seconds, C is the concentration of single-stranded DNA at time t in moles of nucleotides per liter, C_0 is the concentration of single-stranded DNA at zero time in moles of nucleotides per liter, and k is the reassociation rate constant.

Calculate the moles of nucleotides per liter by dividing the milligrams of DNA per milliliter by 331 mg of DNA per millimole of nucleotide. The units (millimoles per milliliter) will be equal to moles per liter (M). Equation 4 is for a hyperbolic curve, for which the general formula is

TABLE 2 Calculations for hypothetical data from HA thermal stability profile

Temp (°C)	Homologous duplexes		Heterologous duplexes	
	cpm eluted	Sum of cpm eluted	cpm eluted	Sum of cpm eluted
50				
1st wash	350	350	2,100	2,100
2nd wash	125	475	750	2,850
3rd wash	25	500	150	3,000
55	300	800	1,800	4,800
60	400	1,200	1,400	6,200
65	1,000	2,200	1,200	7,400
70	2,300	4,500	1,400	8,800
75	2,500	7,000	800	9,600
80	2,200	8,200	200	9,800
85	600	9,800	100	9,900
90	200	10,000	100	10,000

$$y = 1/(1 + ax) \tag{5}$$

When $y = 1/2$ in equation 5, $1/a = x_{1/2}$ and $a = 1/x_{1/2}$. If the logs of the x values are plotted, when $y = 1/2$, $\log a = \log 1/x_{1/2} = -\log x_{1/2}$. A general log C_0t plot is shown in Fig. 10. There is an almost linear region on the curve that extends for about 2 log units. This is because any significant change in C/C_0, occurs between $1/(1 + 0.1)$ and $1/(1 + 10)$, i.e., when kC_0t is between 0.1 and 10. The rate constant $k = 1/C_0t_{1/2}$. The units for the rate constant are reciprocal moles reciprocal hours.

Denatured DNA in free solution may be reassociated for kC_0t values well beyond 10, and it is not critical to know either the exact time of incubation or the exact concentrations of the DNA fragments. When determining $C_0t_{1/2}$ values, however, one must know these parameters. This is because a series of points between $kC_0t = 0.1$ and $kC_0t = 10$ must be determined. In practice, for bacterial DNA preparations, these will be C_0t values from about 0.1 to 50. A particular C_0t value can be reached by altering the time of incubation, the concentration of the unlabeled DNA, or

both. Listed in Table 3 are C_0t values for four DNA concentrations and eight reassociation times.

Set up reassociation mixtures in the usual manner (see section 26.6.2) with duplicate mixtures for each DNA concentration at each temperature. After incubation is completed for a given set of reaction vials, place them into a freezer until all have been incubated. Assay for the extent of duplex formation by either the HA or S1 nuclease procedure.

Plot the fractions of DNA fragments not bound to HA or the fractions which are S1 nuclease sensitive on the y axis and C_0t values on the x axis of four- to five-cycle semilog graph paper. The rate constant, k, can be calculated by using equation 4 when the duplexes are measured by the HA procedure. The results obtained by the S1 nuclease procedure, however, do not reflect second-order kinetics, because the ends of duplexed fragments are hydrolyzed. The relationship between the two procedures is

$$S/C_0 = [1/(1 + kC_0t)]^{0.45} \tag{6}$$

where S is the concentration of S1 nuclease-sensitive DNA (95). Equation 6 can be rearranged as

$$[S/C_0]^{2.222} = 1/(1 + kC_0t) \tag{7}$$

If one plots the S1 nuclease results according to equation 7, one will generate a C_0t curve comparable to that obtained by the HA procedure. Also important to keep in mind is that the reassociation rates are greatly affected by the ionic strengths of the reassociation mixtures, and therefore comparisons should be made by using a single ionic strength.

For additional information about reassociation kinetics and C_0t curves (including optical reassociation), see references 6, 13, 14, 91, 110, and 111.

26.6.3. Membrane Methods

The ability to immobilize DNA on a solid support has been one of the more important developments for the measurement of DNA reassociation. DNA was first successfully immobilized in an agar matrix (45), which was soon replaced by nitrocellulose membranes (38, 108). The immobilization of denatured DNA on nitrocellulose membranes involves slow filtration of a high-salt (e.g., 6× SSC)-denatured DNA solution through the membrane, followed

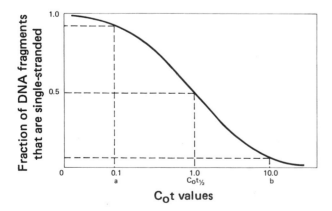

FIGURE 10 Generalized C_0t plot. At C_0t values of less than or equal to a, $1/(1 + C_0t)$ values are 0.9 or greater. At C_0t values equal to or greater than b, $1/(1 + C_0t)$ values are less than 0.1.

TABLE 3 C_0t values for various DNA concentrations and reassociation times

Time in h(s)	Unlabeled DNA concn (μg/110 μl)[a]	C_0t
0.5 (1.8×10^3)	1	0.049
	5	0.247
	205	0.99
	30	1.48
1.0 (3.6×10^3)	1	0.097
	5	0.49
	20	1.98
	30	2.97
2.0 (7.2×10^3)	1	0.194
	5	0.986
	20	3.95
	30	5.93
4.0 (1.44×10^4)	1	0.388
	5	1.97
	20	7.90
	30	11.9
8.0 (2.88×10^4)	20	15.8
	30	23.7
16 (5.76×10^4)	20	31.6
	30	47.5
20 (7.2×10^4)	20	39.5
	30	59.3
24 (8.64×10^4)	20	47.4
	30	71.2

[a]Equivalents: 1 μg/110 μl = 9.1 μg/ml or 0.27×10^{-4} mol/liter; 5 μg/110 μl = 45 μg/ml or 1.37×10^{-4} mol/liter; 20 μg/110 μl = 182 μg/ml or 5.49×10^{-4} mol/liter; 30 μg/110 μl = 272 μg/ml or 8.24×10^{-4} mol/liter.

by air drying and baking in an oven. The chaotropic salt NaI has also been used for immobilizing DNA to nitrocellulose membranes (12). Other membranes developed for immobilizing DNA include diazobenzyloxymethyl paper and membranes made from positively charge-modified nylon 66 (7, 19, 90). All of these products are available from scientific supply companies, together with detailed protocols for their use.

The membrane reassociation methods differ from the free-solution methods in that the unlabeled DNA fragments are unable to reassociate with each other. As a result, the rate of probe DNA reassociation to the immobilized DNA is linear for commonly used reassociation times and the rate of reassociation is dependent on the amount of DNA on the membrane. The advantages of membrane methods are as follows. (i) It is relatively easy to wash non-reassociated DNA—and nonspecifically labeled contaminants—away from the reassociated DNA. (ii) In addition to radioactively labeled probes, nonradioactive probes (i.e., biotin, digoxigenin, and direct enzyme cross-linking) can be used. (iii) Membranes can be probed more than once. The major disadvantage of membrane procedures is that the results are not as quantitative as the solution methods. An indirect membrane procedure (competition method), which is more quantitative (50), requires such large amounts of DNA that it is seldom used. Procedures given here are for the direct binding of radioactively labeled DNA probes to membrane-immobilized DNA and for the determination of the thermal stabilities of the duplexes.

A general scheme for performing the membrane method is given in Fig. 11. The reassociation mixtures may include either a single small membrane in a vial or a larger membrane (dot blot or slot blot formats) containing immobilized DNA from a number of sources. However, the dot and slot blot formats are only qualitative because the DNA in different "dots" compete with each other for the radiolabeled probe.

26.6.3.1. DNA Immobilization on Membranes

The following adaptation of the Gillespie and Spiegelman (38) procedure is for a 15-cm-diameter BA 85 nitrocellulose membrane filter (Schleicher & Schuell Co., Keene, NH) having an effective filtering surface of about 173 cm^2. The immobilization of 4.35 mg of DNA will result in 25 μg/cm^2. For smaller membranes, reduce the amount of DNA and the buffer volumes accordingly.

1. Dilute 4.35 mg of DNA to 50 μg/ml with 0.1$\times$ SSC (approximately 90 ml).

2. Place the solution in a 125-ml Erlenmeyer flask, and denature the DNA by heating in a boiling-water bath for 10 min. Cool the solution quickly by pouring it into 800 ml of ice-cold 6$\times$ SSC (to give about 5 μg of DNA per ml).

3. Float the 15-cm nitrocellulose membrane filter on distilled water so that the pores will be filled with water (if the membrane is submersed immediately, air pockets will form and cause uneven filtration).

4. Place the wet membrane filter on the filtration device (section 26.6.1), and wash the membrane with 500 ml of cold 6$\times$ SSC, using a flow rate of approximately 30 ml/min. Then, pass the denatured DNA through the filter, and wash it again with 500 ml of 6$\times$ SSC.

5. Let the membrane dry initially at room temperature and then overnight at 60°C.

6. Label the membrane on the edge with a pencil, place in an envelope, and store at room temperature over desiccant.

Handle the membrane by the outside edge, and avoid touching the surface on which the DNA is bound. When cutting small membranes out of the large one, avoid getting too close to the edge.

26.6.3.2. Preincubation of DNA-Containing Membranes

The Denhardt preincubation reagent (28), which is used to block additional DNA-binding sites on the DNA-containing membrane, contains the following ingredients dissolved in 6$\times$ SSC.

Bovine serum albumin (fraction V; Sigma
Chemical Co.) 0.02%
Polyvinylpyrrolidone (Calbiochem-Behring,
La Jolla, CA) 0.02%
Ficoll 400 (Pharmacia Fine Chemicals,
Uppsala, Sweden) 0.02%

DIRECT-BINDING METHOD   COMPETITION METHOD

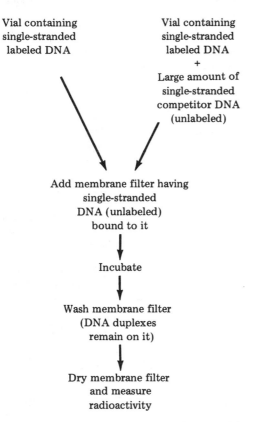

FIGURE 11 Scheme for performing membrane DNA reassociation experiments by using either the direct-binding or competition procedure.

1. Cut small membranes (about 5 μg of DNA per membrane) from the large one with a paper punch.

2. Preincubate the membranes in the Denhardt mixture at the incubation temperature of the experiment for 0.5 to 2 h. Swirl the membranes several times during the preincubation period.

3. Remove the membranes, place on a paper towel to blot off excess liquid, and then place into the reaction vials.

26.6.3.3. Reassociation Measured by Direct Binding

In the direct-binding procedure, the amount of a labeled probe DNA that forms duplexes with DNA preparations immobilized on membrane filters is measured. Reaction mixtures contain the following.

Labeled DNA (1×10^6 to 3×10^6 cpm/ml) ... 10 μl
6.6× SSC .... 100 μl

1. Preincubate the required number of DNA membrane filters. Keep each kind separate.

2. Make up the reaction mixtures, and mix them on a Vortex mixer.

3. Place the membranes in vials, stopper the vials, and incubate them at 25°C below the T_m (as determined in 6.6× SSC) of the labeled DNA. Incubate for 12 to 15 h.

4. Remove the membranes from the vials, and place them in the washing chamber (Fig. 4). Wash them in 2× SSC at the incubation temperature. Use two 300-ml volumes of washing buffer, and wash for 5 min in each. Move the washing chamber back and forth in the beaker several times during the washing period to ensure buffer flow through each compartment. Shake the washing chamber free from buffer. Remove the membrane filters, and dry them on a paper towel under a heat lamp.

5. Skewer the membranes on insect pins (if there are duplicates, place both membranes on the same pin, as shown in Fig. 12).

6. Place into scintillation vials (Fig. 12), and measure the radioactive counts.

The sequence similarity calculations are similar to those used for the S1 nuclease procedure (Table 1) in that the counts bound on the heterologous DNA membrane are divided by the counts bound on the homologous DNA membrane. This ratio is then multiplied by 100 to give the percent reassociation.

Comments

The size of the membranes dictates the size of the reaction vial and the volume of the reaction mixture.

For higher moles percent G+C DNAs, include 20 to 30% deionized formamide so that the reassociation temperature can be lowered (0.6°C/1% formamide).

Although 1× Denhardt reagent (as given in section 26.6.3.2) works well as a blocking agent of nonspecific binding, other hybridization protocols (see chapter 30.4.4.2) call for 5× Denhardt reagent (4) or the milk reagent called BLOTTD (Bovine Lacto Transfer Technique Optimizer) (47). The exact concentrations of blocking agents probably are not critical; their main function is to ensure that there is a low background of probe DNA on membranes containing either no DNA or an unrelated DNA, without causing any interference with the reassociation reaction.

The membrane-binding experiments are also amenable to the use of nonradioactive probes (36, 64–66, 84, 85,

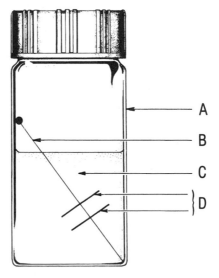

FIGURE 12 Apparatus for counting radioactivity on membrane filters. A, scintillation vial; B, insect pin; C, nonaqueous-based cocktail; D, duplicate filters impaled on pin.

101). The major advantage of these probes is that they are not subject to the safety precautions and regulations required when using radioactive substances; their major disadvantage is that an additional reaction is introduced into the experiment. This will increase the assay time; moreover, the experiment may not be as quantitative as when radioactive probes are used. Nonradioactive probes have the major impact in diagnostic bacteriology for identification of isolates.

26.6.3.4. Thermal Stability Measurement

1. Follow the direct-binding procedure through the washing step (step 4 above).

2. Remove the membrane filters from the washing chamber, skewer them on an insect pin near the head of the pin, insert the pin into the end of a rubber-stoppered tube (Fig. 6), and place into the first elution tube. When using more than one filter, separate them with Teflon disks.

3. Use 13- by 100-mm elution tubes containing 1.2 ml of $0.5\times$ SSC. Place the tubes into a water bath, and perform stepwise elution of the duplexes by increasing the bath temperature 5°C at 10-min intervals. At the end of each period, transfer the filters to the next elution tube and increase the temperature by 5°C. Start the elution profile about 10°C below the incubation temperature used for the direct-binding procedure. Use nine elution temperatures, and count the residual radioactivity on the filters (i.e., 10 scintillation vials per stability profile).

4. Empty the contents of each elution tube into a scintillation vial, and rinse the tube with 0.5 ml of $0.5\times$ SSC, also adding that to the vial.

5. Add 10 ml of a scintillation fluid that is miscible with aqueous samples.

The integral elution profile curves are obtained by summing the radioactivities at each temperature and dividing by the total radioactivity; this is essentially the same as illustrated for HA data (Table 2; Fig. 8) except that one cannot readily measure the unbound DNA, i.e., the washes.

In obtaining the profile curves, place the 1.2 ml of $0.5\times$ SSC in all of the 13- by 100-mm tubes at the same time by using an automatic pipette. Keep one elution tube ahead in the water bath so that each is preheated. Have an extra tube containing a thermometer, and record the exact tube temperature at the middle and the end of each temperature interval.

26.6.4. Fluorometric Plate Method

In the last 10 years, a fluorometric method performed in microdilution plates has become widely used (32). This method is similar to the membrane method except that the reference DNAs are immobilized in the wells of a microdilution plate. Hybridization is then performed with biotinylated DNA so that the extent of hybridization can be measured without use of radioisotopes. A major advantage of this method is the potential for automation using commercially available microdilution plate washers and readers. It can also be readily adapted to dot blot and other qualitative methods (30, 31). The procedure described here is based upon the protocol of Ezaki et al. (32).

Sample Preparation

The reference DNA is prepared by the Marmur procedure (section 26.2.1) and heat denatured prior to being diluted to 20 μg per ml of phosphate-buffered saline (PBS) (8 mM Na_2HPO_4, 1.5 mM KH_2PO_4, 137 mM NaCl, 2.7 mM KCl [pH 7.2]) containing 0.1 M $MgCl_2$. The DNA, 0.1 ml, is then added to the wells of a microdilution plate (MicroFluor type B plates; Dynatech Laboratories, Alexandria, VA) and incubated for 60 min at 37°C. The solutions are then aspirated, and the wells are washed once with PBS containing 0.1 M $MgCl_2$. The plate is then dried

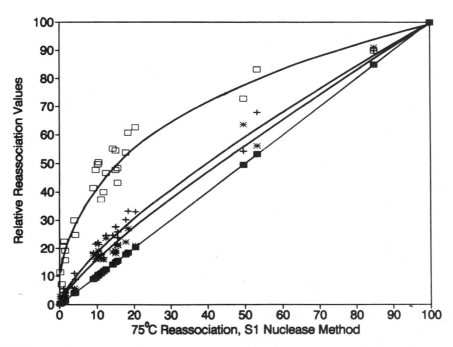

FIGURE 13 Comparison of S1 and HA reassociation methods at different temperatures. Symbols: ■, S1 method, 75°C; *, HA method, 75°C; +, S1 method, 60°C; □, HA method, 60°C.

overnight at 60°C. Each well of the plate is then incubated at 37°C for 60 min with 200 μl of a prehybridization solution composed of 2× SSC (see section 26.2.1), 5× Denhardt reagent (see section 26.6.3.2) and 50% formamide.

The labeled DNA, about 5 μg, is prepared by photobiotinylation as described in section 26.5.2 and used immediately.

Upon removing the prehybridization solution, 100 μl of a hybridization solution composed of 2× SSC, 5× Denhardt reagent, 50% formamide, 3% dextran sulfate, 50 μg of denatured salmon sperm DNA per ml, and 50 ng of biotinylated DNA are added to each well. The microdilution plate is then sealed with vinyl tape and incubated at 55°C below the melting temperature or T_m of the DNA for 2 h. Because of the high concentration of formamide, this condition is equivalent to the common incubation temperature of the $T_m - 25°C$ in the absence of formamide and is utilized to limit evaporation.

After the hybridization, the wells are washed four times with 300 μl of 2× SSC buffer. Streptavidin-β-D-galactosidase, 100 μl of a solution of 10 ng/ml in PBS containing 0.5% bovine serum albumin and 0.1% Triton X-100, is added to the wells and incubated for 30 min at 37°C. The wells are then washed twice with 300 μl of PBS containing 0.1% Triton X-100.

For development of the fluorescence, 100 μl of 3×10^{-4} M 4-methylumbelliferyl-β-D-galactopyranoside in PBS is added, and the plate is incubated at 37°C. The fluorescence is then determined within 15 to 60 min.

26.7. RELIABILITY AND COMPARISON OF RESULTS

One does not have to search the DNA reassociation literature very extensively to realize that there are variations in the published results, many of which are the result of experimental errors (96). Bacteriologists have been criticized for being a bit sloppy in doing and reporting DNA reassociation experiments, compared with those researchers working with eucaryotes. Although some of this criticism is justified, there are additional reasons for greater variability in bacterial DNA experiments. First, the sheer number of bacteriology laboratories involved (often temporarily) in performing DNA reassociation experiments has contributed to the variability of results. Second, the HA procedure has been used for most eucaryotic studies, whereas all of the procedures discussed in this chapter have been used extensively by bacteriologists. Probably the major factor causing experimental variations is the quality of the DNA. Although there are few differences in the isolation of DNA from one mammal to another or one bird to another, the ease of isolation and the quality (i.e., degree of fragmentation and various contaminants) of the DNA from microorganisms are much more variable. Therefore, one cannot overemphasize the importance of pure DNA preparations and the size homogeneity of fragmented DNA preparations. Another variable that is more pronounced with bacteria is the range of moles percent G+C in their DNA; experimental conditions must be adjusted to take this into account. All of these characteristics of bacterial DNA must be considered in order to reduce the amount of experimental error in the experiments.

The choice of reference organisms for use in DNA reassociation experiments can also have a profound effect on the apparent relationships in cluster analysis (41). The ideal situation is one in which every strain is used as a reference organism. If such studies involved more than 10 to 20 organisms, they would soon become quite unwieldy; consequently, fewer reference strains must be carefully chosen. When investigating a group of unidentified organisms, a rule of thumb is that one must first select reference strains rather arbitrarily from each phenotypic cluster of organisms. Organisms whose DNA has 80 to 100% reassociation with a given reference need not be used as references, whereas additional reference DNAs must be selected from each of the clusters that have lower percent reassociation values. This process is repeated until the DNA from each organism used in the study either belongs to a cluster (80 to 100% reassociation) or is unique when used as a reference organism. This type of data can be used for cluster analysis and dendrogram generation. It is also good to have more than one reference DNA preparation within a large cluster for better average reassociation values. When named species are included in a study, use the type strains. If one wants only to demonstrate that an unidentified cluster of organisms has no measurable reassociation with a number of species, it is only necessary to use the DNAs of the type strains of these species as references, but these data will not be suitable for cluster analysis.

26.7.1. DNA Reassociation Values

For the discussion and evaluation of data, it is important to know how one method of determining DNA reassociation values compares with another. Similar values are obtained when the photometric, membrane, and HA reassociation procedures are used. All are affected similarly by the reassociation temperature, with greater specificity at higher ($T_m - 10$) versus lower ($T_m - 25$) temperature. The optical method has higher background values, such that values up to 30% are not considered significant (46). The reassociation temperature has less effect on the S1 nuclease results, because the localized mismatched regions and overhanging ends are degraded. Correlations between procedures and reassociation temperatures (10, 40) are illustrated in Fig. 13.

26.7.2. DNA Thermal Stability Values

Theoretically, the $T_{m(i)}$ values are probably as good a measurement of similarity as any. These values correlate with the degree of base pair mismatches and are for the most part independent of the G+C content of the DNAs (48, 106) and the ionic strength of the elution buffer. The major disadvantage is that the experiments are laborious to perform.

26.8. REFERENCES

26.8.1. General References

1. **Hills, D. M., and C. Moritz (ed.).** 1990. *Molecular Systematics.* Sinauer Associates, Inc., Sunderland, MA.
   Although this useful book emphasizes the systematics of eucaryotic organisms, many of the techniques are directly applicable to bacteria.

2. **Johnson, J. L.** 1985. Determination of DNA base composition, p. 1–31. *In* G. Gottschalk (ed.), *Methods in Microbiology,* vol. 18. Academic Press, Ltd., London, United Kingdom.
   Review of the determination of moles percent G+C of DNA preparations.

3. **Johnson, J. L.** 1989. Nucleic acid hybridization: principles and techniques, p. 3–31. *In* B. Swaminathan and G. Prakash (ed.), *Nucleic Acids and Monoclonal Antibody Probes.* Marcel Dekker, Inc., New York, NY.

Reviews reassociation and hybridization kinetics, effects of experimental variables, duplex and hybrid specificities, and techniques.

4. **Sambrook, J., E. F. Fritsch, and T. Maniatis.** 1989. *Molecular Cloning: A Laboratory Manual,* 2nd ed. Cold Spring Harbor Laboratory Press, Cold Spring Harbor, NY.
This large manual (it is published as three sequential books) includes protocols for nearly all molecular biology techniques.

5. **Stackbrandt, E., and M. Goodfellow (ed.).** 1991. *Nucleic Acid Techniques in Bacterial Systematics.* John Wiley & Sons Ltd., Chichester, England.
Provides an extensive range of molecular methods for studying bacterial classification by the techniques of molecular biology.

6. **Wetmur, J. G.** 1976. Hybridization and renaturation kinetics of nucleic acids. *Annu. Rev. Biophys. Bioeng.* **5:**337–361.
This review deals primarily with free-solution reassociation kinetics.

26.8.2. Specific References

7. **Alwine, J. C., D. J. Kemp, and G. R. Stark.** 1977. Method for detection of specific RNAs in agarose gels by transfer to diazobenzyloxymethyl-paper and hybridization with DNA probes. *Proc. Natl. Acad. Sci. USA* **74:**5350–5354.

8. **Bacon, M. F.** 1965. Analysis of DNA preparations by a variation of the cysteine-sulfuric acid test. *Anal. Biochem.* **13:**223–228.

9. **Beji, A., D. Izard, F. Gavini, H. Leclere, M. Leseine-Delstanche, and J. Krembel.** 1987. A rapid chemical procedure for isolation and purification of chromosomal DNA from Gram-negative bacilli. *Anal. Biochem.* **162:**18–23.

10. **Bouvet, P. J. M., and P. A. D. Grimont.** 1986. Taxonomy of the genus *Acinetobacter* with the recognition of *Acinetobacter baumannii* sp. nov., *Acinetobacter haemolyticus* sp. nov., *Acinetobacter johnsonii* sp. nov., and *Acinetobacter junii* sp. nov. and emended descriptions of *Acinetobacter calcoaceticus* and *Acinetobacter lwoffi*. *Int. J. Syst. Bacteriol.* **36:**228–240.

11. **Brenner, D. J., G. R. Fanning, A. V. Rake, and K. E. Johnson.** 1969. Batch procedure for thermal elution of DNA from hydroxyapatite. *Anal. Biochem.* **28:**447–459.

12. **Bresser, J., J. Doering, and D. Gillespie.** 1983. Quick-blot: selective mRNA or DNA immobilization from whole cells. *DNA* **2:**243–253.

13. **Britten, R. J., D. E. Graham, and B. R. Neufeld.** 1974. Analysis of repeated DNA sequences by reassociation. *Methods Enzymol.* **24E:**363–406.

14. **Britten, R. J., and D. E. Kohne.** 1968. Repeated sequences in DNA. *Science* **161:**529–540.

15. **Britten, R. J., M. Pavich, and J. Smith.** 1969. A new method for DNA purification. *Carnegie Inst. Wash. Year Book* **68:**400–402.

16. **Burton, K.** 1968. Determination of DNA concentration with diphenylamine. *Methods Enzymol.* **12B:**163–166.

17. **Byvoet, P.** 1966. Determination of nucleic acids with concentrated H_2SO_4. II. Simultaneous determination of ribo- and deoxyribonucleic acid. *Anal. Biochem.* **15:**31–39.

18. **Caccone, A., R. DeSalle, and J. R. Powell.** 1988. Calibration of the change in thermal stability of DNA duplexes and degree of base pair mismatch. *J. Mol. Evol.* **27:**212–216.

19. **Cannon, G., S. Heinhorst, and A. Weissbach.** 1985. Quantitative molecular hybridization on nylon membranes. *Anal. Biochem.* **149:**229–237.

20. **Cesarone, C. F., C. Bolognesi, and L. Santi.** 1979. Improved microfluorometric DNA determination in biological material using 33258 Hoechst. *Anal. Biochem.* **100:**188–197.

21. **Chan, H. C., W. T. Ruyechan, and J. G. Wetmur.** 1976. *In vitro* iodination of low complexity nucleic acids without chain scission. *Biochemistry* **15:**5487–5490.

22. **Chassy, B. M., and A. Giuffrida.** 1980. A method for improved lysis of some gram-positive asporogenous bacteria with lysozyme. *Appl. Environ. Microbiol.* **39:**153–158.

23. **Colwell, R. R., R. Johnson, L. Wan, T. E. Lovelace, and D. J. Brenner.** 1974. Numerical taxonomy and deoxyribonucleic acid reassociation in the taxonomy of some gram-negative fermentative bacteria. *Int. J. Syst. Bacteriol.* **24:**422–433.

24. **Commerford, S. L.** 1971. Iodination of nucleic acids *in vitro*. *Biochemistry* **10:**1993–1999.

25. **Crosa, J. H., D. J. Brenner, and S. Falkow.** 1973. Use of single-strand-specific nuclease for analysis of bacterial and plasmid deoxyribonucleic acid homo- and heteroduplexes. *J. Bacteriol.* **115:**904–911.

26. **Cummins, C. S., and J. L. Johnson.** 1971. Taxonomy of the clostridia: wall composition and DNA homologies in *Clostridium butyricum* and other butyric acid-producing clostridia. *J. Gen. Microbiol.* **67:**33–46.

27. **DeLey, J., H. Cattoir, and A. Reynaerts.** 1970. The quantitative measurement of DNA hybridization from renaturation rates. *Eur. J. Biochem.* **12:**133–142.

28. **Denhardt, D. T.** 1966. A membrane-filter technique for the detection of complementary DNA. *Biochem. Biophys. Res. Commun.* **23:**641–646.

29. **Downs, T. R., and W. W. Wilfinger.** 1983. Fluorometric quantification of DNA in cells and tissue. *Anal. Biochem.* **131:**538–547.

30. **Ezaki, T., Y. Hashimoto, N. Takeuchi, H. Yamamoto, S. Liu, H. Miura, K. Matsui, and E. Yabuuchi.** 1988. Simple genetic method to identify viridans group streptococci by colorimetric dot hybridization and fluorometric hybridization in microdilution wells. *J. Clin. Microbiol.* **26:**1708–1713.

31. **Ezaki, T., N. Takeuchi, S. Liu, A. Kai, H. Yamamoto, and E. Yabuuchi.** 1988. Small-scale DNA preparation for rapid genetic identification of *Campylobacter* species with radioisotope. *Microbiol. Immunol.* **32:**141–150.

32. **Ezaki, T., Y. Hashimoto, and E. Yabuuchi.** 1989. Fluorometric deoxyribonucleic acid-deoxyribonucleic acid hybridization in microdilution wells as an alternative to membrane filter hybridization in which radioisotopes are used to determine genetic relatedness among bacterial strains. *Int. J. Syst. Bacteriol.* **39:**224–229.

33. **Feinberg, A. P., and B. Vogelstein.** 1983. A technique for radiolabeling DNA restriction endonuclease fragments to high specific activity. *Anal. Biochem.* **132:**6–13.

34. **Feinberg, A. P., and B. Vogelstein.** 1983. A technique for radiolabeling DNA restriction endonuclease fragments to high specific activity. *Anal. Biochem.* **137:**266–267. (Addendum.)

35. **Ferragut, C., and H. Leclerc.** 1976. Etude comparative des methodes de determination du Tm de L'ADN bactérien. *Ann. Microbiol. (Inst. Pasteur)* **127A:**223–235.

36. **Forster, A. C., J. L. McInnes, D. C. Skingle, and R. H. Symons.** 1985. Non-radioactive hybridization probes prepared by the chemical labeling of DNA and RNA with a novel reagent, photobiotin. *Nucleic Acids Res.* **13:**745–762.

37. **Giles, K. W., and A. Meyers.** 1965. An improved diphenylamine method for estimation of deoxyribonucleic acid. *Nature* (London) **206:**93.

38. **Gillespie, D., and S. Spiegelman.** 1966. A quantitative assay for DNA hybrids with DNA immobilized on a membrane. *J. Mol. Biol.* **12:**829–842.

39. **Gold, D. V., and D. Shochat.** 1980. A rapid colorimetric assay for the estimation of microgram quantities of DNA. *Anal. Biochem.* **105:**121–125.

40. **Grimont, P. A. D., M. Y. Popoff, F. Grimont, C. Coyault, and M. Lemelin.** 1980. Reproducibility and correlation study of three deoxyribonucleic acid hybridization procedures. *Curr. Microbiol.* **4:**325–330.

41. **Hartford, T., and P. H. A. Sneath.** 1988. Distortion of taxonomic structure from DNA relationships due to different choice of reference strains. *Syst. Appl. Microbiol.* **10:** 241–250.

42. **Hill, B. T., and S. Whatley.** 1975. A simple, rapid microassay for DNA. *FEBS Lett.* **56:**20–23.

43. **Holdeman, L. V., E. P. Cato, and W. E. C. Moore.** 1977. *Anaerobe Laboratory Manual,* 4th ed. Virginia Polytechnic Institute and State University, Blacksburg.

44. **Horinouchi, S., T. Uozumi, T. Beppu, and K. Arima.** 1977. A new isolation method of plasmid deoxyribonucleic acid from *Staphylococcus aureus* using a lytic enzyme of *Achromobacter lyticus*. *Agric. Biol. Chem.* **41:**2487–2489.

45. **Hoyer, B. H., B. J. McCarthy, and E. T. Bolton.** 1964. A molecular approach in the systematics of higher organisms. *Science* **144:**959–967.

46. **Huss, V. A. R., H. Festl, and K. H. Schleifer.** 1983. Studies on the spectrophotometric determination of DNA hybridization from renaturation rates. *Syst. Appl. Microbiol.* **4:**184–192.

47. **Johnson, D. A., J. W. Gautsch, J. R. Sportsman, and J. H. Elder.** 1984. Improved technique utilizing nonfat dry milk for analysis of proteins and nucleic acids transferred to nitrocellulose. *Gene Anal. Tech.* **1:**3–8.

48. **Johnson, J. L.** 1973. Use of nucleic acid homologies in the taxonomy of anaerobic bacteria. *Int. J. Syst. Bacteriol.* **23:** 308–315.

49. **Johnson, J. L.** 1978. Taxonomy of the bacteroides. I. Deoxyribonucleic acid homologies among *Bacteroides fragilis* and other saccharolytic *Bacteroides* species. *Int. J. Syst. Bacteriol.* **28:**245–256.

50. **Johnson, J. L.** 1981. Genetic characterization, p. 450–472. *In* P. Gerhardt, R. G. E. Murray, R. N. Costilow, E. W. Nester, W. A. Wood, N. R. Krieg, and G. B. Phillips (ed.), *Manual of Methods for General Bacteriology.* American Society for Microbiology, Washington, DC.

51. **Johnson, J. L.** 1985. DNA reassociation and RNA hybridization of bacterial nucleic acids, p. 33–74. *In* G. Gottschalk (ed.), *Methods in Microbiology,* vol. 18. Academic Press, Ltd., London, United Kingdom.

52. **Johnson, J. L., L. V. H. Moore, B. Kaneko, and W. E. C. Moore.** 1990. *Actinomyces georgiae* sp. nov., *Actinomyces gerencseriae* sp. nov., designation of two genospecies of *Actinomyces naeslundii,* and inclusion of *A. naeslundii* serotypes II and III and *Actinomyces viscosus* serotype II in *A. naeslundii* genospecies 2. *Int. J. Syst. Bacteriol.* **40:**273–286.

53. **Jones, A. S.** 1953. The isolation of bacterial nucleic acids using cetyltrimethylammonium bromide (CETAVLON). *Biochim. Biophys. Acta* **10:**607–612.

54. **Jones, D., and P. H. A. Sneath.** 1970. Genetic transfer and bacterial taxonomy. *Bacteriol. Rev.* **34:**40–81.

55. **Kane, M. D., A. Brauman, and J. A. Breznak.** 1991. *Clostridium mayombei* sp. nov., an H_2/CO_2 acetogenic bacterium from the gut of the African soil-feeding termite, *Cubitermes speciosus*. *Arch. Microbiol.* **156:**99–104.

56. **Kapuciski, J., and B. Skoczylas.** 1977. Simple and rapid fluorometric method for DNA microassay. *Anal. Biochem.* **83:**252–257.

57. **Karsten, U., and A. Wollenberger.** 1977. Improvements in the ethidium bromide method for direct fluorometric estimation of DNA and RNA in cell and tissue homogenates. *Anal. Biochem.* **77:**464–470.

58. **Keswani, J., and W. B. Whitman.** 2001. Relationship of 16S rRNA sequence similarity to DNA hybridization in prokaryotes. *Int. J. Syst. Evol. Microbiol.* **51:**667–678.

59. **Kirby, K. S.** 1957. A new method for the isolation of deoxyribonucleic acids: evidence on the nature of bonds between deoxyribonucleic acid and protein. *Biochem. J.* **66:** 495–504.

60. **Kirby, K. S., E. Fox-Carter, and M. Guest.** 1967. Isolation of deoxyribonucleic acid and ribosomal ribonucleic acid from bacteria. *Biochem. J.* **104:**258–262.

61. **Ko, C. Y., J. L. Johnson, L. B. Barnett, H. M. McNair, and J. R. Vercellotti.** 1977. A sensitive estimation of the percentage of guanine plus cytosine in deoxyribonucleic acid by high performance liquid chromatography. *Anal. Biochem.* **80:**183–192.

62. **Labarca, C., and K. Paigen.** 1980. A simple, rapid, and sensitive DNA assay procedure. *Anal. Biochem.* **102:**344–352.

63. **Lachance, M. A.** 1980. A simple method for determination of deoxyribonucleic acid relatedness by thermal elution in hydroxyapatite microcolumns. *Int. J. Syst. Bacteriol.* **30:**433–436.

64. **Langer, P. R., A. A. Waldrop, and D. C. Ward.** 1981. Enzymatic synthesis of biotin-labeled polynucleotides: novel nucleic acid affinity probes. *Proc. Natl. Acad. Sci. USA* **78:**6633–6637.

65. **Leary, J. J., D. J. Brigati, and D. C. Ward.** 1983. Rapid and sensitive colorimetric method for visualizing biotin-labeled DNA probes hybridized to DNA or RNA immobilized on nitrocellulose: bio-blots. *Proc. Natl. Acad. Sci. USA* **80:**4045–4049.

66. **Leary, J. J., and J. L. Ruth.** 1989. Nonradioactive labeling of nucleic acid probes, p. 33–57. *In* B. Swaminathan and G. Prakash (ed.), *Nucleic Acids and Monoclonal Antibody Probes.* Marcel Dekker, Inc., New York, NY.

67. **Legros, M., and A. Kepes.** 1985. One-step fluorometric microassay of DNA in procaryotes. *Anal. Biochem.* **147:** 497–502.

68. **Le Pecq, J.-B., and C. Paoletti.** 1966. A new fluorometric method for RNA and DNA determination. *Anal. Biochem.* **17:**100–107.

69. **Lutz, L. H., and A. A. Yayanos.** 1985. Spectrofluorometric determination of bacterial DNA base composition. *Anal. Biochem.* **144:**1–5.

70. **Mackey, B. M., C. A. Miles, S. E. Parsons, and D. A. Seymour.** 1991. Thermal denaturation of whole cells and cell components of *Escherichia coli* examined by differential scanning calorimetry. *J. Gen. Microbiol.* **137:**2361–2374.

71. **Mackey, B. M., S. E. Parsons, C. A. Miles, and R. J. Owen.** 1988. The relationship between the base composition of bacterial DNA and its intracellular melting temperature as determined by differential scanning calorimetry. *J. Gen. Microbiol.* **134:**1185–1195.

72. **Mandel, M., L. Igambi, J. Bergendahl, M. L. Dodson, Jr., and E. Schelgen.** 1970. Correlation of melting temperature and cesium chloride buoyant density of bacterial deoxyribonucleic acid. *J. Bacteriol.* **101:**333–338.

73. **Markov, G. G., and I. G. Ivanov.** 1974. Hydroxyapatite column chromatography in procedures for isolation of purified DNA. *Anal. Biochem.* **59:**555–563.

74. **Marmur, J.** 1961. A procedure for the isolation of deoxyribonucleic acid from microorganisms. *J. Mol. Biol.* **3:** 208–218.

75. **Marmur, J., and P. Doty.** 1962. Determination of the base composition of deoxyribonucleic acid from its thermal denaturation temperature. *J. Mol. Biol.* **5:**109–118.

76. **McConaughy, B. L., C. D. Laird, and B. J. McCarthy.** 1969. Nucleic acid reassociation in formamide. *Biochemistry* **8:**3289–3294.

77. **McIntyre, P., and G. R. Stark.** 1988. A quantitative method for analyzing specific DNA sequences directly from whole cells. *Anal. Biochem.* **174:**209–214.

78. **Mesbah, M., U. Premachandran, and W. B. Whitman.** 1989. Precise measurement of G+C content of deoxyribonucleic acid by high-performance liquid chromatography. *Int. J. Syst. Bacteriol.* **39:**159–167.

79. **Mesbah, M., and W. B. Whitman.** 1989. Measurement of deoxyguanosine/thymidine ratios in complex mixtures by high-performance liquid chromatography for determination of the mole percentage guanine + cytosine of DNA. *J. Chromatogr.* **479:**297–306.

80. **Murray, M. G., and W. F. Thompson.** 1980. Rapid isolation of high molecular weight plant DNA. *Nucleic Acids Res.* **8:**4321–4325.

81. **Orosz, J. M., and J. G. Wetmur.** 1974. *In vitro* iodination of DNA. Maximizing iodination while minimizing degradation; use of buoyant density shifts for DNA-DNA hybrid isolation. *Biochemistry* **13:**5467–5473.

82. **Paul, J. H., and B. Myers.** 1982. Fluorometric determination of DNA in aquatic microorganisms by use of Hoechst 33258. *Appl. Environ. Microbiol.* **43:**1393–1399.

83. **Perkin-Elmer Corporation.** 1986. *Fluorometric Determination of DNA Concentration Biotechnology, Technical Report 1-913A (September).* Perkin-Elmer Corp., Norwalk, CT.

84. **Pollard-Knight, D., C. A. Read, M. J. Downs, L. A. Howard, M. R. Leadbetter, S. A. Pheby, E. McNaughton, A. Syms, and M. A. W. Brady.** 1990. Nonradioactive nucleic acid detection by enhanced chemiluminescence using probes directly labeled with horseradish peroxidase. *Anal. Biochem.* **185:**84–89.

85. **Renz, M., and C. Kurz.** 1984. A colorimetric method for DNA hybridization. *Nucleic Acids Res.* **12:**3435–3444.

86. **Rigby, P. W., J. M. Dieckmann, C. Rhodes, and P. Berg.** 1977. Labeling deoxyribonucleic acid to high specific activity in vitro by nick translation with DNA polymerase I. *J. Mol. Biol.* **113:**237–251.

87. **Sanders, C. A., D. M. Yajko, W. Hyun, R. G. Langlois, P. S. Nassos, M. J. Fulwyler, and W. K. Hadley.** 1990. Determination of guanine-plus-cytosine content of bacterial DNA by dual-laser flow cytometry. *J. Gen. Microbiol.* **136:**359–365.

88. **Schindler, C. A., and V. T. Schuhardt.** 1964. Lysostaphin: a new bacteriolytic agent for the *Staphylococcus*. *Proc. Natl. Acad. Sci. USA* **51:**414–421.

89. **Schwinghammer, E. A.** 1980. A method for improved lysis of some Gram-negative bacteria. *FEMS Microbiol. Lett.* **7:**157–162.

90. **Seed, B.** 1982. Attachment of nucleic acids to nitrocellulose and diazonium-substituted supports. *Genet. Eng.* **4:**91–102.

91. **Seidler, R. J., and M. Mandel.** 1971. Quantitative aspects of deoxyribonucleic acid renaturation: base composition, state of chromosome replication, and polynucleotide homologies. *J. Bacteriol.* **106:**608–614.

92. **Selin, Y. M., B. Harich, and J. L. Johnson.** 1983. Preparation of labeled nucleic acids (nick translation and iodination) for DNA homology and rRNA hybridization experiments. *Curr. Microbiol.* **8:**127–132.

93. **Setaro, F., and C. D. G. Morley.** 1977. A rapid colorimetric assay for DNA. *Anal. Biochem.* **81:**467–471.

94. **Sibley, C. G., and J. E. Ahlquist.** 1981. The phylogeny and relationships of the ratite birds as indicated by DNA-DNA hybridization, p. 301–335. *In* G. G. E. Scudder and J. L. Reveal (ed.), *Evolution Today. Proceedings of the Second International Congress of Systematic and Evolutionary Biology.* Carnegie-Mellon University, Pittsburgh, PA.

95. **Smith, M. J., R. J. Britten, and E. H. Davidson.** 1975. Studies on nucleic acid reassociation kinetics: reactivity of single-stranded tails in DNA-DNA renaturation. *Proc. Natl. Acad. Sci. USA* **72:**4805–4809.

96. **Sneath, P. H. A.** 1989. Analysis and interpretation of sequence data for bacterial systematics: the view of a numerical taxonomist. *Syst. Appl. Microbiol.* **12:**15–31.

97. **Stackebrandt, E., W. Frederiksen, G. M. Garrity, P. A. D. Grimont, P. Kämpfer, M. C. J. Maiden, X. Nesme, R. Rosselló-Moro, J. Swings, H. G. Trüper, L. Vauterin, A. C. Ward, and W. B. Whitman.** 2002. Report of the ad hoc committee for the re-evaluation of the species definition in bacteriology. *Int. J. Syst. Evol. Microbiol.* **52:**1043–1047.

98. **Stackebrandt, E., and B. M. Goebel.** 1994. Taxonomic note: a place for DNA-DNA reassociation and 16S rRNA sequence analysis in the present species definition in bacteriology. *Int. J. Syst. Bacteriol.* **44:**846–849.

99. **Steiner, R. F., and H. Sternberg.** 1979. The interaction of Hoechst 33258 with natural and biosynthetic nucleic acids. *Arch. Biochem. Biophys.* **197:**580–588.

100. **Sterzel, W., P. Bedford, and G. Eisenbrand.** 1985. Automated determination of DNA using the fluorochrome Hoechst 33258. *Anal. Biochem.* **147:**462–467.

101. **Takahashi, T., T. Mitsuda, and K. Okuda.** 1989. An alternative nonradioactive method for labeling DNA using biotin. *Anal. Biochem.* **179:**77–85.

102. **Tamaoka, J., and K. Komagata.** 1984. Determination of DNA base composition by reversed-phase high-performance liquid chromatography. *FEMS Microbiol. Lett.* **25:**125–128.

103. **Tereba, A., and B. J. McCarthy.** 1973. Hybridization of ^{125}I-labeled ribonucleic acid. *Biochemistry* **12:**4675–4679.

104. **Thompson, L. M., III, R. M. Smibert, J. L. Johnson, and N. R. Krieg.** 1988. Phylogenetic study of the genus *Campylobacter*. *Int. J. Syst. Bacteriol.* **38:**190–200.

105. **Ullman, J. S., and B. J. McCarthy.** 1973. Alkali deamination of cytosine residues in DNA *Biochim. Biophys. Acta* **294:**396–404.

106. **Ullman, J. S., and B. J. McCarthy.** 1973. The relationship between mismatched base pairs and the thermal stability of DNA duplexes. *Biochim. Biophys. Acta* **294:**416–424.

107. **Van Dilla, M. A., R. G. Langlois, D. Pinkel, D. Yajko, and W. K. Hadley.** 1983. Bacterial characterization by flow cytometry. *Science* **220:**620–622.

108. **Warnaar, S. O., and J. A. Cohen.** 1966. A quantitative assay for DNA-DNA hybrids using membrane filters. *Biochem. Biophys. Res. Commun.* **24:**554–563.

109. **Wayne, L. G., D. J. Brenner, R. R. Colwell, P. A. D. Grimont, O. Kandler, M. I. Krichevsky, L. H. Moore, W. E. C. Moore, R. G. E. Murray, E. Stackebrandt, M. P. Starr, and H. G. Trüper.** 1987. Report of the ad hoc committee on reconciliation of approaches to bacterial systematics. *Int. J. Syst. Bacteriol.* **37:**463–464.

110. **Werman, S. D., M. S. Springer, and R. J. Britten.** 1990. Nucleic acids I. DNA-DNA hybridization, p. 204–249. *In* D. M. Hillis and C. Moritz (ed.), *Molecular Systematics.* Sinauer Associates, Inc., Sunderland, MA.

111. **Wetmur, J. G., and N. Davidson.** 1968. Kinetics of renaturation of DNA. *J. Mol. Biol.* **31:**349–370.

112. **Wiener, S. L., M. Urievetzky, S. Lendval, S. Shafer, and E. Meilman.** 1976. The indole method for determination of DNA: conditions for maximal sensitivity. *Anal. Biochem.* **71:**579–582.

113. **Yamada, K., and K. Komagata.** 1970. Taxonomic studies of coryneform bacteria. III. DNA base composition of coryneform bacteria. *J. Gen. Appl. Microbiol.* **16:**215–224.

114. **Yokogawa, K., S. Kawata, and Y. Yoshimura.** 1972. Bacteriolytic activity of enzymes derived from *Streptomyces* species. *Agric. Biol. Chem.* **36:**2055–2065.

115. **Yokogawa, K., S. Kawata, and Y. Yoshimura.** 1975. Purification and properties of lytic enzymes from *Streptomyces globisporus* 1829. *Agric. Biol. Chem.* **39:**1533–1543.

116. **Zolan, M. E., and P. J. Pukkila.** 1986. Inheritance of DNA methylation in *Coprinus cinereus*. *Mol. Cell. Biol.* **6:**195–200.

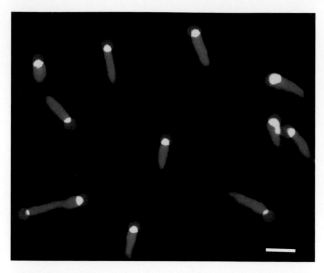

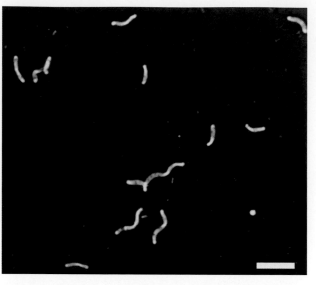

COLOR PLATE 1 (chapter 2) Dually fluorescently labeled *Mycoplasma pneumoniae* cells where the DNA has been stained with DAPI (blue) and the P1 adhesion protein has been stained with anti-P1/Alexa 488 (red) to show their respective cellular positions. Scale bar, 1 μm. (Reprinted from reference 48 with the authors' permission.)

COLOR PLATE 2 (chapter 2) Gray-scale image of *Aquaspirillum serpens* negatively stained with India ink. The spiral shape of this gram-negative bacterium can be seen as well as black aggregates of ink. Scale bar, 5 μm. (Kindly prepared and supplied by R. Van Twest, University of Guelph, Guelph, Ontario, Canada.)

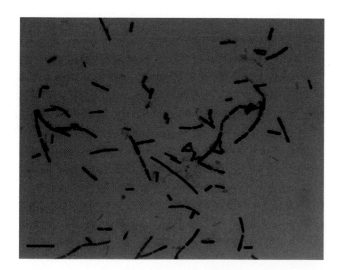

COLOR PLATE 3 (chapter 2) Gram stain of *Bacillus subtilis* that has been incubated with egg white lysozyme to digest the peptidoglycan in the cell wall. Some cells have lysed and therefore stain as gram negative (red), whereas others remain intact and stain as gram positive (purple). Scale bar, 5 μm. (Kindly prepared and supplied by R. Van Twest.)

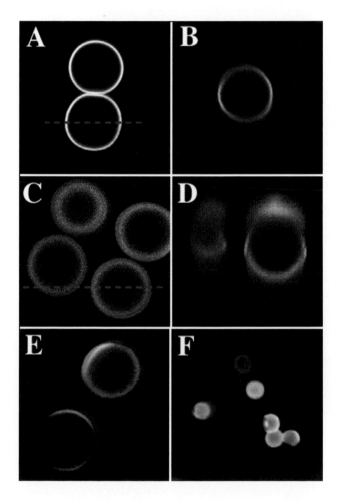

COLOR PLATE 4 (chapter 3) CLSM images of 6-μm-diameter Focal Check beads stained with green, red, and blue fluors. (A) Beads imaged with a 60×, 1.4-NA lens. The image shows a white ring indicating good alignment of the LSM and correction of the lens. (B) Loss of resolution in the xz plane due to light wavelength and point spread function error. (C) Effect of using a 63×, 0.9-NA lens on the appearance of the same beads in the xy plane. (D) Further reduced quality of the xz image (dashed lines indicate location of xz image). (E) Effect of imaging with the outer edge of the 63× lens. (F) Loss of confocality (presence of multiple planes and bead colors) and chromatic correction in a 20×, 0.4-NA lens.

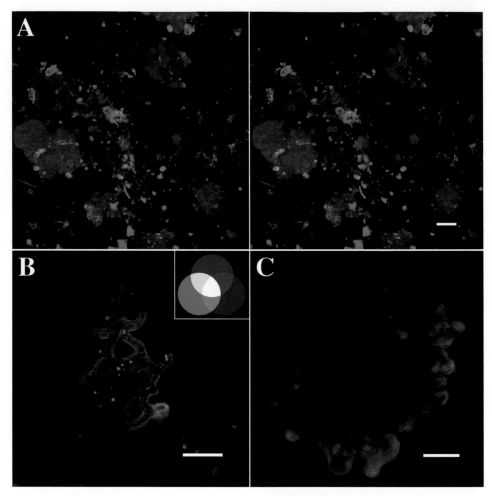

COLOR PLATE 5 (chapter 3) (A) Three-channel 1P-LSM Z-series projected as a stereo pair showing the colocalization of binding sites for three lectins, *Triticum vulgaris*-TRITC (red), *Lycopersicum esculentum*-FITC (green), and *Tetragonolobus purpureas*-Cy5 (blue), in a river biofilm. A variety of microcolonies and individual cells can be seen to bind single or combinations of the lectins as indicated by the color coding of the area. (B) High-resolution imaging of a microcolony stained with the lectin conjugates *Solanum tuberosum*-FITC (green), *Cicer arietinum*-Alexa568 (red), and *Tetragonolobus purpureas*-Cy5 (blue). (C) *Wisteria floribunda*-FITC (green), *Lens culinaris*-TRITC (red), and *Canavalia ensiformis*-Cy5 (blue). These three-channel images show the details of binding and structure of the multicomponent polymer surrounding individual bacterial cells in a microcolony. (A color wheel is included to allow interpretation of stain combinations.)

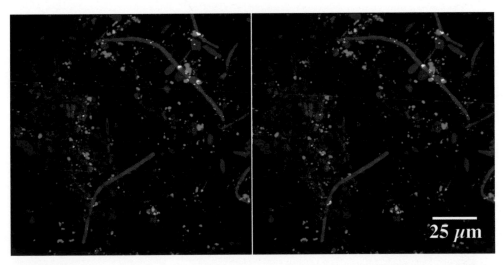

COLOR PLATE 6 (chapter 3) Multiparameter image showing the distribution of bacteria (green), algae (autofluorescence red), and exopolymeric substances stained with *Triticum vulgaris* lectin (blue). The image stack was projected with one version of the stack offset and aligned to form a stereo pair.

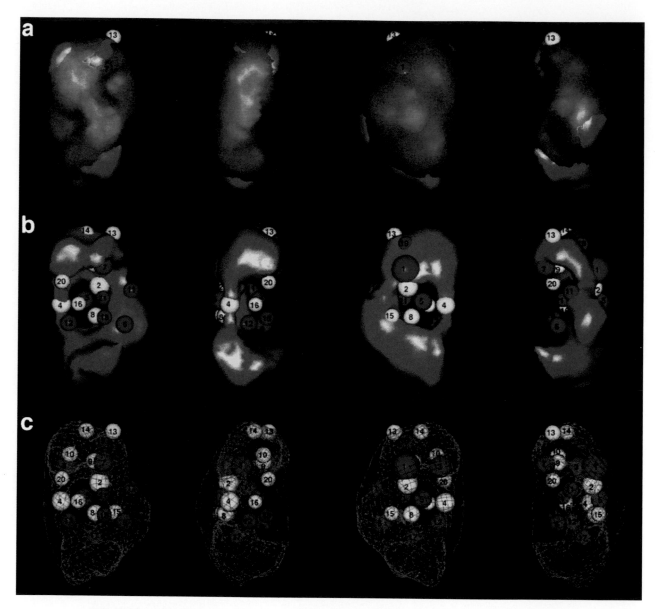

COLOR PLATE 7 (chapter 4)   An example of the use of ESI, TEM, and three-dimensional re-construction (see chapter 5) to determine the location of rRNA (by following the phosphorus line) on the surface of the small (30S) ribosomal subunit of *E. coli*. A 150- ±9-eV loss (green) and NetP (orange) are shown in four views which are related to one another by 90° rotations about the long axis of the subunit. In panel a the reconstructions are represented as solid surfaces, in panel b only the NetP reconstructions are shown, and in panel c both reconstructions are shown as wire mesh models. Spheres with numbers indicate positions of proteins mapped by neutron diffraction in stud-ies by other researchers. In the first column of images, the intersubunit face of the small subunit is facing the viewer. The bottom of the subunit is shown to be phosphorus rich (i.e., rRNA rich). (From D. R. Beniac, G. J. Czarnota, B. L. Rutherford, F. P. Ottensmeyer, and G. Harauz, *J. Microsc.* **188:**24–35, 1997.)

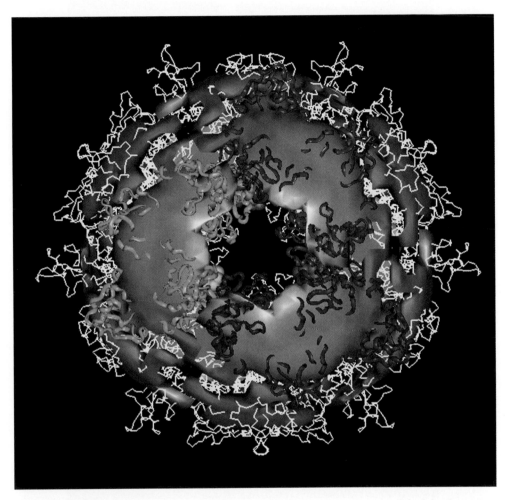

COLOR PLATE 8 (chapter 5) As described in reference 23, an atomic model of the *S. marinus* phosphoenolpyruvate synthase subunit was constructed by using Insight software (i) and 24 subunits were computationally matched to the 3-D reconstruction in Fig. 8d. Here, the view down the three-fold axis of rotational symmetry shows the enzymatically active sites of each subunit clustered around a putative channel. This figure was provided by F. P. Ottensmeyer, Ontario Cancer Institute, Toronto, Canada.

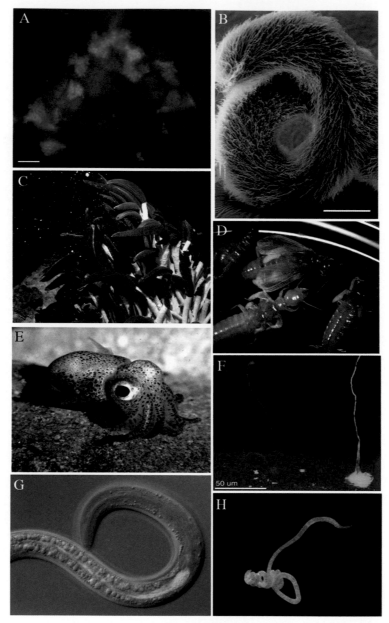

COLOR PLATE 9 (chapter 16) Diversity of symbiosis. (A) Anaerobic methane-oxidizing consortium (FISH image; red, archaeal group; green, sulfate-reducing bacteria (credit: V. Orphan, Caltech). Scale bar, 10 μm. (B) Nematode (*Eubostrichus dianae*) with epibionts (credit: M. Polz, MIT, and M. Bright, University of Vienna). Scale bar, 100 μm. (C) *Riftia pachyptila* tube worms living near a deep-sea hydrothermal vent (credit: S. Goffredi, Monterey Bay Aquarium Research Institute). (D)Termite, *Zootermopsis nevadensis* (credit: Amy Vu, Caltech). (E) Hawaiian bobtail squid, *Euprymna scolopes*. *Vibrio fischeri*-containing light organ is located on the ventral side of mantle cavity (not shown) (credit: M. Mcfall-Ngai, University of Wisconsin at Madison). (F) GFP- and dsRed-labeled *Sinorhizobium meliloti* inside infection threads of alfalfa. Clonal selection during the development of the root nodule is evident due to distinct GFP and dsRed-*S. meliloti* in each infection thread. Scale bar, 50 μm. [credit: D. Gage, University of Connecticut (61)]. (G) GFP-labeled *Photorhabdus luminescens* in the intestine of an infective juvenile nematode, *Heterorhabditis bacteriophora*. The nematode and bacteria cooperate in insect pathogenesis (credit: T. Ciche, Michigan State University), (H) The gutless marine annelid *Inanidrilus leukodermatus* (credit: A. Blazejak, Max Planck Institute).

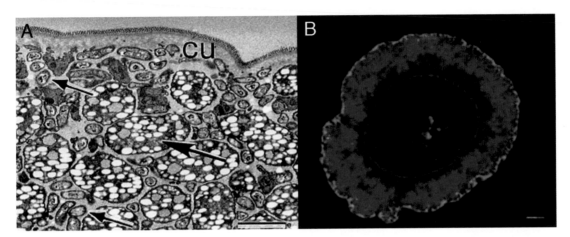

COLOR PLATE 10 (chapter 16)   (A) Transmission electron micrograph of the symbiont-containing region of *Inanidrilus leukodermatus*. Smaller and larger arrows indicate smaller and larger symbiont morphotypes, respectively. Scale bar, 2 μm. cu, cuticle (44). (B) In situ epifluorescence identification of bacterial symbionts in O. *crassitunicatus*. Cross sections through the entire worm. Dual hybridization with the GAM42a and DSS658/DSR651 probes, showing γ-proteobacterial symbionts (red) and δ-proteobacterial symbionts (green). Scale bar, 20 (μm (12).

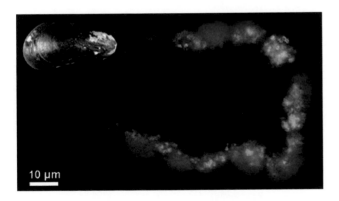

COLOR PLATE 11 (chapter 16)   mRNA FISH on gill filament of the mussel *Bathymodiolus puteoserpentis*. Blue, thiotrophic symbiont labeled with a specific 16S rRNA probe; red, methanotrophic symbiont labeled with a specific 16S rRNA probe; green, activity of particulate methane monooxygenase of the symbiont (subunit A; *pmoA*), a key enzyme of aerobic methane oxidation (credit: Annelie Pernthaler, Max Planck Institute).

COLOR PLATE 12 (chapter 27) DNA sequence chromatogram tracings. Screen shots were taken during the running of Sequencher software (Gene Codes Corp., Ann Arbor, MI). The sequence identified by the computer program is given above each tracing. (A) Normal sequence; (B) double peaks from a mixed clone; (C) failure with dirty template; (D) heterozygote template; (E) failure with high salt contamination; (F) blockage possibly due to hairpin formation.

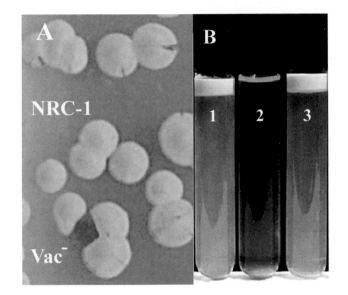

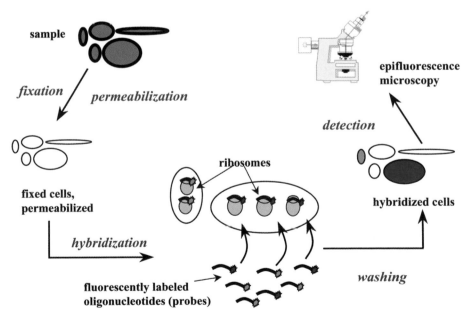

COLOR PLATE 13 (chapter 33) *Halobacterium* sp. NRC-1. (A) Colonies of the model haloarchaeal strain *Halobacterium* sp. NRC-1. Among pink wild-type (NRC-1) colonies, an orange gas vesicle-minus (Vac⁻) mutant is visible. (B) Liquid cultures of NRC-1 (tube 1), Vac⁻ mutant SD109 (tube 2), and a SD109 transformant containing the gas vesicle gene cluster on pFL2 (tube 3). Vac⁺ floating cells at the top of tubes 1 and 3 and Vac⁻ nonfloating cells at the bottom of tube 2 are visible. The meniscus is visible at the top of tube 2.

COLOR PLATE 14 (chapter 39) General protocol of whole-cell fluorescence in situ hybridization.

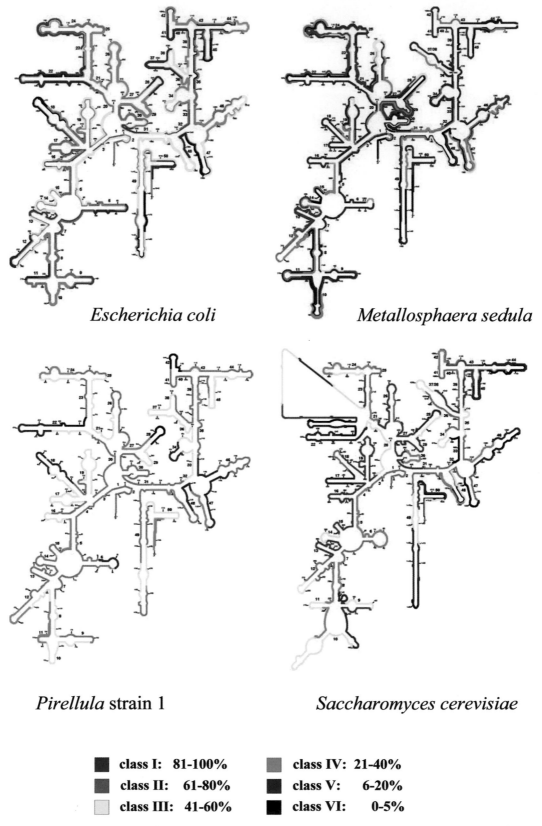

Escherichia coli

Metallosphaera sedula

Pirellula strain 1

Saccharomyces cerevisiae

■ class I: 81-100%	■ class IV: 21-40%	
■ class II: 61-80%	■ class V: 6-20%	
☐ class III: 41-60%	■ class VI: 0-5%	

COLOR PLATE 15 (chapter 39) In situ accessibility of the 16S rRNA of *E. coli* (*Enterobacteriaceae*), *Pirellula* sp. strain 1 (*Planctomycetales*), *M. sedula* (Archaea), and *S. cerevisiae* (*Eukarya*) for fluorescently labeled oligonucleotide probes. Different colors indicate differences in accessibility. From reference 8 with permission.

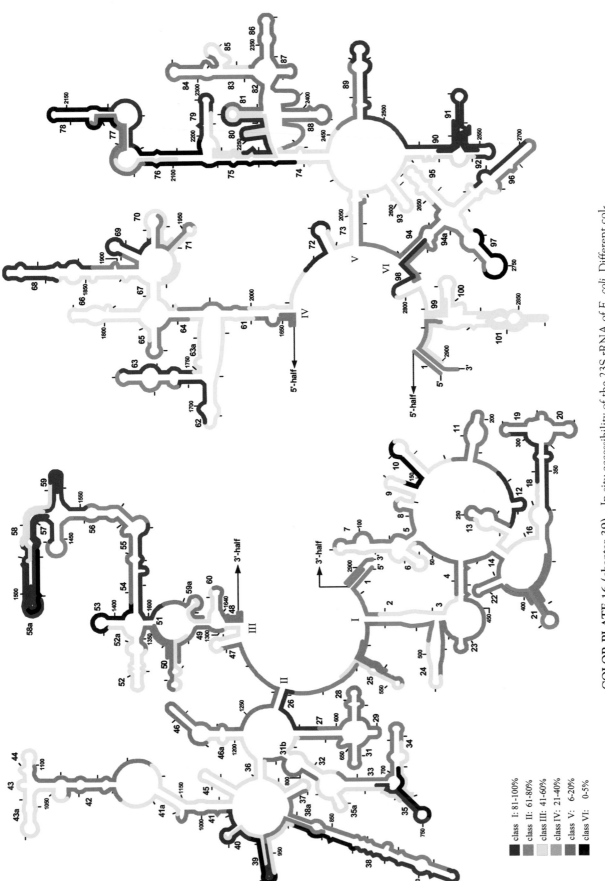

COLOR PLATE 16 (chapter 39) In situ accessibility of the 23S rRNA of *E. coli*. Different colors indicate differences in accessibility for oligonucleotide probes. From reference 17 with permission.

class I: 81-100%
class II: 61-80%
class III: 41-60%
class IV: 21-40%
class V: 6-20%
class VI: 0-5%

A.

◇	A	B	C	D	E	F	G	H	I	J	K	L	M	N	O	P	Q	R	S
	Sample Info	A01 1	A02 1	A03 1	A04 1	A05 1	A07 1	A09 1	A11 1	A12 1	A14 1	A16 1	A18 1	A19 1	A20 1	A22 1	A23 1	A24 1	A25 1
2	H1H		57.39	58.89		61.45		64.87	76.6			√	93.79		95.35	100.57	102.26	110.08	
3	H2H		√	59.01		61.59		64.8		80.62		91.67	93.79		95.35	100.36		110.08	
4	H3H		√	58.79		61.37		64.8				91.88	94		95.35	100.36	√	110.08	
5	H4H	56.79	√	59.16		61.74		64.96	76.7			91.9	94		95.31	100.44	102.32	109.92	
6	N1H		57.43	59.16			63.03	65.17		80.73	83.06	91.27	93.79		95.31				124.31
7	N2H		57.43	59.16		61.53	63.03	64.96		80.73	83.06	91.48	93.79	95.1	95.1				124.28
8	N3H		57.43	59.16			63.03	64.96		80.73	83.06	91.48	93.79		√				124.05
9	N4H	57.22	57.22	60.09			64.04	66.01		81.81	83.69	92.16	94.63		95.89				124.05
10	T1H		57.47	59.92		62.35		66.08	77.05	81.89	83.76	92.79	94.83				109.89		
11	T2H			59.07		61.44		65.11		80.71	82.62	92.11	94					109.9	
12	T3H			59.07	60.15	61.44		65.32		80.71	82.62	92.11	94.22	94.22				109.9	
13	T4H			58.92		61.07		64.95		80.6	√	91.88	94						

B.

◇	A	B	C	D	E	F	G	H	I	J	K	L	M	N	O	P	Q	R	S
1	H1H	0	1	1	0	1	0	1	1	0	0	1	1	0	1	1	1	1	0
2	H2H	0	1	1	0	1	0	1	0	1	0	1	1	0	1	1	1	1	0
3	H3H	0	1	1	0	1	0	1	0	0	0	1	1	0	1	1	1	1	0
4	H4H	1	1	1	0	1	0	1	0	0	0	1	1	0	1	1	1	1	0
5	N1H	0	1	1	0	0	1	1	0	1	1	1	1	0	1	0	0	0	1
6	N2H	0	1	1	0	1	1	1	0	1	1	1	1	1	1	0	0	0	1
7	N3H	0	1	1	0	0	1	1	0	1	1	1	1	0	1	0	0	0	1
8	N4H	1	1	1	0	0	1	1	0	1	1	1	1	0	1	0	0	0	1
9	T1H	0	1	1	0	1	0	1	1	1	1	1	1	0	0	0	1	0	0
10	T2H	0	0	1	0	1	0	1	0	1	1	1	1	0	0	0	0	1	0
11	T3H	0	0	1	1	1	0	1	0	1	1	1	1	1	0	0	0	1	0
12	T4H	0	0	1	0	1	0	1	0	1	1	1	1	0	0	0	0	0	0

C.

	SOIL 1				SOIL 2				SOIL 3			
	H1	H2	H3	H4	N1	N2	N3	N4	T1	T2	T3	T4
H1	1											
H2	0.767	1										
H3	0.86	0.91	1									
H4	0.93	0.85	0.98	1								
N1	0.419	0.56	0.47	0.44	1							
N2	0.488	0.56	0.56	0.5	0.979	1						
N3	0.488	0.58	0.54	0.48	0.958	0.96	1					
N4	0.349	0.47	0.4	0.44	0.867	0.87	0.84	1				
T1	0.5	0.48	0.43	0.45	0.475	0.45	0.43	0.58	1			
T2	0.558	0.67	0.58	0.62	0.533	0.58	0.6	0.47	0.8	1		
T3	0.558	0.66	0.58	0.59	0.479	0.52	0.5	0.42	0.825	0.91	1	
T4	0.558	0.73	0.58	0.61	0.5	0.55	0.55	0.41	0.7	0.86	0.96	1

D.

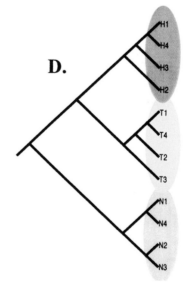

COLOR PLATE 17 (chapter 41)   An example of analysis of T-RFLP data from three soils. Each soil type (H, N, and T) is presented with four replicates. (A) Aligned profiles from 12 samples taken into Excel™. Each profile includes the estimated size for each detected fragment. The checks indicate that a fragment was detected by visual inspection of the electropherogram but was not scored by the software because the peak amplitude value was below the preset limit. These are recorded as "present." (B) The data set from panel A converted into binary format (presence or absence) for similarity comparisons. (C) Similarity analysis of the 12 samples with the T-RFLP function of the RDP. (D) UPGMA analysis of the 12 samples with PAUP (http://paup.csit.fsu.edu/).

COLOR PLATE 18 (chapter 44)   (Top left) Replica plating block used for replication of *A. nidulans* colonies. (Top right) *A. nidulans* cleistothecia appearing on mycelium near the mycelial frontier. (Bottom left) *A. nidulans* cleistothecia. The needle shaft is about 1 mm wide. The cleistothecium to its immediate left is uncleaned, with adhering Hülle cells and other debris. The cleistothecium to the far left has been cleaned by rolling on the agar. (Bottom right) Multipin replicator (to the right) used for rapid transfer of arrays of *A. nidulans* colonies. The colonies on the upper plate have grown after spreading conidia. Colonies are sampled individually from this plate and grown in the array shown in the lower plate. This will be the master plate used for transfers made with the multipin replicator.

27

Nucleic Acid Analysis

WILLIAM HENDRICKSON AND DON WALTHERS

This chapter describes several of the most common nucleic acid analyses performed in vitro to characterize a cloned DNA segment carrying a gene or genes. However, Southern hybridization, which is done to locate a cloned DNA segment within a physical map, is described in chapter 29.2.3.2 and will not be discussed here. DNA sequencing, described in section 27.1, is essential for many other methods and is often one of the first procedures done with a cloned DNA segment. Section 27.2 describes the gel mobility shift assay, DNase protection and chemical protection ("footprinting") assays, and an immunoprecipitation assay used to locate DNA-binding sites for regulatory proteins at promoters and in control regions of genes. The precise 5′ end of the mRNA can be located by S1 nuclease protection or by primer extension methods, as described in section 27.3. The size of an mRNA is determined by doing a Northern blot hybridization, also described in section 27.3. Both the S1 protection and Northern (RNA) blot hybridization methods provide information on the abundance of mRNA for studies of gene regulation.

27.1. DNA SEQUENCING

The most common method used to determine the sequence of nucleotides in a DNA fragment (50) is based on the generation of sequencing ladders by using a DNA oligomer as a primer, a double-stranded target DNA segment as the template, various types of DNA polymerase, deoxynucleoside triphosphates (dNTPs), and 2′,3′-dideoxynucleoside triphosphates (ddNTPs). The polymerase extends the primer from the 5′ to the 3′ direction. The ddNTPs are analogs of dNTPs, but they lack a hydroxyl residue at the

3' position of deoxyribose. When a ddNTP residue is incorporated into the growing chain of the newly synthesized DNA strand, it cannot form a phosphodiester bond with another dNTP and, consequently, the DNA chain terminates. For manual sequencing, one specific ddNTP is added to each of four reaction mixtures and the DNA chain is terminated at specific base positions, generating sequence (oligonucleotide) ladders of random sizes with a fixed 5' terminus. Separate reactions are run for each of the four nucleotides. The DNA oligomers are separated by denaturing polyacrylamide gel electrophoresis. A maximum of 400 to 600 bases can be read accurately from one gel (Fig. 1). The longer reads are achieved by loading additional aliquots of the reaction mixes during the electrophoresis run.

Automated techniques of DNA sequencing that allow for large-scale sequencing projects have been improved over the last decade and make it possible to obtain a first draft of an entire bacterial genome in as little as 1 day in a high-throughput sequencing facility. The average run length is now 700 to 1,000 bases, depending on the quality of the template, and the cost has been reduced to a few dollars per run. For high-throughput projects using templates submitted in 96-well plates, the cost is often reduced an additional 50%. Therefore, it is most cost-effective to have all sequencing projects, whether large or small, performed at a DNA-sequencing facility. Almost every major research university has such a facility, and new, low-cost commercial facilities are opening as automation techniques are improved. Nevertheless, simple manual procedures of nucleotide sequencing are still used in the laboratory for routine molecular biology, for example, promoter mapping by primer extension and footprinting studies.

27.1.1. Manual DNA Sequencing

27.1.1.1. DNA Polymerases

The large fragment of the *Escherichia coli* DNA polymerase (the Klenow fragment), originally used by Sanger et al. (50) for DNA sequencing, has a low level of 3'-to-5' exonuclease activity and an intermediate level of 5'-to-3' polymerase activity. The exonuclease can hydrolyze 3'-terminal deoxy or dideoxy monophosphates, interfering with the analysis of polymerase reactions, and therefore, the enzyme is no longer used for sequencing. Newer polymerases that are used for DNA sequencing include modified T7 phage-derived DNA polymerase (60, 61) and a variety of thermostable DNA polymerases such as that from the thermophilic bacterium *Thermus aquaticus* (11). A genetically engineered form of the enzyme is generally used for automated sequencing (AmpliTaq; Perkin-Elmer, Emeryville, CA). The modified T7 phage-derived polymerase is popular for manual sequencing and is marketed under the trade name Sequenase 2.0 (U.S. Biochemicals, Cleveland, OH; http://www.usbweb.com).

The Klenow fragment polymerase has low processivity (dissociates from and reassociates with the template before 10 nucleoside triphosphates are incorporated) and incorporates ddNTPs at a 1,000-fold-lower rate than dNTPs. These properties result in differential band intensities and a low net level of incorporation of the labeled substrate used and limit the extension of the primer chain to approximately 350 nucleotides in a typical reaction. Sequenase 2.0 lacks exonuclease activity and is highly processive (for several thousand nucleotides), and the rate of the polymerization reaction is very high. *Taq* DNA polymerase lacks 3'-to-5' exonuclease activity, has an intermediate level of processivity, and yields an intermediate-rate polymerization reaction, but it has 5'-to-3' exonuclease activity and works best at 70 to 80°C (33). This enzyme is useful for the sequencing of double-stranded DNA (dsDNA), especially when plasmid dsDNA from "mini" preparations, phage dsDNA, or PCR products are used as templates. There is a high degree of selectivity of priming since the sequencing reaction is carried out at a high temperature. Genetically modified versions of the *Taq* DNA polymerase (Thermo Sequenase for manual sequencing, from U.S. Biochemicals, and AmpliTaq for automated sequencing, from Applied Biosystems Inc.) lack any exonuclease activity and can survive repeated incubations at 90 to 95°C. DNA polymerases from other thermophilic microorganisms are also now used for DNA sequencing. These polymerases have specific characteristics that are suitable for special needs. For a comprehensive update on these enzymes and their use in DNA sequencing, see reference 2.

High-quality commercial sequencing kits are relatively inexpensive, provide quality control, and come with detailed descriptions of protocols and free access to a host of technical tips. For example, kits include the dNTP-ddNTP mixes that have been tested for quality and proper run length for the reaction products in the gel. We have successfully used the kit marketed by U.S. Biochemicals for many years; however, other products may be just as effective. It is strongly recommended that laboratories use kits

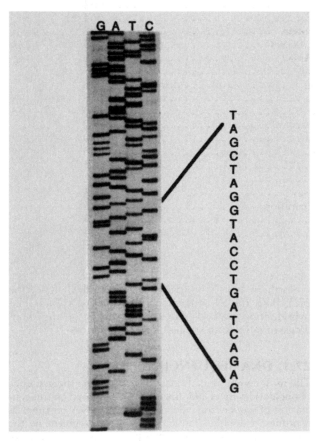

FIGURE 1 Autoradiogram of a sequencing gel. A central portion of the gel approximately 100 bases from the primer is shown. Lanes are identified at the top, and a portion of the sequence is indicated at the side.

from U.S. Biochemicals or competing manufacturers for applications in which they need to do manual sequencing and gel electrophoresis. Since the kits include detailed description of protocols, these descriptions will not be reproduced here.

27.1.1.2. DNA Templates

The most critical aspects of both manual and automated sequencing are the quality and quantity of the template. dsDNA obtained from plasmids is the most common template. Methods for the preparation of plasmid DNA are described in chapter 30. Plasmid DNA preparations should be free from RNA. Plasmid DNA prepared by using mini preparation kits can work very well; however, care must be taken to ensure that RNA is completely digested and that contaminants are minimized.

Several factors should be considered in preparing template DNA.

1. The bacterial strain used to propagate the plasmid can affect the template quality. A number of strains that lack a functional *endA1* gene (encodes DNA endonuclease), including *E. coli* DH5α, DH10B, TOP-10, XL1-Blue, and XL10-Gold, have been shown to produce good results. Other commercially available *E. coli* strains, for example, JM101, HB101, and TB1 (54), should be avoided since they release endonuclease and large amounts of carbohydrates that can greatly reduce the template quality or recovery.

2. Cells should be cultured properly. Growth for 16 to 18 h at 35°C in Luria-Bertani or YT medium works well, and for most plasmids and strains, the yield of DNA from a miniprep is more than sufficient for DNA sequencing. If DNA yields are low, the amount of culture used in the miniprep may be doubled or a "midiprep" can be done. Alternatively, a super-rich medium such as superbroth or 2× YT may be used, but growth should be monitored so that cultures do not overgrow and begin to lyse. When more cells are used, the prep columns may become overloaded or the amount of contaminants may increase and interfere with the sequencing reactions. Resuspending the cells in a salt solution such as SSC (1× SSC is 0.15 NaCl plus 0.015 M sodium citrate) and pelleting a second time will reduce the levels of some contaminants. Low yields may also occur if the proper antibiotic concentration is not maintained. Ampicillin will degrade during storage and after exposure to moisture (such as that from repeated opening when the bottle is cold). In the absence of selection, cells will begin to lose the plasmid.

3. Template preparation methods must produce sufficient DNA with minimal contaminants. Although you may use very inexpensive methods such as boiling preps, to obtain the most consistent, long-sequence reads, it is recommended that proven plasmid kits be used. QIAGEN (http://www1.qiagen.com) and Invitrogen (http://www.invitrogen.com) market miniprep kits that work well for both manual and automated sequencing. For QIAGEN kits, we find that an additional wash step before the elution of the DNA can improve results. Modest amounts of RNA contaminating the sample can be tolerated; however, determining the concentration of DNA in the sample by using a spectrophotometer will be impossible. Salts or ethanol left over from the elution step will inhibit the polymerase, especially in performing cycle sequencing. The presence of contaminants may be due to the improper removal of supernatants, the precipitation of salt by ethanol, the overloading of columns, or failure to fully dry

the sample. Additional washes with binding solution can help remove cellular contaminants, additional wash buffer can reduce salts, and additional 70% ethanol washes can reduce salts and contaminants. Resuspend the DNA in water or a 1/5 dilution of Tris-EDTA (TE) buffer (Table 1). TE can inhibit *Taq* polymerase and cause shortened runs. Store the DNA samples frozen at ≤ −20°C for future use. The average yield of DNA should be 10 to 20 μg for a miniprep.

TABLE 1 Preparation of buffers and solutions

Buffer or solution	Ingredient or step	Amt
NaTMM buffer (10×)	2 M Tris-HCl (pH 8.0)	100 μl
	2 M NaCl	50 μl
	1 M MgCl$_2$	200 μl
	14.2 M 2-mercaptoethanol	14 μl
	Water	1,636 μl
Luria-Bertani medium	Bacto tryptone	10 g
	Bacto yeast extract	5 g
	NaCl	5 g
	Water	1 liter
	Adjust pH to 7.5	
Polyethylene glycol solution	Polyethylene glycol 6000 or 8000 (Carbowax)	15 g
	NaCl	14.6 g
	Water	To 100 ml
S1 buffer (10×)	3 M potassium acetate (pH 4.6)	1.1 ml
	5 M NaCl	5 ml
	Glycerol	5 ml
	ZnSO$_4$	30 mg
Taq sequencing buffer (5×)	1 M Tris-HCl (pH 4.6)	1.1 ml
	1 M MgCl$_2$	0.1 ml
	Water	1.4 ml
TBE buffer (5×)	Tris base	60.55 g
	Boric acid	25.8 g
	Disodium-EDTA	1.85 g
TE buffer	5 M NaCl	0.2 ml
	2 M Tris HCl (pH 8.0)	0.5 ml
	0.2 M disodium-EDTA (pH 8.0)	50 μl
	Water	To 100 ml
TES buffer	5 M NaCl	6 ml
	2 M Tris-HCl (pH 8.0)	5 ml
	0.2 M disodium-EDTA (pH 8.0)	0.5 ml
	Water	To 100 ml

4. PCR products also may be sequenced directly; however, care must be taken to purify the products away from contaminating deoxynucleotides, primers, and other interfering molecules. The contaminating primers can especially interfere with cycle sequencing reactions. Also, the PCR should produce a single product. These problems can be avoided by gel purifying a single band from the PCR. The gel slice containing the band is purified by using a gel extraction kit (QIAGEN and Invitrogen kits work well).

5. Determine the approximate amount of DNA by running samples and standard DNA solutions (DNA quantification standards may be obtained commercially) on an agarose gel and comparing intensities of ethidium bromide-stained bands in the gel (for agarose gel electrophoresis and staining procedures, see chapters 15 and 30). The concentrations of PCR products should also be checked after gel purification. Although spectrophotometer readings can be used for plasmid preps, any RNA or carbohydrates in the sample will make the readings inaccurate. Once the concentration has been determined by using gel electrophoresis, readings from the spectrophotometer can be compared and calculations can be adjusted if necessary. If the minipreps are done consistently by using the same plasmids, strains, and media, spectrophotometer readings can then be used. The spectrophotometer also provides an estimate of protein contamination. An A_{260}/A_{280} ratio of less than 1.7 indicates contamination. Finally, accurate, sensitive measurements of DNA concentrations are made with a fluorometer by using dyes such as Picogreen (Invitrogen) and Hoechst 33258 (Molecular Probes or Invitrogen).

27.1.1.3. Manual DNA Sequencing Reactions

DNA polymerase uses dNTPs as substrates for the extension of the primer that hybridizes with the single-stranded DNA template. ddNTPs are substrate analogs, the incorporation of which results in the termination of the polymerization reaction. If dGTP is used in the sequencing reaction mixture, upon electrophoresis anomalous separations of oligonucleotides occasionally result from the compression of G-C stretches present in the DNA, which forms a localized secondary structure. This problem can be alleviated by replacing dGTP with dITP or with 7'-deaza-2'-dGTP in the polymerization reaction mixture. The concentrations of dNTPs and ddNTPs are critical for the sequencing reaction; hence, they must be handled properly. The manufacturer's recommendations must be followed carefully, reagents must be kept on ice during use, and reagents that have been stored for a long period of time or that produce gradually poorer results should be discarded. For manual sequencing, four reactions are set up for each template and primer pair. Then a short labeling reaction is done to produce labeled DNA fragments from the template and primer, dNTPs, and a labeled dNTP. This step is followed by a termination reaction that is started by adding a mix containing all four dNTPs and only one ddNTP. Each of the four tubes gets either the A, C, G, or T ddNTP. Since the elongation of the DNA chain terminates with the incorporation of one molecule of ddNTP, the oligonucleotides in the sequencing ladder will end with the specific ddNTP used in the reaction. By using four different ddNTPs in separate reactions, and then separating the oligonucleotides extended from the primer in parallel on a polyacrylamide gel, the sequence of bases is read (Fig. 1). The fragments that terminate near the primer appear at the bottom of the gel. A readable sequence adjacent to the primer is usually not obtained, so primers should be designed at least 20 bases away from the sequence to be read.

The radiolabeled dNTP (usually an α-^{35}S-labeled dNTP) should be obtained fresh. ^{35}S labeling is preferred because it produces sharp bands upon exposure of the polyacrylamide gel to an X-ray film, and ^{35}S-labeled dNTPs have long half-lives (the ^{35}S half-life is 87.4 days), thus reducing waste. Overnight exposure of the dried gel to a phosphor screen (used with laser digital imaging systems) or to sensitive X-ray film (Kodak XAR) is sufficient to observe the sequencing bands.

For standard sequencing reactions with T7 DNA polymerase, the following materials and reagents are needed: template DNA (2 to 3 μg for a 3-to-6-kb plasmid), a DNA oligonucleotide primer (an oligomer of at least 18 deoxynucleotides; 10 ng/μl), and the appropriate sequencing kit components. The Sequenase™ version 2.0 DNA sequencing kits from U.S. Biochemicals include Sequenase version 2.0 DNA polymerase. It is highly processive and able to efficiently incorporate nucleotide analogs (ddNTPs, dITP, 7-deaza-dGTP) and is not easily impeded by template secondary structures. Reactions are done in two basic steps. First, the polymerase, primer, and nucleotides are added in a short labeling reaction. Then the ddNTP mixes are added (for each template, four reactions each with a mix including one of the four ddNTPs) to terminate the chains. Occasionally, weak bands may occur with prolonged reaction times or with the use of dITP in the sequencing reaction. The addition of pyrophosphatase can prevent weak band intensities brought on by sequence-specific pyrophosphorolysis catalyzed by the polymerase (U.S. Biochemicals Sequenase protocol). Specialized kits can be used for difficult templates with significant secondary structures. In one kit, the substitution of 7-deaza-dGTP for standard dGTP helps to resolve gel compressions that may arise from secondary structures of the DNA in the gel.

Manual sequencing protocols using *Taq* DNA polymerase (cycle sequencing) require much less of the template (0.1 μg). The reactions are done in two basic steps: thermal cycle labeling-amplification and thermal cycle termination by the ddNTPs. Primers of 18- to 30-nucleotide oligomers are used. Thirty to 35 thermal cycles are performed with the inclusion of the labeled deoxynucleotide. In each cycle, the temperature is first raised to 95°C to denature the template and then it is reduced to 55°C to anneal the primer, followed by an extension step at 70°C. Depending on the melting temperature of the primer and the empirical results of the sequencing run, the annealing step may be adjusted to obtain better annealing or to reduce nonspecific priming. Then the four termination mixes are added individually to four tubes per template-primer pair, and an additional 30 to 35 cycles are performed. These cycles produce a set of single-stranded DNA fragments that are terminated at each respective ddNTP, either A, C, G, or T, in a particular reaction tube. The reaction mixtures are cleaned up and applied to a sequencing gel as described for standard sequencing, producing the same type of gel image (Fig. 1). Cycle sequencing has the advantage that the higher temperature of the reaction reduces the effects of template secondary structure.

27.1.1.4. Gel Electrophoresis

Gel apparatuses of various lengths and widths are commercially available. Larger and wider gels are difficult to handle. Apparatuses from Bio-Rad work well. Gels of 40 cm (length) by 20 cm (width) by 0.3 to 0.4 cm (thickness) can be conveniently handled. Plastic spacers placed between glass plates determine the thickness of the gel. In order to read the maximum number of nucleotides from a gel, wedge-shaped spacers (tapered from 0.3 to 0.8 cm) may be used (1, 34); however, they are more difficult to dry. Polyacrylamide gradient gels (4 to 18%) or buffer gradient gels have also been used (4, 24, 43). Buffer gradient gels with different ionic concentrations of the electrode buffer are the easiest to handle among all different kinds of gradient gels. Two different kinds of combs are used for forming wells in gels: those producing conventional square or rectangular wells that are spaced a few millimeters apart from one another and "shark's tooth" combs forming wells separated from one another by minimum spacing. In order to form conventional wells, the comb is inserted into the gel mold before the gel solution polymerizes. In contrast, the flat end of the shark's tooth comb is inserted into the gel mold prior to the polymerization of the gel solution. The comb is removed after the gel is polymerized, the comb is rotated 180°, and wells are formed by slightly inserting the sharp teeth of the comb into the gel. The loading of samples onto the shark's tooth gel is more difficult, and leaking of samples from one well to another well is a common problem. On the other hand, the reading of sequences from such a gel is improved since the lanes run very close together.

For the preparation of sequencing gels, use a solution containing 38% (wt/vol) acrylamide and 2% (wt/vol) N,N'-methylenebisacrylamide as a stock solution. Acrylamide is a neurotoxin, so wear gloves while handling it, and use a mask while handling powdered acrylamide. Make the solution in warm water (40 to 50°C), deionize by stirring with 5 g of Amberlite MB-1 resin (from Mallinckrodt or Sigma Chemical Co.) or an equivalent monobed resin/liter for a period of 30 min or longer, and then filter through Whatman no. 1 filter paper (preferably fluted). The stock solution can be stored in the dark at 4°C for months (12 months). Ammonium persulfate and TEMED (N,N,N',N'-tetramethylethylenediamine) are used as catalysts for the polymerization reaction. A stock solution of ammonium persulfate (10% [wt/vol] in water) can be stored for up to a week at 4°C.

The volumes of the solutions required for a gel depend on the sizes of the glass plates and the thickness of the spacers used. For casting gels, follow the recommendation of the supplier of the gel apparatus. Polymerized gels can be stored clamped with large binder clamps and wrapped with plastic wrap at room temperature to prevent dehydration from the area where the comb is inserted. Gels should not be stored for more than 24 h under the above-described conditions. Use clean glass plates to prevent the trapping of bubbles in the gel sandwiched between the glass plates. The glass plates are best cleaned with warm water without detergents or acids immediately following electrophoresis. Spray ethanol on the washed glass plates, and wipe with a few Kim wipes. Applying a layer of silicone to one side of one of the plates (with Sigmacoat [Sigma Chemical Co.] or an equivalent chemical) facilitates the separation of the gel that is sandwiched between the glass plates. A protocol for the preparation of a 7% gel (20 by 40 by 0.3 cm) is given below.

1. Materials and solutions required are as follows: glass plates, spacers, combs, large binder clamps, 40% acrylamide-bisacrylamide solution (38:2), 10% ammonium persulfate, TEMED, ultrapure urea, and 5× Tris-borate-EDTA (TBE) (Table 1).

2. Two glass plates (one notched) are put together with spacers in between the plates on the longitudinal edges and at the bottom (optional). The spacers must be placed on the siliconized surface of one of the glass plates. Use clamps to hold the plates and spacers together. The bottom can be sealed by using tape (3M yellow electrical tape or its equivalent) instead of the bottom spacer. The bottom can also be sealed by inserting a piece of filter paper in between the plates and soaking the filter paper with a mixture of 12 to 18% acrylamide solution and high concentrations of catalysts (51).

3. For casting two 7% gels, mix the following in a beaker: urea, 33 g; 40% acrylamide solution, 13.1 ml; 10% ammonium persulfate, 0.38 ml; 5× TBE, 15 ml; and water to 75 ml. The addition of warm water (about 50°C) will help in dissolving urea. After the solutions are thoroughly mixed, filter the solution through a fluted Whatman no. 1 filter paper if you see any suspended particles (some batches of urea contain insoluble particles). Cool the solution on ice, add 38 µl of TEMED, and mix by swirling. Cold solution will polymerize slowly in the presence of TEMED, allowing sufficient time to pour the gel solution into the gel mold.

4. Draw the above-described solution into a 25-ml pipette or a hypodermic syringe. Hold the glass plates at an angle (45°), and pour gel solution through one corner of the gel mold at a uniform rate. If any air bubbles are trapped, the plates are not clean. However, air bubbles usually can be removed by tapping on the upper plate (the wooden handle of a large spatula works well for tapping). When the mold is filled, lay the plates on the top of a rubber stopper in a near-horizontal position. Insert the comb. Place clamps holding the two glass plates together in the area where the comb is inserted. Allow the gel to polymerize over a period of 1 to 2 hours at room temperature before electrophoresis. Efficient polymerization can be achieved in 30 min at 37°C.

5. When ready to use the gel for electrophoresis, remove the clamps, the bottom spacer (or the piece of sealing tape), and the comb. Wash the top part of the gel with 1× TBE. Remove any polymerized gel particles trapped in the wells by forcing 1× TBE through a Pasteur pipette (with one end drawn to a capillary shape). If a shark's tooth comb is used, insert the comb with the teeth just penetrating the acrylamide. Make sure all the teeth have penetrated, but do not insert so far that the surface of the well is bowed.

6. Pour electrode buffer (1× TBE or a different buffer for buffer gradient gel) into the lower buffer reservoir. Place the plates vertically (unnotched plate facing outside), and secure them to the gel apparatus by using clamps or screws. Air bubbles between the lower edge of the gel and the lower electrode buffer should be avoided (forcing of buffer through a Pasteur pipette may help remove any trapped air bubbles). Fill the upper buffer reservoir with 1× TBE, making connection between the upper part of the gel and the upper electrode buffer.

7. Prerun the gel for 30 min to warm up to 40 to 45°C (set the power as recommended by the supplier of the gel apparatus).

8. Disconnect the power, and wash urea from the wells by squirting 1× TBE with a Pasteur pipette or syringe.

9. Load 1 to 3 µl of samples into each well by using a pipette and capillary tips (commercially available as "flat" tips or "gel-loading" tips). Samples should be heated to 70°C before loading. Load in the order G, A, T, and C for each template. Rinse the pipette tip by pipetting buffer from the lower reservoir in and out between the loading of each sample, or use a separate tip for each sample. When loading samples with reactions from many templates, load samples for one to three templates and then electrophorese for 3 min before loading another set of samples. Remove urea from the wells, as described above, between each loading.

10. Turn on the power, and run the gel for about 20 min after the bromophenol blue dye runs out of the bottom of

the gel. Previously loaded samples can be loaded again in adjacent wells at this time, and electrophoresis is continued for an equal amount of time to obtain the maximum sequence information for each template from a single gel. The choice of using multiple gels with different run times or staggered loading of one gel depends on the number of templates and the width of the gel (number of wells) used.

11. After electrophoresis is complete, disassemble the glass plates from the electrophoresis apparatus, wash the plates with tap water, and lay the plates flat on a sheet of absorbent paper (for safe handling of radioactive materials). Pry the plates apart by starting at one corner. Observe carefully and lay down the plate to which the gel remains attached. If desired, cut about the top 3 cm of the gel with a pizza cutter (works best) or a scalpel and discard the top portion of the gel. This part of the gel contains the maximum amount of radioactivity, and sequences cannot be read in this area. Note that the buffer in the lower reservoir is radioactive and should be disposed of properly.

12. Place the gel with the support of the attached glass plate in a liter (or more) of 5% acetic acid bath (a plastic tray can be used for making the acetic acid bath). Soak the gel in acetic acid for a period of 30 min at room temperature without shaking. Soaking with acetic acid helps fix the oligonucleotide bands in the gel and removes urea.

13. Carefully remove the gel with the support of the glass plate from the acetic acid. Soak excess acetic acid from the top of the gel by placing two to three layers of tissue papers (Kim wipes) over the gel and slowly rolling the tissue papers off the gel from one corner. Place a piece of a Whatman 3MM paper on the gel, and press the paper slowly to adhere to the gel surface.

14. Invert the plate with hand support onto the Whatman paper, and slowly peel the plate away from the gel. Note the orientation of the gel to ascertain the proper order of the samples loaded. Place a piece of plastic wrap on the gel, and dry the gel under a vacuum at 70 to 80°C. Gel-drying apparatuses are commercially available.

15. Remove the plastic wrap, and expose the dried gel overnight to an X-ray film (Kodak XAR-5 or its equivalent) for autoradiography or expose it for at least 4 h to a phosphor screen (be sure the screen is sensitive to α-^{35}S) for laser digital imaging.

16. After the image is developed, sequences can be read from the film or computer. Mark the lanes in the appropriate order of loading. Start reading from the bottom of the film in ascending order of bands present in the lanes (G, A, T, and C). Note that the sequence read is complementary to the sequence of the template DNA strand.

27.1.2. Automated DNA Sequencing

27.1.2.1. Overview

Automated fluorescence sequencing utilizes the same general Sanger chain termination methods as manual sequencing; however, the methods are substantially modified to greatly increase speed and reduce the cost per sample. As in manual cycle sequencing, a primer is hybridized to the DNA and the template is amplified by using 35 reaction cycles (note that this is not PCR since only one primer is used). Most sequencing facilities use AmpliTaq and reagents marketed by Applied Biosystems. The enzyme and reagents are premixed in the present ABI BigDye terminator version 3.1 reagent kit. A set of four fluorescently labeled ddNTPs representing each of the four bases, AGCT, is present in each reaction mixture, along with all dNTPs. Unlike in manual sequenc-

ing, all four labeled dideoxynucleotides will be detected in a single tube. The different bases are labeled with fluorescent probes that each emit light at a unique wavelength. As the DNA fragments are synthesized, termination and labeling occur when a labeled ddNTP is encountered. The ratio of ddNTPs to dNTPs is set so that these termination reactions occur evenly across all possible positions. All reactions are done in 96-well plates, greatly increasing the throughput. The reaction mixtures are cleaned up to remove unincorporated ddNTPs and other contaminants by using 96-well clean-up plates and then put into an automated DNA sequencer.

The most popular present capillary sequencing instruments are the ABI 3730 and 3730xl, which are capable of processing hundreds of samples per day with unattended operation. The sequencers are equipped with an autoloader, a sample injection system, and an array of 48 or 96 very thin capillaries. A separation matrix (the latest is POP7) is automatically injected into the capillaries before each sample run. Then an electrokinetic injection system draws samples from the 96-well plate and the DNA is carried by an electric current into the capillary matrix.

Although this remarkable system has greatly reduced labor, it does have certain limitations. Templates must be cleaner than with previous gel-based systems. Any negatively charged contaminants compete with the DNA injection, reducing the signal and possibly broadening bands. TE and salts eluted with the DNA from miniprep kits are common culprits. It is highly recommended to elute with water or a 1/10 dilution of TE. Any silica from miniprep kits carried over with DNA can be especially damaging to the capillaries. Since the entire block of capillaries must be replaced when even a few are blocked, contaminated DNA preps can greatly increase the cost of sequencing for all who use a facility. Even moderate levels of contaminants will reduce the lifespan of the capillaries.

As the fluorescently labeled DNA fragments reach a detection cell at the end of the capillaries, a laser beam excites the fluors. The subsequent emissions are captured by a charge-coupled device camera, and the software interprets which color of dye is passing through at any time. The software evaluates the size, spacing, and other properties of the peaks and identifies the base sequence.

27.1.2.2. Interpreting Chromatograms

The output provided to the user for each run consists of a simple text file with the sequence identified by the software and a much larger file with the chromatogram. ABI supplied the free viewing and editing program Editview for the Macintosh, and later Chromas for personal computers became available. Editview is no longer supported and is not compatible with the Mac OSX system. A new program called FinchTV from Geospiza (http://www.geospiza.com/finchtv) is an excellent program for viewing chromatogram traces on Linux, Mac OSX, Windows, and Solaris systems.

Bases are shown in colors: black, G; red, T; blue, C; and green, A. The first things to notice are the peak heights and spacing and the background. An example of a reasonably good sequence is shown in Color Plate 12A. The part of the sequence close to the primer is often not readable, so primers should be designed 20 to 40 bp away from the region of interest. For 500 to 700 bases, the sequence should be more than 99% accurate, and then as the peaks spread and lose intensity, the error rate increases. Even with 900 to 1,100 bases, a sequence with occasional errors is obtained if the template is of good quality. Due to the higher error rate, it is unadvisable to design primers beyond 700 bp from single-run data. With the use of a pure template, appropriate primers, and good sequencing procedures, the peak heights are relatively even and the background signal is almost non-existent. When the computer program is unable to make an identification, an N is placed at the position in question. Experienced users can often detect patterns in the data and accurately extend the base identification beyond the limits of the software, but as the base-identifying algorithms have improved, the benefit from peering over chromatograms to determine a few additional bases has dwindled.

27.1.2.3. Troubleshooting

Color Plate 12B illustrates a common problem that has nothing to do with the quality of the sequencing reactions. Throughout the entire sequence, two bands at each position are observed. This is a clear indication that the template is a mixture of two plasmids. Excess DNA is often used for bacterial transformations, and this practice often results in multiple plasmids entering the same cell. The remedy is to simply plate the strain and isolate single colonies. Similar results (although usually with more bands observed) may also appear if there are multiple priming events. Either design a new primer or ask if the sequencing reaction can be run with a higher annealing temperature. Most facilities perform all standard cycle sequencing at one temperature, about 55°C; however, they may do runs with problem samples at other temperatures. In Color Plate 12C, there is also a site with multiple peaks, near position 145. In this case, since two bands appear only in two positions, it is likely that the template is derived from a heterozygote. This scenario is most common in the sequencing of PCR products.

A total failure to obtain a sequence is shown in Color Plate 12D. This result may be caused by improper design or degradation of the primer, loss of the primer binding site on the plasmid, a greatly reduced concentration of intact DNA, or the inhibition of the reaction by contaminants. The result in Color Plate 12E is an example of inhibition by a high salt content. Bands are broadened, and base intensities are quite variable. The high level of background signal in Color Plate 12G was not sufficient to interfere with most base identifications in the early part of the run; however, the readable sequence is very short. High levels of background signals occur with multiple priming or insufficient purification of PCR products, poor-quality plasmid templates, or primers containing shortened contaminants from the oligonucleotide synthesis. Very large bands in the 60- to 80-base region are "dye blobs" resulting from contamination with unincorporated labeled ddNTPs. If the sequence is weak due to a low template concentration or contamination, the excess unincorporated nucleotides overwhelm the purification procedure. If the dye blobs persist even with a good sequence signal, the facility should consider alternative purification procedures.

Truncated sequences such as that shown in Color Plate 12F also may result from a number of problems. G-C-rich regions produce a high frequency of hairpin structures when single stranded. During the sequencing reaction, the hairpin forms and may block polymerase, causing an abrupt drop in the intensity of the sequence signal. A gradual loss of the signal may also occur if the blockage is only partial. The ABI BigDye 3.1 formulation has greatly reduced the effects of hairpins, but they are still seen regularly. There are several common remedies, including adding denaturants such as dimethyl sulfoxide to the sequencing reaction mixtures, increasing the temperature of the cycle sequencing reactions, or sequencing from the opposite strand.

27.2. DNA-PROTEIN BINDING

A variety of techniques are available to study the factors that influence the regulation of a gene. If sequence-specific DNA-binding proteins are thought to be involved, it is of interest to determine the locations of the binding sites and the biochemical properties of the protein-DNA interactions. In bacteria, these sites can be close to the transcription start site or they may be located more than a kilobase away. An analysis in vivo of variants with deletion and point mutations can be used to obtain information on the locations of the sites, but ultimately biochemical assays must be used. To study in vitro the factors that influence transcription, several DNA-binding assays can be employed. If the location of the suspected binding site is known within a few hundred base pairs, nuclease or chemical protection (footprinting) assays can be used (see section 27.2.2). If the location of the site is not suspected, or if the protein is of low abundance and purity, use gel mobility shift or chromatin immunoprecipitation (ChIP) assays. The recently developed ChIP assay is capable of simultaneously isolating for one protein all protein-binding sites within a genome. The basic aspects of the gel mobility shift assay using purified protein are described first, and then the cell lysate assay is discussed. Next, protection assays are considered, followed by a review of the ChIP assay. Further discussion of all the assays may be found in reviews (10, 27, 46) and method manuals (2, 7).

27.2.1. Gel Mobility Shift Assay

The gel mobility shift assay (also called the gel retardation assay) is based on the differences in the degrees of electrophoretic mobility between nucleic acid fragments and nucleic acid-protein complexes (14, 19). Although the assay has been used successfully to study binding to RNA (52), this discussion will be limited to the most common use, the study of binding to dsDNA. A solution of protein is mixed with DNA fragments in an appropriate binding buffer. After allowing the protein to bind, the mixture is applied to a nondenaturing gel, and following a period of electrophoresis, the positions of the DNA fragments are determined. If the protein binds the DNA stably, the increased mass and the alteration in the charge of the complex will generally cause it to migrate more slowly than the DNA alone. The relative amounts of free DNA and complex bands can be analyzed quantitatively to obtain information on the concentrations of the reactants, equilibrium and kinetic binding constants, stoichiometry, and some aspects of the structure of the complex (12, 14–17, 19, 41, 52, 59, 67).

The greatest advantage of the gel shift assay over other methods is its versatility. Specific protein-DNA interactions can be detected by using a mixture of DNA fragments, such as a digest of a whole plasmid. This is valuable for quickly locating binding sites within a large segment of DNA. By using a single DNA fragment, interactions with multiple proteins can usually be resolved since different proteins or different stoichiometries of binding yield characteristic shift positions. Thus, the binding of multiple activator or repressor proteins can be distinguished from the binding of one species, which in turn can be distinguished from the binding of a large molecule such as RNA polymerase (32, 59). This property allows for functional assays in studies of protein interactions at promoters. Even in a crude cell lysate, the specific binding of one protein can be detected if appropriate controls are performed to demonstrate specificity. This is perhaps the most powerful aspect of the assay, since it provides a rapid, convenient method to quantify protein from new constructs or to assay fractions from purification schemes. The method can also be used preparatively to obtain the bound and free fractions of the DNA, to isolate bound proteins from a crude mixture, or to obtain the protein-DNA complexes after treatments such as UV cross-linking and chemical modification (27, 46).

All these applications rely on the basic finding that protein-DNA complexes are more stable in the gel than in solution. This observation has been explained on the basis of the "caging" effect of the gel. The complexes are not more stable; dissociated protein reassociates at a greatly increased rate so that the reactants do not have a chance to migrate apart. Since the electrostatic component of the binding interaction is stronger at lower salt concentrations, an additional stability of complexes may be obtained by using low-salt-content gels.

27.2.1.1. DNA Fragments
The most common DNA templates used in mobility shift assays are restriction fragments, synthetic oligodeoxynucleotides, and PCR products. Best results, easily observable shifts, are obtained with fragments of less than 1 kb p, although assays with larger fragments and even whole plasmids have been successful. It should be kept in mind that the larger the DNA fragment, the smaller the shift in the mobilities of the complexes and the greater the amount of nonspecific protein binding. If the fragment of interest can be cut by restriction enzymes, leaving only one or two large vector fragments, the mixture may be used in the assay; otherwise, the fragment should be isolated by gel electrophoresis. Synthetic oligodeoxynucleotides and fragments produced by PCR should also be purified to eliminate any contaminants that may interfere with binding. Mixtures containing fragments that may comigrate with the protein-DNA complexes should be avoided. Unlabeled DNA can be detected by soaking the gel in 1 μg of ethidium bromide/ml after electrophoresis. Labeling the fragments with ^{32}P greatly reduces the necessary amounts of reactants and is the method of choice. One of the standard end-labeling or internal labeling protocols may be used; however, protocols such as nick translation and random primer labeling should be used with caution as they may leave partially filled-in DNA that will significantly alter quantitative measurements and produce smeared bands. For most applications, labeling to the extent of producing high specific activity is not necessary since as little as 200 cpm is sufficient for the detection of bands upon overnight exposure to a phosphor screen and imaging using a laser digital imager or overnight exposure to X-ray film with an intensifying screen. After labeling, the unincorporated nucleotides may be removed by chromatography on a 1-ml Sephadex G-50 column or by similar methods. The purification or removal of unincorporated nucleotides may be done by using DNA clean-up kits (QIAGEN, Invitrogen, or Promega), but do not use kits that employ loose silica beads. Frequently, a small portion of the DNA remains bound with silica and is observed as a band that fails to migrate into the gel. Choose kits that use a membrane-bound matrix. Gel purification can be used to isolate the fragment from vector fragments as well as free nucleotides. The DNA should be stored in a buffer such as 10 mM Tris (pH 7.5)–50 mM NaCl–1 mM EDTA. Storage of small fragments in low-salt-content

buffers such as TE may allow some denaturation of the DNA, resulting in the appearance of a band migrating more slowly than the intact, dsDNA exactly where a protein-DNA complex is anticipated.

27.2.1.2. Gels and Buffers
Polyacrylamide and agarose gels have both been used successfully for mobility shift assays. Polyacrylamide gels may be 4 to 10% with 0.33 to 0.7% bisacrylamide, depending on the sizes of the DNA fragment and the protein. For most regulatory proteins and DNA fragments of fewer than 600 bp, 4 to 6% acrylamide is a good concentration for pilot experiments. NuSieve agarose (Cambrex Corp.) at 3 to 4% may be substituted in most cases since it has sieving properties similar to those of polyacrylamide. Regular agarose is generally useful only when the protein is large or multiprotein complexes are bound.

Many different gel buffers have been used. In the first gel mobility shift experiments, very-low-salt-content TE (10 mM Tris, 1 mM EDTA, pH 7.5) buffer was used to maximize protein binding during electrophoresis (11, 16). The low buffering capacity requires that a circulator be used during electrophoresis; in as little as 30 min without circulation, the pH of the gel can drop to ≤5. Usually, 1× or 2× TBE (45 mM Tris-borate, 1 mM EDTA, pH 8.3) and TAE (40 mM Tris-acetate, 15 mM Na acetate, 1 mM EDTA, pH 8.0) may be substituted; these buffers have the advantage that no circulation is necessary. Care should be taken with TBE, however, since borate may significantly affect the DNA-binding properties of protein. The gel buffer need not be the same as the buffer used for the binding reaction. Some cofactors required for the stable binding of protein to DNA may also be necessary in the gel, such as metal ions like Mg^{2+} or allosteric effectors, for example, cyclic AMP (cAMP) for the E. coli catabolite gene activator protein.

27.2.1.3. DNA Labeling
The following protocol is for filling in the ends of restriction fragments with 5′ overhangs. End labeling with T4 polynucleotide kinase (section 27.3.1.3) may be used for 5′ overhangs or blunt-end fragments. For fragments with 3′ overhangs, T4 polymerase can be used to digest the ends back and then to fill in with labeled nucleotides (49). The incorporation of radioactivity may be measured by placing a portion of the sample in the scintillation counter without using scintillation fluid and using the discriminator setting for tritium or an open setting. This method of counting by using the Cerenkov effect is about 50% as efficient as counting by using ^{32}P with a scintillation fluid, but the method is quick and it does not destroy the counted sample.

1. Use an appropriate restriction enzyme in a volume of 40 μl to digest 5 to 10 μg of plasmid DNA containing the region of interest (see chapter 30 for plasmid purification and handling). Inactivate the enzyme by heating to 70°C for 10 min. In the case of a heat-resistant enzyme, remove by phenol extraction and ethanol precipitation or with a clean-up kit. Tables for the heat inactivation of restriction enzymes may be found in the catalogs of manufacturers such as New England Biolabs.

2. If a low- or moderate-salt-content buffer or a 1× concentration of a universal buffer (50 mM potassium acetate or glutamate) is used for the restriction digestion, proceed to step 3. If a high-salt-content buffer is used or if the sample was treated with phenol, remove the buffer by ethanol precipitation with a 1/10 volume of 3 M Na acetate, pH 5, and 2.5 vol-

umes of ethanol at −20°C overnight or −70°C for 30 min. Alternatively, isolate the DNA by using a silica method. If a silica method is used, follow the manufacturer's instructions but repeat the final elution step and carefully draw off the supernatant to reduce the amount of silica carried over into the sample. Resuspend the DNA in 40 μl of TE (Table 1).

3. Add 50 μCi of an α-^{32}P-labeled dNTP and 1 μl of a 10 mM stock of each of the other dNTPs as necessary to fill in the fragment ends. The nucleotides used will depend on the restriction enzyme, but it is best that the labeled nucleotide not be the last to be incorporated. If the DNA is in TE, add 5 μl of 10× polymerase buffer (500 mM Tris-HCl [pH 8.0], 100 mM MgCl$_2$, 10 mM dithiothreitol [DTT], 500 mg of bovine serum albumin [BSA]/ml). Add 2 U of Klenow fragment DNA polymerase, and incubate at room temperature for 15 min.

4. Add 2 μl of a solution of all four nucleotides (5 mM each), and continue the incubation for 5 min.

5. Ethanol precipitate as described in step 2, and dry. Centrifuge 10 min in a microcentrifuge, remove the supernatant, add 70% ethanol to the pellet (without disturbing it), centrifuge 2 min, remove the supernatant, and dry the pellet in air or in a vacuum concentrator. Resuspend the DNA in 20 μl of TE, and add 2 μl of 10× loading buffer (50% glycerol, 0.05% bromophenol blue).

6. Purify the fragment of interest by electrophoresis on agarose or NuSieve agarose, depending on size. NuSieve agarose (3% NuSieve–1% agarose mixture) will separate small fragments in the range of 50 to 500 bp well. After electrophoresis in the presence of 0.6 μg of ethidium bromide/ml, visualize the band with UV light and make an incision in the gel just ahead of the band. Place a small piece of NA45 membrane (Schleicher & Schuell, Keene, NH) into the incision down to the gel tray. Electrophorese the DNA onto the membrane at two times the voltage used for the gel run; take care not to run additional bands onto the membrane. Remove the membrane, wash in 1 ml of low-salt-content buffer (10 mM Tris [pH 7.5], 100 mM NaCl), and elute for 45 min in 300 μl of high-salt-content buffer (10 mM Tris [pH 7.5], 1 M NaCl) at 70°C according to the manufacturer's directions. Remove the membrane and precipitate the DNA by adding 1.5 μl of 1 M MgCl$_2$ and 2.5 volumes of ethanol, hold at −70°C for 30 min, centrifuge for 10 min, and wash the pellet with 70% ethanol. Resuspend in 100 μl of TES (Table 1) and store frozen. This method is simple and produces quite clean DNA with a specific activity of >5 × 10^6 cpm per μg of fragment.

Clean-up kits (Qiaex or Invitrogen gel extraction kits) also work well to isolate the DNA from NuSieve or standard agarose gels. For fragments of fewer than 100 bp, such as synthetic oligonucleotides, be sure that the clean-up kit is rated for oligonucleotides. The gel purification may be done before labeling, especially if several fragments are produced by the restriction digestion. In this case, remove the free nucleotides by column chromatography after labeling.

27.2.1.4. Gel Preparation

The binding assay can be done most easily on horizontal, "submarine" gels (28); however, since horizontal acrylamide gel apparatuses are not used widely, the following procedure is for a standard vertical system. It is best if the vertical gel apparatus has fittings for buffer circulation, but this is not essential if the running time is limited.

1. Cast a 1.5-mm-thick gel with standard size plates (usually 14 to 16 cm long) and a comb to produce wells 5 to 10 mm wide. Use 6% polyacrylamide–0.1% bisacrylamide in TAE buffer (40 mM Tris base, 15 mM Na acetate, 1 mM EDTA; adjust to pH 8.0 with acetic acid), which is filtered and degassed. A 20% polyacrylamide–0.33% bisacrylamide stock can be stored in the dark at 4°C for up to several months; store TAE as a 10× stock.

2. Polymerize with 20 μl of TEMED and 200 μl of fresh 10% ammonium persulfate for 50 ml of gel.

3. If available, set up a pump to circulate the gel buffer. The gel should be run in the cold for pilot experiments until the stability of the complex is known.

4. Soak the gel in running buffer (the same as the gel buffer) for several hours (preferably overnight) to allow the TEMED and persulfate to diffuse out. Add ligands necessary for the protein (such as cAMP for the cAMP receptor protein) at this time. For horizontal gels, soak for 1 h.

5. Prerun the gel at 5 to 10 V/cm for 1/2 to 1 h. If the buffer is not circulated, mix the buffer portions from the upper and lower tanks or replace the buffer before loading the samples.

27.2.1.5. Binding Reactions and Electrophoresis

1. Make an appropriate binding buffer. Usually, some salt is added to reduce nonspecific binding, components are added to stabilize the protein, and glycerol is added to allow layering on the gel. A typical buffer is as follows: 10 mM Tris, 50 mM KCl, 1 mM EDTA, 5% glycerol, 50 μg of BSA/ml, and 1 mM DTT. NP-40 (0.05%) may be added if the protein is unstable and/or aggregates. An easy method is to make a 5× concentrated stock with the first four items (store them frozen) and add the BSA, DTT, water, any additional factors, and DNA to a single tube just before use. Make up 10% more mix than required to allow for pipetting error.

2. Add enough DNA for 500 cpm (Cerenkov) for each sample (assuming a single labeled DNA fragment). Pipette 20 μl of the complete binding mix into 1.5-ml microcentrifuge tubes for each binding reaction.

3. Make up the binding mix without DNA for diluting the protein. Make 10-fold serial dilutions of the protein, and maintain on ice for a limited time. *Never vortex solutions containing protein;* mix gently by using the pipette tip or by flicking the tube with a finger. For a pilot experiment, alternate adding 1 and 3 μl of each dilution to reaction tubes. For a purified protein, the lowest target concentration can be estimated from the amount of DNA in each sample (generally less than 1 ng) by assuming 100% active protein. A protein at 1 mg/ml should require a dilution of 10^{-4} to 10^{-5}. One tube should have no protein added. Thus, the full range of concentrations can be assayed in 12 reactions.

4. After adding protein, incubate the reaction mixtures for 30 min at room temperature. Many proteins require only 5 to 10 min.

5. Gently add 1 μl of 0.05% bromophenol blue, pH 7.5 (the dye should be neutralized with Tris or other buffer before use). Turn off the power (and circulator if used), and load the samples carefully but quickly onto the gel. A consideration important for this assay is that the binding characteristics may be changed when the sample is mixed with gel-running buffer. To minimize this effect, samples are loaded with a minimum of mixing and frequently the samples are loaded with the current turned on. (*Caution:* take great care to avoid electric shock.) It takes about 1 to 3 min for the complexes to enter the gel. After the dye has entered the gel, the circulator may be turned on.

6. Electrophorese at 5 to 10 V/cm (usually 100 to 150 V) until the dye is one-third of the way to halfway down the

gel, 1.5 to 2 h). The dye should run with a DNA fragment of about 50 bp. The voltage should be adjusted so that minimal heating occurs.

7. Gently pry the plates apart by using a spatula, and pull the top plate away, leaving the gel on one plate. Gently press a piece of filter paper onto the gel, and pull it off the plate. Cover with a piece of plastic wrap, and dry in a vacuum dryer (1 h at 80°C).

8. Expose the dried gel to X-ray film. Expose overnight with an enhancing screen (such as DuPont Cronex) for 300 to 1,000 cpm (Cerenkov) per band. Place the screen in the film holder, then add the film, and finally add the gel (facing down). For shorter exposures, or if intensifying screens are not available, use an amount of DNA corresponding to 2,000 to 5,000 cpm per band. Exposure to a phosphor screen for 4 h to overnight will also provide a sufficient signal for the laser digital imager.

An example of protein titration using the gel mobility shift assay is given in Fig. 2. The relative mobility of the shifted band can vary widely with different combinations of protein and DNA. If only a slight shift is seen, the gel concentration can be increased. Running conditions may also greatly affect the mobility. For *trp* repressor-operator complexes, reducing the pH is necessary to noticeably lessen the mobility of the complexes (9). Thus, running a sample of protein alone (detected by silver staining of the gel) may show whether the protein runs far into the gel at the pH selected, and conditions can be altered to maximize the difference in the degrees of mobility of the protein and the free DNA. If the shifted band is near the top of the gel, repeat the assay with an agarose gel. Sometimes the free-DNA band disappears with increasing amounts of protein, but no complex band appears. In this case, the protein is probably dissociating during the run. A lower-salt-content gel, different binding conditions, or different running buffer may allow for more-stable complexes.

A common observation is that a single shifted band is observed at one protein concentration and multiple, higher bands are seen at higher protein concentrations, until all the DNA is near the gel origin at the highest concentrations. Since weak protein-DNA interactions are stabilized in the gel, nonspecific binding is readily detected. Once one molecule of protein is bound at its specific site, additional protein added to the reaction mixture results in the moderately stable binding of multiple proteins. Even if no specific binding occurs, nonspecifically bound protein-DNA bands may be observed. Therefore, it is important to carefully titrate the DNA and to characterize the binding with proper controls. One simple control is to add equal amounts of a labeled, irrelevant DNA fragment and a specific DNA fragment to the binding mix. The specific fragment should shift in position while the control fragment remains unchanged. This result is easiest to observe if the control fragment is somewhat smaller than the specific fragment.

The gel mobility shift assay can be used to obtain reasonably accurate quantitative information about the binding reaction. Several methods can be employed to determine the equilibrium constant for binding. For any of these methods, the concentration of the protein active in specific DNA binding must be determined accurately. The simplest method to accomplish this is to accurately determine the concentration of the DNA fragment containing a single, specific binding site and then to titrate the protein against the DNA.

27.2.1.6. Protein Concentration

1. Carefully measure the concentration of plasmid DNA containing the binding site of interest by absorbance at 260 nm against a blank with the same buffer. An A_{260} of 1 is equivalent to 50 µg of DNA/ml. The A_{260}/A_{280} ratio should be approximately 1.8. It is important that the DNA be free of contaminating RNA. RNase treatment is not sufficient since mono- and oligonucleotides will still be present even after CsCl purification of the DNA. Use DNA that has been purified two times by a CsCl-ethidium bromide gradient method or some other method that ensures the removal of RNA.

2. Digest 20 µg of the plasmid DNA with the restriction enzyme(s).

3. Label a small portion (1 µg) of the DNA by filling in the ends with the Klenow fragment polymerase as described in section 27.2.1.3.

4. Inactivate the enzymes in the labeled sample by heating to 70°C for 20 min or by phenol extraction. Remove the unincorporated label by purification on a 1-ml Sephadex G-50 column equilibrated with TE buffer or by a silica method.

5. Inactivate the restriction enzyme in the unlabeled sample by heating to 70°C. Do not use procedures such as ethanol precipitation that may cause a loss of DNA during handling.

6. Mix the labeled and unlabeled samples together. The molar concentration of the binding site is calculated from the molecular weight of the plasmid. Since the labeled sample is only 5% of the total, even a significant loss during handling will not affect the calculation. If the affinity of the protein for the specific site is thought to be weak or the nonspecific binding is strong, protein binding to the vector sequences will alter the results. In such cases, at least 1 µg of the fragment must be purified after restriction cutting and the fragment concentration must be determined spectrophotometrically. Do not add carriers such as tRNA or BSA since these will interfere with the readings.

7. Perform the binding assay as described in section 27.2.1.3 by using protein at a level increased by twofold or more for each sample. The DNA concentration should be high so that the measurements are above the dissociation

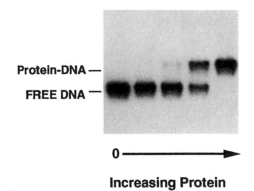

Protein-DNA —
FREE DNA —

0 ⟶

Increasing Protein

FIGURE 2 Gel mobility shift assay. Shown is an autoradiogram from an assay with a 200-bp DNA fragment corresponding to 500 cpm per sample run on a 6% polyacrylamide gel for 1 h at 8 V/cm. Lanes 2 through 5 contain samples with threefold increases in levels of protein (purified *E. coli* AraC protein). Film was exposed overnight at −70°C with an intensifying screen.

binding constant of the protein. For most regulatory proteins, a concentration above 10 nM is sufficient, and 100 nM would be sufficient for even very-low-affinity specific binding reactions. The binding can be quantitatively measured by densitometry of the gel as long as data are within the linear range for the film. Generally, the film is preflashed (cover a camera flash attachment with two layers of 3MM paper and flash from a distance of about 2 ft) so that the absorbance at 450 nm is about 0.1. The film is then tested by exposing a range of labeled standards. Alternatively, the bands may be cut from the gel and measured directly in a scintillation counter. As long as the excised gel fragments are nearly equal in size, quenching of the radiation for ^{32}P-labeled samples is not a problem. The best results are obtained by using a phosphorimager system. The linear range for this system is as much as 5 orders of magnitude, or 100-fold, greater than that for film.

8. For the most accurate calculations, plot the following: (1 − fraction of DNA not bound) versus the level of protein. The fraction of free DNA in each sample should be calculated by dividing the counts for the sample band by the counts for the band from a control lane with DNA but with no protein. Since some of the protein-DNA complex may be lost during the gel run, the fraction of DNA in the complex sometimes is not an accurate reflection of the amount of DNA bound in solution. The plot should be linear to a point well beyond that where 50% of the DNA is bound. At the point that 50% of the DNA is bound, the amount of protein is equal to one-half of the amount of DNA, and the active protein concentration can be calculated. It is quite common for highly purified proteins to be 10 to 50% active in specific binding. This method is also useful for estimating the amount of active protein from partially or unpurified samples (see below) as long as the impurities do not strongly interfere with the binding reaction.

27.2.1.7. Equilibrium Binding Constant

The simplest method to estimate the equilibrium binding constant is to perform a titration step as described in section 27.2.1.6; however, in this assay the DNA concentration is held at a level lower than the equilibrium binding concentration. If the midpoint for binding is obtained under conditions in which the protein is in vast excess relative to the DNA concentration, the equilibrium dissociation constant (k_D) is identical to the protein concentration. This principle follows from the simple binding relationship expressed here: $P_f + D_f = [P \cdot D]$, where P_f is the free-protein concentration, D_f is the free-DNA concentration, and $[P \cdot D]$ is the concentration of the complex. The K_D can be expressed as follows: $K_D = P_f D_f/[P \cdot D]$. At the point where 50% of the DNA is bound, $[P \cdot D]$ equals D_f. Substituting D_f for $[P \cdot D]$ in the equation gives the following: $K_D = 50\%(P_f)$. If $P_t \gg D_t$, where P_t and D_t are the amounts of total protein and total DNA, respectively, then the amount of protein in the complex is negligible and $P_t \approx P_f$. Thus, $K_D \approx [P_t]$. If the DNA concentration is 10-fold lower than that of the protein, P_f is only about 5% lower than the total protein concentration and the K_D obtained will be well within the experimental variation of the assay. The equilibrium constant may also be expressed as the equilibrium association constant, K_a, which is simply the inverse of K_D.

K_Ds of 10^{-7} M to 10^{-13} M for regulatory proteins in *E. coli*, have been obtained, with most falling in the range of 10^{-9} to 10^{-12} (7–10, 16, 17). The constant may be affected strongly by the binding conditions. The electrostatic contribution to the binding free energy is highly dependent on the concentration and the type of salt in the buffer. Many proteins bind with much higher affinity in a lower salt concentration and with anions such as acetate and glutamate than with chloride. All sequence-specific DNA-binding proteins also bind nonspecifically. The nonspecific binding constant is mostly electrostatic and, therefore, more salt sensitive. To maximize specific over nonspecific binding, an optimum salt concentration and type can be determined. A good starting point is to use 50 mM KCl and test increasing concentrations. A buffer with potassium acetate may also be beneficial.

1. Use DNA labeled to an extent corresponding to high specific activity. Since only 100 cpm per reaction is necessary, DNA concentrations of around 10^{-11} M are easily obtained in a 20-μl reaction mixture. For lower concentrations, alternative labeling schemes are needed. T4 polymerase can be used to digest and then fill in one strand of DNA with labeled nucleosides (49), or nick translation may be used if the procedure is optimized so that intact fragments are obtained.

2. Perform a gel mobility shift assay as described in section 27.2.1.5. Allow plenty of time for the reaction to come to equilibrium. The time required for the approach to equilibrium is proportional to the product of the concentrations of the reactants and the association and dissociation rate constants, so at very low concentrations the time needed is greatly extended. Nevertheless, most systems will be at equilibrium within 30 min. Use a wide range of protein concentrations, starting with one equal to the estimated concentration of the DNA. An easy method is to use 1 and 3 μl of each of a series of 10-fold serial dilutions. In this way, a range in protein concentrations of 10^4M can be tested with just eight samples. Very carefully load the gel while it is running to minimize the mixing of the sample with the running buffer. Since protein binding is salt sensitive, the effects of exposure to a low-salt-content running buffer are a major concern; however, experiments with a number of well-characterized proteins suggest that data from carefully performed gel mobility shift assays with low-salt-content running buffer closely match binding data from other methods such as DNase footprinting. At 8 V/cm, a sample in 50 mM KCl will fully enter the gel in about 1 min. Run the gel for sufficient time to resolve the complexes, dry it, and expose it to X-ray film. At the lowest levels of the label, exposures of several days may be necessary without an enhancing screen.

3. Determine the fraction of free DNA as described above, and plot the fraction of free DNA (or the fraction of bound DNA) against the active protein concentration. If necessary, correct for any DNA that is not bound even at very high protein concentrations. This is sometimes a factor when using hybridized, synthetic oligonucleotides. If a densitometry analysis of autoradiograms is performed, the linearity of the response must be checked under the conditions of the binding assay and film exposure. Accurate quantification may also be achieved by assessing the gel with a beta-scanning device for direct counting or by excising the bands and placing them in a scintillation counter. Mincing the gel slices and placing them in an aqueous counting cocktail is sufficient for good counting efficiency for ^{32}P-labeled samples. An alternative, simple method to obtain a reasonable estimate ($\pm$ a factor of two) of the binding affinity is to visually estimate the binding half point. For the initial characterization of a protein, such accuracy is usually sufficient.

For the simple, bimolecular reaction, the binding curve should follow the relation shown here: $K_D = [P_t](1 - f)/f$, where f is the fraction of bound DNA. If the binding data do not fit a theoretical curve assuming this relation, cooperativity, multiple binding sites, interference by partially active protein, or some other factor may be involved (5, 6, 25). The gel mobility shift assay is especially useful in determining the quantitative parameters of cooperative binding, and detailed discussions of such measurements have been published previously (7, 8, 32). It is important to validate such measurements with an independent method, and DNase I protection assays (see section 27.2.2) are frequently useful.

27.2.1.8. Kinetics

Rate measurements are often used to confirm the equilibrium measurements, as well as to compare differences in related binding sites, such as those obtained from mutants. Dissociation rates for many regulatory proteins, with reaction half times of minutes, are easily measured with great precision. Since the dissociation reaction is first order, it is independent of the concentrations of the reactants. The dissociation assay is as follows.

1. Perform a gel mobility shift assay. Add just sufficient protein to bind most of the DNA. Do not add excess. Allow the binding reaction to come to equilibrium.
2. Add a 20- to 100-fold molar excess of unlabeled DNA containing the binding site. This DNA can be synthetic oligonucleotides, a purified restriction fragment, or a whole plasmid. The excess unlabeled site sequesters the protein as it dissociates, so the most precise measurements are made with the maximum level of the competitor. Experiments should also be performed with different levels of the competitor to ensure that there are no concentration-dependent effects.
3. At different time points, load samples onto the gel while it is running. Samples for different time points can be removed from a single reaction tube, or parallel tubes can be used. The dissociation is stopped by loading onto the gel. With moderate-salt-content buffer (50 mM KCl), permeation into the gel adds about 1 min to the effective dissociation time. If samples are removed from a single tube, the samples corresponding to the different time points will run on the gel for different lengths of time. For quick dissociation reactions, this method can work well, but for reactions spanning more than 1 h, it can be problematic. If separate tubes are used, the starting points for each sample can be adjusted so that all the reaction mixtures are loaded onto the gel within a short period (30 s for each sample).
4. Determine the quantity of DNA in the lanes as described in section 27.2.1.5. The dissociation rate constant is equal to the slope of a plot of the log fraction of bound DNA against time. Differences in dissociation rates of less than twofold are easily detected with this assay. Adjustments in the salt concentration or type usually will alter the rate so that it falls within a range that can be measured; however, proteins with very weak binding affinities may dissociate substantially during the time required for the complexes to enter the gel.

Association rates are more difficult to measure and will not be treated in detail here (53). Since the association rate is proportional to the product of the free DNA and free protein concentrations, both concentrations must be determined. The reaction is stopped by adding an unlabeled competitor or by loading onto the gel. An unlabeled competitor can be used only if the dissociation rate is insignificant compared with the time required to process the sample, and quenching by loading onto the gel is useful only for relatively slow reactions that last much longer than 1 min.

27.2.1.9. Stoichiometry

Accurate determination of the stoichiometry of the protein-DNA complex can be difficult. Double-label experiments are performed with ^{32}P-labeled DNA and ^{3}H-, ^{125}I-, or ^{35}S-labeled protein. The relative amounts of the labels in the complexes are determined by cutting out and counting bands after a mobility shift experiment. To calculate the ratios of protein-to-DNA molar concentrations, the specific activities of both the protein and DNA must be known. One method is given below; detailed treatments of several methods have been published previously (15, 19, 28, 68).

1. Label the DNA and calculate the specific activity as described in section 27.2.1.3. High levels of DNA in the mobility shift assay mixture are required to produce sufficient complexes to measure labeled protein; however, the specific activity must be low to minimize overlap with that of the labeled protein during scintillation counting. Adjust the DNA specific activity to approximately 100 cpm/ng of fragment.
2. Label the protein by growing cells in medium lacking methionine with 1 mCi of [^{35}S]-methionine added per 5 ml of culture. Purify the protein. The protein can also be labeled after purification (2). If the extinction coefficient is known, the specific activity can be calculated directly and a correction can be applied for the percentage of inactive protein. The protein activity is measured by running a mobility shift assay with excess, unlabeled DNA and measuring the ratio of labeled protein in specific complexes to input protein. This measuring can be done only if the complexes are stable during the gel run. If the extinction coefficient is not known, the protein specific activity is calculated from the specific activity of the methionine in the growth medium and the number of labeled amino acids in the protein (9, 29). Alternatively, the protein concentration can be deduced from quantitative amino acid analysis (68). Labeling before purification is feasible only if the protein is overproduced. Protein overproduced with the T7 polymerase-promoter system (63) can be specifically labeled by adding rifampicin after the induction of the promoter. This step eliminates the background of labeled contaminants and may allow measurements with partially purified proteins.
3. Perform the gel mobility shift assay with 20 ng of DNA per sample and sufficient protein to bind 90% of the DNA.
4. Expose the wet gel to film or stain the gel for 30 min in ethidium bromide. Use the developed film or illuminate a stained gel with UV light to locate bands containing complexes. Cut out the bands, and place them in scintillation vials.
5. Dissolve the gel slice by adding 1 ml of 21% H_2O_2 and 17% perchloric acid and incubating overnight in a tightly capped scintillation vial at 60°C. Add 19 ml of aqueous scintillation fluid, and store the samples overnight in the dark before counting. Excise blank regions of the gel, and treat similarly to determine backgrounds. It is especially important to perform counting with sections of the gel where complexes migrate by using a sample with protein alone to determine the level of background in the region. Calculate quenching by adding known quantities of DNA or protein to blank gel fragments. Calculate the overlap of ^{32}P into the

other channel of the scintillation counter by using identically treated standards.

6. Convert the counts to moles of DNA and moles of protein by applying the quenching and overlap corrections to the calculated specific activities. Ratios of protein monomer to DNA are then calculated.

27.2.1.10. Assay with Crude Protein

One of the most useful aspects of the gel mobility shift assay is the capability it provides to identify and monitor the purification of sequence-specific DNA-binding proteins. Proteins from mutants can also be screened rapidly for qualitative or quantitative changes in DNA binding. The assay is performed as usual with the exception that the binding of irrelevant proteins to the target DNA must be inhibited. In addition, the specificity of binding must be conclusively demonstrated with competition assays. The simplest assay is to mix cell extract with two labeled fragments simultaneously; one carries the specific protein-binding site, and the other is a nonspecific control. Differential shifting of the specific DNA indicates the level of protein necessary to achieve specific binding. Proteins present in low concentrations or with very low binding affinities may not be detected above the nonspecific background. In such cases, the protein must be overproduced or initial purification steps must be performed prior to using the assay. Small amounts of the protein may be purified by a preparative mobility shift procedure; the yield is governed by the level of DNA in the binding reaction mixture. The following protocol is for detecting binding proteins from *E. coli* extracts. For other bacteria, the lysis procedure may be altered and additional protease inhibitors may be added as needed.

1. Label a DNA fragment containing the specific binding site. Small fragments will have fewer sites for nonspecific binding and are preferred for lysate assays. Use an amount corresponding to 2,000 cpm/reaction so that as little as 5% binding can be detected. Higher levels of DNA can be used, but nonspecific binding becomes more of a problem as the amount of protein needed to bind the DNA is increased. Also label a second DNA fragment, such as another fragment from the plasmid vector, to use as a nonspecific control. The second fragment should be easily distinguished from the target fragment when run on the gel. It is easier to observe binding if the target is larger than the nonspecific fragment so that the protein-DNA complexes are not obscured on the gel.

2. Culture 10 ml of bacteria to late log phase (maximum density) in rich medium. Centrifuge, resuspend in 1 ml of TES buffer (Table 1), and transfer to a 1.5-ml centrifuge tube. Centrifuge again, and resuspend in 1 ml of lysate buffer (50 mM Tris-HCl [pH 7.5], 100 mM KCl, 10% glycerol, 1 mM EDTA, 1 mM DTT, and 160 μg of phenylmethylsulfonyl fluoride/ml [16-mg/ml stock in 95% ethanol, made up fresh]).

3. Sonicate with a microprobe. Set the sonicator on the maximum setting appropriate for the probe, and sonicate the sample on ice for two to four 10-s pulses, allowing 30 s between pulses. It is important to keep the sample cool and to avoid creating foam, which can occur if the probe is too close to the meniscus. If foaming is a problem, increase the volumes in step 2 by twofold. If necessary, the energy level of the probe may be reduced. Check the cell lysis by viewing a few microliters of the sample under a microscope. Use the minimum sonication necessary to lyse 90% of the cells. Following the reduction in absorbance (A_{550}), a spec-

trophotometer will also provide an estimate of cell lysis; look for about 80% reduction.

4. Centrifuge at ≥12,000 × g for 20 min at 4°C. Centrifugation at 40,000 × g can be used to obtain a slightly more pure preparation. Transfer the supernatant to a new tube, and add a 1/5 volume of cold glycerol. Mix gently with a pipette. Remove a portion of the sample for the assay, and store aliquots of the remainder at −70°C. Many proteins will remain active for days at −20°C under these conditions. Some activity is lost upon freezing (up to 50%), but the remaining activity is usually stable for months at −70°C. Thaw aliquots by placing at −20°C, and gently mix after thawing.

5. Make serial dilutions, and perform a gel mobility shift assay. Add amounts of the specific and nonspecific DNA corresponding to equal counts per minute to all samples. If the protein is highly overproduced, to ≥1% of the total protein, the normal binding buffer is used. The chromosomal DNA in the lysate should block the binding of nonspecific proteins. For lysates with lower concentrations of the specific protein, add a competitor to the reaction mixture. Both the concentration and the type of competitor must be determined empirically. Start with 0.1 to 1 μg of poly (dI-dC)/reaction. This competitor has the lowest probability of inhibiting specific binding. Other competitors include calf thymus, salmon sperm, and *E. coli* DNA. A large number of samples are run to test different concentrations of the competitor with various levels of protein. The goal is to find the lowest levels of the competitor and the protein that produce a shift in the specific DNA fragment without shifting the nonspecific fragment. With crude lysates, only a portion of the specific DNA may shift at protein levels below those at which nonspecific binding interferes. Further purification of the protein or its overproduction will increase the signal.

6. Small amounts of column buffers do not strongly affect the assay, so fractions from purification steps can be assayed directly. A total monovalent salt concentration above 200 mM may cause the bands in the gel to smear badly, as will >20 mM divalent cations and >100 mM phosphate. Since the binding of highly labeled DNA requires less than 1 ng of active protein, dilutions of 100- to 1,000-fold are usually required for proteins from lysates of overproducer strains or from fractions obtained during purifications.

7. To confirm the specificity of the binding reaction, perform additional competition assays. In a series of tubes, add the labeled specific DNA and increasing levels of an unlabeled specific DNA competitor. Then add sufficient protein to bind the labeled DNA alone; do not add excess. The competitor concentration should vary from a 0- to 100-fold excess over that of the labeled DNA. A similar set of reactions is done with an unlabeled nonspecific competitor. The nonspecific competitor should be as similar as possible to the specific competitor; do not use calf thymus DNA. If the assay is detecting specific binding, the level of the specific competitor required to block binding should be significantly lower than the level of the nonspecific competitor.

27.2.2. DNA Protection (Footprinting) Assays

An alternative DNA-protein binding assay is to cleave or modify the DNA bound by a protein such that, on average, each DNA molecule is cleaved one time. On a sequencing gel, a ladder of cleavage products is produced much like in a sequencing reaction. The DNA region covered by protein will be protected from cleavage or modification, creating a "footprint" in the cleavage pattern compared with the pattern of an unbound control (Fig. 3). Sequencing reactions with the same DNA are run in parallel to determine the position of the binding site. Both enzymatic and chemical modification techniques have been developed (reviewed in reference 62). An assay using cleavage by DNase I is capable of localizing the protein binding site to within a few base pairs (18). An assay relying on the modification of DNA by dimethylsulfate (DMS) provides additional information beyond the general region covered by the protein (56). This method also identifies the close association of protein with guanine residues in the DNA major groove and adenine residues in the DNA minor groove. Both methods can also provide quantitative data on binding affinity, kinetics, cooperativity, and stoichiometry. Both also have been adapted to obtain protection data by using supercoiled DNA and whole cells (23, 34, 42, 51). For an in

vivo technique, DNA is modified by adding DMS (see below) to cells. After the isolation of the modified DNA, primer extension reactions with a labeled oligonucleotide (section 27.3.2) are performed to create a set of fragments ending at the modified sites. When the fragments are run on sequencing gels, the data are similar to those from the direct method using labeled restriction fragments.

Other methods can provide additional information. The hydroxyl radical footprinting technique provides high-resolution data on the phosphates of the DNA backbone that are protected by bound protein (63, 64), and the hydroxyl radical interference assay indicates specific base interactions with protein (25). The hydroxyl radical is produced by the reaction of hydrogen peroxide with Fe(II)-EDTA. The advantage of this method is that all bases are modified, whereas cleavage by DNase I is not random and gaps may appear in the regions sampled. Since the DNase I molecule is large compared with the hydroxyl radical, the enzymatic approach is also much lower in resolution. Chemical agents such as copper phenanthroline and potassium permanganate have been used to study changes in the DNA structure in the complex (5, 37). Both have been used to detect DNA unwinding at promoters. The iron chelate S-1-(p-bromoacetamidobenzyl) ethylenediaminetetraacetate (FeBABE) can be used to covalently tether Fe(II) to a protein of interest. The bound DNA is cleaved only in proximity to the FeBABE, so it is possible to localize the signal of the tagged protein without interfering footprints from additional untagged proteins. DNase I, DMS, and FeBABE footprinting methods are described here; for the other methods, see references 2, 62, and 64.

27.2.2.1. DNase I Protection Assay

1. DNA must be labeled on one end only. Accomplish this by filling in the end with Klenow fragment polymerase or by adding a labeled phosphate with T4 polynucleotide kinase. The easiest method is to cut a plasmid containing the insert to be studied with one restriction enzyme, label the DNA end, and extract with phenol to inactivate the enzyme. Then cut on the opposite end of the DNA insert with a second enzyme and purify the single end-labeled fragment by electrophoresis onto NA45 membranes or by the use of a DNA clean-up kit. The overall length of the DNA fragment is not critical, but the labeled end must be within 300 bp of the binding site. Beyond this distance, the separation on a sequencing gel is insufficient to obtain a convincing footprint.

2. Perform sequencing reactions with the DNA as described in section 27.1. The 5′ end of the sequencing primer must exactly match the 5′ end of the labeled DNA fragment used for footprinting.

3. Add DNA corresponding to 10,000 to 20,000 cpm (Cerenkov) to each binding reaction mixture. Various buffer conditions are acceptable (0 to 200 mM salt, pH 6 to 8), but all reaction mixtures must contain at least 3 mM Mg^{2+} for the DNase I. $CaCl_2$ is usually added to optimize the enzyme also. A common buffer contains 20 mM Tris-HCl (pH 7.5), 3 mM $MgCl_2$, 1 mM $CaCl_2$, 0.1 mM EDTA, 50 mM KCl, 1 mM DTT, and 50 μg of BSA/ml in a volume of 100 μl per reaction. Calf thymus or salmon sperm DNA can also be added (50 ng/reaction) to avoid variations in the activity of the DNase I caused by differences among the samples in the availability of the DNA substrate.

4. Make serial dilutions of the protein in the binding buffer, and keep on ice. Make a 1-mg/ml solution of DNase

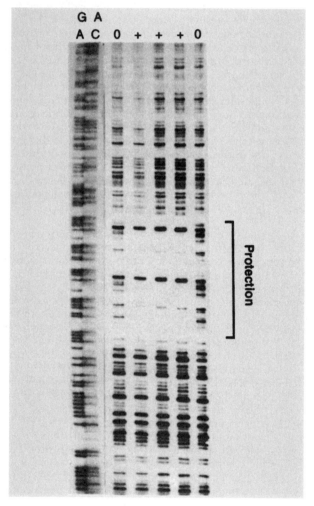

FIGURE 3 DNase I footprint. Shown is an autoradiogram from a footprinting assay of AraC protein binding to a synthetic binding site cloned into the *araFGH* promoter. G+A and A+C chemical sequencing reactions (31) were run in the first two lanes. 0, reactions with no protein; +, reactions with a twofold excess of protein relative to DNA. Each sample was loaded onto an 8% sequencing gel in an amount corresponding to 5,000 cpm. The protected region is indicated by a bracket. Two bands of strongly enhanced cleavage appear within the protected region.

I in 20 mM Tris-HCl (pH 7.5)–3 mM MgCl$_2$–1 mM CaCl$_2$–0.1 mM EDTA–50 mM KCl–50% glycerol. This solution can be stored for months at −20°C. Just before use, dilute the DNase I to 2 μg/ml in binding buffer and keep on ice.

5. Set up reaction tubes with no protein and with increasing concentrations of protein, and incubate at room temperature for 30 min. Add 3 μl of DNase I (final concentration, 0.06 μg/ml) to the first tube, mix gently but quickly, and incubate for 30 s at room temperature. Stop the reaction by adding 50 μl of 4 M ammonium acetate and 5 μg of yeast tRNA (or any other carrier such as DNA or glycogen), vortex, and place the sample on ice. Repeat the procedure for each sample. With a second set of reaction mixtures, repeat the procedure, adding 10 μl of DNase I. If carrier DNA was added, or if the protein is not purified, it may be necessary to add higher concentrations of DNase I (up to a 1-μg/ml final concentration).

6. Add 400 μl of 95% ethanol to each sample, vortex, and incubate at −20°C for several hours or at −70°C for 30 min. Centrifuge for 10 min (preferably at 4°C), carefully discard the supernatant, rinse the pellet with 70% ethanol, and dry. Perform scintillation counting (Cerenkov) with the samples to determine recovery levels, resuspend in 80% formamide dye, and mix, adjusting the volumes so that equal counts are loaded onto the gel. Also check to make sure that the DNA is resuspended in the formamide by removing the sample and performing counting with the empty tube. If much of the DNA remains in the tube, samples can be resuspended in 5 μl of TE and then mixed with 5 μl of formamide loading buffer. If the problem persists, it may be necessary to extract the protein in the samples with phenol before the precipitation step.

7. Denature the DNA by heating the sample tubes to 90°C for 3 min, and load no more than 5 μl onto a sequencing gel as described in section 27.1. Also load the sequencing reaction mixtures. The gel running time will depend on the location of the binding site with respect to the labeled end of the DNA. A second loading of samples after 1.5 h can be used to maximize the amount of readable sequence.

8. Expose the gel to X-ray film or a phosphor screen. If the samples loaded onto the gel contain fewer than 5,000 cpm, an enhancing screen (e.g., DuPont Cronex) can be used to increase the sensitivity by fivefold; however, some resolution will be lost as the bands will be more diffuse. Expose the film overnight at room temperature, and develop. Alternatively, a phosphor screen and a laser digital imager may be used, but resolution should be set to the highest value to avoid indistinct bands. The location of the protein-binding site is determined by comparing the bands in the no-protein sample with the bands in the other samples. Missing bands indicate a footprint. Within the footprint, there may also be one or more darker bands indicating enhanced cleavage caused by an altered DNA structure induced by protein binding.

The two concentrations of DNase should provide a maximum intensity of bands over a broad stretch of DNA. If the bands are too light and most of the DNA is uncut at the top of the gel, increase the DNase concentration and/or increase the incubation time to 1 min. If there is little or no uncut DNA, decrease the DNase concentration. Purified proteins are usually added in such small amounts that the protein will not interfere with the sequencing gel. If the bands are smeared, it may be necessary to phenol extract the DNA before the ethanol precipitation in step 6. It is advisable to perform the phenol extraction for crude lysates or partially purified proteins. At high concentrations of protein or in samples with significant protein contaminants, the DNA may be coated with protein, causing changes in DNA bands over a large stretch of DNA. The apparent activity of the DNase also may be reduced. If necessary, add a competitor such as poly (dI-dC) to reduce background binding and add additional DNase.

27.2.2.2. DMS Protection Assay

The DMS assay is performed much like the DNase protection assay with the exception that the sites modified by the DMS must be cleaved in a subsequent reaction (56). The procedure is as follows.

1. End label and purify the DNA as described in section 27.2.1.3.

2. For each sample, add DNA corresponding to 20,000 cpm to the binding buffer (20 mM Tris-HCl [pH 7.5], 50 mM KCl, 1 mM EDTA, 50 μg of BSA/ml) in a volume of 100 μl in a 1.5-ml centrifuge tube. Set up a sufficient number of tubes for a no-protein standard and the necessary protein dilutions.

3. Just before use, make protein dilutions in binding buffer. Add protein to sample tubes, and incubate at room temperature for 30 min.

4. To the first sample, add 1 μl of DMS, and incubate at room temperature for 1 min. (*Caution*: DMS is a carcinogen! Use it in a hood and take proper precautions.)

5. Stop the reaction by adding 25 μl of DMS stop solution (1 M Tris-HCl [pH 7.5], 1.5 M sodium acetate, 1 M β-mercaptoethanol, and 50 mM Mg acetate, stored at 4°C), 5 μg of yeast tRNA, and 400 μl of 95% ethanol. Place on ice, and repeat steps 4 and 5 with the rest of the samples.

6. Precipitate for several hours at −20°C or 30 min at −70°C. Centrifuge for 10 min, discard the supernatant, wash the pellet with 70% ethanol, dry, and resuspend in 100 μl of 1 M piperidine. The piperidine is highly volatile and should be diluted to 1 M (the stock is 10 M) the day it is used. Place the samples in tightly capped, 0.5-ml centrifuge tubes, and heat in a 90°C water bath for 30 min to cleave the DNA at the modified sites. It may be necessary to place a weight on the caps to prevent them from popping open. This reaction will cleave guanines methylated at the N7 position in the major groove of the DNA, and adenines modified at the N3 position in the minor groove will be partially cleaved. If increased cleavage at adenines is desired, resuspend the pellet in 85 μl of 20 mM sodium acetate instead of piperidine, and then add 15 μl of 1 M NaOH, vortex, and incubate at 90°C for 30 min.

7. After the incubation, neutralize samples containing NaOH with 15 μl of 1 M HCl and then add 5 μg of carrier tRNA. Add 250 μl of 95% ethanol to all samples, precipitate, centrifuge, wash with 70% ethanol, and dry.

8. Measure the counts per minute by using the Cerenkov effect. Resuspend the pellet in 80% formamide loading buffer, adjusting the volumes so that equal counts are loaded for each sample. Heat the samples to 90°C, and load no more than 5 μl on a sequencing gel along with the sequence standard. Process the gel as described above, and expose the film overnight. Reduced and enhanced cleavages are seen at the position of the protein-binding site as in the DNase protection assay. For overnight exposure of DNA corresponding to fewer than 5,000 cpm, an enhancing screen or phosphorimager should be used.

27.2.2.3. FeBABE Footprinting

DNase I and hydroxyl radical protection footprinting yield high-resolution data concerning the sequences at which DNA-binding proteins are bound. These approaches are limited in that they do not yield information concerning the orientation of each monomer bound to DNA. An additional shortcoming is that it may not be possible to distinguish between the binding sites of two or more proteins that bind to the same region of DNA. Both of these questions can be addressed by tethering a site-specific nuclease to a protein of interest. The iron chelate FeBABE reacts with free cysteine sulfhydryls to covalently tether Fe(II) to the protein of interest (24; see reference 30 for a review). In the presence of ascorbic acid and hydrogen peroxide, free radicals are generated by reaction with Fe-EDTA, which results in site-specific cleavage of the DNA backbone at a radius of 12 Å from the point of tethering. Thus, with a reaction mixture containing two or more DNA-binding proteins, it is possible to localize the signal of the tagged protein without interfering footprints from additional untagged proteins. This approach can be used to determine if two or more proteins bind simultaneously to overlapping targets or if binding by one protein is altered in the presence of an additional protein(s). The orientation of monomers within a dimer or higher-order complex can be determined by choosing appropriate locations for FeBABE tagging. If a cysteine residue near the dimer interface is tagged with FeBABE, cleavage directed at the center of a binding site suggests a symmetric dimer. An asymmetric dimer would yield a cleavage pattern with cuts directed at the center and one end of the binding site. The orientation of monomers is then verified by footprinting with FeBABE tagged to different surfaces of the protein.

A limitation of FeBABE footprinting is that some knowledge of the protein motif involved in DNA binding is required. The crystal structure of a related protein may provide a rational basis for choosing positions at which to engineer unique cysteines for subsequent tagging. Otherwise, a cysteine scan library of the DNA-binding motif and flanking regions can be used to identify potential targets for FeBABE labeling. The tagged cysteine should be positioned near the DNA without interfering with binding. Proteins that contain endogenous cysteines need to be tested to determine if labeling at these positions results in site-specific or nonspecific cleavage of target DNA. If denaturing is required to label the desired cysteine, solvent-inaccessible cysteines should be considered potential targets for labeling. Thus, a protein containing a unique engineered cysteine near the DNA-binding motif is the preferred substrate for FeBABE tagging and subsequent footprinting. Since these methods are more complicated and varied than those described in previous sections, readers are referred to descriptions of specific applications of FeBABE footprinting for methodology (6, 20, 21, 24, and 66).

27.2.2.4. ChIP Assay

ChIP is a technique to visualize multiple protein-DNA interactions within living cells (see references 13, 36, and 44 for reviews). ChIP was first employed in studies of eukaryotic gene regulation and chromatin structure. More recently, the technique has been applied to studying protein-DNA interactions in prokaryotes (3, 8, 40, 47, 55). When combined with microarray analysis, ChIP enables a genome-wide identification of transcription factor-binding sites (22, 30, 38). A basic ChIP experiment involves (i) reversible cross-linking of protein-DNA complexes, (ii) shearing of total DNA to fragments of 500 to 1,000 bp in length, (iii) immunoprecipitation of protein-DNA complexes of interest, (iv) reversal of cross-linking and removal of immunoprecipitated protein, and (v) identification of immunoprecipitated (enriched) DNA. The outcome of the assay is the identification of specific DNA targets of the protein of interest. This section reviews ChIP assays from a number of laboratories relevant to prokaryotic systems. Readers are referred to references 33 and 54 for a more detailed discussion on technical considerations of employing ChIP, and reference 58 provides an excellent discussion on the modification of ChIP for particular antibody-antigen combinations.

Commonly used in vitro techniques described elsewhere in this chapter generally require biochemical purification of a DNA-binding protein of interest and some knowledge concerning the location of its DNA target(s). The advantage of ChIP is that it enables the identification of physiologically relevant protein-DNA interactions. The assay can be performed with cells grown under different environmental conditions to determine when the protein-DNA interaction occurs, which may not always correlate with the point when the protein of interest is expressed. The absence of a biochemical fractionation step ensures that the protein will have its DNA-binding capability intact because the loss of a coactivator protein, small-molecule cofactor, or covalent modification required for binding is avoided. In the absence of information concerning the location of a transcription factor-binding site(s), ChIP can be coupled to microarray analysis not only to determine the specific targets of a DNA-binding protein but also to perform a genome-wide characterization of a regulon.

1. Cross-linking. The initial step in a standard ChIP experiment is in vivo cross-linking of protein-DNA complexes by using formaldehyde (57). Cross-linking is initiated by the addition of 1% reagent-grade formaldehyde directly to bacterial cultures. The reaction is carried out at room temperature for 5 to 20 min and quenched by the addition of 125 mM glycine, with further incubation for 5 min at room temperature. Formaldehyde reacts with side chain nitrogens, predominantly those of lysine but also those of arginine and histidine, and the α-amino groups of all amino acids and, in DNA, the side chain nitrogens of adenine, guanine, and cytosine. Cross-linking is limited to a distance of 2 Å (44). Thus, protein-DNA cross-linking occurs between proteins that are directly bound to DNA. Because formaldehyde treatment also results in protein-protein cross-linking, higher-order nucleoprotein complexes can be immunoprecipitated. The exposure time should be calibrated to ensure maximal cross-linking without resulting in nonspecific protein-DNA cross-linking, protein-protein cross-linking, or epitope masking that may lower the yield of immunoprecipitated material. The reactivity of the antibody should be tested by Western analysis in the presence and absence of cross-linking (see step 4).

2. Cell lysis. The cross-linked cell culture is then lysed to expose cross-linked protein-DNA complexes for the subsequent sonication step. Ten milliliters of culture is washed twice with phosphate-buffered saline or Tris-buffered saline (pH 7.5) and suspended in 0.5 to 1 ml of lysis buffer (10 mM Tris-HCl [pH 8.0], 20% sucrose, 50 mM NaCl, 10 mM EDTA). The quantity of lysozyme varies depending on the organism (2 mg/ml for *E. coli* or 20 mg/ml for *Bacillus subtilis*). The final volume of lysis solution should be minimized to facilitate efficient sonication. Allow lysis to proceed for

30 min at 37°C. An optional freeze-thaw cycle may aid lysis, particularly for assays with a larger starting culture volume or for cultures grown to a high optical density at 600 nm (6). The lysate is then suspended in an equal volume of 2× immunoprecipitation buffer (200 mM Tris [pH 7], 300 mM NaCl, 2% Triton X-100, 1 mM polymethylsulfonyl fluoride) and incubated at 37°C for 10 min (50). Modifications to the basic immunoprecipitation buffer may be more suitable to particular antigen-antibody combinations in the immunoprecipitation reaction and subsequent wash steps (3, 22, 55). The protease inhibitor 4-(2-aminoethyl) benzenesulfonyl fluoride at a final concentration of 1 mM can be used in place of polymethylsulfonyl fluoride, and the protease inhibitor can also be included in the lysis step.

3. *Sonication.* Shearing of DNA to fragments of an average size of 500 to 1,000 bp is critical to ensure that only DNA that is bound to the protein of interest is enriched in the subsequent PCR step. The immunoprecipitation step ensures that most DNA not bound by the protein of interest is omitted from subsequent manipulation. However, if the fragments are too long, the primers chosen for subsequent PCR may amplify DNA at a region of the template that is not relevant to the protein-DNA complex occurring elsewhere. Excessive sonication can reduce the amplification efficiency of the PCR because less template will be available. Sonication on ice is performed, typically with 10 s pulses at 30 to 40% output power with pauses in between. Sonication should be performed with a sample volume of no more than 1 ml, and the probe should be placed 1 cm below the sample surface (58). Insoluble debris is then removed by centrifugation at 10,000 × g for 5 min at 4°C. The supernatant is transferred to a new tube, and centrifugation is repeated for 10 min. DNA shearing is examined by agarose gel electrophoresis after the reversal of cross-linking and the removal of protein. DNA shearing can initially be tested directly with the sonicated samples after centrifugation and phenol-chloroform extraction; however, the extent of shearing should be confirmed with experimental samples that have undergone a cross-linking step. A small aliquot of the cleared, sonicated sample is saved for subsequent PCR analysis (input DNA).

4. *Immunoprecipitation.* The choice of antibody for immunoprecipitation is the most critical step to ensure that the desired protein-DNA complex is isolated for subsequent identification. Ideally, antibodies should be affinity purified. Generally, monoclonal antibodies display less cross-reactivity; however, an antibody cannot participate in the immunoprecipitation reaction if its epitope is masked by reaction with formaldehyde or is otherwise occluded in the protein-DNA complex. Antibodies that are highly cross-reactive with other proteins, particularly additional DNA-binding proteins, can yield undesired immunoprecipitates. To determine if an antibody is suitable for use in ChIP, Western analysis of formaldehyde-fixed cells is performed. If an antibody is unavailable or if it is not ChIP grade, it is possible to epitope tag the protein of interest and use a commercially available antibody. This modification requires that the epitope-tagged protein retain its DNA-binding function in vivo. Anti-FLAG and anti-c-Myc antibodies have been successfully employed in ChIP assays (40, 55). Epitopes can be altered by buffer systems, thereby decreasing their reactivity with an antibody. Thus, if an antigen-antibody interaction required optimization under particular buffer conditions (e.g., in an immunoblotting or enzyme-linked immunosorbent assay procedure), it may be necessary to modify the immunoprecipitation reaction. The amount of antibody required for immunoprecipitation will need to be determined empirically, with 0.1 to 25 μg/ml as a recommended starting range (26). The antibody is adsorbed at 4°C overnight with agitation; however, shorter incubations (60 to 90 min) at room temperature have also been reported (3, 22). Shorter incubation times may reduce the nonspecific cross-reactivity of the antibody. The immunoprecipitation is initiated by the addition of protein A-Sepharose (50% slurry; 30 μl per 1 ml of reaction volume). It may be necessary to empirically determine the quantity or choice of immunosorbent required to precipitate the cross-linked protein-DNA complex (58). The use of alternate protein A or G formulations may require a modification of the immunoprecipitation step to minimize nonspecific binding to chromatin, in addition to a modification of the wash steps. Immunoprecipitation complexes are then harvested by gentle centrifugation at 500 × g for 10 min at 4°C in a standard benchtop microcentrifuge. The complexes are washed five times in 1× immunoprecipitation buffer, followed by a single wash in TE (10 mM Tris-HCl [pH 8], 0.1 mM EDTA [pH 8]), and then suspended in 50 μl of TE. The stringency of the wash step for a particular antibody-antigen interaction may require modification to optimize specificity.

5. *Reversal of cross-linking.* The formaldehyde cross-linking is reversed to remove precipitated protein, rendering the DNA suitable for subsequent analysis (e.g., PCR, Southern analysis, or array hybridization). Prior to this step, an aliquot of the immunoprecipitation complex can be removed for immunoblotting. The sample is heated to 65°C for 4 to 6 h or overnight. A protease step can be added prior to heating to facilitate the purification of the DNA. Proteinase K (0.1 mg/ml) can be added directly to the heating reaction mixture (45); otherwise, incubation at 42°C for 2 h with 0.8 mg of pronase/ml is performed prior to heating (3). The reaction mixtures are then extracted with phenol-chloroform, ethanol precipitated, and suspended in TE. Spin columns used to purify PCR mixtures can be used in place of the phenol-chloroform extraction. Precipitation is enhanced in the presence of glycogen (5 μg) or carrier DNA.

6. *PCR and appropriate controls.* The identification of target DNA is typically done by PCR with *Taq* polymerase. Primers (~25-mers with similar G+C contents) are designed to amplify products in the ~200- to 300-bp range. Master reaction mixes for each primer pair are used to minimize variability among the input samples. PCRs with serial dilutions of each input sample are performed at a range of cycle numbers to ensure that each reaction product is within the linear range of amplification. A minimum quantity of the product is necessary for accurate quantification via ethidium bromide-agarose gel electrophoresis, whereas too many PCR cycles result in reaction end points, thereby making a quantitative comparison between input samples impossible. Total (input) DNA is quantified via PCR with experimental primers by using the cleared, sonicated sample as the template. Each experimental or control sample is standardized to the respective input reaction mixture. The following controls are possible: (i) a mock immunoprecipitation reaction with the antibody omitted, (ii) a null mutant background that does not express the protein of interest, (iii) a wild-type strain if an epitope-tagged protein was targeted for immunoprecipitation, and (iv) an experimental condition under which the target protein is not expressed or does not have DNA-binding capability. The degree of enrichment (*n*-fold) for each experimental sample is then determined relative to an appropriate control. A

mock immunoprecipitation reaction without antibody is also used to demonstrate that target-DNA enrichment is due to the immunoprecipitation of the protein in question. It is also necessary to perform an internal control with primers designed to amplify a region of DNA not predicted to be enriched by the immunoprecipitation (e.g., a house-keeping gene or the downstream coding sequence of the gene in question). This control demonstrates that the immunoprecipitation is specifically enriching with DNA targeted by the protein under study. The average shearing size of the DNA should be taken into consideration in choosing an internal control target.

7. *Precise mapping of a protein-binding site.* Lin and Grossman used a modification of the standard ChIP procedure to more precisely define the site of the binding of the *B. subtilis* partitioning protein Spo0J to a 16-bp inverted repeat element (40). They constructed a series of plasmids containing various regions of the predicted Spo0J-binding site, a precise deletion of the binding site, and a derivative of the binding site containing point mutations. ChIP was performed, and target DNA was identified by using plasmid-specific primers. This in vivo approach can be used to validate binding sites determined by standard in vitro techniques (e.g., protection footprinting) that are not necessarily representative of physiological conditions. One limitation of ChIP is that the protein targeted for immunoprecipitation may not reveal the complete story concerning the particular nucleoprotein complex in question. Additional proteins may be required for DNA binding, or the protein under study may simply be an accessory factor not directly involved in DNA binding. In this scenario, an in vitro strategy using purified protein is necessary to demonstrate a direct protein-DNA interaction. Thus, ChIP and the more widely employed in vitro strategies provide complementary information about protein-DNA interactions.

27.3. RNA TRANSCRIPT MAPPING

Once a gene or promoter has been cloned, it is of interest to determine the start site and the size of the RNA transcript. Two common methods for determining the start site are presented; each has advantages and disadvantages. For the S1 nuclease protection assay, a restriction fragment that spans the region near the beginning of the transcript is end labeled and hybridized to cellular RNA. The portion of DNA that is not hybridized is removed by S1 endonuclease digestion, and the position of the end of the RNA is then inferred from the size of the remaining DNA fragment. In the primer extension assay, a labeled oligonucleotide primer is hybridized to the RNA and reverse transcriptase extends the DNA to the 5′ end of the RNA. Again, the start site is inferred from the size of the resulting fragment. Both procedures require that the position (sequence) of the 5′ end of the probe be known accurately, and both require only that the 5′-proximal portion of the RNA be intact. The S1 nuclease method also can be used to map the 3′ end of the RNA. A simple procedure for determining the overall size of the transcript is to perform Northern blot hybridization with cellular RNA. Total cellular RNA is separated by fragment size in a denaturing gel and transferred onto a membrane, and the specific RNA of interest is detected by the hybridization of a labeled probe.

27.3.1. S1 Nuclease Protection Assay

Make sure that all buffers, glassware, and other materials are RNase free. Liquids should be decontaminated by removing

RNase with the addition of a 0.2% final concentration of diethylpyrocarbonate (DEPC; Sigma Chemical Co.). Shake the mixture vigorously to dissolve as much DEPC as possible, incubate overnight, and then autoclave to decompose the DEPC into CO_2. Do not treat Tris solutions with DEPC; instead, make them up with DEPC-treated and autoclaved water. Handle all materials only with clean gloves. Autoclaving generally does not destroy all RNase activity. RNases can be readily removed from plasticware by washing with a strong detergent such as Absolve (DuPont) or soaking in DEPC solution; however, disposables such as microtubes and tips are not contaminated in the manufacturing process, so careful handling should eliminate the need for cleaning. Bake all glassware and spatulas at ≥150°C for ≥2 h. Then proceed with RNA isolation and the assay as follows.

27.3.1.1. Crude-RNA Isolation

Kits utilizing columns that bind the RNA and allow impurities to be washed away can be used to reliably obtain high-quality RNA from cells (QIAGEN). As an inexpensive alternative, a simple method for isolating total crude RNA from gram-negative bacteria (48) is as follows.

1. Grow a 50-ml cell culture in a rich medium to late log phase (an absorbance at 550 nm of to 0.7 to 0.8 for *E. coli*).

2. Harvest the cells by pouring the culture into a centrifuge bottle containing 50 ml of ice, 5 mg of chloramphenicol (50-mg/ml stock solution in 95% ethanol), and 5 mM NaN_3. Pellet the cells by spinning at 8,000 × *g* for 10 min. Discard the supernatant, and resuspend the cells in 2 ml of 10 mM Tris-HCl (pH 7.4)–10 mM KCl–5 mM $MgCl_2$.

3. Add 0.7 ml of lysozyme from a freshly prepared stock of 10 mg/ml in 10 mM Tris, pH 7.4. Mix well, incubate on ice for 15 min, and freeze in a bath of dry ice-ethanol.

4. Add 0.3 ml of 10% sodium dodecyl sulfate, and thaw in a 65°C water bath. Incubate until the sample clears (<5 min). For a cleaner preparation, proteinase K can be added (0.1 ml of 5-mg/ml stock) and the sample can be incubated at 37°C for 20 min.

5. Add 0.1 ml of 3 M sodium acetate, pH 5.2, and then add 1 volume of TE-saturated phenol. Extract for 5 min by placing the tube on a water bath shaker at 65°C and setting on moderate shaking (best mixing is obtained by capping the tube and laying it on its side). Centrifuge the sample briefly to separate the phases, remove the aqueous phase to a new tube, and repeat the phenol extraction. After recovering the aqueous phase again, measure the volume (estimate by filling a similar tube to the same level).

6. Add sufficient 5 M NaCl to bring the concentration to 2 M NaCl and then add 1.5 volumes of ethanol and precipitate at −20°C for 60 min. Centrifuge for 10 min at ≥12,000 × *g*, discard the supernatant, rinse the pellet three times with 70% ethanol, and resuspend in 1 ml of 10 mM Tris (pH 7.4)–10 mM NaCl. Clarify by centrifugation for 5 min in a microcentrifuge, and remove the sample to a new tube.

7. Read the absorbance at 260 nm in a spectrophotometer. Most of the absorbance is due to RNA, the majority of which is rRNA and tRNA. The concentration of RNA corresponds to an A_{260} of approximately 1, or 40 μg/ml. A highly expressed mRNA species will represent no more than 1% of the total RNA.

27.3.1.2. Pure-RNA Isolation

An alternative method for isolating total RNA is adapted from a method for eukaryotic cells described by Sambrook et al. (49). This method removes DNA and gives a much cleaner intact RNA preparation, especially from more difficult cell cultures containing large amounts of RNase and including gram-positive as well as gram-negative bacteria.

1. Pour a late-log-phase culture (50 to 200 ml) over an equal volume of ice-cold ethanol-acetone (1:1), and pellet the cells by centrifugation at 6,000 × g for 10 min.

2. Lyse the cells. For gram-positive bacteria, such as *Staphylococcus aureus*, wash (resuspend and pellet again) once with SMM (1 M sucrose, 40 mM maleic acid, 40 mM MgCl$_2$, pH 7.6) and suspend the cells in 9 ml of SMM. Prepare protoplasts by adding 1 ml of lysostaphin (1 mg/ml in SMM), and incubate in ice for 30 min. Pellet the cells by centrifugation, and resuspend in 2 ml of 4 M guanidinium thiocyanate (in 0.1 M Tris-HCl [pH 7.5], 1% β-mercaptoethanol). Add sodium lauryl sarcosinate (Sarkosyl) to a 0.5% final concentration. For *E. coli* cultures, chill the culture in ethanol-acetone, pellet and wash the cells with 0.8% NaCl, and add 2 ml of the 4 M guanidinium thiocyanate–β-mercaptoethanol solution. The guanidinium thiocyanate should be filtered through Whatman no. 1 filter paper and stored at room temperature, and the β-mercaptoethanol should be added just before use.

3. For all cultures, freeze-thaw twice by alternately plunging into dry ice-ethanol and 45°C baths. Centrifuge at 6,000 × g for 10 min, and recover the supernatant.

4. Pellet the RNA. Prepare a solution of 5.7 M CsCl–10 mM Na$_2$-EDTA (pH 7.5), and add 3 ml to a centrifuge tube for an SW50.1 or similar swinging-bucket rotor. Layer the cleared lysate onto the top of the CsCl solution, and centrifuge at 160,000 × g and 20°C for at least 24 h. With a pipette held at the meniscus so that some air is drawn up along with the liquid, carefully draw off the supernatant and

discard. Rinse the RNA pellet with 75% ethanol to remove the CsCl, allow to dry, and dissolve the pellet in 200 μl of 10 mM Tris, pH 7.5. Remove to a microcentrifuge tube, and precipitate with a 1/10 volume of Na acetate, pH 5.2, and 2.5 volumes of ethanol at −20°C for 1 h. Centrifuge, discard the supernatant, rinse with 75% ethanol, dry, and dissolve the pellet in 100 μl of water. If the RNA is to be stored, add 3 volumes of ethanol and keep at −70°C. Add Na acetate and precipitate aliquots as needed.

27.3.1.3. Assay Procedure

1. Label the DNA probe. The DNA fragment should have one 5' overhang that is labeled with [^{32}P]ATP and T4 polynucleotide kinase. Use 0.2 to 1 μg of an appropriately cut DNA fragment. For quantitative results, the amount of probe must be in at least 10-fold molar excess relative to the mRNA being detected. The labeled end of the fragment should be within the region that hybridizes to the RNA (Fig. 4). A DNA restriction fragment is chosen so that one 5' end that is 50 to 400 bases downstream from the transcription start site can be labeled. After hybridization and digestion of the nonhybridized, single-stranded portion of the DNA fragment with S1 nuclease, the 3' end of the DNA strand will end at the beginning of the transcript. If this DNA is run on a sequencing gel with similar DNA that has undergone sequencing reactions (the 5' end of the sequencing primer must correspond to the 5' end of the labeled probe), the position of the band will correspond to the transcription start site (4, 65). The expected direction of transcription also must be considered in order to determine which strand will hybridize. Separate probes labeled at opposite ends can be used to determine the direction of transcription. To label a restriction fragment, remove phosphates

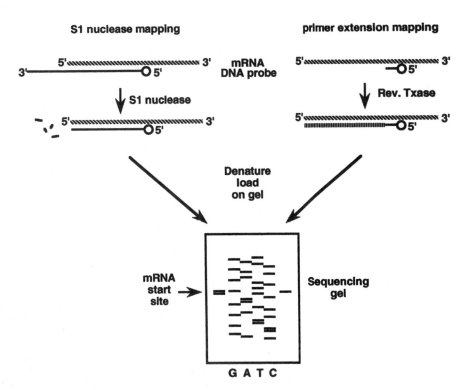

FIGURE 4 S1 and primer extension assays. Cross-hatched line, mRNA; solid line, DNA; circle, labeled 5' end of DNA. The cartoon of a sequencing gel shows sequencing reactions in the center, an S1 reaction to the left, and primer extension to the right.

from the ends of the DNA. Digest with a restriction enzyme, and isolate a DNA fragment if necessary. Do not add tRNA as a carrier for precipitation since it will interfere with subsequent reactions. If a carrier is necessary, use 10 mg of glycogen (Boehringer Mannheim)/ml. Add DNA and 5 μl of 10× phosphatase buffer (0.2 M Tris [pH 8.5], 10 mM MgCl$_2$, 10 mM ZnCl$_2$, 1 mM EDTA) plus water to a final volume of 50 μl. If a low- or moderate-salt-content buffer is used for the restriction enzyme, precipitation is not necessary; just omit the 10× buffer and add 1 μl of 1 M Tris-HCl (pH 9.5)–50 mM ZnCl$_2$. Add 0.5 U of calf intestine alkaline phosphatase, and incubate at 37°C for 60 min. Heat to 65°C for 15 min. Extract the protein by adding 50 μl of 0.6 M Na acetate and 100 μl of TE-saturated phenol. Vortex for 1 min, centrifuge for 1 min, and remove the aqueous (top) phase to a new tube. Repeat the extraction with phenol-chloroform, and precipitate the DNA as usual. Resuspend the DNA in 21 μl of TE and add 3 μl of 10× kinase buffer (0.5 M Tris-HCl [pH 7.6], 0.1 M MgCl$_2$, 50 mM DDT, 1 mM spermidine HCl, 1 mM EDTA) and 5 μl of [γ^{32}P]-ATP (3,000 Ci/mmol, or 10 mCi/ml). Add 10 U of T4 polynucleotide kinase, and incubate at 37°C for 45 min. Heat to 65°C for 10 min, and remove the unincorporated nucleotides by using chromatography or a commercial clean-up kit. Digest the probe with a restriction enzyme to remove one labeled end, and isolate the appropriate fragment by gel electrophoresis.

2. Hybridize the probe. Add the probe in an amount corresponding to 2×10^4 to 2×10^5 cpm (Cerenkov) to 25 μg of RNA in a 1/10 volume of 3 M sodium acetate and then add 2.5 volumes of 95% ethanol and precipitate at −70°C for 30 min. For a low-abundance message, the amount of RNA can be increased by up to eightfold. Centrifuge, remove the supernatant, and wash and air dry the pellet. Do not overdry, or the pellet will be difficult to resuspend. Resuspend the pellet in 20 to 30 μl of hybridization buffer {80% deionized formamide, 40 mM PIPES HCl [Na$_2$ piperazine-N,N′-bis (2-ethanesulfonic acid) adjusted with HCl, pH 6.4], 400 mM NaCl, 1 mM EDTA}. It may be necessary to pipette the sample up and down and incubate it for several hours to completely dissolve it. Heat to 80°C for 10 min, and then quickly place in a water bath at the hybridization temperature. The hybridization temperature depends on the length of the hybrid and its G+C content. The temperature is determined empirically and usually falls within 42 to 60°C. Start by incubating samples at 55°C for hybrids expected to be longer than 100 bases and to have a 50% G+C content. Additional samples can be hybridized at 5°C above and below this temperature to ensure proper hybridization. Since RNA-DNA hybrids will form at about 10°C higher than DNA-DNA hybrids, incubation at the optimal temperature will maximize the desired hybridization and give quantitative results. Allow to hybridize for 6 to 12 h.

3. Dilute the samples with 400 μl of cold S1 buffer (30 mM sodium acetate [pH 4.5], 1 mM ZnSO$_4$, 250 mM NaCl, 5% glycerol). Add 30 U of S1 nuclease and incubate at 37°C for 30 min. Digestion conditions may be varied to obtain maximal digestion. The temperature can be increased to 45°C, although this increase may result in some cleavage of A-T-rich regions. The concentration of S1 may also be increased to 500 U. Stop the reaction with 100 μl of 2.5 M ammonium acetate–50 mM EDTA.

4. Add 500 μl of isopropanol, precipitate for 30 min at 4°C, centrifuge, discard the supernatant, and resuspend in 100 μl of 0.3 M sodium acetate. Precipitate with 2.5 volumes of ethanol, rinse with 70% ethanol, and dry.

5. Resuspend in 5 μl of formamide loading buffer (as described in section 27.1). Heat to 90°C for 3 min, and load onto a sequencing gel along with sequence standards. Run the gel, and expose to X-ray film or use a phosphorimager as with sequencing samples (section 27.1).

6. A single band or, more commonly, a cluster of a few bands in the lanes containing the S1-treated samples indicates the start site for transcription. The position is determined by comparison with the sequencing lanes. The S1 may "nibble" the DNA end, so multiple bands do not necessarily indicate more than one start position. S1 may also cleave the DNA at sites with high A-T contents, giving a false start position. Because of these artifacts, it is often useful to confirm the start site by another means such as the primer extension assay.

27.3.2. Primer Extension Assay

Isolate RNA by one of the above-described methods. The 5′ end of a message can be mapped by hybridizing a primer to the RNA and synthesizing a DNA strand from the 3′ end of the primer extending to the 5′ end of the RNA. Reverse transcriptase is used for the synthesis. The products of the reaction are loaded onto a sequencing gel along with a set of sequencing reaction mixtures as described for S1 mapping. Generally, the yield of DNA is lower than that obtained with the S1 procedure, but some of the artifacts of the S1 method, such as bands appearing at A-T-rich regions, are avoided. Restriction fragments labeled at one end can be used as primers, but synthetic oligonucleotides of 20 to 30 bases are preferable. The oligonucleotide is labeled to generate high specific activity, and there is no competition with a second DNA strand during hybridization with the RNA. Choose a sequence that is 100 to 400 bases from the expected position of the 5′ end of the RNA, and make sure that it is the appropriate strand and has the correct polarity.

1. Label 100 ng of the oligonucleotide with [γ^{32}P]ATP by using T4 polynucleotide kinase as described in section 27.2.2.1. A 5′-phosphate is not present on the oligonucleotide, so treatment with phosphatase is unnecessary. Separate the labeled DNA from the unincorporated label by ethanol precipitation in the presence of 2 M ammonium acetate or by column chromatography.

2. Mix an amount of the probe corresponding to 10^4 to 10^5 cpm with 10 to 100 μg of RNA prepared as described for S1 mapping. If the extensions do not work, a higher-purity RNA prep may help (2). A large excess of probe can increase the number of premature bands during the extension reaction, so it is advisable to try several levels of the probe. Add a 1/10 volume of 3M Na acetate, pH 5.2, and 2.5 volumes of ethanol, precipitate at −20°C for 60 min, and recover the nucleic acids. Air dry the pellet after washing with 70% ethanol.

3. Dissolve the pellet in 30 μl of hybridization buffer, and hybridize as described in step 2 in section 27.3.1.2. Heat to 80°C for 10 min, and then transfer to a water bath at the hybridization temperature. For synthetic oligonucleotides, the optimal temperature is usually between 30 and 42°C; try 32°C first. Incubate for 6 to 12 h. Add 170 μl of 0.3 M Na acetate and 2.5 volumes of ethanol, and precipitate at −20°C for 60 min. Centrifuge, rinse with 70% ethanol, and carefully remove all ethanol. Allow to air dry.

4. Dissolve the sample in 25 μl of reverse transcriptase buffer (50 mM Tris-HCl [pH 7.6], 60 mM KCl, 10 mM MgCl$_2$, 1 mM DTT, 50 μg of actinomycin D/ml, 25 U of RNasin, 0.4 mM [each] dNTPs). Add 40 U of AMV reverse

transcriptase, and incubate at 42°C for 90 min. (The actinomycin D destabilizes DNA-DNA hybrids, and the RNasin inhibits contaminating RNases.) Heat at 70°C for 10 min.

5. Add 1 μl of 0.5 M EDTA and 1 μl of 1-mg/ml RNase A, and incubate for 30 min at 37°C.

6. Add 100 μl of 2.5 M ammonium acetate, extract by vortexing for 1 min with an equal volume of phenol chloroform, centrifuge for 1 min, and remove the aqueous phase to a new tube. Precipitate the DNA with 300 μl of ethanol at −20°C for 60 min, centrifuge, rinse the pellet, and dry.

7. Resuspend in 3 μl of TE buffer and then add 3 μl of formamide loading buffer and heat at 80°C for 3 min. Load on a sequencing gel with DNA sequence reaction mixtures.

27.3.3. Northern Blot Hybridization Assay

Northern blot hybridization is a method to determine the size of an RNA species (39). RNA from a cellular extract is run on a denaturing gel and then transferred onto nitrocellulose paper, much like a Southern blot procedure (chapter 29.2.3.2). The specific RNA is detected by the hybridization of a labeled probe and autoradiography. The most difficult step in the procedure is to obtain relatively intact RNA without degradation by RNases. This is especially important for mRNA from bacteria which has a short half-life. All buffers, water, and glass- and plasticware must be RNase free.

27.3.3.1. Cell Lysate RNA Preparation

The following procedure is for RNA in cell lysates (35), not for purified RNA. It has the advantages that minimal handling will result in the least degradation of the RNA and quantitative recoveries are possible. Thus, the assay can be used to determine relative levels of transcription during the induction of a gene. This procedure has worked well for *S. aureus*, *E. coli*, and *B. subtilis*.

1. Grow a 10-ml culture of bacteria in rich medium at 37°C to late log phase.

2. Rapidly cool in ice water, and immediately collect the cells by centrifugation at 4°C for 3 min in a microcentrifuge. Remove the supernatant, and resuspend the pellet in 100 μl of cold lysis buffer (20% sucrose, 20 mM Tris HCl [pH 7.6], 10 mM Na$_2$-EDTA, 50 mM NaCl, and either 100 μg of lysostaphin/ml for *S. aureus* or 1 mg of lysozyme/ml for *B. subtilis* and *E. coli*). Incubate on ice until the suspension clears (10 to 20 min).

3. Complete the cell lysis by adding 100 μl of 2% sodium dodecyl sulfate and 50 μl of proteinase K (5-mg/ml solution). Shake at room temperature for 15 min.

4. Freeze and thaw the lysate two times by using a dry ice-ethanol bath and a 45°C water bath. Add 50 μl of loading dye (50% glycerol, 200 mM EDTA, 0.1% bromophenol blue, 0.1% xylene cyanol), and mix well.

5. Check the integrity of the RNA by electrophoresis of a 6-μl aliquot on a 1% agarose gel. Run the gel until the blue dye is two-thirds of the distance from the origin. The rRNA should form two distinct bands after the gel is stained with ethidium bromide.

27.3.3.2. Assay Procedure

1. Prepare a 1.2% denaturing agarose gel. Mix 1.2 g of agarose with 73 ml of distilled water, and heat until boiling. After the agarose has dissolved, cool to 65°C and add 10 ml of 10× running buffer (0.2 M MOPS [morpholinepropanesulfonic acid; pH 7.0], 50 mM Na acetate, 1 mM EDTA)

and 16.2 ml of 37% formaldehyde. Stir briefly, and pour into the gel apparatus. After the agarose has hardened, cover the gel with running buffer plus 2.2 M formaldehyde and 1 μg of ethidium bromide per ml.

2. Mix 10 μl of lysate prepared as described above with 20 μl of deionized formamide (50%) and 4 μl of 10× running buffer. Denature the RNA by heating at 65°C for 10 min. After chilling on ice, load the entire 34-μl sample onto the gel. Also load RNA markers: 6 μl of RNA markers (Invitrogen or Fermentas) mixed with 4 μl of loading dye, denatured and chilled.

3. Carry out electrophoresis at room temperature in a hood at 100 V of constant voltage. Circulate the buffer from the cathode to the anode. A peristaltic pump is sufficient. Run the gel until the blue dye is two-thirds of the distance from the origin. The running time and voltage can be adjusted depending on the size of the RNA of interest and the particular apparatus used.

4. After electrophoresis, cut off the lanes of the gel containing the marker RNA and stain in a 1-mg/ml ethidium bromide solution for 20 min followed by a 15-min destaining in water. View the gel with UV light, and measure the migration of the markers from the origin.

5. Transfer the RNA onto a nitrocellulose filter. This procedure is quite similar to that described in section 27.3.1, where buffer formulas and transfer details are given. The same blotting setup is used (section 27.3; Fig. 3). No pretreatment (denaturation or nicking) is necessary; however, the formaldehyde may be removed by soaking the gel in two changes of 500 ml of water for 15 min for each. Blot the RNA from the gel onto the filter (Amersham) overnight by using 20× SSC as the transfer buffer. The origins of the gel lanes can be marked on the filter by using a soft pencil to aid in molecular weight calculations. The transfer should be done in a fume hood if the formaldehyde was not removed.

6. After transfer, place the filter between two pieces of Whatman 3MM paper and bake at 80°C for 2 h in a vacuum oven.

7. Prehybridize the membrane (section 27.3.1) by soaking it with agitation in 10 ml (for a 12-by-12-cm filter) of 2× Denhardt's solution (0.04% each BSA, Ficoll, and polyvinyl pyrrolidone; 10 mM EDTA; 0.2% sodium dodecyl sulfate; 5× SSC) for 3 h at 60°C in a sealed plastic bag.

8. Open a corner of the bag, and add 0.2 ml of denatured sonicated salmon sperm DNA (5-mg/ml stock) and denatured probe in an amount corresponding to 10^7 cpm. The probe can be prepared by random primed synthesis as described in section 27.3.1.

9. Reseal the bag, and incubate with gentle agitation for 18 h at 60°C.

10. Remove the membrane from the bag, and wash two times in 100 ml of 2× SSC–0.1% sodium dodecyl sulfate for 5 min at room temperature with constant agitation. Wash two times with 200 ml of 2× SSC–1% sodium dodecyl sulfate at 60°C for 30 min, and wash twice with 100 ml of 0.1× SSC at room temperature for 30 min.

11. Place the membrane on a sheet of filter paper, and allow to dry. Place on top of a piece of X-ray film. Mark the origins of the gel lanes on the film with a pencil, or mark the filter with an autoradiography pen (a fluorescent dye that will expose the film). Expose overnight. An intensifying screen (DuPont Lightning Plus) can be used if the signal is weak; expose overnight at ≤−70°C with the screen behind the film. Use the distances measured for the marker RNAs to construct a log-linear graph of RNA sizes

(molecular weights or numbers of bases) versus migration distances. A linear relationship should be obtained, which can be used to interpolate the size(s) of the RNA of interest from the developed autoradiogram.

27.4. REFERENCES

1. **Ansorge, W., and S. Labeit.** 1984. Field gradients improve resolution on DNA sequencing gels. *J. Biochem. Biophys. Methods* **10:**237–243.

2. **Ausubel, F. M., R. Brent, R. E. Kingston, D. D. Moore, J. G. Seidman, J. A. Smith, and K. Struhl.** 2003. *Current Protocols in Molecular Biology.* Wiley-Interscience, New York, NY.

3. **Beloin, C., S. McKenna, and C. J. Dorman.** 2002. Molecular dissection of VirB, a key regulator of the virulence cascade of *Shigella flexneri. J. Biol. Chem.* **277:**15333–15344.

4. **Berk, A. J., and P. A. Sharp.** 1977. Sizing and mapping of early adenovirus mRNAs by gel electrophoresis of S1 endonuclease-digested hybrids. *Cell* **12:**721–732.

5. **Borowiec, J. A., L. Zhang, S. Sasse-Dwight, and J. D. Gralla.** 1987. DNA supercoiling promotes formation of a bent repression loop in *lac* DNA. *J. Mol. Biol.* **196:**101–111.

6. **Boucher, P. E., A. E. Maris, M. S. Yang, and S. Stibitz.** 2003. The response regulator BvgA and RNA polymerase alpha subunit C-terminal domain bind simultaneously to different faces of the same segment of promoter DNA. *Mol. Cell* **11:**163–173.

7. **Brenowitz, M., D. F. Senear, M. A. Shea, and G. K. Ackers.** 1986. Quantitative DNase I footprint titration: a method for studying protein-DNA interactions. *Methods Enzymol.* **130:**132–181.

8. **Brenowitz, M., E. Jamison, A. Majumdar, and S. Adhya.** 1990. Interaction of the *Escherichia coli* Gal repressor protein with its DNA operators in vitro. *Biochemistry* **29:**3374–3383.

9. **Carey, J.** 1988. Gel retardation at low pH resolves trp repressor-DNA complexes for quantitative study. *Proc. Natl. Acad. Sci. USA* **85:**975–979.

10. **Carey, J.** 1991. Gel retardation. *Methods Enzymol.* **208:**103–117.

11. **Chien, A., D. B. Edgar, and J. M. Trela.** 1976. Deoxyribonucleic polymerase from the extreme thermophile *Thermus aquaticus. J. Bacteriol.* **127:**1550–1557.

12. **Crothers, D. M., M. R. Gartenberg, and T. E. Shrader.** 1991. DNA bending in protein-DNA complexes. *Methods Enzymol.* **208:**118–146.

13. **Das, P. M., K. Ramachandran, J. vanWert, and R. Singal.** 2004. Chromatin immunoprecipitation assay. *BioTechniques* **37:**961–969.

14. **Fried, M., and D. M. Crothers.** 1981. Equilibria and kinetics of lac repressor-operator interactions by polyacrylamide gel electrophoresis. *Nucleic Acids Res.* **9:**6505–6525.

15. **Fried, M., and D. M. Crothers.** 1983. CAP and RNA polymerase interactions with the *lac* promoter: binding stoichiometry and long range effects. *Nucleic Acids Res.* **11:**141–158.

16. **Fried, M. G., and D. M. Crothers.** 1984. Equilibrium studies of the cyclic AMP receptor protein-DNA interaction. *J. Mol. Biol.* **172:**241–262.

17. **Fried, M. G., and D. M. Crothers.** 1984. Kinetics and mechanism in the reaction of gene regulatory proteins with DNA. *J. Mol. Biol.* **172:**263–282.

18. **Galas, D. J., and A. Schmitz.** 1978. DNase footprinting: a simple method for the detection of protein-DNA binding specificity. *Nucleic Acids Res.* **5:**3157–3170.

19. **Garner, M. M., and A. Revzin.** 1982. Stoichiometry of catabolite activator protein/adenosine cyclic 3′,5′-monophosphate interactions at the lac promoter of *Escherichia coli. Biochemistry* **21:**6032–6036.

20. **Grainger, D. C., T. A. Belyaeva, D. J. Lee, E. I. Hyde, and S. J. Busby.** 2003. Binding of the *Escherichia coli* MelR protein to the melAB promoter: orientation of MelR subunits and investigation of MelR-DNA contacts. *Mol. Microbiol.* **48:**335–348.

21. **Grainger, D. C., T. A. Belyaeva, D. J. Lee, E. I. Hyde, and S. J. Busby.** 2004. Transcription activation at the *Escherichia coli melAB* promoter: interactions of MelR with the C-terminal domain of the RNA polymerase alpha subunit. *Mol. Microbiol.* **51:**1311–1320.

22. **Grainger, D. C., T. W. Overton, N. Reppas, J. T. Wade, E. Tamai, J. L. Hobman, C. Constantinidou, K. Struhl, G. Church, and S. J. Busby.** 2004. Genomic studies with *Escherichia coli* MelR protein: applications of chromatin immunoprecipitation and microarrays. *J. Bacteriol.* **186:**6938–6943.

23. **Gralla, J. D.** 1985. Rapid "footprinting" on supercoiled DNA. *Proc. Natl. Acad. Sci. USA* **82:**3078–3081.

24. **Greiner, D. P., R. Miyake, J. K. Moran, A. D. Jones, T. Negishi, A. Ishihama, and C. F. Meares.** 1997. Synthesis of the protein cutting reagent iron (S)-1-(p-bromoacetamidobenzyl)ethylenediaminetetraacetate and conjugation to cysteine side chains. *Bioconjug. Chem.* **8:**44–48.

25. **Hayes, J. J., and T. D. Tullius.** 1989. The missing nucleoside experiment: a new technique to study recognition of DNA by protein. *Biochemistry* **28:**9521–9527.

26. **Hecht, A., and M. Grunstein.** 1999. Mapping DNA interaction sites of chromosomal proteins using immunoprecipitation and polymerase chain reaction. *Methods Enzymol.* **304:**399–414.

27. **Hendrickson, W.** 1985. Protein-DNA interactions studied by the gel electrophoresis-DNA binding assay. *BioTechniques* **4:**198–207.

28. **Hendrickson, W., and R. Schleif.** 1984. Regulation of the *Escherichia coli* L-arabinose operon studied by gel electrophoresis DNA binding assay. *J. Mol. Biol.* **178:**611–628.

29. **Hendrickson, W., and R. Schleif.** 1985. A dimer of AraC protein contacts three adjacent major groove regions of the *araI* DNA site. *Proc. Natl. Acad. Sci. USA* **82:**3129–3133.

30. **Herring, C. D., M. Raffaelle, T. E. Allen, E. I. Kanin, R. Landick, A. Z. Ansari, and B. O. Palsson.** 2005. Immobilization of *Escherichia coli* RNA polymerase and location of binding sites by use of chromatin immunoprecipitation and microarrays. *J. Bacteriol.* **187:**6166–6174.

31. **Hong, G. F.** 1987. The use of DNase I, buffer gradient gel, and ^{35}S label for DNA sequencing. *Methods Enzymol.* **155:**93–110.

32. **Hudson, J. M., and M. G. Fried.** 1990. Co-operative interactions between the catabolite gene activator protein and the lac repressor at the lactose promoter. *J. Mol. Biol.* **214:**381–396.

33. **Innis, M. A., K. B. Myambo, D. H. Gelfand, and M. A. D. Brow.** 1988. DNA sequencing with *Thermus aquaticus* DNA polymerase and direct sequencing of polymerase chain reaction-amplified DNA. *Proc. Natl. Acad. Sci. USA* **85:**9436–9440.

34. **Khoury, A. M., H. S. Nick, and P. Lu.** 1991. *In vivo* interaction of *Escherichia coli lac* repressor N-terminal fragments with the *lac* operator. *J. Mol. Biol.* **219:**623–634.

35. **Kornblum, J. S., S. J. Projan, S. L. Moghazeh, and R. P. Novick.** 1988. A rapid method to quantitate non-labeled RNA species in bacterial cells. *Gene* **63:**75–85.

36. **Kuo, M. H., and C. D. Allis.** 1999. In vivo cross-linking and immunoprecipitation for studying dynamic protein: DNA associations in a chromatin environment. *Methods* **19**:425–433.

37. **Kuwabara, M., and D. S. Sigman.** 1987. Footprinting DNA-protein complexes in situ following gel retardation assays using 1,10-phenanthroline-copper ion: *Escherichia coli* RNA polymerase-*lac* promoter complexes. *Biochemistry* **26**:7234–7238.

38. **Laub, M. T., S. L. Chen, L. Shapiro, and H. H. McAdams.** 2002. Genes directly controlled by CtrA, a master regulator of the *Caulobacter* cell cycle. *Proc. Natl. Acad. Sci. USA* **99**:4632–4637.

39. **Lehrach, H., D. Diamon, J. M. Wozney, and H. Boedtker.** 1977. RNA molecular weight determinations by gel electrophoresis under denaturing conditions: a critical examination. *Biochemistry* **16**:4743–4749.

40. **Lin, D. C., and A. D. Grossman.** 1998. Identification and characterization of a bacterial chromosome partitioning site. *Cell* **92**:675–685.

41. **Liu-Johnson, H. N., M. R. Gartenberg, and D. M. Crothers.** 1986. The DNA binding domain and bending angle of *Escherichia coli* CAP protein. *Cell* **47**:995–1005.

42. **Nick, H., and W. Gilbert.** 1985. Detection in vivo of protein-DNA interactions within the lac operon of *Escherichia coli*. *Nature* **313**:795–797.

43. **O'Neill, L. P., and B. M. Turner.** 1996. Immunoprecipitation of chromatin. *Methods Enzymol.* **274**:189–197.

44. **Orlando, V.** 2000. Mapping chromosomal proteins in vivo by formaldehyde-crosslinked-chromatin immunoprecipitation. *Trends Biochem. Sci.* **25**:99–104.

45. **Pereira, S. L., R. A. Grayling, R. Lurz, and J. N. Reeve.** 1997. Archaeal nucleosomes. *Proc. Natl. Acad. Sci. USA* **94**:12633–12637.

46. **Revzin, A.** 1989. Gel electrophoresis assays for DNA-protein interactions. *BioTechniques* **7**:346–355.

47. **Rokop, M. E., J. M. Auchtung, and A. D. Grossman.** 2004. Control of DNA replication initiation by recruitment of an essential initiation protein to the membrane of *Bacillus subtilis*. *Mol. Microbiol.* **52**:1757–1767.

48. **Salzer, W., R. F. Gestland, and A. Bolle.** 1967. In vitro synthesis of bacteriophage lysozyme. *Nature* **215**:588–591.

49. **Sambrook, J., E. F. Fritsch, and T. Maniatis.** 1989. *Molecular Cloning: A Laboratory Manual*. Cold Spring Harbor Laboratory Press, Cold Spring Harbor, NY.

50. **Sanger, F., S. Nicklen, and A. R. Coulson.** 1977. DNA sequencing with chain terminating inhibitors. *Proc. Natl. Acad. Sci. USA* **74**:5463–5467.

51. **Sasse-Dwight, S., and J. D. Gralla.** 1988. Probing cooperative DNA-binding in vivo: the *lac* $O_1:O_3$ interaction. *J. Mol. Biol.* **202**:107–109.

52. **Shanblatt, S. H., and A. Revzin.** 1983. Two catabolite activator protein molecules bind to the galactose promoter region of *Escherichia coli* in the presence of RNA polymerase. *Proc. Natl. Acad. Sci. USA* **80**:1594–1598.

53. **Shanblatt, S. H., and A. Revzin.** 1984. Kinetics of RNA polymerase-promoter complex formation: effect of nonspe-cific DNA-protein interactions. *Nucleic Acids Res.* **12:** 5287–5306.

54. **Sheen, J.-Y., and S. Brian.** 1988. Electrolyte gradient gels for DNA sequencing. *BioTechniques* **6**:942–944.

55. **Shin, D., and E. A. Groisman.** 2005. Signal-dependent binding of the response regulators PhoP and PmrA to their target promoters in vivo. *J. Biol. Chem.* **280**:4089–4094.

56. **Siebenlist, U., and W. Gilbert.** 1980. Contacts between *Escherichia coli* RNA polymerase and an early promoter of phage T7. *Proc. Natl. Acad. Sci. USA* **77**:122–126.

57. **Solomon, M. J., and A. Varshavsky.** 1985. Formaldehyde-mediated DNA-protein crosslinking: a probe for in vivo chromatin structures. *Proc. Natl. Acad. Sci. USA* **82**:6470–6474.

58. **Spencer, V. A., J. M. Sun, L. Li, and J. R. Davie.** 2003. Chromatin immunoprecipitation: a tool for studying histone acetylation and transcription factor binding. *Methods* **31**:67–75.

59. **Straney, D. C., S. B. Straney, and D. M. Crothers.** 1989. Synergy between *Escherichia coli* CAP protein and RNA polymerase in the *lac* promoter open complex. *J. Mol. Biol.* **206**:41–57.

60. **Tabor, S., and C. C. Richardson.** 1985. A bacteriophage T7 RNA polymerase/promoter system for controlled exclusive expression of specific genes. *Proc. Natl. Acad. Sci. USA* **82**:1074–1078.

61. **Tabor, S., and C. C. Richardson.** 1987. DNA sequence analysis with a modified bacteriophage T7 DNA polymerase. *Proc. Natl. Acad. Sci. USA* **84**:4767–4771.

62. **Tullius, T. D.** 1989. Physical studies of protein-DNA complexes by footprinting. *Annu. Rev. Biophys. Biophys. Chem.* **18**:213–237.

63. **Tullius, T. D., and B. A. Dombroski.** 1986. Hydroxyl radical "footprinting": high-resolution information about DNA-protein contacts and application to lambda repressor and Cro protein. *Proc. Natl. Acad. Sci. USA* **83**:5469–5473.

64. **Tullius, T. D., B. A. Dombroski, M. E. A. Churchill, and L. Kam.** 1987. Hydroxyl radical footprinting: a high resolution method for mapping protein-DNA contacts. *Methods Enzymol.* **155**:537–558.

65. **Weaver, R. F., and C. Weissman.** 1979. Mapping of RNA by a modification of the Berk-Sharp procedure: the 5′ termini of 15S β-globin mRNA precursor and mature 10S β-globin mRNA have identical map coordinates. *Nucleic Acids Res.* **7**:1175–1193.

66. **Wigneshweraraj, S. R., A. Ishihama, and M. Buck.** 2001. In vitro roles of invariant helix-turn-helix motif residue R383 in sigma(54) (sigma(N)). *Nucleic Acids Res.* **29**:1163–1174.

67. **Wu, H. M., and D. M. Crothers.** 1984. The locus of sequence-directed and protein-induced DNA bending. *Nature* **308**:509–513.

68. **Yang, C.-C., and H. A. Nash.** 1989. The interaction of *Escherichia coli* IHF protein with its specific binding sites. *Cell* **57**:869–880.

28

Measuring Spontaneous Mutation Rates

PATRICIA L. FOSTER

28.1. INTRODUCTION

Spontaneous mutations are mutations that occur in the absence of exogenous causes. They can be due to errors made by DNA polymerases during replication or repair, errors made during recombination, the movement of genetic elements, spontaneous events such as DNA base loss, and DNA damage from reactive compounds produced by normal cellular metabolism. The rate at which spontaneous mutations occur can yield useful information about cellular processes. For example, the occurrence of specific classes of mutations in various mutant backgrounds has been used to deduce the prevalence and importance of specific DNA repair pathways (23). Of particular interest are the spontaneous mutation rates of organisms that live in environments so extreme that the coding properties of their DNA should be destroyed. Do these organisms have extremely efficient mechanisms to protect or repair their DNA, or do they simply tolerate a high rate of genetic change (10, 12)?

Accurately determining the rates of mutations due to spontaneous events is more challenging than determining the rates of mutations due to exogenously applied DNA-damaging agents. If a population of cells is subjected to DNA damage at a specific time point, and if the level of induced mutation is well above the rate of spontaneous mutation, the frequency of mutants among the survivors reflects only the number of mutations induced. (It is important to distinguish a mutation—a heritable change in the genetic material—from a mutant—an individual that carries a mutation.) Importantly, because spontaneous mutations are rare relative to induced mutations, the mutant faction remains a constant as the cells proliferate. In contrast, the events that give rise to spontaneous mutations occur continuously and so the fraction of cells that are mutant due to spontaneous events increases with every division. In addition, as a population proliferates, the time at which a spontaneous mutation occurs has a dramatic effect on the ultimate mutant fraction; mutations that occur early will give rise to "jackpots," which are cultures with abnormally large numbers of mutant individuals.

The mutation rate is the expected number of mutations that an average cell will sustain during its lifetime. Since mutation rates are very low, this number is best understood as a probability, usually of the order of 1 per 10^9 to 10^{10} cell lifetimes for a given mutation. The mutant fraction or frequency is the proportion of cells that are mutant and is typically of the order of 1 per 10^5 to 10^7 cells for a given mutant phenotype. Although mutant frequencies are adequate to assess rates of induced mutations, they are very inaccurate measures of rates of spontaneous mutations. And, unlike most biological phenomena, the accuracy of the measurement does not improve with the number of replicates. This is the fundamental property of spontaneous mutation that was exploited in the famous "fluctuation test" of Luria and Delbrück (21). The theory is as follows.

During the growth of a population, every cell has an equal and constant chance of sustaining a mutation during its lifetime. This probability, which is the mutation rate, is low, and the number of cells is large, so the numbers of mutations sustained among replicate cultures follow the Poisson distribution. However, once a mutation occurs, the cell sustaining it will produce a clone of mutant progeny, all of which will contribute to the number of mutants in that culture. The size of a given mutant clone will depend on when during the growth of the population the mutation occurred; mutations occurring early will produce large clones, whereas mutations occurring late will produce small clones. Thus, the distribution of the numbers of mutants among replicate cultures is far from Poisson and is usually referred to as the Luria-Delbrück distribution. Because of this basic biology, the frequency of mutants in a culture has no simple relationship to the mutation rate, but it can be used to estimate the mutation rate by various mathematical manipulations.

The Luria-Delbrück distribution, which has no closed-form solution, has stimulated the interest of mathematicians since the original paper describing it was published in 1943. Unfortunately, methods to calculate mutation rates from this distribution remain opaque to most investigators. This chapter will provide a guide to these methods in the hope that it will be useful to scientists and students who wish to use mutation rates in their research. In addition, the Luria-Delbrück distribution applies to other cases in which a rare initiating event is amplified in a population. For example, disease genes can be mapped by using the Luria-Delbrück distribution to estimate the rate at which linkage disequilibrium is lost after a founder mutation is introduced into a population (13).

There are two basic methods to determine mutation rates: mutant accumulation and fluctuation analysis. These two methods are described below, with an emphasis on fluctuation analysis.

28.2. TERMINOLOGY

It is unfortunate that many different symbols have been used for the terms that enter into the equations for calculating mutation rates. Most of the terms used here date from the report by Luria and Delbrück (21) and are given in Table 1. It is particularly important to distinguish between m, the mean number of mutations (mutational events) that occurred during the growth of a culture, and μ, the mutation rate, which is the probability that a cell will sustain a mutation during its lifetime. Almost all methods to calculate the mutation rate start by determining m and then obtain μ by dividing m by some measure of the number of cells at risk. For example, Luria and Delbrück (21) divided m by the total number of cells, N_t, to give the mutation rate.

It is also important to distinguish between the number of mutations per culture, m, and the number of mutants per culture, r. The number of mutants derives from both the number of new mutants that arose during the growth of a culture and the clonal expansion of the progeny of each founder mutant. r is what we can actually observe and use to calculate m. r divided by N_t is the mutant fraction or mutant frequency, f.

TABLE 1 Definition of terms

Term	Definition
m	Number of mutations per culture
μ	Mutation rate; probability of mutation per cell per division or generation
N_0	Initial number of cells in a culture; the inoculum
N_t	Final number of cells in a culture
r	Number of mutants in a culture
$\bar{r}$	Mean number of mutants per culture
$\tilde{r}$	Median number of mutants per culture
f	Mutant fraction or frequency; r/N
V	Volume of a culture
C	Number of cultures in an experiment
p_0	Proportion of cultures without mutants
p_r	Proportion of cultures with r mutants
P_r	Proportion of cultures with r or more mutants
z	Dilution factor or fraction of a culture plated
Q_1	Value of r at 25% of the ranked series of r's
Q_2	Value of r at 50% of the ranked series of r's; the median
Q_3	Value of r at 75% of the ranked series of r's
c_r	Number of cultures with r mutants
σ	Standard deviation
CL	Confidence limit

28.3. ASSUMPTIONS ABOUT MUTATIONAL PROCESSES

All methods to estimate spontaneous mutation rates are based on theoretical models of mutational processes and cell growth. The most common model originated with Luria and Delbrück (21) and was further defined by Lea and Coulson (19); it is usually called the Lea-Coulson model. It has the following assumptions.

The population increases exponentially

The initial number of cells is negligible compared to the final number of cells

The probability of mutation is constant per cell lifetime

The probability of mutation per cell lifetime does not vary during the growth of the culture

The growth rates of mutants and nonmutants are the same

The proportion of mutants is always small

The number of reverse mutations is negligible

The death rate is negligible

All mutants are detected

No mutants arise after selection is imposed

An additional implicit assumption is that the probability that a mutation occurs is not influenced by previous mutational events, which is the randomness required for the Poisson distribution. Various theoretical papers have addressed how departures from the Lea-Coulson model affect the distribution of mutant numbers. In general, most of the model's requirements can be met with proper experimental protocols, but some departures reflect real biological phenomena.

28.4. MUTANT ACCUMULATION

After an exponentially growing population reaches a sufficient size so that m is $\gg 1$, the accumulation of mutants becomes a simple function of the cell number (20). As illustrated in Fig. 1, the combination of new mutants and the growth of preexisting mutants results in a constant increase in the mutant fraction; the mutation rate is equal to this increase divided by the number of individuals present. Therefore, the conceptually simplest of all ways to determine a mutation rate is to measure the change in the mutant fraction in a growing population. Considering the population growth to be continuous, μ can be calculated by using equation 1 (9):

$$\mu = \left(\frac{r_2}{N_2} - \frac{r_1}{N_1} \right) \div \left[\ln\left(\frac{N_2}{N_1} \right) \right]$$
$$= (f_2 - f_1) \div (\ln N_2 - \ln N_1) \tag{1}$$

There are important limitations to the use of this method. By the time the population reaches a size large enough to allow the mutant fraction to be measured, mutations have already occurred and their numbers will be subject to the fluctuation effect. In many cases, these fluctuations will make it impossible to accurately measure the accumulation of new mutants. However, the mutant accumulation method can be used successfully when the number of mutations within a large population can be kept low. For example, stable stocks containing large numbers of bacteriophages or viruses can be generated and tested for mutants; the stocks with low mutant frequencies can then be used repeatedly for experiments. Mutant accumulation can also be used if a population can be purged of preexisting mutants by using

Generation = k	Cells = N	New Mutants	Growth of Pre-existing Mutants	Total Mutants = r	Mutant Fraction = r/N	Change per Generation = μ
4	Nt/16	1		1	16/Nt	16/Nt
3	Nt/8	2	2	4	32/Nt	16/Nt
2	Nt/4	4	4 4	12	48/Nt	16/Nt
1	Nt/2	8	8 8 8	32	64/Nt	16/Nt
0	Nt	16	16 16 16 16	80	80/Nt	16/Nt

FIGURE 1 Illustration of the constant increase in the mutant fraction after a population reaches a size sufficiently large that the accumulation of mutants is simply a function of population size. Luria's conventions are followed (20): k, the generation numbered backwards from 0; N, the number of cells present at each generation; N_t, the final number of cells in the population; μ, the mutation rate per cell (assuming a synchronous population). At each generation, there are $N_t/2^k$ individuals that produce $\mu N_t/2^k$ new mutations, which will produce a total of μN_t mutant progeny by the last generation. Adapted from reference 10a with permission of the publisher.

a mutational target that can be selected both for and against (i.e., a counterselectable marker). For example, if a population of Lac$^-$ cells carrying a temperature-sensitive *galE* mutation is incubated in lactose medium at the nonpermissive temperature, Lac$^+$ revertants will die (because one product of β-galactosidase is galactose and the accumulation of UDP-galactose is toxic in the absence of the epimerase encoded by *galE*); if the population is then shifted to the permissive temperature, the accumulation of new Lac$^+$ mutants can be monitored (24). Another possibility is to use a mutational target that allows mutants to be eliminated by cell sorting. For example, mutants that allow green fluorescent protein to be produced can be eliminated by fluorescence-activated cell sorting; new mutants can then be detected by flow cytometry (4).

28.5. FLUCTUATION ANALYSIS

28.5.1. Experimental Design

A normal fluctuation assay begins by inoculating parallel cultures with a population of cells that is small enough to be unlikely to contain any preexisting mutants. The cultures are allowed to grow, usually to saturation, and the total number of cells is determined by plating appropriate dilutions onto nonselective medium. The number of mutants is determined by plating each culture onto a selective medium, where each mutant produces a colony. The mutation rate is calculated from the distribution of the numbers of mutants among the parallel cultures. This basic design can be used for single-celled microorganisms, cultured cells, bacteriophage, and viruses. Selection usually takes place on solid medium so that individual mutants can be enumerated; however, the method utilizing the proportion of cultures without mutants (p_0; see below) can be carried out with liquid medium as well.

Proper design of a fluctuation assay will maximize the precision of the estimate of the mutation rate. Precision is a measure of reproducibility, not accuracy. Accuracy is how well the resulting estimate reflects the actual mutation rate, and that will depend on the applicability of the underlying assumptions. The parameters that are important in designing a fluctuation assay are as follows: m, the number of mutations per culture; r, the number of mutants per culture; N_0, the inoculum; N_t, the final number of cells; V, the culture volume; and C, the number of parallel cultures. Preliminary experiments should be done to determine the range of r's obtained when a given number of cells is plated. These experiments consist simply of growing three to five cultures in the desired medium, plating aliquots of various sizes onto the selective medium, and determining the number of mutants per culture obtained. A provisional m can then be estimated from the median of these experiments (by using method 3 or 4; see below), and this value can be used to adjust the other parameters. In general, m should be between 0.3 and 15; at low values of m, the error is large, but if m is above 15, some of the more convenient methods to calculate the mutation rate are not valid (11, 25). Obviously, m also has to be within a range so that the number of mutants per plate is countable; however, if the number of mutants per plate is in the typical range of 0 to 50, there is little loss in precision if occasionally high colony counts are truncated at a value of about 150 (3, 15).

The desired m can be achieved by adjusting N_t, the final number of cells. When defined medium is used, this adjustment can be done by limiting the carbon source. However, it is not recommended to limit the cells with other growth factors, such as vitamins or amino acids, because the cell physiology may change and mutants that do not require the growth factor may arise. The alternative to limiting cell growth is to adjust the volume of the culture to obtain the desired N_t. It has long been considered the best practice to subject the entirety of each culture to selection because sampling introduces errors. Sampling or low plating efficiency (which is the same thing) narrows the distribution of the number of mutants, increasing the influence of small numbers and decreasing the influence of large numbers (7, 28). However, a recent analysis has concluded that it is better to plate a small aliquot from a large culture than all of a small culture because the large culture contains more information than the small culture (15). In addition, if several mutant phenotypes are to be assayed simultaneously, sampling may be unavoidable. Simple corrections for sampling can be applied when using some, but not all, of the various methods to calculate mutation rates.

It is extremely important that when selection is applied, N_t be the same for each culture. Usually this condition can be ensured by growing cells to saturation, although care should be taken when the strain used has growth defects. Also, in some cases it may be desirable to use exponentially growing cells. Cell numbers can be monitored before mutant selection by measuring the optical density or by counting the cells microscopically (e.g., by using a Petroff-Hausser chamber); however, these measurements should be confirmed by plating to determine the numbers of viable cells. Presently there is no valid method to correct mutant numbers for different N_t's, so deviant cultures must be eliminated from the analysis (11).

To easily calculate m, the initial inoculum, N_0, must be negligible compared to N_t. A ratio of 1/1,000 is usually sufficient if m is ≤ 10 (26), but a smaller inoculum may be needed to ensure that there are no preexisting mutants. Obviously, with a small inoculum, the cultures will take longer to saturate and viability may be a problem, which may increase the variation of N_t's. A reasonable rule of thumb is that N_0 should be approximately equal to (N_t/m) $\times 10^{-4}$ (11).

It is a common practice to begin a fluctuation analysis by inoculating each culture with a single colony of the strain being tested. There are two problems with this procedure. First, N_0's will vary among the cultures. Second, a colony has undergone the same process of mutant accumulation as a culture, and each colony will contain a different number of mutants. Thus, some of the cultures in the fluctuation test may be inoculated with preexisting mutants. A better practice is to dilute one culture or colony into an appropriate volume of fresh medium and then distribute aliquots into the individual cultures for nonselective growth.

Because the precision of the estimate of m depends on the number of parallel cultures, C, the larger C is, the better the estimate is. Most experiments have 10 to 100 parallel cultures, with numbers around 40 being most common. There is little gain in precision if C is increased above this level unless m is less than about 0.3 (11, 15, 25).

In experiments with nonlethal selections, for example, the reversion of an auxotrophy or the utilization of a carbon source, mutants can arise after selection has been applied. These mutants can be the result of mutations occurring as the cells proliferate on the selective medium, or the result of mutations occurring in the nongrowing cells (adaptive mutations). In either case, the m estimated for preplating mutations will be inflated by these postplating mutations. If nonmutant cells can grow on the selective medium because of contaminating nutrients, the problem can be solved by adding an excess of scavenger cells that cannot mutate (because, for example, they have a deletion of the relevant gene) (5). If the cells proliferate because the selection is not stringent, the time it takes for a mutant colony to form on the selective medium can be determined and then colonies can be counted at the earliest possible time after plating.

28.5.2. Analyzing Results of Fluctuation Assays

A fluctuation assay yields a distribution of the number of mutants per culture, r, which is used to calculated m, the mean or most likely number of mutations per culture. It is helpful to remember that although the distribution of mutants is not Poisson the distribution of mutations is, so m is a Poisson parameter. All the methods to calculate m depend on the theoretically described distribution of mutant clone sizes, but the methods differ in the mathematics used to obtain m from the experimentally obtained distributions. The equations and parameters used to calculate m are usually called estimators. Over the last 50 years, the methods have improved but the complexity of the calculations has increased, and access to a computer is required for some of the more complex calculations. The MSS maximum-likelihood method (method 8) is the "gold standard" because it utilizes all of the results of an experiment, it can be used over the entire range of mutation rates, and it yields results that can be statistically evaluated. Of the less complicated methods, the Lea-Coulson method of the median (method 3) and the Jones median estimator (method 4) are reliable if mutation rates are low to moderate; if mutation rates are very low ($m \leq 1$), only the p_0 method (method 1) is applicable (11, 25).

Method 1: the p_0 Method

Since the distribution of the numbers of mutations per culture is Poisson, the mean can be calculated from p_0, the proportion of cultures with no mutants (because if there are no mutants, there were no mutations) (21).

Since

$$p_0 = e^{-m} \tag{2}$$

Then

$$m = -\ln p_0 \tag{3}$$

The great advantage of the p_0 method is its simplicity; N_t is adjusted so that about a third of the cultures have no mutants, and m is calculated from equation 3. Although simple, the method is limited. Obviously, it can be used only when the number of mutants is small, and the error of small numbers is large. The range of usefulness of the p_0 method is: $0.3 \leq m \leq 2.3$ ($0.7 \geq p_0 \geq 0.1$). No method is reliable when m is <0.3 unless hundreds of cultures are used, and if m is ≥ 2.3, there are too few cultures with no mutants to use the p_0 method. In addition, the p_0 method is inefficient (i.e., it requires more cultures for the same precision) compared to other methods (11, 18, 25).

The p_0 estimator is also sensitive to factors other than the mutation rate that increase p_0. For example, if the phenotype being selected is not immediately expressed (i.e., it shows phenotypic lag), mutations that occur in the last few generations will not produce clones and, thus, the p_0 will be inflated. Likewise, if mutants die or have a poor plating efficiency, p_0 will be increased. However, if only a fraction of all cells (not just mutants) is detected and this fraction is known, a correction factor can be applied to m. For example, if only a fraction, z, of a culture is plated, the p_0 will be increased, but the m calculated from this observed p_0 (m_{obs}) can be corrected to give the actual m (m_{act}) by using equation 4 (14, 28).

$$m_{act} = m_{obs} \frac{z - 1}{z \ln(z)} \tag{4}$$

Note that when z is 1, $(z - 1)/[z \ln (z)]$ equals 1 from l'Hôpital's rule and m_{act} equals m_{obs}.

Method 2: the Method of the Mean

The method of the mean was first formulated by Luria and Delbrück (21). Rearranging their equation gives the following:

$$\bar{r} = m \ln(mC) \tag{5}$$

where $\bar{r}$ is simply the mean number of mutants per culture and C is the number of cultures. This transcendental equation can easily be solved by iteration by using the following rearrangement:

$$m \ln(mC) - \bar{r} = 0 \tag{6}$$

This method is very easy to apply but, like all methods relying on the mean, has severe limitations. Because the distribution of mutant numbers is inherently skewed even without obvious jackpots, the mean will be dominated by the few cultures with large numbers of mutants. Luria and Delbrück attempted to eliminate this problem by considering only mutations that occur after the population reaches a size sufficiently large that the accumulation of mutants is deterministic (see above) (Fig. 1). They took this to be the point when NC, the whole population size in the experiment (i.e., all the cells in all the cultures), equals $1/\mu$. Alternatively, Armitage (2) considered the deterministic period to begin when the population of each culture reaches $1/\mu$, in which case the equation is as follows:

$$m \ln(m) - \bar{r} = 0 \tag{7}$$

Obviously, mean estimators should not be used to calculate mutation rates when N_t is $\leq 1/\mu$, i.e., when m is ≤ 1. In addition, the method assumes that no mutations occur before the deterministic period, which is unlikely to be true. As a result, methods using the mean will frequently overestimate mutation rates. If jackpots could be eliminated, then the mean might be useful, but there is no way to nonarbitrarily decide what is or is not a jackpot. The use of the median instead of the mean improves the method somewhat (11, 25).

Method 3: Lea-Coulson Method of the Median

Lea and Coulson (19) observed that for m over the range from 4 to 15, a plot of the probability that a culture contains r or fewer mutants versus $[r/m - \ln(m)]$ gives a smooth curve with a median of 1.24. Rearranging gives the following transcendental equation:

$$\frac{\tilde{r}}{m} - \ln(m) - 1.24 = 0 \tag{8}$$

where $\tilde{r}$ is the median number of mutants per culture. This equation can easily be solved by iteration. The Lea and Coulson method of the median is easy to apply and remarkably accurate as judged by its performance in computer simulations (3, 27). It performs well within the range $1.5 \leq m \leq 15$ ($2.5 \leq \tilde{r} \leq 60$) (11, 25). Because it uses the median, it is relatively insensitive to deviations that affect either end of the distribution. For example, experiments can be designed so the $\tilde{r}$ is large enough to be unaffected by phenotypic lag. However, the method of the median compares unfavorably to the maximum-likelihood method (method 8) because little of the information obtained from the fluctuation test is used, making the method relatively inefficient (11, 25).

Method 4: the Jones Median Estimator

Jones et al. (16) derived an estimator based on the hypothetical dilution that would be necessary to produce a dis-

tribution with its median equal to 0.5 (which also means that p_0 equals 0.5). The two advantages of the Jones estimator are that it is explicit and that it accommodates dilutions. The basic equation is as follows:

$$m = \frac{\tilde{r} - \ln(2)}{\ln(\tilde{r}) - \ln[\ln(2)]} = \frac{\tilde{r} - 0.693}{\ln(\tilde{r}) + 0.3665} \tag{9}$$

If z is the dilution factor (i.e., the fraction of the culture that is plated) or the plating efficiency, then the observed median, $\tilde{r}_{obs}$, can be used in the following equation:

$$m = \frac{\left(\dfrac{\tilde{r}_{obs}}{z}\right) - 0.693}{\ln\left(\dfrac{\tilde{r}_{obs}}{z}\right) + 0.367} \tag{10}$$

The Jones median estimator method was verified with simulated data for the range of m from 1.5 to 10 ($1.5 \leq m \leq 10$; $3 \leq \tilde{r} \leq 40$) and proved to be as reliable as and more efficient than the Lea and Coulson median estimator (16). It also performs well with real data (11, 25).

Method 5: Drake's Formula

Drake's formula (8) provides an easy way to calculate mutation rates from mutant frequencies and is especially useful in comparing data from different sources. Starting from equation 1 above, Drake sets N_1 to be not the initial inoculum, but $1/\mu$, which is the population size when the probability of mutation approaches unity. Setting f_1 as zero and f_2 as the final mutant frequency gives the following:

$$u = f \div \ln(uN_t) \tag{11}$$

which can be solved for μ by iteration. Drake's formula is based on the same assumption as the Luria-Delbrück method of the mean, i.e., that mutations occur only during the deterministic period of mutant accumulation. To minimize the influence of jackpots, the median frequency is used instead of the mean (8). By setting μ equal to m/N_t, Drake's formula can be rearranged into a form similar to that of Lea and Coulson's (see equation 8) as follows (25):

$$\frac{\tilde{r}}{m} - \ln(m) = 0 \tag{12}$$

Because of its derivation, Drake's formula is best used only when mutation rates are high. When m is <4, the estimates of m from equation 12 are significantly higher than those from equation 8. As m becomes larger, values calculated from equation 12 asymptotically approach those calculated from equation 8; the differences are trivial when m is ≥ 30 (11, 25).

Method 6: Accumulation of Clones

Luria (20) pointed out that P_r, the proportion of cultures with r or more mutants, approaches $2m/r$ during the deterministic portion of growth. Formally,

$$P_r = \sum_{i=r}^{i=N_1} p_i = \frac{2m}{r} \tag{13}$$

Taking logarithms gives the following:

$$\ln(P_r) = -\ln(r) + \ln(2m) \tag{14}$$

Thus, a plot of $\ln(P_r)$ versus $\ln(r)$ will yield a straight line with a slope of -1 and an intercept [where $\ln(r) = 0$] equal to $\ln(2m)$ (Fig. 2). Dividing P_r at the intercept by 2 gives m (11, 25).

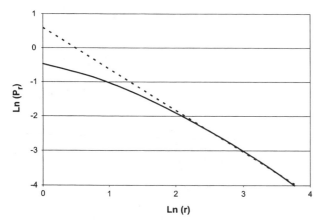

FIGURE 2 $\ln(P_r)$ versus $\ln(r)$ for the Luria-Delbrück distribution. The distribution (solid line) was calculated for m of 1 by using equation 18. A curve (dotted line) has been fitted by the least-squares method to the upper part of the distribution [$\ln(r) \geq 2$]; it has a slope of -1.2 and an intercept of 0.6, giving an m of 0.9.

Method 7: the Quartiles Method

The median is the central (50%) quartile of a distribution and is the parameter used in most of the methods described above. More of the data from a fluctuation assay are incorporated into the calculation of m if the upper (75%) and lower (25%) quartiles are also used. By regressing m versus the theoretical values of r at the quartiles, Koch (17) derived the following empirical equations:

$$m_1 = 1.7335 + 0.4474Q_1 - 0.002755(Q_1)^2 \quad (15)$$

$$m_2 = 1.1580 + 0.2730Q_2 - 0.000761(Q_2)^? \quad (16)$$

$$m_3 = 0.6658 + 0.1497Q_3 - 0.0001387(Q_3)^2 \quad (17)$$

where Q_1, Q_2, and Q_3 are the values of r at 25, 50, and 75% of the ranked series of observed r's. For a perfect Luria-Delbrück distribution, the three m's should be equal. These equations are valid over the range of m from 2 to 14; Koch also gives graphs that can be used up to values of m of 120 (17).

Method 8: the MSS Maximum-Likelihood Method

Sarkar et al. (26) derived the following recursive algorithm that efficiently computes the Luria-Delbrück distribution for a given value of m

$$p_0 = e^{-m}; \quad p_r = \frac{m}{r} \sum_{i=0}^{r-1} \frac{p_i}{(r-i+1)} \quad (18)$$

where p_r is the proportion of cultures with r mutants. This algorithm can be used to estimate m from experimental results by using the MMS maximum-likelihood method (22), the formula for which is as follows:

$$f(r|m) = \prod_{i=1}^{C} f(r|m) \quad (19)$$

where $f(r_i | m)$ equals p_r from equation 18. The procedure is to start with a trial m (obtained from equation 8, for example) and use equation 18 to calculate the probability, p_r, of obtaining each possible r from 0 to the maximum value obtained (even if a given r was not obtained in the experiment). For most experiments, values of r greater than 150

are so rare that they contribute little to the distribution and can be lumped into one category (3). The likelihood function, equation 19, is the product of these calculated p_r's for each culture. The easiest way to do this calculation is to arrange the mutant counts from an experiment in order and count the number of cultures with each r number of mutants. Then the product of the p_r's is as follows:

$$(p_0)^{c_0} \times (p_1)^{c_1} \times (p_2)^{c_2} \times (p_3)^{c_3} \times (p_4)^{c_4} \cdots \quad (20)$$

where each p_r is from equation 18 and c_r is the number of cultures that had r mutants (note that because equation 18 is recursive, r's that were not obtained enter into the calculation of the p_r's, but because these give a value of 1 in equation 19, they do not have to appear in equation 20). Alternatively, the values of c_r times $\ln(p_r)$ for each r can be added. Several m's over a small range are evaluated to find the m that maximizes the likelihood function.

As mentioned above, the MSS maximum-likelihood method is the best method presently available to estimate m. Unlike the other methods discussed above, except methods 6 and 7, the maximum-likelihood method uses all the results from a fluctuation experiment. Because of its derivation, it is valid over the entire range of mutation rates (although the calculations become tedious at high values of r). In addition, simulations have shown it to perform in a manner regular enough to allow for statistical evaluation (27).

28.5.3. Calculating the Mutation Rate

The mutation rate, μ, is calculated by dividing m, which is an estimate of the number of mutations that occurred during the growth of a culture, by some measure of the number of cells at risk for mutation during that growth. There are three such measures, each of which reflects different assumptions about the underlying mutational process.

Luria and Delbrück (21) and most theorists assume that the probability of mutation is constant over the cell division cycle. Since the final number of cells in a culture, N_t, arose from $(N_t - 1)$ divisions, the mutation rate is as follows:

$$u = \frac{m}{(N_t - 1)} \approx \frac{m}{N_t} \quad (21)$$

The same calculation applies if mutations are assumed to occur at the end of the division cycle (i.e., at or during division), so that the number of "opportunities" for mutation is equal to the number of divisions (1).

If mutations are assumed to occur at the beginning of the division cycle (i.e., shortly after division), then the number of cells at risk is equal to the total number of cells that ever existed in the culture, which is $2N_t$ (because N_t cells had $N_t/2$ parents and $N_t/4$ grandparents, etc., and the sum of the series is $2N_t$) . Thus, the mutation rate (1) is as follows:

$$u = \frac{m}{2N_t} \quad (22)$$

Finally, microorganisms rarely grow synchronously. During one generation period for an asynchronously exponentially growing population, there is an average of $N/\ln(2)$ cells. Thus, the total number of divisions during the growth of a culture is $N_t/\ln(2)$ and the mutation rate (1) is as follows:

$$u = \frac{m\ln(2)}{N_t} = \frac{0.6932m}{N_t} \quad (23)$$

For the same m, these three methods will give mutation rates that differ by 1, 0.5, and 0.693.

28.5.4. Statistical Methods To Evaluate Mutation Rates

The estimates of m or μ obtained from fluctuation tests are neither normally distributed nor unbiased; therefore, normal statistics cannot be used to evaluate the results (3, 16, 27). No matter how many times a fluctuation experiment is repeated, it is simply not valid to take the mean and standard deviation of the results. There are two general approaches that can allow reasonable confidence limits to be placed around mutation rate estimates.

If the parameter that was used to calculate m has a defined distribution, that distribution can be used to evaluate m. For example, because a culture either has mutants or does not, p_0 (but not m) can be considered a binomial parameter (19). The confidence limits for a binomial parameter can be found by using standard statistical methods, and the new p_0 at each limit can be used to calculate new m's. These new m's are estimates of the confidence limits for m as determined by the p_0 method. Similarly, when the median is used to estimate m, standard methods can be used to calculate confidence limits for the median and then m can be recalculated using these median limits to estimate the confidence limits for m (11, 25).

The second approach to statistically evaluating m is to find a transforming function that gives m a normal distribution. Using simulated fluctuation tests, Stewart (27) compared the distributions of m's obtained by several methods. These results indicated that the logarithms of m's obtained by using the MSS maximum-likelihood method (equation 19) are approximately normally distributed. From this, Stewart calculated the standard deviation, σ, of the logarithm of $\ln(m)$ to be as follows:

$$\sigma_{\ln(m)} \approx \frac{1.225 m^{-0.315}}{\sqrt{C}} \qquad (24)$$

where C is the number of cultures. Since $\ln(m)$ is normally distributed, the 95% confidence limits for $\ln(m)$ should be

$$\ln(m) \pm 1.96\sigma_{\ln(m)} \qquad (25)$$

While this is a reasonable approximation, the true confidence limits must be calculated from the actual m and σ of the population, not the experimentally determined m and σ. Methods to calculate or estimate the correct confidence limits are given by Stewart (27). A close approximation can be obtained from the following equations (11, 25)

$$CL_{+95\%} = \ln(m) + 1.96\sigma(e^{1.96\sigma})^{-0.315} \qquad (26)$$

$$CL_{-95\%} = \ln(m) - 1.96\sigma(e^{1.96\sigma})^{+0.315} \qquad (27)$$

Once the upper and lower limits for $\ln(m)$ are obtained, the upper and lower limits for m are simply the antilogarithms.

Once confidence limits are obtained for m, these can then be divided by N_t (or $2N_t$ or $N_t/\ln 2$) to estimate the confidence limits for μ, the mutation rate. Of course, this procedure ignores the variance associated with the determination of N_t, which is approximately equal to N_t. Although this variance is nontrivial, most theorists consider it to be negligible, and it is probably justifiable to ignore it as long as N_t is determined accurately.

28.5.5. Departures from the Luria-Delbrück and Lea-Coulson Model

All the methods for calculating mutation rates from fluctuation tests discussed above are dependent for their applica-bility on the model of expansion of mutant clones originally described by Luria and Delbrück (21) and extended by Lea and Coulson (19). If the assumptions of the model are not true for a particular application, then the calculated mutation rate will be wrong. Several meaningful biological departures have been considered by theorists and can, with some care, be accounted for to derive a meaningful mutation rate.

A plating efficiency for all cells (not just mutants) of less than 100% has the same effect as plating only samples of the cultures: all clones will be reduced in size by the same relative amount. The actual m can be calculated by using equation 4 or 10, where z is either the fraction of the culture that is plated or the fraction of the cells that can grow after plating.

Phenotypic lag means that the expression of a mutant phenotype is delayed for several generations. As a result, clones that arise in the last few generations of growth will contribute few mutants, but clones that arise early during growth will contribute a normal number of mutants. Thus, the lower end of the distribution will be affected, but depending on the length of the lag, the upper end will not, resulting in an inflated m. The actual m can be estimated graphically with method 6 by using only the upper part of the curve and eliminating any obvious jackpots (11, 25) (Fig. 2). If the length of the phenotypic lag can be estimated, Koch (17) provides a method for estimating m from the quartiles (method 7). First, the length of the phenotypic lag is estimated in number of doublings, n. Then the three experimental r values that are the points 2^n-fold less than the actual quartiles are used to derive provisional m's from equations 15 through 17 or from the graphs given in reference 17. Different values of n can be tried until the three m's have similar values. Then, the provisional m is multiplied by $(2^n - 1)$ to give the actual m.

If during nonselective growth mutants grow more slowly that nonmutants, mutant clones arising early during the growth of a culture will be reduced in size, whereas clones arising late will be more normal. This occurrence shifts the distribution of mutant numbers from the Luria-Delbrück toward the Poisson (17, 28). Koch (17) gives graphs that can be used to estimate m from the quartiles when the relative growth of mutants ranges from 0.6 to 0.9 of that of the nonmutants. If there is more than one type of mutant and each type has a different growth rate, the distribution can be approximated by assuming that there is only one type whose growth rate is the average (28).

Deviations from the Luria-Delbrück distribution can be detected as distortions of the curve defined by equation 14 (Fig. 2). For example, if mutations occur both during nonselective growth and after selection is applied, a plot of $\ln(P_r)$ versus $\ln(r)$ gives a curve that is a combination of the Luria-Delbrück and Poisson distributions (6). The effects that several other deviations from the Luria-Delbrück and Lea-Coulson assumptions have on the shape of this curve have been modeled previously (28). Experimental data can be fit to these curves to test whether a given factor, such as selection against mutants, is operative. However, all of the deviant curves are rather similar, so any conclusion that a given factor is distorting the distribution would have to be tested experimentally.

28.6. REFERENCES

1. **Armitage, P.** 1952. The statistical theory of bacterial populations subject to mutation. *J. R. Stat. Soc. B* **14:**1–40.
2. **Armitage, P.** 1953. Statistical concepts in the theory of bacterial mutation. *J. Hyg.* **51:**162–184.

3. **Asteris, G., and S. Sarkar.** 1996. Bayesian procedures for the estimation of mutation rates from fluctuation experiments. *Genetics* **142**:313–326.

4. **Bachl, J., M. Dessing, C. Olsson, R. C. von Borstel, and C. Steinberg.** 1999. An experimental solution for the Luria Delbruck fluctuation problem in measuring hypermutation rates. *Proc. Natl. Acad. Sci. USA* **96**:6847–6849.

5. **Cairns, J., and P. L. Foster.** 1991. Adaptive reversion of a frameshift mutation in *Escherichia coli*. *Genetics* **128**:695–701.

6. **Cairns, J., J. Overbaugh, and S. Miller.** 1988. The origin of mutants. *Nature* **335**:142–145.

7. **Crane, B. J., S. M. Thomas, and M. E. Jones.** 1996. A modified Luria-Delbrück fluctuation assay for estimating and comparing mutation rates. *Mutat. Res.* **354**:171–182.

8. **Drake, J. W.** 1991. A constant rate of spontaneous mutation in DNA-based microbes. *Proc. Natl. Acad. Sci. USA* **88**:7160–7164.

9. **Drake, J. W.** 1970. *The Molecular Basis of Mutation.* Holden-Day, Inc., San Francisco, CA.

10. **Fitz-Gibbon, S. T., H. Ladner, U. J. Kim, K. O. Stetter, M. I. Simon, and J. H. Miller.** 2002. Genome sequence of the hyperthermophilic crenarchaeon Pyrobaculum aerophilum. *Proc. Natl. Acad. Sci. USA* **99**:984–989.

10a. **Foster, P. L.** 1999. Sorting out mutation rates. *Proc. Natl. Acad. Sci. USA* **96**:6862–6867.

11. **Foster, P. L.** 2006. Methods for determining spontaneous mutation rates. *Methods Enzymol.* **409**:195–213.

12. **Grogan, D. W.** 2000. The question of DNA repair in hyperthermophilic archaea. *Trends Microbiol.* **8**:180–185.

13. **Hastbacka, J., A. de la Chapelle, I. Kaitila, P. Sistonen, A. Weaver, and E. Lander.** 1992. Linkage disequilibrium mapping in isolated founder populations: diastrophic dysplasia in Finland. *Nat. Genet.* **2**:204–211.

14. **Jones, M. E.** 1993. Accounting for plating efficiency when estimating spontaneous mutation rates. *Mutat. Res.* **292**:187–189.

15. **Jones, M. E., S. M. Thomas, and K. Clarke.** 1999. The application of a linear algebra to the analysis of mutation rates. *J. Theor. Biol.* **199**:11–23.

16. **Jones, M. E., S. M. Thomas, and A. Rogers.** 1994. Luria-Delbrück fluctuation experiments: design and analysis. *Genetics* **136**:1209–1216.

17. **Koch, A. L.** 1982. Mutation and growth rates from Luria-Delbrück fluctuation tests. *Mutat. Res.* **95**:129–143.

18. **Koziol, J. A.** 1991. A note of efficient estimation of mutation rates using Luria-Delbrück fluctuation analysis. *Mutat. Res.* **249**:275–280.

19. **Lea, D. E., and C. A. Coulson.** 1949. The distribution of the numbers of mutants in bacterial populations. *J. Genet.* **49**:264–285.

20. **Luria, S. E.** 1951. The frequency distribution of spontaneous bacteriophage mutants as evidence for the exponential rate of phage reproduction. *Cold Spring Harbor Symp. Quant. Biol.* **16**:463–470.

21. **Luria, S. E., and M. Delbrück.** 1943. Mutations of bacteria from virus sensitivity to virus resistance. *Genetics* **28**:491–511.

22. **Ma, W. T., G. v. H. Sandri, and S. Sarkar.** 1992. Analysis of the Luria-Delbrück distribution using discrete convolution powers. *J. Appl. Prob.* **29**:255–267.

23. **Miller, J. H.** 1996. Spontaneous mutators in bacteria: insights into pathways of mutagenesis and repair. *Annu. Rev. Microbiol.* **50**:625–643.

24. **Reddy, M., and J. Gowrishankar.** 1997. A genetic strategy to demonstrate the occurrence of spontaneous mutations in non-dividing cells within colonies of *Escherichia coli*. *Genetics* **147**:991–1001.

25. **Rosche, W. A., and P. L. Foster.** 2000. Determining mutation rates in bacterial populations. *Methods* **20**:4–17.

26. **Sarkar, S., W. T. Ma, and G. v. H. Sandri.** 1992. On fluctuation analysis: a new, simple and efficient method for computing the expected number of mutants. *Genetica* **85**:173–179.

27. **Stewart, F. M.** 1994. Fluctuation tests: how reliable are the estimates of mutation rates? *Genetics* **137**:1139–1146.

28. **Stewart, F. M., D. M. Gordon, and B. R. Levin.** 1990. Fluctuation analysis: the probability distribution of the number of mutants under different conditions. *Genetics* **124**:175–185.

29

Transposon Mutagenesis

SILVIA ROSSBACH AND FRANS J. DE BRUIJN

Transposable elements are distinct DNA segments that have the unique capacity to move (transpose) to new sites within the genomes of their host organisms. The transposition process is independent of the classical homologous recombination (*rec*) system of the organism. Moreover, the insertion of a transposable element into a new genomic site does not require extensive DNA homology between the ends of the element and its target site. Transposable elements have been found in a wide variety of prokaryotic and eukaryotic organisms, where they can cause null mutations, chromosome rearrangements, and novel patterns of gene expression upon insertion into the coding regions or regulatory sequences of resident genes and operons. For general reviews, see the books edited by Berg and Howe (21) and Craig et al. (47).

Prokaryotic transposable elements can be divided into different groups. One group consists of simple elements such as insertion sequences (IS elements), which are approximately 800 to 1,500 bp in length. IS elements normally consist of a gene encoding an enzyme required for transposition (transposase), flanked by terminally repeated DNA sequences which serve as substrate for the transposase. IS elements were initially identified in the lactose (*lac*) and galactose (*gal*) utilization operons of enteric bacteria, where they were found to cause often unstable, polar mutations upon insertion (198).

Traditionally, structurally more complex transposable elements (transposons, or Tn elements) have been assigned to three different classes. Class I transposons in prokaryotes have been defined as a class of composite transposons that contain IS elements (or parts thereof) as direct or inverted repeats at their termini. Tn5 and Tn10 are members of this class of transposons. Class I transposons behave formally like IS elements but carry additional genes unrelated to transposition functions, such as antibiotic resistance, heavy-metal resistance, or pathogenicity determinant genes (33, 34). Transposons carrying antibiotic resistance genes were first identified in the mid-1970s after transposition from drug resistance transfer plasmids into other replicons in the cell, such as bacteriophage λ (20, 80, 92, 107); see references 21 and 47 for general reviews. A simple italic number is assigned to each independent isolate from nature, e.g., Tn5 (33, 34). An abbreviated list of the most extensively studied transposons is shown in Table 1; for more detailed lists, see references 14 and 15. The insertion of a transposon into a particular genetic locus or replicon (phage) is designated by using a double colon, e.g. lacZ::Tn5 or λ::Tn5 (33, 34). These designations will also be used in this chapter.

Class II transposons (noncomposite transposons) are described as transposons that do not contain IS elements and are related to Tn3. Usually, they contain genes encoding antibiotic (ampicillin) resistance, a transposase, and an enzyme involved in the resolution of the cointegrate transposition intermediate, the resolvase.

Class III transposons include transposable bacteriophages, such as Mu and its relatives. Phage Mu is in fact both a virus and a transposon and has been extremely useful in elucidating the mechanism of transposition of transposable elements (35, 150). It was first discovered in the early 1960s as a novel phage that upon lysogenization could integrate at multiple sites in the host chromosome, thereby frequently causing mutations (mutator phage); see references 35, 150, and 206 for reviews and reference 141 for the complete DNA sequence of phage Mu.

Recently, other mobile elements have been discovered that do not fit into these three categories. They include integrons, conjugative transposons (CTns), mobilizable transposons (MTns), and retrotransposons (213). Integrons usually consist of a gene encoding an integrase enzyme, a strong promoter, and a recombination site (159). The integrase is a site-specific recombinase, which integrates incoming circular gene cassettes (often carrying antibiotic resistance genes) at the recombination site. The expression of the incoming genes is driven by the promoter of the integron. For example, the transposon Tn7 contains an integron with trimethoprim, streptothricin, and streptomycin/spectinomycin resistance genes. CTns have the capability to move themselves from one cell to another (41, 42, 213). First discovered in gram-positive bacteria, they have also been found to be prevalent among gram-negative Bacteroides strains. Tn916 is one of the best-described examples of a CTn. In addition to genes responsible for the integration and excision of the transposon and antibiotic (often tetracycline) resistance genes, the CTns contain so-called tra genes responsible for their conjugal transfer. After excision, CTns form a circular intermediate; however, they are not called plasmids, since they do not form self-replicating structures. Nevertheless, comparable to conjugative plasmids, they do contain an origin of transfer (oriT), a site at which a nick occurs to generate single-stranded DNA, which is then transferred through the conjugation apparatus into another cell. MTns also contain an origin of transfer, but they do not contain tra genes. For their transfer, they rely on the tra gene products provided by either CTns or plasmids. MTns have been identified mainly in Bacteroides species (41, 213). Group II introns have been identified first in eukaryotes, but also recently in prokaryotes (129, 223). They are also called retrotransposons, since they have the ability to move to new, heterologous sites via an RNA intermediate. One

TABLE 1 Common bacterial transposable elements and their characteristics

Designation	Antibiotic resistance phenotype(s)[a]	Length (kb)	Transposition frequency (per cell generation)	Insertion specificity	Reference(s)
Tn3	Ap^r	4.957	10^{-7}–10^{-5}	Prefers A+T-rich sequences; attracted to sites having some similarity to transposon ends	82, 186
Tn5	Km^r Nm^r; Bm^r; Sm^r	5.820	10^{-3}–10^{-2}	Nearly none; hot spot in pBR322	18, 163
Tn7	Tp^r Sm^r Sp^r	14	High frequency if target site attTn7 is present; low frequencies at secondary sites	Very high for target site attTn7	44, 46
Tn9	Cm^r	2.6	2×10^{-5}	Many sites	7, 80
Tn10	Tc^r	9.147	10^{-7}	Insertion hot spots (symmetrical 6-bp sequences); prefers nontranscribed regions	86, 106
Phage Mu		36.717	High frequency for chromosome	Essentially random	35, 150

[a]Abbreviations used for resistance phenotypes are as follows: Ap, ampicillin; Km, kanamycin; Nm, neomycin; Bm, bleomycin; Sm, streptomycin; Tp, trimethoprim; Sp, spectinomycin; Cm, chloramphenicol; Tc, tetracycline. Bm^r and Sm^r phenotypes associated with Tn5 are expressed only in certain nonenteric bacteria.

example is the retrotransposon L1.LtrB of *Lactococcus lactis*. The intron encodes a catalytic RNA and a multifunctional protein, which possesses a maturase activity to stabilize the RNA structure and to promote splicing, a reverse transcriptase activity to synthesize cDNA, and a DNA nuclease activity for the cleavage of the recipient DNA (12, 43, 213).

It was recognized soon after the discovery of transposons that these elements could be used as extremely efficient tools in bacterial genetics (30, 108) because of the following characteristics.

1. The transposition or insertion of a transposon into a gene generally leads to the gene's inactivation, and the resulting null mutations are relatively stable since the frequency of precise excision of the transposon is very low.

2. If a transposon integrates into an operon, it usually exerts a strong polar effect on genes located downstream of its insertion site, allowing transposon mutagenesis to be used to determine operon structure.

3. Upon integration, transposons introduce new genetic and physical markers into the target locus, such as antibiotic resistance genes, new restriction endonuclease cleavage sites, and unique DNA sequences which can be identified by DNA-DNA hybridization, electron microscopic heteroduplex analysis, or PCR-based methods. The genetic markers can be used to rapidly map and transduce the mutated loci.

4. Transposons can generate a variety of genomic rearrangements, such as deletions, inversions, translocations, and duplications, and can be used to introduce specific genes into the genome of a target bacterium.

These characteristics have made transposon mutagenesis a valuable addition to the more classical techniques such as chemical mutagenesis or mutageensis induced by UV irradiation. Transposon mutagenesis has been used to clone genes, to construct reporter gene fusions, to construct correlated physical and genetic maps of cloned DNA segments, to map entire bacterial genomes by pulsed-field gel electrophoresis, to construct conditional mutations with portable promoters, to introduce desired origins of conjugal transfer or replication into chromosomes and plasmids of interest, and to determine the sequence of large DNA regions without the need of sub- or deletion cloning. For additional information on the use of transposons and bacteriophages for gene manipulations in bacteria, see chapters 31, 32, and 34.

Even in the time of the availability of complete bacterial genome sequences, transposon mutagenesis continues to play a role in analyses of the functions of putative genes. Often, when complete bacterial genomes are characterized, no function can be assigned to around 40% of the putative open reading frames. Thus, the technique of transposon mutagenesis can be combined with genomics to analyze the phenotypes of mutants with transposon insertions in specific genes.

In this chapter, a subset of these applications and the corresponding experimental protocols are described. The focus is on one of the transposons, namely, Tn5, which has been extensively used in bacterial molecular genetics, but the principles and most of the experimental strategies described are applicable to a variety of other transposons. For a treatise on the biology of transposons and their transposition mechanisms, the interested reader is referred to the publication by Berg and Howe (21) and the volume edited by Craig et al. (47). For a detailed description of the application of transposon mutagenesis to more classical bacterial

genetics, the reader is referred to the article by Kleckner et al. (108). For an in-depth overview of the use of different transposable elements in the genetic engineering of bacteria, as well as comprehensive lists of transposable elements and the map positions of known insertions in *Escherichia coli* and *Salmonella*, the reader is referred to articles by Berg and Berg (14, 15), Berg et al. (16), Altman et al. (6), Sherratt (186), Craig (44–46), Kleckner (106), Haniford (86), and Hayes (91).

Because of the great diversity of bacterial strains, transposons, delivery vehicles, and gene transfer methods available today, it is not possible to present a complete set of transposon mutagenesis protocols in this chapter. Nevertheless, the methods, examples, and references included here should provide some basic tools.

29.1. Tn5 AS A MODEL TRANSPOSON

Transposon Tn5 is a 5,820-bp composite transposon consisting of two inverted repeats of 1,534 bp (IS50L and IS50R) flanking a unique region which carries three genes that confer on certain bacterial hosts resistance to the antibiotics kanamycin or neomycin (the kanamycin or neomycin phosphotransferase [*nptII*] gene), bleomycin (the *ble* gene), and/or streptomycin (the *str* gene) (18) (Fig. 1). The *nptII*, *ble*, and *str* genes are organized in an operon and are transcribed from a promoter located at the inside end of IS50L (18) (Fig. 1). IS50R genes encode the transposase responsible for Tn5 transposition (the *tnp* gene), as well as a transposition inhibitor (the *inh* gene) (18) (Fig. 1). The first 19 bp of both IS50L and IS50R have been found to be essential for the efficient transposition of Tn5 (18).

Tn5 is capable of transposing at a very high frequency (10^{-2} to 10^{-3} transpositions/cell generation) in *E. coli* by using a conservative "cut-and-paste" mechanism, and it generates a 9-bp target site duplication upon insertion (18). The three-dimensional structure of the transposition intermediate has been determined (49). Tn5 has a low insertional specificity and therefore can insert itself into a large number of locations in bacterial genomes, as well as into multiple positions within single genes, although occasional hot spots for insertion have been reported (18, 19, 50, 52). Tn5-induced insertion mutations are very stable, and mutants only occasionally revert to the wild type via a process known as precise excision, although the reversion frequency is somewhat dependent on the precise location of Tn5 in the genetic locus, the nature of the replicon, and the host genotype (18). These parameters are important for the experiments described below and should be examined when Tn5 mutagenesis experiments are performed with new organisms or when problems are encountered with the protocols presented here. For in-depth discussions of these parameters, see reviews by Berg and colleagues (16, 18, 19) and de Bruijn and Lupski (52).

Tn5 mutagenesis experiments can be divided into two distinct categories: random mutagenesis and region-directed mutagenesis (50). The terms site-directed and site-specific transposon mutagenesis have also been used in the literature to describe the latter category (50). The term region-directed mutagenesis is used here to distinguish this category of transposon mutagenesis from base pair-specific oligonucleotide mutagenesis methods, which are usually referred to as site specific. Random mutagenesis involves the introduction of Tn5 into the bacterial species of interest via transformation, transduction, conjugation, or electroporation by using "suicide" plasmid or phage vectors, followed

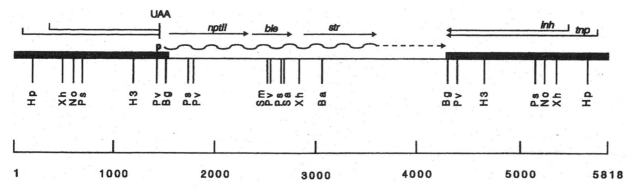

FIGURE 1 Structure of the transposable element Tn5. Stippled bars represent the insertion sequences IS50L (left) and IS50R (right), present as terminal inverted repeats in Tn5. Gene symbols: *tnp*, coding region for the transposase; *inh*, coding region for the inhibitor; UAA, stop codon responsible for the truncated versions of the transposase and the inhibitor in IS50L; p, promoter for the transcript of the operon coding for resistance to three antibiotics (*nptII*, neomycin or kanamycin resistance; *ble*, bleomycin resistance; *str*, streptomycin resistance). Abbreviations for restriction enzymes: Ba, BamHI; Bg, BglII; H3, HindIII; Hp, HpaI; No, NotI; Ps, PstI; Pv, PvuII; Sa, SalI; Sm, SmaI; Xh, XhoI.

by selection for the antibiotic resistance marker(s) carried by Tn5. The transposition and insertion of Tn5 into the genome of the recipient bacterium are detected after the vector has been lost by segregation, and the phenotype associated with the Tn5-induced mutations can be analyzed subsequently. Region-directed mutagenesis usually involves the isolation of Tn5 insertion mutations in genes cloned into (multicopy) plasmids in *E. coli*, followed by the characterization of the mutant phenotype of this (heterologous) host or the organism of origin of the cloned genes after the reintroduction of the Tn5-mutated loci by gene replacement (50). Experimental strategies for these two types of mutagenesis are presented in the following sections and have been reviewed previously (16, 50, 52).

29.2. RANDOM Tn5 MUTAGENESIS

To generate transposon insertion mutations in the genomes of bacteria, the transposon must be introduced into its new host and cells in which transposition has occurred must be identified. These objectives are achieved by the use of so-called suicide vectors, which can be of phage or plasmid origin. In both cases, after the introduction of the suicide vector carrying the Tn5 DNA molecule into the cell, the vector must be incapable of stable replication or integration so that selection for one or more of the antibiotic resistance markers on the transposon element leads to the identification of cells in which the element has transposed from the vector into the genome. A commonly used phage vector for the introduction of transposons into bacterial cells is a modified derivative of bacteriophage λ, which can be used for the mutagenesis of all bacteria that are naturally susceptible to λ adsorption and infection or have been genetically modified to be λ sensitive. The protocol for generating the latter category of engineered organisms has been described previously (121, 149) and is not presented here. For bacteria that are not λ sensitive or cannot be engineered to become λ sensitive, the transposon can be introduced into the cells by using plasmids, which can be conjugally mobilized or electroporated (62) into recipient cells (see chapter 31.4) but lack the appropriate (wide-host-range) origins of repli-

cation for stable maintenance. Phage P1 has also been used to introduce transposons such as Tn5 into different bacterial species, and examples of the use of this vector system have been described previously (11, 114).

The choice of the suicide vector and transposon element to be used depends on the purpose of the random mutagenesis experiment and a number of criteria, including the phage sensitivity and the intrinsic antibiotic resistance of the bacterium to be mutagenized, as well as the availability of a conjugal transfer or direct transformation system. A short list of transposons and their relevant properties is shown in Table 1, but for a discussion of the above-mentioned parameters and an extensive list of the transposons available for specific purposes, the reader is referred to previous reviews (14–16, 52).

This chapter focuses on the use of bacteriophage λ and conjugable narrow-host-range or conditionally replicating plasmid vectors to deliver Tn5 and its derivatives and describes experimental protocols that are commonly used to carry out the random mutagenesis of (predominantly gram-negative) organisms. For a discussion of chemical transformation and high-voltage electroporation techniques (62) to introduce plasmids carrying Tn5 into cells, the reader is referred to chapter 31.4. For a discussion of transposons in gram-positive organisms and their use for mutagenesis experiments, the reader is referred to articles by Murphy (142) and Youngman (222).

29.2.1. Phage λ Suicide Vectors

For λ::Tn5-mediated random mutagenesis experiments, a derivative of phage λ (λ467) is routinely used; it has the following genotype: λ b221 rex::Tn5 cI857 Oam29 Pam80 (52, 108). The b221 mutation is a deletion in the λ genome that removes the phage attachment (*att*) site. This deletion prevents the lysogenization of the target strain. The *rex* gene is a nonessential λ gene and carries the Tn5 insertion. Both the O and P genes are involved in phage replication, and therefore, amber mutations in these genes will prevent phage replication in a suppressor-negative (Su⁻) background. The λ467-type phage can be propagated in a Su⁺ *E. coli* strain (see below), will adsorb to a λ-sensitive strain, and will inject its DNA; however, this DNA will be unable to replicate and will be lost by segregation. Selection for Km^r Nm^r (and, in selected cases, Bm^r or Sm^r) survivors results in the identification of cells in which Tn5 has transposed and integrated into the genome. This type of Tn5 mutagenesis has been successfully used to isolate mutations in many genes of *E. coli* (15, 123, 184).

29.2.1.1. Phage Lysate Preparation

To prepare λ::Tn5 lysates, the Su^+ *E. coli* LE392 strain (F^- *hsdR514 supE44 supF58 lacY1 galK2 galT22 metB1 trpR55* λ^-) and the following protocol (modified from reference 52) are used.

1. Start an overnight culture of *E. coli* LE392 at 37°C in 10 ml of Luria-Bertani (LB) medium (see section 29.5) plus 0.2% maltose.

2. Pellet the cells by centrifugation, wash once with 5 ml of 10 mM MgSO_4, and pellet again.

3. Resuspend the cells in 4 ml of 10 mM MgSO_4.

4. Prepare five sterile glass tubes with 0.1 ml of washed LE392 cells, and add 10 μl of λ::Tn5 lysate or the phages from a single plaque to four of the tubes. Leave one tube uninfected as a control.

5. Mix gently, and let the phages adsorb for 2 h at room temperature without agitating the tubes.

6. Mix the infected cells with 3 ml of soft agar (45°C) and pour the mixture evenly onto λ agar plates (see section 29.5).

7. Incubate the plates for 6 to 8 h (or overnight) at 37°C, and scrape off the clear soft agar into a fresh tube.

8. Mix the agar-phage cell suspension well but gently with 2.5 ml of SM buffer (see section 29.5) and 0.5 ml of chloroform in a sterile (Corex) centrifuge tube. Incubate the mixture for 5 min on ice, and centrifuge it for 15 min at 1,000 × g (e.g., at 3,000 rpm in an SS34 fixed-angle rotor) at 4°C.

9. Transfer the supernatant solution into a small glass tube, add 20 drops of chloroform, and store at 4°C.

10. Determine the titer of the lysate by infecting LE392 cells (prepared as described in steps 1 to 3) with several dilutions of the lysate (from 10^{-2}, 10^{-4}, to 10^{-12}) and counting the resulting single plaques on the indicator plates.

Controls and Comments

The plaques on the titer plates should be small and well defined (pinpoint). No plaques should be observed on the plates with bacterial cell suspensions to which no phage was added. The titer of the lysate should be in excess of 5×10^{10} for successful use in subsequent Tn5 mutagenesis experiments.

29.2.1.2. Infection and Mutant Selection

1. Start an overnight culture of the λ-sensitive Su^- bacterial strain to be mutagenized in 5 ml of complete medium for the respective strain, supplemented with 0.2% maltose.

2. Inoculate 100 μl (1:100) of the saturated culture into 10 ml of complete medium + 0.2% maltose, and incubate the culture for 2 to 4 h.

3. Pellet the cells, wash them with a sterile solution of 10 mM MgSO_4, and pellet them again. Resuspend the pelleted bacteria in 4 ml of 10 mM MgSO_4.

4. Infect 1-ml aliquots of the washed cells with 100 μl of λ::Tn5 lysate (10^8 to 10^9 CFU/ml), and keep 1 ml of uninfected cells as a control.

5. Mix the phage and the cells gently, and allow the phage to adsorb for 2 h at room temperature without agitating the tubes.

6. Plate 200-μl aliquots of the infected cells on complete-medium plates containing kanamycin (or neomycin, bleomycin, and streptomycin) at the desired concentration, and incubate the plates at the appropriate temperature until clearly defined colonies appear on the selection plates (but not on the control plates).

7. Purify the resistant colonies on complete-medium plates with the desired antibiotics, and analyze them as described in sections 29.2.3 and 29.2.4.

Controls and Comments

No Km^r colonies should be observed on the plates onto which the uninfected cells were plated.

29.2.2. Plasmid Suicide Vectors

The first class of suicide plasmids to be used for the random Tn5 mutagenesis of λ-insensitive bacterial strains was based on an IncP-type plasmid carrying a copy of Mu in addition to Tn5 (pJB4JI) (24). The presence of Mu in this replicon was found to prevent its stable maintenance in different gram-negative bacteria, and Tn5 transpositions could be identified after antibiotic resistance selection (24). However, for several bacterial species, it was found that a large percentage of the insertion mutants carried Mu sequences at the insertion site, in addition to Tn5 (130), and pJB4JI was observed to be relatively stable in other species (191), leading to a reduced use of this class of vectors.

A second class of Tn5-containing suicide plasmids is based on replicons carrying a temperature-sensitive origin of replication, allowing the selection for Tn5 transposition at elevated temperatures (115). In addition, derivatives of the broad-host-range conjugative plasmids R388 (pCHR71) and pRK290 (pRK340), carrying temperature-sensitive mutations in plasmid replication functions, have been described and their utility in the Tn5 mutagenesis of, for example, *Shigella* and *Legionella* species has been demonstrated previously (104, 179, 180).

A third class of suicide vectors for Tn5 mutagenesis, based on narrow-host-range IncW- and IncN-type plasmids, has been described previously and used for the mutagenesis of different bacterial species, such as *Rhizobium meliloti* (pGS9) (181), *Bradyrhizobium japonicum* (171), *Azorhizobium caulinodans* (pGS9) (96), *Pseudomonas solanacearum* (pW1281) (140), and *Azotobacter vinelandii* (105), to mention a few examples.

A fourth class of suicide vectors is derived from the R6K replicon and is based on the principle of transcomplementation-dependent maintenance of the R6K replication origin (109). In this system, a plasmid carrying (a fragment of) the R6K replication origin (e.g., pRK703) can be stably maintained only if a *trans*-acting factor (the π protein encoded by the *pir* gene) is provided in the bacterial cell, for example, by a λpir prophage (109). Miller and Mekalanos (139) inserted into plasmid pRK703 an IncP-type origin of conjugal transfer (*mob*, *oriT* of RP4) to allow high-frequency transfer to a variety of gram-negative bacteria with the help of RP4 mobilization functions provided in *trans* (see below), as well as a polylinker containing several cloning sites (pGP704). Transposon Tn5 derivatives (mini-Tn5's), consisting of the 19 bp of the Tn5 inverted repeats required for transposition coupled to antibiotic resistance or other selectable marker genes (see section 29.4.4), have been inserted into this vector, in addition to the Tn5 transposase (*tnp*) gene, to construct an elegant suicide transposon mutagenesis system. The suicide plasmid carrying a mini-Tn5 construct is conjugally mobilized (see sections 29.2.2.1 and 29.2.2.2) from an *E. coli* strain harboring λ*pir* to the gram-negative bacterium to be mutagenized, and selection for one of the mini-Tn5-borne marker genes is applied. This procedure results in the selection of transconjugants in which the mini-Tn5 construct has transposed from the vector, which itself is no longer capable of replicating in the absence of the *pir* gene, into the resident genome. Once inserted into the genome, the mini-Tn5 element is incapable of further transposition, since it is lacking the suicide vector-borne *tnp* gene, resulting in a very stable integration event (95). The stable integration is especially relevant when Tn5-carried genes must be stably maintained in the absence of antibiotic selection in natural environments (95). For a detailed description of the use of these vectors, the reader is referred to the articles by Herrero et al. (95) and de Lorenzo et al. (55, 56).

Another commonly employed class of suicide vectors for Tn5 mutagenesis in gram-negative bacteria other than *E. coli* is based on replicons derived from plasmid RP4 carrying the same broad-host-range conjugal transfer or mobilization functions and sites as those used for the above-described vectors but a narrow-host-range origin of replication (190, 191). These vectors can be mobilized at a high frequency from *E. coli* to other gram-negative bacteria but cannot be stably maintained in the recipient species.

These vectors contain, in addition to Tn5, the IncP-type mobilization (*mob*; *oriT*) site and are based on commonly used *E. coli* cloning vectors, such as pACYC184 (37) and pBR325 (28), which cannot replicate in nonenteric bacteria (pSUP series) (189, 191). pSUP-type plasmids can be mobilized in diparental mating experiments (see section 29.2.2.1) by providing the transfer functions in *trans* from a chromosomally integrated copy of the IncP plasmid RP4 in the donor strain itself, e.g., with strain S17-1 (RP4-2-Tc::Mu Km::Tn7 Tp^r Sm^r *thi pro hsdR*) (190), or in triparental mating experiments (see section 29.2.2.2.) by providing the transfer functions from plasmid pRK2013, harbored by a nondonor, nonrecipient helper strain of *E. coli* (*E. coli*/pRK2013) (59, 60). The use of this vector system will be described here; for a detailed review of the conjugal transfer of pSUP vectors, the reader is referred to other reviews (189, 191). As an alternative to conjugation methods to transfer suicide vectors, high-voltage electroporation can be used (62) (chapter 31.4).

29.2.2.1. Diparental Conjugation and Mutant Selection

1. Start a 10-ml culture of the recipient bacterial strain in complete medium, and grow it until it has reached saturation at the appropriate temperature. Reinoculate the cells 1:10 in fresh medium, and incubate the cultures for another 2 to 5 h (depending on the doubling time of the bacterial strain used).

2. Start a 10-ml overnight culture of the donor strain *E. coli* S17-1 harboring the Tn5 delivery plasmid (189, 190) in LB medium supplemented with the appropriate antibiotics (e.g., usually kanamycin for Tn5 and tetracycline, ampicillin, or chloramphenicol for the pSUP vector), and grow the culture until saturation at 37°C. Reinoculate the cells 1:10 in fresh medium, and grow for 1 to 2 h at 37°C.

3. Pellet the logarithmically growing recipient and donor cultures (1,000 × *g* for 10 min), wash the bacteria with 0.9% NaCl to remove the antibiotics, and pellet again.

4. Resuspend the pelleted bacteria in 1 ml of 0.9% NaCl.

5. Spot 100 μl of the donor and recipient cell cultures separately and 100 μl of a 1:1 mixture of both onto conjugation plates containing the complete medium suitable for the recipient strain, in the absence of antibiotics. Allow the spots to dry, and incubate the plates at the appropriate temperature for the recipient strain for 12 to 36 h.

6. Scrape the mating mixtures and the single spots of donor and recipient cells off the plates with a sterile spatula, resuspend the bacterial cells in 1 ml of 0.9% NaCl (or H$_2$O for osmotically stable recipient bacteria, such as rhizobia), and plate out the suspensions (or appropriate dilutions thereof) onto selective plates containing antibiotics to select for the recipient strain (if the recipient strain harbors a gene conferring resistance to an antibiotic to which the donor strain is sensitive) and for the presence of Tn5 (kanamycin, neomycin, bleomycin, or streptomycin).

7. Purify the single colonies appearing on the selective plates once onto the same selective medium to avoid contamination with donor bacteria, or (when screening large numbers of colonies for mutant phenotypes) characterize them directly.

Km^r derivatives of the recipient strain should be the result of the introduction of the suicide vector, the subsequent (random) transposition of Tn5 into the resident genome, and the loss of the vector. These colonies can now be subjected to further studies (see sections 29.2.3 and 29.2.4).

Controls and Comments

No colonies should appear on the selection plates onto which the donor and recipient cells were plated separately. If colonies do appear repeatedly on the control plates, alternative counterselection protocols should be considered (see below). Care should be taken to treat the cells gently during pelleting and washing and to keep them at the growth temperature of the recipient strain. Instead of or in addition to the antibiotic selection, a specific minimal medium suitable for the recipient bacteria but lacking the appropriate carbon source for *E. coli* (e.g., one containing sucrose) or lacking the amino acid or vitamin requirements of the *E. coli* donor strain (e.g., thiamine and proline for strain S17-1) (190) can be used to counterselect the *E. coli* donor bacteria. The latter method can of course be used only if one is not trying to isolate Tn5-induced auxotrophic mutations in the target bacterial strain (see section 29.2.2.3).

29.2.2.2. Triparental Conjugation and Mutant Selection

1. Prepare a culture of logarithmically growing recipient bacteria as described above. Similarly, prepare a logarithmically growing culture of a donor *E. coli* strain harboring the RP4-derived Tn5 mutagenesis vector. In this case, the donor strain does not contain a chromosomally integrated copy of RP4 to provide transfer functions but can be any *E. coli* strain, preferably a strain carrying one or more auxotrophic markers for counterselection (see above). In addition, prepare a logarithmically growing culture of a third *E. coli* (helper) strain harboring a plasmid (pRK2013) (59, 60), providing the RP4 transfer functions in *trans*, and allowing the conjugal transfer of the Tn5 mutagenesis vector from the donor to the recipient bacteria in this type of triparental mating experiment. Since the pRK2013 plasmid carries a Km^r or Nm^r gene, kanamycin or neomycin should be added to the growth medium to select for the presence of the plasmid.

2. Pellet the cultures (1,000 × *g* for 10 min), wash the bacteria with 0.9% NaCl, and pellet them again. Resuspend the pelleted bacteria in 1 ml of 0.9% NaCl.

3. Spot 100 μl of the donor, helper, and recipient cell cultures separately, in pairwise combinations (1:1), and all three together (1:1:1) onto conjugation plates containing the complete medium suitable for the recipient strain in the absence of antibiotics. Incubate the plates at the appropriate temperature for the recipient strain for 12 to 36 h.

4. Follow steps 6 and 7 of the diparental mating protocol.

Controls and Comments
No colonies should appear on the selection plates onto which the donor, helper, or recipient strain was plated separately or in a pairwise combination. The use of an unsupplemented minimal medium and auxotrophic donor and helper E. coli strains should help reduce background on the selection plates. If colonies do form consistently on the control plates, the individual strains should be checked or other (or modified) counterselection conditions should be explored (e.g., using a different minimal medium, using an alternative antibiotic, or raising the concentration of the antibiotic).

29.2.2.3. Mutagenesis Protocols and Mutant Characterization
The protocols outlined above normally result in approximately 10^{-3} to 10^{-6} Tn5-containing colonies per recipient bacterial cell. If the total number of Tn5-containing strains is too small for the planned mutant-screening program, the mutagenesis protocol should be optimized. For the conjugation protocols, the ratios of donor and recipient cells should be varied (e.g., 2- to 20-fold excess of donor cells compared to recipient cells) and the conjugation time should be extended. In addition, instead of spotting the conjugation mixtures directly, the cell mixtures should be combined onto 0.2-μm-pore-size nitrocellulose filters placed on (prewarmed) conjugation plates, incubated for the desired time, and removed from the filters by vortexing in 0.9% NaCl (or H₂O) before plating onto selection plates. The concentration of the antibiotic(s) used for (counter)selection should also be varied.

When first carrying out random Tn5 mutagenesis experiments with a particular bacterial strain, it is useful to examine the frequency of Tn5-mediated gene inactivation. This is most conveniently done by determining the percentage of auxotrophic mutants resulting from a given Tn5 mutagenesis experiment. This determination requires the availability of a defined minimal medium for the strain under study. Five hundred to several thousand Km^r colonies are picked and streaked or replica plated (117) onto complete and minimal medium, and the frequency of colonies unable to grow on minimal medium is examined. Approximately 0.1 to 3.0% of the colonies should fall into the latter category. The occurrence of different auxotrophic mutants can be examined by using the Holliday test (100). The actual percentage of different auxotrophic mutants observed is an important piece of data that can be used to determine how many random Tn5-containing colonies must be screened to find an insertion in a particular gene of interest, since it reflects the insertional specificity of Tn5 in the bacterial strain under study.

Once it has been established that Tn5 mutagenesis can indeed lead to the identification of different mutations in the bacterial strain of interest, it should be determined whether the Km^r (mutant) colonies isolated carry a simple insertion of a single Tn5 transposon into a particular genomic locus. These events should be distinguished from multiple Tn5 insertions or aberrant cointegration events in-

volving the suicide vector. Cointegration is of particular importance since, in some bacterial species, the predominantly Tn5-mediated formation of cointegrates resulting in the integration of the entire Tn5-carrying suicide vector into the genome is the cause of the observed mutant phenotype (for examples, see references 61 and 96). For a detailed description of this phenomenon, the reader is referred to reference 18.

The cointegration of the entire suicide vector can usually be ruled out conveniently by examining the Km^r colonies resulting from a random Tn5 mutagenesis experiment for the absence of one or more of the antibiotic resistance genes carried by the non-Tn5 part of the vector, e.g., the Tc^r, Ap^r, or Cm^r gene for the pSUP vectors (189, 190). Single versus multiple Tn5 insertions and the cointegration of parts of the suicide vector should also be examined by DNA-DNA hybridization (Southern blotting; see the following sections).

29.2.3. Physical Mapping of Insertions
To physically characterize Km^r colonies resulting from the random Tn5 mutagenesis protocols outlined in sections 29.2.1 and 29.2.2, total genomic DNA is isolated from cultures derived from the respective colonies, digested with restriction enzymes, separated by agarose gel electrophoresis, transferred onto nitrocellulose or nylon membranes, and hybridized with Tn5 and suicide vector-derived DNA probes.

The method for isolating total genomic DNA from bacterial cultures depends strictly on the genus and species being analyzed, and therefore, no universal protocol is available. However, the following method modified from references 50 and 130 is applicable to a variety of gram-negative enteric and soil bacteria.

29.2.3.1. Isolation of Genomic DNA
1. Generate a 1-ml culture of logarithmically growing cells.
2. Pellet the cells (e.g., in a 1.5-ml microcentrifuge tube), and wash them twice with 1 M NaCl to remove (most) extracellular polysaccharides. Resuspend the pelleted cells in 1 ml of TE (10 mM Tris, 1 mM EDTA, pH 8.0).
3. Incubate the cells for 20 min at 37°C in the presence of 0.1 ml of a freshly prepared 2-mg/ml lysozyme solution in TE, with gentle mixing.
4. Add 0.125 ml of a 10% Sarkosyl–5-mg/ml pronase solution in TE (which has been predigested for 1 h at 37°C), and incubate the mixture for 1 h at 37°C.
5. Extract the viscous mixture of lysed cells with TE (pH 8.0)-saturated phenol two or three times, and extract the final aqueous phase with chloroform.
6. Add ammonium acetate to a final concentration of 0.3 M, and precipitate the nucleic acids by the addition of 2.5 volumes of ice-cold ethanol. "Spool out" the DNA strands with a glass rod, or pellet the nucleic acids by centrifugation. Resuspend the spooled-out DNA or the pellet in 0.1 to 0.5 ml of TE by gentle mixing. Leave the DNA solution at 4°C overnight, and mix gently to achieve optimal resuspension.
7. Determine the concentration of the DNA preparations, and load 0.1 to 0.5 μg of undigested DNA onto an agarose gel to verify its high molecular weight and to check for the presence of nucleases in the preparations.

29.2.3.2. Southern Blotting and Hybridization
The total genomic DNA from the Km^r colonies (1 to 2 μg) can now be digested with different restriction endonucle-

ases. For the first analysis, an enzyme which does not cut Tn5 should be used (e.g., EcoRI, KpnI, or ClaI). For modified Tn5 derivatives carrying additional sequences, the restriction maps should be examined to verify that these enzymes do not cut; if necessary, other enzymes should be selected. The insertion of Tn5 into a specific EcoRI or ClaI fragment (X) should give rise to a new, larger EcoRI or ClaI fragment (X + Tn5; Tn5 ≈ 5.8 kb). In the second level of analysis, an enzyme should be used which cuts the transposon twice (in the inverted repeats; e.g., HindIII or BglII) (Fig. 1), thereby generating distinct Tn5 target junction fragments, as well as constant internal Tn5 fragments (50). To visualize the Tn5-containing (junction) fragments, the digested DNA should be separated by agarose gel electrophoresis and transferred onto membranes by Southern blotting (for details, see chapter 30.4.4 and references 174 and 175). To determine the orientation of the Tn5 insertion relative to known restriction sites in the mutated locus, enzymes which cut the transposon once (asymmetrically; e.g., SalI and SmaI) (Fig. 1) should be used. The membranes should be hybridized with labeled Tn5 sequences (e.g., the HpaI internal fragment of Tn5) (Fig. 1), and the hybridizing fragments should be visualized by autoradiography or another appropriate visualization technique (see references 174 and 175).

For the enzymes not cutting Tn5, single hybridizing fragments of different sizes (all larger than 5.8 kb) should be observed in the DNA of different Km^r colonies, suggesting that the fragments are the result of single Tn5 transpositions into the genome. For the enzymes cutting Tn5 once, two distinct hybridizing fragments should be observed in the DNA of each Km^r colony, representing defined transposon-host junction sequences. For the enzymes cutting Tn5 twice within the inverted repeats, two distinct junctions and one constant internal Tn5 fragment should hybridize.

The Southern blots carrying the digested genomic DNA of the Km^r colonies should also be probed with labeled suicide vector sequences (without Tn5). If simple Tn5 transpositions have occurred, fragments hybridizing with vector sequences should not be observed.

The Km^r colonies have now been shown to be the result of the simple insertion of a single copy of Tn5 and can be characterized further, both genetically (see section 29.2.4) and physically, by more extensive restriction mapping, cloning, and DNA sequencing (see sections 27.1, 29.2.5, 29.4.6, and 29.4.7).

29.2.4. Genetic Mapping of Insertions and Tn5-Induced Mutations

Once a Tn5 insertion, a Tn5-induced mutation, or a collection thereof has been isolated and characterized, it can be used for genetic mapping experiments to determine its relative position with reference to previously identified mutations or to create a genetic map of an organism. This procedure is facilitated by the presence of the antibiotic resistance marker on the transposon, which permits selection for the inheritance of the integrated transposon and selection or screening for the mutations caused by the transposon's insertion (108). Transposon-induced insertion mutations behave essentially as standard genetic markers and therefore can be used in two-point crosses to determine genetic distances and in three-point crosses to establish gene order (108). By using standard methods (see chapter 31.1), these crosses can be carried out and genetic map positions can be determined. This process is further facilitated by the use of

Tn5 derivatives carrying an origin of conjugal transfer (see section 29.4.4).

Detailed strategies for carrying out mapping experiments with transposon insertions are described by Kleckner et al. (108), and such descriptions exceed the scope of this chapter. Examples of the mapping of specific Tn5-induced insertion mutations have been described previously (89, 184), and a list of mapped Tn5-induced insertion mutations in *E. coli* and *Salmonella* has been compiled elsewhere (15). Examples of the use of Tn5 mutagenesis to create a genetic map have also been described previously (9, 155). For an example of a fine-structure analysis of Tn5 insertions within a single operon, see Miller et al. (138).

Once the genetic map of a region (operon or regulon) has been established with the help of Tn5-induced insertion mutations and a recombinant plasmid corresponding to the wild-type operon (regulon) has been obtained, a correlated physical and genetic map can be prepared by using restriction mapping and Southern blotting. For a description of the strategy used and an example of this type of analysis, see Riedel et al. (167).

29.2.5. Cloning of Tn5-Mutated Genes

On the basis of hybridization data (see section 29.2.3), a fragment generated by a restriction enzyme that does not cut within the Tn5 sequence is usually chosen. This choice enables the cloning of regions upstream as well as downstream of the Tn5 insertion. The fragment should be large enough to carry, besides Tn5, the coding sequences of the gene(s) of interest, at least with a high probability. Any suitable vector can be used, but of course it will facilitate the cloning procedure if the antibiotic resistance marker of the vector is not kanamycin resistance.

A great advantage at this step in the random mutagenesis procedure is the use of a Tn5 construct that carries an origin of replication (see section 29.4.4) in addition to the Tn5 antibiotic resistance genes. In this case, the chromosomal DNA can be isolated, digested with a restriction enzyme that does not cut within the Tn5 derivative, and self-ligated, and *E. coli* cells can be transformed with the construct. Only fragments containing the Tn5 derivative with the origin will replicate and propagate in *E. coli*.

As soon as the mutated gene region is cloned, it can be used to conduct a gene replacement experiment (see section 29.3.4) to verify that the Tn5 insertion indeed caused the observed phenotype.

Finally, the mutated gene region can be used as a probe to clone the corresponding wild-type region from the original bacterial strain. Different strategies are available for this purpose; they are summarized in the following paragraphs (for detailed protocols, see molecular biology manuals [174, 175]).

1. After the plasmid carrying the Tn5-mutated region is mapped with restriction enzymes, prepare a DNA probe from the cloned and mutagenized bacterial locus. This probe may consist of a restriction fragment carrying DNA sequences immediately adjacent to the Tn5 insertion or a fragment carrying Tn5 plus target sequences. Alternatively, if the DNA sequence of the mutagenized locus has been determined (see also section 29.4.6), a DNA oligonucleotide probe can be synthesized or a specific PCR-generated probe can be prepared for hybridization experiments (see chapter 30.4.3 and reference 175).

2. Hybridize a Southern blot, containing genomic DNA from both the Tn5-mutated and the wild-type strains

digested with enzymes that do not cut the transposon (see section 29.2.3.2), with the DNA probe prepared in step 1. This analysis should lead to the identification of the same single hybridizing fragments found to hybridize with the Tn5 probe (see section 29.2.3.2) in the Tn5-containing mutant strains and corresponding smaller single fragments (minus the length of Tn5) in the wild-type strain.

3. Create a complete clone bank of DNA from the wild-type bacterial strain used to generate the Tn5 mutants in a plasmid or phage cloning vector (see reference 175). Alternatively, a partial clone bank can be constructed by purifying DNA fragments of the expected size (see step 2) from a gel and cloning them into a plasmid vector.

4. Screen the library (step 3) with the probe generated in step 1, and purify hybridizing plaques or colonies (see reference 175).

5. Hybridize a Southern blot containing genomic DNA from the strain and DNA from the recombinant phages or plasmids, digested with the same enzymes as those used in step 2, with the probe (step 1) to verify that the fragment of interest has been cloned.

6. Further analyze the cloned wild-type locus by restriction mapping, complementation studies, DNA sequencing, and expression studies (175). In addition, the cloned region can be subjected to region-directed Tn5 mutagenesis to create a precise, correlated physical and genetic map (see section 29.3.3) or subjected to transposon mutagenesis to create gene fusions for regulation studies (see section 29.4.1).

In selected cases, the wild-type counterpart of a Tn5-mutated locus can also be cloned by direct complementation. A readily screenable or selectable phenotype of the complemented strain would have to be available. In this case, the clone bank of the wild-type strain (step 3) would be introduced into the mutant strain via transformation, transduction, or conjugation and individual transformants, tranductants, or transconjugants would be examined for the restoration of the wild-type phenotype.

29.3. REGION-DIRECTED Tn5 MUTAGENESIS

When a wild-type locus has been cloned, region-directed (Tn5) mutagenesis can be used to delimit the genes and operons carried by the cloned region. The whole procedure can be carried out most efficiently if the presence of the cloned locus in E. coli results in an observable phenotype. Possible scenarios include those in which the cloned locus of interest is able to complement an auxotrophic mutation or a growth defect of a special E. coli strain, endows strains with the ability to catabolize or grow in the presence of novel compounds, or produces protein products which can be detected enzymatically or immunologically. If the cloned locus does not carry genes whose presence results in an observable phenotype, region-directed Tn5 mutagenesis can still be performed as described below. The insertions are mapped, and a subset is chosen for gene replacement experiments (see section 29.3.4) to analyze the effects of the different Tn5 insertions when introduced into the genomic locus of the original bacterial strain.

29.3.1. Vectors

In principle, every plasmid cloning vector carrying a cloned region but not a kanamycin resistance gene can be used as a target for region-directed Tn5 mutagenesis. The length of the vector relative to the length of the insert should be considered. Assuming a random insertional specificity for Tn5

(18, 52), the longer the vector sequence, the more insertions will be made in the vector and not in the cloned locus of interest. In addition, the copy number of the cloning vector is important. High-copy-number vectors will be easier to mutagenize, and it will be more convenient to isolate DNA from strains harboring them. In addition, high-copy-number plasmids can be used for Tn5 mutagenesis experiments with a chromosomal copy of Tn5 as the donor (22, 188). This protocol will not be presented here but is detailed in reference 50.

Some commonly used cloning vectors, such as pBR322 (29), carry "hot" regions for Tn5 insertion (18) and may therefore be less useful. However, the Tn5 mutagenesis of cloned genes has been performed successfully, not only with small, higher-copy-number vectors such as pACYC184 (37) but also with low-copy-number, broad-host-range vectors such as pRK290 and pLAFR1 (60, 66).

The most commonly used Tn5 delivery (mutagenesis) vector for region-directed mutagenesis is λ::Tn5 (see section 29.2.1), and a protocol modified from references 50 and 52 for this type of experiment follows.

29.3.1.1. Phage λ::Tn5

1. Transform any λ-sensitive E. coli host with the plasmid to be mutagenized (see chapter 31.1.2 and reference 175). The plasmid-selectable marker should not be kanamycin or neomycin resistance.

2. Purify a single colony, make a small-scale plasmid DNA preparation (see chapter 30.1.2 and reference 175), and verify the monomeric structure of the recombinant plasmid to be mutagenized. (Determine the size of the undigested plasmid by agarose gel electrophoresis.)

3. Start an overnight culture of the E. coli strain with the target plasmid in 5 ml of LB supplemented with the appropriate antibiotic (corresponding to the plasmid vector marker gene) and 0.2% maltose at 37°C.

4. Inoculate 100 μl (1:100) of the saturated culture into 10 ml of LB–antibiotic–0.2% maltose, and let it grow for 2 h at 37°C.

5. Spin down the cells, wash them with 10 mM $MgSO_4$, and pellet them again. Resuspend the pelleted bacteria in 4 ml of 10 mM $MgSO_4$.

6. Infect l-ml aliquots of cells with 100 μl of λ::Tn5 lysate, prepared as described in section 29.2.1.1 (10^8 to 10^9 CFU/ml), and keep 1 ml of uninfected cells as a control.

7. Mix gently, and let the phage adsorb for 2 h at room temperature without agitating.

8. Plate 200-μl aliquots of infected cells onto LB plates containing kanamycin (25 μl/ml) and the antibiotic selecting for the presence of the plasmid. Incubate the plates for 2 days at 37°C.

9. Wash the colonies off the plates with 1 to 2 ml of LB medium by using a sterile spreading rod, and pellet the cells by centrifugation. Resuspend the pellet in 2 to 5 ml of LB medium. Isolate plasmid DNA from the cell suspension by using a (scaled-up) miniprep plasmid DNA preparation protocol (see chapter 30.1.2).

10. Transform competent E. coli cells (see chapter 31.1.3) with the plasmid DNA isolated in step 9, and select for kanamycin- and plasmid marker-resistant transformants. Store the desired number of transformants on master plates for further analysis. These colonies should harbor plasmids carrying (generally) a single copy of Tn5 inserted into a distinct plasmid location (pXX::Tn5, where pXX is any plasmid vector used), which can be determined by restriction mapping (see section 29.3.2).

29.3.1.2. Mutagenesis Protocol and Mutant Characterization

Generally, step 8 of the above-described protocol should yield plates with at least 500 to 1,000 Km^r colonies, from which plasmid DNA is to be isolated. If fewer than 500 colonies appear, the titer of the λ::Tn5 lysate should be checked. Steps 9 and 10 will select colonies in which Tn5 has inserted into the plasmid and not into the chromosome of the original E. coli/pXX strain infected with λ::Tn5. The effect of the Tn5 insertion on the expression of the gene(s) or operon(s) carried on the cloned DNA of plasmid pXX can now be directly examined if there is a measurable phenotype.

1. If the cloned DNA is capable of complementing an auxotrophy or growth requirement of the E. coli strain used in step 10, plate the E. coli/pXX::Tn5 transformants onto the appropriate minimal medium and select pXX::Tn5 derivatives no longer capable of complementing. These plasmids are expected to carry Tn5 insertions in the locus of interest, either in the structural gene(s) or in the corresponding control region(s). For examples of this type of analysis, see references 53, 54, and 169. A similar strategy would apply in analyzing a cloned DNA region conferring on the host E. coli strain the ability to catabolize certain compounds. For an example of this type of analysis, see reference 70.

2. If the detectable phenotype associated with the cloned region is solely the production of a polypeptide of a known molecular weight, introduce the pXX::Tn5 plasmids and the parental plasmid into minicells (160) or maxicells (176) via transformation and analyze the plasmid-encoded polypeptides by sodium dodecyl sulfate-polyacrylamide gel electrophoresis (see chapter 17.5.1 and reference 175). The insertion of Tn5 into the locus would result in the disappearance of the polypeptide of interest. In many cases, because of the presence of transcriptional as well as translational stop signals on Tn5, truncated polypeptides will appear. The sizes of these truncated polypeptides are indicative of the position of the Tn5 in the gene (or the operon) relative to the start codon, and this information can be used to deduce the direction of the transcription of the gene (operon) once it is combined with the physical mapping data (section 29.3.2). Examples of this type of analysis have been described previously (53, 54, 169).

3. If the cloned region encodes an enzymatic activity detectable in E. coli or produces a protein for which an antibody has been isolated, screen E. coli/pXX::Tn5 transformants appropriately for a loss of function. For antibody-screening methods, see reference 175, and for an example of the latter type of Tn5 mutant screening, see reference 122.

Clearly, there are many variations not listed here on the theme of how to screen for the desired phenotype. For an extensive list of transposon-induced insertion mutations in distinct genes of E. coli and Salmonella (and screening procedures used to identify them), the reader is referred to reference 15. For further discussion, see references 16 and 52.

29.3.2. Physical Mapping of Plasmid-Borne Insertions

1. Isolate plasmid DNA from the desired E. coli/pXX::Tn5 strains by using a rapid, miniprep isolation protocol (see chapter 16.1.2).

2. Digest the DNA with several restriction enzymes. Several convenient enzymes are available for the physical mapping of Tn5 insertions in pXX::Tn5 plasmids, some of which do not cut Tn5 (e.g., EcoRI, KpnI, and ClaI) or cut Tn5 once (e.g., SalI, SmaI, and BamHI) or twice (e.g., HindIII, BglII, NotI, and HpaI) (Fig. 1). Depending on the restriction sites available in the pXX plasmid, different single and double digestions can be carried out to map the position and relative orientation of the Tn5 insertion. As discussed above for the mapping of chromosomal Tn5 insertions (section 29.2.3.2), first use an enzyme which does not cut Tn5 and subsequently use an enzyme which cuts (twice) within the inverted repeats in order to generate specific Tn5-pXX junction fragments. By measuring the lengths of the junction fragments and subtracting the number of base pairs contributed by Tn5 (Fig. 1), the approximate distance of the Tn5 insertion from the nearest cleavage site for the enzyme used can be deduced. To help determine the relative orientation of the transposon within the cloned DNA, an enzyme cutting Tn5 asymmetrically (e.g., SmaI) can be used. For a detailed discussion of mapping strategies, see references 50 and 52.

29.3.3. Correlated Physical and Genetic Maps

The physical mapping data (section 29.3.2) can now be combined with the data obtained by examining the phenotypes corresponding to the different insertions (section 29.3.1.2) to create a correlated physical and genetic map. The Tn5 insertions that disturb the expression of the cloned locus under analysis would be expected to cluster and to be flanked by Tn5 insertions that do not affect the expression of the locus. Thus, the extent of the locus can be mapped to an accuracy of approximately 100 bp, depending on the extent of the saturation mutagenesis experiment and the lack of insertional specificity of Tn5 in the particular cloned region under investigation. For examples of correlated maps thus created, see references 53, 54, and 169, and for discussions of the insertional specificity of Tn5, see references 18, 19, and 52.

The information thus obtained on the precise location of a gene of interest in a large fragment of cloned DNA can now be used to characterize this gene further, for example, by directed DNA sequencing with Tn5-derived primers (see section 29.4.6), characterization of the gene's regulatory regions and expression (see section 29.4.1), and the construction, by directed gene replacement, of well-characterized Tn5 insertion mutants of the bacterial species from which the gene was cloned (section 29.3.4).

Since Tn5 insertion mutations are generally polar on genes located downstream of the mutated gene in the same operon (18), transposon mutagenesis can also be used to determine the structure of an operon carried on a cloned segment of DNA, especially when the method is combined with an analysis of the polypeptides encoded by the (mutagenized) recombinant plasmids (see section 29.3.1.2). However, Tn5-mediated polarity is not always observed (23), especially in analyzing Tn5-mutated operons carried by multicopy plasmids (54), and therefore one should exercise caution when interpreting the results obtained with region-directed Tn5 mutagenesis of cloned DNA segments; for a discussion, see references 18 and 52.

29.3.4. Gene Replacement Vectors and Methods

A powerful extension of the Tn5 mutagenesis protocol is the use of gene replacement techniques to replace the wild-type gene in the original bacterial strain with its well-characterized Tn5-mutated analog carried by a plasmid in E. coli. This technique allows the determination of the phenotypes associated with specific Tn5-induced insertion mutations in the original organism. The general principle is to introduce the Tn5-mutated gene region back into the original organism and force a double-crossover event between the wild-type gene sequences in the bacterial genome and the corresponding sequences flanking the Tn5 insertion, followed by the loss of the wild-type gene.

Two methods widely applicable to gram-negative bacteria other than *E. coli* have been described previously (50). The first method was developed by Ruvkun and Ausubel (172). In this method, the broad-host-range vector pRK290 (60) is used to introduce the Tn5-mutated gene into the original bacterial strain by triparental conjugation (see section 29.2.2.2) with the helper plasmid pRK2013 (60). This step is followed by the introduction of a plasmid incompatible with pRK290, e.g., pPH1JI (97), into the same strain. Selection for the antibiotic resistance marker of Tn5 (kanamycin resistance) and that of the newly introduced plasmid pPH1JI (gentamicin resistance) forces a double-homologous-crossover event and the subsequent loss of the pRK290 replicon carrying the wild-type sequences.

The second method uses narrow-host-range plasmids carrying the ColE1 mobilization (bom) site (such as pBR322). These plasmids are able to replicate only in *E. coli*, but they are transferable to a wide range of other gram-negative bacteria if the ColE1 Mob and Tra functions are provided in *trans* (see section 29.2.2). Van Haute et al. (209) described the use of *E. coli* GJ23 harboring the plasmids pGJ28 (Mob⁺) and R64*drd*11 (Tra⁺) to transfer pBR322 derivatives carrying a Tn5-mutated gene effectively into *Agrobacterium tumefaciens*. The ColE1 replicon cannot be stably maintained, and therefore, upon selection for the resistance marker of Tn5, the mutated region will integrate into the genome via homologous recombination between cloned plasmid-borne and resident sequences. In most cases, a cointegration of the vector will occur via a single crossover, but at a certain frequency (depending on the host and the mutated gene region), a double crossover will occur, leading to a bona fide gene exchange. Alternatively, the above-described vectors of the pSUP series (see section 29.2.2) can be used for the same purpose, since they carry the *mob* region of RP4 cloned into narrow-host-range plasmids such as pACYC184 and pBR325 (190). Therefore, they can be mobilized into other gram-negative bacteria by providing the Tra functions in *trans*. This step is normally achieved by using strain S17-1, which carries a chromosomally integrated copy of RP4 (Tra⁺).

Gene replacement techniques can also be used to replace already existing Tn5 insertions with other Tn5 derivatives carrying alternative selectable marker genes, reporter gene fusions, portable promoters, DNA primers for restriction sites, and origins of transfer or replication (see section 29.4 and references 23, 58, and 154). This replacement allows a large degree of versatility. The only limitation in these types of replacement experiments is the length of the homologous sequences needed for the double-recombination event. This consideration is particularly relevant for transposons used for the construction of gene fusions and those carrying portable promoters, in which very limited Tn5 sequences may be present between the reporter gene or promoter and the end of the transposon.

29.3.4.1. Broad-Host-Range and Incompatible Plasmids

1. Clone the Tn5-mutated region into the broad-host-range vector pRK290.

2. Perform a triparental mating, as outlined in section 29.2.2.2, with the recipient strain (from which the cloned region was derived), an *E. coli* strain harboring the Tn5-mutated gene region cloned into pRK290, and an *E. coli* strain harboring the helper plasmid pRK2013. Select for the recipient strain, Tn5 (Km⁺), and pRK290 (Tc⁺).

3. Purify the obtained exconjugant once on the same selective plates.

4. Perform a diparental mating (section 29.2.2.1) with the recipient strain harboring the Tn5-mutated gene on the pRK290 replicon and an *E. coli* strain harboring pPH1JI. Select for the recipient strain, Tn5 (Km⁺), and pPH1JI (Gm⁺).

5. Pick 100 to 1,000 exconjugants.

6. Screen for the loss of the pRK290 replicon by streaking the exconjugants onto master plates with and without tetracycline.

7. Purify about 10 different tetracycline-sensitive exconjugants once or twice on selective plates.

8. Isolate the total chromosomal DNA from these tranconjugants, digest it with one or two restriction enzymes, separate it by agarose gel electrophoresis, blot it onto a filter, and hybridize it with the cloned wild-type gene region. The wild-type region should be replaced by the Tn5-mutated region. (In a digest with a restriction enzyme that does not cut Tn5, such as EcoRI, the wild-type fragment should be replaced by a fragment 5.8 kb larger [see section 29.2.3.2 and reference 50].)

29.3.4.2. Narrow-Host-Range Plasmids

1. Clone the Tn5-mutated region into a narrow-host-range vector, such as pBR322 or pSUP202.

2. Transform an *E. coli* strain providing the Mob and Tra functions with the clone obtained in step 1. (Use strain GJ23 for pBR322 derivatives and strain S17-1 for pSUP derivatives.)

3. Perform a diparental mating (section 29.2.2.1) with the recipient strain (from which the region was cloned) and the *E. coli* strain harboring the narrow-host-range vector with the cloned Tn5-mutated region. Select for the recipient strain and Tn5 (Km⁺).

4. Pick 100 to 1,000 transconjugants.

5. Screen for loss of the vector sequences by streaking the exconjugants onto master plates with and without the antibiotic resistance marker of the vector used but containing kanamycin for the selection of Tn5.

6. Purify, once or twice on selective plates, about 10 colonies that do not exhibit the antibiotic resistance phenotype conferred by the vector.

7. Isolate the total chromosomal DNA from these transconjugants, digest it with one or two restriction enzymes, separate it by agarose gel electrophoresis, blot it onto a filter, and hybridize it with the wild-type gene region. The wild-type region should have been replaced by the Tn5-mutated region (see section 29.3.4.1, step 8).

29.3.4.3. Screening for Double Crossovers

In some bacterial systems, depending on the host or the special gene region mutagenized, it may be difficult to find double-crossover events. Sometimes more than 1,000 colonies must be screened to find exconjugants that have undergone the double homologous crossover (151). Alternatively, a conditionally lethal gene has been used to improve the gene replacement protocol. This gene allows selection and screening for the bona fide gene exchange event in several bacterial species. The *sacB* gene of *Bacillus subtilis* (coding for levansucrase) confers sensitivity to 5% sucrose; i.e., gram-negative organisms expressing the *sacB* gene cannot survive on solid medium containing 5% sucrose (72). If the vector used for the gene replacement experiment carries the *sacB* gene, all single-crossover events leading to the cointegration of the vector will be lethal. Therefore, only exconjugants in which double-crossover events have occurred will be obtained in the presence of 5% sucrose. Before trying this approach, the sensitivity of the strain of interest, containing the cloned *sacB* gene, to the presence of sucrose in the growth medium should be tested.

29.4. USES OF Tn*5* DERIVATIVES

29.4.1. Reporter Gene Fusions

The utility of transposon Tn*5* mutagenesis has been extended to the generation of gene fusions for regulation studies. An abbreviated list of Tn*5*-based vehicles for the construction of chimeric genes is shown in Table 2. For additional compilations of useful transposon derivatives, see references 14, 16, and 91.

Transposons for the generation of both transcriptional and translational reporter gene fusions are available. In the case of transcriptional fusions, the reporter gene carries its own start codon and DNA sequences recognized by the translational machinery. In translational fusions, the reporter gene lacks its own start codon and ribosome-binding site and its expression is dependent not only on transcription originating from the promoter of the mutagenized gene but also on the simultaneous creation of an in-frame fusion of the open reading frame of the mutagenized gene with the truncated reporter gene.

The reporter genes which are commonly used to construct gene fusions include the β-galactosidase (β-Gal; *lacZ*), β-glucuronidase (Gus; *uidA*), alkaline phosphatase (AP; *phoA*), luciferase (Lux; *lux*), and green fluorescent protein (*gfp*) genes. Convenient dyes to detect the expression of these genes on solid growth media, enzymatic assays, and histochemical staining methods or other methods for the detection of these reporter gene products are available, making these genes very useful to monitor the quantitative

TABLE 2   Tn*5* derivatives used for the creation of gene fusions

Reporter gene(s)	Designation of derivative	Antibiotic resistance phenotype	Length (kb)	Type of fusion	Reference
cat	pTn*5cat*	Km^r Nm^r	5.9	Transcriptional	128
cya	Mini-Tn*5cyaA'*	Km^r Nm^r	~2.9	Translational	207
gfp	Tn*5GFP1*	Tc^r	~6.2	Transcriptional	32
	Mini-Tn*5gfp*	Gm^r Km^r Nm^r	~2.25	Transcriptional	203
	Mini-Tn*5gfp-km*	Gm^r Km^r Nm^r	~2.6	Transcriptional	205
	mTn*5-gfp-pgusA*	Km^r Nm^r	~4.7	Transcriptional	219
gusA	Tn*5-gusA1*	Km^r Nm^r Tc^r	8.3	Transcriptional	183
	Tn*5-gusA9*	Sm^r Sp^r	5.0	Transcriptional	183
	Tn*5-gusA2*	Km^r Nm^r Tc^r	8.3	Translational	183
	Tn*5-gusA10*	Sm^r Sp^r	5.0	Translational	183
	mTn*5SSgusA40*	Sm^r Sp^r	~4.4	Transcriptional	214
	mTn*5gusA-pgfp*	Km^r Nm^r	~4.7	Transcriptional	218
	mTn*5gusA-pgfp11*	Km^r Nm^r	~5.1	Transcriptional	218
	mTn*5gusA-pgfp12*	Km^r Nm^r	~5.4	Transcriptional	218
	mTn*5gusA-pgfp21*	Km^r Nm^r	5.742	Transcriptional	218
	mTn*5gusA-pgfp22*	Km^r Nm^r	~6.0	Transcriptional	218
hlyA	mTn*5hlyAs*	Km^r Nm^r	3.4	Translational	74
lacZ	Tn*5-lac*	Km^r Nm^r	12.0	Transcriptional	113
	Tn*5-B20*	Km^r Nm^r	8.3	Transcriptional	192
	Tn*5-B21*	Tc^r	8.4	Transcriptional	192
	Tn*5-B22*	Gm^r	9.4	*Transcriptional*	192
	Tn*5-OT182*	Tc^r Ap^r Cb^r	10.9	Transcriptional	135
	Mini-Tn*5lacZ1*	Km^r Nm^r	5.4	Transcriptional	55
	Tn*5ORFlac*	Tc^r	~12	Translational	110
	Mini-Tn*5lacZ2*	Km^r Nm^r	5.3	Translational	55
luxAB	Tn*5-1063*	Km^r Nm^r Bm^r Sm^r	7.8	Transcriptional	216
	Mini-Tn*5luxAB*	Tc^r	5.2	Transcriptional	55
nptII	Tn*5-VB32*	Tc^r	~5	Transcriptional	13
	Tn*5-B30*	Tc^r	6.2	Transcriptional	192
phoA	Tn*5-phoA*	Km^r Nm^r	7.7	Translational	125
	Mini-Tn*5phoA*	Km^r Nm^r	3.5	*Translational*	55
phoA and *lacZ*	Mini-Tn*5phoAlac1*	Km^r Nm^r	~2.8	Translational	4
	Mini-Tn*5phoAlac2*	Cm^r	~2.6	Translational	4
xylE	Mini-Tn*5xylE*	Km^r Nm^r	3.7	Transcriptional	55

and in situ expression of chimeric genes. β-Gal, Gus, and Lux activities are cytoplasmic, whereas AP activity can be measured only if the protein is periplasmic or excreted. Thus, Tn5-phoA fusions can be used to identify protein signal sequences required for excretion (124–126). Moreover, saturation mutagenesis of a gene with, for example, Tn5-lac or Tn5-phoA can reveal sequences that are translated into cytoplasmic protein domains or transmembrane domains, respectively. A simplification of this procedure was achieved by the construction of dual pho-lac transposon derivatives that are used in a single mutagenesis step to analyze the topology of membrane proteins (4). Thus, both β-Gal and AP activities can be measured at the same fusion point of the protein to give complementary information.

For detailed descriptions of the use of the lac (187), phoA (126), uidA (71, 102), lux (134, 200), and gfp (36, 202) reporter gene systems, the reader is referred to the respective references. For the detection of β-Gal enzyme activity in vivo on plates or in cell extracts, see chapter 19.3.6.1 and reference 137. For the detection of AP activity, see reference 178. For the detection of Gus activity, see references 71, 102, 116, and 183. For the detection of Lux activity, see references 134, 158, 200, and 216. For the detection of gfp activity, see references 36 and 202.

Recently, reporter gene transposons that contain a promoterless gusA reporter gene and constitutively expressed gfp as an additional marker gene have been constructed (218, 219). These constructs are useful not only for measuring the expression level of a particular gene but also for tracking the bacterial strain in host tissue during symbiotic or pathogenic relationships.

In addition to reporter genes lacking transcriptional and/or translational start signals, a number of other useful DNA sequences and genes have been cloned into Tn5 to facilitate their random or region-directed introduction into bacterial chromosomes or recombinant plasmids. A number of examples of useful Tn5 derivatives are discussed in the following sections. For a more comprehensive list of such derivatives and examples of their use, see references 14 and 16. In most cases, the protocols for the generation and characterization of Tn5-containing replicons are the same as those described above (sections 29.2 and 29.3).

The actual mutagenesis protocols for the generation of Tn5-lac, Tn5-phoA, Tn5-uidA, Tn5-lux, and Tn5-gfp insertions into genomic loci or cloned DNA fragments carried by recombinant plasmids are the same as those described above (sections 29.2 and 29.3). For details of generating gene fusions with the mini-Tn5-reporter gene transposons, see de Lorenzo et al. (55, 56).

The only difference from conventional Tn5 mutagenesis is that, at the stage of selecting Tn5-reporter gene-containing strains or plasmids, the appropriate indicator compounds (e.g., 5-bromo-4-chloro-3-indolyl-β-D-galactopyranoside [X-Gal] for β-Gal activity, 5-bromo-4-chloro-3-indolylphosphate [XP] for AP activity, and 5-bromo-4-chloro-3-indolyl-β-D-glucuronic acid [X-Gluc] for Gus activity) are added to the solid medium and positive colonies containing the active transcriptional or translational fusions are identified under the proper physiological (inducing or repressing) conditions for the gene of interest. For the Tn5-lux transposons, colonies containing proper gene fusions can be identified by their level of light emission by using a photonic detection system (216). Tn5-gfp fusions can be detected by their fluorescence with fluorometers (luminescence spectrometers), fluorescence microscopes, or in extreme cases, the naked eye (202). The positions and ori-

entations of the Tn5-reporter gene insertions in the genome or on plasmids are mapped as described in sections 29.2.3 and 29.3.2.

Once an interesting Tn5-reporter gene-mediated gene fusion on a recombinant plasmid has been generated and physically characterized, it can be integrated into the genome of the host organism by gene replacement (section 29.3.4) or cointegrate formation. In the latter case, the protocols described in sections 29.3.4.1 and 29.3.4.2 can be used, but instead of screening for transconjugants which have lost the antibiotic marker(s) of the plasmid vector carrying the fusion, colonies which retain the vector marker are selected and the presence of the vector is verified by Southern blotting (see section 29.2.3.2).

Tn5 derivatives carrying reporter genes have recently been used as tools in whole-cell environmental monitoring. For example, a mini-Tn5luxCDABE construct was integrated into Pseudomonas strains for ecotoxicity testing of copper, zinc, or toluene via light emission (212). Mini-Tn5's with lux, lac, or gfp genes were joined with promoters inducible by tetracycline or mercury and used to determine the amount of antibiotics in milk or of mercury in soil, respectively (87, 88). In addition, the mini-Tn5luxAB-tet transposon joined with fluorene-inducible promoters was used in the construction of whole-cell biosensor strains for monitoring the amount of the polycyclic aromatic hydrocarbon fluorene in soil (10).

29.4.2. Secondary Transposon Mutagenesis for the Identification of Regulatory Genes

A successful mutagenesis procedure with a reporter gene transposon will result in the identification of a variety of mutants in which the genes targeted by the transposon are differentially expressed under certain environmental conditions. Now, a second round of mutagenesis with another transposon can be used to identify genes that encode trans-acting regulatory proteins regulating the genes identified in the first round of mutagenesis. As an example, this strategy has been employed to identify regulators of genes involved in tyrosine degradation and nutrient deprivation (48, 136). In both cases, the original reporter gene transposon was Tn5 based and contained luxAB as the reporter genes. For the second round of mutagenesis, a transposon carrying a different antibiotic resistance marker (to facilitate selection) and belonging to another class (to avoid homologous recombination) was used. Davey and de Bruijn (48) and Milcamps et al. (136) used Tn1721, which is Tn3 based and carries the tetracycline resistance gene. The genes of interest identified in the first round of mutagenesis showed high levels of gene expression (high levels of light emission) under specific environmental conditions. Thus, in the second round, the investigators screened for mutants that did not show high levels of light emission under these conditions. This approach led to the successful identification of novel regulatory genes (48, 136).

29.4.3. Portable Promoters

Tn5 derivatives carrying strong, regulatable, outward-facing promoters have been constructed in vitro. The synthetic, inducible tac promoter and the constitutive promoter for the neomycin phosphotransferase (nptII) gene have been introduced into Tn5, near one of the ends, to generate Tn5-tac1 and Tn5-B50, respectively (40, 192); additional derivatives are listed in Table 3. Upon the insertion of these transposon elements upstream of the 5' end of the coding sequence of a certain gene, the expression of this gene is

TABLE 3 Tn5 derivatives carrying portable promoters, alternative marker genes, and useful sites

Relevant transposon characteristic	Designation in original reference	Antibiotic resistance phenotype and/or element(s) with novel sequences	Length (kb)	Reference(s) or source
Carry outward-reading promoters	Tn5-tac1	Km^r; ptac	4.5	40
	Tn5-B50	Tc^r; pnptII	5.0	192
	Tn5-B60	Km^r; ptac	5.0	192
	Tn5-B61	Gm^r; ptac	5.5	192
Carry replication and mobilization origins	Tn5V	Km^r; ori pSC101	~6	69
	Tn5-oriT	Km^r; oriT (RK2)	6.5	220
	Tn5-Mob	Km^r; oriT (RP4)	7.8	188
	Tn5-B11	Gm^r; (RP4)	8.4	192
	Tn5-B12	Gm^r; Sp^r; oriT (RP4)	10.2	192
	Tn5-B13	Tc^r; oriT (RP4)	9.4	192
	Tn5-OT182	Ap^r Cb^r Tc^r; oriV (ColE1)	10.9	135
	pTn5cat	Km^r; oriV (ColE1), oriT (RP4)	5.9	128
Carry sequencing primer sites	Tn5seq1	Km^R; pT7, pSP6	3.2	143
	Tn5supF	supF	0.3	153, 154
	EZ::TN <KAN-2>	Km^r Nm^r	1.2	Epicentre, Madison, WI
	EZ::TN <TET-1>	Tc^r	1.7	Epicentre, Madison, WI
	EZ::TN <DHFR-1>	Tp^r	0.9	Epicentre, Madison, WI
Carry novel antibiotic resistance genes	Tn5-Tc2	Tc^r	5.1	180
	Tn5-Cm	Cm^r	3.9	180
	Tn5-Gm	Gm^r	7.5	180
	Tn5-Tp	Tp^r	5.2	180
	Tn5-Sm	Sm^r	5.2	180
	Tn5-Ap	Ap^r	5.0	180
	Tn5.7	Tp^r Sm^r Sp^r	7.6	224
	Tn5-233	Gm^r/Km^r, Sm^r/Sp^r	6.6	58
	Tn5-235	Km^r; lacZY	10.0	58
	Tn5-GmSpSm	Gm^r Sp^r Sm^r	6.8	98
	Tn5-751	Km^r Tp^r	9.0	161
	Tn5-30Tp	Km^r Tp^r	7.9	1
	Tn5-31Tp	Km^r Tp^r	7.9	1
	Mini-Tn5 Sm/Sp	Sm^r Sp^r	2.1	55
	Mini-Tn5 Tc	Tc^r	2.2	55
	Mini-Tn5 Cm	Cm^r	3.5	55
	Mini-Tn5 Km	Km^r	2.3	55
Carry other marker genes	Tn5 Amy^{+a}	Km^r; amy	8.7	211
	Tn5 Bgl^{+b}	Km^r; bgl	11.3	211
	Tn5-Lux	Km^r; luxAB	6.6	27
	Tn5-tox^c	Km^r; tox	10.5	144
	Mini-Tn5 ars^d	ars	5.0	95
	Mini-Tn5 bar^e	bar	2.5	95
	Mini-Tn5 mer^f	mer	4.5	95
	Mini-Tn5 lacZY	lacZY	~5	63
Carry dual-marker systems	Mini-Tn5 PpsbA-gfp luxAB	Km^r Nm^r; gfp, luxAB	~3	208
Useful for restriction mapping	Tn5cos	Km^r Nm^r	6.2	225
Useful for plasmid curing	Tn5-rpsL	Km^r Nm^r	7.6	201

(Continued on next page)

TABLE 3 Tn5 derivatives carrying portable promoters, alternative marker genes, and useful sites (*Continued*)

Relevant transposon characteristic	Designation in original reference	Antibiotic resistance phenotype and/or element(s) with novel sequences	Length (kb)	Reference(s) or source
	Tn5-B12S	Km^r Nm^r; *sacBR*	9.9	101
	Tn5-B13S	Km^r Nm^r Tc^r; *sacBR*	13.2	101
Carry multiple unique restriction sites	Tn5-K20	Km^r Nm^r Tc^r	9.5	103
	Tn5-K28	Sm^r Tc^r	7.4	103
	mTn5 in pBSL118	Km^r	~1.3	5
	mTn5 in pBSL202	Gm^r	~0.7	5
	mTn5 in pBSL203	Cm^r	~1.1	5
	mTn5 in pBSL204	Tc^r	~1.5	5
	mTn5 in pBSL299	Sm^r	~2.1	5
	pTn5cat	Km^r; oriV (ColE1), oriT (RP4)	5.9	128
	pTnMod-OGM	Gm^r	4.7	57

^{a}Amy$^+$, ability to degrade starch.
^{b}Bgl$^+$, ability to ferment cellobiose.
^{c}tox, delta endotoxin gene from *Bacillus thuringiensis*.
^{d}ars, gene for resistance to arsenite; vector designated pUT/Ars by Herrero et al. (95).
^{e}bar, gene for resistance to the herbicide bialaphos; vector designated pUT/PTT by Herrero et al. (95).
^{f}mer, gene for resistance to mercuric salts and organomercurial compounds; vector designated pUT/Hg by Herrero et al. (95).

rendered inducible by isopropyl-β-D-thiogalactoside (for Tn5-*tac*1) or constitutive (for Tn5-B50). This property also applies to genes positioned distally in an operon. When Tn5 derivatives such as Tn5-*tac*1 and Tn5-B50 insert into an operon, the downstream genes come under the control of the Tn5-borne promoters. By using random or region-directed Tn5 mutagenesis protocols, cryptic or normally down-regulated genes can be activated in this fashion. This technique can be very helpful for the overproduction of proteins.

29.4.4. Transfer and Replication Origins

Tn5 derivatives in which the central region of Tn5 has been replaced with the IncP-specific mobilization target site (Tn5-Mob and Tn5-*oriT*) have been constructed previously (188, 220); (Table 3). The insertion of these transposon elements allows the target replicon to be efficiently mobilized via the broad-host-range mobilization functions of RP4 or RK2. If the insertion is in the chromosome, the insertion site becomes the origin of Hfr-like oriented transfer. This technique is therefore very useful for the construction of genetic maps; see references 38, 65, 145, and 220 for examples (see also section 29.2.4). If the insertion is in a (recombinant) plasmid, it becomes F′-like and allows the plasmid to be conjugally mobilized at high frequencies in di- or tri-parental mating experiments (see sections 29.2.2.1 and 29.2.2.2); for reviews, see references 191 and 192.

Tn5 derivatives carrying a narrow-host-range origin of replication have also been constructed previously. Furuichi et al. (69) cloned the origin of replication of the ColE1-type plasmid pSC101 into Tn5 to construct Tn5V (Table 3). Tn5V-mutated chromosomal loci generated by random Tn5 mutagenesis (see section 29.2) can be readily cloned by using the antibiotic resistance marker and the replication origin carried on the transposon by digestion of the total DNA with an enzyme which does not cut Tn5V, religation, and transformation (see section 29.2.5). Wolk

et al. (216) also constructed a Tn5 derivative which carries a P15A (pACYC184) origin of replication in addition to a *lux* reporter gene (Tn5-1063; see section 29.4.1 and Table 2), which is extremely useful for recloning Tn5-1063-mutagenized chromosomal loci with *lux* gene fusions. The mini-Tn5 derivative pTn5cat constructed by Marsch-Moreno et al. (128) combines several of the above-mentioned features: it contains a promoterless chloramphenicol acetyltransferase gene as a reporter, a ColE1 origin of replication to allow easy recloning, the *oriT* region of RP4 for mobilization, and several rare restriction sites to facilitate mapping.

29.4.5. Alternative Marker Genes

Many derivatives of the original Tn5 transposon containing a variety of additional antibiotic resistance and other marker genes have been constructed (16) (Table 3). Therefore, not only the original kanamycin and neomycin resistance markers (which are expressed in all organisms tested thus far) or the streptomycin and bleomycin resistance markers (which are expressed in several nonenteric bacteria), but also genes encoding resistance to tetracycline, ampicillin, chloramphenicol, trimethoprim, and gentamicin, can be used to monitor the presence of Tn5 derivatives (Table 3). These derivatives are especially useful if the target bacterial strain has an intrinsic resistance to the antibiotics normally used to select for Tn5 (kanamycin, neomycin, bleomycin, and streptomycin) or if a secondary mutagenesis of a strain already harboring a Tn5 derivative (e.g., a Tn5-reporter gene fusion; see section 29.4.1) is to be carried out. They can also be used to replace already existing genomic Tn5 insertions via homologous recombination (gene replacement) to exchange antibiotic resistance marker genes.

Mini-Tn5 derivatives consisting of the terminal 19 bp of IS50L and IS50R flanking the genes encoding resistance to the herbicide bialaphos (*bar*), mercury and organomercurial compound resistance (*mer*), or arsenite resistance (*ars*) have been constructed by Herrero et al. (95). The transposase required for the transposition of these derivatives is provided by a (modified) *tnp* gene, cloned immediately adjacent to the mini-Tn5 construct on the suicide vector pGP704 (95) (see section 29.2.2). A mini-Tn5 construct has been modified to carry the *lacZY* genes encoding β-galactosidase as a nonantibiotic marker to track *Pseudomonas* strains in the rhizospheres of plants (63). For similar purposes, e.g., the tracking of bacteria in soil, Unge et al. (208) constructed the dual-marker transposon mini-Tn5 P*psbA*-*gfp luxAB*, in which the *gfp* marker is independent of the metabolic activity of the marked strain but the *luxAB* genes require energy and therefore an active metabolic state of the bacterial cell. Thus, when this construct is inserted into a bacterial strain, it allows the simultaneous monitoring of the metabolic activity and the total cell number in a specific soil sample.

Several additional non-antibiotic resistance genes have been inserted into Tn5 to facilitate their transfer to and maintenance in a range of gram-negative bacteria. For example, genes encoding β-glucosidase and amylase, tryptophan biosynthesis, catechol-2,3-dioxygenase, and delta endotoxin production have been inserted into Tn5 (Table 3). The expression of these genes can be used to monitor the (stable) presence of the modified transposons in different bacterial species. To facilitate the cloning of other genes into Tn5 for stable integration into the genomes of bacteria, Tn5 derivatives carrying many unique restriction sites, such as Tn5-K20 and Tn5-K28, have been constructed (103). Also, the use of mini-Tn5's has been extended by including multiple restriction sites in the vector series constructed by Alexeyev et al. (5) (Table 3).

A derivative of Tn5 carrying the λ *cos* site has been constructed. It allows the rapid restriction mapping of large plasmids by using λ terminase (225). After the transposition of Tn5*cos* into the desired plasmid, the DNA is linearized at the *cos* site and the left and right termini thus generated are labeled selectively in vitro. A partial digestion of the linearized DNA is carried out with the restriction enzyme of choice, and the resulting (partial) fragments are visualized by means of the (radioactive or nonradioactive) label. This method should also be applicable to the establishment of a restriction map of a specific region of a bacterial chromosome after Tn5*cos* mutagenesis (225).

Tn5 derivatives that can be used for direct selection for the curing and deletion of large, indigenous plasmids in bacteria have been constructed by inserting the *Bacillus subtilis* *sacBR* genes and the RP4 *mob* site into Tn5 (Tn5-B12S and Tn5-B13S) (Table 3) (101). When one of these modified transposons inserts into a large, cryptic plasmid, the resulting tagged plasmid can be mobilized to other bacterial strains. However, when the gram-negative bacteria containing the Tn5-B12S- or Tn5-B13S-tagged plasmid are grown in the presence of sucrose, the expression of the *sac* genes is lethal (see also section 29.3.4.3), and therefore, curing of the plasmid or deletion formation events are selected (101).

An additional derivative of Tn5 useful in plasmid-curing experiments has been described previously (201). Tn5-*rpsL* carries the *E. coli rpsL* gene, which confers streptomycin sensitivity on strains carrying the transposon. After a Sm^r derivative of the desired bacterial strain containing a plasmid to be cured has been isolated, it is mutagenized with Tn5-*rpsL* and insertions into the plasmid are isolated. The resulting strains are now Sm^s again. By selecting for Sm^r derivatives of the plasmid- and Tn5-*rpsL*-containing strain,

the loss of this plasmid is selected for, resulting in plasmid curing and deletion formation.

A Tn5 deletion derivative which has the ability to act as a recombinational switch activating distal gene expression when inserted in one orientation (Tn5-112) has also been reported previously (17). Already-existing Tn5 insertions at any site can be replaced with this transposon via homologous recombination and can be inserted to activate (or silence) downstream genes.

29.4.6. Primers for DNA Sequencing and PCR

Tn5 derivatives that facilitate directed DNA sequencing by the chain termination method (177) have been developed previously (also see chapter 27.1). To use this method, it is important to bring the DNA to be sequenced into the immediate vicinity of unique sequences that can be used as oligonucleotide-priming sites. Because of the nature of its inverted repeats, Tn5 does not contain unique primer sites at its ends. Detailed mutational analyses have shown that only the outermost 19 bp are needed for the efficient transposition of Tn5 (152). Nag et al. (143) have constructed Tn5*seq1*, which contains unique primer sites derived from the promoters of the phages SP6 and T7 adjacent to these essential sequences at the subterminal ends of Tn5. These unique sequences, normally not present in the bacterial genome or plasmids, can be used as primer hybridization sites for commercially available oligonucleotides. Because Tn5 transposes in vivo into many different sites, its ends can serve as portable priming sites for DNA sequencing, preventing the need for subcloning or deletion experiments (see chapter 27.1).

Another mini-Tn5-derived tool for the DNA sequencing of phage λ-borne DNA segments, Tn5*supF*, has been constructed by Phadnis et al. (153). This transposon carries a synthetic *E. coli supF* gene cloned between the 19-bp termini of Tn5 and, when inserted after transposition, allows amber-mutant phage λ derivatives (such as the cloning vector Charon 4A) to form plaques on a Su$^-$ strain. By using unique primers hybridizing to the ends of the transposon, DNA sequencing in both directions reading outward from the Tn5*supF* insertion can be carried out (153). This transposon is also very suitable for gene replacement experiments with *E. coli* (154).

With the help of PCR, Rich and Willis (166) have developed a method of isolating DNA sequences adjacent to the ends of a chromosomal Tn5 insertion. They have used a single oligonucleotide complementary to and extending outward from the inverted repeats of Tn5 to amplify the target sequences. By digesting the chromosomal DNA with a restriction enzyme not cutting within Tn5 (EcoRI) and self-ligating the chromosomal DNA at a low concentration to favor intramolecular ligation, the circular DNA molecule containing Tn5 and the adjacent target sequences will be the only plasmid containing the two priming sites for the PCR. The resulting PCR product can then be used as a probe for the cloning of the wild-type region.

A combination of the PCR and the Tn5*supF* mutagenesis protocols (153) (see above) have been used to further facilitate Tn5*supF*-based DNA sequencing of λ phage clones (112).

Avoiding any cloning or ligation steps, Subramanian et al. (204) have used primers based on the ends of Tn5 and the so-called repetitive extragenic palindrome (REP) sequences to map chromosomal Tn5-induced mutations. The REP sequences are highly conserved 35-bp-long palindromic sequences first identified in a number of intercistronic regions of numerous operons in *E. coli* and *Salmonella* but later also found in a variety of other eubacteria (51, 210). The chromosome of *E. coli* contains 581 REP sequences, and these sequences account for 0.54% of the genome (26). REP sequences are very useful for the genomic fingerprinting of bacteria (51, 210). For the PCR, one primer homologous to the end of Tn5 and another degenerate primer homologous to the genomically anchored REP sequence are used. Thus, a unique segment of chromosomal DNA will be amplified and can be used as a probe to identify the corresponding region in a genomic library of the bacterial species of interest, or the PCR product can be sequenced directly.

A similar technique to determine the DNA sequence adjacent to a transposon insertion without a cloning step has been developed by O'Toole and Kolter and O'Toole et al. (146, 147). In the arbitrary PCR method, two rounds of amplification are used to enrich the DNA flanking the transposon insertion site. Two sets of oligonucleotides are used, including two arbitrary primers and two primers that are homologous to the transposon; one of the two anneals a little bit further to the transposon ends than the other. In the first round of PCR, one of the transposon-based oligonucleotides is used as a primer, together with an arbitrary oligonucleotide that has at its 3' end a mixture of all four bases. In the second round, the oligonucleotide that is closer to the ends of the transposon (the nested primer) is used together with an oligonucleotide that is homologous to the known sequence of the 5' end of the arbitrary primer used in the first round. The exact PCR conditions and primers are described in detail by O'Toole and Kolter and O'Toole et al. (146, 147). The template in the first reaction can be chromosomal DNA, a purified colony, or a liquid culture of the insertion mutant. The arbitrary oligonucleotide will anneal to many sites in the genome, but only the annealing at a site close to the transposon-based primer will result in the generation of PCR products. In the second round, 1 μl of the PCR product from the first round is used as a template in the enrichment step. Usually, the PCR products obtained in the second round are between 400 and 600 bases. They can be purified by gel electrophoresis or the use of DNA purification columns and subjected to direct DNA sequencing. This approach has been used successfully to determine the DNA sequences adjacent to a variety of transposon insertions without the need for DNA cloning (120, 146, 147, 168, 170).

29.4.7. Genome Mapping

Tn5 and its derivatives are also useful for genome mapping. Pulsed-field gel electrophoresis is a tool for separating large DNA fragments (193). To obtain large DNA fragments, several new restriction endonucleases that cut only at rare sites in the genome are available. For example, for A+T-rich organisms, the restriction enzyme NotI, which recognizes an octameric sequence consisting only of G and C residues, is used to generate large fragments. For G+C-rich organisms, PacI and SwaI, which recognize octameric sequences consisting of A and T residues only, can be used. For example, before the complete genomic DNA sequence of *E. coli* was known, a genomic map based on the 22 different NotI fragments had been established (194). NotI also cuts within the inverted repeats (IS50) of Tn5 (Fig. 1); therefore, the insertion of Tn5 into the *E. coli* genome will generate two additional NotI sites, which will result in the appearance of two new fragments (plus a small internal Tn5 fragment). By measuring the sizes of the two new (junction) fragments, the position of the Tn5 insertion on the NotI re-

striction map can be determined. Smith and Kolodner (195) have used this method to assign Tn5 insertions to specific NotI fragments of the *E. coli* genome. A similar strategy has been used by Wolk et al. (216) to map Tn5-*lux*-induced mutations in *Anabaena* mutants.

Mini-Tn5 derivatives which carry rare restriction sites such as M·XbaI/DpnI, NotI, SwaI, PacI, and SpeI have been constructed previously (217). These derivatives have been used to construct a map of *Salmonella* via pulsed-field gel electrophoresis and should be extremely useful for the genomic mapping of chromosomes and large replicons of other bacteria, since they introduce a variety of rare cutting sites upon (random) insertion. In addition, already-existing Tn5 insertions can be converted into derivatives carrying these rare sites via gene replacement, further extending the utility of this method for large-scale gene mapping.

29.4.8. In Vitro Transposition

The molecular events of Tn5 transposition have been studied in detail by Reznikoff and coworkers and Steiniger-White et al. (162, 164, 165, 199). A hyperactive mutant transposase facilitated the studies, and it was found that, indeed, only the transposase and the 19-bp outer ends of Tn5 are needed for Tn5 transposition (77, 79). Any DNA sequence located between the outer ends will be transposed. By using the hyperactive transposase and the (slightly modified) 19-bp outer ends of Tn5, an in vitro transposition system was developed and is commercially available (EZ::TN™; Epicentre, Madison, WI). Containing one of three antibiotic resistance markers (a kanamycin, tetracycline, or trimethoprim resistance gene), this transposon is incubated with the target DNA (DNA cloned into any plasmid or cosmid or linear DNA) and the transposase in vitro. After incubation, *E. coli* is transformed with the transposition products and selection for the antibiotic resistance marker of the transposon will yield the transposition products. This method facilitates the random insertion of resistance markers into cloned DNA. In addition, the kit provides sequencing primers. Thus, this system offers a convenient way to introduce random priming sites into any DNA to be sequenced.

An extension of this method is the use of a preformed transposon-transposase complex (Transposome™; Epicentre, Madison, WI) for random mutagenesis. The transposon-transposase complex is stable in the absence of Mg^{2+}, and it can be electroporated into living cells. The Mg^{2+} concentration in the cell is sufficiently high to promote the random insertion of the transposon. Since this method does not use any suicide vectors, it is a convenient approach for creating random gene knockouts in many bacteria, including those for which no delivery vehicles are known. Using this strategy, Goryshin et al. (76) were able to show highly efficient transposition in *E. coli*, *Salmonella*, and *Proteus vulgaris* and even in the yeast *Saccharomyces cerevisiae*. Fernandes et al. (64) used this system to create knockout mutants of *Rhodococcus* species, which were previously not amenable to the delivery of transposons. This system has also been used to successfully mutagenize *Streptomyces coelicolor* (73) and *Xylella fastidiosa* (84). Oligonucleotides that are homologous to the ends of the transposon are also included in this kit (Epicentre, Madison, WI). Combined with the possibility of the direct sequencing of bacterial genomes without cloning (93, 99, 111) and the availability of complete bacterial genome sequences in public databases such as Genbank, it is a powerful and fast approach to easily identify the genotypes of insertional mutants and to correlate them to the observed phenotypes.

Moreover, a strategy to generate deletions in cloned DNA is based on the use of the purified, hyperactive transposase of Tn5 (in vitro deletion strategy; Epicentre, Madison, WI) (221). This method is an alternative to the use of nucleases to generate deletions. A transposon containing an origin of replication, an ampicillin resistance gene, and a multiple cloning site to clone the DNA of interest was constructed. A kanamycin resistance gene is located outside of the ends of the transposon. When the purified transposase is provided in vitro, it cleaves at the transposon ends and releases the kanamycin resistance gene, and since in low concentrations the intramolecular transposition is favored, the transposon ends "attack" themselves. This will lead to the generation of either deletions or inversions. Since the kanamycin resistance gene is lost during these intramolecular events, the correct clones can be identified by their kanamycin sensitivity. Because inversion clones will be the same size as the original clone, the deletion clones are easily recognizable by agarose gel electrophoresis. The deletion clones can be used for DNA sequencing or for the precise mapping of the extensions of open reading frames.

Other in vitro mutagenesis systems are based on transposon Tn7 (25), which is commercially available (New England Biolabs, Beverly, MA); the mariner-family transposon Himar1 (2); or mini-Mu (85). For detailed protocols of the in vitro transposition systems, see the publications of the manufacturers. It is interesting that during in vitro Tn5 transposition, occasionally a 10-bp target duplication instead of the usual 9-bp duplication has been observed (4).

29.4.9. Protein Structure-Function Analysis Using Transposons

Transposon derivatives have been constructed that allow the insertion of short, in-frame linkers to identify sites in proteins that allow these manipulations without a loss of function ("permissive sites"). These transposon derivatives are useful for dissecting regulatory, binding, or catalytic domains of proteins. For an overview of the different transposons available for this use, see Manoil and Traxler (127). In general, first a transposon is used to generate an in-frame translational fusion, for example, with TnlacZ/in (127), Tnpholac1, or Tnpholac2 (4). In a second step, most of the inserted sequence is removed either by restriction cleavage and rejoining or by Cre-lox site-specific recombination. This leaves a short in-frame insertion in the protein of interest, for example, a peptide between 24 and 93 amino acids long (127). These peptide tags can be used to screen for sites that allow an insertion without destroying the function of the protein to analyze structure-function relationships such as protein-protein interactions, multimerizations, and the binding of effector molecules. A peptide tag can also serve as an epitope for detection via immunological methods. Many of the peptides contain a protease cleavage site, e.g., one for trypsin, which can be used to determine the membrane topology by analyzing protease-accessible and -inaccessible sites of membrane proteins (127). For similar purposes, the EZ::TN™ in-frame linker insertion kit is available commercially (Epicentre, Madison, WI).

New tools and new reporter genes have been developed to study protein secretion. A mini-Tn5 derivative (mTn5hlyAs) encoding the C-terminal secretory signals for the E. coli hemolysin secretion pathway has been used to insert secretion signals into proteins that are normally not se-creted (74, 197). Recently, another mini-Tn5 derivative has been constructed (mini-Tn5cyaA') which carries part of the Bordetella pertussis adenylate cyclase toxin gene, encoding the first 400 catalytically active, N-terminal amino acids (207). This transposon was successfully used to identify genes in Bordetella bronchiseptica that encode surface-exposed or secreted proteins. The determination of adenylate cyclase activity is not as simple as the determination of AP activity, but whereas the phoA reporter can be used to detect any protein that is delivered beyond the inner membrane, the cyaA reporter detects only surface-exposed or secreted proteins, excluding inner membrane and periplasmic proteins (207).

29.4.10. Signature-Tagged Mutagenesis

A variation of transposon mutagenesis useful for the identification of bacterial virulence genes was introduced in 1995 (94). This method facilitates the isolation of virulence-attenuated mutants from a large pool of bacterial mutant strains, thereby cutting down on the number of experimental animals being used. Mini-Tn5 transposons are uniquely marked by the introduction of an individual random sequence tag. This tag consists of a central, 40-bp random sequence, flanked by invariable arms of 20 bp that serve as priming sites for oligonucleotides during PCR. After the random transposon mutagenesis of a virulent bacterial species, mutant strains containing the transposon are assembled into a 96-well microtiter dish and pooled for DNA isolation, and a pool of these 96 mutant isolates ("input" pool) is inoculated into an experimental animal. When the mutant pool is reisolated from the animal, those mutants with attenuated virulence will not have survived the passage through the host and will not be represented in the "output" pool (negative selection). By using oligonucleotides homologous to the invariable arms surrounding the central variable sequence, the unique DNA tags of each transposon insertion are amplified via PCR. This amplification is done not only with the input pool, but also with the output pool. The unique tags of the input and output pools are radiolabeled separately and hybridized to colony blots or dot blots obtained from the original microtiter dish of 96 mutant isolates. Those mutant strains that do not survive the passage through the animal will not be represented in the output pool, and their unique DNA tag will not be amplified during the PCR amplification of the output pool. Colonies that hybridize to the probe generated from the input pool, but not from the output pool, represent the bacterial mutants with attenuated virulence. This approach has been successfully used to identify virulence genes in a variety of animal pathogens, including Actinobacillus pleuropneumoniae (68), Brucella melitensis (120), Campylobacter jejuni (81), Legionella pneumophila (156), Listeria monocytogenes (8), Pasteurella multocida (67), Proteus mirabilis (31), Salmonella enterica serovar Gallinarum (182), Salmonella enterica serovar Typhimurium (94), Staphylococcus aureus (133), Streptococcus pneumoniae (157), Streptococcus sanguinis (148), Vibrio cholerae (39), and Yersinia pseudotuberculosis (132). For reviews and more extensive protocols for signature-tagged mutagenesis, see references by Lehoux and Levesque (118, 119), Mecsas (131), Shea et al. (185), and Saenz and Dehio (173). In addition, signature-tagged mutagenesis has been adapted to identify genes relevant for the survival of bacteria in abiotic environments, e.g., sediments (83).

Due to recent progress in the determination of complete bacterial genome sequences, techniques based on transposon mutagenesis have experienced a reemergence (91).

Methods employed for genome-wide analyses include the identification of novel virulence loci through signature-tagged mutagenesis as described above (173), the formation of chromosomal deletions to make minimal genomes (78), and the identification of genes that are important for viability under certain environmental conditions by genetic footprinting (3, 75, 90, 196) or by combining transposon insertion libraries with whole-genome oligonucleotide microarrays (215).

29.5. RECIPES FOR MEDIA

The media listed here are used to prepare the λ::Tn5 lysate for region-directed Tn5 mutagenesis.

LB Medium

Tryptone	10 g/liter
Yeast extract	5 g/liter
NaCl	5 g/liter

SM Buffer

0.1 M NaCl
0.02 M Tris (pH 7.5)
0.01 M $MgSO_4$
0.01% gelatin

λ Agar Medium

Tryptone	10	g/liter
NaCl	2.5	g/liter
Bacto agar (Difco)	11	g/liter

λ Soft Medium

Tryptone	10	g/liter
NaCl	2.5	g/liter
Bacto agar	6	g/liter

For all other experiments, the appropriate growth medium and antibiotic concentration for the target organism should be used. For a general discussion about nutrition and a list of media, see chapters 10.8, 11.6, 15.4, and 31.8.

We thank Lynda King for help with the manuscript.

29.6. REFERENCES

1. **Abe, M., M. Tsuda, M. Kimoto, S. Inouye, A. Nakazawa, and T. Nakazawa.** 1996. A genetic analysis system of *Burkholderia cepacia*: construction of mobilizable transposons and a cloning vector. *Gene* **174:**191–194.
2. **Akerley, B. J., E. J. Rubin, A. Camilli, D. J. Lampe, H. M. Robertson, and J. J. Mekalanos.** 1998. Systematic identification of essential genes by *in vitro* mariner mutagenesis. *Proc. Natl. Acad. Sci. USA* **95:**8927–8932.
3. **Akerley, B. J., E. J. Rubin, V. L. Novick, K. Amaya, N. Judson, and J. J. Mekalanos.** 2002. A genome-scale analysis for identification of genes required for growth or survival of Haemophilus influenzae. *Proc. Natl. Acad. Sci. USA* **99:**966–971.
4. **Alexeyev, M., and H. H. Winkler.** 2002. Transposable dual reporters for studying the structure-function relationships in membrane proteins: permissive sites in *R. prowazekii* ATP/ADP translocase. *Biochemistry* **41:**406–414.
5. **Alexeyev, M. F., I. N. Shokolenko, and T. P. Croughan.** 1995. New mini-Tn5 derivatives for insertion mutagenesis and genetic engineering in Gram-negative bacteria. *Can. J. Microbiol.* **41:**1053–1055.
6. **Altman, E., J. R. Roth, A. Hessel, and K. E. Sanderson.** 1996. Transposons currently in use in genetic analysis of *Salmonella* species, p. 2613–2626. *In* F. C. Neidhardt, R. Curtiss III, J. L. Ingraham, E. C. C. Lin, K. B. Low, B. Magasanik, W. S. Reznikoff, M. Riley, M. Schaechter, and H. E. Umbarger (ed.), Escherichia coli *and* Salmonella: *Cellular and Molecular Biology*, 2nd ed., vol. 2. ASM Press, Washington, DC.
7. **Alton, N. K., and D. Vapnek.** 1979. Nucleotide sequence analysis of the chloramphenicol resistance transposon Tn9. *Nature* **282:**864–869.
8. **Autret, N., I. Dubail, P. Trieu-Cuot, P. Berche, and A. Charbit.** 2001. Identification of new genes involved in the virulence of *Listeria monocytogenes* by signature-tagged transposon mutagenesis. *Infect. Immun.* **69:**2054–2065.
9. **Barrett, J. T., R. H. Croft, D. M. Ferber, C. J. Gerardot, P. V. Schoenlein, and B. Ely.** 1982. Genetic mapping with Tn5-derived auxotrophs of *Caulobacter crescentus. J. Bacteriol.* **151:**888–898.
10. **Bastiaens, L., D. Springael, W. Dejonghe, P. Wattiau, H. Verachtert, and L. Diels.** 2001. A transcriptional *luxAB* reporter fusion responding to fluorene in *Sphingomonas* sp. LB126 and its initial characterisation for whole-cell bioreporter purposes. *Res. Microbiol.* **152:**849–859.
11. **Belas, R., A. Mileham, M. Simon, and M. L. Silverman.** 1984. Transposon mutagenesis of marine *Vibrio* spp. *J. Bacteriol.* **158:**890–896.
12. **Belfort, M., V. Derbyshire, M. M. Parker, B. Cousineau, and A. M. Lambowitz.** 2002. Mobile introns: pathways and proteins, p. 761–783. *In* N. L. Craig, R. Craigie, M. Gellert, and A. M. Lambowitz (ed.), *Mobile DNA II.* ASM Press, Washington, DC.
13. **Bellofatto, V., L. Shapiro, and D. A. Hodgson.** 1984. Generation of a Tn5 promoter probe and its use in the study of gene expression in *Caulobacter crescentus. Proc. Natl. Acad. Sci. USA* **81:**1035–1039.
14. **Berg, C. M., and D. E. Berg.** 1996. Transposable element tools for microbial genetics, p. 2588–2612. *In* F. C. Neidhardt, R. Curtiss III, J. L. Ingraham, E. C. C. Lin, K. B. Low, B. Magasanik, W. S. Reznikoff, M. Riley, M. Schaechter, and H. E. Umbarger (ed.), Escherichia coli *and* Salmonella: *Cellular and Molecular Biology*, 2nd ed., vol. 2. ASM Press, Washington, DC.
15. **Berg, C. M., and D. E. Berg.** 1987. Uses of transposable elements and maps of known insertion, p. 1071–1109. *In* J. L. Ingraham, K. B. Low, B. Magasanik, M. Schaechter, and H. E. Umbarger (ed.), Escherichia coli *and* Salmonella typhimurium: *Cellular and Molecular Biology.* American Society for Microbiology, Washington, DC.
16. **Berg, C. M., D. E. Berg, and E. A. Groisman.** 1989. Transposable elements and the genetic engineering of bacteria, p. 879–926. *In* D. E. Berg and M. M. Howe (ed.), *Mobile DNA.* American Society for Microbiology, Washington, DC.
17. **Berg, D. E.** 1980. Control of gene expression by a mobile recombinational switch. *Proc. Natl. Acad. Sci. USA* **77:** 4880–4884.
18. **Berg, D. E.** 1989. Transposon Tn5, p. 163–184. *In* D. E. Berg and M. M. Howe (ed.), *Mobile DNA.* American Society for Microbiology, Washington, DC.
19. **Berg, D. E., and C. M. Berg.** 1983. The prokaryotic transposable element Tn5. *Bio/Technology* **1:**417–435.
20. **Berg, D. E., J. Davies, B. Allet, and J. D. Rochaix.** 1975. Transposition of R factor genes to bacteriophage lambda. *Proc. Natl. Acad. Sci. USA* **72:**3628–3632.
21. **Berg, D. E., and M. M. Howe (ed.).** 1989. *Mobile DNA.* American Society for Microbiology, Washington, DC.

22. **Berg, D. E., M. A. Schmandt, and J. B. Lowe.** 1983. Specificity of transposon Tn5 insertion. *Genetics* **105:**813–828.

23. **Berg, D. E., A. Weiss, and L. Crossland.** 1980. Polarity of Tn5 insertion mutations in *Escherichia coli. J. Bacteriol.* **142:**439–446.

24. **Beringer, J. E., J. L. Beynon, A. V. Buchanan-Wollaston, and A. W. B. Johnston.** 1978. Transfer of the drug-resistance transposon Tn5 to *Rhizobium. Nature* **276:**633–634.

25. **Biery, M. C., F. J. Stewart, A. E. Stellwagen, E. A. Raleigh, and N. L. Craig.** 2000. A simple *in vitro* Tn7-based transposition system with low target site selectivity for genome and gene analysis. *Nucleic Acids Res.* **28:**1067–1077.

26. **Blattner, F. R., G. Plunkett III, C. A. Bloch, N. T. Perna, V. Burland, M. Riley, J. Collado-Vides, J. D. Glasner, C. K. Rode, G. F. Mayhew, J. Gregor, N. W. Davis, H. A. Kirkpatrick, M. A. Goeden, D. J. Rose, B. Mau, and Y. Shao.** 1997. The complete genome sequence of *Escherichia coli* K-12. *Science* **277:**1453–1462.

27. **Boivin, R., F. P. Chalifou, and P. Dion.** 1988. Construction of a Tn5 derivative encoding bioluminescence and its introduction in *Pseudomonas, Agrobacterium* and *Rhizobium. Mol. Gen. Genet.* **213:**50–55.

28. **Bolivar, F.** 1978. Construction and characterization of new cloning vehicles. III. Derivatives of plasmid pBR322 carrying unique *EcoRI* sites for selection of *EcoRI* generated recombinant molecules. *Gene* **4:**121–136.

29. **Bolivar, F., R. L. Rodriguez, P. J. Greene, M. C. Betlach, H. L. Heyneker, H. W. Boyer, and J. H. Crosa.** 1977. Construction and characterization of new cloning vehicles. II. A multipurpose cloning system. *Gene* **2:**95–113.

30. **Bukhari, A. I., J. A. Shapiro, and S. L. Adhya (ed.).** 1977. *DNA Insertion Elements, Plasmids, and Episomes.* Cold Spring Harbor Laboratory, Cold Spring Harbor, NY.

31. **Burall, L. S., J. M. Harro, X. Li, C. V. Lockatell, S. D. Himpsl, J. R. Hebel, D. E. Johnson, and H. L. Mobley.** 2004. *Proteus mirabilis* genes that contribute to pathogenesis of urinary tract infection: identification of 25 signature-tagged mutants attenuated at least 100-fold. *Infect. Immun.* **72:**2922–2938.

32. **Burlage, R. S., Z. K. Yang, and T. Mehlhorn.** 1996. A transposon for green fluorescent protein transcriptional fusions: application for bacterial transport experiments. *Gene* **173:**53–58.

33. **Campbell, A., D. E. Berg, D. Botstein, E. M. Lederberg, R. P. Novick, P. Starlinger, and W. Szybalski.** 1979. Nomenclature of transposable elements in prokaryotes. *Gene* **5:**197–206.

34. **Campbell, A., D. E. Berg, D. Botstein, R. Novick, and P. Starlinger.** 1977. Nomenclature of transposable elements in prokaryotes, p. 15–22. *In* A. I. Bukhari, J. A. Shapiro, and S. L. Adhya (ed.), *Insertion Elements, Plasmids, and Episomes.* Cold Spring Harbor Laboratory, Cold Spring Harbor, NY.

35. **Chaconas, G., and R. M. Harshey.** 2002. Transposition of phage Mu DNA, p. 384–402. *In* N. L. Craig, R. Craigie, M. Gellert, and A. M. Lambowitz (ed.), *Mobile DNA II.* ASM Press, Washington, DC.

36. **Chalfie, M., Y. Tu, G. Euskirchen, W. W. Ward, and D. C. Prasher.** 1994. Green fluorescent protein as a marker for gene expression. *Science* **263:**802–805.

37. **Chang, A. C. Y., and S. N. Coben.** 1978. Construction and characterization of amplifiable multicopy DNA cloning vehicles derived from the P15A cryptic miniplasmid. *J. Bacteriol.* **134:**1141–1156.

38. **Charles, T. C., and T. M. Finan.** 1990. Genetic map of *Rhizobium meliloti* megaplasmid pRmeSU47b. *J. Bacteriol.* **172:**2469–2476.

39. **Chiang, S. L., and J. J. Mekalanos.** 1998. Use of signature-tagged transposon mutagenesis to identify *Vibrio cholerae* genes critical for colonization. *Mol. Microbiol.* **27:**797–805.

40. **Chow, W.-Y., and D. E. Berg.** 1988. Tn5tac1, a derivative of transposon Tn5 that generates conditional mutations. *Proc. Natl. Acad. Sci. USA* **85:**6468–6472.

41. **Churchward, G.** 2002. Conjugative transposons and related mobile elements, p. 177–191. *In* N. L. Craig, R. Craigie, M. Gellert, and A. M. Lambowitz (ed.), *Mobile DNA II.* ASM Press, Washington, DC.

42. **Clewell, D. B.** 1998. Conjugative transposons, p. 130–139. *In* F. J. de Bruijn, J. R. Lupski, and G. M. Weinstock (ed.), *Bacterial Genomes: Physical Structure and Analysis.* Chapman & Hall, New York, NY.

43. **Cousineau, B., S. Lawrence, D. Smith, and M. Belfort.** 2000. Retrotransposition of a bacterial group II intron. *Nature* **404:**1018–1021.

44. **Craig, N. L.** 2002. Tn7, p. 423–456. *In* N. L. Craig, R. Craigie, M. Gellert, and A. M. Lambowitz (ed.), *Mobile DNA II.* ASM Press, Washington, DC.

45. **Craig, N. L.** 1996. Transposition, p. 2339–2362. *In* F. C. Neidhardt, R. Curtiss III, J. L. Ingraham, E. C. C. Lin, K. B. Low, B. Magasanik, W. S. Reznikoff, M. Riley, M. Schaechter, and H. E. Umbarger (ed.), Escherichia coli *and* Salmonella: *Cellular and Molecular Biology,* 2nd ed., vol. 2. ASM Press, Washington, DC.

46. **Craig, N. L.** 1989. Transposon Tn7, p. 211–225. *In* D. E. Berg and M. M. Howe (ed.), *Mobile DNA.* American Society for Microbiology, Washington, DC.

47. **Craig, N. L., R. Craigie, M. Gellert, and A. M. Lambowitz.** 2002. *Mobile DNA II.* ASM Press, Washington, DC.

48. **Davey, M. E., and F. J. de Bruijn.** 2000. A homologue of the tryptophan-rich sensory protein TspO and FixL regulate a novel nutrient deprivation-induced *Sinorhizobium meliloti* locus. *Appl. Environ. Microbiol.* **66:**5353–5359.

49. **Davies, D. R., I. Y. Goryshin, W. S. Reznikoff, and I. Rayment.** 2000. Three-dimensional structure of the Tn5 synaptic complex transposition intermediate. *Science* **289:**77–85.

50. **de Bruijn, F. J.** 1987. Transposon Tn5 mutagenesis to map genes. *Methods Enzymol.* **154:**175–196.

51. **de Bruijn, F. J.** 1992. Use of repetitive (repetitive extragenic palindromic and enterobacterial repetitive intergeneric consensus) sequences and the polymerase chain reaction to fingerprint the genomes of *Rhizobium meliloti* isolates and other soil bacteria. *Appl. Environ. Microbiol.* **58:**2180–2187.

52. **de Bruijn, F. J., and J. R. Lupski.** 1984. The use of transposon Tn5 mutagenesis in the rapid generation of correlated physical and genetic maps of DNA segments cloned into multicopy plasmids: a review. *Gene* **27:**131–149.

53. **de Bruijn, F. J., S. Rossbach, M. Schneider, P. Ratet, W. W. Szeto, F. M. Ausubel, and J. Schell.** 1989. *Rhizobium meliloti* 1021 has three differentially regulated loci involved in glutamine biosynthesis, none of which is essential for symbiotic nitrogen fixation. *J. Bacteriol.* **171:**1673–1682.

54. **de Bruijn, F. J., I. L. Stroke, D. J. Marvel, and F. M. Ausubel.** 1983. Construction of a correlated physical and genetic map of the *Klebsiella pneumoniae hisDGO* region using transposon Tn5 mutagenesis. *EMBO J.* **2:**1831–1838.

55. **de Lorenzo, V., M. Herrero, U. Jakubzik, and K. N. Timmis.** 1990. Mini-Tn5 transposon derivatives for insertion mutagenesis, promoter probing, and chromosomal insertion of cloned DNA in gram-negative eubacteria. *J. Bacteriol.* **172:**6568–6572.

56. **de Lorenzo, V., and K. N. Timmis.** 1994. Analysis and construction of stable phenotypes in Gram-Negative

bacteria with Tn5- and Tn10-derived minitransposons. *Methods Enzymol.* **235**:386–405.

57. **Dennis, J. J., and G. J. Zylstra.** 1998. Plasposons: modular self-cloning minitransposon derivatives for rapid genetic analysis of gram-negative bacterial genomes. *Appl. Environ. Microbiol.* **64**:2710–2715.

58. **De Vos, G. F., C. C. Walker, and E. R. Signer.** 1986. Genetic manipulations in *Rhizobium meliloti* utilizing two new transposon Tn5 derivatives. *Mol. Gen. Genet.* **204**:485–491.

59. **Ditta, G.** 1986. Tn5 mapping of *Rhizobium* nitrogen fixation genes. *Methods Enzymol.* **118**:519–528.

60. **Ditta, G., S. Stanfield, D. Corbin, and D. R. Helinski.** 1980. Broad host range DNA cloning system for gram-negative bacteria. Construction of a gene bank of *Rhizobium meliloti*. *Proc. Natl. Acad. Sci. USA* **77**:7347–7351.

61. **Donald, R. G. K., C. K. Raymond, and R. A. Ludwig.** 1985. Vector insertion mutagenesis of *Rhizobium* sp. strain ORS571: direct cloning of mutagenized DNA sequences. *J. Bacteriol.* **162**:317–323.

62. **Dower, W. J., J. F. Miller, and C. W. Ragsdale.** 1988. High efficiency transformation of *E. coli* by high voltage electroporation. *Nucleic Acids Res.* **16**:6127–6145.

63. **Fedi, S., D. Brazil, D. N. Dowling, and F. O'Gara.** 1996. Construction of a modified mini-Tn5lacZY non-antibiotic marker cassette: ecological evaluation of *lacZY* marked *Pseudomonas* strain in the sugarbeet rhizosphere. *FEMS Microbiol. Lett.* **135**:251–257.

64. **Fernandes, P. J., J. A. C. Powell, and J. A. C. Archer.** 2001. Construction of *Rhodococcus* random mutagenesis libraries using Tn5 transposition complexes. *Microbiology* **147**:2529–2536.

65. **Finan, T. M., I. Oresnik, and A. Bottacin.** 1988. Mutants of *Rhizobium meliloti* defective in succinate metabolism. *J. Bacteriol.* **170**:3396–3403.

66. **Friedman, A. M., S. R. Long, S. E. Brown, W. J. Buikema, and F. M. Ausubel.** 1982. Construction of a broad host range cosmid cloning vector and its use in the genetic analysis of *Rhizobium* mutants. *Gene* **18**:289–296.

67. **Fuller, T. E., M. J. Kennedy, and D. E. Lowery.** 2000. Identification of *Pasteurella multocida* virulence genes in a septicemic mouse model using signature-tagged mutagenesis. *Microb. Pathog.* **29**:25–38.

68. **Fuller, T. E., S. Martin, J. F. Teel, G. R. Alaniz, M. J. Kennedy, and D. E. Lowery.** 2000. Identification of *Actinobacillus pleuropneumoniae* virulence genes using signature-tagged mutagenesis in a swine infection model. *Microb. Pathog.* **29**:39–51.

69. **Furuichi, T., M. Inouye, and S. Inouye.** 1985. Novel one-step cloning vector with a transposable element: application to the *Myxococcus xanthus* genome. *J. Bacteriol.* **164**:270–275.

70. **Furukawa, K., S. Hayashida, and K. Taira.** 1991. Gene-specific transposon mutagenesis of the biphenyl/polychlorinated biphenyl-degradation controlling *bph* operon in soil bacteria. *Gene* **98**:21–28.

71. **Gallagher, S. R.** 1992. *GUS Protocols: Using the GUS Gene as a Reporter of Gene Expression.* Academic Press, San Diego, CA.

72. **Gay, P., D. L. Coq, M. Steinmetz, T. Berkelman, and C. I. Kado.** 1985. Positive selection procedure for entrapment of insertion sequence elements in gram-negative bacteria. *J. Bacteriol.* **164**:918–921.

73. **Gehring, A. M., J. R. Nodwell, S. M. Beverley, and R. Losick.** 2000. Genomewide insertional mutagenesis in *Streptomyces coelicolor* reveals additional genes involved in morphological differentiation. *Proc. Natl. Acad. Sci. USA* **97**:9642–9647.

74. **Gentschev, I., G. Maier, A. Kranig, and W. Goebel.** 1996. Mini-TnhlyAs: a new tool for the construction of secreted fusion proteins. *Mol. Gen. Genet.* **252**:266–274.

75. **Gerdes, S. Y., M. D. Scholle, M. D'Souza, A. Bernal, M. V. Baev, M. Farrell, O. V. Kurnasov, M. D. Daugherty, F. Mseeh, B. M. Polanuyer, J. W. Campbell, S. Anantha, K. Y. Shatalin, S. A. Chowdhury, M. Y. Fonstein, and A. L. Osterman.** 2002. From genetic footprinting to antimicrobial drug targets: examples in cofactor biosynthetic pathways. *J. Bacteriol.* **184**:4555–4572.

76. **Goryshin, I. Y., J. Jendrisak, L. M. Hoffman, R. Meis, and W. S. Reznikoff.** 2000. Insertional transposon mutagenesis by electroporation of released Tn5 tranposition complexes. *Nat. Biotechnol.* **18**:97–100.

77. **Goryshin, I. Y., J. A. Miller, Y. V. Kil, V. A. Lansov, and W. S. Reznikoff.** 1998. Tn5/IS50 target recognition. *Proc. Natl. Acad. Sci. USA* **95**:10716–10721.

78. **Goryshin, I. Y., T. A. Naumann, J. Apodaca, and W. S. Reznikoff.** 2003. Chromosomal deletion formation system based on Tn5 double transposition: use for making minimal genomes and essential gene analysis. *Genome Res.* **13**:644–653.

79. **Goryshin, I. Y., and W. S. Reznikoff.** 1998. Tn5 *in vitro* transposition. *J. Biol. Chem.* **273**:7367–7374.

80. **Gottesman, M. M., and J. L. Rosner.** 1975. Acquisition of a determinant for chloramphenicol resistance by coliphage lambda. *Proc. Natl. Acad. Sci. USA* **72**:5041–5045.

81. **Grant, A. J., C. Coward, M. A. Jones, C. A. Woodall, P. A. Barrow, and D. J. Maskell.** 2005. Signature-tagged transposon mutagenesis studies demonstrate the dynamic nature of cecal colonization of 2-week-old chickens by *Campylobacter jejuni*. *Appl. Environ. Microbiol.* **71**:8031–8041.

82. **Grindley, N. D. F.** 2002. The movement of Tn3-like elements: transposition and cointegrate resolution, p. 272–302. *In* N. L. Craig, R. Craigie, M. Gellert, and A. M. Lambowitz (ed.), *Mobile DNA II.* ASM Press, Washington, DC.

83. **Groh, J. L., Q. Luo, J. D. Ballard, and L. R. Krumholz.** 2005. A method adapting microarray technology for signature-tagged mutagenesis of *Desulfovibrio desulfuricans* G20 and *Shewanella oneidensis* MR-1 in anaerobic sediment survival experiments. *Appl. Environ. Microbiol.* **71**:7064–7074.

84. **Guilhabert, M. R., L. M. Hoffman, D. A. Mills, and B. C. Kirkpatrick.** 2001. Transposon mutagenesis of *Xylella fastidiosa* by electroporation of Tn5 synaptic complexes. *Mol. Plant-Microbe Interact.* **14**:701–706.

85. **Haapa, S., S. Taira, E. Heikkinen, and H. Savilahti.** 1999. An effective and accurate integration of mini-Mu transposons *in vitro*: a general methodology for functional genetic analysis and molecular biology applications. *Nucleic Acids Res.* **27**:2777–2784.

86. **Haniford, D. B.** 2002. Transposon Tn10, p. 457-483. *In* N. L. Craig, R. Craigie, M. Gellert, and A. M. Lambowitz (ed.), *Mobile DNA II.* ASM Press, Washington, DC.

87. **Hansen, L. H., and S. J. Sorensen.** 2000. Detection and quantification of tetracyclines by whole cell biosensors. *FEMS Microbiol. Lett.* **190**:273–278.

88. **Hansen, L. H., and S. J. Sorensen.** 2000. Versatile biosensor vectors for detection and quantification of mercury. *FEMS Microbiol. Lett.* **193**:123–127.

89. **Harayama, S., E. T. Palva, and G. L. Hazelbauer.** 1979. Transposon-insertion mutants of *Escherichia coli* K12 defective in a component common to galactose and ribose chemotaxis. *Mol. Gen. Genet.* **171**:193–203.

90. **Hare, R. S., S. S. Walker, T. E. Dorman, J. R. Greene, L. M. Guzman, T. J. Kenney, M. C. Sulavik, K. Baradaran, C. Houseweart, H. Yu, Z. Foldes, A. Motzer, M. Walbridge, G. H. Shimer, Jr., and K. J. Shaw.** 2001. Genetic footprinting in bacteria. *J. Bacteriol.* **183**:1694–1706.

91. **Hayes, F.** 2003. Transposon-based strategies for microbial functional genomics and proteomics. *Annu. Rev. Genet.* **37:**3–29.

92. **Hedges, R. W., and A. E. Jakob.** 1974. Transposition of ampicillin resistance from RP4 to other replicons. *Mol. Gen. Genet.* **132:**31–40.

93. **Heiner, C. R., K. L. Hunkapillar, S.-M. Chen, J. I. Glass, and E. Y. Chen.** 1998. Sequencing multi-megabase-template DNA with BigDye terminator chemistry. *Genome Res.* **8:**557–561.

94. **Hensel, M., J. E. Shea, C. Gleeson, M. D. Jones, E. Dalton, and D. W. Holden.** 1995. Simultaneous identification of bacterial virulence genes by negative selection. *Science* **269:**400–403.

95. **Herrero, M., V. de Lorenzo, and K. N. Timmis.** 1990. Transposon vectors containing non-antibiotic resistance selection markers for cloning and stable chromosomal insertion of foreign genes in gram-negative bacteria. *J. Bacteriol.* **172:**6557–6567.

96. **Hilgert, U., J. Schell, and F. J. de Bruijn.** 1987. Isolation and characterization of Tn*5*-induced NADPH-glutamate synthase (GOGAT−) mutants of *Azorhizobium sesbaniae* ORS571 and cloning of the corresponding *glt* locus. *Mol. Gen. Genet.* **210:**195–202.

97. **Hirsch, P. R., and J. E. Beringer.** 1984. A physical map of pPH1JI and pJB4JI. *Plasmid* **12:**139–141.

98. **Hirsch, P. R., C. L. Wang, and M. J. Woodward.** 1986. Construction of a Tn*5* derivative determining resistance to gentamicin and spectinomycin using a fragment cloned from R1033. *Gene* **48:**203–209.

99. **Hoffman, L. M., J. J. Jendriska, R. J. Meis, I. Y. Goryshin, and W. S. Reznikoff.** 2000. Transposome insertional mutagenesis and direct sequencing of microbial genomes. *Genetica* **108:**19–24.

100. **Holliday, R.** 1956. A new method for identification of auxotrophic mutants in microorganisms. *Nature* **178:**987.

101. **Hynes, M., J. Quandt, M. P. O'Connell, and A. Puchler.** 1989. Direct selection for curing and deletion of *Rhizobium* plasmids using transposons carrying the *Bacillus subtilis sacB* gene. *Gene* **78:**111–120.

102. **Jefferson, R. A., S. M. Burgess, and D. Hirsh.** 1986. β-Glucuronidase from *Escherichia coli* as a gene-fusion marker. *Proc. Natl. Acad. Sci. USA* **83:**8447–8451.

103. **Kaniga, K., and J. Davison.** 1991. Transposon vectors for stable chromosomal integration of cloned genes in rhizosphere bacteria. *Gene* **100:**201–205.

104. **Keen, M. G., E. D. Street, and P. S. Hoffman.** 1985. Broad host-range plasmid pRK340 delivers Tn*5* into the *Legionella pneumophila* chromosome. *J. Bacteriol.* **162:**1332–1335.

105. **Kennedy, C., R. Gamal, R. Humphrey, I. Ramos, K. Brigle, and D. Dean.** 1986. The *nifH*, *nifM* and *nifN* genes of *Azotobacter vinelandii*: characterization by Tn*5* mutagenesis and isolation from pLAFR1 gene banks. *Mol. Gen. Genet.* **205:**318–325.

106. **Kleckner, N.** 1989. Transposon Tn*10*, p. 227–268. *In* D. E. Berg and M. M. Howe (ed.), *Mobile DNA.* American Society for Microbiology, Washington, DC.

107. **Kleckner, N., R. K. Chan, B. K. Tye, and D. Botstein.** 1975. Mutagenesis by insertion of a drug resistance element carrying an inverted repetition. *J. Mol. Biol.* **95:**561–575.

108. **Kleckner, N., J. Roth, and D. Botstein.** 1977. Genetic engineering *in vivo* using translocatable drug-resistant elements: new methods in bacterial genetics. *J. Mol. Biol.* **116:**125–159.

109. **Kolter, R., M. Inuzuka, and D. R. Helinski.** 1978. Trans-complementation-dependent replication of a low molecular weight origin fragment from plasmid R6K. *Cell* **15:**1199–1208.

110. **Krebs, M. P., and W. S. Reznikoff.** 1988. Use of a Tn*5* derivative that creates *lacZ* translational fusions to obtain a transposition mutant. *Gene* **63:**277–285.

111. **Krin, E., F. Hommais, O. Soutourina, S. Ngo, A. Danchin, and P. Bertin.** 2001. Description and application of a rapid method for genomic DNA direct sequencing. *FEMS Microbiol. Lett.* **199:**229–233.

112. **Krishnan, B. R., D. Kersulyte, I. Brikun, C. M. Berg, and D. E. Berg.** 1991. Direct and crossover PCR amplification to facilitate Tn*5supF*-based sequencing of λ phage clones. *Nucleic Acids Res.* **19:**6177–6182.

113. **Kroos, L., and D. Kaiser.** 1984. Construction of Tn*5lac*, a transposon that fuses *lacZ* expression to exogenous promoters, and its introduction into *Myxococcus xanthus*. *Proc. Natl. Acad. Sci. USA* **81:**5816–5820.

114. **Kuner, L. M., and D. Kaiser.** 1981. Introduction of transposon Tn*5* into *Myxococcus* for analysis of developmental and other nonselectable mutants. *Proc. Natl. Acad. Sci. USA* **78:**425–429.

115. **Laird, A. J., and I. G. Young.** 1980. Tn*5* mutagenesis of the enterocholin gene cluster of *Escherichia coli*. *Gene* **11:**359–366.

116. **Lambrecht, M., A. Vande Broek, and J. Vanderleyden.** 2000. The use of the GUS reporter system to study molecular aspects of interactions between bacteria and plants, p. 87–99. *In* J. K. Jansson, J. D. van Elsas, and M. J. Bailey (ed.), *Tracking Genetically-Engineered Microorganisms.* Eurekah.com/Landes Bioscience, Georgetown, TX.

117. **Lederberg, L., and E. M. Lederberg.** 1952. Replica plating and indirect selection of bacterial mutants. *J. Bacteriol.* **63:**399.

118. **Lehoux, D. E., and R. C. Levesque.** 2000. Detection of genes essential in specific niches by signature-tagged mutagenesis. *Curr. Opin. Microbiol.* **11:**434–439.

119. **Lehoux, D. E., and R. C. Levesque.** 2002. Polymerase chain reaction-based signature-tagged mutagenesis. *Methods Mol. Biol.* **182:**127–137.

120. **Lestrate, P., R. M. Delrue, I. Danese, C. Didembourg, B. Taminiau, P. Mertens, X. D. Bolle, A. Tibor, C. M. Tang, and J. J. Letesson.** 2000. Identification and characterization of *in vivo* attenuated mutants of *Brucella melitensis*. *Mol. Microbiol.* **38:**548–551.

121. **Ludwig, R. A.** 1987. Gene tandem-mediated selection of coliphage λ-receptive *Agrobacterium*, *Pseudomonas*, and *Rhizobium* strains. *Proc. Natl. Acad. Sci. USA* **84:**3334–3338.

122. **Lupski, J. R., L. S. Ozaki, J. Ellis, and G. N. Godson.** 1983. Localization of the *Plasmodium* surface antigen epitope by Tn*5* mutagenesis mapping of a recombinant cDNA clone. *Science* **220:**1285–1288.

123. **Lupski, L. R., Y. H. Zhang, M. Rieger, M. Minter, B. Hsu, T. Koeuth, and E. R. B. McCabe.** 1990. Mutational analysis of the *Escherichia coli glpFK* region with Tn*5* mutagenesis and the polymerase chain reaction. *J. Bacteriol.* **172:**6129–6134.

124. **Manoil, C., and J. Beckwith.** 1986. A genetic approach to analyzing membrane protein topology. *Science* **233:**1403–1408.

125. **Manoil, C., and J. Beckwith.** 1985. TnphoA: a transposon probe for protein export signals. *Proc. Natl. Acad. Sci. USA* **82:**8129–8133.

126. **Manoil, C., J. J. Mekalanos, and J. Beckwith.** 1990. Alkaline phosphatase fusions: sensors of subcellular location. *J. Bacteriol.* **172:**515–518.

127. **Manoil, C., and B. Traxler.** 2000. Insertion of in-frame sequence tags into proteins using transposons. *Methods* **20:**55–61.

128. **Marsch-Moreno, R., G. Hernandez-Guzman, and A. Alvarez-Morales.** 1998. pTn*5cat*: a Tn*5*-derived genetic

element to facilitate insertion mutagenesis, promoter probing, physical mapping, cloning, and marker exchange in phytopathogenic and other Gram-Negative bacteria. *Plasmid* **39:**205–214.

129. **Martinez-Abarca, F., and N. Toro.** 2000. Group II introns in the bacterial world. *Mol. Microbiol.* **38:**917–926.

130. **Meade, H. M., S. R. Long, G. B. Ruvkun, S. E. Brown, and F. M. Ausubel.** 1982. Physical and genetic characterization of symbiotic and auxotrophic mutants of *Rhizobium meliloti* induced by transposon Tn5 mutagenesis. *J. Bacteriol.* **149:**114–122.

131. **Mecsas, J.** 2002. Use of signature-tagged mutagenesis in pathogenesis studies. *Curr. Opin. Microbiol.* **5:**33–37.

132. **Mecsas, J., I. Bilis, and S. Falkow.** 2001. Identification of attenuated *Yersinia pseudotuberculosis* strains and characterization of an orogastric infection in BALB/c mice on day 5 postinfection by signature-tagged mutagenesis. *Infect. Immun.* **69:**2779–2787.

133. **Mei, J.-M., F. Nourbakhsh, C. W. Ford, and D. W. Holden.** 1997. Identification of *Staphylococcus aureus* virulence genes in a murine model of bacteraemia using signature-tagged mutagenesis. *Mol. Microbiol.* **26:**399–407.

134. **Meighen, E. A.** 1991. Molecular biology of bacterial bioluminescence. *Microbiol. Rev.* **55:**123–142.

135. **Merriman, T. R., and I. L. Lamont.** 1993. Construction and use of a self-cloning promoter probe vector for Gram-negative bacteria. *Gene* **126:**17–23.

136. **Milcamps, A., P. Struffi, and F. J. de Bruijn.** 2001. The *Sinorhizobium meliloti* nutrient-deprivation-induced tyrosine degradation gene *hmgA* is controlled by a novel member of the *arsR* family of regulatory genes. *Appl. Environ. Microbiol.* **67:**2641–2648.

137. **Miller, J. H.** 1972. *Experiments in Molecular Genetics.* Cold Spring Harbor Laboratory, Cold Spring Harbor, NY.

138. **Miller, J. H., M. P. Calos, M. Hofer, D. Buechel, and B. Mueller-Hill.** 1980. Genetic analysis of transpositions in the *lac* region of *Escherichia coli. J. Mol. Biol.* **144:**1–18.

139. **Miller, V. L., and J. J. Mekalanos.** 1988. A novel suicide vector and its use in construction of insertion mutations: osmoregulation of outer membrane proteins and virulence determinants in *Vibrio cholerae* requires *toxR. J. Bacteriol.* **170:**2575–2583.

140. **Morales, V. M., and L. Sequeira.** 1985. Suicide vector for transposon mutagenesis in *Pseudomonas solanacearum. J. Bacteriol.* **163:**1263–1264.

141. **Morgan, G. J., G. F. Hatfull, S. Casjens, and R. W. Hendrix.** 2002. Bacteriophage Mu genome sequence: analysis and comparison with Mu-like prophages in *Haemophilus, Neisseria* and *Deinococcus. J. Mol. Biol.* **317:**337–359.

142. **Murphy, E.** 1989. Transposable elements in gram-positive bacteria, p. 269–288. *In* D. E. Berg and M. M. Howe (ed.), *Mobile DNA.* American Society for Microbiology, Washington, DC.

143. **Nag, D. K., H. V. Huang, and D. E. Berg.** 1988. Bidirectional chain termination nucleotide sequencing: transposon Tn5seq1 as a mobile source of primer sites. *Gene* **64:**135–145.

144. **Obukowicz, M. G., F. J. Perlak, K. Kusano-Kretzmer, E. J. Mayer, and L. S. Watrud.** 1986. Integration of the delta endotoxin gene of *Bacillus thuringiensis* into the chromosome of root colonizing strains of pseudomonads using Tn5. *Gene* **45:**327–331.

145. **Osteras, M., J. Stanley, W. J. Broughton, and D. N. Dowling.** 1989. A chromosomal genetic map of *Rhizobium* sp. NGR234 generated with Tn5-*Mob. Mol. Gen. Genet.* **220:**157–160.

146. **O'Toole, G. A., and R. Kolter.** 1998. Initiation of biofilm formation in *Pseudomonas fluorescens* WCS365 proceeds via multiple, convergent signalling pathways: a genetic analysis. *Mol. Microbiol.* **28:**449–461.

147. **O'Toole, G. A., L. A. Pratt, P. I. Watnick, D. K. Newman, V. B. Weaver, and R. Kolter.** 1999. Genetic approaches to study of biofilms. *Methods Enzymol.* **310:**91–109.

148. **Paik, S., L. Senty, S. Das, J. C. Noe, C. L. Munro, and T. Kitten.** 2005. Identification of virulence determinants for endocarditis in *Streptococcus sanguinis* by signature-tagged mutagenesis. *Infect. Immun.* **73:**6064–6074.

149. **Palva, E. T., P. Liljestroem, and S. Harayama.** 1981. Cosmid cloning and transposon mutagenesis in *Salmonella typhimurium* using phage λ vehicles. *Mol. Gen. Genet.* **181:**153–157.

150. **Pato, M. L.** 1989. Bacteriophage Mu, p. 23–52. *In* D. E. Berg and M. M. Howe (ed.), *Mobile DNA.* American Society for Microbiology, Washington, DC.

151. **Pawlowski, K., P. Ratet, J. Schell, and F. J. de Bruijn.** 1987. Cloning and characterization of *nifA* and *ntrC* genes of the stem nodulating bacterium ORS571, the nitrogen fixing symbiont of *Sesbania rostrata*: regulation of nitrogen fixation (*nif*) genes in the free living versus symbiotic state. *Mol. Gen. Genet.* **206:**207–219.

152. **Phadnis, S. H., and D. E. Berg.** 1987. Identification of base pairs in the outside end of insertion sequence IS50 that are needed for IS50 and Tn5 transposition. *Proc. Natl. Acad. Sci. USA* **84:**9118–9122.

153. **Phadnis, S. H., H. V. Huang, and D. E. Berg.** 1989. Tn5supF, a 264 base-pair transposon derived from Tn5 for insertion mutagenesis and sequencing DNAs cloned in phage λ. *Proc. Natl. Acad. Sci. USA* **86:**5908–5912.

154. **Phadnis, S. H., S. Kulakauskas, B. R. Krishnan, J. Hiemstra, and D. E. Berg.** 1991. Transposon Tn5supF-based reverse genetic method for mutational analysis of *Escherichia coli* with DNAs cloned in λ phage. *J. Bacteriol.* **173:**896–899.

155. **Pischl, D. L., and S. K. Farrand.** 1984. Characterization of transposon Tn5-facilitated donor strains and development of a chromosomal linkage map for *Agrobacterium tumefaciens. J. Bacteriol.* **159:**1–8.

156. **Polesky, A. H., J. T. Ross, S. Falkow, and L. S. Tompkins.** 2001. Identification of *Legionella pneumophila* genes important for infection of amoebas by signature-tagged mutagenesis. *Infect. Immun.* **69:**977–987.

157. **Polissi, A., A. Pontiggia, G. Feger, M. Altieri, H. Mottl, L. Ferrari, and D. Simon.** 1998. Large-scale identification of virulence genes from *Streptococcus pneumoniae. Infect. Immun.* **66:**5620–5629.

158. **Prosser, J. I., A. J. Palomares, M. T. Karp, and P. J. Hill.** 2000. Luminescence-based microbial marker systems and their application in microbial ecology, p. 69–85. *In* J. K. Jansson, J. D. van Elsas, and M. J. Bailey (ed.), *Tracking Genetically-Engineered Microorganisms.* Eurekah .com/Landes Bioscience, Georgetown, TX.

159. **Recchia, G. D., and D. J. Sheratt.** 2002. Gene acquisition in bacteria by integron-mediated site-specific recombination, p. 162–176. *In* N. L. Craig, R. Craigie, M. Gellert, and A. M. Lambowitz (ed.), *Mobile DNA II.* ASM Press, Washington, DC.

160. **Reeve, J.** 1979. Use of minicells for bacteriophage directed polypeptide synthesis. *Methods Enzymol.* **68:**493–503.

161. **Rella, M., A. Mercenier, and D. Hass.** 1985. Transposon insertion mutagenesis of *Pseudomonas aeruginosa* with a Tn5 derivative: application to physical mapping of the *arc* gene cluster. *Gene* **33:**293–303.

162. **Reznikoff, W. S.** 2003. Tn5 as a model for understanding DNA transposition. *Mol. Microbiol.* **47:**1199–1206.

163. **Reznikoff, W. S.** 2002. Tn5 transposition, p. 403–422. *In* N. L. Craig, R. Craigie, M. Gellert, and A. M. Lambowitz (ed.), *Mobile DNA II.* ASM Press, Washington, DC.

164. **Reznikoff, W. S.** 2006. Tn5 transposition: a molecular tool for studying protein structure-function. *Biochem. Soc. Trans.* **34:**320–323.

165. **Reznikoff, W. S., A. Bhasin, D. R. Davies, I. Y. Goryshin, L. A. Mahnke, T. Naumann, I. Rayment, M. Steiniger-White, and S. S. Twining.** 1999. Tn5: a molecular window on transposition. *Biochem. Biophys. Res. Commun.* **266:**729–734.

166. **Rich, J. J., and D. K. Willis.** 1990. A single oligonucleotide can be used to rapidly isolate DNA sequences flanking a transposon Tn5 insertion by the polymerase chain reaction. *Nucleic Acids Res.* **18:**6673–6676.

167. **Riedel, G. E., F. M. Ausubel, and F. C. Cannon.** 1979. Physical map of chromosomal nitrogen fixation (*nif*) genes of *Klebsiella pneumoniae. Proc. Natl. Acad. Sci. USA* **76:**2866–2870.

168. **Rossbach, S., M. L. Kukuk, T. L. Wilson, S. F. Feng, M. M. Pearson, and M. A. Fisher.** 2000. Cadmium-regulated gene fusions in *Pseudomonas fluorescens. Environ. Microbiol.* **2:**373–382.

169. **Rossbach, S., J. Schell, and F. J. de Bruijn.** 1988. Cloning and analysis of *Agrobacterium tumefaciens* C58 loci involved in glutamine biosynthesis: neither the *glnA* (GSI) nor the *glnII* (GSII) gene plays a special role in virulence. *Mol. Gen. Genet.* **212:**38–47.

170. **Rossbach, S., T. L. Wilson, M. L. Kukuk, and H. A. Carty.** 2000. Elevated zinc induces siderophore biosynthesis genes and a *zntA*-like gene in *Pseudomonas fluorescens. FEMS Microbiol. Lett.* **191:**61–70.

171. **Rostas, K., P. Sista, J. Stanley, and D. P. S. Verma.** 1984. Transposon mutagenesis of *Rhizobium japonicum. Mol. Gen. Genet.* **197:**230–235.

172. **Ruvkun, G. B., and F. M. Ausubel.** 1981. A general method for site-directed mutagenesis in prokaryotes. *Nature* **289:**85–88.

173. **Saenz, H. L., and C. Dehio.** 2005. Signature-tagged mutagenesis: technical advances in a negative selection method for virulence gene identification. *Curr. Opin. Microbiol.* **8:**612–619.

174. **Sambrook, J., E. F. Fritsch, and T. Maniatis.** 1989. *Molecular Cloning: a Laboratory Manual,* Cold Spring Harbor Laboratory, Cold Spring Harbor, NY.

175. **Sambrook, J., and D. Russell.** 2001. *Molecular Cloning: a Laboratory Manual,* 3rd ed., Cold Spring Harbor Laboratory Press, Cold Spring Harbor, NY.

176. **Sancar, A., A. M. Hack, and W. D. Rupp.** 1979. Simple method for identification of plasmid-coded proteins. *J. Bacteriol.* **137:**692–693.

177. **Sanger, F., S. Nicklen, and A. R. Coulson.** 1977. DNA sequencing with chain-terminating inhibitors. *Proc. Natl. Acad. Sci. USA* **74:**5463–5467.

178. **Sarthy, A., S. Michaelis, and J. Beckwith.** 1981. Deletion map of the *Escherichia coli* structural gene for alkaline phosphatase. *J. Bacteriol.* **145:**288–292.

179. **Sasakawa, C., K. Kamata, T. Sakai, S. Makino, M. Yamada, and N. Okada.** 1988. Virulence-associated genetic regions comprising 31 kilobases of the 230 kilobase plasmid in *Shigella flexneri* 2a. *J. Bacteriol.* **170:**2480–2484.

180. **Sasakawa, C., and M. Yoshikawa.** 1987. A series of Tn5 variants with various drug-resistance markers and suicide vector for transposon mutagenesis. *Gene* **56:**283–288.

181. **Selvaraj, G., and V. N. Iyer.** 1983. Suicide plasmid vehicles for insertion mutagenesis in *Rhizobium meliloti* and related bacteria. *J. Bacteriol.* **156:**1292–1300.

182. **Shah, D. H., M. J. Lee, J. H. Park, J. H. Lee, S. K. Eo, J. T. Kwon, and J. S. Chae.** 2005. Identification of *Salmonella gallinarum* virulence genes in a chicken infection model using PCR-based signature-tagged mutagenesis. *Microbiology* **151:**3957–3968.

183. **Sharma, S. B., and E. R. Signer.** 1990. Temporal and spatial regulation of the symbiotic genes of *Rhizobium meliloti* in planta revealed by transposon Tn5-*gusA. Genes Dev.* **4:**344–356.

184. **Shaw, K. J., and C. M. Berg.** 1979. *Escherichia coli* K-12 auxotrophs induced by insertion of the transposable element Tn5. *Genetics* **92:**741–747.

185. **Shea, J. E., J. D. Santangelo, and R. G. Feldman.** 2000. Signature-tagged mutagenesis in the identification of virulence genes in pathogens. *Curr. Opin. Microbiol.* **3:**451–458.

186. **Sherratt, D.** 1989. Tn3 and related transposable elements, p. 163–184. *In* D. E. Berg and M. M. Howe (ed.), *Mobile DNA.* American Society for Microbiology, Washington, DC.

187. **Silhavy, T. J., and J. R. Beckwith.** 1985. Uses of *lac* fusions for the study of biological problems. *Microbiol. Rev.* **49:**398–418.

188. **Simon, R.** 1984. High frequency mobilization system for *in vivo* constructed Tn5 Mob transposon. *Mol. Gen. Genet.* **196:**413–420.

189. **Simon, R., M. O'Connell, M. Labes, and A. Puehler.** 1986. Plasmid vectors for the genetic analysis and manipulation of rhizobia and other Gram-negative bacteria. *Methods Enzymol.* **118:**640–659.

190. **Simon, R., U. Priefer, and A. Puehler.** 1983. A broad host range mobilization system for *in vivo* genetic engineering: transposon mutagenesis in Gram-negative bacteria. *Bio/Technology* **1:**784–791.

191. **Simon, R., and U. B. Priefer.** 1990. Vector technology of relevance to nitrogen fixation research, p. 13–49. *In* P. M. Gresshoff (ed.), *Molecular Biology of Symbiotic Nitrogen Fixation.* CRC Press, Boca Raton, FL.

192. **Simon, R., J. Quandt, and W. Klipp.** 1989. New derivatives of transposon Tn5 suitable for mobilization of replicons, generation of operon fusions and induction of genes in Gram-negative bacteria. *Gene* **80:**161–169.

193. **Smith, C. L., and C. R. Cantor.** 1987. Purification, specific fragmentation and separation of large DNA molecules. *Methods Enzymol.* **151:**449–467.

194. **Smith, C. L., J. G. Econome, A. Schutt, S. Klco, and C. R. Cantor.** 1987. A physical map of the *Escherichia coli* K12 genome. *Science* **236:**1448–1453.

195. **Smith, C. L., and R. D. Kolodner.** 1988. Mapping of *Escherichia coli* chromosomal Tn5 and F insertions by pulsed field gel electrophoresis. *Genetics* **119:**227–236.

196. **Smith, V., D. Botstein, and P. O. Brown.** 1995. Genetic footprinting: a genomic strategy for determining a gene's function given its sequence. *Proc. Natl. Acad. Sci. USA* **92:**6479–6483.

197. **Spreng, S., and I. Gentschev.** 1998. Construction of chromosomally encoded secreted hemolysin fusion proteins by use of mini-Tn*hlyAs* transposon. *FEMS Microbiol. Lett.* **165:**187–192.

198. **Starlinger, P., and H. Saedler.** 1976. IS-elements in microorganisms. *Curr. Top. Microbiol. Immunol.* **75:**111–152.

199. **Steiniger-White, M., I. Rayment, and W. S. Reznikoff.** 2004. Structure/function insights into Tn5 transposition. *Curr. Opin. Struct. Biol.* **14:**50–57.

200. **Stewart, G. S. A. B., and P. Williams.** 1992. *lux*-genes and the applications of bacterial bioluminescence. *J. Gen. Microbiol.* **138:**1289–1300.

201. **Stojiljkovic, I., Z. Trgovcevic, and E. Salaj-Smic.** 1991. Tn5*rpsL*, a new derivative of transposon Tn5 useful in plasmid curing. *Gene* **99:**101–104.

202. **Stoltzfus, J. R., J. K. Jansson, and F. J. de Bruijn.** 2000. Using green fluorescent protein (GFP) as a biomarker or bioreporter for bacteria, p. 101–116. *In* J. K. Jansson,

J. D. van Elsas, and M. J. Bailey (ed.), *Tracking Genetically-Engineered Microorganisms*. Eurekah.com/ Landes Bioscience, Georgetown, TX.

203. **Suarez, A., A. Guttler, M. Stratz, L. H. Staendner, K. N. Timms, and C. A. Guzman.** 1997. Green fluorescent protein-based reporter systems for genetic analysis of bacteria including monocopy applications. *Gene* **196:**69–74.

204. **Subramanian, P. S., I. Versalovic, E. R. B. McCabe, and J. R. Lupski.** 1992. Rapid mapping of *Escherichia coli*::Tn5 insertion mutations by REP-Tn5 PCR. *PCR Methods Appl.* **1:**187–194.

205. **Tang, X., B. F. Lu, and S. Q. Pan.** 1999. A bifunctional transposon mini-Tn5*gfp-km* which can be used to select for promoter fusions and report gene expression levels in *Agrobacterium tumefaciens*. *FEMS Microbiol. Lett.* **179:** 37–42.

206. **Taylor, A. L.** 1963. Bacteriophage induced mutation in *Escherichia coli*. *Proc. Natl. Acad. Sci. USA* **50:**1043–1051.

207. **Tu, X., I. Nisan, J. F. Miller, E. Hanski, and I. Rosenshine.** 2001. Construction of mini-Tn5*cyaA*′ and its utilization for the identification of genes encoding surface-exposed and secreted proteins in *Bordetella bronchiseptica*. *FEMS Microbiol. Lett.* **205:**119–123.

208. **Unge, A., R. Tombolini, L. Molbak, and J. K. Jansson.** 1999. Simultaneous monitoring of cell number and metabolic activity of specific bacterial populations with a dual *gfp-luxAB* marker system. *Appl. Environ. Microbiol.* **65:** 813–821.

209. **Van Haute, E., H. Joos, M. Maes, G. Warren, M. V. Montagu, and J. Schell.** 1983. Intergenic transfer and exchange recombination of restriction fragments cloned in pBR322: a novel strategy for the reversed genetics of the Ti-plasmids of *Agrobacterium tumefaciens*. *EMBO J.* **2:** 411–429.

210. **Versalovic, J., T. Koeuth, and J. R. Lupski.** 1991. Distribution of repetitive DNA sequences in eubacteria and application to fingerprinting of bacterial genomes. *Nucleic Acids Res.* **19:**6823–6831.

211. **Walker, M. J., and J. M. Pemberton.** 1988. Construction of transposons encoding genes for beta-glucosidase, amylase and polygalacturonate *trans*-eliminase from *Klebsiella oxytoca* and their expression in a range of Gram-negative bacteria. *Curr. Microbiol.* **17:**69–75.

212. **Weitz, H. J., J. M. Ritchie, D. A. Bailey, A. M. Horsburgh, K. Killham, and L. A. Glover.** 2001. Construction of a modified mini-Tn5 *luxCDABE* transposon for the development of bacterial biosensors for ecotoxicity testing. *FEMS Microbiol. Lett.* **197:**159–165.

213. **Whittle, G., and A. A. Salyers.** 2002. Bacterial transposons: an increasingly diverse group of elements, p. 387–427. *In* U. N. Streips and R. E. Yasbin (ed.), *Modern Microbial Genetics*, 2nd ed. Wiley- Liss, New York, NY.

214. **Wilson, K. J., A. Sessitch, J. C. Corbo, K. E. Giller, A. D. L. Akkermans, and R. A. Jefferson.** 1995. β-Glucuronidase (GUS) transposons for ecological and genetic studies of rhizobia and other Gram-negative bacteria. *Microbiology* **141:**1691–1705.

215. **Winterberg, K. M., J. Luecke, A. S. Bruegl, and W. S. Reznikoff.** 2005. Phenotypic screening of *Escherichia coli* K-12 Tn5 insertion libraries, using whole-genome oligonucleotide microarrays. *Appl. Environ. Microbiol.* **71:** 451–459.

216. **Wolk, C. P., Y. Cai, and J.-M. Panoff.** 1991. Use of a transposon with luciferase as a reporter to identify environmentally responsive genes in a cyanobacterium. *Proc. Natl. Acad. Sci. USA* **88:**5355–5359.

217. **Wong, K. K., and M. McClelland.** 1992. Dissection of the *Salmonella typhimurium* genome by use of a Tn5 derivative carrying rare restriction sites. *J. Bacteriol.* **174:** 3807–3811.

218. **Xi, C., G. Dirix, J. Hofkens, F. C. D. Schryver, J. Vanderleyden, and J. Michiels.** 2001. Use of dual marker transposons to identify new symbiosis genes in *Rhizobium*. *Microb. Ecol.* **41:**325–332.

219. **Xi, C., M. Lambrecht, J. Vanderleyden, and J. Michiels.** 1999. Bi-functional *gfp-* and *gusA*-containing mini-Tn5 transposon derivatives for combined gene expression and bacterial localization studies. *J. Microbiol. Methods* **35:** 85–92.

220. **Yakobson, E. A., and D. G. Guiney.** 1984. Conjugal transfer of bacterial chromosomes mediated by the RK2 plasmid transfer origin cloned into transposon Tn5. *J. Bacteriol.* **160:**451–453.

221. **York, D., K. Welch, I. Y. Goryshin, and W. S. Reznikoff.** 1998. Simple and efficient generation *in vitro* of nested deletions and inversions: Tn5 intramolecular transposition. *Nucleic Acids Res.* **26:**1927–1933.

222. **Youngman, P.** 1990. Use of transposons and integrational vectors for mutagenesis and construction of gene fusions in *Bacillus* species, p. 221–266. *In* C. R. Harwood and S. M. Cutting (ed.), *Molecular Biological Methods for Bacillus*. John Wiley and Sons, Chichester, England.

223. **Zimmerly, S., G. Hausner, and X.-C. Wu.** 2001. Phylogenetic relationships among group II intron ORFs. *Nucleic Acids Res.* **29:**1238–1250.

224. **Zsebo, K. M., F. Wu, and J. E. Hearst.** 1984. Tn5.7 construction and physical mapping of pRPS404 containing photosynthetic genes from *Rhodopseudomonas capsulata*. *Plasmid* **11:**182–184.

225. **Zuber, U., and W. Schumann.** 1991. Tn5cos: a transposon for restriction mapping of large plasmids using phage lambda terminase. *Gene* **103:**69–72.

30

Plasmids

MARCELO E. TOLMASKY, LUIS A. ACTIS, TIMOTHY J. WELCH, AND
JORGE H. CROSA

Plasmids range in size from 1 to more than 200 kbp (31), and even larger megaplasmids were detected in *Rhizobium* spp. (8, 48, 79). Although they replicate separately from the host genome, plasmids do rely on host-encoded factors for their replication. Although not essential for the survival of bacteria, plasmids may encode a wide variety of genetic determinants, which permit their bacterial hosts to survive better in an adverse environment or to compete better with other microorganisms occupying the same ecological niche. Plasmids are found in a wide variety of bacteria, and it is as difficult to generalize about plasmids as it is to generalize about the bacteria that harbor them. The medical importance of plasmids that encode antibiotic resistance and specific virulence traits has been well documented (60, 132, 157). Plasmids have also been shown to influence significantly several properties contributing to the usefulness of bacteria in agriculture and industry (21, 32, 42, 61). The types of genes that encode these properties are frequently located in transposable elements, producing great variation and flexibility in the constitution of plasmids. These extrachromosomal elements are of equal importance, however, for the study of the structure and function of DNA. Plasmids have taken on paramount importance in recombinant DNA technology, particularly as cloning vehicles for foreign genes. Most bacterial plasmids are usually covalently closed circular (CCC) DNA; however, linear DNA plasmids, as well as single-stranded plasmids which are intermediates in replication, have also been described (2, 24, 38, 39, 50, 64, 75, 79, 135, 137, 158). Plasmid-encoded maintenance functions have also been shown to interact with the host cell (72). Plasmids have also been found in eukaryotic microorganisms. Recently, plasmid transfer from bacteria to eukaryotic cell lines maintained in culture has also been demonstrated (30, 55, 148). With the advent of the genomics era, websites that include complete sequences of plasmids as well as a data base for plasmids replicons have been created (http://www.ncbi.nlm.nih.gov/genomes/static/o.html; http://www.essex.ac.uk/bs/staff/osborn/DPR_home.htm [in construction at the time of writing of this chapter]; http://www.genomics.ceh.ac.uk/plasmiddb/ [currently updated]).

This chapter deals with the isolation, purification, and characterization of bacterial plasmids and also provides some information concerning the novel application of plasmid biology in gene therapy. Many of the techniques currently used for the isolation of plasmid DNA from bacteria are based on its supercoiled CCC configuration. All of the techniques require some means of gently lysing bacterial cells so that the plasmid DNA is preserved intact and can be physically separated from the more massive chromosomal DNA. This separation is more easily achieved with smaller plasmids, and the degree of difficulty increases as the size of the plasmid increases, particularly with some megaplasmids. This complication is often compounded by the fact that very large plasmids are normally present in low copy numbers. Even more specialized techniques are needed to isolate linear plasmids. Kits for the isolation of plasmid DNA have been recently developed and are currently being used in many laboratories.

The initial characterization of a bacterial plasmid usually is at the genetic level. If a bacterial trait is suspected of being plasmid mediated, gene transfer experiments often document transmissibility of plasmid determinants independently of chromosomal determinants. Moreover, the elimination (curing) of a genetic trait by treating the bacterial population with chemical or physical curing agents such as acridine dyes, ethidium bromide, sodium dodecyl sulfate (SDS), antibiotics, high temperature, or electroporation (63, 95) indicates that the expression of that genetic trait is linked to the presence of a plasmid. Most curing agents are also mutagenic agents. In most cases, however, it is essential to document that a plasmid is present and unequivocally associated with the genetic trait in question. Clinically, the plasmid content of a cell can also be a useful epidemiological marker (62, 103). If possible, it is best to transfer a plasmid by physical or genetic means into a bacterial host that is known to be devoid of plasmids. The advantage, of course, is that any single plasmid transferred to and subsequently isolated from such a host can be analyzed without fear of contamination by a preexisting plasmid. In some cases, genetic methods are not available, and so one must directly examine a bacterium to determine its plasmid content. Such an analysis usually can be performed to determine simply whether a plasmid is or not present. Subsequently, one can examine a strain "cured" of a trait to determine whether a plasmid is lost concomitantly with a particular host cell function. The compositions of the commonly used reagents mentioned throughout this chapter are indicated in Table 1. In this chapter we give a representative rather than comprehensive account of techniques and references concerning the biology of plasmids and their use as tools in molecular biology and pharmacology. Additional information on bacterial plasmids can be found in references 11, 60, 114, and 132.

30.1. ISOLATION

In this section are described methods of isolating plasmid DNA that do not require the use of commercially available materials such as prepared chromatography columns or reagents. Some characteristics are explained before each procedure. When dealing with a new isolated bacterium, a trial-and-error strategy in selecting the appropriate method is recommended.

30.1.1. Ordinary Plasmids by Large-Scale Methods

Large-scale methods permit the preparation of ordinary plasmid DNA in amounts such that repetitive analyses (restriction endonuclease digestions, preparative isolation of specific DNA fragments, cloning and subcloning, etc.) can be carried out. Generally, these methods are used in conjunction with a subsequent purification step, such as equilibrium density gradient centrifugation in CsCl-ethidium

TABLE 1 Composition of commonly used reagents

Reagent	Composition
Acetate solution	60 ml of 5 M potassium acetate, 11.5 ml of glacial acetic acid, 28.5 ml of water
Boiled RNase	5 mg of RNase per ml in 0.15 M NaCl, heated at 100°C for 20 min to destroy DNase
50× Denhardt's solution	1% Ficoll, 1% polyvinylpyrrolidone, 1% bovine serum albumin
Gel-loading buffer	0.25% bromophenol blue, 0.25% xylene cyanol FF, 30% glycerol
Phenol	Equilibrated with 0.1 M Tris-HCl (pH 8) and 0.2% β-mercaptoethanol. Oxidation is prevented by the addition of 0.1% 8-hydroxyquinoline
20× salt-sodium citrate (SSC)	175.3 g of NaCl and 88.2 g of sodium citrate per liter adjusted to pH 7.8
20× salt-sodium phosphate-EDTA (SSPE)	3.6 M NaCl, 0.2 M sodium phosphate (pH 7.7), 0.02 M EDTA
Sucrose-Triton-EDTA-Tris (STET)	8% Sucrose, 0.5% Triton X-100, 0.05 M EDTA, 0.05 M Tris-HCl (pH 8.0)
TE buffer	0.01 M Tris-HCl, 0.001 M EDTA (pH 8.0)
TES buffer	0.05 M Tris-HCl, 0.005 M EDTA, 0.05 M NaCl (pH 8.0)
Tris-acetate (electrophoresis buffer)	0.04 M Tris, 0.001 M EDTA (pH 8.0), 0.02 M glacial acetic acid
Tris-borate (electrophoresis buffer)	0.089 M Tris, 0.0025 M EDTA (pH 8.0), 0.089 M boric acid
Triton X-100 lytic mixture	1 ml of Triton X-100 in 0.01 M Tris-HCl (pH 8.0), 25 ml of 0.25 M EDTA (pH8.0), 5 ml of 1 m Tris-HCl (pH 8.0), and 69 ml of water

bromide gradients or precipitation with polyethylene glycol. Most of the methods described below are generally used for isolating ordinary plasmid DNA with molecular size no greater than 160 kbp.

30.1.1.1. Cold Triton X-100

The cold Triton X-100 procedure (28, 81) works well for isolating plasmid DNA from *Escherichia coli* K-12 strains, with the cells lysed by the nonionic detergent Triton X-100.

1. Grow the cells at 37°C either in a rich medium or in a suitable minimal medium (under selection, if possible), with gentle aeration achieved by shaking the flask at a rate just sufficient to keep the surface of the medium in motion. The following details are for a 100-ml culture in a 250-ml Erlenmeyer flask, but the method can be scaled up or down proportionally. To ensure optimum lysis, harvest the cells in the mid-logarithmic phase of growth.

2. Harvest the cells by centrifugation in a 250-ml bottle (e.g., for 10 min at 12,000 × g at 5°C). Resuspend the pellet with 15 ml of Tris-EDTA-salt (TES) buffer (Table 1), transfer to a 40-ml tube, and centrifuge for 10 min at 12,000 × g at 5°C. Resuspend the pellet with 1 ml of ice-cold 25% sucrose solution (in 0.05 M Tris–0.001 M EDTA [pH 8.0]), and place the tube in crushed ice for 30 min.

3. Add 0.2 ml of a freshly prepared lysozyme solution (5 mg/ml in 0.25 M Tris-HCl [pH 8.0]). Mix the contents by swirling the tube several times, and then place the tube in ice for 10 min. This step can be skipped when working with some bacterial species.

4. Add 0.4 ml of 0.25 M EDTA (pH 8.0). Swirl the tube, and then place it in ice for another 10 min.

5. Add 1.6 ml of Triton X-100 lytic mixture (1 ml of 10% Triton X-100 in 0.01 M Tris-HCl [pH 8.0], 25 ml of 0.25 M EDTA [pH 8.0], 5 ml of 1 M Tris-HCl [pH 8.0], 69 ml of water). After very gentle swirling to mix the contents thoroughly, place the tube in ice for 20 min.

6. Centrifuge at 35,000 × g at 5°C for 20 min, decant the supernatant, and discard the pellet.

About 95% of the plasmid is separated from the pellet containing the chromosomal DNA and cellular debris and can be used for further analysis. The plasmid-enriched supernatant fraction (3.2 ml) can also be purified and concentrated by centrifugation in a CsCl-ethidium bromide density gradient as described in section 30.2.

30.1.1.2. Hot Triton X-100

The hot Triton X-100 method is a modification of the foregoing procedure, which improves the sharpness of the DNA bands in agarose gels for lysates obtained from certain bacteria, e.g., *Pseudomonas aeruginosa*, *Serratia marcescens*, *Proteus rettgeri*, and *Klebsiella pneumoniae*. This procedure was successfully used in the analysis of plasmid profiles of clinical isolates of *Acinetobacter baumannii* (3, 62). This method is useful for isolation of plasmids of sizes ranging from 2 to 100 kbp.

1. Grow the culture in 40 ml of brain heart infusion broth or any other appropriate rich medium in a 100-ml flask on a shaker, as in step 1 of the method described above.

2. Harvest the cells by centrifugation at 3,000 × g for 10 min.

3. Suspend the cell pellet in 5 ml of TES buffer as in step 2 of the method described above.

4. Centrifuge the cells as in step 2.

5. Suspend the cell pellet in 2 ml of a 25% sucrose solution (in 0.001 M EDTA–0.05 M Tris-HCl [pH 8.0]). Place the tube in ice for 20 min.

6. Add 0.4 ml of a lysozyme solution (10 mg/ml in 0.25 M Tris HCl [pH 8.0]) to the suspension. Place the tube in ice for 20 min.

7. Add 0.8 ml of 0.5 M EDTA (pH 8.0) to the cell suspension.

8. Lyse the cells with 4.4 ml of Triton lytic mixture as in step 5 of the method described above. Mix gently.

9. Heat the tube at 65°C for 20 min.

10. Sediment the cellular debris by centrifugation at 27,200 × g for 40 min.

11. Decant, and then adjust the supernatant solution to 0.5 M NaCl and 10% polyethylene glycol by use of stock solutions of 5 M NaCl and 40% polyethylene glycol (molecular weight, 1,000 to 6,000).

12. Store the tube at 4°C overnight.

13. Sediment the resulting precipitate by centrifugation at 3,000 × g for 10 min, and resuspend the pellet in 1 to 2 ml of 0.25 M NaCl containing 0.001 M EDTA and 0.01 M Tris-HCl (pH 8.0) at 4°C.

14. Precipitate the DNA by adding 2 volumes of 95% ethanol at −20°C and let the tube stand overnight at −20°C. Alternatively, the DNA can be precipitated by letting the tube stand for 30 min at −70°C. Depending on the bacteria and the nature of the plasmid, the ideal length and temperature of incubation may vary.

15. Collect the plasmid DNA pellet by centrifugation at 12,000 × g at 4°C. Resuspend it in TES.

30.1.1.3. Alkaline pH

The alkaline pH method described below is a shorter version of the one described by Birnboim and Doly (13). It is used routinely to prepare plasmid DNA from *E. coli*, *Vibrio anguillarum*, *K. pneumoniae*, and *Enterobacter cloacae* and has been used by others for many other genera. The pH should not be allowed to go above 12.5, to avoid plasmid DNA alteration.

1. Harvest cells from a 50- to 200-ml (depending on the plasmid copy number) culture by centrifugation at 12,000 × g at 4°C. Resuspend the pellet with 5 ml of a solution containing 0.05 M Tris-HCl, 0.010 M EDTA, and 0.050 M glucose (pH 8.0).

2. Add 10 ml of a solution containing 0.2 N NaOH and 1% SDS. Mix gently. The solution should clear and turn viscous.

3. Add 7.5 ml of acetate solution (make the solution by adding 60 ml of 5 M potassium acetate, 11.5 ml of glacial acetic acid, and 28.5 ml of water). Mix thoroughly by shaking. A white precipitate should form.

4. Centrifuge at 30,000 × g for 20 min at 4°C.

5. Discard the pellet. If the supernatant contains pellet particles or is milky, filter it to obtain a clear solution. Add 12.5 ml of isopropanol, and let stand 15 min at room temperature.

6. Sediment the plasmid DNA by centrifugation at 12,000 × g at room temperature.

7. Resuspend the DNA in Tris-EDTA (TE) buffer (Table 1).

The original technique recommends that, after addition of the alkali and acetate solutions in steps 2 and 3, the sample stand in ice for a certain time. However, skipping these waiting steps results in the same yield and quality of plasmid DNA.

30.1.1.4. Sodium Dodecyl Sulfate

The SDS method below is a modification (57) of that described by Hirt (69).

1. Sediment the cells by centrifugation at 12,000 × g for 10 min at 4°C, and then resuspend them in 1.5 ml of 25% sucrose containing 0.05 M Tris-HCl and 0.001 M EDTA (pH 8.0).

2. Add 0.2 ml of lysozyme solution (10 mg of lysozyme per ml in 0.25 M Tris-HCl [pH 8.0]) to the cell suspension, and mix gently. Place the tube in ice for 15 min.

3. Add 0.1 ml of 0.25 M EDTA at pH 8.0, mix gently by slow inversion of the tube, and replace the tube in ice for 10 min.

4. Add 0.1 ml of 20% SDS solution. Mix the suspension gently, and keep in ice for 10 min. During this time, the cells lyse and the solution becomes viscous.

5. Precipitate chromosomal DNA by adding 0.9 ml of 3 M NaCl (to bring the final concentration to 1 M NaCl). Place the tube in ice for at least 2 h to allow complete precipitation.

6. Centrifuge the precipitated chromosomal DNA and any remaining cell debris at 35,000 × g for 30 min at 4°C, and decant the supernatant (enriched for plasmid DNA) into a 15-ml Corex tube.

7. Remove the RNA by adding 1 volume of distilled water and 4 μl of a 5 mg/ml boiled RNase solution in 0.15 M NaCl (the enzyme must be heated at 100°C for 20 min before use to destroy DNases). Incubate the tube at 37°C for an additional 1 h.

8. Add an equal volume of Tris-saturated phenol to deproteinize the mixture (as described in section 30.2.1). Shake the mixture vigorously, and then centrifuge at 3,000 × g for 20 min at 2°C. Remove the aqueous (upper) phase containing the plasmid DNA. Repeat this step once. Extract the aqueous phase once with 1 volume of chloroform-isoamyl alcohol (24:1, vol/vol) to remove residual phenol. Warning: phenol is poisonous and caustic.

9. Transfer the aqueous phase to a 30-ml Corex tube, and add 0.6 ml of 3 M sodium acetate to make the final concentration 0.3 M.

10. Add 2 volumes of cold (−20°C) 95% ethanol. Mix the solution well, and place at −20°C overnight.

11. Sediment the precipitated DNA at 12,000 × g for 20 min at −10°C, decant the ethanol thoroughly, and resuspend the plasmid DNA in 0.2 ml of 0.006 M Tris-HCl (pH 7.5).

30.1.1.5. Boiling

The boiling method described below is a modification by Reddy et al. (110) of the original method of Holmes and Quigley (70), which is commonly used as a small-scale method. A related modification was also reported by Lev (85) and Lev and Seveg (86). The following method has been used with *E. coli* cells, but it is possible that it could be used with other bacteria.

1. Collect cells from a 1-liter culture by centrifugation at 16,000 × g for 10 min.

2. Resuspend the cells in 10 ml of a solution containing 0.050 M Tris-HCl (pH 7.5), 0.062 M EDTA, 0.4% Triton X-100, and 2.5 M LiCl. Add 0.5 ml of 20% lysozyme in water. Mix gently, and incubate at 25°C for 10 min.

3. Transfer the solution to a plastic bag (8 by 8 cm), and seal. Immerse the bag in a boiling-water bath for 1 min.

4. Transfer the contents to a centrifuge tube, let it stand in ice for 5 min, and centrifuge at 27,200 × g at 4°C for 20 min.

5. Transfer the supernatant to a fresh tube, add 2 volumes of ethanol, mix, and keep the tube in dry ice for 10 min.

6. Sediment the DNA by centrifugation as in step 4, wash with 70% ethanol, and let dry.

7. Resuspend the DNA in 0.5 ml of TE buffer (Table 1), and add 10 μl of a boiled RNase solution (10 mg/ml dissolved in 0.15 M NaCl). Incubate at 25°C for 20 min.

8. Extract the plasmid DNA-containing solution with phenol as described in section 30.2.1, and precipitate with 1/9 volume of 5 M potassium acetate (pH 4.8) and 2.5 volumes of ethanol.

9. Collect the plasmid DNA by centrifugation, and resuspend in TE buffer.

30.1.1.6. Lysostaphin

The following procedure has been used successfully to obtain plasmid DNA from gram-positive bacteria, as well as from gram-negative bacteria that are resistant to lysis by Triton X-100 or SDS.

1. Pellet the cells from a 30-ml overnight culture by centrifugation at $16,000 \times g$ for 10 min and suspend them in 1.5 ml of 0.0075 M NaCl–0.050 M EDTA (pH 7.0).

2. Add lysostaphin to a final concentration of 15 µg/ml. (Double the enzyme concentration for *Staphylococcus epidermidis*.) Incubate the suspension at 37°C for 15 min with gentle agitation, and then place the flask in ice.

3. Add 1.5 volumes of a mixture containing 0.4% deoxycholate, 1% Brij 58, and 0.3 M EDTA (pH 8.0) to achieve cell lysis. Mix the viscous contents of the tube gently, and put the tube in ice for 15 min.

4. Pellet the cellular debris by centrifugation at $23,000 \times g$ for 20 min at 4°C, and decant the supernatant fluid into a 15-ml Corex tube.

5. Add 1 volume of distilled water and 4 µl of boiled RNase solution (1 mg/ml) to the supernatant fluid. Incubate the tube at 37°C for 1 h. Proceed with plasmid DNA purification as described for SDS lysis, step 8.

30.1.1.7. Lysostaphin and Lysozyme

The lysostaphin and lysozyme method was adapted from that of Lyon et al. (88).

1. Collect the cells from a 20-ml culture, wash them, and resuspend them with 1 ml of TES buffer (Table 1). Add lysostaphin and lysozyme to a final concentration of 75 µg/ml for each enzyme. Incubate for 1 h at 37°C.

2. Add 1 ml of a solution containing 2.5% Sarkosyl, 0.4 M EDTA (pH 8), and 100 µg proteinase K. Mix by inversion, and incubate at 65°C for 30 min.

3. Add 2 ml of a solution containing 3 M sodium acetate and 0.005 M EDTA (pH 7.0), and keep in ice for 30 min.

4. Centrifuge at $40,000 \times g$ at 4°C for 30 min. Transfer the supernatant to a fresh tube, and add 1 volume of isopropanol. Let stand at room temperature for 10 min.

5. Collect the DNA by centrifugation at $27,000 \times g$ for 10 min.

6. Dissolve the DNA pellet in 0.04 ml of water. Add boiled RNase to a final concentration of 50 µg/ml. Incubate for 30 min at 37°C. This DNA can be phenol extracted and ethanol precipitated.

30.1.1.8. Sucrose

The sucrose method can be used with gram-negative as well as gram-positive bacteria. Bacterial lysis is performed by applying an osmotic shock and then treating with detergent (112).

1. Harvest cells from 50 ml of culture, and wash them with 0.7 ml of 0.05 M Tris-HCl (pH 8.0) containing 25% sucrose.

2. Resuspend the cells in 0.7 ml of 0.05 M Tris-HCl (pH 8.0) containing 100% sucrose. Mix by pipetting, which takes time since 100% sucrose solution is very thick. Incubate for 3 h at 37°C.

3. Add 3 ml of a solution made by mixing 0.5 parts of 0.25 M EDTA (pH 8.1) with 2.3 parts of 1% Brij 58–0.4% sodium deoxycholate–0.05 M Tris-HCl (pH 8.0)–0.06 M EDTA–2 M NaCl. Mix gently, and incubate at room temperature for 1 h.

4. Pellet the cellular debris for 30 min at $10,000 \times g$ at 4°C.

5. Add 0.6 volume of isopropanol to the supernatant, mix, and keep for 15 min.

6. Centrifuge at $10,000 \times g$ at 4°C, and resuspend the pellet in 0.7 ml of TE.

7. This preparation can be phenol extracted and ethanol precipitated.

30.1.1.9. Penicillin

The following procedure can be used for obtaining plasmid DNA from gram-positive and gram-negative organisms that are resistant to Triton and SDS but sensitive to penicillin. If the bacterium under study produces a β-lactamase, however, a penicillinase-resistant antibiotic (e.g., a cephalosporin) may be effective. It is not certain whether other antimicrobial agents active against cell wall biosynthesis would prove successful, although it may be useful to investigate this possibility.

1. Grow bacteria to the mid-logarithmic phase, add 1 mg of penicillin G per ml, and incubate the culture at the optimum growth temperature for an additional 2 h. (Note: It is convenient to perform a preliminary experiment to determine the time and antibiotic concentration that give optimum protoplast formation; the goal is to achieve this maximum without massive lysis. The addition of 1 M sucrose to the growth medium may be beneficial in achieving this goal.)

2. Harvest the penicillin-treated cells by centrifugation, and wash them twice with an equal volume of 0.01 M Tris-HCl (pH 8.2) of the original culture.

3. Suspend the cell pellet in 0.25 volume of 0.02 M Tris-HCl (pH 8.2)–0.5 volume of 1 M sucrose–0.25 volume of lysozyme (4 mg/ml) in 0.02 M Tris-HCl (pH 8.2).

4. Incubate the cell suspension at 3°C for 1 h with shaking. Centrifuge at $27,000 \times g$ for 15 min.

5. Gently suspend the cells in 0.02 M Tris-HCl (pH 8.2) and 0.01 M EDTA, taking care to ensure that the cells are homogeneously dispersed.

6. Add 0.1 volume of 10% SDS to the cells. Lysis should be complete within 10 min.

Proceed from here as described for the SDS lysis method (section 30.1.1.4, step 8).

Among gram-positive bacteria such as streptococci, lactococci, and lactobacilli, addition of 0.01 M L-threonine to the growth medium weakens the cell wall and leads to easier lysis after lysozyme treatment followed by detergent lysis (23, 90). Glycine added at the early to mid-log phase can be just as effective as penicillin but, like the latter and unlike threonine, will stop growth of the culture (90).

30.1.2. Ordinary Plasmids by Small-Scale Methods

Small-scale methods permit the rapid preparation of ordinary plasmid DNA in small amounts, normally enough for a few assays such as restriction endonuclease digestion for screening or preparation of probes. Some of them produce a DNA of high enough quality to perform a few sequencing reactions, and some produce a DNA of inferior quality. These methods do not require a subsequent purification step for the stated purposes.

30.1.2.1. Cold Alkaline pH

The cold alkaline pH method is a scaled-down version of the one described in large scale (section 30.1.1.3) that has been used to prepare *E. coli* and *V. anguillarum* plasmids.

1. Harvest cells from 1.5 ml of culture by centrifugation in a microcentrifuge ($16,000 \times g$) for 10 s. Resuspend the pellet with 100 µl of a solution containing 0.05 M Tris-HCl, 0.010 M EDTA, and 0.050 M glucose (pH 8.0).

2. Add 200 μl of a solution containing 0.2 N NaOH and 1% SDS. Mix gently. The solution should clear and turn viscous.

3. Add 150 μl of acetate solution (make the solution with 60 ml of 5 M potassium acetate, 11.5 ml of glacial acetic acid, and 28.5 ml of water). Mix thoroughly by shaking. A white precipitate should form.

4. Centrifuge in a microcentrifuge (16,000 × g) for 5 min.

5. Transfer 400 μl of the supernatant to a fresh microcentrifuge tube, and add 1 ml of cold ethanol.

6. Recover the plasmid DNA by centrifugation for 5 min, pour off the supernatant, wash once with 70% cold ethanol, and dry.

7. Resuspend the DNA in 20 to 50 μl of TE buffer.

This DNA is suitable for restriction endonuclease analysis and Southern blot hybridization (see section 30.4.4). After RNase treatment and either phenol extraction or purification with silica powder (e.g., GENECLEAN [Q•BIOgene, Carlsbad, CA]), the DNA can be used for probe preparation and cloning experiments. *CAUTION:* (i) If the plasmid DNA is to be used as a probe against genomic DNA from an organism that is homologous to the host from which the plasmid is isolated, hybridization might occur between the blotted DNA and the residual chromosomal DNA from the host present in the plasmid preparation. This situation can be avoided either by purifying the plasmid DNA by dye buoyant-density gradient centrifugation or by preparing plasmid DNA by the acid-phenol method described below. (ii) Sometimes the plasmid DNA is not clean enough at this point for restriction endonuclease analysis. In this case, a phenol extraction usually renders the preparation suitable for this purpose.

30.1.2.2. Hot Alkaline pH
The hot alkaline pH method is a modification (73) of the alkaline lysis method described above (section 30.1.2.1). This rapid miniscreen method has been used to analyze large and small plasmids in many types of gram-positive as well as gram-negative bacteria.

1. Harvest cells from 2 ml of culture by centrifugation, and resuspend the cell pellet in 1 ml of 0.002 M EDTA–0.04 M Tris-acetate (pH 8.0).

2. Add 2 ml of a solution containing 0.05 M Tris and 3% SDS (pH 12.6) (adjust the pH with 2 N NaOH). Mix uniformly, and incubate at 60°C for 30 min.

3. Add 6 ml of phenol-chloroform (1:1), and mix until complete emulsification occurs. Separate the aqueous phase by centrifugation at 10,000 × g for 15 min at 4°C.

4. Transfer the aqueous phase to a fresh tube containing 1 volume of chloroform. Mix, and again separate the phases.

5. Recover the aqueous phase containing the plasmid DNA.

30.1.2.3. Acid Phenol
The acid phenol method (151) is based on the addition of an acid phenol extraction to the cold alkaline lysis preparation of DNA (section 30.1.2.1) and has been used very successfully. The additional use of phenol produces a preparation of quality similar to that obtained after separation in a CsCl-ethidium bromide gradient. This DNA can be used for sequencing and other procedures such as exonuclease III-generated nested deletions. Acid phenol removes nicked and linear DNA, leaving only the supercoiled molecules, by a mechanism still not well understood (159).

1. Proceed through step 4 as described for the alkaline lysis method (section 30.1.2.1). Transfer 400 μl of the supernatant to a fresh microcentrifuge tube, and add 400 μl of acid phenol (made by equilibration of distilled phenol with 0.05 M sodium acetate pH 4.0). Vortex and separate the layers in a microcentrifuge (16,000 × g).

2. Transfer the upper phase to another microcentrifuge tube, and extract with 1 volume of chloroform-isoamyl alcohol (24:1, vol/vol).

3. Transfer the upper phase to another microcentrifuge tube, and add 2.5 volumes of cold ethanol.

4. Recover the plasmid DNA by centrifugation for 5 min, pour off the supernatant, wash once with 70% cold ethanol, and dry. Resuspend the pellet in 50 μl of TE buffer (Table 1) containing 50 μg of boiled RNase per ml.

30.1.2.4. Phenol and Chloroform
The phenol-chloroform method involves the lysis of bacterial cells by phenol-chloroform, which renders a DNA preparation suitable for restriction endonuclease treatment as well as transformation (4, 110, 120). Broth cultures or pooled bacterial growth from solid surfaces can be used for this method.

1. Transfer 1.5 ml of culture or suspended scraped cells from a plate to a microcentrifuge tube, and centrifuge for 3 min.

2. Resuspend the pellet in 0.05 ml of a solution containing 0.01 M Tris-HCl (pH 8.0), 0.1 M NaCl, and 0.001 M EDTA. Add 0.05 ml of phenol-chloroform solution (made by mixing phenol, chloroform, and isoamyl alcohol [25:24:1, vol/vol/vol]), and vortex.

3. Centrifuge for 5 min to yield a clear supernatant. Transfer the upper phase to another tube, add ammonium acetate to a final concentration of 2 M, and add 2 volumes of cold ethanol. Let stand in ice for 15 min.

4. Centrifuge for 10 min to collect the DNA. Wash the pellet with 70% cold ethanol, dry it, and resuspend it in 0.02 ml of TE buffer.

30.1.2.5. Boiling
The boiling method described here is the original method of Holmes and Quigley (70), used effectively for screening of high-copy-number plasmids.

1. Transfer 1.5 ml of culture to a microcentrifuge tube, and pellet the cells.

2. Resuspend in 85 μl of STET buffer (8% sucrose, 0.5% Triton X-100, 0.05 M EDTA, 0.05 M Tris-HCl [pH 8.0]). Add 10 μl of 10 mg/ml lysozyme dissolved in water and let stand on ice for 5 min.

3. Place in a boiling-water bath for 90 s. The optimal time should be determined for different strains, since effectiveness seems to be strain dependent.

4. Immediately centrifuge for 10 min in a microcentrifuge at full speed.

5. Remove the viscous pellet with a pipette tip, and add 85 μl of cold isopropanol. Let stand at −20°C for 20 min.

6. Collect the DNA by centrifugation in a microcentrifuge (16,000 × g), wash the pellet once with 70% cold ethanol, and dry it. Resuspend the pellet in 20 μl of TE buffer. Another method involving lysis by boiling is reported in reference 161.

30.1.3. Large Plasmids
Large plasmids have been found in several bacterial systems, including the tumor-inducing and root-inducing plasmids

of *Agrobacterium tumefaciens* and *A. rhizogenes* (145, 156), the Sym plasmids of *Rhizobium* spp. (6), the H incompatibility group of antibiotic resistance plasmids (31), the camphor-degradative plasmid of *Pseudomonas putida* (22, 102), and the various F′ factors (91, 122, 127). Several methods have been developed which permit the isolation of small as well as large plasmids (5, 36, 58, 149). In addition, several techniques have been developed which permit the use of very small working volumes or even material from a single colony (43). A method similar to those mentioned above has been applied to lactic acid bacteria and several other groups of gram-positive organisms (5).

30.1.3.1. Lysis in Solution

The following method has the advantage of using small working volumes, shares the best features of the previously published methods, and permits the isolation of small as well as large plasmids. It takes advantage of the efficiency of plasmid DNA molecules to renature after treatment with alkali, which may account for the improved yields in plasmid DNA of very high molecular mass (up to 500 kbp).

This method has been used for the isolation of plasmids from gram-negative bacteria (e.g., *A. tumefaciens*, *Yersinia enterocolitica*, *Salmonella enterica* serovar Typhi, *Klebsiella* spp., and *E. coli*) and also gram-positive bacteria (e.g., *Staphylococcus* and *Streptococcus* spp.). Consequently, it is recommended for the routine screening of all types of bacteria for all types of plasmids.

Plasmid DNA obtained by this method from even 2 ml of culture is relatively free from chromosomal DNA and, after a few additional steps, can be used directly for restriction endonuclease analyses. The general procedure is as follows.

1. Grow 2 ml of culture overnight in an appropriate medium (e.g., brain heart infusion broth) on a shaker. The cells may be used directly or, better, can be used to obtain logarithmically growing cells by diluting the overnight culture 1:20 into 2 ml of fresh medium and incubating the subculture for 2 to 3 h. Harvest the cells by centrifugation at about 2,500 × *g* for 10 min.

2. Resuspend the cells in 2 ml of TE buffer (Table 1), recentrifuge, and resuspend the cell pellet thoroughly in 40 μl of TE. For marine vibrios and other halophilic bacteria, it is necessary to carry out this and subsequent washing steps in the presence of high salt concentrations (up to 1 M NaCl).

3. Add 0.6 ml of the lysis buffer (4% SDS in TE [pH 12.4]) to a 1.5-ml microcentrifuge tube, and, with a Pasteur pipette, transfer 40 μl of the cell suspension into the lysis buffer. Mix the suspension well, but avoid vigorous agitation or the use of mechanical mixing devices. It is important to determine the pH accurately (use a high-pH electrode), since values higher than 12.5 will irreversibly denature the plasmid DNA.

4. Incubate the suspension at 37°C for 20 min to achieve full lysis of the cells.

5. Neutralize the solution by adding 30 μl of 2.0 M Tris-HCl (pH 7.0). Slowly invert the tube until a change in viscosity is noted.

6. Precipitate chromosomal DNA by adding 0.24 ml of 5 M NaCl. This step should be done immediately after step 5. For complete removal of the chromosomal DNA, put the tube in ice for 4 h. If a large number of cultures are being screened and maximum plasmid purity is not important, the time in ice can be shortened to 1 h.

7. Centrifuge in a microcentrifuge (16,000 × *g*) for 10 min to sediment debris, and then pour the supernatant fluid into another microcentrifuge tube (do not attempt to recover the small amount of fluid remaining in the bottom of the tube).

8. Add 0.55 ml of isopropanol to the supernatant fluid to precipitate the DNA. After mixing, place the tube at −20°C for 30 min.

9. Centrifuge for 3 min in a microcentrifuge (16,000 × *g*), pour off the supernatant fluid, and invert the tube on a paper towel. Then, dry the tube under vacuum.

10. Resuspend the precipitate in 30 μl of TES. Allow the DNA to dissolve overnight at 4°C. Ordinarily, 10 μl of this solution shows readily visible DNA bands after gel electrophoresis. Note that plasmid DNA obtained by this method should be further purified for restriction endonuclease analysis by the following additional steps.

11. Resuspend the pellet obtained in step 9 in 100 μl of 0.010 M Tris (pH 8.0). Add 100 μl of phenol equilibrated with 0.010 M Tris-HCl (pH 8.0). Mix well. Add 100 μl of chloroform.

12. Centrifuge for 30 s in a microcentrifuge (16,000 × *g*) to separate the aqueous phase from the phenol-chloroform phase. Remove the upper, aqueous phase.

13. Precipitate the plasmid DNA with 2 volumes of ice-cold 95% ethanol, and proceed as in step 9. Resuspend the precipitate in a suitable buffer for the restriction endonuclease reactions.

For lactic streptococci, another lysis protocol is described in reference 5.

30.1.3.2. Lysis in Agarose Gel

The following method is based on bacterial cell lysis directly in the wells of a vertical agarose gel (43). Since most of the chromosomal DNA remains intact after the lysis, it remains in the well, and therefore less than 0.5% of the total chromosomal DNA appears in the gel as a band of linear DNA. Megaplasmids have been successfully detected by this method.

1. Pick cells from a colony with a flat toothpick, and resuspend them in the well of a gel containing 15 μl of a solution consisting of 7,500 U of lysozyme per ml, 0.3 U of boiled RNase per ml, 20% Ficoll 400,000, and 0.05% bromophenol blue in electrophoresis buffer (0.09 M Tris base, 0.012 M EDTA, and 0.09 M boric acid) (for gram-negative bacteria) or 75,000 U of lysozyme per ml, 0.3 U of boiled RNase per ml, 20% Ficoll 400,000, 0.05 M EDTA (pH 8.0), 0.1 M NaCl, and 0.05% bromophenol blue in electrophoresis buffer (for gram-positive bacteria). The gel dimensions recommended are 110 by 140 by 2.5 mm, and the well sizes should be 6.5 by 15 by 2.5 mm. Gel concentrations may vary from 0.75 to 1.2% agarose.

2. On top of the bacterium-lysozyme mixture add 30 μl of a solution containing 0.2% SDS and 10% Ficoll 400,000 in electrophoresis buffer for gram-negative bacteria or 30 μl of the same solution but also containing 2% SDS for gram-positive bacteria.

3. Gently mix the two layers with a toothpick by moving it from side to side (not more than twice for gram-negative bacteria). Complete mixing should be avoided. The two layers must still be distinguishable.

4. On top add 0.1 ml of a solution containing 0.2% SDS and 5% Ficoll 400,000 in electrophoresis buffer without disturbing the viscous DNA lysate.

5. Seal the wells with hot agarose, fill both electrophoretic chambers with electrophoresis buffer, and proceed with electrophoresis for 60 min at 2 mA and then for 60 to 150 min at 40 mA.

30.1.4. Linear Plasmids

Although most plasmids are composed of circular DNA, linear plasmids have also been described. They were found in lankamicin- and lankacidin-producing *Streptomyces rochei* as well as in *Borrelia burgdorferi* and *B. coriaceae*, the etiological agents of Lyme disease and epizootic bovine abortion, respectively (7, 67, 106, 113).

The protocols to isolate linear plasmids involve cell lysis, in some cases followed by CsCl-ethidium chloride gradient purification, and the separation of the linear plasmid from the chromosomal DNA by either sucrose gradient or agarose gel electrophoresis (7, 67, 106). A procedure designed to isolate linear plasmids of *B. coriaceae* is as follows.

1. Pellet cells from a 20-ml culture, and wash them with phosphate-buffered saline containing 0.005 M $MgCl_2$.
2. Resuspend the pellet in 0.4 ml of a solution containing 0.05 M Tris-HCl (pH 8.0) and 0.05 M EDTA.
3. Add 0.1 ml of 10% SDS and 25 μl of 0.02 M proteinase K. Incubate at 37°C for 1 h.
4. Load this mixture onto a 10 to 40% sucrose gradient prepared in 0.05 M Tris-HCl (pH 8.0)–0.05 M EDTA. Run overnight at 100,000 × *g* in an SW41 rotor (Beckman-Coulter Instruments, Palo Alto, CA).
5. Collect 3-drop fractions from the bottom of the gradient.
6. Run the fractions in 0.3% agarose gel electrophoresis, and use UV transillumination to locate those containing the linear plasmid DNA.
7. Pool these fractions, and precipitate the DNA by addition of 2.5 volumes of cold ethanol and centrifugation. Wash with 70% cold ethanol and dry. Resuspend the DNA in TE buffer.

30.1.5. Single-Stranded Plasmids

Several plasmids of gram-positive bacteria may be present as both single- and double-stranded DNA. All these plasmids replicate via a rolling-circle mechanism, and so the single-stranded DNA is a replicative intermediate (56). Single-stranded DNA plasmids have been isolated from gram-positive bacteria: lysozyme or lysostaphin in combination with Sarkosyl were used with *Bacillus subtilis* and *Staphylococcus aureus* (134), and electroporation was recently applied to *Clostridium acetobutylicum* for the same purposes (76). The method developed by te Riele et al. (134) is as follows.

1. Collect cells from a culture of optical density at 650 nm (OD_{650}) of 1, and wash with a solution containing 0.1 M EDTA (pH 8) and 0.15 M NaCl.
2. Resuspend the cells in a solution containing 0.01 M EDTA (pH 8.0) and 0.15 M NaCl.
3. Add 10 mg of lysozyme per ml (for *B. subtilis*) or 50 μg of lysostaphin per ml (for *S. aureus*), and incubate at 37°C.
4. Add Sarkosyl to a final concentration of 1%, and incubate at 65°C for 20 min.
5. Extract the cell lysate with phenol and chloroform.
6. Add boiled RNase to 50 μg/ml, and incubate at 37°C for 20 min.

The single-stranded plasmid can be recognized by performing Southern blot hybridization (as described in section 30.4.4) but skipping the alkali treatment of the gel to impede the binding of the double-stranded DNA to the membrane.

Further purification can be achieved by hydroxyapatite column chromatography. Dialyze the extract obtained in step 6 against 0.075 M sodium phosphate buffer (pH 6.8). Apply this sample to a hydroxyapatite column equilibrated with the same buffer. After washing, elute the single-stranded DNA with a linear gradient consisting of 0.075 to 0.250 M phosphate buffer.

30.1.6. Single-Stranded Phagemids

Although several techniques that used to require that the cloned DNA fragment be in the form of a single-stranded template have been now replaced by other, newer techniques, for some purposes it may still be desirable to purify single-stranded plasmid DNA. Dotto et al. (41) demonstrated that a plasmid carrying the major intergenic region of f1, in the presence of a helper phage, enters the f1 replication pathway by generating a single-stranded DNA that is packaged and exported in phage-like particles. The major intergenic region of f1 contains all of the sequences required in *cis* for initiation and termination of viral DNA synthesis and for morphogenesis of bacteriophage particles. Phagemids are plasmid vectors that have the ColE1 origin of replication, an antibiotic resistance gene for selection, a multiple cloning site, and the origin of replication of one filamentous phage. Most of these phagemids also have the T3 and/or T7 promoter to allow the synthesis of specific RNA by using the clone as a template. The cloned DNA can be propagated as a regular recombinant plasmid. Single-stranded DNA is obtained in the culture supernatant after the *E. coli* cells harboring the phagemid are coinfected with a helper phage. Different phagemids are available from several commercial sources (Stratagene, Bio-Rad, and others). In the following paragraphs a general protocol is described for the generation of single-stranded phagemid DNA.

1. Inoculate 5 ml of 2× YT broth with a single colony of *E. coli* harboring a phagemid, and add 2 × 10^7 PFU of helper phage suspension (e.g., phage Ml3KO7 [Bio-Rad, Richmond, CA], which has a kanamycin resistance gene) per ml.
2. Incubate at 37°C for 30 min with strong agitation (250 to 300 cycles/min), and then add kanamycin to a final concentration of 70 μg/ml. This addition prevents growth of cells not infected with the helper phage.
3. Incubate in a rotary incubator (250 to 300 cycles/min) at 37°C for 18 h.
4. Transfer 1.5 ml of the culture to a microcentrifuge tube, and pellet the cells in a microcentrifuge (16,000 × *g*) for 5 min at 4°C.
5. Transfer 1.2 ml of the supernatant to a new tube, add 7 μg of boiled RNase, and incubate for 30 min at room temperature.
6. Add 240 μl of a solution containing 20% polyethylene glycol 8000 and 15% NaCl (a solution containing 3.5 M sodium acetate instead of NaCl may also be used) to precipitate the phage particles. Let the mixture stand in ice for 30 min.
7. Collect the phage by centrifugation in a microcentrifuge (16,000 × *g*) for 1.5 min. Remove the supernatant. Resuspend the phage pellet in 200 μl of a solution containing 0.3 M NaCl, 0.1 M Tris-HCl (pH 8.0), and 0.001 M EDTA. Transfer to another tube, and let stand in ice for 30 min. Centrifuge again, and transfer the supernatant to a

fresh tube (this step removes insoluble products). This phage suspension can be stored at 4°C for 1 week.

8. To isolate the phagemid DNA, extract the phage suspension with 1 volume of phenol (prepared as described in section 30.2.1). Transfer the aqueous phase to a fresh tube.

9. Extract once with 1 volume of phenol-chloroform-isoamyl alcohol (25:24:1), recover the aqueous phase, and reextract several times with chloroform-isoamyl alcohol (24:1) until no interphase is visible.

10. To the aqueous phase add 1/10 volume of 7.8 M ammonium acetate and 2.5 volumes of cold ethanol. Keep at −70°C for 30 min.

11. Recover the phagemid DNA by centrifugation, wash the pellet with 70% cold ethanol, dry, and resuspend in 20 μl of TE buffer.

30.2. PURIFICATION

The subsequent analysis of plasmid DNA determines the degree of purity required. For example, to analyze and compare plasmid profiles for epidemiological purposes, further purification is normally not required. Conversely, other purposes such as restriction endonuclease analysis, cloning, and probe preparation may require further purification. Furthermore, a higher degree of purification is required for nucleotide sequencing in which all linear or nicked DNA molecules as well as RNA must be eliminated. After purification, the plasmid DNA concentration can be determined by either classical methods such as spectrophotometric determination or ethidium bromide fluorescence or commercial methods such as colorimetric quantitation with Dipstick (Invitrogen, Carlsbad, CA). This is a rapid method that consists of spotting a small amount of the nucleic acid solution on the Dipstick, dipping the stick in a solution provided with the kit, and determining the concentration by matching the color intensity of the sample to a standard chart.

30.2.1. Extraction with Organic Solvents

Extraction with organic solvents is used to remove proteins that are normally present after plasmid DNA isolation or after boiled RNase or restriction endonuclease treatment.

1. Mix the plasmid DNA solution with 1 volume of phenol (Table 1) or phenol-chloroform-isoamyl alcohol (25:24:1, vol/vol/vol) until emulsified.

2. Separate the phases by centrifugation in a microcentrifuge (16,000 × g) for 10 s.

3. Transfer the aqueous upper phase to a fresh tube, add 1 volume of either chloroform-isoamyl alcohol (24:1) or water-saturated ether, and mix until the aqueous phase becomes translucent. Warning: ether is very flammable and tends to form explosive peroxides under the influence of air and light.

4. Separate the phases by centrifugation as in step 2, and recover the aqueous phase containing the plasmid DNA. With chloroform extraction, the aqueous phase is on top; with ether extraction, the aqueous phase is below.

5. Add 1/9 volume of 5 M potassium acetate (pH 4.8) and 2.5 volumes of cold ethanol. Precipitation can also be achieved by addition of 0.5 volume of 7.5 M ammonium acetate and 3 volumes of ethanol.

6. Pellet the DNA by centrifugation in a microcentrifuge (16,000 × g) for 5 min, wash the pellet with 70% cold ethanol, and dry.

7. Resuspend the DNA in TE buffer.

To obtain a higher degree of purification, perform an extraction with phenol followed by another extraction with phenol-chloroform-isoamyl alcohol (25:24:1).

30.2.2. Equilibrium Gradient Centrifugation

Equilibrium gradient centrifugation is a classical method to purify plasmid DNA from large preparations. Ethidium bromide is one in a series of phenathridinium dyes that bind to DNA and RNA and inhibit nucleic acid functions (29, 65). The DNA is sedimented to equilibrium by high-speed centrifugation. Ethidium bromide intercalates into the DNA, thus reducing its density (27, 108). A linear DNA molecule or an open-circular (OC) (nicked) molecule of plasmid DNA does not have the physical constraints that are imposed on a CCC molecule of plasmid DNA. Consequently, the former types of DNA can bind significantly more ethidium bromide molecules and are rendered less dense than the latter type of DNA. This difference in binding permits the separation of the different forms of plasmid DNA. OC and linear plasmid DNA that may form as a result of handling during the preparation, as well as linear chromosomal DNA (also formed during handling), will all appear in the same band unless the G+C contents of plasmid and chromosome are drastically different. Only CCC plasmid DNA and some replicative intermediates will be separated from the linear and OC DNA molecules. In gradients of cesium chloride, these forms can be visualized as discrete bands when the gradient is illuminated with UV light, as a result of the fluorescence of the ethidium bromide intercalated in the DNA.

1. To a polyallomer ultracentrifuge tube (5/8 by 3 in. [1.6 by 7.6 cm]), add about 5 g of cesium chloride, 2 ml of TES buffer, and 3.2 ml of the DNA solution.

2. Mix the contents of the tube until the cesium chloride is in solution, and add 0.2 ml of ethidium bromide solution (10 mg/ml in TES buffer). Add the ethidium bromide after the CsCl, since the relaxation protein present in some plasmids (which leads to nicking of the DNA) can be activated by ethidium bromide (82). The high CsCl concentration (7 M) inactivates the relaxation complex of most plasmids (82), and consequently higher yields of DNA can be obtained. Even at a high salt concentration, visible light (in the presence of ethidium bromide) can induce nicking of plasmid DNA. Therefore, all operations must be conducted under indirect illumination.

3. Adjust the refractive index of the solution to 1.3925 ±0.001 g/ml by adding TES buffer or solid cesium chloride. The refractive index is a suggested value and may vary for each instrument; it must be standardized by doing trial experiments. If necessary, top the tubes off with mineral oil. *Rule of thumb*: To the dried DNA pellet resuspended in 4.0 ml of TE buffer, add 4.5 g of CsCl and 0.1 ml of 10 mg/ml ethidium bromide. Under these conditions, good separation of the bands is usually obtained.

4. Prepare the tubes for ultracentrifugation as specified by the centrifuge manufacturer, and place them in the appropriate rotor. Fixed-angle, vertical, swinging bucket or near-vertical rotors can be used. However, in vertical and near-vertical rotors the equilibrium is reached faster.

5. Centrifuge the tubes at 180,000 × g at 15°C for 4 to 8 h in a vertical or near-vertical rotor or 12 to 16 h in a fixed-angle rotor. After centrifugation, examine the gradients by illuminating the tubes with UV light and looking for the presence of fluorescent bands due to the DNA-ethidium bromide complex. Two bands should be observed,

the upper one corresponding to chromosomal and linear or nicked plasmid DNA, and the lower band corresponding to CCC plasmid DNA. In some cases only the lower band is observed. Collect the plasmid DNA by puncturing the tube at the bottom or just below the plasmid DNA band with a needle.

6. Free the DNA of ethidium bromide by extracting the preparation at least three times with cesium chloride-saturated isopropanol.

7. Dialyze the DNA against TE buffer for 2 to 3 h to eliminate the cesium chloride. Alternatively, the cesium chloride can be eliminated by twofold dilution of the DNA solution with TE buffer and precipitation with 2.5 volumes of ethanol.

30.2.3. Precipitation with Polyethylene Glycol

1. Add 1 volume of 5 M LiCl (to precipitate high-molecular-weight RNA) to the DNA solution. Mix, and centrifuge at 12,000 × g at 4°C for 10 min.

2. Transfer the supernatant to a fresh tube, and add 1 volume of isopropanol. Mix and centrifuge as in step 1 but at room temperature.

3. Discard the supernatant, wash the pellet with cold 70% ethanol, and allow it to dry.

4. Dissolve the DNA pellet in 0.5 ml of TE buffer containing boiled RNase (20 μg/ml). Let stand for 30 min at room temperature.

5. Add 0.5 ml of a solution containing 1.6 M NaCl and 13% polyethylene glycol 6000. Mix.

6. Collect the plasmid DNA by centrifugation in a microcentrifuge (16,000 × g) for 5 min, and discard the supernatant.

7. Dissolve the DNA pellet in 20 μl of TE buffer, extract with phenol, and precipitate the DNA as described above.

30.2.4. Chromatography

Several authors have described the application of high-pressure liquid chromatography and fast protein liquid chromatography to the purification of plasmid DNA (92, 155). These methods can be fast and convenient, although they require special equipment.

30.2.5. Commercial Methods

Many companies have introduced products for nucleic acid purification that are rapid and nontoxic alternatives to organic-solvent extraction. These products can be used to purify and concentrate DNA preparations, to extract-purify DNA fragments from agarose gels, and to separate labeled DNA from unincorporated nucleotides. The methods consist basically of the adsorption of the DNA to glass or silica powder or of minicolumns for adsorption or ion-exchange chromatography.

The principle of glass or silica matrix methods is the selective binding of DNA to these matrixes in the presence of sodium iodide. The contaminants (RNA, proteins, salts, agarose, and polysaccharides normally present in agarose) are eliminated by washing steps. Finally the DNA is recovered from the matrix by elution in low-ionic-strength buffer. A few examples of these commercial products are GENECLEAN kits (Q•BIOgene), which are used to purify DNA fragments of 200 bp or larger; MERmaid kits (Q•BIOgene), which are used to purify oligonucleotides of 10 to 200 nucleotides; or QIAquick DNA cleanup system (QIAGEN, Valencia, CA). An alternative method is based on the use of a hydroxylated silica resin that has high affin-

ity for proteins but low affinity for nucleic acid at neutral pH, such as StrataClean Resin (Stratagene, La Jolla, CA). Quick minicolumn chromatography products are based on ion exchange or adsorption. Some of these columns are prepacked in small devices, such as pipette tips or small columns that can be centrifuged for easier elution, and are components of DNA purification kits. Commercial kits for plasmid DNA purification are available from several companies.

Recently, plasmid DNA isolation and purification has become almost routine in most clinical and research laboratories. Therefore, several companies have developed products to make nucleic acid isolation and purification easier and faster. Most of these products are basically combinations of alkaline or boiling lysis methods followed by a purification step such as those described above. Some examples of these kits are Prep-A-Gene (Bio-Rad), Wizard DNA purification system (Promega), PlasmidQuick (Stratagene), Miniprep Kit Plus (Pharmacia), and QIAGEN plasmid purification systems, including kits for the isolation of plasmids containing large genomic inserts such as those generated using BAC, PAC, and P1 cloning vectors. In addition, most of these systems have been adapted for the isolation of plasmid DNA from a large number of samples such as those processed by high-throughput systems involved in genome sequencing projects. The Concert96 lysis and purification system was developed recently by Invitrogen, in which the bacterial cells of up to 96 samples are collected and lysed directly onto the matrix used for the subsequent purification of plasmid DNA. Some of these kits work with both gram-positive and gram-negative organisms.

30.3. QUANTIFICATION

30.3.1. Quantification of Plasmid Concentration

30.3.1.1. Spectrophotometric Methods

Absorbance at 260 nm is a widely used method to determine the concentration of nucleic acids in aqueous solutions (114). Although very practical and nondestructive, the drawbacks of this method include the requirement of pure nucleic acid preparations, the relatively narrow range at which reliable concentrations can be determined, its relatively low sensitivity, and the effect of pH and ionic strength on the absorbance of UV light by the purines and pyrimidines. A general practice when using this quantitative method is the determination of the absorbance of the sample at different wavelengths to determine the purity of the sample being tested. For example, contamination with phenolate and thiocyanate compounds, which are generally used for the isolation of nucleic acids, increases significantly the absorption at 230 nm, while absorption at 330 nm and higher wavelengths indicates contamination with particulate matter. The contamination of DNA samples with proteins can be tested by determining the OD_{260}:OD_{280} ratio, although it should be kept in mind that this ratio is a better indication of contamination of protein samples with nucleic acids rather than the reverse. In general, the initial data are obtained as the A_{260} or OD_{260} units of appropriate dilutions of DNA or RNA samples. This information is then transformed into relative concentration of nucleic acids by considering that 1 OD_{260} unit is equal to 50 μg/ml of double-stranded DNA or 38 μg/ml of single-stranded DNA or RNA. The newer spectrophotometers, e.g., NanoDrop, obtain concentration and quality data by direct measurement of samples as small as 1 μl.

30.3.1.2. Fluorometric Methods

30.3.1.2.1. Nonintercalating Dyes

The concentration of DNA can be determined using some other approaches that are as simple as direct spectrophotometry but more sensitive. One of these methods is based on the binding of nonintercalating dyes such as Hoechst 33258 by double-stranded DNA larger than 1 kbp, which results in a 60-fold increase in fluorescence at 458 nm when the sample is excited at 365 nm. It is important to consider that this assay is also affected by the pH of the sample, the presence of detergents and salts above 3 M, and the quenching of the fluorescence of the sample by excess of dye in the final reaction mixture.

1. Dilute 50 μl of Hoechst 33258 stock solution (0.2 mg/ml in H_2O) in 100 ml of fluorometry buffer (2M NaCl, 50 mM $NaPO_4$, pH 7.4). Solutions should be filtered through a 0.45-μm-pore-size filter, and the Hoechst 33258 stock solution should be kept protected from the light.

2. Transfer 3 ml of diluted dye to an appropriate number of glass tubes.

3. Construct a standard curve using DNA of known concentration (100 to 500 ng).

4. Mix and read the fluorescence of the standards immediately in a prewarmed fluorometer set to 365 nm and 458 nm excitation and emission wavelengths, respectively.

5. Add 0.1, 1.0, and 10 μl of the DNA sample being tested to individual tubes containing the diluted Hoechst 33258 dye solution. Mix and read immediately as described for the standard samples.

6. Determine the concentration of DNA in the sample by using a curve constructed by plotting the fluorescence of each standard point against its corresponding DNA concentration.

30.3.1.2.2. Intercalating Dyes

The UV-induced fluorescence emitted by ethidium bromide when intercalated with DNA is a practical and widely used alternative to determine as little as 1 ng of double-stranded DNA. In general, the most popular methods are only semiquantitative because they are based on the visual comparison of the fluorescence produced by the test samples with that produced by dilutions of standard DNA, which are all spotted on the surface of UV-transparent material (Saran Wrap). Alternatively, the samples and standard dilutions can be either spotted onto an agarose slab or electrophoretically fractionated through agarose containing 0.5 μg/ml ethidium bromide. These approaches avoid interference due to contaminants. New dyes such as Pico Green and SYBR Green I (Molecular Probes, Eugene, OR) are now being used to quantify DNA because of their low intrinsic fluorescence and other properties that make them more suitable for DNA quantification (16, 126). Affordable fluorometers are now on the market, e.g., Qubit (Invitrogen). Recently, a sensitive method (the microplate method) that allows the rapid and practical quantification of a large number of samples, such as those produced in high-throughput processes, was developed using SYBR Green I (146).

The Saran Wrap Method

1. Stretch a piece of Saran Wrap over a UV transilluminator and spot 5 μl of standard DNA serially diluted covering a concentration range from 0.1 to 20 μg/ml.

2. Spot 1 to 5 μl of sample DNA close to the standard spots.

3. Add to each standard and sample spots 1 volume of TE buffer containing 2 μg/ml ethidium bromide and mix either by pipetting up and down with a micropipette or mixing with a pipette tip.

4. Turn on the UV transilluminator and determine the fluorescence of the samples and standard dilutions immediately either by direct visual inspection or by taking a photograph, using an appropriate gel documentation system equipped for DNA detection.

5. Estimate the plasmid DNA concentration of the samples by comparing their fluorescence intensity with that of the standard dilutions.

The Agarose Slab Method

1. Prepare the samples and standard dilutions as described for the Saran Wrap method.

2. Prepare a 1% agarose slab by pipetting 2 to 3 ml of 1% agarose in TE buffer containing 0.5 μg/ml ethidium bromide on a flat surface such as that of a microscope slide.

3. Spot the samples and standard dilutions on the surface of the agarose after it has solidified and incubate at room temperature from 1 to 2 h.

4. Transfer the gel slab onto the surface of a UV transilluminator carefully, turn it on, and photograph the gel immediately using an appropriate gel documentation system equipped for DNA detection.

5. Estimate the concentration of the samples by comparing their fluorescence intensity with that of the standard dilutions.

The Agarose Minigel Method

This method is as rapid as those described in the two previous sections with the advantage of providing information regarding the presence of different plasmid topoisomers as well as the potential contamination with RNA.

1. Mix 2 to 5 μl of plasmid DNA sample with 2 μl of gel-loading buffer on the surface of a piece of Parafilm or Saran Wrap.

2. Mix 2 to 5 μl of each of a twofold dilution series of plasmid DNA standard, covering a concentration range from 0 to 50 μg/ml, with 2 μl of gel-loading buffer on the surface of a piece of Parafilm or Saran Wrap. Purified plasmids similar in size to those being tested should be used for this purpose.

3. Prepare a 0.8% mini-agarose gel (20 to 30 ml) containing 0.5 μg/ml ethidium bromide.

4. Load the plasmid samples and the standards and run the electrophoresis using standard conditions until the bromophenol blue present in the gel-loading buffer reaches one-third to one-half of the gel length.

5. Destain the gel by immersing it in 50 to 100 ml of distilled water or 0.01 M $MgCl_2$ for 5 min at room temperature. Skip this step if the gel background is low and does not interfere with the quantification of plasmid DNA.

6. Photograph the gel immediately using an appropriate gel documentation system equipped for DNA detection.

7. Estimate the concentration of the samples by comparing their fluorescence intensity with that of the standard dilutions.

The Microplate Method

1. Dilute the dye SYBR Green I 6,250-fold in 10 mM Tris-HCl (pH 7.5)–1 mM EDTA.

2. Pipette up to 250 μl of diluted dye into each well of a standard microtiter plate.

3. Prepare a serial dilution of standard DNA covering a final concentration range from 2 ng/ml to 2 μg/ml of DNA.

4. Add 1 to 5 μl of standard dilutions and plasmid samples to each well containing diluted SYBR Green I.

5. Mix by tapping on the side of the microplate gently, and incubate at room temperature no longer than 15 min.

6. Illuminate the samples by using a 365-nm UV transilluminator and capture the image with a CCD camera (BioPrint gel documentation system; LTF-Labortechnik, Wasserburg, Germany) with a standard yellow filter (485 nm) using an aperture of 1.8. The plate should be located with minimal proximity to the camera objective, and the exposure time should be adjusted to obtain maximal sensitivity.

7. Calculate the fluorescence intensity of each sample by determining their pixel density with a software such as the Image Tool 2.00 (University of Texas Health Science Center, San Antonio, TX).

8. Estimate the concentration of the plasmid samples by comparing their fluorescence intensity with that of the standard dilutions.

30.3.2. Quantification of Plasmid Forms

Plasmids can be isolated in different forms such as circular elements with different degrees of coiling, partially cleaved, as well as linear molecules. In addition, the extrachromosomal elements can exist as monomers and multimers that can be present either as concatemer or catenete complexes. Recently, a fluorescence-based method was developed to determine the presence of supercoiled DNA in plasmid preparations (87). This method is based on the fluorescence enhancement of PicoGreen bound specifically to double-stranded DNA and the fact that supercoiled DNA regains its original double-stranded structure after denaturation with chemical and physical agents such as high temperature. In contrast, OC and linear DNA molecules are irreversibly denatured and remain as single-stranded DNA after these treatments, resulting in the minimal fluorescence of these DNA forms because of the very limited binding of PicoGreen to single-stranded DNA. Although sensitive and practical for the rapid analysis of a large number of samples, this method does not provide information regarding the presence of the different topoisomers and their relative concentration in a particular plasmid DNA preparation. Most of these forms can be detected and quantified by simple methods such as agarose gel electrophoresis and ethidium bromide staining. A similar analysis can also be conducted using capillary gel electrophoresis, which provides more reliable and quantitative data and can be automated to examine a large number of samples. A comparative analysis of these two methods was reported by Schmidt et al. (116), who concluded that capillary gel electrophoresis is an appropriate alternative for the detection and quantification of plasmid forms.

30.3.2.1. Agarose Gel Electrophoresis

1. Isolate plasmid DNA by a standard method. In general the preparation is composed mostly of monomers and dimers in their supercoiled or CCC forms.

2. Incubate an aliquot of the purified plasmid either with excess or limiting amounts of a restriction endonuclease, which has a single recognition site within the plasmid molecule, to produce monomers and multimers in their cognate linear forms.

3. Irradiate an aliquot of the purified plasmid for 30 min at room temperature with UV light (254 nm, 15 W) to create the OC (nicked) forms of plasmid monomers and multimers.

4. Mix each of the samples (up to 500 ng of DNA) with gel-loading buffer.

5. Prepare a 0.8% medium-size agarose gel using the Tris-acetate buffer system.

6. Load the plasmid samples and run the electrophoresis at 5 V/cm (50 V) until the bromophenol blue present in the gel-loading buffer reaches the bottom of the gel.

7. Stain the DNA samples with 1 μg/ml ethidium bromide for 30 min and record the gel image using an appropriate gel documentation system equipped for DNA detection.

8. Comparative analysis of the electrophoretic mobility of the DNA bands detected in each sample provides information regarding the plasmid forms present in each sample. Under the conditions described, the monomeric CCC form is the fastest-migrating plasmid species, which is followed by the plasmid monomers in their nicked and linear forms, respectively. The CCC, nicked, and linear forms of plasmid dimers migrate immediately above the monomers, with the CCC dimer being the fastest species of this second group of plasmid forms.

9. Comparative analysis of the relative fluorescence of each plasmid species present in the untreated sample provides information regarding the relative proportion of each species in the plasmid preparation.

30.3.2.2. Capillary Gel Electrophoresis

1. Equip a capillary electrophoresis system such as a Beckman P/ACE 2050 CE (Beckman) instrument with an argon laser and a 37-cm-long coated capillary (100-μm inner diameter, 0.1-μm coating thickness) (DB-17; J. and W. Scientific, Folsom, CA). Use 488 and 520 nm as the excitation and detection wavelengths, respectively.

2. Prepare the running buffer, consisting of 89 mM Tris, 89 mM boric acid, 2 mM EDTA, pH 8.4, and 0.1% (wt/wt) hydroxypropylmethylcellulose (H-7509; Sigma Chemical Co., St. Louis, MO).

3. Add 1 μl of YOYO stain (Molecular Probes) to 15 ml of running buffer immediately before sample analysis.

4. Stain the samples prepared as described for agarose gel electrophoresis by incubating them at room temperature for 2 min with YOYO at a DNA base pair-to-dye molar ratio of 5:1.

5. Apply samples to the capillary column by pressure injection for 1 s and perform electrophoresis at 100 V/cm and 30°C.

6. Collect the data as suggested by the instrument manufacturer and analyze them using a package such as the System Gold Software (Beckman), recording migration time and peak area for each peak detected in the analyzed samples.

7. Quantify each plasmid form present in a particular sample by determining the peak area:migration time ratio for each peak detected during the electrophoresis of a plasmid sample.

30.4. CHARACTERIZATION

Physical characterization of plasmid DNA is covered in sections 30.4.1 through 30.4.4. For details of genetic and functional characterization, see sections 30.4.5 through 30.4.8.

30.4.1. Size

30.4.1.1. Gel Electrophoresis of Native Plasmid DNA

Simple adaptations of gel electrophoresis methods (43, 96) are suitable for the detection and preliminary size characterization of plasmid DNA present in clinical or environmental isolates and in laboratory strains of gram-negative and gram-positive bacteria.

When plasmid DNA is subjected to electrophoresis in gel, the migration rate of the different DNA species is inversely related to their molecular weights; i.e., the higher the molecular weight, the slower the migration. Procedures involve either a vertical or a horizontal slab of agarose gel. Even partially purified lysates (generally 10 to 20 μl of lysate is sufficient) can be subjected to electrophoresis. This technique has been used for purposes ranging from the simple visualization of a plasmid in a bacterial cell to epidemiological studies made by analysis of plasmid profiles (62).

For electrophoresis, use standard vertical or horizontal gels. Gel concentrations vary from 0.25 to 1.5% agarose in a standard electrophoresis buffer (e.g., 0.04 M Tris-acetate and 0.001 M EDTA [pH 8.0], prepared from a 50× concentrated solution that is made by addition of 242 g of Tris base, 57.1 ml of glacial acetic acid, and 100 ml of 0.5 M EDTA [pH 8.0] for 1 liter). Tris-acetate is more convenient because precipitation of salts in the electrophoresis tanks is avoided. However, it is advisable to try two different buffer systems when examining plasmid DNA for the first time. Plasmids in the 10- to 15-kbp range migrate with the chromosomal DNA band in Tris-borate gels, whereas plasmids in the 20- to 23-kbp range migrate with the chromosomal band in Tris-acetate gels. Therefore, if a plasmid is masked by residual chromosomal DNA in one buffer, it will be revealed by running the same sample in the second buffer. If speed is important, plasmids in Tris-borate migrate much faster than those in Tris-acetate.

The sample (in about 20 μl) is mixed with 2 μl of gel-loading buffer (0.25% bromophenol blue, 0.25% xylene cyanol FF, and 30% glycerol in water). Electrophoresis is ordinarily carried out for 2 h at room temperature, 60 mA, and 120 V (constant voltage) or until the bromophenol blue band (the dark blue band) reaches the end of the gel. The gel is then placed in a solution of ethidium bromide (0.4 to 1 μg/ml) and stained for 15 min. The DNA then can be visualized in a UV light transilluminator.

Plasmids of known molecular weight can be used as standards to determine the molecular weight of an unknown plasmid by agarose gel electrophoresis. A good source of plasmids of different molecular weight is *E. coli* V517 (90) (available through The CABRI Consortium, http://www.cabri.org/guidelines/catalogue/CPdishostplasm.html or The German Collection of Microorganisms and Cell Cultures, http://www.dsmz.de/plasmids/pls08872.htm). By plotting the $\log_{10}$ of the distance migrated from the origin by the plasmid DNA versus the $\log_{10}$ of the molecular mass of the plasmid DNA, a straight-line relationship is obtained from about 1 to 100 kbp. Interpolation permits the determination of molecular weights of the various plasmids. A concentration of agarose of 0.7% is appropriate for most plasmids within this molecular size range.

30.4.1.2. Gel Electrophoresis of Digested Plasmid DNA

After restriction endonuclease treatment, plasmid DNAs (and all forms of DNA, for that matter) have characteristic fragment patterns which depend on the number and spacing of the specific recognition sites within the genome. These patterns can be analyzed by electrophoresis on either agarose or polyacrylamide gels. In general, larger DNA fragments are better resolved in agarose than polyacrylamide gels, whereas smaller DNA fragments are better separated in polyacrylamide gels. The separation of the DNA fragments also depends on the concentration of agarose or polyacrylamide used to cast the gels. For example, 0.7% agarose gels are suitable to resolve fragments within 0.5 to 10 kbp, whereas 1.5% agarose gels resolve fragments from 0.2 to 3 kbp. Normally no more than 2% agarose gels are used for DNA electrophoresis. Instead, for example, 15% polyacrylamide gels are good to separate fragments ranging from 0.025 to 0.250 kbp and 20% polyacrylamide gels are suitable to resolve fragments smaller than 0.1 kbp. Once electrophoresed, the DNA fragments can be transferred from the agarose gel to nitrocellulose or nylon membranes. After hybridization with a radioactive DNA "probe," the fragments containing sequences homologous to those present in the probe can be detected as sharp bands by autoradiography.

Agarose gels are prepared with the same buffer described above, but for linear DNA, submarine horizontal gels (the gel is completely submerged in the buffer) are recommended. These types of gels can also be used for the analysis of undigested plasmid DNAs. The size of these gels, as well as the agarose concentration, varies according to the size of the fragments to be separated. Polyacrylamide gels are commonly used in a vertical apparatus; however, in our laboratories we use submarine horizontal minigels. To prepare these gels, mix 3 ml of solution A (18.1% Tris base, 0.24% N,N,N',N'-tetramethylethylenediamine [TEMED], 0.25 N HCl), 4.2 ml of solution B (28% acrylamide, 0.75% bisacrylamide), and 4.8 ml of 0.14% ammonium persulfate. Degassing is not necessary. This makes a 12% polyacrylamide gel of 5 by 6 by 0.3 cm, which is electrophoresed by using 0.06% Tris–0.28% glycine as the electrophoresis buffer. Under these conditions, the migration of the bromophenol blue dye corresponds to 20-bp oligonucleotides and that of the xylene cyanol corresponds to 70-bp oligonucleotides. Depending on the sizes of the DNA fragments to be analyzed, the porosity of the gel can be adjusted by varying the concentration of acrylamide. The samples for both kinds of gels are prepared as mentioned in the previous paragraph.

For a comparison between agarose concentration and migration of DNA fragments of different molecular sizes, we recommend visiting http://tools.neb.com/NEBcutter2/index.php, where the migration of DNA fragments on gels with different agarose concentrations can be viewed.

30.4.1.3. Pulsed-Field Gel Electrophoresis

Traditional agarose gel electrophoresis techniques such as those described above use an electric field of constant strength and direction. These methods are good for separating linear DNA of up to approximately 50 kbp. To separate larger fragments, pulsed-field gel electrophoresis is the technique of choice (117). It involves the use of pulsed alternating electric fields to separate large DNA fragments. Applications of this technique include, but are not limited to, separation of very large DNA fragments like those commonly present in recombinant derivatives generated using cosmid and BAC cloning vectors. Resolution of restriction fragments in the analysis of restriction fragment length polymorphisms is also achieved by this technique. Simske and Scherer (125) described an application of this method for the analysis of CCC and OC DNA. Many apparatuses for pulsed-field gel electrophoresis currently are on the market, and a profuse literature describing the different designs and protocols is available (14, 20, 59, 83, 114, 117).

In some circumstances, the ability to extract fragments that have been separated on agarose or polyacrylamide gels is highly desirable (49). These fragments can be useful for physical and chemical studies, as well as for fine-structure restriction analysis, superhelical density determination, electron microscopy, DNA sequencing, heteroduplex analysis, subcloning, and hybridization by the S1 endonuclease method (33) or by the method of Southern (128) (section 30.4.4). Below are described techniques that permit the extraction of a variety of physically intact, biologically active DNAs in high yield directly from agarose gel bands. These techniques include electroelution, low-melting-point agarose, silica powder, and treatment with GELase (Epicentre Technologies, Madison, WI) for agarose gels; and "crush and soak" for polyacrylamide gels. The newer silica powder (in a slurry or in columns) and GELase methods are the easiest and fastest and provide good results. However, some shearing can occur when eluting fragments larger than 15 kbp with silica powder. Several different kits are offered by a number of companies. Electroelution has the advantage that no commercial products are required.

Electroelution

1. Visualize DNA fragments separated on an agarose gel by illumination with a UV lamp, and slice out the fragments with a razor blade. Locate the DNA band with a hand-held, long-wavelength UV light to minimize DNA damage.
2. Fill a dialysis bag with agarose electrophoresis buffer, transfer the gel slice containing the DNA to be eluted, and close the bag, avoiding air bubbles. The amount of buffer should be the minimum required to keep the gel in contact with it.
3. Immerse the bag in the electrophoresis buffer in the electrophoresis tank. Apply 4 to 5 V/cm during 2 to 3 h.
4. Reverse the polarity of the current for 1 min.
5. Recover the buffer containing the extracted DNA in solution.
6. The DNA can be purified further by extraction with organic solvents or by the use of glass or silica powder as described in section 30.2.5.

Commercial options are also available for electroelution of DNA. The Analytical Unidirectional Electroeluter (IBI, New Haven, CN) and the SixPac electroelution unit (Hoefer) are examples of these devices.

Low-Melting-Point Agarose

1. Once a regular agarose gel is prepared, cut out a transversal strip with respect to the lane in which the fragment of interest is going to run. Fill this space with low-melting-point agarose, and let it solidify in a cold room.
2. Load and run the DNA sample in a cold room, monitoring the migration of the band of interest, preferably with a hand-held, long-wavelength UV light to minimize DNA damage.
3. Once this band enters the portion of low-melting-point agarose, stop the agarose electrophoresis.
4. Cut out a slice of low-melting-point agarose containing the band, and transfer it to a plastic tube.

5. Add 5 volumes of TE buffer, and incubate for 5 min at 65°C. Then, treat it as a regular DNA solution; i.e., purify the DNA by extraction with organic solvents.

Silica Powder

Here the protocol for extraction of DNA with GENECLEAN is described. Protocols involving other similar products are similar and are provided by the supplier.

1. After electrophoresis, cut out a slice of agarose containing the fragment of interest. Weigh it and add 3 volumes of 5 M sodium iodide, assuming that 1 mg of gel equals 1 ml.
2. Incubate at 55°C for about 5 min, shaking occasionally until the agarose melts.
3. Add about 5 μl of the suspension of silica powder, vortex, and let the solution remain in contact with the powder for 5 min at room temperature, mixing occasionally to prevent sedimentation of the powder.
4. Recover the powder by centrifugation in a microcentrifuge (16,000 × g) for 5 s, and wash three times with the washing solution provided by the supplier.
5. Extract the DNA by incubation at 55°C in the presence of TE buffer. Recover the DNA solution by centrifugation.

GELase

GELase is an enzyme that degrades low-melting-point agarose into small oligosaccharides that are soluble in water and alcohol. The product of this reaction is a clear liquid that does not become viscous even when put on ice. Two protocols have been developed by the supplier.

Fast protocol for GELase

1. Cut out the low-melting-point agarose portion containing the DNA band of interest, and determine the weight of the gel slice.
2. Add 50× GELase buffer (1 μl/50 mg of gel), and completely melt the gel by incubation at 70°C.
3. Equilibrate the temperature of the molten gel at 45°C, add the proper amount of GELase (1 U/300 mg of molten 1% low-melting-point agarose when the agarose electrophoresis buffer was Tris-acetate or morpholinepropanesulfonic acid [MOPS], or 1 U/80 mg when Tris-borate buffer was used). Incubate at 45°C for 1 h.

High-activity protocol for GELase

1. Cut out the low-melting-point agarose portion containing the DNA band of interest, and determine the weight of the gel slice.
2. Replace the electrophoresis buffer by soaking the gel slice in 10 volumes of 1× GELase buffer at room temperature for 1 h.
3. Melt the agarose slice at 70°C for 20 min, and then equilibrate the tube at 45°C.
4. Digest the molten agarose by adding 1 U of GELase per 600 mg of low-melting-point agarose for 1 h at 45°C.

The alcohol precipitation steps that follow are common for both GELase protocols.

1. Add 1 volume of 5 M ammonium acetate and 2 volumes of ethanol, and mix gently and thoroughly.
2. Recover the DNA by centrifugation in a microcentrifuge (16,000 × g) for 5 min.
3. Wash the pellet with 70% ethanol, and dry.
4. Dissolve the DNA in the appropriate buffer.

"Crush and Soak"

1. Locate the band of interest in the polyacrylamide gel, and cut it out.

2. Transfer to a plastic tube, crush the agarose polyacrylamide with a pipette tip, and add 1 to 2 volumes of a solution containing 0.5 M ammonium acetate, 0.01 M magnesium acetate, 0.001 M EDTA (pH 8.0), and 0.1% SDS.

3. Incubate at 37°C on a rotary shaker for 3 to 24 h depending on the size of the DNA fragment (fragments larger than 0.5 kbp take between 12 and 24 h).

4. Centrifuge the sample in a microcentrifuge (16,000 × g) for 1 min at 4°C, and transfer the supernatant to a fresh tube.

5. Reextract the polyacrylamide pellet with 0.5 volume of the same solution.

6. Combine the two supernatants, and precipitate the DNA by addition of 2 volumes of cold ethanol. Recover the DNA by centrifugation.

7. Redissolve the DNA in 0.2 ml of TE buffer. Reprecipitate the DNA by the addition of 1/9 volume of 5 M potassium acetate (pH 4.8) and 2.5 volumes of cold ethanol.

8. Collect the DNA by centrifugation, wash the pellet with cold 70% ethanol, dry, and resuspend in TE buffer.

30.4.2. Copy Number

The number of copies of a plasmid present per chromosome equivalent is a parameter that characterizes a plasmid and also gives information about the nature of its replication. There are several ways to express copy number, e.g., per chromosome equivalent, per cell volume, or per mass. Plasmid copy number can be estimated by centrifugation of a [³H]thymine-labeled total cell lysate in an ethidium bromide-cesium chloride gradient or by agarose gel electrophoresis (34). A method based on in situ colony lysis with known cell volumes was developed by Shields et al. (124) for mini-F, a low-copy-number plasmid. A specific probe for mini-F was used for hybridization with DNA immobilized on nitrocellulose. Although this method offers the advantage of simplicity as well as sensitivity and low cost, a potential drawback is the requirement for a specific probe that recognizes sequences only in the plasmid DNA.

30.4.2.1. CsCl-Ethidium Bromide Gradient Centrifugation

If the bacterial strain requires thymine, use an adequate concentration of thymine for growth plus [³H]thymine (7 μCi/ml). For a strain that does not require thymine, use 1 mg of thymine, 250 μg of deoxyadenosine, and 7 μCi of [³H]thymine per ml. For some gram-positive bacteria, [³H]thymidine is all that need be added to the cultures in rich media to label the DNA.

The procedure for gram-negative bacteria is as follows.

1. Grow a culture (1 ml) of a plasmid-containing strain at 37°C in the presence of 7 μCi of [³H]thymine per ml to a density of 2×10^9 cells per ml.

2. Centrifuge the cells, wash them in TE buffer, and resuspend them in 400 μl of 0.5 M sucrose (ultrapure, RNase-free sucrose in 0.5 M Tris-HCl [pH 8.0]).

3. Add 0.2 ml of 0.2 M EDTA (pH 7.8) to the cell suspension, and mix well.

4. Add 0.2 ml of a fresh solution containing 1% lysozyme (in 0.25 M Tris-HCl [pH 8]) to the cell suspension, and mix well. Shake gently, and let stand in ice for 15 min.

5. Add 1.2 ml of 1.2% Sarkosyl to the suspension, and then immediately add 0.1 ml of pronase (1 mg/ml) in 0.01 M Tris-HCl (pH 8.1) and 0.02 M EDTA. Pronase should be first self-digested for 2 h at 37°C and then heated for 2 min at 80°C before use. Mix gently, and incubate at 37°C until the suspension clears completely.

6. Draw the DNA solution rapidly in and out of a 5-ml pipette (to shear the chromosomal DNA) until the solution is no longer viscous.

7. Centrifuge a sample containing about 500,000 cpm of radioactive DNA in a 5-ml ethidium bromide-cesium chloride gradient (the refractive index must be 0.0025 lower than the refractive index used for Triton X-100 lysate CsCl-ethidium bromide gradient). Also centrifuge a sample in a 5 to 20% linear sucrose gradient. Sucrose gradients can be prepared in a gradient mixing chamber or, alternatively, by making a sucrose solution of an average concentration between the two desired extreme concentrations. Freezing of this solution at −20°C and slow thawing at 4°C (overnight) permits the development of a gradient suitable for analytical purposes. For example, a frozen 12.5% sucrose solution after thawing renders a gradient that is approximately 5 to 20%.

8. Fractionate the gradients directly onto Whatman 3MM filter disks.

9. Precipitate radioactive DNA by immersing the filters in cold 5% trichloroacetic acid containing 50 μg of thymine per ml, wash with 95% ethanol, and dry. Determine radioactivity in a scintillation spectrometer.

Figure 1 shows a cesium chloride-ethidium bromide gradient with the profile obtained for ³H-labeled total-cell DNA from a plasmid-containing strain of E. coli. The ratio between the plasmid and chromosome peaks can be used for the calculation of copy numbers per chromosome equivalent, as follows.

$$\text{Copy number} = \text{cpm of plasmid peak/cpm of chromosome peak} \times \text{MWc/MWp}$$

where MWc and MWp are the molecular weights of the chromosome and the plasmid, respectively.

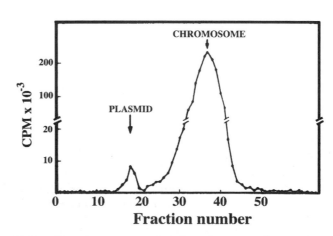

FIGURE 1 Determination of plasmid copy number. A plasmid-containing bacterial strain was grown in a minimal salts medium plus Casamino Acids and glucose. Labeling and lysis were carried out, and the lysate was centrifuged in the CsCl-ethidium bromide gradient. Seven-drop fractions were collected on microtiter trays and assayed for radioactivity.

Comments about Determination of Plasmid Copy Number

Plasmid forms other than CCC DNA are detected in the same cesium chloride-ethidium bromide gradient (or in sucrose gradients) by their banding characteristics. For example, dimers of two CCC DNA molecules coband with CCC DNA in a CsCl-ethidium bromide gradient but sediment ahead of monomeric CCC DNA in a sucrose gradient. Dimers of one CCC and one OC DNA molecule band at a position intermediate between CCC DNA and OC DNA in the CsCl-ethidium, bromide gradient. Dimers of two OC DNA molecules coband with OC DNA in a CsCl-ethidium bromide gradient but band ahead of OC DNA in a sucrose gradient.

30.4.2.2. Agarose Gel Electrophoresis

A faster method to estimate copy number by using agarose gel electrophoresis (150) is as follows.

1. Grow cells until the OD_{600} reaches 0.8. Pellet the cells by centrifugation of 1.5 ml of the culture.
2. Wash the cells twice with 0.85% NaCl. Freeze the cells at −20°C for 1 h.
3. Thaw the cells, resuspend in 30 µl of a solution containing 0.01 M EDTA, 10 µg of boiled RNase per ml, and 2 mg of lysozyme per ml, and incubate for 15 min at 37°C.
4. Add 30 µl of 2% SDS, mix, and incubate for 10 min at 60°C.
5. Add 10 µl of loading solution (40% sucrose, 0.25% bromophenol blue).
6. Load in the well of a horizontal agarose gel (section 30.4.1.2), and after all air bubbles are removed (this can be achieved by holding a flamed loop over the well), fill the well with agarose to prevent the viscous lysate from floating out of the well. Run the electrophoresis.
7. Stain the gel with ethidium bromide, photograph, and scan with a densitometer.
8. Cut out the peak areas from the densitometry plots, and weigh them. Some densitometers have an integrator; if so, the integrated values can be used to calculate the copy number. In the formula shown above, counts per minute can be replaced by the weight or the integrated values of the peaks.

This method can also be performed by using a coresident plasmid of known copy number as an internal control, as follows.

1. Transform E. coli cells already carrying a control plasmid (such as the low-copy-number plasmid pSC101) with the plasmid of unknown copy number.
2. Prepare plasmid DNA from this E. coli strain by any of the methods described above.
3. Subject the plasmid DNA to agarose gel electrophoresis as described in section 30.4.1.2. Stain the gel with ethidium bromide, photograph, and scan the bands corresponding to the control and problem plasmids with a densitometer.
4. Cut out the peak areas from the densitometry plots, and weigh them. Some densitometers have an integrator; if so, the integrated values can be used to calculate copy number by using the copy number and molecular weight of the reference plasmid (rp) in the following formula.

Copy number = weight of plasmid peak/weight of rp peak/rp copy number × MWrp/MWp

If both the control and problem plasmids contain unique restriction sites, the following protocol can be applied.

1. Transform E. coli cells already carrying a control plasmid with the plasmid of unknown copy number.
2. Prepare plasmid DNA from this E. coli strain by using any of the methods described above. Digest the plasmids with a restriction endonuclease that cuts each plasmid once.
3. End label the plasmid DNAs by mixing 5 ml of 10× kinase buffer (0.5 M Tris-HCl [pH 9.51], 0.1 M MgCl₂, 0.05 M dithiothreitol, 50% glycerol), 10 U of T4 polynucleotide kinase, 150 µCi of [γ³²P]ATP, 9.8 µl of DNA (0.1 ng/µl in TE buffer), and 19.2 µl of distilled water. Incubate the reaction mixture for 60 min at 30°C.
4. Precipitate the end-labeled plasmid with 0.5 volume of 6 M ammonium acetate and 2 to 3 volumes of ethanol. Let stand at −20°C for 1 h. Collect the DNA by centrifugation in a microcentrifuge (16,000 × g) for 5 min. Wash the pellet three times with 70% ethanol, and dry under vacuum.
5. Resuspend the DNA in 20 µl of water, add 2 µl of gel-loading buffer (see section 30.4.1.2), mix, and separate the DNAs by agarose gel electrophoresis. Localize the bands by UV transillumination, and excise the gel fragments containing each plasmid. Place each gel fragment in a scintillation vial, add scintillation cocktail, and determine radioactivity in a liquid scintillation counter. The copy number of the plasmid is determined by normalizing to the reference plasmid. In this case the determination is independent of the molecular weights of the plasmids.

30.4.3. Base Composition

Another means of characterizing plasmid DNA is by its base composition. The percentage of guanosine and cytosine (G+C content) in the total plasmid DNA content is a characteristic of a particular molecule of DNA. For example, by comparing the G+C content of plasmids or chromosomes, using software packages such as Artemis (http://www.sanger.ac.uk/Software/Artemis/), it can be determined whether the plasmids are related to each other or to the chromosome of the bacterial host.

30.4.4. DNA "Homology"

The dissociation of double-stranded plasmid DNA followed by specific hybridization (reassociation, reannealing) with single-stranded DNA from a homologous or heterologous source is the basis of all homology studies. The following methods use single-strand-specific endonuclease and immobilized DNA for hybridization. The technique based on the use of single-strand-specific endonuclease is more sensitive and is thus the preferred method of quantitatively determining the homology among DNA molecules. Hybridization with immobilized DNA on nitrocellulose or nylon membranes is easier and faster, but the relationship between the degree of homology and the signal is not linear. Therefore, this last technique is convenient when DNA fragments with a high level of homology (usually more than 70%) must be detected.

DNA homology experiments can be carried out to study the characteristics of a certain DNA, in which case the method that uses single-strand-specific endonuclease is recommended, or as a tool to detect or identify DNA fragments, in which case the membrane-immobilized DNA method is chosen.

30.4.4.1. Single-Strand-Specific Endonuclease

Plasmid homo- and heteroduplexes can be analyzed with the single-strand-specific endonuclease, S1, of *Aspergillus oryzae*. The following procedure (33) permits an accurate and rapid determination of polynucleotide sequence relationships. It is particularly useful for surveys and for other investigations, which require a large number of DNA-DNA hybridization assays.

Purified, radiolabeled plasmid DNA is used as the reference probe and is hybridized with purified total cell DNA from plasmid-containing strains. DNA from a plasmid-free strain serves as a control. DNA-DNA reassociation reactions assayed by the S1 endonuclease method are performed by incubating approximately 5,000 to 10,000 cpm (typically less than 0.001 μg of DNA) of sheared, denatured, purified plasmid DNA with 150 μg of unfractionated, total, sheared, denatured bacterial DNA preparations from plasmid-containing and plasmid-free bacteria in a total volume of 1 ml of 0.42 M NaCl. The DNA mixtures are incubated at a temperature that depends on the G+C content of the plasmid DNA. The time of reassociation is such that essentially complete reassociation for the homologous reaction is achieved. The time required for a given plasmid to reassociate completely is a function of its molecular mass and the ionic concentration. The time of reassociation can be determined by calculating the $C_0t_{1/2}$. C_0 is the initial concentration of DNA; $t_{1/2}$ is the time taken for 50% reassociation of the DNA (at C_0 concentration) at a given temperature, salt concentration, and DNA fragment size. The empirical relationship (17) is as follows.

$$C_0t_{1/2} = \text{molecular weight of plasmid}/3 \times 10^7$$

The approximate time of incubation to get complete reassociation is equivalent to about 8 $C_0t_{1/2}$ in 0.42 M NaCl. Thus,

$$8 C_0t_{1/2} = C_0t_x$$

where C_0 is the initial concentration of plasmid DNA (which can be calculated or estimated from the plasmid copy number) and t_x is the time of reassociation.

Under optimum conditions, less than 10% of the labeled plasmid DNA incubated alone should reassociate with itself, while more than 85% of the labeled plasmid DNA should reassociate with its homologous unlabeled DNA. The reassociation of labeled plasmid DNA with DNA of a plasmid-free bacterial strain is also included in each experiment, and this value (about 10% or less) is subtracted from the values of all reactions.

Hybridization Step

1. Prepare the reassociation mixture to contain 150 μg of unlabeled sonicated total-cell DNA, 5,000 to 10,000 cpm of labeled sonicated plasmid DNA, and sufficient NaCl to bring the solution to 0.42 M in a total volume of 1.0 ml.

2. Denature the DNA by boiling the solution for 10 min. Then, place the DNA solution in ice.

3. Incubate the reassociation mixture at 55 to 70°C depending on the G+C content of the plasmid DNA.

[³H]thymine-labeled plasmid DNA is prepared by isolating CCC DNA by the Triton X-100 lysis technique (see section 30.1.1.1), or it can be labeled in vitro. The specific activity of the [³H]thymine plasmid DNA obtained by this method is usually about 10^6 cpm/pg. The plasmid DNA is subjected to sonic treatment to obtain fragments of an approximate molecular size of 0.5 kbp.

Total-cell DNA is prepared as described previously (71). This DNA is also degraded by sonic treatment to an approximate molecular size of 0.5 kbp.

S1 Endonuclease Reaction Step

1. Prepare a stock solution of S1 reaction mixture (0.125 $\times 10^{-3}$ M ZnSO$_4$, 0.087 M NaCl, 0.038 M sodium acetate buffer [pH 4.5], 25 μg of calf thymus DNA per ml). Store frozen at -20°C in 0.8-ml aliquots. The reaction mixture will then contain 0.8 ml of the S1 reaction mixture and 0.2 ml of the reassociation mixture. The final reagent concentrations are 0.1 $\times 10^{-3}$ M ZnSO$_4$, 0.150 M NaCl (this includes the NaCl added with 0.2 ml of the reassociation mixture), 0.03 M sodium acetate buffer (pH 4.5), and 20 μg of calf thymus DNA per ml. For each reassociation reaction, prepare two S1-treated and two untreated samples.

2. Add 187.5 U of S1 endonuclease to start the reaction. Incubate at 50°C for 20 min.

3. Stop the reaction by placing the tubes in an ice bath; add 50 μg of calf thymus or salmon sperm DNA per ml as carrier, and add 0.3 ml of cold 20% trichloroacetic acid. Collect the trichloroacetic acid precipitate by vacuum filtration on a membrane filter (type HA; Millipore, Bedford, MA). Dry the filters at 70°C, and determine the radioactivity by counting in a liquid scintillation spectrometer. Controls should include native-labeled plasmid DNA, which will detect any double-stranded nuclease activity in the S1 enzyme preparation; denatured and reassociated labeled DNA alone, which is a control for the amount of self-reassociation of the labeled DNA under the conditions of reassociation; and denatured labeled DNA, which is a control for the efficiency of the single-stranded activity of the S1 endonuclease preparations. Table 2 shows how to transform the raw radioactivity results into percent DNA homology. The results are expressed as the percentage of the untreated controls normalized to the values obtained for the homologous reaction.

30.4.4.2. DNA Immobilization and Hybridization

DNA can be immobilized on nitrocellulose or nylon membranes without any isolation or purification steps (colony hybridization) or after agarose gel electrophoresis (Southern blot hybridization).

Colony Transfer

1. Transfer cells from each colony to be tested with a sterile toothpick to two plates, one containing a regular solid medium (master plate) and the other on whose surface a membrane has been laid. Incubate.

2. After incubation, keep the master plate at 4°C and lyse the cells in situ by placing the membrane, colony side up, on top of a filter paper soaked with 0.5 N sodium hydroxide. Let stand for 10 min at room temperature.

3. Transfer the membrane to a filter paper soaked with 0.5 M Tris-HCl (pH 7.5) for 5 min.

4. Transfer the membrane to a filter paper soaked with a solution containing 0.5 M Tris-HCl (pH 7.5) and 1.5 M NaCl for 5 min.

5. Transfer the membrane to a filter paper soaked with 2× SSC solution (20× SSC solution is 175.3 g of NaCl and 88.2 g of sodium citrate per liter adjusted to pH 7.8).

6. Immobilize the DNA by either baking under vacuum for 2 h at 80°C or UV cross-linking. UV cross-linking is performed with commercial equipment. The entire cross-linking process takes only 30 to 60 s, in contrast to the 2 h needed for baking.

7. Remove all cell debris by wiping gently the surface of the filter with paper tissue (Kimwipes) soaked with 2× SSC.

TABLE 2 Determination of homology between heterologous plasmids by use of single-strand-specific endonuclease

Sample	cpm after S1 nuclease[a]		Raw radioactivity (%)	Avg %	Corrected and normalized homology[b]
	+	−			
1. Heat-denatured ^{3}H-labeled plasmid DNA	20	821	2.4	2.3	
	18	832	2.2		
2. Heat-denatured and reassociated ^{3}H-labeled plasmid DNA	50	850	6.0	6.0	
	53	890	6.0		
3. Heat-denatured reassociated mixture of ^{3}H-labeled plasmid DNA and total-cell DNA containing the same plasmid	790	850	9.29	94.8	100
	798	825	96.7		
4. Heat-denatured and reassociated mixture of ^{3}H-labeled plasmid DNA and total-cell DNA containing a heterologous plasmid	45	820	5.5	6.3	<0.3
	63	881	7.1		
5. Heat-denatured and reassociated mixture of ^{3}H-labeled plasmid DNA and total-cell DNA from a plasmidless derivative	60	860	7.0	7.0	<1
	62	880	7.0		

[a]The count-per-minute values were obtained by subtracting a background of 20 cpm. +, positive strand; −, negative strand.
[b]Values in this column were obtained by subtracting the value in sample 2 from the respective values in samples 3, 4, and 5 and then normalizing to the corrected value in sample 3.

8. Keep the membrane in a sealed plastic bag until hybridization. Any commercial bag sealer will be suitable for this purpose. Also, many laboratories store blots between tissue wiper sheets (Kimwipes) in a drawer for long periods.

This technique can be applied to *B. subtilis* and other *Bacillus* species without modification. However, other gram-positive bacteria such as *S. aureus* and streptococci are more difficult to lyse. In these cases it is recommended to soak the nitrocellulose paper in a lysostaphin solution prior to its use.

Southern Blot Transfer

1. Soak the agarose gel containing the DNA in 0.25 N HCl (partial depurination treatment), and gently rock for 15 min. Make sure that the gel is totally covered. Rinse twice with distilled water quickly. This step ensures uniform transfer of a broad range of DNA fragment sizes.

2. Denature the DNA by soaking the gel in a solution containing 1.5 M NaCl and 0.5 M NaOH with gentle rocking for 20 min. Rinse twice with distilled water quickly.

3. Neutralize by soaking the gel in a solution containing 1.5 M NaCl and 1 M Tris-HCl (pH 8.0) with gentle rocking for 20 min. Rinse twice with water.

4. Place the gel on top of a stack of glass plates covered with a strip of filter paper soaked with 6× SSC. The width of the filter paper should be such that it can accommodate the gel, and its length should overhang the plates. Soak a nitrocellulose or nylon membrane the size of the gel with 6× SSC, and place it on top of the gel, avoiding air bubbles between the gel and the membrane.

5. Place this device in a pan containing 6× SSC, and allow the overhanging filter to touch the solution.

6. Place on top of the membrane a piece of filter paper the size of the membrane. Avoid air bubbles between the membrane and the filter paper.

7. Cut a stack of paper towels just smaller than the filter, and place on the filter paper. Add a glass plate and a weight of about 500 g on top of the paper towels.

8. Allow transfer of the DNA for about 10 h.

9. Remove the membrane, immobilize the DNA by baking or UV cross-linking, and keep the membrane as described above under "Colony Transfer" (section 30.4.4.2).

The transfer of DNA from agarose gels to hybridization membranes can also be accomplished using commercially available systems such as the PosiBlot 30-30 pressure blotter (Stratagene), which increase the speed and efficiency of this step of the hybridization protocol. This is achieved by driving the buffer through the gel, which transfers the DNA onto the hybridization membrane, using positive pressure.

Probe Labeling

Two basic methods are available to radioactively label DNA, nick translation and random priming. The latter method provides probes of higher specific activity. Presently, commercial kits to label DNA by either of these two methods are available from many companies. A protocol for the random-priming method is as follows.

1. In a microcentrifuge tube containing 10 μl of oligolabeling buffer (OLB), mix 2 μl of 10 mg/ml gelatin or the same concentration of nuclease-free bovine serum albumin, 2 U of Klenow fragment of DNA polymerase I, and 5 μl of [α-^{32}P]dATP. To make OLB, mix solution A (1.25 M Tris-HCl [pH 8.0], 0.125 M MgCl$_2$, 0.25 M 2-mercaptoethanol, 0.0005 M deoxynucleoside triphosphates except dATP), solution B (2 M HEPES [pH 6.6]), and solution C (90 OD$_{260}$ units/ml random hexanucleotides dissolved in 0.001 M Tris-HCl [pH 7.0], 0.001 M EDTA) in a proportion of 100:250:150. Keep the solutions at −20°C. The hexanucleotides are commercially available from several companies.

2. Add 8 μl of the DNA to a separate microcentrifuge tube and denature by placing it in a heating block at 100°C for 5 min. Transfer the tube to an ice-water bath.

3. Transfer the DNA to the tube containing the reaction mixture, and incubate at room temperature for 2 h.

4. To determine the radioactive incorporation, spot 1 μl of the mixture on a filter paper. Dry, and determine the total counts per minute. Precipitate the labeled DNA by submerging the filter in cold 5% trichloroacetic acid, and let it stand for 10 min on ice. Pour off the liquid, and wash once with cold 5% trichloroacetic acid and once with cold ethanol. Dry, and determine the incorporated counts per minute. The yield is calculated by dividing the incorporated

counts per minute by the total counts per minute. Normally about 90% of labeled nucleotide is incorporated.

Other systems of labeling probes and detection of the hybrid products by different strategies that do not involve radioactivity have been described and can be obtained as kits from several companies. They are based on modified nucleotides that can be recognized by specific ligands such as the complex streptavidin-biotin or antigen-antibody. These complexes can be detected either by enzyme (phosphatase or peroxidase)-specific chromogenic substrates that produce a visible signal or by chemiluminescence. The reagent 4-methoxy-4(3phosphatephenyl)spiro[1,2-dioxetane-3,2′-adamantane] (LUMI-PHOS 530; Lumigen, Detroit, MI) has been used for this purpose and has a sensitivity comparable to that obtained using radiolabeled probes.

Hybridization

1. To the membrane contained in a plastic bag, add enough prehybridization solution to keep it wet. Prehybridize for 2 h at 37°C in prehybridization solution (6× SSC, 0.5% SDS, 50% formamide, 5× Denhardt's solution). 50× Denhardt's solution is 1% Ficoll, 1% polyvinylpirrolidone, and 1% bovine serum albumin. The stringency can be modified by changing the concentration of formamide. Each increase of 1% in the formamide concentration in the hybridization buffer lowers the melting temperature of duplex DNA by 0.7°C (93).

2. Add 10^6 cpm of the radiolabeled probe to a tube containing 200 µl of 100 µg/ml sonicated salmon sperm DNA. Denature by heating at 100°C for 5 min.

3. Hybridize at 37°C for about 3 h.

4. Remove the membrane from the bag, and wash it with a solution containing 2× SSC and 0.1% SDS at 65°C for 1 h, changing the washing solution three times, once every 20 min. Depending on the degree of homology between the DNAs, different temperatures ranging from 45 to 70°C should be tested.

5. Wash the membrane briefly with 2× SSC.

6. Dry the membrane, and expose it to an X-ray film. For nylon membranes, drying should be avoided if reuse with a different probe is intended. To remove the radioactive bound probe from used nylon membranes, wash them at 65°C for 30 min in a solution containing 50% formamide and 6× SSPE (20× SSPE is 3.6 M NaCl, 0.2 M sodium phosphate [pH 7.7], and 0.02 M EDTA) and rinse with 2× SSPE.

Alternatively the hybridization reactions can be conducted with hybridization incubators that are more convenient and practical than the plastic bag-water bath method described above. This type of equipment is available from a variety of commercial sources.

30.4.5. Replication and Maintenance Functions

All plasmids possess essential determinants that constitute the minimal replicon. These determinants, including self-perpetuation (replication and stable inheritance) in dividing bacteria, are known as replication and maintenance functions. Maintenance of a plasmid includes different functions, the main one being the partition system. Others include site-specific recombination systems and killing of cells that have lost the plasmid (12, 89, 132, 133, 136, 140). Plasmids with high copy number under conditions where they can freely diffuse within the cell may not require a partition system, because the probability that one cell gets all the copies and the other gets none during cell division is

very low. It should be said that most plasmids are stably maintained in a limited number of bacterial species. Some, such as RK2, have the characteristic of being stable in many species and are called broad-host-range plasmids (47). Others, such as ColE1, p15A, and pSC101, can exist in only a few species and are designated narrow-host-range plasmids (10). The host range of these three plasmids is limited to *E. coli* and closely related enterobacterial species. Besides these determinants, several large low-copy-number plasmids also encode conjugation functions that promote their transfer to bacterial cells of either the same or different species. In addition to inheritance functions, most plasmids encode a wide range of specialized functions, such as resistance to antimicrobial agents, acquisition of essential nutrients, degradation of organic compounds, and virulence factors. However, there are cryptic plasmids for which none of the latter set of functions have been assigned.

Replication and maintenance functions are studied by introducing the plasmid into the bacterial cell. This can be achieved by using transformation or transduction for naked and packaged DNA, respectively. Electroporation is a good alternative for those cases involving large plasmids that are more difficult to transform using standard transformation methods. Replication and maintenance can be monitored by taking advantage of one of the specialized functions. Isolation of cells expressing the specialized function, for example antibiotic resistance, is an indication that the plasmid replicates in that strain; however, the possibility of a genetic rearrangement must be discounted by isolating the original plasmid from transformants.

Maintenance can be studied as follows (105, 129).

1. Inoculate the appropriate liquid medium without antibiotic with cells grown on selective medium as described previously (140).

2. Take samples of the culture at different times until it reaches stationary phase. The times vary according to the bacterial species used; for *E. coli*, take samples every 3 h.

3. Dilute each sample to allow colony counting, and spread on plates with and without antibiotic.

4. After incubation, count the colonies grown under both conditions and estimate the frequency of plasmid loss by plotting the percentage of plasmid-containing cells versus the number of generations. The number of generations is calculated by dividing the time at which the sample was taken by the doubling time of the bacterial strain. A stable plasmid should display about the same percentage of plasmid-containing cells throughout the growth of the liquid culture. It is recommended to confirm the plasmid presence by screening some of the colonies growing in the presence of the antibiotic.

30.4.6. Visualization

DNA molecules have been visualized inside bacterial cells by using a number of techniques (9, 52, 53, 66, 98–100, 107, 109, 111, 131). Plasmids containing the *E. coli oriC* as well as derivatives of plasmid R1 and F, or plasmids containing the *Bacillus subtilis* partition system have been visualized in fixed cells by using fluorescence in situ hybridization (FISH) (99, 100, 153, 154). Plasmids P1, F, pUC19, and RK2 have been localized in *E. coli* by visualization in living cells (52, 107). Multiple *lacO* sequences were cloned into the plasmid molecules and a green fluorescent protein (GFP)-LacI fusion was expressed to tag the plasmid molecules, which were visualized by fluorescence microscopy. Below we briefly describe an example of each of these techniques, FISH and GFP-tagging.

30.4.6.1. FISH

The following description has been adapted from the methodology used by Niki and Hiraga (98, 99).

1. Grow *E. coli* cells in M9 minimal medium at 37°C and fix by mixing 1 volume of culture with 1 volume of fixing solution (methanol:acetic acid 3:1). Let stand 5 min at room temperature, centrifuge, and resuspend the cells in 1 ml of fixing solution. These treated cells are stable at 4°C for several months.

2. Spread 10 μl of fixed cells onto a poly-L-lysine-coated slide and let dry.

3. Incubate the slide in denaturing solution (70% formamide, 2× SSC) and heat at 75°C for 2 min before transferring the slide to a 70% ethanol solution at −35°C.

4. After 5 min transfer the slide to a 90% ethanol bath and let stand for 5 min.

5. Transfer the slide to a 100% ethanol bath, let stand for 5 min, remove slide, and let dry.

6. Cover the slide with freshly prepared lysozyme solution (25 mM Tris HCl, 10 mM EDTA, 50 mM glucose, 2 mg/ml lysozyme, pH 8) and keep for 10 min at room temperature.

7. Wash the slide by submerging it in PBS (80 mM Na_2HPO_4, 20 mM NaH_2PO_4, 100 mM NaCl, pH 7.5) for 5 min followed by rinsing in 70% and then 100% ethanol. Dry the slide.

8. Incubate the slide overnight at 42°C in the presence of hybridization buffer containing denatured probe labeled with Cy3-dCTP.

9. Wash the slide with a solution containing 50% formamide and 2× SSC at 37°C for 15 min. Wash with a series of SSC solutions (2×, 1×, 4×, 2×) 5 min each at room temperature.

10. Wash the slide in PBS containing 10 mM EDTA for 5 min and then in water for 5 min.

11. Dry the slide and apply mounting medium (90% glycerol, 1 mg/ml *p*-phenylenediamine dihydrochloride, 0.15 mg/ml DAPI [4′,6′-diamidino-2-phenylindole]).

12. Record image with a cooled charge-coupled-device camera using a phase-contrast and fluorescence microscope. Niki and Hiraga used a charge-coupled device camera C5810-01 (Hamamatsu, Japan) and a Nikon microscopy system (98, 99).

30.4.6.2. GFP-LacI Fusion

The methodology applied by Pogliano et al. (107) has been adapted for the following description. The plasmid to be visualized included a 10-kbp *lacO* array, and another plasmid harbored the GFP-LacI fusion under the control of the arabinose promoter.

1. Culture cells overnight at 23 to 30°C, dilute 1/100 on fresh medium, and let grow until the OD_{600} reaches 0.1.

2. Induce GFP-LacI by adding 0.2% arabinose. Culture for 20 min and add 0.2% glucose to repress the promoter and avoid overexpression.

3. Culture until OD_{600} is between 0.4 and 0.6 (it should take about 2.5 to 3 h).

4. Stain the cells with 1 μg/ml FM 4-64 (Chroma Technology Corp, Quebec City, Canada) and 5 μg/ml DAPI (Sigma).

5. Fix to a poly-L-lysine-coated slide and obtain images. Pogliano et al. (107) captured images from six optical sections spaced 0.15 μm apart with an Applied Precision Deconvolution microscope.

30.4.7. Incompatibility

Incompatibility is the inability of two plasmid species that have been introduced or transferred to a single bacterial clone to be stably maintained in the same cell line in the absence of selective pressure for both plasmids (31, 101). Incompatibility is a consequence of relatedness between the plasmids; i.e., incompatible plasmids share some or all functions involved in replication or maintenance. Plasmids have been classified by being assigned to incompatibility groups (plasmids in the same group are incompatible).

A simple method to determine incompatibility between two plasmids, provided they have different markers, is as follows.

1. Transform a strain carrying plasmid A (resident) with plasmid B (incoming). This transfer can also be accomplished by conjugation. Select transformants or transconjugants with antibiotics selective for both plasmids. It is convenient to use a *recA* bacterial host to prevent recombination between the two plasmids.

2. Inoculate broth containing both antibiotics with the strain carrying both plasmids, and allow growth up to the stationary phase.

3. Dilute the culture 10^6-fold into antibiotic-free broth, and incubate until the stationary phase. At this point, plate a suitable dilution on antibiotic-free agar medium to obtain isolated colonies.

4. Toothpick the colonies to two agar plates, each containing one antibiotic, to determine the frequency of cells carrying one or the other of the two plasmids. Growth of colonies in the two plates indicates that the two plasmids are compatible.

Incompatibility testing presents some potential problems: the lack of resistance markers in one or both plasmids or the presence of the same marker in the plasmids, the inhibition of entry of the incoming plasmid (entry or surface exclusion), and the presence of more than one incompatibility function on either plasmid. To overcome these difficulties, a series of miniplasmids (representing six major plasmid incompatibility groups) containing incompatibility functions of reference plasmids and the *gal* operon was designed by Davey et al. (37). The miniplasmids were constructed by cloning plasmid DNA fragments that do not include surface exclusion functions. The use of galactose utilization permits one to determine incompatibility by monitoring the loss of the miniplasmid, revealed by the appearance of Gal⁻ segregants on indicator plates.

Gal⁻ indicator plates are prepared by adding 1% galactose and 25 μg of 2,3,5-triphenyltetrazolium chloride per ml to sterile L-agar medium (141). Colonies that ferment galactose are white, and those that do not ferment the sugar are red.

Despite these improvements, incompatibility tests can still give misleading results because of the presence of more than one replicon in a single plasmid or the presence of some replicons that, as a result of their mechanism of control of replication by using antisense RNAs, either can be detected as compatible even though they differ in just a few bases or can show cross-incompatibility (31).

To address these problems, a classification scheme based on replicon typing was developed for plasmids present in gram-negative bacteria; it involved the use of specific DNA probes to detect the presence of a known basic replicon by DNA-DNA hybridization (31). Before performing replicon typing, the plasmid profile of the strain to be tested should be determined. If only a single plasmid replicon is present, typing can be carried out directly by colony hybridization. If two or more plasmids are present, they must be separated by gel electrophoresis and transferred to nitrocellulose or nylon membrane before the hybridization reaction. The protocol can be summarized as follows.

1. Immobilize the DNA in nitrocellulose or nylon, obtained by colony lysis or plasmid purification and gel electrophoresis as described in sections 30.2 and 30.4.

2. Digest the hybrid plasmid carrying the probe with the appropriate restriction enzyme. Separate the fragments by agarose gel electrophoresis, extract the DNA band to be used as probe, and label the DNA as described in section 30.4.4.2.

3. Hybridize the immobilized DNA with the labeled probe, wash, and detect the radioactive hybrids by using the conditions described in section 30.4.4.2. Plasmids sharing highly related replicons show a positive signal on X-ray films.

30.4.8. Curing

Elimination ("curing") of a plasmid from a bacterial culture is the best test to substantiate the relationship between a genetic trait and carriage of that specific plasmid by the culture. A phenotype linked to the presence of the plasmid will not be expressed in cured cultures. Reintroduction of the plasmid or a recombinant clone derivative carrying the specific genetic trait must be accompanied by regaining of the phenotype.

Curing of plasmids was first reported more than 25 years ago (74). The artificial elimination of F was reported by Hirota more than 35 years ago, before the nature of the "F factor" was even understood (68). Since then, various protocols involving chemical and physical agents have been developed to cure plasmids. Depending on the nature of the bacterial host and/or the plasmid, some methods work better than others. Protocols for curing plasmids consist basically of exposure of a culture to different concentrations of the chemical agent or to higher than normal temperature, followed by plating and analysis of the presence of a trait carried by the plasmid. Colonies that lose the trait are then analyzed for their plasmid content to confirm loss of the plasmid. Some plasmid curing protocols are described below.

30.4.8.1. Elevated Incubation Temperature

The elevated incubation temperature protocol was used to cure *V. anguillarum* (139), which grows normally at 30°C.

1. Inoculate 1 ml of Trypticase soy broth supplemented with 1% NaCl with 0.01 ml of an overnight culture of *V. anguillarum*. Incubate at 37°C for 16 h.

2. Use 0.01 ml of this culture as an inoculum for 1 ml of the same broth, and incubate at 37°C for another 16 h. This procedure can be repeated several times.

3. Spread the cells on Trypticase soy agar plates supplemented with 1% NaCl, and incubate at 30°C.

4. Analyze colonies for the presence of the plasmid. With this protocol, about 3% of the colonies are cured.

Basically the same protocol was used to cure plasmids from *Acinetobacter baumannii* and the *E. coli* formaldehyde-resistant strain VU3695, but L medium and incubation at 42°C were used in these two cases.

30.4.8.2. Intercalating Dyes

1. Inoculate 2 ml of L broth in tubes with 20 μl of an overnight inoculum. Add to each tube increasing concentrations of an intercalating dye (e.g., ethidium bromide or acridine orange) ranging from 10 to 300 μg/ml.

2. Incubate overnight, and select the culture with the highest concentration of the acridine dye at which growth occurred.

3. Dilute the culture, plate on L agar, and incubate to obtain isolated colonies.

4. Analyze the colonies for plasmid content.

30.4.8.3. Sodium Dodecyl Sulfate

1. Inoculate 2 ml of broth containing 10% SDS with 20 μl of an overnight inoculum.

2. Incubate overnight at 37°C.

3. Plate after dilutions to obtain isolated colonies. Incubate the plates at 37°C overnight.

4. Analyze the colonies for the presence of the plasmid.

30.4.8.4. Electroporation

Electroporation was used originally for the transformation of bacteria with DNA (95). This method was used to cure a pBR322 derivative from *E. coli* DH1 (63) as follows.

1. Incubate the cells in 10 ml of L broth.

2. Pellet the cells, and wash three times with distilled water.

3. Let the cells stand in ice for 5 min.

4. Subject the cells to one electrical pulse (Gene Pulser [Bio-Rad]; pulse at 2.5 kV and 25 μF, time constant ca. 4.5 ms).

5. Add 1 ml of SOC (2% tryptone, 0.5% yeast extract, 0.01 M NaCl, 0.0025 M KCl, 0.01 M MgCl$_2$, 0.01 M MgSO$_4$, 0.02 M glucose) immediately after the pulse, and incubate for 20 min at 37°C.

6. Make dilutions in SOC to plate the cells on L agar to obtain isolated colonies.

7. Cells can be analyzed by toothpicking to plates with and without ampicillin (100 μg/ml). Colonies that are sensitive to ampicillin are cured. In this particular case, related to the presence of a plasmid encoding resistance to ampicillin, the frequency of curing was 80 to 90%.

30.4.8.5. Kick-Out

A plasmid can be eliminated by transformation with an incompatible plasmid that encodes resistance to a different antibiotic. Selection in the presence of antibiotic for which the incoming plasmid codes for resistance results in growth of cells that lost the resident plasmid. This method has been used to eliminate a resident plasmid in the mutagenesis by marker exchange technique (138).

30.5. NOVEL USE OF PLASMID DNA

30.5.1. Plasmid DNA from a Pharmaceutical Perspective

Plasmid DNA has recently been proposed as a drug substance in response to the needs created by the advent of gene therapy and polynucleotide-based vaccines (35, 94). These novel technologies led to the necessity of developing new pharmaceuticals that are biological macromolecules rather than small molecules. Plasmid DNA has been used in experiments as a means to transfer genetic material, using a variety of diseases and cell systems such as cystic fibrosis (19), Duchenne muscular dystrophy (18), solid tumors (26, 40, 97), diabetes (1), fibroblasts (44), and others. In the field of DNA vaccination, plasmid DNA has also been utilized recently in a wide variety of systems (77, 78, 80, 84, 119, 123, 147, 152). Although the descriptions of detailed protocols to obtain plasmid DNA formulations that can be used in gene therapy or vaccination are beyond the scope of this article, we briefly describe below the current trends in this new plasmid DNA field.

Naked DNA formulations have been explored for skin and muscle injection with a degree of success (25), but the most effective plasmid DNA-based pharmaceuticals will probably be formulated with some kind of delivery vehicle or special preparation (46, 77, 78, 94, 118, 130, 144). Direct injection of naked plasmid DNA resulted in lower levels of gene expression than what was obtained with liposomal vectors. Naked plasmid DNA must be administered by direct injection into tissues because systemic administration is prevented by the presence of serum nucleases. Plasmid DNA has been used in the form of plasmid-liposome complexes as gene transfer vectors which possess several advantages such as their ability to transfer expression cassettes, their inability to replicate and become infectious (unlike virus vectors), or their reduced capability to evoke inflammatory or immune responses (35, 45). On the other hand, they present some disadvantages such as the need of high numbers of plasmid molecules to achieve a successful gene transfer (35). Plasmid DNA-liposome complexes have been obtained using a variety of compositions but typically include cationic lipids which spontaneously interact with the negatively charged DNA (51). "Classical" liposomes (negatively charged) showed problems with the DNA encapsulation and formation of empty liposomes. Another advantage of cationic liposomes is their interaction with negatively charged cell membranes resulting in delivery of the DNA directly across the plasma membrane. A wide variety of cationic lipids have been used to generate plasmid DNA-liposome complexes. In order to facilitate membrane fusion, a neutral lipid such as dioleoylphosphatidyl-ethanolamine is commonly added (97, 160).

A promising delivery route for plasmid DNA is the lung, via inhalation techniques similar to those used for asthma medicines (104, 121). To be able to utilize this route of administration, as it has been done for proteins (142; R. Sloan, W. Hollowood, W. Ashraf, G. Humphreys, and P. York, presented at the 5th Meeting on Supercritical Fluids, Nice, France, 1998), plasmid DNA-loaded particles in a dry powder could represent a viable formulation (142). Although several methods have been described for powder production, it has been recently postulated that supercritical fluid technology, which facilitates controlled particle size formation at near room temperature (143), may be the methodology of choice to generate plasmid DNA-loaded pharmaceutical powder (142). Tservistas et al. (142) determined conditions for the "solution enhanced dispersion by supercritical fluid" technique to generate powder formulations using the plasmid pSVβ with mannitol as excipient. These assays led to the definition of conditions in which the formulation contained up to 80% supercoiled plasmid DNA capable of transfection. Further development of this technology may lead to the preparation of active pharmaceutical formulations involving plasmid DNA. Other possible formulations that may be used to deliver plasmid DNA involve DNA-peptide and DNA-synthetic polymer complexes (15, 54, 115).

30.6. REFERENCES

1. **Abai, A. M., P. M. Hobart, and K. M. Barnhart.** 1999. Insulin delivery with plasmid DNA. *Hum. Gene Ther.* **10**:2637–2649.

2. **Actis, L. A., M. E. Tolmasky, and J. H. Crosa.** 1999. Bacterial plasmids: replication of extrachromosomal genetic elements encoding resistance to antimicrobial compounds. *Front. Biosci.* **4**:D43–D62.

3. **Actis, L. A., M. E. Tolmasky, L. Crosa, and J. H. Crosa.** 1993. Effect of iron-limiting conditions on growth of clinical isolates of *Acinetobacter baumannii. J. Clin. Microbiol.* **31**:2812–2815.

4. **Alter, D. C., and K. N. Subramanian.** 1989. A one step, quick step, mini prep. *BioTechniques* **7**:456–458.

5. **Anderson, D. G., and L. L. McKay.** 1983. Simple and rapid method for isolating large plasmid DNA from lactic streptococci. *Appl. Environ. Microbiol.* **46**:549–552.

6. **Banfalvi, Z., E. Kondorosi, and A. Kondorosi.** 1985. *Rhizobium meliloti* carries two megaplasmids. *Plasmid* **13**:129–138.

7. **Barbour, A. G., and C. F. Garon.** 1987. Linear plasmids of the bacterium *Borrelia burgdorferi* have covalently closed ends. *Science* **237**:409–411.

8. **Barnett, M. J., R. F. Fisher, T. Jones, C. Komp, A. P. Abola, F. Barloy-Hubler, L. Bowser, D. Capela, F. Galibert, J. Gouzy, M. Gurjal, A. Hong, L. Huizar, R. W. Hyman, D. Kahn, M. L. Kahn, S. Kalman, D. H. Keating, C. Palm, M. C. Peck, R. Surzycki, D. H. Wells, K. C. Yeh, R. W. Davis, N. A. Federspiel, and S. R. Long.** 2001. Nucleotide sequence and predicted functions of the entire *Sinorhizobium meliloti* pSymA megaplasmids. *Proc. Natl. Acad. Sci. USA* **98**:9883–9888.

9. **Barre, F. X., M. Aroyo, S. D. Colloms, A. Helfrich, F. Cornet, and D. J. Sherratt.** 2000. FtsK functions in the processing of a Holliday junction intermediate during bacterial chromosome segregation. *Genes Dev.* **14**:2976–2988.

10. **Barth, P. L., L. Tobin, and G. Sharpe.** 1981. Development of broad host-range plasmid vectors, p. 439–448. *In* S. B. Levy, R. C. Clowes, and E. L. Koenig (ed.), *Molecular Biology, Pathogenicity, and Ecology of Bacterial Plasmids.* Plenum Press, New York, NY.

11. **Berfer, A., and A. Kimmel.** 1987. *Guide to Molecular Cloning Techniques*, vol. 152. Academic Press, San Diego, CA.

12. **Bhattacharyya, A., and D. H. Figurski.** 2001. A small protein-protein interaction domain common to KlcB and global regulators KorA and TrbA of promiscuous IncP plasmids. *J. Mol. Biol.* **310**:51–67.

13. **Birnboim, H. C., and J. Doly.** 1979. A rapid alkaline extraction procedure for screening recombinant plasmid DNA. *Nucleic Acids Res.* **7**:1513–1523.

14. **Boyd, E. F., and D. L. Hartl.** 1998. *Salmonella* virulence plasmid. Modular acquisition of the spv virulence region by an F-plasmid in *Salmonella enterica* subspecies I and insertion into the chromosome of subspecies II, IIIa, IV and VII isolates. *Genetics* **149**:1183–1190.

15. **Brown, M. D., A. G. Schatzlein, and I. F. Uchegbu.** 2001. Gene delivery with synthetic (non viral) carriers. *Int. J. Pharm.* **229**:1–21.

16. **Brunk, C. F., C. K. Jones, and T. W. James.** 1979. Assay for nanogram quantities of DNA in cellular homogenates. *Anal. Biochem.* **92**:497–500.

17. **Burkardt, B., D. Schillik, and A. Puhler.** 1987. Physical characterization of *Rhizobium meliloti* megaplasmids. *Plasmid* **17**:13–25.

18. **Campeau, P., P. Chapdelaine, S. Seigneurin-Venin, B. Massie, and J. P. Tremblay.** 2001. Transfection of large plasmids in primary human myoblasts. *Gene Ther.* **8**:1387–1394.

19. **Caplen, N., E. Alton, P. Middelton, J. Dorin, B. Stevenson, X. Gao, S. Durham, P. Jeffrey, M. Hodson, C. Coutelle, L. Huang, D. Porteus, R. Williamson, and**

D. Geddes. 1995. A phase trial of liposome mediated CFTR gene transfer to the nasal epithelia of patients with cystic fibrosis. *Nat. Med.* **1:**39–46.

20. Carle, G. F., and M. V. Olson. 1985. An electrophoretic karyotype for yeast. *Proc. Natl. Acad. Sci. USA* **82:**3756–3760.

21. Chakrabarty, A. M. 1992. Microorganisms having multiple compatible degradative energy-generating plasmids and preparation thereof. *Biotechnology* **24:**535–545.

22. Chakrabarty, A. M. 1976. Plasmids in *Pseudomonas. Annu. Rev. Genet.* **10:**7–30.

23. Chassy, B. M., and A. Giuffrida. 1980. Method for the lysis of gram-positive, asporogenous bacteria with lysozyme. *Appl. Environ. Microbiol.* **39:**153–158.

24. Chattoraj, D. K. 2000. Control of plasmid DNA replication by iterons: no longer paradoxical. *Mol. Microbiol.* **37:**467–476.

25. Choate, K. A., and P. A. Khavari. 1997. Direct cutaneous gene delivery in a human genetic skin disease. *Hum. Gene Ther.* **8:**1659–1665.

26. Clark, P. R., A. T. Stopeck, M. Ferrari, S. E. Parker, and E. M. Hersh. 2000. Studies of direct intratumoral gene transfer using cationic lipid-complexed plasmid DNA. *Cancer Gene Ther.* **7:**853–860.

27. Clayton, D. A., R. W. Davis, and J. Vinograd. 1970. Homology and structural relationships between the dimeric and monomeric circular forms of mitochondrial DNA from human leukemic leukocytes. *J. Mol. Biol.* **47:**137–153.

28. Clewell, D. B., and D. R. Helinski. 1969. Supercoiled circular DNA-protein complex in *Escherichia coli*: purification and induced conversion to an open circular DNA form. *Proc. Natl. Acad. Sci. USA* **62:**1159–1166.

29. Clowes, R. C. 1972. Molecular structure of bacterial plasmids. *Bacteriol. Rev.* **36:**361–405.

30. Courvalin, P., S. Goussard, and C. Grillot-Courvalin. 1995. Gene transfer from bacteria to mammalian cells. *C. R. Acad. Sci. III* **318:**1207–1212.

31. Couturier, M., F. Bex, P. L. Bergquist, and W. K. Maas. 1988. Identification and classification of bacterial plasmids. *Microbiol. Rev.* **52:**375–395.

32. Crosa, J. H. 1997. Signal transduction and transcriptional and posttranscriptional control of iron-regulated genes in bacteria. *Microbiol. Mol. Biol. Rev.* **61:**319–336.

33. Crosa, J. H., D. J. Brenner, and S. Falkow. 1973. Use of a single-strand specific nuclease for analysis of bacterial and plasmid deoxyribonucleic acid homo- and heteroduplexes. *J. Bacteriol.* **115:**904–911.

34. Crosa, J. H., L. K. Luttropp, F. Heffron, and S. Falkow. 1975. Two replication initiation sites on R-plasmid DNA. *Mol. Gen. Genet.* **140:**39–50.

35. Crystal, R. 1995. Transfer of genes to humans: early lessons and obstacles to success. *Science* **270:**404.

36. Currier, T. C., and E. W. Nester. 1976. Isolation of covalently closed circular DNA of high molecular weight from bacteria. *Anal. Biochem.* **76:**431–441.

37. Davey, R. B., P. I. Bird, S. M. Nikoletti, J. Praszkier, and J. Pittard. 1984. The use of mini-Gal plasmids for rapid incompatibility grouping of conjugative R plasmids. *Plasmid* **11:**234–242.

38. del Solar, G., J. C. Alonso, M. Espinosa, and R. Diaz-Orejas. 1996. Broad-host-range plasmid replication: an open question. *Mol. Microbiol.* **21:**661–666.

39. del Solar, G., and M. Espinosa. 2000. Plasmid copy number control: an ever-growing story. *Mol. Microbiol.* **37:**492–500.

40. Densmore, C. L., E. S. Kleinerman, A. Gautam, S. F. Jia, B. Xu, L. L. Worth, J. C. Waldrep, Y. K. Fung, A. T'Ang, and V. Knight. 2001. Growth suppression of established human osteosarcoma lung metastases in mice by aerosol gene therapy with PEI-p53 complexes. *Cancer Gene Ther.* **8:**619–627.

41. Dotto, G. P., V. Enea, and N. D. Zinder. 1981. Functional analysis of bacteriophage f1 intergenic region. *Virology* **114:**463–473.

42. Dunny, G. M., D. A. Krug, C. L. Pan, and R. A. Ledford. 1988. Identification of cell wall antigens associated with a large conjugative plasmid encoding phage resistance and lactose fermentation ability in lactic streptococci. *Biochimie* **70:**443–450.

43. Eckhardt, T. 1978. A rapid method for the identification of plasmid desoxyribonucleic acid in bacteria. *Plasmid* **1:**584–588.

44. Estruch, E. J., S. L. Hart, C. Kinnon, and B. G. Winchester. 2001. Non-viral, integrin-mediated gene transfer into fibroblasts from patients with lysosomal storage diseases. *J. Gene Med.* **3:**488–497.

45. Farhood, H., X. Gao, K. Son, Y. Y. Yang, J. S. Lazo, L. Huang, J. Barsoum, R. Bottega, and R. M. Epand. 1994. Cationic liposomes for direct gene transfer in therapy of cancer and other diseases. *Ann. N. Y. Acad. Sci.* **716:**23–34, 34–35.

46. Felgner, J. H., R. Kumar, C. N. Sridhar, C. J. Wheeler, Y. J. Tsai, R. Border, P. Ramsey, M. Martin, and P. L. Felgner. 1994. Enhanced gene delivery and mechanism studies with a novel series of cationic lipid formulations. *J. Biol. Chem.* **269:**2550–2561.

47. Figurski, D. H., and D. R. Helinski. 1979. Replication of an origin-containing derivative of plasmid RK2 dependent on a plasmid function provided in trans. *Proc. Natl. Acad. Sci. USA* **76:**1648–1652.

48. Finan, T. M., S. Weidner, K. Wong, J. Buhrmester, P. Chain, F. J. Vorholter, I. Hernandez-Lucas, A. Becker, A. Cowie, J. Gouzy, B. Golding, and A. Puhler. 2001. The complete sequence of the 1,683-kb pSymB megaplasmid from the N2-fixing endosymbiont *Sinorhizobium meliloti. Proc. Natl. Acad. Sci. USA* **98:**9889–9894.

49. Finkelstein, M., and T. H. Rownd. 1978. A rapid method for extracting DNA from agarose gels. *Plasmid* **1:**557–562.

50. Gerdes, K., J. Moller-Jensen, and R. Bugge Jensen. 2000. Plasmid and chromosome partitioning: surprises from phylogeny. *Mol. Microbiol.* **37:**455–466.

51. Gershon, H., R. Ghirlando, S. B. Guttman, and A. Minsky. 1993. Mode of formation and structural features of DNA-cationic liposome complexes used for transfection. *Biochemistry* **32:**7143–7151.

52. Gordon, G. S., D. Sitnikov, C. D. Webb, A. Teleman, A. Straight, R. Losick, A. W. Murray, and A. Wright. 1997. Chromosome and low copy plasmid segregation in *E. coli*: visual evidence for distinct mechanisms. *Cell* **90:**1113–1121.

53. Gordon, G. S., and A. Wright. 2000. DNA segregation in bacteria. *Annu. Rev. Microbiol.* **54:**681–708.

54. Gottschalk, S., J. T. Sparrow, J. Hauer, M. P. Mims, F. E. Leland, S. L. Woo, and L. C. Smith. 1996. A novel DNA-peptide complex for efficient gene transfer and expression in mammalian cells. *Gene Ther.* **3:**48–57.

55. Grillot-Courvalin, C., S. Goussard, and P. Courvalin. 1999. Bacteria as gene delivery vectors for mammalian cells. *Curr. Opin. Biotechnol.* **10:**477–481.

56. Gruss, A., and S. D. Ehrlich. 1989. The family of highly interrelated single-stranded deoxyribonucleic acid plasmids. *Microbiol. Rev.* **53:**231–241.

57. Guerry, P., D. J. LeBlanc, and S. Falkow. 1973. General method for the isolation of plasmid deoxyribonucleic acid. *J. Bacteriol.* **116:**1064–1066.

58. Hansen, J. B., and R. H. Olsen. 1978. Isolation of large bacterial plasmids and characterization of the P2 incompatibility group plasmids pMG1 and pMG5. *J. Bacteriol.* **135:**227–238.

59. **Hardy, D. A., J. I. Bell, E. O. Long, T. Lindsten, and H. O. McDevitt.** 1986. Mapping of the class II region of the human major histocompatibility complex by pulsed-field gel electrophoresis. *Nature* **323:**453–455.

60. **Hardy, K. G. (ed.).** 1993. *Plasmids: a Practical Approach*, 2nd ed. IRL Press at Oxford University Press, Oxford, United Kingdom.

61. **Haro, M. A., and V. de Lorenzo.** 2001. Metabolic engineering of bacteria for environmental applications: construction of *Pseudomonas* strains for biodegradation of 2-chlorotoluene. *J. Biotechnol.* **85:**103–113.

62. **Hartstein, A. I., A. L. Rashad, J. M. Liebler, L. A. Actis, J. Freeman, J. W. Rourke, Jr., T. B. Stibolt, M. E. Tolmasky, G. R. Ellis, and J. H. Crosa.** 1988. Multiple intensive care unit outbreak of *Acinetobacter calcoaceticus* subspecies *anitratus* respiratory infection and colonization associated with contaminated, reusable ventilator circuits and resuscitation bags. *Am. J. Med.* **85:**624–631.

63. **Heery, D. M., R. Powell, F. Gannon, and L. K. Dunican.** 1989. Curing of a plasmid from *E. coli* using high-voltage electroporation. *Nucleic Acids Res.* **17:**10131.

64. **Helinski, D., A. Toukdarian, and R. Novick.** 1996. Replication control and other stable maintenance mechanisms of plasmids, p. 2295–2234. *In* F. Neidhart, R. Curtis III, J. Ingraham, C. Lin, K. Low, B. Magasanik, W. Reznikoff, M. Riley, M. Schaechter, and H. Umbarger (ed.), Escherichia coli *and* Salmonella, 2nd ed., vol. 2. ASM Press, Washington, DC.

65. **Helinski, D. R.** 1973. Plasmid determined resistance to antibiotics: molecular properties of R factors. *Annu. Rev. Microbiol.* **27:**437–470.

66. **Hiraga, S.** 2000. Dynamic localization of bacterial and plasmid chromosomes. *Annu. Rev. Genet.* **34:**21–59.

67. **Hirochika, H., and K. Sakaguchi.** 1982. Analysis of linear plasmids isolated from *Streptomyces*: association of protein with the ends of the plasmid DNA. *Plasmid* **7:**59–65.

68. **Hirota, Y.** 1956. Artificial elimination of the F factor in *Bact. coli* K-12. *Nature* (London) **178:**92.

69. **Hirt, B.** 1967. Selective extraction of polyoma DNA from infected mouse cell cultures. *J. Mol. Biol.* **26:**365–369.

70. **Holmes, D. S., and M. Quigley.** 1981. A rapid boiling method for the preparation of bacterial plasmids. *Anal. Biochem.* **114:**193–197.

71. **Hull, R. A., R. E. Gill, P. Hsu, B. H. Minshew, and S. Falkow.** 1981. Construction and expression of recombinant plasmids encoding type 1 or D-mannose-resistant pili from a urinary tract infection *Escherichia coli* isolate. *Infect. Immun.* **33:**933–938.

72. **Ingmer, H., C. Miller, and S. N. Cohen.** 2001. The RepA protein of plasmid pSC101 controls *Escherichia coli* cell division through the SOS response. *Mol. Microbiol.* **42:**519–526.

73. **Kado, C. I., and S. T. Liu.** 1981. Rapid procedure for detection and isolation of large and small plasmids. *J. Bacteriol.* **145:**1365–1373.

74. **Kawakami, M., and O. Landman.** 1965. Experiments concerning the curing and intracellular site of episomes. *Biochem. Biophys. Res. Commun.* **18:**716–724.

75. **Khan, S. A.** 2000. Plasmid rolling-circle replication: recent developments. *Mol. Microbiol.* **37:**477–484.

76. **Kim, A. Y., A. A. Vertes, and H. P. Blaschek.** 1990. Isolation of a single-stranded plasmid from *Clostridium acetobutylicum* NCIB 6444. *Appl. Environ. Microbiol.* **56:**1725–1728.

77. **Klavinskis, L. S., C. Barnfield, L. Gao, and S. Parker.** 1999. Intranasal immunization with plasmid DNA-lipid complexes elicits mucosal immunity in the female genital and rectal tracts. *J. Immunol.* **162:**254–262.

78. **Klavinskis, L. S., L. Gao, C. Barnfield, T. Lehner, and S. Parker.** 1997. Mucosal immunization with DNA-liposome complexes. *Vaccine* **15:**818–820.

79. **Krum, J. G., and S. A. Ensign.** 2001. Evidence that a linear megaplasmid encodes enzymes of aliphatic alkene and epoxide metabolism and coenzyme M (2-mercaptoethanesulfonate) biosynthesis in *Xanthobacter* strain Py2. *J. Bacteriol.* **183:**2172–2177.

80. **Kumar, M., A. K. Behera, J. Hu, R. F. Lockey, and S. S. Mohapatra.** 2001. IFN-gamma and IL-12 plasmid DNAs as vaccine adjuvant in a murine model of grass allergy. *J. Allergy Clin. Immunol.* **108:**402–408.

81. **Kupersztoch, Y. M., and D. R. Helinski.** 1973. A catenated DNA molecule as an intermediate in the replication of the resistance transfer factor R6K in *Escherichia coli*. *Biochem. Biophys. Res. Commun.* **54:**1451–1459.

82. **Kupersztoch-Portnoy, Y. M., M. A. Lovett, and D. R. Helinski.** 1974. Strand and site specificity of the relaxation event for the relaxation complex of the antibiotic resistance plasmid R6K. *Biochemistry* **13:**5484–5490.

83. **Lai, E., B. W. Birren, S. M. Clark, M. I. Simon, and L. Hood.** 1989. Pulsed field gel electrophoresis. *BioTechniques* **7:**34–42.

84. **Leifert, J. A., J. A. Lindencrona, J. Charo, and J. L. Whitton.** 2001. Enhancing T cell activation and antiviral protection by introducing the HIV-1 protein transduction domain into a DNA vaccine. *Hum. Gene Ther.* **12:**1881–1892.

85. **Lev, Z.** 1987. A procedure for large-scale isolation of RNA-free plasmid and phage DNA without the use of RNase. *Anal. Biochem.* **160:**332–336.

86. **Lev, Z., and O. Seveg.** 1986. The RNA transcripts of *Drosophila melanogaster* src are differentially regulated during development. *Biochim. Biophys. Acta* **867:**144–151.

87. **Levy, M. S., P. Lotfian, R. O'Kennedy, M. Y. Lo-Yim, and P. A. Shamlou.** 2000. Quantitation of supercoiled circular content in plasmid DNA solutions using a fluorescence-based method. *Nucleic Acids Res.* **28:**E57.

88. **Lyon, B. R., J. W. May, and R. A. Skurray.** 1983. Analysis of plasmids in nosocomial strains of multiple-antibiotic-resistant *Staphylococcus aureus*. *Antimicrob. Agents Chemother.* **23:**817–826.

89. **Maas, R.** 2001. Change of plasmid DNA structure, hypermethylation, and Lon-proteolysis as steps in a replicative cascade. *Cell* **105:**945–955.

90. **Macrina, F. L., D. J. Kopecko, K. R. Jones, D. J. Ayers, and S. M. McCowen.** 1978. A multiple plasmid-containing *Escherichia coli* strain: convenient source of size reference plasmid molecules. *Plasmid* **1:**417–420.

91. **Manis, J. J., and H. J. Whitfield.** 1977. Physical characterization of a plasmid cointegrate containing an F'his gnd element and the *Salmonella typhimurium* LT2 cryptic plasmid. *J. Bacteriol.* **129:**1601–1606.

92. **McClung, J. K., and R. A. Gonzales.** 1989. Purification of plasmid DNA by fast protein liquid chromatography on superose 6 preparative grade. *Anal. Biochem.* **177:**378–382.

93. **McConaughy, B. L., C. D. Laird, and B. J. McCarthy.** 1969. Nucleic acid reassociation in formamide. *Biochemistry* **8:**3289–3295.

94. **Middaugh, C., R. Evans, D. Montgomery, and D. Casimiro.** 1998. Analysis of plasmid DNA from a pharmaceutical perspective. *J. Pharm. Sci.* **87:**130–146.

95. **Miller, J. F., W. J. Dower, and L. S. Tompkins.** 1988. High-voltage electroporation of bacteria: genetic transformation of *Campylobacter jejuni* with plasmid DNA. *Proc. Natl. Acad. Sci. USA* **85:**856–860.

96. **Myers, J. A., D. Sanchez, L. P. Elwell, and S. Falkow.** 1976. Simple agarose gel electrophoretic method for the identification and characterization of plasmid deoxyribonucleic acid. *J. Bacteriol.* **127:**1529–1537.

97. **Nabel, G., E. Nabel, Z. Yang, B. Fox, G. Plautz, X. Gao, L. Huang, S. Shu, D. Gordon, and A. Chang.** 1993. Direct gene transfer with DNA-liposome complexes in melanoma: expression, biologic activity, and lack of toxicity in humans. *Proc. Natl. Acad. Sci. USA* **90:**11307–11311.

98. **Niki, H., and S. Hiraga.** 1998. Polar localization of the replication origin and terminus in *Escherichia coli* nucleoids during chromosome partitioning. *Genes Dev.* **12:**1036–1045.

99. **Niki, H., and S. Hiraga.** 1997. Subcellular distribution of actively partitioning F plasmid during the cell division cycle in *E. coli. Cell* **90:**951–957.

100. **Niki, H., and S. Hiraga.** 1999. Subcellular localization of plasmids containing the *oriC* region of the *Escherichia coli* chromosome, with or without the sopABC partitioning system. *Mol. Microbiol.* **34:**498–503.

101. **Novick, R. P.** 1987. Plasmid incompatibility. *Microbiol. Rev.* **51:**381–395.

102. **Palchaudhuri, S.** 1977. Molecular characterization of hydrocarbon degradative plasmids in *Pseudomonas putida. Biochem. Biophys. Res. Commun.* **77:**518–525.

103. **Patton, C. M., I. K. Wachsmuth, G. M. Evins, J. A. Kiehlbauch, B. D. Plikaytis, N. Troup, L. Tompkins, and H. Lior.** 1991. Evaluation of 10 methods to distinguish epidemic-associated *Campylobacter* strains. *J. Clin. Microbiol.* **29:**680–688.

104. **Patton, J.** 1998. Breathing life into protein drugs. *Nat. Biotechnol.* **16:**141–143.

105. **Perez-Casal, J. F., and J. H. Crosa.** 1987. Novel incompatibility and partition loci for the REPI replication region of plasmid ColV-K30. *J. Bacteriol.* **169:**5078–5086.

106. **Perng, G. C., and R. B. LeFebvre.** 1990. Expression of antigens from chromosomal and linear plasmid DNA of *Borrelia coriaceae. Infect. Immun.* **58:**1744–1748.

107. **Pogliano, J., T. Q. Ho, Z. Zhong, and D. R. Helinski.** 2001. Multicopy plasmids are clustered and localized in *Escherichia coli. Proc. Natl. Acad. Sci. USA* **98:**4486–4491.

108. **Radloff, R., W. Bauer, and J. Vinograd.** 1967. A dye-buoyant-density method for the detection and isolation of closed circular duplex DNA: the closed circular DNA in HeLa cells. *Proc. Natl. Acad. Sci. USA* **57:**1514–1521.

109. **Recchia, G. D., M. Aroyo, D. Wolf, G. Blakely, and D. J. Sherratt.** 1999. FtsK-dependent and -independent pathways of Xer site-specific recombination. *EMBO J.* **18:**5724–5734.

110. **Reddy, B. A., P. M. Billingsley, and L. Etkin.** 1990. A quick protocol for plasmid DNA analysis. *BioTechniques* **9:**716–718.

111. **Robinett, C., A. Straight, G. Li, C. Willhelm, G. Sudlow, A. Murray, and A. Belmont.** 1996. *In vivo* localization of DNA sequences and visualization of large-scale chromatin organization using *lac* operator-repressor recognition. *J. Cell Biol.* **135:**1685–1700.

112. **Saha, B., D. Saha, S. Niyogi, and M. Bal.** 1989. A new method of plasmid DNA preparation by sucrose-mediated detergent lysis from *Escherichia coli* (gram-negative) and *Staphylococcus aureus* (gram-positive). *Anal. Biochem.* **176:**344–349.

113. **Sakaguchi, K., H. Hirochika, and N. Gunge.** 1985. Linear plasmids with terminal inverted repeats obtained from *Streptomyces rochei* and *Kluyveromyces lactis. Basic Life Sci.* **30:**433–451.

114. **Sambrook, J., and D. W. Russell.** 2001. *Molecular Cloning. A Laboratory Manual*, 3rd ed. Cold Spring Harbor Laboratory Press, Cold Spring Harbor, NY.

115. **Schatzlein, A. G.** 2001. Non-viral vectors in cancer gene therapy: principles and progress. *Anticancer Drugs* **12:**275–304.

116. **Schmidt, T., K. Friehs, M. Schleef, C. Voss, and E. Flaschel.** 1999. Quantitative analysis of plasmid forms by agarose and capillary gel electrophoresis. *Anal. Biochem.* **274:**235–240.

117. **Schwartz, D. C., and C. R. Cantor.** 1984. Separation of yeast chromosome-sized DNAs by pulsed field gradient gel electrophoresis. *Cell* **37:**67–75.

118. **Sedegah, M., R. Hedstrom, P. Hobart, and S. L. Hoffman.** 1994. Protection against malaria by immunization with plasmid DNA encoding circumsporozoite protein. *Proc. Natl. Acad. Sci. USA* **91:**9866–9870.

119. **Sedegah, M., W. Weiss, J. B. Sacci, Jr., Y. Charoenvit, R. Hedstrom, K. Gowda, V. F. Majam, J. Tine, S. Kumar, P. Hobart, and S. L. Hoffman.** 2000. Improving protective immunity induced by DNA-based immunization: priming with antigen and GM-CSF-encoding plasmid DNA and boosting with antigen-expressing recombinant poxvirus. *J. Immunol.* **164:**5905–5912.

120. **Serghini, M. A., C. Ritzenthaler, and L. Pinck.** 1989. A rapid and efficient 'miniprep' for isolation of plasmid DNA. *Nucleic Acids Res.* **17:**3604.

121. **Service, R. F.** 1997. Drug delivery takes a deep breath. *Science* 277:1199–1200.

122. **Sharp, P. A., M. T. Hsu, E. Otsubo, and N. Davidson.** 1972. Electron microscope heteroduplex studies of sequence relations among plasmids of *Escherichia coli.* I. Structure of F-prime factors. *J. Mol. Biol.* **71:**471–497.

123. **Shi, W., J. Liu, Y. Huang, and L. Qiao.** 2001. Papillomavirus pseudovirus: a novel vaccine to induce mucosal and systemic cytotoxic T-lymphocyte responses. *J. Virol.* **75:**10139–10148.

124. **Shields, M., B. Kline, and J. Tam.** 1986. A rapid method for the quantitative measurement of gene dosage: mini-F plasmid concentration as a function of cell growth rate. *J. Microbiol. Methods* **6:**33–46.

125. **Simske, J. S., and S. Scherer.** 1989. Pulsed-field gel electrophoresis of circular DNA. *Nucleic Acids Res.* **17:**4359–4365.

126. **Singer, V. L., L. J. Jones, S. T. Yue, and R. P. Haugland.** 1997. Characterization of Pico Green reagent and development of a fluorescence-based solution assay for double-stranded DNA quantitation. *Anal. Biochem.* **249:**228–238.

127. **Skurray, R. A., H. Nagaishi, and A. J. Clark.** 1976. Molecular cloning of DNA from F sex factor of *Escherichia coli* K-12. *Proc. Natl. Acad. Sci. USA* **73:**64–68.

128. **Southern, E. M.** 1975. Detection of specific sequences among DNA fragments separated by gel electrophoresis. *J. Mol. Biol.* **98:**503–517.

129. **Stirling, C. J., G. Stewart, and D. J. Sherratt.** 1988. Multicopy plasmid stability in *Escherichia coli* requires host-encoded functions that lead to plasmid site-specific recombination. *Mol. Gen. Genet.* **214:**80–84.

130. **Stopeck, A. T., E. M. Hersh, J. L. Brailey, P. R. Clark, J. Norman, and S. E. Parker.** 1998. Transfection of primary tumor cells and tumor cell lines with plasmid DNA/lipid complexes. *Cancer Gene Ther.* **5:**119–126.

131. **Straight, A., A. Belmont, C. Robinett, and A. Murray.** 1996. GFP tagging of budding yeast chromosomes reveals that protein-protein interactions can mediate sister chromatid cohesion. *Curr. Biol.* **6:**1599–1608.

132. **Summers, D. K.** 1996. *The Biology of Plasmids.* Blackwell Science Ltd., Oxford, United Kingdom.

133. **Summers, D. K., and D. J. Sherratt.** 1988. Resolution of ColE1 dimers requires a DNA sequence implicated in the three-dimensional organization of the *cer* site. *EMBO J.* **7:**851–858.

134. **te Riele, H., B. Michel, and S. D. Ehrlich.** 1986. Single-stranded plasmid DNA in *Bacillus subtilis* and *Staphylococcus aureus. Proc. Natl. Acad. Sci. USA* **83:**2541–2545.

135. **Thomas, C. M.** 2000. Paradigms of plasmid organization. *Mol. Microbiol.* **37:**485–491.

136. **Thomas, C. M., C. A. Smith, V. Shingler, M. A. Cross, A. A. Hussain, and M. Pinkney.** 1985. Regulation of replication and maintenance functions of broad host-range plasmid RK2. *Basic Life Sci.* **30:**261–276.

137. **Tolmasky, M. E., L. Actis, and J. H. Crosa.** 1999. Plasmid DNA replication, p. 2004–2019. *In* M. Flickinger and S. Drew (ed.), *Bioprocess Technology: Fermentation, Biocatalysis, and Bioseparation.* John Wiley & Sons, Inc., New York, NY.

138. **Tolmasky, M. E., L. A. Actis, and J. H. Crosa.** 1993. Virulence plasmids, p. 95–118. *In* K. Hardy (ed.), *Plasmids: a Practical Approach,* 2nd ed. IRL Press at Oxford University Press, Oxford, United Kingdom.

139. **Tolmasky, M. E., L. A. Actis, A. E. Toranzo, J. L. Barja, and J. H. Crosa.** 1985. Plasmids mediating iron uptake in *Vibrio anguillarum* strains isolated from turbot in Spain. *J. Gen. Microbiol.* **131:**1989–1997.

140. **Tolmasky, M. E., S. Colloms, G. Blakely, and D. J. Sherratt.** 2000. Stability by multimer resolution of pJHCMW1 is due to the Tn*1331* resolvase and not to the *Escherichia coli* Xer system. *Microbiology* **146:**581–589.

141. **Tolmasky, M. E., R. J. Staneloni, and L. F. Leloir.** 1982. Lipid-bound saccharides in *Rhizobium meliloti. J. Biol. Chem.* **257:**6751–6767.

142. **Tservistas, M., M. Levy, M. Lo-Yin, R. O'Kennedy, P. York, G. Humphrey, and M. Hoare.** 2001. The formation of plasmid DNA loaded pharmaceutical powders using supercritical fluids technology. *Biotechnol. Bioeng.* **72:**12–18.

143. **Tservistas, M., T. Scheper, and R. Freitag.** 1999. Supercritical fluid extraction (SFE)—novel strategies in the processing of biomaterials, p. 106–113. *In* S. Grabley and R. Thiericke (ed.), *Drug Discovery from Nature.* Springer Verlag, Berlin, Germany.

144. **Vahlsing, H. L., M. A. Yankauckas, M. Sawdey, S. H. Gromkowski, and M. Manthorpe.** 1994. Immunization with plasmid DNA using a pneumatic gun. *J. Immunol. Methods* **175:**11–22.

145. **Van Larebeke, N., G. Engler, M. Holsters, S. Van den Elsacker, I. Zaenen, R. A. Schilperoort, and J. Schell.** 1974. Large plasmid in *Agrobacterium tumefaciens* essential for crown gall-inducing ability. *Nature* **252:**169–170.

146. **Vitzthum, F., G. Geiger, H. Bisswanger, H. Brunner, and J. Bernhagen.** 1999. A quantitative fluorescence-based microplate assay for the determination of double-stranded DNA using SYBR Green I and a standard ultraviolet transilluminator gel imaging system. *Anal. Biochem.* **276:**59–64.

147. **Wang, R., D. Doolan, T. Le, R. Hedstrom, K. Coonan, Y. Charoenvit, T. Jones, P. Hobart, M. Margalith, J. Ng, W. Weiss, M. Sedegah, C. de Taisne, J. Norman, and S. Hoffman.** 1998. Induction of antigen-specific cytotoxic T lymphocytes in humans by malaria DNA vaccine. *Science* **282:**476–480.

148. **Waters, V.** 2001. Conjugation between bacterial and mammalian cells. *Nat. Genet.* **29:**375–376.

149. **Watson, B., T. C. Currier, M. P. Gordon, M. D. Chilton, and E. W. Nester.** 1975. Plasmid required for virulence of *Agrobacterium tumefaciens. J. Bacteriol.* **123:**255–264.

150. **Wegrzyn, G., P. Neubauer, S. Krueger, M. Hecker, and K. Taylor.** 1991. Stringent control of replication of plasmids derived from coliphage lambda. *Mol. Gen. Genet.* **225:**94–98.

151. **Weickert, M., and G. Chambliss.** 1989. Acid-phenol minipreps make excellent sequencing templates. *U.S. Biochemical Corp. Editorial Comments* **16:**5–6.

152. **Weiss, W. R., K. J. Ishii, R. C. Hedstrom, M. Sedegah, M. Ichino, K. Barnhart, D. M. Klinman, and S. L. Hoffman.** 1998. A plasmid encoding murine granulocyte-macrophage colony-stimulating factor increases protection conferred by a malaria DNA vaccine. *J. Immunol.* **161:**2325–2332.

153. **Weitao, T., S. Dasgupta, and K. Nordstrom.** 2000. Plasmid R1 is present as clusters in the cells of *Escherichia coli. Plasmid* **43:**200–204.

154. **Weitao, T., S. Dasgupta, and K. Nordstrom.** 2000. Role of the *mukB* gene in chromosome and plasmid partition in *Escherichia coli. Mol. Microbiol.* **38:**392–400.

155. **Whisenant, E. C., B. K. Rasheed, and Y. M. Bhatnagar.** 1988. Plasmid purification using high-performance gel filtration chromatography. *Nucleic Acids Res.* **16:**5202.

156. **White, F. F., and E. W. Nester.** 1980. Hairy root: plasmid encodes virulence traits in *Agrobacterium rhizogenes. J. Bacteriol.* **141:**1134–1141.

157. **Woloj, M., M. E. Tolmasky, M. C. Roberts, and J. H. Crosa.** 1986. Plasmid-encoded amikacin resistance in multiresistant strains of *Klebsiella pneumoniae* isolated from neonates with meningitis. *Antimicrob. Agents Chemother.* **29:**315–319.

158. **Yarmolinsky, M. B.** 2000. A pot-pourri of plasmid paradoxes: effects of a second copy. *Mol. Microbiol.* **38:**1–7.

159. **Zasloff, M., G. D. Ginder, and G. Felsenfeld.** 1978. A new method for the purification and identification of covalently closed circular DNA molecules. *Nucleic Acids Res.* **5:**1139–1152.

160. **Zhang, Y. P., D. L. Reimer, G. Zhang, P. H. Lee, and M. B. Bally.** 1997. Self-assembling DNA-lipid particles for gene transfer. *Pharm. Res.* **14:**190–196.

161. **Ziai, M. R., C. V. Hamby, R. Reddy, K. Hayashibe, and S. Ferrone.** 1989. Rapid purification of plasmid DNA following acid precipitation of bacterial proteins. *BioTechniques* **7:**147.

31

Gene Transfer in Gram-Negative Bacteria†

JOSEPH E. PETERS

---

†A chapter by the same title was authored by D. L. Provence and Roy Curtiss III for *Methods for General and Molecular Bacteriology*. This chapter has been completely revised and reorganized by the current author.

One realization that has come from comparing multiple bacterial genome sequences, including multiple isolates from the same species, is that gene transfer is an important force in bacterial genome evolution. In the laboratory gene transfer is essential for the study of bacteria and for learning more about all living organisms.

Three processes in bacteria can broadly define the transfer of DNA: transformation, transduction, and conjugation. In each transfer, genetic information from one bacterium, the donor, is acquired by another bacterium, the recipient. Generally an organism which has had its genetic content modified is called a recombinant organism, but specific terms are also used depending on how DNA is introduced. **Transformation** is a gene transfer process in which DNA from the donor is taken up directly by the recipient and incorporated into its genome. A recipient cell that now expresses a donor genetic trait is called a transformant. **Transduction** is a gene transfer process in which a bacterial virus propagating in the donor picks up some of its genetic information and, on infection of the recipient, causes a heritable change in the recipient. A recipient cell that acquires a donor trait by this process is called a transductant. **Conjugation** is a gene transfer process in which the so-called donor bacterium makes physical contact with, and transfers genetic material to, the recipient. A recipient that acquires donor genetic information is called a transconjugant. These three processes have been central in allowing bacterial geneticists to analyze the genetic and biochemical basis of bacterial functions. In turn they have led to the establishment of principles of gene structure, function, and regulation as well as the dissection of more complex processes of macromolecular synthesis, growth, and cell division.

This chapter focuses on the many genetic tools available to manipulate the genetic content of *Escherichia coli*. *E. coli* remains important as a model system and as a workhorse for manipulating DNA prior to introduction into other organisms. In addition to *E. coli*, references are made to a variety of other proteobacteria (Table 1). It would be impossible to definitively cover gene transfer in all gram-negative organisms. However, an attempt is made to cover the basic ideas in gene transfer in organisms other than *E. coli* where there have been successes in a variety of proteobacteria and in cyanobacteria. Plasmids are the topic of another chapter in this book and are mentioned only briefly in this chapter.

31.1. INTEGRATING DNA INTO THE CHROMOSOME

After genetic information is introduced into a cell, it must be replicated to create a recombinant bacterium. A DNA molecule that does not have its own origin of replication must integrate into either the host chromosome or another autonomously replicating element such as an endogenous plasmid. Therefore, one cannot really talk about gene transfer without describing how the genetic information will be maintained after it is introduced.

DNA recombination systems encoded by bacteria and bacteriophages provide important tools for the in vivo modification of chromosomes. Most of these strategies can also be applied to large low-copy-number plasmids, such as the Fertility or F plasmid that has been adapted to make bacterial artificial chromosomes (BACs). Early recombination methods used the host-encoded recombination system as a tool for engineering the chromosome. These strategies can still be useful in many bacteria for generating recombinants in wild-type bacteria via single-crossover and double-crossover events from circular and linear DNA molecules. In *E. coli* and a number of other bacteria, mutations in various host recombination functions are required to allow chromosome modification with linear substrates. Site-specific recombination systems can be utilized for introducing DNA into the chromosome via circular DNA molecules. The greatest recent advance in bacterial genetics is the use of bacteriophage-encoded recombination systems like λ Red-mediated recombination that have been adapted as important laboratory tools.

31.1.1. Homology-Based RecA Systems

The major bacterial recombination protein RecA is conserved in bacteria and plays an important role in DNA repair (72, 73). RecA-mediated recombination was the original source for manipulating bacterial genomes and provided the original evidence for sexual recombination in bacteria (77). Work in *E. coli* indicates that RecA cannot act on its own in recombination. A variety of accessory proteins are required for RecA to function by tailoring DNA substrates and promoting the binding of RecA to a single-stranded DNA in competition with the single-stranded DNA binding protein. The RecBCD machine allows RecA-mediated recombination to initiate at broken double-strand ends by generating a 3′ single-strand end and loading RecA (6). The RecBCD machine is a potent nuclease, and recombination with linear DNA substrates in *E. coli* and many other bacteria is essentially impossible without inactivation of its nuclease activity. Another RecA-dependent system, called the RecF system, requires a larger set of recombination proteins and is predominantly involved in promoting recombination at single-strand interruptions. Once RecA is loaded onto a single-strand DNA substrate, it mediates recombination by searching for homology and carrying out strand invasion and exchange. Other recombination proteins are required for processing recombination intermediates, as well as DNA replication proteins. Bacteriophage recombination proteins can generate recombinant molecules in the absence of RecA and are dealt with later.

31.1.1.1. Single-Crossover Systems

Circular DNA substrates can be integrated into the bacterial chromosome in a RecA-dependent process as a tool for manipulating the bacterial chromosome (104). This strategy is based on selecting for a gene, such as antibiotic resistance, that is carried on a circular DNA substrate in a strain background where the DNA element cannot replicate autonomously (54, 95) (Fig. 1). These experiments utilize a conditional plasmid that cannot replicate in the given host naturally or because it lacks an essential *trans*-acting protein (70, 89) or because it is temperature sensitive for

TABLE 1 Methods for transferring DNA in selected species within the phylum proteobacteria

Organism	Subdivision	Method of DNA introduction[a]	Narrow-host-range vectors[b]	Broad-host-range vectors[c]	Moving markers between strains	Reference(s)
Agrobacterium tumefaciens	Alpha	E, M	No	Yes	None	24
Caulobacter crescentus	Alpha	E, M	No	Yes	Transduction	13, 46
Rhodobacter sphaeroides	Alpha	M, E	No	Yes	None	45
Sinorhizobium meliloti	Alpha	M, E	No	Yes	Transduction	47, 50, 115
Neisseria gonorrhoeae	Beta	N, E, C, M	No	Yes	Natural transformation	51, 111, 123
Helicobacter pylori	Epsilon	N, E	No	No	Natural transformation	
Escherichia coli	Gamma	C, E, M	Yes	Yes	Transduction	90
Salmonella enterica	Gamma	C, E, M	Yes	Yes	Transduction	
Vibrio cholerae	Gamma	E, C, M	Yes	Yes	Transduction	57, 116
Vibrio harveyi	Gamma	E, M	Yes	Yes	None	116
Pseudomonas aeruginosa	Gamma	E, M	No	Yes	Transduction	20, 106
Pseudomonas syringae	Gamma	E, C, M	No	Yes	None	106
Legionella pneumophila	Gamma	E, M	No	Yes	None	
Yersinia enterocolitica	Gamma	E, M	Yes	Yes	None	33
Haemophilus influenzae	Gamma	N, E	Yes	Yes	Natural transformation	11
Geobacter sulfurreducens	Delta	E	No	Yes	None	35

[a]Mechanisms of introduction are abbreviated as follows (in order of preference): E, electroporation; M, mating via conjugation; N, natural competency; and C, chemical competency.

[b]Representatives from the narrow-host-range plasmids are reported to replicate in this species (Table 3).

[c]Representatives from the broad-host-range plasmids are reported to replicate in this species (Table 3).

DNA replication (54). This strategy can be used to inactivate genes as well as to introduce gene fusion to promoters and proteins (120). Promoter fusions can be very useful to assay gene regulation. Despite the ease of this type of construction, it also has several limitations; these constructs may not be genetically stable, and the integrated DNA can be polar on downstream genes. In addition, promoters included in the integrated DNA segment can activate downstream genes.

31.1.1.2. Double-Crossover Systems with Circular Substrates

Circular DNA substrates can be used to isolate constructs where the allele replacement has taken place via a double-crossover event (Fig. 2). In theory this procedure is similar to the single-crossover event, but only the new allele is left behind in the chromosome and not the entire plasmid. These experiments utilize conditional or temperature-sensitive replicons as described above. Presumably a single-crossover event is an intermediate in all double-crossover events. The second crossover event that allows the plasmid to be lost can have one of two outcomes; the plasmid is excised from the replicon and leaves either the original allele or the allele pro-

vided in the vector. A marker in the vector backbone can be screened to determine if the plasmid is lost (for example, by screening for loss of a marker at positions 3 to 5 in Fig. 2). It may sometimes be sufficient to introduce a selective marker that replaces a large portion of the gene (for example, by selecting a marker cloned at position 1 or 2 in Fig. 2). However, this is not always an option because it can alter the expression of downstream genes in an operon. Therefore, integration of a mutant allele with an in-frame deletion is often more appropriate to eliminate the possibility of a polar effect on the transcription of downstream genes. Additionally, a very specific mutation may be required (e.g., a point mutation conferring a specific amino acid change or a conditional mutation such as an amber suppressible mutation). In cases such as these one must have an alternative strategy to identify the recombinant organisms, for example, a phenotype or through the use of DNA sequencing.

One potent limitation in procedures where circular substrates are used to isolate recombinants by double-crossover events is that single-crossover events that allow integration of the entire circular substrate often must be sorted away from the desired replacement event. Therefore, a counterselection or negative selection may be required to select for

a.

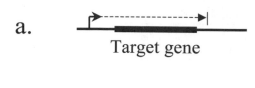

Target gene

b.

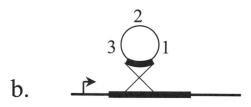

c.

FIGURE 1   Genetic information can be integrated into the chromosome of a bacterium using a single crossover from a circular DNA construct that cannot replicate in the host. (a) A single open reading frame (thick line) to be inactivated is shown with its promoter (solid arrow) and RNA transcript (dashed line with arrow). (b) A plasmid (circle) that cannot replicate in the strain can be maintained only by integrating into the chromosome using homology provided on the plasmid indicated with an "X." Three arbitrary positions are indicated with numbers to show the orientation. (c) Integration of the circular DNA substrate disrupts and fuses the plasmid-borne genes behind the target gene promoter. Depending on the construct, the transcript may be terminated within the integrated DNA segment. Promoters within the plasmid could also activate adjacent genes in the chromosome.

a.

Target gene

b.

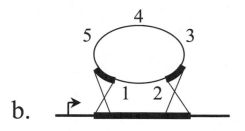

c.

FIGURE 2   Genetic information can be integrated into the chromosome of a bacterium using a double-crossover event from a circular DNA construct that cannot replicate in the host. (a) A single open reading frame (thick line) to be inactivated is shown with its promoter (solid arrow) and RNA transcript (dashed line with arrow). (b) A plasmid (oval) that cannot replicate in the strain can be maintained only by integrating into the chromosome using the homology provided on the plasmid indicated with "X's" (thick lines). Two crossover events ensure that only a portion of the circular DNA is integrated into the gene. Five arbitrary positions are indicated with numbers to show the DNA that is integrated and the orientation. A single crossover event likely occurs at a mid-step in the reaction and is not shown (Fig. 1). (c) Integration of the circular DNA substrate removes all or a portion of the gene and fuses the encoded genes behind the target gene promoter. Depending on the construct, the transcript may be terminated within the integrated DNA segment. Promoters within the plasmid could activate adjacent genes in the chromosome.

loss of the integrating DNA element if the frequency of loss is very low (79). One tool for negative selection capitalizes on the hypersensitivity of elements encoding the *tetAR* genes, like Tn*10*, to lipophilic chelating agents such as fusaric acid (17). Tetracycline-sensitive clones can be selected from an otherwise tetracycline-resistant population by streaking the strain on tetracycline-sensitive-selective (TSS) plates and incubating 1 to 2 days at 37°C. The *sacB* gene from *Bacillus subtilis*, which imparts sensitivity to sucrose in proteobacteria and cyanobacteria, can also be used for negative selection by selecting for colonies capable of growth on rich media containing 5% sucrose (sucrose is filter sterilized and added after the medium cools) (16, 22, 49, 68). An additional strategy involves the use of the *rpsL* gene encoding the S12 subunit of the ribosome. Certain mutations in the *rpsL* gene confer resistance to streptomycin. However, streptomycin resistance is recessive to the wild-type allele when placed in *trans* in the cell allowing negative selection for loss of the wild-type allele. This procedure was initially used in *E. coli* and subsequently in other less tractable bacteria like *Bordetella pertussis* and *Vibrio cholerae* (107, 119, 125).

Negative selection can be used as a mechanism to facilitate and select RecA-mediated recombination events in

the chromosome by inducing double-strand breaks at a specific location in the chromosome (101). In this procedure, the I-SceI meganuclease from yeast, which recognizes a large sequence that is not found in *E. coli* and other bacteria, is used. The desired mutant copy of the gene is introduced on a suicide plasmid. By selecting a plasmid-encoded drug resistance encoded on the plasmid backbone, the suicide plasmid will integrate into the chromosome by homologous recombination. The plasmid also contains the I-SceI recognition site in its backbone such that upon integration the I-SceI recognition site resides between the mutant and wild-type alleles (for example at positions 1 to 3 in Fig. 1). In a second step, expression of the I-SceI endonuclease in *trans* allows for selection of recombinants that lost the plasmid. Following loss of the integrated plasmid a second strategy must be used to identify recombinants that have maintained the altered allele and not the wild-type allele. The double-strand break acts both as a negative selection and also as a means to facilitate recombination. Variations on the use of I-SceI as a mechanism for negative selection have also been used with bacteriophage-encoded recombination systems (see below) (69).

31.1.1.3. Double-Crossover Systems with Linear Substrates

When linear DNA substrates are used to modify the chromosome, a double-crossover event must take place because a single-crossover event would result in a broken chromosome and cell death (Fig. 3). However, as mentioned above, the potent nuclease activity of RecBCD in wild-type *E. coli* cells prevents linear DNAs from residing in the bacterium long enough for recombination to occur. Mutant strain backgrounds have been used to introduce linear DNA, which inactivate the nuclease activity of RecBCD but also allow RecA to load onto the DNA substrate. Mutations abolishing *recB* and/or *recC* functions in a strain background (like JC7623) that also contains a *sbcBC* suppressor allele allow recombination with linear DNAs (63, 133). Mutations in the *recD* gene abolish the RecBCD exonuclease activity without inactivating its ability to load RecA, thereby preserving its recombination function (108). A strain with a Tn*10* insertion inactivating the *recD* gene, such as CAG12135, can be used for modifying the chromo-

some with large linear DNAs (Table 2). Following introduction of a desired mutation in such a genetic background, a second mechanism is needed to move the recombinant allele into a *rec*⁺ background. Some of the limitations of these recombination strategies that abolish RecBCD function are the need for long stretches of homology to the target substrate (~300 to 1,000 bp) and a low frequency of recombinants requiring direct selection for the recombination event. One strategy that appears to circumvent the recombination barrier imposed by the exonuclease activity of RecBCD with linear DNA is through the inclusion of very large stretches of homology flanking the DNA substrate. This allowed large fragments (~20 kb) in linearized BACs to be crossed into RecBCD⁺ *E. coli* strains when the fragment was flanked by ~6-kb segments of DNA that were homologous to the *E. coli* chromosome (105).

31.1.2. Site-Specific Recombination Systems Can Remove Selectable Markers

Another strategy for the removal of a selectable marker that was introduced via a double crossover is the use of various site-specific recombination systems like the Cre and Flp recombinases. The Cre recombinase functions at its cognate *lox* sites, and the Flp recombinase acts at *frt* sites. When the recombinase is expressed in *trans*, a selectable marker situated between the cognate recombination sites is efficiently removed. The Cre system originally came from bacteriophage P1, but this system has been utilized in a number of eukaryotic organisms as well (110). The Cre system has been adapted for use in a variety of hosts as a tool to remove antibiotic resistance determinants (8, 85). The Flp recombinase from the yeast 2μm plasmid can be used in a similar fashion and is described below.

31.1.3. Recombination Systems from Bacteriophages

One of the greatest recent advances in tools for the genetic modification of *E. coli* has been the widespread establishment of the recombination system from bacteriophage λ, called the λ Red recombinase, as well as a system from a related phage (39, 136, 139). These systems now allow recombination efficiencies that were previously possible only in yeast (97, 126). Linear DNA substrates have been used for some time in yeast, capitalizing on its highly efficient recombination-based double-strand break repair system that is stimulated by linear DNA. The segments of homology that are required for recombination in yeast can be as short as 50 bp (12, 76).

Bacteriophage-encoded recombination functions are well suited for manipulating bacterial chromosomes and large low-copy-number plasmids such as BACs. However, there are some limitations for manipulating small high-copy-number cloning vectors (136). Red-mediated recombination is much simpler than the host-encoded RecA-mediated systems. The λ Red recombinase functions reside in three genes, *exo* (*redα*), *bet* (*redβ*), and *gam* (37, 74). The Beta and Exo proteins have functions analogous to those of RecA and the RecBCD proteins, respectively. The Beta protein binds stably to single-stranded DNA greater than 35 nucleotides and promotes pairing or annealing between complementary single-stranded DNAs. The exonuclease function of Exo can tailor double-stranded DNA substrates to make them competent for recombination with Beta. The Exo function is dispensable when single-stranded DNA substrates are used for recombination (136, 137). Homologs of the Exo and Beta proteins, called RecE and RecT, are encoded on a cryptic prophage called Rac that is found in the

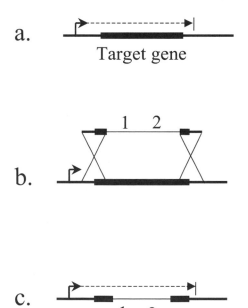

a.

Target gene

b.

1 2

c.

1 2

FIGURE 3 Genetic information can be integrated into the chromosome of a bacterium using a double-crossover event from a linear DNA construct. (a) A single open reading frame (thick line) to be inactivated is shown with its promoter (solid arrow) and RNA transcript (dashed line with arrow). (b) Two crossover events must occur for the linear DNA to be integrated into the gene when selecting for gene products indicated by 1 or 2. Homology provided on the fragment is indicated with "X's." (c) Integration of the DNA substrate removes all or a portion of the gene and fuses the genes carried behind the target gene promoter. Depending on the construct, the transcript may be terminated within the integrated DNA segment. Promoters within the integrated DNA segment could activate adjacent genes in the chromosome.

TABLE 2 *E. coli* strains

Strain or plasmid	Genotype[a]	Comments[b]
E. coli strains AB1157	*thr-1 araC14 leuB6 Δ(gpt-proA)62 lacY1 tsx-33 qsr'-0 glnV44 galK2 Rac-0 hisG4 rfbD1 mgl-51 rpoS396 rpsL31 kdgK51 xylA5 mtl-1 argE3 thi-1*	A common laboratory strain (43); CGSC 1157
BW20767	*Δ3(lac)$_{X74}$ RP4-2-tet:Mu-1kan::Tn7 integrant leu-63::IS10 recA1 creC510 hsdR17 endA1 sbf-5 uidA(ΔMluI)::pir$^+$ thi*	Strain for mobilizing plasmids with RP4 origins of transfer via an integrated RP4 fragment (*RP4-2-tet:Mu-1kan::Tn7*) which imparts resistance to spectinomycin, streptomycin, and trimethoprim. *pir* gene allows replication of conditional R6K origin plasmids (88, 89); ATCC 47084.
BW23473	*Δ(argF-lac)169^c robA1 creC510 hsdR514 endA9 recA1 ΔuidA::pir$^+$*	Chromosome has the wild-type *pir* gene which encodes the π protein required for the R6K origin replication at a copy number of ~15 (88, 89); CGSC 7837.
BW23474	*Δ(argF-lac)169^c robA1 creC510 hsdR514 endA9 recA1 ΔuidA::pir-116*	Chromosome has the mutant *pir-116* gene which encodes a π protein required for the R6K origin replication at a copy number of ~250 (88, 89); CGSC 7838.
CAG12135	*MG1655 recD1901::Tn10*	MG1655 strain with a Tn*10* insertion in *recD*. Allows linear DNA to persist long enough for recombination into the chromosome (118); CGSC 7429.
DH5α	*endA1 hsdR17 glnV44 thi-1 recA1 gyrA96 Δ(argF-lac)169^c deoR (φ80dlacΔ(lacZ)M15)*	Cloning strain that allows α complementation. Available from Invitrogen (134).
ER1793	*fhuA2 Δ(lacZ)r1 glnV44 e14$^-$ trp-31 his-1 rpsL104 xyl-7 mtl-2 metB1 Δ(mcrC-mmr)114::IS10*	Cloning strain that lacks the native *E. coli* K-12 restriction systems. Should be used when cloning DNA directly from non-*E. coli* strains. Available from New England Biolabs.
ER2925	*ara-14 leuB6 fhuA31 lacY1tsx78 glnV44 galK2 galT22 mcrA dcm-6 hisG4 rfbD1 R(zgb210::Tn10)TetS endA1 rpsL136 dam13::Tn9 xylA-5 mtl-1 thi-1 mcrB1 hsdR2*	Cloning strain that is both *dam-* and *dcm*-negative. Useful for propagating DNA that will be introduced into strains that restrict methylated DNA. Available from New England Biolabs.
HB101	*thi-1 hsdS20 glnV44 recA13 ara-14 leuB6 proA2 lacY1 galK2 rpsL20 xyl-5 mtl-1*	A common laboratory strain (75). ATCC 33694.
HME5	*inversion(rrnD-rrnE)1 rph-1 Δ(argF-lac)169^c λcI857 Δ(cro-bioA)*	Temperature-sensitive strain that allows for the induction of the λ Red (Exo Bet Gam) proteins with a short period of growth at 42°C (136).
JC7623	*AB1157 sbcC201 sbcB15 recB21 recC22*	Mutations in *recBC sbcBC* allow linear DNA to persist long enough for recombination into the chromosome (63). CGSC 5188.
JM109	*endA1 recA1 gyrA96 thi hsdR17 relA1 glnV44 Δ(lac-proAB) (F' traD36 proAB lacIqZΔM15*	Reference 135; ATCC 53323.
CAG18642	*MG1655 srlD3131::Tn10*	Can be used to move mutations in *recA* with P1 transduction; CGSC 7423.
MC4100	*araD139 Δ(argF-lac)169^c rpsL150 relA1 flhD5301 deoC1 ptsF25 rbsR22 e14$^-$ Δ(fimB-fimE)632::IS1 Δ(fruK-yeiU)725*	A common laboratory strain. The Δ*(fruK-yeiU)725* allele was formally referred to as *fruA25* (25, 99); CGSC 6152.
MG1655	*rph-1*	A common laboratory strain. The genome has been sequenced (15); CGSC 6300.

(Continued on next page)

TABLE 2 *Continued*

Strain or plasmid	Genotype[a]	Comments[b]
S17-1	*thi pro hsdR recA (RP4-2-tet::Mu aphA::Tn7)*	Used for mobilizing plasmids with RP4 origins of transfer. The strain is sensitive to kanamycin and tetracycline but resistant to spectinomycin, streptomycin, and trimethoprim. A derivative with λ*pir* is also available for maintenance of R6K origin plasmids (117); ATCC 47055.
SM10	*C600 (RP4-2-tet:Mu-1kan)*	Used for mobilizing plasmids with RP4 origins of transfer. The strain is sensitive to tetracycline but resistant to kanamycin. A derivative with λ*pir* is also available for maintenance of R6K origin plasmids (117).
W3110	*inversion (rrnD-rrnE)1 rph-1*	W3110 is a common laboratory strain. The genome has been sequenced (9); CGSC 4474.
XL1-Blue	*F'::Tn10 proA$^+$B$^+$ lacI$_q$ Δ(lacZ)M15/recA1 endA1 gyrA96 thi hsdR17 glnV44 relA1 lac*	Cloning strain that allows α complementation. Available from Stratagene.
Plasmids		
pCP20		Temperature-sensitive plasmid with thermal induction of Flp recombinase (27). CGSC 7629.
pKD46		Temperature-sensitive plasmid with the λ Red proteins under arabinose control (39); CGSC 7669.
pKD3		Plasmid encoding ampicillin resistance that allows PCR amplification of a gene cassette encoding chloramphenicol resistance flanked by *frt* sites recognized by the Flp recombinase (39); CGSC 7631.
pKD4		Plasmid encoding ampicillin resistance that allows PCR amplification of a gene cassette encoding kanamycin resistance flanked by *frt* sites recognized by the Flp recombinase (39); CGSC 7632.
pACBSR		Chloramphenicol-resistant plasmid with λ Red proteins and the I-SceI meganuclease under arabinose control (58).

[a]Unless indicated, all of the strains do not contain bacteriophage λ or the F plasmid.
[b]The catalog numbers for the American Type Culture Collection (ATCC) and the *E. coli* Genetic Stock Center (CGSC) are given when available.
[c]The Δ(*argF-lac*)169 deletion allele actually deletes *ykfD-mhpD* (99).

common laboratory *E. coli* strain K-12 and most other non-laboratory strains. RecE and RecT are normally not expressed except in the presence of activating mutations called *sbcA* mutations (10). The Gam protein is a modifier of two exonucleases in *E. coli*, RecBCD and SbcCD. As mentioned above, RecBCD is a potent nuclease responsible for degrading linear DNAs in *E. coli*. The Gam protein is capable of binding RecBCD, thereby inhibiting many, but not all, activities of RecBCD, and functionally protecting linear DNA.

There are a number of different systems that allow expression of the λ Red recombination functions. In one system using strain HME5 and its derivatives, the phage proteins are expressed at high temperature from the chromosome by a defective λ phage (Table 2) (136). Expression of the Exo, Beta, and Gam proteins is induced to high levels by a 15-min heat shock at 42°C just prior to making the cells electrocompetent (78). An advantage of the chromosomally encoded system is its tight regulation and coordi-

nated expression (37). Constructs made in this background will likely be moved to a new strain background via transduction or conjugation. Other strategies utilize plasmid constructs for expression of the Exo, Beta, and Gam proteins utilizing an arabinose-inducible promoter (39, 58, 69, 139). One advantage of the plasmid-based systems is the ability to move the system into the strain of choice. Indeed, these plasmid constructs have even allowed the system to be adapted to a few other species including *Yersinia* sp., *Shigella flexneri*, and *Salmonella enterica* serovar Typhimurium (42, 87, 102, 113, 127, 128). Anecdotal accounts indicate that the system from bacteriophage lambda has met with little success in nonenterics. The best strategy for organisms unrelated to *E. coli* may be to adapt the recombination proteins from bacteriophages that are endogenous to the target bacterium.

One of the greatest advantages of using the λ Red recombination proteins to recombine substrates into the chromosome is that small regions of homology (30 to 70 bp) are sufficient to allow recombination (Fig. 3). Recombination that is catalyzed with RecA typically requires much larger regions of homology (see above). PCR can be used to quickly synthesize recombination substrates for in vivo recombination with the λ Red recombinase, which works efficiently with such small regions of homology (39, 136); primers can be designed to amplify a selectable marker where a 5′ tail can be included on the primer with homology that flanks the region to be deleted (12, 76). This procedure provides a convenient way to introduce antibiotic-resistance-conferring genes (cassettes) to replace genes in the *E. coli* chromosome. However, it is often desirable to remove the selectable marker since leaving a selectable marker in an operon could produce polar effects on downstream genes or the selectable marker may simply be needed for other purposes. Template plasmids can be used to make recombination substrates with *frt* recombination sites flanking the antibiotic resistance determinant that are recognized by the Flp recombinase (Fig. 4a) (39). When the template plasmids pKD3, pKD4, and pKD13 are used as a source for the antibiotic resistance cassette, the *frt* sites flanking the resistance gene can be used to remove the gene and leave behind a deletion that allows translation of the remaining portion of the gene intact; in-frame deletions will not cause a polar effect on downstream genes. The Flp recombinase that is used to catalyze site-specific recombination between the *frt* sites is thermally induced from the plasmid pCP20 (Table 2 and Fig. 4c). The plasmid also has a temperature-sensitive replication system that is therefore lost (or cured) at 42°C.

31.1.3.1. Other Applications with Bacteriophage Recombination Systems

There are numerous applications for the λ Red recombination system for manipulating chromosomes and plasmids. One consideration when carrying out these procedures is the hazard of unwanted recombination events. Given that the frequency of unselected recombination events is reported to exceed 1% (36, 96, 136), it stands to reason that a number of unintended recombination events will likely occur in the bacterial genome besides the one that was intended. Therefore, it is important that any time λ Red recombination is used, the recombinant allele should be moved into a clean genetic background by using transduction or natural transformation. As with RecA-mediated recombination (see above), I-SceI-mediated double-strand breaks can be used as a form of negative selection that also enhances recombination. In such a procedure the I-SceI recognition site is first introduced into the gene of interest by recombination or transposition. In a second step, a linear recombination substrate is introduced into the cells at the same time as a vector expressing the I-SceI meganuclease and the λ Red recombination proteins. In another efficient procedure called gene-gorging, the DNA substrate that is to be recombined into the genome is linearized in vivo from a plasmid through the use of I-SceI sites that flank the DNA segment (58). The ability to linearize the substrate in vivo addresses the issue of transformation efficiency, the greatest limiting factor in the λ Red system, because all cells will have the linearized substrate. Single-stranded DNA oligonucleotides have also been reported to efficiently recombine into the chromosome in a process that obviates the need for the Gam and Exo proteins (96, 137). As expected, the process of λ Red-mediated recombination is enhanced in strains that lack the methyl-directed repair system (36).

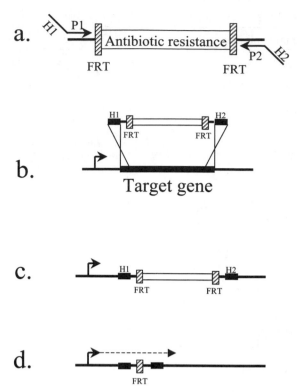

FIGURE 4 Recombination substrates can be made in vitro for use with λ Red recombination. The antibiotic resistance used to select for recombinants can subsequently be removed by expressing the Flp recombinase that will excise the region between the two *frt* sites (FRT). (a) Primers are designed to amplify from a universal antibiotic-resistant cassette with *frt* recombination sites using primer binding sites P1 and P2. The primers contain 5′ "tails" that are complementary to the target gene at regions H1 and H2 for the subsequent crossover event. (b) The PCR-generated substrate is transformed into a cell expressing the Exo, Beta, and Gam proteins (substrate shown miniaturized). (c) Selecting for the antibiotic resistance marker crosses in the cassette with its flanking *frt* recombination sites using the homology at regions H1 and H2. (d) The Flp recombinase is transiently expressed and subsequently lost because replication from the pCP20 vector is temperature sensitive. The Flp recombinase efficiently catalyzes recombination between the *frt* recombination sites removing the antibiotic resistance-conferring cassette. See the text for details.

Other novel strategies for the Red proteins include using them to manipulate large DNA substrates encoded on bacterial artificial chromosome (e.g., for building substrates for mouse genetics), for cloning genes via homology, and as a tool to recombine DNA substrates subjected to in vitro transposition into the chromosome (34, 37, 38).

31.1.4. Introducing DNA at Specific Sites in the Chromosome

There are multiple strategies for introducing DNA at specific locations in the chromosome. One mechanism involves using site-specific recombination systems endogenous to *E. coli* and its bacteriophage. An advantage of integrating DNA directly into the chromosome is that it allows genes to be expressed at one copy per chromosome, thereby avoiding artifacts caused by expressing genes on multiple-copy plasmids. One clever system for introducing DNA into the chromosome makes use of five insertion sites for bacteriophages in *E. coli* and their cognate recombinases that can be supplied on plasmids (53). In this system the genes of interest are cloned into plasmid vectors that are subsequently introduced into an *E. coli* host that allows integration into the chromosomal bacteriophage attachment sites, but not replication of the plasmid. The availability of multiple attachment sites for integration with constructs containing various expression systems and different antibiotic resistances allows many genes to be expressed in a single strain. Integrated vectors can be removed using plasmid constructs encoding excision genes, or the construct can be rescued as a plasmid when transduced into the appropriate host. Presumably, similar systems could be established in other less tractable bacteria using bacteriophage integration sites or integrative and conjugative elements (conjugative transposons) (21). Another system that utilizes bacteriophage attachment sites is based on a λ bacteriophage derivative, called λ InCh, encoding two regions that are homologous to common pBR322-based plasmids encoding ampicillin resistance (19). An advantage of this system is that genes that are already cloned in a pBR322-based vector can be constructed into the chromosome with minimal additional effort. A strain containing the gene of interest cloned into a pBR322-based plasmid is used as a host to grow the λ InCh bacteriophage. Recombination between the plasmid and a truncated copy of the *bla* gene encoded on the bacteriophage results in resistance to ampicillin; however, the same recombination event removes a bacteriophage-encoded kanamycin resistance gene. This procedure allows the isolation of recombinant bacteriophages by selecting strains that have the bacteriophages integrated in the chromosome (lysogens), which are resistant to ampicillin but are kanamycin sensitive. Lysogens can be prevented from excising and locked into the final host by selecting for a second recombination event that removes a large portion of the bacteriophage.

Transposons can be used to introduce DNA into the chromosomes of proteobacteria and cyanobacteria (31, 41). However, one disadvantage of using random transposition to insert cloned genes into the chromosome is that the insertion events can cause problems associated with gene inactivation. An additional consideration is that the transcription of each transposition event will be affected to different degrees depending on where the DNA inserted. A solution for this problem came from the use of the transposon Tn7, which inserts at a single position, called its attachment site (*attTn7*), and only in one orientation (98). The *attTn7* site is found in the transcriptional terminator of the *glmS* gene in proteobacteria, and insertion into this site does not negatively impact the host. A transposition system has been developed whereby the DNA of interest is cloned into a Tn7 derivative that can then be introduced into the *attTn7* site of proteobacteria following conjugation into the host (67).

Another option for introducing DNA directly into the chromosome involves direct selection for insertions by gene inactivation. For instance, examples exist where inactivating a single gene product results in resistance to an antibiotic or other substance (89). This strategy has proved successful at introducing genes into the chromosome of *Helicobacter pylori*. There is only one shuttle plasmid system for *H. pylori*, and an alternative system was therefore needed to express genes in this organism (59). Null mutations in the *rdxA* gene of *H. pylori* impart resistance to metronidazole; therefore, inactivation of this gene could be used as a selection for gene cassettes into this site via double-crossover events (121). Similar strategies can likely be developed for other organisms.

31.2. INTRODUCING DNA ON PLASMIDS

An important strategy for manipulating the genetic content of bacteria is the use of plasmids. Plasmids can be described in two broad families of replicons, narrow- and broad-host-range vectors. Narrow-host-range plasmids are stably maintained in *E. coli* and its close relatives (Table 1). Common examples of narrow-host-range replicons are the ColEI/pMB1, p15A, and pSC101 replicons (Tables 2 and 3). The ColEI/pMB1, p15A, and pSC101 replicons all fall into different incompatibility groups and can therefore reside in the same host at the same time. Broad-host-range vectors are capable of replicating in *E. coli* and a large number of different proteobacteria (Table 3). Some of the broad-host-range vectors are also capable of replicating in various cyanobacteria (68). All of the vectors in Table 3 are in different incompatibility groups and could in theory be maintained in the same permissive host organism. There are multiple considerations when choosing which type of plasmid to use to replicate any given substrate. Many vectors contain controllable promoters to limit the expression of cloned genes that are derived from sugar catabolism systems such as those from lactose or arabinose (52, 65).

Plasmids often contain a region of sequence called a multiple cloning site (MCS) with recognition sites for many different restriction endonucleases to facilitate cloning. A useful MCS often encodes a deletion derivative of the LacZ protein called LacZα as a tool to identify clones. When the Δ(*LacZ*)M15 allele is expressed in a strain with the LacZα mutant protein, it allows LacZ activity to be restored in a process called α-complementation (see "Bacterial Strains" below).

An additional consideration when selecting plasmids is the copy number, the steady-state number of copies of the plasmid that will reside in the cell. Copy number can vary from very low (for example, pSC101 and F plasmid derivatives have a copy number of one or a few per genome) to very high (for example, ColEI/pMB1 derivatives can have a copy number approaching 1,000) (109).

Conditional replicons, in which replication occurs only in certain hosts or at certain permissive temperatures, are important tools for genetics. Multiple temperature-sensitive derivatives of the pSC101 replicon have been adapted for cloning with the addition of a variety of antibiotic resistance selection systems and a useful MCS (100). An extremely useful conditional replication system has been assembled using the R6K plasmid replicon (122). Plasmids that contain the R6K-γ origin are maintained only in cells where the π protein (encoded by the *pir* gene) is expressed (70, 89). *E. coli* normally does not encode the π protein and will not replicate plasmids with the R6K-γ origin. However, R6K-γ origin plasmids will replicate in *E. coli* strains with the *pir* gene on a λ phage or otherwise integrated into the chromosome, such as in the *uidA* gene in strains BW23473 and BW23474 (Table 2) (88, 89). A mutant *pir-116* gene also exists that allows R6K-γ origin plasmids to be maintained at a higher copy number in the host cell (Table 2).

TABLE 3 General classes of plasmids commonly used in gram-negative bacteria

Founding replicon	Common examples	Host range	Comments
pMB1/ColEI	pBR322 (18), pUC vectors (135), pGEM vectors, pBluescript vectors	Narrow	pBR322 is a low-copy-number vector (~20 copies/cell) which has been adapted as a very-high-copy number vector (>300).
p15A	pACYC177, pACYC184 (26)	Narrow	The pACYC vectors are low-copy-number vectors (~15/cell). The p15A is similar to, but compatible with, the pMB1/ColEI replicon.
pSC101	pSC101	Narrow	Low-copy-number vector (~5) good for toxic genes. Temperature-sensitive derivatives exist (100).
F plasmid	pBeloBAC11 (114)	Narrow	The original Fertility (F) plasmid, Replication origin utilized in BACs
RK2 (RP4)	pSP329 (24), pCM62, pCM66 (86)	Broad	IncP incompatibility group
RSF1010	pJRD215 (40), pSUP104, pSUP204 (103)	Broad	IncQ incompatibility group
pSa	pUCD2 (30)	Broad	IncW incompatibility group
R6K	R6K	Broad	IncX incompatibility group
pBBR1MCS	pBBR1MCS-2, pBBR1MCS-3, etc. (71)	Broad	Undefined incompatibility group

31.3. GENERALIZED TRANSDUCTION

In **generalized transduction**, alleles from a donor strain can be moved to a recipient in a bacteriophage-mediated process when host DNA is mistakenly packaged into the bacteriophage. This is contrasted by another form of bacteriophage-mediated transduction called specialized transduction. In **specialized transduction**, genes adjacent to a bacteriophage integrated in the chromosome are packaged into the phage particle replacing a portion of the phage genome. The process of specialized transduction was of important historical significance and is still of great importance for making libraries of large DNA fragments but has previously been covered in detail (124).

31.3.1. Generalized Transduction in *E. coli* with Bacteriophage P1

P1 is a well-studied bacteriophage that has been adapted for generalized transduction (124). P1 is normally a lysogenic bacteriophage that replicates as a plasmid in *E. coli*. However, virulent derivatives of P1 have been isolated for use in generalized transduction. When P1 is grown on an *E. coli* host, 0.3 to 6.0% of the phage particles contain host DNA that can subsequently be moved or transduced into a new host where 10^{-4} to 10^{-5} transductants per infected donor cell can be obtained. P1 packages 110 to 115 kbp of DNA, 2.3% of the genome of standard laboratory strains. Therefore, the longest possible distance where a selected genetic marker can bring along a second marker, in a process called cotransduction, is approximately 2 min according to the genetic map of *E. coli* K-12 originally derived by conjugal mapping (14).

Virtually all strains of *E. coli* can act as donors for P1 transduction. *E. coli* strains that are *recA*⁺ produce more P1 phage particles than *recA*-negative strains (25 to 150 versus 5, respectively) per donor cell. The lower titers of P1 phage found with *recA*-negative cells contribute to lower levels of transducing phage particles. Host recombination is required to recombine the DNA packaged into P1 particles into the recipient chromosome. Therefore, recipient cells must be *recA*⁺. General principles for handling bacteriophages are explained in depth in another chapter in this text. Brief protocols are given here for use in transduction experiments.

31.3.1.1. Protocol for Determining Plaque-Forming Units of P1

1. Prepare an overnight culture with a *recA*⁺ *E. coli* strain in 5 ml of LB media with 5 mM CaCl₂.
2. Serially dilute the P1 stock preparation, called a bacteriophage lysate, with 100-fold dilutions (10^{-2}, 10^{-4}, 10^{-6}, and 10^{-8}).
3. Combine 100 μl of each phage dilution with 200 μl of the overnight culture in a glass test tube.
4. Incubate at 37°C without shaking for 15 min to allow the bacteriophage to adsorb to the cells.
5. Add 3 ml of molten top agar to one test tube containing the mixture of cells and diluted bacteriophage. The top agar should have been prepared in advance and set out of the dry or wet bath so that it is warm to touch, but still molten.
6. Gently mix the agar preparation in the test tube without producing bubbles, pour onto a fresh LB plate, and allow the solution to solidify. Repeat for each bacteriophage dilution and incubate inverted overnight at 37°C.
7. The following day count the plaques from plates bearing between 30 and 300 plaques to calculate the number of plaque-forming units of bacteriophage in the stock lysate.

31.3.1.2. Procedure for P1 Growth from Plaques

1. Prepare an overnight culture of a *recA*⁺ *E. coli* strain in 5 ml of LB media with 5 mM CaCl₂.
2. Use a sterile Pasteur pipette to collect bacteriophage plaques from an LB plate as plugs. Assemble six test tubes with 100 μl of overnight cells. To each tube with cells add 0, 1, 2, 3, 4, or 5 plaques, mix gently, and incubate at 37°C for 5 min without shaking.
3. Add 5 ml of LB media containing 5 mM CaCl₂ and continue incubating at 37°C with vigorous aeration until lysis is visible in one of the tubes (usually 3 to 5 h). If multiple tubes show good lysis, use the tube that was produced with the fewest plaques.

4. Add several drops of chloroform and vortex briefly. Centrifuge at 2,500 × g to remove the cell debris.

5. Store phage lysates at 4°C protected from light with additional chloroform. The chloroform will kill the residual donor bacteria overnight at 4°C or within a couple of hours at room temperature. Phage titers with this procedure should be in the range of 10^9 to 10^{10} PFU per ml.

31.3.1.3. Procedure for P1 Growth on Donor Strains

1. Prepare an overnight culture of a donor strain of *E. coli* in 5 ml of LB media with 5 mM CaCl$_2$ (for example, strain CAG12135 containing a Tn10 encoding tetracycline resistance inserted in the *recD* gene) (Table 2).

2. Add 100 μl of overnight cells to each of four test tubes.

3. To each tube add 0, 50, 100, or 200 μl of stock P1 bacteriophage lysate and incubate at 37°C with vigorous aeration until lysis is visible in one of the tubes (usually 3 to 5 h). Use the tubes that showed the best clearing with the fewest phage as a donor lysate (see below). Use the tube with 0 μl of stock for comparison.

4. Add several drops of chloroform and vortex briefly. Centrifuge at 2,500 × g to remove the cell debris.

5. Store phage lysates at 4°C protected from light with additional chloroform. The chloroform will kill the residual donor bacteria overnight at 4°C or within a couple of hours at room temperature.

31.3.1.4. Procedure for P1 Transduction

1. Prepare an overnight culture of a recipient strain of *E. coli* in 5 ml of LB media with 5 mM CaCl$_2$ (for example, the tetracycline-sensitive strain MG1655) (Table 2).

2. Add 300 μl of overnight cells to each of four test tubes.

3. To each test tube with recipient cells add 0, 50, 100, or 200 μl of P1 bacteriophage grown on the donor cells in a test tube (i.e., the best donor phage lysate above). Incubate without shaking at 37°C for 25 min (do not allow to go over 30 min).

4. Add 5 ml of LB medium with 10 mM sodium citrate to each tube and centrifuge at 2,500 × g to collect the cells.

5. Remove the supernatant and resuspend the cells in 5 ml of LB medium with 10 mM sodium citrate and incubate at 37°C with aeration for 1 h.

6. Centrifuge at 2,500 × g to collect the cells and plate on selective medium. Also plate 200 μl of the donor phage lysate as a control. It is also a good idea to spot all of the reagents on a selection plate.

This procedure should result in hundreds of transductants. It is important to streak purify transductants at least once on selective media to ensure that the bacteria are free from residual bacteriophage (124). Very fresh plates or plates that are wet with condensation can limit the procedure because they contribute to the killing of potential recombinants by residual bacteriophage on the selection plate. Minimal growth media can be used instead of LB where required.

31.3.3. Generalized Transduction in Other Systems

There are many bacteriophages that are suitable for generalized transduction in *E. coli* and in other bacteria (83). In *E. coli* a modified derivative of the bacteriophage T4 offers some advantages for transduction in that it packages twice as much DNA as P1 and also is less sensitive to capsules found on many pathogenic strains of *E. coli*. Special T4 derivatives must be used for transduction in that do not degrade and modify host DNA. P1 can also be used in some other bacteria, such as *Myxococcus xanthus* (64, 94). The bacteriophage P22 was first used to demonstrate the process of generalized transduction in the bacterium *S. enterica* serovar Typhimurium (140). Bacteriophages that are capable of generalized transduction have since been reported in a number of bacteria, including *V. cholerae* (CP-T1) (57), *Vibrio parahaemolyticus* and *Vibrio alginolyticus* (phi VP253 and phi VP143) (91), *Pseudomonas aeruginosa* (DMS3) (20), *Caulobacter crescentus* (Cr30T) (13), *Xanthomonas campestris* pv. *campestris* (XTP1) (131), *Sinorhizobium meliloti* (phage 11 M12) (47, 115), and *Myxococcus xanthus* (phage MX4) (23).

31.3.4. Applications for Generalized Transduction

There are many applications for generalized transduction in genetics for proteobacteria. For example, generalized transduction is an important tool for moving mutations that might be in different backgrounds into a single background; an important consideration when determining the effect of various mutant alleles is the effect of strain background. If a mutant allele has an associated selectable marker, such as a transposon insertion in a given gene, transductants can be selected directly. However, it is important to confirm with a separate assay that the allele has been recombined correctly into the new background by using a phenotype associated with the mutation or by physical means like PCR or Southern hybridization. Plasmids can also be moved by generalized transduction in a mechanism that must still accommodate the headfull packaging mechanism (81, 83). A very clean way to move a given mutation is to utilize linked alleles that can be selected for and selected against. One example involves producing bacteria that are *recA*-negative by using a linked Tn10 insertion in the sorbitol genes, *srlD3131*::Tn10. The tetracycline resistance encoded by Tn10 can be used to introduce the *srlD3131*::Tn10 allele into a target strain which in turn results in a strain incapable of growth on sorbitol as a sole carbon source. A *recA*-negative *srlD*$^+$ strain can then be used as donor for transduction. By selecting for the ability to grow on sorbitol as the sole carbon source with the *recA*-negative *srlD*$^+$ donor lysate, the *recA*-negative allele can be cotransduced. The transductants that have replaced the *srlD3131*::Tn10 allele with the *srlD*$^+$ allele can now grow on sorbitol minimal plates. The transductants that can grow on sorbital are screened for the *recA*-negative allele by using sensitivity to UV light or mitomycin C. Because the strain has lost the *srlD3131*::Tn10 allele, the organisms are now tetracycline sensitive.

Generalized transduction can also be used in mapping mutations and in a variety of mutagenesis strategies. For example, a strain containing a mutation of interest can be used to make a pool of bacteria with random transposition events. The transposons are then transduced to a new genetic background where one can screen for the mutant allele. Strains that contain both the transposon insertion and the mutation from the donor strain must have the two alleles linked. Transposon insertions that are very tightly linked to the mutation can then be mapped using arbitrary PCR to determine the region where the original mutation occurred (28).

Generalized transduction can also be used for localized mutagenesis (60). A strain containing a selectable marker in the region of interest is heavily mutagenized and then used to grow bacteriophage. The region of the chromosome that is linked to the marker is then moved to a clean genetic background, virtually ensuring that only a small region of the chromosome is mutagenized and not the entire genome. A very useful set of transposon insertions at every minute position (about every 46 kb) in *E. coli* is available for mapping and cotransducing mutations (92, 118). The set of transposon insertions have isolates with antibiotic resistances to tetracycline or kanamycin at each position and are available from the *E. coli* genetic stock center (see below). Multiple projects are under way to produce null mutations in every nonessential gene in *E. coli*.

31.4. TRANSFORMATION

Transformation of bacteria by use of either naturally competent organisms, the process of electroporation, or chemical competency relies on the direct uptake of DNA by bacteria. A number of commonly used protocols that are optimized for *E. coli* are given below. The chemical competency procedures have limited success for organisms other than *E. coli*; however, the electroporation procedure is widely applicable to a variety of proteobacteria, cyanobacteria, gram-positive bacteria, and eukaryotic cells, often with some modifications (Table 1) (56).

31.4.1. Chemical Transformation

Chemical transformation was discovered over 30 years ago and provides a convenient strategy for introducing DNA into *E. coli* and some other bacteria (Table 1) (82). While the parameters that affect the process are well understood, the exact mechanism that allows for the uptake of DNA remains obscure (56).

31.4.1.1. Procedure for Calcium Chloride-Competent *E. coli*

This simple procedure should allow transformation efficiencies of 10^6 to 10^7 CFU of bacteria per μg of DNA, which is sufficient for many applications (32).

1. Inoculate 1 liter of LB broth with 20 ml of an overnight culture of *E. coli* at 37°C.
2. Grow the cells with shaking at 300 rpm at 37°C to an overnight density at 600 nm (OD_{600}) of approximately 0.4 to 0.6. Cells should be collected in early to mid-log phase, and therefore, the appropriate OD_{600} and time will vary by strain and growth conditions.
3. Chill cells as quickly as possible by swirling the flask by hand in an ice bath for 15 min. From this point on, keep cells at 0°C. Transfer the cells to sterile prechilled centrifuge bottles and centrifuge in a cold rotor at 4,000 × g for 15 min at 4°C. Discard as much supernatant as possible including residual drops on the sides of the tubes, even if some cells are removed in the process.
4. Gently resuspend the pellets in a total of 300 ml of ice-cold sterile $MgCl_2$-$CaCl_2$ solution (80 mM $MgCl_2$, 29 mM $CaCl_2$). For example, resuspend by swirling instead of pipetting or vortexing the suspension. Centrifuge as above.
5. Gently resuspend the pellets in a total of 100 ml 0.1M $CaCl_2$ 10% glycerol and transfer to prechilled sterile 50 ml (28- by 115-mm) plastic disposable centrifuge tubes. Centrifuge at 2,500 × g at 4°C .
6. Gently resuspend in 40 ml of ice-cold sterile 0.1 M $CaCl_2$–10% glycerol and centrifuge as above.

7. Aliquot the cell suspension in 0.5-ml amounts into prechilled microcentrifuge tubes, freeze in a dry ice/ethanol bath or in liquid nitrogen and store at −70°C. When needed for transformation, the frozen cells can be thawed on ice and used as described below.
8. For transformation, thaw the cells on ice and transfer 200 μl of cell suspension to a prechilled sterile 15-ml (17- by 119-mm) plastic disposable centrifuge tube. Prepare one tube for each transformation plus tubes for the positive and negative controls (see below). Add the DNA in a volume of 5 μl or less, swirl to mix, and incubate on ice for 30 min.
9. Transfer the tubes in a plastic rack to a circulating-water bath equilibrated at 42°C. Allow the tube to stand without shaking for 60 s.
10. Rapidly transfer the tubes back into ice for 2 min.
11. Add 800 μl of SOC medium (see "Media" below) and incubate at 37°C with aeration for 1 h to allow recovery of the transformants and expression of the genes encoding antibiotic resistance.
12. Plate out the cells on selective media in a volume of 100 to 200 μl. It is also possible to collect the cells with centrifugation at 2,500 × g and plate the entire amount in a volume of 100 to 200 μl.

31.4.1.2. Procedure for Competent *E. coli* with the Inoue Method

The Inoue method is not as convenient as making calcium chloride-competent cells but allows for higher-competency cells (10^8 CFU of bacteria per μg of DNA) (61). The procedure requires growing cells at or below standard room temperature (18 to 22°C).

1. Prepare and freeze Inoue transformation buffer in advance.
2. Grow a 25-ml bacterial culture in LB at 37°C for 6 to 8 h from a plate that was streaked out the prior day.
3. In the evening subculture the starter culture into three 1-liter flasks with 250 ml of SOB medium (see "Media" below), adding 10, 4, or 2 ml individually to each flask. Allow the flasks to incubate overnight at 18 to 22°C with moderate shaking (this may need to be adjusted according to the strain used).
4. The following morning, check the OD_{600} for all three flasks to find if any is at the appropriate OD_{600} of 0.55. If none of the flasks is at the appropriate OD_{600}, continue to read the OD every 45 min until one reaches 0.55. Having three cultures helps ensure that one reaches the appropriate OD in the correct time frame. Continue with this flask and discard the other two.
5. Chill cells as quickly as possible by swirling the flask by hand in an ice bath for 15 min. From this point on, keep cells at 0°C. Transfer the cells to prechilled centrifuge bottles and centrifuge in a cold rotor at 4,000 × g for 15 min at 4°C. Discard as much supernatant as possible including residual drops on the sides of the tubes, even if some cells are removed in the process.
6. Gently resuspend the cells in a total volume of 80 ml of ice-cold Inoue transformation buffer (for example by swirling instead of pipetting or vortexing) and transfer the suspension to prechilled sterile 50-ml (28- by 115-mm) plastic disposable centrifuge tubes.
7. Centrifuge at 2,500 × g for 10 min at 4°C. Discard as much supernatant as possible including residual drops on the sides of the tubes.

8. Gently resuspend the cells in a total volume of 20 ml of ice-cold Inoue transformation buffer. Add 1.5 ml of dimethyl sulfoxide (DMSO) with gentle swirling and store on ice for 10 min. Quickly aliquot into microcentrifuge tubes, freeze in liquid nitrogen, and store at −70°C.

9. For transformation, thaw an aliquot of cells on ice. Add up to 2.5 μl of DNA solution to 50 μl of competent cells in a prechilled sterile 15-ml (17- by 119-mm) plastic disposable centrifuge tube. Prepare one tube for each transformation plus tubes for the positive and negative controls (see below). Gently mix the cell-DNA suspension with swirling, and incubate on ice for 30 min.

10. Transfer the tubes in a plastic rack to a circulating-water bath equilibrated at 42°C. Allow the tubes to sit without shaking for 60 s.

11. Rapidly transfer the tubes back into ice for 2 min.

12. Add 800 μl of SOC medium and incubate at 37°C with aeration for 1 h to allow recovery of the transformants and expression of the genes encoding antibiotic resistance.

13. Plate out the cells on selective SOB medium with 20 mM MgSO$_4$ in a volume of 100 to 200 μl. It is also possible to collect the cells with centrifugation at 2,500 × g and plate the entire amount in a volume of 100 to 200 μl.

31.4.1.3. Procedure for Rubidium Chloride-Competent E. coli

This procedure is reported to allow transformation efficiencies of 10^8 CFU of bacteria per μg of DNA or greater (55).

1. Inoculate 100 ml of Psi medium with 1 ml of an overnight culture of *E. coli* at 37°C.

2. Grow the cells with shaking at 300 rpm at 37°C to an OD$_{600}$ of approximately 0.4 to 0.6. Cells should be collected in early to mid-log phase, and therefore, the appropriate OD$_{600}$ and growth time will vary by strain and growth conditions.

3. Chill cells as quickly as possible by swirling the flask by hand in an ice bath for 15 min. From this point on, keep cells at 0°C. Transfer the cell suspension to two prechilled sterile 50-ml (28- by 115-mm) plastic disposable centrifuge tubes. Centrifuge at 2,500 × g at 4°C. Discard as much supernatant as possible including residual drops on the sides of the tubes, even if some cells are removed in the process.

4. Gently resuspend in a total of 40 ml of ice-cold TfbI (see "Buffers" below) and leave on ice for 15 min. Centrifuge at 2,500 × g at 4°C. Discard as much supernatant as possible including residual drops on the sides of the tubes.

5. Gently resuspend in a total of 4 ml of TfbII (see "Buffers" below) and leave on ice for 15 min.

6. Aliquot the cell suspension in 0.5-ml amounts into prechilled microcentrifuge tubes, freeze in a dry ice/ethanol bath or on liquid nitrogen and store at −70°C. When needed for transformation, the frozen cells can be thawed on ice and used as described below.

7. For transformation, thaw an aliquot of cells on ice. Add up to 2.5 μl of DNA solution to 50 μl of competent cells in a prechilled sterile 15-ml (17- by 119-mm) plastic disposable centrifuge tube. Prepare one tube for each transformation plus tubes for the positive and negative controls (see below). Gently mix the cell-DNA suspension with swirling, and incubate on ice for 30 min.

8. Transfer the tubes in a plastic rack to a circulating-water bath equilibrated at 42°C. Allow the tube to stand without shaking for 60 s.

9. Rapidly transfer the tubes back into ice and allow to chill for 2 min.

10. Add 800 μl of SOC medium and incubate at 37°C with aeration for 1 h to allow recovery of the transformants and expression of the genes encoding antibiotic resistance.

11. Plate out the cells on selective LB medium in a volume of 100 to 200 μl. It is also possible to collect the cells with centrifugation at 2,500 × g and plate the entire amount in a volume of 100 to 200 μl.

31.4.2. Electroporation

Electroporation is a broadly successful strategy for introducing DNA into proteobacteria and cyanobacteria (56, 68, 93). In this technique, bacteria are subjected to a high-strength electrical field to form pores in the cells. While the pores are present, DNA in solution can diffuse into the cells. Electroporation has proven to be useful in a variety of *E. coli* strains and, under optimal conditions, results in frequencies of 10^9 to 10^{10} transformants per μg of plasmid DNA. The identification and optimization of electroporation conditions are different for different genera, and often for different species, and require systematic testing of several variables. These variables include, but are not limited to, growth phase, cell density, DNA concentration, field strength, pulse duration, and temperature. Electroporation is the mechanism of choice for making DNA libraries, introducing DNA substrates for λ Red recombination, and difficult ligation reactions. Because of the high frequency of transformation one must also consider the possibility of cotransformation of multiple plasmids.

31.4.2.1. Procedure for Electroporation

The following procedure is a modification of the procedure recommended by Bio-Rad Laboratories (Richmond, CA). It is routinely used to transform *E. coli* and *S. enterica* serovar Typhimurium. The apparatus is a Bio-Rad Gene Pulser with a Pulse Control Unit or equivalent.

1. Inoculate 1 liter of LB broth with 20 ml of an overnight culture of *E. coli* at 37°C.

2. Grow the cells with shaking at 300 rpm at 37°C to an OD$_{600}$ of approximately 0.4 to 0.6. Cells should be collected in early to mid-log phase, and therefore, the appropriate OD$_{600}$ and growth time will vary by strain and growth conditions.

3. Chill cells as quickly as possible by swirling the flask by hand in an ice bath for 15 min. From this point on, keep cells at 0°C. Transfer the cells to prechilled centrifuge bottles and centrifuge in a cold rotor at 4,000 × g for 15 min at 4°C. Discard as much supernatant as possible including residual drops on the sides of the tubes, even if some cells are removed in the process.

4. Resuspend the pellets in a total of 1 liter of ice-cold sterile H$_2$O (use very pure water, ideally Milli-Q water or higher). Centrifuge as above. Resuspend in 0.5 liter of ice-cold sterile H$_2$O and centrifuge as above.

5. Resuspend in a total of 20 ml of ice-cold sterile 10% glycerol in H$_2$O and transfer to prechilled sterile 50 ml (28- by 115-mm) plastic disposable centrifuge tubes. Centrifuge at 2,500 × g for 10 min at 4°C. Resuspend the cells in 3 ml of ice-cold sterile 10% glycerol in H$_2$O. The cell concentration should be about 3 × 10^{10} CFU of bacteria per ml.

6. Aliquot the cell suspension in 0.1- to 0.2-ml amounts into prechilled sterile microcentrifuge tubes, freeze in a dry ice/ethanol bath or on liquid nitrogen and store at −70°C.

When needed for electroporation, the frozen cells can be thawed on ice and used as described below.

7. Prechill all tubes and electroporation cuvettes on ice, preparing enough supplies for each sample and the positive and negative controls (see below). Mix 40 μl of cells with 1 or 2 μl of DNA in a chilled microcentrifuge tube (note that DNA should be in H_2O or a low-ionic-strength buffer such as Tris-EDTA [TE]). Mix well, but gently, and incubate on ice for ~1 min. Transfer the cell-DNA mixture to a cold 0.2-cm cuvette, shake to the bottom, and place in the Gene Pulser cuvette holder. Apply one pulse with the gene pulser set at 2.5 kV and 25 μF and the pulse controller set at 200 Ω. If using a 0.1-cm cuvette, set the apparatus at 1.5 to 1.8 kV.

8. Immediately suspend the cells in 1 ml of SOC. It is important to add SOC immediately, since waiting will result in lower transformation frequencies. Transfer the cells to a tube, and incubate at 37°C with aeration for 1 h to allow recovery of the transformants and expression of the genes encoding antibiotic resistance. The apparatus should allow you to read an expression of the pulse strength for each sample called the "time constant." Ideally the time constant should be between 4 and 5 ms for a 0.2-cm cuvette containing 40 μl of cells.

9. Plate out the cells on selective medium in a volume of 100 to 200 μl.

It should be noted that electroporation of E. coli with ligated DNA results in low transformation frequencies unless the DNA is first transferred to a low-ionic-strength buffer (62, 132, 138). This can be accomplished by precipitating the reaction with ethanol or by diluting the reaction 1/10 to 1/20. For ethanol precipitation add 2.5 volumes of ethanol and 0.1 volume of 3 M sodium acetate (pH 5.2) followed by centrifugation and a 70% ethanol wash (109).

Cuvettes may be recycled by soaking them in bleach for a few minutes and washing extensively with tap water, then in distilled water, and finally with 70% ethanol. Allow the cuvettes to dry before using. Do not use acetone and do not autoclave the cuvettes.

31.4.2.2. Adaptations for Electroporation

There are additional considerations for some strains of bacteria that are unable to withstand the change in conditions associated with preparing the cells (35, 93, 129). For example, with Geobacter sulfurreducens an optimized electroporation buffer was required consisting of 1 mM HEPES (pH 7.0), 1 mM MgCl$_2$, and 175 mM sucrose (35). In G. sulfurreducens the addition of MgCl$_2$ prevents cell lysis, presumably by stabilizing the outer membrane. Sucrose was added to abolish the change in osmolarity from the growth media. Finally, high viability during storage at −70°C required that the cells be frozen in 10% DMSO and not glycerol. Other variations may be required to allow this procedure to be optimized for other nonenteric bacteria (93).

31.4.3. Natural Competency

Natural competence is perhaps the easiest way to introduce DNA into bacteria. Well-studied systems for natural competence exist in H. pylori (130), Haemophilus influenzae (11), and Neisseria sp. (51, 123). Some of the cyanobacteria are also naturally competent (68). Interestingly, the genes encoding natural competence are fairly widespread in many bacteria that are not generally considered naturally competent (29). One possibility is that the appropriate signal or conditions that are needed to allow these organisms to be made naturally competent have not been

identified. Interestingly, a novel genetic strategy was used to isolate a hypercompetent Legionella pneumophila strain from a normal noncompetent background (112). It is also worth noting that natural competency can be of limited value when transforming plasmids because in many of these systems the DNA is imported as single-stranded fragments.

31.4.3.1. Procedure for Chromosomal DNA Isolation

Isolation of plasmid DNA is covered in depth elsewhere in this text. This procedure is appropriate for preparing multiple small samples by using a microcentrifuge (7). Many commercial kits are also available for isolating plasmid and chromosomal DNA.

1. Pellet 1.5 ml of an overnight culture of bacteria for 2 min at top speed in a microcentrifuge. Remove all traces of media.

2. Use a vortex mixer and/or pipetting to completely resuspend the cell pellet in 567 μl of TE buffer.

3. Add 30 μl of 10% sodium dodecyl sulfate (prepared in H_2O) and mix by inversion 10 times.

4. Add 4 μl of proteinase K (prepared at a concentration of 10 mg/ml and stored at −20°C as aliquots that are never used twice) and mix by inversion 10 times.

5. Incubate the samples for 1 h at 37°C, mixing every 15 min by inverting several times.

6. Add 100 μl of 5 M NaCl and mix by extensive mixing (200 times); to avoid shearing of the DNA in this and all subsequent steps, mix by flicking or shaking the tubes; do not vortex or use a pipette for mixing.

7. Add 80 μl of prewarmed (65°C) hexadecyltrimethyl-ammonium bromide-NaCl solution (dissolve 10g of hexadecyltrimethyl-ammonium bromide and 4.1 g of NaCl into 80 ml of H_2O with stirring with low heat. Adjust the final volume to 100 ml and store at room temperature). Mix very well by flicking or shaking the tube (200 times). Incubate at 65°C for 10 min, mixing 10 times at 5 min.

8. Add 780 μl of chloroform and isoamyl alcohol solution (24:1), mix by flicking or shaking the tube 100 times, and centrifuge for 5 min at 9,000 × g in a microcentrifuge.

9. Transfer 700 μl of the aqueous phase (upper layer) to a new tube. Add 700 μl of phenol-chloroform-isoamyl alcohol solution (24:24:1), mix 100 times, and centrifuge for 5 min at 9,000 × g in a microcentrifuge.

10. Transfer 600 μl of the aqueous phase (upper layer) to a new tube and precipitate with 360 μl of isopropanol. High-molecular-weight DNA and RNA should be visible. Pellet briefly in a microcentrifuge and wash once with 70% ethanol.

11. Dry briefly and dissolve in 50 μl of sterile water.

31.4.4. Controls and Precautions for Transformation Strategies

In every transformation experiment it is important to include both positive and negative controls. For the negative control a sample containing water or TE and lacking DNA should be included. The negative control is important to rule out contamination and defective selection plates. A positive control should be included where 10 picograms of circular superhelical plasmid DNA is transformed to determine the transformation efficiency with the electroporation and chemical competency procedures (lower-efficiency competency procedures may require up to 1 nanogram of DNA). An additional control to consider is doping one of the samples with circular superhelical plasmid DNA. This will control for the very real possibility that a contaminant in your DNA preparation is inhibiting transformation. For example, residual salts, agarose, or enzyme present in the DNA sample could be affecting transformation efficiencies in cells that would otherwise be highly competent.

31.5. CONJUGATION

Conjugation remains an important tool to introduce DNA into proteobacteria and cyanobacteria. The process of conjugation is very efficient and can reach 100% in many organisms. The high efficiency of this process likely overwhelms the restriction and modification systems that might otherwise limit other methods for introducing DNA. Historically conjugation was critical in the mobilization of chromosomal genes for genetic mapping. The topic of conjugal mapping has been covered in great detail in another review (80).

Conjugation requires direct cell-to-cell contact between the plasmid-containing cell and the recipient bacterium. During conjugation, one strand of the plasmid is delivered into the recipient cell in a process that resembles rolling-circle DNA replication. Plasmids can be characterized as either self-transmissible or mobilizable. Self-transmissible plasmids encode all of the functions required to move between bacteria. Self-transmissible plasmids can also facilitate the movement of other plasmids (and chromosomes) by a process called mobilization. Mobilizable plasmids contain only a subset of the functions required for transfer, such as a cis-acting origin of transfer (oriT) and perhaps the proteins (sometimes called Mob functions) that specifically recognize this oriT. Mobilizable plasmids can move between bacteria only when specific transfer functions are provided in trans in the cell. Mobilizable plasmids can be produced de novo but also occur naturally.

Conjugation can be used as a tool to deliver plasmids that are capable of being stably maintained in the target host or as a tool to deliver "suicide vectors" that cannot replicate in the target host. Suicide vectors can be used to deliver transposons or they can be integrated into the host chromosome if they contain sequences that are homologous to the chromosome.

Two widely used strains for mobilizing the transfer of plasmids containing RP4 origins of transfer are SM10 and S17-1 (117). To provide the RP4 transfer functions without having the normally self-transmissible RP4 vector itself being transferred into recipient cells, the RP4 vector was integrated into the E. coli chromosome. Strain SM10 was created from E. coli strain C600 by integrating the plasmid RP4-2-Tet::Mu (kanamycin resistant and tetracycline sensitive) into the chromosome by forcing its maintenance in the presence of the incompatible RP4 derivative pSR120 (tetracycline resistant); the cell was later cured of pSR120. The integrated RP4-2-Tet::Mu plasmid imparts kanamycin resistance and can excise and transfer only at a very low frequency ($\sim 10^{-5}$) (117). The same strategy was used to produce S17-1 from the E. coli strain 294, in which an RP4-2-Tet::Mu derivative was used where a Tn7 insertion inactivated the kanamycin resistance gene (RP4-2-Tet::Mu Kan::Tn7), thereby imparting a strain that is kanamycin sensitive, but spectinomycin, streptomycin, and trimethoprim resistant. E. coli strain BW20767 is a more recent construction containing the RP4-2-Tet::Mu Kan::Tn7 region in a strain that is also capable of maintaining plasmids with the R6K origin of replication (Table 2) (88). A convenient

set of strains and plasmids that allow cloning of genes, into conditionally replicating R6K plasmids, and mobilization with the RP4 system is available (88). These vectors are also capable of both positive and negative selection.

Conjugation occurs at highest efficiency on solid media. Recipient and donor cells can be mixed directly on nonselective agar plates and allowed to mate from 1 h to overnight. Transconjugants are isolated on a medium that selects for a marker encoded on the conjugal plasmid and a marker found only in the recipient cells (usually carried on the chromosome). Controls include plating the donor and recipient cells individually. It is sometimes more convenient to do a "triparental" mating. In a triparental mating three strains are used; one donor strain contains a self-transmissible plasmid which can transfer to a second strain and mobilize the resident plasmid to transfer to the final recipient. A strategy is normally used to prevent the self-transmissible plasmid from being maintained in the final recipient (e.g., the use of a narrow-host-range conjugal plasmid). An advantage of triparental mating is that the mobilizable plasmid does not need to be separately introduced into the special donor cell.

31.6. BARRIERS TO DNA TRANSFER

Many bacterial host-encoded endonucleases can inhibit the transfer of DNA between bacteria. Multiple genetic and technical strategies exist for overcoming these barriers.

31.6.1. Restriction of Foreign DNA by E. coli

Many standard laboratory strains of E. coli have multiple systems that allow the destruction of foreign DNA. These systems can form a barrier to introducing DNA. E. coli K-12 strains naturally carry the EcoKI restriction system in the hsdRMS genes. DNA that is not modified by adenine methylation on the appropriate recognition site will be cleaved by the EcoKI system. Strains bearing mutations in the hsdR gene are useful because they will not restrict DNA without the appropriate methylation, but they will methylate DNA at the appropriate recognition sites so that the DNA can subsequently be used to propagate DNA for introduction into $hsdR^+$ hosts. Strains bearing mutations in the hsdS gene will not restrict or methylate DNA at the appropriate recognition site.

E. coli also has a series of systems for restricting DNA that is methylated at specific cytosine and adenine positions. These systems are encoded by the mcrA, mcrBC, and mrr genes. Strains that are mutant in the mcr and mrr systems should be considered when difficulty is encountered in cloning DNA that has not been propagated in E. coli. The mcrA gene is carried on the excisable genetic element e14. Strains that are e14-negative do not have the mcrA gene (14) (Table 2).

31.6.2. Barriers That Inhibit Moving DNA from E. coli into Other Bacteria

As mentioned, most bacteria encode restriction systems that can inhibit the introduction of exogenously replicated DNAs. Various strategies can be used to limit the effects of these systems. In cases where the actual restriction system is not known, genetic strategies can be used to directly select strains with a lower restriction barrier (84, 129). It is also possible to specifically mutate endonucleases that can be identified from an available genome sequence. It was found that deleting the restriction endonuclease YenI vastly increased the transformability of Yersina enterocolitica (66). A similar strategy was utilized for *Vibrio cholerae* strains, whereby deleting multiple DNases allowed for a modest increase in transformation efficiency (48).

Various technical strategies can also deal with barriers that inhibit moving DNA from *E. coli* into other bacteria. One simple strategy is to use more DNA along with the highest-efficiency mechanism to introduce DNA in order to overwhelm the restriction systems. This technique has been found to be successful in *Neisseria* sp. (51). There are anecdotal reports that the addition of sheared salmon sperm DNA can allow a 100-fold increase in transformation efficiency by electroporation in *Pseudomonas* sp., possibly by acting as a nonspecific target for endogenous nucleases. Because methylation can also make DNA a target for degradation, a property that is common in gram-positive bacteria, it might be appropriate in some cases to propagate DNA in strains of *E. coli* that do not methylate DNA (Table 2). Strategies also exist for methylating DNA in vitro, and it is formally possible that commercially available methylases such as the CpG methylase (M.SssI) could protect DNA from restriction. An interesting strategy that was used in *H. pylori* involves the in vitro treatment of DNA with cell-free extracts from the recipient strain to apply the appropriate methylation pattern (44). This strategy could provide an important option when working with strain backgrounds with multiple restriction-modification systems.

31.7. BACTERIAL STRAINS

For historical reasons many laboratories work with different derivatives of the original *E. coli* K-12 bacterium (9). Only two laboratory strains of *E. coli* have been sequenced, MG1655 and W3110 (15). MG1655 is an early derivative of the original *E. coli* K-12 strain that was cured of the endogenous λ lysogen and Fertility plasmid. Many *E. coli* K-12 derivatives continue to be used including AB1157, MC4100, and HB101 (Table 2). However, there is a general move towards doing experiments with MG1655.

31.7.1. Basics of *E. coli* Strain Genotypes

There are a few common genotypic features that should be checked when choosing strains for experimentation. Genes in the lactose operon are often important in strains used for cloning because many cloning vectors have special genetic requirements. As described above in the section on plasmids, the LacZα fragment located in many useful multicloning sites requires that a special deletion derivative of *lacZ* reside in the cell called Δ(*lacZ*)M15. The Δ(*lacZ*)M15 allele is often not carried at its natural position on the *E. coli* chromosome. Most cloning strains have extensive deletions that remove the lactose operon on the chromosome such as the 97-kb Δ(*ykfD-mhpD*)169 deletion allele, which for historical reasons is called Δ(*argF-lac*)169 (99). Typically the Δ(*lacZ*)M15 allele is carried on an F plasmid derivative as in JP109 and XL1-Blue or in the chromosome on the λ-like bacteriophage lysogen φ80 as in the case of DH5α (Table 2). Cloning strains may have mutations in various restriction systems to remove barriers to cloning some DNAs (see above). Ideally cloning strains should also be *endA*-negative, as this gene encodes a stable nonspecific endonuclease that contaminates plasmid DNA preparations. When attempting to use expression plasmids that require arabinose for induction, it is important to note that the *araD139* allele found in strains such as MC4100 causes the strain to be sensitive to arabinose (52, 65).

Many bacterial strains have resistance to various antibiotics. Many mutations in the *rpsL* gene encoding the ribosomal protein, small S12 subunit, can impart resistance to streptomycin (e.g., *rpsL20*, *rpsL31*, *rpsL104*, and *rpsL136*). Mutations in *gyrA* can impart resistance to nalidixic acid (e.g., *gyr29* and *gyr96*). Transposons often encode antibiotic resistance determinants. Commonly used transposons include Tn5 (encoding kanamycin resistance), Tn7 (encoding streptomycin, spectinomycin, and trimethoprim resistance), Tn9 (encoding chloramphenicol resistance), and Tn10 (encoding tetracycline resistance). Transposons that have localized to a certain gene are indicated with a specific nomenclature; for example, a Tn10 insertion in the *srlD* gene would be indicated as *srlD*::Tn10.

31.7.2. How To Get Bacterial Strains

E. coli strains can be obtained from a variety of sources. The most comprehensive collection of strains can be found at the *E. coli* Genetic Stock Center at Yale University (http://cgsc2.biology.yale.edu). Large *E. coli* genome projects are currently under way to isolate selectable null mutations and conditional alleles for every open reading frame in *E. coli*. Plasmid clones for every open reading frame in *E. coli* are also being constructed. These reagents should be available in the near future from the *E. coli* Genome Project at University of Wisconsin—Madison (http://www.genome.wisc.edu/resources.htm) and the *E. coli* National BioResource Project in Japan (http://shigen.lab.nig.ac.jp/ecoli/strain/top/top.jsp). Other sources of bacterial strains include the American Type Culture Collection (http://www.atcc.org), biotechnology companies, and independent investigators.

31.8. MEDIA

31.8.1. LB Medium

Dissolve 10 g of Bacto tryptone, 5 g of Bacto yeast extract, and 10 g of NaCl in 1 liter of deionized H_2O. Adjust to pH 7.0 with 5 N NaOH and sterilize by autoclaving.

31.8.2. SOB Medium

Dissolve 20 g of Bacto tryptone, 5 g of Bacto yeast extract, and 0.5 g of NaCl in 950 ml of deionized H_2O. Add 10 ml of 250 mM KCl, adjust to pH 7.0 with 5 N NaOH, and adjust the volume to 1 liter. Sterilize by autoclaving. Just prior to use add 5 ml of sterile 2 M $MgCl_2$ per liter (sterilized by autoclaving).

31.8.3. SOC Medium

SOC is identical to SOB medium except that it contains 20 mM glucose. Glucose is added to SOB after autoclaving and cooling as 20 ml of a 1 M filter-sterilized solution.

31.8.4. Psi Medium

Dissolve the following into 800 ml of deionized H_2O: 5 g of Bacto yeast extract, 20 g of Bacto tryptone, 5 g of $MgSO_4$. Adjust the pH to 7.6 with 5 M NaOH and adjust volume to 1 liter. Aliquot and sterilize by autoclaving.

31.8.5. M9 Medium

Make a 10× stock of M9 salts by adding the following to 800 ml of deionized H_2O: 60 g of Na_2HPO_4 (anhydrous), 30 g of KH_2PO_4 (anhydrous), 5 g of NaCl, 10 g of NH_4Cl. Adjust the volume to 1 liter and sterilize by autoclaving.

Combine the 10× salts with sterile distilled water to a 1× solution in 1 liter with $CaCl_2$ (10 ml of 0.01 M $CaCl_2$), $MgSO_4$ (1 ml of 1 M $MgSO_4$–$7H_2O$), vitamin B1/thiamine hydrochloride (0.5 ml of a 1% stock), and a carbon source (10 ml of a 20% solution of sugar or glycerol). If needed, add 40 μg/ml of the required L-amino acid.

31.8.6. M63 Medium

Make a 10× stock of M63 salts by adding the following to 800 ml of deionized H_2O: 136 g of KH_2PO_4, 20 g of $(NH_4)_2SO_4$, 5 mg of $FeSO_4$–$7H_2O$. Adjust the pH to 7.0 with KOH, adjust the volume to 1 liter, and sterilize by autoclaving. Combine with sterile distilled water to a 1× solution in 1 liter with $MgSO_4$ (1ml of 1 M $MgSO_4$–$7H_2O$), vitamin B1/thiamine hydrochloride (0.5 ml of a 1% stock), and a carbon source (10 ml of a 20% solution of sugar or glycerol). If needed, add 40 μg/ml of the required L-amino acid.

31.8.7. Supplements

Antibiotics are prepared, stored, and added to media prior to use according to Table 4. Stock solutions of 5-bromo-4-chloro-3-indolyl-β-D-galactopyranoside (X-Gal) and 5-bromo-4-chloro-3-indolyl phosphate (XP) are prepared at 20 mg/ml in dimethylformamide. Each is light sensitive and should be stored in an amber glass bottle or wrapped in aluminum foil at −20°C. A stock solution of isopropyl-β-D-thiogalactopytanoside (IPTG) is 200 mg/ml in H_2O. This solution is filter sterilized and stored in aliquots at −20°C. X-Gal (80 μl) and IPTG (8 μl) can be spread on plates just prior to use, but the solution should be allowed to dry completely. XP and X-Gal are added to agar media at a concentration of 40 mg/l.

31.8.8. Preparation of Agar Media

LB agar plates are prepared by including 15 g of Bacto agar per liter of medium prior to autoclaving. Soft agar solutions for agar overlays are made by adding 8 g of Bacto agar per liter of medium. Soft agar can be made and stored in 100-ml aliquots. Soft-agar aliquots can be carefully melted in the microwave prior to use (with the bottle cap removed) and then stored in a wet or dry bath for many days.

31.8.9. Tetracycline-Sensitive-Selective (TSS) Agar Plates

Combine 2.5 g of Bacto tryptone, 2.5 g of Bacto yeast extract, 5 g of NaCl, 7.5 g of agar, and 25 mg of chlorotetracycline HCl (Sigma) with 400 ml of H_2O and autoclave. Cool to ~45°C and add sterile solutions of 5 g of NaH_2PO_4–H_2O in 100 ml of H_2O, 3 ml of 2 mg/ml fusaric acid, and 2.5 ml of 20 mM $ZnCl_2$. TSS plates must be used within a few days of preparation (17, 88).

31.9. BUFFERS

31.9.1. Inoue Transformation Buffer

Dissolve all of the following buffer constituents in 800 ml of pure H_2O: 55 mM $MnCl_2$ (10.88 g of $MnCl_2$–$4H_2O$), 15 mM $CaCl_2$ (2.20 g of $CaCl_2$–$2H_2O$), 250 mM KCl (18.65 g of KCl). Add 20 ml of 0.5 M PIPES (piperazine-N,N′-bis [2-ethanesulfonic acid]) to give a final concentration of 10 mM (prepare as a 0.5 M solution by dissolving 15.1 g of PIPES in 80 ml of Milli-Q H_2O [with pH adjusted to 6.7 with 5 M KOH] and final volume brought to 100 ml, and stored in 20-ml aliquots at −20°C). Adjust the volume to 1 liter, filter sterilize, aliquot, and store at −20°C.

31.9.2. TfbI

Dissolve all of the following buffer constituents in 150 ml of distilled H_2O; 30 mM CH_3COOK (0.588 g of CH_3COOK), 100 mM RbCl (2.42 g of RbCl), 10 mM $CaCl_2$ (0.294 g of $CaCl_2$–$2H_2O$), 50 mM $MnCl_2$ (0.100 g of $MnCl_2$–$4H_2O$). Add 30 ml of glycerol (15% [vol/vol] final) and adjust the pH to 5.8 with dilute acetic acid. Adjust volume to 200 ml, filter sterilize, aliquot, and store at −20°C.

31.9.3. TfbII

Dissolve all of the following buffer constituents in 70 ml of distilled H_2O: 10 mM MOPS (morpholinepropanesulfonic acid) (0.21 g of MOPS), 75 mM CaCl2 (1.1 g of $CaCl_2$), 10 mM RbCl (1.1 g of RbCl). Add 15 ml of glycerol (15% [vol/vol] final) and adjust the pH to 6.5 with dilute NaOH. Adjust the volume to 100 ml, filter sterilize, aliquot, and store at −20°C.

TABLE 4 Antibiotic concentrations for stock solutions and medium supplementation

Antibiotic	Solvent	Storage temp (°C)	Stock concn (mg/ml)	Plate concn (μg/ml)	Amount (ml/liter)
Ampicillin	Water[a]	−20	100	100	1
Chloramphenicol	EtOH[b]	−20	10	30	3
Gentamicin	Water[a]	−20	10	10	1
Kanamycin	Water[a]	−20	50	50	1
Nalidixic acid	NaOH[c]	4	5	20	4
Rifampicin	MeOH[d]	−20	25	100	4
Spectinomycin	Water[a]	−20	50	100	2
Streptomycin	Water[a]	−20	50	50	1
Tetracycline[e]	MeOH[d]	−20	5	20	4
Trimethoprim	DMSO[f]	−20	50	100	2

[a]Filter sterilize the solution with a 0.2-μm-pore-size filter.
[b]Suspend in 100% ethanol.
[c]Suspend in 0.1 N NaOH, which converts the acid to the sodium salt.
[d]Suspend in 100% methanol; store in the dark.
[e]Tetracycline is reported to be incompatible with magnesium ions and therefore should be tested before being used in minimal media.
[f]Suspend in DMSO.

31.10. FINAL CONSIDERATIONS

Many of the compounds used in these experiments are known to be toxic, caustic, and/or carcinogenic. In all cases, before using a compound the material safety data sheet should be read and all safety recommendations followed. In addition, laboratory personnel should work with their local environmental safety group to ensure proper disposal of all waste. Investigators should consult the latest National Institutes of Health guidelines for work with recombinant DNA before initiating any experiments and work closely with their local institutional biosafety committee.

I owe thanks to Matthew DeLisa and Steve Winans for comments on the chapter and to Jose Carlos Huguet-Tapia, Qiaojuan Shi, and Adam Parks, who trial ran experiments. Special thanks to the many individuals who provided important species-specific information: Andrew Camilli, Maddalena Coppi, Andrew Darwin, Brian Kvitko, Nina Salama, James Shapleigh, Howard Shuman, Daniel Stein, Chris Waters, Patrick Viollier, and Steve Winans. Work in the lab is supported by the National Science Foundation (MCB-0315316).

31.11. REFERENCES

31.11.1. General References

1. Ausubel, F. M., R. Brent, R. E. Kingston, D. D. More, J. G. Seidman, J. A. Smith, and K. Struhl. 1988. *Current Protocols in Molecular Biology*. Greene Publishing Associates Inc. and John Wiley & Sons, Inc., New York, NY.
2. Miller, J. H. 1991. *A Short Course in Bacterial Genetics: A Laboratory Manual and Handbook for* Escherichia coli *and Related Bacteria*. Cold Spring Harbor Laboratory Press, Cold Spring Harbor, NY.
3. Miller, J. H. (ed.). 1991. *Methods in Enzymology*, vol. 204. *Bacterial Genetic Systems*. Academic Press, San Diego, CA.
4. Sambrook, J., and D. W. Russell. 2001. *Molecular Cloning: A Laboratory Manual*, 3rd ed. Cold Spring Harbor Laboratory Press, Cold Spring Harbor, NY.
5. Silhavy, T. J., M. L. Berman, and L. Enquist. 1984. *Experiments with Gene Fusions*. Cold Spring Harbor Laboratory Press. Cold Spring Harbor, NY.

31.11.2. Specific References

6. Anderson, D. G., and S. C. Kowalczykowski. 1997. The translocating RecBCD enzyme stimulates recombination by directing RecA protein onto ssDNA in a *chi*-regulated manner. *Cell* 90:77–86.
7. Ausubel, F. M., R. Brent, R. E. Kingston, D. D. Moore, J. G. Seidman, J. A. Smith, and K. Struhl. 1988. *Current Protocols in Molecular Biology*. Greene Publishing Associates Inc. and John Wiley & Sons, Inc., New York, NY.
8. Ayres, E. K., V. J. Thomson, G. Merino, D. Balderes, and D. H. Figurski. 1993. Precise deletions in large bacterial genomes by vector-mediated excision (VEX). The *trfA* gene of promiscuous plasmid RK2 is essential for replication in several gram-negative hosts. *J. Mol. Biol.* 230:174–185.
9. Bachmann, B. J. 1996. Derivations and genotypes of some mutant derivatives of *Escherichia coli* K-12, p. 2460–2488. *In* F. C. Neidhardt, R. I. Curtiss, J. L. Ingraham, E. C. C. Lin, K. B. Low, B. Magasanik, W. S. Reznikoff, M. Riley, M. Schaechter, and H. E. Umbarger (ed.), Escherichia coli *and* Salmonella: *Cellular and Molecular Biology*, 2nd ed. American Society for Microbiology, Washington, DC.
10. Barbour, S. D., H. Nagaishi, A. Templin, and A. J. Clark. 1970. Biochemical and genetic studies of recombination proficiency in *Escherichia coli*. II. Rec+ revertants caused by indirect suppression of *rec-* mutations. *Proc. Natl. Acad. Sci. USA* 67:128–135.
11. Barcak, G. J., M. S. Chandler, R. J. Redfield, and J.-F. Tomb. 1991. Genetic systems in *Haemophilus influenzae*. *Methods Enzymol.* 204:321–342.
12. Baudin, A., O. Ozier-Kalogeropoulos, A. Denouel, F. Lacroute, and C. Cullin. 1993. A simple and efficient method for direct gene deletion in *Saccharomyces cerevisiae*. *Nucleic Acids Res.* 21:3329–3330.
13. Bender, R. A. 1981. Improved generalized transducing bacteriophage for *Caulobacter crescentus*. *J. Bacteriol.* 148:734–735.
14. Berlyn, M. K., K. B. Low, and K. E. Rudd. 1996. Linkage map of *Escherichia coli* K-12. *In* F. C. Neidhardt, R. I. Curtiss, J. L. Ingraham, E. C. C. Lin, K. B. Low, B. Magasanik, W. S. Reznikoff, M. Riley, M. Schaechter, and H. E. Umbarger (ed.), Escherichia coli *and* Salmonella: *Cellular and Molecular Biology*, 2nd ed. American Society for Microbiology, Washington, DC.
15. Blattner, F. R., G. Plunkett III, C. A. Bloch, N. T. Perna, V. Burland, M. Riley, J. Collado-Vides, J. D. Glasner, C. K. Rode, G. F. Mayhew, J. Gregor, N. W. Davis, H. A. Kirkpatrick, M. A. Goeden, D. J. Rose, B. Mau, and Y. Shao. 1997. The complete genome sequence of *Escherichia coli* K-12. *Science* 277:1453–1474.
16. Blomfield, I. C., V. Vaughn, R. F. Rest, and B. I. Eisenstein. 1991. Allelic exchange in *Escherichia coli* using the *Bacillus subtilis sacB* gene and a temperature-sensitive pSC101 replicon. *Mol. Microbiol.* 5:1447–1457.
17. Bochner, B. R., H. C. Huang, G. L. Schieven, and B. N. Ames. 1980. Positive selection for loss of tetracycline resistance. *J. Bacteriol.* 143:926–933.
18. Bolivar, F., R. L. Rodriguez, P. J. Greene, M. C. Betlach, H. L. Heyneker, and H. W. Boyer. 1977. Construction and characterization of new cloning vehicles. II. A multipurpose cloning system. *Gene* 2:95–113.
19. Boyd, D., D. S. Weiss, J. C. Chen, and J. Beckwith. 2000. Towards single-copy gene expression systems making gene cloning physiologically relevant: lambda InCh, a simple *Escherichia coli* plasmid-chromosome shuttle system. *J. Bacteriol.* 182:842–847.
20. Budzik, J. M., W. A. Rosche, A. Rietsch, and G. A. O'Toole. 2004. Isolation and characterization of a generalized transducing phage for *Pseudomonas aeruginosa* strains PAO1 and PA14. *J. Bacteriol.* 186:3270–3273.
21. Burrus, V., and M. K. Waldor. 2004. Shaping bacterial genomes with integrative and conjugative elements. *Res. Microbiol.* 155:376–386.
22. Cai, Y. P., and C. P. Wolk. 1990. Use of a conditionally lethal gene in *Anabaena* sp. strain PCC 7120 to select for double recombinants and to entrap insertion sequences. *J. Bacteriol.* 172:3138–3145.
23. Campos, J. M., J. Geisselsoder, and D. R. Zusman. 1978. Isolation of bacteriophage MX4, a generalized transducing phage for *Myxococcus xanthus*. *J. Mol. Biol.* 119:167–178.
24. Cangelosi, G. A., E. A. Best, G. Martinetti, and E. W. Nester. 1991. Genetic analysis of *Agrobacterium*. *Methods Enzymol.* 204:384–397.
25. Casadaban, M. J. 1976. Transposition and fusion of the *lac* genes to selected promoters in *Escherichia coli* using bacteriophage lambda and Mu. *J. Mol. Biol.* 104:541–555.
26. Chang, A. C., and S. N. Cohen. 1978. Construction and characterization of amplifiable multicopy DNA cloning vehicles derived from the p15A cryptic miniplasmid. *J. Bacteriol.* 134:1141–1156.
27. Cherepanov, P. P., and W. Wackernagel. 1995. Gene disruption in *Escherichia coli*: TcR and KmR cassettes with the option of Flp-catalyzed excision of the antibiotic-resistance determinant. *Gene* 158:9–14.

28. **Chun, K. T., H. J. Edenberg, M. R. Kelley, and M. G. Goebl.** 1997. Rapid amplification of uncharacterized transposon-tagged DNA sequences from genomic DNA. *Yeast* **13:**233–400.

29. **Claverys, J. P., and B. Martin.** 2003. Bacterial "competence" genes: signatures of active transformation, or only remnants? *Trends Microbiol.* **11:**161–165.

30. **Close, T. J., D. Zaitlin, and C. I. Kado.** 1984. Design and development of amplifiable broad-host-range cloning vectors: analysis of the *vir* region of *Agrobacterium tumefaciens* plasmid pTiC58. *Plasmid* **12:**111–118.

31. **Cohen, M. F., J. C. Meeks, Y. A. Cai, and C. P. Wolk.** 1998. Transposon mutagenesis of heterocyst-forming filamentous cyanobacteria. *Methods Enzymol.* **305:**3–17.

32. **Cohen, S. N., A. C. Chang, and L. Hsu.** 1972. Nonchromosomal antibiotic resistance in bacteria: genetic transformation of *Escherichia coli* by R-factor DNA. *Proc. Natl. Acad. Sci. USA* **69:**2110–2114.

33. **Conchas, R. F., and E. Carniel.** 1990. A highly efficient electroporation system for transformation of *Yersinia*. *Gene* **87:**133–137.

34. **Copeland, N. G., N. A. Jenkins, and D. L. Court.** 2001. Recombineering: a powerful new tool for mouse functional genomics. *Nat. Rev. Genet.* **2:**769–779.

35. **Coppi, M. V., C. Leang, S. J. Sandler, and D. R. Lovley.** 2001. Development of a genetic system for *Geobacter sulfurreducens*. *Appl. Environ. Microbiol.* **67:**3180–3187.

36. **Costantino, N., and D. L. Court.** 2003. Enhanced levels of lambda Red-mediated recombinants in mismatch repair mutants. *Proc. Natl. Acad. Sci. USA* **100:**15748–15753.

37. **Court, D. L., J. A. Sawitzke, and L. C. Thomason.** 2002. Genetic engineering using homologous recombination. *Annu. Rev. Genet.* **36:**361–388.

38. **Court, D. L., S. Swaminathan, D. Yu, H. Wilson, T. Baker, M. Bubunenko, J. Sawitzke, and S. K. Sharan.** 2003. Mini-lambda: a tractable system for chromosome and BAC engineering. *Gene* **315:**63–69.

39. **Datsenko, K. A., and B. L. Wanner.** 2000. One-step inactivation of chromosomal genes in *Escherichia coli* K-12 using PCR products. *Proc. Natl. Acad. Sci. USA* **97:**6640–6645.

40. **Davison, J., M. Heustersprente, N. Chevalier, V. Ha-Thi, and F. Brunel.** 1987. Vectors with restriction site banks. V. pJRD215, a wide-host-range cosmid vector with multiple cloning sites. *Gene* **51:**275–280.

41. **de Lorenzo, V., M. Herrero, J. M. Sanchez, and K. N. Timmis.** 1998. Mini-transposons in microbial ecology and environmental biotechnology. *FEMS Microbiol. Ecol.* **27:** 211–224.

42. **Derbise, A., B. Lesic, D. Dacheux, J. M. Ghigo, and E. Carniel.** 2003. A rapid and simple method for inactivating chromosomal genes in *Yersinia*. *FEMS Immunol. Med. Microbiol.* **38:**113–116.

43. **DeWitt, S. K., and E. A. Adelberg.** 1962. The occurrence of a genetic transposition in a strain of *Escherichia coli*. *Genetics* **47:**577–585.

44. **Donahue, J. P., D. A. Israel, R. M. Peek, M. J. Blaser, and G. G. Miller.** 2000. Overcoming the restriction barrier to plasmid transformation of *Helicobacter pylori*. *Mol. Microbiol.* **37:**1066–1074.

45. **Donohue, T. J., and S. Kaplan.** 1991. Genetic techniques in *Rhodospirillaceae*. *Methods Enzymol.* **204:**459–485.

46. **Ely, B.** 1991. Genetics of *Caulobacter crescentus*. *Methods Enzymol.* **204:**372–384.

47. **Finan, T. M., E. Hartweig, K. LeMieux, K. Bergman, G. C. Walker, and E. R. Signer.** 1984. General transduction in *Rhizobium meliloti*. *J. Bacteriol.* **159:**120–124.

48. **Focareta, T., and P. A. Manning.** 1991. Distinguishing between the extracellular DNases of *Vibrio cholerae* and de-

49. **Gay, P., D. Le Coq, M. Steinmetz, T. Berkelman, and C. I. Kado.** 1985. Positive selection procedure for entrapment of insertion sequence elements in gram-negative bacteria. *J. Bacteriol.* **164:**918–921.

50. **Glazebrook, J., and G. C. Walker.** 1991. Genetic techniques in *Rhizobium meliloti*. *Methods Enzymol.* **204:**398–418.

51. **Gunn, J. S., and D. C. Stein.** 1996. Use of a non-selective transformation technique to construct a multiply restriction/modification-deficient mutant of *Neisseria gonorrhoeae*. *Mol. Gen. Genet.* **251:**509–517.

52. **Guzman, L.-M., D. Belin, M. J. Carson, and J. Beckwith.** 1995. Tight regulation, modulation, and high-level expression by vectors containing the arabinose P_{BAD} promoter. *J. Bacteriol.* **177:**4121–4130.

53. **Haldimann, A., and B. L. Wanner.** 2001. Conditional-replication, integration, excision, and retrieval plasmid-host systems for gene structure-function studies of bacteria. *J. Bacteriol.* **183:**6384–6393.

54. **Hamilton, C. M., M. Aldea, B. K. Washburn, P. Babitzke, and S. R. Kushner.** 1989. New method for generating deletions and gene replacements in *Escherichia coli*. *J. Bacteriol.* **171:**4617–4622.

55. **Hanahan, D.** 1983. Studies on transformation of *Escherichia coli* with plasmids. *J. Mol. Biol.* **166:**557–580.

56. **Hanahan, D., J. Jessee, and F. R. Bloom.** 1991. Plasmid transformation of *Escherichia coli* and other bacteria. *Methods Enzymol.* **204:**63–113.

57. **Hava, D. L., and A. Camilli.** 2001. Isolation and characterization of a temperature-sensitive generalized transducing bacteriophage for *Vibrio cholerae*. *J. Microbiol. Methods* **46:**217–225.

58. **Herring, C. D., J. D. Glasner, and F. R. Blattner.** 2003. Gene replacement without selection: regulated suppression of amber mutations in *Escherichia coli*. *Gene* **311:**153–163.

59. **Heuermann, D., and R. Haas.** 1998. A stable shuttle vector system for efficient genetic complementation of *Helicobacter pylori* strains by transformation and conjugation. *Mol. Gen. Genet.* **257:**519–528.

60. **Hong, J. S., and B. N. Ames.** 1971. Localized mutagenesis of any specific small region of the bacterial chromosome. *Proc. Natl. Acad. Sci. USA* **68:**3158–3162.

61. **Inoue, H., H. Nojima, and H. Okayama.** 1990. High efficiency transformation of *Escherichia coli* with plasmids. *Gene* **96:**23–28.

62. **Jacobs, M., S. Wnendt, and U. Stahl.** 1990. High-efficiency electro-transformation of *Escherichia coli* with DNA from ligation mixtures. *Nucleic Acids Res.* **18:**1653.

63. **Jasin, M., and P. Schimmel.** 1984. Deletion of an essential gene in *Escherichia coli* by site-specific recombination with linear DNA fragments. *J. Bacteriol.* **159:**783–786.

64. **Kaiser, D., and M. Dworkin.** 1975. Gene transfer to *Myxobacterium* by *Escherichia coli* phage P1. *Science* **187:**653–654.

65. **Khlebnikov, A., K. A. Datsenko, T. Skaug, B. L. Wanner, and J. D. Keasling.** 2001. Homogeneous expression of the P(BAD) promoter in *Escherichia coli* by constitutive expression of the low-affinity high-capacity AraE transporter. *Microbiology* **147:**3241–3247.

66. **Kinder, S. A., J. L. Badger, G. O. Bryant, J. C. Pepe, and V. L. Miller.** 1993. Cloning of the YenI restriction endonuclease and methyltransferase from *Yersinia enterocolitica* serotype O8 and construction of a transformable R-M+ mutant. *Gene* **136:**271–275.

67. **Koch, B., L. E. Jensen, and O. Nybroe.** 2001. A panel of Tn7-based vectors for insertion of the *gfp* marker gene or

velopment of a transformation system. *Mol. Microbiol.* **5:** 2547–2555.

for delivery of cloned DNA into Gram-negative bacteria at a neutral chromosomal site. *J. Microbiol. Methods* **45:** 187–195.

68. **Koksharova, O. A., and C. P. Wolk.** 2002. Genetic tools for cyanobacteria. *Appl. Microbiol. Biotechnol.* **58:**123–137.

69. **Kolisnychenko, V., G. Plunkett III, C. D. Herring, T. Feher, J. Posfai, F. R. Blattner, and G. Posfai.** 2002. Engineering a reduced *Escherichia coli* genome. *Genome Res.* **12:**640–647.

70. **Kolter, R., M. Inuzuka, and D. R. Helinski.** 1978. Trans-complementation-dependent replication of a low molecular weight origin fragment from plasmid R6K. *Cell* **15:** 1199–1208.

71. **Kovach, M. E., P. H. Elzer, D. S. Hill, G. T. Robertson, M. A. Farris, R. M. Roop II, and K. M. Peterson.** 1995. Four new derivatives of the broad-host-range cloning vector pBBR1MCS, carrying different antibiotic-resistance cassettes. *Gene* **166:**175–176.

72. **Kowalczykowski, S. C.** 2000. Initiation of genetic recombination and recombination-dependent replication. *Trends Biochem. Sci.* **25:**156–165.

73. **Kowalczykowski, S. C., D. A. Dixon, A. K. Eggleston, S. D. Lauder, and W. M. Rehrauer.** 1994. Biochemistry of homologous recombination in *Escherichia coli*. *Microbiol. Rev.* **58:**401–465.

74. **Kuzminov, A.** 1999. Recombinational repair of DNA damage in *Escherichia coli* and bacteriophage lambda. *Microbiol. Mol. Biol. Rev.* **63:**751–813.

75. **Lacks, S., and B. Greenberg.** 1977. Complementary specificity of restriction endonucleases of *Diplococcus pneumoniae* with respect to DNA methylation. *J. Mol. Biol.* **114:**153–168.

76. **Lafontaine, D., and D. Tollervey.** 1996. One-step PCR mediated strategy for the construction of conditionally expressed and epitope tagged yeast proteins. *Nucleic Acids Res.* **24:**3469–3471.

77. **Lederberg, J., and E. L. Tatum.** 1946. Gene recombination in *Escherichia coli*. *Nature* **158:**558.

78. **Lee, E. C., D. Yu, J. Martinez de Velasco, L. Tessarollo, D. A. Swing, D. L. Court, N. A. Jenkins, and N. G. Copeland.** 2001. A highly efficient *Escherichia coli*-based chromosome engineering system adapted for recombinogenic targeting and subcloning of BAC DNA. *Genomics* **73:**56–65.

79. **Link, A. J., D. Phillips, and G. M. Church.** 1997. Methods for generating precise deletions and insertions in the genome of wild-type *Escherichia coli*: application to open reading frame characterization. *J. Bacteriol.* **179:** 6228–6237.

80. **Low, K. B.** 1991. Conjugational methods for mapping with Hfr and F-prime strains. *Methods Enzymol.* **204:**43–62.

81. **Low, K. B., and D. D. Porter.** 1978. Modes of gene transfer and recombination in bacteria. *Annu. Rev. Genet.* **12:** 249–287.

82. **Mandel, M., and A. Higa.** 1970. Calcium-dependent bacteriophage DNA infection. *J. Mol. Biol.* **53:**159–162.

83. **Margolin, P.** 1987. Generalized transduction, p. 1154–1168. *In* F. C. Neidhardt, J. L. Ingraham, K. B. Low, B. Magasanik, M. Schaechter, and H. E. Umbarger (ed.), *Escherichia coli and* Salmonella typhimurium: *Cellular and Molecular Biology*, vol. 2. American Society for Microbiology, Washington, DC.

84. **Marra, A., and H. A. Shuman.** 1989. Isolation of a *Legionella pneumophila* restriction mutant with increased ability to act as a recipient in heterospecific matings. *J. Bacteriol.* **171:**2238–2240.

85. **Marx, C. J., and M. E. Lidstrom.** 2002. Broad-host-range cre-*lox* system for antibiotic marker recycling in gram-negative bacteria. *BioTechniques* **33:**1062–1067.

86. **Marx, C. J., and M. E. Lidstrom.** 2001. Development of improved versatile broad-host-range vectors for use in methylotrophs and other Gram-negative bacteria. *Microbiology* **147:**2065–2075.

87. **Maxson, M. E., and A. J. Darwin.** 2004. Identification of inducers of the *Yersinia enterocolitica* phage shock protein system and comparison to the regulation of the RpoE and Cpx extracytoplasmic stress responses. *J. Bacteriol.* **186:**4199–4208.

88. **Metcalf, W. W., W. Jiang, L. L. Daniels, S. K. Kim, A. Haldimann, and B. L. Wanner.** 1996. Conditionally replicative and conjugative plasmids carrying *lacZ* alpha for cloning, mutagenesis, and allele replacement in bacteria. *Plasmid* **35:**1–13.

89. **Metcalf, W. W., W. Jiang, and B. L. Wanner.** 1994. Use of the rep technique for allele replacement to construct new *Escherichia coli* hosts for maintenance of R6K gamma origin plasmids at different copy numbers. *Gene* **138:**1–7.

90. **Miller, J. H.** 1992. *A Short Course in Bacterial Genetics.* Cold Spring Harbor Laboratory Press, Cold Spring Harbor, NY.

91. **Muramatsu, K., and H. Matsumoto.** 1991. Two generalized transducing phages in *Vibrio parahaemolyticus* and *Vibrio alginolyticus*. *Microbiol. Immunol.* **35:**1073–1084.

92. **Nichols, B. P., O. Shafiq, and V. Meiners.** 1998. Sequence analysis of Tn*10* insertion sites in a collection of Escherichia coli strains used for genetic mapping and strain construction. *J. Bacteriol.* **180:**6408–6411.

93. **Nickoloff, J.** 1995. *Electroporation Protocols for Microorganisms*, vol. 47. Humana Press, Totowa, NJ.

94. **O'Connor, K. A., and D. R. Zusman.** 1983. Coliphage P1-mediated transduction of cloned DNA from *Escherichia coli* to *Myxococcus xanthus*: use for complementation and recombinational analyses. *J. Bacteriol.* **155:**317–329.

95. **O'Connor, M., M. Peifer, and W. Bender.** 1989. Construction of large DNA segments in *Escherichia coli*. *Science* **244:**1307–1312.

96. **Oppenheim, A. B., A. J. Rattray, M. Bubunenko, L. C. Thomason, and D. L. Court.** 2004. In vivo recombineering of bacteriophage lambda by PCR fragments and single-strand oligonucleotides. *Virology* **319:**185–189.

97. **Orr-Weaver, T. L., J. W. Szostak, and R. J. Rothstein.** 1981. Yeast transformation: a model system for the study of recombination. *Proc. Natl. Acad. Sci. USA* **78:**6354–6358.

98. **Peters, J. E., and N. L. Craig.** 2001. Tn7: smarter than we thought. *Nat. Rev./Mol. Cell Biol.* **2:**806–814.

99. **Peters, J. E., T. E. Thate, and N. L. Craig.** 2003. Definition of the *Escherichia coli* MC4100 genome by use of a DNA array. *J. Bacteriol.* **185:**2017–2021.

100. **Phillips, G. J.** 1999. New cloning vectors with temperature-sensitive replication. *Plasmid* **41:**78–81.

101. **Posfai, G., V. Kolisnychenko, Z. Bereczki, and F. R. Blattner.** 1999. Markerless gene replacement in *Escherichia coli* stimulated by a double-strand break in the chromosome. *Nucleic Acids Res.* **27:**4409–4415.

102. **Price-Carter, M., J. Tingey, T. A. Bobik, and J. R. Roth.** 2001. The alternative electron acceptor tetrathionate supports B12-dependent anaerobic growth of *Salmonella enterica* serovar Typhimurium on ethanolamine or 1,2-propanediol. *J. Bacteriol.* **183:**2463–2475.

103. **Priefer, U. B., R. Simon, and A. Puhler.** 1985. Extension of the host range of *Escherichia coli* vectors by incorporation of RSF1010 replication and mobilization functions. *J. Bacteriol.* **163:**324–330.

104. **Raibaud, O., M. Mock, and M. Schwartz.** 1984. A technique for integrating any DNA fragment into the chromosome of *Escherichia coli*. *Gene* **29:**231–241.

105. **Rong, R., M. M. Slupska, J. H. Chiang, and J. H. Miller.** 2004. Engineering large fragment insertions into the chromosome of *Escherichia coli*. *Gene* **336:**73–80.

106. **Rothmel, R. K., A. M. Chakrabarty, A. Berry, and A. Darzins.** 1991. Genetic systems in *Pseudomonas. Methods Enzymol.* **204:**485–514.

107. **Russell, C. B., and F. W. Dahlquist.** 1989. Exchange of chromosomal and plasmid alleles in *Escherichia coli* by selection for loss of a dominant antibiotic sensitivity marker. *J. Bacteriol.* **171:**2614–2618.

108. **Russell, C. B., D. S. Thaler, and F. W. Dahlquist.** 1989. Chromosomal transformation of *Escherichia coli recD* strains with linearized plasmids. *J. Bacteriol.* **171:**2609–2613.

109. **Sambrook, J., E. F. Fritsch, and T. Maniatis.** 1989. *Molecular Cloning: a Laboratory Manual*, 2nd ed. Cold Spring Harbor Press, Cold Spring Harbor, NY.

110. **Sauer, B.** 1998. Inducible gene targeting in mice using the Cre/*lox* system. *Methods* **14:**381–392.

111. **Seifert, H. S., and M. So.** 1991. Genetic systems in pathogenic *Neisseriae. Methods Enzymol.* **204:**342–357.

112. **Sexton, J. A., and J. P. Vogel.** 2004. Regulation of hypercompetence in *Legionella pneumophila. J. Bacteriol.* **186:**3814–3825.

113. **Shi, Z. X., H. L. Wang, K. Hu, E. L. Feng, X. Yao, G. F. Su, P. T. Huang, and L. Y. Huang.** 2003. Identification of *alkA* gene related to virulence of *Shigella flexneri* 2a by mutational analysis. *World J. Gastroenterol.* **9:**2720–2725.

114. **Shizuya, H., B. Birren, U. J. Kim, V. Mancino, T. Slepak, Y. Tachiiri, and M. Simon.** 1992. Cloning and stable maintenance of 300-kilobase-pair fragments of human DNA in *Escherichia coli* using an F-factor-based vector. *Proc. Natl. Acad. Sci. USA* **89:**8794–8797.

115. **Sik, T., J. Horvath, and S. Chatterjee.** 1980. Generalized transduction in *Rhizobium meliloti. Mol. Gen. Genet.* **178:**511–516.

116. **Silverman, M., R. Showalter, and L. McCarter.** 1991. Genetic analysis in *Vibrio. Methods Enzymol.* **204:**515–536.

117. **Simon, R., U. B. Priefer, and A. Puhler.** 1982. A broad host range mobilization system for in vivo genetic engineering: transposon mutagenesis in gram-negative bacteria. *Biotechnology* **1:**784–791.

118. **Singer, M., T. A. Baker, G. Schnitzler, S. M. Deischel, M. Goel, W. Dove, K. J. Jaacks, A. D. Grossman, J. W. Erickson, and C. A. Gross.** 1989. A collection of strains containing genetically linked alternating antibiotic resistance elements for genetic mapping of *Escherichia coli. Microbiol. Rev.* **53:**1–24.

119. **Skorupski, K., and R. K. Taylor.** 1996. Positive selection vectors for allelic exchange. *Gene* **169:**47–52.

120. **Slauch, J. M., and T. J. Silhavy.** 1991. Genetic fusions as experimental tools. *Methods Enzymol.* **204:**213–248.

121. **Smeets, L. C., J. J. Bijlsma, S. Y. Boomkens, C. M. Vandenbroucke-Grauls, and J. G. Kusters.** 2000. *comH*, a novel gene essential for natural transformation of *Helicobacter pylori. J. Bacteriol.* **182:**3948–3954.

122. **Stalker, D. M., R. Kolter, and D. R. Helinski.** 1979. Nucleotide sequence of the region of an origin of replication of the antibiotic resistance plasmid R6K. *Proc. Natl. Acad. Sci. USA* **76:**1150–1154.

123. **Stein, D. C.** 1991. Transformation of *Neisseria gonorrhoeae*: physical requirements of the transforming DNA. *Can. J. Microbiol.* **37:**345–349.

124. **Sternberg, N. L., and R. Maurer.** 1991. Bacteriophage-mediated generalized transduction in *Escherichia coli* and *Salmonella typhimurium. Methods Enzymol.* **204:**18–43.

125. **Stibitz, S., W. Black, and S. Falkow.** 1986. The construction of a cloning vector designed for gene replacement in *Bordetella pertussis. Gene* **50:**133–140.

126. **Szostak, J. W., T. Orr-Weaver, R. J. Rothstein, and F. W. Stahl.** 1983. The double-strand-break repair model for recombination. *Cell* **33:**25–35.

127. **Trulzsch, K., T. Sporleder, E. I. Igwe, H. Russmann, and J. Heesemann.** 2004. Contribution of the major secreted Yops of *Yersinia enterocolitica* O:8 to pathogenicity in the mouse infection model. *Infect. Immun.* **72:**5227–5234.

128. **Uzzau, S., N. Figueroa-Bossi, S. Rubino, and L. Bossi.** 2001. Epitope tagging of chromosomal genes in *Salmonella. Proc. Natl. Acad. Sci. USA* **98:**15264–15269.

129. **Visick, K. G., and E. G. Ruby.** 1996. Construction and symbiotic competence of a *luxA*-deletion mutant of *Vibrio fischeri. Gene* **175:**89–94.

130. **Wang, Y., K. P. Roos, and D. E. Taylor.** 1993. Transformation of *Helicobacter pylori* by chromosomal metronidazole resistance and by a plasmid with a selectable chloramphenicol resistance marker. *J. Gen. Microbiol.* **139**(Pt. 10)**:**2485–2493.

131. **Weiss, B. D., M. A. Capage, M. Kessel, and S. A. Benson.** 1994. Isolation and characterization of a generalized transducing phage for *Xanthomonas campestris* pv. *campestris. J. Bacteriol.* **176:**3354–3359.

132. **Willson, T. A., and N. M. Gough.** 1988. High voltage *E. coli* electro-transformation with DNA following ligation. *Nucleic Acids Res.* **16:**11820.

133. **Winans, S. C., S. J. Elledge, J. H. Krueger, and G. C. Walker.** 1985. Site-directed insertion and deletion mutagenesis with cloned fragments in *Escherichia coli. J. Bacteriol.* **161:**1219–1221.

134. **Woodcock, D. M., P. J. Crowther, J. Doherty, S. Jefferson, E. DeCruz, M. Noyer-Weidner, S. S. Smith, M. Z. Michael, and M. W. Graham.** 1989. Quantitative evaluation of *Escherichia coli* host strains for tolerance to cytosine methylation in plasmid and phage recombinants. *Nucleic Acids Res.* **17:**3469–3478.

135. **Yanisch-Perron, C., J. Vieira, and J. Messing.** 1985. Improved M13 phage cloning vectors and host strains: nucleotide sequences of the M13mp18 and pUC19 vectors. *Gene* **33:**103–119.

136. **Yu, D., H. M. Ellis, E. C. Lee, N. A. Jenkins, N. G. Copeland, and D. L. Court.** 2000. An efficient recombination system for chromosome engineering in *Escherichia coli. Proc. Natl. Acad. Sci. USA* **97:**5978–5983.

137. **Yu, D., J. A. Sawitzke, H. Ellis, and D. L. Court.** 2003. Recombineering with overlapping single-stranded DNA oligonucleotides: testing a recombination intermediate. *Proc. Natl. Acad. Sci. USA* **100:**7207–7212.

138. **Zabarovsky, E. R., and G. Winberg.** 1990. High efficiency electroporation of ligated DNA into bacteria. *Nucleic Acids Res.* **18:**5912.

139. **Zhang, Y., F. Buchholz, J. P. Muyrers, and A. F. Stewart.** 1998. A new logic for DNA engineering using recombination in *Escherichia coli. Nat. Genet.* **20:**123–128.

140. **Zinder, N. D., and J. Lederberg.** 1952. Genetic exchange in *Salmonella. J. Bacteriol.* **64:**679–699.

32

Genetic Exchange in Gram-Positive Bacteria[†]

CHRISTOPHER J. KRISTICH, CHRISTINE E. SALOMON,
AND GARY M. DUNNY

[†]A chapter titled "Gene Transfer in Gram-Positive Bacteria" in *Methods for General and Molecular Bacteriology* was authored by Simon M. Cutting and Philip Youngman. This chapter is a completely revised, reorganized, and expanded presentation by the current authors.

32.1. INTRODUCTION AND OVERVIEW OF GENE TRANSFER MECHANISMS AND THEIR APPLICATIONS

The gram-positive (gram$^+$) bacteria represent an extremely diverse group of organisms (Fig. 1), many of which are of significant medical, environmental, or industrial importance. From the basic biology perspective, it is noteworthy that both horizontal gene transfer and the existence of DNA as the genetic material were first discovered in an organism from this group, *Streptococcus pneumoniae* (7, 149). The three known mechanisms of gene transfer in bacteria (conjugation, transformation, and transduction) are widespread in gram$^+$ bacteria, although the relative prevalence of each of these processes in different organisms is highly variable, and in most cases not clearly documented. These differences in the "preferred" modes of genetic transfer have significant implications both for the natural evolution of the organisms and for the development of successful approaches to their genetic manipulation in the laboratory. Thus, when designing schemes for manipulation of gram$^+$ bacteria, a useful rule of thumb is that approaches known to be successful in the most closely related gram$^+$ bacteria are the most likely to work (see Fig. 1), although there certainly are exceptions.

There are several important operational constraints strongly affecting natural gene transfer events and the use of genetic transfer methods in the laboratory. The first important consideration is that the gram$^+$ bacteria can be divided into two main groups on the basis of DNA base composition. In terms of gene regulation, promoter structure/function, transcription, and translation, there are major differences between the "high-GC" and "low-GC" groups resulting from codon usage, DNA secondary structure, and other phenomena directly related to DNA base composition. As noted later in this chapter, certain members of the high-GC group carry out forms of conjugation whose mechanisms may be fundamentally different from that observed in other bacteria. The other overriding issue affecting the manner in which natural and engineered genetic changes occur is whether the organism in question is naturally competent for genetic transformation; this affects not only the entry of transferred DNA into a new host but also the subsequent recombination events that ultimately determine whether a heritable change results from intercellular transfer of genetic material. While the ability to undergo genetic transformation is encoded within the core genome of competent organisms, both conjugation and transduction abilities are generally encoded by accessory elements (plasmids, conjugative transposons, and bacteriophages) which have a degree of genetic autonomy from their bacterial hosts. These elements may be viewed as parasitic selfish DNAs whose long-term evolutionary goals may not completely overlap with those of the host bacteria, although a preponderance of evidence clearly indicates that lateral transfer events mediated by mobile elements have had a tremendous influence on microbial evolution and on how phylogenetic relationships are determined (89, 99).

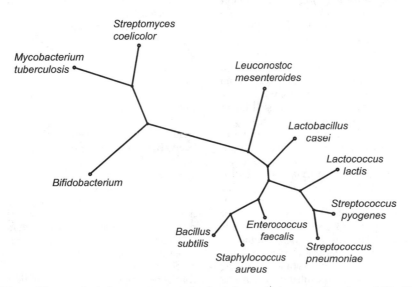

FIGURE 1 Unrooted phylogenetic tree of selected gram$^+$ bacteria based on 16S rRNA sequences. The sequences for each organism were obtained from the NCBI database and subjected to phylogenetic analysis assuming maximum parsimony and bootstrapping (500 times resampling). The length of the lines between the species corresponds to phylogenetic distance.

This chapter presents an overview of gene transfer systems and their applications in several important groups of gram$^+$ bacteria. In the following paragraphs, a summary of each of the known mechanisms of transfer as they have been described for gram$^+$ genera is presented. In the remaining sections of the chapter, transfer systems and their application in various groups of related gram$^+$ bacteria are summarized. We have provided extensive, though not comprehensive, citations both to the primary literature and to review articles. The use of various transfer systems for genetic manipulation of gram$^+$ bacteria is described, and representative protocols are presented as illustrations of general approaches that can be applied to gram$^+$ bacteria. In addition, a table describing selected plasmids (Table 1) that have found use in genetic manipulation of gram$^+$ bacteria is provided. While it is not feasible to provide complete protocols or comprehensive plasmid lists for every individual organism discussed, the material presented provides resources to enable investigators to identify or design the appropriate protocols for transfer studies with their respective organism of choice. In many cases, genetic tools used for manipulation of one gram$^+$ species can be readily adapted for use in others, particularly among the low-GC gram$^+$ species. For example, tools useful for genetic manipulation of the low-GC gram$^+$ *Listeria monocytogenes*, although not discussed explicitly in this chapter, have been derived from tools applicable in other low-GC gram$^+$ species, including some described herein (130).

32.1.1. Conjugation

Conjugation in bacteria is defined as a lateral transfer mechanism where the transmission of genetic material from one cell to another requires direct cell-to-cell contact. During conjugation a channel is formed between the donor and recipient cells. This channel serves as a direct conduit for the transferred DNA, such that the process cannot be disrupted by exogenous nucleases. Transfer is generally unidirectional (although multiple rounds of transfer may give the appearance of bidirectionality) and requires viable, metabolically active donor cells. Conjugation has a tremendous impact on natural evolution of microorganisms (and probably higher organisms as well) because the transfer systems can allow genetic exchange between very distantly related species, because it can occur very efficiently under optimal conditions, and because the number of genes transferred can be very large. While many conjugative transposons and some conjugative plasmids can be maintained stably in a very broad range of bacterial hosts, other conjugative plasmids show a very narrow host range; the latter elements may still effect lateral transfer across species barriers by carrying transposable elements or by mobilizing plasmids that may survive in a new host independently of the conjugative element itself. In gram$^+$ bacteria, there seem to be at least two fundamentally different forms of genetic exchange that fit the definition of conjugation. As described in a recent review (150), a variety of plasmids and conjugative transposons identified in gram$^+$ species encode three essential transfer functions that are also found in well-characterized conjugative elements from gram-negative (gram$^-$) organisms. These include a proteinaceous DNA processing machinery that specifically nicks and unwinds the transferred DNA while remaining attached to the 5' end of the nick. The resulting relaxosome complex interacts with a coupling protein to guide the transferred DNA to a mating channel extending from the inner cytoplasmic membrane through the cell envelopes of both the donor and recipient. Sequence analysis has shown considerable conservation in the protein components of the DNA processing systems and coupling proteins of gram$^+$ and gram$^-$ systems, whereas the channel components seem to be less related and are largely uncharacterized in gram$^+$ bacteria. Conjugative elements transferred by such systems contain an *oriT* (origin of transfer) DNA sequence that serves as a site for binding and strand-specific and nucleotide-specific nicking of the DNA strand destined for transfer. Small nonconjugative plasmids often carry *oriT* and DNA processing components that enable them to be efficiently mobilized by coresident conjugative plasmids whose transfer machinery can interact with the relaxosome complex of the small plasmid. Mobilization of small plasmids in the laboratory can also be effected by insertion of *oriT* sequences using recombinant DNA techniques; these manipulations can be very useful in genetic engineering applications for bacterial species refractory to introduction of DNA by transformation, a common problem for researchers studying gram$^+$ bacteria.

In the past decade, analysis of conjugation systems in the high-GC gram$^+$ genera, particularly the filamentous organisms and the mycobacteria, has suggested that these organisms carry out novel forms of transfer utilizing one or more sets of molecular machinery which show no apparent evolutionary relatedness to the more conventional systems described above. In *Streptomyces* spp., a number of conjugative elements have been identified, including small and large elements, high- and low-copy-number elements, and linear and circular elements; some of these are described later in this chapter. In some cases a single protein encoded by the mobile element has been shown to be sufficient for transfer; these Tra proteins are related to septal DNA translocator proteins of the SpoIIIE/FtsK family. This suggests that the core genome of these organisms encodes functions involved in mycelial fusion events that must be required to form a functional mating contact used by the Tra protein to effect transfer. A small *cis*-acting DNA site called *clt* has been implicated in transfer of certain plasmids (244). Although *clt* would appear to play a role similar to *oriT*s in other systems, attempts to demonstrate strand-specific nicking have been unsuccessful, and it is possible that double-stranded DNA is transferred by these systems. Very recently Derbyshire and colleagues have presented striking evidence for a novel form of conjugation in mycobacterial species (123, 394, 395). As in the *Streptomyces* systems, the evidence suggests that the core genome of these organisms may encode functions that facilitate mating pair formation, and that the DNA processing mechanism(s) is fundamentally distinct from the *oriT*-based systems.

Significantly, the genes encoding the mating pair formation machinery of several gram$^-$ conjugation systems have been shown to belong to the type IV protein secretion superfamily (51). Type IV systems are involved in export of virulence factor proteins from bacterial pathogens into host target cells (61). Even among the better-studied gram$^+$ transfer systems, very little is known about the mechanism of mating pair formation or about whether these systems could be involved in export of virulence factors. This is an unaddressed question that is of great importance relative to the genetic basis of virulence in these organisms. Interestingly, a recent genetic study of mycobacterial conjugation showed that genes important for conjugation mapped to a locus believed to be essential for secretion of virulence factors (123), suggesting that the functional link between conjugative mating pair formation and virulence may extend through all the diverse transfer systems identified in gram$^+$ microbes.

TABLE 1 Representative plasmids of use for genetic manipulation of gram⁺ bacteria[a]

Plasmid name	Gram⁺ origin of replication	Gram⁺ selectable marker	Gram⁻ origin of replication	Gram⁻ selectable marker	Property	Host range	Reference(s)
pAM401	pIP501	Cm	p15A (pACYC184)	Cm, Tc	Shuttle vector for general cloning in LGC	LGC	406
pAT18	pAMβ1	Em	ColE1 (pBR322)	Em	Shuttle vector for general cloning in LGC; carries RK2 *oriT* for mobilization to gram⁺ hosts from *E. coli*	LGC	378
pAT28	pAMβ1	Sp	ColE1 (pBR322)	Sp	Shuttle vector for general cloning in LGC; carries RK2 *oriT* for mobilization to gram⁺ hosts from *E. coli*	LGC	377
pCJK47	Suicide	Em	pWV01 (requires provision of RepA in *trans*)	Em	Conjugative delivery vector for allelic exchange carrying gram⁺ counterselectable marker	*E. faecalis*	210a
pCL52.1	pE194	Tc	pSC101	Sp	TS; delivery vector for allelic exchange in *Staphylococcus*	S	230
pCL84	Suicide	Tc	pSC101	Sp	Site-specific integration into *attB* used by phage L54a in *S. aureus*	S	221
pCN series	pT181 or pI258	Sp, Tc, Cm, Kn, or Em	ColE1 (pUC19)	Ap	Interchangeable cassette-based shuttle vectors for complementation, operon fusions in *Staphylococcus*	S	54
pDG1661–pDG1664 series pDG1728–pDG1731 series	Suicide	Cm, Sp, or Em	ColE1	Ap	Vectors for ectopic single-copy integration into *B. subtilis amyE* locus for complementation, operon fusion analysis	B	152
pDL276	pVA380-1	Kn	ColE1 (pUC18)	Kn	Shuttle vector for general cloning in LGC; RCR replicon	LGC	104
pDL278	pVA380-1	Sp	ColE1 (pUC18)	Sp	Shuttle vector for general cloning in LGC; RCR replicon	LGC	219
pFW series	Suicide	Sp, Tc, Cm, Kn, or Em	ColE1 (pUC18)	Sp, Tc, Cm, Kn, or Em	Delivery vector for gene inactivation or allelic exchange in LGC	LGC	306
pG⁺host4	pWV01	Cm, Em	pWV01	Cm, Em	TS; delivery vector for allelic exchange in LGC	LGC	253
pG⁺host5	pWV01	Cm, Em	ColE1 (pBR322)	Cm, Em	TS; delivery vector for allelic exchange in LGC; contains pBR322 origin for replication at 37°C in *E. coli*	LGC	30
pIJ486	pIJ101	Nm, Kn, Th			Promoter-probe plasmid	A	397
pIJ698	SCP2*	Th, Hy	p15A (RK2)	Vm, Kn	Bifunctional *E. coli/Streptomyces* replacement vector	A	199
pIJ80	SCP2*	Nm			Unstable *Streptomyces* plasmid used to cure SCP2* derivatives from host strains	A	370
pIJ941	SCP2*	Th, Hy			SCP2*-derived *Streptomyces* vector with transfer functions used for cloning and analysis	A	240
pJIR series	pIP404	Cm or Em	ColE1 (pUC18)	Cm or Em	Shuttle vectors for general cloning in *C. perfringens*; some versions carry RP4 *oriT* for mobilization from *E. coli*	C	243
pJOE2577	pSG5	Th, Kn	ColE1 (pUC19)	Ap	TS transposon delivery shuttle vector containing the high-frequency transposase gene *tnpA*	A	388

(Continued on next page)

TABLE 1 Representative plasmids of use for genetic manipulation of gram$^+$ bacteria[a] (*Continued*)

Plasmid name	Gram$^+$ origin of replication	Gram$^+$ selectable marker	Gram$^-$ origin of replication	Gram$^-$ selectable marker	Property	Host range	Reference(s)
pKC1139	pSG5	Am	ColE1 (pBR322)	Am, lacZ	Bifunctional conjugative *E. coli*/*Streptomyces* plasmid	A	28
pKC1218	SCP2*	Am	ColE1 (pUC18)	lacZ	Bifunctional conjugative *E. coli*/*Streptomyces* plasmid	A	28
pMSP3535	pAMβ1	Em	ColE1 (pSP73)	Em	Shuttle vector for inducible gene expression in LGC; nisin-inducible promoter for operon fusions	LGC	41
pMSP3545	pAMβ1	Em	ColE1 (pSP73)	Em	Shuttle vector for inducible gene expression in LGC; nisin-inducible promoter for gene fusions	LGC	41
pMUTIN	Suicide	Em	ColE1	Ap	Insertional inactivation of *Bacillus* genes; operon fusions of inactivated gene to lacZ; IPTG-regulatable expression of downstream genes	B	383
pOJ446	SCP2*	Am	ColE1 (pUC)	Am	Bifunctional conjugative *E. coli*/*Streptomyces* cosmid vector	A	28
pPR27	pAL5000	Investigator determined	p15A (pACYC184)	Gm	TS; delivery vector for allelic exchange or transposon mutagenesis in mycobacteria	M	298
pSET152		Am	ColE1 (pUC18)	lacZ	Conjugative vector that integrates into the chromosome at the φC31 attachment site	A	28
pSG series	pC194 or suicide	Cm or Sp	ColE1 (pSK$^-$)	Ap	Shuttle vectors (or delivery vectors for integration at *amyE* locus) for construction of *gfp*, *cfp*, *yfp*, or *dsRed* fluorescent fusion proteins in *Bacillus*	B	116, 228
pSWEET	Suicide	Cm	ColE1	Ap	Vector for ectopic single-copy integration into *B. subtilis amyE* locus for xylose-inducible gene expression	B	24
pTCV-*lac*	pAMβ1	Em, Kan	p15A (pACYC184)	Em, Kan	Shuttle vector for construction of operon fusions to *lacZ* in LGC; carries RK2 *oriT* for mobilization to gram$^+$ hosts from *E. coli*	LGC	310
pVA838	pAMβ1	Em	p15A (pACYC184)	Em, Cm	Shuttle vector for general cloning in LGC	LGC	249
pVE616		Th	ColE1 (pBR322)	Am	Integrating cosmid vector	A	246
pYUB657	Suicide	Hy	ColE1	Ap	Delivery vector for allelic exchange in mycobacteria carrying *sacB* counterselectable marker	M	295

[a]Abbreviations: LGC, low-GC gram$^+$ bacteria; S, staphylococci; B, bacilli; M, mycobacteria; C, clostridia; A, actinomycetes; Ap, ampicillin resistance; Kn, kanamycin resistance; Cm, chloramphenicol resistance; Sp, spectinomycin resistance; Em, erythromycin resistance; Gm, gentamicin resistance; Tc, tetracycline resistance; Th, thiostrepton resistance; Vm, viomycin resistance; Am, apramycin resistance; Hy, hygromycin resistance; Nm, neomycin resistance; bla, β-lactamase; ds, double stranded.

32.1.2. Transduction

Transduction is a form of lateral genetic transfer mediated by bacteriophage. Traditionally, transduction is envisioned to occur as a result of errors in packaging of phage particles during a lytic phage replication cycle. In generalized transduction, a random segment of the bacterial chromosome is inserted into a nascent phage particle. Upon release from the infected donor bacterium by lysis, this transducing particle can bind to a new bacterial host cell displaying the appropriate surface receptor for phage attachment and inject its DNA, which can recombine into the host chromosome or, in the case of transduced plasmids, can replicate autonomously. Specialized transduction involves an aberrant excision of an integrated prophage from the chromosome during induction of a lysogenic bacterial strain. The excised DNA contains a segment of chromosomal DNA adjacent to the phage attachment site, and a segment of the phage genome is usually left behind in this process, again rendering the transducing particle defective. In specialized transduction, the chromosomal genes linked to the phage attachment site are transmitted at significant frequencies since nearly all transducing particles are generated during the initial excision of the prophage following induction of a lysogen. While numerous examples of transduction of gram$^+$ bacteria have been reported, it is somewhat surprising that refined transduction systems for allelic exchange and chromosomal gene mapping have only been fully developed and utilized with a small number of species, such as *Bacillus subtilis* and *Staphylococcus aureus*. On the other hand, the upswing in bacterial and bacteriophage genome sequencing projects in recent years has sparked a new appreciation for the important role of phages in natural lateral transfer events and in the evolution of many industrially and medically important bacteria, including numerous gram$^+$ species. In addition to transduction of host genes, the related phenomenon of phage conversion (where genes actually carried by the wild-type temperate phage change the phenotype of the host bacterium) plays an essential role in the evolution of bacteria, especially in the case of pathogens. Phage-mediated inversions, integrations, and other forms of DNA recombination can also impact genomic evolution in bacteria; since the numbers of phage species are probably orders of magnitude higher than the number of bacterial species, phage-mediated transfer events have been estimated to occur at rates approaching 10^{15}/s in the marine environment, where the phage titers exceeding 10^7/ml make these elements the most abundant biological entities on the planet (293). Because phage generally exhibit a high degree of specificity for host receptors, one prediction is that natural transfer events mediated by phage might have their largest impact on the evolution of closely related bacterial species or strains. While this is probably true, clearly there is significant genetic exchange between distantly related phages infecting distinct bacterial hosts (4, 93, 163, 164).

32.1.3. Transformation

Lateral genetic transfer mediated by the uptake and recombination of naked DNA into the recipient cell genome (i.e., transformation) was the first form of horizontal transfer documented experimentally (7, 149). Although natural or "competent cell" transformation has been described for more than 40 microbial species, including gram$^+$ and gram$^-$ organisms, investigators have failed to demonstrate transformation for a much larger number of species, and the efficiency with which natural transformation occurs among various strains of competent species varies considerably. However, genome sequencing projects have revealed apparently functional competence genes in organisms which have not been shown to undergo natural transformation (66), and recent reports suggest that even heavily studied organisms like *Escherichia coli* may be naturally transformable under certain conditions (252), so the prevalence of natural transformation may be higher than is currently appreciated.

Given the differences in the cell envelope of gram$^+$ and gram$^-$ bacteria, the existence of significant differences in the transformation machinery utilized by these two groups is perhaps not surprising. However, the competent bacterial species examined to date seem to utilize a core group including binding proteins, nucleases, channel-forming proteins, and nucleoside triphosphate-dependent helicases to move transforming DNA across the cytoplasmic membrane into the recipient cell. Among these DNA transport components there is significant conservation of function and amino acid sequence across both groups of bacteria. In gram$^-$ bacteria, pili are generally parts of the transformation apparatus, and they may serve as the conduit for import of the transforming DNA across the outer membrane into the periplasmic space. Some of the components of the competent cell transformation apparatus also show similarity to conjugative DNA export components and other secretion systems (55). In competent cell transformation of both *B. subtilis* and *Streptococcus pneumoniae*, the two most thoroughly studied systems of gram$^+$ species, the transforming material is bound by a secreted cell wall DNA binding protein and then subjected to both single-stranded and double-stranded nucleolytic attack during its transport into the cytoplasm. Ultimately, a double-stranded cleavage point seems to serve as the substrate for import, and as the DNA is taken into the recipient cell, one strand is completely digested as the opposite strand is imported. In contrast to some gram$^-$ systems, uptake of DNA by competent gram$^+$ strains is not sequence specific, but any heterologous DNA that enters is unlikely to recombine into the chromosome and is subject to rapid degradation. Likewise, autonomously replicating plasmids are taken up efficiently by competent bacteria but are subject to complete degradation of one strand and partial degradation of the other; thus, these elements transform with multihit kinetics, which places constraints on certain recombinant DNA manipulations such as construction of genomic libraries. On the other hand, chromosomal DNA from a strain closely related to the recipient is stably incorporated extremely efficiently, because the induction of competence (see next paragraph) generally results in increased expression of homologous recombination functions. This makes natural transformation the method of choice for allelic exchange and insertion/duplication mutagenesis (83) in competent organisms.

An important aspect of competent cell transformation in gram$^+$ species is that development of competence is generally subject to control by peptide-mediated quorum sensing mechanisms (65). In addition to serving as the first experimental system for demonstration of horizontal genetic transfer by DNA-mediated transformation, *S. pneumoniae* is the organism for which the first direct experimental evidence for release of a bacterial signaling molecule into the growth medium was presented (373). More recently, analyses of cell-cell signaling in the control of competent cell transformation of *B. subtilis* and of several streptococcal species have contributed very significantly to the overall understanding of multicellular microbial activities (15, 65, 83, 154). Interestingly, transformation and its stimulation by peptide-mediated cell-cell signaling occur efficiently in biofilms. Even more striking are recent results suggesting that competence-stimulating peptides themselves are also involved directly in biofilm development (83), and that both biofilm formation and competence development are actually components of global stress response pathways in these organisms (65, 67, 83, 84).

In terms of the effects of natural transformation on evolution via lateral transfer, clearly natural transformation contributes substantially to processes such as antigenic variation, and antibiotic resistance dissemination within species and between closely related species, while being less involved in lateral transfer of large amounts of DNA across substantial phylogenetic gaps. Finally, of note is the fact that the development of physical methods for introduction of DNA into living bacterial cells by techniques such as treatment with divalent cations, protoplast transformation, and especially electroporation has been at least partially successful in many, if not most, strains examined. None of these techniques shows much resemblance to natural transformation. Recently the molecular basis of electroporation was investigated in phospholipid bilayers, with the main conclusion being that the high-voltage pulse across the membrane drives the formation of electric field gradients at the lipid/water interface. This results in penetration of the bilayer by water, generating hydrophilic pores lined with phospholipid head groups (372). In all of these methods, exogenously added DNA enters by diffusion through artificially generated pores. Since competence-associated nuclease and recombinase activities are not induced, there is no specific mechanism operating to promote recombination of added DNA into the genome. On the other hand, recombinant plasmids can potentially be established very efficiently, making artificial transformation the preferred method for many recombinant DNA applications.

32.1.4. Applications

All three naturally occurring forms of genetic exchange have been harnessed for manipulation of gene content and organization of gram⁺ bacteria, although the relative applicability of any particular method is substantially species dependent. Genetic manipulation in gram⁺ bacteria is conceptually very similar to genetic manipulation of gram⁻ bacteria, although in almost every case not nearly as technically advanced as for the familiar gram⁻ genetic workhorse, *Escherichia coli*, or for many other gram⁻ bacteria. Mutant construction, analysis, and complementation in gram⁺ bacteria utilize many of the same procedures frequently employed in gram⁻ bacteria (e.g., the use of conjugation, transduction, and transformation to introduce DNA into strains; the use of replicating and nonreplicating plasmids as vehicles for substrate DNA; and RecA-dependent homologous recombination). Other approaches commonly used for manipulation of gram⁻ bacteria (e.g., lambda red-mediated recombination) are notably absent from the arsenal of genetic tools available for manipulation of gram⁺ bacteria. Chapter 27 of this volume (and references therein) describes the mechanisms of basic genetic manipulations and some general considerations important for their application. We do not recapitulate this information in detail here, but instead we merely reference specific techniques when appropriate and focus on aspects of genetic exchange that are specifically relevant to their application in gram⁺ bacteria. For example, plasmids that are capable of

replication in gram⁻ bacteria generally do not replicate in gram⁺ bacteria. Therefore, common gram⁻ replicons (e.g., pUC) are frequently used as suicide plasmids in gram⁺ bacteria for the delivery of DNA as substrates for homologous recombination. Additionally, shuttle plasmids (i.e., plasmids capable of replication in both a gram⁻ and a gram⁺ species) are typically constructed using molecular techniques to fuse a gram⁺ replicon with a gram⁻ replicon, yielding a recombinant plasmid carrying multiple origins of replication that is therefore capable of replication in both classes of bacteria. This approach allows investigators to use the vast array of tools available for general manipulation of DNA in *E. coli*, prior to insertion of the modified DNA into the gram⁺ organism of interest for analysis. Many such shuttle vectors have been constructed using the broad-host-range gram⁺ replicon pAMβ1. Accordingly, these vectors can often be applied in a wide spectrum of different gram⁺ species, even though their use may have been described only for one or a small number of species. Other tools used for genetic analysis of gram⁻ species can sometimes be adapted for use in gram⁺ bacteria. For example, many reporter genes that are used to analyze gene expression in gram⁻ bacteria can also be used in gram⁺ bacteria by cloning them into shuttle vectors and creating operon fusions with promoters of interest. However, the machinery for gene expression in gram⁺ bacteria (particularly for translation) tends to exhibit somewhat more stringent requirements for efficient function. Therefore, modification of the translational initiation signals present in genes originating from gram⁻ bacteria is often necessary for them to be recognized efficiently by the translational apparatus in gram⁺ species. Conversely, the more "promiscuous" gene expression machinery in gram⁻ bacteria can sometimes cause problems when shuttle vectors bearing cloned gram⁺ genes are propagated in gram⁻ species (i.e., *E. coli*), particularly if the cloned genes encode regulatory or secreted proteins. Aberrant overexpression of the gram⁺ genes can lead to toxicity for the *E. coli* host, which may result in growth inhibition and/or production of deletions or mutations in the cloned gram⁺ genes.

32.2. GENE TRANSFER IN VARIOUS GROUPS OF GRAM-POSITIVE BACTERIA

32.2.1. Streptococci

The streptococci are a heterogeneous and diverse group of low-GC, gram⁺ bacteria. This is manifest in the variety of their habitats, in their roles as human commensals and pathogens, and in the differences in types of disease they cause. Although there are some similarities in the mechanisms of genetic exchange and other genetic approaches that are useful for all streptococci, there also exist substantial differences among different species. For example, genetic exchange by natural transformation is prevalent in members of the mitis and anginosus groups of streptococci (e.g., *S. pneumoniae*, *Streptococcus gordonii*, *Streptococcus mitis*, *Streptococcus infantis*, *Streptococcus cristatus*, *Streptococcus anginosus*) and is also present to various extents in the some members of the mutans group of streptococci (e.g., *Streptococcus mutans*) but not other mutans group streptococci (e.g., *Streptococcus sobrinus*). However, the property of natural competence is apparently absent from members of the salivarius, bovis, and pyogenic groups of streptococci (e.g., *Streptococcus pyogenes*, *Streptococcus agalactiae*, *Streptococcus dysgalactiae*, *Streptococcus salivarius*, and

Streptococcus equinus), although one strain of *Streptococcus bovis* has been found to exhibit natural competence (266). The existence of natural transformation facilitates certain genetic manipulations that are difficult or impossible in bacteria that lack this property. Therefore, we have compared and contrasted two widely studied species of streptococci, with the idea that genetic methods applicable to any given species of *Streptococcus* can be developed by applying concepts from one or both of these diverse groups. Below we discuss genetic exchange in the pyogenic group A streptococci (i.e., *S. pyogenes*; also known as GAS) and in the mitis group streptococcus *S. pneumoniae*.

32.2.1.1. Plasmids

GAS can serve as a host for several classes of plasmids, including the rolling-circle-replication (RCR) plasmids (discussed in more detail in the section on staphylococci) and the θ-replicating plasmids of the Inc18 family (discussed in more detail in the section on enterococci). Members of these plasmid classes were originally identified in other gram$^+$ species, and they are generally capable of replication in a wide variety of gram$^+$ bacteria, but they are mentioned here because they form the basis of a number of tools for genetic analysis in GAS. It is worth noting that the common staphylococcal RCR replicons pC194, pE194, and pUB110 do not replicate reliably in GAS (46), unlike in many other gram$^+$ cocci. However, the lactococcal RCR replicon pWVO1 (227) is functional in GAS, and a temperature-sensitive-for-replication (TS) pWVO1 derivative (known as the pG$^+$host series) (253) has been used for genetic manipulation in this organism. Plasmid vectors based on the θ-replicating Inc18 replicon of pAMβ1 (see the section on enterococci for more details) have also been widely used in GAS (e.g., the shuttle vector pAT28, containing the pUC replicon derived from ColE1 [377]). Plasmids bearing Inc18-type replicons transform GAS at a lower frequency than plasmids with pWVO1-based replicons (46), but the Inc18-based vectors have been found to be useful for obtaining expression of cloned genes that were only poorly expressed by pWVO1-based vectors (124).

In *S. pneumoniae*, the most common indigenous plasmid is pDP1, a cryptic 3-kb RCR plasmid that belongs to the pC194 family (338, 349). However, pDP1 has not been widely exploited for genetic manipulation. The *S. agalactiae* plasmid pMV158 has been used as a cloning vector in *S. pneumoniae* (354). Another broad-host-range plasmid, pVA838, which carries a derivative of the pAMβ1 replicon (248, 249), has been used in *S. pneumoniae* (412).

32.2.1.2. Conjugative Elements

Conjugative elements encode the machinery required for self-transmission to neighboring bacteria by conjugation (see the section on enterococci for more details). Conjugative elements exist in two varieties broadly defined based on their mode of replication: (i) autonomously replicating plasmids and (ii) conjugal elements that depend on integration into the host chromosome for replication. Although no conjugative plasmid indigenous to GAS has been experimentally characterized, the enterococcal conjugative plasmid pAMβ1 (68) exhibits a broad host range and has been transferred into GAS. A relative of pAMβ1, known as pIP501, was originally identified in a clinical isolate of *Streptococcus agalactiae* (group B streptococcus) (170) and has been transferred to GAS. In fact, pIP501 has been transferred by conjugation into a wide variety of gram$^+$ bac-

teria, including a host of streptococci and lactobacilli (371), lactococci (216), *Listeria* spp. (45), staphylococci (335), bacilli (213), and pediococci (143). Recently, the remarkably broad host range of pIP501 was expanded to include the high-GC gram$^+$ bacterium *Streptomyces lividans* and the gram$^-$ bacterium *E. coli* (213). Conjugative plasmids indigenous to *S. pneumoniae* have not been experimentally characterized, but as in the case of GAS, several conjugative plasmids from heterologous streptococci have been successfully transferred into *S. pneumoniae* (e.g., pIP501 [350] and pAMβ1 [111]).

Conjugative transposons belong to a large class of mobile genetic elements now known as integrative and conjugative elements (ICEs) (44) that normally reside integrated in the chromosome of their host (see the section on enterococci for more details). In GAS, one naturally occurring conjugative transposon, Tn1207.3 (carrying the *mefA* macrolide efflux determinant), has recently been described (330). This element has been transferred by conjugation to other streptococcal species and was found to insert at a specific site in the streptococcal chromosome. Other conjugative transposons can also be transferred into GAS. For example, the broad-host-range conjugative transposon Tn916 (see the section on enterococci for more detail on Tn916) can be used for mutagenesis in GAS, even in isolates that are difficult to transform with plasmid delivery vectors bearing transposons. However, the use of Tn916 as a mutagen is subject to certain disadvantages: the element is quite large (>18 kb), it exhibits some degree of site specificity for insertion (the degree of insertion site specificity varies by species and sometimes even by strain), and it frequently generates multiple insertions per host. Despite these limitations, Tn916 has been used to identify a number of mutants of GAS (22, 48, 100, 233, 276, 403). A detailed procedure describing the use of Tn916 as a mutagen in GAS was published by Caparon and Scott (49).

In *S. pneumoniae*, two types of conjugative transposons have been characterized. The first, Tn1545, carries resistance determinants for tetracycline, kanamycin, and erythromycin. Tn1545 is closely related to Tn916 and, accordingly, has been shown to have a similarly broad host range (78). A second type of conjugative transposon, designated Tn5253, is a composite of two unrelated conjugative transposons that are capable of transferring themselves to recipients independently of each other. This complex genetic element is composed of a copy of Tn5252 into which Tn5251 (a Tn916-like conjugative transposon) has inserted (9).

32.2.1.3. Bacteriophage and Transduction

Bacteriophage-mediated generalized transduction is an important method of genetic manipulation that can be accomplished using either lytic or temperate phages. A variety of lytic GAS phages have been described, including phages A6, A12, A25, and A27 (261). A derivative of A25 (255) has been widely used for generalized transduction; a detailed procedure for A25-mediated transduction has been described by Caparon and Scott (49). Temperate phages of GAS have also been described, some of which result in toxigenic conversion of their GAS host upon the establishment of lysogeny. Toxigenic conversion is the introduction of a temperate phage-borne gene that functions as a virulence factor into the genome of its host bacterium, following the establishment of lysogeny. For example, GAS phage T12 carries the gene encoding *speA* pyrogenic exotoxin (186, 401), and phage CS112 carries the gene encoding *speC* pyrogenic exotoxin (187). Thus, genetic exchange by bacteriophage appears to play an important role in the evolution of GAS and in the emergence of virulent pathogenic variants.

It also seems likely that bacteriophage play an important role in the evolution of *S. pneumoniae*, as 70% of clinical *S. pneumoniae* isolates carry putative prophages (235). A number of lytic phage capable of infecting *S. pneumoniae* have been experimentally identified (e.g., phages Cp-1 [258] and Dp-1 [133]), as have temperate phage (e.g., phages MM1 [282] and EJ-1 [323]). However, in contrast to GAS, no system for genetic exchange by generalized transduction using any of these or other phages has been described, perhaps in part due to the ease of performing genetic manipulations by natural transformation in *S. pneumoniae* (see below).

32.2.1.4. Transformation

Competence for transformation, determined by a genetically encoded mechanism, has not been described for GAS. However, a procedure to introduce DNA into GAS based on electroporation has been developed. This procedure relies on growth of the bacteria in the presence of high concentrations of glycine to weaken the cell wall prior to electroporation, an approach that is often used for various gram$^+$ species. A detailed protocol, as reported by Caparon and Scott (49), is described below. An alternative set of conditions for electroporation of GAS has also been described (346) that may be more effective for some strains. Although a mechanistically different method for artificial transformation, known as protoplast transformation, has been developed for some related gram$^+$ cocci, for unknown reasons this approach was not found to be effective for GAS (49). It is worth noting that growth of streptococci in the presence of high concentrations of glycine can sometimes be irreproducible or erratic; an alternative procedure for electroporation of streptococci has been developed (42) that limits glycine exposure of bacteria to 1 h, and it was found to be effective for electroporation of a variety of difficult-to-transform streptococci.

Procedure for Transformation of GAS by Electroporation

The strain to be transformed is cultivated overnight in THY broth supplemented with 20 mM glycine at 37°C. For some strains of GAS, or for other species of streptococci, the concentration of glycine in the growth medium may require empirical optimization. Dilute the overnight culture 20-fold in fresh THY broth plus glycine to yield 100 ml total. Continue incubating as before without aeration until the A_{600} of the culture reaches 0.2 (~2 h). Harvest the bacteria by centrifugation (5,000 × *g*, 10 min, 10°C) and wash twice in 20 ml of ice-cold electroporation buffer. The cell pellet obtained from 100 ml of culture is resuspended in 0.75 ml of ice-cold electroporation buffer and stored on ice. Aliquots of this cell suspension (0.2 ml) are transferred to a chilled 0.2-cm electroporation cuvette. DNA (1 to 5 μg) is added to the experimental samples and the cuvettes are pulsed using the Bio-Rad Gene Pulser apparatus (voltage set at 1.75 kV, capacitance at 25 μF, and pulse controller at 400 Ω). Immediately following the pulse, the cell suspension is withdrawn and placed directly into 10 ml of nonselective THY broth (without glycine) and incubated for 2 h at 37°C to allow for recovery of the cells and expression of the resistance determinants. Finally, the culture is centrifuged as described above, washed once in THY broth, and resuspended in 1 ml of THY broth prior to plating of aliquots on appropriate selective agar plates.

THY broth

Todd-Hewitt broth (Difco) . . . . 30 g/liter
Yeast extract (Difco) . . . . 2 g/liter

Electroporation buffer

Glucose . . . . 49 g/liter
MgCl$_2$ · 6H$_2$O . . . . 0.2 g/liter

Adjust pH to 6.5 with NaOH.

Transformation of *S. pneumoniae*

Natural genetic transformation (or competence) has been extensively investigated in *S. pneumoniae* and exhibits many similarities with that of *B. subtilis*. The competent state is transiently induced when a culture of *S. pneumoniae* achieves a critical cell density. Induction of the competent state is dependent on the accumulation of a competence-stimulating peptide (CSP) in the culture medium; CSP is actively secreted by the pneumococci during growth, using a dedicated processing and export apparatus. When extracellular CSP achieves a suitably high concentration, its presence is detected by a two-component sensory transduction system (encoded by the genes *comDE*), initiating a signal transduction cascade that ultimately results in enhanced transcription of a collection of genes involved in binding, processing, uptake, and recombination of exogenous DNA. Thus, the addition of spent culture medium containing CSP (or addition of synthetic CSP) to a low-density population of *S. pneumoniae* results in the premature development of the competent state. In addition to cell density-dependent control mediated by CSP, development of competence in streptococci also depends on a variety of other factors, some of which are poorly understood. The regulation of competence for transformation has been recently reviewed (65), as has the process of DNA transport by gram$^+$ bacteria (55, 56). From a practical perspective, the ability to easily and routinely obtain reproducibly high frequencies of transformation has been greatly enhanced by the availability of synthetic CSP, which can be added to cultures of *S. pneumoniae* to induce the competent state at will. A protocol for transformation of *S. pneumoniae* using synthetic CSP, as reported by Luo et al. (239), is described below. However, it is important to note that not all strains of *S. pneumoniae* are equally transformable. A modified procedure for reproducible transformation of a type 4 encapsulated strain of *S. pneumoniae* has been described (34).

Procedure for Transformation of *S. pneumoniae*

The *S. pneumoniae* strain of interest (e.g., CP1250 [300]) is cultured at 37°C in CAT medium (269) supplemented with 0.2% bovine serum albumin, 0.5 mM CaCl$_2$, and 10 mM HCl to an optical density at 550 nm (OD$_{550}$) of 0.04. Aliquots of this culture are exposed to transforming DNA (1 μg/ml) at 30°C for 30 min in the presence of CSP (250 ng/ml). The bacteria are then shifted to 37°C and incubated for 1 h before plating in CAT agar with appropriate drugs.

Casein hydrolysate yeast extract medium (CAT medium)

Choline . . . . 5 mg/liter
Tryptone . . . . 5 g/liter
Casein hydrolysate . . . . 10 g/liter
Yeast extract . . . . 1 g/liter
NaCl . . . . 5 g/liter

Autoclave; after cooling, supplement to the following concentrations.

Glucose 0.2%
K₂HPO₄ .... 0.167 M

Synthetic CSP (peptide sequence: EMRLSKFFRD-FILQRKK) can be synthesized by core service facilities at most major research universities. CSP is diluted in sterile water to final concentration 250 μg/ml, filter sterilized, and stored at −20°C.

32.2.1.5. Applications of Gene Transfer

32.2.1.5.1. Construction of Mutants

32.2.1.5.1.1. Transposon mutagenesis.
Use of transposons as agents of insertional mutagenesis is a common and effective means of identifying genes involved in a biological process of interest. In GAS, transposon mutagenesis has been conducted using a variety of transposable elements. As described above, the conjugative transposon Tn916 has been used to identify genes of interest. A more conventional transposon isolated from a gram⁺ bacterium, Tn917, has been cloned into the TS delivery vector pG⁺host4 (109) and used as a tool for transposon mutagenesis (Tn917 is discussed in the section on enterococci). In addition, an engineered derivative of the staphylococcal transposable element Tn4001 (known as TnSpc) has also been used to identify genes of interest in GAS (241), although a significant percentage of transposon insertion mutants generated with this element carry transposon insertions at multiple loci. Thus, mutants generated by TnSpc mutagenesis must be analyzed carefully to determine that the mutant phenotype is due to a particular transposon insertion. Another derivative of Tn4001 carrying the *Enterococcus faecalis* gene encoding alkaline phosphatase (*phoZ*), an enzyme that must be secreted to the extracellular space to become active, has been used to identify secreted proteins of GAS (141).

As noted above, *S. pneumoniae* exhibits natural genetic competence and therefore readily acquires exogenous DNA from its environment. If this DNA contains regions of homology to the streptococcal chromosome, it can be integrated by homologous recombination. This property has been exploited to develop an in vitro strategy for transposon mutagenesis, using a derivative of the *mariner* transposable element (2, 159), which requires only a TA dinucleotide at its site of insertion. The mutagenesis procedure is divided into two parts: high-efficiency insertion of a mini-*mariner* transposable element is performed in vitro with purified pneumococcal chromosomal DNA, purified transposon DNA, and *mariner* transposase; in the second step, the resulting DNA (carrying transposon insertions) is efficiently transformed into *S. pneumoniae* cells and recombined into the genome to generate a pool of transposon mutants.

32.2.1.5.1.2. Directed mutagenesis and allelic exchange.
Insertion mutations to inactivate specific genes in GAS (and most other bacteria) are typically constructed by taking advantage of bacterial homologous recombination systems (see chapter 27 for more details). One common approach is to introduce a circular suicide plasmid, carrying a cloned internal fragment of the target gene, into the strain of interest by electroporation. Subsequent selection for a plasmid-encoded antibiotic resistance determinant yields recombinants in which the plasmid has been integrated by Campbell-type homologous recombination (single-crossover recombination) at the duplicated sequence (thereby inactivating the gene). However, one disadvantage of this approach is that it inherently generates mutations that are polar on the expression of downstream genes. The significance of polar effects can be assessed, to some extent, by constructing a second, analogous strain using a recombinant vector carrying the 3′ end of the target gene, instead of an internal gene fragment. After integration, such an arrangement confers polarity on the expression of downstream genes but does not disrupt the target gene, thereby allowing the investigator to determine if polar effects are resulting in any observed phenotype. However, this approach is often not optimal, as the phenotypic effects of polarity may obscure the effect of inactivation of the gene of interest.

Recently, alternative strategies to inactivate genes have been developed to circumvent such complications. For example, using standard molecular techniques, recombinant plasmids can be constructed that carry cloned chromosomal segments flanking a gene of interest. An antibiotic resistance cassette is then inserted between these two flanking fragments, creating a cloned allele in which the gene sequence has been replaced with a selectable antibiotic marker. If this allele is carried on a suicide plasmid, electroporation of a linear form of this recombinant plasmid into a GAS strain and selection for the appropriate antibiotic resistance will yield viable recombinants only after double-crossover recombination in the flanking DNA on both sides of the antibiotic marker. In such mutants, the wild-type allele will have been exchanged with the plasmid-borne mutant allele. Because no other plasmid-specific sequences are incorporated in the mutant chromosome, this approach can sometimes be used to construct mutations that are not polar on the expression of downstream genes, although this depends on the particular antibiotic resistance cassette and should be verified experimentally. Transforming DNA fragments that carry at least 1 kb of flanking DNA sequence on either side of the gene targeted for mutagenesis yield the highest frequencies of recombination. A rapid approach for PCR-based generation of linear recombinant fragments to be used as substrates for transformation has been developed which eliminates the requirements for extensive in vitro cloning steps (217). The use of linear fragments for transformation is most effective with naturally transformable host bacteria (e.g., *S. pneumoniae* and *S. mutans*), although they can be used with GAS as well (the frequency at which recombinants arise is substantially reduced). For more detail, see the description of long-flanking homology PCR in the section on bacilli (below).

An allelic exchange method to construct mutants of GAS carrying in-frame deletions of target genes has been developed (184). This approach has the advantage that the mutations are stable and nonpolar on the expression of downstream genes, but it suffers from the disadvantage that such mutants are more difficult and time-consuming to construct. This strategy employs a two-step insertion-excision sequence of recombination events to exchange a plasmid-borne allele of a gene of interest with the wild-type chromosomal copy. Initially, a plasmid-borne in-frame deletion allele must be constructed, using standard molecular techniques, in an appropriate conditionally replicating vector (e.g., the TS vector pG⁺host5 [30]). As noted above, clones that contain at least 1 kb of flanking DNA on either side of the deletion yield the highest recombination frequencies. After introducing the recombinant plasmid into the GAS strain of interest, growth under conditions that are nonpermissive for plasmid replication and selection for the plasmid-encoded antibiotic resistance determinant will yield recombinants in which the plasmid has integrated into the host chromosome by single-crossover homologous recombination at either the upstream or the downstream segment of flanking DNA, resulting in duplications of these sequences. A second single-crossover recombination event (during subsequent growth under nonselective conditions) between corresponding duplicated fragments can lead to the excision of the integrated plasmid from the chromosome, either leaving the mutant allele on the chromosome or regenerating the wild-type allele, depending on the site of recombination. The mixed population of cells that results must then be screened to determine which allele has been retained on the host chromosome. Because the recombination event that leads to the excision of the integrated plasmid occurs at a relatively low frequency, it is highly advantageous to incorporate a counterselectable marker into the plasmid that permits a selection to be used to identify such recombinants. However, no such marker is available for GAS. Therefore, either the deletion mutant to be identified must have a known and selectable phenotype or a time-consuming and laborious screening process must be used to identify the appropriate recombinants.

The ability to naturally transform *S. pneumoniae* facilitates the construction of directed mutants but also requires that care be exercised in the interpretation of the results. The frequency of spontaneous mutation increases during transformation with homologous DNA, such that mutations other than those desired are often observed to occur (411). Two methods, based on the same principles described above for GAS, are generally used for gene inactivation in *S. pneumoniae*: (i) insertional mutagenesis via a single-crossover homologous recombination with a circular suicide plasmid and (ii) gene replacement using double-crossover recombination of linear DNA fragments. For insertional mutagenesis using a circular suicide plasmid, a strong dependence of recombination frequency on the length of the cloned segment has been observed (225). By constructing a plasmid library containing random fragments of the pneumococcal genome, this type of insertional-mutagenesis approach was used to create an insertion mutant library of *S. pneumoniae* strains (224). Transformation by linear DNA fragments and double-crossover recombination of these fragments into the genome is highly efficient in *S. pneumoniae* and probably the most common method of directed mutant construction, due to the action of the natural genetic competence system (for more detail on this method, see the description of long-flanking homology PCR, in the section on *Bacillus* below).

32.2.1.5.1.3. Selection and counterselection. As with many other members of the gram$^+$ cocci, nutritional requirements of most streptococci, including GAS and *S. pneumoniae*, are complex. Therefore, nutritional markers (e.g., selection for prototrophy or carbon source utilization) are generally not used for genetic manipulation in these organisms. Instead, a variety of antibiotic resistance determinants serve as selectable markers to facilitate genetic manipulation. Most common resistance determinants that function in other gram$^+$ bacteria are also functional in the streptococci, e.g., *erm* (erythromycin), *cat* (chloramphenicol), *tet* (tetracycline), and *kan* (kanamycin). However, some alleles of these determinants, originally identified on multicopy plasmids, are known to require transcriptional read-through from adjacent chromosomal promoters for ef-

ficient expression. Thus, insertion of such a resistance determinant at a given chromosomal locus will yield viable, antibiotic-resistant recombinants only when inserted in one of the two possible orientations. To circumvent this problem, alleles of some markers have been developed that encode synthetic promoters, rendering their expression independent of chromosomal context (64). It should be noted that DNA encoding a β-lactamase should not be introduced into GAS, because β-lactam antibiotics remain clinically relevant therapeutic agents for these organisms.

A counterselectable marker (i.e., a marker whose absence can be selected for, also known as a negative selection) that can be applied in GAS has not been described. However, a counterselectable marker has been developed for use in negative selections during genetic manipulation of *S. pneumoniae* (362). This strategy is analogous to a similar procedure used in gram$^-$ bacteria and is based on the well-known recessive properties of *E. coli rpsL* mutant alleles that confer streptomycin resistance (Sm^r) (220, 327, 358). When such an allele is present in diploid with a wild-type *rpsL$^+$* allele, the cells exhibit sensitivity to streptomycin (thus, the mutant is recessive to the wild-type allele). Consequently, introduction of *rpsL$^+$* (carried on the backbone of an integrating plasmid) into an *rpsL*-containing (Sm^r) host in the first step of a two-step allelic exchange manipulation (as described above) generates an Sm^s merodiploid strain. The investigator then has the ability to select for the relatively rare Sm^r recombinants that result from the excision of the integrated plasmid (and the concomitant loss of the *rpsL$^+$* allele) in the second step of the procedure, rather than having to screen for such recombinants.

32.2.1.5.2. Analysis of Mutants

32.2.1.5.2.1. Complementation. Complementation analysis requires the introduction of a wild-type copy of a gene of interest into the corresponding mutant genetic background to determine if the mutant phenotype can be rescued. This is accomplished either by introducing a recombinant, autonomously replicating plasmid carrying the gene of interest into the mutant strain or by using recombination to introduce the gene of interest at an ectopic locus in the chromosome of the mutant strain. Plasmid vectors based on the pWV01 or pAMβ1 replicons can be used to express cloned genes in GAS for complementation or analysis (124). In *S. pneumoniae*, plasmids based on pMV158 and pVA838 have been used to express cloned genes. Other plasmid vectors based on the pAMβ1 replicon are also functional. A series of suicide plasmid vectors (the pVIT series) designed to allow the integration of cloned genes or promoter fusions into a resident chromosomal copy of Tn916 have been developed for use in GAS (140). An equivalent approach has been employed in *S. pneumoniae* (311). This approach facilitates the transfer of such recombinant Tn916 elements to other streptococcal strains that may not be readily transformable themselves (because the recombinant Tn916 elements can be transferred by conjugation to such strains), and it provides the advantage of expressing genes at the single-copy level, which may help avoid some complications resulting from overexpression of the cloned gene. However, this approach suffers from the obvious disadvantage that a Tn916-containing derivative of the mutant strain must be constructed and characterized before complementation can be performed.

Often, the ability to regulate the expression of cloned genes in a defined manner is desirable when performing complementation analysis. Expression of cloned genes in a controlled manner in both GAS and *S. pneumoniae* is possible using the lactococcal nisin-inducible promoter system that is functional in a variety of gram$^+$ bacteria (108, 238). In GAS, the isopropyl-β-D-thiogalactopyranoside (IPTG)-inducible P$_{spac}$ promoter (108) that was originally developed for use in *B. subtilis* has also been used. Other options are available for use in *S. pneumoniae*. These include the promoter for α-galactosidase (inducible by raffinose [238, 325]) and a fucose-regulated promoter (53) that has been used for analysis of essential genes in *S. pneumoniae*.

32.2.1.5.2.2. Expression analysis.
Reporter genes that specify easily assayed gene products are routinely used as surrogate indicators of the level of expression for genes of interest. This is usually done by using molecular cloning techniques to fuse the promoter of the gene of interest to a suitable reporter gene and then introducing the recombinant construct into appropriate strains for analysis, either on an autonomously replicating plasmid or after integration into a chromosomal locus. A wide variety of reporter genes are available for use in GAS to monitor promoter activity. These include the genes for chloramphenicol acetyltransferase (47, 264), β-galactosidase (124, 333), β-glucuronidase (31), alkaline phosphatase (147), luciferase (110, 210, 307), and green fluorescent protein (151). Most of these reporters have also been used to monitor gene expression in *S. pneumoniae*, including the genes for chloramphenicol acetyltransferase (96), alkaline phosphatase (296, 297), luciferase (16, 127), and green fluorescent protein (90, 189). The *lacZ* reporter can be used in *S. pneumoniae* but requires a genetic background that carries a mutation reducing the endogenous production of β-galactosidase (64).

32.2.2. Staphylococci

The significance of some species of staphylococci as human pathogens has prompted considerable study of these bacteria, particularly *Staphylococcus aureus*. As a result, it is now clear that genetic exchange plays a significant role in staphylococcal evolution and the emergence of antibiotic-resistant and pathogenic variants. For example, lysogenic staphylococcal phage can endow their hosts with the ability to produce toxins. Furthermore, plasmids are widespread in the staphylococci and are associated with the emergence of antibiotic-resistant strains in health care settings. Members of one class of plasmids discovered in staphylococci, the RCR plasmids, are capable of replication in a wide variety of gram$^+$ bacteria and have thus been developed as tools for genetic manipulation in many species.

32.2.2.1. Plasmids

Staphylococcal plasmids can be grouped into four general classes. (i) Small RCR plasmids utilizing an asymmetric single-stranded replication mechanism are widespread in staphylococci. Four families of RCR plasmids have been identified: the pT181 family, pC194 family, pSN2 family, and pE194 family. These are normally less than 5 kb in size, generally maintain themselves at relatively high copy number (10 to 50 per cell), typically are cryptic or carry only a single antibiotic resistance determinant, and are rarely found to contain transposable elements (278). RCR plasmids can replicate in many heterologous gram$^+$ bacteria (280) and have therefore been widely employed as cloning and expression vectors, but derivatives of RCR plasmids carrying cloned genes can sometimes be structurally unstable, leading to deletions of cloned gene segments. Thus,

when a plasmid based on an RCR replicon is used for cloning or expression purposes, it is important to monitor the integrity of the plasmid throughout the study. Replication of RCR plasmids has been reviewed in detail (193, 194). (ii) Staphylococcal multiresistance plasmids (e.g., pI524, pSK23, and pI258) are typically 15 to 40 kb in size and exist at copy numbers of 4 to 6 per cell. These are thought to replicate via a θ-type mechanism, which confers enhanced stability. They typically carry some combination of resistance determinants for β-lactam antibiotics, macrolides, or heavy metal ions, some of which can be encoded on transposable elements that reside on the plasmids. Although the transposable elements and any associated resistance genes may be able to transpose to heterologous genetic elements, the plasmids themselves in this class are not mobile by conjugation. (iii) Conjugative plasmids are discussed in detail below. (iv) The fourth class of plasmids includes those that do not obviously belong to any of the preceding three classes, including pSK639 (226) and related plasmids. Plasmids of this class have not been characterized in as much detail as those of the other classes.

32.2.2.2. Conjugative Elements

Conjugative plasmids represent the primary type of conjugative element described for staphylococci. This class of plasmids is typically 30 to 60 kb in size. Members encode the machinery required for conjugative self-transfer, as well as some (variable) combination of resistance determinants. For example, plasmid pSK41 encodes resistance to aminoglycosides, gentamicin, tobramycin, and kanamycin as well as neomycin, and to antiseptics and disinfectants (19). The best-studied family of conjugative plasmids contains pSK41 (19, 117), pGO1 (270), and pJE1 (113). Multiple copies of insertion sequences, such as IS257, are typically present in these plasmids, and the IS elements are often associated with antibiotic resistance determinants, suggesting that the plasmids acquired their resistance determinants through a series of transposition events. Conjugative transfer of pSK41 family plasmids occurs on solid surfaces at relatively low frequencies (10^{-5} to 10^{-7} transconjugants per donor cell) and does not involve a pilus-like structure. These plasmids can be transferred to other staphylococcal species.

Although not originally discovered in staphylococci, the broad-host-range conjugative plasmids pAMβ1 and pIP501 can be transferred to *S. aureus* (111, 247, 334).

Conjugative transposons indigenous to *S. aureus* have not been experimentally characterized. However, a putative Tn916-like conjugative transposon has been identified by comparative genomics in the completely sequenced genome of strain Mu50 (214). Whether or not this element retains the ability to transfer itself to other staphylococci or has become inactivated by mutation remains to be determined. The broad-host-range conjugative transposon Tn916 can be transferred into *S. aureus* and has been used as a tool for mutagenesis (188).

32.2.2.3. Bacteriophage and Transduction

Most natural isolates of *S. aureus* carry one or more prophages. The repertoire of prophages present in a given strain influences susceptibility to further phage infection, thereby forming the basis for a historically widely used phage typing scheme (5). This procedure utilizes a standard reference set of phages and has been employed for epidemiological analysis, although it has now been largely supplanted by molecular techniques (5). As with many other species of bacteria, temperate phage integration results in lysogenic conversion for staphylococci. For example, the staphylococcal enterotoxin

A gene is carried by phage PS42-D (21) and staphylokinase is encoded by Pφ-2 (208). Establishment of lysogeny by various phages is also known to abolish production of certain staphylococcal gene products. The genes encoding staphylococcal β-toxin and lipase both contain phage attachment sites, such that lysogenic integration of the cognate phage inactivates the gene (76, 222). Lytic staphylococcal phages have also been identified (317, 390).

Generalized transduction is routinely used for genetic exchange of chromosomal markers or plasmids in staphylococci (60, 177). The best-studied generalized transducing phage is φ11 (277), a temperate phage typically yielding transduction frequencies of 10^{-4} to 10^{-6} for plasmids and 10^{-6} to 10^{-8} for chromosomal markers (280). However, transduction of markers between strains of different phage types may be difficult due to restriction barriers (but see below). During production of φ11 transducing particles, DNA molecules corresponding in size to a single phage genome are encapsulated, truncating DNA fragments that are too large and incorporating linear multimers of smaller DNAs (i.e., plasmids) (107, 281). Incorporation of a small segment of φ11 DNA into a target plasmid has been found to dramatically increase the plasmid transduction frequency (234, 281), thereby overcoming the restriction-related inability to transduce plasmids to heterologous staphylococcal strains.

Procedure for Transduction with φ11

Protocols for preparation of φ11 transducing lysates and for subsequent transduction of recipient strains are described below, as derived from the reports of Novick (280) and Cheung et al. (60).

Preparation of transducing lysates

Dilute a mid-exponential-phase culture of the donor strain in CY broth at approximately 3×10^8 CFU/ml with an equal volume of phage buffer and add φ11 at a multiplicity of infection of 0.1 to 1. Shake the mixture gently at 30°C until lysis occurs (approximately 2 h). Centrifuge the lysate at 10,000 × g at room temperature for 10 min to remove unbroken cells and cellular debris. Filter the supernatant through a 0.2-μm-pore-size membrane filter to remove any remaining bacteria. The phage titer can be determined by dilution in phage buffer and plating in 3 ml of phage top agar on phage plates, using RN450 (280, 281) as an indicator strain (add approximately 10^8 CFU/plate). Using this procedure, phage titers are usually in the range of 10^{10} to 10^{11} PFU/ml.

Transduction

An exponential-phase culture of the recipient strain is prepared in broth (approximately 5×10^8 CFU/ml) or by scraping cells from a fresh overnight plate. One-tenth milliliter of this cell suspension is treated with a low multiplicity of infection of transducing phage (phage-to-recipient ratio of 0.1) at room temperature for 5 min, followed by addition of 0.3GL top agar and plating on the appropriate selectable medium (e.g., 0.3GL agar supplemented with antibiotic). As a control, the phage lysate alone (no recipient cells) should be plated to ensure its sterility.

Materials
CY broth

Casamino Acids	10 g/liter
Yeast extract	10 g/liter
Glucose	5 g/liter
NaCl	5.9 g/liter
1.5 M β-glycerophosphate (added after autoclaving)	40 ml/liter

0.3GL agar

Casamino Acids	3 g/liter
Yeast extract	3 g/liter
NaCl	5.9 g/liter
Sodium lactate (60% syrup)	3.3 ml/liter
25% glycerol	4 ml/liter
Agar	15 g/liter (bottom agar)
Agar	7.5 g/liter (top agar)

Phage buffer

0.1 M MgSO$_4$	10 ml/liter
0.4 M CaCl$_2$	10 ml/liter
2.5 M Tris, pH 7.8	20 ml/liter
NaCl	5.9 g/liter
Gelatin	1 g/liter

Phage agar

Casamino Acids	3 g/liter
Yeast extract	3 g/liter
NaCl	5.9 g/liter
Agar	15 g/liter (bottom agar)
Agar	5 g/liter (top agar)

32.2.2.4. Transformation

Competence for transformation, determined by a genetically encoded mechanism, has not been described for staphylococci. Thus, transformation is not likely a significant contributor to natural genetic exchange. However, methods for transformation via electroporation (190, 337) or protoplast fusion (292) in the laboratory have been developed. An important practical consideration for the application of these methods is that staphylococci possess potent restriction systems that inhibit the introduction of DNA from other species, or even certain other lineages of staphylococci. To circumvent this problem, plasmids destined for *S. aureus* are typically first introduced into a restriction-defective host strain (RN4220) (279) to allow for plasmid modification. The modified plasmids can subsequently be isolated from RN4220 and introduced into the strain of interest (assuming that the strain of interest contains the same restriction system as does RN4220, which is the case for most of the commonly used laboratory strains of *S. aureus*).

32.2.2.5. Applications of Gene Transfer

32.2.2.5.1. Construction of Mutants

32.2.2.5.1.1. Transposon mutagenesis.
Several systems for transposon mutagenesis in *S. aureus* have been developed, including approaches based on the staphylococcal transposon Tn*551*, the enterococcal transposon Tn*917*, and the *mariner* element *Himar1*. Mutagenesis by Tn*551* has employed a delivery plasmid that is TS (60), as has Tn*917* mutagenesis (59). It is worth noting that Tn*551* differs from Tn*917* in only 11 of 5,266 nucleotide positions (409), indicating that the two transposons should be considered essentially the same. A two-plasmid in vivo transposon mutagenesis system for *S. aureus* based on the *mariner* transposable element has been developed (referred to as *bursa aurealis*). In this system, the *mariner* transposase is expressed from one plasmid, and the transposable element is carried on a separate, compatible, and temperature-sensitive plasmid.

Growth of bacteria carrying these two plasmids at nonpermissive temperature and selection for the transposon-encoded antibiotic resistance marker result in loss of the TS replicon and isolation of transposon insertion mutants. Some 10,000 mutants were generated using this system, and evaluation of the locations of the resulting insertions revealed them to be distributed throughout the staphylococcal genome in a more random fashion than insertions generated by Tn917 (11).

32.2.2.5.1.2. Directed mutagenesis and allelic exchange. An allelic exchange approach (i.e., a two-step insertion-excision procedure using a cloned gene fragment encoded on a conditionally replicating plasmid) has been used to transfer mutant alleles to the staphylococcal chromosome by Ince and Hooper (176). In that study, the conditionally replicating plasmid was the temperature-sensitive plasmid pCL52.1 (230). However, no counterselectable marker has been described that would allow the selection of recombinants that have excised the integrated plasmid in the second step of the procedure. In the study by Ince and Hooper (176), the phenotype of the desired mutant was already known and conveniently provided a selection that could be used to identify the desired recombinants. An alternative approach makes use of pCL52.1 to supply a mutant allele in which the gene of interest has been replaced with an antibiotic resistance determinant in the context of flanking chromosomal DNA, as a substrate for homologous recombination. At nonpermissive temperature, a double-crossover recombination event in the flanking homologous fragments can result in the transfer of the antibiotic resistance-marked allele to the chromosome, yielding a mutant that can be identified by selection and replica plating (177).

32.2.2.5.1.3. Selection and counterselection. As with many other members of the gram$^+$ cocci, nutritional requirements of staphylococci are complex. Therefore, nutritional markers (e.g., selection for prototrophy or carbon source utilization) are not widely used for genetic manipulation in these organisms, despite that fact that a number of auxotrophic mutants of *S. aureus* have been identified (292). Instead, a variety of antibiotic resistance determinants serve as selectable markers to facilitate genetic manipulation. Most common resistance determinants that function in other gram$^+$ bacteria are also functional in staphylococci, e.g., *erm* (erythromycin), *cat* (chloramphenicol), *tet* (tetracycline), and *kan* (kanamycin). A counterselectable marker useful in staphylococci has not been described.

32.2.2.5.2. Analysis of Mutants

32.2.2.5.2.1. Complementation. Plasmid vectors have been developed (e.g., pCL84) that allow single-copy integration into the *S. aureus* genome via site-specific recombination at the *attB* site used by phage L54a (221). These suicide vectors do not replicate in *S. aureus* and therefore can only be maintained in the recipient by recombination at the chromosomal phage attachment site, which occurs in the gene encoding staphylococcal lipase. This strategy results in a level of gene dosage that is equivalent to that of the chromosome, thereby eliminating concerns about phenotypic effects that may result from differences in copy number of genes under analysis.

A series of modular *E. coli-S. aureus* shuttle vectors has been developed (54) that permit the combinatorial use of a variety of different selectable markers (Erm, Tet, Cam, Kan, and Spc), staphylococcal replicons (the replicons from pT181, from several pT181 conditionally replicating variants, or from pI258), reporter genes (*blaZ*, *gfp*, and *luxAB*), a choice of a constitutively active promoter or a cadmium-inducible promoter to drive expression of cloned genes, and the option of integrating the construct into a specific chromosomal location (site-specific integration into the staphylococcal pathogenicity island SaPI1). This system is designed in a modular fashion that facilitates combining these elements as needed to produce recombinant plasmid vectors with distinct properties to fulfill disparate experimental requirements.

In addition to the Cd-inducible promoter used in the modular plasmid system described above (54, 77), several other inducible promoter systems have been used to control gene expression in staphylococci. For example, the IPTG-inducible P$_{spac}$ promoter has been successfully used in *S. aureus* (304). A tetracycline-inducible promoter has been used to control expression of antisense RNA in *S. aureus* (183). Plasmid vectors carrying a xylose-inducible promoter that is functional in *Staphylococcus carnosus* have been described (404).

32.2.2.5.2.2. Expression analysis. Most of the common reporter genes have been used to analyze gene expression in staphylococci. β-Galactosidase (231, 265), β-lactamase (280, 396), chloramphenicol acetyltransferase (*cat*) (105, 280), green fluorescent protein (177, 257), and luciferase (77, 126) fusions have all been described.

32.2.3. Enterococci

The enterococci are commensal members of the gastrointestinal microflora of animals and humans. Enterococci have emerged as significant nosocomial opportunistic pathogens over the past two decades; they are now among the top three etiological agents responsible for each of the most frequent types of hospital-acquired infections. The enterococci have earned a reputation as reservoirs for the acquisition and dissemination of mobile genetic elements. The observation that ~25% of the recently sequenced genome of an *Enterococcus faecalis* strain consists of putative mobile genetic elements is consistent with this view (294). Dissemination of antibiotic resistance determinants that accompanies the intra- and interspecies exchange of these mobile elements clearly plays a significant role in the emergence of antibiotic-resistant enterococci as well as emergence of resistant isolates of other genera, such as vancomycin-resistant *S. aureus* (402). Due to the prevalence and significance of genetic exchange, research on enterococci has historically focused on the identification and characterization of enterococcal mobile genetic elements. The end result is that several widely studied and/or used mobile genetic elements were discovered in enterococci, including the conjugative transposon Tn916, the broad-host-range conjugative plasmid pAMβ1, and the enterococcal pheromone-responsive conjugative plasmids.

32.2.3.1. Plasmids

Members of three broad classes of plasmids are competent for replication in enterococci: (i) the RCR plasmids, (ii) the θ-replicating plasmids of the Inc18 family, and (iii) the pheromone-responsive plasmids (discussed below). Plasmids of classes i and ii are, to various extents, broad-host-range plasmids because they are capable of replication in many other gram$^+$ bacteria. In contrast, the pheromone-responsive plasmids are known to replicate only in enterococci. Although no RCR plasmid indigenous to enterococci has been identified, some rolling-circle replicons (e.g., pVA380-1 [250], pE194, and pWVO1) are capable of replication in enterococci and have been exploited as tools for cloning (104, 219) or transposon delivery (398). These plasmids tend to be small (<10 kb) and can be notoriously structurally unstable, often yielding plasmid progeny that have experienced partial or complete deletion of cloned DNA fragments. Thus, the integrity of such plasmids should be carefully monitored throughout their use.

The Inc18 family pAMβ1 replicon is capable of replication in a wide variety of gram$^+$ bacteria, including virtually every species of *Streptococcus* (68) as well as members of the genera *Lactococcus*, *Staphylococcus* (111, 334, 335), *Lactobacillus* (142), *Bacillus* (215), *Clostridium* (286), and probably others. Many members of this family are capable of transferring themselves by conjugation. These plasmids tend to be 25 to 30 kb in size and maintain themselves at a relatively low copy number (<10 copies/cell). Mechanisms of replication and copy control for this plasmid family have been reviewed (181, 400). Inc18-type plasmids replicate using a θ-type mechanism and therefore do not suffer from problems of structural instability that can plague RCR plasmids. Consequently, the Inc18 replicon has been used as the basis of a number of cloning and expression vectors, such as pAM401 (405), pWM401 (405), pAT18 (378), and pMSP3535 (41). Due to their distinct mechanisms of replication, Inc18 plasmids are compatible with RCR plasmids and can therefore be used with the latter plasmids when required.

32.2.3.2. Conjugative Elements

Enterococci possess numerous examples of conjugative plasmids and conjugative transposons. The pheromone-responsive plasmids, as well as some members of the θ-replicating class of plasmids, encode the machinery required for self-transmission to neighboring bacteria by conjugation. The apparatus mediating this transfer appears to comprise a subgroup of type IV secretion systems that has evolved the capacity for intercellular translocation of DNA-protein complexes (61). Putative homologs of type IV secretion system components, such as ATPases, lytic transglycosylases, and coupling proteins, are encoded in the transfer regions of such conjugative plasmids (reviewed in reference 150), but conjugation apparently does not utilize a pilus-like structure. The structure of the actual transfer apparatus or channel has not been described for gram$^+$ bacteria.

The conjugative Inc18 family plasmid pAMβ1 was originally identified in a clinical isolate of *E. faecalis* (75). It is a broad-host-range member of the θ-replicating class of plasmids (like pIP501) that carries the *ermB* erythromycin resistance determinant as well as transfer functions required for conjugation. Conjugative transfer occurs across wide species boundaries (e.g., transfer to *Lactococcus*, *Staphylococcus*, *Clostridium*, *Pediococcus*, and *Listeria* has been observed) and generally requires cocultivation of donor and recipient cells on a solid surface for efficient transfer, due to the requirement for cell-cell contact. Little is known about the regulation of pAMβ1 conjugative transfer.

The enterococcal pheromone-responsive plasmids have been studied in the greatest detail. These are large plasmids (~40 to 90 kb in size) that are maintained at low copy number in their enterococcal hosts. They are capable of self-transfer to plasmid-free recipients at extremely high frequencies by conjugation in liquid matings, due to the production of a plasmid-encoded surface protein that facilitates the formation of large cellular aggregates in which con-

jugative transfer can take place. The pheromone-responsive plasmids are remarkable in that they respond to a specific pheromone signal, secreted by plasmid-free enterococcal recipients, by inducing functions required for high-frequency conjugative transfer. Over 20 pheromone-responsive plasmids have been identified; these are known to respond to at least six different pheromones. Each pheromone signal is a linear, hydrophobic peptide of 7 or 8 amino acids. Once plasmid transfer is complete, the strain that has acquired the plasmid (which then becomes a new plasmid donor strain) no longer produces biologically active cognate pheromone signal (or masks the presence of the cognate pheromone by secreting a specific inhibitor), but it continues to produce the other unrelated pheromones. Thus, such strains can continue to collect (and disseminate) heterologous plasmids and any antibiotic resistance determinants or virulence factors that they might encode. While it is difficult to provide a quantitative assessment of the contribution of pheromone-responsive plasmids to virulence, it is clear that these plasmids enhance virulence in experimental models and are widespread in clinical isolates of enterococci. For more information on pheromone-responsive plasmids, the reader is referred to recent reviews that discuss replication (400) and conjugation of these plasmids (70, 72) in more detail than is possible here. Although the pheromone-responsive plasmids have been studied extensively, they have not been developed as tools for genetic exchange in enterococci thus far.

A third type of putative conjugative plasmid, the gentamicin resistance plasmid pMG1, has been identified in the enterococci (174). pMG1 can efficiently transfer among enterococci in liquid or filter matings. One pMG1-borne gene, designated *traA*, has been shown to be differentially regulated following mixing of donor and recipient cells, and mutation of this gene reduces the frequency of transfer in liquid matings (366). However, hybridization experiments indicate that there is no similarity between pMG1 and either the pAMβ1-like conjugative plasmids or the pheromone-responsive conjugative plasmids, suggesting that pMG1 carries a previously undescribed set of conjugation and regulation genes. pMG1 family plasmids likely play a role in genetic exchange of enterococci, as vancomycin-resistant clinical isolates have been shown to harbor pMG1-like plasmids that have acquired Tn*1546*-like transposons encoding vancomycin resistance (375).

The conjugative transposon Tn*916* belongs to a large class of mobile genetic elements now known as ICEs (44) that normally reside integrated in the chromosome of their host. These elements undergo excision from the chromosome to form a circular intermediate that is transferred by conjugation to recipient cells. These elements carry all the necessary genes for conjugative self-transfer. Upon completion of transfer to a recipient, the element recircularizes and subsequently becomes integrated into the recipient chromosome by the action of its self-encoded integrase. The prototypical gram$^+$ element of this type, Tn*916*, was originally discovered in a clinical isolate of *E. faecalis* (128). Although other large conjugative elements had been previously identified in other streptococci (343), Tn*916* was the first well-characterized conjugative transposon. Subsequent characterization of Tn*916*-like elements has revealed that these elements carry the *tetM* tetracycline resistance determinant and, in some cases, determinants encoding other types of antibiotic resistance as well. Frequencies of transfer of Tn*916* have been found to be moderately enhanced by treatment of donor cells with tetracycline, suggesting that exposure to tetracycline both promotes dissemination of the tetracycline resistance determinant and selects for tetracycline-resistant bacteria. Furthermore, these elements have been transferred to, or found naturally in, more than 50 divergent species of bacteria (71). The broad host range of Tn916-like elements suggests that they may represent significant agents of the dissemination of antibiotic resistance determinants.

In *E. faecalis*, Tn916 integrates into a variety of sites, although it exhibits a clear preference for AT-rich regions (339) and there has been a suggestion that regions of locally "bent" DNA are preferred as well (340). It is common to find that Tn916 recipients acquire several copies of the element that are distributed around the chromosome (138). However, the degree of site specificity of Tn916 integration is dependent on the identity of the recipient species. For example, Tn916 integrates preferentially at a unique site in some strains of *Clostridum difficile* (273, 393). Thus, although a number of attempts have been made to use Tn916 as a mutagen in several bacterial species, its utility in this capacity is limited. The literature on Tn916 is extensive: for a historical perspective of the discovery of Tn916, the reader is referred to an early review by Clewell and Gawron-Burke (73). For more recent reviews, see references 71, 318, 340, and 400.

General Procedure for Conjugative Mating

For conjugative transfer to occur, donors and recipients must establish cell to cell contact. To accomplish this, three variations on a similar theme are typically used for conjugative matings: liquid matings (planktonic cells of donors and recipients mixed in suspension), filter matings (cells of donors and recipients mixed in suspension and subsequently deposited on filter membranes which have been placed on the surface of agar plates; alternatively, the cell mixture is deposited directly on the surface of agar plates), or cross-streak matings (cells of donors and recipients streaked perpendicular to each other on agar plates such that they intersect). It is essential to have strains with the appropriate, stable genetic markers that can be used to differentiate transconjugants from donors and recipients. In *E. faecalis*, markers useful for this purpose include chromosomally borne resistance determinants for rifampin, fusidic acid, streptomycin, and spectinomycin. These markers can be introduced into strains of interest by simply selecting for spontaneous, resistant point mutants on appropriate selective agar medium.

To perform a conjugation experiment, fresh overnight cultures of donor and recipient strains (independent overnight cultures) are diluted 10-fold in broth and incubated at 37°C for 1 h. These cultures are then mixed at the desired ratio (typically a donor-to-recipient ratio of 0.1). At this point, the cell suspension can be incubated as such for the desired length of time (from 1 h to overnight) to perform a liquid mating (103), or the cell mixture can be concentrated and applied to a filter on a nonselective agar plate (69, 87) to perform a filter mating. Two mechanisms for concentrating the cells are common: (i) using a plastic, Luer-Lock syringe to apply the cells to a removable filter carried within a Swinnex-13 Millipore disk filter holder followed by placement of the filter on an agar plate (cells up) and (ii) centrifuging the cells and resuspending the pellet in a small volume (e.g., 25 µl) of fresh broth, which can then be pipetted onto a filter that has been placed on an agar plate. The agar plates are then incubated for the desired length of time. To perform a cross-streak mating (74), simply use a sterile, cotton-tipped applicator to streak the donors and recipients

in perpendicular lines on the surface of a nonselective agar plate such that they intersect with each other. Cross-streak matings generally are not suitable for quantification of transfer frequency but are a convenient means of transferring a conjugative element between strains.

After the desired incubation period, cells are recovered and plated on appropriate selective media for enumeration of donors and transconjugants. For liquid matings, this simply involves vortexing the cell suspension, to ensure it is thoroughly mixed, and removing aliquots for dilution and plating. For filter matings, cells are typically recovered by placing the filter in 1 ml of fresh broth and vortexing vigorously. For cross-streak matings, cells from the intersection of the donor and recipient streaks are recovered, for example, by using a sterile cotton-tipped applicator. To enumerate transconjugants, the medium that is used must present selective challenges that are simultaneously satisfied by a stable marker present only in the recipients (to select against donor cells) and the marker present on the conjugative element. To enumerate donors, the medium must present selective challenges that are simultaneously satisfied by the marker present on the conjugative element and a stable marker present only in the donors (to select against recipient cells and transconjugants). Pure cultures of the donor and recipient strains should also be plated on the selective enumeration media as controls. Frequencies of conjugation are typically presented as a ratio of transconjugants per donor and can vary widely depending on the mobile element that is being transferred, the time allowed for mating, and the method of mating that is employed.

32.2.3.3. Bacteriophage and Transduction

Generalized transduction has not been described for enterococci. However, two observations suggest that phage have played a significant role in genetic exchange in enterococci. First, seven loci encoding putative prophages have been identified in the completely sequenced genome of an *E. faecalis* strain (294), some of which encode apparent homologs of putative virulence factors (*Streptococcus mitis* adhesins PblA and PblB, which have been implicated in binding human platelets). Second, enterococcal phages can be isolated from various environmental niches that enterococci inhabit (10, 50, 288, 320, 379).

32.2.3.4. Transformation

Competence for transformation, determined by a genetically encoded mechanism, has not been described for enterococci. However, electroporation can be used to insert DNA into cells made competent by either of two procedures: growth in high concentrations of glycine (80, 104) or treatment with low concentrations of lysozyme prior to electroporation (13). Both of these treatments serve to weaken the thick cell wall and presumably facilitate the passage of DNA across this barrier. A procedure for protoplast transformation has also been described (406), but it has been largely superseded by the electroporation-based protocols due to technical difficulty and lack of reliability associated with the production and regeneration of protoplasts.

32.2.3.5. Applications of Gene Transfer

32.2.3.5.1. Construction of Mutants

32.2.3.5.1.1. Transposon mutagenesis. The Tn3-like transposon Tn917 was originally discovered in a clinical isolate of *E. faecalis* and has been employed to isolate transposon mutants of *E. faecalis* (374, 399) as well as numerous other species of gram$^+$ bacteria. However, use of Tn917 as a tool for random mutagenesis in *E. faecalis* is ineffective due to the highly biased regional insertion site preference this transposon exhibits (135). Analysis of the insertion sites of 1,923 unique transposon insertion mutations in *E. faecalis* revealed that 65% of Tn917 insertions were found within a 200-kb region of DNA flanking the chromosomal terminus of replication. The biological basis for this regional insertion preference is unknown, but the practical implications indicate that use of Tn917 as an agent of random mutagenesis is ineffective. An efficient alternative system for random transposon mutagenesis in enterococci has not been described thus far.

32.2.3.5.1.2. Directed mutagenesis and allelic exchange.

Inactivation of genes in enterococci has traditionally been accomplished by using electroporation to introduce a suicide plasmid (e.g., pTEX4577 [313]) carrying a cloned internal fragment of the target gene into an appropriate host strain. Selection for plasmid-encoded antibiotic resistance yields recombinants in which the circular plasmid has integrated into the chromosome by Campbell-type (single-crossover) homologous recombination. The main drawback to this approach is that recombinants are difficult to obtain, due to the relatively low frequency of DNA uptake and recombination following electroporation. Recently, a conjugative delivery system based on the enterococcal pheromone-responsive plasmid pCF10 has been developed, enabling highly efficient introduction of conditionally replicating plasmids into target strains to serve as substrates for homologous recombination and allelic exchange (210a). An alternative strategy for allelic exchange based on the use of a TS plasmid (the pG$^+$ host replicon) has also been described (6). As described above (in the section on GAS), this is a multistep procedure that requires a genetic screening step to identify the desired recombinants (which arise at a relatively low frequency), thereby rendering the procedure somewhat inefficient and laborious. The recent development of counterselectable markers for *E. faecalis* has eliminated the need to use laborious screening procedures to identify the desired recombinants, allowing the development of allelic exchange systems in which genetic selections can be used for each of the two steps required to construct a wide variety of types of unmarked mutations, including in-frame deletions (210a, 211). This development makes efficient and sophisticated genetic manipulation of *E. faecalis* possible.

32.2.3.5.1.3. Selection and counterselection.

As with many other members of the gram$^+$ cocci, nutritional requirements of enterococci are complex. Therefore, nutritional markers (e.g., selection for prototrophy or carbon source utilization) are generally not used for genetic manipulation in these organisms. Instead, a variety of antibiotic resistance determinants serve as selectable markers to facilitate genetic manipulation. Most common resistance determinants that function in other gram$^+$ bacteria are also functional in enterococci, e.g., *erm* (erythromycin), *cat* (chloramphenicol), *tet* (tetracycline), and *aad9* (spectinomycin). Determinants conferring resistance to β-lactam antibiotics should not be introduced into enterococci, as certain β-lactam antibiotics remain clinically useful in some contexts.

Recently, two counterselectable markers (i.e., markers useful for negative selection) have been described for use in *E. faecalis*. The first is analogous to a previously described

approach used in *B. subtilis* (114) and is based on the observation that *E. faecalis upp* mutants exhibit 5-fluorouracil (5-FU) resistance (211). Because this 5-FU-based approach requires that the host strain carry a mutation in the resident chromosomal *upp* gene in order for counterselection to be applied (see the section on counterselection in *B. subtilis* for more detail on 5-FU-dependent counterselection), a "second-generation" counterselectable marker was developed that functions independently of the host genotype (210a). This approach exploits a dominant missense mutant allele of the *E. faecalis* phenylalanyl-tRNA synthetase α-subunit (termed *pheS**) that exhibits relaxed substrate specificity, such that it aminoacylates its cognate tRNA with halogenated derivatives of Phe (including *p*-Cl-Phe). In strains expressing *pheS**, subsequent production of nonfunctional proteins containing *p*-Cl-Phe residues (in place of Phe) results in growth inhibition. Thus, only clones that have lost *pheS** are capable of growth in the presence of *p*-Cl-Phe. A functional copy of *pheS** has been incorporated into a conditionally replicating plasmid vector that can be used for allelic exchange, thereby providing the benefit of a selectable marker for both steps in the two-step allelic exchange procedure (210a).

32.2.3.5.2. Analysis of Mutants

32.2.3.5.2.1. Complementation.

A variety of replicating plasmids have been developed that can be used to clone and express genes (or gene clusters) that possess their own promoters in enterococci. These are primarily based on the Inc18 θ-type replicon derived from pAMβ1 or pIP501 or the RCR replicon derived from pVA380-1. These plasmids include pAM401 (406), pWM401 (405), pAT18/19 (378), and pDL276/8 (104, 219). Expression of cloned genes in a controlled manner in enterococci is possible using the lactococcal nisin-inducible promoter system found on the plasmids pMSP3535 and pMSP3545 (41). A set of suicide plasmid vectors have been developed to permit integration of cloned genes into a resident copy of Tn916 in the *E. faecalis* chromosome by recombination, yielding recombinants in which the Tn916 variant is either competent for conjugative transfer or defective as a result of the recombination event (52, 256). One potential advantage of this approach is that the cloned gene is present in single copy, thereby circumventing any potential problems associated with differences in gene dosage from the wild-type situation. However, a disadvantage is that this approach obviously requires that a Tn916-containing mutant be generated to serve as a host for complementation analysis.

32.2.3.5.2.2. Expression analysis.

Several plasmid vectors have been developed for the construction and analysis of promoter fusions to β-galactosidase in enterococci (179, 310). Both the chloramphenicol acetyltransferase (*cat*) and β-glucuronidase (*gusA*) genes have also been used as reporter genes in enterococci (12).

32.2.4. Lactococci and Related Food-Fermenting Bacteria

The common features shared by lactococci and related food-fermenting bacteria include the facts that they are all gram$^+$ bacteria that produce lactic acid from fermentation, and they play essential roles in the preparation of fermented food products. While the lactic acid bacteria (LAB) actually include several pathogenic species discussed elsewhere in this chapter, this section as well as most published literature referring to "LAB" focuses on the food-fermenting organisms. As has been pointed out in previous reviews (94, 106), phylogenetic analysis based on 16S rRNA sequences reveals a considerable degree of variation among these organisms, with one of the most important groups, the lactococci, showing a high level of relatedness to the streptococci, while others, such as the bifidobacteria, are quite distantly related (Fig. 1). These differences correlate generally with the growth properties and with the extent to which genetic principles and methodologies derived from one organism can be applied to others in this large group. In terms of practical applications of gene transfer methods to these bacteria, there are two important constraints. The first is that the genetics of *Lactococcus lactis* are very well developed, and with the possible exception of *B. subtilis*, this species may be the most advanced of all gram$^+$ bacteria in terms of current level of genetic analysis and available approaches for genetic manipulation. One reason that lactococcal genetic transfer technology is so well developed is that the early research on lactococcal genetics revealed that most of the industrially important traits of these organisms are encoded within mobile genetic elements (120); thus, the study of these elements has been the focus of attention of many excellent researchers over the past 30 years. On the other hand, many other industrially important organisms in the LAB group are less well studied and more refractory to genetic analysis. There is much variation in the degree to which genetic tools developed in lactococci can be readily adapted to these other species. In the past few years remarkable progress in both genomic analysis and development of in vivo genetic techniques for nonlactococcal food bacteria and their phages has been made (94, 106, 267), and further development of gene transfer methods in many of these organisms will be accelerated by genomic data.

A second important consideration for researchers studying these organisms is the fact that the objective of this type of research is the development of new strains with improved industrial properties. Since such strains ultimately may be consumed in large quantities by humans, they must have been derived through the use of "food grade" vectors and techniques. Such strains must not carry potentially deleterious genes such as antibiotic resistance determinants, other genes from pathogenic bacteria, or any DNA not native to organisms already used in food production. (A notable exception to this principle is in the development of oral live vaccines, where LAB strains may be used to deliver antigenic determinants from pathogens to humans in order to induce a protective immune response against the pathogenic agent [311a].) Thus, the choice of cloning vectors, selective markers, etc., depends on whether the research being carried out is basic or applied directly to starter culture development. While space limitations prohibit a comprehensive compilation of all publications in this area, several examples of food-grade tools are summarized in Table 2. For more comprehensive treatments of this subject, the reader is referred to several recent reviews (94, 106, 209).

32.2.4.1. Plasmids

Lactococci from industrial starter culture collections typically contain multiple autonomously replicating plasmids (≥6 in a single strain is not uncommon), ranging in size from <2 to >100 kb. Although a few plasmid replicons from LAB have been analyzed in detail, relatively little is known about the molecular mechanisms that allow this vast pool of extrachromosomal DNA to be replicated and maintained stably. In terms of cloning vectors, two types of replicon have been most extensively developed for use in *L. lactis* and also in a number of other LAB. Vectors derived from the Inc18 (pAMβ1 and pIP501) family (Table 1) display two very useful properties of cloning vectors. They tend to be very broad in host range, and their θ-mode of replication generally means that recombinant plasmids constructed using such vectors do not typically suffer deletions and other rearrangements that frequently occur with cloning vectors utilizing RCR mechanisms. Although the naturally broad host range of these plasmids may actually extend to gram$^-$ spp., most of the commonly employed *E. coli*-LAB shuttle vectors derived from the Inc18 family are engineered to contain an *E. coli* replicon to facilitate the use of well-characterized *E. coli* host strains as intermediates in cloning manipulations. A few examples of shuttle vectors in this family used successfully in LAB are shown in Table 1. Although the wild-type Inc18 replicons generally have a relatively low copy number, elevated-copy-number mutants such as pAT18 (378) may also be useful for some applications. Importantly, most widely used vectors in this family may be unsuitable for direct use in many food-grade applications, since they contain antibiotic resistance genes.

TABLE 2 Examples of vectors and other genetic tools for engineering of food-grade LAB strains

Application	Organism	Selection	Reference
Cloning/selection	*L. lactis*	D-Alanine prototrophy	37
Cloning/selection	*Lactobacillus plantarum*	D-Alanine prototrophy	37
Cloning/selection	*Lactobacillus johnsonii*	Lactacin resistance	3
Cloning/selection	*Lactobacillus acidophilus*	Thymidine prototrophy	131
Cloning/selection	*L. lactis*	Nisin resistance	364
Cloning/selection	*L. lactis*	Nisin or heavy-metal resistance	232
Cloning/selection	*Streptococcus thermophilus*	Cd^{2+} resistance	408
Cloning/expression	*Lactobacillus casei*	Lactose (phospho-β-galactosidase)	365
Cloning/expression/ mobilization/integration[a]	*L. lactis*	Antibiotic resistance (markers deleted in final construct)	165
Gene delivery/integration[b]/ excision	*L. casei*	Antibiotic resistance (markers deleted in final construct)	259

[a]Integration based on homologous recombination.
[b]Integration based on site-specific recombination.

A second family of heavily used vector plasmids has been derived from the pWVO1 naturally occurring RCR plasmid. Due to the remarkably wide natural host range of this plasmid (it was originally identified in a strain of *L. lactis* subsp. *cremoris* [207]), it serves as a natural shuttle vector, and its replication mechanism has been analyzed extensively. Studies of the replication initiator protein RepA from pWVO1 have led to the identification of temperature-sensitive mutants which have been employed to generate conditionally replicating derivatives of this plasmid for various applications, as described below and in other parts of this chapter (Table 1).

Since small, cryptic plasmids are common in LAB, a number of investigators have developed cloning vectors for their organism of interest by isolating and sequencing such a cryptic plasmid and then using standard recombinant DNA methods to insert selectable markers and, in most cases, a replication origin functional in *E. coli*. Because so many different pWVO1-derived plasmids have utility in a variety of LAB, development of novel cloning vectors typically involves testing potential replicons for compatibility with pWVO1 at an early stage, as illustrated by the recent reports of Sanchez et al. (207, 329, 331).

32.2.4.2. Conjugative Elements

Some of the seminal studies of lactococcal genetics involved demonstration of horizontal transfer of industrially important traits such as lactose fermentation ability, proteinase production, and phage resistance by conjugation; for reviews of the early literature in this area, see references 120 and 355. Naturally occurring conjugative plasmids carrying useful traits have been used directly in starter strain construction, to generate cultures with improved properties using simple mating protocols (355, 356). In several cases, characterization of the functional origin of transfer (*oriT*) region of conjugative elements has facilitated the construction of mobilizable vectors (18, 166, 329), which could be of considerable use for introduction of DNA into LAB strains refractory to efficient transformation (by using an *L. lactis* intermediate strain to propagate recombinant plasmids, followed by mating into the target strain). Another potential tool of great utility in the genomic era would be the construction of Hfr strains for transfer of large blocks of chromosomal DNA. Gasson and colleagues demonstrated the ability of the lactococcal sex factor to generate Hfr strains by integration into the chromosome (136), but this technique has not been widely exploited. A notable exception to the lack of conjugation-based genetic systems is the report of Henrich et al. (165), who used pG⁺host-derived vectors recombined with the sex factor to deliver cloned genes under the control of inducible promoters into the *L. lactis* chromosome, generating food-grade strains with stably integrated genes whose expression was inducible.

Conjugative transposons in the Tn*916* family have been shown to insert into LAB (20, 35, 36, 119), although they have not been used widely for chromosomal mutagenesis. Interestingly, at least one report indicates that Tn*916* cannot excise and undergo conjugative transposition from the *L. lactis* genome, due to a lack of an unknown host factor (35). The best-studied naturally occurring conjugative transposon in these organisms is Tn*5276*, which encodes both the production of nisin and the ability to ferment sucrose (314). Although considerable information about the excision/insertion mechanism of this element (315) is available, Tn*5276* has not been widely used for random mutagenesis, perhaps due to the large size of the element.

32.2.4.3. Bacteriophage and Transduction

The interactions of LAB with bacteriophage have been studied extensively, largely due to the fact that phage-mediated lysis of starter cultures is a major problem in the industrial setting (40, 204, 209). Both temperate and lytic phages are extremely common in LAB, and a great deal of effort has been made to survey the diversity and host range of LAB phages as well as naturally occurring bacterial mechanisms of resisting phage infection. Notably, one of the most commonly used culture media for cultivation of lactococcal strains is M17, which was originally developed to allow for identification of phages (367). As summarized in a recent review, genomic analysis of LAB phages has revealed an extensive amount of biological diversity and genetic recombination in the evolution of the genes encoding the ability of LAB phage to infect and replicate (40). Some of the earliest literature on LAB gene transfer was comprised of reports of the transfer of industrially important genes of *L. lactis* by transduction (137). Transduction has also been described for other LAB (236). However, in the past 5 to 10 years, transduction has, surprisingly, not been one of the most frequently used genetic tools. In the case of organisms like *L. lactis*, the relative ease of transformation- and conjugation-based transfer methods might account for this trend. However, generalized transduction can be a particularly efficient method for in vivo genetic engineering, particularly in the case of chromosomal loci, and may be especially useful with organisms where genomic data are available but effective genetic tools are limited. Because some recently sequenced LAB strains fall into this category, generalized transduction may be more heavily used in the future.

32.2.4.4. Transformation

Natural competent cell transformation of the LAB used in food fermentations for the most part has not been documented. However, this form of genetic exchange could be taking place with certain strains or under certain conditions that have not yet been identified. In support of this idea, analysis of the genome sequence of *L. lactis* has revealed the presence of a number of competence gene homologs (66), and evidence for natural transformation (albeit at very low frequency) of *Leuconostoc carnosum* has been reported recently (161). Electroporation has been used to introduce DNA into many LAB (237). Many commonly used protocols involve the growth of the organisms in complex medium containing a high concentration of glycine (0.5 to 5%, wt/vol) to weaken the cell wall, and washing and resuspension in an osmotically stabilized (sucrose is often used) buffer solution. A collection of suggested electroporation protocols published as appendices to *Genetics and Molecular Biology of Streptococci, Lactococci and Enterococci* (102a) serves as a reasonable starting point for optimization of a method for electroporation for a variety of LAB. A sample protocol for *L. lactis* based on the work of Holo and Nes (168) is given below.

Protocol for Transformation of *Lactococcus lactis* by Electroporation

The strain to be transformed is grown at 30°C to exponential phase in M17 medium (BBL) with 0.5% glucose (GM17). This culture is used to inoculate M17 medium with 0.5% sucrose (SM17) and 3% glycine; this culture is grown for 12 to 16 h at 30°C. To obtain optimal transformation frequencies, it may be necessary to grow several overnight cultures with glycine concentrations ranging from 0.5 to 5% and use the cells from the highest glycine concentration that still allowed growth of the overnight cultures to OD_{600} levels of 50 to 100% relative to that obtained with a parallel culture lacking glycine. The overnight culture is then diluted into the same medium and grown to an OD_{600} of 0.2 to 0.7. The cells are harvested by centrifugation and washed twice with 1 volume of ice-cold 0.5 M sucrose containing 10% (wt/vol) glycerol. The washed cell pellets are resuspended in 1/100 volume of the washing solution and can then either be frozen at −85°C for storage or used directly for electroporation. Electroporation of 40 μl of bacterial sample mixed with 1 to 5 μl of purified DNA (in 10 mM Tris HCl–1 mM EDTA) is carried out in a Bio-Rad Gene Pulser unit at 10 to 12.5 kV/cm. Immediately following electroporation, the solution is diluted into 0.96 ml of SGM17 containing 20 mM $MgCl_2$ and 2 mM $CaCl_2$ and incubated for 1 to 2 h at 30°C. The cells are then plated on either SR or GM17 agar medium containing appropriate antibiotics to select for transformants and incubated at 30°C.

SR medium

Tryptone	10 g/liter
Yeast extract	5 g/L
Sucrose	200 g/liter
Glucose	10 g/liter
Gelatin	25 g/liter
Agar	15 g/liter
$MgCl_2$	2.5 mM
$CaCl_2$	2.5 mM

Final pH is 6.8.

32.2.4.5. Applications of Gene Transfer

32.2.4.5.1. Construction of Mutants

32.2.4.5.1.1. Transposon mutagenesis and site-specific recombination systems. As described above, the Tn916 family of conjugative transposons can potentially be used for insertional mutagenesis of LAB, and derivatives of enterococcal transposon Tn917 have been employed for mutagenesis of LAB (179, 251, 316). In addition, naturally occurring insertion sequences such as ISS1 have been used to promote formation of cointegrate plasmids (321, 322) and drive random insertion of pG+host derivatives into the chromosome (254). Another cutting-edge approach reported recently involves the high-efficiency integration of cloned genes into the chromosome of LAB using phage-mediated site-specific recombination systems, generally of phage origin (259). This technique may be particularly useful in constructing stable starter culture strains where controlled expression of genes of interest can be used for industrial applications; these systems generally require a naturally occurring or engineered target site in the chromosome for the site-specific recombinase. They offer many potential advantages in terms of genetic stability, as well as the use of a common bacterial host "platform" for expression of a variety of useful genes in industrial settings.

32.2.4.5.1.2. Directed mutagenesis and allelic exchange. A number of different conditionally replicating vectors have been employed for delivery of transposable elements, or for allelic exchange in LAB (see reference 267 for a review). The most widely used systems are the thermosensitive pG+host vector series and the pWVO1-based pORI plasmids (Table 1 and other sections of this chapter).

The pG+host plasmids are temperature-sensitive shuttle vectors (253) and have been very widely used to generate allelic replacements in several species. The pORI series of plasmids are pWVO1 derivatives (218) in which the repA gene, encoding the replication initiator protein, has been deleted. These plasmids can be maintained in specially constructed host strains where the gene is supplied in trans from a helper plasmid (LAB hosts) or chromosomal location (E. coli hosts). In the LAB strain targeted for insertional mutagenesis or allelic exchange, the helper plasmid is temperature sensitive, such that the pORI plasmid can be inserted by electroporation and allowed to establish in the new host. After several generations, the temperature is raised, resulting in curing of both the helper plasmid and the pORI vector, which may then integrate by homologous recombination; subsequently a second recombination event can excise the integrated plasmid and, in some cases, leave behind the desired mutation of the target gene in the chromosome. The key feature of this system is that the helper plasmid allows the investigator to separate the initial transformation event introducing the pORI-derivative from the integration of the plasmid into the chromosome. In addition to the resistance genes that provide selection for the pORI vectors, a β-galactosidase gene has also been included to facilitate screening for excisants in the second step of the process. As described in the section on enterococci, a significant improvement is to modify these vectors such that the excision event can be selected rather than screened.

In addition to methods based on homologous recombination, two novel approaches for targeted integration and insertional mutagenesis have been reported. One of these involves the use of a group II intron called Ll.ltrB (268). This self-splicing mobile element was discovered within the conjugative relaxase gene, ltrB, of L. lactis. In addition to splicing from mRNAs, group II introns can insert themselves into DNA targets by retrohoming or retrotransposition processes (79). Retrohoming involves intron insertion into an intronless allele of the native exon gene, whereas retrotransposition entails insertion into other locations (usually related in sequence to the homing site). In either case, exon-binding site sequences in the intron play a key role in targeting the intron for insertion into complementary intron-binding site sequences. Introns can be retargeted for insertion into new sites, as well as using conditionally splicing defective introns to produce conditional mutations in target genes (129). As reviewed by Mills (267), this system has considerable promise as a genetic engineering tool for bacteria, although the technical issues relating to optimization of the system for high-throughput use in diverse species are not trivial.

32.2.4.5.1.3. Selection and counterselection. While nearly all of the antibiotic resistance markers that can be used in enterococci and streptococci (described elsewhere in this chapter) will function in most LAB, with appropriate adjustment of concentration based on empirical testing, none of these markers can be employed in food-grade strains. Therefore, considerable effort has been made to develop selective markers that utilize native genes from LAB starter culture strains. Because virtually all biosynthetic gene clusters encoding bacteriocin production include one or more immunity genes, a common approach is to insert these genes into cloning vectors or into the chromosome of recipient strains such that these constructs can be selected for by incorporation of the cognate bacteriocin into the growth medium (3, 364). In some cases, heavy metal resistance

determinants originally identified in starter strains have been incorporated into potential food-grade vectors (232, 408). An alternative approach to the use of antimicrobial agents for selection is to manipulate the metabolic capabilities of the organism such that the desired construct may be selected by simply modifying the composition of the growth medium. Bron et al. (37, 38) have shown that deletion of the *alr* (alanine racemase) genes of *L. lactis* and *Lactobacillus plantarum* resulted in auxotrophy for D-alanine, enabling the use of *alr* to generate food-grade cloning vectors. Inactivation of the *thyA* gene, encoding thymidylate synthetase, and the resulting thymidine requirement have been employed as selective strategies in several LAB (131, 332). Variations on this theme include the use of sugar utilization genes (305, 365). The use of the *upp* gene and the *pheS** cassette in development of a counterselection system for *E. faecalis* described elsewhere in this chapter may be adaptable to various LAB strains, although this will need to be tested for each species. Finally, it should be noted that some investigators have generated strains that they consider to be food grade using antibiotic resistance markers to select for integration of recombinant DNA into the chromosome, followed by a secondary excision event that removes all of the "non-food-grade" sequences in the final strain (165).

32.2.4.5.2. Analysis of Mutants

32.2.4.5.2.1. Complementation. The analysis of promoter activity in LAB has been a very active research area. Considerable information is available about the in vivo activity of a large number of promoters, and this has been applied to the development of various constitutive and inducible expression systems (for a review, see reference 94). As is typical with other aspects of LAB genetics, promoter function is best studied in *L. lactis*. The most comprehensive analysis of LAB promoters was carried out by Jensen and Hammer (182), who took selected *L. lactis* promoters of various strengths, mutagenized the spacer regions between the −35 and −10 sequences, and derived a set of 38 constitutive promoters with a range of activity over 400-fold. These serve as a very useful resource for research and industrial applications. Chen and Steele (58) recently tested the ability of cloned genomic fragments containing seven putative constitutive promoters from *Lactobacillus helveticus* to drive expression of a *gusA* reporter gene in the native host, as well as in *Lactobacillus casei* and *L. lactis*. All promoters except *pepN* were stably maintained and functional in all LAB hosts, as well as in *E. coli*. The *pepN* promoter was overly active, leading to toxic levels of reporter protein expression and genetic instability. A variety of inducible systems have also been developed, including those derived from biosynthetic operons for bacteriocins, which are often controlled by quorum sensing. The best characterized of these is the regulatory mechanism for nisin biosynthesis, for which nisin itself serves as the autoinducer. Several expression systems based on nisin induction (Table 1) are quite widely used in LAB and can also be employed in some gram$^+$ pathogens (92, 205, 212). Other inducible promoters are controlled by sugars, such as lactose (146, 365) or xylose (309), or by shifts in pH (179). Finally, "explosive" expression systems based on phage promoters that produce a burst of expression upon phage infection have utility for some applications (359, 392).

32.2.4.5.2.2. Expression analysis. A variety of native and heterologous reporter genes have been employed as tools to monitor transcription, translation, and protein secretion in LAB. Reporters shown to be useful in monitoring expression in the cytoplasm include various β-galactosidase genes (179, 229, 283, 285, 365), *gusA* (58, 146, 351, 368), *cat*-86 (207, 218), *lux* (54), and *gfp* (54, 145). In terms of protein secretion, two of the most frequently used reporter enzymes are staphylococcal nuclease (95, 308, 345) and amylase (155, 332). The former system has been studied in considerable detail and has been used successfully to analyze the topology of membrane-spanning proteins as well as protein secretion. Use of reporter enzymes of gram$^-$ origin, such as alkaline phosphatase, and β-lactamase can be problematic, although the *E. faecalis* alkaline phosphatase gene, *phoZ*, has been adapted for use in streptococci (63, 223).

32.2.5. Bacilli

Investigation of genetic exchange in the genus *Bacillus*, particularly in *B. subtilis*, has a long and rich history. As a result, the literature on genetic analysis of *B. subtilis* is vast. These studies have traditionally been driven, in part, by the practical advantage afforded by the phenomenon of natural genetic transformation exhibited by laboratory strains of *B. subtilis* (although some wild strains of *B. subtilis* apparently do not exhibit natural transformation) and, more generally, by the extensive use of the endospore-forming *B. subtilis* as a model for microbial development. Any attempt at comprehensive review of *B. subtilis* genetics would easily fill an entire chapter by itself. Several reviews have covered this material in various amounts of detail (81, 82, 101, 102). The primary focus of the present chapter is on advances that have been developed for genetic analysis of *B. subtilis* since these reviews appeared, although we do include some basic protocols for particularly important types of genetic manipulation. We refer the interested reader to the older reviews for more detailed discussion of the previously described methods. We also note that many strains of *Bacillus*, as well as plasmids available for use in bacilli, can be obtained from the repository at the *Bacillus* Genetic Stock Center (located at Ohio State University), which maintains an online, searchable catalog of materials that are available for distribution.

32.2.5.1. Plasmids

Although some *Bacillus* strains carry cryptic plasmids, laboratory strains of *B. subtilis* in common use do not contain endogenous plasmids. As a result, most plasmids that have found use in *B. subtilis* were derived from staphylococcal plasmids. Plasmids that replicate in *B. subtilis*, as in other gram$^+$ bacteria, broadly fall into two classes: the RCR plasmids (these consist primarily of the broad-host-range RCR replicons, and their derivatives, isolated from staphylococci) and the broad-host-range θ-replicating plasmids. Plasmids useful in *B. subtilis* have been extensively reviewed (39, 181). Much information about available plasmids, as well as the plasmids themselves, is also available from the *Bacillus* Genetic Stock Center (http://www.bgsc.org/); a few plasmids useful for specific applications in *Bacillus* are included in Table 1.

32.2.5.2. Conjugative Elements

Broad-host-range conjugative plasmids, such as pIP501, are known to be able to transfer into bacilli (213), and conjugative transposons have been transferred into and out of *B. subtilis* (271). However, possibly due to the ease of genetic manipulation by natural transformation, conjugative elements indigenous to bacilli have not been extensively characterized. Some exceptions include the 95-kb plasmid p19 of *B. subtilis* (307a) and the *Bacillus thuringiensis* plasmid pHT73, encoding the Cry1Ac toxin (171a). Other conjugative *B. thuringiensis* plasmids include the pXO11-pXO16 collection (15a, 316a, 367a).

32.2.5.3. Bacteriophage and Transduction

Bacteriophage have been extensively used as tools for genetic manipulation of bacilli. Transduction using recombinant derivatives of phage SPβ has been used for site-specific integration of recombinant constructs into the *B. subtilis* chromosome, allowing for complementation or expression from a single, stable genomic copy of the gene (see below). Generalized transduction can be performed using the large phage PBS1, which encapsulates approximately 300 kb of host chromosomal DNA in its transducing particles. The use of these phage for genetic manipulation of *B. subtilis* has been reviewed (82). Below, we describe methods for PBS1-mediated generalized transduction, taken from Cutting and Vanderhorn (81). PBS1 infects cells of *B. subtilis* by adsorbing to the flagella; therefore, successful use of this phage requires highly motile recipient cells. Some practical approaches that may help produce motile recipient cells include allowing the cells to grow well into stationary phase or removing the culture flask from the incubator and placing it on the bench for ~20 min.

Liquid Culture Method for Preparing a PBS1 Transducing Lysate

1. Streak the donor strain onto a Luria-Bertani (LB) or tryptose blood agar base (TBAB) (Difco) agar plate and incubate overnight at 37°C.
2. Inoculate 5 ml of Penassay broth (PAB) in a test tube with a single colony and incubate at 37°C with aeration.
3. When the culture reaches mid-log phase (OD_{600}, ~0.7), check a sample for motility under the microscope. Continue to check the growing culture every 10 to 15 min for motility.
4. When it is clear that the majority of the cells are motile, prepare the following two tubes. In tube 1, add 0.8 ml of motile culture and 0.2 ml of PBS1 lysate. In tube 2, add 0.9 ml of motile culture and 0.1 ml of PBS1 lysate.
5. Incubate these tubes at 37°C for 30 min with aeration. Then add 8.5 ml of prewarmed PAB to each tube and incubate for a further 30 min.
6. Add 20 μl of chloramphenicol (2 mg/ml) and incubate for a further 2 h.
7. Leave the tubes upright in a 37°C incubator overnight. After overnight incubation, it is likely that only partial lysis of the culture will be observed, but this does not necessarily mean that no transducing particles are present.
8. Centrifuge the lysed culture (8,000 × *g*, 15 min, room temperature). Sterilize the supernatant (lysate) by filtration or by addition of 0.5 ml of chloroform and store at 4°C. Verify the sterility of the supernatant prior to use.

Method for Transducing Cells with PBS1 Bacteriophage

1. Streak the recipient strain on an LB or TBAB agar plate and incubate overnight at 37°C.
2. The following day, inoculate 10 ml of brain heart infusion (BHI) broth (Difco) in a 250-ml flask and incubate overnight at 30°C with slow aeration.
3. If the cells are motile, inoculate 10 ml of fresh BHI broth with 2 ml of motile culture. Incubate at 37°C with aeration.

4. After 3 to 4 h, check a sample for motility under the microscope. If there is no motility, remove the flask from the incubator, leave on the bench top for 15 to 20 min, and check again.
5. Add 1 ml of motile culture to each of two sterile tubes. One will serve as a control; to the other, add 0.1 ml of PBS1 donor transducing lysate. Incubate with slow aeration for 30 min at 37°C.
6. Centrifuge the culture (8,000 × *g*, 10 min, room temperature), and wash the cells at least twice with 10 ml of T base plus 0.5% glucose (see the section on transformation below for composition). Resuspend the cell pellet in 1 ml of T base plus 0.5% glucose and plate aliquots onto selective agar medium.

32.2.5.4. Transformation

Like *S. pneumoniae*, *B. subtilis* possesses a genetically encoded program of natural competence development, the activation of which is cell density dependent. The enzymes involved in binding, processing, uptake, and recombination of exogenous DNA are highly similar between the two organisms. However, the molecular details of the regulatory pathway that links the cell density signal to expression of competence genes in *B. subtilis* differs from that of *S. pneumoniae* (102, 376).

Importantly, not all species of *Bacillus* exhibit a genetic program for natural transformation competence. For example, *Bacillus anthracis* and its close relatives have not been reported to be naturally transformable. For such species, electroporation has been used to introduce plasmid DNA into cells (206).

Two-step Transformation Procedure for *B. subtilis*

The procedure described below is taken from Cutting and Vanderhorn (81) and is an efficient method of obtaining high yields of competent cells. Note that starting with a low-cell-density inoculum in SpC medium will extend the time needed to reach stationary phase. If a heavy inoculum in SpC is used, typically it may take 4 to 5 h for the cells to reach stationary phase, equivalent to an OD_{600} of ~2.5 to 3.0. For strains that grow poorly, cultivating the bacteria on TBAB plates overnight may yield the best results.

1. Prepare a heavy streak of the strain to be made competent on an LB or TBAB agar plate and incubate overnight at 30°C.
2. Scrape cells and inoculate prewarmed SpC medium (20 ml) to an OD_{600} of about 0.5.
3. Incubate at 37°C with vigorous aeration and monitor the OD_{600} to assess growth.
4. When exponential growth ceases (i.e., no significant change in cell density over 20 to 30 min), inoculate 200 ml of prewarmed SpII medium with 2 ml of the initial culture. Incubate at 37°C with slower aeration.
5. After 90 min, pellet the cells by centrifugation (8,000 × *g*, 5 min) at room temperature.
6. Decant the supernatant and save.
7. Gently resuspend the cell pellet in 18 ml of the supernatant and add 2 ml of sterile glycerol; mix gently.
8. Aliquot the competent cells (0.5 ml) into sterile tubes and use for transformation or freeze at −70°C. Freezing competent cells may result in a 10-fold drop in yield of competent cells, although frozen cells can remain competent for several years.
9. Thaw competent cells rapidly by immersing frozen tubes in a 37°C water bath.

10. Immediately add 1 volume of SpII medium plus EGTA and mix gently.

11. Add competent cells (0.2 to 0.5 ml) to the DNA solution (<0.1 ml) in a sterile test tube. Incubate at 37°C to allow phenotypic expression of the selectable marker (only 5 to 10 min is required for DNA uptake).

12. Dilute the transformed cells as appropriate in T base containing 0.5% glucose and plate onto selective agar medium.

Media required for two-step transformation procedure:

T base

$(NH_4)_2SO_4$	2 g/liter
$K_2HPO_4 \cdot 3H_2O$	18.3 g/liter
KH_2PO_4	6 g/liter
Trisodium citrate · $2H_2O$	1 g/liter

Autoclave.

SpC medium (made fresh the day of use from the following sterile solutions)

T base	20 ml
50% (wt/vol) glucose	0.2 ml
1.2% (wt/vol) $MgSO_4 \cdot 7H_2O$	0.3 ml
10% Bacto yeast extract	0.4 ml
1% Casamino Acids	0.5 ml

Growth supplements as needed to satisfy auxotrophic requirements

SpII medium (made fresh the day of use from the following sterile solutions)

T base	200 ml
50% (wt/vol) glucose	2 ml
1.2% (wt/vol) $MgSO_4 \cdot 7H_2O$	14 ml
10% Bacto yeast extract	2 ml
1% Casamino Acids	2 ml
0.1 M $CaCl_2$	1 ml

Growth supplements as needed to satisfy auxotrophic requirements

SpII medium plus EGTA

SpII medium (200 ml) with 4 ml of EGTA (0.1 M, pH 8.0) but without $CaCl_2$. SpII medium plus EGTA can be frozen in small aliquots at −20°C. Avoid repeated freeze-thawing. SpII medium plus EGTA should be prepared without any additional growth supplements.

32.2.5.5. Applications of Gene Transfer

32.2.5.5.1. Construction of Mutants

32.2.5.5.1.1. Transposon mutagenesis. Transposons have been used extensively as tools for genetic manipulation in bacilli, especially *B. subtilis*. Early approaches used the enterococcal transposon Tn*917* (412a, 412b), although anecdotal reports of biased insertion sites for Tn*917* are common in *B. subtilis* (383a), just as in other gram-positive bacteria. Derivatives of Tn*10* in which the transposable element lacks a copy of the transposase have been widely used (301a, 357a) and provide the important benefit that, once transposition has occurred, the inserted elements are stable and cannot hop to other sites in the genome during subse-

quent growth of the strain. Such mini-Tn*10* derivatives have also been used in other bacilli (86a). Recently, several versions of the *mariner* transposable element have been successfully applied to various species of bacilli as well (219a, 365a).

32.2.5.5.1.2. Directed mutagenesis and allelic exchange. The pMUTIN series of suicide vectors (383) can be used to insertionally inactivate genes in *B. subtilis* by single-crossover homologous recombination. All the pMUTIN plasmids carry a *lacZ* reporter gene and an inducible P_{spac} promoter, which is reported to be tightly regulated and can be induced about 1,000-fold. Integration of a pMUTIN vector into a target gene has three consequences: (i) the target gene is inactivated; (ii) *lacZ* becomes transcriptionally fused to the target gene, allowing its expression pattern to be monitored; and (iii) the P_{spac} promoter controls the transcription of downstream genes in an IPTG-dependent fashion, thereby reducing the possibility of polar effects on their expression.

As an alternative to plasmid-based strategies for gene inactivation, a no-cloning-required approach known as long-flanking homology PCR (391) has been applied to analysis of *B. subtilis* (as well as to the competent streptococci *S. pneumoniae* and *S. mutans*). This approach is particularly effective for gene disruption in organisms that exhibit natural genetic competence, as they are capable of importing exogenous linear DNA fragments (including PCR amplicons) and recombining this DNA into their genomes by double-crossover recombination at high efficiencies. DNA fragments to be used as substrates for transformation are constructed using a multistep PCR approach such that a selectable antibiotic resistance marker is sandwiched between two pieces of chromosomal DNA that flanks the gene to be deleted (see references 217 and 391 for detailed descriptions of this process; a variant of the procedure is described by Fabret et al. [114]). For optimal recombination of the PCR amplicon, approximately 1 kb of flanking DNA should be included on both sides of the resistance marker.

32.2.5.5.1.3. Selection and counterselection. A counterselection strategy has been developed for *B. subtilis* (114). Wild-type *B. subtilis* is susceptible to growth inhibition when cultivated in the presence of the pyrimidine base analog 5-FU. This phenotype is dependent on the function of the *B. subtilis* uracil phosphoribosyl transferase, encoded by the *upp* gene, which converts 5-FU to toxic derivatives inside the cell. Accordingly, *upp* mutants of *B. subtilis* are resistant to the inhibitory effects of 5-FU, thereby providing a selection for the loss of *upp* function. To create a deletion mutant of a specific gene, a PCR amplicon is assembled carrying DNA segments that flank the gene of interest with an intervening cassette bearing a functional copy of *upp*, adjacent to an antibiotic resistance determinant. Thus, such an amplicon encodes a deletion of the target gene. The amplicon is constructed in such a way that it contains a pair of direct repeats flanking the *upp* antibiotic resistance cassette. This amplicon is transformed into a competent *B. subtilis* strain carrying a deletion in its chromosomal copy of *upp*, with selection for the amplicon-encoded antibiotic resistance marker. Once the amplicon is established in the chromosome by homologous recombination, a secondary recombination event between the cassette-encoded direct repeats leads to the excision of the *upp* antibiotic resistance cassette, leaving the deleted allele of the target gene in the chromosome. Such recombinants can be selected using medium supplemented with 5-FU.

32.2.5.5.2. Analysis of Mutants

32.2.5.5.2.1. Complementation. For complementation analysis, genes can be integrated into a lysogenic copy of phage SPβ by homologous recombination and subsequently moved between strains by transduction (described in reference 82). Alternatively, a set of plasmid vectors have been developed that allow ectopic integration of recombinant constructs at either the *B. subtilis amyE* locus or the *thrC* locus by double recombination, resulting in a loss of function of the corresponding gene. One important advantage of this strategy is that the recombinant constructs are present in single copy in the host genome, thereby avoiding problems related to gene dosage effects during complementation analysis. Some of these integration vectors provide the ability to construct operon fusions to a *lacZ* reporter for expression analysis (152). Integration of these vectors confers a detectable phenotype on the recipient strain (i.e., loss of the ability to hydrolyze starch [for *amyE* integrants] or threonine auxotrophy [for *thrC* integrants]), facilitating the identification of desired recombinants.

Expression of cloned genes in a controlled manner in *B. subtilis* has been accomplished using the IPTG-inducible P_{spac} hybrid promoter (410). An alternative xylose-inducible system has been developed and is carried on the vector pSWEET; this system facilitates integration by double recombination of cloned constructs into the *amyE* chromosomal locus in single copy (24). This *xyl* expression system was effective in achieving very low levels of expression in the absence of an inducer (it is worth noting that a common problem with the P_{spac} promoter is significant residual promoter activity in the uninduced state), was capable of a wide range of induction levels, and had the capacity for modulated expression over a broad scale of inducer concentrations.

32.2.5.5.2.2. Expression analysis. In addition to the typical β-galactosidase reporter, fluorescent proteins have been widely used in *B. subtilis* to monitor gene expression and subcellular protein localization. A series of plasmid vectors facilitating the construction of *gfp* fusions and their integration into the *B. subtilis* chromosome has been developed (228), as well as an expanded series of vectors encoding spectral variants of *gfp* (116). Luciferase (*lux*) fusions have also been used to analyze gene expression in *B. subtilis* (173).

32.2.6. Clostridia

The clostridia are a diverse collection of anaerobic, gram+ endospore-forming organisms, some of which are commensal members of the gastrointestinal microflora, some of which are important producers of industrial solvents, and others of which are pathogenic toxin producers. The genetics of the genus *Clostridium* have been recently reviewed in more detail than is possible here (243, 324). Many types of plasmids have been found in clostridia, although most of these have not been characterized in detail. A few of the plasmids isolated from clostridia have been modified for use as shuttle vectors in genetic manipulation. A number of conjugative elements (both plasmids and ICEs) have also been described.

32.2.6.1. Plasmids

Plasmids identified in clostridia are known to encode determinants of antibiotic resistance, toxins, and other virulence factors, but most of these plasmids have not been characterized in detail. The best-studied plasmid is the bacteriocin-encoding pIP404, which has been completely sequenced (134). The origin of replication of this plasmid has been defined, and copy control elements have been studied as well. Shuttle vector derivatives of pIP404, such as the pJIR series (243), have been developed for genetic analysis in clostridia.

32.2.6.2. Conjugative Elements

Two conjugative plasmids have been identified in *Clostridium perfringens*, one of which (pIP401, encoding tetracycline and chloramphenicol resistance) appears to be a derivative of the other (pCW3, encoding tetracycline resistance) carrying a transposon. Not surprisingly, heterologous broad-host-range conjugative plasmids are also capable of replication in clostridia, as pAMβ1 has been transferred into *Clostridium acetobutylicum* (286).

A family of mobilizable transposons (represented by Tn4451) have been identified in *C. perfringens*. These elements are not themselves conjugative, but conjugative transfer of such mobilizable transposons can be facilitated by other self-conjugative elements that reside in the same host. If a mobilizable transposon is located on a nonmobilizable plasmid, transfer of the mobilizable transposon can result in cotransfer of the plasmid as well (1). At least one conjugative transposon indigenous to clostridia has been identified, Tn5397, encoding tetracycline resistance (271), which is capable of conjugative transfer to *B. subtilis* and subsequent transfer back to *Clostridium difficile*. The broad-host-range conjugative transposon Tn916 can transfer into and throughout the clostridia and has been used as an agent of mutagenesis in *C. perfringens*. However, as in other species, Tn916 is not optimal for this purpose due to its propensity to insert at multiple loci in the same strain. Furthermore, in *C. difficile* Tn916 inserts preferentially at specific sites (273, 393), rendering it useless as a tool for mutagenesis. However, this insertion site preference appears to be strain dependent, as insertion in other strains of *C. difficile* occurs at multiple sites (319).

32.2.6.3. Bacteriophage and Transduction

Clostridial bacteriophage have been identified (148), and at least some phage are known to result in toxigenic conversion of the host upon establishment of lysogeny. For example, the botulinum neurotoxins are known to be encoded on lysogenic bacteriophage in *Clostridium botulinum* (162, 185). Thus, as with many other bacteria, bacteriophage appear to play a significant role in the evolution of clostridia and emergence of pathogenic strains. However, the use of clostridial bacteriophage as agents of generalized transduction has not been described.

32.2.6.4. Transformation

Competence for transformation, determined by a genetically encoded mechanism, has not been described for clostridia. However, several investigators have developed approaches for artificial transformation by electroporation (324). The success of electroporation is strain dependent, as some strains cannot be transformed using current electroporation protocols. In such cases, an alternative approach to deliver mobilizable shuttle plasmids by conjugation from an *E. coli* donor may be useful (242).

32.2.6.5. Applications of Gene Transfer

32.2.6.5.1. Construction of Mutants

32.2.6.5.1.1. Transposon mutagenesis. Transposon mutagenesis has been attempted in *C. perfringens* using the broad-host-range conjugative transposon Tn916. However, as noted above and as has been observed with other gram$^+$ bacteria, Tn916 was found to generate multiple insertions per strain, thereby limiting its usefulness as an agent of mutagenesis.

32.2.6.5.1.2. Directed mutagenesis and allelic exchange. Procedures based on homologous recombination have been used to inactivate genes in *C. perfringens* (8, 342). These approaches are conceptually similar to those used for gene inactivation in other gram$^+$ bacteria: they involve using electroporation to introduce suicide plasmids carrying either a cloned internal fragment of the target gene (for single-crossover recombination) or a gene replacement allele in which the gene has been replaced with an antibiotic resistance determinant (for double-crossover recombination in flanking segments). Approximately 1 kb of cloned, homologous DNA appears to be necessary in the suicide plasmid to achieve reasonable frequencies of recombination (324).

32.2.6.5.1.3. Selection and counterselection. Several antibiotic resistance determinants have been identified in *C. perfringens*, including those encoding resistance to chloramphenicol, erythromycin, and tetracycline. These determinants are similar or identical to equivalent genes from various other gram$^+$ bacteria (243). There have been no reports on the development of counterselectable markers in the clostridia.

32.2.6.5.2. Analysis of Mutants

32.2.6.5.2.1. Complementation. A suicide vector was developed to integrate cloned genes into a chromosomal copy of Tn916 residing in *B. subtilis*, which could then be transferred by conjugation into *C. difficile* (272) for purposes of complementation or expression analysis. The recombinant Tn916 elements could be further transferred to other strains of *C. difficile* for analysis in other genetic backgrounds, circumventing the difficulties often associated with transformation of this group of bacteria. Shuttle plasmids based on the clostridial pIP404 replicon have been developed (summarized by Lyras and Rood [243]) as well and can be used for cloning of target genes.

32.2.6.5.2.2. Expression analysis. Reporter genes that have been used in clostridia include *cat*, *luxAB*, and *gusA* (243). However, *gfp* cannot be used in the anaerobic clostridia due to the requirement for oxygen during the maturation of the *gfp* chromophore.

32.2.7. Mycobacteria

In contrast to the gram$^+$ bacteria discussed so far in this chapter, members of the genus *Mycobacterium* belong to the cluster of so-called high-GC gram$^+$ bacteria based on the nucleotide composition of their genomes. Some members of this genus are highly pathogenic for humans; *Mycobacterium tuberculosis*, for example, infects one-third of the world's population and resulted in an estimated two million deaths in 2002 according to the World Health Organization.

Various aspects of the biology of *M. tuberculosis*, including its extremely slow growth (3 to 6 weeks are required for colonies to arise from single cells on agar plates), its tendency to grow as clumps of cells, and the associated substantial biological safety risk make this a particularly difficult organism to study in the laboratory. However, the past 10 to 15 years have seen the development of a suite of tools for genetic analysis of *M. tuberculosis* and other mycobacteria. We summarize those developments in this section. The reader is referred to several excellent reviews of mycobacterial genetics (33, 91, 156–158, 180, 263, 290, 380) for more detail than can be presented here.

32.2.7.1. Plasmids

Clinical isolates of *M. tuberculosis* do not commonly carry naturally occurring plasmids. However, plasmids of various sizes and numbers are commonly found in the related species *Mycobacterium avium*. Many such plasmids are cryptic, and most have not been characterized in detail. Recently, an in vitro transposition-based approach designed to rescue cryptic plasmids from cytoplasmic extracts was described, thereby facilitating characterization of these elements (203). Three main classes of mycobacterial plasmids have been characterized to various degrees; most of the plasmids developed thus far for genetic manipulation of mycobacteria are based on the low-copy-number replicon pAL5000. More recently, two additional plasmid classes, both compatible with pAL5000-derived replicons, have been described in some detail. These include pVT2 (203) and the members of the pMSC262 family of related replicons (290). Many of these plasmids can stably replicate in a variety of mycobacterial species. A few useful mycobacterial plasmids are listed in Table 1. For an extensive listing of mycobacterial plasmids that can be used for cloning, gene expression, delivery of transposons, expression analysis, and integration into the mycobacterial chromosome, see the review by Pashley and Stoker (290). Linear plasmids, which are common in high-GC members of the actinomycetes, have been identified in some mycobacterial species but have not been adapted for use as genetic tools.

32.2.7.2. Conjugative Elements

Although genetic exchange by transduction in mycobacteria has been studied extensively (see below), and procedures for transformation of mycobacteria by electroporation are now routine, genetic exchange by conjugation in mycobacteria has received relatively little attention. Consequently, no conjugative plasmids or conjugative transposons analogous to those that are so prevalent in other gram$^+$ bacteria have been reported for mycobacteria. However, recently a novel mechanism for conjugal exchange of mycobacterial DNA has been reported (395), but this conjugation system has not yet been developed as a tool for genetic manipulation in these bacteria.

32.2.7.3. Bacteriophage and Transduction

Mycobacterial bacteriophage and their biology have been extensively investigated over the last several decades (reviewed by Hatfull [158]). Studies of phage biology played an important role in the initial development of methods for the introduction of DNA into mycobacteria, and phages or their derivatives have subsequently formed the basis for numerous vectors and other genetic tools and strategies that have revolutionized genetic analysis of mycobacteria. Complete genome sequences are now available for several mycobacterial phages, and at least two segments of phage-related genes are present in the completely sequenced genomes of M. *tuberculosis* strains, raising questions about the roles of these elements in mycobacterial physiology.

Mycobacteriophage I3 is a widely used generalized transducing phage of mycobacteria. Approximately 145 kb of DNA can be packaged into its head, and I3 has been shown to transduce both auxotrophic and antibiotic resistance markers (157). A number of widely used mycobacterial genetic tools based on bacteriophages have been developed. Chimeric molecules, known as phasmids, capable of growth both as a phage (in mycobacteria) and as a plasmid (in *E. coli*), are highly efficient means of introducing foreign DNA into mycobacteria. Phasmids are typically constructed by making a random library of the phage genome in an *E. coli* plasmid vector; these allow in vitro manipulation to form recombinant derivatives. A variety of modified phages (reviewed by McAdam et al. [263] and Derbyshire and Bardarov [91]) have been developed for use as tools for delivery of DNA, taking advantage of the highly efficient manner in which phage particles deliver their DNA cargo to essentially 100% of target cells when infected at the appropriate multiplicity of infection. Thus, even strains resulting from low-frequency events can be recovered. Phage derivatives have also been developed as tools for single-copy integration into the mycobacterial chromosome (see below for more detail).

32.2.7.4. Transformation

Competence for transformation, determined by a genetically encoded mechanism, has not been described for mycobacteria. However, procedures for reproducible transformation by electroporation have been developed. Although electroporation is now relatively straightforward, the variation in mycobacterial growth rates significantly affects the timing of the process. For example, transformant colonies of the fast-growing *Mycobacterium smegmatis* can be observed after approximately 3 days of growth at 37°C postelectroporation. In contrast, transformants of the slow-growing M. *tuberculosis* require about 3 to 4 weeks at 37°C to appear. Detailed protocols for electroporation of mycobacteria are described by Jacobs et al. (180) and Braunstein et al. (33).

32.2.7.5. Applications of Gene Transfer

32.2.7.5.1. Construction of Mutants

32.2.7.5.1.1. Transposon mutagenesis. Transposon mutagenesis has been used to isolate a variety of mutants in various mycobacterial species (263). Most of these transposons are composite elements based on insertion sequences that were originally identified in mycobacteria, including transposons based on IS*6100* and IS*1096*. For example, Tn*5367* was engineered from IS*1096* to carry a gene encoding kanamycin resistance; it transposes at a frequency of 10^{-5} (33). Other investigators have taken advantage of the impressive broad host range of the *mariner* transposable element to successfully mutagenize mycobacteria (326). Because conjugation in mycobacteria has not yet been extensively developed as a genetic tool, transposons are typically delivered using either a temperature-sensitive plasmid or phage derivative. These approaches allow relatively low-frequency transposition events to be recovered in significant numbers, due to the fact that most (or all) cells in a population carry (or receive) a copy of the donor element carrying the transposon that can serve as a substrate for transposition. For example, vectors such as pPR27, bearing

a TS derivative of the pAL5000 replicon, have been used to deliver transposons derived from IS*1096* (298).

32.2.7.5.1.2. Directed mutagenesis and allelic exchange. Procedures based on homologous recombination have been used to inactivate genes in mycobacteria. These approaches are conceptually similar to those used for gene inactivation in other gram$^+$ bacteria; they involve the introduction of nonreplicating recombination substrate DNA into the target cell, bearing either a cloned internal fragment of the target gene (for single-crossover recombination) or a gene replacement allele in which the gene has been replaced with an antibiotic resistance determinant (for double-crossover recombination in flanking segments). However, caution is warranted when these approaches are used in slow-growing mycobacteria such as M. *tuberculosis*, as illegitimate recombination (i.e., recombination at nonhomologous sites) has been observed to occur at relatively high frequencies, particularly if the substrate DNA is introduced in the linear form. Thus, candidate recombinants must be evaluated carefully to ensure the accuracy of the recombination event. The most significant advance that has facilitated allelic exchange in mycobacteria has been the development of phasmid vectors that permit highly efficient delivery of recombination substrates to target cells. For example, a conditionally replicating derivative of the broad-host-range TM4 phage phAE87 was used to transfer a mutant allele of the *lysA* gene to the chromosome of M. *smegmatis* mc^{2}155 (295). Transductants were obtained at a frequency of 10^{-6} per input cell, 95% of which exhibited a LysA$^-$ phenotype.

32.2.7.5.1.3. Selection and counterselection. A variety of selectable markers have been employed in mycobacteria. Many of these are antibiotic resistance determinants derived from plasmids or transposons originally isolated in other bacteria. The most commonly used marker is the *aph* gene, encoding kanamycin resistance. An important advantage of the antibiotic kanamycin is that the molecule is stable over the extended periods of incubation required for growth of slow-growing mycobacteria; other less stable antibiotics, such as tetracycline, can be used only with fast-growing mycobacterial species such as M. *smegmatis*. Other antibiotic resistance determinants that have found use in mycobacteria include those encoding resistance to chloramphenicol, streptomycin, sulfonamide, and hygromycin (156). β-Lactam antibiotics are generally not useful, because most laboratory mycobacterial strains contain an endogenous β-lactamase conferring resistance to these agents.

In some situations, antibiotic-based selectable markers are not suitable. For example, recombinant mycobacterial strains intended for use as live human vaccines should not contain antibiotic resistance determinants, to avoid the possible dissemination of such determinants to other pathogenic bacteria. In these situations, an alternative scheme for selection utilizes the bacteriophage L5 immunity gene to confer protection from phage L5 infection. This strategy has been shown to function well in M. *smegmatis* (98) and also appears to work in the vaccine strain bacillus Calmette-Guérin.

A counterselection scheme based on the *sacB* gene of B. *subtilis*, encoding levansucrase, has been successfully applied in mycobacteria. Although this approach has found wide use in various gram$^-$ bacteria, it generally cannot be used in low-GC gram$^+$ bacteria. To implement this strategy, bacteria are cultured on growth medium containing sucrose.

The *sacB* gene product catalyzes hydrolysis of sucrose and polymerization of the resulting fructose monosaccharides, inhibiting the growth of the bacteria. Thus, cells that have lost the *sacB* gene do not produce fructose polymers and are therefore able to grow in the presence of sucrose. The *sacB* determinant has been incorporated into a variety of mycobacterial plasmid vectors for use in allelic exchange, such as pYUB657 (295).

32.2.7.5.2. Analysis of Mutants

32.2.7.5.2.1. Complementation.
Perhaps the most valuable of the genetic tools derived from mycobacteriophages are the integration-proficient vectors, which utilize the phage *attP* and integrase genes to promote site-specific integration of the plasmid into the host chromosome. A wide variety of different phages have been used to construct these vectors, such as L5, Ms6, and D29 (reviewed by Hatfull [158]). Among the advantages of this strategy for the introduction of genes into the mycobacterial chromosome, the placement of genes in single copy is perhaps the most important, as this approach helps to eliminate complications that can arise from overexpression effects when genes are expressed from multicopy plasmids. Furthermore, the integration sites for these vectors typically are located in conserved tRNA genes, which are likely to be present in diverse species of mycobacteria, allowing their use across species boundaries. Finally, these vectors are extremely stable in the absence of selection once integrated into the chromosome, in the absence of the corresponding phage excisionase.

32.2.7.5.2.2. Expression analysis.
Reporter genes applicable to the study of gene expression in mycobacteria include most of those available for other bacteria, such as *cat*, *lux*, *lacZ*, *xylE*, *phoA*, and *gfp* (reviewed by Tyagi et al. [380]). However, the use of *lacZ* in mycobacterial strains carrying IS*1096* can be problematic, because *lacZ* contains at least one hot spot for insertion of this insertion sequence (62). Thus, repeated culturing of mycobacterial strains that simultaneously carry *lacZ* and IS*1096* should be avoided to reduce the possibility of unintended insertional inactivation of the reporter gene.

32.2.8. Actinomycetes
The actinomycetes are a diverse group of high-GC (70 to 74 mol%) sporulating organisms that generally develop a filamentous or branching form at some stage in their life cycle. Members of the genus *Streptomyces* are ubiquitous in soil habitats and are some of the most important sources of antibiotics and other biologically active chemicals. The actinomycetes have only a limited role as human pathogens (with the significant exception of M. *tuberculosis*), and the study of genetics in this group has focused primarily on their morphologically complex life cycles and secondary metabolic biosynthetic pathways. One of the most notable differences in gene transfer techniques used for many actinomycetes versus most other bacteria is the use of spores and subsequent protoplast formation for strain fusions, transformations, and transfections. Additionally, the molecular mechanisms of interspecies (and intergeneric) conjugation are fundamentally different than for all other groups of bacteria (150). Most of the techniques have been developed for specific strains of *Streptomyces*, with some applicability to other species and members of related genera. Some of the most well-established general procedures for genetic exchange in *Streptomyces* are presented below, but for more in-depth information regarding techniques, protocol, and theory, the reader is directed to the comprehensive manual *Practical Streptomyces Genetics* by Kieser et al. (197).

32.2.8.1. Plasmids
A wide variety of different extrachromosomal elements have been isolated from *Streptomyces* spp., including small and large circular as well as linear plasmids. Most encode replication or fertility traits, although several of the large linear plasmids have been shown to encode antibiotic biosynthetic pathways and resistance functions (202, 289, 363). The latter are the only known *Streptomyces* plasmids to encode a phenotypic trait (150). Almost all naturally occurring *Streptomyces* plasmids contain transfer functions that allow them to be transferred by conjugation to other related species (197).

The host ranges of most *Streptomyces* phages and plasmids are generally limited to the genus or related genera, with a few exceptions. The naturally occurring broad-host-range IncQ plasmid RSF1010, originally isolated from gram⁻ bacteria, has been shown to replicate in *Streptomyces* (144, 160). This non-self-transmissible plasmid can be mobilized between E. *coli* and a large number of gram⁺ strains, including *Streptomyces* and *Mycobacterium* spp., if the transfer gene (*tra*) functions are supplied in *trans* by the donor host (144). The Inc18 plasmid pIP501 (8.6 kb), originally isolated from *Streptococcus agalactiae*, has been shown to have a broad host range within gram⁺ bacteria and was recently shown to transfer conjugatively from E. *faecalis* to S. *lividans* (213). Within the genus *Streptomyces*, the high-copy-number RCR plasmid pIJ101 (8.9 kb) has the broadest host range, is conjugative, and promotes high-frequency chromosomal recombination (192, 198). pSG5 (12.2 kb) is an RCR temperature-sensitive plasmid that replicates in many species of *Streptomyces* (274). This plasmid cannot replicate at temperatures above 34°C, and this feature has been utilized to engineer several useful vectors for transposon delivery and insertional mutagenesis (28, 275). Another RCR plasmid that has a relatively broad host range within *Streptomyces* as well as related genera such as *Saccharopolyspora* is the multicopy plasmid pJV1 (11.1 kb) (14, 386). pSAM2 (11.2 kb) is a *Streptomyces* RCR plasmid that is usually integrated into a specific chromosomal tRNA gene, but it has also been isolated as a covalently closed circular plasmid (299, 348).

32.2.8.2. Conjugative Elements
There are no known naturally occurring conjugative transposons in *Streptomyces* species, although several transposons have been engineered into delivery vectors which can be transferred by conjugation, such as Tn*5096*, Tn*5099*, and Tn*5100* derived from IS*493* (see the section below on transposon mutagenesis) (352, 353).

Although there has been significant progress in the development of transformation systems in *Streptomyces* spp., protocols are often species specific and depend on laborious preparations of protoplasts and development of regeneration conditions. An alternative method for introducing DNA into *Streptomyces* is either interspecific or intergeneric conjugation, described below.

32.2.8.2.1. Interspecies Conjugation
Genetic transfer between strains of *Streptomyces* and closely related genera in the natural environment appears

to be quite frequent and has been attributed to the efficient conjugation of plasmids. Perhaps one of the most striking aspects of actinomycete biology is the unique mode of conjugal DNA transfer which does not require any genes for pilus formation or aggregation factors and depends on the physical contact of mycelia in the mating strains. Most *Streptomyces* conjugating elements require only a single transfer protein (Tra), related to the septal DNA translocator proteins in the SpoIIIE/FtsK family, and repressor (150, 197). Since inactivation of the repressor usually results in overexpression of *tra* and subsequent cell death, the repressor gene is known as *kill override* (*kor*) (191). However, when the *tra* genes from pSA1.1 from *Streptomyces azureus* and pSG5 from *Streptomyces ghanaensis* are cloned without their corresponding repressors and expressed in *Streptomyces* strains, lethality is not induced, suggesting that in some plasmids Tra is not a kill function (97, 244). Some plasmids, such as pIJ101 and pJV1, also require the presence of a *cis*-acting locus (*clt*) for mobilization, but the mechanism of its action is not yet known (150, 301). Unlike almost all other known conjugative systems in bacteria, conjugation among *Streptomyces* does not seem to require the nicking of one plasmid strand to allow the transfer of single-stranded DNA, supporting the hypothesis that DNA may be transferred as a double strand (150, 197).

Interspecies transfer of conjugative plasmids in *Streptomyces* is performed on solid media by replica plating the plasmid donor (as mycelia, spores, or protoplasts) onto a lawn of plasmid-free culture (197). Once the initial transfer of the plasmid has occurred, the plasmid is then spread to adjacent hyphae through the cross walls of the mycelium via the products of spread (*spd*) and *tra* genes (150, 198, 302). Detection of exconjugants is straightforward in many cases due to lethal zygosis (Ltz) which is manifested as visible areas (0.5 to 2 mm) of growth retardation or "pocks" in mycelial lawns (27, 198). The recipient mycelia carrying the plasmid can then be streaked out onto a new plate one or more times to purify the strain.

32.2.8.2.2. Shuttle Vectors for Intergeneric Conjugation

Conjugation between *E. coli* and *Streptomyces* was first demonstrated by Mazodier et al. (262) by modifying the mobilization system found in gram⁻ bacteria (347). Shuttle plasmids were constructed by incorporating origins of replication from natural *Streptomyces* plasmids (pIJ101 and pBR322) and *oriT* from the IncP plasmid RK2. Conjugation with these plasmids requires the presence of transfer gene function (*tra*) supplied in *trans* by the donor *E. coli* strains (262). Two commonly used *E. coli* strains are S17-1 with an integrated RP4 derivative and the methylation-deficient ET12567/pUB307 or pUZ8002, used to circumvent many *Streptomyces* restriction systems (122, 246, 262, 347). The following general conjugation protocol for *S. coelicolor* is taken from Kieser et al. (197) and can be used with *Streptomyces* spores or mycelia. Many modifications of the protocol exist, and conditions may need to be optimized for each new conjugation system. Some variations include different liquid and solid media, the choice of *E. coli* strain (methylation proficient or deficient), introduction of cosmid DNA, and the use of spores versus mycelial fragments (28, 122, 144, 260, 262, 387).

Procedure for Plasmid Conjugation from *E. coli*
Prepare competent cells of *E. coli* strain ET12567/pUB307 in the presence of kanamycin and chloramphenicol to

maintain the plasmid and *dam* mutation, respectively. These cells are transformed with a vector containing *oriT* and selected only for the incoming plasmid. Inoculate a colony into 10 ml of LB medium containing chloramphenicol, kanamycin, and the antibiotic to select for the *oriT* vector and grow overnight. Dilute the culture 1:100 in LB medium with antibiotics and grow to an OD_{600} of 0.4 to 0.6. Wash the cells twice with LB medium to remove traces of antibiotics and resuspend in 0.1 volume of LB medium. During this washing step, add recipient spores ($\sim 10^8$) to 500 µl of 2× YT broth (for each conjugation), heat shock at 50°C to pregerminate the spores, and allow to cool. If mycelia are used instead of spores, harvest from a 3- to 4-day culture growing on MS agar using 2 to 3 ml of 20% glycerol. Vortex the suspension and use 0.5 ml per conjugation without the heat shock step. Add 0.5 ml of the *E. coli* cells to 0.5 ml of the prepared spores or mycelial fragments, mix, and spin. Pour off the supernatant and take up the remaining cell pellet in the residual liquid. Dry an MS agar plate containing 10 mM $MgCl_2$ in a laminar-flow hood for 1 h to assist in absorbing liquid during subsequent steps. Plate the cells onto the dried MS plate and incubate at 30°C for 16 to 20 h. Overlay the plate with 1 ml of water containing 0.5 mg of nalidixic acid and the appropriate selective antibiotic for the plasmid, using a spreader to coat the surface evenly. Continue to incubate plates at 30°C and transfer exconjugants to selective media containing nalidixic acid (25 µg/ml).

Materials
2× YT medium

Difco Bacto tryptone	16 g
Difco Bacto yeast extract	10 g
NaCl	5 g
Water . . . . to 1,000 ml

Mannitol Soya (MS) medium

Agar	20 g
Mannitol	20 g
Soya flour	20 g
Water . . . . to 1,000 ml

Dissolve mannitol in the water and pour 200 ml into 250-ml Erlenmeyer flasks each containing 2 g of agar and 2 g of soya flour. Close/cover the flasks and autoclave twice, with gentle shaking between runs.

32.2.8.3. Bacteriophage and Transduction

The first natural transducing phage described from a *Streptomyces* strain is SV1 from *S. venezuelae* (360). Due to the narrow host specificity of this phage, it has found only limited use in the transduction of chromosomal markers. Four additional generalized transducing phage (DAH2, DAH4, DAH5, and DAH6), which have a much broader host range, have been isolated from *S. coelicolor* (43). DAH5 and DAH6 produced plaques on *S. lividans*, *S. avermitilis* 31272, *S. verticillus* 15003, and *S. venezuelae* 10712 and may prove to be useful tools for genetic manipulations of otherwise intractable species (43).

32.2.8.4. Transformation

There is no general natural competence system known in actinomycetes, although there is at least one published example demonstrating the transfer of plasmids from *S. lividans* to *M. smegmatis* by spontaneous transformation (23). The two primary artificial methods of introducing DNA via transformation are the use of polyethylene glycol (PEG) with protoplast preparations and the use of electroporation of mycelia when protoplasts prove to be recalcitrant to DNA uptake. Due to the restriction systems of many *Streptomyces* species, it may be necessary to produce plasmid or cosmid DNA using a nonmethylating *E. coli* host (i.e., ET12567). The following protocols for protoplast formation and transformation were adapted from Bibb et al. (27), Kieser et al. (197), and Thompson et al. (370).

Protoplast Preparation

Add 0.1 ml of spore suspension (10^9 spores ml^{-1} as a fresh suspension or frozen stock in 20% glycerol) and required growth factors to 25 ml of YEME medium (see below) in a baffled flask. Incubate for 36 to 40 h at 30°C in an orbital incubator. Pellet mycelia in 20-ml bottles (1,000 × g, 10 min) and discard the supernatant. Resuspend the pellet in 15 ml of 10.3% sucrose, centrifuge again, and discard the supernatant. Repeat the wash and centrifugation step. Resuspend the pellet in 4 ml of lysozyme solution (1 mg/ml in P buffer, filter sterilized). Incubate at 30°C for 15 to 60 min (formation of protoplasts can be monitored using a phase-contrast microscope). Gently mix cells by pipetting up and down three times with a 5-ml pipette and incubate for 5 min. Add 5 ml of P buffer and repeat gentle pipette mixing and incubation step. Filter protoplasts through a cotton wool plug (prepare filters by loosely tamping cotton wool into the ends of 5-ml plastic pipette tips and autoclaving). Spin protoplasts gently (e.g., 1,000 × g, 7 min). Discard the supernatant and resuspend in 1 ml of P buffer by gentle tapping or pipetting. Protoplasts can be used immediately or frozen for storage. To freeze protoplast aliquots, place tubes in ice in a plastic beaker and leave in a −70°C freezer overnight. Frozen protoplasts should be thawed quickly by shaking tubes under warm running water.

PEG-Mediated Protoplast Transformation

For each transformation add 50 μl of freshly washed protoplasts (~10^{10} ml^{-1}) to a tube. Add up to 5 μl of DNA in TE buffer to protoplasts and mix immediately by gently tapping on the tube. Add 200 μl of PEG 1000 (25%, wt/vol) in P buffer and mix by pipetting up and down four times. Spread approximately half of the suspension on each of two dried R2YE plates, using additional P buffer to facilitate spreading if necessary. Incubate plates at 30°C. Plates can be left to sporulate and then replicated to selective media if pock-forming plasmids are used (causing small visible patches of resistant growth). Alternatively, after 14 to 20 h to allow for expression of resistance, the regeneration plates can be overlaid with the appropriate antibiotic.

Materials
YEME medium

Difco yeast extract	3 g
Difco Bacto peptone	5 g
Oxoid malt extract	3 g
Glucose	10 g
Sucrose	340 g (34% final)
Distilled water	to 1,000 ml

After autoclaving, add MgCl$_2$ · 6H$_2$O (2.5 M), 2 ml/liter (5 mM final). For preparing protoplasts, also add glycine (20%), 25 ml/liter (0.5%). Different strains may require different concentrations of glycine. For *S. lividans* and *S. coelicolor*, 0.5% is the standard concentration.

R2YE medium

Sucrose	103 g
K$_2$SO$_4$	0.25 g
MgCl$_2$ · 6H$_2$O	10.12 g
Glucose	10 g
Difco Casamino Acids	0.1 g
Distilled water	800 ml

Combine the above and pour 80 ml of the solution into 250-ml Erlenmeyer flasks each containing 2.2 g of Difco Bacto agar. Cover the flasks and autoclave. For each time of use, melt the media and add to each flask the following autoclaved solutions in the order listed:

KH$_2$PO$_4$ (0.5%)	1 ml
CaCl$_2$ · 2H$_2$O (3.68%)	8 ml
L-Proline (20%)	1.5 ml
TES buffer (5.73%, pH 7.2)	10 ml
Trace element solution (see below)	0.2 ml
NaOH (1 N)	0.5 ml
Difco yeast extract (10%)	5 ml

P (protoplast) buffer

Sucrose	103 g
K$_2$SO$_4$	0.25 g
MgCl$_2$ · 6H$_2$O	2.02 g
Trace element solution (see below)	2 ml
Distilled water	to 800 ml

Dispense into 80-ml aliquots and autoclave. Before use, add the following autoclaved solutions in the order listed to each aliquot.

KH$_2$PO$_4$ (0.5%)	1 ml
CaCl$_2$ · H$_2$O (3.68%)	10 ml
TES buffer (5.73%, pH 7.2)	10 ml

Trace element solution

ZnCl$_2$	40 mg
FeCl$_3$ · H$_2$O	200 mg
CuCl$_2$ · 2H$_2$O	10 mg
MnCl$_2$ · 4H$_2$O	10 mg
NaB$_4$O$_7$ · 10H$_2$O	10 mg
(NH$_4$)$_6$Mo$_7$O$_{24}$ · 4H$_2$O	10 mg

32.2.8.4.1. Transformation by Electroporation

Some *Streptomyces* strains cannot be transformed efficiently using the PEG-mediated protoplast method. MacNeil (245) successfully transformed *S. lividans* using electroporation of protoplasts. Additionally, a growing number of publications demonstrate the uptake of DNA by *Streptomyces* spp. using electroporation of mycelia, which avoids the development of protoplast and regeneration conditions (121, 303). There are also reports of increased transformation efficiency using germinated spores instead of mycelia (112, 381). The following mycelium electroporation protocol is taken from Pigac and Schrempf (303) and involves the treatment of mycelia with lysozyme to weaken the cell walls and facilitate DNA uptake.

Transformation by Electroporation Protocol

Preparation of electrocompetent *Streptomyces* mycelia

Inoculate the actinomycete into 100 ml of CRM medium and incubate at 30°C for 24 h. Pellet the mycelia (4°C, 10,000 rpm) and resuspend cells in 100 ml of ice-cold 10% sucrose. Recentrifuge and resuspend in 50 ml of ice-cold 15% glycerol. Centrifuge and resuspend in 10 ml of lysozyme solution (100 μg ml^{-1} lysozyme in 15% glycerol) and incubate at 37°C for 30 min. Centrifuge and wash the pellet twice with 10 ml of ice-cold 15% glycerol. Resuspend mycelia in 1 to 5 ml of 30% PEG 1000 and 6.5% sucrose in 10% glycerol. Make 50 μl aliquots of cell suspension and freeze quickly in a dry ice-methanol bath or liquid nitrogen and store at −80°C.

Electroporation of *Streptomyces* mycelia

For each transformation, thaw one tube of cells at room temperature, place on ice, and add 0.01 to 1 μg of plasmid DNA in 1 to 2 μl of water and gently mix by pipetting. Transfer suspension to a chilled electrocuvette and apply pulse (2 kV, 25 μF capacitor, 400 Ω). Immediately add 0.75 of ml ice-cold CRM medium to cells and incubate at 30°C for 3 h with shaking. Add 200 μl of CRM medium and plate dilutions onto TSB plates containing appropriate antibiotics.

CRM medium

Glucose	10 g
Sucrose	103 g
MgCl$_2$ · 6H$_2$O	10.12 g
Tryptic soy broth	15 g
Yeast extract	5 g
Water	to 1,000 ml

TSB medium

Oxoid tryptone soya broth powder (CM129)	30 g
Water	to 1,000 ml

32.2.8.5. Applications of Gene Transfer

32.2.8.5.1. Construction of Mutants

32.2.8.5.1.1. Transposon mutagenesis. Transposable elements have proven to be a vital tool for mutational analysis of many *Streptomyces* strains. Many natural transposons have been described from *Streptomyces* spp. and engineered into useful derivatives with a variety of resistance and indicator genes because they lack innate selectable markers. Introduction of transposons to both plasmids and the chromosome is often accomplished using suicide vectors that are unstable (such as temperature-sensitive derivatives of pSG5), allowing the elimination of the vector after transposition. Some of the most commonly used transposons are derivatives of the Tn3-like Tn4556, which transposes efficiently (0.1 to 1%) and can contain *vph* and *neo* resistance markers and a variety of reporter genes (175, 344). Another efficient transposon used in *Streptomyces* is the minitransposon Tn5493 (1.1 kb) derived from the gram⁻ transposon

Tn5 (388). This transposon is delivered by the temperature-sensitive shuttle vector pJOE2577, which also contains the *tsr* resistance gene. High-frequency transposition is facilitated by a mutant transposase encoded by *tnpA*, with the *S. lividans mer* (mercury resistance operon) promoter in the plasmid. More recently, several efficient in vitro transposition systems have been developed in *Streptomyces* that utilize purified high-efficiency transposase, derivatives of Tn5, and the target DNA (29, 139). These simple reactions circumvent many of the limitations of in vivo systems because they do not require the expression of transposase in the host and eliminate introduction of the transposon on a suicide vector and subsequent curing step.

32.2.8.5.1.2. Directed mutagenesis and allelic exchange. The simplest method for generating insertional mutants is to introduce an antibiotic resistance marker into the gene of interest by a single-crossover event. Although mutants can be made relatively quickly and easily, the major disadvantage to this method is the possibility that the strain will revert to the wild type by excision of the marker during another single crossover. However, the technique has been particularly successfully in systems that lack more developed genetic recombination tools such as cloning vectors and the ability to transform or transfect the strain (172). More commonly, a double-crossover strategy is employed to generate stable mutations. The vector is constructed by cloning a mutant allele with an internal selectable marker as well as a different, external resistance gene. The interrupted gene is then used to replace the wild-type allele in the chromosome via two homologous recombination events in the upstream and downstream regions flanking the marker. Loss of the vector can be tested by demonstrating the absence of the external resistance gene (sensitivity to the antibiotic by replica plating) or by using another counterselection method (see item 32.2.8.5.1.3 below). The resulting mutant allele containing the internal resistance gene is then stable and cannot easily revert to the wild type. A disadvantage to this method is the difficulty in obtaining double- versus single-crossover events, sometimes requiring many additional rounds of growth and selection. Some techniques developed to increase the frequency of recombination are the use of single-stranded circular (167) or denatured double-stranded circular or linear DNA (284). An additional consideration when choosing insertional inactivation methods is the possibility of polar effects on downstream genes potentially cotranscribed with the gene of interest (197). In-frame deletion or gene/cluster replacement strategies can be used to avoid polar effects and are carried out in a number of different ways. One deletion method utilizes PCR to generate a resistance gene flanked by both FRT (FLP recombinase recognition sites) and small (30- to 40-nucleotide) homologous regions of the chromosome target (85). This cassette is then introduced into an *E. coli* strain carrying an expression plasmid for λ Red recombinase and a cosmid containing the cloned target gene. The resulting recombinant cosmid containing the interruption cassette is then integrated into the *Streptomyces* chromosome by a double crossover, and the resistance gene is subsequently removed by FLP recombinase-mediated recombination (153). This method and variations of it utilizing *recE* and *recT* ("ET cloning") to recombine linear and circular DNA (414) have been particularly useful for analyzing and manipulating secondary metabolic pathways in *Streptomyces* and related spp. (291).

32.2.8.5.1.3. Selection and counterselection. Many *Streptomyces* species possess natural resistance to common antibiotic markers, and therefore each new strain should be tested for sensitivity. The inhibitory concentrations of some antibiotics can also highly depend on the composition of the medium; sensitivity to viomycin and the aminoglycosides hygromycin, kanamycin, and neomycin tends to be enhanced on low-salt media (197). Some of the resistance genes that have been successfully used for selection include *aac(3)IV* (apramycin) (32), *ermE* (lincomycin and erythromycin) (382), *aaaC1* (gentamicin) (407), *hyg* (hygromycin) (413), *neo* (kanamycin and neomycin) (17), *aadA* (spectinomycin and streptomycin) (312), and *vph* (viomycin) (26). Because *tsr* (thiostrepton resistance) is used so frequently as a selectable marker in many *Streptomyces* vectors, it is generally not useful for gene disruptions (197).

The use of a counterselectable marker is advantageous when screening for loss of the vector during double-crossover experiments. One system developed for *S. coelicolor* utilizes a plasmid containing an antibiotic resistance marker and the glucose kinase gene, *glkA* (118, 384). The presence of this gene product renders the strain sensitive to 2-deoxyglucose and can be used for counterselection to ensure plasmid excision after the second crossover. One of the significant limitations of this system is the need to conduct the gene exchange in a *glkA* deletion mutant strain. A similar counterselection system utilizing the *sacB* gene, encoding levansucrase, from *B. subtilis* has been successfully used in *S. venezuelae* (200). Expression of this exoenzyme causes the hydrolyzation of sucrose and subsequent polymerization of the resulting fructose molecules, causing cell death (88). An additional marker used for counterselection in *Streptomyces roseosporus* is the *rpsL* gene, which encodes ribosomal protein S12 and confers streptomycin sensitivity (Sm^s) in Sm^r background strains (171). However, *rpsL* is an essential gene, and experiments were performed on spontaneous Sm^r mutants (rather than deletion strains) which could be then be counterselected by the dominant wild-type *rpsL* allele (171).

32.2.8.5.2. Analysis of Mutants

32.2.8.5.2.1. Complementation. A variety of vectors based on the *Streptomyces* plasmid SCP2* have been used for complementation cloning and analysis (240). Relatively simple derivatives such as pIJ941 have stability and transfer functions and have been used for in *trans* complementation of deletion mutants of antibiotic biosynthetic clusters with corresponding genes from heterologous pathways (201, 240, 341). Some of the most useful SCP2* derivatives are transmissible bifunctional vectors, such as pIJ698, which have the advantage of maintaining low copy number to avoid adverse regulatory effects, are relatively stable, can be propagated in *E. coli*, and contain *oriT* (RK2) for conjugation from *E. coli* to *Streptomyces* (199). This vector has been successfully used to complement morphological mutants using a plate mating strategy (86). A library of wild-type *Streptomyces* DNA is constructed with the vector and transformed into an indicator strain, such as a nutritional auxotroph. The resulting transformants are first grown on nonselective media and then replica plated onto the mutant strain to permit plate mating. Colonies are then replicated onto selective media and subsequently analyzed for the complemented phenotype (86, 328). As a test for in *trans* complementation, SCP2* derivatives can be cured from the host strain using the displacement plasmid pIJ80, a deriva-

tive of SCP103 lacking stability functions (341, 369). Derivatives of the integrating temperature-sensitive *Streptomyces* plasmid pSG5 can be used for in *cis* complementation analysis (195, 196, 274). Strategies employing these shuttle vectors can be used to complement deletion strains with wild-type genes and gene fragments upon curing of the vector and chromosomal integration. A related technique frequently used for domain replacement in biosynthetic gene clusters is to construct mutant strains by in *cis* functional complementation with heterologous genes/domains from other strains (195, 196). Bierman et al. (28) described several useful bifunctional plasmids, such as pKC1139, pKC1218, and pOJ446, which can each replicate in both *E. coli* and *Streptomyces* and are useful for mutational complementation. pOJ446 is a cosmid vector and can accept large DNA fragments up to 35 kb. pSET152 is one of the most commonly used conjugating integrating vectors for both mutational cloning and complementation and integrates into the chromosome at the φC31 attachment site (28). Another convenient integrating plasmid is the *E. coli* λ cosmid vector pVE616, which has the advantages of being small (4.2 kb), having a medium copy number, being able to be selected for in *E. coli* by *apr* and in *Streptomyces* by *tsr*, and containing λ *cos* sites for phage packaging (246). Interestingly, complementation studies with this integrating vector showed that the majority of transformants resulted from in *trans* complementation rather than homologous recombination (246).

32.2.8.5.2.2. Expression analysis. Several useful reporter systems have been developed for genetic analysis in *Streptomyces* and related species. In *S. lividans*, multicopy promoter-probe plasmids such as pSLP124 and pIJ486, containing the antibiotic resistance genes *cat* (chloramphenicol acetyltransferase) and *neo* (aminoglycoside phosphotransferase), respectively, have been used to screen for promoter activity (25, 397). Other reporters include genes encoding colored or fluorescent compounds such as melanin (*melC* operon from *Streptomyces glaucescens*) (57, 287), spore pigment (*whiE* operon from *S. coelicolor*) (115, 169), undecylprodiosin (*redD* from *S. coelicolor*) (385), and enhanced green fluorescence protein (*egfp* from *Aequorea victoria*) (361). Additional reporter systems depend on the production of specific enzymatic activities such as catechol 2,3-dioxygenase, which converts colorless catechol to a yellow compound (*xylE* from a *Pseudomonas* TOL plasmid) (178); alkaline phosphatransferase (*phoA* from *E. coli*) (132); luciferase (*luxAB* genes from *Vibrio harveyi*) (336); malate dehydrogenase (*mdh* from *Thermus flavus*) (389); β-lactamase (*ampC* from *E. coli*) (125); and β-galactosidase (β-gal from *S. lividans*) (357).

We thank Joanne Bartkus for assistance with preparation of Fig. 1. C.J.K. was supported by NRSA fellowship F32-AI56684 from the National Institutes of Health (NIH). C.E.S. was supported by MinnCrest training grant T32-DE07288 from the NIH. Research in the laboratory of G.M.D. was supported by the NIH.

32.3. REFERENCES

1. **Adams, V., D. Lyras, K. A. Farrow, and J. I. Rood.** 2002. The clostridial mobilisable transposons. *Cell. Mol. Life Sci.* **59:**2033–2043.
2. **Akerley, B. J., E. J. Rubin, A. Camilli, D. J. Lampe, H. M. Robertson, and J. J. Mekalanos.** 1998. Systematic identification of essential genes by in *vitro mariner* mutagenesis. *Proc. Natl. Acad. Sci. USA* **95:**8927–8932.

3. **Allison, G. E., and T. R. Klaenhammer.** 1996. Functional analysis of the gene encoding immunity to lactacin F, *lafI*, and its use as a *Lactobacillus*-specific, food-grade genetic marker. *Appl. Environ. Microbiol.* **62:**4450–4460.

4. **Anderson, T., P. Brian, P. Riggle, R. Kong, and W. Champness.** 1999. Genetic suppression analysis of non-antibiotic-producing mutants of the *Streptomyces coelicolor* *absA* locus. *Microbiology* **145:**2343–2353.

5. **Arbeit, R. D.** 1997. Laboratory procedures for epidemiologic analysis, p. 253–286. *In* K. B. Crossley and G. L. Archer (ed.), *The Staphylococci in Human Disease.* Churchill Livingstone, New York, NY.

6. **Arbeloa, A., H. Segal, J.-E. Hugonnet, N. Josseaume, L. Dubost, J.-P. Brouard, L. Gutmann, D. Mengin-Lecreulx, and M. Arthur.** 2004. Role of class A penicillin-binding proteins in PBP5-mediated β-lactam resistance in *Enterococcus faecalis. J. Bacteriol.* **186:**1221–1228.

7. **Avery, O. T., C. M. Macleod, and M. McCarty.** 1944. Studies on the chemical nature of the substance inducing transformation of pneumococcal phenotypes. Induction of transformation by a deoxyribonucleic acid fraction isolated from Pneumococcus Type III. *J. Exp. Med.* **79:**137–158.

8. **Awad, M. M., A. E. Bryant, D. L. Stevens, and J. I. Rood.** 1995. Virulence studies on chromosomal alpha-toxin and theta-toxin mutants constructed by allelic exchange provide genetic evidence for the essential role of alpha-toxin in *Clostridium perfringens*-mediated gas gangrene. *Mol. Microbiol.* **15:**191–202.

9. **Ayoubi, P., A. O. Kilic, and M. N. Vijayakumar.** 1991. Tn*5253*, the pneumococcal omega (*cat tet*) BM6001 element, is a composite structure of two conjugative transposons, Tn*5251* and Tn*5252. J. Bacteriol.* **173:**1617–1622.

10. **Bachrach, G., M. Leizerovici-Zigmond, A. Zlotkin, R. Naor, and D. Steinberg.** 2003. Bacteriophage isolation from human saliva. *Lett. Appl. Microbiol.* **36:**50–53.

11. **Bae, T., A. K. Banger, A. Wallace, E. M. Glass, F. Aslund, O. Schneewind, and D. M. Missiakas.** 2004. *Staphylococcus aureus* virulence genes identified by *bursa aurealis* mutagenesis and nematode killing. *Proc. Natl. Acad. Sci. USA* **101:**12312–12317.

12. **Bae, T., S. Clerc-Bardin, and G. M. Dunny.** 2000. Analysis of expression of *prgX*, a key negative regulator of the transfer of the *Enterococcus faecalis* pheromone-inducible plasmid pCF10. *J. Mol. Biol.* **297:**861–875.

13. **Bae, T., B. Kozlowicz, and G. M. Dunny.** 2002. Two targets in pCF10 DNA for PrgX binding: their role in production of Qa and *prgX* mRNA and in regulation of pheromone-inducible conjugation. *J. Mol. Biol.* **315:**995–1007.

14. **Bailey, C. R., C. J. Bruton, M. J. Butler, K. F. Chater, J. E. Harris, and D. A. Hopwood.** 1986. Properties of in vitro recombinant derivatives of pJV1, a multi-copy plasmid from *Streptomyces phaeochromogenes. J. Gen. Microbiol.* **132:**2071–2078.

15. **Bartilson, M., A. Marra, J. Christine, J. S. Asundi, W. P. Schneider, and A. E. Hromockyj.** 2001. Differential fluorescence induction reveals *Streptococcus pneumoniae* loci regulated by competence stimulatory peptide. *Mol. Microbiol.* **39:**126–135.

15a. **Battisti, L., B. D. Green, and C. B. Thorne.** 1985. Mating system for transfer of plasmids among *Bacillus anthracis*, *Bacillus cereus*, and *Bacillus thuringiensis. J. Bacteriol.* **162:**543–550.

16. **Beard, S. J., V. Salisbury, R. J. Lewis, J. A. Sharpe, and A. P. MacGowan.** 2002. Expression of *lux* genes in a clinical isolate of *Streptococcus pneumoniae*: using bioluminescence to monitor gemifloxacin activity. *Antimicrob. Agents Chemother.* **46:**538–542.

17. **Beck, E., G. Ludwig, E. A. Auerswald, B. Reiss, and H. Schaller.** 1982. Nucleotide sequence and exact localization of the neomycin phosphotransferase gene from transposon Tn*5. Gene* **19:**327–336.

18. **Beltramo, C., M. Oraby, G. Bourel, D. Garmyn, and J. Guzzo.** 2004. A new vector, pGID052, for genetic transfer in *Oenococcus oeni. FEMS Microbiol. Lett.* **236:**53–60.

19. **Berg, T., N. Firth, S. Apisiridej, A. Hettiaratchi, A. Leelaporn, and R. A. Skurray.** 1998. Complete nucleotide sequence of pSK41: evolution of staphylococcal conjugative multiresistance plasmids. *J. Bacteriol.* **180:**4350–4359.

20. **Bertram, J., M. Stratz, and P. Durre.** 1991. Natural transfer of conjugative transposon Tn*916* between gram-positive and gram-negative bacteria. *J. Bacteriol.* **173:**443–448.

21. **Betley, M. J., and J. J. Mekalanos.** 1985. Staphylococcal enterotoxin A is encoded by phage. *Science* **229:**185–187.

22. **Betschel, S. D., S. M. Borgia, N. L. Barg, D. E. Low, and J. C. De Azavedo.** 1998. Reduced virulence of group A streptococcal Tn*916* mutants that do not produce streptolysin S. *Infect. Immun.* **66:**1671–1679.

23. **Bhatt, A., H. M. Kieser, R. E. Melton, and T. Kieser.** 2002. Plasmid transfer from *Streptomyces* to *Mycobacterium smegmatis* by spontaneous transformation. *Mol. Microbiol.* **43:**135–146.

24. **Bhavsar, A. P., X. Zhao, and E. D. Brown.** 2001. Development and characterization of a xylose-dependent system for expression of cloned genes in *Bacillus subtilis*: conditional complementation of a teichoic acid mutant. *Appl. Environ. Microbiol.* **67:**403–410.

25. **Bibb, M. J., and S. N. Cohen.** 1982. Gene expression in *Streptomyces*: construction and application of promoter-probe plasmid vectors in *Streptomyces lividans. Mol. Gen. Genet.* **187:**265–277.

26. **Bibb, M. J., J. M. Ward, and S. N. Cohen.** 1985. Nucleotide sequences encoding and promoting expression of 3 antibiotic resistance genes indigenous to *Streptomyces. Mol. Gen. Genet.* **199:**26–36.

27. **Bibb, M. J., J. M. Ward, and D. A. Hopwood.** 1978. Transformation of plasmid DNA into *Streptomyces* at high frequency. *Nature* **274:**398–400.

28. **Bierman, M., R. Logan, K. Obrien, E. T. Seno, R. N. Rao, and B. E. Schoner.** 1992. Plasmid cloning vectors for the conjugal transfer of DNA from *Escherichia coli* to *Streptomyces* spp. *Gene* **116:**43–49.

29. **Bishop, A., S. Fielding, P. Dyson, and P. Herron.** 2004. Systematic insertional mutagenesis of a streptomycete genome: a link between osmoadaptation and antibiotic production. *Genome Res.* **14:**893–900.

30. **Biswas, I., A. Gruss, S. D. Ehrlich, and E. Maguin.** 1993. High-efficiency gene inactivation and replacement system for gram-positive bacteria. *J. Bacteriol.* **175:**3628–3635.

31. **Biswas, I., and J. R. Scott.** 2003. Identification of *rocA*, a positive regulator of *covR* expression in the group A streptococcus. *J. Bacteriol.* **185:**3081–3090.

32. **Brau, B., U. Pilz, and W. Piepersberg.** 1984. Genes for gentamicin-(3)-N-acetyltransferases-III and gentamicin-(3)-N-acetyltransferase-IV. I. Nucleotide sequence of the *aac(3)-IV* gene and possible involvement of an IS*140* element in its expression. *Mol. Gen. Genet.* **193:**179–187.

33. **Braunstein, M., S. S. Bardarov, and W. R. Jacobs, Jr.** 2002. Genetic methods for deciphering virulence determinants of *Mycobacterium tuberculosis. Methods Enzymol.* **358:**67–99.

34. **Bricker, A. L., and A. Camilli.** 1999. Transformation of a type 4 encapsulated strain of *Streptococcus pneumoniae. FEMS Microbiol. Lett.* **172:**131–135.

35. **Bringel, F., G. L. Van Alstine, and J. R. Scott.** 1991. A host factor absent from *Lactococcus lactis* subspecies *lactis*

MG1363 is required for conjugative transposition. *Mol. Microbiol.* **5:**2983–2993.

36. **Bringel, F., G. L. Van Alstine, and J. R. Scott.** 1992. Transfer of Tn*916* between *Lactococcus lactis* subsp. *lactis* strains is nontranspositional: evidence for a chromosomal fertility function in strain MG1363. *J. Bacteriol.* **174:**5840–5847.

37. **Bron, P. A., M. G. Benchimol, J. Lambert, E. Palumbo, M. Deghorain, J. Delcour, W. M. de Vos, M. Kleerebezem, and P. Hols.** 2002. Use of the *alr* gene as a food-grade selection marker in lactic acid bacteria. *Appl. Environ. Microbiol.* **68:**5663–5670.

38. **Bron, P. A., S. M. Hoffer, S. Van II, W. M. De Vos, and M. Kleerebezem.** 2004. Selection and characterization of conditionally active promoters in *Lactobacillus plantarum*, using alanine racemase as a promoter probe. *Appl. Environ. Microbiol.* **70:**310–317.

39. **Bron, S.** 1990. Plasmids, p. 75–174. *In* C. R. Harwood and S. M. Cutting (ed.), *Molecular Biological Methods for Bacillus.* John Wiley & Sons, Chichester, United Kingdom.

40. **Brussow, H.** 2001. Phages of dairy bacteria. *Annu. Rev. Microbiol.* **55:**283–303.

41. **Bryan, E. M., T. Bae, M. Kleerebezem, and G. M. Dunny.** 2000. Improved vectors for nisin-controlled expression in gram-positive bacteria. *Plasmid* **44:**183–190.

42. **Buckley, N. D., C. Vadeboncoeur, D. J. LeBlanc, L. N. Lee, and M. Frenette.** 1999. An effective strategy, applicable to *Streptococcus salivarius* and related bacteria, to enhance or confer electroporation competence. *Appl. Environ. Microbiol.* **65:**3800–3804.

43. **Burke, J., D. Schneider, and J. Westpheling.** 2001. Generalized transduction in *Streptomyces coelicolor. Proc. Natl. Acad. Sci. USA* **98:**6289–6294.

44. **Burrus, V., G. Pavlovic, B. Decaris, and G. Guedon.** 2002. Conjugative transposons: the tip of the iceberg. *Mol. Microbiol.* **46:**601–610.

45. **Buu-Hoi, A., G. Bieth, and T. Horaud.** 1984. Broad host range of streptococcal macrolide resistance plasmids. *Antimicrob. Agents Chemother.* **25:**289–291.

46. **Caparon, M.** 2000. Genetics of group A streptococci, p. 53–65. *In* V. A. Fischetti, R. P. Novick, J. J. Ferretti, D. A. Portnoy, and J. I. Rood (ed.), *Gram-Positive Pathogens.* ASM Press, Washington, DC.

47. **Caparon, M. G., R. T. Geist, J. Perez-Casal, and J. R. Scott.** 1992. Environmental regulation of virulence in group A streptococci: transcription of the gene encoding M protein is stimulated by carbon dioxide. *J. Bacteriol.* **174:**5693–5701.

48. **Caparon, M. G., and J. R. Scott.** 1987. Identification of a gene that regulates expression of M protein, the major virulence determinant of group A streptococci. *Proc. Natl. Acad. Sci. USA* **84:**8677–8681.

49. **Caparon, M. G., and J. R. Scott.** 1991. Genetic manipulation of pathogenic streptococci. *Methods Enzymol.* **204:**556–586.

50. **Caprioli, T., F. Zaccour, and S. S. Kasatiya.** 1975. Phage typing scheme for group D streptococci isolated from human urogenital tract. *J. Clin. Microbiol.* **2:**311–317.

51. **Cascales, E., and P. J. Christie.** 2003. The versatile bacterial type IV secretion systems. *Nat. Rev. Microbiol.* **1:**137–149.

52. **Casey, J., C. Daly, and G. F. Fitzgerald.** 1991. Chromosomal integration of plasmid DNA by homologous recombination in *Enterococcus faecalis* and *Lactococcus lactis* subsp. *lactis* hosts harboring Tn*919. Appl. Environ. Microbiol.* **57:**2677–2682.

53. **Chan, P. F., K. M. O'Dwyer, L. M. Palmer, J. D. Ambrad, K. A. Ingraham, C. So, M. A. Lonetto, S.** Biswas, M. Rosenberg, D. J. Holmes, and M. Zalacain. 2003. Characterization of a novel fucose-regulated promoter (P*fcsK*) suitable for gene essentiality and antibacterial mode-of-action studies in *Streptococcus pneumoniae. J. Bacteriol.* **185:**2051–2058.

54. **Charpentier, E., A. I. Anton, P. Barry, B. Alfonso, Y. Fang, and R. P. Novick.** 2004. Novel cassette-based shuttle vector system for gram-positive bacteria. *Appl. Environ. Microbiol.* **70:**6076–6085.

55. **Chen, I., and D. Dubnau.** 2003. DNA transport during transformation. *Front. Biosci.* **8:**s544–556.

56. **Chen, I., and D. Dubnau.** 2004. DNA uptake during bacterial transformation. *Nat. Rev. Microbiol.* **2:**241–249.

57. **Chen, L. Y., W. M. Leu, K. T. Wang, and Y. H. W. Lee.** 1992. Copper transfer and activation of the *Streptomyces* apotyrosinase are mediated through a complex formation between apotyrosinase and its *trans*-activator MelC1. *J. Biol. Chem.* **267:**20100–20107.

58. **Chen, Y. S., and J. L. Steele.** 2005. Analysis of promoter sequences from *Lactobacillus helveticus* CNRZ32 and their activity in other lactic acid bacteria. *J. Appl. Microbiol.* **98:**64–72.

59. **Cheung, A. L., J. M. Koomey, C. A. Butler, S. J. Projan, and V. A. Fischetti.** 1992. Regulation of exoprotein expression in *Staphylococcus aureus* by a locus (*sar*) distinct from *agr. Proc. Natl. Acad. Sci. USA* **89:**6462–6466.

60. **Cheung, A. L., C. Wolz, M. R. Yeaman, and A. S. Bayer.** 1995. Insertional inactivation of a chromosomal locus that modulates expression of potential virulence determinants in *Staphylococcus aureus. J. Bacteriol.* **177:**3220–3226.

61. **Christie, P. J.** 2001. Type IV secretion: intercellular transfer of macromolecules by systems ancestrally related to conjugation machines. *Mol. Microbiol.* **40:**294–305.

62. **Cirillo, J. D., R. G. Barletta, B. R. Bloom, and W. R. Jacobs, Jr.** 1991. A novel transposon trap for mycobacteria: isolation and characterization of IS1096. *J. Bacteriol.* **173:**7772–7780.

63. **Clancy, A., M. H. Lee, A. L. Jones, and C. E. Rubens.** 2004. Construction and characterization of transposon Tn*phoZ* for the identification of genes encoding exported proteins in *Streptococcus agalactiae. FEMS Microbiol. Lett.* **241:**257–264.

64. **Claverys, J. P., A. Dintilhac, E. V. Pestova, B. Martin, and D. A. Morrison.** 1995. Construction and evaluation of new drug-resistance cassettes for gene disruption mutagenesis in *Streptococcus pneumoniae*, using an *ami* test platform. *Gene* **164:**123–128.

65. **Claverys, J. P., and L. S. Havarstein.** 2002. Extracellular-peptide control of competence for genetic transformation in *Streptococcus pneumoniae. Front. Biosci.* **7:**1798–1814.

66. **Claverys, J. P., and B. Martin.** 2003. Bacterial "competence" genes: signatures of active transformation, or only remnants? *Trends Microbiol.* **11:**161–165.

67. **Claverys, J. P., M. Prudhomme, I. Mortier-Barriere, and B. Martin.** 2000. Adaptation to the environment: *Streptococcus pneumoniae*, a paradigm for recombination-mediated genetic plasticity? *Mol. Microbiol.* **35:**251–259.

68. **Clewell, D. B.** 1981. Plasmids, drug resistance, and gene transfer in the genus *Streptococcus. Microbiol. Rev.* **45:**409–436.

69. **Clewell, D. B., F. Y. An, B. A. White, and C. Gawron-Burke.** 1985. *Streptococcus faecalis* sex pheromone (cAM373) also produced by *Staphylococcus aureus* and identification of a conjugative transposon (Tn*918*). *J. Bacteriol.* **162:**1212–1220.

70. **Clewell, D. B., and G. M. Dunny.** 2002. Conjugation and genetic exchange in enterococci, p. 265–300. *In* M. S. Gilmore, D. B. Clewell, P. Courvalin, G. M. Dunny, B. E. Murray, and L. B. Rice (ed.), *The Enterococci: Pathogenesis,*

Molecular Biology, and Antibiotic Resistance. ASM Press, Washington, DC.

71. **Clewell, D. B., S. E. Flannagan, and D. D. Jaworski.** 1995. Unconstrained bacterial promiscuity: the Tn916-Tn1545 family of conjugative transposons. *Trends Microbiol.* **3:**229–236.

72. **Clewell, D. B., M. V. Francia, S. E. Flannagan, and F. Y. An.** 2002. Enterococcal plasmid transfer: sex pheromones, transfer origins, relaxases, and the *Staphylococcus aureus* issue. *Plasmid* **48:**193–201.

73. **Clewell, D. B., and C. Gawron-Burke.** 1986. Conjugative transposons and the dissemination of antibiotic resistance in streptococci. *Annu. Rev. Microbiol.* **40:**635–659.

74. **Clewell, D. B., P. K. Tomich, M. C. Gawron-Burke, A. E. Franke, Y. Yagi, and F. Y. An.** 1982. Mapping of *Streptococcus faecalis* plasmids pAD1 and pAD2 and studies relating to transposition of Tn917. *J. Bacteriol.* **152:**1220–1230.

75. **Clewell, D. B., Y. Yagi, G. M. Dunny, and S. K. Schultz.** 1974. Characterization of three plasmid deoxyribonucleic acid molecules in a strain of *Streptococcus faecalis:* identification of a plasmid determining erythromycin resistance. *J. Bacteriol.* **117:**283–289.

76. **Coleman, D., J. Knights, R. Russell, D. Shanley, T. H. Birkbeck, G. Dougan, and I. Charles.** 1991. Insertional inactivation of the *Staphylococcus aureus* beta-toxin by bacteriophage phi 13 occurs by site- and orientation-specific integration of the phi 13 genome. *Mol. Microbiol.* **5:**933–939.

77. **Corbisier, P., G. Ji, G. Nuyts, M. Mergeay, and S. Silver.** 1993. luxAB gene fusions with the arsenic and cadmium resistance operons of *Staphylococcus aureus* plasmid pI258. *FEMS Microbiol. Lett.* **110:**231–238.

78. **Courvalin, P., and C. Carlier.** 1986. Transposable multiple antibiotic resistance in *Streptococcus pneumoniae. Mol. Gen. Genet.* **205:**291–297.

79. **Cousineau, B., D. Smith, S. Lawrence-Cavanagh, J. E. Mueller, J. Yang, D. Mills, D. Manias, G. Dunny, A. M. Lambowitz, and M. Belfort.** 1998. Retrohoming of a bacterial group II intron: mobility via complete reverse splicing, independent of homologous DNA recombination. *Cell* **94:**451–462.

80. **Cruz-Rodz, A. L., and M. S. Gilmore.** 1990. High efficiency introduction of plasmid DNA into glycine treated *Enterococcus faecalis* by electroporation. *Mol. Gen. Genet.* **224:**152–154.

81. **Cutting, S. M., and P. B. Vanderhorn.** 1990. Genetic analysis, p. 27–74. *In* C. R. Harwood and S. M. Cutting (ed.), *Molecular Biological Methods for* Bacillus. John Wiley & Sons, Chichester, United Kingdom.

82. **Cutting, S. M., and P. Youngman.** 1994. Gene transfer in gram-positive bacteria, p. 348–364. *In* P. Gerhardt, R. G. E. Murray, W. A. Wood, and N. R. Krieg (ed.), *Methods for General and Molecular Bacteriology.* ASM Press, Washington, DC.

83. **Cvitkovitch, D. G.** 2001. Genetic competence and transformation in oral streptococci. *Crit. Rev. Oral Biol. Med.* **12:**217–243.

84. **Dagkessamanskaia, A., M. Moscoso, V. Henard, S. Guiral, K. Overweg, M. Reuter, B. Martin, J. Wells, and J. P. Claverys.** 2004. Interconnection of competence, stress and CiaR regulons in *Streptococcus pneumoniae:* competence triggers stationary phase autolysis of ciaR mutant cells. *Mol. Microbiol.* **51:**1071–1086.

85. **Datsenko, K. A., and B. L. Wanner.** 2000. One-step inactivation of chromosomal genes in *Escherichia coli* K-12 using PCR products. *Proc. Natl. Acad. Sci. USA* **97:**6640–6645.

86. **Davis, N. K., and K. F. Chater.** 1990. Spore color in *Streptomyces coelicolor* A3(2) involves the developmentally regulated synthesis of a compound biosynthetically related to polyketide antibiotics. *Mol. Microbiol.* **4:**1679–1691.

86a. **Day, W. A., Jr., S. L. Rasmussen, B. M. Carpenter, S. N. Peterson, and A. M. Friedlander.** 2007. Microarray analysis of transposon insertion mutations in *Bacillus anthracis:* global identification of genes required for sporulation and germination. *J. Bacteriol.* **189:**3296–3301.

87. **De Boever, E. H., D. B. Clewell, and C. M. Fraser.** 2000. *Enterococcus faecalis* conjugative plasmid pAM373: complete nucleotide sequence and genetic analyses of sex pheromone response. *Mol. Microbiol.* **37:**1327–1341.

88. **Dedonder, R.** 1966. Levan sucrase from *Bacillus subtilis. Methods Enzymol.* **8:**500–505.

89. **de la Cruz, F., and J. Davies.** 2000. Horizontal gene transfer and the origin of species: lessons from bacteria. *Trends Microbiol.* **8:**128–133.

90. **De Las Rivas, B., J. L. García, R. López, and P. García.** 2002. Purification and polar localization of pneumococcal LytB, a putative endo-β-N-acetylglucosaminidase: the chain-dispersing murein hydrolase. *J. Bacteriol.* **184:**4988–5000.

91. **Derbyshire, K. M., and S. S. Bardarov.** 2000. DNA transfer in mycobacteria: conjugation and transduction, p. 93–110. *In* G. F. Hatfull and W. R. Jacobs, Jr. (ed.), *Molecular Genetics of Mycobacteria.* ASM Press, Washington, DC.

92. **de Ruyter, P. G., O. P. Kuipers, and W. M. de Vos.** 1996. Controlled gene expression systems for *Lactococcus lactis* with the food-grade inducer nisin. *Appl. Environ. Microbiol.* **62:**3662–3667.

93. **Desiere, F., S. Lucchini, C. Canchaya, M. Ventura, and H. Brussow.** 2002. Comparative genomics of phages and prophages in lactic acid bacteria. *Antonie Leeuwenhoek* **82:**73–91.

94. **de Vos, W. M.** 1999. Gene expression systems for lactic acid bacteria. *Curr. Opin. Microbiol.* **2:**289–295.

95. **Dieye, Y., S. Usai, F. Clier, A. Gruss, and J. C. Piard.** 2001. Design of a protein-targeting system for lactic acid bacteria. *J. Bacteriol.* **183:**4157–4166.

96. **Dillard, J. P., and J. Yother.** 1994. Genetic and molecular characterization of capsular polysaccharide biosynthesis in *Streptococcus pneumoniae* type 3. *Mol. Microbiol.* **12:**959–972.

97. **Doi, K., Y. Ono, E. Yokoyama, Y. Tsukagoe, and S. Ogata.** 1998. Whole sequence of spoIIIE-like, sporulation-inhibitory, and transfer gene (spi) in a conjugative plasmid, pSA1.1, of *Streptomyces azureus* and detection of spi-like gene in the actinomycete chromosome. *Biosci. Biotechnol. Biochem.* **62:**1597–1600.

98. **Donnelly-Wu, M. K., W. R. Jacobs, Jr., and G. F. Hatfull.** 1993. Superinfection immunity of mycobacteriophage L5: applications for genetic transformation of mycobacteria. *Mol. Microbiol.* **7:**407–417.

99. **Doolittle, W. F.** 1999. Phylogenetic classification and the universal tree. *Science* **284:**2124–2129.

100. **Dougherty, B. A., and I. van de Rijn.** 1992. Molecular characterization of a locus required for hyaluronic acid capsule production in group A streptococci. *J. Exp. Med.* **175:**1291–1299.

101. **Dubnau, D.** 1993. Genetic exchange and homologous recombination, p. 555–584. *In* A. L. Sonenshein, J. A. Hoch, and R. Losick (ed.), Bacillus subtilis *and Other Gram-Positive Bacteria: Biochemistry, Physiology, and Molecular Genetics.* American Society for Microbiology, Washington, DC.

102. **Dubnau, D., and C. M. Lovett.** 2002. Transformation and recombination, p. 453–471. *In* A. L. Sonenshein, J. A. Hoch, and R. Losick (ed.), Bacillus subtilis *and Its Closest Relatives: from Genes to Cells.* ASM Press, Washington, DC.

102a.**Dunny, G. M., P. P. Cleary, and L. L. McKay (ed.).** 1991. *Genetics and Molecular Biology of Streptococci, Lactococci, and Enterococci.* American Society for Microbiology, Washington, DC.

103. **Dunny, G. M., and D. B. Clewell.** 1975. Transmissible toxin (hemolysin) plasmid in *Streptococcus faecalis* and its mobilization of a noninfectious drug resistance plasmid. *J. Bacteriol.* **124:**784–790.

104. **Dunny, G. M., L. N. Lee, and D. J. LeBlanc.** 1991. Improved electroporation and cloning vector system for gram-positive bacteria. *Appl. Environ. Microbiol.* **57:**1194–1201.

105. **Duvall, E. J., D. M. Williams, P. S. Lovett, C. Rudolph, N. Vasantha, and M. Guyer.** 1983. Chloramphenicol-inducible gene expression in *Bacillus subtilis. Gene* **24:** 171–177.

106. **Duwat, P., K. Hammer, A. Bolotin, and A. Gruss.** 2000. Genetics of lactococci, p. 295–306. *In* V. A. Fischetti, R. P. Novick, J. J. Ferretti, D. A. Portnoy, and J. I. Rood (ed.), *Gram-Positive Pathogens.* ASM Press, Washington, DC.

107. **Dyer, D. W., M. I. Rock, C. Y. Lee, and J. J. Iandolo.** 1985. Generation of transducing particles in *Staphylococcus aureus. J. Bacteriol.* **161:**91–95.

108. **Eichenbaum, Z., M. J. Federle, D. Marra, W. M. de Vos, O. P. Kuipers, M. Kleerebezem, and J. R. Scott.** 1998. Use of the lactococcal *nisA* promoter to regulate gene expression in gram-positive bacteria: comparison of induction level and promoter strength. *Appl. Environ. Microbiol.* **64:**2763–2769.

109. **Eichenbaum, Z., and J. R. Scott.** 1997. Use of Tn*917* to generate insertion mutations in the group A streptococcus. *Gene* **186:**213–217.

110. **Elsner, A., B. Kreikemeyer, A. Braun-Kiewnick, B. Spellerberg, B. A. Buttaro, and A. Podbielski.** 2002. Involvement of Lsp, a member of the LraI-lipoprotein family in *Streptococcus pyogenes*, in eukaryotic cell adhesion and internalization. *Infect. Immun.* **70:**4859–4869.

111. **Engel, H. W., N. Soedirman, J. A. Rost, W. J. van Leeuwen, and J. D. van Embden.** 1980. Transferability of macrolide, lincomycin, and streptogramin resistances between group A, B, and D streptococci, *Streptococcus pneumoniae*, and *Staphylococcus aureus. J. Bacteriol.* **142:** 407–413.

112. **English, R. S., J. S. Lampel, and T. J. Vanden Boom.** 1998. Transformation of *Saccharopolyspora erythraea* by electroporation of germinating spores: construction of propionyl Co-A carboxylase mutants. *J. Ind. Microbiol. Biotechnol.* **21:**219–224.

113. **Evans, J., and K. G. Dyke.** 1988. Characterization of the conjugation system associated with the *Staphylococcus aureus* plasmid pJE1. *J. Gen. Microbiol.* **134**(Pt. 1):1–8.

114. **Fabret, C., S. D. Ehrlich, and P. Noirot.** 2002. A new mutation delivery system for genome-scale approaches in *Bacillus subtilis. Mol. Microbiol.* **46:**25–36.

115. **Feitelson, J. S.** 1988. An improved plasmid for the isolation and analysis of *Streptomyces* promoters. *Gene* **66:** 159–162.

116. **Feucht, A., and P. J. Lewis.** 2001. Improved plasmid vectors for the production of multiple fluorescent protein fusions in *Bacillus subtilis. Gene* **264:**289–297.

117. **Firth, N., K. P. Ridgway, M. E. Byrne, P. D. Fink, L. Johnson, I. T. Paulsen, and R. A. Skurray.** 1993. Analysis of a transfer region from the staphylococcal conjugative plasmid pSK41. *Gene* **136:**13–25.

118. **Fisher, S. H., C. J. Bruton, and K. F. Chater.** 1987. The glucose kinase gene of *Streptomyces coelicolor* and its use in selecting spontaneous deletions for desired regions of the genome. *Mol. Gen. Genet.* **206:**35–44.

119. **Fitzgerald, G. F., and D. B. Clewell.** 1985. A conjugative transposon (Tn*919*) in *Streptococcus sanguis* [*sic*]. *Infect. Immun.* **47:**415–420.

120. **Fitzgerald, G. F., and M. J. Gasson.** 1988. In vivo gene transfer systems and transposons. *Biochimie* **70:**489–502.

121. **Fitzgerald, N. B., R. S. English, J. S. Lampel, and T. J. Vanden Boom.** 1998. Sonication-dependent electroporation of the erythromycin-producing bacterium *Saccharopolyspora erythraea. Appl. Environ. Microbiol.* **64:**1580–1583.

122. **Flett, F., V. Mersinias, and C. P. Smith.** 1997. High efficiency intergeneric conjugal transfer of plasmid DNA from *Escherichia coli* to methyl DNA-restricting streptomycetes. *FEMS Microbiol. Lett.* **155:**223–229.

123. **Flint, J. L., J. C. Kowalski, P. K. Karnati, and K. M. Derbyshire.** 2004. The RD1 virulence locus of *Mycobacterium tuberculosis* regulates DNA transfer in *Mycobacterium smegmatis. Proc. Natl. Acad. Sci. USA* **101:**12598–12603.

124. **Fogg, G. C., and M. G. Caparon.** 1997. Constitutive expression of fibronectin binding in *Streptococcus pyogenes* as a result of anaerobic activation of *rofA. J. Bacteriol.* **179:**6172–6180.

125. **Forsman, M., and B. Jaurin.** 1987. Chromogenic identification of promoters in *Streptomyces lividans* by using an *ampC* β-lactamase promoter-probe vector. *Mol. Gen. Genet.* **210:**23–32.

126. **Francis, K. P., D. Joh, C. Bellinger-Kawahara, M. J. Hawkinson, T. F. Purchio, and P. R. Contag.** 2000. Monitoring bioluminescent *Staphylococcus aureus* infections in living mice using a novel *luxABCDE* construct. *Infect. Immun.* **68:**3594–3600.

127. **Francis, K. P., J. Yu, C. Bellinger-Kawahara, D. Joh, M. J. Hawkinson, G. Xiao, T. F. Purchio, M. G. Caparon, M. Lipsitch, and P. R. Contag.** 2001. Visualizing pneumococcal infections in the lungs of live mice using bioluminescent *Streptococcus pneumoniae* transformed with a novel gram-positive *lux* transposon. *Infect. Immun.* **69:**3350–3358.

128. **Franke, A. E., and D. B. Clewell.** 1981. Evidence for a chromosome-borne resistance transposon (Tn*916*) in *Streptococcus faecalis* that is capable of "conjugal" transfer in the absence of a conjugative plasmid. *J. Bacteriol.* **145:** 494–502.

129. **Frazier, C. L., J. San Filippo, A. M. Lambowitz, and D. A. Mills.** 2003. Genetic manipulation of *Lactococcus lactis* by using targeted group II introns: generation of stable insertions without selection. *Appl. Environ. Microbiol.* **69:**1121–1128.

130. **Freitag, N. E.** 2000. Genetic tools for use with *Listeria monocytogenes*, p. 488–498. *In* V. A. Fischetti, R. P. Novick, J. J. Ferretti, D. A. Portnoy, and J. I. Rood (ed.), *Gram-Positive Pathogens.* ASM Press, Washington, DC.

131. **Fu, X., and J. G. Xu.** 2000. Development of a chromosome-plasmid balanced lethal system for *Lactobacillus acidophilus* with *thyA* gene as selective marker. *Microbiol. Immunol.* **44:**551–556.

132. **Gandlur, S. M., L. Wei, J. Levine, J. Russell, and P. Kaur.** 2004. Membrane topology of the DrrB protein of the doxorubicin transporter of *Streptomyces peucetius. J. Biol. Chem.* **279:**27799–27806.

133. **Garcia, P., A. C. Martin, and R. Lopez.** 1997. Bacteriophages of *Streptococcus pneumoniae*: a molecular approach. *Microb. Drug Resist.* **3:**165–176.

134. **Garnier, T., and S. T. Cole.** 1988. Complete nucleotide sequence and genetic organization of the bacteriocinogenic plasmid, pIP404, from *Clostridium perfringens. Plasmid* **19:**134–150.

135. Garsin, D. A., J. Urbach, J. C. Huguet-Tapia, J. E. Peters, and F. M. Ausubel. 2004. Construction of an *Enterococcus faecalis* Tn917-mediated-gene-disruption library offers insight into Tn917 insertion patterns. *J. Bacteriol.* **186:**7280–7289.

136. Gasson, M., J.-J. Godon, C. J. Pillidge, T. J. Eaton, K. L. Jury, and C. A. Shearman. 1995. Characterization and exploitation of conjugation in *Lactococcus lactis. Int. Dairy J.* **5:**757–762.

137. Gasson, M. J. 1990. In vivo genetic systems in lactic acid bacteria. *FEMS Microbiol. Rev.* **7:**43–60.

138. Gawron-Burke, C., and D. B. Clewell. 1982. A transposon in *Streptococcus faecalis* with fertility properties. *Nature* **300:**281–284.

139. Gehring, A. M., J. R. Nodwell, S. M. Beverley, and R. Losick. 2000. Genomewide insertional mutagenesis in *Streptomyces coelicolor* reveals additional genes involved in morphological differentiation. *Proc. Natl. Acad. Sci. USA* **97:**9642–9647.

140. Geist, R. T., N. Okada, and M. G. Caparon. 1993. Analysis of *Streptococcus pyogenes* promoters by using novel Tn916-based shuttle vectors for the construction of transcriptional fusions to chloramphenicol acetyltransferase. *J. Bacteriol.* **175:**7561–7570.

141. Gibson, C. M., and M. G. Caparon. 2002. Alkaline phosphatase reporter transposon for identification of genes encoding secreted proteins in gram-positive microorganisms. *Appl. Environ. Microbiol.* **68:**928–932.

142. Gibson, E. M., N. M. Chace, S. B. London, and J. London. 1979. Transfer of plasmid-mediated antibiotic resistance from streptococci to lactobacilli. *J. Bacteriol.* **137:**614–619.

143. Gonzalez, C. F., and B. S. Kunka. 1983. Plasmid transfer in *Pediococcus* spp.: intergeneric and intragenic transfer of pIP501. *Appl. Environ. Microbiol.* **46:**81–89.

144. Gormley, E. P., and J. Davies. 1991. Transfer of plasmid rsf1010 by conjugation from *Escherichia coli* to *Streptomyces lividans* and *Mycobacterium smegmatis. J. Bacteriol.* **173:**6705–6708.

145. Gory, L., M. C. Montel, and M. Zagorec. 2001. Use of green fluorescent protein to monitor *Lactobacillus sakei* in fermented meat products. *FEMS Microbiol. Lett.* **194:**127–133.

146. Gosalbes, M. J., C. D. Esteban, J. L. Galan, and G. Perez-Martinez. 2000. Integrative food-grade expression system based on the lactose regulon of *Lactobacillus casei. Appl. Environ. Microbiol.* **66:**4822–4828.

147. Granok, A. B., D. Parsonage, R. P. Ross, and M. G. Caparon. 2000. The RofA binding site in *Streptococcus pyogenes* is utilized in multiple transcriptional pathways. *J. Bacteriol.* **182:**1529–1540.

148. Grant, R. B., and H. P. Riemann. 1976. Temperate phages of *Clostridium perfringens* type C1. *Can. J. Microbiol.* **22:**603–610.

149. Griffith, F. 1928. The significance of pneumococcal types. *J. Hyg.* (London) **27:**113–159.

150. Grohmann, E., G. Muth, and M. Espinosa. 2003. Conjugative plasmid transfer in gram-positive bacteria. *Microbiol. Mol. Biol. Rev.* **67:**277–301.

151. Gryllos, I., C. Cywes, M. H. Shearer, M. Cary, R. C. Kennedy, and M. R. Wessels. 2001. Regulation of capsule gene expression by group A Streptococcus during pharyngeal colonization and invasive infection. *Mol. Microbiol.* **42:**61–74.

152. Guerout-Fleury, A. M., N. Frandsen, and P. Stragier. 1996. Plasmids for ectopic integration in *Bacillus subtilis. Gene* **180:**57–61.

153. Gust, B., G. L. Challis, K. Fowler, T. Kieser, and K. F. Chater. 2003. PCR-targeted *Streptomyces* gene replacement identifies a protein domain needed for biosynthesis of the sesquiterpene soil odor geosmin. *Proc. Natl. Acad. Sci. USA* **100:**1541–1546.

154. Hamoen, L. W., G. Venema, and O. P. Kuipers. 2003. Controlling competence in Bacillus subtilis: shared use of regulators. *Microbiology* **149:**9–17.

155. Hashiba, H., R. Takiguchi, K. Jyoho, and K. Aoyama. 1992. Establishment of a host-vector system in *Lactobacillus helveticus* with beta-galactosidase activity as a selection marker. *Biosci. Biotechnol. Biochem.* **56:**190–194.

156. Hatfull, G. F. 1993. Genetic transformation of mycobacteria. *Trends Microbiol.* **1:**310–314.

157. Hatfull, G. F. 1996. The molecular genetics of *Mycobacterium tuberculosis. Curr. Top. Microbiol. Immunol.* **215:**29–47.

158. Hatfull, G. F. 2000. Molecular genetics of mycobacteriophages, p. 37–54. *In* G. F. Hatfull and W. R. Jacobs, Jr. (ed.), *Molecular Genetics of Mycobacteria.* ASM Press, Washington, DC.

159. Hava, D. L., and A. Camilli. 2002. Large-scale identification of serotype 4 *Streptococcus pneumoniae* virulence factors. *Mol. Microbiol.* **45:**1389–1406.

160. Heffron, F., C. Rubens, and S. Falkow. 1975. Translocation of a plasmid DNA sequence which mediates ampicillin resistance: molecular nature and specificity of insertion. *Proc. Natl. Acad. Sci. USA* **72:**3623–3627.

161. Helmark, S., M. E. Hansen, B. Jelle, K. I. Sorensen, and P. R. Jensen. 2004. Transformation of *Leuconostoc carnosum* 4010 and evidence for natural competence of the organism. *Appl. Environ. Microbiol.* **70:**3695–3699.

162. Henderson, I., T. Davis, M. Elmore, and N. Minton. 1997. The genetic basis of toxin production in *Clostridium botulinum* and *Clostridium tetani*, p. 261–294. *In* J. I. Rood, B. A. McClane, J. G. Songer, and R. W. Titball (ed.), *The Clostridia: Molecular Biology and Pathogenesis.* Academic Press, Inc., London, United Kingdom.

163. Hendrix, R. W. 2003. Bacteriophage genomics. *Curr. Opin. Microbiol.* **6:**506–511.

164. Hendrix, R. W., M. C. Smith, R. N. Burns, M. E. Ford, and G. F. Hatfull. 1999. Evolutionary relationships among diverse bacteriophages and prophages: all the world's a phage. *Proc. Natl. Acad. Sci. USA* **96:**2192–2197.

165. Henrich, B., J. R. Klein, B. Weber, C. Delorme, P. Renault, and U. Wegmann. 2002. Food-grade delivery system for controlled gene expression in *Lactococcus lactis. Appl. Environ. Microbiol.* **68:**5429–5436.

166. Hickey, R. M., D. P. Twomey, R. P. Ross, and C. Hill. 2001. Exploitation of plasmid pMRC01 to direct transfer of mobilizable plasmids into commercial lactococcal starter strains. *Appl. Environ. Microbiol.* **67:**2853–2858.

167. Hillemann, D., A. Puhler, and W. Wohlleben. 1991. Gene disruption and gene replacement in *Streptomyces* via single-stranded DNA transformation of integration vectors. *Nucleic Acids Res.* **19:**727–731.

168. Holo, H., and I. F. Nes. 1995. Transformation of *Lactococcus* by electroporation. *Methods Mol. Biol.* **47:**195–199.

169. Horinouchi, S., and T. Beppu. 1985. Construction and application of a promoter-probe plasmid that allows chromogenic identification in *Streptomyces lividans. J. Bacteriol.* **162:**406–412.

170. Horodniceanu, T., D. H. Bouanchaud, G. Bieth, and Y. A. Chabbert. 1976. R plasmids in *Streptococcus agalactiae* (group B). *Antimicrob. Agents Chemother.* **10:**795–801.

171. Hosted, T. J., and R. H. Baltz. 1997. Use of *rpsL* for dominance selection and gene replacement in *Streptomyces roseosporus. J. Bacteriol.* **179:**180–186.

171a. Hu, X., B. M. Hansen, J. Eilenberg, N. B. Hendriksen, L. Smidt, Z. Yuan, and G. B. Jensen. 2004. Conjugative transfer, stability and expression of a plasmid encoding a cry1Ac gene in Bacillus cereus group strains. *FEMS Microbiol. Lett.* **231:**45–52.

172. **Hu, Z. H., D. Hunziker, C. R. Hutchinson, and C. Khosla.** 1999. A host-vector system for analysis and manipulation of rifamycin polyketide biosynthesis in *Amycolatopsis mediterranei*. *Microbiology* **145:**2335–2341.

173. **Hutter, B., C. Fischer, A. Jacobi, C. Schaab, and H. Loferer.** 2004. Panel of *Bacillus subtilis* reporter strains indicative of various modes of action. *Antimicrob. Agents Chemother.* **48:**2588–2594.

174. **Ike, Y., K. Tanimoto, H. Tomita, K. Takeuchi, and S. Fujimoto.** 1998. Efficient transfer of the pheromone-independent *Enterococcus faecium* plasmid pMG1 (Gmr) (65.1 kilobases) to *Enterococcus* strains during broth mating. *J. Bacteriol.* **180:**4886–4892.

175. **Ikeda, H., Y. Takada, C. H. Pang, H. Tanaka, and S. Omura.** 1993. Transposon mutagenesis by Tn*4560* and applications with avermectin-producing *Streptomyces avermitilis*. *J. Bacteriol.* **175:**2077–2082.

176. **Ince, D., and D. C. Hooper.** 2000. Mechanisms and frequency of resistance to premafloxacin in *Staphylococcus aureus*: novel mutations suggest novel drug-target interactions. *Antimicrob. Agents Chemother.* **44:**3344–3350.

177. **Ingavale, S. S., W. Van Wamel, and A. L. Cheung.** 2003. Characterization of RAT, an autolysis regulator in *Staphylococcus aureus*. *Mol. Microbiol.* **48:**1451–1466.

178. **Ingram, C., M. Brawner, P. Youngman, and J. Westpheling.** 1989. *xylE* functions as an efficient reporter gene in *Streptomyces* spp: use for the study of *galP1*, a catabolite-controlled promoter. *J. Bacteriol.* **171:**6617–6624.

179. **Israelsen, H., S. M. Madsen, A. Vrang, E. B. Hansen, and E. Johansen.** 1995. Cloning and partial characterization of regulated promoters from *Lactococcus lactis* Tn*917-lacZ* integrants with the new promoter probe vector, pAK80. *Appl. Environ. Microbiol.* **61:**2540–2547.

180. **Jacobs, W. R., Jr., G. V. Kalpana, J. D. Cirillo, L. Pascopella, S. B. Snapper, R. A. Udani, W. Jones, R. G. Barletta, and B. R. Bloom.** 1991. Genetic systems for mycobacteria. *Methods Enzymol.* **204:**537–555.

181. **Janniere, L., A. Gruss, and D. Ehrlich.** 1993. Plasmids, p. 625–644. *In* A. L. Sonenshein, J. A. Hoch, and R. Losick (ed.), Bacillus subtilis *and other Gram-Positive Bacteria: Biochemistry, Physiology, and Molecular Genetics.* American Society for Microbiology, Washington, DC.

182. **Jensen, P. R., and K. Hammer.** 1998. The sequence of spacers between the consensus sequences modulates the strength of prokaryotic promoters. *Appl. Environ. Microbiol.* **64:**82–87.

183. **Ji, Y., A. Marra, M. Rosenberg, and G. Woodnutt.** 1999. Regulated antisense RNA eliminates alpha-toxin virulence in *Staphylococcus aureus* infection. *J. Bacteriol.* **181:**6585–6590.

184. **Ji, Y., L. McLandsborough, A. Kondagunta, and P. P. Cleary.** 1996. C5a peptidase alters clearance and trafficking of group A streptococci by infected mice. *Infect. Immun.* **64:**503–510.

185. **Johnson, E.** 1997. Extrachromosomal virulence determinants in the clostridia, p. 35–48. *In* J. I. Rood, B. A. McClane, J. G. Songer, and R. W. Titball (ed.), *The Clostridia: Molecular Biology and Pathogenesis.* Academic Press, Inc., London, United Kingdom.

186. **Johnson, L. P., and P. M. Schlievert.** 1984. Group A streptococcal phage T12 carries the structural gene for pyrogenic exotoxin type A. *Mol. Gen. Genet.* **194:**52–56.

187. **Johnson, L. P., P. M. Schlievert, and D. W. Watson.** 1980. Transfer of group A streptococcal pyrogenic exotoxin production to nontoxigenic strains of lysogenic conversion. *Infect. Immun.* **28:**254–257.

188. **Jones, J. M., S. C. Yost, and P. A. Pattee.** 1987. Transfer of the conjugal tetracycline resistance transposon Tn*916* from *Streptococcus faecalis* to *Staphylococcus aureus* and

189. **Kadioglu, A., J. A. Sharpe, I. Lazou, C. Svanborg, C. Ockleford, T. J. Mitchell, and P. W. Andrew.** 2001. Use of green fluorescent protein in visualisation of pneumococcal invasion of broncho-epithelial cells in vivo. *FEMS Microbiol. Lett.* **194:**105–110.

190. **Katayama, Y., T. Ito, and K. Hiramatsu.** 2000. A new class of genetic element, staphylococcus cassette chromosome *mec*, encodes methicillin resistance in *Staphylococcus aureus*. *Antimicrob. Agents Chemother.* **44:**1549–1555.

191. **Kendall, K. J., and S. N. Cohen.** 1987. Plasmid transfer in *Streptomyces lividans*: identification of a *kil-kor* system associated with the transfer region of pIJ101. *J. Bacteriol.* **169:**4177–4183.

192. **Kendall, K. J., and S. N. Cohen.** 1988. Complete nucleotide sequence of the *Streptomyces lividans* plasmid pIJ101 and correlation of the sequence with genetic properties. *J. Bacteriol.* **170:**4634–4651.

193. **Khan, S. A.** 1997. Rolling-circle replication of bacterial plasmids. *Microbiol. Mol. Biol. Rev.* **61:**442–455.

194. **Khan, S. A.** 2000. Plasmid rolling-circle replication: recent developments. *Mol. Microbiol.* **37:**477–484.

195. **Khosla, C., S. Ebertkhosla, and D. A. Hopwood.** 1992. Targeted gene replacements in a *Streptomyces* polyketide synthase gene cluster: role for the acyl carrier protein. *Mol. Microbiol.* **6:**3237–3249.

196. **Khosla, C., R. McDaniel, S. Ebertkhosla, R. Torres, D. H. Sherman, M. J. Bibb, and D. A. Hopwood.** 1993. Genetic construction and functional analysis of hybrid polyketide synthases containing heterologous acyl carrier proteins. *J. Bacteriol.* **175:**2197–2204.

197. **Kieser, T., M. J. Bibb, M. J. Buttner, K. F. Chater, and D. A. Hopwood.** 2000. *Practical* Streptomyces *Genetics.* The John Innes Foundation, Norwich, England.

198. **Kieser, T., D. A. Hopwood, H. M. Wright, and C. J. Thompson.** 1982. pIJ101, a multi-copy broad host-range *Streptomyces* plasmid: functional analysis and development of DNA cloning vectors. *Mol. Gen. Genet.* **185:**223–238.

199. **Kieser, T., and R. E. Melton.** 1988. Plasmid pIJ699, a multi-copy positive-selection vector for *Streptomyces*. *Gene* **65:**83–91.

200. **Kim, B. S., D. H. Sherman, and K. A. Reynolds.** 2004. An efficient method for creation and functional analysis of libraries of hybrid type I polyketide synthases. *Protein Eng. Des. Sel.* **17:**277–284.

201. **Kim, E. S., D. A. Hopwood, and D. H. Sherman.** 1994. Analysis of type-II polyketide β-ketoacyl synthase specificity in *Streptomyces coelicolor* A3(2) by *trans* complementation of actinorhodin synthase mutants. *J. Bacteriol.* **176:**1801–1804.

202. **Kinashi, H., M. Shimaji, and A. Sakai.** 1987. Giant linear plasmids in *Streptomyces* which code for antibiotic biosynthesis genes. *Nature* **328:**454–456.

203. **Kirby, C., A. Waring, T. J. Griffin, J. O. Falkinham III, N. D. Grindley, and K. M. Derbyshire.** 2002. Cryptic plasmids of *Mycobacterium avium*: Tn*552* to the rescue. *Mol. Microbiol.* **43:**173–186.

204. **Klaenhammer, T. R., and R. B. Sanozky.** 1985. Conjugal transfer from *Streptococcus lactis* ME2 of plasmids encoding phage resistance, nisin resistance and lactose-fermenting ability: evidence for a high-frequency conjugative plasmid responsible for abortive infection of virulent bacteriophage. *J. Gen. Microbiol.* **131:**1531–1541.

205. **Kleerebezem, M., M. M. Beerthuyzen, E. E. Vaughan, W. M. de Vos, and O. P. Kuipers.** 1997. Controlled gene

expression systems for lactic acid bacteria: transferable nisin-inducible expression cassettes for *Lactococcus*, *Leuconostoc*, and *Lactobacillus* spp. *Appl. Environ. Microbiol.* **63:**4581–4584.

206. **Koehler, T. M., Z. Dai, and M. Kaufman-Yarbray.** 1994. Regulation of the *Bacillus anthracis* protective antigen gene: CO_2 and a trans-acting element activate transcription from one of two promoters. *J. Bacteriol.* **176:**586–595.

207. **Kok, J., J. M. van der Vossen, and G. Venema.** 1984. Construction of plasmid cloning vectors for lactic streptococci which also replicate in *Bacillus subtilis* and *Escherichia coli*. *Appl. Environ. Microbiol.* **48:**726–731.

208. **Kondo, I., and K. Fujise.** 1977. Serotype B staphylococcal bacteriophage singly converting staphylokinase. *Infect. Immun.* **18:**266–272.

209. **Konings, W. N., J. Kok, O. P. Kuipers, and B. Poolman.** 2000. Lactic acid bacteria: the bugs of the new millennium. *Curr. Opin. Microbiol.* **3:**276–282.

210. **Kreikemeyer, B., M. D. Boyle, B. A. Buttaro, M. Heinemann, and A. Podbielski.** 2001. Group A streptococcal growth phase-associated virulence factor regulation by a novel operon (Fas) with homologies to two-component-type regulators requires a small RNA molecule. *Mol. Microbiol.* **39:**392–406.

210a.**Kristich, C. J., J. R. Chandler, and G. M. Dunny.** 2007. Development of a host-genotype-independent counterselectable marker and a high-frequency conjugative delivery system and their use in genetic analysis of *Enterococcus faecalis*. *Plasmid* **57:**133–144.

211. **Kristich, C. J., D. A. Manias, and G. M. Dunny.** 2005. Development of a method for markerless genetic exchange in *Enterococcus faecalis* and its use in construction of a *srtA* mutant. *Appl. Environ. Microbiol.* **71:**5837–5849.

212. **Kuipers, O. P., H. S. Rollema, R. J. Siezen, and W. M. De Vos.** 1995. Lactococcal expression systems for protein engineering of nisin. *Dev. Biol. Stand.* **85:**605–613.

213. **Kurenbach, B., C. Bohn, J. Prabhu, M. Abudukerim, U. Szewzyk, and E. Grohmann.** 2003. Intergeneric transfer of the *Enterococcus faecalis* plasmid pIP501 to *Escherichia coli* and *Streptomyces lividans* and sequence analysis of its *tra* region. *Plasmid* **50:**86–93.

214. **Kuroda, M., T. Ohta, I. Uchiyama, T. Baba, H. Yuzawa, I. Kobayashi, L. Cui, A. Oguchi, K. Aoki, Y. Nagai, J. Lian, T. Ito, M. Kanamori, H. Matsumaru, A. Maruyama, H. Murakami, A. Hosoyama, Y. Mizutani-Ui, N. K. Takahashi, T. Sawano, R. Inoue, C. Kaito, K. Sekimizu, H. Hirakawa, S. Kuhara, S. Goto, J. Yabuzaki, M. Kanehisa, A. Yamashita, K. Oshima, K. Furuya, C. Yoshino, T. Shiba, M. Hattori, N. Ogasawara, H. Hayashi, and K. Hiramatsu.** 2001. Whole genome sequencing of meticillin-resistant *Staphylococcus aureus*. *Lancet* **357:**1225–1240.

215. **Landman, O. E., D. J. Badkin, C. W. J. Finn, and R. A. Pepin.** 1980. Presented at the Proceedings of the Fifth European Meeting on Bacterial Transformation and Transfection.

216. **Langella, P., and A. Chopin.** 1989. Effect of restriction-modification systems on transfer of foreign DNA into *Lactococcus lactis* subsp. *lactis*. *FEMS Microbiol. Lett.* **59:**301–306.

217. **Lau, P. C., C. K. Sung, J. H. Lee, D. A. Morrison, and D. G. Cvitkovitch.** 2002. PCR ligation mutagenesis in transformable streptococci: application and efficiency. *J. Microbiol. Methods* **49:**193–205.

218. **Law, J., G. Buist, A. Haandrikman, J. Kok, G. Venema, and K. Leenhouts.** 1995. A system to generate chromosomal mutations in *Lactococcus lactis* which allows

fast analysis of targeted genes. *J. Bacteriol.* **177:**7011–7018.

219. **LeBlanc, D. J., L. N. Lee, and A. Abu-Al-Jaibat.** 1992. Molecular, genetic, and functional analysis of the basic replicon of pVA380-1, a plasmid of oral streptococcal origin. *Plasmid* **28:**130–145.

219a.**Le Breton, Y., N. P. Mohapatra, and W. G. Haldenwang.** 2006. In vivo random mutagenesis of *Bacillus subtilis* by use of TnYLB-1, a *mariner*-based transposon. *Appl. Environ. Microbiol.* **72:**327–333.

220. **Lederberg, J.** 1951. Streptomycin resistance: a genetically recessive mutation. *J. Bacteriol.* **61:**549–550.

221. **Lee, C. Y., S. L. Buranen, and Z. H. Ye.** 1991. Construction of single-copy integration vectors for *Staphylococcus aureus*. *Gene* **103:**101–105.

222. **Lee, C. Y., and J. J. Iandolo.** 1986. Lysogenic conversion of staphylococcal lipase is caused by insertion of the bacteriophage L54a genome into the lipase structural gene. *J. Bacteriol.* **166:**385–391.

223. **Lee, M. H., A. Nittayajarn, R. P. Ross, C. B. Rothschild, D. Parsonage, A. Claiborne, and C. E. Rubens.** 1999. Characterization of *Enterococcus faecalis* alkaline phosphatase and use in identifying *Streptococcus agalactiae* secreted proteins. *J. Bacteriol.* **181:**5790–5799.

224. **Lee, M. S., B. A. Dougherty, A. C. Madeo, and D. A. Morrison.** 1999. Construction and analysis of a library for random insertional mutagenesis in *Streptococcus pneumoniae*: use for recovery of mutants defective in genetic transformation and for identification of essential genes. *Appl. Environ. Microbiol.* **65:**1883–1890.

225. **Lee, M. S., C. Seok, and D. A. Morrison.** 1998. Insertion-duplication mutagenesis in *Streptococcus pneumoniae*: targeting fragment length is a critical parameter in use as a random insertion tool. *Appl. Environ. Microbiol.* **64:**4796–4802.

226. **Leelaporn, A., N. Firth, I. T. Paulsen, and R. A. Skurray.** 1996. IS257-mediated cointegration in the evolution of a family of staphylococcal trimethoprim resistance plasmids. *J. Bacteriol.* **178:**6070–6073.

227. **Leenhouts, K. J., B. Tolner, S. Bron, J. Kok, G. Venema, and J. F. Seegers.** 1991. Nucleotide sequence and characterization of the broad-host-range lactococcal plasmid pWVO1. *Plasmid* **26:**55–66.

228. **Lewis, P. J., and A. L. Marston.** 1999. GFP vectors for controlled expression and dual labelling of protein fusions in *Bacillus subtilis*. *Gene* **227:**101–110.

229. **Lin, M. Y., S. Harlander, and D. Savaiano.** 1996. Construction of an integrative food-grade cloning vector for *Lactobacillus acidophilus*. *Appl. Microbiol. Biotechnol.* **45:**484–489.

230. **Lin, W. S., T. Cunneen, and C. Y. Lee.** 1994. Sequence analysis and molecular characterization of genes required for the biosynthesis of type 1 capsular polysaccharide in *Staphylococcus aureus*. *J. Bacteriol.* **176:**7005–7016.

231. **Lithgow, J. K., E. Ingham, and S. J. Foster.** 2004. Role of the *hprT-ftsH* locus in *Staphylococcus aureus*. *Microbiology* **150:**373–381.

232. **Liu, C. Q., P. Su, N. Khunajakr, Y. M. Deng, S. Sumual, W. S. Kim, J. E. Tandianus, and N. W. Dunn.** 2005. Development of food-grade cloning and expression vectors for *Lactococcus lactis*. *J. Appl. Microbiol.* **98:**127–135.

233. **Liu, S., S. Sela, G. Cohen, J. Jadoun, A. Cheung, and I. Ofek.** 1997. Insertional inactivation of streptolysin S expression is associated with altered riboflavin metabolism in *Streptococcus pyogenes*. *Microb. Pathog.* **22:**227–234.

234. **Löfdahl, S., J. E. Sjöström, and L. Philipson.** 1981. Cloning of restriction fragments of DNA from staphylococcal bacteriophage φ11. *J. Virol.* **37:**795–801.

235. **Lopez, R.** 2004. *Streptococcus pneumoniae* and its bacteriophages: one long argument. *Int. Microbiol.* **7:**163–171.

236. **Luchansky, J. B., E. G. Kleeman, R. R. Raya, and T. R. Klaenhammer.** 1989. Genetic transfer systems for delivery of plasmid deoxyribonucleic acid to *Lactobacillus acidophilus* ADH: conjugation, electroporation, and transduction. *J. Dairy Sci.* **72:**1408–1417.

237. **Luchansky, J. B., P. M. Muriana, and T. R. Klaenhammer.** 1988. Application of electroporation for transfer of plasmid DNA to *Lactobacillus, Lactococcus, Leuconostoc, Listeria, Pediococcus, Bacillus, Staphylococcus, Enterococcus* and *Propionibacterium. Mol. Microbiol.* **2:**637–646.

238. **Luo, P., H. Li, and D. A. Morrison.** 2003. ComX is a unique link between multiple quorum sensing outputs and competence in *Streptococcus pneumoniae. Mol. Microbiol.* **50:**623–633.

239. **Luo, P., H. Li, and D. A. Morrison.** 2004. Identification of ComW as a new component in the regulation of genetic transformation in *Streptococcus pneumoniae. Mol. Microbiol.* **54:**172–183.

240. **Lydiate, D. J., F. Malpartida, and D. A. Hopwood.** 1985. The *Streptomyces* plasmid SCP2*: its functional analysis and development into useful cloning vectors. *Gene* **35:**223–235.

241. **Lyon, W. R., C. M. Gibson, and M. G. Caparon.** 1998. A role for trigger factor and an *rgg*-like regulator in the transcription, secretion and processing of the cysteine proteinase of *Streptococcus pyogenes. EMBO J.* **17:**6263–6275.

242. **Lyras, D., and J. I. Rood.** 1998. Conjugative transfer of RP4-*oriT* shuttle vectors from *Escherichia coli* to *Clostridium perfringens. Plasmid* **39:**160–164.

243. **Lyras, D., and J. I. Rood.** 2000. Clostridial genetics, p. 529–539. *In* V. A. Fischetti, R. P. Novick, J. J. Ferretti, D. A. Portnoy, and J. I. Rood (ed.), *Gram-Positive Pathogens.* ASM Press, Washington, DC.

244. **Maas, R.-M., J. Gotz, W. Wohlleben, and G. Muth.** 1998. The conjugative plasmid pSG5 from *Streptomyces ghanaensis* DSM 2932 differs in its transfer functions from other *Streptomyces* rolling-circle-type plasmids. *Microbiology* **144:**2809–2817.

245. **Macneil, D. J.** 1987. Introduction of plasmid DNA into *Streptomyces lividans* by electroporation. *FEMS Microbiol. Lett.* **42:**239–244.

246. **Macneil, D. J., K. M. Gewain, C. L. Ruby, G. Dezeny, P. H. Gibbons, and T. Macneil.** 1992. Analysis of *Streptomyces avermitilis* genes required for avermectin biosynthesis utilizing a novel integration vector. *Gene* **111:**61–68.

247. **Macrina, F. L., and G. L. Archer.** 1993. Conjugation and broad host range plasmids in streptococci and staphylococci, p. 313–329. *In* D. B. Clewell (ed.), *Bacterial Conjugation.* Plenum Press, New York, NY.

248. **Macrina, F. L., J. A. Tobian, K. R. Jones, and R. P. Evans.** 1982. Molecular cloning in the streptococci, p. 195–220. *In* A. Hollaender, R. DeMoss, S. Kaplan, J. Konisky, D. Savage, and R. Wolfe (ed.), *Genetic Engineering of Microorganisms for Chemicals.* Plenum Press, New York, NY.

249. **Macrina, F. L., J. A. Tobian, K. R. Jones, R. P. Evans, and D. B. Clewell.** 1982. A cloning vector able to replicate in *Escherichia coli* and *Streptococcus sanguis. Gene* **19:**345–353.

250. **Macrina, F. L., P. H. Wood, and K. R. Jones.** 1980. Genetic transformation of *Streptococcus sanguis* (Challis) with cryptic plasmids from *Streptococcus ferus. Infect. Immun.* **28:**692–699.

251. **Madsen, S. M., J. Arnau, A. Vrang, M. Givskov, and H. Israelsen.** 1999. Molecular characterization of the pH-inducible and growth phase-dependent promoter P170 of *Lactococcus lactis. Mol. Microbiol.* **32:**75–87.

252. **Maeda, S., A. Sawamura, and A. Matsuda.** 2004. Transformation of colonial *Escherichia coli* on solid media. *FEMS Microbiol. Lett.* **236:**61–64.

253. **Maguin, E., P. Duwat, T. Hege, D. Ehrlich, and A. Gruss.** 1992. New thermosensitive plasmid for gram-positive bacteria. *J. Bacteriol.* **174:**5633–5638.

254. **Maguin, E., H. Prevost, S. D. Ehrlich, and A. Gruss.** 1996. Efficient insertional mutagenesis in lactococci and other gram-positive bacteria. *J. Bacteriol.* **178:**931–935.

255. **Malke, H.** 1969. Transduction of *Streptococcus pyogenes* K 56 by temperature-sensitive mutants of the transducing phage A 25. *Z. Naturforsch. B* **24:**1556–1561.

256. **Manganelli, R., R. Provvedi, C. Berneri, M. R. Oggioni, and G. Pozzi.** 1998. Insertion vectors for construction of recombinant conjugative transposons in *Bacillus subtilis* and *Enterococcus faecalis. FEMS Microbiol. Lett.* **168:**259–268.

257. **Manna, A., and A. L. Cheung.** 2001. Characterization of *sarR*, a modulator of *sar* expression in *Staphylococcus aureus. Infect. Immun.* **69:**885–896.

258. **Martin, A. C., R. Lopez, and P. Garcia.** 1996. Analysis of the complete nucleotide sequence and functional organization of the genome of *Streptococcus pneumoniae* bacteriophage Cp-1. *J. Virol.* **70:**3678–3687.

259. **Martin, M. C., J. C. Alonso, J. E. Suarez, and M. A. Alvarez.** 2000. Generation of food-grade recombinant lactic acid bacterium strains by site-specific recombination. *Appl. Environ. Microbiol.* **66:**2599–2604.

260. **Matsushima, P., M. C. Broughton, J. R. Turner, and R. H. Baltz.** 1994. Conjugal transfer of cosmid DNA from *Escherichia coli* to *Saccharopolyspora spinosa*: effects of chromosomal insertions on macrolide A83543 production. *Gene* **146:**39–45.

261. **Maxted, W. R.** 1955. The influence of bacteriophage on *Streptococcus pyogenes. J. Gen. Microbiol.* **12:**484.

262. **Mazodier, P., R. Petter, and C. Thompson.** 1989. Intergeneric conjugation between *Escherichia coli* and *Streptomyces* species. *J. Bacteriol.* **171:**3583–3585.

263. **McAdam, R. A., S. Quan, and C. Guilhot.** 2000. Mycobacterial transposons and their applications, p. 69–84. *In* G. F. Hatfull and W. R. Jacobs, Jr. (ed.), *Molecular Genetics of Mycobacteria.* ASM Press, Washington, DC.

264. **McIver, K. S., A. S. Heath, and J. R. Scott.** 1995. Regulation of virulence by environmental signals in group A streptococci: influence of osmolarity, temperature, gas exchange, and iron limitation on *emm* transcription. *Infect. Immun.* **63:**4540–4542.

265. **McKinney, T. K., V. K. Sharma, W. A. Craig, and G. L. Archer.** 2001. Transcription of the gene mediating methicillin resistance in *Staphylococcus aureus* (*mecA*) is corepressed but not coinduced by cognate *mecA* and beta-lactamase regulators. *J. Bacteriol.* **183:**6862–6868.

266. **Mercer, D. K., C. M. Melville, K. P. Scott, and H. J. Flint.** 1999. Natural genetic transformation in the rumen bacterium *Streptococcus bovis* JB1. *FEMS Microbiol. Lett.* **179:**485–490.

267. **Mills, D. A.** 2001. Mutagenesis in the post genomics era: tools for generating insertional mutations in the lactic acid bacteria. *Curr. Opin. Biotechnol.* **12:**503–509.

268. **Mills, D. A., L. L. McKay, and G. M. Dunny.** 1996. Splicing of a group II intron involved in the conjugative transfer of pRS01 in lactococci. *J. Bacteriol.* **178:**3531–3538.

269. **Morrison, D. A., S. A. Lacks, W. R. Guild, and J. M. Hageman.** 1983. Isolation and characterization of three new classes of transformation-deficient mutants of *Streptococcus pneumoniae* that are defective in DNA

transport and genetic recombination. *J. Bacteriol.* **156:** 281–290.

270. **Morton, T. M., D. M. Eaton, J. L. Johnston, and G. L. Archer.** 1993. DNA sequence and units of transcription of the conjugative transfer gene complex (*trs*) of *Staphylococcus aureus* plasmid pGO1. *J. Bacteriol.* **175:**4436–4447.

271. **Mullany, P., M. Wilks, I. Lamb, C. Clayton, B. Wren, and S. Tabaqchali.** 1990. Genetic analysis of a tetracycline resistance element from *Clostridium difficile* and its conjugal transfer to and from *Bacillus subtilis. J. Gen. Microbiol.* **136:**1343–1349.

272. **Mullany, P., M. Wilks, L. Puckey, and S. Tabaqchali.** 1994. Gene cloning in *Clostridium difficile* using Tn*916* as a shuttle conjugative transposon. *Plasmid* **31:**320–323.

273. **Mullany, P., M. Wilks, and S. Tabaqchali.** 1991. Transfer of Tn*916* and Tn*916* delta E into *Clostridium difficile*: demonstration of a hot-spot for these elements in the *C. difficile* genome. *FEMS Microbiol. Lett.* **63:**191–194.

274. **Muth, G., M. Farr, V. Hartmann, and W. Wohlleben.** 1995. *Streptomyces ghanaensis* plasmid pSG5: nucleotide sequence analysis of the self-transmissible minimal replicon and characterization of the replication mode. *Plasmid* **33:**113–126.

275. **Muth, G., B. Nussbaumer, W. Wohlleben, and A. Puhler.** 1989. A vector system with temperature-sensitive replication for gene disruption and mutational cloning in streptomycetes. *Mol. Gen. Genet.* **219:**341–348.

276. **Nida, K., and P. P. Cleary.** 1983. Insertional inactivation of streptolysin S expression in *Streptococcus pyogenes. J. Bacteriol.* **155:**1156–1161.

277. **Novick, R.** 1967. Properties of a cryptic high-frequency transducing phage in *Staphylococcus aureus. Virology* **33:** 155–166.

278. **Novick, R. P.** 1989. Staphylococcal plasmids and their replication. *Annu. Rev. Microbiol.* **43:**537–565.

279. **Novick, R. P.** 1990. The staphylococcus as a molecular genetic system, p. 1–40. *In* R. P. Novick (ed.), *Molecular Biology of the Staphylococci.* VCH, New York, NY.

280. **Novick, R. P.** 1991. Genetic systems in staphylococci. *Methods Enzymol.* **204:**587–636.

281. **Novick, R. P., I. Edelman, and S. Lofdahl.** 1986. Small *Staphylococcus aureus* plasmids are transduced as linear multimers that are formed and resolved by replicative processes. *J. Mol. Biol.* **192:**209–220.

282. **Obregon, V., J. L. Garcia, E. Garcia, R. Lopez, and P. Garcia.** 2003. Genome organization and molecular analysis of the temperate bacteriophage MM1 of *Streptococcus pneumoniae. J. Bacteriol.* **185:**2362–2368.

283. **Obst, M., E. R. Meding, R. F. Vogel, and W. P. Hammes.** 1995. Two genes encoding the beta-galactosidase of *Lactobacillus sake* [*sic*]. *Microbiology* **141:**3059–3066.

284. **Oh, S. H., and K. F. Chater.** 1997. Denaturation of circular or linear DNA facilitates targeted integrative transformation of *Streptomyces coelicolor* A3(2): possible relevance to other organisms. *J. Bacteriol.* **179:**122–127.

285. **O'Sullivan, D. J., and T. R. Klaenhammer.** 1993. High- and low-copy-number *Lactococcus* shuttle cloning vectors with features for clone screening. *Gene* **137:**227–231.

286. **Oultram, J. D., and M. Young.** 1985. Conjugal transfer of plasmid pAMbeta1 from *Streptococcus lactis* and *Bacillus subtilis* to *Clostridium acetobutylicum. FEMS Microbiol. Lett.* **27:**129–134.

287. **Paget, M. S. B., G. Hintermann, and C. P. Smith.** 1994. Construction and application of streptomycete promoter probe vectors which employ the *Streptomyces glaucescens* tyrosinase-encoding gene as reporter. *Gene* **146:**105–110.

288. **Paisano, A. F., B. Spira, S. Cai, and A. C. Bombana.** 2004. In vitro antimicrobial effect of bacteriophages on human dentin infected with *Enterococcus faecalis* ATCC 29212. *Oral Microbiol. Immunol.* **19:**327–330.

289. **Pandza, S., G. Biukovic, A. Paravic, A. Dadbin, J. Cullum, and D. Hranueli.** 1998. Recombination between the linear plasmid pPZG101 and the linear chromosome of *Streptomyces rimosus* can lead to exchange of ends. *Mol. Microbiol.* **28:**1165–1176.

290. **Pashley, C., and N. G. Stoker.** 2000. Plasmids in mycobacteria, p. 551–567. *In* G. F. Hatfull and W. R. Jacobs, Jr. (ed.), *Molecular Genetics of Mycobacteria.* ASM Press, Washington, DC.

291. **Patel, K., M. Piagentini, A. Rascher, Z. Q. Tian, G. O. Buchanan, R. Regentin, Z. H. Hu, C. R. Hutchinson, and R. McDaniel.** 2004. Engineered biosynthesis of geldanamycin analogs for Hsp90 inhibition. *Chem. Biol.* **11:** 1625–1633.

292. **Pattee, P. A.** 1981. Distribution of Tn*551* insertion sites responsible for auxotrophy on the *Staphylococcus aureus* chromosome. *J. Bacteriol.* **145:**479–488.

293. **Paul, J. H., M. B. Sullivan, A. M. Segall, and F. Rohwer.** 2002. Marine phage genomics. *Comp. Biochem. Physiol. B* **133:**463–476.

294. **Paulsen, I. T., L. Banerjei, G. S. Myers, K. E. Nelson, R. Seshadri, T. D. Read, D. E. Fouts, J. A. Eisen, S. R. Gill, J. F. Heidelberg, H. Tettelin, R. J. Dodson, L. Umayam, L. Brinkac, M. Beanan, S. Daugherty, R. T. DeBoy, S. Durkin, J. Kolonay, R. Madupu, W. Nelson, J. Vamathevan, B. Tran, J. Upton, T. Hansen, J. Shetty, H. Khouri, T. Utterback, D. Radune, K. A. Ketchum, B. A. Dougherty, and C. M. Fraser.** 2003. Role of mobile DNA in the evolution of vancomycin-resistant *Enterococcus faecalis. Science* **299:**2071–2074.

295. **Pavelka, M. S., Jr., and W. R. Jacobs, Jr.** 1999. Comparison of the construction of unmarked deletion mutations in *Mycobacterium smegmatis, Mycobacterium bovis* bacillus Calmette-Guerin, and *Mycobacterium tuberculosis* H37Rv by allelic exchange. *J. Bacteriol.* **181:**4780–4789.

296. **Pearce, B. J., A. M. Naughton, and H. R. Masure.** 1994. Peptide permeases modulate transformation in *Streptococcus pneumoniae. Mol. Microbiol.* **12:**881–892.

297. **Pearce, B. J., Y. B. Yin, and H. R. Masure.** 1993. Genetic identification of exported proteins in *Streptococcus pneumoniae. Mol. Microbiol.* **9:**1037–1050.

298. **Pelicic, V., M. Jackson, J. M. Reyrat, W. R. Jacobs, Jr., B. Gicquel, and C. Guilhot.** 1997. Efficient allelic exchange and transposon mutagenesis in *Mycobacterium tuberculosis. Proc. Natl. Acad. Sci. USA* **94:**10955–10960.

299. **Pernodet, J. L., J. M. Simonet, and M. Guerineau.** 1984. Plasmids in different strains of *Streptomyces ambofaciens*: free and integrated form of plasmid pSAM2. *Mol. Gen. Genet.* **198:**35–41.

300. **Pestova, E. V., L. S. Havarstein, and D. A. Morrison.** 1996. Regulation of competence for genetic transformation in *Streptococcus pneumoniae* by an auto-induced peptide pheromone and a two-component regulatory system. *Mol. Microbiol.* **21:**853–862.

301. **Pettis, G. S., and S. N. Cohen.** 1994. Transfer of the pIJ101 plasmid in *Streptomyces lividans* requires a *cis*-acting function dispensable for chromosomal gene transfer. *Mol. Microbiol.* **13:**955–964.

301a. **Petit, M. A., C. Bruand, L. Janniere, and S. D. Ehrlich.** 1990. Tn10-derived transposons active in *Bacillus subtilis. J. Bacteriol.* **172:**6736–6740.

302. **Pettis, G. S., and S. N. Cohen.** 2000. Mutational analysis of the *tra* locus of the broad-host-range *Streptomyces* plasmid pIJ101. *J. Bacteriol.* **182:**4500–4504.

303. **Pigac, J., and H. Schrempf.** 1995. A simple and rapid method of transformation of *Streptomyces rimosus* R6 and

other streptomycetes by electroporation. *Appl. Environ. Microbiol.* **61:**352–356.

304. **Pinho, M. G., and J. Errington.** 2003. Dispersed mode of *Staphylococcus aureus* cell wall synthesis in the absence of the division machinery. *Mol. Microbiol.* **50:**871–881.

305. **Platteeuw, C., I. van Alen-Boerrigter, S. van Schalkwijk, and W. M. de Vos.** 1996. Food-grade cloning and expression system for *Lactococcus lactis. Appl. Environ. Microbiol.* **62:**1008–1013.

306. **Podbielski, A., B. Spellerberg, M. Woischnik, B. Pohl, and R. Lutticken.** 1996. Novel series of plasmid vectors for gene inactivation and expression analysis in group A streptococci (GAS). *Gene* **177:**137–147.

307. **Podbielski, A., M. Woischnik, B. A. Leonard, and K. H. Schmidt.** 1999. Characterization of *nra*, a global negative regulator gene in group A streptococci. *Mol. Microbiol.* **31:**1051–1064.

307a.**Poluektova, E. U., E. A. Fedorina, O. V. Lotareva, and A. A. Prozorov.** 2004. Plasmid transfer in bacilli by a self-transmissible plasmid p19 from a Bacillus subtilis soil strain. *Plasmid* **52:**212–217.

308. **Poquet, I., S. D. Ehrlich, and A. Gruss.** 1998. An export-specific reporter designed for gram-positive bacteria: application to *Lactococcus lactis. J. Bacteriol.* **180:**1904–1912.

309. **Posno, M., P. T. Heuvelmans, M. J. van Giezen, B. C. Lokman, R. J. Leer, and P. H. Pouwels.** 1991. Complementation of the inability of *Lactobacillus* strains to utilize D-xylose with D-xylose catabolism-encoding genes of *Lactobacillus pentosus. Appl. Environ. Microbiol.* **57:**2764–2766.

310. **Poyart, C., and P. Trieu-Cuot.** 1997. A broad-host-range mobilizable shuttle vector for the construction of transcriptional fusions to beta-galactosidase in gram-positive bacteria. *FEMS Microbiol. Lett.* **156:**193–198.

311. **Pozzi, G., R. A. Musmanno, E. A. Renzoni, M. R. Oggioni, and M. G. Cusi.** 1988. Host-vector system for integration of recombinant DNA into chromosomes of transformable and nontransformable streptococci. *J. Bacteriol.* **170:**1969–1972.

311a.**Pozzi,, G., and J. M. Wells (ed.).** 1997. *Gram-Positive Bacteria: Vaccine Vehicles for Mucosal Immunization.* Springer-Landes Bioscience, Georgetown, TX.

312. **Prentki, P., and H. M. Krisch.** 1984. In vitro insertional mutagenesis with a selectable DNA fragment. *Gene* **29:**303–313.

313. **Qin, X., K. V. Singh, Y. Xu, G. M. Weinstock, and B. E. Murray.** 1998. Effect of disruption of a gene encoding an autolysin of *Enterococcus faecalis* OG1RF. *Antimicrob. Agents Chemother.* **42:**2883–2888.

314. **Rauch, P. J., and W. M. De Vos.** 1992. Characterization of the novel nisin-sucrose conjugative transposon Tn*5276* and its insertion in *Lactococcus lactis. J. Bacteriol.* **174:**1280–1287.

315. **Rauch, P. J., and W. M. de Vos.** 1994. Identification and characterization of genes involved in excision of the *Lactococcus lactis* conjugative transposon Tn*5276. J. Bacteriol.* **176:**2165–2171.

316. **Ravn, P., J. Arnau, S. M. Madsen, A. Vrang, and H. Israelsen.** 2000. The development of Tn*Nuc* and its use for the isolation of novel secretion signals in *Lactococcus lactis. Gene* **242:**347–356.

316a.**Reddy, A., L. Battisti, and C. B. Thorne.** 1987. Identification of self-transmissible plasmids in four *Bacillus thuringiensis* subspecies. *J. Bacteriol.* **169:**5263–5270.

317. **Rees, P. J., and B. A. Fry.** 1981. The morphology of staphylococcal bacteriophage K and DNA metabolism in infected *Staphylococcus aureus. J. Gen. Virol.* **53:**293–307.

318. **Rice, L. B.** 1998. Tn*916* family conjugative transposons and dissemination of antimicrobial resistance determinants. *Antimicrob. Agents Chemother.* **42:**1871–1877.

319. **Roberts, A. P., C. Hennequin, M. Elmore, A. Collignon, T. Karjalainen, N. Minton, and P. Mullany.** 2003. Development of an integrative vector for the expression of antisense RNA in *Clostridium difficile. J. Microbiol. Methods* **55:**617–624.

320. **Rogers, C. G., and W. B. Sarles.** 1963. Characterization of *Enterococcus* bacteriophages from the small intestine of the rat. *J. Bacteriol.* **85:**1378–1385.

321. **Romero, D. A., and T. R. Klaenhammer.** 1990. Characterization of insertion sequence IS946, an Iso-ISS1 element, isolated from the conjugative lactococcal plasmid pTR2030. *J. Bacteriol.* **172:**4151–4160.

322. **Romero, D. A., and T. R. Klaenhammer.** 1991. Construction of an IS946-based composite transposon in *Lactococcus lactis* subsp. *lactis. J. Bacteriol.* **173:**7599–7606.

323. **Romero, P., R. Lopez, and E. Garcia.** 2004. Genomic organization and molecular analysis of the inducible prophage EJ-1, a mosaic myovirus from an atypical pneumococcus. *Virology* **322:**239–252.

324. **Rood, J. I.** 1997. Genetic analysis in C. *perfringens*, p. 65–72. *In* J. I. Rood, B. A. McClane, J. G. Songer, and R. W. Titball (ed.), *The Clostridia: Molecular Biology and Pathogenesis.* Academic Press, Inc., London, United Kingdom.

325. **Rosenow, C., M. Maniar, and J. Trias.** 1999. Regulation of the alpha-galactosidase activity in *Streptococcus pneumoniae*: characterization of the raffinose utilization system. *Genome Res.* **9:**1189–1197.

326. **Rubin, E. J., B. J. Akerley, V. N. Novik, D. J. Lampe, R. N. Husson, and J. J. Mekalanos.** 1999. In vivo transposition of *mariner*-based elements in enteric bacteria and mycobacteria. *Proc. Natl. Acad. Sci. USA* **96:**1645–1650.

327. **Russell, C. B., and F. W. Dahlquist.** 1989. Exchange of chromosomal and plasmid alleles in *Escherichia coli* by selection for loss of a dominant antibiotic sensitivity marker. *J. Bacteriol.* **171:**2614–2618.

328. **Ryding, N. J., G. H. Kelemen, C. A. Whatling, K. Flardh, M. J. Buttner, and K. F. Chater.** 1998. A developmentally regulated gene encoding a repressor-like protein is essential for sporulation in *Streptomyces coelicolor* A3(2). *Mol. Microbiol.* **29:**343–357.

329. **Sanchez, C., and B. Mayo.** 2004. General and specialized vectors derived from pBM02, a new rolling circle replicating plasmid of *Lactococcus lactis. Plasmid* **51:**265–271.

330. **Santagati, M., F. Iannelli, C. Cascone, F. Campanile, M. R. Oggioni, S. Stefani, and G. Pozzi.** 2003. The novel conjugative transposon Tn*1207.3* carries the macrolide efflux gene *mef*(A) in *Streptococcus pyogenes. Microb. Drug Resist.* **9:**243–247.

331. **Sanz, Y., F. C. Lanfermeijer, M. Hellendoorn, J. Kok, W. N. Konings, and B. Poolman.** 2004. Two homologous oligopeptide binding protein genes (*opp*A) in *Lactococcus lactis* MG1363. *Int. J. Food Microbiol.* **97:**9–15.

332. **Sasaki, Y., Y. Ito, and T. Sasaki.** 2004. ThyA as a selection marker in construction of food-grade host-vector and integration systems for *Streptococcus thermophilus. Appl. Environ. Microbiol.* **70:**1858–1864.

333. **Savic, D. J., W. M. McShan, and J. J. Ferretti.** 2002. Autonomous expression of the *slo* gene of the bicistronic *nga-slo* operon of *Streptococcus pyogenes. Infect. Immun.* **70:**2730–2733.

334. **Schaberg, D. R., D. B. Clewell, and L. Glatzer.** 1981. Cell-to-cell transfer of R-plasmids from *Streptococcus faecalis* to *Staphylococcus aureus*, p. 658. *In* S. B. Levy, R. C. Clowes, and E. L. Koenig (ed.), *Molecular Biology, Patho-*

genicity, and Ecology of Bacterial Plasmids. Plenum Publishing Corp., New York, NY.

335. **Schaberg, D. R., D. B. Clewell, and L. Glatzer.** 1982. Conjugative transfer of R-plasmids from *Streptococcus faecalis* to *Staphylococcus aureus. Antimicrob. Agents Chemother.* **22:**204–207.

336. **Schauer, A., M. Ranes, R. Santamaria, J. Guijarro, E. Lawlor, C. Mendez, K. Chater, and R. Losick.** 1988. Visualizing gene expression in time and space in the filamentous bacterium *Streptomyces coelicolor. Science* **240:**768–772.

337. **Schenk, S., and R. A. Laddaga.** 1992. Improved method for electroporation of *Staphylococcus aureus. FEMS Microbiol. Lett.* **73:**133–138.

338. **Schuster, C., M. van der Linden, and R. Hakenbeck.** 1998. Small cryptic plasmids of *Streptococcus pneumoniae* belong to the pC194/pUB110 family of rolling circle plasmids. *FEMS Microbiol. Lett.* **164:**427–431.

339. **Scott, J. R., F. Bringel, D. Marra, G. Van Alstine, and C. K. Rudy.** 1994. Conjugative transposition of Tn*916:* preferred targets and evidence for conjugative transfer of a single strand and for a double-stranded circular intermediate. *Mol. Microbiol.* **11:**1099–1108.

340. **Scott, J. R., and G. G. Churchward.** 1995. Conjugative transposition. *Annu. Rev. Microbiol.* **49:**367–397.

341. **Sherman, D. H., E. S. Kim, M. J. Bibb, and D. A. Hopwood.** 1992. Functional replacement of genes for individual polyketide synthase components in *Streptomyces coelicolor* A3(2) by heterologous genes from a different polyketide pathway. *J. Bacteriol.* **174:**6184–6190.

342. **Shimizu, T., W. Ba-Thein, M. Tamaki, and H. Hayashi.** 1994. The *virR* gene, a member of a class of two-component response regulators, regulates the production of perfringolysin O, collagenase, and hemagglutinin in *Clostridium perfringens. J. Bacteriol.* **176:**1616–1623.

343. **Shoemaker, N. B., M. D. Smith, and W. R. Guild.** 1980. DNase-resistant transfer of chromosomal *cat* and *tet* insertions by filter mating in *Pneumococcus. Plasmid* **3:** 80–87.

344. **Siemieniak, D. R., J. L. Slightom, and S. T. Chung.** 1990. Nucleotide sequence of *Streptomyces fradiae* transposable element Tn*4556:* a class-II transposon related to Tn*3. Gene* **86:**1–9.

345. **Simoes-Barbosa, A., H. Abreu, A. Silva Neto, A. Gruss, and P. Langella.** 2004. A food-grade delivery system for *Lactococcus lactis* and evaluation of inducible gene expression. *Appl. Microbiol. Biotechnol.* **65:**61–67.

346. **Simon, D., and J. J. Ferretti.** 1991. Electrotransformation of *Streptococcus pyogenes* with plasmid and linear DNA. *FEMS Microbiol. Lett.* **66:**219–224.

347. **Simon, R., U. Priefer, and A. Puhler.** 1983. A broad host range mobilization system for *in vivo* genetic engineering: transposon mutagenesis in gram-negative bacteria. *Bio/Technology* **1:**784–791.

348. **Simonet, J. M., F. Boccard, J. L. Pernodet, J. Gagnat, and M. Guerineau.** 1987. Excision and integration of a self-transmissible replicon of *Streptomyces ambofaciens. Gene* **59:**137–144.

349. **Smith, M. D., and W. R. Guild.** 1979. A plasmid in *Streptococcus pneumoniae. J. Bacteriol.* **137:**735–739.

350. **Smith, M. D., N. B. Shoemaker, V. Burdett, and W. R. Guild.** 1980. Transfer of plasmids by conjugation in *Streptococcus pneumoniae. Plasmid* **3:**70–79.

351. **Solem, C., and P. R. Jensen.** 2002. Modulation of gene expression made easy. *Appl. Environ. Microbiol.* **68:**2397–2403.

352. **Solenberg, P. J., and R. H. Baltz.** 1991. Transposition of Tn*5096* and other IS*493* derivatives in *Streptomyces griseofuscus. J. Bacteriol.* **173:**1096–1104.

353. **Solenberg, P. J., and R. H. Baltz.** 1994. Hypertransposing derivatives of the streptomycete insertion sequence IS*493. Gene* **147:**47–54.

354. **Stassi, D. L., P. Lopez, M. Espinosa, and S. A. Lacks.** 1981. Cloning of chromosomal genes in *Streptococcus pneumoniae. Proc. Natl. Acad. Sci. USA* **78:**7028–7032.

355. **Steele, J. L., and L. L. McKay.** 1989. Conjugal transfer of genetic material by *Lactococcus lactis* subsp. *lactis* 11007. *Plasmid* **22:**32–43.

356. **Steenson, L. R., and T. R. Klaenhammer.** 1985. *Streptococcus cremoris* M12R transconjugants carrying the conjugal plasmid pTR2030 are insensitive to attack by lytic bacteriophages. *Appl. Environ. Microbiol.* **50:** 851–858.

357. **Stein, D. S., K. J. Kendall, and S. N. Cohen.** 1989. Identification and analysis of transcriptional regulatory signals for the *kil* and *kor* loci of *Streptomyces* plasmid pIJ101. *J. Bacteriol.* **171:**5768–5775.

357a. **Steinmetz, M., and R. Richter.** 1994. Easy cloning of mini-Tn*10* insertions from the *Bacillus subtilis* chromosome. *J. Bacteriol.* **176:**1761–1763.

358. **Stibitz, S., W. Black, and S. Falkow.** 1986. The construction of a cloning vector designed for gene replacement in *Bordetella pertussis. Gene* **50:**133–140.

359. **Sturino, J. M., and T. R. Klaenhammer.** 2004. Bacteriophage defense systems and strategies for lactic acid bacteria. *Adv. Appl. Microbiol.* **56:**331–378.

360. **Stuttard, C.** 1979. Transduction of auxotrophic markers in a chloramphenicol-producing strain of *Streptomyces. J. Gen. Microbiol.* **110:**479–482.

361. **Sun, J. H., G. H. Kelemen, J. M. Fernandez-Abalos, and M. J. Bibb.** 1999. Green fluorescent protein as a reporter for spatial and temporal gene expression in *Streptomyces coelicolor* A3(2). *Microbiology* **145:**2221–2227.

362. **Sung, C. K., H. Li, J. P. Claverys, and D. A. Morrison.** 2001. An *rpsL* cassette, *janus,* for gene replacement through negative selection in *Streptococcus pneumoniae. Appl. Environ. Microbiol.* **67:**5190–5196.

363. **Suwa, M., H. Sugino, A. Sasaoka, E. Mori, H. Fujii, H. Shinkawa, O. Nimi, and H. Kinashi.** 2000. Identification of two polyketide synthase gene clusters on the linear plasmid pSLA2-L in *Streptomyces rochei. Gene* **246:** 123–131.

364. **Takala, T. M., and P. E. Saris.** 2002. A food-grade cloning vector for lactic acid bacteria based on the nisin immunity gene *nisI. Appl. Microbiol. Biotechnol.* **59:**467–471.

365. **Takala, T. M., P. E. Saris, and S. S. Tynkkynen.** 2003. Food-grade host/vector expression system for *Lactobacillus casei* based on complementation of plasmid-associated phospho-beta-galactosidase gene *lacG. Appl. Microbiol. Biotechnol.* **60:**564–570.

365a. **Tam, C., E. M. Glass, D. M. Anderson, and D. Missiakas.** 2006. Transposon mutagenesis of *Bacillus anthracis* strain Sterne using Bursa aurealis. *Plasmid* **56:**74–77.

366. **Tanimoto, K., and Y. Ike.** 2002. Analysis of the conjugal transfer system of the pheromone-independent highly transferable *Enterococcus* plasmid pMG1: identification of a tra gene (*traA*) up-regulated during conjugation. *J. Bacteriol.* **184:**5800–5804.

367. **Terzaghi, B. E., and W. E. Sandine.** 1975. Improved medium for lactic streptococci and their bacteriophages. *Appl. Environ. Microbiol.* **29:**807–813.

367a. **Thomas, D. J., J. A. Morgan, J. M. Whipps, and J. R. Saunders.** 2000. Plasmid transfer between the *Bacillus thuringiensis* subspecies *kurstaki* and *tenebrionis* in laboratory culture and soil and in lepidopteran and coleopteran larvae. *Appl. Environ. Microbiol.* **66:**118–124.

368. **Thompson, A., and M. J. Gasson.** 2001. Location effects of a reporter gene on expression levels and on native protein synthesis in *Lactococcus lactis* and *Saccharomyces cerevisiae*. *Appl. Environ. Microbiol.* **67:**3434–3439.

369. **Thompson, C. J., J. M. Ward, and D. A. Hopwood.** 1980. DNA cloning in *Streptomyces*: resistance genes from antibiotic producing species. *Nature* **286:**525–527.

370. **Thompson, C. J., J. M. Ward, and D. A. Hopwood.** 1982. Cloning of antibiotic resistance and nutritional genes in *Streptomyces*. *J. Bacteriol.* **151:**668–677.

371. **Thompson, J. K., and M. A. Collins.** 1988. Evidence for the conjugal transfer of the broad host range plasmid pIP501 into strains of *Lactobacillus helveticus*. *J. Appl. Bacteriol.* **65:**309–319.

372. **Tieleman, D. P.** 2004. The molecular basis of electroporation. *BMC Biochem.* **5:**3.

373. **Tomasz, A.** 1965. Control of the competent state in *Pneumococcus* by a hormone-like cell product: an example for a new type of regulatory mechanism in bacteria. *Nature* **208:**155–159.

374. **Tomich, P. K., F. Y. An, and D. B. Clewell.** 1980. Properties of erythromycin-inducible transposon Tn917 in *Streptococcus faecalis*. *J. Bacteriol.* **141:**1366–1374.

375. **Tomita, H., K. Tanimoto, S. Hayakawa, K. Morinaga, K. Ezaki, H. Oshima, and Y. Ike.** 2003. Highly conjugative pMG1-like plasmids carrying Tn1546-like transposons that encode vancomycin resistance in *Enterococcus faecium*. *J. Bacteriol.* **185:**7024–7028.

376. **Tortosa, P., and D. Dubnau.** 1999. Competence for transformation: a matter of taste. *Curr. Opin. Microbiol.* **2:**588–592.

377. **Trieu-Cuot, P., C. Carlier, C. Poyart-Salmeron, and P. Courvalin.** 1990. A pair of mobilizable shuttle vectors conferring resistance to spectinomycin for molecular cloning in *Escherichia coli* and in gram-positive bacteria. *Nucleic Acids Res.* **18:**4296.

378. **Trieu-Cuot, P., C. Carlier, C. Poyart-Salmeron, and P. Courvalin.** 1991. Shuttle vectors containing a multiple cloning site and a *lacZ* alpha gene for conjugal transfer of DNA from *Escherichia coli* to gram-positive bacteria. *Gene* **102:**99–104.

379. **Trotter, K. M., and G. M. Dunny.** 1990. Mutants of *Enterococcus faecalis* deficient as recipients in mating with donors carrying pheromone-inducible plasmids. *Plasmid* **24:**57–67.

380. **Tyagi, A. K., S. K. Das Gupta, and S. Jain.** 2000. Gene expression: reporter technologies, p. 131–148. *In* G. F. Hatfull and W. R. Jacobs, Jr. (ed.), *Molecular Genetics of Mycobacteria*. ASM Press, Washington, DC.

381. **Tyurin, M., L. Starodubtseva, H. Kudryavtseva, T. Voeykova, and V. Livshits.** 1995. Electrotransformation of germinating spores of *Streptomyces* spp. *Biotechnol. Tech.* **9:**737–740.

382. **Uchiyama, H., and B. Weisblum.** 1985. N-methyl transferase of *Streptomyces erythraeus* that confers resistance to the macrolide-lincosamide-streptogramin B antibiotics: amino acid sequence and its homology to cognate R-factor enzymes from pathogenic bacilli and cocci. *Gene* **38:**103–110.

383. **Vagner, V., E. Dervyn, and S. D. Ehrlich.** 1998. A vector for systematic gene inactivation in *Bacillus subtilis*. *Microbiology* **144:**3097–3104.

383a. **Vandeyar, M. A., and S. A. Zahler.** 1986. Chromosomal insertions of Tn917 in *Bacillus subtilis*. *J. Bacteriol.* **167:**530–534.

384. **van Wezel, G. P., and M. J. Bibb.** 1996. A novel plasmid vector that uses the glucose kinase gene (*glkA*) for the positive selection of stable gene disruptants in *Streptomyces*. *Gene* **182:**229–230.

385. **van Wezel, G. P., J. White, G. Hoogvliet, and M. J. Bibb.** 2000. Application of *redD*, the transcriptional activator gene of the undecylprodigiosin biosynthetic pathway, as a reporter for transcriptional activity in *Streptomyces coelicolor* A3(2) and *Streptomyces lividans*. *J. Mol. Microbiol. Biotechnol.* **2:**551–556.

386. **Vara, J., M. Lewandowskaskarbek, Y. G. Wang, S. Donadio, and C. R. Hutchinson.** 1989. Cloning of genes governing the deoxysugar portion of the erythromycin biosynthesis pathway in *Saccharopolyspora erythraea* (*Streptomyces erythreus*). *J. Bacteriol.* **171:**5872–5881.

387. **Voeykova, T., L. Emelyanova, V. Tabakov, and N. Mkrtumyan.** 1998. Transfer of plasmid pTO1 from *Escherichia coli* to various representatives of the order Actinomycetales by intergeneric conjugation. *FEMS Microbiol. Lett.* **162:**47–52.

388. **Volff, J. N., and J. Altenbuchner.** 1997. High frequency transposition of the Tn5 derivative Tn5493 in *Streptomyces lividans*. *Gene* **194:**81–86.

389. **Vujaklija, D., K. Ueda, S. K. Hong, T. Beppu, and S. Horinouchi.** 1991. Identification of an a-factor-dependent promoter in the streptomycin biosynthetic gene cluster of *Streptomyces griseus*. *Mol. Gen. Genet.* **229:**119–128.

390. **Vybiral, D., M. Takac, M. Loessner, A. Witte, U. von Ahsen, and U. Blasi.** 2003. Complete nucleotide sequence and molecular characterization of two lytic *Staphylococcus aureus* phages: 44AHJD and P68. *FEMS Microbiol. Lett.* **219:**275–283.

391. **Wach, A.** 1996. PCR-synthesis of marker cassettes with long flanking homology regions for gene disruptions in *S. cerevisiae*. *Yeast* **12:**259–265.

392. **Walker, S. A., and T. R. Klaenhammer.** 2000. An explosive antisense RNA strategy for inhibition of a lactococcal bacteriophage. *Appl. Environ. Microbiol.* **66:**310–319.

393. **Wang, H., A. P. Roberts, and P. Mullany.** 2000. DNA sequence of the insertional hot spot of Tn916 in the *Clostridium difficile* genome and discovery of a Tn916-like element in an environmental isolate integrated in the same hot spot. *FEMS Microbiol. Lett.* **192:**15–20.

394. **Wang, J., and K. M. Derbyshire.** 2004. Plasmid DNA transfer in *Mycobacterium smegmatis* involves novel DNA rearrangements in the recipient, which can be exploited for molecular genetic studies. *Mol. Microbiol.* **53:**1233–1241.

395. **Wang, J., L. M. Parsons, and K. M. Derbyshire.** 2003. Unconventional conjugal DNA transfer in mycobacteria. *Nat. Genet.* **34:**80–84.

396. **Wang, P. Z., S. J. Projan, K. R. Leason, and R. P. Novick.** 1987. Translational fusion with a secretory enzyme as an indicator. *J. Bacteriol.* **169:**3082–3087.

397. **Ward, J. M., G. R. Janssen, T. Kieser, M. J. Bibb, and M. J. Buttner.** 1986. Construction and characterization of a series of multi-copy promoter-probe plasmid vectors for *Streptomyces* using the aminoglycoside phosphotransferase gene from Tn5 as indicator. *Mol. Gen. Genet.* **203:**468–478.

398. **Weaver, K. E., and D. B. Clewell.** 1987. Transposon Tn917 delivery vectors for mutagenesis in *Streptococcus faecalis*, p. 17–21. *In* J. J. Ferretti and R. I. Curtiss (ed.), *Streptococcal Genetics*. ASM Press, Washington, DC.

399. **Weaver, K. E., and D. B. Clewell.** 1988. Regulation of the pAD1 sex pheromone response in *Enterococcus faecalis*: construction and characterization of *lacZ* transcriptional fusions in a key control region of the plasmid. *J. Bacteriol.* **170:**4343–4352.

400. **Weaver, K. E., L. B. Rice, and G. Churchward.** 2002. Plasmids and transposons, p. 219–263. *In* M. S. Gilmore, D. B. Clewell, P. Courvalin, G. M. Dunny, B. E. Murray, and L. B. Rice (ed.), *The Enterococci: Pathogenesis, Molec-*

ular Biology, and Antibiotic Resistance. ASM Press, Washington, DC.

401. **Weeks, C. R., and J. J. Ferretti.** 1984. The gene for type A streptococcal exotoxin (erythrogenic toxin) is located in bacteriophage T12. *Infect. Immun.* **46:**531–536.

402. **Weigel, L. M., D. B. Clewell, S. R. Gill, N. C. Clark, L. K. McDougal, S. E. Flannagan, J. F. Kolonay, J. Shetty, G. E. Killgore, and F. C. Tenover.** 2003. Genetic analysis of a high-level vancomycin-resistant isolate of *Staphylococcus aureus. Science* **302:**1569–1571.

403. **Wessels, M. R., A. E. Moses, J. B. Goldberg, and T. J. DiCesare.** 1991. Hyaluronic acid capsule is a virulence factor for mucoid group A streptococci. *Proc. Natl. Acad. Sci. USA* **88:**8317–8321.

404. **Wieland, K. P., B. Wieland, and F. Gotz.** 1995. A promoter-screening plasmid and xylose-inducible, glucose-repressible expression vectors for *Staphylococcus carnosus. Gene* **158:**91–96.

405. **Wirth, R., F. An, and D. B. Clewell.** 1987. Highly efficient cloning system for *Streptococcus faecalis:* protoplast transformation, shuttle vectors, and applications, p. 25–27. *In* J. J. Ferretti and R. I. Curtiss (ed.), *Streptococcal Genetics.* American Society for Microbiology, Washington, DC.

406. **Wirth, R., F. Y. An, and D. B. Clewell.** 1986. Highly efficient protoplast transformation system for *Streptococcus faecalis* and a new *Escherichia coli-S. faecalis* shuttle vector. *J. Bacteriol.* **165:**831–836.

407. **Wohlleben, W., W. Arnold, L. Bissonnette, A. Pelletier, A. Tanguay, P. H. Roy, G. C. Gamboa, G. F. Barry, E. Aubert, J. Davies, and S. A. Kagan.** 1989. On the evolution of Tn*21*-like multi-resistance transposons: sequence analysis of the gene (*aacC1*) for gentamicin acetyltransferase-3-I(AAC(3)-I), another member of the Tn*21*-based expression cassette. *Mol. Gen. Genet.* **217:**202–208.

408. **Wong, W. Y., P. Su, G. E. Allison, C. Q. Liu, and N. W. Dunn.** 2003. A potential food-grade cloning vec-tor for *Streptococcus thermophilus* that uses cadmium resistance as the selectable marker. *Appl. Environ. Microbiol.* **69:**5767–5771.

409. **Wu, S. W., H. de Lencastre, and A. Tomasz.** 1999. The *Staphylococcus aureus* transposon Tn*551*: complete nucleotide sequence and transcriptional analysis of the expression of the erythromycin resistance gene. *Microb. Drug Resist.* **5:**1–7.

410. **Yansura, D. G., and D. J. Henner.** 1984. Use of the *Escherichia coli lac* repressor and operator to control gene expression in *Bacillus subtilis. Proc. Natl. Acad. Sci. USA* **81:**439–443.

411. **Yother, J.** 2000. Genetics of *Streptococcus pneumoniae*, p. 232–243. *In* V. A. Fischetti, R. P. Novick, J. J. Ferretti, D. A. Portnoy, and J. I. Rood (ed.), *Gram-Positive Pathogens.* ASM Press, Washington, DC.

412. **Yother, J., L. S. McDaniel, and D. E. Briles.** 1986. Transformation of encapsulated *Streptococcus pneumoniae. J. Bacteriol.* **168:**1463–1465.

412a. **Youngman, P. J., J. B. Perkins, and R. Losick.** 1983. Genetic transposition and insertional mutagenesis in *Bacillus subtilis* with *Streptococcus faecalis* transposon Tn*917. Proc. Natl. Acad. Sci. USA* **80:**2305–2309.

412b. **Youngman, P., J. B. Perkins, and R. Losick.** 1984. Construction of a cloning site near one end of Tn*917* into which foreign DNA may be inserted without affecting transposition in *Bacillus subtilis* or expression of the transposon-borne *erm* gene. *Plasmid* **12:**1–9.

413. **Zalacain, M., A. Gonzalez, M. C. Guerrero, R. J. Mattaliano, F. Malpartida, and A. Jimenez.** 1986. Nucleotide sequence of the hygromycin B phosphotransferase gene from *Streptomyces hygroscopicus. Nucleic Acids Res.* **14:**1565–1581.

414. **Zhang, Y. M., F. Buchholz, J. P. P. Muyrers, and A. F. Stewart.** 1998. A new logic for DNA engineering using recombination in *Escherichia coli. Nat. Genet.* **20:**123–128.

33

Genetics of Archaea

KEVIN R. SOWERS, PAUL H. BLUM, AND SHILADITYA DASSARMA

The Archaea represent a phylogenetic lineage of microorganisms that are distinct from both the Bacteria and Eukarya. Although this phylum of microorganisms exhibits a wide range of morphological and physiological diversity, which includes sulfur-metabolizing hyperthermophiles, methanogens, and extreme halophiles, all Archaea have one characteristic in common: the requirement for growth conditions that have periodically redefined the limits for extant life as we know it. These include saline concentrations up to saturation, highly reduced anoxic environments poised below −350 mV, and extremely high temperatures of up to 113°C. Numerous reports have been presented on the physiology and biochemistry of all three groups of Archaea, but several unique characteristics of this phylum have precluded the ability to apply standard bacterial or eukaryal genetic protocols. Traditional colony growth of many of the fastidious anaerobes in roll tubes is not as practical as plating for screening large numbers of clones; unique cell wall structures prevent the use of commonly used antibiotic genetic markers that target cell wall synthesis; bacterial gene promoters associated with many of the commonly used genetic markers are not recognized by the archaeal transcriptional apparatus; bacterial plasmids and phages will not replicate in archaeal species. In recent years several laboratories have overcome these difficulties by developing effective plating techniques, identifying genetic markers that do not target cell wall synthesis, fusing archaeal promoters with recombinant genes, and isolating native vectors and promiscuous nonnative vectors.

This chapter does not attempt to describe historical aspects or gene transfer systems under development, but focuses instead on tractable systems that are currently available for the Archaea. Because of fundamental differences between gene transfer systems for each archaeal branch, the chapter is divided into three inclusive sections covering the halophilic and methanogenic Euryarchaeota and the hyperthermophilic Crenarchaeota. All three Archaea require methodologies that are specialized to varying degrees. The halophilic Euryarchaeota require only media with high concentrations of salt, utilizing otherwise standard protocols for mesophilic, aerobic bacteria. The hyperthermophilic Crenarchaeota require incubators adapted for higher temperatures (80°C) and sealed containers to prevent evaporation of medium, but otherwise utilize standard protocols for aerobic bacteria. Additional information on methods for growing halophilic Euryarchaeota and hyperthermophilic Crenarchaeota can be found elsewhere (31, 95). The methanogenic Euryarchaeota require specialized apparatus and methods to maintain anaerobic media at a low redox potential and prevent exposure of cultures to oxygen. Because details of these methods are beyond the scope of this chapter, the reader should refer elsewhere for a comprehensive description of the growth of methanogenic Euryarchaeota before beginning genetic experiments with this system (108). Despite varying degrees of difficulty growing Archaea, all three systems are routinely used by laboratories conducting research on archaeal genetics and can be mastered by anyone with a fundamental knowledge of microbial genetic techniques. For reviews on archaeal genetics see references 2, 69, 98, and 110.

33.1. HALOPHILIC EURYARCHAEOTA

33.1.1. Introduction

Halophilic Archaea grow optimally in hypersaline environments containing 2 to 5 M NaCl, usually under aerobic conditions, with normal atmospheric temperature and pressure (29, 42). In contrast to methanogens and thermophiles, both of which require specialized culturing facilities, haloarchaea can be grown in any microbiology laboratory by using standard media supplemented with 12 to 25% salt. As a result, many methods for genetic manipulation of haloarchaea have been developed and standardized during the past 20 years (see reference 31). With the recent availability of complete or nearly complete genome sequences for several haloarchaea, these microorganisms have come to represent very tractable model systems for genetic studies.

Haloarchaea are found in evaporatic brine pools in the tropical, temperate, and arctic regions of the world, usually at the surface, but also under the sea in submarine pools and underground in ancient salt deposits (29). Many hypersaline environments are dynamic with respect to temperature, pH, oxygen, sunlight, etc.; therefore, haloarchaea display a wide variety of biological responses (28, 59). To balance the high sodium chloride concentration in the medium, haloarchaea accumulate potassium chloride as a compatible solute, and all of their metabolic activities take place in a hypersaline cytoplasm. Haloarchaea also exhibit a high level of resistance to UV light and γ-radiation as protection against the intense solar radiation usually found in their environment. Many species are highly motile, displaying phototaxis, chemotaxis, and gas-vesicle-mediated flotation. Some are capable of phototrophic growth, and most display facultative anaerobic capabilities.

All of the haloarchaea described thus far belong to the *Halobacteriaceae* family (42). Of the 19 different genera of haloarchaea that have been described, genetic and genomic analyses have been concentrated on a few species and strains. The most intensively studied are *Halobacterium* spp., which are well known for their production of purple membrane, and *Haloferax volcanii*, which can grow on minimal medium. *Halobacterium* spp. and *H. volcanii* have well-developed genetic tools, including a facile transformation system, selectable markers, cloning and expression vectors, reporter genes, and gene replacement and knockout systems (31).

Genome sequences are available for some of the common laboratory strains of haloarchaea. The first complete genome sequence became available in 2000 for *Halobacterium* sp. NRC-1 (78, 80). Several additional genome sequences are under investigation (Table 1), including *H. volcanii* and *Haloarcula marismortui* and also *Halorubrum lacusprofundi*, *Natronomonas pharaonensis*, and *Haloquadratum walsbyi*. The genomes of haloarchaea are relatively G+C-rich (58 to 68%), with the exception of some large and small extrachromosomal species that are rich in insertion sequence (IS) elements. These extrachromosomal species have been studied in *Halobacterium* spp. and shown to rearrange at high frequency. This is in contrast to the *Halobacterium* chromosome, which is quite stable and in fact contains a smaller number of transposable IS elements than some other archaea, e.g., *Sulfolobus solfataricus* (28).

The combination of easy culturing, highly developed genetic tools, and interesting biology makes haloarchaeal systems attractive for study. Past studies have been extremely fruitful, providing a deeper understanding of prokaryotic genome structure and dynamics, lateral gene transfer, genetic regulatory mechanisms, and DNA replication, transcription, and translation. With the recent availability of complete genome sequences, systematic knockout strategies, DNA arrays, and proteomics, this system has the potential to provide tremendous opportunities for advancing the biology of haloarchaea.

TABLE 1 Archaeal genomic sequences

Archaeal strains	Website	Reference
Halophilic Euryarchaeota		
Halobacterium sp. NRC-1	http://zdna2.umbi.umd.edu/	80
Haloferax volcanii DS2	http://zdna2.umbi.umd.edu/~haloweb/hvo.html	Unpublished
Haloarcula marismortui	http://zdna2.umbi.umd.edu/~haloweb/hma.html	Unpublished
Methanogenic Euryarchaeota		
Methanococcus maripaludis LL	http://www.genome.washington.edu/UWGC/methanococus/Methanococcus.html	50
Methanosarcina acetivorans C2A	http://www.broad.mit.edu/annotation/microbes/methanosarcina/	39
Methanosarcina barkeri Fusaro	http://genome.jgi-psf.org/finished_microbes/metba/metba.home.html	64a
Hyperthermophilic Crenarchaeota		
Sulfolobus solfataricus P2	http://www-archbac.u-psud.fr/projects/sulfolobus/	103

33.1.2. Strain Characteristics

33.1.2.1. *Halobacterium* sp. NRC-1

Halobacterium sp. strain NRC-1 (ATCC 700922) is a common, rod-shaped strain which grows well in complex media containing 4.5 M NaCl with peptone, yeast extract, or Casamino Acids as carbon and energy sources (Color Plate 13). Optimal growth (6-h generation time) is observed at 42°C. Growth on defined medium containing 15 amino acids and vitamins has also been reported, and cell lysis is minimized in the presence of trace quantities of metal ions. Under low oxygen tension, *Halobacterium* sp. NRC-1 induces purple membrane patches in the cell membrane and buoyant gas vesicles intracellularly, which increases the availability of light and oxygen and allows a period of light-driven proton pumping and phototrophic growth (13).

One of the most useful mutant strains of NRC-1 has a deletion of the *ura3* gene (NRC-1 [Δ*ura3*]) and is auxotrophic for uracil and 5-fluoroorotic acid (5-FOA) resistant (Ura⁻, Foaʳ) (86). The NRC-1 (Δ*ura3*) strain is used for generating knockouts using either selection for mevinolin resistance (Mevʳ) and counterselection against *ura3* (Foaʳ) or both selection and counterselection for *ura3* (62, 86). Several single and multiple gene mutants have been successfully constructed by using these procedures (e.g., Δ*brp*, Δ*blh*, Δ*brp/blh*, and Δ*crtY*, coding for enzymes of retinal biosynthesis, and Δ*arsADRC*, Δ*arsB*, and Δ*arsM*, coding for heavy metal resistance), indicating that knockout of any nonessential gene is possible (87, 88). Numerous studies have been conducted on the purple membrane protein, bacteriorhodopsin (BR), and the *bop* gene. The commonly used BR overproducer (Pum⁺⁺⁺) strain S9 is not a derivative of or isogenic with NRC-1 (13). It was isolated from another natural *Halobacterium* strain by extensive chemical mutagenesis. S9 likely contains dozens of mutations and allelic differences from NRC-1. The mutation responsible for the Pum⁺⁺⁺ (deep purple) phenotype of S9 has been shown to be a double frameshift in the *bat* gene ("*bat*" allele), which results in a 4-amino-acid change in the encoded protein. Mutants of S9 lacking BR, e.g., SD23, which is purple membrane and bacterioruberin deficient (Pum⁻, Rub⁻) as a result of an ISH1 insertion in *bat*, and SD20, which is Pum⁻ as a result of an ISH1 insertion in *bop*, have been characterized extensively (13, 27). Other strains lacking BR have also been isolated and characterized (e.g., R1mR and L33) (27).

Numerous genetic studies have also been conducted on *gvp* genes with a gas vesicle-deficient (Vac⁻) NRC-1 strain deleted for the pNRC100 *gvp* gene cluster (30, 77). The most common strain, SD109 (Color Plate 13), has a 67-kbp deletion of pNRC100 resulting from intramolecular transposition of ISH8. A second strain, SD112, has a 59-kbp deletion resulting from an ISH2 transposition.

33.1.2.2. *H. volcanii*

H. volcanii DS2 (ATCC29605), isolated from the Dead Sea, is disk or cup shaped and grows optimally at 45°C in 1.5 to 2.5 M NaCl. It is tolerant to high concentrations of Mg²⁺, up to 1.5 M, reflecting the composition of its natural environment. *H. volcanii* can grow in minimal medium on glucose as sole carbon and energy source. However, the growth rate is slow compared with rich medium, where the generation time is 4 h. A derivative of DS2, *H. volcanii* WFD11, is cured of the natural cryptic miniplasmid pHV2 (63). Several pHV2 derivatives are popular cloning and shuttle vectors. An *H. volcanii* Δ*pyrE1*Δ*pyrE2* mutant, WR479, can be used for gene knockout with a method similar to that developed for *Halobacterium* sp. NRC-1 (16).

Genotypic characteristics for some haloarchaeal strains are summarized in Table 2.

33.1.3. Genetic Markers for Selection

One of the most commonly used selective markers for the haloarchaea confers resistance to mevinolin (62). Mevinolin is an inhibitor of 3-hydroxy-3-methylglutaryl coenzyme A reductase, an enzyme involved in the production of sterols in eukaryotes and isoprenoid side chains of lipids in archaea. Selection of mevinolin-resistant (Mevʳ) mutants of *H. volcanii* resulted in the finding of up-promoter mutations, an allele that has been used for selection in *H. volcanii*, *Halobacterium* spp., and other haloarchaea. The background mutation rate to Mevʳ is relatively low ($<10^{-7}$), making this an excellent selection system for the haloarchaea.

Selectable markers for halophilic archaea allowing both selection and counterselection have been developed recently. These are the *Halobacterium* sp. NRC-1 *ura3* gene, encoding orotidine-5'-monophosphate decarboxylase, and *H. volcanii pyrE2* genes, coding for orotate phosphoribosyltransferase (16, 86). These markers are extremely useful for both gene replacements and gene knockouts. Forward selection can be carried out in a Δ*ura3* or Δ*pyrE2* background by selection for uracil prototrophy (Ura⁺), using uracil-dropout medium for *Halobacterium* sp. and Casamino Acids for *H. volcanii*. Counterselection can be carried out using resistance to 5-FOA (Foaʳ), which is metabolized to a toxic intermediate in a Ura⁺ background.

A variety of other selectable markers have also been described. The *gyrB* gene allele has been used for selection of

TABLE 2 Archaeal strain characteristics and sources[a]

Archaeal strains	Characteristics	Reference
Halophilic Euryarchaeota		
Halobacterium sp. NRC-1	Wild-type ATCC 700922, JCM 11081	80
Halobacterium sp. SK400/MPK414	NRC-1 Δ*ura3*	123
Halobacterium sp. S9	Purple membrane overproducer	13, 113
Halobacterium sp. R1	Gas-vesicle-deficient mutant ATCC 19700	113
Haloferax volcanii DS2	Wild-type ATCC 29605, DSMZ 3757, JCM 8879	24
Halobacterium sp. SD20	Purple-membrane-deficient mutant of S9 *bat*::ISH1	9
Halobacterium sp. SD23	Purple-membrane-deficient mutant of S9 *bop*::ISH1	9
Halobacterium sp. SD109	NRC-1 with 67-kb deletion of *gvp* gene cluster	61
Halobacterium sp. SD112	NRC-1 with 59-kb deletion of *gvp* gene cluster	61
Halobacterium sp. R1mR	Purple membrane mutant of R1 *bop*::ISH2	24
Halobacterium sp. L33	Purple membrane mutant of S9 *bop*::ISH2	24
Haloferax volcanii WFD11	DS2 cured of pHV2	63
Haloferax volcanii WR479	WR341 Δ*pyrE1*Δ*pyrE2*	16
Methanogenic Euryarchaeota		
Methanococcus maripaludis JJ	Wild-type ATCC 43000, DSMZ 2067, JCM 10722, OCM 175	55
Methanosarcina acetivorans C2A	Wild-type ATCC 35395, DSMZ 2834, JCM 12185, OCM 95	105
Methanosarcina barkeri Fusaro	Wild-type ATCC 29787, DSMZ 804, OCM 83	58
Methanosarcina mazei Gö1	Wild-type ATCC BAA-159, DSMZ 3647, JCM 11833, OCM 88	57
Hyperthermophilic Crenarchaeota		
Sulfolobus solfataricus 98/2	Wild type	96
Sulfolobus solfataricus PBL2025	98/2 ΔSSO3004–SSO3050, *lacS* deletion	101
Sulfolobus solfataricus PBL2030	98/2 ΔSSO3013-SSO3037, ΔSSO3051, *malA* deletion	Hoang and Blum, unpublished
Sulfolobus solfataricus Gtheta	Wild-type derivative of MT3	23
Sulfolobus solfataricus P2	Wild-type ATCC 35092, DSMZ 1617	130
Sulfolobus solfataricus P1	Wild-type ATCC 35091, DSMZ 1616	130
Sulfolobus acidocaldarius	Wild-type ATCC 33909	22

[a]ATCC, American Type Culture Collection (http://www.atcc.org/); DSMZ, Deutsche Sammlung von Mikroorganismen und Zellkulturen GmbH (http://www.dsmz.de/); JCM, Japan Collection of Microorganisms (http://www.jcm.riken.go.jp/); OCM, Oregon Collection of Methanogens (http://methanogens.pdx.edu/).

resistance to the DNA gyrase inhibitor novobiocin (53). Several protein synthesis inhibitors, including anisomysin, thiostrepton, and chloramphenicol, have also been used as selectable markers. These are particularly useful for gene replacements of rRNA genes, especially for *Halobacterium* spp., which have a single copy of the rRNA operon (65).

33.1.4. Vectors for Gene Transfer

33.1.4.1. Shuttle and Cloning Vectors

Halophilic archaea are rich in plasmid diversity, and many recombinant vectors have been constructed (31). Some of the most common vectors are derivatives of *H. volcanii* miniplasmid pHV2 and *Halobacterium* spp. minichromosomes or megaplasmids, such as pNRC100 and pHH1 or miniplasmids pGRB and pHSB. In general, these vectors replicate in all commonly used strains of halophilic archaea tested.

Vectors derived from the *H. volcanii* miniplasmid pHV2 are in wide use and can be selected and maintained in *Haloferax* spp., *Halobacterium* spp., or *Escherichia coli*. These vectors were constructed by first cloning the *hmg* gene, encoding resistance to the 3-hydroxy-3-methylglutaryl coenzyme A reductase inhibitor mevinolin, into the pHV2 replication region to form pWL2. The resistance determinant, portions of pHV2, and an ampicillin- and tetracycline-resistance-conferring pBR322 derivative, pAT153, were ligated together to form the shuttle vector pWL102 (63). This plasmid was reduced in size by minimizing the mevinolin resistance region, and a multiple cloning region with unique ApaI, EcoRI, SmaI, SacI, SphI, KpnI, XbaI, HindIII, and NotI sites was inserted to form the 8.8-kbp shuttle vector pWL104.

A *Halobacterium* pNRC100-derived vector, pNG168, 8.9 kbp (Fig. 1), contains the pTZ18 vector, including the Amp[r] determinant for selection in *E. coli*; multiple cloning region with unique ApaI, HindIII, EcoRI, PstI, SmaI, BamHI, SpeI, and NoI restriction sites; the pNRC100 minimal replicon pNG101; and the minimal *H. volcanii* Mev[r] marker for selection in halophiles. pNG168 is capable of replication in both *Halobacterium* sp. NRC-1 and *H. volcanii* and has been used for a variety of genetic experiments (30, 79).

A vector similar to pNG168, pUBP2, is somewhat larger (12.3 kbp) and was constructed from the related *Halobacterium* sp. PHH1 plasmid pHH1 (18). This vector contains the *E. coli* vector pIBI, with the *H. volcanii hmg* marker for Mev[r] selection in halophiles, and pHH9, a natural deletion derivative of pHH1. Several unique restriction sites are available for cloning in this vector.

Three related *Halobacterium* spp. miniplasmids, pGRB1, pHSB1, and pHGN1, which replicate by a single-stranded intermediate, have been used for vector construction by inserting the Mev[r] marker. Another vector, pMSD20, was

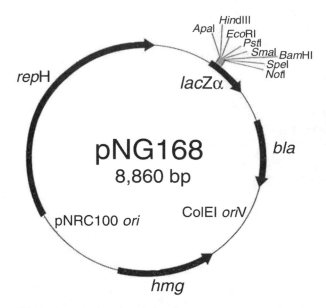

FIGURE 1 Haloarchaeal shuttle vector pNG168. This plasmid contains the *E. coli* pTZ19r replicon and the *Halobacterium* NRC-1 pNRC100 minimal replication region. The *bla* gene provides selection with ampicillin in *E. coli*, and the *hmg* gene provides selection with mevinolin in haloarchaea. The multiple cloning site is located in the *lacZα* fragment gene and permits blue-white screening in *E. coli*. The plasmid is available from ATCC (catalog no. MBA-77) and the sequence is available in GenBank (accession no. AY291460).

constructed from cryptic *Haloferax* sp. plasmid pHK2 by cloning the novobiocin resistance (Nov^r) *gyrB* gene from *H. volcanii*. This plasmid was used to transform *H. volcanii* at high frequency.

33.1.4.2. Gene Replacements and Knockouts

An exciting recent development is a new gene replacement and knockout method in the sequenced strain *Halobacterium* sp. NRC-1 using the *ura3* gene, which can be both selected for and counterselected against (86). This approach is diagrammed in Fig. 2. A gene allele of interest (e.g., a precise deletion) is first cloned into an *E. coli* plasmid, and the suicide plasmid is introduced into the haloarchaeon host by transformation. Integrants are selected by uracil prototrophy (Ura^+, using commercially available uracil-dropout media components available from Sigma-Aldrich Corp., St. Louis, MO). Alternatively, mevinolin resistance (Mev^r) may also be used. Subsequently, excisants are selected by counterselecting for 5-FOA resistance (Foa^r), giving rise to both the replaced allele and the original, which can be distinguished by PCR or phenotypic analysis. A similar approach has proved successful for *H. volcanii*, but substituting the *pyrE2* gene (in place of *ura3*).

In the past, gene deletions and replacements were constructed, but they required extensive manual screening. The largest number of mutants was characterized for the purple membrane *bop* gene, employing the mevinolin resistance marker (e.g., reference 60). Another interesting application of gene replacement was for the single rRNA operon of *Halobacterium* spp., where dominant mutations resulting in antibiotic resistance could be selected (65).

33.1.4.3. Insertion and Transposon Mutagenesis

Insertion mutagenesis is conducted by employing natural IS element insertions, gene cassettes and oligonucleotide linkers, and transposons constructed with haloarchaeal IS elements. Natural phenotypic variants were used in early studies to identify *bop* and *gvp* regulatory genes and coordinately regulated genes. IS element insertions identified the *brp* and the *bat* genes necessary for purple membrane production upstream of *bop* (27). Similarly, IS element insertions identified *gvpD*, *gvpE*, and other genes necessary for wild-type gas vesicle synthesis (30). More recently, IS element insertions have been found in the *ura3* gene in Foa^r mutants (86). These studies represent a valuable natural transposon mutagenesis system in haloarchaea that has provided significant insights into gene regulation.

Gene disruptions made by using gene cassettes and oligonucleotide linkers are also useful for genetic analysis. To study the functions of the *gvp* gene cluster on pNRC100 in gas vesicle formation, a kanamycin resistance (κ) gene cassette was used for scanning mutagenesis of the gene cluster in pFL2, a 24.5-kb *E. coli*-*Halobacterium* shuttle plasmid (30). Transformation of *Halobacterium* sp. SD109, deleted for the entire *gvp* gene cluster, with pFL2 and mutated pFL2 derivatives showed that, while the unmutated gene cluster successfully programmed gas vesicle formation (Color Plate 13), derivatives with insertion of the κ cassette in nearly all of the *gvp* genes lacked normal gas vesicles. In most cases, the block in gas vesicle synthesis did not result from polar effects, since similar results were obtained for derivatives of the insertion mutants in which most of the internal portion of the κ cassette was deleted and only small (15- to 54-bp) insertions remained.

Similarly, one pNRC100 minireplicon, pNG11Δ12, was analyzed by linker-scanning mutagenesis by using a short oligonucleotide (79). Insertions in the *repH* gene knocked out the capability of the minireplicon for autonomous replication, while a second insertion at the same site, restoring the *repH* reading frame, resulted in reversion to replication proficiency.

Some effort has been directed toward construction of recombinant transposons with selectable markers from natural IS elements (33, 122). Synthetic transposons were constructed consisting of haloarchaeal ISH elements (ISH2, ISH26, or ISH28) flanking a mevinolin resistance determinant. Introduction of an ISH28-based transposon (ThD28) into *Haloarcula hispanica* on nonreplicating plasmids produced numerous non-site-specific transformants, with some bias toward low G+C content regions. However, when transposons were introduced into haloarcheal strains with perfectly homologus DNA (e.g., *H. volcanii*), plasmids integrated at high frequency by homologous recombination.

33.1.4.4. Gene Reporters

Several gene reporters are used for haloarchaea, including the *Aequorea victoria* green fluorescent protein (GFP) (83). The *gfp* gene was fused to the *bop* gene and expressed as a BR-GFP fusion in a *Halobacterium* sp. The fusion protein preserved the intrinsic function of each component under conditions with an extremely high-salt concentration and was shown to be properly localized in the plasma membrane. The results indicate that GFP can be used as a versatile reporter of gene expression in *Halobacterium* for investigations of halophilic membrane proteins.

A β-galactosidase (*bgaH*) gene from *Haloferax alicantei* has also become available as a reporter in haloarchaea (84). When introduced into a *Halobacterium* sp. on a plasmid

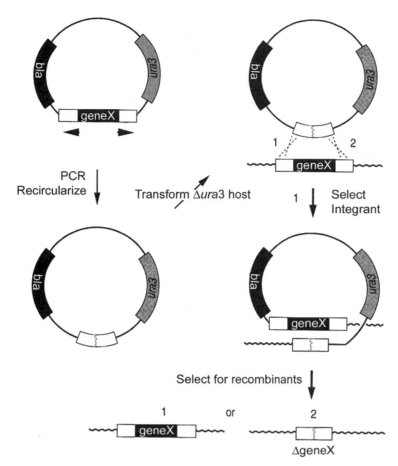

FIGURE 2 Gene knockout and replacement in the halophilic Euryarchaeota. The example shown is for selection and counterselection with *ura3*. A cloned haloarchaeal target gene (geneX) in a plasmid vector, which does not replicate in haloarchaea, is used for PCR amplification (primers designated by arrowheads) and recircularization to provide for a precisely deleted gene. The plasmid is introduced into a Δ*ura3* haloarchaeon by transformation. Integrants are selected by uracil prototrophy using uracil dropout plates. Excisants are selected for by plating on plates containing 5-FOA. Depending on the site of the recombination (1 or 2), different outcomes are possible. Alternatively, mevinolin selection can also be used for integration and the *pyrE2* gene can be used for selection and counterselection.

vector-carrying *bgaH*, the enzyme activity in cell lysates can be determined by a colorimetric assay and colonies screened for activity on plates containing X-Gal substrate. Expression of *bgaH* under the control of various halobacterial promoters of known strength leads to different specific β-galactosidase activities in the lysates. In one study, the *bgaH* gene was used as a reporter to investigate three different haloarchaeal promoter regions derived from *gvpA* (43).

33.1.4.5. Promoter Analysis and Expression Vectors

The two main promoters commonly utilized for heterologous or homologous gene expression in haloarchaea are the ferredoxin (*fdx*) promoter, mainly for soluble proteins, and the *bop* promoter, usually for membrane proteins. Both promoters have been characterized by saturation mutagenesis and up-promoter mutations have been reported (11, 12, 26). Vectors containing 200 bp upstream of the *fdx* coding region are sufficient for constitutive expression. For *bop*, only 53 bp upstream of the coding region is sufficient, but regulation is complex, and expression is highly dependent on the specific gene and construct.

Fusions of the *bop* promoter and N-terminal coding region have been used successfully for overproduction of bacteriorhodopsin mutants, halorhodopsin (the chloride pump), several sensory rhodopsins (the phototactic receptors), and two halotransducers (e.g., references 51, 56, 60, 81). Additionally, foreign membrane proteins, e.g., mammalian G-coupled protein receptors (GPCRs), such as the human muscarinic acetylcholine (M_1) and adrenergic (a2b, β_2) receptors, have been expressed in haloarchaea (14, 85). However, both proteolysis and uncharacterized determinants at the 3′ end of the *bop* gene were found to limit the usefulness of this heterologous gene expression system.

The *fdx* promoter has been used to express a variety of soluble proteins (e.g., references 56 and 89). These expression systems have sometimes been coupled to His-tagged proteins, which have been successfully purified using metal affinity chromatography in a variety of systems. However, difficulties have also been encountered in haloarchaea due to binding of endogenous proteins with metal affinity (e.g., *Halobacterium* sp. NRC-1 protein VNG2021 is a frequent contaminant).

In addition to the *fdx* and *bop* promoter vectors, a few other specialized expression systems have also been used successfully. The gas vesicle gene cluster has been modified to incorporate a cloning site in the *gvpC* gene in the pFM104D shuttle vector and used to produce fusion proteins attached to the surface of floating vesicles (114). This system has been used for antigen expression and delivery and for development and testing of vaccine candidates. Also, pWL204, a derivative of pWL102 with a *H. volcanii* tRNA^Lys gene promoter cloned in, has been used for expression of tRNA genes to study RNA processing (82).

33.1.5. Transformation

Both *Halobacterium* and *Haloferax* spp. and several other haloarchaea can be transformed using a procedure involving spheroplast formation, by chelating Mg^{2+} with EDTA, generating competent cells with PEG treatment, DNA uptake, cell regeneration, and plating (31). The routine efficiency of the procedure is about 10^5 to 10^6 transformants per microgram of common shuttle plasmid DNA. The main procedural variations for transformation of different halophiles reflect the specific ionic concentrations of the media. An alternate procedure for *Halobacterium* sp. has been reported by using freeze-thaw of cells, but the efficiency is relatively low in comparison with the EDTA-PEG procedure (128).

Most of the haloarchaeal species tested have a restriction system, which reduces the efficiency of transformation with DNA from a heterologous host by a factor of 10^3 or greater (31). The modification is evident by the inability of some restriction enzymes to digest chromosomal DNA from haloarchaea, and corresponding haloarchaeal restriction-modification activities have been reported (16, 51, 62). Purification of transforming DNA from an *E. coli* Dam^+ strain can reduce or eliminate the restriction problem.

33.1.5.1. Transformation with PEG

1. Grow a 50-ml culture of *Halobacterium* sp. NRC-1 at 42°C to an OD_{600} of 1.0 in a flask with good aeration.

2. Gently pellet cells by centrifugation at 6,000 × *g* for 10 min.

3. Gently resuspend cells in 5 ml of spheroplasting solution (2 M NaCl, 27 mM KCl, 50 mM Tris-Cl [pH 8.75], 15% [wt/vol] sucrose).

4. After 5 min, examine cells in a phase-contrast microscope to verify spheroplast formation, which is characterized by cell morphology with perfectly circular spheres.

5. Add 10 μl of DNA in spheroplasting solution or 10 mM Tris–HCl–1 mM EDTA to 200 μl of spheroplasts. Examine cell morphology under a phase-contrast microscope to check for cell lysis. Cells should appear spherical or pleomorphic, and the number of visible cells should not change substantially after addition of DNA. Cell lysis results in increased viscosity and greatly reduces the transformation efficiency.

6. Add 200 μl of PEG solution (6 ml of PEG600 in 4 ml of spheroplasting solution) and mix gently by tapping.

7. After 20 min, add 1 ml of spheroplast dilution medium (4.3 M NaCl, 27 mM KCl, 80 mM MgSO₄, 10 mM sodium citrate, 1.4 mM CaCl₂, 50 mM Tris-Cl [pH 7.4], 15% [wt/vol] sucrose, 1% [wt/vol] Oxoid Peptone) to the mixture and gently invert tubes to mix.

8. Centrifuge at 6,000 × *g* for 4 min and remove supernatant. Gently resuspend cell pellets in 1 ml of spheroplast dilution medium, transfer to sterile test tubes, and incubate the tubes with shaking at 42°C overnight.

9. After overnight incubation, add 1 ml of CM^+ medium (see below) and grow with shaking overnight.

10. Check cultures under a phase-contrast microscope for regeneration. When >80% of cells have returned to rod shape (in general, 1 to 2 days after addition of CM^+ media), inoculate transformants on selective plates by spreading and incubate at 42°C. Visible transformant colonies develop over 5 to 10 days.

NOTE. Similar procedures can also be used to transform *H. volcanii*, *H. mediterranei*, *H. hispanica*, and *H. vallismortis*, with changes in the spheroplasting and regeneration media to reflect the different ionic optima for the different species (21). Many of the details may be obtained from *Halophiles*, volume 1 of *Archaea: A Laboratory Manual* (31).

33.1.6. Media and Growth

In general, culturing of halophiles is relatively simple and conducted in the same way as for common aerobic bacteria. Liquid cultures are grown at 37 to 42°C in test tubes or Erlenmeyer-Fernbach flasks with shaking at 200 to 300 rpm. For phototrophic strains, fluorescent lighting may be used but is not essential. Plates contain 2% agar, which is required for solidifying under hypersaline conditions, and are placed in airtight containers or humidified environment to avoid crystallization of salt.

33.1.6.1. Medium Preparation

Descriptions are provided here for rich media for the three model haloarchaea, *Halobacterium* sp. NRC-1, and *H. volcanii*.

Halobacterium sp. NRC-1 CM^+ Complex Medium (per Liter)

NaCl	250 g
MgSO₄ · 7H₂O	20 g
KCl	2 g
Sodium citrate	3 g
Oxoid peptone (Oxoid L34)	10 g
Trace metal stock (2,000 ×)	0.5 ml

pH to 7.2 with NaOH.

Trace Metal Stock (2,000 ×)

FeSO₄ · 7H₂O	3.50 mg/ml
ZnSO₄ · 7H₂O	0.88 mg/ml
MnSO₄ · H₂O	0.66 mg/ml
CuSO₄ · 5H₂O	0.02 mg/ml

NOTE. Oxoid peptone should be used for optimum purple membrane production. Peptone may be substituted with 7.5 g of Casamino Acids and 10 g of yeast extract. Trace metal stock is filter sterilized and added to cooled autoclaved medium.

Halobacterium sp. NRC-1 Uracil-Dropout Medium (per Liter)

NaCl	250 g
MgSO₄ · 7H₂O	20 g
Sodium citrate	3 g
KCl	2 g
Nitrogen base (Sigma Y0626)	10 g
Dropout formula (Sigma Y1501)	1.92 g

pH 7.0 with NaOH

H. volcanii Complex Medium (per Liter)

```
NaCl ............................... 206 g
MgSO₄ · 7H₂O ...................... 37 g
KCl ................................ 3.7 g
Yeast extract ....................... 3 g
Tryptone ........................... 5 g
CaCl₂ · 2H₂O (10%) ................. 5.0 ml
MnCl₂ (75 mg/ml) .................. 1.7 ml
1 M Tris-HCl (pH 7.2) .............. 50 ml
```

NOTE.

- For agar plates, add 20 g of Bacto-Agar per liter of medium.
- Autoclave at 15 lb/in² for 20 min.
- Cool agar media to 65°C before addition of antibiotics and/or trace metals.

33.1.6.2. Storage of Strains

Cultures may be stored for weeks to several months without extreme loss of viability, although extended storage may result in accumulation of mutants from IS element transpositions and other DNA rearrangements. For long-term storage, cultures may be frozen at −70°C after addition of glycerol to 15% (vol/vol) in liquid medium. Culturing conditions and growth media for additional haloarchaeal strains may be found in *Halophiles* (31).

33.1.7. Troubleshooting

High salinity of halophile media presents some unique problems for genetic studies. Salts inhibit gelling of agar, making a relatively high concentration essential to obtain a solid surface. Salts also reduce surface tension of the medium, making removal of bubbles more difficult. Extra care must be taken to avoid formation of bubbles when pouring plates since their removal by flaming is less effective. Another problem in culturing halophiles arises from the contamination of some batches of peptone with bile acids, leading to growth inhibition and even cell lysis. Synthesis of purple membrane is especially sensitive to bile acids. For suspension of cells and transformation, the ionic strength of the medium is critical to maintain integrity of spheroplasts. Even a small amout of lysis will result in significant reduction in transformation efficiency through the release and precipitation of DNA.

The selections for transformation work well, in general, although rare cultures may contain large numbers of drug-resistant mutants and high background, resulting from Luria-Delbruck fluctuations. This is true for both mevinolin and 5-FOA resistance. The presence of homologous DNA will lead to the integration of transforming plasmids at high frequency, with a greater problem for larger regions of identity. Finally, the high mutation rate of some species and strains can lead to genetic variation among populations. Strict microbiological practice, including purification of strains and their long-term storage in frozen stocks, is essential for genetic studies.

33.1.8. Recent Developments

The genome sequences of *H. marismortui* and *N. pharaonensis* have recently been published (10, 36). Also, the number of interesting mutants is increasing, e.g., mutants of dimethyl sulfoxide reductase and its regulator *dmsR* (76); photolyase (*phr1* and *phr2*) gene mutants (9, 67) of *Halobacterium* sp. NRC-1; a *radB* mutant of *H. volcanii* (46); and rRNA operon deletions of *H. marismortui* (115). An inter-

esting application for the GFP reporter system has been to study proteolysis in *H. volcanii* (94). Finally, several microarray studies have been published for *Halobacterium* sp. NRC-1, including one for anaerobic growth and another for UV response (68, 76).

33.2. METHANOGENIC EURYARCHAEOTA

33.2.1. Introduction

The methanogenic Archaea have a significant role in the global carbon cycle as microbial catalysts in anaerobic degradative processes in habitats where O_2 is not available. The methanogens are a phylogenetically coherent group that currently includes more than 60 described species in five orders within the archaeal kingdom *Euryarchaeota*. They range from psychrophilic species from Antarctica that grow at 1.7°C to extremely thermophilic species from deep submarine vents that grow at 113°C; acidophiles from marine vents that grow at pH 5.0 to alkaliphiles from alkaline lake sediments that grow at pH 10.3; species from freshwater lake sediments that grow at saline concentrations less than 0.1 M to extreme halophiles from solar salterns that grow at nearly saturated NaCl concentrations; autotrophs that use only CO_2 for cell carbon and methylotrophs that utilize methylated carbon compounds. Despite the range and diversity of growth habitats where methanogens are found, methanogens have one common attribute: they all generate CH_4 during growth.

Two tractable genetic systems are currently available for the methanogenic Archaea within the *Methanococcales* and *Methanosarcinales*. The order *Methanococcales* includes marine autotrophs that grow exclusively by CO_2 reduction with H_2. Morphologically, these species form irregularly shaped cocci. Instead of a rigid cell wall typical of the *Bacteria*, these species form an S-layer composed of a paracrystalline array of glycoproteins and are subject to osmotic lysis at NaCl concentrations below seawater. This order includes several mesophilic species, the moderate thermophile *Methanothermococcus thermolithotrophicus*, and the extreme thermophiles *Methanotorris igneus*, *Methanocaldococcus jannaschii*, and *Methanocaldococcus infernus*. A genetic system has been developed for the mesophilic species *Methanococcus maripaludis*. Although genetic approaches were first developed with *Methanococcus voltae*, later investigators focused on *M. maripaludis* after the discovery of a native plasmid. As the genome of this species will soon be completed (http://www.genome.washington.edu/UWGC/), this has become the preferred obligate hydrogen-utilizing species for genetic research.

The order *Methanosarcinales* is the most catabolically diverse species of methanogens. In addition to growth and methanogenesis by CO_2 reduction with H_2, some species grow by the dismutation or "splitting" of acetate and by methylotrophic catabolism of methanol, methylated amines, pyruvate, and dimethyl sulfide. Species in most genera are either obligate methylotrophs or, in the case of *Methanothrix* spp., obligate acetotrophs. However, species of *Methanosarcina* can grow by all three catabolic pathways. All species have a protein S-layer cell wall and grow as irregularly shaped cocci, but in low-saline medium some species also grow as multicellular aggregates embedded in a heteropolysaccharide matrix synthesized adjacent to the S-layer.

The discovery of a native plasmid in *Methanosarcina acetivorans* led to the development of a genetic system for this metabolically diverse genus of hydrogen-utilizing, aceticlastic, and methylotrophic species. Although optimized for

M. *acetivorans*, the genetic system can be used with most species. A slightly modified transformation system has been developed for *Methanosarcina mazei* (34). Genomes for M. *acetivorans* C2A (39), M. *mazei* Gö1 (32), and *Methanosarcina barkeri* Fusaro have been completed (http://www.jgi.doe.gov). For overviews of genetic systems for the methanogenic Archaea, see Lang and Ahring (64) Sowers and Schreier (110), and Tumbula and Whitman (118).

33.2.2. Strain Characteristics

33.2.2.1. *M. acetivorans* C2A
M. *acetivorans* C2A is an irregularly shaped, coccus-shaped cell isolated from a submarine canyon; it grows in a defined marine mineral medium with acetic acid, methanol, or methylated amines as sources of energy for growth and methanogenesis (105). No exogenous growth factors are required. This species generates a protein S-layer rather than a rigid cell wall. In addition to an S-layer, several other *Methanosarcina* spp. synthesize a rigid chondroitin-like (methanochondroitin) outer layer, causing them to grow as multicellular aggregates, which would normally preclude genetic manipulation. However, when the methanochondroitin-producing species are adapted and grown at higher salt concentrations, they no longer generate methanochondroitin and grow as individual cells with only an S-layer (106). For transformation of *Methanosarcina* spp., spheroplasts are generated by suspending cells in a Mg^{2+}-free sucrose buffer, which disrupts the integrity of the S-layer (37). After transformation, the S-layer is regenerated by resuspension of the cells in medium that contains Mg^{2+}. Although a functional restriction system has not been reported in this species, a putative restriction modification has been identified in the annotated genome sequence. M. *acetivorans* contains a 5.4-kb plasmid open reading frame encoding a putative RecA replication initiation protein associated with a rolling-circle mechanism of replication and a putative site-specific recombinase (71). This plasmid has been recombinantly modified to serve as a shuttle vector for genetic transformation of most *Methanosarcina* spp (Table 3).

TABLE 3 Archaeal genetic vectors

Plasmids	Features	Reference
Halophilic Euryarchaeota		
pWL104	Haloarchaeal-*E. coli* shuttle plasmid for *Halobacterium* and *Haloferax* spp.: has multiple cloning sites and confers mevinolin resistance in haloarchaea and ampicillin resistance in *E. coli*	63
pNG168	Haloarchaeal-*E. coli* shuttle plasmid for *Halobacterium* and *Haloferax* spp.: has multiple cloning sites and confers mevinolin resistance in haloarchaea and ampicillin resistance in *E. coli*	31
pUBP2	Haloarchaeal-*E. coli* shuttle plasmid for *Halobacterium* and *Haloferax* spp.: confers mevinolin resistance in haloarchaea and ampicillin resistance in *E. coli*	18
pMSD20	Haloarchaeal-*E. coli* shuttle plasmid for *Haloferax* spp.: confers mevinolin-resistance in haloarchaea and ampicillin resistance in *E. coli*	53
pHRZH	Haloarchaeal rRNA gene replacement vector for *Halobacterium* spp.: confers anisomycin and thiostrepton resistance	65
pMPK408/pSK400	Haloarchaeal *ura3* suicide vector for *Halobacterium* gene knockouts for *Halobacterium* spp.: integrates into genome and confers uracil prototrophy and 5-FOA sensitivity	86, 119
pGB70	Haloarchaeal *pyrE* suicide vector for *Haloferax* spp.: integrates into genome and confers uracil prototrophy and 5-FOA sensitivity	16
pXLNov	Haloarchaeal *bop* gene expression vector for *Halobacterium* spp. used for green fluorescent protein (GFP) fusion: confers resistance to novobiocin	83
pSO7	Haloarchaeal sensory rhodopsin SRI (*sop1*) gene expression vector for *Halobacterium* spp.: confers resistance to mevinolin	61
pBBEV1	Haloarchaeal *bop* promoter expression vector for *Halobacterium* spp.: confers resistance to mevinolin	Berquist & DasSarma, unpublished data
pKJ408*sfdx*	Haloarchaeal HtrI expression vector using super *fdx* promoter for Halobacterium spp.: confers resistance to mevinolin	56
pSE1	Haloarchaeal *fdx*-DHFR fusion gene for promoter analysis in *Haloferax volcanii*: confers resistance to trimethoprim	26
pSD1	Haloarchaeal GPCR expression plasmid for *Haloferax* spp.: confers resistance to trimethoprim	85
pENDS	Haloarchaeal *bop* promoter-based GPCR expression plasmid in *Halobacterium* spp.: confers resistance to mevinolin	14
pFL2	Haloarchaeal gas vesicle expression plasmid for *Halobacterium* spp.: confers resistance to mevinolin	30

(Continued on next page)

TABLE 3 *Continued*

Plasmids	Features	Reference
pFM104D	Antigen display vector on gas vesicles for *Halobacterium* spp.: confers resistance to mevinolin	114
pWL204	Haloarchaeal tRNA gene expression vector for Haloferax spp.: confers resistance to mevinolin	82
Methanogenic Euryarchaeota		
pJK 28 through -41	Methanoarchaeal-*E. coli* shuttle plasmids for site-directed gene disruption in *Methanosarcina* spp.: do not replicate in *Methanosarcina*; have different *lacZα* polylinker cloning sites for cloning DNA homologous to the target gene; confer puromycin resistance in methanoarchaea and ampicillin resistance in *E. coli*	71
pDLT44	Methanoarchaeal-*E. coli* shuttle plasmid for gene expression/complementation in M. *maripaludis*: autonomously replicates in M. *maripaludis*, has unique *XbaI* and *SacI* restriction sites for cloning heterologous DNA, and confers puromycin resistance in methanoarchaea and ampicillin resistance in *E. coli*	117
pWM309 through -321	Methanoarchaeal-*E. coli* shuttle plasmids for gene expression-complementation in *Methanosarcina* spp.: autonomously replicate in *Methanosarcina* spp., have different *lacZα* polylinker restriction sites for cloning heterologous DNA, and confer puromycin resistance in methanoarchaea and ampicillin resistance in *E. coli*	71
pJK60	Methanoarchaeal-*E. coli* shuttle plasmid for in vivo random gene disruption in *Methanosarcina* spp.: contains the mariner transposon and confers puromycin resistance in methanoarchaea and kanamycin resistance in *E. coli*	126
pEA105	Methanoarchaeal-*E. coli* shuttle plasmid for gene expression-complementation in *Methanosarcina* spp.: derivative of pWM315 with greater transformation efficiency in *E. coli*	Apolinario and Sowers, unpublished
pJK5	Methanoarchaeal-*E. coli* shuttle plasmid for site-directed gene disruption in *Methanosarcina* spp.: *pac-oriR6K-aph* cassette plasmid with symmetrical *XbaI* and *EcoRI* flanking restriction sites for cloning DNA homologous to the target gene.	127
pPB12	Methanoarchaeal-*E. coli* shuttle plasmid for gene expression-complementation in *Methanosarcina* spp.: autonomously replicates in *Methanosarcina* spp. and confers pseudomonic acid resistance in methanoarchaea and ampicillin resistance in *E. coli*	19
pMipuid	Methanoarchaeal-*E. coli* shuttle plasmid for assaying gene expression in M. *maripaludis*: integration vector with EcoRI and NdeI restriction sites upstream of *uidA* for detecting archaeal promoter strength by β-glucuronidase activity; confers puromycin resistance in methanoarchaea and ampicillin resistance in *E. coli*	15
pMudpur	Methanoarchaeal-*E. coli* shuttle plasmid for in vitro directed gene disruption in M. *maripaludis* genomic DNA: contains the Mu transposon and confers puromycin resistance in methanoarchaea and chloramphenicol resistance in *E. coli*	17
pWLG30+*lacZ*	Methanoarchaeal-*E. coli* shuttle plasmid for assaying gene expression in M. *maripaludis*: autonomously replicating reporter vector with NsiI restriction site upstream of *lacZα* for detecting archaeal promoter strength by β-galactosidase activity	40
Hyperthermophilic Crenarchaeota		
pAG1	Ap[r] *E. coli* shuttle vector: butanol[r] benzyl alcohol[r] (alcohol dehydrogenase gene from *S. solfataricus*) *Pyrococcus* cryptic plasmid (pGT5) for *S. acidocaldarius*	5
pEXSs	Ap[r] *E.coli* shuttle vector: hygromycin B[r] (mutant hygromycin phosphotransferase gene from *E. coli*), SSVI viral origin for *S. solfataricus* Gtheta	20
pNOB8-lacS	A self-transmissible plasmid derived from pNOB8 with the β-glycosidase gene (*lacS*) fused to ribosomal protein S12 promoter from *S. acidocaldarius* and inserted at unique *Sal* I site; for use in *S. solfataricus* P1	35
pKMSD48	Ap[r] *E. coli* shuttle vectors: SSVI virus hybrid. For use in *S. solfataricus*	112

33.2.2.2. *M. maripaludis* JJ

M. maripaludis JJ is an irregularly shaped, coccus-shaped cell isolated from marine sediments; it utilizes hydrogen as the exclusive source of energy for growth and methanogenesis (55). Wild-type M. maripaludis can be grown in a defined minimal medium with the addition of isoleucine and calcium pantothenate (105). This species generates a glycoprotein S-layer rather than a rigid cell wall, and spheroplasts are generated as described above. M. maripaludis JJ harbors a cryptic 8.3-kb plasmid, which has been recombinantly modified to serve as a shuttle vector for genetic transformation of M. maripaludis JJ (117). This species has a Pst-like restriction system, which reduces transformation efficiency by about fourfold when transformed with non-PstI methylated DNA (55).

33.2.3. Genetic Markers for Selection

Mesophilic methanogenic Archaea are sensitive to puromycin, pseudomonic acid, fusidic acid, neomycin, and bacitracin at reversion frequencies of $\leq 10^{-7}$ (20, 47, 90). Bacterial genes that encode resistance to three of these antibiotics have been modified for expression as genetic markers in the methanogenic Archaea. The structural gene from *Streptomyces alboniger* that encodes puromycin transacetylase (*pac*) has been used for genetic selection in M. voltae, M. maripaludis, and Methanosarcina spp. As the S. alboniger bacterial promoter is not recognized and transcribed in Archaea, the pac gene is flanked with the promoter and terminator for the gene encoding constitutively expressed methyl coenzyme M reductase (*mcr*) from M. voltae (pMEB.2 and pIJA03) or M. barkeri (pJK3) (41, 71, 111). The pac cassette is flanked by multiple restriction sites in plasmids pMEB.2 and pJK3, enabling it to be excised for insertion into other vectors. A second antibiotic resistance marker has been developed for M. maripaludis utilizing aminoglycoside phosphotransferase genes APH3′I and APH3′II from bacterial transposons Tn903 and Tn5, respectively, engineered with flanking archaeal mcr promoter and terminator sequences (6). When introduced into M. maripaludis on plasmids, these genes confer resistance to the protein synthesis inhibitor neomycin. APH3′I and APH3′II cassettes are flanked by EcoRI sites in plasmids pJLA6 and pJLA5, respectively, for ready excision and insertion into other vectors. However, neomycin is not an effective growth inhibitor of *Methanosarcina* spp. (19). A second antibiotic resistance marker developed for *Methanosarcina* spp. utilizes a mutagenized *ileS* gene from M. barkeri, which confers resistance to pseudomonic acid A. pPB12 is an autonomously replicating E. coli/Methanosarcina shuttle vector that carries modified ileS flanked on both sides by polylinkers. The genetic marker can be readily excised and inserted into genes for genomic disruption by homologous recombination (see below) (19). pPB32 carries modified ileS and a lacZα with a multiple cloning site for blue-white screening of recombinant clones. With a wild-type gene or operon inserted in the polylinker, the construct will complement mutagenized genomic genes that are disrupted with pac (19). However, pseudominic acid is not commercially available. The Reference Materials Group of SmithKline Beecham Pharmaceuticals (Philadelphia, PA) has provided small amounts to investigators upon request, or it can be purified from *Pseudominas fluorescens* (NCIB 10586) by acidic isobutylmethyl ketone extraction (38).

33.2.4. Vectors for Gene Transfer

33.2.4.1. Shuttle and Cloning Vectors

Plasmid-based shuttle vectors have been constructed by combining origins of replication (*ori*) from indigenous bacterial and archaeal plasmids and selectable genetic markers fused to bacterial and archaeal promoters (see Fig. 3). Bacterial and eukaryal transposable elements have been adapted for gene disruption in vitro and in vivo, respectively. By combining genetic elements from bacterial, eukaryal, and archaeal microbes, tractable systems for gene disruption, expression, and reporting have been developed for the methanogenic Archaea (Table 3).

33.2.4.2. Transposon-Mediated Random Gene Disruption

A powerful tool for use in identifying new roles for genes is random mutagenesis, whereby an entire genome is randomly mutagenized and specific phentotypes are identified by selection or screening. A very effective approach employed for random mutagenesis in vivo is to use a transposon-mediated vector. Although transposons have not been detected in the methanogenic Archaea, the *mariner* transposable element from the hornfly *Hematobia irritans*, which contains a host-independent transposase, has been incorporated into a transformable suicide vector for M. acetivorans (126). The construct includes the mini-*mariner* inverted repeats flanking the *pac* and *aph* genes and the E. coli R6K origin of replication, and the mariner transposase Tnp transcribed from the M. barkeri methyl-CoM reductase *mcrB* promoter (see Fig. 4). The plasmid also contains a high-copy-number pMB1 *ori* for replication of the plasmid in E. coli and the β-lactamase gene (*bla*) for selection of E. coli transformants with ampicillin. E. coli is transformed by the transposable vector using heat shock or electroporation and isolated using standard protocols. For random mutagenesis in the archaeon, M. acetivorans is transformed by the puri-

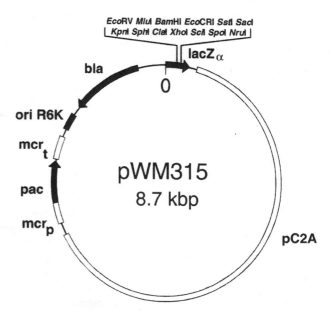

FIGURE 3 Recombinant plasmid showing construction typical for an *E. coli-Methanosarcina* shuttle vector. The construct includes the *pir*-dependent R6K *ori* for replication of the plasmid in *pir*⁺ E. coli and the bla gene for selection of E. coli transformants with ampicillin; pC2A *ori* and *repA* for replication in *Methanosarcina* spp.; and *pac* under transcriptional control of the archaeal methyl-coenzyme M reductase *mcrB* gene for selection of methansarcinal recombinants on puromycin. Filled and open elements represent genes from the Bacteria and Archaea, respectively.

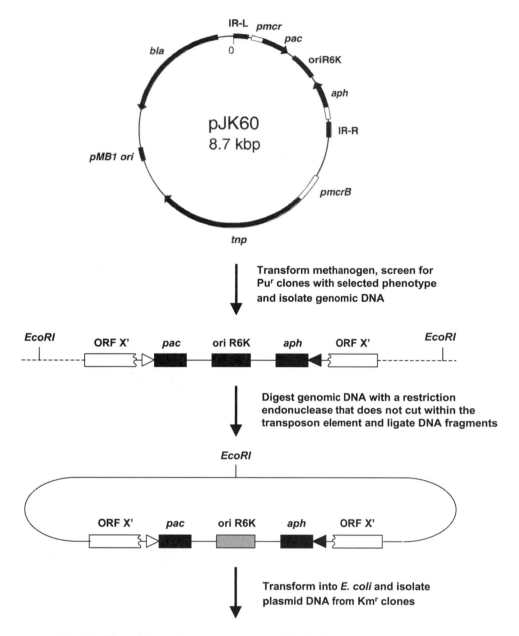

FIGURE 4 Random gene mutagenesis in *Methanosarcina* with an *E. coli-Methanosarcina* shuttle vector containing a transposable element. pJK60 is a modified *E. coli-Methanosarcina* shuttle vector that contains the puromycin resistance *pac* gene, the R6K *E. coli* plasmid origin of replication, and the kanamycin resistance *aph* gene flanked by the transposable elements of the insect *mariner*-family *Himar1* transposon. The transposase is expressed in the methanogen from the methyl-coenzyme M reductase gene (*mcrB*) from *Methanosarcina barkeri*. The vector is transformed into *Methanosarcina* spp. and transposed into random sites in the genome, and then puromycin-resistant colonies with the desired phenotype are selected. The transposed DNA is purified and digested with EcoRI, and the fragments are closed by treatment with T4 ligase. The circular DNA is transformed into *E. coli*, the DNA is repurified from kanamycin-resistant clones, and then DNA flanking the transposable element is sequenced to identify the disrupted gene.

fied plasmid, then inoculated onto solidified medium containing puromycin as described above. The *pac* gene transcribed by the archaeal *mcr* promoter encodes puromycin resistance to clones with the *mariner* element incorporated

into their genomes. The expected frequency of tranposition for pJK60 in M. *acetivorans* is 208 ± 169 per µg of DNA or a transposition frequency of ca. 2.5×10^{-5} per CFU (126). Using conditions optimized for M. *acetivorans*, the vector will also transpose in M. *maripaludis*, but at a lower frequency of ca. 1×10^{-7} per CFU (J. Leigh, personal communication). To identify the gene disrupted by the transposon, genomic DNA isolated from the recombinant clone is digested with a restriction endonuclease that does not digest the transposon (Fig. 4). The DNA fragment is circularized and ligated with T4 DNA ligase, then transformed into *pir*$^+$ *E. coli* (e.g., DH5α/λpir), where replication is mediated by the λpir-dependent R6K *ori* (70). The *aph* gene confers resistance of transformants to kanamycin, and the purified plasmid can be sequenced with upstream- and downstream-directed, transposon-specific primers to identify the site of insertion.

33.2.4.3. Directed Gene Disruption and Complementation

A method developed for directed mutagenesis in *Methanosarcina* spp. and *Methanococcus* spp. employs homologous recombination-mediated gene replacement. The target gene is disrupted in vitro with the *pac-ori-aph* cassette, which confers puromycin resistance in the methanogen and kanamycin resistance and R6K *ori*-mediated autonomous replication of circularized DNA in *E. coli*. The disrupted target gene is ligated into the multiple cloning site of an *E. coli* cloning vector such as pBluscript SK (Stratagene) at sites that flank the respective symmetric XbaI or EcoRI restriction sites, thereby deleting one of the sites. The *pac-ori-aph* cassette in plasmid pJK5 is digested at one of the respective symmetrical XbaI or EcoRI flanking sites and ligated into the target gene. The disrupted gene is digested at restriction sites that flank the *pac-ori-aph* cassette by approximately 500 bp. The methanogen is transformed with the linearized fragment by protocols described below (see section 33.2.5), and transformants are selected by resistance to puromycin. To confirm disruption of the target gene, total DNA from the methanogen clone is digested with a restriction endonuclease that does not digest the disrupted gene fragment, and the fragments are circularized by ligation and then transformed into competent *pir*$^+$ *E. coli* cells by standard protocols. Plasmid DNA isolated from kanamycin resistance clones is sequenced upstream and downstream of the *pac-ori-aph* by using primers specific for the cassette. This approach has been used successfully for disruption of the *proABC* locus for proline biosynthesis in *M. acetivorans* and the *echABCDEF* locus for hydrogenase in *M. barkeri* (73, 127). A similar approach has been used for disrupting the genes in *M. maripaludis* (111, 121). In this case, cassettes containing *pac* from pJK3 or *neo* from pRCN230, each under control of the *mcr* promoter, were flanked with selected restriction sites by PCR and ligated into the target formate dehydrogenase-encoding *fdhAB* genes. Gene disruptions were selected by growing transformed *M. maripaludis* with puromycin or neomycin, respectively.

Two transposon-mediated systems have been developed for directed gene disruption. Zhang et al. (127) employed a commercially available transposon and hyperactive transposase, EZ::TN™ (Epicenter), to disrupt proline-encoding *proABC* in *M. barkeri*. The *pac-ori-aph* or *ileS12* (19) cassettes, which encode resistance to kanamycin in *E. coli* and to puromycin or pseudomonic acid, respectively, in *Methanosarcina* spp., were ligated into the EZ::TN™ transposon, then used to disrupt plasmid-borne *proABC* in *E. coli* DH10B by the manufacturer's recommendations. Mutagenized plasmids were recovered from ampicillin- and kanamycin-resistant transformants. The *proABC* was linearized by restriction digestion, inserted into the *M. ace-*

tivorans genome by liposome-mediated transformation and homologous recombination, and then screened for puromycin or pseudomonic acid resistance. Two micrograms of linearized DNA yielded 10^2 to 10^3 antibiotic-resistant colonies with 80 to 100% recovery of Pro$^-$ transformants. Confirmation of mutations in the correct locus was confirmed by transforming the mutants with a copy of the wild-type *proABC* locus ligated into the autonomously replicating shuttle vector pWM321 (73).

Blank et al. (17) developed an efficient phage-based ex vivo gene disruption system for *M. maripaludis* based on *mu*-mediated transposition. The vector pMudpur was constructed with the left and right ends of *mu*, Ap and *cat* for selection of ampicillin- and chloramphenicol-resistant transformants of *E. coli*, and a cassette from pMip1 (41) containing *pac*, under control of the methyl-CoM reductase archaeal promoter, for selection of puromycin-resistant transformants of *M. maripaludis*. The target DNA is ligated into a multiple cloning site and transformed into *E. coli* MH132 by standard protocols. One to two milliliters of culture grown with ampicillin and chloramphenicol is infected with 10^9 PFU of Mmpλ-1 for 20 min without shaking at room temperature, then transferred to 50 ml of LB medium with chloramphenicol preheated to 42°C and shaken gently for 20 min to induce transposition. The culture is then incubated at 37°C for 3 to 24 h to induce lysis. Phages are harvested by chloroform treatment, and plaques are isolated from infected *E. coli* P2392 on modified N-Z-Amine yeast medium (NZY) (17) with chloramphenicol. Phage DNA is isolated by polyethylene glycol treatment, confirmed by restriction analysis or sequencing, and transformed into the methanogens as described above.

33.2.4.4. Gene Reporters

Gene reporters for assaying promoter strength in the methanogenic Archaea have been developed for both *Methanosarcina* and *Methanococcus* using bacterial genes *uidA* and *lacZ*, encoding β-glucuronidase and β-galactosidase, respectively. Beneke et al. (15) developed an integration vector, pMipuid, which contains the *mcr*-mediated *pac* cassette from pMIP (41) and *E. coli uidA* under control of the archaeal promoter for the hydrogenase *hmvA*. The *pac-uidA* region is flanked on both sides by *M. voltae hisA*, which encodes a histonelike protein, for homologous recombination into the archaeal genome. Different promoters can be substituted for *hmvA* at the EcoRI/NdeI site upstream of the *uidA*. Since the p*hmvA-uidA* insert is flanked by EcoRI restriction sites, it can be readily inserted into other vectors to create β-glucuronidase reporter constructs for *M. maripaludis* and *Methanosarcina* spp. β-Galactosidase-based promoter expression has been reported for studies with *nifH* (25), encoding nitrogenase, and *fdhABC* in *M. maripaludis* (25, 121) and with *cdhA* in *Methanosarcina thermophila* (3). pWLG30+*lacZ* and pJK104 are shuttle vectors with both *E. coli* and methanogen origins of replication (Fig. 5). Both have an *mcr*-mediated *pac* cassette and *bla* gene for selection in the methanogen and *E. coli*, respectively. Archaeal promoters are ligated into the restriction sites upstream of *lacZ*, purified from *E. coli*, then transformed into their respective methanogen host. Cells grown under desired test conditions are assayed for gene expression based on β-galactosidase activity by using standard protocols (74). Screening for expression by colorimetric detection of colonies is also possible, but transformed colonies or a colony lift must be exposed to air to develop the color in the chromophore (15). Alternatively, pWLG30+*lacZ* has been used to express het-

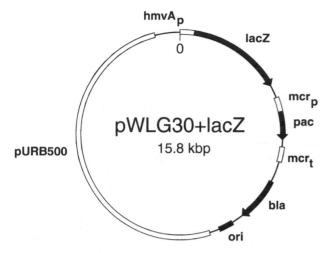

FIGURE 5 Reporter vector pWLG30+*lacZ* for detecting archaeal promoter strength based on β-galactosidase activity. This *Escherichia-Methanococcus* shuttle vector includes a bacterial *ori* for replication of the plasmid in *E. coli* and the *bla* gene for selection of *E. coli* transformants with ampicillin; methanococcal pURB500 for replication in *Methanococcus maripaludis*; and *pac* under transcriptional control of the archaeal methyl-coenzyme M reductase *mcrB* gene for selection of methanococcal recombinants on puromycin. The methanococcal *hmvA* hydrogenase promoter is fused upstream of *lacZ* for measuring expression of hydrogenase by β-galactosidase activity. Filled and open elements represent genes from the Bacteria and Archaea, respectively.

erologous genes in *Methanococcus* (99). In this case, the *lacZ* is replaced by the heterologous gene using the blue-white screening in *E. coli* (99). Upon transformation into *Methanococcus*, the cloned gene is expressed by the *hisA* promoter of *M. voltae*, which is inserted just upstream of a multiple cloning site in the vector.

33.2.5. Transformation

Two transformation approaches have been developed and optimized for the methanogenic Archaea. Liposome-mediated transformation is highly efficient for *Methanosarcina* spp. with an efficiency of $\leq 10^8$ transformants per μg of DNA for M. *acetivorans* and 1 to 2 orders of magnitude lower efficiency for other species. Polyethylene glycol-mediated transformation of M. *maripaludis* yields efficiencies of $\leq 10^7$ transformants per μg of DNA, but is specific for this species.

33.2.5.1. Transformation of *M. acetivorans* with Liposomes

1. In an anaerobic glove box inoculate 10 ml of M medium containing 0.05 M trimethylamine or methanol in a 18- by 150-mm anaerobe tube (Bellco) with a single colony transferred by removing a plug of agar with a sterile Pasteur pipette. Seal tube with a butyl rubber stopper secured with an aluminum crimp seal (Bellco).

2. Transfer tube to a 35°C incubator for 5 to 7 days to mid-exponential growth (OD$_{550}$, 0.3 to 0.5).

3. For each transformation, inoculate 10 ml of fresh MT medium with 1 to 2 ml of culture and incubate to mid-exponential growth (OD$_{550}$, 0.3 to 0.5). Estimate incubation time based on a doubling rate of 7.3 h.

4. Harvest cells in the culture tubes by centrifugation at 3,200 × *g* in a Beckman GPA swinging bucket rotor or equivalent for 15 min. For other rotors, the culture can be transferred to a sterile aluminum seal culture tube that has been shortened to 10 cm at the glass shop. These tubes can be pressurized and withstand centrifugation at low speeds. Alternatively, culture can be transferred to a sterile Oak Ridge tube, sealed in an anaerobic glove box, then centrifuged outside of the glove box in any standard centrifuge.

5. Transfer culture to an anaerobic glove box, open culture, and gently decant medium from cell pellet. Discard supernatant.

6. Add 1 ml of 0.85 M sucrose to cell pellet and resuspend cells by gently tapping the tube.

7. To a sterile, disposable 13-by-100-mm borosilicate glass tube add: 0.85 M sucrose, buffer, 25 μl of DOTAP liposome reagent (Roche Diagnostics), 2 μg of DNA, and distilled water to a final volume of 150 μl. Cap tube and mix by gentle tapping.

8. Incubate at room temperature in a glove box for 1 h.

9. Add 1 ml of resuspended culture to DNA/liposome mixture. Mix by gently tapping tube.

10. Incubate 2 to 4 h at room temperature.

11. Transfer entire transformation mix to 10 ml of M medium in an 18-by-150-mm anaerobe tube by syringe.

12. Incubate 12 to 16 h at 35°C. Culture can be incubated outside of glove box.

33.2.5.2. Transformation of *M. maripaludis* with PEG

1. In an anaerobic glove box, inoculate 5 ml of McC medium in an 18-by-150-mm anaerobe tube with a single colony transferred by removing a plug of agar with a sterile Pasteur pipette. Seal tube with a butyl rubber stopper secured with an aluminum crimp seal.

2. Incubate in a shaking incubator or water bath at 35°C until the early-stationary phase of growth (OD$_{600}$, 0.8) is obtained, usually 1 to 2 days.

3. For each transformation, inoculate 5 ml of fresh McC medium with 0.1 to 0.2 ml of culture and incubate overnight. The next morning, repressurize the culture with H$_2$ + CO$_2$ and allow growth to continue to early-stationary growth (OD$_{600}$, 0.8). Estimate incubation time based on a doubling rate of 5 h.

4. Harvest cells in the culture tubes as described in step 4 of section 33.2.5.1.

5. After centrifugation, the spent medium is removed by inverting the tube and inserting a 22-gauge Vacutainer Blood Collection needle through the stopper and into the bottom corner. Because the tube is pressurized, the medium is rapidly expelled. Alternatively, transfer the culture to an anaerobic glove box, open culture, and gently decant medium from cell pellet. Discard supernatant.

6. Wash cells by resuspending the cell pellet in 5 ml of transformation buffer (50 mM Tris-HCl [pH 7.5], 0.35 M sucrose, 0.38 M NaCl, 0.05% [wt/vol] cysteine-HCl, 1 mM dithiothreitol, 0.00001% resazurin, 1 mM MgCl$_2$) followed by centrifugation as described above. If the transformation buffer is in a pressurized culture tube, it can be conveniently added to the cell pellet using a 22-gauge Vacutainer Blood Collection needle already in the tube holding the cell pellet. Simply invert the tube with the transformation buffer and insert the free end of the Blood Collection needle through the stopper. The pressure in the tube will force the buffer into the pellet. The resuspended cells are then centrifuged as described above and the supernatant is decanted.

7. Add 0.375 ml of transformation buffer to cell pellet and resuspend cells by gently tapping the tube.

8. Add 0.8 μg of DNA in 0.1 ml of TE buffer, followed by 0.225 ml of PEG solution (40% [wt/vol] PEG8000).

9. Seal tube, exchange the gas with N_2, and incubate at 37°C for 1 h.

10. Add 4 ml of McC medium, mix by gently tapping the tube, and harvest cells by centrifugation.

11. Resuspend cell pellet in 5 ml of McC medium, and exchange the headspace with 275 kPa H_2/CO_2 (4:1).

12. Incubate with shaking at 37°C for 5 h.

13. Harvest cell by centrifugation and resuspend cell pellet in 0.6 ml of McC medium.

33.2.5.3. Colonization of Transformants on Agar Plates

1. Inside an anaerobic glove box, transfer 100-μl aliquots of transformed culture with a 1-ml tuberculin syringe to the surface of agar-solidified selective medium in petri plates.

2. Spread culture evenly on surface with a sterile culture spreader.

3. Load inoculated plates, inverted, into a glass or stainless steel anaerobic jar.

4. Immediately before closing jar, place a 23-by-47-mm serum vial (Wheaton) containing an Na_2S solution in bottom of jar. Estimate amount of Na_2S based on total jar volume without plates: 0.8 ml of 2.5% Na_2S (wt/vol) per 1-liter total jar volume for M. acetivorans; 6.4 ml of 2.5% Na_2S (wt/vol) per 1-liter total jar volume for M. maripaludis. Alternatively, if the jar is equipped with an inlet valve, H_2S can be added directly to the jar to a final volume of 0.2% (vol/vol). Adequate precautions should be observed when using H_2S, which is an extremely toxic gas.

5. For M. acetivorans, seal jar and incubate at 35°C for 5 to 7 days. For M. maripaludis, purge and pressurize with H_2/CO_2 (4:1 [vol/vol]) at maximum pressure recommended for jar and incubate at 37°C for 5 to 7 days.

6. To observe plates, purge jar with N_2/CO_2 in a fume hood to remove H_2S. Jar can now be transferred to an anaerobic glove box and opened.

7. Plates can be resealed in the jar for further incubation. It is not necessary to add H_2S to the jars for reincubation.

33.2.6. Media and Growth

Unlike many obligately anaerobic bacteria, which can tolerate brief exposure to oxygen, the methanogenic Archaea are fastidiously anaerobic, requiring complete exclusion of oxygen and a redox potential below −330 mV for growth (104). All procedures including medium preparation, culture transfer, and colonization on plates require specialized equipment to maintain anaerobic conditions (109). Medium is made anaerobic by boiling or sparging with anaerobic gas to exclude oxygen, chemically reduced with Na_2S and cysteine, then dispensed into culture tubes, serum vials, or bottles. Media are sealed with rubber stoppers under an atmosphere of nitrogen and carbon dioxide before sterilization by autoclaving. Solidified medium is prepared by adding agar to liquid medium (116). After autoclaving, the melted medium is transferred to an anaerobic glove box (Fig. 6) and poured into petri plates under an atmosphere of nitrogen and carbon dioxide. The medium is inoculated by spreading the culture on the surface in molten agar or directly with a sterile rod, then incubated in an anaerobic jar (Fig. 6) under an atmosphere of nitrogen, carbon dioxide, and hydrogen sulfide. The long incubation periods of 2 to 14 days require that the jars selected should not be wholly composed of plastics or other polymers, which will permit the gradual permeation of oxygen. Metal culture jars are available from commercial sources (e.g., Equitron, Medica Instrument Mfg. Co., Mumbai, India; DW Scientific Ltd., West Yorkshire, UK). Alternatively, they can be constructed from modified pressure cookers, paint pressure canisters, or glove box airlocks (8, 72, 116) or custom-made by a machine shop (7). Inexpensive glass containers can be constructed from a modified canning jar (4). Several commercially manufactured glass and/or metal anaerobe jars suitable for growing methanogens from TORBAL, Oxoid, and BBL are no longer available, but they often can be found in storage or through used-equipment distributors. Since the plates are incubated in a sealed environment, the plates must be predried to a critical water content; too much water causes confluence of colonies and overdrying inhibits growth. In general, predrying medium solidified with 1.5% agar in plates in an anaerobic glove box in 40% relative humidity for 48 h is optimal for Methanosarcina spp. and Methanococcus spp. (4).

For M. acetivorans medium, add Na_2S and substrate stock solutions at least 5 min before inoculation. Inoculate sterile medium with 5 to 10% (vol/vol) of culture using a syringe that has been purged with N_2. For M. maripaludis, add the Na_2S stock solution at least 5 min before inoculation. Inoculate sterile medium as described above, then pressurize the inoculated culture to 275 kPa with H_2/CO_2 (4:1 [vol/vol]). Incubate cultures at 37°C. NOTE: Whenever working with pressurized glass vessels, wear safety glasses as a precaution against accidental explosion. Additional information on the preparation of anaerobic medium and media for additional methanoarchaeal strains may be found in Methanogens, volume 2 of Archaea: A Laboratory Manual (108).

33.2.6.1. Media Preparation

M. acetivorans Growth Medium M (107) (per Liter)

NaCl	23.38 g
$MgSO_4 \cdot 7H_2O$	12.32 g
NH_4Cl	0.50 g
KCl	0.50 g
$CaCl_2 \cdot 2H_2O$	0.14 g
Resazurin stock solution (100 ×)	10.0 ml
Trace mineral stock solution (100 ×)	10.0 ml
Vitamin stock solution (100 ×)	10.0 ml
$Na_2HPO_4 \cdot 7H_2O$	1.12 g
$NaHCO_3$	3.0 g
Cysteine-HCl · H_2O	0.25 g
NaS · $9H_2O$ stock solution (2.5% w/v)	10.0 ml
Sodium acetate, methanol, or trimethylamine, 5 M substrate stock solution	10.0 ml

M. maripaludis Growth Medium McC (116) (per Liter)

NaCl	21.97 g
$MgSO_4 \cdot 7H_2O$	3.45 g
$MgCl_2 \cdot 6H_2O$	2.75 g
NH_4Cl	0.50 g
KCl	0.34 g
$CaCl_2 \cdot 2H_2O$	0.14 g

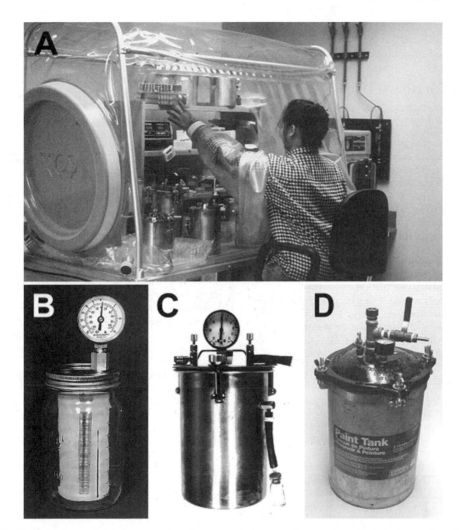

FIGURE 6 Specialized apparatus for plating methanogenic Euryarchaeota. (A) Anaerobic glove box used for plating methanogenic Archaea. The gas phase is composed of a mixture of N_2, CO_2, and H_2 in a volume ratio of 75:20:5. The CO_2 maintains the equilibrium of the carbonate buffer at a neutral pH, the H_2 combined with palladium catalyst pellets located in the glove box reduces any oxygen that may diffuse into the chamber, and the N_2 is inert. Lower figures show anaerobic jars for incubation of colony clones on solidified medium. Examples include a modified glass canning jar (B), a commercial jar manufactured by TORBAL (C), and a modified paint pressure tank (D). Photo in panel D is courtesy of W. B. Whitman.

$Fe(NH_4)_2(SO_4)_2 \cdot 6H_2O$	0.01 g
Yeast extract (exclude from solidified medium)	2.0 g
Sodium acetate (exclude from solidified medium)	1.36 g
Resazurin stock solution (100 ×)	10.0 ml
Trace mineral stock solution (100 ×)	10.0 ml
Vitamin stock solution (100 ×)	10.0 ml
K_2HPO_4	0.14 g
$NaHCO_3$ (2.0 g for solidified medium)	5.0 g
Cysteine-HCl $\cdot$ H_2O	0.25 g
$NaS \times 9H_2O$ stock solution (2.5% w/v)	10.0 ml

Resazurin Stock Solution (100 ×) (per Liter)

Resazurin (Sigma)	0.01 g

Trace Mineral Stock Solution (100 ×) (120) (per Liter)

Nitrilotriacetic acid	1.5 g
$MgSO_4 \cdot 7H_2O$	3.0 g
$MnSO_4 \cdot H_2O$	0.5 g
NaCl	1.0 g
$FeSO_4 \cdot 7H_2O$	0.1 g
$CoSO_4$ or $CoCl_2$	0.1 g
$CaCl_2 \cdot 2H_2O$	0.1 g
$ZnCl_2$	0.08 g
$CuSO_4 \cdot 5H_2O$	0.01 g
$AlK(SO_4)_2$	0.01 g
H_3BO_3	0.01 g
$NiCl_2 \cdot 6H_2O$	0.024 g
$Na_2MoO_4 \cdot 2H_2O$	0.02 g

Dissolve nitrilotriacetic acid in H_2O containing ca. 8 pellets of KOH, adjust pH to 6.5 with additional KOH, and then proceed to add additional ingredients (120). Store 50-ml aliquots at $-20°C$.

Vitamin Stock Solution (100 ×) (120) (per Liter)

Biotin . . . . 0.002 g
Folic acid . . . . 0.002 g
Pyridoxine-HCl . . . . 0.010 g
Riboflavin . . . . 0.050 g
Thiamine . . . . 0.050 g
Nicotinic acid . . . . 0.050 g
Pantothenic acid . . . . 0.050 g
Vitamin B_{12} . . . . 0.0001 g
p-Aminobenzoic acid . . . . 0.050 g
Thioctic acid (lipoic acid) . . . . 0.050 g

Store 50-ml aliquots at $-20°C$.

Na_2S and Substrate Stock Solutions

Under a stream of nitrogen, add water boiled to remove dissolved oxygen to a premeasured amount of $NaS \cdot 9H_2O$ or substrate in an anaerobic culture tube or serum vial. Seal vessel with a butyl rubber stopper secured with an aluminum crimp seal, and autoclave at 121°C for 20 min.

33.2.6.1.1. Liquid Medium Preparation

1. Add half the total volume of H_2O to a round-bottom flask containing a stir bar.

2. Dissolve ingredients of the vitamin stock solution in water. Use some of the remaining H_2O to wash powder off the walls of the flask.

3. Insert a gassing stone and gas solution with a mixture of N_2/CO_2 (4:1 [vol/vol]) while stirring, and dissolve ingredients completely.

4. Add $Na_2HPO_4 \cdot 7H_2O$ and dissolve completely.

5. Add $NaHCO_3$ and dissolve completely.

6. Use the remaining H_2O to wash powder off the walls of the flask.

7. Continue to gas for 15 min to displace dissolved oxygen.

8. Add cysteine-HCl $\cdot$ H_2O and dissolve *completely*.

9. Adjust pH to 6.8 to 7.0 with concentrated HCl.

10. Insert a gassing cannula into the flask, then remove the gassing stone and turn off the stir plate.

11. Dispense medium with a pipette or syringe into tubes or bottles under N_2/CO_2.

12. After dispensing medium into tubes or bottles, secure septa with aluminum crimp seals. For M. *maripaludis*, purge and pressurize the medium gas phase to 125 kPa with H_2/CO_2 (4:1 [vol/vol]).

13. Autoclave media for 20 min under fast exhaust (gravity). Bottles should be autoclaved in wire baskets to protect from risk of explosion upon removal from the autoclave, and safety glasses should be worn.

NOTE: Precipitates will form during autoclaving because dissolution of CO_2 increases the pH, but it may be redissolved by briefly shaking the medium after it has cooled and allowing it to stand for a time. Medium can be stored without the addition of Na_2S for up to 6 months.

33.2.6.1.2. Solidified Medium Preparation

1. Prepare media as described above, and dispense up to 100 ml into 160-ml serum vials containing premeasured quantities of agar (1% [wt/vol]).

2. After sterilization by autoclaving, transfer media precooled to 60°C into an anaerobic glove box and pour into petri plates. For M. *acetivorans*, add substrate and antibiotic stock solutions to each vial and mix by gentle swirling before pouring. For preparing large quantities of medium, solidification can be prevented by holding the medium in the glove box at 60°C in an incubator or water bath filled with glass beads until ready for use.

3. After solidification of the medium, dry plates by storing in an anaerobic glove box with ca. 40% relative humidity at room temperature for 48 h. Thereafter, plates can be stored inverted in a sealed container such as an anaerobe jar or canning jar until ready for use.

4. Inoculate media by spreading inocula onto the surface with a sterile spreading rod. The rod should be stored in the anaerobic chamber overnight before use to ensure that all O_2 is desorbed from it.

5. Transfer inoculated plates, inverted, into an anaerobe jar.

6. H_2S is introduced into the gas phase by including a small vial or empty petri plate containing ca. 0.5 ml of Na_2S solution. As the Na_2S solution is acidified by CO_2 in the gas phase, H_2S is released into the vessel. Optimal amounts of Na_2S are 0.02 g/liter volume of vessel and 0.16 g/liter volume of vessel for M. *acetivorans* and M. *maripaludis*, respectively (4). Remove sealed anaerobe jars from the glove box. Incubate M. *acetivorans* at 37°C. For M. *maripaludis*, purge and pressurize the gas phase of the jar to 200 kPa with H_2/CO_2 (4:1), then incubate at 37°C. Alternatively, H_2S can be added to the jars directly as a gas phase by purging and pressurizing the vessels with premixed 79.9% N_2, 20% CO_2, 0.1% H_2S or 79.9% H_2, 20% CO_2, 0.1% H_2S (72). Pressurization should not exceed value recommended for canister. NOTE: Take precautions necessary for protection from accidental explosion when pressurizing vessels.

33.2.6.2. Storage of Strains

Cultures may be stored in liquid medium at room temperature for up to 2 months without transfer. For long-term storage, cultures are frozen at $-70°C$ in liquid medium prepared with 50% glycerol (vol/vol).

33.2.7. Troubleshooting

The most common problem encountered when transforming *Methanosarcina* spp. and *Methanococcus* spp. is low transformation efficiency. Assuming that the recombinant DNA being introduced does not generate a lethal product, the most common reason for low efficiency is the exposure of cells to oxygen during the plating procedure. As methanogens are much more vulnerable to oxygen during the plating process because they are no longer suspended in a redox-buffered liquid medium, even a trace amount of oxygen can be lethal to the cells. Before plating, palladium catalyst used to reduce oxygen in the anaerobic glove box should be freshly charged, and all glassware and plasticware should be equilibrated in the glove box overnight. As an additional precaution, the airlock should not be used to transfer anything into the glove box 1 h prior to plating transformed cells, to ensure that the palladium catalyst reduces all traces of oxygen. All anaerobic incubation jars should be regularly checked for leaks to prevent exposure to oxygen during incubation. For additional information on plating *Methanosarcina* spp. and *Methanococcus* spp., refer to Apolinario and Sowers (4).

33.2.8. Recent Developments

A markerless gene disruption system has been developed for the methanogens using a mutant defective for the purine salvage pathway, which is resistant to inhibition by the purine base analog 8-*azo*-2,6-diamino-purine (8-ADP) (21, 75, 92). Wild-type *htp* and *pac* flanked by target DNA is inserted into the wild-type gene by homologous recombination; transformants are then selected for puromycin resistance and screened for resistance to 8-ADP. Selected transformants are grown under nonselective conditions without puromycin to promote plasmid excision, followed by screening for the mutant allele by Southern hybridization or PCR. A more recent modified markerless disruption method eliminates the requirement for screening by utilizing an Ftp recombinase to remove the genetic marker (98).

33.3. HYPERTHERMOPHILIC CRENARCHAEOTA

33.3.1. Introduction

Hyperthermophilic Crenarchaeota comprise 23 genera, represented by 50 type species. These organisms inhabit a range of terrestrial geothermal and marine hydrothermal environments. Among the Crenarchaeota, only those species belonging to the genus *Sulfolobus* have genetic systems, and they inhabit highly acidic geothermal pools. *Sulfolobus* species with genetic systems include various strains of *S. solfataricus* and the more distantly related *S. acidocaldarius* (Table 2). Genetic studies in *S. acidocaldarius* have focused primarily on chromosomally mediated conjugation (93) and spontaneous mutation (44, 45). With one exception (5), genetic systems for DNA delivery into this organism remain undeveloped (Table 2). In contrast, genetic manipulation of *S. solfataricus* is at a more advanced state. This reflects the occurrence of conjugative plasmids and viruses native to this organism (129), as well as the availability of a completed genome sequence (Table 1). Several genetic systems derived from extrachromosomal elements have been developed (23, 35, 112) (Table 3). In addition, this organism harbors numerous insertion elements that have enabled recovery of spontaneous loss of function (disruption) mutations (48, 66, 102, 125).

Targeted manipulation of the chromosome by directed recombination was recently added to the growing list of approaches for the genetic analysis of *S. solfataricus* (125). This subsection on hyperthermophilic archea will focus on the directed recombination method because of its utility for conducting functional genomics. Directed recombination of *S. solfataricus* involves the transformation by electroporation of prepared cells using nonreplicating (suicide) plasmids encoding selected genes. Recombinants are recovered by enrichment in liquid culture with selection for one of several genes. In the present version of the method, enrichment in liquid culture is required rather than direct plating on solid medium. This is because the marker genes encode glycosyl hydrolases whose selection depends on the use of a defined medium that can be realized in liquid culture but not on plates because of the presence of other carbon and energy sources in the plating support called gellan gum (Gelrite®, Merck & Co.). Recombinants occur as a result of two successive crossover events between the cloned target locus, in this case disrupted by *lacS*, and the chromosomal copy of the target locus. A transient duplicated intermediate can form after only the initial crossover. Allele replacement results when the two crossovers occur on either side of the selectable maker. Cultures that have been

successfully enriched are then diluted and plated to purify individual recombinants. At this stage, phenotypic screens can be used to confirm the identity of putative recombinants. The targeted locus of individual recombinants is then characterized by PCR and restriction analysis of genomic DNA. Isolates with the desired genomic alteration can then be studied immediately or stored for later use.

33.3.2. Strain Characteristics

33.3.2.1. *S. solfataricus* Strains for *lacS* Selection

Several *S. solfataricus* strains are used as hosts for directed recombination (Table 2) and their use depends on the nature of the selectable marker for genomic manipulation (Table 4). When the *lacS* gene is used as a selectable marker, strain PBL2025 and its derivatives are used (101). Strain PBL2025 was selected from a collection of deletion strains lacking regions covering *lacS* and neighboring genes because it retains the *malA* gene (97), which is necessary for rapid growth on sucrose minimal medium. PB2025 harbors a 58-kb deletion spanning open reading frames SSO3004 to SSO3050 as annotated by She and coworkers (103). The deleted region in PBL2025 includes the *lacS* gene and the flanking hypothetical lactose transporter (SSO3017) (91). PBL2025 is unable to utilize β-linked disaccharides, and reintroduction of *lacS* as a selectable marker restores β-linked disaccharide utilization.

Directed recombination has been conducted with PBL2025 for many loci. Whether the target locus is nonessential or is a locus in which disruption confers a growth defect (remedial phenotype), PBL2025 is the appropriate strain to use. Elimination of the *lacS* locus in PBL2025 permits recovery only of recombinants at the target locus, preventing recovery of recombinants at *lacS* that otherwise dominate recombinant populations. If the target gene is essential, the transient intermediate harboring a duplication of the target locus will be recovered, resulting from a single crossover event. Such intermediates will necessarily segregate the disrupted second copy of the target locus (and vector) to reproduce the original wild-type locus. This process can be monitored (see section 3.3.2) and frequencies can be determined to provide direct genetic evidence of the essentiality of the target gene.

33.3.2.2. *S. solfataricus* Strains for *malA* Selection

Strain PBL2030 (Table 2) is appropriate for use with *malA* as a selectable marker (Table 4). This strain harbors two chromosomal alterations, a deletion of the *lacS* region spanning 26 kb and open reading frames SSO3013 to SSO3037 as well as a disruption of the *malA* gene (SSO3051) resulting from insertion of *lacS* (Hoang and Blum, unpublished data). The deleted region removes *xylS* (SSO3022), an α-xylosidase, which confers a portion of this organism's ability to utilize α-linked oligosaccharides. Because *S. solfataricus* has additional genes encoding proteins with α-glucosidase activity, loss of *malA* renders this strain incapable of growing on starch and glycogen but not maltose.

Figure 7 shows the enrichment time course for six transformations, two selecting for *lacS* integrants and growth on lactose and four selecting for *malA* integrants and growth on either glycogen or starch. Differences in the time of recovery of positive enrichment cultures are evident; recovery of *lacS* integrants is fastest. Two transformations use PBL2025 and pLacS while four employ PBL2030 and plasmid pMalA. The two plasmids contain marker genes that are flanked by chromosomal DNA derived from other loci.

TABLE 4 Selectable genetic markers for Archaea

Selection agent	Target species	Mode of action	Resistance gene	Gene product	Working concn	Reference
Halophilic Euryarchaeota						
Mevinolin	*Halobacterium* and *Haloferax* spp.	Inhibits lipid synthesis	*hmg/mev*	3-Hydroxy-3-methylglutaryl-coenzyme A reductase	13–16 μg/ml	63
5-Fluoroorotic acid (5-FOA)	*Halobacterium* and *Haloferax* spp.	Inhibits pyrimidine biosynthesis	*ura3/pyrF* and *ura5/pyrE*	Orotidine-5′-monophosphate decarboxylase and orotate phospho-ribosyltransferase	0.15–0.25 mg/ml	16, 86
Novobiocin	*Haloferax* spp.	Inhibits DNA gyrase	*gyrB*	DNA gyrase subunit	0.1–0.2 μg/ml	54
Methanogenic Euryarchaeota						
Puromycin	*Methanosarcina* spp. and *M. maripaludis*	Inhibits protein synthesis	*pac*	Puromycin transacetylase	1–2 μg/ml	71, 100
Neomycin	*M. maripaludis*	Inhibits protein synthesis	APH3′I APH3′II	Aminoglycoside phosphotransferase	250–1,000 μg/ml	6
Pseudomonic acid A	*Methanosarcina* spp.	Inhibits protein synthesis	*ileS*	Pseudomonic resistant isoleucyl-tRNA synthetase	35 μg/ml	19
Hyperthermophilic Crenarchaeota						
Lactose	*S. solfataricus*	SCES[a]	*lacS*	β-Glycosidase	0.2% (wt/vol)	125
Starch or glycogen	*S. solfataricus*	SCES	*malA*	β-Glucosidase	0.2% (wt/vol)	This work

[a]SCES, sole carbon and energy source.

Though selections based on *malA* require significantly longer enrichment times than those based on *lacS*, the availability of an alternative selection maker can be useful for the construction of strains with multiple lesions.

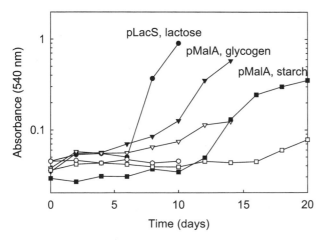

FIGURE 7 Transformation enrichment culture time courses. *S. solfataricus* PBL2025 (circles) and PBL2030 (triangles and squares) were transformed with plasmids pLacS (filled circles) or pMalA (filled triangles and squares) to growth on lactose (circles), glycogen (triangles), or starch (squares). Untransformed control cultures were PBL2025 (open circles) and PBL2030 (open triangles and open squares).

33.3.3. Genetic Markers for Selection

33.3.3.1. *lacS* and *malA* as Selectable Markers

The *S. solfataricus lacS* gene encodes a β-glycosidase with broad substrate specificity including activity against β-galactosides and their chemical analogs such as the colorimetric indicator X-Gal (125) (Table 3). *S. solfataricus* forms blue colonies upon exposure to X-Gal, similar to the behavior of *E. coli* β-galatosidase. Strains lacking this gene are unable to grow on β-linked disaccharides as sole carbon and energy sources but regain this ability (and a blue colony color) when *lacS* is reintroduced into the chromosome. β-Linked disaccharides such as lactose undergo spontaneous hydrolysis when heated. Because *S. solfataricus* cultivation depends on the use of a high-temperature environment, *lacS*-based selections must accommodate substrate hydrolysis. Salacin [2-(hydroxymethyl)phenyl-β-D-glucopyranoside], lactose, and then cellobiose, in this order, exhibit decreasing relative resistance to hydolysis upon heating and have all been used for *lacS*-based selections. Confirmation of the presence of the *lacS* gene in recombinant strains can be verified by several tests. In addition to genomic analysis using PCR and restriction digestion, these include detection of LacS specific activity in colonies by X-Gal hydrolysis, colorimetric enzyme assays of cell lysates using *p*-nitrophenyl-β-glucoside (β-PNPG), and Western blot analysis (48, 49).

The *malA* gene encodes an α-glucosidase that hydrolyzes maltose, maltose oligomers, and dextrin and its branched forms, including starch and glycogen (96, 97) (Table 3). The occurrence (or absence) of MalA activity can be detected by spraying colonies with X-Glu (5-bromo-

4-chloro-3-indoyl-β-D-glucopyranoside), an analog of X-Gal, by colorimetric enzyme assay of cell lysates using *p*-nitrophenyl-α-glucoside (α-PNPG), and by Western blot analysis (96, 97).

33.3.3.2. Screening for Recombinants

The following protocol describes the procedure for identification of *lacS* recombinants derived from enrichment cultures produced as described below (section 33.3.5). An analogous procedure is performed for identification of *malA* recombinants.

1. Plate 10-fold serial dilutions of the enrichment culture on replicate rich medium (0.2% [wt/vol] tryptone) plates and incubate at 80°C until colonies form (4 to 6 days).

2. Recombinant strains are detected by spraying colonies with an X-Gal solution using an atomizer (10 mg ml^{-1} in *N,N*-dimethyl formamide, stored at −20°C protected from light exposure). Incubate plates an additional 20 min at 80°C until blue colonies are observed.

3. Pick and patch blue colonies on rich medium plates, incubate at 80°C for 3 days, and retest the *LacS* phenotype by treating with X-Gal.

4. Inoculate confirmed patches into 25-ml shake-flask cultures in rich medium and grow to mid log phase (e.g., OD$_{540}$ = 0.5).

5. Using 10 ml of culture, isolate genomic DNA (97) for PCR and restriction analysis. With an additional 10 ml of culture, prepare a frozen stock.

33.3.4. Vectors for Gene Transfer

Plasmids that do not replicate in *S. solfataricus* can be used to introduce DNA into the genome. It is easiest to use conventional cloning vectors readily propagated in *E. coli* such as the pUC vectors and their commercial analogs (e.g., pBluescript, pGEM, pNEB). Linear DNA is also suitable but retards recovery of recombinants, as indicated by an increase in enrichment time of 12%.

33.3.4.1. DNA Methylation

Methylation of plasmid DNA improves recovery of recombinants and decreases enrichment time by 15%. The procedure for DNA methylation is as follows:

1. Prepare the following DNA methylation mixture in a sterile 1.5-ml microcentrifuge tube and incubate overnight at 37°C:

HaeIII 10 × methylase buffer (New
  England Biolabs, Inc) . . . . 2.0 μl
S-Adenosylmethionine 16 mM stock . . . . 0.1 μl
Plasmid DNA . . . . 5.0 μg
HaeIII methylase enzyme . . . . 1.5 μl

Adjust volume to 20 μl using double-distilled water.

2. Purify methylated plasmid DNA using a column purification method and elute the DNA with approximately 10 μl of sterile double-distilled water to achieve a final DNA concentration of 0.5 to 1.0 μg/μl (total, 10 μg).

33.3.5. Transformation

Transformation of *S. solfataricus* employs electroporation. Factors affecting the transformation frequency include the type and amount of DNA, electroporation conditions, and cell preparation procedure. Frozen electrocompetent cell preparations can also be used.

33.3.5.1. Transformation of *S. solfataricus* by Electroporation

1. Prepare a cell pellet from 20 ml of a mid-exponential-phase culture (see section 33.3.6.2) by centrifugation for 10 min at 3,000 × *g* in a Sorval SS34 rotor at room temperature.

2. Add 1 ml of 20 mM sucrose and gently tap tube to resuspend cells.

3. Add an additional 4 ml of 20 mM sucrose and recover the cells as in step 1.

4. Repeat steps 2 and 3, two additional times.

5. Resuspend the cell pellet in 20 mM sucrose to a final volume of 1 ml, which is sufficient for up to 20 transformations.

6. For each transformation, transfer 0.05 ml of cells to a 1.5-ml polypropylene sterile microcentrifuge tube and preheat at 50°C for 10 min in a heating block.

7. Add DNA (typically 1 μg), then transfer the mixture into an electroporation cuvette (0.1-cm electrode) and incubate at 50°C for an additional 3 min in a heating block.

8. Electroporate at 100 ohms, 2.0 kV, 25 μF (2.5 to 5 ms).

9. Transfer mixture into a flask containing preheated (80°C) medium and incubate with shaking for 8 h at 80°C.

10. Recover cells by centrifugation (step 1). Decant supernatant, resuspend in 1 ml of Brock Salts, and subculture into 50 ml of prewarmed lactose selection medium.

11. When growth is apparent (typically 6 days), process culture (see section 33.3.3.2).

33.3.6. Media and Growth

Cells are cultivated aerobically at 80°C in glass Ehrlenmeyer flasks tightly sealed with polypropylene screw caps in rotary water bath shakers with fitted Lucite lids to minimize evaporation. Rotation speed is 100 rpm using flasks at least five times the volume of the culture (e.g., 250-ml flask for 50-ml culture). Glycerin rather than water is used to maintain flask temperature. Growth in liquid cultures is monitored spectrophotometrically at a wavelength of 540 nm. Growth on solid medium employs polystyrene plastic plates incubated at 80°C in plastic containers with sufficient hydration to prevent desiccation. Hydration is ensured by adding water to the bottom of the container to a 5-cm depth.

33.3.6.1. Media Preparation

33.3.6.1.1. Liquid Medium Preparation

S. solfataricus is grown in Allen's medium (1) as modified by Brock (22) (referred to as "Brock Salts") or as modified by Worthington (124). The major salts of Brock Salts include (per liter):

$(NH_4)_2SO_4$ . . . . 1.3 g
KH_2PO_4 . . . . 0.28 g
$MgSO_4$ . . . . 0.12 g
$CaCl_2 \cdot 2H_2O$ . . . . 0.072 g
$FeCl_3 \cdot 6H_2O$ . . . . 0.02 g
$Na_2B_4O_7 \cdot 10H_2O$ . . . . 0.0045 g

The minor salts of Brock Salts are prepared fresh in water and are added volumetrically. These salts, their stock solution concentrations, and the amounts added per liter of medium are:

$MnCl_2 \cdot 4H_2O$ . . . . 100 μl of 18 mg/ml
$ZnSO_4 \cdot 7H_2O$ . . . . 10 μl of 22 mg/ml

CuCl$_2$·2H$_2$O .... 10 μl of 5.0 mg/ml
NaMoO$_4$·2H$_2$O .... 10 μl of 3.0 mg/ml
VOSO$_4$·xH$_2$O .... 10 μl of 3.0 mg/ml
CoSO$_4$·7H$_2$O .... 10 μl of 1.0 mg/ml

The pH of Brock Salts is adjusted to 3.0 by addition of sulfuric acid; the final volume is adjusted to 1 liter and sterilized by autoclaving. Defined media contain carbon sources added after autoclaving from concentrated stock solutions to final concentrations of 0.2% (wt/vol). Sucrose is used as the carbon source for a general minimal medium, while lactose, starch, or glycogen is used for enrichment cultures to obtain recombinants. Rich (complex) medium contains tryptone, added as a powder prior to autoclaving to a final concentration of 0.2% (wt/vol). In all cases, the effect of carbon source addition on medium pH must be accommodated by prior adjustment.

33.3.6.1.2. Solidified Medium Preparation

A solid medium is prepared using 0.6% (wt/vol) gelrite (Kelco) and Brock Salts containing tryptone 0.2% (wt/vol) and adjusted to pH 3.0 with sulfuric acid. Magnesium chloride is added at a final concentration of 8.0 mM to solidify the medium.

33.3.6.2. Storage of Strains

The following protocol is used for long-term storage of strains.

1. Prepare a cell pellet by centrifugation at room temperature by using 10 ml of a mid-exponential-phase culture at 3,000 × g in an SS34 Sorvall rotor. Mid-exponential-phase growth is a culture density of 0.5 OD$_{540}$.
2. Remove supernatant and resuspend in 0.93 ml of Brock Salts.
3. Add 0.07 ml of dimethyl sulfoxide (DMSO, spectral grade).
4. Transfer to a 1.5-ml polypropylene microcentrifuge tube, label, and flash freeze for 5 min in an ethanol–dry ice mixture.
5. Store in a −80°C freezer.

33.3.7. Troubleshooting

The most common problem yet encountered precluding recovery of chromosomal recombinants is the lack of sufficient homologous flanking DNA. In some cases it is apparent that 200 bp of flanking DNA is not sufficient to enable recovery of recombinant strains and that longer stretches are required. Other typical problems pertaining to the general cultivation of hyperthermophilic archaea include the need to use clean glassware untainted by detergents, and sufficient aeration to support their obligate aerobic metabolism. It is important also to avoid using cultures that have entered stationary phase. *Sulfolobus* is an extreme acidophile and cannot withstand periods of nutrient deprivation. Without an energy source, proton extrusion is compromised and the organism undergoes lysis.

33.3.8. Recent Developments

For the hyperthermophiles, further developments include creation of recombinant strains with chromosomal deletions and strains with integrated promoter gene fusions (52). Newer but as yet unpublished methods have been developed for *S. solfataricus*. They include: (i) markerless exchange, a method that allows reuse of a selectable marker to create multiple mutations in a single strain; (ii) linear DNA recombination, a method using PCR or other linear DNA molecules rather than circular forms and in which recombinants arise by double-crossover events; (iii) topological mapping, a method for localizing chromosomal mutations using large insert libraries; and (iv) protein tagging and expression, a suite of methods for high-throughput purification and shotgun functional screening of heterologous genomic libraries.

Preparation of this chapter subsection was supported in part by National Science Foundation grant MCB-0110762 and U.S. Department of Energy grant DE-FG02-93-ER20106 (to K.S.), National Science Foundation grants MCB-0235167 and MCB-0085216 (to P.B.), and National Science Foundation grants MCB-0296017, 0135595, 0196502, and 0296019 (to S.D.). Thanks are due to the many students and collaborators who contributed to the development of techniques and execution of experiments in the laboratory, and to Ethel Apolinario, Brian Berquist, Viet Hoang, Kwan-Hwan Kevin Jung, and James Schelert for providing data prior to publication. The authors also thank W. Metcalf and W. Whitman for helpful suggestions.

33.4. REFERENCES

1. **Allen, M. B.** 1959. Studies with *Cyanidium caldarium*, an anomalously pigmented chlorophyte. *Arch. Mikrobiol.* **32:** 270–277.
2. **Allers, T., and M. Mevarech.** 2005. Archaeal genetics—the third way. *Nat. Rev. Genet.* **6:**58–73.
3. **Apolinario, E. E., K. M. Jackson, and K. R. Sowers.** 2005. Development of a plasmid-mediated reporter system for in vivo monitoring of gene expression in the archaeon Methanosarcina acetivorans. *Appl. Environ Microbiol.* **71:** 4914–4918.
4. **Apolinario, E. E., and K. R. Sowers.** 1996. Plate colonization of *Methanococcus maripaludis* and *Methanosarcina thermophila* in a modified canning jar. *FEMS Microbiol. Lett.* **145:**131–137.
5. **Aravalli, R. N., and R. A. Garrett.** 1997. Shuttle vectors for hyperthermophilic archaea. *Extremophiles* **1:**183–191.
6. **Argyle, J. L., D. L. Tumbula, and J. A. Leigh.** 1996. Neomycin resistance as a selectable marker in *Methanococcus maripaludis*. *Appl. Environ. Microbiol.* **62:**4233–4237.
7. **Balch, W. E., G. E. Fox, L. J. Magrum, C. R. Woese, and R. S. Wolfe.** 1979. Methanogens: reevaluation of a unique biological group. *Microbiol. Rev.* **43:**260–296.
8. **Balch, W. E., and R. S. Wolfe.** 1976. New approach to the cultivation of methanogenic bacteria: 2-mercaptoethanesulfonic acid (HS-CoM)-dependent growth of *Methanobacterium ruminantium* in a pressurized atmosphere. *Appl. Environ. Microbiol.* **32:**781–791.
9. **Baliga, N. S., S. J. Bjork, R. Bonneau, M. Pan, C. Iloanusi, M. C. Kottemann, L. Hood, and J. DiRuggiero.** 2004. Systems level insights into the stress response to UV radiation in the halophilic archaeon Halobacterium NRC-1. *Genome Res.* **14:**1025–1035.
10. **Baliga, N. S., B. R. Bonneau, M. T. Facciotti, M. Pan, G. Glusman, E. W. Deutsch, P. Shannon, Y. Chiu, R. S. Weng, R. R. Gan, P. Hung, S. V. Date, E. Marcotte, L. Hood, and W. V. Ng.** 2004. Genome sequence of Haloarcula marismortui: a halophilic archaeon from the Dead Sea. *Genome Res.* **14:**2221–2234.
11. **Baliga, N., and S. DasSarma.** 1999. Saturation mutagenesis of the TATA-box and upstream activator sequence in the haloarchaeal *bop* gene promoter. *J. Bacteriol.* **181:** 2513–2518.
12. **Baliga, N. S., and S. DasSarma.** 2000. Saturation mutagenesis of the haloarchaeal *bop* gene promoter: identification of DNA supercoiling sensitivity sites and absence of

TFB recognition element and UAS enhancer activity. *Mol. Microbiol.* **36:**1175–1183.

13. **Baliga, N. S., S. P. Kennedy, W. V. Ng, L. Hood, and S. DasSarma.** 2001. Genomic and genetic dissection of an archaeal regulon. *Proc. Natl. Acad. Sci. USA* **98:**2521–2525.

14. **Bartus, C. L., V. P. Jaakola, R. Reusch, H. H. Valentine, P. Heikinheimo, A. Levay, L. T. Potter, H. Heimo, A. Goldman, and G. J. Turner.** 2003. Downstream coding region determinants of bacterio-opsin, muscarinic acetylcholine receptor and adrenergic receptor expression in *Halobacterium salinarum. Biochim. Biophys. Acta* **1610:**109–123.

15. **Beneke, S., H. Bestgen, and A. Klein.** 1995. Use of the *Escherichia coli uidA* gene as a reporter in *Methanococcus voltae* for the analysis of the regulatory function of the intergenic region between the operons encoding selenium-free hydrogenases. *Mol. Gen. Genet.* **248:**225–228.

16. **Bitan-Banin, R. O. G, and M. Mevarech.** 2003. Development of a gene knockout system for the halophilic archaeon *Haloferax volcanii* by use of the *pyr*E gene. *J. Bacteriol.* **185:**772–778.

17. **Blank, C. E., P. S. Kessler, and J. A. Leigh.** 1995. Genetics in methanogens: transposon insertion mutagenesis of a *Methanococcus maripaludis nif*H gene. *J. Bacteriol.* **177:**5773–5777.

18. **Blaseio, U., and F. Pfeifer.** 1990. Transformation of *Halobacterium halobium:* development of vectors and investigation of gas vesicle synthesis. *Proc. Natl. Acad. Sci. USA* **87:**6772–6776.

19. **Boccazzi, P., K. J. Zhang, and W. W. Metcalf.** 2000. Generation of dominant selectable markers for resistance to pseudomonic acid by cloning and mutagenesis of the *ile*S gene from the archaeon *Methanosarcina barkeri* Fusaro. *J. Bacteriol.* **182:**2611–2618.

20. **Bock, A., and O. Kandler.** 1985. Antibiotic sensitivity of archaebacteria. *In* C. R. Woese and R. S. Wolfe (ed.), *The Bacteria. A Treatise on Structure and Function*, vol. VIII. Academic Press, Inc., New York.

21. **Bowen, T. L., and W. B. Whitman.** 1987. Incorporation of exogenous purines and pyrimidines by *Methanococcus voltae* and isolation of analog-resistant mutants. *Appl. Environ. Microbiol.* **53:**1822–1826.

22. **Brock, T. D., K. M. Brock, R. T. Belly, and R. L. Weiss.** 1972. *Sulfolobus:* a new genus of sulfur-oxidizing bacteria living at low pH and high temperature. *Arch. Mikrobiol.* **84:**54–68.

23. **Cannio, R., P. Contursi, M. Rossi, and S. Bartolucci.** 1998. An autonomously replicating transforming vector for *Sulfolobus solfataricus. J. Bacteriol.* **180:**3237–3240.

24. **Charlebois, R. L., L. C. Schalkwyk, J. D. Hofman, and W. F. Doolittle.** 1991. Detailed physical map and set of overlapping clones covering the genome of the archaebacterium *Haloferax volcanii* DS2. *J. Mol. Biol.* **222:**509–524.

25. **Cohen-Kupiec, R., C. Blank, and J. A. Leigh.** 1997. Transcriptional regulation in Archaea: *In vivo* demonstration of a repressor binding site in a methanogen. *Proc. Natl. Acad. Sci. USA* **94:**1316–1320.

26. **Danner, S., and J. Soppa.** 1996. Characterization of the distal promoter element of halobacteria *in vivo* using saturation mutagenesis and selection. *Mol. Microbiol.* **19:**1265–1276.

27. **DasSarma, S.** 1989. Mechanisms of genetic variability in *Halobacterium halobium:* the purple membrane and gas vesicle mutations. *Can. J. Microbiol.* **35:**65–72.

28. **DasSarma, S.** 2003. Genome sequence of an extremely halophilic archaeon. *In* T. R. C. M. Fraser and K. E. Nelson (ed.), *Microbial Genomes.* Humana Press, Inc, Totowa, NJ.

29. **DasSarma, S., and P. Arora.** 2002. Halophiles, p. 458–466. *In Encyclopedia of Life Sciences.* Macmillan Press, London, England.

30. **DasSarma, S., P. Arora, F. Lin, E. Molinari, and L. R. Yin.** 1994. Wild-type gas vesicle formation requires at least ten genes in the gvp gene cluster of *Halobacterium halobium* plasmid pNRC100. *J. Bacteriol.* **176:**7646–7652.

31. **DasSarma, S., and E. M. Fleischmann (ed.).** 1995. *Halophiles,* vol. 1. *Archaea: A Laboratory Manual.* Cold Spring Harbor Laboratory Press, Plainview, NY.

32. **Deppenmeier, U., A. Johann, T. Hartsch, R. Merkl, R. A. Schmitz, R. Martinez-Arias, A. Henne, A. Wiezer, S. Bäumer, C. Jacobi, H. Brüggemann, T. Lienard, A. Christmann, M. Bömeke, S. Steckel, A. Bhattacharyya, A. Lykidis, R. Overbeek, H.-P. Klenk, R. P. Gunsalus, H. J. Fritz, and G. Gottschalk.** 2002. The genome of *Methanosarcina mazei:* evidence for lateral gene transfer between Bacteria and Archaea. *J. Mol. Microbiol. Biotechnol.* **4:**453–461.

33. **Dyall-Smith, M. L., and W. F. Doolittle.** 1994. Construction of composite transposons for halophilic Archaea. *Can. J. Microbiol.* **40:**922–929.

34. **Ehlers, C., K. Weidenbach, K. Veit, U. Deppenmeier, W. W. Metcalf, and R. A. Schmitz.** 2005. Development of genetic methods and construction of a chromosomal glnK1 mutant in Methanosarcina mazei strain Go1. *Mol. Genet. Genomics* **273:**290–298.

35. **Elferink, M. G., C. Schleper, and W. Zillig.** 1996. Transformation of the extremely thermoacidophilic archaeon *Sulfolobus solfataricus* via a self-spreading vector. *FEMS Microbiol. Lett.* **137:**31–35.

36. **Falb, M., F. Pfeiffer, P. Palm, K. Rodewald, V. Hickmann, J. Tittor, and D. Oesterhelt.** 2005. Living with two extremes: conclusions from the genome sequence of Natronomonas pharaonis. *Genome Res.* **15:**1336–1343.

37. **Firtel, M., G. B. Patel, and T. J. Beveridge.** 1995. S layer regeneration in *Methanococcus voltae* protoplasts. *Microbiology* **141:**817–824.

38. **Fuller, A., G. Banks, G. Mellows, K. Barrow, M. Woolford, and E. Chain.** 1971. Pseudomonic acid: an antibiotic produced by *Pseudomonas fluorescens. Nature* **234:**416–417.

39. **Galagan, J. E., C. Nusbaum, A. Roy, M. G. Endrizzi, P. Macdonald, W. FitzHugh, S. Calvo, R. Engels, S. Smirnov, D. Atnoor, A. Brown, N. Allen, J. Naylor, N. Stange-Thomann, K. DeArellano, R. Johnson, L. Linton, P. McEwan, K. McKernan, J. Talamas, A. Tirrell, W. J. Ye, A. Zimmer, R. D. Barber, I. Cann, D. E. Graham, D. A. Grahame, A. M. Guss, R. Hedderich, C. Ingram-Smith, H. C. Kuettner, J. A. Krzycki, J. A. Leigh, W. X. Li, J. F. Liu, B. Mukhopadhyay, J. N. Reeve, K. Smith, T. A. Springer, L. A. Umayam, O. White, R. H. White, E. C. de Macario, J. G. Ferry, K. F. Jarrell, H. Jing, A. J. L. Macario, I. Paulsen, M. Pritchett, K. R. Sowers, R. V. Swanson, S. H. Zinder, E. Lander, W. W. Metcalf, and B. Birren.** 2002. The genome of *Methanosarcina acetivorans* reveals extensive metabolic and physiological diversity. *Genome Res.* **12:**532–542.

40. **Gardner, W. L., and W. B. Whitman.** 1999. Expression vectors for *Methanococcus maripaludis:* overexpression of acetohydroxyacid synthase and beta-galactosidase. *Genetics* **152:**1439–1447.

41. **Gernhardt, P., O. Possot, M. Foglino, L. Sibold, and A. Klein.** 1990. Construction of an integration vector for use in the archaebacterium *Methanococcus voltae* and expression of a eubacterial resistance gene. *Mol. Gen. Genet.* **22:**273–279.

42. **Grant, W. D., and H. Larsen.** 1984. Extremely halophilic archaeobacteria, p. 2216–2233. *In* S. Staley (ed.), *Bergey's Manual of Systematic Bacteriology*, vol. 3. The Williams & Wilkins Co., Baltimore, MD.

43. **Gregor, D., and F. Pfeifer.** 2001. Use of a halobacterial *bga*H reporter gene to analyse the regulation of gene expression in halophilic archaea. *Microbiology* **147:**1745–1754.

44. **Grogan, D. W., G. T. Carver, and J. W. Drake.** 2001. Genetic fidelity under harsh conditions: analysis of spontaneous mutation in the thermoacidophilic archaeon *Sulfolobus acidocaldarius. Proc. Natl. Acad. Sci. USA* **98:**7928–7933.

45. **Grogan, D. W., and J. E. Hansen.** 2003. Molecular characteristics of spontaneous deletions in the hyperthermophilic archaeon *Sulfolobus acidocaldarius. J. Bacteriol.* **185:**1266–1272.

46. **Guy, C. P., S. Haldenby, A. Brindley, D. A. Walsh, G. S. Briggs, M. J. Warren, T. Allers, and E. L. Bolt.** 2006. Interactions of RadB, a DNA repair protein in Archaea, with DNA and ATP. *J. Mol. Biol.* **358:**46–56.

47. **Harris, J. E., and P. A. Pinn.** 1985. Bacitracin-resistant mutants of a mesophilic *Methanobacterium* species. *Arch. Microbiol.* **143:**151–153.

48. **Haseltine, C., R. Montalvo-Rodriguez, A. Carl, E. Bini, and P. Blum.** 1999. Extragenic pleiotropic mutations that repress glycosyl hydrolase expression in the hyperthermophilic archaeon *Sulfolobus solfataricus. Genetics* **152:**1353–1361.

49. **Haseltine, C., R. Montalvo-Rodriguez, E. Bini, A. Carl, and P. Blum.** 1999. Coordinate transcriptional control in the hyperthermophilic archaeon *Sulfolobus solfataricus. J. Bacteriol.* **181:**3920–3927.

50. **Hendrickson, E. L., R. Kaul, Y. Zhou, D. Bovee, P. Chapman, J. Chung, E. C. de Macario, J. A. Dodsworth, W. Gillett, D. E. Graham, M. Hackett, A. K. Haydock, A. Kang, M. L. Land, R. Levy, T. J. Lie, T. A. Major, B. C. Moore, I. Porat, A. Palmeiri, G. Rouse, C. Saenphimmachak, D. Soll, S. Van Dien, T. Wang, W. B. Whitman, Q. Xia, Y. Zhang, F. W. Larimer, M. V. Olson, and J. A. Leigh.** 2004. Complete genome sequence of the genetically tractable hydrogenotrophic methanogen Methanococcus maripaludis. *J. Bacteriol.* **186:**6956–6969.

51. **Heymann, J. A., W. A. Havelka, and D. Oesterhelt.** 1993. Homologous overexpression of a light-driven anion pump in an archaebacterium. *Mol. Microbiol.* **7:**623–630.

53. **Hoang, V., E. Bini, V. Dixit, M. Drozda, and P. Blum.** 2004. The role of cis-acting sequences governing catabolite repression control of lacS expression in the archaeon *Sulfolobus solfataricus. Genetics* **167:**1563–1572.

53. **Holmes, M. L., and M. L. Dyall-Smith.** 1990. A plasmid vector with a selectable marker for halophilic archaebacteria. *J. Bacteriol.* **172:**756–761.

54. **Holmes, M. L., and M. L. Dyall-Smith.** 1991. Mutations in DNA gyrase result in novobiocin resistance in halophilic archaebacteria. *J. Bacteriol.* **173:**642–648.

55. **Jones, W. J., W. B. Whitman, F. D. Fields, and R. S. Wolfe.** 1983. Growth and plating efficiency of methanococci on agar media. *Appl. Environ. Microbiol.* **46:**220–226.

56. **Jung, K. H., and J. L. Spudich.** 1998. Suppressor mutation analysis of the sensory rhodopsin I-transducer complex: insights into the color-sensing mechanism. *J. Bacteriol.* **180:**2033–2042.

57. **Jussofie, A., F. Mayer, and G. Gottschalk.** 1986. Methane formation from methanol hydrogen by protoplasts of new methanogenic isolates and inhibition on by dicyclohexylcarbodiimide. *Arch. Microbiol.* **146:**245–249.

58. **Kandler, O., and H. Hippe.** 1977. Lack of peptidoglycan in the cell walls of Methanosarcina barkeri. *Arch. Microbiol.* **113:**57–60.

59. **Kennedy, S. P., W. V. Ng, S. L. Salzberg, L. Hood, and S. DasSarma.** 2001. Understanding the adaptation of *Halobacterium* species NRC-1 to its extreme environment through computational analysis of its genome sequence. *Genome Res.* **11:**1641–1650.

60. **Krebs, M. P., R. Mollaaghababa, and H. G. Khorana.** 1993. Gene replacement in *Halobacterium halobium* and expression of bacteriorhodopsin mutants. *Proc. Natl. Acad. Sci. USA* **90:**1987–1991.

61. **Krebs, M. P., E. N. Spudich, H. G. Khorana, and J. L. Spudich.** 1993. Synthesis of a gene for sensory rhodopsin I and its functional expression in *Halobacterium halobium. Proc. Natl. Acad. Sci. USA* **90:**3486–3490.

62. **Lam, W. L., and W. F. Doolittle.** 1992. Mevinolin-resistant mutations identify a promoter and the gene for a eukaryote-like 3-hydroxy-3-methylglutaryl-coenzyme A reductase in the archaebacterium *Haloferax volcanii. J. Biol. Chem.* **267:**5829–5834.

63. **Lam, W. L., and W. F. Doolittle.** 1989. Shuttle vectors for the archaebacterium *Halobacterium volcanii. Proc. Natl. Acad. Sci. USA* **86:**5478–5482.

64. **Lange, M., and B. K. Ahring.** 2001. A comprehensive study into the molecular methodology and molecular biology of methanogenic Archaea. *FEMS Microbiol. Rev.* **25:**553–571.

64a. **Maeder, D. L., I. Anderson, T. Brettin, D. Bruce, P. Gilna, C. S. Han, A. Lapidus, W. W. Metcalf, E. Saunders, R. Tapia, and K. R. Sowers.** 2006. The *Methanosarcina barkeri* genome: comparative analysis with *Methanosarcina acetivorans* and *Methanosarcina mazei* reveals extensive rearrangement within methanosarcinal genomes. *J. Bacteriol.* **188:**7922–7931.

65. **Mankin, A. S., I. M. Zyrianova, V. K. Kagramanova, and R. A. Garrett.** 1992. Introducing mutations into the single-copy chromosomal 23S rRNA gene of the archaeon *Halobacterium halobium* by using an rRNA operon-based transformation system. *Proc. Natl. Acad. Sci. USA* **89:**6535–6539.

66. **Martusewitsch, E., C. W. Sensen, and C. Schleper.** 2000. High spontaneous mutation rate in the hyperthermophilic archaeon *Sulfolobus solfataricus* is mediated by transposable elements. *J. Bacteriol.* **182:**2574–2581.

67. **McCready, S., and L. Marcello.** 2003. Repair of UV damage in Halobacterium salinarum. *Biochem. Soc. Trans.* **31:**694–698.

68. **McCready, S., J. A. Muller, I. Boubriak, B. R. Berquist, W. L. Ng, and S. Dassarma.** 2005. UV irradiation induces homologous recombination genes in the model archaeon, Halobacterium sp. NRC-1. *Saline Syst.* **1:**3.

69. **Metcalf, W. W.** 1999. Genetic analysis in the domain Archaea, p. 277–326. *In* M. C. Smith and R. E. Sockett (ed.), *Genetic Methods for Diverse Prokaryotes,* vol. 29. Academic Press, New York.

70. **Metcalf, W. W., W. Jiang, and B. L. Wanner.** 1994. Use of rep technique for allele replacement to construct new *Escherichia coli* hosts for maintenance of R6K g origin plasmids at different copy numbers. *Gene* **138:**1–7.

71. **Metcalf, W. W., J. K. Zhang, E. Apolinario, K. R. Sowers, and R. S. Wolfe.** 1997. A genetic system for Archaea of the genus *Methanosarcina*: liposome-mediated transformation and construction of shuttle vectors. *Proc. Natl. Acad. Sci. USA* **94:**2626–2631.

72. **Metcalf, W. W., J. K. Zhang, and R. S. Wolfe.** 1998. An anaerobic, intrachamber incubator for growth of *Methanosarcina* spp. on methanol-containing solid media. *Appl. Environ. Microbiol.* **64:**768–770.

73. **Meuer, J., H. C. Kuettner, J. K. Zhang, R. Hedderich, and W. W. Metcalf.** 2002. Genetic analysis of the archaeon *Methanosarcina barkeri* Fusaro reveals a central role for Ech hydrogenase and ferridoxin in methanogenesis and carbon fixation. *Proc. Natl. Acad. Sci. USA* **99:**5632–5637.

74. **Miller, J. H.** 1992. *A Short Course in Bacterial Genetics.* Cold Spring Harbor Laboratory Press, Plainview, NY.

75. **Moore, B. C., and J. A. Leigh.** 2005. Markerless mutagenesis in Methanococcus maripaludis demonstrates roles for alanine dehydrogenase, alanine racemase, and alanine permease. *J. Bacteriol.* **187:**972–979.

76. **Muller, J. A., and S. DasSarma.** 2005. Genomic analysis of anaerobic respiration in the archaeon Halobacterium sp. strain NRC-1: dimethyl sulfoxide and trimethylamine N-oxide as terminal electron acceptors. *J. Bacteriol.* **187:**1659–1667.

77. **Ng, W.-L., P. Arora, and S. DasSarma.** 1994. Large deletions in class III gas-vesicles deficient mutants of *Halobacterium halobium.* Syst. Appl. Microbiol. **16:**560–568.

78. **Ng, W.-L., S. A. Ciufo, T. M. Smith, R. E. Bumgarner, D. Baskin, J. Faust, B. Hall, C. Loretz, J. Seto, J. Slagel, L. Hood, and S. DasSarma.** 1998. Snapshot of a large dynamic replicon from a halophilic archaeon: Megaplasmid or minichromosome? *Genome Res.* **8:**1131–1141.

79. **Ng, W. L., and S. DasSarma.** 1993. Minimal replication origin of the 200-kilobase *Halobacterium* plasmid pNRC100. *J. Bacteriol.* **175:**4584–4596.

80. **Ng, W. V., S. P. Kennedy, G. G. Mahairas, B.Berquist, M. Pan, H. D. Shukla, S. R. Lasky, N. S. Baliga, V. Thorsson, J. Sbrogna, S. Swartzell, D. Weir, J. Hall, T. A. Dahl, R. Welti, Y. A. Goo, B. Leithauser, K. Keller, R. Cruz, M. J. Danson, D. W. Hough, D. G. Maddocks, P. E. Jablonski, M. P. Krebs, C. M. Angevine, H. Dale, T. A. Isenbarger, R. F. Peck, M. Pohlschroder, J. L. Spudich, K. W. Jung, M. Alam, T. Freitas, S. Hou, C. J. Daniels, P. P. Dennis, A. D. Omer, H. Ebhardt, T. M. Lowe, P. Liang, M. Riley, L. Hood, and S. DasSarma.** 2000. Genome sequence of Halobacterium species NRC-1. *Proc. Natl. Acad. Sci. USA* **97:**12176–12181.

81. **Ni, B. F., M. Chang, A. Duschl, J. Lanyi, and R. Needleman.** 1990. An efficient system for the synthesis of bacteriorhodopsin in *Halobacterium halobium.* Gene **90:**169–172.

82. **Nieuwlandt, D. T., and C. J. Daniels.** 1990. An expression vector for the archaebacterium *Haloferax volcanii.* J. Bacteriol. **172:**7104–7110.

83. **Nomura, S., and Y. Harada.** 1998. Functional expression of green fluorescent protein derivatives in *Halobacterium salinarum.* FEMS Microbiol. Lett. **167:**287–293.

84. **Patenge, N., A. Haase, H. Bolhuis, and D. Oesterhelt.** 1995. The gene for a halophilic beta-galactosidase (*bga*H) of *Haloferax alicantei* as a reporter gene for promoter analyses in *Halobacterium salinarum.* EMBO J. **14:**667–673.

85. **Patenge, N., and J. Soppa.** 1999. Extensive proteolysis inhibits high-level production of eukaryal G protein-coupled receptors in the archaeon *Haloferax volcanii.* FEMS Microbiol. Lett. **171:**27–35.

86. **Peck, R. F., S. DasSarma, and M. P. Krebs.** 2000. Homologous gene knockout in the archaeon *Halobacterium* with ura3 as a counterselectable marker. *Mol. Microbiol.* **35:**667–676.

87. **Peck, R. F., C. Echavarri-Erasun, E. A. Johnson, W. V. Ng, S. P. Kennedy, L. Hood, S. DasSarma, and M. P. Krebs.** 2001. *brp* and *blh* are required for synthesis of the retinal cofactor of bacteriorhodopsin in *Halobacterium.* J. Biol. Chem. **276:**5739–5744.

88. **Peck, R. F., E. A. Johnson, and M. P. Krebs.** 2002. Identification of a lycopene beta-cyclase required for bacteriorhodopsin biogenesis in the archaeon *Halobacterium salinarum.* J. Bacteriol. **184:**2889–2897.

89. **Piatibratov, M., S. Hou, A. Brooun, J. Yang, H. Chen, and M. Alam.** 2000. Expression and fast-flow purification of a polyhistidine-tagged myoglobin-like aerotaxis transducer. *Biochim. Biophys. Acta* **1524:**149–154.

90. **Possot, O., P. Gernhardt, A. Klein, and L. Sibold.** 1988. Analysis of drug resistance in the archaebacterium *Methanococcus voltae* with respect to potential use in genetic engineering. *Appl. Environ. Microbiol.* **54:**734–740.

91. **Prisco, A., M. Moracci, M. Rossi, and M. Ciaramella.** 1995. A gene encoding a putative membrane protein homologous to the major facilitator superfamily of transporters maps upstream of the beta-glycosidase gene in the archaeon *Sulfolobus solfataricus.* J. Bacteriol. **177:**1614–1619.

92. **Pritchett, M. A., J. K. Zhang, and W. W. Metcalf.** 2004. Development of a markerless genetic exchange method for Methanosarcina acetivorans C2A and its use in construction of new genetic tools for methanogenic archaea. *Appl. Environ. Microbiol.* **70:**1425–1433.

93. **Reilly, M. S., and D. W. Grogan.** 2001. Characterization of intragenic recombination in a hyperthermophilic archaeon via conjugational DNA exchange. *J. Bacteriol.* **183:**2943–2946.

94. **Reuter, C. J., and J. A. Maupin-Furlow.** 2004. Analysis of proteasome-dependent proteolysis in *Haloferax volcanii* cells, using short-lived green fluorescent proteins. *Appl. Environ. Microbiol.* **70:**7530–7538.

95. **Robb, F. T., and A. R. Place (ed.).** 1995. *Thermophiles,* vol. 3. *Archaea: A Laboroatory Manual.* Cold Spring Harbor Laboratory Press, Plainview, NY.

96. **Rolfsmeier, M., and P. Blum.** 1995. Purification and characterization of a maltase from the extremely thermophilic crenarchaeote *Solfataricus solfataricus.* J. Bacteriol. **177:**482–485.

97. **Rolfsmeier, M., C. Haseltine, A. C. E. Bini, and P. Blum.** 1998. Molecular characterization of the a-glucosidase gene (*malA*) from the hyperthermophilic archaeon *Sulfolobus solfataricus.* J. Bacteriol. **180:**1287–1295.

98. **Rother, M., and W. W. Metcalf.** 2005. Genetic technologies for Archaea. *Curr. Opin. Microbiol.* **8:**745–751.

99. **Rother, M., A. Resch, W. L. Gardner, W. B. Whitman, and A. Böck.** 2001. Heterologous expression of archaeal selenoprotein genes directed by the SECIS element located in the 3′ non-translated region. *Mol. Microbiol.* **40:**900–908.

100. **Sandbeck, K. A., and J. A. Leigh.** 1991. Recovery of an integration shuttle vector from tandem repeats in *Methanococcus maripaludis.* Appl. Environ. Microbiol. **57:**2762–2763.

101. **Schelert, J., V. Dixit, V. Hoang, J. Simbahan, M. Drozda, and P. Blum.** 2004. Occurrence and characterization of mercury resistance in the hyperthermophilic archaeon *Sulfolobus solfataricus* using gene disruption. *J. Bacteriol.* **186:**427–437.

102. **Schleper, C., R. Roder, T. Singer, and W. Zillig.** 1994. An insertion element of the extremely thermophilic archaeon *Sulfolobus solfataricus* transposes into the endogenous β-galactosidase gene. *Mol. Gen. Genet.* **243:**91–96.

103. **She, Q., R. K. Singh, F. Confalonieri, Y. Zivanovic, G. Allard, M. J. Awayez, C. C. Chan-Weiher, I. G. Clausen, B. A. Curtis, A. D. Moors, G. Erauso, C. Fletcher, P. M. Gordon, I. H.-d. Jong, A. C. Jeffries, C. J. Kozera, N. Medina, X. Peng, H. P. Thi-Ngoc, P. Redder, M. E. Schenk, C. Theriault, N. Tolstrup, R. L. Charlebois, W. F. Doolittle, M. Duguet, T. Gaasterland, R. A. Garrett, M. A. Ragan, C. W. Sensen, and J. V. d. Oost.** 2001. The complete genome of the Crenarchaeote *Sulfolobus solfataricus* P2. *Proc. Natl. Acad. Sci. USA* **98:**7835–7840.

104. **Smith, P. H., and R. E. Hungate.** 1958. Isolation and characterization of *Methanobacterium ruminatium* n. sp. *J. Bacteriol.* **75:**713–718.

105. **Sowers, K. R., S. F. Baron, and J. G. Ferry.** 1984. *Methanosarcina acetivorans* sp. nov., an acetotrophic methane-producing bacterium isolated from marine sediments. *Appl. Environ. Microbiol.* **47:**971–978.

106. **Sowers, K. R., J. E. Boone, and R. P. Gunsalus.** 1993. Disaggregation of *Methanosarcina* spp. and growth as single cells at elevated osmolarity. *Appl. Environ. Microbiol.* **59:**3832–3839.

107. **Sowers, K. R., and R. P. Gunsalus.** 1988. Adaptation for growth at various saline concentrations by the archaebacterium *Methanosarcina thermophila*. *J. Bacteriol.* **170:**998–1002.

108. **Sowers, K. R., and H. J. Schreier (ed.).** 1995. *Methanogens*, vol. 2. *Archaea: A Laboratory Manual.* Cold Spring Harbor Laboratory Press, Cold Spring Harbor, NY.

109. **Sowers, K. R., and H. J. Schreier.** 1995. Techniques for anaerobic growth, p. 15–55. *In* F. T. Robb, K. R. Sowers, S. DasSharma, A. R. Place, H. J. Schreier, and E. M. Fleischmann (ed.), *Archaea: A Laboratory Manual.* Cold Spring Harbor Laboratory Press, Plainview, NY.

110. **Sowers, K. R., and H. J. Schreier.** 1999. Gene transfer systems for the archaea. *Trends Microbiol.* **7:**212–219.

111. **Stathopoulos, C., W. Kim, T. Li, I. Anderson, B. Deutsch, S. Palioura, W. Whitman, and D. Soll.** 2001. Cysteinyl-tRNA synthetase is not essential for viability of the archaeon *Methanococcus maripaludis*. *Proc. Natl. Acad. Sci. USA* **98:**14292–14297.

112. **Stedman, K. M., C. Schleper, E. Rumpf, and W. Zillig.** 1999. Genetic requirements for the function of the archaeal virus SSV1 in *Sulfolobus solfataricus*: construction and testing of viral shuttle vectors. *Genetics* **152:**1397–1405.

113. **Stoeckenius, W., R. H. Lozier, and R. A. Bogomolni.** 1979. Bacteriorhodopsin and the purple membrane of halobacteria. *Biochim. Biophys. Acta* **505:**215–278.

114. **Stuart, E. S., F. Morshed, M. Sremac, and S. DasSarma.** 2001. Antigen presentation using novel particulate organelles from halophilic archaea. *J. Biotechnol.* **88:**119–128.

115. **Tu, D., G. Blaha, P. B. Moore, and T. A. Steitz.** 2005. Gene replacement in Haloarcula marismortui: construction of a strain with two of its three chromosomal rRNA operons deleted. *Extremophiles* **9:**427–435.

116. **Tumbula, D. L., T. L. Bowen, and W. B. Whitman.** 1995. Growth of methanogens on solidified medium, p. 49–55. *In* F. T. Robb, A. R. Place, K. R. Sowers, H. J. Schreier, S. DasSarma, and E. M. Fleischmann (ed.), *Archaea: A Laboratory Manual.* Cold Spring Harbor Laboratory Press, Plainview, NY.

117. **Tumbula, D. L., T. L. Bowen, and W. B. Whitman.** 1997. Characterization of pURB500 from the archaeon *Methanococcus maripaludis* and construction of a shuttle vector. *J. Bacteriol.* **179:**2976–2986.

118. **Tumbula, D. L., and W. B. Whitman.** 1999. Genetics of Methanococcus: possibilities for functional genomics in Archaea. *Mol. Microbiol.* **33:**1–7.

119. **Wang, G., S. P. Kennedy, S. Fasiludeen, C. Rensing, and S. DasSarma.** 2004. Arsenic resistance in Halobacterium sp. strain NRC-1 examined by using an improved gene knockout system. *J. Bacteriol.* **186:**3187–3194.

120. **Wolin, E. A., M. J. Wolin, and R. S. Wolfe.** 1963. Formation of methane by bacterial extracts. *J. Biol. Chem.* **238:**2882–2886.

121. **Wood, G. E., A. K. Haydock, and J. A. Leigh.** 2003. Function and regulation of the formate dehydrogenase genes of the methanogenic Archaeon *Methanococcus maripaludis*. *J. Bacteriol.* **185:**2548–2554.

122. **Woods, W. G., K. Ngui, and M. L. Dyall-Smith.** 1999. An improved transposon for the halophilic archaeon *Haloarcula hispanica*. *J. Bacteriol.* **181:**7140–7142.

123. **Woodson, D., R. Peck, M. Krebs, and J. Escalante-Semerena.** 2003. The *cobY* gene of the Archaeon *Halobacterium* sp. strain NRC-1 is required for *de novo* cobamide synthesis. *J. Bacteriol.* **185:**311–316.

124. **Worthington, P., P. Blum, F. Perez-Pomares, and T. Elthon.** 2003. Large scale cultivation of acidophilic hyperthermophiles for recovery of secreted proteins. *Appl. Environ. Microbiol.* **69:**252–257.

125. **Worthington, P., V. Hoang, P. Perez-Pomares, and P. Blum.** 2003. Targeted disruption of the a-amylase gene in the hyperthermophilic archaeon *Sulfolobus solfataricus*. *J. Bacteriol.* **185:**482–488.

126. **Zhang, J. K., M. A. Pritchett, D. J. Lampe, H. M. Robertson, and W. W. Metcalf.** 2000. *In vivo* transposon mutagenesis of the methanogenic archaeon *Methanosarcina acetivorans* C2A using a modified version of the insect *mariner*-family transposable element *Himar1*. *Proc. Natl. Acad. Sci. USA* **97:**9665–9670.

127. **Zhang, J. K., A. K. White, H. C. Kuettner, P. Boccazzi, and W. W. Metcalf.** 2002. Directed mutagenesis and plasmid-based complementation in the methanogenic archaeon *Methanosarcina acetivorans* C2A demonstrated by genetic analysis of proline biosynthesis. *J. Bacteriol.* **184:**1449–1454.

128. **Zibat, A.** 2001. Efficient transformation of *Halobacterium salinarum* by a "freeze and thaw" technique. *Biotechniques.* **31:**1010–1012.

129. **Zillig, W., H. P. Arnold, I. Holz, D. Prangishvili, A. Schweier, K. Stedman, Q. She, H. Phan, R. Garrett, and J. K. Kristjansson.** 1998. Genetic elements in the extremely thermophilic archaeon *Sulfolobus*. *Extremophiles* **2:**131–140.

130. **Zillig, W., K. O. Stetter, S. Wunderl, W. Schulz, H. Priess, and J. Scholz.** 1980. The *Sulfolobus*-"Caldariella" group: taxonomy on the basis of the structure of DNA-dependent RNA polymerases. *Arch. Microbiol.* **125:**259–269.

34

Genetic Manipulations Using Phages

GRAHAM F. HATFULL, DEBORAH JACOBS-SERA, MICHELLE H. LARSEN,
AND WILLIAM R. JACOBS, JR.

34.1. INTRODUCTION AND OVERVIEW OF PHAGE GENETICS AND ITS APPLICATIONS

34.1.1. The World of Bacteriophages

Bacteriophages were discovered by Frederick Twort (1916) and Felix d'Herelle (1917) as agents that have the ability to kill bacteria (49). The nature of the killing property remained unclear for many years, although the dilution experiments of d'Herelle provided compelling evidence that they are particulate in nature. Thus at the appropriate concentration, individual areas of killing or "plaques" appear on bacterial lawns. Each plaque is derived from a single submicroscopic entity that had the ability to replicate on its bacterial host. The next few decades saw continued advances in understanding these viruses, although it was not until the development of the electron microscope in the 1940s that the morphologies of phages became apparent (44). With the description of the structure of DNA following shortly after

(51), it became clear that the ability to quickly and simply propagate phages at high concentrations would help to unravel the secrets of the genetic code and provide invaluable tools for manipulating DNA molecules.

More recently, the vastness of the phage population has come into sharper focus (15). Determination of the concentrations of phages in a multitude of environmental samples shows that they are commonly present at 10^6 to 10^7/ml, and usually about 10-fold more prevalent than bacteria (15). Extrapolation of these numbers suggests that there are about 10^{31} phage particles in the biosphere, making them an absolute majority of all biological forms (52). Individual phages show specificity for particular hosts, and the spectrum of available hosts is defined as the "host range"; some phages have extremely narrow host ranges, such as a single serovar, while others are broad and can infect many different bacterial species. The genetic complexity of this huge population remains unclear, although the genomic characterization of about 300 phage genomes shows that they are extremely diverse and typically constructed with mosaic architectures (10, 12, 20, 37). This unusual genetic texture suggests that phages have existed for several billion years and that horizontal genetic exchange has played a major role in their evolution (18–20). Particular phage types thus cannot easily be classified within any generally accepted species concept, and their taxonomy is fraught with complications (31).

While phage morphologies are diverse, the population is dominated by doubled-stranded DNA (dsDNA) tailed phages. Typically, these have a DNA-containing head with icosahedral symmetry attached to a tail, and the viruses can be grouped into three main classes depending on the type of the tail: *Podoviridae* contain short stubby tails, *Siphoviridae* have long flexible tails, and *Myoviridae* have contractile tails (1). Most phages can also be divided into those that are temperate and those that are lytic. Lytic (or virulent) phages usually have only a single outcome upon infection of a bacterial host: reproduction of the virus and lysis of the bacterium. As a consequence, plaques appear as perfectly clear areas of infection, within which all of the bacterial cells are dead. In contrast, there are two alternative outcomes of infection by a temperate phage. One is the same process as occurs with lytic phages, but the second is the establishment of lysogeny, in which the phage genome becomes established and maintained within the cell for many generations (and may either integrate or replicate extra-chromosomally) and the lytic phage functions are switched off. Temperate phages form turbid plaques that arise from a combination of killed cells and lysogens that are now immune to subsequent reinfection by the phage. A variety of plaque morphologies are shown in Fig. 1. In practice, a significant number of phages form "hazy" plaques (intermediate between turbid and clear) and cannot readily be described as either truly temperate or lytic. For example, many mycobacteriophages form hazy plaques and encode integration functions in their genomes. A plausible explanation is that such phages are competent to form stable lysogens, but perhaps not necessarily in all bacterial hosts. Thus phages may exhibit different growth cycles depending on which bacterial host they are infecting.

34.1.2. The Power of Phage Genetics

The term "phage genetics" comes with multiple meanings depending on the perspectives of the reader. In the more traditional and narrower application of the term, it is used to describe the isolation, mapping, and functional analysis of bacteriophage mutations. In the broader context, it refers to the use of bacteriophages to genetically manipulate a bacterial host, and as sources for the development of genetic tools. It is in this broader context that we will use the term "phage genetics" in this chapter.

The role of phage genetics in microbiology has expanded with the realization that we understand only a small fraction of bacterial diversity. Microbes such as *Escherichia coli* and *Salmonella* have been deeply studied and to great effect, but as our attention has turned to the broader bacterial community, new bacteria have brought new genetic challenges. Fortunately, phages have proven to be great friends of the microbiologist for a multitude of reasons. First, phages can be isolated for virtually any bacterial species that can be propagated in the laboratory and from which smooth lawns can be grown on agar plates. Second, these phages can be propagated to high titer (~10^{10}/ml) without any detailed knowledge of replication systems, antibiotic markers, or transformation methods. Third, DNA can be extracted from these particles and characterized genomically. Fourth, phage genomes are replete with genetic features that are of great utility to the genetic engineer, and

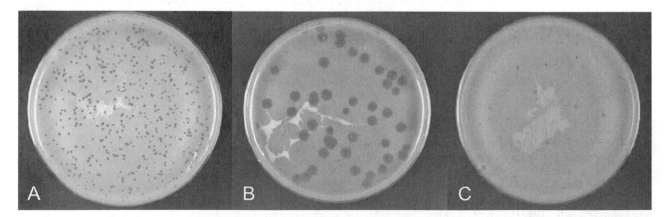

FIGURE 1 Mycobacteriophage plaque morphologies on M. *smegmatis* lawns. (A) Small clear plaques (Mycobacteriophage Oasis). (B) Large clear plaques (Mycobacteriophage Hammer). (C) Turbid plaques (Mycobacteriophage RedRock).

their modular nature means that these can be harvested from the genome and employed to develop genetic tools. Last, the high efficiency of phage infection provides a means of introducing recombinant molecules to every cell within a bacterial population.

In this chapter we will describe some common phage methods and approaches that can be applied to virtually any bacterial host. However, we will utilize an example that we are familiar with—the development of mycobacterial genetics—to illustrate this awesome power of phage genetics.

34.1.3. The Role of Phages in Mycobacterial Genetics: A Case Study

The predominant role of the mycobacteria in human health at the global scale demands the development of a facile genetic system for understanding and manipulating them; thus it is an obvious system for phage genetics. The World Health Organization estimates that one in three of all people are infected with *Mycobacterium tuberculosis*, and each year more people die of these infections than because of any other single infectious agent (11, 13). However, these numbers also illustrate what a formidable opponent *M. tuberculosis* is; most people that it infects do not die of the disease or even get serious symptoms. Instead, a lifelong latent infection is established, a time bomb that can explode into full-blown tuberculosis whenever the immune system loses its full proficiency.

Twenty years ago, two main challenges faced the mycobacterial geneticist. The first was the pathogenicity of *M. tuberculosis* that warrants growing it under strict biosafety conditions (BSL-3), and the second was its extremely slow growth rate with a doubling time of 24 h. (Consider that a typical colony has 10^8 bacterial cells, which would require 30 generations of growth from a single cell, thereby requiring 30 days to obtain a *M. tuberculosis* colony.) The good news is that it grows on defined medium and there are a variety of possible surrogate bacteria that grow much faster and are nonvirulent (such as *Mycobacterium smegmatis*). But in 1986 no simple and efficient methods existed for introducing and propagating recombinant DNA molecules or for the efficient isolation of mutants. Bacteriophages played a central role in overcoming these hurdles (22).

Several collections of mycobacteriophages had been assembled previously with a view to using their host-range specificities for the typing of clinical isolates (26, 27). One of these phages had been shown to mediate generalized transduction in *M. smegmatis* (40) (although none in *M. tuberculosis*), and Tokunaga and colleagues achieved uptake of phage DNA (i.e., transfection) by *M. tuberculosis* (50). However, these did not immediately facilitate the development of methods for introducing foreign DNA into mycobacteria.

This important advance was accomplished through the creation of novel chimeric molecules termed "phasmids," which have the ability to grow in *E. coli* as large extrachromosomal plasmid molecules and as bacteriophages in mycobacteria (25). These can be generated from any phage genome that is appropriately sized (40 to 50 kbp) by ligation into an *E. coli* cosmid vector, packaging into phage lambda particles in vitro, and infection of *E. coli*; DNA can then be prepared from this library of recombinants and used to transfect mycobacteria (see Fig. 2) (24). Phage lysates can then be prepared and used to infect any suitable mycobacterial host. The key advantage of these phasmids is that they can simply be genetically manipulated in *E. coli* and used as vehicles to introduce foreign DNA into mycobacte-

ria. Furthermore, genes can be introduced this way and their expression and phenotypic consequences evaluated by screening without any need for selection, especially when temperate phages are used (47). A notable example is that antibiotic resistance genes can be introduced and their utility can be determined without any prior knowledge of whether and how they work in that bacterial host. Shuttle phasmids thus helped in showing how DNA can be efficiently introduced into mycobacteria and what drug resistance genes might be useful and how. Further details in shuttle phasmid construction are provided below.

The general utility of shuttle phasmids is illustrated by three key applications. First, they can be used to efficiently introduce transposons, taking advantage of the high efficiency of infection to overcome the inherent low frequency of transposition (4). Second, they can be used to introduce reporter genes such as the firefly luciferase, where the efficiency of infection has the potential to enable the simple diagnosis of mycobacteria (and drug susceptibilities) within a clinical sample (23, 45). Third, the efficiency of infection can be utilized to overcome the inherently low frequencies of homologous recombination when constructing mutants by allelic exchange (3).

While mycobacteriophage shuttle phasmids have proven critical to the development of mycobacterial genetics, they represent only one of many examples of phage genetics in the mycobacterial system. The field of mycobacteriophage genetics has been significantly fueled by the isolation and genomic characterization of more than 30 individual phages, and these have served as a genetic reservoir for tool development (17). For example, phage-derived integration vectors are useful for constructing single-copy stable recombinants (28, 32), and recombinase proteins can be identified and adapted for mycobacterial recombineering (50a).

In this chapter, we will describe how phages can be isolated, purified, and genomically characterized, how shuttle phasmids can be constructed and utilized, and how phages can be used to construct a variety of genetic tools. The mycobacterial context provides specific examples, but the approaches are adaptable to many other bacterial systems. Finally, it will become apparent that the potential for phage genetics in any bacterial host is vast, and the approaches and methods described here represent just a small part of what bacteriophages have to offer to the microbial genetic engineer.

34.2. PHAGE ISOLATION

34.2.1. Sample Collection

The process used to isolate phages from environmental samples naturally depends on the ecology of the bacterial host and the environments where the host and its phages usually reside. To isolate phages, simply collect samples from a variety of places where the bacterial host is likely to be found, extract the phages using a simple buffer, and plate a portion of this directly on lawns of the bacterial host. An amplification step can be included if desired, but we typically avoid it to prevent biasing the collection of phages toward those that flourish under laboratory conditions and out-compete all others. Detailed protocols for isolating phages by amplification have been described elsewhere and will not be presented here in any detail.

A typical procedure for isolating mycobacteriophages is as follows. First, gather a small sample from the environment (such as from soil, compost, ponds, etc.) into a 15-ml screw-cap conical tube, typically enough to fill about one-third of the tube. If the sample is solid, add an approximately equal volume of phage buffer (make sure that the buffer includes 1 mM CaCl$_2$). It is helpful to avoid adding too much liquid since this will dilute the sample too much; if too little buffer is added, it will be difficult to subsequently clarify the sample. Thoroughly mix the sample by shaking and then let it sit for about 20 min; most of the solid material will have settled to the bottom of the tube. Remove about 1 ml of liquid from the liquid phase by drawing up the sample into a 1-ml syringe, attach a 0.22-μm-pore-size filter to the syringe, and collect the filtered sample in a 1.5-ml microcentrifuge tube. The sample is now clear and sterile.

34.2.2. Identification of Plaques

The specific conditions that are used to grow a bacterial host will obviously depend on what host is used. The following description details the process of isolating mycobacteriophages by using M. *smegmatis* as a host, but it can be readily adapted to other hosts.

In sterile tubes, add about 50 μl of sterile phage sample to 0.5 ml of a culture of saturated M. *smegmatis*. Set up a parallel sample to act as a negative control to which no phage sample is added; simply add 50 μl of phage buffer. Let these sit on the bench for 15 to 30 min to allow the phage to adsorb to the bacteria. During this time, melt some top agar by microwaving a 50-ml bottle of 0.7% top agar until boiling, then adding an equal volume (50 ml) of medium to bring the final agar concentration to about 0.35%; maintain the temperature at about 55°C; we typically add CaCl$_2$ to a final concentration of 1 mM after the agar has been boiled and diluted. Approximately 4.5 ml of this top agar is then added to each sample tube, and the entire contents is poured onto a standard agar plate. For this step to work well, it is important that the plates are relatively freshly poured, are prewarmed to either room temperature or 37°C, and are not too wet (i.e., fresh but not too fresh!). Immediately after pouring the contents out, swirl the plate to make sure it is evenly covered and then let stand until the top agar layer hardens (usually about 15 to 30 min). Although neither of these actions appears to be difficult, they are critical to success. A minimal amount of swirling will distribute the agar and sample over the entire agar surface, although it is imperative that the plates are not moved during cooling, to ensure that the top agar solidifies uniformly. Turn the plate over carefully, and incubate at 37°C overnight or until a bacterial lawn has grown across the entire plate.

When the bacterial lawn is fully grown, examine it for plaques. Since amplification has not been done in this procedure, the number of plaques is usually small and highly variable. Frequently, only about 3 to 10% of all the samples give any plaques at all, and when they do, it is usually only one or two plaques; occasionally, a sample gives a couple of dozen plaques or more. It is usual to also see a few bubbles in the top agar, and these can easily be mistaken for plaques! If several plaques appear to have different plaque morphologies (such as large, small, clear, turbid) (Fig. 1), then these may be different and can be picked for further testing.

To confirm that plaques are real (rather than bubbles), use a plastic pipette tip to touch the putative plaque, and then twirl the tip in 100 μl of phage buffer and perform a spot test on a fresh lawn of bacteria. To do the spot test, prepare the lawn as described above for the no-phage control, let the agar harden, and then spot with a 5-μl drop of phage sample. Allow to dry and incubate at 37°C overnight. If the sample contains phage particles, then either complete killing or individual plaques within the spot will be observed. This is convenient if testing large numbers of samples, since many can be added onto a single agar plate. A positive spot test will usually show complete cell killing in the entire spotted drop.

34.2.3. Plaque Purification

If the preceding procedures are successful, then it is necessary to ensure the homogeneity of the phage sample by performing at least three rounds of plaque purification—more won't do any harm! To do this, pick a single plaque from a plate by using a plastic pipette tip (or from a spot from a spot test) and twirl into 100 μl of phage buffer. (Be cautious if the spot test plate is used for this step. It may be best to go back to the sample used to make the spot test to avoid contamination from other spots). Prepare serial dilutions of the sample in phage buffer—typical dilutions are 10^{-2}, 10^{-3}, 10^{-4}—and add 10 μl of each dilution to 0.5 ml of plating cells. After allowing 15 to 30 minutes for adsorption, add 4.5 ml of top agar and pour onto plates. Allow agar to harden, invert plates, and incubate until a lawn has grown and plaques are visible. When starting with a single plaque, between 10^4 and 10^6 particles are transferred into the original tube. One reason why it is important to do many rounds of plaque purification is that the phage particles can continue to diffuse through the top agar layer after the lawns have fully grown and the plaques do not appear to get any larger. Thus, there is always a risk of picking up particles other than those that are in the intended plaque. Avoid extended incubation or wait times between rounds of purification to prevent this problem.

34.2.4. Preparation of a Working Stock Lysate

To prepare a lysate as a working stock, use a freshly prepared plate from the final plaque purification process that contains just enough phage to have cleared the entire plate. To this add 5 ml of phage buffer and swirl it so that the whole plate is covered; let this stand at room temperature for 2 to 3 h (or overnight in the cold). Collect the liquid using a Pasteur pipette and filter through a 0.22-μm-pore-size filter into a clean sterile tube. Determine the titer of this by making and plating serial dilutions as described above.

34.2.5. Growth of High-Titer Stocks

For isolation and characterization of phage particles and phage DNA, it is usually necessary to prepare a high-titer stock and then to purify and concentrate the phage. While there are many possible ways to do this, a typical approach is to prepare a plate lysate as described above but using 30 large petri plates. For each of these large plates, use 1 ml of host bacterial culture, mixed with just enough phage particles to give complete clearing of the bacterial lawn, and 9 ml of top agar. After the plates have grown, flood each plate with 10 ml of phage buffer and collect the phage as described above, typically generating about 300 ml of lysate. Transfer this to a centrifuge bottle, spin for 5 min at 3,500 × g, and collect the clear supernatant.

34.2.6. Precipitation with PEG 8000

To precipitate the phage particles, add polyethylene glycol 8000 (PEG 8000) to a final concentration of 10% and NaCl to a final concentration of 1 M. To do so, measure the amount of the lysate (e.g., 275 ml) and then add the appropriate amount of solid PEG 8000 (e.g., 27.5 g in this example) and solid NaCl. Add a stirrer bar and stir overnight in the cold.

The following morning, collect the phage pellet by centrifugation at 3,500 × g for 5 min and pour off the supernatant; drain as well as possible. Resuspend the pellets in 10 ml of phage buffer by running the sample up and down a pipette. When as much phage sample is resuspended as possible, clarify by spinning the sample again at $3,500 \times g$ for 10 min. There is always a significant amount of insoluble debris, but the phage usually suspend well.

34.2.7. CsCl Equilibrium Density Centrifugation

Once the precipitated phage is resuspended, it can be further purified and concentrated using CsCl equilibrium density centrifugation, although there are several reasonable alternative approaches (such as step gradients and high-speed centrifugation). To do this, add solid CsCl to the lysate and dissolve, adding about 8.5 g of CsCl for each 10 ml of lysate, and check the density using a refractometer. Alternatively, the density can be determined by simply weighing 1 ml of the sample. Adjust the density to 1.5 by adding either a small amount of a saturated CsCl solution (to increase the density) or phage buffer (to reduce it).

Transfer the solution to an appropriate centrifuge tube (such as heat-sealable tubes) and spin at 38,000 rpm at 10°C for 16 to 24 h. The phage should appear as an opaque band positioned approximately one-third or one-half of the way up the tube; there is usually a small pellet of debris at the bottom. Collect the phage band by carefully placing the tube in a clasp on a ring stand, excising the top of the tube by using a hot scalpel, and inserting a needled syringe (18 gauge) through the side of the tube about 4 to 5 mm below the band. With the needle turned so that the bevel is facing upward and the needle of the syringe is pointing upward, carefully draw off the band from beneath. It is helpful to try to remove as small a sample as possible to avoid diluting it. Phage particles usually store well in CsCl at 4°C, and this is a reasonable option for long-term storage. But for further processing, dialyze away the CsCl by transferring a portion to a dialysis tube and dialyze against two changes of cold phage buffer for several hours.

34.3. PHAGE GENOMIC CHARACTERIZATION

34.3.1. Preparing Phage DNA

DNA can be recovered from a concentrated phage stock by several rounds of phenol extraction. Typically, begin with about 0.5 ml of concentrated phage, add an equal volume of buffer-saturated phenol, and mix by inversion. The phases can then be separated by centrifugation in a microfuge for about 2 min at top speed (~14,000 K rpm). Remove the top aqueous phase to a clean tube; avoid as much of the white particulate material at the interface as possible. Repeat the phenol extraction several more times, and also backextract the phenol layers. All aqueous phase samples are then combined, and the DNA is precipitated by adding 0.1 volume of 3 M sodium acetate and 2.5 volumes of cold ethanol. Mix well and freeze at −20°C for several hours. Collect the DNA by centrifugation, remove the supernatant, and wash at least once with 70% ethanol. Allow the pellet to air dry and resuspend the DNA in a small volume (100 to 200 μl) of Tris-EDTA (TE). The DNA can be slow to dissolve, and if necessary you can speed it up by a brief incubation at 60°C. Determine the DNA concentration by using a spectrophotometer.

34.3.2. Restriction Analysis

Restriction enzyme analysis is useful for evaluating the uniqueness of the phage genome. Phage DNA can be cut with restriction enzymes according to the manufacturer's recommended conditions, and the products can be analyzed by agarose gel electrophoresis. Note several points. First, many phage genomes contain short single-stranded DNA complementary ends, which can anneal and remain annealed during electrophoresis. Thus different restriction patterns may be seen depending on how you treat the sample prior to electrophoresis. If you specifically want to see the end fragments, then after restriction, heat the sample to 75°C for 10 min and snap cool it on wet ice. Second, some or many enzymes may not digest phage DNA if the DNA is modified, and using a variety of different enzymes is helpful. Last, since phage genomes can be quite large, it is helpful to use agarose gel electrophoresis conditions that can separate relatively large DNA fragments.

34.3.3. Construction of Genomic Libraries

For genomic sequencing, prepare a plasmid library of DNA fragments derived by hydrodynamic shearing of phage genomic DNA. We use a HydroShear instrument (GeneMachines Inc., San Carlos, CA) that has been calibrated to shear the DNA predominantly into 1- to 3-kbp pieces. After the shearing, the DNA is repaired by incubation with both T4 DNA polymerase and Klenow according to the manufacturer's instructions, and ethanol precipitated. After washing (70% ethanol), drying, and resuspending in a small volume of TE, the entire sample is loaded onto a wide-slot 0.7% agarose gel along with a size marker. After electrophoresis, a gel slice containing 1- to 3-kbp fragments is excised and the DNA is recovered.

For generating genomic libraries, we typically use one of the pBluescript vectors, such as pBluescriptII SK+, that has been cleaved to completion with EcoRV and dephosphory-lated with CIAP. Ligations are performed in small volumes under conditions favoring blunt-end ligation, using different quantities of the size-fractionated DNA (such as 0.5 μl, 1 μl, or 2 μl from a total of 15 to 20 μl) and then transformed by electroporation into a suitable E. coli strain such as E. coli XL1-Blue. The degree of success can be determined by the ratio of white and black colonies on selective plates containing S-Gal indicator (or blue colonies using X-Gal). With a good preparation of vector DNA, self-ligation in the absence of phage DNA generates only a few black colonies and no white ones. In the presence of phage DNA, a similar number (or a small increase) of black colonies and a significant number (dozens or hundreds) of white colonies is desired. If one particular amount of phage DNA yields optimal numbers of white colonies, the remainder of the ligation mix can be electroporated to generate sufficient white colonies to complete a genome-sequencing project, typically 300 to 600 depending on the size of the genome. The presence of recombinants among the white colonies can be determined by growing up a small number (e.g., 12), isolating plasmid DNA, and analyzing by restriction and agarose gel electrophoresis.

34.3.4. Sequencing and Assembly

Large numbers of recombinant plasmids can be prepared by growing them in "blocks" of 96 samples, and the DNA preparation is greatly simplified by using a robot such as the Qiagen BioRobot 9600. Blocks of DNA samples can then be used to perform standard sequencing reactions as described by the manufacturer, cleaned up to remove excess dyes and/or primers, and analyzed using an automating DNA sequencer such as ABI3730. Each block can be sequenced using both forward and reverse primers. It is important to sequence sufficient blocks to provide good coverage and to minimize the number of weak areas that need to be worked on subsequently. When the sequence coverage is good (~sevenfold redundant), oligonucleotide primers extended on phage DNA template is normally the quickest route to completion of the project.

A variety of programs are available for the assembly of sequence fragments into a final single contiguous sequence. The Phred/Phrap/Consed package is one that we have found to be simple and effective to use; besides providing a reliable assembly, it can be used to simply design primers for completion of the weaker areas as well as the genome termini.

34.3.5. Genome End Analysis

Some phage genomes contain defined ends, while others have multiple possible ends such that the genome assembles into a circle. If the assembly circularizes, then it is helpful to organize the final sequence into a linear array with an arbitrarily defined left end; if a terminase gene can be identified, then its left end is a suitable choice for this. If the genome has defined ends, then this often becomes apparent from the assembly, since the ends tend to be overrepresented (as there are two fixed ends among the sheared DNA population). The simplest way to define the ends precisely is to design primers that are close to the ends (50 to 100

bases away) and pointing outward; use these to extend on phage DNA templates.

34.4. SHUTTLE PHASMIDS AND THEIR CONSTRUCTION

34.4.1. Overview

Shuttle phasmids (part phage, part plasmid) are phage-cloning vectors for mycobacteria that also shuttle to *E. coli* (Fig. 2). They are chimeric molecules made up of a truncated mycobacteriophage genome cloned into an *E. coli*-bacteriophage cosmid. They can be packaged into bacteriophage heads as cosmids, thereby allowing for the use of bacteriophage in in vitro packaging mixes. Upon transfection of the fast-growing M. *smegmatis*, these chimeric DNA molecules become packaged into mycobacteriophage particles that can then infect M. *tuberculosis*, *Mycobacterium bovis*, BCG, and many other mycobacteria. TM4 mycobacteriophage shuttle phasmids enabled the introduction of foreign DNA into both fast- and slow-growing mycobacteria for the first time (25). Moreover, the generation of temperate shuttle phasmids from mycobacteriophage L5 allowed a

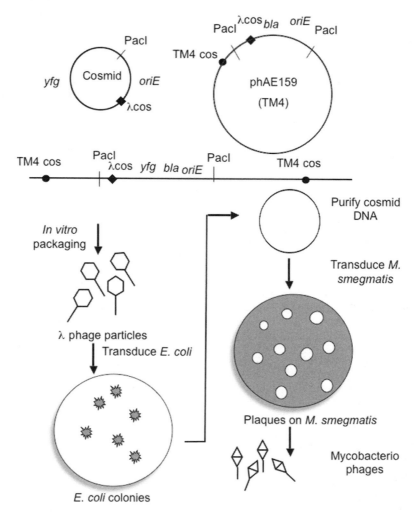

FIGURE 2 Construction of shuttle phasmids. Shuttle phasmids are chimeric molecules that replicate as extrachromosomal plasmids in *E. coli* and as phages in another bacterial host. Construction details are provided in the text.

functional selectable marker to be identified for mycobacteria, which led to the development of the first transformation system for mycobacteria (47). For organisms for which genetics is nonexistent or limited, shuttle phasmid construction provides a powerful approach.

Shuttle phasmids cannot be made from all phages. For example, they cannot be easily made with a phage whose genome is readily expressed in *E. coli*, such as *E. coli* bacteriophage or *Salmonella* phage P22, since the multicopy plasmid is usually toxic to *E. coli* due to the expression of phage genes. Mycobacterial chromosomal DNA and most mycobacteriophage genomes are not readily expressed in *E. coli* because of the inability of the *E. coli* transcriptional machinery to recognize high G+C% mycobacterial promoters (21). Based on this attribute, it should be possible to make shuttle phasmids from any high G+C% phage, including those from any *Actinomycetes* family members including *Mycobacterium*, *Streptomyces*, *Rhodococcus*, *Norcardia*, and *Corynebacterium*. In addition, this might also work for organisms from the high-G+C% gram-negative bacteria like *Pseudomonas* or *Caulobacter*. In contrast to a similar phasmid φC31 vector (42), shuttle phasmids also incorporate a bacteriophage *cos* sequence, which allows for their packaging in vitro into bacteriophage heads. This addition facilitates the genetic manipulation of shuttle phasmids in *E. coli*. Shuttle phasmids have been successfully constructed from the lytic mycobacteriophages TM4 (25) and D29 (36) and the temperate mycobacteriophages L5 and L1 (47).

34.4.2. Generation of a Library of Cosmid Insertions in a Mycobacteriophage Genome

The successful generation of a shuttle plasmid from a mycobacteriophage genome requires the insertion of an *E. coli* cosmid in a nonessential region of the phage genome (Fig. 2). This can be achieved by generating a library of phage genome insertions into a cosmid vector in *E. coli*, followed by selection for those recombinants that yield viable phage particles following transfection of the library (25). A suitable phage genome must contain cohesive ends (a headfull packaging phage would be difficult to use) and a genome size of 40 to 60 kb (capable of being packaged into bacteriophage lambda heads in vitro). The steps to construct a library of cosmid insertions in a mycobacteriophage genome include: (i) ligation of purified mycobacteriophage genomic DNA to form concatemers, (ii) partial digestion of concatemers with Sau3A to generate 40- to 50-kb-long fragments, (iii) ligation of large Sau3A fragments to arms of a double *cos* vector, (iv) package ligated DNA with lambda in vitro packaging mix, (v) transduction of packaging mix into lambda-susceptible *E. coli* strain, and (vi) characterization of DNA recovered from individual transductants to confirm the presence of phage DNA. A good shuttle phasmid library will have two key features: most of the transductants will contain phage DNA, and cosmid insertions into the mycobacteriophage genome will occur around the entire genome. Each insertion point for an *E. coli* cosmid on the mycobacteriophage genome should also be accompanied by a small deletion of nonessential phage sequences. See Jacobs et al. (23) for further explanation. The end result of this genomic cloning is a library of insertions of an *E. coli* cosmid around the mycobacteriophage genome (Fig. 2).

34.4.3. Identification of Shuttle Phasmids Following Transfection of *M. smegmatis*

In section 34.4.2, we described a means of generating a library of cosmid insertions with simultaneous small deletions

in a mycobacteriophage genome. An easy way to think about this is to imagine the phage genome as representing the outline of a clock, with the mycobacteriophage cohesive end site (*cos*) positioned at 12 o'clock. Then imagine that each cosmid is inserted at some minute position around the clock. Cosmid A may have the cosmid inserted at 2 o'clock. Cosmid B may have the cosmid inserted at 7 o'clock on the genome. A viable shuttle phasmid will be one in which the cosmid is inserted in a nonessential region of the phage, and this should produce plaques upon transfection of *M. smegmatis*. If cosmid A has an insertion in a DNA replication gene, the phage will not likely yield viable phages. However, if cosmid B has the cosmid inserted at a nonessential accessory region, it should yield plaques. From a library of thousands of insertions, one can simply identify those insertions in the nonessential region by transfecting the cosmids into *M. smegmatis* and looking for plaques (Fig. 2).

A general protocol for characterizing shuttle phasmids after transduction of *M. smegmatis* includes: (i) isolate cosmid DNA from pooled *E. coli* transductants as described in section 34.4.2, (ii) electroporate pooled cosmid DNA into *M. smegmatis* and mix electroporated cells with *M. smegmatis* cells on a phage plaque lawn, (iii) pick and purify resulting plaques, (iv) make phage stocks and isolate the resulting phage DNA, (v) ligate the putative shuttle phasmid DNA to form concatemers and in vitro package with lambda mix, and (vi) transduce *E. coli* and select for ampicillin-resistant colonies. Ampicillin-resistant colonies in *E. coli* indicate that the DNA fragment isolated from a phage particle can replicate in *E. coli* as a cosmid.

Surprising differences occurred in the plaques obtained from TM4 compared with L5 and D29. The library of TM4 cosmids yield hundreds of plaques, but most of these plaques resulted from the propagation of wild-type phage. Plaque hybridization with the cosmid revealed that only 1 in 400 plaques had cosmid DNA. It is likely that the wild-type TM4 phages had arisen by recombination, since it was highly likely that any one of the transfected *M. smegmatis* cells got more than one cosmid molecule. Consider if a cell received two cosmids, each with an insert in an essential region. If the nonessential regions did not overlap and recombination occurred, one recombination product would be a wild-type TM4 phage. This possibility has been rigorously demonstrated (J. C. van Kessel and G. F. Hatfull, unpublished communication). In contrast, if L5 or D29 are used, the number of plaques obtained is less, but more than 95% have cosmid inserts. The working hypothesis is that TM4 possesses a phage-mediated recombination system. This system may contribute to the success of allelic exchange in *M. tuberculosis*, though these studies are ongoing. Regardless of the phage recombination system, shuttle phasmids have been obtained from three different mycobacteriophages to date.

34.4.4. Cloning Genes into Shuttle Phasmids

Phage delivery of DNA represents the "absolute" most efficient way to introduce DNA into any bacterial cell. Based on the ease of cloning genes into shuttle phasmids in *E. coli*, and the ability to generate TM4 phage particles of packaged shuttle phasmid DNA, shuttle phasmids are ideal for delivery of DNA to mycobacterial cells. Shuttle phasmids have been useful for delivering reporter genes, transposons, or allelic exchange substrates. The cloning of any gene into a shuttle phasmid is a straightforward task that involves the replacement of the existing cosmid with a cosmid of choice. The steps involved in cloning genes into shuttle phasmids include: (i) preparation of the plasmid form of the shuttle phasmid in *E. coli*, (ii) digestion of both the shuttle phasmid and the "new" cosmid with PacI, (iii) ligation of equimolar amounts of phasmid and new cosmid, (iv) packaging of ligated molecules using lambda in vitro packaging mix, (v) transduction of lambda-sensitive *E. coli* strain with selection for the antibiotic resistance gene on the "new" cosmid, (vi) analysis of *E. coli* propagated shuttle phasmid by restriction enzymes to verify exchange of previous cosmid for the "new" cosmid, and (vii) transfection of *M. smegmatis* and preparation of TM4-packaged shuttle phasmids.

34.5. GENE DELIVERY USING PHAGES

34.5.1. Reporter Gene Delivery

Luciferase reporter phages were first constructed with the goal of rapidly assessing drug susceptibilities in clinical isolates of *M. tuberculosis* (23). Drug resistance could be distinguished from drug susceptibility by treating *M. tuberculosis* cells with or without different concentrations of drugs and infecting with a luciferase reporter phage. Drug-sensitive cells would fail to generate any luciferase activity compared with the non-drug-treated control. Drug-resistant cells would glow like untreated cells. TM4 phages are particularly attractive since hundreds of *M. tuberculosis* isolates have been infected and no strain has yet been found to be resistant to TM4 phage. Luciferase reporter phages have also been constructed from both phages D29 and L5 (36, 45) and several studies have used this methodology (2, 41).

34.5.2. Conditionally Replicating Mycobacteriophages and Transposon Delivery

Historically, conditionally replicating phages, usually amber mutants, have been used to efficiently deliver transposons (29). To achieve such a goal with mycobacteriophages, a battery of mutants of TM4 or D29 shuttle phasmids were isolated that permit replication of the phage in *M. smegmatis* at 30°C, but not at 38 or 39°C (4, 5). It is noteworthy that phasmid phAE87, derived from the TM4 phage ph101 (5), has a reversion frequency of 10^{-8} to 10^{-9} in contrast to most thermosensitive mutants, which have a reversion frequency of 10^{-4} to 10^{-5}; this low reversion frequency is likely the result of two independent thermosensitive mutations. This low level of reversion makes this phage the vehicle of choice for delivery of transposons to *M. tuberculosis* (4), *M. smegmatis* (4), and *Mycobacterium avium* (16). Initially this phage was used to deliver IS1096-derived transposons (4, 6, 35). Such transposon libraries have been useful in identifying many previously unknown virulence genes including *pcaA* (14), PDIM biosynthetic genes (8), and PE PGRS (9). More recently, by incorporating the Himar *mariner* transposon, a transposon that displays little sequence specificity (30, 43), newer transposon delivery phages have been constructed.

34.5.3. Allelic Exchange by Specialized Transduction

The ability to generate null mutants in *M. tuberculosis* is critical for rigorous genetic analysis. In any knockout scheme, an allelic exchange substrate containing the modified gene of interest needs to be delivered to the cells, followed by selection for recombinants that have the knockout allele. Several methods have been used to successfully generate knockout strains of *M. tuberculosis* (3), including transformation with linear DNA substrates, suicide plas-

mids, chemical modification of allelic exchange substrates, and specialized phage transduction (3). Both suicide plasmids and chemical modifications of plasmids have the advantage that only a plasmid construction and subsequent *M. tuberculosis* strain electroporation is necessary. The disadvantage of using suicide plasmids or chemically modified plasmids is that electroporation can be very inefficient and electroporation efficiencies vary widely between *M. tuberculosis* strains. Specialized phage transduction has the disadvantage of more preparation steps, but the advantage of higher efficiencies of substrate delivery than electroporation as well as less variability in efficiency for many different strains of *M. tuberculosis*.

Generation of knockout mutants with specialized transduction entails five steps (see Fig. 3): (i) construction of an allelic exchange cosmid; (ii) phasmid transduction; (iii) phage lysate preparation; (iv) phage transduction; and (v) generation of double-crossover alleles (Fig. 3). In brief, the plasmid for constructing allelic exchange substrates (AES) has multiple cloning sites (MCS) upstream and downstream of an antibiotic resistance marker (hygromycin), as well as a counter-selectable marker (*sacB*); also on the AES plasmid is *oriE*, an *E. coli* plasmid origin of replication, lambda-*cos* site for lambda packaging, and a unique PacI restriction site for eventual ligation to a modified temperature-sensitive mycobacteriophage such as TM4. DNA segments (typically 500 bp) derived from each side of the gene to be replaced are generated by PCR amplification and cloned into convenient restriction sites in the AES plasmid on either side of the resistance gene (3). Detailed protocols for these manipulations can be found at www.tbvaccines.org.

34.6. OTHER PHAGE-DERIVED GENETIC TOOLS

34.6.1. Construction of Integration-Proficient Plasmid Vectors

Since many phages carry an apparatus for integrating their DNA into the host chromosome, it is normally not hard to find this feature for use in the construction of integration-proficient plasmid vectors. Conceptually, this is straightforward and requires the incorporation of a segment of phage DNA containing the integrase gene and the *attP* site into a plasmid vector; this vector should then transform the host by integrase-mediated integration into the host chromosome. The advantages of such vectors are that they should be stably maintained and they provide single-copy recombinants.

Although conceptually easy, there are several considerations when constructing integration-proficient vectors. First, it is important to carefully define the segment of phage DNA to be used. Unless experimental data are available, the integrase gene can usually be identified bioinformatically and usually belongs to one of two groups, the tyrosine integrases (such as phage lambda) or the serine integrases (such as φC31 or Bxb1). While these can be readily identified, the phage attachment sites (*attP*) are often more difficult, although in almost every known case the *attP* site is located adjacent to the integrase gene, either to the 5′ or 3′ side. Since members of the tyrosine integrase family predominantly utilize chromosomal attachment sites (*attB*) that overlap tRNA genes, which are reconstituted following integration, then the corresponding *attP* sites include portion of tRNA genes corresponding to their 3′ ends. Thus, if the genome sequence of the host is available

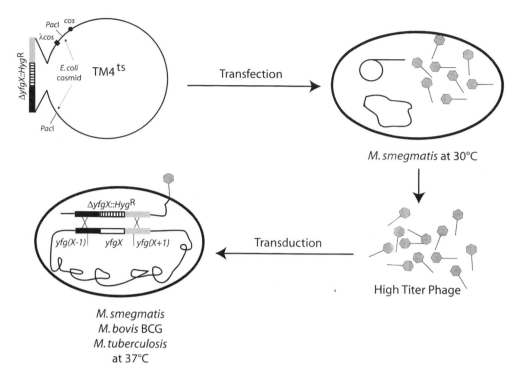

FIGURE 3 Specialized transduction. Regions flanking the upstream and downstream regions of "Your favorite gene" (*yfg*) are cloned into a hygromycin-containing cassette on a plasmid containing both lambda *cos* sites and mycobacteriophage *cos* sites. M. *smegmatis* is transfected at 30°C to allow for production of a high-titer phage lysate containing the Δ*yfgX::Hyg*^R^ allele. The high-titer phage lysate is then used to transduce M. *tuberculosis* at 37°C (nonpermissive temperature for phage lysis) to deliver Δ*yfgX::Hyg*^R^ substrate to the cell for subsequent allelic exchange.

(or a near relative) then a BLAST search with sequences flanking the integrase genes can reveal 20 to 40-bp regions of near-sequence identity. If this corresponds to an annotated host tRNA gene, then this almost certainly indicates the position of *attP*. Unfortunately, this approach normally does not work for serine integrases because they do not use tRNA genes as integration sites and the extent of sequence identity with the host is small, often just 3 to 8 bp (46).

The second important consideration involves the location of the excise gene. Most phage genomes that integrate usually also encode a recombination directionality factor (RDF) that facilitates an integrase-mediated excision reaction; sometimes this is encoded nearby the integrase gene, but several examples exist where it is not (34). Unfortunately, these RDFs are highly varied in their sequences and they can be difficult to identify if relying solely on bioinformatic approaches. When defining a segment of phage DNA for constructing integration-proficient vectors, it is helpful to try to avoid the RDF gene, since it will contribute to either low transformation frequencies or instability of the integrated DNAs (32). If the RDF gene can be identified, it is usually simple to exclude it; if not, it is best to use as small a segment as possible that includes both the integrase gene and the *attP* site.

A third consideration is the expression of the integrase gene. While it is possible that a promoter may be needed to promote integrase expression, this has not been necessary for any of the mycobacteriophage-derived vectors that have been constructed; presumably the gene is expressed either from vector sequences or from closely associated phage signals (28, 32, 39). Since it is possible that integrase can promote RDF-independent excision at low frequency include and that this frequency will be elevated with high integrase expression (48), it may be desirable to avoid the use of strong heterologous promoter systems. It is also important to avoid the inclusion of plasmid replication origins that function in the host, since these are likely to be toxic to the host when integrated into the chromosome.

An alternative approach to avoiding undesirable excision is to express the integrase transiently from a nonreplicating plasmid that is coelectroporated with an *attP*-containing plasmid (38). This generates far fewer transformants, but those recovered are no longer expressing integrase and the integrated sequences are maintained without any significant loss. If serine integrase systems are used, then the extent of sequence identity shared by the two junction sites *attL* and *attR* flanking the integrated sequence is short (3 to 8 bp), and therefore undesirable excision by the host recombination system is infrequent.

34.6.2. Recombineering: Phage-Mediated Recombination for Allelic Exchange

Two phage-derived recombination systems—encoded by the lambda Red system and the Rac prophage—have been developed for high-frequency allelic exchange and mutagenesis in E. *coli*. This process is referred to as recombineering (7). While these systems work well in E. *coli*—and can be used to construct point mutations in the absence of selection—they may not function as well in other bacteria, especially gram-positive bacteria. While the reasons for this are not clear, it could be due to differences in G+C% content, the activities of small DNA-binding proteins, or alternative modes of DNA replication. However, a reasonable solution to this problem is to develop host-specific recombineering systems using analogous (or homologous) systems encoded by phages that infect the host of interest. This approach has been successful for mycobacterial recombineering and should be applicable in general to other bacteria (50a).

Homologues of the RecET proteins encoded by the Rac prophage are rare among some groups of phages. For example, only one (Che9c) of 30 completely sequenced mycobacteriophage genomes encodes homologues of both the RecE exonuclease and the RecT DNA-paring enzyme. However, these can be adapted for recombineering by expressing both proteins from an inducible acetamidase promoter and then preparing electrocompetent cells. These can then be transformed using a linear DNA substrate that contains approximately 500 bp of chromosomal DNA flanking a drug resistance gene. If these 500-bp segments correspond to sequences flanking gene X, then selection for the drug marker generates transformants in which gene X is now replaced by that marker. In the mycobacterial system, approximately 100 transformants can be recovered using only 100 ng of DNA in both fast-growing (e.g., *M. smegmatis*) and slow-growing (e.g., *M. tuberculosis*) mycobacterial strains; approximately 90% or more of these are the products of allelic exchange. Such a recombineering system could likely be further adapted to function with shorter segments of DNA homology or with single-stranded DNA substrates that require only the pairing enzyme.

34.6.3. Generalized Transduction

Generalized transduction is an essential component of any well-developed system for bacterial genetics, since it provides a simple method for constructing isogenic strains. Generalized transducing phages have thus been identified and described for a wide variety of bacterial hosts, and they can be used similarly. Methods have long been established for generalized transduction and usually involve the preparation of a phage lysate on the donor bacterial strain, followed by infection of a recipient strain with this lysate and selection for the desired genetic marker. Transduction frequencies vary greatly, but even if they are fairly low and produce only a small number of colonies, this is usually useful provided that the marker does not have a high reversion frequency. A primary concern is that there may be significant levels of killing by the phage lysate, which may need to be minimized following initial adsorption by addition of antiphage antisera or chelating agents that lower the concentration of available divalent cations such as calcium. In the mycobacteria, several transducing phages have been described (e.g., I3 and Bxz1), but none for *M. tuberculosis* (33, 40).

34.7. RECIPES

34.7.1. Culture of *M. smegmatis*

M. smegmatis mc²155 is a suitable host bacterium for phage propagation. It has a doubling time of approximately 3 h at 37°C and grows to a colony on solid medium in 3 to 4 days; bacterial lawns typically grow in about 24 h, and plaques can be seen after about 16 h. It grows in defined media as described below and is resistant to carbenicillin (CB), which is typically included in media (50 μg/ml) to help re-

duce bacterial contamination; it is also common to include cycloheximide (CHX) at 10 (g/ml to eliminate fungal contamination. Since it has a strong tendency (like most mycobacteria) to clump when grown in liquid media, addition of 0.05% Tween® 80 is typically added to help form a dispersed culture. However, this causes a serious problem for phage infections, since mycobacteriophages are frequently sensitive to these detergents. Since mycobacteria hydrolyze Tween as they grow, this problem can usually be circumvented either by using mid- or late-logarithmic cultures or by subculturing a stationary-phase culture into fresh medium lacking detergent. *M. smegmatis* cultures are grown in standard test tubes placed at an angle on a shaker, but for bigger volumes baffled flasks with agitation are appropriate. Since many phages require calcium for infection, $CaCl_2$ at a final concentration of 1 mM is routinely included in all media. We typically use 0.35% top agar to screen for new phages (since large phages do not produce visible plaques on thicker agar), but 0.7% agar can be used in subsequent steps.

34.7.2. Preparation of *E. coli* Competent Cells

To prepare *E. coli* HB101 for transduction, grow the HB101 strain overnight in Luria Broth (LB) supplemented with 10 mM $MgSO_4$ and 0.2% maltose. Inoculate 25 ml of fresh medium with 0.5 ml of overnight culture. Shake culture at 200 rpm at 37°C until OD_{600} reaches 0.8 to 1.0. Transfer cells to a 50-ml conical tube and pellet cells in a tabletop centrifuge at 3,000 rpm for 10 min at 4°C. Decant supernatant and resuspend the cell pellet in 12.5 ml of 10 mM $MgSO_4$. Store cells at 4°C until ready for use.

34.7.3. Transformation of *M. smegmatis*

1. Grow fresh 50 ml of culture of *M. smegmatis* mc²155 in either Middlebrook 7H9 broth or LB with Tween (0.05%) to mid-log phase (OD_{600} 0.5 to 1.0).

2. Incubate cells on ice for 10 min but no longer than 2 h.

NOTE: For the following steps, keep cells as close to 0°C (ice bath) as possible. The electroporation cuvettes, 50-ml conical tubes, and 10% glycerol should be prechilled to 4°C.

3. Transfer chilled cells to a 50-ml prechilled conical tube. Spin down cells in a tabletop centrifuge at 2,000 × *g* for 10 min. Decant supernatant from cell pellet.

4. Wash cells two times with 40 ml of ice-cold 10% glycerol.

NOTE: For optimal results, prepare 10% glycerol fresh weekly with distilled water; sterilize by filtration through 0.2-μm membrane pore; do not autoclave.

5. Resuspend the washed cell pellet in 5 ml of ice-cold 10% glycerol (1/10 culture volume) and store on ice (will use 0.4 ml per sample).

6. Aliquot transforming DNA (1 to 10 μl) in a prechilled GenePulser electroporation cuvette with a 0.2-cm electrode gap (Bio-Rad, catalog no. 165-2086) and incubate on ice. To each aliquot of DNA, add 400 μl of prepared cell suspension and mix by gently pipetting up and down (use prechilled Pipetman tips for cell transfer).

7. Pulse one time with GenePulser apparatus (Bio-Rad catalog no. 1652076) set at 2.5 V: resistance, 1,000 Ohms; capacitance, 25 μFD. The time constant reading should be about 19 to 21 for control cells with no DNA added.

8. Immediately dilute transformed cell mixture with 2 ml of medium. Transfer cell suspension to a 15-ml snap-cap Falcon tube (Falcon catalog no. 2059).

9. Incubate at 37°C (with or without shaking) for 2 h to allow cell recovery and expression of antibiotic resistance.

10. Spin down cells in tabletop centrifuge at 2,000 × g for 10 min. Aspirate supernatant from cell pellet and resuspend cell pellet in 0.5 ml of medium.

11. Plate desired volume of cell resuspension on plates containing the antibiotic of choice (kanamycin at 10 to 20 μg/ml; hygromycin B at 50 μg/ml).

34.7.4. Media and Reagents

ADC (NO HEAT, filter sterilize)

Dextrose	60 g
NaCl	25.5 g
Albumin	150 g
ddH$_2$O	2,850 ml

Be sure to add water before adding albumin. Mix on hot plate on the lowest setting (or not at all) to dissolve. Filter sterilize using a 0.22-μm-pore-size filter.

0.1 M CaCl$_2$ (Stock Solution)

Calcium is added to mycobacterial growth medium to ensure adequate calcium is available for necessary cellular metabolic processes. Note that, for the most part, calcium is added to obtain a **0.1 mM concentration** in the final solutions of each medium or reagent. Therefore, add 1ml of the 1 mM CaCl$_2$ stock solution to phage buffer, but 2 ml of CaCl$_2$ to 7H9, because it will be used to dilute top agar by an additional 50%.

Calcium chloride (anhydrous)	5.55 g
dH$_2$O	to 500 ml

Carbenicillin (CB) (50 μg/ml)

Carbenicillin	2.5 g
dH$_2$O	to 50 ml

Filter sterilize. Store in the refrigerator.

Cycloheximide (CXH) (10 μg/ml)

Cycloheximide	1 g
dH$_2$O	to 100 ml

Filter sterilize. Store in the refrigerator.

Hygromycin (100 μg/ml)

Hygromycin	1g
dH$_2$O	to 10 ml

Filter sterilize. Store at 4°C.

Kanamycin (50 mg/ml)

Kanamycin	500 mg
dH$_2$O	to 10 ml

Filter sterilize. Store at 4°C.

Luria Broth

Tryptone	10 g
Yeast extract	5 g
NaCl	5 g
dH$_2$O	to 1 liter

Autoclave to sterilize.

NOTE: For LB Agar, add 15 g of agar. For L-Kan drug plates, add 0.8 ml of kanamycin per liter after temperature is less than 60°C. For L-CB drug plates, add 1 ml of carbenicillin after temperature is less than 60°C.

Middlebrook 7H9 Liquid Medium

(Autoclave and add antibiotics, ADC, and calcium prior to use.)

7H9 broth base	4.7 g
40% glycerol	5 ml
ddH$_2$O	900 ml

Middlebrook 7H10 Solid Medium

7H10 agar	19 g
40% glycerol	12.5 ml
ddH$_2$O	890 ml

Autoclave, cool to 55°C, and add 100 ml of ADC and 10 ml of 0.1 mM CaCl$_2$, CB, CHX, and then pour.

MBTA (Middlebrook Top Agar)

7H9	4.7 g
Agar	7 g
H$_2$O	to 900 ml

Prepared at 0.7% concentration, melted, diluted to 0.35% with 7H9 (plus 2 ml of 0.1 M CaCl$_2$).

Phage Adsorption Medium

10% albumin dextrose saline	2% glycerol

Phage Buffer

1 M Tris, pH 7.5	10 ml
1 M MgSO$_4$	10 ml

Autoclave or filter sterilize. Add 0.1 mM CaCl$_2$ prior to use.

20% Tween®

Tween® 80	20 ml
ddH$_2$O	80 ml

Warm to 50°C to dissolve and filter sterilize.

We wish to thank all former and current members of the Hatfull and Jacobs laboratories for developing the methods and approaches described in this chapter.

34.8. REFERENCES

1. **Ackermann, H. W.** 2001. Frequency of morphological phage descriptions in the year 2000. Brief review. *Arch. Virol.* **146:**843–857.

2. **Banaiee, N., M. Bobadilla-del-Valle, P. F. Riska, S. Bardarov, Jr., P. M. Small, A. Ponce-de-Leon, W. R. Jacobs, Jr., G. F. Hatfull, and J. Sifuentes-Osornio.** 2003. Rapid identification and susceptibility testing of *Mycobacterium tuberculosis* from MGIT cultures with luciferase

reporter mycobacteriophages. *J. Med. Microbiol.* **52:**557–561.

3. **Bardarov, S., S. Bardarov, Jr., M. S. Pavelka, Jr., V. Sambandamurthy, M. Larsen, J. Tufariello, J. Chan, G. Hatfull, and W. R. Jacobs, Jr.** 2002. Specialized transduction: an efficient method for generating marked and unmarked targeted gene disruptions in *Mycobacterium tuberculosis*, M. *bovis* BCG and M. *smegmatis*. *Microbiology* **148:**3007–3017.

4. **Bardarov, S., J. Kriakov, C. Carriere, S. Yu, C. Vaamonde, R. A. McAdam, B. R. Bloom, G. F. Hatfull, and W. R. Jacobs, Jr.** 1997. Conditionally replicating mycobacteriophages: a system for transposon delivery to *Mycobacterium tuberculosis*. *Proc. Natl. Acad. Sci. USA* **94:**10961–10966.

5. **Carriere, C., P. F. Riska, O. Zimhony, J. Kriakov, S. Bardarov, J. Burns, J. Chan, and W. R. Jacobs, Jr.** 1997. Conditionally replicating luciferase reporter phages: improved sensitivity for rapid detection and assessment of drug susceptibility of *Mycobacterium tuberculosis*. *J. Clin. Microbiol.* **35:**3232–3239.

6. **Cirillo, J. D., R. G. Barletta, B. R. Bloom, and W. R. Jacobs, Jr.** 1991. A novel transposon trap for mycobacteria: isolation and characterization of IS1096. *J. Bacteriol.* **173:**7772–7780.

7. **Court, D. L., J. A. Sawitzke, and L. C. Thomason.** 2002. Genetic engineering using homologous recombination. *Annu. Rev. Genet.* **36:**361–388.

8. **Cox, J. S., B. Chen, M. McNeil, and W. R. Jacobs, Jr.** 1999. Complex lipid determines tissue-specific replication of *Mycobacterium tuberculosis* in mice. *Nature* **402:**79–83.

9. **Delogu, G., C. Pusceddu, A. Bua, G. Fadda, M. J. Brennan, and S. Zanetti.** 2004. Rv1818c-encoded PE_PGRS protein of *Mycobacterium tuberculosis* is surface exposed and influences bacterial cell structure. *Mol. Microbiol.* **52:**725–733.

10. **Desplats, C., and H. M. Krisch.** 2003. The diversity and evolution of the T4-type bacteriophages. *Res. Microbiol.* **154:**259–267.

11. **Dye, C., S. Scheele, P. Dolin, V. Pathania, and M. C. Raviglione.** 1999. Consensus statement. Global burden of tuberculosis: estimated incidence, prevalence, and mortality by country. WHO Global Surveillance and Monitoring Project. *JAMA* **282:**677–686.

12. **Edwards, R. A., and F. Rohwer.** 2005. Viral metagenomics. *Nat. Rev. Microbiol.* **3:**504–510.

13. **Frieden, T. R., and C. R. Driver.** 2003. Tuberculosis control: past 10 years and future progress. *Tuberculosis (Edinb)* **83:**82–85.

14. **Glickman, M. S., J. S. Cox, and W. R. Jacobs, Jr.** 2000. A novel mycolic acid cyclopropane synthetase is required for cording, persistence, and virulence of *Mycobacterium tuberculosis*. *Mol. Cell.* **5:**717–727.

15. **Hambly, E., and C. A. Suttle.** 2005. The viriosphere, diversity, and genetic exchange within phage communities. *Curr. Opin. Microbiol.* **8:**444–450.

16. **Harris, N. B., Z. Feng, X. Liu, S. L. Cirillo, J. D. Cirillo, and R. G. Barletta.** 1999. Development of a transposon mutagenesis system for *Mycobacterium avium* subsp. *paratuberculosis*. *FEMS Microbiol. Lett.* **175:**21–26.

17. **Hatfull, G. F., M. L. Pedulla, D. Jacobs-Sera, P. M. Cichon, A. Foley, M. E. Ford, R. M. Gonda, J. M. Houtz, A. J. Hryckowian, V. A. Kelchner, S. Namburi, K. V. Pajcini, M. G. Popovich, D. T. Schleicher, B. Z. Simanek, A. L. Smith, G. M. Zdanowicz, V. Kumar, C. L. Peebles, W. R. Jacobs, J. G. Lawrence, and R. W. Hendrix.** 2006. Exploring the mycobacteriophage metaproteome: phage genomics as an educational platform. *PLoS Genetics* **2:**e92.

18. **Hendrix, R. W.** 1999. The long evolutionary reach of viruses. *Curr. Biol.* **9:**914–917.

19. **Hendrix, R. W.** 2003. Bacteriophage genomics. *Curr. Opin. Microbiol.* **6:**506–511.

20. **Hendrix, R. W., M. C. Smith, R. N. Burns, M. E. Ford, and G. F. Hatfull.** 1999. Evolutionary relationships among diverse bacteriophages and prophages: all the world's a phage. *Proc. Natl. Acad. Sci. USA* **96:**2192–2197.

21. **Jacobs, W. R., M. A. Docherty, R. D. Curtiss, and J. E. Clark-Curtiss.** 1986. Expression of *Mycobacterium leprae* genes from a *Streptococcus mutans* promoter in *Escherichia coli* K-12. *Proc. Natl. Acad. Sci. USA* **83:**1926–1930.

22. **Jacobs, W. R., Jr.** 2000. *Mycobacterium tuberculosis*: a once genetically intractable organism, p. 1–16. *In* G. F. Hatfull and W. R. Jacobs Jr. (ed.), *Molecular Genetics of the Mycobacteria*. ASM Press, Washington, DC.

23. **Jacobs, W. R., Jr., R. G. Barletta, R. Udani, J. Chan, G. Kalkut, G. Sosne, T. Kieser, G. J. Sarkis, G. F. Hatfull, and B. R. Bloom.** 1993. Rapid assessment of drug susceptibilities of *Mycobacterium tuberculosis* by means of luciferase reporter phages. *Science* **260:**819–822.

24. **Jacobs, W. R., Jr., S. B. Snapper, M. Tuckman, and B. R. Bloom.** 1989. Mycobacteriophage vector systems. *Rev. Infect. Dis.* **11** (Suppl 2):S404–S410.

25. **Jacobs, W. R., Jr., M. Tuckman, and B. R. Bloom.** 1987. Introduction of foreign DNA into mycobacteria using a shuttle phasmid. *Nature* **327:**532–535.

26. **Jones, W. D., Jr.** 1975. Phage typing report of 125 strains of "*Mycobacterium tuberculosis*." *Ann. Sclavo* **17:**599–604.

27. **Jones, W. D., Jr., and J. Greenberg.** 1978. Modification of methods used in bacteriophage typing of *Mycobacterium tuberculosis* isolates. *J. Clin. Microbiol.* **7:**467–469.

28. **Kim, A., P. Ghosh, M. A. Aaron, L. A. Bibb, S. Jain, and G. F. Hatfull.** 2003. Mycobacteriophage Bxb1 integrates into the *Mycobacterium smegmatis groEL1* gene. *Mol. Microbiol.* **50:**463–473.

29. **Kleckner, N., J. Bender, and S. Gottesman.** 1991. Uses of transposons with emphasis on Tn10. *Methods Enzymol.* **204:**139–180.

30. **Lamichhane, G., M. Zignol, N. J. Blades, D. E. Geiman, A. Dougherty, J. Grosset, K. W. Broman, and W. R. Bishai.** 2003. A postgenomic method for predicting essential genes at subsaturation levels of mutagenesis: application to *Mycobacterium tuberculosis*. *Proc. Natl. Acad. Sci. USA* **100:**7213–7218.

31. **Lawrence, J. G., G. F. Hatfull, and R. W. Hendrix.** 2002. Imbroglios of viral taxonomy: genetic exchange and failings of phenetic approaches. *J. Bacteriol.* **184:**4891–4905.

32. **Lee, M. H., L. Pascopella, W. R. Jacobs, Jr., and G. F. Hatfull.** 1991. Site-specific integration of mycobacteriophage L5: integration-proficient vectors for *Mycobacterium smegmatis*, *Mycobacterium tuberculosis*, and *Bacille Calmette-Guerin*. *Proc. Natl. Acad. Sci. USA* **88:**3111–3115.

33. **Lee, S., J. Kriakov, C. Vilcheze, Z. Dai, G. F. Hatfull, and W. R. Jacobs, Jr.** 2004. Bxz1, a new generalized transducing phage for mycobacteria. *FEMS Microbiol. Lett.* **241:**271–276.

34. **Lewis, J. A., and G. F. Hatfull.** 2001. Control of directionality in integrase-mediated recombination: examination of recombination directionality factors (RDFs) including Xis and Cox proteins. *Nucleic Acids Res.* **29:**2205–2216.

35. **McAdam, R. A., S. Quan, D. A. Smith, S. Bardarov, J. C. Betts, F. C. Cook, E. U. Hooker, A. P. Lewis, P. Woollard, M. J. Everett, P. T. Lukey, G. J. Bancroft, W. R. Jacobs, Jr., and K. Duncan.** 2002. Characterization of a *Mycobacterium tuberculosis* H37Rv transposon library reveals insertions in 351 ORFs and mutants with altered virulence. *Microbiology* **148:**2975–2986.

36. **Pearson, R. E., S. Jurgensen, G. J. Sarkis, G. F. Hatfull, and W. R. Jacobs, Jr.** 1996. Construction of D29 shuttle phasmids and luciferase reporter phages for detection of mycobacteria. *Gene* **183:**129–136.

37. **Pedulla, M. L., M. E. Ford, J. M. Houtz, T. Karthikeyan, C. Wadsworth, J. A. Lewis, D. Jacobs-Sera, J. Falbo, J. Gross, N. R. Pannunzio, W. Brucker, V. Kumar, J. Kandasamy, L. Keenan, S. Bardarov, J. Kriakov, J. G. Lawrence, W. R. Jacobs, R. W. Hendrix, and G. F. Hatfull.** 2003. Origins of highly mosaic mycobacteriophage genomes. *Cell* **113:**171–182.

38. **Peña, C. E., M. H. Lee, M. L. Pedulla, and G. F. Hatfull.** 1997. Characterization of the mycobacteriophage L5 attachment site, attP. *J. Mol. Biol.* **266:**76–92.

39. **Pena, C. E., J. Stoner, and G. F. Hatfull.** 1998. Mycobacteriophage D29 integrase-mediated recombination: specificity of mycobacteriophage integration. *Gene* **225:**143–151.

40. **Raj, C. V., and T. Ramakrishnan.** 1970. Transduction in *Mycobacterium smegmatis*. *Nature* **228:**280–281.

41. **Riska, P. F., and W. R. Jacobs, Jr.** 1998. The use of luciferase-reporter phage for antibiotic-susceptibility testing of mycobacteria. *Methods Mol. Biol.* **101:**431–455.

42. **Rodicio, M. R., C. J. Bruton, and K. F. Chater.** 1985. New derivatives of the Streptomyces temperate phage phi C31 useful for the cloning and functional analysis of Streptomyces DNA. *Gene* **34:**283–292.

43. **Rubin, E. J., B. J. Akerley, V. N. Novik, D. J. Lampe, R. N. Husson, and J. J. Mekalanos.** 1999. In vivo transposition of mariner-based elements in enteric bacteria and mycobacteria. *Proc. Natl. Acad. Sci. USA* **96:**1645–1650.

44. **Ruska, H.** 1940. Die Sichtbarmachung der BakteriophagenLyse im Ubermikroskop. *Naturwissenschaften* **28:**45–46.

45. **Sarkis, G. J., W. R. Jacobs, Jr., and G. F. Hatfull.** 1995. L5 luciferase reporter mycobacteriophages: a sensitive tool for the detection and assay of live mycobacteria. *Mol. Microbiol.* **15:**1055–1067.

46. **Smith, M. C., and H. M. Thorpe.** 2002. Diversity in the serine recombinases. *Mol. Microbiol.* **44:**299–307.

47. **Snapper, S. B., L. Lugosi, A. Jekkel, R. E. Melton, T. Kieser, B. R. Bloom, and W. R. Jacobs, Jr.** 1988. Lysogeny and transformation in mycobacteria: stable expression of foreign genes. *Proc. Natl. Acad. Sci. USA* **85:**6987–6991.

48. **Springer, B., P. Sander, L. Sedlacek, K. Ellrott, and E. C. Bottger.** 2001. Instability and site-specific excision of integration-proficient mycobacteriophage L5 plasmids: development of stably maintained integrative vectors. *Int. J. Med. Microbiol.* **290:**669–675.

49. **Summers, W. C.** 1999. *Felix d'Herelle and the Origins of Molecular Biology.* Yale University Press, New Haven, CT.

50. **Tokunaga, T., and R. M. Nakamura.** 1967. Infection of *Mycobacterium tuberculosis* with deoxyribonucleic acid extracted from mycobacteriophage B1. *J. Virol.* **1:**448–449.

50a. **van Kessel, J. C., and G. F. Hatfull.** 2007. Recombineering in *Mycobacterium tuberculosis*. *Nat. Meth.* **4:**147–152.

51. **Watson, J. D., and F. H. Crick.** 1953. Molecular structure of nucleic acids; a structure for deoxyribose nucleic acid. *Nature* **171:**737–738.

52. **Wommack, K. E., and R. R. Colwell.** 2000. Virioplankton: viruses in aquatic ecosystems. *Microbiol. Mol. Biol. Rev.* **64:**69–114.

COMMUNITY AND GENOMIC ANALYSIS

Introduction to Community and Genomic Analysis

THOMAS M. SCHMIDT

Remarkable advances in DNA-sequencing technology and the computational capacity to analyze sequences has had a tremendous impact on our understanding of microbes and microbial communities. No longer is microbiology constrained by the absolute requirement for cultivation of microbes since genome-based approaches have extended the reach of microbiologists to include microbes that have not yet been cultured. While physiological and genetic studies of cultivated microbes and biochemical analyses of cloned genes continue to provide the "ground truth" necessary to interpret DNA sequences, the capacity to determine complete genome sequences and meta-genomes from microbial communities has helped usher in a new era of microbiology that has been dubbed "The Third Golden Age" of microbiology (1). In this era, the subdisciplines of microbiology are being integrated, due in part to the common use of nucleic acid sequences to study the genetics, evolution, physiology, and ecology of diverse microbes.

The advances in genomics underlying this new era of microbiology have warranted the creation of a new section in this manual that is focused on nucleic acid-based methods for community and genome analysis. This section includes descriptions of the methods to analyze gene sequences, beginning with effective strategies for searching genome databases for homologs (chapter 35). Although it is common practice to search databases using nucleic acid sequences, Pearson describes more effective approaches for finding homologs that include a concise explanation of the advantages of searching with protein sequences. Once homologs have been identified, additional information can be extracted from the sequence data through phylogenetic analyses. Douady and Nesbø provide a thoughtful overview of molecular phylogenetics and specific recommendations for constructing and interpreting phylogenetic trees in chapter 36.

While the capacity to study complex microbial communities is expanding rapidly, basic questions still frequently top the list when studies of a new microbial community are initiated. Those questions include "How much biomass is present?" and "What is the metabolic activity of the total microbial community?" In chapter 37, Karl provides a review of the methods for answering these questions, including detailed methodology for using nucleotide fingerprints to assess biomass and metabolic activity. A more detailed look at the structure of microbial communities is made possible by analysis of the genes and genomes present in a community. While genome sequencing and sequence assembly is conducted primarily at specialized sequencing facilities, probing the genomic content of microbial communities through large insert libraries requires the construction of the clone libraries. DeLong (chapter 38) provides a discussion of the cloning vectors appropriate for this task, along with the "tricks of the trade" resulting from his years of experience as a pioneer of this approach.

DNA sequences are the foundation for the design of nucleic acid probes that serve as pivotal tools in a series of techniques that provide a window into the structure and activities of microbial communities. Microbial ecologists have the choice of probing intact cells with fluorescent or radioactive probes, or sometimes in combination to provide phylogenetic and physiological information simultaneously. This collection of techniques is presented in chapter 39 by Fuchs, Pernthaler, and Amann, who have developed and applied these approaches to numerous microbial communities. Community analysis can also be accomplished through the probing of purified RNA to target metabolically active populations. Buckley and Schmidt present this method (chapter 40) as one strategy to avoid the potential bias introduced by the amplification of nucleic acids. However, phylogenetically informative genes or functional genes can be successfully targeted for analysis through PCR amplification and then separation of the amplicons on denaturing gels or analysis of the terminal fragments of the amplicons (chapter 41). Nakatsu and Marsh provide a useful explanation of these approaches and discuss their potential pitfalls.

Each of the methods presented in this section offers opportunities and limitations in the analysis of genomes and community structure, and the authors provide the thoughtful commentary necessary for readers to make an appropriate selection of the most reasonable approaches to their research questions. Since methods that utilize nucleic acid sequences are rapidly expanding, the collection of chapters in this section provides a strong foundation of methods for the microbiologists planning to probe the complexity of microbial communities and genomes. The references provided in each chapter provide the reader with useful entry points into this burgeoning literature, including many complementary laboratory manuals for more specialized approaches to community and genomic analyses.

REFERENCE

1. **Schaechter, M.** 2003. Integrative microbiology—the Third Golden Age. *J. Biosci.* **28:**149–154.

35

Characterization of Bacterial Genome Sequences by Similarity Searching

WILLIAM R. PEARSON

35.1. INTRODUCTION

With more than 100 fully sequenced microbial genome sequences publicly available from many of the deep divisions of bacteria and archaea, protein and DNA sequence similarity searching is one of the most widely used methods for analyzing bacterial genes and genomes. Most molecular bacteriologists have used similarity searches to identify individual genes, and with the decreasing cost of DNA sequencing, it is becoming more routine to consider large-scale sequencing of diverse bacterial populations. While the common practice of "BLAST"-ing bacterial sequences against the NCBI's comprehensive NR (nonredundant) database can provide important new scientific insights, it is rarely the most effective strategy for characterizing bacterial sequences. In this chapter, I focus on more effective methods for similarity searching with protein and DNA sequences, with the goal of identifying distantly related genes in bacteria and bacterial populations.

While strategies for similarity searching are often presented as a process of selecting a sequence comparison program, and possibly the scoring matrix and gap penalties, this chapter provides a different perspective. The inference of homology is based on parsimony and statistics. If two sequences share much more similarity than is expected by chance, the most parsimonious explanation for the excess similarity is that they derived from a shared ancestor. Homologous sequences have similar three-dimensional structures, and often, but not always, they have similar functions. The goal of sequence similarity searches is to find homologous sequences, and the inference of homology is statistical. Thus, this chapter summarizes the statistical

basis for inferring homology; from this statistical perspective, several strategies emerge for improving the effectiveness of similarity searches.

Indeed, the four most common errors made when searching with BLAST and FASTA involve statistics: (i) comparing DNA sequences to a DNA sequence database, instead of using protein or translated DNA sequence comparisons; (ii) focusing on percent identity, rather than expectation value, when inferring homology; (iii) searching databases that are much larger than necessary; and (iv) using expectation value thresholds that are too stringent. The sections below present the philosophical basis for the relationship between homology and statistical significance and then present the statistical equations that describe local similarity scores. From these equations, several strategies for improving statistical significance, and thus the ability to recognize homologs, emerge. Finally, potential failures in the statistical model are discussed.

35.2. SIMILARITY, HOMOLOGY, AND STATISTICS

Effective strategies for sequence similarity searching are best understood by exploring the statistical basis for inferring homology. Sequence comparison programs like BLAST and FASTA are designed to find homologous sequences—sequences that share excess similarity because they descended from a common ancestor. Over the past 20 years, the term "homology" has emerged from evolution journals to common parlance, but the term is still frequently misused (two sequences are never 50% homologous, though they may be

50% identical [14]). Homology is a qualitative inference—two proteins (or protein domains) either are homologous or are not—and it is a statement about the evolutionary history of a protein that existed hundreds of millions or billions of years in the past—the common or shared ancestor of the homologous proteins.

The great success of computational genome analysis reflects the fact that for many protein families, their sequences change so slowly that statistically significant alignments can be found after billions of years of divergence. For example, the human mismatch repair *mutS* homolog MSH2 shares more than 30% protein sequence identity with most bacterial *mutS* proteins; alignments between the human and prokaryotic proteins have similarity scores that are expected less than once in 10^{40} database searches. The most parsimonious explanation for this high level of sequence similarity is descent from a common ancestor. An alternative explanation would require that the bacterial and eukaryotic mismatch repair proteins arose independently in at least two separate events; it is much simpler to assume, when presented with two sequences that are much more similar than expected by chance, that they were copied from a single common ancestor. Thus, the inference of homology is statistical; we need to have confidence that we know how often a similarity score might occur by chance, so that we can recognize excess similarity that is significantly greater than that expected by chance.

35.2.1. Homology and Statistical Significance

Most biologists are comfortable concluding that when two sequences are 70 to 100% identical over their entire length, they must be homologous. But many investigators are less comfortable with this inference when the level of similarity drops below 40% identity. Because of the complex relationship between sequence similarity, sequence identity, alignment length, and statistical significance, statistical intuition often fails us. Fortunately, modern sequence databases are large and very comprehensive, with hundreds of thousands of unrelated sequences. Thus, when analyzing the results from a sequence similarity search, we need not speculate whether a score is much greater than expected by chance; we can examine the distribution of scores, particularly the high-scoring sequences, to evaluate the accuracy of the statistical estimates. With rare exceptions, these estimates are accurate, so that statistically significant similarity, even between sequences that share less than 20% identity, can be used to infer homology.

Thus, to infer homology reliably, we need accurate estimates of how likely an alignment score will occur by chance. The probability that two random sequences will produce local similarity score S is given by the extreme value distribution (2; reviewed in reference 13):

$$P\left(S' > x\right) = 1 - \exp(e^{-x}) \tag{1}$$

where

$$S' = \lambda S_{raw} - \ln(Kmn) \tag{2}$$

Equation 1 is a simplified form of the extreme value distribution; equation 2 reflects the fact that sequence raw similarity scores for random alignments depend on the scoring matrix and the lengths of the two sequences. The role of the scoring matrix is apparent from the following argument: if all the values in the scoring matrix were doubled, then alignment scores using the doubled matrix would be twice as high, yet their statistical significance would be unchanged. Thus, raw similarity scores S_{raw} are scaled with the

l parameter; if the scoring matrix values are doubled, l is halved so that lS remains the same. The K parameter corrects for the lengths of the sequences, and there are more places to start an alignment between two long sequences than between two short sequences, and the normalized score S' must be adjusted to correct for this effect. Once the "raw" alignment score S_{raw} has been normalized to S' using equation 2, the probability of the score can be calculated using equation 1.

Equation 1 can also be rewritten as

$$P(S_{bit} > x) = 1 - \exp(mn2^{-x}) \tag{3}$$

where

$$S_{bit} = [\lambda S_{raw} - \ln(K)]/\ln(2) \tag{4}$$

S_{bit} is the "bit" score reported by current versions of the BLAST and FASTA programs. For statistically significant bit scores ($P < 0.01$), equation 3 can be rewritten as

$$P(S_{bit} > x) = mn2^{-x} \tag{5}$$

S_{bit} similarity scores have the convenient property that increasing the score by 1 bit halves the probability (a 3-bit increase improves it eightfold), and the probability of the score can be calculated knowing only m and n, the lengths of the two sequences, and without knowing λ or K.

The extreme value distribution (equations [REF: extv-eqn]1 and [REF: s-norm]2) reports the probability that two sequences, picked independently, will produce a given pairwise similarity score. However, the high-scoring sequences from a similarity search are not independent; the sequences of interest are usually examined because they were the highest-scoring sequences in a comprehensive similarity search. Thus, one must correct for the fact that tens or hundreds of thousands of sequences were examined to find the highest-scoring sequences. (While the odds of a "fair" coin coming up heads 10 times in exactly 10 tosses is about 1/1,000, one expects to see 10 heads in a row after about 1,000 tosses, and one would be very surprised not to see 10 heads in a row after 3,000 tosses.) As a result, modern similarity searching programs report the expectation that a score would occur by chance:

$$E(S) = P(S)D \tag{6}$$

where D is the size of the database searched (the number of alignment scores calculated). [The BLAST programs calculate the expectation using a slightly different form: $E(S) = P(S)N/n$, where N is the total number of residues in the database and n is the number of residues in the library sequence.]

To illustrate the use of the various statistical measures of similarity and the effects of different search strategies on search sensitivity, I will use two example query sequences from *Methanococcus jannaschii*: (i) hypothetical protein MJ0050 (SwissProt Y050_METJA, NCBI RefSeq NP_247014), a group II decarboxylase homolog (Table 1A), and (ii) hypothetical protein MJ1633 (SwissProt YG33_METJA, RefSeq NP_248643), which shares a potassium efflux antiporter domain with several *Escherichia coli* proteins (Table 1B). The MJ0050 decarboxylase homolog is an example of a protein that shares global similarity with most of its homologs, i.e., similarity that extends from the beginning of the query sequence and extends to the end, so that the alignment length is similar to the query sequence length. In contrast, only a portion of MJ1633 aligns with the potassium transport domain; here the alignment length is only about 40% of the query sequence length.

For the alignment between MJ0050 and glutamate decarboxylase (NP_416010) in Table 1A, one can calculate the probability that the 66-bit alignment score would occur by chance in a single pairwise comparison using equation 5:

$$P(S_{bit} \geq x) = mn2^{-x}$$
$$P(S_{bit} \geq 66) = 396 \times 466 \times 2^{-66}$$
$$= 184{,}536 \times 1.36 \times 10^{-20}$$
$$= 2.4 \times 10^{-15}$$

However, this pairwise probability does not reflect the chance of seeing a 66-bit match after examining each of the 4,311 sequences in this *E. coli* proteome, which is

$$E(S_{bit} \geq x) = P(S_{bit} \geq x)D$$
$$E(S_{bit} \geq 66) = 2.4 \times 10^{-15} \times 4{,}311$$
$$= 1.1 \times 10^{-11}$$

This 66-bit similarity score is expected to occur by chance only once in 10^{11} database searches; the most parsimonious explanation for this strong similarity is descent from a shared ancestor.

35.2.2. Homology and Nonhomology

One of the most confusing aspects of homology inference is its nonreversibility. Sequences that share statistically significant similarity are almost always homologous (some pathological cases are considered in section 35.7), but the inverse is not true: sequences that do not share statistically significant similarity are not always nonhomologous. For example, in Table 1B, the alignment between MJ1633 and the putative transport protein (NP_415011) with an $E()$ of <0.025

[the parentheses in the $E()$ notation indicate that the expectation depends on the number of sequences, in this case the 4,311 *E. coli* proteins] would not generally be considered statistically significant (similar scores will occur by chance every 40 database searches; a more reliable significance threshold is 0.001 to 0.01), but MJ1633 and NP_415011 do share a homologous domain, which can be readily identified by searches with members of the family from *E. coli*.

Homologous sequences that do not share significant pairwise similarity are generally identified using one of three strategies. (i) The first is significant tertiary structural similarity (when three-dimensional structures are available for the sequences or their clear homologs). (ii) The second is significant similarity to intermediate homologs (transitive homology)—in Table 1B, MJ1633 does not share significant similarity to NP_415011, but it does share significant similarity with NP_414589 [$E() < 10^{-4}$] which in turn shares very strong similarity with NP_415011 [$E() < 10^{-20}$]. Homology is transitive; if A is homologous to B and B is homologous to C *over the same domain*, then A can be inferred to be homologous to C, even when they do not share significant pairwise similarity. (iii) The third strategy is significant similarity to a profile, a position-specific scoring matrix (4), or Hidden Markov model (16) for the protein family. Thus, conventional pairwise similarity searches can establish homology based on statistically significant similarity, but they cannot be used to establish nonhomology.

35.2.3. Expectation and Percent Identity

The MJ0050 glutamate decarboxylase alignment illustrates a common error in inferring homology from sequence simi-

TABLE 1 Sample search with query sequences from *M. jannaschii*[a]

Sequence	Length	S-W	Bits	E(4,311)	% ID	AL
A. MJ0050, 396 aa, best scores						
NP_416010 glutamate decarboxylase	466	250	66.0	1.1×10^{-11}	24.9	401
NP_417379 glycine decarboxylase	957	169	46.2	2×10^{-5}	22.1	420
NP_417025 putative aminotransferase	404	122	35.3	0.016	23.6	254
NP_414772 aminoacyl-histidine	485	110	32.3	0.15	23.4	188
NP_415139 alkyl hydroperoxide	531	99	29.6	1.0	26.9	156
NP_417345 putative transcription	592	92	27.9	4.0	25.2	107
B. MJ1633, 478 aa, best scores						
NP_417809 KefB (*E. coli*)	601	196	49.0	2.2×10^{-6}	28.2	177
NP_414589 K$^+$ efflux antiporter	620	175	44.4	5.4×10^{-5}	25.4	142
NP_415011 putative transport protein	558	133	35.4	0.025	23.2	142
NP_417748 TrkA (*E. coli*)	458	128	34.4	0.041	23.7	135
NP_416807 putative sugar nuclease	297	103	29.2	0.98	26.1	92
NP_418046 putative oxidoreductase	382	104	29.3	1.2	22.2	171
NP_416911 PEP-protein phosphotransferase	575	106	29.6	1.5	22.2	234
NP_418023 xylose binding protein	330	100	28.5	1.8	21.3	253
NP_417468 putative hydrogenase	567	104	29.2	1.9	23.4	316
NP_417367 peptide chain release	365	99	28.3	2.3	22.3	260

[a]"Typical" sequence similarity searching results for comparison of MJ0050, SwissProt Y050_METJA, or MJ1633, SwissProt YG33_METJA, with the *E. coli* proteome. Comparisons used the Smith-Waterman algorithm (4) as implemented by the SSEARCH program from the FASTA package (5). Results include the description of the *E. coli* library sequence, the length of the library sequence, and four measures of sequence similarity: the raw Smith-Waterman alignment (S-W) score using the BLOSUM50 matrix and a penalty of -10 for a gap open and -2 for each residue in the gap; the bit score (bits) as defined in equation 5; $E(4,311)$, the expectation, as defined in equation 6; the percent identity (% ID); and alignment length (AL). aa, amino acids.

larity—focusing on the *percent identity* and *alignment length* rather than the *bit score* or *expectation*. In this example, the closest homolog (NP_416010) has a very significant alignment score [$E() < 10^{-11}$], yet shares less than 25% identity over 400 amino acids, a level of identity often considered "marginal." For example, the nonhomologous NP_415139 shares higher percent identity (26.9%) with MJ0050 (though over a shorter aligned region, 156 residues). And the alignment between MJ0050 and glycine decarboxylase has even lower identity (22%), but it is also very statistically significant [$E() < 2 \times 10^{-5}$]. While the widely used rule of thumb—that two sequences that share more than 30% identity over their entire lengths are homologous—is usually correct in predicting homologs, it is much too conservative. In a large-scale comparison of human proteins to *E. coli* proteins, the median percent identity for alignments with $10^{-12} < E() < 10^{-6}$ was 28.5%, one-fifth of the alignments shared less than 25% identity, and one 19.2% identical alignment has an $E() < 10^{-10}$. While high identity (>30%) over long regions is a reliable indicator of homology, many biologically interesting homologs are readily found with very significant expectation values despite sharing less than 25% identity. Statistical significance—the expectation value—is the most sensitive and most reliable indicator of homology.

35.2.4. Improving Sensitivity with Smaller Databases—Part I

Equation 6 provides a simple strategy for dramatically improving the sensitivity of a similarity search: search smaller, genome-sized, databases. For example, an alignment with an S_{bit} of 40 between two 200-residue proteins would not be significant in a search of the NCBI NR database, with about 1.6 million sequences:

$$
\begin{aligned}
E(S_{bit} \geq 40) &= P(S_{bit} \geq 40)D \\
&= 200 \times 200 \times 2^{-40} \times 1.6 \times 10^{6} \\
&= 3.64 \times 10^{-8} \times 1.6 \times 10^{6} \\
&= 0.06
\end{aligned}
$$

However, in a search of either the *E. coli* or human genome alone, the expectation would be

$$
\begin{aligned}
E(S_{bit} \geq 40) &= P(S_{bit} \geq 40)D \\
&= 3.64 \times 10^{-8} \times 4,300 \ (E.\ coli) \\
&= 1.6 \times 10^{-4} \\
&= 3.64 \times 10^{-8} \times 40,000 \ (human) \\
&= 1.5 \times 10^{-3}
\end{aligned}
$$

The comprehensive NR database has almost 400 times as many sequences as most bacterial genomes, so any search against the NR database is 400 times less sensitive than a search against a single prokaryotic genome. Thus, while the most common BLAST search strategy—searching the default NR database—provides the most comprehensive search, it also provides the least sensitive search. A taxonomically diverse selection of sequenced genomes can provide a much more sensitive search strategy (e.g., 10 prokaryotic genomes might contain 40,000 sequences) with little loss in diversity (section 35.5.3).

35.2.5. Homology and Function—Orthology

While similarity searching is a very sensitive and reliable technique for characterizing new sequences, it is important to remember that it is essentially a structural method. Homologous proteins have similar structures, but not necessarily similar functions. Homology can be subdivided into two types: orthology and paralogy. Orthologous sequences differ because of speciation events; for example, cytochrome *c* in humans is both homologous and orthologous with cytochrome *c* in fish, insects, yeasts, and plants. Orthology is most easily established for genes like the cytochrome *c* gene which are found once per haploid genome, and are found in most organisms. The differences in the cytochrome *c* sequences of humans, insects, plants, and fungi can be used to produce an accurate "species" phylogenetic tree that reflects the evolutionary history of the organisms.

In contrast, paralogous sequences (which are also homologous) have diverged from one another after a gene duplication event. Thus, hemoglobin α, β-globin, and myoglobin are all paralogous sequences; they appeared after successive gene duplication events in the vertebrate lineage. Evolutionary trees of paralogous sequences do not reflect the evolutionary history of the organisms where they are found; they reflect the evolutionary history of the gene family.

While *homologous* sequences need not share a common function, *orthologous* sequences are very likely to have very similar or identical functions. In the case of proteins like cytochrome *c*, which are present only once per haploid genome, the critical function of the gene must be preserved in other organisms. Thus, databases like COGS (Clusters of Orthologous Groups [18]) can be used to provide more reliable functional inferences than similarity searches alone. Alternatively, paralogous sequences, because they are duplicated genes, are more likely to acquire different functions. Reliably distinguishing orthology from paralogy over long evolutionary distances can be very difficult. A common strategy for assigning orthology is by finding the "reciprocal best BLAST hit"; sequence b in genome B is orthologous to sequence a in genome A if b is the best match to a in genome B, and a is the best match to b in genome A. More robust strategies for inferring orthology build evolutionary trees (chapter 36) for the protein families of interest (20).

35.2.6. Statistical Thresholds and Large-Scale Similarity Searching

The expectation value of an alignment score is the most sensitive and most reliable indicator of homology (section 35.2.3). The expectation in equation 6 reports the number of times a score is expected in a *single* similarity search. For protein sequences, the expectation value calculated by the BLAST and FASTA programs is quite accurate; random sequences will receive expectation values of 0.02 about 1 time in 50, and real, unrelated protein sequences will typically receive expectation values between 0.004 and 0.01 about 1 time in 50 (6, 10). Thus, with real sequences, reported expectation values sometimes overestimate statistical significance by a small factor, of 2 to 5. To adjust for this error, conservative statistical significance thresholds between 0.001 and 0.01 are generally used to infer homology.

The relationship between statistical significance and homology is not symmetric; while proteins with pairwise alignment scores with an $E()$ of <0.001 are almost always homologous, proteins with worse $E()$ values are not necessarily nonhomologous. Proteins with alignment scores between $0.01 < E() < 0.05$ are probably homologous as well, but unrelated sequences will obtain scores this good by chance in 5 to 10% of searches. Indeed, since very distantly related homologous sequences need not share statistically significant pairwise similarity, homologous sequences sometimes have an $E()$ of >1.0 (and homology must be inferred from other analyses). The $E()$ ~ 1 criterion is useful for evaluating statistical estimates because, on average, the highest-scoring *unrelated* sequence should have $E()$ ~ 1. In Table 1, the highest scoring sequences that are clearly unrelated to the query have an $E()$ of <0.15 (MJ0050 versus NP_414772, aminoacyl-histidine dipeptidase) and an $E()$ of <0.98 (MJ1633 versus NP_416807, a putative sugar nucleotide epimerase). Although one cannot infer nonhomology based on a lack of statistical significance in a single similarity search, additional searches can confirm that library sequences with expectations around 1 are probably not homologous. The decarboxylases in Table 1A and the transport proteins in Table 1B belong to large protein families; the *E. coli* glutamate decarboxylase (NP_416010) shares statistically significant similarity $[E() < 0.01]$ with more than 20 glutamate decarboxylases in SwissProt (5) from a broad range of bacteria, archaea, yeasts, plants, and animals. None of the members of this family share significant similarity with NP_414772, strongly suggesting nonhomology. Identification of the highest-scoring candidate nonhomolog, and confirming the nonhomology with additional sequence database searches, provides empirical confirmation of the accuracy of the statistical estimates and allows a less stringent significance threshold $[E() < 0.01]$.

The statistical thresholds discussed in this section apply to one or a few sequence similarity searches. When a very large number of searches are performed, more conservative thresholds must be used. An accurate $E()$ of <0.01 states that the alignment score will occur one time in 100 searches by chance; thus, in a large-scale genome analysis that searches with 2,000 candidate open reading frames, an $E()$ of <0.01 should occur $2,000 \times 0.01 = 20$, i.e., times by chance. A conservative correction for the large number of searches reduces the statistical significance threshold by a factor based on the number of searches. It is common to use a statistical thresholds of an $E()$ of $<10^{-6}$ when thousands of searches are done. This conservative approach seeks to reduce the number of *false positives*—sequences tentatively identified as homologous but that actually obtain their high scores by chance because of the large number of searches. Unfortunately, particularly in bacterial genome analysis, where the evolutionary distances to the nearest phylogenetic neighbor can be quite large, increasing the statistical stringency from 10^{-3} to 10^{-6} can also dramatically increase the number of false negatives, sequences that are homologous but no longer meet the statistical criterion. For example, in a comparison of 1,731 *M. jannaschii* sequences with the *E. coli* proteome using the Smith-Waterman algorithm, there were 3,251 alignments with an $E(4,311)$ of $<10^{-3}$ but only 2,227 with an $E()$ of $<10^{-6}$. Because there were 1,711 queries, we might expect that 2 of the 1,024 alignment scores with $10^{-6} < E() < 10^{-3}$ occurred by chance, but more than 1,000 of these alignments probably reflect genuine distant homologs that become false negatives at the stricter statistical threshold. The statistical threshold for definitively annotating sequences based on significant similarity is very different from the appropriate threshold for tentatively identifying a sequence as unique, or unrelated to known proteins.

35.3. SIMILARITY SEARCHING PROGRAMS

As this chapter is written in late 2003, there are two widely used sets of programs for searching protein and DNA sequence databases: BLAST (4) and FASTA (11). BLAST is by far the more popular; BLAST is considerably faster than FASTA and integrated into the comprehensive sequence, protein family, and structure databases at the NCBI website (http://www.ncbi.nih.gov). The FASTA programs are available at the European Bioinformatics Institute website http://www.ebi.ac.uk/fasta; source code for the FASTA package is available from ftp://ftp.virginia.edu/pub/fasta.

Both the BLAST and FASTA packages provide a variety of different programs for protein, DNA, and translated-DNA searches (Table 2). As discussed in the next section, all similarity searching programs are most effective when comparing sequences at the protein or translated-protein level, and modern versions of the BLAST and FASTA programs provide efficient algorithms for translated searches with sequences that may include frameshift errors and termination codons.

For protein sequence comparison, the major difference between BLASTP and FASTA is the strategy used to estimate statistical significance. (The two programs also differ in their default amino acid scoring matrices; BLASTP uses BLOSUM62, while FASTA uses BLOSUM50; see section 35.5.1). The original BLAST programs (3) used analytical formulas to calculate the λ and K parameters for equation 2. Unfortunately, these equations applied only to local alignments without gaps, and alignments with gaps are significantly more sensitive. With the introduction of *gapped*-BLAST (4), λ and K were calculated for specific combinations of scoring matrices and gap penalties using a simulation strategy. Several recent improvements to BLAST involve improvements to the statistical estimates.

The FASTA programs calculate statistical estimates from the distribution of unrelated-sequence scores that are produced in any comprehensive database search (10). (When unrelated scores are unavailable, fasta can use shuffled sequences to estimate the statistical parameters.) Thus, FASTA effectively calculates different λ and K parameters for each query sequence in each database search. This strategy can be more robust to unusual properties in query sequences.

For DNA sequence comparison, the empirical statistical estimation approach used by FASTA is often more accurate than BLASTN estimates because unrelated DNA sequences are highly nonuniform and differ considerably from simple random sequences. (In contrast, real protein sequences are difficult to distinguish from random sequences.) In addition, the default BLASTN parameters for DNA alignments, $+1$ for a match and -3 for a mismatch, optimally identify sequences that are more than 99% identical. DNA comparison can identify significant alignments between sequences that share as little as 70% identity; the FASTA program uses $+5/-4$ match/mismatch values, which are targeted for 70% identical alignments (17). For more sensitive comparison with the BLAST package, the TBLASTX program, which compares six-frame translations of both the DNA query and DNA library sequence, is sometimes used. In general, FASTA is both more sensitive and more selective (because of more accurate statistics) than BLASTN for DNA sequence comparison. Unfortunately, FASTA is also considerably slower.

For translated-DNA sequence comparison, both BLASTX/TBLASTN and FASTX/TFASTX use sophisticated alignment algorithms that provide almost as much sensitivity as protein-protein sequence comparison. The major difference between the two programs is in the treatment of frame shifts; FASTX and TFASTX use a more sophisticated alignment algorithm that allows alignments to continue across nucleotide insertions and deletions that

TABLE 2 BLAST and FASTA similarity search programs

Query	Library	BLAST	Matrix	FASTA	Matrix
Protein	Protein	BLASTP	BL62	FASTA	BL50
				SSEARCH	BL50
DNA	DNA	BLASTN	+1/−3	FASTA	+5/−4
DNA	Protein	BLASTX	BL62	FASTX	BL50
Protein	DNA	TBLASTN	BL62	TFASTX	BL50
DNA	DNA	TBLASTX	BL62		
Protein Fragments	Protein			FASTS	MD20
Protein Fragments	DNA			TFASTS	MD10

cause translation frameshifts. Frameshifts cause BLASTX/ TBLASTX to break alignments into separate segments, which can have incorrect boundaries.

The FASTA package also includes SSEARCH, an implementation of the optimal Smith-Waterman algorithm (15). Although it is as much as 100 times slower than BLASTP, the Smith-Waterman algorithm can be more sensitive than the heuristic BLAST and FASTA algorithms, and on modern gigahertz computers, searches against bacterial genomes take only tens of seconds. In addition, network parallel versions of the FASTA programs (including SSEARCH) are available that allow large-scale bacterial genome comparisons using the Smith-Waterman algorithm to finish in a few hours.

35.4. PROTEIN VERSUS DNA SEQUENCE COMPARISON

The most *common* sequence similarity search performed at the NCBI website, using BLASTN to compare a DNA sequence to a DNA sequence database, is also the *least effective* search strategy. On average, DNA-DNA sequence comparison is 5 to 10-fold less sensitive than the equivalent protein or translated-DNA similarity search. In practice, the most sensitive DNA searches for protein coding sequences provide a 200 million- to 500 million-year lookback time; similar protein or translated-DNA searches routinely identify homologs that diverged 1 billion to 2 billion years ago.

The dramatic difference in search effectiveness with DNA versus translated-DNA or protein searches is illustrated in Table 3. Here, the taxonomic results from three searches are shown, each with exactly the same DNA query (the DNA sequence encoding *E. coli* glutamate decarboxylase): a BLASTX translated-DNA search against the proteins in the bacterial division of Genpept, and two DNA searches, one with relatively sensitive BLASTN parameters, +3 for a match and −3 for a mismatch, and a second with the default (+1/−3) BLASTN scoring values. With the translated BLASTX search, the DNA query can identify homologs in a broad variety of organisms, including several that are very distantly related to the proteobacteria. In contrast, when BLASTN is used with its default scoring parameters, only very closely related homologs are found. These genes probably diverged within the past 100 million years (the α-proteobacterial homolog is likely to have arisen by horizontal gene transfer; the *E. coli* and *Brucella melitensis* sequences are 77% identical at the protein sequence level).

BLASTN searching is far more sensitive when less stringent scoring parameters are used (+3/−3), but even then, the most distant DNA matches correspond to translated-

DNA–protein alignments with bit scores of >330, or expectation values of an $E()$ of $<10^{-89}$ with 40% identity. The BLASTX alignments with bit scores between 45 and 200 correspond to much more distant homologs, with amino acid percent identities as low as 25%, which are missed with DNA sequence comparison. Indeed, using a translated-DNA search against archaeal proteins in Genpept, homologs with bits scores of >50 are found in 12 euryarchaeotal organisms and *Aeropyrum pernix*, a crenarchaeota. None of these homologs are found with BLASTN using the more sensitive +3/−3 scoring parameters.

Protein or translated-DNA sequence comparison is far more sensitive than DNA sequence comparison for several reasons. (i) Protein substitution matrices distinguish conservative (e.g., lysine to arginine) and nonconservative (e.g., leucine to glutamate) amino acid replacements. (ii) This distinction is lost in DNA scoring matrices, and the problem is compounded by the fact that some DNA substitutions are silent at the codon level, while others cause dramatic amino acid replacements. Protein-based comparison is also more sensitive because of the larger alphabet and the shorter sequences. (iii) In addition, protein alignment statistics are far more accurate than DNA statistics. An $E()$ of $<10^{-3}$ is a very safe threshold for statistical significance for protein alignments; for DNA comparisons, thresholds of an $E()$ of $<10^{-6}$ or even 10^{-9} are required to avoid false positives. (iv) Moreover, low-complexity regions, which are the most common cause of statistical errors, are easily removed from protein sequences; many DNA sequences that encode low-complexity protein sequences are not easily recognized.

A common excuse for searching with DNA sequences, rather than protein sequences, is that the sequences of interest are of relatively low quality and may contain frameshifts or codon errors that would produce incorrect translations. However, it is not necessary to first translate a DNA sequence before searching a protein database; both FASTX and BLASTX use algorithms that are robust to both frameshifts and codon errors. These errors can decrease the sensitivity of the search substantially, but not nearly to the level of a DNA-DNA search.

The most effective strategy for improving sequence similarity search effectiveness is to search with protein or translated-DNA sequences. Protein-based alignments provide 5- to 10-fold-deeper evolutionary look-back time. For studies of *Bacteria* sequences, where the most closely related organism may have diverged more than 2 billion years ago, DNA sequence comparison is impractical for all but the most highly conserved sequences. Even in those cases, the more sensitive (+3/−3) BLASTN parameters should be used.

TABLE 3 DNA and translated-DNA similarity searches[a]

Taxonomic group	BLASTX	BLASTN +3/−3	BLASTN +1/−3	TAXON
Bacteria eubacteria				
Proteobacteria				
γ-Proteobacteria				
Enterobacteriaceae				
Shigella enterobacteria				
Shigella flexneri 2a	979	2,165	2,595	Enterobacteria
Escherichia coli CFT073	976	2,130	2,508	Enterobacteria
Escherichia coli O157:H7	959	2,184	2,642	Enterobacteria
Escherichia coli	758	2,253	2,817	Enterobacteria
Edwardsiella tarda	784	1,102	180	Enterobacteria
Brucella melitensis 16M	496	854	113	α-proteobacteria
Mesorhizobium loti	60			α-proteobacteria
Bordetella bronchiseptica RB	330	217		β-proteobacteria
Geobacter metallireducens	53			δ-proteobacteria
Geobacter sulfurreducens PCA	53			δ-proteobacteria
Prochlorococcus marinus MIT	517	458		Cyanobacteria
Synechocystis sp. strain PCC 6803	466	284		Cyanobacteria
Clostridium perfringens strain 13	427			Eubacteria
Streptomyces coelicolor A3(2)	417			High-GC gram[+]
Mycobacterium tuberculosis	414	311		High-GC gram[+]
Listeria innocua	414	257		Eubacteria
Listeria monocytogenes	414	234		Eubacteria
Enterococcus faecium	411			Eubacteria
Streptomyces avermitilis MA4680	409			High-GC gram[+]
Lactococcus lactis	405	183		Eubacteria
Lactobacillus plantarum WCFS1	390	231		Eubacteria
Bacteroides thetaiotaomicron VPI	387	233		CFB group bacteria
Chloroflexus aurantiacus	72			GNS bacteria
Gloeobacter violaceus PCC 7421	48			Cyanobacteria
Streptomyces viridifaciens	45			High-GC gram[+]
Clostridium tetani E88	45			Eubacteria

[a]Bit scores from BLASTX and BLASTN searches presented using the BLAST taxonomy summary option. The DNA sequence (M84025) encoding *E. coli* glutamate decarboxylase was used to search the bacterial division of GenBank or Genpept. Species that contain a homolog with a bit score of ≥45 [E() < 10⁻³ for BLASTX] are shown. The numbers under the BLASTX and BLASTN columns indicate the highest bit score obtained for that taxonomic group. gram[+], gram positive; CFB, *Cytophaga-Flavobacterium-Bacteroides*; GNS, green nonsulfur.

35.5. IMPROVING SEARCH SENSITIVITY

Because homology is a statistical inference, the terms in equations 1, 2, and 6 are the basis for any strategy to improve search sensitivity. From equations 1 and 2, the probability of a pairwise score can be increased by increasing $S' = \lambda S - \ln(Kmn)$; from equation 6, one can improve the expectation by searching smaller databases.

35.5.1. Improving Pairwise Similarity Scores

Improving S' largely involves increasing λS, the similarity score. [The $\ln(Kmn)$ term in equation 2 is usually much smaller than λS, and m and n are fixed.] Similarity scores can be improved by selecting a scoring matrix and gap penalties that are most appropriate for the evolutionary distance and length of the homologous domain. For example, the BLOSUM62 matrix and gap penalties (−11 to open a gap; −1 for each residue in the gap) used by BLASTP are more effective at identifying shorter domains than the default matrix (BLOSUM50) and gap penalties used by FASTA and SSEARCH; the latter scoring parameters are more effective for long alignments (>100 amino acids). The relationship between scoring matrix, evolutionary distance, and alignment length was described in detail in reference 1; for the BLOSUM matrices (8), matrices with higher numbers (BLOSUM62–BLOSUM80) are more effective at finding shorter homologous domains while the BLOSUM45–BLOSUM50 matrices can be used for longer, more distant, relationships.

The effects of different scoring matrices and gap penalties are shown in Table 4. The MJ0050 query shares global similarity with most of its homologs, and the BLOSUM50 matrix is the most effective at identifying homologs. When the default BLASTP scoring parameters (BLOSUM62, −11/−1) are used, the statistical significance and the align-

TABLE 4 Changing scoring parameters[a]

Sequence	BLOSUM50 −10/−2				BLOSUM62 −7/−1				BLOSUM62 −11/−1			
	S-W	E()	% ID	AL	S-W	E()	% ID	AL	S-W	E()	% ID	AL
A. MJ0050, best scores												
NP_416010 glutamate decarboxylase	250	10^{-11}	24.9	401	216	10^{-7}	25.3	415	137	10^{-8}	22.9	332
NP_417379 glycine decarboxylase	169	10^{-5}	22.1	420	163	0.001	23.3	430	88	0.004	22.1	331
NP_417025 aminotransferase	122	0.02	23.6	254	119	0.12	24.5	257	76	0.04	23.7	118
NP_414772 aminoacyl-histidine	110	0.15	23.4	188	108	0.74	23.2	311	57	6.9	23.4	188
NP_415139 alkyl hydroperoxide	99	1.1	26.9	156	104	1.5	24.5	233	62	2.0	28.9	97
B. MJ1633, best scores												
NP_417809 KefB	196	10^{-6}	28.2	177	162	0.02	27.3	176	143	10^{-8}	34.4	96
NP_414589 K^{+} antiporter	175	10^{-4}	25.4	142	141	0.2	24.7	166	131	10^{-7}	25.4	142
NP_415011 transport protein	133	0.03	23.2	142	113	4.4	23.2	142	89	0.005	23.2	142
NP_417748 TrkA	128	0.04	23.7	135	114	2.9	22.2	176	99	10^{-3}	21.8	133
NP_416807 NAD(P) binding	103	0.98	26.1	92				92	70	0.29	26.1	92

[a]Searches were performed as for Table 1 using different scoring matrices and gap penalties. Three different combinations of BLOSUM scoring matrices and gap penalties are shown: BLOSUM50, with a gap open penalty of −10 and a gap extension penalty of −2 (the default for FASTA); BLOSUM62, with a gap open penalty of −11 and a gap extension penalty of −1 (the default for BLASTP), and BLOSUM62, with a gap open penalty of −7 and a gap extension penalty of −2 (gap penalties that are more effective than the BLASTP values when used with FASTA). For abbreviations, see Table 1, footnote a.

ment length of both homologs decrease substantially. When lower gap penalties (−7/−1) are used with BLOSUM62, the alignment length increases to that found with BLOSUM50, presumably the entire length of the homologous domain.

For the local alignments of MJ1633 with its homologous domains, BLOSUM62, −11/−1, performs dramatically better than the other scoring parameters, even though sometimes a shorter aligned domain is found. Indeed, it seems likely that the alignment between MJ1633 and *kefB* produced by BLOSUM50 is too long; other alignments of this domain are ~140 residues.

Thus, for weak or *marginally significant* alignments [10^{-3} < E() < 0.1], the length of the alignment can serve a guide for the most appropriate scoring matrix. If the alignments are short (<200 residues), then the BLOSUM62 matrix with stringent gap penalties (−11/−1) may improve statistical significance; if the alignments are even shorter (<100 residues), BLOSUM80 may be more appropriate. BLOSUM62 and BLOSUM80 are particularly effective when searching with short query sequences, as when searching for unannotated proteins in short intergenic regions. However, these examples were selected to highlight differences in performance with different scoring matrices. In most cases, particularly for less distant relationships, the scoring matrix does not matter; the homologs will have very similar bit scores and E() values with either BLOSUM50 and BLOSUM62.

35.5.2. Adjusting Scoring Matrices for Evolutionary Distance

The default scoring matrices and gap penalties used by BLASTP and FASTA are designed for maximum sensitivity—detecting the most distant evolutionary relations possible. While this is certainly the most common task, in some cases we seek to restrict our focus to a particular evolutionary distance. This is particularly true when trying to identify orthologs of a protein family in relatively closely related species (<500 My). Table 5 shows a typical case, where I seek to identify the ortholog of the *E. coli* KefB protein in *Vibrio cholerae* and *Yersinia pestis*.

While the widely used BLOSUM (8) scoring matrices are the most effective at finding distant homologs, the PAM series of matrices (7) are built using an evolutionary model that can produce matrices appropriate for any evolutionary distance. (A set of PAM matrices based on a more modern and comprehensive protein set is described in reference 9.) The number associated with each PAM matrix, e.g., PAM20 or PAM120, indicates the amount of evolutionary change for which the matrix is targeted; PAM20 is targeted for 20% change, or sequences that are 80% identical; 120% change produces sequences that are about 40% identical. (The 120% change, or PAM120, denotes 120 changes per 100 residues; after 120 changes in 100 positions, the two sequences will be 40% identical, on average, because of multiple hits at some sites and reversions at other sites. After 250% change [PAM250], homologous sequences remain 20% identical [7].) Table 5 shows that the scoring matrix that most closely approximates the correct evolutionary distance produces the most statistically significant alignment (1). For example, *E. coli* KefB and the *E. coli* paralog NP_414589 are about 45% identical, and PAM120 produces the most statistically significant alignment between these two sequences. While the BLOSUM50 alignment is about the same length and identity, it is much less statistically significant; BLOSUM50 is targeted to much more distant (<25%) identical alignments.

TABLE 5 Identifying orthologs[a]

Sequence	BLOSUM50			PAM120			PAM20		
	$E(4,311)$	% ID	AL	$E(4,311)$	% ID	AL	$E(4,311)$	% ID	AL
E. coli orthologs and paralogs, best scores									
NP_417809 KefB (*E. coli*)	5×10^{-173}	100	601	0	100.0	601	0	100.0	601
NP_414589 K$^+$ efflux antiporter	4×10^{-72}	44.2	608	10^{-143}	45.3	563	10^{-95}	47.8	502
NP_415011 putative transport protein	6×10^{-31}	30.5	525	2×10^{-43}	30.2	493	0.045	47.1	51
NP_415709 hypothetical protein	0.013	22.5	427						
NP_418489 hypothetical protein	0.12	22.5	306						
NP_417959 arsenical pump	0.22	23.9	364						
NP_417364 putative permease (*E. coli*)	0.24	24.2	397						
NP_417559 putative transport protein	0.3	25.9	263						
Y. pestis orthologs and paralogs, best scores									
NP_671265 K$^+$ efflux; NEM-act	3×10^{-138}	77.9	602	0	77.9	602	0	78.1	603
NP_668414 putative K$^+$ efflux	7×10^{-30}	31.1	534	2×10^{-40}	30.4	493	3.5	49.0	51
NP_668178 arsenical pump	0.12	23.7	375						
NP_669263 putative transport protein	0.31	25.3	340						
NP_668704 putative permease	0.56	25.7	296						
NP_671240 permease (major)	0.71	23.6	296						
NP_671313 potassium uptake protein	0.76	29.3	99	0.059	28.3	99			
NP_671412 putative membrane	0.85	24.0	329						
V. cholerae orthologs and paralogs, best scores									
NP_232234 glutathione regulator	4×10^{-107}	58.3	597	6×10^{-216}	58.9	589	0	59.6	589
NP_230638 glutathione regulator	3×10^{-16}	25.2	527	3×10^{-5}	30.7	137	9.6	80.0	10
NP_232330 conserved hypothetical	3×10^{-5}	24.5	335	2.1	33.3	48			
NP_229702 potassium uptake protein	0.024	28.9	114						
NP_230338 conserved hypothetical	0.051	20.6	451						
NP_231264 Na$^+$/H$^+$ antiporter protein	0.29	26.2	294						
NP_229728 multidrug resistance	0.38	23.7	354						
NP_232555 NADH dehydrogenase	0.55	23.0	357						

[a]High-scoring sequences found in searches of the *E. coli*, *Y. pestis*, and *V. cholerae* genomes using the *E. coli* KefB protein sequence (NP_417809). Results with three different scoring matrices are shown: BLOSUM50 (the FASTA default), PAM120 (appropriate for sequences that are about 40% identical), and PAM20 (appropriate for sequences that are about 80% identical). For abbreviations, see Table 1, footnote *a*.

The *E. coli* KefB alignment with *Y. pestis* NP_668414 shows how a "shallow" scoring matrix can focus on more recent evolutionary events. The most significant alignment uses the PAM120 matrix, and because the sequences share about 30% identity, the PAM20 matrix misses the homology altogether. Conversely, the *V. cholerae* KefB ortholog, NP_232234, shares 60% identity and has the best expectation value with PAM20. Targeted scoring matrices, e.g., PAM120 for sequences that have diverged in the past 400 million years or PAM20 for sequences that have diverged in the past 100 million years allow one to focus on recent evolutionary events.

35.5.3. Improving Sensitivity with Smaller Databases—Part II

A second strategy for improving search sensitivity is to reduce the size of the database being searched (equation 6). The most comprehensive database at the NCBI website, the NR database, contained about 1.6 million sequences at the end of 2003. In contrast, bacterial genomes typically contain 2,000 to 4,000 proteins, and metazoan genomes typically contain 15,000 to 30,000 protein coding genes. When the taxonomy of a bacterial sequence is approximately known, the most efficient and sensitive characterization strategy would initially search completed genomes from the same class (e.g., γ-*Proteobacteria*) or phylum (*Actinobacteria*, *Proteobacteria*, etc.).

The NCBI BLAST server (http://www.ncbi.nlm.nih.gov/BLAST) provides a powerful facility for limiting database searches to a phylogenetic subset. Under the "options for advanced blasting," any taxonomic division can be specified, e.g., "Actinobacteria[orgn]" or "txid68525[Organism:exp]" (the latter selects δ/ε *proteobacteria*; the txid can be found with the NCBI taxonomy browser; many NCBI/Entrez search term strategies can be found by selecting the Details option). By selecting *Actinobacteria*, one reduces the size of the database search from 1.6 million entries and 520 million residues to about 20 million residues. Thus, a search for *Actinobacteria* sequences from the taxonomic subset is about 26 times more sensitive than searching all of NR.

While the NCBI BLAST/Entrez taxonomy subset options can reduce the database size 10- to 20-fold, the taxonomic subset is still very redundant. The NR database does not contain any identical sequences, but it contains many sequences that differ by a single residue, and thus the NR database is very redundant. For example, while the annotated *E. coli* genome contains 4,311 protein sequences, the NR database contains 9,522 *E. coli* K-12 sequences (simply restricting the search to *Escherichia coli* would select 87,754 sequences).

A more sensitive strategy searches against the proteomes of individual genomes. The NCBI provides a very powerful facility for searching individual genomes, and combinations of genomes, through their Genomic BLAST pages, available from the NCBI Genome pages. Using http://www.ncbi.nlm.nih.gov/sutils/genomtable.cgi, one can select taxonomic sets of genomes, and include or exclude specific genomes (for example, when fully sequenced *Actinobacteria* proteomes are selected, two strains of *Mycobacterium tuberculosis* and *Tropheryma whipplei* are included). When Genomic BLAST is used to search the *Actinobacteria*, only 35,000 sequences, with 10 million residues, from 10 genomes are examined. Indeed, one can select a set of 70 bacterial genomes that represents every different genus only once. This set of genomes includes 221,223 sequences (70 million residues), about 40% of the number of sequences

produced by selecting all bacterial sequences from the NR database.

This strategy can also be implemented on local computers by downloading the appropriate microbial proteome databases from the NCBI and searching with BLASTP (or BLASTX), FASTA (FASTX), or SSEARCH. Complete bacterial genomes and protein sets are available from ftp://ftp.ncbi.nih.gov/genomes/bacteria. Both the BLAST and FASTA programs can be downloaded and run on local computers. Versions of the programs are available for Unix, Linux, Windows, and MacOS operating systems, but it is typically much easier to manage the various downloading and searching processes on a Unix/Linux platform.

35.6. IDENTIFYING ANONYMOUS SEQUENCES FROM BACTERIAL POPULATIONS

rRNA sequences are widely used to study bacterial diversity from uncultured samples. While this has proved to be a very powerful strategy, and more than 7,000 rRNA sequences are available in current databases, an alternative strategy involves simply cloning and sequencing DNA from uncultured, anonymous bacterial samples. The current taxonomic distribution of sequenced bacteria and archaea is sufficiently complete that, in general, one would expect that most bacterial DNA sequences longer than 5,000 nucleotides should contain at least one gene, and typically several, that is recognizable based on similarity to sequenced prokaryotic genomes.

To test the ability of translated-DNA similarity searching to identify anonymous DNA sequences, I selected either 10 10,000-nucleotide sequences or 100 1,000-nucleotide sequences at random from the completed *Archaeoglobus fulgidus*, *E. coli* O157:H7, or *Streptococcus pyogenes* genome. These DNA sequences were used to search the NR database, excluding all sequences from either the phylum (e.g., *Proteobacteria*, *Firmicutes*, *Euryarchaeota*, etc.) or superkingdom (*Bacteria*, *Archaea* [Table 6]). Searches were also performed using DNA sequences with random mutations (substitutions, insertions, or deletions) introduced at 5% of the positions. This analysis sought to simulate the analysis of anonymous DNA sequences, with sequencing errors that might be found in an environmental sample.

Anonymous DNA sequences with average lengths of 10,000 or 1,000 nucleotides were selected at random from *A. fulgidus* (RefSeq NC_000917), *E. coli* O157H7 (NC_002655), and *S. pyogenes* (NC_004070). These sequences were also randomly mutated at 5% of sites, with replacements twice as likely as insertions or deletions. Using the FASTX program, both the correct and mutated sequences were compared with subsets of the NR database that excluded the taxonomic divisions indicated. The fraction of the anonymous query sequences that had a statistically significant alignment with an NR sequence, the median length of the fraction of the query that aligned for queries that matched the NR sequence, and the median number of NR sequences that matched the query (coverage depth) are shown.

The data in Table 6 illustrate the effectiveness of characterizing anonymous DNA sequences by similarity searching. If long (~10,000-nucleotide) sequences are available with 5% errors, then 50 to 60% will align with sequences from different phyla in the NR database, with 20 statistically significant alignments per open reading frame. This depth of coverage—20 homologs per query—should

TABLE 6 Identification of anonymous DNA sequences at different evolutionary distances

Unknown DNA	Excluded sequences	Query length	$E()$ threshold	Fraction found	Length	Coverage (X)
A. fulgidis	Euryarchaeota	10,000	10^{-6}	1.00	0.785	632
			10^{-12}	1.00	0.781	344
		1,000	10^{-6}	0.64	0.811	30
			10^{-12}	0.64	0.748	21
A. fulgidis, 5% mutated	Euryarchaeota	10,000	10^{-6}	1.00	0.657	260
			10^{-12}	1.00	0.648	148
		1,000	10^{-6}	0.64	0.811	30
			10^{-12}	0.64	0.748	21
A. fulgidis	Archaea	10,000	10^{-6}	1.00	0.725	607
			10^{-12}	1.00	0.781	344
		1,000	10^{-6}	0.57	0.746	33
			10^{-12}	0.52	0.733	21
A. fulgidis, 5% mutated	Archaea	10,000	10^{-6}	1.00	0.553	240
			10^{-12}	1.00	0.781	344
		1,000	10^{-6}	0.57	0.746	33
			10^{-12}	0.52	0.733	21
E. coli	Proteobacteria	10,000	10^{-6}	1.00	0.727	336
			10^{-12}	1.00	0.679	281
		1,000	10^{-6}	0.57	0.788	28
			10^{-12}	0.52	0.746	17
E. coli, 5% mutated	Proteobacteria	10,000	10^{-6}	1.00	0.636	256
			10^{-12}	1.00	0.619	171
		1,000	10^{-6}	0.57	0.788	28
			10^{-12}	0.52	0.746	17
E. coli	Bacteria	10,000	10^{-6}	1.00	0.430	102
			10^{-12}	0.90	0.392	109
		1,000	10^{-6}	0.44	0.665	17
			10^{-12}	0.36	0.682	11
E. coli, 5% mutated	Bacteria	10,000	10^{-6}	0.90	0.375	92
			10^{-12}	0.70	0.396	61
		1,000	10^{-6}	0.44	0.665	17
			10^{-12}	0.36	0.682	11
S. pyogenes	Firmicutes	10,000	10^{-6}	1.00	0.723	570
			10^{-12}	1.00	0.695	377
S. pyogenes, 5% mutated	Firmicutes	10,000	10^{-6}	0.90	0.628	332
			10^{-12}	0.90	0.524	195
S. pyogenes	Bacteria	10,000	10^{-6}	1.00	0.480	150
			10^{-12}	0.90	0.475	89
S. pyogenes, 5% mutated	Bacteria	10,000	10^{-6}	0.90	0.433	72
			10^{-12}	0.80	0.433	43

provide considerable phylogenetic detail about the likely origin of the sequence. When shorter (1,000-nucleotide) "reads" were examined, 40 to 60% of the sequences produced significant alignments to sequences from other phyla or other superkingdoms.

As one might expect, the effectiveness of this strategy depends on the origin of the "anonymous" DNA sample, but the taxonomic exclusions used in Table 6 are very challenging. Because of the large number of diverse bacterial sequences available, all of the 10,000-nucleotide archaeal sequences samples shared a statistically significant match with an NR sequence with an $E()$ of $<10^{-12}$, even with 5%

error. And 30% of mutated 1,000-nucleotide sequences found a homolog, with about 20 significant alignments. The A. fulgidus case is perhaps the easiest; the S. pyogenes data are probably more representative, but even in this case, if long sequences are available, many homologs are found.

Naturally, over the next several years, the phylogenetic sampling of microbial genomes will become much more informative. Today, one expects to be able to identify more than half of long anonymous DNA sequences simply by translated-DNA–protein similarity searching. This number will only increase as the microbial kingdoms are better sampled.

35.7. PATHOLOGICAL CASES

35.7.1. Low-Complexity Regions—SEG

This chapter has focused on exploiting our understanding of local similarity statistics to develop more sensitive and selective sequence comparison strategies. This strategy assumes, however, that the similarity statistics are accurate: that an alignment with an $E()$ of $<10^{-3}$ will occur, by chance, about once in 1,000 searches. While statistical estimates for pairwise alignment scores are generally accurate, alignments that appear to be "significant" can occur between nonhomologs when the statistical assumptions fail. The most common failure is caused by low-complexity domains in proteins—regions with a highly biased amino acid composition (19). Figure 1 shows an example of a protein with several low-complexity domains, as identified by the SEG program. These small domains of biased amino acid composition can produce unexpectedly high similarity scores, not because of homology but because of shared composition bias. The BLASTP and TBLASTN programs automatically applies SEG filtering to protein query sequences, where it is extremely effective in excluding significant alignments caused by low-complexity domains.

BLAST and FASTA use slightly different strategies for excluding low-complexity domains from alignment scores. By default, BLASTP and TBLASTN run the SEG (19) program on all query sequences before the search begins. BLAST uses a SEG option that masks the low-complexity regions by converting the residues to X's. The FASTA programs use a slightly different strategy. SEG can be run on a protein library using a mode (pseg -q -z 1) that converts low-complexity domains into lowercase letters, rather than X's. Then, FASTA or FASTX can be run with the -S option, and the lowercase residues in library sequences are masked out for the initial score calculation but are included in the final alignment so that homologous low-complexity regions are displayed. (The same strategy can be used to lowercase low-complexity regions in query sequences with TFASTX.) Figure 2 shows similarity scores obtained with the NR database when low-complexity domains are excluded as well as when they are included.

Removing low-complexity domains is especially important when comparing translated-DNA sequences to protein sequence databases with BLASTX or FASTX. Out-of-frame translations of DNA sequences frequently produce low-complexity regions (12), which show up as statistically significant matches to low-complexity protein domains. Unfortunately, the current web interface to the BLAST programs at the NCBI does not offer this option, though one can search a SEG'ed database with a local BLAST installation.

35.7.2. Confirming Statistical Significance with Shuffled Comparisons

The statistical estimates reported by the BLAST and FASTA programs report the number of times an alignment score is expected by chance, from sequences with compositions and lengths similar to those in the sequence database. For sequences with "average" compositions, these estimates are very accurate, but as discussed above, for proteins with low-complexity domains having very biased composition, statistical estimates may be off by many orders of magnitude. However, there are intermediate cases as well; some sequences may have only slightly biased regions (the most common examples are transmembrane proteins). These proteins do not have average compositions, and their high alignment scores may reflect shared composition bias rather than shared ancestry. (In my experience, nonhomologous membrane proteins never share statistically significant sequence similarity, though often the high-scoring nonsignificant alignments include other membrane proteins.)

The PRSS and PRFX programs in the FASTA package can be used to confirm that statistical estimates do not simply reflect shared composition bias. PRSS is a derivative of the SSEARCH (Smith-Waterman) program that compares two sequences to produce an optimal local alignment score and then shuffles the second sequence 200 to 1,000 or more times, calculating Smith-Waterman alignment scores between the first sequence and each shuffled sequence. The 200 to 1,000 shuffled sequence scores are used to estimate a λ and K (equation 1) that are specific for the length and composition of the sequence pair of interest. In addition, the shuffling method can be modified to increase the stringency

```
>gi|21909550|ref|NP663818.1| putative secreted protein
                             1-46   MKKRILSAVLVSGVTLGAATTVGAEDLSTK
                                    IAKQDSIISNLTTEQK
        aaqnqvsalqaqvsslqs   47-64  EQDKLTARNTELEALSKRFEQEIKALTSQI
                                    VARNEKLKNQARSAYKNNETSGYINALLNS
                                    KSISDVVNRLVAINRAVSANAKLLEQQKAD
                                    KVSLEEKQ
     aanqtaintiaanmamaeenqn  163-184
                             185-187 TLR
        tqqanleaatanlalqlasat 188-208
                             209-213 EDKAN
lvaqkeaaekaaaealaqeqaakvkaqeqa 214-288
aqqaasveaaksaitpapqatpaaqssnai
        epaaltapaapsarp
                             289-310 QTSYDSSNTYPVGQCTWGAKSL
        apwagnnwgnggqwa      311-325
                             326-354 YSAQAAGYRTGSTPMVGAIAVWNDGGYGH
        vavvvevqsass         355-366
                             367-398 IRVMESNYSGRQYIADHRGWFNPTGVTFIY
                                    PH
```

FIGURE 1 Low-complexity regions identified in a putative secreted protein from *S. pyogenes* (NP_663818) by the SEG program (19). Low-complexity regions are shown to the left of the residue numbers.

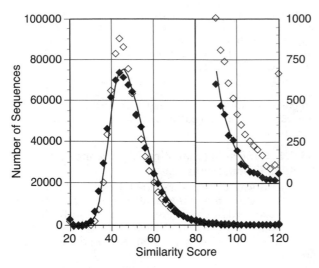

FIGURE 2 Low-complexity domains increase similarity scores. Distribution of scores obtained searching the NR database excluding *Firmicutes* sequences with a putative secreted protein from *S. pyogenes* (NP_663818), a 398-amino-acid protein that is about 30% low complexity. The filled symbols show the distribution of scores obtained when low-complexity regions are excluded from the alignment scores; the open diamonds show the scores obtained when low-complexity domains are included. The inset (upper right) shows that about 1,000 additional sequences have high scores if low-complexity domains are included; three of these sequences have an $E()$ of $< 10^{-3}$, and six more have an $E()$ of <0.01. There are no statistically significant matches when low-complexity domains are excluded.

of the analysis: the second sequence can be shuffled uniformly, so that each residue from the unshuffled sequence is placed anywhere in the shuffled copy; or the second sequence can be shuffled in segments (window shuffle), so that residues 1 to 20 of the unshuffled sequence are found only in positions 1 to 20 of the shuffled sequences, residues 21 to 40 remain in positions 21 to 40, etc. The segmental or window shuffling preserves not only the global but also the local composition in the shuffled sequences, so that transmembrane domains in the unshuffled sequence will be placed in the same position, but in shuffled order, in the shuffled sequences.

For proteins with average amino acid composition, there is rarely more than a twofold difference between statistical estimates reported from similarity search and estimates based on PRSS shuffles. For proteins containing multiple transmembrane domains (e.g., G protein-coupled receptors or ion channels), the "uniform" shuffle estimates differ only slightly from the similarity search statistical estimates, but segmental shuffle estimates may differ 10- to 100-fold (the difference is generally greater for more significant scores).

While protein-protein statistical estimates are generally very accurate, translated-DNA–protein estimates can be more misleading because of the unusual properties of out-of-frame alignments. The PRFX program performs a FASTX translated-DNA–protein alignment and then shuffles the protein sequence 200 to 1,000 or more times to es-

timate λ and K for the translated sequence comparison. As with PRSS, either a uniform or segmental shuffle can be used, but a uniform shuffle is usually sufficient to identify "significant" scores that reflect shared composition bias, rather than homology. FASTX statistical estimates for alignments that exclude low-complexity regions (SEG'ed proteins) are generally very accurate; 5 to 10 per thousand will have significance estimates that drop 10^3- to 10^6-fold when PRFX shuffles are done. For large-scale translated-DNA sequence comparison, PRFX analysis should be used routinely to confirm biologically unexpected observations.

Statistical estimates from shuffled protein sequences are very reliable because, with the exception of composition effects, shuffled protein sequences appear to be very similar to unrelated protein sequences. This is not true for DNA sequences; DNA sequences have higher-order dependencies (e.g., codon biases) that are usually disrupted by sequence shuffling, so that 1,000 shuffled DNA sequences may have alignment properties that are very different from 1,000 real, unrelated sequences. Thus, in general, the DNA-DNA statistical significance reported by FASTA are probably more accurate than those produced by PRSS.

35.8. SUMMARY

BLAST and FASTA similarity searches against comprehensive sequence databases are among the most sensitive and reliable tools available for characterizing bacterial genes and genomes. Similarity searches are most effective when used to identify homologous genes; with the exceptions noted above, protein and translated-DNA sequences that share statistically significant similarity are homologous, with similar three-dimensional structures.

When characterizing bacterial genomes, a few strategies can improve dramatically the effectiveness of similarity searches. (i) Whenever possible, compare protein sequences or perform translated-DNA comparisons; DNA-DNA comparisons are 5- to 10-fold less sensitive than protein comparisons. (ii) Search the smallest comprehensive database that is likely to contain the sequences of interest. If homologs are not found in the smaller database, a larger database can be searched, but searches against comprehensive databases like the NR database are 100- to 1,000-fold less sensitive than searching bacterial genomes. (iii) Focus on the statistical estimates when inferring homology; inferences based on percent identity and alignment length are either much less sensitive or much less reliable. (iv) Sequence alignments with expectation values between 10^{-3} and 10^{-6} are almost always homologous, except for low-complexity domains or out-of-frame translations. Low-complexity domains can usually be removed with the SEG program, and statistical estimates can be confirmed with PRSS and PRFX.

The inference of homology from statistically significant similarity is a powerful strategy for characterizing sequences. Current versions of the BLAST and FASTA comparison programs produce very reliable statistics. But homology reflects the evolutionary history of a sequence, preserved through constraints on structure. Today, the most common errors attributed to similarity searches and homology inference involve assertions about biological function. Nevertheless, searching for homologs remains the most informative analysis available for sequences and promises to remain so as more genomes are sequenced and sequence databases expand.

35.9. REFERENCES

1. **Altschul, S. F.** 1991. Amino acid substitution matrices from an information theoretic perspective. *J. Mol. Biol.* **219**:555–565.

2. **Altschul, S. F., M. S. Boguski, W. Gish, and J. C. Wootton.** 1994. Issues in searching molecular sequence databases. *Nat. Genet.* **6**:119–129.

3. **Altschul, S. F., W. Gish, W. Miller, E.W. Myers, and D. J. Lipman.** 1990. Basic local alignment search tool. *J. Mol. Biol.* **215**:403–410.

4. **Altschul, S. F., T. L. Madden, A. A. Schaffer, J. Zhang, Z. Zhang, W. Miller, and D. J. Lipman.** 1997. Gapped BLAST and PSI-BLAST: a new generation of protein database search programs. *Nucleic Acids Res.* **25**:3389–3402.

5. **Boeckmann, B., A. Bairoch, R. Apweiler, M. Blatter, A. Estreicher, E. Gasteiger, M. J. Martin, K. Michoud, C. O'Donovan, I. Phan, S. Pilbout, and M. Schneider.** 2003. The Swiss-Prot protein knowledgebase and its supplement trembl in 2003. *Nucleic Acids Res.* **31**:365–370.

6. **Brenner, S. E., C. Chothia, and T. J. Hubbard.** 1998. Assessing sequence comparison methods with reliable structurally identified distant evolutionary relationships. *Proc. Natl. Acad. Sci. USA* **95**:6073–6078.

7. **Dayhoff, M., R. M. Schwartz, and B. C. Orcutt.** 1978. A model of evolutionary change in proteins, p. 345–352. *In* M. Dayhoff (ed.), *Atlas of Protein Sequence and Structure*, vol. 5, suppl. 3. National Biomedical Research Foundation, Silver Spring, MD.

8. **Henikoff, S., and J. G. Henikoff.** 1992. Amino acid substitutions matrices from protein blocks. *Proc. Natl. Acad. Sci. USA* **89**:10915–10919.

9. **Jones, D. T., W. R. Taylor, and J. M. Thornton.** 1992. The rapid generation of mutation data matrices from protein sequences. *Comput. Appl. Biosci.* **8**:275–282.

10. **Pearson, W. R.** 1998. Empirical statistical estimates for sequence similarity searches. *J. Mol. Biol.* **276**:71–84.

11. **Pearson, W. R.** 1999. Flexible similarity searching with the fasta3 program package, p. 185–219. *In* S. Misener and S. A. Krawetz (ed.), *Bioinformatics Methods and Protocols.* Humana Press, Totowa, NJ.

12. **Pearson, W. R., T. Wood, Z. Zhang, and W. Miller.** 1997. Comparison of DNA sequences with protein sequences. *Genomics* **46**:24–36.

13. **Pearson, W. R., and T. C. Wood.** 2001. Statistical significance in biological sequence comparison, p. 39–65. *In* D. J. Balding, M. Bishop, and C. Cannings (ed.), *Handbook of Statistical Genetics.* Wiley, London, United Kingdom.

14. **Reeck, G. R., C. de Haen, D. C. Teller, R. F. Doolittle, W. M. Fitch, R. E. Dickerson, P. Chambon, A. D. McLachlan, E. Margoliash, T. H. Jukes, and E. Zuckerland.** 1987. "Homology" in proteins and nucleic acids: a terminology muddle and a way out of it. *Cell* **50**:667.

15. **Smith, T. F., and M. S. Waterman.** 1981. Identification of common molecular subsequences. *J. Mol. Biol.* **147**:195–197.

16. **Sonnhammer, E. L., S. R. Eddy, and R. Durbin.** 1997. Pfam: a comprehensive database of protein domain families based on seed alignments. *Proteins* **28**:405–420.

17. **States, D. J., W. Gish, and S. F. Altschul.** 1991. Improved sensitivity of nucleic acid database searches using application-specific scoring matrices. *Methods Companion Methods Enzymol.* **3**:66–70.

18. **Tatusov, R. L., E. V. Koonin, and D. J. Lipman.** 1991. A genomic perspective on protein families. *Science* **278**:631–637.

19. **Wootton, J. C., and S. Federhen.** 1993. Statistics of local complexity in amino acid sequences and sequence databases. *Comput. Chem.* **17**:149–163.

20. **Zmasek, C. M., and S. R. Eddy.** 2002. Rio: analyzing proteomes by automated phylogenomics using resampled inference of orthologs. *BMC Bioinformatics* **3**:14.

36

Reconstructing and Interpreting Evolutionary Relationships

CHRISTOPHE J. DOUADY AND CAMILLA L. NESBØ

36.1. INTRODUCTION

Phylogenetics is the study of evolutionary relationships among organisms or genes, and phylogenetic analyses aim at estimating or reconstructing the evolutionary relationships among such "operational taxonomic units" (OTUs). These relationships are usually visualized as a phylogenetic tree, which portrays the evolutionary history of the OTUs. In this chapter the OTUs of interest are represented either by DNA or amino acid sequences. We will give a general introduction to methods for reconstructing phylogenetic trees from a set of homologous DNA or amino acid sequences (see section 36.5), obtained for example by doing database searches. Given space limitations, it is impossible to guide inexperienced users across all or even most of the steps of tree reconstruction. Additionally, this task has already been accomplished by several books and book chapters (28, 47, 52). A reader with little or no prior experience, however,

should be able to use this chapter as a guide to the options available for different types of data as well as to software available for doing the analysis. Such software is becoming increasingly user friendly. Our goal, then, is to provide some background and understanding of what methods may be useful for what purpose.

All phylogenetic tree-building methods inherently rely on a common set of assumptions. For example, all analyses assume that the sequences to be analyzed are homologs, that they are not a mixture of different sets of paralogs (see section 36.5), and that gene flow between taxa is nonexistent or extremely low (see below). There are four steps in the phylogenetic analysis of molecular sequence data: (i) building an alignment, (ii) determining an appropriate substitution model, (iii) tree building, and (iv) tree evaluation. It is important to remember that all subsequent steps in the analysis are highly dependent on the assumption that the initial alignment is correct, i.e., that each position in the alignment is composed of homologous residues (positional homology; see section 36.5). Even the most sophisticated phylogenetic analysis software cannot correct for bad input data. Similarly, obtaining the best model for the data is also important, and a phylogenetic analysis should be viewed as a search for the optimal evolutionary model as well as for the optimal tree.

36.2. CONSTRUCTION OF PHYLOGENETIC TREES

36.2.1. Alignment

The primary input required by phylogenetic software is an alignment. An alignment is a matrix (usually an electronic file) where each sequence of interest occupies a line and each homologous site or position in these sequences occupies the same column. Selecting the relevant sequences to be studied is an important process, and pertinent information can be found in recent literature (see chapter 35, this volume, and reference 28, p. 8–20). The next step consists of generating columns of homologous sites and would be trivial if not for insertion and deletion mutations (indels). These indels can be compensated for in the alignment by introducing "gaps" (Fig. 1a, b). However, as shown in Fig. 1, multiple options for introducing gaps usually exist. Hence, generating an optimal alignment can be summarized as reducing the number of mismatches (putatively homologous positions that do not share the same base or nucleotide) and adding as few (in terms of both number and size) gaps as possible.

There are several different alignment file formats; FASTA, NEXUS, and PHYLIP formats are the most commonly used. The simplest format is the FASTA format, shown in Fig. 2. Information on alignment format options for different phylogenetic analysis programs is available in their respective user manuals.

36.2.1.1. Automated Alignment

When only a few relatively similar sequences are investigated, visual inspection in a text editor or sequence editor such as BIOEDIT, Se-Al, or MacClade (see Table 1) generally allows the unambiguous introduction of gaps. When the number and divergence of the sequences increases, the operation can become extremely tedious. Phylogeneticists usually use software such as ClustalX, which applies a heuristic algorithm to generate an alignment (for more information see p. 86–98 in reference 27 and the ClustalX

documentation [63, 64]). ClustalX can accept raw data in FASTA format and other formats (Fig. 2) and lets the operator decide the penalty that should accompany the opening and extension of a new gap, compared with creating a mismatch. ClustalX starts by generating pairwise alignments between all the sequences to create a dendrogram that is used during the final, multiple alignment step. It should be noted that the dendrogram produced during the second phase is treelike but is *not* a valid phylogenetic tree. Furthermore, no objective reason exists for choosing one set of penalties over another, and often many should be tried and the alignment inspected manually to determine the most suitable set for the data. This step is crucial for all subsequent inference. More practical information can be found in reference 28 (p. 21–37).

For rRNA analyses, alignments based on structural data can be downloaded from several databases, such as the European Ribosomal RNA Database (66, 67) (http://www.psb.ugent.be/rRNA/), the Ribosomal Database Project II (RDP II) (7) (http://rdp.cme.msu.edu), and the Comparative RNA Web Site (5) (http://www.rna.ccbb.utexas.edu/). Sequences can be added to these alignments online, manually (see below) or in ClustalX, and should be inspected and edited by eye (see below). These databases also offer other services such as BLAST analyses, primer construction, or distance-based phylogenetic analyses using clustering algorithms such as neighbor joining (see section 36.2.4). For the online phylogenetic analyses at RDPII, it should be noted that the user is less free to manipulate the different parameter settings than in, for instance, PAUP*, and trees should be built using other phylogenetic software as well. The trees are useful as a starting point, however. The "Quick phylogeny search" at the European Ribosomal database, which implements the BIBI program (http://pbil.univ-lyon1.fr/bibi/), is based on Clustal alignments (not structural alignments), and as stated on the web page, the tool should be used with care. However, the tree produced is a true neighbor-joining (NJ) tree (even if Clustal produced it) and not the dendrogram produced during the alignment process. Alignments can also be built using ARB (http://www2.mikro.biologie.tu-muenchen.de/arb/about.html). This program package also offers phylogenetic analyses and probe design. The trees are useful as a starting point.

36.2.1.2. Alignment Editing

Automatic alignments in general require subsequent human editing. Most common actions include (i) correction of obvious alignment errors, (ii) edge trimming of the alignment (it is unlikely that all the sequences in the file will start and end at the same position), and (iii) removal of inherently ambiguous sections. Ambiguous portions occur when sequences share zones of weak similarity and much length variation, forcing an arbitrary choice between equally likely alignments (i.e., the gap could be introduced at several places within a small range of sites). Recently developed software can help perform this final editing (e.g., G blocks [6]; SOAP [42]), but even if such software represent progress toward objectivity, caution should prevail and a final check by eye is advisable.

36.2.1.3. Codon Alignment

When aligning protein-coding DNA sequences, gaps should not only be introduced to decrease the number of mismatches, but they should also try to respect the triplet structure of the code (i.e., usually gaps are introduced by

```
A  Sequence1    A C G C T A C T T G A C T A G
   Sequence2    A C G C A C A T G

B  Sequence1    A C G C T A C T T G A C T A G
   Sequence2    A C G C - A C A T G

C  Sequence1    A C G C T A C T T G A C T A G
   Sequence2    A C G C - A C - - - A - T - G
```

FIGURE 1 Three possible alignments between the same two sequences, showing the compromises between the number of gaps and the number of mismatches. Positions in bold indicate matching sites.

```
>gi|6468294|emb|AJ251082.1|HMA251082 Haloarcula marismortui 5S rRNA
gene, rrnB operon
TCATTCATACGCACTGTGACTCATTCACCGACGATTTAACTCGTCGCTGAACGAGTCCAGGCGCAAACTG
GATCGCACGTAATCACACGGTGGAAGAGTTAATCGAGACTGGTACTATCGCGGTTCGATTCCGTGACTCG
ACGTTAGGCGGCCACAGCGGTGGGGTTGCCTCCCGTACCCATCCCGAACACGGAAGATAAGCCCACCAGC
GTTCCGGGGAGTACTGGAGTGCGCGAGCCTCTGGGAAACGCGGTTCGCCGCCACCATTCATACCTTTCAT
AGCCCACTCAGGAGAGATATCTCTCCCGAGTGGGCTTTCCGTATTTAAACAGAGCCGAACCACTCAGTAA
ATGACCGGTTCTCGCACTCTGTGGAATACGGCTTCAATCGGTGAGATCAGACGTGCGACTAGCGATCGTG
ATCGAGTCGTTGA
>gi|439198|gb|L27163.1|HACRRA Haloarcula vallismortis 5S ribosomal RNA
GTAGCGGCCACAGCGGTGGGGTTCCTCCCGTACCCATCCCGAACACGGAAGATAAGCCCACCAGCGTTCC
GGGGAGTACTGGAGTGCGCGACCCTCTGGGAAACCGGGTTCGCCGCTACYY
>gi|43554|emb|X03407.1|HHRRNA Halobacterium halobium rRNA operon with
genes for 16S rRNA, 23S rRNA, 5S rRNA and tRNA-Ala (UGC)
GGTACCACTCGGCCCGACCGAACGCACTCGCGCGGGATGACCGGCCGACCTCCGCCTACGCAATACGCTG
TGGCGTGTGTCCCTGGTGTGGGCCGCCATCACGAAGCGCTGCTGGTTCGACGGTGTTTTATGTACCCCAC
CACTCGGATGAGATGCGAACGACGTGAGGTGGCTCGGTGCACCCGACGCCACTGATTGACGCCCCCTCGT
CCCGTTCGGACGGAACCCGACTGGGTTCAGTCCGATGCCCTTAAGTACAACAGGGTACTTCGGTGGAATG
CGAACGACAATGGGGCCGCCCGGTTACACGGGTGGCCGACGCATGACTCCGCTGATCGGTTCGGCGTTCG
GCCGAACTCGATTCGATGCCCTTAAGTAATAACGGGTGTTCCGATGAGATGCGAACGACAATGAGGCTAT
CCGGTTCGTCCGGGTGGCTGATGCATCTCTTCGACGCTCTCCATGGTGTCGGTCTCACTCTCAGTGAGTG
TGATTCGATGCCCTTAAGTAATAACGGGCGTTACGAGGAATTGCGAACGACAATGTGGCTACCTGGTTCT
>Hb_halobi
GGTACCACTCGGCCCGACCGAACGCACTCGCGCGGGATGACCGGCCGACCTCCGCCTACGCAATACGCTG
TGGCGTGTGTCCCTGGTGTGGGCCGCCATCACGAAGCGCTGCTGGTTCGACGGTGTTTTATGTACCCCAC
CACTCGGATGAGATGCGAACGACGTGAGGTGGCTCGGTGCACCCGACGCCACTGATTGACGCCCCCTCGT
CCCGTTCGGACGGAACCCGACTGGGTTCAGTCCGATGCCCTTAAGTACAACAGGGTACTTCGGTGGAATG
CGAACGACAATGGGGCCGCCCGGTTACACGGGTGGCCGACGCATGACTCCGCTGATCGGTTCGGCGTTCG
GCCGAACTCGATTCGATGCCCTTAAGTAATAACGGGTGTTCCGATGAGATGCGAACGACAATGAGGCTAT
CCGGTTCGTCCGGGTGGCTGATGCATCTCTTCGACGCTCTCCATGGTGTCGGTCTCACTCTCAGTGAGTG
TGATTCGATGCCCTTAAGTAATAACGGGCGTTACGAGGAATTGCGAACGACAATGTGGCTACCTGGTTCT
```

FIGURE 2 Sequences in FASTA format. Each sequence is composed of two parts. The first part includes the first line only, which always starts with a "greater than" sign (>) and is a more or less complex description of the sequence. To avoid any downstream problems, we recommend using a nine-character descriptor with spaces represented by an underscore sign (e.g., the last sequence). The second part is the actual string of nucleotides or amino acids. By default this part will start after the descriptor paragraph mark and will end just before the next "greater than" sign.

multiples of three between two codons). This task can be performed relatively easily by using CodonAlign (see comment in reference 28 p. 146–149) or ARB. These softwares use a protein alignment as a guide to generate an alignment of the corresponding DNA sequence.

36.2.2. Tree Building Options

When starting a phylogenetic analysis, numerous choices have to be made. Of these choices, none is straightforward and many still divide the phylogenetic community. Therefore, the following section should be read critically and with the understanding that the perfect analysis, if such is possible, has yet to be developed. The users of phylogenetic analysis programs should remember that compromises are often made among at least three parameters: (i) power (the amount of data that will produce a result similar to an infinite data set), (ii) consistency (assurance that the correct answer will be found given that enough data are provided), and (iii) efficiency (the amount of time used to deliver a tree). Efficiency might seem to be an irrelevant parameter to enter into this equation, but at first sight only. The number of solutions (tree topologies) grows factorially with the number of OTUs considered.

Hence, to be sure that the optimal tree is recovered for 10 taxa, more than 2 million trees have to be inspected, and for 20 taxa more than 2×10^{20} trees will have to be inspected.

36.2.2.1. What Type of Tree Can Be Expected from the Data?

The first question one has to answer is not strictly phylogenetic but one that regards the amount of gene flow between the OTUs present in the data set. The tree reconstruction methods usually used in phylogenetic analyses assume that the data are dichotomous (i.e., that the gene flow has stopped). In the remaining part of this chapter, we will focus on this type of reconstruction. However, if the data have been sampled from a population, or if one can expect very high levels of lateral gene transfer (LGT) and /or recombination, the gene flow should be considered in the analysis, and readers can turn themselves toward population genetic methods (if the flow is likely to be very high) or toward network analysis. Network analyses do not assume that the data are dichotomous and therefore allow for a taxon to be connected with more than another. Programs that can be used to do such analyses include TCS

TABLE 1 Useful software[a]

Software	Where to get it	Platform	Price
ARB	http://www2.mikro.biologie.tu-muenchen.de/arb/documentation.html	Mac, Unix, Linux	Free
Bioedit	http://www.mbio.ncsu.edu/BioEdit/bioedit.html	PC	Free
Clustal X	http://bips.u-strasbg.fr/fr/Documentation/clustalX	PC, Mac, Unix, Linux	Free
CodonAlign	http://homepage.mac.com/barryghall/CodonAlign.html	PC, Mac	Free
Consel	http://www.is.titech.ac.jp/~shimo/prog/consel/	PC, Unix, Linux	Free
CoVARES	http://hades.biochem.dal.ca/Rogerlab/Software/software.html	Unix	Free
Gblocks	http://molevol.ibmb.csic.es/Gblocks.html	PC, Mac, Unix, Linux	Free
GDE	ftp://ftp.bio.indiana.edu/molbio/unix/GDE	Mac, Unix, Linux,	Free
MacClade	http://macclade.org/	Mac	$125
Mega	http://www.megasoftware.net/	PC	Free
Modeltest	http://darwin.uvigo.es/software/modeltest.html	PC, Mac, Unix	Free
Molphy	http://www.ism.ac.jp/software/ismlib/softother.e.html#molphy	Unix	Free
MrBayes	http://mrbayes.csit.fsu.edu/	PC, Mac, Unix	Free
Must	herve.philippe@umontreal.ca	PC, Linux	Free
PAML	http://abacus.gene.ucl.ac.uk/software/paml.html	PC, Mac, Unix, Linux	Free
PAUP*	http://paup.csit.fsu.edu/	PC, Mac, Unix, Linux	<$150
PHYLIP	http://evolution.genetics.washington.edu/phylip.html	PC, Mac, Unix, Linux	Free
Se-Al	http://evolve.zoo.ox.ac.uk/software.html?id=seal	Mac	Free
SOAP	http://ueg.ulb.ac.be/SOAP/	PC, Mac, Unix, Linux	Free
Tree-puzzle	http://www.tree-puzzle.de/	PC, Mac, Unix	Free
Treeview	http://taxonomy.zoology.gla.ac.uk/rod/treeview.html	PC, Mac, Unix, Linux	Free

[a]This list is far from being exhaustive, but presents a few of the most widely used software packages. A more complete list can be found at Joe Felsenstein's web page (http://evolution.genetics.washington.edu/phylip/software.html).

(http://darwin.uvigo.es/software/tcs.html) and Splitstree (http://www.splitstree.org).

36.2.2.2. Tree Search: Exhaustive versus Heuristic Searches

As a direct consequence of the size problem (with 20 taxa, more than 2×10^{20} trees will have to be inspected; see above), users will have to choose a way to search the tree space, comprising all possible topologies (the haystack) in which the optimal topology (the needle) is found. There are two approaches: (i) picking and inspecting each piece of "straw," one after another (exhaustive search), and (ii) using a "magnet" to find the "needle" (heuristic searches). Unfortunately, phylogeneticists cannot rely on precise heuristic searches to solve their problem, as the "needle" has many of the same properties as all the "straws." In fact, this problem belongs to a class of mathematical problem, called NP-complete, for which no efficient algorithms are known that provide a solution (26). Exhaustive searches are usually not an option, and thus imperfect heuristic searches have been developed.

The most typical heuristics are generally called "hill climbing." They resemble the process of a hiker lost in the fog who is seeking the highest peak in a landscape. Like a hiker who moves forward, avoiding downhill and choosing uphill paths, hill-climbing heuristics start from any given tree and move toward a different and "higher" topology (in a tree landscape, each location represents a tree). The analogy is quite easy to visualize, but there is still a need to define what the heuristic starting point will be, how it will move, and what defines a higher topology. The starting point of a search can be chosen at random by using stepwise addition, or "star decomposition." A description of these procedures can be found in reference 61, but more adequately an NJ (neighbor-joining) tree can be used (see below; 51). The search will move using topological perturbation. These rearrangements can affect any part of the tree and operate in various processes presented as choices to the user of PAUP or PHYLIP, for instance: nearest-neighbor interchanges (NNI, local rearrangements in PHYLIP), subtree pruning and regrafting (SPR, global rearrangements in PHYLIP), and tree bisection and reconnection (TBR). Unless time becomes a limitation, the operator should favor TBR to SPR and SPR to NNI, as NNI represent only a subset of the SPR results that are themselves only a subset of the TBR rearrangements (60, 61). It is important to recall that these searches are imperfect. Thus, even if one is sure of having found a high topology, one cannot be sure of having found the highest one(s), a notion known as "local optimum entrapment." The relative height of topologies will be discussed in section 36.2.2.4.

Heuristic searches are not the only alternative to exhaustive searches. Hendy and Penny (29) applied a "branch-and-bound" method to phylogenetic problems. In simple terms, this method offers a unique way to find the best (highest) tree without actually evaluating each tree (29, 61). However, for large data sets the time required for this type of search may still force one to choose a heuristic approach.

36.2.2.3. Type of Data

Another crucial choice that the user must make is the form of the data to be analyzed. One must first choose between DNA and amino acid sequences. It is often argued that to avoid saturation, slowly evolving characters are typically favored to investigate deep phylogenies and faster-evolving characters are used at more shallow levels. This translates into using amino acid and nucleotide alignments for deep and shallow levels, respectively. The second set of choices having to do with type of data concerns the utilization of raw data or its reduction into a distance matrix. Again this choice is far from trivial, since it might greatly affect the outcome and efficiency of the analyses. The use of distance matrices results in information loss, since the sequence data are reduced to a single distance measure for each pairwise comparison, and will prevent using parsimony or maximum-likelihood methods. Therefore this conversion is not advisable per se. However, the gain in efficiency from distance matrices can allow analyses of data that would not otherwise be feasible, or permit more realistic assumptions when selecting the optimal (highest) topology.

36.2.2.4. Tree Selection: Optimality Criterion

In section 36.2.2.2, we stated that heuristic searches help select higher topologies, but what defines a higher or more optimal topology? This tree selection process relies on the concept of optimality criteria. In choosing a criterion, the user chooses what should be optimized. For distance data, the optimality criterion can seek the tree that minimizes the difference between the observed distance (the distance matrix) and the tree distance for each pair of taxa. Another option would be to look for a tree that is the shortest (minimizes the sum of tree distances). The first criterion belongs to a family of optimality criteria known as goodness-of-fit measures, which includes criteria such as Farris's f statistic, least-squares, or Fitch-Margoliash methods, while the second is called minimum evolution.

Maximum parsimony is a site-based (raw data) method that searches for the tree that requires the fewest changes or mutations (i.e., that is the most parsimonious solution). In its simplest form, changes of all types and at all sites are regarded as equally parsimonious (equally weighted parsimony). However, different positions or types of changes can be considered as having different cost (generalized parsimony; see, e.g., reference 8).

The maximum-likelihood method is another approach that utilizes raw data. In this case, the criterion to be optimized is the likelihood of observing the data given the considered topology and a model of evolution (see below). Finally, a newcomer in phylogeny is the Bayesian category of methods (33, 50). These approaches seek the most probable tree given the data. Without entering into any mathematical considerations, the main difference between maximum likelihood and Bayesian approaches is the way they deal with the parameters of the model of evolution. Maximum likelihood will seek parameter values that maximize the likelihood function, whereas a Bayesian inference approach integrates over all possible values of those parameters (see MrBayes documentation).

In Table 2 we have listed some of the most commonly used phylogenetic computer software and the types of analyses that are available in each of the packages.

TABLE 2 Function implemented in some of the most common phylogenetic software[a]

Software	Alignment		Tree building								Model selection	Statistical test	Tree viewing
	Building	Editing	Distance		Parsimony		Maximum likelihood		Bayesian				
			nt	aa	nt	aa	nt	aa	nt	aa			
ARB	√	√	(√)		(√)		(√)					(√)	√
Bioedit	(√)	√	√*	√*		√*	√*					(√)	
Clustal X	√		√	√									
CodonAlign	√												
Consel												√	
Gblocks		√											
GDE	√	√	√*	√*	√*	√*	√*					(√*)	√*
MacClade	(√)	√											
Mega			√	√	√	√						√	√
Modeltest											√		
Molphy							√	√					
MrBayes									√	√		(√)	
Must	√	√	√	√								√	√—
PAML							√	√	√				
PAUP*			√	(√)	√	√	√	(√)				√	(√)
PHYLIP			√	√	√	(√)	√	√					
Se-Al	(√)	√											
SOAP	√	√											
Tree-puzzle							√	√				(√)	
Treeview													√

[a]Absence of a check mark does not mean that the given software cannot perform the task, but that we do not find it advisable to carry out the task. When between parentheses, a check mark indicates a weak possibility. *, via other software. nt, nucleotides; aa, amino acids.

36.2.2.5. Models of Evolution

36.2.5.1. Models of Substitution and State Frequency

Models of substitution are ways to define the probability of a change or mutation. They can be seen as the distance, likelihood, and Bayesian equivalent of a parsimony-weighting scheme. For nucleic acid sequences, by allowing six independent rates of change for the six types of nucleotide substitutions without assuming that bases are present in equal frequency, the General Time Reversible (GTR) model (38) is the most complex "stationary" (see section 36.2.2.5.3) model commonly implemented. Other classical models such as Tamura-Nei, HKY85, Kimura-2-Parameter, or Jukes-and-Cantor are simpler versions of this general model (see Fig. 11 in reference 61). All these models are available in most of the commonly used software for estimating DNA phylogenies (PAUP*, Table 2).

Most models for protein analysis are based on a compilation of empirical data and estimation of probability of amino acid change. Dayhoff and JTT are the two most classical models and were both inferred from nuclear encoded proteins (10, 36). The BLOSUM 62 is also a nuclear matrix but was deduced from very divergent proteins (30). The mtREV24 was estimated, as its name indicates, from mitochondrial data (1). Simpler but analytical models include Poisson and proportional model (respectively, Jukes-and-Cantor and F81 counterpart for protein). The commonly used software for building protein phylogenies, PHYLIP and TREE-PUZZLE, together cover most of these models (Table 2).

36.2.2.5.2. Across-Site Rate Variation

Models of evolution leave the user free to determine the probabilities of various types of change in the data. However, it is very unlikely that any rate of change will remain constant across the complete length of the sequences. Consider an rRNA data set: who would assume that both loops and stems evolve at the same rate? This heterogeneity in rate of evolution can be taken into account by introducing a new parameter in the model of evolution. This parameter, known as α, will determine the shape of a Γ distribution of the rate (68). When α is close to infinity, the substitution rates are equal for all sites, and when α is near 0, then most of the sites do not evolve and a few evolve extremely rapidly. In addition, a proportion of sites that are invariable can often be implemented. These invariable sites correspond to only a subset of the constant sites, since some can change but have not done so yet. Several software packages can be used to estimate these parameters; PAUP* and TREE-PUZZLE are commonly employed for DNA alignments and TREE-PUZZLE is used for amino acid alignments (Table 1).

The site-specific option is another way to take rate variation into account. Unlike the Γ-distributed rates, users specify which sites belong to which category. This approach is useful, for instance, for giving different rates to the three positions of the codon or for stems versus the loops of RNA.

36.2.2.5.3. Over-Time Rate Variation and Further Refinements

Most of the commonly used algorithms assume that the substitution rate, regardless of its variation across sites, has been constant over time (i.e., that the model is stationary). This assumption, however mathematically convenient, has often been shown to be violated, especially for data sets involving long diverged organisms (see, for instance, references 39 and 41). To correct for this bias, covarion or heterotachous models (i.e., that do not assume stationarity) have been proposed (e.g., references 19, 21, 39, 40, and 48; also see section 36.4). However, they are seldom used. Other standard simplifications include the assumption of reversibility (the rate of change from A to B equals that from B to A), Markovian mutation (no memory is kept of the former states), and the independence of the sites (change in site X does not affect changes at any other position). These simplifications are clearly no more sound than assuming a stationary model, but they are often necessary compromises, since adding a parameter to a model is not without cost (see section 36.2.3.5).

36.2.3. How To Decide?

36.2.3.1. Need for Compromise

In the previous sections we saw that there are at least four choices that have to be made before estimating a phylogenetic tree (type of search, data, optimality criterion, and model of evolution). When considered individually, some of these choices seem rather trivial. For example, exhaustive searches are superior to the branch-and-bound method in describing the complete universe of possible trees, and both of these are superior to heuristic searches in always finding the best tree (global optimum). However, all choices have an influence on the time required to complete the tree search, often forcing the investigator to make difficult decisions. Depending on the size of the data set, an exhaustive search may only be performed for overly simplistic models of evolution, while an imperfect heuristic search would allow usage of more realistic models. Unfortunately, no objective reason for favoring one set of choices over another exists. Often analysis methods are chosen because they are implemented in the user's favorite phylogenetic software package. These sorts of choices should be avoided, however, and Table 2 should guide the less experienced readers toward sound decisions. We also provide a brief recap of relevant factors below.

36.2.3.2. Type of Search

As mentioned in the previous section, exhaustive searches are preferable to branch-and-bound searches, which are, in turn, superior to heuristic searches. However, experienced phylogeneticists often sacrifice exhaustiveness first. In fact, congruence between two or three independent heuristic searches (starting from different trees) is often regarded as a strong indicator that the heuristic was not trapped in local optima and is usually obtained in a fraction of the time required by a full search.

36.2.3.3. Type of Data

The choice of data is much more controversial within the phylogenetic community. As we have already pointed out, amino acid and nucleotide alignments are generally used for deep and shallow levels, respectively. However, concerns have been raised recently, indicating that DNA alignments might also be useful at deeper levels (57). Therefore, the most pragmatic position one can take is to analyze both data sets (as long as the proteins show some divergence and the DNA sequences can be aligned) and compare the results. With few exceptions, distance methods are regarded as less desirable than maximum-likelihood and Bayesian ones, since distance matrices imply a reduction of the information. Nonetheless, distance matrices often permit a much better modeling of sequence evolution and can end up being more reliable and/or more robust than a data-based analysis. Relative performance of distance-based over maximum-parsimony methods is far less clear and often seems to depend on the data (e.g., distance methods often underperform for sequence collections with much missing data).

36.2.3.4. Type of Optimality Criterion

Choice between the two families of distance optimality criteria (goodness of fit or minimum evolution) is usually of modest importance. Similarly, Bayesian and maximum-likelihood searches usually produce very similar phylogenies. Parsimony, on the other hand, can sometimes be at odds with other methods, particularly when data contain highly divergent homologous sequences. In this context, parsimony is more likely to artificially group such divergent sequences (a phenomenon better known as long-branch attraction [17]). This is not to say that parsimony always will be erroneous when differing from likelihood, but that this known bias should be kept in mind when analyzing the results. As a rule of thumb, it is always better to run more than one type of analysis. In the best-case scenario, all methods will agree (be congruent in phylogenetic jargon) and the operator will have confidence in the results. On the other hand, when some analyses differ, it may be possible to identify the artifact.

36.2.3.5. Model of Evolution

When choosing between models of evolution, the investigator needs to search for the most realistic model allowed by the data. From a statistical point of view, adding parameters to the model (e.g., increase the number of different rates of substitution, take into account base frequencies, or allow across-site rate variation) has a cost, and the cost is paid for by the data. When the data do not contain enough information, adding parameters will increase stochastic error and therefore reduce the chance of finding the correct topology (61).

Posada and Crandall (49) have developed software to help determine which model fits nucleotide data best by maximizing model realism without overparameterizing the analysis (Modeltest; see Table 1). For proteins, where models are mostly empirically based, it seems sensible to prefer the model that was defined on similar data. For example, if the data are of mitochondrial origin, using the mtREV24 model is a good idea, while one would use the PMB matrix (65) for a set of distant nuclear proteins and the JTT model in most other cases.

36.2.4. Clustering Algorithms: a Shortcut

In their war against time, phylogeneticists have developed clustering algorithms that can find a reasonable topology in only a fraction of the time required by criterion optimization methods. What these clustering methods have in common is that the data sets are first reduced to a distance matrix and that trees are constructed by sequentially grouping the sequences that are more similar until a tree is completely resolved. Unlike previous methods, no rearrangement of the full tree is performed, and the tree recovered is not compared with other possible topologies.

Depending on the technique, different rules will govern the clustering process. We will limit ourselves to the most commonly used: the NJ (neighbor-joining) methods of Saitou and Nei (51). Using these procedures, all sequences (OTUs) are first clustered in a single starlike tree. Then, a first step is initiated in which all topologies having a pair in one side and the remaining starlike tree on the other are formed. The topology that displays the shortest sum of branch lengths is then selected, and the two OTUs are paired. A new distance matrix is then calculated and the process is reiterated. The procedure stops when the starlike component is resolved.

This method is extremely popular due to its extreme efficiency and the fact that it produces one tree, whereas parsimony can support hundreds of equally optimal trees. However, it is less accurate than maximum-likelihood-based methods (22) and is therefore often seen only as a good starting point for topological rearrangement performed by the optimality criterion-based approach.

Many software packages include the option to construct NJ trees, for example, ClustalX, PHYLIP, PAUP*, and Mega (Table 1).

36.3. INTERPRETATION OF PHYLOGENETIC TREES

36.3.1. Reading of the Tree

36.3.1.1. Rooting and Polarizing the Tree

Polarization of phylogenetic trees is an extremely important step and is often perceived as the most difficult one (61). At this stage the investigator defines the directionality of evolution and the topology will pass from an unrooted (Fig. 3A) to a rooted one (Fig. 3B, C). To root a tree, external information needs to be included. For example, if the sequences under investigation are known to fall into two monophyletic (see below) groups (e.g., the Crenarchaeota and the Euryarchaeota in Fig. 3), then it is possible to use this information to place the root between these two independent entities. However, most of the time such a priori information is not known and the investigator will have to rely on outgroups. Outgroups are sequences that are assumed not to belong to the group investigated but to be relatively close (e.g., other taxonomic units, paralogous genes). Given this assumption, one can use these sequences (usually at least two) to a posteriori polarize the tree, by placing the root between them and those from the group under study. For example, if we are interested in the relationships within Euryarchaeota, we can add three crenarchaeal sequences so that we will be able to root the tree. Importantly, if the rooting aspect of a tree is crucial to resolve a question, it is also likely to be the most unreliable part of the analysis. By not being part of the studied group, outgroups often display the longest branches and are most likely to attract long branches within the studied group. Therefore, extra care should be taken when discussing inferences based on outgroup position. Note that removing an outgroup can often reveal weak but true linkages that are masked by the phylogenetic noise inherent to distant sequence usage.

36.3.1.2. Phylogenetic Terminology

Once polarization has been addressed, it is possible to further define taxonomic relationships. A group of taxa that include all the descendants of a common ancestor, as well as the ancestor, is said to be *monophyletic*. Two such examples are highlighted by ovals in Fig. 3B. These monophyletic groups, or clades, are the only units recognized by cladist classification, since they are the only ones that are defined in terms of sharing of derived (evolved) characters: *synapomorphies*. When a monophyletic group is composed of two entities, generally these entities are referred to as *sister*

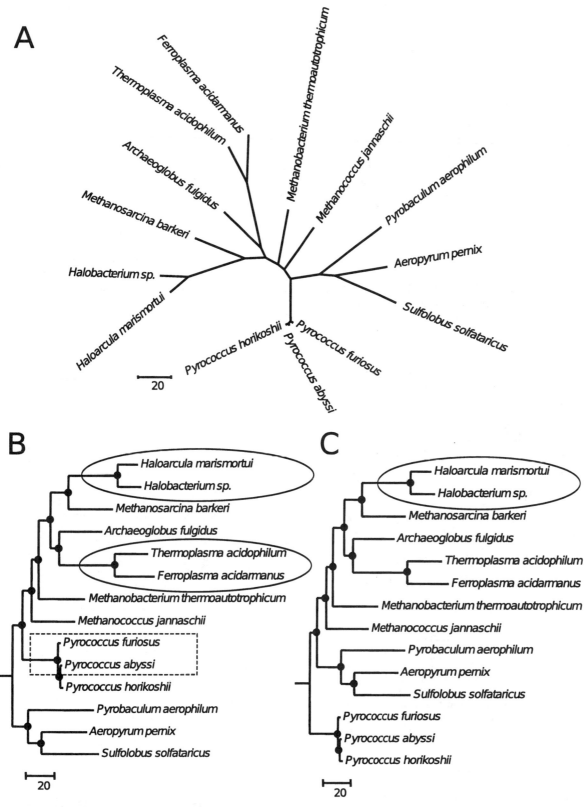

FIGURE 3 Unrooted (A) versus rooted (B and C) trees. Ovals enclose monophyletic groups and rectangles enclose paraphyletic groups. Underlined taxa are members of a polyphyletic group. The black dots on the tree represent each group's ancestor. Note that when the outgroup is changed, some monophyletic groups may become paraphyletic (e.g., the *Euryarchaeota* are monophyletic in tree B but not in C). The scale bar represents substitutions per 100 sites.

groups. A *paraphyletic* group, on the other hand, would unite some but not all of the descendents of a common ancestor as well as the ancestor. Such a group is represented in Fig. 3B by the dotted rectangles. Paraphyletic groups were recognized, in addition to monophyletic groups, by other schools of classification (e.g., Linnaean and evolutionary systematics) and are generally supported by *synplesiomorphies* (shared ancestral characters). Fishes or Reptilia are classical examples of paraphyletic groups, as they would only turn monophyletic if tetrapods and birds were to be included, respectively. In Fig. 3B, the third *Pyrococcus* sequence would have to be included. Finally, a *polyphyletic* group is a group of taxa that share similarities that were presumably not present in their last common ancestor. For example, in Fig. 3B the three underlined taxa share a methanogenic ability, but it seems unlikely that the last common ancestor of these three orders, plus the *Halobacteriales*, *Archaeglobales*, and *Thermoplasmatales*, had this function. In other words, polyphyletic groups have not inherited nor retained their similarity, but obtained the similarity through convergent evolution.

All these definitions imply that the tree has been rooted. However, the term monophyletic is commonly used if the investigator feels that it is unlikely that the root will fall within the group of interest. Fig. 3C, for example, shows that if the real root of our tree lay along the *Pyrococcus* lineage, instead of between Crenarchaeota and Euryarchaeota, then Halobacteria would remain monophyletic (oval) but the Euryarchaeota would become paraphyletic.

36.3.2. Robustness

When phylogenetic relationships have been identified, it is important to critically evaluate them. In other words, is the observed relationship the result of a strong phylogenetic signal (a sharp peak in the phylogenetic landscape), or is it only very slightly better than a multitude of different arrangements (a rock on a plateau)? To address this very important question, phylogeneticists have developed a large number of tools. These tools can be classified as methods that do or do not resample the data.

36.3.2.1. Initial Matrix without Resampling

36.3.2.1.1. Decrease in Value of the Optimality Criterion
This approach is highly intuitive and basically states that the cost, in terms of optimality criterion value, of tearing apart a tree or breaking a given clade will depend on the robustness of the tree or clade. This technique is computationally efficient and can be used for any criterion-based methods (4, 44). Software such as PAUP, PAML, Molphy, and MacClade (see Table 1) implement possibilities to assess the optimality criterion for a tree that is chosen a priori. However, this method does not allow the determination, statistically or objectively, of what is a reliable hypothesis.

36.3.2.1.2. Nonparametric and Parametric Statistical Tests
Statistical tests have been developed to circumvent the previously mentioned problem. The Templeton test (62) is classically used with parsimony-based approaches, whereas the KH test (37) has long been regarded as the reference for likelihood-based studies. However, the KH test is not valid when there is a priori reason to believe that one topology is

better than the other (e.g., test the best tree versus the next best one). Other tests such as the SOWH (61) or the SH (54) have been proposed to be valid in such cases, but often appear too liberal and too conservative, respectively (25). More recently, Shimodaira (53) introduced a new Approximately Unbiased (AU) test that seems to be gaining acceptance within the community. The AU test and numerous others are implemented in the CONSEL software package (55).

36.3.2.1.3. Posterior Probabilities
The major difference between Bayesian inferences and maximum likelihood lies in the way they deal with parameters. Likelihood parameter optimizations are very computer intensive, while Bayesian integrations are analytically infeasible (34). Procedures that are able to approximate posterior probability not only of the tree but also of each node have been implemented (e.g., MrBayes [32]). These local posteriors have been proposed to replace bootstrap values (see below) (34), but this early optimism needs to be revised as posterior probabilities are more likely to represent reliability's upper bound and bootstrap its lower bound (14).

36.3.2.2. Resampling
Resampling-based methods attempt to estimate the variance of sampling (variance associated with taking, at best, only a few genes from few species into consideration) by generating new data sets by resampling from the original data set. The number of resamplings should ideally be greater than 1,000, but one or a few hundred are often all that can be reasonably computed.

36.3.2.2.1. Bootstrap Percentages (15, 18)
Bootstrapping generates new data sets by performing multiple drawings with replacement. If the initial data contain 150 sites, then the resampled data set (pseudoreplicate) will also be 150 sites long, but the first site could be present three times while the second may not be included at all. In subsequent steps, optimal trees for each pseudoreplicate are computed and a consensus tree is drawn. For example, when any given clade is recovered by two thirds of the pseudoreplicates, then this node will have a bootstrap support of 66.6%. At or above 75%, the nodes are often regarded as "supported," and those above 95% are "strongly supported." However, evidence seems to suggest that bootstrap values are overly conservative when in the high range, meaning that a node receiving 75% of bootstrap would be accurate more than 75% of the time (conversely, low bootstrap support would be overestimates [31]).

36.3.2.2.2. Jackknife (46)
The jackknife procedure shares many features with bootstrapping, but instead of generating full-length pseudoreplicates, jackknife produces truncated ones. Second, the resampling is done without replacement so a site can only be present one or zero times. It is unclear how many sites should be dropped, but somewhere between Felsenstein's 50% and Farris's e^{-1} is likely to be reasonable.

36.4. SUMMARY AND CONCLUSION: WHY DO WE CONSTRUCT PHYLOGENETIC TREES?
The classical and principal reason for constructing phylogenetic trees is to determine how different sequences are related. This can be then extrapolated as reflecting the evo-

lutionary history of the organisms from which the sequences were sampled. In molecular microbiology, trees estimated from 16S rRNA genes have classically been used for this purpose (11). However, with the increasing amount of sequences available, mainly from the genome-sequencing projects, it has become increasingly evident that many (if not most) genes from prokaryotic genomes have different evolutionary relationships due to transfer of genes between lineages or species, a phenomenon termed lateral or horizontal gene transfer (LGT) (12, 24). The phylogenetic trees estimated from all the genes in a genome will, in most cases, represent gene trees and not organismal trees. However, phylogenetic analysis can be used to identify the laterally transferred genes (some examples are given in Table 1 in reference 13). For instance, if a bacterial gene is found nested within a eukaryotic clade, it is likely that the gene was transferred from a eukaryote to the bacterium. For discussions on how to detect LGT from phylogenetic analyses, see, for example, references 2 and 59.

Phylogenies are not only used for inferring the evolutionary relationship between sequences per se; phylogenetics can also be used to help in the annotation of genomes (16, 56). Most genes from genome-sequencing projects are annotated based on results from pairwise comparisons from database searches. However, paralogous genes often do not have the same function, and such genes cannot be distinguished by simple pairwise comparisons. Hence, tools for automated phylogenetic analyses of complete genome data have recently been developed which can help facilitate gene annotation as well as flag interesting genes for more detailed phylogenetic analyses (see, e.g., reference 56).

Phylogenetic analyses can also be used to identify functionally and structurally constrained residues in proteins or RNA (see, e.g., references 5, 9, and 35). For instance, using maximum-likelihood phylogeny, it is possible to assign relative rates to each site and use this quantity as a measure of constraints (3). This type of analysis relies on the hypothesis that functionally important sites usually evolve more slowly than nonfunctional sites. Conversely, if a protein family has been well studied, phylogenetic analyses can also be used to investigate whether "new" sequences also possess functionally important residues. This can provide valuable information on how proteins evolve new substrate specificities or how generally applicable previously identified functional residues may be across taxonomic diversity (see references 9 and 43 for further details).

Another related type of comparative analysis is prediction of functional divergence by identifying site-specific rate shifts (see reference 23 for a review). These analyses map the evolutionary rates of amino acid replacement at sites along the protein sequence, using the gamma distribution described in section 36.2.2.5.2. If a functional shift has occurred in the evolution of a protein family, then the distribution of evolutionary rates along the protein sequence can be expected to vary in different subtrees. When data from structural analyses are included, these comparative approaches can be extremely powerful. Recently, methods for characterizing rate-independent functional divergence have been described (3) and can be found in the software package covARES. One advantage of this package is that it provides both evolutionary and structural parameter mapping and analysis tools.

Prediction of functionally important sites and detection of molecular adaptation can also be done by measuring the ratio of synonymous to nonsynonymous mutations (see sec-

tion 36.5), and methods have been developed to detect positive selection along branches on a phylogeny as well as detecting amino acid sites under positive selection (see reference 69 for a review, as well as the PAML package in Table 1). Since these methods are based on DNA alignments and synonymous mutations reach saturation rapidly, they are best suited for relatively closely related sequences, whereas the methods described above can be used for more divergent sequence data.

Phylogenetic trees can also be used to reconstruct ancestral sequences. These ancestral sequences can provide structural and functional information on protein families, both when used in computational studies (45) and when used to reconstruct ancestral sequences through recombinant DNA technology (58). This approach has also been explored to detect covariation signals in proteins (20). Such signals are detected as patterns of compensatory amino acid replacements. For example, the replacement of an amino acid at a given position with a residue that reverses the change may consistently be associated with the replacement of an amino acid at a second position with a residue that restores the charge. Fukami-Kobayashi et al. (20) showed that reconstructing ancestral sequences gives a stronger covariation signal than comparisons of contemporary sequences.

It should be apparent from this chapter that the range of phylogenetic techniques is considerable, as are the questions that can be addressed using these techniques. It should also be clear that phylogenetic analysis cannot be regarded as a "black box" and that each analysis will be "custom made." We hope that this chapter has provided some guidelines to what methods and choices are available.

36.5. SOME DEFINITIONS

Homology Two genes are **homologous** if they descend from a common ancestral gene. When working with sequence alignments, it is important to remember that even if two sequences are homologous, insertions and deletions (indels) may mix up the **positional homology** of the nucleotide or amino acid sites. Note that homology is commonly *misused* to mean similarity.

Orthology Two homologous sequences are **orthologs** if their common history can be traced back to a speciation event.

Paralogy Two homologous sequences are **paralogs** if their common history can be traced back to a duplication event.

Synonymous and nonsynonymous substitutions Synonymous mutations are mutations that do not result in an amino acid replacement, whereas nonsynonymous substitutions do result in an amino acid replacement.

Xenology Two homologous sequences are **xenologs** if their common history can be traced back to a lateral gene transfer event.

36.6. FREQUENTLY ASKED QUESTIONS

Why is there no single recipe?

Unlike DNA or RNA extractions, phylogenetic analyses are "custom made" and therefore cannot be accurately summarized in a single protocol. The analyses will need to be adapted depending on the number of taxa (OTUs), the length of the sequences analyzed, the degree of divergence, the percentage of missing data, and the computing facilities at hand.

Is there an easy way to make a tree?

No, but if you really need a tree quickly and do not want to put too much time into the analysis:

1. Create a file in FASTA format (Fig. 2) that includes all sequences of interest.

2. Download and install the latest version of ClustalX on your system (http://www.bips.u-strasbg.fr/fr/Documentation/ClustalX).

3. Start ClustalX.

4. Click on "File," "Load sequences."

5. Select your FASTA file.

6. Click "open."

7. Click on "Alignment," "Do Complete Alignment," "Align."

8. Wait until completion (look for a "CLUSTAL-Alignment file created []" message at the bottom left of the window).

9. Click on "Trees."

10. Check "Exclude Positions with Gap," "Correct for Multiple Substitution."

11. Draw NJ tree.

12. Download (http://taxonomy.zoology.gla.ac.uk/rod/treeview.html) and install the latest version of Treeview on your system to visualize, present, and print your tree.

What is maximum parsimony?

The most parsimonious topology is the tree, or set of trees, that requires the smallest number of evolutionary changes. When doing an exhaustive search (see section 36.2.2.2.), the minimum number of changes for all sites in the alignment is computed for all possible topologies, and the tree requiring the fewest changes (or "steps," in parsimony jargon) is selected. Under a heuristic search, the minimum number of steps is determined for the two topologies to be compared (the best topology so far in the search and the rearranged one; see section 36.2.2.2) and the more parsimonious of the two is selected and itself rearranged for the next step. Over the years, several variants of parsimony have been developed that, for example, forbid convergence (Dollo parsimony), take into account that some changes are "easier" than others (generalized parsimony), or assume some sites are better phylogenetic markers than others (weighted parsimony).

What is maximum likelihood?

Not unlike maximum parsimony, a maximum-likelihood procedure will look for the tree that maximizes the likelihood of observing the data given the topology, its branch lengths, the model of evolution, and parameters, such as the shape of the gamma distribution, that account for the among-site rate variation. In this way, searches will compare topologies by multiplying the likelihood of each individual site and selecting the tree having the highest product.

What is a neighbor joining?

Neighbor joining (NJ) is an extremely efficient, distance-based algorithm (see section 36.2.4). Unlike maximum likelihood or parsimony, this method is not based on the maximization of any criterion and does not do any comparison of trees. On the contrary, this algorithmic procedure will carry out an iterative hierarchization of the OTUs. All sequences (OTUs) are first clustered in a single starlike tree. Then, a first step is initiated in which all topologies having a pair in one side and the remaining starlike tree on the

other are formed. The topology that displays the shortest sum of branch lengths is then selected and the two OTUs are paired. A new distance matrix is then calculated and the process is reiterated. The procedure stops when the starlike component is resolved.

What method is the best?

This question is the most frequent and yet the most debated in the field. Based on simulated data, maximum likelihood is often regarded as the best tool (32). Bayesian methods, if we consider only the most probable topology (and not the still highly questionable posterior probabilities of the nodes), are likely to be just as useful as maximum likelihood. Depending on the data, parsimony- or distance-based methods will then be the next best choice. For example, if a data set contains much missing data, a parsimony-based reconstruction will be more likely to provide a sensible topology, but this method will become highly variable when extremely diverged taxa are under study. However, given that likelihood and Bayesian inference can be significantly more time consuming and that bias can interfere with simulation results, we recommend one more time that each phylogenetic analysis should be treated as a unique case and that custom-made decisions should be made before each study.

We are grateful to Andrew Roger, Rebecca Case, W. Ford Doolittle, and two anonymous referees for comments and suggestions. This work was supported by Genome Atlantic (an affiliate of Genome Canada), the Killam Foundation, and the Canadian Institutes for Health Research.

36.7. REFERENCES

36.7.1. Further Reading

1. **Felsenstein, F.** 2004. *Inferring Phylogenies.* Sinauer Associates, Sunderland, MA.
   The most complete. THE reference.

2. **Graur, D., and W.-H. Li.** 2000. *Fundamentals of Molecular Evolution.* Sinauer Associates, Sunderland, MA.
   As indicated by its title, the primary focus of this book is molecular evolution. However, numerous notions introduced mostly in chapters 3 to 5 will be of great help to any apprentice phylogeneticists. The help will nevertheless remain on the theoretical side, as few if any practical considerations are discussed.

3. **Hall, B. G.** 2001. *Phylogenetic Trees Made Easy: A How-To Manual for Molecular Biologists.* Sinauer Associates, Sunderland, MA.
   By author's own admission, this book should be taken as a cookbook and not as a textbook. It will prove most useful for inexperienced Mac users who are interested in generating neighbor-joining and Bayesian trees. This book also includes a very useful section on generating and improving alignments.

4. **Page, R. D. M., and E. C. Holmes.** 1998. *Molecular Evolution: a Phylogenetic Approach.* Blackwell Science, Oxford.
   This book can definitively be regarded as the textbook in phylogenetic reconstruction. With the exception of Bayesian inference, which was being developed at the time, most of the major topics in the field are nicely presented. This work is the most adequate introduction to the Swofford et al. (1996) publication.

5. **Page, R. D. M., O. Gascuel, J. C. Wilgenbush, D. Swofford, D. Posada, H. A. Schmidt, A. von Haeseler, and D. H. Huson.** 2003. Inferring evolutionary relationships. *In* A. D. Baxevanis, D. B. Davison, R. D. M. Page, G. A. Petsko, L. D. Stein, and G. D. Stormo (ed.), *Current Protocols in Bioinformatics,* vol. 1. John Wiley & Sons, New York, NY.
   This section of the Current Protocols in Bioinformatics *provides classic and yet accessible protocols for choosing models of*

evolution and finding most parsimonious and neighbor-joining trees as well as visualizing them. The section is completed by an introduction to phylogenetic inference, usage of quartet puzzling to approximate the most likely tree, and an alternative approach for non-tree-like data.

6. **Swofford, D. L., G. P. Olsen, P. J. Waddell, and D. M. Hillis.** 1996. Phylogenetic inference, p. 407–514. *In* D. M. Hillis, C. Moritz, and B. K. Mable (ed.), *Molecular Systematics.* Sinauer Associates, Sunderland, MA.

This book chapter is essential for any tree builder who wants to know more. A few of the newer techniques are missing, but it still remains valuable. It is likely to be a bit cryptic to any newcomer, but reading Page and Holmes's book first will make things clearer.

36.7.2. General References

1. **Adachi, J., and M. Hasegawa.** 1996. Model of amino acid substitution in proteins encoded by mitochondrial DNA. *J. Mol. Evol.* **42:**459–468.

2. **Andersson, J. O., W. F. Doolittle, and C. L. Nesbo.** 2001. Genomics. Are there bugs in our genome? *Science* **292:**1848–1850.

3. **Blouin, C., Y. Boucher, and A. J. Roger.** 2003. Inferring functional constraints and divergence in protein families using 3D mapping of phylogenetic information. *Nucleic Acids Res.* **31:**790–797.

4. **Bremer, K.** 1994. Branch support and tree stability. *Cladistics* **10:**295–304.

5. **Cannone, J. J., S. Subramanian, M. N. Schnare, J. R. Collett, L. M. D'Souza, Y. Du, B. Feng, N. Lin, L. V. Madabusi, K. M. Muller, N. Pande, Z. Shang, N. Yu, and R. R. Gutell.** 31 May 2002. The Comparative RNA Web (CRW) Site: an online database of comparative sequence and structure information for ribosomal, intron, and other RNAs. *BMC Bioinformatics* **3:**2. [online.] http://www.rna.ccbb.utexas.edu/.

6. **Castresana, J.** 2000. Selection of conserved blocks from multiple alignments for their use in phylogenetic analysis. *Mol. Biol. Evol.* **17:**540–552.

7. **Cole, J. R., B. Chai, T. L. Marsh, R. J. Farris, Q. Wang, S. A. Kulam, S. Chandra, D. M. McGarrell, T. M. Schmidt, G. M. Garrity, and J. M. Tiedje.** 2003. The Ribosomal Database Project (RDP-II): previewing a new autoaligner that allows regular updates and the new prokaryotic taxonomy. *Nucleic Acids Res.* **31:**442–443.

8. **Cunningham, C. W.** 1997. Is congruence between data partitions a reliable predictor of phylogenetic accuracy? Empirically testing an iterative procedure for choosing among phylogenetic methods. *Syst. Biol.* **46:**464–478.

9. **Dacks, J. B., and W. F. Doolittle.** 2002. Novel syntaxin gene sequences from *Giardia, Trypanosoma* and algae: implications for the ancient evolution of the eukaryotic endomembrane system. *J Cell Sci.* **115:**1635–1642.

10. **Dayhoff, M. O., R. M. Schwartz, and B. C. Orcutt.** 1978. A model of evolutionary change in proteins, p. 345–352. *In* M. O. Dayhoff (ed.), *Atlas of Protein Sequence Structure,* vol. 5. National Biomedical Research Foundation, Washington, DC.

11. **DeLong, E. F.** 2002. Microbial population genomics and ecology. *Curr. Opin. Microbiol.* **5:**520–524.

12. **Doolittle, W. F.** 1999. Phylogenetic classification and the universal tree. *Science* **284:**2124–2129.

13. **Doolittle, W. F., Y. Boucher, C. L. Nesbø, C. J. Douady, J. O. Andersson, and A. J. Roger.** 2003. How big is the iceberg of which organellar genes in nuclear genomes are but the tip? *Philos. Trans. Biol. Sci.* **358:**39–58.

14. **Douady, C. J., F. Delsuc, Y. Boucher, W. F. Doolittle, and E. J. P. Douzery.** 2003. Comparison of Bayesian and maximum likelihood bootstrap measures of phylogenetic reliability. *Mol. Biol. Evol.* **20:**248–254.

15. **Efron, B.** 1979. Bootstrap methods: another look at the jackknife. *Ann. Stat.* **7:**1–26.

16. **Eisen, J. A., and M. Wu.** 2002. Phylogenetic analysis and gene functional predictions: phylogenomics in action. *Theor. Popul. Biol.* **61:**481–487.

17. **Felsenstein, J.** 1978. Cases in which parsimony or compatibility methods will be positively misleading. *Syst. Zool.* **27:**401–410.

18. **Felsenstein, J.** 1985. Confidence limits on phylogenies: an approach using the bootstrap. *Evolution* **39:**783–791.

19. **Fitch, W. M.** 1971. Rate of change of concomitantly variable codons. *J. Mol. Evol.* **1:**84–96.

20. **Fukami-Kobayashi, K., D. R. Schreiber, and S. A. Benner.** 2002. Detecting compensatory covariation signals in protein evolution using reconstructed ancestral sequences. *J. Mol. Biol.* **319:**729–743.

21. **Galtier, N.** 2001. Maximum-likelihood phylogenetic analysis under a covarion-like model. *Mol. Biol. Evol.* **18:**866–873.

22. **Gascuel, O.** 2003. Getting a tree fast: neighbor joining and distance based methods, p. 6.3.1–6.3.18. *In* G. D. Stormo (ed.), *Current Protocols in Bioinformatics,* vol. I. John Wiley & Sons, Inc., Hoboken, NJ.

23. **Gaucher, E. A., X. Gu, M. M. Miyamoto, and S. A. Benner.** 2002. Predicting functional divergence in protein evolution by site-specific rate shifts. *Trends Biochem. Sci.* **27:**315–321.

24. **Gogarten, J. P., W. F. Doolittle, and J. G. Lawrence.** 2002. Prokaryotic evolution in light of gene transfer. *Mol. Biol. Evol.* **19:**2226–2238.

25. **Goldman, N., J. P. Anderson, and A. G. Rodrigo.** 2000. Likelihood-based tests of topologies in phylogenetics. *Syst. Biol.* **49:**652–670.

26. **Graham, R. L., and L. R. Foulds.** 1982. Unlikelihood that minimal phylogenies for a realistic biological study can be constructed in a reasonable computational time. *Math. Biosci.* **60:**133–142.

27. **Graur, D., and W.-H. L. Li.** 2000. *Fundamentals of Molecular Evolution,* 2nd ed. Sinauer Associates, Sunderland, MA.

28. **Hall, B. G.** 2001. *Phylogenetic Trees Made Easy: A How-to Manual for Molecular Biologists.* Sinauer Associates, Sunderland, MA.

29. **Hendy, M. D., and D. Penny.** 1982. Branch and bound algorithms to determine minimal evolutionary trees. *Math. Biosci.* **59:**277–290.

30. **Henikoff, S., and J. G. Henikoff.** 1992. Amino acid substitution matrices from protein blocks. *Proc. Natl. Acad. Sci. USA* **89:**10915–10919.

31. **Hillis, D. M., and J. J. Bull.** 1993. An empirical test of bootstrapping as a method for assessing confidence in phylogenetic analysis. *Syst. Biol.* **42:**182–192.

32. **Huelsenbeck, J. P.** 1995. The robustness of two phylogenetic methods: four-taxon simulations reveal a slight superiority of maximum likelihood over neighbor joining. *Mol. Biol. Evol.* **12:**843–849.

33. **Huelsenbeck, J. P., and F. Ronquist.** 2001. MRBAYES: Bayesian inference of phylogenetic trees. *Bioinformatics* **17:**754–755.

34. **Huelsenbeck, J. P., F. Ronquist, R. Nielsen, and J. P. Bollback.** 2001. Evolution: Bayesian inference of phylogeny and its impact on evolutionary biology. *Science* **294:**2310–2314.

35. **Inagaki, Y., C. Blouin, W. F. Doolittle, and A. J. Roger.** 2002. Convergence and constraint in eukaryotic release factor 1 (eRF1) domain 1: the evolution of stop codon specificity. *Nucleic Acids Res.* **30:**532–544.

36. **Jones, D. T., W. R. Taylor, and J. M. Thornton.** 1992. The rapid generation of mutation data matrices from protein sequences. *CABIOS* **8:**275–282.

37. **Kishino, H., and M. Hasegawa.** 1989. Evaluation of the maximum likelihood estimate of the evolutionary tree topologies from DNA sequence data, and the branching order in hominoidea. *J. Mol. Evol.* **29:**170–179.

38. **Lanave, C., G. Preparata, C. Saccone, and G. Serio.** 1984. A new method for calculating evolutionary substitution rates. *J. Mol. Evol.* **20:**86–93.

39. **Lockhart, P. J., M. A. Steel, A. C. Barbrook , D. H. Huson, M. A. Charleston, and C. J. Howe.** 1998. A covariotide model explains apparent phylogenetic structure of oxygenic photosynthetic lineages. *Mol. Biol. Evol.* **15:** 1183–1188.

40. **Lopez, P., D. Casane, and H. Philippe.** 2002. Heterotachy, an important process of protein evolution. *Mol. Biol. Evol.* **19:**1–7.

41. **Lopez, P., P. Forterre, and H. Philippe.** 1999. The root of the tree of life in the light of the covarion model. *J. Mol. Evol.* **49:**496–508.

42. **Loytynoja, A., and M. C. Milinkovitch.** 2001. SOAP, cleaning multiple alignments from unstable blocks. *Bioinformatics* **17:**573–574.

43. **Lucas, P., C. Otis, J. P. Mercier, M. Turmel, and C. Lemieux.** 2001. Rapid evolution of the DNA-binding site in LAGLIDADG homing endonucleases. *Nucleic Acids Res.* **29:**960–969.

44. **Meireles, C. M., J. Czelusniak, M. P. Schneider, J. A. Muniz, M. C. Brigido, H. S. Ferreira, and M. Goodman.** 1999. Molecular phylogeny of ateline new world monkeys (Platyrrhini, atelinae) based on gamma-globin gene sequences: evidence that brachyteles is the sister group of lagothrix. *Mol. Phylogenet. Evol.* **12:**10–30.

45. **Messier, W., and C. B. Stewart.** 1997. Episodic adaptive evolution of primate lysozymes. *Nature* **385:**151–154.

46. **Mueller, L. D., and F. J. Ayala.** 1982. Estimation and interpretation of genetic distance in empirical studies. *Genet. Res.* **40:**127–137.

47. **Page, R. D. M., O. Gascuel, J. C. Wilgenbush, D. Swofford, D. Posada, H. A. Schmidt, A. von Haeseler, and D. H. Huson.** 2003. Inferring evolutionary relationships. *In* L. D. Stein and G. D. Stormo (ed.), *Current Protocols in Bioinformatics*, vol. 1. John Wiley & Sons, Hoboken, NJ.

48. **Penny, D., B. J. McComish, M. A. Charleston, and M. D. Hendy.** 2001. Mathematical elegance with biochemical realism: the covarion model of molecular evolution. *J. Mol. Evol.* **53:**711–723.

49. **Posada, D., and K. A. Crandall.** 1998. MODELTEST: testing the model of DNA substitution. *Bioinformatics* **14:** 817–818.

50. **Rannala, B., and Z. H. Yang.** 1996. Probability distribution of molecular evolutionary trees: a new method of phylogenetic inference. *J. Mol. Evol.* **43:**304–311.

51. **Saitou, N., and M. Nei.** 1987. The neighbor-joining method: a new method for reconstructing phylogenetic trees. *Mol. Biol. Evol.* **4:**406–425.

52. **Salemi, M., and A.-M. Vandamme.** 2003. *The Phylogenetic Handbook. A Practical Approach to DNA and Protein Phylogeny.* Cambridge University Press, Cambridge, United Kingdom.

53. **Shimodaira, H.** 2002. An approximately unbiased test of phylogenetic tree selection. *Syst. Biol.* **51:**492–508.

54. **Shimodaira, H., and M. Hasegawa.** 1999. Multiple comparisons of log-likelihoods with applications to phylogenetic inference. *Mol. Biol. Evol.* **16:**1114–1116.

55. **Shimodaira, H., and M. Hasegawa.** 2001. CONSEL: for assessing the confidence of phylogenetic tree selection. *Bioinformatics* **17:**1246–1247.

56. **Sicheritz-Ponten, T., and S. G. Andersson.** 2001. A phylogenomic approach to microbial evolution. *Nucleic Acids Res.* **29:**545–552.

57. **Simmons, M. P., H. Ochoterena, and J. V. Freudenstein.** 2002. Amino acid vs. nucleotide characters: challenging preconceived notions. *Mol. Phylogenet. Evol.* **24:**78–90.

58. **Stackhouse, J., S. R. Presnell, G. M. McGeehan, K. P. Nambiar, and S. A. Benner.** 1990. The ribonuclease from an extinct bovid ruminant. *FEBS Lett.* **262:**104–106.

59. **Stanhope, M. J., A. Lupas, M. J. Italia, K. K. Koretke, C. Volker, and J. R. Brown.** 2001. Phylogenetic analyses do not support horizontal gene transfers from bacteria to vertebrates. *Nature* **411:**940–944.

60. **Swofford, D. L. (ed.).** 2001. *PAUP*. Phylogenetic Analysis Using Parsimony (*and Other Methods)*, Version 4. Sinauer Associates, Sunderland, MA.

61. **Swofford, D. L., G. P. Olsen, P. J. Waddell, and D. M. Hillis.** 1996. Phylogenetic inference, p. 407–514. *In* B. K. Mable (ed.), *Molecular Systematics*. Sinauer Associates, Sunderland, MA.

62. **Templeton, A. R.** 1983. Phylogenetic inference from restriction endonuclease cleavage site maps with particular reference to the evolution of humans and the apes. *Evolution* **37:**221–244.

63. **Thompson, J. D., D. G. Higgins, and T. J. Gibson.** 1994. CLUSTAL W: improving the sensitivity of progressive multiple sequence alignment through sequence weighting, position-specific gap penalties and weight matrix choice. *Nucleic Acids Res.* **22:**4673–4680.

64. **Thompson, J. D., F. Plewniak, and O. Poch.** 1999. A comprehensive comparison of multiple sequence alignment programs. *Nucleic Acids Res.* **27:**2682–2690.

65. **Veerassamy, S., A. Smith, and E. R. M. Tillier.** 2003. A transition probability model for amino acid substitutions from BLOCKS. *J. Comput. Biol.* **10:**997–1010.

66. **Wuyts, J., P. De Rijk, Y. Van de Peer, T. Winkelmans, and R. De Wachter.** 2001. The European Large Subunit Ribosomal RNA database. *Nucleic Acids Res.* **29:**175–177.

67. **Wuyts, J., Y. Van de Peer, T. Winkelmans, and R. De Wachter.** 2002. The European database on small subunit ribosomal RNA. *Nucleic Acids Res.* **30:**183–185.

68. **Yang, Z.** 1996. Among-site rate variation and its impact on phylogenetic analyses. *Trends Ecol. Evol.* **11:**367–372.

69. **Yang, Z.** 1996. Phylogenetic analysis using parsimony and likelihood methods. *J. Mol. Evol.* **42:**294–307.

37

Microbial Nucleotide Fingerprints in Nature

DAVID M. KARL

37.1. INTRODUCTION

Microorganisms are the key to Earth's habitability. They harvest light energy, produce organic matter, and facilitate the turnover of key bioelements like nitrogen (N), phosphorus (P), and sulfur (S). During the process of microbial cell metabolism and growth, inorganic nutrients are assimilated into new biomass that fuels a complex series of interactions between various groups of microorganisms and among microbes and multicellular organisms. In summary, microbes make things happen.

Microbial communities in nature are diverse and complex. With the advent of modern techniques in molecular biology, one can now identify and track even the "virtual" microbes—those species that have successfully evaded pure culture isolation. It is also possible to interrogate ecosystems for the presence of specific genes and gene products, and even determine in situ expression of target genes. The present level of analytical sophistication and the continued improvement of techniques are impressive and hold great promise to yield new, fundamental information regarding the phylogenetic and physiological structure of microbial assemblages in nature. By comparison with the development of novel autecological approaches, progress in the development and refinement of methods that target the entire microbial community, specifically estimation of total living biomass, metabolic activity, growth, and reproduction, has been less dramatic.

For many ecological studies, accurate estimates of total microbial biomass, metabolic activity, and productivity are essential; they are the key master variables for a complete ecological assessment. Despite the recognized importance of total microbial community assessments, however, few methods currently exist, and those that have been developed have unique limitations that derive from the fact that the assay targets are generally diverse, partially uncharacterized, and in variable states of metabolism and growth. A forward-looking review article by Madsen (40), "Epistemology of environmental microbiology," presents an up-to-date progress report of the constraints imposed by the complexity of natural ecosystems and by limited methodology.

In pure culture studies, the measurement of biomass is fairly easily satisfied. However, as stated by Koch (36) in his authoritative review on growth measurements, "the methods for measuring biomass seem obvious and straightforward, but in fact they are complicated if accuracy is sought." He goes on to review the methods that are most often applied to laboratory studies of pure culture isolates, namely wet and dry weight, light scattering and turbidimetry, and various enumeration techniques. These last include light and electron microscopy and flow cytometry for direct counts, as well as spread plates, most probable number, and other culture methods for viable cell counts. If the average dimensions (or average mass or carbon content) of the cells in culture are known, or measured, then the direct or viable colony count can be extrapolated to biomass. However, even with pure culture isolates, cell size, mass, and elemental composition may vary depending on composition of the growth medium, culture conditions (temperature, light, pressure), and growth rate. These laboratory-based methods for direct and indirect estimation of biomass are not easily adapted to the complex mixed cultures of microorganisms that one generally finds in nature.

The direct measurement of carbon, nucleic acids, protein, or other similar cell-derived compounds is generally unacceptable due to the relatively long residence times in the environment after cell death and lysis. For example, in most open-ocean habitats there is more particulate carbon (PC) in nonliving than in living materials, and the proportion varies considerably with water depth (approximately 50% of total carbon is living near the surface, decreasing to about 5% at 1,000 m) (29). There is even more nonliving particulate DNA in most natural ecosystems than there is inside all of the cells that are present. Nevertheless, measurements of total PC or DNA can be used to set upper bounds on total biomass and, as such, they have been used extensively in ecological studies.

An effective biochemical or molecular biomarker of microbial biomass should satisfy the following criteria (22): (i) the biomarker selected must be present in all living cells or in all organisms of a specific population; (ii) it must be readily metabolized, hydrolyzed, or otherwise decomposed following cell death; (iii) it must exist as a uniform and constant percentage of total biomass regardless of environmental or physiological conditions; (iv) a convenient method must exist to extract and purify (if necessary) the compound from environmental samples; and (v) a sensitive, quantitative assay procedure must be available. Microbial biomass in many natural ecosystems is often low, so assay sensitivity is an important criterion.

Only a few common features are shared by all of the diverse members of natural microbial assemblages that might serve as the basis for an environmental biomass assay system, and for this reason, progress in these important areas of microbial ecology has been slow. Two promising candidate classes of molecular biomarkers for ecological investigation are membrane lipids and cellular nucleotides. Other compounds have been used in environmental studies to detect the biomass of a portion of the total microbial assemblage. These include chlorophyll *a* for the sum of all photosynthetically active bacteria and eukarya, lipopolysaccharide for gram-negative bacteria, and muramic acid for gram-positive plus gram-negative bacteria (29).

Determination of total lipid phosphate (LP) as a quantitative measure of microbial biomass was originally described by White et al. (51) in their application to marine sediments. The general method was subsequently expanded to include (i) LP production rate, using radioactive phosphate (^{32}P or ^{33}P) as a precursor, as a measure of total microbial production (50); (ii) the pattern of signature ester-linked phospholipids fatty acids (PLFA) and hydroxyl fatty acids in the lipopolysaccharide of gram-negative bacteria as measures of microbial community structure (46, 52); and (iii) changes in the concentrations of poly-β-hydroxyalkanoic acid (PHA) in bacteria or triglyceride in microeukaryotes as measures of nutritional-physiological status (13). These diagnostic lipid-based profiles, or environmental microbial lipid fingerprints, have proven invaluable in a variety of ecological studies (11). Major limitations, according to White et al. (52), are the nontraditional units that result from these analyses (e.g., picomoles of PLFA per sample, rather than total cell mass or carbon), the relative insensitivity (detection limit of approximately 10^{10} bacteria per sample for biomass by LP analysis; for oligotrophic ocean seawaters that would equate to ~10 liters), and the requirement for specialized analytical equipment (gas chromatograph–mass spectrometer [GC-MS]) for the lipid profile analyses. Furthermore, it now appears that certain ubiquitous marine microorganisms, e.g., *Synechococcus* and *Prochlorococcus*, may have reduced cell quotas of membrane phospholipids (3, 9), so they would not be accurately represented in the environmental microbial biomass assessment.

The second and, perhaps, more broadly applicable class of cellular biomarkers is nucleotides and their derivatives. All living cells studied to date have an identical suite of more than 100 different nucleotides that are essential for viability, energy transduction, metabolic regulation, and biosynthesis. Although it is impossible to be absolutely certain on this point, it is highly unlikely that any microorganisms will be discovered with alternative pathways; the universal role of nucleotides and their derivatives appears to be well established (21).

The central role of the adenine nucleotides, and especially ATP, as intermediate carriers of chemical energy linking catabolism and biosynthesis has been known since the publication of Lipmann's classic paper (39). Nucleotides have at least three other crucial cellular functions: (i) synthesis of DNA and RNA; (ii) activation and transfer of precursors for cellular biosynthesis; and (iii) control and regulation of cellular metabolism. By measuring the intracellular concentrations of key nucleotides and ratios thereof, the intracellular pool turnover rates, and nucleic acid (DNA, RNA) synthesis rates, an environmental nucleotide fingerprint can be compiled that provides valuable information concerning in situ biomass and metabolic activity of the total microbial community.

Over 40 years ago, methods were developed for quantitative extraction of ATP from cells and for its detection in cell extracts. Initially, the ATP detection method was promoted as a life detection system for possible use on the Martian spacecraft *Viking* in 1974 (38), but a few years later the method was expanded into a total microbial biomass assay by Holm-Hansen and Booth (15). The steady-state intracellular concentration of ATP (referred to as the "ATP pool") in metabolically active microorganisms (prokaryotes and eukaryotes) appears to be well regulated at a value of 2 to 6 nmol of ATP mg^{-1} dry weight (1 to 3 mM) regardless of growth rate, culture condition, or mode of nutrition (21). This served as the theoretical basis for its use in total microbial biomass estimation.

Subsequent investigation led to the development of the adenylate energy charge (EC$_A$) concept and its application to microbial assemblages in nature (53). The EC$_A$ is equal to one-half of the number of anhydride-bound phosphate groups per adenine moiety, or EC$_A$ = ([ATP] + [ADP])/([ATP] + [ADP] + [AMP]), and is equivalent to a linear measure of the total amount of chemical potential energy momentarily stored in the adenine nucleotide pool (1). In vitro rate responses of several "key" enzymes in cellular metabolism to variations in EC$_A$ have provided the background data for this control hypothesis. The well-established correlation between cellular EC$_A$ and metabolic or growth potentials of individual organisms, in theory, provides a framework for estimating metabolic potentials of naturally occurring microbial populations (6).

In addition to ATP, the measurement of non-adenine NTPs may provide relevant ecological information. Many important cellular reactions, notably those associated with microbial biosynthesis and growth, are coupled to energy derived from NTPs other than ATP (20). For example, guanosine-5′-triphosphate (GTP) and uridine 5′-triphosphate (UTP) are both required for the activation and interconversion of carbohydrate precursors for polysaccharide biosynthesis; cytidine 5′-triphosphate (CTP), GTP, and UTP are required for RNA transcription; and the deoxyribose derivatives of CTP, GTP, ATP, and thymidine 5′-triphosphate are required for DNA replication. In addition, GTP is an obligate requirement for the initiation, the aminoacyl-tRNA binding, and the translocation processes of protein synthesis. Unlike the ATP pool, which is maintained at a relatively constant level independent of growth rate, the intracellular concentrations of non-adenine NTPs fluctuate in direct proportion to their requirements for biosynthesis (12, 20, 47). Consequently, quantitative measurements of GTP and the estimation of total GTP:ATP ratios of microbial communities in nature constitute a key component of the environmental nucleotide fingerprint.

Some previous studies of ATP in microorganisms and in environmental samples have emphasized the futility of extrapolating these static pool measurements to estimates of metabolic energy flux. It is the turnover rate of the ATP pool, rather than the steady-state concentration, that varies in proportion to cellular metabolic energy requirements. It follows, then, that direct measurements of ATP pool turnover, when coupled with independent estimates of ATP pool size, should provide useful information on biological energy flux in cells, populations, or natural microbial assemblages (26). Finally, the measurement of rates of RNA and DNA synthesis, using the uptake and incorporation of specific radiolabeled precursors (e.g.,^{3}H-adenine, ^{3}H-thymidine, ^{32}P/^{33}P-phosphate) provides invaluable information to the environmental nucleotide fingerprint (24, 27, 43). The correlations between nucleic acid synthesis, protein synthesis, and cell growth are so universally accepted that they lend themselves well to the study of complex microbial assemblages in nature.

This chapter will focus on the most basic and most widely used aspect of the environmental microbial nucleotide fingerprint, namely, the measurement of cellular ATP as a biomass indicator. The stepwise methodology that is presented will focus on aquatic habitats. However, ATP measurements have also been used for the analysis of soils and in wastewater treatment and clinical applications (21). The ATP extraction and analysis protocols will vary considerably with sample type and expected microbial biomass. Other features, including the EC$_A$ ratio, GTP:ATP ratio, total energy flux determination by cellular ATP turnover, and measurement of rates of RNA and DNA synthesis, will be mentioned, but the detailed analytical procedures published elsewhere (16, 23–25, 31, 32) will not be repeated here.

37.2. SAMPLING AND SUBSAMPLING THE ENVIRONMENT

Sampling is one of the most important aspects of quantitative microbial ecology, but sampling design, including frequency of sampling, replication, and other relevant issues, is sometimes overlooked. All naturally occurring microbial communities are variable in space and in time, and the number of samples typically collected in any single study is usually small and, therefore inadequate for a complete description of the habitat under investigation. To the extent possible, multiple samples should be collected, covering as many microenvironments of the bulk habitat as possible to obtain a true representation of the ecosystem under investigation. For relatively homogeneous aquatic environments this is straightforward, but for sediments and soils it is not. The nature of the sampler itself is also important. For ex-

ample, when is a sterile sampler absolutely necessary, and when can that requirement be relaxed?

The number of samples that are necessary to accurately describe microbial processes in a given ecosystem depends to a certain extent on the structure of the habitat. Because most natural habitats are actually mosaics of many microhabitats with a heterogeneous distribution of microorganisms, variance between multiple samples generally exceeds variance between subsamples of a single collection. Therefore, replication for microbial biomass and activity estimation is most meaningful when performed at the highest level (35). Unfortunately, this is rarely achieved. Automation in sample collection and, especially, in sample analysis—including remote sensing—should greatly improve the current situation with respect to the overall accuracy of ecosystem measurement.

37.3. BIOMASS DETERMINATION BY ATP

ATP has several unique characteristics that make it a reliable indicator of microbial biomass in aquatic environments. It is ubiquitous in all living cells, has a relatively short half-life following cell death and autolysis, and is present at a fairly constant intracellular concentration regardless of nutritional mode (e.g., photoautotroph, chemoheterotroph, chemolithoautotroph, phagotroph) or growth rate. Furthermore, particulate ATP (P-ATP) can be rapidly and efficiently extracted from cells and stabilized in solution using boiling buffers, cold mineral or organic acids, or a variety of organic solvents (21). The preferred method of ATP quantification is the firefly bioluminescence reaction, but a variety of analytical techniques are available for either discrete sample or continuous flow analyses. Data on the P-ATP content of a water, sediment, or soil sample can be extrapolated to total microbial biomass using C:ATP relationships derived from either laboratory or field studies. Other advantages include the low detection limit (less than 10^{-12} mol; roughly equivalent to $\sim 10^3$ *Escherichia coli* cells) and acceptable level of precision for field replicates (typically 1 to 10%, depending on ATP concentration and operator), the high degree of objectivity compared with methods requiring operator recognition of "live" cells and estimation of cell dimensions, and the potential for near "real time" analyses and continuous flow applications. Finally, compared with most other biomass assays, there is also an extensive laboratory and field database for comparisons, conclusions, and ecological interpretations.

37.4. MATERIALS REQUIRED

37.4.1. Equipment

- Sampling gear (water column): Niskin bottles (or equivalent), Kevlar line or equivalent, subsampling bottles, 202–μm Nitex mesh
- Sampling gear (sediment column): corer device, core tubes, syringe samplers, glove box (optional)
- Filtration gear: polyvinyl chloride (PVC) or stainless steel manifold (3- or 6-place), equipped with glass filter bases with stainless steel screens for 25–mm-diameter filters and large-volume funnels (100– to 200–ml capacity)
- Vacuum pump with gauge (compressed N_2 or air can be used as an alternative to vacuum filtration)
- Block heater (Sybron Thermolyne Type (16500 Dri-bath, or equivalent), capable of heating extraction menstruum

in test tubes to $100 \pm 1°C$ (1 atm) and maintaining boiling temperatures throughout the extraction period

- Storage freezer ($-20°C$)
- pH meter and reference buffers
- Magnetic stirrer and stir bars
- Light-detection instrument: Any one of many commercially available general instruments, such as fluorometers, spectrophotometers, or liquid scintillation counters, is suitable for measuring light emission. However, to obtain efficient and reliable data with maximum sensitivity, a specially designed ATP photometer is required. Instruments specifically designed for ATP analyses are marketed by several manufacturers. Whatever light-measuring device is selected, it is imperative to have either a strip-chart recorder, integrator, or suitable computer-assisted data station to quantify light emission.

37.4.2. Supplies

- Filters (24/25–mm-diameter Whatman GF/F, Millipore HA/GS, or equivalent)
- Filter forceps
- Extraction and assay glassware: test tubes for ATP extraction (approximately 15–mm diameter by 125– to 160–mm length); assay vials (18–mm diameter by 40–mm height). Size requirements may vary depending on equipment/instruments used.
- Adjustable automatic pipettes (0.1– to 1–ml and 1– to 5–ml capacities)

37.4.3. Solutions and Reagents

(Unless otherwise indicated, catalog numbers or information refer to items sold by Sigma Chemical Company.)

- Firefly lantern extract (FLE-50)
- Potassium arsenate buffer, 100 mM, pH 7.4
- Tris(hydroxy)aminomethane-HCl (TRIS) buffer, 20 mM, pH 7.75
- ATP, sodium salt, 2 μM solution in Tris buffer, in 1– to 5–ml aliquots, and stored frozen
- Magnesium sulfate solution, 40 mM
- Potassium phosphate buffer, 60 mM, pH 7.4
- D-Luciferin, optional (L-9504)
- H_3PO_4 (1 to 1.5 M)
- NaOH (1 and 0.1 M)

37.5. PROCEDURES

37.5.1. Sampling, Subsampling, Extraction, and Sample Storage

37.5.1.1. Water Column

Typically, oceanic or lake samples are collected at predetermined depths throughout the water column using standard commercially available PVC bottles (e.g., General Oceanics Niskin or Go-Flo bottles, or equivalent) mounted on a CTD-rosette sampler. In addition to obtaining complementary information on physical and chemical characteristics of the water column, this sampling protocol allows interactive, directed sampling at specific regions of interest (e.g., particle, fluorescence, or oxygen maxima or minima, density discontinuities, etc.). Alternatively, water can be obtained using submersible pumps, manually operated

evacuated bottles, syringe samplers, or any other effective means. Selected habitats such as the sea-surface microlayer or high-temperature hydrothermal vents require the use of specialized samplers.

1. Prior to use, the samplers are cleaned with dilute HCl (0.5 to 1 M) or ethyl alcohol (95%) and rinsed thoroughly with distilled water; sterilization of the samplers is neither required nor practical for most field studies.

2. To measure microbial ATP, it is necessary to remove metazoans and other nonmicrobial ATP prior to sample extraction. This is done most conveniently by passing water samples through a 202–μm Nitex mesh as part of the subsampling procedures (e.g., during subsampling from Niskin bottles). It is conceivable that in certain aquatic environments this procedure may also remove large detrital particles to which microorganisms are attached. In these extreme cases, metazoans can be hand sorted from the respective water samples before analysis.

3. Following collection and subsampling, the particulate matter is concentrated and extracted as soon as possible. Although numerous methods have been described for the extraction of ATP from microorganisms (21), the most commonly used method for aqueous samples involves P-ATP concentration by vacuum or pressure filtration and extraction into boiling Tris (0.02 M, pH 7.4 to 7.7) or boiling phosphate (60 mM, pH 7.4) buffers. The latter is recommended for samples suspected of containing alkaline phosphatase, as would be expected in phosphorus-depleted habitats (28). After concentration of microbes onto a filter, the filter is immersed into the boiling buffer as quickly as possible after the final portion of liquid passes through the filter. If left on the filtration manifold for an extended period, measured in seconds, the cells desiccate, causing a loss of cellular ATP. It is imperative that the extraction buffer be at boiling point at the time of filter insertion. At temperatures below approximately 90°C, inefficient extraction occurs due to enzyme-catalyzed ATP hydrolysis.

4. After the filter is placed into the boiling buffer, the sample is heated for an additional 5 min, during which time the test tubes are partially covered to minimize evaporation and resultant volume changes.

5. Following extraction, the samples are removed from the heating block or temperature-controlled bath, cooled, then stored frozen ($-20°C$) until assayed. At this point the sample extracts are extremely stable, with ATP losses of less than 1% per year in properly buffered solutions.

37.5.1.2. Sediment Column

Many aquatic sediments are well stratified and characterized by steep depth gradients in microbial biomass. Consequently, it is imperative that the sampling and subsampling methods used to collect sediment for microbial ATP analysis preserve the unique depth distribution. For intertidal or shallow subtidal habitats, sediment cores can be collected manually by inserting PVC or acrylic tubes (10– to 15–cm diameter, 30– to 50–cm length) into the sediment and placing stoppers on both ends prior to retrieval. Deeper samples require the use of a spade box corer (or equivalent device), which is designed to minimize both sediment disruption during sampling and winnowing during sample recovery. Unfortunately, most gravity corers, which are easier to operate than box corers, create an unacceptable bow wave prior to penetration into the sediment. This results in a disruption of the microbial biomass gradients that are of greatest interest to the microbial ecologist. As for water column samples, the greatest variability is expected to occur at the level of sample (i.e., sediment core) replication. This variability is also of greatest relevance to the microbial ecologist.

Once collected, the core samples should be processed as quickly as possible and, as discussed above, with care taken to minimize changes in environmental conditions. For most anoxic sediments, this requires the use of a N_2-filled glove box to prevent sample oxidation and subsequent transitions in intracellular ATP pools or potential loss of obligate anaerobe viability.

A variety of procedures have been described for core subsampling, including techniques for the collection of the water-sediment interface, where microbial biomass and activity are expected to be the highest (45), and for the preservation of millimeter-scale habitat variability (8). If required, the sediment samples can be screened to remove ATP-containing metazoa, macroalgae, or higher plant rhizomes prior to analysis. The replicated subsamples from the different depth strata are then mixed thoroughly with a spatula in preparation for final subsampling for ATP extraction.

1. Triplicate subsamples (1 to 2 cm^3) of each sediment fraction are collected using a 3–ml plastic syringe barrel (luer lock end removed). Additional replicates are also taken for the determination of wet-volume–to–dry-weight conversion and for other bulk chemical parameters (e.g., total carbon).

2. The plugs of sediment are immediately discharged into test tubes containing 10 ml of cold H_3PO_4 (0.5 M, 4°C), capped, and thoroughly mixed. Additional subsamples should be prepared with a known amount of ATP added as an internal standard to assess and correct for ATP losses by the combined effects of adsorption, hydrolysis, and various potential sources of chemical interference. Representative sediment porewaters should also be collected (by centrifugation or pressure filtration through a 0.2–μm-pore-size filter) to assess the potential interference due to dissolved ATP (D-ATP). If the sediment samples contain $CaCO_3$, occasional venting to release accumulated CO_2 gas may be required.

3. After an extraction period of 15 to 20 min at 4°C, the extracted nucleotides are separated from the solid phase by centrifugation or vacuum filtration.

4. The pH of the acid extracts is adjusted to 7.4 by titration with NaOH (1.0 and 0.1 M). At this point, the ATP is stable in samples stored at −20°C. If the ATP concentration in the acid extract is >100 nM, the sample can be diluted with 60 mM PO_4 buffer as an alternative to base titration.

5. If the ATP concentration is ≤1 nM, the acid extract must be concentrated by either the activated charcoal procedure (14) or brucite coprecipitation (4, 34) prior to analysis. At this point, the ATP is relatively stable and the samples may either be stored at 4°C for up to 2 to 3 weeks or processed immediately.

37.5.2. Detection of ATP by Firefly Bioluminescence

Although ATP (and other adenine and nonadenine nucleotides) can be measured using any one of a variety of analytical detection systems, the firefly bioluminescence assay and high-performance liquid chromatography (HPLC) are most commonly used in ecological studies. A major advantage of HPLC is the ability to separate complex nucleotide mixtures (i.e., cell extracts) into individual components (ATP, ADP, AMP, etc.) that can be quantified during a single sample run. This additional information on the concentrations of non-ATP nucleotides can provide useful data on the metabolic states and in situ growth rates of microbial communities in nature (21, 22). The more commonly used bioluminescence assay, however, has a much lower ATP detection limit, is straightforward and inexpensive to perform, has a high level of precision, and requires less-specialized instrumentation. Furthermore, non-ATP nucleotides (e.g., ADP, AMP, GTP, etc.) can also be measured by the firefly bioluminescence reaction following stoichiometric generation of ATP from other nucleotide triphosphates via specific transphosphorylation reactions, as discussed later in this chapter.

Several reviews have been published concerning the specificity, kinetics, and mechanism of the firefly bioluminescence reaction. The postulated steps are (10):

$$LH_2 \text{ (luciferin)} + ATP \xrightarrow{Mg^{2+}-\text{luciferase}} E\text{-}LH_2\text{-}AMP + PP_i \quad (1)$$

$$E\text{-}LH_2\text{-}AMP + O_2 \xrightarrow{\text{neutral pH}} \text{oxyluciferin} + E + CO_2 + AMP + \text{light} \quad (2)$$

When all necessary reactants are present in excess, the in vitro light emission is directly proportional to the concentration of ATP in solution. Reaction kinetics and specificity depend on the purity of the enzyme preparation; sensitivity is controlled by luciferin concentration (30). The measurement of either the initial rise of the luminescence curve (0 to 2 s), the peak height of luminescence (0 to 5 s), or a predetermined integrated portion (e.g., 15 to 75 s) of the light emission decay curve can be used to relate ATP concentrations in reference standards to those in the unknown sample extracts. The major advantages of the integrated mode are increased sensitivity, ease and reliability of mixing, and nonreliance on the peak-height response, which is difficult or impossible to measure with certain instruments. However, a major disadvantage of the integrated mode is the nonspecificity of light emission with certain crude enzyme preparations (20). Reliability of peak-height analyses depends on a very rapid and complete mixing of all reactants. This is best accomplished using an automatic injection system, which ensures consistent mixing velocities for all samples. The peak-height mode of analysis offers the advantages of speed of assay and minimum interference from other enzymes or substrates (e.g., non-ATP nucleotides) that may affect the rate of the luciferase-catalyzed reaction.

1. Firefly lantern extract (catalog FLE-50 or FLE-250; Sigma Chemical Co., St. Louis, MO) is stored with desiccant at −20°C. To activate lyophilized FLE, a 50–mg vial is hydrated in 5 ml of distilled water. This enzyme preparation is allowed to "age" at room temperature for a minimum of 4 to 6 h to a maximum of 24 h, depending on the desired sensitivity. During the aging process, endogenous ATP initially present in the crude extract is consumed, thereby decreasing the background light emission. Because of variations among individual enzyme preparations, it is imperative that only a single batch of enzyme be used for the analysis of a given set of samples and ATP standards.

2. Next, the hydrated, aged FLE is further diluted using equal volumes (generally 10 ml each for a single 50–mg vial of FLE) of $MgSO_4$ (0.04 M) and $KHAsO_4$ buffer (0.1 M,

pH 7.4), and the mixture is incubated at room temperature for 1 h.

3. If desired, the FLE preparation can be diluted to a greater final volume depending on the required sensitivity of the assay. However, if a single vial of FLE-50 is diluted to a working volume of >50 ml, the bioluminescence reaction can become limiting for luciferin, which must be added back to maintain first-order reaction kinetics with respect to ATP (30).

4. Immediately before use, the insoluble residue is removed by vacuum filtration (Whatman GF/F filter) or centrifugation (1500 × g, 5 min).

37.5.3. ATP Standards

For each enzyme preparation, an ATP standard curve is prepared and analyzed with the sample extracts.

1. A stock ATP solution containing 1 to 2 μM ATP is prepared in TRIS buffer (0.02 M, pH 7.4 to 7.7) and stored at −20°C in 1-ml aliquots until needed. Under these conditions, ATP hydrolysis is <1% per year. The exact concentration of the stock solution is determined by absorption spectrophotometry at 259 nm, using the relationship

$$A = cle \qquad (3)$$

where c = ATP (M); l = absorption pathlength (cm); and e = 15.4×10^3 (ATP molar extinction coefficient at pH 7.4).

2. A working ATP standard solution is prepared on the day of the assay by diluting the stock ATP preparation with the appropriate extraction menstruum (TRIS or phosphate buffer depending on sample type, and preferably the same batch as used for sample extraction).

3. Between seven and eight ATP standards (including a buffer blank) covering the expected range of the sample concentrations (approximately 1 to 100 nM ATP) are prepared. If ≥100 samples are analyzed or if the analysis time exceeds approximately 1 h, it is desirable to measure a set of standards at the beginning and at the end of the analysis to monitor temporal changes in the response of the FLE preparation. Otherwise, a single standard curve measured midway through the experiment is sufficient. In general, no significant hydrolysis of the diluted ATP standard solutions occurs during a typical 3- to 4-h working period.

4. Following a single use, all thawed ATP stock standards and serial dilutions thereof should be discarded.

37.5.4. Data Reduction and Extrapolations of ATP to Biomass

The relationship between ATP concentration and light emission is linear provided substrate (O_2 and luciferin) limitation does not occur. Peak heights or integrated areas are regressed on ATP standard concentrations. From a model I linear regression analysis of these data, the ATP concentrations in sample extracts can be calculated. By correcting for the proportion of the sample actually assayed and the volume of medium originally extracted, the ATP per liter of the original water or sediment sample can be determined.

The C:ATP ratio in microorganisms varies considerably, although somewhat predictably, among taxa and even for a given species when grown under different nutritional constraints (data summarized by Karl [21]). Among the most conspicuous differences in C:ATP ratios are those observed between unicellular microorganisms (i.e., bacteria and microalgae: C:ATP ≈ 200 to 350, by weight) and micrometazoa (C:ATP = 50 to 150) and the large increases in the

C:ATP ratio when microorganisms are starved for phosphorus (21, 22). However, under most conditions found in nature, the C:ATP ratio of the microbial community is about 250:1. Although originally developed to estimate total microbial "biomass," ATP concentrations are, in theory, more closely coupled to "protoplasm" biomass and, more specifically, total biovolume. Because of the obligate role of ATP in cellular bioenergetics, intracellular ATP levels are carefully maintained at a concentration of approximately 1 to 2 mM (6). Unfortunately, it is difficult to use biovolume estimates directly, however accurate, in most studies of microbial ecology. Consequently, one must rely on empirically determined C:ATP ratios to extrapolate P-ATP determinations to estimates of total microbial biomass. In so doing, both the level of precision and the accuracy of the initial ATP determination are decreased. Furthermore, in habitats where copious amounts of capsular materials, extracellular secretions, or slimes occur (7), ATP-based values of total biomass probably provide only minimum estimates.

37.6. NOTES AND COMMENTS

37.6.1. Sampling and Subsampling

Exposure of metabolically active microorganisms to environmental conditions that are substantially different from the collection site should be avoided, to minimize short-term transitions in intracellular ATP pools. However, this becomes nearly impossible during the collection of many samples, for example, abyssal water samples from the equatorial ocean. Decreases in pressure and increases in temperature, even during the time required for the samples to reach the ocean's surface, are almost certain to alter microbial ATP concentrations and perhaps even cell viability. However, at the present time these effects have not been systematically evaluated. In the future, a technique that provides for the in situ extraction of ATP from microorganisms needs to be developed and compared with conventional sampling procedures in order to provide quantitative constraints on the potential changes in P-ATP during routine sample recovery.

37.6.2. Extraction

The concentration step, which is necessary for most low-biomass environments (i.e., <1 g of C m^{-3}), must be performed with extreme care; sample size is especially important. It has been shown that the recovery of microbial ATP from certain high-particulate-load samples is volume dependent (49). Initially, the observed P-ATP losses were thought to be the result of cell lysis during prolonged filtration. However, it is now known to be caused by a filtration-induced metabolic stress that results in the hydrolysis of ATP to ADP and AMP (31). Consequently, if the total adenine nucleotide pool (the sum of ATP + ADP + AMP) rather than ATP is measured, filtration volume becomes less critical.

In certain eutrophic habitats where total microbial biomass exceeds 1 g of C m^{-3}, water samples can be extracted directly, thereby eliminating the preextraction concentration step (17). It is imperative, however, that the sample volume injected does not exceed 5% (by volume) of the extractant volumes; otherwise, temperature changes may affect the efficiency of ATP extraction. If desired, multiple injections can be made, allowing time between single injections for the extraction menstruum to return to its boiling temperature. Furthermore, if direct injection is employed for P-ATP extraction, it is necessary to measure, and correct for, dissolved ATP (operationally defined as passing through a 0.2–μm-pore-size filter), which is also present in most marine (2, 44) and freshwater (41) ecosystems.

37.6.3. Sensitivity, Precision, and Accuracy

The sensitivity of the ATP assay is determined by the instrumentation used to detect light emission and by the purity of luciferase and luciferin concentration in the enzyme preparation. Using the crude FLE-50 luciferase mixture prepared as described in this chapter and commercially available ATP photometers, the lower limit of ATP detection (i.e., twice the background light emission) is about 0.2 nM ATP. For greater sensitivity, exogenous luciferin is added to the enzyme preparations, enabling the detection of 1 pM ATP (30). The precision of the peak height assay procedure as routinely performed is ±1 to 2% of the mean (n = 8) throughout the entire range of ATP standards. Accuracy is estimated by analyzing diluted ATP standards and treating them as unknown samples. At the present time, no commercially available certified reference materials are available for independent determination of accuracy.

37.6.4. Analytical Interferences and Use of Internal Standards

In addition to the potential problems discussed in the previous sections, several sources of analytical interference are possible. These include (i) the presence of inorganic and organic ions in the sample extracts, resulting in loss of ATP in solution (i.e., through chelation) or in decreased luciferase activity; (ii) the presence of humic acid-like substances in the sample extracts that impart a yellow color to the solution, thereby resulting in attenuation of the emitted light; (iii) turbidity of the final extracts, resulting in light scattering and absorption; (iv) the presence of a high concentration of inorganic particulate material in the final extracts, resulting in loss of ATP through adsorption; and (v) the presence of contaminating enzymes, in either the sample extracts or the luciferase preparation, that compete with luciferase for the ATP in solution (e.g., ATPase or adenylate kinase) or that result in the production of ATP through transphosphorylase reactions (e.g., nucleoside diphosphate kinase or pyruvate kinase).

Most of the preceding sources of error are detected, and corrected for, through the use of an ATP internal standard, as discussed by Strehler (48). The internal standard may be added in the form of an ATP salt solution, as live or lyophilized bacterial cells, or as radiolabeled ATP. To minimize the effects of ionic interference, it is imperative that the standard ATP solutions be prepared in an ionic medium identical with that of the samples. Peak-height measurements significantly decrease the analytical interference due to the presence of nonadenine nucleotide triphosphates and therefore are strongly recommended. A review of analytical issues concerned with ATP extraction efficiency from soils has recently appeared (42).

37.7. ENHANCING THE MICROBIAL NUCLEOTIDE FINGERPRINT

The basic method of ATP measurement as an indicator of total microbial biomass has given rise to the more comprehensive protocol of environmental nucleotide fingerprinting (21). The development of specific and sensitive techniques for quantitative measurements of selected non-ATP intracellular nucleotides has enabled researchers to estimate the rates of protein and nucleic acid biosyntheses, nucleotide metabolism, metabolic activity, and growth. Furthermore, Karl and Bossard (26) developed a method to estimate the turnover rates of ATP to quantify energy flux in natural populations of marine microorganisms. This expanded approach of nucleotide fingerprinting is strongly recommended for field investigations to obtain corroborative data relating to the in situ physiological and growth states of microbial assemblages in nature.

To calculate the mean cellular EC_A ratio, measurements of ADP and AMP must be made in addition to ATP determination, as described above. ADP and AMP are both quantitatively coextracted with ATP, so the methods described above for sample collection and initial processing are identical to those for ATP determination. After the extraction of cellular nucleotides, ADP and AMP are quantitatively converted to ATP by stepwise enzymatic reactions involving pyruvate kinase and adenylate kinase plus pyruvate kinase, respectively. ATP is measured again in each sample, and ADP and AMP concentrations are calculated by difference (16).

For GTP concentration measurements, the ATP and UTP present in the nucleotide extracts are first destroyed by specific enzymatic reactions involving the addition of hexokinase, glucose-6–phosphate dehydrogenase, and UDP-glucose pyrophosphorylase. The remaining NTPs, mainly GTP and CTP, are separated kinetically in the light emission reaction involving firefly luciferase (18–20).

For the determination of ATP and total adenine nucleotide (TAN) pool turnover rates as estimates of energy flux and specific growth rate of microbial populations in nature, an environmental sample is incubated with radioactive phosphate (^{32}P- or ^{33}P-PO_4), as a tracer for P-flux through the acid anhydride-bound P groups of ATP (β-P and γ-P) and into the α-P position as adenine nucleotide molecules are removed for net growth (27). With increasing incubation time, the three phosphate moieties of the "community" ATP pool (or isolated fraction thereof) become labeled, albeit at different rates. Experimental work has demonstrated that the kinetic constants of γ-P and β-P labeling are indistinguishable, both in laboratory cultures and in field samples, because of the rapid intracellular interconversions of ADP and ATP. Because both anhydride bonds have identical free energies of hydrolysis, this simplifies the measurement of total metabolic energy flux (5). By comparison, the α-P labeling lags behind the γ-P and β-P moieties because the α-P position of ATP becomes labeled only as a result of the net removal of adenine nucleotides for biosynthesis, the sum of salvage and de novo synthesis processes. The exact turnover time of the ATP or the TAN pool is determined by an analysis of the time rate of change in the specific activity of either the γ-P, β-P, or α-P positions of ATP. Labeling of the individual positions is determined by selective hydrolysis of ATP to ADP + P_i. Since the γ-P radioactivity is equal to the β-P radioactivity, α-P can be calculated by difference: (α-P* = [total ATP*] − [2 × γ-P*]).

As predicted by radiotracer theory, the change in the ATP or TAN pool-specific radioactivity (SA) (nanocuries per picomole) following the addition of an appropriate radioactive precursor is an exponential function of incubation time (t). The initial SA (at t = 0) is zero, and the relative SA (at isotopic equilibrium with precursor pool) is 1.000. Under these conditions, the decimal equivalent of SA at any time (SA_t) can be described by the equation

$$SA_t = 1 - (2^{-N}) \qquad (1)$$

where N is equal to the number of turnover cycles (N is dimensionless) observed during the incubation period. ATP or TAN pool turnover time (T) can then be calculated from the expression

$$T = t/N \qquad (2)$$

This relationship is applicable for all incubation periods (t) less than or equal to $5N$, at which time the pools are expected to be near isotopic equilibrium and, in theory, would not change until the exogenous radioactive precursor pool is exhausted. For oligotrophic surface seawaters, a 3- to 4-h incubation period is sufficient to reach isotopic equilibrium (5).

Once the turnover time (T) of the ATP or TAN pool has been determined, it is straightforward to extrapolate these data to the more meaningful ecological parameters of energy flux and specific growth rate. Assuming a value of -11 (± 1) kcal mol^{-1} for the free energy of ATP hydrolysis for either the β-P and γ-P positions under in vivo conditions (54), then the total microbiological energy flux (EF, expressed in units of kilocalories per liter of sample per hour) can be calculated as

$$EF \text{ (kcal liter}^{-1} \text{ h}^{-1}) = -22([ATP]/T_{ATP}) \qquad (3)$$

where [ATP] is equal to the total particulate ATP pool in moles per liter^{-1} of sample and T_{ATP} is ATP pool turnover time (h). This value for total available free energy can be compared directly to the solar energy flux, to the energy stored by photosynthesis, or to respiration. In this way, carbon and energy fluxes can be directly compared.

Extrapolation of the measured TAN turnover time (T_{TAN}) to specific growth rate (μ) is based on the theoretical predictions (6) and empirical observations (33) that the T_{TAN} is equivalent to a value that is 2 to 3% of the generation time (i.e., TAN pool turns over, on average, 40 times per generation). Consequently, the doubling time (T_d, expressed in hours) of the substrate-responsive population is

$$T_d = T_{TAN} \times 40 \qquad (4)$$

and specific growth rate (μ, expressed in units per hour) is

$$\mu = 1/T_d \times \ln(2) \qquad (5)$$

These kinetic model formulations will provide "ideal" results when all of the radiotracer-responsive microorganisms in a given habitat are growing at identical rates. A novel, statistical treatment of field data, described in detail by Laws et al. (37), provides a method for determining the coefficient of variation among the individual estimates of T_{ATP} and T_{TAN}. Consequently, one can quantify the variability among individual subcomponents of the microbial community. Although applied here for the turnover of ATP and TAN pools, this mathematical formulation can be used to assess heterogeneity in the turnover rate of any intracellular pool for which a suitable radioactive or stable isotopic precursor exists.

37.8. CONCLUSION

No single approach to the study of microorganisms in their natural environments is universally accepted, or acceptable (22). Early in the development of microbiology as a scientific discipline, specific groups of microorganisms were isolated from nature and studied in the laboratory. This led to major advances in microbiology, especially physiology, genetics, and metabolism. Now, armed with new information about the phylogenetic and metabolic diversity of microorganisms in nature and the probable presence of novel, as yet uncultivatable microbes, there is a renewed interest in microbial ecology and in whole-community assays. Despite recent and significant progress, the field of microbial ecology is still "methods-limited" with regard to the most fundamental properties of natural microbial assemblages, namely, biomass and metabolic activity estimation of the total population. One of the most significant challenges for the future is to link microbial phylogeny to biogeochemistry and ecology to gain a better understanding of material and energy flow in the biosphere.

37.9. REFERENCES

1. **Atkinson, D. E.** 1969. Regulation of enzyme function. *Annu. Rev. Microbiol.* 23;47–68.
2. **Azam, F., and R. E. Hodson.** 1977. Dissolved ATP in the sea and its utilization by marine bacteria. *Nature* 267:696–698.
3. **Bertilsson, S., O. Berglund, D. M. Karl, and S. W. Chisholm.** 2003. Elemental composition of marine *Prochlorococcus* and *Synechococcus*: implications for the ecological stoichiometry of the sea. *Limnol. Oceanogr.* 48:1721–1731.
4. **Björkman, K., and D. M. Karl.** 2001. A novel method for the measurement of dissolved adenosine and guanosine triphosphate in aquatic habitats: applications to marine microbial ecology. *J. Microbiol. Meth.* 47:159–167.
5. **Bossard, P., and D. M. Karl.** 1986. The direct measurement of ATP and adenine nucleotide pool turnover in microorganisms: a new method for environmental assessment of metabolism, energy flux and phosphorus dynamics. *J. Plankton Res.* 8:1–13.
6. **Chapman, A. G., and D. E. Atkinson.** 1977. Adenine nucleotide concentrations and turnover rates. Their correlation with biological activity in bacteria and yeast. *Adv. Microb. Physiol.* 15:253–306.
7. **Costerton, J. W., R. T. Irvin, and K. J. Cheng.** 1981. The bacterial glycocalyx in nature and disease. *Annu. Rev. Microbiol.* 35:299–324.
8. **Craven, D. B., R. A. Jahnke, and A. F. Carlucci.** 1986. Fine-scale vertical distributions of microbial biomass and activity in California Borderland sediments. *Mar. Biol.* 83:129–139.
9. **Cuhel, R. L., and J. B. Waterbury.** 1984. Biochemical composition and short term nutrient incorporation patterns in a unicellular marine cyanobacterium, *Synechococcus* (WH7803). *Limnol. Oceanogr.* 29:370–374.
10. **DeLuca, M., and W. D. McElroy.** 1978. Purification and properties of firefly luciferase. *Methods Enzymol.* 57:3–15.
11. **Dobbs, F. C., and R. H. Findlay.** 1993. Analysis of microbial lipids to determine biomass and detect the response of sedimentary microorganisms to disturbance, p. 347–358. *In* P. F. Kemp, B. F. Sherr, E. B. Sherr, and J. J. Cole (ed.), *Handbook of Methods in Aquatic Microbial Ecology.* Lewis Publishers, Boca Raton, FL.
12. **Franzen, J. S., and S. B. Binkley.** 1961. Comparison of the acid-soluble nucleotides in *Escherichia coli* at different growth rates. *J. Biol. Chem.* 236:515–519.
13. **Guckert, J. B., C. P. Antworth, P. D. Nichols, and D. C. White.** 1985. Phospholipid, ester-linked fatty acid profiles as reproducible assays for changes in prokaryotic community structure of estuarine sediments. *FEMS Microbiol. Ecol.* 31:147–158.
14. **Hodson, R. E., O. Holm-Hansen, and F. Azam.** 1976. Improved methodology for ATP determination in marine environments. *Mar. Biol.* 34:143–149.

15. **Holm-Hansen, O., and C. R. Booth.** 1966. The measurement of adenosine triphosphate in the ocean and its ecological significance. *Limnol. Oceanogr.* **11:**510–519.

16. **Holm-Hansen, O., and D. M. Karl.** 1978. Biomass and adenylate energy charge determination in microbial cell extracts and environmental samples. *Methods Enzymol.* **57:** 73–85.

17. **Jones, J. G., and B. M. Simon.** 1977. Increased sensitivity in the measurement of ATP in freshwater samples with a comment on the adverse effect of membrane filtration. *Freshwater Biol.* **7:**253–260.

18. **Karl, D. M.** 1978. A rapid sensitive method for the measurement of guanine ribonucleotides in bacterial and environmental extracts. *Anal. Biochem.* **89:**581–595.

19. **Karl, D. M.** 1978. Determination of GTP, GDP and GMP in cell and tissue extracts. *Methods Enzymol.* **57:**85–94.

20. **Karl, D. M.** 1978. Occurrence and ecological significance of GTP in the ocean and in microbial cells. *Appl. Environ. Microbiol.* **36:**349–355.

21. **Karl, D. M.** 1980. Cellular nucleotide measurements and applications in microbial ecology. *Microbiol. Rev.* **44:**739–796.

22. **Karl, D. M.** 1986. Determination of *in situ* microbial biomass, viability, metabolism and growth, p. 85–176. *In* J. S. Poindexter and E. R. Leadbetter (ed.), *Bacteria in Nature,* vol. 2. *Methods and Special Applications in Bacterial Ecology.* Plenum Press, New York, NY.

23. **Karl, D. M.** 1993. Adenosine triphosphate (ATP) and total adenine nucleotide (TAN) pool turnover rates as measures of energy flux and specific growth rate in natural populations of microorganisms, p. 483–494. *In* P. F. Kemp, B. F. Sherr, E. B. Sherr, and J. J. Cole (ed.), *Current Methods in Aquatic Microbial Ecology.* Lewis Publishers, Boca Raton, FL.

24. **Karl, D. M.** 1993. Microbial RNA and DNA synthesis derived from the assimilation of [2-^{3}H] adenine, p. 471–481. *In* P. F. Kemp, B. F. Sherr, E. B. Sherr, and J. J. Cole (ed.), *Current Methods in Aquatic Microbial Ecology.* Lewis Publishers, Boca Raton, FL.

25. **Karl, D. M.** 1993. Total microbial biomass estimation derived from the measurement of particulate adenosine-5′-triphosphate, p. 359–368. *In* P. F. Kemp, B. F. Sherr, E. B. Sherr, and J. J. Cole (ed.), *Current Methods in Aquatic Microbial Ecology.* Lewis Publishers, Boca Raton, FL.

26. **Karl, D. M., and P. Bossard.** 1985. Measurement and significance of ATP and adenine nucleotide pool turnover in microbial cells and environmental samples. *J. Microbiol. Meth.* **3:**125–139.

27. **Karl, D. M., and P. Bossard.** 1985. Measurement of microbial nucleic acid synthesis and specific growth rate by ^{32}PO$_4$ and [^{3}H]adenine: field comparison. *Appl. Environ. Microbiol.* **50:**706–709.

28. **Karl, D. M., and D. B. Craven.** 1980. Effects of alkaline phosphatase activity on nucleotide measurements in aquatic microbial communities. *Appl. Environ. Microbiol.* **40:**549–561.

29. **Karl, D. M., and F. C. Dobbs.** 1998. Molecular approaches to microbial biomass estimation in the sea, p. 29–89. *In* K. E. Cooksey (ed.), *Molecular Approaches to the Study of the Ocean.* Chapman & Hall, London, United Kingdom.

30. **Karl, D. M., and O. Holm-Hansen.** 1976. Effects of luciferin concentration on the quantitative assay of ATP using crude luciferase preparations. *Anal. Biochem.* **75:** 100–112.

31. **Karl, D. M., and O. Holm-Hansen.** 1978. ATP, ADP and AMP determinations in water samples and algal cultures, p. 197–206. *In* J. A. Hellebust and J. S. Craigie (ed.),

Handbook of Phycological Methods, Vol. III. *Physiological and Biochemical Methods.* Cambridge University Press, New York, NY.

32. **Karl, D. M., and O. Holm-Hansen.** 1978. Methodology and measurement of adenylate energy charge ratios in environmental samples. *Mar. Biol.* **48:**185–197.

33. **Karl, D. M., D. R. Jones, J. A. Novitsky, C. D. Winn, and P. Bossard.** 1987. Specific growth rates of natural microbial communities measured by adenine nucleotide pool turnover. *J. Microbiol. Meth.* **6:**221–235.

34. **Karl, D. M., and G. Tien.** 1992. MAGIC: a sensitive and precise method for measuring dissolved phosphorus in aquatic environments. *Limnol. Oceanogr.* **37:**105–116.

35. **Kirchman, D., J. Sigda, R. Kapuscinski, and R. Mitchell.** 1982. Statistical analysis of the direct count method for enumerating bacteria. *Appl. Environ. Microbiol.* **44:**376–382.

36. **Koch, A.** 1994. Growth measurement, p. 248–277. *In* P. Gerhardt, R. G. E. Murray, W. A. Wood, and N. R. Krieg (ed.), *Methods for General and Molecular Bacteriology.* ASM Press, Washington, DC.

37. **Laws, E. A., D. Jones, and D. M. Karl.** 1986. Method for assessing heterogeneity in turnover rates within microbial communities. *Appl. Environ. Microbiol.* **52:**866–874.

38. **Levin, G. V., J. R. Clendenning, E. W. Chappelle, A. H. Heim, and E. Rocek.** 1964. A rapid method for detection of microorganisms of ATP assay. *BioScience* **14:**37–38.

39. **Lipmann, F.** 1941. Metabolic generation and utilization of phosphate bond energy. *Adv. Enzymol. Relat. Areas Mol. Biol.* **1:**99–162.

40. **Madsen, E. L.** 1998. Epistemology of environmental microbiology. *Environ. Sci. Tech.* **32:**429–439.

41. **Maki, J. S., M. E. Sierszen, and C. C. Remsen.** 1983. Measurements of dissolved adenosine triphosphate in Lake Michigan. *Can. J. Fish. Aquat. Sci.* **40:**542–547.

42. **Martens, R.** 2001. Estimation of ATP in soil: extraction methods and calculation of extraction efficiency. *Soil Biol. Biochem.* **33:**973–982.

43. **Moriarty, D. J. W.** 1986. Measurement of bacterial growth rates in aquatic systems using rates of nucleic acid synthesis. *Adv. Microbiol. Ecol.* **9:**245–264.

44. **Nawrocki, M. P., and D. M. Karl.** 1989. Dissolved ATP turnover in the Bransfield Strait, Antarctica, during a spring bloom. *Mar. Ecol. Prog. Ser.* **57:**35–44.

45. **Novitsky, J. A.** 1983. Heterotrophic activity throughout a vertical profile of seawater and sediment in Halifax Harbor, Canada. *Appl. Environ. Microbiol.* **45:**1761–1766.

46. **Pinkart, H. C., D. B. Ringelberg, Y. M. Piceno, S. J. Macnaughton, and D. C. White.** 2002. Biochemical approaches to biomass measurements and community structure analysis, p. 101–113. *In* C. J. Hurst, R. L. Crawford, G. R. Knudsen, M. J. McInerney, and L. D. Stetzenbach (ed.), *Manual of Environmental Microbiology,* 2nd ed. ASM Press, Washington, DC.

47. **Smith, R. J.** 1979. Increasing guanosine 3′-diphosphate 5′-diphosphate concentration with decreasing growth rate in *Anacystis nidulans. J. Gen. Microbiol.* **113:**403–405.

48. **Strehler, B. L.** 1968. Bioluminescence assay: principles and practice. *Meth. Biochem. Anal.* **16:**99–181.

49. **Sutcliffe, W. H., Jr., E. A. Orr, and O. Holm-Hansen.** 1976. Difficulties with ATP measurements in inshore waters. *Limnol. Oceanogr.* **21:**145–149.

50. **White, D. C., R. J. Bobbie, S. J. Morrison, D. K. Oosterhof, C. W. Taylor, and D. A. Meeter.** 1977. Determination of microbial activity of estuarine detritus by relative rates of lipid biosynehtsis. *Limnol. Oceanogr.* **22:**1089–1099.

51. **White, D. C., W. M. Davis, J. S. Nickels, J. D. King, and R. J. Bobbie.** 1979. Determination of sedimentary microbial biomass by extractible lipid phosphate. *Oecologia (Berlin)* **40:**51–62.

52. **White, D. C., H. C. Pinkart, and D. B. Ringelberg.** 1997. Biomass measurements: Biochemical approaches, p. 91–101. *In* C. J. Hurst, G. R. Knudsen, M. J. McInerney, L. D. Stetzenbach, and M. V. Walter (ed.), *Manual of Environmental Microbiology.* ASM Press, Washington, DC.

53. **Wiebe, W. J., and K. Bancroft.** 1975. Use of adenylate energy charge ratio to measure growth rate of natural microbial communities. *Proc. Natl. Acad. Sci. USA* **72:**2112–2115.

54. **Wilson, D. F., M. Stubbs, R. L. Veech, M. Erecinska, and H. A. Krebs.** 1974. Equilibrium relations between the oxidation-reduction reactions and the adenosine triphosphate synthesis in suspensions of isolated liver cells. *Biochem. J.* **140:**57–64.

38

Construction of BAC and Fosmid Libraries from Naturally Occurring Microbial Populations

EDWARD F. DeLONG

38.1. CULTIVATION-INDEPENDENT APPROACHES IN MICROBIAL ECOLOGY AND POPULATION BIOLOGY

Most definitive microbiological studies have been conducted in laboratories using pure cultures. This approach has been essential for the development of the science of microbiology and remains a methodological mainstay. No methodology is perfect, however, and it has long been recognized that many microbial species still resist isolation in pure culture, for many and varied reasons. One interest of contemporary microbiologists has therefore been to devise better methods to identify, characterize, and understand those naturally occurring microbes that are difficult or impractical to cultivate and study by traditional methods. This chapter discusses one potential approach involving genomic characterization of the naturally occurring microbial genomes, without the requirement of isolating individual microbes one by one in pure culture.

Norman Pace and colleagues invented molecular phylogenetic cultivation-independent approaches to study natural microbial populations in the mid-1980s (8–10). Pace's strategy exploits nucleic acid-based cloning and sequencing to access naturally occurring microbial diversity. The development of these approaches has led to the now-common practice of cloning rRNA genes from mixed microbial assemblages. These cultivation-independent rRNA gene surveys have revealed a significant number of new phylogenetic lineages, including abundant types of bacteria and archaea previously unknown to science. Our current understanding of the extent and nature of natural microbial diversity has improved dramatically, in part because of the development and application of the cultivation-independent rRNA approach.

The past decade also witnessed an amazing acceleration in genome-sequencing capabilities. Just seven years after the first report of a complete microbial genome sequence, more than 100 full microbial genome sequences have now become available for study. A large percentage of the genomes sequenced (about 75%) originate from clinically important microbes (4). Of the total prokaryotic sequences, about 90% are from the domain *Bacteria*. To date, most full microbial genome sequences have originated from pure cultures, although a few are derived from obligately symbiotic or parasitic bacteria.

A powerful extension of the tremendous genome-sequencing capacity available today aims to describe genomically the natural world of indigenous microorganisms. Using the same logic as was applied in rRNA gene surveys (9), it has now become theoretically possible, using genomics, to access large portions of the genomes of all naturally occurring microbes, cultivated or not. This really is just an extension of Norm Pace's approach, but with a twist. Instead of analyzing just a single genetic locus like rRNA, however, it has now become possible to recover and analyze large portions of the genomes of naturally occurring microbes, cultivated or not (Fig. 1) (3, 11, 13, 14, 18). There are a variety of potential strategies for developing and applying the "environmental genomics" (14, 18) or "metagenomics" (12) approach. This chapter will describe some of the methodological options and considerations and present a few of the protocols that have been used successfully for preparing genomic libraries from natural microbial assemblages.

38.2. CONSTRUCTION OF LARGE INSERT GENOMIC LIBRARIES FROM ENVIRONMENTAL SAMPLES: GENERAL CONSIDERATIONS

38.2.1. Microbial Sampling, Still a Challenging Enterprise

Sampling strategies are always context dependent and influenced by the nature of the microbial community in question, the environment that it occupies, spatial constraints, population numbers, and other contaminating organisms

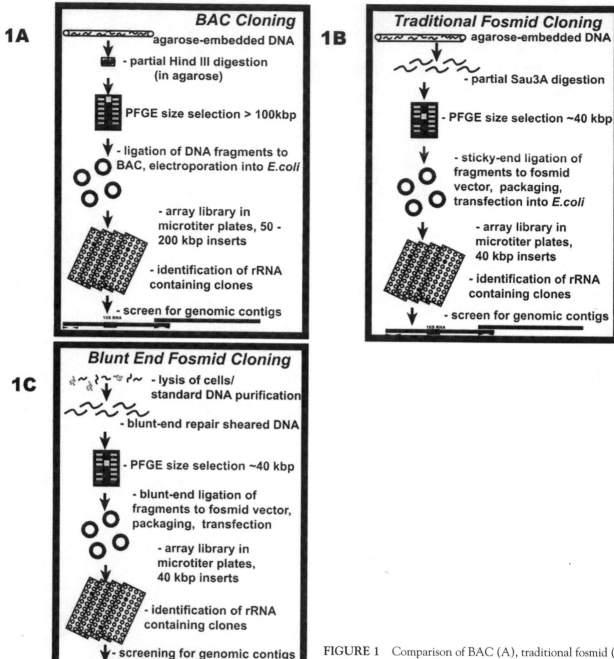

FIGURE 1 Comparison of BAC (A), traditional fosmid (B), and blunt-end fosmid (C) cloning of DNA fragments for environmental genomic studies.

and substances. Several up-front questions need to be addressed when designing sampling strategies. Will the cells need to be purified away from a soil, sediment, or rock matrix? The interference of contaminating materials in DNA purification and downstream steps requiring enzymatic modifications are major considerations. Do the cells need to be concentrated in some fashion before DNA extraction? For some procedures, like bacterial artificial chromosome (BAC) libraries, a minimum of 10^9 cells or more may be required, which can sometimes impart severe limitations on

the types of communities amenable to the approach. Other important questions are: What size of DNA fragments are required? What cloning strategy will be used? The answers to these questions will dictate whether the cells need to be embedded in agarose to maintain high-molecular-weight, intact chromosomes during DNA preparation. (For inserts of 40 kb or less, embedding cells in agarose is not an absolute requirement.) What are the complexity, richness and evenness, and specific composition of the population/sample in question? For example, it would take only a small proportion of eukaryotic genomes in a sample, each having 10 to 50 times greater genome size, to dramatically affect the representation of prokaryotic genome sequences in any given library. All these questions and considerations are important to keep in mind before embarking on studies in environmental genomics. Prior knowledge of the nature, complexity, and composition of the original sample, both matrix and microbes, will largely influence experimental design and the final outcome of library construction and downstream analyses. The possibilities and permutations of sample preparation from diverse microbial communities are too numerous to cover adequately here. But some of these considerations can help guide in sampling design and preparation strategies.

38.2.2. Large-Insert DNA Libraries: Contemporary Approaches

Some of the earliest large-fragment genome libraries from natural microbial populations made use of bacteriophage lambda cloning protocols. Today, advanced vector systems such as fosmids and bacterial artificial chromosomes (BACs) are now in wide use. BACs and fosmids are plasmid vectors whose multiplication is regulated by the F_1 origin of replication (6, 15). The development of these new vectors, which stably maintain and propagate extremely large DNA inserts, was an important milestone in the whole-genome-sequencing revolution. BAC vectors, developed in the early 1990s for use in the Human Genome Project, have been particularly useful. By virtue of their lowcopy number, BACs can be used to stably propagate very large DNA fragments that are otherwise unstable (15). One version of the BAC vector (originally termed fosmid, for F1 origin-based cosmid-sized vector [6]) is introduced into *Escherichia coli* via transfection, after the DNA has been packaged as phage particles using bacteriophage lambdapackaging extracts. Transfection of the packaged fosmid DNA is extremely efficient, facilitating construction of BAC libraries with uniform insert sizes closely bracketing 40 kbp from very small amounts of DNA (6). Another approach to construct BAC libraries relies on the embedding of cells in an agarose plug, followed by in-gel lysis using detergent and proteinase K. The intact, agarose-embedded DNA is then subjected to partial restriction enzyme digestion and size fractionated via pulsed-field gel electrophoresis (PFGE). The size-separated fragments are excised from the gel and ligated into a BAC vector. Recombinant BAC clones are then introduced into *E. coli* cells via electroporation and archived in an ordered library. Depending on the source DNA, this approach can yield cloned inserts exceeding 300 kbp (15), representing as much as one quarter of an entire bacterial genome.

Fosmids (for F-factor cosmids) are essentially identical to BACs in sequence and their other properties, once packaged via lambda extracts and transfected into *E. coli* (6). The main difference between BACs and fosmids is the method used to introduce them into *E. coli*. BACs are introduced into host cells via electoporation, whereas fosmids are packaged in vitro in phage lambda capsids and transfected into *E. coli* host cells. These differences in methodology influence the maximum size of recombinant DNA inserts recovered. BACs recovered from natural microbial populations can be as large as 200 kb (1), whereas fosmid sizes range from 32 to 45 kb (6, 18).

The general strategy for BAC cloning used today is essentially identical in strategy to original protocols introduced in the early 1990s (6, 15, 16) (Fig. 1A). In brief, cells are first embedded in agarose to prevent physical shearing during downstream lysis and purification steps. Although this method is very efficient, any other material copurifying with cells will also be trapped in the agarose and may inhibit downstream enzymatic modification (e.g., restriction digestion or ligation). Therefore, cells to be embedded in agarose need to be relatively free of contaminating materials. Cells concentrated from relatively clean, dilute aqueous environments may be more amenable to embedding than those derived from more challenging environments like soils or sediments. After proteinase K/detergent digestion of the agarose-embedded cells, the liberated DNA is subjected to partial digestion by restriction endonucleases (BamHI and HindIII are commonly used). Partial digestion is adjusted to generate a maximum amount of DNA in the appropriate size range, approximately 100 to 300 kb. Often there is sheared DNA in the original preparation, so partial digestion conditions need to be carefully monitored above and below the size regions of interest. After purification via PFGE, the DNA fragments are recovered from the agarose, ligated into the BAC vector, and electroporated into *E. coli*. Improvements in blue-white selection and optimization of electroporation conditions have elevated the efficiency of the entire process. BAC libraries with average insert sizes approaching 100 kb and maximum insert sizes near 200 kb have been successfully recovered from natural microbial populations (1). Importantly, these general procedures need to be tuned to optimize them for any given sample examined. For standard BAC libraries, it is particularly important to ensure that there is sufficient material to work with, in general a minimum of 10^{10} cells/ml in the original agarose plug.

Although insert size is smaller on average (ca. 40 kb), fosmid cloning approaches have several potential advantages compared with standard BACs for constructing high-quality libraries from environmental samples. In general, less biomass, lower amounts, and smaller-sized DNA are required for successful fosmid library construction in comparison with BACs. This alleviates the stringent requirement in BAC cloning for isolation of largely intact chromosomes from many (10^{10}) cells (see above). The original fosmid cloning procedure (6) is outlined in Fig. 1B. In the original protocol, the vector "arms" are prepared, and Sau3A partial digests of the DNA are prepared, generally with a size selection step. After ligation, lambda extract packaging, and transfection, the libraries are arrayed in microtiter dishes, and standard plasmid purification and screening procedures are used to evaluate the clones.

Newer approaches have greatly improved the efficiencies in fosmid library preparation. These new techniques incorporate random physical shearing of DNA inserts, followed by end repair and blunt-end ligation into the circularized fosmid (Fig. 1C), in contrast to the partial restriction endonuclease digestion and "sticky end" cloning required with standard BAC and fosmid cloning approaches. In general, DNA prepared by standard purification procedures produces DNA of the required size (40 kb), eliminating the agarose embedding of cells, partial restriction endonuclease digestion, and PFGE steps. In addition, more rigorous DNA purification protocols (for instance, CsCl equilibrium density gradient centrifugation) can be used successfully with the blunt-end fosmid cloning approach. These advantages allow the preparation of high-quality fosmid libraries from environmental samples (11). Such blunt-end fosmid libraries may even be more representative than standard BAC libraries (H. Shizuya, personal communication). This is because regions with few HindIII or BamHI sites are expected to be less well represented in BAC or fosmids prepared using partial digests that are not required in the blunt-end fosmid cloning approach. With this streamlined procedure, it is now possible to successfully prepare high-quality fosmid libraries using as little as 1 μg or less of natural population DNA (unpublished).

38.2.3. Library Screening and Sequencing

One of the advantages of BACs and fosmids is that standard screening approaches developed for conventional plasmids are all applicable. High-density colony blots (macroarrays) that allow screening of tens of thousands of clones on a 20-cm-by-20-cm blot are one typical approach (18). Multiplex screening by PCR is another rapid and very useful way to screen for specific recombinant clones. Both approaches have been used successfully to identify clones of interest in environmental BAC and fosmid libraries (1, 18). For screening of rRNA genes, other rapid multiplexing approaches that exploit automated capillary electrophoresis include length heterogeneity polymorphism (19) or terminal restriction fragment PCR (7) screening. Additionally, "functional" gene screening can be very useful for gaining information about specific metabolic pathways and processes (2). Another screening method that has proven very useful is random end sequencing, which can be used for library quality assessment, gene surveys, contig identification, and comparative community analyses (see, for example, http://www.tigr.org/tdb/MBMO/BAC_end_ann_info .shtml). Although end sequencing of single-copy BACs and fosmids was once challenging, improvements in sequencing technologies have now removed technical barriers to high-throughput BAC end sequencing. This is probably the first pass method of choice for gaining an in-depth qualitative and quantitative assessment of any given environmental library.

A significant point to consider in library assessments is the fact that even in low-copy-number BAC vectors (17) there may be differential recovery of different genomic sequences. These artifacts can arise from the presence of highly repetitive, tandem sequence motifs (17), or due to gene expression of recombinant BAC DNA in *E. coli* (especially if it is of bacterial origin) (13), which may sometimes prove lethal. Assessing the extent and fidelity of genome recovery from complex assemblages is important in many contexts, but remains difficult to quantify accurately. Screening for phylogenetic representation within the library (via rRNA genes) is one useful approach to assess the diversity of genomes recovered in libraries (3, 12) relative to the representation in the starting DNA. Statistical methods to estimate total genome recovery in populations may also prove useful (5), but may require a fairly extensive sampling regime. The type of coverage desired in recombinant libraries also largely depends on the purpose of the investigation. For genome-oriented population biology studies, sample "normalization" is not necessarily desirable, since quantitative information about genome representation is a major goal. Bioprospectors, on the other hand, may wish to maximize the diversity of recovered types and therefore seek methods to amplify low-abundance genome types in the libraries. In either case, both the "richness" (species number) and "evenness" (relative representation of each species) of microbial genomes in the original sample will greatly influence the end result. Prior knowledge of this representation may guide the strategies used for any given environmental sample or microbial assemblage from which libraries are constructed.

38.3. PROTOCOLS

38.3.1. BAC Libraries from Mixed Microbial Populations: Protocol

Supplementary protocols for BAC library preparation and related procedures can be found on the web at http://informa .bio.caltech.edu/ and http://www.genome.clemson.edu. The following protocols are based on several previously published protocols (1–3) and the advice and suggestions of Hiroaki Shizuya. The success in library construction depends on many variables, in particular, cell concentration, absence of contaminating inhibitory materials in the agarose-embedded sample, and the quality of the vector preparation. The following protocol has been used to create BAC libraries with average insert sizes approaching 100 kb from naturally occurring marine picoplankton populations concentrated by tangential flow filtration.

BAC Library Construction Protocol

Cell embedding and lysis

1. Resuspend concentrated bacterioplankton cells in 0.5 ml of sterile seawater to a final concentration of 10^{10} to 10^{11} cells/ml. Mix the cell suspension with an equal volume of melted and prewarmed (42°C) 1% SeaPlaque agarose. Draw the cells in agarose into a 1ml syringe, seal with parafilm, and chill on ice to solidify the agarose.

2. Place the agarose plug in 15 ml of STE (500 mM NaCl, 0.1 M EDTA [pH 8], 10 mM Tris-HCl [pH 8]), and place the plug on a rotating incubator (room temperature) for 1 h.

3. Exchange the buffer with 15 ml of lysis buffer (50 mM NaCl, 0.1 M EDTA [pH 8], 10 mM Tris-HCl [pH 8.0], 0.2% [wt/vol] sodium deoxycholate, 1% [wt/vol] Sarkosyl, 1 mg/ml lysozyme). Place the agarose plug in a rotating incubator and incubate at 37°C for 0.5 h.

4. Transfer the agarose plug to a 50-ml Falcon tube and add 45 ml of ESP buffer (0.5 M EDTA [pH 8.0], 1% Sarkosyl, 1 mg/ml proteinase-K). Incubate in a rotating incubator overnight at 50°C. After overnight incubation, resuspend the agarose plug in freshly prepared ESP buffer and incubate at 50°C for another 16 to 24 h.

5. Following ESP buffer incubations, transfer the agarose plug into 50 mM EDTA (pH 8.0) and store at 4°C.

6. Before restriction endonuclease digestion, incubate the agarose plug in 15 ml of PMSF solution (10 mM Tris-HCl [pH 7.5], 0.1 mM EDTA, 1 mM phenylmethylsulfonyl fluoride) for 2 h at 37°C. Replace the PMSF solution with fresh PMSF solution, and incubate overnight at 37°C. Finally, wash the plug twice for 2 h each wash in 10 mM Tris-HCl (pH 7.5) + 0.1 mM EDTA.

Pilot-scale partial digestion of chromosomal DNA

1. Incubate a 1-mm slice of the DNA-containing agarose plug with 100 μl of a mixture of 1× HindIII reaction buffer, 0.1 mg/ml acetylated bovine serum albumin, and 4 mM spermidine on ice for 30 min.

2. Replace the solution with fresh solution plus HindIII and allow to diffuse into the plug for 60 min on ice. Typically, HindIII concentrations for pilot titrations range from 0.5 to 100 U per 100-μl reaction.

3. Transfer the reaction mixture to a 37°C water bath for partial digestion and incubate for 20 min.

4. Stop the reaction by adding 1/10 volume of 0.5 M EDTA (pH 8.0) and placing the tubes on ice.

5. Load the partially digested DNA on a 1% agarose gel in 0.5× TBE with a wide-bore tip and seal the wells with the same molten agarose as the gel.

6. Perform PFGE on a CHEF Mapper (Bio-Rad) under the conditions of 6.0 V/cm, 90-s pulse, 0.5× TBE buffer, 12°C, 18 h.

7. After checking the ethidium bromide-stained gel from step 6, choose the enzyme concentration giving a majority of DNA fragments ranging from about 300 to 600 kb. Due to DNA trapping during the step 6 running conditions, we expect the actual size of the DNA fragments obtained from the 300-to-600-kb fraction to be between 50 and 300 kb in size. Environmental samples often contain considerable amounts of low-modecular-weight DNA, which masks the partially digested DNA of interest. It is sometimes necessary, therefore, to "guesstimate" the appropriate concentration range of restriction enzyme to use in partial digests. The results are then evaluated by ligation, electroporation, and assessment of actual BAC clone sizes.

Preparing partially digested DNA for ligation

1. Select the amount of enzyme giving the optimum digest and perform the digestion reaction in a large scale by carrying out several reactions (10 to 20 reactions) under exactly the same conditions as above (same volumes, same tubes, same dilution of enzyme, same enzyme tube, etc.).

2. Load the partially digested DNA on a 1% LMP agarose gel (Seaplaque GTG, FMC) in 1× TAE and seal with the same molten agarose.

3. Perform PFGE on a CHEF Mapper under the conditions of 6.0 V/cm, 20 to 40 s pulse, 1× TAE buffer, 12°C, 18 h.

4. Cut DNA fragments ranging from about 300 to 600 kb from the gel and use for ligation.

Ligation

1. Melt the dialyzed LMP DNA piece from the size selection at 65°C for 5 min and transfer to a 45°C water bath.

2. Add 1 unit of GELase (Epicentre) for every 100 mg of excised, melted agarose gel slice and incubate at 45°C for 1 h.

3. Check the concentration of the DNA solution by loading 10 μl on a 1% agarose minigel with ethidium bromide and running a quick electrophoresis at 60 V for 1 h in 0.5× TBE with uncut lambda DNA standard solutions (e.g., lambda 5 ng, 10 ng, 20 ng, 40 ng, and 80 ng/well). Usually the concentration of the size-selected DNA is between 0.5 and 2 ng/μl.

4. Combine 50 to 200 ng of the size-selected DNA with the HindIII-digested, dephosphorylated BAC vector in a molar ratio of 1 to 10 and incubate at 55°C for 10 min. Cool to room temperature for 15 min and only then add the ligase and buffer in a total volume of 100 μl. Use 4 to 6 units of T4 DNA ligase and incubate at 14°C for 16 h.

5. Drop-dialyze the ligation solution by pipetting into a Millipore filter (filter type VS, 0.025 μm) floating on Tris-EDTA buffer (TE). Incubate for 1 h at room temperature to remove the salts present in the ligation buffer. Carefully remove the ligation from the top of the floating filter by pipetting. If necessary, store at 4°C, or proceed to the next step.

Transformation by electroporation

1. Transform 1 to 2.5 μl of the ligation material into 20 to 25 μl of E. coli DH10B competent cells (Gibco BRL) by using the Bio-Rad Gene Pulser II (Bio-Rad) using settings of 100 Ohms, 13 kV/cm, and 25 μFa.

2. Transfer the electroporated cells to 15-ml culture tubes with 0.4 to 1 ml of SOC and shake at 220 rpm for 55 min at 37°C.

3. Spread the SOC medium from step 2 on one or two LB plates containing 12.5 μg/ml chloramphenicol, 50 μg/ml X-Gal (stock, 20 mg/ml in dimethylformamide), and 25 μg/ml isopropyl-β-D-thiogalactopyramoside (IPTG) and incubate at 37°C for 20 to 36 h.

38.3.2. Fosmid Library Construction by Blunt-End Cloning: Protocol

The following protocol is a modified version of a blunt-end repair and cloning strategy for preparing fosmid libraries that is conveniently available as a commercial kit from EpiCentre. There are several advantages to the general approach of blunt-end fosmid cloning, as discussed previously in this chapter. The method has been used successfully to prepare libraries from soil (11) as well as from seawater and hydrothermal vent sulfide samples (unpublished), using less than 1 μg of DNA. Standard detergent lysis and phenol/chloroform extraction procedures can be used to prepare the DNA initially, which yields DNA in the appropriate size range (40 kb) for blunt-end fosmid cloning. Small-scale cesium chloride purification can then be used in a final purification step before size fractionation by gel electrophoresis. These steps greatly facilitate the removal of contaminants and potentially inhibitory substances that often contribute to cloning failures. Additionally, potential biases introduced by partial restriction endonuclease digestion of the genomic DNA template are entirely avoided, further enhancing cloning efficiency and library representation.

Fosmid Library Construction Protocol

Small-scale CsCl DNA purification

1. Purify DNA by standard detergent lysis and phenol chloroform extraction procedures. This generally produces DNA in the appropriate size range of ≥40 kb.

2. In a 1.5-μl epitube, combine 160 mg of CsCl, 178 μl of DNA, and 10 μl of 10 μg/μl ethidium bromide.

3. After mixing and dissolving the CsCl, transfer to a 200-μl polycarbonate ultracentrifugation tube (Beckman catalog no. 342775).

4. Place the tubes in a Beckman TLA 100 rotor, ensuring that the rotor is balanced.

5. Centrifuge the samples at 100,000 rpm, 20°C, overnight.

6. Remove the tubes and visualize the DNA bands with a long-wavelength UV lamp.

7. Using a sterile 1-ml syringe and needle, remove the DNA band and place in an epitube.

8. Extract the DNA with an equal volume of water-saturated butanol. Remove the upper, ethidium bromide- and butanol-containing layer. Repeat the extraction three times.

9. Bring the volume of the DNA to 2 ml with sterile TE, and place the sample in an Amicon Centricon 100 spin dialysis unit. Centrifuge at 500 × g at room temperature until the DNA volume is reduced to approximately 50 μl. Resuspend the DNA to 2 ml again with sterile TE. Repeat this washing step two more times to remove residual CsCl. In the final centrifugation, concentrate the DNA to about 50 μl final volume.

End repair and PFGE gel purification

1. Place 52 μl of the CsCl-purified DNA (≥40 kb DNA) in a 0.5-ml sterile epitube.

2. Add 8 μl of 10× end-repair buffer, 8 μl of 2.5 mM dNTP mix, and 8 μl of 10 mM ATP. Add 4 μl of Epicentre End-Repair Enzyme Mix.

3. Incubate at room temperature for 45 min.

4. Heat inactivate at 70°C, 10 min.

5. Add 10 μl of 10× gel loading buffer (50% glycerol, xylene cyanol, bromphenol blue in TE).

6. Prepare a 0.8% SeaPlaque agarose gel in 1× TAE buffer.

7. Load sample and also 100 ng of T7 DNA size marker in adjacent lanes. Run gel overnight on a PFGE apparatus (CHEF-DR II (Bio-Rad), or the equivalent, in 1× TAE Buffer, under the following conditions: 6 V/cm, 5 to 15 s pulse switch time, 13 h run time, 12°C buffer temperature.

8. The next morning, remove the gel and stain with 1× SybrGold (10,000× stock SybrGold diluted 1/10,000 in sterile water) for 30 min, per the manufacturer's instructions.

9. Visualize DNA on DarkReader Transilluminator (EpiCentre). Excise DNA in the 40- kb size range, and place in a 1.5-ml epitube.

10. Melt gel slice at 70°C, 10 min.

11. Place at 45°C, 10 min.

12. Add an appropriate volume of 50× Gelase Buffer to the melted gel slice to achieve a 1× final Gelase Buffer concentration. Add 1 μl of Epicentre GELase for every 100 μl of melted gel slice. Mix gently by inverting the tube several times, and incubate for 1 to 2 h at 45°C.

13. Inactivate Gelase by incubation at 70°C for 10 min.

14. Bring the volume of DNA to 500 μl with sterile H2O. Wash the DNA by spin dialysis by centrifuging at 4,000 × g in an Amicon Microcon-100 unit. Resuspend to 500 μl with sterile H2O. Repeat this washing step two more times. Bring the final volume of the DNA down to about 50 μl final volume. The DNA is now ready for ligation.

Ligation and packaging

1. Place 15 μl of end-repaired, gel-purified (now 40-kb-sized) DNA in a 0.5-ml epitube.

2. Add 2 μl of 10× Fast-link Ligase buffer, 1 μl of 10 mM ATP, 1 μl of pEPIFOS plasmid, and 1 μl of Fast-link ligase. Mix by gentle pipetting, avoiding bubbles.

3. Incubate at room temperature for 2 h.

4. Inactivate ligase at 70°C for 10 min. (Can now store the ligation at 4°C before proceeding to the packaging step.)

5. Thaw MaxPlax packaging mix on ice. Remove 25 μl into prechilled epitube. Place this aliquot at −80°C for later use in step 8.

6. To remaining packaging mix, add 10 μl of ligation (avoid bubble formation and mix by gentle pipetting).

7. Incubate 90 min at 30°C .

8. Remove remaining 25 μl of packaging mix from −80°C. Add the entire 25 μl to the packing reaction at 30°C. Incubate another 90 min at 30°C.

9. Add 100 μl of phage dilution buffer (10 mM Tris [pH 8.3], 100 mM NaCl, 10 mM MgCl2), then add 5 μl of CHCl3. Mix gently. Centrifuge briefly. The packaged library may be stored at 4°C for several weeks before transduction and plating.

Transduction

1. Inoculate 50 ml of LB/10 mM MgSO4 with E. coli EPI100 plating strain. Grow overnight with shaking at 37°C.

2. In the morning, inoculate 50 ml of LB/10 mM MgSO4 with 3 ml of the overnight culture of E. coli EPI100 plating strain.

3. Grow to an OD600 of 0.8 to 1.0 (do not allow culture to exceed OD600 = 1.0).

4. Cells may be stored at 4°C until needed, for up to 3 days.

5. For typical titers in environmental libraries, mix 25 μl of packaged library with 500 μl of E. coli EPI100 plating strain (from above). This dilution can be bracketed with serial dilutions to more accurately determine the titer.

6. Incubate at room temperature 20 min. To titer, plate 25 μl and 250 μl onto LB plates (150-mm petri dish) containing 12.5 μg/ml chloramphenicol. Incubate plates for 24 h at 37°C and count colonies to determine titer in the packaging mix.

38.4. CONCLUSION AND FUTURE PROSPECTS

The applications of BAC libraries for environmental microbiology are many and varied. Bioprospecting for antibiotics and other natural products has been one of the earlier and more obvious applications to exploit their novel genetic content. Tremendous potential exists for genomic approaches using these libraries to answer many questions about evolution, ecology, and population biology in the context of natural microbial ecosystems. BAC and fosmid libraries are useful tools that can be used to sidestep many thorny problems that still inhibit the accurate characterization of naturally occurring microbial communities. They represent useful tools for conducting sequence surveys to determine the phylogenetic and functional content and properties inherent in native microbial populations. They can also serve as reagents for gene discovery, gene expression, and functional genomic characterization of naturally occurring microbes, some still resistant to cultivation techniques. BAC and fosmid libraries can also serve as useful reagents in microarray construction and analyses of naturally occurring microbial dynamics and variability.

Environmental genomic approaches are cultivation independent and theoretically can provide equal access to the collective genomes contained in natural microbial assemblages. Applied microbial population genomics can now facilitate predictions about the relationships between genotype and ecological distribution and function. These approaches can broaden our knowledge of natural environmental processes and extend our abilities to exploit natural microbial products and processes. Perhaps most importantly, application of environmental genomic approaches has the potential to foster a better understanding of microbial genome evolution, ecology, and population biology and to contribute significantly to theoretical models and predictive capabilities in environmental microbiology and microbial ecology.

38.5. REFERENCES

1. Béjà, O., L. Aravind, E. V. Koonin, M. T. Suzuki, A. Hadd, L. P. Nguyen, S. B. Jovanovich, C. M. Gates, R. A. Feldman, J. L. Spudich, E. N. Spudich, and E. F. DeLong. 2000. Bacterial rhodopsin: evidence for a new type of phototrophy in the sea. Science 289:1902–1906.

2. Béjà, O., M. T. Suzuki, J. F. Heidelberg, W. C. Nelson, C. M. Preston, T. Hamada, J. A. Eisen, C. M. Fraser, and E. F. DeLong. 2002. Unsuspected diversity among marine aerobic anoxygenic phototrophs. Nature 415:630–633.

3. **Béjà , O., M. T. Suzuki, E. V. Koonin, L. Aravind, A. Hadd, L. P. Nguyen, R. Villacorta, M. Amjadi, C. Garrigues, S. B. Jovanovich, R. A. Feldman, and E. F. DeLong.** 2000. Construction and analysis of bacterial artificial chromosome libraries from a marine microbial assemblage. *Environ. Microbiol.* **2:**516–529.

4. **Doolittle, R. F.** 2002. Microbial genomes opened up. *Nature* **392:**339–342.

5. **Hughes, J. B., J. J. Hellmann, T. H. Ricketts, and B. J. Bohannan.** 2001. Counting the uncountable: statistical approaches to estimating microbial diversity. *Appl. Environ. Microbiol.* **67:**4399–4406.

6. **Kim, U.-J., H. Shizuya, P. Dejong, B. Birren, and M. Simon.** 1992. Stable propagation of cosmid sized human DNA inserts in an F-factor based vector. *Nucleic Acids Res.* **20:**1083–1185.

7. **Marsh, T. L., P. Saxman, J. Cole, and J. Tiedje.** 2000. Terminal restriction fragment length polymorphism analysis program, a web-based research tool for microbial community analysis. *Appl. Environ. Microbiol.* **66:**3616–3620.

8. **Olsen, G. J., D. J. Lane, S. J. Giovannoni, N. R. Pace, and D. A. Stahl.** 1986. Microbial ecology and evolution: a ribosomal RNA approach. *Annu. Rev. Microbiol.* **40:**337–365.

9. **Pace, N. R.** 1997. A molecular view of microbial diversity and the biosphere. *Science* **276:**734–740.

10. **Pace, N. R., D. A. Stahl, G. J. Olsen, and D. J. Lane.** 1985. Analyzing natural microbial populations by rRNA sequences. *ASM News* **51:**4–12.

11. **Quaiser, A., T. Ochsenreiter, H. P. Klenk, A. Kletzin, A. H. Treusch, G. Meurer, J. Eck, C. W. Sensen, and C. Schleper.** 2002. First insight into the genome of an uncultivated crenarchaeote from soil. *Environ. Microbiol.* **4:**603–611.

12. **Rondon, M. R., P. R. August, A. D. Bettermann, S. F. Brady, T. H. Grossman, M. R. Liles, K. A. Loiacono, B. A. Lynch, I. A. MacNeil, C. Minor, C. L. Tiong, M.** Gilman, M. S. Osburne, J. Clardy, J. Handelsman, and R. M. Goodman. 2000. Cloning the soil metagenome: a strategy for accessing the genetic and functional diversity of uncultured microorganisms. *Appl. Environ. Microbiol.* **66:**2541–2547.

13. **Rondon, M. R., S. J. Raffel, R. M. Goodman, and J. Handelsman.** 1999. Toward functional genomics in bacteria: analysis of gene expression in Escherichia coli from a bacterial artificial chromosome library of Bacillus cereus. *Proc. Natl. Acad. Sci. USA* **96:**6451–6455.

14. **Schleper, C., E. F. DeLong, C. M. Preston, R. A. Feldman, K. Y. Wu, and R. V. Swanson.** 1998. Genomic analysis reveals chromosomal variation in natural populations of the uncultured psychrophilic archaeon Cenarchaeum symbiosum. *J. Bacteriol.* **180:**5003–5009.

15. **Shizuya, H., B. Birren, U. J. Kim, V. Mancino, T. Slepak, Y. Tachiiri, and M. Simon.** 1992. Cloning and stable maintenance of 300-kilobase-pair fragments of human DNA in Escherichia coli using an F-factor-based vector. *Proc. Natl. Acad. Sci. USA* **89:**8794–8797.

16. **Shizuya, H., and H. Kouros-Mehr.** 2001. The development and applications of the bacterial artificial chromosome cloning system. *Keio J. Med.* **50:**26–30.

17. **Song, J., F. Dong, J. W. Lilly, R. M. Stupar, and J. Jiang.** 2001. Instability of bacterial artificial chromosome (BAC) clones containing tandemly repeated DNA sequences. *Genome* **44:**463–469.

18. **Stein, J. L., T. L. Marsh, K. Y. Wu, H. Shizuya, and E. F. DeLong.** 1996. Characterization of uncultivated prokaryotes: isolation and analysis of a 40-kilobase-pair genome fragment from a planktonic marine archaeon. *J. Bacteriol.* **178:**591–599.

19. **Suzuki, M., M. S. Rappe, and S. J. Giovannoni.** 1998. Kinetic bias in estimates of coastal picoplankton community structure obtained by measurements of small-subunit rRNA gene PCR amplicon length heterogeneity. *Appl. Environ. Microbiol.* **64:**4522–4529.

39

Single Cell Identification by Fluorescence In Situ Hybridization

BERNHARD M. FUCHS, JAKOB PERNTHALER, AND RUDOLF AMANN

39.1. INTRODUCTION

This chapter presents an updated collection of protocols for the identification of individual microbial cells by fluorescence in situ hybridization (FISH) with rRNA-targeted oligonucleotide probes. This "phylogenetic staining" (14) method allows the enumeration, biomass determination, and localization of microorganisms in their environmental context. FISH is an integral part of the rRNA approach to microbial ecology and evolution (38).

Several reviews are available on the principles of hybridization (5, 58) and on methodological aspects and applications of FISH with rRNA-targeted oligonucleotides (1, 4). Therefore, we will first outline the background in brief and then focus on protocols.

The rRNAs are among the most conserved macromolecules. The three prokaryotic rRNAs, 5S, 16S, and 23S rRNA, are found in abundance in every cell. Therefore, the following protocols are not readily applied to the detection of molecules that have low cellular abundance, such as mRNA, plasmids, or chromosomal genes. Comparative 16S rRNA sequence analysis has become a standard method in microbial taxonomy. Consequently a 16S rRNA database has been compiled that comprises to date more than 400,000 (2007) sequences covering not only most pure cultures but also many 16S rRNA genes of so far uncultured bacteria (9, 30, 59). The comparison of 16S rRNA sequences allows for the reconstruction of microbial phylogeny (65) and the identification of short signature sequences that are unique for different groups of microorganisms. These signatures are used as targets for short complementary oligonucleotides,

called probes, with a typical length of about 15 to 30 nucleotides. For standard FISH applications, oligonucleotides are typically monolabeled with a fluorochrome.

FISH of whole cells starts with a fixation of the sample containing the target cell types (Color Plate 14). Fixation stabilizes macromolecules and cytoskeletal structures, thus preventing lysis of the cells during hybridization, and at the same time permeabilizes the cell walls for fluorescently labeled oligonucleotide probes. The fixed cells are incubated in a buffer containing the specific probe at a temperature near but below the melting point of the probe-rRNA hybrid. The subsequent washing step will remove unbound probe and leave only those probe-rRNA pairs intact that have no mismatches in the hybrid. Consequently, only target cells that contain the full signature sequence on their rRNA will be stained. Finally, hybridized cells can be enumerated by epifluorescence microscopy or by flow cytometry.

39.2. rRNA DATABASE

The reliability of FISH for the identification of cells in complex environments strongly depends on the quality of the 16S rRNA database. A comprehensive and well-maintained database is vital for rational probe design. Powerful software tools are required to manage such a database. We recommend the highly integrated software package ARB released by the Technical University of Munich (60). This program runs on different UNIX-based operating systems (including Linux), it is capable of maintaining several hundred thousand aligned 16S rRNA sequences, and it allows the easy import of additional sequences from various sources. In an ideal world, probe design would be solely based on high-quality, full-length sequences. In practice, the public domain databases are filled with partial sequences and sequences that contain sequencing errors. The sequence gaps in partial sequences can limit probe design by reducing the number of potential target regions and by failing to provide information about the specificity of existing probes (Fig. 1).

39.3. PROBE DESIGN

For FISH, a probe length of 15 to 25 nucleotides is most commonly used. Sequence signatures serving as suitable target sites for nucleic acid probing can be convenient and automatically searched with the PROBE_DESIGN tool within the ARB software package. First, a target group of organisms must be specified, e.g., in a phylogenetic tree contained in ARB. The PROBE_DESIGN tool searches for possible signature sequences that are diagnostic for the selected species. The G+C content of probe sequences influences their melting behavior. By default this parameter is set in PROBE_DESIGN between 50 and 100% to ensure a tight binding, but usually it ranges between 50 and 70%. Sometimes the PROBE_DESIGN tool cannot find a suitable probe. However, the program provides options for modifying the search parameters to look for signatures in subsets of the group originally selected, or by choosing to allow for the signature to be found in a defined number of species outside the target group. By the combinatorial use of probes with overlapping specificity, the selected group of organisms may be fully targeted.

A problem that should be considered during the design of FISH probes is target site accessibility. The higher-order structure of the ribosome may hinder the binding of the probe to its target site. The 16S rRNA in situ accessibility for oligonucleotide probes has recently been studied for two members of the domain Bacteria (*Escherichia coli*, *Pirellula* sp. strain 1), one eukaryote (*Saccharomyces cerevisiae*), and one archaeon (*Metallosphaera sedula*) (8, 18) (Color Plate 15). In 2001 a complete accessibility map for the 23S rRNA of *E. coli* was published (17) (Color Plate 16). These color-coded maps clearly demonstrate dramatic differences in the binding of different fully complementary probes of approximately identical length to one batch of fixed target cells. Even though the secondary structure of the ribosome is highly conserved, and even though a consensus map could be developed from the 16S rRNA accessibility studies (8), each probe should be checked on its respective target group of organisms to ensure high probe signals.

It has recently been shown that inaccessible target regions can be made accessible by the use of unlabeled oligonucleotides called helpers (16). These bind adjacent to the diagnostic probe, thereby "opening" the target region for the probe. Helpers should be a few nucleotides longer than the diagnostic probe; that is, if the probe is an 18-mer, the helper should be a 21-mer to ensure a tight binding beyond the melting point of the diagnostic probe.

Another strategy of probe selection is searching for already published probes. Loy and coworkers established a probe database with currently more than 890 entries, not all developed for FISH (www.microbial-ecology.net/probebase) (29).

39.4. SELECTION OF FLUORESCENT LABEL

A range of fluorochromes is available for the labeling of nucleic acids, but not all fluorochromes are equally suited as labels for oligonucleotides. Newly developed dyes, in particular, should be checked for nonspecific staining. Standard labels for in situ hybridization are the green fluorescein and the red tetramethylrhodamine derivatives (Table 1). These dyes are well suited for standard applications when the ribosome content of target cells is high. In addition, these dyes can be used in conjunction to label different probes in double-staining experiments. CY3 and CY5 are members of the indocarbocyanine family. The high signal intensity of these two dyes makes them the fluorochromes of choice when trying to detect small cells with low ribosome content, such as bacterioplankton cells. These fluorochromes exhibit high fluorescence because of their high quantum yield and high molar extinction coefficient (Table 1). It is important to note that the CY5 derivative that fluoresces in the near infrared is detectable only with a CCD camera or on a confocal laser scanning microscope with red laser excitation. Concomitant to the selection of the fluorescent label, care should be taken that the right optical filters are chosen for the detection of the respective dyes. Well-adapted optical filters with a high transmission in the emission spectrum of the dye and a strong and sharp blocking of the excitation light are essential for a confident detection of weak signals.

In environmental samples, single oligonucleotides carrying only one fluorochrome may not provide enough fluorescence signal to detect cells with low ribosome content (43). Polynucleotide probes with a length of more than 100 nucleotides, carrying several fluorochromes, are an alternative (13, 62). However, these probes lack specificity for narrow target groups that are at the level of species and genera. An alternative labeling technique that increases fluorescence signal intensity uses horseradish peroxidase (HRP)-labeled oligonucleotides. When using HRP-labeled probes, fluorescent staining results from a secondary incubation with fluorescently labeled tyramide. The specifically bound peroxidase molecules catalyze the deposition of these labeled reporter compounds within cells targeted by the HRP-tagged

```
Probe     3'-5'                              GGAAGCCCGGAGAACGGT
Target    5'-3'  UCGCAAGACCAAAGAGGG  CCUUCGGGCCUCUUGCCA  CGGAUGUG
                 UCGCAAGACCAAAGAGGG  CCUUCGTGCCUCUUGCCA  CGGAUGUG
Non-target?      ..................  ........CCUCUUGCCA  CGGAUGUG
                 UCGCAAGACCNNNGAGGG  CCUUCGGSCCUCUUGYCA  CGGAUGUG
```

FIGURE 1 Example of an rRNA alignment showing the probe sequence, the target sequence, and possible false identities.

TABLE 1 Dye labels frequently used for oligonucleotide probes and their characteristics[a]

Label	Excitation (± 10 nm)	Emission (± 10 nm)	Mol mass (Da)	ϵ[b] (mol^{-1} cm^{-1})
CY3[c]	512/552	565/615	766	150,000
CY5[c]	625–650	670	792	250,000
Carboxyfluorescein (FAM)[d]	492	518	376	79,000
Fluorescein[d]	490	520	389	77,000
Tetramethylrhodamine[d]	543	571	444	99,000
Carboxytetramethylrhodamine (TAMRA)[d]	542	568	430	91,000

[a]Fluorescein and derivatives are pH sensitive and exhibit maximum fluorescence at pH ≥ 9.
[b]ϵ, molar extinction coefficient.
[c]Data compiled from Amersham Biosciences, San Francisco, CA.
[d]Data compiled from reference 50 and Invitrogen, Carlsbad, CA.

probe. FISH signals are up to 20-fold brighter with HRP-labeled probes than with conventional single-labeled probes (52). However, cell permeabilization protocols need to be adjusted to enable the larger enzyme-labeled oligonucleotides to penetrate into cells (42).

39.5. QUALITY CHECK OF PROBES

Oligonucleotide probes are custom made by solid-phase synthesis. In the last step of synthesis a fluorochrome is added to the 5′ end of the oligonucleotide. Purified probe stocks are frequently delivered lyophilized. Upon reconstitution with 100 μl of sterile water, a probe synthesis at 0.02-μmol scale yields approximately a 1,500 ng μl^{-1} stock solution. To determine the exact probe concentration, the absorbance of the 1:100-diluted stock solution at 260 nm should be measured, giving 1 OD$_{260\,nm}$ ≈ 20 ng μl^{-1} DNA. Furthermore, the labeling of the oligonucleotide should be checked. For a pure monolabeled oligonucleotide, the ratio of absorption of the dye (A_{dye}) versus the absorption of the nucleic acids at 260 nm (A_{260}) of a monolabeled oligonucleotide should match the ratio of the extinction coefficients (ϵ) of the dye and oligonucleotide. The extinction coefficient ϵ at 260 nm (ϵ_{260}) of an oligonucleotide can be estimated from its nucleotide composition as the sum of the extinction coefficients of the individual nucleotides (dATP = 15.4 cm^3 μmol^{-1}, dCTP = 7.3 cm^3 μmol^{-1}, dGTP = 11.7 cm^3 μmol^{-1}, dTTP = 8.8 cm^3 μmol^{-1}, from reference 50). Taking into account the extinction coefficient of the dye (ϵ_{dye}; see also Table 2), the quality of labeled oligonucleotide can be estimated by calculating a ratio k according to the following formula:

$$k = \frac{\epsilon_{260} / \epsilon_{dye}}{A_{260} / A_{dye}}$$

Values of $k < 1$ indicate an incomplete labeling of a probe, whereas values >1 point to the presence of additional, potentially unbound dye. Considering inaccuracies in the estimation of the extinction coefficients of oligonucleotides, k values between 0.7 and 1.3 are acceptable.

Working solutions are prepared at concentrations of 50 ng μl^{-1} and stored in the dark at −20°C. Prepare only small portions of probe working solutions (50 to 100 μl). Repeated freeze-thawing may damage the probe and might result in a precipitation of the probe, which in turn would lead to weak hybridization signals and high background.

Do not freeze working solutions of HRP-labeled probes, but store them in the refrigerator at 4°C. Portions of stock solution can be stored in the freezer, but once an aliquot is thawed to prepare working solutions, keep it at 4°C. Lyophilized HRP-labeled probes are suspended in sterile H$_2$O, too. For calculating the concentration, it has to be taken into account that the enzyme itself contributes to the measured absorbance at 260 nm. Therefore the measured A_{oligo} has to be lowered by a correction factor (cf) of 0.276. Presuming optimal labeling, the peak ratio (A_{260}/A_{HRP}) should be about 3.

39.6. CELL FIXATION

FISH of whole fixed microbial cells with rRNA-targeted oligonucleotide probes is based on the binding of a short, fluorescently labeled oligonucleotide to the intracellular target nucleic acids (Color Plate 14). Therefore the fixation of the sample is one of the most critical steps in the protocol. A good fixative should preserve the cell morphology while concomitantly permeabilizing all cells for the labeled oligonucleotide. Standard fixatives are aldehydes and alcohols. For many microorganisms, good results are achieved by fixation at a final concentration of 1% formaldehyde overnight at 4°C. The formaldehyde solution should be freshly prepared from paraformaldehyde.

Preparation of a 4% Paraformaldehyde Fixative

1. Pour 2 g of paraformaldehyde (PFA) powder in 50 ml of phosphate-buffered saline (PBS; 130 mM NaCl, 10 mM Na$_2$HPO$_4$/NaH$_2$PO$_4$ [pH 7.4]) or a similar buffer (use mask for weighing PFA; irritant if inhaled).

2. Heat to ca. 60°C (must not boil!) until suspension is clear (ca. 0.5 h); if not, add some drops of 1N NaOH.

3. Check pH and adjust to pH 7.0.

4. Filter through 0.2-μm-pore-size filter and place on ice.

5. Usable for up to 1 week, and up to 6 months if stored in the dark under nitrogen atmosphere at room temperature (RT).

Commercially available 35 to 37% formalin solution is often stabilized with methanol, which decreases FISH signal intensity and tends to precipitate upon longer storage. After the fixation, a dehydration series of 50, 80, and 96% ethanol may help to permeabilize cells for FISH.

Fixation of Pure Cultures with Gram-Negative Cell Wall

1. Harvest cells during logarithmic growth by centrifugation of an aliquot (ca. 2 ml) in microcentrifuge (10 min at 4,000 × g).

2. Discharge supernatant and resuspend cells in 750 μl of PBS (145 mM NaCl, 1.4 mM NaH$_2$PO$_4$, 8 mM Na$_2$HPO$_4$ [pH 7.4]).

3. Fix by adding 250 μl of a 4% PFA fixative (1% final concentration).

4. Incubate for 1 h (for thick-walled cells) to 24 h (for fragile cells) at 4°C.

5. Pellet cells by centrifugation (10 min at 4,000 × g); discharge supernatant.

6. Thoroughly resuspend fixed cells in 500 μl of PBS.

7. Repeat steps 5 and 6.

8. Add 500 μl of absolute ethanol and resuspend cells thoroughly.

9. At this stage samples can be stored at −20°C for several months.

Fixation of Pure Cultures with Gram-Positive Cell Wall

1. Harvest cells during logarithmic growth by centrifugation of an aliquot (ca. 2 ml) in microcentrifuge (10 min at 4,000 × g).

2. Discharge supernatant and wash cells in PBS.

3. Pellet cells by centrifugation (10 min at 4,000 × g); discharge supernatant.

4. Add 500 μl of PBS and resuspend cells thoroughly.

5. Add 500 μl of cold, absolute ethanol and mix.

6. At this stage samples can be stored at −20°C for several months.

Fixation of Planktonic Samples

1. Add freshly prepared PFA fixative to water sample to a final concentration of 1 to 3% and fix for 1 to 24 h at 4°C.

2. Place a moistened support filter (0.45 μm pore size, cellulose nitrate, 47 mm diameter; Sartorius, Germany) and a membrane filter (0.2 μm pore size, white polycarbonate, 47 mm diameter; Millipore, Eschborn, Germany) into a filtration tower.

3. Filter an appropriate volume of the fixed sample by applying gentle vacuum. Support filters may be utilized for several samples. For cell numbers of approximately 10^6 ml^{-1}, 10 ml of sample is generally sufficient.

4. After complete sample filtration, wash with 10 to 20 ml of sterile H$_2$O; remove H$_2$O by filtration.

5. Put membrane filter in a plastic petri dish, cover, and allow air drying.

6. Store at −20°C until processing; filters can be stored frozen for several months without apparent loss of hybridization signal.

Fixation and Preparation of Sediment and Soil Samples

1. Suspend 0.5 ml of freshly collected sediment or soil in 1.5 ml of 4% PFA fixative in a 2-ml screw-top microcentrifuge tube; fix for 1 to 24 h.

2. Centrifuge at 10,000 rpm for 5 min; pour off supernatant.

3. Add 1.5 ml of PBS and resuspend sample.

4. Repeat steps 2 and 3.

5. Centrifuge at 10,000 rpm for 5 min; pour off supernatant.

6. Add 1.5 ml of a 1:1 mix of PBS/ethanol and store sample at −20°C until further processing.

7. Resuspend sample and transfer 20 to 100 μl of aliquot to 500 μl of a 1:1 mix of PBS/ethanol in a 2-ml microcentrifuge tube.

8. Sonicate aliquot for 20 to 30 s at low intensity using 1-s sonication pulses (Sonopuls HD70, Bandelin, Berlin, Germany); if required, the sonicated sample can be further diluted.

NOTE: Depending on the type of sediment or soil, it might be necessary to also adapt the aliquot size prior to sonication. If too much sediment is suspended, sonication will lead to incomplete detachment of cells from particles.

9. Place cellulose nitrate support filters beneath the membrane filters to improve the distribution of cells; add 15 to 20 μl of aliquot from the sonicated sample to 2 ml of distilled water, and filter this volume onto the membrane filters.

10. Air-dry filtered preparations and store in petri dishes at −20°C until hybridization.

39.7. PROBE TESTING; OPTIMIZATION OF HYBRIDIZATION CONDITIONS

When using probes that have been previously described, it is important to always check the probe against a recently updated 16S rRNA database (e.g., by a NCBI BLAST search; www.ncbi.nlm.nih.gov/BLAST/). Ideally, the probe of interest still has more than one base mismatch with all nontarget microorganisms. For the design of new probes, it is important to keep discriminatory positions central since a mismatch between probe and nontarget rRNA at the 3′ or 5′ end of the oligonucleotide is only weakly destabilizing. Competitor oligonucleotides (32) have been shown to strongly enhance single-mismatch discrimination. Competitors are unlabeled oligonucleotides, which are fully complementary to the mismatch-containing nontarget sequence. They suppress unspecific probe binding to a defined one-mismatch sequence.

If possible, new probes should be tested by FISH of isolates that have zero, one, and more mismatches with the oligonucleotide. A series of hybridizations is performed at increasing stringency either by increasing the temperature of hybridization or by increasing the concentration of a denaturing agent like formamide in the hybridization buffer. The changes in the fluorescence intensities of individual cells can be quantified by computer-assisted image analysis (37) or by flow cytometry (18). The most desirable hybridization stringency often occurs at the point immediately before the target cell fluorescence begins to decrease (Fig. 2). At this formamide concentration, hybridization to the nontarget organism should be low or absent. As a rule of thumb, an 18-mer oligonucleotide with a G+C content between 50 and 60% will start to dissociate from its fully complementary rRNA target at a formamide concentration of approximately 30 to 40% in our standard buffer at 46°C. Alternatively, to determine their temperature of dissociation (T_d), probes can be radiolabeled and hybridized to extracted rRNAs that have been blotted onto nylon membranes (58).

Frequently, new probes are designed to target yet-uncultured microorganisms that are only known from their rDNA sequences. In this case it is not possible to test these probes on isolates. Two strategies are available to optimize the hybridization conditions. (i) The cloned 16S rRNA gene of interest can be transcribed in vitro into RNA, which is then blotted on a nylon membrane and hybridized with a radiolabeled oligonucleotide at increasing levels of formamide (e.g., references 45 and 47). This quite laborious method is based on the assumption that the temperature of dissociation from isolated rRNA is the same as from rRNA in fixed cells. (ii) Recently a new method was published using clones for adjusting the hybridization conditions. Clones carrying the target sequence of the new probe are

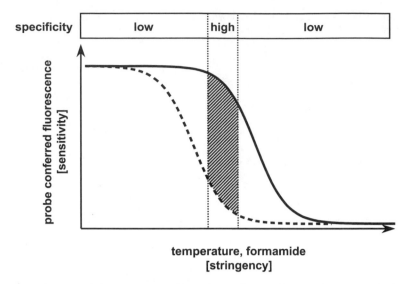

FIGURE 2 Theoretical dissociation profiles of a nucleic acid probe from a perfectly matched (bold line) and an imperfectly matched immobilized target nucleic acid (dashed line). The probe-conferred signal (Y axis), which is directly proportional to the sensitivity, is shown over temperature and formamide concentration (x axis), which represents the hybridization stringency. Note that the temperature of dissociation of the perfect hybrid is higher than that of the imperfect hybrid. The primary goal of the probe design is to maximize the difference between the temperature of dissociation of a probe from target and nontarget nucleic acid. The bar above the dissociation profiles indicates hybridization stringencies with high and low specific discrimination of target and nontarget nucleic acid, respectively.

grown with chloramphenicol and isopropyl-β-D-thiogalactopyranoside (IPTG). If a vector with an inducible promoter upstream of the multiple cloning site is chosen, then this treatment leads to an in vivo transcription of the cloned 16S rDNA and accumulation of 16S rRNA of the uncultured organism inside the *E. coli* cell. After standard fixation of these clones, they could be used as analogs to cultured organisms for determining the melting point of probes (54).

39.8. HYBRIDIZATION OF CULTURED ORGANISMS ON GLASS SLIDES

In a quick but nonquantitative protocol, fixed cells are transferred onto gelatin-coated slides and incubated in a moisture chamber with a buffer containing an oligonucleotide probe. After a brief washing, the cells are embedded in antifading reagent for microscopic visualization.

1. Heat a 0.01% $CrK(SO_4)_2$ (prevents fouling of gelatin)–0.1% gelatin solution to 65°C; dip precleaned multiwell slides in this solution; let air dry.

2. Spot 2 to 20 μl of fixed cell suspension (depending on cell density) in the wells of the gelatin-coated slide; let air dry, then dehydrate for 3 min each in 50, 80, and 100% ethanol.

3. Prepare hybridization buffer (Tables 2 and 3).

NOTE: The formamide concentration depends on the probe used and determines the stringency of the hybridization. Hybridization stringency may also be adjusted by temperature rather than by the chemical compositon of buffers. We find that it is more convenient to keep the incubator and water bath at one set temperature and to modulate the stringency by adding formamide.

4. For the hybridization mixtures, add 1 volume of probe working solution (50 ng μl^{-1}) to 9 volumes of hybridization buffer in a 0.5-ml microcentrifuge tube; keep probe solutions dark and on ice.

5. Prepare hybridization vessels from 50-ml polyethylene tubes: insert a piece of blotting paper into a polyethylene tube and soak it with the remaining hybridization buffer; use separate tubes for each concentration of formamide.

6. Add 10 μl of hybridization mix to the samples in each well and place the slide into the polyethylene tube (in a horizontal position).

7. Incubate at 46°C for at least 90 min (maximum, 3 h).

8. Prepare 50 ml of washing buffer in a polyethylene tube and preheat in a 48°C water bath (see Tables 4 and 5).

TABLE 2 Standard hybridization buffer

Stock reagent	Volume	Final concn in hybridization buffer
5 M NaCl	360 μl	900 mM
1 M Tris-HCl	40 μl	20 mM
Formamide	% depending on probe (see Table 3)	
Distilled H$_2$O	Add to 2 ml	
10% SDS (added last to avoid precipitation)	2 μl	0.01%

TABLE 3 Formamide concentration of some frequently applied probes[a]

Probe	Specificity	Sequence	Formamide %	Target	Competitor	Reference
ALF968	*Alphaproteobacteria*	GGTAAGGTTCTGCGCGTT	35	16S		35
ARCH915	Archaea	GTGCTCCCCCGCCAATTCCT	0	16S		58
BET42a	*Betaproteobacteria*	GCCTTCCCACTTCGTTT	35	23S	Gam42a	32
CF319a	*Cytophaga/Flavobacterium* cluster	TGGTCCGTGTCTCAGTAC	35	16S		31
EUB338	Most Bacteria, no *Planctomycetales*	GCTGCCTCCCGTAGGAGT	0	16S		3
EUK516	Eukarya	ACCAGACTTGCCCTCC	0	16S		2
GAM42a	*Gammaproteobacteria*	GCCTTCCCACATCGTTT	35	23S	Bet42a	32
HGC69a	Many *Actinobacteria*	TATAGTTACCACCGCCGT	20	23S		49
LGCa354	Some *Firmicutes*	TGGAAGATTCCCTACTGC	20	16S		34
Non338	Nonsense probe, for detection of nonspecific binding	ACTCCTACGGGAGGCAGC	0	16S		58
Pla886	*Planctomycetales*	GCCTTGCGACCATACTCCC	35	16S	cPla	36

[a]For horseradish-labeled probes, use 20% more formamide in hybridization buffer. See text for details.

NOTE: The stringency in the washing buffer is achieved by adjusting the NaCl concentration. This avoids the use of excess amounts of formamide.

9. Quickly rinse the slide carefully with a bit of washing buffer, transfer slide into preheated washing buffer, and incubate for 25 min at 48°C (water bath).

10. Rinse slide with distilled H_2O and let air dry.

11. For counterstaining, cover each well with 10 μl of a 1 μg ml^{-1} 4′,6′-diamidino-2-phenylindole (DAPI) solution and incubate for 3 min; rinse slide with distilled H_2O and let air dry.

12. Samples are mounted in a 4:1 mix of Citifluor (Citifluor Ltd, London, United Kingdom) and Vecta Shield (Vector Laboratories, Inc., Burlingame, CA). Vecta Shield contains a superior antibleaching reagent, but quenches DAPI fluorescence. The wells have to be completely dry before embedding; otherwise a fraction of cells will detach during inspection.

13. Double-stained and air-dried preparations as well as mounted slides can be stored in the dark at −20°C for several days without substantial loss of probe fluorescence.

14. Probe-conferred fluorescence fades much more rapidly than DAPI fluorescence in the microscopic image, and

TABLE 4 Standard washing buffer

Stock reagent	Volume	Final concn in hybridization buffer
5 M NaCl	Concentration depending on probe (see Table 5)	
1 M Tris-HCl	1 ml	20 mM
0.5 M EDTA	500 μl	5 mM
Distilled H_2O	Add to 50 ml	
10% SDS (added last to avoid precipitation)	50 μl	0.01%

UV excitation will also bleach the probe signal; for counting, it is therefore safer to first quantify probe-stained cells and subsequently all cells from the same field of vision in UV excitation.

39.9. HYBRIDIZATION OF FIXED CELLS ON MEMBRANE FILTERS

1. Cut sections from membrane filters with a razor blade. A 47-mm-diameter filter should allow the preparation of 16 to 20 individual hybridizations; a 25-mm-diameter filter yields eight sections. Label filter sections with a pencil, e.g., by numbering them.

2. Put filter sections on glass slides (facing up). Several filter sections can be placed on one slide and hybridized simultaneously with the same probe.

3. Prepare 2 ml of hybridization buffer in a microcentrifuge tube (see above).

4. Per filter section, mix 20 μl of hybridization buffer with 2 μl of probe working solution.

5. Put a piece of blotting paper into a polyethylene tube and soak it with the remaining hybridization buffer.

6. Carefully cover the filter section with the hybridization mix and place the slide with filter sections into the polyethylene tube (in a horizontal position).

7. Incubate at 46°C for at least 90 min (maximum, 3 h).

8. Prepare 50 ml of washing buffer in a polyethylene tube (see above).

9. Quickly transfer filter sections into preheated washing buffer and incubate for 15 min at 48°C (water bath).

10. Pour washing buffer with filter sections into a petri dish. Pick filter sections and rinse them by placing them into a petri dish with distilled H_2O for several seconds, then let them air dry on blotting paper.

11. For counterstaining, put filter sections on a glass plate, cover with 50 μl of DAPI solution, and incubate for 3 min; afterward, wash filter sections for several seconds in 80% ethanol to remove unspecific staining, followed by rinsing in distilled H_2O and air drying.

12. Samples are mounted in a 4:1 mix of Citifluor and Vecta Shield. The filter sections have to be completely dry

TABLE 5 Corresponding NaCl concentration in 50 ml of washing buffer[a]

% Formamide in hybridization	Washing at 35°C		Washing at 48°C	
	[NaCl] in mol	μl 5 M NaCl	[NaCl] in mol	μl 5 M NaCl
0	–	–	0.900	8,900
5	–	–	0.636	6,260
10	–	–	0.450	4,400
15	–	–	0.318	3,080
20	0.145	1,350	0.225	2,150
25	0.105	950	0.159	1,490
30	0.074	640	0.112	1,020
35	0.052	420	0.080	700
40	0.037	270	0.056	460
45	0.026	160	0.040	300
50	0.019	90	0.028	180
55	0.013	30	0.020	100
60	0.009	0	0.014	40
65	0.007	0	–	–
70	0.005	0	–	–

[a]Calculated using the formula from Lathe (26). Note that the addition of EDTA contributes to the Na^+ concentration. Therefore the required volume of 5 M NaCl solution in the washing buffer is reduced by 100 μl.

before embedding; otherwise part of the cells might detach during inspection.

13. Double-stained and air-dried preparations as well as filters mounted on slides can be stored in the dark at −20°C for several days without substantial loss of probe fluorescence.

NOTE: At least 500 DAPI-stained cells should be counted in plankton samples to obtain a counting error of less than ±5%. Do not attempt to determine absolute cell counts from filters after hybridization, but only the percentage of hybridized cells. The distribution of cells on sections of a 47-mm-diameter membrane filter is never as even as on a small filter, resulting in a higher error of the total DAPI counts. We find that, following our procedure, 80 to 90% of the initial bacterial cell numbers are recovered after hybridizations of bacterioplankton on membrane filters. This fraction, however, may depend on the type of sample and should be verified experimentally.

39.10. HYBRIDIZATION IN LIQUID FOR FLOW CYTOMETRY

1. Prepare hybridization buffer (Table 2).
2. Distribute subsamples of 100 μl of hybridization buffer in a 1.5-ml centrifuge tube.
3. Add approximately 10^6 to 10^7 cells.
4. Add 4 μl of probe working solution (2.5 ng $μl^{-1}$) and mix well.
5. Incubate at 46°C for 2 to 3 h.
6. Cells are pelleted by centrifugation for 2 min at 4,000 × g and resuspended in 100 μl of prewarmed hybridization buffer containing no probe for washing.
7. After washing for 25 min at 46°C, cells are pelleted by centrifugation for 2 min at 4,000 × g.
8. Cells are resuspended with 300 μl of 1 × PBS (pH 7.2 for CY3 and pH 9.0 for FLUOS), immediately placed on ice, and analyzed within 3 h in the flow cytometer.

39.11. HYBRIDIZATION FOLLOWED BY TYRAMIDE SIGNAL AMPLIFICATION (CARD-FISH)

Fixation

1. Sample is concentrated on a membrane filter and air dried.
2. Dip filter in 0.2% low-gelling-point agarose, place filters face-up onto glass slides, and let air dry.
3. Dehydrate in 96% ethanol (1 min, RT) and let air dry.

NOTE: The embedding in agarose prevents the lysis of cells during subsequent permeabilization.

Inactivation of Endogenous Peroxidases

4. Incubate membrane filters with 0.01 M HCl for 10 min at RT.
5. Wash twice in distilled H_2O.
6. Let air dry.

Permeabilization

7. Incubate in lysozyme (10 mg ml^{-1}, dissolved in 0.05 M EDTA, 0.1 M Tris-HCl [pH 7.5]) at 37°C for 60 min.
8. Wash twice in distilled H_2O for 1 min at RT.
9. Wash in 96% ethanol (1 min, RT).
10. Let air dry.

Hybridization

11. Cut filters in sections.
12. Place sections in reaction vial (0.5 ml, 10 to 20 sections per vial).

NOTE: The hybridization and washing buffers used for HRP-oligos are different from the standard hybridization mixtures described above (see Table 6). First, because of the instability of the enzyme at higher temperatures, the tem-

TABLE 6 Hybridization buffer for CARD-FISH

Stock reagent	Volume	Final concn in hybridization buffer
5 M NaCl	360 μl	900 mM
1 M Tris-HCl	40 μl	20 mM
Dextran sulfate	0.2 g	10% (wt/vol)
20% SDS	2 μl	0.02% (wt/vol)
Formamide	% depending on probe (see Table 3, plus 20%)	
10% blocking reagent	400 μl	2% (wt/vol)
Distilled H₂O	Add to 2 ml	

perature for hybridization is lowered to 35°C and that for washing is lowered to 37°C. The amount of formamide is therefore 20% higher than for the standard hybridization conditions to compensate for the lower incubation temperatures. Second, the buffer contains dextran sulfate, which enhances the binding of the probe to the target, and blocking reagent to block potential unspecific binding sites of the HRP. Therefore the amount of water is adjusted accordingly. If the concentration of formamide should exceed 60%, the volume of blocking reagent is reduced.

NOTE: The hybridization buffer is prepared as follows: distilled H₂O, dextran sulfate, NaCl, sodium dodecyl sulfate (SDS), and Tris-HCl are mixed, brought into solution at 60°C, and subsequently cooled on ice. Then add formamide and blocking reagent. The hybridization buffer can be stored at −20°C for up to 3 months without any apparent effects on FISH detection rates.

13. Mix 400 μl of hybridization buffer and 4 μl of probe working solution (50 ng μl⁻¹) and add to filter sections.

14. Incubate at 35°C for 2 h.

15. Wash filters in prewarmed washing buffer (5 min, 37°C) (Table 7).

16. Do not let filter run dry.

Tyramide Signal Amplification

17. Dab filter on blotting paper to remove excess solution, but do not blot to dryness.

18. To equilibrate the probe-delivered HRP, filter sections are placed in 10 ml of 1× PBS amended with 0.05% of Triton X-100 for 15 min at room temperature.

TABLE 7 Washing buffer for CARD-FISH

Stock reagent	Volume	Final concn in hybridization buffer
5 M NaCl	Concentration depending on probe (see Table 5)	
1 M Tris-HCl	1 ml	20 mM
0.5 M EDTA	500 μl	5 mM
Distilled H₂O	Add to 50 ml	
10% SDS (added last to avoid precipitation)	50 μl	0.01%

19. Prepare substrate mix: 1 part of tyramide plus 100 to 300 parts of PBS (pH 7.6; important for proper functioning of HRP) containing 0.0015% H₂O₂.

20. Dab filter on blotting paper to remove excess solution, but do not blot to dryness; incubate in substrate mix at 37°C, 10 min, in the dark.

21. Dab filter on blotting paper to remove excess solution, but do not blot to dryness.

22. Wash in 10 ml of 1× PBS amended with 0.05% of Triton X-100 (RT, 10 min, in the dark, mild agitation).

23. Wash in 10 ml of distilled H₂O (RT, 1 min).

24. Wash in 10 ml of 96% ethanol (RT, 1 min).

25. Let air dry.

26. Cells can be counterstained with DAPI as described above or directly embedded in antifading mountant.

NOTE: At this point filters can be stored at −20°C.

39.12. TROUBLESHOOTING

Some common failure symptoms during FISH, their possible causes, and suggestions for improvements are listed in Table 8.

39.13. APPLICATIONS AND FURTHER READING

39.13.1. Methodological Aspects

Behrens and coworkers compare the native three-dimensional structure of the ribosome (6) with the data from FISH on fixed ribosomes within cells (7). Apparently the FISH procedure strongly denatures the three-dimensional structure of the ribosome (7).

39.13.2. Marine Environments

In marine planktonic research, several manuscripts deal with FISH alone or in combination with a complementary technique. A comprehensive FISH study on coastal waters was done by Eilers et al. in 2000 (15). This study explored the distribution of different phylogenetic groups in the North Sea. Fuchs and coworkers describe the combination of flow cytometric sorting and FISH for a dilution culture and point out the potential of such an approach for the open ocean (19). Karner and coworkers found a surprising dominance of Archaea in deep ocean water masses (24). Amino acid uptake by Archaea was shown by Ouverney and Fuhrman in 2000 (41). The ubiquitously found SAR11 clade was successfully brought into pure culture by a FISH-guided cultivation approach (46). A multidisciplinary approach including FISH was used to show that anaerobic ammonium oxidation (51) occurs also in anoxic sea basins like the Black Sea (25). Fuller and coworkers developed probes for the *Synechococcus* clade, a dominant prokaryotic primary producer in the marine environment (20). Cottrell and Kirchman demonstrated with their study the necessity of FISH for proving the abundance of bacterioplankton groups in the environment (11).

39.13.3. Sediment

In a pioneering work, the composition of the microbial community in the sediments of tidal mud flats (Wadden Sea) was analyzed by Llobet-Brossa et al. (28). FISH was also used to study permanently cold marine sediments (48). FISH played an important role in the discovery that structured microbial consortia perform the anaerobic oxidation of methane (10, 39).

TABLE 8 Common failure symptoms during FISH, their possible causes, and suggestions for improvements

Symptom	Cause
No cells stained	1. Probe is damaged; check probe in photometer, whether oligonucleotide and label are present in equimolar amounts; check probe on a polyacrylamide gel: only a single band representing intact probe should be visible.
	2. Cells are "overfixed": excess fixing in PFA cross-links cell wall components until cells are completely impermeable to probes; this may occur when cells are stored for a long period in PFA; try an increasing enzymatic digestion of the cells with, e.g., lysozyme, proteinase K, or achromopeptidase or incubate in 0.1 M Hcl.
	3. No SDS in hybridization buffer; SDS denatures the native ribosome structure by removing the ribosomal proteins; it helps the probe penetrating into the cell.
	4. CARD-FISH: check age of reagents, especially dextran sulfate, HRP probes, tyramides.
Unspecific staining of cells	1. Probe is damaged; check probe in photometer, whether oligonucleotide and label are present in equimolar amounts; check probe on a polyacrylamide gel: only a single band representing intact probe should be visible.
	2. Check sequence of the probe ordered.
	3. Use a pencil (not a marker) to label slides and filters; marker dyes are often fluorescent.
	4. Prepare fresh formamide stock.
FISH signal low	1. Probe target site is inaccessible; try to use helper probes (16).
	2. For probes labeled with fluorescein derivatives: check if pH of mountant/resuspension buffer exceeds pH 9.0 (fluorescence maximum of fluorescein is pH > 9.0).
	3. Try to use CARD-FISH.
	4. Use alternative mounting reagent.
No cells visible	1. Fixation was too short and cells lysed; try to fix longer or fix the remaining sample again.
	2. Lysing procedure was too harsh; reduce lysozyme concentration and incubation time.
	3. Slides were not covered with gelatin; cells were washed off.

39.13.4. Activated Sludge

FISH has been widely applied to activated sludge samples (33, 63, 64); in recent years, Crocetti et al. (12) used FISH to study polyphosphate-accumulating bacteria in a phosphate removal reactor. Filamentous activated sludge bacteria Eikelboom type 021N were further phylogenetically affiliated and characterized by FISH from the Kanagawa group (23). Schramm and coworkers coupled microelectrodes and FISH for structure–function analysis in biofilms (53, 55). A direct link of activity and identity could be achieved by the combination of microautoradiography and FISH (27, 40).

39.13.5. Limnology

Some of the earliest community studies with FISH were conducted by Pernthaler and coworkers on Lake Gossenkölle (44, 45). The complicated trophic interactions in a eutrophic reservoir were studied by a group around Simek in 1997 (57). A FISH-based comparison of dominant groups of bacterioplankton in oceans and lakes was done by Glöckner et al. (21). Grossart studied river snow by a combined approach of microsensor measurements, radiotracer experiments, and FISH (22). Community analysis of sulfate reducers in a high mountain lake were done by Tonolla and coworkers (61). Recently, the protocol for the abundant limnic *Actinobacteria* cluster was improved by Sekar and coworkers (56).

The original method development by our group reviewed in this manuscript was supported by the Max Planck Society, the German Science foundation, the Federal Ministry of Education and Research, and the European Community (BASICS, EVK3–CT-2002–00078). We acknowledge Annelie Pernthaler and Sebastian Behrens for valuable comments on earlier versions of the manuscript.

Note in proof. The protocols and references herein represent the state of the art in 2003. For updates and further information, the reader is referred to publications by the authors and others on FISH, and to www.arb-silva.de.

39.14. REFERENCES

1. **Amann, R., B. M. Fuchs, and S. Behrens.** 2001. The identification of microorganisms by fluorescence in situ hybridisation. *Curr. Opin. Biotechnol.* **12:**231–236.
2. **Amann, R. I., B. J. Binder, R. J. Olson, S. W. Chisholm, R. Devereux, and D. A. Stahl.** 1990. Combination of 16S rRNA-targeted oligonucleotide probes with flow cytometry for analyzing mixed microbial populations. *Appl. Environ. Microbiol.* **56:**1919–1925.
3. **Amann, R. I., L. Krumholz, and D. A. Stahl.** 1990. Fluorescent-oligonucleotide probing of whole cells for determinative, phylogenetic, and environmental studies in microbiology. *J. Bacteriol.* **172:**762–770.
4. **Amann, R. I., W. Ludwig, and K. H. Schleifer.** 1995. Phylogenetic identification and in situ detection of individual microbial cells without cultivation. *Microbiol. Rev.* **59:**143–169.
5. **Amann, R. I., and K.-H. Schleifer.** 2001. Nucleic acid probes and their application in environmental microbiology, p. 67–82. *In* G. Garrity, D. R. Boone, and R. W. Castenholz (ed.), *Bergey's Manual of Systematic Bacteriology,* 2nd ed, vol. 1. Springer-Verlag, New York, NY.
6. **Ban, N., P. Nissen, J. Hansen, P. B. Moore, and T. A. Steitz.** 2000. The complete atomic structure of the large ribosomal subunit at 2.4 angstrom resolution. *Science* **289:**905–920.

7. **Behrens, S., B. M. Fuchs, F. Mueller, and R. Amann.** 2003. Is the in ditu accessibility of the 16S rRNA of *Escherichia coli* for Cy3-labeled oligonucleotide probes predicted by a three-dimensional structure model of the 30S ribosomal subunit? *Appl. Environ. Microbiol.* **69:**4935–4941.

8. **Behrens, S., C. Ruhland, J. Inacio, H. Huber, A. Fonseca, I. Spencer-Martins, B. M. Fuchs, and R. Amann.** 2003. In situ accessibility of small-subunit rRNA of members of the domains bacteria, Archaea, and Eucarya to Cy3-labeled oligonucleotide probes. *Appl. Environ. Microbiol.* **69:**1748–1758.

9. **Benson, D. A., I. Karsch-Mizrachi, D. J. Lipman, J. Ostell, B. A. Rapp, and D. L. Wheeler.** 2002. GenBank. *Nucleic Acids Res.* **30:**17–20.

10. **Boetius, A., K. Ravenschlag, C. J. Schubert, D. Rickert, F. Widdel, A. Gieseke, R. Amann, B. B. Jorgensen, U. Witte, and O. Pfannkuche.** 2000. A marine microbial consortium apparently mediating anaerobic oxidation of methane. *Nature* **407:**623–626.

11. **Cottrell, M. T., and D. L. Kirchman.** 2000. Community composition of marine bacterioplankton determined by 16S rRNA gene clone libraries and fluorescence in situ hybridization. *Appl. Environ. Microbiol.* **66:**5116–5122.

12. **Crocetti, G. R., P. Hugenholtz, P. L. Bond, A. Schuler, J. Keller, D. Jenkins, and L. L. Blackall.** 2000. Identification of polyphosphate-accumulating organisms and design of 16S rRNA-directed probes for their detection and quantitation. *Appl. Environ. Microbiol.* **3:**1175–1182.

13. **DeLong, E., L. Taylor, T. Marsh, and C. Preston.** 1999. Visualization and enumeration of marine planktonic archaea and bacteria by using polyribonucleotide probes and fluorescent in situ hybridization. *Appl. Environ. Microbiol.* **65:**5554–5563.

14. **DeLong, E. F., G. S. Wickham, and N. R. Pace.** 1989. Phylogenetic stains: ribosomal RNA-based probes for the identification of single cells. *Science* **243:**1360–1363.

15. **Eilers, H., J. Pernthaler, F. O. Glockner, and R. Amann.** 2000. Culturability and in situ abundance of pelagic bacteria from the North Sea. *Appl. Environ. Microbiol.* **66:**3044–3051.

16. **Fuchs, B. M., F. O. Glöckner, J. Wulf, and R. Amann.** 2000. Unlabeled helper oligonucleotides increase the in situ accessibility to 16S rRNA of fluorescently labeled oligonucleotide probes. *Appl. Environ. Microbiol.* **66:**3603–3607.

17. **Fuchs, B. M., K. Syutsubo, W. Ludwig, and R. Amann.** 2001. In situ accessibility of *Escherichia coli* 23S rRNA to fluorescently labeled oligonucleotide probes. *Appl. Environ. Microbiol.* **67:**961–968.

18. **Fuchs, B. M., G. Wallner, W. Beisker, I. Schwippl, W. Ludwig, and R. Amann.** 1998. Flow cytometric analysis of the in situ accessibility of *Escherichia coli* 16S rRNA for fluorescently labeled oligonucleotide probes. *Appl. Environ. Microbiol.* **64:**4973–4982.

19. **Fuchs, B. M., M. V. Zubkov, K. Sahm, P. H. Burkill, and R. Amann.** 2000. Changes in community composition during dilution cultures of marine bacterioplankton as assessed by flow cytometric and molecular biological techniques. *Environ. Microbiol.* **2:**191–202.

20. **Fuller, N. J., D. Marie, F. Partensky, D. Vaulot, A. F. Post, and D. J. Scanlan.** 2003. Clade-specific 16S ribosomal DNA oligonucleotides reveal the predominance of a single marine Synechococcus clade throughout a stratified water column in the Red Sea. *Appl. Environ. Microbiol.* **69:**2430–2443.

21. **Glöckner, F. O., B. M. Fuchs, and R. Amann.** 1999. Bacterioplankton composition in lakes and oceans: a first comparison based on fluorescence in situ hybridization. *Appl. Environ. Microbiol.* **65:**3721–3726.

22. **Grossart, H. P., and H. Ploug.** 2000. Bacterial production and growth efficiencies: direct measurements on riverine aggregates. *Limnol. Oceanogr.* **45:**436–445.

23. **Kanagawa, T., Y. Kamagata, S. Aruga, T. Kohno, M. Horn, and M. Wagner.** 2000. Phylogenetic analysis of and oligonucleotide probe development for Eikelboom type 021N filamentous bacteria isolated from bulking activated sludge. *Appl. Environ. Microbiol.* **66:**5043–5052.

24. **Karner, M. B., E. F. DeLong, and D. M. Karl.** 2001. Archaeal dominance in the mesopelagic zone of the Pacific Ocean. *Nature* **409:**507–510.

25. **Kuypers, M. M. M., A. O. Sliekers, G. Lavik, M. Schmid, B. B. Joergensen, J. G. Kuenen, J. S. S. Damste, M. Strous, and M. S. M. Jetten.** 2003. Anaerobic ammonium oxidation by anammox bacteria in the Black Sea. *Nature* **422:**608–611.

26. **Lathe, R.** 1985. Synthetic oligonucleotide probes deduced from amino acid sequence data: theoretical and practical considerations. *J. Mol. Biol.* **183:**1–12.

27. **Lee, N., P. Nielsen, K. Andreasen, S. Juretschko, J. Nielsen, K. Schleifer, and M. Wagner.** 1999. Combination of fluorescent in situ hybridization and microautoradiography—a new tool for structure-function analyses in microbial ecology. *Appl. Environ. Microbiol.* **65:**1289–1297.

28. **Llobet-Brossa, E., R. Rosselló-Mora, and R. Amann.** 1998. Microbial community composition of Wadden Sea sediments as revealed by fluorescence in situ hybridization. *Appl. Environ. Microbiol.* **64:**2691–2696.

29. **Loy, A., M. Horn, and M. Wagner.** 2003. probeBase—an online resource for rRNA-targeted oligonucleotide probes. *Nucleic Acids Res.* **31:**514–516.

30. **Maidak, B. L., J. R. Cole, T. G. Lilburn, C. T. J. Parker, P. R. Saxmann, R. J. Farris, G. M. Garrity, G. J. Olsen, T. M. Schmidt, and J. M. Tiedje.** 2001. The RDP II (Ribosomal Database Project). *Nucleic Acids Res.* **29:**173–174.

31. **Manz, W., R. Amann, W. Ludwig, M. Vancanneyt, and K.-H. Schleifer.** 1996. Application of a suite of 16S rRNA-specific oligonucleotide probes designed to investigate bacteria of the phylum *Cytophaga-Flavobacter-Bacteroides* in the natural environment. *Microbiology* **142:**1097–1106.

32. **Manz, W., R. Amann, W. Ludwig, M. Wagner, and K.-H. Schleifer.** 1992. Phylogenetic oligodeoxynucleotide probes for the major subclasses of proteobacteria: problems and solutions. *Syst. Appl. Microbiol.* **15:**593–600.

33. **Manz, W., M. Wagner, R. Amann, and K.-H. Schleifer.** 1994. In situ characterization of the microbial consortia active in two wastewater treatment plants. *Water Res.* **28:**1715–1723.

34. **Meier, H., R. Amann, W. Ludwig, and K. Schleifer.** 1999. Specific oligonucleotide probes for in situ detection of a major group of Gram-positive bacteria with low DNA G+C content. *Syst. Appl. Microbiol.* **22:**186–196.

35. **Neef, A.** 1997. Application of in situ identification of bacteria to population analysis in complex microbial communities. Ph.D. dissertation. Technical University of Munich, Munich, Germany.

36. **Neef, A., R. Amann, H. Schlesner, and K.-H. Schleifer.** 1998. Monitoring a widespread bacterial group: in situ detection of planctomycetes with 16S rRNA-targeted probes. *Microbiology* **144:**3257–3266.

37. **Neef, A., A. Zaglauer, H. Meier, R. Amann, H. Lemmer, and K. H. Schleifer.** 1996. Population analysis in a denitrifying sand filter: conventional and in situ identification of *Paracoccus* spp. in methanol-fed biofilms. *Appl. Environ. Microbiol.* **62:**4329–4339.

38. **Olsen, G. J., D. J. Lane, S. J. Giovannoni, N. R. Pace, and D. A. Stahl.** 1986. Microbial ecology and evolution:

a ribosomal rRNA approach. *Annu. Rev. Microbiol.* **40:** 337–365.

39. **Orphan, V. J., C. H. House, K. U. Hinrichs, K. D. McKeegan, and D. F. Delong.** 2001. Methane-consuming archaea revealed by directly coupled isotopic and phylogenetic analysis. *Science* **293:**484–487.

40. **Ouverney, C. C., and J. A. Fuhrman.** 1999. Combined microautoradiography-16S rRNA probe technique for determination of radioisotope uptake by specific microbial cell types in situ. *Appl. Environ. Microbiol.* **65:**1746–1752.

41. **Ouverney, C. C., and J. A. Fuhrman.** 2000. Marine planktonic Archaea take up amino acids. *Appl. Environ. Microbiol.* **66:**4829–4833.

42. **Pernthaler, A., J. Pernthaler, and R. Amann.** 2002. Fluorescence in situ hybridization and catalyzed reporter deposition for the identification of marine bacteria. *Appl. Environ. Microbiol.* **68:**3094–3101.

43. **Pernthaler, A., C. M. Preston, J. Pernthaler, E. F. DeLong, and R. Amann.** 2002. Comparison of fluorescently labeled oligonucleotide and polynucleotide probes for the detection of pelagic marine Bacteria and Archaea. *Appl. Environ. Microbiol.* **68:**661–667.

44. **Pernthaler, J., A. Alfreider, T. Posch, S. Andreatta, and R. Psenner.** 1997. In situ classification and image cytometry of pelagic bacteria from a high mountain lake (Gossenköllesee, Austria). *Appl. Environ. Microbiol.* **63:** 4778–4783.

45. **Pernthaler, J., F. O. Glöckner, S. Unterholzner, A. Alfreider, R. Psenner, and R. Amann.** 1998. Seasonal community and population dynamics of pelagic bacteria and archaea in a high mountain lake. *Appl. Environ. Microbiol.* **64:**4299–4306.

46. **Rappe, M. S., S. A. Connon, K. L. Vergin, and S. J. Giovannoni.** 2002. Cultivation of the ubiquitous SAR11 marine bacterioplankton clade. *Nature* **6898:**630–632.

47. **Raskin, L., B. E. Rittmann, and D. A. Stahl.** 1996. Competition and coexistence of sulfate-reducing and methanogenic populations in anaerobic biofilms. *Appl. Environ. Microbiol.* **62:**3847–3857.

48. **Ravenschlag, K., K. Sahm, C. Knoblauch, B. B. Jorgensen, and R. Amann.** 2000. Community structure, cellular rRNA content, and activity of sulfate-reducing bacteria in marine Arctic sediments. *Appl. Environ. Microbiol.* **66:**3592–3602.

49. **Roller, C., M. Wagner, R. Amann, W. Ludwig, and K.-H. Schleifer.** 1994. *In situ* probing of Gram-positive bacteria with high DNA G+C content using 23S rRNA-targeted oligonucleotides. *Microbiology* **140:**2849–2858.

50. **Sambrook, J., E. F. Fritsch, and T. Maniatis.** 1989. *Molecular Cloning: a Laboratory Manual*, 2nd ed, vol. 3, p. 11.21. Cold Spring Harbor Laboratory Press, Cold Spring Harbor, NY.

51. **Schmid, M., U. Twachtmann, M. Klein, M. Strous, S. Juretschko, M. Jetten, J. W. Metzger, K. H. Schleifer, and M. Wagner.** 2000. Molecular evidence for genus level diversity of bacteria capable of catalyzing anaerobic ammonium oxidation. *Syst. Appl. Microbiol.* **23:**93–106.

52. **Schönhuber, W., B. Fuchs, S. Juretschko, and R. Amann.** 1997. Improved sensitivity of whole-cell hybridization by the combination of horseradish peroxidase-labeled oligonucleotides and tyramide signal amplification. *Appl. Environ. Microbiol.* **63:**3268–3273.

53. **Schramm, A., D. de Beer, J. C. van den Heuvel, S. Ottengraf, and R. Amann.** 1999. Microscale distribution of populations and activities of *Nitrosospira* and *Nitrospira* spp. along a macroscale gradient in a nitrifying bioreactor: quantification by in situ hybridization and the use of microsensors. *Appl. Environ. Microbiol.* **65:**3690–3696.

54. **Schramm, A., B. M. Fuchs, J. L. Nielsen, M. Tonolla, and D. A. Stahl.** 2002. Fluorescence in situ hybridization of 16S rRNA gene clones (Clone-FISH) for probe validation and screening of clone libraries. *Environ. Microbiol.* **4:**713–720.

55. **Schramm, A., L. H. Larsen, N. P. Revsbech, N. B. Ramsing, R. Amann, and K.-H. Schleifer.** 1996. Structure and function of a nitrifying biofilm as determined by in situ hybridization and the use of microelectrodes. *Appl. Environ. Microbiol.* **62:**4641–4647.

56. **Sekar, R., A. Pernthaler, J. Pernthaler, F. Warnecke, T. Posch, and R. Amann.** 2003. An improved protocol for quantification of freshwater actinobacteria by fluorescence in situ hybridization. *Appl. Environ. Microbiol.* **69:**2928–2935.

57. **Simek, K., P. Hartman, J. Nedoma, J. Pernthaler, D. Springmann, J. Vrba, and R. Psenner.** 1997. Community structure, picoplankton grazing and zooplankton control of heterotrophic nanoflagellates in a eutrophic reservoir during the summer phytoplankton maximum. *Aquat. Microb. Ecol.* **12:**49–63.

58. **Stahl, D. A., and R. Amann.** 1991. Development and application of nucleic acid probes, p. 205–248. *In* E. Stackebrandt and M. Goodfellow (ed.), *Nucleic Acid Techniques in Bacterial Systematics.* John Wiley & Sons Ltd., Chichester, United Kingdom.

59. **Stoesser, G., W. Baker, A. van den Broek, E. Camon, M. Garcia-Pastor, C. Kanz, T. Kulikova, R. Leinonen, Q. Lin, V. Lombard, R. Lopez, N. Redaschi, P. Stoehr, M. A. Tuli, K. Tzouvara, and R. Vaughan.** 2002. The EMBL Nucleotide Sequence Database. *Nucleic Acids Res.* **30:**21–26.

60. **Strunk, O., R. Westram, H. Meier, G. Jobb, L. Richter, M. May, S. Hermann, N. Stuckmann, O. Gross, B. Nonhoff, M. Lenke, R. Jost, B. Reichel, W. Foerster, T. Ginhart, A. Vilbig, S. Gerbers, and W. Ludwig.** 2002. ARB: a software environment for sequence data. [Online.] www.arb-home.de.

61. **Tonolla, M., A. Demarta, S. Peduzzi, D. Hahn, and R. Peduzzi.** 2000. In situ analysis of sulfate-reducing bacteria related to *Desulfocapsa thiozymogenes* in the chemocline of meromictic Lake Cadagno (Switzerland). *Appl. Environ. Microbiol.* **66:**820–824.

62. **Trebesius, K., R. Amann, W. Ludwig, K. Mühlegger, and K.-H. Schleifer.** 1994. Identification of whole fixed bacterial cells with nonradioactive 23S rRNA-targeted polynucleotide probes. *Appl. Environ. Microbiol.* **60:**3228–3235.

63. **Wagner, M., R. Amann, H. Lemmer, and K.-H. Schleifer.** 1993. Probing activated sludge with oligonucleotides specific for proteobacteria: inadequacy of culture-dependent methods for describing microbial community structure. *Appl. Environ. Microbiol.* **59:**1520–1525.

64. **Wallner, G., R. Erhart, and R. Amann.** 1995. Flow cytometric analysis of activated sludge with rRNA-targeted probes. *Appl. Environ. Microbiol.* **61:**1859–1866.

65. **Woese, C. R.** 1987. Bacterial evolution. *Microbiol. Rev.* **51:**221–271.

40

Measurement of rRNA Abundance by Hybridization with Oligodeoxynucleotide Probes

DANIEL H. BUCKLEY AND THOMAS M. SCHMIDT

40.1. BACKGROUND INFORMATION

40.1.1. Introduction

The past 20 years have been witness to a phenomenal revolution in our understanding of microbial diversity. During this time, the number of recognized bacterial groups has more than quadrupled and newly discovered genera have become legion (15). Two primary driving forces have been behind this revolution. The first was the discovery that the sequences of rRNA genes could be used to infer the evolutionary history of microorganisms, and the second was the invention of PCR. With these tools, it has been possible to recover rRNA genes directly from natural environments and determine the phylogenetic diversity of the microorganisms present. Numerous studies have used this approach to discover novel groups of microorganisms that had previously escaped detection because they had not been cultivated in the laboratory (for review see references 15 and 29). Many of these novel microbial groups are widespread and numerically abundant in our biosphere, and their ac-

tivities may have far-reaching impacts on ecosystems at local, regional, or even global scales. As a result, numerous techniques have evolved that allow for the study of microorganisms as they occur in their natural habitats and without the need for cultivation.

Two fundamental questions that need to be answered when examining a natural microbial community are: "Who is there?" and "What is the relative abundance of microbial populations?" More formally, these questions refer to two components of community diversity: richness and evenness, where richness is the total number of different species present and evenness represents the relative proportions of those species (16). Molecular techniques for studying microbial communities may be more or less well suited to answering these two questions. For example, techniques that utilize PCR provide qualitative data on the richness of microbial communities, but because of several types of PCR artifacts that can cause the relative proportion of rRNA genes to change during PCR amplification (10, 13, 48, 51), these techniques may not be suitable for quantitative analyses of the evenness of communities. Quantitative measures of microbial abundance can be obtained by designing DNA probes that target either rRNA or the rRNA-encoding gene (rDNA) sequences from particular microorganisms or microbial groups. These probes can be hybridized to nucleic acids in intact cells by fluorescent in situ hybridization (FISH) as described in chapter 39, or they can be hybridized to nucleic acids extracted from a microbial community as described in this chapter. In both approaches, hybridization of probes generates quantitative data on microbial abundance that can provide valuable insight on the ecology of microorganisms and microbial communities.

This chapter provides background information and basic methodologies required to carry out the measurement of rRNA abundance by hybridization with oligonucleotide probes. Much of the discussion in this chapter focuses on hybridization to RNA, but many of the methods presented are equally applicable to DNA. This chapter addresses the extraction of RNA from environmental samples, the design of oligonucleotide probes, probe labeling and hybridization, and methods for calculating rRNA abundance from hybridization data. Several methodological studies have evaluated the use of RNA hybridization in microbial ecology and have contributed to many of the protocols and methodological caveats presented in this chapter (1, 7, 9, 21, 32–34, 46). Wherever possible, we have tried to provide information on the assumptions and potential biases that exist in the methods, to allow the reader to evaluate the suitability of these methods toward their individual research objectives.

40.1.2. Factors That Influence rRNA Abundance

40.1.2.1. rRNA and Growth Rate

It is important to note that the measurement of rRNA abundance does not provide a measure of microbial cell numbers in the same manner as would direct cell counts made by FISH. rRNA abundance does not correspond to cell numbers, because the number of ribosomes per cell is regulated in response to cellular growth rate (18, 39, 43). As a result, rapidly growing cells tend to have higher numbers of ribosomes than the same cells when growing at a slower rate. In addition, different species tend to carry different numbers of ribosomes at any given growth rate depending on their particular needs for protein synthesis. The amount of rRNA present in an environment is therefore a function of both the number of cells present in the environment and their mean growth rate at the time of sampling. rRNA abundance of each sequence type therefore provides a measure of the relative contribution of each actively growing community member to the protein synthetic capacity of the entire community (50).

40.1.2.2. Measurement of rRNA versus rDNA

Although the number of rRNA molecules can vary from 10^3 to 10^5 per cell depending on species and growth rate, the number of rRNA genes (rDNA) present per genome varies little for a given species. The genes for the 5S rRNA, the 16S rRNA, and the 23S rRNA tend to be organized into operons. Different microbial species have different characteristic numbers of rRNA operons. Although many microbial species have only one or a few rRNA operons per genome, others have as many as 15 (19). As a result, the amount of rDNA present in an environment cannot be

used to directly estimate cell number unless the number of rRNA genes has previously been determined for the target organisms. The rRNA operon copy number for an organism can be determined by performing a Southern hybridization while using a probe specific for the rRNA genes; alternatively, a compilation of rRNA operons previously determined is provided by Klappenbach et al. (20).

rRNA rather than rDNA is the target of choice for most hybridization studies primarily because there are thousands or tens of thousands more copies of a given rRNA molecule per cell than there are copies of the corresponding rDNA. As a practical concern, this increased number of target molecules greatly enhances the ability to detect a particular molecular signature in an environmental sample. The use of rDNA as a target for hybridization in microbial ecology occurs most frequently in attempts to determine the in situ growth rate of specific microorganisms (21, 26, 30). As the cellular concentration of rRNA is correlated with growth rate, there is a direct relationship between the rRNA/rDNA ratio in a cell and cellular growth rate that is characteristic for the growth of an organism over a wide range of growth rates. Provided that it is possible to determine the rRNA/rDNA ratio for a particular species over a range of growth rates in the laboratory, it should be possible to estimate the in situ growth rate of the species by extracting community nucleic acids and using hybridization with species-specific probes to determine the amount of rRNA and rDNA from the given species that is present in the environment.

40.1.3. Basic Approach to Measuring rRNA Abundance

40.1.3.1. Overview

rRNA hybridization is a method that is useful for determining the relative physiological activity of microorganisms in the environment (3, 38, 50). The sequence conservation of rRNA genes makes it possible to design oligonucleotide probes that target the rRNA molecules from specific groups of organisms. These probes can be used to quantify the abundance of specific rRNAs in a microbial community. Once an oligonucleotide probe for a specific microbial group has been designed and tested empirically, rRNA abundance can be determined by radioactively labeling the probe, allowing the probe to hybridize to an RNA sample, and then measuring the amount of radioactivity associated with the RNA sample.

40.1.3.2. Quantity of RNA Needed for Probe Hybridization Assays

Probe hybridization assays require considerably more nucleic acids than PCR-based assays. While under ideal circumstances the detection limit of hybridization should be less than 1 pg of target nucleic acid, environmental nucleic acid extracts are rarely ideal samples. For nucleic acids extracted from environmental samples, actual detection limits will vary depending on sample characteristics. The presence of inhibitors and nonhomologous nucleic acids can raise the actual detection limit of end-labeled oligonucleotide probes to approximately 100 pg of 16S rRNA. As a result, typical amounts of community RNA added to individual slots in slot-blot hybridizations range from 10 ng to 1 to 3 μg. The total amount of RNA needed to perform hybridization analyses on a given environmental sample will ultimately depend on the number of different probes to be tested and the number of replications per probe.

40.1.3.3. Potential Pitfalls

The two most important considerations when measuring rRNA abundance by hybridization are the quality of RNA extracted from environmental samples and the hybridization conditions used to ensure that probes achieve their desired specificity. Issues associated with probe specificity are covered in depth in section 40.3, so only issues related to the quality of rRNA will be discussed in this section.

Several problems can arise during the extraction of RNA from environmental samples. The first problem that may be encountered is the inability to obtain sufficient quantities of RNA to perform the desired analyses. Increasing the size or the number of samples used to generate RNA best solves this problem. A second problem associated with RNA extracts results from RNA degradation. Degradation of RNA (as described in sections 40.1.4 and 40.2.1.3) can reduce the availability of targets for hybridization, but a more insidious problem can result if selective degradation of certain portions of rRNA molecules occurs. As described in section 40.4, it is usually advisable to relate rRNA abundance measurements made with a probe that targets a specific microbial group to measurements made with a probe that targets all rRNA molecules in a community. The two probes will usually target different regions of the rRNA molecule. Thus, if selective degradation of these probe target regions occurs, then measurements of rRNA relative abundance can be either increased or decreased artificially. It is therefore necessary to take precautions to prevent RNA degradation (as described in sections 40.1.4 and 40.2.1.3). The last problem associated with RNA extracts is caused by the presence of inhibitory compounds that copurify with RNA. Discussion of the potential problems associated with inhibitors of hybridization and methods for overcoming these problems are presented in section 40.4.1.

40.1.4. RNA Handling Considerations

Very stable RNase enzymes that rapidly degrade RNA are everywhere, on hands, lab counters, and other contact surfaces. Care must be taken to ensure that all materials that might come in contact with extracted RNA are free of RNases. All glassware and heat-resistant materials (e.g., glass wool, spatulas) should be baked in an oven at greater than 150°C for 5 h before use. Disposable plasticware should be used wherever possible, and all pipette tips and microcentrifuge tubes should be handled with gloved hands prior to autoclaving. During sample handling, care should be taken to prevent contamination of any materials with bare hands and untreated laboratory surfaces. Water used during extraction and purification can be treated by adding 0.2% diethylpyrocarbonate (DEPC), stirring for at least 2 h to disperse DEPC droplets, and autoclaving. All solutions and buffers should be made up with DEPC-treated water (36).

40.2. RNA EXTRACTION

40.2.1. Basic Considerations

The techniques used to isolate RNA from environmental samples share many of the concerns associated with the extraction of DNA from environmental samples, as described in chapter 41. The primary difference between RNA and DNA extractions stems from the fact that RNA is much more labile than DNA and so RNA extraction protocols must be designed to prevent RNA degradation. The extraction of RNA from environmental samples has four primary considerations: cell lysis, RNA stabilization, yield, and purity. Relative to techniques depending on DNA and RNA amplification, RNA hybridization requires a large amount of nucleic acid (up to 1 to 3 µg of community RNA per hybridization), and so extraction protocols need to be designed to provide large quantities of RNA. In addition, when extracting RNA from an environment such as soil, humic acids in the soil tend to copurify with nucleic acids. These contaminating humic acids can inhibit nucleic acid hybridization (2, 49), so extraction protocols must be designed to eliminate or at least minimize the presence of humic acids. The environmental RNA extraction protocol described in this section was designed to yield RNA of sufficient quantity and quality to allow multiple RNA hybridization experiments. Though the protocol was designed for soil, it should be easily adaptable to provide RNA from other environmental samples.

40.2.1.1. Comparison between RNA and DNA Extraction

Fewer methods for RNA extraction from environmental samples have been published than for DNA extraction. The differences between RNA extraction protocols and DNA extraction protocols are necessitated by the need for steps to both inactivate and prevent contamination from RNA-degrading RNases. RNases are both ubiquitous and notoriously hardy enzymes, and special precautions need to be taken to ensure that laboratory equipment and solutions are free of RNase contamination (see section 40.1.4). Solutions used for RNA extraction frequently contain powerful chaotropic agents such as guanidine isothiocyanate or guanidine hydrochloride to inactivate RNases either present in the environmental matrix or originating from lysed bacteria (17). Another difference between RNA and DNA extraction protocols occurs when an organic extraction (with phenol or phenol/CHCl$_3$) is required, as a neutral or alkaline pH will cause both RNA and DNA to partition into the aqueous phase during an organic extraction, while a more acidic pH (as used in most RNA extractions) will cause DNA to be localized to the interface between the aqueous and organic phase (17). When DNA-free RNA or RNA-free DNA is needed, it is common practice to include either a DNase or an RNase step, respectively. Methods for simultaneous extraction of DNA and RNA from soils have also been described (12, 31, 41). The choice of a particular method will depend on the experimental objectives, the amount and type of nucleic acid needed for analysis, the chemical and physical properties of the samples, and characteristics of the microorganisms under study.

40.2.1.2. Cell Lysis

Most extraction methods combine chemical, enzymatic, and/or physical treatments to achieve efficient cell lysis. An ideal extraction buffer causes all microorganisms in the sample to become detached from soil colloids, renders cells susceptible to breakage, maintains the integrity of nucleic acids released from cells by inactivating all nuclease enzymes, and prevents adsorption of nucleic acids to the environmental matrix. Typically, soils or sediments are suspended in alkaline buffers (pH 8) containing high concentrations of Na salts to promote osmotic disruption and disperse soil colloids. Phosphate buffers (100 to 200 mM) are commonly used for direct extraction, as the excess phosphate is thought to compete with nucleic acid binding to clays in soil (28), though higher phosphate concentrations may lead to the formation of insoluble phosphate precipitates. Detergents such as sodium dodecyl sulfate (SDS) or sarcosine are used to destabilize cell membranes and render cells more susceptible to breakage, while chaotropic agents such as guanidine isothiocyanate or guanidine hydrochloride are used to inactivate the enzymes that degrade RNA. Once the soil or sediment particles are suspended and dispersed, they are subjected to one or more physical treatments to achieve further cell lysis. Physical treatments include heating at 80°C, repeated freeze-thawing, sonication, and bead-beating, either with glass or silica-zirconium beads. Methods that rely on mechanical lysis through bead mill homogenization provide higher rates of cell disruption than most other methods (25).

Cell lysis efficiency affects RNA yield, but more importantly, the cell lysis procedure must be designed to minimize the possibility of differential cell lysis. If certain microbial groups are less susceptible to lysis than others and the lysis technique used is not sufficiently rigorous, certain microbial groups may be consistently underrepresented in extracted RNA. To ensure that the RNA present in extracts is representative of the indigenous microbial community, high cell disruption efficiencies must be achieved. The most effective way to assess cell lysis efficiency is by comparing direct microscopic counts of soil smears before lysis treatments to direct counts of the extracted pellets after treatment. In preparing soil smears, do not apply more than 0.25 mg of dry soil per cm^2 of slide area, to prevent masking of cells by soil particles (5). Suitable cell stains are acridine orange, 4',6'-diamidino-2-phenyindole (DAPI), or DTAF 5-[4,6-dichlorotriazin-2-yl]amino fluorescein (DTAF) (5).

The extraction protocol described in section 40.2.3 relies on bead mill homogenization in the presence of a strong chaotropic agent. Microscopic cell counts of DTAF-stained cells made before and after homogenization revealed that 97.3% ± 1.6% (standard error) of cells in selected soil samples were disrupted (7). The observed lysis efficiency is consistent with previous measurements of lysis efficiency obtained for bead mill homogenization of soil samples (25). Thus, while the lysis protocol employed does not provide complete cell lysis, the disruption rate is high enough to ensure that any bias caused by differential cell lysis will be kept below a few percent of the total rRNA abundance.

40.2.1.3. RNA Yield and Purity

The removal of humic acids from RNA extracts requires extensive purification steps including nucleic acid precipitation with polyethylene glycol and purification on both hydroxyapatite and Sephadex G-75 columns. Assessment of humic acid contamination of RNA extracts requires measuring sample light absorbance at 230 nm and 260 nm. As humics absorb light strongly at 230 nm and nucleic acid absorbs light at 260 nm, increasing A_{260}/A_{230} ratios tend to indicate increasing sample purity (52). Total RNA concentrations in extracts can be estimated by using the orcinol reaction to determine ribose concentrations (11). Sizes of extracted RNA can also be evaluated by denaturing agarose gel electrophoresis (37). If total RNA has not been sheared, distinct bands for 23S and 16S subunit RNAs can be visualized.

40.2.2. RNA Extraction from Cultures

The following is a fairly rapid protocol suitable for the extraction of total RNA from pure cultures. To maximize the yield of rRNA, cultures should be harvested late in the exponential phase of growth. All solutions should be prepared with RNase-free water, and standard precautions should be taken at all steps to prevent RNase contamination of samples (see section 40.1.4).

1. Pellet cells from 2 to 5 ml of culture by spinning at 5,000 × g for 15 min at 4°C (50 to 100 mg wet weight of cells is required).

2. Resuspend cell pellet in 0.7 ml of acetate buffer.

3. In a 2.2-ml conical tube on ice, combine the 0.7 ml of cells suspension from above with 0.7 ml of phenol, (pH 5.1), 50 μl of 20% SDS, and 0.7 ml of silica-zirconium beads (0.1 mm) (beads should be baked as glassware, to destroy RNase enzymes).

4. Place in a bead beater (Mini-Beadbeater1 [no. 3110BX] or Mini-Beadbeater8 [no. 693] from BioSpec Products [www.biospec.com], or FastPrep Instrument [FP120A] from QBiogene [previously Bio101; www.qbiogene.com] and shake for 2 min at the highestspeed setting (it may be necessary to repeat this step for organisms that are difficult to lyse).

5. Spin tube at 500 × g in a centrifuge for 10 min (CAUTION: spinning at higher speeds may cause beads to rupture tube).

6. Transfer the aqueous phase to a fresh 1.5-ml Eppendorf tube containing 0.7 ml of phenol/chloroform/isoamyl alcohol (pH 5.1), mix well, and then spin at 14,000 × g for 5 min. Perform this step a total of two times.

7. Transfer the aqueous phase to a fresh 1.5-ml Eppendorf tube containing 0.7 ml of chloroform, mix well, and then spin at 14,000 × g for 5 min. Perform this step for a total of two times.

8. Transfer the aqueous phase to a fresh 1.5-ml Eppendorf tube and precipitate RNA by adding 0.1 volume of 3 M sodium acetate and 2 volumes of 100% ethanol, mix well, and then spin at 14,000 × g for 20 to 30 min in a centrifuge at 4°C.

9. Decant carefully and resuspend in 1 ml of 70% ethanol, mix well, and then spin at 14,000 × g for 5 to 10 min.

10. Decant carefully and dry briefly under a vacuum. RNA will not form a pellet like DNA, but rather will be distributed on the lower wall of the tube. Resuspend RNA in 250 μl of RNase-free water.

Solutions

Acetate Buffer. 50 mM sodium acetate, 10 mM EDTA, pH to 5.1.

Phenol pH 5.1. Add acetate buffer to an equal volume of phenol and shake well. Allow phases to separate and check the pH of the buffer. Remove the buffer and repeat until the buffer pH after mixing is 5.1.

Phenol/chloroform/isoamyl alcohol (pH 5.1). Mix volumes of phenol, chloroform, and isoamyl alcohol at a 25:24:1 ratio. Equilibrate at pH 5.1 with acetate buffer as for phenol.

40.2.3. Extraction of RNA from Soil

Using the protocol below, RNA can be extracted from eight 10-g samples in approximately 8 to 10 h (7). All solutions should be prepared using RNase-free water, and standard precautions should be taken to prevent RNase contamination of samples at all steps.

Extraction of RNA

1. Add the following to the chamber of a beadbeater (Beadbeater [1107900] with small chamber kit [110803] from BioSpec Products [www.biospec.com]): 10 g of soil, 20 g of 0.1-mm silica/zirconium beads (BioSpec Products), and 20 ml of soil homogenization buffer (SHB).

2. Bead beat for two 1-min cycles on ice.

3. Decant soil slurry to a sterile 50-ml tube on ice.

4. Centrifuge at 3,000 × g for 5 min.

5. Decant supernatant fluid into another sterile 50-ml tube on ice.

6. Resuspend soil pellet in 10 ml of SHB.

7. Repeat steps 4 and 5, pooling the supernatant fractions on ice.

8. Add 0.1 volume of 5 M NaCl and 0.5 volumes of 50% polyethylene glycol (PEG 8000).

9. Mix well and incubate on ice for 2 h.

Sample Concentration and Phenol Extraction

10. Centrifuge sample for 30 min, 4°C, at 12,000 × g.

11. Discard supernatant fluid and carefully rinse pellet with cold 70% ethanol.

12. Discard ethanol and resuspend pellet in 2 ml of 120 mM sodium phosphate buffer (pH 7.2) containing 0.7 M NaCl.

13. To each of two 2-ml microcentrifuge tubes, add 1 ml of extract and 1/10 volume of 10% hexadecyltrimethylammonium bromide (CTAB) solution containing 0.7 M NaCl.

14. Heat at 60°C for 5 min.

15. Add 0.7 ml of phenol/chloroform/isoamyl alcohol (see section 40.2.2 for details on preparing solution) and mix well.

16. Centrifuge for 5 min at 15,000 × g, and collect the aqueous phase.

Hydroxyapatite Column Purification

17. Add pooled sample gently to the top of a hydroxyapatite (HPT) column.

18. Centrifuge for 2 min at 100 × g, or until sample completely passes through the column.

19. Wash the column three times with 1 ml of 120 mM sodium phosphate (pH 7.2).

20. Place a 1.5-ml tube with cap removed under the syringe barrel. Add 1 ml of 300 mM K$_2$HPO$_4$ and elute at 100 × g for 4 min.

Sephadex Desalting and Purification

21. Gently add eluent from HPT column to the top of the packed Sephadex column.

22. Place a 1.5-ml tube under the column and spin at 1,400 × g for 6 min at room temperature.

23. Precipitate the RNA overnight at −20°C with 1/10 volume of 3 M sodium acetate and 0.6 volume of isopropyl alcohol.

24. Resuspend the RNA in 200 μl of DEPC-treated water.

Preparing Hydroxyapatite (HPT) Columns

Prepare a slurry of HTP (Bio-Rad HTP BioGel, not DNA-grade) in 10 mM sodium phosphate (pH 7.2). Add the slurry to a 3-ml syringe barrel that has been plugged with baked glass wool until the bed volume is 2.5 ml. Centrifuge the syringe in a 15-ml Corex glass tube at 100 × g for 2 min to pack the column, which should have a final volume of 1.5 to 2 ml. Wash the column three times with 1 of ml 120 mM sodium phosphate (pH 7.2). Columns are ready to use and may be stored at 4°C for several days, if needed.

Preparing Sephadex G-75 Columns

Prepare a slurry of Sephadex G-75 by mixing 4 g of Sephadex with 50 ml of DEPC-treated water. Allow the Sephadex to swell overnight or by autoclaving for 15 to 30

min. Plug 3-ml syringe barrels with baked glass wool, and slowly add Sephadex until the syringe is completely filled. Wash the column three times with 1 ml of DEPC-treated water by centrifugation at 1,400 × g for 4 min. Columns are ready for use and may be stored hydrated (do not spin after adding water the last time) at 4°C for several days.

Solutions

Soil homogenization buffer (SHB). 4 M guanidine isothiocyanate, 200 mM sodium phosphate (pH 8), 25 mM sodium citrate, and 0.5% N-lauryl sarcosine.

40.3. OLIGONUCLEOTIDE PROBES

40.3.1. Probe Selection

rRNA molecules consist of a mosaic of conserved and variable domains whose nucleotide sequences evolve at different rates. As a result, it is possible to design rRNA-based probes that have different levels of phylogenetic specificity. For example, probes can be designed to target a single species, a particular microbial group, or an entire domain. In general, rRNA molecules have a very slow rate of evolution and so different strains of the same species commonly have few if any differences in their rRNA sequences. Thus, rRNA-targeted probes cannot distinguish organisms below the species level. A large selection of phylogenetic probes can be found in the Oligonucleotide Probe Database (1) or in a review by Amann et al. (3).

There are some general considerations to keep in mind when preparing to use oligonucleotide probes to measure rRNA abundance. An oligonucleotide probe is designed using the rRNA sequence database that is available at a given point in time. At any given time in the foreseeable future, it is safe to imagine that there will be more species in the world than there are represented in our species databases. As a result, a possibility remains that a probe targeting a particular group may also target other organisms present in an environment that are unknown in sequence databases. Whenever possible, it is advisable to use multiple probes that target a particular group or to use nested probes that will allow independent corroboration of hybridization data (3, 45). In addition, when choosing a probe reported in the literature, it is advisable to check the specificity of that probe against the sequences available in the current ribosomal database (22).

40.3.2. Probe Design

There are two critical stages in the process of designing a new oligonucleotide probe. The first stage involves determining the actual probe DNA sequence that will provide selectivity for the organisms that you desire to target. The second stage requires an empirical determination of the hybridization conditions required to provide probe specificity and validation that the probe actually functions as designed. Many of these issues have been reviewed elsewhere (3, 44, 45, 50).

The basic idea behind designing an oligonucleotide for a species or a group of species is to find a stretch of sequence from 15 to 25 bases long that is conserved in all of the target organisms but is not present in any other sequence. The difficulty of this enterprise will vary considerably depending on the particular group of organisms being targeted, in a manner that is difficult to predict a priori. Perhaps the best way to design phylogenetic probes is to use the ARB phylogenetic software package (47). This software package has sophisticated but simple-to-use tools that streamline the probe design process. The ARB package, installation instructions, and user manual are available free over the Internet (www.mpi-bremen.de/molecol/arb). Without ARB, the most basic way to design a probe is to bring all of the target sequences into a DNA sequence editor and to make a consensus sequence from the group of sequences. This consensus sequence will reveal the regions of conservation between the target organisms. These conserved regions should then be compared with homologous sequences from closely related nontarget organisms to identify putative probe regions that are specific to the targeted organisms. A single base pair mismatch with nontarget organisms can be sufficient to provide specificity; however, it should be noted that the character and location of the mismatch in the probe sequence will have an effect on the ability of the probe to discriminate between target and nontarget sequences. For example, a mismatch located in the middle of a probe sequence will tend to be more destabilizing than a mismatch on the termini of the probe and will thus make for a better probe. It is also important to keep in mind that for RNA-RNA or RNA-DNA hybrids, guanine is able to base-pair to uracil to form G-U base pairs. In addition, mispairing between adenine and guanine is only weakly destabilizing relative to all other base pair mismatches. Putative probe targets should be checked by performing a BLAST search of GenBank to determine whether the probe's complement sequence is present in organisms outside your target group. (As many regions of rRNA molecules are highly conserved, it may be necessary to screen many putative sequence targets until a successful probe sequence is found.)

Once a probe sequence has been identified, it is necessary to test the performance of the probe empirically. The objective of this exercise is to determine the hybridization conditions that provide the greatest amount of hybridization between probe and target sequences while minimizing hybridization between probe and nontarget sequences. When using oligonucleotide probes, it should be possible to achieve discrimination between sequences that differ in a single nucleotide position. The affinity with which nucleic acids bind to one another is influenced by the degree of sequence complementarity, buffer composition, and temperature. Thus, the hybridization conditions required to achieve probe specificity are determined by hybridizing a probe to both target nucleic acids and nontarget nucleic acids and then performing wash steps of increasing stringency by raising either the temperature or changing the composition of the wash buffer. For example, formamide can be added at various concentrations to both hybridization and wash buffers to increase the specificity of hybridization in lieu of altering temperature (this alternative is commonly used in FISH, reviewed in chapter 39). The temperature of dissociation (T_d) is a standard measure of the affinity between nucleic acid molecules in solution and those on a membrane (as opposed to the T_m, which is a measure of the affinity between different nucleic acid molecules in solution). The T_d of a probe for a given nucleic acid immobilized on a membrane is defined as the temperature at which one half of the probe becomes dissociated from that nucleic acid. Note that formulas used to estimate the T_d of a probe very rarely agree with empirical measurements of T_d.

40.3.3. Probe 5' End Labeling

End labeling of oligonucleotide probes is accomplished by using T4 polynucleotide kinase to transfer and exchange the $^{32}P_i$ from the γ position of [γ-^{32}P]ATP to the 5' hydroxyl terminus of an oligonucleotide. Probes need to be purified following labeling to remove unincorporated [γ-^{32}P]ATP. There are several different protocols for separating unincorporated nucleotides from oligonucleotides. Most of these methods require some sort of size exclusion or ion exchange resin. Methods for end labeling and probe purification are provided. Noted that ^{32}P-labeled probes have a limited shelf life, as the half-life of ^{32}P is approximately 14 days. To achieve maximum sensitivity, hybridization experiments should be carried out soon after probe labeling.

5' End Labeling of Oligonucleotides

1. Mix together 25 pmol of oligonucleotide, 50 pmol of [γ-^{32}P]ATP (3,000 Ci/mmol), 5 μl of 10× T4 polynucleotide kinase buffer (supplied with enzyme), and 20 units of T4 polynucleotide kinase (New England Biolabs) and add RNase-free water to a final volume of 50 μl.

2. Incubate reaction at 37°C for 30 min.

3. Stop reaction by heat inactivation at 65°C for 20 min, or by adding EDTA to a final concentration of 10 mM.

Oligonucleotide Purification

4. Combine probe-labeling reaction with 12 μl of 250 mM ammonium acetate (pH 8.5 with ammonium hydroxide) and add this mixture to the top of a TSK-DEAE column (see preparation of TSK-DEAE columns below).

5. Wash off unincorporated [γ-^{32}P]ATP by adding 1 ml of 50 mM ammonium acetate (pH 8.5) followed by 1 ml of 250 mM ammonium acetate (pH 8.5) and then 1 ml of 500 mM ammonium acetate (pH 8.5). Allow solutions to pass through column by the force of gravity. Do not allow the meniscus to reach the level of the resin in the column at any time.

6. Elute labeled probe from the column with 1 ml of 1 M ammonium acetate (pH 8.5).

7. Dry down the probe in a SpeedVac (alternatively, the probe can be precipitated with 0.1 volume of sodium acetate and 2.5 volumes of 100% ethanol).

8. Resuspend the probe in 50 μl of 50% methanol, then repeat step 7 (this step is not needed if ethanol precipitation is used).

9. Resuspend the probe in 100 μl of RNase-free water and determine the activity of 1 μl by using a liquid scintillation counter.

Preparing TSK-DEAE Columns

Use a small quantity of glass wool to plug the bottom of a 1-ml plastic pipette tip. Slide a 4-cm-long piece of silicon tubing (0.1 to 0.2 mm diameter) over the narrow end of the pipette tip. Clamp the tubing and suspend the pipette tip on a ring stand. Add 250-μl of water to the pipette tip and mark the location of the meniscus. Remove the clamp and allow the water to drain from the column. Add TSK-DEAE resin (Supelco, DEAE-650M or equivalent) as a slurry in water to the column until the bed volume reaches the 250-μl mark. Wash the column two times with 1 ml of 50 mM ammonium acetate (pH 8.5). To ensure the settled integrity of the column, do not allow the meniscus to go below the top of the gel bed; use the clamp to stop the flow through the column.

40.3.4. T_d Determination

Before determining the T_d of an oligonucleotide probe, it is necessary to isolate RNA from an organism that has a nucleic acid sequence complementary to the probe. The RNA from this organism is then immobilized on a nylon membrane, and hybridization between ^{32}P-labeled probe and target is carried out as described in section 40.4.2. To determine the T_d, it will be necessary to have at least four slots containing 1 μg of RNA on the membrane.

1. After hybridization, wash the membrane twice in 50 ml of wash buffer (see section 40.4.2 for details) for 30 min at room temperature.

2. Cut out each of the slots that contain target RNA on the membrane and place into separate vials.

3. Add scintillation fluid to one of the membrane pieces and use a liquid scintillation counter to determine the total counts per min (cpm) of the probe on the membrane.

4. Add 10 ml of wash buffer to each of the other membrane pieces and place in a shaking incubator at 30°C and 100 rpm.

5. After the samples have reached the desired temperature, incubate for exactly 10 min.

6. Remove 500 μl from each vial and place in scintillation fluid for liquid scintillation counting. Replace the removed buffer by adding 500 μl from a fresh supply of wash buffer kept at temperature in the incubator.

7. Raise the temperature in 2°C increments and repeat steps 5 and 6 at each interval.

8. At the end of the experiment, place each membrane piece in liquid scintillation fluid to determine the amount of probe remaining on the membrane.

9. Plot the percentage of cpm released as a function of the wash temperature (Fig. 1). The T_d is the temperature at which one half of the probe has been released from the membrane.

40.4. RNA HYBRIDIZATION

40.4.1. Basic Considerations

In hybridization experiments it is important to account for the presence of contaminants that may affect rRNA quan-

tification. Compounds such as humic acids have been observed to inhibit hybridization (2, 49). The quantity of humic acids present, and hence the degree of inhibition, can vary depending on the characteristics of the parent sample, presenting the possibility for a systematic bias in hybridization experiments with environmental nucleic acid extracts. Fortunately, it is possible to design RNA hybridization experiments to account for the presence of inhibitory compounds. Inhibitors of hybridization, such as humic acids, most commonly act by competing with nucleic acids for membrane occupancy during the immobilization of nucleic acid samples (2, 6). Prior to cross-linking, positively charged nylon membranes bind and retain negatively charged nucleic acids. Negatively charged compounds like humic acids can saturate the positive charge on the membrane and prevent nucleic acids from associating with membranes (Fig. 2). It is possible to detect the presence of hybridization inhibitors by blotting a sample at a range of concentrations and observing the amount of probe bound following hybridization as a function of sample dilution. A nonlinear relationship characterized by a logarithmic decline in signal response is characteristic of inhibition (6). If humic acids or other inhibitors are present in any degree, then any attempt at absolute quantitation of the number of rRNA molecules present by hybridization results will result in an underestimate. In addition, since the amount of humic acids present in a sample may vary, it would be difficult to distinguish changes in rRNA absolute abundance that result from changes in the microbial community from apparent changes that result from changes in the humic acid content of samples. A simple way to compensate for the presence of

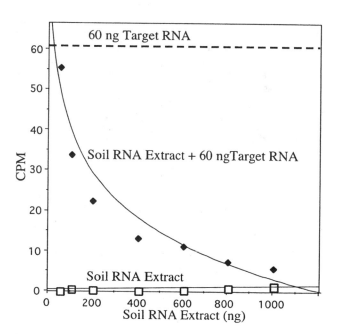

FIGURE 2 Effect of humic acid contamination on rRNA hybridization. A constant amount of target RNA (60 ng) was added to increasing amounts of a soil RNA extract that was heavily contaminated by humic acids. Hybridization with a radiolabeled oligonucleotide probe that bound only to the target RNA was performed on the target RNA spiked with soil RNA extract, soil RNA extract alone, and target RNA alone. Hybridization of probe to target RNA decreased with increasing amounts of the contaminated soil RNA extract.

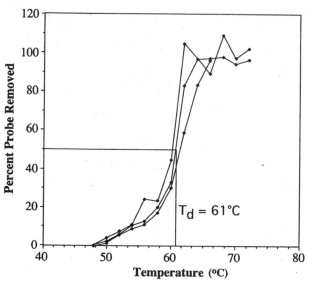

FIGURE 1 The temperature of dissociation (T_d) for an oligonucleotide probe calculated from three replicate experiments. Data are obtained by hybridizing the probe to 1 μg of complementary RNA on a nylon membrane. The probe is then removed from the RNA by gradually increasing the wash temperature. The probe removed is expressed as a percentage of the probe bound to the RNA prior to washing.

inhibitors when quantifying rRNA abundance is to normalize hybridization results to the total amount of rRNA present in a sample as measured by a "universal" probe that targets the rRNA of all known organisms (as described in section 40.5.1). Provided that the same amount of sample is immobilized on the membranes used for hybridization with both group-specific and universal probes, then calculating the ratio of group-specific probe signal to universal probe signal will cancel out the effects of hybridization inhibitors.

40.4.2. Hybridization Methodology

The general procedure for performing RNA hybridization requires that RNA be denatured and subsequently immobilized on a nylon membrane (46). The membrane is bathed in hybridization solution containing a labeled probe and then a series of wash steps is used to remove probe that is nonspecifically bound to the membrane. The temperature of the final wash step is set to ensure that probe is washed off noncomplementary nucleic acid sequences. The temperature of this stringent wash step needs to be determined empirically for each probe (see section 40.3). The stringent wash temperatures for a variety of different probes are given in Table 1.

It is important to include both positive and negative controls on all membranes used in hybridization experiments. Controls consist of RNA extracted from pure cultures of organisms that have either a nucleic acid sequence that is complementary to the probe being used (in the case of a positive control) or a nucleic acid sequence that differs from the probe target in only a few base positions (in the case of negative controls). In the case that rRNA from a pure culture is not available to use as a control, as is frequently the case with noncultivated organisms, it is possible to use rRNA that has been transcribed in vitro from a cloned sequence (7). Methods for in vitro transcription are described elsewhere (37). The best negative controls will be those organisms that have nucleic acid sequences that are as similar as possible to the probe target sequence. To verify that the probe provides specificity under the hybridization

conditions used, it is advisable to select several negative controls that vary in the number and type of mismatches to the probe.

Preparation of RNA

1. Mix each RNA sample with 3 volumes of glutaraldehyde solution and let nucleic acids denature for 15 min at room temperature. (RNA samples should contain between 300 ng and 3,000 ng of RNA; the lower value is recommended for controls while the upper value is recommended for mixed community samples. Note that exact quantification of RNA is not crucial for measurement of rRNA relative abundance.)

2. Add poly(A) water to a final volume of 1,300 μl and place samples on ice.

Blotting Denatured RNA on a Membrane

3. Wet Magna Charge nylon membrane (Micron Separation, Inc.) and several pieces of absorbent filter paper with RNase-free water. (Note: Membranes other than Magna Charge can be used as described in reference 32).

4. Place membrane and filter paper into a clean blotting apparatus so that the membrane is facing up. Use enough pieces of filter paper so that the wells of the blotting apparatus do not leak when assembled (this can be checked before by running a dye through the wells of the blotting apparatus).

5. Add 400 μl, 320 μl, 240 μl, 160 μl, and 80 μl of each sample to five adjacent wells on the membrane. Repeat for all samples and controls.

6. Apply a slight vacuum to the blotting apparatus to pull the samples onto the nylon membrane.

7. Disassemble blotting apparatus and immobilize RNA on the membrane by UV cross-linking with a total energy of $1,200 \times 100$ μJ cm^{-2}. Alternatively, RNA can be immobilized on the membrane by baking for 30 min at 80°C in a vacuum oven, or heating for 2 to 3 min at full power in a microwave oven (750 to 900 W) as described elsewhere (37).

Hybridization with ^{32}P-labeled Oligonucleotide Probes

8. Place the membrane in a hybridization bottle (35 by 150 mm) with the RNA side facing the center of the bottle. Add 10 ml of hybridization buffer and incubate in a rolling incubator at 45°C for at least 30 min.

9. Pour out buffer and replace with 10 ml of fresh hybridization buffer. Add ^{32}P-labeled oligonucleotide probe (see section 40.3.3 for labeling instructions) so that the final hybridization buffer contains approximately 10^6 cpm/ml hybridization buffer.

10. Hybridize the membrane in a rolling incubator at 45°C overnight.

Washing the Membrane

11. Pour hybridization buffer into a radioactive waste container and replace with 50 ml of wash buffer prewarmed to 45°C; incubate for 30 min in the rolling incubator at 45°C.

12. Pour wash buffer into a radioactive water container and replace with 50 ml of fresh wash buffer prewarmed to 45°C; incubate for 30 min in the rolling incubator at 45°C.

13. Pour wash buffer into a radioactive water container, replace with 50 ml of fresh wash buffer prewarmed to the stringent temperature specified for the probe, and incubate

TABLE 1 Final stringent wash temperatures for a variety of oligonucleotide probes[a]

Probe	Final wash temp (°C)	Reference
Univ1390	45	46
Eub338	45	45
Arc915	45	45
Euk1195	45	14
Acd31	53	4
Alf1b	55	23
Bet42a	62	23
CF319	55	23
Cren745	61	7
Gam42a	62	23
HGC69a	50	35
Nso1225	47	24
Pla46	55	27
Ver49	45	8

[a]Stringent wash temperatures for a selection of commonly used rRNA-targeted oligonucleotide probes in buffer lacking formamide as described in section 40.4.2.

in the rolling incubator for 30 min at the desired temperature. (Wash temperatures for many probes are listed in Table 1; otherwise it may be necessary to determine this temperature empirically as described in section 40.3.4.)

14. Pour wash buffer into a radioactive water container, remove membrane from hybridization bottle using forceps, and rinse membrane briefly with Milli-Q water in a shallow dish.

15. Allow membrane to air dry and then wrap in Mylar or Saran Wrap.

16. Determine the amount of probe bound to each spot on the membrane by using a system that is capable of detecting the emission of beta particles, either directly or through the exposure of phosphor plates.

Solutions

Glutaraldehyde solution. 2% glutaraldehyde, 50 mM sodium phosphate, pH 7.0. Filter sterilize and store at $-20°C$.

Poly(A) water. 1 μg ml^{-1} polyadenylic acid, 0.06 μg ml^{-1} bromphenol blue, 0.5% glutaraldehyde. Filter sterilize and store at $-20°C$.

Hybridization buffer. 900 mM NaCl, 120 mM Tris-HCl (pH 8.0), 6 mM disodium EDTA, 0.5% SDS, 1× Denhardt's solution, 100 μg ml^{-1} polyadenylic acid. Make fresh daily.

Wash buffer. 300 mM NaCl, 40 mM Tris-HCl (pH 8.0), 2 mM disodium EDTA, 0.5% SDS.

Denhardt's solution. 1% Ficoll, 1% polyvinylpyrrolidone, 1% bovine serum albumin (fraction V). Filter sterilize and store at $-20°C$.

40.5. DETERMINATION OF rRNA ABUNDANCE

40.5.1. Calculation of rRNA Relative Abundance

RNA molecules extracted from pure cultures of microorganisms are placed on every hybridization membrane to provide the two levels of control. These RNA samples are both from organisms that are targeted by the rRNA probe being used and from several microorganisms that are closely related but outside of the target group. The positive controls are used to account for variations in the labeling efficiency of different oligonucleotide probes, while the negative controls are used to account for any hybridization signal that results from nonspecific interactions. Calculations of relative abundance are made by taking the ratio of signal intensities obtained for specific and universal probe binding to an RNA sample as $R = \Sigma^n_{i=1}[G_i(U_i)^{-1}]n^{-1}$, where G_i and U_i represent, respectively, the corresponding signal intensities obtained for group-specific and universal probe binding to each aliquot representing the sample, and n equals the total number of aliquots representing the RNA sample. The value R is then calculated for each soil RNA sample (R_s), and a mean value of R is determined for all positive (R_p) and negative (R_n) controls present on each membrane. The relative abundance (expressed as a percentage) of rRNA from a specific microbial group is then defined as $(R_s - R_n)(R_p - R_n)^{-1} \times 100$. An example calculation of rRNA relative abundance is shown in Table 2.

The calculation of rRNA relative abundance from RNA hybridization data involves multiple levels of control that account completely for any possible problems caused by

TABLE 2 Example calculation of rRNA relative abundance for a single RNA sample[a]

Name	Volume (μl)	Universal probe signal	Group-specific probe signal	$G_i(U_i)^{-1}$	Equation
Positive control	400	12080000	3450000	0.2856	$\Sigma^n_{i=1}[G_i(U_i)^{-1}] = 1.4065; n = 5$
	320	9800000	2800000	0.2857	
	240	7360000	2100000	0.2853	$R_P = \Sigma^n_{i=1}[G_i(U_i)^{-1}]n^{-1}$
	160	5000000	1410000	0.2820	$R_P = (1.4065)(5)^{-1}$
	80	2650000	710000	0.2679	$R_P = 0.2813$
Negative control	400	25590000	100000	0.0039	$\Sigma^n_{i=1}[G_i(U_i)^{-1}] = 0.0214; n = 5$
	320	22130000	80000	0.0036	
	240	17270000	60000	0.0035	$R_N = \Sigma^n_{i=1}[G_i(U_i)^{-1}]n^{-1}$
	160	10420000	50000	0.0048	$R_N = (0.0257)(5)^{-1}$
	80	5360000	30000	0.0056	$R_N = 0.0043$
Sample	400	63670000	930000	0.0146	$\Sigma^n_{i=1}[G_i(U_i)^{-1}] = 0.0807; n = 5$
	320	55200000	790000	0.0143	
	240	36450000	700000	0.0192	$R_S = \Sigma^n_{i=1}[G_i(U_i)^{-1}]n^{-1}$
	160	25050000	400000	0.0160	$R_S = (0.0807)(5)^{-1}$
	80	11470000	190000	0.0166	$R_S = 0.0161$

$$(R_S - R_N)(R_P - R_N)^{-1} \times 100 = \text{percent rRNA abundance}$$
$$(0.0161 - 0.0043)(0.2813 - 0.0043)^{-1} \times 100 = \textbf{4.3\%}$$

[a]Data are from one hybridization experiment with a universal probe and a second hybridization experiment with a group-specific probe. The hybridization experiments were carried out on duplicate membranes that contained RNA from a positive control organism, a negative control organism, and a soil sample. Five different volumes of each RNA sample were blotted on each membrane. With a 96-well blotting apparatus it is possible to create a single membrane that contains 19 different samples and controls.

impurities in soil RNA extracts and variability in soil sample RNA abundance. To control for differences in RNA quantity and quality, all determinations of rRNA relative abundance are based on the ratio of the results from two hybridization experiments carried out with two distinct probes. In such hybridization experiments, it is important that the same amount of RNA (and humic acids as the case may be) is present in each corresponding well on the blots used for specific probe hybridization and on the blots used for universal probe hybridization. Thus, the amount of inhibition for a given sample will be the same on both blots, and dividing the hybridization signal for the sample on a specific blot by its signal on a universal blot will cancel out any inhibition effects, giving an accurate measure of rRNA relative abundance.

There are two caveats to keep in mind when using the above technique to calculate rRNA relative abundance. First, a change in rRNA relative abundance can result from a change in the amount of a target rRNA in a community (specific probe signal) or from a change in the total number of rRNA molecules present in a community (as detected by a universal rRNA probe). Thus, it is important to consider the hypotheses that are being tested to determine whether relative abundance measures are appropriate. Comparisons of universal probe signal intensity can be used to determine the degree to which changes in community rRNA abundance influence measurements of group-specific rRNA relative abundance, though inhibitors such as humic acids, if present, may have a significant impact on total signal intensity. If inhibitory substances are verified to be absent, then it is possible to calculate the absolute abundance of rRNA present from a specific group by relating hybridization signal intensity directly to known amounts of positive control rRNA included on hybridization membranes. A second caveat to using this method of calculating rRNA relative abundance is that the negative controls used to correct for nonspecific hybridization tend to be overly conservative. For example, a typical negative control may contain only a single mismatch in the probe target region, while a microbial community will contain an array of sequences in the probe target region, most of which will have more than a single mismatch. As a result of this conservative use of negative controls, it is difficult to detect signal from target groups that are present at low relative abundance (generally <0.5%).

40.5.2. Data Analysis

Measurements of rRNA relative abundance are commonly made to examine whether some component of a microbial community is affected by a particular factor or group of factors. In the simplest case, it might be necessary to test whether the activity of a particular microbial group is influenced by a single experimental treatment. In a more complicated analysis, it may be necessary to examine whether microbial community structure varies in relation to a range of experimental treatments administered over time. These are ultimately ecological questions about microbial communities. A suite of analytical tools has been developed for the analysis of ecological experiments that are useful for the analysis of rRNA relative abundance data. An in-depth review of these techniques is out of the scope of this chapter, but brief descriptions of some techniques that may be of interest are provided below. More information on the design and analysis of ecological experiments can be found elsewhere (40, 42).

The particular type of hypothesis being addressed will inform the choice of analytical method. Decisions about the type of analysis to be performed should be made prior to the design of experiments. A regression analysis is appropriate for determining the likelihood of a linear relationship between a continuous variable and the rRNA abundance of a particular microorganism. When it is necessary to examine the degree to which multiple continuous variables correlate with rRNA abundance, then either multiple regression analysis or canonical correspondence analysis will be necessary. In many cases, an investigator manipulates one or more factors to determine their effect on the rRNA abundance of a given organism. An example would be an experiment in which the rRNA abundance of the autotrophic ammonia oxidizers is examined in response to fertilizer addition in a set of fields that received different amounts of fertilizer. In these cases, an analysis of variance (ANOVA) is the analytical method of choice. ANOVA is an extremely powerful analytical technique for examining the relationship between one or more factors and a single dependent variable, such as rRNA abundance. It is also possible to examine variation in the rRNA abundance of multiple microbial groups simultaneously in response to one or more experimentally manipulated factors through the use of multiple analysis of variance (MANOVA) or correspondence analysis. MANOVA is useful for determining whether there is an overall change in the structure of a microbial community in response to experimental manipulation. All of these analytical techniques can be carried out on computer packages such as StatView v 5.0 (SAS Institute, Inc.) and SAS v 7.0 (SAS Institute, Inc.).

40.6. REFERENCES

1. **Alm, E. W., D. B. Oerther, N. Larsen, D. A. Stahl, and L. Raskin.** 1996. The oligonucleotide probe database. *Appl. Environ. Microbiol.* **62:**3557–3559.

2. **Alm, E. W., D. D. Zheng, and L. Raskin.** 2000. The presence of humic substances and DNA in RNA extracts affects hybridization results. *Appl. Environ. Microbiol.* **66:** 4547–4554.

3. **Amann, R. I., W. Ludwig, and K.-H. Schleifer.** 1995. Phylogenetic identification and in situ detection of individual microbial cells without cultivation. *Microbiol. Rev.* **59:**143–169.

4. **Barns, S. M., S. L. Takala, and C. R. Kuske.** 1999. Wide distribution and diversity of members of the bacterial Kingdom *Acidobacterium* in the environment. *Appl. Environ. Microbiol.* **65:**1731–1737.

5. **Bloem, J.** 1995. Fluorescent staining of microbes for total direct counts, p. 1–12. *In* A. D. L. Akkermans, J. D. VanElsas, and F. J. deBruijn (ed.), *Molecular Microbial Ecology Manual.* Kluwer Academic Publishers, Dordrecht, The Netherlands.

6. **Buckley, D. H.** 2000. The diversity and dynamics of microbial groups in soils from agroecosystems. Ph.D. dissertation. Michigan State University, East Lansing, MI.

7. **Buckley, D. H., J. R. Graber, and T. M. Schmidt.** 1998. Phylogenetic analysis of nonthermophilic members of the kingdom *Crenarchaeota* and their diversity and abundance in soils. *Appl. Environ. Microbiol.* **64:**4333–4339.

8. **Buckley, D. H., and T. M. Schmidt.** 2001. Environmental factors influencing the distribution of *Verrucomicrobia* in soil. *FEMS Microbiol. Ecol.* **35:**105–112.

9. **Buckley, D. H., and T. M. Schmidt.** 2001. The structure of microbial communities in soil and the lasting impacts of cultivation. *Microb. Ecol.* **42:**11–21.

10. **Chandler, D. P., J. K. Fredrickson, and F. J. Brockman.** 1997. Effect of PCR template concentration on the composition and distribution of total community 16S rDNA clone libraries. *Mol. Ecol.* **6:**475–482.

11. **Daniels, L., R. S. Hanson, and J. A. Phillips.** 1994. Chemical analysis, p. 536. *In* P. Gerhardt, R. G. E. Murray, W. A. Wood, and N. R. Krieg (ed.), *Methods for General and Molecular Bacteriology.* American Society for Microbiology, Washington, DC.

12. **Duarte, G. F., A. S. Rosado, L. Seldin, A. C. Keijzer-Wolters, and J. D. van Elsas.** 1998. Extraction of ribosomal RNA and genomic DNA from soil for studying the diversity of the indigenous bacterial community. *J. Microbiol. Meth.* **32:**21–29.

13. **Farrelly, V., F. A. Rainey, and E. Stakebrandt.** 1995. Effect of genome size and *rrn* gene copy number on PCR amplification of 16S rRNA genes from a mixture of bacterial species. *Appl. Environ. Microbiol.* **61:**2798–2801.

14. **Giovannoni, S. J., E. F. DeLong, G. J. Olsen, and N. R. Pace.** 1988. Phylogenetic group-specific oligonucleotide probes for identification of single microbial cells. *J. Bacteriol.* **170:**720–726.

15. **Hugenholtz, P., B. M. Goebel, and N. R. Pace.** 1998. Impact of culture-independent studies on the emerging phylogenetic view of bacterial diversity. *J. Bacteriol.* **180:**4765–4774.

16. **Huston, M. A.** 1994. *Biological Diversity.* Cambridge University Press, Cambridge, United Kingdom.

17. **Johnson, J.** 1994. Similarity analysis of rRNAs, p. 685. *In* P. Gerhardt, R. G. E. Murray, W. A. Wood, and N. R. Krieg (ed.), *Methods for General and Molecular Bacteriology.* American Society for Microbiology, Washington, DC.

18. **Kemp, P. F., S. Lee, and J. LaRoche.** 1993. Estimating the growth rate of slowly growing marine bacteria from RNA content. *Appl. Environ. Microbiol.* **59:**2594–2601.

19. **Klappenbach, J. A., J. M. Dunbar, and T. M. Schmidt.** 2000. rRNA operon copy number reflects ecological strategies of bacteria. *Appl. Environ. Microbiol.* **66:**1328–1333.

20. **Klappenbach, J. A., P. R. Saxman, J. R. Cole, and T. M. Schmidt.** 2001. rrndb: the ribosomal RNA operon copy number database. *Nucleic Acids Res.* **29:**181–184.

21. **Lepp, P. W., and T. M. Schmidt.** 1998. Nucleic acid content of *Synechococcus* spp. during growth in continuous light and light/dark cycles. *Arch. Microbiol.* **170:**201–207.

22. **Maidak, B. L., J. R. Cole, T. G. Lilburn, C. T. Parker, P. R. Saxman, R. J. Farris, G. M. Garrity, G. J. Olsen, T. M. Schmidt, and J. M. Tiedje.** 2001. The RDP-II (Ribosomal Database Project). *Nucleic Acids Res.* **29:**173–174.

23. **Manz, W., R. Amann, W. Ludwig, M. Wagner, and K.-H. Schleifer.** 1992. Phylogenetic oligonucleotide probes for the major subclasses of proteobacteria: problems and solutions. *Syst. Appl. Microbiol.* **15:**593–600.

24. **Mobarry, B. K., M. Wagner, V. Urbain, B. E. Rittman, and D. S. Stahl.** 1996. Phylogenetic probes for analyzing abundance and spatial organization of nitrifying bacteria. *Appl. Environ. Microbiol.* **62:**2156–2162.

25. **More, M. J., J. B. Herrick, M. C. Silva, W. C. Ghiorse, and E. L. Madsen.** 1994. Quantitative cell lysis of indigenous microorganisms and rapid extraction of microbial DNA from sediment. *Appl. Environ. Microbiol.* **60:**1572–1580.

26. **Muttray, A. F., and W. W. Mohn.** 1999. Quantitation of the population size and metabolic activity of a resin acid degrading bacterium in activated sludge using slot-blot hybridization to measure the rRNA:rDNA ratio. *Microb. Ecol.* **38:**348–357.

27. **Neef, A., R. Amann, H. Schlesner, and K. H. Schleifer.** 1998. Monitoring a widespread bacterial group: in situ detection of Planctomycetes with 16S rRNA-targeted probes. *Microbiology* **144:**3257–3266.

28. **Ogram, A., G. S. Sayler, and T. Barkay.** 1987. The extraction and purification of microbial DNA from sediments. *J. Microbiol. Meth.* **7:**57–66.

29. **Pace, N. R.** 1997. A molecular view of microbial diversity and the biosphere. *Science* **276:**734–740.

30. **Poulsen, L. K., G. Ballard, and D. A. Stahl.** 1993. Use of ribosomal-RNA fluorescence in situ hybridization for measuring the activity of single cells in young and established biofilms. *Appl. Environ. Microbiol.* **59:**1354–1360.

31. **Purdy, K. J., T. M. Embley, S. Takii, and D. B. Nedwell.** 1996. Rapid extraction of DNA and rRNA from sediments by a novel hydroxyapatite spin-column method. *Appl. Environ. Microbiol.* **62:**3905–3907.

32. **Raskin, L., W. C. Capman, M. D. Kane, B. E. Rittmann, and D. A. Stahl.** 1996. Critical evaluation of membrane supports for use in quantitative hybridizations. *Appl. Environ. Microbiol.* **62:**300–303.

33. **Raskin, L., W. C. Capman, R. Sharp, and D. A. Stahl.** 1997. Molecular ecology of gastrointestinal ecosystems, p. 243–298. *In* R. I. Mackie, B. A. White, and R. E. Isaacson (ed.), *Ecology and Physiology of Gastrointestinal Microbes*, vol. 2. Chapman and Hall, New York, NY.

34. **Raskin, L., L. K. Poulsen, D. R. Noguera, B. E. Rittmann, and D. A. Stahl.** 1994. Quantification of methanogenic groups in anaerobic biological reactors by oligonucleotide probe hybridization. *Appl. Environ. Microbiol.* **60:**1241–1248.

35. **Roller, C., M. Wagner, R. Amann, W. Ludwig, and K.-H. Schleifer.** 1994. *In situ* probing of Gram-positive bacteria with high DNA G + C content using rRNA-targeted oligonucleotides. *Microbiology* **140:**2849–2858.

36. **Sambrook, J., E. F. Fritsch, and T. Maniatis.** 1989. *Molecular Cloning: a Laboratory Manual*, 2nd ed. Cold Spring Harbor Laboratory Press, Cold Spring Harbor, NY.

37. **Sambrook, J., and D. W. Russell (ed.).** 2001. *Molecular Cloning: a Laboratory Manual*, 3rd ed. Cold Spring Harbor Laboratory Press, Cold Spring Harbor, NY.

38. **Sayler, G. S., and A. C. Clayton.** 1990. Environmental application of nucleic acid hybridization. *Annu. Rev. Microbiol.* **44:**625–648.

39. **Schaechter, M. O., O. Maaloe, and N. O. Kjeldgaard.** 1958. Dependency on medium and temperature of cell size and chemical composition during balanced growth of *Salmonella typhimurium. J. Gen. Microbiol.* **19:**592–606.

40. **Scheiner, S. M., and J. Gurevitch (ed.).** 1993. *Design and Analysis of Ecological Experiments.* Chapman and Hall, New York, NY.

41. **Selenska-Pobell, S.** 1995. Direct and simultaneous extraction of DNA and RNA from soil, p. 1–17. *In* J. D. v. E. D. L. Akkermans, and E. M. H. Wellington (ed.), *Molecular Microbial Ecology Manual.* Kluwer Academic Publishers, Dordrecht, The Netherlands.

42. **Sokal, R. R., and F. J. Rohlf.** 1995. *Biometry: the Principles and Practice of Statistics in Biological Research.* W. H. Freeman & Co., New York, NY.

43. **Srivastava, A. K., and D. Schlessinger.** 1990. Mechanism and regulation of bacterial ribosomal RNA processing. *Annu. Rev. Microbiol.* **44:**105–129.

44. **Stahl, D. A.** 1997. Molecular approaches for the measurement of density, diversity, and phylogeny, p. 102–114. *In* C. J. Hurst, G. R. Knudsen, M. J. McInerney, L. D. Stetzenbach, and M. V. Walter (ed.), *Manual of Methods in Environmental Microbiology.* ASM Press, Washington, DC.

45. **Stahl, D. A., and R. Amann.** 1991. Development and application of nucleic acid probes in bacterial systematics,

p. 205–248. *In* E. Stackebrandt and M. Goodfellow (ed.), *Nucleic Acid Techniques in Bacterial Systematics*. John Wiley & Sons Ltd., Chichester, United Kingdom.

46. **Stahl, D. A., B. Flesher, H. R. Mansfield, and L. Montgomery.** 1988. Use of phylogenetically based probes for studies of ruminal microbial ecology. *Appl. Environ. Microbiol.* **54:**1079–1084.

47. **Strunk, O., and W. Ludwig.** 1997. *ARB: A Software Environment for Sequence Data*, version 2.5. Department of Microbiology, Technical University of Munich, Munich, Germany.

48. **Suzuki, M. T., and S. J. Giovannoni.** 1996. Bias caused by template annealing in the amplification of mixtures of 16S rRNA genes by PCR. *Appl. Environ. Microbiol.* **62:**625–630.

49. **Tebbe, C. C., and W. Vahjen.** 1993. Interference of humic acids and DNA extracted directly from soil in detection and transformation of recombinant DNA from bacteria and yeast. *Appl. Environ. Microbiol.* **59:**2657–2665.

50. **Ward, D. M., M. M. Bateson, R. Weller, and A. L. Ruff-Roberts.** 1992. Ribosomal RNA analysis of microorganisms as they occur in nature. *Adv. Microb. Ecol.* **12:**219–268.

51. **Wintzingerode, F. V., U. B. Gobel, and E. Stackebrandt.** 1997. Determination of microbial diversity in environmental samples: pitfalls of PCR-based rRNA analysis. *FEMS Microbiol. Rev.* **21:**213–229.

52. **Zhou, J., M. A. Bruns, and J. M. Tiedje.** 1996. DNA recovery from soils of diverse composition. *Appl. Environ. Microbiol.* **62:**316–322.

41

Analysis of Microbial Communities with Denaturing Gradient Gel Electrophoresis and Terminal Restriction Fragment Length Polymorphism

CINDY H. NAKATSU AND TERENCE L. MARSH

41.1. INTRODUCTION

41.1.1. Microbial Communities

Microbial communities can exhibit an enormous range of complexity, from those with a mere handful of populations (38) to those with thousands of species derived from all three domains of life (11, 18, 42, 47). In order to better understand these communities, microbial ecologists seek a detailed description of the community that would include (i) a phylogenetic census of all populations as a function of time, (ii) a description of all food webs and the participating populations, (iii) the identification of all sexual cohorts,

(iv) the nature of all inter- and intraspecies interactions (quorum sensing and chemical warfare/defense), and (v) the life cycles that influence population dynamics (e.g., sporulation and encystment). Such a complete description of a community is indeed a challenging task. As a first step, many of us labor simply to identify the collection of species present. This in itself is difficult inasmuch as the majority of the microbes present in most communities have not yet been cultivated (1, 46). This so-called plating paradox has led to the development of culture-independent approaches for describing diversity within a community. Critical to this approach is the use of phylogenetic markers (16S–23S rRNA, *rpoC*, and gyrases) that are present in most, if not all, populations. Within these markers, there are both conserved and variable regions of sequence. The former permit PCR amplification directly from community DNA that has been extracted from environmental samples, while the latter provide a view of diversity. Thus, an early strategy of community analysis included the comparative sequence analysis of clone libraries that established the single-gene phylogenetic diversity of the community (21, 35). While this approach can provide a richly detailed view of diversity, such detail comes at an expense that renders ecological questions addressing temporal or spatial changes in ecosystems much too dear. In recent years, the solution to this dilemma has been to sacrifice detail for high-throughput capabilities. For community analysis, genetic fingerprinting techniques allow the comparative profiling of many environmental samples and thus facilitate the spatial and temporal analysis of microbial communities in ecosystems. Two of these approaches, denaturing gradient gel electrophoresis (DGGE) and terminal restriction fragment length polymorphism (T-RFLP), are described herein.

41.1.2. Microbial Community Analysis Methods

Microbial community analysis methods fall into two broad categories, those that interrogate the community without PCR amplification and those that are dependent on PCR. Examples of PCR-independent methods include fluorescence in situ hybridization (see chapter 39), reverse genome probing, DNA:DNA reassociation kinetics (chapter 26), nucleic acid hybridization (chapter 30), phospholipid fatty acid analysis (chapter 15), and community level metagenome analysis (chapter 38). PCR-dependent approaches include DGGE, T-RFLP, single-strand conformational polymorphism, and quantitative PCR. One should note that PCR-dependent analytical techniques are subject to all of the potential pitfalls of PCR, including inhibition from compounds that copurify with nucleic acids, the formation of chimeras, preferential priming of select targets, nonspecific priming of unexpected targets (49), and the production of single-stranded products (10) and formation of heterologous duplexes. The investigator is forewarned that many of these pitfalls are likely to arise in any experiment, underscoring the need for controls.

Analytical approaches like DGGE and T-RFLP have been referred to as genetic fingerprinting. By targeting a robust phylogenetic marker such as 16S rRNA, the investigator hopes that all populations in a community will be revealed in the fingerprint. In reality, complex microbial communities with population richness exceeding several hundred species are beyond single-primer pair PCR analysis. The best gels and capillaries resolve 20 to 30 operational taxonomic units in DGGE and 75 to 125 in T-RFLP. Clearly, this is a relatively small fraction of a complex community. Thus, at the level of current community finger-

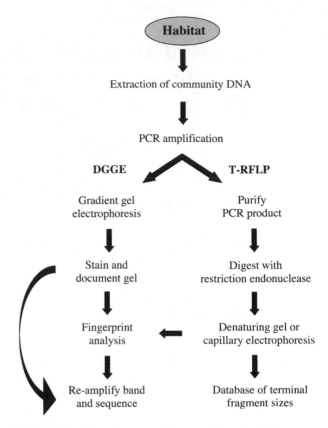

FIGURE 1 Flow diagram of the steps for microbial community analysis using either DGGE or T-RFLP.

printing technology with DGGE and T-RFLP, the investigator plays a game of chance regarding the possibility that the critical population in a changing community will be detected with the experimental technique. Targeting a community with multiple primer pairs of different specificities is one approach to increasing the number of detectable populations and hence the odds of identifying a relevant community shift.

The general protocols for DGGE and T-RFLP are presented in Fig. 1. Note that the first two steps, extraction of nucleic acid and PCR amplification, are common to both protocols. These steps, along with data analysis approaches common to both, are discussed in section 41.2 below. Details specific to DGGE and T-RFLP follow in sections 41.3 and 41.4, respectively. The intent is to provide the investigator with sufficient information such that he/she can avoid the potential pitfalls, regardless of the habitat being investigated.

41.2. STRATEGIES AND TECHNIQUES SHARED BY DGGE AND T-RFLP

41.2.1. General Design Features of an Experiment

As mentioned above, the power of DGGE and T-RFLP lies in their high-throughput features. This high throughput makes robust comparative analysis possible at the community level. An essential prerequisite for all comparative analyses is the standardization of protocols. For both DGGE and T-RFLP, this means that all steps of the protocol, including DNA extraction, PCR amplification, primer design and construction, electrophoresis systems, size standards, and data analysis approaches, must be identical for each sample or community in the experiment. Because of the sensitivity and highly technical nature of the protocols, both experimental and technical replicates are required. The investigator should assure himself/herself that all steps in the protocol are reproducible.

41.2.2. Community Nucleic Acid Extraction

The extraction of community DNA from environmental samples has been intensively investigated, and many technical variations exist in the literature (see chapter 38). In general, nucleic acids can be extracted directly from the substratum (e.g., soil and sediment) or from cells that have been stripped off of the substratum. The latter approach decreases contamination with substratum-bound components but may also be biased towards cells that are separated more easily from substratum. Most researchers isolate DNA because it is more stable than RNA; however, protocols have been developed for the isolation of RNA from various environments (see chapter 40). Commercial kits that provide a rapid and standardized approach are also available (see, for example, MoBio and Q-BIOgene). An important factor when extracting nucleic acids from environmental samples is to ensure that the approach achieves equivalent lysis of all cell types so that the analysis is representative of the community. To this end, robust physical techniques such as bead-beating and freeze-thaw have been coupled to harsh chemical treatment to provide efficient, reasonably unbiased cell lysis. The reader is encouraged to read reviews in this collection and elsewhere to identify a lysis protocol suitable for his/her specific needs.

41.2.3. PCR and DNA Quantification

Although it is emphasized in other chapters, it is important to note here that PCR conditions must be optimized each time DNA extracts from a new community are being examined. DNA extraction methods have advanced considerably, but some chemical contamination may still occur and interfere with PCR. Also, DNA concentrations must be determined before and after PCR. Nucleic acid concentrations in a sample can be estimated using any of the common approaches: spectroscopy, fluorimetry, and agarose gel electrophoresis (see chapters 26 and 30). Use of suboptimal PCR conditions and DNA concentrations can lead to PCR artifacts that can contribute to the over- or underestimation of community diversity. Equivalent concentrations of PCR products need to be used for DGGE or T-RFLP analyses if profile comparisons are going to be made.

41.2.4. Data Analysis

In general, microbial ecologists are interested in the structure and function of specific communities, as in the gastrointestinal tract, or large-scale trends in the structure and function of communities across spatial or temporal zones. The distribution of phylogenetic markers across communities, as determined by DGGE and T-RFLP, provides a rapid assessment of microbial diversity. Phylogenetic markers derived from different populations are separated by sequence-determined helix stability (DGGE) or terminal restriction fragment length (T-RFLP). Each unique band or terminal fragment represents an operational taxonomic unit (OTU), and the distribution of these OTUs constitutes the profile. By comparing profiles derived from communities, one can quantify the degree of relatedness across communities. The

most obvious data from these techniques is a simple count of OTUs (or phylotypes), which is an estimate of species richness. The evenness of detectable sequence phylotypes is obtained from the measurement of band intensities. In DGGE this requires a densitometric scan of the gel image, while in T-RFLP the amplitude of each fragment signal is automatically recorded. While this can be a highly quantitative measurement of species evenness *within the PCR product pool*, because the techniques are PCR dependent and hence one uncontrolled step removed from the original community structure, the investigator cannot extrapolate back to the original community and assert that peak amplitudes reflect population sizes. For this reason, the application of biodiversity indices (13) to the species richness and evenness estimates from DGGE and T-RFLP are not informative regarding community structure.

More statistically robust comparisons of communities can be made by treating the DGGE and T-RFLP profiles as collections of characters that can be aligned using electrophoretic mobility values associated with each band. The resulting multiple alignments can be used to produce a matrix for statistical comparisons. For example, using common similarity indices, such as that of Dice (5) [$C_s = 2j/(a + b)$, where a is the number of bands in sample A, b is the number of bands in sample B, and j is the number of bands common to A and B], a large number of communities can easily be compared. These data can be represented by cluster analysis, e.g., unweighted-pair group method with arithmetic means (UPGMA) (44), in which a dendrogram is generated illustrating the relationship between communities (32). An example of such an analysis is presented in Color Plate 17. Color Plate 17A shows T-RFLP data taken into a spreadsheet program and aligned based on terminal fragment size. The complete data set includes four replicates of three different soils (labeled H, N, and T). Each profile consists of fragment lengths for each detectable phylotype (only a fraction of each profile is shown). The next panel shows the same data set after being converted to binary form (1's or 0's, representing presence or absence). The similarity of these profiles can be determined using similarity indices such as that of Dice (described above) or the T-RFLP analysis function available at the Ribosomal Database Project (RDP) (4). The results from the latter are shown in Color Plate 17C. The data are also amenable to cluster analysis with, for example, UPGMA. Color Plate 17D presents a dendrogram of the UPGMA analysis of the data in Color Plate 17B from the three soil data sets. In Color Plates 17C and D, the three soils are easily distinguished and the replicate samples form a statistical cluster.

Alternatively, multidimensional scaling (15) can be used to obtain a graphical representation of these (dis)similarities. As data sets increase in size, sources of variation increases and multivariate analysis methods, such as principal-component analysis (37), become more appropriate. This method also provides a visual representation of the relationship between the profiles. Principal-component analysis calculates and ranks the contribution of each variable in a profile, and the approach can be used to identify the main sources of variation observed between profiles (52). For example, in DGGE profiles, the source (band) contributing the greatest variability can be statistically determined, and then the bands can be extracted from the gel and its nucleotide sequence can be determined to identify the population. A limitation of DGGE is that a complex community (e.g., soil) may be comprised of numerous populations (>100) in relatively equivalent proportions, thus resulting in a smear of bands which makes it difficult to identify individual populations (34). It is still possible to qualitatively state that two communities are different if the smear of bands looks different; however, the converse may not be true. Even better resolution of very complex data sets may be obtained using methods such as self-organizing map neural networks (6). Many would consider the current analytical methods more exploratory in nature, just the beginning to quantitatively address the complex relationships between microbial communities.

There are commercial software programs available specifically for community fingerprint analysis, for example, Bionumerics (Applied Maths, Belgian) and Fingerprinting II Informatix (Bio-Rad). Many other statistical packages are available that can be adapted to DGGE and T-RFLP data analysis, and some, like Phylip (http://evolution.genetics .washington.edu/phylip.html), are available at no cost.

41.3. DGGE

The premise of DGGE is that DNA strands of the same length will migrate differentially when electrophoresed through a linear gradient of denaturant if their sequence composition differs. Denaturation occurs when hydrogen bonds holding together the double strands of DNA are disrupted by exposure to denaturants. The strands partially separate when the lowest melting temperature within the sequence is reached, at which point the separation of the strands retards migration in the gel. To ensure that the strands do not entirely separate, a high-GC sequence (called the GC clamp) is added to one primer (43). There are two types of denaturants used: chemicals such as urea and formamide in DGGE and temperature in thermal gradient gel electrophoresis (TGGE). The concentration of denaturant determines the distance the DNA molecules migrate through the gel. Higher denaturant concentrations are needed to disrupt the three hydrogen bonds formed by guanine and cytosine base pairing. Therefore, sequences with higher

G+C content will migrate through the gel to higher concentrations of denaturant. Separation of a mixture of PCR products of ribosomal DNA (rDNA) amplified from a community produces a profile, which represents the different populations present in that community.

Typically, polyacrylamide gels are used for DGGE, but denaturing high-performance liquid chromatography (dHPLC) is being examined as an alternative (50, 53). This approach may overcome some of the resolution and reproducibility challenges faced when using polyacrylamide gradient gels. Specialized dHPLC instruments (WAVE System; Transgenomics Inc., Omaha, NE) have been tested for microbial analysis (7, 12, 14).

41.3.1. DGGE PCR

Typically, PCR products of specific gene sequences are used for DGGE and TGGE analysis. For microbial community analysis, the most commonly used target has been the 16S rRNA gene, but it has also been used to determine genetic polymorphisms in other genes, e.g., (Ni/Fe) hydrogenase genes (51). Some commonly used primers for small-subunit (SSU) rDNA DGGE analysis are listed in Table 1. The primers chosen for PCR can be used to determine the level of discrimination between groups in the community. For example, primers can be used to amplify all Bacteria, or only the β subdivision ammonia oxidizer (17). As the number of submissions to genetic databases increases, all primers should be reassessed to ensure that the desired specificity is conserved. Primers for DGGE and TGGE that yield a relatively short PCR product length are preferred in order to minimize the number of melting domains and to reduce chances of producing PCR artifacts. Some researchers also use touchdown PCR to improve specificity (33). In this approach the first few cycles are performed at temperatures above the standard annealing temperature and then the temperature is lowered with subsequent cycles until the annealing temperature is reached. Nested PCR has been used to amplify target DNA that constitutes a very small fraction

TABLE 1 Examples of PCR primers for DGGE and T-RFLP

Primer	Position[a]	Target rRNA	Synthesized primer sequence[b]	Reference
BA338F	338–358	Bacteria, 16S	5'[c] ACT CCT ACG GGA GGC AGC AG 3'	20
UN518R	534–518	Universal, 16S	5' ATT ACC GCG GCT GCT GG 3'	33
ARC344	344–363	Archaea, 16S	5'[c] ACG GGG IGC AGC AGG CGC GA 3'	40
ARC915	915–934	Archaea, 16S	5' GTG CTC CCC CGC CAA TTC CT-3'	45
F1427	1427–1452	Eukarya, 18S	5'[d] TC TGT GAT GCC CTT AGA TGT TCT GGG 3'	48
R1616	1616–1595	Eukarya, 18S	5' GCG GTG TGT ACA AAG GGC AGG G 3'	48
1511R	1526–1511	Universal	5' YGC AGG TTC ACC TAC	30
1492R	1506–1492	Universal	5' ACC TTG TTA CGA CTT	1
1392R	1406–1392	Universal	5' ACG GGC GGT GTG TRC	1
21F	2–21	Archaea, 16S	5' TTC CGG TTG ATC CYG CCG GA	1
25F	9–25	Eukarya, 18S	5' CTG GTT GAT CCT GCC AG	30
27F	8–27	Bacteria, 16S	5' AGA GTT TGA TCM TGG CTC AG	1
63F	42–63	Bacteria, 16S	5' CAG GCC TAA YAC ATG CAA GTC	26
1387R	1404–1387	Bacteria, 16S	5'-GGG CGG WGT GTA CAA GGC	26
927R	942–927	Bacteria, 16S	5' ACC GCT TGT GCG GGC CC	1

[a]For prokaryotes relative to 16S rRNA gene sequence of Escherichia coli and for eukaryotes 18S rRNA gene sequence of Saccharomyces cerevisiae.
[b]A GC clamp must be added to the 5' end of one of the primer pairs used for PCR.
[c]GC clamp used 5' CGC CCG CCG CGC GCG GCG GGC GGG GCG GGG GCA CGG GGG G 3'.
[d]GC clamp used 5' CGC CCG CCG CGC CCC GCG CCC GGC CCG CCG CCC CCG CCC C 3'.

of the community (34). In this method, PCR is performed twice, first with a set of primers that may also amplify DNA from populations outside of the target group. Then the second PCR is performed using primers within the first amplicon and specific for the target group. As mentioned earlier, one of the primers used for PCR amplification for DGGE and TGGE analysis must include a GC clamp (43) to stabilize strand melting.

41.3.2. Denaturing Gradient Gel Preparation

A vertical polyacrylamide gel electrophoresis apparatus and gradient former are needed to run the chemical type of denaturing gradient. Bio-Rad Laboratories Inc. (DCode mutation detection system) and C.B.S. Scientific Company Inc. (Del Mar, CA) market equipment for DGGE analysis. Alternatively, a standard vertical protein gel apparatus can be adapted for DGGE. The system is adapted by using a constant-temperature bath to maintain the buffer temperature at 60°C and a magnetic stirrer to keep the buffer well mixed during electrophoresis. A specialized cam-operated gradient former is supplied with the Bio-Rad system; otherwise a standard two-cylinder gradient former can be used. In the latter case, as the different denaturant gel solutions are being mixed, they can be delivered into the gel plates using gravity flow or, for better reproducibility, using a peristaltic pump.

The appropriate denaturing gradient for the analysis of a community can be determined using either perpendicular or parallel gels. These gel types are differentiated by orientation of the gradient with respect to the direction of electrophoresis; the electrical current is applied either parallel or perpendicular to the denaturing gradient. In parallel gels the higher denaturant concentration is towards the bottom of the gel, whereas in perpendicular gels it is towards the right side. Perpendicular gels are poured with the plates on their side and then rotated to the correct orientation for electrophoresis. Using parallel gels to determine the appropriate gradient, communities are first examined using a broad gradient range (e.g., 20 to 80% denaturant). The denaturant concentrations are then adjusted to include all the bands while maximizing their separation. The advantage of this approach is that optimal gradients for many different communities can be assessed simultaneously, whereas the denaturing gradient for only one community at a time can be determined using perpendicular gels. In this method, a well spanning the width of the gel is loaded with the community PCR product. After electrophoresis, the gel is examined to determine the approximate denaturant concentrations at which the DNA strands melted (Fig. 2). These denaturant concentrations are then used for subsequent analysis using parallel gels. The advantage of this method is that the melting behavior of the DNA can be observed, which is particularly useful if previously untested primers are being used. Regardless of the method chosen to determine the appropriate gradient, community analysis is performed using parallel gels. In some cases, it may be necessary to characterize communities using two or more different gradients to resolve bands with similar melting characteristics.

41.3.3. Running the Gel

The other important variable that must be determined is the time required for electrophoresis to maximize resolution between bands. Once the denaturant gradient to be used is established, the optimal electrophoresis time can be determined using a time interval experiment. For example, a

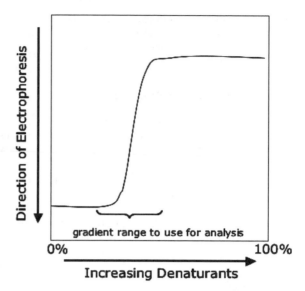

FIGURE 2 Diagram of a melting profile of a single PCR product separated using perpendicular DGGE. The approximate denaturing gradient to use would be 25 to 50% denaturant.

mixture of PCR products is loaded into the gel at 20-min intervals. At the end of the experiment, the minimum time for maximum separation of the PCR products can be observed on the gel (Fig. 3).

Standard markers are used to monitor gel-to-gel variability. The marker should be composed of DGGE PCR products, and under no circumstances should size markers be used. Enough organisms with different sequence compositions should be chosen for the markers to span the denaturing gradients typically used in one's laboratory. In addition, the most commonly studied organisms in your laboratory should be included to readily detect potential contamination in your PCR. To minimize intensity differences between

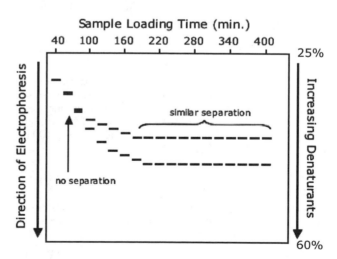

FIGURE 3 Diagram of the separation of a two-member community inoculated over 6.6 h at 20-min intervals using parallel DGGE. This example uses a 25 to 60% denaturing gradient and electrophoresis at 200 V and 60°C. The optimal separation time would be 4.5 to 5 h.

strains, PCR should be performed independently on each strain chosen and then the products should be mixed in equal concentrations. Standard markers must be run on all gels to assess gel variations and to make comparisons between gels.

41.3.4. Detailed DGGE Protocol

41.3.4.1. PCR Amplification for DGGE

A basic protocol for amplification of the V3 region (bases 338 to 534) of the 16S rRNA gene from community DNA extract is as follows.

1. Make a PCR master mix for the total number of reactions (final volume, 50 μl per reaction) to be performed. When calculating the required volume, include additional reactions for negative (no DNA) control, positive control, and liquid loss during handling. The final concentrations in the PCR mixture are:

- 1× PCR buffer
- 2.0 μM MgCl$_2$
- 0.8 mM deoxynucleoside triphosphates (dNTPs)
- a 0.5 μM concentration of each primer (BA338F and UN518R)
- 0.1% bovine serum albumin (BSA) and 5% dimethyl sulfoxide
- 1 U of *Taq* polymerase

2. Dispense 49 μl (less if more than 1 μl of DNA template is needed) into each PCR tube.

3. Add 1 μl of DNA (5 to 50 ng) template. The optimal template DNA concentrations to maximize PCR products and minimize PCR artifacts may vary with the source and should be determined empirically.

4. Place tubes into a thermal cycler and run the following PCR program.

- Denaturation at 94°C for 5 min
- 25 cycles of:

  ■ Denaturation at 94°C for 30 s
  ■ Annealing at 55°C for 30 s (this temperature will vary depending on the primer pair chosen)
  ■ Extension at 72°C for 90 s

- Final extension at 72°C for 10 min
- Hold at 4°C until retrieved

5. Check PCR products for size and quantity on an agarose gel. Load 1 to 5 μl onto a 1 to 1.5% agarose gel (agarose percentage is dependent on PCR product size; buffer is 1× TAE where a liter stock of 50× TAE contains 242 g of Tris base, 57.1 ml of glacial acetic acid, and 100 ml of 0.5 M EDTA of pH 8.0 (41).

41.3.4.2. Preparing Gels Using the Bio-Rad DCode Mutation Detection System

A basic protocol using the DCode mutation detection system (Bio-Rad Laboratories Inc.) is given below. See the Bio-Rad manual for more details.

1. Prepare two stocks of 8% bis-polyacrylamide (from a 40% stock made with 37.5:1 acrylamide/bisacrylamide) in 0.5× TAE gel solution with one containing 0% and the other 100% (7 M urea, 40% formamide) denaturants.

2. Wash glass plates, spacers, and combs thoroughly with warm water and detergent. Assemble the plates to-

gether using the 1.5-mm-thick spacer. The spacers can be lightly coated with silicone grease to reduce "smiling" of bands in the outer lanes of the gel.

3. Before locking the plates into the pouring stand, put a drop of melted 1.5% agarose at each end of the gasket under the position where the plate spacers will fall to prevent leaking. Clamp in the plates, and then seal around the bottom edges of the plates if leaking has been a problem. If leaking becomes excessive, it may be necessary to replace the gasket.

4. For each gel, make 15 ml of desired low-denaturing and 15 ml of high-denaturing solutions. For example, make a low denaturant of 25% (mix 11.25 ml of 8% bis-polyacrylamide gel solution containing 0% denaturant and 3.75 ml of gel solution containing 100% denaturant) and a high denaturant of 60% (mix 6.0 ml of 0% denaturant gel solution and 9.0 ml of 100% denaturant gel solution). If gels with the same denaturant concentrations are to be repeatedly used, make stock solutions of 8% bis-polyacrylamide gel solution with those denaturant concentrations. The volumes for different gel systems may vary and can be determined by measuring the volume of water needed to fill the gel plates.

5. The cam-operated gradient former uses disposable 30-ml syringes to deliver the denaturant gel solutions. The levels of the syringes on the cam must be adjusted for the volume of the gel. Before filling syringes, make sure they move smoothly when the cam is turned and that the syringes will completely extrude the solutions. A "Y" tubing system mixes the two denaturant solutions and delivers it between the gel plates using a syringe needle.

6. Once everything is properly assembled, add 120 μl of APS (ammonium persulfate; 10% fresh or frozen stock) and 5 μl of N, N, N', N"-tetramethylenediamine (TEMED) to the low- and high-denaturant solutions and swirl gently to mix.

7. Take up denaturing and acrylamide solutions into the syringes. For parallel gels, the syringe with the low denaturant is placed in front and the high denaturant is placed on the backside of the gradient maker. Make sure all the air is removed from the syringes before placing them into the gradient maker. Place the delivery syringe needle between the two plates.

8. Slowly turn the wheel until the syringes are empty and the gel is completely poured. For reproducibility, each operator must pour gels at the same rate each time.

9. Let the gel polymerize for about 10 to 15 min and then place comb at a slight angle between the plates.

10. To form wells, add 40 μl of APS and 2 μl of TEMED to 5 ml of 8% bis-polyacrylamide containing 0% denaturant and mix by swirling. Using a pipette, slowly add solution to the top of the gel and then gently push in comb, avoiding making bubbles. Let gel polymerize for at least 2 h to overnight.

41.3.4.3. Running the Gel

1. To run the gel, make approximately 7 liters of 0.5× TAE (must be the same concentration as the gel) and fill the buffer chamber. Reserve about 0.5 liter for use later.

2. Preheat buffer until the temperature is about 50°C, interrupt heating, and attach the gel plates to the core apparatus. Two sets of plates must be attached. If only one gel is required, then two plates with no spacers or gel should be used as the other set.

3. Place plates into the buffer chamber, fill top reservoir with remaining buffer, and continue heating until 65°C is reached (about 2 h). The buffer will cool to the desired 60°C while loading.

4. Flush wells with a syringe to expel dissolved acrylamide. Load 5 to 50 μl of PCR product with loading dye into the wells. The volume loaded will depend on the number of expected bands; as the number of bands increases, the volume loaded must increase.

5. Reset the temperature to 60°C, run at 20 V for 10 min and then at 200 V for 10 min, and then turn on the recirculation pump (this minimizes sample washout from wells). Continue running at 200 V for the predetermined optimal time (typical time is 5 h). It is possible to vary the voltage and electrophoresis time to accommodate your workday. However, comparisons should be done to ensure that bands remain sharp for easier analysis.

6. When the run is complete, take apart the gel apparatus and remove the gel onto plastic wrap for easier manipulation.

7. Leave the gel on the plastic wrap and place into a no. 5 plastic container that just fits the gel. Glass trays should not be used because glass binds the SYBR Green dye.

8. Stain the gel for 10 to 15 min with SYBR Green I (FMC) diluted 5,000 times (4 μl in 20 ml of 0.5× TAE for a full gel) with slight shaking and minimal light exposure.

9. Pour off the stain and rinse gel briefly with 0.5× TAE to remove excess dye.

10. Visualize on a transilluminator and photograph. Alternatively, bands can be visualized after silver staining or ethidium bromide staining. Silver staining is as sensitive as SYBR Green I but is more time-consuming, whereas ethidium bromide takes a similar amount of time but is less sensitive. Figure 4 illustrates different community profiles observed using DGGE.

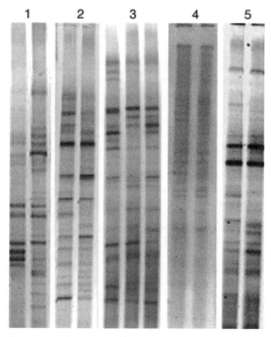

FIGURE 4 DGGE profiles of PCR products of 16S rDNA communities from bioreactors (lanes 1), lakes (lanes 2), corn rhizosphere (lanes 3), bulk agricultural soils (lanes 4), and soils contaminated with heavy metals and organic solvents (lanes 5). With the exception of the bulk agricultural soils, intense bands are observed in each profile. Shown is an inverted image of DGGE gels stained with SYBR Green I.

41.3.4.4. DGGE Band Sequence Determination

One of the advantages of this method is that bands from the gel can be sampled and the nucleotide sequence determined (36). However, only a fraction of the rRNA gene is determined from DGGE bands; therefore, identification of the population represented by the band is tentative.

1. Locate the band(s) of interest after staining and then collect a sample by touching a sterile 1-ml pipette tip to the center of the band.

2. Remove core and place into 20 μl of sterile PCR-grade H₂O in a 1.5-ml microcentrifuge tube. Use a new pipette tip for each band.

3. Incubate tube at 4°C overnight to allow the PCR product to diffuse from the gel into the water.

4. Confirm that the correct band has been extracted by amplifying 5 to 10 μl of the water extract by PCR and compare the product by DGGE with the original sample.

5. If a mixture of bands is found, the band picking process can be repeated until a single band is achieved. Alternatively, the mixture can be PCR amplified using the same primers but without the GC clamp. The PCR products can then be cloned and the band of interest identified by DGGE after amplification of each clone using the primers with GC clamp.

6. Either the PCR product or clone can then be used for nucleotide sequence determination. The primer used for amplification can also be used for sequencing.

7. Since it is possible that a band represents more than one population, multiple clones or PCR products must be sequenced.

41.4. T-RFLP

T-RFLP has been used as an effective tool in the dissection of microbial communities (3, 16, 19, 23–25, 27, 28, 31). The overall protocol is depicted in Fig. 1 and includes (i) the isolation of community DNA (or RNA), (ii) the PCR amplification of the target, (iii) restriction digestion, and (iv) the size separation of the restriction fragments. The essential technical features of the procedure include the use of fluorescently tagged primers in the amplification step and the size separation of fragments on a high-resolution automated sequencing system. This combination of technical protocols yields a system with high throughput capabilities and sensitivity, making it ideal for microbial ecological studies where extensive samplings and replicates are necessary.

41.4.1. The Target

Similar to the case with DGGE, the 16S rDNA has been the most frequently used marker for T-RFLP analysis. The tremendous phylogenetic utility of this marker and the robust sequence database have contributed to this. Moreover, when dissecting or comparing communities, one of the first attributes of interest is the biodiversity, or what species are present? T-RFLP and DGGE of the SSU rRNA genes provide a reasonable first draft estimate of biodiversity, although as pointed out previously, neither DGGE nor T-RFLP of SSU rRNA genes can describe the enormous diversity of the microbial world.

Other markers have been targeted using the T-RFLP approach. These include mercury resistance genes (3), nitrate reductase (39), *nifH* genes, and methyl-coenzyme M reductase, alpha-subunit (*mcrA*/*mrtA*), in methanogenic archaea. In the case of translated genes, codon degeneracy and the wobble base can make primer design considerably more challenging. However, judicious degeneracy within the primer is tolerated in T-RFLP, where the final separation is based on fragment length and not thermal stability, as with DGGE or TGGE. Care and rigor in primer design must be exercised nonetheless, particularly in the case of markers with known paralogs. Paralogs that are amplified along with the intended target will confound estimates of diversity. In this regard, the robustness of one's results resides in the specificity of one's primers.

41.4.2. Primer Selection

In the following discussion, rDNA is used as an example for primer selection. In the case of 16S rDNA, there are currently over 78,000 rRNA sequences in the RDP (4). Considerable effort has been invested in identifying universal, domain-specific, group-specific, and even species-specific probes and primers (1, 22). There are several highly conserved regions in 16S rRNA that have served as targets for probes. Table 1 presents the more commonly employed primers for the bacteria, archaea, and eukaryotes. With gel and capillary electrophoresis, the limits of resolution for fragment size are approximately 700 and 1,000 bases, respectively. Optimally, the primers selected for T-RFLP should amplify nearly the entire gene so that all possible restriction sites within the target are included. However, with the stated resolution of the electrophoresis systems, not all fragments will be detectable if only one labeled primer is employed. This can lead to losing entire phylogenetic groups from the analysis. Thus, complete coverage of the molecule through the use of both forward and reverse labeled primers is recommended. It should be noted that the collection of fragments derived from both forward and reverse labeled primers are a partially overlapping set; hence, diversity cannot be estimated by a simple addition of data sets. The type of attached fluorescent label is usually dictated by the detection technology employed in the automated sequencers. All automated sequencers with multiple fluor detection capabilities allow for the option of running three differently labeled PCR products as well as labeled size markers in the same lane or capillary. This increases the cost-effectiveness of the procedure. For the ABI systems, carboxyhexachlorofluorescein (HEX), 6-carboxyfluorescein (FAM), 6-carboxy-2,4,7,7-tetrachlorofluorescein (TET), 6-carboxytetramethylrhodamine (TAMRA), and carboxyrhodamine (ROX) have been routinely employed. Table 1 lists primers commonly used in T-RFLP of 16S rRNA genes.

41.4.3. Primer Specificity

Primer design has been a particularly productive experimental field. The use of a phylogenetic marker in the analysis means that the sequence differences that distinguish phylogenetic groups can be utilized in the construction of primers with phylogenetic specificity. Well-characterized primers and probes have been collected in several reviews (1, 22) and databases (http://www.microbial-ecology.de/probebase/).

41.4.3.1. In Vitro Specificity

Once primers have been selected and synthesized, it is important to test for the expected specificity. A thermocycler that can form a temperature gradient across the block is invaluable in this step. PCR amplifications are set up with genomic DNA from representative isolates from inside and outside of the specifically targeted group. Testing PCR am-

plification across a temperature range for these isolates will establish the level of stringency required for the advertised specificity.

41.4.3.2. In Silico Specificity

In addition to in vitro tests of specificity, it is also worthwhile to check the specificity in silico, realizing that the specificity of primers is related to the robustness of the database at the time of design. With sequences being added to databases at an increasing rate each year, phylogenetic diversity is a changing entity, and a primer perceived as highly specific 5 years ago may be only moderately specific with today's database. The most familiar method is undoubtedly a BLAST search against the nonredundant sequence database (BLAST at http://www.ncbi.nlm.nih.gov/BLAST/). The RDP (4) also provides a "Probe Match" function that searches the updated database for matches. In addition, it is worth mentioning ARB, a freely distributed software written and maintained by the Technical Institute of Munich (http://www.arb-home.de/). This software provides not only a systematic probe/primer analysis function that reports on the number and phylogenetic distribution of hits (matching sequences) in an rRNA database but also alignment and phylogenetic functions and a variety of features that make maintaining a curated database within the laboratory possible. ARB runs on Unix/Linux and is a powerful and sophisticated sequence analysis environment with a commensurate learning curve.

41.4.4. PCR

PCR is an exquisitely sensitive technology, and for this reason, great care must be taken to avoid contamination. PCR amplification from community DNA is frequently more difficult than amplification from isolates. This is certainly due, in part, to the copurifying contaminants that can accompany DNA through the extraction protocols (as discussed above). PCR adjuvants such as BSA and T4 gene 32 protein can suppress the actions of inhibitors. Thus, a combination of effective DNA isolation procedures and PCR additives will give the most consistent results.

41.4.4.1. The Reaction

After isolating community DNA, pilot PCRs are run to establish that the DNA is of sufficient purity to serve as a productive template. Pilot reactions as small as 12.5 μl can be performed with thermal cyclers designed to handle 200- to 250-μl tubes. Several template concentrations are used to test a 10-fold concentration range (5 to 50 ng for a 25-μl reaction), and unlabeled primers are used to reduce costs. The resulting PCR products are visualized on an agarose gel (0.8 to 1.0%) for product abundance (PCR efficiency) and priming specificity. If needed, PCR efficiency can be improved through a systematic optimization of component concentrations. Focusing on the template and Mg^{2+} concentrations will bring most amplifiable templates to an acceptable level of productivity in PCRs. After optimizing the PCRs, fluor-derivatized primers are substituted in the reaction for unlabeled primers and the reaction volume is increased to 100 μl. Depending on amplification efficiency, from one to three 100-μl reactions may be required to provide enough labeled product. The detection system used in the sequencing apparatus dictates the fluors that can be used. In the ABI systems, the detectable dyes are 6-FAM, HEX, TET, TAMRA, and ROX. Because these fluors are subject to photobleaching, the primer stocks and the reactions should be protected from direct and prolonged exposure to light. If two labeled primers are to be used for full coverage of the target as mentioned above, different fluors should be selected for the forward and reverse primers so that both signals can be detected in the same lane/capillary.

41.4.4.2. Specificity of the PCR

Specificity in PCR amplification is a critical issue for comparative analysis. The fluors used in T-RFLP provide high sensitivity in the gel and capillary systems; hence, any spurious PCR products will be detected and possibly scored as a restriction fragment. If PCR products of unanticipated sizes are detected in the analytical agarose gel, the PCR must be optimized, usually through manipulation of the annealing temperature or concentrations of $MgCl_2$, primer, or template. If secondary products remain problematic after optimization attempts, the PCR product of correct size can be gel purified. This is an unsavory option because of the additional preparative step and the highly variable recovery of DNA from the gel.

41.4.4.3. PCR Components

Standard PCR conditions are presented in Table 2. Observing standard aseptic techniques throughout the setup of PCRs will prevent contaminating templates from contributing to one's experiments. Many commercially available Taq polymerases are suitable for T-RFLP. In most cases, the vendor supplies 10× buffer and a separate MgCl solution. In practice, the PCR is usually set up in a two-step process that serves to minimize chances of contamination. A master mix is assembled in a clean UV-irradiated area. The master mix contains, in order of addition, water, buffer, $MgCl_2$, dNTPs, primer, and Taq. The templates are then added to the reaction tubes and a uniform volume of master mix is pipetted into each reaction tube. Reproducibility will be increased if the stock concentrations are set up such that the volumes pipetted are not smaller than 2 to 3 µl. It is imperative to include both a negative and positive PCR in every set of PCR amplifications.

Either a hot-start or touchdown PCR technique (8) is recommended to diminish nonspecific priming events at the outset of PCR. An initial soak for 3 to 5 min at 94°C is followed by 25 to 30 cycles of 1 min of denaturation at 94°C, 45 s of annealing at 45 to 60°C (temperature determined by primer melting temperature), and 45 s at 72°C for

TABLE 2 Standard PCR conditions

Component	Stock concn	Final concn
Buffer	10×	1×
dNTPs	2.0 mM	200 µM
$MgCl_2$	25 mM	1.5–2.5 mM
Forward primer	~50–100 pmol/µl	~5–10 pmol/25-µl reaction
Reverse primer	~50–100 pmol/µl	~5–10 pmol/25-µl reaction
BSA	1%	0.05%
Template	10–500 ng/µl	5–50 ng/25-µl reaction
H_2O		To volume

extension. A final extension at 72°C for 5 min completes the amplification.

41.4.4.4. Purification of PCR Products

PCR products should be purified away from residual Taq activity and ionic conditions that may be suboptimal for restriction digestion. Commercially available PCR cleanup kits have proved effective for this step (e.g., Promega and Qiagen).

41.4.4.5. Quantitation of PCR Products

As mentioned above, comparative analysis requires that all samples be equivalent in concentration. Thus, PCR products are quantitated after purification by A_{260} or fluorometry (see above). Because sequencing machines vary in sensitivity, determination of the optimal concentration loaded onto the system is derived empirically.

41.4.5. Restriction Digestion

Digestion with restriction endonucleases is conducted according to the manufacturer's recommendations. While this would seem to be a trivial aspect of the procedure, given the extensive use made of restriction enzymes, a number of pitfalls face the investigator at this step.

41.4.5.1. Selection of Enzymes

The selection of an enzyme for T-RFLP analysis is not trivial. For many investigators, the objective is to reveal as many phylotypes in the community as possible. In this way, community diversity can be estimated and/or followed as a function of time or amendment. If 16S or 23S rDNA is the target, then a sufficiently large sequence database exists for identifying enzymes appropriate to the task. In general, four-base cutters have been used in order to maximize the likelihood of a restriction target within the boundaries of the amplified product. However, some four-base target sequences are more conserved within the 16S rRNA sequence than others. Figure 5 displays the frequency distribution of 16S rDNA terminal fragments generated from the in silico digestion of 20,070 16S rDNA sequences (1,200 bases minimum in length) with two restriction enzymes, HaeIII and RsaI. The digestions were carried out assuming that PCR products were generated with a labeled 63F bacterial domain-specific primer. Hence, all fragments are derived from the 5′ terminus of bacterial rRNA genes. In the top panel, showing the HaeIII digestion, there is a skewed distribution with two highly conserved positions at approximately positions 32 and 171 (Escherichia coli numbering) and 503 unique terminal fragment sizes. In contrast, digestion with RsaI revealed a broader distribution of terminal fragments, though still clustered at some level, with a greater number of unique fragments (730) derived from the same database. Thus, the phylotype resolving power of RsaI was substantially better than that of HaeIII. While HaeIII is clearly suboptimal for detecting the greatest diversity, under circumstances where a targeted phylotype has a HaeIII restriction site different from the highly conserved site, it may prove useful by separating most of the community from the target of interest. Using restriction enzymes in this way can only be achieved with reference to the database. Figure 6 displays the number of unique fragments that are generated with 19 different restriction enzymes in an in silico restriction site analysis of 27,060 rRNA sequences. The data were derived using PatScan (9) and the 63F primer with a single restriction site as the queried sequence motifs. Note that

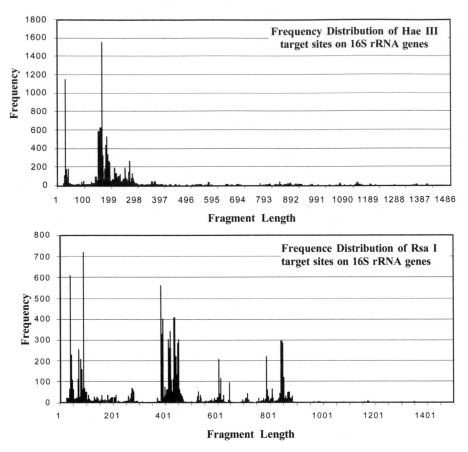

FIGURE 5 Terminal fragment size distribution. An rRNA database of 20,070 sequences was digested in silico with HaeIII (upper panel) and RsaI (lower panel). The frequency (ordinate) of terminal fragment sizes (abscissa) appearing in the database is presented.

two restriction enzymes result in the absence of unique fragments because there is a conserved restriction site within the primer (CviRI and NlaIII). Based on this analysis, the enzymes most suited for estimating phylotypes diversity are AvaII, DpnI, HhaI, NciI, RsaI, and TaqI. Table 3 summarizes the same digestions but also shows the distribution of restriction sites across the 16S gene. The table is sorted in order of increasing number of unique fragments derived from the listed enzymes. While NciI and MaeII look attractive based on the number of unique fragments generated, note that there are a significant number of sequences with target sites more than 1,000 bp from the 5′ terminus.

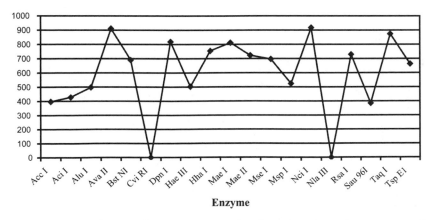

FIGURE 6 Number of unique fragments generated with commonly used restriction endonucleases. A 20,070-sequence rRNA database was digested in silico with 19 restriction enzymes (abscissa). The number of unique terminal fragment sizes is reported for each enzyme (ordinate).

TABLE 3 Number of 5′ proximal restriction target sequences on 16S rDNA[a]

Restriction enzyme	Recognition sequence	No. of sequences primed (63F) and digested[b]	Distribution of terminal fragments		
			<500 bp	500–1,000 bp	1,000–1,500 bp
NlaIII	CATG	18,929	0	0	0
CviRI	TGCA	18,928	0	0	0
Sau96I	GGNCC	18,923	18,828	87	8
AccI	CGCG	18,929	18,864	65	0
AciI	CCGC	18,929	18,886	43	0
AluI	AGCT	18,928	18,463	411	54
HaeIII	GGCC	18,917	18,249	483	185
MspI	CCGG	18,914	18,205	699	10
TspE1	AATT	18,926	10,989	7,935	2
BstNI	CCWGG	18,913	6,149	12,752	12
MseI	TTAA	18,928	9,515	9,387	26
MaeII	ACGT	18,923	16,588	796	1,539
RsaI	GTAC	18,927	14,305	4,569	53
HhaI	GCGC	18,913	14,854	2,934	1,125
MaeI	CTAG	18,888	13,903	4,332	653
DpnI	GATC	18,795	16,358	1,219	1,218
TaqI	TCGA	18,865	7,086	11,655	124
AvaII	GGACC	15,365	10,695	4,297	373
NciI	CCCGG	18,478	7,462	6,932	4,084

[a]Based on an RDP library of 27,060 sequences, each of which is at least 1,200 bases long (REF). The database was screened with PatScan (REF) for primer and restriction sites.

[b]A total of 18,932 out of 27,060 sequences were recognized by the 63F primer when the criteria required a perfect match in the four bases at the 3′ terminus and allowed up to three mismatches in the 5′ proximal 17 bases.

Hence, if only the forward primer is labeled, neither gel nor capillary electrophoresis will resolve these large terminal fragments. Thus, in selecting restriction enzymes for T-RFLP, both the number of unique targets and the distribution of target sites should be considered.

41.4.5.2. The Digestion Reaction

The community itself is a variable, and some empirical assessment of enzyme effectiveness will inevitably be required for each community. In our experience, digesting from 200 to 600 ng of PCR product in a 10- to 15-μl reaction provides sufficient fluorescent material for subsequent detection on an ABI gel or capillary, assuming that approximately 1 to 3 μl of the digest is loaded onto gels. If the resulting profiles appear to be underloaded, the amount of material loaded should be increased. In a complex community, we have found that a total fluorescence of 5,000 to 10,000 U is optimal (2).

41.4.5.3. Incomplete Digestions, Nonspecific Cuts, and Conformational Twists

Restriction enzymes are usually well behaved if used as recommended by the vendor. However, because the fluorescent detection is quite sensitive, controls should be employed to ensure that the digestions are complete. Clearly, partial digestions will produce fragments that will be misinterpreted as terminal. In the same spirit, overly long digestions (>4 h) should be eschewed so as to avoid the formation of illegitimate digestion products. Recently, single-stranded PCR products and/or malformed duplexes have been implicated

as potential problems (10) because kinetically rather than thermodynamically formed secondary structures may block proximal restriction sites and reconstitute distal restriction sites, leading to pseudoterminal fragments. Hence, running controls is imperative. Included should be undigested samples for each PCR and a digested amplified product derived from an isolate or clone to test for spurious PCR products and partial or nonspecific digestions, respectively. In addition, pretreating the PCR products with mung bean nuclease may be necessary (10). Any inexplicable band detected in these controls serves as an indicator of undesirable events.

41.4.6. Loading the Gels or Capillaries—Size Selection of Restriction Fragments

After digestion with restriction endonucleases and inactivation of the enzyme, the products are mixed with labeled size markers and denaturant (formamide), loaded, and electrophoresed under denaturing conditions in an automated sequencing apparatus. The general format among the different manufacturers of this equipment is roughly the same. In general, a slab gel format will provide reliable reads to 600 to 700 bases, while capillary setups extend the reliability of runs out to 800 to 1,000 bases. The former are relatively unaffected by the ionic conditions of digestion. However, capillaries are loaded by electrophoretically injecting the DNA into the capillary; therefore, the ionic conditions of the restriction digest may, if possessing high ionic strength, reduce the efficiency of DNA loading. In this case a desalting step is recommended after digestion. This can be as simple as performing a quick ethanol precipitation on the digestion products. In many cases, excellent results can be obtained without desalting.

41.4.6.1. Electrophoresis Conditions

The conditions employed for electrophoresis vary from one automated system to another. We provide two examples here, derived from experience with ABI gel and capillary systems. For acrylamide gels, restriction digests of PCR-amplified community DNA are denatured at 94°C, chilled, and loaded onto a 36-cm, 6% denaturing polyacrylamide gel. Size standards (e.g., ABI Tamara 2500) are included in each lane. Electrophoretic runs were 16 to 20 h in an ABI automated sequencer (model 373A; Applied Biosystems Instruments, Foster City, CA) run in GeneScan mode with limits of 2,500 V and 40 mA. For capillary systems (e.g., ABI 3100 Genetic Analyzer) the denatured sample containing size standards (e.g., MM1000; Bioventures) is loaded with injection times of 10 to 60 s into a 36-cm capillary containing POP4 polymer. Electrophoresis is for approximately 2 h at 50 kV.

41.4.6.2. Size Markers and Internal Controls

Of paramount importance for fragment size determination is the inclusion of size standards in each electrophoretic run. Because multiple detection channels can be operated on many systems, size markers can be included in every lane, making size comparisons across lanes highly reproducible. In general, each automated electrophoresis manufacturer includes software for determining fragment sizes based on the electrophoretic mobility of size standards included in the run. In most cases, this provides sizing accuracy of less than one base. However, in both gels and capillaries we have seen anomalies with respect to fragment size estimate errors that are larger and range from 1 to 4 bases. Such anomalies underscore the need for technical replicates.

41.4.7. Viewing and Collecting the Data

Viewing and retrieving the data are largely dependent on vendor-supplied software. In the case of ABI, GeneScan, the software used to run the gel, is used to view and collect tabulated data as well. Figure 7 shows a T-RFLP profile derived from a soil sample in full range (Fig. 7A, 50 to 600 bases) and expanded range (Fig. 7B, 100 to 200 bases). Note that by expanding the horizontal scale (fragment size), one can easily discern peaks at a resolution of less than a base in length. From a GeneScan view of the profiles, the tabulated data are extractable in an ASCI formatted file. The viewing screen of GeneScan is limited to eight panels of data; thus, comparisons within a large data set are problematic. Genotyper software (ABI) overcomes this limitation by permitting the user to import as many GeneScan files (a single GeneScan file corresponds to one lane or capillary) as is desired. Moreover, it also facilitates the alignment of T-RFLP profiles by controlling the extraction of the tabulated data from all selected scans in a way that allows the user to bin all fragments within user-specified ranges. Thus, all fragments that are within a window of, for example, ±0.5 base are grouped together in the tabulated data. The data are formatted with community samples (lanes or capillaries) in rows and fragments (size and peak height) in columns. The shortcoming of this software is the column limitation of 122. Thus, if your data set is derived from a community with high diversity and therefore many terminal fragments, sev-

eral extractions of data may be required to collect all of the tabulated data from the Genotyper project file. T-RLFP has another advantage in that the sequence database can be used to determine the possible genus and species of the detected populations (29). It should be clear that one cannot identify the genus and species present with the size of a single terminal fragment. Hence, the information derived from comparison of detected terminal fragment sizes with the database supplies guidance for confirmatory experiments, not identity.

41.4.8. Detailed Protocol for T-RFLP

The following specific protocols have been productive for broad surveys of phylogenetic domains using domain-specific primers.

1. PCR amplification for T-RFLP. After extracting community DNA samples, small-volume pilot PCRs are performed to determine optimal conditions followed by labeling reactions with increased volumes. As with DGGE, a master mix containing all components except for the template is prepared under aseptic conditions. The template is aliquoted to reaction tubes, and the reaction is initiated with the addition of master mix to the tubes. The following PCR conditions are for 25-μl pilot reactions. The volume is scaled up to 100 μl for the labeling reactions.

Sterile deionized water	15.25 μl
10× PCR buffer	2.5 μl
dNTP mix (2 mM each)	2.5 μl
MgCl$_2$ (50 mM)	0.75 μl
BSA (10 mg/ml)	0.25 μl
27F primer (10 μM)	0.5 μl
1492R or 1392R primer (10 μM)	0.5 μl
Taq DNA polymerase (10 U/μl)	0.25 μl
Template DNA	2.5 μl

2. When employing the general domain-specific primers described in Table 1, the following thermal cycling parameters have proven successful.

- 94°C for 5 min
- 25 to 30 cycles of:
  - 94°C for 40 s for denaturation
  - 55 to 57°C for 40 s for annealing
  - 72°C for 1.5 min for polymerization
- 72°C for 10 min (final extension)
- 4°C until retrieval

3. Check PCR products (2 to 5 μl) on a 1% agarose gel (1× TAE buffer; see section 41.3.4.2) and optimize PCR conditions if necessary.

4. Scale up volume to two 100-μl PCRs/sample for labeling and check reaction products on an agarose gel as described above.

5. Combine sample duplicates and purify the PCR products with a commercially available kit (e.g., Promega or Qiagen). We have noted that elution of the DNA from some commercial matrices is influenced by the source of the DNA (presumably by copurifying salts).

6. Quantitate DNA concentration spectrophotometrically (A$_{260}$) or by fluorometry with PicoGreen (Molecular Probes, Inc.).

7. Incubate 200 to 400 μg of PCR product in a restriction digestion according to the manufacturer's recommen-

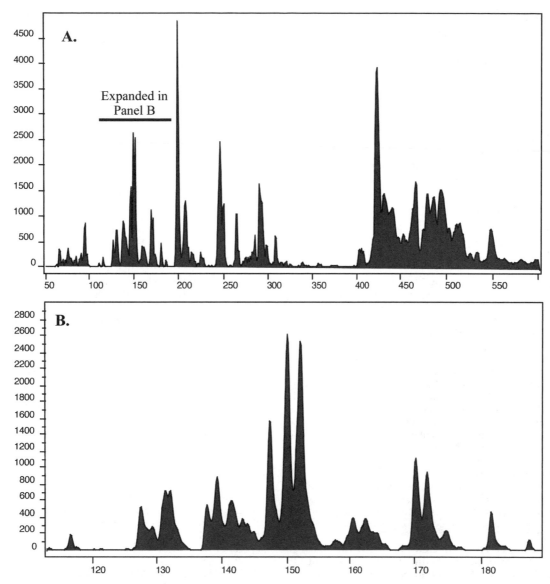

FIGURE 7 T-RFLP profile of a soil community. Samples were amplified with HEX-labeled 27F primer matched with unlabeled 1492R targeting the 16S rRNA genes. PCR products were digested with HhaI and run on an ABI 273A gel system. T-RFLP profiles were viewed with GeneScan. (A) A 50- to 600-base scale on abscissa; (B) expanded base scale of data from panel A.

dation. Generally this is performed in 10- to 20-μl reactions as described below.

- 200 to 400 ng of labeled PCR product
- 5 to 10 U of restriction endonuclease
- 1.0 μl of 10× buffer
- H₂O to 10 μl

8. Terminate digestion by heating to 75°C for 10 min, and freeze samples until electrophoresis.

9. Load 1 to 2 μl of each sample onto a sequencing gel or inject for 10 to 30 s onto a capillary for electrophoretic separation of fragments. Each lane or capillary also contains size markers for determination of fragment sizes. Electrophoresis is performed as described above.

10. Terminal fragment sizes are estimated with several algorithms that are usually part of software for the automated gel system. For example in the ABI systems, the GeneScan software provides five algorithms for interpolating fragment sizes from known standards (second- and third-order least squares, cubic spline, and local and global southern).

11. The fragment data, including fragment length and peak amplitude, are binned or grouped by fragment length across samples or communities. These data are extracted from the sequencer software and imported into a spreadsheet (e.g., Excel™) for comparative community analysis (see above). Fragment sizes of apparent interest, as judged by distribution across communities or peak amplitudes, are compared to a terminal restriction fragment database to determine possible phylogenetic affiliation.

41.5. SUMMARY

It has been only through the window of molecular techniques that we have come to appreciate the great diversity of the microbial world. DGGE and T-RFLP are imperfect methods for assessing this microbial diversity because at present they are incapable of distinguishing every population within a community and every genomic variant within a population. Nonetheless, these techniques provide the most rigorously tested and cost-effective approaches for the ecological assessment of a large number of communities. Used judiciously, comparative appraisals of complex communities with DGGE and T-RFLP can reveal even subtle changes in community structure.

41.6. REFERENCES

1. **Amann, R. I., W. Ludwig, and K.-H. Schleifer.** 1995. Phylogenetic identification and in situ detection of individual microbial cells without cultivation. *Microbiol. Rev.* **59**:143–169.
2. **Blackwood, C. B., T. Marsh, S. H. Kim, and E. A. Paul.** 2003. Terminal restriction fragment length polymorphism data analysis for quantitative comparison of microbial communities. *Appl. Environ. Microbiol.* **69**:926–932.
3. **Bruce, K. D., and M. R. Hughes.** 2000. Terminal restriction fragment length polymorphism monitoring of genes amplified directly from bacterial communities in soils and sediments. *Mol. Biotechnol.* **16**:261–269.
4. **Cole, J. R., B. Chai, T. L. Marsh, R. J. Farris, Q. Wang, S. A. Kulam, S. Chandra, D. M. McGarrell, T. M. Schmidt, G. M. Garrity, and J. M. Tiedje.** 2003. The Ribosomal Database Project (RDP-II): previewing a new autoaligner that allows regular updates and the new prokaryotic taxonomy. *Nucleic Acids Res.* **31**:442–443.
5. **Dice, L. R.** 1945. Measures of the amount of ecologic association between species. *Ecology* **26**:297–302.
6. **Dollhopf, S. L., S. A. Hashsham, and J. M. Tiedje.** 2001. Interpreting 16S rDNA T-RFLP data: application of self-organizing maps and principal component analysis to describe community dynamics and convergence. *Microb. Ecol.* **42**:495–505.
7. **Domann, E., G. Hong, C. Imirzalioglu, S. Turschner, J. Kuhle, C. Watzel, T. Hain, H. Hossain, and T. Chakraborty.** 2003. Culture-independent identification of pathogenic bacteria and polymicrobial infections in the genitourinary tract of renal transplant recipients. *J. Clin. Microbiol.* **41**:5500–5510.
8. **Don, R. H., P. T. Cox, B. J. Wainwright, K. Baker, and J. S. Mattick.** 1991. 'Touchdown' PCR to circumvent spurious priming during gene amplification. *Nucleic Acids Res.* **19**:4008.
9. **Dsouza, M., N. Larsen, and R. Overbeek.** 1997. Searching for patterns in genomic data. *Trends Genet.* **13**:497–498.
10. **Egert, M., and M. W. Friedrich.** 2003. Formation of pseudo-terminal restriction fragments, a PCR-related bias affecting terminal restriction fragment length polymorphism analysis of microbial community structure. *Appl. Environ. Microbiol.* **69**:2555–2562.
11. **Friedrich, M. W., D. Schmitt-Wagner, T. Lueders, and A. Brune.** 2001. Axial differences in community structure of *Crenarchaeota* and *Euryarchaeota* in the highly compartmentalized gut of the soil-feeding termite *Cubitermes orthognathus*. *Appl. Environ. Microbiol.* **67**:4880–4890.
12. **Goldenberg, O., S. Herrmann, G. Marjoram, M. Noyer-Weidner, G. Hong, S. Bereswill, and U. B. Gobel.** 2007. Molecular monitoring of the intestinal flora by denaturing high performance liquid chromatography. *J. Microbiol. Methods* **68**:94–105.
13. **Hill, T. C. J., K. A. Walsh, J. A. Harris, and B. F. Moffett.** 2003. Using ecological diversity measures with bacterial communities. *FEMS Microbiol. Ecol.* **43**:1–11.
14. **Hurtle, W., D. Shoemaker, E. Henchal, and D. Norwood.** 2002. Denaturing HPLC for identifying bacteria. *BioTechniques* **33**:386–391.
15. **Johnson, A. R., and D. W. Wichern.** 2003. *Applied Multivariate Statistical Analysis*, 5th ed. Prentice Hall, Upper Saddle River, NJ.
16. **Kitts, C. L.** 2001. Terminal restriction fragment patterns: a tool for comparing microbial communities and assessing community dynamics. *Curr. Issues Intest. Microbiol.* **2**:17–25.
17. **Kowalchuk, G. A., J. R. Stephen, W. Deboer, J. I. Prosser, T. M. Embley, and J. W. Woldendorp.** 1997. Analysis of ammonia-oxidizing bacteria of the β subdivision of the class *Proteobacteria* in coastal sand dunes by denaturing gradient gel electrophoresis and sequencing of PCR-amplified 16S ribosomal DNA fragments. *Appl. Environ. Microbiol.* **63**:1489–1497.
18. **Kudo, T., M. Ohkuma, S. Moriya, S. Noda, and K. Ohtoko.** 1998. Molecular phylogenetic identification of the intestinal anaerobic microbial community in the hindgut of the termite, *Reticulitermes speratus*, without cultivation. *Extremophiles* **2**:155–161.
19. **Kuske, C. R., L. O. Ticknor, M. E. Miller, J. M. Dunbar, J. A. Davis, S. M. Barns, and J. Belnap.** 2002. Comparison of soil bacterial communities in rhizospheres of three plant species and the interspaces in an arid grassland. *Appl. Environ. Microbiol.* **68**:1854–1863.
20. **Lane, D. J.** 1991. 16S/23S rRNA sequencing, p. 115–175. *In* M. Goodfellow (ed.), *Nucleic Acid Techniques in Bacterial Systematics*. John Wiley & Sons, New York, NY.
21. **Lane, D. J., B. Pace, G. J. Olsen, D. A. Stahl, M. L. Sogin, and N. R. Pace.** 1985. Rapid determination of 16S ribosomal RNA sequences for phylogenetic analyses. *Proc. Natl. Acad. Sci. USA* **82**:6955–6959.
22. **Lipski, A., U. Friedrich, and K. Altendorf.** 2001. Application of rRNA-targeted oligonucleotide probes in biotechnology. *Appl. Microbiol. Biotechnol.* **56**:40–57.
23. **Liu, W. T., T. L. Marsh, H. Cheng, and L. J. Forney.** 1997. Characterization of microbial diversity by determining terminal restriction fragment length polymorphisms of genes encoding 16S rRNA. *Appl. Environ. Microbiol.* **63**:4516–4522.
24. **Liu, W. T., T. L. Marsh, and L. J. Forney.** 1998. Determination of the microbial diversity of anaerobic-aerobic activated sludge by a novel molecular biological technique. *Water Sci. Technol.* **37**:417–422.
25. **Lukow, T., P. F. Dunfield, and W. Liesack.** 2000. Use of the T-RFLP technique to assess spatial and temporal changes in the bacterial community structure within an agricultural soil planted with transgenic and non-transgenic potato plants. *FEMS Microbiol. Ecol.* **32**:241–247.
26. **Marchesi, J., T. Sato, A. Weightman, T. Martin, J. Fry, S. Hiom, and W. Wade.** 1998. Design and evaluation of useful bacterium-specific PCR primers that amplify genes coding for bacterial 16S rRNA. *Appl. Environ. Microbiol.* **64**:795–799.
27. **Marsh, T. L.** 1999. Terminal restriction fragment length polymorphism (T-RFLP): an emerging method for characterizing diversity among homologous populations of amplification products. *Curr. Opin. Microbiol.* **2**:323–327.
28. **Marsh, T. L., W.-T. Liu, L. J. Forney, and H. Cheng.** 1998. Beginning a molecular analysis of the eukaryal community in activated sludge. *Water Sci. Technol.* **37**:455–460.
29. **Marsh, T. L., P. Saxman, J. Cole, and J. Tiedje.** 2000. Terminal restriction fragment length polymorphism analysis program, a web-based research tool for microbial community analysis. *Appl. Environ. Microbiol.* **66**:3616–3620.

30. **Medlin, L., H. J. Elwood, S. Stickel, and M. L. Sogin.** 1988. The characterization of enzymatically amplified eukaryotic 16S-like rRNA-coding regions. *Gene* **71:**491–499.

31. **Moeseneder, M. M., J. M. Arrieta, G. Muyzer, C. Winter, and G. J. Herndl.** 1999. Optimization of terminal-restriction fragment length polymorphism analysis for complex marine bacterioplankton communities and comparison with denaturing gradient gel electrophoresis. *Appl. Environ. Microbiol.* **65:**3518–3525.

32. **Morgan, C. A., A. Hudson, A. Konopka, and C. H. Nakatsu.** 2002. Analyses of microbial activity in biomass-recycle reactors using denaturing gradient gel electrophoresis of 16S rDNA and 16S rRNA PCR products. *Can. J. Microbiol.* **48:**333–341.

33. **Muyzer, G., E. C. de Waal, and A. G. Uitterlinden.** 1993. Profiling of complex microbial populations by denaturing gradient gel electrophoresis analysis of polymerase chain reaction-amplified genes coding for 16S rRNA. *Appl. Environ. Microbiol.* **59:**695–700.

34. **Nakatsu, C. H., V. Torsvik, and L. Øvreås.** 2000. Soil community analysis using DGGE of 16S rDNA polymerase chain reaction products. *Soil Sci. Soc. Am. J.* **64:**1382–1388.

35. **Olsen, G. J., N. R. Pace, M. Nuell, B. P. Kaine, R. Gupta, and C. R. Woese.** 1985. Sequence of the 16S rRNA gene from the thermoacidophilic archaebacterium Sulfolobus solfataricus and its evolutionary implications. *J. Mol. Evol.* **22:**301–307.

36. **Ovreås, L., L. Forney, F. L. Daae, and V. Torsvik.** 1997. Distribution of bacterioplankton in meromictic Lake Saelenvannet, as determined by denaturing gradient gel electrophoresis of PCR-amplified gene fragments coding for 16S rRNA. *Appl. Environ. Microbiol.* **63:**3367–3373.

37. **Pielou, E. C.** 1969. *An Introduction to Mathematical Ecology.* Wiley-Interscience, New York, NY.

38. **Preston, C. M., K. Y. Wu, T. F. Molinski, and E. F. DeLong.** 1996. A psychrophilic crenarchaeon inhabits a marine sponge: *Cenarchaeum symbiosum* gen. nov., sp. nov. *Proc. Natl. Acad. Sci. USA* **93:**6241–6246.

39. **Priemé, A., G. Braker, and J. M. Tiedje.** 2002. Diversity of nitrite reductase (*nirK* and *nirS*) gene fragments in forested upland and wetland soils. *Appl. Environ. Microbiol.* **68:**1893–1900.

40. **Raskin, L., J. M. Stromley, B. E. Rittmann, and D. A. Stahl.** 1994. Group-specific 16S rRNA hybridization probes to describe natural communities of methanogens. *Appl. Environ. Microbiol.* **60:**1232–1240.

41. **Sambrook, J., and D. W. Russell.** 2001. *Molecular Cloning: a Laboratory Manual*, 3rd ed. Cold Spring Harbor Laboratory Press, Cold Spring Harbor, NY.

42. **Schmitt-Wagner, D., M. W. Friedrich, B. Wagner, and A. Brune.** 2003. Axial dynamics, stability, and interspecies similarity of bacterial community structure in the highly compartmentalized gut of soil-feeding termites (*Cubitermes* spp.). *Appl. Environ. Microbiol.* **69:**6018–6024.

43. **Sheffield, V. C., D. R. Cox, L. S. Lerman, and R. M. Myers.** 1989. Attachment of a 40-base-pair G + C-rich sequence (GC-clamp) to genomic DNA fragments by the polymerase chain reaction results in improved detection of single-base changes. *Proc. Natl. Acad. Sci. USA* **86:**232–236.

44. **Sokal, R., and P. H. A. Sneath.** 1963. *Principles of Numerical Taxonomy.* Freeman Publishers, San Francisco, CA.

45. **Stahl, D. A., and R. I. Amann.** 1991. Development and amplification of nucleic acid probes, p. 205–248. *In* M. Goodfellow (ed.), *Nucleic Acid Techniques in Bacterial Systematics.* Wiley, New York, NY.

46. **Staley, J. T., and A. Konopka.** 1985. Measurement of in situ activities of nonphotosynthetic microorganisms in aquatic and terrestrial habitats. *Annu. Rev. Microbiol.* **39:**321–346.

47. **Torsvik, V., J. Goksoyr, and F. L. Daae.** 1990. High diversity in DNA of soil bacteria. *Appl. Environ. Microbiol.* **56:**782–787.

48. **van Hannen, E. J., M. P. van Agterveld, H. J. Gons, and H. J. Laanbroek.** 1998. Revealing genetic diversity of eukaryotic microorganisms in aquatic environments by denaturing gradient gel electrophoresis. *J. Phycol.* **34:**206–213.

49. **von Wintzingerode, F., U. B. Gobel, and E. Stackebrandt.** 1997. Determination of microbial diversity in environmental samples: pitfalls of PCR-based rRNA analysis. *FEMS Microbiol. Rev.* **21:**213–229.

50. **Wagner, T., D. Stoppa-Lyonnet, E. Fleischmann, D. Muhr, S. Pages, T. Sandberg, V. Caux, R. Moeslinger, G. Langbauer, A. Borg, and P. Oefner.** 1999. Denaturing high-performance liquid chromatography detects reliably BRCA1 and BRCA2 mutations. *Genomics* **62:**369–376.

51. **Wawer, C., and G. Muyzer.** 1995. Genetic diversity of *Desulfovibrio* spp. in environmental samples analyzed by denaturing gradient gel electrophoresis of [NiFe] hydrogenase gene fragments. *Appl. Environ. Microbiol.* **61:**2203–2210.

52. **Wilbur, J. D., J. K. Ghosh, C. H. Nakatsu, S. M. Brouder, and R. W. Doerge.** 2002. Variable selection in high-dimensional multivariate binary data with application to the analysis of microbial community DNA fingerprints. *Biometrics* **58:**378–386.

53. **Xiao, W. Z., and P. J. Oefner.** 2001. Denaturing high-performance liquid chromatography: a review. *Hum. Mutat.* **17:**439–474.

MYCOLOGY

Introduction to Mycology

GEORGE A. MARZLUF

It is fitting that fungi have been included in this new edition of *Methods for General and Molecular Microbiology*. Filamentous fungi and yeasts represent a huge and diverse group of organisms that impact human life in many ways, including their use in the production of various foods and of life-saving antibiotics. Many fungal plant pathogens cause enormous loss of food production, and others contribute to spoilage. Pathogenic fungi which attack animals, including humans, cause severe diseases and are difficult to control. In another vein, both filamentous fungi and yeasts have been employed as model organisms in research that has provided major new scientific insights. The first auxotrophic mutants isolated in any organism were obtained in *Neurospora* by Beadle and Tatum in their ground-breaking work that represented the first step in the molecular revolution of genetics. *Neurospora*, yeasts, *Aspergillus*, *Sordaria*, *Coprinus*, *Ascobolus*, and other fungi have made many fundamental contributions, e.g., to understanding the mechanism of recombination, chromosome behavior, complementation, and eukaryotic gene regulation. With the advent of recombinant DNA approaches, interest in the fungi has increased dramatically in recent years, and major developments in understanding complex phenomena such as circadian rhythms and light responses have resulted. DNA-mediated transformation is now routine with yeasts and multiple species of filamentous fungi; indeed, it appears that most fungal species can be transformed. The Fungal Genetics Stock Center located at the University of Kansas maintains and distributes thousands of wild-type and mutant strains of *Aspergillus*, *Neurospora*, *Fusarium*, *Podospora*, *Sordaria*, *Ascobolus*, and other fungal species, as well as vectors and genomic libraries. The Fungal Genetics Meeting held biannually has grown in attendance so greatly that space at the Asilomar Conference grounds in Monterey, CA, is limited. A new, rigorous American Society for Microbiology-sponsored journal, *Eukaryotic Cell*, is devoted to eukaryotic microorganisms and highlights many of the most exciting new developments with fungi. The entire genomes of *Neurospora crassa*, *Aspergillus nidulans*, *Saccharomyces cerevisiae*, and other fungi have completely revolutionized molecular approaches and allow meaningful comparisons of entire genomes and proteomes.

The chapters devoted to the fungi are designed to provide information to allow new investigators to initiate work with these fascinating organisms but also to be of assistance to workers already in the field. The ecology of fungi and the methods used to study this challenging area are addressed in an authoritative chapter by Thorn et al. in which they describe the various groups of fungi and other fungus-like organisms and point out that no single method will allow identification and isolation of fungi from vastly different situations. A series of detailed methods are then described to study fungi and yeasts from a diversity of environmental settings, including sampling methods, microscopic examination, and culturing. Complete formulations to prepare eight different common media for fungal isolation are provided. The chapter by Davis and Clutterbuck describes practical microbial and genetic techniques for filamentous fungi, including basic minimal and complete media as well as a special synthetic crossing medium. They present several methods for measuring growth rate and describe procedures for extraction of low-molecular-weight molecules and macromolecules, followed by a description of methods to isolate mutants and to analyze them by various genetic approaches. The next chapter describes the physiology and special molecular aspects of filamentous fungi. This includes a description of the biochemistry and the genetic regulation of nitrogen, carbon, sulfur, and phosphorus metabolism, and assays for representative enzymes in each of these areas. After a description of pH control of enzyme synthesis and iron acquisition, various special molecular techniques for filamentous fungi are presented, including nucleic acid isolation, cloning, transformation protocols, and methods to prepare knockout strains. In the chapter, the strategies and power of microarray analysis are presented in sufficient detail to allow new investigators to adopt it in their work.

To gain more insight into the topics presented in these chapters and to explore many other important areas relating to filamentous fungi, one can consult *The Mycota*, a series of eight volumes that represent a comprehensive treatise on fungi as experimental systems for basic and applied research (2). A number of diverse and exciting subjects are covered in an earlier work, *Molecular Biology of Filamentous Fungi* (6), and also in a very recent volume, *Molecular and Cellular Biology of Filamentous Fungi* (7). A superb book entitled *Neurospora, Contributions of a Model Organism*, by Rowland Davis (coauthor of the second chapter in the section devoted to fungi), although focused on *Neurospora*, presents a wealth of information for anyone interested in filamentous fungi (1). A recently published volume deals with fungal genomics and addresses issues that arise with complete genomic sequences (5). Finally, three extensive volumes by C. Guthrie and G. R. Fink in the *Methods in Enzymology* series present a comprehensive set of sophisticated genetic and molecular biology methods for yeasts (3, 4).

REFERENCES

1. **Davis, R. H.** 2000. Neurospora, *Contributions of a Model Organism*. Oxford University Press, Oxford, United Kingdom.
2. **Esser, K., and P. A. Lemke (ed.).** 1995. *The Mycota*, vol. 1 to 8. Springer-Verlag, Berlin, Germany.
3. **Guthrie, C., and G. R. Fink.** 1991. *Methods in Enzymology*, vol. 194. *Guide to Yeast Genetics and Molecular Biology*. Academic Press, San Diego, CA.

4. **Guthrie, C., and G. R. Fink.** 2002. *Methods in Enzymology*, vol. 350 and 351. *Guide to Yeast Genetics and Molecular and Cell Biology*. Academic Press, San Diego, CA.

5. **Prade, R. A., and H. J. Bohnert (ed.).** 2003. *Genomics of Plants and Fungi*. Marcel Dekker, Inc., New York, NY.

6. **Stahl, U., and P. Tudzynski (ed.).** 1992. *Molecular Biology of Filamentous Fungi*. VCH, Weinheim, Germany.

7. **Talbot, N. (ed.).** 2001. *Molecular and Cellular Biology of Filamentous Fungi—a Practical Approach*. Oxford University Press, Oxford, United Kingdom.

42

Methods for Studying Terrestrial Fungal Ecology and Diversity

R. G. THORN, J. SCOTT, AND M. A. LACHANCE

42.1. INTRODUCTION

42.1.1. Importance of Terrestrial Fungi

Fungi are of fundamental importance in terrestrial ecosystems, and their roles and importance are usually overlooked or underestimated by ecologists who study plants or animals. Fungi drive terrestrial nutrient cycles through their abilities to decompose the complex carbohydrates—cellulose, hemicelluloses, and lignin—that make up the majority of plant biomass, which in turn accounts for approximately 90% of the total biomass in most terrestrial ecosystems (34, 37, 89). Other fungi, the mycorrhizal fungi, form symbiotic associations with the roots of living plants. These mutualistic symbioses are vital to the survival of most green plants in natural ecosystems. Mycorrhizal fungi help plants obtain nutrients and water in environments where these are chronically or periodically lacking, and

they may also protect plant roots from attack by pathogens, including other fungi (4, 102). Fungi living within aboveground plant parts, the endophytic fungi, may protect these plants from herbivory or attack by pathogens, sometimes at a cost of flowering and sexual reproduction (33, 90, 92, 96). The filamentous growth form of many fungi and their ability to simultaneously or successively perform more than one nutritional role (decomposer, symbiont, or predator) allow them to act as bridges across time, space, and trophic levels in ecosystems (5, 76, 109, 111). In contrast, predominantly unicellular, nonfilamentous yeasts are adapted for rapid response to rich, moist environments and may convert sugar solutions such as nectar or plant sap into more nutritious—or alcoholic—food for associated insects or other animals, including humans (70, 115). Filamentous fungi with a habit of exuding adhesive extracellular polysaccharides or mucopolysaccharides are important in soil stabilization through the formation of microaggregates and the binding of aggregates and particles (29, 122). In the human environment, fungi are important in many food and industrial fermentations, but they also cause food spoilage and grow unwanted in our living and working spaces, creating problems of environmental health (7, 41, 87, 100). This chapter can only touch on the fundamentals of discovering and identifying fungi in these diverse environments.

42.1.2. Groups of Fungi and Funguslike Organisms

42.1.2.1. Nonfungi: *Actinomycetes, Oomycota* and *Hyphochytriomycota,* and Slime Molds

Several groups of organisms that are not true fungi have traditionally been studied by mycologists or thought of as fungi. *Actinomycetes* are prokaryotic and are members of the G+C-rich gram-positive *Bacteria* (85, 119). They resemble fungi because of their filamentous growth, reproduction by spores, and production of extracellular enzymes (46, 62, 88; http://www.nih.go.jp/saj/DigitalAtlas/index.htm). Although mycologists and soil biologists will encounter actinomycetes in their studies, they are not treated here. The *Oomycota* and *Hyphochytriomycota* are "water molds" with motile, flagellated spores. The anterior flagella in both groups are called tinsel flagella and have two rows of fine, tubular hairs resembling those of the golden-brown algae. The natural relationships of the water molds, as determined by DNA sequence analyses and other lines of evidence, are with the chrysophyte algae in the kingdom *Stramenopila* (113, 114) (also called *Chromista* or *Heterokonta* [13]). Slime molds (the *Mycetozoa* or "fungus animals") are relatives of amoebae and have an ameboid feeding stage and a spore-bearing reproductive stage (11–13, 56). None of these groups belongs in the kingdom *Fungi*.

42.1.2.2. Kingdom *Fungi: Chytridiomycota, Zygomycota, Glomeromycota, Ascomycota,* and *Basidiomycota*

Most current classifications of fungi recognize four phyla: *Chytridiomycota, Zygomycota, Ascomycota,* and *Basidiomycota* (3, 60). There is good evidence from molecular phylogenetic studies, fossil record, and ecology for a fifth fungal phylum, *Glomeromycota* (98). The Assembling the Fungal Tree of Life group (http://aftol.org/) has spearheaded recent phylogenetic analyses of the fungi based on sequences of ribosomal and other genes. These have led to substantial changes to classification and naming of fungal groups (52a) compared to the conservative approach taken here. In addition, sequencing of ribosomal DNAs (rDNAs) isolated from soils has led to the recent discovery of several unknown and uncultured lineages with very deep roots in the fungal kingdom (112). The truly innovative investigator may develop techniques to see, cultivate, and describe these fungi and discover their role in nature, but our discussion is limited to groups known as whole organisms.

Like the *Oomycota*, chytrids are aquatic fungi with flagellated zoospores that require free water for motility (Fig. 1c). Zoospores of most chytrids have one posterior flagellum, but members of one group found in the rumen of herbivorous mammals are multiflagellate. Soil-inhabiting chytrids include decomposers of cellulose, chitin, and keratin, as well as parasites of soil algae, invertebrates, and vascular plants (15). The chytrids in rumen contribute to the digestion of cellulosic plant cell walls in the animal's food (81). The four other phyla consist of predominantly filamentous forms, plus secondarily unicellular yeasts belonging to both *Ascomycota* and *Basidiomycota*.

The *Glomeromycota* (Fig. 1d) are the fungi that form mutualistic symbioses called arbuscular mycorrhizae with the roots of approximately 90% of the vascular plants in the world; these symbioses enable plants in natural environments to survive drought and nutrient stress (102). *Glomeromycota* are coenocytic, with tubular cells containing hundreds of nuclei. Although they form large spores, of 100 to 1,000 μm in diameter, and networks of hyphae that may extend a meter or more, they are otherwise only microscopically visible. Recognition of the monophyletic phylum *Glomeromycota* resolves the paraphyly of the formerly broad *Zygomycota* including *Glomales* with respect to the *Chytridiomycota* (98). The remaining *Zygomycota* (Fig. 1e) are a diverse group of fungi in terms of morphology and ecology and include soil saprobes, parasites of soil fungi or algae, and parasitoids of insects and soil invertebrates, but few significant plant pathogens (20).

The *Ascomycota* (Fig. 1a) are the largest group of fungi and include most of the asexual or "mitosporic" fungi that were formerly classified in the *Deuteromycetes* or "Fungi Imperfecti" (60, 93, 99). The *Ascomycota* include soil-borne pathogens of crop plants, endophytes that live within plant tissues, usually without causing disease symptoms, most fungi that cause human and animal diseases, plus groups that are mycorrhizal, parasitic on other soil fungi, or predatory or parasitic on insects or other invertebrates (16, 19, 30, 33, 45, 58, 86, 92, 94, 96). In addition, the majority of lichens are in the *Ascomycota* (60). Lichens are symbiotic associations between fungi and green algae or cyanobacteria. The fungal partner (mycobiont) forms the characteristic structure that we call a lichen, which encloses and protects the alga or cyanobacterium (photobiont) (2, 50). Finally, most yeasts (Fig. 1f), including the most economically important ones such as *Saccharomyces cerevisiae*, belong in the *Ascomycota* (65). Yeasts are fungi adapted to life in aqueous environments by growth as separate, usually elliptical cells that divide by budding or fission (14, 63) and are discussed further in section 42.2.3 below.

The *Basidiomycota* (Fig. 1b) are the second largest group of true fungi and may be divided into four major groups, the *Ustilaginomycetes* (smuts), *Urediniomycetes* (rusts), *Heterobasidiomycetes* (jelly fungi), and *Homobasidiomycetes* (mushrooms and relatives) (60). The thousands of species of *Homobasidiomycetes* whose mycelial phase occurs in soil have mostly been overlooked in surveys of soil fungi (55, 110). The *Homobasidiomycetes* include important crop pathogens, including *Rhizoctonia* (sexual state *Thanatephorus* [6, 103, 105]) and timber pathogens *Armillaria, Phellinus,* and *Ganoderma* (108), saprotrophic leaf- and wood-decomposing fungi (53, 121), and the majority of fungi that form ectomycorrhizal symbioses with woody vascular plants in 30 families (47, 54, 102).

The fundamentally different biologies of different groups of fungi mean that no single method will work to discover or isolate all fungi in any material or area. For this reason, some general methods are presented here, which users may need to modify to study their fungi of particular interest. *Biodiversity of Fungi* (81a) presents an extensive review of this topic.

42.2. METHODS

42.2.1. Terrestrial Filamentous Fungi

42.2.1.1. Collecting and Culturing Macrofungi

Macrofungi are those fungi with fruiting bodies large enough to see with the naked eye, usually defined as being

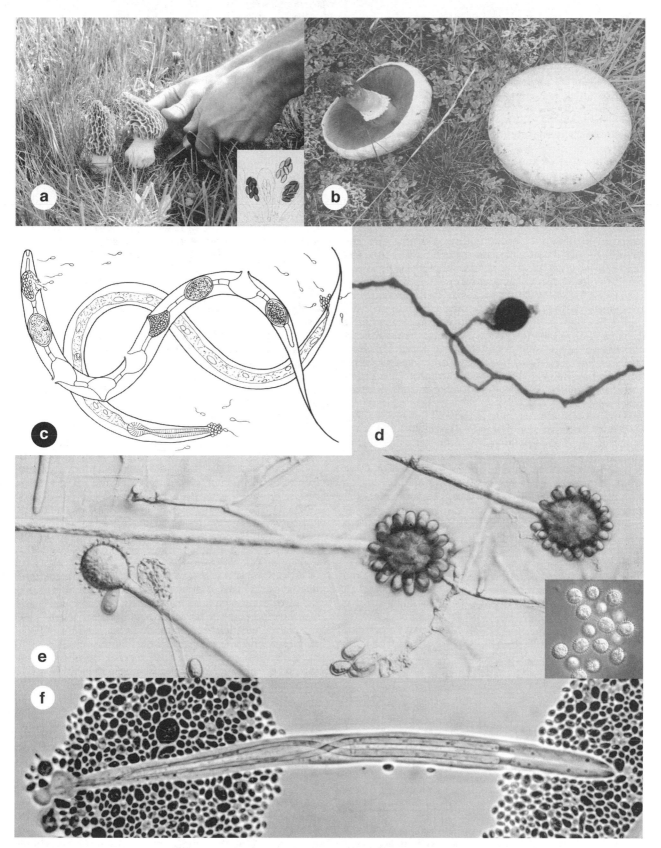

FIGURE 1 The major types of fungi. (a) *Ascomyota: Morchella esculenta*, showing the correct method of collecting macrofungi; inset, eight ascospores are produced in each sac-like ascus in this *Saccobolus* (photo by G. L. Barron, University of Guelph). (b) *Basidiomycota: Agaricus*, associated with a fairy ring in Wyoming. (c) *Chytridiomycota: Catenaria anguillulae* attacking nematodes (drawing by G. L. Barron, University of Guelph, from reference 18). (d) *Glomeromycota*: a spore and hyphae of an endomyorrhizal fungus washed from soil. (e) *Zygomycota: Cunninghamella echinulata*, a striking and common fungus from soil. The inset shows a close-up of the sporangiospores (both photos by G. L. Barron, University of Guelph). (f) Vegetative cells and ascus of *Metschnikowia hawaiiensis*, an ascomycetous yeast.

at least 1 cm in one dimension. These fungi are relatively easy to collect, but a systematic approach and sampling design are required for studies intended to yield results that are comparable between areas or years. Sampling plots should be based on a stratified random or regular transect design for comparability between studies (Fig. 2), but for discovery of maximum diversity they should be combined with opportunistic sampling of known or suspected "good habitats" (91). Plots should be repeatedly sampled during the fruiting season or seasons, once a week being desirable but once per month probably more practical in terms of collection time and effort.

Care should be taken in collecting fruiting bodies of macrofungi to obtain the entire fruiting body (Fig. 1a). The base of the stem, if one is present, may be buried in the soil or other substratum. Collections should immediately be assigned a collection number and recorded in a field notebook. Collections should be packaged in the field to prevent damage during transit back to the lab, including drying or, in rainy areas, saturation. Waxed paper packets or Kraft paper bags are suitable for relatively robust specimens and dry collecting areas, aluminum foil for rainy areas, and hard plastic fishing tackle boxes for minute or fragile specimens. The collection number should be written on the packet, or a slip of paper with the collection number should be packed with the specimens.

Many macrofungi can only be reliably identified if detailed notes are made of characters detectable only when fresh (Fig. 3a). A good photograph, in situ or in the lab, can reduce the need for description of colors, textures, and size, but notes on taste, odor, staining reactions, and the color of a fresh and dried spore print are often vital for final identification. Too often a mycologist is asked to identify the voucher specimens from a detailed ecological study in some remote locale, but finds that the specimens are without notes and, because they were poorly dried, are useless for study of microscopic features as well. For details and terminology of macroscopic features of macrofungi, see reference 72. To obtain a spore print, lay a fresh, moist portion of the fruiting body, hymenium (spore-bearing surface) down on a piece of white paper or a microscope slide, and protect it

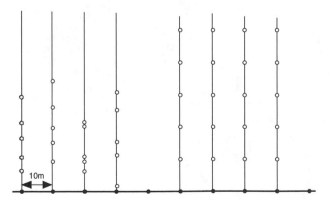

FIGURE 2 Sampling designs: stratified random (left side) and regular grid (right). The heavy horizontal line represents the baseline, filled circles represent 10-m intervals from which sampling transects are struck at 90° to the baseline, and open circles represent sampling plots 4 m² in area, with a radius of 1.14 m (91). In stratified random sampling, sample plots are located on the transects at distances taken from a random numbers table, e.g., in the left line, with their centers at 6.33 m from the baseline, and then 2.17, 5.94, 4.67, and 8.07 m from the center of the previous plot.

from drying by covering with a cup or small collecting jar (Fig. 3c). Keep in a cool place overnight, and a pattern of thousands or millions of spores will be deposited (Fig. 3d), enabling the detection of spore color en masse and microscopic measurement of mature spores.

Once a spore print is set up and notes are taken, including color annotations with reference to some color standard such as in the work of Kornerup and Wanscher (61), the decision is made whether to culture the specimen, preserve a small piece of tissue for future DNA studies, or dry it for the herbarium. A good field preservative for tissues intended for DNA studies is 2× CTAB buffer (see section 42.2.4.11 and reference 43). A piece of tissue approximately 3 by 3 by 3 mm, not including any contaminated external surface, can be cut out with a flamed and cooled scalpel and preserved in 0.5 to 1.0 ml of 2× CTAB in a 1.5- to 2.0-ml screw-cap vial and then kept at ambient temperatures until it is practical to refrigerate. Freezing is unnecessary and may not preserve the DNA as well as temperatures of 4 to 5°C. Specimens, and their field packets if Kraft paper bags were used, should be dried at 40 to 50°C with ample air flow. In many situations, a portable food dehydrator with an electric heating element and fan is ideal. Most specimens will dry overnight, and very large specimens may be cut into slices to speed drying. In humid areas, dried specimens should be packed immediately into airtight plastic bags such as Ziploc™ bags to prevent rehydration and attack by molds or insects.

Culturing from specimens of macrofungi is often very simple and adds a whole new biological dimension to the collection beyond the preserved, dead herbarium specimen. The spores of many saprotrophic macrofungi germinate readily on simple agar media, such as ME agar (see section 42.2.4.3). The isolation medium may be supplemented with antibacterial antibiotics and, for isolation of basidiomycetes, with benomyl. To obtain a polyspore culture, affix a small portion of fruiting body, approximately 2 by 2 mm, hymenium side down, to the lid of a petri dish using a dot of petroleum jelly. When spores on the agar surface are visible by eye or using a dissecting or compound microscope, the piece of tissue may be removed and a small piece of agar bearing spores may be transferred aseptically to a fresh plate of isolation medium. To obtain monospore, usually haploid, cultures for mating studies, make a dilution series of spores in sterile distilled water (sdH₂O). Spores deposited on agar should be resuspended and plated immediately after deposit, whereas spores collected on sterile microscope slides may be resuspended and plated up to several weeks after deposit. The correct dilution must be chosen so that germinating spores are well separated on the agar surface, and germinated spores should immediately be transferred to separate plates or tubes. For robust specimens, and for fungi such as ectomycorrhizal species whose spores do not readily germinate, it is easier or better to make a tissue culture, which will usually be dikaryotic. Break open the fruiting body and extract a small tissue piece from this freshly exposed surface using flamed and cooled forceps or scalpel, and transfer it directly to a plate of isolation medium (Fig. 3b). To be certain that you are growing what you intended, you should follow the germination and growth of spores or tissue pieces at frequent intervals using a dissecting and compound microscope, and immediately transfer the target fungus out of any plate that becomes contaminated with molds. Once cultures are growing cleanly, they should be transferred to plates (short-term working cultures) or slants (stock cultures) of medium without antibiotics. Slant cultures may be preserved by aseptically filling the vials with sterile, heavy

FIGURE 3 Collecting and culturing macrofungi (a to d) and microfungi (e to f). (a) Taking notes on a fresh collection of *Amanita*, *Basidiomycota*, using a color guide (61); (b) making a tissue culture from a fruiting body of *Agaricus*, *Basidiomycota*; (c) the procedure for obtaining a spore print; (d) a spore print; (e) a soil sprinkle plate for isolation of nematophagous fungi; (f) particle washing to remove spores of abundantly sporulating molds to selectively recover more recalcitrant fungi from soil, including *Basidiomycota*. In this example, 5 g of soil was first dispersed by shaking for 1 h at 4°C in 0.1 M sodium pyrophosphate decahydrate and then poured through sieves of 250- and 53-μm mesh. Particles remaining on the 53-μm mesh sieve were washed with a stream of tap water. The remaining organic materials may be picked up in a 1-ml, broad-bore pipette and inoculated onto agar or in liquid medium or used in DNA extractions that are enriched for filamentous fungi.

mineral (paraffin) oil and storing at room temperature or with cryoprotectant such as 15% (vol/vol) glycerol and storing in the vapor phase of liquid nitrogen or in a freezer at −80°C (59).

42.2.1.2. Detection and Isolation of Microfungi

Many of the most interesting and important fungi cannot be seen with the naked eye and collected. Instead, one must collect the material on which or in which they occur and either examine this microscopically to observe the fungi or process it in some way to isolate the fungi into culture. What follows is a discussion of methods suitable for recovery of many filamentous microfungi, particularly those belonging to the *Ascomycota* and *Zygomycota*. Although many *Chytridiomycota* are members of terrestrial soil ecosystems, methods for their isolation into culture are essentially those for aquatic fungi and are not included here (23, 42). However, amplifications of genomic DNA from soil using fungus-specific primers yield diverse sequences in the *Chytridiomycota* and related basal lineages of fungi (112).

One of the best and easiest methods to detect microfungi is periodic observation of organic materials kept in moist chambers, using a dissecting microscope at ×5 to ×50 magnification. Fresh or moistened herbivore dung, leaf litter, tree bark, living or dead roots, and decorticated dead wood are all excellent materials for study by the moist-chamber method. A moist chamber may be as small as a petri dish or as large as a 10-liter plastic crisper. Depending on the material, a layer of moistened sand, perlite, or vermiculite covered with filter paper or paper towels may be required in the bottom of the chamber to maintain humidity. The soil sprinkle plate method for isolation of nematode-destroying fungi (18) is a variation on the moist-chamber technique, in which a small amount of soil or organic matter is sprinkled onto a plate of water agar (15 g of agar per liter of distilled water) and approximately 200 nematodes from a stock culture are added as bait to stimulate germination, trap formation, and development of nematode-destroying fungi (Fig. 3e). Astute fungal biologists may develop their own particular moist-chamber technique to detect or discover their fungi of interest. Observations are made periodically to note the appearance of novel fungi, and these may often be obtained in culture by picking single spores with a Pasteur pipette that has been drawn to a hair tip in a small flame and transferring these to plates of isolation medium.

In addition to the moist-chamber technique, two basic methods for culturing fungi from soil or organic materials are dispersion plating (also called dilution plating) and particle plating (with soil, also called soil washing). For dispersion plating of soil, a small amount such as 5 g (fresh weight) of soil is dispersed in a series of volumes of sdH$_2$O to obtain a dispersion that yields 25 to 30 colonies on a petri plate of isolation medium. The series of 5 g of soil in 125 ml of sdH$_2$O, 1 ml dispersion 1 in 39 ml of sdH$_2$O, and 1 ml of dispersion 2 in 9 ml of sdH$_2$O yields dispersion ratios of 1:25, 1:1,000, and 1:10,000. The last two are useful for many soils, with 0.5 to 1.0 ml spread evenly over the surface of a 100-mm petri dish. It is important that the soil be well dispersed and evenly suspended when taking the aliquot for the subsequent dispersion or for plating. Unused portions of soil samples are weighed, dried, and reweighed to back-calculate the dry weight of soil used in the dispersion. A low-nutrient medium supplemented with antibacterial antibiotics is generally used, and we recommend Martin's soil extract medium with rose bengal and chloramphenicol (see section 42.2.4.1). Dispersion plating yields a diverse subset of the soil fungal community, but it generally misses most basidiomycetes and nutritional specialists such as the nematode-destroying fungi, mycoparasites, and mycorrhizal fungi (32, 110). Comparability between studies requires a systematic approach to picking colonies to be identified: either all or the first 25 to 30 encountered randomly per plate should be chosen (32). Particle plating starts by dispersing the material, or grinding it to fine particles if necessary, and washing particles (Fig. 3f) through sieves of a chosen size range, such as 50 and 250 μm, with tap water or sdH$_2$O to remove the spores of heavily sporulating molds such as *Penicillium* and *Aspergillus* (10, 22). Washed particles are plated on standard or selective isolation medium, depending on the group of fungi desired (21, 110). Although a few principal species may be recovered by such techniques, there may also be an enormously long "tail" to the species accumulation curve, with 8 to 15 previously unseen species added for every 100 incremental isolates past 1,100 (31, 32).

42.2.1.3. Methods for Mycorrhizae and Endophytes

Fungi growing symbiotically within plant tissues require specific techniques for isolation into culture. These include fungi involved in a variety of mycorrhizal associations and also those growing endophytically within aerial tissues of plants, usually without causing disease symptoms. Among the mycorrhizal fungi are *Glomeromycota* forming arbuscular mycorrhizae (also called vesicular-arbuscular mycorrhizae or endomycorrhizae), *Ascomycota* and *Basidiomycota* forming ectomycorrhizae, and several other forms. As mentioned above, the basidiospores of most ectomycorrhizal basidiomycetes do not germinate readily on standard media, although they may be stimulated to do so using activated charcoal or filter-sterilized root extracts (25, 84). Many of these fungi form large fruiting bodies from which tissue cultures may be derived, using more complex media than might be used for saprotrophic fungi. Two agar media commonly used for isolation of ectomycorrhizal basidiomycetes are MMN medium and BAF agar (sections 42.2.4.4 and 42.2.4.5), usually supplemented with antibacterial antibiotics and with benomyl to suppress molds (57). These media may also be used for isolation of ectomycorrhizal basidiomycetes from mycorrhizal root tips that have been scrupulously washed and surface sterilized, although many attempts will still yield nonmycorrhizal ascomycetous molds that were on or in the root tip (107). MMN medium and BAF agar without benomyl may be used to isolate ascomycetous mycorrhizal fungi from root tips or fruiting bodies. Ectomycorrhizal root tips may be characterized and identified to some extent using morphological features (1, 49), and the fungal associate may be identified by selective amplification followed by sequencing of fungal ribosomal genes (27, 43, 44). Glomalean mycorrhizal fungi (GMF) have generally resisted attempts at axenic culture but can be cultivated from infected root pieces or spores recovered from soil in coculture with a host plant such as leeks (*Allium porrum* [25]). Selective primers are also available to amplify rDNA of GMF from roots or soil (101). Sequence-based identification of GMF promises to open the field of research into their ecology, previously inhibited by the difficult and imprecise identification using spore shape and wall structure (20). One difficulty with molecular identification of GMF, however, is that they are coenocytic. Their spores may each contain over 1,000 genetically distinct nuclei, with several different alleles for rDNA (95), and frequently also contain sequences of unrelated contaminant or endosymbiotic fungi as well (97).

Fungi in plant parts may be observed by direct microscopy following clearing of the plant tissues and staining of fungal structures. For nonpigmented roots and leaves, clear for 1 h in 1 M KOH at 60 to 80°C or 15 min at 121°C, neutralize in one to several rinses of 0.1 M HCl and then distilled water, and stain in one of the following: 0.05% (wt/vol) trypan blue in lactic acid, 0.1% (wt/vol) acid fuchsin in lactic acid, or 0.01% (wt/vol) chlorazol black E in water. For highly pigmented, tannin-rich roots, clear in 3% (vol/vol) hydrogen peroxide prior to KOH clearing (24–26).

Culturable endophytes may be isolated onto appropriate media containing antibiotics following surface sterilization and serial washings of the plant tissues to remove superficial contaminants. Surface sterilization using 75% ethanol and 3% bleach (sodium hypochlorite) may work for thick plant parts, but it may kill endophytic fungi in thinner tissues. For these samples, repeated washings using sdH$_2$O containing 0.1% (wt/vol) detergent such as Tween-20™ are recommended. Portions of the plant tissue may then be aseptically chopped or macerated, and these fine particles, 1 mm^2 or less, may be spread on media containing antibiotics to permit the endophytic fungi to grow out (91).

42.2.2. Sampling and Analysis of Indoor Environmental Fungi

There is increasing recognition that exposures to indoor environmental fungi and their emissions are associated with a range of human illnesses. As a result, the detection, identification, and quantification of indoor fungi, particularly molds, has become an important part of the testing mandate of many environmental testing laboratories, particularly commercial laboratories. In contrast to bacteriology and food microbiology, the commercialization of analytical environmental mycology has posed a number of challenges, primarily due to the lack of standard testing procedures and evaluation criteria. This section discusses several sampling methods that are used commonly in commercial mycology laboratories for the assessment of populations of fungi in home and indoor work environments.

42.2.2.1. Sampling Methods

A wide range of sampling procedures exist for surveying environmental fungi. Primarily these methods are used (i) to evaluate hazards and (ii) to assess control and cleanup measures. Because of the highly episodic nature of spore release, short-term air sampling methods cannot be used to evaluate human exposures to fungi. In addition, the health hazards posed by fungal exposures are complex and may vary in terms of allergenic, toxic, and pathogenic effects, depending upon the species present, growth conditions, and individual human susceptibility. Therefore, the establishment of "safe exposure limits" is widely agreed to be untenable. However, the presence of indoor fungal growth has been strongly correlated to elevated levels of indoor airborne spores and increased symptom reporting by occupants. Thus, there has been widespread consensus that the elimination of indoor fungal growth sources is a prudent measure. Based on this guiding principle, most air sample interpretation guidelines have been developed in order to query the likelihood of indoor fungal growth rather than to recommend safe limits for human exposure.

42.2.2.1.1. Bulk Samples

Bulk samples refer to (i) solid materials such as wood; (ii) fibrous materials, including broadloom and dust; and (iii) friable materials like gypsum wallboard. The assessment of bulk samples for fungal colonization is considered central because the presence of fungal growth on normally barren materials indicates a deficiency and is not acceptable. Furthermore, the examination of bulk samples is of fundamental value in permitting conclusive identification of the source of contamination.

42.2.2.1.1.1. Direct microscopic examination. Direct microscopy is considered the "gold standard" in environmental mycology for establishing the presence of fungal growth. Together, the observations of fungal spores, spore-bearing structures, and vegetative hyphae provide an unequivocal indication of past or current growth. All methods involving direct microscopy should first examine the bulk specimen in reflected light at a magnification of ×10 to ×40 for gross evidence of colonization. Subsequently, preparations from one or more areas of the specimen should be made in a suitable mounting fluid (discussed below) for microscopic evaluation in transmitted light. Two principal methods exist for the examination of bulk specimens by transmitted light microscopy: (i) adhesive tape lift and (ii) scraping/shaving.

42.2.2.1.1.1.1. ADHESIVE TAPE LIFTS. Methods for the collection of fungal growth from surfaces using shellac peels to preserve the arrangement of spores and spore-bearing structures have long been used to study fungi in situ (for example, see reference 106). Currently, the simplest method to do this is by using clear adhesive tape (28, 38, 39). Tape-based methods are more practical than liquid shellacs because they do not require setting time and they can be used easily on highly porous materials. Adhesive tape methods are rapid and practical for both field collection and laboratory analysis.

A number of genera of important indoor fungal contaminants may be recognized directly from spores alone (e.g., *Chaetomium* and *Stachybotrys*). Many more genera may be identified if both the spores and spore-bearing structures are preserved intact. Tape lift sampling allows for the identification and semiquantification of fungal colonists on a surface.

Indications

This sampling method is useful for (i) field collection of surface samples suspected of fungal colonization, (ii) laboratory examination of bulk specimens, and (iii) microscopic examination of delicate structures from cultures where minimal disruption is desirable.

Procedure

1. Press the adhesive side of a 2- to 5-cm segment of optically clear tape (e.g., 3M Scotch "Red Tartan" brand tape) to the sample to be tested. The collection of superficial debris from rough or irregular surfaces can be improved by gently rubbing the back of the tape.

2. For transport, adhere the sampled tape strip to a glass microscope slide or the interior surface of a sturdy closable plastic bag such as a zippered freezer bag.

3. Cut a piece of sampled tape small enough to fit beneath a glass coverslip.

4. Using a clean coverslip, gently scrape away any large surface debris (e.g., greater than 0.2 to 0.5 mm in diameter) from the tape adhesive such as soil particles, glass fibers, etc., as these may cause air bubbles to form in the preparation.

5. Place a small drop of 85% clear lactic acid (~10 µl) on a glass microscope slide, and mount the tape segment adhesive side up. Capillary action will help to draw the lactic acid throughout the space between the tape backing and the glass slide. Any air bubbles that form can be eliminated by gently lifting and repositioning the tape.

6. Place a small drop of mounting fluid such as lactofuchsin or lactophenol cotton blue (60) (~10 µl) in the center of a 22-mm no. 1 glass coverslip. Invert the coverslip to hang the droplet of mounting fluid.

7. Place the coverslip and hanging drop of mounting fluid on the adhesive side of the tape. Slide mounts should be prepared 5 to 10 min prior to examination in order to allow adequate time for staining.

8. Scan two or three microscopic transects of the tape specimen at ×400 magnification to examine a minimum of 20 to 50 mm² of tape area. Results should be collected on fungus type and category of structural element observed (i.e., spores versus vegetative hyphae/mycelia/ spore-bearing organs).

Interpretation

Fungal spores produced on outdoor plant debris are common in indoor settled dusts. Therefore, the observation of fungal spores alone on a tape lift sample is insufficient to characterize the sample site as a fungal growth site. A minimum standard of evidence for the interpretation of active fungal growth requires the microscopic observation of fungal filaments (vegetative hyphae/mycelia/spore-bearing structures) together with spores.

Notes

Tape used for sampling should be stored in a clean zippered plastic bag and not be used for any other purpose. Prior to sampling, the leading 2 to 5 cm of tape should be stripped off and discarded to guard against cross-contamination from earlier samples. Furthermore, to control against cross-contamination, one field blank (unsampled tape lift handled similarly to actual field samples) should be taken and analyzed for every 10 field samples.

Tape mounts prepared in the manner described above cannot be retained as a permanent record because lactic acid causes the tape adhesive to deteriorate within 12 to 24 h of preparation. If a longer-term specimen record is required, a sufficient portion of unmounted specimen should be retained in original packaging in a cool, dark, dry area.

Standard recipes for lactofuchsin and lactophenol cotton blue contain more stain than is necessary for this procedure. For tape lift mounts, these reagents are better prepared at 1/4- to 1/8-strength stain.

Confusion of small starch granules and fungal spores can be avoided by exposing the tape lift (adhesive side up) for 30 to 60 s in a chamber saturated with iodine vapor, which causes any starch granules to appear uniformly bluish, allowing them to be differentiated from fungal spores of similar size and shape. This method is similar to a standard procedure for visualizing chemicals separated by thin-layer chromatography. A suitable chamber can be made by placing several iodine crystals inside a closable wide-mouth amber glass jar set on its side. Once closed, the air inside the container will become saturated with iodine vapor after several hours. The chamber can be maintained by periodically replacing iodine crystals that have been lost to sublimation.

42.2.2.1.1.1.2. SCRAPINGS AND SHAVINGS. The simplest way to examine a bulk specimen for fungal colonization is by direct visualization of small particles of the specimen by transmitted light microscopy. This method has the advantage of allowing the analyst to observe both the sample surface and the subsurface aspect of the bulk material. Thus, this method may facilitate the detection of residual fungal contamination in specimens from which loose surface debris may have been cleaned or wiped. A second advantage of scrapings and shavings is that permanent microscopic mounts can be prepared from these sample formats as necessary. Thirdly, cultures can be prepared readily from scraping material should this be necessary. However, the scraping or shaving of substrates in the field may be time-consuming and requires the availability of appropriate sampling tools (e.g., disposable scalpels) as well as a means of safe disposal of sharps.

Indications

Surface scraping and shaving is an appropriate sampling method in the field for samples with little evident surface irregularity. It can also be used as a fallback sampling method for those materials from which tape lift sampling fails to remove appreciable surface debris.

Procedure

1. Scrape or shave the surface to be sampled, using a clean, sharp scalpel or other suitable tool.

2. Collect the scrapings in a clean tube, closable plastic bag, or paper packet. Plastic tubes are undesirable for the collection of scrapings due to their potential for dispersion of sample by static electric charge. For paper packets it is best to collect scrapings or shavings on gloss-coated heavy black paper because fine specimen fragments are easily seen and unlikely to adhere.

3. Place a small amount of scrapings or shavings on a clean glass microscope slide. Remove or further grind any large fragments, e.g., >200 µm in diameter, because particles of this size will cause uneven distribution of staining fluid. If scrapings or shavings are prepared in the laboratory directly from a bulk sample, they may be collected directly on a glass microscope slide.

4. Place a small drop of mounting fluid such as lactofuchsin or lactophenol cotton blue (60) (~10 µl) in the center of a coverslip, invert the coverslip to hang the droplet of mounting fluid, and place on the scraped material. Woody specimens or those containing alkaline minerals (e.g., wallboard samples containing gypsum dust) should be mounted instead in a solution of 10 to 15% potassium or sodium hydroxide with 10% glycerol to promote softening and avoid foaming. Hydroxide preparations incubated at 30 to 40°C for 30 to 60 min may be useful for wood shavings or other tough organic materials. The addition of a drop of 0.025% phloxine to hydroxide mounts may help differentiate hyaline fungal structures, especially those of basidiomycetes (74). Slide mounts should be prepared 5 to 10 min prior to examination in order to facilitate staining. An optional prewetting step may be useful for hydrophobic materials; this involves adding 1 or 2 drops of 95% ethanol to the specimen and allowing the ethanol to evaporate almost completely prior to the addition of mounting fluid.

5. Scan 20 to 50 microscopic fields at ×400 magnification. Record the fungus type and category of structural element observed, e.g., spores versus vegetative hyphae/ mycelia/spore-bearing organs).

Interpretation

As with adhesive tape lift sampling, the visualization of fungal filaments en masse with or without the observation of fungal spores provides the best indication of fungal growth on a substrate. Scrapings or shavings are not readily quantifiable, and thus this technique is best suited to qualitative analysis.

Notes

Materials containing significant amounts of friable alkaline mineral salts (e.g., gypsum wallboard) for which microscopic analysis is required are usually best sampled on-site (by either adhesive tape lift or scraping) since transport of the entire bulk specimen to the laboratory for testing tends to cause powdering of the surface with gypsum dust, impeding microscopy.

42.2.2.1.1.2. Culture. Assay of bulk samples by direct microscopic examination is not always feasible, particularly for samples that are fibrous (e.g., dust, broadloom, and insulation) or granular (e.g., soil, plaster, and concrete). For these matrices, culture-based methods are preferred in conjunction with direct microscopy. A number of methods exist for culture-based analysis of bulk materials, largely based on techniques developed for soil fungi (17). The most commonly used methods are (i) elution, (ii) particle plating, and, occasionally, (iii) swabs and wipes (however, swabs and wipes offer little valuable information in environmental fungal sampling and should not be used as a routine sampling method). The suitability of elution or particle plating depends on the composition of the sample, as discussed below.

The validity of culture results depends on accurate identification of fungi to as detailed a taxonomic level as practical—in most cases to species level. Certain groups of fungi (notably *Fusarium* and *Penicillium*) require considerable expertise and extensive testing on an isolate-by-isolate basis to provide accurate identification to the species level. For these fungi, species identification is beyond the scope of routine practice, although identification to subgenus or morphological group may provide a practical alternative. If multiple cultures derived from a single material or site yield high numbers of what appears to be a single species of a taxonomically difficult genus, referral of one or more isolates to a relevant expert for accurate species identification may be useful.

42.2.2.1.1.2.1. ELUTION METHODS. Elution methods are best used for samples that are soluble or insoluble but readily dispersible in liquid. Typically, elution methods measure an amount of sample (mass, volume, or area) and disperse the sample in a known volume of liquid eluent, usually a sterile isotonic broth. However, sdH$_2$O is also satisfactory because fungal cell walls provide considerable protection against plasmolysis. Aliquots of eluent may be either plated directly or diluted serially prior to plating. Following incubation, growing colonies are enumerated and the culturable fungal burden of the specimen may be determined in CFU per unit of sample.

Indications

Unlike other culture-based analytical methods, elution methods yield quantitative results. This permits evaluation of the test results against qualitative as well as quantitative interpretation criteria. One disadvantage of elution-type methods over other culture-based techniques is that they tend to overrepresent heavily sporulating fungi, producing results that do not necessarily reflect the true or active fungal populations of the specimen in terms of relative biomass. Even with this limitation, these methods remain widely used because they permit the comparison of specimens of similar composition using numerical data that are well suited to statistical analysis. Usually the elution results are expressed according to mass units (e.g., CFU per gram or CFU per milligram). However, area-based measures (e.g., CFU · 100 cm^{-2} or CFU · cm^{-2}) may be better suited for specific applications such as duct interiors. Two procedural variations are commonly used for the plating of eluted samples: (i) serial dilution and (ii) direct plating. Serial dilution is a more cumbersome method than direct plating due to the need for dispersion of aliquots of eluent in molten agar. Also, there is potential for reduction of fungal viability due to heat shock. Compared to direct plating, however, serial dilution provides more even dispersion of sample and improved quantitation.

Procedure

1. Disperse a known quantity of specimen in a volume of eluent such as sdH$_2$O or 2% peptone broth. The addition of 10 to 200 ppm of a wetting agent such as Tween 20 is helpful for wetting hydrophobic specimens. The ratio of specimen to eluent is arbitrary, but a good starting point is in the range of 20 to 100 mg of specimen per ml of eluent. Higher elution factors may be needed if the test material contains fungicidal additives.

OPTION A—*Serial dilution*

2a. Transfer 1 ml of well-mixed eluent to another tube containing 9 ml of sterile eluent and mix. Transfer 1 ml from this tube to another containing 9 ml of eluent broth and mix; repeat two or three times.

3a. Transfer 1 ml from each of the dilution tubes separately to an empty sterile petri plate.

4a. Disperse the elution aliquot by gentle swirling in 20 to 30 ml of sterile molten agar cooled to 45°C. Let plates cool to room temperature overnight to allow excess moisture to evaporate from the agar surface.

OPTION B—*Direct plating*

2b. Using a sterile disposable micropipette or calibrated loop, dispense aliquots of 100, 10, and 1 µl separately on the surfaces of agar plates.

3b. Disperse the aliquot on the medium surface by first spreading the fluid in a central streak across the diameter of the plate and then rotating the plate 90° and streaking back and forth across the entire length of the initial central streak. For small aliquot volumes (e.g., 1 µl) it may be useful to dispense the eluent in a larger volume of sterile fluid previously transferred to the plate.

4b. Allow inoculated plates to sit overnight to let excess moisture evaporate.

5. Incubate the plates inverted, under appropriate conditions, for 7 to 21 days. The inclusion of a diurnal light/dark cycle during incubation may promote sporulation in some fungi and facilitate identification. Alternatively, exposure to blue-black (near-UV) light may be helpful in inducing sporulation in some fungi, especially dematiaceous phylloplane molds.

6. Select one plate from the prepared concentration series to identify and enumerate colonies. Ideally, the number of colonies on the enumerated plate should be between 30 and 60. Analyzing plates with higher colony densities may lead to underrepresentation of slow-growing fungi (e.g., *Stachybotrys*) due to competitive inhibition or overgrowth by more rapidly growing taxa (73 [section 5.2.3.1]).

Interpretation

Fungi and their propagules are ubiquitous and are likely to be recovered during any investigation employing culture-based methods, even in the absence of indoor fungal growth.

Without direct microscopic evidence of the presence of fungal growth, the simple isolation of a fungus from a bulk sample does not indicate colonization. The interpretation of elution culture results is one of several areas in environmental mycology where objective interpretive criteria are almost entirely lacking and subjectivity dominates.

There has been long-standing recognition that these methods do not provide reliable measures of the biological activity of environmental samples. Excellent discussion of the limitations of these methods is given by Barron (17) and Miller (78). For the purposes of indoor mycology, the quantitative burden of culturable fungi does not correlate well with the presence of fungal growth. This is especially true of indoor dust. For example, in strict quantitative terms a concentration of 10 to 100 CFU of culturable fungi mg^{-1} in quiescent surface dust is at most an ambiguous finding and arguably an expected finding in the absence of a moisture or mold problem. In contrast, the qualitative composition of the fungal community of environmental specimens is often relevant to the evaluation of a problem.

The following general guidelines are proposed for culture-based results.

The following apply if fungal growth is confirmed by direct microscopy of a specimen.

- The recovery of the same fungus from culture of a specimen is confirmatory. This procedure is indicated only where species level identification or any other information not available by means of direct microscopy is relevant to remedial, protective, or control measures.

- The failure to recover fungi from a specimen may be related to reduced viability of propagules or vegetative hyphae or unsuitability of the culture conditions for cultivating the taxa concerned. Nevertheless, the result of direct microscopy is definitive in establishing the presence of fungal colonization, and this conclusion is not weakened by the lack of a positive culture result.

The following apply if fungal growth is not confirmed by direct microscopy of a specimen.

- Failure to recover viable fungi from a specimen supports but does not confirm the absence of contamination or colonization.

- Recovery of viable fungi from a microscopically "clean" specimen presents one of the greatest interpretive challenges in environmental mycology. The extent to which a positive culture result indicates fungal growth on a specimen or surface contamination from a local source usually requires establishing that the recovered flora differs from the expected "normal flora" of the material. Normal flora in this case refers to the expected surface load of viable fungi on the material. Because fungal propagules are ubiquitous, the recovery of low levels of environmentally common species from otherwise clean surfaces is an unremarkable finding. Several examples of expected microfungal flora of various materials under normal conditions compared to fungal contamination follow. These examples are not intended as safe limits.

  - *Wood and wood products*—A wide range of fungi are known to inhabit wood, ranging from basidiomycetes to molds. Recovery of basidiomycetes from wood by dilution methods is poor since many of the common wood rot basidiomycetes that are likely to be encountered as indoor contaminants bear their spores on a highly differentiated fruiting body and do not typically produce asexual spores on the surface of their vegetative mycelium. Thus, the isolation of low levels of wood rot basidiomycetes by dilution methods is strongly suggestive of colonization. In contrast, most molds are copious sporulators. For example, a microcolony of *Cladosporium* or *Penicillium* on a wood surface shaving may yield very high levels per unit mass in a culture-based assay (e.g., >100 or 1,000 $CFU \cdot mg^{-1}$) but have little relevance to the condition of the material.

- *Fibrous insulation*—Fibrous insulation products are used for thermal and sound insulation. These materials consist of fibrous batts or panels (usually fiberglass) containing a resin binder sometimes affixed to a backing membrane. Inorganic insulation materials are fairly resistant to fungal attack. Under conditions of persistent moisture, however, these materials are prone to mold growth (often yielding high numbers of a single or a few species of *Aspergillus* or *Penicillium*, e.g., >100 $CFU \cdot mg^{-1}$). Fibrous insulation on the interior of ducts or wall cavities can act as a filter to trap spores and other airborne debris. In such cases, insulation becomes discolored or dirty, yielding moderate to high levels of culturable fungi (e.g., up to 100 CFU of mostly phylloplane molds mg^{-1}) even though the materials themselves may not exhibit fungal growth.

- *Broadloom and broadloom dust*—As with insulation, the fibrous nature of broadloom and broadloom dust imparts a high capacity to trap and retain ambient airborne fungal spores. In addition, the common North American habit of wearing street footwear indoors further contributes to the content of fungi in these materials. Thus, by culture methods alone it is extraordinarily difficult to distinguish conclusively a carpet that is actively growing fungi from one that is simply dirty. In both cases, levels of total culturable fungi in vacuum-collected broadloom dust are typically up to 500 $CFU \cdot mg^{-1}$. In dirty carpets, phylloplane fungi (e.g., *Alternaria*, *Cladosporium*, *Epicoccum*, and *Ulocladium*) and soil fungi (e.g., *Mucor*, *Rhizopus*, and *Trichoderma*) predominate. In contrast, the fungal flora of actively moldy carpets usually includes at least several of the following taxa in a dominant rank proportion: *Aspergillus*, *Aureobasidium*, *Chaetomium*, *Eurotium*, *Paecilomyces*, *Penicillium* subgenus *Penicillium*, *Scopulariopsis*, and *Wallemia*.

- *Wallboard*—Paper-faced gypsum wallboard is the most commonly used wallboard product in North America today. This product category has been in widespread use for nearly 40 years as a rigid-sheet finishing panel for interior walls. More recently, gypsum wallboard has gained notoriety for the propensity of its paper facing layers to serve as substrates for hazardous molds, notably *Stachybotrys chartarum*. In the absence of moisture under normal conditions of use, gypsum wallboard should be devoid of fungal growth. The interior-facing surfaces of gypsum wallboard are painted or finished, greatly reducing their porosity and surface irregularities that might otherwise trap moisture or airborne particles. Gypsum wallboard paper with more than 10 $CFU \cdot mg^{-1}$ suggests growth on the material itself, particularly if the taxa recovered comprise only one or a few nonphylloplane molds. Levels below 1 to 10 $CFU \cdot mg^{-1}$ may represent passive deposition of spores or hyphal fragments on the paper surface (especially if the taxa recovered are phylloplane molds). However, low levels of nonphylloplane molds (e.g., *Acremonium, Aspergillus, Chaetomium, Paecilomyces, Penicillium, Scopulariopsis,* or *Stachybotrys*) may indicate a nearby mold growth site.

- *Quiescent settled dust (swab or wipe collected, not vacuum collected)*—Like broadloom dust, settled dust is a complex mix of inorganic and organic particles and fibers, most of which at one time were airborne. Because quiescent dusts remain dry under normal circumstances, their microfungal content tends to remain intact and ungerminated. For this reason, sampling passively settled dusts has been suggested as a good way to measure the long-term indoor airborne spore load. Thus, the common analytical goal in testing settled dust is the determination of a fungal growth source in the area where the dust was taken rather than investigating the dust itself for growth. Negative results should be interpreted cautiously because the presence of fungal growth in a building interior does not always cause an increase in the fungal content of settled dust.

Notes

Most laboratories express the results of quantitative culture of bulk specimens as a function of the wet mass of the material. Although the expression of viable fungal content by wet mass is an underestimate of the same calculation by dry mass, the difference is likely to be insignificant with most bulk materials. Thus, adjustment for dry weight of sample is not indicated unless there is considerable variation in the moisture content of a sample set or there is a need to compare results with existing data expressed by dry mass. For quality control, one sterility control should be performed on the elution buffer for each batch of 10 samples processed.

For nonporous or very dense materials such as linoleum flooring or solid wood, quantitative culture expressed by unit mass may be less informative than an expression based on surface area since the interior of the specimen is likely to be barren yet may account for the majority of the sample's mass.

42.2.2.1.1.2.2. PARTICLE PLATING. Particle plating, also called the "soil crumb" method, involves the direct transfer of small fragments of material to a suitable fungal growth medium (78, 117). This method can provide a more reliable qualitative determination of active fungal flora than elution-based methods, which tend to overrepresent heavily sporulating fungi and underrepresent mycelial taxa. The plating of multiple particles from a single test material can provide a semiquantitative assessment of fungal colonization by expressing the recovered fungi in terms of the proportion of particles colonized.

Indications

Particle plating favors the isolation of fungi that are present as actively growing hyphae as opposed to dormant spores. Thus, it is a useful method where mycelial fungi are the suspected agents of biocontamination, such as wood rot basidiomycetes.

Procedure

1. Remove a fine particle of material from the test specimen using sterilized fine forceps or a fine needle. Ideally, the particle should be smaller than 1 mm in diameter. It may be helpful to use a stereomicroscope for the selection the particle. Rolling the tip of the transfer tool in sterile mineral oil may assist in particle recovery.

2. Transfer the particle to the surface of a petri plate containing a suitable growth medium. If the particle is very fine, it is a good idea to circle the area on the underside of the plate where the particle was placed in order that resulting fungal colonies are properly attributed. Up to 10 or 12 particles can be inoculated on a single petri plate as long as they are from the same specimen.

3. Incubate the plates upright under appropriate conditions and examine them under a stereo microscope for signs of growth at 24 and 48 h. Subculture emergent hyphal tips individually for identification.

Interpretation

The first fungal filaments to emerge from a plated particle may represent the active fungal flora of the particle. This is based on two assumptions: (i) actively growing fungi exist in a state of log-phase growth, and (ii) most fungal spores possess a brief lag phase between exposure to suitable growth conditions and germination. Comparing the results of particle plating to those of other culture-based methods or direct microscopy may help to establish the fraction of fungal flora that is biologically active.

Notes

This procedure complements direct microscopic examination by assisting in the determination of active fungal colonists. However, not all fungi present and actively growing can be recovered by culture methods, particularly if complex physiological requirements are not met by the isolation conditions.

Based on the notion that actively growing fungi are more likely than dormant spores to adhere to particles, Warcup (116) suggested that particle plating be preceded by gentle washing of the particles in water to remove loose spores and enhance the recovery of mycelial fungi. Similarly, surface sterilization may be used on larger solid particles to eliminate exterior contaminants.

42.2.2.1.2. Volumetric Air Sampling

Air sampling for fungi was first employed as a deliberate sampling method by Pasteur in the 19th century. Although the principle remains the same, a large number of sophisticated modern fungal air sampling methods exist that are used widely. These techniques can be divided into two categories according to detection method: (i) spore trap sampling and (ii) viable sampling. Spore trap sampling involves the collection of airborne particles by filtration or impaction. The analysis and quantitation of fungal content are done by light microscopy. In contrast, viable sampling collects airborne particles by impaction or centrifugation on the surface of a suitable growth medium. Following incubation, the resulting fungal colonies are counted and identified. A third technique, liquid impinging, is suitable for culture or direct microscopy, but is cumbersome and seldom used.

It is important to note that for several reasons air sampling data derived by either technique cannot be used to assess human exposure. (i) Airborne spore loads vary greatly over time, changing by orders of magnitude during the course of a single day. Brief "grab sample" techniques characteristic of most air sampling methods cannot account for these fluctuations. (ii) Results of air sampling methods cannot readily be converted to standard units of exposure assessment (milligrams of contaminant per cubic meter of air). (iii) The airborne fungal spores are a complex, ever-changing mixture where each taxon may contribute a unique hazard set. Despite these limitations, air sampling methods remain widely used primarily for the detection of indoor fungal growth sites.

42.2.2.1.2.1. Spore traps. The spore trap method involves collecting total airborne particulate on a filter membrane or adhesive-coated slide. Analysis is by direct examination by light microscopy. A number of collection devices are used for this method. The most commonly used devices are (i) mixed cellulose ester membrane filters (MCEMs) and (ii) slit impactors such as the Allergenco MK-3 (Allergenco, Charleston, SC) and the convenient, disposable Air-O-Cell® cassette (Zefon International, St. Petersburg, FL).

The high variability in airborne spore content greatly reduces the importance of variation in counting method as a significant source of error. The most critical aspect of spore trap analysis is accuracy in identification. Spore trap analysis is the most technically challenging task in environmental mycology, requiring considerable skill and experience of the analyst for accurate identification of spores and other particles. It literally requires years of experience, not just analyzing spore trap samples but also examining whole specimens from field collections, in order to develop the depth of knowledge required to reliably and accurately analyze these kinds of samples. This level of experience cannot be trained in a short time, nor can it be acquired easily by individuals lacking advanced training in mycology or botany. Therefore, laboratories engaging in the analysis of spore trap samples are strongly encouraged to ensure that spore trap analysts possess a minimum of a graduate level university degree in mycology with an emphasis on fungal taxonomy and identification as well as field experience. Analyst certification programs will undoubtedly help to promote uniformity in these analytical procedures (e.g., Pan-American Aerobiology Certification Board [www.paacb.org]).

Indications
Spore trap methods have an advantage over culture-based techniques in that they enumerate fungal particles regardless of viability. Because of this, spore trap methods are sometimes incorrectly termed "nonviable" techniques when they might more appropriately be called "total" sampling methods since both viable and nonviable propagules are counted. The major limitation of these techniques is their inability to provide low-level identifications of spores because many species level and even genus level fungal identifications require the examination of spores as well as spore-producing structures. The identification of certain fungi additionally requires knowledge of physiological characteristics. All of this information can be derived from a growing culture but not from spore trap specimens. As a result, spores in spore trap samples are only identified presumptively based on size, shape, and color.

Procedures
MCEM
MCEM spore traps typically use longer-term sample periods than slit impactor methods (e.g., 1 to 8 h versus 0.5 to 30 min). However, the fact that particles tend to be collected in an even distribution on the membrane in a single focal plane makes this sampling method appealing from an analytical standpoint. In addition, breakthrough (re-entrainment of sampled particles) is not a problem with this technique because the size of fungal cells greatly exceeds the pore size of the membrane filters used for collection.

Generally, a 25-mm MCEM is used with a 0.8-μm pore size supported by a backup pad housed in a three-piece electrically conductive cassette with a 50-mm extension cowl to reduce the dispersion of collected particles due to static electric effects. This is a standard sampling device used for airborne asbestos sampling as set out by National Institute for Occupational Safety and Health Analytical Method No. 7400. The laboratory analysis of MCEM samples requires the use of a "hot block" vaporizer to generate acetone vapor for clearing the filter. Preassembled cassettes and acetone vapor generators are available from health and safety supply houses.

Sample acquisition
1. Calibrate a high-volume air sampling pump using an equivalent calibration cassette in-line to draw between 2 and 15 liters of air min^{-1}.
2. Connect the pump hose to the outlet nipple of the test cassette and remove the end cap from the cowl. Samples should be collected with the cassette facing downward to reduce oversampling due to particle deposition by gravity. A total of 1.0 to 1.2 m^3 of air should be collected. In dusty areas a lower sampling volume may be used.
3. One blank MCEM should be analyzed for every 10 field samples. Blank MCEMs should be handled exactly as test samples with the exception that air should not drawn through them.
4. Completed samples and blanks should be re-capped, labeled, and placed in a clean box or bag away from light and moisture until analysis.

Laboratory analysis
1. Remove the MCEM from the cassette and backup pad and place it, sampled surface upward, on a clean glass microscope slide.
2. Insert the slide and filter into the vapor chamber of a hot block set at 70 ± 2°C.
3. Inject 100 to 200 μl of acetone into the injection port. Wait 3 to 5 s before removing the slide and filter. If the filter did not clear completely, repeat the procedure. A minimum volume of acetone should be used to avoid dislodging sampled particles.
4. Place a small drop of mounting fluid in the center of a square 22-mm no. 1 glass coverslip (e.g., 5 to 10 μl). Suitable mounting fluids include lactophenol cotton blue and lactofuchsin.
5. Place the coverslip on the cleared filter and allow to stain for 5 to 10 min prior to examination. Using a darkly colored fine-point permanent marker, trace a line corresponding to the sampled area of the filter on the reverse of the microscope slide.
6. Using a microscope with a calibrated stage micrometer, examine sufficient microscopic fields to evaluate a minimum of 3.5% of the sample area of the filter. For densely sampled filters, it may be sufficient to stop counting once 100 fungal elements have been counted. Warner and Emberlin (118) suggested that counting a minimum of 1.4% of the area of a Hirst spore trap provides a satisfactory representation of the total sample in terms of taxonomic diversity; however, it is unclear if this threshold can be applied to other sample formats. In any case, microscopic fields should be enumerated in tandem along three or four adjacent transects of the full diameter of the filter. Each transect should be separated from its neighbor by at least

one field diameter. Care must be taken not to count beyond the sampled area of the filter. Identify and count all fungal elements that lie within the field of view, including floating elements present initially. As a general rule, elements lying partially outside of the field of view should be counted on only one half of the field (e.g., right side) and ignored on the other half of the field (e.g., left side).

7. Extrapolated airborne concentration in counts per cubic metre of air (C) is given by

$$C = \frac{A_T}{n \times A_f} \times \frac{C_R}{Q \times t} \times 10^3 \ \text{liters} \cdot \text{m}^{-3} \quad (1)$$

where C is extrapolated airborne counts (per cubic meter), A_f is area of one microscopic field (in square millimeters),

A_T is total sampled area of the MCEM (in square millimeters), n is number of microscopic fields evaluated, C_R is raw count, Q is flow rate of sampler (in liters per minute), and t is duration of sample (in minutes).

Air-O-Cell®

Air-O-Cell® is a sticky slide packaged in a single-use slit impactor cassette. Airborne particles are collected on a glass wafer coated with an optically clear gel. The cassettes are individually serial numbered and stamped with an expiry date (Fig. 4).

Sample acquisition

1. Calibrate a high-volume air sampling pump using an equivalent calibration cassette in-line to draw 15 to

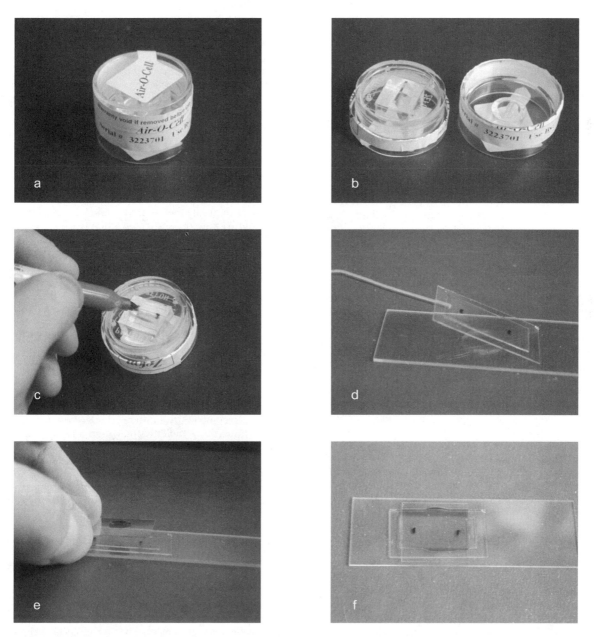

FIGURE 4 (a) Preparation of the Air-O-Cell® spore trap. (b) Separate the cassette by breaking the shrink seal. (c) Mark the ends of the inlet slit. (d) Remove the gel-coated glass wafer and gently lower it, gel side up, onto a drop of clear lactic acid. Using the placement marks, adjust the wafer so that the sampled area lies parallel to the sides of the microscope slide. (e) Hang a drop of mounting fluid on a coverslip and gently lower it onto the gel surface. (f) Completed mount.

20 liters · min^{-1}. Lower flow rates reduce the impaction efficiency of the device and broaden the particle trace, whereas higher flow rates may produce "bounce off" and breakthrough of sampled particles.

2. Note that the shelf life of the cassette is current, and connect the outlet of the cassette to the sampling pump. Suspend the cassette so that the sample slit faces downward. Retract the vinyl slit cover.

3. The manufacturer-recommended sampling volumes are 150 liters for generally clean indoor environments and 75 liters for dusty environments or outdoors. Lower sample volumes (e.g., 10 to 25 liters) may be appropriate for very dusty environments such as remediation enclosures or industrial processes.

4. A minimum of one field blank should be collected for every 10 field samples. Field blanks should be handled exactly as test samples with the exception that air is not drawn through them.

5. Completed samples and blanks should be resealed, labeled, and placed in a clean box or bag away from light and moisture until analysis.

Laboratory analysis

1. Separate the upper and lower halves of the sample cassette using a coin or a fork.

2. Mark the location of the sample trace on the underside of the glass wafer using a fine-point indelible marker.

3. Place a small drop (e.g., 5 µl) of clear lactic acid on a microscope slide. Remove the glass sample wafer from the plastic cassette housing and place it, gel side up, in the lactic acid. Using the pen marks for reference, adjust the sample so that the trace is parallel to the long axis of the microscope slide.

4. Place a drop of mounting fluid (e.g., lactophenol cotton blue or lactofuchsin) in the center of a 22-mm square coverglass. Invert to hang the drop of mounting fluid and carefully lower onto the sticky surface of the Air-O-Cell® sample.

5. Allow to stain for 5 to 10 min prior to examination.

6. The field diameter of the microscope lens used to analyze spore trap samples must be calibrated using a stage micrometer. Identify and enumerate the fungal elements present in a minimum of one-third of the trace by counting along transects perpendicular to the trace. The microscopic transects should be enumerated in the central portion of the trace and separated from each other by at least one microscopic field diameter. In samples showing an elevated density of fungal elements the number of transects evaluated may be reduced to the number of complete transects required to count a total of 100 or more fungal elements. Note that all of the transects counted must be counted in their entirety (i.e., from one side of the slide to the other).

7. Extrapolated airborne concentration in counts per cubic metre of air (C) is given by

$$C = \frac{L}{n \times \Phi} \times \frac{C_R}{Q \times t} \times 10^3 \text{ liters} \cdot \text{m}^{-3} \quad (2)$$

where C is extrapolated airborne counts (per cubic meter), L is 14.4 mm (length of Air-O-Cell® sampling slit), n is number of transects evaluated, Φ is diameter of microscopic field (in millimeters), C_R is raw count, Q is flow rate of sampler (in liters per minute), and t is duration of sample (in minutes).

Interpretation

In the absence of indoor fungal growth, outdoor air can be expected to be the largest contributor to the indoor air-borne spore load. Thus, an indoor fungal growth source should be suspected if the taxonomic profile of fungi in the indoor air varies significantly from that of the outdoor air. One of the major difficulties in interpreting the results of any air sampling method is the problem of accounting for the enormous quantitative fluctuations in airborne spore load that occur over time. There has been general acknowledgment that the numerical data generated by these methods are unreliable and should not be used as a primary determinant in interpretation (78). A more meaningful index of interior condition as reflected by the results of air sampling can be determined by a comparison of the organisms recovered indoors relative to outdoors. When the predominant organisms indoors are not present outdoors or vice versa, the interpretation of spore trap results is straightforward. However, where an objective interpretation is required, the comparative assessment of the biota of indoor versus outdoor air can be accomplished statistically using Spearman's rho or rank correlation (36, 104).

Taxa observed in all samples are assigned a rank according to airborne concentration from highest to lowest. For each taxon observed, the absolute value of the difference between indoor and outdoor rank is summed and divided by the total number of taxa seen. Spearman's coefficient (r_s) is given by

$$r_s = 1 - \frac{6\Sigma(d_i)^2}{n(n^2 - 1)} \quad (3)$$

where d_i = difference between indoor and outdoor rank for the same taxon, and n = number of taxa observed.

An example dataset is shown in Table 1. The average value of tied ranks is substituted when ties occur. In the case where a taxon is absent from one site yet present in the other, that taxon is assigned the lowest rank where it is absent. In the example data set shown in Table 1, the sum of the squared rank differences is

$$(0)^2 + (-4)^2 + (+2)^2 + (-2)^2 + (+3.5)^2 + \\ (-2)^2 + (+2.5)^2 = 46.5$$

and Spearman's coefficient for this data set is

$$r_s = 1 - \frac{6 \times 46.5}{7(7^2 - 1)} = 0.170$$

The null hypothesis of Spearman's test is that there is no correlation between the sample sets (i.e., the biodiversities in the two compared are different). Degrees of freedom (df) for Spearman's coefficient is calculated as $n - 2$, or 5 for this data set, and the corresponding critical value, r_s^*, at α = 0.05 with 5 degrees of freedom is 0.900. Critical values for Spearman's coefficient are given in many general statistics textbooks (e.g., reference 83). In our example, the computed value r_s is well below the critical value, r_s^*, and the null hypothesis cannot be rejected. In other words, the taxonomic profiles of indoor and outdoor air are statistically different, implying the existence of an indoor growth source.

This test is useful but does not account for nominal variation between indoor and outdoor airborne taxa that might directly suggest an indoor growth site. For example, "indicator fungi" such as Chaetomium and Stachybotrys are sufficiently rare in outdoor air samples that their presence indoors individually or collectively even at detectable levels is decisively symptomatic of indoor mold growth. A good discussion on limitations of Spearman's rank correlation in the interpretation of air sample results is given by Miller (78).

TABLE 1 Ranking of indoor versus outdoor air spora

Taxon	Indoor		Outdoor		d_i
	cts^a · m^{-3}	Rank^b	cts · m^{-3}	Rank	
Basidiospores	660	1	1,980	1	0
Aspergillus/Penicillium	506	2	22	6	−4
Ascospores NOC	220	4	990	2	+2
Fusarium	440	3	44	5	−2
Alternaria	<LOD^c	6.5	264	3	+3.5
Scopulariopsis	110	5	<LOD	7	−2
Cladosporium	<LOD	6.5	154	4	+2.5

^{a}cts, counts.
^{b}1 = most abundant.
^{c}LOD, limit of detection.

Notes

Spore trap specimens should be examined at a minimum of ×600 magnification in bright-field microscopy or ×500 magnification if an oil immersion objective is used. Other light microscopy formats such as phase contrast or, still better, Nomarski differential interference contrast microscopy offer improvements over basic bright-field microscopy for the analysis of spore trap samples. Regardless of the illumination method, planachromat or planapochromat objectives should be used in the analysis of spore trap samples to ensure an optically flat field of view free of chromatic aberration. This will greatly help to reduce the possibility of misidentifying or miscounting peripheral particles. The use of a graticule may improve counting by permitting more accurate measurement of area assessed and help to reduce eye fatigue.

Standard recipes for microscopic mounting media often include much more dye than is desirable for the preparation of spore trap samples. Preparation of mounting fluids with 10 to 20% of the recommended dye content is usually adequate for staining and makes the preparations easier to examine. Iodine vapor staining (discussed above) may help to distinguish small hyaline spores from starch granules. As well, many fungal spores have diagnostically relevant surface features that can be revealed by iodine staining.

While a number of resources exist as aids to spore identification, there has been little literature discussion of the standardization of criteria for counting particles on spore traps. Although variability is normally quite high as a function of the nonhomogeneous dispersion and episodic nature of spore aerosols, several common procedural variations in spore counting may increase imprecision and thus warrant discussion.

- *Single spores, spore chains, and clustered elements*—In general, most analysts count the total number of spores or other elements observed with little regard to grouping or distribution. In the context of a naturally produced bioaerosol, the propagules of many aerially dispersed fungi exist commonly as aggregates rather than as single spores. This is certainly true of many plant-pathogenic conidial fungi and ascomycetes for which spore aggregation is a beneficial property, increasing the terminal velocity of the infective particle and enhancing its penetration of the boundary layer of air surrounding plant leaf surfaces. In contrast, a culturable, multispored particle will yield a single colony if deposited on a growth medium and thus will be counted only once by viable sampling methods (viable sampling is discussed in the next section). This methodological difference contributes to the unjustified notion that a vast majority of airborne fungal spores are nonviable or at least nonculturable. Given that (i) single and aggregated airborne spores both behave as single particles and (ii) measurement units used for air sampling are not relevant to exposure measures (e.g., counts per cubic meter versus milligrams per cubic meter), it is recommended for comparative purposes that airborne fungal particles on spore trap samples be tallied by impaction event rather than spore number.

- *Elements floating freely in the mount*—The use of viscous mounting media and removal of excess mounting fluid by gently blotting will help to reduce the movement of spores in microscopic examination of a spore trap preparation. Nevertheless, with all spore trap sampling methods there are occasional "floaters." One way to deal with drifting particles is to count only those particles present in the field initially and ignore those that enter the field during counting.

- *Elements within but extending beyond the field of view*—An approach based on that of Lacey and Venette (66) and adopted by many analysts is to count such elements on only one half of the field of view and to ignore them on the other half.

- *Taxa observed during setup but not encountered during analysis proper*—The low-level presence of certain problem indicator taxa (e.g., *Chaetomium* and *Stachybotrys*) should be reported if observed both formally during counting and informally during setup. If these taxa are only observed informally, then they should be reported but at a concentration below the limit of detection (LOD). The inclusion of a deliberate final raster scan of the complete slide may help to formalize the reporting of rare taxa below the LOD.

42.2.2.1.2.2. Viable sampling. The mainstay of fungal air sampling is viable sampling. This method involves the collection of airborne particulate matter on the surface of a culture medium by impaction or centrifugation. Airborne fungal content is extrapolated from the colony counts obtained following an appropriate incubation period. The results of viable air sampling are usually expressed in CFU per cubic meter of air. As is the case with spore trap air sampling, viable sampling methods cannot provide measurements relevant to the assessment of exposure. Even so, these techniques are useful in detecting the existence of indoor fungal growth. Viable air sampling offers the practical benefit of basing identifications on actively growing colonies of fungus rather than solitary spores, permitting a low level of taxonomic discrimination (e.g., species level identifications may be determined directly for many genera of microfungi). Viable sampling methods have several limitations.

- Short-duration sampling is necessary to prevent excessive drying of growth media. The use of short sampling times increases the extent to which sample results are influenced by transient airborne spore bursts.
- Standard growth media do not support the growth of all fungi present at a given time in the air. A great number of commonly occurring environmental fungi have specific nutritional requirements that collectively cannot be met by the use of a single growth medium. Sometimes these fungi are referred to as nonculturable. Thus, the media that are typically recommended for viable sampling procedures are intended to optimize recovery of the most relevant subset of the problem indoor aeromycota.
- An incalculable proportion of the fungal component of air is nonviable and thus cannot be detected by viable sampling methods.
- Lengthy incubation times are often required for sufficient growth and sporulation to perform accurate identifications (e.g., 7 to 21 days). Reducing incubation time or increasing the incubation temperature subselects fast-growing or thermotolerant taxa, respectively, biasing results.

Indications

Viable sampling is useful as a frontline investigative technique where indoor fungal biocontamination is suspected but not directly observed during a walk-through inspection. Viable sampling is also the preferred air sampling method when low-level identifications (e.g., species level) are required. This method is indicated where excessive nonfungal background particulate present in the air may interfere with spore trap analysis.

A number of commercially available devices exist for viable air sampling such as centrifugal samplers (standard and high-flow RCS; Biotest, Dreieich, Germany), Anderson N6 and two-stage samplers (Thermo Anderson, Smyrna, GA), and the Surface Air Sampler (Bioscience International, Rockville, MD).

Procedure

1. The sampling device should be wiped with a disinfectant solution such as 70% isopropyl alcohol and allowed to dry. This will help to prevent carryover contamination from previous work. As well, a sampling plan should be clearly laid out prior to beginning.

2. Insert the sampling medium into the air sampler by following the manufacturer's recommendations. Wear latex or nitrile gloves during this process to prevent handling contamination.

3. Switch on the sampler or sampling pump to take the air sample. The flow rates of most units are fixed according to the manufacturer in the range of 25 to 60 liters $\cdot$ min^{-1}. Air samples taken in the summer months should be taken with a reduced sample volume to avoid oversampling due to the normally elevated outdoor levels of plant leaf surface fungi. For example, a final sample volume of 80 to 100 liters

is usually sufficient for an Anderson sampler or Biotest RCS during the summer months. Accordingly, the total sampling volume of samples taken during the wintertime may be doubled. The intent of varying the sample volume according to the expected level of culturable airborne fungi is to ensure reliability of counting. Specifically, the sample volume should be roughly calibrated to deliver a colony density on agar plate or strip samples in the range of 10 to 60 fungal colonies (e.g., roughly 1 colony $\cdot$ cm^{-2}) (73 [section 5.2.3.1]). It is not recommended to sample volumes larger than 200 liters with any viable sampling instrument, in order to avoid desiccation of the surface of the sampling medium, reducing spore viability by hyperosmolar action. Impaction samplers are especially susceptible to these effects (79).

4. Incubate samples inverted under appropriate conditions for 7 to 21 days. The inclusion of a diurnal light/dark cycle during incubation may promote sporulation in some fungi and facilitate identification.

5. Colonies should be counted and identified to as low a level as practical. Care should be taken to notice patterns of nonhomogeneous distribution of colonies on sampling medium. For instance, confluent or patchy distribution of a single species restricted to the margin is suggestive of handling contamination. Similarly, patchy distribution of small colonies around or in a plume extending from a single larger colony may indicate satellite colony formation, particularly where dry-spored molds such as *Aspergillus* and *Penicillium* are present. These artifacts result in spurious elevation of airborne counts and should be noted in the laboratory report whenever they are observed. Generation of satellite colonies can be avoided by incubating sampling media in a still environment and not handling samples until the time of analysis.

6. Level of airborne culturable fungi (C) can be determined from the raw colony count such that

$$C = \frac{C_R}{Q \times t} \times 10^3 \text{ liters} \cdot \text{m}^{-3} \quad (4)$$

where C is extrapolated airborne concentration (CFU per cubic meter), C_R is raw count, Q is flow rate of sampler (in liters per minute), and t is duration of sample (in minutes).

Interpretation

Several qualitative and semiquantitative guidelines have been widely used for interpreting the results of viable air samples (36, 51, 73). These guidelines use air samples to predict indoor fungal growth. It bears repeating that neither these nor any other currently available guidelines are intended to give safe limits for exposure. Many interpretive guidelines for viable air sampling use the rank order criteria set out by Health Canada (51) based on a set of more than 3,000 viable air samples taken from more than 100 Canadian government buildings in the decade between 1986 and 1996 with an RCS taking 160-liter samples on rose bengal agar. The nine-point Health Canada interpretive criteria are given below.

- Significant numbers of certain pathogenic fungi should not be present in indoor air (e.g., *Aspergillus fumigatus*, *Histoplasma*, and *Cryptococcus*). Bird or bat droppings near air intakes, in ducts, or in buildings should be assumed to contain these pathogens. Action should be taken accordingly. Some of these species cannot be measured by air sampling techniques.
- The persistent presence of significant numbers of toxigenic fungi (e.g., *Stachybotrys chartarum*, toxigenic

Aspergillus, *Penicillium*, and *Fusarium* species) indicates that further investigation and action should be taken accordingly.

- The confirmed presence of one or more fungal species occurring as a significant percentage of a sample in indoor air samples and not similarly present in concurrent outdoor samples is evidence of a fungal amplifier. Appropriate action should be taken.
- The "normal" air mycoflora is qualitatively similar to and quantitatively lower than that of outdoor air. In federal government buildings, the 3-year average has been approximately 40 CFU · m^{-3} for *Cladosporium*, *Alternaria*, and nonsporulating basidiomycetes.
- More than 50 CFU of a single species present per cubic meter (other than *Cladosporium* or *Alternaria*) may be reason for concern. Further investigation is necessary.
- Up to 150 CFU · m^{-3} is acceptable if there is a mixture of species reflective of the outdoor air spores. Higher counts suggest dirty or low-efficiency air filters or other problems.
- Up to 500 CFU · m^{-3} is acceptable in summer if the species present are primarily *Cladosporium* or other tree and leaf fungi. Values higher than this may indicate failure of the filters or contamination in the building.
- The visible presence of fungi in humidifiers and on ducts, moldy ceiling tiles, and other surfaces requires investigation and remedial action regardless of the airborne spore load.
- There are certain kinds of fungal contamination not readily detected by the methods discussed in this chapter. If unexplained (sick building syndrome) symptoms persist, consideration should be given to collecting dust samples with a vacuum cleaner and having them analyzed for fungal species.

While very useful, it is important to note that air spora vary considerably by geography. Thus, the numerical and to some extent the species composition data upon which the above guidelines are based are likely to be less relevant at southern latitudes. Dillon et al. (36) suggested the use of Spearman's rho coefficient for detecting differences between the indoor and outdoor air spora for a particular location (as discussed previously in the interpretation of spore trap samples). Other authors have also supported this approach (78), acknowledging that the presence of certain indicator taxa (e.g., *Cladosporium sphaerospermum*, *Aspergillus sydowii*, *Aspergillus versicolor*, *Paecilomyces variotii*, *Scopulariopsis* spp., *Stachybotrys* spp., *Chaetomium* spp., and *Wallemia sebi*) in indoor air samples is strongly suggestive of indoor fungal growth.

Notes
The validity of viable sample results is dependent on the accurate identification of the taxa present. Certain groups of fungi (notably *Fusarium* and *Penicillium*) require considerable expertise and extensive testing on an isolate-by-isolate basis to provide accurate identification to the species level. For these fungi, species identification is beyond the scope of routine practice, although identification to subgenus or morphological group may provide a practical alternative. However, if multiple cultures derived from a given material or site yield high numbers of what appears to be a single species of a taxonomically difficult genus, referral of one or more isolates to a relevant expert for accurate species identification may be warranted.

Rapidly growing fungi pose a particular problem on cultured air samples by overgrowing less competitive fungi and biasing sample counts. Often it is possible to observe overgrown colonies by examining the colony reverse. Overgrown samples should be retested at a lower sample volume. Additionally, the use of restrictive growth media (e.g., rose bengal agar and DG18) may help to prevent sample overgrowth.

42.2.3. Principles and Methods of Yeast Community Analysis

42.2.3.1. Introduction
Information on the ecology of yeast communities is scattered across a profuse taxonomic literature and otherwise overshadowed by studies of industrially or medically important species. The following is therefore based largely on our own experience and practice. For a recent review of the subject, see reference 70.

The position of yeasts in nature is founded on the intrinsic properties of the yeasts themselves (fundamental niche) and the environmental factors (realized niche) that ultimately determine where and how yeasts grow in nature. Yeasts are a highly polyphyletic collection of ascomycetous and basidiomycetous fungi (40, 64). Because they share a predominantly unicellular growth cycle, specifically lacking aerial hyphae, yeasts behave very much like heterotrophic bacteria in the laboratory. Consequently, the methods used in the study of yeasts are typical of those used in bacteriology and quite different from those used with the filamentous fungi. The unicellular habit defines both the methodology used in exploring yeast communities and the habitats that should be explored for the occurrence of yeasts.

42.2.3.1.1. The Fundamental Niche
Nutritionally, yeasts share a strictly organotrophic habit (for a comprehensive discussion of yeast physiology, see reference 115). Suitable carbon sources may include many carbohydrates and their polyol derivatives, simple alcohols, organic acids, ketones, esters, alkanes, or more complex compounds, but only a small number of sugars, including D-glucose, are universally assimilated by all known species. Likewise, the nitrogen requirement of yeasts may be satisfied by a variety of inorganic and organic molecules, but ammonia is a quasiuniversal nitrogen source. Most other nutrients may be supplied in the inorganic form, except for vitamins, which are frequently required. A few species require the presence of some amino acids. Various yeasts differ in their growth temperature responses, but most grow well at 25°C. A few species are psychrophilic and only grow at temperatures below 18 to 25°C, and in rare cases, growth is limited to a narrow range from ca. 30 to 40°C. No truly thermophilic yeast has yet been identified. Only a handful of species grow at 45°C or slightly above. The vast majority of yeasts tolerate a broad range of pH values between ca. 3.5 and 8.5. Most are moderately osmotolerant.

Among the yeasts, considerable variation exists with respect to nutritional breadth and metabolic versatility within the limits set above. As a very broad generalization, and with many exceptions, basidiomycetous species tend to be polyphagous with respect to carbon sources and strictly oxidative in the utilization of carbohydrates. The growth of ascomycetous yeasts is usually limited to a narrower range of substrates, and carbohydrates often can be used both oxidatively and fermentatively. Most yeasts appear to be copiotrophic with respect to carbon, and many appear to be oligotrophic with respect to nitrogen.

42.2.3.1.2. The Realized Niche

The characteristics of the fundamental niche not only serve as a basis for the design of culture media but also define the potential range of habitats where yeasts abound in nature (the realized niche). Unicellularity favors submerged growth in liquid substrates or surface growth on moist, solid substrates, although several species are dimorphic and produce invasive hyphae or pseudohyphae. Yeast communities thrive in habitats that are relatively rich in organic carbon, and where other biotic or abiotic factors may restrict the intensity of competition with bacteria and molds. Such selective conditions are most frequently met in the immediate vicinity of other living organisms or their early decay products. Yeasts are considered saprotrophic, although some are intimately associated with other living organisms. Some yeasts are found in association with vertebrates as opportunistic pathogens, but very few have an absolute dependency on their animal hosts and most act as epizootics. A small group of yeasts are able to produce specialized structures that allow them to penetrate other yeasts or even filamentous fungi, resulting in their death (invasive necrotrophy). This may result in a nutritional benefit to "predacious" yeasts, although all are capable of growth at the expense of inert material (68).

Communities consisting of large populations of ascomycetous yeasts commonly occur in tree sap flows, nectar, necrotic tissue of succulents, or damaged fruit. Insects and other invertebrates that utilize these substrates as feeding or breeding grounds are important vectors. In contrast, basidiomycetous yeasts are more often found in the phylloplane. Many are thought to benefit from the formation of carotenoid pigments as protective agents against toxic photoreaction products. Basidiomycetes appear to be less dependent on invertebrates for their movement. Many are ostensibly wind or water dispersed, and some even discharge their propagules actively. As living beings invariably die and find their way to soil and water, populations of yeasts also exist in soils and aquatic habitats. It is not clear, however, whether soil and seawater are habitats that harbor dynamic yeast communities.

42.2.3.1.3. The Place of Yeasts in Nutrient Cycling and the Food Chain

Compared to bacteria, yeasts have rather modest physiological abilities, as all are heterotrophs that require fairly abundant carbon sources. In terms of carbon cycling, yeasts can be regarded as decomposers, although their role in that capacity is unquestionably minor in comparison with that of filamentous fungi. Only a fairly small proportion of yeasts produce extracellular enzymes capable of hydrolyzing polysaccharides, proteins, or fats, and the degradation of the most naturally abundant polymers such as cellulose, lignin, or chitin seems to be confined to the filamentous fungi. Yeasts may effect a reversal of mineralization with respect to nitrogen, thereby enriching the diet of insects that feed on plant exudates, nectar, sap, or decaying tissue (71).

42.2.3.2. Collection Methods

Although in situ identification and enumeration of species with molecular probes may become practicable in the future, ecological studies of yeasts currently require that pure cultures be isolated on culture media and purified prior to characterization in the laboratory. The methods described below were developed for the study of natural communities and may differ from those used in clinical or industrial contexts.

42.2.3.2.1. Media

Isolation is generally done on plates. Common media contain various mixtures of glucose and peptones. YM agar is adequate for most situations. Media that do not contain yeast extract should be avoided. As the nutritional requirements of yeasts also satisfy those of many bacteria, the addition of antibiotics (e.g., 50 to 100 ppm of chloramphenicol) is usually necessary for isolation purposes. Acidification of the medium to pH 3.5 (after autoclaving) has been used also, but it is less effective. Molds such as *Mucor* or *Penicillium* also grow well on yeast media, and even a single colony can rapidly invade the surface of a plate. Molds cannot be suppressed selectively without harming some yeasts. Young mold colonies may be excised from agar plates with a sterile instrument, but in cases where they are abundant, a liquid enrichment step may be necessary prior to plating. A small amount of sample is inoculated into a tube of liquid medium (YM broth with antibiotics) and incubated on a Rollordrum™ (New Brunswick Scientific). The rotational mixing delays the formation of conidia and causes the mold to grow as a mycelial ball that is easily removed from the tube with a bent inoculation needle. A loopful of the remaining yeast suspension is then streak inoculated onto agar. Because of the potential for enrichment of antibiotic-resistant bacteria, it is prudent not to use the same antibiotic in both the broth and the agar medium. Kanamycin, streptomycin, and doxicycline have been used.

42.2.3.2.2. Substrate Type

Liquids containing relatively low cell numbers (e.g., seawater) should be passed through a filter membrane (0.45-μm pore size) that is then plated. If cell numbers are high, a liquid substrate may be streak inoculated directly. The streak pattern should ensure that a broad dilution gradient is formed on the plate so that single colonies will arise regardless of the cell concentration in the sample. Colony separation may be facilitated by diluting the liquid substrate with water. Solid materials are mixed vigorously with a suitable amount of sterile water, followed by streak-inoculation of a loopful of suspension. Addition of a small amount (ca. 10 mg/liter) of Tween 80 to the water improves the dispersion of yeast cells stuck to solids. Small insects are conveniently sampled by allowing them to walk over the agar for a few minutes. Larger insects may require dissection or homogenization. Surfaces of solid substrates may be swabbed or scraped with an inoculation loop and sterile water. Ballistosporogenous yeasts can be sampled by affixing the substrate, usually a leaf, to the lid of the petri dish to allow for spore discharge.

Isolation of yeasts from natural substrates is most effective when the material is fresh. Once harvested, yeast-rich materials, especially moist solids, tend to be overcome by molds. As it is rarely convenient to plate in the field, samples can be carried to the laboratory in sterile Whirl-Packs or in small vials containing sterile water. Refrigeration may or may not be advantageous, but protection from excessive heat is essential.

42.2.3.2.3. Isolation

Inoculated plates must be examined periodically. As yeast colonies develop, usually after 2 days, cells from representative colonies are picked with a sterile inoculation needle, suspended thinly in sterile water, and streak inoculated onto a plate of nonselective medium to ascertain the absence of contaminants. Direct plating of yeast cells without resuspension is less effective. The use of dilution blanks and purification plates in the field, away from the laboratory, may be impractical. In that case, selected colonies may be transferred to small agar slants (1 ml in 2-ml vials). Purified cultures are stored on agar slants or in sdH$_2$O. Permanent preservation requires cryogenic storage or lyophilization. Liquid nitrogen storage in small vials containing porcelain beads (Microbank™ [ProLab Diagnostics]) is particularly convenient.

The details of strain characterization exceed the scope of this discussion, and the reader is referred to reference 123. However, it is useful to mention that standard growth tests are conducted with two chemically defined media available commercially. Yeast nitrogen base may be used in combination with individual carbon sources in searching for yeasts that have the ability to utilize specific carbon compounds. Similarly, yeast carbon base (1.17%) may be supplemented with a nitrogen source to look for yeasts with specific nitrogen assimilation abilities. Until recently, standard growth tests were essential for identification. They now serve a descriptive function. Accurate identification is based on DNA sequencing.

42.2.3.3. Sampling and Data Management, Sample Size, and Numbers

Samples should be assigned simple consecutive reference numbers, to which is attached information about the locality, type of substrate, and voucher specimens. As individual yeast isolates are recovered, they should be assigned a unique number, preferably a decimal extension of the sample number, which serves to cross-reference information on colony characteristics, approximate abundance in the sample, microscopic features, photographs, physiological responses, and any other information. The use of identification codes or acronyms based on the isolation locality, the type of substrate, or any other descriptor inevitably leads to loss of data due to irretrievability in the long term. The information as a whole can be managed using database or spreadsheet software.

If required, the determination of cell numbers in natural substrates may follow any of the procedures used in bacteriology, such as the determination of CFU or most-probable-number approaches. However, in studies of natural yeast communities, establishing the magnitude of species diversity and determining which species are autochthonous members of each community are the most important objectives. Cell numbers vary enormously from one sample to the next and contribute little information relevant to the pursuit of those objectives. The strongest indication that a species is an important member of a community is its repeated occurrence in multiple, independent samples. Indeed, many common species diversity indices are based on presence or absence data and not on relative numbers of individuals (8). Such measures typically stabilize once ca. 15 samples have been screened (69). Accordingly, effort should be spent on examining a sufficient number of representatives of the community, and not on determining cell numbers.

The primary agar plate obtained from inoculation of a sample invariably contains replicates of each species. A decision must therefore be made about how many representatives of each colony type should be picked and purified for further study. When large numbers of samples from the same habitat are examined, the chances of overlooking distinct species because their colonies are similar are minimal, and one representative of each type is sufficient. The cost of identifying large numbers of replicates from a few samples is not justified if many independent samples are available.

42.2.3.4. Identification

Correct identification of species is probably the most important step of an ecological study and once was the most difficult. The so-called conventional approach is based on morphological traits that may be difficult to interpret, even by experienced workers, plus a battery of onerous growth tests (63). As our knowledge of species diversity continues to expand, the discriminatory power of those tests is increasingly insufficient. Identification keys based on morphology and physiology are becoming less and less satisfactory and have a high probability of ambiguous results or error. This is not to say that the knowledge of the reproductive and nutritional features of yeast species is not essential to the characterization of yeast communities. Indeed, they provide important insights on the fundamental niche.

At present, rapid and correct identification is best achieved by determining the sequence of the D1 and D2 variable domains of the large rDNA subunit. The target sequence of an unknown is conveniently amplified by suspending a small amount of a pure yeast culture directly in a PCR mixture containing suitable primers (67). The purified amplification product is sequenced and the result is compared with known sequences. The D1/D2 region offers three important advantages. Most importantly, a comprehensive database for all known species is available in GenBank. Other popular targets, such as the internal transcribed spacer or the small subunit rDNA (SSU) regions, have not been determined systematically for all species. Secondly, for identification purposes, the complete D1/D2 sequence can be obtained in a single reaction, as opposed to three for the SSU. Last, the phylogenetic signal of D1/D2 sequences is normally strong at the species level and the sequences are sufficiently conserved to allow assignment of new species to a genus. In most cases a perfect match or the presence of fewer than two or three substitutions provides a conclusive species identification (64). Exceptions do exist, and some cases may not be resolved without the intervention of an experienced yeast systematist. Better alternatives are not available at this time, although they no doubt will be in the future. The cost of sequencing still precludes its indiscriminate application to very large collections of strains unless one has continuous access to a high-throughput sequencing facility. The present solution is to subject isolates to a preliminary screening procedure based on colony or cell morphology, selected growth tests, or simple molecular fingerprinting. The last should be used with the utmost caution (52).

42.2.3.5. Ecological Interpretation

Yeast community ecology is still in its infancy. Considerable effort is still devoted to the physiological and biochemical characterization of yeasts, but our understanding of their interactions among themselves and with other species, microbial or otherwise, is limited. It would be presumptuous to propose analytical methods purported to lead to genuinely new insights. Great advances will no doubt be made when the methodology is such that identification can be performed in the field.

42.2.4. Common Media and Reagents

42.2.4.1. Rose Bengal Agar

Rose bengal agar is a good, widely available medium for general-purpose isolations that permits the growth of most commonly occurring indoor fungal contaminants and rapidly sporulating soil fungi. Although there are a number of variations, the basic formula uses a base of peptone (5 g/liter), glucose (10 g/liter), mineral salts (KH$_2$PO$_4$, 1 g/liter; MgSO$_4$ · 7H$_2$O, 0.5 g/liter), and agar (15 g/liter) to which antibiotics are added to prevent the overgrowth of bacterial contaminants and rose bengal (25 mg/liter) is added to restrict and delimit the growth of fungal colonies as an aid to enumeration. For soil fungi, the nutrients are omitted and half of the water is replaced with 500 ml of boiling water extract of 75 to 200 g of soil. Dichloran may be added to reduce growth of zygomycetous molds, and prepared dichloran rose bengal chloramphenicol (DRBC) agar is available from Oxoid. Rose bengal is known to become somewhat toxic following exposure to light mainly due to the liberation of singlet oxygen species (^{1}O$_2$). For this reason, these plates should be incubated in the dark.

42.2.4.2. DG18 Agar

Dichloran 18% glycerol (DG18) agar is a good medium for xerophiles such as *Aspergillus* and *Penicillium*, but it does not generally support the growth of *Stachybotrys*. The formula, per liter, includes the following: KH$_2$PO$_4$, 1 g; MgSO$_4$ · 7H$_2$O, 0.5 g, dichloran, 1.0 ml of a 0.2% solution in ethanol; peptone, 5 g; glucose, 10 g; and agar, 15 g. These ingredients are steamed to dissolve and melt the agar, and then 220 g of glycerol is added prior to autoclaving and 0.1 g of chloramphenicol in ethanol is added after autoclaving.

42.2.4.3. ME Agar

Malt extract (ME) agar is an excellent medium for culturing wood-inhabiting and litter-decomposing basidiomycetes, but it is not recommended for sampling fungi of human or indoor environments. Several standard formulations exist in the literature, of which the best is that of Nobles (82): 12.5 g of ME and 15 g of agar per liter of distilled water. Antibacterial agents and antifungal agents may be added to improve isolations of basidiomycetes. Addition of sugars increases production of toxic metabolites by the growing cultures and is not recommended. A weak ME broth, with 2 to 5 g of ME per liter of dH$_2$O, is excellent for preparing mycelia of many saprotrophic fungi for DNA isolation.

42.2.4.4. MMN Medium (9, 75)

Modified Melkin-Norkrans (MMN) medium and BAF agar are excellent for isolating cultures of fastidious ectomycorrhizal fungi. Benomyl and dichloran may be added to reduce growth of ascomycetous and zygomycetous molds, respectively, and antibacterial antibiotics are recommended to prevent the growth of bacterial contaminants. The formula, per liter, includes the following: CaCl$_2$, 0.05 g; NaCl, 0.025 g; KH$_2$PO$_4$, 0.5 g; (NH$_4$)$_2$HPO$_4$, 0.25 g; MgSO$_4$ · 7H$_2$O, 0.15 g; FeCl$_3$, 1.2 ml of a 1% aqueous solution; thiamine HCl, 100 µg; ME, 3 g; sucrose, 10 g; and agar (optional), 15 g.

42.2.4.5. BAF Agar (9, 80)

BAF agar has threefold more sugar than MMN medium and contains a few additional micronutrients and growth factors; BAF agar and MMN medium seem to work equally well for isolating ectomycorrhizal fungi. The formula, per liter, includes the following: CaCl$_2$ · 2H$_2$O, 0.1 g; NaCl, 0.025 g; KH$_2$PO$_4$, 0.5 g; MgSO$_4$ · 7H$_2$O, 0.5 g; FeCl$_3$ · 6H$_2$O, 10 mg; MnSO$_4$, 5 mg; ZnSO$_4$ · 7H$_2$O, 1 mg; thiamine HCl, 50 µg; folic acid, 100 µg; inositol, 50 µg; biotin, 1 µg; yeast extract, 0.2 g; peptone, 2 g; glucose, 30 g; and agar (optional), 15 g. As with MMN medium, benomyl and dichloran may be added to reduce growth of ascomyce-

tous and zygomycetous molds, respectively, and antibacterial agents are recommended to prevent the growth of bacterial contaminants.

42.2.4.6. YM Agar

Available commercially (Difco), "yeasts and molds" agar contains 1% glucose, 0.5% peptone, 0.3% ME 0.3% yeast extract, and 2.0% agar. It can be used for isolation and maintenance of most yeast species.

42.2.4.7. Yeast Nitrogen Base

Yeast nitrogen base is a synthetic medium available from Difco in three formulations. All contain a complete array of defined nutrients, including several vitamins. The standard medium is supplemented with 0.5% ammonium sulfate and three amino acids, some of which are required by a few yeast species. One variant does not contain the amino acids and can be used to isolate prototrophs. Either medium is used at a concentration of 0.67%. Another variant contains neither supplement so that an alternative nitrogen source can be chosen. The normal concentration is 0.17%. All three formulations require the addition of an organic carbon source at the rate of 0.5 to 1.0%.

42.2.4.8. Yeast Carbon Base

Similar to yeast nitrogen base, yeast carbon base contains glucose as a carbon source, but no nitrogen source. Trace amounts of the three amino acids are included to ensure the growth of yeasts that have such a requirement. The normal concentration is 1.17%. Nitrogen sources can be added at a concentration of 0.05% or less in the case of compounds such as sodium nitrite.

42.2.4.9. Antibacterial Agents

Rose bengal, used at 25 mg/liter, is light sensitive and may be toxic to fungi as well as bacteria, but it may be added to media prior to autoclaving (see section 42.2.4.1). Streptomycin sulfate, tetracycline hydrochloride, penicillin G, or chloramphenicol may be added at 50 to 100 mg/liter to media after autoclaving. Addition of a 5-ml/liter concentration of a solution of 20-mg/ml chloramphenicol in 70 to 95% ethanol is convenient and effective for reducing bacterial contamination in isolations from soil or contaminated tissues.

42.2.4.10. Antifungal Agents

Various fungicides may be used to make an isolation medium more selective for the fungi of interest. Fast-growing ascomycetous and zygomycetous molds are sometimes problematic when isolating basidiomycetes, and these may be combated with benomyl and dichloran, respectively (120). Add 1 ml per liter of medium of a solution containing 2 to 5 mg of active ingredient/ml of benomyl or dichloran in 1:1 acetone/ethanol. Cycloheximide (Actidione) has broad toxicity to fungi but may be used at 2 mg/liter to isolate certain fungi that have some tolerance, such as *Ophiostoma* (48, 77). Cycloheximide is added as 2 ml per liter of a 1-mg/ml aqueous solution, before or after autoclaving.

42.2.4.11. 2× CTAB for Preserving Samples for DNA Studies

CTAB is a useful preservative composed of 2% (wt/vol) hexadecyl trimethylammonium bromide (also known as cetyl trimethylammonium bromide or CTAB), 0.1 M Tris-HCl (pH 8.0), 20 mM EDTA (pH 8.0), and 1.4 M NaCl in distilled water (43). The mixture may require heating to bring the CTAB into solution. For convenience, you may choose to buy ready-made 1 M Tris (pH 8) and 0.5 M EDTA (pH 8) from your chemical supplier.

42.3. REFERENCES

1. **Agerer, R. (ed.).** 1987–1993. *Colour Atlas of Ectomycorrhizae.* Einhorn-Verlag, Schwäbische Gmünd, Germany.

2. **Ahmadjian, V.** 1993. *The Lichen Symbiosis.* John Wiley, New York, NY.

3. **Alexopoulos, C. J., C. W. Mims, and M. Blackwell.** 1996. *Introductory Mycology*, 4th ed. John Wiley, New York, NY.

4. **Allen, M. F.** 1991. *The Ecology of Mycorrhizae.* Cambridge University Press, Cambridge, United Kingdom.

5. **Amaranthus, M. P. and D. A. Perry.** 1994. The functioning of ectomycorrhizal fungi in the field: linkage in space and time. *Plant Soil* **159:**133–140.

6. **Andersen, T. F.** 1996. A comparative taxonomic study of *Rhizoctonia sensu lato* employing morphological, ultrastructural and molecular methods. *Mycol. Res.* **100:**1117–1128.

7. **Arora, D. K., R. P. Elander, and K. G. Mukerji.** 1992. *Fungal Biotechnology.* Marcel Dekker, New York, NY.

8. **Atlas, R. M., and R. Bartha.** 1993. *Microbial Ecology: Fundamentals and Applications*, 3rd ed. Addison Wesley, Reading, United Kingdom.

9. **Atlas, R. M., and L. C. Parks.** 1996. *Handbook of Microbiological Media*, 2nd ed. CRC Press, Boca Raton, FL.

10. **Bååth, E.** 1988. A critical examination of the soil washing technique with special reference to the effect of the size of the soil particles. *Can. J. Bot.* **66:**1566–1569.

11. **Baldauf, S. L.** 1999. A search for the origins of animals and fungi: comparing and combining molecular data. *Am. Nat.* **154:**178–188.

12. **Baldauf, S. L., and W. F. Doolittle.** 1997. Origin and evolution of the slime molds (Mycetozoa). *Proc. Natl. Acad. Sci. USA* **94:**12007–12012.

13. **Baldauf, S. L., A. J. Roger, I. Wenk-Siefert, and W. F. Doolittle.** 2000. A kingdom-level phylogeny of eukaryotes based on combined protein data. *Science* **290:**972–977.

14. **Barnett, J. A., R. W. Payne, and D. Yarrow.** 1990. *Yeasts: Characteristics and Identification*, 2nd ed. Cambridge University Press, Cambridge, United Kingdom.

15. **Barr, D. J. S.** 2001. Chytridiomycota, p. 93–112. *In* D. J. McLaughlin, E. G. McLaughlin, and P. A. Lemke (ed.), *The Mycota*, vol. VII. *Systematics and Evolution, Part A.* Springer-Verlag, New York, NY.

16. **Barr, M. E., and S. M. Huhndorf.** 2001. Loculoascomycetes, p. 283–306. *In* D. J. McLaughlin, E. G. McLaughlin, and P. A. Lemke (ed.), *The Mycota*, vol. VII. *Systematics and Evolution, Part A.* Springer-Verlag, New York, NY.

17. **Barron, G. L.** 1971. Soil fungi. *Methods Microbiol.* **4:**405–427.

18. **Barron, G. L.** 1977. *The Nematode-Destroying Fungi.* Canadian Biological Publications, Guelph, Ontario, Canada.

19. **Barron, G. L.** 1981. Parasites and predators of microscopic animals, p. 167–200. *In* G. T. Cole and B. Kendrick (ed.), *Biology of Conidial Fungi*, vol. 2. Academic Press, New York, NY.

20. **Benny, G. L., R. A. Humber, and J. B. Morton.** 2001. Zygomycota: Zygomycetes, p. 113–146. *In* D. J. McLaughlin, E. G. McLaughlin, and P. A. Lemke (ed.), *The Mycota*, vol. VII. *Systematics and Evolution, Part A.* Springer-Verlag, New York, NY.

21. **Bills, G. F., M. Christensen, M. Powell, and G. Thorn.** Saprobic soil fungi, 271–302. *In* G. M. Mueller, G. F. Bills, and M. S. Foster (ed.), *Biodiversity of Fungi: Standard*

22. **Bills, G. F., and J. D. Polishook.** 1994. Abundance and diversity of microfungi in leaf litter of a lowland rain forest in Costa Rica. *Mycologia* **86:**187–198.

23. **Booth, T., and P. Barrett.** 1976. Taxonomic and ecological observations of zoosporic fungi in soils of a high-arctic ecosystem. *Can. J. Bot.* **54:**533–538.

24. **Brundrett, M., N. Bougher, B. Dell, T. Grave, and N. Malajczuk.** 1996. *Working with Mycorrhizas in Forestry and Agriculture.* Monograph 32. Australian Centre for International Agricultural Research, Canberra, Australia.

25. **Brundrett, M., L. Melville, and L. Peterson.** 1994. *Practical Methods in Mycorrhiza Research.* Mycologue Publications, Sidney, British Columbia, Canada.

26. **Brundrett, M. C., Y. Piché, and R. L. Peterson.** 1984. A new method for observing the morphology of vesicular-arbuscular mycorrhizae. *Can. J. Bot.* **62:**2128–2134.

27. **Bruns, T. D., T. M. Szaro, M. Gardes, K. W. Cullings, J. J. Pan, D. L. Taylor, T. R. Horton, A. Kretzer, M. Garbelotto, and Y. Li.** 1998. A sequence database for the identification of ectomycorrhizal basidiomycetes by phylogenetic analysis. *Mol. Ecol.* **7:**257–272.

28. **Butler, E. E., and M. P. Mann.** 1959. Use of cellophane tape for mounting and photographing phytopathogenic fungi. *Phytopathology* **49:**231–232.

29. **Caesar-Ton That, T. C., and V. L. Cochran.** 2000. Soil aggregate stabilization by a saprophytic lignin decomposer basidiomycete fungus. I. Microbiological aspects. *Biol. Fertil. Soils* **32:**374–380.

30. **Calderone, R. A., and R. L. Cihlar.** 2002. *Fungal Pathogenesis: Principles and Clinical Applications.* Marcel Dekker, New York, NY.

31. **Christensen, M.** 1981. Species diversity and dominance in fungal communities, p. 201–232. *In* D. T. Wicklow and G. C. Carroll (ed.), *The Fungal Community.* Marcel Dekker, New York, NY.

32. **Christensen, M.** 1989. A view of fungal ecology. *Mycologia* **81:**1–19.

33. **Clay, K.** 1990. Fungal endophytes of grasses. *Annu. Rev. Ecol. Syst.* **21:**275–297.

34. **Cooke, R. C., and A. D. M. Rayner.** 1984. *Ecology of Saprotrophic Fungi.* Longman, London, United Kingdom.

35. **Davies, R. R.** 1971. Air sampling for fungi, pollen and bacteria. *Methods Microbiol.* **4:**367–404.

36. **Dillon, H. K., P. A. Heinsohn, and J. D. Miller.** 1996. *Field Guide for the Determination of Biological Contaminants in Environmental Samples.* AIHA Press, Fairfax, VA.

37. **Dix, N. J., and J. Webster.** 1995. *Fungal Ecology.* Chapman and Hall, London, United Kingdom.

38. **Dring, D. M.** 1971. Techniques for microscopic preparation. *Methods Microbiol.* **4:**95–111.

39. **Endo, R. M.** 1966. A cellophane tape-cover glass technique for preparing microscopic slide mounts of fungi. *Mycologia* **58:**655.

40. **Fell, J. W., T. Boekhout, A. Fonseca, G. Scorzetti, and A. Statzell-Tallman.** 2000. Biodiversity and systematics of basidiomycetous yeasts as determined by large-subunit rDNA D1/D2 domain sequence analysis. *Int. J. Syst. Evol. Microbiol.* **50:**1351–1371.

41. **Flannigan, B., R. A. Samson, and J. D. Miller (ed.).** 2001. *Microorganisms in Home and Indoor Work Environments: Diversity, Health Impacts, Investigation and Control.* Taylor Francis, London, United Kingdom.

42. **Fuller, M. S., and A. Jaworski (ed.).** 1987. *Zoosporic Fungi in Teaching and Research.* Southeastern Publishing, Athens, GA.

43. **Gardes, M., and T. D. Bruns.** 1993. ITS primers with enhanced specificity for basidiomycetes: application to the

identification of mycorrhizae and rusts. *Mol. Ecol.* **2**:113–118.

44. **Gardes, M., T. J. White, J. A. Fortin, T. D. Bruns, and J. W. Taylor.** 1991. Identification of indigenous and introduced symbiotic fungi in ectomycorrhizae by amplification of nuclear and mitochondrial ribosomal DNA. *Can. J. Bot.* **69**:180–190.

45. **Geiser, D. M., and K. F. Lobuglio.** 2001. The monophyletic Plectomycetes: Ascosphaeriales, Onygenales, Eurotiales, p. 201–220. *In* D. J. McLaughlin, E. G. McLaughlin, and P. A. Lemke (ed.), *The Mycota*, vol. VII. *Systematics and Evolution, Part A.* Springer-Verlag, New York, NY.

46. **Goodfellow, M., M. Mordarski, and S. T. Williams.** 1984. *The Biology of the Actinomycetes.* Academic Press, New York, NY.

47. **Halling, R. E.** 2001. Ectomycorrhizae: co-evolution, significance, and biogeography. *Ann. Mo. Bot. Gard.* **88**:5–13.

48. **Harrington, T. C.** 1981. Cycloheximide sensitivity as a taxonomic character in *Ceratocystis. Mycologia* **73**:1123–1129.

49. **Haug, I., and K. Pritsch.** 1992. *Ectomycorrhizal Types of Spruce (Picea abies (L.) Karst.) in the Black Forest: a Microscopical Atlas.* Kernforschungazentrum Karlsruhe, Universität Tübingen, Tübingen, Germany.

50. **Hawksworth, D. L.** 1988. The variety of fungal-algal symbioses, their evolutionary significance, and the nature of lichens. *Bot. J. Linn. Soc.* **96**:3–20.

51. **Health Canada.** 1995. *Indoor Air Quality in Office Buildings: a Technical Guide.* 93-EHD-166 rev. 1995. Environmental Health Directorate, Ottawa, Ontario, Canada.

52. **Herzberg, M., R. Fischer, and A. Titze.** 2002. Conflicting results obtained by RAPD-PCR and large-subunit rDNA sequences in determining and comparing yeast strains isolated from flowers: a comparison of two methods. *Int. J. Syst. Evol. Microbiol.* **52**:1423–1433.

52a. **Hibbett, D. S., et al.** A higher-level phylogenetic classification of the fungi. *Mycol. Res.*, in press.

53. **Hibbett, D. S., and M. J. Donoghue.** 2001. Analysis of character correlations among wood decay mechanisms, mating systems, and substrate ranges in *Homobasidiomycetes. Syst. Biol.* **50**:215–242.

54. **Hibbett, D. S., L.-B. Gilbert, and M. J. Donoghue.** 2000. Evolutionary instability of ectomycorrhizal symbioses in basidiomycetes. *Nature* **407**:506–508.

55. **Hibbett, D. S., and R. G. Thorn.** 2001. *Homobasidiomycetes*, p. 121–168. *In* D. J. McLaughlin, E. G. McLaughlin, and P. A. Lemke (ed.), *The Mycota*, vol. VII. *Systematics and Evolution, Part B.* Springer-Verlag, New York, NY.

56. **Hu, S.-N., and L.-F. Yan.** 1999. Actin and eukaryotic evolution. *Acta Zool. Sin.* **45**:440–447.

57. **Hutchison, L. J.** 1991. Description and identification of cultures of ectomycorrhizal fungi found in North America. *Mycotaxon* **42**:387–504.

58. **Jacobs, P. H., and L. Nall (ed.).** 1997. *Fungal Disease: Biology, Immunology, and Diagnosis.* Marcel Dekker, New York, NY.

59. **Jong, S.-C., and J. M. Birmingham.** 2001. Cultivation and preservation of fungi in culture, p. 193–202. *In* D. J. McLaughlin, E. G. McLaughlin, and P. A. Lemke (ed.), *The Mycota*, vol. VII. *Systematics and Evolution, Part B.* Springer-Verlag, New York, NY.

60. **Kirk, P. M., P. F. Cannon, J. C. David, and J. . Stalpers.** 2001. *Dictionary of the Fungi*, 9th ed. CABI Publishing, Wallingford, United Kingdom.

61. **Kornerup, A., and J. H. Wanscher.** 1978. *Methuen Handbook of Colour*, 3rd ed. Eyre Methuen, London, United Kingdom.

62. **Korn-Wendisch, F., and H. J. Kutzner.** 1992. The family Streptomycetaceae, p. 921–995. *In* A. Balows, H. G. Truper, and M. Dworkin (ed.), *The Prokaryotes*, 2nd ed., vol. 1. Springer-Verlag, New York, NY.

63. **Kurtzman, C. P., and J. W. Fell (ed.).** 1998. *The Yeasts, a Taxonomy Study*, 4th ed. Elsevier, Amsterdam, The Netherlands.

64. **Kurtzman, C. P., and C. J. Robnett.** 1998. Identification and phylogeny of ascomycetous yeasts from analysis of nuclear large subunit (26S) ribosomal DNA partial sequences. *Antonie Leeuwenhoek* **73**:331–371.

65. **Kurtzman, C. P., and J. Sugiyama.** 2001. Ascomycetous yeasts and yeastlike taxa, p. 179–200. *In* D. J. McLaughlin, E. G. McLaughlin, and P. A. Lemke (ed.), *The Mycota*, vol. VII. *Systematics and Evolution, Part A.* Springer-Verlag, New York, NY.

66. **Lacey, J., and J. Venette.** 1995. Outdoor air sampling techniques, p. 407–471. *In* C. S. Cox and C. M. Wathes (ed.), *Bioaerosols Handbook.* CRC Press, Boca Raton, FL.

67. **Lachance, M. A., J. M. Bowles, W. T. Starmer, and J. S. F. Barker.** 1999. *Kodamaea kakaduensis* and *Candida tolerans*, two new yeast species from Australian *Hibiscus* flowers. *Can. J. Microbiol.* **45**:172–177.

68. **Lachance, M. A., and W. M. Pang.** 1997. Predacious yeasts. *Yeast* **13**:225–232.

69. **Lachance, M. A., and W. T. Starmer.** 1986. The community concept and the problem of nontrivial characterization of yeast communities. *Coenoses* **1**:21–28.

70. **Lachance, M. A., and W. T. Starmer.** 1998. Ecology and yeasts, p. 21–30. *In* C. P. Kurtzman and J. W. Fell (ed.), *The Yeasts, a Taxonomy Study*, 4th ed. Elsevier, Amsterdam, The Netherlands.

71. **Lachance, M. A., W. T. Starmer, C. A. Rosa, J. M. Bowles, J. S. F. Barker, and D. H. Janzen.** 2001. Biogeography of the yeasts of ephemeral flowers and their insects. *FEMS Yeast Res.* **1**:1–8.

72. **Largent, D. L.** 1986. *How To Identify Mushrooms to Genus. I. Macroscopic Features.* Mad River Press, Eureka, CA.

73. **Macher, J. (ed.).** 1999. *Bioaerosols: Assessment and Control.* ACGIH Press, Cincinnati, OH.

74. **Malloch, D. W.** 1981. *Moulds: Their Isolation, Cultivation and Identification.* University of Toronto Press, Toronto, Canada.

75. **Marx, D. H.** 1969. The influence of ectotrophic mycorrhizal fungi on the resistance of pine roots to pathogenic infections. I. Antagonism of mycorrhizal fungi to root pathogenic fungi and soil bacteria. *Phytopathology* **59**:153–163.

76. **Massicotte, H. B., R. Molina, L. E. Tackaberry, J. E. Smith, and M. P. Amaranthus.** 1999. Diversity and host specificity of ectomycorrhizal fungi retrieved from three adjacent forest sites by five host species. *Can. J. Bot.* **77**:1053–1076.

77. **Micales, J. A., and R. J. Stipes.** 1986. The differentiation of *Endothia* and *Cryphonectria* species by exposure to selected fungitoxicants. *Mycotaxon* **26**:99–117.

78. **Miller, J. D.** 2001. Mycological investigations of indoor environments, p. 231–246. *In* B. Flannigan, R. A. Samson, and J. D. Miller (ed.), *Microorganisms in Home and Indoor Work Environments: Diversity, Health Impacts, Investigation and Control.* Taylor Francis, London, United Kingdom.

79. **Morris, K. J.** 1995. Modern microscopic methods of bioaerosol analysis, p. 285–316. *In* C. S. Cox and C. M. Wathes (ed.), *Bioaerosols Handbook.* CRC Press, Boca Raton, FL.

80. **Moser, M.** 1960. *Die gattung Phlegmacium.* Julius Klinkhardt, Bad Heilbrunn, Austria.

81. **Mountfort, D. O., and C. G. Orpin (ed.).** 1994. *Anaerobic Fungi: Biology, Ecology and Function.* Marcel Dekker, New York, NY.

81a. **Mueller, G. M., G. F. Bills, and M. S. Foster (ed.).** 2004. *Biodiversity of Fungi: Inventory and Monitoring Methods.* Elsevier Academic Press, Amsterdam, The Netherlands.

82. **Nobles, M. K.** 1948. Studies in forest pathology. VI. Identification of cultures of wood-rotting fungi. *Can. J. Res. C* **26:**281–431.

83. **Norman, G. R., and D. L. Streiner.** 2000. *Biostatistics: the Bare Essentials*, 2nd ed. Decker, Hamilton, Ontario, Canada.

84. **Norris, J. R., D. J. Read, and A. K. Varma.** 1994. *Techniques for Mycorrhizal Research*. Academic Press, London, United Kingdom.

85. **Olsen, G. J., C. R. Woese, and R. Overbeek.** 1994. The winds of (evolutionary) change: breathing new life into microbiology. *J. Bacteriol.* **176:**1–6.

86. **Pfister, D. H., and J. W. Kimbrough.** 2001. Discomycetes, p. 257–282. *In* D. J. McLaughlin, E. G. McLaughlin, and P. A. Lemke (ed.), *The Mycota*, vol. VII. *Systematics and Evolution, Part A.* Springer-Verlag, New York, NY.

87. **Pitt, J. I., and A. D. Hocking.** 1997. *Fungi and Food Spoilage*, 2nd ed. Blackie Academic and Professional, New York, NY.

88. **Ramachandra, M., D. L. Crawford, and G. Hertel.** 1988. Characterization of an extracellular lignin peroxidase of the ligninolytic actinomycete *Streptomyces viridosporus. Appl. Environ. Microbiol.* **54:**3057–3063.

89. **Rayner, A. D. M., and L. Boddy.** 1988. *Fungal Decomposition of Wood: Its Biology and Ecology.* Wiley, Chichester, United Kingdom.

90. **Redlin, S. C., and L. M. Carris.** 1996. *Endophytic Fungi in Grasses and Woody Plants: Systematics, Ecology, and Evolution.* APS Press, St. Paul, MN.

91. **Rossman, A. Y., R. E. Tuloss, T. E. O'Dell, and R. G. Thorn (ed.).** 1997. *Protocols for an All Taxa Biodiversity Inventory of Fungi in a Costa Rican Conservation Area.* Parkway Publishers, Boone, NC.

92. **Saikkonen, K., S. H. Faeth, M. Helander, and T. J. Sullivan.** 1998. Fungal endophytes: a continuum of interactions with host plants. *Annu. Rev. Ecol. Syst.* **29:**319–343.

93. **Samuels, G., and K. Seifert.** 1995. The impact of molecular characters on systematics of filamentous Ascomycetes. *Annu. Rev. Phytopathol.* **33:**37–67.

94. **Samuels, G. J., and M. Blackwell.** 2001. Pyrenomycetes—fungi with perithecia, p. 221–256. *In* D. J. McLaughlin, E. G. McLaughlin, and P. A. Lemke (ed.), *The Mycota*, vol. VII. *Systematics and Evolution, Part A.* Springer-Verlag, New York, NY.

95. **Sanders, I. R., M. Alt, K. Groppe, T. Boller, and A. Wiemken.** 1995. Identification of ribosomal DNA polymorphisms among and within spores of the Glomales: application to studies on the genetic diversity of arbuscular mycorrhizal fungal communities. *New Phytol.* **130:**419–427.

96. **Schardl, C. L.** 1996. *Epichloe* species: fungal symbionts of grasses. *Annu. Rev. Phytopathol.* **34:**109–130.

97. **Schüßler, A.** 1999. Glomales SSU rRNA gene diversity. *New Phytol.* **144:**205–207.

98. **Schüßler, A., D. Schwarzott, and C. Walker.** 2001. A new fungal phylum, the Glomeromycota: phylogeny and evolution. *Mycol. Res.* **105:**1413–1421.

99. **Seifert, K. A., and W. Gams.** 2001. The taxonomy of anamorphic fungi, p. 307–348. *In* D. J. McLaughlin, E. G. McLaughlin, and P. A. Lemke (ed.), *The Mycota*, vol. VII. *Systematics and Evolution, Part A.* Springer-Verlag, New York, NY.

100. **Shelton, B. G., K. H. Kirkland, W. D. Flanders, and G. K. Morris.** 2002. Profiles of airborne fungi in buildings and outdoor environments in the United States. *Appl. Environ. Microbiol.* **68:**1743–1753.

101. **Simon, L., M. Lalonde, and T. D. Bruns.** 1992. Specific amplification of 18S fungal ribosomal genes from vesicular-arbuscular endomycorrhizal fungi colonizing roots. *Appl. Environ. Microbiol.* **58:**291–295.

102. **Smith, S. E., and D. J. Read.** 1997. *Mycorrhizal Symbiosis*, 2nd ed. Academic Press, San Diego, CA.

103. **Sneh, B., L. Burpee, and A. Ogoshi.** 1991. *Identification of* Rhizoctonia *Species.* APS Press, St. Paul, MN.

104. **Sokal, R. R., and F. J. Rohlf.** 1995. *Biometry*, 3rd ed., W. H. Freeman, New York, NY.

105. **Stalpers, J. A., T. F. Andersen, and W. Gams.** 1998. Two proposals to conserve the names *Rhizoctonia* and *R. solani* (Hyphomycetes). *Taxon* **47:**725–727.

106. **Stevens, F. L.** 1916. A convenient, little-known method of making micromounts of fungi. *Phytopathology* **6:**367–368.

107. **Summerbell, R. C.** 1989. Microfungi associated with the mycorrhizal mantle and adjacent microhabitats within the rhizosphere of black spruce. *Can. J. Bot.* **67:**1085–1095.

108. **Tainter, F. H.** 1996. *Principles of Forest Pathology.* Wiley, New York, NY.

109. **Thorn, R. G.** 2001. Soil fungi: nature's nutritional network, p. 2910–2918. *In* G. Bitton (ed.), *Encyclopedia of Environmental Microbiology.* Wiley, New York, NY.

110. **Thorn, R. G., C. A. Reddy, D. Harris, and E. A. Paul.** 1996. Isolation of saprophytic basidiomycetes from soil. *Appl. Environ. Microbiol.* **62:**4288–4292.

111. **Trappe, J. M., and D. L. Luoma.** 1992. The ties that bind: fungi in ecosystems, p. 17–27. *In* G. C. Carroll and D. T. Wicklow (ed.), *The Fungal Community*, 2nd ed. Marcel Dekker, New York, NY.

112. **Vandenkoornhuyse, P., S. L. Baldauf, C. Leyval, J. Straczek, and J. P. W. Young.** 2002. Extensive fungal diversity in plant roots. *Science* **295:**2051.

113. **Van de Peer, Y., and R. De Wachter.** 1997. Evolutionary relationships among the eukaryotic crown taxa taking into account site-to-site rate variation in 18S rRNA. *J. Mol. Evol.* **45:**619–630.

114. **Van der Auwera, G., R. De Baere, Y. Van de Peer, P. De Rijk, I. Van den Broeck, and R. De Wachter.** 1995. The phylogeny of the Hyphochytriomycota as deduced from ribosomal RNA sequences of Hyphochytrium catenoides. *Mol. Biol. Evol.* **12:**671–678.

115. **Walker, G. M.** 1998. *Yeast Physiology and Biotechnology.* Wiley, Chichester, United Kingdom.

116. **Warcup, J. H.** 1955. On the origin of colonies of fungi developing on soil dilution plates. *Trans. Br. Mycol. Soc.* **38:**298–301.

117. **Warcup, J. H.** 1965. Growth and reproduction of soil microorganisms in relation to substrate, p. 52–68. *In* R. F. Baker and W. C. Snyder (ed.), *Ecology of Soil-Borne Plant Pathogens.* University of California Berkeley Press, Berkeley.

118. **Warner, F. E., and J. Emberlin.** 1998. An evaluation of fungal spore counting techniques for the Hirst-type spore trap tape. International Microscopy Conference and Exhibition (MICRO 98), Royal Microscopical Society London.

119. **Woese, C. R.** 1987. Bacterial evolution. *Microbiol. Rev.* **51:**221–271.

120. **Worrall, J. J.** 1991. Media for the selective isolation of Hymenomycetes. *Mycologia* **83:**296–302.

121. **Worrall, J. J., S. E. Anagnost, and R. A. Zabel.** 1997. Comparison of wood decay among diverse lignicolous fungi. *Mycologia* **89:**199–219.

122. **Wright, S. F., and A. Upadhyaya.** 1998. A survey of soils for aggregate stability and glomalin, a glycoprotein produced by hyphae of arbuscular mycorrhizal fungi. *Plant Soil* **198:**97–107.

123. **Yarrow, D.** 1998. Methods for the isolation and identification of yeasts, p. 77–100. *In* C. P. Kurtzman and J. W. Fell (ed.), *The Yeasts, a Taxonomy Study*, 4th ed. Elsevier, Amsterdam, The Netherlands.

43

Physiology, Metabolism, and Molecular Aspects of Filamentous Fungi

GEORGE A. MARZLUF

43.1. INTRODUCTION

The filamentous fungi display remarkable and diverse metabolic pathways and show great versatility in the utilization of sources of carbon, nitrogen, phosphorus, sulfur, and other metabolites and in acquiring essential elements, e.g., iron and potassium. Many of the filamentous fungi, particularly the well-studied *Aspergillus* and *Neurospora* species, can grow on defined media. Other than inorganic salts, e.g., sodium nitrate and ammonium sulfate, *Neurospora crassa* requires only a carbon source and one vitamin, biotin. The ability to grow on defined media coupled with a haploid genome has facilitated the isolation of mutants which have been instrumental in studying fungal metabolism (see chapter 44). Mutants which lack a specific enzyme are very useful in metabolic studies. Many mutant strains of *Aspergillus nidulans* and other *Aspergillus* species, *Neurospora crassa* and additional *Neurospora* species, *Fusarium*, and *Gelasinospora* are available at the Fungal Genetics Stock Center, University of Kansas Medical Center (http://www.fgsc.net). The Stock Center also maintains and distributes an excellent selection of cloned genes from *Neurospora* and *Aspergillus* as well as useful cloning vectors for fungi and

several fungal genomic libraries. The Stock Center oversees the annual publication of the *Fungal Genetics Newsletter*. The genome sequences of *N. crassa*, *A. nidulans*, and *Magnaporthe grisea* have been completed, and sequencing of the genomes of several additional filamentous fungi will be finished soon. All of this sequence information is available via online links with the Fungal Genetics Stock Center website.

Three entire volumes of *Methods in Enzymology* (41–43) are devoted to the methods of yeast genetics and molecular and cell biology, with articles ranging from "Getting Started with Yeast" to detailed descriptions of genomics and protocols specialized for yeasts. Researchers who wish to carry out any work with the yeast *Saccharomyces cerevisiae* should consult these authoritative volumes.

The filamentous fungi secrete many enzymes into the extracellular environment, which allows them to hydrolyze polysaccharides, nucleic acids, and proteins and to utilize the small-molecular-weight products for growth. Many fungi also carry out a pattern of secondary metabolism, following the exponential growth phase, which results in the synthesis of unusual compounds, e.g., penicillin (7, 14, 91). Entire areas of metabolism appear to be regulated in an integrated fashion, most often at the level of transcription, and involve the action of regulatory DNA binding proteins, metabolic inducers, and repressors. Feedback inhibition, compartmentation, and protein turnover also contribute significantly to metabolic controls. Extensive and authoritative reviews of many aspects of fungal metabolism are available (5, 17, 28, 50, 70, 71, 74, 96). Two recent books, *The Neurospora Compendium* by Perkins et al. (92) and *Neurospora* by Davis (22), are invaluable for *Neurospora* workers. *Neurospora* contains a wealth of information concerning fungal cell biology and genetics and describes experimental approaches and techniques which should be generally applicable to various fungi. An article by Goldman and Morris (39) describes methods for cell and molecular biology studies of *Aspergillus nidulans*, many of which can provide insight for work with any fungal species. Consult chapter 44 in this volume for a list of websites useful for fungal researchers. This chapter emphasizes practical aspects, methods, and simple assays for investigating fungal physiology and metabolism and introduces some of the specialized molecular techniques that have proved invaluable in studying this fascinating group of eukaryotic microorganisms.

43.2. NITROGEN METABOLISM

43.2.1. Nitrogen Regulation

Nitrogen metabolism and its genetic regulation are important topics since most of the macromolecules of living cells are rich in nitrogen, most notably nucleic acids and proteins. The filamentous fungi, particularly *Aspergillus nidulans* and *Neurospora crassa*, have contributed much to the field of nitrogen metabolism (17, 70, 93). These fungi must assimilate nitrogen from a wide range of potential nitrogenous compounds and utilize it to derive all the precursors required for DNA, RNA, and protein synthesis. Both *Aspergillus* and *Neurospora* preferentially utilize certain nitrogen sources, e.g., ammonium and glutamine, a phenomenon known as nitrogen catabolite repression. Under derepression conditions, when such primary nitrogen sources are lacking or available only in limited amounts, global positive-acting regulatory proteins, AREA in *Aspergillus*

and NIT2 in *Neurospora*, turn on the expression of multiple sets of structural genes which encode enzymes of various nitrogen catabolic pathways (11, 17, 33, 35, 70, 77, 95, 114). AREA and NIT2 are members of the GATA family of transcription factors, which have a Cys2/Cys2-type of zinc finger DNA binding motif. Interestingly, filamentous fungi contain multiple GATA factors; e.g., five GATA factors have been characterized in *Neurospora crassa*, each of which is responsible for controlling a specific set of downstream genes (Table 1); the *Neurospora* genome sequence suggests the presence of a sixth GATA factor which has not yet been characterized. Nitrogen metabolite signaling involves an interaction of a negative-acting protein, NMR, with both the C terminus and GATA domain of the AREA or NIT2 global activator (90, 93). Expression of each specific set of nitrogen catabolic genes also usually requires induction by a substrate or intermediate of the pathway and is mediated by a pathway-specific regulatory protein. The promoters of the candidate structural genes contain binding sites for the global activator (AREA or NIT2) and for the pathway-specific activator, most of which have a single binuclear Cys6/Zn2 DNA binding motif (Table 2).

43.2.2. Nitrate Reductase

Investigators often focus on a key enzyme within a pathway of interest, although several additional enzymes should also be examined to ensure that correct conclusions are obtained. Most filamentous fungi can utilize nitrate as an excellent nitrogen source, and its use requires a nitrate transporter and the enzymes nitrate reductase and nitrite reductase. Synthesis of these activities is highly regulated by nitrogen catabolite repression, induction, and major and minor control genes. Nitrate reductase is regarded as the key activity for the pathway of nitrate assimilation and has been studied extensively (12, 17, 34). Nitrate reductase is a very large homodimeric enzyme and possesses three separate domains, an amino-terminal domain that contains an unusual molybdopterin cofactor (Mo-Co), a smaller central heme-containing domain, and a carboxy-terminal domain which contains a flavin (FAD). Electrons derived from NADPH are transferred stepwise to the flavin domain, to the heme domain, and to the molybdopterin-containing domain and, finally, used to reduce nitrate to nitrite.

TABLE 1 GATA factors of filamentous fungi

GATA factor	Regulatory function[a]	No. of zinc fingers	Organism
AFΛREA	Nitrogen +	1	*Aspergillus fumigatus*
AREA	Nitrogen +	1	*Aspergillus nidulans*
AREA	Nitrogen +	1	*Gibberella fujikuroi*
AREA	Nitrogen +	1	*Metarhizium anisopliae*
NRE	Nitrogen +	1	*Penicillium chrysogenum*
NIT2	Nitrogen +	1	*Neurospora crassa*
NUT1	Nitrogen +	1	*Magnaporthe grisea*
SRE	Siderophore −	2	*Neurospora crassa*
SREP	Siderophore −	2	*Penicillium chrysogenum*
URBS1	Siderophore −	2	*Ustilago maydis*
WC-1	Light +	1	*Neurospora crassa*
WC-2	Light +	1	*Neurospora crassa*
ASD4	Ascospore formation +	1	*Neurospora crassa*

[a]+, positive regulation; −, negative regulation.

TABLE 2 Pathway-specific regulatory factors which mediate induction of catabolic genes in filamentous fungi

Pathway-specific factor	Inducer	Pathway	DNA binding domain	Species
ALCR	Alcohol	Alcohol	Cys_6/Zn_2	*Aspergillus nidulans*
AMDA	Acetate	Acetate	Cys_2/His_2	*Aspergillus nidulans*
AMDR	T-amino acid	Acetamide	Cys_6/Zn_2	*Aspergillus nidulans*
FACB	Acetate	Acetate	Cys_6/Zn_2	*Aspergillus nidulans*
NIRA	Nitrate	Nitrate	Cys_6/Zn_2	*Aspergillus nidulans*
NIT4	Nitrate	Nitrate	Cys_6/Zn_2	*Neurospora crassa*
PRNA	Proline	Proline	Cys_6/Zn_2	*Aspergillus nidulans*
UAY	Purine	Purine	Cys_6/Zn_2	*Aspergillus nidulans*
PCO1	Purine	Purine	Cys_6/Zn_2	*Neurospora crassa*

Nitrate reductase activity can be readily determined using a convenient colorimetric assay. The candidate fungus should be grown on contrasting media containing as nitrogen sources 25 mM ammonium or glutamine (repression), 5 mM xanthine or proline (derepression), 5 mM nitrate (induction), and 25 mM ammonium or glutamine plus 5 mM nitrate (repression and induction). Alternatively, the organism can be grown on a single medium and then divided into aliquots which are incubated for 2 to 4 h on the various contrasting media before the cells are harvested for assay. The fungal cells are collected and homogenized in a small volume of buffer using any of several convenient methods, one of the simplest being to grind the mycelia to a milky-like paste in an ice-cold mortar with acid-washed sand and a small amount of buffer (see chapter 44). The homogenate is centrifuged to remove cell debris, and the supernatant fluid is used for assay. Another popular method for making protein extracts utilizes a Mini-BeadBeater (Biospec Products, Bartlesville, OK). This instrument holds 2-ml tubes, each of which is filled one-third to one-half full with zirconia/silica 0.5-mm beads, to which is added approximately 0.3 g of a fungal cell pad plus sufficient extraction buffer to fill the tube. The tubes are pulsed at maximum speed for 1 min and then put on ice for 4 min; this is repeated four times. The tubes are then centrifuged at 14,000 rpm in a microfuge, and the supernatant is used for enzyme assays. This procedure allows rapid preparation of multiple samples.

Nitrate reductase activity is assayed in a mixture containing 20 mM phosphate buffer (pH 7.75), 20 mM $NaNO_3$, 5 mM sodium sulfite, 100 μM FAD, and 0 to 0.2 ml of the enzyme extract (38). The reaction is carried out in a total volume of 0.5 ml and is started by adding NADPH to a final concentration of 200 μM. The samples are incubated at 25°C for 0 to 30 min. Zero-time blanks should be included. The reaction is stopped by addition of 1% sulfanilamide in 20% (vol/vol) HCl. After mixing well, add 0.5 ml of 0.129% (wt/vol) naphthylethylene-diamine-diHCl, then add 1.5 ml of distilled water, mix well, and determine the absorbance of samples at 540 nm. Subtract background for each sample from an appropriate zero-time blank. The samples can be centrifuged before their absorbance is taken if they appear cloudy. A standard curve can be constructed with freshly prepared sodium nitrite.

One of the reasons that nitrate reductase is an attractive enzyme in the study of nitrogen metabolism is that a two-way selection system is highly efficient to obtain mutants and revertants in its structural gene and in multiple cofactor genes required for the synthesis and assembly of Mo-Co, the molybdopterin cofactor, which is also a component of

xanthine dehydrogenase (Table 3). Moreover, mutations that define regulatory genes which control nitrate reductase are also obtained in the same procedure (Table 3). When plated on solid medium containing a derepressing nitrogen source plus chlorate, wild-type cells express nitrate reductase and are killed, presumably due to conversion of chlorate to chlorite by the enzyme. Mutants lacking nitrate reductase are chlorate resistant and survive and grow in this medium, thus providing a powerful selective system for nitrate reductase-negative mutants. Since such mutants cannot grow with nitrate as a sole nitrogen source, revertants and suppressors can easily be obtained by plating the mutant cells on nitrate medium. This system has been utilized to provide selectable markers for introduction of recombinant DNAs into various fungal species (110).

The expression of the structural gene for nitrate reductase can also be examined by Northern blot analysis of RNA isolated from wild-type and mutant cells grown under contrasting conditions. In *Neurospora*, upon induction nitrate reductase mRNA is synthesized very rapidly and reaches a steady-state level within 15 min (84). Turnover of nitrate reductase message also occurs quickly upon repression, with a half-life of approximately 5 min (84).

43.2.3. Other Nitrogen Pathways

The filamentous fungi are extremely versatile in their ability to use a wide range of metabolites as nitrogen sources. These include nitrate, purines, urea, amino acids, peptides, amides, and extracellular proteins (45, 49, 50, 102, 115). In the area of amino acid metabolism, assay of transport activity with radioactively labeled basic, acidic, neutral, and aromatic amino acids is easily carried out (115). L-Amino acid oxidase is an enzyme involved in amino acid metabolism that is readily assayed. The candidate fungus should be grown or incubated in a medium to induce the enzyme, e.g., using phenylalanine as the sole nitrogen source. A useful control to include is a parallel culture grown or incubated with ammonium, which would be expected to repress the activity. A cell-free homogenate is prepared in phosphate buffer, pH 6, as described above and dialyzed overnight against the extraction buffer. The assay mixture contains 0.5 ml of 0.2 M phosphate buffer, pH 7.5, with 2 mM L-leucine plus 10 μl of catalase to which is added 0.0 to 0.5 ml of the crude enzyme plus sufficient water to obtain a total volume of 1 ml. Individual samples can be incubated for 0, 30, and 60 min at 37°C; the reaction is stopped by adding 1 ml of 0.1% 2,4-dinitrophenylhydrazine in 2 M HCl. After mixing and standing for 5 min, 2 ml of absolute ethanol and 5 ml of 2.5 M NaOH are added. The tubes are mixed well and allowed to stand for 5 min, centrifuging if

TABLE 3 *A. nidulans* and *N. crassa* genes which function in nitrate assimilation

Activity	Genetic locus		Growth of mutant upon[a]			
	Aspergillus	*Neurospora*	NH$_4^+$	Nitrate	Nitrite	Xanthine
Encodes nitrate transporter	*crnA*	*nit-10*	+	−	+/−	+
Encodes nitrate reductase	*niaD*	*nit-3*	+	−	+	+
Encodes nitrite reductase	*niiA*	*nit-6*	+	−	−	+
Gene for synthesis or assembly of a molybdenum cofactor	*cnxABC*	*nit-1*	+	−	+	−
	cnxE	*nit-7*	+	−	+	−
	cnxF	*nit-8*	+	−	+	−
	cnxG	*nit-9*	+	−	+	−
	cnxH		+	−	+	−
Pathway-specific control gene (mediates induction)	*nirA*	*nit-4*	+	−	+	−
Global nitrogen regulatory gene (activator during N derepression)	*areA*	*nit-2*	+	−	−	−
Negative-acting regulatory gene (mediates nitrogen repression)	*nmrA*	*nmr*	+	+	+	+

[a]Wild-type strains grown on all of these nitrogen sources. Wild-type and all mutant strains grow using glutamine as the sole nitrogen source. Nitrate transporter mutants (*crnA* and *nit-10* mutants) cannot grow with low concentrations of nitrate.

necessary to clarify, when the absorbance at 550 nm is determined. An alternative procedure involves grinding cells in approximately 20 volumes of buffer in an omnimixer, precipitating the enzyme by adding ammonium sulfate to 85% saturation, and then dialysis. The dialyzed sample is used for assay. A standard curve can be derived using pyruvic acid.

Fungi secrete proteases into the extracellular medium under conditions when their growth depends upon the use of an exogenous protein as a source of nitrogen, sulfur, or carbon (45, 48). A mixture of proteases may be present and their relative abundance may depend upon the pH of the growth medium; e.g., acidic proteases are expected to predominate when fungi are grown in medium at pH 5, whereas alkaline proteases will be the major species when the growth medium is buffered at a more alkaline pH (4). A simple sensitive protease assay is achieved by incubating a 1-ml sample of the cell-free growth medium with 1-ml of 2% casein for various times, after which 2 ml of 5% trichloroacetic acid is added; this stops the reaction and precipitates any undigested protein, which is removed by centrifugation. The soluble products released by protease activity are determined with 1 ml of the supernatant using the Folin phenol reagent as described by Lowry et al. (68), recording the absorbance at 750 nm. By the judicious use of specific inhibitors, e.g., phenylmethylsulfonyl fluoride to inhibit alkaline serine protease activity, and determining the activity at different pH values, the contribution of distinct proteases can be ascertained.

Purines serve as relatively good nitrogen sources for the filamentous fungi. In *Aspergillus nidulans*, synthesis of the various purine catabolic enzymes requires nitrogen derepression and induction and is subject to precise genetic control. A pathway-specific regulatory gene, *uaY*, mediates induction of at least eight unlinked structural genes which encode permeases and enzymes involve in purine transport and degradation. The UAY protein contains a GAL4-like Cys6/Zn2 binuclear DNA binding domain and recognizes promoter elements with the sequence TCGG-N6-CCGA (102). The *uaY* gene itself is not subject to regulation but is

expressed at a very low abundance in a constitutive fashion. Assays are available for various purine catabolic enzymes, e.g., uricase, allantoinase, and allantoicase, but a favorite enzyme to follow in this pathway appears to be xanthine dehydrogenase, whose activity is easily determined using a fluorometric assay (40). The reaction is carried out in a cuvette with 1 ml of 0.1 M phosphate buffer, pH 8, containing 10 nmol of 2-amino-4-hydroxypteridine, 200 nmol of NAD$^+$, and 50 μl of a fungal crude extract. The excitation wavelength is set at 326 nm, and the emission wavelength is set at 410 nm. Xanthine dehydrogenase reduces the substrate, 2-amino-4-hydroxypteridine to isoxanthopterin, which is fluorescent (40), and enzyme activity is readily followed as an increase in fluorescence. Expression of each of the genes encoding purine catabolic enzymes can also be directly determined using Northern blot analysis of total RNA isolated from fungal cells grown under contrasting repressing versus inducing conditions. Additional enzyme activities are of interest, several important ones being glutamate dehydrogenase (22), glutamine synthetase, and various transaminases, as well as specific activities such as proline oxidase and phenylalanine hydroxylase, but none of these can be described here.

43.3. CARBON METABOLISM

43.3.1. Carbon Sources and Catabolite Repression

The filamentous fungi are capable of utilizing a variety of compounds as sources of carbon and energy. These include monosaccharides such as arabinose, glucose, fructose, galactose, and mannose; disaccharides such as sucrose, trehalose, lactose, and maltose; and polysaccharides, sorbitol, mannitol, ethanol, and quinic acid, to name some of the more commonly used carbon sources. Carbon catabolite repression refers to a mechanism that allows certain carbon sources, e.g., glucose, to be used preferentially to less readily utilized compounds. It is well known that in *Escherichia coli*, when glucose is limiting, cyclic AMP serves to signal a lifting of catabolite repression, but its involvement, if any, in fungal regulation is unclear. Pioneering research in carbon metabolism with filamentous fungi has been carried out primarily with *Aspergillus nidulans* (28). A hierarchy among carbon sources for *Aspergillus* has been determined based upon their strength of repressors (6). Detailed studies of the genetic regulation of enzymes subject to specific induction in addition to carbon catabolite repression have been quite instructive. These studies have been conducted at several levels, including growth, enzyme activity, and RNA content. Two of the most noteworthy systems are the *alc* genes, which control alcohol dehydrogenase activity (ADH) and ethanol utilization in *A. nidulans*, and the quinate (*qa*) utilization cluster of genes in *N. crassa* and in *A. nidulans* (28). Carbon catabolite repression in *Aspergillus* is mediated by CREA, a DNA binding protein with two Cys2/His2 type zinc fingers that acts negatively to prevent gene expression (28, 57). During conditions of repression, CREA apparently competes with positive activators by binding at elements that overlap the elements for the positive factors, thereby preventing the latter's function.

ADH and its regulation have been extensively studied in *Aspergillus*. A cluster of closely linked genes includes, in order, *alcR*, *alcX*, *alcA*, *alcM*, and *aldA* (28, 30, 57). Expression of *alcA*, which encodes ADH, and of *aldA*, which encodes aldehyde dehydrogenase, requires induction by ethanol and is mediated by ALCR, a positive-acting transactivator protein that contains a Cys6/Zn2 DNA binding motif (31, 56, 57). Induction by ALCR of both *alcA* and *aldA* is inhibited by CREA in the presence of repressing amounts of glucose.

43.3.2. Complex Regulatory Systems

Certain compounds can serve as both a carbon source and a nitrogen source for *Aspergillus*, and their utilization is highly regulated by both carbon and nitrogen catabolite repression in addition to specific induction (28). Proline serves as both a nitrogen and carbon source for *Aspergillus*, and its utilization involves a cluster of five closely linked genes, arranged in the order *prnA*, -*X*, -*D*, -*B*, and -*C* (100, 101). The *prnA* gene codes for a positive-acting regulatory protein which mediates proline induction; *prnD*, -*B*, and -*C* encode the proline catabolic enzymes and permeases, proline oxidase, proline permease, and pyrroline-5-carboxylate dehydrogenase, respectively. Expression of the *prn* genes requires a functional PRNA protein, induction with proline, and a lifting of either carbon or nitrogen catabolite repression. Nitrogen derepression is signaled by the AREA protein, whereas the CREA protein mediates carbon catabolite repression, as described above. In *Neurospora* and *Aspergillus*, an extracellular protein can serve as the sole source of carbon, nitrogen, or sulfur and a single alkaline protease is synthesized and secreted in response to a limitation for any of these required elements (45). Acetamide also serves as both a carbon and nitrogen source for *Aspergillus*. The genetic regulation of acetamidase and several related enzymes has been the subject of elegant research by Hynes and Davis (50). Control of expression of the *amdS* gene, which encodes acetamidase, is very complex and involves the positive action of AREA and the negative signaling of CREA plus interactions with a number of pathway-specific effectors (50, 78).

43.3.3. Enzyme Assays

The utilization of sucrose requires its hydrolysis by invertase, whose synthesis is subject to catabolite repression exerted by glucose. Invertase activity is determined with fungal cells that have been grown with different carbon sources, including glucose, which should strongly repress invertase synthesis. In *Neurospora*, sucrose is not transported into the cells but must be hydrolyzed extracellularly, after which the fungus can import the monosaccharide products. Thus, in *Neurospora* invertase is largely an extracellular enzyme and can be assayed using a suspension of conidia or young mycelia without a need to prepare a homogenate. Although a number of assays for invertase are available, a simple colorimeteric assay which depends upon the release of reducing sugars is one of the most convenient. The cells or enzyme extract is incubated at 37°C in 0.05 mM acetate or succinate buffer, pH 5.5, with 0.5 mM sucrose in a total volume of 1.5 ml for various times, and then the reaction is stopped by placing in a boiling water bath for 1 min. After cooling and centrifugation, if necessary, to clarify, to 1 ml of the mixture add 2 ml of DNS reagent, mix, and heat the capped tubes in a boiling water bath for 5 min. The tubes are cooled quickly in an ice bucket and then allowed to reach room temperature, and then the absorbance at 560 nm is determined. In some cases it will be more convenient to stop the reaction by adding 3 ml of the DNS reagent, followed by the heat step. The DNS reagent is prepared as follows: 10 g of dinitrosalicylic acid, 400 ml of 1 M NaOH, and 300 g of $KNaC_4H_4O_6 \cdot 4H_2O$ are mixed and diluted to 1 liter. Trehalose is also a nonreducing sugar, and thus trehalase activity can be assayed in a similar fashion, using trehalose as the substrate. Chromogenic substrates such as *O*-nitrophenyl-α-glucoside, *O*-nitrophenyl-β-galactoside, and *O*-nitrophenyl-β-*N*-acetylglucosamine are useful to follow specific hydrolase activities of enzymes, e.g., β-galactosidase, that are subject to carbon catabolite repression. ADH activity is determined by following the increase in absorbance at 340 nm, which reflects the reduction of NAD^+ to NADH. Cell extracts derived from fungal cells that have been induced with ethanol should be compared with extracts from cells grown on repressing medium with glucose as the carbon source. The reaction is carried out in a total volume of 3 ml in a cuvette at 25°C, and the reaction mixture contains 15 mM pyrophosphate buffer (pH 8.8), 0.3 M ethanol, and 7.8 mM NAD^+. The enzyme extract (0 to 0.2 ml) is added at zero time and the absorbance at 340 nm is recorded at 15-s intervals.

43.4. SULFUR METABOLISM

43.4.1. The Sulfur Regulatory Circuit

Sulfur is an essential nutrient and is required for the synthesis of cysteine and methionine plus many other sulfur-containing compounds. Thus, it is not surprising that the fungi possess a sophisticated genetic system to ensure that a steady supply of sulfur can be assimilated from a wide variety of possible environmental sources. Methionine and inorganic sulfate are preferred sulfur sources, and both strongly repress the expression of multiple genes that encode enzymes necessary for use of various secondary sulfur sources, e.g., tyrosine-*O*-sulfate and choline-*O*-sulfate (71). During sulfur derepression conditions, which occur when favored sulfur sources are missing or present in small amounts, the genes encoding an entire set of sulfur catabolic enzymes are activated (71). Unlike the situation with the nitrogen and carbon metabolic circuits, induction by substrates is not required and derepression alone leads to expression of the sulfur-related structural genes. Activation of this set of structural genes in *Neurospora crassa* is mediated by CYS3, a positive-acting DNA binding protein with a basic leucine zipper (bZip) motif (36, 37, 89). The CYS3 protein binds specifically at elements in the promoter region serving the various downstream structural genes (63). The cys-3 gene itself is controlled by positive autogenous regulation, and both the cys-3 mRNA and the CYS3 protein are subject to relatively rapid turnover under sulfur-repressing conditions (104). Additional sulfur regulatory factors, SCON1 and SCON2, appear to act by controlling the synthesis, activity, or stability of CYS3 (58, 88). Recently, Sizemore and Paietta (99) have characterized a new sulfur regulatory factor, SCON3, a Skp-1 homolog which occurs in a complex with SCON2. Skp-1 is an important component of ubiquitin conjugating systems in S. cerevisiae, suggesting an important role for SCON3 in controlling turnover of CYS3.

43.4.2. Enzymes of Sulfur Metabolism

In Neurospora, the uptake of inorganic sulfate is highly regulated (51, 54, 71). The unlinked genes which encode two distinct sulfate permease species, cys-13 and cys-14, are both controlled as members of the sulfur regulatory circuit, and their expression requires activation by CYS3 and only occurs during conditions of sulfur derepression. Sulfate transport assays are easily conducted with ^{35}S-labeled sulfate using cells that have been grown with derepressing conditions, e.g., with 0.25 mM methionine as the sole sulfur source, versus cells growing with 5 mM methionine, which completely represses the uptake systems (69). Assays can be carried out with young mycelia or with conidia. Other fungi are expected to have active transport systems for the uptake of sulfate. The cells are incubated in medium or buffer with ^{35}S-labeled sulfate for various times, and then the cells are collected on a filter and washed several times with a solution of nonradioactive sulfate prior to determining the amount of radioactivity by scintillation counting. Nitrate strongly inhibits sulfate transport and should not be present in the assay mixture.

Aryl sulfatase, an enzyme which hydrolyzes aromatic sulfate esters, is subject to a high degree of regulation and is simple to assay, and thus, it is a valuable activity to examine and can be used to monitor repression/derepression conditions of fungal cells (72, 76). Fungal cells which have been grown under contrasting conditions of repression and derepression are homogenized using acid-washed sand in 0.05 M Tris-HCl buffer, and the clear supernatant obtained after centrifugation is used as the crude enzyme. Various amounts (0.05 to 0.2 ml) of a cell extract are incubated with 0.3 M Tris-HCl buffer, pH 8.1, and 6 mM p-nitrophenyl-sulfate in a total volume of 1 ml at 37°C for various times (0 to 60 min), after which 2 ml of 0.5 M NaOH in 90% ethanol is added and, after centrifugation if necessary to clarify, the absorbance at 405 nm is determined. If the substrate has a significant yellow color due to the presence of free p-nitrophenol, it can be treated with norite and filtered before use. Paietta isolated the Neurospora structural gene encoding arylsulfatase and demonstrated that it is strongly regulated at transcription (87).

43.5. PHOSPHORUS METABOLISM

43.5.1. Genetic Control of Phosphorus Metabolism

Phosphorus is another element that is an important component of nucleic acids, coenzymes, and many other biological molecules. The fungi possess complex regulatory mechanisms to maintain a sufficient intracellular pool of phosphate for continuous growth. Inorganic phosphate is an excellent source of phosphorus, but when it is not available the fungi can utilize a variety of secondary phosphorus sources, including nucleic acids and various phosphate esters. In Neurospora, in which regulation of phosphorus metabolism has been extensively studied, the synthesis of a number of enzymes involved in phosphorus metabolism is repressed as a group by high levels of inorganic phosphate. These activities include a high-affinity phosphate permease, O-phosphorylethanolamine permease, an alkaline phosphatase, an acid phosphatase, nucleotidases, and one or more nucleases (47, 62, 67, 81). Three distinct, interrelated regulatory genes, nuc-1, nuc-2, and preg, control the synthesis of this entire family of phosphorus catabolic enzymes (61, 66). The regulatory factors act sequentially, such that when phosphorus is limiting, NUC2 protein inhibits the expression of preg, whose product in turn inhibits the expression of nuc-1. The NUC1 protein, which is expressed only during phosphorus limitation conditions, is a positive-acting DNA binding protein that activates the expression of the entire set of phosphorus-related genes (53).

43.5.2. Phosphorus-Related Permeases and Enzymes

Permeases for phosphate transport can be assayed using ^{32}P-labeled phosphate with cells which have been grown with high repressing levels of inorganic phosphate in comparison with cells grown with derepressing levels of phosphate or a poorly utilized phosphorus source (10, 67). Alkaline phosphatase and acid phosphatase are convenient enzymes to examine with respect to phosphorus regulation, and their activities can be determined in cells grown under repressing and derepressing conditions, as well as with cells with mutations in regulatory or structural genes. Convenient assays for phosphatases utilize the release of p-nitrophenol from p-nitrophenyl phosphate (61, 81). To assay alkaline phosphatase, various amounts (0 to 0.2 ml) of a cell extract are incubated with 3 ml of 1 M Tris buffer, pH 8, containing 1 mM p-nitrophenyl phosphate. The enzyme reaction can be followed directly by determining the absorbance at 410 nm at various time intervals. Alternatively, the assay can be conducted in a 1-ml total volume; individual reaction mixtures are incubated for various times (0 to 30 min), and then 2 ml of 0.5 M NaOH in 90% ethanol is added and, after centrifugation if necessary to clarify, the absorbance at 410 nm is determined. Acid phosphatase activity can be assayed similarly in 0.1 M sodium acetate buffer, pH 5, and 1 mM p-nitrophenyl phosphate. In this case, it is necessary to stop the reaction with NaOH in ethanol before determining the absorbance at 410 nm. Some fungi may secrete phosphatases into the extracellular growth medium under phosphorus derepression conditions, and these activities can be assayed in the spent growth medium after the cells are removed by filtration or centrifugation.

43.6. pH CONTROL OF EXTRACELLULAR ENZYMES AND PERMEASES

Pioneering research carried out by Arst and his colleagues with Aspergillus nidulans has revealed that fungi contain a remarkable regulatory system that controls the synthesis of specific extracellular enzymes and permeases depending upon the pH of the growth medium (4). Thus, under acidic conditions, an extracellular acid phosphatase and other enzymes and permeases with acidic pH optima are synthesized and secreted. In sharp contrast, an alkaline phosphatase and other activities with alkaline pH optima are synthesized and secreted when the external pH is basic (4). Expression of a fungal penicillin biosynthetic gene is strongly influenced by pH (26). The regulatory mechanism involves a series of six *pal* genes which appear to sense the ambient pH and control the activity of PacC, a DNA binding protein (4, 25, 79). The full-length PacC protein, which predominates during acid pH conditions, is not functional, and genes expressed under acidic conditions are actively expressed, but those expressed under alkaline conditions are silent (106). Under alkaline pH conditions, the PacC protein is specifically proteolyzed to yield its active form (86). At alkaline ambient pH, PacC activates expression of genes expressed under alkaline conditions and also inhibits transcription of those expressed under acidic conditions (25, 106). It can be expected that most fungi have a similar pH control mechanism, which implies that the expression of various extracellular enzymes may be strongly dependent upon the ambient pH. A PacC homolog appears in the annotated *Neurospora* genome sequence, suggesting the presence of a similar system for pH control.

43.7. ACQUISITION OF IRON

43.7.1. Mechanisms for Iron Acquisition

In addition to the obvious requirement for substantial amounts of carbon, nitrogen, sulfur, and phosphorus, the fungi and other organisms need smaller amounts of numerous other metabolites, including Mg^{2+}, K+, Ca^{2+}, Na^+, Zn^{2+}, Cu^{2+}, MoO_4^-, and Fe^{2+}. Fungi possess specific mechanisms to acquire sufficient levels of such ions for optimal growth. Iron is needed to complete synthesis of a number of enzymes and coenzymes, most notably the cytochromes. Pioneering work by Leong and her colleagues with *Ustilago* has contributed much to our understanding of iron homeostasis in fungi (2, 3, 73, 111). Fe^{2+} is present in the environment at extremely low concentrations due to its very limited solubility. Bacteria and fungi secrete low-molecular-weight compounds known as siderophores which chelate Fe^{2+} for transport into the cells. On the other hand, too great an intracellular level of Fe^{2+} is extremely toxic due to its role in the oxidation of macromolecules, membrane lipids, proteins, and DNA. Thus, the fungi have evolved a system for genetic control of siderophore synthesis to ensure that intracellular Fe^{2+} is maintained within a safe concentration range. When the level of Fe^{2+} available is very low, the enzymes in the pathway responsible for siderophore synthesis are expressed, but under conditions of high Fe^{2+}, siderophore synthesis is strongly repressed, thereby precluding additional uptake of iron. When iron is present at repressing levels, a negative-acting GATA factor—URBS1 in *Ustilago* and its homologs, SRE in *Neurospora* and SREP in *Penicillium*—which contains two zinc finger motifs represses the expression of L-ornithine-N^5-oxygenase, the first enzyme unique to siderophore formation(44, 111, 116). When Fe^{2+} is available only in limited amounts, this enzyme and others of the synthetic pathway are expressed and siderophores are made and secreted into the extracellular medium.

43.7.2. Assays

After the candidate fungus has been grown in an iron-depleted liquid medium (low iron) and a parallel culture with a repressing level (0.01 mM $FeSO_4$) of iron, the cells are removed by centrifugation. To determine siderophore

concentrations, an equal volume of a solution of 5 mM $Fe(ClO_4)_3$–0.1 M $HClO_4$ is added to the cell-free supernatant. After mixing, the absorbance at 495 nm is determined, using fresh medium for a background reading.

L-ornithine-N^5-oxygenase activity is assayed with fungal cell homogenates prepared in potassium phosphate buffer, pH 8, after centrifugation to remove cell debris. A total of 0.4 ml of a cell extract is added to an assay mixture consisting of 0.1 mM phosphate buffer (pH 8), 1.5 mM L-ornithine, 0.5 mM NADPH, and 5 μM FAD in a total volume of 1 ml, and the samples are incubated for various times from 0 to 2 h. The reaction is stopped by adding 0.5 ml of 0.2 M perchloric acid. After centrifugation, 1 ml of the supernatant fluid is subjected to the iodine oxidation test (108) by adding 4 ml of 5% sodium acetate and 0.5 ml of 1% sulfanilic acid in 25% acetic acid and 0.2 ml of 1.3% iodine in glacial acetic acid. After 5 min at room temperature, 0.2 ml of 0.1 M sodium thiosulfate and 0.1 ml of 0.6% α-naphthylamine in 30% HCl are added and the samples are held at room temperature for 30 min. After an additional centrifugation, if needed to remove any denatured protein, the absorbance at 520 nm is recorded.

43.8. RNA AND PROTEIN SYNTHESIS

In studies of metabolism and gene expression, it is often necessary to examine RNA and protein synthesis. Moreover, the ability to specifically inhibit either RNA synthesis or protein synthesis helps to determine whether the appearance of a new activity or a significant increase in an activity requires transcription and/or translation. Neither actinomycin D nor α-amanitin inhibits RNA synthesis in *Aspergillus* or in *Neurospora*, and both probably are ineffective in most fungi. Proflavin has been used with some success to block transcription in fungi (19, 105). Proflavin at a final concentration of 100 μg/ml completely inhibited the induction of nitrate reductase in *Neurospora* (105). To ensure that the block imposed occurs at the intended level, it is desirable to show that proflavin, or any potential inhibitor, actually inhibits RNA synthesis and does not act at an unexpected level, e.g., protein synthesis. In *Aspergillus* and *Neurospora*, RNA synthesis can be examined with the use of ^3H- or ^{14}C-labeled uridine, preferably with uridine labeled with tritium in the 6 position. Uridine will be incorporated primarily into RNA, although a small amount may label DNA due to metabolic conversions of uridine to other pyrimidines. [^3H]uridine is added to a final concentration of 0.1 mM to a fungal culture, either one newly inoculated with conidia or one with young rapidly growing mycelia. After various times, homogeneous samples are removed, collected on filter papers, and washed with at least five aliquots of ice-cold 10% trichloroacetic acid, with a final wash with distilled water. The filters are dried and radioactivity is determined by scintillation counting.

The increase in enzyme activity which occurs upon addition of an inducer or lifting of catabolic repression often requires de novo protein synthesis, rather than simply activating a preexisting, inactive protein. Cycloheximide at a final concentration of 0.1 mM is very effective at inhibiting cytoplasmic protein synthesis in at least the majority of fungi and can be used to support the concept that expression of a specific enzyme is dependent upon de novo protein synthesis. To ensure that cycloheximide is inhibiting protein synthesis, incorporation of [^3H] leucine or of [^{35}S]-methionine can be examined. After a culture has been incubated in the presence of the radioactive amino acid for various times, homogeneous samples are collected on filter paper, washed with trichloracetic acid, and processed as described above. In the case of *Neurospora*, 0.1 mM cycloheximide nearly completely inhibits any incorporation of amino acids. Use of a mutant strain that requires leucine (or methionine) can improve the incorporation of the radioactive precursor.

43.9. ISOLATION OF NUCLEIC ACIDS

A number of different methods have been described to isolate DNA from filamentous fungi, all of which are useful (59, 75, 82, 83, 97). A method devised by Metzenberg and Baisch works very well and uses the chaotropic agent ethanolic perchlorate, which is easier (and safer) to utilize than phenol (75). Schechtman (97) combined steps from several different procedures to develop a simple protocol that yields 200 to 300 μg of pure DNA of greater than 50 kb in length from a single 40-ml culture. At least a dozen samples can be processed in 1 day, and the procedure can readily be scaled up. The mycelial pad harvested from a 40-ml culture is washed with water and ground to a fine powder in a prechilled mortar with liquid nitrogen. Transfer the powder to a 50-ml polypropylene tube, add 15 ml of 50 mM EDTA (pH 8.5)–0.2% sodium dodecyl sulfate (SDS)–15 μl of diethylpyrocarbonate (DEP), and after extracting by vigorous shaking, incubate the sample at 70°C for 15 min, then chill on ice for at least 10 min. Add 0.95 ml of potassium acetate (pH 4.3) and incubate on ice for 1 h, then centrifuge at 4°C for 15 min. Then add 15 ml of isopropanol to the supernatant with gentle mixing, which will immediately precipitate DNA. Rinse the DNA pellet with 70% ethanol, drain, and air dry, then resuspend in 4 ml of 1 mM EDTA (pH 8.0). Add 2 ml of high-salt buffer plus 15 μl of 10 mg/ml RNase A (pre-boiled), and incubate the sample at 37°C for 30 min. Add 180 μl of 0.1 M spermidine-HCl, mix, incubate on ice for 20 min, and pipette off the liquid from the clotted DNA. Rinse the pellet three times for 30 min with 2 ml of 75% ethanol–10 mM Mg acetate–0.3 M sodium acetate (pH 6.0), pipetting off the rinse each time, then rinse with 70% ethanol and drain and air dry the pellet. Resuspend the pellet overnight in 1 ml of 10 mM Tris-HCl (pH 8.0)–1 mM EDTA–0.1 M NaCl. After this, reprecipitate the DNA with 2 ml of ethanol at room temperature. Pipette off the supernatant, rinse the pellet again with 70% ethanol, air dry, and dissolve in 1 ml of TE (10 mM Tris-HCl, 1 mM EDTA [pH 8.0]). The DNA obtained is of high molecular weight and is readily cut by all restriction enzymes tested. It can be used for cloning, Southern blotting, DNA footprinting, gel shift assays, in vitro transcription, and other procedures.

Methods are available to make minipreps of DNAs from multiple fungal samples (59, 83).

The isolation of high-quality RNA from any organism, and particularly from fungi, requires vigilance to prevent its degradation by ever-present nucleases. Reagents and glassware for RNA preparation should be kept separate from general laboratory supplies, and glassware should be baked overnight at 250°C prior to each use. Buffers should be treated with 0.1% DEP overnight and then autoclaved. A number of useful methods for RNA isolation from filamentous fungi have been devised, only one of which will be described here. Total RNA can be isolated with a modification of the procedure described by Weaver et al. (113). Mycelia are ground to a fine powder with a mortar and pestle in the presence of liquid nitrogen and then suspended in

lysis buffer (50 mM sodium acetate buffer [pH 5.3], 10 mM EDTA, 1% SDS) using a ratio of 1 g/5 ml. An equal volume of acidic phenol–chloroform—equilibrated with the lysis buffer (pH 5.3), omitting SDS, at 65°C—is added to the lysate, and the mixture is incubated at 65°C for 30 min with shaking. After centrifugation, the aqueous phase is recovered, and this extraction process is repeated several additional times until no protein interface is visible. RNA is then precipitated by adding 0.6 vol of isopropyl alcohol and recovered after centrifuging. The RNA pellet is dissolved in a small volume of DEP-treated water or buffer and can be used for Northern blot analysis, reverse transcription-PCR, in vitro translation, or preparation of cDNA libraries. Various alternative methods for RNA isolation are available, including one which employs guanidinium isothiocyanate and was originally devised to obtain RNA from mammalian tissues rich in RNase (16) and another which uses aurintricarboxylic acid, which irreversibly and tightly binds RNA (65), have proved successful for isolation of fungal RNA. Poly(A) RNA can be isolated from total RNA by use of oligo(dT) cellulose chromatography.

43.10. IDENTIFICATION AND CLONING OF FUNGAL GENES

Many hundreds of fungal genes, including structural, regulatory, rRNA, and tRNA genes, have now been isolated and characterized, leading to a revolution in our understanding of molecular, genetic, and cellular processes. A number of fungal genes have been isolated by approaches in widespread use, e.g., differential screening of cDNA libraries (27), using heterologous probes, PCR approaches, or chromosome walking. The *am* gene of *Neurospora crassa* was isolated using a synthetic DNA probe based upon the amino acid sequence of the encoded product, glutamate dehydrogenase (55). Paietta employed chromosome walking to isolate the *ars* gene, which encodes arylsulfatase (87). However, in many cases, fungal genes can be isolated using more specialized techniques that are applicable to fungi.

The wealth of mutants of *A. nidulans*, *N. crassa*, and other fungi makes possible gene isolation by complementation. However, the recovery of a cloned gene following complementation of a given mutant by simply recovering a recombinant plasmid is usually not possible because of the lack of suitable autonomously replicating vectors. Transforming DNA usually integrates into random chromosomal locations in filamentous fungi, making recovery of the complementing DNA problematic. One approach which has been employed is marker rescue, in which genomic DNA of the transformed strain is cut with a restriction enzyme and the fragments are ligated to form circular molecules; this complex mixture is then used to transform *E. coli* to recover the vector which carries the desired gene. Marker rescue was used to obtain the *pyrG* gene of *Aspergillus nidulans* (82). Akins and Lambowitz (1) devised a more generally applicable technique, known as sib selection, to isolate genes without a requirement for any knowledge of the nature of the gene product. In sib selection, a gene is isolated from a genomic library by successive rounds of transformation of a mutant fungal strain with pools of plasmid or cosmid DNAs which are progressively reduced in complexity until a single candidate clone is obtained. In the case of *Neurospora*, a cosmid with a dominant selectable marker for benomyl resistance was used to generate libraries which contain large DNA inserts. Individual cosmid clones are ordered in the wells of microtiter dishes, allowing rapid and efficient screening of libraries. This approach has been responsible for the isolation of many different genes which were identified by complementation. An important virtue of sib selection is that it allows the isolation of genes identified only by mutation, including those that affect complex cellular processes for which nothing is known of the nature of the gene product. Cosmid-based genomic libraries for *Aspergillus nidulans* as well as lambda-based and YAC libraries are available from the Fungal Genetics Stock Center. The *Aspergillus* and *Neurospora* genome sequences have revolutionized gene cloning in these fungi. Primers can be used to clone any gene of interest using PCR. In the case of *Neurospora*, the ends of all clones in the Orbach/Sachs cosmid library have been sequenced, which has allowed the Whitehead annotation group to place genes within individual cosmids. It is far easier to clone from a cosmid DNA, available from the Fungal Genetics Stock Center, than from genomic DNA. Work is progressing to allow extensive transcriptional profiling using microarrays with both *Aspergillus* and *Neurospora*.

43.11. TRANSFORMATION WITH EXOGENOUS DNA

43.11.1. Transformation Procedures

The ability to readily transform filamentous fungi with exogenous DNA has been revolutionary and provided the essential step required in cloning, characterization, and sophisticated manipulation of individual genes. Transformation now is a standard procedure with *Aspergillus* and *Neurospora* (85, 107, 112) and can be accomplished with many other fungi, e.g., *Gibberella fujikuroi* (109), *Penicillium chrysogenum* (13), and the rice pathogen *Magnaporthe grisea* (23). A typical protocol used with *Neurospora* involves treating germinated conidia suspended in 1 M sorbitol with filter-sterilized Novozyme (2.5-mg/ml final concentration) for 30 to 60 min at 30°C with gentle shaking in order to produce spheroplasts. The extent of spheroplasting is checked by placing a 5-μl sample onto a slide, adding a coverslip, and then adding water. When a sufficient amount of the cell wall has been digested, the spheroplasts will swell and burst. The Novozyme digestion should not be prolonged because the viability of the spheroplasts will be significantly reduced. The resulting spheroplasts can be used immediately or suspended in Tris buffer containing sorbitol, calcium chloride, polyethylene glycol, and dimethyl sulfoxide and dispensed in aliquots into microtubes and frozen at −80°C; they will remain competent for transformation for at least several months (112). Preparation of competent spheroplasts from some mutant strains may prove considerably more difficult than with wild-type ones. Each batch of spheroplasts must be tested for competence; in the case of *Neurospora*, this is easily done using plasmid pBT6, which carries the benomyl resistance gene (85).

For transformation, after thawing on ice, 100 μl of competent spheroplasts is incubated with 1 μg of DNA plus 5 μl of heparin (5 mg/ml) in STC (1 M sorbitol, 50 mM Tris-HCl [pH 8.0], 50 mM CaCl$_2$) on ice for 30 min. Then 1 ml of PTC (40% PEG 4000, 50 mM Tris-HCl [pH 8.0], 50 mM CaCl$_2$) is added and incubated for 20 min at room temperature, after which samples are placed in 50-ml screw-cap tubes to which 15 ml of molten but cooled top agar (2% sorbose, 1 M sorbitol, 0.05% glucose, 0.05% fructose, Vogel's salts, and 2.8% agar) is added. After gentle mixing, the sample is poured onto petri plates containing 25 ml of bottom agar (2% sorbose, 0.05% glucose, 0.05% fructose, Vogel's salts, and 1.5% agar plus an appropriate selective agent, e.g., benomyl [5 μg/ml] or lacking an essential metabolite, e.g., histidine). The plates are incubated at the desired temperature until transformed colonies appear.

Transformation can also be carried out by electroporation, using a protocol devised by K. Borkovich. Conidia from 8- to 10-day-old cultures are harvested, washed, resuspended in 25 ml of ice-cold 1 M sorbitol, and pelleted in a clinical centrifuge. This procedure is repeated twice, after which the final pellet is resuspended in 0.5 ml of 1 M sorbitol. The concentration of conidia is determined with a hemacytometer and adjusted to 2.5 × 10^9 spores/ml; 40-μl aliquots are added to microtubes on ice. One tube serves as a negative control with no added DNA; 1 to 2 μg of DNA is added to the experimental tube. After electroporation (e.g., with an Eppendorf model 2510 electroporator), samples are plated on selective media as described for the spheroplast transformation protocol.

43.11.2. Selectable Markers

The structural gene which encodes nitrate reductase (*niaD*$^+$ in *Aspergillus*; *nit-3* in *Neurospora*) can be used as a selective marker for many different filamentous fungi. Mutants lacking this enzyme are readily selected via chlorate resistance. The Neurospora *pyr-4*$^+$ gene, which codes for orotidine-5′-phosphate decarboxylase, is frequently used to select for transformants of *Aspergillus nidulans* using *pyrG* mutant strains as the recipient (82). The *his-3*$^+$ gene of *Neurospora* has been used to target exogenous DNA to a precise genomic site as described below. Orbach et al. (85) developed a very useful dominant selectable marker, a mutant *N. crassa* β-tubulin gene that confers resistance to benomyl, which makes it possible to transform virtually any wild-type or mutant *Neurospora* strain. The bacterial hygromycin resistance gene, *hph*, and the bleomycin resistance gene, *ble*, from transposon Tn5 both work well as selectable markers in *Aspergillus* and *Neurospora*, and in other fungi.

It is frequently desired to introduce a manipulated gene for which no selective condition is available but a lack of suitable restriction sites may make it difficult to move it into a vector with a proven selectable marker. In these cases, cotransformation with two vectors, one carrying the gene being studied and the other with a convenient selectable marker, works quite well because a high percentage of the transformants selected will contain both vectors. Fungal transformations can be carried out with linear DNA as well as closed circular plasmid DNAs and even very large cosmid DNAs, 30 to 40 kb in size.

43.11.3. Nature of Transformed Fungal Strains

Most transformants of *N. crassa*, *A. nidulans*, and probably most filamentous fungi have one or more copies of the transforming DNA integrated into ectopic genomic sites, and only rarely will the exogenous DNA be incorporated at the homologous site unless special procedures are employed (107). *N. crassa* conidia and thus spheroplasts prepared from them are multinucleate, with an average of two or three nuclei, but usually only one of the nuclei is actually transformed by integration of one or more copies of the exogenously supplied DNA. Thus, the transformed strains are heterokaryotic, which makes them unsuitable for analysis of the function or regulation of the inserted gene of interest. Homokaryons can be obtained by plating microconidia, which have a single nucleus, obtained from the heterokaryon. It should be noted that genetic crosses cannot be used to obtain homokaryons in *Neurospora* because of repeat-induced point mutations (RIP), as described briefly below.

In some cases, although an exogenous gene has been successfully inserted into a fungal host, it may not be expressed, possibly because of DNA methylation or its incorporation in a region of the genome which is not accessible for transcription.

In examining various manipulated forms of a cloned gene for function via transformation of a fungal host strain, a serious consideration is the possibility of significant variability in expression due to incorporation of the transforming DNA at different genomic sites. To examine a series of transformants, each of which possesses a different manipulated version of a gene of interest, but all present at the same genomic location, the constructs can each be coupled with a truncated selectable marker which can function only upon homologous recombination with the host's endogenous mutated gene. Such a strategy has proved invaluable with *Neurospora*, in which selection for *his*⁺ colonies identifies transformants in which a truncated *his-3* gene has been targeted to the *his-3* locus, and the vector carrying this truncated selectable marker includes the gene of interest, whose presence can be verified by Southern blotting. A similar approach should be applicable to any fungus which can be transformed with exogenous DNA.

43.11.4. Constructing Gene Disruptions in Fungi

Gene disruption, a strategy in which a specific gene is partially or totally deleted or otherwise rendered completely inactive, is one extremely valuable use of transformation. Gene disruptions, or "knockouts," allow determination of the phenotype resulting from a complete loss of function of any cloned gene and also provide ideal host strains for analysis of various manipulated forms of the gene being investigated. The most direct procedure for disruption involves replacing all or a central region of the cloned gene with a dominant selectable marker, e.g., hygromycin resistance, leaving long regions of 5′ and 3′ flanking DNA for homologous recombination. The construct is transformed into a wild-type host, selecting for the dominant marker and against the function of the gene being disrupted, if possible. Transformants with the desired properties must be screened via PCR or Southern blotting to identify those with the desired disruption. A more powerful version of this approach involves the use of two dominant selectable markers in the construct, one of which is selected for as described above and the other selected against because it will be absent in the desired transformants where a double crossover has occurred and disrupted the target gene. The second dominant marker will usually be present if the DNA has integrated at ectopic sites; thus, such transformants can quickly be eliminated from consideration, allowing the investigator to focus only on those which may have the desired knockout. If the function of a cloned gene is uncertain, it is quite possible that disrupting it will be lethal and thus no simple knockouts can be obtained. When the nature of the fungus being studied makes it possible, one can use a diploid host strain for the disruption technique, allowing one copy of the gene of interest to be disrupted with the other copy remaining functional to ensure viability.

RIP is an intriguing phenomenon which occurs in *Neurospora crassa*. When a second copy of a cloned gene is introduced into a wild-type strain by transformation, both copies undergo multiple GC-to-AT mutations throughout the length of the duplicated segments during a genetic cross

(98). This occurs whether the duplicated copies are closely linked or even are situated on different chromosomes. The multiple point mutations introduced by RIP can be so extensive that the DNA fragment cannot be recognized by Southern blot hybridization when probed with the original wild-type sequence (29, 98). RIP can be used to obtain mutant strains in which the desired gene has been completely inactivated (29, 52, 98). A special technique, "sheltered RIP," can be utilized when disruption of the target gene might be lethal (46). See chapter 44 for additional information concerning RIP.

43.12. OTHER METABOLIC PATHWAYS

Additional metabolic pathways and closely related subjects beyond those discussed above represent important topics and research areas, some of which are especially relevant to fungi. These include polyamine biosynthesis (21), pathways and regulatory mechanisms of amino acid biosynthesis (22, 96), compartmentation of metabolic reactions (8, 20, 22), light-regulated gene expression (15, 64, 103), mating type determination and sporulation (9, 80), heat shock proteins and stress responses (94), the biological clock and circadian rhythms (18, 24, 32, 60), and the virulence factors of plant and animal fungal pathogens (23). The fungi have many remarkable characteristics and promise a wealth of exciting new phenomena and discoveries as their biology and genetics are further explored.

43.13. REFERENCES

1. **Akins, R. A., and A. M. Lambowitz.** 1985. General method for cloning *Neurospora crassa* nuclear genes by complementation of mutants. *Mol. Cell. Biol.* **5:**2272–2278.
2. **An, Z. Q., B. G. Mei, W. M. Yuan, and S. A. Leong.** 1997. The distal GATA sequences of the sid1 promoter of *Ustilago maydis* mediate iron repression of siderophore production and interact directly with Urbs1, a GATA family transcription factor. *EMBO J.* **16:**1742–1750.
3. **An, Z. Q., Q. Shao, J. McEvoy, W. M. Yuan, J. L. Markley, and S. A. Leong.** 1997. The second finger of Urbs1 is required for iron-mediated repression of sid1 in *Ustilago maydis. Proc. Natl. Acad. Sci. USA* **94:**5882–5887.
4. **Arst, H. N.** 1996. Regulation of gene expression by pH, p. 235–240. *In* R. Brambl and G. A. Marzluf (ed.), *The Mycota: Biochemistry and Molecular Biology*. Springer, Berlin, Germany.
5. **Arst, H. N., and C. Scazzocchio.** 1985. Formal genetics and molecular biology of the control of gene expression in *Aspergillus nidulans*, p. 309–343. *In* J. Bennett and L. Lasure (ed.), *Gene Manipulations in Fungi*. Academic Press, Orlando, FL.
6. **Bailey, C. R., and H. N. Arst.** 1975. Carbon catabolite repression in *Aspergillus nidulans. Eur. J. Biochem.* **51:**573–577.
7. **Barredo, J. L., P. van Solingen, B. Diez, E. Alvarez, J. M. Cantoral, A. Kattevilder, E. B. Smaal, M. A. Groenen, A. E. Venstra, and J. F. Martin.** 1989. Cloning and characterization of the acyl-coenzyme A:6-amino-penicillanic acid acyltransferase gene of *Penicillium chrysogenum. Gene* **83:**291–300.
8. **Bernhardt, S. A., and R. H. Davis.** 1972. Carbamoyl phosphate compartmentation in *Neurospora*: histochemical localization of aspartate and ornithine transcarbamoylases. *Proc. Natl. Acad. Sci. USA* **69:**1868–1872.
9. **Bolker, M., R. Dahl, R. Schlesinger, B. Bergemann, B. Gillissen, R. Schauwecker, M. Urban, B. Schroer, and**

R. Kahman. 1991. Mating type genes of *Ustilago maydis*, p. 231–240. *In* U. Stahl and P. Tudzynski (ed.), *Molecular Biology of Filamentous Fungi*, VCH, Weinheim, Germany.

10. Burns, D. J., and R. E. Beever. 1979. Mechanisms controlling the two phosphate uptake systems in *Neurospora crassa*. *J. Bacteriol.* 139:195–204.

11. Caddick, M. X., H. N. Arst, L. H. Taylor, R. I. Johnson, and A. G. Brownlee. 1986. Cloning of the regulatory gene areA mediating nitrogen metabolite repression in *Aspergillus nidulans*. *EMBO J.* 5:1087–1090.

12. Campbell, W. H., and J. R. Kinghorn. 1990. Functional domains of assimilatory nitrate reductases and nitrite reductases. *Trends Biochem. Sci.* 15:315–319.

13. Cantorai, I. M., B. Diez, J. L. Barredo, E. Avarez, and J. F. Martin. 1987. High frequency transformation of *Penicillium chrysogenum*. *Biotech* 5:494–497.

14. Car, L. G., P. L. Skatrud, M. E. Scheetzii, S. W. Queener, and T. D. Ingolia. 1986. Cloning and expression of the isopenicillin N synthetase gene from *Penicillium chrysogenum*. *Gene* 48:257–266.

15. Chamberlain, N. L., E. D. Driver, and R. L. Miesfeld. 1994. The length and location of CAG trinucleotide repeats in the androgen receptor N-terminal domain affect transactivation function. *Nucleic Acids Res.* 22:3181–3186.

16. Chirgwin, J. M., A. E. Przybyla, R. J. MacDonald, and W. J. Rutter. 1979. Isolation of biologically active ribonucleic acid from sources enriched in ribonuclease. *Biochemistry* 18:5294–5299.

17. Crawford, N. M., and H. N. Arst. 1993. The molecular genetics of nitrate assimilation in fungi and plants. *Annu. Rev. Genet.* 27:115–146.

18. Crosthwaite, S. K., J. C. Dunlap, and J. J. Loros. 1997. *Neurospora* wc-1 and wc-2: transcription, photoresponses, and the origins of circadian rhythmicity. *Science* 276:763–769.

19. Cybis, J., and P. Weglenski. 1972. Arginase induction in *Aspergillus nidulans*. *Eur. J. Biochem.* 30:262–268.

20. Davis, R. H. 1986. Compartmental and regulatory mechanisms in the arginine pathways of *Neurospora crassa* and *Saccharomyces cerevisiae*. *Microbiol. Rev.* 50:280–313.

21. Davis, R. H. 1996. Polyamines in fungi p. 347–356. *In* R. Brambl and G. A. Marzluf (ed.), *The Mycota*, vol. III. *Biochemistry and Molecular Biology*. Springer, Berlin, Germany.

22. Davis, R. H. 2000. *Neurospora*. Oxford University Press, New York, NY.

23. Dobinson, K. F., and J. E. Hamer. 1991. *Magnaporthe grisea*, p.67–86. *In* U. Stahl and P. Tudzynski (ed.), *Molecular Biology of Filamentous Fungi*. VCH, Weinheim, Germany.

24. Dunlap, J. C. 1999. Molecular bases for circadian clocks. *Cell* 96:271–290.

25. Espeso, E. A., and H. N. Arst. 2000. On the mechanism by which alkaline pH prevents expression of an acid-expressed gene. *Mol. Cell. Biol.* 20:3355–3363.

26. Espeso, E. A., J. Tilburn, H. N. Arst, and M. A. Penalva. 1993. pH regulation of a major determinant in expression of a fungal penicillin biosynthetic gene. *EMBO J.* 12:3941–3958.

27. Exley, G. E., J. D. Colandene, and R. H. Garrett. 1993. Molecular cloning, characterization, and nucleotide sequence of nit-6, the structural gene for nitrite reductase in *Neurospora crassa*. *J. Bacteriol.* 175:2379–2392.

28. Felenbok, B., and J. M. Kelly. 1996. Regulation of carbon metabolism in mycelial fungi, p. 357–368. *In* R. Brambl and G. A. Marzluf (ed.), *The Mycota*, vol. 3, *Biochemistry and Molecular Biology*. Springer, Berlin, Germany.

29. Feng, B., H. Haas, and G. A. Marzluf. 2000. ASD4, a new GATA factor of *Neurospora crassa*, displays sequence-specific DNA binding and functions in ascus and ascospore development. *Biochemistry* 39:11065–11073.

30. Fillinger, S., and B. Felenbok. 1996. A newly identified gene cluster in *Aspergillus nidulans* comprises five novel genes localized in the alc region that are controlled both by the specific transactivator AlcR and the carbon-catabolite repressor CreA. *Mol. Microbiol.* 20:475–488.

31. Fillinger, S., C. Panozza, M. Mathieu, and B. Felenbok. 1995. The basal level of transcription of the alc genes in the ethanol regulon in *Aspergillus nidulans* is controlled by the specific transactivator AlcR and the general carbon catabolite repressor CreA. *FEBS Lett.* 368:547–550.

32. Froehlich, A. C., Y. Liu, J. J. Loros, and J. C. Dunlap. 2002. White Collar-1, a circadian blue light photoreceptor, binding to the frequency promoter. *Science* 297:815–819.

33. Fu, Y.-H., and G. A. Marzluf. 1987. Characterization of nit-2, the major nitrogen regulatory gene of *Neurospora crassa*. *Mol. Cell. Biol.* 7:1691–1696.

34. Fu, Y. H., and G. A. Marzluf. 1987. Molecular cloning and analysis of the regulation of nit-3, the structural gene for nitrate reductase in *Neurospora crassa*. *Proc. Natl. Acad. Sci. USA* 84:8243–8247.

35. Fu, Y. H., and G. A. Marzluf. 1990. nit-2, the major positive-acting nitrogen regulatory gene of *Neurospora crassa*, encodes a sequence-specific DNA-binding protein. *Proc. Natl. Acad. Sci. USA* 87:5331–5335.

36. Fu, Y. H., and G. A. Marzluf. 1990. cys-3, the positive-acting sulfur regulatory gene of *Neurospora crassa*, encodes a sequence-specific DNA binding protein. *J. Biol. Chem.* 265:11942–11947.

37. Fu, Y.-H., J. V. Paietta, D. G. Mannix, and G. A. Marzluf. 1989. cys-3, the positive-acting sulfur regulatory gene of *Neurospora crassa*, encodes a protein with a putative leucine zipper DNA-binding element. *Mol. Cell. Biol.* 9:1120–1127.

38. Garrett, R. H., and A. Nason. 1969. Further purification and properties of *Neurospora* nitrate reductase. *J. Biol. Chem.* 244:2870–2882.

39. Goldman, G. H., and N. R. Morris. 1995. *Aspergillus nidulans* as a model system for cell and molecular biology studies, p. 48–65. *In* K. W. Adolph (ed.), *Microbial Gene Techniques*. Academic Press, New York, NY.

40. Griffith, A. N., and R. H. Garrett. 1987. Xanthine dehydrogenase expression in *Neurospora crassa* does not require a functional nit-2 regulatory gene. *Biochem. Genet.* 26:37–52.

41. Guthrie, C., and G. R. Fink (ed.). 1991. *Methods in Enzymology*, vol. 194. *Guide to Yeast Genetics and Molecular Biology*. Academic Press, San Diego, CA.

42. Guthrie, C., and G. R. Fink (ed.). 2002. *Methods in Enzymology*, vol. 350. *Guide to Yeast Genetics and Molecular and Cell Biology, Part B*. Academic Press, San Diego, CA.

43. Guthrie, C., and G. R. Fink (ed.). 2002. *Methods in Enzymology*, vol. 351. *Guide to Yeast Genetics and Molecular and Cell Biology, Part C*. Academic Press, San Diego, CA.

44. Haas, H., K. Angermayr, and G. Stoffler. 1997. Molecular analysis of a *Penicillium chrysogenum* GATA factor encoding gene (sreP) exhibiting significant homology to the *Ustilago maydis* urbs1 gene. *Gene* 184:33–37.

45. Hanson, M. A., and G. A. Marzluf. 1975. Control of the synthesis of a single enzyme by multiple regulatory circuits in *Neurospora crassa*. *Proc. Natl. Acad. Sci. USA* 72:1240–1244.

46. Harkness, T. A., R. L. Metzenberg, R. Schneider, R. Lill, W. Neupert, and F. E. Nargang. 1994. Sheltered Rip in *Neurospora crassa*. *Genetics* 136:107–118.

47. Hasunuma, K., and T. Ishikawa. 1977. Control of the production and partial characterization of repressible ex-

tracellular 5'-nucleotidase and alkaline phosphatase in *Neurospora crassa*. *Biochim. Biophys. Acta* **480**:178–193.

48. **Hensel, M., C. M. Tang, H. N. Arst, and D. W. Holden.** 1995. Regulation of fungal extracellular proteases and their role in mammalian pathogenesis. *Can. J. Bot.* **73** (Suppl. 1):S1065–S1070.

49. **Hull, E. P., P. M. Green, H. N. Arst, and C. Scazzocchio.** 1989. Cloning and characterization of the L-proline catabolism gene cluster of *Aspergillus nidulans*. *Mol. Microbiol.* **3**:553–560.

50. **Hynes, M. J., and M. A. Davis.** 1996. Regulation of acetamide catabolism, p. 381–393. *In* R. Brambl and G. A. Marzluf (ed.), *The Mycota*, vol. III. *Biochemistry and Molecular Biology*, Springer, Berlin, Germany.

51. **Jarai, G., and G. A. Marzluf.** 1991. Sulfate transport in *Neurospora crassa*: regulation, turnover, and cellular localization of the CYS14 protein. *Biochemistry* **30**:4768–4773.

52. **Jarai, G., and G. A. Marzluf.** 1991. Generation of new mutants of *nmr*, the negative-acting nitrogen regulatory gene of *Neurospora crassa*, by repeat induced mutation. *Curr. Genet.* **20**:283–288.

53. **Kang, S., and R. L. Metzenberg.** 1990. Molecular analysis of *nuc-1*⁺, a gene controlling phosphorus acquisition in *Neurospora crassa*. *Mol. Cell. Biol.* **10**:5839–5848.

54. **Ketter, J. S., and G. A. Marzluf.** 1988. Molecular cloning and analysis of the regulation of *cys-14*⁺, a structural gene of the sulfur regulatory circuit of *Neurospora crassa*. *Mol. Cell. Biol.* **8**:1504–1508.

55. **Kinnaird, J. H., M. A. Keighren, J. A. Kinsey, M. Eaton, and J. R. S. Fincham.** 1982. Cloning of the *am* (glutamate dehydrogenase) gene of *Neurospora crassa* through the use of a synthetic DNA probe. *Gene* **20**:387–396.

56. **Kulmburg, P., N. Judewicz, M. Mathieu, F. Lenouvel, D. Sequeval, and B. Felenbok.** 1992. Specific binding sites for the activator protin, ALCR, in the *alcA* promoter of the ethanol regulon of *Aspergillus nidulans*. *J. Biol. Chem.* **267**:21146–21153.

57. **Kulmburg, P., M. Mathieu, C. Dowzer, J. Kelly, and B. Felenbok.** 1993. Specific binding sites in the *alcR* and *alcA* promoters of the ethanol regulon for the CREA repressor mediating carbon catabolite repression in *Aspergillus nidulans*. *Mol. Microbiol.* **7**:847–857.

58. **Kumar, A., and J. V. Paietta.** 1995. The sulfur controller-2 negative regulatory gene of *Neurospora crassa* induces a protein with β-transducin repeats. *Proc. Natl. Acad. Sci. USA* **92**:3343–3347.

59. **Leach, J., B. Finkelstein, and J. A. Rambosek.** 1986. Rapid miniprep of DNA from filamentous fungi. *Neurospora Newsl.* **33**:32–33.

60. **Lee, K., J. J. Loros, and J. C. Dunlap.** 2000. Interconnected feedback loops in the *Neurospora* circadian system. *Science* **289**:107–110.

61. **Lehman, J. F., M. K. Gleason, S. K. Ahlgren, and R. L. Metzenberg.** 1973. Regulation of phosphate metabolism in *Neurospora crassa*. Characterization of regulatory mutants. *Genetics* **75**:61–73.

62. **Lehman, J. F., and R. L. Metzenberg.** 1976. Regulation of phosphate metabolism in *Neurospora crassa*: identification of the structural gene for repressible alkaline phosphatase. *Genetics* **84**:175–182.

63. **Li, Q., and G. A. Marzluf.** 1996. Determination of the *Neurospora crassa* CYS3 sulfur regulatory protein consensus DNA-binding site: amino-acid substitutions in the CYS3 bZip domain that alter DNA-binding specificity. *Curr. Genet.* **30**:298–304.

64. **Linden, H., and G. Macino.** 1997. White collar 2, a partner in blue light signal transduction, controlling expression of light-regulated genes in *Neurospora crassa*. *EMBO J.* **16**:98–107.

65. **Lindgren, K. M., A. Lichens-Park, J. L. Loros, and J. C. Dunlap.** 1990. A quick RNA mini-prep for *Neurospora* mycelial cultures. *Fungal Genet. Newsl.* **37**:21–22.

66. **Littlewood, B. S., W. Chia, and R. L. Metzenberg.** 1975. Genetic control of phosphate-metabolizing enzymes in *Neurospora crassa*: relationships among regulatory mutants. *Genetics* **79**:419–434.

67. **Lowendorf, H. D., G. F. Bazinet, and C. W. Slayman.** 1975. Phosphate transport in *Neurospora*. Derepression of a high-affinity transport system during phosphorus starvation. *Biochim. Biophys. Acta* **389**:541–549.

68. **Lowry, O. H., N. J. Rosebrough, A. L. Farr, and R. J. Randall.** 1951. Protein measurement with the Folin phenol reagent. *J. Biol. Chem.* **193**:265–275.

69. **Marzluf, G. A.** 1970. Genetic and biochemical studies of distinct sulfate permease species in different developmental stages of *Neurospora crassa*. *Arch. Biochem. Biophys.* **138**:254–263.

70. **Marzluf, G. A.** 1997. Genetic regulation of nitrogen metabolism in the fungi. *Microbiol. Mol. Biol. Rev.* **61**:17–32.

71. **Marzluf, G. A.** 1997. Molecular genetics of sulfur assimilation in filamentous fungi and yeast. *Annu. Rev. Microbiol.* **51**:73–96.

72. **Marzluf, G. A., and R. L. Metzenberg.** 1968. Positive control by the *cys-3* locus in regulation of sulfur metabolism in *Neurospora*. *J. Mol. Biol.* **33**:423–437.

73. **Mei, B., A. D. Budde, and S. A. Leong.** 1993. *sid1*, a gene initiating siderophore biosynthesis in *Ustilago maydis*: Molecular characterization, regulation by iron, and role in phytopathogenicity. *Proc. Natl. Acad. Sci. USA* **90**:903–907.

74. **Metzenberg, R. L.** 1979. Implications of some genetic control mechanisms in *Neurospora*. *Microbiol. Rev.* **43**: 361–383.

75. **Metzenberg, R. L., and T. J. Baisch.** 1981. An easy method for preparing *Neurospora* DNA. *Neurospora Newsl.* **28**:20–21.

76. **Metzenberg, R. L., and J. W. Parson.** 1966. Altered repression of some enzymes of sulfur utilization in a temperature-conditional lethal mutant of *Neurospora*. *Proc. Natl. Acad. Sci. USA* **55**:629–635.

77. **Muro-Pastor, M., R. Gonzalez, J. Strauss, F. Narendja, and C. Scazzocchio.** 1999. The GATA factor AreA is essential for chromatin remodelling in a eukaryotic bidirectional promoter. *EMBO J.* **18**:1584–1597.

78. **Narendja, F. M., M. A. Davis, and M. J. Hynes.** 1999. AnCF, the CCAAT binding complex of *Aspergillus nidulans*, is essential for the formation of a DNase I-hypersensitive site in the 5' region of the *amdS* gene. *Mol. Cell. Biol.* **19**:6523–6531.

79. **Negrete-Urtasun, S., W. Reiter, E. Diez, S. H. Denison, J. Tilburn, E. A. Espeso, M. A. Penalva, and H. N. Arst.** 1999. Ambient pH signal transduction in *Aspergillus*: completion of gene characterization. *Mol. Microbiol.* **33**:994–1003.

80. **Nelson, M. A., and R. L. Metzenberg.** 1992. Sexual development genes of *Neurospora crassa*. *Genetics* **132**:149–162.

81. **Nelson, R. E., J. F. Lehman, and R. L. Metzenberg.** 1976. Regulation of phosphate metabolism in *Neurospora crassa*: identification of the structural gene for repressible acid phosphatase. *Genetics* **84**:183–192.

82. **Oakley, B. R., J. E. Rinehart, B. L. Mitchell, C. E. Oakley, C. Carmona, G. L. Gray, and G. S. May.** 1987. Cloning, mapping and molecular analysis of the *pyrG* (orotidine-5'-phosphate decarboxylase) gene of *Aspergillus nidulans*. *Gene* **61**:385–399.

83. **Oakley, C. E., C. F. Weil, P. L. Kretz, and B. R. Oakley.** 1987. Cloning of the *riboB* locus of *Aspergillus nidulans*. *Gene* **53**:293–298.

84. **Okamoto, P. M., Y. H. Fu, and G. A. Marzluf.** 1991. *nit-3*, the structural gene of nitrate reductase in *Neurospora crassa*: nucleotide sequence and regulation of mRNA synthesis and turnover. *Mol. Gen. Genet.* **227**:213–223.

85. **Orbach, M. J., E. B. Porro, and C. Yanofsky.** 1986. Cloning and characterization of the gene for β-tubulin from a benomyl-resistant mutant of *Neurospora crassa* and its use as a dominant selectable marker. *Mol. Cell. Biol.* **6**:2452–2461.

86. **Orejas, M., E. A. Espeso, J. Tilburn, S. Sarkar, and H. N. Arst.** 1995. Activation of the *Aspergillus* PacC transcription factor in response to alkaline ambient pH requires proteolysis of the carboxy-terminal moiety. *Genes Dev.* **9**:1622–1632.

87. **Paietta, J. V.** 1989. Molecular cloning and regulatory analysis of the arylsulfatase structural gene of *Neurospora crassa*. *Mol. Cell. Biol.* **9**:3630–3637.

88. **Paietta, J. V.** 1990. Molecular cloning and analysis of the *scon-2* negative regulatory gene of *Neurospora crassa*. *Mol. Cell. Biol.* **10**:5207–5214.

89. **Paietta, J. V., R. A. Akins, A. M. Lambowitz, and G. A. Marzluf.** 1987. Molecular cloning and characterization of the *cys-3* regulatory gene of *Neurospora crassa*. *Mol. Cell. Biol.* **7**:2506–2511.

90. **Pan, H. G., B. Feng, and G. A. Marzluf.** 1997. Two distinct protein-protein interactions between the NIT2 and NMR regulatory proteins are required to establish nitrogen metabolite repression in *Neurospora crassa*. *Mol. Microbiol.* **26**:721–729.

91. **Perez-Esteban, B., M. Orejas, E. Gomez-Pardo, and M. A. Penalva.** 1993. Molecular characterization of a fungal secondary metabolism promoter: transcription of the *Aspergillus nidulans* isopenicillin N synthetase gene is modulated by upstream negative elements. *Mol. Microbiol.* **9**:881–895.

92. **Perkins, D. D., A. Radford, and M. S. Sachs.** 1981. *The Neurospora Compendium*. Academic Press, New York, NY.

93. **Platt, A., T. Langdon, H. N. Arst, D. Kirk, D. Tollervey, J. M. Sanchez, and M. X. Caddick.** 1996. Nitrogen metabolite signaling involves the C-terminus and the GATA domain of the *Aspergillus* transcription factor AREA and the 3′ untranslated region of its mRNA. *EMBO J.* **15**:2791–2801.

94. **Plesofsky-Vig, N.** 1996. The heat shock proteins and the stress response, p. 171–190. *In* R. Brambl and G. A. Marzluf (ed.). *The Mycota*, vol. III. *Biochemistry and Molecular Biology*. Springer, Berlin, Germany.

95. **Punt, P. J., J. Strauss, R. Smit, J. R. Kinghorn, C. A. M. J. J. van den Hondel, and C. Scazzocchio.** 1995. The intergenic region between the divergently transcribed *niiA* and *niaD* genes of *Aspergillus nidulans* contains multiple NirA binding sites which act bidirectionally. *Mol. Cell. Biol.* **15**:5688–5699.

96. **Sachs, M. S.** 1996. General and cross-pathway controls of amino acid biosynthesis, p. 315–345. *In* R. Brambl and G. A. Marzluf (ed.). *The Mycota*, vol. 3, *Biochemistry and Molecular Biology*. Springer, Berlin, Germany.

97. **Schechtman, M.** 1986. A moderate-scale DNA prep for *Neurospora*. *Fungal Genet. Newsl.* **33**:45–46.

98. **Selker, E. U., and P. W. Garrett.** 1988. DNA sequence duplications trigger gene inactivation in *Neurospora crassa*. *Proc. Natl. Acad. Sci. USA* **85**:6870–6874.

99. **Sizemore, S. T., and J. V. Paietta.** 2002. Cloning and characterization of *scon-3+*, a new member of the *Neurospora crassa* sulfur regulatory system. *Eukaryot. Cell* **1**:875–883.

100. **Sophianopoulou, V., and C. Scazzocchio.** 1989. The proline transport protein of *Aspergillus nidulans* is very similar to amino acid transporters of *Saccharomyces cerevisiae*. *Mol. Microbiol.* **3**:705–714.

101. **Sophianopoulou, V., T. Suárez, G. Diallinas, and C. Scazzocchio.** 1992. Operator derepressed mutations in the proline utilisation gene cluster of *Aspergillus nidulans*. *Mol. Gen. Genet.* **236**:209–213.

102. **Suarez, T., M. V. de Queiroz, N. Oestreicher, and C. Scazzocchio.** 1995. The sequence and binding specificity of UaY, the specific regulator of the purine utilization pathway in *Aspergillus nidulans*, suggest an evolutionary relationship with the PPR1 protein of *Saccharomyces cerevisiae*. *EMBO J.* **14**:1453–1467.

103. **Talora, C., L. Franchi, H. Linden, P. Ballario, and G. Macino.** 1999. Role of a white collar-1-white collar-2 complex in blue-light signal transduction. *EMBO J.* **18**:4961–4968.

104. **Tao, Y., and G. A. Marzluf.** 1998. Synthesis and differential turnover of the CYS3 regulatory protein of *Neurospora crassa* are subject to sulfur control. *J. Bacteriol.* **180**:478–482.

105. **Tao, Y., and G. A. Marzluf.** 1999. The NIT2 nitrogen regulatory protein of *Neurospora*: expression and stability of *nit-2* mRNA and protein. *Curr. Genet.* **36**:153–158.

106. **Tilburn, J., S. Sarkar, D. A. Widdick, E. A. Espeso, M. Orejas, J. Mungroo, M. A. Penalva, and H. N. Arst.** 1995. The *Aspergillus* PacC zinc finger transcription factor mediates regulation of both acid- and alkaline-expressed genes by ambient pH. *EMBO J.* **14**:779–790.

107. **Tilburn, J., C. Scazzocchio, G. G. Taylor, J. J. Zabicky-Zissman, R. A. Lockingham, and R. W. Davies.** 1983. Transformation by integration in *Aspergillus nidulans*. *Gene* **26**:205–221.

108. **Tomlinson, G., W. H. Cruickshank, and T. Viswanatha.** 1971. Sensitivity of substituted hydroxylamines to determination by iodine oxidation. *Anal. Biochem.* **44**:670–679.

109. **Tudzynski, B., V. Homann, B. Feng, and G. A. Marzluf.** 1999. Isolation, characterization and disruption of the *areA* nitrogen regulatory gene of *Gibberella fujikuroi*. *Mol. Gen. Genet.* **261**:106–114.

110. **Unkles, S. E., E. I. Campbell, D. Carrez, C. Grieve, R. Contreras, W. Fiers, and C. A. M. van den Hondel.** 1989. Transformation of *Aspergillus niger* with the homologous nitrate reductase gene. *Gene* **78**:157–166.

111. **Voisard, C., J. Wang, J. L. McEvoy, P. Xu, and S. A. Leong.** 1993. *urbs1*, a gene regulating siderophore biosynthesis in *Ustilago maydis*, encodes a protein similar to the erythroid transcription factor GATA-1. *Mol. Cell. Biol.* **13**:7091–7100.

112. **Vollmer, S. J., and C. Yanofsky.** 1986. Efficient cloning of genes of *Neurospora crassa*. *Proc. Natl. Acad. Sci. USA* **83**:4869–4873.

113. **Weaver, P. L., C. Sun, and T.-H. Chang.** 1997. Dbp3p, a putative RNA helicase in *Saccharomyces cerevisiae*, is required for efficient pre-rRNA processing predominantly at site A_3. *Mol. Cell. Biol.* **17**:1354–1365.

114. **Wilson, R. A., and H. N. Arst, Jr.** 1998. Mutational analysis of AREA, a transcriptional activator mediating nitrogen metabolite repression in *Aspergillus nidulans* and a member of the "streetwise" GATA family of transcription factors. *Microbiol. Mol. Biol. Rev.* **62**:586–596.

115. **Wolfinbarger, L., Jr.** 1976. Mutations in *Neurospora crassa* which affect multiple amino acid transport systems. *Biochim. Biophys. Acta* **436**:774–788.

116. **Zhou, L. W., H. Haas, and G. A. Marzluf.** 1998. Isolation and characterization of a new gene, *sre*, which encodes a GATA-type regulatory protein that controls iron transport in *Neurospora crassa*. *Mol. Gen. Genet.* **259**:532–540.

44

Microbiological and Genetic Methods for Filamentous Fungi

ROWLAND H. DAVIS AND A. JOHN CLUTTERBUCK

44.1. INTRODUCTION

The study of filamentous fungi has a long history, covering investigations of their taxonomy, life cycles, physiology, and nutrition. These studies led to the genetic, biochemical, and molecular investigations initiated by Dodge, Beadle and Tatum, and Pontecorvo between 1927 and 1953 (20, 36). The area loosely called "fungal genetics and biology" is now a formal scientific field and includes studies of all major groups of filamentous fungi. The field originated, in a sense, in the first *Neurospora* Information

Conference in 1961. Investigators working on *Aspergillus nidulans* soon joined this group, which expanded to all workers on filamentous fungi in the mid-1980s under the heading of the newly named biennial Fungal Genetics Conferences. The current field is united by discourse based on genetics, cell biology, development, and the many areas of molecular biology.

The filamentous growth habit of the fungi and their conspicuous, differentiated sexual structures set them apart from other microbial taxa. Methods for cultivating filamentous fungi and for study of their classical molecular genetic mechanisms have also developed in distinct ways from those used for bacteria and yeast. This article describes methods drawn largely from the study of the ascomycetes *Neurospora crassa* and *A. nidulans*. We add, where appropriate, comments and references to other species now used in similar research. Certain features of filamentous fungi require special technical attention in any methodological treatment. These are (i) a filamentous habit, restricting studies of steady-state liquid cultures; (ii) a tough cell wall, requiring harsh cell disruption methods for biochemical and molecular work; (iii) the use of heterokaryons, mycelia having genetically different nuclei in the same cell; and (iv) in many cases, the lack of a natural diploid vegetative phase, for which heterokaryons, partial diploids, or artificial diploids may often substitute.

The life cycles of two model ascomycetes, *N. crassa* and *A. nidulans*, are shown in Fig. 1, with some of the structural terminology used in this article. The biology and life cycles of other fungi, in particular the basidiomycetes such as *Coprinus cinereus* and *Schizophyllum commune*, may be found in the treatise of Fincham et al. (25). Many other general references may be consulted (7, 8, 15, 16, 20, 21, 36, 42, 43). A recent, comprehensive analysis of the genome of *N. crassa* (9) contains much relevant information of use to modern workers.

A World Wide Web site of use to fungal geneticists, with links to many other fungal websites, includes the Fungal Genetics Stock Center (FGSC) (http://www.fgsc.net). This site, initially serving only the *Neurospora* community, is now a clearinghouse for general information on filamentous fungi. Other URLs include, for *A. nidulans*, http://www.genetics .unimelb.edu.au/hdlab/ (Hynes), http://www.ibls.gla.ac.uk/ aspergillus/index.html (linkage maps), http://www.genome .broad.mit.edu/annotation/fungi/aspergillus (Broad Institute; genome), and http://www.aspergillus.man.ac.uk/index .html. For *N. crassa*, http://www.fgsc.net (FGSC) and http:// www.genome.broad.mit.edu/annotation/fungi/neurospora (Broad Institute; genome) are available. A general site for fungi, having many other useful links, is http://mycology .cornell.edu/fgenetic.html.

This article concentrates most heavily on *A. nidulans* and *N. crassa* as model organisms, allowing us to give specific information on culture and techniques rather than exhaustive variations for all filamentous fungi that have contributed to the growth of the field. The reader is referred to the references given and to the websites above for up-to-date information about other fungi.

44.2. MEDIA

Ingredients are given per liter final volume unless otherwise indicated.

44.2.1. *Aspergillus nidulans* (15)

44.2.1.1. Minimal Medium

This medium is a better-buffered version of the minimal medium described by Pontecorvo et al. (44) and Cove (17).

Salts: KH_2PO_4, 1.4 g; $K_2HPO_4 \cdot 3H_2O$, 0.9 g; $MgSO_4 \cdot 7H_2O$, 0.1 g; KCl, 0.1 g. (These salts can be kept as a 100× concentrate.)

To these are added, per liter: $NaNO_3$, 2 g; trace elements, 1 ml (e.g., the solution described by Cove [17]: 40 mg of $Na_2B_4O_7 \cdot 10H_2O$, 400 mg of $CuSO_4 \cdot 5H_2O$, 800 mg of $FePO_4 \cdot 2H_2O$, 800 mg of $MnSO_4 \cdot 2H_2O$, 800 mg of $Na_2MoO_4 \cdot 2H_2O$, 8 g of $ZnSO_4 \cdot 7H_2O$), 10 g of D-glucose, and 15 g of agar (if required).

A slight, but readily dispersed, precipitate may develop on autoclaving for 10 min at 15 lb. Longer autoclaving may also cause caramelization. This medium supports both vegetative growth and cleistothecium production. Other nitrogen sources (generally 10 mM) may replace sodium nitrate and alternatives to glucose as carbon source are used at 1%.

44.2.1.2. Complete Media

A simple yeast extract medium (31) consists of 5 g of yeast extract and 25 g of glucose per liter. For a more complex medium, add to minimal medium 2 g of Difco peptone, 1 g of yeast extract, 1.5 g of casein hydrolysate (if acid-hydrolyzed, this lacks tryptophan), and 1 ml of vitamin solution (containing 100 mg of riboflavin, 100 mg of nicotinamide, 10 mg of *p*-aminobenzoic acid, 5 mg of pyridoxine, 5 mg of thiamine, and 1 mg of biotin per liter).

44.2.2. *Neurospora crassa* (20, 21)

44.2.2.1. Minimal Medium

Vogel's "Medium N" Formula for 50× Concentrate
To 750 ml of H_2O add, with stirring in the following order: 150 g of Na_3 citrate $\cdot$ 5.5 H_2O, 250 g of KH_2PO_4, 100 g of NH_4NO_3, 10 g of $MgSO_4 \cdot 7H_2O$, 5 g of $CaCl_2 \cdot 5H_2O$, 5 ml of trace elements (see below), 5 ml of D-biotin stock (see below), and a few milliliters of chloroform as preservative. Predissolving the $CaCl_2$ in ca. 20 ml of H_2O aids greatly in introducing it into the solution. Store at room temperature. For use, dilute 50× and add 15 to 20 g of sucrose and 15 g of agar per liter if required. All ingredients may be autoclaved together.

Trace Elements
For *trace elements*, dissolve successively, with stirring, in 95 ml of water: citric acid $\cdot$ H_2O, 5 g; $ZnSO_4 \cdot 7H_2O$, 5 g; $Fe(NH_4)_2(SO_4)_2 \cdot 6H_2O$, 1 g; $CuSO_4 \cdot 5H_2O$, 0.25 g; $MnSO_4 \cdot 1H_2O$, 0.05 g; HBO_3, 0.05 g; $Na_2MoO_4 \cdot 2H_2O$, 0.05 g; and chloroform as preservative, 1 ml. Store at room temperature. Any other formulation that assures similar amounts of the metal ions is allowable.

Biotin Solution
For biotin solution, dissolve 5 mg of D-biotin in 100 ml of 50% ethanol. Store at 4°C.

44.2.2.1.1. Plating and Spotting Media
If colonial growth on agar is required, replace the sucrose by 20 g of sorbose, 0.5 g of glucose, and 0.5 g of fructose. (Platings are better if the synthetic crossing salts, below, are used; 10 g rather than 20 g of sorbose per liter is sufficient in that medium.) For spot tests of growth on agar, use 8 g of sorbose and 4 g of sucrose per liter. All ingredients may be autoclaved together; sorbose will darken the medium with autoclaving but remains effective.

FIGURE 1 Life cycles. (Left) *Neurospora crassa*; (right) *Aspergillus nidulans*. For scale, note that the ascus of *N. crassa* is about 150 µm long, and ascospores are about 20 µm long. By contrast, *A. nidulans* asci are only 15 µm in diameter, and the ascospores are correspondingly smaller.

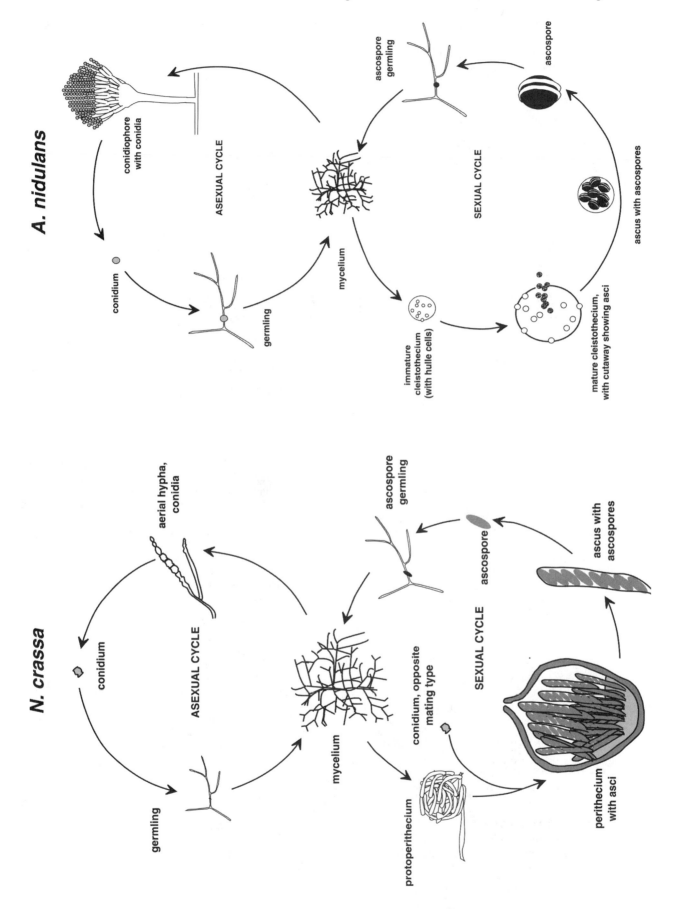

A. nidulans

conidiophore with conidia

ASEXUAL CYCLE

conidium

germling

mycelium

ascospore germling

ascospore

ascus with ascospores

SEXUAL CYCLE

immature cleistothecium (with hulle cells)

mature cleistothecium, with cutaway showing asci

N. crassa

aerial hypha, conidia

conidium

ASEXUAL CYCLE

germling

mycelium

protoperithecium

conidium, opposite mating type

ascospore germling

ascospore

ascus with ascospores

SEXUAL CYCLE

perithecium with asci

44.2.2.2. Synthetic Crossing Medium

Per liter, use KNO₃, 1 g; K₂HPO₄, 0.7 g; KH₂PO₄, 0.5 g; MgSO₄ · 7H₂O, 0.5 g; NaCl, 0.1 g; CaCl₂, 0.1 g; D-biotin, 5 μg; and trace elements, as for Vogel's at the same final concentration. Can be kept as 2× strength stock. Add for use: nutritional supplements, carbon source (see below), and agar, 15 g. Heat to melt agar, distribute to culture tubes, and insert Whatman no. 1 filter strips as carbon source; autoclave, and slant. Grow one mating type at 25°C for 6 days to form protoperithecia before fertilizing by barely wetting the surface of the slant or plate with a suspension of conidia of the other mating type. A few crosses proceed better with 10 g of sucrose per liter in place of filter paper. An alternative to synthetic medium is corn meal agar (without glucose; Difco). For plating of conidia, the salts above may be used with 0.5 g of fructose, 0.5 g of glucose, and 10 g of D-sorbose per liter.

44.2.2.3. Complete Media

Addition of 0.5% yeast extract and 0.5% casein hydrolysate is sometimes used for *N. crassa*. Other complex additives may be made (20, 21) but are now rarely used. Many additives, particularly those that serve as nitrogen sources, may inhibit crossing.

44.2.3. *Podospora anserina* (24)

44.2.3.1. Minimal Medium

Use *N. crassa* synthetic crossing medium salts, trace elements, and biotin, to which 20 g of fructose and 200 μg of thiamine are added. However, modifications are required for some purposes, such as mating, microconidial formation, and spore germination. These may be found in Esser (24).

44.2.4. *Sordaria* spp. (32)

44.2.4.1. Minimal Medium

Vogel's minimal medium (see above for *Neurospora*). For ascospore germination, add 7 g of sodium acetate per liter.

44.2.4.2. Crossing Medium

Cornmeal agar (17%) with 0.2% glucose; alternatively, 10 g of sucrose, 7 g of glucose, 1 g of yeast extract, 0.1 g of KH₂PO₄, and 17 g of Difco cornmeal agar.

44.2.5. *Coprinus* spp. and *Schizophyllum commune* (46)

44.2.5.1. Minimal Medium

L-Asparagine (1.5 g) or 1.5 g of (NH₄)₂HPO₄, 0.46 g of KH₂PO₄, 1 g of K₂HPO₄, 0.5 g of MgSO₄ · 7H₂O, trace elements (as for *N. crassa*, above), 120 μg of thiamine, 20 g of glucose, and 15 g of agar.

44.2.5.2. Complete Medium (Supports Fruiting)

Two grams of yeast extract, 2 g of peptone, 0.46 g of KH₂PO₄, 1 g of K₂HPO₄, 0.5 g of MgSO₄ · 7H₂O, 20 g of glucose, and 15 g of agar. *Coprinus* fruits much better on sterilized horse dung.

44.2.6. Nutritional Supplements

In general, work with auxotrophic strains requires additions of vitamins (5 μg/ml), amino acids (50 to 200 μg/ml), and pyrimidines and purines or their nucleosides (100 μg/ml) as appropriate. The concentrations of each required nutrient should be established specifically in each case lest those rec-

ommended above be inhibitory or insufficient. For *Aspergillus*, add vitamins, 0.1 to 100 μg/ml; purines and pyrimidines, 50 to 100 μg/ml; amino acids, 4 mM for the most demanding mutants. Better uptake may be obtained if an amino acid (e.g., arginine or proline), rather than nitrate or ammonium, is supplied as sole nitrogen source. Alternative nitrogen sources are supplied at 5 to 10 mM, alternative carbon sources at 1% (wt/vol).

44.3. GROWTH

44.3.1. Measurement of Growth

Fungal strains may be grown to test for the amount of growth or the rate of growth in given conditions. Strains may be grown for isolation of small molecules, macromolecules, organelles, or permeabilized cells. Finally, some species may be grown in exponential cultures for kinetic measurements of enzyme regulation, short-term isotope incorporation, and the like.

44.3.1.1. Mycelial Yield

Small inocula are introduced into media in small flasks and harvested when growth is complete, or at a specified time if rates of growth are quite different. Such tests are often done to test for the efficiency with which strains use a supplement to the medium or display a response to an inhibitor.

44.3.1.2. Growth Rate

Multiple cultures such as those just described can be harvested at different times to gauge roughly the rate of growth. In many fungi, growth rate can be ascertained by the rate of increase of colony radius in petri dish cultures. For *A. nidulans*, which grows naturally as a dense colony, the degree of conidiation and the visual diameter and bulk of the colony are good qualitative indicators of the efficiency of use of various carbon and nitrogen sources. In rapidly spreading species like *Neurospora*, one may measure daily progress down a "race tube," a 300- to 500-cm glass tube (bent upward at the ends) half-filled with solid medium and inoculated at one end (Fig. 2). A simpler arrangement is the use of disposable 25-ml plastic pipettes partially filled with molten medium, allowed to solidify in a horizontal position, and later broken and capped for use (53).

44.3.1.3. Mass Culture

Heavy inocula must be used for large liquid cultures. For *Neurospora*, these are prepared by growing crops of conidia for inocula for 7 days (the first two at 32°C, the last at room temperature) in 25 or 50 ml of solidified medium in 125-ml or 250-ml flasks, respectively, followed by suspension of the resulting conidia in sterile water introduced into the growth flask. Conidia are then filtered through two layers of cheesecloth, held by a rubber band as a bag in the mouth of a sterile, dry flask. The conidia are centrifuged and resus-

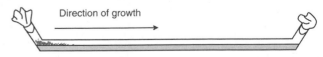

FIGURE 2 *Neurospora* "race tube." The tube (300 to 500 mm long) is supported on a rack in the position shown and is inoculated at one end. The position of the mycelial frontier is marked at intervals and distances are recorded thereafter.

pended in a small volume of water for use. Collection of *A. nidulans* conidia in water may be accomplished by flooding a mature plate culture or slant with 0.1% Tween 80 and scraping the surface. Large cultures (200 to 2,000 ml) of liquid culture will yield large amounts of mycelium if they are well aerated by shaking or by bubbling filtered air through them. In such cultures, a large inoculum (ca. 10^6 conidia per ml for *Neurospora*) will begin to grow rapidly, and growth will remain relatively homogeneous and well dispersed for several doublings. Many fungi clump or grow in tight buttons, leading to poor aeration in the centers of growth and a consequent heterogeneity of cells. Therefore, if experiments permit it, large inocula are better to maintain a young, rapidly growing culture. In *A. nidulans*, growth at 37°C in 100 ml of medium in 250-ml flasks or 300 ml in 1,000-ml flasks with vigorous shaking assures a well-dispersed culture, which can be harvested through nylon net muslin. In general, methods for mass growth are best for short-term work, owing to the dangers of contamination. Subcultures of such cultures, if they must be made, should be checked carefully for contamination before further use. Continuous culture is important in industrial mycology and is therefore the subject of considerable experiment, but can only be touched on here. Typical of experimental studies with a genetical basis are those considering the effect of regulatory mutants on secondary metabolite production (e.g., reference 1) and studies of the stability of fungi transformed for overexpression of native or exogenous genes (22).

44.3.1.4. Induction of Development

In *A. nidulans*, conidial and sexual development occur spontaneously in surface cultures, but submerged cultures are usually sterile. Submerged conidiation is, however, possible under certain conditions (35). For developmental studies, conidiation can be induced by exposing liquid-grown mycelium to air by plating onto agar or filter paper. Cultures must reach a certain age (~16 h at 37°C) before they are competent to conidiate (4), the first signs of conidiophore development becoming evident ca. 8 h after induction. Wild-type strains do not conidiate well in the dark, but most laboratory strains carry the *veA1* ("velvet") mutation, which overcomes the requirement for light. Sexual development is encouraged by partial anaerobiosis (see section 44.5.4.2) and occurs more slowly than conidiation, starting at about 50 h at 37°C (14). The first ascospores are formed at 90 h, and cleistothecia are fully mature in 3 weeks, by which time each contains up to 5×10^5 ascospores.

In *N. crassa*, formation of macroconidia occurs on the surface of cultures and is much more profuse when cultures are exposed to light. Starvation for carbon and nitrogen also elicits rapid conidiation. Macroconidial development can be synchronized by transferring mycelial mats to filter paper and exposing them to light. Macroconidia do not form in submerged culture of wild-type strains. *N. crassa* also forms uninucleate microconidia, a quite different form of propagule. They are used occasionally to isolate pure strains of transformants, which are usually heterokaryotic when first isolated (23). Conditions for induction of protoperithecia of *N. crassa* are given above (section 44.2.2.2).

44.4. COLLECTION AND EXTRACTION

44.4.1. Dry-Weight Measurement

Mycelia can be collected on paper or membrane filters, using suction flasks, or by filtration through muslin. If the mycelium is coherent, small cultures may be collected with a simple spatula. Dry-weight measurements can be made on material dried on towels or filters at 80°C or dried with acetone while the material is collected on a suction filter. These weights may be slightly different owing to acetone leaching of lipids and some small molecules. Therefore, a consistent method should be decided on at the outset.

44.4.2. Extraction Methods

44.4.2.1. Small Molecules

The small molecules of mycelia may be extracted in 5% trichloroacetic acid or 5% perchloric acid, and more selectively with acetone, chloroform, hot alcohol, and the like, depending on the experiment. The methods differ little from those used in animal and plant cell cultures.

44.4.2.2. Macromolecules

For extraction of macromolecules and organelles, the tough cell wall of fungi presents a problem. Three methods generally get around this. The first is to wash the mycelium, as it is collected, in the buffer to be used, draw off excess liquid or press the mycelium dry, and grind the pad in a cold mortar with sand or glass beads. One should add very little buffer until the macerate is collected. The second is to suspend the washed mycelium in buffer and subject it to 10- to 20-s periods of agitation with glass beads in a Bead-Beater® (Biospec Products, Bartlesville, OK). This instrument, with a selection of chambers, has been used widely to disrupt cells in a large range of volumes (18). The chamber is half-filled with buffer-wetted beads (usually 0.3 to 0.5 mm). The buffer (e.g., 10 mM TES [pH 7.2] with 1 mM EDTA) is osmotically stabilized with 1 M sorbitol if the goal is to isolate organelles. For proteins, the outcome is a rather dilute suspension, inappropriate for many studies. The third method is to collect cells in cold acetone and grind them to a powder in a spark-free blendor (Omnimixer). The powder is air dried and then used, at high density (>100 mg/ml buffer), to extract proteins. Many enzymes and other proteins survive this very well and may therefore be extracted with little difficulty in a relatively lipid-free solution.

Lyophilization of pressed-dry mycelium is often used to prepare materials for extraction. Powders can be made by vortexing the pad in a plastic tube with several large glass beads. Finally, for DNA and RNA extractions, many investigators use mycelium that is pressed dry and frozen. The pad is ground to a fine powder in liquid nitrogen with a mortar and pestle, and the macerate is mixed with an appropriate extraction mixture while still frozen (e.g., guanidine thiocyanate buffer for RNA). See chapter 43 by Marzluf, this volume, for handling nucleic acids.

44.4.2.3. Permeabilized Cells

Permeabilized cells are extremely useful for rapid enzyme assay (6). Most enzymes can be monitored in this way, assuming that specificity can be assured. The following method is based on *Neurospora*. Five-milliliter samples are withdrawn from growing, dispersed mass cultures and placed in chilled 12-ml centrifuge tubes. The tubes are centrifuged, the supernatant medium is poured off, and tubes are drained well by inversion on a paper towel. The cells are resuspended in 3 ml of 10 mM EDTA in 50 mM K^+PO_4 (pH 7.2). A toluene-ethanol mixture (0.2 ml; 1:4 [vol/vol]) is added, and the suspension is mixed. The suspension is centrifuged after 1 or 2 min, the supernatant is poured off, and the cells are frozen at −80°C. Freezing of cells is essential for full permeabilization, even if they are to be used immediately after collection. Small molecules are largely removed by the toluene treatment and centrifugation prior to freezing. They therefore do not complicate the ingredients of enzyme reaction mixtures when they are added. After thawing, the cells are brought to 1 ml with approximately 0.85 ml of a buffer compatible with the enzyme assay. Once suspended, they may be sampled for enzyme assay and for protein determination. The latter may require extraction overnight in 1 N NaOH to solubilize all proteins.

Treatment of *Aspergillus* mycelium with toluene vapor is sufficient for X-Gal detection or assay of β-galactosidase fusion proteins, while protoplasting enzymes are used to make cells permeable to antibodies (29).

44.5. GENETIC METHODS

44.5.1. Mutagenesis and Mutant Selection

Suspend single-celled units (conidia, oidia, etc.) in water and irradiate with UV light at 260 nm to 50% survival. Chemical mutagens such as ethylmethane sulfonate, hydroxylamine, ICR-170, and nitrosoguanidine may also be used, although their mutant/survivor ratios may be different, depending on the dose, than with UV. Nitroquinoline oxide (5) is recommended as an effective but readily inactivated mutagen. The 50% survival level with UV light is sufficient to induce many mutations, but not so heavy as to yield a large fraction of multiple mutants. Nevertheless, outcrossing the initial mutants to wild type should be done to remove cryptic, additional mutations from the background and to establish that all phenotypic attributes of the primary mutant are associated with a single gene. UV light is also suspected of inducing more chromosomal mutations (translocations and inversions) than chemical mutagens.

In direct selection, mutants are plated at high density and picked from a selective medium, as one would do for antibiotic-resistant variants or prototrophic revertants of an auxotroph. These must be separated from nonmutant cells by successive conidial isolation or by crossing. The latter is necessary in species in which vegetative propagules are multinucleate and sexual spores are uninucleate.

In indirect selection, several enrichment schemes may be applied. For auxotrophic mutants of *Neurospora*, which has a spreading mode of growth, the filtration-concentration method is suitable. This is done by suspending mutagenized conidia in minimal liquid medium, shaken continuously over 48 to 60 h, with filtration of the culture through cheesecloth at close enough intervals so that little growth remains in the flask. Wild-type cells grow and are removed; auxotrophs pass through the cheesecloth and survive. The culture is finally concentrated by brief centrifugation and plated on media appropriate to the mutant phenotype desired (see references 20 and 21 for details). Other indirect methods include inositol-less (*Neurospora*) or biotinless (*Aspergillus*) death, in which the starting material bears a nonrevertible inositol or biotin mutation. Such strains will die if allowed to germinate and grow without their supplements. However, if additional mutations are induced that block growth early in minimal medium, these derivatives will prevent the suicidal growth and thereby cells with new mutations will become enriched among the survivors after 24 to 36 h of growth. Plating of the final population on media with inositol or biotin and the appropriate supplement for the desired new mutants follows. Germlings are then picked into individual tubes or onto a grid pattern on plates and tested by spot testing or replication onto selective media to identify mutants.

Two hundred or more distinct *Aspergillus* microcolonies can be obtained on a 9-cm petri dish by adding 0.01% Triton X-100 to the medium before pouring. Mutants can be identified after replica plating such colonies from complete, or other supplemented medium, to minimal medium using damp velvet or filter paper (Color Plate 18). In *Neurospora*, a similar density of colonies is obtained on plating medium (see above), but they must be picked to individual tubes and spot tested when mature.

Gene disruption by random integration of plasmids or transposable elements has the advantage that the target DNA is tagged for subsequent gene isolation, e.g., in *Magnaporthe grisea* (51). Plasmid integration can be facilitated by restriction enzyme-mediated integration (REMI), in which a restriction enzyme is included in the transformation mix, e.g., in *Aspergillus fumigatus* (13) and *Coprinus cinerea* (19), while *Agrobacterium*-mediated transformation has been used for insertional mutagenesis in *Fusarium* (39). In *A. nidulans*, transposon-induced mutagenesis is less effective than had been hoped, possibly because disruption of functional genes seems to be infrequent (33).

Targeted mutagenesis—that is, the mutation of particular genes—takes advantage of cloned DNA. Two general methods are available in *N. crassa* (20) and *A. nidulans* (38), each dependent on a different mode of stable integration of DNA during transformation. In one mode, the DNA integrates by recombination with the homolog in the genome. In the other, the DNA integrates quasirandomly in a nonhomologous point or points in the genome. A linear, incomplete fragment of the gene to be disrupted may be introduced into cells. Cells that integrate it (by a double crossover) into the homologous gene will suffer a disruption of the gene in question, and the resulting mycelium should display a mutant phenotype, if known or predicted. The transformant, once purified by serial conidial isolation, may then be crossed (see below) to verify its normal genetic behavior and to purify the mutant nuclear type of other, nonhomologous integration events. In *Neurospora*, this technique has been greatly simplified by the use of strains unable to integrate DNA into nonhomologous locations in the genome (41).

In *A. nidulans*, recessive lethal gene disruptions can be identified by analysis of heterokaryons or diploids in which only one of the two copies in the cell has been disrupted (e.g., reference 34).

In *N. crassa*, a peculiar phenomenon named repeat induced point mutation (RIP) has been discovered (48). The RIP phenomenon takes advantage of the frequent nonhomologous integration of transforming DNA in normal strains. Such integration creates a strain with a genetic duplication. Both copies of the duplicated DNA, resident and exogenous, are frequently and multiply mutated during the sexual cycle. This mutagenic technique is explained in section 44.5.4.

44.5.2. Heterokaryons and Diploids: Complementation Tests

Heterokaryons and (in *Aspergillus*) diploids are used for determination of dominance and complementation. In *Aspergillus*, diploid formation is also the first step in parasexual analysis (see section 44.5.3), which is routinely used to locate new mutants to chromosomes and which can be used for further mapping.

Finally, in any species in which hyphal fusion can be arranged, extranuclear replicating elements such as mitochondrial DNA, viruses, and plasmids may be transferred by making a heterokaryon and then resolving it, after growth, into the homokaryons. In one of these homokaryons, the extranuclear element will be associated with another nuclear type. A great many extranuclear variants are infective and may prove hard to remove without resorting to sexual crosses, as indicated below. In *A. nidulans*, extranuclear inheritance is readily demonstrated using mitochondrially encoded markers (11, 12), e.g., *csA* (cold sensitivity), *oliA*, and *camA* (resistance to oligomycin and chloramphenicol, respectively). Heterokaryons between mutant and wild-type strains for these markers show vegetative segregation, independently of nuclear markers. Recombination can also occur between mitochondrial markers, and a simple linkage map has been drawn up for four loci (52). Similar work in *N. crassa* has been reviewed (20).

Complementation tests enable one to group mutants into allelic classes. For *Neurospora*, heterokaryons of all possible pairs of mutants are made and tested for growth on a medium rigorously selective for wild-type growth (see references 20 and 21 for more detailed procedures). This is done conveniently for *Neurospora* in half-grids of tubes (13 × 100 mm) with 1 ml of liquid medium, introducing suspensions of conidia of each mutant into rows and columns of tubes. Even in minimal medium, conidia of auxotrophs will germinate enough to fuse at the bottom of tubes. Smaller-scale tests are made on doubly inoculated spots on agar medium. All such tests should be accompanied by tests of the homokaryons so that positive responses, where they occur in a heterokaryon, can be compared with possible "leaky" growth of one or the other mutant. *Note:* the mating-type locus (*matA/mata*, or *A* and *a*) of *N. crassa* blocks the formation of heterokaryons, so mutants derived from different mating types cannot be assessed directly for complementation.

A. nidulans heterokaryons are unstable unless the heterokaryotic condition is forced by the use of additional, complementing auxotrophies. Therefore, heterokaryons are made by germinating mixed conidia of the parent strains on liquid or solid enriched medium and transferring the resulting growth (i) to medium permissive for the mutants to be tested, but selective for complementation between the forcing markers, then (ii) to the test medium, allowing growth only for pairs showing complementation between the mutants to be tested.

Since *Aspergillus* spp. heterokaryons vary in the proportions of the two constituent nuclei, the use of diploids will give clearer results in both complementation and dominance tests. *Aspergillus* conidia are uninucleate, so rare diploids can be selected by plating conidia from a heterokaryon on media selective for complementation between the forcing markers. Heterozygous diploid conidia form by fusion of unlike nuclei in heterokaryons at a frequency of approximately 1×10^{-7}. They are best selected by dispersing conidia of a heterokaryon (at various concentrations) into selective agar medium cooled to 40°C and still molten. A valuable aid to the identification of diploids is the use of different spore color markers in the two parents, e.g., *yA* (yellow) and *wA* (white), which complement in individual, uninucleate diploid spores to give the wild-type green color. (Heterokaryotic colonies will yield a variegated spore mass.) Another visible characteristic of diploid conidia is their volume, which is twice that of haploid conidia.

Heterokaryon incompatibility in many fungal species limits formation of heterokaryons and diploids between wild strains. In *A. nidulans*, incompatibility is determined by at least eight loci (3), but most laboratory strains are derived from a single wild isolate, so incompatibility is not a problem. The same considerations apply to *N. crassa*, but some lineages in use require that investigators be sure of compatibility before embarking on complementation tests.

44.5.3. Parasexual Analysis

Aspergillus diploids are moderately stable. Diploid colonies grow like haploids, but give rise to genetically variant sectors by two processes: haploidization and mitotic recombination. Both events are rare and occur independently, but the combined result is a messy, stepwise equivalent to meiosis after diploid formation, a series of events that is known as the parasexual cycle (30).

Haploidization of diploids formed between new mutants and a master strain carrying markers for all eight chromosomes is routinely used to locate new loci to chromosomes (37). Haploidization can be induced with the mitotic poison benomyl (2 µg/ml in complete medium), haploid sectors being identified by segregating spore color markers and conidial diameter. Since haploidization is independent of crossing over, classification of a small number of haploids is usually sufficient to show complete linkage of the new mutation with one of the chromosomal markers, and independent assortment with the remainder. Linkage to markers on two different chromosomes is indicative of a translocation in the mutant strain.

Mitotic crossing over can be used to order markers on one chromosome with respect to the centromere and to each other. During or after duplication of two homologous chromosomes in mitosis, a crossover may occasionally occur between nonsister chromatids of a homologous pair, resulting in exchange of distal chromosomal segments. Because thereafter the two chromatids of each homolog go to opposite poles at nuclear division, a recombinant chromatid may end up in the same daughter nucleus as a nonrecombinant chromatid of the other homolog. The result will be homozygosity for all markers distal to the crossover. The crossover may take place at any point, and therefore selection for such events, facilitated by selection for homozygosity of a distal, visible, or drug-resistance marker, allows ordering of all more proximal markers with respect to the centromere. This technique has great value in *Aspergillus*, since meiotic chromosomal linkage maps are long (some more than 400 map units) and loci on the same chromosome may therefore not show meiotic linkage. However, it is not a routinely employed procedure because it requires test strains carrying all the markers to be mapped in coupling with a distal selectable marker.

Haploidization is believed to be precipitated by mitotic nondisjunction of single chromosomes (30). Rarely, however, some nondisjunctions are followed by loss or gain of chromosomes to restore the diploid state. If in this process one homolog is lost and replaced by a copy of the other, the result will be homozygosity for whole chromosomes, to be distinguished from homozygosity of chromosome arms resulting from mitotic crossing over.

Neurospora spp. do not form diploids suitable for parasexual analysis.

44.5.4. Sexual Crosses and Genetic Analysis

44.5.4.1. General

Genetic analysis can be performed by rationales described in a variety of chapters, books, and manuals (10, 12, 16, 20,

21, 25, 32, 37, 44) focused on various species. In general, almost all genetic analysis can be done with random meiotic products if the latter derive from a single nucleus and are easily separable in suspension. Platings of sexual spores on nonselective medium, followed by picking the germlings to individual tubes and testing them thereafter, will allow the ratio of phenotypic classes to be ascertained. Germination percentages should be recorded in the event that certain classes of progeny are inviable and might be accounted for in the nongerminated fraction.

Segregation, independent assortment, and linkage can be recognized crosses by characteristic ratios among random meiotic products (ascospores or basidiospores). If parental strains differ in one characteristic (e.g., having the phenotypes Arg^+ and Arg^-), a cross can reveal whether they bear alternative alleles at a single gene. Thus the cross $Arg^+ \times Arg^-$ may yield 65 Arg^+ and 70 Arg^-. The approximate 1:1 ratio among the progeny signifies a single gene difference between alleles arg^+ and arg^-. Notice that the notation differs for phenotype (Arg^-) and genotype (arg^-). In what follows, the nomenclatural conventions of *N. crassa* are used (20).

If the parental strains have two phenotypic differences (e.g., $Pyr^+ Arg^-$ and $Pyr^- Arg^+$), the genes responsible may be unlinked and show independent assortment. Thus the cross of $Pyr^+ Arg^- \times Pyr^- Arg^+$ might yield 103 $Pyr^+ Arg^-$, 99 $Pyr^- Arg^+$, 89 $Pyr^+ Arg^+$, and 110 $Pyr^- Arg^-$, that is, about 25% of each possible class. By the convention of the previous cross, we would assign gene and allele symbols pyr^+/pyr^- and arg^+/arg^- to the two genes.

If the two differences are due to linked genes, linkage is detected by observing more parental phenotypes than recombinant phenotypes among the meiotic products. Thus a cross of strains having genotypes $arg^+ inl^- \times arg^- inl^+$ may yield 345 $arg^+ inl^-$, 320 $arg^- inl^+$, 23 $arg^+ inl^+$ and 30 $arg^- inl^-$. Such a ratio, with the recombinant genotypes in the minority, is the criterion for linkage of the genes *arg* and *inl*. In this example, the map distance between the two genes is simply the percentage of recombinants among the total meiotic products: $53/718 = 7.4\%$, or 7.4 map units.

If two mutants have the same phenotype (e.g., Arg^-), the outcome of crosses can be interpreted with these rules in mind. First, if two strains carry mutations for different *arg* loci, *arg-1* and *arg-2*, the cross can best be understood with the following notation: $arg\text{-}1^- arg\text{-}2^+ \times arg\text{-}1^+ arg\text{-}2^-$. Among the progeny of this cross, we will usually see an auxotrophic class, Arg^-, comprising three genotypes ($arg\text{-}1^- arg\text{-}2^+$, $arg\text{-}1^+ arg\text{-}2^-$, $arg\text{-}1^- arg\text{-}2^-$) and a prototrophic class, Arg^+ ($arg\text{-}1^+ arg\text{-}2^+$). If the genes are unlinked, the percentage of Arg^+ phenotypes among the progeny will be approximately 25% (assuming viability among the four genotypes is equal). This will also be true of a cross between an Arg^- strain of the genotype $arg\text{-}1^- arg\text{-}2^-$ and a prototroph ($arg\text{-}1^+ arg\text{-}2^+$), the result differing quantitatively and diagnostically from the case of segregation of two alleles of a single gene, above. If linkage prevails, the prototroph percentage will be even lower, allowing one to calculate the linkage distance. In both of these crosses, one must double the percentage of prototrophs, because it must be assumed that there are an equal number of double-mutant progeny among the Arg^- auxotrophs. In the case of independent assortment of *arg-1* and *arg-2*, 25% prototrophs is a maximum, because in either cross (mutant × mutant or double mutant × wild type) it indicates 50% recombinant genotypes.

It is good practice to cross new mutants, after isolating them, to wild-type or standard strains to ascertain whether the phenotypic features associated with the new mutant are determined by one or more genes, and also to obtain a mutant stock that is derived from a single nucleus. In haploid fungi, such crosses are expected to yield a 1:1 ratio of mutant and wild-type progeny if all features of the mutant are determined by a single gene, but the results will be more complex if the mutant phenotype is due to mutation of more than one gene. In determining progeny ratios, it is important to be aware that deleterious mutants often reduce spore viability, so that mutant phenotypes may be underrepresented.

In classical genetic practice, the next step is to determine the position of the new mutant locus on a linkage map. This procedure has several functions. First, it relates the mutant locus to other genes, sometimes revealing close linkage or possible allelism to mutants of the same phenotypic class. In other cases, close linkage may indicate a cluster of genes of related function and possibly coordinate control. Other features of map position, e.g., proximity to centromeres or telomeres, where heterochromatin is often located, may also be of interest. Finally, an ascertained map position can facilitate cloning that requires chromosome walking.

Recombination frequencies are subject to both genetic and environmental variation, so linkage map distances may vary considerably from one cross to another or among different laboratories where standard conditions and the genetic background of fungal strains might differ. Because of this variability, reliable linkage maps require data drawn from three-point crosses, in which the position of any new gene is related to two other markers in the same cross. These crosses at least yield unambiguous gene order, whatever the source of the information. Methods for efficient localization and mapping of fungal genes with three-point crosses may be found in several accessible sources (e.g., reference 25). In *N. crassa*, rapid localization of new mutations to linkage groups may be made by crossing the mutants to special translocation strains (20, 21) and assessing the outcome on the basis of random-spore analysis, whereas in *A. nidulans* initial localization of mutants to linkage groups is done via the parasexual cycle (see section 44.5.3).

In work with cloned DNA, detailed linkage maps can be obtained without a collection of already mapped mutants. Such maps are based on polymorphism for restriction fragment length (RFLP) or other DNA markers. In *N. crassa*, a species in which considerable polymorphism is known, crosses of a mutant-marked standard strain with an exotic strain have provided the community with sets of progeny (maintained at the Fungal Genetics Stock Center) recombinant for numerous polymorphic sites, determined with DNA probes. These sites have been mapped with respect to one another and with respect to standard genetic markers. A DNA probe that recognizes a polymorphism between the parents can then be used to screen the progeny set. The progeny will each display one or the other parent's variant when probed; distribution among the progeny will be unique to that probe and other probes used to make the RFLP map for that chromosomal region. The identity (or near identity) of the pattern revealed by the new probe is thus determined by reference to the RFLP map.

Detailed RFLP maps are also available for other fungi, e.g., *Cochliobolus heterostrophus* (49). Surprisingly little DNA variation has been detected between wild *A. nidulans* strains, but hybrids between related species can be used to provide molecular markers (50).

In fungi, centromeres can be mapped genetically without reference to cytology. For species forming vegetative diploids, this can be achieved by analysis of mitotic recombination (see section 44.5.3). In species such as *Neurospora*, meiosis distributes the first-division products cleanly to the upper and lower halves of the ascus. Examination of the resulting eight-spored ascus demonstrates this as a "first-division segregation" (Fig. 3). However, a crossover between the marker gene and the centromere is detected by the formation of a "second division segregation" ascus. In such asci, both spore phenotypes appear in both the top and bottom halves of the ascus (Fig. 3). Since each ascus contains the products of a single meiosis, the frequency of crossing over in the interval between a scorable marker and its centromere can be measured readily. *Podospora* differs here, owing to the overlap of spindles at the second meiotic division, and this must be taken into account (25).

Crosses may also be used in allele testing: if two mutations are unlinked, they must be at different loci, while if they are alleles of the same gene, they will be completely linked except in cases where intragenic crossing over might take place. Because lack of observable recombination is not a secure criterion of allelism, complementation tests should be done to decide the matter. Intragenic crosses, i.e., those used to map sites within a gene, are only practicable in microorganisms, including fungi. Such crosses usually employ pairs of allelic mutants in strains carrying "outside markers" that are used to orient crossover events, effectively giving a four-point map. Such crosses invariably demonstrate the phenomenon of "gene conversion." This term was given to nonreciprocal recombination events first detected dependably in fungi (see reference 25). The phenomenon can be most clearly demonstrated by means of tetrad analysis, in which one isolates an entire ascus and observes or dissects out the ascospores. For instance, in *Sordaria* spp. segregation patterns of viable ascospore color mutants can be analyzed by visual inspection (Fig. 3). Gene conversion in its simplest form is seen as a 6:2 or 5:3 ratio of alleles, rather than the orthodox 4:4 expected in normal segregation. The term gene conversion derives from the impression that one of the alleles had "converted" the other one during pairing at meiosis.

Intragenic recombination is normally a rare event, so recombinants are usually obtained by selective plating for wild types. In such an experiment, conversion is detected as an event that replaces a mutant site with the wild-type version, but without crossing over of outside markers (Fig. 4).

In most crosses between two strains with different mutations in the same gene, conversion at one site is usually more common than the other, and if enough mutational sites are studied, a gradient of conversion frequency will be found along the gene; this is termed conversion polarity. In *A. nidulans*, intragenic crossing over is sufficiently frequent to allow ordering of the sites by examination of outside

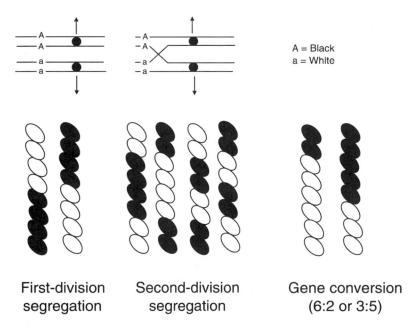

A = Black
a = White

First-division   Second-division   Gene conversion
segregation   segregation   (6:2 or 3:5)

FIGURE 3 Meiotic tetrads of various types. In first-division segregation, no crossover occurs between the gene A/a and the centromere of the chromosome. Segregation of centromeres (arrows) places all copies of the A allele in one half and all copies of the a allele in the other half. In second-division segregation, crossing over exchanges the distal parts of the two central chromatids; segregation of A and a alleles is delayed until the chromatids (lines) separate at the second meiotic division. The patterns shown are all equally probable, but in all cases, both alleles of the A/a gene are found in both halves of the ascus. In gene conversion, molecular interaction of chromatids leads to replacement of information of one chromatid by transfer from the other. One or both DNA strands of the recipient chromosome may be replaced, leading to 3:5 and 6:2 segregations, respectively. Conversions may occur in either direction, from A to a (as shown in the 6:3 example) or from a to A (as shown in the 3:5 example).

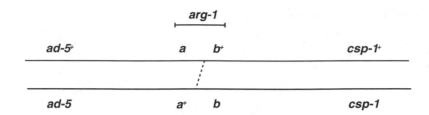

Orthodox crossing over: **ad-5 arg-1⁺ csp-1⁺**

Gene conversion **a** to **a⁺** : **ad-5⁺ arg-1⁺ csp-1⁺**

Gene conversion **b** to **b⁺** : **ad-5 arg-1⁺ csp-1**

FIGURE 4   Intragenic crossing over. Rare cases of intragenic crossing over must be detected by selective means, such as plating ascospores of a cross between two auxotrophs on minimal medium. Here, two different sites of mutation, *a* and *b* in the *arg-1* gene, are shown. Prototrophic progeny (*arg-1⁺*) recovered in plating may arise in three ways. In the first, orthodox crossing over leads to reconstruction of the wild-type allele. (In such a process, a double mutant will form as the reciprocal product.) The crossover leads to the recombination of the markers on either side, as shown. *arg-1⁺* prototrophs may, however, frequently form by gene conversion of either the *a* mutational site or the *b* mutational site without any recombination of the flanking markers.

markers (45). For instance, if in the equivalent of the cross shown in Fig. 4, the most common crossover genotype was *ad-5 arg⁺ csp⁺*, it would be concluded that the **b** site was to the right of the **a** site. However, in many fungi, including *N. crassa*, almost all intragenic recombinants arise by conversion, and sites can only be ordered in terms of the conversion gradient, e.g., if in a cross of **a** × **b**, site **a** is converted more frequently than site **b**, then **a** is at the higher end of the polarity gradient. Descriptions of the molecular events underlying gene conversion and its association with meiotic recombination may be found in most sophisticated genetics texts.

N. crassa offers a novel mutagenic technique based on the RIP phenomenon (48). As mentioned above (see section 44.5.1), *N. crassa* DNA fragments or fragments carried by plasmids integrate most frequently into nonhomologous (ectopic) locations of the genome, creating a duplication (or higher multiple) of the resident gene(s). Once purified, the behavior of the transformant in sexual crosses is irregular. During the several divisions of the transformant nucleus just *prior* to nuclear fusion, both the exogenous (if more than about 1 kb) and the resident copies of the gene or DNA segment are mutated, often massively, by GC-to-AT transitions. The segments are often heavily methylated as well. Therefore, the ascospore products of a cross of an ectopic transformant × wild type will include many that display the mutant phenotype for the gene in question. If viable, the mutants may then be studied physiologically thereafter, although screening the mutant ascospore progeny is desirable to recognize those in which the ectopic copy has been lost by recombination or independent assortment. If the mutant phenotype is inviable or unpredictable, more sophisticated means of recovering "RIPped" nuclei in heterokaryons may have to be used. The "sheltered RIP" method devised by Harkness et al. (27) may be consulted for details. The active mutagenic RIP process has been demonstrated so far only in *Neurospora* and *Podospora* (26), although evidence of RIPped sequences have been found in

many other fungi (e.g., references 2, 28, and 40). So far, the RIP process has been used as a mutagenic system only in *N. crassa*.

Extranuclear variants, such as the respiratory deficiency caused by the mitochondrial [*poky*] mutation, have a distinct pattern of inheritance. First, they are rarely or never transmitted by the male (conidial) parent in sexual crosses. Second, the ascospore progeny display a lack of segregation for the variant; they all have the phenotype of the maternal parent. In *A. nidulans*, heterokaryon-incompatible strains have been used to show that mitochondria are inherited, as in other fungi, from the parent taking the female role in the cross, the latter being identified by maternally determined cleistothecial pigment markers (47).

44.5.4.2. Procedures for Particular Fungi

44.5.4.2.1. N. crassa (20, 21)

Neurospora spp. shoot their spores from perithecia as they develop, and these collect in culture tubes on the wall opposite the agar slant. They are collected no less than 21 days after fertilization to ensure their maturity. This is done by scraping a sample from the wall of the tube with a wet inoculation loop and suspending the spores in 0.5 ml of sterile water in a culture tube. The tubes are heated to 60°C for 30 min and cooled. The heating activates the ascospores and kills any conidia that may have been collected with them. The activated spores are plated on nonselective plating medium (see section 44.2.2.1.1) if the ratios of all progeny types are to be determined. (Selective platings may be appropriate, particularly if percentages of rare recombinants are of interest.) Colonies are picked at a very early germling stage (using a dissection microscope with substage illumination and a condenser) with a spear-point needle or similar device to individual tubes. This avoids colony overlap with further growth and ensures that only one germling is being isolated each time. (The dark spore wall is a good indicator of whether the germling is a single meiotic product.) When mature, the cultures are spot tested in 5 × 5 arrays on petri dishes of spotting medium (see section 44.2.2.1.1). In spot testing, a straight inoculation needle is used to stab very small inocula to spots premarked on a dish. The inoculations are done as the dish remains inverted, to minimize contamination. The results are scored after overnight growth at 32°C, with confirmation later in the day if necessary. Different temperatures may be used if necessary or convenient.

44.5.4.2.2. A. nidulans

Sexual interaction is achieved by mixing conidia in a small area on thick (40 ml per 9-cm dish) minimal agar and, after germination, sealing the plates with cellophane tape to produce partial anaerobiosis (44). Crosses are normally made between strains bearing complementing auxotrophies and spore colors. Germination and fusion of the parent strains on minimal medium is encouraged by addition of a few drops of complete medium at the point of inoculation. Cleistothecia develop within 10 days and are fully mature in 3 weeks. By this stage the asci, which are very fragile, have broken down, leaving up to 5×10^5 free ascospores in each cleistothecium. Ascospores germinate without activation, but high germination rates are only achieved with spores from fully mature fruiting bodies.

A. nidulans is homothallic, so cleistothecia developing from a mixed inoculum may result from a self-cross or from a cross between the parents (hybrid cleistothecium). For this reason, strains used as parents should be marked with conidial color mutants so that hybrid cleistothecia and their contents can be recognized as such. Where analysis is by selective plating, massed, uncleaned hybrid cleistothecia taken from the surface of the mycelium (Color Plate 18A) can be crushed and the resulting ascospore suspension is plated as desired. Alternatively, single cleistothecia are cleaned of adhering conidia and Hülle cells by rolling on sterile agar (Color Plate 18B), then each is crushed in 10 ml of sterile water and the suspension is test-plated on one dish, usually containing Triton X-100 (see section 44.5.1). Hybrid cleistothecia can be identified by the mix of spore colors among the progeny, and suspensions from selected cleistothecia are then replated on a larger scale, usually omitting the Triton X-100, which may select against some phenotypes. Occasionally, "twinned" cleistothecia are found; these contain a mixture of ascospores of hybrid and selfed origin.

Ascospore-derived colonies are classified by transferring them individually to a master plate containing a 5 × 5 colony array, which is then replicated to test media using an appropriately spaced multipin replicator on a translucent plastic base (Color Plate 18). Master plates should be poured some hours before inoculation, or else the open, inverted plates may be briefly dried at 37°C. Inoculation of master plates should be done in still air, holding the recipient plate inverted so that stray spores fall away from it. Similarly, when replicating from a master plate, place it, open, on the bench, stab the multipin replicator down onto it, then invert the replicator and bring recipient plates, in turn, down onto its pins.

44.5.4.2.3. Coprinus and Schizophyllum

The tetrapolar basidiomycetes Coprinus and Schizophyllum require the recognition and use of monokaryotic mycelia of compatible mating type (differences at the unlinked A and B loci) before mating. Confrontation of monokaryons on complete medium in petri dishes leads, after growth and anastomosis, to the formation of fruiting bodies, which form basidiospores, the haploid products of meiosis. Plating of the spores on suitable media is followed by germination; the germlings can be picked and analyzed genetically as a random population of meiotic products. With respect to analysis of vegetative cultures, Coprinus forms uninucleate oidia from monokaryons and chlamydospores from dikaryons; in both cases these are suitable for plating and, in the former case, for resolving heterokaryons. Schizophyllum does not form single-celled spores except via meiosis. Therefore, macerates must be used to resolve heterokaryons and for work involving mutagenesis. More detailed methods may be found in the articles in the volume by King (32).

44.5.5. Stock Management

Any use of fungi requires good management of stocks, in particular in genetic studies. Most stocks should be stored in a manner that minimizes growth or aging of cultures. Such methods include freezing ($-20°C$), lyophilization, and silica gel storage (for N. crassa, see reference 20). In this way, the accumulation of spontaneous mutants in heterokaryotic association with the wild type, or selective survival of mutants over wild type, during prolonged storage in the refrigerator can be avoided. Stocks should have a formal designation and recorded source, and this designation should be distinct from the name of any mutant loci in the strains. The names of loci, in turn, should be distinct from the particular allele of the locus used. Standard rules of nomenclature may be found at many of the websites (for links, see www.fgsc.net) and books cited and may be inferred from recent publications.

Note: *Although minor changes in this chapter have been made more recently, no attempt has been made to incorporate the many rationales and techniques that have appeared since 2003. Readers should consult relevant websites, reviews, and technical literature as they embark on the procedures described here.*

44.6. REFERENCES

1. **Agger, T., A. B. Spohr, and J. Nielsen.** 2001. [alpha]-Amylase production in high cell density submerged cultivation of *Aspergillus oryzae* and *A. nidulans*. *Appl. Microbiol. Biotechnol.* **55**:81–84.

2. **Aleksenko, A., and A. J. Clutterbuck.** 1996. The AMA1 plasmid replicator in *Aspergillus nidulans* is an inverted duplication of a low-copy-number dispersed genomic repeat. *Mol. Microbiol.* **19**:565–574.

3. **Anwar, M. M., J. H. Croft, and R. G. B. Dales.** 1993. Analysis of heterokaryon incompatibility between heterokaryon-compatibilty (h-c) groups R and GL provides evidence that at least eight *het* loci control somatic incompatibility in *Aspergillus nidulans*. *J. Gen. Microbiol.* **139**:1599–1603.

4. **Axelrod, D. E.** 1972. Kinetics of differentiation of conidiophores and conidia by colonies of *Aspergillus nidulans*. *J. Gen. Microbiol.* **73**:181–184.

5. **Bal J., E. M. Kajtaniak, and N. J. Pieniazek.** 1977. 4-Nitroquinoline-1-oxide: a good mutagen for *Aspergillus nidulans*. *Mutat. Res.* **56**:153–156.

6. **Basabe, J. R., C. A. Lee, and R. L. Weiss.** 1979. Enzyme assays using permeabilized cells of Neurospora. *Anal. Biochem.* **92**:356–360.

7. **Bennett, J. W., and L. L. Lasure (ed.).** 1985. *Gene Manipulations in Fungi.* Academic Press, Inc., Orlando, FL.

8. **Bennett, J. W., and L. L. Lasure (ed.).** 1991. *More Gene Manipulations in Fungi.* Academic Press, Inc., San Diego, CA.

9. **Borkovich, K. A., L. A. Alex, O. Yarden, M. Freitag, G. E. Turner, N. D. Read, S. Seiler, D. Bell-Pedersen, J. Paietta, N. Plesofsky, M. Plamann, M. Goodrich-Tanrikulu, U. Schulte, G. Mannhaupt, F. E. Nargang, A. Radford, C. Selitrennikoff, J. E. Galagan, J. C. Dunlap, J. J. Loros, D. Catcheside, H. Inoue, R. Aramayo, M. Polymenis, E. U. Selker, M. S. Sachs, G. A. Marzluf, I. Paulsen, R. Davis, D. J. Ebbole, A. Zelter, E. R. Kalkman, R. O'Rourke, F. Bowring, J. Yeadon, C. Ishii, K. Suzuki, W. Sakai, and R. Pratt.** 2004. Lessons from the genome sequence of *Neurospora crassa*: tracing the path from genomic blueprint to multicellular organism. *Microbiol. Mol. Biol. Rev.* **68:**1–108.
10. **Bos, C. J. (ed.).** 1996. *Fungal Genetics, Principles and Practice.* Marcel Dekker, New York, NY.
11. **Brown, T. A.** 1993. Mitochondrial genome of *Aspergillus nidulans*, p. 3.85–3.86. *In* S. J. O'Brien (ed.). *Genetic Maps*, 6th ed. Cold Spring Harbor Press, New York, NY.
12. **Brown, T. A.** 1994. Mitochondria, p. 505–524. *In* S. D. Martinelli and J. R. Kinghorn (ed.), *Aspergillus 50 Years On.* Elsevier, Amsterdam, The Netherlands.
13. **Brown, J. S., A. Aufauvre-Brown, and D. W. Holden.** 1998. Insertional mutagenesis of *Aspergillus fumigatus*. *Mol. Gen. Genet.* **259:**327–335.
14. **Champe, S. P., D. L. Nagle, and L. N. Yager.** 1994. Sexual sporulation, p. 429–454. *In* S. D. Martinelli and J. R. Kinghorn (ed.), *Aspergillus 50 Years On.* Elsevier, Amsterdam, The Netherlands.
15. **Clutterbuck, A. J.** 1974. *Aspergillus nidulans*, p. 447–510. *In* R. C. King (ed.), *Handbook of Genetics*, 1. *Bacteria, Bacteriophages and Fungi.* Plenum Press, New York, NY.
16. **Clutterbuck, A. J.** 1995. Genetics of fungi, p. 239–253. *In* N. A. R. Gow and G. M. Gadd, (ed.), *The Growing Fungus.* Chapman and Hall, London, United Kingdom.
17. **Cove, D. J.** 1966. The induction and repression of nitrate reductase in the fungus *Aspergillus nidulans*. *Biochim. Biophys. Acta* **113:**51–56.
18. **Cramer, C. L., J. L. Ristow, T. J. Paulus, and R. H. Davis.** 1983. Methods for mycelial breakeage and isolation of mitochondria and vacuoles of *Neurospora*. *Anal. Biochem.* **128:**384–392.
19. **Cummings, W. J., M. Celerin, J. Crodian, L. K. Brunick, and M. E. Zolan.** 1999. Insertional mutagenesis in *Coprinus cinereus*: use of a dominant selectable marker to generate tagged, sporulation-defective mutants. *Curr. Genet.* **39:**371–382.
20. **Davis, R. H.** 2000. *Neurospora: Contributions of a Model Organism.* Oxford University Press, Oxford, United Kingdom.
21. **Davis, R. H., and F. J. de Serres.** 1970. Genetic and microbiological research techniques for *Neurospora*. *Methods Enzymol.* **17A:**79–143.
22. **Dunn-Coleman, N. S., E. A. Brodie, G. L. Carter, and G. L. Armstrong.** 1992. Stability of recombinant strains under fermentation conditions, p. 152–174. *In* J. R. Kinghorn and G. Turner (ed.), *Applied Molecular Genetics of Filamentous Fungi.* Chapman & Hall, London, United Kingdom.
23. **Ebbole, D., and M. S. Sachs.** 1990. A rapid and simple method for isolation of *Neurospora crassa* homokaryons using microconidia. *Fungal Genet. Newsl.* **37:**17–18.
24. **Esser, K.** 1974. *Podospora anserina*, p. 531–551. *In* R. C. King (ed.), *Handbook of Genetics*, vol. 1. *Bacteria, Bacteriophages and Fungi.* Plenum Press, New York, NY.
25. **Fincham, J. R. S., P. R. Day, and A. Radford.** 1979. *Fungal Genetics.* Univ. of California Press, Berkeley, CA.
26. **Graïa, F., O. Lesprinet, B. Rimbault, M. Deqard-Chablat, E. Coppin, and M. Picard.** 2001. Genome quality control: RIP (repeat-induced point mutation) comes to *Podospora*. *Mol. Microbiol.* **40:**586–595.
27. **Harkness, T. A. A., R. L. Metzenberg, H. Schneider, R. Lill, W. Neupert, and F. E. Nargang.** 1994. Inactivation of the *Neurospora crassa* gene encoding the mitochondrial protein import receptor MOM19 by the technique of "sheltered RIP." *Genetics* **136:**107–118.
28. **Hua-Van, A., F. Héricourt, P. Capy, M. J. Daboussie, and T. Langin.** 1998. Three highly divergent subfamilies of the impala transposable element coexist in the genome of the fungus *Fusarium oxysporum*. *Mol. Gen. Genet.* **259:**354–362.
29. **Jung, M. K., N. Ovechkina, C. E. Oakley, and B. Oakley.** 2000. The use of beta-D-glucanase as a substitute for Novozyme 234 in immunofluorescence and protoplasting. *Fungal Genet. Newsl.* **47:**65–66.
30. **Käfer, E.** 1961. The processes of spontaneous recombination in vegetative nuclei of *Aspergillus nidulans*. *Genetics* **46:**1581–1609.
31. **Käfer, E.** 1977. Meiotic and mitotic recombination in *Aspergillus* and its chromosomal aberrations. *Adv. Genet.* **19:**33–131.
32. **King, R. C. (ed.).** 1974. *Handbook of Genetics*, vol. 1. *Bacteria, Bacteriophages and Fungi.* Plenum Press, New York, NY.
33. **Li Destri Nicosia, M. G., C. Brocard-Masson, S. Demais, A. H. Van, M.-J. Daboussi, and C. Scazzocchio.** 2001. Heterologous transposition in *Aspergillus nidulans*. *Mol. Microbiol.* **39:**1330–1344.
34. **Marhoul, J. F., and T. H. Adams.** 1996. Aspergillus *fabM* encodes an essential product that is related to poly(A)-binding proteins and activates development when overexpressed. *Genetics* **144:**1463–1470.
35. **Martinelli, S. D.** 1976. Conidiation of *Aspergillus nidulans* in submerged culture. *Trans. Br. Mycol. Soc.* **67:**121–128.
36. **Martinelli, S. D., and J. R. Kinghorn (ed.).** 1994. *Aspergillus: 50 Years On.* Elsevier, Amsterdam, The Netherlands.
37. **McCully, K. S., and E. Forbes.** 1965. The use of master strains of *Aspergillus nidulans* for assigning genes to linkage groups. *Genet. Res.* **6:**352–359.
38. **Miller, B. L., K. Y. Miller, and W. E. Timberlake.** 1985. Direct and indirect gene replacements in *Aspergillus nidulans*. *Mol. Cell. Biol.* **5:**1714–1721.
39. **Mullins, E. D., X. Chen, P. Romaine, R. Raina, D. M. Geiser, and S. Kang.** 2001. *Agrobacterium*-mediated transformation of *Fusarium oxysporum*: an efficient tool for insertional mutagenesis and gene transfer. *Phytopathology* **91:**173–180.
40. **Nakayashiki, H., N. Nishimoto, K. Ikeda, Y. Tosa, and S. Mayama.** 1999. Degenerate MAGGY elements in a subgroup of *Pyricularia grisea*: a possible example of successful capture of a genetic invader by a fungal genome. *Mol. Gen. Genet.* **261:**958–966.
41. **Ninomiya, Y., K. Suzuki, C. Ishii, and H. Inoue.** 2004. Highly efficient gene replacements in *Neurospora* strains deficient for nonhomologous end-joining. *Proc. Natl. Acad. Sci. USA* **101:**12248–12253.
42. **Perkins, D. D., and R. H. Davis.** 2000. Neurospora at the millennium. *Fungal Genet. Biol.* **31:**153–167.
43. **Perkins, D. D., A. Radford, and M. S. Sachs.** 2001. *The Neurospora Compendium. Chromosomal Loci.* Academic Press, New York, NY.
44. **Pontecorvo, G., J. A. Roper, L. M. Hemmons, K. D. MacDonald, and A. W. J. Bufton.** 1953. The genetics of *Aspergillus nidulans*. *Adv. Genet.* **5:**141–238.
45. **Pritchard, R. H.** 1955. the linear arrangement of a series of alleles in *Aspergillus nidulans*. *Heredity* **9:**343–371.
46. **Raper, J. R., and R. M. Hoffman.** 1974. *Schizophyllum commune*, p. 597–626. *In* R. C. King (ed.), *Handbook of Genetics*, vol. 1. *Bacteria, Bacteriophages and Fungi.* Plenum Press, New York, NY.

47. **Rowlands, R. T., and G. Turner.** 1976. Maternal inheritance of extranuclear markers in *Aspergillus nidulans*. *Genet. Res.* **285:**281–290.

48. **Singer, M. J., and E. U. Selker.** 1995. Genetic and epigenetic inactivation of repetitive sequences in *Neurospora crassa*: RIP, DNA methylation and Quelling. *Curr. Top. Microbiol. Immunol.* **197:**165–177.

49. **Tzeng, T.-H., L. K. Lyngholm, C. F. Ford, and C. R. Bronson.** 1992. A restriction fragment length polymorphism map and electrophoretic karyotype of the fungal maize pathogen *Cochliobolus heterostrophus*. *Genetics* **130:**81–96.

50. **Varga, J., and J. H. Croft.** 1994. Assignment of RFLP, RAPD and isozyme markers to *Aspergillus nidulans* chromosomes, using chromosome-substituted segregants of hybrid *A. nidulans* and *A. quadrilineatus*. *Curr. Genet.* **25:**311–317.

51. **Villalba, F., M.-H. Lebrun, A. Hua-Van, M.-J. Daboussi, and M.-C. Grosjean-Cournoyer.** 2001. Transposon *impala*, a novel tool for gene tagging in the rice blast fungus *Magnaporthe grisea*. *Mol. Plant-Microbe Interact.* **14:**308–315.

52. **Waring, R. B., and C. Scazzocchio.** 1983. Mitochondrial four-point crosses in *Aspergillus nidulans*: mapping of a suppressor of a mitochondrially inherited cold-sensitive mutation. *Genetics* **103:**297–311.

53. **White, B., and D. O. Woodward.** 1995. A simple method for making disposable race tubes. *Fungal Genet. Newslet.* **42:**79.

45

Principles and Practice of DNA Microarray Technology

KRISHNAMURTHY NATARAJAN, MATTHEW J. MARTON,
AND ALAN G. HINNEBUSCH

45.1. INTRODUCTION

The availability of whole-genome sequences has made it possible to identify the full complement of genes and begin a comprehensive assignment of gene functions. However, there are serious limitations to functional analysis of sequenced genomes. Several organisms are not amenable to genetic analysis, and mutant construction and phenotype analysis cannot be carried out routinely. Even for genetically tractable organisms, such as budding yeasts, reliable and sensitive assays are not available for numerous biological processes and pathways. Thus, one-third of the yeast genome has not been characterized experimentally and no function can be ascribed to this large gene set, as there are no known homologs in other organisms. The postgenome era has heralded new technologies that alleviate these difficulties and accelerate gene discovery and functional analysis. Valuable clues about the function of the uncharacterized genes can be obtained using microarray technology to assess expression of the entire genome under numerous experimental conditions in parallel. Microarray technology has evolved rapidly, and indeed this powerful technology is not intimidating and can now be established in most laboratories. This has been possible in large part due to the ready availability of reagents in the form of kits, prearrayed whole-genome microarray slides, data analysis software, and instrumentation from several vendors.

In this chapter we describe the principles and detailed methodology for carrying out microarray experiments using the yeast *Saccharomyces cerevisiae*. These protocols can be adapted to other microorganisms, including different fungi, after suitable modifications to the protocols for growing, harvesting, and breaking cells. Although microarray technology is widely used to study the transcriptome and its differential expression, this technology has also been applied in novel ways to study various aspects of cell biology, and these innovations are reviewed briefly in the latter part of this chapter.

45.2. MICROARRAY TECHNOLOGY

Central to the microarray technology is probe-target hybridization as practiced in Southern and Northern blot hybridization experiments. A microarray or a chip contains thousands of cDNAs or oligonucleotides printed on a substrate glass slide in a high-density array with defined x-y coordinates. It is important to note that in the microarray nomenclature, immobilized DNA on a glass slide is the probe on the microarray, and fluorescently labeled cDNA or cRNA synthesized from total or poly(A)$^+$ RNA is the target nucleic acid.

Several microarray platforms are available, including the short oligonuecleotide chips from the company Affymetrix, spotted cDNA, spotted long oligonucleotides (50- to 60-mer sequences), and in situ-synthesized long oligonucleotides using either an ink-jet technology (26) from the company Agilent or a maskless photodeprotection technology (65) from the company Nimblegen. In this chapter, we discuss some of the most commonly used microarray platforms. The Affymetrix microarray platform is made of 25-mer oligonucleotides (43) and contains perfect-match probes and corresponding mismatched probes for each transcript. Several such pairs of oligonucleotides spanning each transcript are used, with the perfect-match probe being used as a reporter and the mismatch probe used as a control for hybridization specificity (43). The cDNA microarray platform contains PCR-amplified DNA representing either the complete open reading frame (ORF) or relatively large portions of the ORF, e.g., 500 to 600 bp from the 3' end of each ORF. Microarrays containing 50- to 70-mer oligonucleotides representing each ORF are increasingly favored because cumbersome PCR amplification steps are not required and ready-to-spot oligonucleotides can be custom-synthesized or purchased, albeit at a high cost. The PCR-amplified ORF probes suffer from defects due to cross-hybridization between homologous or orthologous genes, whereas the long oligonucleotide probes can be designed to minimize such cross-hybridization and therefore have higher selectivity (26). More recently, high-density long-oligonucleotide microarrays have been developed with up to 385,000 oligonucleotide features synthesized in situ on glass slides. High-density genomic tiling microarrays cover a complete genome or a large fraction of it with densely tiled oligonucleotide probes. These arrays are used for interrogation of transcriptome profiles, DNA-protein binding (section 45.3.6), and comparative genome hybridization (section 45.3.7). Of course, the use of oligonucleotide arrays is limited to sequenced genomes. In cases where genome sequence is not available, DNA amplified from expressed sequence tag libraries (5) can be spotted to generate microarrays. Moreover, for smaller fungal genomes such as that of *Aspergillus terreus*, it has been possible to use a short restriction fragment library (average size, ~2 kb) as probes deposited on glass microarrays (3). However, a major limitation of the arrays made from expressed sequence tag and genomic DNA libraries is probe redundancy, such that complete genome coverage requires a severalfold increase in the number of clones that must be spotted to the microarray.

Two approaches are used for assessing differential gene expression using the microarrays. The short-oligonucleotide platform employs hybridization of a biotin-labeled cDNA or cRNA derived from a single test RNA sample to a single glass microarray. The hybridized target is then stained with phycoerythrin (43), referred to as single-color hybridization, and the intensity of the signal provides a direct estimate of the concentration of the transcript. The reference RNA sample is treated similarly and hybridized to a duplicate microarray. The ratio between the fluorescence intensities emanating from experimental and the reference target hybridizations is calculated for each feature on the microarray.

In this chapter, we provide detailed methodology for expression profiling using the cDNA platform and dual-color hybridization. The two RNA samples, one derived from experimental RNA and the other from the control RNA sample, are converted to cDNA and labeled individually using fluorescent dyes such as cyanine 5 (Cy5) and cyanine 3 (Cy3). Next, the two labeled samples (targets) are combined in equal proportions and hybridized to a single microarray slide (Fig. 1), therefore referred to as dual-color hybridization, involving competitive hybridization, i.e., simultaneous hybridization of the reference and experimental targets to a single microarray (12, 60). The fluorescent signal emanating from each feature on the microarray slide is quantified by scanning the slide in a microarray scanner. The ratio of the fluorescence intensities emanating from the experimental versus control samples is then subjected to extensive statistical analyses to obtain the relative expression level of each gene on the microarray (Fig. 1).

45.2.1. Microarray Design and Fabrication

Microarray technology is exquisitely sensitive such that often changes in gene expression as small as 1.2-fold can be reliably detected. Therefore, it is very important to devote careful attention to the experimental design in order to minimize systematic noise emanating from cultures and handling of samples. For instance, replicate hybridizations exhibited smaller-magnitude fluctuation than the biological variability of separate cultures (see reference 48, Fig. 1). It has also been observed that two seemingly identical wild-type cultures with no treatments and subjected to identical medium and growth conditions displayed small but significant differences in transcript levels of certain genes (see reference 27, Fig. 2). In addition, DeRisi et al. showed that relatively small differences in cell densities can have dramatic consequences, such as in expression of genes involved in sugar metabolism (12). Therefore, experiments have to be devised carefully when comparing strains that have different growth rates or when comparing different culture conditions. Although the procedures below may seem excessive to those unfamiliar with the extreme sensitivity of microarrays, they are designed to minimize variability of results obtained in nominally identical replicate experiments. There are several other important considerations for optimal design of microarray experiments, including adequate technical and biological replication of experiments and correct choice of pair of samples for hybridization, and they are discussed in detail elsewhere (9, 78).

A critical issue with the microarray technology is the choice of experimental controls. Both positive and negative

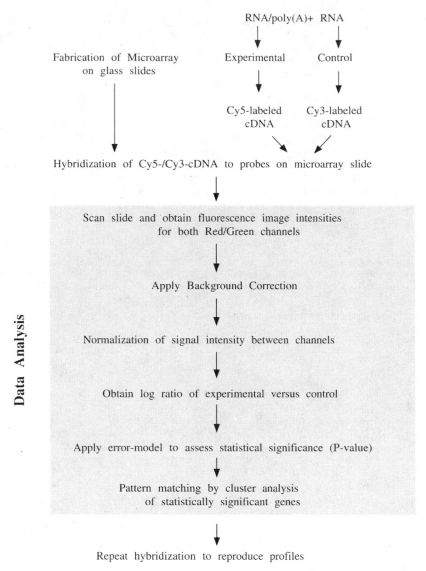

FIGURE 1 Overview of DNA microarray scheme.

hybridization controls have to be carefully chosen ahead of time and printed alongside the probe spots during fabrication of the microarrays. Positive controls on the microarray represent housekeeping genes and often serve as normalization controls (see also section 45.2.8.2). Typically, negative controls are either synthetic artificial sequences or sequences derived from heterologous organisms. Thus, a yeast microarray could contain control spots representing plant or vertebrate genes with no sequence homology to any yeast genes. Randomized versions of oligonucleotide sequences derived from the same organism may also serve as negative controls in the oligonucleotide microarrays.

The PCR-amplified DNA or oligonucleotides can be deposited on glass slides by robotic spotter instruments. These can be fabricated in-house, as pioneered in the Pat Brown laboratory, or purchased from commercial sources. A detailed comparison of the features of the commercial spotter instruments can be found elsewhere (25). Details pertaining to various aspects of the fabrication of a robotic microar-

rayer are described at Pat Brown's lab website (http://cmgm.stanford.edu/pbrown/mguide/index.html). For our studies, a custom-built spotter with 16 quill pins was used to deposit PCR products corresponding to >6,000 complete ORFs from the *S. cerevisiae* genome within a 2- by 2-cm^2 area on standard laboratory glass slides treated with poly-L-lysine. The spotted slides were postprocessed using standard protocols (reference 5 and http://cmgm.stanford.edu/pbrown/protocols/index.html). However, to accelerate research using microarrays, ready-to-use prearrayed microarray slides containing PCR-amplified DNA or synthetic long oligonucleotides can be purchased from several companies or from nonprofit microarray facilities worldwide.

45.2.2. Growth of Yeast Cells
If the microarray experiment involves comparison of two different strains, both experimental and control strains should be revived in parallel from a −80°C freezer stock. The strains can be stored on plates for up to 1 week at 4°C.

For strains harboring growth defects, it is advisable to determine the generation time prior to actual experiments. It is important to also note that the apparent optical density at 600 nm (OD_{600}) and cell density (cells per milliliter) can be different for strains of different ploidy. In our laboratory, 1 OD_{600} unit corresponds to 2×10^7 cells/ml for haploid wild-type *S. cerevisiae* BY4741. For each strain or condition under examination, a control strain should be assigned in advance that is processed strictly in parallel with the experimental sample, from cell growth and harvesting to the hybridization steps.

1. Streak out strain(s) to obtain single colonies on a YPD plate (1% yeast extract, 2% peptone, 2% glucose, 2% Bacto agar) or the most nutrient-rich medium permissible, and grow for 2 days at 30°C.

2. Resuspend a single colony of about a 3-mm diameter in 1 ml of synthetic complete (SC) medium (48).

3. Inoculate 5 ml of SC medium with 50 μl of the resuspended colony, and grow overnight for 10 to 14 h such that this culture is still in logarithmic phase of growth. However, for strains with growth defects, 250 μl of the resuspended colony suspension may be used to achieve higher initial cell density such that the final cell densities of wild-type and mutants are comparable.

4. Measure the OD_{600} of the overnight culture on the following day, and use only those cultures with an OD_{600} of less than 1.0 to inoculate cultures for microarray analysis. The overnight cultures are diluted in 50 ml (for small scale) or 250 ml (for large scale) of prewarmed SC medium to a starting OD_{600} of 0.05 to 0.1 and incubated in an air shaker at 30°C for 6 to 7 h at 220 rpm. Cells are harvested when cultures reach a final cell density of about 0.6 OD_{600} unit (1.2×10^7 cells/ml), as outlined in the next section.

45.2.3. Cell Harvesting

When cells are ready to be harvested, try to work as quickly as possible to minimize the time during which any changes in gene expression may occur. Therefore, before the cultures reach the required cell density, label the requisite number of 50-ml conical tubes to collect cells and obtain liquid nitrogen for freezing the cell pellets. It is also critical to maintain the temperature for cell harvest as close to the culture conditions as possible in order to minimize gene expression changes. Remove flasks from the shaker in pairs that will end up later in the same hybridizations. The number of cultures to be processed in parallel will also depend on the type and capacity of the centrifuge or the harvesting method to be adopted. We provide two methods for cell harvesting. For experiments involving time course studies, such as cell cycle analysis, drug treatments, or temperature shifts, we highly recommend the membrane filtration method for harvesting cells. For all of our single-time-point experiments, we have harvested cells by centrifugation.

To harvest cells by filtration, we use the Millipore polyvinylidene difluoride membrane (0.45-μm pore size) in a membrane filtration device (Millipore). Both the membrane filter and the filtration device are to be presterilized by autoclaving for 15 min. If multiple strains are to be handled in parallel, it is helpful to use a multiple-filtration manifold, also available from Millipore.

After the cultures reach the appropriate cell density or at defined time intervals, cells are collected by quickly passing each culture through a membrane filter connected to a suitable vacuum source. When all of the liquid is removed, apply vacuum for one additional minute, and then place the membrane filter in a polypropylene conical centrifuge tube, flash-freeze it briefly in liquid nitrogen, and store at −80°C until RNA isolation.

To harvest cultures by centrifugation, transfer small-scale cultures to 50-ml conical tubes and immediately spin at 3,000 to 4,000 × *g* in a tabletop swinging-bucket centrifuge for 2 min or the minimum time required to pellet cells. When the centrifuge has stopped, remove up to six tubes at a time from the centrifuge and process them to completion (three experimental-control pairs). For each tube, decant the supernatant and invert on Parafilm or paper towels and proceed to the next tube. After the first six tubes have been drained, and while maintaining the inverted orientation, flick each tube twice to remove any remaining liquid and re-cap the tube. Freeze six tubes at a time by immersing the bottoms of the tubes in liquid nitrogen for about 5 s. Immediately transfer the tubes to a rack in the −80°C freezer. Repeat for the next six tubes.

For large-scale cultures, harvest the cells by centrifugation in 250-ml bottles in the GSA rotor (in a Sorvall RC5C centrifuge) for 2 min at 7,000 rpm at ambient temperature. Decant the supernatant, place the bottle at a slight angle to facilitate drainage of liquid from the pellet, and aspirate residual liquid with a micropipette. Resuspend cells in 4 ml of AE buffer (see section 45.2.4), transfer to a 50-ml conical centrifuge tube and pellet cells as described above. The cell pellets can be stored frozen as described above or can be used immediately for RNA isolation.

45.2.4. Preparation of Total RNA by the Hot-Phenol Method

The following protocol is essentially derived from the method described previously (35), and it relies on the principle of selective fractionation of DNA into the phenol phase and interphase at acidic pH while retaining RNA in the aqueous phase. For isolation of RNA from actively budding yeast cells, glass beads are not required in this protocol. However, for ascospore or mycelial cultures, 3.5 to 4 volumes of glass beads are added along with phenol. The following protocol is for 50-ml cultures. For each sample, combine 400 μl of AE buffer (50 mM sodium acetate [pH 5.3], 10 mM EDTA, 1% sodium dodecyl sulfate [SDS]) with 400 μl of phenol equilibrated with sodium acetate buffer (50 mM sodium acetate and 10 mM EDTA, adjusted to pH 5.3) in a set of microcentrifuge tubes. Prepare also a second set of tubes with 400 μl of phenol-chloroform (1:1) mixture. Hold all tubes in a water bath at 65°C.

1. To the cell pellets (in 50-ml tubes, either fresh or frozen) immediately add 800 μl of hot phenol-AE buffer mixture, vortex briefly, and incubate in a 65°C water bath. If cells were harvested by membrane filtration, wash the cells from the membrane with the 800 μl of hot phenol-AE buffer, rinsing three or four times, to completely strip the membrane filter of all cells. Remove the filter from the tube and discard. Incubate resuspended cells for 10 min at 65°C. Vortex for 30 s several times (typically three or four times) during the incubation period, and return tubes to 65°C. Transfer contents from large tubes to new microcentrifuge tubes. Vortex briefly to keep the cell-phenol mixture resuspended.

2. Cool down tubes to room temperature by placing in an ice-water bath for 2 min. Avoid cooling the tubes below 15°C, as this will lead to precipitation of SDS. Spin in a microcentrifuge (16,000 × *g*) for 10 min.

3. Transfer aqueous supernatant to the microcentrifuge tube containing 400 μl of phenol-chloroform mixture (prewarmed to 65°C). Vortex for 10 s, and spin in the microcentrifuge for 10 min. Collect aqueous supernatant and process as described in section 45.2.4.1.

45.2.4.1. Purification and Concentration of RNA

The following procedure is employed to remove residual phenol-chloroform by repeated exchange with 10 mM Tris-HCl–1 mM EDTA (TE) buffer (pH 7.5) and to concentrate the RNA without the need for an ethanol precipitation step.

1. Transfer aqueous supernatant directly to a prewetted Microcon YM-30 concentrator. Spin in a microcentrifuge (16,000 × g) for 20 min, or until the volume is reduced to <100 μl. The volume remaining in the filter tube can be estimated by monitoring the level of the liquid above the white section of the Microcon filter. For reference, use a filter tube containing 100 μl of water. If the volume is not reduced to <100 μl after 20 min, the filter is probably clogged. In that event, transfer the contents to a new Microcon filter. Because most of the RNA will be in a gelatinous clump, be sure to resuspend it by repeated pipetting before transferring it to a new filter tube.

2. Add 400 μl of TE buffer to the Microcon concentrator and spin for 15 min as described above. The volume should now be ~20 to 50 μl.

3. Repeat step 2 twice and concentrate the sample to ~20 to 30 μl. At this point the samples should be devoid of phenol.

4. Invert the Microcon concentrator into a microcentrifuge tube and spin for 10 s at full speed to collect the RNA. If samples were divided into different Microcon tubes, collect them all in the same tube. Estimate the volume recovered and adjust it to about 40 μl with TE buffer and mix well.

5. The concentration of RNA may be estimated by measuring the OD_{260} (58). The sample will be viscous but can be pipetted accurately after brief heating to 65°C. At this stage, the RNA can be immediately carried to the labeling step or stored at −80°C. The RNA yield should be 400 to 500 μg from 50 ml of cell culture.

45.2.5. Isolation of Poly(A)$^+$ RNA

For studies with *S. cerevisiae*, total RNA can be used directly as a template for cDNA synthesis. However, additional purification steps during poly(A)$^+$ RNA isolation have the benefit of eliminating contaminants that inhibit cDNA synthesis. In addition, random primers can be used for cDNA synthesis from poly(A)$^+$ RNA.

1. Preparation of oligo(dT) columns. Weigh out 0.7 g of oligo(dT)-cellulose (New England Biolabs) (sufficient for 12 to 14 0.25-ml columns) and resuspend in 0.1 N NaOH. Mix on a rotary mixer such as the Nutator (Clay Adams) for a few minutes. To a 0.8- by 4-cm sterile polypropylene column (such as from Bio-Rad), pack the slurry by gravity flow until a 0.25-ml bed volume is achieved. Wash the column with 4 ml of diethyl pyrocarbonate (DEPC)-treated water (58), followed by washing with 2 to 4 ml of 1× column loading buffer (20 mM Tris-HCl [pH 7.6], 0.5 M NaCl, 1 mM EDTA, 0.1% *N*-lauryl sarcosine, sodium salt), or until the pH of the eluate is ~8.0.

2. To denature the RNA, heat sample at 65°C for 5 min, followed by quick chilling on ice for 3 min. An aliquot

of total RNA may be saved at this stage to examine by gel electrophoresis.

3. Add an equal volume (40 μl) of 2× column loading buffer to the RNA sample prepared as described in section 45.2.4.1, and apply to the column by gravity flow. When all of the liquid has entered the column, wash the column with 1 column volume of 1× loading buffer by gravity flow. Collect flowthrough and reload on the column two additional times. Prior to each reloading, reheat the samples at 65°C for 5 min, followed by cooling for 1 min on ice, to improve the yield.

4. Wash with 8 ml of 1× column loading buffer. Discard flowthrough.

5. Wash column with 2 column volumes of middle-wash buffer (20 mM Tris-HCl [pH 7.6], 150 mM NaCl, 1 mM EDTA, 0.1% sodium lauryl sarcosine) and allow buffer to drain from the column completely.

6. Elute poly(A)$^+$ RNA with 1.2 ml of elution buffer (10 mM Tris-HCl [pH 7.6], 0.1 mM EDTA, 0.005% sodium lauryl sarcosine), preheated to 65°C, by adding 400 μl of elution buffer to the column in three aliquots and collecting in the same tube. The resulting poly(A)$^+$ RNA preparation is then purified on the same oligo(dT) column for a second time as follows to provide a substantial enrichment for mRNA.

7. Wash the column with 4 ml of DEPC-treated water and with 2 to 4 ml of 1× loading buffer. Meanwhile, the eluate from the first purification is heat denatured as in step 2. Mix with an equal volume of 2× loading buffer and load as in step 3. Collect flowthrough and reload as in step 3. Remember to heat samples as in step 2 except for cooling on ice for 2 min. Wash with 5 ml of 1× loading buffer, followed by washing with middle-wash buffer as in step 5.

8. Elute mRNA with two 400-μl aliquots of elution buffer in separate microcentrifuge tubes. To each tube of eluate, add 44 μl of 3 M sodium acetate (pH 5.2), 4 μl of 2.5-μg/μl linear acrylamide, and 1.1 ml of ethanol and mix well. Store at −80°C for at least 30 min. Note: Linear acrylamide is used as a neutral coprecipitating agent (18) and can be prepared in the laboratory (5) or purchased (Ambion).

9. Spin for 60 min in a cold microcentrifuge at top speed, and aspirate the supernatant carefully.

10. Wash pellet with 75% ethanol as follows. Add 500 μl of cold (−20°C) 75% ethanol, being careful not to dislodge the pellet. Aspirate the supernatant and discard. Remove the residual liquid with a micropipette tip and allow to air dry for 5 to 10 min. Resuspend in 10 μl of DEPC-treated water or 2× TE buffer (20 mM Tris [pH 7.6], 0.2 mM EDTA). To quantify by spectrophotometry, dilute 1 μl in 60 μl of TE buffer, and read in a low-volume (e.g., 50-μl) cuvette.

11. Regeneration and storage of columns. Wash column with 4 to 6 ml of DEPC-treated water and store in 0.5 ml of water at 4°C. Columns can be used up to three times after being regenerated with 2 ml of 0.1 N NaOH, followed by equilibration in column loading buffer, as in step 1 above. The typical yield of poly(A)$^+$ RNA from *S. cerevisiae* is 1.5 to 2.5% of total RNA.

45.2.6. Synthesis of Fluorescently Labeled cDNA

There are two methods for producing cDNA labeled with fluorescent dyes. In the direct-labeling method, Cy3- or Cy5-deoxynucleoside triphosphate (Cy5-dNTP) is incorporated into cDNA during reverse transcription using total

RNA or poly(A)$^+$ as a template. If the biological material is in limiting amounts, one or two RNA amplification steps could be introduced as described by Van Gelder et al. (72), and the amplified RNA is reverse transcribed prior to labeling. The direct-labeling method is relatively simple and was extensively used in microarray labeling. However, it is now clear that there are several limitations with this method. Because some reverse transcriptase enzymes are sluggish to incorporate the bulky fluorescent dyes into cDNA, there is low fluorescence intensity of labeled cDNA molecules. Another limitation is that Cy3-dNTP and Cy5-dNTP are not incorporated with equal efficiencies by the reverse transcriptase. Therefore, dye-swap experiments are routinely practiced now, wherein the Cy dyes are interchanged between experimental and control RNA samples so that the dye bias can be corrected by averaging the data from the two independent hybridizations. Although this increases the cost for microarray experiments, it is an effective way to perform a replicate measurement while correcting a known systematic bias. The second labeling method is a two-step indirect-labeling procedure in which aminoallyl-dUTP is first incorporated into cDNA during reverse transcription and then the resulting primary amine groups of the amino allyl residues are nonenzymatically conjugated to N-hydroxysuccinimidyl (NHS)-esterified Cy3 or Cy5 (26). This method overcomes both of the limitations associated with the direct-labeling method. The reagents for microarray target labeling are available in kit formulations from several vendors. Indeed, the indirect-labeling method is being adopted as the preferred method in several laboratories, despite being a more demanding technique due to the additional labeling steps (see section 45.2.6.3.2).

45.2.6.1. Direct Fluorescent Dye Labeling during Reverse Transcription

1. Prepare two reaction premixes, one for Cy3 and one for Cy5. For n number of RNA samples, the size of the premix is 11.5 μl × (n + 0.5). Each reaction premix of 11.5 μl is prepared as follows:

6 μl of 5× first-strand buffer (Superscript II reverse transcriptase kit)

0.5 μl of 50× dNTPs (10 mM dTTP and a 25 mM concentration each of dATP, dGTP, and dCTP)

3 μl of 0.1 M dithiothreitol

2 μl of Superscript II enzyme

2. For each RNA to be assayed, dispense 1 to 2 μg of poly(A)$^+$ RNA into two microcentrifuge tubes. Add 2 μl of 2.5-μg/μl oligo(dT)$_{20}$ and adjust the volume to 15.5 μl with DEPC-treated water (58) for each tube. Incubate at 70°C for 2 min to denature the RNA, and chill on ice for 10 min.

3. To one tube, add 11.5 μl of premix followed by 3 μl of Cy3-dUTP, mix, and incubate for 2 h at 42°C. Prepare an identical reaction mixture in the other tube using Cy5-dUTP.

4. Stop the reaction by adding 10 μl of a solution of 0.5 N NaOH–0.25 M EDTA. Mix well. Incubate at 65°C for 20 to 30 min. Neutralize the reaction by adding 20 to 25 μl of 1 M Tris-Cl (pH 7.5).

5. The quality of labeled cDNA can be judged by resolving 1 to 2 μl of the Cy5-labeled sample by electrophoresis in a 2% agarose gel (prepared in 1× TAE buffer) followed by fluorescent imaging on a fluorescence scanner equipped for 630-nm excitation and 650-nm emission (e.g., GE

Typhoon). Successful labeling would yield a smear of labeled cDNA ranging from about 400 to at least 2,000 nucleotides long. The concentration of cDNA and the amount of incorporated dye can be quantified by spectrophotometry (5).

45.2.6.2. Purification of the Labeled cDNA

Add 400 μl of TE buffer to the reservoir of a Microcon YM-30 concentrator (Amicon/Millipore). Transfer the Cy3 and Cy5 reaction products to separate Microcon filter tubes, and mix briefly by repeated pipetting up and down. Spin the tube in a microcentrifuge at full speed for 8 min or until the volume is reduced to ~20 μl. Add 400 μl of TE buffer again and spin as described above to reduce the sample volume to ~20 μl. Recover sample by centrifuging the filter tube in an inverted position into a sample collection tube. Store at 4°C in the dark, or proceed directly for hybridization.

45.2.6.3. Two-Step Labeling of cDNA Using Aminoallyl-dUTP and Coupling to Cy Dyes

45.2.6.3.1. Reverse Transcription

1. Prepare 12.2 μl of the reaction premix as follows:

6 μl of 5× first-strand buffer (Superscript reverse transcriptase kit)

1.2 μl of 25× dNTP–aminoallyl-dUTP labeling mix (a 12.5 mM concentration each of dATP, dCTP, and dGTP; 10 mM aminoallyl-dUTP and 2.5 mM dTTP)

3 μl of 0.1 M dithiothreitol

2 μl of Superscript II enzyme

2. Set up one reaction each for sample RNA and a separate reaction for control RNA. To anneal primer, mix 1 to 4 μg of poly(A)$^+$ RNA with 5 μg of oligo(dT)$_{20}$ (2 μl of 2.5 μg/μl) in a total volume of 17.8 μl. Heat to 70°C for 10 min. Cool on ice for 10 min.

3. Combine the reaction premix with the RNA sample (step 1), mix well, and incubate the reaction mixture for 2 h at 42°C.

4. To each reaction mixture, add 3 μl of 0.1 M EDTA. Mix thoroughly and add 3.5 μl of 0.1 M NaOH. Incubate the reaction mixture for 10 min at 65°C to hydrolyze the RNA.

5. Neutralize the reaction mixture by the addition of HEPES (pH 7.0) to a final concentration of 500 mM.

6. At this stage, the aminoallyl-cDNA is to be purified from unincorporated aminoallyl-dUTP, hydrolyzed RNA, and free amines. For this purification, a Microcon YM-30 device is used essentially as described above (section 45.2.6.2), with a critical difference of using water instead of TE in all steps. This is essential because free amines left in the sample would interfere with the subsequent coupling step. The samples are concentrated to a volume of ~10 μl.

45.2.6.3.2. Nonenzymatic Covalent Coupling and Purification of cDNA

The Cy5-NHS ester and Cy3-NHS ester postlabeling reactive dye packs are available from GE Healthcare (product no. RPN 5661). These are available in single-use aliquots and therefore are more stable and convenient.

1. To the aminoallyl-UTP-labeled cDNA samples from step 45.2.6.3.1, add 1.5 μl of freshly made 1 M sodium bicarbonate (pH 9.0). Remove an aliquot each of NHS-Cy3 and NHS-Cy5, add the bicarbonate-buffered cDNA, and

resuspend the pellet vigorously. Gently but thoroughly mix by repeated pipetting. Incubate for 1 h at ambient temperature in the dark and wrapped with foil, with intermittent mixing by pipetting every 15 min.

2. Stop the reaction by quenching the unreacted NHS-esterified Cy dye molecules by addition of 3.6 μl of 4 M hydroxylamine (Sigma) for a final concentration of ~1.5 M. Mix well and allow reaction to continue for 15 min in the dark.

3. At this point, the efficiency of cDNA synthesis and labeling can be judged by gel electrophoresis as outlined in section 45.2.6.1, step 5.

4. To remove unincorporated/quenched Cy dye, purify the DNA using QIAquick columns available as part of the PCR purification kit (Qiagen). The method described below is as specified by the manufacturer.

5. To the Cy3 and Cy5 reaction mixtures in separate tubes, add 70 μl of water and mix.

6. Add 500 μl of PB buffer, apply to a QIAquick column, and spin at 12,000 × g in a microcentrifuge for 1 min.

7. Decant flowthrough. Add 750 μl of PE buffer and spin at 12,000 × g for 30 to 60 s. Decant flowthrough. Spin as described above to dry the filter unit, and transfer to a fresh microcentrifuge tube.

8. Add 30 μl of EB buffer to the center of the filter and incubate for 3 min at room temperature. Spin at 12,000 × g for 1 min. Repeat elution step again with another 30 μl of EB buffer.

9. Pool eluates and add 20 μl (20 μg) of herring sperm DNA. Add 420 μl of TE buffer (pH 8) and apply to a fresh Microcon YM-30 filter to concentrate the dye-labeled cDNA. Spin at 12,000 × g to reduce the volume to 20 μl.

10. The labeled cDNA can be used for hybridization to microarrays as described in section 45.2.7. We note that other laboratories use slightly different versions of the indirect-labeling protocol. The labeled cDNA and amount of dye incorporation can be quantified by spectrophotometry (5).

45.2.7. Hybridization and Processing of Microarrays

Microarray hybridization can be performed either in an automated hybridization instrument or in a simple manual device. The automated hybridization instruments are available from many vendors and seem to offer many advantages, such as good reproducibility, low hybridization volume, the option of reusing hybridization solutions, and high-throughput capability of handling 12 to 24 slides in parallel. Detailed instructions come with the instruments. Hybridization can also be performed manually either in a humidified box or using a hybridization cassette device, which can be purchased from many vendors, including Telechem International. Below we describe a procedure for manual hybridization carried out in the hybridization cassette. It is of utmost importance to prevent dust contamination. The microarray slide should not be allowed to dry during any of the steps after initiating the hybridization experiment, since the labeled target may stick irreversibly to the microarray. A constant humidity has to be maintained in the chamber since high humidity tends to flood liquid around the coverslip and reduces signal strength, whereas low humidity will evaporate the hybridization fluid. Suppression of bubble formation between slide and cover glass is also very critical for the success of hybridization. For hybridization, equal amounts of Cy3 and Cy5 incorporated dyes are combined. Typically, about 40 pmol of each dye-labeled DNA is used for hybridization. The hybridization is performed in 3× SSC (1× SSC is 0.15

M NaCl plus 0.015 M sodium citrate)–0.75-mg/ml poly(dA) (GE Healthcare)–0.2% SDS for 12 to 16 h at 63°C as follows.

1. To dye-labeled cDNA (from sections 45.2.6.2 and 45.2.6.3), add 4 μl of 20× SSC and 2 μl of 10-mg/ml poly (dA) and make up volume to about 25 μl. Poly (dA) of 40 to 60 dATP bases in length promotes specific hybridization between the labeled cDNA and probe DNA on the chip by annealing to the poly(T) tract in the labeled cDNA. Note that SDS is added only after the filtration step.

2. Remove all particulate material from the hybridization buffer, by filtration through a Millipore 0.22-μm-pore-size centrifugal membrane filter tube as follows. Wet the membrane with 20 μl of sterile water. Spin briefly in a microcentrifuge (up to 60 s) at 8,000 × g, then decant water from the bottom collection tube. Add labeled target to the side of the filter tube (avoid extended contact of probe with membrane). Spin briefly (30 to 60 s) at 8,000 × g. Typically the recovery is approximately 26 μl.

3. Transfer flowthrough to a fresh PCR tube. Add 0.4 μl of 10% SDS. (Do not put on ice after adding SDS.) To denature labeled cDNA, heat to 94°C for 2 min, and transfer to a water bath at room temperature for 3 to 4 min.

4. The hybridization cassette is prepared as follows. Ensure that the clamps fit snugly to the frame. While the labeled target is cooling, add a 10-μl drop of water to the bottom well of the hybridization chamber and load the microarray slide in the hybridization chamber. Add the entire labeled target to the top of the microarray slide and with Millipore forceps quickly apply at an angle the 22- by 30-mm coverslip (no. 1 thickness), avoiding bubbles. Close the hybridization chamber and incubate in a water bath for 12 to 16 h at 63°C.

Following hybridization, slides are washed essentially as reported previously (46). Briefly, remove the hybridization chamber from the water bath and blot dry. Lay the cassette (top of slide facing up) on a clean bench surface and disassemble the hybridization cassette. Care is required so that the printed spots on the slide are not damaged by the forceps. Remove slide from chamber with forceps and immediately immerse in wash solution I (1.1× SSC, 0.03% SDS) at room temperature until the cover glass comes off. If only a few slides are being processed, this can be done individually by hand; otherwise, we recommend placing the slides (up to 20), with coverslip still attached, into a slide holder. Dip them all simultaneously into the wash solution until all the coverslips come off. To minimize carryover of SDS and salt to the secondary wash, it may be better to transfer slides one by one into a new rack. Transfer the rack containing slides to a slide-staining dish containing wash solution II (0.06× SSC). Rinse vigorously by moving the slide holder up and down several times (up to 50 times). Keeping the slides submerged in the secondary wash, carry the slides to the centrifuge, remove them from the wash solution, and spin dry the slides at 600 rpm for 4 min at room temperature (centrifugation at speeds in excess of 600 rpm could break the slides). Hybridized arrays are extremely sensitive to environmental ozone (17), and once dried, the arrays are stable for several weeks if stored protected from light and ozone.

45.2.8. Image Processing and Data Analysis

45.2.8.1. Feature Extraction and Quantification

Several types of microarray scanners, along with image extraction software, are available from commercial sources. The type of instrument most commonly used is a confocal laser scanner with the ability to detect at least two fluorophores and with the additional capability to adjust laser power, photomultiplier tube (PMT) voltage, and scan resolution in the software. Optimization of laser power is important, as high laser intensity could induce photobleaching of the hybridized slide and high PMT settings may saturate pixel intensity. In practice, with a fixed laser power, the microarray is scanned using multiple PMT voltage settings independently for each channel. The PMT voltage is set so as to have the brightest pixels just below the level of saturation.

Obtaining high-quality data from microarray experiments is hampered by several technical limitations, including imperfect microarrays (especially the spotted arrays) such as nonuniform spot sizes, abnormal spot morphology (not perfectly round), often less-efficient labeling and hybridization steps, and uneven hybridization patterns (doughnut or crescent moon shapes). Thus, robust feature extraction methods are needed to minimize noise, extract the best-quality data, and identify authentic differentially hybridized array elements. In the past, feature extraction methods relied on manual techniques for flagging improper spots, but automated techniques are now available (45). There are several steps for feature extraction that form part of microarray data analysis (Fig. 1) and are briefly described here. First, a grid is superimposed over the scanned images. Upon input of vital information about the coordinates of the microarray spots on a glass slide by the user, all image extraction software can automatically determine and provide a tentative grid bearing the spot location on a slide; the user may only confirm the correct grid location and the center of each spot. Prior to quantification, the user can edit the grid to exclude spots with slide artifacts such as dust, abnormal spot morphology (i.e., pixel heterogeneity), salt residue, and spots with saturated intensities in either channel. Once the features have been identified and correctly gridded, image analysis software measures the intensities in each channel for each of the features on the microarray. Typically, we calculate the ratio of fluorescence intensities for each feature in the two channels for the Cy3-Cy5 two-color hybridization. Quantification of ratios and intensities is carried out as follows. First, the background is determined either from the signal emanating from the interfeature intensity in each of the Cy3 and Cy5 channels or from the signal values derived from a set of negative control spots outlined in section 45.2.1. Second, the intensity of each spot in the foreground is measured in each of the channels. Finally, the spot intensity is determined by subtracting the background signal value derived from step 1 above. Image processing and normalization routines have been described in detail before (7, 66, 67).

In two-color ratio measurements, one would like to calculate Cy5/Cy3 ratio value for each spot such that the ratio is a measure of RNA abundance in the two channels. Because ratios are not normally distributed, it is necessary that the ratios be log transformed. For example, we would see a normal distribution if we take log (to base 2) of ratio 2 and 0.5; we would get 1 and −1, respectively. The most common data visualization tool is a scatter plot of the log ratio data on the y axis and sum of medians on a log x axis that would reveal the performance of the microarray experiment. A scatter plot can be easily constructed in a Microsoft Excel spreadsheet or in other dedicated microarray data analysis software. For instance, the induced and repressed genes and the low-intensity spots would be clearly revealed graphically in the scatter plot. Also, the log (base 2) Cy5/Cy3 ratio of two microarray hybridization data can be plotted on a scatter plot and would reveal the behavior of the genes in the two experiments. For two hybridization experiments that are highly reproducible, the scatter plot would yield a correlation coefficient (r^2) close to ~1.0. An alternative and very useful approach to visualizing microarray data is to construct a scatter plot of the log (to base 2) ratio against the average intensity of each feature on the microarray. These plots are called an R-I plot or M-A plot because average intensity is referred to as A and log ratio as M. If the two channels behave similarly, then all points would be centered around a horizontal line passing through zero.

45.2.8.2. Data Normalization

The next step in data analysis (Fig. 1) is data normalization, which helps in removing systematic variation potentially affecting the measured level of gene expression. For example, for a self-versus-self experiment, the spot signals should be equal in both channels across the dynamic range of the intensities. However, differences in reverse transcriptase performance during incorporation of Cy dyes into cDNA in separate labeling reactions can potentially cause a differential change of signals in the two channels. Other sources for such systematic variation between signal intensities include inherent chemical differences in dyes with respect to their quantum efficiencies, variability in the amount of mRNA used for labeling, substrate fluorescence, propensity of nonspecific binding to the glass surface, possible biases introduced by microarray scanner response, i.e., differences in the power settings of the two lasers and PMT voltages, gradient effects arising from hybridization and washing steps and array printing processes.

The normalization process involves an initial probe selection step that is subsequently used for computation and optimization of dye normalization parameters. This is broadly of two types: a signal-independent, nonlocal linear regression performed across the entire normalization data set, essentially resulting in a global scaling of the data, and a signal-dependent, locally weighted linear regression method or LOWESS, which essentially normalizes the data, taking into account the consequences of any local perturbation. Normalization using mean signal intensity in each channel was practiced during the early phase of implementation of the microarray technology. It is now clear that this approach will not be successful if, for example, hybridizing signals from the bulk of the spots changed as a result of a mutation that reduced expression of the majority of the genome. Therefore, other methods have been developed that utilize a minimum set of housekeeping genes because their expression is constant across a range of samples in selected experiments. The selection of normalization probes is very important and involves use of either preselected housekeeping genes or virtual housekeeping genes that are generated from the body of microarray data itself (59, 69), and is used to select a set of invariant genes that have same the intensities in both Cy5 and Cy3 channels. However, the use of a small set of housekeeping genes is not always ideal due to the unforeseen changes in their differential expression under certain conditions.

Another approach is to spot genomic DNA samples of increasing concentrations from the same organism on the microarray. This has been employed for yeasts but may also be useful for other organisms with a low proportion of noncoding sequences in the genome. In cases where this is not feasible, a concentration series of probes comprising a pool of the cDNAs or oligonucleotides can be incorporated as control spots on the same microarray. However, a limitation of this approach is the possibility of overrepresenting a few highly expressed genes that could confound the signal values. The use of external standards, or spike-ins, is efficient and a good option, since they serve as a basis to correct for RNA recovery, dye-labeling efficiency, and other technical limitations; importantly, this method does not assume a priori that the total RNA content is the same in the reference and experimental populations (2, 26, 60, 71). For instance, spots representing five to seven *Bacillus subtilis* genes that do not cross-hybridize to the yeast genome were spotted on the microarray as normalization control probes. Then their cognate mRNAs were transcribed in vitro and known amounts were added to the reference and test RNA populations prior to mRNA isolation, cDNA synthesis, and labeling (2, 71). Although a single recommendation cannot be made as to the choice of the normalization method, it is often helpful and prudent to consider various options and test the most suitable ones before settling with a normalization method.

45.2.8.3. Statistical Tools Accelerate the Identification of True Expression Changes

The core DNA microarray technology was established several years ago, but methods for microarray data analysis are still evolving. Several new software tools have been developed recently that aim to increase the reliability of detecting true expression changes based on robust statistical criteria. In this section, we briefly review some of the steps in microarray data analysis.

A key initial step in handling microarray data is to assess the reproducibility of the data by repeating each experiment at least three times (i.e., from start to finish, from the same glycerol stock of the yeast strain) and examining the reproducibility of profiles in the multiple experiments using either scatter plot or clustering algorithms. The minimum number of repeat experiments is often determined by the quality of data and the magnitude of expression change. Next, the question becomes, what is a statistically meaningful change? That depends on the level of "noise" in the assay. The noise is broadly composed of biological noise reflected in RNA preparation, which can be assayed by replicates, and platform noise that emanates from processing of RNA samples. The individual sources of noise can be sorted out with analysis of variance (ANOVA), a statistical tool to determine variability in multiple data (11). The ANOVA model is applied to transformed intensity data obtained from a logarithmic transformation of raw intensity data. It allows one to account for sources of variation in the data that are attributable to factors other than differential expression of genes; thus, it effectively normalizes the data. A software implementation of ANOVA for microarray data analysis is also available (76).

Significance analysis of microarray (SAM) is another statistical approach to determine how likely it is that the gene is expressed differently between the two cell populations (70). For instance, SAM assigns each gene a Δ score based on both its level of expression and reproducibility. All genes whose scores are higher than a selected threshold are then deemed significant. SAM also calculates a false-discovery rate for each threshold of significance. This value estimates the percentage of the genes with scores higher than a given threshold that are likely to be false positives. SAM is a very popular statistical tool, and the software is available for download in the public domain (70).

A rigorous way to deal with inherent noise in the microarray data is by examination of control "same-versus-same" experimental data as reported earlier (27). In a seminal study, Hughes et al. (27) carried out 63 control pairs of experiments performed in epoch with the actual experiments, wherein expression profiles from two untreated isogenic wild-type yeast cultures grown in parallel were compared to each other. Remarkably, at least one gene exhibited ≥2-fold change (up- or down-regulated) in 55 of the 63 experiments (27). Examination of these genes revealed that many were known to be regulated by nutritional quality or stress. These fluctuations were considered a form of biological noise, and this information was used to derive an error model so as to reduce the significance of expression changes for these groups of genes. Indeed, data analysis of the 300 compendium experiments showed that expression of these groups of genes fluctuated in most of these experiments, indicating that the same biological noise was inherent in the 300 experiments (27). These results emphasize the importance of negative control experiments involving comparisons of identically treated cells and the need for a gene-specific error model. The goal in the implementation of an error model is to determine variance when "truth" is known, i.e., all expression ratios are equal to 1.0. Deviations from 1.0 are noise. The error model assigns a statistical significance numerically expressed as a P value for the expression ratio of each spot on the array, which is applied to all similarly performed experiments with the assumption that those experiments will experience similar errors. A range of P value cut is employed, often with a P value of ≤0.01 (a confidence level of 99% or more). If a given gene produces a ratio greater than seen in same-versus-same control experiments, the change is assigned a very small P value, indicating a very low likelihood that the change was due to chance and this represented a true expression change. Many commercially available feature extraction software packages apply error models, and a comparative study of the performance of the different error models is also available (55). A detailed comparison of the features of some of the commercial software suites for microarray data analysis is also available (13).

The above sections described a numerical basis to obtain statistically significant data that minimized systematic errors and biological variability. But true expression change can often be verified by cleverly designed follow-up experiments. Numerous microarray studies have revealed that yeast cells, upon exposure to a variety of perturbations, elicit large changes in the mRNA expression profile (for examples, see references 8, 12, 19, 20, 27, 33, and 48). Such expression changes could be a direct effect of the perturbation or due to secondary effects generated in the experimental situation. It is therefore necessary to identify what constitutes a true biologically relevant expression change. Although not infallible, time course studies have the potential to minimize secondary effects by focusing on early events during gene expression reprogramming. In addition, use of a genetic strategy such as mutant analysis or regulated gene expression to validate the results would help to authenticate the biological relevance of the microarray data. The magnitude of expression change is not the sole indicator of a biologically relevant phenomenon. Upon obtaining statistically significant data, the next task is to assess the expression pattern of known genes and infer the functions of unknown genes from their expression patterns. These analyses were pioneered using cluster analysis (16), wherein genes are sorted and arranged on the basis of their expression patterns across a range of experiments. The rationale for this approach is that genes that function in a common pathway are often co-regulated in different experimental conditions. Cluster analysis has three steps: (i) applying significance cuts to the expression data matrix, (ii) defining a similarity (or dissimilarity) measure matrix, and (iii) performing clustering based on the similarity matrix. In two-dimensional cluster analysis, gene clustering and experiment clustering are performed independently, without interference between the two dimensions. It is noted that formalisms for gene clustering and experiment clustering are symmetric between genes and experiments. A detailed description of cluster analysis can be found elsewhere (16, 54, 63).

45.2.8.4. Microarray Data Reproducibility, Standards, and Databases

The reproducibility of gene expression measurements using the different microarray platforms has been a matter of concern. Several attempts were made to compare data obtained using one platform with another using the same sample RNA. For example, using the 60-mer ink-jet-synthesized oligonucleotide microarray and spotted cDNA microarray platforms, transcriptome differences were studied in yeast cells grown in sporulation medium versus synthetic complete medium (26). Remarkably, transcript abundance ratios correlated exceptionally well between measurements in the two platforms. However, in a different study, same pairs of RNA samples were analyzed in three different platforms involving single-color or dual-color hybridizations. While the results of the two single-color platforms largely correlated well, little overlap was found between genes that showed significant gene expression change between one-color and the dual-color hybridization platforms (68). Systematic comparisons were carried out on a monumental scale by several laboratories both outside and within the framework of the Microarray Quality Control Project. The reproducibility of microarray data emanating from multiple microarray platforms and laboratories was evaluated (4, 30, 38, 51, 64). In general, it was concluded that the use of standardized protocols, RNA reference samples, and careful design of experiments analysis methods yielded quite comparable data across different microarray platforms and across different laboratories.

As microarray data are accumulated at a rapid pace, archival expression data become important. New insights could be obtained by analyses of the new microarray data alongside previous expression data. Moreover, accumulation of expression profiles from a large number of separate conditions facilitates the use of clustering algorithms to reveal coordinately regulated genes (16). It was realized that for reliable comparison of microarray data, there was a need for standardized methods to describe and communicate microarray experiments and to develop centralized databases to facilitate meta-analysis. One published standard and guideline was developed by the Microarray Gene Expression Data Group (6). Minimum Information About a Microarray Experiment (MIAME) provides a framework for a comprehensive specification of the details of the microarray data, including experimental and array design, sample preparation, hybridization protocols, actual quantitative results, and normalization controls (6). This initiative has been accepted widely, and most major journals now require compliance with MIAME for any new submissions employing microarray data. Details of the microarray experiments and raw microarray data may be deposited with one of the two public repositories: Gene Expression Ominbus (http://www.ncbi.nlm.nih.gov/geo), and ArrayExpress (http://www.ebi.ac.uk/arrayexpress).

45.3. APPLICATIONS

45.3.1. Uncovering Genetic Circuits Governing Metabolic Pathways

Numerous microarray studies have been carried out to explore differential gene expression. As an example of how one can dissect metabolic pathways, we briefly review the experimental setup and salient findings from a few studies (12, 48). DeRisi et al. (12) used microarrays to assess global changes in gene expression during diauxic shift, a natural transition of yeast cells from fermentative metabolism to respiration. They demonstrated that genes displaying similar expression patterns were functionally related and regulated by upstream sequence motifs common to the promoters of the genes. Indeed, expression of about one-third of the yeast genome was altered as cells encountered glucose limitation during the diauxic shift. Interestingly, expression patterns of many uncharacterized genes provided clues to their possible functions.

Yeast cells respond to starvation for one of the many amino acids and other forms of stress by inducing the translation of Gcn4p, a bZIP transcriptional activator (23, 24). Gcn4p binds to general control response elements (UAS_{GCRE}) in the promoter regions to activate transcription of cognate target genes. To determine the genome-wide changes in gene expression elicited by histidine starvation, microarray hybridizations were carried out using fluorescently labeled cDNA from wild-type cells treated with 3-aminotriazole (3AT), an inhibitor of histidine biosynthesis, and from untreated cells as a control. We employed a P value criterion (see section 45.2.8.3) of ≤ 0.01 or ≤ 0.05 to filter genes whose expression was less statistically probable with a confidence level of 95 to 99%. Microarray data analysis revealed that 3AT treatment led to up-regulation of about 24% of the yeast genome and down-regulation of about an equal fraction (~28%) of the genome (48). Multiple hybridization experiments were carried out with a second non-isogenic strain, and the data revealed that the vast majority of the genes that were induced or repressed in the first experiment also exhibited similar behavior in the different strain background. This suggested that true expression changes are likely to be reproduced in multiple strain backgrounds and, therefore, such comparisons could be adopted as a general practice. To identify what fraction of the 3AT-regulated genes is dependent on Gcn4p for regulation, expression profiling was conducted in isogenic wild-type and $gcn4\Delta$ mutant strains both exposed to 3AT. Agglomerative two-dimensional hierarchical clustering was carried out as described previously (27). The results showed that the bulk of the genes up- or down-regulated by 3AT treatment of wild-type strains also showed Gcn4p dependence for their expression in the $GCN4/gcn4\Delta$ experiment, indicating that Gcn4p function was required for both up-regulation and down-regulation of genes in amino acid-starved cells. Furthermore, to comprehensively identify the set of genes regulated by Gcn4p, we applied Boolean logic, which required that genes up-regulated upon 3AT treatment in a wild-type strain should also yield a twofold change in the $GCN4/gcn4\Delta$ experiment. Applying these criteria, a group of 539 genes representing about 9% of the genome were identified to be induced by Gcn4p in response to amino acid starvation (48). Searching the upstream promoter regions of the Gcn4p target genes revealed that only 235 genes (44%) had one or more consensus UAS_{GCRE} elements in its promoter, indicating that they are directly regulated by Gcn4p. The rest of the Gcn4p target genes may have other cis elements, such as the cyclic AMP response element recognized by Gcn4p (49, 61). Alternatively, they may be regulated by another transcription factor which, in turn, is regulated by Gcn4p. The majority of genes negatively regulated by Gcn4p lacked UAS$_{GCRE}$, leading us to propose that their repression results from squelching by Gcn4p, wherein general transcription factors or cofactors are sequestered by Gcn4p at its target genes and less available for other transcriptional activators. In any case, these results revealed that Gcn4p is a pervasive transcriptional activator in *S. cerevisiae*.

Comparative expression profiling is a powerful approach to infer new connections between regulatory pathways. Central to this approach is the two-dimensional clustering of expression data from different experimental situations. Cluster analysis of the above-described data from 3AT experiments with the microarray data obtained by treatment of yeast cells with methyl methanesulfonate (MMS) (32, 33) showed that the bulk of genes induced by Gcn4p in 3AT-treated cells are also induced in MMS-treated cells. Indeed, it was shown that Gcn4p expression was induced by MMS in a manner dependent on the upstream regulator Gcn2p (48).

45.3.2. Unsupervised Identification of Gene Function: a Compendium Approach

Uncovering the function of novel genes is an important aspect of functional genomics projects. Hughes et al. (27) proposed that a compendium or a comprehensive database of diverse expression profiles could serve as a reference to which a query expression profile could be compared to obtain clues as to the pathway(s) that is affected by a perturbation or a mutation in a novel gene. In a landmark study, a compendium of 300 expression profiles associated with a multiplicity of diverse mutations, regulated expression of genes, and chemical treatments was created (27). The following example illustrates the utility of this compendium for gene discovery. Inhibition of ergosterol biosynthesis leads to transcriptional induction of the genes encoding the pathway enzymes. As expected, microarray analysis of mutants and treatments affecting ergosterol biosynthesis showed similar expression profiles. Profiling of the *yer044cΔ* mutant strain and two-dimensional cluster analysis showed that the expression profile of this mutant was clustered with those elicited by the known sterol pathway mutants *erg2Δ* and *erg3Δ*, and upon overexpression of *ERG11*. These findings suggested that *YER044C* (renamed *ERG28*) functions in ergosterol biosynthesis, and indeed follow-up experiments provided confirmation (27).

45.3.3. Translational Arrays

DNA microarrays have also been used to assess genome-wide profiles of mRNA translation or mRNA stability. For example, Wang et al. determined that yeast mRNAs have half-lives between 3 min and 90 min (73). In this study, a temperature-sensitive allele of a gene encoding an RNA polymerase II subunit (*RPB1*) was used to rapidly initiate transcription at the nonpermissive temperature. Following the temperature shift, total RNA was isolated at short time intervals and fluorescently labeled cDNA was synthesized and hybridized to ORF microarrays. Interestingly, it was found that mRNAs that encode proteins involved in a common function or that are constituents of the same macromolecular complex, such as ribosomal protein mRNAs or the nucleosomal core histone mRNAs, have similar half-lives (73). These results add a new dimension to gene regulation wherein fine control of gene regulation is

achieved by coordination of mRNA transcription rates and stability.

An additional level of gene regulation is achieved by differential efficiency of mRNA translation. Microarrays have been used to assess mRNA translation efficiency on a genome-wide scale by comparing mRNA to protein levels (2, 34, 36, 62). The ratio between the proportions of an mRNA associated with polyribosomes versus monosomes is indicative of its translation initiation rate. Accordingly, mRNA was isolated from sucrose gradient fractions containing polyribosomes or from monoribosome fractions, and the total unfractionated mRNA, used as a reference, was reverse transcribed to prepare fluorescently labeled cDNAs. The cDNAs from polysomes and monosomes were each combined with the reference cDNA and hybridized to all of the ORFs in the yeast genome (2). It was found that 43 mRNAs were not associated with either monosomes or polysomes and thus could potentially be regulated at the translational level by associating with the translation machinery only under certain special conditions. An additional group of 53 mRNAs, including *GCN4* and *CPA1*, both known to be regulated by translational control mechanisms mediated by upstream ORFs, was predominantly associated with the monoribosome fraction. These results provided a first genome scale view of posttranscriptional gene regulatory events.

45.3.4. mRNA Splicing Arrays

Splicing of mRNA has important consequences in synthesis of functional mRNA in different genomes. In higher eukaryotes, alternative splicing leads to novel forms of mRNA from a primary transcript and has also been implicated in producing proteins of distinct or even antagonistic function. A detailed analysis of the use of splicing arrays for discovery of alternative splice forms can be found elsewhere (39). For most organisms, however, splicing of mRNA leads to intron removal to produce a functional mRNA. The functional genomics and genome annotation project largely depends on bioinformatics tools to predict mRNAs that are processed by the alternate splicing pathways. Therefore, a systematic genome-wide analysis is required to accurately assess the role of splicing in genome function.

Splicing occurs in the nucleus in a large multiprotein complex called the spliceosome, which contains essential and several nonessential subunits. To discriminate between spliced and unspliced RNAs for each intron-containing yeast gene, a novel microarray was employed (10). Oligonucleotides were designed to detect the splice junction (specific to spliced RNA and not found in the genome), the intron (present in unspliced RNA), and the second exon (common to spliced and unspliced RNA) for each of the 244 known or predicted intron-containing genes. The oligonucleotides were printed on glass slides to create splicing-sensitive microarrays for yeasts. As normalization controls, oligonucleotides corresponding to seven intron-lacking genes were spotted in the same microarray slide. Fluorescently labeled cDNA was prepared from total RNA by reverse transcription with a mixture of oligo(dT) and random hexamer primers and hybridized to the microarray. To validate this approach, cDNA made from a wild-type strain was labeled with Cy3, whereas Cy5-labeled cDNA was made from RNA isolated from temperature-sensitive mutant *prp4-1*, defective in a subunit of the spliceosome. Hybridization revealed that intron-containing probes accumulated, whereas splice junction probes declined, indicating a reduction in splicing in the mutant grown at a nonpermissive temperature. To evaluate the contribution of different mRNA processing factors for splicing, strains bearing mutations in 18 different nonessential subunits were employed (10). Their analysis showed a requirement for a specific mRNA processing factor(s) vis-à-vis a particular intron-containing gene, rather than a general requirement for all or most mRNA processing factors. It was previously thought that Prp17 and Prp18 factors were required for splicing of mRNA with short distance of branch point to the 3′ splice site using *ACT1* in vitro. In contrast, results from the splicing arrays involving almost all intron-containing genes showed no such strict distance specificity. The splicing array could be employed for genome annotation and validation of alternatively spliced genes, but perfect designing of splicing array would depend on the strength of the bioinformatic tools to accurately predict all potential intron-containing genes.

45.3.5. Fitness Testing in Competitive Growth Assays To Identify Gene Function and Drug Targets

Subtle phenotypes associated with mutations are often difficult to score by classical techniques, and therefore, microarray technology has been used to determine phenotypes of mutants in a mixed population of yeast strains (21, 27). Briefly, a library of yeast deletion mutants corresponding to each of the approximately 4,400 nonessential genes was first constructed by individually replacing every ORF with a G418 resistance selectable marker gene and short sequence tags called UP tag and DOWN tag that are unique to each deletion. These two unique sequence tags serve as "molecular bar codes" with which to identify the abundance of each mutant strain. Pools of mutant strains were cocultured under specific growth conditions in a competitive growth assay, and cells were harvested at zero time and subsequently at defined time intervals. The rationale of this assay is that if a mutation impaired growth of the mutant cell in a particular growth condition, then the relative abundance of such a mutant strain would diminish in the population upon extended incubation time. On the other hand, deletion mutations that do not impair growth would be selectively enriched in the population. Thus, the rate of decrease of the mutant cells from a population in a competitive growth assay is indicative of the requirement for its cognate gene for growth. As a proof-of-principle experiment, the competitive growth assay was employed to identify all genes required for growth in galactose media (21). In addition to genes already known to be involved in galactose utilization, about 9 or 10 new genes required for galactose metabolism were identified by this approach. Microarray coupled with competitive growth assay has also been used for drug target identification (44). Here pools of heterozygous mutants were exposed to different compounds and the fitness of the individual mutants was scored essentially as described above for the galactose regulon. Drug targets for several known compounds were validated, and new targets for several compounds were discovered. Thus, the competitive growth test in combination with the microarray technology is a valuable tool to identify gene function and to identify gene targets for pharmaceutical compounds and drugs.

45.3.6. Use of Intergenic DNA Microarrays To Detect Protein-DNA Interactions

The microarray technology has been used in conjunction with the chromatin immunoprecipitation assay in so-called ChIP-array experiments to detect genome-wide protein-DNA association in vivo. Briefly, at an appropriate stage of cell growth, cells are fixed with formaldehyde, which cross-links protein-protein and protein-DNA interactions. Chromatin is isolated and sonicated to shear chromatin fragments to an average size of about 300 to 1,000 bp. Antibodies to either epitope-tagged proteins or highly specific antibodies are used to coimmunoprecipitate genomic regions associated with proteins of interest. The immunoprecipitated DNA is amplified and labeled by either of two methods. In the first method, a two-step PCR is employed wherein degenerate primers are used to nonspecifically prime two rounds of DNA synthesis using the immunoprecipitated DNA. In the second step, a specific primer is used to prime exponential amplification (31, 37, 41, 57). In the second method, ligation-mediated PCR was carried out to incorporate fluorescent dye into the immunoprecipitated and control DNAs samples (40, 56, 77). In both the approaches, labeled DNAs from experimental and control immunoprecipitated samples were mixed and hybridized to intergenic array consisting of spots corresponding to the DNA sequences located between two ORFs.

In a systematic large-scale study, the ChIP-array technology has been employed to investigate genome-wide promoter occupancy of 106 transcriptional regulators encoded by the yeast genome (40). Based on these studies, Lee et al. (40) provided evidence that alteration in gene expression profiles is executed by a network of interactions among transcriptional regulators. Several types of interactions were uncovered by their study. In the most common situation, one or more transcription factors bound directly to the promoters of a target group of coregulated genes. In another interesting situation, one transcriptional activator bound to the promoter of a second transcriptional activator and both activators in turn bound to the promoter of the same target gene. In this way, a given transcriptional activator could regulate expression of the target gene directly and also indirectly by controlling the expression of the second transcription factor working at the same target gene. Lee et al. (40) showed that, depending on the choice of the P value cutoff, the number and reliability of identified targets can vary dramatically. Only a few of the expected binding sites for the transcriptional activator Gcn4p could be identified despite using a relaxed P value filter (reference 39 and our unpublished data), indicating that the experimental and statistical methods need to be refined for comprehensive identification of genome-wide transcription factor binding in vivo. Extending this study, Pokholok et al. (53) employed a modified tiling microarray that contained 44,290 features consisting of 60-mer oligonulceotide probes that covered 12 Mb of the yeast genome (85%), with an average probe density of 266 bp, barring the highly repeated regions. By employing further modifications to the experimental protocol, Pokholok et al. obtained substantially higher resolution and accuracy than previous methods. Analysis of Gcn4 binding with the new method revealed a total of 210 genes whose promoters are bound at high confidence values within an optimal p value threshold of 6×10^{-6} (53).

45.3.7. Comparative Genome Hybridization

The microarray technology has also been adopted for comparative genome hybridization (CGH) studies. Briefly, CGH involves isolation of genomic DNA from a sample strain and the reference strain and then labeling by Cy3 and Cy5, followed by Cy3/Cy5 hybridization to a standard cDNA microarray containing whole-genome probes from the reference strain. This strategy would enable detection of the relative abundances of the chromosome(s) or chromosomal segments in the sample versus the reference strain. In one of the early implementations of this methodology, Hughes et al. (28) reported that several mutant yeast strains used in microarray experiments in various laboratories were aneuploid for whole chromosomes or chromosomal segments that presumably emanated from strain construction steps. The rationale behind this discovery was their observation that a substantial fraction of the mutants exhibited chromosome-wide aneuploidies. Therefore, it has now become imperative that each expression profile be judged for artifactual expression biases by plotting the expression data on a chromosomal map. More recently, CGH has been applied to phylogenetic studies by comparing the relatedness of all genes between two organisms (14, 15, 47, 75). Once the CGH technology is improved to be able to use small amounts of clinical material, CGH could be used routinely for pathogenic strain or species typing in a rapid and robust manner.

45.3.8. Integration of Technologies for Comprehensive Assessment of Genome Function

For comprehensive assessment of the function(s) of novel genes and that of a newly sequenced genome, technologies are now available for measurement of mRNA expression (transcriptome), the protein expression (proteome), and the metabolites (metabolome), leading to a systems biology approach integrating all of this information (Fig. 2). In this scheme, the microarray technology may be the first level of analysis leading to assessment of gene expression under specific conditions. Next, leads obtained from the microarray data can be pursued by phenotype testing using appropriate mutant/genetic strategies (27, 48), to narrow down specific pathway(s) or regulators responsible for alteration in gene expression as discussed earlier (see section 45.3.1). Given that proteins are effectors of programmed gene expression, assessment of their abundance and activity is critical to understand genome function. Also, posttranscriptional regulatory mechanisms can offset the measured abundance of the transcriptome of a cell. Mass spectrometry (MS)-based pro-

teomics technologies can be used at the whole-genome level to assess either the proteome abundance or relative proteome differences between two conditions or cell types (1, 50). Although these technologies have not yet reached the robustness and sensitivity achieved in transcriptome analysis, the current estimates show that at least 1,000 proteins can be identified in a single liquid chromatography-tandem MS experiment (1, 42).

Several reports have now provided insights into the correlation between mRNA and protein levels (22, 29, 74). For instance, Washburn et al. studied the impact of culturing yeast cells in rich versus minimal media (74). They showed that almost the entire amino acid and purine biosynthetic pathways were induced in minimal medium at both the transcriptome and proteome levels (74). Increases of mRNA and protein expression ratios of several amino acid and purine biosynthetic pathways were well correlated at each locus and at each pathway level, and also largely correlated well with an earlier transcriptome study carried out under histidine starvation conditions (48). This suggests that culturing yeast cells in minimal medium elicits a physiological response similar to that of inducing Gcn4p expression upon severe histidine starvation. Interestingly, several proteome level changes did not correlate with the transcriptome level change. For example, in rich medium the Tup1 repressor protein was eightfold overexpressed, but this change was not detectable at the mRNA level; the RNA polymerase II holoenzyme subunits Anc1p and Sin4p were overexpressed in minimal medium, but these changes were not detectable at the transcript level. Moreover, the overall correlation between mRNA and protein expression was weak, with a Spearman rank correlation coefficient of 0.45 for 678 loci. These results highlight the importance of application of both transcriptome and proteome measurements to assess gene expression changes (Fig. 2).

There are also examples where a transcriptional profile generated from microarrays has been integrated into metabolite profiles obtained using liquid chromatography-

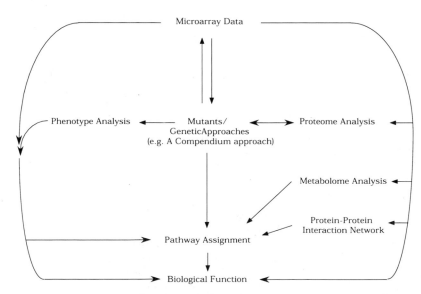

FIGURE 2 Suggested workflow for comprehensive assessment of genome function. Integration of results from microarray studies into the results of proteome analysis, metabolome analysis, and protein-protein interaction data would provide a broad platform for assignment of genes to respective pathways and inference of the biological function of novel genes.

MS technology for the lovastatin-producing fungal strain *Aspergillus terreus* (see reference 3, Fig. 2). Askenazi et al. devised an association approach linking the microarray data, and the metabolite profiling data and several interesting leads were obtained (3). Lovastatin and (+)-geodin are synthesized from polyketide-derived secondary metabolites. Genes involved in catabolism of fatty acids leading to the synthesis of polyketide precursors acetyl coenzyme A and malonyl coenzyme A, positively correlate with lovastatin and (+)-geodin production and negatively correlate with genes involved in anabolic steps of fatty acid metabolism. Genes encoding enzymes that catabolize alternative nutrient sources (such as glycogen phosphorylase, lysophospholipase, and alkaline protease) are expressed in late growth phase, and this positively correlated with lovastatin production and negatively correlated with genes involved in energy generation via the glycolytic and tricarboxylic acid pathways active during early growth phase (3).

Several studies have found coregulation of genes encoding proteins that are integral subunits of multiprotein complexes. For example, mRNA levels for 90 genes encoding 60S and 40S ribosomal subunits were coordinately downregulated upon severe amino acid starvation (48). Moreover, proteome analysis showed that eight loci encoding subunits of the 60S ribosomal particle were overexpressed in rich medium (YPD) compared to minimal medium (74). Indeed, association analysis of microarray data of transcript levels using two-dimensional clustering approaches revealed that loci that function as a multisubunit complex are more often clustered together (27, 48). Thus, integration of protein-protein interaction assays (52) with complementary mRNA and protein expression analysis (Fig. 2) would provide vital clues about function of novel genes. In summary, we have provided detailed methodology for carrying out the DNA microarray technology and briefly reviewed its attendant applications. A careful interpretation of the microarray data is important so that meaningful hypotheses can be generated for designing follow-up experiments. Further refinement of the statistical tools for microarray data analysis, including standardization of data normalization methods, would yield more robust and reliable data and enable comparison of microarray data generated in different laboratories and possibly find use in clinical practice.

45.4. REFERENCES

1. Aebersold, R., and M. Mann. 2003. Mass spectrometry-based proteomics. *Nature* **422:**198–207.
2. Arava, Y., Y. Wang, J. D. Storey, C. Liu, P. O. Brown, and D. Herschlag. 2003. Genome-wide analysis of mRNA translation profiles in *Saccharomyces cerevisiae*. *Proc. Natl. Acad. Sci. USA* **100:**3889–3894.
3. Askenazi, M., E. M. Driggers, D. A. Holtzman, T. C. Norman, S. Iverson, D. P. Zimmer, M. E. Boers, P. R. Blomquist, E. J. Martinez, A. W. Monreal, T. P. Feibelman, M. E. Mayorga, M. E. Maxon, K. Sykes, J. V. Tobin, E. Cordero, S. R. Salama, J. Trueheart, J. C. Royer, and K. T. Madden. 2003. Integrating transcriptional and metabolite profiles to direct the engineering of lovastatin-producing fungal strains. *Nat. Biotechnol.* **21:**150–156.
4. Bammler, T., R. P. Beyer, S. Bhattacharya, G. A. Boorman, A. Boyles, B. U. Bradford, R. E. Bumgarner, P. R. Bushel, K. Chaturvedi, D. Choi, M. L. Cunningham, S. Deng, H. K. Dressman, R. D. Fannin, F. M. Farin, J. H. Freedman, R. C. Fry, A. Harper, M. C. Humble, P. Hurban, T. J. Kavanagh, W. K. Kaufmann, K. F. Kerr, L. Jing, J. A. Lapidus, M. R. Lasarev, J. Li, Y. J. Li, E. K. Lobenhofer, X. Lu, R. L. Malek, S. Milton, S. R. Nagalla, P. O'Malley J, V. S. Palmer, P. Pattee, R. S. Paules, C. M. Perou, K. Phillips, L. X. Qin, Y. Qiu, S. D. Quigley, M. Rodland, I. Rusyn, L. D. Samson, D. A. Schwartz, Y. Shi, J. L. Shin, S. O. Sieber, S. Slifer, M. C. Speer, P. S. Spencer, D. I. Sproles, J. A. Swenberg, W. A. Suk, R. C. Sullivan, R. Tian, R. W. Tennant, S. A. Todd, C. J. Tucker, B. Van Houten, B. K. Weis, S. Xuan, and H. Zarbl. 2005. Standardizing global gene expression analysis between laboratories and across platforms. *Nat. Methods* **2:**351–356.
5. Bowtell, D., and J. Sambrook. 2003. *DNA Microarrays: a Molecular Cloning Manual.* Cold Spring Harbor Laboratory, Cold Spring Harbor, NY.
6. Brazma, A., P. Hingamp, J. Quackenbush, G. Sherlock, P. Spellman, C. Stoeckert, J. Aach, W. Ansorge, C. A. Ball, H. C. Causton, T. Gaasterland, P. Glenisson, F. C. Holstege, I. F. Kim, V. Markowitz, J. C. Matese, H. Parkinson, A. Robinson, U. Sarkans, S. Schulze-Kremer, J. Stewart, R. Taylor, J. Vilo, and M. Vingron. 2001. Minimum Information About a Microarray Experiment (MIAME)—toward standards for microarray data. *Nat. Genet.* **29:**365–371.
7. Causton, H. C., J. Quackenbush, and A. Brazma. 2003. *Microarray Gene Expression Data Analysis: a Beginner's Guide*, p. 40–69. Blackwell Publishing Limited, Malden, MA.
8. Causton, H. C., B. Ren, S. S. Koh, C. T. Harbison, E. Kanin, E. G. Jennings, T. I. Lee, H. L. True, E. S. Lander, and R. A. Young. 2001. Remodeling of yeast genome expression in response to environmental changes. *Mol. Biol. Cell* **12:**323–337.
9. Churchill, G. A. 2002. Fundamentals of experimental design for cDNA microarrays. *Nat. Genet.* **32**(Suppl.):490–495.
10. Clark, T. A., C. W. Sugnet, and M. Ares, Jr. 2002. Genomewide analysis of mRNA processing in yeast using splicing-specific microarrays. *Science* **296:**907–910.
11. Cui, X., and G. A. Churchill. 2003. Statistical tests for differential expression in cDNA microarray experiments. *Genome Biol.* **4:**210.
12. DeRisi, J. L., V. R. Iyer, and P. O. Brown. 1997. Exploring the metabolic and genetic control of gene expression on a genomic scale. *Science* **278:**680–686.
13. Dresen, I. M., J. Husing, E. Kruse, T. Boes, and K. H. Jockel. 2003. Software packages for quantitative microarray-based gene expression analysis. *Curr. Pharm. Biotechnol.* **4:**417–437.
14. Dunham, M. J., H. Badrane, T. Ferea, J. Adams, P. O. Brown, F. Rosenzweig, and D. Botstein. 2002. Characteristic genome rearrangements in experimental evolution of *Saccharomyces cerevisiae*. *Proc. Natl. Acad. Sci. USA* **99:**16144–16149.
15. Edwards-Ingram, L. C., M. E. Gent, D. C. Hoyle, A. Hayes, L. I. Stateva, and S. G. Oliver. 2004. Comparative genomic hybridization provides new insights into the molecular taxonomy of the *Saccharomyces* sensu stricto complex. *Genome Res.* **14:**1043–1051.
16. Eisen, M. B., P. T. Spellman, P. O. Brown, and D. Botstein. 1998. Cluster analysis and display of genome-wide expression patterns. *Proc. Natl. Acad. Sci. USA* **95:**14863–14868.
17. Fare, T. L., E. M. Coffey, H. Dai, Y. D. He, D. A. Kessler, K. A. Kilian, J. E. Koch, E. LeProust, M. J. Marton, M. R. Meyer, R. B. Stoughton, G. Y. Tokiwa, and Y. Wang. 2003. Effects of atmospheric ozone on microarray data quality. *Anal. Chem.* **75:**4672–4675.
18. Gaillard, C., and F. Strauss. 1990. Ethanol precipitation of DNA with linear polyacrylamide as carrier. *Nucleic Acids Res.* **18:**378.

19. Gasch, A. P., M. Huang, S. Metzner, D. Botstein, S. J. Elledge, and P. O. Brown. 2001. Genomic expression responses to DNA-damaging agents and the regulatory role of the yeast ATR homolog Mec1p. *Mol. Biol. Cell* **12**:2987–3003.

20. Gasch, A. P., P. T. Spellman, C. M. Kao, O. Carmel-Harel, M. B. Eisen, G. Storz, D. Botstein, and P. O. Brown. 2000. Genomic expression programs in the response of yeast cells to environmental changes. *Mol. Biol. Cell* **11**:4241–4257.

21. Giaever, G., A. M. Chu, L. Ni, C. Connelly, L. Riles, S. Veronneau, S. Dow, A. Lucau-Danila, K. Anderson, B. Andre, A. P. Arkin, A. Astromoff, M. El-Bakkoury, R. Bangham, R. Benito, S. Brachat, S. Campanaro, M. Curtiss, K. Davis, A. Deutschbauer, K. D. Entian, P. Flaherty, F. Foury, D. J. Garfinkel, M. Gerstein, D. Gotte, U. Guldener, J. H. Hegemann, S. Hempel, Z. Herman, D. F. Jaramillo, D. E. Kelly, S. L. Kelly, P. Kotter, D. LaBonte, D. C. Lamb, N. Lan, H. Liang, H. Liao, L. Liu, C. Luo, M. Lussier, R. Mao, P. Menard, S. L. Ooi, J. L. Revuelta, C. J. Roberts, M. Rose, P. Ross-Macdonald, B. Scherens, G. Schimmack, B. Shafer, D. D. Shoemaker, S. Sookhai-Mahadeo, R. K. Storms, J. N. Strathern, G. Valle, M. Voet, G. Volckaert, C. Y. Wang, T. R. Ward, J. Wilhelmy, E. A. Winzeler, Y. Yang, G. Yen, E. Youngman, K. Yu, H. Bussey, J. D. Boeke, M. Snyder, P. Philippsen, R. W. Davis, and M. Johnston. 2002. Functional profiling of the *Saccharomyces cerevisiae* genome. *Nature* **418**:387–391.

22. Griffin, T. J., S. P. Gygi, T. Ideker, B. Rist, J. Eng, L. Hood, and R. Aebersold. 2002. Complementary profiling of gene expression at the transcriptome and proteome levels in *Saccharomyces cerevisiae*. *Mol. Cell. Proteomics* **1**:323–333.

23. Hinnebusch, A. G. 1997. Translational regulation of yeast GCN4: a window on factors that control initiator-tRNA binding to the ribosome. *J. Biol. Chem.* **272**:21661–21664.

24. Hinnebusch, A. G., and K. Natarajan. 2002. Gcn4p, a master regulator of gene expression, is controlled at multiple levels by diverse signals of starvation and stress. *Eukaryot. Cell* **1**:22–32.

25. Holloway, A. J., R. K. van Laar, R. W. Tothill, and D. D. Bowtell. 2002. Options available—from start to finish—for obtaining data from DNA microarrays II. *Nat. Genet.* **32**(Suppl.):481–489.

26. Hughes, T. R., M. Mao, A. R. Jones, J. Burchard, M. J. Marton, K. W. Shannon, S. M. Lefkowitz, M. Ziman, J. M. Schelter, M. R. Meyer, S. Kobayashi, C. Davis, H. Dai, Y. D. He, S. B. Stephaniants, G. Cavet, W. L. Walker, A. West, E. Coffey, D. D. Shoemaker, R. Stoughton, A. P. Blanchard, S. H. Friend, and P. S. Linsley. 2001. Expression profiling using microarrays fabricated by an ink-jet oligonucleotide synthesizer. *Nat. Biotechnol.* **19**:342–347.

27. Hughes, T. R., M. J. Marton, A. R. Jones, C. J. Roberts, R. Stoughton, C. D. Armour, H. A. Bennett, E. Coffey, H. Dai, Y. D. He, M. J. Kidd, A. M. King, M. R. Meyer, D. Slade, P. Y. Lum, S. B. Stepaniants, D. D. Shoemaker, D. Gachotte, K. Chakraburtty, J. Simon, M. Bard, and S. H. Friend. 2000. Functional discovery via a compendium of expression profiles. *Cell* **102**:109–126.

28. Hughes, T. R., C. J. Roberts, H. Dai, A. R. Jones, M. R. Meyer, D. Slade, J. Burchard, S. Dow, T. R. Ward, M. J. Kidd, S. H. Friend, and M. J. Marton. 2000. Widespread aneuploidy revealed by DNA microarray expression profiling. *Nat. Genet.* **25**:333–337.

29. Ideker, T., V. Thorsson, J. A. Ranish, R. Christmas, J. Buhler, J. K. Eng, R. Bumgarner, D. R. Goodlett, R. Aebersold, and L. Hood. 2001. Integrated genomic and proteomic analyses of a systematically perturbed metabolic network. *Science* **292**:929–934.

30. Irizarry, R. A., D. Warren, F. Spencer, I. F. Kim, S. Biswal, B. C. Frank, E. Gabrielson, J. G. Garcia, J. Geoghegan, G. Germino, C. Griffin, S. C. Hilmer, E. Hoffman, A. E. Jedlicka, E. Kawasaki, F. Martinez-Murillo, L. Morsberger, H. Lee, D. Petersen, J. Quackenbush, A. Scott, M. Wilson, Y. Yang, S. Q. Ye, and W. Yu. 2005. Multiple-laboratory comparison of microarray platforms. *Nat. Methods* **2**:345–350.

31. Iyer, V. R., C. E. Horak, C. S. Scafe, D. Botstein, M. Snyder, and P. O. Brown. 2001. Genomic binding sites of the yeast cell-cycle transcription factors SBF and MBF. *Nature* **409**:533–538.

32. Jelinsky, S. A., P. Estep, G. M. Church, and L. D. Samson. 2000. Regulatory networks revealed by transcriptional profiling of damaged *Saccharomyces cerevisiae* cells: Rpn4 links base excision repair with proteasomes. *Mol. Cell. Biol.* **20**:8157–8167.

33. Jelinsky, S. A., and L. D. Samson. 1999. Global response of *Saccharomyces cerevisiae* to an alkylating agent. *Proc. Natl. Acad. Sci. USA* **96**:1486–1491.

34. Johannes, G., M. S. Carter, M. B. Eisen, P. O. Brown, and P. Sarnow. 1999. Identification of eukaryotic mRNAs that are translated at reduced cap initiation complex eIF4F concentrations using a cDNA microarray. *Proc. Natl. Acad. Sci. USA* **96**:13118–13123.

35. Kohrer, K., and H. Domdey. 1991. Preparation of high molecular weight RNA. *Methods Enzymol.* **194**:398–405.

36. Kuhn, K. M., J. L. DeRisi, P. O. Brown, and P. Sarnow. 2001. Global and specific translational regulation in the genomic response of *Saccharomyces cerevisiae* to a rapid transfer from a fermentable to a nonfermentable carbon source. *Mol. Cell. Biol.* **21**:916–927.

37. Kurdistani, S. K., D. Robyr, S. Tavazoie, and M. Grunstein. 2002. Genome-wide binding map of the histone deacetylase Rpd3 in yeast. *Nat. Genet.* **31**:248–254.

38. Larkin, J. E., B. C. Frank, H. Gavras, R. Sultana, and J. Quackenbush. 2005. Independence and reproducibility across microarray platforms. *Nat. Methods* **2**:337–344.

39. Lee, C., and M. Roy. 2004. Analysis of alternative splicing with microarrays: successes and challenges. *Genome Biol.* **5**:231.

40. Lee, T. I., N. J. Rinaldi, F. Robert, D. T. Odom, Z. Bar-Joseph, G. K. Gerber, N. M. Hannett, C. T. Harbison, C. M. Thompson, I. Simon, J. Zeitlinger, E. G. Jennings, H. L. Murray, D. B. Gordon, B. Ren, J. J. Wyrick, J. B. Tagne, T. L. Volkert, E. Fraenkel, D. K. Gifford, and R. A. Young. 2002. Transcriptional regulatory networks in *Saccharomyces cerevisiae*. *Science* **298**:799–804.

41. Lieb, J. D., X. Liu, D. Botstein, and P. O. Brown. 2001. Promoter-specific binding of Rap1 revealed by genome-wide maps of protein-DNA association. *Nat. Genet.* **28**:327–334.

42. Lin, D., D. L. Tabb, and J. R. Yates III. 2003. Large-scale protein identification using mass spectrometry. *Biochim. Biophys. Acta* **1646**:1–10.

43. Lockhart, D. J., H. Dong, M. C. Byrne, M. T. Follettie, M. V. Gallo, M. S. Chee, M. Mittmann, C. Wang, M. Kobayashi, H. Horton, and E. L. Brown. 1996. Expression monitoring by hybridization to high-density oligonucleotide arrays. *Nat. Biotechnol.* **14**:1675–1680.

44. Lum, P. Y., C. D. Armour, S. B. Stepaniants, G. Cavet, M. K. Wolf, J. S. Butler, J. C. Hinshaw, P. Garnier, G. D. Prestwich, A. Leonardson, P. Garrett-Engele, C. M. Rush, M. Bard, G. Schimmack, J. W. Phillips, C. J. Roberts, and D. D. Shoemaker. 2004. Discovering modes of action for therapeutic compounds using a genome-wide screen of yeast heterozygotes. *Cell* **116**:121–137.

45. Luscombe, N. M., T. E. Royce, P. Bertone, N. Echols, C. E. Horak, J. T. Chang, M. Snyder, and M. Gerstein. 2003. Expressyourself: a modular platform for processing and visualizing microarray data. *Nucleic Acids Res.* **31:** 3477–3482.

46. Marton, M. J., J. L. DeRisi, H. A. Bennett, V. Iyer, M. R. Meyer, C. J. Roberts, R. Stoughton, J. Burchard, D. Slade, H. Dai, D. E. Bassett, L. H. Haratwell, P. O. Brown, and S. H. Friend. 1998. Drug target validation and identification of secondary drug target effects using DNA microarrays. *Nat. Med.* **4:**1293–1301.

47. Murray, A. E., D. Lies, G. Li, K. Nealson, J. Zhou, and J. M. Tiedje. 2001. DNA/DNA hybridization to microarrays reveals gene-specific differences between closely related microbial genomes. *Proc. Natl. Acad. Sci. USA* **98:**9853–9858.

48. Natarajan, K., M. R. Meyer, B. M. Jackson, D. Slade, C. J. Roberts, A. G. Hinnebusch, and M. J. Marton. 2001. Transcriptional profiling shows that Gcn4p is a master regulator of gene expression during amino acid starvation in yeast. *Mol. Cell. Biol.* **21:**4347–4368.

49. Pascual-Ahuir, A., R. Serrano, and M. Proft. 2001. The Sko1p repressor and Gcn4p activator antagonistically modulate stress-regulated transcription in *Saccharomyces cerevisiae*. *Mol. Cell. Biol.* **21:**16–25.

50. Patterson, S. D., and R. H. Aebersold. 2003. Proteomics: the first decade and beyond. *Nat. Genet.* **33**(Suppl):311–323.

51. Patterson, T. A., E. K. Lobenhofer, S. B. Fulmer-Smentek, P. J. Collins, T. M. Chu, W. Bao, H. Fang, E. S. Kawasaki, J. Hager, I. R. Tikhonova, S. J. Walker, L. Zhang, P. Hurban, F. de Longueville, J. C. Fuscoe, W. Tong, L. Shi, and R. D. Wolfinger. 2006. Performance comparison of one-color and two-color platforms within the MicroArray Quality Control (MAQC) project. *Nat. Biotechnol.* **24:**1140–1150.

52. Phizicky, E., P. I. H. Bastiaens, H. Zhu, M. Snyder, and S. Fields. 2003. Protein analysis on a proteomic scale. *Nature* **422:**208–215.

53. Pokholok, D. K., C. T. Harbison, S. Levine, M. Cole, N. M. Hannett, T. I. Lee, G. W. Bell, K. Walker, P. A. Rolfe, E. Herbolsheimer, J. Zeitlinger, F. Lewitter, D. K. Gifford, and R. A. Young. 2005. Genome-wide map of nucleosome acetylation and methylation in yeast. *Cell* **122:** 517–527.

54. Quackenbush, J. 2001. Computational analysis of microarray data. *Nat. Rev. Genet.* **2:**418–427.

55. Rajagopalan, D. 2003. A comparison of statistical methods for analysis of high density oligonucleotide array data. *Bioinformatics* **19:**1469–1476.

56. Ren, B., F. Robert, J. J. Wyrick, O. Aparicio, E. G. Jennings, I. Simon, J. Zeitlinger, J. Schreiber, N. Hannett, E. Kanin, T. L. Volkert, C. J. Wilson, S. P. Bell, and R. A. Young. 2000. Genome-wide location and function of DNA binding proteins. *Science* **290:**2306–2309.

57. Robyr, D., Y. Suka, I. Xenarios, S. K. Kurdistani, A. Wang, N. Suka, and M. Grunstein. 2002. Microarray deacetylation maps determine genome-wide functions for yeast histone deacetylases. *Cell* **109:**437–446.

58. Sambrook, J., and D. W. Russel. 2001. *Molecular Cloning: a Laboratory Manual*, 3rd ed. Cold Spring Harbor Laboratory, Cold Spring Harbor, NY.

59. Schadt, E. E., C. Li, B. Ellis, and W. H. Wong. 2001. Feature extraction and normalization algorithms for high-density oligonucleotide gene expression array data. *J. Cell. Biochem. Suppl.* **37:**120–125.

60. Schena, M., D. Shalon, R. W. Davis, and P. O. Brown. 1995. Quantitative monitoring of gene expression patterns with a complementary DNA microarray. *Science* **270:**467–470.

61. Sellers, J. W., A. C. Vincent, and K. Struhl. 1990. Mutations that define the optimal half-site for binding yeast GCN4 activator protein and identify an ATF/CREB-like repressor that recognizes similar DNA sites. *Mol. Cell. Biol.* **10:**5077–5086.

62. Serikawa, K. A., X. L. Xu, V. L. MacKay, G. L. Law, Q. Zong, L. P. Zhao, R. Bumgarner, and D. R. Morris. 2003. The transcriptome and its translation during recovery from cell cycle arrest in *Saccharomyces cerevisiae*. *Mol. Cell. Proteomics* **2:**191–204.

63. Shannon, W., R. Culverhouse, and J. Duncan. 2003. Analyzing microarray data using cluster analysis. *Pharmacogenomics* **4:**41–52.

64. Shi, L., L. H. Reid, W. D. Jones, R. Shippy, J. A. Warrington, S. C. Baker, P. J. Collins, F. de Longueville, E. S. Kawasaki, K. Y. Lee, Y. Luo, Y. A. Sun, J. C. Willey, R. A. Setterquist, G. M. Fischer, W. Tong, Y. P. Dragan, D. J. Dix, F. W. Frueh, F. M. Goodsaid, D. Herman, R. V. Jensen, C. D. Johnson, E. K. Lobenhofer, R. K. Puri, U. Schrf, J. Thierry-Mieg, C. Wang, M. Wilson, P. K. Wolber, L. Zhang, S. Amur, W. Bao, C. C. Barbacioru, A. B. Lucas, V. Bertholet, C. Boysen, B. Bromley, D. Brown, A. Brunner, R. Canales, X. M. Cao, T. A. Cebula, J. J. Chen, J. Cheng, T. M. Chu, E. Chudin, J. Corson, J. C. Corton, L. J. Croner, C. Davies, T. S. Davison, G. Delenstarr, X. Deng, D. Dorris, A. C. Eklund, X. H. Fan, H. Fang, S. Fulmer-Smentek, J. C. Fuscoe, K. Gallagher, W. Ge, L. Guo, X. Guo, J. Hager, P. K. Haje, J. Han, T. Han, H. C. Harbottle, S. C. Harris, E. Hatchwell, C. A. Hauser, S. Hester, H. Hong, P. Hurban, S. A. Jackson, H. Ji, C. R. Knight, W. P. Kuo, J. E. LeClerc, S. Levy, Q. Z. Li, C. Liu, Y. Liu, M. J. Lombardi, Y. Ma, S. R. Magnuson, B. Maqsodi, T. McDaniel, N. Mei, O. Myklebost, B. Ning, N. Novoradovskaya, M. S. Orr, T. W. Osborn, A. Papallo, T. A. Patterson, R. G. Perkins, E. H. Peters, R. Peterson, et al. 2006. The MicroArray Quality Control (MAQC) project shows inter- and intraplatform reproducibility of gene expression measurements. *Nat. Biotechnol.* **24:**1151–1161.

65. Singh-Gasson, S., R. D. Green, Y. Yue, C. Nelson, F. Blattner, M. R. Sussman, and F. Cerrina. 1999. Maskless fabrication of light-directed oligonucleotide microarrays using a digital micromirror array. *Nat. Biotechnol.* **17:**974–978.

66. Stekel, D. 2003. *Microarray Bioinformatics*, p. 62–72. Cambridge University Press, Cambridge, United Kingdom.

67. Stekel, D. 2003. *Microarray Bioinformatics*, p. 73–99. Cambridge University Press, Cambridge, United Kingdom.

68. Tan, P. K., T. J. Downey, E. L. Spitznagel, Jr., P. Xu, D. Fu, D. S. Dimitrov, R. A. Lempicki, B. M. Raaka, and M. C. Cam. 2003. Evaluation of gene expression measurements from commercial microarray platforms. *Nucleic Acids Res.* **31:**5676–5684.

69. Tseng, G. C., M. K. Oh, L. Rohlin, J. C. Liao, and W. H. Wong. 2001. Issues in cDNA microarray analysis: quality filtering, channel normalization, models of variations and assessment of gene effects. *Nucleic Acids Res.* **29:**2549–2557.

70. Tusher, V. G., R. Tibshirani, and G. Chu. 2001. Significance analysis of microarrays applied to the ionizing radiation response. *Proc. Natl. Acad. Sci. USA* **98:**5116–5121.

71. van de Peppel, J., P. Kemmeren, H. van Bakel, M. Radonjic, D. van Leenen, and F. C. P. Holstege. 2003. Monitoring global messenger RNA changes in externally controlled microarray experiments. *EMBO Rep.* **4:**387–393.

72. Van Gelder, R. N., M. E. von Zastrow, A. Yool, W. C. Dement, J. D. Barchas, and J. H. Eberwine. 1990. Amplified RNA synthesized from limited quantities of heterogeneous cDNA. *Proc. Natl. Acad. Sci. USA* **87:**1663–1667.

73. Wang, Y., C. L. Liu, J. D. Storey, R. J. Tibshirani, D. Herschlag, and P. O. Brown. 2002. Precision and functional specificity in mRNA decay. *Proc. Natl. Acad. Sci. USA* **99:**5860–5865.

74. Washburn, M. P., A. Koller, G. Oshiro, R. R. Ulaszek, D. Plouffe, C. Deciu, E. Winzeler, and J. R. Yates III. 2003. Protein pathway and complex clustering of correlated mRNA and protein expression analyses in *Saccharomyces cerevisiae. Proc. Natl. Acad. Sci. USA* **100:** 3107–3112.

75. Watanabe, T., Y. Murata, S. Oka, and H. Iwahashi. 2004. A new approach to species determination for yeast strains: DNA microarray-based comparative genomic hybridization using a yeast DNA microarray with 6000 genes. *Yeast* **21:**351–365.

76. Wu, H., K. M. Kerr, X. Cui, and G. A. Churchill. 2003. MAANOVA: a software package for the analysis of spotted cDNA microarray experiments, p. 313–341. *In* G. Parmigiani, E. S. Garett, R. A. Irizarry, and S. L. Zeger (ed.), *The Analysis of Gene Expression Data: Methods and Software.* Springer, New York, NY.

77. Wyrick, J. J., J. G. Aparicio, T. Chen, J. D. Barnett, E. G. Jennings, R. A. Young, S. P. Bell, and O. M. Aparicio. 2001. Genome-wide distribution of ORC and MCM proteins in *S. cerevisiae*: high-resolution mapping of replication origins. *Science* **294:**2357–2360.

78. Yang, Y. H., and T. Speed. 2002. Design issues for cDNA microarray experiments. *Nat. Rev. Genet.* **3:**579–588.

APPENDICES

46

Laboratory Safety†

W. EMMETT BARKLEY AND CLAUDIA A. MICKELSON

†We acknowledge the contributions of the late John H. Richardson, coauthor of the original chapter, whose words, guidance, and spirit remain in this revision.

997

Laboratory safety requires an awareness of the possible risks associated with the handling of hazardous materials, knowledge of mechanisms by which exposures may occur, use of safeguards and techniques that reduce the potential for exposures, and vigilance against compromise and error. Microorganisms, radioisotopes, and hazardous chemicals are invariably found in bacteriology, microbiology, and biomedical laboratories. The risk of occupational infection in these laboratories is associated with the use of pathogenic microorganisms or the handling of materials contaminated with them. Biological safety and the prevention of laboratory-acquired infection is the principal subject addressed in this chapter. When pathogenic microorganisms are not present in the laboratory, the risk of physical injury and accidental exposure to hazardous chemicals and radioactive materials should be the primary concern of the laboratory worker, and a brief discussion of these hazards is also included.

Laboratory safety is the subject of several federal health and safety regulations and guidelines. The Nuclear Regulatory Commission has regulated the use of radioisotopes in laboratories since 1946 (47). The National Institutes of Health (NIH) established federal guidelines for the safe conduct of research involving recombinant DNA molecules in 1976 (42). In 1984 the Centers for Disease Control and Prevention (CDC) and NIH established consensus guidelines for protecting the health of laboratory workers who handle pathogenic microorganisms (44). The Occupational Safety and Health Administration (OSHA) established rules in 1990 and 1991 to protect the health and safety of laboratory workers from occupational exposures to hazardous chemicals and blood-borne pathogens, respectively (48, 49). The CDC and the U.S. Department of Agriculture (USDA) issued two new rules in 2002 as part of the nation's priority homeland security initiative to protect animal, plant, and public health from acts of bioterrorism (39, 40, 45). These rules contain provisions for controlling the possession, security, and use of select agents and toxins to safeguard the public health and to protect the health and safety of laboratory workers who will handle these agents in conducting medical and biodefense research.

Biological safety is an important concern for people who work in bacteriology, microbiology, and biomedical laboratories. It is natural to want to protect oneself from harm, and thoughtful people would never knowingly do something that would increase the risk for a colleague to contract an occupational illness. The ability to prevent laboratory-acquired infection, however, requires certain skills and knowledge that can best be acquired through training and careful guidance provided by experienced colleagues. A study of the literature is not sufficient to prepare a laboratory worker for safely handling pathogenic microorganisms. It is necessary to develop proficiency in microbiological techniques through practice with nonpathogenic microorganisms before higher-risk microorganisms are introduced into the laboratory routine (2).

Several books provide comprehensive coverage of the practice of biological safety (9, 19, 21, 44). The consensus guidelines published by the CDC and the NIH (44) should be consulted before pathogenic bacteria are introduced into the laboratory. Authoritative books on chemical safety (22), radiation safety (35), physical safety (16), and disinfection and sterilization (3, 31) are available. Technical handbooks on protocols in molecular biology (6, 33) and microbiology (13) are an excellent sources of laboratory safety information. Also a safety manual (10) is available that can serve as an excellent example of how one university met the intent of regulatory standards that require written safety plans. These references provide detailed information, which is beyond the basic scope of this chapter.

46.1. FEDERAL BIOSAFETY GUIDELINES AND REGULATIONS

46.1.1. NIH Recombinant DNA Guidelines

The NIH Recombinant DNA Guidelines (NIH Guidelines) were developed in 1976 as a response to the public and scientific concern that "gene splicing" technology might give rise to new biological agents with unknown and possibly dangerous characteristics. The NIH Guidelines described four levels of physical containment—referred to at that time as P1, P2, P3, and P4—each providing a level of protection greater than the one preceding it. The assessment of potential risks used to define classes of experiments assigned to each containment level, and the selection of safe laboratory practices, containment equipment, and facility safeguards that describe the safety requirements for each level were based on the scientific knowledge and experience acquired in bacteriological and microbiological laboratories conducting research and diagnostic studies involving human pathogens. The NIH Guidelines establish a multilevel review and approval process that is designed to ensure active and robust oversight of the uses of recombinant DNA technology. The process involves the participation of the principal investigator, the Institutional Biosafety Committee, the NIH Office of Biotechnology Assessment, the NIH Recombinant DNA Advisory Committee (RAC), and the NIH Director. The NIH Guidelines are mandatory for institutions that receive any federal funding for research involving recombinant DNA technologies. The NIH also encourages other institutions that do not receive NIH funding to adopt the NIH Guidelines voluntarily.

Recombinant DNA technology provides a powerful set of biological "tools" for identifying and isolating genes and discerning gene function. These "tools" are well characterized host/vector systems and restriction enzymes and include genetically engineered bacterial hosts and plasmids, debilitated bacteriophages and viruses, and enzymes with the capability of cutting DNA at specific sequences. The development and use of host-vector systems that lack virulence genes, require enriched laboratory media for growth, and cannot colonize the human gut (9, 17) demonstrated that scientists could exploit this new technology to enhance the safety of infectious disease research by studying the virulence and pathogenicity of infectious agents at the molecular level rather than by performing experiments that involve handling the infectious agent. The capability to develop debilitated vectors also made it possible to consider applying this technology to the conduct of human gene transfer experiments for the purpose of evaluating the safety and potential therapeutic benefits of the emerging field of molecular medicine.

The NIH Guidelines were frequently revised as knowledge and experience from the use of this technology in research was acquired. The rapid gain in knowledge quickly allayed the original concerns that led to the development of the initial NIH Guidelines. NIH delegated compliance responsibility to the institutions in 1978. In subsequent revisions, NIH tasked the RAC to develop detailed and explanatory appendices that would aid institutions in carrying out their responsibilities. For example, Appendix B presents a classification of human etiological agents on the basis of hazard. The classification schema uses prevalence of infection and severity of disease as the primary indicators of hazard. Investigators should review Appendix B when determining the appropriate biosafety level for experiments involving the introduction of recombinant DNA into a human pathogen or the cloning of DNA from a human pathogen into a nonpathogenic prokaryotic or lower eukaryotic host-vector system. The current primary role of the RAC is to provide guidance to investigators and IBCs on the scientific and ethical issues associated with human gene transfer experimentation and to maintain the relevancy and applicability of the NIH Guidelines to modern biological research. NIH also changed the P designations for physical containment to correspond with the biosafety level (BL) nomenclature adopted by CDC and NIH in 1984 for new consensus guidelines for the safe use of human pathogens in microbiological and biomedical laboratories (44).

46.1.2. Biosafety in Microbiological and Biomedical Laboratories

Occupational health and safety, environmental, and community protection must be integral components of the routine practices and procedures of research laboratories. It is the responsibility of institutions, investigators, and staff to encourage safety and promote the safe conduct of research. Following the development of the NIH Guidelines, the CDC and NIH recognized the value in promoting biosafety more broadly among diagnostic and research laboratories that handle human pathogens and zoonotic agents. CDC and NIH formed a partnership to develop a voluntary code of practice that would provide authoritative guidance for protecting the health and safety of laboratory workers and the public health from hazards associated with the possession and use of infectious materials in microbiological and biomedical laboratories. CDC and NIH invited scientists and laboratory directors, experienced laboratory workers, and health and safety professionals to work collaboratively in developing consensus practices for the code. CDC and NIH published the first edition of the code of practice, Biosafety in Microbiological and Biomedical Laboratories (BMBL) in 1984 (44). They published the fourth edition in 1999 and the fifth edition in 2007.

This code of practice is an excellent resource. It describes four biosafety levels that parallel the four containment levels of the NIH Guidelines. Recommendations for the laboratory control and containment of specific agents and their use in small laboratory animals is based on the potential hazard of the agent, the mode of transmission, the severity of the disease, and the laboratory activity. The BMBL provides consensus recommendations on laboratory practices and techniques, safety equipment and primary barriers, and facility features for each of the four biosafety levels. The BMBL encourages laboratory directors and in-

vestigators to conduct risk assessments in selecting the appropriate practices for use in their laboratories.

46.1.3. OSHA Occupational Exposure to Bloodborne Pathogens Rule

The Occupational Safety and Health Administration (OSHA) of the Department of Labor issued voluntary guidelines in 1983 to reduce the risk of occupational exposure to hepatitis B virus in the health care industry. In 1987 the Department of Labor and the Department of Health and Human Services published a Joint Advisory Notice on the protection against occupational exposure to hepatitis B virus (HBV) and human immunodeficiency virus (HIV). OSHA initiated rulemaking in 1987 after determining that over 2,000,000 health care workers in the United States, including 500,000 laboratory workers, were at risk for occupational exposure to hepatitis B virus (46). These agencies estimated that 12,000 cases of HBV infection occur in health care workers each year, that an estimated 10% of these patients become chronic HBV carriers, and that 200 to 300 fatalities occur as a result of HBV-related disease. Other blood-borne pathogens (such as human immunodeficiency virus and the causative agents of syphilis, malaria, and hepatitis C) have similar transmission patterns and may pose an occupational hazard for health care workers. The OSHA blood-borne pathogens standard was published on December 6, 1991, and became effective on March 6, 1992 (49).

The OSHA blood-borne pathogens standard has had a far-reaching effect on most of the biological research laboratories in the United States. Careful consideration was given to the possible risks inherent in a wide range of human-derived materials, including primary cells, tissues, organs, and established human cell lines. The inclusion of established human cell lines in the regulatory framework in 1994 has extended the requirements of the OSHA Standard into basic research and molecular biology laboratories, far from the health care and clinical diagnostic settings originally considered in the preamble to the regulation. The OSHA blood-borne pathogens standard requires institutions, investigators, and laboratory safety personnel to work together to assess the risks inherent in the research; develop control and containment practices and procedures, including the offer of a very safe vaccine for individuals at risk from their work; and ensure awareness and expertise through training. The concept of universal precautions, the safety practices and procedures outlined in the OSHA blood-borne pathogens standard for work with human materials, cells, and cell lines are consistent with the CDC and NIH definition of biosafety level 2 practices and procedures. The 4th edition of Biosafety in Microbiological and Biomedical Laboratories contains a section on containment for working with human cells and tissues.

46.1.4. HHS and USDA Rules for the Possession, Use, and Transfer of Select Agents and Toxins

The September 11, 2001 terrorist attack on the United States and the subsequent acts of bioterrorism involving the dissemination of Bacillus anthracis exposed the vulnerability of the United States to terrorism, and the risk to humans, animals, and plants from the use of dangerous biological agents in domestic or international terrorism. Two new laws strengthen the nation's ability to protect the public health and safety from future bioterrorism threats. The USA Patriot Act of 2001 strengthens the criminal laws against terrorism and makes it illegal to possess any biological agent or toxin, including any genetically engineered organism created using recombinant DNA technology, for any purpose not reasonably justified by a prophylactic, protective, bona fide research, or other peaceful purpose. The act also defines who may transfer and possess dangerous pathogens and toxins. The Public Health Security and Bioterrorism and Response Act of 2002 improves the ability of the United States to prevent, prepare for, and respond to bioterrorism and other public health emergencies. The primary goals of this law are to ensure the prompt reporting to the federal government of entities and individuals who possess select agents, to increase the security over such agents, and to establish a comprehensive and detailed national database of the location and characterization of such agents and the identities of individuals who have authorization to possess them.

The Secretary of the Department of Heath and Human Services (HHS) has the authority and responsibility for regulating activities regarding select agents and toxins to protect the public health and safety. The Secretary of the Department of Agriculture (USDA) has the authority and responsibility for regulating activities regarding select agents and toxins to protect animal and plant health and animal and plant products. The list of regulated bacterial, rickettsial, and fungal select agents and toxins is accessible on the HHS web page www.cdc.gov/od/sap/docs/salist.pdf. Several select agents come under the jurisdiction of both agencies. Interagency coordination assures that implementing regulations regarding these "overlap" agents are not in conflict. Clinical and diagnostic laboratories are exempt from most provisions of the HHS and USDA select agents and toxins rules provided that all exemption criteria stated in the rules apply.

Implementing regulations were issued in December 2002 (39, 40, 45). Laboratory directors and principal investigators have a moral and legal responsibility to obtain authorization before taking possession of a select agent and to ensure that they meet all of the requirements of the implementing regulations for the possession and use of select agents, such as the provisions addressing safety plans, security plans, emergency response plans, training, record keeping, inspections, and notifications. The HHS rule encourages entities to consider the consensus guidelines in *Biosafety in Microbiological and Biomedical Laboratories* in preparing the required select agent safety plan (44).

46.2. PRINCIPLES

46.2.1. Directors' Responsibilities

Laboratory directors have the primary responsibility for setting the expectations and means for protecting the health and safety of all laboratory workers under their supervision. This involves ensuring that (i) hazards associated with laboratory activities are recognized, (ii) routine and special practices that are helpful in controlling or eliminating those hazards are selected for use in the laboratory, and (iii) laboratory workers are provided with adequate instruction and training so that they understand the relative significance of the hazards that relate to them personally and acquire the proficiency necessary to protect themselves and their colleagues from these hazards. The laboratory director is also responsible for deciding who should have access to the laboratory and for establishing additional requirements aimed at reducing potential exposure of persons who may be at increased risk of infection. The preparation of a safety manual

for the laboratory is helpful in clearly defining safety expectations and setting a positive attitude that will sustain a high level of safety awareness. The references cited above emphasize the importance and value of safety manuals that are relevant to the actual work activities of the laboratory and provide excellent guidance for developing such manuals.

The laboratory director has a legal obligation to prepare and use a safety manual or plan if the laboratory is involved in the handling of hazardous chemicals, human blood, blood-borne pathogens, or select agents and toxins. Hazardous chemicals, as defined by the OSHA standard on occupational exposure to hazardous chemicals in laboratories (48), are found in virtually every general or molecular bacteriology laboratory, and therefore the standard will apply to them all. Among the provisions of the standard is the requirement to develop and follow a written Chemical Hygiene Plan. The plan must describe the specific measures that are to be taken to protect employees from health hazards associated with hazardous chemicals in the laboratory. Laboratories that handle human blood and blood-borne pathogens are required by the OSHA blood-borne pathogens standard (49) to establish a similar document called an Exposure Control Plan. The purpose of this plan is to describe the emergency controls, laboratory practices, personal protective equipment, and housekeeping practices that will eliminate or minimize employee exposure to human blood and blood-borne pathogens.

A healthful and safe laboratory environment cannot be sustained without every laboratory worker sharing in the responsibility for safety. All persons in the laboratory must be vigilant toward their own safety and the safety of their colleagues and neighbors. They must know and follow safety guidelines, regulations, and procedures that apply to their work. They should actively seek information and advice to improve their safety performance. They should always report accidents, injuries, and unsafe conditions to their supervisor or safety officer. It is imperative that all laboratory employees know how to respond to emergencies. Directors can promote good work habits among all members of the laboratory by encouraging them to participate in the selection of safe practices and corresponding proficiency standards that all should meet to ensure safe working conditions for everyone (2). Ideally, the methods and attitudes that enhance laboratory safety will become a conscious part of all work activity in the laboratory.

46.2.2. Biosafety Principles

Awareness of the potential risks, infectious and otherwise, associated with the procedure or activity being conducted is the safety cornerstone in clinical and research laboratories. To be aware of these risks, the individual worker must be knowledgeable about the biological agents and materials, reagents, chemicals, equipment, and facilities being used. Knowledge implies appropriate training in risk recognition and prevention, in the proper execution of procedures, and in the use of equipment.

The concept of universal precautions that is used in the clinical setting is a prudent standard (41). It is based on the realization that all clinical specimens, body fluids, tissues, and cultures are potentially infectious and should be handled with common sense practices and barrier precautions. Standard and special microbiological practices, containment equipment, and facilities recommended in the CDC and NIH consensus guidelines (44) for biosafety level 2 (see section 46.5) constitute universal precaution for the laboratory and should be adopted as a minimal standard for work with any potentially infectious materials.

Containment of infectious agents within laboratory and research animal work areas is provided by appropriate combinations of laboratory practices, safety equipment, and facility design features. Primary containment protects the worker and the immediate work area from infection or contamination. It is provided by consistent use of accepted good laboratory practices and appropriate use of safety equipment, such as the use of biological safety cabinets (43), for activities involving known or potentially infectious materials. The use of safe and efficacious vaccines may provide additional reductions in infection risks of workers.

Secondary containment, i.e., the protection of the environment external to the laboratory, is provided by a combination of facility design features (e.g., directional airflow, sealed penetrations, double doors), management of wastes and other infectious materials, and access control.

46.2.3. Facility Features

Safety features of laboratory facilities provide architectural and mechanical barriers that serve to segregate activities performed within a laboratory module from nonlaboratory areas such as administrative offices and cafeterias. The ventilation system provides for correct operations of safety cabinets and, in containment laboratories, can prevent the migration of airborne contaminants between occupied spaces. The design of entry areas to laboratories can be helpful in maintaining proper access control. It is also important that an autoclave for decontaminating infectious laboratory wastes be available within the facility.

The basic safety features of individual laboratory modules within a biosafety level 2 facility are floor and wall surfaces that are easily cleaned, a hand-washing sink, and bench tops that are impervious to water and resistant to acids, alkalis, organic solvents, and moderate heat. Spaces between and beneath casework and equipment should be accessible for cleaning. A safety shower and eye wash fountain should also be easily accessible to all laboratory occupants.

Additional safety features are required for biosafety level 3 facilities. The laboratory is physically separated from nonlaboratory areas by entry corridors that require passage through two sets of doors. Floor, wall, and ceiling surfaces are water resistant. Penetrations through these surfaces should be sealed to allow space decontamination. Foot-operated, elbow-operated, or automatically operated hand-washing sinks are located in each laboratory near the exit door. The ventilation system creates directional airflow by drawing air into the laboratory through the entry area. Exhaust air is discharged directly to the outside and dispersed away from occupied areas and air intakes of buildings.

46.3. HISTORICAL PERSPECTIVE

Occupational infections of laboratory workers have been documented for a century. Although infections with specific agents have varied by time and place, occupational infections have typically involved agents prevalent in the community and, consequently, likely to be present in materials received and handled in laboratories involved in diagnostic, production, reference, and research activities.

In 1941, the first survey of occupational infections in laboratory workers in the United States was published by Meyer and Eddy (18). In 1949, Sulkin and Pike published the first in a series of surveys of laboratory-associated infec-

tions listing 222 viral infections, 21 of which were fatal (37). A second survey, involving 5,000 laboratories and published in 1951, summarized 1,342 cases, only one-third of which had previously been reported (38). Brucellosis was the most common occupational infection; together, brucellosis, tuberculosis, tularemia, typhoid, and streptococcal infections accounted for 72% of all bacterial infections and 319 of the total number of cases caused by all categories of infectious agents. The overall case fatality rate was 3%.

The survey data were updated in 1965 and again in 1976, summarizing a cumulative total of 3,921 reported occupational infections worldwide (27, 28). Brucellosis was again the most commonly reported infection; brucellosis, typhoid fever, tularemia, tuberculosis, hepatitis, and Venezuelan equine encephalitis were the most frequently recorded occupational infections at that time.

Of 703 cases in which the source of infection was identified, 80% were due to contact with infectious materials from spills and splashes, accidents involving the use of needles and syringes, mouth pipetting, and injury from contaminated broken glass or other sharp objects. Bites and scratches by experimentally or naturally infected animals or ectoparasites was the fifth major cause of known accidents resulting in laboratory-acquired infection. The most frequently reported cases were 18 cases of ratbite fever (*Streptobacillus moniliformis*), 13 cases of leptospirosis (*Leptospira interrogans*), and 9 cases of B virus (herpesvirus simiae) infections. Awareness of occupational risks, good practices, and modest barrier precautions might have effectively prevented these occupational infections.

From these cited reports, as well as others, a consistent pattern and trend characterizing occupational infections in laboratory workers have emerged. Pertinent issues include the following.

1. Occupational infections are underreported. The Sulkin and Pike studies (37, 38) indicated that, historically, only one-third of the occupational infections that were brought to their attention were reported in the scientific literature. Furthermore, no national focus exists for the reporting or active surveillance of occupational infections.

2. Fewer than 20% of occupational infections are associated with an identified exposure, accident, or incident. The most plausible explanation is that diseases such as brucellosis, tuberculosis, and tularemia, and those caused by certain arboviruses occur as a result of exposure to respirable airborne infectious particles (droplet nuclei). Diseases such as typhoid, cholera, and viral hepatitis are most likely to be associated with poor laboratory practices, poor personal hygiene, and failure to use simple barrier precautions.

3. The majority of cases in which the source of infection is identified could have been prevented by regular use of good laboratory safety practices.

Harding and Byers (12) analyzed data from more than 200 reports on laboratory-associated infections found in the literature worldwide between 1979 and 1999. Their work finds that the pattern and trend of laboratory-associated infections observed in earlier studies in general have not changed. The top 10 laboratory-acquired diseases in both periods included brucellosis, Q fever, and tuberculosis, all of which are diseases that can result from inhalation exposures. Although hepatitis remains on this list a decline in the number of infections has been noted. There were no reports of laboratory-associated infections with HBV in the United States after 1987, suggesting that the OSHA blood-borne pathogens standard has had a positive impact on worker protection. The authors noted that the distribution of laboratory-associated infections among diagnostic and research laboratories has changed. In the earlier studies, diagnostic and research laboratories accounted for 17% and 59%, respectively, of all reported laboratory-associated infections, whereas the distribution observed between 1979 and 1999 for diagnostic and research laboratories was 45% and 51%, respectively. These trends reinforce the importance of the safe practices, good habits, training, and proficiency testing in contemporary bacteriology, microbiology, and biomedical laboratories for preventing laboratory-associated infections.

46.4. BIOSAFETY LEVELS

Although the number of laboratories in the United States is unknown, it is reasonable to assume that clinical laboratories (including physicians' office laboratories) constitute the most common type of laboratory operation. Clinical laboratory activities typically involve the handling, manipulation, and, sometimes, collection of blood, urine, feces, and other body fluids, excretions, or secretions. These clinical materials are all potentially infectious and may contain a variety of bacterial, fungal, viral, and protozoal agents. The concept of universal blood and body fluid precautions (universal precaution) should apply to all clinical specimens since it is not realistically possible to rule out all infectious agents on the basis of medical history and signs and symptoms at the time of medical examination or clinical diagnosis (41). It is also difficult to determine, with reasonable confidence, whether certain body fluids typically not associated with the transmission of blood-borne diseases may contain visible blood. Consequently, it is prudent to handle all body fluids, tissues, and cultures as potentially infectious. The estimated prevalence of HBV infection in the United States (approximately 300,000 new cases annually) would suggest that this viral disease is the occupational infection hazard most likely to be present in clinical materials submitted for laboratory examination.

HBV shares a common transmission pattern in the laboratory occupational setting with most bacterial pathogens that are regularly present in a community. Such agents may be transmitted parenterally, by ingestion, via broken skin, and following exposure of the mucous membranes of the mouth, eyes, or anterior nose. Characteristically, these pathogens are not transmitted in the laboratory via infectious aerosols.

Consequently, it is reasonable to use laboratory practices and barrier precautions that prevent exposure of laboratory workers from the infectious agent most likely to be present in specimens handled on a regular basis. It is especially important that clinical chemists, geneticists, and other scientists without specific training or experience in working with potentially infectious materials are made aware of such hazards and informed about appropriate preventive measures. These practices and precautions are succinctly summarized in the description of standard and special microbiological practices recommended for biosafety level 2 (44) and are discussed in detail in section 46.5.

A higher level of occupational hazard occurs with the handling of primary pathogens that can cause disease through the inhalation of aerosols as well as routes of direct exposure. Diseases such as brucellosis, tularemia, tuberculosis, anthrax, Q fever, and diseases caused by certain indigenous arboviruses are transmitted by aerosols. All of these diseases, except for tuberculosis and diseases caused by the deliberate release or use of a select agent as a weapon of mass destruction (e.g., *B. anthracis*), are currently diseases of low incidence and sporadic distribution. The standard and special practices and containment equipment described for biosafety level 3 are recommended for the handling of pathogenic organisms presenting an aerosol risk of infection (44); these are discussed further in section 46.6.

Tuberculosis, historically important as an occupational infection, continues to represent an especially significant infection hazard to laboratory workers, who may encounter this agent in sputum samples, biopsy materials, and cultures. *Mycobacterium bovis* BCG vaccine, which is licensed and readily available, is not customarily used in the United States, so susceptibility to tuberculosis infection is essentially universal in this country. The adult infectious dose is estimated to be 10 or fewer organisms (51). Although the incidence of occupationally acquired tuberculosis is presently unknown, reports over several years attest to the continuing risk of infection with this classical and still important pathogen (G. P. Kubica, personal communication). A survey in 1976 indicated that the incidence of tuberculosis among medical laboratory workers in England was five times greater than that for the general community (11). Tuberculosis is an appropriate representative of occupational infections that may be transmitted via infectious aerosols.

Biosafety level 1 is not appropriate for work with pathogenic bacteria. This level was described in guidelines developed by CDC and NIH (44) to indicate basic practices appropriate for undergraduate and secondary educational training and teaching laboratories for which introductory microbiological protocols would involve only defined and characterized strains of viable microorganisms not known to cause disease in healthy adult humans. Emphasis is placed on the use of mechanical pipetting aids; hand-washing; not eating, drinking, or smoking in the work area; and daily decontamination of work surfaces. This minimal level of containment has been adopted by molecular biology laboratories involved in research activities for which the host-vector systems in use have been exempted from the NIH Guidelines for handling recombinant DNA molecules (42).

Most human pathogens encountered in diagnostic and clinical laboratories and used in biomedical and microbiological research are safely handled using practices described for biosafety levels 2 and 3. These practices are the subject of this chapter. The safe conduct of research on exotic lethal pathogens like the hemorrhagic viruses requires biosafety level 4 practices and sophisticated biocontainment facilities. A discussion of biosafety level 4 practices, containment equipment, and facilities is beyond the scope of this chapter. However, investigators and technicians who will join the nation's scientific initiative to conduct research to protect the public health from the adverse consequences of bioterrorism will have to gain a thorough understanding of this area of biological safety, acquire proficiency in the use of safe technique, and develop the habits necessary to sustain a high level of performance.

46.5. LEVEL 2 BIOSAFETY PRACTICES

The consistent use of good practices for the manipulation of potentially infectious materials is the single most important element in preventing occupational infections. It is prudent to consider that all bacteria are potential hazards to human health. Consequently, biosafety level 2 practices should be adopted even in general and molecular bacteriology laboratories.

Biosafety level 2 practices are recommended for most laboratory activities involving known pathogens and for experiments involving either the introduction of recombinant DNA into pathogenic bacteria or the introduction of DNA from pathogenic bacteria into nonpathogenic procaryotes or lower eucaryotes. *Brucella* spp., *Francisella tularensis*, M. *bovis*, and *Mycobacterium tuberculosis* require level 3 containment because of the serious hazard of aerosol infection. However, DNA from these bacteria can be cloned in nonpathogenic procaryotic or lower-eucaryotic host-vector systems under level 2 containment (42). The deliberate formation of recombinant DNAs containing genes for the biosynthesis of toxic molecules lethal for vertebrates, and the deliberate transfer of drug resistance genes to microorganisms that are not known to acquire the trait naturally, may require RAC review or NIH approval before experiments are initiated.

Use the following biosafety level 2 practices routinely when handling even potentially pathogenic bacteria. The bacteria capable of causing serious disease when inhaled in infectious aerosols require level 3 containment.

1. Keep laboratory doors closed when work is in progress.

2. Post a hazard warning sign incorporating the universal biohazard symbol (Fig. 1) on the laboratory door when infectious materials are in use and where special provisions for entry are required, such as an immunization requirement to handle hepatitis B virus. The OSHA blood-borne pathogens standard requires the sign to be orange or orange-red with symbol and lettering in a contrasting color (49). Also display the biohazard symbol on freezers, refrigerators, and other units used to store infectious materials.

3. Decontaminate work surfaces daily and immediately after spills of viable material.

4. Decontaminate infectious liquid or solid wastes before disposal. If wastes are decontaminated at a site away from the laboratory, place them in a durable leak-proof container and close it before removal from the laboratory.

5. Always use a mechanical pipetting device when pipetting.

6. Do not permit eating, drinking, smoking, storage or preparation of food for human consumption, or cosmetics application in the laboratory.

7. Wear protective gloves when contact with infectious material may be unavoidable. Remove gloves carefully when they become soiled with contaminated material, and wash hands immediately. Put on a new pair of gloves before continuing the procedure. Do not wash or reuse soiled gloves.

8. Wash hands after handling infectious materials and always before leaving the laboratory.

FIGURE 1 Biohazard warning sign for biomedical facilities.

9. Avoid the use of syringes, needles, and other sharp instruments whenever possible. When they are required, handle them with caution to avoid accidental punctures or cuts. Use only needle-locking syringes or disposable syringe-needle units for injection or aspiration of infectious fluids.

10. Discard used needles and disposable cutting instruments into a puncture-resistant container that has a lid.

11. Wear laboratory coats, gowns, smocks, or uniforms while in the laboratory.

12. Exercise care in all procedures and manipulations to minimize aerosol formation. Use biological safety cabinets or physical containment devices such as centrifuge safety cups to contain operations that will generate aerosols capable of causing significant contamination or hazardous exposure.

13. Wear face protection, such as goggles, mask, or face shield, when performing outside a biological safety cabinet a procedure that may splash or spray infectious materials on the face.

14. A biological safety manual that is appropriate for the work of the laboratory should be available in the laboratory for staff to use for reference as necessary.

Several routine operations in the laboratory can easily become the source of laboratory-acquired infections. They require vigilance to guard against compromise and error. The following sections emphasize the hazards associated with these operations and underscore the precautions necessary to conduct them safely.

46.5.1. Personal Hygiene

Appropriate personal-hygiene practices for laboratory workers have the simple objective of reducing personal risks of self-infection by avoiding ingestion, skin, or mucous-membrane exposure to infectious materials. Wash hands with soap and water following the completion of bench activities, the removal of protective gloves, and the overt contamination of skin and when leaving the laboratory. Hand washing is among the most important actions that workers can take in reducing the risk of infection by agents commonly handled in biomedical laboratories. A 10-second hand wash with soap and water reduces the number of contaminating transient microorganisms to levels below the infective threshold. Use powdered soap from a dispenser rather than bar or liquid soap, which may harbor bacteria (see section 46.8.7). Follow hand washing by drying with disposable towels. Hand washing may also have the added benefit of reducing the level of cross-contamination between specimens or activities.

Wash hands promptly after removing protective gloves. Unrecognized small holes, abrasions, or tears, an opening at the wrist, or solvents can enable bacteria to pass the glove barrier and contaminate the skin. A disinfectant wash or dip may be desirable, but this procedure can cause roughening, desiccation, or sensitization of the skin.

Work with bacteria should not be performed by anyone with a new or old cut, an abrasion, a lesion of the skin, or any open wound, including that resulting from a tooth extraction.

Food, candy, gum, and beverages for human consumption should not be stored or consumed in the laboratory. Smoking should not be permitted. A beard is undesirable because it retains particulate contamination and because a clean-shaven face is essential to the adequate fit of a face mask or respirator. Develop the habit of keeping hands away from the mouth, nose, eyes, face, and hair to prevent self-inoculation. Keep personal items such as coats, hats, storm rubbers or overshoes, umbrellas, and purses out of the laboratory. Laboratory design that allows personal items to remain outside the work area plays an important role in overall safety.

46.5.2. Pipetting

Mouth pipetting and suctioning have been identified as practices that are significant and consistent causes of occupational infections (20, 21). Pipetting aids prevent accidental aspiration of cultures or other infectious fluids into the mouth and eliminate the hazard of transferring infectious agents from contaminated surfaces to the pipette mouthpiece by bare or gloved hands. Other hazards are associated with pipetting infectious materials. Forcibly expelling residual contents from pipettes, bubbling air through liquid cultures, and falling drops of liquid culture from pipette tips can create infectious aerosols in the breathing zone of the worker. These hazards are eliminated by good technique. Use pipettes that do not require expulsion of the last drop. Take care to avoid dropping cultures from the pipette. When pipetting materials containing pathogenic bacteria, place a disinfectant-soaked towel on the work surface. Discharge liquids from pipettes as close as possible to the fluid or agar level of the receiving vessel, or allow the contents to run down the wall of the receiving vessel. Avoid dropping the contents from a height into vessels. Place contaminated pipettes into a container with enough suitable disinfectant for complete immersion.

A variety of pipetting aids are available, ranging from simple bulb- and piston-actuated devices to sophisticated devices that contain their own vacuum pumps (19). In selecting a pipetting aid, consider the procedure, the ease of operation, and the accuracy needed. Use several pipetting aids; none is available that is appropriate for all procedures.

46.5.3. Hypodermic Syringes and Needles

Accidents involving the use of syringes and needles are among the most common cause of occupational infections in laboratories and health care facilities. They account for an estimated one-fourth of laboratory-acquired infections that are caused by accidents (29). Needle "sticks" (punctures) account for an estimated one-third of work-related accidents in health care workers (14). In some health care facilities, annual injury rates from being accidentally stuck by a needle may exceed 100 per 1,000 employees, and these injuries may be underreported by 40 to 70% (1, 14). Accidental parenteral inoculation of infectious fluids during phlebotomy procedures is the most obvious and common means of occupational infection associated with needles. Infections have also been associated with sprays or aerosols of infectious materials when needles separate from syringes and with direct or indirect skin contact with infectious fluids from leaking syringes or needles.

Of seven categories of phlebotomy devices used in a university hospital, disposable syringes are the devices most commonly associated with occupational needle sticks (14). However, when the data comparing the number of injuries with the number of devices purchased and presumably used were reviewed, disposable syringes had the lowest rate of associated injuries among all categories of devices used. Injuries associated with the use of disposable syringes most commonly occurred during the recapping of used needles.

In contrast to health care activities, when needles and syringes or other phlebotomy devices are commonly used to collect blood and other fluids from patients, there are relatively few indications for the use of such devices in most ordinary laboratory activities. Restrict the use of needles and syringes to the collection of blood or fluids from patients or experimental animals, the injection of materials into patients or experimental animals, and the aspiration of reagents or other materials from containers with diaphragm stoppers. Conduct all other manipulations of potentially infectious or hazardous fluids with mechanical pipettes or devices other than needles and syringes.

In response to the continuing problem of needle sticks, manufacturers of phlebotomy devices have developed a variety of "safety syringes" (e.g., a retractable needle). The use of these devices may provide additional safeguards during phlebotomy procedures. The wearing of gloves may also further reduce the risk of occupational exposures of skin to body fluids during phlebotomy and other procedures involving the collection or handling of potentially infectious body fluids.

The following practices are recommended for hypodermic needles and syringes when used for parenteral injections.

1. Avoid quick and unnecessary movements of the hand holding the syringe.

2. Examine syringes for chips and cracks, and examine needles for barbs and plugs. Do this before sterilization and before use.

3. Use needle-locking (Luer-Lok type) syringes only, and be sure that the needle is locked securely.

4. Wear surgical or other rubber gloves.

5. Fill syringes carefully to minimize air bubbles and frothing.

6. Expel excess air, liquid, and bubbles vertically into a cotton pledget moistened with a suitable disinfectant or into a small bottle of sterile cotton.

7. Do not use a syringe to forcefully expel infectious fluid into an open vial or tube for the purpose of mixing. Mixing with a syringe is appropriate only if the tip of the needle is held below the surface of the fluid in the tube.

8. If syringes are filled from test tubes, take care not to contaminate the hub of the needle, since this may result in transfer of infectious material to the fingers.

9. When removing a syringe and needle from a rubber-stoppered bottle, wrap the needle and stopper in a cotton pledget moistened with a suitable disinfectant. If there is danger that the disinfectant may contaminate sensitive experimental materials, a sterile dry pledget may be used and discarded immediately into a disinfectant solution.

10. Inoculate animals with the hand positioned behind the needle to avoid punctures.

11. Be sure that the animal is properly restrained prior to inoculation, and be alert for any unexpected movements. Consider anesthetizing the animal during injections if approved by the Institutional Animal Care and Use Committee (IACUC).

12. Before and especially after injecting an animal, swab the injection site with a disinfectant.

13. Do not bend, shear, recap, or remove the needle from the syringe by hand. Place used needle-syringe units directly into a puncture-resistant container, and decontaminate before disassembly, reuse, or disposal.

46.5.4. Opening Containers

The opening of vials, flasks, petri dishes, culture tubes, embryonated eggs, and other containers of potentially infectious materials poses potential but subtle risks of creating droplets, aerosols, or contamination of the skin or the immediate work area.

Because the level of energy applied in such activities is relatively low, the efficiency for creating aerosols is also relatively low and the droplet is typically large. The potential for creating respirable droplets (in the 1- to 5-μm range) is relatively low. However, the potential for creating larger droplets, which may contaminate hands and immediate work surfaces, is relatively high.

The most common opening activity in most health care laboratories is the removal of stoppers from containers of clinical materials (e.g., blood, urine, spinal fluid, feces) submitted for examination. Seventeen percent of blood samples received in a clinical laboratory have blood on the container label, 6% of stool specimens have feces on the outside of the container, and 4 to 5% of laboratory specimen forms are visibly blood-stained (5). Specimens should be received and opened only by personnel who are knowledgeable about occupational infection risks and of the concept of Universal Precautions. Open clinical specimen containers in well-lighted designated areas with impervious and easily cleanable work surfaces. Wear a laboratory coat and suitable gloves. The use of plastic-backed absorbent paper towels on work surfaces facilitates cleanup and minimizes the creation of aerosols in the event of overt or unobserved spills of infectious materials. Some laboratories have found it to be advantageous to open all incoming specimens in the work chamber of a class I or class II biological safety cabinet, which is especially useful if incoming specimens are broken or leaking.

Appropriate techniques for opening sealed ampoules, dry powders, sealed containers under relatively negative pressure, embryonated eggs, and the like are described in detail elsewhere (20). The use of appropriate personal protection (coats, gloves), primary containment (biological safety cabinet), and physical barriers (e.g., acrylic "beta" shields) in general addresses the ubiquitous problems of exposure of personnel and contamination of work surfaces during the opening of containers.

Tubes containing bacterial suspensions should be manipulated with care. Simple procedures such as removing a tube cap or transferring an inoculum can create aerosols. Clearly mark tubes and racks of tubes containing biohazardous material. Use safety test tube trays in place of conventional test tube racks to minimize spillage from broken tubes. A safety test tube tray is one that has a solid bottom and sides deep enough to hold the liquid should the test tube break.

Vigorous shaking of liquid cultures creates a heavy aerosol. In general, swirling action will create a homogeneous suspension with a minimum of aerosol. When resuspending a liquid culture, allow a few minutes to elapse before opening the container, to reduce the aerosol.

The insertion of a sterile hot loop or needle into a liquid or slant culture can cause spattering and release of aerosol. To minimize aerosols, allow the loop to cool in the air or be cooled by touching to the inside of the container or to the agar surface where no growth is evident prior to contact with the culture or colony. After use of the inoculating loop or needle, sterilize it in an electric or gas incinerator specifically designed for this purpose rather than heating it in an open flame. These small incinerators have a shield to contain any material that may spatter from the loop. Disposable inoculating loops are available commercially; rather than decontaminating them with heat, discard them into a disinfectant.

Streaking an inoculum on rough agar results in aerosol production, created by the vibrating loop or needle. This generally does not occur if the operation is performed on smooth agar. Discard all rough agar-poured plates intended for streaking with a wire loop.

Water of syneresis in petri dish cultures may contain viable bacterial and can form a film between the rim and lid of the inverted plate. Aerosols result when this film is broken by opening the plate. Vented plastic petri dishes, whose lid touches the rim at only three points, are less likely to create this hazard. Filter papers fitted into the lids reduce but do not prevent aerosols. If plates are obviously wet, open them in a safety cabinet.

Less obvious is the release of aerosols when screwcap bottles or plugged tubes are opened. This happens when a film of contaminated liquid between the rim and the liner is broken during opening. Removing cotton plugs or other closures from flasks, bottles, and centrifuge tubes immediately after shaking or centrifugation can release aerosols. Removal of wet closures, which can occur if the flask or centrifuge tube is not held upright, is also hazardous. In addition, when using the centrifuge, there may be a small amount of foaming and the closure may become slightly moistened. Because of these possibilities, open all liquid cultures of infectious material in a safety cabinet and wear gloves and a long-sleeved laboratory garment while doing so.

When a sealed ampoule containing a lyophilized or liquid culture is opened, an aerosol may be created. Open ampoules in a safety cabinet. When recovering the contents of an ampoule, take care not to cut the gloves or hands or to disperse broken glass into the eyes, face, or environment. In addition, the material itself should not be contaminated with foreign organisms or with disinfectants. To accomplish this, work in a safety cabinet and wear gloves. Nick the ampoule with a file near the neck. Wrap the ampoule in disinfectant-wetted cotton. Snap the ampoule open at the nick, being sure to hold it upright. Alternatively, apply a hot wire or rod at the file mark to develop a crack. Then wrap the ampoule in disinfectant-wetted cotton and snap it open. Discard the cotton and ampoule tip into disinfectant. Reconstitute the contents of the ampoule by slowly adding fluid to avoid aerosolizing the dried material. Mix the contents without bubbling, and withdraw them into a fresh container. Commercially available ampoules prescored for easy opening are available. However, prescoring may weaken the ampoule and cause it to break during handling and storage. Open ampoules of liquid cultures in a similar way.

Harvesting cultures from embryonated eggs is a hazardous procedure and leads to heavy contamination of the egg trays and shells, the environment, and the hands of the operator. Conduct operations of this type in a safety cabinet. A suitable disinfectant should be at hand and used frequently.

46.5.5. Centrifuging

Centrifugation is a common procedure in laboratories where infectious materials are handled. The demonstrated hazards are relatively low, however, and are typically confined to diseases that have low infective doses and are efficiently transmitted by means of aerosols (e.g., tuberculosis, tularemia, brucellosis, Q fever, and certain arbovirus infections). Historically, infections associated with centrifugation typically have occurred as sporadic clusters of cases and are often associated with broken or leaking specimens in open, low-speed clinical centrifuges (20, 50).

Blood-borne pathogens (e.g., *Treponema pallidum*, hepatitis B virus) are not transmitted by infectious aerosols in the community and have not been documented to be transmitted by infectious aerosols in the laboratory. The viscosity of blood and other body fluids and the use of low-speed centrifugation may further reduce the production of infectious droplet nuclei and minimize occupational infection risks.

The use of centrifuge safety cups and enclosed or safety centrifuges (50) for low-speed centrifugation and rotors with O-ring seals for high-speed centrifugation may effectively reduce the hazards associated with these centrifugation procedures. Mechanical failure of centrifuges can be minimized by appropriate servicing and maintenance and careful attention to good technique. Infectious or sensitizing materials, concentrated by centrifugation, should be handled by using appropriate primary containment and personal protection devices.

Use safety centrifuge trunnion cups when centrifuging infectious bacterial suspensions. Fill and open centrifuge tubes and trunnion cups in a biological safety cabinet. Decontaminate the outside surface of the cups before removing them from the cabinet. Since some disinfectants are corrosive, rinse with clean water after an appropriate contact time has elapsed.

Before centrifuging, eliminate tubes with cracks and chipped rims, inspect the inside of the trunnion cups for rough walls caused by erosion or adhering matter, and carefully remove bits of glass and other debris from the rubber cushion.

Add a disinfectant between the tube and trunnion cup to disinfect the materials in case of accidental breakage. This also provides an excellent cushion against shocks that might break the tube. Take care not to contaminate the culture material with the disinfectant. Also, if the tube breaks, the disinfectant may not completely inactivate the infectious material because of the dilution of the disinfectant and the high concentration of packing of the cells.

Avoid pouring supernatant fluid from centrifuge tubes. If this must be done, wipe off the outer rim with a disinfectant afterward; otherwise, in a subsequent step, infectious fluid may be spun off as droplets and form an aerosol. Use of vacuum system with appropriate in-line safety reservoirs and filters is preferable to pouring from centrifuge tubes or bottles.

Avoid filling the centrifuge tube to the point that the rim, cap, or cotton plug becomes wet with culture. Screw caps or caps that fit over the rim are safer than plug-in closures. Some fluid usually collects between a plug-in closure and the rim of the tube. Even screw-cap bottles are not without risk; if the rim is soiled and sealed imperfectly, some fluid will escape down the outside of the tube.

Do not use aluminum foil to form centrifuge tube caps, because they often become detached or rupture during handling and centrifuging.

The balancing of buckets and trunnion cups is often improperly performed. Take care to ensure that matched sets of trunnions, buckets, and plastic inserts do not become mixed. If the components are not inscribed with their weights by the manufacturer, apply colored stains for identification.

High-speed rotor heads are prone to metal fatigue, and if there is a chance that they may be used on more than one machine, each rotor should be accompanied by its own log book indicating the number of hours run at top or derated speeds. Failure to observe this precaution can result in dangerous and expensive disintegration. Frequent inspection, cleaning, and drying are important to ensure the absence of corrosion or other damage that may lead to the development of cracks. If the rotor is treated with a disinfectant, rinse it with clean water and dry it as soon as the disinfectant has adequately decontaminated the rotor. Examine rubber O-rings and tube closures for deterioration, and keep them lubricated as recommended by the manufacturer. When tubes of different materials (e.g., celluloid, polypropylene, stainless steel) are provided, take care that tube closures designed specifically for the type of tube are used. Caps are often similar in appearance but are prone to leak if applied to tubes of the wrong material. Properly designed tubes and rotors, when well maintained and handled, should never leak.

46.5.6. Mixing and Disruption

Manual or mechanical mixing, grinding, or disruption of infectious materials has a high potential for producing infectious aerosols. The more vigorous the manipulation, the more likely that respirable (1- to 5-μm) droplet nuclei may be created and liberated into the laboratory. Droplet nuclei, containing infectious agents, may remain suspended indefinitely and pose infection risks long after completion of the activity. Infectious droplet nuclei may also be transported to other laboratory and nonlaboratory areas by personnel or via the building ventilation system. Larger droplets tend to settle rapidly on horizontal surfaces and may serve as direct and indirect sources of infection or contamination.

All manipulations of known or potentially infectious materials should be conducted in such a manner that the contamination of hands, clothing, and work surfaces and the production of aerosols are avoided. These occupational hazards are best addressed by the routine use of leak-proof equipment (safety blenders) in combination with protective clothing and primary containment devices (biological safety cabinets).

It is important that the selection of devices and containment used for grinding, mixing, or disruption be based on the assessed infection risks or the consequences of infection of the agent being handled. For example, although the recommended barrier and containment precautions for grinding brain tissue infected with rabies virus or yolk sacs infected with Q-fever rickettsia are the same, the demonstrated infection risks are remarkably different. As few as 10 viable rickettsia can produce infection via the respiratory route. On the other hand, occupational infection with rabies virus via the respiratory route has not been demonstrated in personnel working with clinical materials and conducting diagnostic activities.

When using blenders, mixers, ultrasonic disintegrators, colloid mills, jet mills, grinders, and mortars and pestles with pathogenic bacteria, do the following.

1. Use a biological safety cabinet always.
2. Use safety blenders designed to prevent leakage from the rotor bearing at the bottom of the bowl. In the absence of a leak-proof rotor, inspect the rotor bearing for leakage prior to operation. Test it in a preliminary run with sterile water, saline, or methylene blue solution.
3. Use a towel moistened with disinfectant over the top of the blender. Sterilize the device and residual contents promptly after use.
4. Do not use glass blender bowls with infectious materials.
5. Blender bowls sometimes require cooling to prevent destruction of the bearings and to minimize thermal effects on the contents.

6. Before opening the blender bowl, wait for at least one minute to allow the aerosol to settle.

46.5.7. Handling Animals

Exposure to experimental animals with natural or induced infections is among the most frequently identified sources of occupational infections in laboratory workers. Of 703 occupational infections for which the source of infection was identified, almost 14% were associated with bites or scratches by experimental animals or exposure to their ectoparasites (44). The current use of research animals, however, is largely restricted to facilities involved in research and other specialty activities, rarely in clinical and diagnostic laboratories. The practices, safety equipment, and facilities recommended for animal biosafety level 2 constitute the basic and minimal conditions for working with naturally or experimentally infected vertebrate animals (44).

Airflow in experimental-animal rooms should be inward and nonrecirculating. Ideally, only one species of animal and one infectious agent should be used for experiments within the same room. Access should be restricted to those working with the animals or providing services to the area. All personnel entering the animal room should wear appropriate protective clothing (e.g., coats and gloves) and meet the requirements for entry, including immunization(s). Physical and chemical restraints appropriate to the species involved in the study should be used routinely. Appropriate physical barriers and primary containment caging should be provided to minimize cross-contamination and exposure of personnel working with agents that may be transmitted by infectious aerosols. Cages for animals used for infectious-disease studies should be autoclaved or chemically decontaminated before cleaning and reuse. A biohazard warning sign indicating the agent(s) in use and conditions for entry should be posted on or near the animal room door.

These combined safety practices supplement the applicable requirements of the Department of Agriculture Animal Welfare Act and the recommendations of the National Research Council Guide for the Care and Use of Laboratory Animals (23).

Animals are capable of shedding pathogens in saliva, urine, or feces. In the absence of information to the contrary, all animals should be regarded as shedders. Careful handling procedures, including changing small cages used in biological safety cabinets, will minimize the dissemination of dust from animal and cage refuse. Sterilize dirty cages by autoclaving. Refuse, bowls, and watering devices should remain in the cage during sterilization.

Wear heavy gloves when feeding, watering, handling, or removing infected animals. Do not insert bare hands in the cage to move any object.

When injecting bacteria into animals, wear protective gloves. Ensure that animals are properly restrained (e.g., use a squeeze cage for primates) or tranquilized to avoid accidents that might disseminate infectious material or cause injury to the animal or personnel.

The oversized canine teeth of large monkeys present a biting hazard and are important in the transmission of naturally occurring, very dangerous monkey virus infections. Such teeth should be blunted or surgically removed by a veterinarian. Many zoonotic diseases, including infectious hepatitis and tuberculosis, can be transmitted from nonhuman primates to humans. Newly imported animals may be naturally infected with these or other diseases, and persons in close contact with such animals may become infected.

Personal protective equipment or cage systems designed to contain infectious material should be used.

Keep the doors to animal rooms closed at all times, except for necessary entrance and exit. Do not permit unauthorized persons in animal rooms.

Keep a container of disinfectant, prepared fresh each day, in each animal room for disinfecting gloves and hands and for general decontamination, even though no infectious animals are present. Wash hands, floors, walls, and cage racks with an approved disinfectant at the recommended strength at regular and frequent intervals.

Fill floor drains in animal rooms with water or disinfectant periodically to prevent backup of sewer gases. Do not wash shavings and other refuse on floors down the floor drain, because such refuse clogs sewer lines.

Maintain an insect and rodent control program in all animal rooms and in animal food storage areas. Take special care to prevent live animals, especially mice, from finding their way into disposable trash.

Carry out necropsy of infected animals in a safety cabinet. Wear surgeon's gowns over laboratory clothing, and wear rubber gloves. Wet the fur of animals with a suitable disinfectant.

On completion of a necropsy, place all potentially infectious materials in suitable containers and sterilize immediately. Place contaminated instruments in a horizontal tray containing a suitable disinfectant. Disinfect the inside of the safety cabinet as well as other potentially contaminated surfaces. Clean grossly contaminated rubber gloves in disinfectant before removal from the hands, preparatory to sterilization. Place dead animals in proper leak-proof containers, autoclave when this is practical, and tag properly before removing for incineration.

46.5.8. Vacuum Systems

The use of vacuum suctioning devices for aspiration and collection of biological fluids (e.g., tissue culture media, suspension fluids) for discard is a convenient and common laboratory practice. Vacuum systems are of two general types: small individual pumps serving a single laboratory or activity and centralized "house" systems serving a building or multiple users. There are obvious safety advantages to the former system. Small systems, with short distances between the point of suction and the point of discard, can be effectively protected against aspiration of infectious aerosols or fluids into the vacuum system.

Protect the collection flask for discarded liquids by using a downstream overflow flask. The use of antifoaming agents and disinfectants is recommended for both of these containers. Distal to the collection and overflow containers, use an in-line cartridge-type filter to protect the vacuum pump (20). Empty the collection flask and overflow flask periodically or when the contents constitute approximately two-thirds of the container's capacity. Clamp off the flexible vacuum tubing during transport to the autoclave. Then remove the clamp and autoclave the entire apparatus. Following autoclaving, discard the treated fluids into the sanitary sewer. Use heavy-walled, autoclavable, nonglass (Nalgene) sidearm flasks for the primary and overflow containers. Autoclave the in-line filter, and replace it periodically or whenever it becomes overly contaminated with fluids from the overflow flask.

46.5.9. Miscellaneous Operations

Water baths used to inactivate, incubate, or test bacteria should contain a disinfectant. For cold-water baths, 70% propylene glycol is recommended. CAUTION: Sodium azide creates an explosive hazard and should not be used as a disinfectant.

Freezers, liquid nitrogen, dry-ice chests, and refrigerators should be checked, cleaned, and decontaminated periodically. Use rubber gloves and face and respiratory protection during cleaning. Label all infectious material stored in refrigerators or freezers. Develop a standard labeling format for long-term storage of materials. Hazardous fluid cultures or viable powdered infectious materials in glass vessels should be transported, incubated, and stored in easily handled, nonbreakable, leak-proof containers large enough to contain all the fluid or powder in case of leakage or breakage of the glass vessel.

Inoculated petri plates or other inoculated solid media should be transported and incubated in leak-proof pans or leak-proof containers.

Exercise care when using membrane filters to obtain sterile filtrates of infectious materials. Because of the fragility of the membrane and other factors, consider the filtrates to be infectious until culture tests have proved their sterility.

Examine shaking machines carefully for potential breakage of flasks or other containers being shaken. Use durable plastic or heavy-walled screw-cap glass flasks, and securely fasten them to the shaker platform. As an additional precaution, enclose each flask in a plastic bag with or without an absorbent material.

Never work alone on a hazardous operation.

46.6. LEVEL 3 BIOSAFETY PRACTICES

Biosafety level 3 practices are used for handling microorganisms that are capable of causing serious or lethal infection by the inhalation route. Two principal practices distinguish biosafety level 3 from biosafety level 2 and result in a higher level of containment. At biosafety level 3, the laboratory director restricts entry into the laboratory to persons whose access is required for program or support purposes, and the use of infectious materials is conducted in biological safety cabinets or other physical containment devices. This is important to reduce of the risk of airborne transmission. In addition, the laboratory facility in which biosafety level 3 practices are used meets higher standards than those normally associated with conventional laboratories to safeguard against the accidental dissemination of airborne pathogens to adjacent and surrounding areas.

46.7. BIOSAFETY CABINETS

Biological safety cabinets are used to reduce the spread of contamination and to eliminate exposures to microbial aerosols. Activities involving pathogens that cause infection through inhalation should be conducted in safety cabinets or other containment devices since most laboratory techniques produce inadvertent aerosols. Class I and II biological safety cabinets are recommended for use at biosafety levels 2 and 3, respectively. Protection against aerosols generated by work activities is provided by the movement of room air into the front opening of the cabinet. Safety can be compromised if this inward airflow is disrupted by other room air currents. For this reason, biological safety cabinets are located in laboratory areas away from doorways, supply air diffusers, and spaces of high activity. In general, the best location for the cabinet is on a side wall at the position farthest from the door.

Cabinet users should consider wearing long-sleeved gowns with knit cuffs and gloves to protect the hands and arms from contact contamination. It is good practice to decontaminate the work surface of the cabinet before equipment and material are placed into the cabinet. Ideally, everything needed for the procedure should be placed in the cabinet before work is started. In class II cabinets, nothing should be placed over the exhaust grills at the front and rear of the work surface.

The arrangement of equipment and material on the work surface is an important consideration. Segregate contaminated items from clean items and locate them to ensure they are not passed over clean items. Discard trays should be located at the rear. The use of aseptic techniques is important to prevent contact contamination. Contain or disinfect all infectious items before removing them from the cabinet. Trays of discarded pipettes and glassware should be covered before they are removed. After all items have been removed from the cabinet, the work surface should be decontaminated.

Biosafety cabinets may be obtained from various manufacturers (e.g., LABCONCO, Kansas City, MO; NuAire, Inc., Minneapolis, MN; Baker Co., Inc., Sanford, ME).

46.7.1. Class I Cabinets

The class I cabinet (Fig. 2) may be used in one of three modes: with a full-width open front, with an installed front panel without gloves, or with the panel equipped with arm-length rubber gloves. Materials may be introduced and removed through the panel opening and, if provided, through a hinged front-view panel or a side airlock. Room air enter-

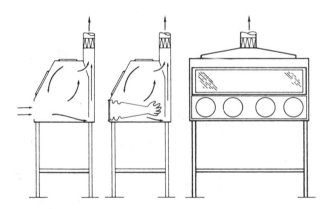

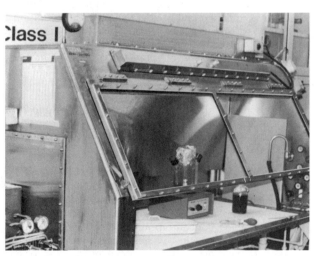

FIGURE 2 Class I biological safety cabinet.

ing the cabinet prevents the escape of airborne contaminants. The air flows across the work space, over and under a back wall baffle, and out through a high-efficiency particulate air (HEPA) filter and blower in an overhead duct to the building air exhaust system or outdoors. When operated with the front open, a minimum inward face velocity of 75 ft (22.9 m)/min is needed.

The protection afforded by a class I cabinet can be compromised by sudden withdrawal of the hands, rapid opening and closing of the room door, or rapid movements past the front of the cabinet. Aerosols created forcefully and in large quantities may escape despite the inward flow of air. Also, the cabinet does not protect the hands and arms from contact with hazardous materials. Such protection depends on technique and on the use of gloves and other protective clothing.

46.7.2. Class II Cabinets

Class II cabinets provide worker safety and product protection, an advance in cabinetry design that has significantly increased the air quality of the cabinet's work area. Class II cabinets (Fig. 3) utilize nonturbulent (laminar) airflow and have front openings for access to the work space and for introduction and removal of materials. Airborne contaminants in the cabinet are prevented from escaping by a curtain of air formed by unfiltered air flowing from the room into the cabinet and by HEPA-filtered air supplied from an overhead grill. The air curtain prevents airborne contaminants in the room air from entering the work space of the cabinet. The air curtain is drawn through a grill at the forward edge of the work surface into a plenum, where it is HEPA filtered and recirculated through the overhead grill down into the cabinet. A portion of the air is used to maintain the air curtain, and the remainder passes down onto the work surface and is drawn out through a grill at the back edge of the work surface. The filtered air from the overhead grill flows in a uniform downward movement to minimize turbulence. This air provides and maintains a filtered-air work environment. A percentage of air drawn through the front and back grills of the work surface, which is equal to the flow of room air into the cabinet, is also filtered and exhausted from the cabinet. Several configurations of class II cabinets are available; they vary primarily in the percentage of filtered airflow that is recirculated within the cabinet. An excellent reference article is available on class I and II biological safety cabinets and other safety cabinets designed to contain hazardous materials (36).

Consensus standards have been developed by the National Sanitation Foundation for the design, construction, and performance of class II biological safety cabinets (24). It is prudent to require that the purchase and on-site certification of class II cabinets conform to these standards. It is important to recertify class II cabinets annually and when moving them to another location. This practice will ensure that the cabinets function correctly.

46.8. CHEMICAL DISINFECTION

Disinfection is the removal or destruction of pathogenic microorganisms from inanimate objects or surfaces, usually by use of a chemical agent, and is needed to protect the integrity of bacteriological test results or to prevent the occurrence and spread of disease. Chemical disinfection is necessary because the use of pressurized steam, the most reliable method of sterilization, and other physical methods are not normally feasible for disinfection of large spaces, surfaces, and

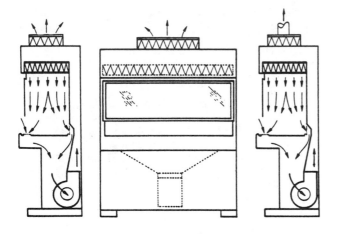

FIGURE 3 Class II biological safety cabinet.

stationary equipment. Moreover, high temperatures and moisture often damage delicate instruments, in particular, those with complex optical and electronic components.

Recommendations for the selection and use of chemical disinfectants in health care facilities have recently been compiled comprehensively by Rutala for the Association for Practitioners of Infection Control (APIC) (32). While focused primarily on patient care items and environments, these recommendations have generic application in bacteriological laboratories, including general ones. Recommendations for high-, intermediate-, and low-level disinfection methods described in the Guidelines for Prevention of Transmission of HIV and HBV in Healthcare and Public-Safety Workers (41) may be interpolated and applied from the health care to the laboratory setting. For further general information, see references 3 and 31.

46.8.1. General Guide

The information summarized in Table 1 provides a general guide for the selection and use of chemical disinfectants in bacteriological laboratories based on experience gained in health care facilities. The APIC recommendations (32) and manufacturer claims should be carefully reviewed in the selection and use of chemical disinfectants in laboratory workplaces.

As a general principle and to ensure the maximum efficacy of the chemical decontamination process selected, the potentially contaminated item or surface must first be cleaned of visible soil. Appropriate precautions, including the wearing of protective gloves, should be taken to prevent exposure of personnel during this predisinfection process. The chemical disinfectant selected should be of demonstrated efficacy against target organisms and prepared and used in accordance with the written instructions of the manufacturers.

The registration of a chemical disinfectant with the Environmental Protection Agency (EPA) as a tuberculocidal hospital disinfectant is based on information provided to this agency by the manufacturer. Verification testing of manufacturer claims of antimicrobial or antiviral efficacy is not conducted by the EPA. The effectiveness of chemical disinfectants in meeting label claims may be significantly influenced by the presence of organic materials on items or surfaces subjected to disinfection, the age of the use solution, the concentration of the chemical used, and the time of exposure or contact.

The combining of two or more chemical disinfectants or disinfectant chemicals and cleaning chemicals may reduce rather than enhance the effectiveness of the cleaning or decontamination effort. Additionally, the intentional or inadvertent mixing of certain compounds (e.g., household bleach and mineral or organic acids or ammonia compounds) may liberate reaction products into the work environment that are toxic, corrosive, or caustic to personnel.

46.8.2. Disinfectant Types

Many disinfectants are available under a wide variety of trade names. In general, disinfectants are classified as acids or alkalis, halogens, heavy metals, quaternary ammonium compounds, phenolic compounds, aldehydes, ketones, alcohols, amines, and peroxides. Unfortunately, the more active the disinfectant, the more likely it is to possess undesirable characteristics, such as toxic or corrosive properties. No disinfectant is equally useful or effective under all conditions. Resistance to the action of chemical disinfectants can be substantially altered by such factors as concentration of active ingredient, duration of contact, pH, temperature, humidity, and presence of organic matter.

46.8.2.1. Alcohols

Ethyl or isopropyl alcohol in a concentration of 50 to 70% is often used to disinfect surfaces. However, these agents are slow in their germicidal action (several minutes being required) and ineffective against spores.

46.8.2.2. Formaldehyde

For use as a disinfectant, formaldehyde is usually marketed as a 37% concentration of the gas in water solution (formalin) or as a solid polymerized compound (paraformaldehyde). Formaldehyde in a concentration of 5% is an effective liquid disinfectant and is an effective space disinfectant for sterilizing rooms or buildings. However, formaldehyde loses considerable disinfectant activity at refrigeration temperatures, and

TABLE 1 Some considerations in the selection and use of chemical disinfectants in biomedical laboratories

Disinfection	Effect	Method	Use	Examples
High-level disinfection	Destroys all microbial life forms except bacterial spores present in large numbers	Exposure of contaminated items to an EPA-registered "sterilant" chemical in accordance with manufacturer use recommendations	Reusable instruments and devices that cannot be subjected to autoclave temperatures	Chlorine dioxide; glutaraldehyde; solutions containing 1,000 ppm of free available chlorine (1:50 dilution of household bleach); stabilized hydrogen peroxide
Intermediate-level disinfection	Destroys M. *tuberculosis*, other vegetative bacteria, and most viruses and fungi, but not bacterial spores	EPA-registered "hospital disinfectants" (label claim for tuberculocidal activity)	Items or surfaces that are visibly contaminated with blood, body fluids, or other potentially infectious materials; surfaces must be pre-cleaned before application of chemical disinfectant	Solutions containing 500 ppm of free available chlorine (1:100 dilution of household bleach); iodophor; phenolics
Low-level disinfection	Destroys most bacteria, some viruses and fungi, but not M. *tuberculosis* or bacterial spores	EPA-registered "hospital disinfectants" (no label claim for tuberculocidal activity)	For routine housekeeping or removal of soils from environmental or work surfaces (e.g., bench work, biological safety cabinets) in the absence of visible blood contamination	Ethyl or isopropyl alcohol; solutions containing 100 ppm of free dilution of household bleach); quaternary ammonium, iodophor, or phenolic germicidal detergent solutions prepared and used as specified on the product label

its pungently irritating odor requires that care be taken when using it in solution in the laboratory. The use of formaldehyde is also regulated by OSHA, and strict practices for protection of human health are mandatory.

46.8.2.3. Phenolic Compounds

Phenol is not often used by itself as a disinfectant. The odor is unpleasant, and a gummy residue remains on treated surfaces. Phenolic compounds, however, are basic to several popular disinfectants that are effective against vegetative bacteria at high dilutions and are essentially odorless. They are not effective in ordinary usage against bacterial spores.

46.8.2.4. Quaternary Ammonium Compounds

Quaternary ammonium compounds are strongly surface active, which makes them good surface cleaners. They attach to proteins, and so in dilute solutions they lose effectiveness. They clump bacteria and are neutralized by anionic detergents, such as soap. They are bactericidal at medium concentrations, but they are not tuberculocidal or sporicidal even at high concentrations. They have the advantages of being odorless, nonstaining, noncorrosive to metals, stable, inexpensive, and relatively nontoxic.

46.8.2.5. Chlorine

Chlorine is a universal disinfectant that is active against all bacteria, including bacterial spores, and is effective over a wide range of temperatures. It combines readily with protein, so that an excess of chlorine must be used if proteins are present. Free, available chlorine is the active element. It is a strong oxidizing agent and is corrosive to metals. Chlorine solutions gradually lose strength, so that fresh solutions must be prepared frequently. Sodium hypochlorite is

usually used as a base for chlorine disinfectants. An excellent disinfectant can be prepared from household or laundry bleaches, which usually contain 5.25% (52,500 ppm) available chlorine. If one dilutes them 1:100, the solution will contain 525 ppm of available chlorine, and if 0.7% of a nonionic detergent is added, a very good disinfectant is created. This is an effective disinfectant against blood-borne pathogens and M. *tuberculosis*.

46.8.2.6. Iodophors

Iodophors, organic iodine compounds, constitute one of the most popular groups of disinfectants used in the laboratory. Recommended-use concentrations for commercial products provide 75 ppm of free iodine. This small amount can be rapidly taken up by the extraneous proteins present. Clean surfaces or clear water can be effectively treated by 75 ppm of available iodine, but difficulties may be experienced if an appreciable amount of protein is present. A concentration of 1,600 ppm of available iodine is required to achieve effective sporicidal properties. An alcoholic solution (tincture) of iodine is effective as a disinfectant but is irritating.

46.8.2.7. Heavy Metals

Heavy metals are not recommended. Mercury and other heavy-metal preparations are toxic, and they are more bacteriostatic than bactericidal.

46.8.3. Disinfectant Characteristics

Although the ideal disinfectant does not exist, it is useful to consider what characteristics it should have, as follows.

- High activity: effectiveness at high dilutions in the presence of organic matter.

- Broad spectrum of antimicrobial activity: effectiveness against gram-positive, gram-negative, and acid-fast bacteria and against spores, viruses, and fungi.
- Stability: retention of potency after storage for prolonged periods.
- Homogeneity: no settling out of active ingredients.
- Adequate solubility: solubility in water, fats, and oils for good penetration into microorganisms.
- Low surface tension: penetration into cracks and crevices.
- Minimum toxicity: lack of acute and chronic toxicity, mutagenicity, carcinogenicity, teratogenicity, allergenicity, irritability, and photosensitization.
- Detergent activity: ability to solubilize and remove dirt and debris.
- Minimum material effects: low or acceptable effects on metals, wood, plastics, and paint.
- Odor control: pleasant odor, odorless, or having deodorizing properties.
- Cost: inexpensive in relation to efficiency.

The effectiveness of a disinfection process can be maximized by attention to the following factors.

1. Select a disinfectant appropriate against the microorganism to be inactivated. If its identity is unknown, select an agent with as broad a spectrum of activity as possible.

2. Reduce the bioburden and the organic content to a minimum on surfaces or objects to be disinfected.

3. Take into account the fact that most chemical disinfectants have limited effectiveness against bacterial spores, and allow additional exposure time for the effective ones.

4. Use the disinfectant in the proper concentration. Inadequate concentrations may result in lack of disinfection, and excessively concentrated solutions may pose problems of toxicity and effects on materials. The concentration may determine whether the action is static or cidal.

5. Carefully consider the exposure or contact time necessary for disinfection.

6. Be sure that the chemical concentration and contact time are compatible with the temperature of disinfection. In general, lower temperatures require longer contact times and higher temperatures increase efficiency two- to threefold per 10°C rise in temperature.

7. Ensure that the water used for preparing dilutions of disinfectants for use is of the proper quality. With certain chemicals, water hardness (e.g., calcium ions) in excess of 300 to 400 ppm destroys disinfecting ability.

8. Consider the amount of organic matter present on the object being treated. Organic matter reacts with disinfecting chemicals and, in effect, removes the active ingredient of the solution.

9. No single chemical disinfectant or method will be effective or practical for all situations. When selecting a chemical disinfectant, the following questions should be considered.

a. What is (are) the target organism(s)?

b. What disinfectants in what form are known to, or can be expected to, inactivate the target organism(s)?

c. What degree of inactivation is required?

d. In what menstruum is the organism suspended (i.e., simple or complex, on solid or porous surfaces, or airborne)?

e. What is the highest concentration of cells anticipated to be encountered?

f. Can the disinfectant (as either a liquid, a vapor, or a gas) be expected to contact the organisms, and can effective duration of contact be maintained?

g. What restrictions apply with respect to compatibility of materials?

h. What is the stability of the disinfectant in use concentrations, and does the anticipated use situation require immediate availability of the disinfectant or will sufficient time be available for preparation of the working concentration shortly before its anticipated use?

10. Actual concentrations and contact times used in the laboratory may differ from the recommendations of the manufacturers. The efficacy of the selected disinfectant procedure should be validated by the individual user. This is particularly important for laboratory activities that involve serious pathogens, such as M. tuberculosis.

46.8.4. Environmental Surfaces

Routine disinfection of floor surfaces is often a function of the housekeeping department. For laboratories and other areas where infectious material is handled, clean floors daily with a suitable disinfectant-detergent at the specified use dilution. Use a broad-spectrum disinfectant-detergent that is active against bacteria, fungi, viruses, and acid-fast organisms and is not materially affected by hard water or organic matter.

Disinfect bench tops, tables, and large equipment by using a disinfectant-detergent at the use dilution with a clean cloth, sponge, or disposable towel. Allow surfaces to remain damp, moist, or wet. If 0.5% hypochlorite solutions are considered, it must be remembered that they are easily neutralized by excess organic matter, may discolor surfaces, and are highly corrosive to metals.

46.8.5. Bacterial Spills

A spill that is confined to the interior of the biological safety cabinet should present little or no hazard to personnel in the area. However, initiate chemical disinfection procedures at once, while the cabinet ventilation system continues to operate, to prevent the escape of contaminants. Spray or wipe walls, work surfaces, and equipment with a disinfectant. A disinfectant with a detergent will help clean the surfaces by removing both dirt and bacteria. A suitable disinfectant is a 3% solution of an iodophor or a 1:100 dilution of household bleach with 0.7% nonionizing detergent. Wear gloves during this procedure. Use sufficient disinfectant solution to ensure that the drain pans and catch basins below the work surface contain the disinfectant. Lift the front exhaust grill and tray, and wipe all surfaces. Wipe the catch basin, and drain the disinfectant into a container. Discard the disinfectant, gloves, wiping cloth, and sponges into a pan, and autoclave them. This procedure will not disinfect the filters, blower, air ducts, or other interior parts of the cabinet.

If the entire interior of the cabinet is to be disinfected, formaldehyde gas should be used. The procedure is potentially hazardous and requires special knowledge and equipment. It should be conducted only by experienced personnel. For this procedure, most institutions hire professional companies that certify biological safety cabinets.

If an organism that can transmit disease by infectious aerosols is spilled outside a biological safety cabinet, laboratory occupants can be placed at considerable risk. The first step in response to such an occurrence is to avoid inhaling airborne material by holding the breath and leaving the laboratory. Warn others in the area, and go directly to a washroom or changeroom area. If clothing is known or suspected to be contaminated, remove it with care and fold the contaminated area inward. Discard clothing into a bag, or place it directly into an autoclave. Wash all potentially contaminated areas of the body as well as the arms, face, and hands. Delay reentry into the laboratory for a period of 30 min to allow reduction of the aerosols generated by the spill. When entering the laboratory to clean the spill area, wear protective clothing (rubber gloves, autoclavable footwear, an outer garment, and a respirator). If the spill was on the floor, do not wear a surgical gown that may trail on the floor when bending down. Place a discard container near the spill, and (using forceps) transfer large fragments of material into it; replace the cover. Cover the spill area with paper towels or other absorbent materials. Using a freshly prepared 1:10 dilution of household bleach or other appropriate disinfectant, carefully pour the disinfectant around and into the visible spill. Avoid splashing. Allow a contact time of 20 min. Use paper or cloth towel to wipe up the disinfectant and spill, working toward the center of the spill. Discard towels into a discard container as they are used. Wipe the outside of the discard container, especially the bottom, with a towel soaked in disinfectant. Place the discard container and other materials in an autoclave, and sterilize them. Remove shoes, outer clothing, respirator, and gloves, and sterilize by autoclaving. Wash hands, arms, and face; if possible, shower. The preceding decontamination procedure can be begun immediately for spills of pathogens that do not present inhalation risks.

46.8.6. Instruments

Contaminated syringes, instruments, pipettes, thermometers, and other glassware should be decontaminated before being reprocessed. Whenever possible, carry out final processing with steam for heat-stable items or with ethylene oxide for heat-labile, moisture-sensitive materials. When decontamination is essential before handling, use containers of liquid disinfectant. Completely immerse the items in the solution, preferably for as long as 20 to 30 min. This allows a safety factor if the item is not extremely clean or if the solution has been used for some period previously. Before reuse, rinse the items carefully with distilled water and, if they are to be used in or on the body, with sterile distilled water.

Decontaminate laboratory pipettes by placing them in a vertical or horizontal container with disinfectant. Glassware should be immersed completely. Before glassware is handled and reprocessed, steam sterilization is recommended.

46.8.7. Hands

Vigorous washing with soap and water is the most practical means of removing organisms from the hands. When working with pathogenic bacteria, the hands should be washed after the removal of gloves and before exiting the laboratory.

Bar soap is discouraged, not only because of the inherent sloppiness of the soap dish but also because some organisms survive on the soap surface. Liquid soaps, unless they contain a preservative, may gradually develop large populations of organisms in the reservoir, which should be cleaned out routinely and new soap added. Powdered soaps and leaf soaps have the advantage of not being contaminated or allowing growth of organisms.

46.9. STERILIZATION (adapted from reference 16)

The objective of a sterilization process is to destroy or remove all living organisms in or on an item. In general, in microbiological laboratories, steam is used for the sterilization of culture media, equipment, and glassware; dry heat is used for glassware and metallic equipment; gas is used for instruments; and filtration is used for solutions. Sterilization is an evolving technology, with constant improvements in equipment design and control. For example, hydrogen peroxide is a promising sterilant for many laboratory, industrial, and hospital applications.

Sterilization is a probability function. Bacterial cells usually die at an exponential (logarithmic) rate similar to that of a first-order chemical reaction. Sterilization efficiency is expressed as the statistical probability that the process will yield a survivor (25). Terminal processes (moist heat, dry heat, ionizing radiation, and ethylene oxide) should have a probability of 10^{-6} or less of a survivor being present, based on extrapolation of the rate determined from measurable survivors.

With the exception of filtration, sterilization methods can be characterized by the rate of the killing process. A D-value is used to characterize sterilization and, for heat, represents the time required at a given temperature to cause a decimal (90%) reduction in a bacterial population; the temperature is often indicated as a subscript (e.g., D_{100}). The D-value can be derived graphically by plotting the logarithm of the number of survivors against time of exposure and is measured as the time on the abscissa corresponding to a decade of reduction in numbers on the ordinate (Fig. 4).

Deviations from exponential killing occur frequently, resulting in patterns different from the idealized rectilinear plot shown in Fig. 4. A common deviation is for a short lag or even an increase in initial cell numbers ("shoulder") to occur before the onset of an exponential regression. Another common deviation is for a long, concave, curvilinear decrease in cell numbers ("tailing") to occur after the rectilinear exponential regression. Other deviations, such as rectilinear biphasic or sigmoid ones, occur less frequently. Clearly, a D-value has meaning only in the primary rectilinear portion of the plot, and extrapolation from this portion is invalid if tailing occurs. These deviations seem especially prevalent in the killing of bacterial spores, which may

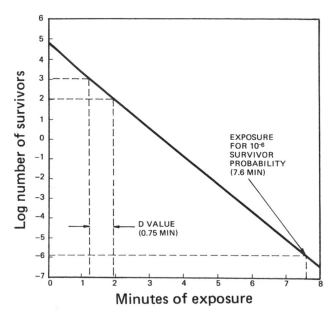

FIGURE 4 Exponential death plot, showing the graphic determination of a D-value and a survivor probability.

be affected by population heterogeneity, clumping, activation, germination, adaptation, and other factors (30).

The efficacy of the sterilization process used in the laboratory should be validated by using biological indicators. These are bacterial spores that exhibit a predictable death rate when exposed to a defined treatment. Highly resistant species of bacterial spores such as *Bacillus subtilis*, *B. stearothermophilus*, and *B. pumilus* are normally used in validation studies. Spore suspensions should be evaluated periodically to ensure that their resistance properties persist. Validation studies should include the use of at least three replicate units for each specific exposure time and condition that is to be evaluated. The accuracy of the result is a function of the number of replicate samples used. The equipment should be thoroughly evaluated for performance to ensure consistent sterilization results.

The authoritative compendium of Block (3) should be consulted for guidance on specific sterilization processes and validation methods, and Quesnel (29) provides basic methods for sterilization in biotechnology. For further general information, see references 3 and 31.

46.9.1. Moist Heat

Steam sterilization is accomplished by using an autoclave with saturated steam under pressure for 15 min at a temperature of 121°C. Other time and temperature relationships can also result in effective sterilization.

Most laboratory autoclaves are gravity or downward-displacement devices that depend on the difference in density between air and steam. Problems occur with moisture penetration, superheating, entrainment or removal of air, and heat and/or moisture damage. Correct preparation of materials and their proper loading is important. The manufacturer of the autoclave can provide directions for installing thermocouples in materials to be sterilized to determine loading patterns within the autoclave.

Automatic autoclaves, equipped with cycle timers and automatic temperature controls, are most suitable for laboratory use. Several other moist-heat systems are available that reduce trapped air by high vacuum, steam and vacuum pulsations, and hot-water cascade or immersion. Air causes the chamber to heat more slowly and results in lower temperatures. Table 2 shows the lower temperature of several air-steam mixtures compared with the temperatures of pure steam. Laboratory autoclaves should be controlled by temperature readings (which depend on saturated steam pressures).

Contrary to commonly held views, carefully controlled studies have shown that air (oxygen) at 100% relative humidity significantly enhances the lethality of moist heat on bacterial spores, despite the practical effect of slower heating when air is present in an autoclave (34).

It is important to know the effects on steam penetrability of loading patterns, viscosity of solutions, size of containers, and densities. Also, the chemistry of the solution may influence microbial killing. Cool zones within loads of materials must be identified. Packages should be loosely arranged to facilitate steam penetration. Articles should be arranged to permit the downward displacement of the heavier air.

Table 3 illustrates the effect of the liquid volume and the number of containers within a load on the time required for sterilization. The time is shorter for small volumes. Thus, sterilization cycle times must be adjusted for the sizes or number of containers.

Precautions must be taken when sterilizing solutions in glass containers. Use a controlled rate of cooling and slow release of pressure to prevent glass shock when the cycle is completed. Some sterilizers use controlled water cool-down systems to avoid glass breakage. Take care to ensure that an autoclave has returned to ambient pressure before opening it. Most modern autoclaves prevent this problem through automatic control. When sterilizing flasks or bottles, loosen container closure fittings to permit air venting and prevent container breakage. Also place the containers in shallow metal or autoclavable plastic pans to capture liquid in the event of breakage. When the container is removed from the autoclave, a vacuum may result and cause an influx of air into the container, which can contaminate the solution unless the closure is made tight or other preventive measures are taken.

46.9.2. Dry Heat

Materials that cannot be sterilized by moist heat (e.g., greases, lubricants, mineral oil, waxes, and powders) are often sterilized by dry heat. Dry heat is also frequently used to sterilize laboratory glassware. Hot forced air and infrared energy have been used for dry-heat sterilization. There are a variety of commercial gas and electric dry-heat sterilizers, including batch ovens and continuous-heating tunnels.

Dry heat is less efficient than moist heat and requires higher temperatures for longer durations. The principal advantages of dry heat are that it is not corrosive for metals and instruments, does not affect glass surfaces, and is suitable for sterilizing powders and nonaqueous, nonvolatile viscous substances. Disadvantages include slow heat penetration and longer sterilization times. In dry-heat steriliza-

TABLE 2 Influence of incomplete air discharge on autoclave temperatures[a]

Gauge pressure (lb/in²)	Temp (°C)			
	Saturated steam, complete air discharge	Two-thirds air discharge	One-third air discharge	No air discharge
5	109	100	90	72
10	115	109	100	90
15	121	115	109	100
20	126	121	115	109
25	130	126	121	115
30	135	130	126	121

[a]Data from reference 24a.

TABLE 3 Effect of liquid volume and number of containers on time required for liquid to reach 121°C in an autoclave[a]

Liquid vol/container (liters)	No. of containers/ load	Liquid temp at initiation of cycle (°C)	Time for liquid to reach 121°C (min)	Total time of cycle (min)
0.5	30	29	19	29
1.0	20	26	34	44
2.0	10	27	37	47
3.0	8	26	43	53
4.0	5	26	52	62
5.0	5	26	60	70
6.0	4	26	62	72

[a]Data from reference 27.

tion, the higher temperatures may adversely affect some materials, and temperature stratification can occur unless the air is circulated.

46.8.3. Gases

Gases are used in sterilization because some materials and supplies cannot be sterilized by other means without damage or destruction. Gaseous processes are more difficult to control than other modes of sterilization because of the various process parameters. Factors such as gas stratification, temperature stratification, availability of moisture, chemically reactive barriers, physical diffusion barriers (such as packaging materials), devaporization, and polymerization must be considered when selecting a gaseous process (6–8). Sterilant gases are extremely hazardous to human health, and their use should be limited to persons who have been carefully trained in carrying out the sterilization process and are knowledgeable about the intrinsic hazards.

Safety and health regulations have been established for most gases that are commonly used as sterilants. These standards require that employers limit occupational uses to established concentration limits.

46.9.3.1. Ethylene Oxide

Ethylene oxide (EtO) has been widely used in the sterilization of devices because of its wide compatibility with various materials. It is used in specially designed chambers or vessels. EtO is highly flammable and explosive by itself. Detailed instructions by the manufacturer of EtO sterilization equipment must be closely followed.

Mixtures of dichlorodifluoromethane and EtO are frequently used to minimize flammability. Although 100% EtO processes are used in the pharmaceutical industries, they are rarely used for laboratory purposes.

The use of EtO as a sterilization agent requires an understanding of the parameters affecting its activity (3). The diffusion of EtO, moisture, and heat into materials can be a limiting factor. Prehumidification overcomes diffusion barriers in packaging materials and permits better penetration of gas. Humidification prior to sterilization also ensures hydration and thus killing of bacterial spores.

46.9.3.2. Formaldehyde

Gaseous formaldehyde is a good space and surface sterilant and is the agent of choice for sterilizing biological safety cabinets. It is effective against bacteria and bacterial spores, fungi, viruses, and rickettsiae, as well as insects and other animal life (26). Formaldehyde gas for sterilization purposes can be generated by heating paraformaldehyde, which is in a mixture of polyoxymethylene glycols containing 90 to 99% formaldehyde. Paraformaldehyde depolymerizes to release gaseous formaldehyde when heated. Sterilization by formaldehyde gas generated from paraformaldehyde is more effective than sterilization by vaporized formaldehyde solution.

46.9.3.3. Hydrogen Peroxide

Vapor-phase hydrogen peroxide (VPHP) is a promising surface sterilant that does not possess the toxic and carcinogenic properties of EtO or formaldehyde (15). The use of VPHP as a gaseous disinfectant is a relatively new development and may offer a less toxic alternative for decontamination of rooms, sealed biological safety cabinets, and heat-sensitive medical equipment once the sterilization and equipment monitoring process are fully developed. VPHP is an antimicrobial pesticide registered with the EPA to kill bacterial spores on environmental surfaces in enclosed areas. It is one of the seven chemical pesticides that may be used in the "Federal Anthrax Decontamination Efforts" under carefully controlled conditions. At present it is registered for use in commercial, institutional, and industrial settings to decontaminate or sterilize sealed enclosures such as isolators, workstations, and pass-through rooms such as air locks. A VPHP process is being adopted for the decontamination of biological safety cabinets with research focusing on methods for ensuring rapid circulation of the gas to offset the rapid decomposition of H_2O_2 vapor and establishing appropriate sterilization cycle times.

46.9.3.4. Chlorine Dioxide

Chlorine dioxide is an antimicrobial pesticide recognized for its disinfectant properties since the early 1900s. Chlorine dioxide is a strong oxidizing agent and kills microorganisms by disrupting transport of nutrients across the cell wall through denaturation of cell wall proteins. Chlorine dioxide gas is an efficient and effective sterilant as long as the relative humidity is kept above 50% and appropriate contact and exhaust or degassing times are ensured. Pesticide products containing either sodium chlorite or stabilized chlorine dioxide are mixed with another "reactive" chemical, usually an acid, to produce chlorine dioxide in a liquid or gaseous state. Chlorine dioxide gas was used in the decontamination of the Hart Senate Office Building in 2001 and 2002 and is one of a series of chemicals that the

EPA has approved for use in the "Federal Anthrax Decontamination Efforts" under carefully controlled conditions.

46.9.4. Filtration

Filtration is one of the oldest methods used to sterilize solutions and is frequently used to remove microorganisms or particulate matter, or both, from solutions and gases. As a terminal process, it is less desirable than the use of moist heat because there is a higher probability that a microorganism will pass through a filter.

Filters function by entrapping microorganisms within the porous structure of the filter matrix. Vacuum or pressure is required to move solutions through the filter. There are two basic types: depth filters and membrane filters. Depth filters consist of fibrous or granular materials that are pressed, wound, fired, or bonded into a maze of flow channels. In these, retention of particulates is a matter of a combination of absorption and mechanical entrapment in the filter matrix. Membrane filters have a continuous structure, and entrapment occurs mainly on the basis of particle size.

Filtration is particularly applicable for oil emulsions or solutions that are heat labile. Membrane filtration is widely used to sterilize oils, ointments, ophthalmic solutions, intravenous solutions, diagnostic drugs, radiopharmaceuticals, tissue culture medium, and vitamin and antibiotic solutions.

For further information, the reader is referred to the excellent monograph by Brock (4).

46.10. NONBIOLOGICAL SAFETY CONSIDERATIONS

Numerous physical and chemical hazards are present in general and molecular bacteriology laboratories. Lacerations are common laboratory injuries; they are most frequently caused by accidents involving the handling of razor blades, scalpels, scissors and other cutting instruments, glassware, and Pasteur pipettes. The microwave oven has become a common piece of laboratory equipment, but its use has introduced a significant new hazard to the laboratory. Careless operating procedures have resulted in the violent release of superheated fluids and explosions caused by rapid pressure buildup in accidentally sealed containers. Serious injury is always a potential consequence of the misuse of this equipment.

Although most commercially available electrophoresis units are designed with safeguards to interrupt the power source when electrodes are exposed or the fluid evaporates, older units and those fabricated locally may not include these safety devices. Such units present serious electrical shock and fire hazards. The considerable use of flammable solvents, in particular, ethanol, in molecular bacteriology laboratories creates potential fire hazards associated with storage, use, and disposal practices. The storage of samples in liquid nitrogen creates the potential for injury from accidental skin contact and from explosion when improperly sealed vials are removed from storage.

Sulfur-35, phosphorus-32, iodine-125, carbon-14, and tritium are common isotopes in the molecular bacteriology laboratory. Good personal-hygiene habits are essential to prevent internal contamination, since, in general, the effect of internal radioactivity is more serious than that of external exposure. Plan experiments carefully so that exposure to the isotopes will be limited to the shortest time possible. To reduce exposure, use forceps and tongs when handling high

activity, and always use appropriate shielding for storage locations, source containers, and waste containers. Use a low-density material such as plastic to shield phosphorus-32 and other high-energy beta emitters to minimize production of bremsstrahlung. It is important that the properties of isotopes used in the laboratory be known and that the radiation safety practices required under the laboratory's license are rigorously followed. See Shapiro (35) for more detailed information on radioactivity safety.

The use of hazardous chemicals in the laboratory can cause serious injury to workers if they are not instructed and trained in how to handle them safely. Overt exposure can result in serious tissue damage and adverse acute effects. A major hazard is the potential for splashing of chemicals into the eyes. Corrosive chemicals such as phenol, for example, can have devastating effects. Safety goggles should be worn when there is a potential hazard of chemical splashes. Special eye protection is also important when working with lasers and UV-light sources, glassware under reduced or elevated pressure, and other activities that could cause objects to strike the eye. See National Research Council (21) for detailed information on chemical hazards..

The brief discussion above is intended only to describe an array of hazards that are constantly present in the general or molecular bacteriology laboratory. Awareness of these hazards and biohazards that may be present is the first step toward establishing a safe laboratory environment. The attitudes and skill necessary for controlling these hazards and the continual reinforcement of good laboratory practice are essential to sustaining a safe and healthful general and molecular bacteriology laboratory.

46.11. SECURITY

Security of hazardous materials in laboratories is not a new issue. For example, most laboratories have established procedures to prevent the theft of radioactive materials and controlled substances. It is also appropriate for laboratories to consider measures to prevent the theft of infectious agents that cause serious disease in humans. This is a federal requirement (39, 45) for institutions that possess select agents. Security provisions should address the need to prevent unauthorized entry to areas where human pathogens are used and stored. This is part of the laboratory director's responsibility to limit or restrict access to laboratories when work with infectious agents is in progress. It is also important to physically secure dangerous pathogens when they are not in use, such as by locking the laboratory. The security provisions should be practical so that they do not unnecessarily constrain laboratory activities. They should also be commensurate with the risk such agents pose to human health. Severity of disease and transmissibility are two important risk characteristics of pathogens for determining both an appropriate level of security and the appropriate biosafety level for use of a particular pathogen. Routine security measures to protect occupants, equipment, and supplies may be appropriate. A card-key access system would provide a higher level of security that may be appropriate for biosafety level 3 laboratories. A staff that is attentive to who is in the laboratory and what is going on in the laboratory would improve security in both biosafety level 2 and 3.

46.12. REFERENCES

1. **Ausubel, F. M., R. Brent, R. E. Kingston, D. D. Moore, J. G. Seidman, J. A. Smith, and K. Struhl.** 2003. *Current*

Protocols in Molecular Biology. John Wiley & Sons, Inc., Hoboken, NJ.

2. **Barkley, W. E. and G. P. Kubica.** 1994. Biological safety in the experimental tuberculosis laboratory, p. 61–71. *In* B. R. Bloom (ed.), *Tuberculosis Pathogenesis, Protection, and Control.* American Society for Microbiology, Washington, DC.

3. **Block, S. S. (ed.).** 2001. *Disinfection, Sterilization and Preservation,* 5th ed. Lippincott Williams & Wilkins, Philadelphia, PA.

4. **Brock, T. D.** 1983. *Membrane Filtration: A User's Guide and Reference Manual.* Science Tech, Inc., Madison, WI.

5. **Centers for Disease Control and Prevention.** 1987. Recommendations for prevention of HIV transmission in healthcare settings. *Morb. Mortal. Wkly. Rep.* **36**(28):1S–17S.

6. **Collins, C. H.** 1983. *Laboratory-Acquired Infections—History, Incidence, Causes and Prevention.* Butterworths, London, United Kingdom.

7. **Ernst, R. R.** 1969. *Federal Regulations and Practical Control Microbiology for Disinfectants, Drugs, and Cosmetics,* special publication no. 4, p. 55–60. Society for Industrial Microbiology, Linden, NJ.

8. **Ernst, R. R., and J. E. Doyle.** 1968. Sterilization with gaseous ethylene oxide-a review of chemical and physical factors. *Biotechnol. Bioeng.* **10:**1–31.

9. **Fleming, D. O., and D. L. Hunt (eds).** 2000. *Biological Safety: Principles and Practices.* American Society for Microbiology, Washington, DC.

10. **Gershey, E. L., E. Party, and A. Wilkerson.** 1991. *Laboratory Safety in Practice: A Comprehensive Compliance Program and Safety Manual for Industrial, Diagnostic and Clinical, Research and Development, Academic (School, College, University), Quality Control, Analytical and Testing Laboratories.* Van Nostrand Reinhold, New York.

11. **Harrington, J. M., and H. S. Shannon.** 1976. Incidence of tuberculosis, hepatitis, brucellosis and shigellosis in British medical laboratory workers. *Br. Med. J.* **1:**759–762.

12. **Harding, L. A., and K. B. Byers.** 2000. Epidemiology of laboratory-associated infections, p. 35–54. *In* D. O. Fleming and D. L. Hunt (ed.), *Biological Safety: Principles and Practices.* American Society for Microbiology, Washington, DC.

13. **Isenberg, H. D. (ed.)** 1995. *Clinical Microbiology Procedures Handbook.* American Society for Microbiology, Washington, DC.

14. **Jagger, J., E. H. Hunt, J. Brand-Elnagger, et al.** 1988. Rates of needlestick injury caused by various devices in a university hospital. *N. Engl. J. Med.* **319:**284–288.

15. **Klapes, N. A., and D. Vesley.** 1990. Vapor-phase hydrogen peroxide as a surface decontaminant and sterilant. *Appl. Environ. Microbiol.* **56:**503–506.

16. **Mahn, W. J.** 1991. *Fundamentals of Laboratory Safety: Physical Hazards in the Academic Laboratory.* Van Nostrand Reinhold, New York.

17. **Marshall, B., et al.** 1981. Survival and transfer in the human gut of poorly mobilizable (pBR322) and of transferable plasmids from the same carrier *E. coli. Gene* **14:** 145–154.

18. **Meyer, K. F, and B. Eddy.** 1941. Laboratory infections due to Brucella. *J. Infect. Dis.* **68:**24–32.

19. **Miller, B. M., D. H. M. Grdschel, J. H. Richardson, D. Vesley, J. R. Songer, R. D. Housewright, and W. E. Barkley (ed.).** 1986. *Laboratory Safety: Principles and Practices.* American Society for Microbiology, Washington, DC.

20. **National Institutes of Health.** 1978. *Laboratory Safety Monograph.* Office of Research Safety, National Cancer Institute, and the Special Committee of Safety and Health Experts, Bethesda, MD.

21. **National Research Council.** 1989. *Biosafety in the Laboratory: Prudent Practices for the Handling and Disposal of Infectious Materials.* Committee on Hazardous Biological Substances in the Laboratory, Board on Chemical Sciences and Technology, Commission on Physical Sciences, Mathematics, and Resources. National Academy Press, Washington, DC.

22. **National Research Council.** 1995. *Prudent Practices in the Laboratory—Handling and Disposal of Chemicals.* Committee on Prudent Practices for Handling, Storage, and Disposal of Chemicals in Laboratories, Commission on Physical Sciences, Mathematics and Applications. National Academy Press, Washington, DC.

23. **National Research Council.** 1996. *Guide for the Care and Use of Laboratory Animals.* Committee to Revise the Guide for the Care and Use of Laboratory Animals, Institute of Laboratory Animal Resources, Commission on Life Sciences. National Academy Press, Washington, DC.

24. **National Sanitation Foundation.** 2002. *Class II (Laminar Flow) Biohazard Cabinetry,* Standard 49. National Sanitation Foundation, Ann Arbor, MI.

24a.**Perkins, J. J.** 1976. *Principles and Methods of Sterilization in Health Sciences,* 2nd ed. Charles C Thomas, Publisher, Springfield, IL.

25. **Pflug, I. J., and J. Bearman.** 1974. *Treatment of Sterilization Process Microbial Survivor Data in Environmental Microbiology as Related to Planetary Quarantine,* progress report 9, NASA grant NGL 24-005-160. University of Minnesota Department of Food Science and Nutrition and School of Public Health, University of Minnesota, Minneapolis, MN.

26. **Phillips, G. B., and W. S. Miller.** 1975. *Remington's Pharmaceutical Sciences,* 15th ed., p. 1389–1404. Mack Publishing, Easton, PA.

27. **Pike, R. M.** 1976. Laboratory-associated infections-summary and analysis of 3921 cases. *Health Lab. Sci.* **13:**105–114.

28. **Pike, R. M., S. E. Sulkin, and M. L. Schultz.** 1965. Continuing importance of laboratory-acquired infections. *Am. J. Public Health* **55:**190–199.

29. **Quesnel, L. B.** 1987. Sterilization and sterility, p. 197–215. *In* J. Bullock and B. Kristiansen (ed.), *Basic Biotechnology.* Academic Press Ltd., London, United Kingdom.

30. **Russell, A. D.** 1982. *The Destruction of Bacterial Spores.* Academic Press Ltd., London, United Kingdom.

31. **Russell, A. D., W B. Hugo, and G. A. J. Ayliffe.** 1992. *Principles and Practice of Disinfection, Preservation and Sterilization,* 2nd ed. Blackwell Scientific Publications, Oxford, United Kingdom.

32. **Rutala, W. A.** 1990. APIC guidelines for selection and use of disinfectants. *Am. J. Infect. Control* **18:**99–117.

33. **Sambrook, J., E. R. Fritsch, and T. Maniatis (ed.).** 1989. *Molecular Cloning: A Laboratory Manual,* 2nd ed. Cold Spring Harbor Laboratory, Cold Spring Harbor, NY

34. **Seruton, M. W.** 1989. The effect of air on the moist-heat resistance of Bacillus stearothermophilus spores. *J. Hosp. Infect.* **14:**339–350.

35. **Shapiro, J.** 2002. *Radiation Protection: A Guide for Scientists, Regulators, and Physicians,* 4rd ed. Harvard University Press, Cambridge, MA.

36. **Stuart, D. G.** 2000. Primary barriers: biological safety cabinets, fume hoods, and glove boxes, p. 313–330. *In* D. O. Fleming and D. L. Hunt (ed.). *Biological Safety: Principles and Practices.* American Society for Microbiology, Washington, DC.

37. **Sulkin, S. E., and R. M. Pike.** 1949. Viral infections contracted in the laboratory. *N. Engl. J. Med.* **241:**205–213.

38. **Sulkin, S. E., and R. M. Pike.** 1951. Survey of laboratory acquired infections. *Am. J. Public Health* **41:**769–781.

39. **U.S. Department of Agriculture.** 2002. *U.S. Code of Federal Regulations, Title 7*, U.S. Department of Agriculture, Part 331 (7 CFR 331). U.S. Government Printing Office, Washington, DC.

40. **U.S. Department of Agriculture.** 2002. *U.S. Code of Federal Regulations, Title 9*, U.S. Department of Agriculture, Part 121 (9 CFR 121). U.S. Government Printing Office, Washington, DC.

41. **U.S. Department of Health and Human Services.** 1989. Guidelines for prevention of transmission of human immunodeficiency virus and hepatitis B virus to healthcare and public-safety workers. U.S. Public Health Service, Centers for Disease Control, National Institute for Occupational Safety and Health. *Morb. Mortal. Wkly. Rep.* **38**(5–6):1–36.

42. **U.S. Department of Health and Human Services.** 1994. Guidelines for research involving recombinant DNA molecules (NIH Guidelines). National Institutes of Health. *Fed. Regist.* **59**:34496.

43. **U.S Department of Health and Human Services.** 1995. *Primary Containment for Biohazards: Selection, Installation and Use of Biological Safety Cabinets.* Centers for Disease Control and Prevention and National Institutes of Health. U.S. Government Printing Office, Washington, DC.

44. **U.S. Department of Health and Human Services.** 1999. *Biosafety in Microbiological and Biomedical Laboratories*, 4th ed. HHS Publication No. (CDC) 93-8395, U.S. Public Health Service, Centers for Disease Control and Prevention and the National Institutes of Health. U.S. Government Printing Office, Washington, DC.

45. **U.S. Department of Health and Human Services.** 2002. *U.S. Code of Federal Regulations, Title 42*, U.S. Department of Health and Human Services, Part 73, (42 CFR 73). U.S. Government Printing Office, Washington, DC.

46. **U.S. Department of Labor, Department of Health and Human Resources.** 1987. Joint Advisory Notice: Protection against occupational exposure to hepatitis B virus (HBV) and human immunodeficiency virus (HIV). *Fed. Regist.* **52**(210):41818–41824.

47. **U.S. Nuclear Regulatory Commission.** 1946. *U.S. Code of Federal Regulations, Title 10*, U.S. Nuclear Regulatory Commission, Part 20, (10 CFR 20). U.S. Government Printing Office, Washington, DC.

48. **U.S. Occupational Safety and Health Administration.** 1990. *U.S. Code of Federal Regulations, Title 29*, Occupational Safety and Health Administration, Part 1910.1450 (29 CFR 1910.1450). U.S. Government Printing Office, Washington, DC.

49. **U.S. Occupational Safety and Health Administration.** 1991. *U.S. Code of Federal Regulations, Title 29*, Occupational Safety and Health Administration Part 1910.1030 (29 CFR 1910.1030). U.S. Government Printing Office, Washington, DC.

50. **Wedum, A. G.** 1973. Microbiological centrifuging hazards, p. 5–16. *In Centrifuge Biohazards: Proceedings of a Cancer Research Safety Symposium*, DHEW Publication no. (NIH) 78-373. Public Health Service, National Institutes of Health. U.S. Government Printing Office, Washington, DC.

51. **Wedum, A. G., W E. Barkley, and A. Hellman.** 1972. Handling of infectious agents. *J. Am. Vet. Med. Assoc.* **161**:1557–1567.

47

Culture Preservation

ROBERT L. GHERNA AND C. A. REDDY

Most bacteriology laboratories maintain stock cultures for educational, research, bioassay, industrial, or other purposes. Biopreservation is the process of preserving the integrity and functionality of cells, tissues, and organs held outside the native environment for extended storage times (1). The proper preservation of cultures is extremely important and should be (but often is not) given the attention it deserves. Numerous programs have been hampered by the loss or variation of a stock culture that is critically needed for research or for the production of useful products.

The primary aim of culture preservation is to maintain the organism alive, uncontaminated, and without variation or mutation, that is, to preserve the culture in a condition that is as close as possible to the original isolate (1, 4, 18, 22, 36). Many methods have been used to preserve bacteria, but not all species respond similarly to a given method. In fact, some strains of the same species give variable results with the same procedure. It should be emphasized that the success of any preservation method depends on the use of the appropriate medium and cultivation procedure and on the age of the culture at the time of preservation. This is particularly true when working with bacteria that contain plasmids or recombinant DNA or that exhibit growth phases such as morphogenesis or spore formation. The availability of equipment, storage space, and skilled labor often dictates the method employed.

Record keeping is an important and often neglected aspect of stock culture maintenance. Too often, isolates are not given strain designations or, if they are, the designations are not documented in strain data files. A strain designation (e.g., by number or letter) refers to a specific strain with distinct phenotypic characteristics which should be recorded. Culture collections, for example, use an identifying acronym and a unique numbering system (such as ATCC 12301, NCTC 34567, and NRRL 23456) to identify their cultures. Another method is to use the initials of the institution followed by the day, month, year, and isolate number (such as UCLA 090892:2).

It is also extremely important to obtain as much documentation as possible on the history of the strain. Data such as the identity of the culture, the individual who isolated it, the source, geographical information, and the date of isolation are important, especially if the organism or its utility will be the subject of a patent application. If the culture is received from another investigator, the documentation must also include the name of the investigator, history if available, date of acquisition, and notation on whether a new strain designation is being used, since the same number may be assigned to other strains by different workers. Some strain variations that have been attributed to inadequate preservation conditions have turned out to be due to poor strain documentation; e.g., the wrong number was noted on the data card or on the containers.

47.1. SHORT-TERM METHODS

47.1.1. Subculturing

The traditional method of preserving bacterial cultures is by periodic serial transfer to fresh medium. The interval between such transfers varies with the organism, the medium used, and the external conditions. Some bacteria must be transferred every other day, whereas others need be transferred only after several weeks or months. The major disadvantages of the serial-transfer technique are the risks of contamination, transposition of strain numbers or designations (mislabeling), selection of variants or mutants, and possible loss of culture, as well as the required storage space.

Three conditions must be determined when using this method for the preservation of cultures: suitable maintenance medium, ideal storage temperature, and frequency between transfers. *Subculturing is not recommended as a long-term preservation method.*

47.1.1.1. Maintenance Medium

Minimal media are preferred for subculturing because they lower the metabolic rate of the organism and thus prolong the period between transfers. However, some bacteria require a complex medium for growth, or their retention of specific physiological properties necessitates the presence of complex compounds in the medium. When a complex medium is used, more frequent transfers may be necessary as a result of accelerated growth or metabolite accumulation. A variety of bacteria, including members of the *Archaea* (11, 19, 21, 29), have been preserved by subculturing in biphasic (liquid-solid) medium (also see sections 10.8 and 33.2).

47.1.1.2. Storage

Storage in a refrigerator in duplicate is the preferred method for subcultures; however, if refrigeration presents a problem, the simplest method is storage at room temperature in a test tube rack or in a specially constructed storage box containing shelves with holes for test tubes. Cultures stored in this fashion require constant care since they tend to dry quickly and, unless the laboratory has a controlled environment, are subject to temperature fluctuations.

To minimize dehydration, use screw caps with rubber liners, wrap the tube caps in Parafilm (VWR Scientific, Inc.; see section 47.5 for addresses of suppliers), or place the tubes in a plastic bag. To reduce the metabolic rate of the organism, place the culture in a refrigerator at about 4 to 8°C. Most bacteria can be kept for 1 to 5 months between transfers if these precautions are used.

47.1.1.3. Transfer Schedule

Determine the subculturing interval by experience. *Keep subculturing to a minimum to avoid the selection of variants.* Maintain duplicate tubes as a precaution against loss. Examine the cultures for purity after each transfer, and perform an abbreviated characterization check periodically to monitor any changes in phenotypic characteristics. *Do not select single colonies in transferring cultures*, since the chances of selecting a mutant are greater when this technique is used.

47.1.2. Immersing in Oil

Many bacterial species can be successfully preserved for months (or years) simply by immersing in sterile medicinal-grade mineral oil (33). Paraffin oil with a specific gravity of 0.865 to 0.890 is also satisfactory (16).

Sterilize 5 ml of oil per Pyrex test tube (18 by 150 mm), sealed with a metal or other heat-resistant closure by dry heat at 180°C for 2 h; autoclaving is not recommended (9). Allow sufficient time for the hot-air oven and the contents of the test tubes to reach sterilization temperature before beginning the timing of the cycle. Use commercially available strips (section 47.5) impregnated with *Bacillus subtilis* spores to check the efficacy of the sterilization equipment and cycle. Do not exceed the recommended sterilization temperature, since the oil will smoke and may ignite. Contamination of cultures when this method is used is often due to improperly sterilized mineral oil. Each lot of sterilized oil must be tested for sterility by streaking on an appropriate medium such as nutrient agar or Trypticase soy agar.

Grow the culture in the appropriate medium as an agar slant or stab or as a broth culture. After growth reaches late logarithmic phase or sporulation occurs, add the sterile mineral oil aseptically to a depth of at least 2 cm (a slant must be entirely covered) to prevent dehydration and to reduce metabolic activity and growth of the culture (16, 33).

Store the oil-covered culture upright in a refrigerator. Perform viability tests periodically by removing a loopful of growth and touching it to a sterile filter paper strip to remove the oil. Streak the growth on agar plates containing a suitable growth medium such as nutrient agar or Trypticase soy agar. Incubate the plates at the appropriate growth temperature for the strain, and assess the resultant growth. Confluent growth represents good survival, but plates which show few colonies indicate a low viability, and new stocks must be prepared.

Use an inoculating needle to transfer the culture under oil to fresh medium, and overlay the subsequent growth with sterile oil. Take care to cool the needle after flaming, since otherwise the oil will splatter and contaminate the surrounding area and personnel. Keep the original culture for at least several weeks to enable the recovery of the culture in the event that the subculture is contaminated or shows aberrant characteristics. The disadvantages of this method are the same as those of ordinary subculturing (section 47.1.1). In addition, the oil is messy.

47.1.3. Ordinary Freezing

The preservation of bacteria in the freezing compartment of a refrigerator or in an ordinary freezer with a temperature range of 0 to −20°C produces variable results, and its success depends on the bacterial species (8, 10, 22, 24). Some bacteria can be kept by ordinary freezing for 6 months to 2 years. In general, ordinary freezing is not recommended for preservation because the freezing process damages the cells.

When the temperature of a cell suspension drops below 0°C, the extracellular liquid water begins to freeze and form external ice crystals; the actual temperature at which this occurs depends on the supercooling properties of the suspension and can be depressed by the addition of a cryoprotective agent such as glycerol or dimethyl sulfoxide (DMSO) or some other protectant to the cell suspensions (3, 17, 25, 39). Ice crystals continue to form in the menstruum, the solute concentration increases, and the internal cellular water begins to migrate out of the cell because of the difference in osmotic pressure. Removal of too much of the internal water results in cell damage. Eventually, all the free water crystallizes as pure ice, leaving only the concentrated solute and its water of hydration. The concentrated solute eventually also freezes at what is called the eutectic temperature. Too rapid a drop in temperature below 0°C often results in the formation of internal ice crystals, which damages the cells. A balance between damage caused by solute concentration and internal ice formation is achieved by controlling the rate of cooling. A rate of 1 to 10°C/min is satisfactory for all bacteria. *CAUTION:* Most modern refrigerators that have a freezer compartment in the range of 0 to −20°C are frost free. The frost-free condition is achieved by an alternating warming and cooling cycle, which can result in further damage to cells with a concomitant loss of viability.

In addition, the concentration of solutes during freezing can alter medium components by the formation of complexes or the lowering of pH in poorly buffered media. The eutectic temperature of a solution will vary depending on its composition (18, 27, 36).

47.1.4. Deep-Freezing

Preservation of bacteria by deep-freezing at −70°C in a mechanical freezer is superior to ordinary freezing at −20°C. The deep-freezing procedure involves the freezing of a bacterial suspension in the presence of a cryoprotective agent such as glycerol or DMSO to prevent cellular damage during the freezing process. Many bacterial species have been successfully preserved for several years by using this method (Table 1).

Grow the cells in an appropriate medium. Harvest the bacteria as described in section 47.2.2.4. Prepare the cell suspensions in broth containing 30 to 50% (vol/vol) glycerol. Dispense the cell suspensions into vials as described below (section 47.2.2) and place the vials in a mechanical freezer set at −70°C. Recover the cells by rapidly thawing the frozen cell suspensions in a 37°C water bath.

Freezing at −70°C in a mechanical freezer with glass beads has also been used to preserve a wide variety of bacteria with 15% glycerol as the cryoprotectant (10). Prepare the glass beads as described in section 47.1.5. Place approximately 30 beads in 2-ml glass screw-cap vials (Wheaton no. 224841) and sterilize for 15 min at 121°C. The vials must be labeled with ink that will not come off easily at ultralow temperatures. Prepare a cell suspension containing approximately 10^8 cells per ml in sterile 15% glycerol, and dispense 0.5 ml into each vial. Gently shake the vial to ensure that all of the beads are wetted with the bacterial suspension, and then remove the excess from the vial with a sterile Pasteur pipette. Place the labeled vials in a storage box or tray and then into a mechanical freezer at −70°C. Recover the bacteria by aseptically removing a bead from the vial with a sterile spatula or forceps and placing it in a suitable broth medium. *CAUTION:* Do not allow the contents of the vial to thaw; withdraw the vial from the freezer, and place it in a box containing crushed dry ice or in a frozen paraffin block containing holes for the vials.

47.1.5. Drying

Effective microbial preservation procedures for long-term storage of bacteria in a viable condition were well explored through experimentation (32). Most cultures die if left to dry in laboratory media. Some cultures, however, especially bacterial spores, can be preserved for years by being dried on a suitable menstruum (17, 26, 32, 40). Bacterial spores have been shown to survive most preservation procedures, including drying (38). Indeed, spores are resistant mainly because of their dehydrated protoplast, and most spores become more resistant if they are dried further in a dried and more hygroscopic medium (such as soil). It is greatly advantageous to grow spore-forming bacteria under conditions which favor spore formation (chapter 7.4.10).

47.1.5.1. Paper

A relatively simple and inexpensive method of preserving various bacteria involves drying them on sterile filter paper (Whatman no. 4; VWR Scientific) strips or disks. The technique is ideally suited for the maintenance of quality control cultures. Many disks containing the same culture can be stored in a single test tube or screw-cap vial. A single disk can be removed aseptically with sterile forceps as needed and inoculated into a suitable broth. Members of the *Enterobacteriaceae*, as well as a variety of other bacteria, are successfully preserved for several years in this manner.

The procedure for this method is as follows. Place the sterile filter paper strips or disks into a sterile petri plate, and add a bacterial suspension containing approximately 10^8 cells or more per ml. Cell concentrations can be determined by direct plate counts of cell dilutions and correlated with optical densities (OD) or nephelometric readings. A faint turbidity is about 10^7 cells per ml, an OD of 1.0 is about 10^9 cells per ml, and an OD of 0.1 is about 10^8 bacteria per ml. Add the bacterial suspension dropwise onto each strip or disk until saturated, and dry in air or a vacuum desiccator (dry under vacuum for greater survival). Store the strips or disks sealed in tubes in desiccators or between

TABLE 1 Expected shelf life of 33 representative bacterial genera preserved by various methods

Genus	Subculturing[a] (mo)	Immersing in oil (yrs)	Deep freezing (yrs)	Freeze-drying (yrs)	Ultra freezing/liquid nitrogen (yrs)
	Shelf life of bacteria preserved by:				
Acetobacter	1–2	1	1–3	>40	>40
Achromobacter	1	1–2	1–3	>40	>40
Acinetobacter	1 wk			>40	>40
Actinobacillus	1 wk	2–3		>40	>40
Actinomyces	1		2–3	>40	>40
Agrobacterium	1–2	1–2		>40	>40
Arthrobacter	1–2		1–2	>40	>40
Bacillus	2–12	1	2–3	>40	>40
Bacteroides	1 wk		1	>40	>40
Bifidobacterium	1 wk			>40	>40
Chromatium	1			>5	>40
Clostridium	6–12	1–2	2–3	>40	>40
Corynebacterium	1–2	1	1–2	>40	>40
Enterobacter	1–4	1–2		>40	>40
Erwinia	1–4	1–2		>40	>40
Escherichia	1–4	1–2		>40	>40
Flavobacterium	1	2		>40	>40
Gluconobacter	1			>40	>40
Haemophilus	1 wk	1 mo (37°C)		>40	>40
Klebsiella	1–4	1	1–2	>40	>40
Lactobacillus	1 wk			>40	>40
Methanobacterium[b]	1			>5	>40
Methanomonas[b]	1			>5	>40
Micromonospora	1		1	>40	>40
Neisseria	1			>40	>40
Nocardia	1–4	1	1–2	>30	>40
Proteus	1–2	1	1–2	>40	>40
Pseudomonas	1–3			>40	>40
Spirillum[c]	1 wk				>40
Staphylococcus	1–2		1	>40	>40
Streptococcus	1–2	1		>40	>40
Streptomyces	1–8	1–2	1–3	>40	>40
Xanthomonas			1–2	>40	>40

[a]The transfer schedule depends on the medium used. The times listed are approximations, and species variation occurs within a genus.
[b]Depends on species.
[c]Spirillum volutans.

two layers of sterile clear plastic. Store the desiccators in a refrigerator, because this extends the shelf life of the culture.

47.1.5.2. Gelatin

Many heterotrophic bacteria can be preserved in dried gelatin drops or disks (18, 22). The storage temperature is important for successful preservation. In general, −20°C is superior to refrigerator or room temperature.

For preparing the gelatin cultures, grow the culture in the appropriate medium, harvest by aseptic centrifugation, resuspend the pellet in a small amount of broth, and inoculate a tube containing 2.0 to 5.0 ml of melted sterile nutrient gelatin held at 30°C to yield a density of 10^8 to 10^9 cells per ml. Using a sterile Pasteur pipette or syringe, place a drop of the bacterial suspension on the bottom of a sterile petri plate.

Place the petri plate in a desiccator containing phosphorus pentoxide, and evacuate with a vacuum pump. After the gelatin drops have dried, aseptically transfer them to sterile screw-cap tubes and store in a refrigerator or, better, in a freezer. Propagate the cultures by aseptically transferring a gelatin drop into a tube containing a suitable medium.

47.1.5.3. Silica Gel

A variety of bacteria have been successfully preserved by drying on sterile silica gel granules (15). The procedure is simple and inexpensive. Sterilize screw-cap test tubes (16 by 125 mm) half filled with silica gel (6 to 12 mesh, grade 40, desiccant activated; W. R. Grace and Co., Baltimore, MD) in an oven at 180°C for 2 h. Prepare spore or cell suspensions by suspending in 1 to 2 ml of sterile 10% (vol/vol) skim milk (no. 0032; Difco, Detroit, MI). Add the cell or spore suspension dropwise (0.5 ml) to prechilled silica test tubes (which must be kept in an ice bath to diminish the effect of heat generated by absorption of the culture to the anhydrous granules), and hold at 0°C for approximately 10 min. Place the tubes with cap loosened half a turn in a desiccator containing activated silica gel for 1 week at room temperature, or dry at 25°C for 2 days. Seal the tubes, and store in a close container over indicator silica gel (grade 42; 6 to 16 mesh, Tel-Tale brand; Fisher Scientific Co., Pittsburgh, PA).

47.1.5.4. Porous Glass and Porcelain Beads

A diverse group of heterotrophic bacteria have been successfully preserved by drying under vacuum on perforated glass beads (28) or porous porcelain beads (20). Only 13 of a total of 202 cultures were nonviable after 21 years of storage on glass beads.

Wash glass or porcelain beads (diameter, 2 to 5 mm) in distilled water containing a mild detergent, and then rinse thoroughly in distilled water. Soak the beads in 2% hydrochloric acid for 2 h to eliminate the alkalinity of the beads; then rinse the beads in distilled water until the pH is neutral, and dry them in an oven. Place the beads in screwcap tubes and sterilize in a dry-heat oven at 180°C for 2 h. Tighten the caps securely while hot to prevent entry of moist air into the tubes. Do not use rubber or plastic closures since these will melt in the oven; instead, use metal caps or other suitable closures. Sterilize test tubes half filled with blue indicator silica gel and a wad of glass wool on top of the gel in a dry-heat oven at 180°C for 2 h.

Prepare a cell or spore suspension containing approximately10[8] cells per ml in sterile 10% (vol/vol in water) skim milk. Aseptically transfer 20 to 40 sterile glass beads into a sterile test tube, and add 1 ml of the cell suspension to the tube. Gently shake the tube to ensure that all the beads are wetted with the suspension. Remove the excess suspension from the tube with a Pasteur pipette or syringe, and transfer the beads into sterile test tubes containing the indicator silica gel and glass wool plug. Tighten the caps securely to prevent entry of moist air. Store the test tubes at room temperature for 1 week and then in a refrigerator.

47.2. LONG-TERM METHODS

47.2.1. Freeze-Drying (Lyophilization)

Freeze-drying (lyophilization) is one of the most economical and effective methods for long-term preservation of bacteria and other microorganisms. Many physiologically diverse bacterial species and bacteriophages have been successfully preserved by this technique and have remained viable for more than 50 years. The method enables large numbers of vials to be produced, and the small size of the vial facilitates storage. Although the freeze-drying procedure is relatively simple, the theoretical aspects are complex. The reader is referred to several excellent reviews for in-depth discussions (4, 6, 18, 27, 30, 31).

Freeze-drying involves the removal of water from frozen bacterial suspensions by sublimation under reduced pressure; that is, the water is evaporated without going through a liquid phase. The dried cells can be stored for long periods if kept away from oxygen, moisture, and light. They can, at any time, be easily rehydrated and restored to their previous state.

Lyophilization can be performed in several ways, for which various types of apparatus have been devised. A simple lyophilization system can be assembled at a reasonable cost. The system consists of a vacuum pump capable of an ultimate pressure of less than 10 μm Hg (a pump rated at 35 to 50 liters/min is usually adequate), a small stainless steel condenser for cooling with dry ice, a thermocouple vacuum gauge to monitor the vacuum system, and heavy-walled vacuum/pressure tubing to connect the components (Fig. 1 illustrates component assembly). Details of other procedures and equipment are given in several reviews (2, 6, 18, 22).

Two of the most common methods used are centrifugal freeze-drying and prefreezing. In centrifugal freeze-drying, the cell suspension is first centrifuged to eliminate frothing due to the removal of dissolved gases. The bacterial suspension is then frozen by the loss of heat during evaporation of the water by vacuum. After primary freeze-drying, the vials are constricted with a narrow flame and then placed on a manifold for secondary freeze-drying. For a detailed description of centrifugal freeze-drying, see references 22 and 24.

The more frequently used prefreezing method is as follows.

47.2.1.1. Equipment

Freeze-drying equipment varies from a simple and relatively inexpensive condenser-vacuum pump system to a complex commercial freeze-dryer which costs thousands of dollars. Reproducibility and shelf life depend on the system used. Excellent results can be obtained, however, with a simple system consisting of a high-vacuum pump, a condenser, and a chamber or manifold (Fig. 1 and 2).

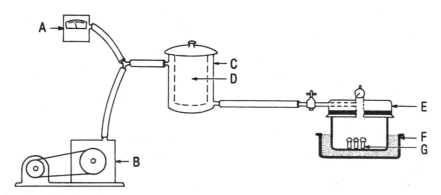

FIGURE 1 Double-vial method of freeze-drying. (A) Vacuum gauge; (B) vacuum pump; (C) VirTis condenser; (D) reservoir filled with dry ice and ethylene glycol; (E) acrylic top plate; (F) stainless steel pan filled with crushed dry ice and ethylene glycol; (G) specimen vial.

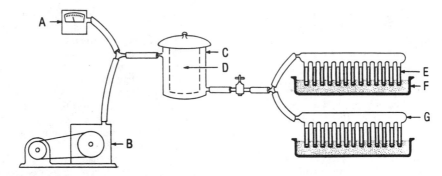

FIGURE 2 Manifold method of freeze-drying. (A) Vacuum gauge; (B) vacuum pump; (C) VirTis condenser; (D) reservoir filled with dry ice and ethylene glycol; (E) specimen vial; (F) stainless steel pan filled with crushed dry ice and ethylene glycol; (G) manifold.

47.2.1.2. Vials

Although dimensions and shapes of vials are variable, there are two basic designs used for freeze-drying. When the cells are to be freeze-dried in a chamber (Fig. 1), double vials are recommended (Fig. 3). Such vials are easier to handle, less susceptible to contamination, and safer for preserving pathogens than are single manifold vials, which are dried and sealed while directly attached to a manifold (Fig. 2) (4, 22, 24, 31).

For double vials, prepare the outer vials (14.25 by 85.0 mm; Glass Vials, Inc., Baltimore, MD) by adding a small quantity of indicator silica gel granules (grade 42, 6 to 16 mesh, Tel-Tale brand) into the vial to cover approximately half of the bottom, and then add a small wad of cotton on top of the granules to cushion the inner vial. Heat the vials in an oven at 100°C for 6 to 8 h, and then cool them in dry box (<10% relative humidity). The silica gel should be dark blue after heating and should remain blue during sealing and storage of the cultures.

For the single-vial manifold procedure, two styles of vials are used. A 1.0-ml bulb-shaped vial (outside diameter, 8 mm; Kimble/Kontes, Vineland, NJ) is used for freezing a shell of the cell suspension, and it is generally used when sucrose is the cryoprotective agent. A tubular type of vial (outside diameter, 8 mm; Bellco Glass Co., Vineland, NJ) is used when a pellet is desired, and this is used when the culture is suspended in skim milk. Rinse the manifold vials by soaking them in distilled water, and autoclave them for 15 min at 121°C to ensure penetration of the water into areas where air is trapped. Dry the vials in an oven and then plug them lightly with non-oil-absorbent cotton (USP grade). Sterilize the manifold vials for 30 min at 121°C.

Prepare the inner vials (11.5 by 35 mm; Glass Vials, Inc.) by rinsing them in distilled water and drying them in an oven. Plug the vials with non-oil-absorbent cotton (USP grade) which extends about 4 cm into the vial. Allow a sufficient amount of cotton to extend above the vial to facilitate removal for dispensing the bacterial suspension. Sterilize the vials by autoclaving for 30 min at 121°C, with the vials on their sides to allow steam penetration.

Vials must be properly labeled with ink that does not come off easily. A labeling machine (model 135A; Markem Machine Co., Keene, NH) with specially formulated ink (no. 7224; Markem Machine Co.) is recommended. This ink will withstand ultralow temperatures and brief exposure to solvents such as alcohol. The letters and numbers are assembled on a master plate and imprinted on a pad. The inner vials are rolled over the pad to pick up the print, and then the vials are heated at 160°C in an oven for 20 min to set the ink. After labeling, sterilize the inner vials again by autoclaving for 30 min at 121°C. Manifold vials are labeled in the same manner as the inner vials.

47.2.1.3. Culture Preparation

Successful freeze-drying depends on using healthy cells grown under optimum conditions in a medium of choice for each strain to ensure the retention of the desired features of the bacterium. Grow a sufficient number of cells to provide a suspension of at least 10^8 cells per ml (OD = 01). Harvest the culture in the late logarithmic or early stationary phase.

47.2.1.4. Cryoprotective Agents

Cryoprotective agents (also see section 47.2.2.3) are chemical compounds such as glycerol or DMSO that are added to a bacterial suspension to help reduce the damage caused during freezing. These compounds must possess certain properties such as being nontoxic, able to penetrate the cell membrane easily, and able to bind either the electrolytes that increase in concentration during freezing or the water molecules to delay freezing (8, 27, 30). Prepare the cells for lyophilization by suspending them in a cryoprotective agent. The American Type Culture Collection (ATCC) has experienced considerable success in long-term preservation of physiologically diverse bacteria by using either 20% (wt/vol) skim milk for the double-vial method or a 24% (wt/vol) sucrose solution diluted equally with growth medium to yield a 12% (wt/vol) sucrose solution (final concentration) for the single-vial manifold procedure. Others have used 10% (wt/vol) dextran, horse serum, inositol, raffinose, trehalose, methylcellulose, glycine betaine, charcoal, and other cryoprotective chemical agents (4, 7, 25, 27, 35, 39).

47.2.1.5. Double-Vial Chamber Method

The following procedures are used for the double-vial method. Prepare a 20% (wt/vol) solution of skim milk (no. 0032; Difco), and sterilize it in small volumes (5 ml) at 116°C for 20 min. Avoid overheating, which can cause caramelization of the milk. For cultures grown on agar surfaces, harvest by aseptically washing the growth off with the 20% skim milk solution. Harvest broth cultures by aseptic centrifugation, and suspend the pellet with the sterile skim milk to yield a cell suspension containing at least 10^8 cells per ml. This can be standardized by resuspending the pellet in a small amount of sterile 0.01 M phosphate buffer (pH

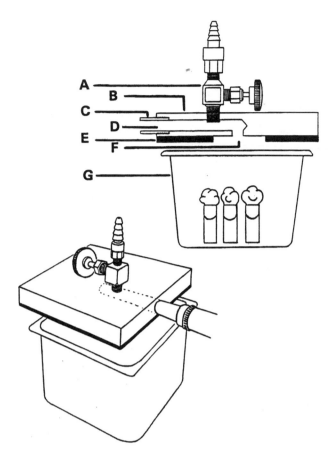

FIGURE 3 Sectional view (top) of acrylic top plate, chamber double-vial method. (A) Vacuum release valve; (B) acrylic plastic top (7 in. wide by 7 in. long by 1 in. deep); (C) stainless steel inlet port (outer diameter, 11/16 in.; inner diameter, 1/2 in.); (D) hole (inner diameter, 1/2 in.) drilled in acrylic plastic to accept the stainless steel tube; (E) rubber gasket (7 in. wide by 7 in. long by 1/4 in. deep) with 2 3/8-in. hole drilled in the center; (F) center hole in bottom of acrylic plastic plate (diameter, 0.5 in.); (G) stainless steel specimen pan (5 in. long by 5 in. wide by 4 in. deep). Top view (bottom) of acrylic top plate, showing Tygon pressure tubing attached to the stainless steel tube and the pressure release valve on top.

7.0) and measuring the OD (an OD of 0.1 equals 10^8 cells per ml). Recentrifuge the cells aseptically, and resuspend the pellet with the same volume of sterile 20% skim milk to yield the desired cell concentration. *CAUTION: Do not use skim milk for bacteria that are inhibited by milk; instead, use sucrose and the manifold single-vial procedure for these bacteria.*

As soon as the cell suspensions are prepared, dispense 0.2 ml into each vial, and trim the cotton plugs with scissors to the tops of the vials. The interval between dispensing and the freeze-drying process should be kept to a minimum to avoid possible alteration of the culture.

Figure 1 shows a typical chamber double-vial freeze-drying system. All vacuum lines are made of Tygon tubing (inner diameter, 3/8 in. [0.95 cm]; outer diameter, 7/8 in. [2.22 cm]; Norton Co., Granville, NY). The system is monitored by a thermistor vacuum gauge (model 10-324; The VirTis Co., Gardiner, NY) and should be evacuated below 30 μm Hg. When the vacuum sensor is placed between the product and the condenser, it will show an increase in pres-

sure as drying occurs. However, when the drying is complete, the pressure should return to below 30 μm Hg.

Place the filled "inner" vials upright in a stainless steel pan. For efficient drying, use only a single layer of vials. Freeze the cells by placing the stainless steel pan containing the vials on the bottom of a mechanical freezer maintained at −60 to −70°C for 1 h.

Prepare a moisture trap by placing chunks of dry ice into a cylindrical condenser (The VirTis Co.), and add ethylene glycol (no. E-180; Fisher Scientific Co.). After the condenser has cooled for approximately 30 min, turn on the vacuum pump. Close the stopcock in the vacuum line between the acrylic plate and the condenser. The plastic plate has an inlet port and a vacuum release valve on the top surface and a rubber gasket on the bottom surface. These plates seal the stainless steel pans to enable a vacuum to be drawn and can be easily made by a machine shop (Fig. 3). Evacuate the system below 30 μm Hg.

At the end of 1 h at −60 to −70°C, place the stainless steel pan containing the vials into a larger shallow pan containing sufficient crushed dry ice to cover the bottom and surround the sides of the stainless steel container. Let stand for 3 h. Take care to avoid getting dry ice into the pan containing the vials, since this will interfere with the evacuation of the system. Attach the acrylic plate to the pan immediately, and hold it firmly in place while opening the stopcock in the vacuum line. Monitor the stopcock to ensure that no pinhole leaks occur. Drying proceeds under a vacuum of 20 to 30 μm Hg for 18 h. For convenience, a run can be started late in the afternoon and allowed to proceed overnight.

Remove the outer vials containing the silica gel from the oven, place them in a dry box (maintained below 10% relative humidity), and allow them to cool. Close the stopcock between the acrylic plate and the condenser. Connect one end of a rubber tube to the inlet port on the plate, which leads to a column of silica gel located inside the dry box. Open the stopcock on the inlet port to allow air through the silica gel column; the pressure in the stainless steel pan will reach atmospheric. This procedure minimizes the introduction of moisture into the samples. Transfer the vials to the dry box, and insert each inner vial into a soft-glass outer vial. Tamp a 0.6-cm plug of fiberglass paper (no. 1821-915; Whatman, Inc.) above the cotton-plugged inner vial.

Remove the double vials from the dry box, and heat the outer vial just above the fiberglass paper with a compressed-air/gas (natural, mixed, propane, or butane) torch. Rotate the vial and flame just above the fiberglass paper until the glass begins to constrict. Pull the vial slowly with a forceps until the constriction is a narrow capillary.

Allow the vials to cool in a dry box, and then attach them to a vacuum manifold (Fig. 2) by using single-hole rubber stoppers which fit the open end of the vials. Evacuate them to less than 50 μm Hg, and seal them with a double-flame air/gas torch at the capillary constriction.

12.2.1.6. Single-Vial Manifold Method

In the single-vial manifold method, dispense 0.2 ml of the bacterial suspension into each sterile vial. Push the cotton plug about 1.3 cm below the rim of the vial with a sterile probe, and flame the rim of the vial to eliminate protruding cotton fibers that would interfere with the integrity of the vacuum system.

Attach a 2.5-cm piece of sterile nonpowdered amber latex IV tubing (no. 17610-163; Baxter Diagnostics Inc., Scientific Products Division) to the rim of each filled vial by using a tube stretcher. When the tubular vials are used, freeze the cell suspension by direct immersion in a dry-ice–ethylene glycol bath and attach the vials to the manifold while they are still immersed in the dry-ice bath. When the bulb-type vials are used, immerse them in the dry-ice–ethylene glycol bath and rotate them to shell-freeze the suspension before placing the vials on the manifold.

Prepare a moisture trap as described for the double-vial freeze-drying method. The vials attached to the manifold remain in the dry-ice–ethylene glycol bath during the freeze-drying process. Dry with a vacuum below 30 μm Hg for 18 h. During the freeze-drying cycle, the bath will warm to ambient temperature.

After completion of the drying process, seal the vials below the cotton plug by using a dual-tipped air/gas torch and continuously moving the flame up and down the vials within a 2.5-cm area. Allow the vials to cool.

47.2.1.7. Storage

Freeze-dried cultures, in either double or single vials, are stored at 2 to 8°C. Extended shelf life has been obtained when cultures are stored at −30 or −70°C in a mechanical freezer. Room temperature storage of freeze-dried cultures should be avoided. Since most modern refrigerators have frost-free freezer compartments, which use a heating and cooling cycle, vials must be stored in the refrigerator compartment only. Walk-in coolers or mechanical freezers set for a maximum temperature of 4°C can also be used for storage.

47.2.1.8. Recovery

Open the double vials by heating the tapered end of the outer container vigorously in a Bunsen burner flame, and quickly add a drop of water to crack the hot glass (Fig. 4). Remove the broken tip of the container carefully (preferably in a hood) with a sharp blow of a forceps, and then withdraw the fiberglass plug and inner vial. Use a sterile forceps to gently remove the cotton plug, and rehydrate the culture with 0.3 to 0.4 ml of appropriate broth medium.

Open the single vials by first scoring the vial with the edge of a file approximately 2.5 cm from the tip. Disinfect the vial with a piece of gauze dampened with 70% alcohol (70% alcohol will kill most vegetative cells, but it is not effective against mycobacteria; for additional information on disinfectants, the reader is referred to reference 9). Wrap sterile gauze around the vial, and break at the scored area. Flame the opened end gently, prior to rehydration of the culture. Open the vials in a laminar-flow hood.

Rehydrate the culture by adding the broth with a sterile needle, and remove with the same needle. Open vials containing pathogenic bacteria only in a closed safety cabinet. Some investigators use a tungsten needle, heated in an oxygen gas flame to white heat, to poke a small hole in the tip of the ampoule; this allows air to enter the vial more slowly and sterilizes the area around the hole.

Rehydrate a freeze-dried culture immediately after opening by aseptically adding 0.3 to 0.4 ml of suitable sterile broth to the contents of the vial. Mix well so that the pellet dissolves completely. Reuse the syringe to transfer the cell suspension to a test tube containing 5 ml of the rehydrating broth. After thorough mixing, transfer 0.2 ml of the rehydrated culture to an agar slant or a semisolid medium of the same composition.

Pre- and postlyophilization purity checks must always be done. To accomplish this, serially dilute the prelyophilization cell suspension or the reconstituted culture and streak on solid medium. Incubate the tubes and plates at the optimum growth temperature, and subculture onto fresh medium as soon as growth appears and purity is ascertained. Freeze-dried cultures often exhibit a prolonged lag period and should be allowed to incubate for an extended period before the culture is considered dead.

47.2.1.9. Monitoring Viability

Although freeze-drying has facilitated the long-term preservation of bacteria, viability checks must be done before and after freeze-drying to determine the effectiveness of the process. In addition, periodic viability tests must be done to ascertain the shelf life of the cultures. Characterization tests should also be performed on cultures that have been lyophilized to determine whether any changes have occurred as a result of the freeze-drying procedures or during storage.

47.2.2. Ultrafreezing

Long-term preservation of bacterial species not amenable to freeze-drying has been achieved through storage in the frozen state at the temperature of liquid nitrogen (−196°C) or above the liquid (vapor phase, −150°C, well-insulated tanks). The ATCC has successfully preserved many fastidious bacteria in liquid nitrogen for over 30 years without the loss of phenotypic properties.

Long-term preservation of bacterial cultures also can be achieved by using ultralow-temperature mechanical freezers (e.g., from Revco, Inc., West Columbia, SC) with a temperature of −70°C. This method of preserving cultures is fairly successful for a large variety of bacteria. Precautions must be taken, however, against electrical shutdowns or compressor malfunction. Adequate backup systems (such as alarms, backup freezers, or an electrical generator) will help to prevent the loss of a valuable collection. Commercial freezers are now equipped with battery-operated alarm systems as standard equipment. However, additional alarm systems are recommended, such as the sound/off power-temperature monitor (Thomas Scientific, Swedesboro, NJ), which has a temperature range of −75 to +200°C (±0.3°C). The unit is battery powered and can be mounted on a wall. Divided storage of lots in different freezers is an added precaution.

47.2.2.1. Equipment

Liquid nitrogen has been considered an expensive method for the long-term preservation of microorganisms. However, the recent history of the successful storage of physiologically diverse bacteria by this method, along with a decreased need for handling, makes liquid nitrogen feasible to use, especially when one considers the cost of labor.

A variety of liquid-nitrogen refrigerators are now commercially available, with a wide assortment of features and storage capacity (e.g., MVE, Linde, Union Carbide Corp., Piscataway, NJ). The sizes range from 10 to 1,000 liters, permitting storage of 300 to 40,000 vials.

47.2.2.2. Vials

Various kinds of liquid-nitrogen vials are available commercially. Types that have been used successfully are heavy-walled, prescored, and unscored borosilicate Cryule vials and also special wide-mouth gold-band cell and tissue culture dryule vials (Wheaton). These vials have a capacity of 1.2 ml and an indentation groove to facilitate opening. The ATCC uses presterilized polypropylene tubes (no. 3-68632 and 3-75353; Nunc, Naperville, IL) with screw caps and

DOUBLE-VIAL PREPARATIONS

- TIP
- INSULATOR
- COTTON PLUG
- OUTER VIAL (soft glass)
- INNER VIAL
- FREEZE-DRIED PELLET
- COTTON
- DESICCANT WITH INDICATOR

SINGLE-VIAL PREPARATIONS

- BOROSILICATE GLASS
- FREEZE-DRIED CELLS

OPENING THE VIAL

1 HEAT THE TIP OF THE OUTER VIAL IN A FLAME

2 SQUIRT A FEW DROPS OF WATER ON THE HOT TIP TO CRACK GLASS

3 STRIKE WITH FILE OR PENCIL TO REMOVE TIP

4 REMOVE INSULATION AND INNER VIAL

Vials prepared before 1974 have a small amount of asbestos tamped above the inner vial. Remove carefully so that fibers do not become airborne.

5 WITH FORCEPS, GENTLY RAISE COTTON PLUG

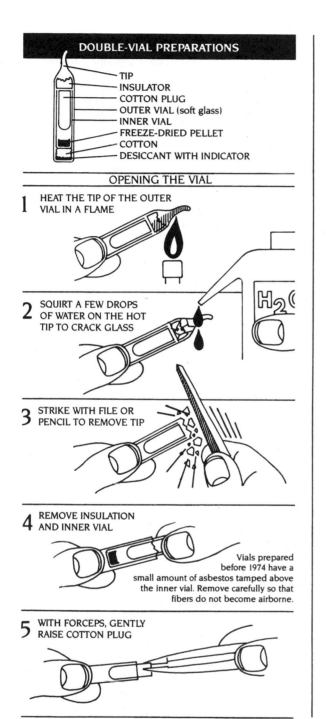

OPENING THE VIAL

1 These preparations may be enclosed in a thin skin of cellulose; this skin must be removed (either with a sharp blade or by soaking in water for a few minutes), score the ampule once briskly with a sharp file about one inch from the tip.

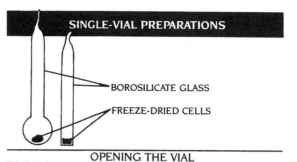

2 Disinfect the ampule with alcohol-dampened gauze.

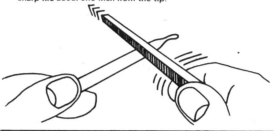

3 Wrap gauze around the ampule, and break at the scored area. Care should be taken not to have the gauze too wet, or alcohol could be sucked into the culture when the vacuum is broken. Rehydrate material at once.

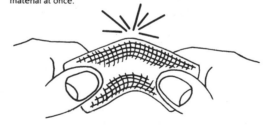

USE A STERILE PASTEUR PIPET TO REHYDRATE THE FREEZE-DRIED CULTURE AS DESCRIBED BELOW.

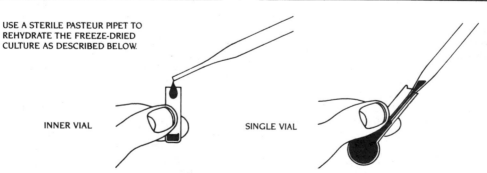

INNER VIAL   SINGLE VIAL

FIGURE 4 Opening of chamber double vials and manifold single vials. (Courtesy of the ATCC.)

silicone washers in place of the glass vials (see *CAUTION* note in section 47.2.2.6).

47.2.2.3. Cryoprotective Agents

Cryoprotective compounds (also see section 47.2.1.4) fall into two types: agents such as glycerol and DMSO, which readily pass through the cell membrane and appear to provide both intracellular and extracellular protection against freezing; and agents such as dextran, glucose, lactose, mannitol, polyglycol, polyvinylpyrrolidone, sorbitol, and sucrose, which appear to exert their protective effect external to the cell membrane. The former type has proven to be more effective, and glycerol and DMSO appear to be equally effective in preserving a wide range of bacteria. The exact choice of cryoprotective agent, however, depends on the bacterial species. A tolerance test must be done when freezing new species to ascertain whether the cryoprotective agent is toxic or beneficial.

Glycerol and DMSO are routinely used at concentrations of 10% (vol/vol) and 5% (vol/vol), respectively, in an appropriate growth medium. Glycerol is usually prepared in double strength (20%, vol/vol) and then mixed with an equal amount of the cell suspension. Sterilize the glycerol by autoclaving at 121°C for 15 min, and store in 6-ml volumes at 2 to 8°C. DMSO (reagent grade) is sterilized with 0.22-μm-pore-size Teflon (polytetrafluoroethylene) membrane filters (Gelman or Millipore) that have been prewashed with methanol and DMSO, collected in 10- to 15-ml quantities in sterile test tubes, and then stored in the frozen state at 5°C (DMSO freezes at 18°C) and protected from light. Because of the accumulation of oxidative breakdown products, an opened bottle of DMSO should not be used for more than 1 month.

47.2.2.4. Culture Preparation

The physiological condition of the culture plays an important part in the ability of the bacterium to survive liquid-nitrogen freezing. In general, use actively growing cells at the mid-logarithmic to late logarithmic phase of growth (38).

Grow the cells in an appropriate medium. For broth cultures, harvest by aseptic centrifugation and resuspend the pellet with sterile fresh medium containing either 10% (vol/vol) glycerol (prepared by adding sterile 20% glycerol to an equal volume of sterile broth) or 5% (vol/vol) DMSO (prepared by adding the appropriate amount of sterile 100% DMSO to the sterile broth). For agar cultures, wash the growth from the agar surface with sterile broth containing the suitable cryoprotective agent. Dispense 0.4 ml of the cell suspension containing at least 10^8 cells per ml into each sterile, prelabeled vial. Plastic presterilized screw-cap vials (Nunc) are recommended for bacteria. Since the screw caps can leak, store the vials in the vapor phase (see *CAUTION* note in section 47.2.2.6). For glass vials, precool the vials for a minimum of 30 min at 4°C before sealing. Then heat the vials above the prescored area by rotating in the torch flame for several seconds to remove moisture. This procedure avoids pressure inside the vial, which could result in bubbling during sealing. Hold the vial at an angle to prevent the culture suspension from contacting the hot glass, and partially withdraw the cotton plug, stopping just short of complete removal. Seal the vials, preferably with a semiautomatic sealer equipped with an oxygen gas torch (e.g., from Kahlenberg Globe Equipment Co., Sarasota, FL) which is designed to pull and seal, thus minimizing the formation of pinhole leaks.

Carefully remove the hot vial with a forceps, and place the vial in a test tube rack immersed in a cool water bath deep enough to immerse the vial to a depth of 0.6 cm. Avoid getting water on the hot portion of the vial; otherwise, the glass will crack. When the ampoules have cooled, they are ready for freezing. If this process is performed satisfactorily, the temperature of the specimen should not rise above 25°C. If the semiautomatic sealer is unavailable, the vials can be sealed with a gas torch by hand, but caution must be used to prevent the presence of pinhole leaks. To detect the presence of leaks, allow the sealed glass ampoules to stand for approximately 30 min at 5°C in a 0.05% methylene blue solution. Improperly sealed vials are detected by penetration of the dye into the vial.

47.2.2.5. Freezing

Much has been written about the effect on survival exerted by the cooling rate during freezing (27, 36). In general, the best results in the preservation of bacterial cultures have been obtained with slow cooling (e.g., 1°C/min).

The procedures for this method are as follows. Place the filled vials onto aluminum canes (no. A545; Nasco, Fort Atkinson, WI), and insert the canes into boxes (open-end cartons with rectangular holes, 2.5 by 2.5 by 11 in. [6.4 by 6.4 by 27.9 cm]; Murray and Heister, Inc., Beltsville, MD). Place the boxes into the freezing chamber of a programmable freezer (Linde BF 3.2; Union Carbide Corp.) which has an adjustable cooling rate. Freeze the cells at a controlled rate of 1°C/min to −40°C and then at a more rapid drop of 10°C/min to −90°C. After this temperature is achieved, transfer the vials to a liquid-nitrogen tank and store them immersed in the liquid phase at −196°C or above in the vapor phase at −150°C.

If a programmable freezer is not available, slow cooling can be achieved by either of the following procedures. (i) Place filled glass or plastic vials in a stainless steel pan on the bottom of a mechanical freezer at −60°C for 1 h, and then plunge them into a liquid-nitrogen bath for 5 mm. The rate of cooling to −60°C with this method is approximately 15°C/min. (ii) Canale-Parola (5) recommended immersing the sealed vials (placed on a cane) in 95% ethanol contained in a graduated cylinder. The cylinder containing the vials is placed in a mechanical freezer set at −85°C and allowed to reach that temperature before the canes and ampoules are placed in liquid nitrogen.

47.2.2.6. Thawing

It has been reported that rapid thawing of frozen cultures results in the greatest recovery of bacteria (27). This has also been the experience with bacterial cells frozen at the ATCC.

To recover frozen cultures, rapidly thaw them with moderate agitation in a 37°C water bath until all the ice melts. This usually takes about 50 s for glass vials and about 90 s for polypropylene ones. Immediately after thawing, remove the vial from the water bath and wipe with 70% ethanol to disinfect. Open and flame the vial, and aseptically transfer the culture to fresh medium.

CAUTION: Because of its extremely low temperature, liquid nitrogen can be hazardous if not used correctly. Improperly sealed glass vials can explode when retrieved from liquid nitrogen because of the rapid expansion of the liquid nitrogen that enters the vials through microscopic holes. Wear protective gloves and a face shield when handling frozen glass vials. Plastic screw-cap vials that are not properly sealed can fill with liquid nitrogen if stored in the liquid phase. Retrieval of the vials to warmer temperatures can result in spraying of the contents as the liquid nitrogen expands. Cell suspensions in plastic vials should be stored in the vapor phase.

Check the viability of the culture to determine the effectiveness of the procedure for a given species. The expected shelf lives of various bacteria are shown in Table 1.

47.3. PRESERVATION OF REPRESENTATIVE GENERA AND SPECIFIC GROUPS

47.3.1. Representative Genera
The reader should consult Table 1 and the various volumes of *Bergey's Manual of Systematic Bacteriology* (11–14) for information on the maintenance and preservation of representative bacterial genera. Certain groups of bacteria have special requirements, as follows.

47.3.2. Anaerobes
It is essential to maintain strictly anaerobic conditions during growth, harvesting, dispensing, and freezing of anaerobic bacteria. The cryoprotective agents must also be prereduced (also see section 14.6 in this volume).

Grow anaerobes such as *Ruminococcus* species in prereduced broth in Hungate test tubes (no. 2047 and 2048; Bellco Glass, Inc.) with rubber stoppers. Harvest the cells in the late logarithmic phase by centrifugation. Conduct all subsequent manipulations under oxygen-free gas flow with a sterile cannula. Remove the supernatant with a sterile Pasteur pipette connected to a filtration flask. Flush all pipettes, syringes, and the freezing or freeze-drying vials with oxygen-free gas before use. Harvest anaerobes grown on slants by washing down the growth with sterile prereduced medium containing the appropriate cryoprotective agent. Pool the pellets or slants to form a cell suspension of about 10^8 cells per ml (OD = 1.0). For freezing, the cryoprotectant is glycerol (final concentration, 10% [vol/vol]). For freeze-drying, the ATCC recommends using cryoprotectant reagent 18, which contains the following: Trypticase soy broth, 1.5 g; sucrose, 10.0 g; bovine serum albumin fraction V, 5.0 g; and distilled water to 100.0 ml. Filter sterilize the solution through a 0.22-μm-pore-size filter.

Flush the freezing or freeze-drying vials with oxygen-free gas, and dispense 0.2 ml of the cell suspension into each vial. Freeze immediately in a dry-ice–ethylene glycol bath prior to freeze-drying, or store in the liquid-nitrogen vapor phase if freeze-drying is delayed. Bacteria that will be frozen in liquid nitrogen are placed immediately in a rack that is partially immersed in liquid nitrogen and then transferred to a liquid-nitrogen tank.

For more details on the isolation and cultivation of anaerobic bacteria, see section 14.6 and references 5, 10, 21, 25, 26, and 29.

47.3.3. Cyanobacteria
The most successful method of preservation of cyanobacteria is freezing and storage in liquid-nitrogen vapor. The cryoprotectant of choice for most strains is DMSO (final concentration, 5% [vol/vol]) in maintenance broth. Some cyanobacteria have been freeze-dried, notably the strains that produce heterocysts and/or akinetes. In these instances, the cells are harvested at the early stationary phase by centrifugation and the pellet is resuspended in sterile cryoprotectant reagent 20 mixed equally with fresh sterile growth broth. The formulation for reagent 20 is as follows: bovine serum albumin fraction V, 10.0 g; sucrose, 20.0 g; and distilled water to 100.0 ml. Filter sterilize the solution through a 0.22-μm-pore-size filter.

Reagent 20 is used instead of reagent 18 for cyanobacterial cultures because they may be adversely affected by

Trypticase soy broth. Dispense 0.2 ml of the cell suspension into sterile glass inner vials (section 47.2.1), and process as described above (section 47.2.1). The freezing is done in a commercial freezer instead of by the chamber method.

47.3.4. Methanogens
The methanogens are a diverse group of the class *Archaea* that require stringent anaerobic conditions for growth. The cultures can be maintained by periodic subculturing in an appropriate medium and periodic replenishment of the appropriate gaseous atmosphere (19). Freezing and storage at low temperature is the most successful method of preservation of this group of bacteria (19). It has been reported that cultures of obligate anaerobes such as methanogens can be successfully stored in sterile 25% glycerol at $-70°C$ for up to 3 years (3).

Most methanogens grow on an H_2-CO_2 gas mixture, and special heavy glass-walled test tubes or serum bottles must be used to accommodate the pressurized atmosphere of the gas. Hungate test tubes (no. 2047-16125; Bellco Glass, Inc.), Balch serum tubes (no. 2048-18150; Bellco Glass, Inc.), and serum bottles are ideal containers for growing small volumes of the methanogenic bacteria.

A modified heavy-walled bottle with a neck that can be closed with a screw cap has been described by Hippe (19). This bottle facilitates the handling of methanogens because it fits in normal laboratory centrifuges and can be used for both the cultivation and harvesting of the cultures. Grow the cultures to an OD of 0.300 at 600 nm in 20 ml of the appropriate medium under H_2-CO_2 in the Hippe container. Harvest the cells by centrifugation directly in the unopened bottle; after removal of the closure, insert a sterile cannula and flush the bottle with oxygen-free gas. Resuspend the pellet with 2 ml of sterile fresh culture medium containing 10% (vol/vol) glycerol or 5% (vol/vol) DMSO (final concentrations). This suspending medium must be prepared just prior to use and must be gassed to ensure anaerobiosis. Dispense 0.2 ml of the cell suspension into sterile glass Cryules (no. 12523; Wheaton) while flushing the vials with oxygen-free gas. It is important to flush all pipettes and syringes with oxygen-free gas prior to use. *CAUTION*: Use an N_2-CO_2 gas mixture instead of H_2-containing mixtures for flushing the vials. After dispensing, carefully move the gassing cannula toward the top of the vial and continue gassing. Seal the middle of the vial by using an air-gas torch, and then place the vial immediately in a rack immersed (only the bottoms of the vials are immersed) in an ice bath to prevent heating of the suspension from the heated vial. Transfer the vials to aluminum canes (Shur-Bend Manufacturing Co., St. Paul, MN), and store them in the vapor phase of a liquid-nitrogen container. Other members of the *Archaea* can also be preserved by similar methods (10).

47.3.5. *Neisseria, Haemophilus, Campylobacter,* and *Helicobacter* Species
Neisseria, Haemophilus, Campylobacter, and *Helicobacter* species have been shown to be sensitive to the preservation process during different stages in the growth phase. It is important, therefore, to harvest these cells at the proper stage in the growth curve before subjecting them to the preservation process. Preservation experiments should be conducted with cells harvested at different stages of growth to obtain optimum conditions. For example, *Campylobacter* cells should be harvested after 48 h of growth on Brucella Albimi broth (MO8600; GIBCO, Grand Island, NY) containing 2% agar under an atmosphere of 5% CO_2, 6% O_2, and 89% N_2. Prepare bacterial suspensions in sterile 10% (vol/vol)

glycerol to yield approximately 10^9 cells per ml (OD = 1.0), and dispense into presterilized plastic screw-cap vials (30). Ultrafreeze the bacteria as described in section 47.2.2.

47.3.6. Plasmid-Containing Bacteria

Preservation of organisms containing plasmids has become an important part of biotechnology programs. These organisms are less stable genetically and are more likely to lose their foreign genetic material (34).

Successful preservation of plasmid-containing bacteria depends on the stability of their plasmids (34). Stable *Escherichia coli* plasmids can be freeze-dried or frozen by standard preservation methods with excellent recovery of both the host and plasmids. The recovery of unstable *E. coli* plasmids can be enhanced by growing the cells in the presence of the appropriate antibiotic, both before and after preservation. In general, bacteria containing unstable plasmids should be grown on antibiotic-containing liquid medium at 30°C, frozen with 10% (vol/vol) glycerol, and stored in the vapor over liquid nitrogen.

47.3.7. Sporeformers

Endospore formers and a few other bacteria have been successfully preserved for years by use of a mixture of sterile, air-dried soil (6, 18, 40, 41). Sterilize 1 g of soil per screw-cap tube (16 by 125 mm) by autoclaving at 121°C for 3 h on three consecutive days. Inoculate 1 ml of spore suspension into each tube, and allow the tubes to stand at room temperature or, preferably, under vacuum over Drierite or P_2O_5, with the cap loosened half a turn until visibly dry. Remove the cap aseptically, replace it with a sterile rubber stopper, and store the tube in the refrigerator.

47.4. CULTURE COLLECTIONS AND THEIR FUNCTIONS

Culture collections represent valuable resources of microbial germplasm (23). Without culture collections, there would be no rare, crucial, or biotechnologically important microbes (37). Culture collections range from those that serve specific research aims to those that house a wide diversity of microorganisms, cell lines, viruses, and genetic material. In general, there are three kinds of culture collections: specialized, reference, and national collections.

47.4.1. Specialized Collections

Specialized collections are usually of a personal nature and are not intended to be permanent. They may be narrow in the variety of microorganisms they hold but extensive in the number of strains of each species. These collections are indispensable as resources in studies of taxonomy, genetics, and strain variabilities. An example of a specialized collection was the bacterial plant pathogen collection of the late Mortimer Starr at the University of California at Davis.

47.4.2. Reference Collections

Reference collections are involved in collaborative efforts of scientists, industry, academe, and government agencies and are concerned mainly with one major group or function. These collections may be located in government agencies, universities, or industries associated with the production of goods, chemicals, and drugs. Examples of these collections are as follows.

Bacillus Genetic Stock Center (BGSC)
D. H. Dean, Curator

The Department of Biochemistry
College of Biological Sciences
The Ohio State University
484 W. 12th Ave.
Columbus, OH 43210
Telephone: 614-422-5550
URL: http://www.bgsc.org/

The Ohio State University maintains genetically characterized strains, cloning vectors, and bacteriophage for the genus Bacillus *and related organisms.*

The *Escherichia coli* K-12 Stock Culture Collection
B. Bachmann, Curator
Department of Biology
Yale University 255 Omi
P.O. Box 666
New Haven, CT 06511-7444
Telephone: 201-449-9395
URL: http://cgsc.biology.yale.edu/top.html

Pseudomonas Genetic Stock Center
Paul V. Phibbs, Professor and Director of Biotechnology Program
Department of Microbiology and Immunology
Brody School of Medicine
East Carolina University
Greenville, NC 27858-4354
Telephone: 252-744-3129
E-mail: phibbspa@mail.ecu.edu
URL: http://www.pseudomonas.med.ecu.edu/

Oregon Collection of Methanogens
David Boone
Department of Biology Portland State University
P.O. Box 751
Portland, OR 97207-0751
Telephone: 503-725-3865
E-mail: BooneD@pdx.edu or methanogens@lists.pdx.edu
URL: http://methanogens.pdx.edu/

The Lancefield-Streptococcus Collection
Vincent A. Fischetti
The Laboratory of Bacterial Pathogenesis and Immunology
The Rockefeller University
New York, NY 10021
Telephone: 212-327-8166
E-mail: vaf@mail.rockefeller.edu
URL: http://www.rockefeller.edu/vaf/catalog.htm

Agricultural Research Service Culture Collection
Cletus Kurtzman, Curator
National Center for Agricultural Utilization Research
1815 N. University St.
Peoria, IL 61604-3999
Telephone: 309-681-6561 or 309-685-4011
E-mail: kurtzman@mail.ncaur.usda.gov
URL: http://nrrl.ncaur.usda.gov/cgi-bin/usda.gov

Fungal Genetics Stock Center
School of Biological Sciences
University of Missouri, Kansas City
Kansas City, MO 64110
Telephone: 816-235-6485
Telephone of curator: 816-235-6484
URL: http://www.fgsc.net/index.html

Fungal Genome Resources
URL: http://www.fgsc.net/outlink.html

47.4.3. National Collections

The national collections, such as the ATCC and the National Collection of Type Cultures (NCTC) in England, are intended to be permanent national resources. They are often called national service collections, and their main function is to acquire, authenticate, preserve, and distribute authentic reference and type cultures. A mission-critical function of the culture collections is to provide continuity with the past through the preservation and distribution of authentic microbial strains of significance to the microbiology community for purposes of taxonomic or other research applications and for biotechnological exploitation of these microbes. These collections attempt to be comprehensive; they often are broad in the variety of species held, but they may be narrow in the number of strains maintained within a species. Many also serve as national and international patent depositories. Many provide a wide variety of services such as identification of unknown strains, proprietary safe deposit of microorganisms, and specialized training. They also represent an invaluable source of strain information, including history, strain data, and special applications. These databases are being made available through computer networking (URL: http://www.wfcc.info/index.html) and provide valuable information to the scientific community through a variety of computer networks. National service collections regularly publish catalogs of their holdings. Furthermore, many culture collections serve as centers of expertise for preserving the microbial germplasm and are very useful in training others in the do's and don'ts of culture preservation and maintenance.

The policy for acquisition of cultures varies among the service collections. A detailed description of their policies can be found in their respective catalogs. Individuals wishing to deposit strains should write to the collection to determine whether there is an interest in acquiring the strains.

Given below is a partial list of national service collections. A directory of nearly 500 national culture collections is available from the *World Data Centre for Culture Collections of Microorganisms* (WDCM) maintained at the National Institute of Genetics in Japan (see below for the address). The world culture collections database at the WDCM is useful for deciphering the acronyms for a given collection such as the ATCC, the Indian Type Culture Collection, DSMZ, etc. (see below).

World Federation of Culture Collections, National Institute of Genetics, WFCC-MIRCEN World Data Centre for Microorganisms, 1111 Yata, Mishima, Shizuoka, 411-8540 Japan
Telephone: 81-55 981 6895
Fax: 81-(0)55-981-6896
E-mail: hsugawar@genes.nig
URL: http://www.wfcc.info/index.html

The WDCM website for the Home Pages of Culture Collections in the World gives on-line access to most culture collections around the globe.
http://wdcm.nig.ac.jp/hpcc.html.

The Culture Collections of Prokaryotes is another useful website that provides information on bacterial culture collections worldwide.
http://www.bacterio.cict.fr/collections.html

American Type Culture Collection (ATCC), P.O. Box 1549, Manassas, VA 20110
Telephone: 703-365-2700 (United States) or 800-638-6597 (United States, Canada, and Puerto Rico)
http://www.atcc.org/Home.cfm

National Collection of Type Cultures (NCTC), L. R. Hill, PHLS Central Public Health Laboratory, 61 Colindale Avenue, London NW9 5HT, United Kingdom
Telephone: 044-81-2004400
Fax: 044-81-2007874

National Collections of Industrial and Marine Bacteria. (NCIMB), Dr. I. Bousfield, 23 St. Machar Drive, Aberdeen AB2 1RY, United Kingdom
Telephone: 044-224-273332
Fax: 044-224-487658

National Collection of Food Bacteria (NCFB), Jackman, AFRC Institute of Food Research, Reading Laboratory, Shinfield, Reading RG2 9AT, United Kingdom
Telephone: 044-734-883103
Fax: 044-734-884763

National Collection of Plant Pathogenic Bacteria (NCPB), E. Stead, Central Science Laboratory, Hatching Green, Harpenden, Hertfordshire AL5 2BD, United Kingdom
Telephone: 044-5827-5241
Fax: 044-5827-62178

German Collection of Microorganisms and Cell Cultures (DSMZ), Hans G. Drexler, Human and Animal Cell Cultures, German Collection of Microorganisms and Cell Cultures, Mascheroder Weg 1b, 38124 Braunschweig, Germany
Telephone: 49-531-2616.161
E-mail: dsmzmutz@gbf-braunschweig.de
URL: http://www.dsmz.de/

Belgian Co-ordinated Collections of Micro-organisms (BCCM) bacterial collection, curator; Ir Claudine Vereecke, Dr. Danielle Janssens, contact; BCCM/LMG, Laboratorium voor Microbiologie, Universiteit Gent (UGent), K.L. Ledeganckstraat 35, B-9000 Gent, Belgium
Telephone: 32-(0)9-264.51.08
E-mail: BCCM.LMG@UGent.be
URL: http://bccm.belspo.be/about/lmg.php#contact

BCCM (Agro) Industrial Fungi and Yeast Collection, Dr. Cony Decock, curator; Dr. Stéphane Declerck, contact; Mycothèque de l'Universite catholique de Louvain, Place Croix du Sud 3, B-1348 Louvain-la-Neuve, Belgium
Telephone: 32-10-47.37.42
Fax: 32-10-45.15.01
E-mail: BCCM.MUCL @mbla.ucl.ac.be
URL: http://bccm.belspo.be/about/mucl.php

BCCM Plasmid and DNA Libraries, Lic. Martine Vanhoucke, promoter, curator, and contact; Department of Molecular Biology, Ghent University, Fiers-Schell-Van Montagu Building, Technologiepark 927, B-9052 Zwijnaarde, Belgium
Telephone: 32-9-33.13.843
E-mail: bccm.lmbp @dmbr.UGent.be
URL: http://bccm.belspo.be/about/lmbp.php

Japan Collection of Microorganisms/Microbe Division (JCMM), RIKEN, BioResource Center, 2-1 Hirosawa, Wako, Saitama 351-0198, Japan
Telephone: 81 48 467 9560
E-mail: inquiry@jcm.riken.jp
URL: http://www.jcm.riken.go.jp/

The Netherlands Culture Collection of Bacteria (NCCB), Centraalbureau voor Schimmelcultures, Uppsalalaan 8, P.O. Box 85167, 3584 CT, The Netherlands
Telephone: 31-30-2122634 (E. Abeln)
Fax (international): 31 30 2512097
E-mail: nccb@cbs.knaw.nl
URL: http://www.cbs.knaw.nl

Collection Nationale de Cultures de Microorganismes (CNCM), Institut Pasteur, 25 rue du Docteur Roux, F-75724 Paris Cedex 15, France
Telephone: 033-1-45688251
Fax: 033-1-45688236
URL: http://www.pasteur.fr/recherche/unites/Cncm/index-en .html

Russian Collection of Microorganisms (VKM), L. V. Kalakoutskii, Institute of Biochemistry and Physiology of Microorganisms, Russian Academy of Sciences, Puschino, Moscow Region 142297, Russia
Telephone: 07-095-231-6576
E-mail: vkm@ibpm.pushchino.ru
URL: http://www.vkm.ru/

Colección Española de Cultivos Tipo (CECT), F. Uruburu, curator; Departamento de Microbiología, Facultad de Ciencias Biologicas, Universidad de Valencia, Es-46100 Burjasot, Spain
Telephone: 034-6-3864300 (outside Spain) or 963 54 46 12 (Spain)
Fax: 963 54 31 87
E-mail: info@cect.org
URL: http://www.uv.es/cect/

VTT Collection of Industrial Microorganisms, Maija-Liisa Suihko, VTT Biotechnical Laboratory afrc. frin. Finland, Tietotie, SF-02150 Espoo, Finland
Telephone: 0358-0-4565133
Fax: 0358-0-4552028
URL: http://www.vtt.fi/services/cluster4/topic4_7/ mikrobikokoelma.jsp?lang=en

Center for General Microbiological Culture Collection (CGMCC),

China Committee for Culture Collections of Microorganisms, Institute of Microbiology, Academia Sinica, Beijing, China
Telephone (86) 10-62555614
Fax: (86) 10-62560912
URL: http://www1.im.ac.cn/typecc/junzhong/en.html

Microbial Type Culture Collection and Gene Bank (MTCC), Institute of Microbial Technology; Tapan, Dr. Chakrabarti, IMTECH, Sector 39-A, Chandigarh, U.T., 160-036
Telephone: (91) 172-690004 or (91) 172-690632
Fax: (91) 172-690632 or (91) 172-690585
URL: http://www.imtech.ernet.in/mtcc/

National Collection of Industrial Microorganisms (NCIM), D. V. Gokhale, National Chemical Laboratory, Dr. Homi Bhabha Road, Pune 411 008, India
Telephone: 91-20-2590 2670
Fax: +91-20-2590 2671
E-mail: dv.gokhale@ncl.res.in

Indian Type Culture Collection (ITCC), J. N. Kapoor, Division of Mycology and Plant Pathology, Indian Agricultural Research Institute Mycology Div, I.A.R.I., New Delhi 110012

Korean Federation of Culture Collections (KFCC), Yu Juhyun, director; Shinchondong Sodaemunku, Laboratorium voor Tropische Hygiene, Seoul 120-749, Korea
Telephone: (82) 2-392-0950
URL: http://www.bacterio.cict.fr/collections.html

National Collection of Agricultural and Industrial Microorganisms, T. Deak, Department of Microbiology and Biotechnology, University of Horticulture and Food Industry, Somloi ut 14-16, H-1118 Budapest, Hungary
Telephone: 036-1-665411
Fax: 036-1-666220

47.5. COMMERCIAL SOURCES

AMSCO Scientific, 2424 West 23rd St., Erie, PA 16514

Baxter Diagnostics Inc., Scientific Products Division, 1430 Waukegan Rd., McGaw Park, IL 60085

Bellco Glass, Inc., 340 Edrudo Rd., P.O. Box "B," Vineland, NJ 08360

Difco Laboratories, P.O. Box 1058A, Detroit, MI 48232

Fisher Scientific Co., 2000 Park Lane Dr., Suite 2, Pittsburgh, PA 15275-1104

Gelman Sciences, Inc., 600 S. Wagner Rd., Ann Arbor, MI 48106

GIBCO Laboratories, Division of Life Technologies, Inc., 3175 Staley Rd., Grand Island, NY 14072

Glass Vials, Inc., 1352 James St., Baltimore, MD 21223

W. R. Grace and Co., Davison Chemical Division, 10 E. Baltimore St., Baltimore, MD 21202

Kahllenberg-Globe Equipment Co., Sarasota, FL 33577

Kimble/Kontes, Spruce St., P.O. Box 739, Vineland, NJ 03860

Markem Machine Co., 150 Congress St., Box 480, Keene, NH 03431

Millipore Filter Corp., 800 Ashby Rd., Bedford, MA 01730

Minnesota Valley Engineering, Inc., 303 7th St. NW, New Prague, MN 56071

Murray & Heister, Inc., 10738 Tucker St., Beltsville, MD 20705

NASCO—Fort Atkinson, 901 Janesville Ave., P.O. Box 901, Fort Atkinson, WI 53538-0901

Norton Performance Plastics, Granville, NY 12832

Nunc, Inc., 2000 North Aurora Rd., Naperville, IL 60563

Revco, Inc., 1100 Memorial Dr., West Columbia, SC 29169

Shur-Bend Manufacturing Co., 4612 N. Chatsworth, St. Paul, MN 55126

Thomas Scientific, P.O. Box 99, Swedesboro, NJ 08085-1780

Union Carbide Corporation, a Subsidiary of The Dow Chemical Company, 171 River Rd., Piscataway, NJ 08854

The VirTis Co., Inc., 815 Route 208, Gardiner, NY 12525

VWR Scientific, Inc., 3202 Race St., Philadelphia, PA 19104

Whatman Inc., 200 Park Ave., Ste. 210, Florham Park, NJ 07932

Wheaton Science Products, 1501 N. 10th St., Millville, NJ 08332-2093

47.6. REFERENCES

1. **Acker, J. P.** 2007. Biopreservation of cells and engineered tissues. *Adv. Biochem. Eng. Biotechnol.* **103:**157–187.

2. **Bjerketorp, J., S. Hakansson, S. Belkin, and J. K. Jansson.** 2006. Advances in preservation methods: keeping biosensor organisms alive and active. *Curr. Opin. Biotechnol.* **17:**43–49.

3. **Briukhanov, A. I., and A. I. Netrusov.** 2006. Long-term storage of obligate anaerobic microorganisms in glycerol. *Prikl. Biokhim. Mikrobiol.* **42:**200–203.

4. **Cabasso, V. J., and R. H. Regamy (ed.).** 1977. *Developments in Biological Standardization,* vol. 36. *International Symposium on Freeze Drying of Biological Products.*

5. **Canale-Parola, E.** 1973. Isolation, growth and maintenance of anaerobic free-living spirochetes. *Methods Microbiol.* **8:**61–73.

6. **Chang, L. T., and R. P. Elander.** 1986. Long-term preservation of industrially important microorganisms, p. 49–55. *In* A. L. Domain and N. A. Solomon (ed.), *Manual of Industrial Microbiology and Biotechnology.* American Society for Microbiology, Washington, DC.

7. **Cleland, D., P. Krader, C. McCree, J. Tang, and D. Emerson.** 2004. Glycine betaine as a cryoprotectant for prokaryotes. *J. Microbiol. Methods* **58:**31–38.

8. **Fahy, G. M.** 1986. The relevance of cryoprotectant "toxicity" to cryobiology. *Cryobiology* **23:**1–13.

9. **Favero, M. S., and M. J. Arduino.** 2006. Decontamination and disinfection, p. 373–381. *In* D. O. Fleming and D. L. Hunt (ed.), *Biological Safety: Principles and Practices,* 4th ed. ASM Press, Washington, DC.

10. **Feltham, R. K. A., A. K. Power, P. A. Pell, and P. H. A. Sneath.** 1978. A simple method for storage of bacteria at −76°C. *J. Appl. Bacteriol.* **44:**313–316.

11. **Garrity, G. M., D. R. Boone, and R. W. Castenholz (ed.).** 2001. *Bergey's Manual of Systematic Bacteriology,* 2nd ed., vol. 1. *The Archaea and the Deeply Branching and Phototrophic Bacteria.* Springer, New York, NY.

12. **Garrity, G. M., D. J. Brenner, N. R. Krieg, and J. T. Staley (ed.).** 2005. *Bergey's Manual of Systematic Bacteriology,* 2nd ed., vol. 2. *The Proteobacteria, Part A: Introductory Essays.* Springer, New York, NY.

13. **Garrity, G. M., D. J. Brenner, N. R. Krieg, and J. T. Staley (ed.).** 2005. *Bergey's Manual of Systematic Bacteriology,* 2nd ed., vol. 2. *The Proteobacteria, Part B: The Gammaproteobacteria.* Springer, New York, NY.

14. **Garrity, G. M., D. J. Brenner, N. R. Krieg, and J. T. Staley (ed.).** 2005. *Bergey's Manual of Systematic Bacteriology,* 2nd ed., vol. 2. *The Proteobacteria, Part C: The Alpha-, Beta-, Delta-, and Epsilonproteobacteria.* Springer, New York, NY.

15. **Grivell, A. R., and J. F. Jackson.** 1969. Microbial culture preservation with silica gel. *J. Gen. Microbiol.* **58:**423–425.

16. **Hartsell, S. E.** 1956. Maintenance of cultures under paraffin oil. *Appl. Microbiol.* **4:**350–355.

17. **Hays, H. C., P. A. Millner, J. K. Jones, and M. H. Rayner-Brades.** 2005. A novel and convenient self-drying system for bacterial preservations. *J. Microbiol. Methods* **63:**29–35.

18. **Heckly, R. J.** 1978. Preservation of microorganisms. *Adv. Appl. Microbiol.* **24:**1–53.

19. **Hippe, H.** 1991. Maintenance of methanogenic bacteria, p. 101–113. *In* B. E. Kirsop and A. Doyle (ed.), *Maintenance of Microorganisms and Cultured Cells,* 2nd ed. Academic Press Ltd., London, England.

20. **Hunt, G. A., A. Gourevitch, and J. Lein.** 1958. Preservation of cultures by drying on porcelain beads. *J. Bacteriol.* **76:**453–454.

21. **Imprey, C. S., and B. A. Phillips.** 1991. Maintenance of anaerobic bacteria, p. 71–80. *In* B. E. Kirsop and A. Doyle (ed.), *Maintenance of Microorganisms and Cultured Cells,* 2nd ed. Academic Press Ltd., London, England.

22. **Kirsop, B. E., and A. Doyle (ed.).** 1991. *Maintenance of Microorganisms and Cultured Cells,* 2nd ed. Academic Press Ltd., London, England.

23. **Labeda, D. P.** 2005. Culture collections, p. 11–113. *In* G. M. Garrity, D. J. Brenner, N. R. Krieg, and J. T. Staley (ed.), *Bergey's Manual of Systematic Bacteriology,* 2nd ed., vol. 2. *The Proteobacteria, Part A: Introductory Essays.* Springer, New York, NY.

24. **Lapage, S. P., J. E. Shelton, T. G. Mitchell, and A. R. Mackenzie.** 1970. Culture collections and the preservation of bacteria. *Methods Microbiol.* **3A:**135–228.

25. **Malik, K. A.** 1990. Use of activated charcoal for the preservation of anaerobic phototrophic and other sensitive bacteria by freeze-drying. *J. Microbiol. Methods* **12:**117–124.

26. **Malik, K. A.** 1990. A simplified liquid-drying method for the preservation of microorganisms sensitive to freezing and freeze-drying. *J. Microbiol. Methods* **12:**125–132.

27. **Meryman, H. T.** 1966. Freeze-drying, p. 609–663. *In* H. T. Meryman (ed.), *Cryobiology.* Academic Press, Inc., New York, NY.

28. **Miller, R. E., and L. A. Simmons.** 1962. Survival of bacteria after twenty-one years in the dried state. *J. Bacteriol.* **84:**1111–1113.

29. **Miller, T. L., M. J. Wolin, and E. A. Kusel.** 1986. Isolation and characterization of methanogens from animals. *Syst. Appl. Microbiol.* **8:**234–238.

30. **Mills, C. K., and R. L. Gherna.** 1988. Cryopreservation studies of *Campylobacter. Cryobiology* **25:**148–152.

31. **Miyamoto-Shinohara, Y., J. Sukenobe, T. Imaizumi, and T. Nakahara.** 2006. Survival curves for microbial species stored by freeze-drying. *Cryobiology* **52:**27–32.

32. **Morgan, C. A., N. Herman, P. A. White, G. Vesey.** 2006. Preservation of micro-organisms by drying: a review. *J. Microbiol. Methods* **66:**183–193.

33. **Nadirova, I. M., and V. L. Zemliakov.** 1970. A study on the degree of stability of typical properties of bacteria preserved under mineral oil. *Mikrobiologiya* **39:**1106–1109.

34. **Nierman, W. C., and T. Feldblyum.** 1985. Cryopreservation of cultures that contain plasmids. *Dev. Ind. Microbiol.* **26:**423–434.

35. **Redway, K. F., and S. P. Lapage.** 1974. Effect of carbohydrates and related compounds on the long term preservation of freeze-dried bacteria. *Cryobiology* **11:**73–75.

36. **Rinfret, A. P., and B. LaSalle (ed.).** 1975. *Round Table Conference on the Cryogenic Preservation of Cell Cultures,* p. 1–78. National Academy of Sciences, Washington, DC.

37. **Smith, D., and M. J. Ryan.** 2001. Culture collections in the twenty-first century. *Biologist* **48:**125–128.

38. **Speck, M. L., and R. A. Cowan.** 1970. Preservation of microorganisms by drying from the liquid state, p. 241–250. *In* H. Iizuka and Y. Hasegawa (ed.), *Proceedings of the First International Conference on Culture Collections.* University of Tokyo Press, Tokyo, Japan.

39. **Suslov, T. V., and M. N. Schroth.** 1981. Bacterial culture preservation in frozen and dry-film methyl cellulose. *Appl. Environ. Microbiol.* **42:**872–877.

40. **Vela, G. R.** 1974. Survival of *Azotobacter* in dry soil. *Appl. Microbiol.* **28:**77–79.

41. **Williams, S. T., and T. Cross.** 1971. Actinomycetes. *Methods Microbiol.* **5:**295–334.

Author Index

Subject Index